The Cambridge
Aerospace Dictionary

The Cambridge
Aerospace Dictionary

Bill Gunston OBE, FRAeS
Editor, Jane's Information Group

PUBLISHED BY THE PRESS SYNDICATE OF THE UNIVERSITY OF CAMBRIDGE
The Pitt Building, Trumpington Street, Cambridge, United Kingdom

CAMBRIDGE UNIVERSITY PRESS
The Edinburgh Building, Cambridge CB2 2RU, UK
40 West 20th Street, New York, NY 10011-4211, USA
477 Williamstown Road, Port Melbourne, VIC 3207, Australia
Ruiz de Alarcón 13, 28014 Madrid, Spain
Dock House, The Waterfront, Cape Town 8001, South Africa

http://www.cambridge.org

This edition adapted and updated from Aerospace Dictionary
published by Jane's Information Group Ltd 1980, 1986, 1988
Cambridge edition first published 2004

Printed in the United States of America

Typefaces Monotype Times New Roman and Adobe Caslon 8/9 pt.   *System* QuarkXPress [PT]

A catalogue record for this book is available from the British Library.

Library of Congress Cataloguing in Publication Data
Gunston, Bill.
   The Cambridge aerospace dictionary / Bill Gunston.
      p. cm – (Cambridge aerospace series)
   Includes bibliographical references and index.
   ISBN 0-521-84140-2
      1. Aerospace engineering – Dictionaries. 2. Aeronautics – Dictionaries.
   3. Astronautics – Dictionaries. I. Title. II. Series.

TL509.G88 2004
629.1'03–dc22   2004043530

ISBN 0 521 84140 2 hardback

Foreword

Gathering terms for an aerospace dictionary is harder than it looks. I recently studied a list of terms used by the US Air Force to describe the status of each of its component organizations. They explained 'These actions are defined in ways that may seem arcane to the non-specialist, but each term has a specific meaning.' The terms are: Activate, Active list, Assign, Attach, Consolidate, Constitute, Designate, Disband, Disestablish, Establish, Establishment, Inactivate, Inactive list, Organize, Provisional organizations, Re-designate, Re-establish, Relieve from active duty, and Unit. I read their meanings through several times and decided not to include any in these pages.

In a previous edition I was criticised by a reviewer for using words 'which have no relevance to aerospace'. He cited as an example 'barrier pattern', a term which BAe Manchester had asked me to define! *My sole objective is to create a useful product.* To this end I have included brief entries on such words as 'generic', 'oxygen' and 'gasoline', which are not aerospace terms. Incidentally, while 'gasoline' is clearly now a preferred spelling, I have had to write quite an essay on 'kerosene/kerosine'.

I once had to defend myself against an air marshal who was offended by such rubbish (as he saw it) as 'hardware' and 'software'. Today the explosion of home computing has opened up millions to such previously unfamiliar language. Indeed, in recent years the number of software terms has begun to get out of hand. The JSF programme alone involves more than 40 software acronyms, and I have omitted most of them.

Partly for this reason, this dictionary is centred (centered) at least in mid-Atlantic, if not further west, so we have 'Petrol *Gasoline*', the brief definition appearing under the latter. Cross-references are italicised. I have used US spellings wherever they are appropriate, and

in this field they tend to predominate. Note: USA means US Army.

I have attempted to include a brief explanation of aerospace materials, even if they are known by a registered tradename. Also included are the names of many organizations, but, with a few exceptions, not armed forces, airlines or flying clubs, and certainly not the names of manufacturers or particular types of aircraft, though such acronyms as TSPJ, Tornado self-protection jammer, are tempting. On the other hand, there is a grey area where a company product appears to merit inclusion, an example being Zero Reader. I have had particular trouble with the names of spacecraft and their payloads.

Entries are in *strict alphabetical order*, thus MW50 appears in the place for MW-fifty. The exception is where an entry has a single alphabetical character followed by a numeral. In such cases it appears immediately after other entries featuring that single character. With a subject as complicated as aerospace, where one finds C, c, c^1, $\bar{c}$, $\bar{\bar{c}}$, (c), C* and a host of C+numeral entries, it is difficult to decide which sequence to adopt. Greek terms are listed in Appendix 1, but some – such as Alpha and Beta – merit a place in the body of the dictionary.

On a lighter note, I read an article by Col. Art Bergman, USAF, explaining how to manage the temperamental F100 engine. I had no difficulty with his EECs, UFCs and Plaps, but was defeated by 'The F100 needs a lot more TLC than the J79 . . .' I asked several certified F-15 drivers, and they were all mystified. I called the 527th TFTS, then the European Aggressor outfit. A charming female voice instantly said "Ever think of tender loving care?" On reflection, I put this meaning in the dictionary. The criterion is whether or not an aerospace person might be confused without it.

One obvious problem area is at what point

one should give up trying to include foreign terms. Some may think I have been over-generous to our Gallic friends, while other countries may think themselves harshly treated by being ignored. It is impossible to say 'Leave out all foreign terms and acronyms', because many have become part of the English language. Nobody would expect 'aileron' to be omitted, and before long 'Fenestron' will be just as universally accepted, probably as fenestron.

At a rough count the number of new entries this time is in excess of 15,000. Almost all the additions are acronyms. There is little point in again saying that acronyms are an infectious disease, especially in the world of aerospace. Whilst admitting that the incentive to abbreviate is often strong, it is self-defeating if the reader has a choice of more than 20 interpretations and does not know which one to pick.

Some acronyms, such as Cardsharp, appear contrived. Another is Tiger, Terrifically Insensitive to Ground-Effect Radar; I had to force myself to include it. In general, I have omitted acronyms which include the name of a company, an example being Caps, Collins adaptive processor system. I have attempted to indicate whether the spoken acronym or spelt-out version predominates. Thus, we have Papi before PAPI. The oustanding exception is NATO. This is always spoken as a word, but the hierarchy in Brussels still insist that it is not written Nato.

Some acronyms bear little resemblance to the actual initial letters of the original words, while a few are quite a mouthful. We have been in particular trouble with the Joint Strike Fighter. This soon spawned JSF-E&MD and JSFPO-AEP, whilst Boeing were awarded a \$28,690,212 contract to perform the JSFPICPTD. This means the Joint Strike Fighter Program Integrated Core Processing Technical Demonstration, and is something I have omitted. Another non-starter has to be Direct, which the US Air Force tell me stands for Defense IEMATS REplacement Command and Control Terminal, which would be

fine were it not for the fact that IEMATS stands for Improved Emergency MEssage Automated Transmission System. Roger Bacon, the sage of *Flight International*, has drawn attention to Boeing's 'no-tail advanced theater transport, tilt-wing super-short takeoff and landing', which creates the handy name NTATTTW/SSTOL. Clearly, we need acronyms within acronyms.

It is often difficult to decide when the name of a specific item has become a more general term which has to be included. In the 1970s the AAH (Advanced Attack Helicopter) meant the AH-64 Apache. This is a particular type of helicopter, so it had no place in these pages. However, over the years AAH has become a term applied to several of the AH-64's later competitors, so exclusion is no longer justified. In the same way Awacs is now a class of aircraft, while, even though there is only one type of AABNCP, that designation is so important it would be unhelpful to omit it.

Both the AAH and AABNCP begin with 'Advanced'. This is a mere pointless buzz-word. Presumably it is intended to imply that something is the very latest, 'state of the art' and better than the competition, but – in aero-space at least – I have seldom heard of anybody designing something that was not 'advanced'. Can these items still be 'advanced' after 40 years? To me, another *bête noire* is 'integrated'. Already we have a zillion AIAs (advanced integrated acronyms). This is an advanced integrated dictionary.

There is an obvious need for a body with the clout to decree what things shall be called, because the present situation is ludicrous. Did you know that the acronym ATAC can mean 'Advanced Target Acquisition and Classification'? Fine, but ATDC stands for 'Assisted Target Detection and Classification' and also for 'Automatic Target Detection and Classification' and also for 'Automated Target Detection and Classification'. Clearly that is not enough, because ATRC stands for 'Aided Target Recognition and Classification' and 'Automatic Target Recognition and Classification'. I did not myself invent these.

And I have just noticed that the USAF, the world's leading offender, has become dissatisfied with the mere ERT (extended-range tank). It has changed it to ERFCS, extended-range fuel-containment system. Feeble! The name could be made *far* more complicated!

In the same way, it should be simple to have an agreed abbreviation for an airspace control zone, but we are now confronted by CTLZ, CTR, CTRZ, and CTZ. In the first edition of this work I included FMEA, for which two elucidations were (and are) current: failure modes and effects analysis and failure-mode effects analysis. I now have to add FMECA, failure-mode effects and criticality analysis, and FMETA, failure-mode effects and task analysis. It is inconceivable that the authors of the two new letter-jumbles were unaware of FMEA, and I cannot comprehend the need for the two new identities. If we go on like this I fear for the sanity of whoever takes over this work when I collapse through exhaustion.

Many of the acronyms in these pages already have more than 20 meanings, and are gathering fresh ones all the time. This trend is leading to texts which, even to most aerospace people, must appear mere gobbledegook. There is no more clearly written periodical than *Aerospace*, published by the august Royal Aeronautical Society, and it strives to remain one of the few bastions of good English. They published an article which told us 'Currently, BASE is developing a Terprom SEM-E standard card for use in the H764G, a high-accuracy INS with embedded GPS. It has two slots, the second being used by an Arinc, MIL-1553A/B or PANIL interface.' Many readers were doubtless happy with this, and one was impelled to respond with 'May I add something to your characterisation of AQP as 'an upgrade of CRM' . . . The human factors elements had to be injected into non-jeopardy Loft and LOE . . . With converging developments in CPL NVQ and recurrent CRM, the AQP may be the shape of things to come in the UK.'

A speaker at a recent conference 'has sat on EUROCONTROL, ICAO, EUROCAE, RTAC and AEEC. In his current position as Programme Manager CNS/ATM he is involved in the CLAIRE and ISATIS using ACARS, a development study of VDL Mode 2 in France. He is evaluation manager of EOLIA and ASD manager in ProATN.' And an advertisement tells me 'Group IV faxes and PCMCIA cards are only supplied with an ISDN S-Bus interface. The ISDN integration provided by the LES means that a SODA is only required at the mobile end'. I think I need a whisky with my SODA.

Preface to the Cambridge edition
This updated and enlarged new edition is the first to be published by Cambridge University Press. I would like to thank Phoenix Typesetting for doing a masterful job with mathematics and Greek symbols, and everyone at Cambridge for their diligence and infectious enthusiasm – all too rare these days in book publishing.

Bill Gunston, Haslemere, 2004

A

A 1 General symbol for area (see S).

2 Aspect ratio (see *As*).

3 Amperes.

4 Atomic weight.

5 Moment of inertia about longitudinal axis, rolling mode.

6 Anode.

7 Amplitude.

8 Degrees absolute.

9 Amber airway.

10 IFR flight plan suffix, fitted DME and 4096-code.

11 JETDS code: piloted aircraft, IR or UV radiation.

12 Airborne Forces category aircraft (UK, 1944–46).

13 Atomic (as in A-bomb).

14 Sonobuoy standard size class, c 1 m/3 ft.

15 Air Branch (UK Admiralty).

16 Calibration (USAF role prefix 1948–62).

17 US military aircraft basic mission or modified mission: attack (USAS, USAAC, USAAF, 1924–48; USN 1948–62; USAF/USN since 1962).

18 Aircraft category, ambulance (USAAS 1919–24, USN 1943).

19 Powered target (USAAC 1940–41).

20 Amphibian (USAF 1948–55).

21 Availability.

22 Aeroplane (PPL).

23 Altitude, followed digits indicated hundreds of feet.

24 Arm, as distinct from safe.

25 Antarctic (but Tor Bergeron's classification = Arctic).

26 Alternate [airport].

27 Weather: hail

28 Accepted (EFIS or nav. display).

29 Arrival chart.

30 Sport-parachuting certificate: 10 jumps, no accuracy demanded.

31 Autotuned (navaid).

32 Magnetic-vector potential.

Å Angström (10^{-10}m), very small unit of length, contrary to SI.

a 1 Velocity of sound in any medium.

2 Structural cross-section area.

3 Anode.

4 (Prefix) atto, 10^{-18}.

5 (Suffix) available (thus, LD_a = landing distance available).

6 Ambient.

7 Acceleration.

A0, A_0 Unmodulated (steady note) CW radio emission.

A0A1 Unmodulated (steady note) radio emission identified by Morse coding in a break period.

A0A2 Unmodulated emission identified by Morse coding heard above unbroken carrier (eg an NDB).

A1, A_1 1 Unmodulated but keyed radio emission, typically giving Morse dots and dashes.

2 Military flying instructor category: two years and 400 h as instructor.

a_1 Lift-curve slope for wing or other primary aerodynamic surface, numerically equal to $dC_L/d\alpha$.

A2 Military flying instructor category; 15 months and 250 h.

a_2 Lift-curve slope for hinged trailing-edge control surface, numerically $dC_L/d\epsilon$.

$A2C^2$ Army airborne [or airspace] command and control [S adds system] (USA).

A2C2 Airborne airstrike command and control (GTACS).

A3 AM radio transmission with double SB.

A^3 Affordable acquisition approach (USAF).

A3H AM, SSB transmission with full carrier.

A^3I Army/NASA aircrew/aircraft integration (USA/US).

A3J AM, SSB transmission with suppressed carrier.

A^3M Advanced air-to-air missiles.

A^3TC Advanced automated air traffic control.

A8-20 Airworthiness approval for classic (usually ex-military) aircraft (CAA, UK).

A-25 Royal Navy form for reporting aircraft accidents.

A-battery Electric cell to heat cathode filament in valve (tube).

A-bomb Atomic bomb, see *nuclear weapon*.

A-check S-check plus routine inspection of flight-control system.

A-class 1 Airspace = 18,000+ ft [5486 m] AMSL and controlled.

2 Aircraft accident = involving loss of life or damage exceeding US$1 million.

A-frame hook Aircraft arrester hook in form of an A; hook at vertex and hinged at base of each leg.

A-gear Arrester gear.

A-Licence Basic PPL without additions or endorsements.

A-line Airway.

A-mode Transponder sends a/c ident code only.

A-sector Sector of radio range in which Morse A is heard, hence **A-signal**.

A-station In Loran, primary transmitting station.

A-Stoff Liquid oxygen (G).

A-type entry Fuselage passenger door meeting FAA emergency exit requirements; typical dimensions 41 in × 76 in.

AA 1 Anti-aircraft.

2 Airship Association (UK).

3 Acquisition Aiding, technique for matching EM waveforms (esp for ECM).

4 Air-to-air (ICAO code).

5 Alert annunciator.

6 Antenna array.

7 Airbrokers Association (UK, 1949, became AAB).

A/A Air-to-air (radar mode).

AAA 1 Airport advisory area.

2 Army Aviation Association (USA), now AAAA.

3 Antique Airplane Association (US).

4 American Airship Association.

5 Anti-aircraft artillery (triple-A).

6 Affordable acquisition approach (usually A^3, USAF).

7 Associazione Arma Aeronautica (I).

AAAA *1* Australian Aerial Agricultural Association.
2 Army Aviation Association of America Inc.
3 Antique Aeroplane Association of Australia.
4 Advanced architecture for airborne arrays.
5 American Aviation Aerospace Alliance.
6 Arizona Antique Aircraft Association.
AAAC Australian Army Aviation Corps.
AAACF Airline Aviation & Aerospace Christian Fellowships (UK charity).
AAAD *1* Airborne anti-armour defence.
2 All-arms air-defence (UK).
AAAE American Association of Airport Executives.
AAAF Association Aéronautique et Astronautique de France.
AAAI American Association for Artificial Intelligence (Menlo Park, CA).
AAAM Advanced air-to-air missile.
AAAS American Association for the Advancement of Sciences.
AAASS Australian airborne acoustic systems strategy [sonobuoy system, also rendered A^3S^2].
AAAV Azienda Autonoma Assistenza al Volo, agency for air navigation and air traffic service (Italy).
AAAW Air-launched anti-armour weapon.
AAB Association of Air Brokers, now BACA.
AABM Air-to-air battle management.
AABNCP Advanced airborne (national) command post (DoD).
AAC *1* Army Air Corps (UK, from 1 September 1957).
2 Army Air Corps (US, 1926– March 1942).
3 Army Aviation Centre (Middle Wallop, UK).
4 Air Armament Center (AFMC).
5 Aviation Advisory Commission (US).
6 Alaskan Air Command (from 1945).
7 All-aspect capability.
8 Advance-acquisition contract (US).
9 Aeronautical, or airline, administrative control, or communications (Satnav).
10 Airborne Analysis Center.
AACA Alaska Air Carrier's Association Inc.
AACAS Auto air-collision avoidance system.
AACC *1* Airport Associations Co-ordinating Council (Int.).
2 See A2C², A2C2.
AACE Aircraft alerting communications EMP.
AACI Aircraft and Accident Commission of Indonesia.
AACMI Autonomous air-combat manoeuvring instrumentation; S adds system, T training.
AACO Arab Air Carriers Association.
AACPP Airport access control pilot program (TSA; note: pilot means initial or preliminary).
AACR Airborne analog cassette recorder.
AACS *1* Army Airways Communications Service [to 1946], Airways and Air Communications Service [1946–51], subsequently AF Com. Service.
2 Airborne advanced communications system.
AACT Air-to-air combat test (USN).
AACU Anti-Aircraft Co-operation Unit (UK, various dates 1937–47).
AAD *1* Aging Aircraft Division, (WPAFB).
2 Assigned altitude deviation.
AADC *1* Area Air-Defense Commander (USN).
2 Analytical air-defence cell (NATO).
AADGE Allied air-defence ground environment.

AADI Advanced area-defence interceptor.
AADP Advanced-architecture display processor.
AADRM Advanced air-breathing dual-range missile.
AADS *1* Advanced air-data system.
2 Airborne active dipping sonar.
3 Airspeed and director sensor.
4 Aircraft activity display system (program).
AADV Autonomous aerial, or air, delivery vehicle.
AAE *1* Above aerodrome/airport/airfield elevation.
2 Army Acquisition Executive (USA).
3 Asociación de Aviación Experimental (homebuilders, Spain).
4 Agrupación Astronáutica Española (Spain).
AAED Advanced airborne expendable decoy.
AAEE, A&AEE Aeroplane & Armament Experimental Establishment (UK, Martlesham Heath 1924–39, then at Boscombe Down to the present but from 1959 under different titles).
AAEEA Association des Anciens Elèves de l'Ecole de l'Air (F).
AAES *1* Association of Aerospace Engineering Societies (US).
2 American Association of Engineering Societies.
AAExS Army/Air Force Exchange Service (US, became AAFES).
AAF Army Air Force[s], full title USAAF, (June 1941–1947).
AAFARS Advanced aviation forward-area refuelling system.
AAFBU AAF Base Unit.
AAFCE Allied Air Forces Central Europe (NATO).
AAFEA Australian Airline Flight Engineers' Association.
AAFES Army and Air Force Exchange Service (US).
AAFIF Automated air-facility information file, compiled by DMA.
AAFRA Association of African Airlines.
AAFSS Advanced aerial fire-support system.
AAG, A/AG *1* Air-to-air gunnery.
2 Air Adjunct General (USAF, ANG).
AAGE Association of Aeronautical Ground Engineers (UK, 1935).
AAGF Advanced aerial gun, far-field.
AAH Advanced attack helicopter.
AAHM Alaska Aviation Heritage Museum.
AAHS American Aviation Historical Society.
AAI *1* Angle-of-approach indicator, or indication (see VASI).
2 Angle-of-attack indicator.
3 Airline Avionics Institute (US).
4 Air aid to intercept (AI was more common).
5 Air-to-air interrogator; see AAICP.
6 Arrival, or arriving, aircraft interval.
7 Airports Authority of India.
AAIB Air Accident Investigation Board (DETR, UK).
AAIC Air Accidents Investigation Commission (US).
AAICP Air-to-air interrogator control panel.
AAII Accelerated accuracy improvement initiative (GPS Navstar).
AAILS Airmedical airborne information for lateral spacing.
AAIM Aircraft autonomous integrity monitoring.
AAIP Analog autoland improvement programme.
AAIR AmSafe aviation inflation restraint.

AAIRA   Assistant Air Attaché (US).
AAL   *1* also **a.a.l.,** Above airfield level.
  2 Australian Air League.
  3 Aircraft approach limitations, UK service usage specifying minima for aircraft type in association with specified ground aids.
AALAAW   Advanced air-launched anti-armour weapon.
AALAE   Association of Australian Licensed Aircraft Engineers.
AALB   Ailes Anciennes Le Bourget (F).
AAM   *1* Air-to-air missile.
  2 Azimuth-angle measuring [unit] (Madge).
  3 Archive Air Museum (BAA).
AAMA   Association des Amis du Musée de l'Air (F).
AAME   Association of Aviation Medical Examiners (UK).
AAMP   *1* Advanced-architecture microprocessor.
  2 Advanced aircraft maneuvering program.
AAMPV   Advanced anti-materiel/personnel/vehicles (US).
AAMRL   Harry G Armstrong Aerospace Medical Research Laboratory (USAF).
AAMS   Association of Air Medical Services (US).
AAN   Airworthiness approval note.
AANCP   Advanced airborne national command post (US).
A&D   *1* Arrival and departure chart.
  2 Aerospace and defense (industry sector).
A&E   Airframe and Engine, qualified engineer.
A&F   Arming and fuzing (ICBM).
A&P   Airframe and Powerplant qualified mechanic (US).
A&R   Assemble and recycle.
AAO   *1* Air-to-air operation[s].
  2 Airborne area of operation.
  3 Air Attack Officer (firefighting).
AAP   *1* Apollo Applications Program (NASA).
  2 Acceptable alternative product (NATO).
  3 Aircraft Acceptance Park (RFC/RAF, to 1918).
AAp   Angle of approach lights.
AAPA   Association of Asia-Pacific Airlines.
AAPP   Airborne auxiliary powerplant.
AAPS   Advanced aviation protection system (EW).
AAR   *1* Aircraft accident report.
  2 Air-augmented rocket.
  3 Air-to-air refuelling.
  4 Antenna azimuth rate.
  5 Airport acceptance rate.
  6 Airport arrival rate.
  7 Active-array radar.
  8 After-action review.
AARA   Air-to-air refuelling area.
AARB   Advanced aerial refuelling boom.
AARF   Aircraft accident report form.
AARGM   Advanced anti-radiation guided missile.
AARL   Advanced applications rotary launcher (S adds system).
AAR points   Ground position of intended hookups.
AARS   *1* Automatic altitude-reporting system.
  2 Attitude/altitude retention system.
  3 Advance [not advanced] airborne reconnaissance system (BAE Systems).
AAS   *1* Airport Advisory Service (FAA).

  2 Army Aviation School (USA).
  3 American Astronautical Society.
  4 Air Armament School (UK).
  5 Advanced automation system (NAS 2).
  6 Aerospace Audiovisual Service (USAF, previously APS, APCS, 1981).
  7 Alternative access to [space] station (NASA).
AASE   Advanced aircraft survivability equipment.
AASF   *1* Advanced Air Striking Force (RAF, 1939–40).
  2 Alaskan Aviation Safety Foundation.
AASM   *1* Armement air-sol modulaire (F).
  2 Advanced ASM (1).
AASU   Aviation Army (or Armies) of the Soviet Union.
AAT   *1* Airworthiness approval tag.
  2 Airports Authority of Thailand.
AATA   *1* Associación Argentina de Transportadores Aéreos.
  2 Animal Air Transport Association (Int.).
AATC   *1* American Air-Traffic Controllers' Council.
  2 ANG/Afres Test Centre (USAF).
AATD   Aviation Applied Technology Directorate (USA).
AATF   Airport and airway trust fund.
AATG   Average annual traffic growth.
AATH   Automatic approach to hover (anti-submarine helicopters).
AATMS   Airborne air-traffic management system (Euret).
AATS   *1* Alternate aircraft takeoff systems.
  2 Access-approval test set [or system].
  3 Aviation and Air-Traffic Services.
AATT   Advanced aviation and transportation technology.
AATSR   Advanced along-track scanning radiometer.
AATTC   Advanced Airlift Tactics Training Center (USAF).
AAU   *1* Aircrew Allocation Unit (UK).
  2 Aircraft Assembly Unit (UK, WW2).
  3 Association of Aerospace Universities, 21 plus 5 commercial organisations (UK).
  4 Audio amplifier unit.
  5 Antenna adaptor unit (IFF).
  6 Articulated audio unit (threat warning).
AAv   Army Aviation (UK).
AAV   Autonomous aerial vehicle.
AAVS   Aerospace Audio-Visual Service (USAF).
AAW   *1* Anti-air warfare.
  2 Active aeroelastic wing.
  3 Aeromedical Airlift Wing (USAF).
AAWEX   Anti-air warfare exercise.
AAWG   Airworthiness Assurance Working Group.
AAWS   *1* Automatic Aviation Weather Service.
  2 Advanced anti-tank weapon system.
AAWWS   Airborne adverse-weather weapon system.
AB   Air base (USAF).
A/B, AB, a/b   *1* Afterburner.
  2 Airbrake.
ABA   American Bar Association; IPC adds International Procurement Committee.
ABAA   Australian Business Aircraft Association.
ABAC   *1* Conversion nomogram, eg for plotting great-circle bearings on Mercator projection.
  2 Association of British Aviation Consultants.
  3 Association of British Aero Clubs and Centres,

formed 1926 as Associated Light Aeroplane Clubs, re-constituted as ABAC 1946, became BLAC 1966.

ABAG   Associação Brasileira de Aviação Geral (Brazilian NBAA).

ABB   Automated beam-builder (space).

ABBCC   Airborne battlefield control center.

Abbey Hill   ESM for British warships, tuned to hostile air (and other) emissions.

ABC   *1* Advance-booking charter.
  2 Advancing-blade concept (Sikorsky).
  3 Automatic boost control.
  4 Airborne commander (SAC).
  5 See Airborne Cigar.
  6 After bottom [dead] centre.

ABCA   American, British, Canadian, Australian Standardization Loan Programme.

ABCCC   Airborne Battlefield Command and Control Center (USAF), upgraded to II and III.

ABCU   Alternate [ie alternative] braking control unit.

ABD   *1* Airborne broadband defence (ECM).
  2 See next.

ABDR   Aircraft battle damage repair.

Abeam   Across the borders European ATM(7) systems effects (Euret).

abeam   Bearing approximately 090° or 270° relative to vehicle.

Aberporth   Chief UK missile test centre, formerly administrated by RAE, on Cardigan Bay.

aberration   Geometrical inaccuracy introduced by optical, IR or similar electromagnetic system in which radiation is processed by mirrors, lenses, diffraction gratings and other elements.

ABE   *1* Air-breathing engine [S adds system].
  2 Aerodrome beacon.
  3 Arinc 429 bus emulator.

ABF   *1* Annular blast fragmentation (warhead).
  2 Auto beam forming (passive sonobuoys).
  3 Advanced bomb family (USN).

ABFAC   Airborne forward air controller.

ABFI   Association of Belgian Flight Instructors.

ABG   Air Base Group (USAF).

ABGS   Air Bombing and Gunnery School (RAF).

ABI   *1* Advanced[d] boundary information.
  2 Airborne broadcast intelligence.

ABIA   Associação Brasileira das Industrias Aeronauticas.

ABICS, Abics, Ada-based interception [or integrated] control system.

ABIHS   Airborne broadcast intelligence hardware system (hazard avoidance).

ABILA   Airborne instrument landing approach.

ab initio   Aircraft or syllabus intended to train pupil pilot with no previous experience.

ABIS   All-bus instrumentation system.

ABITA   Association Belge des Ingénieurs et Techniciens de l'Aéronautique et de l'Astronautique.

ABL   *1* Airborne laser.
  2 Atmospheric boundary layer.
  3 Armoured box launcher.

ablation   Erosion of outer surface of body travelling at hypersonic speed in an atmosphere. An ablative material (ablator) chars or melts and is finally lost by vaporisation or separation of fragments. Char has poor thermal conductivity, chemical reactions within ablative layer

may be endothermic, and generated gases may afford transpiration cooling. Main mechanism of thermal protection for spacecraft or ICBM re-entry vehicles re-entering Earth atmosphere.

AB/LD   Airbrakes/lift dumpers.

ABM   *1* Apogee boost motor.
  2 Anti-ballistic missile, with capability of intercepting re-entry vehicle(s) of ICBM.
  3 Abeam (ICAO code).
  4 Air-burst munition.
  5 Aviation business machine.
  6 Asychronous balanced mode.

abm   Abeam.

ABMA   US Army Ballistic Missile Agency, 1 February 1956, Huntsville.

ABMD   Anti- [or advanced] ballistic missile defense; I adds initiative, P program, S system and T treaty (US).

ABN   Airborne.

ABn   Aerodrome beacon.

Abney level   A spirit-level clinometer (obs.).

abnormal spin   Originally defined as spin which continued for two or more turns after initiation of recovery action; today obscure.

A-bomb   Colloquial term for fission bomb based upon plutonium or enriched uranium (A = atomic).

abort   *1* To abandon course of action, such as takeoff or mission.
  2 Action thus abandoned, thus an *.

abort drill   Rehearsed and instinctive sequence of actions for coping with emergency abort situation; thus, RTO sequence would normally includes throttles closed, wheel brakes, spoilers, then full reverse on all available engines consistent with ability to steer along runway.

above-wing nozzle   Socket for gravity filling of fuel tanks.

ABP   Aerodynamic balance panel.

AB/PM   Air-base protective measure (US).

ABPNL   Association Belge des Pilotes et Navigants techniciens de Ligne.

A-BPSK   Aeronautical binary phase-shift keying.

ABR   *1* Amphibian bomber reconnaissance.
  2 Agile-beam radar.

abradable seal   Surface layer of material, usually non-structural, forming almost gas-tight seal with moving member and which can abrade harmlessly in event of mechanical contact. Some fan and compressor-blade** are silicone rubber with 20% fill of fine glass beads.

ABRC   Advisory Board for the Research Councils (UK).

Abres, ABRES   Advanced ballistic re-entry system[s].

ABRU   Advanced bomb rack unit.

ABRV   *1* Advanced ballistic re-entry vehicle.
  2 Abbreviation.

ABS   *1* Acrylonitrile butadiene styrene, strong thermosetting plastic material.
  2 Anti-blocking system.
  3 Anti-skid brake, or braking, system.

abs   Absolute scale of units.

ABSA   Advanced base support aircraft.

abscissa   *1* In co-ordinate geometry, X-axis.
  2 X-axis location of a point.

ABSL   Ambient background sound level.

absolute aerodynamic ceiling   Altitude at which maximum rate of climb of aerodyne, under specified conditions, falls to zero. Usually pressure altitude amsl, atmosphere ISA, aircraft loading 1 g, and weight must be specified. Except for zoom ceiling, this is greatest height attainable.

absolute alcohol　Pure ethyl alcohol (ethanol) with all water removed.

absolute altimeter　Altimeter that indicates absolute altitude; nearest approach to this theoretical ideal is laser altimeter, closely followed by instruments using longer EM wavelengths (radio altimeter).

absolute altitude　Distance along local vertical between aircraft and point where local vertical cuts Earth's surface.

absolute angle of attack　Angle of attack measured from angle for zero lift (which with cambered wing is negative with respect to chord line).

absolute ceiling　Usually, absolute aerodynamic ceiling.

absolute density　Theoretical density (symbol ρ) at specified height in model atmosphere.

absolute fix　Fix (2) established by two or more position lines crossing at large angles near 90°.

absolute humidity　Humidity of local atmoshere, expressed as gm^{-3}.

absolute inclinometer　Inclinometer reading attitude with respect to local horizontal, usually by precise spirit level or gryo.

absolute optical shaft encoder　Electromechanical transducer giving coded non-ambiguous output exactly proportional to shaft angular position.

absolute pressure　Gauge pressure plus local atmospheric pressure.

absolute system　Of several ** of units, or for calculating aerospace parameters, most important is reduction of aerodynamic forces to dimensionless coefficients by dividing by dynamic pressure head $\frac{1}{2}\rho V^2$.

absolute temperature　Temperature related to absolute zero. Two scales in common use: absolute (°A) using same unit as Fahrenheit or Rankine scale (contrary to SI), and Kelvin (K) using same unit as Celsius scale.

absolute zero　Temperature at which all gross molecular (thermal) motion ceases, with all substances (probably except helium) in solid state. 0K = –273.16°C.

absorbed dose　Energy imparted by nuclear or ionising radiation to unit mass of recipient matter; measured in rads.

absorption band　Range of frequencies or wavelengths within which specified EM radiation is absorbed by specified material; narrow spread(s) of frequencies for which absorption is at clear maximum.

absorption coefficient　*1* In acoustics, percentage of sound energy absorbed by supposed infinitely large area of surface or body.

2 In EM radiation, percentage of energy that fails to be reflected by opaque body or transmitted by transparent body (in case of reflection, part of radiation may be scattered). Water vapour is good absorber of EM at long wavelengths at which solar energy is reflected from Earth's surface, so ** for solar energy varies greatly with altitude.

absorption process　Chemical production of petrols (gasolines) by passing natural gas through heavy hydrocarbon oils.

absorption cross-section　Absorption coefficient of radar target expressed as ratio of absorbed energy to incident energy.

ABT　*1* About (ICAO).

2 Air-breathing threat[s].

ABTA　Association of British Travel Agents, usually pronounced Abta.

ABTJ　Aferburning turobjet.

ABU　Aviation bird unit (airport).

ABV　*1* Air-bleed valve.

2 Above (ICAO).

3 Alternative boost vehicle (BMDS).

ABW　Air Base Wing (USAF).

AC　*1* Aligned continuous (FRP1).

2 Aircraft commander.

3 Army co-operation (UK).

4 Aerodynamic centre (a.c. is preferred).

5 Or A_c, acceleration command.

6 Acquisition cycle.

7 Advisory circular.

8 Aircraft characteristic (JAR).

9 Automated circumferential (riveting).

10 Airworthiness circular.

11 Air carrier.

12 Air conditioner.

13 Airman certification (US).

14 Active component.

15 Area coverage (Satcoms).

16 Analyst console.

17 Approach control.

Ac　Alto-cumulus cloud.

A/C　Approach Control (FAA style).

a.c.　*1* Alternating current (electricity).

2 Aerodynamic centre of wing or other surface.

a/c　Aircraft (FAA = acft).

ACA　*1* Air Crew Association (UK).

2 Aerobatic Club of America.

3 Advanced cargo aircraft.

4 Ammunition-container assembly.

5 Airspace coordination area (GFS).

6 Arms Control Association (US).

7 Address compression algorithm.

ACAA　*1* Air-Carrier Association of America.

2 Australian Civil Aviation Authority.

3 Academic Center for Aging Aircraft (universities + DoD).

ACAAI　Air Cargo Agents Association of India.

ACAAR　Aircraft communications addressing and reporting [s adds system].

ACAB　Air Cavalry Attack Brigade (USA).

ACAC　*1* Arab Civil Aviation Council (Int).

2 Aircooled air cooler.

ACAMS　Aircraft communications and management system.

ACAN　Amicale des Centres Aéronautiques Nationaux (F).

ACAP　*1* Aviation Consumer Action Project (US, 1971–).

2 Advanced composite aircraft (helicopter) program (US).

ACARE　Advisory Council for Aeronautical Research in Europe.

Acars, ACARS　*1* Aircraft communications and automatic reporting system; most common interpretation.

2 Airborne communication and recording system.

3 Arinc communications addressing and reporting system.

4 Airline communication and reporting system (Rockwell Collins).

ACAS   *1* Air-cycle air-conditioning system.

2 Assistant Chief of the Air Staff.

3 Aluminium core, aluminium skin.

4 Airborne collision-avoidance system.

5 Aircraft collision-avoidance system (ICAO is currently *II; 2002).

6 Advisory, Conciliation and Arbitration Service (UK).

7 Airfield chemical-alarm system.

ACASS   Advanced close air support system.

Acat, ACAT   Association of Colleges of Aerospace Technology (UK).

Acatt   Army combined arms team trainer (Cobra/Apache/Scout).

ACAVS   Advanced cab and visual system.

ACBM   Additional conventional-bomb module.

ACC   *1* Area (or aerodrome) control centre.

2 Active clearance control.

3 Air Combat Command (USAF, from 1 June 1992, HQ Langley AFB).

4 Air Co-ordinating Committee (US, military/civilian, 1945–60).

5 Axis-controlled carrier.

6 Avionics computer control.

7 Aeronautical Chamber of Commerce (US, 1921 on).

8 Automatic code change (IFF).

AcC   Alto cumulus castellanus.

ACCA   *1* Air Courier Conference of America.

2 Air Charters Carriers' Association.

ACCC   Australian Competition and Consumer Commission.

ACCE   Air Command and Control Element (RAF).

accelerated flight   Although aircraft that gains or loses speed is accelerating in horizontal plane, term should be used only for acceleration in plane perpendicular to flight-path, esp. in vertical plane.

accelerated history   Test record of specimen subjected to overstress cycling, overtemperature cycling or any other way of 'ageing' at abnormally rapid rate.

accelerated stall   Stall entered in accelerated flight. As common way of inducing stall is to keep pulling up nose, it might be thought all stalls must be accelerated, but in gradual entry flight path may be substantially horizontal. "High-speed stall" is possible in violent manoeuvre because acceleration in vertical plane requires wing to exceed stalling angle of attack. Stall-protection systems are generally designed to respond to rate of change of angle of attack close to stalling angle, so stick-pusher (or whatever form system takes) is fired early enough for critical value not to be reached.

accelerate-stop   Simulation of RTO by accelerating from rest to V_1 or other chosen speed and immediately bringing aircraft to rest in shortest possible distance; hence * distance.

accelerating pump   In piston engine carburettor, pump provided to enrich mixture each time throttle is opened, to assist acceleration of engine masses.

accelerating well   Originally receptacle for small supply of fuel automatically fed into choke tube by increased suction when throttle was opened. Later became small volume connected by bleed holes to mixture delivery passage. Usually absent from modern engines.

acceleration   Rate of change of velocity, having dimensions LT^{-2} and in SI usually measured in $ms^{-2} = 3.28084$ fts^{-2}. As velocity is vector quantity, * can be imparted by changing trajectory without changing speed, and this is meaning most often applied in aerospace.

acceleration control unit   Major element in engine fuel control unit, usually a servo sensing compressor delivery pressure to make fuel flow keep pace with demand for extra fuel to accelerate engine as throttle is opened.

acceleration datum   Engine N_1 corresponding to typical approach power, used in engine type testing for 2½ min. rest period before each simulated overshoot acceleration (repeated 8 or 15 times).

acceleration errors   Traditional direct-reading magnetic compass misreads under linear acceleration (change of speed at constant heading) and in turn (apparent vertical acceleration at constant speed); former is a maximum on E–W headings, increasing speed on W heading in N hemisphere indicating apparent turn to N; Northerly Turning Error (N hemisphere) causes simple compass to lag true reading, while Southerly Turning Error results in over-reading. Simple suction horizon misreads under all applied accelerations, most serious under linear positive acceleration (t-o or overshoot), when indication is falsely nose-up and usually right-wing down (with clockwise rotor, indication is diving left turn).

acceleration manoeuvre   High-speed yo-yo.

acceleration-onset cueing   Simulator technique in which real acceleration is initially imparted and then reduced, usually to zero, at a rate too low for body to notice; thus trainee can even believe in sustained afterburner takeoff.

acceleration stress   Physical deformation of human body caused by acceleration, esp. longitudinal.

acceleration tolerance   See *g-tolerance*.

accelerator   Device, not carried on aircraft, for increasing linear acceleration on takeoff; original name for *catapult*.

accelerator pump   Accelerating pump.

accelerometer   Device for measuring acceleration. INS contains most sensitive * possible. Usually one for each axis, arranged to emit electrical signal proportional to sensed acceleration. Recording * makes continuous hard-copy record of sensed acceleration, or indicates peak. Direct reading * generally fitted in test flying but not in regular aircraft operation.

Accept   Automated cargo clearance enforcement processing technique, computerised inspection of selected items only, to help identify high-risk items (US customs).

acceptable alternative product   One which may be used in place of another for extended periods without technical advice (NATO).

acceptance   One meaning is agreement of air-traffic control to take control of particular aircraft. Hence * rate is (1) actual rate in one-hour period, or (2) the maximum that can safely be handled.

acceptance test   Mainly historic, test of hardware witnessed by customer or his designated authority to demonstrate acceptability of product (usually military). Schedule typically covered operation within design limits, ignoring service life, fatigue, MTBF, MMH/FH and fault protection.

acceptance trials   Trials of flight vehicle carried out by eventual military user or his nominated representative to determine if specified customer requirement has been met.

access door　Hinged door openable to provide access to interior space or equipment.

access light　Until about 1940, light placed near airfield boundary indicating favourable area over which to approach and land.

access panel　Quickly removable aircraft-skin panel, either of replaceable or interchangeable type, removed to provide access to interior.

accessories　Replaceable system components forming functioning integral part of aircraft. Except in general aviation, term is vague; includes pumps, motors and valves, excludes such items as life-rafts and furnishing. In case of fuel system (for example) would include pumps, valves, contents gauges and flowmeters, but not tanks or pipelines.

accessory drive　Shaft drive, typically for group of rotary accessory units, from main engine, APU, EPU, MEPU or other power source.

access time　*1* Time required to access any part of computer program (typically 10^{-3} to 10^{-9} s).

2 Time required to project any desired part of film or roller map in pictorial cockpit display (typically about 3 s).

3 Time necessary to open working section of tunnel and reach model installed (typically about 1,000 s, but varies greatly).

ACCID　Notification of aircraft accident (ICAO).

accident　Incident in life of aircraft which causes significant damage or personal injury (see *notifiable*).

accident-protected recorder　Flight recorder meeting mandatory requirements intended to ensure accurate playback after any crash.

accident rate　In military aviation most common parameter is accidents per 100,000 flying hours; other common measures are fatal accidents, crew fatalities and aircraft write-offs on same time basis, usually reckoned by calendar year. In commercial aviation preferred yardsticks are number of accidents (divided into notifiable and fatal) per 100 million passenger-miles (to be replaced by passenger-km) or per 100,000 stage flights, either per calendar year or as five-year moving average. In General Aviation usual measure is fatal accidents per 100,000 take-offs.

accident recorder　Device, usually self-contained and enclosed in casing proof against severe impact, crushing forces and intense fire, which records on magnetic tape, wire, or other material, flight parameters most likely to indicate cause of accident. Typical parameters are time, altitude, IAS, pitch and roll attitude, control-surface positions and normal acceleration; many other parameters can be added, and some ** on transports are linked with maintenance recording systems. Record may continuously superimpose and erase that of earlier flight, or recorder may be regularly reloaded so that record can be studied.

ACCIS　Automated command and control information system (NATO).

ACCISRC　See AC²ISRC [alphabetically, AC two . . .].

acclrm　Accelerometer.

accompanying cargo/supplies　Cargo and/or supplies carried by combat units into objective area.

ACC-R　Area control centre radar.

accredited medical conclusion　Decision by licensing authority on individual's fitness to fly, in whatever capacity.

accredited sortie　One that puts bomb on target.

ACCS　*1* Airborne Command and Control Squadron (USAF, NATO).

2 Air command and control system (NATO).

3 Air-cycle [modular] cooling system.

4 Airborne computing and communications system.

ACCSA　Allied Communications and Computer Security Agency (NATO).

ACCTS　Aviation Co-ordinating Committee for telecommunications Services (US).

accumulator　*1* Electrical storage battery, invariably liquid-electrolyte and generally lead/acid.

2 Device for storing energy in hydraulic system, or for increasing system elasticity to avoid excessive dynamic pressure loading. Can act as emergency source of pressure of fluid, damp out pressure fluctuations, prevent incessant shuttling of pressure regulators and act as pump back-up at peak load.

3 Device for storing limited quantity of fuel, often under pressure, for engine starting, inverted flight or other time when normal supply may be unavailable or need supplementing.

4 Portion of computer central processor or arithmetic unit used for addition.

accuracy jump　Para-sport jump in which criterion is distance from target.

accuracy landing　In flying training or demonstration, dead-stick landing on designated spot (= spot landing).

accuracy of fire　Linear distance between point of aim and mean point of strikes.

Accu-Time　Magnetron circuit capable of being precisely tuned to different wavelengths.

ACD　*1* Automatic [or automated] chart display.

2 Aeronautical Charting Division (NOAA).

ACDA　Arms Control and Disarmament Agency.

ACDAC　Associación Colombiana de Aviadores Civiles.

ACDB　Airport characteristics data-bank (ICAO).

AC/DC　Air refuelling tanker able both to dispense and receive fuel in flight (colloq).

ACDO　Air-carrier district office (US).

ACDP　Armament control and display panel.

ACDS　*1* Automatic countermeasures [or computer-controlled] dispenser [or dispensing] system.

2 Assistant Chief of the Defence Staff (UK).

3 Air-, or advanced, combat direction system.

ACDTR　Airborne central data tape recorder (now generally called RSD).

ACE　*1* Automatic check-out equipment.

2 Association of Consulting Engineers (UK).

3 Air combat evaluator (CIU software).

4 Aircrew (or accelerated copilot) enrichment.

5 Allied Command Europe (NATO).

6 Association des Compagnies Aériennes de la Communauté Européenne.

7 Advanced crew-station evaluator (helicopter).

8 Automated center for electronics, computer control of all phases of circuit design, development, assembly and test (Lockheed).

9 'Technical acknowlegement' (ACARS code).

10 Actuator control electronics.

11 Advanced-certification equipment.

12 Aerospace Committee (BSI).

13 Avionics capabilities enhancement.

14 Analysis [and] control element.

15 Agile control experiment.

16 Aerobatic certification evaluator.

17 Aviation Career Education, or Educator (US).

18 Aviation Combat Element of MEU

19 Autonomous combat [manoeuvres] evaluation.

20 Air-combat emulator.

ace   Combat pilot with many victories over enemy aircraft. WW2 USAAF scores included strafing (air/ground) "victories". Number required to qualify has varied, but in modern world is usually five confirmed in air combat.

ACEA   Action Committee for European Aerospace (international shop-floor pressure group).

ACEBP   Air-conditioning engine bleed pipe.

ACEC   Ada-compiler evaluation capability.

ACEE   Aircraft energy efficiency (NASA).

ACEL   Air Crew Equipment Laboratory (USN).

ACEM   Aerial camera electro-optical magazine.

ACER   Air Corps Enlisted Reserves (USA).

ACES   *1* Advanced-concept escape system.

2 Advanced-concept ejection seat.

3 Air-carrier engineering support.

4 Aerial combat enhanced [or evaluation, or evaluator] simulation.

5 Advanced carry-on Elint/ESM suite.

6 Adaptation controlled environment system (ATC).

ACeS   Asia cellular satellite system.

ACESNA   Agence Centrafricaine pour la Sécurité Navigation Aérienne.

ACESS   Aircraft computerized equipment support system.

ACET   *1* Air-cushion equipment transporter, for moving aircraft and other loads over soft surfaces, especially over airbase with paved areas heavily cratered (S adds 'system').

2 Automatic cancellation of extended [radar] target[s].

acetate   Compound or solution of acetic acid and alkali. * dope is traditionally based upon acetic acid and cellulose; was used for less inflammable properties (see *nitrate dope*).

ACETEF   Air-combat environmental test and evaluation facility (USA).

acetone $CH_3.CO.CH_3$, inflammable, generally reactive chemical, often prepared by special fermentation of grain, used as solvent. Basis of many 'dopes' and 'thinners'.

ACETS, Acets   Air-cushion equipment transportation system (for post-attack airfields).

acetylene   CH.CH or C_2H_2, colourless gas, explosive mixed with air or when pressurized but safe dissolved in acetone (trade name Prestolite and others). Burns with oxygen to give 3,500°C flame for gas welding; important ingredient of plastics.

Aceval   Air-combat evaluation.

ACEX   Air-coupled electronic transducer.

ACF   *1* Aircraft Components Flight (RAF).

2 Advanced common flightdeck.

3 Area control facility.

ACFC   Aircooled flight-critical.

AC⁴ISR   Adaptive C⁴ISR.

ACFR   Australian Centre for Field Robotics.

acft   Aircraft (ICAO), also loosely ACFT.

ACG   *1* Austro Control GmbH (Austria).

2 Airfield Construction Group (RAF, WW2).

ACGF   Aluminium-coated glassfibre (chaff).

ACGS   Aerospace Cartographic and Geodetic Service (USAF, formerly MAC).

ACH   *1* Advanced Chain Home (UK WW2).

2 Advanced compound helicopter.

ACH/GD   Aircraft-hand, General Duties, "lowest form of life" in RAF (WW2).

achieved navigation performance   The measure of uncertainty in the position element.

achromatic   Transmitting white light without diffraction into special colours; lens system so designed that sum of chromatic dispersions is zero.

ACI   *1* Air Council Instruction (UK).

2 Airports Council International; suffixes denote regions, thus – NA = North America.

3 Avionics caution indicator.

4 Armament control indicator.

ACID   Aircraft identification.

acid engine   Rocket engine in which one propellant is an acid, usually RFNA or WFNA.

acid extraction   Stage in production of lubricating oils in which sulphuric acid is used to extract impurities.

Acids, ACIDS   *1* Automated communications and intercom distribution system.

2 Air conformal ice detection system.

ACI-E   Airports Council International – Europe.

ACINT, Acint   Active acoustic intelligence.

Acips   Airfoil and cowl ice protection system.

ACIS   *1* Advanced CCD imaging spectrometer.

2 Armament, or advanced, control/indicator set.

3 Advanced cabin interphone system.

ACJ   Advisory circular, Joint.

ACK   Acknowlegement of uplink (Acars).

ack   Acknowlegement (ICAO).

Ack-ack   Anti-aircraft (UK WW1, became passé in WW2).

Ackeret formula   There are many, most important being, for thin wing above M_{DET}, regardless of camber, $C_L = 4\alpha / \sqrt{M^2 - 1}$.

Ackeret theory   First detailed treatment [1925] for supersonic flow past infinite wing, suggesting sharp leading and trailing edges and low t/c ratio; favoured profiles were biconvex or trapezium (parallel double wedge).

acknowledged program   A special-access program whose existence is admitted.

acknowledgement   Confirmation from addressee that message has been received and understood.

ACL   *1* Anti-collision light.

2 Allowable cabin load.

3 Aeronautical-chart legend.

4 Altimeter check location.

5 Air Cadet League of Canada.

Aclaim   Airborne coherent lidar for air inflight measurement.

Aclant   Allied Command, Atlantic (NATO).

ACLD, ACld   Above cloud[s].

ACLG   Air-cushion landing gear; underside of aircraft is fitted with inflatable skirt to contain ACV type cushion, suitable for all land, marsh, sand or water surfaces.

Aclics   Airborne communications location, identification and collection system (USA).

aclinic line   Isoclinic line linking all points whose angle of dip is zero.

Aclos, ACLOS   Automatic command to line of sight.

ACLS   *1* Automatic carrier landing system (Bell/USN).

    2 Air-cushion landing system.

ACLT   *1* Aircraft-carrier landing training.

    2 Actual calculated landing time.

ACM   *1* Air-combat manoeuvring, or manoeuvre [US maneuver]; EST adds expert-systems trainer, I instrumentation, R range and S simulator.

    2 Air-cycle machine.

    3 Anti-armour cluster munition.

    4 Air Chief Marshal (not normally abbrev.).

    5 Air-conditioning module.

    6 Advanced cruise missile (USAF).

    7 Aircraft-condition monitoring.

    8 Aircraft manual.

    9 Attitude-control module.

    10 Air Commercial Manual (US Bureau of Air Commerce).

    11 Aircraft-cabin mattress.

ACMA   Advanced concepts and material applications (MoD, UK).

ACME   Advanced-core military engine.

ACMF   Aircraft-condition monitoring function.

ACMG   Air-Cargo Management Group (US).

ACMI   *1* Air-combat manoeuvring instrumentation, or installation.

    2 Aircraft, crew, maintenance and insurance.

ACMP   Alternating-current motor/pump.

ACMR   Air-combat manoeuvring range.

ACMS   *1* Avionics, or advanced, control and management system.

    2 Aircraft, also airport, condition monitoring system.

    3 Armament control and monitoring system.

ACMT   Advanced cruise-missile technology.

ACN   *1* Aircraft Classification Number (ICAO proposal for pavements).

    2 Airborne communications node, C4ISR, now called *AJCN*.

    3 Academia Cosmologica Nova (G).

ACNDT   Advisory Committe for Non-Destructive Testing.

ACNIP   Auxiliary, or advanced, CNI panel.

ACNS   Assistant Chief of the Naval Staff (UK).

ACNSS   Advanced com/nav/surveillance system.

ACO   *1* Airspace control, or coordination, order.

    2 Airborne Control, or Communications, Officer.

    3 Aerosat coordination office.

    4 Advanced concepts of applications.

    5 Air-combat order.

ACOC   Aircooled oil cooler.

acorn   *1* Streamlined body or forebody added at intersection of two aerodynamic surfaces [e.g. fin/tailplane] to reduce peak suction.

    2 Streamlined body introduced at intersection of crossing bracing wires to prevent chafing.

    3 Streamlined fairing over external DF loop.

acorn valve   Small thermionic valve (radio tube) formerly added to VHF or UHF circuit to improve efficiency.

ACOS   Assistant Chief of Staff.

Acost   Advisory Committee on Science and Technology (UK).

ACostE   Association of Cost Engineers (UK).

Acoubuoy   Acoustic sensor dropped by parachute into enemy land area.

acoustic   Associated with sound, and hence material vibrations, at frequencies generally audible to human beings.

acoustic absorption factor   Rate at which acoustic energy is incident on a surface divided by that measured on inner face of material. Varies greatly with frequency.

acoustic delay line   In computer or other EDP device, subsystem for imparting known time delay to pulse of energy; typically closed circuit filled with mercury in which acoustic signals circulate (obs.).

acoustic feedback   Self-oscillation in radio system caused by part of acoustic output impinging upon input.

acoustic impedance   Resistance of material to passage of sound waves, measured in acoustic[al] ohms.

acoustics   In ASW, sonar and other sensing systems relying on underwater sound; thus * operators, * displays.

acoustic splitter   Streamlined wall introduced into flow of air or gas, parallel to streamlines, for acoustic purposes. Usually inserted to reduce output of noise, for which purpose both sides are noise-absorbent. Many are radial panels and concentric long-chord rings (open-ended cylinders).

acoustic tube   Miniature acoustic/electric transducer which has replaced carbon or other types of microphone in aircrew headsets.

ACP   *1* Airborne [or airlift] command post.

    2 Anti-Concorde Project.

    3 Altimeter check point.

    4 Armament control panel.

    5 Africa, Caribbean, Pacific.

    6 Audio control panel, or convertor processor.

    7 Aerosol collector and pyroliser.

    8 Aluminised composite propellant.

Acp   Acceptance.

ACPA   Adaptive-controlled phased array.

ACP(C)   Automatic communications processor (control).

ACPL   Atmospheric Cloud Physics Laboratory.

AC-plonk   AC2 (derogatory reference to this low rank in RAF, 1941–50).

ACPMR   Automatic communications processor and multiband radio.

ASPC   (See *CSPA*.

ACPT, Acpt   Accepted.

acquisition   *1* Act of visually identifying, and remembering location of, object of interest (specific ground or aerial target).

    2 Detection of target by radar or other sensor (plus, usually, automatic lock-on and subsequent tracking).

    3 Detection and identification of desired radio signal or other broadcast emission.

    4 Act of reaching desired flight parameter, such as heading, FL or IAS, or desired point or axis in space such as ILS G/S or LOC (see *capture*).

acquisition round   AAM (1) without propulsion, and usually without wings or fins, carried to provide practice in homing head lock-on.

acquisition scan window   3-D block of airspace into which a VAV can easily be guided, wherein CARS or UCARS acquires it and feeds it to the RIW.

ACR   *1* Aerial combat reconnaissance.

2 Air [or airfield, or approach] control radar.

3 Advanced cargo rotorcraft.

4 Active cockpit rig.

5 Avionics communication[s] router.

ACRA, Acra   Airlift Concepts and Requirements Agency (USA/USAF).

ACRC   Aircrew Reception Centre, (UK, WW2).

acre   Old Imperial (FPS) unit of land surface area, equal to 0.40469 ha (1 ha = 2.47105 acres). For covered area (factory buildings etc) usual SI unit is m^2 (= 0.000247105 acre, so 1-acre plant = 4,047 m^2).

acreage   Superficial area of flight vehicle, especially spacecraft or aerospace craft, as distinct from nose and other parts that need ablative or other special protection.

Ac-Rep   Representative, usually of country of manufacture, accredited to accident investigation.

Acris   Air control recording and information system.

ACRM   Aircrew resource management.

acrobatics   Usual term is aerobatics.

ACRR   Airborne communications restoral relay.

ACRS   Air Crew Refresher School (RAF WW2).

ACRT   Additional cross-reference table.

ACRV   Assured crew-rescue vehicle.

ACRW   Aircraft [aeroplane] with circular rotating wing.

acrylics[s]   Thermosetting plastic[s], usually transparent, based on polymerised esters of * acid; original tradename Perspex (ICI, UK) and Plexiglas (Rohm & Haas, US). Since 1950 improved transparencies result from stretching moulded part prior to setting.

ACS   *1* Attitude, or armament, or active, or audio, or auxiliary, control system.

2 Aeroflight control system, for use by spacecraft within atmosphere.

3 Air-conditioning system.

4 Air Commando Squadron (USAF).

5 Air Control Squadron.

6 Aircraft Certification Service (FAA).

7 Airframe consumable spares.

8 Advanced crew station.

9 Aerial [ie, airborne] common sensor (USA, USN).

10 Air-combat simulator.

11 Assembly & Command Ship (Sea Launch).

ACSA   Allied Communications Security Agency (NATO).

ACSC   Air Command and Staff College (USAF, Maxwell AFB).

ACSE   Access control and switching, or signalling, equipment (Aerosat ground station).

ACSG   *1* Armament computer symbol generator.

2 Aeronautical communications sub-group.

AcSl.   Altocumulus standing lenticular.

ACSM   *1* Advanced conventional standoff missile.

2 Assemblies, components, spare parts and materials (NATO).

ACSR   Active control of structural response.

ACSS   African Centre for Strategic Studies.

ACSSB   Amplitude-commanded single-sideband.

ACT   *1* Actual temperature; ISA ± deviation.

2 Active-control technology.

3 Air-combat tactics.

4 Anti-communications threat.

5 Atlas composing terminal.

6 Airborne crew trainer.

7 Advanced composite technology.

8 Additional centre tank.

9 Advanced-coverage tool.

10 ASR crew trainer.

11 Active, activated, activity.

12 Analysis control team.

13 Allied Command Transformation, strategic force created 2003 in NATO with HQ in US.

ACTC   Air Commerce Type Certificate (US 1934–38).

ACTD   Advanced-concept technology demonstrator, or demonstration.

ACTEW, Actew   Acoustic charged transport electronic warfare, low-cost decoy system in which signals are slowed as they pass across GaAs.

ACTI   Air-combat tactics instructor.

ACTIFT, Actift   Advanced cockpit technology and instrument-flying trainer.

actinic ray   EM radiation, such as short-wave length end of visible spectrum and ultraviolet, capable of exerting marked photochemical effect.

actinometer   Instrument measuring radiation intensity, esp. that causing photochemical effects, eg sunlight; one form measures degree of protection afforded from direct sunlight, while another (see *pyrgeometer*) measures difference between incoming solar radiation and that reflected from Earth.

action   Principal moving mechanism of automatic weapon; in gun of traditional design typically includes bolt, trigger, sear, bent, striker, extractor and ammunition feed.

Actions   Air-combat training interoperable with NATO systems, integrated with *Raids* (see *Units*).

action time   Duration in seconds of significant thrust imparted by solid-propellant or hybrid rocket. Several definitions, most commonly the period between the point at which thrust reaches ten per cent of maximum (or average maximum) and that at which it decays through same level. This period is always shorter than actual duration of combustion, but longer than burn time. Symbol t_a.

action time average chamber pressure, or thrust   Integral of chamber pressure or thrust versus time taken over the action time interval divided by the action time; symbols P_c, F_a.

Actis   Advanced compact thermal-imaging system.

Activ   Air-combat training instrumented virtual range.

ACTIVE, Active   Advanced control technology for integrated vehicles.

activate   To translate planned organisation or establishment into actual organisation or establishment capable of fulfilling planned functions.

activated carbon   Organically derived carbon from which all traces of hydrocarbons have been removed; highly absorbent and used to remove odours and toxic traces from atmospheres; also called activated (or active) charcoal.

active   *1* General adjective for a device emitting radiation (as distinct from passive). Also see * munition.

2 The runway(s) in use.

active aerodynamic braking   Reversed propulsive thrust.

active aeroelastic wing   Instead of trying to prevent flexure and twist the AAW seeks to exploit it. Special F/A-18 works by LE flap control.

active air defence   Direct action against attacking aircraft, as distinct from passive AD.

active clearance control Technique for maintaining an extremely small gap between fixed and rotating components of a machine (for example, by blowing bleed air around a turbine casing in a gas-turbine engine).

active controls Flight-control surfaces and associated operative system energised by vertical acceleration (as in gust) and automatically deflected upwards and/or downwards, usually symmetrically on both sides of aircraft, to alleviate load; thus active ailerons or tailplanes operate in unison to reduce vertical acceleration.

active countermeasures Countermeasures requiring friendly emissions. Subdivisions include microwave, IR and electro-optical.

active decoy round Rocket-launched parawing carrying an EW jammer.

active electronically scanned array Radar, especially for fighter, whose antenna is fixed; scanning is achieved by a progressive phase-shift from one side of the antenna to the other (or from bottom to top), the greater the shift the larger the steering angle θ. Normally slight upward tilt deflects head-on main-lobe reflection to enhance stealth characteristics.

active guidance Active homing guidance.

active homing guidance Guidance towards target by sensing target reflections of radiation emitted by homing vehicle.

active jamming ECM involving attempted masking or suppression of enemy EM signals by high power radiation on same wavelengths.

active landing gear One in which the full suspension force is subject to control.

active loading LO (Stealth) generates signal to cancel that detected by hostile radar.

active magnetic bearing One which holds shaft in position by electro-magnetic field.

active material Many meanings, eg: 1, phosphor, such as zinc phosphate or calcium tungstate, on inner face of CRT; 2, parts of electric storage battery that participates in electrochemical reaction.

active missile Fire-and-forget missile carrying its own active guidance.

active munition One having immediate effect (as distinct from a mine, which is passive).

active noise control Noise-suppressing or countering systems triggered by noise itself and using sound energy against itself.

active pilot On long-haul, the pilot fully alert to FMGS, navigation and other inputs.

active runway Runway currently in use (implied that flying operations are in progress).

active satellite Satellite with on-board electrical power sufficient to broadcast or beam its own transmissions.

active visual camouflage See *counter-illumination*.

activity factor See *blade activity factor*.

ACTP Advanced Computer Technology Project (UK).

ACT-R Air-combat training, rangeless.

Actram Advisory committee on transport of radioactive material.

ACTS Advanced communications technology satellite.

ACT-TO Actual time and fuel state at takeoff.

actual ground zero Point on surface of Earth closest to centre of nuclear detonation.

actuator Device imparting mechanical motion, usually over restricted linear or rotary range and with intermittent duty or duty cycle.

actuator remote terminal Connects the powered flight-control unit in a distributed flight-control system, databus feeding through digital processor to close pilot analog loop and provide redundancy.

AC2 Aircraftman, 2nd Class (RAF, most numerous WW2 rank).

AC²ISRC Aerospace Command and Control Intelligence, Surveillance, Reconnaissance Center (USAF).

ACU *1* Gas-turbine acceleration control unit.

   2 Avionics [or autopilot, or audio, or auxiliary, or acceleration, or apron, or airborne, or adaptive, or annotation, or antenna] control unit.

   3 Airborne computer unit.

acute dose Total radioactive dose received over period so short that biological recovery is not possible.

ACV *1* Air-cushion vehicle.

   2 Escort (or auxiliary) aircraft carrier (CVE from 1943).

   3 Achieved coverage volume (satellite antenna).

ACVC Ada compiler validation capability.

ACW *1* Aircraft control and warning.

   2 Air Control Wing (USAF Awacs).

ACWAR, Acwar Agile continuous-wave acquisition radar.

ACyc Anti-cyclone, anti-cyclonic.

ACZ Airfield, or aerodrome, control zone.

AD *1* Airworthiness Directive (national certifying authorities).

   2 Advisory route (FAA).

   3 Aligned discontinuous (FRP).

   4 Aerodrome (ICAO).

   5 Air defence.

   6 Area-denial munition.

   7 Aerial delivery (ramp-door position).

   8 Autopilot disconnect.

   9 Air diagram, followed by number.

   10 Armament Division (AFSC).

   11 Air Division (USAAF, USAF).

   12 Accidental damage.

   13 Aerodynamic disturbance (which see).

   14 Ashless dispersant.

   15 Assistant Director (UK).

   16 Administrative domain.

A/D *1* Air defence.

   2 Alarm and display.

   3 Aerodrome (common UK usage).

   4 Analog/digital.

Ad Aerodrome (DTI, CAA).

ADA *1* Advisory area.

   2 Air-defence alert [or artillery].

   3 Aeronautical Development Agency (India).

   4 Air-defended area.

   5 Business-aviation association (R).

   6 Americans with disabilities act.

   7 Avion de détection aéroportée (Awacs, F).

   8 Association de Documentation Aéronautique (F).

Ada Standard common high-order language for US DoD software (trademark).

ADAAM Air-directed air-to-air missile.

ADAAPS Aircraft data acquisition, analysis and processing system.

ADAC *1* STOL (F).

   2 Active-radar seeker (F).

Adacs *1* Alarm distributed-access control system.

2 Airborne digital automatic collection system.

ADAD Air-defence alerting device, horizon-scanning IR surveillance system.

ADAE Air Display Association Europe (UK-based).

ADAIRS Air-data and inertial-reference system.

Adam, ADAM *1* Air deflection and modulation.

2 Automated deposition of advanced (or aircraft) material (filament winding).

3 Automated data for aerospace maintenance.

4 Advanced dynamic anthropomorphic manikin.

5 Aerospace data miner.

6 Air-defence air-mobile [or air defense anti-missile] [in each case, S adds system].

7 [Also ADaM] aerostat design and manufacture [J-lens].

ADAMD Air Defence and Aerospace Management Directorate (NATO).

Adams Aircraft dispatch and maintenance safety (int.).

ADAP *1* Aircraft Development Aid Program (US DoT).

2 Air-defence air picture.

Adaps Automatic data acquisition and processing system.

Adapt Air traffic [services] data acquisition, processing and transfer (ATC Switzerland).

adapter Interstage device to mate and then separate adjacent stages of multi-stage vehicle. Often called skirt, especially when lower stage has larger diameter.

adaptive bus Digital data highway to which (almost) any number of inputs and outputs may be connected.

adaptive control system Control system, esp. of vehicle trajectory, capable of continuously monitoring response and changing control-system parameters and relationships to maintain desired result. Adapts to changing environments and vehicle performance to ensure given input demand will always produce same output.

adaptive logic Digital computer logic which can adapt to meet needs of different programs, environments or inputs.

adaptive nulling See *Adars*.

adaptive optical camouflage Active, self-variable form of camouflage which, chameleon-like, alters emitted wavelengths to suit varying background tones.

adaptive radar Usual [not only] meaning is antenna automatically alters gain, sidelobes and directivity according to received signal.

Adapts Adaptive diagnostics and personalised technical support.

ADAR Air-deployed active receiver (ASW).

Adario Analog/digital adaptive recorder input/output.

Adars Adaptive antenna receiver system; antenna (aerial) provides gain towards desired signals arriving from within a protected angle while nulling those arriving from outside that angle.

ADAS *1* Airborne data-acquisition system.

2 Auxiliary (or airborne) data-annotation system (for reconnaissance film, linescan or other hard-copy print-out of reconnaissance or ECM mission).

3 Airfield damage assessment system (USAF).

4 Air-deliverable acoustic sensor.

5 Aeronautical-data access station (AFTN).

6 Advanced digital avionics system (STA.6).

7 Automated weather-observing system data-acquisition system.

ADat-P3 Automatic data-processing [standard]-3 (NATO).

Adats *1* Air-defense [and] anti-tank system (US).

2 Airborne digital avionics test system.

ADAU Air-, or auxiliary-, data acquisition unit.

ADAV VTOL (F).

ADAWS Action-data automated weapon system.

ADAZ Air-defence zone.

ADB *1* Automatic drifting balloon.

2 Apron-drive bridge.

ADC *1* Air-data computer.

2 Air Defense Command (USAAF, 27 March 1946), see next.

3 Aerospace Defense Command (USAF, 15 January 1968, later called Adcom, inactivated 31 March 1980).

4 Advanced design conference.

5 Analog/digital convertor.

6 Aircrew dry coverall (helicopter sea rescues).

ADCA Advanced-design composite aircraft (USAF).

Adcap Advanced capability, or capabilities.

ADCC *1* Air Defence Cadet Corps (UK 1939–41, became ATC).

2 Air-defence, or direction, control centre.

ADCF Aligned discontinuous carbon fibre.

ADCIS Air-defence command information system (UK).

ADCN Aeronautical data communications network.

Adcock aerial Early radio D/F; avoided errors due to horizontal component by using two pairs of veritcal conductors spaced ½-wavelength or less apart and connected in phase opposition to give a figure-8 pattern.

ADCOM, Adcom Aerospace Defense Command (USAF, inactivated 31 March 1980).

ADCoPP Air-defence command-post processor.

ADCP Advanced-display core processor.

ADCS Air-data computer system.

ADCTS Advanced distributed combat training system.

Adcus Advise Customs.

Adcuts, ADCUTS Advanced computerised ultrasonic test system.

ADCV Active destination-coded vehicle (baggage).

ADD *1* Airstream direction detector (stall protection).

2 Long-range aviation (USSR VVS strategic bombing force).

3 Allowable, or acceptable, deferred deficiency, or defect).

ADDA American Design Drafting Association.

ADDC Air-defence data centre (UK).

ADDD Air-defence data dictionary (UK, a mathematical model).

Addison-Luard Large hand-held aluminium-body computers, Type B for triangle of velocities and D for adding fourth vector, eg motion of aircraft carrier (c1928–40).

additive Substance added to fuel, propellant, lubricant, metal alloy etc to improve performance, shelf life or other quality.

ADDL Aerodrome (or airfield) dummy deck landings; pronounced 'addle'.

add-on contract Extension of existing contract to cover new work in same programme.

ADDPB Automatic diluter-demand pressure breathing.

ADDR Aeroklub der Deutschen Demokratischen Republik.

address   Electronic code identifying each part of computer memory, each bit or information unit being routed to different *.

address selective   Adsel.

ADDS   *1* Airborne-decoy [or advanced digital] dispensing system.

   2 Aerial delivery dispersal system.

add time   Time required for single (binary) addition operation in computer arithmetic unit.

ADE   *1* Automated draughting (drafting) equipment.

   2 Aeronautical Development Establishment (India).

   3 Ada development environment.

   4 Australian Defence Executive.

Adecs, ADECS   Advanced digital engine control system.

ADEG   Air traffic services data-exchange requirements group (ICAO).

Adela   Aircraft directed-energy laser applications (AFRL).

Adèle   Alerte detection et localisation des emitteurs (F).

ADELT   Automatically deployable emergency locator transponder, or transmitter.

Adem   Advanced diagnostic and engine monitoring.

Aden   Armament Development, Enfield.

ADEN   Augmented deflector [or deflecting] exhaust nozzle.

Adeos   Advanced Earth-observing satellite.

ADEU   Automatic data-entry unit (punched card input for STOL transport navigation).

Adews   Air-defense EW system (USA).

Adex, ADEX   Air defence exercise.

ADEXP   ATS (1) data-exchange presentation message format.

ADF   *1* Automatic direction-finding or finder. Airborne radio navaid tuned to NDB or other suitable LF/MF broadcast source. Until 1945 aerial was loop mounted in vertical plane and rotated by motor energised by amplified loop current to rest in null position, with plane of loop perpendicular to bearing of ground station. Modern receivers fed by two fixed coils, one fore-and-aft and the other transverse, suppressed flush with aircraft skin (usually on underside).

   2 Australian Defence Force; (A adds Academy).

   3 Air-dominance fighter.

   4 Anti-icing/de-icing fluid.

   5 Airline Dispatcher Federation (office, DC).

ADF sense aerial   Rotatable loop null position gives two possible bearings of ground station;** is added to give only one null in each 360° of loop rotation.

ADFC   Aligned discontinuous fibre composite.

ADG   *1* Auxiliary drive generator.

   2 Air-driven generator.

   3 Accessory drive gear.

   4 Aircraft delivery group (USAF).

ADGB   Air Defence of Great Britain (1943–44).

ADGE, Adge   Air defence ground environment, or equipment.

ADGS   *1* Air-defence gunsight.

   2 Aircraft docking guidance system.

ADI   *1* Attitude director (rarely, display) indicator. 3-D cockpit display forming development of traditional horizon and usually linked with autopilot and other elements forming flight-director system. Most can function in at least two modes, en route and ILS, and in former can provide navigational steering indications.

   2 Anti-detonant injection, such as cylinder-head injection of methanol/water, for high-compression piston engine.

   3 Air Defense Initiative, partner ideas for SDI-type international joint ventures (US DoD).

   4 Aerospace and Defence Industries Directorate [1 to 4] (DTI, UK).

   5 Azimuth display indicator.

adiabatic   Thermodynamic change in system without heat transfer across system boundary. In context of Gas Laws, possible to admit of exact* processes and visualise them happening; shockwave, though not isentropic, is not* in classical sense because thermodynamic changes are not reversible.

adiabatic flame temperature   Calculated temperature of combustion products within rocket chamber, assuming no heat loss. Symbol T_c.

adiabatic lapse rate   Rate at which temperature falls (lapses) as height is increased above Earth's surface up to tropopause (see *DALR, ELR, SALR*).

ADIB   Air-deployable ice beacon.

ADID   Aircraft-data interface device.

ADIMP   Ada improvement programme (UK).

Adints, ADINTS   Automatic depot inertial navigation test system.

ADIR   Air-data [and] inertial-reference (see next and *ADIRU*).

ADIRS   *1* ADIR system.

   2 Airfield damage information and reporting system.

ADIRU   Air-data inertial reference unit.

ADIS   *1* Airport-data information system, or source.

   2 Automatic data-interchange system (FAA, from 1961).

   3 ADS(5) datalink interim system (Australia).

   4 Airport display information system (NAS/LATCC).

ADIT   Automatic detection, identification and tracking (USA).

ADITT   Aerially deployable ice-thickness transponder.

ADIVS   Air-defense interoperabilty validation system.

ADIZ   Air-defence, or defense, identification zone.

ADJ   Adjacent.

adjacent channel   Nearest frequency above or below that on which a radio link is working; can interfere with carrier or sidebands, but ** simplex minimises this.

adjustable propeller   One whose blades can be set to a different pitch on ground, with propeller at rest.

adjustable tailplane [horizontal stabilizer]   Surface which can be reset to different incidence only on ground.

adjuster   Mechanical input (manual, powered or remote-control) for altering a normally fixed setting, such as engine idling speed.

Adkem   Advanced kinetic-energy missile.

ADL   *1* Automatic drag limiter [S adds system].

   2 Arbeitsgemeinschaft Deutscher Luftfahrt-Unternehmen (G).

   3 Armament datum line.

   4 Authorised data list.

   5 Aeronautical data-link.

   6 Advanced distributed learning.

ADLFP   Air-deployed low-frequency projector [Adsid].

ADLGP   Advanced data-link for guided platforms.

ADLP   *1* Aircraft data-link processor.

   2 Airborne data-link protocol.

ADLS   Aeronautical data-link services.

ADLT   Advanced discriminating, or discriminatory, laser technology.

ADLY   Arrival delay.

ADM   *1* Air-decoy [or defense] missile (USAF).

2 Atomic demolition munition.

3 Airport Duty Manager.

4 Air-data module.

5 Asynchronous data modem.

6 Advanced development, or demonstration, model.

7 Aeronautical data management.

8 Admiral (not UK usage).

ADMA   Aviation Distributors & Manufacturers Association (US).

ADMC   Actuator drive and monitor computer.

Ad-Me   Advanced metal evaporated.

administrative loading   Loading transport vehicle (eg, aircraft) for best utilisation of volume or payload, ignoring tactical need or convenience.

Admiral's barge   Aircraft assigned to Flag Officer (FAA, colloq.).

Admit, ADMT   Air distributed mission trainer.

admittance   In a.c. circuit, $1/Z$, reciprocal of impedance, loosely 'conductivity'; made up of real and imaginary parts; symbol Y, unit siemens.

ADMS   Airline data-management system.

ADMU   Air-distance measuring unit.

ADNC   Air-defence notification centre.

ADNL   Additional.

ADNS   Arinc data-network service.

ADO   *1* Advanced development objective.

2 Automatic delayed opening (parachute).

3 Assistant Deputy for Operations.

ADOC   Air Defence Operations Centre (UK).

Adocs, ADOCS   *1* Advanced digital optical control system.

2 Automated deep-operations co-ordination system (DoD, especially used by Norad).

Adora   Analysis and definition of operational requirements for ATM(7) (Euret).

ADP   *1* Acoustic, or air, data processor.

2 Automatic, or airport, data processing.

3 Air-driven pump.

4 Engine aerodynamic design point; determined by cycle parameters.

5 Altitude delay parameter.

6 Aéroports de Paris.

7 Airport development program.

ADPA   American Defense Preparedness Association.

ADPCM   Adaptive differential pulse-code modulation.

ADPE   Automated [radar] data-processing equipment.

ADPG   Air Defence Planning Group.

ADPS   *1* Asars deployable processing station (USAAF).

2 Aeronautical data-processing system [SO replaces system by Selection Office] (USAF).

ADR   *1* Accident, or acoustic, data recorder.

2 All-purpose data-stream replicator, or simplified RMCDE.

3 Air defence region (UK).

4 Advisory route.

5 Airfield damage repair.

6 Air-data reference.

7 Active decoy round.

Adram   Advanced dielectric radar absorbent material (Plessey).

Adras   Aircraft data-recovery and analysis system.

ADRC   Air defence radar controller.

ADRD, ADR/D   Air-data reference disagree.

ADRDE   Air Defence Research & Development Establishment (UK).

Adrep   Accident/incident data report (ICAO).

Adres   Aircraft documentation retrieval system.

ADRG   Arc digital master graphics.

ADR/Hum   Accident data recorder and health/usage monitor, installed as single integrated package with common inputs.

Adries   Advanced digital radar imagery exploitation system, low-level target recognition.

ADRIS   Airport Doppler weather radar information system.

ADRS   Airfield Damage Repair Squadron (RAF).

ADRU   Air defence radar unit.

ADS   *1* Accessory drive system, self-contained yet integrated package.

2 Autopilot disengage switch.

3 Audio distribution system (Awacs).

4 Air-data system.

5 Automated, or automatic, dependent surveillance [-A adds address, -B adds broadcast, -C contrct, -P panel, -PDLC controller/pilot data-link communications, -S system, -U unit].

6 Aviation data server.

7 Airborne data service[s].

8 Aircraft, or airborne, data sensor.

9 Air-defence ship, study or studies.

10 Air Defense Squadron (USAF).

11 Active dipping sonar.

12 Acoustic detection system.

13 Advanced deployable system[s] (USN).

14 Automatic drilling system.

15 Airlifter defense systems (USAF).

16 Aufklärungsdrohnen (UAV) system (Switzerland).

17 Area-denial submunition.

Adsam, ADSAM   Air-directed SAM (USA/USN).

Adsams   Advanced SAM systems.

ADSC   Air-defence siting computer (UK).

Adsel, ADSEL   Address Selective. Improved SSR system in which saturation in dense traffic is avoided by interrogating each aircraft (once acquired) only once on each aerial rotation instead of about 20 times. Transponders reply only when selected by discrete address code, reducing number of replies and mutual interference and opening up space for additional information (such as rate of turn) helpful to ATC computers (see *DABS*).

ADSI   Air-defense systems integrator (UAV).

ADSIA   Allied Data-Systems Interoperability Agency (NATO).

Adsid   Air-delivered seismic detection sensor.

ADSK   Air-droppable, or air-dropped, survival kit.

ADSM   Air defence suppression missile.

adsorption   Removal of molecules of gas or liquid by adhesion to solid surface; activated carbon has very large surface area and is powerful adsorber.

ADSP   *1* Advanced digital signal processor.

2 Automatic dependent surveillance panel.

ADSS   *1* Aeronautical decision support system,

providing instant paperless access to manuals, maps and emergency procedures.

2 Automatic dependent surveillance system.

ADSU *1* Air-data sensor unit.

2 Automatic dependent surveillance unit.

ADT *1* Approved departure time.

2 Automatically deployable transmitter.

3 Air-data transducer.

4 Air-data [or advanced, or alphanumeric, display] terminal.

5 Automatic detection and tracking.

6 Active-denial technology.

ADT³, ADT3 Air defence tactical training theatre.

ADTC Armament Development Test Center (USAF, Eglin AFB).

ADTN Administrative data-transmission network.

ADTS *1* Air-defence threat simulator.

2 Approved departure times.

ADTU Air-data transfer unit.

ADU *1* Alignment display unit (INS).

2 Auxiliary display unit.

3 Avionics [or annotation] display unit.

4 Air-data unit.

5 Actuator drive unit, in digital FCS.

6 Audio distribution unit.

7 Activity display unit (ESM).

8 Air Disarmament Unit (RAF).

ADV *1* Arbeitsgemeinschaft Deutscher Verkehrs-flughafen eV (Federal German Airports Association).

2 Air-defence variant.

ADV, Adv Advise, or advisory area.

advance *1* To * throttle = to open throttle, increase power.

2 In piston engine, to cause ignition spark to occur earlier in each cycle.

3 Forward movement of propeller (see *propeller pitch*).

advance, angle of See *propeller pitch*.

advanced Generalised (overworked) adjective meaning new, complicated and typifying latest technology.

advanced aerobatics Flight manoeuvres with no limits apart from airframe/pilot limits.

advanced common flightdeck Retrofit, initially on FedEx DC-10s, based on MD-11.

advanced airfield, base Base or airfield, usually with minimal facilities, in or near objective area of theatre of operations.

advanced flow-control procedure Any of six theoretical or experimental techniques for ATC in high-traffic airspace.

advanced high-frequency material New coatings [currently classified] for LO aircraft which eliminate the need for laborious maintenance between missions.

advance/diameter ratio Ratio between distance aircraft moves forward for one revolution of propeller(s), under specified conditions, and propeller diameter. Expressed as

$$J = \frac{V}{nD}$$

where V is TAS, n rotational speed and D diameter.

advanced stall Stall allowed to develop fully, yet usually with some lateral control. Many definitions claim longitudinal control must remain, but nose-down rotation is invariably automatic (see *g-break, stall*).

advanced tactial targeting Air-to-air system using Link-16, SADL and other links from TTNT to share information about surface emitters (USAF).

advanced trainer Former military category, more powerful and complicated than ab initio/primary/basic trainer and capable of simulating or performing combat duties when fitted with armament.

advancing blade In rotary-winged aircraft in translational flight, any blade moving forward against relative wind. Each blade advances through 180° of its travel, normally from dead-astern to dead-ahead.

ADVCAP Advanced capability.

advection Generally, transfer by horizontal motion, particularly of heat in lower atmosphere. On gross scale, carries heat from low to high latitudes.

advection fog Fog, generally widespread, caused by horizontal movement of humid air mass over cold (below dew point) land or sea.

adversary aircraft Fighter specially purchased and configured to act role of enemy in dissimilar air-combat training.

adverse rudder Inputs rolling moment opposite to that commanded by lateral-control system.

adverse yaw Negative yawing moment due to roll at high C_L, problem with sailplanes.

Adviser Airborne dual-channel variable-input severe envrionment recorder (RCA).

Advisor Annotated digital video for intelligent surveillance and optimised retrieval (EC aviation security).

advisory Formal recorded helpful message repeatedly broadcast from FAA AAS centre to all local aircraft. Abb: ADVY, ADZ, ADZY.

advisory circular The printed form of information for pilots (FAA).

advisory light Displayed by aircraft (esp. carrier-based) to show LSO status (gear, hook, wing, speed and AOA).

advisory route Published route served by Advisory Service, but not necessarily by ATC (1) or separation monitoring and usually without radar surveillance.

Advisory Service FAA facility to provide information on request to all pilots, and advice to those who need it. Abb: ADVS.

Advon Advanced echelon.

ADW *1* Area-denial weapon.

2 Agent-defeat warhead.

ADWC Air Defense Weapons Center (Tyndall AFB).

ADWES Air-defence weapons-effects simulation, or simulator.

ADZ Advise (ICAO).

AE *1* Aviacion del Ejército [army aviation] (Peru, etc).

2 Augmentor ejector.

A$_e$ Effective area of antenna aperture.

AEA *1* Aeronautical Engineers Association.

2 Aircraft Electronics Association Inc (US).

3 Association of European Airlines.

4 Aircrew equipment assembly.

5 Aerial Experiment Association (US, 1907–09).

6 Airborne electronic attack (V adds Variant).

7 All-electric airplane/aeroplane.

AEAF Allied Expeditionary Air Force (WW2).

AEB *1* Avionics equipment bay.

2 Air Efficiency Board (UK).

AEC *1* Atomic Energy Commission (USA, 1946–74).

2 Automatic exposure control.

AECB Arms Export Control Board (UK).

AECC   Aeromedical Evacuation Control Centre
AéCF   Aéro Club de France.
AECM   Active ECM.
AECMA   Association Européenne des Constructeurs de Matériel Aérospatial (Int.).
AéCS   Aéro Club de Suisse.
AECU   Audio [or advanced] electronic control unit.
AED   *1* Alphanumeric entry device.
   2 Air Engineering Department (TAG).
   3 Automated [or automatic] external defibrillator.
   4 Algol extended for design.
   5 Aviation Environmental Divisions [1 to 4] (DETR, UK).
AEDC   Arnold Engineering Development Center (USAF, mainly air-breathing propulsion systems, at Tullahoma, Tenn).
AEDO   Aeronautical engineering duty officer (USN).
AEDS   Atomic energy detection system (global, run by AFTAC).
AEEC   *1* Airlines Electronic Engineering Committee (US and Int.).
   2 Association of European Express Carriers.
AEELS   Auto Elint emitter-locator system.
AEF   *1* American Expeditionary Force (WW1, WW2).
   2 Air Expeditionary Force (USAF).
   3 Aerospace Education Foundation (US).
   4 Air Experience Flight (RAF).
   5 Armament Engineering Flight (RAF).
   6 Aviation Environment Federation (UK), or Airfields ** (UK).
AEFB   AEF (2) Battlelab.
AEFT   Auxiliary external fuel tank[s].
Aegis   *1* Advanced engine/gearbox integrated system.
   2 CGN ship class (USN).
   3 Airborne early-warning ground-integration segment.
AEH   Airborne emergency hospital.
AEHF   Advanced e.h.f. (S adds satellite).
AEHP   Atmospheric-electricity hazards protection.
AE-I   Aircraft Engineers International (Int.).
AEIS   Aeronautical en-route information service (ICAO).
AEJPT   Advanced European [military] jet-pilot training (proposal).
AEL   *1* Aeronautical Engine Laboratory (S Philadelphia, USN).
   2 Advanced Engineering Laboratory (Australia).
AELS   *1* Augmentor/ejector lift system.
   2 Airborne electronic-library system.
AEM   *1* Air Efficiency Medal (UK).
   2 Automatic emergency mode.
AEMB   Airborne electromagnetic bathymetry.
AEMCC   Air and Expedited Motor Carriers (trade association, US).
AENA   Aeropuertos Españoles y Navegación Aérea (Spain).
AEO   *1* Air Electronics Officer (aircrew trade, RAF).
   2 Air Engineer Officer (USAF).
   3 All engines operating.
AEOSOP   See *Aesop.*
AEP   *1* Airports Economic Panel (ICAO).
   2 Autometric edge product.
   3 Alternate (ie, alternative) engine program (JSF).
   4 Audio-entertainment player.
   5 Autopilot engage panel.

AEPDS   Advanced electronic processing and distribution system (satellite).
AEPO   Aeronautical Enterprise Program Office (WPAFB).
AEPT   Air engineer procedures trainer.
AER   Area expansion ratio in wind tunnel; ratio of cross-sections at start of diffuser to down-stream end at first bend.
AERA   *1* Association pour l'Etude et la Recherche Astronautique et Cosmique (F).
   2 Automated en-route ATC.
Aerad   Commercially published but universally used flight guide and chart system (UK).
Aerall   Association d'Etudes et de Recherches sur Aéronefs Allégés (F).
AERC   Aviation Education Resource Center (US).
AERE   Atomic Energy Research Establishment (UKAEA, Harwell).
aerial   *1* Pertaining to aircraft, aviation or atmosphere.
   2 Part of radio or radar system designed to radiate or intercept energy, with size and shape determined by wavelength, directionality and other variables (US = antenna).
aerial array   Assembly of aerial elements, often identical, usually excited from same source in phase and dimensioned and positioned to radiate in pencil beam or other desired pattern (not necessarily phased-array).
aerial common sensor   Next-generation airborne sensor for tactical reconnaissance, Imint and Sigint (USA).
aerial delivery system   Complete system for air transport and delivery to surface recipient (usually without aircraft landing).
aerial supervision module   Aircraft housing both air attack and leadplane pilot.
aerial survey   Use of aerial cameras and/or other photogrammetric instruments for the making of maps, charts and plans.
aerial swimming vehicle   A micro air vehicle with major dimensions not exceeding 150 mm (c6 in), able to cruise at $c10$ ms^{-1} propelled by aft-mounted reverse-camber flapping wings. Generally synonymous with delphinopter.
aerial work   General aviation for hire or reward other than carriage of passengers or, usually, freight; includes agricultural aviation, aerial photography, mapping and survey, cable and pipeline patrol and similar duties usually not undertaken to full-time fixed schedule.
aerial work platform   Small railed platform for one or two occupants, mounted on vehicle by Z-type [less often scissors] elevating linkage and often providing electric or hydraulic power for occupants.
AERO   Air Education and Recreation Organisation (UK).
aero   Concerned with atmospheric flight.
aeroacoustics   Science and technology of acoustics caused by, and effect upon, aerospace systems. A more general definition is interaction between sound and gas flow, esp. sound generated by the flow.
aeroballistics   Science of high-speed vehicles moving through atmosphere in which both ballistics and aerodynamics must be taken into account. Often asserted aerodynamics and ballistics are applied separately to different portions of flight path, but both act as long as there is significant atmosphere present.
Aerobatic catalogue   Derived from *Aresti,* simplified

scheme for planning and scoring aerobatic routines (FAI).

aerobatic oil system In modern combat aircraft liable to experience prolonged zero-g, meaning is lube system with multiple scavenge ports round all engine bearing chambers leading back to tank in which synthetic gravity is maintained by rapid rotation.

aerobatics Precise and largely standardised manoeuvres, unnecessary in normal flight, executed to acquire or demonstrate mastery over aircraft, for entertainment, or for competition (US = acrobatics). BS: "Evolutions voluntarily performed other than those required for normal flight", which would include a gentle 360.

aerobic propulsion Requiring oxygen.

aerobiology Study of distribution and effects of living matter suspended in atmosphere (small insects, spores, seeds and micro-organisms).

aerobrake _1_ Aerodynamic brake for use in extremely low-density atmospheres at Mach numbers of 5 to 25. Typically can be deployed as a saucer shape, concave side facing direction of travel.

_2_ Deceleration by holding nose high after landing.

Aero-C Message and data-reporting satellite service for satcom aircraft.

aerocapture Technique harnessing drag of atmosphere of planet (especially Mars) to slow spacecraft to planetary orbital speed.

Aeroclinoscope Instrument with semaphore-like arms for indicating wind direction [and, roughly, atmospheric pressure] (obs.).

aeroconical canopy Form of parachute canopy suitable for use at all aerospace Mach numbers.

aerocryptography Representation of aerobatic manoeuvres by 2-D symbols.

aerodone Basic aerodyne, glider relying upon natural stability and having no moving control surfaces. Examples are paper dart and chuck glider, most simple free-flight models, and aeroplanes which continue to fly after being abandoned by their crews.

aerodonetics Science of gliding flight, with or without use of control surfaces.

aerodontalgia Toothache caused by major changes in ambient atmospheric pressure.

aerodontia, aerodontology Branch of dentistry dealing with problems of flying personnel.

aerodrome BS.185, 1940: 'A definite and limited area of ground or water (including any buildings, installations and/or equipment) intended to be used, either wholly or in part, in connexion with the arrival, departure and servicing of aircraft.' Becoming archaic (see _airfield, airport, air base, strip_, etc).

aerodrome traffic zone Airspace up to 2,000 ft (609 m) a.a.l. and within 2.5 nm of centre or 2,000 ft/609 m of boundary (general aviation).

aerodynamic axis Imaginary line through aerodynamic centres of every longitudinal element in solid body moving through gaseous medium. In wing, runs basically from tip to tip, but in swept or slender delta can be an acutely curved, kinky line often having little practical application.

aerodynamic balance _1_ Method of reducing control-surface hinge moment by providing aerodynamic surface ahead of hinge axis (see _Frise aileron, horn balance_).

_2_ Wind-tunnel balance for measurement of aerodynamic forces and moments.

aerodynamic-balance panel Shelf fixed to control surface ahead of the hinge axis, contained inside fixed structure.

aerodynamic braking _1_ Use of atmospheric drag to slow re-entering spacecraft or other RV.

_2_ Use of airbrakes or parachute (drag chute) in passive **.

_3_ Use of reversed propulsive thrust (propeller or jet) in active **.

aerodynamic centre In two-dimensional wing section, point about which there is no change of moment with change in incidence; point about which resultant force appears to rotate with change of incidence. In traditional sections about one-quarter back from leading edge (25% chord) and in symmetrical section lies on chord and thus coincident with CP. Also called axis of constant moments. Abb: a.c. ac or (incorrectly) AC.

aerodynamic chord Reference axis from which angle of attack of two-dimensional aerofoil is measured. Line passing through (supposed sharp) trailing edge and parallel to free-stream flow at zero lift at Mach numbers appreciably below 1 (see _chord, geometric chord, MAC_).

aerodynamic coefficient Aerodynamic force (lift or drag, or moment) may be reduced to dimensionless coefficient by dividing by characteristic length (which must be same parameter for all similar bodies, and in a wing is invariably area) and by dynamic head (symbol q). Traditional divisor is $\frac{1}{2}\rho V^2 S$, where ρ is air density, V velocity and S area, ensuring that units are consistent throughout; if area is m^2 then V must be ms^{-1}. The $\frac{1}{2}\rho V^2$ term, difference between pitot and static pressure, is accurate only at low speeds; if M^2 (square of Mach number) is too large to ignore, a different expression must be found for dynamic head, such as H-p (pitot minus static). (See _force coefficients, moment coefficients, units of measurement_).

aerodynamic damping In flight manoeuvres rotation of aircraft (about c.g. or close to it) changes direction of relative wind to provide restoring moment which opposes control demand and arrests manoeuvre when demand is removed. As altitude increases, combination of increasing TAS (for given EAS) and reduced airflow deflection angles results in ** being progressively decreased, although control demand moment and aircraft inertia do not change. Thus at high altitude pilot must apply greater opposite control movements to arrest rotation.

aerodynamic disturbance Generalised euphimistic term in SR-71 and similar flight reports covering inlet unstarts and related phenomena.

aerodynamic efficiency Most common yardstick is lift/drag ratio (L/D). In general ** maximised when resultant forces are as nearly as possible perpendicular to direction of motion; tend to be reduced as speed is increased, since lift-type forces may be presumed to remain substantially constant while drag-type forces may be presumed to increase in proportion to square of speed.

aerodynamic force Force on body moving through gaseous medium assumed to be proportional to density of medium (ρ), square of speed (u^2 or V^2), characteristic dimension of body (such as length L^2 or area S) and R_n (Reynolds number raised to power n). This broad relationship sometimes called Rayleigh formula. Body assumed to be wholly within homogenous gas, reasonably compact and streamlined (eg not a sheet of paper) and to have smooth surface.

aerodynamic heating As speed of body through gaseous medium is increased, surface temperature increases roughly in proportion to square of speed. Effect due variously to friction between adjacent molecules in boundary layer, to degradation of kinetic energy to heat and to local compression of gases. Maximum temperature is reached on surfaces perpendicular to local airflow where oncoming air or gas molecules are brought to rest on surface. At Mach 2 peak stagnation temperature is about 120°C and at Mach 3 about 315°C; at hypersonic speeds temperature can swiftly rise to 3,000 or 5,000°C in intense shockwave around nose and other stagnation points, causing severe radiation heating, ionisation and dissociation of flow. Adiabatic temperature rise is approximately given by $\Delta T = (^V/_{100})^2 °C$, where V is speed in mph; alternatively a poorer approximation is 41 M^2 where M is Mach number.

aerodynamic mean chord, AMC, MAC, c Chord that would result in same overall force coefficients as those actually measured. Essentially, but not necessarily exactly, same as mean of aerodynamic chords at each station; very nearly same as geometric mean chord.

aerodynamic overbalance Excessive aerodynamic balancing of control surface such that deflection will immediately promote runaway to hard-over position

aerodynamic twist Variation of angle of incidence from root to tip of aerodynamic surface, to obtain desired lift distribution or stalling characteristic (see *wash-in*, *wash-out*).

aerodynamics Science of interactions between gaseous media and solid surfaces between which there is relative motion. Classical * based upon Bernoulli's theorem, concept of boundary layer and circulation. Reynolds number, Kármán vortex street, turbulent flow and stagnation point. High-speed * (M^2 too large to be ignored) assumes gas to be compressible and introduces critical Mach number, shockwave, aerodynamic heating, and relationships and concepts of Prandtl, Glauert, Ackeret, Busemann, Kármán-Tsien and Whitcomb. At Mach numbers above 4, and at heights above 80 miles (130 km), even high-speed * must be modified or abandoned because of extreme aerodynamic heating, violent changes in pressure and large mean free path (see *superaerodynamics*).

aerodyne Heavier-than-air craft, sustained in atmosphere by self-generated aerodynamic force, possibly including direct engine thrust, rather than natural buoyancy. Two major categories are aeroplanes (US = airplanes) and rotorplanes, latter including helicopters.

aeroelasticity Science of interaction between aerodynamic forces and elastic structures. * deflections are increased by raising aerodynamic forces, varying them rapidly (as in gusts and turbulence) and increasing aspect ratio or fineness ratio. All * effects tend to be destabilising, wasteful of energy and degrading to structure.

aeroembolism Release of bubbles or nitrogen into blood and other body fluids as a result of too-rapid reduction in ambient pressure. May be due to return to sea-level from much increased pressure ('caisson sickness', 'the bends') or from sea-level to pressure corresponding to altitude

greater than 30,000 ft (about 10,000 m). Potentially fatal if original, increased, pressure is not rapidly restored.

aeroflight mode Atmospheric flight, by aerospace vehicle (eg, Space Shuttle).

aerofoil (US = airfoil) *1* Solid body designed to move through gaseous medium and obtain useful force reaction other than drag. Examples: wing, control surface, fin, turbine blade, sail, windmill blade, Flettner rotor, circular or elliptical rotor blade with supercirculation maintained by blowing. Some authorities maintain * must be essentially 'wing-shaped' in section.

2 A specific meaning is a gas-turbine rotor blade, without root, for fusion to a ring or disc.

aerofoil section Traditionally, outline of section through aerofoil parallel to plane of symmetry. This must be modified to 'parallel to aircraft longitudinal axis' in variable-sweep and slew wings, and 'perpendicular to blade major axis' in blades for rotors, turbines and propellers. None of these sections may lie even approximately along direction of relative wind, although usually assumed to. Also called *profile*.

aerofoil boat Wing-shaped surface-effect marine craft (or low-altitude aircraft).

aerogel Colloid comprising solution of gaseous phase in solid phase or coagulated sol (colloidal liquid).

aerograph Airborne meteorological recording instrument; aerometeograph.

Aero-H Long-haul cockpit and pax communications, telephony (9.6 kbps), fax (4.8) and data (2.4). H + offers voice codes and better multichannel performance.

Aero-I Short-/medium-haul and corporate communications, telephony (4.8 kbps) and fax/data (2.4).

aero-isoclinic wing Aerofoil which, under aeroelastic distortion, maintains essentially uniform angle of incidence from tip to tip.

aerojumble Aeronautical artefacts in jumble sale.

Aero-L Low-gain satcom service, two-way data exchange at 0.6 kbps.

Aerolite Trade name, low-density bonded sandwich structure based on phenolic-resin-bonded flax fibres (Aero Research, later CIBA).

aerolite Stony meteorite, richer in silicates than metals.

aerology Study of atmosphere (meteorology) other than lower regions strongly influenced by Earth's surface.

aerol strut Early oleo strut relying for energy absorption and damping upon both air and oil.

aerometeograph, aerometeorograph Airborne instrument making permanent record of several meteorological parameters such as altitude, pressure, temperature and humidity.

aerometer Instrument used in determining density of gases, esp. atmosphere.

aeronaut Pilot of aerostat.

Aero Mini-M Service for small corporate and GA, 2.4 kbps data, fax and voice.

aeronautica Aeronautical artefacts, esp. those in auction sale.

aeronautical Pertaining to aeronautics.

aeronautical chart Chart prepared and issued primarily for air navigation. Chief categories include Sectional (plotting), Regional, Radio, Flight Planning, en-route low and high altitude, SID, STAR, TMA, IAP, Great Circle and Magnetic. Most are to conformal projection (esp. Lambert's); non-aeronautical topographic features generally excluded.

aeronautical Earth station   Satcom or navsat station in an aircraft.

aeronautical fixed service   Network of ground radio stations.

aeronautical fixed telecoms, network   National teleprinter network outputting weather forecasts, airfield status and flight plans.

aeronautical ground station   Satcom or navsat station on Earth's surface (conceivably, on ship).

Aeronautical Information Circular   Official publication printed on white paper [admin. matters], green [maps and charts], pink [safety issues], yellow [operations and facilities] and mauve [temporary changes, esp. airspace restrictions] (UK CAA).

Aeronautical Information Documents Unit   Produces all flight-planning documents for UK military aircrew (RAF Northolt).

aeronautical information overprint   Overprint on military or naval map or chart for specific air navigation purposes.

Aeronautical Information Publication   Periodically issued for all civil pilots by national aviation authorities, UK being titled A.I.Circular, see above.

aeronautical light, beacon   Illuminated device approved as aid to air navigation.

aeronautical mile   Nautical mile (British Admiralty standard 6,080 ft = 1.85318 km); defined as length of arc of 1° of meridian at Equator.

aeronautical mobile service   Voice radio linking aircraft and ground stations. AMS(R) serves routes [generally means airways] while AMS(OR) serves off-route airspace.

aeronautical multicommunications   See *multicommunications service*.

aeronautical satellite   Satellite provided to assist aircraft by improving navigation, communications and traffic control. Abb. aerosat.

aeronautical topographic chart/map   Chart or map designed to assist visual or radar navigation and showing features of terrain, hydrography, land use and air navigation facilities.

aeronautics   Science of study, design, construction and operation of aircraft.

Aéronavale   Air arm of French Navy.

aéronef   Aircraft, any species (F).

Aeronet   Secure closed-community information net, not linked to Internet (SITA).

aeroneurosis   Chronic disorder of nervous origin caused by prolonged flying stress.

aeronomy   Study of upper atmosphere of planets with especial reference to effects of radiation, such as dissociation and ionisation.

aeropause   Vague boundary between atmosphere useful to aircraft, and space where air density is too low to provide lift, or air for air-breathing engines, or aerodynamic forces for stability and control. One definition suggests boundary is layer from "12 to 120 miles"; upper limit meaningful for hypersonic aircraft only.

aeroplane (US = airplane)   BS.185, 1940: 'A flying machine with plane(s) fixed in flight'. Modern definition might be 'mechanically propelled aerodyne sustained by wings which, in any one flight regime, remain fixed'. Explicitly excludes gliders and rotorplanes, but could include MPAs, VTOLs and convertiplanes that behave as * in translational flight.

aeroplane effect   Error in radio DF caused by horizontal component of fixed aerial or trail angle of wire (arch.).

Aeropp   Aeronautical message switching system.

aeropulse   Air-breathing pulsejet.

aeroresonator   Resonant air-breathing pulsejet.

aeros   Aerobatics, plural of aero. (colloq.).

aerosat   Aeronautical satellite.

aeroservoelasticity   Study of aeroelasticity in aircraft with automatic control systems.

aeroshell   High-drag aerodynamic-braking heatshield for returning spacecraft or planetary lander.

aerosol   Colloid of finely divided solid or liquid dispersed in gaseous (esp. air) continuous phase. Natural examples: smokes, dustclouds, mist, fog. In commercial product active ingredient is expelled as aerosol by gaseous propellant.

aerospace   *1* Essentially limitless continuum extending from Earth's surface outwards through atmosphere to farthest parts of observable universe, esp. embracing attainable portions of solar system.

2 Pertaining to both aircraft and spacecraft, as in 'aerospace technologies'.

3 Activity of creating and/or operating hardware in aerospace, as in 'the US leads in aerospace'.

aerospace craft   Vehicle designed to operate anywhere in aerospace, and especially both within and above atmosphere.

aerospace data miner   Analyses fleet performance (eg of all aircraft of one type).

aerospace forces   National combat armoury capable of flying in atmosphere or rising into space, including all satellite systems and strategic ballistic missiles.

aerospace medicine   Study of physiological changes, disorders and problems caused by aerospace navigation. Among these are high accelerations, prolonged weightlessness, vertigo, anoxia, ionising radiation, Coriolis effects, micrometeorites, temperature control, recycling of material through human body, and possibility of developing closed ecological systems to support human life away from Earth.

aerospace plane   Colloquial term for space vehicle which can re-enter, manoeuvre within atmosphere and land in conventional way on Earth's surface. Generally assumed to be manned and to include some air-breathing propulsion.

aerospace relay mirror system   Mirror [s] suspended under airship at 65,000 ft (19.8 km) to relay beam from ground-based laser to track and possibly kill objects in space (AFRL).

aerospace warfare   Conflict within and above atmosphere.

aerostat   Lighter-than-air craft, buoyant in atmosphere at a height at which it displaces its own mass of air. Major sub-groups are balloons and airships. In airships aerodynamic lift from hull can be significant, but not enough to invalidate classification under this heading.

aerostatics   The mechanics of gases at rest, in mechanical equilibrium.

aerostation   Operation of aerostat.

aerostructure   *1* The wing [s], engine [s] and tail of a flying boat (term now rare).

2 The supporting and controlling surfaces of an aeroplane (also rare).

3 Today the term is usually synonymous with airframe.

aerothermal flow   Slipstream past hypersonic vehicle in upper atmosphere.

aerothermodynamic border   Region, at height around 100 miles (160 km), at which Earth atmosphere is so attenuated that even at re-entry velocity aerodynamic heating is close to zero and can be neglected, as also can drag.

aerothermodynamic duct   Athodyd, early name for ramjet. Aerothermodynamic in this context does not necessarily conform to definition given below.

aerothermodynamics   Study of aerodynamic phenomena at velocities high enough for thermodynamic properties of constituent gases to become important.

aerothermoelasticity   Study of structures subject to aerodynamic forces and elevated temperatures due to aerodynamic heating.

aerotitis   Pain in the ear caused by pressure difference.

aero-tow   Tow provided for glider by powered aircraft. Not normally applicable to banner or other towing.

Aeroweb   Trade name, range of structural and noise-attenuating bonded honeycomb sandwich materials.

Aerozine   Trade name, family of liquid fuels based on MMH or UDMH.

AES   *1* Aeromedical evacuation system (USAF, MAC).

2 Auger electron spectroscopy.

3 Air Electronics School.

4 Aircraft (or aeronautical) Earth station (satellites).

5 Air EuroSafe, dedicated non-profit.

6 Armament Experimental Station (RFC).

7 Aerodrome emergency service.

8 Atmospheric Environmental Service (Canada).

9 Airborne emitter system.

AESA   Active electronically scanned array/antenna/aperture.

AESC   Aft equipment service centre (on aircraft).

Aescon   Aerospace and Electronics Systems Conference (Int.).

AESF   Avionics electrical systems flight (RAF).

Aesop   *1* Automated engineering and scientific optimisation program (multivariable design tool).

2 Airborne electro-optical special-operations payload.

AESS   *1* Airborne electronic-surveillance system.

2 Aircraft-environment surveillance system (radar, TCAS, EGPWS).

AESU   Aerospace Executive Staff Union (Singapore).

AET   *1* Airfields Environmental Trust (UK).

2 Aerosol explosive thermobaric.

AETA   Association des Anciens Elèves de l'Ecole d'Ensignement Technique de l'Armée de l'Air (F).

AETB   Alumina-enhanced thermal barrier.

AETC   Air Education & Training Command (USAF, Randolph AFB, established 1 July 1993).

AETE   Aerospace Engineering and Test Establishment (Cold Lake, Alberta).

AETMS   Airborne electronic terrain mapping system (3-D colour-coded in real time, plan or elevation).

AETW   AET (2) washed.

AEU   *1* Airborne electronics unit.

2 Auxiliary equipment unit.

AEW   *1* Airborne early warning.

2 Air [or airborne] electronic warfare.

3 Air Expeditionary Wing (USAF).

AEWC, AEW&C, AEW + C   Airborne early warning and control.

AEW/EW   Airborne early warning and electronic warfare.

AEWF   Airborne Early-Warning Force (NATO).

AEWTF   Aircrew Electronic-Warfare Tactics Facility (NATO).

AF, a.f.   *1* Audio frequency, sounds audible to average human ear (20 to 16,000 Hz). In simple radio communications RF carrier is modulated so that it "carries" AF superimposed upon basic waveform.

2 Aerodynamic force.

3 Auto-flight.

4 Airway facilities.

a/f   *1* Airfield.

2 Airframe.

AF³   Anti-fire fighting foam.

AFA   *1* Air Force Association (US).

2 Air Force Act (UK).

3 Aircraft Finance Association (US).

4 Association of Flight Attendants (US).

5 Audio-frequency amplifier.

6 Air Force Academy (Colorado Springs, established 1 April 1954).

7 Academia de Força Aérea (Brazil).

AFAA   Air Force Audit Agency (1 July 1948, Norton AFB, later Washington DC).

AFAC   Airborne forward air controller.

AFADS   Air Force Air Demonstration Squadron (USAF Thunderbirds).

AFAES   Aviation facilities and aircraft engineering support (MoD, UK).

AFAFC   Air Force Accounting and Finance Center (Lowry AFB).

AFAITC   Armed Forces Air Intelligence Training Center (Denver).

AFAL   Air Force Armament Laboratory (Eglin AFB).

AFALC   Air Force Acquisition Logistics Center (Wright-Patterson AFB).

AFAMC   See *AMC*(5).

AFAMRL   See *AAMRL*.

AFAP   Australian Federation of Air Pilots.

AFAPD   Air Force application[s] program, or protocol, development.

Afarmade   Asociación Española de Fabricantes de Armamento y Material de Defensa y Seguridad.

AFARPS   See *ARPS*.

AFASD   Air Force Aeronautical Systems Division, formerly part of AFSC, HQ Wright-Patterson AFB, now Aeronautical Systems Center.

AFATC   Air Force Air Transport Command (1942–47).

AFATL   See *AFAL*, Tadded Testing.

AFAvL   Air Force Avionics Laboratory, now Wright Laboratory (Wright-Patterson AFB).

AFB   *1* Air Force Base (USAF).

2 Air Force Board (RAF).

AFBCA   Air Force Base Conversion Agency (Arlington, Va, established 15 November 1991).

AFBMD   Air Force Ballistic Missile Division, became BMO.

AFC   *1* Air Force Cross (decoration).

2 Aircraft flyaway cost.

3 Aramid-fibre composite.

4 Automatic-feedback control, for interlocking coherent systems.

5 Audio, or automatic, frequency control, or compensation.

6 After [1973] fuel crisis.

AFCAA   *1* Air Force Cost Analysis Agency (Arlington, Va, established 1 August 1992).

2 Air Force Computer Acquisition Center (Hanscom AFB, part of AFCC).

AFCAC   African Civil Aviation Commission (Int., established 1969).

AFCAS, Afcas   Automatic flight control augmentation system.

AFCC   *1* Air Force Communications Command (HQ Scott AFB, formed from AFCS 15 November 1979, became AFC^{4}A 28 May 1993).

2 Office of the Chief of Staff (USAF).

AFCCCCA   See *AFC^{4}A*.

AFCE   Automatic flight control equipment, linked Norden bombsight to autopilot.

AFCEA   Armed Forces Communications and Electronics Association (US).

AFCEE   Air Force Center for Environmental Excellence (Brooks AFB, established 23 July 1991).

Afcent, AFCENT   Allied Forces, Central Europe (Brunssum, Netherlands).

AFCESA   Air Force Civil Engineer Support Agency (Tyndall AFB, established 1 August 1991).

AFC^{4}A   Air Force Command, Control, Communications and Computer Agency (Scott AFB, established 28 May 1993).

AFCI   Arc-fault circuit interruption.

AFCLC   Air Force Contract Law Center (Wright-Patterson AFB).

AFCMC   Air Force Contract Maintenance Center (Wright-Patterson AFB).

AFCMD   Air Force Contract Management Division (Kirtland AFB, was part of AFSC).

AFCMR   Air Force Court of Military Review.

Afcoms, AFCOMS   Air Force Commissary Service, became Defense Commissary Agency (Kelly AFB).

AFCP   Advanced flow-control procedure (ATC).

AFCPMC   Air Force Civilian Personnel Management Center (Randolph AB).

AFCRL   Air Force Cambridge Research Laboratoriesm became part of ESC.

AFCS   *1* Automatic flight-control system.

2 Air Force Communications Service, became AFCC, then part of AFC^{4}A.

3 Active-facility control system.

AFD   *1* Air Force Department (UK, MoD).

2 Adaptive flight display.

3 Advanced flight deck.

4 Autopilot flight director.

A/FD   Airport/facility directory (US).

AFDAS   Aircraft fatigue-data analysis system.

AFDC   *1* Auto flight-director computer; S adds system.

2 Automatic formation drone control (USN).

3 Air Force Doctrine Center (Langley AFB, Va, established 21 July 1993).

AFDK   After dark.

AFDMR   Director of Military Requirements (USAAF).

AFDPS   Automated flight-processing system.

AFDS   *1* Autopilot [and] flight-director system.

2 Air Fighting Development Squadron (RAF, WW2).

3 Advanced flight-deck simulator.

4 Autonomous flight, or freeflight, dispenser system.

AFDTC   Air Force Development Test Center (Eglin AFB).

AFDX   *1* Andio-frequency digital bus [Ethernet].

2 Avionics full duplex Ethernet.

AFEE   Airborne Forces Experimental Establishment (UK, June 1942 Sherburn-in-Elmet, January 1945 Beaulieu Heath, to September 1950).

AFEI   Association for Enterprise Investigation (US).

AFEP   Air Force education plan (USAF).

AFEPS, Afeps   Acars front-end processing system.

AFESA   Air Force Engineering and Services Agency (USAF).

AFEWC   Air Force Electronic Warfare Center (USAF).

AFEWES   Air Force electronic-warfare evaluation simulator.

AFFA   Department of Agriculture, Fisheries and Forestry – Australia.

AFFDL   Air Force Flight Dynamics Laboratory (USAF, at AFWAL).

AFF   *1* Autonomous formation flight.

2 Airmet fax forecast.

3 Automatic flight following.

AFFF, AF^3   Aqueous film-forming foam.

affinity group   Collection of people having a common interest, that interest often being solely an ** charter, at an attractive fare.

affirmative   R/T response meaning 'yes'.

affordable moving surface target engagement   Fuses multiple GMTI and SAR to give accurate direction to inexpensive air/ground munitions.

AFFS   Airborne firefighting system [replaces Maffs].

AFFSA   Air Force Flight Standards Agency (Andrews AFB, established 1 October 1991).

AFFMA   Air Force Frequency Management Agency (Arlington, Va, established 1 October 1991 by renaming AFFM Center).

AFFSCE   Air Forces Flight Safety Committee, Europe (Int.).

AFFTC   Air Force Flight Test Center (Edwards AFB from 1948).

AFG   *1* Aerospace focus group.

2 Airfoil group (LGB).

3 Arbitrary-function generator.

AFGE   American Federation of Government Employees.

AFGL   Air Force Geophysics Laboratory (Hanscom AFB).

AFGS   Autonomous flight-guidance system.

AFGWC   Air Force Global Weather Central.

AFH   *1* Above field height.

2 Airframe flight hours.

3 Advanced fibre heater.

AFHF   Air Force Historical Foundation (Andrews AFB).

AFHRA   Air Force Historical Research Agency (at AFHR Center, Maxwell AFB, established 12 September 1949).

AFHSO   Air Force History Support Office (Washington DC, established 30 September 1994).

AFI   *1* Assistant flying instructor.

2 Authorised Flying Instructor, proposed by FAA 1995 to succeed CFI(2).

3 Authority format identifier.

AFIA   *1* Air Force Inspection Agency (Kirtland AFB, established 1 August 1991).

2 Aerial Firefighting Industry Association (US).

AFIC   AFI(1) course.

AFIL   Air-filed [after takeoff] flight plan.

AFIO   Association of Former Intelligence Officers (US).

AFIRMS   Air Force integrated readiness measurement system.

AFIS   *1* Airfield/aerodrome/airport automatic flight-information service; O adds officer.

2 Air Force Intelligence Service (Washington DC).

3 Airborne flight-information system [= VHF datalink].

AFISC   Air Force Inspection and Safety Center (Norton AFB).

AFISDO   Air Force Information Systems Doctrine Office (Keesler AFB).

AFISQ   AFIS(1) officer.

AFIT   Air Force Institute of Technology (Wright-Patterson AFB, administered by AU, Maxwell).

AFITAE   Association Française des Ingéieurs et Techniciens de l'Aéronautique et de l'Espace (F).

AFIWC   Air Force Information Warfare Center.

AfK   Aramid-fibre composite (G), also called **SfK**.

AFL   . Above field level.

AFLC   Air Force Logistics Command (Wright-Patterson AFB).

AFL-CIO   American Federation of Labor, Congress of Industrial Organizations.

AFLM   Air Force Logistics Management Center (AU, Gunter Annex).

AFLMA   Air Force Logistics Management Agency (Maxwell AFB, established 30 September 1975).

AFLO   Airborne force liaison officer, stationed at departure airfield.

AFLSA   Air Force Legal Services Agency (Bolling AFB, established 1 September 1991).

AFLSC   Air Force Legal Services Center (Washington DC, became LSA).

AFM   *1* Air Force Medal (RAF and Commonwealth air forces).

2 Air Force Manual (USAF).

3 Aircraft/airplane/approved flight manual.

4 Atomic-force microscopy.

5 Airfield friction meter.

6 Affirmative.

7 Note, USAF Museum is The Air Force Museum (TAFM).

AFMA   *1* Armed Forces Management Association (US).

2 Anti-fuel-misting additive.

AFMC   *1* Aluminium-filled metal ceramic.

2 Air Force Materiel Command (Wright-Patterson AFB, activated 1 July 1992).

3 Auxiliary fuel-management computer.

AFMEA   Air Force Management Engineering Agency (Randolph AFB, established 1 November 1975).

AFML   Air Force Materials Laboratory.

AFMOA   Air Force Medical Operations Agency (Bolling AFB, established 1 July 1992).

AFMPC   Air Force Military Personnel Center (Randolph AFB, TX).

AFMS   Automatic, or advanced, flight-management system.

AFMSA   Air Force Medical Support Agency (Brooks AFB, established 1 July 1992).

AFMSS   Air Force mission-support system (aircraft, UAVs, guided munitions, many armed forces worldwide).

AFN   ATS(1) facilities notification.

AFNA   Air Force News Agency (Kelly AFB, established 1 June 1978).

Afnor   Association Française de Normalisation [standardization] (F).

Afnorth   Allied Forces Northern Europe.

Afnorthwest   Allied Forces NW Europe (High Wycombe, UK).

AFNWSP   Air Force nuclear-weapons surety plan.

AFO   *1* Aerodrome/airport fire officer.

2 Announcement of [space] flight opportunities.

AFOAR   Air Force Office of Aerospace Research.

A-FOD tyre   Tyre designed to avoid picking up material causing FOD.

AFOG   Air Force Operations Group (Washington DC, established 26 July 1977).

AFOLTS   Automatic fire overheat logic test system.

AFOMS   Air Force Office of Medical Support (Brooks AFB).

AFOR   Aviation forecast, a visual flight (GA) weather service (Europe, not UK).

AFOSI   Air Force Office of Special Investigations (Bolling AFB, established 1 August 1948).

AFOSP   Air Force Office of Security Police (Kirtland AFB).

AFOSR   Air Force Office of Scientific Research (Bolling AFB).

AFOTEC   Air Force Operational Test and Evaluation Center (Kirtland AFB, three [previously five] detachments, established 1 January 1974).

AFOVRN   Air Force over-run, standard 1,000 ft of approach lights (USAF).

AFP   *1* Alternative flight plan.

2 Air Force Publication.

3 Acceleration along flight path.

4 Area forecast panel.

AFPA   Automatic flight-plan association (an electronic system).

AFPC   Air Force Personnel Center (Randolph AFB, established 1 October 1995).

AFPCA   Air Force Pentagon Communications Agency (Washington DC, established 1 October 1984).

AFP costs   All flight personnel.

AFPEO   Air Force Program Executive Office (Washington DC, established 1 November 1990).

AFPM   Association Française des Pilotes de Montagne.

AFPOA   Air Force Personnel Operations Agency (Washington DC, established 15 August 1993).

AFPRO   Air Force Plant Representatives Office.

AFPSS   Airborne-force protection surveillance system (USAF).

AFP turn   After passing fix.

AFQ   Association Française des Qualiticiens.

AFQI   Air Force Quality Institute (Air University, Maxwell).

AFR   *1* Air-fuel ratio.

2 Air Force Regulation (USAF).

3 AF Reserve is called Afres.

AFRAA   Association of African Airlines (Int.).

AfrATC   African Air Traffic Conference (Int.).

AFRBA   Air Force Review Boards Agency (Andrews AFB, established 1 June 1980).

AFRC   Air Force Reserve Command: (U adds Unit).

AFRCC   Air Force Rescue Co-ordination Center.

AFRCU   Air/fuel ratio control unit.

AFREA   Air Force Real-Estate Agency (Bolling AFB, established 1 August 1991).

Afres, AFRES   Air Force Reserve (Robins AFB, established 14 April 1948).

AFRFW   Air Force Research Flying Wing[s].

AFRL   Air Force Research Laboratory.

AFROC   Air Force Requirements Oversight Council.

AFROTC   Air Force Reserve Officers Training Corps.

AFRP   Aramid-fibre reinforced plastic[s].

AFRPL   Air Force Rocket Propulsion Laboratory (Edwards AFB).

AFRS   Auxiliary flight-reference system.

AFRSI   Advanced flexible reusable surface insulation (Shuttle).

AF/RTL   Auto-flight rudder-travel limit.

AFS   *1* Aeronautical fixed service (ICAO).

2 Auxiliary Fire Service (UK, WW2).

3 Aerodrome/airport/airfield fire service.

4 Air Force Station, or specialty (USAF).

5 Advanced Flying School (RAF).

6 Automatic, or automated, flight system (ie, AFCS[1]).

7 Automatic frequency selection.

8 Air Facilities Service (FAA to 1962).

9 Airborne file server.

AFSA   Air Force Services Agency (San Antonio, TX, established 5 February 1991).

AFSAA   Air Force Studies and Analyses Agency (Washington DC, established February 1991).

AFSAC   Air Force Security Assistance Center (Wright-Patterson AFB).

Afsarc   Air Force Systems Acquisition Review Council.

Afsat   Association Française des Sociétés d'Assurance Transports (F).

Afsatcom   Air Force satellite communications.

AFSATS   Air Force Security Assistance Training Squadron (AETC, Randolph AFB).

AFSC   *1* Air Force Systems Command (ARDC retitled 1 April 1961; inactivated 1 July 1992 on formation of AFMC).

2 Air Force Safety Center (Kirtland AFB, AF Safety Agency renamed 1 January 1996).

3 Air Force specialty code.

4 Aggregate friction surface coat (runway).

5 See *AFSPC*.

AFSCF   Air Force Satellite Control Facility (global network).

AFSCO   Air Force Security Clearance Office (Washington DC).

AFSD   *1* Air Force Space Division, usually called SD, formed as unit of AFSC 1960, incorporated into AFSPC 1982.

2 Airframe, or aircraft, full-scale development.

Afsinc, AFSINC   Air Force Service Information and News Center (Kelly AFB).

AFSK   Audio frequency-shift keying.

AFSOC   Air Force Special Operations Command, (Hurlburt Field, Fla., established 22 May 1990).

Afsouth, AFSOUTH   Allied Forces, Southern Europe (Naples).

AFSPA   Air Force Security Police Agency (Kirtland AFB, established February 1991).

AFSPC   Air Force Space Command (Peterson AFB, established 1 September 1992).

AFSS   *1* Association Française des Salons Spécialises.

2 Active flutter-suppression system.

3 Advanced fire-support system.

4 Air Force Security Service.

5 Automated flight service station.

AFSTC   Air Force Space Technology Center (Kirtland AFB).

AFT   *1* Advanced flying training.

2 Airframe fatigue test.

3 Aft-fuselage trainer.

AFTA   Avionics fault-tree analyser.

Aftac, AFTAC   Air Force Technical Applications Center (Patrick AFB, established 1 May 1960).

AFTC   Aerobatics Flight Training School (Compton Abbas, UK).

AFTENC   Air Force tactical exploitation of national capabilities.

afterbody   *1* Rear part of body, esp of transonic supersonic or hypersonic atmospheric vehicle.

2 Portion of flying-boat hull or seaplane float aft of step, immersed at rest or taxiing.

afterbody angle   In side elevation, acote angle between keel of *afterbody (2)* and (a) undisturbed water line or (b) longitudinal axis.

afterburn   Undesired, irregular combustion of residual proellant in rocket engine after cut-off.

afterburner   Jetpipe equippped for afterburning; in the case of a turbofan, reheat in the core flow only, see *augmentor*.

afterburning   Injection and combustion of additional fuel in specially designed jetpipe (afterburner) of turbojet to provide augmented thrust. Fuel, usually same as in engine, burns swiftly in remaining free oxygen in hot exhaust gas. Downstream of turbine, combustion can reach temperature limited only by radiant heat flux on afterburner wall and rate at which fuel can be completely burned before leaving nozzle. Nozzle must be opened out in area, with con-di profile to give efficiently expanded supersonic jet. Can be applied esp. effectively to turbofan or leaky turbojet, with greater proportion of available oxygen downstream of turbine. Very effective in supersonic flight, but less efficient at low speeds and very noisy (UK = reheat).

afterchine   Rear chine along afterbody (2).

aftercooling   Cooling of gas after compression, esp. of air or mixture before admission to cylinders of highly blown PE designed for operation at high altitudes.

after-flight inspection   Post-flight inspection.

afterglow   *1* Persistence of luminosity from CRT screen, gas-discharge tube or other luminescent device after excitation removed (of importance in design of many radar displays).

2 Pale glow sometimes seen high in western sky well after sunset due to scattering of sunlight by fine dust in upper atmosphere.

3 Transient decay of plasma after switching off EM input power.

Afterm   AFTN terminal.

aft-fan engine   Turbofan in which the fan is a free-running assembly behind the core, driven by a turbine linked only by the gas flow.

aft flap   Auxiliary curved flap mounted behind USB (Coanda) flap to complete turning of USB flow to beyond 90°.

aft flight deck   Rear area of aircraft flight-deck floor where this is at upper level above main floor. Not necessarily occupied by aircrew.

AF³, AF3   See *AFFF*.

AFTI   Advanced fighter technology integration.

AFTIL   Airways Facilities Tower Integration Laboratory (FAA).

aft limit of CG   Rearmost position of CG permitted in flight manual, pilot's notes, certification documentation or other authority. That at which stability in yaw and/or pitch, and static and manoeuvre margins, are still sufficiently good for average pilot to handle most adverse combination of circumstances in safety. CCV concept is leading to revolution in which much reduced, or negative, natural stability is held in check by AFCS.

aft-loaded wing   Supercritical wing, in which centre of pressure is exceptionally far aft because of lift generated by cambered trailing edge.

AFTN   Aeronautical Fixed Telecommunications Network (Int. from 1970).

AFTNS   Aircraft flight-track and noise system, displays 3-D position and noise of all aircraft near airport.

AFTRCC   Aerospace Flight-Text Radio Co-ordination Council.

AFTS   *1* Advanced Flying Training School.
2 Air/fuel test switch.

aft wing   In oblique-(slew-)wing aeroplane, wing pointing rearward.

AFU   Advanced Flying Unit.

AFUC   Average flyaway unit cost [see unit cost].

AFV   Automatic flyback vehicle.

AFVA   Association Française de la Voltige Aérienne.

AFW   Active flexible wing.

AFWA   Air Force Weather Agency (Offutt AFB, Nebraska).

AFWAL   Air Force Wright Aeronautical Laboratory.

AFWC   Air Force Wargaming Center (established 1986 at Maxwell AFB).

AFWL   Air Force Weapons Laboratory.

AFWR   Atlantic Fleet Weapons Range.

AG   *1* Air gunner.
2 Availability guarantee.
3 Assault glider (US, WW2).
4 Reconnaissance Wing (G).
5 Antenna group.
6 Adjutant-General (USA).
7 Arrester gear.

A/G   Air-to-ground.

A_g   Minimum resolvable area within patch illuminated by radar.

AGA   Airfields, ground aids and routes, main output of AIP(1).

AGACS   Automatic ground-to-air communication system [= data-link], not yet achieved.

Agard, AGARD   Advisory Group for Aerospace (formerly Aeronautical) Research and Development (NATO).

AGAQ   Association des Gens de l'Air du Québec.

Agars   Advanced general-aviation research simulator (at CAI [1]).

Agate   Advanced general-aviation transport experiments (NASA/industry).

Agatha   Air/ground anti-jam transmission from helicopter or aeroplane (F).

AGATMS   Action Group for Air Traffic Management Safety (Europe).

AGBR   Affordable ground-based radar.

AGC   *1* Automatic gain control, property of radio receiver designed to vary gain inversely with input signal strength to hold approximately constant output.
2 Affinity-group charter.
3 Adaptive gate centroid (radar tracking algorithms).
4 Active generalised control, digital protected FBW system of Rafale.
5 Active geometry [or geometric] control.

AGCAS   Automatic ground collision-avoidance system.

AGCS   *1* Advanced guidance and control system[s].
2 Air/ground communication system.

AGD   *1* Axial-gear differential.
2 Air generator drive (ie, windmill).

AGE   *1* Aerospace ground equipment (military inventory category).
2 Auxiliary ground equipment (Sigint).
3 Automated ground equipment (space).

A-gear   Arrester gear.

age-hardening   Many metal alloys, especially high-strength aluminium alloys, need time to harden after heat treatment, usually in order that partial precipitation may take place; preferably accomplished at room temperature or chosen higher value.

ageing (US = aging)   Time-dependent changes in microstructure of metal alloys after heat treatment. Some merely relieve internal stress but most improve mechanical properties.

Agent Defeat   Programme to create an air-delivered weapon able to destroy chemical and biological agents without causing their dispersal (DoD).

AGEPL   Association Général des Elèves Pilotes de Ligne (F).

AGES   Air/ground engagement system [also AGES II]; (AD adds air defense).

AGETS, Agets   Automated ground engine-test system.

AGFS   Aviation gridded forecast system (demo 1995).

AGI   *1* Advanced [or, post 1995, authorized] ground instructor (FAA).
2 ADNS/GDSS interface.

Agiflite camera   Hand-held, for photographing surface targets, especially ships (RAF).

AGIFORS, Agifors   Airline Group, International Federation of Operational Research Societies (Int., UK-based).

AGIL   Airborne general illumination light.

Agile   *1* Aircraft ground-induced loads excitation (simulates rough runways).
2 Airborne gyrostablized IR light equipment.
3 Advanced garment integrated life-support ensemble.

agile manufacturing   Rapid response to fluctuation in demand.

Agility Agile information transfer ability, active Satcom antennas.

agility *1* Loosely, manoeuvrability, esp. of air-combat fighter.

2 In particular, ability of fighter to change state quickly, to fly different mission.

AGIMS Air/ground information-management system.

AGINT, Agint Advanced GPS inertial-navigation technology.

AGIS Air/ground intermediate system.

AGL, agl *1* Above ground level.

2 Airborne gun-laying (radar).

3 Airfield ground lighting.

4 Automatic grenade launcher.

AGLT Airborne, or aircraft, gun-laying turret.

AGM *1* Air-to-ground guided missile (inventory category, USAF, USN).

2 Missile range instrumentation ship (US code).

AGMC Aerospace Guidance and Metrology Center (AFSC).

AGN Again.

Agnis Azimuth guidance for nose-in stands; also rendered as approach guidance nose-in to stand or aircraft guidance nose-in system.

AGO *1* Air-to-ground operator.

2 Andes Geophysical Observatory, Santiago.

agonic line Line joining all points on Earth's surface having zero magnetic variation. Two ** exist, one sweeping in curve through Europe, Asia and W Pacific and other roughly N–S through Americas.

AGOS Department of aviation, seaplanes and experimental construction (USSR).

Agpanz Agricultural Pilots' Association of New Zealand.

agplane Agricultural aircraft (colloq.).

AGPO Angle gate pull-off (radar).

AGPPE Advanced general-purpose processor element, a USAF common module.

AGR Air/ground router.

Agra Automatic-gain ranging amplifier.

agravic Hypothetical environment without gravitational field. Unrelated to weightless free-fall in gravitational field, or to possible points where net gravitational field of all mass in universe is zero.

agravic illusion Apparent movement of human visual field in weightless flight due to minute displacements of structure in inner ear.

AGREE Advisory Group on Reliability of Electronic Equipment (DoD/NATO).

AGRI Air/ground radar imaging.

agricultural Colloq., of hardware, essentially primitive and crude, but not necessarily ineffective or obsolescent.

agricultural aircraft Aircraft designed or converted for agricultural aviation.

agricultural aviation Branch of general aviation concerned with agriculture, specif. crop spraying, dusting, top dressing, seeding, disease inspection and, apart from transport, work with livestock.

AGRMS Air/ground router management system.

AGRRM Air/ground router regional manager.

AGS *1* Airborne ground, or air-to-ground, surveillance.

2 Aeronautical ground station (satcom).

3 Aircraft Generation Squadron (POMO).

4 Air Gunnery School.

5 Aircraft General Stores (spare parts).

6 Alliance [airborne radar] ground surveillance (NATO).

AGSS *1* Aerial gunner and scanner simulator.

2 Acars ground-system standard)AEEC).

AGTA Airline Ground Transportation Association Inc. (US).

AGTC Airport ground-traffic control.

AGTFT Anti-jam GPS technology flight test.

AGTS *1* Automated guideway transit system (airport terminal).

2 Air [or aerial] gunnery target system.

3 Air/ground test station.

AGTY Frequency agility.

AGU *1* American Geophysical Union.

2 Airlink gateway unit (satcoms).

AGV *1* Automated guided vehicle, part of most FMS (6); S adds system.

2 Avion à grande vitesse [= hypersonic] (F).

AGZ Actual ground zero.

Ag-Zn, Ag/Zn Silver/zinc electrical storage battery.

AH *1* Artificial horizon.

2 Or A_H, attitude hold.

Ah, A-h Ampere-hour.

AHA Aviation and hazard analysis.

AHB Attack Helicopter Battalion (USA).

AHC *1* Assault Helicopter Company (USA).

2 Attitude/heading computer.

AHD Ahead.

AHE Aerospace Hardware Exchange.

Ahead *1* Attitude, heading and rate of turn indicating system.

2 Advanced hit efficiency and destruction, programmable gun submunition.

AHF Aircooled, heavy fuel.

AHFM Alternate [ie, alternative] or advanced h.f. material (USAF).

AHI Aviation Health Institute (UK).

AHIP Army helicopter improvement program (USA).

AHM *1* Anti-helicopter mine.

2 Airplane health management.

AHMR Aircraft health-monitor recorder.

AHP Army heliport (USA).

AHQ Air headquarters.

AHRS Attitude/heading reference system.

AHS International American Helicopter Society.

AHSA The Aviation Historical Society of Australia.

AHSNZ Aviation Historical Society of New Zealand, Inc.

AHSW Aural high-speed warning.

AHT Automated hover trainer.

AHTR Auto horizontal-tail retrimming after landing.

AHU Aircraft Holding Unit, for military aircraft temporarily surplus to requirements; also said to mean Aircrew HU.

AHWG Aviation Health Working Group (UK Parliamentary Committee 2002).

AI *1* Airborne interception (radar).

2 Artificial intelligence.

3 Air-data/inertial.

4 Attitude indicator.

5 Alternative interrogator.

AIA *1* Aerospace Industries Association of America Inc.

2 Associazione Industrie Aerospaziali (Italy).

3 Associazione Italiana di Aerotechnica.

4 Académie Internationale d'Astronautique.

5 Atelier Industriel de l'Air (F).

6 Air Intelligence Agency (USAF, Kelly AFB, established 1 October 1993).

7 Advanced-information architecture.

AIAA   *1* American Institute of Aeronautics and Astronautics (office Reston, Va).

2 Aerospace Industry Analysts Association (US).

3 Area of intense air activity.

AIAC   Aerospace [previously Air] Industries Association of Canada.

AIAD   Associazione Industrie Aerospaziali e Difesa (Italy).

AIAEA   All-India Aircraft Engineers Association.

AIAH   Advanced integrated avionic, or aircrew, helmet.

AIANZ   The Aviation Industry Association of New Zealand, Inc.

AIB   *1* Accidents Investigation Branch [from 1988 AAIB] (UK).

2 Aeronautical Information Bureau.

3 Airfield, or aerodrome, identification beacon.

AIBU   Advanced interface blanker unit.

AIC   *1* Aeronautical Information Circular (CAA).

2 Advanced industrial countries.

3 Air-inlet control[ler].

4 Aluminium/iron/cerium.

5 Automatic integrity check (MLS).

AICAA   Associazione Italiana Costruttori Amatori d'Aerei (homebuilders).

AICBM   Anti-ICBM.

AICCA   Australian International Cabin-Crew Association.

AICES   Association of International Air Courier and Express Services (office UK).

AICGS   Advanced imagery common ground station.

AICMA   Association Internationale des Constructeurs de Matériel Aérospatial.

AICS   *1* Automatic inlet, or intake, control system.

2 Airborne integrated communications system.

AID   *1* Aeronautical Inspection Directorate (UK, 1913–).

2 Altered-item drawing.

3 Agency for International Development (US).

4 Airport information desk (FAA).

5 Accident/incident/deficiency.

AIDA   *1* Associazione Italiana di Aerofilatelia (philately).

2 Aeronautical integrated data-exchange (EL adds economy line).

3 Artificial-intelligence discrimination architecture.

AIDAA   Associazione Italiana di Aeronautica e Astronautica.

AIDC   Air-traffic services interfacility data communications.

Aidews   Advanced integrated defensive EW system.

Aids, AIDS   *1* Airborne/aircraft/automatic/advanced integrated data system/suite.

2 Acoustic-intelligence data system.

3 Aircraft intrusion detection system (temporary parking area).

AIDU   Aeronautical Information Documents Unit (RAF).

Aieda, AIEDA   Association Internationale des Avocats et Experts en Droit Aérien.

AIEM   Airlines International Electronics Meeting.

AIEU   Aircraft integrated electronics unit.

AIEWS   Advanced integrated EW suite (USN).

AIFF   Advanced identification friend or foe.

AIFS   Advanced indirect fire system[s].

AIG   Acas implementation group.

Aigasa   Associazione Italiana Gestori Aeroporti e Aeroportuali.

AII   Anti-icing/anticing inhibitor.

AIIDED   Active integrated inlet-duct engine demonstration.

AIIS, AI²S   Advanced IR imaging seeker.

AIL   *1* Airworthiness Information Leaflet.

2 Aeronautical Instrument Laboratory (USN, from 1943).

ail   Aileron.

AILA   Airborne, or automatic, instrument landing, or instrumented low, approach; S adds system.

ailavator   Control surface functioning as aileron and elevator (see *elevon*).

aileron   *1* Control surface, traditionally hinged to outer wing and forming part of trailing edge, providing control in roll, ie about longitudinal axis. Seldom fitted to aircraft other than aeroplanes and gliders, and in recent years supplemented or replaced by spoilers, flaperons, elevons and tailerons, while in some high-speed aircraft contentional * are mounted inboard to counter * reversal.

2 Effectiveness of lateral-control system, as in phrase 'to run out of *'.

aileron centring device   Another name for a wing leveller. Typically incorporates two springs, able to overcome friction and air loads.

aileron drag   Asymmetric drag imparted by aileron deflection, greater on down-going aileron (see *differential* *, *Frise* *).

aileron droop   Rigging of manual ailerons so that in neutral position both are at a positive angle relative to the wing.

aileron reversal   As aircraft speed increases, deflection of aileron can twist wing sufficiently to reduce, neutralise and finally reverse rolling moment imparted to aircraft. Many aircraft designed for Mach numbers higher than 0.9 either have no traditional outboard ailerons or else lock these except at low speeds.

aileron reversal speed   That at which pilot input is reduced to zero.

aileron roll   See *slow roll*.

aileron wedge   See *wedge*.

AILS   *1* Airborne information for lateral spacing.

2 Automatic ILS.

AIM   *1* Aeronautical [until 1995 Airmen's] Information Manual (FAA).

2 Air intercept missile (inventory category, USAF, USN).

3 Aerospace industrial modernisation programme (US).

4 Aluminium/iron/molybdenum.

5 Automatic inflation modulation.

6 Advanced intelligent management.

7 ATC/IFF beacon Mk XII.

8 Accelerated introduction of materials.

9 See *AIMD*.

AIMAS Académie Internationale de Médecine Aéronautique et Spatiale.

aim-bias Error between aiming point and centre of dispersion area of statistically valid number of projectiles.

AIMD Aircraft intermediate maintenance department.

aim dot Basic command reference symbol in gunsight or HUD: can show where bullets would hit if gun fired, but usually also gives other indications.

AIMDS Aircraft integrated monitoring and diagnostic system.

AIME Autonomous integrity monitored extrapolation.

AIMES Avionics integrated maintenance expert system (McDonnell Douglas).

Aim/Far Aeronautical Information Manual, Federal Aviation Regulations.

AIMIS, Aimis Advanced integrated modular instrumentation systems (USN).

aim-off Angular allowance when firing at moving target with unguided projectile, usually because of sightline spin resulting in target changing apparent position during projectile's flight, but in air-to-air combat possibly because of lateral air drag (eg in firing at aircraft abeam at same speed and heading, when sightline spin is zero).

AIMS *1* Attitude indicator measurement system.

2 ATCRBS, IFF, Mk 12 transponder, System.

3 Aircraft identification monitoring system (DoD, interceptors).

4 Automated integrated manufacturing system.

5 Airplane, or aircraft, information-management system [reports problems to maintenance staff].

6 Airborne integrated management system.

7 Airport information management system [vast range of data].

8 Advanced imaging multispectral system.

9 Aircraft integrated monitoring system [accident-related and life data].

10 Advanced integrated MAD system.

11 Airspace information monitoring system [major airports, G].

AIMV Aluminium/iron/molybdenum/vanadium.

Aimval Air intercept missile evaluation (USAF/USN).

AIN Airline identification number.

AINS *1* INS with prefix advanced, aided, area or airborne.

2 Associazione Internationale Nomo nello Spazio (I).

AINSC Aeronautical industry service communications.

AIO Action information organization (mainly warships, in relation to aircraft).

AIOA Aviation Insurance Offices Association (UK).

AIP *1* Aeronautical Information Publication[s].

2 Asars, or airport, improvement program.

3 Anti-surface-warfare improvement program[me].

4 Air-independent propulsion.

5 Australian industrial participation.

6 Associação Industrial Portuguesa.

AIPA Australian and International Pilots' Association.

AIPPI Association Internationale pour la Protection de la Propriété Intelectuelle (Int.).

AIPT Advanced image-processing terminal.

AIR *1* French aerospace material specification code.

2 Air intercept[or] rocket (inventory category, USAF).

3 Air-inflatable retarder (similar to ballute).

4 Air-intercept radar.

5 Advanced integrated recorder.

air Air near Earth's surface usually taken to be (% by volume): nitrogen 78.08; oxygen 20.95; argon 0.93; other gases (in descending order of concentration, carbon dioxide, neon, helium, methane, krypton, hydrogen, xenon and ozone) 0.04. In practice also contains up to 4% water vapour. ISA SL pressure at 16.6°C is 10.332 $kg\ m^{-2}$(= 761.848 mm Hg) and density 1.2255 $kg\ m^{-3}$.

air abort Abort after take-off.

Airac Aeronautical information regulation and control, system for disseminating air navigation information (Notams).

Airad Airmen advisory (local).

AI radar Airborne interception radar, carried by fighter for finding and tracking aerial targets.

air attack An experienced firefighter who not only provides the IC(8) with an overview but also knows how best to allocate resources.

air-augmented rocket Usual form of this propulsion system is for first stage of combustion, or primary rocket propellant or gas-generator, to yield fuel-rich range of products which then combine in second stage of combustion with atmospheric air (normally induced through ram intake). Objective is to increase specific impulse, by using oxygen from atmosphere, and also burn time and vehicle range.

air bag Rapidly inflated flexible bag to cushion VL of UAV or other object.

airband Those frequencies used for aeronautical voice communications.

air base *1* Loosely, military or general-aviation airfield (term used mainly by popular media).

2 In photogrammetry, line joining two air stations.

3 Length of (2).

4 Scale distance between adjacent perspective centres as reconstructed in plotting instrument.

air bearing Gas bearing using air as working fluid.

air-bearing table Table supported on single spherical air bearing and thus free to tilt, without sensible friction, to any attitude within design constraints.

air-blast switch Electrical circuit-breaker in which arc formed on breaking circuit is blown away by high-velocity air jet.

air-blast transformer In this context, as in some other electric and electronic equipment dissipating large heat flux, air-blast signifies forced air cooling.

air bleed See *Bleed* (2).

air block Rectilinear volume of atmosphere between designated FLs over published geographical area.

airblown seal Fed with air at pressure slightly higher than surroundings, thus excluding oil or other contaminants.

airborne Sustained by atmosphere or vertical component of propulsive thrust. Implication is that vehicle is not above sensible atmosphere; term not normally used in connection with spaceflights not involving aerodynamically supported vehicles, but applicable to wingless jet-lift devices.

airborne alert Generally, long-duration mission flown by strategic bomber, in all respects ready to make real attack, to reduce reaction time and remove possibility of destruction by ICBM or SLBM attack on its base. Until World War 2 'air alert' was method of deploying interceptor (pursuit) forces, keeping them on sustained flight in likely combat area under ground vector control.

Airborne Cigar   Powerful transmitters on which RAF bombers broadcast misleading instructions to german night fighters in WW2.

airborne early warning, AEW   Use of aircraft to lift powerful search radar to greatest possible height to extend line-of-sight coverage (very approximately, LOS radius in statute miles is square root of 1.5 times observer's height in feet). Modern AEW can give a PPI covering 170,000 sq miles, throughout which two low-level aircraft in close formation can be individually distinguished against ground clutter.

airborne fog blind   Translucent blind or hood admitting light to cockpit or flight deck whilst removing external visual cues.

airborne force   Force constituted for airborne operations.

airborne gunlaying turret, AGLT   Bomber-defence gun turret incorporating automatic provisions for aim-off and other corrections when engaging aerial targets.

airborne interception, AI   Use of aircraft to find, and close with, another aircraft; specifically, use of fighter to intercept, challenge by IFF and, if dissatisfied, destroy another aircraft.

airborne operation   Movement of combat forces and logistic support into combat zone by air.

airborne radio relay   Use of airborne relay stations to increase range, flexibility or security of communications.

airborne target handover system   Coded data-link enabling aircraft to hand over target (usually on ground) to a friendly station, without voice.

Air Box   Air Ministry (RAF, colloq.).

airbrake   Passive device extended from aircraft to increase drag. Most common form is hinged flap(s) or plate(s), mounted in locations where operation causes no significant deterioration in stability and control at any attainable airspeed. Term not normally applied to flaps, drag chute or thrust-reverse systems.

air-breathing   Aspiring air, specifically aircraft propulsion system which sustains combustion of fuel with atmospheric oxygen. Imposes constraints on vehicle speed and height, but invariably offers longer range than rocket system for same vehicle size or mass.

airbridge   *1* Elevated metal 'bridges' linking logic gates on an integrated circuit chip.

2 See *bridge.*

Air Britain   Despite name, international enthusiast body, formed 1948, now has (Historians) added to title.

airburst   Detonation of explosive device well above Earth's surface. Almost all nuclear weapons are programmed for optimised airburst height, which varies with weapon and target.

AIRC   Airlines Industrial Relations Conference (US).

air carrier   Organization certificated or licensed to carry passengers or goods by air for hire or reward.

air cartography   Aerial survey, esp. aerial photography for purpose of mapmaking.

Air Cavalry   Helicopter-borne attack/reconnaissance ground troops (USA).

AIRCMM   Advanced infra-red countermeasures munition.

aircom   Traffic on an Acars link (SITA).

AIRCON   Air communications network, specif. serving US air carriers. Characterised by wide geographical extent, very large information flow, 'on-line, real time,

full-time' storage, and computer-compatible electronic switching.

air conformal ice detection system   Measures thickness and characteristics by scattering of light from fibre optics.

air controller   In military operations, an individual trained for and assigned to traffic control of particular air forces assigned to him within a particular sector.

air control team   Team organised to direct CAS3 strikes in the vicinity of forward ground elements.

air-cooled   Heat-generating device, esp. piston engine, maintained within safe limits of temperature by air cooling. Invariably cooling is direct, in case of piston engine by radiating heat to air flowing between fins around cylinder head and barrel, or around hot rotor casing(s) of RC engine.

air corridor   *1* Defined civil airway crossing prohibited airspace.

2 Restricted air route in theatre of military operations intended to afford safe passage for friendly air traffic.

aircraft   Device designed to sustain itself in atmosphere above Earth's surface, to which it may be attached by tether that offers no support. Two fundamental classes are aerodynes and aerostats. Aircraft need have no means of locomotion (balloons are borne along with gross motion of atmosphere, while kites are tethered and lifted by motion of atmosphere past them), or any control system, nor means for aerodynamic or aerostatic lift (eg, jet VTOL aircraft need be no more than jet engine arranged to direct efflux downwards). Free-falling spacecraft qualifies as aircraft if, after re-entry, its shape endows it with sufficient L/D ratio to glide extended distance, irrespective of whether or not it can control its trajectory.

aircraft cabin mattress   Unpacked from storage bag, converts two facing seats + intermediate table into foam bed.

aircraft cable   Specially designed tensile cable, usually either solid wire or any of eight built-up constructions, used for operating flight control and other mechanical systems.

aircraft carrier   Marine craft, traditionally large surface vessel, designed to act as mobile base for military aircraft.

aircraft categories   *1* For genealogical purposes, family tree of possible classifications.

2 For certification purposes, subdivision of aeroplanes (most important family of aircraft) on basis of performance. In UK aeroplanes certificated before 1951 are categorised as No Performance Group Classification; after 1951 subdivided into Performance Group A, large multi-engined; Performance Group C, light multi-engined; and Performance Group D. Also Group X for large multi-engined aeropanes built outside UK before specified date.

aircraft certificate   In US all aeroplanes (airplanes) and most other aircraft except models are categorised and licensed according to four classes of certificate, each having status of legal document: airworthiness, production, registration and type.

aircraft commander   See *commander.*

aircraft communications and automatic reporting system   Monitors and records many parameters, mainly engine data.

aircraft container   See *container.*

aircraft dispatcher   In US air transport, official charged with overseeing and expediting dispatch of each flight.

Traditional post analogous to train dispatcher of US railroads. Today duties include provision of met. information, flight planning, arranging unloading and loading, stocking with consumables, apron servicing and other turnaround tasks, calling for large staff.

aircraft dope See *dope*.

aircraft fabric See *fabric*.

aircraft fuel See *gasoline*, *kerosene*.

Aircraft Holding Unit Accepted new aircraft off production, or in-service aircraft after major overhaul or repair, and tested them before allocation to operating unit (RAF, RN).

aircraft integrated data system Supplements 'black box' by monitoring and recording many additional engine and system parameters.

aircraft lifting bag Usually made of stout neoprene woven fabric, inflated to assist recovery of disabled or crashed aircraft; among other terms are pneumatic elevator and pneumatic aircraft jack.

aircraft log One or more volumes recording detailed operating life of individual aircraft, listing daily and cumulative flight time, notifiable irregularities or transient unserviceability of any part, all inspections, overhauls, parts replacement, modification and repair.

Aircraftman, Aircraftwoman RAF/WRAF non-commissioned rank, with junior and senior grades, having no bearing on trade in which rank-holder is qualified.

aircraft management simulator Essentially the same as a pre-1960 simulator, equivalent to a modern FFS but without 6-axis motion or synthetic external scenes; capable of training on all cockpit instruments and systems.

aircraft missile Missile launched from aircraft.

aircraft mover Apron vehicle for towing or pushback.

aircraft network interface unit Provides link between aircraft satcoms system and passenger [or possibly crew] PCs.

aircraft pallet See *pallet*.

aircraft performance monitoring Software calculates deviation(s) from specific range caused by aerodynamic deterioration of airframe.

aircraft prepared for service See *weight*.

aircraft rocket Missile launched from aircraft.

aircraft system controller Avionics subsystem performing flight engineering control and monitoring functions to automate hydraulic or electric or fuel or ECS or other system.

aircraft unit-load device See *Unit load*.

aircrew Crew required to operate aircraft, esp. crew, numbering more than one, of military aircraft. Large civil aircraft normally operated by flight crew and cabin crew; * is not used.

aircrew equipment assembly Standard modular fitting incoporating PEC and various other items carried on flying clothing, forming single 'umbilical' for military flight-crew member.

air cushion Volume of air at pressure slightly above local atmospheric, trapped or constantly replenished by suitably arranged air jets (possibly issuing from base of flexible skirt) to support ACV.

air data Parameters derived from measurements of the air mass surrounding the aircraft.

air-data computer Digital computer serving as central source of information on surrounding atmosphere and flight of aircraft through it. Typical ADC senses, measures, computes or transmits (to AFCS and other aircraft systems) pressure altitude, OAT and total temperature, Mach number, EAS, angle of attack, angle of yaw and dynamic pressure. All are corrected for known errors and converted into signals of form required by supplied systems. ADC may have 60 to 90 output channels, most used throughout each flight.

air defence Defence against aerial attack, ie attack by aircraft, atmospheric missiles and RVs entering atmosphere from space.

air defence identification zone Defined airspace within which all traffic must be identified, located and controlled (*ADIZ*).

air defence operations area Geographical area, usually large, within which air and other operations are integrated.

air defence region Geographical subdivision of an AD area.

air defence sector Geographical subdivision of an AD region.

Air-Dek PSP (1), a US registered name.

air despatcher Person trained to supervise release or ejection of cargo from aircraft in flight.

air distance Distance flown through the air, ie with respect to atmosphere.

air distributed mission trainer Features Ro-Ro cockpit to enable aircrew to receive individual or networked training on various aircraft types.

Air Division Largest administrative unit in USAF below Air Force.

air dominance Unquestioned military supremacy in aerospace.

air drag Drag.

air drill *1* Training or display by group of military aircraft which repeatedly change formation or perform manoeuvres.

2 Drill driven by high-pressure air.

air-driven horizon Artificial horizon in which gyro is driven by one or more high-velocity air jets, usually arranged to impinge on cups machined in periphery. In most, instrument case is connected to vacuum line, often generated by venturi, and jets are atmospheric air. Performance reduced at altitude and by contamination by foreign matter blocking or penetrating filter.

airdrome Incorrect corruption of aerodrome.

airdrop Delivery of personnel or cargo from aircraft in flight, usually by parachute.

airdrop platform Platform designed to carry large indivisible loads for airdrop or LAE.

Airep Air report, either spoken weather report by airborne aircrew or written air weather report.

Airex Patented low-density polyetherimide foam.

air exchange Release of a proportion of the air in closed-circuit tunnel on each pass and its replacement by fresh cooler air from atmosphere; hence, * system, control doors, cooling etc, and * rate expressed as % of tunnel airflow. Purpose is to regulate tunnel temperature and overall static pressure level.

air expeditionary force Multi-arms force quickly assembled to meet needs of a local commander and sent to a crisis point within hours (US).

airfield Land area designated and used, routinely or in emergency, for takeoff and landing by full-scale aerodyne(s). Definition excludes aerostats and model aircraft, but admits VTOLs and RPVs. No facilities need be provided.

airfield elevation   Height above MSL, usually of highest point on runway or other used surface.

airfield surface movement indicator   Airfield surveillance radar.

airfield surveillance radar   Radar on or near airfield with scanner well above ground level rotating continuously to give fine-definition PPI display, especially showing aircraft on ground and vehicles.

air-filed   Flight plan sent by aircraft in flight.

airflow   Air flowing past or through body. For immersed solids moving through air, major factors are speed (IAS, EAS, CAS, TAS), angle of attack or yaw, dynamic head, OAT and total temperature. For turbine engine, * normally mass flow, ie mass per unit time passing through engine.

airfoil   *Aerofoil*.

air force station   Usually means location of an air force unit where there is no airfield.

airframe   BS.185/1940: 'A flying machine without the engines', today BS.185 has added 'power driven'. Better definition is: assembled structure of aircraft, together with system components forming integral part of structure and influencing strength, integrity or shape. Includes transparencies, flush aerials, radomes, fairings, doors, internal ducts, and pylons for external stores. In case of ballistic rocket vehicle would not include thrust chambers of liquid-propellant engines, nor separable solid motors, but could include payload fairings. Items where argument exists include: RVs; MAD booms; rigid refuelling booms; mission equipment carried demonstrably outside structure proper (eg, AWACS aerial); and podded engine cowlings. Airframe usually includes landing gear, but not systems, equipment, armament, furnishings and other readily removable items.

airframe attributable   Accident or notifiable incident caused by defect or malfunction in airframe.

airframe parachute   Large parachute deployed from aeroydyne in emergency to provide ground impact at not over 30 ft/s.

airframer   Loosely, company or other organization whose primary business is manufacture of aircraft. Arguably, includes assembler of aircraft from major sections manufactured by partners.

air/fuel ratio   Ratio by mass of air to fuel in air-breathing engine or other combustion system. With hydrocarbon fuels ratio usually in neighbourhood of 16:1.

air gap   *1* Clearance between stationary and moving portions of electrical machine, crossed by magnetic flux.

2 Air space between poles of magnet.

3 Gap left in core of chokes and transformers used in radio or radar circuits to prevent saturation by d.c.

air-gap   Traditional type of piston-engine spark plug; or gas-turbine igniter in which c25,000V is required to make a spark jump through fuel/air mixture from electrode to body.

airglow   Quasi-steady radiation, visible at night, due to chemi-luminescence in upper atmosphere energised by solar radiation. Today often taken to embrace radiation outside visible range.

air gunner   Member of aircrew assigned, for whole or part of mission, to manning guns to defend aircraft [today's nearest equivalent = DSO (2)].

air hardening   Age hardening at room temperature.

airhead   *1* Designated area in hostile or disputed territory needed to sustain air landing; normally objective of airborne assault.

2 Air supply and evacuation base in theatre of operations.

3 In undeveloped region, nearest usable airfield.

4 CTOL base for support of dispersed VTOL operation.

air hostess   Stewardess.

air inlet, air intake   Admits air to duct inside aircraft, esp. to engine.

air interception   Radar or visual contact between a friendly and another aircraft.

air interdiction   Air attack on enemy forces sufficiently far from friendly forces for integration with the latter not to be required; esp. attack on enemy supply routes rather than theatre forces. Differs from BAI in that it interferes with enemy's major operational movement and prevents movement of forces into battle area.

air lag   See *lag*.

air/land warfare   Simultaneous warfare on land and in airspace above.

air launch   Release from aircraft of self-propelled or aerodynamically lifted object: missile, target or other aircraft (manned or RPV) previously attached to it (not towed).

Air League   UK air-minded association; not abb, founded 1909, in 1932–70 added : of the British Empire.

air lever   On early aeroplanes, hand throttle governing engine airflow (not fuel).

air liaison officer   Tactical air force or naval aviation officer attached to surface forces.

airlift   *1* Carriage by air of load, esp. by means other than routine airline operation.

2 Transport operation (usually military) in which aircraft make round-trip flights to transport large load such as army division or refugee population.

3 Continuing, open-ended logistic supply operation, such as Berlin *, 1948–49.

airlifter   Aeroplane [usually large] designed primarily for cargo, esp. military.

airline   *1* Certificated air carrier.

2 Public image of air carrier, created by house logo, aircraft livery and advertising, even where no such single carrier may exist.

3 Any great-circle route.

4 Ground supply pipe conveying air at typically 80 lb/in^2.

airliner   Not defined, but generally applied to large passenger aircraft operated by scheduled carrier; usage UK rather than US and becoming dated.

air load   Aggregate force exerted on surface by relative airflow. In case of aerofoil or control surface, force exerted on three-dimensional entity, not on just one of its surfaces.

air lock   *1* Small chamber through which personnel must pass to enter or leave larger chamber maintained at atmospheric pressure significantly different from ambient. Provided with two doors in series, never more than one door being open at a time.

2 Unwanted volume of air or other gas trapped at high

point in liquid system in such a way as to prevent or degrade proper operation.

air log Instrument for measuring air distance flown.

air mail Mail prepaid and sent by air where there is an alternative, cheaper surface route. European letters and postcards travel by air if this speeds delivery; no separate ** service.

Airman Aircraft maintenance analysis.

airman *1* Loosely, any aviator, aeronaut or man who navigates by air.

2 Tradesman certificated by appropriate licensing authority to work on aircraft.

3 Air force rank category (but not a rank) below NCO, equivalent to Army 'other ranks'.

4 (Capital A) lowest uniformed rank in USAF, with class subdivisions.

Air Maneuver Buzzword suggesting helicopters can have an effect in rapid battlefield movement (UK but US spelling).

airmanship Skill in piloting aircraft. Embraces not only academic knowledge but also qualities of common sense, quick reaction, awareness and experience.

Air Marshal Armed Federal officer riding incognito as ordinary passenger on flights by US carriers to deter terrorism. In view of prior existence of RAF rank, confusion would be reduced by standardizing on Sky-Marshal.

air mass Very large parcel of atmosphere which at lower levels exhibits almost uniform characteristics of temperature and humidity at any given level. According to Bergeron classification, grouped according to origin (Arctic, Polar, Tropical, Equatorial), subdivided into Continental or Maritime within each group, and then again into warm (w) or cold (k).

Airmec International Aircraft Maintenance and Engineering Exhibition and Conference.

Airmen Advisory Notice to Airmen (see *NOTAM*) normally issued locally, often verbally during pre-flight or in-flight briefing.

AIRMET, Airmet *1* In-flight weather advisory category less severe than SIGMET but potentially hazardous to simple aircraft flown by inexperienced pilot (US).

2 Telephone weather service (CAA UK).

air meter *1* Instrument on testbed [e.g., for engines] for measurement and recording of airflow mass per unit time.

2 Confusingly, a 1935 dictionary offers "A type of portable anemometer," which meassures velocity.

air mile *1* Aeronautical mile = nautical mile.

2 One mile flown through the air, following Hdg. at TAS; wind must be added to give distnce along Tr. at G/s. Hence * per gallon.

air mileage unit, AMU Mechanical calculating instrument, 1942–55, to derive continuous value for air distance flown. Output, more accurate and reliable than air log, was fed to air mileage indicator (AMI) and often other instruments.

air-minded Of general public, concerned to further aviation for prosperity, defence or sport.

air ministry In many countries, national department charged with administering military (sometimes all) aviation. In UK, replaced by MoD (RAF).

Airmis Airline management information system, EDP for smaller airlines.

airmiss Incident reported by at least one member of aircrew who considers there was "definite risk of collision" between two airborne aircraft (US). See *Airprox*.

airmobile, air-mobile Ground troops equipped and trained for insertion by air, making conventional landing (fixed or rotary wing).

air-mobile band Band of communications frequencies assigned to air-mobile forces.

air-mobile operation Operation by ground forces carried in air vehicles.

air movement Military air transport operation involving landing and/or airdrop.

air movement table Detailed schedule of utilisation of aircraft load space, numbers and types of aircraft, and departure places and times.

AIMS Airport unterference monitoring system.

Air National Guard, ANG Part-time voluntary auxiliary to USAF equipped with fighter, tactical strike and transport aircraft, organised as self-contained arm by each state.

air navigation Art of conducting aircraft from place of departure to predetermined destination, or along intermediate routes (eg to follow precise tracks in surveying). Originally pure pilotage (contact flying); by 1918 moved into nautical realm of dead reckoning and celestial observation (astro-nav); by 1960 all ** relied upon ground and airborne aids, except in gliders and simple light aircraft.

Air Navigation Commission Body charged with setting standards and operating practice. Reports to the Council, see next.

Air Navigation Council Governing body of ICAO.

air navigation facility Navaid; surface facility for air navigation including 'landing areas, lights, any apparatus or equipment for disseminating weather information, for signalling, for radio direction-finding, or for radio or other electronic communication . . .' (FAA). Today add 'for electronic position-finding.'

Air Navigation Orders Statutory instruments decreeing laws of civil air operations, including flight-crew licensing (UK).

AIRO Airborne IR observatory.

air officer, Air Officer *1* Loosely, officer commissioned in an air force.

2 Specif., officer of Air Rank in RAF.

air/oil strut Telescopic member utilising properties of air and oil to absorb compressive shocks (rarely, tensile) with minimal or controlled rebound.

air operator One who engages in flying for hire or reward, hence * Certificate.

air phone HF air/ground telephony.

airplane Aeroplane (N America).

air plot *1* Continuous air navigation graphic plot constructed (usually on board aircraft) by drawing vectors of true headings for lengths equivalent to air distances flown, today archaic.

2 Similar plot constructed for airborne object derived from visual or radar observation of its flight.

3 Automatic or manually constructed display showing position and movements of airborne objects (if in a ship, relative to the ship).

air pocket Sudden and pronounced gust imparting negative vertical acceleration; down-draught. Suggest archaic.

airport Airfield or marine base designated and used for public air service to meet needs of quasi-permanent community. Need be no facilities for aircraft replenishment or repair, customs facilities, nor scheduled service; but must be facilities for passengers and/or cargo. Community served can be mainly or even exclusively employees of one company (eg at oilfield).

airport advisory area   Area within 5 miles of geographical centre of uncontrolled airport on which is located FSS so depicted on appropriate sectional aeronautical chart (FAA)

airport advisory service   Terminal service provided by FSS located at airport where control tower is not operating.

airport code   Three-letter code identifying all commercial airports (eg, LHR, JFK, LAX).

airport commission   Board of management of most US airports.

Airport-G   Airport integrated research and development project for operational regulation of traffic guidance (Euret).

airport information desk   Unmanned facility at local airport provided for pilot self-service briefing, flight planning and filing of flight plans (FAA).

airport marker   See *marker*.

airport movement area safety system   Uses surface and airspace radar linked to predictive software to warn of future conflict on runway, taxiway or apron.

airport of entry   Airport provided with customs facilities through which air traffic can be cleared before or after international flight.

airport runway configuration   Current runways in use for takeoffs and landings, changes notified in advance.

airport surface detection/movement   See *ASDE, ASMI*.

airport surveillance radar   Approach-control radar used to display position of all traffic in TMA [up to 60 miles/100 km] providing range/azimuth but not height (FAA).

airport traffic area   Unless otherwise designated (FAR Pt 93), airspace within 5 miles of geographical centre of airport with TWR operating, extending up to, but not including, 3,000 ft AAL (FAA).

airport traffic control service   ATC service provided by airport TWR for aircraft operating on movement area and in vicinity (FAA).

airport traffic control tower   Facility providing airport ATC service (FAA).

air position   Georgraphical position airborne aircraft would occupy if entire flight was made in still air; point derived by plotting Hdg. and TAS.

Air Proving Ground Command   USAAF/USAF establishment at Eglin AFB for testing weapons, became AFDTC June 1957.

air-position indicator, API   Instrument which continuously senses Hdg. and TAS (usually not allowing for compressibility error) to indicate current air position.

airprox   Unintended near-miss by two airborne aircraft, considered sufficiently dangerous to be reported. (UK term).

air rage   Anti-social behaviour [usually caused by alcohol or drugs] by airline passenger.

air raid   Aerial attack on surface target, esp. against civil population.

Air Rank   Senior to Group Captain; Air Officer, equivalent to naval Flag Officer.

air refuelling control point   Location in space at which

boom-type tanker is 1,000 ft higher than receiver, heading on reciprocal 9 to 11 miles away laterally and 22 to 29 miles away longitudinally, whereupon 180° turn inwards is started.

air report, AIREP   Meteorological report sent by aircraft in flight.

air route   Defined airspace between two geographical points, subject to navigational regulations. See *airway*.

air route surveillance   Surface radar giving display(s) showing geographical position and height of all traffic along designated civil route (usually airway).

air route traffic control centre, ARTCC   Facility providing ATC service to aircraft operating on IFR flight plan in controlled airspace and principally during en route phase (FAA).

air-run landing   Final deceleration in ground effect followed by vertical landing (fixed-wing V/STOL).

air-run take-off   Vertical take-off followed by horizontal acceleration in ground effect (fixed-wing V/STOL).

AIRS, Airs   *1* Airborne integrated reconnaissance system (USN).

  2 Advanced inertial reference sphere, or system.

  3 Airline inventory redistribution surplus.

  4 Aircrew incident-reporting system.

  5 Advanced IR seeker.

  6 Atmospheric IR sounder (on Aqua EOS).

  7 Airborne IR surveillance.

  8 Alliance icing research study (Int.).

Air Safety Report   Filed by crew after a flight in which they encounter an untoward or potentially dangerous situation, which may be partly or entirely of their own making.

Airsar   Airborne synthetic-aperture radar.

airscape   Broad vista of sky, not necessarily including Earth's surface, from aerial viewpoint.

air scoop   Colloq. ram intake, esp. projecting from exterior profile of aircraft.

airscrew   BS.185, 1951: 'Any type of screw designed to rotate in air'. Word never common in US, but in UK used in early days of powered flight to denote rotary aerodynamic device intended to impart thrust. From about 1920–50 explicitly denoted tractor device ('propeller' being 'an airscrew joined to the engine by a shaft in compression'). Today redundant. See *fan, propeller, rotor, windmill*.

airscrew-turbine engine   Turboprop.

air/sea rescue, ASR   Use of aircraft to rescue life in danger at sea, esp. permanently established service for this purpose (UK, RAF: US, USCG).

airship   BS.185 1951: "A power-driven lighter-than-air aircraft". Thus need not be provided with means for controlling its path, though if * is to be of use such means must be provided. Traditional classes are: blimp, a small non-rigid; non-rigid, in which envelope is essentially devoid of rigid members and maintains shape by inflation pressure; semi-rigid, non-rigid with strong axial keel acting as beam to support load; and rigid, in which envelope is itself stiff in local bending or supported within or around rigid framework.

airside   *1* All parts of airport containing aircraft.

  2 For passengers, beyond departure customs, prior to arrival customs.

air snatch   *1* Recovery of passive body from atmosphere by passing powered aircraft, esp. recovery of space payload descending by parachute.

2 Recovery of human being from hostile territory or sea by passing aircraft unable to hover (see *Fulton*).

air sounding *1* Measurement of atmospheric parameters from sea level to specified upper level by transmitting or recording instruments lifted by rocket or aircraft (esp. balloon).

2 Record thus obtained.

airspace Volume of atmosphere bounded by local verticals and Earth's surface or given flight levels. May be controlled or uncontrolled, but always an administrative unit defined by precise geographical or Earth-referred locations.

airspace denial Military mission flown by fighter to destroy all hostile aircraft entering particular airspace, usually that above friendly troops.

airspeed, air speed Relative velocity between tangible object, such as raindrop or aircraft, and surrounding air. In most aircraft measured by pitot-static system connected to airspeed indicator (ASI) to give airspeed indicator reading (ASIR). When corrected for instrument error (IE), result is indicated airspeed (IAS). When corrected for position error (PE), result is rectified airspeed (RAS). Most ASIs calibrated acccording to ideal incompressible flow ($\frac{1}{2}\rho V^2$), so from RAS subtract compressibility correction to give equivalent airspeed (EAS). Finally density correction, proportional to difference between ambient air density and calibration density (1,225 gm^{-3}), applied to give true airspeed (TAS). This sequence ignores errors, usually transient, due to major changes in angle of attack (eg, in manoeuvres). Some ASIs calibrated to allow for compressibility according to ISA SL, indicating calibrated airspeed (CAS). Confusion caused by fact most authorities now use 'calibrated airspeed' to mean ASIR corrected for IE and PE; CAS thus defined would have to be corrected for density and compressibility. Thus since 1980 CAS must be regarded as ASIR + IE + PE; if allowing for compressibility then at ISA/SL CAS = TAS.

airspeed indicator, ASI Instrument giving continuous indication of airspeed.

airspeed transducer In flight testing, or performance measurement of unmanned vehicle in atmosphere, transducer giving electrical signal proportional to airspeed. In simple systems signal is d.c. voltage.

airsplint Lightweight splint, inflated for rigidity.

air spotting Correcting adjustment of friendly surface bombardment based on air observation.

airspray nozzle Fuel burner in gas-turbine engine which itself mixes fuel spray with primary air, avoiding smoke from fuel-rich combustion and incidentally reducing required fuel feed pressure.

AIRSS Advanced IR suppressor system.

air staging Gas-turbine combustion chamber having variable geometry to redistribute air under different engine operating conditions.

air staging unit Military unit stationed at airfield to handle all assigned air traffic calling at that airfield.

airstairs Passenger and/or crew stairway forming integral part of aircraft and, after use, folded or hinged up and stowed on board.

Airstar Airborne surveillance and target-acquisition radar.

airstart, air start Action of starting or re-starting aircraft main propulsion or lift engines in flight.

air-starter unit Apron vehicle or trailer providing air at 2.8–3.5 bar.

airstream, air stream *1* Moving air mass, esp. that penetrates and divides more stationary mass.

2 Loosely, any localised airflow.

airstrip Prepared operating platform for aeroplanes, usually from STOL to CTOL, distinguished from airfield by either: hasty construction under battlefield conditions; lack of permanent paved surfaces; lack of permanent accommodation for personnel or hardware; or lack of facilities, other than temporary fuel supply or ATC.

air superiority Degree of airspace dominance sufficient to prevent prohibitive enemy interference with one's own operations.

air-superiority fighter Combat aircraft designed specif. to clear airspace of hostile aircraft.

air supremacy Degree of air superiority sufficient to prevent effective enemy interference with one's own operations.

air surface zone Restricted area established to protect friendly surface vessels and aircraft and permit ASW operations unhindered by presence of friendly submarines.

air surveillance Systematic observation of airspace by visual electronic or other means to plot and identify all traffic.

AIRT Air-intercept radar training.

air taxi Aircraft below 12,500 lb TOGW, licensed to ply for hire for casual passenger traffic.

air taxiing Positioning helicopter or other VTOL or STOVL aircraft by short translational flight at very low altitude. Standardised ** manoeuvres form part of VTOL flying insruction.

air terminal Facility in city centre at which passengers can check in for flights and board coach to airport.

air time Elapsed time from start of takeoff run to end of landing run.

Airto Association of Independent Research & Technology Organisations (UK).

air-to-air From one aerial position to another, esp. between one airborne aircraft and another.

air-to-ground Between aerial position, esp. airborne aircraft, and land surface.

air-to-surface Between aerial position, esp. airborne aircraft, and any part of Earth's surface or target thereon.

air-to-underwater Between aerial position, esp. airborne aircraft, and location below water surface, esp. flight profile of ASW weapons and operating regime of ASW detection systems.

Air Track Landing Early form of ILS developed by NBS and Washington Institute of Technology.

air traffic Aircraft operating in air or on airport surface, exclusive of loading ramps and parking areas (FAA); aircraft in operation anywhere in airspace and on manoeuvring area of aerodrome (BSI). To air carriers 'traffic' has entirely different meaning, but this is never qualified by 'air'.

air traffic clearance See *clearance (1)*.

air traffic control radar beacon system Beacons along airways which trigger responses from airborne transponders providing identity, location and [usually] FL of equipped traffic. See *secondary radar*.

air traffic control centre   Unit combining functions of area control centre and flight information centre.

air traffic control service   Service provided for promoting safe, orderly and expeditious flow of air traffic, including airport, approach and en route ATC service.

air train   Aerial tug towing two or more gliders in line-ahead.

air-transportable hangar   Modular lightweight hangar erected over temporary site, such as crashed repairable aircraft.

air-transportable units   Military units, other than airborne, whose equipment is all adapted for air movement.

air transportation oversight system   Method of checking air-carrier safety procedures and programmes (FAA).

air transport operation   BS.185 1951: 'The carriage of passengers or goods for hire or reward'; this eliminates all military air transport, business flying and several other classes of ***.

air trooping   Non-tactical air movement of personnel.

air-tube oil cooler   Oil cooler in which air passes through tubes surrounded by oil.

air turnback   *1* Point at which mission already airborne is abandoned, for any reason.

2 Specif., point at which non-Etops aircraft has to abandon planned flight.

air umbrella   Massive friendly air support over surface operation or other air activity at lower level.

air vane   *1* Small fin carried on pivoted arm to respond to local changes in incident airflow; arm usually drives potentiometer pick-off sending signal of angle of attack or yaw.

2 Powered surface to control trajectory of ballistic vehicle in atmosphere (see *jet vane*).

air vector   In DR navigation, Hdg. and TAS (air plot).

air volume   In aerostat, volume of air displaced by solid body having same size and shape as envelope or outer cover. Volume used in airship aerodynamics.

air ward system   Aircraft used for surveillance, fisheries or customs patrol, police duties, reconnaissance and similar tasks.

airway   BS.185, 1951: 'An air route provided with ground organisation'. Most civil air routes are flown along ICAO IFR airways, typically 10 nm wide with centreline defined by point-source radio navaids spaced sufficiently close for inherent accuracy to be less than half width of airway at midpoint. Each airway has form of corridor, of rectangular cross-section well above Earth. Airspace within is controlled, and traffic separated by being assigned different levels and from ATC having position reports and accurate forecasts of future position (typically, by ETA at next reporting point). In general, made up of a series of route segments each linking two waypoints.

airway beacon   Light beacon located on or near airway (see *NDB*).

airways-equipped   Equipped with functioning statutory avionics and instruments (eg, two pressure altimeters) to satisfy ICAO requirements for flight in controlled airspace.

airways flying   Constrained to dogleg along centrelines of airways instead of flying direct to destination.

airway traffic control   Civil air traffic control formerly exercised on 'airways' basis; today no separate system for designated airways.

Air Wheel   Wheel/tyre combination introduced by Goodyear after World War 1, characterised by small wheel and fat tyre to absorb landing shocks.

air work   In flying instruction, student air time as distinct from classroom time.

airwork   Today usually one word, to explore aircraft's handling or perform tests or demonstrations in flight.

airworthiness   Fitness for flight operations, in all possible environments and foreseeable circumstances for which aircraft or device has been designed.

Airworthiness Directive, AD   Message from national certifying authority requiring [often immediate] mandatory inspection and/or modification.

Airworthiness Notice, ND   Not mandatory, but strong recommendation or advice.

airworthy   Complying with all regulations and requirements of national certifying authority.

airybuzzer   Aeroplane (colloq.).

AIS   *1* Aeronautical Information Service [AG adds automation group] (ICAO, UK).

2 Advanced [or airborne] instrumentation subsystem (ACMR).

3 Avionics intermediate shop.

4 Academic instructor school.

5 Aircraft indicated (air) speed (IAS is preferred).

6 Automated information system.

7 Airport information service.

AI²S   Advanced IR imaging seeker.

AISA   Ada instruction set architecture.

AISAP   AIS(1) Automation panel.

AISD   Airlift Information Systems Division (Scott AFB).

AISFS   Avionics integration support facilities.

aisle   Longitudinal walkway between seats of passenger aircraft.

aisle height   Headroom along aisle.

aisle stand   Pilot's instrument panel mounted on pillar in all-glazed nose [as in B–29].

AIT   *1* Alliance Internationale de Tourisme (Int.).

2 Assembly, integration and test.

3 Airborne integrated terminal [G adds group].

4 Atmospheric intercept [or] testbed [or] technology].

5 Automated information transfer.

6 Avanced intelligence tape.

AITAL, Aital   Asociación Internacional de Transportes Aéreos Latinoamericanos (Int.).

AITFA   Association des Ingénieurs et Techniciens Français des Aéroglisseurs (hovercraft) (F).

AIU   *1* Analog, or aircraft, or armament, interface unit.

2 Astro-inertial unit.

3 Airborne-installation unit.

4 Auto-ignition unit.

5 Audio integration unit.

AIV   *1* Accumulator isolation [or isolator] valve.

2 Aviation-impact variables (program).

AIVSC   Aviation Industry Vocational Standards Council (UK).

AIWS   Advanced interdiction weapon system, a stand-off ASM.

AIX   Advanced interactive executive.

AIZ   Aerodrome/airfield information zone.

AJ   Anti-jam.

A$_j$ Nozzle throat area (occasionally AJ).

AJA   Aft jamming antenna.

Ajax unit   Device for providing artificial feel in pitching plane as function of stick displacement, altitude and airspeed.

AJB   Audio junction box.

AJCN   Adaptive, later advanced, joint C4ISR node.

AJE   Augmented-jet ejector; VL adds vertical lift.

AJJ   Adaptive-jungle jammer, sophisticated ECM self-adapting to variable enemy transmissions.

AJM   Anti-jam modem.

AJPAE, Ajpae   Association des Journalistes Professionnels de l'Aéronautique et de l'Espace (F).

AJPO   Ada Joint Program Office.

AJPS   Afeps journal processing system.

AJS   Attack, Jakt, Spaning, attack, fighter, recon (Sweden).

AJT   Advanced jet trainer.

A/K   Aluminium/Kevlar (armour).

AKU   Avionic[s] keyboard unit.

AL   *1* Approch and landing chart (FAA).

    2 Spec prefixes for methanol or water/methanol (AL-24 piston engines, AL-28 gas turbines).

Al   Aluminium [US aluminum].

A/L   *1* Approach/land, operative mode for airborne system.

    2 Airline.

    3 Autoland.

ALA   *1* Alighting area (ICAO, marine aircraft).

    2 Asociación de Lineas Aéreas (Spain).

    3 Approach and landing accident[s].

ALAA   Aviation Légère de l'Armée de l'Air (F).

ALADC   Australian Light Aircraft Development Council.

Aladdin   Algorithm adaptive and diminished dimension.

ALAE   Association of Licensed Aircraft Engineers (trade union, UK, 1981, Australia).

ALAFS   Advanced lightweight aircraft fuselage structure [or system].

ALAM   Advanced land-attack missile.

ALAR   Aircraft [or approach and] landing accident reduction.

Alarm   *1* Air-launched anti-radiation missile.

    2 Automatic light-aircraft readiness monitor.

Alarms   Airborne laser-radar mine sensor.

ALARR   Air-launched, air-recoverable rocket.

ALAT   Aviation Légère de l'Armée de Terre (F).

Alats   Advanced laser targeting system.

ALAVIS   Advanced low-altitude IR reconnaissance system.

ALB   *1* Air/land battle.

    2 Aircraft lifting bag, for recovery after belly landing, etc.

albedo   *1* Percentage of EM radiation falling on un-polished surface that is reflected from it, esp. percentage of solar radiation (particularly in visible, or other specified, range) reflected from Moon or Earth.

    2 Radiation thus reflected.

ALBM   *1* Air-launched ballistic missile.

    2 Air/land battle management (Lockheed).

ALBT   Air-launched ballistic target.

ALC   *1* Air Logistics Command [now Air Mobility Command] (USAF).

    2 Air Logistics Center.

    3 Automatic level, or levelling, control (radio).

    4 Air logic control, for automated systems.

ALCA   Advanced light combat aircraft.

ALCAC   Air Lines Communications Administrative Council (US).

Alcam   Air-launched conventional attack missile.

ALCC   *1* Airborne launch control centre.

    2 Airlift control centre.

ALCE   Airlift control element (MAC).

Alclad   Trade name (Alcoa) of high-strength light alloys (usually sheet) coated with corrosion-resistant high-purity aluminium. Originally developed for marine aircraft.

ALCM   Air-launched cruise missile.

alcohol   Large family of hydrocarbons containing hydroxl groups, esp. methyl. alcohol CH_3OH (toxic) and ethyl alcohol (ethanol, C_2H_5OH, potable), both used as fuels, anti-detonants and rocket propellants.

Alcoseal   Range of film-forming foam compounds for extinguishing fires involving water-miscible solvents; * VSA has vapour-suppressing additive.

ALCS   *1* Active lift control system, to reduce peak wing stresses in gusts.

    2 Airborne launch-control system.

Alcusing   Light alloy (aluminium, copper, silicon).

ALD   *1* Arbitrary landing distance, standard comparison distance along runway, from touchdown to stop, using specified landing technique; used in determining field-length requirements.

    2 Available landing distance.

    3 JETDS code: piloted aircraft, countermeasures, combination of purposes.

ALDCS   Active-lift distribution control system.

Alder   Advanced laser devices and effects research.

Aldis   *1* Patented hand signalling lamp with optical sight, trigger switch (kept on throughout use) and second trigger to tilt mirror to deflect light beam intermittently down to target.

    2 Airport land-dues information system.

ALDP   Airborne laser designator pod.

ALDS   Airborne laser defensive system.

ALE   *1* Aviazione Leggera Esercito (Italian army aviation).

    2 Automatic [radio] link establishment.

ALEA   Airborne Law Enforcement Association (US).

Aleastrasyl   Refractory material for re-entry heat shields, a resin-impregnated silica fabric.

ALEK   See *anchor-line extension kit*.

Alerfa   Alert phase of SAR operation.

Alert   *1* Attack [and] launch early response [or reporting] to theater (USAF).

    2 Air-launched extended-range transporter.

alert   *1* Specified condition of readiness for action, esp. of military unit.

    2 Warning of enemy air attack.

    3 ATC action taken after 30 min "uncertainty" period (5 min in case of aircraft previously cleared to land) when contact cannot be established.

    4 Response by manufacturer and/or certifying authority to unacceptable incidence of service failures by hardware item.

alert area   Airspace which may contain high volume of pilot-training activities or unusual type of aerial activity (FAA).

Alerte   Anti-aircraft laser enemy ranging and targeting equipment.

alerting centre   Centre designated by appropriate authority to perform functions of RCC where none exists (BS.185).

alerting service   Service provided to notify and assist all appropriate organisations capable of aiding aircraft in need of search or rescue.

alert-level standard   Agreed reliability performance below which special and urgent action must be taken (eg 0.3 IFSD per thousand engine hours).

alerting unit   Encoding-altimeter device which, in potentially dangerous flight conditions, triggers a warning.

alert phase   Aircraft seriously overdue.

Alerts   Airborne-laser EW receiver training system.

ALES   Autonomous Link-Eleven system.

Alex   Automated launching of expendables (EW).

ALF   *1* Auto/lock-follow (target tracker).

2 Auxiliary landing field.

3 Aloft.

4 Airborne, or adaptive, or advanced, low frequency; EAS adds electro-acoustic sonar, S sonar.

Alfa, ALFA   Aéroports de langue Française associés.

Alfens   Automatic, or advanced, low-flying entry, or enquiry, notification system (UK NATS/MoD).

Alfensops   Automated low-flying and flight-planning enquiry and notification system operations centre (UK).

ALFH   Advanced lightweight flying helmet.

Alflex   Automatic-landing flight experiment (J).

ALFS   Airborne, or air-dropped, low-frequency sonar.

ALG   *1* Autonomous landing guidance.

2 Advanced landing ground.

3 Along.

4 Also Ali, altimeter.

AlGaAs   Aluminium/gallium arsenide.

algae   Primitive plants (thallophytes) which elaborate food by photosynthesis, investigated as human food for extended space travel.

algal corrosion   Degradation caused by algae and other microorganisms, especially those dwelling at fuel/air interface.

Algol   Algorithmic language.

algorithm   *1* Established method of computation, numeric or algebraic.

2 Computation with steps in preassigned order, usually involving iteration, for solving particular class of problem.

ALGP   Aviation Loan Guarantee Program (US, post 11–9).

ALH   *1* Advanced light helicopter.

2 Active laser homing.

ALI   *1* Advanced land imager.

2 Automatic line integration.

3 Aegis Leap intercept.

4 Aerospace Lighting Institute (US).

aliasing   Wide variety of errors possible when breaking down image into pixels, such as irregular edges or tendency of small polygons to blink on/off; hence anti-*.

ALIC   *1* Aerodrome/airfield/airport locator indicator code.

2 Aircraft launcher interface computer.

Alicat   Advanced long-wave IR circuit and array technology.

Alice   *1* Air-launched integrated countermeasure[s], expendable.

2 Alcatel integrated control environment.

alidade   Optical sight [microscope or telescope] used to read linear scale on mensuration system.

alien interference   On weather radar, by other radars in scanning area.

alight   To land, esp. of marine aircraft on water.

alighting channel   Part of water aerodrome navigable and cleared for safe alighting or taking off.

alighting gear   See *landing gear*.

align   *1* In INS, to rotate stable platform before start of journey until precisely aligned with local horizontal and desired azimuth.

2 In radio, radar or other equipment having resonant or tuned circuits, to adjust each circuit with signal generator to obtain optimum output at operating frequencies.

3 Normal meaning of word is relevant to erection of airframe jigging, lasers often being used when structures are large.

aligned mat   Intermediate semi-prepared composite structure in which strong and/or stiff reinforcing fibres (rarely whiskers) are arranged substantially parallel in two-dimensional mat.

alignment time   In INS or guidance system, minimum time required to spool-up gyros and align platform, preparatory to allowing significant movement of vehicle.

ALIMS   Automatic laser inspection and measurement system.

ALIS, Alis   *1* Airline interactive services.

2 Airport luggage identification system.

3 Autonomic logistics information system.

Alithalite   Range of medium-density general-purpose Al-Li alloys (Alcoa).

ALJEAL, Aljeal   Association of Lawyers, Jurists and Experts in Air Law (Int.).

ALJS   Airborne-laser jamming system.

alkali metal   Group of metals in First Group of Periodic Table characterised by single electron in outermost shell which they readily lose to form stable cation (thus, strongly reactive). Lithium, sodium, potassium and caesium (cesium) are important in electrical storage batteries and as working fluids in closed-circuit space power generation.

alkylation   Addition of alkyl group (generally, radical derived from the aliphatic hydrocarbons); important in manufacture of gasolines (petrols) having high anti-knock (octane) rating.

ALL   Airborne laser laboratory.

ALLA, Alla   Allied Long Lines Agency (NATO).

all-burnt   Rocket propulsion system which has consumed all propellant (where there are two, which has consumed all of either); specif., time and flight parameters when this occurs.

all-call   Transponder Mode-S broadcast interrogation, thus * address, * reply.

ALLD   Airborne laser locator designator.

alleviation   Reduction of structural loads (eg wing bending moment) in vertical gusts by active controls.

alleviation factor   Numerical multiplier of calculated vertical acceleration or structural load on encountering gust, taking into account fact gust is not sharp-edged and aircraft is already rising before peak intensity is reached. Later refined by making it a function of the ratio of mean chord to gradient distance, aspect ratio and mass parameter.

alleviation lag   Time difference between actual and ideal response of a GAC active control system.

alleviation technique   Method of reducing heat flux on atmospheric re-entry by controlling plasma sheath surrounding vehicle.

all-flying tail   Term formerly used to describe variable-incidence tailplane used as primary control surface in pitch, separate elevators serving merely as additional part of surface or as a means of increasing camber.

Al-Li   Aluminium-lithium alloys.

Alliance ground surveillance   Programme for reconnaissance aircraft (NATO 1980, approved 1993, operational possibly 2010).

allithium   Generic name for aluminium-lithium alloys; also (capital A) trade name.

ALLM   Aft lower-lobe module (AEW radar).

all-moving tail   All-flying tail.

'all out'   Signal signifying glider or other towrope taut, towed vehicle ready for takeoff.

allowable deficiency   Missing, damaged, inoperative or imperfectly functioning item which does not invalidate C of A and does not delay scheduled departure (eg, rudder bias system, fuel flowmeter and almost any item not part of structure or aircraft system). In US called despatch deviation or MEI (1).

allowance   *1* Intentional difference between dimensions (with tolerances permitted on each) of mechanically mating parts, to give desired fit.

2 Calculated quantity of fuel beyond minimum needed for flight carried to comply with established doctrine for diversion, holding and other delays or departures from ideal flight plan.

3 In sheet-metal construction, extra material needed to form bend of given inside radius and angle.

alloy   Mixture of two or more metals, or metal-like elements, often as solid solution but generally with complex structure. Small traces of one element can exert large good or bad influence. Most aircraft made principally of alloys of aluminium (about 95%, rest being copper, magnesium, manganese, tin and other metals) or titanium (commercially pure or alloyed with aluminium, vanadium, tin or other elements), with steels (alloys of iron) at concentrated loads.

all-shot   Aerial target hit by every round from one gunnery pass.

all-speed aileron   Lateral-control surface operable throughout flight envelope, as distinct from second aileron group on same aircraft operable at low speeds only.

'all systems go'   Colloq., absence of mechanical malfunction; not authorised R/T procedure.

ALLTV   All light-levels TV.

all-up round   Munition, especially guided missile, complete and ready to fire.

all-up weight   See *AUW*.

all-ways fuze   Fuze triggered by acceleration in any direction exceeding specified level.

all-weather   Former category of interceptor which could not, in fact, fly in **. Strictly true only for aircraft with triplex or quad AFCS and blind landing system plus ground guidance.

all-wing aircraft   Aerodyne consisting of nothing but wing. Some aeroplanes of 1944–49 were devoid of fuselage, tail or other appendages, and approached this closely.

ALM   *1* Air loadmaster.

2 Air-launched munition[s]; IPT adds integrated project team.

ALMDS   Airborne laser mine-detection system.

ALMS   *1* Aircraft landing measurement system, typically using IR beams and geophones to produce hard-copy print-out of final approach and touch-down.

2 Air lift management system (software).

ALMV   Air-launched miniature vehicle.

Alnico   Permanent-magnet materials (iron alloyed with Al, Ni, Co – hence name – and often Cu) showing good properties and esp. high coercive force.

Alnot   Alert notice.

ALNZ   The Air League of New Zealand Inc.

ALO   Air Liaison Officer.

ALOC   Air line of communication, airlift for spare parts (USA).

Alochrome   Surface treatment for light-alloy structure to ensure good key for paint: chemical cleaning, light etching and final passivating.

Alodine   Proprietary treatment similar to Alochrome.

Aloft   Airborne light-optical-fibre technology.

Alofts   Active low-frequency towed sonar [H adds helo].

AlON   Aluminium oxynitride.

along and across   Configuration of track position display unit in which separate windows show continuous reading of distance to go (along) and distance off track (across), usually driven by Doppler.

ALOS   Advanced land-observing, or observation, satellite.

ALOTS   Airborne lightweight optical tracking system [precision photograph of ballistic vehicles].

ALP   *1* Aircraft landing permit.

2 Aegis Leap program.

AlP   Aluminium powder.

ALPA   Airline Pilots' Association (US, from 1931), now name adds International, to include Canada.

Alpas   Air Line Pilots' Association of Singapore.

Alpax   Aluminium alloy containing c13 per cent Si, for intricate castings (1932).

alpha   Angle of attack of main wing (α, AOA).

alpha exit   The first available runway turnoff.

alpha floor   Mainly to protect against windshear, system which automatically applies full power if AOA exceeds preset value, and earlier if rate of change of TAS/GS passes preset thresholds.

alpha hinge   *1* Crossed-spring pivot (eg in tunnel balance).

2 Confusingly, drag hinge.

alpha max   The maximum attainable AOA with stick fully back.

alphanumeric   Character representing capital letter or numeral portrayed in precisely repeatable stylised form either by electronic output (computer peripheral or display) or printed in same form for high-rate reading by OCR system.

alpha particle   Nucleus of He atom: 2 protons, 2 neutrons, positive charge.

Alpha prot   Short for protection, the maximum attainable stick-free AOA. Auto trim stops there, because there is no reason to maintain this condition.

alpha speed, α-SPD Safe stall-margin speed (auto-throttle mode setting).

alpha vane Transducer measuring AOA.

Alply Trade name (Alcoa) of sandwich comprising polystyrene foam between two sheets of aluminium.

ALQA Automatic link quality analysis.

ALQDS All quadrants.

ALR *1* Alerting message [S adds Service].
 2 Aircraft (or airborne) liferaft.
 3 JETDS code: piloted, countermeasure, passive.
 4 Arbeitsgruppe für Luft und Raumfahrt (Switzerland).

ALRAAM Advanced long-range air-to-air missile.

Alrad Airborne laser ranger and designator.

ALRI Airborne long-range radar input.

ALS *1* Approach light system (FAR Pt 1).
 2 Alert-level standard.
 3 Automatic takeoff and landing system (RPV).
 4 Advanced launch system, HLLV for SDI (US).
 5 Air [or airborne] launch system.
 6 All-weather landing system, or (Matcals) subsystem.
 7 Augmented logistics support.
 8 Application layer structure.

ALSC Advanced logistic systems center (ALFC).

ALSCC Apollo lunar-surface close-up camera.

ALSCU Auxiliary level sense control unit [fuel transfer].

ALSEP Apollo lunar-surface experiment package.

ALSF Approach lights, sequenced flashing.

ALSIP Clear.

ALSL Alternative landing ship, logistic (UK RAF).

AL/SL Weapon capable of being air-launched or surface-launched (ie from surface vessel) (USN).

ALSS *1* Air-launched saturation-system [missile].
 2 Advanced logistic support site[s].

ALSTG Altimeter setting (ICAO).

alt, ALT *1* Altitude, or altimeter.
 2 Alternate [i.e., alternative] airfield or destination].
 3 Automatic altitude hold.
 4 Attack light torpedo.
 5 Approach and landing tests (Shuttle).
 6 Automatic, or airborne, link terminal.
 7 Airborne laser tracker.

ALTA *1* Association of Local Transport Airlines (US).
 2 Associación Latinoamericana de Transportadores Aéreos [C adds cargo].
 3 Airborne [or advanced] lightweight tactical antenna.

Altair ARPA long-range tracking and instrumentation radar.

alternate *1* Incorrectly, has come to mean "alternative" in flight operations; alternative destination, designated in flight plan as chosen if landing not possible at desired destination.
 2 As applied to landing gear, flaps, etc, means using emergency power such as electrically driven pump(s).

alternate hub airport Secondary civil airport at large traffic centre.

alternating light Intermittent light of two or more alternate [correct usage] colours.

alternator A.c. generator.

ALTF Automatic launch test facility, carries out confidence check on XPDR as aircraft taxies to runway.

ALTG Air and Land, or Air/Land, Technology Group (UK MoD).

Alt Hold Altitude-hold mode.

altigraph Recording altimeter; generally aneroid barograph, and thus subject to inaccuracy in pressure/height relationship assumed in calibration.

altimeter Instrument for measuring and indicating height. Pressure * is aneroid barometer or atmospheric pressure gauge calibrated to give reading in height. Sensitive * has stack of aneroid capsules, refined drive mechanism to multiply capsule movements with minimal friction or free play, and setting knob to adjust to different SL or airfield pressures (or to read zero at airfield height). In servo-assisted * mechanism is replaced with more accurate electrical one (see *engine* *, *radio* *).

altimeter errors Apart from servo-assisted altimeter, all pressure altimeters suffer significant lag, so rapid reversal of climb and descent will give a reading up to 200–300 feet in arrears; called lag or hysteresis. There are errors in drive friction and lost motion. Static pressure sensed is subject to PE(2) and compressibility errors, and to transient excursions during manoeuvres. Most significant, parameter measured depends on atmospheric pressure variation, temperature variation and variation in lapse rate between departure airfield and aircraft height (see *altimeter setting*).

altimeter fatigue Supposed tendency of aneroid system to become 'set' in distorted position in long flight at high altitude; this error, not confirmed by most authorities, is called fatigue or, confusingly, hysteresis.

altimeter lag See *altimeter errors*.

altimeter setting For safe vertical separation all altimeters in controlled airspace must be set to uniform datum. Standard is 1013.25 mb (see *ISA*) throughout most en route flying. Instrument then registers vertical separation between aircraft and pressure surface 1013.25 mb, usually below local ground level and may be below local SL or MSL. Second common setting is QNH, at which reading is difference between aircraft height and MSL. Third common setting is QFE, at which reading is difference between aircraft height and appropriate airfield height AMSL; thus at that airfield instrument reads zero. Two other settings, QFF and QNL, seldom necessary.

altimeter switch Triggered by reaching preset altitude, one application to trigger explosive charge.

altimetric valve Device sensitive to increasing cabin altitude (ie, falling pressure) and set to release drop-out oxygen at given level.

altitude *1* Vertical distance of level, point or object considered as point, measured from MSL (normally asociated with QNH) (DTI, UK). In this dictionary meanings are given for 17 other measures of *.
 2 Arc of vertical circle, or corresponding angle at centre of Earth, intercepted between heavenly body and point below it where circle cuts celestial horizon.
 3 In spaceflight, distance from spacecraft to nearest point on surface of neighbouring heavenly body (in contrast to "distance", measured from body's centre).
 4 In aircraft performance measurement and calculation, pressure * shown by altimeter set to 1013.25 mb.

altitude acclimatisation Gradual physiological adaptation to reduced atmospheric pressure.

altitude chamber Airtight volume evacuated and temperature-controlled to simulate any atmospheric level.

altitude clearance Clearance for VFR flight above smoke, cloud or other IFR layer.

altitude datum   Local horizontal level from which heights or altitudes are measured (see *true altitude, pressure altitude, height*).

altitude delay   *1* Deliberate time-lag between emission of radar pulse and start of indicator trace, to eliminate altitude hole or slot.

    2 See next entry.

altitude-delay parameter   Time delay which elapses between pilot nose-up command and establishment of positive climb, esp. during landing approach. ADP is serious in large aircraft with pitch control by tail surfaces, not canard, and without DLC.

altitude hole   Blank area at centre of radial (eg PPI) display.

altitude line   On environmental plot, line joining points of minimum range at which WX main beam intersects ground.

altitude power factor   In piston engine ratio of power developed at specified altitude to power at same settings at ISA SL.

altitude parallax   In altitude (2), angle between LOS from body to observer (assumed on Earth's surface) and LOS from body to centre of Earth.

altitude recorder   Altigraph.

altitude reservation, ALTRV   Airspace utilisation under prescribed conditions, normally employed for mass movement or other special-user requirements which cannot otherwise be accomplished (FAA).

altitude ring   Continuous return across WX display at range equivalent to aircraft's altitude.

altitude sensing unit   Capsule-based unit in engine fuel system sensing aircraft speed/altitude.

altitude sickness   Malaise, nausea, depression, vomiting and ultimate collapse, caused by exposure to atmospheric pressure significantly lower than that to which individual is acclimatised.

altitude signal   In airborne radar operating in forward search mode, unwanted return signal reflected by Earth directly below.

altitude slot   Blank line at origin of SLAR display.

altitude switch   Barometric instrument which makes or breaks electric circuit at preset pressure altitude; contacting altimeter.

altitude tunnel   Wind tunnel whose working section can simulate altitude conditions of pressure, temperature and humidity. In view of advantages of high pressure and driest possible air, conditions chosen usually compromise.

altitude valve   In some carburettors, progressively closed by aneroid to reduce fuel flow at high altitudes.

ALTN   *1* Alternate airfield.

    2 Alternating (two-colour light).

altocumulus, Ac   Medium cloud, about 12,000 ft in groups, lines or waves of white globules.

altostratus, As   Stratiform veil 6,000–20,000 ft with ice-crystal content of variable thickness (giving mottled appearance) but usually allowing Sun/Moon to be seen.

ALTP   Air Line Transport Pilot licence; confusingly, now often called ATPL.

ALTR   Approach/landing thrust reverser.

ALTRV   Altitude reservation.

ALTS   Altimeter setting.

ALTV   Approach and landing test vehicle.

ALU   Arithmetic and logic unit.

Alumel   Ni-Al alloy or coating.

Alumigrip   Trade name of paint used on airframe exterior.

Alumilite   Trade name for sulphuric-acid anodizing process for aluminium and alloys.

alumina   Aluminium oxide Al_2O_3, occurring naturally but also manufactured to close tolerance in various densities. Hard, refractory, white or transparent ceramic.

aluminium (N America, aluminum), Al   Metal element, density about 2.7, MPt 661°C, BPt 2,467°C, most important structural material in aerospace, commercially pure and, esp., alloyed with other metals (see *duralumin*).

aluminium-cell arrester   Lightning arrester/conductor in which insulating film of aluminium plates breaks down and conducts at high applied voltage.

aluminium dip brazing   Method of metallising printed-circuit boards, by closely controlled dipping in molten aluminium.

aluminum   Aluminium (N America).

Alvin   Air-launched vehicle investigation (MALV).

ALVRJ   Air-launched low-volume ramjet.

ALW   Air/land warfare.

ALWS   Airborne laser warning system.

ALWT   Airborne (or advanced) light weight torpedo.

AM   *1* Air Ministry (defunct in UK).

    2 Aircraft mover.

    3 Airspace management.

    4 Or a.m., amplitude modulation.

    5 Air Marshal.

    6 Confusingly, ambient.

    7 Asynchronous machine.

    8 Airlock module.

am   *1* Ambient.

    2 Attometre, 10^{-18}m.

a.m.   Ante-meridian, before noon.

AM-2   Standard military prefab airstrip or landing mat, of 0.16 in aluminium (US).

AM³   Affordable multi-missile manufacturing [program] (US).

AMA   *1* Air materiel area (USAF).

    2 Aerospace Medical Association (US).

    3 Advanced mobility aircraft.

    4 Adaptive multifunction antenna.

    5 Area minimum altitude.

    6 American Management Association.

AMAC   Airborne multi-application computer.

AMACH   Mach number (data-processing).

AMACUS   Automatic microfilm aperture card updating system (Singer).

AMAD   Airframe-mounted accessory [or auxiliary] drive.

AMAGB   Airframe-mounted accessory gearbox.

amagnetic   Having no magnetic properties.

AMAL   Air Medical Acceleration Laboratory (USN).

AMARC   Aerospace Maintenance and Regeneration Center [the Boneyard] (US DoD).

AMARV   Advanced manoeuvring re-entry vehicle.

AMAS   Automated manoeuvring attack system.

Amascos   Airborne maritime situation control system.

Amass   Airport movement-area safety system (FAA).

Amatol   High explosive (**AM**monium nitrate **A**nd **TOL**uene).

AMB   *1* Air Mobile Brigade.

    2 Airwarp Modernization Board (1957–58).

3 Agile multi-beam.

Ambac   R-Nav system (see *Mona*).

AMBE   Advanced multi-band excitation.

Amber   *1* Colour identifying global groups of airways aligned predominantly N–S.

2 Colour = caution, also called yellow.

3 Day/night training equipment, often called two-stage amber, in which pupil pilot is denied visual cues outside cockpit by wearing blue-lens glasses while cockpit transparency is amber; two stages together cut off 99% of light, while allowing pupil to see blue instruments and instructor to see amber outside world (obsolete).

ambient   Characteristic of environment (eg that around aircraft but unaffected by its presence).

ambit   Radar search by missile for target.

AMBL   Air Maneuver Battle Lab (USAF).

AMC   *1* Aerodynamic mean chord.

2 Acceptable means of compliance.

3 Avionics Maintenance Conference (US and Int.).

4 Air Materiel Command (USAF).

5 Air Mobility Command (USAF, 1 June 92).

6 Automatic modulation control.

7 Air management computer.

8 Authorised maintenance centre.

9 Advanced mission computer.

10 Advanced microelectronics converter.

AMCA   Airborne mission control aircraft (USAF).

AMC-C²IPS   AMC (5) C² information-processing system.

AMCD, AMC&D   Advanced mission computer[s] and displays [program] (USN).

AMCM   Airborne mine countermeasures.

AMCOM, Amcom   Aviation and Missile Command (USA); MAT adds multimode airframe technology.

AMCP   Aeronautical mobile communications panel.

AMCS   *1* Adaptive microprogrammed control systems (IBM).

2 Airborne missile control system (aircraft-mounted).

AMD   *1* Aerospace Medical Division (USAF, Brooks AFB).

2 Automatic map display.

3 (rare) Air mileage distance.

4 Angular-momentum desaturation.

5 Amend[ed].

6 Archway metal detector.

7 Aerospace, Maritime and Defence (industrial association, S. Africa).

AMDA   Airlines Medical Directors Association.

Amdar, AMDAR   Automated mission data airborne recording.

AMDB   Airport-mapping database.

AMDP   Air Member for Development and Production (UK, WW2)

AMDS   *1* Automatic manoeuvre device system.

2 Anti-missile discarding sabot.

AMDSS   Airborne mine-detection and survival system (USN).

amdt   Amendment (FAA).

AMDWS   Air and missile defense work station (USA, USAF).

AME   *1* Authorised medical examiner.

2 Alternate [alternative is meant] mission equipment.

3 Air Mobility Element (USAF).

4 Angle-measuring equipment.

5 Amplitude-modulation equivalent, or equipment.

6 Aircraft, multi-engine [L adds land, S sea].

AMEA   Aircraft Maintenance Engineers Association (US, Can.).

AMEC   Advanced multifunction embedded computer.

AMeDAS   Automated met data-acquisition system (J).

AMEL   *1* Active-matrix electroluminescent [D adds display].

2 Aircraft maintenance engineer's licence.

amended clearance   Clearance altered by ATC while flight en route, typically requesting change of altitude or hold, to avoid future conflict unforeseen when clearance filed.

American National   Family of 60° screw (bolt) threads, basically divided into National Coarse (NC), National Fine (NF) and National Special (N), which have in part superseded SAE and ASME profiles.

AMES   *1* Airborne mission-equipment subsystem.

2 Advanced multiple-environment simulator.

Ames   Major NASA laboratory, full title Ames Research Center, Moffett Field, Calif., mainly associated with atmospheric flight (from 1939).

AMET   Advanced military engine(s) technology.

AMF   *1* Allied [Command Europe] Mobile Force [-A adds Air].

2 Armé Marin & Flygfilm (Sweden).

AMFA   Aircraft Mechanics Fraternal Association (US).

AMFI   Aviation Maintenance Foundation International (US).

AMFP   Adaptive matched-field processor, or processing.

AMG   Antenna mast group.

AMGCS   Advanced movements guidance control system (airport).

AMHF   American Military Heritage Foundation.

AMHMS   Advanced magnetic helmet-mounted system.

AMHS   Aeronautical message handling system.

AMI   *1* Airline modifiable information.

2 Avionics midlife improvement.

A/MI   Airspeed/Mach indicator.

AMICS   Adaptive multidimensional integrated control system.

AMID   *1* Airborne mine[field] detection [ARS adds and reconnaissance system, S adds system].

2 Airport management and information display [S adds system].

Amids, AMIDS   *1* Advanced missile-detection system.

2 Airport management information and display system.

AMIK   Automtic target-recognition system (Sweden).

AMIMU   Advanced multisensor inertial measurement unit.

amino   Group –NH₂ which can replace hydrogen atom in hydrocarbon radical to yield amino acids; these play central role in metabolic pathways of living organisms; thus of interest in space exploration.

AMIPS   Adaptive multiple-image projector system.

AMIR   *1* Air Mission Intelligence Report, detailed and complete report on results of air mission.

2 Anti-missile infra-red.

AMIS   *1* Anti-materiel incendiary submunition.

2 Aircraft-movement identification section.

AMK   *1* FAA-approved airplane modification kit.

2 Anti-misting kerosene.

AML *1* Adaptive manoeuvring logic.

2 Aeronautical Materials Laboratory (USN, established 1935).

3 Admiralty Materials Laboratory (Holton Heath).

AMLCD Active-matrix [or advanced multifunction] liquid-crystal display.

AMM *1* Aircraft maintenance manual.

2 Aircraft maintenance and modification.

3 Anti-missile missile.

AMMC Aeronautical materiel management center.

AMMCS Airborne multiservice/multimedia communications system.

ammeter Instrument for measuring electric current (d.c. or a.c., or both in case of 'universal testers') with reading usually given in amperes.

AMMM Affordable multi-missile manufacturing [program].

ammo Ammunition (UK colloq.).

Ammonal High explosive (**AMMON**ium nitrate + **AL**uminium, and often finely divided carbon).

ammonia NH_3, gas at ISA SL, pungent, toxic, present in atmosphere of Jupiter (ice crystals and vapour) and more distant planets (frozen solid). Ammonium chloride in 'dry batteries', nitrate in many explosives, perchlorate plasticised propellants in large solid rocket motors, sulphate soldering and brazing flux and in dry cells, and several compounds in fireproofing.

ammunition Projectiles and propellants for guns; increasingly, guided weapons are logistically treated as * but term normally excludes them.

ammunition quality Coefficient of, symbol r, $= \frac{P}{mw}$ where P is hit probability of each shot, m is mass of projectile and W is required average hits for kill.

ammunition tank Compartment or container housing ammunition for airborne automatic weapon, usually in form of belt arranged in specified way; reloadable, usually when removed from aircraft.

AMNS Airborne mine-neutralization system.

AMNTK Aircraft engine design office (R).

AMO *1* Air Ministry Order[s].

2 Air mass zero, test condition for solar arrays and other space hardware.

3 Approved maintenance organization.

AM1 Single-crystal material for HPT blades.

AMOC Air & Marine Operations Center (US Customs Service at March AFB).

Amors Airborne multifunction optical radar system.

amortisation Fiscal process of writing-down value of goods and chattels over specified period. Typically, transport aircraft * over five, seven or ten years, after which book value is zero. Rate exerts major influence on DOC.

AMOS Automatic meteoroligical observation system, or observing station.

AMOSS, Amoss Airline maintenance and operations support system [self-diagnoses in flight and tells ground computer].

AMP *1* Assisted maintenance period (aircraft carrier).

2 Avionics master plan (USAF).

3 Advanced mission planning, update of CPGS [A adds aid, S adds system].

4 Audio management panel.

5 Aerospace materials program.

6 Advanced modular processor.

7 Avionics modernization program.

8 Atomic materialization process.

9 Application message protocol.

10 Advanced manoeuvre program [me].

11 Air Member for Personnel (UK).

12 Accelerated maturation program.

amp Ampere[s].

AMPA Advanced mission-planning aid.

ampere SI unit of electric current (quantity per unit time), symbol A; named for A.M. Ampère but no accent in unit except in F. Hence: ampere-hour (1 A flowing for 1 h); ampere-turn (unit magnetising force, 1 A flowing round 1 turn of coil).

AMPG, a.m.p.g. Air-miles per gallon, air distance flown per gallon of fuel consumed. UK gallon was Imp; distance was statute miles (not "air miles"). In US, st. mi. and US gal (0.83267 Imp gal). New unit must be found; SI suggests air metres per litre, unless fuel measured by mass (see *AMPP, NAMP*).

amphibian Aerodyne capable of routinely operating from land or water.

amplidyne D.c. generator whose output voltage governed by field excitation; formerly used as power amplifier in airborne systems.

amplification factor In thermionic valve (radio vacuum tube), ratio of change in plate voltage to change in grid voltage for constant plate current (UK, plate = anode).

amplifier Device for magnifying physical or mechanical effect, esp. electronic circuit designed to produce magnified image of weak input signal whilst retaining exact waveform.

amplitude *1* Maximum value of displacement of oscillating or otherwise periodic phenomenon about neutral or reference position.

2 Angular distance along celestial horizon from prime vertical (ie due N–S) of heavenly body, generally as it rises or sets at horizon.

amplitude modulation MCW or A_2 emission in which AF is impressed on carrier by varying carrier amplitude at rate depending on frequency, and depth of modulation depending on audio amplitude.

amplitude modulation equipment [or equivalent] Processes info. and carrier separately and reconstructs them to make equivalent AM signal.

AMPP, a.m.p.p. Air miles per pound (of fuel), air distance flown for each pound avoirdupois of fuel consumed, former measure of specific range (see *AMPG, NAMP*).

AMPS *1* Advanced multi-sensor payload system.

2 Aviation mission-planning system (USA).

3 Automatic message-processing system.

AMPSS Advanced manned precision strike system.

AMPT *1* Air miles per tonne [of fuel].

2 Advanced missile propulsion technology [generally means airbreathing].

AMPTE Active Magnetospheric Particle Tracer Explorer(s).

AMR *1* Airport movement radar.

2 Atlantic Missile Range, military (DoD) range originally run by Pan Am and RCA from Patrick AFB and also serving NASA's KSC at Cape Canaveral.

Amraam, AMRAAM Advanced medium-range AAM.

AMRC Aerospace Maintenance and Regeneration Center [previously MASDC].

AMRDEC Aviation and Missile RD&E Center (USA, Redstone Arsenal).

Amrics Automatic management radio and intercom system.

AMRL Aerospace Medical Research Laboratory (USAF).

AMRS Advanced maintenance-recorder system.

AMS *1* Aeronautical Mobile Service, radio-communication service between aircraft or between aircraft and ground stations.

2 Air Maintenance Squadron (USAF).

3 Aircraft management simulator [or system].

4 Academy of Military Science (USA).

5 Air [or aerospace, or aircraft] material specification.

6 Advanced missile system (USN).

7 Apogee and manoeuvring stage.

8 Automated message switch.

9 Airborne maintenance subsystem.

10 Avionics, or airspace, management system, or service.

11 Altitude management [and alert] system.

12 American Material Standard.

13 American Meteorological Society (office Boston).

14 Automatic marking system, or subsystem.

15 Automatic meteorological system.

16 Apron management service.

17 Automated manifest system.

18 Aviation Manpower and Support (USMC).

19 Aircraft Maintenance Standards (CAA).

AMSA Advanced manned strategic aircraft.

Amsam Anti-missile SAM (1).

Amsar, AMSAR *1* Airborne multimode, or multirole, or multipurpose, solid-state, active-array radar.

2 Airborne multifunction steerable-array radar.

AMSC Automatic message-switching centre.

AMSD Aircraft Maintenance Standards Department (CAA).

AMSE Automatic message-switching equipment.

AMSG Air Mobility Support Group (USAF).

AMSL, a.m.s.l. Above mean sea level.

AMSO Air Member for Supply and Organisation (UK, WW2).

AMSR Advanced microwave scanning radiometer.

AMS(R)S Aeronautical mobile satellite (route) service.

AMSS *1* Advanced multi-sensor system (EW, SOJ, AEW).

2 Aeronautical mobile satellite service [P adds panel].

3 Airborne mission-support system.

4 Automatic message-switching system (AFTN).

AMST Advanced medium STOL transport.

AMSTE Affordable moving surface-target engagement.

AMSU *1* Aircraft-motion sensor unit [digital FCS].

2 Air-motor servo unit.

3 Advanced microwave sounding unit.

AMT *1* Accelerated mission test[ing].

2 Aircraft [or aviation] maintenance technician.

3 Advanced metal-tolerant tracker [or tracking system].

4 Air maneuver transport, C–130/C–17 capability plus V/STOL (USA).

5 Association for Manufacturing Technology (US).

amt Amount.

AMTC Aerospace Medicine Training Centre.

AMTD *1* Aircrew maintenance-training device.

2 Adaptive moving-target detector, detection, or device.

AMTE Adjusted megaton equivalent.

AMTI Airborne, or air, moving-target indicator, or indication.

Amtorg Former organisation for importing and licensing US products (USSR).

Amtos, AMTOS Aircraft-maintenance task-oriented support [second S adds system].

AMTS *1* Adaptive marked-target simulator.

2 Aeronautical message transfer service.

AMTT See *AMT (3)*.

AMU *1* Air mileage unit.

2 Astronaut maneuvering unit.

3 Aircraft maintenance unit; F adds facility.

4 Audio-, or avionics-, management unit.

5 Antenna-matching unit.

6 Auxiliary memory unit.

7 Air Mobility Unit (USAF).

AMUST Airborne manned/unmanned system technology; -D adds demonstration.

AMUX Audio multiplexer.

AMW Air Mobility Wing (USAF).

AMWD Air Ministry Works Department [airfield and building construction, formerly] (UK).

AMWM Aircraft maintenance wiring manual.

amyl Family of univalent hydrocarbon radicals, all loosely C_5H_{11}, esp.: amyl acetate (banana oil) solvent and major ingredient of aircraft dopes; and amyl alcohol, in lacquers.

AN *1* Air navigation.

2 Prefix to designation codes of US military hardware denoting "Army-Navy"; now rare, though code system remains.

3 Airworthiness Notice.

4 [or A/N] alphanumeric.

A_n Acceleration normal to flight path, usually along OZ axis.

AN^2 Product of gas-turbine annulus area at turbine rotor-blade mid-length and square of rotational speed.

ANA *1* Air Navigation Act.

2 Association of Nordic Aeroclubs (Int).

3 Association of Naval Aviation Inc. (US, 53 chapters).

4 Aeroportos e Navegacao Aerea (Portugal).

anabatic wind Wind blowing uphill as result of insolation heating slope and adjacent air more than distant air at same level.

ANAC, Anac Automatic nav/attack control(s).

Anacna Associazione Nazionale Assistenti e Controllori della Navigazione Aerea (I).

anacoustic region Extreme upper level of atmosphere (say, 100 miles above Earth) where mean free path too great for significant propagation of sound.

ANAE Académie Nationale de l'Air et de l'Espace (F).

ANAEM Aircraft noise and aviation emissions mitigation.

anaglyph Picture, generally photographic but often print-out from some other system, comprising stereoscopic pairs of images, one in one colour (eg red) and other in second colour (eg blue). Viewed through corresponding (eg blue/red) spectacles, result appears three-dimensional.

analog computer *1* Computational device functioning by relating or operating upon continuous variables (in contrast to digital computers, which operate with discrete parcels of information). Simplest example, slide-rule.

2 Electronic computer in which input data are continuously variable values operated upon as corresponding electrical voltages. Actual hardware can be coupled directly in so that, for example, control response, angular movement and aeroelastic distortion of control surface can be investigated in situ and in real time.

analog/digital converter, ADC   Device for converting analog output into discrete digital data according to specified code of resolution; also called digitiser and, esp. for linear and rotary movement, encoder.

analog output   Transducer signal in which amplitude (typically quasi-steady voltage) is continuously proportional to function of stimulus.

analyser   In piston engine installatons, device intended to indicate mixture ratio by sampling composition of exhaust gas (hence EGA, exhaust-gas *). Some for static-test purposes depend on chemical absorption of carbon dioxide, but airborne instrument uses Wheatstone bridge to measure variation in resistance due to proportion of carbon dioxide in gas.

analysis   Stress analysis.

ANAO   Australian National Aerospace Organization.

anaprop   Anomalous propagation.

ANASA   Azerbaijan National Aerospace Agency.

Anasics   Alaska National Airspace System Interfacility Com. System.

Anazot   A resin foam for fuel-tank protection.

ANB   *1* Air Navigation Bureau (ICAO).

     2 Adaptive narrow beam(s).

ANC   *1* Air navigation charges.

     2 Air Navigation Commission, or Council (both ICAO).

     3 Army/Navy/civil (US).

     4 Aviate/navigate/communicate.

     5 Active noise cancellation, or control.

Ancat   Abatement of nuisance caused by air transport.

ANCB   Association Nationale Contre les Bangs (supersonic) (F).

anchor-centred   Also anchor charge, anchor grain: solid rocket propellant charge in which initial combustion surface has cross-section resembling radial array of anchors, flukes outward.

anchor light   Riding light.

anchor line, cable   Cable running along interior of airdrop aircraft to which parachute static lines (strops) are secured.

anchor-line extension kit   Assembly arranged to extend anchor line to allow airdropping through rear clamshell doors or aperture with such doors removed.

anchor nut   Large family of nuts positively securable by means of screwed or bolted plate projecting from base (see *nut, nutplate, stiffnut, stopnut*).

ANCOA   Aerial Nurse Corps of America.

AND   *1* Active-nutation damping.

     2 Aircraft nose-down.

     3 Army/Navy drawing.

ANDA   Associazione Nazionale Direttori di Aeroporto (I).

ANDAG   Associazione Nazionale Dipendenti Aviazione Generale (I).

ANDB   Air Navigation Development Board (US 1948–57).

Anderson shelter   Small air-raid shelter assembled from sheet galvanized-steel pressings in pit and covered with deep layer of earth (UK, WW2).

AND gate   Bistable logic function triggered only when all inputs are in ON state; in computers used as addition circuit, performing Boolean function of intersection. Hence AND/OR gate, AND/NOR, AND/NOT.

AND pad   Standard Army/Navy drive accessory pad.

ANDR   Air, or airborne, navigation data recorder.

androgynous   Mating portions of docking system which are topologically identical (eg US and Soviet docking faces).

ANDS   *1* Automatic navigation differential station.

     2 Accelerate N, decelerate S, mnemonic for NTE (2).

ANDVT   Advanced narrow-band digital voice terminal.

anechoic   Without echoes; thus, * facility, * room, in which specially constructed interior walls reduce reflections to infinite number of vanishingly small ones. Chamber can be designed to operate best at given wavelength, with sound, ultrasound, ultrasonic energy, microwaves and various other EM wavelengths. Mobile facilities used for boresighting nose radar of combat aircraft.

anemogram   Record produced by anemograph.

anemograph   Instrument designed to produce permanent record of wind speed (ie recording anemometer) and, usually, direction. Dines * incorporates weathercock vane carrying pitot and static tubes.

anemometer   Instrument for measuring speed of wind, usually 10 m (32.8 ft) above ground level. Robinson Cup * has free-rotating rotor with three or four arms each terminating in a hemispherical or conical cup.

anemoscope   Instrument for checking existence and direction of slow air currents.

ANEPVV   Association Nationale d'Entraide et de Prévoyance du Vol à Voile (gliding) (F).

anergolic   Not spontaneously igniting; thus, most rocket-propellant combinations comprising two or more liquids. Opposite of hypergolic.

aneroid   Thin-walled airtight compartment designed to suffer precisely predictable and repeatable elastic distortion proportional to pressure difference between interior and exterior. Most are evacuated steel capsules in form of disc with two corrugated faces which can approach or recede from each other at centre. To increase displacement a stack can be used linked at adjacent centres. Common basis of pressure altimeter, ASI and Machmeter.

aneroid altimeter   Pressure altimeter; aneroid barometer calibrated to read pressure altitude.

aneroid altitude   Pressure altitude.

ANFCMA   Associazione Nazionale Famille Caduti e Mutilati dell'Aeronautica (I).

ANF   Anti-navire futur (F).

ANG   Air National Guard (US); B adds base.

angels   *1* Historic military R/T code word for altitude in thousands of feet; thus, 'angels two-three' = 23,000 ft.

     2 Distinct, coherent and often strong (40 dB above background) radar echoes apparently coming from clear sky. Probable cause strong pressure, temperature or humidity gradient in lower atmosphere giving even sharper gradient in refractive index.

Angit   Aircraft next-generation identification transponder.

angled deck Aircraft-carrier deck inclined obliquely from port (left) bow to starboard (right) stern to provide greater deck space, greater catapult capacity and unobstructed flight path further from island than with axial deck, with safe parking area towards bow.

angle of . . . In general, see under operative word.

angle of attack indicator Instrument served by *** sensing system.

angle of attack sensing system Incorporated in aircraft, esp. aeroplane, to trigger stall-warning, stall-protection system or other desired output, and possibly serve an indicator. Sensing unit (SU) comprises freely pivoted vane or series of pitot tubes set at different angles of incidence and each connected to different supply pipe to give dP output. SU on wing leading edge, to sense movement of stagnation point, or on side of fuselage, repeated on opposite side to eliminate error due to sideslip. SU anti-iced and must allow for changes in aircraft configuration.

angle of depression Acute angle between axis of oblique camera and horizontal.

angle off Acute angle between own-fighter sightline and longitudinal axis of target aircraft.

angle of incidence indicator Instrument giving continuous reading of angle of foreplane, horizontal tail (especially tailplane where not primary pitch control) or wing, where incidence variable.

angle of view Angle subtended at perspective centre of camera lens by two opposite corners of format.

Anglico Air/naval gunfire liaison company (USMC).

ANGR Air navigation (general) regulation[s] (UK).

ANGRC Air National Guard Readiness Center.

ANGSA Air National Guard support aircraft.

Ångström Unit Å or AU, unit of length equal to 10^{-10} m, formerly used to express wavelengths of light; nearest SI is nanometre; 1 nm = 10 Å.

angular acceleration Time rate of angular velocity of body rotating about axis which need not pass through it; unit rad/s².

angular displacement *1* Angular difference between two directions or axes, esp. between reference axis of hinged or pivoted body and same axis in neutral or previous position.

2 In magneto, angular difference between neutral position of rotor pole and later position giving highest-energy spark (colloq., E-gap).

angular distance *1* Angular displacement.

2 Smaller arc of great circle joining two points expressed in angular measure.

3 In all sine-wave phenomena (radio, radar, astronomy, etc), number of waves of specified frequency between two points (numerically multiplied by 360 or 2π depending on whether unit is degree or radian).

angular measure *1* SI unit of plane angle is radian (rad), angle subtended by arc equal in length to radius of circle on which arc centred. Thus one revolution = 2π rad, and 1 rad = 57.296°. Degree (°) defined as 1/360th part of one revolution, itself subdivided into 60 minutes (') each subdivided into 60 seconds ("); pedantically distinguished from units of time by calling them arc-minutes and arc-seconds. Thus 1 rad = 57° 17' 45". For small displacements milliradian (mrad) to be used; roughly 3' 26¼"; thus 1' = 0.2909 mrad.

angular momentum For rigid body of significant mass (not elementary particle), product of angular velocity and moment of inertia; or, if axis of rotation at some distance from it (as in axial turbine blade), mass, instantaneous linear velocity and radial distance of CG from axis. Thus, L = Iω = mvr.

angular resolution Angular distance between LOS from radar, human eye or other "seeing" system to target and LOS from same system to second target which system just distinguishes as separate object; usually only a few mrad, esp. if targets are pinpoints of light against dark background.

angular speed *1* Loosely, angular velocity.

2 Rate of change of target bearing, esp. as seen on PPI.

angular velocity Symbol ω, time rate of angular displacement of body rotating about axis which need not pass through it. Preferred measure is rad s^{-1} or mrad s^{-1}; in traditional engineering most common is rpm. Multiplied by radius gives *tip speed*, or peripheral speed.

anharmonic Not harmonic, irregular.

anhedral *1* Negative dihedral, smaller angle between reference plane defining wing (such as lower surface or locus of AMCs) which slopes downward from root to tip, and horizontal plane through root. In early aircraft dihedral considered desirable as means to natural stability, esp. in roll; in some, and many modern gliders, wing flexure converts static * into dihedral under 1 g in flight. Tendency to design modern wing with * to counter excessive roll response to sideslip or side gusts, esp. in high-wing or supersonic aircraft. VG aeroplane angle may be varied with sweep.

2 Some authorities define as 'absence of dihedral' (from Greek root of prefix *an* = not), and suggest "cathedral" for downward-sloping wing.

ANIAF Associazione Nazionale Imprese Aerofotogrammetriche (I).

Anics Alaskan NAS(2) interfacility communications system.

ANIE Associazione Nazionale Industrie Elettrotechniche ed Elettroniche (I).

aniline Phenylamine, aminobenzene, $C_6H_5NH_2$, colourless, odorous amine, MPt –6°C (thus, normally liquid), BPt 184°C, turns gradually brown on exposure to air, reacts violently with RFNA or other strong nitric acids with which often used as rocket propellant.

anion Negative charged ion or radical, travels towards anode in electrolytic cell.

anisotropic Exhibiting different physical properties along different axes, esp. different optical properties or, in structural material, different mechanical properties, esp. tensile strength and stiffness.

ANIU Aircraft network interface unit.

ANK Automatic navigation kit.

ANL *1* Auto noise limiter (communications).

2 Automatic noise levelling.

ANLP Arinc network layer protocol.

ANLS Automatic navigation launch station.

ANM AFTN notification message.

ANMI Air navigation multiple indicator.

ANMPG Air nautical miles per gallon.

ANMS *1* Aircraft navigation and management system.

2 Automatic navigation mission station (UAV).

3 Automated noise monitoring system.

ANN Applied neural network.

ann Annunciator.

annealing Heat treatment for pure metal and alloys to

obtain desired physical properties by altering crystalline microstructure. Usually involves heating to above solid-solution or critical temperatures, followed by gentle cooling in air. General aim to make metal less brittle, tougher, more ductile and relieve interior stress.

annotation Identifying and coding reconnaissance ouputs, such as visual-light photographs, IR print-outs, ECM records, etc, with digital data: date, time, place, unit, altitude, flight speed, heading and data specific to reconnaissance system.

annual Annual mandatory inspection of aircraft.

annual variation Amount by which magnetic variation at specified place on Earth varies in calendar year. In UK about –7'(min), reducing local variation to zero in year 2140.

annular combustion chamber In gas-turbine engine, chamber [including flame tube(s) and liners] entirely in form of body of revolution, usually about major axis of engine.

annular gear Ring gear or annulus, gearwheel in which teeth project inwards from outer periphery. In annulus gear there is no centre, teeth being carried on open ring (which in turboprop/turbofan reduction gears may be resiliently mounted and torque-reacted by torque-signalling system). When shaft-mounted, teeth usually on one side only of flat or conical disc.

annular injector Rocket (or possibly other engine) injector in which liquid fuel and/or oxidant is sprayed from narrow annular orifice. In bipropellant engine numerous such orifices spaced around chamber head, alternately for fuel and oxidant.

annular radiator Cooling radiator shaped as body of revolution to fit around axis of aircraft engine, esp. between propeller and piston engine having circular cowl.

annular spring Has form of ring distorted radially under load. Sometimes called ring spring, esp. when given tapered cross-section and used in multiple, in an inter-meshing stack, to resist load along axis of symmetry.

annular wing Wing in form of body of revolution, designed to operate in translational flight with axis of symmetry almost horizontal.

annulus drag Most often refers to base drag of annular periphery around propulsive jet nozzle which incompletely fills base of vehicle, esp. ballistic rocket rising through atmosphere.

annunciator *1* In gyrocompass (remote compass), indicator flag visible through window of cockpit instrument; with a.c. supply on and gyro synchronized with compass detector, indication should hover between dot and cross, never settle on either.

2 In aircraft system, esp. on aircraft having flight deck rather than cockpit, panel or captioned warnings often distributed on schematic diagram of system.

3 In a CDU, the alpha and numeric keys providing part of the operator interface.

ANO Air Navigation Order, UK statutory instrument for enactment of ICAO policy defining laws, licensing and similar fundamental issues regarding aerial navigation (see *ANR*).

anode In electrical circuit (electrolytic cell, valve, CRT), positive pole, towards which electrons flow; that from which "current" conventionally depicted as emanating.

anodising Electrolytic (electrochemical) treatment for aluminium and alloys, magnesium and alloys and, rarely,

other metals, coated with inert surface film consisting mainly of oxide(s) as protection against corrosion. Electrolyte usually weak sulphuric or chromic acid.

ANOE Automated nap of the Earth.

anomalistic period Time between successive passages of satellite through perigee.

anomalistic year Earth's orbital period round Sun, perihelion to perihelion: 365 d 6 h 13 m 53.2 s, increasing by about 0.26 s per century.

anomalous dispersion Local reversal in rule that medium transparent to EM radiation diffracts it with refractive index that falls as wavelength increases; discontinuities in absorption spectrum make index increase as wavelength increases.

anomalous propagation Of wave motions, esp. EM radiation of over 30 kHz frequency or sound, by route(s) grossly different from expected, usually because of atmospheric reflection and/or refraction, sharp humidity gradients and temperature inversion.

anomaly *1* Difference between mean of measured values of meteorological parameter at one place and mean of similar values at all other points on same parallel (in practice, mean of similar values at other stations near same parallel).

2 In general, deviation of observed geodesic parameter from norm or theoretical value.

3 Specif. local distortion of terrestrial magnetic field caused by local concentrations of magnetic material used in aerial geophysical surveys and ASW (see *MAD*).

Anoms Airport noise and operations monitoring system.

ANORAA Association Nationale des Officiers de Réserve de l'Armée de l'Air (F).

Anova Analysis of variance, especially in monitoring flight-crew performance under different adverse or stress conditions.

anoxaemia Hypoxaemia, deficiency in oxygen tension (loosely, concentration) of blood.

anoxia Absence of oxygen available for physiological use by the body (see *hypoxia*).

ANP *1* Aircraft nuclear propulsion; general subject and defunct DoD programme.

2 Air navigation plan (ICAO).

3 Actual [or achieved] navigation performance.

ANPA Aircraft Nuisance Prevention Association (J).

ANPAC Association Nazionale Piloti Aviazione Commerciale (I); other Italian associations include ANPAV (Assistenti di Volo), ANPCAT (Professionale Controllori & Assistenti Traffico Aereo), ANPI (Paracadutisti d'Italia), ANPIC (Piloti Istruttori Civili), ANPiCo (Piloti Collaudatori), and ANPSAM (Piloti Servizi Aerei Minori).

A-NPR, ANPRM Advance notice of proposed rule-making (FAA).

ANR *1* Air Navigation Regulation.

2 Active [or acoustic] noise reduction.

ANRA Association Nationale des Résistants de l'Air (F); résistant, a tough hard worker.

ANRS Automatic navigation relay station.

ANRT Association Nationale de la Recherche Technique (F).

ANS *1* Air Navigation School.

2 Airborne [or air, or area] navigation system.

3 Artificial neural system.

4 Ambient-noise sensor.

5 Answer.

ANSA Advisory group, air navigation services (G).

Anser Autonomous navigation sensing experimental research.

ANSI *1* American National Standards Institution, vital for software.

2 Air navigation services institute [CAA] (Finland).

Ansir Awareness of national security issues and responses (FBI, US).

ANSP Air navigation service provider.

AN² AN-squared, a fundamental gas-turbine parameter in which A is total cross-section area of the gas path through the rotor blades and N is ppm.

Ansyn Analysis by synthesis.

ANT *1* Autonomous negotiation [or negotiating] team.

2 Air-navigation trainer (simulator).

3 Antrieb [powerplant] neuer Technologie (G).

ant Antenna.

ANTAC Association des Navigants Techniciens de l'Aviation Civile (Belg.).

Antares Antenna advanced intertial reference for enhanced sensors.

ANTC Advanced Networking Test Center (US).

antenna US term for aerial (2), portions of broadcasting EM system used for radiating or receiving radiation. Plural antennas, not antennae.

antenna azimuth rate Rotational speed, rpm or rad/s.

anthophyllite Crystalline mineral, essentially (Mg, Fe) Si O₃.

anti-aircraft, AA Surface-based defence against aerial attack. Suggested historic word; better to introduce SA, surface-to-air, as prefix for guns as well as missiles, together with associated radars and other peripherals.

anti-aircraft artillery, AAA Guns and unguided-rocket projectors dedicated to surface-to-air use with calibre 12.7 mm (0.5 in) or greater.

anti-air warfare, AAW All operations intended to diminish or thwart hostile air power, eg air defence and interdiction against enemy airfields.

anti-aliasing filter Inserted in ouput of analog-to-digital converter to screen out multiples of the digital sampling frequency.

anti-balance Opposing, counteracting or reducing balance, esp. in dynamic system.

anti-balance tab Tab on control surface mechanically constrained to deflect in same sense as parent surface to increase surface hinge-moment (ie, to make it more difficult to move in airstream); opposite of servo tab.

anti-ballistic missile, ABM System designed to intercept and destroy hypersonic ballistic missiles, esp. RVs of ICBMs. Speed of such targets, smaller radar cross-section, possible numbers, use of ECM and decoys, nuclear blanketing of large volumes of sky, ability to change trajectory, enormous distances, and need for 100% interception, make ABM difficult.

anti-blocking system Prevents aircraft from making simultaneous or conflicting radio transmissions.

anti-buffet Describes measures adopted on atmospheric vehicles, esp. high-speed aircraft, to reduce or eliminate aerodynamic buffet. Almost always auxiliary or locally hinged surface moved out into airflow to reduce buffet which would otherwise be caused by configuration

change, eg opening weapon-bay doors. Thus * flap, panel, comb, rake, slot.

anticing See *anti-icing*.

anti-clutter Any of many techniques intended to reduce *clutter*.

anti-collision beacon High-intensity [so-called strobe] flashing red light[s] carried by most aircraft, to be visible at great distance from any aspect.

anti-coning In most helicopters, and some other rotorcraft, ** device fitted to prevent main rotor blades from reaching excessive coning angle (being blown upwards, eg by high wind at zero or low rotor rpm on ground), from which they could fall and exceed root design stress when suddenly arrested. Usually fixed range of angular coning permissible between ** device and droop stop.

anti-corrosion Measures taken depend on environment (see *marinising*), and working stress and temperature of hardware; apart from choice of material and surface coating (Alclad, anodising, painting with epoxy-based paint, etc), special agents can be introduced to fuels, lubricants, seals, hydraulic fluids and interior of stored device (inhibiting).

anti-cyclone, anticyclone Atmospheric motion contrary to Earth's rotation. Large area of high pressure, generally with quiet, fine weather, with general circulation clockwise in N hemisphere and anticlockwise in S; divided into 'cold' and 'warm' each group being subdivided into 'permanent' and 'temporary'.

anti-dazzle panel *1* Rearwards extension around top of instrument panel or cockpit coaming to improve instrument visibility and at night prevent reflection of instruments from windscreen.

2 On aircraft with natural metal finish, areas of exterior painted with non-reflective black or dark blue to prevent bright reflections being visible to crew.

anti-drag wire Structural bracing filament, usually incorporated within wing, intended to resist forwards ('anti-drag') forces, usually from trailing edge at root to leading edge at tip.

anti-freeze Most important agent ethylene glycol (CH₂OH.CH₂OH), usually used as aqueous solution with minor additives. Neat 'glycol' (there are many) remains liquid over more than twice temperature range of water, and freezes not as solid ice but as slush.

anti-friction bearing Loose term applicable to bearing suffering only rolling friction (ball, needle, roller) but esp. signifying advanced geometry and high precision.

anti-frosting Measures taken to prevent frost (ice condensing from atmosphere and freezing as layer of fine crystals), esp. on windscreen; typically raise temperature by hot air or fine electrical resistance grids or conductive films.

anti-g Measures counteracting adverse effects of severe accelerations in vertical plane.

anti-glare Against optical glare, generally synonymous with anti-dazzle but esp. dull non-reflective painted panels on airframe and propeller blades.

anti-gravity Yet to be invented mechanism capable of nullifying local region in gravitational field.

anti-g suit See *g-suit*.

anti-g valve See *inverted flight valve*.

anti-icing Measures to prevent formation of ice on aircraft; required on small vital areas where ice should not be allowed to form even momentarily (see *icing, de-icing*).

anti-icing correction   Applied to aircraft, esp. advanced aeroplane, performance with various forms of ice-protection operative. Esp necessary when engine air bleed exerts significant penalty in degraded take-off, climb-out, overshoot and en route terrain-clearance calculations. Required numerical values usually given in flight manual as percentage for each flight condition.

anti-icing inhibitor   Fuel additive preventing freezing of water precipitated out of fuel at high altitude.

anti-knock rating   Measure of resistance of piston engine fuel to detonation (1) (see *octane rating*).

anti-lift   In direction opposite to lift forces, eg loads experienced by wing on hard landing with lift dumpers in use. Thus ** wire (landing wire), structural bracing filament, usually within wing, to resist downloads.

anti-missile   Against missiles, specif system intended to intercept and destroy hostile missiles (which may or may not include guided devices, artillery shells, bullets, mortar bombs and other flying hardware). In large-scale defence against ICBM attack, anti-ballistic missile.

anti-misting kerosene   Jet fuel chemically and physically tailored so that, on sudden release to atmosphere (from ruptured tanks in a crash), it spreads in the form of droplets too large to form an explosive mixture with air.

antimony   Element, abb. Sb, existing in several allotropic forms, most stable being grey metal with brittle crystalline structure. Widely used in aerospace in small quantities: with tin and other metals in bearings and applications involving sliding friction, with lead in storage batteries, as acceptor impurity in semiconductors, in type metals and electronic cathodes.

antinode   *1* Points on wave motion where displacement (amplitude) is maximum.

2 Locations in aircraft structure where flexure (due to vibration or aeroelastic excitation) is maximum.

3 Either of two points in satellite orbit where line in orbit plane perpendicular to line of nodes, and passing through focus, intersects orbit (thus, essentially, points in orbit midway between nodes).

anti-oxidant   Fuel additive which prevents formation of oxides and, esp. peroxides during long storage.

antipode   Point on Earth, or other body, as far as possible from some other point or body; specif. point on Earth from which line through centre of Earth would pass through centre of Moon.

antipodes   Regions on Earth diametrically opposite each other.

anti-radiation missile   Missile designed to home on to hostile radars.

anti-rolling   In rigid airship, measures intended to prevent rolling of any part relative to hull or envelope.

anti-rumble panel   Small anti-buffet panel necessary on grounds of noise.

anti-snaking strip   In early high-subsonic aircraft, strip of cord or metal attached to one side of rudder or elsewhere to prevent snaking (yawing oscillations).

anti-solar point   Point on celestial sphere 180° from Sun; projection to infinity of line from Sun through observer.

anti-sound   Sound generated to cancel out unwanted noise.

anti-spin parachute   Streamed from extremity of aeroplane or glider to assist recovery from spin; most common location is extreme tail.

antistatic   Measures taken to reduce static interference with radio communications, traditionally by trailed ** wire, released from ** cartridge, which serves as pathway for dissipation of charge built up on aircraft. See next.

anti-static additive   Fuel additive which increases electrical conductivity and thus speeds up dissipation of static electricity built up during refuelling.

anti-submarine warfare   See *ASW*.

anti-surface improvement program   Combines sensors, datalinks and displays presenting integrated precision tactical picture.

anti-surge measures   *1* To prevent aerodynamic surging in axial compressor, eg redesign further from surge line, use of variable stators, blow-off valves and interstage bleeds.

2 Valves and baffles in oil cooler to maintain steady oil flow.

anti-torque drift   Inherent lateral drift of helicopter due to side-thrust of tail rotor; often countered by aligning main rotor so that tip-path plane is tilted to give cancelling lateral component.

anti-torque pedals   Common name for foot pedals of helicopter.

anti-torque rotor   Tail rotor of helicopter, or any other rotor imparting thrust (moment) neutralising that of main rotor.

antitrade wind   Semi-permanent winds above surface trades, generally at height of at least 3,000 feet, especially in winter hemisphere, moving in opposite direction (ie westerly).

anti-transmit/receive   Pulsed-radar circuit which isolates transmitter during periods of reception.

antivibration loop   Closed-loop servo system designed to suppress structural or system vibration.

Antle   Affordable near-term low emissions.

ANTMS   Airport noise/track monitoring system.

ANU   Aircraft nose-up.

ANUA   Associazione Nazionale Ufficiali Aeronautica (I).

anvil cloud   Cumulonimbus.

Anvis   Aviator's night-vision system, see next.

Anvis/Hud   Adds head-up display for safe NOE helicopter flight at night.

ANVR   Association of travel agents (Neth.).

ANZUK, Anzuk   Australia, New Zeland, UK, and SE Asia defence.

ANZUS, Anzus   Australia, New Zealand and US (1951 defence pact).

AO   *1* Administrative Operations, major US Federal budget heading.

2 Artillery observation.

3 Aircraft operator.

4 Anti-oxidant.

5 Airplane, observation (USA 1956–62).

6 Announcement of opportunity (NAS, NASA).

AOA   *1* Aerodrome Owners' Association (UK 1934–, became next).

2 Airport Operators Association (US, UK).

3 Angle of attack (units, thus "6 AOA").

4 Angle of arrival (ECM).

5 Air Officer i/c Administration (RAF).

6 Airborne optical adjunct (ABM).

7 Amphibious operating [or operations] area (DoD).

8 At or above (FAA).

9 Abort once around, ie after one orbit.

10 Analysis of alternatives.

AOAC Autonomous operation from aircraft carrier (UAV).

AOB   *1* At or below.

2 Angle of bank.

3 Automatic optical bench (for testing optically tracked missiles).

4 Air Observer (Bombardment).

5 Angle off boresight.

6 Air order of battle.

7 Air-dropped oceanic [or Arctic Ocean] buoy.

AOC   *1* Air Officer Commanding.

2 Air Operator's Certificate (CAA UK).

3 Autopilot omni-coupler.

4 Aerodrome, or airport, obstruction, or obstacle, chart.

5 Assumption of control message (ICAO).

6 Adaptive optical camouflage.

7 Association of Old Crows (EW, US).

8 Air, or airline, Operations Center (US).

9 Air Operations Command (Vietnam AF).

10 Airport [also Airline] Operational Commission (US).

11 Air/oil cooler.

12 Aeronautical operational control (Acars).

13 All other configurations.

14 Acceleration-onset cueing.

15 Attitude and orbit control [S adds system].

16 Airline, or aircraft, operational control, or communications.

AOCC Airline Operation Control Center (US).

AOCI Airport Operators Council International, Inc (Int.).

AOCM Airborne, or advanced, optical countermeasures.

AOCP Airborne operational computer program.

AOCS   *1* Attitude and orbit control system.

2 Airline Operational Control Society (US).

AOD   *1* Aft of datum (c.g.).

2 Airport Operations Director.

3 Airport operational data [B adds base].

4 Age of data [C adds clock, E adds ephemeris] (GPS).

5 Area of display.

6 Above Ordnance datum (Newlyn, see *sea level*).

7 Audio on demand.

AOE Airport of entry.

AOFR Aluminium oxide fibre-reinforced.

AOG Aircraft on ground, code inserted in message (eg for spare parts) indicating aircraft unable to operate until remedial action taken.

AOHE Air/oil heat exchanger.

AOI   *1* Arab Organization for Industrialization.

2 Area of interest.

3 Airborne, or aircraft, overhead interoperability [O adds office, TF adds task force].

AOK "All OK" [Astronaut].

AOM   *1* Annual operational maintenance.

2 Aircraft operating manual.

AOO Analysis of options.

AO1 Automated observation without precipitation discriminator.

AOP   *1* Airborne (or air) observation post.

2 All other persons (airline costings).

3 Airline operational procedure.

4 Aeronautical OSI profile.

5 Advanced onboard processor.

AOPA Aircraft Owners & Pilots Association (US from 1939); LA adds legislative action.

AOPAA Aircraft Owners & Pilots Association of Australia.

AOPF Active optical proximity fuze.

AOPG Aerodrome Operators' Group (UK).

AOPT Accurate [or advanced] optical position transducer.

AOPTS Air Operations Planning and Tasking System[s].

AOR   *1* Atlantic Oceanic Region, [suffix -E or -W].

2 Average operational reliability.

3 Area of responsibility.

AOS   *1* Acquisition of signal (telecommunications, telemetry).

2 Airborne optical sensor [A adds adjunct, P processor, T telemetry].

AOSC Asset-optimization service contract.

AOSP Advanced on-board signal processor.

AOSU Airfield Operations Safety Unit.

AOT   *1* All-operators Telex [issued by prime supplier].

2 Air Officer, Training (RAF).

AOTD Active optical target detector.

AO2 Automated observation from unattended ASOS.

AO2A AO-augmented, from an attended ASOS.

AOU Area of uncertainty.

AOW All-operators wire [AOT is preferred].

AOY Angle of yaw.

AP   *1* Armour-piercing [–DS adds discarding sabot, –E explosive, –F finned, –I incendiary, –T tracer; there are other suffixes].

2 Ammonium perchlorate (solid rocket fuel).

3 Air Publication (UK).

4 Airport (ICAO, Acars).

5 Autopilot.

6 Aviation regiment (USSR).

7 Allied publication (NATO).

8 Advance[d] procurement.

9 Airframe parachute [S adds system].

10 Action Panel [materials R&D].

11 Array processor.

12 Anti-personnel.

13 Anomalous propagation.

14 Approach [apch preferred].

15 Assessment Phase (UK).

16 Automotive [i.e., automatic] picture transmission.

Ap Approach light[s].

A/P   *1* Autopilot.

2 Airplane(s).

3 Airport.

4 Aim point.

APA   *1* Airline Passengers Association (US).

2 Airport (or airfield) pressure altitude.

3 Allied Pilots' Association (US).

4 Army Parachute Association (UK).

5 Autopilot amplifier.

6 Automatic plotting aid.

7 Aerobatic practice area.

8 Accidents to private aviation [now Saga].

9 Altitude preselect/alert[er].

10 Aviation Policy Area.

11 Aerodromes Protection Agency, found necessary to fight closures (UK).

12 Airline performance analyses.

APAC Association of Professional Aviation Consultants (UK).

APACCS Aerial-port command and control systems (USAF).

Apacs Atlas prompting and checking system.

APADS Advanced precision air-delivery system.

APAG Allied Policy Advisory Group (NATO).

Apals Autonomous precision-approach and landing system.

AP/AM Anti-personnel/anti-material (last word often spelt materiel).

APAMA Asia/Pacific Aviation Media Association.

APAP Approach-path alignment panels.

Apapi Abbreviated Papi.

APAR Active phased-array radar (Canada, G, Neth.).

APATC Allied publication air traffic control.

Apatsi Airport/air-traffic system interface.

APAW Air-portable avionic workshop (RAF).

APB *1* Auxiliary power-breaker.

2 Aviation Policy Board (US Congress 1947–48).

APC *1* Association of Parascendng Clubs.

2 Approach power control (or compensator).

3 Armament practice camp (RAF).

4 Avionics planning conference.

5 Aviation Press Club (Belg.).

6 Area positive control.

7 Aeronautical passenger communications.

8 Aeronautical public correspondence [public telephone].

9 Autopilot computer [also A/PC].

10 Adaptive packet compression.

APCA Association Professionnelle de la Circulation Aérienne (F).

APCB Advanced plenum-chamber burning.

APCC Air Pollution Control Center (EPA, US).

apch, apchg Approach, approaching (FAA).

APCO Air Pollution Control Office (EPA, US).

APCR Armour-piercing, composite rigid.

APCS *1* Air Photo and Charting Service (USAF, Orlando AFB).

2 Approach-power compensation system.

APD *1* JETDS code: piloted aircraft, radar, DF/reconnaissance/surveillance (usually SLAR).

2 Aerial position, digital (usually 4,096 pulses per 360°).

3 Avalanche photo-diode.

4 Amplifyng photo-diode.

5 Airports Policy Divisions [1 to 3] (DETR UK).

6 Air Procurement District.

7 Air Passenger Duty (UK).

APDMC, A/PDMC Aircraft and products, or aircraft-propulsion, data-management computer.

APDS Air-picture display system; multi-radar C^2.

APDZ Active parachute drop zone.

APE *1* Airline pallet extender.

2 Airborne polar experiment, study of ozone depletion (R).

APEC Asia-Pacific Economic Co-operation.

APEP Armour-piercing, enhanced penetration.

aperiodic Of any dynamic and potentially oscillatory system, so heavily damped as to have no period; unable to

accomplish one cycle of oscillation; thus * magnetic compass, * electric circuit.

aperture *1* Diameter of objective of optical instrument, either direct length or function of it; also angular *, minor angle subtended at principal focus by extremes of objective diameter; numerical *, n sin u, where n is refractive index between lens and object and u is objective angular radius (half angular aperture); and relative * (f-number) relating focal length to objective diameter.

2 In radio or radar aerial, either greatest dimension; or, with undirectional aerial, greatest length across plane perpendicular to direction of maximum radiation, close to aerial, through which all radiation is intended to pass (ie all except diffuse stray radiation).

aperture card Standardised unit in microfilm filing, comprising frame of microfilm mounted in card border; stored, retrieved and projected automatically.

aperture management Design of radar cavities and apertures to eliminate multiple reflections.

Apex *1* Advanced project for European information exchange, linking all major EC airframe companies.

2 Advanced passenger, or purchased, excursion fare, one of many forms of air carrier fare and flight coupon.

apex Highest point in canopy of parachute in vertical descent.

APF *1* Association des [female] Pilotes Françaises.

2 Adhesive polymer film.

APFA Association of Professional Flight Attendants (US).

APFC Air-portable fuel container [or cell].

APFD Autopilot flight director.

APFSDS Armour-piercing fin-stabilized discarding sabot.

APG *1* JETDS code: piloted aircraft, radar, fire control.

2 Automatic program generator, requires only component and netlist input.

3 Aberdeen Proving Ground, MD (USA, but used by other US services).

4 Air Platforms Group (DSTL, UK).

APGC Air Proving Ground Command (USAF, defunct).

APGM Autonomous precision-guided missile [or munition].

APHAZ Aircraft proximity hazard[s], panel investigating airproxes.

aphelion Point in solar orbit furthest from Sun.

API *1* Air-position indicator.

2 Armour-piercing incendiary.

3 American Petroleum Institute.

4 Associazione Pilote Italiane.

5 Application program[ming] interface.

6 Air-photo interpreter.

7 Airframe/propulsion integration [also A/PI].

8 Aim-point initiative.

9 Ascent-phase intercept.

APICS Automatic pressure indication and control system.

APIHC Armour-piercing incendiary hard core.

Aphids Advanced panoramic helmet interface demo system.

apiquage According to a 1935 authority "Rotation of an aircraft about its lateral axis in the sense which decreses its angle of incidence (there is no English equivalent)."

APIRG African region Planning and Implementation Regional planning Group (ICAO).

APIRS Aircraft piloting inertial reference system, or strapdown sensor.

APIS *1* Apogee/perigee injection system.

2 Automatic priority interrupt system, for large computer systems with multi-programming.

3 Aircraft parking information system [docking guidance].

4 Air passenger information system.

APKWS Advanced precision-kill weapon system (USA).

APL *1* Acceptance performance level.

2 Applied Physics Laboratory of JHU.

APLA Asociación de Pilotos de Lineas Aéreas (Arg.).

aplanatic Free from spherical aberration.

APLS Automated ply laminating system.

APM *1* Aluminium powder metallurgy.

2 Aircraft performance monitoring.

3 Aviation de Patrouille Maritime (F).

4 Assistant programme manager.

APMS Automatic performance-management system (also, in US, rendered as advanced power management system) or automated performance measurement system.

APN *1* JETDS code: piloted aircraft, radar, navaid.

2 Aircraft procurement, Navy (US).

3 Arinc packet network.

APNA Association des Professionnels Navigants de l'Aviation (F).

A.P.970 See *Av.P.970*

apoapsis Point in orbit furthest from primary; apocentre.

apoastron Furthest point in orbit round star.

APOB, Ap Ob Airplane observation [of weather].

apocynthion Point in lunar orbit furthest from Moon.

APOE Air (or aerial) port of embarkation.

apogalacticon Furthest point in orbit round galaxy.

apogee Point in geocentric (Earth) orbit furthest from centre of Earth (in near-circular polar orbit equatorial bulge could result in satellite being closer to surface at equatorial apogee than at polar perigee).

apogee motor Apogee kick motor or kick motor, small rocket designed to impart predetermined (sometimes remotely controllable) velocity change (delta-V) to satellite or spacecraft to change orbit from an apogee position.

Apollo Applications Program, AAP Much altered and largely defunct NASA programme intended to make maximum and earliest post-Apollo use of Apollo technology. Major portion evolved into Skylab; other AAP being built in to various plans for future manned and unmanned spaceflight, include Shuttle missions.

apolune Apocynthion of spacecraft departed from Moon into lunar orbit.

APOR Automated purchase order rescheduling system.

apostilb Non-SI unit of luminance equal to $1/\pi$ international candle (candela) m^{-2} or 10^{-4} lambert (see *luminance*).

Apota Automatic-positioning telemetering antenna.

APP *1* Approach (DTI, UK).

2 Approach control office (ICAO).

3 Approach pattern.

4 Association des Pilotes Privés (F).

5 Association of Priest Pilots (US).

6 Autopilot panel.

APPA *1* Associação de Pilotos e Proprietarios de Aeronaves (Braz.).

2 Association des Pilotes Privés Avions (F).

apparent precession Apparent tilt of gyro due to rotation of Earth; vertical component = topple, horizontal = drift.

apparent solar day Length of Earth day determined by two successive meridian passages of apparent Sun; longer than sidereal day by time taken by Earth to turn additional increment to nullify distance travelled in solar orbit during this day. Basis of most human timescales, being divided into 24 h, hour being thus defined.

apparent wander Apparent precession.

APPC Advanced program-to-program communications.

APP CON Approach control (FAA).

Apple *1* American pilots participating in local education.

2 Aircraft precise-position location equipment.

Appleton layer F layer (F_1 and F_2) of ionosphere, most useful for reflection of EM radiation (see *F-layer*).

Appleyard scale Circular slide-rule.

appliqué Adhesive in the form of thin foils or polymer-based film, usually on aircraft external surface.

APPP, AP3 Airport Privatization Pilot Program (FAA from 1996).

APPR Approach.

approach BS.185: 'To manoeuvre an aircraft into position relative to the landing area for flattening-out and alighting'. Now subdivided into various categories, each of which needs pages of explanation defining circumstances, clearances and procedure. Following are brief notes. VFR * may be made with no radio at uncontrolled airport or airfield. Visual * may be made in IFR by pilot in contact with runway either not following other traffic or else in visual contact with it, with ceiling at least 500 ft above minimum vectoring altitude and visibility at least three miles. Various types of instrument * are admissible in IFR with radio TWR authorization: straight-in, circling, precision (with g/s and runway centreline guidance) and parallel (two parallel ILS runways, or, in military aviation, two parallel runways, each with PAR). In certain circumstances pilot may receive clearance for contact *, even in IFR. ILS * is most important IFR precision *. If required and available, pilot can be 'talked down' in GCA or RCA, his only necessary equipment being primary instruments and operative R/T.

approach area Airspace over designated region of terminal area controlled by approach control unit (in some cases serving two or more airfields).

approach beacon *1* Historically, short-range track beacon (see *BABS*).

2 Today, beacon giving fix before or after approach gate (rare).

approach control BS.185: 'A service established to provide ATC for those parts of an IFR flight when an aircraft is arriving at, or departing from, or operating in the vicinity of, an aerodrome'. DTI (Air Pilot): 'ATC service for arriving or departing IFR flights'. FAA adds 'and, on occasion, VFR aircraft'.

approach control radar ACR, radar at approach control facility displaying PPI positions (and, in advanced models, height or alphanumeric data) of all aircraft within its range (which is not less than radius of furthest point in the controlled airspace).

approach coupler Electronic linkage between aircraft ILS receiver and autopilot and hence to AFCS; thus aircraft can make 'hands off' approach.

approach fix From or over which final approach (IFR) to airport is executed (FAA). On projected centreline 3–5 miles from threshold.

approach glide "A glide preliminary to alighting" (B.S., 1940).

approach gate Point on final-approach course 1 mile beyond approach fix (ie further from airport) or 5 miles from landing threshold, whichever is greater distance from threshold (FAA).

approach indicator Ambiguous: could mean ILS or other cockpit instrument or any of several visual systems on ground indicating angle of approach.

approach lights *1* In modern large airfields, any of several systems of lights extending along projected centre-line of runway in use towards approaching aircraft to provide visual indication of runway location, distances, alignment, glide path slope, and, probably, transverse horizontal.

2 In smaller or older airfields, one or more lights (often green) at, or extending from, downwind end of landing area to show favourable direction of approach.

approach noise Measured on extended runway centre-line 1 nm (one nautical mile = 6,080 ft = 1,853 m) from downwind end of runway, with aircraft at height of 370 ft (112.58 m). [See *Noise*].

approach operations Flight operations within approach area, esp. those of aircraft arriving or departing, designated as IFR or VFR.

approach plane Approach surface, sloping plane below which no aircraft should penetrate; in UK ** to grass airfield extends at inclination of 1:30 in all directions from periphery of landing area.

approach plate Flight-planning document relevant to specific airfield, giving details of minimum heights, safe headings and weather minimums (UK = minima), and including horizontal map and often also vertical profile for approach to each instrument runway.

approach power That used on landing approach, often about 58 percent MTO.

approach power compensator Autothrottle, esp. on combat aircraft. The APCS [S adds system] was devised to hold constant AOA (3) during carrier landings.

approach radar See *PAR, GCA, SRE*.

approach receiver *1* ILS receiver.

2 Historically, radio receiver 'capable of interpreting the special indications given by an approach beacon installation'.

approach sequence Order in which aircraft are placed while awaiting landing clearance and in subsequent approach. In busy TMA traffic drawn in blocks from alternate landing stacks.

approach speed Usually means IAS.

approach surveillance radar Approach control radar.

appropriation Act of Congress enabling a Federal agency to spend money for a specific purpose.

approval *1* Of manufacturer of aerospace hardware, approval by delegated national authority to design, manufacture, repair or modify such hardware, subject to specified conditions and inspection.

2 Of item of aerospace hardware, certificate issued by delegated national authority that item is correctly designed and manufactured and will thus be likely to perform within its design limits satisfactorily. In case of complete aircraft, C of A, or Type Certificate.

3 Of flight plan, signature by ATC officer or other responsible person that proposed plan does not conflict with pilot's qualifications, aircraft equipment, expected met conditions and expected air traffic, and that flight may proceed.

approval flight Required to authorise historic aircraft [usually military] to do one more year flying at airshows.

approval note Issued by importing country to cover aircraft with foreign C of A.

APPSS Association of Police and Public Security Suppliers.

APPT Air-platform propulsion technology; R adds research.

APQ JETDS code: piloted aircraft, radar, combination of purposes.

APR *1* JETDS code: piloted aircraft, radar, passive detection.

2 Airman performance report.

3 Automatic [or auxiliary] power reserve.

4 Air-photo reader.

5 Actual performance reserve.

APRA Air Power Association, previously the Air Public Relations Association (UK, 1947–).

APRL ATN(1) profile requirement list.

APRO Airlines Public Relations Organization (UK).

Aprodeas Association pour la Promotion et le Développement d'actions de formation pour les Entreprises Aéronautiques et Spatiales (F).

apron *1* Large paved area of airfield for such purposes as: loading and unloading of aircraft; aircraft turnaround operations; aircraft modification, maintenance or repair; any other approved purpose other than flight operations.

2 In engine cowling, any portion hinged down to act as walkway or servicing stand.

3 In ejection seat, lower forward face behind occupant's lower legs.

4 In vehicle fuelled with corrosive liquid, corrosion-resistant panel surrounding, and especially beneath, relevant supply hose coupling.

5 Fairing round front of main landing gear, forming underside of nacelle in flight.

apron capacity Nominated number of transport aircraft to be accommodated on particular apron area in designated positions.

apron-drive bridge Passenger loading bridge comprising telescopic sections pivoted to terminal, extended and positioned by steerable powered chassis supporting free end. See *bridge* and next.

apron-drive unit Self-propelled vehicular support for free end of pasenger jetty (jetway), usually provided with two heavy-duty wheels steering through at least 180°.

APRX Approximate[ly].

APS *1* Aircraft prepared for service; standard weighing condition, or condition at which weight is calculated: comprises aircraft in all respects ready to take off on mission of type for which it was designed, complete with all stores, equipment (such as passenger reading material), fuel, crew and all consumable items, but with no revenue load.

2 Appearance-potential spectroscopy.

3 JETDS code: piloted aircraft, radar, search and detection.

4 Adaptive-processor system, or sonar, or sonobuoy.

5 Armament practice station (UK).

6 Auxiliary power system.

7 Aerial Post Squadron (USAF).

8 Advanced planning and scheduling.

9 Airborne-platform subsystem.

10 Armament, or air-vehicle, planning system.

11 Advanced fighter-crew protection system.

12 AIS(1) processing system.

13 Airframe/propulsion/steering.

14 Air Pictorial Service, formed 1951, now AAS(6).

15 Airline Pilots Security Alliance (US, 2002–).

16 Aircraft/altitude/attitude position sensor.

17 Autopilot system [also A/PS].

APSA   Airline Pilots' Security Alliance (gun lobby, US).

APSE, Apse   Ada programming support environment.

APSG   After passing.

APSI   Aircraft, or airframe, propulsion-system integration.

apsides   Plural of apsis.

apsis   Extreme point of orbit, apocentre (furthest) or pericentre (nearest).

APSP   Advanced programmable signal-processor.

APST   Aircraft propulsion systems trainer.

APT   *1* Automatically programmed [machine] tool.

2 Automatic picture transmission, datalink from satellite vidicon.

3 Automatic, or automated, powerplant test [U adds unit].

4 Advanced passive technology.

5 Airport.

6 Aircrew procedure[s] trainer.

APTA   American Public Transportation Association.

APTS   Automatic picture-transmission system.

APTT   Aircrew part-task trainer.

APTU   Aerodynamic and Propulsion unit.

AP-25   C of A (R).

APU   *1* Auxiliary power unit; /GCU adds generator control unit.

2 Weapon release unit (R).

3 Automatic pull-up.

APUC   *1* APU controller.

2 Average program unit cost.

APV   *1* Autopiloted vehicle.

2 Accumulated project value.

3 Approved.

APVO   See *IA-PVO*

APW   *1* Automatic pitch warning, required on aircraft [eg, SR-71] with possible low or negative pitch stability.

2 Aircraft, or airborne, proximity warning; I adds indicator, S system.

APXS   Alpha-proton, or particle, X-ray spectrometer.

AQAD   Aeronautical Quality-Assurance Directorate (MoD-PE, now Qinetiq, UK).

Aquabrasive 330   Sand/water mix for high-velocity stripping of markings or sealant from airfield paved surfaces.

AQAP   Allied quality-assurance publication (NATO).

AQB   Advanced quadrature band.

AQC   Aviator Qualification Course [or Certificate] (USA).

AQD   Operational airline quality determination programme.

AQF   Avionics qualification facility.

AQL   *1* Agreed quality level [material specifications, dimensional tolerance, etc].

2 Advanced quick look (Guardrail).

AQP   *1* Advanced qualification program [US commercial pilots].

2 Avionics qualification procedure.

A-QPSK   Aeronautical quadrature phase-shift keying.

AQR   Airline quality rating.

AQS   Advanced quality system.

AQ-SAP   Acquisition special-access program.

aquaplane   To run wheeled vehicle, esp. landing aircraft, over shallow standing water at so high a speed that weight is supported wholly by dynamic reaction of water; tyres, out of ground contact, unable to provide steering or braking. An empirical formula is $V_a = 9\sqrt{p}$ where V_a is aquaplaning speed in knots and p is tyre pressure in lb/in^2.

aqueous   Pertaining to water, thus * solution.

Aqueous film-forming foam   Typically 3 to 6 per cent halon/Halotron/BCF or other firefighting agent and 97 to 94 per cent water.

Aquila   Code address of EQD.

AQZ   Area QNH zone.

AR   *1* Air [aerial, airborne] refuelling.

2 Air receive.

3 Aspect ratio.

4 Alternative route.

5 Army [or Air] Regulation.

6 or A/R, altitude reporting.

A/R   Approach/reverse (nozzle mode).

Æ̶R   Aspect ratio (US).

ARA   *1* Aircraft Research Association (UK, 1953–).

2 Airborne-radar approach.

3 Airspace restricted area.

4 Airborne Research Australia.

5 Alternative reference area.

6 Air-refuelling area (RAF).

7 Avanced Ram analysis [M adds method].

8 Anti-radar attack (UAV).

9 Advisory radio area.

10 Atmospheric research aircraft (FAAM 3).

ARAC   Aviation Rulemaking Advisory Committee (FAA).

Arades   Automatic radar evaluation system and jammer test set.

Araldite   Trade name (Ciba) of two-component (resin + hardener) epoxy-based adhesive used in airframe structural bonding.

ARALL, Arall   Aramid/aluminium laminate(s).

aramid fibre   Man-made fibre of extraordinary tensile strength, so named because of its chemical and physical similarity to spider-web fibre; see *Kevlar*.

Aramis   *1* Area multiple intercept system (radar).

2 Advanced runway arrivals management to improve airport safety and efficiency (Euret).

AR&M   Availability, reliability and maintainability.

AR&TF   Aircraft Repair and Transportation Flight (RAF).

ARAS   Auto refuelling assembly system.

ARASP   Advanced radar airborne signal processor.

ARATCC   Air Route Air-Traffice-Control Center (FAA).

ARB   *1* Air Registration Board, then Airworthiness Requirements Board (UK).

2 Air Research Bureau (BRA, ICAO).

3 Air Reserve, or Rescue, Base (USAF).

4 Arbitrary waveform generator.

arbitrary landing distance See *ALD*.

ARBS *1* Angle/rate bombing system [or set].

2 Airline Representative Board Sweden.

ARC *1* Aeronautical Research Council (UK).

2 JETDS code: piloted aircraft, radio, communications.

3 Ames Research Center (NASA).

4 Area reprogramming capability (EW).

5 Arc, position of compass rose on EFIS.

6 Air Race Classic (US from 1977, for female pilots only).

7 Airport runway configuration.

8 Automatic radial centring (VOR).

arc Ground track of aircraft flying constant DME distance from navaid.

Arcads Armament control and delivery system.

Arcal Aircraft, or airborne, radio control of airfield lighting.

Arcan Aeronautical Radio of Canada.

arc and plug See *plug aileron*.

ARCC *1* Airworthiness Requirements Co-ordinating Committee (UK).

2 Aeronautical Rescue Co-ordination Centre (RAF Kinloss, UK).

ARCH Agricultural remotely controlled helicopter.

Archie Colloq., anti-aircraft gunfire, 1915–18 (allegedly from 'Archie! Certainly not', music-hall song; archaic).

archway Airport-gate detector requiring passenger to pass through sensitive magnetic field, usually alongside baggage screening; also called AMD or WTMD.

ARCM Anti-radiation countermeasures.

Arcmas Automatic real-time cable monitoring and analysis system.

arc-minute, arc-second See *angular measure*.

ARCO, Arco Airborne remote control officer (RPVs).

ARCP *1* Air refuelling control point.

2 Aerodrome Reference Code Panel (ICAO).

ARCS *1* Acquisition radar and control system.

2 Aerial rocket control system.

3 Airline request communication system.

Arctic air mass Major class of air mass most highly developed in winter over ice and snow, although surface temperature may be higher than that for Polar masses.

Arctic minimum Densest of standard model atmospheres assumed in aircraft performance calculation.

Arctic smoke Surface fog essentially caused by very cold air drifting across warmer water.

ARCTS Automated radar-controlled terminal system.

ARD *1* Anti-radar drone.

2 Atmospheric re-entry demonstrator.

3 ATC-related delay.

4 Advanced requirement[s] definition.

ARDC Air Research and Development Command (USAF, established 1 February 1950, became Systems Command 1 April 1961).

ARDC model atmosphere Devised by ARDC, published 1956 (see *model atmosphere*).

Ardec Armament Research Development and Engineering Center (USA).

ARDF Airborne radio direction-finding.

Ardhan Association pour la Recherche de Documentation sur l'Histoire de l'Aéronautique Navale (F).

ARDS Airborne radar demonstration system, links J-Stars, Astor, Orchidée.

ARDU Aircraft Research & Development Unit (RAAF).

ARE *1* Airborne radar extension (surveillance C-130).

2 Altitude-reporting equipment (towed target).

3 Admiralty Research Establishment.

area SI unit of plane area is square metre (m^2); to convert from ft^2 multiply by 0.092903; from hectares by 10^4; from sq yd by 0.836127.

are Non-SI unit of area = 10^{-2}m^2.

area, aerospace surfaces See *gross wing* *, *net wing* *, *disc* *, *equivalent flat-plate* *, *control-surface* *.

area bombing Bombing in which target occupies large area, such as built-up area of city, with aiming point loosely defined near centre (when expression was current, WW2, marked at night by TIs or TMs).

area defence system In general, anti-aircraft or AAW system capable of providing effective defence over large area (dispersed battlefleet, task force, ground battlefield or large tract of country containing several cities) rather than point target.

area-denial munition Explosive device, usually dispensing cluster bombs each with time-delay fuze, to deny area to enemy ground forces.

area-increasing flap Wing flap which in initial part of travel moves almost directly rearwards to increase wing chord, without significant angular movement.

area loading Mass divided by gross projected area W/S [lifting-body aircraft].

area navigation, R-nav, RNAV Navaid that permits aircraft operations on any desired course within coverage of station-referenced navigation signals or within limits of self-contained system capability (FAA); thus, does not constrain aircraft to preset pathways.

area-navigation route Established R-nav route, predefined route segment, arrival or departure route (including RNAV SIDs and STARs). Route, based on existing high-altitude or low-altitude VOR/DME coverage, which has been designated by Administrator and published (FAA).

area ratio *1* In rocket thrust chamber, usually ratio of idealised cross-section area at nozzle to minimum cross-section area at throat; also called expansion ratio. In general, chambers designed to expand products of combustion into atmosphere have ** 10:1 to 25:1; those for use in upper atmosphere may exceed 50:1; SSME for Space Shuttle has ** 157:1.

2 For a wing, S/b^2, area divided by span squared, reciprocal of aspect ratio.

area rule Formulated by Richard T. Whitcomb at NACA in 1953. For minimum transonic drag at zero lift aircraft should be so shaped that nose-to-tail plot of gross cross-section areas should approximate to that of ideal body for chosen flight Mach number. Thus, addition of wing should be compensated for by reduction in section of body (which gave some early area-ruled aircraft "wasp waists", which are generally undesirable). Obviously, streamlines cannot be sharply deflected; it is not possible to have perfect area-ruling both with and without bulky external stores. In 1954–55 rules extended to Mach 2 by plotting cross-section area distributions on sloping axes approximately aligned with Mach angle.

area sterilization   Seeding part of sky with chaff of such extent and density that radar operation is impossible.

AREF, ARef   Air Refueling Squadron (USAF), also ARS.

Arens   A remote control system in which push/pull commands are transmitted by a steel cable tightly surrounded by a guiding coil spring, the whole sliding in a tube.

Arento   National telecommunications organisation (Egypt).

ARES   *1* Adaptable radar-environment simulator.

2 Airborne radar emitter simulator.

3 Aerial regional scale environmental survey.

Aresa   Association des Radio-Electroniciens de la Sécurité Aérienne (F).

Aresti   International procedures governing competitive aerobatics to set formula stipulating competing aircraft, set and free manoeuvres, judging and marking, now replaced by Aerobatic Catalogue.

ARF   *1* Air Reserve Forces; suffix PDS adds personnel data systems (US).

2 Airborne relay facility, or facilities.

3 Air Reconnaissance Facility.

4 Airlink risk factor.

ARFA, Arfa   Allied Radio-Frequency Agency (NATO).

Arfab   Allied radio frequency allocation board (NATO).

ARFAC   Australian Royal Federation of Aero Clubs.

ARFC   Aerospace Reconstruction Finance Corporation (US Government).

ARFF   Airfield [or airport or aircraft] rescue and fire-fighting (vehicle).

AR5   NBC hood and respirator (UK, RAF, RN).

ARFOR   Area forecast (Int. Met. Figure Code, ICAO).

ARG   *1* Amphibious Ready Group, with air assets (USMC).

2 Aeronautics Research Group (EREA).

ARGMA   Army Rocket and Guided Missile Agency (USA, now Miradcom).

argon   Most widespread inert gas in atmosphere; typical sea-level about 0.934%. Used in some fluorescent lamps and filament bulbs and as constituent of breathing mixtures, but most important in steel and titanium production and to prevent oxidation in welding.

Argos   National space-based navigation system (F).

ARGS   Anti-range-gate stealing.

argument   Angle or arc (astronomy). Thus * of perigee, angular arc traversed from ascendng node to perigee as seen by observer at near focus, measured in orbital plane in satellite's direction of travel.

Argus   Advanced remote ground unattended sensor.

ARH   *1* Active radar homing.

2 Anti-radar homing.

3 Anti-radiation homing.

4 Armed reconnaissance helicopter.

ARI   *1* Aileron/rudder interconnect.

2 Airborne, or aircraft, radio installation (UK).

3 Airpower Research Institute (US).

4 Aviation restructure initiative.

5 Additional requirement[s] for import.

6 Azimuth/range indicator.

7 Aviation Research Institute (USA, not connected with 3).

ARIA   Apollo (later Advanced) range instrumentation aircraft.

ARIARDA, Ariarda   Army Research Institution Aviation R&D Activity (USA).

ARIC   Airborne radio and intercom control.

Aries, ARIES   *1* Airborne research integrated experimental system [flight instrumentation].

2 Airborne recorder for IR and EO sensors.

3 Aeronautical reporting and information-exchange system.

4 Airborne reconnaissance integrated electronics suite.

Arinc   Aeronautical Radio Inc, with subsidiary Arinc Research. Non-profit research organisation responsible for aeronautical radio standards and widespread ground aids, esp. communications, across Pacific and in other regions.

Arinc communications and reporting system   VHF link between aircraft systems and ground-based computer, plus messages generated by menu-driven CDU.

Arinc (ARINC) 429   Initial standard digital data highway for civil aircraft; now 42913.

Arinc 629   Multiplexed bus for up to 120 subsystems at 2 Mbps.

ARIP   Air refuelling initial point.

ARIS   Anti-resonance isolation system.

arithmetic unit   Heart of typical digital computer; portion of central processor where arithmetical and logic functions are performed; invariably contains accumulator(s), shift and sequencing circuitry and various registers.

ARJA   Association Suisse Romande des Journalistes Aéro & Astronautique (Switz.).

ARJS   Airborne radar jamming system.

ARL   *1* Aeronautical   Research   Laboratories (Australia).

2 Aerospace research laboratories (US, OAR).

3 Air Resources Laboratory (NOAA).

AR-L   Air [or airborne] reconnaissance-low.

ARM   *1* Anti-radiation missile.

2 Atmospheric radiation measurement (US Dept of Energy).

3 Advanced radar mode[s].

arm   *1* To prepare explosive or pyrotechnic device so that it will operate when triggered.

2 Horizontal distance from aircraft or missile reference datum to c.g. of particular part of it.

Armak   Aeronautical certification authority (R).

armament   Carried on combat aircraft specifically to cause injury, by direct action, to hostile forces. Excludes radars, laser rangers (unless they cause optical injury by intent), illumination devices, detection or tracking devices, defoliating sprays, smokescreen generators and, unless filled with napalm, drop tanks.

Armat   Anti-radar missile, Matra.

armature reaction   In d.c. generator or alternator, distortion of main field by armature current, factor fundamental to machine design, and speed/voltage regulation.

ARMC   Area Regional Maintenance Center.

ARM/CM   Anti-radiation missile countermeasures.

ARMD   Anti-radiation missile, decoy.

armed   System switched to function upon command; thus, eg, when pneumatic escape chute (slide) is * it extends immediately passenger door is opened.

armed decoy Aerodynamic vehicle launched by penetrating bomber to generate additional target for hostile radars, send out its own countermeasures and, if it finds hostile aerial target, home on it and destory it.

armed reconnaissance Mission flown with primary purpose of finding and attacking targets of opportunity.

ARMET Area forecast, upper winds and temperatures (ICAO).

arming Closing an electrical circuit [or in any other way] enabling a device or system to function when required. A typical system is a thrust reverser.

arming vanes One name for slipstream-driven windmill used in some aerial bombs, mines and other stores to unscrew safety device or in some other way arm device as it falls.

ARMMAC Active remote maintenance monitoring and control (ILS, VOR etc).

armour Typical materials used in airborne armour include thick light alloys, titanium alloys, boron carbide and several filament-reinforced composites.

armourer Trade for military ground crew specialising in armament.

arm restraint In some types of ejection seat, automatic straps or arms energised during firing sequence to hold occupant's arms securely against aerodynamic forces until he is released from seat.

ARMS, Arms *1* Aircraft reporting and monitoring system, combines DMU, FDIU, FDR and AIDS.

2 Airport remote monitoring system.

3 Aviation reconfigurable manned simulator.

4 Airborne reconnaissance and marine surveillance.

5 Aerospace relay mirror system.

Armstrong line Pressure equivalent to about 63,000 ft, 19,200 m, at which human blood boils.

ARMTS Advanced radar maintenance training set.

ARN Active reduction of noise.

Arnd Around.

ARNG, ArNG *1* Army National Guard (USA).

2 Arrange.

ARNO Azimuth/range not operating.

Arnold Engineering Development Center, AEDC Large USAF installation in Tennessee charged with aerodynamic development, especially of air-breathing propulsion.

ARO *1* Air Traffic Services reporting office.

2 Aspheric reflective optics.

3 Airport reservation office, for arranging GA traffic slots.

4 Aerial refueling operator.

5 Aircraft Recovery Officer.

6 Army Research Office (US).

7 Airfield, or airport, reporting office.

AROD *1* Aerodrome runway and obstruction data.

2 Airborne remotely operated device.

AROG Auto roll-out guidance (after blind Cat 3b landing).

aromatic Hydrocarbon petrol (gasoline) fuels containing, in addition to straight-chain paraffins (kerosenes), various linked or ring-form compounds such as toluenes, benzenes and xylenes. Some cause rapid degradation of natural or synthetic rubbers.

AROS African Regional Organization for Standardization.

Arosys Adaptive-rotor system.

ARP *1* Air report (written).

2 Aero-Rifle Platoon (infantry section of Air Cavalary).

3 Air raid precautions (UK, WW2).

4 Aerospace, or aeronautical, recommended practice.

5 Aerodrome reference point (ICAO).

6 Aluminium-reinforced polyimide.

7 Attack reference point (appears in practice to mean IP) (US).

8 Applied research programme.

9 Anti-runway penetrator.

10 Airworthiness review programme.

11 Antenna rotation period.

12 Aviation regulatory proposal (Australian).

13 Anti-rotation period (radar).

14 Air-data reference panel.

ARPA, Arpa Advanced Research Projects Agency, created 1958, became Darpa.

ARPC Air Reserve Personnel Center (USAF, Denver, Colorado).

ARPS *1* USAF Aerospace Research Pilots' School, Edwards AFB, Calif.

2 Advanced radar processing subsystem (AEW).

ARPT Airport.

ARPTT Air-refueling part-task trainer (USAF).

ARQ Automatic error correction [repeat request].

ARR *1* Arrival message.

2 Airborne radio relay.

3 Air-refuelling receiver.

4 Air-traffic-control radar recording.

array Transmitting or receiving aerial (antenna) system made up of two or more (often 20 or more) normally identical aerials positioned to give enormously multiplied gain in desired direction.

ARRC Allied Command Europe Rapid-Reaction Corps (NATO).

ARRCOS Arrival co-ordination system.

arrested landing Normal fixed-wing landing on aircraft carrier, engaging arrester cable.

arrested-propeller system In aircraft with free-turbine turboprop engines, system for bringing one or more propellers to rest and holding them stationary while gas generator continues to run. Should not cause turbine overheat condition; speeds up turn-round and sustains on-board power without causing danger to passengers or others near aircraft.

arresting barrier Runway barrier.

arresting gear, arrester gear Fixed to aeroplane landing area to halt arriving aircraft within specified distance. Many systems qualified for use on aircraft carriers, rough battlefield airstrips (in this case, mainly by light STOL machines) and major military runways. In nearly all cases involves one or more transverse cables traversed by hook on arriving aircraft. Kinetic energy of aircraft dissipated by cable pulling pistons through hydraulic cylinders or rotary brakes, driving fan through step-up gears or towing heavy free chains.

arresting hook, arrester hook Strong hook hinged to some land-based and most carrier-based aeroplanes for engagement of arresting gear; usually released by pilot from flight position to free-fall or be hinged under power to Engage position.

arresting unit Energy-absorbing device on one, or usually both, ends of arrester wire.

ARRGp Aerospace Rescue and Recovery Group (USAF).

arrival *1* In flight planning, calculated time when destination should be reached (see *ETA*); may be determined by plotting straight line from last waypoint to overhead destination, but professional pilots refine this to take account of approach procedures.

2 Inbound unit of traffic (ie one aircraft approaching destination airfield).

3 Colloq., derogatory description of a particular landing.

arrival runway At major airfield, runway currently being used only by arrivals (2).

arrival stall Caused by attempting to line up on landing approach by rudder alone, without bank, causing inner wing to stall. Trying to recover by aileron aggravates situation.

arrival time Time at which inbound aircraft touches down (BS, FAA); airlines sometimes use different definitions, esp. time at which first door opened.

ARRM Affordable rapid-response missile [D adds demonstrator].

ARROW *1* Aircraft routing right of way (not spoken as word) (US).

2 Checklist for documents carried: Airworthiness, Registration, Radio license, Operational limitations, Weight/balance. Suffix-C adds Charts [outside local area].

arrow engine piston engine having three, or multiples of three, cylinders arranged with one (or one row) vertical and others equally inclined on either side. Also called broad-arrow or W (if inverted, M or inverted-arrow).

arrow stability Weathercock stability stemming from simple distribution of mass and side areas.

arrow wing *1* Markedly swept wing; in his Wright Brothers lecture in 1946 von Kármán used ** exclusively, and 'swept wing' did not become universal until 1948.

2 Modern meaning is wing with inboard section [with subtly curved profile] with LE sweep close to 80° and outer panels of more conventional form, eg sweep 30°–50°.

ARRS Aerospace Rescue & Recovery Service (USAF).

ARRW Aerospace Rescue & Recovery Wing (USAF).

ARS *1* Special air report (written).

2 Atmosphere revitalisation subsystem.

3 Auto-relight system.

4 Attack radar set.

5 Air Rescue Service (from 1996 ARRS).

6 American Rocket Society (became AIAA in 1962).

7 Attitude retention system (XV-15 FCS).

8 Automated retrieve system.

9 Aeroplane Repair Shop (RAF 1918–45).

10 Automated radar summary, charts issued hourly showing local echoes.

ARSA *1* Airport radar service area (in US reclassified 1993 as Class C airspace).

2 Apron, or advisable, radar service area.

3 Aeronautical Repair Station Association (US).

ARSAG, Arsag Aerial Refueling Systems Advisory Group (US).

Arsis Aircraft rotation, scheduling and information system.

ARSR Air-route surveillance radar, ARTCC radar to detect and display aircraft en route between TMAs.

ART *1* Actuator remote terminal.

2 Air Reserve Technician (Afres).

3 Airborne-radar technician.

4 Adaptive-resonance theory.

5 Auto reserve thrust.

Artac The Alliance of Independent Travel Agents (UK).

Artads Army tactical data system (USA).

Artas, ARTAS Air-traffic control radar tracker and server.

ARTCC Air-route traffic control center (FAA).

ARTCS Advanced radar traffic control system (FAA).

ARTD Applied research technology demonstrator.

ARTF *1* Alkali-removable temporary finish.

2 Aircraft Recovery and Transportation Flight (USAF).

Arthur Any AFCS (F, colloq.).

ARTI, Arti Advanced rotorcraft technology integration.

article Generalised term for one aircraft, especially one operated by the CIA.

articulated blade In rotorplane, rotor blade connected to hub through one or more hinges or pivots.

articulated rod In radial piston engine, any connecting rod pivoted to piston at one end and master rod at other.

artificial ageing Ageing of alloy at other than room temperature, esp. at elevated temperature.

artificial feel In aircraft control system, esp. AFCS, forces generated within system and fed to cockpit controls to oppose pilot demand. In fully powered system there would otherwise be no feedback and no "feel" of how hard any surface was working. Simulates ideal response while giving true picture of surface moments insofar as response curve of each surface and their harmonisation are concerned. System invariably strongly influenced by dynamic pressure q. Generates force for each surface according to optimised law [not necessarily same for all axes] and prevents pilot from damaging aircraft by primary control (but rarely takes into account rapid trimmer movements).

artificial gravity Simulated gravitational effects (but not field) in space environment. Obvious method involves rotation about axis, which introduces Coriolis forces.

artificial horizon *1* Primary cockpit flight instrument which, often in addition to other functions, indicates aircraft attitude with respect to horizon ahead.

2 Simulation of Earth horizon (planet's limb) for use as uniform and accurate sensing reference in near-Earth spaceflight (Orbital Scanner is programme generating data for this ideal).

artificial satellite Man-made satellite of planetary body.

ARTIP, Artip Advanced radar-technology insertion program (US).

ARTS *1* Automated, or automatic, radar terminal system (FAA, from 1966).

2 Aircraft-recovery transport system (after belly landing, etc.).

3 Automated remote tracking station.

4 All-round thermal surveillance.

5 All-purpose remote transport system (USAF).

ARTT Above real-time training.

ARU *1* Attitude retention unit (helicopter height-hold).

2 Auxiliary readout unit (ECM).

3 Aviation Research Unit, part of ARI (7).

ARV Air recreational vehicle.

ARW *1* Air Refueling Wing (USAF).

2 Advanced radar warning: E adds equipment, S system.

AS *1* Aerospace Standard; major standards are 9001 and 9010

2 Air Station, HQ of unit[s] not directly equipped with aircraft (USAF).

3 Anti-submarine; less often, anti-ship (both also A/S).

4 Airlift Squadron (USAF).

5 Anti-skid (also A/S).

6 Air start, pneumatic service vehicle.

7 Altimeter setting.

8 Anti-spoofing.

As *1* Structural aspect ratio.

2 Altostratus.

A$_s$ Web average burning surface area of solid rocket.

AS3, AS^3 Aviation Services and Suppliers (US).

ASA *1* Anti-static additive [fuel].

2 American Standards Association.

3 Airline Suppliers Association (US).

4 Army Security Agency (USA).

5 Advanced system architecture.

6 Air-services agreement, between operators and/or governments on particular routes.

7 Autoland status annunciator.

8 Aircraft, or airborne, separation assurance.

9 Airborne shared-aperture.

10 Aviation Safety Authorities; SC adds Steering Committee.

ASAA *1* American Society of Aviation Artists (founded 1986).

2 Acars system access approval

ASAAC Allied Standard Avionics Architecture Council.

ASAC *1* Asian Standards Advisory Committee.

2 Aviation Security Advisory Committee.

3 Airborne surveillance, airborne control, comprises autonomous radar, datalink, C^2, navigation/guidance.

4 Australian Sport Aviation Confederation, Inc.

5 Air, later (6–03) Aviation, Security Advisory Committee (TSA).

ASAG Air Security Advisory Group (TSA).

ASAI Automated subsystem of aeronautical information, linked with FDR(1)/RDP(1).

ASALM Advanced strategic air-launched missile.

AS&C Airborne surveillance and control.

ASAP *1* Airborne shared-aperture program (fighter radar).

2 Australian Society for Aero-historical Preservation.

3 Advanced survival avionics program (USAF).

4 Aviation Security Action Plan, audits national standards for counter-terrorism (ICAO).

5 Aviation Safety Action Partnership.

6 Aggressor space applications project.

7 Aerospace Safety Advisory Panel (NASA).

8 Association of Star Alliance Pilots.

ASAR Advanced synthetic-aperture radar.

ASARG Autonomous synthetic-aperture-radar guidance; P adds program.

ASARS *1* Advanced synthetic-aperture-radar system.

2 Airborne search and rescue system.

ASAS *1* All-source analysis system; RW adds remote workstation.

2 Airborne, or aircraft, separation assistance system.

ASAT *1* Aviation situation-awareness trainer.

2 Advanced subsonic aerial target.

A-sat Anti-satellite.

Asata Asociación de Aviadores de Trabajos Aéreos (Spain).

ASB *1* Air Safety Board (UK, and US 1938–40).

2 American Standard Beam (structural sections).

3 Assembly section breakdown (airframe).

4 Alert service bulletin.

5 Airline Stabilization Board (US).

asbestos Fibrous silicate minerals once used for thermal insulation and as reinforcement in composites such as Durestos.

ASBM Air-to-surface ballistic missile, = ALBM.

ASBS Automated self-briefing system.

ASC *1* American Standard Channel (structural sections).

2 Aeronautical Systems Center (Wright-Patterson AFB).

3 Ascend, ascending (ICAO).

4 Aviation Statistic[s] Center (Canada).

5 American Society for Cybernetics.

6 Aircraft-system controller.

7 Air Service Command (USAAF, defunct).

8 Air Support Command (RAF, defunct).

9 Aviation Safety Council (Taiwan).

10 Airborne strain counter.

11 Aerospace planning chart.

12 Airborne surveillance and control.

Ascap Automatic SSR-code assignment procedure.

Ascas Automated security-clearance approval system.

ASCB *1* Avionics synchronized control bus; commercial counterpart to MIL-1553B (Sperry).

2 Avionics, or aircraft, standard communications bus.

ASCC Air Standards Coordinating Committee (US, UK, Canada, Australia, NZ).

ASCCA, ASC^{2}A Air and Space Command and Control Agency (USAF).

ASCE American Society of Civil Engineers.

ascending node Point at which body crosses to north side of ecliptic.

ascent Rise of spacecraft from body other than Earth.

ASCIET All-Service Combined Identification Evaluation Team (US).

ASCII American standard code for information interchange, 7-bit plus 8th bit for parity in serial transmissions.

ASCL Advanced sonobuoy communications link.

ASCM *1* Anti-ship cruise missile.

2 Advanced spaceborne computer module.

Ascon Airport surveillance and control [system] (Alcatel).

Ascot, ASCOT *1* Airspace control and operations training [simulator].

2 Aerial-survey control tool.

ASCPC Air-supply and cabin-pressure control[ler].

ASCS Aircraft-systems central system.

ASCSII Aerospace sensor component and subsystem investigation and innovation (USAF).

ASCTU Air-supply controller/test unit.

ASCU Armament station control unit.

ASD *1* Accelerate/stop distance.

2 Aeronautical Systems Division (AFSC, USAF).

3 Average sortie duration.

4 Aircraft, or airborne, or air, situation display.

ASDA, ASDa *1* Accelerate/stop distance available.

2 Association Suisse de Droit Aérien (now ASDAS) (Switzerland).

Asdar Aircraft-to-satellite data relay.

ASDAS Association Suisse de Droit Aérien et Spatial (Switzerland).

ASDC *1* Alternative Space Defence Centre.

2 Armament signal data converter.

ASDCS Airspace surveillance display and control system (WSMR).

ASDE Airport surface, or airfield surveillance, detection equipment (from 1952).

ASDF Air command and control Simulation and Demonstration Facility.

ASDI Aircraft situation display indicator.

Asdic Armed Services Documents Intelligence Center (US).

asdic Sonar, from "Anti-Submarine Detection Investigation Committee"; UK term originally (1925) applied to high-energy sound systems carried in surface vessels.

ASDL *1* Airborne self-defence laser.

2 Aeronautical satellite data-link.

ASDR *1* Avionic systems demonstrator rig.

2 Airport surface detection [ie, surveillance] radar.

ASDS Aircraft-sound description system.

ASE *1* Airport support equipment.

2 Amalgamated Society of Engineers (US).

3 Autostabilization equipment.

4 Allowable steering error (HUD).

5 Auto slat extension (DC-10).

6 Aircraft survivability equipment.

7 Aero servo-elastic [mode].

8 Altitude setting, or altimetry-system, error.

9 Airplane, single-engine; see *ASEL*, *ASES*.

Asean, ASEAN Association of SE Asia Nations.

Asecna Agence Pour la Sécurité de la Navigation Aérienne (F).

ASEDP Army Space Exploration Demonstration Program (USA).

ASEE American Society for Engineering Education.

ASEG All-Services Evaluation Group (USN).

ASEL Airplane, single-engine, land.

Asema Advanced special electronic mision aircraft (USAF).

ASEP ACCS [Allied command and control system] surveillance exploratory prototype (NATO).

ASES Airplane, single-engine, sea.

ASET *1* Aircraft survivability equipment trainer (threat simulator).

2 Automatic scoring electronic target; A adds aircraft.

Asetma Asociación Sindical Española de Tecnicos de Mantenimiento de Aeronaves.

ASETS Airborne seeker evaluation test system.

ASF *1* Air Safety Foundation.

2 Aviation Sans Frontières (humanitarian, non-profit).

3 Aircraft Servicing Flight (RAF).

4 Aeromedical staging facility/facilities.

ASFA Aviation Safety Foundation of Australia.

AS4 Structural graphite/epoxy composite (Hercules).

ASG *1* Airborne system, gun (DoD hardware code).

2 Piloted aircraft, special/combination, fire control (JETDS).

3 Air Safety Group (UK).

4 Arinc signal gateway.

5 Acoustic-signal generator.

ASGC Airborne-surveillance ground control.

ASH *1* Advanced, or assault, support helicopter (USA).

2 Active seeker homing.

ASHC Assault Support Helicopter Company (USA).

ashless dispersant Lubricating-oil additive which stops the oil from ashing when overheated.

ASI *1* Airspeed indicator.

2 Air Staff Instruction (usually plural, ASIs).

3 Aviation Safety Institute (US).

4 Agenzia Spaziale Italiana.

5 Aviation Society of Ireland (1963–).

6 Arinc standards interface.

7 The Aeronautical Society of India.

8 Augmented spark igniter.

9 Air-system interrogator.

10 Avionics system integration.

a-Si Amorphous silicon, atoms arranged haphazardly, not in lattice.

ASIA Association Suisse de l'Industrie Aéronautique.

ASIC *1* Application-specific integrated circuit.

2 Australian Securities and Investment Commission.

ASICC Australian Space Industry Chamber of Commerce.

ASID American Society for Information Display.

ASIG *1* Airports Special-Interest Group (UK local-government association).

2 Aircraft Service International Group.

Asims Advanced Sita message server.

ASIP Aircraft structural-integrity program (USAF).

ASIR Airspeed indicator reading.

ASIS *1* Air Safety Investigating System (Australia).

2 Abort-sensing and implementation system.

Asist, ASIST Aircraft/ship integrated secure and traverse.

ASIT Adaptable surface-interface terminal.

Asival Assessment of the ATM(7) system configuration subject to validation (Euret).

A6013 A weldable Si/Al alloy (Alcoa).

ASJ Automatic search jammer.

ASK *1* Automatic shift-keying.

2 Available seat-km.

ASL *1* Above sea level (UK = AMSL).

2 Authorised service life.

3 Atmospheric Sciences Laboratory (WSMR).

ASLA Air Services Licensing Authority (NZ).

Aslar Aircraft surge launch and recovery (USAF).

Aslib Association of Special Libraries and Information Bureau [not plural]; (UK 1951–).

ASLO Accident-Site Liaison Officer.

ASLP Air-sol longue portée (long-range ASM) (F).

ASLR Air/surface, or air-to-surface, laser ranger.

ASLS Air Surveillance Liaison Section.

ASLV Augmented satellite launch vehicle (India).

ASM *1* Air-to-surface missile; NZ has used * to mean anti-ship missile.

2 Available, or aircraft, seat-miles (often a.s.m.).

3 Advanced systems monitor (in cockpit).

4 American Society for Metals.

5 Autothrottle servo-motor.

6 Apron-services management.

7 Aircraft schematic[s] manual.

8 AirSpace management.

9 Air separation module.

10 Aerial supervision module.

ASMA   *1* Aerospace Medical Association, often **AsMA** (US).

2 Air Staff management aid (UK).

ASMD   Anti-ship missile defense (USN).

ASME   The American Society of Mechanical Engineers (office New York).

ASMET   Accelerated simulated missile endurance test.

A-SMGCS   Airport, or airfield, or avanced, surface-movement guidance and control system.

ASMI   Airfield surface movement indicator.

ASMK   Advanced serpentine manoeuvrability kit.

ASMO   Arab Standardization and Metrology Organization.

ASMP   Air-sol moyenne portée (medium-range ASM); A adds amélioré = improved.

ASMR   Advanced short/medium-range.

ASMS   *1* Advanced strategic-missile system (previously Abres).

2 Airport surface-management system.

ASMT   Air-supply mission technology.

ASMU   Avionics system management unit.

ASN   Abstract syntax notation, protocol for message handling; can be followed by identity number.

ASNI   Ambient sea-noise indication.

ASNT   American Society for Non-destructive Testing (office Columbus, OH).

ASO   *1* Air-support operations, see *ASOC*.

2 Acoustic-systems operator.

ASOC   *1* Air-Support Operations Centre (RAF), Center (GTACS).

2 Air Sovereignty Operations Centre.

Asops   Airport-security operational process simulation.

ASOS   Automated surface observation, or observing, system; at uncontrolled airfields gives voice readout of useful data.

ASP   *1* Armament status panel.

2 Aircraft servicing platform [or pan].

3 Automated small-batch production.

4 Audio selector panel.

5 Airbase survivability program (US).

6 Adaptive, or advanced, or aircraft, or airborne, signal processing, or processor.

7 Aircrew services package.

8 Application service provider.

9 Altimeter setting panel.

10 Airborne surveillance platform.

11 Airfield[s] systems planning (ICAO).

12 Arrival sequencing program.

13 Aircraft systems processor.

14 Antenna scan[ning] period.

15 AFS(1) planning.

ASPA   Asociación Sindical de Pilotos Aviadores (Mexico).

Asparcs   Air-surveillance and precision-approach radar control system (USMC).

ASPE   Speed of sound in EDP.

aspect change   Changing appearance or signature of reflective target as seen by radar, caused by attitude changes.

aspect ratio   General measure of slenderness of aerofoil in plan. For constant-section rectangular surface, numerical ratio of span divided by chord, discounting effects due to presence of body or other parts of aircraft. For most

wings ** A is defined as b^2/S, where b is span measured from tip to tip perpendicular to longitudinal axis (slew or VG wing as nearly as possible transverse) and S is gross area. Structural ** A_S generally defined as $b^2 \sec^2 \Lambda$ or $\dfrac{A}{\cos^2} \Lambda$, where Λ is ¼-chord sweep angle. Effective or equivalent ** increased by fitting end-plates (eg horizontal tail surface on top of fin), but there is no universally applicable formula. Optimum ** usually means that giving minimum wing weight, but this is seldom the best overall. Generally, faster aircraft have lower **; shorter field length demands higher **. Sailplanes and man-powered aircraft need extremely high **, from 20 to 40.

Aspen   Aerospace planning and execution network (USAF).

Asph   Asphalt.

Aspire   Advanced supersonic propulsion integration and research (NASA).

Aspis   Advanced self-protection integrated suite.

ASPJ   Advanced, or airborne, self-protection jamming, or jammer.

ASPO   *1* Army Space Program Office (USA).

2 Avionics Systems Project Officer (USAF).

ASPP   Airfield [originally aeronautical fixed] Service Planning Panel (ICAO).

ASPR   Armed Services Procurement Regulations (US).

Aspro   Airborne associative [from "advanced signal"] processor.

ASPS   Airborne self-protection system.

ASQC   American Society for Quality Control.

ASRAAM, Asraam   Advanced short-range AAM.

ASR   *1* Air/sea rescue [see *ASR apparatus*].

2 Aerodrome, airfield, airport, or approach, surveillance radar.

3 Air Staff Requirement (UK).

4 Altimeter setting region.

5 Automatic send/receive.

6 Acceleration slip reduction.

7 Air Safety Report, of hazardous event.

ASR apparatus   Term for that dropped to survivor(s), including 3 canisters containing dinghy, food, radio etc.

ASRC   Alabama Space and Rocket Center, Huntsville.

ASRgn   Altimeter setting region.

ASRM   Advanced solid rocket motor (NASA).

Asroc   Anti-submarine rocket.

ASRP   Aviation Safety Reporting Program (FAA/NASA, from 1976).

ASRS   *1* Aviation safety [&] reporting system (NASA).

2 or **AS/RS** Automatic storage/retrieval system.

ASRT   *1* Air support radar team.

2 Autonomous scout rotorcraft testbed (DoD).

ASRTE   Avion station relais de transmissions exceptionelles, also redered Astarte (F).

ASRWSP   Airport surveillance radar with weather systems processor.

ASS   *1* Attitude-sensing system.

2 Airlock support subsystem.

3 Air Signallers' School (RAF).

4 Anti-shelter submunition.

5 Aviation support ship.

6 Atmospheric sounding system.

ASSA   *1* Aeronautical Society of South Africa, more usually in Afrikaans: LKV.

2 Aviation system of systems architecture (US).

ASSAD, Assad   Union of aviation-engine producers (R, CIS).

assault aircraft   Aeroplanes and/or helicopters which convey assault troops to their objective and provide for their resupply.

assembly   *1* Completed subsystem, portion of airframe or other part of larger whole which itself is assembled from smaller pieces.

2 Process of putting together parts of functioning item of equipment, engine, subsystem, portion of airframe or other aerospace hardware (other than complete aircraft, buildings, docks, large launch complexes and similar major structures, for which preferred term is *erection*).

assembly drawing   Engineering drawing giving no information on manufacture but necessary for correct assembly; shows geometric relationships, assembly sequence, necessary tooling (jigging), fits and tolerances, and operations required during assembly.

assembly line   Essentially linear arrangement of work stations in manufacturing plant for assembly or erection of finished product or major component (eg wing). Modern aircraft produced in such small numbers optimum erection-shop layout often not linear. High-rate production (eg car engines or radio sets) parts travel on belt or overhead conveyor from station to station (see *transfer machines*).

ASSET   Aero-Space Structure Environmental Test; USAF research programme.

Asset   Airborne sensor system for evaluation and test (Northrop).

asset   Item or group of items from a nation's military inventory, especially those serving front-line function.

ASSG   Acoustic-sensor signal generator.

assigned amount   In emissions legislation, the maximum permitted.

Assist   Affordable space systems intelligent synthesis technology (USAF/NASA).

assisted takeoff, ATO   Aerodyne takeoff with linear acceleration augmented by accelerator, by self-propelled trolley, by rockets attached to aircraft or by other means not forming part of normal flight propulsion. Criterion is use of external force; mere downward slope, giving component of weight in takeoff direction, does not qualify.

ASSM   Anti-ship supersonic missile.

associated (VOR, Tacan)   VOR and Tacan/DME facilities either co-located or situated as closely as possible; subject to maximum aerial (antenna) separation of 100 ft in TMAs when used for approach or other purposes requiring maximum accuracy, or 2,000 ft elsewhere.

associative processor   Digital computer processor operating wholly as ancillary to another, usually larger, installation. Normally has no parent computer. Future ATC computer installations may use powerful large-memory sequential machine to resolve conflicts and exercise overall control, supplemented by ** working in parallel seeking potential conflicts.

ASSR   *1* Airport surface-surveillance radar.

2 Approach-control secondary surveillance radar.

3 Air-security screening records (TSA).

ASSRP   Air and Space Scientific Research Program (AFRL).

AS³, ASSS   *1* Active-search sonobuoy system.

2 Airport surface-surveillance system.

ASST   *1* Anti-ship surveillance and tracking, or targeting.

2 Advanced supersonic transport.

ASSTC   Aerospace Simulation and Systems-Test Center.

assured destruction   Concept of measurable inevitable damage inflicted on enemy heartlands for purposes of deterrence and arms limitation.

ASSV   Alternate-source select[or] valve.

ASSW   *1* Anti-surface-ship warfare.

2 Associated with (met. refort).

AST   *1* Advanced supersonic transport (or technology).

2 Advanced simulation technology.

3 Atmospheric surveillance technology.

4 Air Staff Target (UK).

5 Accelerated service test.

6 Airborne surveillance testbed.

7 Avionics system trainer.

8 Atlantic Standard Time.

9 Applied signal technology.

10 Air surveillance terminal.

ASTA   *1* American Society of Travel Agents.

2 Airport surface-traffic automation [S adds system] (FAA).

3 Aircrew synthetic-training aid[s].

astable   Not having a stable state.

AST&L   American Society of Transportation and Logistics (Lock Haven, PA).

Astamids   Airborne standoff minefield-detection system (passive IR).

Astar   Airborne search/track, or target, attack radar.

Astarte   Avion station relais de transmissions exceptionelles (F).

astatic   Without specific orientation or direction.

ASTE   *1* Association pour le Développement des Sciences et Techniques de l'Environnement (F).

2 Advanced strategic/tactical expendables [mainly IR sources].

Astec   *1* Automation system[s] for terminal and en-route control.

2 Advanced small turbine-engine core.

3 Advanced strategic/tactical expendable[s] (IR decoy).

Aster   Advanced spaceborne thermal emission and reflection radiometer.

Asterix   All-purpose structure Eurocontrol radar info. exchange.

asteroid   Minor planet, esp. fragments (mostly much less than 50 miles across) orbiting between Mars and Jupiter; thus, small body, such as artificial satellite, in solar orbit.

ASTF   *1* Aeropropulsion Systems Test Facility (at AEDC, commissioned September 1985).

2 Airspace system task force.

ASTI   Airport surface-traffic indicator.

Astia   Armed Services Technical Information Agency (US).

ASTM   American Society for Testing & Materials.

ASTO   Arab Satellite Telecommunications Organization (Int.).

ASTOL   Alternate (alternative is meant) STOL.

Astor   Airborne stand-off radar.

ASTOS   Association of Specialist Technical Organisations [SMEs] for Space (UK).

Astovl, ASTOVL   Advanced Stovl.

ASTP   *1* Apollo-Soyuz Test Project (US, USSR).

2 Advanced Space Transportation Program (NASA).

ASTR Attack store; usually (and very confusingly) means which type of AAM is selected, not air/ground weapon.

Astra, ASTRA *1* Applications of Space Technology to Requirements of Civil Aviation (ICAO panel); 1986 renamed Application of Space Techniques Relating To Aviation, later still Space replaced by Satellite.

2 Air staff training programme.

3 Association in Scotland to Research into Astronautics.

4 Advanced systems training aircraft (ETPS).

5 Attitude steering turn-rate azimuth.

Astral Air-surveillance and targeting radar L-band.

Astras Airport surface-traffic awareness system (ICAO), confusion with ASTA[2].

Astrid Airborne system for target recognition, identification and designation.

astrionics Space electronics.

Astro *1* Air support to regular operations (police helicopters).

2 Autonomous space transport robotic operations, or -and robotic orbiter (Darpa).

astrobiology Science of possible life on planets other than Earth, or elsewhere in space (note, in R means 'space medicine').

astrocompass Non-magnetic instrument, gives direction of true North relative to celestial body which must emit light and be of known direction (see *astronavigation*).

astrodome Optically transparent dome in roof of large aircraft 1935–50 through which navigator could take astro fixes using sextant.

astro fix Fix obtained by sighting two or more stars of known direction using sextant or astrocompass.

astrogation Navigation in space, suggest colloq.

astro-inertial Navigation by means of inertial system updated or corrected by astro fixes.

astronaut One who navigates in space; ie who travels in space. Specif., one selected for space flight by NASA.

astronautics Science of study, design, construction and operation of spacecraft.

astronavigation *1* Navigation of aircraft or spacecraft by measuring declination, right ascension and/or other angular positions of stars and other celestial bodies whose location on celestial sphere is known.

2 Navigation of spacecraft by any means (usage ambiguous).

astronics Astrionics.

astronomical twilight Period between day and night when Sun's centre between 12° and 18° below sea-level horizon (see *civil twilight*, *nautical twilight*).

Astronomical Unit, AU, A.U. Unit of linear distance based on mean distance between Earth and Sun; accepted value was 149,598,500 km, but IAU definition is now $1,496{\times}10^{11}$ m.

astronomy Science of celestial bodies other than Earth. Not included in this definition are celestial phenomena, such as polarisation of stellar light and other measures concerned more with radiation than with 'bodies'. Thus, subdivision radar *, X-ray *, IR *, UV * etc.

astrophysics Physics of observable universe, esp. states of matter and energy generation and transfer.

astroseismology Study of earthquakes on bodies other than Earth.

astro-tracker Automatic sextant capable of searching celestial sphere for particular luminous body, identifying it and determining orientation in terms useful for navigation, and of repeating sequence with same and at least one other celestial body. Corrects and updates INS in long-range aircraft.

ASTS Association Suisse pour les Techniques Spatiales (Switz).

ASU *1* Aeromedical staging unit.

2 Altitude (or, confusingly, attitude) sensing unit.

3 Approval for service use.

4 Aircraft storage unit (UK).

5 Aircraft starting unit.

6 Acoustic simulation unit.

7 Avionics switching unit.

Asupt Advanced simulator for undergraduate pilot training.

ASUW, ASuW Anti-surface [or anti-surface-unit] warfare aircraft category (USN).

ASV *1* Air-to-surface vessel. WW2 airborne search radar. Incorrectly rendered as anti-surface vessel. See *ASVW*.

2 Aerial swimming vehicle.

ASVEH Air-surveillance vehicle.

ASVS Airborne separation video system.

ASVW Anti-surface vessel warfare, today's term.

ASW *1* Anti-submarine warfare [AC adds analysis center, AS area system, DS data system, SOW stand-off weapon and T trainer].

2 Aft-swept wing.

3 Confusingly, also used in US for anti-ship warfare, and in UK (WW2) for air/sea warfare.

4 Acquisition scan window (UAV).

ASWAC Airborne surveillance, warning and control [S adds system] (India).

ASWDU Air/Sea Warfare Development Unit (RAF/RN).

ASWE Admiralty Surface Weapons Establishment (Portsdown, UK).

ASWOC ASW [1] operations centre.

ASWT Air-to-surface weapons technology.

ASXV Air-launched expendable sound velocimeter.

asymmetric flight Flight by aerodyne in sustained grossly asymmetric condition of lift, weight, thrust or drag, esp. flight by multi-engined aircraft in which at least one engine at substantial distance from axis of symmetry is inoperative.

asymmetric loading Flight by aircraft, esp. aerodyne, in which c.g. is located at substantial distance from vertical line through centre of lift with aircraft in level attitude and trimmed for normal horizontal flight (eg strike aircraft unable to release one of two heavy stores carried on outer-wing pylons).

asymmetric warfare Conflict between a high-tech nation and a primitive one.

asymptote Limiting position of tangent to curve, where lines meet at infinity. Thus, asymptotic, where slope of plotted curve becomes parallel to either x or y axis.

asynchronous Not synchronised, not in frequency or phase.

asynchronous computer Electronic computer, usually digital, in which operations do not proceed according to timing clock but are signalled to start by completion of preceding operation.

ASZ   Air surface zone (NATO, USAF).
AT   *1* Advanced trainer (USAAF category, 1924–48).
   2 Anti-tank.
   3 Autogenic training.
   4 Autothrottle [also A/T].
   5 Air transmit.
   6 Armament trainer (F-22).
   7 Air transport (role of tanker).
   8 All traffic.
   9 Advanced targeting.
AT[3]   See *ATTT*.
ATA   *1* Air Transport Auxiliary, UK ferry organization 1940–45; also the Association, 1946–.
   2 Air Transport Association of America (scheduled carriers).
   3 Actual time of arrival.
   4 Advanced tactical aircraft.
   5 Air Transport Association (UK).
   6 Automatic target-acquisition.
   7 Aero Testing Alliance (F, G, Neth.).
   8 Aviation Training Association (UK).
   9 Airport traffic area.
   10 Airline-tariff analysis.
   11 Advanced testbed[s] for avionics.
ata   Atmosphere[s] pressure; 1 ata [atm in UK] = 101.325 kPA = 14.6959 lb/in^2.
ATAAC   Anti-torpedo air-launched countermeasure[s].
ATAAS   Advanced terminal area approach spacing.
ATAB   Air Transport Allocations Board, joint agency in theatre of operations which assigns priorities to loads.
ATAC   *1* Air Transport Association of Canada.
   2 Applied-technology advanced computer (airborne EW).
   3 Air-transportable acoustic communication[s], expandable buoy.
ATACC   Advanced tactical air command center (USMC).
ATACMS   Army tactical missile system[s] (USA).
Ataco   Air Tactical Control Officer.
ATAF   *1* Allied Tactical Air Force (NATO).
   2 Association Internationale des Transporteurs Aériens (acronym from previous title).
ATAFCS   Airborne target acquisition and fire-control system.
ATAG   Air Transport Action Group, coalition pressing for better infrastructure.
Atags   Advanced-technology anti-g suit (USAF).
ATAL, Atal   *1* Automatic test application language.
   2 Appareillage de TV sur aéronef léger (F).
ATALS   Army Transportation and Logistics School (USA).
ATAM   Air-to-air Mistral.
ATAOS   Autonomous tactical attack and observation system.
ATAP   Advanced tactical-, or target-, attack penetrator.
ATAR   *1* Air-to-air recognition (device).
   2 Air-to-air recovery.
   3 Advanced threat-alert and response; CE adds critical experiment (USAF).
   4 Association des Transporteurs Aériens Régionaux (F).
Atares   Air-transport and air-refuelling exchange of services (European Air Group).

Atars   *1* Automatic traffic-advisory and resolution service (UK).
   2 Advanced tactical air-reconnaissance system.
ATAS, Atas   *1* Air Traffic Advisory Service, provides separation between known aircraft in IFR on certain routes (UK DTI).
   2 Advanced target-acquisition sensor.
   3 Air-to-air Stinger.
   4 Automated talking advisory system (FAA).
ATATS   Automatic target-acquisition and tracking-system.
Ataws   Advanced tactical air-warfare system.
ATB   *1* Advanced-technology bomber.
   2 Air-Transport Bureau (ICAO).
   3 Aerospace Technology Board.
   4 Automated ticket and boarding pass.
   5 Advanced-technology blade (HP turbine).
ATBM   Anti-tactical, or anti-theater, ballistic missile.
ATC   *1* Air traffic control; C adds centre.
   2 Air Training Corps (UK, replaced ADCC in 1941).
   3 Air Training Command (USAF, from 15 April 1946).
   4 Air Transport Command, formed from Air Ferrying Command 1 July 1942, became MATS.
   5 Approved Type Certificate, first issued (by DoC) in 1927.
   6 Advanced-technology component (DoD).
   7 Aerospace Technical Council (AIAA).
   8 Air transport conference (travel agencies).
   9 Automatic [usually EW] threat-countering.
   10 After top centre.
   11 Acoustic-torpedo countermeasures.
   12 Automatic tuning control.
   13 Astronomy Technology Centre (Edinburgh, UK).
   14 Airport traffic control (US DoC 1938).
ATCA   *1* Air Traffic Control Association (US, organised 1955).
   2 Air Traffic Conference of America.
   3 Allied Tactical Communications Agency (NATO).
   4 ATC Assistant (UK).
   5 Advanced tanker/cargo aircraft (USAF).
ATCAC   ATC Advisory Committee (US Congress).
Atcap   *1* ATC Automation Panel (ICAO).
   2 Army Telecommunications Automation Program (USA).
Atcare   ATC analysis and recording environment.
Atcas   *1* ATC administration system.
   2 ATC automation system.
ATCC   ATC centre/center (UK/US).
ATC clearance   Authorization by ATC for purpose of preventing collision between known aircraft for aircraft to proceed under specified conditions in controlled airspace (FAA).
ATCCC, ATC3   ATC Command Center (FAA).
ATCCTS   *1* ATC communications training system.
   2 ATC control-tower simulator.
ATCE   Air training centre of excellence.
ATCEU   ATC Evaluation Unit (Hurn, UK).
ATCGS   ATC ground segment [of satellite link].
ATCI   ATC investigation of airprox.
ATCMFT   ATC multifunction trainer.
Atco, ATCO   *1* ATC officer.
   2 Air-taxi and commercial operator.
Atcom   Aviation and Troop Command (USA).

ATCOMS, Atcoms   ATC operations, or and operational, management system[s].

ATCPS   ATC procedures simulator.

ATCPT   ATC procedural trainer.

ATCR   *1* Air Training Command Regulation[s] (USAF).

   2 ATC room[s].

ATCRBS   ATC radar-beacon system (FAA).

ATCRS   ATC radar simulator.

ATCRU   ATC radar unit (UK).

ATCS   *1* ATC Service (UK), or simulator.

   2 Active thermal-control subsystem.

   3 Automated tower control system.

ATCSCC   ATC System (originally Services) Command Center (FAA, Herndon, VA).

ATCSS   ATC signalling system (air/ground datalink tested 1958 as alternative to voice).

ATCT   ATC (14) tower; S adds simulator.

ATD   *1* Actual time of departure.

   2 Airline-, or aviation-, transmitted disease.

   3 [translated] Aviation technical division (USSR, R).

   4 Automatic threat detection; S adds system.

   5 Applied Technology Directorate (USA).

   6 Along-track distance.

ATDA   Augmented target docking adaptor.

ATDC   Automatic, or assisted, target detection and classification.

ATDL   Army tactical, or air-transport, datalink (USA).

ATDMA   Advanced time-division multiple access.

ATDS   *1* Airborne tactical data system.

   2 Air-turbine drive system.

ATDU   Air Torpedo Development Unit (RAF Gosport 1940–46).

ATE   *1* Automatic test equipment.

   2 Aircraft test and evaluation (UK).

   3 Actual time en-route.

   4 Advanced technology and engineering (also AT&E).

ATEC   *1* Aviation Technician Education Council (US).

   2 Automatic test-equipment complex.

Atecma, ATECMA   Agrupación Técnica Española de Constructores de Material Aeroespacial (Spain).

Ategg, ATEGG   Advanced turbine engine gas-generator.

Atems, ATEMS   Advanced threat-emitter simulator.

Atepsa   Asociación Técnicos y Empleados de Protección e Securidad a la Aeronavegación (Argentina).

ATER   Advanced triple ejector rack.

ATES   Aircraft Test and Evaluation Sector (Qinetic/DRA, Boscombe Down).

ATESS   *1* Aerospace and telecommunications engineering support services.

   2 Advanced tactics and engagement simulation subsystem.

ATF   *1* Advanced tactical fighter.

   2 Aviation turbine fuel.

   3 Actual time of fall.

   4 Amphibious task force.

   5 Adaptive terrain-following.

   6 Altitude test facility.

   7 Air traffic flow.

   8 Aerodrome, or airport, traffic frequency.

   9 Air-transport force.

Atfero   Atlantic Ferry Organisation (UK, 1940–42).

Atflir, Atfir, ATFLIR, ATFIR   Advanced targeting forward-looking IR.

ATFM   Air traffic flow management; U adds unit.

ATFPS   Air taffic flow planning system.

ATFS   Authentic tactical flight simulator, or simulation.

ATG   *1* Amphibious task group.

   2 Air-traffic generator.

ATGS   Air Tactical Group Supervisor, an experienced firefighter who provides aerial C^2 and also feeds information to the IC(8).

ATGW   Anti-tank guided weapon.

ATH   *1* Autonomous terminal homing.

   2 Air-transportable hospital.

   3 Automatic target handoff.

athodyd   Aero-thermodynamic duct, ie *ramjet*.

Athos   Airport tower harmonised controller system (Euret).

A/THR   Autothrottle mode.

ATHS   Airborne, or automatic, target handover, or handoff, system.

ATI   Air, or airline, transport instrument [standard panel sizes].

ATIF   *1* Aeronautical telecommunications network trials infrastructure (ICAO).

   2 All-source track and identification fusion.

ATIG   Air Technical Intelligence Group, of FEAF (1945–6).

Atigs   Advanced tactical inertial guidance system.

ATILO   Air technical-intelligence liaison officer.

ATIMS   Airborne target-information management system.

ATIMU   Advanced tactical inertial-measurement unit.

ATIR   Air-traffic incident report.

ATIRCM   Advanced threat, or theatre, IR countermeasures.

ATIS   *1* Automatic, or automated, terminal information service (ICAO, FAA); continuous broadcast of recorded non-control information in selected high-activity terminal areas (to improve controller effectiveness, and relieve congestion by automating repetitive transmission of routine information).

   2 Airfield-terminal information system.

   3 Air-traffic information service, or server, or system.

ATISD   Air Training Information Systems Division (Randolph AFB).

ATITA   Air Transport Industry Training Association (UK).

ATITB   Aviation and Travel Industry Training Board (NZ).

ATK   Aviation turbine kerosene.

ATKHB   Attack Helicopter Battalion (USA).

ATk   *1* Available tonne-kilometres.

   2 Anti-tank.

ATL   *1* Airborne, or advanced, tactical laser.

   2 Auto-trim loop.

   3 Acquisition, technology and logistics (DoD).

ATLA   Air Transport Licensing Authority (Hong Kong, formerly).

Atlantic   Airborne targeting low-altitude navigation thermal imaging and cueing.

Atlas   *1* Abbreviated test language for avionics systems.

   2 Advanced tactical low arresting system (overrun barrier).

3 Advanced tactical light arresting system (Aerazur cable/drum).

4 Advanced-technology ladar system.

5 Airborne topography and land-use assessment system.

6 Azimuth target-intelligence and acquisition system (Israel).

7 Antenna-testing laboratory automated system (NAS Patuxent).

8 Aircraft total lightning advisory system.

ATLB   Air Transport Licensing Board (UK).

Atlis   Automatic tracking laser illumination system.

ATLND   Automatic takeoff and landing (UAV).

ATM   *1* Air-turbine motor.

2 Air transport movement.

3 Air tasking [previously base] message, request for a particular combat mission to be flown (RAF).

4 Anti-tactical missile.

5 Anti-tank missile, or mine, or munition.

6 Asynchronous transfer, or transmission, mode.

7 Airspace, or air, traffic management [C adds centre, MG management group].

atm   Atmospheres pressure (UK usage, see *ata*).

ATMA   Association Technique Maritime et Aéronautique (F).

ATMG   Arms Transfer Management Group (US DoD).

Atmos   Ammunition, toxic material open space.

atmosphere   *1* Gaseous envelope surrounding Earth, subdivided into layers (see *atmospheric regions*, *model atmosphere*). For composition see *Air*.

2 Gaseous or vaporous envelope surrounding other planets and celestial bodies.

3 Theoretical model atmosphere providing standard basis for performance and other calculation.

4 Any of group of units of pressure all approximately equal to pressure of atmosphere on Earth at sea level. Most important is Standard * (abb. ata on European continent, atm in UK) equal to $101,325 \ Nm^{-2} = 101,325 \ Pa = 1,013.25 \ mb = 1.01325$ bars or hectopièze $= 14.6959 \ lbf \ in^{-2} = 761.848$ mm (29.994 in) Hg at 16.6°C. Second is Metric * (also ata) equal to 0.98642 Standard * and defined as 0.981117 bars (981.117 mb, ie acceleration due to 1 g) or $14.223 \ lbf \ in^{-2}$. Third is Technical * (at), usually identical with Metric. Fourth is bar (b), 1000 mb $= 750.07$ mm Hg $= 14.5038 \ lbf \ in^{-2}$ (see *pressure*).

atmospheric absorption   Absorption of EM radiation due to ionisation in atmosphere. Apparent loss of signal or beam power may be much greater, as result of diffraction and dispersion by vapour and particular matter.

atmospheric boundary layer   Generally defined as Earth's surface up to 5,000 ft or 1.5 km.

atmospheric braking   Use of air drag, esp. of upper atmosphere on re-entering spacecraft or RV, converting very high kinetic energy into heat.

atmospheric circulation   Gross quasi-permanent wind system of Earth, based on bands between parallels of latitude.

atmospheric constituents   See *air*.

atmospheric diffraction   Of importance chiefly with sound waves, which can be substantially changed in direction and intensity distribution by changes in air velocity and density. Effect with most EM radiation is small.

atmospheric duct   Almost horizontal layer or channel in troposphere apparently defined by values of refractive index within which EM radiation, esp. in microwave region, is propagated with abnormal efficiency over abnormally great distances.

atmospheric electric field   Intensity of electrostatic field of Earth varies enormously, but on fine day may be about $100 \ V \ m^{-1}$ at SL falling to around $5 \ V \ m^{-1}$ at 10 km height. Air/Earth current continuously degrades ***, believed that thunderstorms reinforce it.

atmospheric entry   Re-entry, or entry of extraterrestrial bodies such as meteors.

atmospheric filtering   Use of upper zones of ionosphere and mesosphere to filter out ICBM decoys from true warheads, it being supposed that latter will have higher density, better thermal protection and increasingly divergent trajectories, while decoys decelerate more violently, fall behind and burn up.

atmospheric pressure   See *atmosphere (4)*, *atmospheric regions*.

atmospheric refraction   Bending of EM radiation as it passes through different layers of atmosphere, esp. obliquely. Affects radio and radar, esp. when directionally beamed; visibly manifest in air over, say, hot roadway in sunshine when objects seen through this air 'shimmer'; in astronavigation ** makes apparent altitude of celestial bodies falsely great.

atmospheric regions   Layers of Earth's atmosphere differ in different model atmospheres; following notes are based on ISA. Lowest layer, troposphere, extends from SL to about 8 km (26,000 ft) at poles, to 11 km (36,090 ft) in temperate latitudes, and to 16 km (52,000 ft) over tropics. Throughout this region ISA characteristics of temperature, pressure and relative density are precisely plotted. Assumed lapse rate is $6.5°C \ km^{-1}$ and at tropopause, taken in ISA to be 11 km (36,090 ft), temperature is –56.5°C. From tropopause stratosphere extends at almost constant temperature but falling pressure to 30 km, stratopause, above which is mesosphere. Here there is reversed lapse rate, temperature reaching peak of about 10°C at 47.35–52.43 km, thereafter falling again to minimum of 180.65°K at mesopause (79.994–90.000 km). Above this is ionosphere, extending to at least 1,000 km, where temperature again rises through 0°C (273°K) at about 112 km and continues to over 1,000°C at 150 km and to peak of about 1,781°C at 700 km. Between 100–150 km lies E (Kennelly-Heaviside) layer; at 200–400 km is F (Appleton) layer, which at night is single band but by day divides into F1 and F2, F2 climbing to 400–500 km on summer day. E and F layers are reflective to suitable EM radiation striking at acute angle. Above ionosphere is open-topped exosphere, from which atmospheric molecules can escape to space and where mean free path varies with direction, being greatest vertically upward. Other ** are based upon composition, electrical properties and other variables.

atmospheric tides   Produced by gravitational attraction of Sun and Moon. Latter exerts small influence, equal to equatorial pressure difference of 0.06 mb, but solar ** has 12 h harmonic component (apparently partly thermal) of 1.5 mb in tropics and 0.5 in mid-latitudes.

atmospheric turbulence   See *gust*, *CAT*.

ATMS   *1* Air-traffic management system[s].

2 Advanced-technology microwave sounder.

ATN   *1* Aeronautical telecommunications network.

2 Air-traffic network.

Atnavics   Air-traffic navigation integration and co-ordination system (USA).

ATNM   Air-traffic network management.

ATNP   ATN(1) panel.

ATNS   Air-traffic [and] navigation services.

ATO   *1* Air task, or tasking, order.

2 Assisted takeoff.

3 Abandoned takeoff.

4 Abort to orbit, ie cannot avoid making a complete orbit.

5 Airborne tactical officer.

6 Auto [reporting of] time over[head].

7 Authorization to offer.

ATOA   Air Taxi Operators Association (UK).

Atoc(s)   Allied tactical operations centre(s) (NATO).

A to F   Authority to fly.

ATOL   Air travel organiser's (or operator's) licence (UK, a major function is to protect passenger after bankruptcy of carrier).

Atol   Advanced trainer on localizer.

Atoll   Assembly/test oriented launch language.

ATOM   Aileron trim offset monitor.

atom bomb   Colloq., fission bomb; very loosely, any NW (see *nuclear weapon*).

atomic materialization   Growth of thin-film coating by bombardment with ions and clusters [IR stealth].

atomic number   Symbol Z, number of protons in atomic nucleus or number of units of positive electronic charge it bears.

atomic weight   Mass of atom of element in units each 1/12th that of atom of carbon 12 (refined to 12.01115 on 1961 table). Numerical value for each element is same as atomic mass.

atomising   Continuous conversion of solid or liquid, esp. high-pressure jet of liquid, into spray of fine particles. Also called atomisation.

ATOP   Airline training orientation program (US).

Atops   Advanced transport operating systems (NASA).

ATOS   *1* Automated technical-orders system.

2 Air Transportation Oversight System (FAA).

ATP   *1* At time, place.

2 Actual track pointer.

3 Authority to proceed.

4 Aviation technical regiment (USSR).

5 Application transaction program (SNA).

6 Airborne [USAF, Advanced] targeting pod.

7 Air-transport pilot [ALTP is preferred].

8 Advanced tow placement, manufacturing process for precise composite structure.

9 Air-turbine pump.

10 Allied technical publication.

11 Attack plot.

12 Acceptance-test procedure.

ATPAC   Air-Traffic Procedures Advisory Committee (US).

ATPCS   Automatic takeoff power control system.

ATPL   Airline [or air] transport pilot's licence; ALTP licence; /H adds endorsement for helicopters [required to be PIC of civil aircraft ⩾20,000 kg MTOW].

ATR   Air, or airline, transport radio, Arinc system of standardizing dimensions of airborne electronics boxes, thus ATR, ½ATR, ¼ATR etc; broadly defined by Arinc 404, has also been said to mean air-transport[able] rack[ing].

2 Air Transport Rating.

3 Automatic, or aided, target recognition.

4 Airport terminal resources.

5 Air-traffic requirements.

6 Anti-transmit receive.

7 Analog tape recorder.

8 Advanced tactical radar.

9 Armed turn[a]round.

10 Attained turn-rate.

11 Advanced threat resolution, a baggage-screen work-station.

12 Automated time-recording.

Atran   Automatic terrain recognition and navigation, cruise-missile guidance, Goodyear from 1949.

ATRB   Advanced Technology Review Board.

ATRC   *1* Air transport, or traffic, regulation center (US).

2 See next.

ATR/C   Automatic, or aided, target recognition and classification.

ATRD   Active towed radar decoy.

Atrel   Air-transportable reconnaissance exploitation laboratory (RAF).

Atrif   Air Transportation Research International Forum (Int.).

ATRJ   Advanced threat radar jammer.

ATRP   Air-Transport Regulation Panel (ICAO).

ATS   *1* Air Traffic Services, thus * route (ICAO).

2 Applications technology satellite, wide research programme.

3 Suomen Avaruustutkimusseura Ry (Finnish astro-nautical society).

4 Automatic throttle system.

5 Armament training station (UK, WW2).

6 Aircrew training system (USAF).

7 Automatic test system (or station) for LRU check away from aircraft.

8 Aviation training ship (RN).

9 Air-turbine starter.

10 Acoustic tracking system.

11 Accelerator test stand [U adds upgrade].

12 Agile target system.

13 Auxiliary Territorial Service (UK WW2, became WRAC).

14 Advanced tracking system (radar extractor/tracker).

ATSA   *1* Aviation and Transportation Security Act (US, 19 November 2001).

2 Airline and travel services architecture.

3 Air-Traffic Services Agency (Bulgaria).

ATSB   *1* Air Transportation Stabilization Board (FAA).

2 Australian Transport Safety Board.

ATSC   *1* Air Technical Service Command (USAAF).

2 Air Traffic Services Cell, or comunication (FAA/DoD).

AT/SC   Autothrottle/speed control.

ATSCC   Air Traffic Service Command Center.

ATSD   Air-traffic situation display.

ATSG   Acoustic test signal generator.

ATSGF   ATS(1) geographic filter.

ATSM   ATS(1) message [P adds processor].

Atsora   Air traffic service[s] outside regulated airspace.

Comprises RAS, RIS, FIS and non-radar procedural services.

ATSS   Air Transport Security School (UK, RAF).

ATSU   ATS(1) unit.

ATSy   Air Transport Security Section (RAF).

ATT   *1* Automatic attitude hold (AFCS).

2 Advanced tactical [or theater] transport.

3 Advanced theater threat.

4 Automatic target tracking.

Aᴛᴛ   Tail-on-tail aerodynamic influence coefficient.

attach   To place temporarily in a military unit.

attached shockwave   Caused by supersonic body having leading edge or nose sufficiently sharp or pointed not to cause shock to detach and move ahead of it. Critical values of M at which shock will just remain attached; eg for cone of 30° included angle shock will detach below 1.46; for 30° wedge M is 2.55 because wedge exerts larger obstructing effect on airflow. In most supersonic aircraft aim is to keep most shocks attached, especially at engine inlets.

attack aircraft   Combat aircraft, usually aeroplane but sometimes helicopter, designed for attacking surface targets of tactical nature; missions include CAS (3) and interdiction.

attack, angle of   Angle α between wing chord or other reference axis and local undisturbed airflow direction. There are several ways of measuring this crucial parameter. One is absolute angle of attack. Another is *** for infinite aspect ratio, which assumes two-dimensional flow. Effective *** varies greatly with aspect ratio; modern wings of low aspect ratio have no stall in conventional sense even at $\alpha = 40°$. Some authorities in UK cause confusion by using 'angle of incidence', which already has clear meaning unconnected with angle of incident airflow.

attack avionics   Navigation and weapon-aiming systems, often integrated into single 'fit' for particular attack-aircraft type.

ATTC   *1* Automatic takeoff thrust control.

2 Aviation Technical Test Center (USA, Ft Rucker).

3 Aircraft Tactics Training Center (USAF).

attention-getter   Prominently positioned caption in cockpit, or esp. flight deck, triggered by onboard malfunction or hazardous situation (eg potential mid-air collision) immediately to flash bright amber or red. Less strident caption, warning light or other visual and/or aural circuit also triggered, enabling crew to identify cause and, if possible, take remedial action. Where CWS is fitted first task is to trigger ** while routing appropriate signal to captioned warning panel or other more detailed information.

attenuation   Loss of signal strength of EM radiation, esp. broadcast through atmosphere, due to geometric spread of energy through volume increasing as cube of distance, loss of energy to Earth, water vapour, air and possibly ionised E and F layers.

attenuation factor   Ratio of incident dose or dose rate to that passing through radiation shield.

ATTG   Automated tactical target graphic[s].

AT³, AT3, AT- three   See ATTT.

Attinello flap   Blown flap.

ATTITB   Air Transport and Travel Industry Training Board (UK).

attitude   Most, if not all, aircraft * described by relating to outside reference system three major axes OX/OY/OZ (see *axes*); * of flight relates these mutually perpendicular co-ordinates to relative wind; * with respect to ground relates axes to local horizontal.

attitude control system, ACS   Control system to alter or maintain desired flight attitude, esp. in satellite or spacecraft to accomplish this purpose in Earth orbit or other space trajectory. Typical *** uses sensing system, referred to Earth's limb, star or other 'fixed' point or line, and imparts extremely small turning moments to structure by means of gas jets or small rocket motors. In some cases passive *** used (PACS), or vehicle stabilized by spin about an axis, with portions despun if necessary.

attitude gyro   Loosely, gyro instrument designed to indicate attitude of vehicle. Specif., instrument similar to artifical horizon but with 360° freedom in roll and preferably 360° freedom in pitch. Also applicable to conventional horizon with restricted indications of movement in aircraft not intended for aerobatics.

attitude jet   *1* Reaction jet imparting control moments to aircraft at low airspeed (see *RCS*).

2 Sometimes applied to small thrusters or attitude motors used for same purpose on spacecraft.

attitude motor   Small rocket motor used to control attitude of space vehicle (see *thruster, reaction control engine*).

attitude reference symbol   Usually an inverted T giving heading and pitch attitude in HUD symbology.

ATTLA   Air transportability test-loading agency (US).

ATTMA   Advanced transport technology [or tactical transport] mission analysis.

ATTN   Attention.

atto   Prefix, 10^{-18}, symbol a; thus, 1 am (attometre) = 0.000000000000000001 m.

attrition   Wastage of hardware in operational service, esp. of combat or other military aircraft.

attrition buy   Additional increment of production run ordered to make good anticipated attrition over active life of system.

attrition rate   Usually means average [actual or predicted] loss per year.

attrition ratio   Many meanings, none of which compare losses with those of enemy.

ATTS   Air-transportable towed system.

ATTT   Advanced tactical targeting technology.

ATTU   *1* Atlantic to the Urals (NATO).

2 Advanced Tactics and Training Unit

ATU   *1* Antenna, or automatic, tuning unit.

2 Aerial [or aircraft] target unit.

ATUA   Air Transport User's Association (UK).

ATUC   See *AUC*.

Atugs   Armed tactical unattended ground sensor.

ATV   *1* Associazione Tecnici di Volo Aviazione Civile (I).

2 Automated transfer vehicle, also called space tug.

3 Atmospheric test vehicle.

4 Aircrew training vessel (UK MoD).

AT-Vasi   Abbreviated T-Vasi, ten light units on one side of runway in single 4-unit wing bar plus 6-unit bisecting longitudinal line.

ATVC   Ascent thrust-vector control.

ATVS   Advanced TV seeker.

ATW   Advanced tactical workstation.

Aᴛᴡ   Wing-on-tail aerodynamic influence coefficient.

ATWGS   Advanced tactical weapon guidance system.

A2I2　Accelrated-accuracy improvement initiative (Navstar/GPS).

ATWS　Active tail-warning system.

ATZ　Aerodrome traffic zone.

AU, a.u.　Astronomical Unit.

AU　*1* Air University (Maxwell AFB, USAF, founded 15 March 1946).

2 Utility aircraft (USA, USAF, 1956–62).

AUC　Air Transport Users' Council (CAA, UK).

audio box, audio control　Governs voice communication and broadcast throughout aircraft and by telephone to ground crew.

audio frequency　Frequency within range normally heard as sound. Limits vary widely with individuals but normally accepted 15 Hz to 20 kHz, upper limit being depressed with advancing age.

audio integration unit　Interlinks 'classic' cockpit with TCAS, EGPWS and other newer audio systems.

audiometer　Instrument for measuring subject's ability to hear speeches and tones at different frequencies.

audio oscillator　Multi-valve (tube) or multi-transistor stage in superheterodyne receiver serving as local oscillator and amplifier (detector).

audio speed signal　Aural indication of vehicle speed, either on board or at ground station; usually has pitch proportional to sensed velocity, indication being qualitative only (see *aural high-speed warning*).

audio warning　Loud warning by horn, buzzer, bell or voice tape in headphones or cockpit loudspeaker indicating potential danger. Examples: incorrect speed (usually sensed airspeed) for regime, configuration or other flight condition; potentially dangerous excursion of angle of attack; ground proximity; incorrect configuration (gear up, wings at maximum sweep, etc).

Audist　Agence Universitaire de Documentation et d'Information Scientifiques et Techniques (F).

audit　Comprehensive detailed examination of aiframe structure of aircraft in line service, esp. after fatigue problems encountered on type.

AUEW　Amalgamated Union of Engineering Workers (UK).

AUF　*1* Airborne use of force (USCG).

2 Australian Ultralight Federation.

AUFP　Average unit flyaway price.

auger in　To crash, esp.. to fly into ground.

augmentation　*1* Boosting propulsive thrust by auxiliary device, esp. by afterburning in both core and bypass flows of turbofan.

2 Percentage of thrust added by (1).

3 Increasing inadequate natural flight stability of aerodyne by on-board system driving control surfaces (rarely, ad hoc auxiliary surfaces).

4 Enhancing target signature of radar, optical, IR or other radiation by means of corner reflectors, Luneberg lenses or other * devices.

5 Enhancement of fluid flow by ejector effect, esp.. of lift airflow in powered-lift aircraft.

augmentation choke　Pilot-controlled modulating valve in blowing system of CCW for roll control.

augmentation ratio　Ratio of total fluid flow to mass flow of primary ejector flow of hot gas in ejector-lift system.

augmented deflector (or deflected) exhaust nozzle　Nozzle of V/STOL jet engine downstream of augmentor and capable of deflecting flow through at least 90°C.

augmented turbojet, turbofan　Engine equipped with afterburning (reheat) to augment thrust, esp. at transonic or supersonic flight speed. Turbofan augmentation may be in hot and/or cold flows, latter offering greater density of free oxygen.

augmentor　Afterburner for turbofan, with burning in hot and cold flows; US often augmenter.

augmentor ejector　Main lifting system in *ejector lift*.

augmentor wing　General term for STOL aeroplane wing in which engine thrust is directly applied to augment circulation and thus lift. Countless variations, but most fundamental division is into external and internal blowing. Former typically uses engine bleed air (rarely, total efflux) discharged at sonic speed through one or more narrow slits ahead of large double or triple-slotted flaps which, because of blowing, can be depressed to unusually sharp angle. Second method uses either engine bleed air or total efflux to blow through flap system itself (or propulsion engines are distributed across main flap, in some schemes there being as many as 48 small engines). Internal blowing makes flow separation impossible and gives large downwards component of thrust, but is difficult to apply and may severely compromise aircraft in cruise (see *jet flap*, *blown flap*, *externally blown flap*, *upper-surface blowing*).

AUM　Air-to-underwater missile (USA DoD weapon category).

AUP　*1* Advanced unitary penetrator.

2 Avionics upgrade program.

AUR　*1* All-up round.

2 Airplane [aircraft, aeroplane] upset recovery.

AURA　*1* Advanced UHF radar.

2 Autonomous unmanned reconnaissance aircraft.

aural acquisition　Acquisition of target by IR seeker head as confirmed by pilot's headset.

aural high-speed warning　System triggered by sensed flight speed, usually presented as EAS, significantly above allowable maximum. In transport aircraft typically triggered 10 kt above V_{mo} and 0.01 Mach above M_{mo}. Not usually made to do more than warn crew.

aural null　Condition of silence between large regions where sound is heard, eg in early radio DF system, in some types of beacon passage ('cone of silence') and several ground-test procedures.

Aurora　Automatic recovery of remotely-piloted aircraft.

aurora　Luminescence in upper atmosphere, esp. in high latitudes, associated with radiation and/or particles travelling along Earth's magnetic field and at least partly coming from Sun. Exact mechanism not yet elucidated, but 12 classes identified, based on appearance and structure.

AUS　Airspace utilisation section, part of ATC service.

AUSA　Association of the US Army.

Ausrire　Anglicization of 'all-union scientific research institute of radio equipment' (R).

Austaccs　Australian automatic command and control system.

austenitic steel　Ferrous alloys with high proportions of alloying elements and with microstructure transformed by heat treatment to consist mainly of solid solution of austenite (iron carbide in iron, face-centred). Used to make highly stressed aerospace parts, such as turbine discs.

AuTC   Autothrottle control.

AUTEC   Atlantic Undersea Test and Evaluation Center (USN with UK help, in British territory).

AUTH   Authorized.

authority   *1* Organisation empowered to pronounce hardware properly designed, constructed and maintained, and to issue appropriate certificates.

2 Extent to which functioning system is permitted to control itself and associated systems. Greater * demands greater inherent or acquired reliability, typical means of acquiring reliability being to increase redundancy and provide alternative control channels or in some other way provide for failure-survival.

Authorization   Act of Congress establishing a Federal agency or procedure (US).

Auto-Acas   Automatic air-collision avoidance system.

Autob   Automatic observation and reporting of weather.

Autocarp, Auto-CARP   Automatic computed air release point.

autoclave   Pressure chamber which can be heated; oven which can be pressurized. Large * can accept major structural parts of aircraft for adhesive bonding operations.

auto coarse pitch   System used on a few multi-piston engine aircraft to minimise drag after engine failure.

Autodin   Automatic digital network (USAF).

autodyne oscillator   Multi-electrode valve or transisor stage of superheterodyne receiver serving as both local oscillator and amplifier or detector (demodulator).

autofeathering   System for automatically and swiftly feathering propeller when engine fails to drive it; usually triggered by NTS.

autoflare   Flare [1] commanded by autopilot.

autogenic training   Psycho-physiological technique carried out by subject according to prescription by qualified therapist, to reduce stress.

Autogiro   Registered name of Juan de la Cierva autogyros, 1924–45.

autogyro   Rotorplane in which propulsion is effected by horizontal-thrust system, eg propeller, and lift by rotor free to spin under action of air flowing through disc from below to above (ie, autorotating). Some can achieve VTOL by driving rotor in vertical phases of flight, but true * is STOL.

auto-hover   Automatic hover, usually at low altitude, by helicopter or other VTOL aircraft, using radio altimeter and AFCS.

auto-igniting propellant   Hypergolic.

auto-ignition   *1* Of gas turbine engine, auxiliary system which senses angle of attack or other aerodynamic parameter and switches on igniter circuits before engine is fed grossly disturbed airflow which would otherwise pose combustion-extinguishing hazard. In some aircraft flight in rough air with full flap, or flight at high AOA, can cause intermittent or total flame-extinction, and there are other flight conditions (eg violent manoeuvre) when turbulent flow across intake triggers *, indicated by cockpit lights.

2 Specific meaning, ignition of premix fuel/air because of high compression (OPR 45+).

3 Of combustible material, spontaneous combustion.

auto-ignition temperature   At which auto-igniting materials spontaneously combust in air; design factor in some rocket engines and gas generators.

autokinesis   Sensation of movement of a distant light that is stared at.

Autoklean   Patented (UK) lubricating-oil filters based on compressed stack of strainer discs and spacers.

Autoland   *1* Loosely, AFCS capable of landing aeroplane hands-off and qualified to do so in total absence of pilot visual cues.

2 Specific systems developed in UK by Smiths Industries and GEC-Marconi-Elliott.

Autolycus   ASW detection system operating by sensing minute atmospheric concentrations of material likely to have come from diesel submarine running submerged (UK).

Automap   Trade name for map and/or track guide projection system, esp. for single-seat combat aircraft.

Automated Radar Terminal System   Shows terminal controller basic information on all collaborating traffic. * I largely overtaken by * II, and by III which tracks/predicts secondary-radar targets. * IIIA adds primary targets. All forms now modular and programmable. (FAA).

automatic boost control, ABC   On piston engine servo system which senses induction pressure and so governs boost system that permissible limits of boost pressure cannot be exceeded. These were among first airborne closed-loop feedback systems.

automatic coarse pitch   See *Automatic pitch coarsening*.

automatic dependent surveillance   Global system to compensate for lack of radar coverage over oceans and remote areas, involving automatic regular polling of navaids of each aircraft so that ATC can always monitor its position and ensure safe separation. Satellites appear to be essential for implementation.

automatic-direction finder   See *ADF*.

automatic extension gear   Landing gear which extends by itself, typically by sensing airspeed and engine rpm, should pilot omit to select DOWN.

automatic feathering   Autofeathering.

automatic flagman   Electronic installation in ag aircraft to provide precise track guidance on each run in conjunction with ground beacons.

automatic flyback vehicle   Unmanned shuttle between Earth and spacecraft.

automatic frequency control   *1* Radio receiver which self-compensates for small variations in received signal or local oscillator.

2 By different method, self-governing of a time base.

automatic gain control   See *AGC*.

automatic landing   Safe, precisely repeatable landing of advanced aeroplane, helicopter or other aerodyne in visibility so restricted that external visual cues are of no assistance to pilot. Basis of present systems is high-precision ILS, approach coupler, triplexed or quad AFCS, autothrottle, autoflare and ground guidance after touchdown.

automatic manoeuvre device system   Automatically schedules high-lift devices, especially on variable-sweep aircraft; usually governed by AOA.

automatic manual reversion   Fully powered flight-control system may be so designed that no ordinary pilot could control aircraft manually; if not, *** on one, two or three axes allows pilot to drive surfaces giving control in those axes after malfunction of powered system can no longer be accommodated by failure-survival.

automatic mixture control   In piston engine, subsystem which automatically adjusts flow rate of fuel to counteract changes in air density, or which controls intake airflow by restricting carburettor air-intake duct by amount inversely proportional to altitude until wide open at height usually around 15,000 ft.

automatic observer   Self-controlled group of sensors for recovering parameters during flight test.

automatic parachute   *1* Parachute pulled from its pack by static line [usual meaning].

    2 Parachute opened by barometric device at preset pressure altitude[s].

automatic pilot   See *autopilot*.

automatic pitch-coarsening   Facility built into propeller control system causing it to increase pitch automatically, normally from fine-pitch setting to typical cruise angle, when called for by operating regime. This is normally another name for a CSU, and quite distinct from auto coarse pitch.

automatic power reserve   Special increased-thrust rating available on commercial turbofan engines only in emergency, and triggered automatically by loss of power in other engine in same aircraft.

automatic pull-up   Preprogrammed steep climb by (1) aircraft in TFR flight following failure of one pitch channel, or (2) aerial target at start of parachute recovery.

automatic RDF   See *ADF*.

automatic reverse pitch   Facility built into propeller control system causing it to reduce pitch automatically past fine-pitch stop through zero to reverse (braking) position, upon receipt of signal from microswitch triggered when main gears compress shock struts on landing. Very rare for reverse pitch to be obtainable without deliberate selection by pilot, though with *** pilot may select in air, leaving auto system to send operative signal.

automatic riveter   Machine for drilling holes through parts to be joined, inserting rivet and closing (heading) it.

automatic roll-out guidance, AROG   Steering guidance after automatic landing in blind conditions (ideally extended from runway turnoff to terminal parking).

automatic search jammer   ECM intercept receiver and jamming transmitter which searches for and jams all signals having particular signatures or characteristics.

automatic selective feathering   *1* Airborne subsystem which, in the event of engine failure in multi-engined aircraft (invariably aeroplane), decides which engine has failed and takes appropriate action to shut down and feather propeller.

    2 Similar system which, when pilot presses single feathering button, routes signal automatically to failed engine and propeller.

automatic slat   Leading-edge slat pulled open automatically at high angle of attack by aerodynamic load upon it. All slats were originally of this type.

automatic synchronization   In multi-engined aircraft (invariably aeroplane), subsystem which electrically locks rpm governors of all engines to common speed.

automatic terminal information service   See *ATIS*.

automatic threat countering   Ability of EW system to detect, identify, locate and respond to each hostile emission without human intervention.

automatic touchdown release   Device incorporated in sling system for external cargo carried below helicopter or other VTOL aircraft which releases load as soon as sling tension is released.

automatic tracking   Although this could have meaning in system constraining aircraft to follow preset tracks over Earth's surface, universal meaning is property of directionally aimed system to follow moving target through sensing feedback signal from it. Applications found in (1) air-superiority fighter, in which essential to lock-on and track aerial target automatically, by means of radar, IR, optics or other system; and (2) ground tracking station, which may need to follow satellite, aircraft, drone target or other moving body in order to sustain command system, interception system, data-transmission system or other directionally beamed link.

automatic VHF D/F   Ground D/F system in which, instead of requiring manual turning of aerial array to find null position, signal from aircraft causes aerial to rotate automatically to this position, direction being at once displayed as radial line on CRT of the equipment. Also called CRT D/F, CRT/DF.

automatic voice advice   In terminal ATC, computer-generated voice message broadcast to two or more aircraft warning of potential conflict. Intended primarily for VFR traffic and for all traffic not under immediate control, and intended to relieve controller of function that appears to be safely automated.

automatic voice alerting device   Uses digitized human voice to warn of impending or hazardous situation, warnings being arranged in order of priority.

autonomous   *1* Of aircraft or other vehicle, not needing GSE.

    2 Of an SSR, not co-located.

    3 Of airborne equipment, not needing external sensors; not linked to other aircraft systems (though possibly under pilot control), eg * reconnaissance pod.

autonomous formation flight   Saving up to 20% fuel by copying geese and using accurate GPS to place at least one wingtip in tip vortex of preceding aircraft (NASA/Boeing/UCLA).

autonomous landing guidance   Based upon sensors or other devices in the aircraft.

autonomous logistics information system   Monitors all significant functioning parts of an organism, predicts expected life and gives advance warning of failure.

autonomous vehicle   Vehicle, especially unmanned aircraft, which completes mission without external help.

autopilot   Airborne electronic system which automatically stabilizes aircraft about its three axes (sometimes, in light aircraft, only two, rudder not being served), restores original flight path following any upset, and, in modern *, preset by pilot or remote radio control to cause aircraft to follow any desired trajectory. In advanced aircraft * is integral portion of AFCS and can be set by dial, push-button or other control to capture and hold any chosen airspeed, Mach, flight level or heading. In avanced combat aircraft * receives signals from sensing and weapon-aiming systems enabling it to fly aircraft along correct trajectories to fire guns or other ordnance at aerial target or lay down unguided bombs on surface target.

autopilot-disconnect   Advanced autopilots are automatically disconnected by control overloads generated within aircraft and by certain other disturbances likely to reflect wish of pilot, eg triggering of stall-protection stick-pusher.

Autoplan   Portable EDP which digitises navigation plan and combines output with other CPGS data to provide attack-aircraft pilot with complete nav/ECM/weapon-aiming information.

auto power reserve   See *automatic* **.

autopsy   Searching examination of crashed aircraft to discover cause, esp. to detect fatigue failure.

autorotation   *1* Loosely, condition in which airflow past aircraft causes whole aircraft or significant part of it to rotate. Propeller in this context is not significant, though windmilling propeller is autorotating.

2 In helicopter, descent with power off, air flowing in reverse direction upwards through lifting rotor(s), causing it to continue to rotate at approximately cruise rpm. Pilot preserves usual control functions through pedals, cyclic and collective, but cannot grossly alter steep 'glide path'. Rate of descent may exceed design ROD for landing gear, but is reduced just before ground impact by sudden increase in collective pitch; this increases lift, trading stored rotor kinetic energy for increased aerodynamic reaction by blades, and should result in gentle touchdown.

3 In aeroplane, descent in stalled condition, with general direction of airflow coming from well beyond stalling angle of attack but in grossly asymmetric condition (see *spin*).

4 In aeroplane, descent in unstalled condition under conditions apparently not greatly different from straight and level but with stabilized spiral flight path (see *spiral dive, spiral stability*). Distinct case of * which purist might argue is incorrect usage.

5 In helicopter flying training, range of manoeuvres designed to increase confidence and remove fear of power failure; all power-off descents, but differ in whether they are NPR (no power recovery) or terminated at height well above ground by restoring at least partial power. In latter case * terminated either by run-on-landing (running *), run-on climb-out or moderate-flare climb-out.

auto-separation   Automatic (often barometric) release of occupant from ejection seat.

Autosevocom   Automatic secure-voice communications (DoD).

autostabilizer   Loose term for autopilot, esp.. with authority on pitch axis only.

Autosyn   Trade name for remote-indicating system in which angular position of indicator needle precisely follows rotary sensing device moved by fluid level, mechanical displacement or other parameter which must be remotely measured. Sensor and indicator are essentially synchronous electric motors.

autosynchronization   Automatic synchronisation.

autothrottle   Power control system for main propulsion engines linked electro-mechanically to AFCS and automatic-landing system so that thrust is varied automatically to keep aircraft on glide path and taken off at right point in autoflare; in general * will also call for reverse thrust at full power in conjunction with automatic track guidance (roll-out guidance), though this may be left to discretion of pilot.

autotracking   Signal processing technique that enables a target to be automatically acquired and tracked by means of its own image, which can be received at any operating wavelength (usually microwave or optical).

autotrim   Aircraft trim system automatically adjusted by autopilot or other stabilising system to alter or maintain aircraft attitude according to pilot demand or changed distribution of weight or aerodynamic load. Usually

governs pitch only, the autopilot commanding the elevator and the autotrim the tailplane (stabilizer).

Autovon   Automatic voice network (USAF Communications Service).

AUVS   Association for Unmanned Vehicle Systems [I adds International] (US).

AUW, a.u.w.   All-up weight; actual aggregate weight of particular laden aircraft at moment of weighing. For generalized equivalent, more precise term should be used (MRW, MTOW, etc), but AUW has advantage of being not explicit and thus can be used to mean 'total weight of aircraft, whatever that happens to be'. Should never be used to mean MRW or MTOW. If latter are not known, preferred term meaning 'maximum allowable weight' is gross weight.

AUWE   Admiralty Underwater Warfare Establishment (UK).

AUWG   Airspace Users Working Group.

AUX   Auxiliary.

auxiliary bus   Secondary electrical bus serving one or more devices and often maintained at voltage different from that of main bus.

auxiliary fin   Generally, small additional fixed fin carried, not necessarily to enhance directional stability, well outboard on tailplane.

auxiliary fluid ignition   In rocket engine, use of limited supply of hypergolic fluid(s) to initiate combustion of main propellants.

auxiliary inlet, auxiliary intake   Extra inlet to [invariably gas-turbine] engine to admit extra air when needed [normally only on takeoff]. Also called supplementary air inlet and, most commonly, suction-relief door.

auxiliary parachute   Pilot parachute.

auxiliary power unit, APU   Airborne power-generation system other than propulsion or lift engines, carried to generate power for airborne system (electrics, hydraulics, air-conditioning, avionics, pressurization, main-engine starting, etc). In general, term restricted to plant deriving energy from on-board source and supplying constant-speed shaft power plus air bleed. This would exclude a RAT or primitive windmill-driven generator. Some *** can provide propulsive thrust in emergency (see *MEPU*).

auxiliary rigging lines   Branching from main parachute rigging lines to distribute load more evenly around canopy.

auxiliary rotor   In a classical helicopter [one main rotor], the rotor provided to counter drive torque and control fuselage azimuth. This is preferably called the tail rotor.

auxiliary tank   Fuel tank additional to main supply, esp. that can readily be removed from aircraft (see *reserve tank, external tank, drop tank*).

AV   *1* Air vehicle.

2 Audio-visual.

3 Aft vectoring (nozzle mode).

AVA   *1* Automatic voice advice.

2 Applied vector analysis.

AVAD   Automatic voice alert[ing] device.

AVADS   Autotrack Vulcan air-defense system (USA).

aval, Aval   Available.

availability   *1* Symbol A, proportion of time aircraft is serviceable and ready for use, expressed as decimal fraction over period or as number of hours per day or days per month. Also expressed as $\dfrac{\text{uptime}}{\text{uptime + downtime}}$.

2 Period which must elapse between purchase of aircraft, usually second-hand, and handover to customer.

avalanche *1* Any of several processes involving ions or electrons in which collisions generate fresh ions or electrons which in turn go on to have their own collisions. In * tube electrons or other charged particles are accelerated in electric field to generate additional charged particles through collisions with neutral gas atoms or molecules. In semiconductor devices * effect occurs when potential in excess of critical voltage is applied across p–n junction, enormously multiplying liberation of charge carriers.

2 Aerobatic manoeuvre devised by Ranald Porteous involving rapid rotation about all axes in combined stall turn and flick roll.

AVAS Air-vehicle avionics suite.

Avasi Abbreviated Vasi.

AvBatt, AVBATT Aviation Battalion (USA).

AvBM Aviation business machine, PC + com. terminal.

AVC *1* Automatic volume control (see AGC).

2 Automatic variable camber.

3 Active visual camouflage.

4 Attitude/velocity/control subsystem.

Avcat Originally Aviation Carrier [ship] Turbine, kerosene tailored to raise flashpoint above 60°C, freeze ≤–48°C; US = JP-5, NATO = F43 (F44 with FS11 additive), see fuels.

Avcatt Aviation combined-arms tactical trainer; -A adds aviation reconfigurable manned simulator.

AVCR Airborne video-cassette recorder.

AVCS Advanced vidicon camera system (carried by satellites).

AVD *1* Air Vehicles Directorate (AFRL).

2 Atmospheric-vehicle detection.

AVDA Asociación Venezolana de los Deportes Aeros; air sport, Venezuela.

Avdas Airborne-vehicle data-acquisition system.

Avdel Trade-name for large range of rivets/fasteners (Textron).

A-VDV Aviation of airborne forces (USSR).

AVE *1* Airfield visitor enthusiast (PFA).

2 Airborne-vehicle equipment.

3 Aéronef de validation expérimentale (UCAV, F).

AVEN, Aven Axi-symmetric vectoring engine, or exhaust, nozzle.

Aveppa Asociacón Venezolana de Pilotos Privados y Proprietarios de Aeronaves (Venezuela).

average flyaway unit cost Recurring manufacturing cost, all GFE costs, avionics, production support services and cost of tooling, manufacturing and pre-delivery maintenance.

average procurement cost Average flyaway unit cost plus GSE, training and technical aids, spares support and data, handbooks, technical representatives and logistic support.

AVD *1* Air Vehicles Directorate (AFRL).

2 Atmospheric-vehicle detection.

Avdas Airborne-vehicle data-acquisition system.

AVG *1* Average (ICAO).

2 American Volunteer Group (China, WW2).

3 Aircraft escort vessel (later ACV, then CVE).

Avgard Additive to JP-1 and other jet fuels to produce anti-misting kerosene (ICI trade name).

Avgas Aviation gasoline, range of piston engine petrols today being narrowed to 100LL (blue, max. 2.4 ml/Imp gal TEL) and 115 (purple, max. 5.52 ml/Imp gal TEL).

AVHRR Advanced very-high resolution radiometer.

AVI Air-vehicle integration; D adds design.

Aviaregister List of all civil aircraft (R, CIS).

aviation Operation of aerodynes (but only pedant would insist on "aerostation" for aerostat).

aviation medicine Study of all effects of aviation on the human body.

Aviation Policy Area Proposed rules for residential and commercial density in environs of airports with 2,300+ ft runway (US).

Aviation Security Advisory Committee Headed by V-P Al Gore, convened 1996 to improve security on US commercial flights, made numerous recommendations to combat terrorism, all rejected by US carriers.

aviator Operator of an aerodyne, esp. pilot. As term archaic, difficult to define; nearest modern equivalent is aircrew member.

aviatrix Female aviator.

Aviatrust Original (1923) central aviation industry management organisation (USSR).

Aviavnito Aviation department of Vnito, all-union amateur scientific/technical research organisation (USSR).

Avics, AVICS Air-vehicle interface and control system.

Avid Air-vehicle integration design.

aviette Man-powered aircraft.

avigation Aerial navigation (suggest undesirable word). Hence also avigator.

Avim Aviation intermediate-level maintenance (USAF).

Avimid Thermoplastic (strictly 'pseudo'-) polymer composite, marketed as K-III fabric and tape (Du Pont).

avionics Aeronautical (not aviation) electronics, not necessarily by definition restricted to aerodynes. Term implies equipment intended for use in air; purely ground-based equipment could be argued to be outside category.

Aviox Registered name for high-solidity polyurethane surface coatings.

AVIP Avionics integrity program (USAF).

AVIRIS, Aviris Airborne visible and IR imaging spectrometer.

Avlan Avionics local-area network.

AVLC Aviation VHF link control.

AVLF Airborne very low frequency.

AVM *1* Airborne vibration monitor.

2 Air Vice-Marshal.

A-VMF Naval air force (USSR, R).

AVN Aviation system standards (FAA).

AVNDTA Aviation development test activity (USA).

AVNF Air-vehicle near field.

AVNIR, Avnir Advanced visible and near-IR radiation.

AVNL Automatic video noise limiter, or limiting.

AVO Avoid verbal orders.

AVOD Audio/video on demand.

Avogadro number The number of molecules of a substance in one *mole* (2).

avoid curve Plot of TAS/height below which a helicopter may not survive total engine failure.

AVOL Aerodynamic visual, or aerodrome visibility, operational level.

Avometer Pioneer hand-held tester for electrical systems.

Avoss Aircraft vortex spacing system, to permit reduced landing intervals.

Avpac Aviation packt communication.

AVPH Air-vehicle prognostics and health.

Avpin Aviation-specification isopropyl nitrate.

Avplex Avionics planning and execution.

Av.P.970 Aviation Publication 970, *Manual of Design Requirements for Aircraft,* now called DAR.1 (UK).

Avpol Aviation-specification petrol/oil/lubricant.

AVR *1* Active vibration reduction.
2 Additional validation requirement[s].

AVRA Automatic visual-range assessor.

Avradar Aeronautical, or aviation, R&D activity (USA).

AVRS Airborne video-recording system.

AVS *1* Advanced vertical strike.
2 Advanced vision, or visionics, system.
3 Air-vehicle specification.

Avsat Aviation satcom service.

Avsec, AVSEC Aviation Security Panel (ICAO).

AvSP Aviation safety project (NASA).

AVSS *1* Analog voice switching system.
2 See *Avoss.*

AVT *1* Automatic video tracker.
2 Augmented, or avanced, vectored thrust.
3 Analog voice terminal.

Avtag Aviation turbine-engine gasoline, see *fuel.*

Avtoc Aviation tactical-operations center, if A2C²C unavailable (USA).

Avtops Aviation tactical-operations center (USA, USAF).

AVTR Airborne video-tape recorder.

Avtur Aviation turbine-engine fuel, see *fuel.*

AVUM *1* Aviation unit-level maintenance (USA).
2 Air-vehicle unit maintenance.

AVVI Altitude and vertical-velocity indicator.

AW *1* All-weather (usually not literally true).
2 Airway; often A/W, but AWY is preferred.
3 Automatic [gun] weapon.
4 Confusingly, airborne early warning.
5 Airlift Wing (USAF).
6 Airworthiness.

Aw Aircraft total wetted area.

AWA Aviation/Space Writers Association (US/Int.).

Awacs, AWACS Airborne warning and control system.

AWADC Advanced wideband analog-to-digital converter; T adds technology.

Awads All-, or adverse-, weather aerial delivery system.

AWAM Association of Women in Aviation Maintenance (US).

Awans Aviation weather and notice-to-airmen system (FAA, from 1976).

Awards *1* Aircraft wide-angle reflective display system.
2 All-weather airborne reconnaissance drone sensor.

Aware Advanced warning of active-radar emission[s].

Awas Automated weather advisory system.

AWB *1* Airway bill, air waybill.
2 Above-water battlespace.

AWBA, Awba Automated wing-box assembly.

AWC *1* Accumulated water condensate.
2 Air Warfare Centre (RAF Waddington; see AWFC).

AWCCV Advanced-weapon-carriage configured vehicle.

AWCLS All-weather carrier landing system.

AWCS Automatic-weapons control system.

AWD *1* Airworthiness Division (CAA, UK).
2 Air-warfare destroyer (ship).

Awdats Automatic-weapon, or air-warfare, data-transmission system.

Awdrey Atomic-weapon detection, recognition and estimation of yield.

AWDS *1* All-weather delivery, or distribution, system.
2 Automated weather [forecast] distribution system

AWE *1* All-up weight, equipped; generally = OWE.
2 Atomic Weapons Establishment (Aldermaston, UK).
3 Aircraft/weapon/electronic (AFMSS module).

AWES Atomic-weapons effects simulator; AWESS = signature simulator.

AWFC Air Warfare Center (USAF).

AWG *1* JETDS Code: piloted aircraft, armament, fire-control.
2 Airlines Working Group (CAA, UK).
3 Arbitrary waveform generator.
4 Aural-warning generator.
5 American Wire Gauge.

AWI *1* Aircraft weight indicator, on-board system also often indicating c.g. position, usually by sensing deflections of landing-gear shock struts.
2 All-weather intercept.
3 Air weapons instructor.
4 Airframe/weapons integration.

Awiator Aircraft wing with advanced-technology operation (Airbus).

AWIGWG Aircraft wire and inert-generator working group.

AWIM Airport weather information manager.

AWIN, Awin Aviation weather information [S adds system] (NASA/Boeing).

Awips Advanced weather interactive processing system (FAA).

AWIS Automatic weather information system.

AWL Above water level.

AWLS All-weather landing system.

AWM *1* Average working man (in defining aircrew sleep patterns).
2 Aircraft wiring manual.
3 Audio warning mixer.

AWMDS Automatic wing-sweep and manoeuvre devices system.

AWNIS Allied worldwide naval information system.

AWO All-weather operations.

AWOP All-weather Operations Panel (ICAO).

AWOS Automated, or airport, weather-observing, or automatic weather-observation, system.

AWP Aerial work platform.

AWPA Australian Women Pilots' Association.

AWPG Advanced weather products generator.

AWR Airborne weather radar.

AWRA Augmentor-wing research aircraft.

AWRE Atomic Weapons Research Establishment [Aldermaston, now AWE] (UK).

AWRS *1* Airborne weather reconnaissance system (USAF).
2 Automatic weather reporting system.

AWS *1* Air Weather Service (USAF, formerly part of MAC, now an FOA, Scott AFB).
2 Audible (or advanced) warning system.
3 Automatic wing sweep.

4 Air ward system.

5 Area weather system.

6 Aircraft Warning Service (USA/USN, 1942 – 46).

Awsacs, AWSACS   All-weather stand-off aircraft (or attack) control system (USN).

AWSAS   All-weather stand-off attack system.

AWSO   Aviation-warfare systems operator.

AWT   *1* Airborne wideband terminal.

2 Atmospheric wind tunnel.

Aw$_T$ Tail-on-wing aerodynamic influence coefficient.

AWTA   Advise what time available.

AWTSS   All-weather tactical strike system.

AWW   All-, or alert, weather watch.

Aww   Wing-on-wing aerodynamic influence coefficient.

AWX   All-weather interceptor.

AWY, awy   Airway (ICAO, FAA).

AX   *1* Avionics, avionics control (USAF).

2 Int. classes of hot-air balloons, from AX3 [20,000 cu ft, 566 m^3] to AX9 [140,000 cu ft, 3,935 m^3].

AXAF   Advanced X-ray astrophysics facility.

AXBT   Air- [or aircraft-] launched bathythermograph.

axes   Aircraft attitude is described in terms of three sets of *. First set are reference *, also called *body* *,three mutually perpendicular directions originating at c.g. (point O and defined as longitudinal (roll) axis OX, measured positive forwards from O and negative to rear; transverse (pitch) axis OY, measured positive to right and negative to left; and vertical (yaw) axis OZ, measured positive downwards and negative upwards. Position of O defined at design stage in what is considered most likely location for real c.g. in practice. Second, single, * is wind *: direction of relative wind, drawn through O, has angle determined by flight velocity, angle of attack and angle of sideslip. Third * are those known as inertia * and are imaginary lines about which aircraft would actually rotate in manoeuvres. These need not be same as reference *, although OY and OZ inertia axes are usual closely co-incident unless aircraft is asymmetrically loaded. Principal inertia *, however, may often depart substantially from geometrically drawn fore-and-aft axis OX (see inertia coupling). Purists also distinguish between stability axes for aircraft (a special set of body axes) and those for tunnel testing.

AX-5   Space suit for post-1989 Shuttle operations (NASA).

axial cable   *1* In non-rigid airship, main longitudinal member linking supporting cables, in framework carrying crew and engines.

2 In rigid airship, essentially straight cable sometimes linking extreme nose and tail of hull and central fittings of radial or diametral wires.

axial compressor   Compressor for air or other fluid with drum-shaped rotor carrying one or more rows of radial blades in form of small aerofoils (airfoils) arranged to rotate around central axis, with row of stationary stator blades (vanes) between each moving row. Compressed fluid moves through alternate fixed and moving blading in essentially axial direction, parallel to axis of rotation, temperature and pressure increased at each stage.

axial cone   In rigid airship, fabric cone at front and rear of each gas cell providing flexible gas-tight connection between cell and axial cable.

axial cord   In parachute, central rigging line joining apex to eyes formed at lower extremities of rigging lines.

axial deck   In carrier (1) or other ship carrying or serving as operating platform for aircraft, flight deck aligned fore and aft.

axial engine   Usually, piston engine in which axes of cylinders are parallel to crankshaft and/or main output shaft. Also, loosely, axial-flow engine.

axial firing   Fixed to fire directly ahead (usually on helicopters).

axial-flow engine   Gas-turbine engine having predominantly axial compressor, esp. one in which airflow is essentially axial throughout (ie, not reverse-flow).

axial focusing   In supersonic wind tunnel, focusing of shockwaves reflected from tunnel wall on to principal axis. Usually condition to be avoided, typically by minimising such reflections or so shaping working section that they are dispersed in different planes.

axial velocity ratio   In an axial-flow engine, ratio of axial flow velocity Va to turbine rotor blade velocity U, Va/U. Also called flow coefficient.

axis of rotation   In rotorplane, apparent axis about which main lifting rotor rotates: line passing through centre of tip-path circle and perpendicular to tip-path plane. May be widely divergent from mechanical axis on which hub is mounted, especially in articulated rotor.

axis of symmetry   Usually aeronautical *** determined by geometrical form, but in some cases dictated by mass distribution.

Ay   *1* Direct sideforce, changing heading without bank or sideslip.

2 Any lateral acceleration.

AYY   A half-width cargo container for upper deck of a narrow-body aircraft.

Az   Generalised symbol, azimuth.

Az-El, Azel, Az/El   Radar presentation giving separate pictures of azimuth (PPI display, or chosen sector) and elevation (such as side view of glide path).

azication   Azimuth indication.

azimuth   *1* Horizontal bearing or direction; thus * angle.

2 Rotation about vertical axis (yaw is preferred term where motion is that of whole aircraft).

3 Bearing of celestical body measured clockwise from true North, often called * angle and qualified true, compass, grid, magnetic or reference, depending on measure used.

azimuth aerial   Ground radar aerial rotating about vertical axis, or sending out phased-array emission rotating about such axis, intended to measure target azimuth angles.

azimuth compiler   Portion of SSR system, often optional or absent, which provides accurate azimuth information more accurately than the normal plot extractor.

azimuth control   In rotorplane, cyclic pitch.

azimuth error   Radar bearing error due to horizontal refraction.

azimuth marker   Scale used on PPI display to indicate bearing, including electronically generated references when display is offset from central position.

azimuth stabilized   PPI display which does not rotate, despite changes in heading of vehicle.

AZM   Azimuth.

az-ran, azran   General term for target tracking or navigational fixing by means of azimuth and range; more commonly called Rθ or rho-theta.

AZRN   Azimuth range.

Azusa   C-band tracking system operating on short baseline and giving continuous signals of two direction cosines plus slant range (and thus giving 3-D fix and instantaneous velocity). From Azusa, Calif.

B

B *1* Pitching moment of inertia.

2 Blue (ICAO).

3 Base (of semiconductor device).

4 Aircraft category, bomber (USAS, USAAC, USAAF, USAF 1924→, USN 1941–43 and 1962→; UK role prefix).

5 Bel, or [Neth.], total aircraft noise rating..

6 Boron.

7 Magnetic flux density, or induction.

8 Prefix for nuclear bombs (US).

9 Degrees Baumé.

10 Beginning [precipitation].

11 Hourly cost.

12 Airspace near airport up to 1,000 ft AGL (FAA).

13 Beacon.

14 Rotorcraft category: cannot maintain flight after failure of one engine.

15 Receiver bandwidth.

16 Sport-parachuting vertificate: 25 jumps, 10 landing ≤50m of target.

17 Byte[s].

18 Susceptance.

19 Luminance [B for brightness].

b *1* Wing span.

2 Bars (unit allowed within SI).

3 Barns (unit allowed within SI).

4 Engine bleed mass flow.

5 Number of blades in helicopter main motor.

6 Bit[s].

7 Propeller axial slipstream factor.

B_1, B_2 Graduation ratings from CFS.

b_1 Control-surface hinge moment.

b_2 Rate of change of surface hinge moment $dCH/d\epsilon$.

B2B Business to business.

B2C Business to consumer.

B_2H_6 Diborane rocket propellant, usually combined with OF_2.

B4 Aviation petrol (G, WW2).

B-category Aircraft used as non-flying trainer.

B-class *1* Military and civil prototype or experimental aircraft, not certificated but flown by manufacturer under special rules and with SBAC numerical registration (UK).

2 Terminal or control area near large airport (ICAO 1990, and US 1993).

B-code In flight plan, have DME and transponder with 64-code without encoding altitude.

B-display CRT or other display in which horizontal axis is bearing and vertical axis is range.

B-licence Commercial pilot's licence (not ALTP).

B-line 90° to the runway.

B-power supply Plate circuit that generates electron current in CRT or other electron tube.

B-rating Twin-engine pilot rating.

B-slope B-display.

B-station In Loran, transmitter in each pair whose signals are emitted more than half a repetition period after next succeeding signal and less than half an r.p. before next preceding signal of other (A station).

B-Stoff Hydrazine hydrate (G).

B-vehicles Non-flying vehicles in RAF service.

BA *1* Braking action (ICAO).

2 Budget authority.

3 Base Aérienne (air base, F).

4 Breathing apparatus.

B-A gauge Bayard-Alpert ionisation gauge.

b.a. Buffer amplifier.

BAA *1* British Airports Authority.

2 Bombardiers' Alumni Association (US).

3 Broad Agency Announcement (Darpa).

BAAC *1* British Association of Aviation Consultants.

2 British Aviation Archaeological Council, concerned with aircraft relics and documents, not with studying archaeological sites from the air; office Oulton Broad, Suffolk.

BAAEMS British Association of Airport Equipment Manufacturers and Services.

BAAHS Bay Area Airline Historical Society (San Francisco region).

BAAI Balloon and Airship Association of Ireland.

BAAS Broad-area aerial surveillance.

Babbitt (incorrectly, babbit) Family of soft tin-based alloys used to make liners for plain bearings.

babble Incoherent cross-talk in voice communications system.

Babinet point One of three points of zero polarisation of diffuse sky radiation.

BABOV Bureau Aanleg Beheer en Onderhoud van Vliegvelden (airfield plans and maint.) (Neth.).

Babs, BABS Beam-approach beacon system. Outmoded secondary radar system which provided fixed-wing aircraft with lateral guidance and distance information during landing approach.

BAC *1* Blood alcohol content.

2 Bureau of Air Commerce (US, 1934–38).

BACA Baltic, later British, Air Charter Association (UK 1946, became BIATA).

BACE Basic automatic checkout equipment (USN).

BACEA British Airport Construction and Equipment Association.

Bacimo Battlespace atmospheric and cloud impacts on military operations (USAF).

back *1* Of drag curve, aeroplane flight below V_{IMD}, in which reduction in speed results in increased drag.

2 Of propeller or rotor blade, surface corresponding to upper surface of wing.

3 Rear cockpit of tandem two-seat aircraft, especially combat type (hence GIB).

back beam In any beam system, especially ILS localizer, reciprocal beam on other side of transmitter.

back bearing Direction observed from aircraft holding steady course of fixed object over which it has recently passed; reciprocal of track.

backboard *1* Multilayer circuit board.

2 Specially narrow stretcher (litter) for carrying injured passengers along aisles.

back burner To be on * = not urgent, temporarily shelved.

back contamination   Contamination of Earth by organisms introduced by spacecraft and crews returning from missions.

back course   Course flown along back beam, on extended centreline of runway away from airfield.

backdrive   Where controlled device drives control input, eg thrust levers in Autothrottle mode, or when active motor [e.g. for tailplane] drives failed unit.

backed-off   Jet engine slid out of fuselage on rails for maintenance, ready to be at once re-installed.

backfire   Premature ignition of charge in piston engine cylinder such that flame travels through still-open inlet valve(s) and along induction manifold.

backfit   Retrofit (US usage).

background   Ambient (usually supposed steady-state) level of intensity of a physical phenomenon against which particular signal is measured. If signal amplitude never exceeds that of background, it cannot be detected. Thus: * clutter (radar), * count (radiation), * luminance, * noise (this has two meanings: noise in electronic circuit, and ambient level of aural noise at airport or elsewhere). See *microwave* *.

background check   Post – 9/11 investigation of personnel (US Dept. of Justice).

backing   *1* General effort by hardware manufacturer to support his products after they reach customer.

2 Change in direction of prevailing wind in counter-clockwise direction viewed from above; thus, from S to SE.

backing up   Rearwards taxiing of freight aircraft to loading dock, using reverse thrust.

backlash   In any mechanism, lost motion due to loose fitting or wear.

backlog   In manufacturing programme, items sold but not yet delivered.

backout   Any reversal of countdown, usually due to technical hold or fault condition.

backpack   *1* Personal parachute pack worn on back (usually thin enough for wearer to sit comfortably).

2 Any life-support system worn on back of user.

backpacker   Passenger in cheapest seat(s).

backplane   Standard STD or STE bus mounting board on which computer boards and other modules are attached by multiple push-in connectors. Most * accommodate 10 boards and have 20 edge connectors.

backplate   Fixed disc behind single-sided centrifugal compressor.

back porch   In electronic display, esp. TV, brief (eg 6 μ) interval of suppressed video signal at end of each line scan.

back-pressure   *1* Pressure in closed fluid system opposing main flow.

2 Ambient pressure on nozzle of rocket or other jet engine or any other discharge from fluid system.

backscatter   *1* Backward scatter.

2 Signal received by backward scattering.

back-shop   In manufacturing plant, first shop to close on rundown of programme.

backside   Moon face turned away from Earth.

backstagger   Backward stagger.

backswept   Swept.

backswing   Linking manoeuvre between tailslide from zero airspeed with fuselage vertical and start of [upright or inverted] down-45 line.

backtell   Transfer of information from higher to lower echelon.

backtrack   *1* In aircraft operation, to turn through about 180° and follow same track in reverse direction (as allowance must be made for wind, not same as turn on reciprocal).

2 Having landed on runway in use, to turn through 180° and proceed along runway in reverse direction.

back-up   *1* Complete programme, hardware item or human crew funded as insurance against failure of another.

2 Type of hardware item which could, even with degraded system performance, replace new design whose technical success is in doubt.

3 System funded to augment one already in operation (thus, BUIC).

4 Information printed on reverse of map or other sheet to supplement marginal information.

backward compatibility   New or modified device cleared to operate in old system.

backward extrusion   Extrusion by die so shaped that material being worked flows through or around it in reverse direction to that of die.

backward scatter   Electromagnetic energy (eg, radio, radar or laser) scattered by atmosphere back towards transmitter. In some cases whole hemisphere facing towards transmitter is of interest; in radar, attention usually confined to small amount of energy scattered at very close to 180° and detected by receiver.

backward stagger   Stagger such that upper wing is mounted further back than lower.

backward tilt   Tilt such that blade tips are to rear of plane of rotation through centroids of blade roots.

backward wave   In a TWT, any wave whose group velocity is opposite to direction of electron travel.

backwash   Slipstream.

BACM   Bootstrap air-cycle machine.

Bacon cell   Hydrogen/oxygen fuel cell.

BACS   Bleed-air control system.

BAD   *1* Boom-avoidance distance; thus * (A) is boom-avoidance distance measured along track on arrival and * (D) is same on departure.

2 Bomber Division (R).

3 Biological agent defeat.

BADD   Battlefield awareness and data dissemination.

BADGE, Badge   Base air-defence ground environment (Japan).

BAeA   British Aerobatic Association.

BAF   Bleed-air failure.

BAFF   British air forces in France (1939–40).

baffle   *1* Loosely, any device intended to disturb and impede fluid flow.

2 Shaped plates fixed around and between cylinders of air-cooled piston engine to improve cooling.

3 Surface, usually in form of a ring, plate or grating, arranged inside liquid container to minimise sloshing.

4 In two-stroke piston engine, deflector incorporated in crown of piston.

5 Partial obstruction inside pitot tube to minimise ingress of liquid or solid matter.

BAFO   *1* British Air Forces of Occupation.

2 Best and final offer.

BAG   British Airports Group (SBAC).

bag   See *body bag*.

baggage Checked-in possessions of a passenger, normal limits ≤900×700×400 mm (36×28×15 in), ≥350×230×150 mm (14×9×6 in), ≤34 kg (75 lb), no trailing cord or loose binding nor sharp projections.

bag tank Liquid container, especially fuel tank, constructed of flexible material not forming part of airframe.

BAH Belgian Aircraft Homebuilders.

BAI *1* Battlefield air interdiction.

2 Board of Auditors, international (NATO).

bail To loan aircraft or other possession, freely but under contract, to facilitate accomplishment of specific objective; in particular, loan by owner government of military hardware to industrial contractor engaged in particular development programme for that government.

Bailie beam Extra-precise Lorenz beam for Babs.

bail out To abandon dangerously unserviceable aircraft, esp. in midair, by parachute. Not yet used in connection with spacecraft.

bailout bottle Emergency personal oxygen supply, usually high-pressure gox, attached to aircrew harness or ejection seat.

Bairstow number Mach number.

bakes Back-course (ILS).

Bakelite Trade name for a phenol-formaldehyde resin plastic.

bake out In high-vacuum technology, heating to promote degassing.

Baker-Nunn Large optical camera used for tracking objects in space.

BAL *1* Office of air force training (Switzerland).

2 Bombe à guidage laser (F).

balance *1* State of equilibrium attained by aircraft or spacecraft.

2 Mechanism for supporting object under test in wind-tunnel and for measuring forces and moments experienced by it due to gas flow.

3 Mass or aerodynamic surface intended to reduce hinge moment of control surface.

balance area In aerodynamically balanced control surface, projected area ahead of hinge axis.

balance circuit In a WCS, subsystem which prevents, or warns of impending lateral asymmetry due to unbalanced weapon load.

balanced approach Optimum approach path referred to ground, taking in such factors as noise, ATC routing, cutbacks, land-use planning and preferential-runway rules.

balanced field length *1* Hypothetical length of runway for which TODa = EMDa (and sometimes, in addition, TORa).

2 Under CAR.4b, unfactored TOD to 50 ft following failure of one engine at V_1 = EMD to and from V_1, on dry surface.

balanced modulator Modulator whose output comprises sidebands without carrier.

balanced signature One which achieves optimal matching of IRS and RCS.

balanced support Logistic supply based on predicted consumption of each item.

balanced surface Control surface whose hinge moment is wholly or partially self-balanced (usually by means of mass or area ahead of hinge axis or by tabs).

balance rod Mass distributed along or within leading edge of helicopter rotor blade.

balance station zero Imaginary reference plane perpendicular to longitudinal axis of aircraft and at or ahead of nose, used in determinations of mass distribution and longitudinal balance.

balance tab Tab hinged to, and forming part of, trailing edge of control surface, and so linked to airframe that it is deflected in opposition to main surface, and thus reduces hinge moment. Action is thus similar to that of servo tab.

bale out See *bail out*.

ball *1* Small spheroid, or other laterally symmetric shape, in lateral glass tube of ball-type slip indicator.

2 Arbitrary unit of slip, equal to one ball-width.

ball ammunition Bullets of solid metal, containing no explosive, pyrotechnic or AP core.

ballast *1* In aerodynamics, mass carried to simulate payload, and permit c.g. position to be varied (usually in flight).

2 In aerostats, mass carried for discharge during flight to change vertical velocity or adjust trim.

ballast carrier In transport aircraft, holder for metal ballast weights with locking plungers mating with floor rails.

ball bearing Any shaft bearing in which inner race is supported and located by hardened spheres.

ball inclinometer Ball turn-and-slip.

ballistic camera Photographic camera which, by means of multiple exposures on same plate or frame of film, records trajectory of body moving relative to it.

ballistic capsule Capsule enclosing environment suitable for human crew or other payload and moving in * trajectory.

ballistic flight Ballistic trajectory; arguably, not flight.

ballistic galvanometer Undamped galvanometer which, when an electrostatic charge is switched through it, causes large initial swing, taken to be proportional to quantity of electricity passing.

ballistic missile Wingless rocket weapon which, after burnout or cutoff, follows * trajectory.

ballistic parachute canopy spreader Device for accelerating deployment of drag canopy in certain types of ejection seat.

ballistic range Research facility for investigation of behaviour of projectiles or of bodies moving through gaseous media at extremely high Mach numbers; usually comprises calibrated range along which test bodies can be fired, sometimes into gas travelling at high speed in opposite direction.

ballistic recovery system Large parachute deployed [sometimes fired by rocket] from above c.g. of [usually small] G A aircraft.

ballistic re-entry Non-lifting re-entry.

ballistic trajectory Trajectory of wingless body, formerly propelled but now subject only to gravitational forces and, if in atmosphere, aerodynamic drag.

ballistic tunnel High-Mach tunnel into which free projectiles are fired in opposition to gas flow (see * *range*).

ballistic vehicle Vehicle, other than missile, describing a * trajectory. Term usually not applied to spacecraft but to vehicles used in proximity of Earth and at least mainly within atmosphere.

Ballistic Wind[s]   Research into variation of wind with altitude (USAF).

ballistic wind   Theoretical constant wind having same overall effect on a * projectile as varying winds actually encountered.

ballizing   Forcing a hard oversized ball through a hole with a small fatigue crack, to bring crack region into compression.

ball lightning   Rare natural phenomenon, materialising during electrical storms, having appearance of luminous balls which often appear to spin, eject sparks, travel slowly and eventually disappear (sometimes with explosion).

ball mat   Standard squares or rectangles of tough flooring containing pattern of protruding freely-rotating balls facilitating movement of containers or pallets.

ballonet   Flexible gastight compartment inside envelope of airship (rarely, balloon) which can be inflated by air to any desired volume to compensate for variation in volume of lifting gas and so maintain superpressure and alter trim.

ballonet ceiling   Maximum altitude from which pressure aerostat with empty ballonet(s) can return to sea level without loss of superpressure.

balloon sonde   See *registering balloon*.

balloon   Aerostat without propulsion system.

balloon barrage   Protective screen of balloons moored by steel cables around target likely to be attacked by enemy aircraft.

balloon bed   Area of ground prepared for mooring of inoperative captive balloon.

balloon fabric   Range of fabrics of mercerised cotton meeting specifications for covering lightplanes and, impregnated with rubber, aerostats.

ballooning   Colloq., sudden unwanted gain in height of aeroplane on landing approach due to lowering flaps, GCA instruction or, most commonly, flare at excessive airspeed.

balloon master   Person in charge of launch of free-flight balloons, including liaison with ATC.

balloon reflector   In electronic warfare, confusion reflector supported by balloon(s).

balloon tank   Tank for containing liquid or gas constructed of metal so thin that it must be pressurized for stability. Such tanks have formed airframe of large ICBMs.

balloon tyre [tire]   Not defined but generally taken to mean tyre of larger than normal section (profile) and less than normal pressure.

ball screwjack   Screwjack in which friction is reduced by system of recirculating bearing balls interposed between fixed and rotating members.

balls-out   Maximum possible power (colloq.).

ball turn-and-slip   Flight instrument whose means of indicating slip is a ball free to move within liquid-filled curved tube having its centre lower than its ends.

ball turret   Gun turret on certain large aircraft of 1942–45 having part-spherical shape.

ballute   Balloon-parachute; any system of inflatable aerodynamic braking used for upper-atmosphere retardation of sounding rockets or slowing of spacecraft descending into planetary atmospheres.

BALO   Brigade Air Liaison Officer.

Balpa, BALPA   The British Air Line Pilots' Association; 1937, trade union.

balsa   Wood of extremely low density (s.g. about 0.13), originally grown in W. Indies and Central America.

Balt, BALT   Barometric altitude.

Baltnet   Monitors airspace over Estonia, Latvia, Lithuania.

BAM   *1* Bundesanstalt für Material Prüfung (G).
 2 Bird avoidance module.

Bambi basket   Deployed under helicopter, esp. for recovery [esp. from water] of exhausted or injured.

BAMS   *1* Brassboard airborne multispectral sensor.
 2 Broad-area maritime surveillance (USN).

BAMTRI   Beijing Aeronautical Manufacturing Technology Research Institute.

BAN   Beacon alphanumerics (part of FAA ARCTS).

band   *1* Designated portion of EM spectrum, usually bounded by frequencies used for radio communication.
 2 A (usually small) portion of EM spectrum containing frequencies of absorption or emission spectra.
 3 Strip of stronger material built into non-rigid aerostat envelope to distribute stress from mooring line, car or other load.
 4 Group of tracks on magnetic disc or drum.

band-elimination filter   Filter that eliminates one band of EM frequencies, upper and lower limits both being finite.

B&GS   Bombing and gunnery school.

bandit   Air contact (5) known to be hostile.

band of error   Band of position.

band of position   Band of terrestrial position, usually extending equally on each side of position line, within which, for given level of probability, true position is considered to lie.

B&P   Bid and proposal.

bandpass   Width, expressed in Hz, of band bounded by lower and upper frequencies giving specified fraction (usually one-half) of maximum output of amplifier.

bandpass filter   EM wave filter designed to reduce or eliminate all radiation falling outside specified band of frequencies.

B&S   Brown and Sharpe.

bandwidth   *1* Number of Hz between limits of frequency band.
 2 Number of Hz separating closet lower and upper frequency limits beyond which power spectrum of time-variant quantity is everywhere less than specified fraction of its value at reference frequency between limits.
 3 Range of frequencies between which aerial performs to specified standard.
 4 In EDP and information theory, capacity of channel.
 5 Band of position, usually expressed in nm on each side of PL (1) or track.

bang   Sound caused by passage of discontinuous pressure wave in atmosphere (*sonic bang*).

bang-bang   Any dynamic system, especially one exercising control function, which continually oscillates between two extreme 'hard-over' positions. Also called flicker control.

bang out   To eject (colloq.).

bang valley   Land or sea area under track where supersonic flight is permitted (colloq.).

banjo   Structural member having form of banjo, with open ring joined to linear portion (on either or both sides) projecting radially in same plane. Typically used to link spar booms on each side of jet engine or jetpipe aperture.

bank   *1* Attitude of aerodyne which, after partial roll, is flown with wings or rotor not laterally level. Held during any properly executed turn.

2 To roll aerodyne into banked position.

3 Linear group of cylinders in piston engine.

4 Pool of trained aircrew, esp. pilots, for whom no jobs are immediately available.

bank and turn indicator　Turn and slip indicator.

banner cloud　Cloud plume extending downwind of mountain peak, often present on otherwise cloudless day.

banner　*1* Large fabric strip towed behind aircraft, usually bearing advertising statement readable from both sides.

2 Fabric sleeve placed over propeller blade, usually saying FOR SALE.

banner sleeve　Tow target in form of long tube inflated by slipstream, usually called sleeve target.

banner target　Air-to-air or ground-to-air firing target in form of towed strip of flexible fabric like elongated flag.

BAO　Battlefield air operations, Afsoc field kit.

BAOR　British Army of the Rhine.

BAP　*1* Bomber aviation regiment (USSR).

2 Bank-angle projection.

3 Be a pilot (US GA initiative).

BAPA　British Aeromedical Practicioners Association.

BAPC　*1* British Aircraft Preservation Council (1967–).

2 British Association of Parascending Clubs.

3 Bombe[s] anti-personnel (F).

4 Broad Area Review.

Bapta　Bearing and power-transfer assemblies (comsat).

BAQ　Base allowance for quarters (US armed services).

BAR　See * *UK*.

bar　*1* Unit of pressure, allowed within SI and standard in meteorology and many other sciences. Equal to 10^5 Nm^{-2} = 14.5037 lb/in^2. In units contrary to SI, 10^6 dynes cm^{-2} or 750.08 mm; 29.53 in of Hg. One bar (1,000 mb) is Normal Atmospheric Pressure.

2 Metal [usually gold] bar(s) across ribbon of medal to show decoration has been awarded twice (thrice); in written form, each bar appears as a star *.

baralyme　Trade name for mixture of barium and calcium hydroxides used to absorb CO_2.

Barb　Boosted anti-radar bomb (S Africa).

barbecue manoeuvre　Deliberate intermittent half-rolls performed by spacecraft to equalise solar heating on both sides.

barber chair　Chair capable of gross variation in inclination.

barber-pole instrument　Indicator using rotating spirals, such as PVD (2).

barbette　Defensive gun position on large aircraft projecting laterally and providing field of fire to beam.

BARC　Bhaba Atomic Research Centre [NW warheads] (India).

Barcap　Barrier combat air patrol between naval strike force and expected aerial threat.

Barcis　British airports rapid control and indication system.

bar code　Geometric patterns, no two alike, printed on document (eg flight coupon) and read by light pen connected to computer storing data (eg stolen ticket numbers).

bare base　Airfield comprising runways, taxiways and supply of potable water.

bare engine　Generally means engine without fuel/oil/water or any auxiliary devices such as APU, cabin compressor or vectored nozzle(s).

barf bag　Sick bag.

Barif　Bureau of Airlines Representatives in Finland.

barn　*1* Unit of area for measuring nuclear cross-sections. Equal to 10^{-28} m^2 or 10^{-22} mm^2. Symbol b.

2 Non-hardened hangar for single aircraft, eg SR–71.

barnstormer　Formerly, itinerant freelance pilot who would operate from succession of unprepared temporary airfields giving displays and joyrides.

baro　Barometric.

baro-corrected　Pressure altitude corrected to local atmosphere.

barogram　Hard-copy record made by barograph.

barograph　Barometer giving continuous hard-copy record.

barometer　Instrument for measuring local atmospheric pressure.

barometric altimeter　Pressure altimeter.

barometric altitude　Pressure height.

barometric element　Transmitting barometer carried in radiosonde payload, variation in aneroid capsules causing shift in frequency of carrier wave.

barometric fuze　Fuze set to trigger at preset pressure height.

barometric pressure　Local atmospheric pressure.

barometric pressure control　Automatic regulation of fuel flow in proportion to local atmospheric pressure.

barometric pressure gradient　Change in barometric pressure over given distance along line perpendicular to isobars.

barometric tendency　Change in barometric pressure within specified time, usually the preceding three hours.

barometric wave　Any short-period meteorological wave in atmosphere.

barosphere　Atmosphere below critical level of escape.

barostat　Device for maintaining constant atmospheric pressure in enclosed volume.

barostatic relief valve　Automatic regulation of fuel flow by spilling back surplus through relief valve sensitive to atmospheric pressure.

barothermograph　Instrument for simultaneously recording local temperature and pressure.

barotrauma　Bodily injury due to gross or sudden change in atmospheric pressure.

barotropy　Bulk fluid condition in which surfaces of constant density and constant pressure are coincident.

barrage　AA artillery fire aimed not at specific targets but to fill designated rectilinear box of sky.

barrage balloon　Captive balloon forming part of balloon barrage.

barrage jamming　High-power electronic jamming over broadest possible spread of frequencies.

barrel　*1* Of piston engine cylinder, body of cylinder without head or liner.

2 Of rocket engine, thrust chamber of nozzle, esp. of engine having multiple chambers.

3 Any portion of airframe of near-circular section, even if tapering; thus Canadair made the CF–18 nose *.

4 Highly dangerous area with strong defences against attacking aircraft.

5 Non-SI unit of volume, = 42 US gal – 0.15899m³.

barrel engine　piston engine having cylinders with axes

disposed parallel to engine longitudinal axis and output shaft.

barrelling *1* Various methods of using rotating drum filled with abrasive [usually powder] to remove burrs and other surface imperfections from workpiece.

2 Colloq., to make fast low pass over ground location; to barrel in.

barrel roll Manoeuvre in which aerodyne is flown through 360° roll while trajectory follows horizontal spiral such that occupants are always under positive acceleration in vertical plane relative to aircraft.

barrel section *1* Portion of transport aircraft fuselage added in *stretching*. Alternatively called plug section (but see *barrel (3)*).

2 Parallel length of control rod.

barrel wing Wing in form of duct open at both ends and with longitudinal section of aerofoil shape.

barrette Array of closely spaced ground lights that appear to form a solid bar of light.

barricade Barrier (USN).

barrier Net mounted on carrier (1) deck or airfield runway to arrest with minimal damage aircraft otherwise likely to overrun. Normally lying flat, can be raised quickly when required.

barrier crash Incident involving high-speed entry to * with or without damage. Usually applied to carrier operations.

barrier pattern Geometrical pattern of sonobuoys so disposed as to bar escape of submerged submarine in particular direction.

barring Slowly turning gas-turbine engine by hand.

bar stock Standard form of metal raw material: solid rolled or extruded with round, square or hexagonal (rarely, other) section.

Barstur, BARSTUR Barking Sands Tactical Underwater Range (USN, Kauai, Hawaii).

BAR UK Board of Airline Representatives in the UK.

barycentre Centre of mass of system of masses, such as Earth/Moon system (barycentre of which is inside Earth).

barye Unit of pressure in CGS system. Equal to 10 Nm^{-2} = 1 dyne cm^{-2} or 1.4503×10^{-3} lb/in^2 or 10^{-6} bar (hence alternative name of microbar).

BAS *1* Bleed-air system.

2 Base allowance for subsistence (US armed forces).

BASA Bilateral Aviation Safety Agreement (FAA + foreign government).

Basar Breathing air, search and rescue (for diver in helo crew).

BASE Cloud base height AMSL (ICAO).

BaSE Battlespace synthetic environment.

base *1* Locality from which operations are projected or supported.

2 Locality containing installations to support operations.

3 In an object moving through atmosphere, any unfaired region facing rearwards (eg rear face of bullet or shell, trailing-edge area of wedge aerofoils, and projected gross nozzle area of rocket engine after cutoff).

4 Substance constructed of ions or molecules having one or more pairs of electrons in outer shells capable of forming covalent bonds.

5 Loosely, substance that neutralises an acid.

6 In transistor, region of semiconductor material into which minority carriers are emitted (hence, usually, between emitter and collector).

7 Underside of cloud. With cloud having grossly irregular undersurface, surface parallel to local Earth surface below which not more than 50 per cent of cloud protrudes, and which can be taken as upper limit of VFR. In mist or fog * intersects Earth.

Basea, BASEA British Airport Services & Equipment Association (77 members, office Bournemouth).

base area Aggregate area of unfaired rearward-facing surface of aerodynamic body.

base check Examination in flight of crew on completion of conversion to new type.

base drag Drag due to base area experiencing reduced pressures.

base height Minimum height AGL authorised on lo-level training sortie, in peacetime typically 200 ft, 91 m.

base leg Extends from end of downwind leg to start of turn on to finals.

base line, baseline *1* Yardstick used as basis for comparison, specif. known standard of build for functioning system, such as combat aircraft, against which developed versions can be assessed in numerical terms. Hence, * aircraft.

2 Geodesic line between two points on Earth linked by common operative system, eg between two Loran, Decca or Gee stations.

3 In many types of visual display and pen recorder, line dislayed in absence of any signal.

base metal *1* Major constituent of an alloy.

2 Metal of two parts to be joined by welding (as distinct from metal forming joint itself, which is modified or added during welding process).

base pressure Local aerodynamic pressure on base area of body moving through atmosphere.

base surge Expanding toroid surrounding vertical column in shallow underwater nuclear explosion.

base/timing sequencing Automatic sharing of transponder between several interrogators or other fixed stations by use of coded timing signals.

BASF Boron-augmented solid fuel.

BASH, Bash Bird/aircraft strike hazard (USAF team).

BASI Bureau of Air Safety investigation (Australia).

Basic British American Security Information Council (office DC).

basic aircraft Simplest usable form of particular type of aircraft, from which more versatile aircraft can be produced by equipment additions. In case of advanced aircraft, such as combat and large transports, ** includes IFR instruments, communications, and standard equipment for design mission.

basic cloud formations Subdivision of cloud types into: A, high; B, middle; C, low; D, clouds having large vertical development (International Cloud Atlas, 1930).

basic commercial pilot's licence Awarded after 220 h including 100 as P1, allows holder to do aerial work including VFR pleasure passenger flights in aircraft up to 5,700 kg. Suffix (A), aeroplanes; (H), helicopters.

basic cover Aerial reconnaissance coverage of semi-permanent installation which can be compared with subsequent coverage to reveal changes.

basic encyclopedia Inventory of one's own or hostile places or installations likely to be targets for attack.

basic flight envelope Graphical plot of possible or

permissible flight boundaries of aerodyne of particular type. Cartesian plot with TAS or Mach number as horizontal and altitude or ambient pressure as vertical. Boundaries imposed by insufficient lift, thrust or structural strength (and sometimes by social and other considerations); see *basic gust **, *basic manoeuvring **.

basic gross weight   Operating weight empty.

basic gust envelope   Specified form of graphical plot for each new aerodyne design showing permissible limits of speed for passage through vertical sharp-edged gusts of prescribed strength (traditionally ± 25, 50 and 66 ft sec^{-1}). Result is V–n diagram, with EAS as horizontal and gust load factor n as vertical. See *gust envelope*.

basic load   *1* Load (force) transmitted by structural member in condition of static equilibrium, usually in straight and level flight (1 g rectilinear), at a specified gross weight and mass distribution.

2 Aggregate quantity of non-nuclear ammunition, expressed in numbers of rounds, mass or other units, required to be in possession of military formation.

basic manoeuvring envelope   V–n diagram with EAS as horizontal and manoeuvring load factor as vertical. See *manoeuvring envelope*.

basic operating platform   See *bare base*.

basic operating weight   Operating weight empty.

basic research   See *pure research*.

basic runway   Runway without aids and bearing only VFR markings: centreline dashes or arrows, direction number and, if appropriate, displaced threshold.

basic 6   The instruments on a blind-flying panel.

basic supplier   Nominated supplier of hardware item in absence of specific customer option.

basic T   In traditional cockpit instrument panel, primary flight instruments (ASI, horizon, turn/slip and VSI) arranged in a standard T formation.

basic thermal radiation   Thermal radiation from Quiet Sun.

basic trainer   American military aeroplane category used for second stage in pilot training (after primary), with greater power and flight performance. Formerly also 'basic combat' (BC) category, which introduced armament and closely paralleled flight characteristics of operational type; called 'scout trainer' by Navy and redesignated 'advanced trainer' (obs).

basic weight   Superseded term formerly having loose meaning of mass of aircraft including fixed equipment and residual fluids.

basic wing   Aerofoil of known section used as starting point for modified design, often with wholly or partly different section.

basket   *1* Radar-defined horizontal circular area of airspace into which dispensed payloads (eg anti-armour bomblets) are delivered by bus (5). More loosely, volume of sky designated to receive free-fall object(s).

2 Car suspended below aerostat for payload, not necessarily of wickerwork construction.

3 Drogue on a flight-refuelling tanker hose.

4 Across-board sample of aircraft types in calculation of airport charges.

basket tube   Form of construction of liquid-propellant rocket thrust chamber in which throat and nozzle is formed by welded tubes, usually of nickel or copper, through which is pumped liquid oxygen or other cryogenic propellant for regenerative cooling.

BASO   Brigade air support officer.

Bassa   British Airlines Stewards and Stewardesses Association.

BAT   *1* Boom-avoidance technique.

2 Beam-approach training.

3 Bureau of Air Transportation (Philippines).

4 Brilliant anti-tank (submunitions).

5 Bombe[s] d'appui tactique (F).

6 Blind-approach technique (WW2).

7 See *Bat-Cam*.

BATA   British Air Transport Association, formerly BCASC.

Batap   B-type application to application protocol.

Bat-Cam   Battlefield air targeting-camera autonomous micro air vehicle.

Batco   The British Air Traffic Controllers' Association (UK, 1961–).

BATDU   Blind-Approach Training and Development Unit (UK became WIDU).

Bates   Battlefield artillery target-engagement system, UAV-integrated (UK).

bathtub   *1* Bath-shaped structure of heavy plate or armour surrounding lower part of cockpit or other vital area in ground-attack aircraft.

2 Temporary severe recession in production in manufacturing plant or programme, or between programmes; named from appearance on graphical plot (US, colloq.).

3 Graphical plot of equipment's service life: burn-in, useful life, wearout.

bathtub fitting   Fishplate (US).

bathythermograph   Sonobuoy dropped ahead of others to measure water data at various depths.

BATOA   British Air Taxi Operators' Association.

batonet   Tubular or rod-like toggle forming link between rigging line and band on fabric aerostat envelope.

BATR   Bullets at target range (shows location in HUD).

BATS   *1* Ballistic aerial target system.

2 Bathymetric and topographic survey.

batsman   Member of crew of aircraft carrier [rarely, other landing place] charged with guiding landing aircraft by hand signals.

batt   Ceramic filler formed from chopped-strand mat.

batten   *1* Wood or metal strip used in interlinked pairs as ground control lock.

2 Wood or metal strips arranged radially from nose of non-rigid airship to stiffen fabric against dynamic pressure, or, where applicable, mooring loads.

3 Flexible strips used in lofting drawing.

battery   Enclosed device for converting chemical energy to electricity. Most aerospace batteries are secondary (rechargeable), principal families being Ni/Cd (nickel, cadmium), Ag/Zn (silver, zinc) and lead/acid. Fuel cells are batteries continuously fed with reactants.

battery booster   Starting coil.

battle damage   In-flight damage caused directly by enemy (not, eg, by collision with friendly aircraft).

battlefield air interdiction   Air/ground sortie(s) tasked with restricting enemy's tactical movement and preventing him bringing up reserves to reinforce battle in progress.

battle formation   Any of several formations characterised by open spacing and flexible interpretation.

battleship model   Any model or rig used for repeated development testing, usually statically, in which major elements not themselves under test are made quickly and cheaply from 'boilerplate' material to withstand repeated use. Thus battleship tank.

battleship tank   Tank for liquid propellant for static testing of rocket engines having same capacity and serving same feed system as in flight vehicle but made of heavy steel or other cheap and robust material for repeated outdoor use.

battlespace   *1* Earth atmosphere and near space, region of human conflict.

2 3-d space in which actual conflict is taking place.

bat turn   Maximum-rate turn in air combat (colloq.).

bat wing   Becoming common term for BWB projects.

BAUA   Business Aircraft Users' Association (Kinloch Rannoch, UK).

BAUAG   British Airports Users' Action Group.

baud   Unit of telegraphic signalling speed, equivalent to shortest signalling pulse or code element. Thus speed of 7 pulses per second is 7 bauds (pronounced 'boards'); abbreviation Bd.

baulk   *1* Aborted landing due to occurrence in final stages of approach (typical causes would be aircraft on airfield taxiing on to runway or a runaway stick-pusher in landing aircraft).

2 To obstruct landing of an approaching aircraft and cause it to overshoot.

baulked landing   See *baulk (1)*.

Baumé   Density scale of petroleum products; for S.G. less than 1, $°B = \dfrac{140}{S.G.} -1$.

BAWS   Biological aerosol, or agent, warning system, or sensor.

bay   *1* In aerodyne fuselage or rigid airship hull, portion between two major transverse members such as frames or bulkheads.

2 In biplane or triplane, portion of wings between each set of interplane struts; thus, single-* aircraft has but one set of interplane struts on each side of centreline.

3 In any aircraft or spacecraft, volume set aside for enclosing something (eg, engine *, undercarriage *, bomb *, cargo *, lunar rock *).

bayonet   Electronic subsystem permitting radar of strike aircraft to home on designated ground target and providing azimuth for release of weapons equipped with radiation sensors.

bayonet exhaust   Formerly, form of piston engine exhaust stack designed to reduce noise.

BAZ   *1* Bundesamt für Zivilluftfahrt Zentrale (Austria).

2 Back azimuth unit, MLS addition used for departure and overshoots.

BAZL   Federal office for civil aviation (Switzerland).

BB   *1* Back bearing.

2 Battleship (USN).

3 Bulletin board [S adds system] (US).

4 Base band.

BBAC   British Balloon and Airship Club. (1965–).

BBC   *1* Broad-band chaff.

2 Before bottom centre.

BBI   Bit bus interface.

BBL   *1* Billion barrels liquid.

2 Bring-back load.

BBM, BBm   Back beam.

BBMF   Battle of Britain Memorial Flight (RAF).

BBN   Basic backup network for future civil navaids (FRN).

BBOE   Billion barrels oil equivalent.

BBSU   British Bombing Survey Unit.

BBU   Battery back-up unit.

BBW   *1* Brake by wire.

2 Bring-back weight.

BC   *1* Boron carbide armour.

2 Back course.

3 Bus (2) controller.

4 Bomber Command (RAF, USAAF).

5 Basic combat trainer category (USAAC, 1936–40).

6 Patches [of cloud] (ICAO).

7 Bottom of cylinder.

8 Become, becoming.

BCA   *1* Board of Civil Aviation (Sweden).

2 Baro-corrected altitude.

3 Best cruise altitude.

4 Belgian Cockpit Association, also called ABPNL.

5 British Cargo Alliance [carrier pressure group].

BCAA   British Cargo Airline Alliance.

BCAM   Best cruise altitude/Mach.

BCAOC   Balkan Combined Air Operations Centre (NATO).

BCARs   British Civil Airworthiness Requirements.

BCAS   Beacon-based collision-avoidance system.

BCASC   British Civil Aviation Standing Conference, now BATA.

BCATP   British Commonwealth Air Training Plan (1939–45).

BCC   British Chambers of Commerce, representing 126,000+ businesses.

BCD   *1* Binary coded decimal, also bcd.

2 Bulk chaff dispenser.

BCE   Battlefield control element.

BCF   Bromochlorodifluoromethane (fire extinguishant).

BCFG   Fog patches.

BCH   Binary coded hamming.

BC³I   Battlefield command and control communications and intelligence.

BCIU   Bus control and interface unit.

BCKG   Backing (ICAO).

BCL   Braked conventional landing (V/STOL).

BCM   *1* Basic combat manoeuvring, series of manoeuvres simulating interceptions and close dogfights.

2 Best cruise Mach number.

3 Background clutter matching.

4 Back-course marker.

BCMG   Becoming.

bcn   Beacon.

BCO, bco   Binary coded octal.

BCOB   Broken cloud[s] or better.

BCP   *1* Battery command post.

2 BIT control panel.

3 Break[ing] cloud procedure.

BCPL   See *Basic commercial pilot's licence*.

BCPR   Broad – coverage photo-reconnaissance (USAF satellites).

BCPT   Basic communications procedures trainer, for AEOs.

BCR   *1* European Community Bureau of Reference (co-ordinates R & D).

2 Battle casualty replacement.

3 Bombing, combat and reconnaissance.

B/CRS　Back course.

BCS　*1* British Computer Society.

2 Beam control system, esp. for laser on platform subject to jitter and atmospheric turbulence.

3 Block check sequence.

4 Buoy communication system.

BCSG　Bus-computer symbol generator.

BCST, bcst　Broadcast.

BCSV　Bearing-compartment scavenge valve.

BCU　Bird control unit (on airfield).

BCV　Belly cargo volume.

BCW　Binary chemical warhead.

BD　*1* Blowing dust.

2 Baud (correct abbreviation is Bd).

B/D　Bearing and distance.

Bd　Baud.

bd　Candle (unit of luminous intensity), abb.

BDA　Bomb (or battle) damage assessment.

BDAC　Bilateral Defence/Defense Acquisition Committee [being negotiated from 2002 to resolve problems caused by US refusal to disclose information to UK defence partners].

BDC, b.d.c.　*1* Bottom dead centre.

2 Bomb detection chamber.

BDFA　British Disabled Flying Association.

BDHI　Bearing/distance/heading indicator.

B/DHSI, BDHSI　Bearing/distance horizontal-situation indicator.

BDI　Bearing/distance indicator.

BDL　Bistable-diode laser.

BDLI　Bundesverband der Deutschen Luft- und Raumfahrtindustrie (W German aerospace industry association).

BDM　Buhr design method.

BDMIPG　British Defence Manufacturers Industrial Participation Group.

BDMIS　Business data management and invoicing.

BDOE　Barrels per day oil equivalent.

BDP　Bitsync descrambler preprocessor.

BDR　Battle-damage repair.

BDRY　Boundary.

BDS　*1* Comm-B designation subfield.

2 Boost defence segment.

3 Bypass-duct splitter.

BDX　Beacon-data extractor, or extraction.

Be　*Beryllium.*

BEA　*1* British European Airways 1946–72.

2 Bureau Enquêtes Accidents (F).

BEAB　British Electrotechnical Approvals Board.

beaching　Pulling marine aircraft up sloping beach, out of water to position above high tide.

beaching gear　Wheels or complete chassis designed to be attached to marine aircraft in water to facilitate beaching and handling on land.

beacon　*1* System of visual lights marking fixed feature on ground (see *aeronautical light*).

2 Radio navaid (see *fan marker, homing beacon, NDB, LFM, marker beacon, Z marker*).

3 Radar transceiver which automatically interrogates airborne transponders (see *radar beacon, ATCRBS*).

4 Portable radio transmitter, with or without radar reflector or signature enhancement, for assisting location

of object on ground (see *crash locator beacon, personnel locator beacon*).

beacon buoy　Self-contained radio beacon carried in emergency kit. Floats on water.

beacon characteristic　Repeated time-variant code of some visual light beacons, esp. aerodrome beacons emitting Morse letters identifying airfield.

beacon delay　Time elapsed between receipt of signal by beacon of transponder type (eg, in DME) and its response.

beacon identification light　Visual light, emitting characteristic signal, placed near visual light beacon (pre-1950) to identify it.

beacon skipping　Fault condition, due to technical or natural causes, in which interrogator beacon fails to receive full transponder pulse train.

beacon stealing　Interference by one radar resulting in loss of tracking of aerial target by another.

beacon tracking　Tracking of aerial target by radar beacon, esp. with assistance from transponder carried by target.

bead　*1* Corrugation or other linear discontinuity rolled or pressed into sheet to stiffen it, esp. around edges.

2 Thickened edge to pneumatic tyre shaped to mate with wheel rim and usually containing steel or other filament reinforcement.

3 Unwanted blob of weld metal.

beading　Rolling or pressing sheet to incorporate beads.

beadseat　Profiled seating on wheel for bead (2), hence * life, on expiry of which wheel must be reprofiled.

bead sight　Ring and bead sight.

Be/Al　Beryllium-aluminium.

beam　*1* Structural member, long in relation to height and width and supported at either or both ends, designed to carry shear loads and bending moments.

2 Quasi-unidirectional flow of EM radiation.

3 Quasi-unidirectional flow of electrons of particles, with or without focusing to point.

4 Loosely, on either side of aircraft; specif. direction from 45° to 135° on either side measured from aircraft longitudinal axis and extending undefined angle above and below horizontal. Hence, * guns (firing on either side), or surface object described as 'on the port *' (90° on left side).

BEAMA　British Electrical and Allied Manufacturers' Association.

beam approach　Early landing systems in which final approach was directed by beam (2) from ground radio aid (see *BABS, ILS, SBA*).

beam attack　Interception terminating at crossing angle between 45° and 135°.

beam bracketing　Flying aircraft alternately on each side of equisignal zone of radio range or similar two-lobe beam.

beam capture　To fly aircraft to intercept asymptotically a beam (2), esp. ILS localizer and glide path.

beam compass　Drawing instrument based on beam parallel to drawing plane having centre point and carrier for pen or other marker.

beam direction　In stress analysis, direction parallel to both plane of spar web, or other loadbearing member, and aircraft plane of symmetry.

beam-index display　Full-colour CRT using single gun and no shadow mask, computer switching to illuminate spots of red, blue or green phosphor according to instantaneous beam position.

beam jitter   Continuous oscillation of radar beam through small conical angle due to mechanical motion and distortion of aerial.

beam rider   Missile or other projectile equipped with beam-rider guidance.

beam-rider guidance   Radar guidance system in which vehicle being guided continuously senses, and corrects for, deviation from centre of coded radar or laser beam which is usually locked on to target. Accuracy degrades with distance from emitter.

BEAMS   British Emergency Air Medical Service.

beam slenderness ratio   Length of structural beam divided by depth (essentially, divided by transverse direction parallel to major applied load).

beam softening   Progressive reduction in gain of ILS demand signal.

beam width   Angle in degrees subtended at aerial between limiting directions at which power (DoD states "RF power", NATO states "emission power") of radar beam has fallen to half that on axis. Often defined for azimuth and elevation. Determines discrimination.

bean counting   Notional procedure of accountants whose sole interest is the balance sheet.

Bear   Electronic-warfare officer, usually in defence-suppression aircraft.

beard radiator   piston engine radiator mounted under the engine.

bearer   Secondary structure supporting removable part such as fuel tank or engine.

bearing   *1* Angular direction of distant point measured in horizontal plane relative to reference direction.

2 Angular direction of distant point measured in degrees clockwise from local meridian, or other nominated reference. Such measure must be compass, magnetic or true. True * is same as azimuth angle.

3 Mechanical arrangement for transmitting loads between parts having relative motion, with minimum frictional loss of energy or mechanical wear.

bearing chamber   Annular chamber surrounding shaft bearing, in gas-turbine engine as far as possible in cool location and incorporating low-friction sealing.

bearing compass   Portable and hand-held, used for determining magnetic bearing of distant objects.

bearingless rotor   Helicopter rotor in which all control is effected by flexibility in the blade attachments.

bearing-only launch   Missile is launched along approximate known bearing of target and seeker is switched on to search sector ahead, thereafter following various commands depending on whether or not target is acquired.

bearing plate   Simple geometric instrument for converting bearings of distant objects into GS (1) and drift.

bearing projector   Powerful searchlight trained from landmark beacon or other point towards nearby airfield (obs).

bearing selector   See *omni-*.

bearing stress   In any mechanical bearing, with or without relative movement, load divided by projected supporting area.

bear pads   Horizontal plates added to prevent helicopter skid sinking into snow. See *ski pad*.

bear paws   Short skis [fixed-wing aircraft].

beat   *1* Vibration of lower frequency resulting from mutual interaction of two differing higher frequencies. Often very noticeable in multi-engined aircraft.

2 One complete cycle of such interference.

beat frequency   Output from oscillator fed by two different input frequencies which has frequency equal to difference between applied frequencies. (Other outputs have higher frequencies, such as sum of applied frequencies.)

beat-frequency oscillator   Oscillator generating signals having a frequency such that, when combined with received signal, difference frequency is audible. Such ** is heterodyne, used in CW (1) telegraphy. Another is super-heterodyne, in which local ** produces intermediate frequency by mixing with received signal.

beat reception   Heterodyne reception, as used in CW (1) telegraphy.

beat-up   *1* Aggressive dive by aircraft to close proximity of surface object.

2 Repeated close passes in dangerous proximity to slower aircraft.

Beaufort notation   System of letters proposed by Rear-Admiral Sir Francis Beaufort (1805) to signify weather phenomena.

Beaufort scale   System of numbers proposed by Beaufort to signify wind strength, ranging from 0 (calm) to 12 (hurricane).

beaver tail   Tail of fuselage or other body which has progressively flattened cross-section.

beavertail aerial   Radar aerial emitting flattened beam having major beam width at 90° to major axis of aerial.

BEC   Boron-epoxy composite.

BeCA   Belgian Cockpit Association.

BECMG   Becoming.

becquerel   SI unit of radioactivity; Bq = 1 disintegration /s, = 2.7×10^{-11} Ci.

Be/Cu   Beryllium-copper.

Beddown   Process of introducing major new weapon system to combat duty, esp. aircraft (eg B–2) at first operating base.

BEEF, Beef   Base emergency engineering force (USAF).

beefing up   Strengthening of structural parts, either by redesign for new production or by modification of hardware already made. Thus, beef (= added material), beefed (US colloq.).

beehive   Small formation of bombers with close fighter escort (RAF pre-1945).

beep box   Station for remote radio control of activity or vehicle, such as RPV (colloq.).

beeper   *1* Personal radio alerting receiver.

2 RPV pilot.

3 Manual two-way command switch, eg electric trimmer or cyclic-stick thumbswitch for fuel valve control.

BEES, Bees   Battle-force electromagnetic interface evaluation system.

before-flight inspection   Pre-flight inspection (NATO).

behavioural science   Study of behaviour of living organisms, especially under stress or in unusual environments.

Beier gear   Infinitely variable mechanical transmission accomplished by stacks of convex discs intermeshing with stacks of concave discs, drive ratio being varied by bringing parallel shafts closer together, for extreme ratios, or further apart.

bel   Unit for relative intensity of power levels, esp. in relation to sound. One bel is ratio of power to be expressed divided by reference power, expressed as logarithm to base 10. Numerically equal to 10 decibels.

BELF   Breakeven load factor (BLF also common).

Belleville washer   Washer, usually of thin elastic metal in form of flat or convex/concave disc, which offers calibrated resistance to linear deflection of centre, perpendicular to plane of washer.

Bell-Hiller stabilizer   Two masses on short arms attached to hub of two-blade main rotor of helicopter in plane of blades, crossing at 90°.

Bellini-Tosi   First directional radio stations, using triangular loop antennas.

bellows   Aneroid capsule(s).

belly   Underside of central portion of fuselage.

belly-in, belly landing   To make premeditated landing with landing gear retracted or part-extended.

below minima   Weather precludes takeoff or landing.

bench   Static platform or fixture for manual work, system testing or any other function requiring firm temporary mounting (need not resemble a workshop *).

bench check   Mandatory manual strip, inspection, repair, assembly and recalibration of airborne functional parts, made at prescribed intervals and by station and staff certificated by airworthiness authority and/or hardware manufacturer.

bench engine   For bench testing, not cleared for flight.

bench test   Test of complete engine or other functional system on static testbed or rig.

bend   To damage an aircraft, especially in a crash, hence bent (UK colloq.)

bend allowance   Additional linear distance of sheet material required to form bend of specified radius.

bending brake   Workshop power tool for pressing metal sheet without dies.

bending moment   Moment tending to cause bending in structural member. At any section, algebraic sum of all moments due to all forces on member about axis in plane of section through its centroid.

bending relief   Design of aircraft to alleviate aeroelastic deflection, especially of main wing (eg, by distributing mass of fuel and engines across span).

bending stress   Secondary stresses (eg, in wing skins and spar booms) which resist deflection due to applied bending moments.

bends   Acute and potentially dangerous or lethal discomfort caused by release of gases within a mammalian body exposed to greatly reduced ambient pressure. Thus, a hazard of high-altitude fliers and deep-sea divers (see *aeroembolism, decompression sickness*).

bent   *1* Feature of many gun mechanisms, engaging in cocked position with sear.

2 Signal code indicating that facility is inoperative (DoD).

3 Transverse frame capable of offering vertical support and transmitting bending moment.

4 Damaged in an accident (colloq.).

bent beam, bent course   Radio or radar beams significantly diverted from desired rectilinear path by topographic effects, hostile ECM or other cause.

benzene   Liquid hydrocarbon, C_6H_6 with characteristic

ring structure forming base of large number of derivatives. Used as fuel or fuel additive, as solvent in paints and varnishes, and for many other manufactures.

benzine   Mixture of hydrocarbons of paraffin series, unrelated to benzene. Volatile cleaning fluid and solvent.

benzol   Benzene C_6H_6.

BEP   Back-end processor; MS adds management system.

BER   *1* Beyond economic repair.

2 Bit error rate, also b.e.r.

BERD   Business enterprise research and development.

Berline   Single-engined transport aircraft (F).

Bermuda triangle   Region at base of stiffener bonded to composite sheet subject to high stress.

Bernoulli's theorem   Statement of conservation of energy in fluid flow. Basis for major part of classical aerodynamics, and can be expressed in several ways. One form states an incompressible, inviscid fluid in steady motion must always and at all points have uniform total energy per unit mass, this energy being made up of kinetic energy, potential energy and (in compressible fluid) pressure energy. Making assumptions regarding proportionality between pressure and density, and ignoring gravity and frictional effects, it follows that in any small parcel of fluid or along a streamline, sum of static and dynamic pressure is constant, expressed as $p + \frac{1}{2}\rho V^2 = k$. Thus, if fluid flows subsonically through a venturi, pressure is lowest at throat; likewise, pressure is reduced in accelerated flow across a wing.

BERP, Berp   British experimental helicopter main rotor programme.

Berp rotor   RAE 9648 profile inboard, 9645 outboard, broad tip 9634.

beryllium   Hard, light, strong and corrosion-resistant white metal, m.p. about 1,278°C, density about 1.8. Expensive, but increasingly used for aerospace structures, especially heat sinks and shields.

BES   Best estimate.

b.e.s.   Best-endurance speed.

bespoke software   Designed for a specific application.

best-climb speed   Usually V_{IMD}.

best-economy altitude   Narrow band of altitudes where specific range is maximum.

best-endurance speed   Always V_{IMD}.

best-fit parabola   Profile of practical parabolic aerial capable of being repeatedly manufactured by chosen method. In many cases *** is made up of numerous flat elements.

best-range speed V_{IMR}; speed at which tangent to curve of speed/drag is a minimum; hence where speed/drag is a maximum and fuel consumption/speed a minimum.

BET   Best estimate of trajectory.

BETA, Beta   Battlefield exploitation and target acquisition.

beta   *1* Sideslip (β).

2 Angle of sideslip.

beta-1, $\beta1$ Yaw pointing angle (zero sideslip).

beta-2, $\beta2$ Lateral translation (constant yaw angle).

Bet-AB   Rocket-accelerated free-fall deep-penetration bomb (USSR, R).

beta blackout   Communications interference due to beta radiation.

beta control   Control mode for normally automatic propeller in which pilot exercises direct command of pitch for braking and ground manoeuvring. Also called beta mode.

beta lines, β-lines   Arbitrary lines drawn roughly parallel to compressor surge line to assist off-design performance calculations.

beta mode   *See Beta control.*

beta particle   Elementary particle emitted from nucleus during radioactive decay, having unit electrical charge and mass 1/1,837th that of proton. With positive charge, called positron; with negative, electron. Biologically dangerous but stopped by metal foil.

beta strips   Low-intensity strip lights on skin of military aircraft to assist close formation flying at night.

beta target   The trapezoid on a PFD.

beta vane   Transducer measuring yaw (sideslip) angle.

BETT, Bett   Bolt extrusion thrust-terminator.

bev   Billion electron volts. No longer proper term; 10^9ev is correctly 1 Gev.

bevel gear   Gearwheel having teeth whose straight-line elements lie along conical surface (pitch cone); thus, such gears transmit drive between shafts whose axes intersect.

bevelled control   Flight-control surface whose chordivise taper in thickness is increased close to the trailing edge.

BEW   Bare engine weight.

BEXR   Beacon extractor and recorder.

bezel   Sloping part-conical ring that retains glass of watch or instrument; esp. rotatable outer ring of pilot's magnetic compass.

BF   *1* Block fuel.
 2 Blind-flying; thus * instrument, * panel.
 3 Base Flight (USN)
 4 Below freezing.
 5 Bomber/fighter category (USN, 1934–37).

BFA   Balloon Federation of America.

BFAANN   British Federation Against Aircraft Noise Nuisance.

BFCU   Barometric fuel control unit.

BFDAS   Basic flight-data acquisition system.

BFDK   Before dark.

BFE   *1* Buyer-furnished equipment [MS adds management system].
 2 Basic flight envelope.

BFL   *1* Basic field length.
 2 Balanced field length.

BFM   Basic fighting/fighter/flight manoeuvring (or manoeuvres).

BFN   Beam-forming network [satcoms].

BFO   *1* Beat-frequency oscillation, or oscillator.
 2 Battlefield obscuration.
 3 Bits falling off.

BFOM   Basic flight-operations management.

BFP   *1* Blind-flying panel.
 2 British Flying Permit (ultralights).
 3 Best-fit parabola.
 4 Blown fuse-plug (tyre).

BFR   *1* Biennial flight review, for renewal of pilot licence (FAA).
 2 Before.

BFRP   Boron-fibre reinforced plastics.

BFS   *1* Bundesanstalt für Flugsicherung (= ATC, G).
 2 Back-up flight system.

BFT   *1* Basic fitness test[ing].
 2 Blue Force Tracking; I adds Initiatives.

BFTS   *1* Basic flying training school.

 2 British Flying Training School (US 1941–44); A adds Association (from 1948).
 3 Bomber/Fighter Training System (USAF).

BFU   Accident-investigation office (G).

BG   *1* Bomb, or Bombardment, Group (USAAC, USAAF, USAF).
 2 Bomb glider, aircraft category (USAAF 1943–46).
 3 See *lighting.

BGA   The British Gliding Association (1929–).

BGAN   Broadband global area network.

BGFOO   British Guild of Flight Operations Officers.

BGI   *1* Basic ground instructor.
 2 Bus grant inhibit.

BG lighting   Blue/green.

BGM   Designation code; multiple launch environment surface-attack missile, = cruise missile.

BGN   Begin, begun.

BGP   Border gateway protocol.

BGR   Best glide-ratio.

BGS   Blasting grit, soft, such as Carboblast.

BGW   Basic gross weight; not normally defined.

B/H   Curves of magnetic flux density plotted against magnetising force.

BHA   *1* Brazilian Helicopter Association.
 2 Bird-hit area, dangerous during migrations.

BHAB   British Helicopter Advisory Board.

BHGA   British Hang Gliding Association, now BHPA.

BHI   Bureau Hydrographique International.

BHN   Brinell hardness number.

BHO   Black-hole ocarina (tactical IR suppressor).

b.h.p., bhp   Brake horsepower.

BHPA   British Hang-gliding and Paragliding Association.

BHRA   British Hydromechanics Research Association.

BHS   Baggage-handling system.

BI   *1* Burn-in.
 2 Basse intensité.

BIA   Bomb-impact assessment; M adds modification.

BIAM   Beijing Institute of Aeronautical Materials.

BIAS   Battlefield-illumination airborne system.

bias   Voltage applied between thermionic valve (vacuum tube) cathode and control grid.

biased fabric   Multi-ply fabric with one or more plies so cut that warp threads lie at angle (in general, near 45°) to length.

bias error   Any error having constant magnitude and sign.

bias force   Output of accelerometer when true acceleration is zero.

bias ply   Tyre [tire] construction with alternate layers of rubber-coated cord extending under the bead at alternate angles; tread usually circumferentially ribbed.

bias temperature effect   Rate of change of bias force, usually in g °C^{-1}.

BIATA   British Independent Air Transport Association; formed 1946 as BACA.

BIBA   British Insurance Brokers Association.

Bibby coupling   Drive for transmitting shaft rotation without vibration, using multiple flexural cantilevers linking adjacent discs.

Bicep   Battlefield integrated-concept emulation program.

Bices, BICES   Battlefield information collection and exploitation systems.

biconvex Presenting convex surface on both sides. Such wings usually have profile formed from two circular arcs, not always of same radius, intersecting at sharp leading and trailing edges. Inefficient in subsonic flight.

bicycle Form of landing gear having two main legs in tandem on aircraft centreline.

BID Baggage information display.

bidding *1* Phase in procurement process in which rival manufacturers submit detailed proposals with prices.

2 Competitive procedure within air carrier's flying staff for licence endorsement on new type of aircraft.

BIDE Blow-in door ejector (engine nozzle).

BIDS, Bids Battlefield [or baggage] information distribution system.

BIFA British International Freight Association.

BIFAP Bourse Internationale de Fret Aérien de Paris.

BiFET Bistable field-effect transistor (gate).

BIFF, Biff *1* Battlefield identification friend or foe.

2 British Industrial Fasteners Federation.

bifilar Suspension of mass by two well separated filaments; mass normally swings in plane of filaments.

bi-fuel *1* Bipropellant.

2 More rarely, heat engine which can run on either of two fuels but not both together.

bifurcated Rod, tube or other object of slender form which is part-divided into halves; fork-ended.

bifurcation *1* Point at which duct [eg jetpipe] splits into two, usually left/right.

2 Analysis of steady states, a * occurring when stability changes from one state to another as an input [eg, control-surface angle] is altered.

Big BLU Proposed large [30,000-lb, 13.6-tonne] deep-penetration bomb (USAF).

big bone Very large indivisible part of airframe structure, such as a spar or monolithic bulkhead.

Big Chop, the Killed (RAF, colloq.).

Big Ear Battle group exploitation airborne radio.

Big F Commander (Flying) on carrier (RN).

bigraph Two-letter code for airfield name painted on local buildings, eg gasholders.

Bigs Bilingual ground station (Acars).

Bigsworth Chartboard, transparent overlay, Douglas protractor, parallel rules, all integrated (obs).

big-ticket item Subject of a high-value contract (US).

Bihrle See *CAP (7)*.

bilateral Agreement between two parties.

bilateration Position determination by use of AF signal beamed to vehicle, which re-radiates it to original ground station and to second at surveyed location giving known delay path. Can be used for control of multiple vehicles beyond LOS.

Bill, BILL Beacon illuminator laser (ABL).

billet Rough raw material metal form, usually square or rectangular-section bar, made by forging or rolling ingot or bloom.

billow Inflation of each half of Rogallo wing.

BIM *1* Ballistic intercept missile.

2 Blade inspection method (helicopters).

3 Blade integrity monitor.

4 British Institute of Management.

bimetallic joint Joint between dissimilar metals.

bimetallic strip Strip made of sandwich of metals, usually two metals chosen for contrasting coefficients of thermal expansion. As halves are bonded together any

change in temperature will tend to curve the strip. Principle of bimetallic switch and temperature gauge.

bi-mono aircraft Monoplane having detachable second wing for operation as biplane.

bin *1* Electronic three-dimensional block of airspace. All airspace in range of SSR or other surveillance radar is subdivided into *, size of which is much greater than expected dimensions of aircraft. Thus presence of an aircraft cannot load more than one (transiently two)* at a time.

2 Upright cylindrical receptacle, esp. some early gun turrets.

binary code Binary notation.

binary munition One whose filling is composed of two components, mixed immediately before release or launch.

binary notation System of counting to base of 2, instead of common base of 10. Thus 43 (sum of 2^5, 2^3, 2^1 and 2^0) is written 101011. All binary numbers are expressed in terms of two digits, 0 and 1. Thus digital computer can function with bistable elements, distinction between a 0 or 1 being made by switch being on or off, or magnetic core element being magnetised or not.

binary phase modulation Radar pulse-compression technique in which the phases of certain echo segments are reversed.

binary switch Bistable switch.

binaural Listening with both ears.

bind Noun, boring duty; verb, to complain incessantly (RAF, traditional).

BINDT British Institute of Non-Destructive Testing.

bingo *1* As an instruction, radioed command to aircrew (usually military) to proceed to agreed alternative base.

2 As information, radioed call from aircrew (usually military) meaning that fuel state is below a certain critical level (usually that necessary to return to base). Thus calls "*fuel" or "Below*". See next entries.

3 In some single-seat aircraft, alarm [usually bell] in headset at * fuel point.

Bingo 1 External tanks are empty.

Bingo 2 At start of what must be last practice engagement.

Bingo 3 Must break off combat area and recover to base, but with margins not present at Chicken.

binor Binary optimum ranging. Binary code modified for range measurement and minimising receiver acquisition time.

BINOVC, BinOvc Break[s] in overcast.

BIO Biotechnology Industry Organization (US).

bioastronautics Study of effects of space travel on life forms.

biochemical engineering Technology of biochemistry.

biochemistry Chemistry of life forms.

biodegradable Of waste material, capable of being broken down and assimilated by soils and other natural environments.

biodynamics Study of effects of motion, esp. accelerations, on life forms.

biological agent Micro-organism causing damage to living or inanimate material and disseminated as a weapon.

biological decay Long-term degradation of material due to biological agents.

biological warfare Warfare involving use as weapons of

biological agents, toxic biological products and plant-growth regulators.

biomedical monitoring Strictly 'biomedical' is tautological, but term has come to denote inflight monitoring of heartbeat, respiration and sometimes other variables, esp. of astronauts in space.

biometric recognition Identification of individual humans by their unique features, eg eye iris, finger/hand-print.

biometry Geometric measurement of life forms, esp. humans.

bionics Study of manufactured systems, esp. those involving electronics, that function in ways intended to resemble living organisms.

bio-pak Container for housing and monitoring life forms, usually plants, insects and small animals, in space payload. May be recoverable.

biophage Literally, destructive of life : CBW defence payload.

BIOS British Intelligence Objectives Sub-committee (1944 – 46).

biosatellite Artificial satellite carrying life forms for experimental purposes.

biosensor Sensor for measuring variables in behaviour of living systems.

biosphere Location of most terrestrial life: oceans, surface and near-surface of land, and lower atmosphere.

bioterrorism Loosely, germ warfare; hence bioweapons.

biowaste All waste products of living organisms, esp. humans in spacecraft.

BIP borescope inspection port.

biplace Two-seat (US, F usage).

biplane Aeroplane or glider having two sets of wings substantially superimposed (see *tandem-wing*).

biplane interference Aerodynamic interference between upper and lower wings of biplane or multiplane.

biplane propeller Propeller having pairs of blades rotating together in close proximity, resembling biplane wings.

BIPM Bureau International des Poids et Mesures.

bipropellant *1* As adjective, rocket which consumes two propellants – solid, liquid or gaseous – normally kept separate until introduction to reaction process. In most common meaning, propellants are liquid fuel and liquid oxidiser, stored in separate tanks.

2 As noun, rocket propellant comprising two components, typically fuel and oxidiser.

BIR Biennial infrastructure review.

BIRD Banque Internationale pour la Reconstruction et le Développement.

bird Any flight vehicle, esp. aeroplane, RPV, missile or ballistic rocket (colloq.).

bird gun Gun, usually powered by compressed air, for firing real or simulated standard birds at aircraft test specimens to demonstrate design compliance.

BIRDiE Battery integration and radar display equipment (USA).

birdie Spurious radar echo, usually from PRF harmonics.

bird impact Birdstrike; design case for all aerodynes intended for military or passenger carrying use (see *standard bird*).

bird ingestion Swallowing of one or more birds by gasturbine or other air-breathing jet engine, with or without

subsequent damage or malfunction (see *ingestion certification*).

birdnesting Tendency of chaff to stick together in tight bundles, hence forming **bird(s) nests**.

bird strike, birdstrike Collision between aerodyne and natural bird resulting in significant damage to both. Certification requirements for large turbofan engines include the ability to continue to give useful thrust after ingesting a single large bird [in 2002, one weighing 3.629 kg/8.0 lb, but with more severe demands being discussed] or various numbers of smaller birds, whilst running at maximum takeoff power.

Birmabright Trade name of many British alloys of Al, Mg and Mn.

BIRMO British IR Manufacturers' Organisation.

birotative Having two components on the same axis rotating in opposite directions, or in the same direction at different speeds.

BIS *1* British Interplanetary Society (1933–).

2 Board of Inspection and Survey (US Navy).

3 Burn-in screening.

4 Boundary intermediate system.

5 Biometric identification system.

bis Second version of product, equivalent to Mk 2 (F,I,R).

BISA Battlefield Information System Application (GBAD, UK).

bi-signal zone Portions of radio-range beam on either side of centreline where either A or N signal can be heard against monotone on-course background.

BISMS BIS(4) management system.

BISPA, Bispa British Iron and Steel Producers Association.

BISS Base and installation security system.

BIST, Bist Built-in self-test.

bistable Capable of remaining indefinitely in either of two states: thus, bistable switch is stable in either on or off position (see *flip-flop*).

BIT *1* Built-in test.

2 Bureau International du Travail (ILO).

bit Unit of data or information in all digital (binary) systems, comprising single character (0 or 1) in binary number. Thus, capacity in bits of memory is $\log_2$ of number of possible states of device. From 'binary digit'.

BITD Basic- instruments training device.

BITE, bite Built-in test equipment.

BIT/Fi Built-in test and fault isolation.

bit rate Theoretical or actual speed at which information transmission or processing system can handle data. Measured in bits, kbits, Mbits or Gbits per second.

bitube Single automatic weapon with two barrels.

biz Business; hence, **bizjet** (colloq.).

bizav Business aviation.

b$_J$ Span of CC blowing slit.

BJA Baseline jamming assets.

BJSD British Joint Service Designation.

BJSM British Joint Services Mission.

BK, bk Break [in transmission].

BKEP Boosted kinetic-energy penetrator.

BKN Broken [clouds].

BKSA British Kite Soaring Association.

BL *1* Buttock line.

2 Base-line (hyperbolic nav).

3 Bulk loader.

4 Between layers.

5 Blowing [DU adds dust, SA sand, SN snow].

blabbermouth   Foam monitor.

BLAC   British Light Aviation Centre.

Black   *1* SAO security classification.

2 Code, runway is unusable.

black   Big error of judgement or faux pas (RAF, as in "he put up a *").

black aluminium   Carbon-fibre composite (colloq.).

black ball   *1* Traditional form of artificial horizon based on black hemisphere or bi-coloured sphere.

2 Suspended on mast in signals area: TO and landing directions may differ.

black body   Theoretical material or surface which reflects no radiation, absorbs all and emits at maximum rate per unit area at every wavelength for any given temperature. In some cases wavelength is restricted to band from near-IR through visible to far-UV.

black-body radiation   EM spectrum emitted by black body, theoretical maximum of all wavelengths possible for any given body temperature.

black box   *1* Avionic equipment or electronic controller for hardware device or system, removable as single package. Box colour is immaterial (colloq.).

2 In particular, flight-data or accident recorder, usually Day-Glo scarlet or orange.

blacker-than-black   In TV system, portion of video signal containing blanking and synchronizing voltages, which are below black level (with positive modulation) and prevent electrons from reaching screen.

black hole   *1* IRCM design giving engine efflux (esp. of helicopter) or other heat source greatest protection.

2 Approach to land in near-absence of external visual cues.

black hot   TV or IR display mode giving negative picture (warm ship looks black against white sea).

blackjack   Hand tool for manually forming sheet metal; has form of flexible (often leather) quasi-tubular bag filled with lead shot too small to cause local indentations.

black level   In TV system, limit of video (picture) signal black peaks, below which signal voltages cannot make electrons reach screen.

black   men   Ground   crew   [Schwarze   Männer] (Luftwaffe).

black metal   See *black aluminium*.

blackout   *1* In war, suppression of all visible lighting that could convey information to enemy aircraft.

2 Fadeout of radio communications, including telemetry, as result of ionospheric disturbances or to sheath of ionised plasma surrounding spacecraft re-entering atmosphere.

3 Fadeout of radio communications caused by disruption of ionosphere by nuclear explosions.

4 Dulling of senses and seemingly blackish loss of vision in humans subjected to sustained high positive acceleration. In author's experience, more a dark red, not very unlike red-out.

blackout block   Block of consecutive serial numbers deliberately left unused.

black picture   See *black hot*.

bladder tank   Fluid tank made of flexible material, especially one not forming part of the airframe.

Blade   Bristol Laboratory for Advanced Dynamics Engineering.

blade   *1* Radial aerofoil designed to rotate about an axis, as in propeller, lifting rotor, axial compressor rotor or axial turbine. Also see stator *.

2 Rigid array of solar cells, especially one having length much greater than width.

3 Operative element of windscreen wiper.

blade activity factor   Non-dimensional formula for expressing ability of blade (1) to transmit power; integral between 0.1 and 0.5 of diameter of chord and cube of radius with respect to radius. Loosely, low aspect ratio means high activity factor.

blade aerial   Radio aerial, eg for VHF communications, having form of vertical blade, either rectilinear, tapered or backswept.

blade angle   *1* In propeller or fan blade, angle (usually acute) between chord at chosen station and plane of rotation.

2 In helicopter lifting rotor, acute angle between no-lift direction and plane perpendicular to axis of rotation.

blade angle of attack   Angle between incident airflow and blade (1) tangent to mean chord at leading edge at chosen station.

blade articulation   Attachment to hub of helicopter lifting rotor blades by hinges in flapping and/or drag axes.

blade back   Surface of blade (1) corresponding to upper surface of wing.

blade beam   Hand tool in form of beam incorporating padded aperture fitting propeller or rotor blade; for adjusting or checking blade angle.

blade centre of pressure   Point through which resultant of all aerodynamic forces on blade (1) acts.

blade damper   Hydraulic, spring or other device for restraining motion of helicopter lifting rotor blade about drag (lag) hinge.

bladed spinner   Zero stage of part-height blades added to spinner ahead of fan of turbofan; proposed by Rolls-Royce, * is unshrouded and not separated from fan by stators, otherwise similar in principle to TF39 of 1966.

blade element   Infinitely thin slice (ie having no spanwise magnitude) through blade (1) in plane parallel to axis of rotation and perpendicular to line joining centroid of slice to that axis. Thus, blade is made up of infinity of such elements from root to tip, usually all having different section profile and blade angle.

blade face   Surface of blade (1) corresponding to under-side of a wing. With propellers, called thrust face.

blade inspection method   Spars of helicopter main-rotor blades are pressurized, loss of pressure warning of crack (Sikorsky).

blade loading   Of helicopter or autogyro, gross weight divided by total area of all lifting blades (not disc area).

blade passing noise   Component of internally generated noise of turbomachinery, caused by interaction between rotating blades and wakes from inlet guide vanes and stationary blades. Generates distinct tones at blade-passing frequencies, which in turn are product of number of blades per row and rotational speed.

blade root   *1* Loosely, inner end of blade (1).

2 Where applicable, extreme inboard end of blade incorporating means of attachment (see *blade shank*).

Blades   Battlespace laser detection system (AEFB/AFRL).

blades!   Verbal call to pull piston engine through three or four propeller blades before start.

blade section Shape of blade element.

blade shank Where applicable, portion of blade of non-aerofoil form extending from root to inboard end of effective aerofoil section. Unlike root, * of propeller is outside spinner.

blade span axis *1* Axis, defined by geometry of root pitch-change bearings, about which blade is feathered.

2 Axis through centroids of sections at root and tip.

blade station Radial location of blade element, expressed as decimal fraction of tip radius (rarely, as linear distance from axis of rotation, from root or from some other reference).

blade sweep Deviation of locus of centroids of all elements of blade from radial axis tangential to that locus at centre. Was marked in early aircraft propellers, usually towards trailing edge (ie, trailing sweep); leading sweep, in which tips would be azimuthally ahead of hub, is rare.

blade tilt Deviation of locus of centroids of all elements of blade from plane of rotation. Again a feature of early aircraft propellers, more common form being backward tilt, visible in side view as propeller flat at back and tapered from boss to tip in front.

blade twist *1* Unwanted variation in pitch from root to tip caused by aerodynamic loads.

2 Natural twist which reduces blade angle from root to tip.

blade/vortex interaction Between each helicopter main-rotor blade and the vortex created by its predecessor, a principal cause of slap.

blade width ratio Ratio of mean chord to diameter.

BLAM Barrel-launched adaptive munition(s).

blank *1* Workpiece sheared, cut, routed or punched from flat sheet before further shaping.

2 Action of cutting part from flat sheet, esp. by using blanking press and shaped die.

3 Round of gun ammunition without projectile.

4 All-weather cover tailored to engine inlet or other aperture, forming part of AGE for each aircraft type.

blanket *1* Layer of thermally insulating material tailored to protect particular item, typically refractory fibre housed in thin dimpled stainless steel. Term is not normally used for noise insulation.

2 Layer of heating material supplied with electrical or other energy.

blanket cover Fabric cover for aircraft machine-sewn into large sheet, draped over structure, pulled to shape and sewn by hand.

blanketing *1* Supression, distortion or other gross interference of wanted radio signal by unwanted one.

2 In long-range radio communication, prevention of reflection from F layers by ionisation of E layer.

blanketing frequency Signal frequency below which radio signals are blanketed (2).

blank-gore parachute Parachute having one gore left blank, without fabric.

blanking *1* Using press and blanking die to cut blanks (1).

2 In electron tube or CRT, including TV picture tubes, suppression of picture signal on fly-back to make return trace invisible.

blanking cap Removable cap fitted to seal open ends of unused pipe connections or other apertures in fluid system.

blanking plate Removable plate fitted to seal aperture in sheet, such as unused place for instrument in panel.

blanking signal Regular pulsed signal which effects blanking (2) and combines with picture signal to form blanked picture signal. Sometimes called blanking pulse.

Blasius flow Theoretically perfect laminar flow.

blast *1* Loosely, mechanical effects caused by blast wave, high-velocity jet or other very rapidly moving fluid.

2 Rapidly expanding products from explosion and subsequent blast wave(s) transmitted through atmosphere.

blast area Region around launch pad which, before final countdown of large vehicle, is cleared of unnecessary personnel and objects.

blast cooling In rotating electrical machines and other devices, removal of waste heat by airflow supplied under pressure.

blast deflector Structure on launch pad or captive test stand to turn rocket or jet engine efflux away from ground with minimal erosion and disturbance.

blast fence Large barrier constructed of multiple horizontal strips of curved section, concave side upwards, which diverts efflux behind parked jet aircraft upwards and thus reduces annoyance and danger at airfields.

blast/fragmentation Warhead, common on AAMs and SAMs, whose effect combines blast of HE charge and penetration of fragments of rod(s) or casing.

BlastGard Proprietary honeycomb materials in which compartments are part-filled with various foams or expanding materials which attenuate blast and serve as a flame barrier.

blast gate See *waste gate*.

blast line Chosen radial line from ground zero along which effects of nuclear explosion (esp. blast effects) are measured.

blast-off Launch of rocket or air-breathing jet vehicle; usually, from ground or other planetary surface (colloq.).

blast pad Area immediately to rear of runway threshold across which jet blast is most severe. Constructed to surface standards higher than overrun or stopway beyond.

blast pen Small pen, enclosed by strong embankments on three sides, but open above, for ground running jet or rocket aircraft or firing missile engines.

blast pipe see *blast tube*.

blast tube Refractory tube linking rocket combustion chamber or propellant charge with nozzle, where these have to be axially separated.

blast valve Valve in air-conditioning and other systems of hardened facilities which, upon sensing blast wave, swiftly shuts to protect against nuclear contamination.

blast wave Shock wave (N-wave) of large amplitude and followed by significant (4 ata; 40 kPA or more) overpressure. Travels at or above velocity of sound and causes severe mechanical damage. Centred on explosions (local ** caused by lightning); attentuation and effective radius depend on third or fourth power of released energy.

blast-wave accelerator Concept for launching small payloads into space by accelerating them along an evacuated tube incorporating a long series of circumferential shaped charges pointed towards the muzzle.

Blaugas German gas used for airship lift and fuel; mixture of ethylene, methylene, propylene, butylene, ethane and hydrogen; literal meaning, blue gas.

BLC Boundary-layer control, especially gross control of airflow around lifting wing to increase circulation and prevent flow breakaway.

BLD *1* DSB, double sideband.

2 Berufsverband Luftfahrt-Personal in Deutschland eV (G).

BLDG Building (cloud).

BLE Boundary-layer energiser.

bleed *1* To allow quantity of fluid to escape from closed system.

2 To extract proportion (usually small) of fluid from continuously flowing supply; eg compressed air from gas-turbine engine(s).

3 To allow fluid to escape from closed system until excess pressure has fallen to lower level or equalised with surroundings.

4 To remove unwanted fluid contaminating system filled with other fluid, eg * air from hydraulic brakes.

5 To allow speed or height of aircraft to decay to desired lower level, thus *-off speed before lowering flaps.

6 To allow electronic signal or electrical voltage to decay to zero (eg, * glide path during flare in automatic landing).

bleed air Compressed air bled from main engines of gas-turbine aircraft.

bleed and burn Vertical lift system in which fuel is burned in a vertical combustion chamber fed with air from main engine (extra airflow may or may not be induced by ejector effect).

bleeder resistor Resistor permanently coupled across power supply to allow filter capacitor charge to leak away after supply is disconnected (see *bleed-off relay*).

bleeder screw Small screw in tapped hole through highest point of hydraulic or other liquid system to facilitate bleeding (4) air or vapour.

bleeding edge Edge of map or chart where cartographic detail extends to edge of paper.

bleed-off relay In laser, discharges capacitors when switched off, to render accidental firing impossible.

blended Aerodynamic (arguably, also hydrodynamic) shape in which major elements merge with no evident line of demarcation. Thus, aeroplane having * wing/body (see next two entries).

blended-hull seaplane Marine aeroplane, generally called in English flying boat, in which planing bottom is blended into fuselage. Involves dispensing with chine, sacrificing hydrodynamic behaviour in order to reduce aerodynamic drag.

blended wing/body Aircraft in which wing/fuselage intersections are eliminated. Today important for reasons of *stealth*.

blend point In aerodynamic shapes having rigid and flexible surfaces mutually attached (eg Raevam, variable inlet ducts, flexible Krügers), point in section profile at which flexibility is assumed to start.

BLEU Blind Landing Experimental Unit (Bedford, UK).

BLF Breakeven load factor.

BLG *1* Laser-guided bomb (F).

2 Body (-mounted) landing gear.

BLI Belgische Luchtvaart Info.

blimp Non-rigid airship (from 'Dirigible Type B, limp', colloq. until made official USN term in 1939).

blind *1* Without direct human vision.

2 Without external visibility, e.g. in dense cloud.

3 Of radar, incapable of giving clear indication of target (eg see * *speed*).

blind bombing Dropping of free-fall ordnance on surface target unseen by aircrew.

blind bombing zone Restricted area (strictly, volume) where attacking aircraft know they will encounter no friendly land, naval or air forces.

blind fastener See *blind rivet* [though need not have rivet-like form].

blindfire Weapon system able to operate without visual acquisition of target.

blind flying Manual flight without external visual cues.

blind-flying panel Formerly, in British aircraft, separate panel carrying six primary flight instruments: ASI, horizon, ROC (today VSI), altimeter, DG and TB (turn/slip).

blind landing *1* Landing of manned aircraft, esp. aerodyne, with crew deprived of all external visual cues.

2 Landing of RPV unseen by remote pilot except on TV or other synthetic display.

blind nut Nut inserted or attached on far side of sheet or other member to which there is no access except through bolt hole.

blind rivet Rivet inserted and closed with no access to far side of joint. Apart from explosive and rare magnetic-pulse types, invariably tubular.

blind speed effect Characteristic of Doppler MTI systems used with radars having fixed PRF which makes them blind to targets whose Doppler frequencies are multiples of PRF (see *staggered PRF*).

blind spot *1* Not reached by radio or radar, for whatever reason.

2 Region of airfield hidden from tower.

blind toss Programmed toss without acquisition of target (eg on DR from an offset).

blind transmission Station called cannot talk back.

bling Monolithic bladed ring [gas-turbine].

blink *1* Of light or other indicator, to be illuminated and extinguished, or to present black/white or other contrasting colour indication, more than 20 times per minute.

2 In aircraft at night in VFR, manually to switch off navigation lights (typically, twice in as many seconds) as acknowledgement of message.

blinker Light or indicator that blinks (1), eg to confirm oxygen feed.

Blip, blip Background-limited IR performance.

blip *1* Visible indication of target on radar display. Due normally to discrete target such as aircraft or periscope of submarine; in ground mapping mode, term used only to denote strong echo from transponder.

2 Spot, spike or other indication on CRT due to signal of interest.

3 To control energy input to early aeroplane by switching ignition on and off as necessary (normally, on landing approach).

4 To operate bang/bang control manually (eg, electric trim).

blip driver Operator of synthetic trainer for SSR or other surveillance radar with rolling ball or other means of traversing system co-ordinates to give desired blip (1) position and movement (colloq.).

blip/scan ratio   Also written blip: scan, an expression for probability of detection of a target by radar.

blisk   Axial turbine rotor stage (rarely, compressor stage) in which disc [US = disk] and blades are fabricated as single piece of material.

blister   *1* Streamlined protuberance on aerodynamic body, usually of semicircular transverse section.

    2 See *blister spray*.

blister aerial   Aerial projecting from surface of aircraft and faired by dielectric blister.

blister hangar   Prefabricated and demountable hangar having arched roof and fabric covering.

blister spray   Arching sheet of water thrown up and outwards above free water surface on each side of planing hull or float. Compared with ribbon spray, has lower lateral velocity, rises higher, is clear water rather than spray, and is much more damaging.

blister spray dam   Strong strip forming near-vertical wall projecting downwards along forebody chine of hull or float.

blivet   Flexible bag for transporting fuel, usually as heli-copter slung load.

BLK   *1* Block.

    2 Black.

BLM   *1* Background luminance monitor.

    2 Bureau of Land Management, Federal agency responsible for firefighting in wild regions (US).

BLN   Balloon.

BLO   Below clouds (ICAO).

blob   Local atmospheric inhomogeneity, produced by turbulence, with temperature and humidity different from ambient. Can produce angels (2).

block   *1* In quantity production, consecutive series of identical products having same * number. In World War 2 aircraft production a * might number several hundred; with large spacecraft and launch vehicles, fewer than ten. In general, products of two * normally differ as result of incorporation of engineering changes.

    2 In research, groups of experimental items subjected to different treatment for comparative purposes.

    3 In EDP, group of machine words considered as a unit.

    4 In aircraft (usually commercial) operation, chocks (real or figurative) whose removal or placement defines the beginning and end of each flight.

blockbuster   Large thin-case conventional bomb (colloq.).

block check sequence   Cyclic code used as reference bits in error-detection procedure.

block construction   Arrangement of gores of parachute such that fabric warp threads are parallel to peripheral edge.

block diagram   Pictorial representation of system, other than purely electrical or electronic circuit, in which lines show signal or other flows between components, depicted as blocks or other conventional symbols.

blocker   See *inlet* *.

blocker door   In installed turbofan engine, hinged or otherwise movable reverser door (normally one of per-ipheral ring) which when closed blocks fan exit duct and opens peripheral exits directing airflow diagonally forward.

block fuel   Fuel burned during block time.

block-hour cost   DOC for one hour of block time.

blockhouse   Fortified building close to launch pad for potentially explosive vehicles, from which human crew manage launch operations or perform other duties (eg photography).

block in   To park transport aircraft at destination. Term spread from commercial to military transport use.

blocking   *1* In wind tunnel, gross obstruction to flow caused by shockwaves at Mach numbers close to 1, unless throat and working section designed to avoid it (see *choking*).

    2 Use of struts and wedges to prevent movement of loose cargo or cargo inside container.

    3 Use of form block.

blocking capacitor   Capacitor inserted to pass AC and block DC.

blocking layer   Barrier layer in photovoltaic (ie solar) cell.

blocking oscillator   Any of many kinds of oscillator which quench their output after each alternate half-cycle to generate sawtooth waveform.

blocking up   To use shaped masses behind sheet metal being hammered.

block letter   Suffix to aircraft serial number, equivalent to USAF *block number* (USN).

block number   See *block* (1).

block out   To move off blocks, esp. at start of scheduled flight.

block shipment   Rule-of-thumb logistic supply to provide balanced support to round number of troops for round number of days.

block speed   Average speed reckoned as sector distance divided by block time.

block stowage   Loading all cargo for each destination together, for rapid off-loading without disturbing cargo for subsequent destinations.

block template   Template for making form block.

block time   Elapsed period from time aircraft starts to move at beginning of mission to time it comes to rest at conclusion. Historically derived from manual removal and placement of blocks (chocks). Normally used for scheduled commercial operations, either for intermediate sectors or end-to-end.

block upgrade plan, programme   Introduction of succes-sive groups of modifications, each in a particular FY or production block.

Bloctube   Patented mechanical control in which command is transmitted by push/pull action of steel cable on which are threaded guidance rings running inside a tube.

blood-albumen glue   Adhesive used in aircraft plywood, made from dry cattle blood albumen.

blood chit   *1* Form signed by aircraft passenger before flight for which no fare has been paid (eg, in military aircraft) indemnifying operator against claims resulting from passenger's injury or death (colloq.).

    2 Plastic or cloth message, usually in several languages, promising reward if bearer is helped to safety. Often includes representation of bearer's national flag (colloq.).

blood wagon   Ambulance.

bloom   *1* Ingot from which slag has been removed, some-times after rough rolling to square section.

    2 Of ECM chaff, to burst into large-volume cloud after being dispensed as compact payload.

blooming   *1* In surface-coating technology, to coat optical glass with layer a few molecules thick which improves optical properties (by changing refractive index and/or reducing external reflection).

2 In CRT, defocusing effect caused by excessive brightness and consequent mushrooming of beam.

3 In atmospheric laser operations, defocusing or undesired focusing of pulses or beam caused by lens-like properties in atmosphere (see *blob*).

BLOS   Beyond, or below, line of sight.

blossom effect   Sudden apparent growth in size of aircraft on collision course as distance approaches zero.

blow   *1* Rupture in case of solid rocket during firing.

2 To activate blowing system.

blowback   *1* Type of action of automatic gun in which bolt, or other sliding breech-closure, is never locked and is blown back by combustion pressure on its face, or by rearwards motion of empty case, inertia sufficing to keep breech closed until projectile has left muzzle.

2 Improper escape of gas through breech during firing of gun, due to ruptured case, faulty breech mechanism or other malfunction.

3 Closure, or partial closure, of spoiler or speed brake due to aerodynamic load overcoming force exerted by actuation system.

blowback angle   Maximum angle to which a spoiler can be extended under given q (dynamic pressure) without blockback (3).

blowby   Loss of gas leaking past piston engine piston ring.

blowdown   *1* Pilot input or pfcu force overcome by aerodynamic load on control surface, the latter being either prevented from moving or returned to neutral position.

2 In captive firing of liquid rocket, expulsion of residual propellant[s] by gas [usually nitrogen] after burnout or cutoff.

blowdown period   Period in cycle of reciprocating IC engine in which exhaust valve or port is open prior to BDC.

blowdown tunnel   Open-circuit wind tunnel in which gas stored under pressure escapes to atmosphere, or into evacuated chamber, through working section.

blowdown turbine   Turbine driven by piston engine exhaust gas in such a way that kinetic energy of discharge from each cylinder is utilised.

blower   *1* Centrifugal compressor with output (from NTP input) at between 1 and 35 lb/sq in gauge (6.9–240 kN m^{-2}).

2 Centrifugal fan used on piston engine, esp. on radial engine, to improve distribution of mixture among cylinders.

blower pipe   In an airship, duct through which propeller slipstream is rammed to pressurize ballonets.

blow-in door   Door free to open inwards against spring upon application of differential pressure (eg, in aircraft inlet duct).

blowing   Provision for discharging high-pressure, high-velocity bleed air from narrow spanwise slit along wing leading edge, tail surface or ahead of flap or control surface. Greatly increases energy in boundary layer, increases circulation and prevents flow breakaway.

blowing coefficient   For jet or blown-flap or blown aerofoil, $\dfrac{MVj}{qS}$

blown flap   Flap to which airflow remains attached, even at sharp angles, as result of blowing sheet of high-velocity

air across its upper surface (see *boundary-layer control*, *supercirculation*).

blown periphery   Parachute in which part of peripheral hem becomes blown between two rigging lines in another part of canopy, and attempts to inflate inside-out.

blown primer   Percussion cartridge primer which blows rearward out of its pocket, allowing primer gases to escape.

blow off   *1* Explosive or other enforced separation of instrument pack or other payload from rocket vehicle or other carrier. Hence, ** signal.

2 Controlled reduction [if necessary to zero] of deflection of flap, tab or [rare] control surface because of aerodynamic load.

blow-off valve   Safety valve set to open at predetermined dP at chosen point in axial compressor casing to allow escape of part of air delivered by stages upstream. Common in early axial gas turbines to prevent surging and other malfunctions, especially during starting and acceleration.

blowout   *1* Flameout in any fuel/air combustion system caused by excessive primary airflow velocity.

2 In particular, afterburner flameout.

blowout disc   Calibrated disc of thin metal used to seal fluid system pipe, rocket combustion space or any other device subject to large dP. If design dP limit is exceeded, ** ruptures. Also called safety diaphragm.

BLR   *1* Bundesverband der Luftfahrt Zubehör und Raketenindustrie eV (G).

2 Beyond local repair.

3 Bomber, long-range category (USAAC 1934–36).

BLS   Blue-line speed.

BLSN   Blown snow (ICAO).

BLSS   Base-level supply sufficiency.

BLU   *1* Bande latérale unique, = SSB.

2 Bomb, live unit (US).

BLUE   Friendly.

Blue airway   Originally in US, N–S civil airway.

Blue angels   Principal formation aerobatic team (USN 1953–).

blue box   Pre-1945 Link trainer (colloq.).

Blue Core   Versatile wireless technology which, with Bluetooth, enables almost anything to be done remotely.

Blue Flag   Command post exercise emphasising tac-air warfare C^3 management (USAF).

Blue Force   In military exercise, forces used in friendly role.

Blue Force Tracking   Methods devised after 9–11 of combining space and airborne sensors, GPS and advanced com. to give real-time picture of every individual on battle area.

blue ice   Formed when water or lavatory fluids leak at high altitude; on descent, large pieces can cause damage to the aircraft or on ground.

blue key   Blue annotation or image on original map or diagram which does not show in subsequent reproduction.

blue-line speed   For multi-engined aircraft, best rate of climb airspeed after failure of one engine (usually thus marked on ASI), usually called V_{YSE}.

Blue List   Schedule of standard units of measurement (ICAO).

blue-on-blue   Mistaken engagement between friendly fighters.

Blue Paper   Notice of proposed amendment (BCARs).

blue-pole   S-seeking pole of magnet.

blueprint   Drawing reproduced on paper by ammonium ferric citrate or oxalate and potassium ferrocyanide to give white lines on blue background. Seldom used today, but word has common loose meanings: any drawn plan; any written plan of campaign or course of action.

blue room   Toilet, esp. on commercial transport.

Blue sector   That half of ILS localiser beam modulated at 150 Hz (right of centreline).

blue-sky   Research considered [perhaps in ignorance] to have no goal.

blue suit   US Air Force; usually in contradistinction to white suit = contractor personnel (colloq.).

Bluetooth   Low-power short-range radio link for mobile [eg pax] devices and for WAN/LAN access points.

blue water   Oceanic, far from land, as distinct from brown.

bluff body   Solid body immersed in fluid stream which experiences resultant force essentially along direction of relative motion and promotes rapidly increasing downstream pressure gradient. Causes flow breakaway and turbulent wake. Broadly, bluff is opposite of streamlined.

BLUH   Battlefield light utility helicopter (UK).

blunt   A * trailing edge or rear face of body causes turbulence immediately downstream, but main airflow cannot detect that body or aerofoil has come to an end and thus continues to behave as if in passage over surface of greater length or chord.

blushing   Spotty or general milkiness or opacity of doped or varnished surface, caused by improper formulation, too-rapid solvent evaporation or steamy environment.

BLW   Below (ICAO).

BLZD   Blizzard.

BM   *1* Bus monitor.

   2 Bubble memory.

   3 Battle management.

   *4*1 Back marker.

   5 Ballistic missile.

B/M   Boom/mask (switch).

BMAA   British Microlight Aircraft Association.

BMC   Basic mean chord.

BM/C^2   Battle management/command and control.

BM/C^3   Battle management/command, control and communications [I adds intelligence].

BM/C^{4}I   As above, plus computers.

BMCE   Base maintenance certifying engineer.

BMD   *1* US Air Force Ballistic Missile Division (later SAMSO).

   2 Ballistic-missile defence.

BMDS   Ballistic-Missile Defense System (DoD).

BMDO   Ballistic Missile Defense Organisation (USAF).

BME   *1* Basic mass empty.

   2 Bulk memory element.

BMEC   Battlespace Management Evaluation Centre, BAE Systems facility at Farnborough.

BMEP   Brake mean effective pressure.

BMEW   Basic mass empty weight.

BMEWS   Ballistic-missile early-warning system.

BMF   Ministy of Finance (G).

BMFA   British Model Flying Association.

BMFT   Ministry for research and technology (G).

BMH   Basic mechanical helmet.

BMI   Bismaleimide, high-temperature-resistant resin adhesive.

BMR   Bearingless main rotor.

BMS   *1* Bureau Militaire de Standardization (F).

   2 Building, budget or battle management system.

   3 Ballistic-missile sensor.

   4 Battle-management shelter (RAF).

BMTC   Basic Military Training Center (USAF Lackland AFB).

BMTOGW   Basic mission takeoff gross weight.

BMTS   Ballistic-missile target system.

BMUP   Block Modification Upgrade Program (USN).

BMV   Brake metering valve.

BMV$_g$ Bundesministerium der Verteidigung (MoD, G).

BN   *1* Night bomber (F, obs).

   2 Boron nitride.

B/N   Bombardier/navigator.

B$_n$ Receiver noise bandwidth.

BNAE   Bureau de Normalisation de l'Air [or Aéronautique] et de l'Espace (F).

BNASC   Belgian National AIS (1) Centre.

B/NB   Bid or no bid.

BNEA   British Naval Equipment Association.

BNG   Boosted, not guided.

BNH battery   Bipolar nickel/hydrogen.

BNK   Bureau of new construction (USSR).

B$_{NN}$ Null-to-null bandwidth.

BNR   Binary.

BNRID   Basic net radio interface device.

BNS   Boundary notation system.

BNSC   British National Space Centre, formed 1985 as successor to British Space Development Co.

BO   Boom operator.

Bo   *1* Boundary lights.

   2 Boron.

BOA   *1* Basic ordering agreement.

   2 Bulle Operationnelle Aéroterrestre (F).

boarding   Noun, one passenger.

boarding card, boarding pass   Document issued at check-in which admits passenger to aircraft.

boarding status   Current stage reached at gate, ending with 'closing'.

boardroom bomber   Former WW2 or similar warplane converted for executive use.

boards   Speed brakes (colloq.).

boat seaplane   Flying boat (US).

boat-tail   Rear portion of aerodynamic body, esp. body of revolution, tapered to reduce drag. Taper angle must be gentle to avoid breakaway.

BOB   Bureau of the Budget (US).

bobbing   Rare fluctuation in strength of radar echoes allegedly due to alternate attenuation and reinforcement of successive pulse waves.

BOBS, Bobs   Beacon-only bombing system.

bobweight   Mass inserted into flight-control system, usually immediately downstream of pilot's input, to impart opposing force proportional to aircraft linear or angular acceleration.

BOC   *1* Bottom of climb.

   2 Binary offset carrier.

BOD   *1* Biochemical, or biological, oxygen demand.

   2 Beneficial occupancy date.

bod   Male of lowly rank (RAF WW2).

Bode plot   Gain and phase angle against system frequency.

Bodie   Severe test of gas-turbine engine: soak to maximum carcase temperature, slam deceleration to flight idle, then slam to MTO.

body   *1* Any three-dimensional object in fluid flow.

2 In most aircraft, central structure: hull of marine aircraft or airship, fuselage of aeroplane or helicopter, * of missile.

3 Any observable astronomical object, esp. within solar system.

body axes   Outlined by G.H. Bryan in 1903, orthogonal reference axes, fore/aft or longitudinal [called X], transverse or lateral [Y] and vertical [Z]. Problem: they have their origin at the c.g., which has no fixed location. See other axes: *principal*, *stability*, *wind*.

body bag   *1* Occupied by pilot of hang glider, instead of open harness: reduces drag and keeps occupant warm.

2 Container for corpse in transit.

body burden   Aggregate radioactive material (not dose received) in living body.

body English   Guiding the flight of an aerodyne, usually a glider, by shifting the c.g. of one's body.

body gear, body landing gear   Main landing gear retracting into fuselage.

body lift   Lift from fuselage of supersonic aircraft or missile at AOA other than zero.

body of revolution   Body (2) having circular section at any station and surface shape described by rotating side elevation about axis of symmetry. Ideal streamlined forms are generally such.

body plan   Full-scale elevations and sections of aircraft body in lofting.

body sensor   Biomedical sensors worn by astronauts or aircrew to measure parameters such as body temperature, pulse rate and respiration.

body stall   Gross flow breakaway from core engine and afterbody in installed turbofan.

BOE   Black-out exit; predicted time in manned re-entry at which communications will be resumed.

boe   Barrels of oil equivalent; thus boe/d = boe per day.

boffin   Research scientist, esp. senior worker on secret defence project (colloq.).

bog   To taxi across ground so soft that landing gear sinks in and halts aircraft (see *flotation*).

bogey   Air contact (5) not yet identified, usually assumed to be enemy (UK spelling often bogy).

bogie   Landing gear having multi-wheel truck on each leg.

bogie beam   Pivoted beam linking front and rear axles of bogie to each other and to leg.

bog in   To become stuck in soft airfield surface.

bogy   See *bogey*.

BOH   Break-off height.

BOI   *1* Board of Inquiry.

2 Basis of issue.

boiler   Gas-turbine used as a core engine in high-ratio turbofan, as source of hot gas for tip-drive rotor or fan-lift system or any other application calling for central power source (colloq.).

boilerplate   Non-flying form of construction where light weight is sacrificed for durability and low cost (see *battle-ship*).

boiloff   Cryogenic propellant lost to atmosphere through safety valves as result of heat transfer through walls of container (which may be static storage or tank in launch vehicle).

BOK   Bureau of special designers (USSR).

BOL   *1* Bearing-only launch.

2 Bottom of loop [engine s.f.c.].

bold-face procedures   Emergency procedures, written in flight manual in bold-face type.

bollard   Mooring attachment in form of short upright cylinder on marine aircraft hull or float bow.

bollock   APFC (colloq.).

bolometer   Sensitive instrument based on temperature coefficient of resistance of metallic element (usually platinum); used to measure IR radiation or in microwave technology (see *radiometer*).

BOLT   Build, operate, lease, transfer.

bolt   *1* In advanced airframe structure, usually precision fitted major attachment device loaded in shear.

2 In firearm, approximately cylindrical body which oscillates axially behind barrel feeding fresh rounds, closing breech and extracting empty cases (see **breech-block**).

bolter   *1* In carrier (1) flying, aircraft which fails to pick up any arrester wire and overshoots without engaging barrier.

2 Verb, to perform 1.

Boltzmann constant   Ratio of universal gas constant to Avogadro's number; 1.380546×10^{-23} J $°K^{-2}$.

Boltzmann equation   Transport equation describes behaviour of minute particles subject to production, leakage and absorption; describes distribution of such particles acted upon by gravitation, magnetic or electrical fields, or inertia. Boltzmann-Vlasov equations describe high-temperature plasmas.

BOM   Bill of material.

bomb   *1* Transportable device for delivery and detonation of explosive charge, incendiary material (including napalm), smoke or other agent, esp. for carriage and release from aircraft.

2 Streamlined body containing pitot tube towed by aircraft and stabilized by fins to keep pointing into relative wind in region undisturbed by aircraft.

bomb aimer   Aircrew trade in RAF (formerly) and certain other air forces.

Bomb Alarm System   Automatic system throughout Conus for detecting and reporting nuclear bursts.

bombardier   Aircrew trade. bomb aimer, in USA and USAF (formerly).

bombardment ion engine   Rocket engine for use in deep space which produces ion beam by bombarding metal (usually mercury or caesium) with electrons.

bomb bay   In specially designed bomber aircraft, internal bay for carriage of bombs (in fuselage, wings or streamlined nacelles).

bomb-burst   Standard manoeuvre by formation aerobatic teams in which entire team commences vertical dive, usually from top of loop, in tight formation; on command, trailing smoke, members roll toward different azimuth directions and pull out of dive, disappearing at low level 'in all directions'.

bomb damage assessment   Determination of effects on enemy targets of all forms of aerial attack.

bomb detection chamber   Explosion-containment chamber in which objects, such as cargo containers, can be subjected to a complete simulated air-travel environment.

bomb door   Door which normally seals underside of bomb bay. Can slide rearwards, sideways and upwards, open to each side or rotate through 180° about longitudinal axis to release stores from its upper face.

bombed out   Forced to leave home because of serious damage caused by air attack (UK).

bomber   Aircraft designed primarily to carry and release bombs. Term today reserved for strategic aircraft.

bomber-transport   Former category of military aircraft capable of being used for either type of mission.

bomb fall line   Bright line on HUD along which free-fall bombs would fall to the ground if they were released.

bomb impact plot   Graphical picture of single bombing attack by marking all impact of detonation centres on pre-strike vertical reconnaissance photograph(s).

bombing angle   Angle between local vertical through aircraft at bomb release point and line from that point to target.

bombing errors   *1* 50% circular error: radius of circle, with centre at desired mean point of impact, which contains half missiles independently aimed to hit that point (see *CEP (1)*).

2 50% deflection error: half distance between two lines drawn parallel to track and equidistant from desired mean point of impact which contain between them half impact points of missiles independently aimed to hit that point.

3 50% range error: half distance between two lines drawn perpendicular to track and equidistant from desired mean point of impact which contain between them half missiles independently aimed to hit that point.

bombing height   Vertical distance from target to altitude of bombing aircraft.

bombing run   Accurately flown pass over target attacked with free-fall stores.

bombing teacher   Primitive classroom rig in which pupil uses actual bombsight in simulated environment.

bombing up   Loading one or more bombers with bombs.

bomblet   Small bomb, usually of fragmentation type, carried in large clusters and released from single stream-lined container.

bomb line   Forward limit of area over which air attacks must be co-ordinated with ground forces; ahead of ** air forces can attack targets without reference to friendly ground troops.

bomb rack   Formerly, attachments in bomb bay or externally to which bombs were secured; provided with mechanical or EM release, fuzing and arming circuits and sometimes other services. Replaced by universal store carriers tailored to spectrum of weapons.

bomb release line   Locus of all points (often a near-circle) at which aircraft following prescribed mode of attack must release particular ordnance in order to hit objective in centre.

bomb release point   Particular point in space at which free-fall ordance must be released to hit chosen target.

bombsight   Any device for enabling aircraft to be steered to bomb release point, esp. one in which aimer sights target optically and releases bombs by command.

bomb site   Urban area completely cleared of rubble after WW2 (UK).

bomb trolley   Low trolley for carriage of ordnance from airfield bomb stores to aircraft (and often equipped to raise bombs into position on bomb racks).

bomb truck   Originally [1943–45], bomber engaged in carpet or non-precision bombing. Today, deliverer of ordnance to a target marked by a laser in another aircraft or on ground.

bomb winch   Manual or powered winch for hoisting bombs from trolleys into or beneath racks.

bonding   *1* Structurally joining parts by adhesive, esp. adhesives cured under elevated temperature and/or pressure.

2 Joining together all major metal parts of an aircraft, especially an aircraft not of all-metal construction, to ensure low-resistance electrical continuity throughout. Even where metal structures are squeezed together by bolts or rivets a bond of copper strip or braided wire must link them reliably. Bonding is necessary for Earth–return systems and to dissipate lightning strikes and other electrical charges safely with no tendency to arcing or spark formation.

3 Legal agreement linking a pilot to an airline who pays for his tuition.

bonding noise   In older aircraft, radio interference caused by relative movement between metal parts bonded (2) together.

Bondolite   Low-density sandwich of balsa faced with aluminium.

Bone   B-one next enhancement.

bonedome   Internally padded rigid protective helmet worn by combat aircrew (colloq.).

boneyard   Graveyard of unwanted aircraft, usually stripped of potential spares (colloq.). Particularly refers to AMARC, Arizona.

bonker   Small rocket giving high thrust for a fraction of a second designed to impart powerful disturbing blow to extremity of airframe in investigation of aerodynamic/structural damping.

bonnet   Valve hood in aerostat envelope.

BOO   Build, own, operate.

boob   Noun, error; verb, to make mistake (RAF 1935–).

Boolean algebra   Powerful and versatile algebra compatible with binary system and with functions AND, OR and NOT.

boom   *1* Any long and substantially tubular portion of structure linking major parts of an aircraft (esp. linking the tail to the wing or to a short body).

2 Longitudinal structural members forming main compression and tension members of a wing spar, having large section modulus as far as possible from wing flexural axis.

3 Device used in some air-refuelling tanker aircraft in form of pivoted but rigid telescopic tube steered by aerodynamic controls until its tip can be extended into a fuel-tight receptacle on receiver aircraft.

4 Sound heard due to passage of shockwaves from distant supersonic source, such as SST flying high overhead.

5 Spanwise pipe conveying ag-liquid to spraying nozzles; hence * width, * pivots.

boom avoidance   Technique of flying aircraft for minimal boom (4) disturbance on ground.

boom avoidance distance   Distance along track over which ** technique is enforced. Hence, BAD (departure) and BAD (arrival).

boom carpet   Strip of Earth's surface along which observers hear sonic boom from supersonic aircraft.

boomer   *1* Operator of refuelling boom (5) in tanker aircraft (colloq.).

2 USN submarine armed with ballistic missiles [FBMS] (colloq.).

boom microphone   Voice microphone carried on cantilever boom (slender structural beam) pivoted at side of headset so that it can be moved away from the mouth for avoiding unwanted speech broadcasting.

boom receptacle   Flight refuelling socket on military aircraft with which a boom (3) forms a fuel-tight connection despite motion and changes of orientation relative to tanker.

boom throw-forward   Distance along track from origin of shockwave to ground impact.

boom trough   See *boom well*.

boom well   Recess in deck plating of marine-aircraft float or hull to take end-fitting of struts or booms (1).

boondocks   Open land area remote from habitation, esp. when site of forced landing.

boost   *1* Any temporary augmentation of thrust or power in a mechanical or propulsive system.

2 Excess pressure, over and above a datum, in induction manifold of piston engine as result of super-charging. Datum is usually one standard atmosphere.

3 Jettisonable booster rocket for unmanned vehicle (UK, colloq.).

4 Fast-burning portion of a boost/sustain motor.

boost control   Control system, today invariably automatic, for maintaining suitable boost pressure in aircraft piston engine and, in particular, for avoiding excessive boost.

boost/cruise motor   Rocket motor having very large but brief thrust for vehicle launch followed by lower but long-duration thrust for aerodynamic cruise.

boosted controls   Aeroplane flying control surfaces in which balance is reduced and pilot input augmented by brute force, usually by hydraulic jacks. Much simpler and cruder than a powered flying-control system (see *servo*).

booster   *1* Boost rocket.

2 Sensitive high-explosive element detonated by fuze or primer and powerful enough to detonate a larger main charge.

3 LP compressor, with from one to five stages, downstream of an HBPR fan and rotating with it to supercharge the core airflow into the HP spool.

booster APU   APU capable, usually in emergency only, of augmenting aircraft propulsion.

booster coil   Battery-energised induction coil to provide a spark to assist piston engine starting.

booster magneto   Auxiliary magneto, often turned by hand, for supplying hot sparks during piston engine starting.

booster pump   *1* Centrifugal pump, often located at lowest point of a liquid fuel tank, to ensure positive supply and maintain above-ambient pressure in supply line.

2 Auxiliary impeller in cryogenic propellant system to maintain system pressure and prevent vaporization upstream of main pump.

booster rocket, booster stage   See *boost rocket*, though * *stage* implies a large long-burning stage for a large vehicle.

boost/glide vehicle   Aerodyne launched under rocket thrust and accelerated to hypersonic speed in upper fringes of atmosphere, thereafter gliding according to any of various predetermined trajectories over distances of thousands of miles.

boost motor   See *boost rocket*.

boost phase   Initial phase of launch and rapid acceleration of missile or other short-range aerodynamic vehicle fitted with boost rockets.

boost pressure   See *boost (2)*, *international* *, *override* *.

boost pump   Booster pump.

boost rocket   Rocket motor, usually solid propellant and sometimes used in multiple, used to impart very large thrust during stages of launch and initial acceleration of missile or other vehicle launched from ground or another aerial vehicle. Almost all kinds of ** burn for a few seconds only, and in some cases for only a fraction of a second. Sometimes case and chamber forms part of vehicle, but most boost rockets are separate and jettisoned after burnout.

boost rocket impact area   Area within which all ** should fall during launches on a given range.

boost separation   Process by which boost rocket thrust decays and becomes less than drag, causing rearward motion relative to vehicle and subsequent progressive unlocking, possibly relative rotation, and detachment.

boost/sustain motor   Rocket comprising fast-burning high-thrust portion followed by slow-burning low-thrust portion.

boost vehicle   SDI term for space or long-range missile launcher.

boot   *1* Flat array of flexible tubes bonded to leading edge of wings, fins and other aircraft surfaces to break up ice. Fluid pressure is alternately applied to different sets of tubes in each boot to crack ice as it forms.

2 Shroud or vizor enabling cockpit radar to be viewed in bright sunlight.

bootie   *1* Protective cover for pitot tube, usually with streamer.

2 Soft fabric overshoe warn before walking on aircraft skin or entering engine duct.

bootstrap   Noun, hoisting gear to remove disabled engine or lift replacement engine to pylon or [trijet] to tail engine position; and verb, to perform lifting operation. Can also be applied to modules.

bootstrap exploration   Using each space mission to bring back information to help subsequent missions, esp. in lunar exploration (colloq.).

bootstrap operation   Dynamic system operation in which once cycle has been started by external power, working fluid maintains a self-sustaining process. A gas turbine, once started, sustains bootstrap operation because the turbine keeps driving the compressor which feeds it. Thus, * cycle \(cold-air unit), * mainstage engine pump (turbine being fed by propellants delivered by pump), etc.

Boozer   Code name for British ECM [two-colour warning lights] carried by Mosquito aircraft in 1944.

BOP   *1* Balance of payments, esp. with regard to national participation in multinational programme.

2 Basic operating platform (bare base airfield).

3 Bit-oriented protocol.

BOPS   *1* Burn-off per sector; fuel burned on each sector, or segment, or stage (all three words are used in flight-planning documents) in commercial transport operation.

2 Beam-offset phase shifter (Awacs).

bops   Billions of operations per second.

BOR   Basic operational [or operating] requirement.

Boram   Block-orientated random-access memory.

boresafe fuze   Projectile fuze rendered safe by interrupter until projectile has cleared gun muzzle.

borescope   Slender optical periscope, usually incorporating illumination, capable of being inserted into narrow apertures to inspect interior of machinery.

borescope port   Circular ports, fitted with openable caps, through which borescopes may be inserted (esp. in aircraft engines).

boresight   *1* Verb, to align gun or other device by means of optical sighting on a target.

2 Noun, precise aim direction of gun, directional aerial/antenna, camera, etc.

boresight camera   Optical camera precisely aligned with tracking radar and used to assist in alignment of aerial [antenna].

boresight coincidence   Optical alignment of different adjacent devices, such as radar waveguide, reflector, passive interferometer and IR or optical camera.

boresight line   Optical reference line used in harmonising guns and other aircraft weapon launchers.

boresight mode   Radar is locked at one chosen angle between dead ahead and –2° or –3°.

boresight test chamber   Anechoic chamber containing movable near-field test targets and aerials, capable of being wheeled over nose of radar-equipped fighter aircraft.

BORG   Basic Operational Requirements Group (ICAO).

boring   Process of accurately finishing already-drilled hole to precise dimension, usually by using single-point tool.

boron, Bo   Element, either greenish-brown powder, density 2.3, or brown-yellow crystals [2.34], MPt 2,300°C. Used as alloying element in hard steels, as starting point for range of possible high-energy fuels, and above all as chief constituent of boron fibre.

boron doping   Addition, finely divided, to increase I_{sp} of solid rocket motor.

boron/epoxy   Composite plastic materials comprising fibres or whiskers of boron in matrix of epoxy resin.

boron fibre   High-strength, high-modulus structural fibre made by depositing boron from vapour phase on white-hot tungsten filament. Used as reinforcement in aerospace composite materials.

Borsic   Boron fibre coated with silicon carbide. Structurally important as reinforcement in matrices of aluminium and other metals.

bort   Side, hence * number = serial painted on fuselage (R).

Boscombe   Aeroplane & Armament Experimental Establishment, Boscombe Down, Wiltshire, England.

bosom tank   Detachable (usually jettisonable) fuel tank scabbed on under fuselage.

Boss, BOSS   *1* Ballistic offensive suppression system (EW).

2 Battlefield optical surveillance system.

3 Bureau of State Security (S Africa).

4 Bio-optic synthetic systems (inspired by optics of life forms).

boss   *1* In traditional wooden or one-piece metal propeller, the thickened, non-aerodynamic central portion.

2 In a casting, locally thickened area to provide support for shaft bearing, threaded connection or other load.

3 Squadron commander, often initial cap (RAF).

BOT   *1* Boom-operator trainer.

2 Brakes-off time.

3 Build, operate [or own], transfer.

BoT   Board of Trade.

BOTB   British Overseas Trade Board.

Bottlang   Commercially produced loose-leaf binder describing European VFR airfields.

bottle   A JATO rocket (filled or empty case) (colloq.).

bottom dead centre   Position in piston engine in which centre of crankpin is precisely aligned with axis of cylinder, with piston at bottom limit of stroke; at BDC piston cannot exert a turning moment on crankshaft.

bottom rudder   In aeroplane in banked turn, applying rudder towards lower side of aircraft; among other things this will lower the nose.

bottom shock   On underside of supersonic aerofoil.

bought it   Killed (RAF colloq.).

bounce   *1* In air combat, to catch enemy aircraft unawares; to intercept without being seen.

2 In piston engine poppet valve gear, elastic bounce of valves on their seats.

bounce table   Vibration test machine.

boundary layer   Layer of fluid in vicinity of a bounding surface; eg, layer of air surrounding a body moving through the atmosphere. Within the ** fluid motion is determined mainly by viscous forces, and molecular layer in contact with surface is assumed to be at rest with respect to that surface. Thickness of ** is normally least distance from surface to fluid layer having 99 per cent of free-stream velocity. ** can be laminar or, downstream of transition point, turbulent.

boundary-layer bleed   Pathway for escape of ** adjacent to engine inlet mounted close beside fuselage wall. Bleed is either open or ducted, and removes stagnant ** which would otherwise reduce ram pressure recovery and propulsion system performance.

boundary-layer control   Control of ** over aircraft surface to increase lift and/or reduce drag and/or improve control under extreme flight conditions. BLC can be effected by: passive devices, such as vortex generators; ejecting high-velocity bleed air through rearward-facing slits; sucking ** away through porous surfaces; use of engine slipstream to blow wings or flaps.

boundary-layer duct   Duct to carry ** from ** bleed to point at which it can advantageously be dumped overboard.

boundary-layer energiser   Low sharp-edged wall normal to airflow across aerodynamic surface (eg immediately upstream of aileron).

boundary-layer fence   Shallow fence fixed axially across swept wing to reduce or check spanwise drift of ** and its consequent thickening and proneness to separation.

boundary-layer noise   Major source of noise inside aircraft.

boundary-layer scoops   Forward-facing inlets designed to remove thick ** upstream of engine inlet or other object.

boundary-layer separation   Gross separation of ** from boundary surface, space being filled by undirected, random turbulence.

boundary light   Visible steady light defining boundary of landing area.

boundary marker   Markers, often orange cones, defining boundary of landing area.

bound vortex   *1* Circulation round a wing.

2 Vortex embracing any solid body or touching a surface.

Bourdon tube   Flat spiral tube, either glass filled with alcohol or metal filled with mercury, whose radius increases (rotating the centre) with increasing temperature.

Boussinesq   Formula giving distance along uniform tube necessary for laminar, viscous, incompressible flow to become fully developed.

$$x = \frac{0.26 \; Um \; r^2}{v} = 0.25 \; rR.$$

where Um is mean velocity, r tube radius, *v* kinematic viscosity and R Reynolds number.

BOV   Blow-off valve.

BOVC   Base of overcast.

BOW   Basic operating weight [OWE is more precise].

bow   Rhyming with cow, nose of airship or marine hull or float.

bow   Rhyming with go:

1 Curvature along length of turbine blade or other slender forging, or curvature due to instability in structural compression member.

2 Curved member forming tip of wing or other aerofoil, esp. one with fabric covering.

bow cap   Structure forming front end of airship hull or envelope. Alternatively, nose cap.

Bowen-Knapp camera   High-speed strip-film camera used in vehicle flight testing.

bowser   *1* Airfield fuel truck, roadable and self-propelled; unpropelled, * trailer.

2 Used as adjective, specially modified to contain overload or ultra-long range fuel, hence * wing, * fuselage.

Bow's notation   Conventional system of representing structural forces and stresses by letters and/or numbers in graphical stress analysis (rhymes with slow).

bow stiffeners   Longitudinal stiffeners arranged radially around nose of aerostate envelope (esp. blimp or kite balloon) to prevent buckling under aerodynamic pressure. Alternatively called battens.

bow wave   *1* Shockwave from nose of supersonic body, esp. one not having sharply pointed nose.

2 Shockwave caused by motion of planetary body through solar wind.

3 Form of wave caused by bows of taxiing marine aircraft.

4 Form of wave caused by landplane nosewheels running through standing water.

box   *1* Tight formation of four aircraft in diamond (leader, left, right, box).

2 Structural heart of a wing comprising all major spars, ribs and attached skins (often forming integral tankage), but usually excluding leading and trailing edges, secondary structure and movable surfaces.

3 Major sections of fuselage, especially where these are of rectilinear form and thus not describable as *barrel sections*.

4 Aircraft structure formed from two or more lifting planes (wing or tail), linked by struts and bracing wires.

In early aviation no other form could compete for lightness and strength.

5 Airlifter cargo compartment, simplified to basic rectilinear form and dimensioned overall.

6 Above-floor removable cargo container, with various standard dimensions.

7 Container of lights in visual ground guidance system, thus 4 * VASI.

box connector   Multi-circuit connector having four-sided box sockets, with linear pin engagements (2 or 3 rows, up to 240 circuits).

boxer   piston engine having two crankshafts and [e.g. four or six] parallel cylinders in rectilinear formation.

box girder   See *box spar*.

boxing   Process of assembling major airframe sections in erection jig; includes fuselage *box* (3) sections.

boxkite   Kite in form of rectangular- (often square-) section box, open at mid-section and at ends. Structurally related to early biplanes in form of *box* (4).

box position   Rear aircraft in *box* (1).

box rib   Rib assembled from left and right sides separated by a peripheral member following profile of aerofoil.

box sizing   Part of GAMM in which an aircraft fuselage cross-section is selected and optimum cargo box length determined for groupings of vehicles or other large loads.

box spar   Spar assembled from front and rear webs separated by upper and lower booms.

box tool   Tangential cutting tool incorporating its own rest, used on automatic turning machines.

box wing   *Diamond wing.*

Boyle's law   In an ideal gas at constant temperature, pressure and volume are inversely proportional, so that $PV = f(T°)$ and $P/\rho = RT$.

BP   *1* Bite processor.

2 Bottom plug.

3 Braided pultruded.

4 Boron phosphide.

b.p.   *1* Bypass (jet engine); thus, * ratio.

2 Band-pass filter.

BPA   1 British Parachute Association, successor to British Parachute Club (1956).

2 Blanket purchase agreement.

BPC   *1* Gas-turbine barometric pressure control.

2 Benchmark pricing guide.

3 British Purchasing Commission (WW2).

4 Basic primer concept (paint).

BPCU   Bus power control unit.

BPDMS   Base, or basic, point-defense missile system.

BPE   *1* Best preliminary estimate.

2 Bomber penetration evaluation.

B Per T   Squadron for testing heavy aircraft (AAEE, WW2).

BPF   *1* Band-pass filter.

2 Blade-passing frequency.

3 British Pacific Fleet (WW2).

BPI   *1* Boost-phase intercept[or].

2 or **bpi**, bits per inch (EDP).

BPL   Band-pass limiter.

BPLI   Boost phase launch[er] intercept.

BPM   Binary phase-modulation.

BPP   Breakthrough propulsion physics.

BPPA   British Precision Pilots Association; affiliated to

FAI, orienteering without navaids, assisting handicapped/paraplegic pilots.

BPR　Bypass ratio.

BPS　Balanced pressure system: buried glycol pipes trigger alarm if stepped on.

　2 Bistable phosphor storage.

　3 Bytes per second.

　4 See next.

bps　Bits per second (EDP).

BPSK　Binary phase-shift keying.

BPt　Boiling point.

BQ　Ground-launched controllable bomb [ie, SSM] (USAAF 1942–45).

Bq　Becquerel[s].

BR, Br　*1* Bomber-reconnaissance.

　2 Mist (Metar code).

　3 Bearing.

　4 Baggage reconciliation.

　5 Bridge.

br　Reference length.

BRA　*1* Bureau des Recherches Aériennes (ICAO).

　2 Barrel-roll attack.

　3 Bomb-rack assembly.

　4 Braking action; F,G,N,P add fair, good, nil, poor.

　5 Beam rotational, or reference axis.

　6 Basewide remedial assessment.

BRAAT　Base recovery after attack training.

BRAC　Base realignment and closure.

brace position　Adopted for ditching or crash-landing: shoes removed, bent forward with arms protecting head.

Bracis　Biological, radiological and chemical information system (UK).

bracket　Limits of time/distance/altitude for one pre-planned in-flight refuelling.

bracketing　*1* Obsolete method of flying a radio range in order to establish correct quadrant and hence direction to next waypoint.

　2 In flak or other artillery, establishing a short and an over along desired line and then successively splitting resulting 'bracket' in half.

bracket propeller　Variable-pitch (usually two-position) propeller in which each blade is pivoted and automatically adopts a fine or coarse setting according to the position of a sliding mass at the root.

Braduskill　'Slow kill' technique proposed in SDI for gradual closure on hostile satellites.

BRAG　Batteries Research Advisory Group.

Brahms　Baggage reconciliation and handling management system.

brake　*1* Device for removing energy from a moving system to reduce its speed or bring it to rest. Energy withdrawn may be rejected to atmosphere (airbrake, speed brake) or absorbed in heat sinks (wheel brake, propeller or rotor brake).

　2 In sheet-metalwork, a power press for edging and folding.

brake horsepower　Power available at output shaft of a prime mover. Generally synonymous with shaft horsepower [preferred] and torque horsepower.

brake mean effective pressure　A measure of MEP in an operating piston engine cylinder calculated from known BHP (BMEP is to be expressed in kPa or bars and derived from power in kW).

brake-up　Aeroplane nose-over caused by harsh braking (colloq.).

braking action　Measure of likely adhesion of tyres to runway, thus * advisory warns of snow, ice or other hazard.

braking coefficient　Braking force coefficient.

braking ellipse　Elliptical orbit described by spacecraft on entry to a planetary atmosphere. If continued, successive ellipses are smaller, due to drag; purpose of manoeuvre is to dissipate re-entry or entry energy over a much greater time and distance and thus reduce heat flux.

braking force　Linear force exerted by a braked vehicle (ie, aircraft) wheel.

braking force coefficient　Coefficient of friction between wheel and fixed surface (whether rolling or sliding).

braking nod　Nose-down pitch of aircraft when wheel-braked, or nose-up pitch when brakes released at full power, esp. maximum angular movement thus imparted.

braking parachute　Parachute streamed from aircraft to increase drag, increase dive angle or reduce landing run.

braking pitch　Predetermined propeller pitch to give maximum retardation, either windmilling drag or reverse thrust under power.

braking rocket　Retrorocket.

branch　*1* In electrical system, portion of circuit containing one or more two-terminal elements in series.

　2 In computer program, point at which * instructions are used to select one from two or more possible routines.

　3 In crystal containing two or more kinds of atom, either of possible modes of vibration, termed acoustic * or optical *.

branch pipe　Pipe conveying exhaust gas from piston engine cylinder to manifold or collector ring.

brass　Alloys of copper with up to 40 per cent zinc and small proportions of other elements.

brassboard　Functioning breadboard model of avionic system; also adjective and verb.

brassed off　See *browned off*.

BRAT　Benchtop reconfigurable automatic tester.

Brat　*1* Bomb-responsive anti-terrorist.

　2 Graduate of RAF Apprentice School, Halton (colloq.).

Bratt-DaRos　Method of solving problems of inertial coupling.

Bravo exercise　Combat mission called off at point when aircraft ready to taxi.

Brayton cycle　Thermodynamic cycle used in most gas turbines: diagram comprises compression and expansion curves joined by straight lines representing addition or rejection of heat at constant pressure. The so-called 'open cycle'.

brazier-head rivet　Light-alloy rivet having head shallower but of larger diameter than round-head.

brazing　Joining metals by filling small space between them with molten non-ferrous metal having a melting point above a given arbitrary value (originally 1,000°F = 538°C).

breadboard　Preliminary assembly of hardware to prove feasibility of proposed system, without regard to packaging, reliability or, often, safety. May be laboratory rig or flyable system. Often adjective or verb.

break　*1* Point at which pilot senses stall of wing.

　2 Breakaway (1) (colloq.).

　3 Chief meaning in modern air combat: to make

maximum instantaneous turn to destroy hostile fighter's tracking solution. Used as noun or as verb.

4 In carrier flying, point at which aircraft turns sharply left across bows and on to downwind leg.

5 Word inserted by harassed controller to indicate that following words are for a different recipient.

breakaway   *1* Point at which aircraft breaks off trajectory directed against another object, such as stern attack on enemy aircraft or gun-firing run on ground target.

2 Altitude at which pilot abandons approach in bad weather.

3 In nuclear explosion, point in space or time at which shockfront moves ahead of expanding fireball.

breakaway thrust   Engine power needed to initiate movement and reach taxiing speed.

breakdown book   Record of physical changes introduced during maintenance, servicing or repair.

breakdown drawing   Isometric or perspective drawing showing parts separated from each other by being displaced along one or more axes. Often called exploded drawing.

breakdown potential   Dielectric strength.

breaker strip   *1* Linear narrow de-icing element, either thermal or mechanical, arranged along leading edge (eg of wing or engine inlet strut) to split ice accretion into two parts.

2 See *stall strip*.

break-even load factor   Load factor at which a particular flight, service, aircraft type or overall airline operation shows a net profit.

break-even point   *1* In any commercial aircraft operation, load factor at which total revenue equals total cost.

2 In manufacture, number of sales required to cover investment.

breakin, break-in   First bench run of new type of engine or other device.

break lock   To use ECM or other counter measure to make hostile tracking system (eg IR or radar) cease to track friendly or own aircraft.

breakoff phenomenon   Mental state experienced by crews of high-altitude aircraft and spacecraft of being divorced from other humanity.

break-out   Point at which flight crew receive first forward visual cues afer an approach through cloud.

breakout force   Minimum force required to move pilot's flying controls (each axis considered separately). If not measured at zero airspeed, airspeed must be quoted.

breakout panel   One panel in aircraft canopy through which occupant[s] can escape in emergency.

breakpoint   *1* In system responding to high-frequency input, the corner frequency (as seen on a Bode plot) where $f = 1/2T\pi$.

2 In EDP, point in program or routine at which, upon manual insertion of * instruction, machine will stop and verify progress.

3 Sudden change in slope of graphic plot, eg point at which payload has to fall from maximum value in plot against range. Also called *knee*.

break price   Quantity at which unit price changes.

breakthrough propulsion physics   NASA project searching for a way to travel at a significant fraction of the speed of light.

break-up   Separation of single radar blip into discrete parts each caused by a target.

break-up circuit   Electrical circuit linking airborne portions of break-up system.

break-up shot   Artillery shot designed to break into small fragments upon leaving muzzle and thus travel only a short distance.

break-up system   System designed to break unmanned vehicle, such as ballistic missile, space launcher or RPV, into fragments sufficiently small to cause minor damage if they should fall on inhabited area.

break X   To break (2) at point of minimum range for launch of own AAM, where X symbol appears on cockpit display.

breather   Open pipe connecting interior of device, such as piston engine to atmosphere to dissipate moisture or oil vapour.

breathing   *1* Flow of air and exhaust gas through piston engine, esp. the way this is limited by constraints of flow path.

2 Flow of air and/or gas into and out of aerostat in course of flight.

3 Very small air and oil vapour flow through breather holes or ducts provided to equalise pressure inside and outside an engine.

4 Generally, escape of air from all volumes as aircraft climbs, to be replaced by [usually more humid] air on descent.

breech   In a gun, end of barrel opposite muzzle, through which shot is normally loaded.

breech-block   Rigid metal block normally serving to insert and withdraw shell cases, resist recoil force on fired case and seal breech during firing; usually oscillates in line with barrel axis.

breeches piece   Tubular assembly, often shaped like pair of shorts, serving to bifurcate jetpipe or join two jetpipes into common pipe.

Breguet formula   Rule-of-thumb formula for giving flight range of classical aeroplane, and reasonably accurate for all types of aerodyne; range given by multiplying together L/D ratio, ratio of cruising speed divided by sfc, and $\log_e$ of ratio of aircraft weight at start and finish expressed as decimal fraction greater than 1. Units must be compatible throughout.

BREMA, Brema   British Radio and Electronic-equipment Manufacturers Association.

bremsstrahlung   Electromagnetic radiation emitted by fast charged particles, esp. electrons, subjected to positive or negative acceleration by atomic nuclei. Radiation has continuous spectra (eg, X-rays).

brennschluss   Cessation of operation of rocket, for whatever cause.

BREO   On-board avionics (R).

Brétigny   Location, southwest of Paris, of Centre des Essais en Vol (CEV).

brevet   Flying badge worn on uniform, especially denoting qualification as flight-crew member.

BRF   Short approach is required or desired (ICAO).

BRG, Brg   Bearing.

BRH   Bundesrechnungshof [Federal audit office] (G).

BRHZ   Mist, haze.

BRI   Basic-rate interface.

brick   Portable data store which transfers mission data to aircraft systems (colloq.).

bridge   Permanently installed pedestrian connector linking terminal with aircraft. Almost all are covered, and provided with powered movement controlled from the free end. The movement usually includes vertical elevation, lateral traverse and telescopic linear extension, and the terminal end may be able to interface with one level for arrivals or another for departures. Apron-drive bridges are controlled by steerable powered wheels at the free end, running over the apron surface. A variant is the over-the-wing bridge, which with the more common ADB enables a rear-fuselage door to load/unload from the same terminal walkway. So-called glass walls are becoming popular. Noseloaders are parallel to the parked fuselage and have a fixed outer end provided with a short section at 90° to mate with the aircraft door. Commuter bridges provide covered access at ground level. Other names are passenger-boarding or passenger-loading *, airbridge or jetlink; Jetway is a tradename.

bridgehead   End of apron-drive bridge which abuts aircraft; hence * cab.

bridge-type stick   In side-by-side cockpit, control columns linked by pivoted connector.

bridle   *1* Towing linkage, other than expendable strop, transmitting pull of catapult to two hard points on aircraft.

2 Assembly of electric cables or fluid system pipes which, after disconnection, can be removed from supporting structure (eg, landing gear) as a unit.

3 Rigging attached to two or more points on aerostat, esp. blimp, to distribute main mooring pull.

brief   To issue all relevant instructions and information in advance of flying mission (not necessarily military), static test, war game or other operation involving human decision-taking.

bright display   Normally, display which can be viewed clearly without a hood in brightest daylight.

brightness control   Facility provided in radar, TV and other display systems for adjusting CRT bias to control average brightness.

brilliant   Describes munition having both guidance (smart) and programmable software; in practice also means with ability to guide itself to target without external help.

Brinell hardness   Measure of relative hardness of solids, expressed as numerical value of load (either 500 kg or 3,000 kg) and resulting area of indentation made by hard 10 mm ball.

bring-back weight   Weight at which combat aircraft recovers to airbase or carrier, with remaining fuel and unexpended ordnance.

BRITE   *1* Broadcast request imagery technology experiment (satellites).

2 Basic research in industrial technologies for Europe [Int.].

3 Boston rocket ionospheric tomography experiment.

4 Bright radar indicator tower equipment.

BritGFO   British Guild of Flight Operations Officers.

British Parachute Association,   controlling the sport in the UK (office Leicester).

British Thermal Unit   Obsolete [from 1995] measure of heat: quantity required to raise temperature of 1 lb of water from 63° to 64°F; six definitions, all close to 1 Btu = 1,055 J.

brittle fracture   Fracture in solid, usually metal, in which plastic deformation and energy dissipated are close to

zero. Contrasts with ductile frcture, and is rare except at very low temperatures.

BRKG, BRKS   Breaking, breaks (ICAO).

B-RNAV, B-RNav   Basic area navigation, accurate to 5 nm (9.3 km); was VOR/DME, now increasingly GPS.

Broach   Bomb, Royal Ordnance, augmented charge.

broach   Cutting tool having linear row(s) of teeth, each larger than its predecessor.

broad-arrow engine   In-line piston engine having three banks of cylinders with adjacent banks spaced at less than 90°; W-engine.

broadband aerial   Aerial [antenna] capable of operating efficiently over spread of frequencies of the order of ten per cent of centre frequency.

broadband GAN   Worldwide secure shared IP service providing 144 kbit/s.

broadcast   Radio transmission not directed at any specific station and to which no acknowledgement is expected.

broadcast control   Air interception in which interceptors are given no instructions other than running commentary on battle situation.

broad goods   Carbon-fibre and other fibre-reinforced sheet as delivered in the bale.

broadside array   Aerial array in which peak polarization is perpendicular to array plane.

Broficon   Broadcast flight-control (originally fighter control) management of tac-air warfare (USAF).

broken clouds   Sky coverage between 'scattered' and 'continuous', defined by ICAO as five-tenths to nine-tenths, UK usage 5–7 oktas.

broken field   Covered by craters and debris, especially on runway.

brolly   Parachute.

bromine   Br, toxic liquid, density 3.1, BPt 59°C, used in a range of aerospace products.

bronze   Alloys of copper and tin and/or aluminium.

Bronze C   First qualification for glider pilot (BGA).

browned off   Discouraged, bored (RAF WW2).

brown job   Member of friendly army (RAF, WW2).

brownout   Often hypenhated, near-zero surface visibility because of blown sand or topsoil.

brown water   Littoral, close inshore.

BRP   Braked retarded parachute [S adds super].

BRS   *1* Best-range speed.

2 Baggage reconciliation system.

3 Ballistic recovery system.

BRSL   Bomb-release safety lock.

Br-Stoff   Avgas or benzole (G).

BRT   *1* Brightness.

2 Bomb retarding tail.

BRTF   Battery repair and test facility (artillery, guided weapons).

BRU   Bomb release unit, normally complete interface beween hardpoint and munition.

brush discharge   See *Corona discharge.*

brush seal   Ring of fine wire bristles continuously rubbing on erosion-resistant [ceramic] sleeve on rotating shaft.

BRW   Brake release weight, ie at start of takeoff run.

BS   *1* Commercial broadcasting station.

2 British Standard, thus * parts.

3 Bomb (or bombardment) squadron (USAAC, USAAF, USAF).

4 Blowing snow.

b/s   Bits per second.

BSB   British Satellite Broadcasting.

BSC   Beam-steering computer (EW).

BSCU   Brake-system control unit.

BSDH   Bus shared-data highway.

BSI   *1* British Standards Institution (formed 1901 as Engineering Standards Committee).

2 Bus system interface [U adds unit].

BSIN   Alternative for BSI (2).

BSL   *1* British Standard family of light alloys.

2 Base second level (servicing).

BSM   Breakaway support mast.

BSN   Backbone subnetwork.

BSP   *1* Barra side processor.

2 Board support package.

BSPL   Band sound pressure level; sound pressure level in bands each one-third of an octave wide from 50 to 10,000 Hz.

BSPR   Boost/sustainer pressure ratio (rocket).

BSPS   Beam-steering phase-shifter (Awacs).

BSS   *1* British Standard Specification.

2 British Standard family of steels.

BST   British Summer Time.

bst   Boresight.

BS/TA   Battlefield surveillance and target attack.

BSTS   Boost-phase surveillance and tracking satellite (or system), for detection of enemy launches, tracking of BVs and PBVs and kill assessment (SDI).

BSU   *1* Beam-steering unit.

2 Bypass switch unit.

3 Baggage-screening unit.

BSV   Burner staging valve.

BSW   British Standard Whitworth [screwthreads].

BT   *1* Burn time (rocket).

2 Basic trainer (USAAF, USAF category 1930–47).

3 Bomber/torpedo (USN category, 1942–45).

4 Bathythermograph.

BTB   Bus tie breaker.

BTC   *1* Bus tie connector.

2 Before top centre.

3 Belgocontrol Training Centre, Brussels.

4 Business Travel Coalition (US).

BTH   Beyond the horizon (radar).

B$_{3dB}$ 3-decibel bandwidth.

BTID   Battlefield target identity (US).

BTL   Between cloud layers.

BTM   *1* Bromotrifluoromethane (extinguishant).

2 Burn, then mix.

BTMU   Brake-temperature monitor unit.

BTN   Between [also BTW, BTWN].

BTO   Bombing through overcast (WW2).

BTP   Bureau Trilatéral de Programmes (Eur).

BTR   *1* Bus tie relay.

2 Better (ICAO).

BTS   1 Bureau of Transportation Statistics (US).

2 Border and Transportation Security (DHS, US).

BTsVM   On-board [digital] computer (R).

BTT   Basic training target [aircraft].

BTU   *1* British Thermal Unit (alternatively, Btu, BThU).

2 Bus, or basic, terminal unit.

BTV   Boost [rocket motor] test vehicle.

BTVOR   Weather broadcast terminal VOR.

BTW, BTWN   Between.

BU   *1* Break-up, thus a guided-weapon * unit.

2 Back-up.

3 Broken up.

BuA, BuAer   Bureau of Aeronautics (USN, 1921–59).

BUB   Back-up battery/batteries.

bubble   *1* Continuous ovate-blister film of fuel from burner at low flow rate.

2 Region of continuous EW protection.

bubble horizon   Bubble turn and slip.

bubble memory   Computer memory whose bits are distributed among microscopic voids (bubbles) in a 3-D volume of solid.

bubble sextant   Sextant in which local horizontal is established by a bubble device. Often called bubble octant, because arc is usually not greater than 45°, restricting altitude to 90°.

bubble turn and slip   Primitive flight instrument in which lateral acceleration is indicated by sideways displacement of bubble in arched glass tube of liquid.

BUCD   Back-up command destruct.

buck   Dolly or transport frame, with or without wheels and usually making no provision for inverting (rolling over) contents, tailored to carry complete engine or other major equipment item.

bucket   *1* In US, a turbine rotor blade.

2 Principal member of most types of thrust reverser, two buckets normally rotating and translating to block path of efflux and divert it diagonally forwards. Alternative (UK) = clamshell.

3 Graphical plot having basic U shape resembling *, notably produced by adding one plot of negative slope (eg operating and servicing cost against MTBF) to a related plot of positive slope (eg capital cost against MTBF).

bucket brigade   Integrated-circuit device, comprising MOS transistors connected in series, serving as shift register by transferring analog signal charge from one storage node to next.

bucket shop   Retail outlet (shop) offering non-IATA passenger tickets at cut prices.

bucking   Repeated succession of stalls and recoveries, deliberate or otherwise.

bucking bar   Shaped bar held against shank in manual riveting.

buckling   Lateral deflection of structural member under compressive load; state of instability or unstable equilibrium, but may be purely elastic.

BUCS   Back-up control system.

buddy   Aircraft providing preplanned in flight assistance to another, specifically by providing fuel to an aircraft of similar type [as distinct from normal tanker] or by laser-marking a target.

buddy lasing   Designation of a target by one aircraft for attack by another.

buddy pack   Flight-refuelling hose reel and drogue packaged in streamlined container for carriage by standard weapon rack; thus, aircraft A can refuel buddy B flying identical aircraft (formerly colloq.).

BUF   Back-up facility.

buffer   *1* In radio, low-gain amplifier inserted to prevent interaction between two circuits.

2 Amplifier stage hvaing several inputs, any of which may be connected to output.

3 In EDP, temporary store used to smooth out information flow between devices, esp. between I/O and main processor or core store.

buffer distance   In nuclear warfare, horizontal distance (expressed in multiples of delivery error) which, added to radius of safety, will give required acceptable risk to friendly forces; alternatively, vertical distance (expressed in multiples of vertical error) added to fallout safe height to ensure that no fallout will occur.

buffet   Irregular rapid oscillation of structure caused by turbulent wake. * in aeroplanes may be caused by excessive angle of attack (due to low airspeed, extreme altitude or excessive g) or, in subsonic aircraft, an attempt to fly at too high a Mach number.

buffet boundary   For any given aircraft and environment, plot of limiting values of speed and altitude beyond which buffet will be experienced in unaccelerated flight. Also defined as condition at which a 'significant' region of separated flow appears.

buffet boundary parameter M^2C_L, product of lift coefficient and square of Mach number, at which buffet becomes unacceptable.

buffet inducer   Small projection, usually in form of strake, intended to induce buffet (usually as warning in advance of dangerous buffet affecting major part of aircraft).

buffet margin   For any given aircraft and environment, highest vertical acceleration (g) which can be sustained without exceeding given buffet severity (in some cases severity is zero).

buffet threshold   For any given aircraft and environment, point at which buffet is first perceptible, expressed in terms of speed, altitude and vertical acceleration (g).

buffing   Process for polishing sheet metal by rotary tool of soft fabric impregnated with fine abrasive.

BUFR   Binary universal format.

bug   *1* Heading marker on navigational instrument.

2 Fiducial index, esp. on flight instrument, having appearance of *. Can be painted, Chinagraph or removable by peeling.

3 Clandestine monitoring device, esp. for audiosurveillance.

4 To install and conceal (3).

5 System malfunction or other fault, esp. one not yet traced and rectified; hence, to debug (colloq.).

Bug-E   Battlefield universal gateway equipment, translator between SADL and other tactical datalinks (USAF).

bug-eye canopy   Small canopy (usually two, left and right) over each projecting pilot's head in large aircraft (1942–50).

bugged   Value marked by bug (2).

buggying   Riding a ground vehicle, eg trike unit, pulled by kite.

bugle bag   Sick bag (colloq. among cabin crew).

bug out   Eject (colloq.).

bug speed   Speed at which ASI needle passes *bug* (2), usually V_{REF}.

BUIC   Back-up interceptor control; add-on to SAGE system.

build   Growth in received radio signal; opposite of fade.

build standard   Detailed schedule of all possible variable or unresolved items in aircraft or other complex hardware in stage of development or pre-production. Original ** may list features of airframe, development state of

engines, system engineering and, esp. equipment fitted or absent; altered as aircraft is modified.

built-in hold   Pre-planned hold during countdown to provide time for defect correction or other activity without delaying liftoff.

built-up section   Structural members having section assembled from two or more parts, rather than rolled, extruded, hogged from solid or forged. Reinforced composites are not regarded as built-up; essentially parts should be assembled by joints and could be unfastened.

bulb angle, bulb flange   Structural sections, usually used as booms, having circular or polyhedral form. Nearly all were rolled from strip and had eight to 12 faces after assembly, complete ** being built up from one to five segments.

bulk cargo   Homogenous cargo, such as coal. Today also means cargo carried loose; has come to mean cargo or baggage not contained in standard container or pallet.

bulk erasure   Erasure of complete magnetic tape by powerful field.

bulkhead   Major transverse structural member in fuselage, hull or other axial structure, esp. one forming complete transverse barrier. Certain *, such as pressure * in aircraft fuselage and tank * in rocket vehicle, must form presure-tight seal.

bulk-injection   In piston engine injection of fuel into induction airflow upstream of distribution to individual cylinders.

bulk loader   Self-drive belt conveyor vehicle for loading bulk cargo.

bulkmeter   Instrument, esp. in refuelling of aircraft, for measuring liquid flow, typically as mass per second or as summing indication of total mass passed. Some * measure volume and require density correction for each liquid handled.

bulk modulus   Elastic modulus of solid under uniform compressive stress over entire surface, as when immersed in fluid under pressure; numerically, stress multiplied by original volume divided by change in volume.

bulk out   To run out of cargo space while still within allowed weight.

bulk petroleum products   Liquid products carried in tankcars or other containers larger than 45 gal (55 US gal).

bullet   *1* Gun-fired projectile intended to strike target, having calibre less than 20 mm (0.7874 in).

2 Streamlined fairing having form of quasiconical nose or forepart of body of revolution. If rotating, called spinner.

3 Aluminium or steel peg at top of hot-air-balloon rip line.

bull gear   Largest gear in train, esp. large gear on which aerial of surveillance radar is mounted.

bull session   Informal discussion on serious aviation topics, between engineers and/or aviators.

bull's eye   *1* Circular thimble.

2 Ring used to guide or secure rope.

3 Cockade having concentric rings (colloq.).

bump   *1* See *Gust (1)*.

2 Sensation experienced in flight through gust (1). See *bumps*.

3 To form sheet metal on bumping hammer.

4 Thrust bump.

5 See *bumping*.

6 Confusingly, in view of 5, to upgrade a passenger to a higher class.

bumper bag   Padded or inflated bag beneath lowest point of aerostat to absorb shock of ground impacts.

bumper rocket   Pre-1955, first stage of two-stage launch vehicle.

bumper screen   On spacecraft, protective screen intended to arrest micrometeorites and other macroscopic solids.

bumper wheel   Wheeling machine.

Bumpf   One of earliest of 35 English-language safety mnemonics: brakes, undercarriage, mixture, pitch, fuel/flaps; G added gills/gyros.

bumping   Practice of denying a fully booked and confirmed passenger the right to board an overbooked flight; officially called IBR (involuntary boarding refusal). Bumped pax qualify for DBC.

bumping hammer   Power hammer for bumping.

bump rating   Increased engine thrust rating (beyond GM or average TO) cleared for short periods; fixed-wing equivalent of contingency.

bumps   Repeated uncommanded excursions in the vertical plane caused by atmospheric turbulence. Term particularly applies to passage through gusts (2) of exceptional severity.

Buna   A synthetic rubber mass-produced in G, WW2 (tradename).

bunching   *1* In traffic control, tendency of vehicles (esp. aircraft) to reduce linear separation, esp. to dangerous degree.

2 In klystron, separation of steady electron stream into concentrated bunches to generate required very high frequency in oscillatory circuit.

bund   *1* Earth bank constructed at airfield to reduce environmental noise.

2 Fuel/ammo dump sufficient for one day's operations from a STOVL hide, replenished daily from Logspark (RAF).

bungee   Elastic cord comprising multiple strands of rubber encased in braided (usually cotton) sheath.

bunny suit   Electrically heated [usually blue] suit for high-altitude unpressurized aircraft (US, colloq.).

bunt   *1* Severe negative-g manoeuvre comprising first half of outside loop followed by half roll or second half of inside loop.

2 In surface-attack missile trajectory, negative-g pushover from climb to dive in terminal phase near target.

BuOrd   Bureau of Ordnance (USN)

buoyancy   Upthrust due to the displaced surrounding fluid just sufficient to support a mass. Thus, in aerostat, condition in which aircraft mass equals mass of displaced air. In marine aircraft at rest, mass of aircraft equals mass of water displaced (in this case, as with all aerodynes, displaced air mass is usually ignored).

buoyant spacecraft   Spacecraft designed to operate as aerostat in planetary atmosphere.

Buoy communication system   VHF links via satellite from buoys to improve com. over Gulf of Mexico (FAA).

BUP   Block upgrade programme.

burble   *1* Turbulent eddy in fluid flow, esp. in proximity to, or caused by, a bounding surface.

2 Brakedown of unseparated flow (not necessarily with laminar boundary layer) across aircraft surface, esp. across top of wing. First region of separated flow due to

excessive angle of attack or to formation of shockwaves at M_{crit}.

burble point   *1* Angle of attack at which wing first suffers sudden separation of flow.

2 Mach number at which subsonic wing first suffers sudden separation of flow due to shockwave formation.

Bureau Veritas   International organisation for certifying companies [eg, to ISO 9000] and surveying and underwriting vessels, including aircraft.

buried engine   Engine contained within airframe, esp. without causing significant protuberance. Normally applied to jet engine inside wing root.

burn   *1* Operation of rocket engine, esp. programmed operation for scheduled time. Thus, first *, second *.

2 Operation of main flame in burner of hot-air aerostat, Thus, a 20-second *.

3 Authorized destruction of classified material, by whatever means.

burner   *1* In gas turbine, device for mixing fuel or fuel vapour with swirling primary airflow with minimal axial velocity to sustain stable combustion; generally synonymous with fuel nozzle, fuel injector.

2 Afterburner (colloq., R/T).

3 Incorrectly, though common in US, gas-turbine combustion chamber.

4 In liquid or hybrid rocket, device for injecting and/or mixing liquid propellants to sustain primary combustion; more usually called injector.

5 Stainless-steel vaporizing coil and jet of hot-air balloon.

burner can   Combustion chamber in engine of can-annular type [UK = flame tube].

burn in   *1* To enter data in EDP core store so that it will subsequently resist nuclear explosion effects and other hostile action.

2 To operate avionic and other electronic equipment under severe overload conditions to stabilize it before operational service and reduce incidence of faults.

burnish   To smooth and polish metal surface by rubbing (usually with lubricant) with convex surface of harder metal.

burn-off   In aeroplane (esp. commercial transport) operation, fuel burned between takeoff and critical position for establishing terrain clearance.

burnout   *1* Termination of rocket operation as result of exhaustion of propellants. Also called all-burnt.

2 Mechanical failure of part subject to high temperature as result of gross overheating, esp. rocket case or chamber.

burnout plug   In certain rocket motors, esp. storable liquid and hybrid types, combustible plug which, when ignited, releases and fires liquid propellant.

burnout velocity   Vehicle speed at burnout (theoretically, highest attainable for given trajectory).

burnout weight   Mass of vehicle at burnout, including unusable fuel.

burn rate   In solid rocket, linear velocity of combustion measured (usually in millimetres per second) normal to burning surface. Symbol r.

burn-rate constant   Factor applied to ** calculations dependent upon initial grain temperature.

burn-rate exponent   Pressure exponent n in burn rate law $r = aP_c^n$.

burn table   Refractory surface in tail of plume target on which jet fuel is burnt.

burn-through   Operation of radar in face of jamming and similar ECM, esp. by virtue of high transmission power to overcome interference.

burn-through range   Limit of range at which *-* operation can yield useful information, if necessary in emergency short-life operation at abnormally high power.

burn time   Duration in seconds of rocket motor burn. Burn time starts when chamber pressure has risen to 10% of maximum (or averaged maximum during level portion of thrust curve) and ends when pressure drops to 75%. Alternative criterion is to draw tangents to level portion and descending portion of thrust curve and measure time to point at which curve is cut by bisector of angle between tangents. Symbol t_b.

burn-time average chamber pressure   Integral of chamber pressure versus time taken over burn time interval divided by burn time. Symbol $\bar{P}_c$.

burn-time average thrust   Integral of thrust versus time taken over burn time interval divided by burn time. Symbol $\bar{F}_b$.

burnup   *1* On entry to atmosphere from space, partial or complete destruction due to kinetic heating.

2 In nuclear reactor, esp. thermal fission reactor, percentage of available fissile atoms that have undergone fission.

Burro   Broad-area unmanned responsive resupply operations.

burst   In colour TV, transmission of small number of cycles of chroma sub-carrier in back-porch period (in military systems this signal not always used).

2 One round (payload) of ECM dispensed from attacking vehicle; can be active jammer or flare.

3 Period of fire by automatic gun[s].

4 See *turbulent* *.

burst controller   *Burst limiter.*

burst diaphragm   Diaphragm sealing fluid system and designed to rupture either upon command or at predetermined dP.

burst height   Height at which nuclear weapon is programmed to detonate.

burst limiter   Preset control of number of rounds to be fired in burst (3).

burst order   Detonation of missile warhead by command.

BUS   *1* Break-up system (of unmanned vehicle upon command).

2 Backscatter ultraviolet spectrometer.

bus   *1* Spacecraft carrier vehicle for various payloads.

2 In EDP, main route for power or data. Alternatively, trunk.

3 Busbar.

4 In ICBM, carrier vehicle for MIRV payloads.

5 In any delivery system, carrier vehicle for multiple warheads or submunitions.

6 A particular aeroplane (affectionate usage, UK c 1910–30).

busbar   In elecrical system, main conductor linking all generators and/or batteries and distributing power to operative branches.

Busemann biplane   Aeroplane, so far not built, in which

at supersonic speed shockwaves and flows around upper and lower wings would react favourably.

Busemann theory   First theory for two-dimensional supersonic wing to take into account second-order terms (1935).

bush   Open-ended drum tailored to fit inside hole, eg to reduce its diameter, act as shaft bearing or serve as electrical insulation (see *grommet*).

bush aircraft   Aerodyne tailored to utility service in remote (eg Canadian Arctic) regions.

bushie   Bush pilot (Australia).

bush pilot   Operator (often also owner) of bush aircraft, usually freelance jobbing professional.

business aircraft   Aerodyne tailored to needs of business management and executives of government and other organisations.

Business Class   Airline passenger category between Tourist and First; no universal definition of seat pitch or services provided.

bust   A failure to comply with instruction to fly at assigned FL.

buster   R/T command [usually said at least twice] "Fly at maximum continuous power", normally to effect interception.

butterfly   Distorted figure-8 pattern, looking like a butterfly, flown by orbiting combat aircraft on weapon-guidance, EW or, rarely, AWACS duty.

butterfly tail   Comprises two oblique (dihedfal in region 25°–45°) fixed stabilizer surfaces each carrying a hinged surface, the latter operating in unison as elevators or in opposition as rudders; also called V tail.

butterfly valve   Fluid-flow valve in form of pivoted plate, usually having circular form to close a pipe.

butt joint   Sheet joint with edge-to-edge contact without overlap, with jointing strip along either or both sides.

buttock lines   Profiles of intersection of longtitudinal vertical planes with surface of solid bodies, esp. aircraft fuselages and marine floats. Zero ** is that on axis of symmetry. Used in lofting, these lines do not correspond with structural members.

button   Extreme downwind end of usable runway (colloq.).

buttonhead rivet   Rivet with approximately hemispherical head; used where tensile load may be high.

butt rib   Compression rib at joint between outer and inner wing, or wing and fuselage.

butts   Facility for testing aircraft guns [rare, singular].

BUV   Backscatter ultraviolet.

BuWeps   Bureau of Weapons, combined BuAer and BuOrd in 1959 (USN).

Buys Ballot's law   Professor Buys Ballot postulated that an observer with back to wind in N hemisphere has lower pressure to left (in S hemisphere, to right). True for any isobar pattern.

buy the farm   To be killed in a crash, not excluding military action (colloq.).

buzz   *1* Oscillation of skin or other structure at frequency high enough to sound as a note.

2 Oscillation of control surface at high frequency.

3 Loosely, any single-direction-of-freedom vibration at audible frequency.

4 Wake-interaction noise generated by turbomachinery, esp. large fans, at 900–4,000 Hz.

5 High-frequency, often violent, pulsation of airflow at supersonic air-breathing engine inlet.

6 To fly aircraft, esp. one of high performance and manoeuvrability, in way designed to harass another aircraft or ground target. Transitive, thus "to * the control tower".

7 Collective noun for micros.

buzz liner   Sound-absorbent liner to fan duct or other surface bounding wake-interaction noise.

buzz number   Extra-large individual aircraft number (can be unit number or aircraft serial), readable from a distance.

buzz-saw noise   Buzz (3) from shock system of fan with supersonic flow over blades, composed of discrete tones at multiples of N_1.

BV   *1 Bureau Veritas.*

2 Bleed valve.

3 Boost vehicle.

4 Present visibility.

BVA   Bleed-valve actuator.

BVCU   Bleed-valve control unit.

BVD   Battlespace visualization display.

BVI   Blade/vortex interaction (helo).

BVID   Barely visible impact damage.

BVIS   Baggage vector interface server, provides routing system for host systems of all carriers at one airport.

BVOR/BVortac   Weather broadcast VOR or Vortac.

BVQI   Bureau Veritas Quality International.

BVR   *1* Beyond visual range.

2 Best-value rate.

BVTRU   Bleed-valve transient reset unit (controls BVs during transients).

BVU   On-board computer (R).

BW   *1* Biological warfare [or weapons].

2 Bomb (or bombardment). Wing (USAAC, USAAF, USAF).

3 Bandwidth; also B/W.

BWA   Blast-wave accelerator.

BWAN, B-Wan   Back-up WAN.

BWB   *1* Blended wing/body.

2 Bundesamt für Wehrtechnik und Beschaftung [MoD procurement office] (G).

BWC   Biological and Toxin Weapons Convention, 1972.

BWER   Bounded, or boundary of weak echo region [of thunderstorms].

BWFT   Ministry of research and technology (G).

BWO   Backward-wave oscillator.

BWPA   British Women Pilots' Association (1955–).

BWRA   British Welding Research Association.

BX   Base Exchange, today AAFES.

BXA   Bureau of Export Administration (US).

BY   Blowing spray.

BYD   Beyond.

BY$ Base-year dollars.

BYG   Blue/yellow/green.

Bygrave   Slide rule for solving vector triangles, esp. from sextant readings (obs.).

bypass   *1* Capacitor connected in shunt to provide low-impedance alternative path.

2 Alternative flow path for fluid system.

bypass duct   Annular space surrounding engine core through which bypass air flows; in modern turbofans usually called fan duct. May be short, or extended back to a mixer.

bypass engine   Air-breathing jet engine in which air admitted at inlet may take either of two flow paths (see *bypass turbojet*).

bypass ratio   In bypass turbojet or turbofan, numerical ratio of mass flow in bypass duct divided by that through core, ie cold jet divided by hot. Some have defined as total mass flow divided by core mass flow; this is incorrect, and would always yield numbers greater than 1. BPR is normally measured at TO power at S/L.

bypass turbojet   Turbojet in which mass flow through LP compressor stages is slightly greater than that through HP stages, excess being discharged along bypass duct. Also called leaky turbojet. In principle difference between this and turbofan is purely of degree; turbofan has much higher bypass ratio (greater than 1) and probably at least two shafts. In general subsonic engines may be considered turbofans and supersonic engines bypass turbojets.

byte   *1* Group of bits normally processed as unit.

2 Sequence of consecutive bits forming an EDP word, thus an 8-bit *, which gives $2^8 = 256$ possible combinations.

BZ   Benactyzine.

C

C *1* Degrees Celsius.

2 Coulomb[s].

3 Yawing moment of inertia.

4 Compass heading/bearing/course.

5 Capacitance, capacitor, capacity (electrical).

6 Ceiling, or bottom of cloud layer.

7 Thermal conductance.

8 Any constant.

9 Aggregate fuel consumption.

10 Carrier-wave power in watts.

11 Basic mission, cargo (USAS, USAAC, USAAF, USAF since 1925, USN since 1962, UK mission prefix since 1941).

12 JETDS code: air-transportable, carrier-wave, common use.

13 Fighter category (F).

14 Prefix: ground service connection (BSI).

15 Viscous-damping coefficient.

16 Council (ICAO).

17 Customs available.

18 Clear, clears, clearance delivery (ATC).

19 Cell of storm.

20 Continental (air mass).

21 Circling landing minimum.

22 Dirigible class (USN 1914–16).

23 Sport-parachuting certificate; 50+ jumps, 20 landing ≤20m of target.

24 Heat capacity per mole.

25 Chemical concentration.

c *1* Chord.

2 Speed of light in vacuum, $= 2.997925 \times 10^8 \text{ms}^{-1}$.

3 Collector of semiconductor device.

4 Prefix, centi (10^{-2}, non-SI).

5 Compass.

6 Prefix, circa, = approximate.

7 Subscript, convective, convection.

8 Specific heat.

c' Thrust specific fuel consumption.

$\bar{\mathbf{c}}$ Geometric mean chord; sometimes $\bar{C}$.

$\bar{\bar{\mathbf{c}}}$ Aerodynamic mean chord; sometimes $\bar{\bar{C}}$.

(c) Astronomical Unit, see AU.

C* *1* Characteristic exhaust velocity of a rocket.

2 Weighted linear combination of pilot's pitch-control input to aircraft pitch-rate and normal acceleration.

C0 to C9 See *cloud types.*

C1, C2 Avion de chasse [fighter] with 1 or 2 seats (F).

C1A Cr/Al oxidation/oxysulphuration coating.

C² *1* Also called C-squared = command and control; I adds interface or intelligence, IPS information-processing subsystem, ISR intelligence, surveillance and reconnaissance [to which C adds center], IT interoperability trial, MC mobile-capable, P processing, S status, SIM simulation, SS or S² switching system, V vehicle, and W warfare.

2 Camouflage and concealment, thus C2D (usually not C²D) is camouflage, concealment and deception.

C³ *1* Also called C-cubed = command, control and communications; CM adds countermeasures, I intelligence, and ISRSS intelligence, surveillance, reconnaissance and space systems.

2 Crash-crew chart[s], detailed airfield plan[s] carried on RIV.

3 Coated carbon/carbon.

C³D, C3D Cross-cockpit collimated display.

C4 Plastic explosive based on RDX/PETN.

C⁴ Command, control, communications and computers; addition of I and ISR as above, C4ISR previously being called ACN(3); IFTW adds information for the Warrior.

C-band EM frequencies 3.9–6.2 GHz, now covered by Bands S and X (see *Appendix 2*)

C-certificate Highest category for glider pilot.

C-channel C = circuit-mode, provides full duplex, voice 9.6 kbit/s, data 10.5, assigned in pairs uplink/downlink.

C-check A-check plus thorough inspection of structure, removing fairings where necessary, plus test of systems.

C-clamp Headset.

C-class Controlled airspace near busy airport, usually a radar service area.

C-code IFR flight-plan suffix: no-code transponder and approved area navigation.

C-cycle One complete flight simulated in engine development or test.

C-display Rectangular display in which horizontal axis is target bearing and vertical axis is its angle of elevation.

C-duct Half a fan duct forming part of an engine pod cowl, usually pivoted at the top for access to the core.

C-licence Permits ground engineer to inspect and rectify engines.

C-Lite Small polycarbonate fin on wingtip for guidance on crowded airfields and showing if wingtip lights are illuminated.

C-mode Transponder transmits altitude.

C power supply Between cathode and grid, for grid bias.

C-scope C-display.

C-spar Structural member along helicopter rotor blade between D-nose and main spar or I-beam, closed at front, open at rear.

C-stoff Rocket propellant (fuel+coolant), hydrazine hydrate plus methyl alcohol, often plus water, usual percentages 30/57/13 (G).

C-wing Blended wing/body.

CA *1* Controller Aircraft; holder of this office is also Deputy Chief of Defence Procurement (UK).

2 Controlled airspace.

3 Cabin attendant[s] (airline costings).

4 Conversion angle.

5 Circuit analog.

6 Cruiser, gun armed [can have SAM secondary armament] (USN).

7 Cetyl alcohol, a lubricant.

8 Control advises.

9 Conflict alert.

C/A *1* Coarse acquisition (GPS).

2 Course acquisition.

CAA *1* Civil Aviation Authority (UK, from 1972, said to mean 'campaign against aviation').

2 Civil Aeronautics Authority (US, 1938–40).

3 Civil Aeronautics Administration (US, 1940–58, part of Department of Commerce).

4 Civil Aviation Administration (Israel).

5 Conformal-array antenna, or aerial.

6 Chromic-acid anodizing.

7 Cargo Airline Association (US).

8 Component application architecture.

CAAA *1* CAA(1) Approved.

2 Commuter Airlines Association of America, also called C3A or C-triple-A, and now the RAA.

CAAC Civil Aviation Administration of China, from 1964.

CAACU Civilian Anti-Aircraft Co-operation Unit (UK, various dates 1950–71).

CAADRP Civil Aircraft Airworthiness Data Requirements [originally Recording] Programme (UK).

CAAFI CAA(1) of the Fiji Islands.

CAAFU CAA(1) Flying Unit, originally for navaid calibration, now examines candidates for licences.

CAAG CFIT/ALAR Action Group (FSF).

CAARC Commonwealth Advisory Aeronautical Research Council (Int., office in London).

CAARP Cooperatives des Ateliers Aéronautiques des la Région Parisienne (formed 1965).

CAAS *1* Computer-assisted approach sequencing.

2 Common avionics-architecture system.

3 Civil Aviation Authority of Singapore.

CAASA Commercial Aviation Association of Southern Africa.

CAASD Center for Advanced Aviation System Development (Mitre Corpn.).

CAASP Common avionics architecture system program (USSOF).

CAASS Computer-aided aircrew scheduling system (USAF).

CAATER Co-ordinated access to aircraft for transitional environmental research (EU).

CAATS *1* Computer-assisted aircraft trouble-shooting.

2 Canadian automated ATC system.

3 Computer-aided aerodrome training suites.

CAAV CAA of Vietnam.

CAAVTS Compact airborne automatic video tracking system.

CAAZ CAA of Zimbabwe.

CAB *1* Civil Aeronautics Board (US, 1940–84).

2 Common avionics baseline.

cab *1* Structure containing cockpit of large aircraft, often excluding nose.

2 Airport tower, especially workplace of controllers.

CABA Consolidated agile-beam antenna.

cabane Structure of braced struts used to carry load above fuselage or wing, such as upper wing of biplane or parasol wing, engine nacelle or bracing wires to wingtips.

cabin *1* Enclosure for aircraft occupants; today excludes cockpit or flight deck.

2 Occupied part of simulator, with or without motion.

cabin altitude Altitude corresponding to pressure inside cabin.

cabin blower In some pressurized aircraft, shaft-driven cabin blower, also called cabin supercharger.

CabinCall First certificated system allowing use of mobile phone, initially on business aircraft.

cabin crew Staff who attend to passengers in flight.

cabin distribution system Links CTV with passenger and crew telephone and data terminals.

cabin file server Links CMU to ISVSs and POSTs.

cabin fog Caused when cold dry input hits warmer humid cabin air.

cabin pressure Ambiguous; can mean absolute pressure inside aircraft or pressure differential (dP) between cabin and surrounding atmosphere.

cabin supercharger See *cabin blower*.

cabin telecommunications unit A complex PBX for satcoms, to Arinc 746.

cable *1* Non-SI unit of length, = 2.19456×10^{-2}m.

2 Traditional term for filament for winch or other surface launch of glider, irrespective of material.

cable cutter One-shot guillotine, powered by cartridge, on wing leading edge to cut barrage cables.

cable-drag drop Low-level airdrop with load extracted and arrested by ground cable installation.

cable hover Design requirement for ASW helicopter autopilot while dunking.

cable-strike protection See *wire-strike protection system*.

cabotage Freedom of air transport operator to pick up or set down traffic in (usually foreign) country for hire or reward.

cabrage Rotation of aeroplane about lateral axis to increase angle of attack (obs.).

cab-rank patrol Close air-support technique in which instead of striking designated targets or targets of opportunity, aircraft loiter awaiting assignments from surface forces.

cabs, CABS Cockpit air-bag system [USA/Simula].

CAC *1* Combat Air Command.

2 Combat-assessment capability.

3 Computer acceleration control.

4 Centralised approach control.

5 Caution advisory computer.

6 Ciurse acquisition code.

CACA National certification authority (Poland).

Cacas Civil Aviation Council of the Arab States.

CAC³, CACCC Combat air command and control center.

C/A code Coarse acquisition code (GPS).

CACP Cabin-air, or area, control panel.

CAC²S Common aviation command and control system (USMC).

CACTCS Cabin-air conditioning and temperature control system.

CACU Coast Artillery Co-operation Unit (RAF c1926–).

CAD Computer-assisted design [/CAM adds computer-assisted manufacturing, D adds drafting].

2 Cushion-augmentation device [US = LID].

3 Cartridge-activated device.

4 Close-in air defence.

5 Computer-aided dispatch.

6 Component advanced development.

7 Civil Aviation Department (Hong Kong).

8 Chemical agent defeat.

Cadal Communications automation and data-link.

CAD$ Canadian dollars.

CADE Computer-aided design evaluation.

CADEA Confederación Argentina de Entidades Aerodeportivas.

cadence braking   Rapidly repeated jabs on pedals or other input.

cadensicon   Measures fuel density and permittivity as it enters aircraft tank.

Cades, CADES   Computer-aided design and evaluation system.

Cadets   Computer-assisted documentation education tutorial system.

CADF   *1* Commutated-aerial (or antenna) direction-finder, or finding.

2 China Aviation Development Foundation (Taiwan).

Cadin   Czech aeronautical data-interchange network.

Cadiz   Canadian air-defense identification zone.

CADM   Clustered airfield defeat, or dispensed, munition(s).

Cadmat   Computer-assisted, or augmented, design, manufacture and test.

cadmium   Symbol Cd, soft white metal, density 8.7, MPt 321°C, major uses electroplating, NiCd batteries, CdS IR detectors and fusible alloys.

CADO   Central Air Documents Office (USA).

CADP   Central annunciator display panel.

CADRE, Cadre   *1* Communications-actuated data-retrieval equipment.

2 Center for, now College of, Aerospace Doctrine, Research and Education (Air University, Maxwell AFB).

Cadre   Adjective: formed from regular and reserve personnel (obs. in UK).

Cads, CADS   *1* Cushion-augmentation device[s] to increase jet lift near ground.

2 Concept and design study/studies.

3 Computer-aided debriefing system.

4 Controlled aerial delivery system [without aircraft having to land].

CADWS   Close air-defence weapon system[s].

CAE   *1* Computer-aided, or assisted, engineering.

2 Component-application engineer.

3 Control-area extension.

CAé   Commission d'Aérologie (WMO).

CAEDM   Community/airport economic-development model.

CAEE   Committee on Aircraft Engine Emissions (ICAO).

CAEM   Cargo-airline evaluation model.

CAeM   Commission for Aeronautical Meteorology (WMO).

CAEP   Committee on Aviation Environmental Protection, or Protocol (ICAO).

Caepe, CAEPE   Centre d'Achèvement et d'Essais de Propulseurs d'Engins (F).

CAER   See *EARC.*

Caerat, CAERAT   Common American/European Reference Aeronautical Telecommunications [F adds facility, NF network facility].

Caesar, CAESAR   *1* Component and engine structural assessment and research.

2 Coalition aerial surveillance and reconnaissance (NATO).

caesium   In N America cesium, symbol Cs, gold-colour soft metal, density 1.9, MPt 28°C, used in glasses (but the hydroxide dissolves glass), highly reactive and toxic.

CAF   *1* Canadian Armed Forces.

2 Citizen Air Force (South Africa).

3 Confederate Air Force (US, from 1961, now called

Commemmorative AF).

4 Cleared as filed.

CAFAC   CAF(1) Air Command.

Cafac   Commission Africaine de l'Aviation Civile (Int.).

CAFATC   Canadian Air Transport Command.

CAFD   Collection, analysis, fusion and dissemination.

Cafda   Commandement Air des Forces de Défense Aérienne (F).

CAFH   Cumulative airframe flight hours.

CAFI   Commander's annual facilities inspection.

CAFMS   Computer-assisted force management system.

CAFT   Combined advanced field team (evaluate new captured hardware).

C/Aft   CNS/ATM focus team.

CAFU   Civil Aviation Flying Unit (UK).

CAG   *1* Carrier air group (USN).

2 Civil Air Guard (UK, 1937–39).

3 Circulation aérienne générale (F).

CAGE   *1* Commercial and governmental entity.

2 Commercial avionics GPS engine.

cage   *1* To orientate and lock gyro into fixed position relative to its case.

2 Housing for bearing balls/rollers/needles.

caged switch   Protected against inadvertent operation by spring-loaded hinged box.

CAGR   Compound average, or annual, growth rate.

CAGS   Central attention-getting system.

CAH   Cabin-attendant handset.

CAHI   Central Aerodynamics and Hydrodynamics Institute (Moscow, founded 1 December 1918).

CAHS   Canadian Aviation Historical Society.

CAI   *1* Civil Aeromedical Institute (FAA).

2 Computer-aided instruction (see *CMI*).

3 Close approach indicator (STOVL carrier landing).

4 Czech Astronomical Institute.

5 Caution annunciator/indicator.

6 Component analysis and integration.

7 Composites affordability initiative (USAF).

CAIG   Cost analysis improvement group (DoD).

CAIMS   Central aircraft information management, or maintenance, system.

Cains   Carrier aircraft (since 1982, also aligned) inertial navigation system.

CAIP   Civil aircraft inspection procedure[s].

CAIR   Confidential aviation incident reporting [P adds programme].

CAIRA   See *IAARC.*

CAIS   *1* Common Ada [or APSE] interface set.

2 Common airborne instrument, or instrumentation, system.

CAIV   Cost as an independent variable [each decision taken on basis f (cost of program)].

cAk   Continental Arctic air mass, very cold.

cal   Calorie.

CALCM   Conventional [i.e., not nuclear] air-launched cruise missile.

calculated altitude   Celestial altitude calculated but not observed.

Calda   Canadian Airline Dispatchers Association.

Cale gear   Shock-absorbing system in carrier arrester wire anchors.

CALF, calf   Common affordable lightweight fighter.

Calfab   Computer-aided layout and fabrication.

Calfax Patented quick-release panel fastener, latch opened or closed by 540° rotation.

calf-garters Automatic leg-restraint straps in certain ejection seats.

caliber US unit of length, = 10^{-2}in.

calibrated airspeed IAS corrected for ASI system errors; 'true indicated airspeed', but see *airspeed*.

calibrated altitude Not normally used, but signifies pressure altitude or radar height corrected for instrument errors.

calibrated club propeller Club propeller whose drive torque has been measured and plotted against rpm; thus, can serve as dynamometer.

calibrated focal length Equivalent focal length adjusted to equalise positive and negative distortion over view field.

calibration card Graphical or tabular plot of instrument errors (other than compass); usually displayed near instrument.

calibration test Static run of bipropellant rocket engine to check propellant mixture ratio and performance.

calibrator Device for measuring instrument errors.

calibre Bore (ID) of tube, esp. diameter of largest cylinder that fits inside (thus, in rifled barrel, touches highest points of opposing lands).

caliper Instrument for measuring or checking thickness, diameter or gap; with internal or external measuring points on tips of pivoted or sliding arms.

calipher Caliper.

Callback Confidential reporting system to attempt to record civil (especially air carrier) incidents caused by human failures (FAA via NASA).

call fire Fire against specific target delivered as requested.

call for fire Request, by FAC or other observer, for fire on specific target and containing target data.

calling See next.

calling out Spoken data readout by crew member or ground observer to assist pilot or other crew member; thus, co-pilot's speed/altitude checks on instrument approach.

call mission CAS (3) mission at short notice by pre-briefed pilot with pre-armed aircraft, target assigned after take-off.

call number In EDP, number code identifying sub-routine and containing data relevant to it.

callout notes On engineering drawing, written notification of special features (eg material, process, tolerance or equipment installation).

Calls Computer-aided language learning system.

callsign, call sign Pronounceable word(s), sometimes with suffix number, serving to identify a communications station (such as an aircraft). Civil aircraft ** are ICAO phonetic letters and numbers derived from international registration; ground station ** are name of airport followed by type of station (tower, departure, clearance delivery, etc).

calm No sensible wind.

CALNS, Calns Common air-launched navigation system.

calorie Unit of quantity of heat, contrary to SI; International *, cal_{IT} = 4.1868 J by definition, 15°. * = 4.1855 J, thermochemical * = 4.184 J.

calorific value Quantity of heat released by burning unit mass of fuel; kJ/kg = 0.429923 Btu/lb; kJ/m^3 = 0.200784 Btu/Imp. gal.

Calorizing Heating steel part surrounded by aluminium (liquid or granules); gives protection in high-temperature use.

Calow Contingency and limited objective warfare.

Calpa Canadian Air Line Pilots' Association.

Calrod Electric heater [many types] fitting inside shafts of FCS or other mechanisms.

Cals, CALS *1* Computer-aided logistics support.
2 Computer-aided acquisition and logistic [or lifetime] support.
3 Continuous acquisition and life-cycle support.
4 Carrier aircraft-landing system.

Calsel Proposed *Selcal* modification in which signal is combined with a gating tone to produce automatic receiver function.

CALT China Academy of Launch-vehicle Technology, Beijing.

Caltech California Institute of Technology, Pasadena, name since 1920, founded 1891 as Throop Polytechnic Institute.

CalVer Calibration verification.

Calvert lighting Original system of crossbar approach lighting.

CAM *1* Cockpit angle measure (flight deck vision limitations expressed as angles).
2 Catapult armed merchantman (UK ships, 1941–43).
3 Chemical-agent munition (or monitor).
4 Circulation aérienne militaire (F).
5 Computer-assisted manufacture, or computer-aided manufacturing.
6 Conventional attack missile.
7 Content-addressable memory.
8 Counter-air missile.
9 Centre of Aviation Medicine (RAF).
10 Commercial Air Mail routes (US, from 1926).
11 Cockpit audio monitoring.
12 Cabin assignment module (CIDS, later FAP).
13 Civil Aeronautics Manual (US).
14 Control-actuator mechanism.

cam Rotating or oscillating member having profiled surface to impart linear motion to second member in contact with it.

CAMA Civil Aviation Medical Association (US, office Oklahoma City).

camber *1* Generally, curvature of surface in airflow.
2 Curvature of aerofoil section, measured along centre-line or upper or lower surface, positive when centreline is arched in direction of lift force (see *upper* *, *lower* *, *centre-line* *, *conical* *, *reflex* *, *mean* *).
3 Centreline of aerofoil.
4 Inclination of landing wheels away from vertical plane.

cambered Krüger Krüger having flexible profile to increase camber when open.

cambered wing Wing section whose centreline is not coincident with chord.

CAMBS Command active multi-beam sonobuoy.

Camden Co-operative air and missile defense exercise network.

CAMDS Chemical agent munitions disposal system.

CAMEA Canadian Aircraft Maintenance Engineers' Association.

Camel   Cartridge-activated miniature electromagnetic

Camera axis   Perpendicular to film plane through optical centre of lens system.

camera gun   Camera, usually colour ciné, aimed at target with aircraft gun and operated by gun-firing circuit; used to provide combat confirmation, intelligence information and, with unloaded gun(s), as training aid.

camera obscura   Dark room equipped with lens projecting image of external scene on to wall or floor (formerly used as bombing target with roof lens for recording of bomb release position).

camera recorder   One or more cameras arranged to provide continuous film of instrument panel or similar data source.

camera tube   TV converter of optical scan into electrical video signals (Orthicon etc).

CAMF   Christian Airmen's Missionary Fellowship, became MAF).

camfax   Camera facsimile, especially for synoptic charts.

cam follower   Driven member in sliding or rolling contact with cam.

CAMI   *1* Computer-aided (or -assisted) manufacturing and inspection.

2 Civil Aeromedical Institute (previously CARI, renamed 1965).

CAML   Cargo-aircraft minelayer.

cam lobe   Profiled projection from straight or circular baseline.

Camloc handle   Patented self-tightening latch for cowlings and skin access panels.

CAMMS   Co-operative aggregate mission-management system, for multiple UAVs.

CAMM2   Computer-aided maintenance-management system, Version 2.

Camos   Computer-aided meteorological observing system.

camouflage   Attempt to change appearance to mislead enemy, esp. by concealment with portable material or painting to reduce visual contrast with background.

camouflage detection photography   Use of film whose spectral response differs from that of human eye (eg IR-sensitive).

campaign fire   An enormous forest or other fire calling for the assembly of large resources.

Campbell diagram   Plot of natural frequencies against rpm for rotating part.

cam ring   Ring inside crankcase of radial piston engine geared to crankshaft and having sequence of lobes to operate inlet and exhaust valves of all cylinders in that row.

CAMS   *1* Combat aviation management system (USA).

2 Control and monitoring system.

camshaft   Shaft equipped with cams aligned with valve gear of cylinders of in-line piston engine.

Camsim   Canadian airspace management simulator.

CAMU   Central avionics management unit (databus).

CAN   Committee on Aircraft Noise (ICAO).

can   *1* Individual flame tube of can-annular combustion chamber.

2 Complete combustion chamber of multi-combustor engine.

3 Five-sided box projecting into integral wing tank to accommodate slat track.

4 Controlled-environment weapon container.

Canadian break   Max-rate 360° turn.

can-annular   See *cannular*.

canard   *1* Tail-first aerodyne, usually with auxiliary horizontal surface at front (foreplane) but vertical surface (fin, rudder) at rear.

2 Foreplane or nose yaw control fitted to * (1).

cancel   *1* To terminate complete R&D or hardware programme.

2 To countermand order.

3 To deactivate activity (eg, * reverse thrust).

C&C   Command and control.

C&DH   Communications and data-handling.

candela   SI unit of luminous intensity 1/683 W/sr [mono-chromatic at 5.4×10^{14} Hz], abb. cd; [in US sometimes called candle].

candle   *1* Former unit of luminous intensity (based on Harcourt pentane lamp).

2 Of parachute canopy, to become so constrained by rigging lines as to fail to deploy.

candlepower   Former measure of rate of emission of light by source, usually in given direction, abb. cp; see *candela*, *luminous flux/intensity*.

C&M   Care and maintenance.

C&W   Control, or caution, and warning.

CANES   Codes addresses numériquement à nombre d'entrées sélectionnable, = PCM (F).

CAN 5   Committee on Aircraft Noise, rules for new aircraft types (ICAO).

canister   Preferred term for tubular TEL(4) for SAMs.

canned cycle   Complete routine for particular com-puterised process, such as drilling and reaming a hole in NC machining.

cannibalise   To dismantle aircraft or other hardware to provide spare parts.

cannibalisation rate   Actual number of item in given period.

canning   Loading gun ammunition into magazine tank, box or drum.

cannon   Generally, gun of calibre 20 mm or greater.

Cannon plug   Vast range of electrical connectors (trade name, ITT Industries).

cannular   Annular combustion chamber containing separate flame tubes, each of which may have a ring of burners.

canoe radar   Aircraft radar whose radome has canoe-like shape.

canonical time unit   Time required for hypothetical satel-lite in geocentric equatorial orbit, with centre of satellite coincident with surface of Earth, to move distance subtending 1 radian at centre: 13.447 minutes.

canopy   *1* Fairing, usually transparent, over flight crew or, in lightplane, all occupants, which does not form part of airframe and slides or pivots for entry and exit.

2 Rarely, transparent fairing over flight crew which does form part of airframe and is not used for entry/exit.

3 Main deployable body of parachute.

4 Another name for the envelope of a balloon.

CANP   *1* Collision-avoidance notification procedure.

2 Civil air [or aircraft] notification procedure, tells mili-tary of low-level civil activity (UK).

CANS   Civil air navigation school.

Canso   Civil Air-Navigation Services Organization (Int.).

cant angle   *1* Angle between centreline of winglet and local vertical [OZ axis], seen from head-on.

  2 Angle between biplane interplane strut and local vertical, seen from head-on.

canted deck   Angled deck.

canted nozzle   Nozzle of jet engine, usually turbojet or rocket, whose axis is fixed and not parallel to centreline of engine or motor or to line of flight, but passes close to vehicle c.g.

cantilever   Structural member, such as beam, rigidly attached at one end only. Thus * wing is monoplane without external struts or bracing wires.

cantilever ratio   Semi-span divided by maximum root depth.

Canukus   Standards agreed by Canada, Australia, New Zealand, UK, US.

CAO   Combined Air Operations [C adds Center, CC Command Centre, CS Center for Space.

C-X   Centre-Experimental (USAF).

CAP   *1* Civil Air Patrol (US, from 1941).

  2 Combat air patrol or carrier air patrol.

  3 Continuing Airworthiness Panel, and study group on ** problems (ICAO).

  4 Chloroacetophenone (tear gas, also called CN).

  5 Civil Air Publication (various countries).

  6 Contractor assessment program (US).

  7 Consolidation by atmospheric pressure, an advanced powder metallurgy process (Cyclops).

  8 Combat ammunition production (USAF).

  9 Crew alert[ing] panel.

  10 Control anticipation parameter [which see].

  11 Capacity.

  12 Contact approach.

cap   *1* Extreme nose structure of aerostat (see *bow* *).

  2 See *flight* *.

  3 Various portions of parachute system (see *petal* *, *tear-off* *, *vent* *).

  4 Tension boom in form of flat strip attached along top or bottom edge or spar or around rib.

  5 Upper limit on a proposed budget.

CAPA   *1* Central airborne performance analyser.

  2 Coalition of Airline Pilots' Associations (US).

capability insertion   Improvement.

Capa cartridge   Rocket-propelled explosive bird-scarer.

capacitance   In electrical system, ratio of charge to related change in potential. Basis of fuel measurement system which gives readout of fuel mass irrespective of aircraft attitude.

capacity   In an EDP installation, number of bits storable in all cores, registers and other memories.

capacity payload   Payload limited by volume, number of seats or other factor apart from mass.

capacity safety valve   Device on carrier catapult (dial-to-aircraft weight) to prevent overloading cylinder or strop/tow bridle.

capacity ton-mile, CTM   Unit of work performed by transport aircraft with capacity payload.

Capas, CAPAS   Computer-assisted performance-analysis system.

CAPE, Cape   *1* Convective available potential energy.

  2 Computer-aided parametric engineering.

Cape Canaveral   On E Florida coast (US), at one time called Cape Kennedy; site of KSC.

capillary drilling   ECM technique using nitric acid fed

through glass tube containing platinum cathode; can rapidly etch precision holes as small as 0.19 mm (0.0075 in).

CAPP   Computer-aided programme planning, or planning project.

capped   Subjected to legal upper limit (slot[3] allocation).

capping membrane   Flexible sheet hastily placed over filled bomb crater.

capping strip   See cap (4).

CAPPS, Capps   Computer-assisted passenger pre-screening system. CAPPS-2, or -II is new-generation replacement (TSA).

Capri   Compact all-purpose range instrument, made by RCA.

CAPRS   Community air-passenger reporting system (EC7).

CAPS, Caps   *1* Computer antenna pointing system (Satcom).

  2 Conventional armaments planning system (NATO).

  3 Computer-assisted passenger screening.

  4 Civil-aviation purchasing service.

  5 Civil, or commercial, airliner protection system, against terrorist SAM.

Capsin   Civil-aviation packet-switching integrated network.

Cap/strike   Mission is primary CAP (2), secondary strike; strike ordnance jettisoned to engage in air combat.

capstrip   See cap (4).

capsule   *1* Small hermetically sealed compartment (eg aneroid).

  2 Sealed compartment or container for instrumentation in space or other adverse environment.

  3 Small manned spacecraft.

capt.   Captain.

captain   *1* Very loosely, any PIC.

  2 More correctly, officer in charge of military, naval or commercial aircraft having flight crew numbering more than one; not normally used with two-seat combat aircraft. Usually * is PIC but in RAF in WW2 could have any aircrew trade, in today's MR [e.g. Nimrod] * is Tac Nav, and in AAC helo * is Gunner [missile operator] regardless of rank. Airline * has status of rank with four gold stripes.

captain's bars   Parallel lights under tanker fuselage to assist receiver's station-keeping (USAF).

captain's discretion   Undefinable. Most airline captains are permitted to ignore some rules, such as number of hours on duty, or limitations on flight time.

caption   Small rectangular display bearing name of airborne system or device, visible when lit from behind.

caption panel   Array of 20–60 captions, usually serving as CWP or alerting device giving indication of failure on broad system basis.

captive balloon   Balloon secured by cable to surface object or vehicle.

captive firing   Firing of complete vehicle, normally unmanned rocket, while secured to test stand.

CAPTS   Co-operative area passive tracking system, provides air and surface surveillance with target ident and precise position (with ASDE, ASTA, ATCRBS, TCAS).

capture   *1* In flying aircraft or space vehicle, to control trajectory to acquire and hold given instrument reading.

  2 In flying aircraft, to control trajectory to intercept and then follow external radio beam (eg ILS, radio range).

3 In ATC or air-defence system, to acquire and lock-on to target.

4 In interplanetary (eg Earth-Moon) flight, eventual dominant gravitational pull of destination body.

5 In automatic or self-governing system not always operative (eg yaw damper), limits of aircraft attitude and angular velocity within which its authority is complete.

capture rate Data rate in kbits/s or Mbits/s.

CAR *1* Civil Air Regulations (FAA, US).

2 Civil Airworthiness Regulations (CAA, UK etc).

3 Caribbean (ICAO).

4 Conformal-array radar.

5 Crew/aircraft ratio.

car *1* Nacelle housing crew and/or engines suspended beneath aerostat.

2 Loosely, payload container attached to or within aerostat, esp. airship.

CARA *1* Computer-aided requirements analysis.

2 Combined-altitude radar altimeter.

3 Cargo and rescue aircraft.

Carabas Coherent all-radio band sensing.

carabiner Karabiner.

Carad *1* Civil aerospace R & D.

2 Civil aircraft research and demonstration programme (UK).

CA RAM Circuit analog radar absorbent material.

Carat Cargo agents reservation airway-bill insurance and tracking system.

carat Non SI unit of mass, = 0.2g.

carboblast Lignocellulose blasting abrasive (BGS); used to clean gas path of running gas turbine and made from crushed apricot stones.

carbobronze Copper alloys containing 8 per cent tin (and trace of phosporous).

carbon Possibly most important element in aerospace, primary constituent of hydrocarbons [e.g., fuels], carbon/graphite fibre and even diamond [heat sink], symbol C, density [graphite] 2.3, MPt various to 3,600°C.

carbon/carbon Composite material: pyrolised carbon fibres in pyrolised carbon matrix.

carbon fibre Range of fine fibres pyrolised from various precursor materials (eg PAN) and exhibiting outstanding specific strength and modulus. Used as reinforcement in CFRP.

carbonising, carbonitriding See *Nitriding*.

carbon microphone Contains packed carbon granules whose resistance, and hence output signal, is modulated by variable pressure from vibration of sound diaphragm.

carbon oxides Gases produced upon combustion of carbon-containing fuel: principally carbon monoxide CO, carbon dioxide CO_2 and wide range of valence bond variations. CO_2 present in Earth atmosphere (3 parts in 10^4); percentage much higher in exhaled breath.

carbon seal Sliding seal between moving machinery (eg turbine disc) and fixed structure; oil removes heat.

carbureter See *carburettor*.

carburetion Mixing of liquid fuel with air to form optimum mixture for combustion.

carburettor Device for continuously supplying engine, esp. Otto-cycle piston engine, with optimum combustible mixture. Many forms exist, some with choke tube and others injecting liquid fuel direct into cylinders (in which case injection pump can assume * function). Not fitted to most steady-burning devices such as gas turbines and heaters.

carburettor air Induced, usually via ram intake, along separate duct; intake normally anti-iced.

carburettor icing Caused by depression in venturi of choke tube giving local reduction in temperature.

carburising Prolonged heating of fully machined steel part in atmosphere rich in CO or hydrocarbon gases to give hard, tough outer layer.

carburising flame Oxy-acetylene flame having excess acetylene.

Carcinotron Backward wave oscillator; TWT for generating microwaves in which electron beam opposes direction of travel of a wave guided by a slow-wave structure.

CARD Civil aviation research and development (NASA, DoT).

Card Aerobatic-team formation resembling 4 (* 4) or 5 (* 5) playing cards.

card compass Simple compass with magnets attached to pivoted card on which bearings are marked.

CARDE Canadian Armament Research and Development Establishment.

Cardec Civil-Aviation R & D Executive Committee.

cardinal altitudes, cardinal FLs Altitudes or FLs forming an odd or even multiple of 1,000 ft.

cardinal point effect Increased intensity of radar returns when target surface is most nearly perpendicular to LOS, esp. in case of surface features.

cardinal points Bearings 09 (E), 18 (S), 27 (W) and 36/0 (N).

cardioid Heart-shaped; profile of cam or plot of radio signal strength against bearing.

cardioid reception Obtained by combining dipole and reflector or vertical aerial with loop in correct phase.

CARDP Civil-Aviation R & D Program Board.

Cards, CARDS *1* Computer-assisted radar-display system.

2 Computer-aided reporting and diagnosis system.

Cardsharp Capabilities and requirements demo *CBASS* ['sea bass'] high-alt. recon. program.

CARE Center for Aviation Research and Education (US).

carefree handling AFCS provides reliable protection against stall, departure or overstress.

Cares Cratering and related effects simulation.

caret inlet Left and right engine inlets each in front view having form of parallelogram with centrelines meeting below a/c.

CARF Central altitude reservation facility; ATC facility for special users under altitude reservation concept (FAA, from 1956).

cargo Useful load other than passengers or baggage, but including live animals. In military aircraft, all load other than human beings and personal kit and weapons.

cargo conversion Passenger or other non-cargo aircraft permanently converted to carry freight (see QC [*2*]).

cargo net Webbing or rope net for restraining cargo on pallet or igloo.

CARI Civil Aeromedical Research Institute (FAA 1959, became CAMI 1965).

car launch Use of towing motor vehicle to launch glider.

car lines Steel wires or multistrand cables passing from hot-air-balloon load ring down and under basket and back up to load ring.

CARMS   Civil Aviation Radio Measuring Station (DoT, UK).

carnet   Document facilitating crossing frontier by air without customs dues on aircraft; credit card valid for aviation fuel and certain services; sometimes includes medical certificates.

CARNF   Charges for airports and route navigation facilities (ICAO).

Carnot cycle   Ideal reversible thermodynamic cycle: isothermal compression, adiabatic compression, isothermal expansion, adiabatic expansion.

Carousel   Pioneer family of civil INS.

carousel   *1* Circulatory conveyor system to which baggage is delivered in arrival terminal.

2 Large structural ring on which rotating wing of tilt-rotor aircraft is mounted.

CARP   Computed air release point.

carpet   *1* Graphical plot of three variables having appearance of flexible two-dimensional surface viewed obliquely.

2 Strip of Earth's surface subjected to sonic boom.

carpet bombing   Level bombing, using one or more aircraft, to distribute bombs uniformly over target area.

Carquals   Carrier qualification tests (USN).

carrel   Computer-based pilot-training aid: cockpit, keyboard interface and instructor system displays.

carrier   *1* Aircraft carrier.

2 EM wave, usually continuous and constant amplitude and frequency, capable of being modulated to transmit intelligence.

3 Electronic charge *, either so-called hole or mobile electron.

4 Substance chosen to carry trace element or trace of radioactive material too small to handle conveniently.

5 Operator of commercial aircraft engaged in transport of passengers and/or freight for hire or reward.

carrier air group   Two or more aircraft squadrons operating from same carrier (1) under unified command.

carrier-on-board delivery   Air delivery of personnel, mail and supplies to carrier (1) at sea.

carrier suppression   Communications system in which intelligence is transmitted by sidebands, carrier being almost suppressed.

carrier task force   One or more carriers (1) and supporting ships intended to be self-sufficient in prolonged campaign.

carrier vehicle   Parent body or bus of SBI, can be equipped with mid-course sensors independent of SBI (SDI).

carry-on baggage   Brought on board by passenger. Some airlines [eg, Aeroflot] make passenger carry all baggage.

carry the can   Accept, or be awarded, blame for something (RAF).

carry through   Wing spars and other linking structure inside fuselage (esp. of mid-wing aircraft).

carry trials   Programme intended to prove carriage and release of fired, dropped or jettisoned stores.

CARS   *1* Coherent antistrokes Raman spectroscopy.

2 Crew-awareness rating scale.

3 Community aerodrome radio station.

4 Contingency airborne reconnaissance system [MISU adds as in *Carsmisu*].

5 Common automatic recovery system.

Carsmisu   Contingency airborne reconnaissance systems [and] mission intelligence systems upgrade.

CART   Combat aircraft repair team.

Cartesian co-ordinates   System of three mutually perpendicular planes to describe any position in rectilinear space.

cartridge   Portable container of solid fuel or propellant, with self-ignition system, for propulsion of projectile or supplying pressure to one-shot system.

cartridge starter   Main-engine starting system energised by reloadable cartridges.

cartwheel   *1* Aerobatic manoeuvre involving rotation about Z (yaw) axis, at very low airspeed, with that axis approximately horizontal.

2 Crash on ground involving rotation with wings near vertical plane.

carve-out   Removal of black or otherwise classified program from oversight by security or contract-oversight organization (DoD).

CAS   *1* Chief of the Air Staff.

2 Collision avoidance system.

3 Close air support.

4 Calibrated, or computed, airspeed (see *airspeed*).

5 Corrected airspeed (obs.).

6 Control [or command] augmentation system (or sub-system).

7 Controlled airspace.

8 Commission for Atmospheric Sciences.

9 Crisis action system (US JCS).

10 Cockpit avionics system (FSA[3]/CAS).

11 Control actuation section (missiles).

12 Contract Administration Service (or standard[s]).

13 Cost allocation schedule.

14 Combined antenna system.

15 Controlled-access service (satnav).

16 Crashworthy armoured seat.

17 Ceramic abradable seal.

18 Computer-aided software, or support.

19 Cable arresting system.

20 Control actuation system (Goodrich).

CASA, Casa   Civil-Aviation Safety Authority (Australia).

CASB   Canadian Aviation Safety Board.

CASC   Combined acceleration and speed control.

Cascad   Close air support cargo dispenser.

Cascade   *1* Combat air surveillance correlation and display system.

2 Contribution for assessment of common ATM(7) development in Europe (Euret).

cascade   Array of numerous (eg six or more) sharply cambered aerofoils superimposed to handle large gas flow (eg to turn flow round corner of tunnel circuit).

cascade reverser   Thrust reverser incorporating one or more cascades to direct efflux diagonally forwards.

CASCC   Close air support coordination and control.

Case, CASE   *1* Computer-assisted, or aided, software engineering.

2 Controlled-airspace synthetic environment.

case   *1* Outer layer of carburised, nitrided or otherwise case-hardened steel part.

2 Cartridge or shell case housing propellant.

3 Envelope containing solid rocket propellant and withstanding structural and combustion loads.

case-bonded   Solid propellant poured as a liquid into motor case (3) and cast in situ.

case chute   See *case ejection*.

case ejection   Method of disposing of non-consumable case (2), usually stored on board or discharged through chute under assumed positive acceleration.

casein   Cold-water glue manufactured from dehydrated milk curd.

caseless ammunition   Gun ammunition in which case (2) is consumed upon firing.

casevac   Casualty evacuation.

Casex   Combined anti-submarine exercise.

CASF   Composite Air Strike Force (TAC).

CASI   *1* Canadian Aeronautics and Space Institute.
  2 Commission Aéronautique Sportive Internationale (Int.).

Casid   Committee for Aviation and Space Industry Development (Taiwan).

Casits   Close air support integrated targeting system[s].

cask   Container for transport and storage of nuclear fuel or radioactive material.

CASM   Cost per a/c, or available, seat-mile.

Casom   Conventionally armed stand-off missile.

Casp   *1* Canada/Atlantic storms program.
  2 Commercial airborne security patrol[s] (at KSC).

Casper   Composite-aircraft spare parts with enhanced reliability.

CASS   *1* Command active sonobuoy system.
  2 Crab-angle sensing system.
  3 Consolidated automated support system (USN).
  4 Commercial air service standards.
  5 Close air support system.
  6 Continuing analysis and surveillance system (FAA).

Cassegrain   Optical telescope using two parabolic mirrors in series.

cassette   Standard tape container, eg for recording mission data.

CASST   Civil aviation safety strategic team.

CAST   *1* Civil Aircraft Study Team.
  2 Commercial aviation safety team.
  3 Chinese Academy of Space Technology (People's Republic).
  4 Complete aircraft static test.
  5 Command and staff trainer.
  6 Conformal-array seeker technology.
  7 Cyclic auto self-test.
  8 Commercial Aviation Safety Team (US).

cast-block engine   Piston engine with each linear row of cylinders arranged in a single cast block.

castellated nut   Typically, hexagon nut with six radial slots for split-pin or other lock.

Castigliano   Fundamental structural theorems relating loads, deflections and deformation energy.

casting   Forming material by pouring molten into shaped mould.

Castor   Corps airborne stand-off radar (UK).

cast propellant   Solid rocket propellant formed into grain by casting.

CASU   Compact air-supply unit.

casual pay parade   One-time payment to a group of personnel for a specific purpose (RAF).

casualty   Aircraft which suffers damage or sudden severe unserviceability, hence * action, * maintenance.

CASWS   Close air support weapon system.

CAT   *1* Clear-air turbulence.
  2 Comité de l'Assistance Technique.
  3 Combat aircraft technology.
  4 Computer automatic [or aided] testing.
  5 Crisis action team.
  6 Cockpit automation technology.
  7 Compressed-air [wind] tunnel.
  8 Computerized axial tomography (baggage screen).

Cat   *1* Aircraft-carrier catapult (colloq.).
  2 See *Categories*.

CATA   *1* Canadian Air Transportation Administration.
  2 Control automation and task allocation.

Catac   Commande Aérienne Tactique (F fighter command).

Catalin   Cast phenolic thermosetting plastic.

catalyst   Substance whose presence permits or accelerates chemical reaction, itself not taking part.

catalyst bed   Porous structure through which fluid passes and undergoes chemical reaction (eg HTP motor).

cat & trap   Catapult launch and arrested landing.

Cataphos   Chlorinated phosphate rubber-based paint for marking taxiways, aprons, etc, yellow or white.

catapult   Device for externally accelerating aeroplane or other vehicle to safe flying speed in short distance. Those on ships, especially surface warships, were originally operated by compressed air, then hydraulic, and now by steam pressure from main ship propulsion. See *LEM*.

catastrophic instability   Irrecoverably divergent loss of stability at dynamic head sufficient to break primary structure.

CATC   *1* College of Air Traffic Control (Hurn, UK).
  2 Commonwealth Air Transport Council (UK/Int.).

Catca   Canadian Air Traffic Control Association.

CATCC   Carrier air traffic control centre.

catcher   Small fence-like strips around leading edge of sharply swept or delta-wing naval aircraft; designed to engage barrier.

CATCS   Central Air Traffic Control School (RAF Shawbury).

CATE   Conference on Co-ordination of Air Transport in Europe.

Cat E   *1* Category E for an aircraft, a write-off [now Cat 5].
  2 For repaired runway crater, profile allows 4.5 in (114 mm) rise in first 12 ft (3.66 m).

categories   *1* For flight crew, licence authorisation to qualify on all aircraft within broad groups (eg light aircraft, glider, rotary-wing).
  2 For certification of aircraft, grouping based on usage (eg transport, experimental, aerobatic).
  3 In bad-weather landings, operational performance * are defined by runway visible range and decision height:
    Cat 1 or I: DH 60 m [200 ft], RVR 800 m [2,600 ft].
    Cat 2 or II: DH 30 m [100 ft], RVR 400 m [1,300 ft].
    Cat 3a or IIIa: DH 0, RVR 200 m [700 ft but 650 ft is closer conversion].
    Cat 3b or IIIb: DH 0, RVR 50 m [150 ft].
    Cat 3c or IIIc: DH 0, RVR 0.
  4 For aircraft damage and repairability: Cat 1, undamaged; Cat 2, repairable on unit; Cat 3, repairable by 2nd echelon or MU; Cat 4, by manufacturer; Cat 5, a write-off.
  5 For runway repairs, see *rough field*.
  6 For airfield conflicts, A = near-collision demanding

extreme action, B = significant potential of collision; C = ample time to avoid collision (FAA).

Category 2 box Rectangular box or window in HUD defining permissible Cat 2 deviation of localiser and G/S.

catenary Curve described by filament supported at two points not on same local vertical.

catering vehicle Removes and replenishes galley; conventional truck with scissor lift, elevating body usually not including cab.

Caterpillar Club Private club formed by Irvin Air Chute Co and open to all who have saved their lives by using parachute of any kind.

CAT-EVS CAT(1) enhanced vision system.

CATH, Cath Compact air-transportable hospital.

cathedral Anhedral (pronounced cat-hedral).

cathode *1* Positive terminal of source of EMF (eg battery).
2 Electrode at which "positive current" leaves solid circuit.
3 Negative terminal of electroplating cell.
4 In CRT and similar tube, source of electron stream.

cathode-ray oscillograph, cathode-ray oscilloscope, CRO CRT built into device including amplifier, power pack and controls for graphic examination of waveforms and other research.

cathode-ray tube, CRT Vacuum tube along which electrons (cathode rays) are projected, deflected by pairs of plates creating electric field (deflected toward positive) and impact on screen coated with electroluminescent phosphor.

cathode tuning indicator Triode amplifier and miniature CRT giving visual indication, by closure of well-defined shadow area, of changes in carrier amplitude too small to detect aurally. Also called Magic Eye.

CATIA, Catia Computer-aided 3-D interactive analysis [anglicized from next].

Catia Conception assistée tridimensionelle interactive d'applications (F).

Catic (CATIC) China National Aero Technology Industrial Corporation.

cation Positively charged ion, travelling in nominal direction of current (pronounced cat-ion).

CATM Captive air training missile.

CATO Civil air traffic operations (or officer).

catoptric beam Visible light concentrated into parallel beam by accurate reflector.

CATP Commonwealth Air Training Plan, originally *BCATP*.

Cats, CATS *1* Combined Aerial Target Services (MoD, UK).
2 Corporate air travel survey.
3 Contracted Airborne Training Services (Canada).
4 Consequence assessment tool set.

Catsa, CATSA Canadian Air Transport Security Agency.

CATSE Capacity of the air-transport system in Europe (ECAC).

cat shot Launch of naval aircraft by catapult.

cat's whisker Delicate current pickoff, eg from gyro.

CATTS Central Air Traffic Training School (RAF Shawbury, UK).

catwalk Narrow footway along keel of airship.

CAU *1* Cold-air unit.
2 Canard actuation unit.

3 Communication[s] access unit.

CAUFN Caution advised until further notice.

Caul Thin sheet defining surface of composite structure during ATP(8).

caution note Written on approach chart, usually warning of high ground.

caution speeds Speeds published in Flight Manual as limit for each configuration.

Cautra, CAUTRA Co-ordinateur automatique du trafic aérien Français.

CAUWG Commercial Ada users working group.

CAV *Common aero vehicle.*

cavalry charge Aural-warning clarion call usually signifying autopilot disconnect.

cavitation Transient formation and shedding of vapour-pressure bubbles at surface of body moving through non-degassed liquid, or of vacuum cavities at surface of body moving through other fluid. Collapse of cavities is violent, causing extreme implosive pressures.

cavity magnetron Magnetron having resonant cavities within cylindrical anode encircling central cathode.

cavity resonator Precisely shaped volume bounded by conducting surface within which EM energy is stored at resonant frequency.

Cav-OK Ceiling and visibility OK (for VFR), or better than predicted [said as written].

CAVU Ceiling and visibility unlimited.

cAw. Continental Arctic air mass, warmer than land surface.

CAWC Combined air-warfare course.

CAWDSG Combat-aircraft wing-design steering group.

CAWP *1* Central annunciator warning panel.
2 Cockpit assessment working party.

CB *1* Circuit breaker.
2 Chlorobromo-type fire extinguishants.
3 Citizens' band radio.
4 Centre of balance (or c.b., C–B).
5 Chemical/biological.
6 Centre/center barrel (major portion of fuselage).
7 Construction Battalion (USMC 'Seabees', USN).
8 Chaff block.

C_B Lift proportionality constant for circulation-controlled wing.

Cb Cumulonimbus.

CBAA Canadian Business Aircraft Association.

CBACS C-band airborne communications system.

CBASS Common broadband advanced sonar system ['sea bass'].

CBAST Computer-based advanced skills, or systems, trainer.

CBC *1* Carbon/BMI composite.
2 Common booster core.

CBD *1* Central business district of city.
2 Chemical/biological defence.

CBDC Chemical and Biological Defence Centre (Porton Down, UK).

CBDP Chemical and Biological Defense Program (DoD).

CBDS Chemical and biological detection system.

CBE CAIS(2) bus emulator.

CBERS China/Brazil Earth-resources satellite programme.

CBG Carrier Battle Group (USN).

117

CBH   Chemical/biological hardening.
CBI   *1* Computer-based instruction.
2 Component burn-in.
3 Confederation of British Industries.
4 China/Burma/India (theatre WW2).
5 Chemical/biological incident; RF adds Response Force (USMC).
CBIM   Cloudbase information manager.
C-bite   Continuous built-in test equipment.
CBL   *1* Control by light.
2 Conveyor-belt loader (cargo).
3 Crowd barrier line.
CBLS   Carrier, bomb, light store.
CBM   *1* Chlorobromomethane.
2 Confidence-building measure[s].
3 Chronological bus monitor[ing].
C_bM, C_bMAM   Cumulonimbus mammatus.
CBMS   Chemical and biological mass spectrometer.
CBN   Cubic boron nitride.
CBO   Congressional Budget Office (US).
CBP   Contact-burst preclusion.
CBR   *1* California Bearing Ratio; system for assessing ability of soft (ie unpaved) surfaces to support aircraft operations; contains terms for aircraft weight, tyre characteristics, landing gear configuration and rutting after given numbers of sorties.
2 Chemical, biological, radiological warfare.
3 Common bomb rack.
4 Center-barrel [3] replacement.
5 Cloud-base recorder.
CBRN   Chemical, bioogical, radiological and nuclear.
CBS   Cavity-backed spiral.
CBSIFTCB   Common preflight cockpit check for glider: controls, ballast, straps, instruments, flaps, trim, canopy, brake [airbrakes].
CBT   Computer-based training.
CBTE   Carrier, (or conventional) bomb triple ejector.
CBU   Cluster bomb unit.
CBW   *1* Chemical and biological (or bacteriological) warfare.
2 Combat Bombardment Wing (USAAF).
CBX   Control, or copilot control, box (UAV).
CC   *1* Central or countermeasures, computer.
2 Critical crack.
3 Coastal Command (RAF).
4 Communications (UK role prefix).
5 Composite command (USAF).
6 Circulation controlled (wing or rotor).
7 Compass course.
8 Co-ordinating committee.
9 Counterclockwise.
10 C-check.
11 Cape Canaveral.
C/C   Carbon/carbon [also rendered as C-C].
C_c   *1* Equivalent centreline chord.
2 Cirrocumulus.
CC^3   Counter-C^3.
CCA   *1* Cooled cooling air.
2 Current cost accounting.
3 Carrier-controlled approach.
4 Continental Control Area (US + Alaska at 14,500+ ft AMSL).
CCAFS   Cape Canaveral Air Force Station.
CC&D   *1* Camouflage, concealment and deception.

2 Common command and decision.
CCAOU   Central Counties Air Operations Unit (UK).
CCAQ   Consultative Committee on Administrative Questions (ICAO).
CCAS   Centralised crew-alerting system.
CCAT   Carrier control approach trainer.
CCATE   Common-core automatic test equipment.
CCB   *1* Configuration-change, or -control, board (software).
2 Converter circuit-breaker.
3 Common-core booster.
CCC   *1* See C^3 with suffixes.
2 Customs Co-operation Council (ICAO).
3 Combat Control Centre.
CCCA   Conference of city-centre airports (UK, 1996–).
CCCD, C^3D   *1* Cross-cockpit collimated display.
2 Counter-C^3 (Italy).
CCD   *1* Camouflage, concealment and deception.
2 Charge-coupled device, or diode.
3 Cursor-control device.
4 Computerised current density.
CCDA   Cockpit-control driver actuator.
CCDR   Contractor cost data report (US).
CCE   *1* Communications control equipment.
2 Commission des Communautés Européennes (Int.).
3 Cryogenically cooled electronics.
4 Change compositive explorer.
CC89   Cabin Crew 89 (trade union, UK).
CCF   *1* Central control function.
2 Combined Cadet Force (UK).
CCFG   Compact constant-frequency generator.
CCFL   Cold-cathode fluorescent lamp.
CCFP   Collaborative convective forecast product, from Aviation Weather Center, esp. concerned with severe weather (US).
CCG   *1* C-code generator.
2 Computer control and guidance.
3 Communications control group.
CCH   Close-combat helicopter.
CCI   *1* Commission for Climatology (WMO).
2 Chambre de Commerce Internationale (Int.). See *ICC*.
3 Continuous capability improvement.
CCIA   Comitato Coordinazione Industria Aerospaziale (I).
CCID   Composite combat identification (JIADS).
CCIG   Cold-cathode ion gauge.
CCII   Command, control and information infrastructure (UK).
CCIL   Continuously computed impact line; HUD snap-shoot presentation with fully predicted bullet flight profile with various range assessments.
CCIP   *1* Continuously computed impact point; HUD display for air-to-ground weapon delivery with impact point indicated for any manual release from laydown to steep dive. See also *Delayed**.
2 Critical-component improvement program.
3 Common-configuration implementation program (USAF).
CCIP/IP   With preliminary designation of an initial point.
CCIR   *1* Comité Consultatif International des Radiocommunications (UIT), assigns wavebands, frequencies.

2 Commission Consultatif Internationale pour la Radio.

CCIRM Collection, co-ordination and intelligence requirements management.

CCIS Command and control information system[s].

CCISR Command and control intelligence, surveillance and reconnaissance.

CCITT Comité Consultantif International pour Télégraphie et Téléphone.

CCL Climate-change levy.

CCLRC Council for the Central Laboratory of the Research Councils (UK, by Royal Charter 1995–).

CCM *1* Conventional cruise missile.
2 Counter-countermeasures.

CCMA Comite de Compradores de Material Aeronautico de America Latina (Int.).

CCMS *1* Communication control and management system (Scope Command).
2 Content compilation management system.

CCN Contract change notice.

CCO *1* Chief corporate officer.
2 Chief of Combined Operations.

CCOA Centre de Conduite des Operations Aériennes (Taverny, F).

CCOC Combustion-chamber outer casing.

CCP *1* Cutter-centre path (machining).
2 Coherent countermeasures processor.
3 Combat correlation parameter (simulator).
4 Corrosion-control programme.
5 Control and correlation processor.

CCPC Civil Communications Planning Conference (NATO).

CCPDS Command and control processing and display system.

CCPR Cruise compressor pressure ratio.

CCQ Cross-crew qualification.

CCR *1* See *ACC. (1)*.
2 Circulation-controlled rotor.
3 Configuration-change report, required each time a Part or Data-base No. changes.
4 Constant-current regulator.

CCRA Canadian Customs and Revenue Agency.

CCRI Climate Change Research Initiative (NOAA).

CCRP Continuously computed release point; HUD display for air-to-ground weapon delivery with steering command and auto weapon release in any attitude from laydown to OTS, system controlling entire firing sequence and triggering release or firing mechanism automatically.

CCS *1* Communications control system (aircraft R/T and i/c selection and audio routeing).
2 Conformal countermeasures system.
3 Computer and computation subsystem (ACMI).
4 Combat Control Squadron (USAF).
5 Cargo community system, electronically links shippers, airports, forwarders and carriers.
6 Cabin-communication system.
7 Common carriage system [external weapons].

CCSL Cirrocumulus, standing lenticular wave.

CCSS Command and control switching system.

Cct Airfield circuit.

CCTS *1* Co-ordinating Committee for Telecommunications by Satellite.
2 Cabin cordless-telephone system.
3 Combat Crew Training Squadron (USAF).

CCTV *1* Closed-circuit TV.
2 Colour cockpit TV;' S adds sensor.
3 Crew/cargo transfer vehicle.

CCTW Combat Crew Training Wing (USAF).

CCTWT Coupled-cavity travelling-wave tube.

CCU *1* Cockpit, communications, central or common control unit; TSD adds tactical-situation display.
2 Control and compensation unit.

CCV *1* Control-configured vehicle.
2 Command and control vehicle.
3 Chamber coolant valve.

CCW *1* Counter-clockwise.
2 Circulation-controlled wing.

CCWS Common controller workstation (CAATS/ MAATS).

CD *1* Certification demonstration.
2 Concept demonstration.
3 Clearance delivery (US, not UK).
4 Controlled-diffusion.
5 Convergent/divergent.
6 Cycle-dependent.
7 Capacitor-discharge.
8 Cold.
9 Compact disk.
10 Circular dispersion.
11 Civil Defence (UK, WW2).
12 Chrominance difference.
13 Carrier detect.
14 Coast Defence (RAF 1926 - c40).

Cd Cadmium.

cd Candela(s).

Cd, C_d Total drag coefficient.

Cd_0 Zero-lift drag coefficient.

CD-2 Common digitiser 2 (FAA).

CDA *1* Centre for Defence Analyses (UK).
2 Controlled-diffusion aerofoil.
3 Concept-demonstration aircraft.
4 Continuous-descent approach.
5 Cognitive decision-aiding.
6 Co-ordinating design authority.

C/DA Climb and dive angle.

CDAA Circularly disposed aerial (antenna) array.

CDAI Centre de Documentation Aéronautique Internationale.

CDAT Critical-defect assessment technology.

CDB *1* Cast double-base rocket propellant, allows case-bonding, varied formulation and charge configuration, low smoke emission etc.
2 Central [ATC] data bank.

CDBP Command Data Buffer Program (USAF).

CDBR Cabin databus repeater.

CDC *1* Course and distance calculator.
2 Concorde Directing Committee (Comité Direction Concorde).
3 Cour des Comptes [general accounting office] (F).
4 Cabin-display computer.

CDCN Controller of Defence Communications Network (UK).

CDD Credible delicious decoy.

CDDT Countdown demonstration test.

CDE Chemical Defence Establishment (UK).

CDF *1* Clearance delivery frequency.
2 Core-driven fan [s adds stage].
3 Core-distributed interactive-simulation facility.

CDF   Drag coefficient for zero lift.

CDFNT   Cold front.

CD_{fric}   Frictional drag coefficient, usually close to C_{DF}.

CDG   Configuration-database generator.

CDH   Constant delta height.

CDI   *1* Course-deviation indicator.

   2 Collector diffusion isolation.

   3 Capacitor-discharge ignition.

   4 Compass director indicator.

   5 Collateral duty inspector (US).

   6 Chief of Defence Intelligence (UK).

   7 Classification, discrimination and identification.

CDi   Coefficient of induced drag.

CDIP   Continuously displayed impact point (HUD).

CDIRRS   Cockpit display of IR reconnaissance system.

CDIS   Central control function display.

CDIU   Circuit-mode data interface unit.

CDL   *1* Configuration data list (JARs).

   2 Chief of Defence Logistics (UK).

   3 Common datalink.

   4 Cabin-discrepancy log.

CDL**, C**DL   Lift-dependent drag coefficient, sub-critically equal to induced-drag coefficient.

CDM   Collaborative decision-making.

CDMA   Code-division multiple-access.

CDMATC   Collaborative decision-making ATC.

CDMC   Cranfield Disaster-Management Centre (UK).

CDmin**, C**D_{min}   Minimum drag-coefficient.

CDMLS   Commutated-Doppler microwave landing system.

CDMS   CDM system.

CDMT   Central design and management team.

CDMVT   'Cadbury's Dairy Milk, very tasty', one of countless mnemonics, in this case meaning course→deviation→magnetic→variation→true (RAF WW2).

CDN   *1* Co-ordinating message (ICAO).

   2 Certificat de Navigabilité (C of A, F).

CDNU   Control [and] display navigation unit (helicopter).

Cdo   Commando (Unit).

CDOPS, C-dops   Coherent-Doppler scorer.

CDP   *1* Central data processor.

   2 Countermeasures dispenser pod.

   3 Critical decision point.

   4 Chief of Defence Procurement (UK).

   5 Contract-definition phase.

   6 Concept-demonstration phase, or program[me].

   7 Cockpit-display player (F-22).

   8 Continuous-data program.

CDp   Coefficient of profile drag.

CDPI   Crash data position indicator.

CDR   Critical design review.

CDRA   Carbon dioxide removal assembly.

CDRB   Canadian Defense Research Board.

CDRL   Contract data requirements list.

CD-ROM   Compact-disk read-only memory.

CDRS   *1* Control and data retrieval system.

   2 Container-design retrieval system (USAF).

   3 Cockpit display and recording system.

CDS   *1* Controls and displays, or control/display, subsystem.

   2 Chief of Defence Staff (UK).

   3 Coefficient of slush drag, derived from spray

impingement drag, basic precipitation drag and wheel geometry.

   4 Container, or covert, delivery system (USAF).

   5 Cabin distribution system (Satcom).

   6 Concept-definition study.

   7 Cockpit-design simulator.

   8 Component documentation status.

   9 Common display system.

   10 Cockpit development station.

   11 Countermeasure[s] dispensing system.

CdS   Cadmium sulphide/sulfide, IR detector.

CdSe   Cadmium selenide.

CDSS   Center for Defence/Defense and Security Studies (Univ. of Manitoba).

CDT   *1* Central [US] Daylight Time.

   2 Crew duty time.

   3 Controlled departure time.

CDTC   Controlled descent through cloud.

CDTF   Coast Defence Training Flight (RAF to 1935).

CDTI   Cockpit display of traffic (ATC) information.

CD trainer   Cockpit-drill trainer.

CD-TR-TV   Con-di, thrust-reverser thrust vector.

CDU   *1* Control and display unit.

   2 Cockpit, or console, or combined, display unit.

CDVE   Commandement de vol electrique = fly by wire (F).

CDVS   Cockpit-door video surveillance; S adds system.

CDWS   Common defense weapon system (USN/USMC helicopters).

CD0   Drag coefficient at zero lift.

CE   *1* Concurrent engineering.

   2 Computing element.

   3 Chemical energy.

Ce   Specific fuel consumption (R).

CEA   *1* Code of European Airworthiness.

   2 Commissariat à l'Energie Atomique (F).

   3 Circular-error average.

   4 Combined electronics assembly.

CEAA   Commandement des Ecoles de l'Armée de l'Air (F).

CEAC   *1* Commission Européenne de l'Aviation Civile (Int.).

   2 Committee for European Aerospace Cooperation (Int.).

CEAM   Centre d'Expériences Aériennes Militaires (F, Mont de Marsan).

CEAS   Confederation of European Aeronautical Societies (Int.).

CEAT   Centre d'Essais Aéronautiques de Toulouse (F).

CEATS   Central European Air-Traffic Service (Vienna).

CEB   *1* Combined-effects bomblet.

   2 Curve of equal bearings, from NDB.

CEC   *1* Cooperative-engagement capability.

   2 Communications and Electronics Command (USA).

   3 Continental entry chart.

   4 Centre d'Entrainement au Combat (F).

   5 Crew ejectable cabin.

CECAI   Conference of European Corporate Aviation Interests (Int.).

CECC   Cenelec Electronic Components Committee (Int.).

Cecom   Communications and Electronics Command (USA).

Ceconite   Weatherproof fabrics for skinning light aircraft.

CED   *1* Continued engineering development.

2 Competitive engineering definition.

Cedam   Combined electronic display and map; derived from Comed.

Cedap   Cockpit-emergency directed-action program.

Cedocar   Centre de Documentation de l'Armement (F).

CEE   *1* Cabin emergency evacuation.

2 Commission on rules for Electrical Equipment (Int.).

Ceesim   Combat electromagnetic-environment simulator.

CEF   *1* Cost-effectiveness factor (materials).

2 Contrast-enhanced filter.

CEFA   Cooperation for Environmentally Friendly Aviation (EU).

CEFH   Cumulative engine flight-hours.

CEI   *1* Critical engine inoperative.

2 Council of Engineering Institutions (UK).

3 Commission Electrotechnique Internationale.

4 Combat efficiency improvement.

5 Cabin equipment interface.

ceil   Ceiling.

ceiling   *1* Of aircraft, greatest pressure height that can be reached (see *absolute* *, *absolute aerodynamic* *, *service* *, *zoom* *).

2 Of cloud, height above nearest Earth's surface of lowest layer of clouds or obscuring phenomena that is reported as 'broken,' 'overcast' or 'obscuration' and not 'thin' or 'partial' (FAA).

3 Height of cloud base below 6000 m/20,000 ft covering 50+ per cent of sky (ICAO).

4 Amount above which FFP contract cannot be implemented.

ceiling balloon   Small free balloon, whose rate of ascent is known, timed from release to give measure of ceiling (2).

ceiling climb   Aircraft flight authorised for express purpose of measuring ceiling (1).

ceiling height indicator   Device for measuring height of spot produced by ceiling projector.

ceiling light   See *ceiling projector*.

ceiling projector   Source of powerful light beam projected vertically to form bright spot on underside of cloud.

ceiling unlimited   Sky clear or scattered cloud, or base above given agreed height (in US 9,750 ft, 2,970 m).

ceiling zero   Fog.

ceiliometer   Device for measuring ceiling (2), esp. ceiling projector, in later types a lidar, whose beam oscillates about horizontal axis like metronome, linked with photocell with readout in tower.

CEL   *1* Centre d'Essais des Landes (F).

2 Capacitatively-enhanced logic, speed enhanced by incorporation on chip of capacitor(s).

3 Component evolution list.

Celar   Centre Electronique de l'Armement (F).

celestial altitude   Altitude (2) of heavenly body or point on celestial sphere.

celestial body   Meaning arguable, but normally all bodies visible or supposed other than Earth and man-made objects. Diffuse bodies (eg nebulae) often called 'structures'.

celestial equator   Great circle formed by extending Earth equatorial plane to celestial sphere.

celestial fix   See *astro fix*.

celestial guidance   Guidance of unmanned vehicle by automatic star tracking.

celestial horizon   Great circle formed on celestial sphere by plane passing through centre of Earth normal to straight line joining zenith and nadir.

celestial mechanics   Science of motion of celestial bodies.

celestial meridian   Meridian on celestial sphere.

celestial navigation   See *astronavigation*.

celestial pole   Terrestrial pole projected upon celestial sphere.

celestial sphere   Imaginary hollow sphere of infinite radius centred at centre of Earth (for practical purposes, at eyes of anyone on Earth).

celestial triangle   Spherical triangle on celestial sphere, esp. one used for navigation.

cell   *1* Combination of electrodes and electrolyte generating EMF; basic element of battery.

2 Portion of structure having form of rigid box, not necessarily completely enclosed.

3 In biplane or other multi-wing aircraft, complete assembly of planes, struts and wires forming structural box.

4 Gasbag of aerostat, esp. in airship having multiple bags in outer envelope.

5 In EDP, elementary unit of storage.

6 Self-contained air mass of violent character (eg TRS).

7 In military operations, smallest tactical aircraft element flying together (often three); another definition is a small unit of airborne military aircraft which can if necessary operate independently.

Cellophane   Early cellulose-base transparent plastics film.

cell-textured   See *textured visuals*.

cellular logic image processing   Each pixel has its own dedicated logic element to which are attached its eight adjacent pixels in parallel architecture. Thus each instruction is executed by all processing elements simultaneously.

cellule   Cell (3) or assembly of two or more wings on either side of centreline.

Celluloid   Early transparent plastics film from nitrocellulose treated with camphor.

cellulose   Carbohydrate forming major structural constituent of plants; precursor of many aeronautical materials.

cellulose acetate   Thermoplastic from cellulose and acetic acid; used in rayon, film, lacquer, etc.

cellulose dope   See *dope*.

cellulose nitrate   Thermoplastic from cellulose and nitric acid; used in explosives, dopes and structural plastics.

Celsius   Scale of temperature, symbol °C. Unit is same as SI scale K, but numbers are lower by 273. Until 1948 called Centigrade.

CELT   Combined emitter location (or locator) testbed.

CELV   Complementary expendable launch vehicle, ie in addition to Shuttle (US).

CEM   *1* Centre d'Essais de la Méditerranée (F).

2 Combined-effects munition (cluster dispenser).

3 Conventional enhancement modification.

4 Concept evaluation model.

5 Core exhaust mixer.

6 Control-, or contract, equipment manufacturer.

7 Computational electromagnetics.

Cementite   Iron carbide; carbon form in annealed steel.

Ceminal   Cost/effective manufacturing in new aluminum (aluminium) alloys (US).

CEMS   Centre d'Etudes de la Météorologie Spatiale (F).

CEMT   Conférence Européenne des Ministres des Transports.

CEN   *1* Centre d'Etudes Nucléaires (F).

2 Comité Européen Normalisation [Int. standards].

CENA   Centre d'Etudes (formerly d'Expérimentation) de la Navigation Aérienne (F).

CENC   China-Europe Global Navigation Satellite System Technical Training and Co-operation Centre (from September 2003).

Cenelec   Comité Européen pour la standardisation de l'Électrotechnique (Int.).

Cenipa   Centro de Investigaçao e Prevençao de Acidentes Aeronáuticos (Braz.).

cenospheres   Microscopic hollow ceramic spheres.

Cenrap   Center radar processing (terminal ATC).

Centag   Centre Army Group (NATO, formerly).

centi   Prefix, $\times 10^{-2}$ (one hundredth), symbol c (non-SI).

Centigrade   See *Celsius*.

centilitre   Cl, 0.01 litre, measure of volume contrary to SI.

centimetre   0.01 metre; measure of length contrary to SI.

centimetre Hg   Used as unit of pressure = 1.33 kPa.

centimetric radar   Radar operating on wavelengths around 0.01 m, with frequencies 3–30 GHz.

centipoise   See *viscosity*.

centistoke   See *viscosity*.

CENTO, Cento   Central Treaty Organization (1955–80, Iran, Iraq, Pakistan, Turkey, UK, US).

central altitude reservation   See *CARF*.

centralised fault display system   Avionics system accessing all on-board BITE systems to extract and display data and initiate maintenance tests.

centralised servicing   Establishment of one unit and site for all routine maintenance on station, breaking previous intimate relationship between crew chief and each aircraft (RAF).

central warning panel   See *CWP (I)*.

centrebody   Streamlined body in centre of circular, semi-circular or quasi-circular supersonic intake (inlet) to cause inclined shock.

centre controls   To move primary flight controls from deflected to neutral position.

centre engine   Engine on centreline of multi-engined aircraft.

centreline   *1* Principal longitudinal axis; usually also axis of symmetry (eg of aircraft, missile, runway).

2 In aerofoil section, line joining leading and trailing edges and everywhere equidistant from upper and lower surfaces, all measures being normal to line itself.

centreline aircraft   See *centreline engine[s]*.

centreline camber   Ratio of maximum distance between chord and centreline (2) to chord.

centreline engine[s]   Any engine on longitudinal centre-line of multi-engine aircraft [engines hung on side of fuselage or in wing roots are excluded].

centreline gear   Main landing gear on aircraft centreline.

centreline lighting   On runway, flush lights at 50 ft (15 m) intervals terminating 75 ft (23 m) from each threshold.

centreline noise plot   Plot of aircraft noise, usually EPNdB, along runway centreline extended (usually 6 st

miles, 9.6 km) in each direction covering approach and climb-out.

centre of area   See *centroid*.

centre of buoyancy   Point through which upthrust of displaced fluid acts; e.g. in case of aerostats and marine aircraft afloat.

centre of burst   Mean point of impact.

centre of dynamic lift   In aerostat, point on centreline through which lift force due to motion through atmosphere acts.

centre of gravity, c.g.   Point through which resultant force of gravity acts, irrespective of orientation; in uniform gravitational field, centre of mass. For two-dimensional forms, centroid.

centre of gravity limits   In nearly all aircraft, esp. aero-dynes, published fore and aft limits for safe c.g. position; in case of aeroplanes expressed as percentages of MAC.

centre of gravity margin H_n, distance along aircraft major axis from c.g. to neutral point, expressed as % SMC.

centre of gravity travel   Fore and aft wander of c.g. in course of flight due to consumption of fuel, release of loads, etc.

centre of gross lift   In aerostat, usually centre of buoyancy.

centre of gyration   For solid rotating about axis, point at which all mass could be concentrated without changing moment of inertia about same axis.

centre of lift   Resultant of all centres of pressure on a wing or other body.

centre of mass   Point through which all mass of solid body could act without changing dynamics in translational motion; loosely, but not always correctly, called c.g. Alternative title, centre of inertia.

centre of pressure, c.p.   On aerofoil, point at which line of action of resultant aerodynamic force intersects chord. Almost same as aerodynamic centre, but latter need not lie on chord. In general, c.p. is resultant of all aero-dynamic forces on surface of body.

centre of pressure coefficient   Ratio of distance of c.p. from leading edge to chord.

centre of pressure moment   Product of resultant force on wing (or section) and distance from c.p. to leading edge (or leading edge produced at aircraft centreline). Inapplicable to very slender wings.

centre of pressure moment coefficient   As above but divided by dynamic pressure; not same as coefficient of moment.

centre of pressure travel   *1* Linear distance through which c.p. travels along chord over extreme negative to positive operating range of angles of attack, ignoring compressibility in subsonic flow.

2 Linear distance through which c.p. travels along chord over complete aircraft operating range of Mach numbers (supersonic aircraft only).

centre of thrust   Thrust axis, for one or multiple engines.

centre punch   Hand tool for making accurate conical depressions.

centre section   In most winged aircraft, centre portion of wing extending symmetrically through or across fuselage and carrying left and right wings on its tips. Certain aircraft have wing in one piece, or in left and right halves joined at centreline; such have no **, though some author-ities suggest it is then wing inboard of main landing gear.

centrifugal breather   Centrifuge filter for removing oil from air vented overboard from interior of engine, often after passage through porous segments.

centrifugal clearance   Radial clearance between rotating mass and surrounding fixed structure at peak rotating speed.

centrifugal clutch   Freewheels at low speed, but takes up drive as speed is increased. At full power slip tends towards zero. Main purpose is to prevent excessive load on gearteeth.

centrifugal compressor   Rotary compressor in form of disc carrying radial vanes to accelerate working fluid radially outward to leave periphery at very high speed, this being converted to pressure energy in fixed diffuser.

centrifugal separator   See *centrifuge filter*.

centrifugal twisting moment   Moment tending to rotate propeller blades towards zero pitch (opposing coarsening of pitch).

centrifuge   *1* Device for whirling human and other subjects about vertical axis, mainly in aerospace medicine research.

2 Device for imparting high unidirectional acceleration to hardware under test.

3 Device for imparting high unidirectional acceleration to mixtures of fluids with or without particulate solids to separate constituent fractions.

centrifuge filter   Action, inherent in turbofans and some other machines, which in rotating fluid flow causes unwanted particulate solids to be centrifuged outwards away from core.

centring, centering   Radial and, usually, also axial constraint of floated gyro to run equidistant at all points from enclosure.

centring control   System which, on demand, centres a variable in another system.

centripetal   Acceleration toward axis around which body is rotated; usually equal and opposite to centrifugal (see *coriolis*).

centrisep   Centrifugal separator (*centrifuge filter*).

centroid   In geometrical figure (one, two or three dimensional), point whose co-ordinates are mean of all co-ordinates of all points in figure. In material body of uniform composition, centre of mass.

Centrospas   [also rendered *Tsentro-*] Ministry for defence, emergencies and natural disasters (R).

CEO   *1* Chief executive officer of corporation or company.

2 Crew Earth observation.

CEOA   Central Europe Operating Agency (NATO pipelines).

CEOC   Colloque Européen des Organisations de Contrôle (Int.).

CEOI   Common electronic operating instructions.

CEP   *1* See *circle of equal probabilities*; also called circular error probable.

2 Concurrent evaluation phase.

3 Chromate-enriched pellet.

4 Common engine program.

CEPA   *1* Commission d'Evaluation Pratique d'Aéronautique (F).

2 Common European Priority Area.

Cepana   Commission d'Examen Permanent des Matériels Nouveaux d'Aéronautique (F).

CEPME   College for Enlisted Professional Military Education (USAF AU).

CEPr   Centre Essais de Propulseurs (Saclay, F).

CEPS   Central European pipeline system (NATO).

CEPT   *1* Conférence Européenne des Postes et Télécommunications (Int.).

2 Cockpit emergency procedures trainer.

CER   Cost-estimating relationship.

ceramal   See *cernet* (from ceramic + alloy).

ceramel   See *cernet* (from ceramic metal).

Cerap   Combined Center Radar Approach Control.

CERCA   Commonwealth and Empire conference on Radio for Civil Aviation.

Ceres   *1* Clouds and Earth's radiant-energy system.

2 Computer-enhanced radio emission surveillance (UK).

ceria glass   Amorphous semiconductor glass doped with cerium dioxide.

cerium   Reactive metal, Ce, density 8.2, MPt 799°C.

Cerma, CERMA   Centre d'Etudes et de Recherches de Médecine Aéronautique (F).

cermet   *1* Composite material attempting to combine mechanical toughness of metal with hardness and refractory qualities of ceramics. Early examples cemented carbides, eg tungsten carbide sintered with cobalt.

2 Incorrectly, part made of ceramic bonded to metal.

CERNAI   Study Commission on air navigation (Brazil).

CERP   Centre-Ecole Régional de Parachutisme.

Cerrobase   US lead-bismuth alloy, MPt 125°C.

Cerrobend   US bismuth-tin-lead-cadmium alloy, MPt 68°C.

Cerromatrix   US lead-tin-antimony alloy, MPt 105°C.

CERS   *1* Centre Européen de Recherche Spatiale (see *ESRO*).

2 Carrier evaluation and reporting system.

CERT   *1* Centre d'Etudes et de Recherches de Toulouse (ONERA).

2 Committee on Energy Research and Technology (European Community).

Certico   Committee on Certification (ISO).

Certificate of Airworthiness   Issued to confirm each individual aircraft is airworthy, renewed at intervals (CAA).

Certificate of Compliance   Issued to confirm functioning part of aircraft has been made/overhauled/repaired correctly (CAA).

Certificate of experience   Document required by private pilot showing flying as PIC in each preceding 13-month period (CAA UK).

Certificate of Maintenance   Issued upon completion of major overhaul or other routine work affecting airworthiness (CAA).

Certificate of Release [for Service]   Issued after heavy maintenance.

certification   For aircraft, issue of ATC (4) for US, C of A for UK or equivalent by national certifying authority, stating type meets all authority's requirements on grounds of safety. Other aircraft certificates include Production and Registration.

certification pilot   Test pilot employed by national certification authority to evaluate all types of aircraft proposed for use by operator in that country, from whatever source.

certification test   Test conducted by certification authority prior to issue of certificate.

CES   *1* Catapult-end speed.

   2 Civil Engineering Squadron (USAF).

Cesar   Computing environment StralCom architecture.

CESE   Communications equipment support element.

cesium, caesium   Extremely soft silver metal, highly reactive, used as working fluid as jet of charged ions in space thrusters.

CESR   Centre d'Etude Spatiale des Rayonnements (space radiation).

CEST   *1* Convertible-engine system technology.

   2 Centre for the Exploitation of Science and Technology (UK).

CESTA   *1* Continuous engineering services and technical assistance.

   2 Centre d'Etudes des Systèmes et des Technologies Avancées (F).

CET   *1* Combustor exit temperature.

   2 Core-engine test.

   3 Calculated estimated time [surely tautological]; A adds arrival, D departure, O overflight.

cetane rating   Numerical scale for ignition quality of diesel fuels.

CETO   Center for Emerging Threats & Operations (2002, USA).

CETPS   Cooperative engagement transmission processing set.

CETS   Contractor engineering technical services.

CETT   Core engine to test (date).

CEU   Checklist entry unit.

CEV   Centre d'Essais en Vol (Brétigny, Istres, Cazaux, F).

CEWI   Combat electronic warfare [and] intelligence.

CF   *1* Centrifugal force.

   2 Concept formulation.

   3 Carbon-fibre.

   4 Charge [or change] field.

   5 Centrifuge filter.

   6 Controlled fragmentation.

   7 Change frequency.

Cf   *1* Measured thrust coefficient of rocket.

   2 Skin friction coefficient (see $C_{D_{fric}}$).

C_f Cubic feet.

cf   Flap chord.

C_f^0 Theoretical thrust coefficient of rocket.

CFA   *1* Cross- (or crossed-) field amplifier.

   2 Chief flight attendant.

   3 Caribbean Federation of Aeroclubs (Int.).

CFAC   *1* Constant-frequency alternating current.

   2 Combined-Forces Air Component [C adds Commander] (USAF).

CFAO   Conception et fabrication assistée par ordinateur (F).

CFAR   Constant false-alarm rate (ECM).

CFAS   *1* Commandement des Forces Aériennes Stratégiques (F).

   2 Federal office for space (Switzerland).

C-Fast   Counter-force autonomous surveillance and targeting (USAF).

CFB   *1* Cruise fuel burn.

   2 Canadian Forces Base.

CFC   *1* Cycles to first crack (structures).

   2 Cold fog clearing.

   3 Carbon-fibre composite.

   4 Central fire control.

   5 Core-failure clutch.

   6 Centennial of Flight Commission (US).

   7 Chlorofluorocarbon.

CFCF   Central flow control facility (FAA 1).

CFD   *1* Chaff/flare dispenser; C adds computer, CU control unit, IU interface unit, S system.

   2 Centralized fault display (IU adds indicator or interface unit, S adds system).

   3 Computational fluid dynamics.

   4 Centralised fault display [IU adds interface unit, S adds system].

CFE   *1* Contractor-furnished equipment.

   2 Central Fighter Establishment (RAF).

   3 Conventional forces in Europe.

CFES   Continuous-flow electrophoresis in space.

CF_{ex} Excess-thrust coefficient.

CFF   *1* Cost plus fixed fee, sometimes CPFF.

   2 Critical flicker frequency (electronic displays).

CFFT   *1* Critical flicker fusion threshold.

   2 Cockpit and forward-fuselage trainer (F-22).

CFFTS   Canadian Forces Flying Training School.

CFG   Customer focus group.

CFI   *1* Chief flying instructor (UK).

   2 Certificated flight instructor (US).

   3 Call for improvement[s] (DoD).

CFII   Civilian, or certificated, flight instrument instructor (US).

C/FIMS   Contaminant and fluid-integrity measuring system.

CFIT   Controlled flight into terrain.

CFK, CfK   CFRP (G), usually KfK.

CFL   Cleared flight level.

CFM   Common functional module.

cfm   Cubic feet per minute.

CFMO   Command flight medical officer.

CFMT, C/F/M/T   Celsius, fahrenheit, magnetic, true.

CFMV   Central Flow, or Flight, Management Unit (Eurocontrol).

CFO   *1* Coherent-fibre optics.

   2 Chief Financial Officer.

   3 Central Forecast Office.

C433   Very strong al-alloy for lower wing skin (Alcoa).

CFP   *1* Cost plus fixed price.

   2 Communications-failure procedure.

   3 Cold-front passage.

CFPD   Command flight-path display.

CFR   *1* Co-operative Fuel Research.

   2 Crash fire rescue.

   3 Code of Federal Regulations.

   4 Contact flight rules.

   5 Call for release.

CFROI   Cash-flow return on investment.

CFRP   Carbon-fibre reinforced plastics.

CFS   *1* Central Flying School (RAF).

   2 Cabin file server.

   3 Chlorofluorosulphonic acid, suppresses contrails.

   4 Customer fleet service.

CFSO   Command flight safety officer.

CFSP   Common foreign and security policy (EU).

CFT   *1* Conformal fuel tank.

   2 Contractor flight test.

CFTR   Cold-fan thrust reverser.

CFTS   Contracted Flying Training and Support (Canadian DnD).

CFU *1* Colony-forming unit, measure of bacteria per cubic metre of cabin air.

2 Cartridge-firing unit (countermeasures).

CF weight Contractor-furnished.

CFWS Central Flow Weather Service; U adds Unit.

CG *1* Cargo glider, USAAF 1941–47.

2 Lethal gas phosgene.

3 Guided-missile cruiser (USN).

4 Commanding General.

5 Coast Guard.

C_g Ratio of actual/ideal propulsive thrust.

cg, c.g. Centre of gravity.

CGAC Combination, or combined, generator and air-conditioner (trailer mounted).

CGAO Conference of General-Aviation Organizations (UK).

c.g. arm Arm (2) obtained by adding all individual moments and dividing sum by aircraft total mass.

CGAS Coast Guard Air Station (US).

CGB Central gearbox.

CGCC Centre of gravity control computer.

CGF Computer-generated [usually armed hostile] forces.

CGH Guided-missile/helicopter cruiser (USN).

CGI *1* Chief ground instructor.

2 Computer-generated image, or imagery.

CGIVS CGI (2) visual system.

c.g. limits Forward and aft limits, usually expressed as percentage MAC, within which aircraft c.g. must fall for safe operation.

CGM Computer-graphics metafile[s] [sometimes **CGMF**].

CGM load Combined gust and manoeuvre load.

CGN Guided-missile cruiser, nuclear (USN).

CGP Coal-gasification plant, eg for jet fuel.

CGPM Conference Général Poids et Mésures [Int.].

CGRO Compton gamma-ray observatory.

CGS *1* Centimetre, gramme, second system of units, superseded by SI.

2 Central Gunnery, or [later] Gliding, School (RAF).

3 Computer-generated simulation.

4 Common ground segment, or station.

5 Chief of the General Staff (UK).

CGT Consolidated ground terminal.

CGV Computer-generated voice.

CGW Combat gross weight.

CH *1* Channel (ICAO).

2 Compass, or course, heading.

3 Critical height.

4 Chain home.

C_H *1* Hinge coefficient.

2 High cloud.

Ch *1* Channel.

2 Hinge-moment coefficient, hinge moment divided by control-surface area and chord aft of hinge axis and by q. $Ch\delta$ and $Ch\propto$ are derivatives with respect to control deflection and AOA of main surface.

Chaals, CHAALS Communications high-accuracy airborne location system (USAF).

chaff Radar-reflective particulate matter or dipoles sized to known or suspected enemy wavelengths (ECM).

chaff dispenser Tube, gun, projector or other system for releasing chaff either in discrete bursts or in measured stream, manually or upon command by RWS.

chafing patch Local reinforcement of aerostat envelope where likely to suffer abrasion.

Chag Chain arrester gear.

chain *1* Geographical distribution of radio navaid stations linked together and emitting synchronised signals (eg Decca, Loran).

2 Obsolete unit of length = 22 yd = 20.1168 m.

chain arrester Aircraft hook picks up wire attached to heavy chain on each side of runway.

chain gun High-speed automatic cannon with external power supply and open rotating bolt.

Chain Home The original string of interlinked radar stations constructed round east and south England from 1937. Chain Home Low added radars able to detect low-altitude targets.

chain radar beacon Beacon having fast recovery time, thus can be interrogated by many targets up to prf.

chain reaction In fissile material or other reactant, process self-perpetuating as result of formation of materials necessary for reaction to occur. Nuclear chain reaction arranged in weapons to have optimum uncontrolled (accelerating) form.

CHAIR, chair Control handling aid for increased range (target aircraft).

chairborne Condemned to desk job (originally said of RAF pilots).

chair chute Parachute permanently incorporated into seat (not necessarily ejection seat, but removable).

chalk Loosely, one load, especially stick or squad of troops, often 6–10.

chalk number In military logistics, number assigned one complete load and carrier vehicle; hence chalk commander, chalk troops.

challenge and response Basic method of working through any procedural cockpit check, nav (mil) or first officer (civil) reading each item after satisfactory response to previous item.

Chals Communications high-accuracy locating system; -x adds exploitable.

chamber In liquid rocket engine, enclosed space where combustion takes place, between injectors and throat. In solid rocket, enlarging volume in which combustion takes place, varying in form with design of motor.

chamber pressure See *burn time average, action time average, MEOP.*

chamber volume In liquid rocket, total volume as defined; solid motor varies during burn.

chamfered Bevelled (edge or corner, eg of sheet).

champ Cargo-handling and management planning [s adds system].

Champs Common helicopter aviation mission planning system (USN/USMC).

Chance Complete helicopter advanced computational environment.

Chance light Formerly, mobile airfield floodlight illuminating landing area and apron.

chandelle Flight manoeuvre (see *stall turn*); another definition, not necessarily synonymous with stall turn, is a manoeuvre in which speed is traded for altitude whilst reversing flight direction (see also *Immelmann*).

changeover point Ground position and time at which aircraft switches from using one ground-based navaid to another, not necessarily at midpoint of leg.

channel *1* Band of frequencies in EM spectrum, esp. at

radio nav/com frequencies (thus 20 * allotted to ILS in all countries, each with published frequency for loc and g/p).

2 Single end-to-end 'route' in dynamic system, esp. one exerting control authority (eg collective pitch in helicopter AFCS).

3 Structural member of channel form (eg top-hat or U).

4 In EDP, several meanings: any information or data highway, one or more parallel tracks treated as unit, portion of store accessible to given reading station.

5 In semiconductor device (eg transistor), flow bypassing base.

6 Pathways for energetic ions or atoms along crystal lattices.

7 In airport terminal, single routing for departing or arriving passengers.

8 Takeoff and alighting path at marine airport.

channel nut   One forming integral part of channel (3).

channel patch   Channel-shaped reinforcement to aerostat envelope to anchor rigid spar.

channel section   See *channel (3)*.

channel wing   Wing curved in front elevation to fit closely round lower half of propeller disc.

Chapi, CHAPI   Carrier, colour or compact helicopter approach path indicator.

char   Ablative material charred and eroded during re-entry (see *ablation*).

characteristic   Sense (up or down) in which barometric pressure changes in preceding 3 h.

characteristic curve   *1* Curve of atmospheric sounding results plotted on Rossby diagram.

2 Curve of primary characteristic of aerofoil when plotted (see *characteristics*).

characteristic exhaust velocity, C*   Measure of rocket performance, numerically $g_c A_t / W_p$ multiplied by integral of chamber pressure over action time.

characteristic length   *1* In rocket, ratio of chamber volume to nozzle throat area.

2 In rocket, length of cylindrical tube of same diameter as chamber, having same volume as chamber.

3 Convenient reference length (eg chord).

characteristics   *1* Of aerofoil, primary * are: coefficients of lift and drag, L/D ratio, cp position and coefficient of moment, each plotted for all operating AOAs.

2 In electronics, relationships between basic variables (eg anode current/voltage, anode current/grid voltage) for valves (tubes) and correspondingly for transistors.

characteristic velocity   *1* Sum of all changes in velocity, positive and negative all treated as positive, in course of space mission.

2 Velocity required for given planetary (esp. Earth) orbit.

charge   *1* Total mass of propellant in solid rocket.

2 Propellant of semi-fixed or separate-loading ammunition.

3 To fill high-pressure gas or cryogenic (eg Lox) container.

4 Quantity of electricity, measured (SI) in Coulombs.

chargeable   Malfunction (eg IFSD) clearly due to fault in design, workmanship, material or technique by supplier.

charging point   Standard coupling through which aircraft fluid system is replenished or pressurized.

Charles' law   Perfect gas at constant pressure has volume change roughly proportional to absolute temperature change.

Charlie   Preplanned landing-on time in carrier operations.

Charlière   Common term for a gas balloon (pre-c1850).

charm   Composite high-altitude radiation model.

Charme   Concept d'helice pour avions rapides en vue d'une meilleure économie (12-blade single rotation propfan).

Charpy   Destructive test of impact resistance of notched test bar.

chart   Simplified map, typically showing coasts, certain contours, woods, water, and aeronautical information (symbols vary and not yet internationally agreed).

chart board   Rigid board, suitably sized for chart or topographic map; often provided with protractor on parallel arms.

charted approach   Visual flight to destination authorised to radar-controlled aircraft on IFR flight plan.

charted delay   In Loran, published delay.

charts   Compact hydrographic airborne rapid total survey.

Chase   Coronal helium abundance Spacelab experiment.

chase   To accompany other aircraft, esp. one on test, to observe behaviour and warn of visible malfunction; hence * plane, * pilot.

chassis   *1* Rigid base on which electronics are mounted; for airborne equipment, mates with racking.

2 Landing gear. In WW2 the accepted UK term was undercarriage, but the author recalls many cockpits where * was used because there was no room for the longer word.

Chats   Counter-intelligence Humint automated tool system.

chatter   *1* Multiple conversations or signatures, most being of no interest, all heard on same frequency.

2 High-frequency vibration energised by intermeshing gear teeth.

3 [also **chattering**] any small uncommanded repeated operations, e.g. by hydraulic jack.

CHB   Common high bandwidth.

CHC   Chance, traditionally 30–40% likelihood of precipitation.

CHD   Crutching heavy-duty (stores carrier).

cheatline   Bold line[s], usually horizontal, painted at artistically pleasing level along side of aircraft.

check   *1* To verify flight progress (eg ETA *).

2 To examine pilot for proficiency.

3 Examination conducted in (2).

4 Programmed investigation of aircraft for malfunction (before-flight *, after or post-flight *, etc).

5 Programmed procedural routine from entering cockpit to start of flight (cockpit *).

6 To reduce in-flight trajectory variable (* rate of descent).

Checkered Flag   Provides designated bases and procedures for overseas training deployments (USAF, TAC).

checker team   Responsible for maintenance of all runway/taxiway/apron markings.

check finals!   Last brief vital actions before landing (items not standardized).

check in   *1* Several meanings, esp. to call up tower to establish frequencies and procedures at start of cockpit check.

2 For airline passenger, to report at designated * desk, have page of flight coupon torn off, hand over hold baggage and be assigned seat.

checking station   *1* Designated and marked location used in rigging airframe. Also called checking point.

2 Reference station on propeller blade, typically at 42 in radius on lightplanes.

checklist   Written list of sequenced actions taken in check (5), usually printed on plastics or laminated pages. Usually also includes emergency actions, V-speeds and short- or soft-field procedures.

check-out   *1* Programmed test by user immediately before operational use of missile, radar or other system.

2 Sequenced tasks to familiarise operator with new hardware.

check point, checkpoint   *1* See *waypoint*.

2 Geographical point happening to offer precise fix.

3 Planned strategic review of major industrial programme for government customer, esp. collaborative.

4 Predetermined geographical location serving as reference for fire adjustment or other purpose.

5 Mean point of impact.

check six!   Look astern for hostile fighters.

check valve   One-way valve in fluid line.

cheese aerial (antenna)   Shaped roughly like D in side elevation with quasi-circular or parabolic reflector and flat parallel sides nodding in elevation; propagates more than single mode.

chemical fuel   Has ambiguous connotation excepting common fuels; hence, 'exotic' fuel, esp. for air-breathing engines.

chemical laser   Strictly, one in which chemical reaction occurs, eg perfluoromethane/hydrogen.

chemical milling   Use of acid or alkaline bath to etch metal workpiece in controlled manner.

chemical munition   Non-explosive ordnance operating by chemical reaction: incendiary, smoke, irritant or lethal gas, defoliant, flare or flash, dye marker etc.

chemical propulsion   Propulsion by energy released by chemical reaction, eg fuel + oxygen, with or without air breathing.

chemical rocket   Rocket operating by chemical reaction; not ion, photon or nuclear.

chemical warfare   Use of major chemical munitions, esp. irritant or lethal gases.

chemosphere   Region of upper atmosphere (say 15–120 miles, 24–190 km) noted for photochemical reactivity.

chequered flag   *1* Black/white, various meanings including race winner.

2 Red/yellow, do not move off blocks until ATC permits.

3 See *Checkered Flag*.

cherl   Change in Earth's rate (tangential velocity) of rotation with latitude.

Cherry rivet   Tubular rivet inserted blind and closed by internal mandrel (shank) which then breaks and is removed.

chest-type parachute   Pack stored in aircraft separate from harness, to which it is secured by quick clips on chest.

cheval du bois   'Wooden horse' = swing [1] (F).

cheval vapeur   Metric horsepower, 1 cv = 0.98632 hp = 0.7355 kW; reciprocals 1.01387, 1.35962.

chevron mixer   Sawtooth nozzle.

Cheyenne Mountain   Location of USAF/Norad Space Command HQ.

CHF   Commando helicopter force.

CHG   *1* Change, ie modifying previous message; CHGD, changed.

2 Charge.

CHI   Computer/human interface.

Chicken   See *State Chicken*.

chicken bolts   Temporary fasteners used in metal airframe assembly.

chicks   *1* Fighters, especially airborne group under unified local command (USN).

2 Fighters in group round air-refuelling tanker.

chiefy   Flight sergeant in charge of erks (RAF, colloq.).

chilldown   Pre-cooling of tanks and system hardware before loading cryogenic propellant.

chilled   Stage at which design is almost frozen [change accepted reluctantly].

chilled casting   Made in mould of metal, usually ferrous, giving rapid cooling and thus surface hardness.

chime   Melodious warning of imminent announcement to passengers, typical output 120 W.

chin   Region under aircraft nose; hence * blister, * intake, * radome, * turret.

china-clay   Technique for distinguishing regions of laminar and turbulent boundary layer from changed appearance of thin coating of * suspension.

China Lake   California desert home of NAWS.

chine   In traditional marine aircraft hull or float, extreme side member running approximately parallel to keel in side elevation. In supersonic aircraft, sharp edge forming lateral extremity of fuselage, shedding strong vortex and merging into wing.

chined tyre   Tyre, esp. for nosewheels, with chines to depress trajectory of water or slush.

chin fairing   On centreline on underside of leading edge [T-tail].

chin fin   Fixed destabilizing fin under nose.

Chinook   Warm dry wind on E side of Rocky Mountains.

chip   *1* Single completed device separated from slice, wafer or other substrate of single-crystal semiconductor.

2 Metal fragment, visible to eye, broken from engine or other machinery.

chip chart   Rectangles of paint showing colours available.

chip detector   Device, often permanent magnet, for gathering every chip (2), usually from lube oil.

Chips   Cosmic hot interstellar plasma spectrometer.

chip width   Length of PN code bit, T_c.

Chirp   Confidential human-factors incident reporting procedure (or programme) (CAA).

chirp   *1* Radar/communications pulse compression or expansion (colloq.).

2 Particularly, pulse compression by linear FM.

CHIS   Center hydraulic isolation system.

chisel window   Small oblique nose window for LRMTS or camera.

CHL   Chain Home Low.

ch.lat.   Change of latitude (pronounced sh-lat).

ChLCD   Cholesteric liquid-crystal display.

ch.long.   Change of longitude (pronounced sh-long).

chlorine   Cl, toxic green/yellow gas, density 3.2, MPt -101°C, present in vast range of aerospace products.

Chobert   Tubular rivet inserted blind and closed by withdrawal of re-usable mandrel.

chock   Portable obstruction placed in front of and/or behind landplane wheel(s) to prevent taxiing.

chock-to-chock   See *block time*.

choke   *1* Inductance used to offer high reactance at chosen frequency to pass d.c. or lower a.c. frequencies only.

2 In typical car (auto) engine, manual control for reducing inlet airflow to enrich mixture when cold, rare in aviation.

choked flow   Flow of compressible fluid in duct [eg tunnel or jet-engine nozzle] in which local Mach number has reached 1 and velocity cannot be increased significantly by increasing upstream pressure.

choked inlet   Containing normal shock and suffering choked flow.

Chol, CHOL   Common [or Collins] high-order language.

cholesteric   LCD with layers each aligned in preferred, different direction.

chomp   Changeover (from one waypoint VOR, NDB etc to next) at midpoint (of leg).

Chop   Countermeasures hands-on program[me].

chop   *1* Changeover point.

2 Change of operational control, precisely promulgated time.

3 To get the *, sudden termination (of human life, project, place on flying-training course, etc. colloq.).

4 To close throttle(s) completely and suddenly.

5 Atmospheric turbulence, esp. CAT, categorised as mild (also called light), moderate (or medium) and severe (or heavy).

chopped fibre   Reinforcing fibre chopped into short lengths.

chopped random mat   Chopped fibre made into mat (two-dimensional sheet) with random orientation.

chopper   *1* Rotary-wing aircraft, esp. helicopter (colloq.).

2 Mechanical device for periodically interrupting flow, esp. light beam, or switching it alternately between two sources.

3 Device for modulating signal by making and breaking contacts at frequency higher than frequencies in signal.

chop rate   Rate of aircraft or crew loss on operations, or wastage rate in flying training (colloq.).

chord   *1* Straight line parallel to longitudinal axis joining centres of curvature of leading and trailing edges of aerofoil section.

2 Loosely, breadth of wing or other aerofoil from front to rear.

3 Boundary members of structural truss.

chord direction   In stress analysis, usually parallel to chord at aircraft centreline (of wing, or wing produced to centreline).

chord length   Length of chord (1), not measured round profile.

chord line   See *chord (1)*. Ambiguously, sometimes line tangent at two points to lower surface (see *geometric chord*).

chord plane   Plane containing chord lines of all sections forming three-dimensional aerofoil (assuming no twist).

chord position   Defined by location of quarter-chord point and inclination to aircraft x-y plane, point being defined on primary centreline co-ordinates.

chord wire   Wire tying vertices of airship frame.

chordwise   Parallel to chord (normally also to longitudinal axis).

chosen instrument   Carrier selected as national monopoly [can be private company].

CHP   Controlled-humidity preservation.

CHR   Cooper-Harper rating.

Christmas tree   Aircraft temporarily set aside as source of spare parts, but to be eventually returned to service.

chromate enriched pellet   Small source of antifungal chemical in integral-tank low point (possible water trap).

chromate primer   Anti-corrosive, antimicrobiological surface treatment, esp. for water traps in airframe.

chromatic aberration   Rainbow effect caused by simple lens having different focal length for each wavelength.

chrome steels   Steels containing chromium; often also with vanadium, molybdenum etc.

chromic acid   Red crystalline solid, $H_2 Cr O_4$, used in solution as cleaner and etchant, as electrolyte [eg, Alocrome) for Cr plating and anodising, and for crack detection.

chromic paint   Coating which changes colour (usually white-grey-blue-black) as temperature increases.

chromium   Cr, hard silvery metal taking brilliant polish, density 7.2, MPt 1,860°C.

Chromoly   Alloy steels containing chromium and molybdenum.

chromosphere   Thin (under 15,000 km) layer of gas surrounding Sun's photosphere.

chronograph   Device for producing hard-copy readout of variable against time, with particular events recorded, typically by pen and disc or drum chart.

chronometer   Accurate portable clock with spring drive and escapement.

CHS, CH/S   Common hardware and software.

CHT   Cylinder-head temperature.

chuck   Rotating vice (US "vise") for gripping tool or workpiece.

chuck glider   Small model glider [not necessarily a toy] launched by hand.

chuffing   Pulsating irregular rocket combustion.

chum   Chart update module.

chutai   Squadron (J).

chute   *1* Parachute (colloq.).

2 Axial/radial ducts around engine periphery to guide fan air into afterburner or engine airflow through silenced nozzle.

3 Duct for ventral crew escape or discharge of ECM, leaflets etc. Not to be confused with *escape slide*.

CI   *1* Chief instructor.

2 Compression ignition.

3 Catalytic ignition [see *CIS (6)*].

4 Certificate of Interest, signed by intending purchaser, unless rescinded becomes binding contract after stipulated period.

5 Cubic inches [strongly deprecated].

6 Configuration item.

7 Cabin interphone.

8 Counter-intelligence.

9 Chief inspector.

10 Competitive intelligence, primarily for multinational corporations.

11 Control indicator.

12 Continuous improvement.

Ci   *1* Cirrus.

2 Curie[s].

ci   Cubic inches (not recommended).

CIA   *1* Contractor interface agreement.

2 Central Intelligence Agency (US, 1947 on).

3 Commission Internationale d'Aérostation (FAI).

4 Captured in action.

CIAA   *1* Consorzio Italiano di Assicurazioni Aeronautiche.

2 Centre International d'Aviation Agricole.

CIACA   Commission International des Aéronefs de Construction Amateur (FAI).

CIAM   *1* Computerised integrated and automated manufacturing.

2 Commission Internationale d'Aéromodélisme (FAI).

CIANA   Latin American commission for air navigation (office Madrid, from October 1926).

CIAP   Climatic Impact Assessment Program (US DoT).

CIARA   See *IAARC*.

CIAS   Comitato Interministeriale Attività Spaziali (I).

CIASE   China Institute of Aeronautic Systems Engineering.

CIB   Controlled image base.

CIC   *1* Combat information center (USA, USAF).

2 Commander-in-Chief [UK usage, C-in-C].

3 Corrosion-inhibiting compound.

Cicas   Check-in counter allocation system, displays logo and flight details at each counter.

C(I)CT   Completion of interim certification testing.

CID   *1* Combat identification.

2 Check-in display.

3 Category-interaction.

CIDA   Co-ordinating installation design authority.

Cidef   Conseil des Industries de Défense Français.

CIDIN, Cidin   Common ICAO data-interchange network (Int.).

CIDS   *1* Cabin interface [previously intercommunication or interphone] data, or distribution, system, with microprocessor control for speakers, lamps, PA, entertainment systems, safety signs, crew intercom, etc, to facilitate changes in interior layout.

2 Check-in display system.

CIE   *1* Commission Internationale de l'Eclairage (= ICI).

2 Centro de Investigaciones Espaciales (Arg.).

CIEA   Commission Internationale d'Enseignement Aéronautique (FAI).

CIEM   Computer-integrated engineering and manufacturing.

CI-F   Control indicator, front.

CIFMS   Computer-integrated flexible manufacturing systems.

CIFRR   Common IFR room (New York, civil/military).

CIFS   Computer-interactive flight simulation.

CIG   *1* Computer image-generation.

2 Control/indicator group.

3 Commission Internationale de Giraviation (FAI).

4 Ceiling.

CIGAR   Many US mnemonics begin: Controls, instruments, gas, attitude [trim/flaps], run-up [mags., carb. heat, etc].

Cigar   *Airborne Cigar.*

cigar   Aeroplane from which wings have been removed (colloq.).

cigarette-burning   Solid propellant ignited at one end across entire section and burning towards other end.

Cigars   Mnemonic reminding pilot of vital actions before takeoff: controls, instruments, gas, attitude indicator, run up, seat belt (US, WW2).

CIGFTPR   Controls-instruments-gas-flaps-trim-prop-run-up (US).

CIGS   Chief of the Imperial General Staff [now CGS] (UK).

CIGSS   Common imagery ground/surface standards.

CIGTF   Central Inertial Guidance Test Facility (USAF).

CIIP   Critical information infrastructure protection.

CIJ   Close-in jamming.

CIL   *1* Candidate-items list.

2 Command Information Library (NIMA).

CIM   Computer-integrated manufacturing.

CIMA   *1* Chartered Institute of Management Accountants (UK).

2 Commission Internationale de Micro Aviation (FAI).

CIME   Commission Intergouvernementale des Migrations Européennes (Int., arranges mass flights, eg for refugees).

CIMT   Configuration integration management team.

Cimtic C^4I and munitions test improvement contract.

CINA   Commission Internationale de la Navigation Aérienne (Int., office Paris, 1922–).

Cincat   Capacity increase through controller assistance tools (Euret).

Cinch   Compact inertial navigation combining HUD.

CINS   Compact INS.

CIO   *1* Central Imagery Office (US).

2 See *AFL-CIO*.

3 Chairman in office.

4 Chief information officer.

CIOD   Counterspace and Information Operations Division (USAF).

CIOS   Combined Intelligence Objectives Sub-committee (US/UK, WW2).

CIP   *1* Cold iso-pressing (beryllium).

2 Component [or communications] improvement program(me).

3 Commission Internationale de Parachutisme (FAI).

4 Commercially important passenger.

5 Command input potentiometer.

6 Capital investment plan (FAA).

7 Critical infrastructure protection.

8 Common imagery, or core-integrated, processor.

CIPR   Cubic inches per revolution.

CIR   *1* Constant, or continuous, infra-red (heat source on target).

2 Cockpit image recorder.

CIRA   *1* Cospar International Reference Atmosphere.

2 Centro Italiane Ricerche Aerospaziali (I).

CIRC   Central information reference and control.

circ   Circling or circulating.

circadian rhythm   Change in physiological activity on approximate 24 hour cycle.

Circe   Cossor interrogation and reply cryptographic equipment, enabling all participating nations to have own secure cryptographic IFF.

Circle   Circuit (US); can also be an instruction to join stack.

circle marker On unpaved airfield, white circle indicating centre of landing area.

circle of confusion Image of any distant point on lens focal plane (eg on film).

circle of equal probabilities, CEP Radius of circle within which half the strikes (eg bullet impacts on single aiming point) fall or within which probability is equal that one bullet, bomb or RV will fall inside or outside. Also called circular error probable.

circle of latitude On celestial sphere through ecliptic poles, perpendicular to ecliptic.

circle of longitude On celestial sphere parallel to ecliptic.

circle of origin Normally, Equator or Prime Meridian.

circle of position Circle on Earth's surface centred on line joining centre of Earth to heavenly body, from which altitude of body is everywhere equal. Sometimes called circle of equal altitude.

circuit *1* Basic element in pilot training, short flight comprising takeoff and precisely executed * back to landing, if necessary ready for repeated *. In US often circular, but in UK and most other countries rectilinear, comprising takeoff, straight climbout [upwind leg] to [typically] 300 m/1,000 ft, turn 90° [in either direction, but usually L] on to crosswind leg, second 90° turn on to downwind leg, passing airfield parallel to runway, third 90° turn with power off on to descending second crosswind leg [often called base leg] followed by 90° turn on to finals. Depending on circumstances, * may be followed by immediate takeoff called touch-and-go [if the aircraft is brought to rest, called stop-and-go] for second * or by taxi back to downwind end of runway for next takeoff; * can be L-hand or R-hand. US term pattern, circle or traffic circle.

2 List of airshows [usually annual] at which particular aircraft regularly appear.

3 Closed loop of electrical conductors.

circuit analog absorber Large family of RAM (2) in which outer resistive sheet is given an imaginary part to its admittance, by laying it down in form of many discrete elements such as dipoles, crosses and meshes. See *frequency-selective*.

circuit breaker Switch for opening circuit (3) while carrying large electrical load.

circuit length Length round closed-circuit wind tunnel traversed by streamline always equidistant from walls.

circuits and bumps Repeated circuits (1) with landing at end of each (colloq.).

circular approach Precision training manoeuvre, chiefly associated with carrier flying, where circuit (1) is circular.

circular dispersion Diameter of smallest circle within which 75 per cent of projectiles strike.

circular error probable See *bombing errors (1)*.

circular isotropic Aerial (antenna) radiation polar equal in all directions (normally in azimuth).

circularisation Refinement of satellite orbit to approach perfect circle, usually at given required height.

circular mil Area of circle of one mil (1/1,000 inch) diameter = $5.067 \times 10^{-10} \text{m}^2$ ($7.85 \times 10^{-7} \text{in}^2$).

circular mil foot Unit of resistivity; resistance of one foot of wire of one circular mil section, equal to ohm-mm $\times$ 6.015×10^5.

circular milliradian Conical beam or spread of fire having angle (not semi-angle) of one milliradian.

circular nose Control surface whose leading-edge section is semicircle about hinge axis.

circular velocity At given orbital height, velocity resulting in circular orbit, $V_c = \sqrt{Rg}$ where R is radius from Earth centre.

circulation *1* Rotary motion of fluid about body or point; vortex.

2 Ideal flow around (not past) circular body, with streamlines concentric circles and velocity inversely proportional to radius (body needed to avoid infinite V at centre).

3 Streamline flow around body of any form, defined as integral of component of velocity along closed circuit with respect to distance travelled around it. Wing lift created by * superimposed on rectilinear flow past surface (see *bound vortex, Magnus, Zhukovsky*).

4 Gross motions of planetary (eg Earth) atmosphere.

circulation controlled Wing, rotor blade or other aerofoil in which external power is used to enhance lift, typically by high-velocity tangential blowing of various kinds.

circulator Non-reciprocal device in microwave circuit to produce phase-shift as function of direction of wave flow (see *duplexer*).

circulatory flow Rectilinear flow past lifting body inducing circulation (3) (see *Zhukovsky*).

circumaural Fitting around the ear.

Circus Small formation of bombers with much larger fighter escort, objective being to lure enemy fighters to combat (RAF, WW2).

circus *1* Large loose formation of fighters, usually with distinctive individual markings, flown by aces (G, WW1).

2 Loosely, group of itinerant aircraft entertaining public and offering rides (1919–39).

CIRF Consolidated intermediate repair facility.

CIRM Comité International de Radio Maritime.

CIRO Centre Interarmée de Recherches Opérationelles (F).

Cirpas Centre for interdisciplinary remotely-piloted aircraft studies (USN).

Cirris Cryogenic infrared radiance instrument for Shuttle.

cirrocumulus Cc, layer of globular cloud masses at about 6,000 m/20,000 ft. Also known as mackerel sky.

cirrostratus Cs, high milky-white or grey sheet cloud, 7,000 m/23,000 ft.

cirrus Ci, high white cloud; detached, fibrous, silky, 7,500–12,000 m/25,000–40,000 ft.

Cirstel Combined IR suppression and tail-rotor elimination.

CIRTEVS, CIRTVS Compact IR TV system.

CIS *1* Combat identification system.

2 Computer interface system.

3 Communications, or command, or combat, or corporate, information systems.

4 Co-operative independent surveillance.

5 Cluster ion spectrometer.

6 Chemical ignition system.

7 Control indicator set, or suite.

8 Commonwealth of Independent States, of former USSR; CST adds Collective Security Treaty.

9 Cargo-inspection system, usually PFNA or X-ray.

cislunar Between Moon's orbit and Earth.

CISPR Comité International Spécial des Peturbations Radiophoniques [radio interference].
CIT *1* Compressor inlet temperature (flight envelope limit).
2 Central integrated testing.
3 Cranfield Institute of Technology.
4 Control in turbulence [mode].
5 Critical-item test.
6 The Chartered Institute of Transport (UK, 1919, received Charter 1926).
7 Near or over a city.
8 Combined interrogater and transponder.
9 Commission for Integrated Transport (UK think tank).
CITA *1* Commission Internationale de Tourisme Aérien.
2 Confederación Interamericana de Transportadores Aéreos.
CITE *1* Computer integrated test equipment (USAF).
2 Compression-ignition and turbine engine (fuel).
CITEJA, Citeja Comité International Technique d'Experts Juridiques Aériens (1925–47, now part of ICAO).
Citeps Central integrated test experimental parameter subsystem.
CITES Convention on International Trade in Endangered Species.
CITIS Contractor integrated technical information system.
CITS *1* Central integrated test subsystem (eg Shuttle).
2 CAS (3) integrated targeting system.
3 Combat information transport system.
city pair Pair of cities studied from viewpoint of mutual passenger/cargo traffic.
city pair ranking Lists of ** in order of current or projected traffic generation.
CIU *1* Computer, central, cockpit, coupler, communications, or control interface unit.
2 Central Interpretation Unit (RAF, WW2).
3 Control-information unit (cartridge dispensing).
CIV *1* Crossbleed isolation valve.
2 Coannular inverted-velocity (nozzle).
3 Civil.
CIVA Commission International de Vol Aérobatique.
Civil Aeronautics Administration Since 1958 FAA (1).
Civil Aeronautics Board, CAB US Government (DoC) agency responsible for civil aviation, including CARs, licensing, routes and US mail rates.
civil aircraft Not in government [including military] service.
Civil Air Patrol, CAP US para-military organization using pilot and lightplane resources of general aviation for national ends.
civil day Day of constant 24 hours (sometimes counted as two periods of 12 hours); mean solar day.
Civil Reserve Air Fleet US airline transport aircraft and flight crews predesignated as available at any time for reasons of national emergency.
civil time See *mean solar time*.
civil twilight Period at sunrise or sunset when Sun's centre is between 0° 50' and 6° below horizon.
CIVL Commission International de Vol Libre (FAI hang-gliding organization).

CIVRES Congrès International des Techniques du Vide et de la Recherche Spatiale.
Civs, Civils CAA [1], (UK, colloq.).
CIVV Commission International de Vol à Voile (gliding).
CIWS Close-in weapon system.
C$_j$ Blowing coefficient, or thrust coefficient of jet engine.
CJAA Classic Jet Aircraft Association (US).
CJAP Commonwealth Joint Air Training Plan (1939–45).
CJCS Chairman of the Joint Chiefs of Staff.
CJO Chief of Joint Operations (UK MoD).
CJTF Combined [or commanders] joint task force.
CK Cape Kennedy.
CK, Ck Check.
CKD Component, or completely, knock-down, parts imported for assembly in importing country.
CKEM Compact kinetic-energy missile.
CL *1* Centreline of aircraft.
2 Checklist.
3 Chemical laser.
4 Catapult-launched.
5 Charge limit, ie limit payload (RAF).
6 Creeping landing.
7 Centre of lift.
8 Compass locator.
9 Centreline lights of runway.
C$_L$ *1* Coefficient of lift.
2 Low cloud.
Cl Rolling moment coefficient (BSI).
cl Centilitre.
c$_l$ Section lift coefficient.
CLA *1* Clear ice formation.
2 Centreline average (surface roughness).
3 Collective labor agreement (US).
4 Consortium of Lancashire Aerospace, Became NWAA.
CLAC Comisión Latino Americana de Civil Aviación (Int.).
clack, clacking Aural warning, esp. of Mach limit.
clack valve Fluid one-way valve having freely hinged flap seated on one side.
Clads Common large-area display set.
CLAES Cryogenic limb array etalon spectrometer.
CLAEX Air-force flight test centre (Spain).
clag Widespread low cloud, mist and/or rain (colloq.).
CL/AL Catapult-launched, arrested landing.
CLAMP Closed-loop aeronautical management programme.
clamp Weather unfit for flight (colloq.).
clamping To hold either or both peaks of waveform or signal at desired reference potential (d.c. restoration). Increasingly used in processing sensor images; black-level * references all black levels to darkest point of image.
clamshell *1* Cockpit canopy hinged at front or rear.
2 Nose or tail of cargo aircraft hinged into lower and upper or left and right halves.
3 Reverser opening in upper and lower halves meeting on jet centreline behind nozzle [US = bucket].
clandestine aircraft Aircraft designed to overfly without detection, having minimal noise, IR and radar signatures.
clang box Jet-engine switch-in deflector for V/STOL comprising an internal valve and side nozzle with deflecting cascade.

Clansman   Army tactical radio communication system (UK).

CLAP   Centre Laïque d'Aviation Populaire (F).

clapper   Part-span shroud.

Clara   Carbon-dioxide-laser radar, for obstacle avoidance.

CLASB   Citizens' League Against the Supersonic Boom (US).

Class   Coherent laser airborne shear sensor.

class action   Litigation in US courts in which plaintiffs represent a class, eg airline passengers, or passengers of a particular carrier.

classic   Term merited by aircraft produced for many years, esp. to distinguish from later versions of same type.

classical aeroplane   Aeroplane having clearly defined fuselage, nacelles and aerodynamic surfaces, not necessarily with all tail surfaces at rear. Opposite of integrated aeroplane.

classical flutter   Occurring because of coupling – aerodynamic, inertial or elastic – between two degrees of freedom.

classify   *1* To protect official information from unauthorised disclosure [UK and US have numerous classification grades].

2 In ASW to sort sonar returns according to types of source.

claw   *1* Accelerator hook.

2 Operative part of arrester hook.

Claws   Complementary low-altitude weapon system (USMC).

CLB   *1* Crash locator beacon.

2 Climb, helicopter autopilot mode.

C$_{L\beta}$ Dihedral effect, the rolling moment due to sideslip.

CLBR   Calibration.

CLC   *1* Command launch computer.

2 Course-line computer.

CLD   *1* Cloud (ICAO).

2 Crutching light-duty (stores carrier).

CLDP   Convertible laser-designation pod.

CLDS   *1* Cockpit laser-designation system.

2 Clouds.

CLE   Central Landing Establishment, RAF Ringway 1941, pioneer paratroop/glider school.

Clean   Component validator for environmentally friendly aero engine.

clean   *1* Of aircraft design: streamlined, devoid of struts and other excrescences.

2 Of aircraft condition: landing gear, high-lift systems and other extendible items retracted, and not carrying drop tanks, external ordnance or other drag-producing bodies.

3 Nuclear weapon designed for reduced, or minimal, residual radioactivity compared with normal weapon of same yield.

cleaning   In prolonged glide with piston-engined aircraft, to open up engine briefly to high power to clear over-rich mixture and gummy or carbon deposits.

clean room   Sealed airlock-entrance facility for manufacture [eg, of inertial gyro] or examination of space samples, with rigid rules on humans admitted.

clean up   To retract gear and flaps, and other high-lift devices, after takeoff.

clear   *1* To authorise hardware as fit for use.

2 To authorise person to receive classified information.

3 To rectify stoppage in automatic weapon.

4 To unload weapon and demonstrate no ammunition remains.

5 To empty core store, register or other memory device.

6 In flight operations, authorised to take off, land or make other manoeuvre under ground control.

7 En route, to pass over waypoint.

8 To destroy all hostile aircraft in given airspace.

9 Of local sky, devoid of clouds ("the *"), but may be above or between cloud layers.

10 To clean piston engine; see *cleaning*.

11 To fly out of a local area, eg a flying display.

12 Not secure [communications].

clear air turbulence, CAT   Significant turbulence in sky where no clouds present, normally at high altitude in high windshear near jetstream.

clearance   *1* Authorisation by ATC (1), for purpose of preventing collision between known aircraft, for aircraft to proceed under specified conditions within controlled airspace (see *abbreviated *, SIDS, STARs, * delivery, * items, * limits*).

2 Minimum gap between portions of hardware in relative motion (eg fan blade and case).

3 Transport of troops and material from beach, port or airfield using available communications.

4 Approval for publication of written text, image or film concerning sensitive subject, after excision of offending parts.

clearance amendment   Change in clearance (1) made by controller to avoid foreseeable conflict.

clearance delivery   ATC service, with assigned frequency, for issuing pre-taxi, taxi and certain other pre-flight clearances.

clearance function   Clearance delivery (UK).

clearance limit   Fix or waypoint to which outbound flight may be cleared, there to receive clearance to destination.

clearance void   Automatic cancellation if takeoff not made by specified time.

clearance volume   Minimum volume remaining in piston engine cylinder at TDC.

cleared flight level   FL to which flight is cleared, though possibly not yet reached.

cleared through   Valid to clearance limit, including intermediate stops.

clear ice   Glossy, clear or translucent accretion from slow freezing of large supercooled water droplets.

clearing manoeuvre   Change of aircraft attitude, on ground or in flight, to give better view of other traffic.

clearing procedure   Clearing manoeuvre, often combined with vocal callouts (esp. when pupil under instruction) before takeoff or any other flight operation (eg scrutiny of airspace beneath prior to spin).

clearing turn   Turn in which pilot checks local airspace, especially below, before stall or spin.

clear-vision panel   See *DV panel*.

clearway   *1* Rectangular area at upwind end of runway or other takeoff path devoid of obstructions and prepared as suitable for initial climbout.

2 Specif., area beyond runway, extending not less than 250 ft/76 m wide on each side of centreline, no part of which (other than threshold lights away from centreline and not over 26 in/660 mm high) projects above * plane.

clearway plane　Plane extending from upwind end of runway at slope positive and not exceeding 1.25 per cent.

cleat　In airframes, a triangular brace at a junction.

clevis joint　Fork and tongue joint (eg between solid motor cases) secured by large-diameter pin.

CLF　Carbon-loaded foam, common single-layer RAM.

CLFA　Centre de Laser Franco-Allemagne.

CLG　*1* Ceiling (ICAO).
　2 Calling.

C$_L\gamma$, C$_L$ gamma　Circulation lift coefficient.

CLGE　Cannon-launched guidance electronics.

CLGP　Cannon-launched guided projectile.

CLI　Common languages interactions.

Climate Change Levy　Financial penalty imposed [in absence of precise numerical values] on users of energy from non-renewable sources (EC).

climatic test　Static test in simulated adverse environments (rain, ice, temperature extremes, salt, sand, dust) to demonstrate compliance with requirements.

climb　*1* Any gain in height by aircraft (verb or noun).
　2 More commonly, deliberate and prolonged gain in height by appropriate trajectory and power setting (ie not zoom).

climb corridor　Positive controlled military airspace of published dimensions extending from airfield.

climb gradient　Vertical height gained expressed as percentage of horizontal distance travelled.

climb indicator　See *VSI*.

climbing cruise, climb cruise　Compromise between speed and range, typically at 1.15 V_{md} planned from published tables for peak efficiency higher than attainable in constant-height cruise.

climbing shaft　Access hatch and ladder leading from bottom to top of airship hull.

climb out　*1* Loosely, flight from unstick to setting course (lightplane in VFR).
　2 Specif., flight from screen height (35 ft/11 m) to 1,500 ft/460 m. Comprises six segments: 1, 35 ft to gear up (V_2); 2, gear up to FRH (V_2); 3, level (accelerate to FUSS); 4, FRH to 5-minute power point (FUSS); 5, level (accelerate to initial ERCS); 6, to 1,500 ft/460 m (ERCS) (see *NFP*).

clinker-built　Marine hull or float constructed from diagonal or longitudinal planks overlapping at edges.

clinodromic　Holding constant lead angle.

clinometer　*1* Instrument for measuring angle of elevation, used in some ceilometers.
　2 Pre-1935, a lateral-level flight instrument.
　3 Several authorities use * as synonymous with *inclinometer*.

clip, CLIP　*1* Cellular logic image processor.
　2 Pack of air-launched missiles loaded as a unit.

clipped wing　Aircraft having wing modified by removal of tips or outer portions (eg for racing).

clipper　Clipping (1) circuit.

clipping　*1* Limiting positive and/or negative parts of waveform to chosen level.
　2 Mutilation of communications by cutting off or distorting beginnings and/or ends of words or syllables.
　3 Limitation of frequency bandwidth.
　4 Reduction of amplification below given frequency.

Clircm　Closed-loop IRCM.

CLK　Clock, clock time.

CLL　Centreline lighting provided.

C$_t$M　Centreline (major axis) of missile.

CLMA　Contact localization and mission analysis (ASW).

CL$_{max}$　Maximum attainable lift coefficient.

Clnc, CLNC　Clearance (UK), hence Clnc Del, for delivery.

CLNP　Connectionless network protocol.

CLNS　Connectionless network service.

CLNTS　China Lake Naval Testing Station (CA, USN).

CLO　*1* Counter-LO (low-observables).
　2 Logistics and training command (KL, RNethAF).

CLOAR　Common low-observable[s] autorouter (AFMSS).

clobber　To knock out a ground or air target (colloq.).

clocking　Precisely aligning groups of rotating airfoils, especially of turbine stages.

clock rate　Precise frequency at which pulses are generated to control computer arithmetic unit, digital chip or other device.

CLOS　Command to line of sight; can be prefaced by A = automatic, M = manual or SA = semi-automatic.

close air support, CAS　Air attack on targets close to friendly surface force, integrated with latter's fire and movement.

close-controlled interception　One in which interceptor is under continuous ground control until target is within visual or AI radar range.

closed-circuit tunnel　Wind tunnel which recirculates given mass of working fluid.

closed-circuit TV　Camera/microphone linked to TV receiver/speaker by wires.

closed competition　Procurement competition in which prices, performances and design details are not disclosed to rival bidders.

closed-jet tunnel　Tunnel, not necessarily closed-circuit, in which working section is enclosed by walls.

closed-loop system　Dynamic system in which controlled variables are constantly measured, compared with inputs or desired values and error signals generated to reduce difference to zero.

closed thermodynamic cycle　Cycle which can transfer energy but not matter across its boundary.

close flight plan　To report safe arrival to appropriate ATC authority and thus terminate flight plan. (Failure to close may trigger emergency.)

close hangar doors!　Stop talking shop (RAF, colloq.).

close out　*1* To seal spacecraft, esp. manned; task performed by ad hoc ** crew who are last to leave pad area.
　2 To complete manufacturing programme.

close parallel operation　Runways less than 200 m [656 ft] apart.

closest approach　*1* Time, location or separating distance at which two planets are closest.
　2 Same for fly-by spacecraft.

close support　See *close air support, CAS*.

closet　Above-floor bay or compartment for carry-on baggage or folded wheelchairs.

closure　Relative closing velocity between two air or space vehicles.

clot　Idiot (RAF colloq.).

cloud   Large agglomeration of liquid droplets (water in case of Earth) or ice crystals suspended in atmosphere.

cloud absorption   Absorption of EM radiation by planetary cloud depends on cloud structure, size and EM wavelength, long waves reflected from planet surface being strongly absorbed even by thin layers.

cloud amount   Estimated as apparent coverage of celestial dome, as seen by observer; expressed in oktas and written in symbolic form on met chart.

cloud attenuation   Reduction in strength of microwave or IR radiation by cloud, usually due to scattering rather than absorption.

cloud banner   See *banner cloud.*

cloud break approach   Final approach beginning in cloud and ending in visual contact (though possibly with precipitation).

cloud chamber   Sealed chamber filled with saturated gas which, when cooled by sudden expansion, gives visible track of fog droplets upon passage of ionising radiation or particle.

cloud/collision warning   See *weather radar.*

cloud cover   See *cloud amount.*

cloud-cover satellite   Satellite equipped to measure by spectral response cloud cover on Earth or planet below.

cloud deck   Cloud layer, esp. visibly dense, seen from above.

cloud droplet   Water or ice particle with diameter ≤0.2 mm.

cloud 9   To be on * = feeling of elation and/or haziness.

cloud point   Temperature at which cooling liquid becomes cloudy.

cloud seeding   Scattering finely divided particles into cloud to serve as nuclei for precipitation (rainmaking).

cloud types   Each type has its own entry. They are classified by numbers giving an indication of danger: cirrus 0, cirrocumulus 1, cirrostratus 2, altocumulus 3, altostratus 4, nimbostratus 5, stratocumulus 6, stratus 7, cumulus 8, and cumulonimbus 9.

clovers   Common low-observables verification system (USAF).

CLP   Club der Luftfahrtpublizisten (Austria).

CLR   *1* Clearance, or cleared to (given height).

 2 Clear sky [≤10% cloud].

 3 Compact, long-range (Flir).

CLRC   Central Laboratory of the Research Councils (UK).

CLRS   Weather clear and smooth.

CLS   *1* Contingency landing site.

 2 Cargo loading system [M adds manual] (JARS).

 3 Computer loading system.

 4 Contractor [or co-operative] logistic system [or support].

 5 Central logging system.

 6 Capsule launch system.

C_{L_x} Lift coefficient at stall.

C/LS   Cruising/loiter speed.

CLSD   Closed.

CLSU   Culham Lightning Studies Unit.

CLT   *1* Centreline tracking (ILS/ILM).

 2 Customised lead time.

 3 calculated landing time.

CLTF   Closed-loop transfer function.

$C_{L_{to}}$ Takeoff lift coefficient.

CLTP   Connectionless mode transport protocol.

C_{L_U} Lift coefficient, unblown.

club layout   Pairs of seats facing each other, often with table between.

club propeller   Propeller having stubby coarse-pitch blades for bench-testing engine with suitable torque but reduced personnel danger and slipstream.

clue   Piece of information, hence: clued up, well informed; clueless, ignorant (RAF colloq.).

cluster   *1* A group of off-the-shelf computers linked together to create a high-performance (e.g. over 10 teraflops) computing system.

 2 Two or more parachutes linked to support single load.

 3 Several bombs or other stores dropped as group.

 4 Several stars or other pyrotechnic devices fired simultaneously from single container.

 5 Several engines forming group controlled by single throttle.

 6 Several rocket motors fired simultaneously to propel single vehicle.

cluster joint   Structural joint of several members not all in same plane.

cluster munition   Container which, after release from aircraft, opens to dispense numerous bomblets (rarely, ECM or other payloads).

cluster weld   See *cluster joint.*

clutter   Unwanted indications on display, esp. radar display, due to atmospheric interference, lightning, natural static, ground/sea returns or hostile ECM.

CLX   Combat leadership exercise.

CM   *1* Command module.

 2 Configuration, or context, management (EDP, software).

 3 Crew member, thus *1, *2, etc.

 4 Cluster [or cratering] munition.

 5 Cruise missile.

 6 Comsec module.

 7 Countermeasure[s].

 8 Classified message.

 9 Capability Manager (MoD UK).

C_M Coefficient of pitching moment about half-chord.

C_m Coefficient of pitching moment about quarter-chord.

cm   Centimetre[s].

C_{mac} Coefficient of pitching moment about aerodynamic centre.

CMA   Central[ized] maintenance application.

CMAG   Cruise-missile advanced guidance.

CMATZ   Combined military air, aerodrome, traffic zones.

CMB   *1* Continuous monofilament, braided.

 2 Concorde Management Board.

 3 Cosmic microwave background.

 4 Central Medical Board (RAF).

 5 Ceiling-mounted bin.

 6 Climb, climbing.

CMBRE   Common munitions built-in test reprogramming equipment.

CMC   *1* Cruise-missile carrier (A adds aircraft).

 2 Ceramic-matrix composite(s).

 3 Central maintenance computer [F adds function, S system].

 4 Cheyenne Mountain Complex (USAF).

C_{mcg} Coefficient of pitching moment about c.g.

CMD   *1* Command, ie total autopilot authority.
   2 Countermeasures dispenser, or duties.
   3 Cruise-missile defense.
   4 Colour [or common] multipurpose [or multifunction] display [S adds system, U unit].
CMDR   *1* Coherent monopulse Doppler radar.
   2 Card maintenance data recorder.
CMDS   Countermeasures dispensing system.
CME   *1* ECM (1) (F).
   2 Coronal mass ejection.
   3 Central Medical Establishment (RAF).
CMEA   Council for Mutual Economic Assistance.
CMF   *1* Conceptual military framework (NATO).
   2 Central maintenance function.
   3 Common message format.
CMFT   Canadian Museum of Flight and Transportation, Surrey BC.
CMG   Control-moment gyro.
CMH   Center for Military History (US).
CMI   *1* Computer-managed instruction (see *CAI* [2]).
   2 Cruise-missile interface.
   3 Catia Metaphase Interface.
CMIK   Cruise-missile integration kit.
CMIS   *1* Command management information system.
   2 Conical microwave image/sounder.
CMISE   Combat management integration support environment.
CMIV   Cabin management and interactive video.
CML   Consumable materials list.
CMLP   Cruise-missile launch point.
CMLS   Commercial microwave landing system [A adds avionics].
CMM   *1* Computerised modular monitoring (of health of hardware).
   2 Condition-monitored maintenance.
   3 Co-ordinate measuring machine.
   4 Common-mode monitor (AFCS).
   5 Component maintenance manual.
   6 Capability maturity model; I adds integration (SEI4).
   7 Common modular missile.
   8 Command memory management.
CMMI   See *CMM(6)*.
CMMCA   Cruise-missile mission control aircraft.
CMMS   Congressionally mandated monthly study (US).
CMN   Control-motion noise (MLS).
CMO   Certificate Management Office (FAA).
C$_{mo}$ Coefficient of pitching moment (¼-chord) at zero lift.
CMOS   *1* Complementary metal-oxide silicon, or semiconductor.
   2 Cockpit maintenance operations simulation, or simulator.
CMP   *1* Countermeasures precursor (aircraft penetrating hostile airspace ahead of attacking force).
   2 Counter-military potential (strategic balance).
   3 Central maintenance panel.
   4 Configuration management plan.
CMPL, cmpl   Completion, completed.
CMR   Central[ised] maintenance record.
CMRA   Cruise-missile radar altimeter.
CMRB   Composite main-rotor blade.
CMRS   *1* Countermeasures receiver system.
   2 Crash/maintenance recorder system.

CMS   *1* Continuous monofilament, spun.
   2 Commission de Météorologie Synoptique.
   3 Cockpit, cabin or circuit [electric/electronic, not ATC] management system.
   4 Constellation maintenance system [unrelated to that a/c].
   5 Common modular, or combat-mission, simulator.
   6 Computer module system.
   7 Cassette memory system.
   8 Component-management support.
CMSAF   Chief master sergeant of the Air Force (USAF).
CMT   *1* Cadmium mercury telluride (IR detector).
   2 Communications management terminal.
   3 Certificate management team (ATOS).
CMTC   Committee for Military-Technical Co-operation.
Cµ, C$_{mu}$ Blowing coefficient of circulation-controlled aerofoil.
CMU   *1* Communications, or central, management unit.
   2 Control and monitor unit (Hums).
CMUP   Conventional-mission upgrade program.
CMW   Compartmented mode workstation.
CMWS   Common missile warning system.
CN   Consigne de navigabilité [= AD(1)] (F).
Cn   Directional stability, yawing moment coefficient due to sideslip.
c/n   Constructor's number.
CNA   *1* Computer network attack.
   2 Center for Naval Analyses.
   3 Common-nozzle assembly.
   4 Cast nickel alloy.
CNAD   Conference of National Armaments Directors (NATO).
CNATRA, Cnatra   Chief of Naval Air Training (USN).
CNATS   Controller of National Air Traffic Services (UK).
CNC   *1* Computer numerical control (NC machining).
   2 Com/nav controls.
CNCE   Communications nodal control element.
CNCS   Central Navigation and Control School (RAF).
CND   *1* Computer network, defense.
   2 Campaign for nuclear disarmament (UK).
CNDB   Customised navigation database.
CND/RTOK   Could not duplicate, retest OK.
CNEIA   Comité National d'Expansion pour l'Industrie Aéronautique (F).
CNEL   Community noise equivalent level.
CNES   Centre National d'Etudes Spatiales (F).
CNF   Central notice-to-airmen facility.
CNG   *1* Compressed natural gas.
   2 Chief of [State] National Guard.
CNI   *1* Communications, navigation, identification.
   2 Chief navigational instructor.
   3 Continuous nitrogen inerting.
CNIE   Comision Nacional de Investigaciones Espaciales (Arg.).
CNIEW   CNI (1) electronic warfare.
CNII   Central research institute (R).
CNIMS   CNI (1) management system.
CNIR   Comunication, navigation, identification and reconnaissance.
CNITI   Central scientific institute for radiotechnical measurement; often rendered TsNITI (R).

CNIU   CNI (2) unit.

CNK   Cause not known.

CNL   Cancel, cancelled.

CNMA   Communications network for manufacturing applications, search for ISO standards complementary to MAP6 and TOP (EEC).

CNO   *1* Chief of Naval Operations (USN).

2 Computer network operations; JTF adds Joint Task Force.

C/NO, C/No   Carrier-to-noise density ratio.

C/N/P   Com./nav./pulse.

CNPI   Communication(s), navigation and position(ing) integration.

CNR   *1* Community noise rating.

2 Consiglio Nazionale Ricerche (I).

CNRA   Certificat de Navigabilité Restreint (homebuilts, F).

CNRE   Centre National de Recherches de l'Espace (F).

CNRI   Combat net radio interface.

CNRS   Centre National de la Recherche Scientifique (F).

CNS   *1* Continuous.

2 Communications network simulator.

3 Communications, navigation, surveillance; ATM adds air-traffic management (ICAO).

4 Common nacelle system, able to accept different types of engine.

5 Chief of Naval Staff [First Sea Lord] (UK).

CNSAC   Comité National de Sûreté de l'Aviation Civile (F).

CNST   Center for NanoSpace Technologies.

CNT   Certificat de Navigabilité de Type (F).

CNTR   Centre.

CN₂D   Coefficient of usable lift (variable aerofoil profile).

CN²H   Conduit nuit 2nd generation helicopters.

CNVTV   Convective.

CO   *1* Commanding officer.

2 Crystal oscillator.

3 Checkout.

4 Aerodynamic mean chord.

5 Corps observation (USA, 1919–24).

6 Carbon monoxide.

COA   *1* Course of action.

2 Corps observation, amphibian (USA 1919–24).

CoA   Circle of ambiguity

coach   Formerly, US domestic high-density seating configuration.

coalescing filter   Works by coalescing finely divided liquid droplets (eg water in fuel) into removable masses.

coaming   *1* Edge of open-cockpit aperture, often padded.

2 In flight deck, along top of main instrument panel.

Coanda effect   Tendency of fluid jet to adhere to solid wall even if this curves away from jet axis.

Coanda flap   Flap relying on Coanda effect for attachment of flow to upper surface even at extreme angles.

coannular inverted nozzle   Nozzle of variable-cycle jet engine with low-velocity core and high-velocity surrounding jet.

coarse pitch   Making large angle between blade chord and plane of disc, thus giving high forward speed for given rotational speed.

coarse-pitch stop   Mechanical stop to prevent inefficient over-coarse setting (removed when feathering).

coast   *1* Radar memory technique tending to slave to original target trajectory and avoid lock-on to stronger target passing same LOS.

2 Unpowered phase of trajectory, esp. in atmosphere (usually verb).

coastal refraction   Change in direction of EM radiation in crossing coast; also called shoreline effect, land effect.

coast-boost   Period of coasting followed by rocket burn.

coasted track   Continued on basis of previous characteristics in absence of surveillance data (TCAS).

coastline refraction   See *coastal refraction*.

COAT   Corrected outside air temperature (OAT minus TAS/100).

co-axial   Propeller or rotor having two or more sets of blades on same axis rotating in opposite senses independently. Not same as contra-rotating.

co-axial cable   Comprises central conductor wire and conducting sheath separated by dielectric insulator.

COB   *1* Co-located operating base.

2 Certificated operational base.

3 Catenary obstruction beacon, mounted on pylons supporting power cables .

cobalt   Hard, silver-white metal, density 8.9, MPt 1,495°C, important in steels and in high-temperature engine alloys. Co-60 is dangerous radioisotope theoretically producible in large amount by nuclear weapons.

cobblestone turbulence   *1* High frequency * due to large mass of randomly disturbed air without significant gross air movement.

2 Buffet experienced by jet V/STOL descending into ground effect.

COBE   Cosmic-origin background explorer.

Cobol   Common business-oriented language.

cobonding   Manufacture of composite aerofoil, esp. wing, in which entire surface is assembled and cured, but with one skin (usually upper) separated by debonding agent. This skin is then attached by removable bolts.

Cobra   *1* Manoeuvre in which from level flight at moderate airspeed pilot applies maximum symmetric nose-up command, reaching AOA 90° up to possibly 130°, when control neutralised for flip-down recovery to level flight about 5 s later. Modest gain in height, large loss in airspeed [energy].

2 Co-optimized booster for reusable applications.

3 Coastal battlefield reconnaissance and analysis (USMC).

COBY   Current operating budget year.

COC   *1* Common (or combat) operations centre, for tactical control of all arms in theatre.

2 Catalytic ozone converter.

3 Copper on ceramic.

4 Chamber of Commerce.

COCC   Contractor's operational control centre.

cockade   National insignia worn by military aircraft, esp. one of concentric rings.

cocked   Aircraft, especially combat type, preflighted through all checklists to point of starting engines.

cocked hat   Triangle formed by three position lines that do not meet at a point.

cockpit   Space occupied by pilot or other occupants, esp. if open at top. Preferably restricted to small aircraft in which occupants cannot move from their seats; most * contain only one seat. Term could arguably be applied to all aerodyne pilot stations, but flight deck preferred for large aircraft.

cockpit alert State of immediate readiness with combat aircrew fully suited, in * and ready to start engine.

cockpit audio monitoring Activated by flight crew, continuously transmits live audio via satellite from aircraft [jet airliner] experiencing emergency.

cockpit cowling Aircraft skin around cockpit aperture.

cockpit television sensor Solid-state CCD camera recording what the pilot sees during each flight.

cockpit voice recorder Automatic recycling recorder storing all crew radio and intercom traffic, plus background noise, during previous several missions.

Coco exercise Combat mission exercise called off when aircraft are lined up on runway.

Cocomo Constructive cost model (software).

Cocraly Anti-oxidation coating for hot metal, from Co, Cr, Al, Yttrium.

COD *1* See *carrier on-board delivery*.
2 Component operating data.
3 Cash on delivery.
4 Chemical oxygen demand.

CODA, Coda Centre Opérationnel de Défense Aérienne (Taverny, F).

Codamps Coupled ocean/atmosphere mesoscale prediction system.

Codan Carrier-operated device anti-noise.

Codar Correlation detection and recording, or ranging (ASW).

Code *1* Two capital [upper-case] letters assigned to airline [any public carrier]; sometimes shared, thus Cronus aircraft operate on * of Aegean. These letters preface the three- or four-digit number identifying a particular time-tabled flight.
2 See * *letters*.
3 Another meaning is the series of pulses from a transponder.

Code Bambini Literally 'child's talk', multi-lingual tactical radio language (Switz.).

code block Standardised format of data identifying each frame in visual, IR or SLAR film, with provision for high-speed computer recall.

Codec Coder/decoder.

code letters Pairs of letters [from 1944 often letter + number] identifying unit of aircraft in WW2. Each aircraft also assigned individual letter (RAF, USAAF).

code light Surface light giving signal, usually Morse; if at airfield could be called beacon.

codem Coded modulator/demodulator.

coder Part of DME transponder which codes identity into responses.

code rate Ratio of actual data bits to total information digits transmitted in radar or communications system having deliberate redundancy. Symbol R.

Coderm Committee for Defence Equipment R & M (UK).

Codes Common digital exploitation system.

codes Numbers assigned to multiple-pulse reply signals transmitted by ATCRBS and SIF transponders.

Codib Controlled-diffusion blade (or blading).

coding Arrangement of problem-solving instructions in format and sequence to suit particular computer.

CODSIA, Codsia Council of Defense and Space Industry Associations (US).

COE *1* Certification of equivalency (USAF).
2 Co-operative emitter.
3 Common operating environment.

COEA Cost and operational effectiveness analysis.

coefficients Except for next four entries, see under appropriate characteristics.

Coefficient A In simple magnetic compass, deviations on cardinal and quadrantal points summed and divided by 8.

Coefficient B In simple magnetic compass, deviation E minus deviation W divided by 2.

Coefficient C In simple magnetic compass, deviation N minus deviation S divided by 2.

coefficient conversion factor Formerly, multiplier 0.00256 required to convert absolute to engineering coefficients.

COEIA Combined operational effectiveness and investment appraisal (UK 2001).

COF Centrifugal oil filter.

C of A Certificate of Airworthiness.

COFAS Centre d'Opérations des Forces Aériennes Stratégiques (Taverny, F).

CoFAS Commandement des FAS, same address.

C of C Certificate of Compliance.

COFDM Code orthogonal frequency-division multiplexing (helicopters).

C of E Certificate of Experience

C of F Construction of facilities.

coffin *1* Missile (ICBM) launcher recessed into ground but not hardened.
2 Symbol which appears in place of a downed aircraft (ACMR).

C of G See *c.g* .

C of M Certificate of Maintenance.

C of P See *centre of pressure.*

C of T Certificate of test.

C of R Certificate of Registration of aircraft.

cogbelt Flexible belt incorporating teeth to prevent slip.

COGT Centre-of-gravity towing.

coherent Radiation in which, over any plane perpendicular to direction of propagation, all waves are linked by unvarying phase relationships (common simplified picture is of waves 'marching in step' with all peaks in exact alignment).

coherent echo Radar return whose amplitude and phase vary only very slowly (from fixed or slowly moving object).

coherent pulse radar, coherent radar Incorporates circuitry for comparing phases of successive echo pulses (one species of MTI).

coherent transponder Transmitted pulses are in phase with those received.

coherer RF detector in which conductance of imperfect part of circuit (eg iron filings) is improved by received signal.

Cohoe Computer-originated holographic optical elements.

COI Co-ordinator of Information (US, WW2).

CoI Central Office of Information (UK).

Coil, COIL Chemical oxygen iodine laser.

coin, Co-In, CO-IN Counter-insurgency; aircraft designed for guerilla war.

Coincat   Community of Interests in Civil Air Transport (G).

coincidence circuit   Gives output signal only when two or more inputs all receive signals simultaneously or within agreed time.

COINS, Coins   Computer-operated instrument system.

COIS   Coastal ocean imaging spectrometer.

Cojas   Coherent jammer simulator.

coke   Verb, to modify aircraft with Küchemann 'Coke bottle' fuselage.

col   In atmosphere isobar field, saddle-shaped region separating two highs on opposite sides and two lows on remaining sides.

colander   In some ramjet engines, perforated shell controlling secondary airflow into combustion chamber. Generally equivalent to gas-turbine flame tube.

cold   Without using afterburner.

cold air mass   Colder than surrounding atmosphere.

cold-air unit   Air-cycle machine, usually in an ECS, which greatly reduces temperature of working fluid by extracting mechanical energy in expansion through a turbine.

Coldama   Co-ordination of loads data acquisition management.

cold bucket   In aft fan with double-deck blades, outer blades handling cold air.

cold cathode   Highly emissive coating and operating at ambient temperature.

cold-cockpit alert   Combat aircraft has no ground power supplies and is 'cold' until pilot enters and initiates start sequence for engine, gyros and systems.

cold cordite charge   Does not detonate but burns to give high-pressure flow of gas.

cold drawing   Drawing workpiece at room temperature.

cold flow test   Static test of liquid rocket propulsion system to verify propellant loading and feeding but without firing engine(s).

cold front   Front of advancing cold air mass moving beneath and lifting warmer air, esp. intersection of this front with Earth's surface.

cold gas   Reaction-control jet or rocket using as working fluid gas released from pressure or monopropellant decomposed without combustion.

cold launch   *1* Launch of missile or other ballistic vehicle under external impulse, usually from tube (in atmosphere, in silo or on sea bed) with vehicle's propulsion fired later.

2 Takeoff of aircraft with INS not aligned.

cold mission   Mission or test judged non-hazardous, thus not interfering with other activities.

cold plate   In high-vacuum technology, refrigerated plate used to condense out last molecules of gas in chamber.

cold plug   Spark plug having short insulated electrode keeping relatively cool (because rate of carbon deposit from oil or fuel is very low).

cold rating   Cold thrust; rated output of jet engine without afterburning. Can be MIL.

cold rocket   Operating on pressurized gas or monopropellant, without combustion.

cold rolling   Performed on steels to harden and increase strength, at expense of ductility.

cold round   Test missile launched without active propulsion.

Colds   Common opto-electronic laser detection system (detects laser beams and measures angle of arrival).

cold shut   Porosity due to premature surface freezing in casting, or formation of gas bubble in weld.

cold soak   *1* Test of complete aircraft by prolonged exposure to lowest terrestrial temperature available before flying a mission.

2 Test of cryogenic propulsion system by prolonged passage of propellant.

cold stream   Fan airflow; hence * reverser, one not affecting core.

cold test   Determines lowest temperature at which oil or other liquid will flow freely.

cold thrust   Maximum without afterburner.

cold wave   Sudden major fall in surface ambient temperature in winter.

cold working   Forming metal workpiece at room temperature; increases hardness and often strength but reduces ductility (increases brittleness).

Coleman theory   Derived by NACA's R. P. Coleman and A. M. Feingold, basic explanation of ground resonance of helicopters with articulated rotors; hence such resonance called Coleman instability.

coleopter   Aircraft having annular wing with fuselage at centre; usually tail-standing VTOL.

collaborative programme   Undertaken by industrial companies in two or more countries as result of legal agreements between those companies or between their national governments.

collar   Impact-absorbent ring around bottom of balloon gondola (usually lightweight foamed polystyrene).

collateral damage   *1* Refers esp. to injury to friendly eyes from clumsy use of powerful lasers in warfare.

2 Damage caused to anything other than the intended target.

collation   Selection in correct sequence and stacking in exact register of pre-cut piles to make part in composite material.

collective pitch   Pilot control in rotary-wing aircraft directly affecting pitch of all blades of lifting rotor(s) simultaneously, irrespective of azimuth position. Main control for vertical velocity. Colloq. = 'collective'.

collective stick   Collective-pitch lever (colloq.).

collector   *1* Bell-mouth intake downstream of working section of open-jet tunnel.

2 Region of transistor between * junction and * connection carrying electrons or holes from base.

collector ring   Circular manifold collecting exhaust from cylinders of radial piston engine engine.

collimate   To adjust optical equipment to give parallel beam from point source or vice versa.

collimating mark   A short line or cross at the mid-point of each edge of a reconnaissance photo.

collimating tower   Carries visual and radio/radar target for establishing axes of aerials (antennas) with minimal interference from other electrical fields. Alternatively collimation tower.

collision-avoidance system   Provides cockpit indication of all conflicting traffic, without latter carrying any helpful equipment or co-operating in any way, and increases intensity of warning as function of range and rate of closure.

collision beacon   Powerful rotating visual light, normally flashing Xenon tube, carried by IFR-equipped aircraft (normally one dorsal, one ventral).

collision-course interception   Aimed at point in space

which target will occupy at a selected future time; interceptor may approach this point from any direction.

collision-warning radar　See *weather radar*.

collision-warning system　See *collision-avoidance*.

colloidal propellant　Having colloidal structure, with particles never larger than 5×10^{-3} mm and apparently homogeneous to unaided eye.

co-located　Two ground navaids, usually VOR and DME, at the same site.

colours of the day　Particular combination, changed daily, of [usually two] Very [or similar] signal cartridges, fired to confirm aircraft as friendly to suspicious ground forces.

colour stripping　Removal of all MES (6) colours except those indicating organic substances such as plastic explosives.

Colpar　Confederacion Latino Americana de Paracaidismo (sport parachuting, office Argentina).

COLT　CO_2 laser technology.

COM　*1* Computer output on microfilm (direct recording).

　2 Company operations manual.

　3 Acronyms based on Command[er] or communications [over 50].

　4 Cockpit operating manual.

com　Communications (FAA = comm).

Comac　Cockpit-management computer.

ComAO, COMAO　Composite air operation.

comb　*1* Rake, usually linear, of pressure heads.

　2 IFF aerial (antenna) with linear array of dipoles often sized to match spread of wavelengths.

combat aircraft　Aircraft designed to use its own armament for destruction of enemy forces; thus includes ASW but not AEW or transport (definition controversial).

combat air patrol, CAP　Maintained over designated area for purpose of destroying hostile aircraft before latter reach their targets.

combat camera　Colour ciné camera aligned with fighter armament to film target.

combat control team　Air force team tasked with establishing and operating navaids, communications, landing aids and ATC facilities in objective area of airborne operation.

combat fuel tank　*Combat tank*.

combat gross weight　See *weight*.

combat load　Aggregate of warlike stores carried (includes guns/ammunition but excludes radars, lasers/receivers and drop tanks carried for propulsion).

combat mission　Mission flown by balloon, airship, kite, aeroplane, helicopter or other aircraft such that it may expect to encounter enemy land, sea or air forces.

combat persistence　Ability of fighter aircraft to engage numerous successive targets, by virtue of large number of AAMs carried.

combat plug　Manual control of fighter engine permitting TET limit to rise to new higher level for period of emergency (typically 30 sec to 3 min).

combat radius　*Radius of action*.

combat spread　Variable loose formation affording best visual lookout.

combat tank　External jettisonable fuel tank used on combat missions; possibly smaller than ferry tank.

combat thrust loading　Thrust loading assumed for fighter in typical combat.

combat trail　Combat aircraft, usually interceptors, in loose trail formation, maintaining position visually or by radar.

combat wing loading　Wing loading assumed for fighter in typical combat.

combat zone　*1* Geographic area, including airspace, required by combat forces for conduct of operations.

　2 Territory forward of army rear boundary.

combi, Combi　Transport aircraft with main deck furnished for both passenger and freight (from 'combination'). Proportion devoted to freight usually variable.

combination　Tug and glider, before separation.

combination aircraft　Combi.

combination propulsion　See *mixed-power aircraft*.

combination slide　Escape slide designed for subsequent use as life raft.

Combined　Involving armed forces of two or more allied nations. Thus * common user item, * forces, * staff etc.

combined display　Presents information from two or more sources, usually radar superimposed on moving-map display.

combined-effects munition　One having anti-armour, anti-personnel and incendiary effects.

combined sight　Weapon-aiming device able to operate in more than one mode, eg optical and thermal imaging.

combined stresses　Two or more simple stresses acting simultaneously on same body.

combiner　Optical element in HUD for aligning, collimating or focusing at infinity all displayed elements on single screen.

combining gearbox　Reduction gearbox driven by two or more engines or [e.g.] surface power units, and driving single or contra-rotating propeller or lifting rotor.

Combre　See *CMBRE*.

COMBS　Contractor-operated and managed base supply, ie manufacturer of major system manages and maintains government-owned GSE and spare parts and carries out heavy maintenance.

combustion　Chemical combination with oxygen (burning).

combustion chamber　*1* In piston engine, space above piston(s) at TDC, arguably extended over part of stroke depending on progress of flame front.

　2 In gas turbine, entire volume in which combustion takes place, including that outside flame tube(s) occupied by dilution air.

　3 In liquid rocket or ramjet, entire volume in which combustion takes place, bounded by injector face, walls of chamber and plane of nozzle throat (not nozzle exit).

　4 In solid or hybrid rocket, inapplicable.

combustion efficiency　Ratio of energy released to potential chemical energy of fuel, both usually expressed as a rate.

combustion ratio　Ratio of fuels or propellants actually achieved; in case of fuel/air usually termed mixture ratio.

combustion ring　Combustion chamber of annular (eg Aerospike) liquid rocket engine.

combustion space　See *combustion chamber (1)*.

combustion starter　Engine-start energised by burning fuel, either fuel/air, monopropellant (eg Avpin) or solid cartridge.

combustion test vehicle　Free-flight vehicle (RPV or missile) whose purpose is test or demonstration of propulsion performance.

combustor   *1* See *combustion chamber (2)*.

2 Combustion chamber (2) together with fuel manifolds, injectors, flameholders and igniters.

3 Rarely, afterburner burning region, with fuel spray bars, flameholders and ignition system.

combustor loading   Expressed as a function of mass flow, chamber volume, and inlet pressure and temperature.

Comdac   Command, display and control (USCG).

Comecon   Council for Mutual Economic Assistance.

Comed   Combined map and electronic display (pronounced co-med).

Comeds   Conus meteorological data system (DoD).

Comest   European colour-TV satellite management consortium.

Comfile   Expandable network connecting ATC data, voice and radar to digital recorders.

comfort chart   Plot of dry-bulb T° against humidity (sometimes modified to include effect of air motion).

Comint   Communications intelligence.

comlo   Compass locator.

comm   Communications (FAA).

command   *1* Intentional control input by flight crew or remote pilot.

2 Electrical or radio signal used to start or stop action.

3 In EDP, portion of instruction word specifying operation to be performed.

4 Authority over precise flight trajectory exercised by ATC or military authority (hence * altitude, * height, * heading, * speed etc.).

command airspeed   A target airspeed displayed as a *command parameter*.

command augmentation system   Compares pilot demand with aircraft response, FCS receiving the difference; latest CAS have full authority and often high gain.

command bars   Principal reference index on flight director instruments, giving attitude in pitch and roll.

command destruct   System which, at range safety officer's discretion, can explode malfunctioning missile, RPV or other unmanned vehicle, or trigger BUS, thereby averting hazard to life or property.

command dot   *Command marker* in form of bold dot or small disc.

command ejection   Ordered [not necessarily triggered] by captain of aircraft.

commander   Used only in military aviation, aircraft * has authority over everyone on board even though he may not be a member of flight crew. Not synonymous with PIC or with civil term captain.

command guidance   Steering by remote human operator.

command marker   Reference index (line, bug, arrow or other shape) indicating target value, set by pilot on tape (sometimes dial) instrument and then flown to centre reference line. (See *command reference symbol*.)

command parameter   Variable subject to command (1), (2), (4) and thereafter displayed as target value on instrument or display.

command reference symbol   HUD symbology in form of ring or other shape showing a point at which to aim ahead of aircraft, eg landing touchdown point or an aerial point for optimum AOA on overshoot (go-around).

comma rudder   Rudder shaped like comma, with balance area ahead of hinge axis, used without a fixed fin.

commercial   In military use, purchasable from civilian source (eg aircraft rivet).

commercial aircraft   Aircraft flown for hire or reward.

commercial electrics   Electrical systems serving passenger functions only (eg steward call circuits, PA system, cabin lighting).

commercial support   Assistance to operator of civil aircraft given or sold by original manufacturer or dealer.

com./met./ops.   Communications, meteorology, operations.

commitment   Announced decision to purchase an aircraft type, usually commercial transport.

committal height   See *decision height*.

commodity loading   All cargo of one kind grouped together, without regard to destination.

commodity rate   Price charged to fly specified kind of cargo, typically per kilogramme over particular route.

common aero vehicle   *1* Originally this was a standard design of RV [to house different payloads] for ICBMs.

2 Today, a common vehicle structure for deploying a variety of customised payloads, including weapons, into the atmosphere (MSP).

3 Capitalized, "an unpowered, manoeuvrable hypersonic glide vehicle carrying c1,000lb of munitions" launched from space to hit within 10ft (USAF).

commonality   *1* Hardware quality of being similar to, and to some degree interchangeable with, hardware of different design.

2 Objective of using one basic design of aircraft, or other major system, to meet needs of more than one user service in more than one role (with economies in training, spares and other areas).

common automatic recovery system   To retrieve UAVs on surface ship: electronic guidance to system of nets and cables on LPD quarterdeck.

common configuration   Numerous plans, mainly USAF, to bring as many aircraft of one type as possible to uniform standard, usually by upgrades.

common display system   Standardised glass cockpit.

common-flow afterburner   Augmented turbofan in which fan and core flows mix upstream of afterburner.

common infrastructure   Financed by two or more allies, eg by all members of NATO.

common mark   Marking assigned by ICAO to aircraft of international agency (eg UN) on other than national basis. Hence ** registering authority.

common module(s)   Use of identical "black box" subsystems as building blocks for different major equipments, eg * IR components to build night-vision, recon., weapon guidance and other systems for different armed forces or civilian customers.

common route   Portion of N American route west of coastal beacon.

common sensor   The principal meaning is a sensor that intercepts both communications and Elint.

common servicing   Performed by one military service for another without reimbursement.

common-user airlift   In US, provided on same basis for all DoD agencies and, as authorised, other Federal Government agencies.

communication deception   Interference with hostile communications (including ATC and navaids) with intent to confuse or mislead.

communication language   Complete language structure for linking otherwise completely separate (and possibly dissimilar) EDP (1) systems.

communications intelligence Gained by listening to hostile communications.

communications satellite Vehicle, normally man-made, orbiting planetary body, usually Earth, for purpose of relaying intercontinental telecommunications (telephone, telex, radio, TV, online etc.) (see *active* **, *passive* **, *synchronous* **).

communications security Made up of physical security of transmitter and receiver, emission security of transmitter, transmission security en route and cryptosecurity of message.

community Clearly defined group, usually of aircrew, eg all who fly particular aircraft type or particular type of mission.

community boundary Drawn around inhabited or urban areas surrounding airport or airfield.

community noise level Flyover, sideline and approach NLs measured at designated points on or beyond community boundary (see *noise*).

commutated Doppler Form of MLS in which beam is frequency-coded and/or linearly commutated instead of scanned in azimuth and elevation.

commutation *1* Repeated reversal of current flow in winding of electrical machine, esp. to change output from a.c. to d.c.

2 Transfer of current between elements of polyphase rectifier to produce unidirectional output.

commutator Typically, radially separated series of conductors forming ring round rotating generator shaft, opposite pairs of which are touched by brushes in external circuit to give d.c. output by commutation (1).

commuter aircraft See *feederliner*.

commuter airline In theory, air carrier operating between outlying regions and major hub(s). In practice, applied to anything from air-taxi operator to – in undeveloped regions – national carrier (see *third-level*).

com/nav Communications and navigation aids; usually means complete avionic fit.

Comlo Compass locator (usually comlo).

Comos Common Mode S (Eur ATC).

comp *1* Component of W/V along Tr (strictly, along flight-plan track between check points).

2 Compressor.

Compacta tyre Landing wheel tyre of reduced diameter and greater than normal width (Dunlop).

companion body Hardware from launch system accompanying space vehicle or satellite on its final trajectory.

comparative cover Reconnaissance coverage of same scene at different times.

comparative vacuum monitoring Potentially very important method of detecting even the smallest cracking in structures by measuring any flow of air into a volume maintained as partial vacuum.

compartment marking Stencilled subdivisions of cargo aircraft interior to assist compliance with floor loading and c.g. position limits.

Compas Computer-oriented metering, planning and advising system.

Compass Compact multipurpose advanced stabilized system.

compass acceleration error See *acceleration errors*.

compass base Area on airfield, usually paved disc, on which aircraft can conveniently be swung.

compass calibration pad Compass base.

compass compensation See *compensating magnets*.

compass course See *heading*.

compass deviation Deviation (2).

compass error *1* Vector sum of variation E plus variation W.

2 Sum of deviation, variation and northerly turning error.

compass heading See *heading*.

compass locator Low-power beacon used with ILS, 2-letter ident.

compass points 32 named directions comprising cardinal points, quadrantal points and 24 intermediate points.

compass rose Disc divided into 360°, either on simple magnetic compass or on compass base.

compass swing See *swing*.

compass testing platform See *compass base*.

compass variation See *variation*.

compatibility Ability of materials (solids, liquids and gases) and dynamic operating systems to interface for prolonged periods without interference under prescribed environmental conditions.

compatible *1* Colour TV transmission capable of being received as monochrome by monochrome receiver.

2 Language and software capable of being used in given computer.

compensated gyro Incorporates correction for apparent wander.

compensating magnets Two pairs of bar magnets carried on arms rotatable about axis of magnetic compass to correct or minimise deviation.

compensation manoeuvres Aircraft manoeuvres required for accurate use of compensator (2), always involving four orthogonal headings, and sometimes circle or cloverleaf.

compensator *1* Instrument for measuring phase difference between components of elliptically polarised light (Babinet * has pair of quartz wedges with optical axes perpendicular).

2 Device, manually or computer-controlled, carried in ASW aircraft to eliminate false readings caused by permanent (airframe and equipment hardware), induced and eddy-current interference signals.

Compglas Low-density composite of graphite fibres in ceramic matrix, offering strength at very high temperatures (United Technologies).

compiler ECP (1) program more powerful than assembler for translating and expanding input instructions into correctly assembled sub-routines.

complementary shear Induced in tension field (eg aircraft skin) at right angles to applied shear, in plane of field.

completion business Process of taking green airframes from manufacturer and equipping and furnishing to each customer's specification (principally in field of executive or commuter transports). Hence, a completion = one aircraft ready for customer.

complex See *launch complex*.

compliance Demonstrated fulfilment of requirements or certificating authority.

compliance limit Time (usually GMT) by which compliance must be demonstrated.

compliant member Capable of substantial elastic or otherwise recoverable deflection.

compliant volume Trapped body of fluid, usually oil, having predetermined stiffness resulting from fluid's bulk modulus. Often sealed by diaphragm or piston having small bleed, to even out pressures over a period (see *stiffness*).

component   *1* One of assemblage of structural members.

2 One of assemblage of parts used to build hardware system.

3 Major subdivision of prime mover, esp. gas turbine (eg fan, compressor, combustor, turbine, afterburner, nozzle); hence * efficiency.

4 Force, velocity or other vector quantity along reference axis, such that components along two mutually perpendicular axes sum vectorially to actual vector. Thus, crosswind * on landing.

5 Major portion of aircraft that can be separated in flight, esp. if this leaves two complete aircraft able to proceed independently.

component efficiency   Measure of performance of part of machine, normally on basis of energy output × 100 divided by energy input. Thus overall efficiency of gas turbine is product of ** of each part, considered on both mechanical and thermodynamic basis.

component life   Authorised period of usage without attention, as stipulated by manufacturer or other authority. At expiry may be discarded or overhauled. Period may be extended from time to time.

components tree   Notional "tree" formed by interlinking of aircraft systems, highlighted in CBT by ability to strip aircraft layer by layer.

composite aircraft   *1* Comprising two aircraft joined together at take-off [see *component* (5)] but separated later in flight.

2 Aircraft made principally of composite material(s).

composite air picture   Fed from many sources to give giant hi-resolution monitor with many overlays controlled by keyboards, mice and trackballs.

composite beam   Composed of dissimilar materials bonded together.

composite cloud   Combination of, or intermediate between, basic forms, eg cirro-cumulus.

composite cooling   Evaporative cooling.

composite double-base   Solid rocket filling of combined double-base and composite types (eg AP (2) + AlP in matrix of NC + NG).

composite flight plan   One specifying VFR for one or more portions and IFR for remainder.

composite flying   Long-range navigation along great circle but modified (eg to avoid high mountains) by inserting sectors using other methods.

composite launch   Single launch vehicle carrying two or more distinct payloads.

composite material   Structural material made up of two or more contrasting components, normally fine fibres or whiskers in a bonding matrix. Unlike an alloy, usually anisotropic.

composite power   See *mixed power*.

composite propellant   Solid rocket filling comprising separate fuel and oxidiser intimately mixed.

composite route   One where composite separation is authorised.

composite separation   Reduction [usually to half normal] of lateral and vertical minima on oceanic routes meeting criteria.

compound aerofoil   Not defined, but has been applied to wing whose trailing edge comprises separately hinged upper and lower sub-aerofoils leaving controllable gaps.

compound aircraft   Having wing(s) and lifting rotor(s).

compound balance   Compound shelf.

compound curvature   Sheet or surface curved in more than one plane, thus not formable by simple bending.

compound die   Performs two or more sheet-forming operations on single stroke of press.

compound engine   Expands working fluid two or more times in two or more places, eg in HP and LP cylinders or in piston engine followed by gas turbine or blow-down exhaust turbine.

compound helicopter   Having propulsion (usually turbofan or turbojet) in addition to thrust component of lifting rotor.

compound shelf   Control surface comprising two [rarely, three] spanwise sections hinged together one behind the other and moving in opposition. LE of main [front] section normally has fabric seal to fixed surface.

compound stress   Not simple tension/compression, torque, bending or shear but combination of two or more of these.

compound taper   Outer wing is tapered more or less sharply than inboard.

compound wing   Wing made up of major fixed portion and upper/lower rear foils, with or without blowing between them. Also called multi-foil section. T/c up to 30% has been achieved at high M_D.

compressed-air starter   Expands HP airflow through piston engine cylinders or ATM or turbine-blade impingement jet. In multi-engined aircraft cross-bleed can start second and subsequent engines.

compressed-air tunnel   Closed-circuit tunnel filled with gas or air under pressure; can be smaller, and cheaper to run, than one at atmospheric pressure for given M and R.

compressibility   In aerodynamics, phenomena manifest at speeds close to local sonic speed, when air can no longer be regarded as incompressible. Loosely, behaviour of airflow subject to pressure/density changes of 50 per cent or more of free-stream values.

compressibility correction   From RAS to EAS (see *airspeed*).

compressibility effects   Manifest as local speed, at peak suctions, exceeds that of sound in surrounding flow; include abnormally rapid increase in drag, rearward shift of CP (2) on lifting wings, appearance of shockwaves, tendency to boundary-layer breakaway and, in improperly designed aircraft, control buzz and other more severe losses of stability and control.

compressibility error   Manifest in all instrument readings derived from simple pitot/static system at high subsonic Mach numbers; typically, progressive under-reading until pressure and static orifices have penetrated bow shock.

compression   Control of signal gain, esp. to increase it for small signal voltages and reduce it for large.

compression ignition   Combustion of fuel/air mixture triggered by high temperature due to compression in diesel cylinder or in highly supersonic ramjet with suitable internal profile.

compression lift   Lift gained at supersonic speed by favourable flow field by forcing flow to accelerate beneath wing (accentuated by down-turned wingtips).

compression pressure   Gauge pressure in piston engine cylinder at TDC (in absence of combustion).

compression ratio Ratio of entrapped volume above piston at BDC to volume at TDC.

compression rib Provided inside fabric-covered wing to withstand tension of drag bracing.

compression ring(s) Top ring(s) on piston, of plain rectangular section, serving to seal mixture into combustion space on compression stroke.

compression wave See *blast wave*.

compressor Machine for compressing working fluid (see *axial* *, *centrifugal* *, *skew* *, *Roots* *, *positive-displacement* *). In general, term used for device handling large mass flow at moderate pressure (say, up to 40 ata, 400 kPa); small flow at high pressure = pump.

compressor blade *1* Loosely, rotor blade or stator vane in axial compressor.

2 Precisely, operative aerofoil from axial compressor rotor.

compressor casing Fixed casing closely surrounding compressor rotor.

compressor diffuser Passage for working fluid immediately downstream of compressor wherein pressure is increased at expense of flow velocity.

compressor efficiency Useful work done in delivering fluid at higher pressure, in assumed adiabatic operation, expressed as percentage of power expended in driving rotor.

compressor map Fundamental graphical plot of compressor performance showing variation of pressure ratio (ordinate) against mass flow (abscissa) for each rpm band.

compressor pressure ratio Ratio of total-head pressure at delivery to that at inlet (if ratio is 24:1, conveniently written as 24, for example).

compressor rotor Main moving part in compressor of rotary form (ie, not reciprocating type).

compressor stator Stationary part of axial compressor carrying fixed vanes.

compressor vane Stationary blade attached to stator (case), one row of such vanes preceding each row of rotor blades.

compromised *1* Classified information known or suspected to have been disclosed to unauthorised persons.

2 Of serial number or civil registration, one inadvertently applied to two aircraft.

Comptuex Composite training unit exercise (USN).

Compu-Scene Add-on visual system for existing simulators (General Electric).

computational fluid dynamics Representation of a surface by a fine grid, enabling program to determine fluid flow over it in terms of velocity, pressure, force, moment, temperature and possibly other variables. Impossible before powerful computers.

computed air release point Air position at which first paratrooper or cargo item is released to land on objective.

computed approach MLS approach to a runway not aligned with an MLS radial.

computer *1* Machine capable of accepting, storing and processing information and providing results in usable form; function may be direct control of one or more operating systems.

2 Simple mechanical device for solving problems (eg Dalton *).

computer acceleration control Use of airborne computer linked to AFCS to limit (close to zero) unwanted flight

accelerations, esp. in vertical plane, on aeroplanes and helicopters.

computer-assisted approach sequencing Use of one, or several interlinked, computers in ATC system to solve problem of feeding arrivals automatically into optimised trajectories so that each arrives at destination runway at correct spacing and with minimal delay.

computer board Component part of a computer or similar device, each being a driver, RAM, EPROM, A/D converter, video interface or similar self-contained unit which can be assembled with others on to a bus (eg, backplane) to form a purpose-designed EDP system.

computer-programmable Capable of being controlled by digital computer without additional interfacing (typical item would be microwave signal generator for radar testing).

computing gunsight Automatically compensates for most predictable or measurable variables in weapon aiming.

comsat See *communications satellite*.

Comsec Office of Communications Security (US, NSA).

comsnd Commissioned (of facilities on airfield charts).

COMSS Coastal/oceans monitoring satellite system.

CON, con *1* Consol beacon.

2 Continuous.

3 Console.

4 Control.

Conac Continental Air Command (1 December 1948, became part of ADC).

Conaero Consorzio Italiano Compagnie Lavoro Aereo (1).

Conar Continental Norad Region (US).

Conc Concrete surfaced runway (ICAO).

concentrated force, load See *point force, load*.

concentration ring *1* In balloon, ring, usually rigid, attached to envelope or (if applicable) surrounding net, and from which basket is suspended.

2 In airship, ring to which several mooring lines may be secured (sometimes also helping support car, if this is suspended below hull).

concentric Having common centre or central axis.

concession *1* Allowable departure from drawing in manufacture of part (eg on material spec., surface finish or manufacturing tolerance).

2 Allowable non-compliance with certification or other requirement, esp. in emergency (eg take-off permitted with one engine or one altimeter inoperative).

concurrence Policy adopted for reasons of national emergency in which most, or all, parts of major system programme are implemented simultaneously, even though several large portions may need to be grossly modified or updated (eg Atlas ICBM hurriedly deployed above ground, then in surface shelters and finally in silos).

concurrent engineering Consideration of market, design, manufacture [and tooling], test and life support, from outset.

concurrent forces Acting through common point.

Cond, Conds Condition[s].

condensation Physical change from gaseous or vapour state to liquid.

condensation level Height at which rising parcel of air reaches saturation; cools at DALR and reached 100% RH at ** at intersection of DALR and DPL.

condensation nuclei Minute particles, solid or liquid,

upon which nucleation begins in process of condensation; most effective ** are hygroscopic.

condensation shock   Sudden condensation of super-saturated air in passage through normal or inclined shock, rendering shock field visible, often showing elliptic lift distribution around transonic aircraft.

condensation trail   Visible trail, usually white but sometimes darker than sky background, left by winged or propelled vehicle when flying above condensation level. May be due to reduced pressure (eg in tip vortices), but nearly all persistent ** due to condensation (and probable freezing) of water vapour formed by combustion of fuel.

condenser   *1* Capacitor.

2 Device for changing flow of vapour to liquid by removing latent heat of evaporation. Essential feature of closed-cycle space power systems in which working fluid must be used repeatedly.

condenser-discharge light   Gives very short flashes of great intensity caused by capacitor discharge through low-pressure gas tube (eg collision beacon).

con-di nozzle   Jet-engine nozzle having cross section which converges to throat and then diverges; subsonic flow accelerates to throat, becomes supersonic and then accelerates in divergent portion.

conditionally unstable   Unsaturated air above or through which temperature falls with height faster than SALR but less than DALR; thus if air becomes saturated it will be unstable.

condition monitoring   Health inspection of operative hardware, eg engine, using intrascope, X-ray photography, oil sampling and BITE.

Condo   Contractors on deployed operations.

Condor   *1* Confidential direct occurrence reporting, system for non-attributably ensuring that nothing having a direct bearing on flight safety is kept hidden (RAF, CAB, etc).

2 Electronic 'sniffer' which by mass spectometry identifies traces of vapour or particles emitted by explosives and drugs (from contraband detector, British Aerospace).

3 Covert night and day operations for rotorcraft.

conductance   *1* Real part of admittance in electric circuit; symbol Λ.

2 In circuit having no reactance, ratio of current to potential difference, ie reciprocal of resistance. Symbol G, unit siemens, $= {}^1/_\Omega$.

3 In vacuum system, throughput Q divided by difference in p between two specified cross-sections in pumping system.

4 Several meanings in electrolytes (little aerospace relevance).

5 See *thermal* *.

conduction   Transfer of heat from hotter to colder material or of electrons from higher to lower potential.

conduction band   Band of electron energies corresponding to free electrons able to act as carriers of negative charges.

conductivity   Measure of ability of material to transmit energy, eg heat or electricity. Thermal *, symbol k or λ, measured in $Jm/M^2s°C$. Electrical *, symbol δ, measured in mhos/m (per cube); reciprocal of resistivity.

conductor   Material having very low electrical resistivity, esp. such material fashioned in form useful for electric circuits.

cone   *1* Drag and stabilizing member trailed on end of HF aerial wire (trailing *) or on end of air-refuelling hose.

2 Drag and stabilizing member incorporating pressure and/or static heads trailed beneath aircraft under test in supposed undisturbed air.

cone angle   Semi-angle of right circular cone having same increase in surface area per unit length as diffuser; hence diffuser **.

coned   Caught in beams of two or more searchlights.

cone of confusion   Inverted cone of airspace with vertical axis centred on VOR or other point navaid.

cone of escape   Volume in exosphere with vertex pointing directly to Earth centre through which atom or molecule could theoretically escape to space without collision. Opens out in angle to infinity at critical level of escape.

cone of silence   Inverted cone of airspace with vertical axis centred on certain marker beacons, NDBs and other point navaids within which signal strength reduces close to zero.

cone passage   Flight through cone (of confusion or of silence) above point navaid.

cone yawmeter   Cone flying point-first, with pitot holes spaced at 90° intervals, to obtain yaw indication at supersonic speeds (avoids averaging effect of wing-type yawmeter).

confidence level   Used in statistical sense, eg as percentage probability that an actual MTBF will exceed estimated or published MTBF. Value of ** increases with number of samples. Sometimes called confidence limit.

confidence manoeuvres   Set pattern of ground and air tasks easily mastered by new and inexperienced pupil pilot (eg, swinging propeller, letting aircraft recover from unnatural flight attitude hands-off); devised to ease problem of apprehension and tension. Sometimes called confidence actions.

configuration   *1* Gross spatial arrangement of major elements, eg in case of aircraft disposition of wings, bodies, engines and control surfaces.

2 Aerodynamic shape of aircraft where variable by pilot command, eg position of landing gear, leading/trailing-edge devices and external stores. Thus high-lift *, clean *.

3 Standard of build or equipment for task. Thus helicopter in dunking ASW *, passenger transport converted to all-cargo *.

4 Apparent positions of heavenly bodies, esp. in solar system, as seen from Earth at particular time.

5 A new (1990–) usage: the number of seats in a passenger airliner, thus '*220'.

6 Used, incorrectly, to mean 'application', eg 'Chaparral is the Sidewinder missile in ground-to-air *'. This would be correct if hardware was physically changed in *.

configuration bias   Channel or subsystem in stall protection or stick-pusher system allowing for changes in configuration (2).

configuration deviation list   Comprehensive schedule of all variable parts of a/c, such as door panels and seals.

conflict   In ATC (1), two aircraft proceeding towards potentially dangerous future situation. Hence, * alert, * resolution, * situation.

conflicting traffic   With respect to one aircraft, other traffic at or near same FL heading towards future conflict.

conformal-array aerial   Electronically scanned, fits exterior surface of vehicle.

conformal-array radar   Having plurality of small or light ES aerials covered by radomes fitting vehicle shape (eg wing or rotor leading and trailing edges, etc).

conformal gears   Having teeth whose mating profiles conform, both sets having instantaneous centres of curvature on same side of contact. Usually applied to W-N gears.

conformal projection   Having all angles and distances correct at any point, but with scale changing with distance from point.

conformal tank   Removable [not necessarily jettisonable] fuel tank shaped to fit precisely against skin of aircraft.

confusion reflector   Designed to reflect strong echo to confuse radar, proximity fuze, etc. Form of passive ECM.

conical camber   Applied to wing leading edge so that, from root or intermediate station to tip, it is progressively drooped, centreline of profile following surface of cone with vertex at root (or at start of ** if this is some distance along semi-span).

conical flow   Theory for supersonic flow over thin flat plate having corner (apex), with flow perpendicular to rear edge: constant pressure, velocity, density and temperature along any radius (to infinity) from apex.

conical scanning   Common search mode for radar, esp. AI radar, in which beam is mechanically or electronically scanned in cone extending ahead of aerial, often using beam-switching to give az/el data.

conical sleeve   Cone-shaped flexible sleeve extending inwards into gas cell of airship from aperture for line, providing near gas-tightness with freedom for line to move axially through envelope.

conic apogee   Apogee of satellite if all mass of primary were at its centre.

conic perigee   Perigee of satellite if all mass of primary were at its centre.

conic sections   Perpendicular to axis = circle; parallel to axis = parabola (eccentricity 1); eccentricity less than 1 = ellipse; eccentricity greater than 1 = hyperbola. All are found in trajectories of bodies moving in space.

Conie   Comision Nacional de Investigacion de Espacio (Spain).

coning   *1* Tunnel test in which model is rotated whilst held at constant AOA and sideslip by rotary balance.

2 Capturing hostile aircraft in beams of several searchlights.

coning angle   *1* Angle between longitudinal axis of blade of lifting rotor and tip-path plane (assuming no blade bending). Symbol β.

2 Incorrectly, sometimes given as average angle between blade and plane perpendicular to axis of rotation.

conjugate   Many specialised meanings in theory of groups, complex numbers and geometry of curved surfaces.

conjugate beam   Hypothetical beam whose bending moment assists determination of deflection of real beam.

conjugate foci   In optics, interdependent distances object/lens and lens/image.

conjunction   Alignment of two heavenly bodies sharing same celestial longitude or sidereal hour angle.

connecting rod   Joins reciprocating piston to rotary crank in piston engine, reciprocating pump, etc.

connector   Standard mating end-fitting for fluid lines, multi-core cables, co-ax. cables and similar transmission hardware, providing automatic coupling of all circuits. Term preferred for multipin electric *; with fluid systems prefer "pipe coupling".

Conops   Concept[s] of operations (USN, now all-US).

conplan   Contingency plan.

Conradson   Standard test apparatus and procedure for determining carbon residue left after combustion of hydrocarbon oils, especially lubricating oils.

conrod   Piston[s] engine connecting rod (colloq.).

consensus   Majority vote concept in logic systems, multichannel redundant systems etc; thus, * can command landing flare against presumed failed channel.

Consequence assessment tool set   Central program used by Federal and local agencies in responding to domestic emergencies, now part of ECHO (DoD).

Consol   Simple long-range navaid providing PLs (within range of two * stations, a fix) over N Atlantic. LF/MF receiver is tuned to identified * station and operator counts dots and dashes in repeated 'sweep' lasting about 30 seconds; PL is then obtained by reading off * chart.

Consolan   Consol-type system radiating daisy pattern at c300 kHz, formerly based at Nantucket (US).

console   *1* Control station for major device or system, normally arranged for seated operator.

2 Control and instrument installation for pupil navigator, esp. when such * repeated along fuselage (but not used for pilot station on flight deck).

3 Single bank of controls and/or instruments on flight deck, eg roof *, left side *.

4 Station for manual input/output interface with large system, eg air defence, ATC, EDP (1).

5 Tailored box for storage of maps, cameras and other items, eg 'The Cessna 210 has centre-aisle * as an option'; misleading and ambiguous.

consolidation   Period between first solo and issue of PPl or other ab initio licence; hence * exercise, * flight.

consolute   Of two or more liquids, miscible in any ratio.

constantan   Alloy of copper with 10–55 per cent nickel; resistivity essentially unchanged over wide range of temperature.

constant-colour   Philosophy for cockpit warning systems, usually: no caption illuminated = no fault, all buttons normal; blue = normal-temperature operation; white = button abnormal, either from mis-select or to rectify/suppress fault; red or amber = fault.

constant duty cycle   Device or system whose rate of operation is unvarying despite variable demand; eg DME ground transponder beacon has *** behaving as though continuously interrogated by 100 aircraft.

constant-energy line   Plots taken in steep dive at terminal velocity, when increase in dive angle has no effect on V.

constant-flow oxygen   Crew-breathing system in which gox is fed at steady rate, in contrast to demand-type supply.

constant-g re-entry   RV uses aerodynamic lift in skip trajectory to impose constant total acceleration down to relatively low velocity.

constant-heading square   Helicopter pilot training manoeuvre: large square described at low level with helicopter constantly facing into wind (so one leg forwards, one backwards and two sideways).

constant-incidence cruise   Transport aircraft flight plan calculated on basis of constant angle of attack over major portion, angle being chosen for best L/D or other optimised point between time and fuel consumption.

constant-level balloon Designed to float at constant pressure level.

constant of gravitation See *gravitational constant.*

constant-pressure chart Plot of contours showing height above MSL of selected isobaric surfaces.

constant-speed drive CSD, infinitely-variable-ratio gear between two rotating systems, esp. variable speed aircraft engine and constant-frequency alternator; output maintained invariant despite variation in input speed and output torque.

constant-speed propeller, c/s propeller Propeller whose control system incorporates governor and feedback which automatically adjusts pitch to maintain selected rpm.

constant speed unit CSU, engine-driven governor controlling c/s propeller, maintaining rotational speed by varying pitch according to airspeed and engine power.

constant torque on takeoff Turboprop electronic unit which modifies DECU voltage according to pilot's torque command.

constant wind *1* W/V assumed for navigational purposes, until updated or refined.

2 Used in contradistinction to gust (2).

constellation *1* Traditional conspicuous group of fixed stars having supposed resemblance to Earth object.

2 Arbitrary portion of celestial sphere containing a * (1) bounded by straight lines, whole sphere being thus divided for use as reference index.

constituent day Period of Earth rotation with respect to hypothetical fixed star.

constrictor *1* Obstruction in pipe or other fluid flow constraint pierced by small hole giving precisely known mass flow per unit pressure difference.

2 Annular or distributed constriction in nozzle of air-breathing jet engine, esp. ramjet or pulsejet.

consumables Materials aboard spacecraft which must undergo once-only irreversible change during mission, eg propellants, foods (in present state of art) and some other chemicals such as in SPS.

consumables update Regular housekeeping chore, reporting to Earth mission control exact quantities (usually masses) remaining.

cont Continuous, continuously, or continue.

contact *1* Visual link between pilot (rarely, other aircrew) and ground or other external body. Thus, in * = seen, * flying = by reference to ground.

2 Unambiguous radar link (radar *).

3 Single positive mechanical hook-up between FR tanker and receiver aircraft (dry * if no fuel to be transferred).

4 Shouted by pilot of simple aircraft to person swinging propeller of piston engine, indicating ignition about to be switched on.

5 Unidentified target appearing on radar or other surveillance system (rarely, seen visually).

contact altimeter See *contacting altimeter.*

contact approach Visual approach to airfield requested by, and granted to, pilot making IFR flight.

contact-burst preclusion Nuclear-weapon fuzing system which, in the event of failure of desired air burst, prohibits unwanted surface burst.

contact flying Aircraft attitude and navigation controlled by pilot looking at Earth's surface. (Certain

authorities, questionably, include clouds as source of visual cues.)

contact height That at which runway is first glimpsed during landing approach.

contacting altimeter Makes or breaks electrical circuit (eg warning or radio transmission) at chosen reading(s).

contact ion engine Space thruster stripping electrons from caesium or other supply material infiltrated in substrate (eg tungsten). Bombardment ion engine more common.

Contact Judy AAM firing mode: target is within correct parameters.

contact lights White lights on either side of runway in use, parallel to centreline (obsolescent, see *runway edge lights*).

contact lost Situation in which contact (5) can no longer be seen, though target believed still present.

contactor Electric switch having remote (usually electromagnetic) control.

contact patrol Patrol beyond front line with intention of encountering hostile a/c (WW1).

contact point In CAS (3), geographical or time point at which leader established R/T contact with FAC or ground ATC.

contact print Photograph made from negative or diapositive in contact with sensitised material; optical, radar or IR.

contact race Competitors are required to land at several intermediate points where their logbooks are signed by a marshal.

contagious failure One likely to transmit to an adjacent item.

container *1* Standard rigid box for baggage or cargo: maindeck *, ISO 96 in × 96 in × 10, 20, 30 or 40 ft; SAE 10, 96 in × 96 in × 125 in; SAE 20, 96 in × 96 in × 238.5 in; underfloor *, IATA A1 (LD3) 92 in × 60.4 in × 64 in; A2 (LD1) 79 in × 60.4 in × 64 in.

2 Standard ASR package dropped to aircrew in dinghy.

container delivery Standard military airdrop supply of from one to 16 bundles of 1,000 kg (2,200 lb) each.

containment Demonstrated ability to retain every part within machine, following mechanical breakup of portion or whole of moving machinery. Applies particularly to gas-turbine engines, certification of which usually prohibits ejection of fragments even through inlet or nozzle.

contaminate Aerospace meanings include transfer of terrestrial germs and other organisms to spacecraft sterilized for mission, transfer of unwanted atoms to single-crystal (eg semi-conductor) materials, and deposit and/or absorption and/or adsorption of any NBC material on friendly surfaces.

contaminated runway Surface all or partly covered with water, snow, slush, blown sand or foreign objects capable of causing damage.

Contap Consol Technical Advisory Panel.

Conticell Proprietary low-density sandwich structure.

contingency air terminal Mobile air-transportable unit providing all necessary functions to handle air transport at combat airfield.

contingency plan Drawn up and implemented by military commander or civil manager in event of failure of original plan, for anticipated reasons.

contingency power Exceptional power available from

engine[s] of multi-engined aircraft after failure of another; in Concorde * was 5 per cent above normal reheat T-O rating. See next.

contingency rating Power levels required of helicopter and VTOL engines in emergency conditions, time-limited [usually to from 1.5 to 30 min] and normally requiring subsequent special inspection (see *maximum* **, *intermediate* **).

contingency retention item Surplus to requirements but authorised for retention to meet unpredicted contingency.

contingent effects Those of nuclear detonation other than primary effects.

continuation trainer Trainer aircraft for experienced pilots, esp. those in desk jobs.

continuity line Portion of line system diagram in cockpit or other human interface superimposed on push-button or magnetic indicator.

continuous beam Single structural member having more than two supports.

continuous-descent approach Especially important at night, philosophy of eliminating stacking and enabling every arrival to avoid power settings for level flight.

continuous-element system Fire-detection system comprising either electrical circuit or gas-filled tube; heating any part sends signal.

continuous-flow system See *constant-flow oxygen*.

continuous half rolls Display/competition manoeuvre in which numerous half rolls are made, marking being on accuracy of intermediate wings-level positions, which are held very briefly.

continuous-path machining Shaping of workpiece by cutter traversing unbroken path, esp. this form of NC control and machine program.

continuous strip Film produced by ** photography, using ** camera, in which ** film passes at constant speed, related to speed of aircraft, past slit in optical focal plane.

continuous wave, CW, c.w. EM waves repeated without breaks indefinitely, usually with constant amplitude and length (frequency); ie, not pulsed.

continuum Spectral region in which absorption or emission is continuous, with no discrete lines.

continuum flow See *free-molecule flow*.

Contour Comet-nucleus tour (NASA).

contour *1* On topographic map or chart, line joining all points of equal surface elevation above datum (eg MSL).

2 On * chart, line joining all points of equal elevation (height above or below datum, eg MSL, and above or below ground or sea surface) of selected pressure surfaces. Thus can plot * of 1,000 mb surface at -120 ft, MSL and +120 ft.

3 On weather radar, area blanked out in centre of display of storm cell, or whenever return level exceeds given threshold.

contour capability *1* Of mapping radar, ability to display all ground above selected height above MSL or other datum.

2 Of weather radar, ability to make contour display.

contour display Radar display in which all echoes above given strength are cancelled. Normally used in viewing storm clouds. With CONTOUR operative cloud echo has black centre showing region of greatest precipitation (and assumed greatest gust severity). With colour radar each contour has distinctive hue.

contour flying Normally denotes holding constant small

height AGL, ie not following contours (1) but terrain profile (see *NOE*).

contour interval Difference in height between adjacent contours (1, 2).

contour template Hard copy of profile of 2-D or 3-D shape, eg for tunnel throat, press tool, form block.

contract definition phase Important process in procurement linking end of feasibility study and other conceptual phases with full hardware development, CDP involves collaboration with one or more industrial contractors and can involve detailed computer study and hardware test to establish what is to be bought and on what terms.

Contracting State Sovereign country party to an international agreement.

contraction Duct of diminishing cross-section through which fluid is flowing; eg front part of venturi.

contraction ratio *1* In subsonic tunnel, ratio of maximum cross-section to that at working section.

2 In supersonic tunnel, ratio of cross-sectional area just ahead of contraction to that at throat (can be variable).

contractor-furnished equipment, CFE Hardware, software or, rarely, specialist knowledge or experience, supplied by contractor to support programme; esp. items normally GFE, bought-out or supplied from other source.

contractor-furnished weight, CF weight Total mass of aircraft in precise state in which ownership is transferred to customer.

contract oversight Ongoing monitoring of contracts, with particular attention to finance and national security (DoD). Does not mean to fail to notice an irregularity.

Contrafan Registered name for studies of advanced direct-drive shrouded propeller engines in Mach 0.9 class (Rolls-Royce).

contra-flow engine Loosely, any engine involving fluid flow in opposite directions; specif., gas turbine having compressor and turbine back-to-back, with flows (1) axially towards each other and radially out together, (2) radially out from compressor and radially in through turbine, or (3) forward through compressor and back through ducts to turbine.

contrail *Condensation trail* (abbn. not admitted by NATO).

contra-injection Upstream injection of fuel droplets into airflow or of one liquid rocket propellant against another.

contra-orbit defence Supposed technique of defending area by launching missile along predicted trajectory or orbit of hostile weapon.

contrapop Contra-rotating propeller.

contra-rotating *1* Two or more propellers rotating at equal speed in opposite directions on common shaft axis, and sharing common drive.

2 Installation of similar tandem piston engine/propeller combinations back-to-back on opposite ends of common nacelle. (Not to be used for propellers rotating in opposite directions but not on common axis. See *handed*, *co-axial*.)

3 Of any rotating assembly, turning in opposite directions, possibly at different speeds.

contrast Difference in luminous intensity between different parts of picture (photograph, radar display, synthetic display or TV).

control *1* Exercise of civil or military authority, eg over air traffic.

2 In hardware system, device governing system operation.

3 In man/machine system, device through which human command is transmitted across interface.

4 In photogrammetry, points of known position and elevation.

5 In research experiment, unmodified test subject used as yardstick.

control airport   See *tower airport*.

control and reporting centre   Subordinate air-control element of tactical air control centre from which radar control and warning operations are conducted within its area of responsibility (USAF, NATO etc).

control and reporting system   Organisation set up for (1) early warning, tracking and identification of all air and sea traffic, and (2) control of all active air defence.

control anticipation parameter   In a sudden large nose-up command, ratio of initial to steady-state normal acceleration [in simple manual aircraft].

control area   Controlled airspace extending upwards from specified height (ICAO prefers 'limits') above Earth (NATO adds 'without upper limit unless specified').

control augmentation system   See *command augmentation*.

control bar   Main pilot's input to hang glider.

control cable   Physical connection between human control (3) and operating system, esp. between pilot's flying controls and control surfaces.

control car   Housing pilot or coxswain of airship.

control centre   See *launch* **.

control column   Aerodyne trajectory control (flight control input) normally exercising authority in pitch and roll. May be stick, wheel, miniature sidestick or spectacles (see *yoke*).

control-configured vehicle   See *CCV*.

contrôle auto généralisé   Voice + computer (F).

control feel   See *feel*.

control jet   See *reaction control jet*.

controllable-pitch propeller   Capable of having blade pitch manually altered in flight, either to set positions or over infinite range (but not c/s).

controllable rocket   Having rate of combustion of liquid, solid or hybrid propellants capable of being varied at will during burn.

controllable twist   Helicopter rotor blade capable of changing angle of incidence in predetermined manner from root to tip in course of flight.

controlled aerodrome   One at which ATC service is supplied to aerodrome traffic (does not imply existence of control zone).

controlled airspace   Airspace of defined dimensions within which ATC service is provided (ICAO adds 'to controlled flights'). Can be IFR only, IFR/VFR or visual exempted [no control provided].

controlled attack   Bombing target with Master Bomber in attendance (RAF WW2).

controlled environment   One in which such variables as temperature, pressure, atmospheric composition, ionising radiation and humidity are maintained at levels suitable for life or hardware.

controlled flight   Provided with ATC service.

controlled flight into terrain   Unexpectedly encountering terra firma (land or water, but usually hills or mountains),

the No 1 killer in commercial aviation. The flight need not be controlled (see previous).

controlled interception   One in which interceptors are under positive control (from ground, ship or AWACS).

controlled leakage   Environment for life or hardware in which harmful products (eg carbon dioxide) are allowed to leak away and be replaced by fresh oxygen or other material.

controlled mosaic   One in which distances and directions are accurate.

controlled response   Chosen from range of options as being that giving best all-round result.

controlled torque tightening   Use of special adjustable tool to tighten bolts/nuts etc according to material, diameter, plating and lubricant.

control line   *1* Connection between operator and ** aircraft.

2 Connection between control car of airship and controlled item.

control-line aircraft   Model aircraft whose trajectory is controlled by varying tensions or signals in two or more filaments linking it with ground operator.

control lock   Physical lock preventing movement of control surface, either built into aircraft or brought to it and fastened in place.

control-motion noise   Sufficient to cause small surface movement in coupled ILS, but not affecting trajectory.

control panel   Self-contained group of controls, indicators, test connections and other devices serving whole or portion of aircraft system, either accessible in flight or only during ground maintenance.

control pattern   In SSR/IFF, governs reply code for each mode selected.

control point   Fixed position, marked by geographic feature, electronic device, buoy, aircraft or other object, used as designated aid to navigation or traffic control (NATO, USAF).

control reversal   In aircraft flight control system, dangerous state in which pilot demand causes response in opposite sense. Normally caused by either mechanical malfunction (eg crossed controls) or aeroelastic distortion of airframe.

control rocket   Usually small and intermittently fired thruster for changing spacecraft attitude and refining velocity.

controls   As 'the *', primary flight control input devices, esp. in aerodyne; typically stick and rudder pedals.

control sector   Defined block of airspace within which one controller, or group of controllers, has authority [normally feature of civil ATC].

control stick   Control column (colloq.).

control-stick steering   Control of aircraft trajectory by input to AFCS by means of primary flight controls. Not same as * -*wheel* *.

control surface   Aerofoil or part thereof hinged near extremities of airframe so that, when deflected from streamwise neutral position, imparts force tending to change aircraft attitude and thus trajectory.

control surface angle   Measured between reference datum on control surface and chord of fixed surface or aircraft longitudinal axis.

control system   In missile, RPV or aircraft flying on AFCS, serves to maintain attitude stability and correct deflections (NATO, USAF). Also, not included in this definition, translates guidance demands into changes in trajectory.

control tower ATC organization, normally located on tower or near airfield, providing ATC service for airfield traffic and possibly within other airspace.

control vane Refractory surface, usually small, pivoted in jet of rocket or other propulsion system to control attitude, and hence trajectory, of vehicle when deflected from neutral setting.

control warfare Information warfare.

control-wheel steering Autopilot mode giving manual control of heading while holding velocity and/or attitude.

control zone Controlled airspace extending upwards from Earth's surface (NATO, USAF). SEATO has long and involved definition including 'and including one or more airdromes' (*sic*). ICAO adds 'to a specified upper limit'.

Conus Continental US, ie US and its territorial waters between Mexico and Canada plus Alaska, but excluding overseas states.

convection *1* In fluid dynamics, transfer of fluid property by virtue of gross fluid motion.

2 In atmosphere, transfer of properties by vertical motion, normally thermally induced.

convection cooling Method of cooling hot hardware, esp. gas turbine rotor blades, by removing heat from within bulk of material by flow of cooler air passing through system of holes or passages (see *film cooling, transpiration cooling*).

convective cloud Cumuliform, CuF, triggered by convection; normal vertical development fair-weather cumulus; extreme form is cumulonimbus. Bottom lies at condensation level; top can be in stratosphere.

Convective Sigmet Issued for convective weather posing potential danger.

convenience bag Sick bag [despite name, not for urine].

conventional Not nuclear, ie HE.

conventional enhancement Modifies B-52H for electrical and software interfaces for future weapons, using MIL-STD-1760.

conventional stores Free-fall HE devices.

conventional take-off and landing, CTOL Aeroplanes other than STOL, VTOL and other short-field forms.

convergence *1* Condition in which, at least reckoned on surface winds, there is net inflow of air into region.

2 Of mathematical series, one having a limit.

3 Of vector field, contraction.

4 Of terrestrial meridians, angular difference between adjacent pair at particular position.

convergence factor Ratio of convergence (4) and change of latitude (zero at Equator, max. at poles).

convergent Of oscillation – eg sinusoidal motion, phugoid or structural vibration – tending to die out to zero within finite (possibly small) number of cycles.

convergent/divergent See *con-di nozzle*.

converging flight rule Aircraft approaching from right has right of way.

conversion angle That between great-circle and rhumb-line bearings.

convertible aircraft *1* Transport aircraft designed for rapid conversion from passenger to all-cargo configuration or vice versa.

2 Generally unsuccessful aircraft which can change their configuration [eg. from rotor to fixed wing] in flight.

convertible brake Able to make quick change anywhere between carbon/compo/steel.

convertible engine One capable of giving either fan thrust or shaft power.

convertible laser designation pod Any 'convertible' pod usually offers a choice of LWIR or TV.

converticar One term for a roadable VTOL.

convertiplane Aerodyne capable of flight in at least two distinct modes, eg vertical flight supported by lifting rotor and forward translational flight supported by wing.

convertor Among many other meanings;

1 Rotary machine for changing alternating into direct current.

2 Self-regulating boiler for drawing on Lox storage and supplying flow of Gox.

convo Convolution response algorithm.

COO *1* Chief operating officer of company or corporation.

2 Cost of ownership.

cookie HC bomb 4,000 lb or over (RAF colloq.).

cook-off Inadvertent firing of automatic weapon due to round being detonated by residual heat in breech.

coolant Liquid circulated through closed circuit to remove excess heat, eg from piston engine.

cooldown See *chilldown*.

cooled cooling air Use of a fuel/air heat exchanger to cool [hot] compressor-bleed air used to cool the turbine and nozzle, permitting higher TGT.

cooling drag That due to need to dump excess heat to atmosphere (with skill can be made negative).

cooling effectiveness Expressed as $\dfrac{T_{gas} - T_{metal}}{T_{gas} - T_{coolant}}$.

cooling gills Hinged flaps forming partial or complete ring around rear edge of cowling of air-cooled piston engine to control airflow.

co-operative aircraft In ATC, one carrying transponder for SSR.

co-operative emitter Any friendly emitter, esp. those provided for surveillance and tracking of hostile targets.

Co-operative Fuel Research Permanent committee of SAE including fuel and engine representatives with special brief to measure and improve anti-knock ratings.

co-operative independent surveillance Monitoring aircraft position, beyond radar range, by satellite tracking; co-operative because aircraft emits a signal, and independent because aircraft's navaids are not used.

Cooper-Harper Refined scale of flying qualities, broad bands being: up to 3.5 satisfactory, 3.5–6.5 adequate, improvement warranted, over 6.5 inadequate, improvement required.

Cooper scale Scale for quantified Pilot Opinion Rating.

Co-ops CO_2 observational platform system.

co-ordinated-turn One in which controls about three axes are used to avoid slip or skid.

co-ordinates Inter-related linear and/or angular measures by which the position of a point may be defined with reference to fixed axes, planes or directions.

co-ordination In a pilot, ability to control simultaneous unrelated motions, by left and right hands and feet.

COP *1* Changeover point from one navaid to next (US = chop).

2 Common operating [or operational] picture; 21 adds 21st century.

3 Cab over [snow] plough.

4 Character-oriented protocol.

COPA Canadian Owners and Pilots Association.

copal Natural resin from tropical trees used in some varnishes.

co-pilot Licensed pilot serving in any piloting capacity other than (1) PIC or (2) being on board solely to receive instruction.

copper Malleable metal of distinctive red-gold colour, Cu, density 9.0, MPt 1,084°C.

Copper Flag Air-defence equivalent of Red Flag, held at Tyndall AFB (USAF).

COPR Cruise overall pressure ratio.

Cops, COPS *1* Common operational performance specification.

2 Common operating procedures.

copter Helicopter (approach procedure).

Copy "I read you" (radio voice code).

copy machining Using machine tool having means for copying shape of template or master part.

copy milling See *copy machining*.

COR, cor *1* Correct, corrected, correction.

2 Certificate of Registration.

Coral British computer language very similar to Jovial.

cord US measure of volume, = 3.6246 m^3.

cord, corded From pioneer era rigging of flight-control surfaces was adjusted by doping on length [guessed from experience] of cord. Even today trim can be improved by cord on one side of trailing edge, and overbalance by adding cord on both sides.

cordite Gun propellant prepared mainly from nitro-celulose (gun-cotton) dissolved in nitroglycerine.

Cords *1* Coherent on-receive Doppler system

2 Centre for Orbital and Re-entry Debris Studies (US).

Cordtex Blasting or cutting cord comprising high explosive in flexible filament form.

Cordwood Electronic technology (1948–55) designed to achieve maximum packing density of discrete components in pre-semiconductor era.

CORE, Core Controlled requirements expression, discipline which defines software design (BAe).

core *1* Gas-generator portion of turbofan, term especially when * small in relation to fan; less relevant to bypass or 'leaky turbojet' engines.

2 Central part of launch vehicle boosted by lateral or wrap-round rockets.

3 Low-density stabilizing filling inside honeycomb, foam-filled or other two-component structure.

4 High-density penetrative filling in armour-piercing projectile.

5 Magnetic circuit of transformer or inductor.

6 Central portion of nuclear reactor in which reaction occurs.

7 Solid shape(s) which make casting hollow.

8 Loosely, EDP (1) memory of magnetic type, from * (5).

9 Interior of carburised or nitrided part unaffected by surface treatment.

core booster *Booster 3.*

core deposits Solids deposited on metal surfaces of core (1).

core exhaust mixer In engine of ejector-lift STOVL, core nozzle capable of inflight limited vectoring and, in jet-lift mode, of deflecting at least 90° while entraining fresh air from above.

core-failure clutch Upon major mechanical failure of core (1), disconnects drive to tilting rotors (rarely, to helicopter transmission).

corel Combined omnidirectional runway/taxiway edge light[ing].

coring Uneven flow of oil through oil cooler due to reduced viscosity of oil in hot central core.

coriolis acceleration Acceleration of particle moving in co-ordinate system which is itself accelerating, eg by rotating. In Earth-referenced motion, ** is experienced in all motion parallel to local surface except for that on Equator.

coriolis correction Applied to all celestially derived fixes to allow for coriolis acceleration.

coriolis effect *1* Physiological response (eg vertigo, nausea) felt by persons moving inside rotating container (eg space station with rotation-induced gravity) in any direction other than parallel to axis.

2 According to AGARD: 'The acceleration, due to an aircraft flying in a non-linear path in space, which causes the displacement of the apparent horizon as defined by the bubble in a sextant'. This definition is inadequate.

coriolis force Apparent inertial force acting on body moving with radial velocity within a rotating reference system. Such a force is necessary if Newtonian mechanics are to be applicable. On Earth, ** acts perpendicular to direction of travel, towards right in N hemisphere and towards left in S hemisphere. Also called deflecting force, compound centrifugal force, geostrophic force.

coriolis parameter Twice component of Earth's angular velocity about local vertical, ie twice Earth rate multiplied by sin lat.

coriolis rate sensor Instrument based on beam vibrating in plane of aircraft-referenced vertical, sensing any disturbance about longitudinal axis.

corkscrew Evasive manoeuvre, esp. when subjected to stern attack by fighter; interpretation variable but * axis basically horizontal.

corncob Descriptive generic name for multi-row radial or multi-bank in-line piston engine (colloq.).

corner point Instantaneous change in slope of graph; eg kink in payload/range curve, esp. limiting range for max payload.

corner reflector Passive device for giving strong radar echo, based on three mutually perpendicular metal plates or screens which automatically send back radiation directly towards source.

corner speed Lowest airspeed at which a fighter can pull structure- or aerodynamic-limiting g.

Corogard Vinyl-modified polysulphide paint resistant to hydraulic fluid, usually silver from added aluminium powder.

Corona Radio countermeasure: issuing misleading voice commands to enemy fighters (RAF Bomber Command WW2).

corona discharge Electric discharge occurring when potential gradient around conductor is sufficient to ionise surrounding gas. Unlike point discharge, can be luminous and audible, but unlike spark discharge there are an infinity of transmission paths carrying continuous current. Also called brush discharge, St Elmo's fire (see *static wick*).

co-rotating wheels Landing-gear wheels on live axle and thus constrained to rotate together.

Co-Route Company route.

CoRP Common radar processor, partner to MoRE.

corpuscular cosmic rays Cosmic rays are primary particles (protons, alpha particles and heavier nuclei) which react with Earth atmosphere to yield particles and EM radiation. Term corpuscular is redundant.

CORR Corridor.

corrected advisory Resolution advisory that instructs pilot to change vertical speed [ROC]. [TCAS].

corrected airspeed No defined meaning [see *airspeed*, *SSEC*].

corrected altitude No defined meaning, other than "true height above SL" (see *altimeter errors*).

corrected gyro Normally taken to be one corrected (by latitude nut) for apparent wander due to Earth rotation.

correction Many, such as SSEC.

corrective advisory Resolution advisory commanding changes in ROC, vertical speed.

correlation Confirmation that aircraft or other target seen visually or on radar display or plotting table is same as that on which information is being received from other source(s).

correlation criterion Statistical basis for defruiting or decoding raw IFF, typically on ** of 2/7, ie 2 valid synchronous replies detected within any 7 successive interrogations.

correlation factor In nuclear warfare, ratio of ground dose-rate reading taken at approximately same time as one at survey height over same point.

correlation protection Development by RAE with industry of method of avoiding false ILS indications caused by spurious signals reflected from large objects near runway; localizer and glide-path aerials duplicated (respectively horizontally and vertically) and emit signals which, if not received almost simultaneously at aircraft, are suppressed.

corridor *1* Geographically determinate path through atmosphere, typically curved-axis cone with apex at surface, along which space vehicle must pass after launch.

2 Path through atmosphere, geographically determinate for given entry point, along which space vehicle must pass during re-entry; has precisely defined upper and lower limits, above which vehicle will skip back into space and below which it will suffer severe deceleration and risk injuring occupants or burnup through heating.

3 Assumed safe track in LO penetration of hostile territory.

4 Path through atmosphere, usually at low level, along which defences are assumed handicapped by prior seeding with chaff and decoys.

5 Region of any shape on graph within which solution to problem is possible.

6 In Europe pre-1960, nominated tracks along which aircraft were permitted to cross a frontier.

corrosion A normally used word, but see *exfoliation*.

corrugated mixer Turbofan core nozzle of deep multi-lobe form to promote rapid mixing with fan airflow.

corrugated skin Stabilized against local bending by uniform rolled corrugations which, when used as external skin of aircraft, are aligned fore and aft (incorrectly assumed parallel to local airflow).

corrugated strip Interposed between welded sections of gas-turbine flame tube, admits film of cooling air; colloq. wiggly-strip.

corruption Degradation of EDP (1) memory, typically from severe EM interference or, with volatile memory, from switching off power.

CORS Continuously operating reference station (NGS).

Corsaire Co-ordination of research for the study of aircraft impact on the environment (EU).

COS Corporation for open systems, software improvement concept.

CoS Chief of Staff.

Cosac Computing system[s] for air cargo.

Cosim Variometer (colloq., obs.).

Coslane Constant [lateral] separation lane.

Coslettising Anti-corrosion treatment involving a wet deposition of Zn.

cosmetic RFP Issued for sake of appearance, contract award being already decided.

cosmic speeds Those sufficiently high for interstellar exploration, similar to that of light; even allowing for relativistic time effects these are wholly unattainable at present.

cosmodrome Space launching site (USSR).

cosmology Science of the Universe.

cosmonaut Member of spacecraft crew (USSR, R).

cosmonautics See *astronautics*.

Cospar Committee on Space Research (Int., office in Paris).

Cospas Anglicised form of space system for search for distressed vessels, in conjunction with *Sarsat* (R).

Cosro Conical scan, receive only, i.e. only during reception.

Cossi Commercial Operations and support savings initiative.

cost In procurement main elements may include R&D, T&E, flyaway, spares provisioning, ground equipment, base, crew and publications. Operating adds fuel and other consumables, depreciation and various indirect *.

costa Rib, translated in aviation not as wing rib but as fuselage frame.

costal Pertaining to frames or ribs; hence intercostal.

Costar Correcting optics space telescope axial replacement (Hubble).

cost/economical Cruise conditions for minimum trip cost.

cost-effectiveness Measure of desirability of product, esp. a weapon system, in which single quantified figure for capability (including reliability, survivability and other factors) is divided by various costs (total ownership, acquisition etc).

cost plus fixed fee Reward invariant with actual costs but fee may be renegotiated.

cost plus incentive fee Reward covers actual costs plus a fee which depends on contractor performance and possibly costs.

cost-sharing No fee, contractor merely reimbursed agreed percentage of costs.

COT *1* Compressor outlet temperature.

2 At the coast.

Cotal Confederación de Organizaciones Turisticas de la America Latina (Int.).

CoTAM Commandement du Transport Aérienne Militaire (F).

Cotim Compact thermal-imaging module.

COTP Connection-oriented transport protocol.

COTS, Cots   Commercial off-the-shelf [item already available, esp. for military a/c].

cottage loaf   Fuselage with smaller-section upper deck and unfaired sides [almost figure 8].

cotter pin   *1* Wedge-shaped pin used in joining parts.
   2 In US, often split pin.

CO₂   Carbon dioxide.

Cougar   Co-operative unmanned ground-attack robot (USA).

coulomb   SI unit for quantity of electricity or electric charge, = 1 As, symbol C.

coulomb damping   That due to opposing force independent of distance or velocity; also called dry friction damping.

coulomb excitation   Raising of energy level as a result of charged particle passing outside range of nuclear interactions.

Coulomb's law   Force between two magnetic or electric charges is proportional to product of charges and inversely to square of distance apart: $F = \dfrac{Q_1 Q_2}{4\pi\epsilon r^2}$.

countdown   Oral telling-off of time, usually at first in minutes, then in seconds, remaining before launch of vehicle or other event.

counter   *1* Portion of ship hull from stern overhanging water; thus applicable to undersurface of rear fuselage above and behind jet nozzles or other lower section.
   2 Electronic circuit which counts bits, impulses, waves or other repeated signals.

counter air   Defensive and offensive actions against enemy air power.

counterfeit part   An unapproved part knowingly installed.

counterforce   Attack directed against enemy ICBMs and SLBMs or other strategic forces.

counter-illumination   Challenging LO technology in which appearance of an object is changed or [in theory] eliminated by nullifying incident illumination; also called active visual camouflage.

counter-insurgent, Coin   Directed against supposed primitive guerrilla forces.

countermeasures   All techniques intended to confuse or mislead hostile sensors such as radar, IR, visual, TV or noise.

counter-pointer   Dial indication comprising rotating pointer(s) and counter readout in same instrument.

counter readout   Numerical display generated by numerals on adjacent rotating drums, also called veeder.

counter-reflector   Metal mesh or other radio reflector arranged in pattern under VOR or other ground station to nullify interference and give radiation as from perfect level-surface site.

counter-rotating   See *contra-rotating*.

countersilo   Counterforce attack against ICBM silos.

countersink   To form or cut conical depression in workpiece to receive rivet or bolt head flush with surface.

countersurveillance   All active or passive measures to prevent hostile surveillance.

countertrade   Trade in reverse direction generated to assist high-tech (eg defence) exports by an industrialised country; in no sense barter.

countervalue   Attack directed against enemy homeland society and industry.

country cover diagram   Small-scale map and index showing availability of air reconnaissance information of whole country for planning purposes.

countup   Oral telling-off of time, usually in seconds, elapsed since liftoff.

coupé   Aircraft, normally with open cockpit[s], fitted with * top, generally synonymous with canopy forming integral part of fuselage.

couple   Two parallel opposing forces not acting through same point, producing rotative force equal to either force multiplied by perpendicular distance separating axes. SI unit Nm (newton-metre), = 0.748604 lb-ft. Also called moment, turning moment, torque.

coupled engines   Geared to same propeller(s) but not necessarily mechanically joined.

coupled flutter   In which energy is transferred through distorting structure linking two fluttering masses to augment flutter of either or both.

coupling   *1* Inertia *, tendency for inertia forces in manoeuvres to overcome stabilizing aerodynamic forces, esp. in long, dense aircraft having large inertia in pitch and yaw; eg, rapid roll results in violent cyclic oscillation in pitch about principal inertial axis, increasingly marked with altitude owing to divergence of this axis from relative wind.
   2 Connection (electrical, electronic or mechanical) between flight-control system and other onboard system such as ILS or TFR.
   3 Unwanted connection or interference between two radiating elements in a planar-array antenna.

coupon   *1* Small extra piece formed on casting or, rarely, forging or extrusion, to provide metallurgical test specimen.
   2 See *flight**.

courier   See *delayed repeater comsat*.

course   UK term for *heading*.

course and distance calculator   Aluminium disc with pivoted arms for solving three- [even four-] vector navigation problems (1917–40).

course and speed calculator   More advanced yet compact mechanical computer for solving vector problems (UK 1935–50).

course corrections   Allowances for deviation and variation.

course deviation indicator   Vertical needle of VOR display.

course light   Visual beacon on airway, or light indicating course [track] of airway (both obs.).

course line   Locus of points nearest to runway centreline in any horizontal plane along which DDM is zero.

course sector   Horizontal sector in same plane as course line limited by loci of nearest points having DDM of 0.155.

course selector   See *OBS*.

courtesy vehicle   Battery electric car providing up to 6 seats for elderly or disabled passengers.

courtyard   Space in centre of closed-circuit tunnel; hence * wall, inner wall of tunnel.

COV   *1* Common operational value (RAF).
   2 Covered, cover, covering (ICAO).

cove   Local concave curved region where two structures meet, eg wing/pylon or pylon/pod.

cover   *1* Protection of friendly aircraft by fighters or EW platforms at higher level.
   2 Ground area shown in imagery, mosaics etc.

3 To maintain continuous EM receiver watch.

4 To use fighters to shadow hostile contact from designated BVR distance.

coverage diagram   Plot of air-defence radar performance against target of particular cross-section for different elevation angles, plotted on altitude (ordinate) and slant range.

coverage index   See *covertrace*.

Coverage Level   In aerial firefighting, quantity of retardant per unit area, in US usually USG per 100 sq ft.

cover mod   Paperwork [documents] by which DA accepts SEM (3) or STF (UK).

cover search   To select best cover (2) for air reconnaissance for particular requirement.

covertrace   Map overlay listing all air reconnaissance sorties over that ground area, marking tracks and exposures.

covert search   Patrol using advanced sensors from high level so that aircraft's presence is undetected from ground, esp. in offshore patrol for customs, immigration or fishery protection.

Covos   Comité d'Etudes sur les Conséquences des Vols dans la Stratosphère (Int.).

cowl   Covering over installed engine or other device, normally mainly of hinged or removable panels.

cowl flaps   See *gills*.

cowling   See *cowl*.

CP   *1* Critical point.
   2 Centre of pressure (often c.p.).
   3 Controllable pitch (not constant speed).
   4 Chlorinated paraffin.
   5 Circularly polarised.
   6 General call to several specified stations (ICAO).
   7 Co-pilot.
   8 Cadet pilot (or C/p).
   9 Cathode protection.
   10 Command post.
   11 Centre-perforate (rocket grain).
   12 Computer, or communications, processor.
   13 Constant power.
   14 Control panel.
   15 Conflict probe.

C_p   *1* Pressure coefficient.
   2 Specific heat at constant pressure.

cP   *1* Centipoise.
   2 Continental polar air mass.

cp   Candlepower.

CP^3   CPPP.

CPA   *1* Critical-path analysis (see *critical path*).
   2 Continuous patrol aircraft.
   3 Closest point of approach.
   4 Cabin public address.
   5 Certified public accountant.
   6 Civilian Production Administration (succeeded WPB, US).

C/PA   Cost/performance analysis.

CPACS   Coded-pulse anti-clutter system.

CPAM   *1* Committee of Purchasers of Aviation Materials (Int.).
   2 Cabin-pressure acquisition module.

CPC   *1* Cabin-pressure control[ler]; S adds system.
   2 Cursor-position control.
   3 Controller/pilot communication[s].

CPCI   Computer program-configuration item.

CPCP   Corrosion prevention and control programme.

CPCS   Cabin-pressure control system.

CPD   *1* Command planning and direction (GTACS).
   2 Continuing professional development.

CPDL   Controller/pilot data link [C adds communications].

CPE   *1* Central Photographic Establishment (RAF).
   2 Circular position error.

CPF   *1* Complete power failure.
   2 Central processing facilities.

CPFF   Cost plus fixed fee.

CPG   Co-pilot/gunner.

CPGS   Cassette-preparation ground station.

CPI   *1* Cost plus incentive (F adds 'fee').
   2 Chief pilot instructor.
   3 Crash position indicator.

CPIF   Cost plus incentive fee.

CPIFT   Cockpit procedures and instrument flight trainer (Pacer Systems).

CPILS   Correlation-protected ILS.

cPk   Continental polar, colder than surface.

CPL   *1* Commercial pilot's licence.
   2 Current flight plan message (ICAO).

CPL/A   Commercial pilot's licence, aeroplanes.

CPL/H   Commercial pilot's licence, helicopter.

CPL/IR   Commercial pilot's licence, instrument rating.

CPL/SEL   Commercial pilot's licence, single-engine limitation.

CPM   *1* Capacity passenger-miles.
   2 Critical-path method.
   3 Core [or control, or central] processor module.
   4 Certification program manager.
   5 Command-post modem; P adds processor.

CPMIEC   China Precision Machinery Import and Export Corporation, Beijing.

CPO   Close parallel operation.

CPP   *1* Cost per passenger.
   2 Critical parts plan (ECPP).
   3 Crossfeed phasing parameter [μ is preferred].

CPPC   Cost plus percentage of cost.

CPR   *1* Coherent-pulse radar.
   2 Crack-propagation rate.
   3 Contract (or contractor, or cost) performance report.
   4 Covert penetration radar.

CPRSR   Compressor.

CPRTM   Cents per revenue ton-mile.

CPS   *1* Central processing system (or site).
   2 Cabin-pressure sensor.
   3 Covert penetration system.
   4 Characters per second (also cps).
   5 Control power supply.
   6 Conventional[ly] profiled sortie.

cps   Cycles per second (Hz is preferred).

CPT   *1* Cockpit procedure[s] trainer.
   2 Central passenger terminal complex.
   3 Civilian pilot training program (US, 1939–46).
   4 Clearance, pre-taxi.

CPTA   Civilian Pilot Training Act (1939).

CPTP   CPT Program (US 1939–42), became WTS.

CPTR   Command-post terminal replacement.

CPU   *1* Contractor payment unit.
   2 Central [or communications] processing unit.
   3 Control-panel unit; -F adds front, -S side.

cPw   Continental polar, warmer than surface.

153

CPX   Command-post exercise.
CQ   *1* Carrier (ship) qualification.
  2 General message to all stations.
  3 Target control (remotely piloted target), USAAF 1942–47.
CqS   Constant-q stagnation trajectory.
CR   *1* Compression ratio.
  2 Credit (aerial victory).
  3 Cost-reimbursable.
  4 ATC request (FAA).
  5 Change request.
  6 Fighter-reconnaissance (F).
  7 Contrast ratio.
  8 Root chord, also C_R.
  9 Canard/rotor, also C/R.
  10 Countermeasures receiver.
  11 Close-range.
  12 Component repair.
  13 See next.
C/R   *1* Counter-, or contra-, rotating, or rotation, usually refers to handed engines driving single-rotation propellers in opposite directions.
  2 Command/response.
  3 See CR (9).
C$_R$ *1* Resultant-force coefficient.
  2 Range constant, velocity × wt/fuel flow.
Cr   Chromium.
Cr$_2$O$_3$ One of the three chromium oxides.
CRA   Centro Ricerche Aerospaziali, Rome.
crab   *1* To fly with wings level but significant drift due to crosswind.
  2 To fly with wings level but significant yaw due to asymmetric thrust.
  3 To fly with wings level but significant yaw imparted by rudder to neutralise effect of crosswind.
  4 Miniature trolley driven by Link trainer and certain other simulators which reproduces aircraft track on map on instructor's desk.
crab angle   *1* Drift angle.
  2 In landing, angle between runway axis and aircraft heading.
  3 Angle between fore/aft camera axis and track.
crab list   List of snags after flight test (US, WW2).
crab-pot   Fabric non-return valve in circular duct in airship, controlled by bidirectional pull of cord attached to centre.
crack   *1* Microscopic rupture in stressed metal part which under repeated loads progressively grows longer, without deformation of structure, until remaining material suddenly breaks.
  2 To break down hydrocarbons by cracking. Originally done continuously in giant cat-crackers in refineries, this is becoming a procedure necessary in JP7-fuelled hypersonic ramjets.
cracking   Application of heat and usually pressure, sometimes in presence of catalysts, to break down complex hydrocarbons, esp. petroleums, into desired products.
crack-stopper   Structural design feature, such as assembly of part from several components with joints perpendicular to expected crack directions, to prevent crack progressing right across.
CRAD, Crad   Critical R&D.
CRADA, Crada   Co-operative R&D agreement.

CRAF   *1* Civil Reserve Air Fleet (US, from 1951).
  2 Comet rendezvous/asteroid flyby.
  3 Committee on Radio Astronomy Frequencies.
crafted   Made (US usage).
Crag, CRAG   From Pacer -*, compass, radar, GPS.
Cram   Conditional route-availability message.
crane helicopter   Designed for local lifting and positioning of heavy or bulky items rather than normal transport; characterised by vestigial fuselage with payload attached externally or slung.
Cranfield   Formerly College of Aeronautics, now Cranfield University (UK).
crank   Apart from familiar meanings, a single rotation of crank handle [human inceptor], thus full flap may need 12 cranks.
cranked wing   Has acute anhedral inboard, dihedral outboard, usually with abrupt change at about 30 per cent semi-span.
cranking   *1* Turning engine (any type) by external power.
  2 Making a max-rate turn away from the target immediately upon launching an AAM, hence: a crank.
C-Rap   Condensed recognized air picture.
crash   Unpremeditated termination of mission at any point after start of taxi caused by violent impact with another body, with or without pilot in control, usually causing severe damage to aircraft. Term never used in official language.
crash arch   Strong structure above or behind pilot(s) head(s), esp. in open cockpit or small cabin aircraft, able to bear all likely loads in overturning and sliding inverted on ground.
crash barrier   See *barrier*.
crash gate   Gate in airfield periphery through which crash/fire/rescue teams can most quickly reach nearby crashed aircraft.
crash landing   Emergency forced landing with severe features such as rugged terrain or incapacitated pilot, resulting in more than superficial damage to aircraft.
crash locator beacon   Automatic radio beacon designed to be ejected from crashing aircraft, thereafter to float and survive all predictable impacts, crushing forces or fire while broadcasting coded signal.
crash pan   Secondary structure under para-dropped load, esp. vehicle or artillery, which absorbs landing shock by plastic deformation.
crashproof tank   Euphemistic, denotes fuel or other tank designed not to rupture, leak or catch fire in all except most severe crash.
crash pylon   Structure having same purpose as crash arch.
crash switch   Electrical switch triggered by various crash symptoms to shut off fuel, activate fire/explosion suppression, release CLB, etc.
crashworthiness   Generally unquantifiable ability of aircraft to crash without severely injuring occupants or preventing their escape.
crate   Aerodyne, esp. aeroplane (colloq., derogatory, archaic).
CRAW   Carrier replacement air wing (USN).
CRB   Chlorinated rubber-based; P adds paint [airfields].
CRC   *1* Control and reporting centre.
  2 Carbon-fibre reinforced composite.
  3 Communications Research Center (Canada).
  4 Central [ised] radio control.

5 Cyclic redundancy check, or code.

6 Cassegrain Ritchey/Chretien.

CRCO   Central Route Charges Office (ICAO).

CRD   *1* Controller, Research & Development (MAP, WW2).

2 Current routing domain.

CRDA   Cooperative research and development agreement (FAA).

CRD/F   Cathode-ray direction-finding; ground D/F receiver in which aerial automatically rotates to null azimuth as soon as pilot transmits, bearing being instantly shown on circular display.

CRE   *1* Command-readiness exercise.

2 Communications radar exciter.

3 Control and reporting element.

4 Central Reconnaissance Establishment (RAF, formerly).

creamed   Shot down, destroyed (colloq.).

creamer   A perfect landing.

credible   Of deterrent, demonstrably capable of being used and having desired effect; depends on its ability to penetrate and on government's resolution.

credit   Unit of aerial victory scores made up of air-combat plus strafing (aircraft destroyed on ground), with fractions for targets shared (US).

creep   *1* Slow plastic deformation under prolonged load, greatly accelerated by high temperatures.

2 Gradual rotation of tyre around wheel; hence * marker, white index marks on wheel and tyre initially in alignment.

3 See next.

creepback   Tendency of bombs to fall progressively further back in front of target (RAF Bomber Command).

creeping landing   Landing by jet-lift STOVL aircraft with just enough forward speed to avoid reingestion of hot gas or debris from unpaved surface.

creep life   Safe service life of turbine rotor blades, normally set at or near point at which elongation ceases to be proportional to time.

creep strength   Stress that will produce specified elongation over given period (typically 0.1% over 1,000 h) at given temperature.

CRES, Cres   Corrosion-resistant steel.

crescent wing   Has progressive reduction in both t/c radio and sweep angle from root to tip, usually in discrete stages.

Crest   *1* Comprehensive radar effects simulator trainer.

2 Consolidated reporting and evaluating subsystem, tactical.

3 Crew escape technology.

crevice corrosion   Initiated by presence of crevice in structure in which foreign material may collect; eliminated by modern structural coating and assembly methods.

crew   Divided into flight * to fly aircraft, mission * to carry out other duties in flight, cabin * to minister to passengers and, arguably, instructors; all assigned to these duties by appropriate authority.

crew duty time   Measured from reporting for duty to completion of all post-flight duties.

crewing   *1* Make-up of flight crew by trade or appointment.

2 Make-up of flight crew by individual rostered names.

crew ratio   Number of complete air crews authorised per

line aircraft (civil) or per aircraft in unit complement (military).

crew resource management   Ever-refined improvement in in-flight [airline] crew behaviour, esp. in flight-deck and cabin communications, esp. in crisis.

crew return vehicle   Lifting-body vehicle, with final descent by inflatable wing, to bring ISS crew of six back to Earth.

crew room   Room reserved for (usually military) flight crews, some on standby and others relaxing after a mission, where publications are kept and notices promulgated.

crew trainer   Aircraft designed to train whole flight crew, esp. of traditional military aircraft requiring several flight-crew trades: pilot/navigator/bombardier/signaller/engineer/gunner.

CRG   Contingency Response Group (USAF).

crib   Shop-floor container for small tools, parts or material other than scrap.

CrIMSS   Cross-track IR microwave sounder system.

CRIP   Coat-rod inches per passenger.

crisis management   Management of military (war or near-war) situations or of civil crises such as major accidents or natural disasters.

CrIS   Cross-track IR sounder.

Crisp, CRISP   *1* Contra-rotating integrated shrouded propfan.

2 Computer-reconstructed images from space photographs.

3 Compact reconfigurable interactive signal processor.

Crista   Cryogenic infrared spectrometer telescope for the atmosphere.

CRIT   Centre de Recherches Industrielles et Techniques (F).

critical altitude   *1* The highest density altitude which a supercharged piston engine can maintain its maximum continuous rated power.

2 See *decision height*.

critical angle   *1* Angle from local vertical at which radio signals of given frequency do not escape through ionosphere but just return to Earth.

2 Incorrectly used to mean stalling angle of attack.

critical case   That combination of failures (of propulsion, flight controls or systems) giving worst performance (see *critical engine*).

critical crack   One of * length.

critical engine   Engine, the failure of which is most disadvantageous, due to asymmetric effects, loss of system power or other adverse factors; failure of ** at V_1 is basis of takeoff certification in most multi-engine aircraft.

critical frequency   *1* That corresponding to natural resonance of blade, control surface or other structure.

2 Helicopter main-rotor blade-passing frequency at which whole machine resonates on landing gear.

3 Frequency at which critical angle becomes zero; highest at which vertical reflection is possible.

critical-length crack   Crack of length at which application of limit load causes failure.

critical line   Locus of critical points (when track is not known precisely).

critical Mach number   *1* M_{crit}; Mach number at which most-accelerated flow around a body first becomes locally supersonic; for thin wing might be M 0.9 while thick wing may have * below 0.75.

2 Mach number at which compressibility effects significantly influence handling.

critical mass   Mass of fissile material in which chain reaction becomes self-sustaining.

critical path   That traced through number of tasks proceeding both consecutively and concurrently (as during turn-round of aircraft) that determines minimum total elapsed time.

critical-path technique   Minimisation of total elapsed time by concentrating on those elements that form critical path.

critical point   That from which two fixed bases, such as departure airfield and destination, are equidistant in time.

critical position   That over large city or mountain range at which propulsion failure would be most serious.

critical pressure   In fluid flow through nozzle, that final pressure below which no further reduction results in increase in flow from fixed initial pressure; usually rather more than 50% of initial pressure (fixed ratio for any given medium and temperature).

critical pressure coefficient C_{pc}; pressure coefficient at critical Mach number, approximately given by Prandtl-Glauert.

critical pressure ratio   That at which particular axial compressor suddenly ceases to operate efficiently due to choking, stall or other flow breakdown.

critical speed   *1* See V_1.

2 That rotational speed at which machinery (eg, engine) suffers dangerous resonance or whip of shafting.

critical static pressure   That at critical Mach number; symbol P_c.

critical temperature   That below which gas or vapour may be liquefied by pressure alone.

critical velocity   Speed at which fluid flow becomes sonic, ie locally reaches Mach number of unity;

$$V_{cr} \text{ or } V_{crit} = a_o \sqrt{\frac{(\gamma - 1)\, M_0^2 + 2}{\gamma - 1}}$$

CRL   *1* Common rail launcher.

2 Cambridge [Massachusetts] Research Laboratory (USAF).

CRLCN   Circulation.

CRM   *1* Originally cockpit resource management, now crew resource management.

2 Collision-risk model.

3 Customer relationship, or resource, management.

CRN   Common random number.

CRNA   Centre Régional de la Navigation Aérienne (F).

CRO   *1* Civilian Repair Organization.

2 Cathode-ray oscilloscope.

3 Community relations officer (RAF).

Crocco   Luigi Crocco (1932) derived equation:

$$T = Tw - (Tw - Tf)\,\frac{u}{Uf} + \frac{u(Uf - u)}{2Cp}$$

where T is temperature within boundary layer, Tw temperature of adjacent solid surface, Tf free-stream temperature, u local velocity, Uf free-stream velocity, and Cp specific heat at constant pressure.

crocodile   *1* Control surface, usually aileron, which can split apart into upper and lower halves as airbrake; see *deceleron*.

2 Covered gangway to protect passengers from slipstream, c 1920–40.

cropped-fan engine   Turbofan whose fan has been reduced in diameter to match reduced thrust requirement and permit LP turbine and other parts to be simplified.

cropped surface   Wing, tail or other surface whose tip is cut off diagonally at Mach angle appropriate to particular supersonic flight condition.

cropped tip   *Cropped surface.*

cross   To pass over a fix under ATC at a specified altitude, or a specified maximum or minimum altitude.

cross-bar   System of approach lighting using straight rows of white lights perpendicular to runway centreline. Calvert and some other systems use several bars decreasing in width to threshold while US practice is single white bar followed by red undershoot zone.

cross-beam rotor   Helicopter (usually tail) rotor comprising two two-blade assemblies superimposed; usually set at 90° but in AH-64 at 55°/125°.

cross-bleed   Pneumatic pipe system connecting all engines so that bleed from one can start, or drive accessories on, any other.

Crossbow   Code for air attacks on flying-bomb launch sites, 1944.

cross-bracing   Use of crossed diagonal wires, cables or struts/ties to achieve a rigid structure.

crosscheck   Brief message from one pilot to another, or another crew member, in same aircraft giving or confirming situation, eg "Inner marker" or "crosscheck, I have the yoke".

cross-cockpit collimated display   Simulator display providing large visual scene on back-projected screen viewed in curved concave mirror, giving correct perspectives with no discontinuities.

cross-country   Flight to predetermined destination, where landing may or may not be made, esp. one to gain practice in map-reading and navigation.

cross-crew qualification   Training course for mixed-fleet flying.

cross-deck   Operations by two or more aircraft carriers, not necessarily of same navy, in which aircraft operate from unfamiliar decks on exchange basis; hence * ing.

crossed controls   Application of flight-control movements in opposite sense to those in normal turns or manoeuvres, eg right stick and left rudder; rarely required.

crossed-spring balance   Wind-tunnel balance whose pivots are made up of two or more leaf springs crossing diagonally and giving virtually frictionless flexure through defined axes.

cross-fall   Transverse slope of runway surface, to ensure sufficiently quick run-off of water to avoid aquaplaning except in particular adverse crosswinds.

cross-feed   *1* Feeding items (eg, engines) on one side of aircraft from supply (eg, fuel) on opposite side; abnormal condition under pilot control.

2 Often crossfeed, use of rudder to minimise sideslip in roll or in sustained very steep turn. See next.

crossfeed phasing parameter   Not quantifiable-value μ derived from ratios of transfer-function numerators of rudder: sideslip and aileron: sideslip, with profound effect on pilot rating.

cross-flow   Having two fluids flowing past each other at 90° while separated by thin metal walls.

cross-level   Lateral clinometer, instrument formerly used to indicate direction of local vertical as aircraft manoeuvres in rolling plane.

cross-modulation   Unwanted modulation from one

carrier being impressed on another in same receiver, usually resulting from inability to filter out certain sidebands.

cross-needle Instrument display based on two pivoted needles which pilot attempts to keep crossed at 90° in centre of display.

cross-over exhaust Gas from inboard cylinders of multi-piston engine aircraft is piped to discharge on outboard side to reduce noise in fuselage.

cross-over model That model of compensatory operation in a powered flight-control system at which the open-loop frequency response has a gain of unity, the same as the closed-loop bandwidth Ω_c, i.e. at which open-loop amplitude response crosses 1.0 [zero db] line.

cross-over struts Inclined radial gas paths in one form of coannular inverted-flow engine to convey high-V core jet to outer periphery and low-V fan flow to centre.

crossover turn Fighter battle formation in which left aircraft move across to right.

cross-qualification Among other meanings, qualification of pilots on a type of aircraft with characteristics and flight deck similar to that habitually flown, but (except on simulator) not actually flown; eg, A300B/A310, B737-300/B757.

cross-radial navigation Routeing not on a radial constituting a promulgated airway; ie, RNav using VOR and/or DME to fly direct from A to B (see *GNav*).

cross-range Approximately at 90° to axis of missile or space launch range. See next.

cross-range limit Maximum lateral distance to either side of re-entry trajectory which can be reached by a lifting body on a particular re-entry.

cross-section *1* Transverse section through object, eg fuselage or structural member.

2 Measure of radar reflectivity of object, usually expressed as area of perfect reflector perpendicular to incident radiation; depends on structural materials, incident angles, physical size of target, radar wavelength and possibly other factors.

3 In nuclear or atomic reactions, area (expressed in barn) giving measure of probability of process occurring.

cross-servicing Between-flights routine maintenance and replenishment of aircraft at base of different armed force or different nation; * guide is manual facilitating operational turnround at locations where relevant documents are not available.

crosstalk Unwanted signals generated in one set of circuits in communications or EDP (1) system by traffic in another.

cross-trail Distance bomb or other free-fall object falls downwind measured perpendicular to track (or track at release point projected ahead).

cross-trail angle Angle in horizontal plane measured at release point between track and line to point of bomb impact.

crosstube Transverse tube forming main spar of wing in most microlight and similar aircraft.

cross-turn Rapid 180° in which each half of formation turns towards remainder.

crosswind One blowing more or less at right angles to track, to runway direction, or to other flown direction.

crosswind axis Straight line through c.g. perpendicular to lift and drag axes.

crosswind component Velocity of wind component at 90°

to runway, track or other direction; = WV sin A where A is angle between WV and direction concerned.

crosswind force Component along crosswind axis of resultant force due to relative wind.

crosswind landing gear One whose wheels can be castored or prealigned with runway while aircraft crabs on to ground with wings level.

crosswind leg In landing circuit, that made at 90° to landing direction from end of downwind leg to start of approach.

crosswind testing Testing of engine with high-velocity wind (simulated at known speed) blowing across inlet.

crowbar Unswept wing (c 1950 colloq.).

crowd-line Often one word, line defining front edge of airshow crowd, parallel to runway.

crown *1* Upper part of fuselage, above cabin ceiling, of passenger transport, especially large pressurized aircraft.

2 Loosely, upper part of any fuselage.

3 Top of canopy [envelope] of balloon.

CRP *1* Carbon-fibre reinforced plastics.

2 Control and reporting point (or post).

3 Counter-rotation propfan.

4 Compulsory reporting point.

CRPA Controlled reception-pattern antenna.

CRPAE Cercle des Relations Publiques de l'Aéronautique et de l'Espace (F).

CRPM Compressor rpm.

CRPMD Combined radar and projected-map display.

CRR Cutover readiness review.

CRRA Capabilities review and risk assessment.

CRS *1* Container release system.

2 Control and reporting squadron (or section).

3 Component repair squadron (US).

4 Computer reservation system.

5 Child restraint system.

6 Congressional Research Service (US).

7 Cosmic-ray subsystem.

crs *1* Course.

2 Cruise.

CRT Cathode-ray tube.

CRTE Combat rescue training exercise.

CRTS CRT scope; term not recommended.

CRU *1* Control routing unit [MIL-1553B].

2 Chemical-resistant urethane.

3 Computer receiver unit.

CRUAV Communications-relay UAV.

crucible Hot source designed for radiating IR, for decoy or training.

cruciform Having approximate form of a cross, in aerospace usually when viewed from front; thus * wing missile has four wings arranged radially (often at 90°) at same axial position round body.

cruise *1* In any flight from one place to another, that portion of flight from top of climb to top of descent en route to destination, usually at altitudes, engine settings and other factors selected for economy and long life.

2 Verb, to perform (1).

3 Tour of operations by naval air unit aboard carrier.

cruise configuration Describes not only aerodynamic (normally fully clean) status but also systems status and possibly location and duties of flight crew, during cruise (1).

cruise missile Long-range pilotless delivery system whose flight is wing-supported within atmosphere.

cruise motor   Propulsion, of any kind, used to sustain speed of missile from boost burnout onwards.

cruising altitude   That assigned to or selected by captain for flight from top of climb to top of descent; varies with type of aircraft, sector distance, take-off weight, ATC rules and other traffic, winds and other factors.

cruising boost   With piston engine, that available in weak mixture for continuous operation giving best time or lowest fuel burn.

cruising ceiling   Formerly, greatest height at which 1.35 V_{i-mp} could be maintained at max WM cruise power.

cruising speed   That selected for cruise (1).

cruising threshold   1.35 V_{i-mp}, considered (1935–50) practical lower limit to cruising speed.

crutches   Lateral arms carrying pads which are screwed down on upper sides of bomb, missile or other store to prevent movement relative to rack, pylon or other carrier.

CRV   *1* Centre-reading voltmeter.
     2 Crew rescue vehicle.
     3 Crew-return vehicle.

CR/W   Canard rotor/wing.

CRW   Circular rotating wing; spinning wing provides gyrostabilization as well as [in fast forward flight] adequate lift.

cryogen   See *refrigerant*.

CryoGenesis   Cleaning by blast of air + solid CO_2 [dry ice].

cryogenic   Operating at extremely low temperatures.

cryogenic materials   Limited range of highly specialized materials suitable for sustained structural or other use at below -180°C.

cryogenic propellants   Gases used in liquid state as oxidants and/or fuels in rocket engines, esp. Lox, LH_2, Fl and various Fl compounds or mixtures.

cryopump   High-vacuum pump operating by cooling chamber walls so that residual gas molecules are condensed on to them, leaving vapour pressure below that required.

cryostat   Usually small lab rig for experiments at ultra-low temperatures, eg NMR, superconductivity etc.

Cryotech   Range of deicer materials, used alone or in combination with sodium or potassium acetate.

crystal laser   One whose lasing medium is a perfect-lattice crystal, eg ruby.

crystal lattice   Three-dimensional orthogonal space lattice whose intersections locate the atoms of a perfect crystal (except on small scale, most crystals contain important imperfections).

crystal oscillator   One with added subcircuit containing piezo-electric crystal (eg quartz) whose extremely rigid response gives high frequency stability.

crystal transducer   Transducer containing piezo-electric crystal which translates mechanical strain into electrical voltage.

CRZ   Cruise.

CS   *1* Constant-speed (c/s is preferred).
     2 Cassegrain system.
     3 Chemical harassing agent OCBM.
     4 Certification standard.
     5 Or C/S, callsign.
     6 Communications subsystem.
     7 Control subsystem (ECM).
     8 Colour stripping (baggage screening).
     9 Cargo system.

     10 Common service.

Cs   *1* Cirro-stratus.
     2 Caesium.

C_s Specific fuel consumption (F).

c/s   *1* Constant-speed.
     2 Course/airspeed.
     3 Cycles per second (Hz is preferred).
     4 Centre-section of wing.
     5 Callsign.
     6 Characters per second.

CSA   *1* Canadian Space Agency.
     2 Configuration status accounting (software).
     3 Control-stick assembly.
     4 Chief Scientific Advisor (UK MoD).
     5 Customer Service Agent.

CSAA   Chinese Society of Aeronautics and Astronautics (Beijing).

CSAF   Chief of Staff (USAF).

CSAM   Conseil Supérieur de l'Aviation Marchande (F).

CSAR   Combat search and rescue.

CSAS   *1* Command stability augmentation system.
     2 Common-service airlift system.

CSATC   Central School for Air Traffic Control (RAF).

CSAV   Academy of Sciences (Czech).

CSAW   Commander's situational-awareness workstation.

CSB   *1* Closely spaced basing (ICBM).
     2 Carrier and sidebands (ILS).

CSBM   Confidence and security building measures (MBFR treaty).

CSBPC   Control-stick boost and pitch compensator.

CSBS   Course-setting bombsight.

CSC   *1* Course and speed calculator.
     2 Constant symbol contrast (HUD).
     3 Centreline stowage cabinet.
     4 Chief sector controller.
     5 Cargo- or communication-, or compass-system controller.
     6 Cataloging and Standardization Center (USAF Battle Creek, Mich.).
     7 Carbon/silicon carbide.

CSCE   *1* Conference on Security and Co-operation in Europe.
     2 Communication systems control element.

CSCG   Communications system control group.

CSCI   Computer-software configuration item.

CSCP   *1* Computer software change proposal (usually SCP).
     2 Cabin-system control panel.

CSCS   Contractor['s] satellite control site.

C/SCSC   Cost/schedule control systems criteria.

CSD   *1* Constant-speed drive.
     2 Common strategic Doppler.
     3 Critical-sector detector.

CSDB   Commercial standard data-bus, or digital bus.

CSDC   Computer signal data converter.

CSDE   Central Servicing Development Establishment (RAF).

CSDS   *1* Constant-speed drive starter.
     2 Cargo smoke-detector system.

CSE   *1* Central Studies Establishment (Australia).
     2 Central Signals Establishment (UK).

CS/EL   Combat survivor/evader locator.

CSELT Centro Studi e Laboratori Telecomunicazioni (I).

CSET Certification, standardization and evaluation team.

CSEU Control-system electronics unit.

CSF Command/status frame.

CSFIR Crash-survivable flight-information recorder.

CSG *1* Centre Spatial Guyanais (F).
2 Constant-speed generator.
3 Counterterrorism Security Group (NSC).
4 Computer signal [or symbol] generator.

CSH Combat support helicopter, basically transport role.

CSI *1* Combined speed indicator, displaying airspeed and Mach.
2 Computer-synthesised image, or imagery, in hybrid simulator blended with CGI.
3 Commercial satellite imagery.

CSiC Carbon/silicon carbide composite.

CSII Centre for Study of Industrial Innovation (UK).

CSINA Conseil Supérieur de l'Infrastructure et de la Navigation Aérienne (F).

CSIRO Commonwealth Scientific and Industrial Research Organization (Int.).

CSIRS Covert survivable in-weather reconnaissance/strike.

CSIS *1* Cabin sensor indicating system.
2 Canadian Security Intelligence Service.
3 Center for Strategic and International Studies (US).

CSLC Coherent side-lobe canceller.

CSM *1* Command/service module.
2 Customer support manager.
3 Crash-survivable memory [M adds module, U unit].
4 Cabin-systems management [U adds unit].

CSMA Carrier-sense multiple access.

CSMM Crash-survivable memory module (CSMU, see above).

CSN Catalogue sequence numbers[s].

CSO Command Signals Officer (RAF).

CSOC Combined Space (or Satellite) Operations Center (pronounced C-sock).
2 Consolidated space operations contract.

C/SOIT Communications/surveillance operational implementation team.

CSP *1* Common signal-processor.
2 Capability-sustainment programme (UK), or plus.
3 Comprehensive surveillance plan[s] (ATOS).

CSPA Canadian Sport Parachuting Association.

CSR Covert strike radar.

CSRDF Crew-station research and development facility.

CSRL Common strategic rotary launcher.

CSRP Cabin-safety research program.

CSRS Counter-surveillance and reconnaissance system.

CSS *1* Control-stick steering.
2 Cockpit system(s) simulator.
3 Clean stall[ing] speed.
4 Communications subsystem.
5 Complementary satellite system.
6 Computer support [or sighting] system.

CSSD Strategic Defense Command (USA).

C/SSR Cost/schedule status report.

CSSS Combat service support system.

CST *1* Combined station and tower.
2 Centre Spatial de Toulouse (F).
3 Commercial space transportation.
4 Central Standard Time.
5 Coast, coastal.

cSt Centistoke[s].

CsTe Caesium telluride, photocathode material.

CSTF Cross-scan terrain following.

CSTI Control-surface tie-in.

CSTM Centro Studi Trasporti Missilistici (I).

CSTMS Customs.

CSU *1* Constant-speed unit.
2 Central suppression unit, prevents mutual interference in complex avionics.
3 Command sensor unit.
4 Cabin service unit.
5 Configuration stopping [or strapping] unit.
6 Communications switching unit.
7 Crew-station unit.
8 Control-status unit.
9 Control selection unit.
10 Categorization and status unit (ILS).
11 Cross-strap unit.

CSV Capacity (or catapult) safety-valve.

CSVR Crash-survivable voice recorder.

CSVTS Scientific and technical society (Czech).

CSW *1* Conventional standoff weapon.
2 Combat Support Wing.

CSWIP Certification scheme for weldment inspection personnel.

CSWS Corps-support weapon system.

CT *1* Carry trials.
2 Counter-terrorist.
3 Current transformer.
4 Tip chord [also C$_\text{T}$].
5 Clearance time.
6 Contact.
7 Computerized, or computed, tomography.
8 Crew-member terminal.
9 Control transmitter, or tower.
10 Cockpit trainer; /IPS adds interacting pilot station, /IPS-E further adds enhanced.

C$_\text{T}$ Thrust coefficient, eg of forced flow through a slit in CC aerofoil.

cT Continental tropical airmass.

c$_t$ Equivalent tip chord.

CTA *1* Control area.
2 Companion trainer aircraft.
3 Centro Técnico Aeroespacial (Braz.).
4 Controlled, or calculated, time of arrival.
5 Cryogenic telescope assembly.
6 Canadian Transportation Agency (issues licences).

CTAA Commandement des Transmissions de l'Armée de l'Air (F).

CTAF *1* Comité des Transporteurs Aériens Français.
2 Common-traffic advisory frequency areas.

CTAGS Co-operative transatlantic air-ground surveillance; S adds system.

CTAI Cowl thermal anti-icing.

CTAM Climb to and maintain.

CTAPS Contingency theater [or tactical] air [or aircraft, or automated] planning system (USAF).

CTAR Comité des Transporteurs Aériens Complémentaires (F).

CTAS   Center Tracon automation system.

CTB   *1* Comprehensive Test Ban (T adds 'treaty'), 1996.

  2 Central terminal building.

CTC   *1* Canadian Transport Commission.

  2 Command track counter.

  3 Cabin-temperature control[ler].

  4 Counter-Terrorism Committee (UN).

  5 Central telemetry control.

ctc   Contact.

CTD   *1* Compound turbo diesel.

  2 Colour tactical display.

CTDC   Civil Transport Development Corporation (J).

CTF   *1* Combined test force.

  2 Central Test Facility.

  3 Conventional turbofan.

CTFE   Chlorotrifluoroethylene, advanced non-flam hydraulic fluid.

CTHA   Contrawound toroidal helical antenna.

CTI   *1* Costings technology integration [O adds office] (USAF).

  2 Computer/telephone, or telephony, integration.

  3 Continuous technology insertion.

  4 Commission Technique et Industrielle (F).

  5 Combined threat image.

CTIPS   See CT (10).

CTK, ctk   *1* Capacity tonne-kilometres.

  2 Command track.

cTk   Continental tropical air, colder than surface.

CTL   *1* Control.

  2 Tactical air command (Netherlands).

CTLA   Control area.

CTLZ   Control zone.

CTM   *1* Capacity ton-miles (unless otherwise stated, short tons, statute miles).

  2 Centrifugal twisting moment.

  3 Cost per ton-mile.

CTMO   Central[ized] air-traffic flow management organization.

CTN   *1* Caution.

  2 Case, throat, nozzle.

CTO   *1* Chief technical officer.

  2 Conventional takeoff.

  3 Crypto operator.

CTOL   Conventional takeoff and landing, ie ordinary aeroplane.

CTOT   *1* Calculated takeoff time.

  2 Constant torque on takeoff.

CTP   *1* Chief test pilot.

  2 Command track pointer.

  3 Common technology programme (ASTOVL).

  4 Critical technology project[s].

CTPA   Comité Technique de Programmes d'Armement (F).

CTPB   Carboxy-terminated polybutadiene rocket propellant.

CTR   *1* Controlled airspace, control zone.

  2 Controllable-twist rotor.

  3 Common Type Rating.

  4 Co-operative Threat Reduction (US, R).

  5 Click to refresh.

  6 Continuous technology refreshment.

  7 Common, or configuration, test requirements; D adds document.

  8 Civil tilt-rotor.

  9 Center, centre.

  10 Conversion to role; T adds training.

CTRD   Configuration test requirements document.

CTRDAC   Civil tilt-rotor development advisory committee.

CTRL   Control.

CTS   *1* Central tactical system.

  2 Cockpit television sensor.

  3 Common termination system.

  4 Clear to send.

C^T/s   Helicopter blade-loading coefficient.

CTSS   Commercial training simulator, or simulation, services (USAF).

CTT   *1* Capital-transfer tax (UK).

  2 Commander's tactical terminal; H/R adds hybrid-receive only.

  3 Controlled-torque tightening.

  4 Conversion to type; T adds training.

CTTO   Central Tactics & Trials Organization (RAF).

CTTTF, CT^{3}F   Combating Terrorism Technology Task Force (DoD).

CTU   *1* Control terminal unit.

  2 Cabin telecommunications, ot telephone, unit (satcom).

CTV   *1* Curved trend vector.

  2 Crew transfer vehicle.

CTVS   Cockpit TV sensor [or system].

CTZ   *1* Control zone [this is preferred].

  2 Corps tactical zone.

CU   *1* Conversion unit.

  2 Cage/uncage; gyro system control.

  3 Common use[r]: BES adds baggage enterprise system [cubes], PS passenger self-service, SS self-service and TE [cute] terminal equipment.

  4 Control unit (HMS).

  5 Combiner unit (HUD).

  6 Channel utilization.

Cu   Cumulus.

CUAV   Clandestine UAV.

cubage   Total volume of rectilinear cargo that can be accommodated; typically 0.7 of pressurized above-floor cargo volume.

Cuban eight   Manoeuvre in vertical plane normally comprising ¾ loop, half-roll, ¾ loop, half-roll.

cube out   To run out of payload volume (either pax, cargo or both) at less than MSP (5).

cubic foot   Non-SI measure of volume, 1* = 28,316.7 cm^3.

cubic inch   Non SI measure of volume, 1 cu in = 16.387 cm^3 = 0.0164 litre; reciprocals 0.06102, 60.9756.

CUDS   Common-user data services.

CUE   Computer update equipment.

cue   *1* Glimpse of Earth's surface through cloud or darkness giving helpful attitude and distance information.

  2 To slave homing seeker of missile to target, using information from other source.

CuF   Cumuliform cloud.

cuff   *1* Secondary structure added around propeller blade root, usually for aerodynamic reasons.

  2 Structure added ahead of wing LE extending chord 3–5%, with sharp inboard end.

  3 Heated muff round drain valve or drain mast.

CuFra   Cumulus fractus.

CUG   Computer Utilization Group (OECD).

CUGF Counter underground facilities [weapons against caves].

CUGR Cargo utility GPS receiver.

CUI Committee on Unlawful Interference.

cu in Cubic inch, 16.387 cm^3.

culture Man-made terrestrial features.

cumulonimbus Cb, extremely large cumuliform clouds whose tops reach stratosphere and spread in form of fibrous ice-crystal anvil. Extreme vertical velocities and turbulence make them dangerous.

cumulus Cu, dense white clouds with almost horizontal base and large vertical development, domeshaped tops (cauliflower) showing growth in strong upcurrents.

cumulus mammatus Cumuliform clouds having pendulous protuberances on underside.

CUP Capabilities upkeep program (USN).

cup Non-SI unit of volume, = 2.3659 × 10^{-4}m^3.

Cupid Common, or combat, upgrade plan integration details (USAF).

cupola *1* Turret-like structure projecting from aircraft with windows giving maximum field of view.

2 Manned viewing station on space station for observation of docking, truss construction and placement of antennas and other equipment.

cupping Re-rigging to increase angle of incidence.

curie, Ci Unit of radioactivity equal to 3.7 x 10^{10} disintegrations or transformations per second, replaced by becquerel.

curie point Critical temperature, different for each material, above which ferromagnetic materials lose permanent or spontaneous magnetisation.

curing Process by which most synthetic rubbers, plastics and solid-propellant binders are converted to compositions of higher molecular weight; may involve heating (condensation polymerization), chain reaction via free-radical or ionic mechanism (addition polymerisation), or use of catalysts. In solid rocket motors semi-liquid is often cured in case, solidifying and becoming case-bonded.

curl Vector resulting from action of operator del (differential operator in vector analysis) on vector; sometimes called rotation.

curling die Used with curling punch to bend sheet edges to tubular form.

curlover Possibly dangerous downdraft and turbulence downwind of trees or buildings.

current *1* Pilot is qualified on particular type and routinely flying it.

2 Civil aircraft is on active register and in routine operation.

3 Flight plan is that being followed.

cursive writing Rounded, flowing writing with strokes joined; hence formed in display by actual strokes rather than TV-type raster scanning.

curtain *1* Non-gastight partition in aerostat.

2 Electronic barrier formed by wall of chaff.

curved approach *1* Adopted by some aircraft, notably WW2 fighters, because of inadequate forward view straight ahead at low airspeeds.

2 Any of numerous possible quasi-elliptical paths followed when using MLS or other system offering such approach paths on either side of straight centreline.

curved trend Turn information imparted by three future track-lines on EHSI terminating 30 sec, 60 sec and 90 sec hence; these are straight with wings level but in banked turn show *, in extreme case linking in 360° circle (does not allow for drift).

curve of pursuit Followed by any aircraft chasing another and continuously steering towards latter's present position; with non-manoeuvring target curve soon becomes asymptotic with target straight-line course.

curvic coupling Joint between driving and driven shaft systems which transmits torque perfectly; allows for small errors in alignment or angle but does not secure one to other. In simplest form comprises two sets of meshing radial teeth of smooth curving profile.

curvilinear flight Accelerated flight, ie not straight and level.

Cus, CUS Customs available.

cushion See *ground cushion*.

cushion creep Use of ground cushion for gradual helicopter takeoff.

CUSRPG Canada/US Regional Planning Group.

Cuss, CUSS Common-use self-service, for check-in desks which automatically identify and process passenger. CUSS is an electronic standard [v1.0], produced by SITA but specified by IATA.

customer Usually purchaser and operator are synonymous; where purchaser is government agency and operator an air force, or purchase is finance company or bank, * normally applies chiefly to operator.

customer base Total list of customers (term usually refers to civil air carriers) committed to purchasing or leasing new type.

customer mock-up Exact reproduction of aircraft interior, or part thereof, furnished with materials, fabrics, colours, seats and other equipment as specified by customer.

customer supplies Bleed-air or shaft power, other than that required for propulsion, needed for aircraft services.

customised lead time No spares supplied until needed, * typically 2h–2 years.

customising *1* Finishing GA aircraft to customer's spec., eg furnishing, avionics kit, external paint.

2 Finishing avionics or instruments for particular task with chosen language, labels, IC chips and self-test.

cusum Cumulative sum [suggest: tautology].

cut Sudden complete shutdown of power [noun and verb]; hence: the *, command given by batsman on carrier.

C/UT Code/unit test.

cutaway General term for detailed perspective drawing showing maximum detail of 3-D object.

cutback *1* Sudden partial closure of throttles at end of first climb segment for noise-abatement reasons.

2 Reduction in existing or planned procurement.

3 Reduction in manufacturing rate.

cut-back nozzle Normally, one (not of con-di type) shortened in length and terminating obliquely to give thrust slightly inclined to pipe axis.

cutback speed ASIR at top of first segment.

Cute Common-user terminal equipment.

Cutlass Combat UAV target locate and strike system.

cutlet Cutlet-shaped flattened outer arm of hub forging of rigid rotor.

cutoff *1* Termination of rocket propulsion before burnout because desired trajectory and velocity have been reached.

2 Flying shortest track to intercept an air target.

cutoff ports Circular apertures in forward face of solid motor which can swiftly be blown open to terminate combustion.

cut-out   *1* Aperture in pressurized fuselage, for door, window, hatch or other purpose.

2 Absence of rear inner part of elevator, terminating in diagonal edge, to allow full rudder to be applied.

cut-out switch   One isolating or inactivating circuit or subsystem.

CV   *1* Fleet [aircraft] carrier.

2 Carrier vehicle (SDI).

3 Compiler vendor.

4 Cryptographic variable.

C_v Specific heat at constant volume.

cv   Cheval vapeur, metric horsepower = 0.98632 hp = 0.7355 kW; reciprocals 1.01387, 1.35962.

CV(A), CVA   Fleet carrier, attack.

CVBG   Carrier battle group.

CVD   Chemical vapour deposition.

CVDR   Cockpit voice data recorder.

CVE   *1* Escort carrier.

2 Combat value enhancement.

CVF, cv(f)   Future aircraft carrier.

CVFD   Cockpit voice and flight data [R adds recorder].

CV/DFDR   Cockpit voice and digital flight data recorder.

CVFP   Charted visual flight procedure.

CVFR   Controlled visual flight rules.

CVH   Helicopter carrier.

CVI   Counter-flow virtual impactor.

CVID   Clearly visible impact damage.

CVL   *1* Controlled-vortex lift.

2 Light aircraft carrier.

CVM   Comparative vacuum monitoring.

CV(N), CVN   Fleet carrier, nuclear-propelled [X adds next-generation]

CVOR   Commutated VOR.

CVR   *1* Cockpit voice recorder [CP adds control panel].

2 Crystal video receiver.

CVRS   Computerized voice reservation system.

CV(S), CVS   Carrier, escort/ASW.

CVV   *1* Compressor variable vane.

2 Combined validation and verification.

CVW   Carrier air wing (USN).

CV-WST   Carrier-based weapon-system trainer.

CW   *1* Continuous wave.

2 Ambiguously, carrier wave.

3 Clockwise.

4 Chemical warfare.

5 Composite Wing (USAAF, USAF).

CWA   *1* Center weather advisory (inflight, unscheduled).

2 Civil Works Administration (US, 1933).

3 Communications Workers of America.

CWAN   Coalition wide-area network.

CWAR   CW (1) acquisition radar.

CWC   *1* Crosswind component.

2 Chemical Weapons Convention, 15 January 1993.

3 Comparator warning computer.

CWCS   Common weapons control system.

CWD   Chemical warfare defence.

CWDS   Clean-wing detection system, senses thickness of contaminant, eg ice.

CWFS   Crashworthy fuel system.

CWG   *1* Charges Working Group (IATA).

2 Capability Working Group (MoD and NATO).

CWI   Continuous-wave illuminator [or interference].

CWIN   *1* Cockpit weather information system.

2 Cyber warfare integration network.

CWM   Comparator warning monitor.

CWN   Call when needed, short-term contract prevalent in firefighting.

CWP   *1* Central warning panel.

2 Contractor's working party.

3 Compact when packed (antennas).

4 Controller, or controlled, work[ing] position[s].

5 Central West Pacific (ICAO).

CWR   Continuous-wave [or colour weather] radar.

CWS   *1* Caution/warning system.

2 Central [or collision] warning system.

3 Control-wheel steering.

4 Container weapon system.

CWSG   Civil Wing Study Group.

CWSU   Central [or Center] Weather Service Unit (US).

cwt   Hundredweight, archaic unit of mass, = 112 lb = 50.8032 kg; US short * = 100 lb = 45.3592 kg.

CWU-45P   Classic USAF leather flight jacket.

CWV   Crest working voltage.

CWW   Cruciform-wing weapon.

CWY   Clearway.

C_x *1* Longitudinal force coefficient (= C_d cos A minus C_L sin A, where A is angle of attack).

2 Controlled expansion, for fatigue enhancement of holes.

CXO   Chandra X-ray Observatory.

CXR   Helicopter configuration, co-axial rotor(s).

CXRS   Coherent X-ray scattering (baggage screening).

CY   Calendar year.

cyan   Bright blue colour, e.g. marking FLOT/FEBA on radar.

cyaniding   Surface hardening of steel by immersion in bath of cyanide (and other) salts producing nitrogen as chief agent.

cyber   Prefix, concerned with information, computers and the internet, hence * attack, * crime, * defence, * security, * strategy, * tools.

Cyc, CYC   Cyclonic.

cycle   One complete sequence of events making up portion of life of machine; thus for piston engine, four strokes of Otto *, while for aircraft usually start-up, taxi, takeoff, climb, cruise, possibly combat, descent, landing, thrust-reverse, taxi, shutdown.

cycle-dependent costs   Directly proportional to usage, eg reverser.

cycle efficiency   Measure of performance of heat engine derived from PV or entropy diagram; usually synonymous with thermal efficiency.

cycle parameters   For gas turbine, primary * are EPR and TET.

cyclic pitch   In most helicopters main rotor blade pitch progressively increases from minimum (a very small) angle when head-on to airstream (momentarily occupying position of wing) to maximum 180° later (when in position of wing trailing-edge-on to airstream); this makes blade fall on advancing side of rotor and rise on retreating side, effectively decreasing and increasing angle of attack to even-out lift on both sides. This is also called feathering, and results in blade flapping (see * *control*).

cyclic-pitch control   Primary helicopter flight control.

Usually governed by stick, similar to aeroplane control column, which in central position causes basic cyclic variation as described above by tilting stationary and rotating stars on rotor hub. Pilot demand is passed through mixing unit and output tilts fixed star in desired direction to superimpose additional cyclic variation causing disc to tilt in desired direction to cause helicopter to rotate about pitch or roll axis.

cyclic rate   Rate at which automatic gun fires, expressed in shots per minute, measured after maximum rate has been attained and not necessarily attainable except for brief periods.

cyclic stick   Cyclic-pitch control stick.

cyclic testing   Repeated application of supposed operating cycle, usually of exceptionally severe nature, under arduous environmental conditions to prove endurance or life of hardware.

cycling   Cyclic testing.

cyclogenesis   Development of a cyclone.

cyclogiro   Aerodyne, never successfully achieved, lifted and propelled by pivoted blades rotating about substantially horizontal transverse axes as in paddle steamer.

cyclone   Tropical revolving storm.

Cyclonite   See *RDX*.

cyclostrophic force   That experienced by wind following curved isobars acting in addition to geostrophic force to give resultant wind along isobar according to Buys Ballot's law. At Equator geostrophic force vanishes, leaving pure cyclostrophic wind.

cyclostrophic wind   As explained above, wind near Equator with strong circular motion, such as a tornado.

Cyclotol   Specially formulated high explosive, used alone or with PBX-9500 series, to cause implosion trigger (NW).

cyclotron   Family of magnetic resonance particle accelerators, many extremely large.

cyclotron resonance   Motion of moving charged particle in magnetic field on which is superposed alternating electric field normal to magnetic field.

cyc/sec   Cycles per second, SI unit is Hz.

CYI   Canary Islands (AMR).

cylinder   One unit of piston engine, or, specif., surrounding cylinder enclosing combustion space and guiding piston.

cylinder block   Single unit enclosing row of liquid-cooled in-line piston engine cylinders.

cylinder head   Usually removable top of piston engine cylinder containing plugs, inlet/exhaust connections and (except with sleeve valve) valves.

cylinder liner   Hard abrasion-resistant lining inserted into cylinder of light alloy or other soft material.

Cytac   Loran-C.

CY2KSS   Center for year-2000 strategic stability (US/R).

CZ   Control zone.

C_z   Normal force coefficient, $C_L \cos A + C_d \sin A$, where A is angle of attack; rarely called C_N.

cZ   Fore/aft magnetic VSI component.

CZCS   Coastal zone colour scanner.

CZI   Compressor-zone inspection.

D

D *1* Total aerodynamic drag.

2 Danger area (ICAO).

3 Duration of phenomenon in seconds, eg D_{60}.

4 Drift.

5 Diameter (rarely, d); for tyre [tire], at rim ledge.

6 For airspace, see *-class.

7 Departure chart.

8 Pavement bending strength for dual-wheel landing gear.

9 Drone (UK).

10 Drone director (US modified-mission prefix).

11 Electric flux density.

12 Fuze delay time.

13 PPL Group for microlights (CAA).

14 Sport-parachuting certificate: 200 free-falls, 20 landing $\leqslant$15m of target.

15 Other meanings include Doppler, downward, distance, day, dust, delete, designated, delay, displacement, differential coefficient, chemical diffusion coefficient and decision.

d *1* Distance.

2 Differential.

3 Deci, prefix, multiply by 10^{-1} (not recommended).

4 Clear distance between contact areas of landing wheels (can include axial or transverse distance between wheels of bogie).

5 Thickness of RAM surface-wave absorber.

6 Usually as subscript, design.

7 Diameter of jet or propeller (alternative to D).

8 Diode.

9 Relative density.

D^3 Data download and display.

D^3S Dynamic data-display subsystem.

D8PSK Differential-8 phase-shift keying.

D_{100} Drag at 100 ft/s.

D-check Major overhaul carried out every 3–5 years.

D-class Airspace up to 2,500 ft (762 m) AGL above airfield with operating tower; 2-way dialogue radio required.

D-code In flight plan, have DME.

D-factor Actual, or true, altitude divided by pressure altitude.

D-gun Detonation gun, firing suspended particles of hard surface coating by detonation of oxy-acetylene.

D-layer Region of increasing electron and ion density in ionosphere, existing in daytime only and merging with bottom of E-layer.

D-licence For inspection of engines after overhaul.

D-nose Strong leading edge of aerofoil, often forming principal structural basis of wing or helicopter rotor blade.

D-notice Issued regularly to advise [esp. Press, broadcasters] of changes in classification status of defence or other sensitive subjects (UK MoD).

D-nozzle Propulsive nozzle of jet engine on centreline of fuselage or nacelle vectoring to give lift or thrust (from cross-section).

D-ring Steel handle with which parachutist pulls ripcord.

D-spar D-nose.

D-tube Leading edge of lightplane or micro comprising spar and load-bearing skin, generally simpler than D-nose.

D-value Departure from pressure altitude.

DA *1* Drift angle.

2 Diplomatic authorization.

3 Double attack.

4 Long-range (bomber) aviation, predecessor of ADD (USSR).

5 Delayed-action (bomb).

6 Dual-alloy (turbine disc).

7 Direct action (fuze).

8 Deck alert.

9 Decision [or density] altitude.

10 Danger area.

11 Development aircraft.

12 Design authority.

13 Duplex aluminide.

14 Air defence (F).

15 Defence advisory.

16 Drought area.

17 Direct access (telecoms).

18 Descent advisor, or advisory.

19 Display Authorization.

da Deca, prefix, multiplied by 10, non-SI.

d_α Radar resolution in azimuth.

D/A Digital/analog.

DAA *1* Directorate of Air Armament (UK).

2 Digital/analog adaptor.

DAACM Direct airfield-attack cluster munition.

DAAIS Danger area activity information service (CAA, UK).

DAAS Defense advanced automation system (ATC)

DAAT Digital angle-of-attack transmitter.

DAB *1* Defense Acquisitions Board (US).

2 Digital audio broadcast[ing].

DABF Digital adaptive beam-forming; N adds network

DABM Defence against ballistic missile(s).

DABRK Daybreak.

Dabs Discrete-address, or addressable, beacon system; can address individual aircraft via transponder, pointing a narrow beam at it to transmit messages via data-link.

Dabsef Dabs Experimental Facility (Lincoln Laboratories, US).

DAC *1* Deployable ACCS component (USAF).

2 Design aperture card.

3 Dual annular combustor.

4 Dangerous air cargo.

5 Defensive-aids computer.

6 Duplex aluminide coating.

7 DSMC (2) analysis code.

dac Digital-to-analog converter.

DACC Dangerous Air Cargo Committee (UK, RAF).

Dacota Dispositif d'association, de correlation et de traitement radar pour les approches (F).

Dacron A commercial polyethylene glycol terephthalate fibre and woven fabric related to Terylene and Mylar.

DACS *1* Directorate of Aerospace Combat Systems (Canada).

2 Danger area crossing service (CAA, UK).

3 See next.

Dacs Divert and attitude control system.

DACT Dissimilar air-combat training (or tactics).

DAD *1* Deep air defence.

2 Dual-alloy disc.

3 Density - altitude display.

Dadacs Danger-area divert and attitude-control system.

DADC Digital air-data computer.

dadopanel Cabin wall just above floor [one word]

DADR Deployable [not fixed-base] air-defence radar.

DADS *1* Deployable air-data sensor.

2 Digital air-data system.

DAeC Deutsche Aero Club (G).

Daedalians National fraternity of military pilots (US).

DAES Directorate of Avionic Equipment and Systems (UK).

DAFCS Digital automatic flight-control system.

DA/FD Digital autopilot/flight director.

Dafics Digital automatic flight inlet control system.

Dafusa Data-fusion airports (Euret).

DA fuze Direct-action fuze; designed to explode on impact.

DAG Deutsche Angestellten-Gewerkschaft Bundes-gruppe Luft- und Raumfahrt.

Dagmar Faired shape [a body, not a blister] in front of a bluff projection.

DAGR Defense advanced GPS receiver, probably to replace PLGR.

DAI *1* Direction des Affairs Internationales (F, MoD).

2 **DCMS** Audio interface.

DAIR Direct altitude and identity readout (FAA/USAF).

DAIRS Distributed-architectures, or aperture, IR system, or sensing.

DAIS *1* Digital avionics information system.

2 Distributed airport information system.

Daisy Decision-aid for interpretation of air situation display (Alcatel from 1994).

daisy chain Several helpers link arms to swing large propeller.

DAIW Danger area infringement warning.

DALGT Daylight.

Dallenbach layer Pioneer form of RAM(2) coating consisting of homogeneous lossy layer backed by metallic plate (eg, aircraft skin); if lossy layer has same impedance as free space there will be no surface reflection.

Dalmatian effect Increase in number of spots in map of V/STOL air bases compared with airfields.

DALO Divisional Air Liaison Officer (UK).

DALR Dry adiabatic lapse rate.

D-alt Density altitude.

Dalton computer Family of pocket-size mechanical calculators for navigation [esp triangle of velocities] problems.

Dalton's Law Empirical generalization that, for many so-called perfect gases, a mixture will have pressure equal to sum of partial pressures each would have as sole component within same volume and temperature, provided there is no chemical interaction.

DAM Dollars per aircraft-mile.

Dama, DAMA. Demand-assigned, or assignment, multiple access.

damage assessment Determination of effect of attacks on targets.

damage cycle Loss of life of engine or other hardware.

damage limitation Ability to limit effects of nuclear destruction by using offensive and defensive measures to reduce weight of enemy attacks.

damage-tolerant Structure so designed as to continue to bear normal in-flight loads after failure (through fatigue, external damage or other cause) of any member (see *fail-safe*).

Damask Direct-attack munition affordable seeker.

DAME Designated aviation medical examiner.

damped natural frequency Frequency of free vibration of damped linear system; decreases as damping increases.

damped wave Wave whose amplitude decreases with time or whose total energy decreases by transfer to other frequencies.

damper *1* Mass[es] attached to crankweb, either rigidly or free to oscillate, to eliminate dangerous vibration at critical frequencies of crankshaft.

2 Snubber or part-span shroud on fan blade.

3 See *flame* *.

4 See *roll* *.

5 See *shimmy* *.

6 See *yaw* *.

damping factor Ratio of peak amplitudes of successive oscillations.

damping moment Proportional to rate of displacement; tends to restore aircraft to normal flight attitude after upset.

DAMS *1* Dynamic airspace management system(s).

2 Drum auxiliary memory subunit.

Damsl Dictionary and message specification language.

daN Decanewton, unit of force = 2.248 lbf.

Danac Decca area-navigation airborne computer (1984 also appeared as digital air-navigation control).

D&C *1* Design and clearance.

2 Diagnostic and conditioning.

D&D *1* Distress and diversion (ATC).

2 Diesel and dye (smoke-making).

D&F Determination and findings.

D&O Description and operation.

D&P Development and Production (MoD contracts).

D&V Demonstration and validation.

danger area Airspace of defined dimensions in which activities dangerous to flight may exist at specific times.

dangle Angle between local horizontal at glider and end of tow-rope (usually air tow).

DAO Defence Attaché Office.

DAP *1* Distortion of aligned phases (LCD).

2 Director(ate) of Aircraft (originally Aeroplane) Production (UK, WW2).

3 Distributed-array processor.

4 Directorate of Airspace Policy (CAA, UK)

5 Digital service accept product.

DAPU Data acquisition and processing unit.

DAR *1* Design and Airworthiness Requirements (UK)

2 Drone, anti-radar.

3 Direct-access recorder.

4 Design assurance review.

5 Defense Acquisition Regulation.

6 Diffuser area ratio.

7 Digital archive recorder.

Dara, DARA   *1* Defence Aviation Repair Agency (UK).

2 National space agency (G).

Darc, DARC   Direct-access radar channel.

dark   Switched off.

dark burst   Gamma-ray burst that fades very rapidly.

dark cockpit   All lights out, ie correct configuration and all systems normal.

darkfire   Missile system operable at night in clear visibility.

dark-trace   Display phosphor creating image through reflection/absorption of light instead of light emission from phosphor (see *skiatron*).

Darlington pair   High-gain amplifier stage using two transistors in which base of second is fed from emitter of first.

DArmRD   Directorate of Armament Research and Development (UK).

DARO, Daro   Defense Airborne Reconnaissance Office[r] (DoD).

Darp   *1* Dynamic rerouting procedures, or programs.

2 Digital audio recording and playback.

Darpa, DARPA   Defense Advanced Research Projects Agency (DoD).

Darps   Dynamic aircraft route-planning study.

DARS, Dars   *1* Digital attitude-reference system.

2 Dynamic-assist retargeting system.

3 Drogue air-refuelling system.

4 Digital audio radio service.

Dart, DART   *1* Defensive-avionics receive, or receiver, transmitter; jams hostile IR-seekers.

2 Directional automatic realignment of trajectory (ejection seat).

3 Dual-axis rate transducer.

4 Deployable automatic relay terminal.

dart   *1* Unpowered aerodynamic vehicle with stabilizing tailfins.

2 Any freeflight vehicle with stabilizing tailfins but no trajectory control.

3 Guided missile accelerated to speed by rocket[s] jettisoned at burnout, thereafter coasting to target.

Darts   *1* Diversified aircrew readiness training support.

2 Digital airborne radar threat simulator.

DARU, Daru   Data acquisition and recording unit.

Darwin   Design assessment of [engine] reliability with inspection (FAA).

DAS   *1* Defensive-aids suite, or subsystem.

2 Defensive avionics system.

3 Director, Air Staff (UK).

4 Directorate of Aerodrome Standards (UK).

5 Distributed-aperture system.

6 Designated alteration station.

DASA   *1* Defense Atomic Support Agency (US, MoD).

2 Defence Analytical Services Agency (UK).

Dasals   Distributed-aperture semi-active laser seeker.

DASC   *1* Direct air support centre.

2 Defence Aviation Safety Centre (MoD, UK).

DASD   Deputy Assistant Secretary for Defense.

DASE   Digital autostabilizer equipment.

DA718   Advanced (1995) DS compressor-blade alloy.

DASH   *1* Drone anti-submarine helicopter.

2 Differential airspeed hold.

3 Display and sight[ing] equipment.

dash   Portion of attack mission through defended hostile

territory at full afterburning power at low level, ignoring high fuel burn.

DASP   Discrete analog signal processing.

DASR   *1* Digital airspace, or airport, surveillance radar.

2 Direct air-to-satellite relay.

DASS   *1* Defensive-aids subsystem.

2 Dynamic-assembly scheduling system.

DAST, Dast   Drone[s] for aerodynamic and structural testing (NASA).

DAT   *1* Damage-to-aircraft trials.

2 Digital audio tape.

DATA   Defense Air Transportation Association (US).

databus   Highway for digital data, most common linking aircraft sensors and other air or ground systems being MIL-STD-1553B or Arinc 419 (one-way) or Arinc 619 (two-way).

DATACOM, Datacom   Data compilation, large handbook and CD which attempts to give designers complete knowledge of effect on lift and moment of changes in design (USAF).

Datacs   Digital autonomous terminal access communications system (Boeing).

data fusion   Integration and management of possibly billions per second of bits of information from recon sensors, C^3 and battle management systems.

datalink   *1* Any highway or channel along which messages are sent in digital form.

2 Communications channel or circuit used to transmit data from sensor to computer, readout device or storage.

data-logger   Short-term store for digital or analog information, eg for one flight or one week, periodically read back to build up service history of system, engines or other devices.

data plate   Permanently fixed to aircraft, engine or other product, giving basic data, serial numbers and dates.

Datar, DATAR   Detection and tactical alert of radar (helicopter RWR).

data reconstruction   Assembling correct bar-codes from brief any-angle glimpses (mainly in checking baggage).

data recorder   Device, usually electronic, for recording data [previously analog, now mainly digital] for subsequent playback and analysis (see *flight recorder, maintenance recorder*).

Datas   Data-link and transponder analysis system.

Data-3   Inmarsat system enabling aircraft to link direct to ground networks.

Datco   Duty air-traffic-control officer.

DATF   Deployable air task force.

DATIS, D-Atis, Datis   *1* Digital air-traffic information service, or system.

2 Digital automated, or automatic, terminal information service.

DATM   Dummy air-training missile.

Datmas   Danish air-traffic management system (2007-).

DATS   Data-acquisition and telemetry system.

Datsa   Depot automatic test system for avionics.

DATT, DAtt   Defense Attaché (US).

DATTS   Data acquisition, telecommand and tracking station.

datum   *1* Numerical, geometric or spatial reference or base for measurement of other quantities.

2 Vertical (rarely, horizontal or other) reference line from which all structural parts are measured and identified. Most * lines are exactly at, or close in front of or behind, nose; thus, frame 443 is a nominal 443 in or mm behind*; wing * is often aircraft centreline.

DAU *1* Directly Administered Unit[s] (RAF).

  2 Digital amplifier unit.

  3 Data-acquisition unit.

DAUG Danger-area users group (NATS).

Da Vinci Departure and arrival integrated management system for co-operative improvement of airport traffic flow (Euret).

Davis barrier Retractable crash barrier across carrier (1) deck.

Davis tables List altitude and azimuth of astro-navigation targets.

Davis wing High-aspect ratio wing designed by David R. Davis; intended to cruise at low angle of attack with low drag.

DAVSS Doppler/acoustic vortex sensing system.

DAVVL Birdstrike committee, with several sub units (G).

DAW (A) Dedicated all-weather (aircraft).

day Mean solar * is defined at 8.64×10^4s; sidereal * is approximately 8.616×10^4s.

Day-Glo Family of dyes and paints with property of converting to visible light wavelengths outside normal visible spectrum, thus giving unnaturally bright hues.

day/night Equipment giving cheap and convenient IFR training, using tinted pilot goggles and complementary tinted cockpit transparency (eg blue goggles and amber canopy or red + green); pilot sees clear but tinted cockpit while outside world appears black.

DB *1* Development batch.

  2 Direct broadcast. [S adds satellite, service or system].

  3 Database.

  4 Databus.

  5 Double base (rocket propellant).

  6 Diffusion bonding.

  7 Day bombardment category (USA 1919–24).

dB Decibel, see *noise*.

DBA *1* Dominant battlespace awareness.

  2 long-range bombing aviation (USSR, R).

dBA Decibels absolute, or adjusted, see *noise*.

d.b.a. Doing business as.

DBC *1* Denied boarding compensation for bumped passengers; for pax reaching their destination within 4 h of original booked time 50% of flight-coupon value in Europe, 200% in US. No compensation for aircraft under 60 seats.

  2 DCMS bus coupler.

  3 Data-bank, Comecon.

DBE Data bank, Eurocontrol.

DBF *1* Doppler beat frequency.

  2 Digital beam forming.

  3 Destroyed by fire.

  4 Doppler blade flash.

DBGS Data-base generation system.

DBFM Defensive basic flight manoeuvres.

DBI *1* DCMS bus interface.

  2 Downlink block identifier.

dBi Decibels referenced to isotropic antenna or above isotropic circular.

dBm Decibel meter, unit of power referenced to $1 \text{ mW} = \text{dB} \times 10^{-3}$.

DBM/C Data bus monitor/controller.

DBMS Database management system; also rendered **DBMX**.

DB/N Data-base No.

DBNS Doppler bombing/navigation system.

DBPS Digital [electron] beam-positioning system.

DBR Dual-band radar.

DBS *1* Doppler beam-sharpening.

  2 Direct-broadcast satellite, or service.

  3 Database storage.

DBSA Directorate for Broadening Smart Acquistion (MoD, UK).

dBsm Decibel unit of radar beam cross-section referenced to 1 m^2.

DBT Diffusion-bonded titanium.

DBTE Data-bus test equipment.

DBTF Duct-burning turbofan.

DBU Database unit.

DBUF Defence buildup plan, or programme (J).

DBV Diagonally braked vehicle, for runway friction measures.

DBVOR Doppler VOR with weather broadcast. Hence DBVortac.

DBW Differential ballistic wind.

dBw Decibels referenced to 1 Watt.

DBWS Database work-station.

DC *1* Depth charge.

  2 Departure control.

  3 Direct cycle.

  4 Display controller.

  5 Directionally cast.

  6 Drag control.

  7 Dead centre.

  8 Detection centre (homing).

  9 Dry chemical.

  10 Direct cost.

  11 Digital compass.

dc, d.c. Direct current.

DCA *1* Defense Communications Agency (US).

  2 Directorate of Civil Aviation.

  3 Defense Contre Avions (F, 1935–40).

  4 Department of Civil Aviation (A, Braz.).

  5 Dual-capable aircraft.

  6 Document content architecture (IBM).

  7 Defensive counter-air.

  8 Design Chain Accelerator.

  9 Drift correction angle.

  10 Defense Certification Authority.

  11 Defence Codification Agency (UK).

DCAA Defense Contract Audit Agency (US).

DCACMRM Defense and control airspace configuration/manufacturing resource management.

D-carts Decoy cartridges.

DCAS *1* Deputy Chief of the Air Staff (UK).

  2 Digital core avionics system.

  3 Defense Contract Administration Service (DoD).

  4 Digitally controlled audio system.

DCAV STOVL (F).

DCC *1* Drone Control Center.

  2 Direct computer control.

  3 Digital computer complex.

DCCA Direction Centrale du Commissariat de l'Air (F).

DCCR Display-channel complex rehost.

DCD *1* Data collector and diagnoster.

2 Data-collection device (helicopter).

3 Double-channel duplex.

DCDI   Digital course deviation indicator.

DCDS   Deputy Chief of the Defence Staff (UK).

DCDU   *1* Datalink control and display unit.

2 Digital, or data, communications and display unit.

DCE   *1* Data communications equipment.

2 Data-circuit terminating equipment.

DCEE   Distributed continuous experimentation environment.

DCFS   Digitally controlled frequency service.

DCGA   Deck closed, go-around.

DCGF   Data-conversion gateway function.

DCGS   *1* Distributed common ground system.

2 Distributed common group station.

DCH   Destination change (input button).

DCI   *1* DCMS crew-member interface; replaced by PCC (5).

2 Defence-capabilities initiative (NATO), or interface.

DCIC   Defence Capability Investment Committee (Australia).

DCIU   Digital control and interface unit.

DCKG   Docking.

DCL   Defence Contractors List (UK).

DCM   *1* Defense Contract Management; A adds Agency, C Command.

2 Data-conversion management; F adds function.

3 Diagnostic and condition monitoring.

DCMAA   Direction Centrale du Matériel de l'Armée de l'Air (F).

DCMF   Data communication management function.

DCMS   Digital, or data, communications management system.

dcmsnd   Decommissioned.

DCMU   Digitally coloured map unit.

DCN   *1* Drawing, or design, or document, change notice.

2 The Defence Contrators' Network (UK)

3 Diplomatic clearance number.

DCNO   *1* Deputy Chief of Naval Operations.

2 Duty not carried out.

DCO   Duty carried out.

DCOM   Distributed-component object model.

DCoS   Deputy Chief of Staff.

DCP   *1* Distributed communications processor.

2 Display control panel.

3 Data collection and processing, or pack.

DCPG   Defense Communications Planning Group (DoD).

DCPS   Data collection and processing system.

DCPU   Display control power unit (IFF).

DCR   *1* Digitally-coded radar.

2 Direct-conversion receiver (radio).

3 Digital-cartridge recording; S adds system.

4 Previously DCRJ, dual-combustion ramjet.

5 Decrease.

DCRD   Dynamic-component repair and development [helicopter and V/STOL].

DCRF   Director, or Directorate, for Construction and Research Facilities (UK, 1935–57).

DCS   *1* Deputy Chief of Staff (DCoS) preferred).

2 Defense Communications System (US).

3 Decompression sickness.

4 Double-channel simplex.

5 Departure control system.

6 Double-correlated sampling.

7 Digital or duplex communications system, or suite.

dcs   Data collection system.

DCSA   Defence Communications [and] Services Agency (UK).

DCSO   Deputy Commander for Space Operations (USAF).

DCS/R&D   Deputy Chief of Staff for Research and Development.

DCSU   Dual crew-station unit.

DCSWI   Deputy Chief of Staff for Warfighting Integration (USAF).

DCT   *1* Digital communications terminal.

2 Direct.

DCTE   Data circuit terminating equipment.

DCTL   From DC to light, ie entire usable FM spectrum.

DCTN   Defense Commercial Telecommunications Network, subset of DISN.

Dctr   Director.

DCTS   Data communications terminal system.

DCTU   Digital calibration trim unit.

DCU   *1* Digital [engine] control unit.

2 Data concentration, or collection, unit.

DCV   *1* Demonstrated crosswind velocity.

2 Directional control valve.

3 Destination-coded vehicle (baggage handling).

Dcw   Crosswind drag component.

DD   *1* Direct drive (servo valve).

2 Dewpoint depression.

3 Distress and Diversion.

4 Destroyer ship class (USN).

5 Differential Doppler.

6 Data delivery.

dD   Any component of drag; thus dD_N could be drag of a nacelle; -dD is a forward thrust, such as negative drag of a winglet.

DDA   *1* Design-deviation authorization, or authority.

2 Defence Diversification Agency (UK).

3 Digital differential analyser.

4 Distance data adapter.

DDAFCS   Dual digital automatic flight-control system.

DDAU   Digital-data acquisition unit.

DDB   Dynamic database [fusion of input] (Darpa).

DDBS   Distributed database system.

DDC   *1* Digital display console.

2 Distributed digital computer.

3 Distress and Diversion call.

4 Deducted damage computation.

DDCA   Dual-designated complex architecture.

DDCU   Data display and control unit.

DDD, D^3   *1* Dynamic data display; S adds subsystem.

2 Detail data display.

3 Dual disk drive.

DDF, DD/F   Digital direction-finding.

DDG   Guided-missile [-armed] destroyer (USN).

DDI   *1* Digital display indicator; C adds control.

2 Data display indicator (Awacs).

3 Direct-dial indicator.

DDL   *1* Down datalink.

2 Dummy deck landing; see *Addls*.

3 Drag due to lift.

DDM   *1* Difference in depth of modulation.

2 Distributed data management.

3 Display diagonal measure [corner to corner].

DDMS DoD Manager for Space Shuttle (support) operations.

DDN Defense data network.

DDNP Diazodinitrophenol.

DDOR Deputy Director of Operational Requirements.

DDP *1* Declaration of Design and Performance, formal statement by manufacturer accompanying each functioning item or modification, (required by CAA).

2 Digital data, or Doppler, processor.

DDPE Digital data-processing equipment.

DDPS Digital display, or data, processing system.

DDR *1* Depressed-datum reheat.

2 Draft document review.

DDR&E Director of Defense Research and Engineering (DoD).

DDRMI DME/VOR radio-magnetic indicator, usually duplicated and partnered by two ADF/RMIs.

DDS *1* Display and debriefing subsystem.

2 Digital-data set.

3 Data distribution system.

4 Dynamic directional stability.

5 Direct digital synthesizer.

6 Digital debrief[ing] station.

DDT *1* Direct digital targeting.

2 Detail data display.

3 Runway strength for double dual tandem [landing gear].

4 Downlink data transfer.

DDTC Data-link delivery of [expected] taxi clearance[s].

DDT&E Design, development, test and evaluation.

DDU *1* Diagnostic display unit.

2 Disk drive unit.

DDV Direct-drive valve (hydraulics/brakes).

DDVR Displayed-data video recorder.

DE *1* Directed energy.

2 Direct-entry (RAF).

3 Weather map.

4 Deflection error.

D_E Effective drag.

d_e *1* Diameter of single jet or nozzle with area equal to total of system of multiple nozzles.

2 Distance between elements of an array antenna.

DEA *1* Delegated engineering authority.

2 Data encryption algorithm.

3 Drug Enforcement Agency [OAO adds Office of Air Operations].

DEAD Destruction of enemy air defence[s].

dead centre *1* In piston engine, with conrod aligned with cylinder axis [in normal designs], piston at end of stroke.

2 Precisely on target [esp. sport parachuting].

dead engine One that cannot be operated after IFSD.

deadeye Circular block pulled by surrounding cable or rope to exert tension on other cables passing through transverse holes.

deadface To cut off all system power by circuit interrupters at interface between modules, stages or spacecraft, prior to separation.

dead foot Failed engine of twin- [rarely more] engined aircraft.

deadhead *1* To fly to maintenance base off-route.

2 Of aircrew, to ride as passenger(s) while on duty.

dead men Masses [not necessarily anthropomorphic] simulating passengers.

dead reckoning Plotting aircraft position by calculations of speed, course, time, effect of wind, and previous known position.

dead-rise Difference in height from keel to chine of float or flying-boat hull.

dead-rise angle That between line joining keel and chine and transverse horizontal through keel.

dead side *1* Side away from aircraft formation, eg left seat when in echelon to right.

2 Side of airfield or active runway away from that of circuit [pattern] in use, or from which arrivals join circuit.

dead spot In a system, region centred about neutral position where small inputs produce no response.

dead-stick landing Landing of powered aircraft with all engines inoperative.

dead vortex Remnants of vortex after breakup and decay.

dead zone *1* Surface area within maximum range of weapon, radar or observer which cannot be covered by fire or observation because of obstacles, nature of ground, or trajectory characteristics or pointing limitations of weapon.

2 Zone within range of radio transmitter in which signal is not received.

3 Region above gun or missile into which weapon cannot fire because of mechanical or electronic limitations.

4 Area(s) next to surfaces of aircraft plate for integrally machined parts which cannot be ultrasonically inspected and for which ultrasonic-inspection thickness allowances can be removed.

de-aerator Static or centrifugal screen for removing air from circulating lubricating oil.

deal Bad error by ATC controller.

dealer plate No Issued temporarily to a/c for export, often to several in succession, to avoid need for proper US registration (FAA).

Deatac Directed-energy applications in tactical airborne [or air] combat.

DEB Digital European backbone (major NATO programme).

deboost Retrograde or braking manoeuvre which lowers either perigee or apogee of orbiting spacecraft.

debrief To interrogate aircrew or astronauts after mission to obtain maximum useful information.

debris *1* Remains from catastrophic accident.

2 In particular, fragments from exploded engine.

3 Loosely, BFO(3).

debug To isolate, correct or remove faults or malfunctions, especially from computer program.

DEC *1* Data-Exchange Committee.

2 Digital engine [or electronic] control [S adds system, U unit].

3 Decrease.

4 Declination.

5 Decommissioned.

6 Digital electronic clock.

7 Defence; or Directorate of, equipment capability (UK).

deca Prefix, multiplied by 10, symbol ∝a (non-SI).

Decade DFS(4) Eurocontrol ATM(7) development.

decal Insignia or other mark applied by transfer, usually to a model.

decalage   Difference in angles of incidence of wings of biplane or multiplane; angle between chord of upper plane and that of lower plane in section parallel to plane of symmetry. Negative when angle of lower plane is greater.

decalescence point   Temperature, characterised by sudden evolution of heat, at which definite crystalline transformation takes place when heating steel.

decametric   Having wavelengths in the order of 10 m.

decanewton   10 N of force or thrust, = 2.24809 lbf.

decant   Drain dregs of fuel from lowest point of integral or other tank, hence *hole, *assembly.

decarburising   Heating iron or carbon steel to temperature sufficiently high to burn out or oxidise carbon.

decay   *1* Progressive, accelerating reduction in orbital parameters, esp. apogee and perigee, of body in orbit affected by an atmosphere.

2 Progressive reduction in intensity of many natural processes, eg radioactivity, phosphorescence.

decay curve   Plot of radiation intensity against time.

decay orbit   Usually, final orbit terminating in re-entry.

decay product   Usually, radioactive nuclides.

decay rate   Rate of disintegration with respect to time.

decay time   *1* Time for electronic pulse to fall to 0.1 of peak.

2 Time for charge in storage tube to fall to given fraction of initial, usually 1/e where e is e(5).

3 Estimated lifetime of satellite in low orbit.

Decca chain   Single system of master and three slave Decca Navigator stations giving guidance over one geographic region.

Decca Flight Log   Pictorial presentation of Decca Navigator inputs on roller-map display.

Decca lane   In original Navigator, any hyperbolic region between two adjacent position lines.

Decca Navigator   Pioneer hyperbolic navaid using CW.

Decca Omnitrac   Airborne digital computer which eliminates Flight Log chart distortion, sets pen accurately after chart change and enables system to be coupled to autopilot.

DECD   Digital expandable color display.

deceleration limit   That sustained value allowed for fully equipped astronauts or aircrew, normally – 10 g.

deceleron   Aileron which splits into upper/lower halves to serve as speed brake (originally Northrop patent).

decentralised control   In air defence, normal mode whereby higher echelon merely monitors unit actions, making direct target assignments only when necessary.

deception   Measure designed to mislead enemy by manipulation, distortion or falsification of evidence, eg by DECM (1).

deci   Prefix, one-tenth, symbol d (non-SI).

DECIBE   Defence equipment capability indirect battle-field effect (UK).

decibel   Fundamental unit of sound pressure (see *noise*).

decimetre 10^{-1} m = 3·937 in (contrary to SI).

decimetric   Having wavelengths in the order of 10^{-1} m (not recommended).

decimillimetric   Having wavelengths in the order of 10^{-1} mm (not recommended).

decision height   Specified height AGL at which missed approach must be initiated if the required visual reference to continue approach to land has not then been established; normally but not exclusively ILS, PAR or MLS approach.

decision height abuse   For test purposes, landing from points deliberately offset laterally or longitudinally at DH.

decision speed   Usually, V_1.

deck   *1* Any ground or water surface (colloq.).

2 From 1966, FL just above such surface.

deck alert   Ready for immediate takeoff from ship, normally by fighter.

deck-edge elevator   Lift built into side of aircraft carrier for moving aircraft between decks.

decking   Top surface of fuselage.

deck letter   Identifying letter painted on flight deck of aircraft carrier (USN: *number*).

deck park   Parking area for aircraft or other vehicles on aircraft carrier flight deck.

deck plate[s]   Electroluminescent panel[s] recessed flush with deck of ship or oil rig or other platform marking helicopter landing area.

deck run   Distance run along ship deck in free (non-cat.) takeoff.

deck spot   Area allocated to, or occupied by, one aircraft on deck.

declaration   Size of force committed by government to special purpose, esp. to support multinational alliance; hence declared force.

declarative language   So-called 'fifth generation' computer language used for AI(2) which requires merely that programmer describes problem and declares facts and parameters necessary for solution; * then decides how this information will be used in solution process.

declared   Numerical or factual data published or filed before flight.

declared alternate   Airfield specified in flight plan to which flight may proceed, should landing at original airfield become inadvisable.

declared destination   Airfield specified in flight plan at which flight is intended to terminate.

declared distance   See ASDA, ED, LDA, TODA, TORA.

declared thrust   Generally, those ratings published by manufacturers.

declared weight   Generally, that filed in flight plan.

declassification   *1* Removal of item from security classification.

2 At public exhibition or display, removal of sensitive items prior to opening.

declination   *1* Angular distance to body on celestial sphere measured north or south through 90° from celestial equator along hour circle of body. Comparable to latitude on terrestrial sphere.

2 Magnetic variation.

DECM   *1* Deceptive ECM.

2 Defensive ECM.

decode   To translate into plain language aeronautical telecommunications and other signals from ground to air, esp. in Notam and Q-code.

Decometer   Original dial-type phase-meter display, one per lane, in Decca Navigator before 1953.

decompression chamber   Capsule or chamber in which human beings or hardware can undergo process of decompression.

decompression sickness   See *aeroembolism*.

decompression stress   Human stress arising from decompression syndrome.

deconfliction   In air display, arranging each slot to avoid scheduled traffic from same or nearby airfield[s].

decoration   Extra manoeuvres, such as flick rolls, inserted into aerobatic routine while aircraft is being repositioned.

decoupler   Large-amplitude elastic connection separating two systems of masses which, if rigidly linked, would be prone to dangerous flutter; hence * pylon for separating vibration of wing and heavy stores hung below it.

decoy   Device or technique used to simulate attacking aircraft and their defensive systems. Usually operates at radar or IR wavelengths.

DECR   Decrease.

decrab   To yaw crabbing aircraft landing in crosswind to align wheels with track.

decrement   Quantified decrease in value of variable.

DECS   Defence Economic-Commerce Service (UK).

Dectrac   Decca Navigator display for GA aircraft.

DECU   Digital engine (or electronic) control unit.

ded   Dedicated.

DED   *1* Data entry display.

    2 Directed-Energy Directorate (AFRL).

dedicated   Available only for one declared application; thus a * dock is tailored to one type of aircraft.

dedicated runway   That permanently assigned as main instrument runway.

DEE   *1* Di-ethyl ether.

    2 Department of Education and Employment (UK).

DEEC   Digital electronic engine control.

DEEP   Dangerous-environment electrical protection system.

deep   Far down the runway [said of a landing].

deep cycling   Pre-delivery test of electronics (rarely, other hardware) by subjecting circuits to slightly excessive voltages.

deepening   Decreasing pressure in centre of existing low.

deep overhaul   Major overhaul, not normally performable by user.

deep space   Not in vicinity of Earth.

deep stall   Condition associated with T-tail, rear-engined configuration characterised by rapid increase in angle of attack to point where effectiveness of horizontal tail is inadequate for longitudinal control, and stable longitudinal trim point is reached with AOA up to 90°. Apparent synonyms are locked-in stall, superstall.

DEEU   Data-entry electronics unit.

DEFA   Direction des Etudes et Fabrications d'Armement (F).

Defamm   Development of demonstration facilities for airport movements guidance control and management (Euret).

Defcon, DEFCON   Defence contracting (UK, MoD.).

DEFCS   Digital electronic flight-control system.

Defdars   Digital expandable flight-data acquisition and recording system.

defence suppression   Secondary objective of air attack on enemy territory, to reduce or eliminate anti-aircraft defences.

defense VFR   Filed by aircraft cruising at 180 kt or above intending to penetrate an ADIZ in VFR (US).

defensive combat spread   Loose pair 1–2 miles (1.6–3.2 km) apart and slightly separated fore/aft and vertically.

defensive electronics   Airborne ECM and EW equipment used to protect aircraft against hostile defences.

defensive spiral   Accelerating high-g continuous-roll dive to negate attack.

defensive split   Controlled separation of target element into different planes to force enemy interceptors to commit to one of them.

defensive turn   Basic defensive manoeuvre to prevent an attacker from achieving a launch or firing position.

deferment   Agreed postponement of delivery.

deficiency   Fault condition (known or suspected), equipment shortage or other imperfection which may or may not render aircraft unairworthy.

definition   *1* Clarity and sharpness of image in display.

    2 In contract proposal, complete description by contractor of product offered.

defl   Deflection.

deflation port   Aperture in top of balloon gas envelope through which contents are [Ed. opinion, foolishly] dissipated after landing.

deflected slipstream   Horizontal slipstream from propulsion system deflected downwards by mechanical means to augment lift for STOL or V/STOL performance.

deflecting yoke   Mutually perpendicular coils around neck of CRT which control position of electron beam, enabling it to scan screen.

deflection   *1* Bending or displacement of neutral axis of structural member due to external load.

    2 Change in radius of pneumatic tyre, expressed as percentage.

deflection angle   *1* In supersonic flight, that between longitudinal axis and outer surface of bow [nose] of body.

    2 That between longitudinal axis and surface [esp. trailing edge of airfoil] determining angle of top and bottom shocks.

deflection crash switch   One triggered by impact significantly changing shape of structure.

deflection error   Lateral artillery error, as distinct from range error, usually problem with land rather than air targets.

deflectometer   Instrument for measuring deflection under load of airfield surface. There are several species.

defruiting   Elimination of fruiting by rejecting all non-synchronous replies; PRFs varying by $2.5\mu s$ can be eliminated.

Defstan, DefStan   Defence standard (UK).

DEFT   Directorate of Elementary Flying Training (UK, until 2003).

DEFTS   Defence Elementary Flying Training School (replaced JEFTS 2003).

defueller   Unit, usually vehicle-mounted, for draining fuel and condensate from aircraft.

DEG   *1* Dressed engine gearbox.

    2 Degree[s].

degarbling   Elimination of garbling by trying to extract interleaved replies, differentiating between the exact leading and trailing edges of the pulses.

degraded flight control   Usually means failure of surface power unit.

degraded performance   Performance reduced by internal shortcomings, eg airframe tiredness, engine gas-path deposits, etc. Not normally used for external influences, eg hot-and-high conditions.

degraded surface   Airfield covered with snow, ice or standing water.

degreasing   Removal of grease, oil or related residue by solvent, either liquid such as naphtha or vapour such as trichlorethylene.

degree   Non-SI unit of plane angle, = 1.745329 × 10^{-2} rad.

degree of freedom   *1* Mode of motion, angular or linear, with respect to co-ordinate system; free body has six possible ***, three linear and three angular.

2 Specif., of gyro, number of orthogonal axes about which spin axis is free to rotate.

3 In unconstrained dynamic system, number of independent variables required to specify state at given moment. If system has constraints, each reduces *** by one.

4 Of mechanical system, minimum number of independent generalised co-ordinates required to define positions of all parts at any instant. Generally, *** equals number of possible independent generalised displacements.

DEI   Development engineering inspection.

de-icing   Removal of ice accretion by thermal, mechanical or chemical means. Note: anti-icing prevents accretion.

de-icing fluid   Glycol/alcohol mixtures are common for removing frost from parked aircraft. Fluids for use in flight include ethyl or isopropyl alcohol and ethylene or propylene glycols.

DEIMS   Defense economic-impact modelling system (DoD).

de-ionisation time   Time for gas-discharge tube to return to neutral condition after interruption of anode current.

del   Delete.

DEL   Delay, delay message; also DLA (ICAO code).

de Laval nozzle   Con-di nozzle used in steam turbines, certain rockets, and some tunnels.

delay   *1* Distance from point directly beneath aircraft to beginning of area visible to its radar.

2 Electronic delay at start of time base used to select particular segment of total.

3 Difference in phase between two EM waves of same frequency.

delayed automatic gain control   Applied only to received signals above predetermined level, so permitting only weak signals to be fully amplified.

delayed CCIP   CCIP for highly retarded bomb.

delayed drop   Live parachute descent begun by prolonged free fall. Controlled by wearer, unlike delayed opening.

delayed flap approach   Otherwise conventional landing approach, usually by commercial jet, in which AFCS (TCC) or FMS is programmed to postpone final configuration until very late stage, typically near airport perimeter. Reduces noise and fuel burn.

delayed opening   Opening of parachute canopy automatically delayed by barostat to allow rapid fall through stratosphere to safer altitude, usually 10,000–15,000 ft, 3050–4570 m.

delayed repeater   Comsat which stores messages and retransmits later, usually at high rate in brief period of time.

delay indefinite   ATC cannot yet estimate duration of delay; usually followed by Expect further clearance.

delay line   Passive network, such as closed loop of mercury, capable of delaying signal without introducing distortion.

delay parameter   Also called altitude *, time that elapses between sharp nose-up command and start of climb or arrest of sink. Significant in large aircraft on landing, when sudden download at tail has opposite to desired effect [except canard aircraft].

delay rate   Unusual measure of airline punctuality: number of delays (usually 5 min) per 100 scheduled departures.

delivery error   Overall inaccuracy of weapon system resulting in dispersion of shots about aiming point.

delphinopter   Class of micro air vehicles weighing c4·5g combining tail-flapping propulsion with a forward wing which twists for trajectory control. Most alternate between flapping and gliding.

Delrin   UV-stabilized acetyl-resin adhesives.

DELSC   Defence Electrical and Electronic Standardization Committee.

delta (δ)   *1* Surface deflection angle, thus eδ = elevator deflection angle.

2 Difference; thus * 1700–1745 is 45 min.

3 Delta wing, or delta-wing aircraft.

Delta Gold   Top FAI rating for glider pilot, requiring flight of 300 km or closed circuit (landing back at start) of 200 km.

delta h, δh   Quantified change in altitude or height above ground.

delta hinge   Helicopter main-rotor flapping hinge, giving blade freedom to flap up/down vertically. Thus, * is perpendicular to both blade axis and axis of rotation.

Delta Silver   FAI qualification for glider pilot requiring distance flight 50+ km and [can be same flight] 5+ h duration.

Delta-3   Helicopter tail rotor with two pairs of blades not crossing at 90°.

delta-V, δV   Quantified change in velocity, usually airspeed.

delta wing   Wing of basically triangular plan-form with one apex at front and transverse trailing edge, usually with sharp leading-edge sweep giving low aspect ratio.

deluge pond   Facility at site for testing or launching large vertically mounted rockets into which cooling water is flushed; also called skimmer basin.

DEM   Digital elevation model.

demand breathing   See demand mask.

demand mask   Mask through which oxygen or other therapeutic gas flows only on inspiration of wearer.

demand mode   Acars mode initiated by either aircraft or ground processor.

demand oxygen   See demand mask.

demijohn   Fluid container of cylindrical form (F).

Demiz   Distant early-warning military identification zone.

demodulation   Detection of received signal by extracting modulating signal from carrier.

Demon   Demodulation of noise.

demonstrate   *1* To display new hardware according to detailed test schedule before certificating authority or sponsoring military customer.

2 More specifically, to show compliance with numerical performance values, reliability or maintainability.

demonstration flight   Made for potential customer [on

board], normally not forming part of an airshow programme.

demonstrator programme *1* Showing of new civil aircraft in visits to potential customers.

2 Agreed schedule of tests of new hardware, including complete aircraft, before military customer in advance of any decision on procurement and often to establish what is possible.

demounting One meaning is to remove tyre from wheel; the wheel may include a demountable flange.

Demoval, Demval Demonstration and validation.

Dempi [pronounced dimpy]. Designated mean point of impact.

Denalt Density altitude.

DENEB, Deneb Fog-dispersal operation.

DEngRD Directorate of Engine Research and Development (UK).

denitrogenation Removal of nitrogen dissolved in blood and body tissues, usually by breathing pure oxygen for extended period, to minimise aeroembolism. Also called preoxygenation.

densified wood Multiple laminates bonded under high pressure.

densitometer *1* Instrument for measurement of optical density, generally of photographic image.

2 Instrument for measuring fuel density, usually part of fuel measurement system.

density *1* Mass per unit volume; SI unit $kg/m^3 = 0.062428$ $lb/ft^3 = 0.01002$ lb/Imp gal; $1g/cm^3 = 0.036127$ lb/in^3; $1lb/in^3 = 27.6799$ g/cm^3; 1 $lb/ft^3 = 16.0185$ kg/m^3. Often needs qualifying for temperature and pressure (see *absolute**, *relative**).

2 Of aircraft, MTOW divided by total aircraft volume calculated from external envelope, or divided by both wing area and mean chord.

density altitude Pressure altitude corrected for non-ISA temperature.

density error Correction to EAS to give TAS (see *airspeed*).

deorbit Deliberately to depart from spacecraft orbit, usually to enter descent phase or change course.

DEOS Digital engine operating system.

DEP *1* Department of Employment and Productivity (UK).

2 ICAO code for depart, departure, departure message.

3 Departure airfield.

4 Design eye position (usually of pilot).

5 Direct-entry pilot.

6 Data-entry panel.

departure *1* Any aircraft taking off from airport (as distinct from other airfields) under departure control.

2 In air navigation, distance made good in E/W direction, usually expressed in nm.

3 General term for uncontrolled flight beyond the stall; see *divergence* (1) or *disturbance* (2).

departure alternate Alternate airfield specified in flight plan filed before takeoff.

departure control Function of approach control providing service for departing IFR aircraft and, on occasion, VFR aircraft in such matters as runway clearances, vectors away from congested areas and radar separations, all at nominated time.

departure pattern That flown in 3-D by departure (1).

departure point Navigational check point, such as VOR or visual fix, used as a marker for setting course.

departure procedures ATC procedures (usually SID) flown by departing aircraft during climb-out to minimum en route altitude.

departure profile Flight profile flown by departure (1) to suit needs of vertical and horizontal separation, noise abatement, obstacle clearance, etc.

departure runway That from which departures (1) are cleared.

departure stall On attempted takeoff from small field, pilot avoids obstacle ahead by steep bank and sharp turn, then applying top rudder, stalling upper wing.

departure strip Flight progress strip recording callsign, ETD and route of departure.

departure tax Imposed by most states at flat rate per passenger.

departure track That followed by departure (1).

departure traffic Total number of departures, scheduled and non-scheduled, from one airport, usually expressed in movements per hour or per day.

DEP CON Departure control.

Depcos Departure co-ordination system (Airsys); see Depos.

depigram Plot of variation of dewpoint with pressure for given sounding on tephigram.

deplane Normally transitive verb, to ask all occupants to leave aircraft, especially because of fault or potential danger.

depleted uranium Dense metal [see uranium] removed in spent fuel rods from nuclear reactors, used in flight-control surface balances and gun ammunition.

depletion layer In semiconductor, region in which mobile carrier charge density is insufficient to neutralise net fixed charge density of donors (N-type) and acceptors (P-type). Also known as depletion barrier or zone.

deployable simulator Installed at front-line airbase or aboard carrier.

deployment *1* Strategic relocation of forces to desired area of operation.

2 Extension or widening of front of military unit.

3 Change from cruising approach, or contact disposition, to formation for battle.

4 Process from pulling ripcord to fully opened parachute.

5 Extension of solar panels from spacecraft.

6 Basic meaning of word: to use a military or naval aircraft operationally.

deploy range Range of combat aircraft on transfer from one theatre to another, if necessary with internal or external auxiliary fuel.

DEPM Data evaluation program manager[s] (ATOS).

Depos Departure co-ordination system (ATC).

depot-level maintenance Performed at a specialized overhaul facility, remote from user unit.

depreservation run Test run of machinery after storage, to validate performance.

depressed-datum reheat Engine control mode for jet STOVL giving reheat operation at low (dry) thrust levels, giving smooth auto control to max. thrust.

depressed-sightline attack Shallow dive.

depressed trajectory Flight profile of ballistic missile, esp. SLBM, fired over relatively short range with altitude kept low to reduce exposure to defending radars.

depression *1* Region of relatively low barometric pressure, also known as cyclonic area or low; secondary * is small low accompanying primary.

2 Negative altitude, angular distance below horizon.

depth Aerospace meanings include * of depression, * of modulation and distance down runway; * of wing profile is *thickness*.

DER *1* Designated engineering representative.

2 Departure end of runway.

DERA Defence Evaluation and Research Agency, in 2001 renamed *DRA* (UK).

derated engine One whose maximum power is governed at a lower than normal value. Hence derating.

Derd, DERD *1* Display of extracted radar data.

2 Incorrectly used to mean DEngRD.

deregulation Removal of rules regarding admission to air-transport industry of new carriers, routes and equipment.

derivative Not precisely defined, but taken by certifying authority to mean that new aircraft or engine is so similar to the original version that no new certification programme is needed.

derivatives, resistance *1* Lateral ** give variation of forces and moments caused by small changes in lateral, rolling and yawing velocities.

2 Longitudinal ** give variation of forces and moments caused by small changes in longitudinal, normal and pitching velocities.

derivatives, stability Quantities expressing variation of forces and moments on aircraft due to any disturbance to steady motion.

derotation To put nose gear on runway after landing.

DES *1* Design environmental simulator (USAF).

2 Design engineering support.

3 Data encryption standard.

4 Descend, descent.

DESC Defense Electronic Support Center (US).

descending node Longitude or time at which satellite crosses Equator from N to S.

descent fuel Fuel burned from TOD until either hold or approach.

descent idle Engine setting to optimise parameters in near-glide.

descent indicator See *VSI*.

descent orbit insertion Start of lunar or planetary landing procedure from orbit, with retrograde thrust into descent transfer orbit.

descent propulsion That providing trajectory control for soft lunar or planetary landing.

descent stage Lower part of two-way lunar or planetary lander which, when mission is completed, acts as launch pad for ascent stage.

descent transfer orbit Highly elliptical around Moon, can be circular around planet, in which soft lander is placed before descent to surface.

Descr. Dscription.

deselect *1* To switch off.

2 Eliminate contender from competition.

desensitization Reduction in TCAS threat volume.

design Entire process of translating hardware requirement or specification into final production drawings and NC tapes.

designated flying course Prior to carrier landing, 15 seconds before turning downwind.

designated target One at which friendly designator (2) is pointed.

designation marking Use of laser or other designator.

designator *1* Letter/number code identifying each flight by a scheduled carrier.

2 Number/letter code identifying each runway, thus 26L = 260° left runway of pair.

3 Laser or other device pointed at target to make latter emit signals on which missile can home.

Design Chain Accelerator One of the first commercially offered clusters (1) for simulating complex systems (Intel/MSC/HP).

design gross weight Anticipated MTOW used in design calculations; design takeoff weight.

design landing weight Anticipated MLW used in design calculations.

design leader *1* Individual leading design team.

2 Nation in collaborative project said to have political dominance.

design load Specified load below which structural member or part is designed not to fail, usually expressed as probable maximum limit load, unfactored.

design load factor Maximum repeated vertical acceleration which an aircraft structure is designed to withstand without accretion of damage. Typical values for a jet transport are +2·5/-1g (with flaps extended reduced to +2), and for a fighter +12/-6.

design maximum weight Assumed weight used in stressing structure for flight loads.

design office That in which design takes place, and authority vested therein.

design points Specific combinations of variables upon which design process is based; together these cover every combination of air density, airspeed, Mach, dynamic pressure, structural loads (including free or accelerated take-off and normal or arrested landing) and system demands aircraft can encounter.

design verification First item built to new design to prove compliance with drawings and demonstrate correct functioning (see *DVA*).

design weight No standard meaning, but with most design/certification authorities is less than MTOW.

design wing area Area enclosed by wing outline (including flaps in retracted position and ailerons, but excluding fillets or fairings) on surface containing wing chords, extended through nacelles and fuselage to plane of symmetry.

Desir Direct English statement information retrieval (EDP).

desmodromic Mechanical drive giving perfect to/fro action, esp. of cam drive to piston-engine valve.

Deso, DESO Defence Export Services Organization (UK).

despatch *1* To supervise exit of parachutists, or to unload stores with parachutes attached, from aircraft in flight.

2 Process of supervising readiness of civil transport for next flight, with departure on schedule.

despatch deficiency Malfunction, failure, breakage, missing equipment item or other irregularity which does not prohibit on-time departure.

despatch delay Any notifiable delay, measured variously from either 5 or 15 min, in departure of scheduled flight.

despatch deviation   Any reportable irregularity other than deficiency which does not prohibit on-time departure.

despatcher   One who is responsible for despatching an airline flight.

despatch reliability   Percentage of all scheduled flights by particular aircraft or all aircraft of that type, often over specified period or for particular operator, that departed on time (measured as within 5 or 15 min).

de-spin   To rotate part or whole of satellite or other spacecraft to neutralise spin previously imparted (see *next*).

despun antenna   One mounted on satellite spun for reasons of stability which, because it must point continuously towards an Earth station, must rotate relative to satellite.

dessyn   Synchro (trade name).

dest   Destination airport.

destage   To redesign an engine by removing one stage of blading from an axial compressor (usually the last stage).

destretch   To produce new version of transport aircraft with fuselage of reduced length.

destruct   To destroy vehicle after launch because of guidance or other failure making it dangerous.

destructive test   One which destroys specimen.

destruct line   Map boundaries which vehicle must not cross; any which does is immediately destructed.

destructor   *1* Device, explosive or incendiary, for intentionally destroying all or part of vehicle such as wayward missile or aircraft down in enemy territory.

    2 NW for undersea use.

DESU   Digital electronic sequence unit (APU).

DET   Direct energy transfer.

DETA   Di-ethylene triamine.

detachable   Capable of being removed from aircraft with normal hand tools.

detached shockwave   One proceeding ahead of body causing it.

detail   *1* To design small part such as attachment bracket.

    2 Drawing (can be inset on main design drawing) giving graphical representation of features.

    3 Small military detachment for particular task.

    4 To assign to special task or duty.

detail part   One not normally broken down during service or storage.

Detasheet   Plastic explosive based on RDX/PETN.

detectable crack   Nominal length 100 mm, 4 in.

detector   Sensitive receiver for observing and measuring IR.

detent   A spring-loaded catch permitting linear movement in one direction only.

deterrence   Prevention of aggression through fear of consequences.

DETF   Data-exchange test facility.

detolerancing   The principal meaning is to open out (relax) dimensional limits on airframe structure.

detonating cord   Flexible explosive [usually shaped-charge] pipe for emergency severing of doors, canopies, etc.

detonation   *1* Violent and irregular combustion in piston engine cylinder resulting from excessive compression ratio or supercharging, or using inferior fuel; also known as knocking or pinking.

    2 Correct triggering of explosive.

detonator   Explosive device usually sensitive to mechanical or electrical action and employed to set off larger charge of explosive.

detotalizing counter   Indicates total remaining of substance being measured, such as rounds for a gun or kg of fuel.

DETR   Department of the Environment, Transport and the Regions (UK)

DEU   *1* Display electronic unit.

    2 Decoder/encoder unit (CIDS).

deuteron   Nucleus of deuterium.

deuterium   Isotope of hydrogen (heavy hydrogen) whose nucleus contains a neutron as well as a proton; used as projectile in nuclear processes. Forms heavy water (D_2O) with oxygen.

deuteride   Compound of deuterium. Lithium-6 deuteride is a standard fusion material in NW.

DEV   Deviation.

Devco   Development Committee (ISO).

development   *1* Process of converting first flight article into mature product ready for delivery.

    2 Ongoing process of improving production aircraft to carry heavier load, fly farther, accomplish new tasks, etc.

    3 Determining by mathematical calculation, computer graphics or drafting methods, size, shape and other pertinent characteristics of non-flat parts.

    4 Opening of parachute canopy.

    5 Generally not precisely quantifiable, process in which aircraft becomes locked-in to stall, superstall or spin.

development contract   Calls for development (1) of particular hardware item.

development stage   Begins as soon as hardware to new design is available; main phase complete at service (production) release or certification, but continues throughout active life of aircraft.

deviation   *1* Distance by which impact misses target.

    2 Angular difference between magnetic and compass headings caused by magnetic fields other than that of Earth.

    3 In statistics, difference between two numbers (also known as departure), difference of variable from its mean (esp. standard *), or difference of observed value from theoretical.

    4 In meteorology, angle between wind and pressure gradient.

    5 In radio, apparent variation of frequency above and below unmodulated centre frequency.

    6 In flying, sudden excursion from normal flightpath.

    7 Any significant variation from plan.

deviation card   Records compass courses corresponding to desired magnetic headings.

devil   Dust devil.

deviation light[s]   Warn pilot or ground controller of excessive departure from ILS beam.

DEW   *1* Distant early warning.

    2 Directed-energy weapon.

    3 Dressed engine weight.

dew   Atmospheric moisture condensed upon cold objects, esp. at night.

dewar   Thermally insulated container, eg for cryogenics.

DEWD   Dedicated electronic-warfare display.

DEWIZ   Distant early-warning identification zone, extends from surface north of DEW line and around Alaska.

DEW Line   Distant early-warning radar stations at about 70th parallel across North American continent (1955–58).

dewpoint   Temperature at which, under ordinary conditions, condensation begins in cooling mass of air.

Dews   Digital electronic-warfare simulator.

DF, D/F   *1* Direction-finding (or finder).

    2 Digital filter.

    3 Directed-flow (reverser).

    4 Methylphosphonic difluoride, component of GB Sarin.

    5 Data fusion.

    6 Deutsche Flugsichering [ATC] (G), also DFS.

    7 Dong Feng = east wind, family designations of strategic ballistic missiles (China).

    8 Double-fuselage.

    9 Diesel fuel.

    10 Defensive [or direct].

    11 Downlink format.

D_f Zero-lift drag, usually of whole aircraft.

DF-1, DF-2, DF-A   GA specifications for diesel fuel.

DFA   *1* Deutsche Flug-Ambulanz Gemeinnützige GmbH (G).

    2 Delayed-flap approach.

    3 Direction-finding antenna.

DFAD   Digital-feature analysis data.

DFAR   Defense Federal Acquisition Regulations (US).

DFAS   Defense Finance and Accounting Service (US).

DFBW   Digital fly-by-wire.

DFC   *1* Distinguished Flying Cross.

    2 Direct force control, eg on F-16.

    3 Digital fuel control [s adds system].

    4 Duty-free confederation.

DFCC   Digital flight-control computer.

DFCL   Director(ate) of flight-crew licensing.

DFCS   Digital flight – [or fuel] – control system.

DFCT   Directorate of Foreign and Commonwealth Training (UK, multiservice).

DFD   *1* Digital frequency discriminator.

    2 Data flow diagram (real time).

    3 Digital flight data.

DFDA   Defence Force Discipline Act (UK).

DFDAF   Digital flight-data acquisition function.

DFDAMU   DFD(3) acquisition management unit.

DFDAU   Digital [or distributed] flight-data acquisition unit.

DFDR   Digital flight-data recorder.

DFDS   Digital fire-detection system.

DFDU   DFD(3) unit.

DFF   Display failure flag.

DFG   Digital flight guidance [C adds computer, S system, U unit].

DF/GA   Day fighter/ground attack.

DFI   Direction-finding interferometer.

DFIC   Duty-free import certificate.

DFIDU   Dual-function interactive display unit.

DFIR   Deployable flight-incident recorder.

DFIU   Digital flight-instrument unit.

DFL   Dry-film lubricant.

DFLCC   *DFCC.*

DF loop   Direction-finding aerial consisting of one or more turns of wire on vertically pivoted frame, giving maximum response in plane of frame, and thus PL through ground station (see *Adcock, ADF*).

DFLS   Day Fighter Leaders' School (RAF).

DFM   *1* Distinguished Flying Medal.

    2 Digital frequency measurement.

    3 Distortion-factor meter.

    4 Direct-force mode.

DFMS   Digital fuel-, or flight-, management system.

D_{form} Form drag.

DFQI   Digital fuel-quantity indicator.

DFR   *1* Dynamic flap restraint.

    2 Departure flow regulation.

    3 Digital flight recorder.

DFRC   Dryden Flight Research Center [previously Facility] (NASA, at Edwards).

D_{fric} Total frictional drag, at low speeds almost equal to D_f.

DFRR   Data-fusion risk reduction.

DFS   *1* Directorate of Flight Safety (UK).

    2 Digital frequency synthesis, or select.

    3 Deutsche Forschungsanstalt für Segelflug [glider research institute] (G).

    4 Deutsche Flugsicherung [air-traffic control] (G).

    5 Dynamic fuel-slosh measure.

    6 Detailed functional specification.

DFSC   Defense fuel supply center (DoD).

DFT   *1* Distance from threshhold.

    2 Discrete Fourier transform.

    3 Demand flow technology.

    4 Defence fixed telecommunications [S adds system] (UK).

DfT   Department for Transport (UK).

DFTDS   Data-fusion technology demonstrator system.

DFTI   Distance-from-touchdown, or threshold, indicator.

DFU   *1* Digital function unit.

    2 Deployable flotation unit.

DFV   Deutsche Flugdienstberater Vereinigung (G).

DFVLR   Deutsche Forschungs- und Versuchsanstalt für Luft- und Raumfahrt, now DLR (G).

DFWD   Discrete flight warning display.

DFWF   Direct-fire weapons effects; S adds simulator or simulation.

DG   *1* Directional gyro, = DI (2).

    2 Dichromated gelatin.

Dg   Maximum growth of tyre [tire] outside diameter.

DGA   *1* Dispersed ground alert.

    2 Délégation Générale pour l'Armement (F).

    3 Director-General Aircraft, (N) adds Navy (UK).

    4 Displacement gyro assembly.

DGAA   Director-General for Aeronautical Armaments (I, for NATO).

DGAC   *1* Direction Générale de l'Aviation Civile (F).

    2 Directorate-General of Air Communications (Indonesia).

    3 Direzione Generale dell'Aviazione Civile (I).

    4 Direccao-General da Aviacao Civil (Portugal).

    5 Dirección General de Aviación Civil (Spain etc).

DGES   Director-General Equipment Support; can have suffix (Air). (UK).

DGI   Directional gyro instrument or indicator.

DGIA   Dirección General de Infraestructura Aeronautica (Uruguay).

DGIIA   Defence Geographic and Imagery Intelligence Agency (UK).

DGLR   Deutsche Gesellschaft Für Luft- und Raumfahrt (G).

DGLRM   Deutsche Gesellschaft Für Luft- und Raumfahrtmedizin (G).

DGM   *1* Distance-gone meter (Doppler).

2 Digital-group multiplexer.

DGMS   Director-General of Medical Services (RAF).

DGNS   Differential global navigation system.

DGNSS   Differential global nav-sat system; U adds unit.

DGON   Deutsche Gesellschaft Für Ortung and Navigation (G).

DGPS   Differential GPS.

DGR   Dangerous-goods requirements (1ATA).

DGRR   Deutsche Gesellschaft Für Raketentechnik und Raumfahrt (G).

DGSM   Director-General of Support Management (RAF).

DGRST   Direction Générale à la Recherche Scientifique et Technique (F).

DGS   *1* Disc-generated signal.

2 Digital-generation subsystem (ECM).

3 Docking guidance system.

DGSI   Drift and groundspeed indicator (Doppler).

DGT   Digital GPS translator.

DGTA   Dirección General de Transporte Aéreo (Peru, etc).

DGTE   Director-General Test & Evaluation (UK).

DGU   Display generator unit.

DGVS   Doppler ground velocity system.

DGW   Design gross weight.

DGZ   Desired ground zero; point on Earth's surface nearest to centre of planned nuclear detonation (see *actual ground zero, ground zero*).

DH   *1* Decision height.

2 Dataflash Header.

DHB   Dynamic hot bench.

DHDA   Digicon header diode array.

DHFS   Defence Helicopter Flying School (Shawbury, UK).

DHHKHH   Aeronautic Association (S. Korea).

DHMI   Airports authority (Turkey).

DHS   *1* Data-handling system.

2 Department of Homeland Security (US, 2002).

DHSA   Defence Helicopter Support Authority (UK).

DHUD   Diffraction-optics HUD.

DHV   *1* Deutscher Hubschrauber Verband, helicopter association (G).

2 Deutscher Hängergleiterverband, hang gliding and sport parachuting (G).

DI   *1* Daily inspection.

2 Direction indicator.

3 Director of Intelligence.

4 Duty instructor.

5 Direct-injection.

6 Data interrupt.

D_i Induced drag.

DIA   *1* Documentation Internationale des Accidents (DocIntAcc).

2 Defense Intelligence Agency (US).

3 Document interchange architecture (IBM).

4 Data-interaction architecture; DEM adds demonstrator.

5 Digital interface adaptor card.

diabatic process   Process in thermodynamic system with transfer of heat across boundaries.

diabolo   Landing gear with two wheels side-by-side on centreline of aircraft [esp. MLG].

DIAC   Data-Interpretation Analysis Center (US).

diagnostic routeing equipment   Automatic or semi-automatic fault-isolating tester with ability progressively to narrow down location of fault.

diagonal-flow compressor   One in which air flows diagonally to plane of rotation, centrifugal with axial component.

Dial   Differential-absorption lidar.

dial-a-STOL   Notional method of operating CTOLs from bomb-damaged runways in which weapon/fuel load is selected according to length of undamaged runway available.

Dialmet   Automated Metar and TAF service.

Dials   Digital integrated automatic landing system.

dial your weight   Small computer on whose keyboard is manually inserted all fuel, crew, payload and other on-board items, displaying MTOW and c.g. position (colloq.).

diamagnetic   Reacting negatively to magnetic field, developing magnetic moment opposed to it, with permeability less than 1; includes aluminium, non-ferrous alloys and corrosion and heat-resistant steels.

diameter   *1* That of any circular arcs making up fuselage external cross-section.

2 In optics, unit of linear measurement of magnifying power.

3 Of parachute canopy, that while fully spread out on flat surface.

diametral pitch   Ratio of number of teeth on gearwheel divided by pitch diameter.

Diamond C   Highest proficiency award for which sailplane pilots can qualify.

diamond landing gear   Tandem centreline mainwheels, and outriggers.

diamonds   See *shock diamonds.*

diamond-wing aircraft   Has swept-back front wing merged at tips into forward-swept rear wing; also called *twin-wing.*

Diane   *1* Digital integrated attack navigation equipment.

2 Détection identification analyse des nouveaux emetteurs [helo threat warning] (F).

DIAP   Defense Information Assurance Program (DoD).

diaphragm   Fabric partition within aerostat; may be gastight (ballonet *) or non-gastight (stabilizer *).

diathermy   Generation of heat by HF power, usually at 0.5/1.5 MHz.

Diatms   DISN interim asynchronous transfer services.

DIB   De-icer boot.

dibber   Weapon intended to penetrate concrete runway before exploding.

DIC   Defence Industries Council (UK member of EDIG).

Dicarps   Digital cassette recorder for passive sonar.

Dicass   Directional command-activated sonobuoy system.

dice   *1* Semiconductor chips or IC after scribing and separation.

2 To fly, esp. in exciting manner or on operations

(collog. RAF, WW2). Hence, **dicing**: 'op' is on, not scrubbed.

dichroic mirror   One coated with molecular-thickness layer of reflector, usually metal, so as to transmit some EM wavelengths (esp. visible colours) and reflect others.

DICU   Display interface control unit.

DID   *1* Data-item description [S adds sheet(s)].

   2 Digital-image design.

   3 Data-insertion device.

diddler   CRT auxiliary electrostatic plates which can collapse elongated blips to sharp spots.

DIE   Defense [defence] information environment.

die   *1* Press tool, often in mating male/female halves, which cuts sheet or imparts three-dimensional shape to workpiece.

   2 Shaped tools used in *-casting.

   3 Shaped tools used in *-forging.

   4 Shaped female mould used in explosive or magnetic forming.

   5 Shaped male tool used in ultrasonic, ECM and related mechanical, chemical or electrochemical shaping.

   6 Tool with shaped aperture used in extrusion.

dielectric   Substance capable of supporting electric stress, sustaining electric field and undergoing electric polarization; includes all insulators and vacuum.

dielectric constant   Ratio of capacitance of material to same condenser using air or vacuum, or of ratio of flux densities in the two media. Also called **permittivity**. Symbol ϵ, but Δ, χ and other symbols can be found.

dielectric heating   Generated in dielectric subjected to HF field, resulting from molecular friction due to successive reversals of polarization; power dissipated is **dielectric loss**.

dielectric strength   Measure of resistance of dielectric to electrical breakdown under intense electric field; SI unit is Vm^{-1}; also known as breakdown potential.

dielectric tape camera   TV recording camera (Vidicon) giving output on tape in form of varying electric field.

DIELI   Direction des Industries Electroniques et de l'Informatique (F).

DIEPS   Digital imagery exploitation and production system.

diergolic   Non-hypergolic, thus requiring an igniter system.

diesel   IC engine utilising heat of compression to ignite fuel oil injected in highly atomised state direct into cylinder, with piston nearly at TDC.

dieseling   Any spontaneous ignition of combustible gaseous mixture due solely to temperature caused by compression.

diesel ramjet   Ramjet operating at Mach number high enough for fuel to ignite by heat of air compression.

DIF, dif   Diffuse weather.

Difar   Directional acoustic frequency analysis and recording (ASW).

difference in depth of modulation   Modulation of stronger [usually ILS] signal minus that of weaker, both expressed as percentages, divided by 100.

differential ailerons   Ailerons interconnected so that upgoing aileron travels through larger angle than downgoing. This increases drag of wing with upgoing aileron and minimises extra drag of other wing.

differential ballistic wind   In bombing, hypothetical wind equal to difference in velocity between ballistic and actual winds at release altitude.

differential controls   Control surfaces on opposite sides of body or fuselage which move in opposition to cause or arrest roll.

differential fare   Difference in airline fare levels usually reflecting time-saving and passenger appeal of new aircraft.

differential GPS   Operates by placing a receiver at a point precisely referenced to a point on a runway. It then makes satellite measurements, from which error signals are transmitted to the airborne receiver which then corrects the signals received from the satellite, esp. for precision approach.

differential laser gyro   Two lasers of opposite polarization operate in same cavity; comparison of outputs gives twice angular-measure sensitivity of normal laser gyro.

diffrential pitch   Original term for *cyclic pitch*.

differential positioning   See *differential GPS*.

differential pressure   Pressure difference between two systems or volumes (abb. dP). That of fuselage or cabin is maximum design figure for pressurization system, beyond which point spill valves open.

differential spoilers   Wing spoilers used as primary or secondary roll control.

differential tailplane   See *taileron*.

differential tracker   Radar that can simultaneously measure angular separation of target and friendly missile, so that guidance system can reduce this value to zero.

differentiating circuit   Circuit delivering output voltage in approximate proportion to rate of change of input voltage or current.

diffraction   Phenomena which occur when EM wave train, such as beam of light, is interrupted by opaque object(s). Rays passing through narrow slit, or a grating made of slits, are bent slightly as they pass edges; thus waves can 'bend' around obstacle.

diffraction grating   Several forms of grating with lines so close that they diffract EM wavelengths.

diffraction-optics HUD   Uses a precise 3-D array of microminiature grids, or light apertures, to create a volume hologram which makes possible a wide-angle HUD suitable for all-weather low-level navigation and weapon-aiming.

diffuser   Expanding profiled duct or chamber, sometimes with internal guide vanes, that decreases subsonic velocity of fluid, such as air, and increases its pressure, downstream of compressor or supercharger, upstream of afterburner, and in some wind tunnels. In contrast, supersonic flow through a * is reduced in pressure and increased in velocity, hence con/di nozzle.

diffuser area ratio   *1* Ratio of outlet to inlet cross-section area of diffuser, esp. of ramjet.

   2 Ratio of area of jet-lift mixed-flow nozzle divided by that of primary jet.

diffuser, compressor   Ring of fixed vanes or expansion passages in compressor delivery of gas turbine to assist in converting velocity of air into pressure.

diffuser efficiency   Ratio of total energy at exit to entry or achieved/theoretical pressure rise.

diffuser tunnel   Wind tunnel containing section in which velocity is converted into pressure.

diffuser vanes   Guide vanes inside diffuser that assist in converting velocity into pressure.

diffusion   *1* In atmosphere or gaseous system, exchange of molecules across border between two or more concentrations so that adjacent layers tend towards uniformity of composition.

2 Of stress, variation along length of structure of transverse distribution of stress due to axial loads.

3 In materials, movement of atoms of one material into crystal lattice of adjoining material.

4 In ion engines, migration of neutral atoms through porous structure prior to ionisation at emitting surface.

5 Of light, scattering of rays, either when reflected from rough surface or during transmission through translucent medium.

6 In electronic circuitry, method of making p-n junction in which n- or p-type semiconductor is placed in gaseous atmosphere containing donor or acceptor impurity.

7 Of uranium, repeated gaseous-phase concentration of fissile U-235.

diffusion bonding   Use of diffusion (3) to join solids with high surface finish in uncontaminated intimate contact.

diffusion coefficient   Absolute value of ratio of molecular flux per unit area to concentration gradient of gas diffusing through gaseous or porous medium, evaluated perpendicular to gradient.

DIFM   Digital instantaneous frequency-measurement; R adds receiver.

DIFOT   Duty involving flight operations and training.

DIG   *1* Display/indicator group.

2 Directional gyro (usually DG).

3 Digital image generator (or generation).

Digest   Digital international geographic[al] exchange standard.

digibus   Any digital multiplex data highway.

digicon   Diode array giving a light input to electrical signals which are then amplified and analysed.

Digilin   Digital plus linear functions on one chip (trade name).

Digitac   Digital tactical aircraft control (USAF).

digital   Operating on discrete numbers, bits (0s and 1s) or other individual parcels of data.

digital/analog converter   Device which converts analog inputs (eg. varying voltages) into digits. Also known as **digitiser**.

digital display   Usually means numbers instead of needle/dial.

digital phase coding   Basic radar technique for hi-PRF and LPI.

DIGS   Digital generation simulator.

DIH   Department of Information Handling (ESOC).

dihedral   Acute angle between left (port) and right (starboard) mainplanes or tailplanes (measured along axis of centroids) and lateral axis.

dihedral effect   *1* Roll due to sideslip.

2 Extra dihedral due to flexure of wing [esp. of sailplane] under load.

3 Sideslip effect in variable-sweep aircraft that causes change in rolling moment; too much augments roll response while too little (adverse sideslip) opposes it.

DII   *1* Defense information infrastructure [COE adds common operating environment, IC adds integration contract] (UK MoD, US DoD).

2 DCMS interphone interface.

DIL   *1* Digital integrated logic.

2 Dual in-line.

Dilag   Differential laser gyro.

DILS   Doppler ILS.

dilution holes   Precisely arranged air holes in combustion-chamber liner or flame tube.

dilution of precision   Caused by the often very small angle at which a GPS customer sees a satellite; GDOP adds geometric *.

DIM   Dispense interface microprocessor.

dim   RAF slang, 1, stupid; 2 to disagree, as 'to take a * view of'.

Dime   *1* Dynamic IR missile evaluation.

2 Distributed integrated modular electronics.

dimensional similarity   Of physical quantities, made up of same selection of fundamental M (mass), L (length) and T (time) raised to same indices.

Dimes   Descent image-motion estimation subsystem (planetary landers).

diminishing manufacturing source[s]   Redesign of obsolete parts [esp. avionics] to ensure continued procurement.

DIMM   Dual-part integrated memory monitor.

Dimpi *Dempi.*

dimpled tyre   Landing-wheel tyre whose contact surface is covered with small recesses, mainly to provide visual index of wear.

dimpling   Countersinking thin sheet metal by tool which dimples (recesses) without cutting, so that rivethead is flush with surface.

DIMSS   Dynamic interface modeling and simulation system.

DIN   *1* Deutsches Institut für Normung eV (G equivalent of BSI, NBS).

2 Digital inertial navigation.

DINA   Direct noise amplification.

DINAS, Dinas   Digital inertial nav/attack system.

DINFIA   Direction Nacional de Fabricaciones e Investigaciones Aeronauticas (Arg.).

dinghy   Small boat, usually of inflatable rubberised fabric, for use by crew and passengers after aircraft has ditched. Correct term is liferaft.

dinghy drill   Procedure for unpacking, inflating and entering dinghy.

dining-in night   Formal dinner, usually once per month, attended by all members of mess and invited guests (RAF).

dinking   Use of thin blade-like shaped die(s) to cut soft sheet materials such as leather, cloth, rubber or felt, and to cut lightening holes in thin sheet-metal; inexpensive die is used, and cutting action is by steady pressure or hand hammer.

DINS   Digital inertial navigation system.

diode   Two-electrode thermionic valve containing cathode and anode, or semiconductor device having unidirectional conductivity.

diode lamp   Semiconductor diode which, when subject to applied voltage, emits visible light. Smaller than most switchable light sources. Also known as light-emitting diode (LED).

DIOT & E   Dedicated initial operational test and evaluation, requires four primary aircraft plus backup, all close to production configuration.

DIP   *1* Digital image processing.

2 Defense industrial plant (US); EC adds 'equipment center'.

3 Defense industrial participation.

4 Debtor in possession.

5 Dual inline package [ICs].

6 Data-interrupt program.

7 Diplexer.

dip   *1* Angle between magnetic compass needle perfectly poised or on horizontal axis and local horizontal plane. Also known as magnetic inclination.

2 Vertical angle at eye of observer between astronomical horizon and apparent line of sight to visible horizon.

3 Angle between local horizontal and lines of force of terrestrial magnetic field (indicated by [1]).

4 Salutation by briefly rolling aircraft towards observer, to * wing in salute.

DIPA   Defence Industrial Program Authorization.

diplexer   Device permitting antenna (aerial) system to be used simultaneously or separately by two transmitters.

DIP/LNA   Diplexer and low-noise amplifier.

diplomatic authorization   Authority for over-flight or landing obtained at government level.

diplomatic cleatance   The number and callsign allocated to a military aircraft to permit it to overfly foreign territory.

dipole   *1* System composed of two separated and equal electric or magnetic charges of opposite sign.

2 Antenna (aerial) composed of two conductors in line, fed at mid-point. Total length equal to one half wavelength.

DIPP   Defense/industry partnership program (Canada).

dipper   See *fuel dipper*.

dipping sonobuoy   One designed to be suspended but not released from helicopter and immersed in selected places in sea.

DIPR   Directorate of Intellectual Property Rights (UK).

DIPS   Dipole inches per second (chaff dispenser).

dipstick   Graduated quasi-vertical gauge of fluid level in container, usually disconnected for reading.

dipsydoodle   Official term for rollercoaster manoeuvre performed by SR-71 and some other supercruise aircraft following inflight refuelling, comprising dive to supersonic speed followed by accelerating climb back to operating height.

DIQAP   Defence Industries Quality-Assurance Panel (UK).

DIR   *1* Diagnostic imaging radar.

2 Distributed IR.

3 Direct, direction, director.

4 Digital instant recall.

5 Dwell illumination region.

Dircen   Direction des Centres d'Experimentations Nucléaires (F).

Dircm   Directional, or directed, IR countermeasures (said as a word, USAF).

direct-action fuze   See *DA fuze*.

direct approach   Unflared landing.

direct broadcast   Satellite powerful enough to transmit TV direct to terrestrial recipient or subscriber.

direct coupling   Association of two circuits by having an inductor, condenser or resistor common to both.

direct-cranking starter   Hand crank or starter geared to crankshaft to start engine.

direct current   Electric current constant in direction and magnitude.

direct damage assessment   Examination of actual strike area by air or ground observation or air photography.

directed-energy weapon   One whose effect is produced by a high-power beam, normally of EM radiation, having essentially instantaneous effect at a distance. Most important are lasers and HPM.

directed-flow reverser   Reverser whose discharge in the reverse mode is confined within limited angular limits to avoid the airframe or FOD/reingestion problems.

directed mode   DME mode allowing FMCS to select one to five DMEs for interrogation.

directed slipstream   Means of achieving STOL in which slipstream created by propellers or fans is blown over entire wing. Also known as deflected slipstream.

direct flight   *1* Portions of flight not flown on radials or courses of established airways.

2 Point-to-point space flight, without rendezvous, docking or other manoeuvre.

direct force control   Control of aeroplane trajectory by application of force normal to flightpath without prior need to rotate to different attitude; eg lateral force by combined rudder and chin fin, vertical by tailerons/flaps/spoilers or vectored thrust.

direct frontal   Air-combat tactic for double attack in which one interceptor closes head-on on each side of enemy force.

direct injection   Precise metered doses of fuel sprayed directly into cylinder combustion space, not into eye of supercharger.

direction   See *azimuth*, *bearing*, *course*, *heading*, *track*.

directional aerial, antenna   Aerial which radiates or receives more efficiently in one direction than in others.

directional beacon   Transmitter emitting coded signals automatically to enable aircraft to determine their bearing from the beacon with a communications receiver.

directional gyro   Free-gyro instrument for indicating azimuth direction.

directional instability   Tendency to depart from straight flight by a combination of sideslipping and yawing.

directional marker   Ground marker indicating true north and direction and names of nearest towns.

directional solidification   Casting metal alloys in such a way that all transverse grain boundaries are eliminated, leaving long columnar crystals aligned with direction of principal stress.

directional stability   Tendency of an aircraft to return at once to its original direction of flight from a yawing or sideslipping condition; also known as weathercock stability.

direction-finder   Automatic or manually operated airborne receiver designed to indicate bearing of continuous-wave ground radio beacon (see *ADF*).

direction indicator   See *directional gyro*.

direct lift control   Use of aerodynamic surfaces, esp. symmetric spoilers, to provide instantaneous control of rate of descent without need to rotate aircraft in pitch.

direct operating cost   Costs of operating transport aircraft, usually expressed in pence or cents per seat-mile, per US ton-mile or per mile, and including crew costs, fuel and oil, insurance, maintenance and depreciation. Excluding indirect expenses, such as station costs or advertising; usually taken as 100 per cent of direct costs.

director   *1* Aircraft equipped to control RPV or missile.

2 In air traffic control, a radar controller.

3 Fire-control tower in warships.

direct shadow photo   Simplest and oldest shadow photography: bright point source of light (in former days, spark) throws shadow of body and shockwaves on to photographic plate.

direct side-force control   DFC (2) flight-control system in which aircraft (heavier than air) can be translated sideways without yaw or change of heading by application of direct lateral force.

direct transit   Special rules under which aircraft may pause [eg, to refuel] in a Contracting state.

direct-view storage tube   CRT storage tube needing no visor in bright sunshine.

direct-vision panel   Flight-deck window or part of window that can be opened.

direct voice input   Control of function [eg. panel display, weapon selection, radio channel] by spoken command.

dirigible   Capable of being guided or steered; thus an airship but not a balloon.

Dir/Intc   Direct intercept.

DIRP   *1* Defense industrial reserve plant (US).

2 Defense Industrial Research Program (Canada).

DIRS   *1* Damage information reporting system.

2 Distributed IR system.

dirty   *1* Aircraft configuration in which aerodynamic cleanness is spoilt by extension of drag-producing parts, eg landing gear, flaps, spoilers, airbrakes.

2 NW whose detonation releases large quantity of toxic radiological material or emissions.

dirty bird   Stealth aircraft coated [especially freshly] with ferrite paint.

DIS   *1* Distributed-intelligence system (MMI).

2 Defense Investigative Service (US DoD).

3 Defence Intelligence Service (UK MoD).

4 Distributed interactive simulation.

5 Data-intensive system[s].

6 Drawing-introduction schedule.

DISA   Defense Information Systems Agency.

Disc   Disconnect.

DISC   Defence Intelligence and Security Centre (Chicksands, UK).

disc   *1* Ring on which one stage of compressor blades is carried.

2 Hub carrying blades of fan or turbine.

3 Circular area swept by lifting rotor.

disc area   Of propeller or helicopter rotor, area of circle described by tips of blades.

Disch   Discharge.

dischargeable weight   *1* All masses which may be jettisoned overboard in emergency.

2 Of airship, total weight that can be consumed or jettisoned and still leave ship in safe condition with specified reserves of fuel, oil, ballast and provisions.

discharge correction factor   Of rocket nozzle, ratio of mass flow to that of ideal nozzle which expands identical working fluid from the same initial conditions to same exit pressure.

discharge valve   Manually operated and opened sparingly to release hot air from balloon envelope. Generally = dump valve (3).

discing   Operation of propeller in ground fine pitch to cause aerodynamic drag.

disc loading   Helicopter weight divided by main-rotor disc area.

Disco   Directional composite whose resin-impregnated fibres can slip past each other, giving highly deformable product which retains directional strength properties.

Discon   Defence integrated secure communications network (Australia).

disconnect   Inadvertent or deliberate severance of flow during boom-type air refuelling.

discontinuity   *1* Sudden break in the continuity of mathematical variable.

2 In meteorology, zone within which there is rapid change, as between two air masses.

discontinuous fibre   Chopped roving as distinct from yarn or tow.

discount carrier   Despite next entry, one that [often in partnership with another] legally offers permanent low-cost travel, principally for tourists.

discounting   Illegal selling of airline tickets, for affinity group and other promotional fares, at below agreed tariffs.

Discr   Discrepancy.

discrete code   Any of the 4096 xpdr codes available to ATCRBS except those ending with a zero.

discretion   Flight time outside normal crew duty limits but legally permitted with concurrence of captain or PIC.

discrimination   *1* Of radar, minimum angular separation at which two targets can be seen separately.

2 Precision with which satellite antenna can focus in particular direction.

discriminator   Stage of FM receiver which converts frequency deviations of input voltage into amplitude variations.

discus   Of variable-geometry wing, part-circular portion of upper surface of fixed glove on which swinging portion can slide.

disembark   To step down from COD aircraft.

dish   Reflector for centimetric radar waves whose surface forms part of paraboloid or sphere.

dishing   *1* Pressing regular depressions in thin sheet to increase stability and resistance to bends.

2 In formation aerobatics, unwanted distortion of planar formation into dish shape [e.g. in formation roll].

disk   Disc, except for compact *.

dismounted flight training   Hands-off training on ground using hand-held model aircraft, particularly for air-combat tactics.

DISN   Defense Information System Network (DoD).

Disney bomb   Armour-piercing free-fall bomb weighing 4,500 lb (2041 kg) finally accelerated to 2,400 ft/s (1089 ms^{-1}) by rocket (UK WW2).

DISOSS, Disoss   Distributed office support systems (SNA).

DISP   Displaced.

dispatchable   Cleared to fly despite deficiencies [e.g. engines in N_1 instead of EPR mode].

dispatcher   See *despatcher*.

dispensation   Agreement to waive a rule without affecting safety.

dispenser   *1* Container from which objects [e.g., ECM chaff cartridges, flares and active emitters] can be ejected in predetermined sequence.

2 Externally carried container for bomblets or other small multiple munitions.

dispensing *1* Release of ECM payloads in controlled manner.

2 Supply of fuel to aircraft via hydrant.

dispensing sequence Graphical or tabular plan for ECM to meet expected threats.

dispersal *1* Geographical spreading out of aircraft, material, establishments or other activities to reduce vulnerability to enemy action.

2 Dispersal area.

3 Parking area, usually paved, accessible from perimeter track, on which one aircraft could be parked. Some WW2 airfields had over 100.

dispersal area Area usually on remote parts of airfield to which aircraft and support equipment can be dispersed in wartime.

dispersant oil Lubricating oil with additives which slow or even prevent formation of sludge and other solid deposits.

dispersion *1* Average distance from aim point of bombs dropped under identical conditions or by projectiles fired from same weapon or group of weapons with same firing data.

2 In AAA, scattering of shots about target.

3 In chemical operations, dissemination of agents in liquid or aerosol form from bombs and spray tanks.

4 In rocketry and AAM testing, deviation from prescribed flight path; circular dispersion.

5 Measure of scatter of data points around mean value or around regression curve, usually expressed as standard-deviation estimate or standard error.

6 Process in which EM radiation is separated into its components.

7 Measure of resolving power of spectroscope or spectrograph, usually expressed in A/mm.

8 Tendency over long period of commercial traffic to move from primary to secondary airports.

9 Scatter of actual touchdown points on runway over a period.

dispersion error Distance from aim point to mean point of impacts.

dispersion hardening Scattering of fine particles of different phase within metallic material, resulting in overall strengthening.

dispersion pattern Distribution of series of rounds fired from one weapon or group of weapons on fixed aim under conditions as nearly identical as possible.

dispersion warhead Discharging bomblets, FAE or other multiple or dispersed payloads.

displaced threshold Threshold not at downwind end of full-strength runway pavement. It is usually beyond it, and is available for takeoff or for end of landing roll, but not for touchdown.

displacement *1* In air interception, separation between target and interceptor tracks to provide interceptor acquisition space.

2 Distance from standard point (usually origin) measured in given direction.

3 Of IC engine, total volume swept by pistons during crankshaft rotation from BDC to TDC. Also known as swept volume.

4 Of airship or balloon, mass of air displaced by gas, expressed as weight or volume.

5 Lateral, vertical or angular * of any point of zero DDM from localizer or glidepath.

displacement thickness Dimension characteristic of all boundary layers and equal to thickness of completely stagnant fluid having same overall effect. Equal to distance through which each streamline is displaced from position it would have assumed had fluid been inviscid.

$$\delta* = \int_0^\infty \left(1 - \frac{u}{V}\right) dy \triangleq \sqrt{\frac{3vl}{V}}$$

where u is local boundary layer velocity, V free-stream velocity, y distance from solid surface, v kinematic viscosity and 1 characteristic length; actual boundary-layer thickness is nearly three times $\delta*$.

display Graphic presentation of data for human study.

Display Authorisation Document required from national aviation authority before pilot can take part in airshow.

display datum Also called display centre, the mid-point of the crowd-line.

disposable lift Gross lift less fixed weight of an aerostat.

disposable load Maximum ramp weight minus OEW.

DISR *1* Descent imager spectral radiometer.

2 Department of Industry, Science and Resources (Australia).

disreef system Timing system for automatically releasing reef of parachute.

disrupter-type spoiler Maximises local turbulence.

dissimilar air-combat training Mock air combat with friendly fighters of different type(s) acting part of enemy aircraft, chosen for performance similar to that of enemy types and usually painted to resemble them.

dissipation trail Rift in clouds caused by passage of [jet] aircraft. Abb. **distrail**.

Dist Distance or district (ICAO).

distance Standard airline unit is nm (contrary to SI); up to 1,200 nm airline * calculated as D (great-circle distance) + (7 + 0.015D); above 1,200 nm measure is D + 0.02D.

distance bar Rigid bar linking tow vehicle with aircraft.

distance marker *1* Numbers painted on runway side to indicate thousands of feet to upwind end.

2 Reference marker on radar display; usually one of series of concentric circles. Also known as range marker.

distance-measuring equipment Airborne secondary radar sending out paired pulses (interrogation) received at ground transponder; time for round trip is translated into distance. DME offers 252 frequencies from 962 to 1,213 MHz at 1 MHz spacing, providing 126 channels each comprising two frequencies 63MHz apart.

Distant Marshal At gliding championship, official charged with arranging tugs and gliders in correct start sequence.

distillate *1* Any petroleum product.

2 Fuel oil, eg for diesels.

distortion *1* Undesired change in shape.

2 Undesired change in waveform.

3 In radio or sound reproduction, failure exactly to transmit or reproduce received waveform.

4 Variation of flow velocity or temperature across transverse plane through gas turbine.

distraction ECM mode in which hostile missile locks-on to decoy before it can see real target.

Distress & Diversion ATC cells [or in UK RAF units] which maintain 24-h monitor on VHF/HF emergency frequencies to offer assistance.

distress frequency   Internationally 121.5 kHz.

distress signal   Signal transmitted by vehicle in imminent danger.

distributed-aperture system   EO sensors providing a protective sphere around aircraft for missile warning, navigation support and night operations.

distributed data-processing   Distribution of EDP (1) capability among a number of positions in a geographically large system.

distributed jet system   Any arrangement in which a power source is arranged to augment lift along the length of an aerofoil, examples being the jet flap, augmentor wing, EBF, IBF, CCW and USB.

distributed load   One which has no single point of application but is distributed over a line or area, such as air load on a surface.

distributed mass-balance   One distributed along span of control surface.

distributed mission training   Creating realistic battle-space for aircrew by using networked simulators.

distributor   *1* Rotary switch feeding HT in sequence to spark plugs.

2 Circumferential gallery connecting engine fuel manifold[s] to burner nozzles, probably incorporating a * valve, to compensated for gravity head and ensure all burners receive same supply.

disturbance   *1* Upset to normal flight involving uncommanded change in AOA (α), normally quantified as change in $C_L = \Delta\alpha$.

2 Situation involving unpremeditated loss of control, eg pitch-up or stall/spin.

3 Local departure from normal wind conditions; often used to mean cyclone or depression.

disturbance motion   Uncommanded movement of cockpit caused by turbulence, vibration or other input beyond pilot's capacity to counter.

disturbed-state concept   Advanced yet simplified modelling of the mechanics of materials and interfaces.

disturbing moment   Moment which tends to rotate aircraft about an axis.

Ditacs   Digital tactical system.

DITC   Department of Industry, Trade and Commerce (Canada).

ditching   Emergency alighting of aircraft, especially landplane, on water; thus verb to ditch.

ditching characteristic   Way in which aircraft behaves on being ditched, dynamically and structurally.

ditching device   Causes RPV to land or crash land when control has been lost.

ditching drill   Emergency procedures for aircraft crew and passengers, performed before and after ditching.

Ditco   Defense Information Technology Contracting Organisation.

dither   Signal applied to keep servo motor or valve constantly quivering and unable to stick in null position.

DITS   Data information transfer system [or set], centralised control of military aircraft communications.

dits   Digital information transport standard.

ditty bag   Container for AC(2)'s personal items and mission documents carried aboard combat aircraft.

DITU   De-icer timer unit.

DIU   Data interface unit.

diurnal   Adjective generally meaning daily, or in 24h cycles.

DIV   Divert, diverting.

DIVC   Digital imagery and video compression.

DIVADS   Division air-defense system (USA).

dive   Steep descent with or without power.

dive bomber   Aircraft designed to release bombs at end of steep dive towards objective.

dive brake   Extensible and retractable surface designed to enable aircraft to dive steeply at moderate airspeed.

dive-recovery flap   Simple plate flap hinged at leading edge on underside of wing at about 30% chord and opened to assist recovery from dive by changing pitching moment, removing local compressibility effects and increasing drag. Common c1942–50.

divergence   *1* Disturbance which increases without oscillation.

2 Expansion or spreading out of vector field; considered to include convergence, or negative divergence.

3 Aeroelastic instability which results when rate of change of aerodynamic forces or couples exceeds rate of change of elastic restoring forces or couples.

divergence, lateral   Divergence in roll, yaw or sideslip; tends to a spin or spiral descent with increasing rate of turn.

divergence, longitudinal   Non-periodic divergence in plane of symmetry; leads to nose dive or stall.

divergence Mach No   Value higher than M_{crit} beyond which there is rapid drag rise.

divergence speed   Lowest EAS at which aeroelastic divergence occurs.

divergent oscillation   One whose amplitude increases at accelerating rate.

diversion   *1* Change in prescribed route or destination made because of weather or other operational reasons.

2 Traffic diverted or claimed to be diverted from one airline by another, or to non-scheduled, charter or supplemental operators on same route. Frequently called material *.

diversity receiver   See *spaced diversity*.

divided landing gear   Traditional fixed main gear but with no axle or horizontal member linking wheels.

divided shielding   Nuclear radiation shield in two or more separated layers.

divider   Logic circuit which performs arithmetical division.

dividing streamline   That which eventually separates a flow into two parts, such as that which impacts the dividing line along the leading edge of a wing.

division, air   *1* Air combat organization normally consisting of two or more wings of similar type units (US).

2 Tactical unit of naval aircraft squadron, consisting of two or more sections.

DIWS   Digital-imagery workstation.

Dixie cup   Simple continuous-supply drop-down oxygen mask for passengers.

DJ   Detector-jammer.

DJE   Deception-jamming equipment.

DJTF   Deployable Joint Task Force (NRF).

DK   Docked.

DKATMS   Danish air-traffic management system.

DL   *1* Delay line.

2 Deck landing.

3 Downlink, or datalink.

DLA   *1* Delay message.

2 Defense Logistics Agency (US).

3 Dedicated lease agreement.
DLAD　Delayed.
DLAIND　Delay indefinite (DLI more usual).
DLAND　Development of Landing Areas for National Defense (US, 583 airfields 1941–44)
DLAP　Downlink application processor.
DLB　Datalink buffer.
DLC　*1* Direct-lift control.
　2 Downlink communication (sonobuoy).
　3 Datalink control; DU adds display unit, I identifier.
　4 Diamond-like carbon.
DLCO　Deck-landing control officer.
DLCRJ　Detect, locate, classify, record and jam.
DLE　Datalink entry.
DLF　Design load factor.
DLFA　Deutsche Luftundraumfahrt ForschungsAnstalt (G).
DLGF　Data-load gateway function.
DLGS　Datalink ground station, for reconnaissance pods (RAF).
DLI　*1* Deck-launched intercept.
　2 Delay indefinite.
　3 Datalink interface, or interpreter.
DLIR　Downward-looking IR.
DLJ　Downlink jammer, or jamming.
DLK　Datalink (AEEC).
DLL　*1* Design limit load.
　2 Datalink library.
DLLR　Deutsche Liga für Luft- und Raumfahrt (G).
DLM　*1* Declarative language machine.
　2 Depot-level maintenance (US, NATO).
DLME　*1* Direct lift and manoeuvre enhancement.
　2 Datalink and messsage engineering.
DLMS　Digital land-mass simulation, common though superseded by MIL-STD protocols.
DL/MSU　Data-loader mass storage unit.
DLMU　Datalink management unit.
D/LNA　Diplexer/low-noise amplifier.
DLO　Defence Logistics Organization (UK); ES(A) adds Equipment Support (Air).
DLOC　Datalink operations centre (RAF).
DLODS　Duct leak and overheat detection system.
DLP　*1* Deck-landing practice.
　2 Datalink processor.
　3 Directional lethal package.
DLPP　Datalink pre-processor.
DLPU　Datalink processor unit.
DLQ　Deck-landing qualification.
DLS　*1* DME landing system.
　2 Datalink splitter.
　3 Data-load[er] system.
　4 Defect-location system.
DLST　Datalink surface terminal[s].
DLT　Digital linear tape.
DLTS　*1* Datalink and tracking system.
　2 Datalink test set.
DLU　*1* Data-logger unit.
　2 Download unit.
　3 Dual laser unit.
DLV　Deutsche Luftsport Verband (1919–45).
DLW　Design landing weight.
DLY　Daily.
DM　*1* Data management.
　2 Disconnected mode.

3 Docking module.
dm　Decimetre.
DMA　*1* Defense Mapping Agency, now part of NIMA (US).
　2 Délégation Ministérielle pour l'Armement (F).
　3 Direct memory access, or addressing.
　4 Defence Manufacturers' Association of Great Britain.
　5 Dimethylamine.
　6 Descent-mode annunciator.
　7 Directorate of Military Aeronautics (UK 1914–17).
DMAAC　Defense Mapping Agency Aerospace Center (US).
DMAB　Defended modular-array basing.
DMB　Digital multi-broadcasting.
DMC　*1* Direct manufacturing cost[s].
　2 Dynamic metal compaction, by EM pulses.
　3 Disaster-monitoring constellation.
　4 Display[s] and mission computer.
DMD　*1* Deployment manning document[s] (USAF).
　2 Digital message device.
　3 Digital map display; G adds generator.
DMDR　Digital mission-data receiver.
DME　*1* Distance-measuring equipment; suffixes /N, /P, /T and /W signify normal, precision, Tacan or time, and wide-spectrum.
　2 Designated maintenance examiner.
DMEA　Defect-mode and effect[s] analysis.
DMED　Digital-message entry device.
DME distance　Slant range.
DME fix　Geographical position determined by reference to navaid which provides distance and azimuth information.
DMEP　Data-management and entry panel.
DME-P　DME, precision.
DMES　Deployable mobility execution system.
DME separation　Spacing of aircraft on airway in terms of distance determined by DME.
DMET　DME with respect to time.
DMF　*1* Digital matched filter.
　2 Département Militaire Fédéral (Switz.).
　3 Database, menu, function (software).
　4 Date of manufacture.
DMFV　Deutscher Modellflieger Verband eV (G).
DMG　*1* Deutsche Meteorologische Gesellschaft eV (G).
　2 Digital map generator.
DMI　*1* Department of Manufacturing Industry (Australia).
　2 Meteorological Institute [and service] (Denmark).
DMICS　Design methods for integrated control systems.
DMIF　Dynamic multi-user information fusion.
DMIR　Designated manufacturer [or manufacturing] inspection representative.
DML　Decision and modeling/modelling language.
DMLS　Doppler microwave landing system.
DMM　*1* Dama modem module.
　2 Data memory, or management, module.
　3 Digital multimeter.
DMMF　Developmental manufacturing and modification facility.
DMMH/FH　Direct maintenance man-hours per flight hour.
DMN　Data multiplexing network.
DMO　*1* Dependent meteorological office.

2 Development Manufacturing Organization (modifies aircraft as system development vehicles).

3 Defense Materiel Organization (US).

4 Defence Material Organization (Australia).

5 Defence Management Office (central procurement body, Australia).

DMOR   Digest of mandatory occurrence reports.

DMOS   *1* Double-diffused MOS.

2 Diffusive mixing of organic solutions (spaceflight).

DMP   *1* Display management panel.

2 Direct-manning personnel.

DMPI   Desired mean point of impact.

DMPP   Display and multi-purpose processor.

DMR   Delayed multipath replica.

DMS   *1* Defensive management system, or subsystem.

2 Defense Mapping School.

3 Data, or database; management system. [**DMSS**, subsystem].

4 Data multiplexer subunit.

5 Domestic military sales.

6 Display mode selector.

7 Diminishing manufacturing service [or support], obsolete spare parts.

8 Diminishing materiel shortage[s].

9 Debris monitoring sensor.

10 Dual-mode seeker.

11 Digital-map system.

12 Defense Message Service (US).

DMSH   Diminishing.

DMSK   Differential-minimum shift keying.

DMSO   Defense Modeling and Simulation Office.

DMSP   Defense Meteorological Satellite Program (DoD).

DMSS   *1* Deployable mission support system [can go overseas on detachment].

2 See *DMS*[3].

DMT   Distributed mission trainer [or training].

DMTI   Digital moving-target indicator, or indication.

DMU   *1* Distance-measurement unit.

2 Data-management unit.

3 Digital master unit.

DMV   See *DMFV*.

DMWH   Diret-maintenance working hour[s].

DMZ   Demilitarized zone.

DNA   *1* Defense Nuclear Agency (US).

2 Do not approach area.

3 Direction de la Navigation Aérienne (SGAC,F).

4 Does not apply.

DNAC   Direction Nationale de l'Aviation Civile (Mali).

DNAPS   Day/night adverse piloting system.

DNAW   Director[ate] of Naval Air Warfare (UK).

D/NAW   Day/night adverse weather.

DNC   Direct (or digital, or direction[al], or distributed) numerical control (NC machining).

DNCO   Duty not carried out.

DND   Department of National Defense (Canada).

DNG   Danger[ous].

DNI   Director of National Intelligence.

DNIA   Duty not involving alert.

DNIF   Duty not involving flying.

DNMI   Meteorological Institute [and service] (Norway).

DNS   *1* Direct numerical simulation.

2 Dense.

DNSARC   Department of the Navy System Acquisition Review Council (US).

DNSLP   Downslope.

DNTAC   Dirección Nacional de Transporte Aéreo Civile (Arg.).

DNV   Audit [certification] bureau (Neth.).

DNVT   Digital non-secure voice terminal.

DNW   Deutsch-Niederländischer Windkanal [wind tunnel] (G/Neth.).

DO   *1* Drawing office.

2 Drop-out (mask).

3 Director of Operations.

D$_o$ *1* Wing profile drag.

2 Maximum outside diameter of tyre [tire].

DOA   *1* Delegation option authorization; FAA document authorizing company to do its own aircraft type-certification.

2 Defence Operational Analysis Establishment (UK).

3 Direction of arrival (ECM).

4 Dominant obstacle allowance.

5 Department of Aviation (several countries, e.g. Thailand).

6 Design organization approval [-JA adds joint airworthiness].

DOB   Dispersed [or deployment] operating base.

DOC   *1* Direct operating cost[s].

2 Delayed-opening chaff.

3 Designed operational capability.

DoC   Department of Commerce (US 1926–38).

Doc   Document [ation].

DOCC   Deep-operations co-ordination cell.

DOCCT/S   Dama orderwire channel controller trainer/simulator.

dock   *1* Structure surrounding whole or portion of aircraft undergoing maintenance, to provide easy access for ground crew to reach all parts.

2 Large volume in factory, usually extending well below floor level, for installation of giant tools (jigs) and master tools.

docking   *1* Mechanical linking of two spacecraft or payloads.

2 Forward movement of airliner to nose-in stand at terminal.

3 Process of manoeuvring airship into its shed after landing.

documentation   *1* In EDP (1), formal standardized recording of detailed objectives, policies and procedures.

2 Any hard-copy media, such as aircraft servicing manual, illustrated parts catalogue, repair manual, servicing diagram manual, cross-servicing guide, aircrew or flight manual (usually three volumes), flight reference cards, equipment manuals, servicing cards, and possibly checklists. In aircraft restoration * is required to authenticate every part.

DoD   Department of Defense (US, from 1947).

Dodac   DoD ammunition code.

DODIIS   DoD intelligence information system.

DoE   *1* Department of the Environment, now DETR (UK).

2 Department of Energy (US).

DOETR   See *DETR*.

Dof, DOF   Degree[s] of freedom.

DofA   Department of Aviation (Australia).

dog   *1* Bad individual example of particular aircraft type.

2 Aircraft type all examples of which exhibit bad flying qualities (both meanings colloq.).

dogbone Bone-shaped tie, eg linking rigid-rotor blade to hub, or reacting engine thrust on testbed.

dogfight Air-to-air combat at close visual range.

doghouse *1* Fairing for instrumentation, esp. on rocket (colloq.).

2 Balloon landing which results in basket being over-turned.

dogleg *1* Track over several waypoints away from direct route.

2 Directional turn in space launch trajectory to improve orbit inclination.

dogship Repeatedly modified developmental prototype a/c [no reflection on handling qualities].

dogtooth Discontinuity at inboard end of leading-edge chordwise extension, generating strong vortex.

DOHC Double overhead camshaft.

DOI Descent (or docking) orbit insertion.

DoI Department of Industry (UK, now DTI).

DOL Dispersed operating location.

DOLE Detection of laser emitters.

doll's eye Cockpit magnetic indicator which when trig-gered clicks to a white warning aspect.

dolly *1* Airborne data-link equipment.

2 Metal back-up block used in hand riveting or hammering out dents in sheet.

3 Pneumatic-tyred truck tailored to elevate and grasp engine, skid-equipped helicopter, radar or other item, and transport it on ground.

4 Vehicle or trolley equipped with ballmats, rollers [can be powered] or other interfaces for ULDs. Other names: pallet transporter or trailer, container trailer or carrier or even cargo trailer.

5 Each truck in airport baggage train.

dolly roll Dolly with payload carried on two roller-supported rings for rotation to any desired angle to facilitate inspection and maintenance.

Dolram, DOLRM Detection of laser, radar and millimetre (millimetric) waves.

DOM Domestic, within US.

dome Flight simulator [esp. for combat a/c] with replica cockpit at centre of hemisphere on which images projected.

dome rivet Rivet with deep head, curved top and almost parallel sides.

domestic Involving one's own country only.

domestic brief Before combat or training mission, portion of briefing which allocates aircraft (and explains where they are parked) and call-signs.

domestic reserves Fuel reserves for scheduled domestic flight.

domestic service Airline service within one country.

domicile Country in which air carrier is registered.

Domsat Domestic (usually communications) satellite.

donk Aircraft engine[s] or power (colloq. noun or transative verb).

DO-160 'Environmental conditions and test procedures for airborne equipment' (RTCA).

DO-178 'Software considerations in airborne systems and equipment certification' (RTCA).

DO-178B Certification procedure for software [civil aviation].

door bundle Para-dropped load immediately preceding stick of parachutists.

door-hinge rotor Articulated blades on flapping hinges visually similar to door hinge.

Doors Dynamic-object oriented requirements system.

DOP *1* Dioctyl phthalate [air-filter measures].

2 Defence and overseas policty (UK).

3 Detailed operation[al] procedure.

4 Dilution of precision.

5 Digital on-board processor.

DoPAA Description of proposed actions and alternatives.

dope *1* Liquid applied to fabric to tauten it by shrinking, strengthen it and render it airtight by acting as filler. Usually compounded from nitrocellulose or cellulose acetate base, and soluble in thinners.

2 Ingredient added to fuel in small quantities to prevent premature detonation (colloq.).

doping *1* Treatment of fabric with dope.

2 Addition of impurities to semiconductor to achieve desired electronic characteristics.

3 To prime piston engine with spray of neat fuel prior to starting from cold.

doploc Doppler phase lock; active tracking system which determines satellite orbit by measuring Doppler shift in radio signals transmitted by satellite.

Doppler *Doppler effect.*

Doppler beam sharpening As aircraft radar aerial points anywhere other than dead-ahead, computer breaks each reading into small pieces and reassembles them as high-resolution map using Doppler correction to eliminate background clutter.

Doppler blade flash Transient bright spots on radar display caused by returns from rotating helicopter rotor.

Doppler correction Numerical correction to observed frequency or wavelength to eliminate effect of relative velocity of source and observer (eg removal of sea wave velocity from Doppler groundspeed).

Doppler effect Increase or decrease in frequency of wave motion, such as EM radiation or sound, sensed by observer or receiver having relative speed with respect to source. Thus, police-car siren seems to drop in pitch as it passes stationary observer at high speed. Approximate figures for X (I/J)-band: 34.3 Hz per kt, 30 per mph, 19 per km/h, 20 per ft/s. Also known as Doppler shift.

Doppler error In Doppler radar, error in measurement of target radial velocities due to atmospheric refraction.

Doppler groundspeed Groundspeed output from Doppler.

Doppler hover VTOL, esp. helicopter, hover controlled over desired geographical spot by Doppler coupler to AFCS.

Doppler navigation Dead reckoning by airborne navaid which gives continuous indication of position by inte-grating along-track and across-track velocities derived from measurement of Doppler effect of radar signals sent out (usually in four diagonal directions) and reflected from ground.

Doppler radar Radar which measures Doppler shift to distinguish between fixed and moving targets, or serve as airborne navaid by out-putting groundspeed and track.

Doppler ranging (Doran) CW trajectory-measuring system which uses Doppler effect to measure velocities between transmitter, vehicle transponder and several receiving stations; obviates necessity of continuous recording by making simultaneous measurements with four different frequencies.

Doppler shift   *1* See *Doppler effect*.

2 Magnitude of Doppler effect measured in Hz or (astronomical) in terms of visible-light spectrum.

Doppler spectrum   Output of Doppler radar with finite beam width.

Doppler velocity and position (Dovap)   CW trajectory-measuring using Doppler effect; ground transmitter interrogates a frequency-doubling transponder and output is received at three or more sites for comparison with interrogation frequency, intersection of ellipsoids formed by transmitter and each receiver providing spatial position.

DOR   *1* Directorate of Research (previously of Operational Research), under Chief Scientist (CAA, UK).

2 Dynamic observation report (ATOS).

DORA   Directorate of Research and Analysis (UK).

Dora   New technology for aerospace digital computers (R).

Doran   *Doppler ranging.*

Dorca   Directive on occurrence-reporting in civil aviation (CAA, UK).

dorsal   *1* Pertaining to the back, interpreted as upper surface of vehicle body.

2 Structural member running longitudinally along centreline at top of flying boat hull.

dorsal fin   Shallow vertical surface on upper centreline sloping gradually upwards to blend with main fin.

dorsal spine   Ridge running along top of fuselage from cockpit to fin for aerodynamic or system-access purposes.

dorsal turret   Powered gun turret on top of fuselage, normally able to cover upper hemisphere.

DOS   *1* Disk operating system.

2 Denial of service (cyber attack).

DoS   *1* Department of Supply (Australia).

2 Department of Space (India).

DOSAAF   Voluntary society for support of Army, Air Force and Navy (USSR).

DOSAV   Voluntary society for assisting Air Force (USSR, 1948–51).

DOSC   Direct oil-spray cooled.

dose rate   Incident rate of ionising radiation, measured in röntgens or mrem per hour.

dose rate contour line   Line joining all points at which dose rates at given times are equal.

dosimeter   *1* Instrument for measuring ultra-violet in solar and sky radiation.

2 Device worn by persons which indicates dose to which they have been exposed (each Apollo astronaut wore four passive * and carried a fifth personal-radiation * in sleeve pocket).

DoT   *1* Department of Transportation (US, 1967), Canada and UK.

2 Designating optical tracker.

3 Day of training.

dot   Electronic dot displayed on CRT for cursive writing, providing steering guidance or other information.

DOTE   Director, operational test and evaluation.

DoTI   Department of Trade and Industry (UK, DTI is preferred).

Dotram   Domain-tip random-access memory.

DOTS   Dynamic ocean[ic] track system [flexible routing responsive to PDWC].

double attack   Co-ordinated air/air operation by two partners making repeated synchronized yoyos (lo and hi-speed) and BRA(2)s as an effective ACM system.

double-base propellant   Solid rocket propellant using two unstable compounds, such as nitrocellulose and nitro-glycerine, which do not require a separate oxidiser.

double blank   Parachute with two gores removed.

double-bubble   Fuselage cross-section consisting of two intersecting arcs [almost complete circles] with floor forming common chord.

double-channel simplex   Two RF channels, one being disabled while the other is used to transmit.

double curvature   Curvature in more than one plane; also known as compound curvature.

double delta   Delta wing with sharply swept leading edge inboard changing at about mid-semi-span to less sharply swept outer section.

double designation   Nomination by a national government of two of that country's airlines as national flag-carriers operating scheduled service on same international route.

double drift   Method of determining wind velocity by observing drift on three true headings flown in specific pattern.

double engine   Power unit containing two engines driving co-axial propellers; usually one half can be shut down for cruising flight.

double-entry compressor   Centrifugal or radial-flow compressor that takes in fluid on both sides of impeller.

double farval   Aerobatic routine by section of four in diamond with 1 (box) and 4 (lead) inverted with respect to 2 and 3; thus at times it would be 2 and 3 that were inverted, 1 and 4 then being upright.

double-flow engine   Usually, bypass turbojet or turbofan.

double-fluxe   Bypass or turbofan engine (F).

double-headed   Warship having SAM systems both fore and aft of central superstructure.

double-hinged rudder   Rudder with additional hinge near mid-chord for maximum effectiveness at low speeds with large camber and total angle.

double horn   Dangerous ice accretion on LE forming two LEs, with channel between them.

double lift   Winchman and rescuee together.

double manned   Two crews per aircraft.

double modulation   Carrier wave of one frequency is first modulated by signal wave and then made to modulate second carrier wave of another frequency.

double notch   Flight-control system giving reduced surface movement for given input.

double propulsion   Use of two independent sources of thrust, one for lift and the other for propulsion, in an aerodyne. Historical examples include Rotodyne and Mirage III-V.

double-protection honeycomb   Honeycomb structure protected by chemical surface conversion process and then varnish dip.

doubler   Additional layer of sheet or strip to reinforce structural joint.

double-root blade   Rotating blade or vane with root fitting at each end to enable it to be fitted into disc either way.

double-row aircraft   Cargo airlifter able to carry unit loads (military vehicles, pallets, etc) side-by-side.

double-row engine   See *twin-row*.

double sideband   AM signal with carrier removed [still needs some bandwidth].

double slot   Passive-suction system in upper surface of wing of high-subsonic aircraft in which air is continuously extracted immediately downstream of shockwave (to stabilize and weaken it) and discharged immediately upstream of it. Improves C_L and buffet boundary.

double-slotted flap   Flap with vane and thus two slots, one between wing and vane, and the second between vane and flap.

double-T   Tail comprising a vertical surface above each of two booms joined at top by a horizontal.

doublet   *1* In fluid mechanics, source and sink of equal strength whose distance apart is zero.

2 Violent uncommanded stop-to-stop rudder movement.

doublet antenna   Antenna (aerial) composed of two similar elements in line but separated and fed in centre, and having total length equal to half wavelength; half a dipole.

double taper   Taper of aerofoil which incorporates change in angle at part-span.

double-wedge aerofoil   Section suitable for straight supersonic wings and blades of supersonic axial compressors, characterised by sharp wedge-like taper and sharp leading and trailing edges.

double wing   See Junkers.

doughnut   *1* Common shape for plasma contained in toroidal bottle.

2 Low-pressure tyre on small wheel for soft airfields.

3 Figurative representation of a thermal, in which central upflow can rise no further and turns over into surrounding downflow.

Douglas protractor   Transparent square covered with precise rectilinear grid and degrees around the edge.

Douglas scale   Table of numerical values of sea state [nine, from calm to confused] as one axis and swell [again nine] as the other.

DOV   *1* Data over voice.

2 Discrete operational vehicle (low-profile, does not appear to be military or para-military), usually non-flying.

Dovap   Doppler velocity and position.

Dover control   Patented linkage permitting all engines of multi-engine aircraft to vary power in unison or with any desired differences [obs.].

DOW   Dry operating weight.

Dowgard   Aviation ethylene glycol (Dow Chemicals).

down   Faulty and unusable.

downburst   Local but potentially dangerous high-velocity downward movement of air mass, eg when arrested by sea-breeze front; chief cause of windshear.

down-conversion   To lower EM frequency-band.

downcutting   Milling so that teeth enter upper surface of workpiece.

downdraft   Bulk downward movement of air such as commonly found on lee side of mountain or caused by descending body of cool air.

downdraft carburettor   One in which air is taken in at top and travels downwards; reduced fire hazard, and less risk of foreign-object ingestion.

downed aircraft   Aircraft that has made forced landing or ditching, esp. through battle damage.

down-45 line   Straight sustained dive at inclination of 45°.

downgrade   To reduce security classification of a document or item of classified material.

downlink   Radio transmission from air- or spacecraft to Earth.

downlink data   Transmissions to Earth from spacecraft giving such information as astronaut respiration or cabin temperature; computers alert flight controllers to any deviation (7).

download   *1* Any load acting downwards, eg on wing at negative angle of attack.

2 To remove unexpended ordnance or camera magazines from aircraft after operational sortie or aborted mission.

downlock   Locks landing gear in extended position.

down-look radar   Radar capable of detecting targets close to ground when seen from above.

downrange   Away from launch site in direction of target or impact area.

downselect   To select [Pentagonese].

downsize   *1* Of manufacturer, to reduce payroll.

2 Of air-carrier, to switch to smaller aircraft.

downspring   Long-travel coil spring imparting near-constant force on lightplane elevator control which at low airspeeds makes pilot pull stick back, effect fading as airspeed rises.

downstage   From upper stages of multi-stage vehicle downwards (signals, vibration, fluid flow, pressure, etc).

downstairs   At a lower altitude (UK colloq.).

downtilt   Downward tilt of thrust axis, normally of single-engine tractor propeller aircraft, to maintain adequate longitudinal stability, esp. in high-power climb.

downtime   Period during which hardware is inoperative following technical failure or for maintenance.

downward identification light   White light on underside of aircraft for identification, with manual keying for Morse transmission.

downwash   *1* Angle through which fluid stream is deflected down by aerofoil or other lifting body, measured in plane parallel to plane of symmetry close behind trailing edge; directly proportional to lift coefficient.

2 Angle through which fluid stream is deflected by rotor of rotary-wing aircraft, measured parallel to rotor disc.

3 Some authorities consider it not as an angular measure but as a rate of change of momentum, equal but opposite to lift.

4 Not least, often taken to mean linear velocity of flow through helicopter main rotor in hovering flight.

downwind   Direction away from source of wind; G/S = TAS + W/V.

downwind leg   Leg of circuit flown downwind, 180° from landing direction.

downwing side   Inner side of aeroplane in banked turn.

DP   *1* Direction des Poudres (F).

2 Data (or display) processor.

3 Differential protection.

4 Dual-purpose (in case of artillery, against air and surface targets).

5 Dew point.

6 Departure, or departure procedure.

7 Deep.

8 Deep penetration.

D$_p$ Parasite drag.

dP   Differential pressure.

DPA   *1* Defence Procurement Agency (UK).

2 Data Protection Act.

3 Diphenylamine.

4 Digital pressure altimeter.

5 Digital pre-assembly.

DPAC   Direction des Programmes Aéronautiques Civiles (F).

DPAO   Defence Public-Affairs Organization (Australia).

DPB   Dual-purpose bomblet, for soft- or hard-skinned targets.

DPBAC   Defence Press and Broadcasting Advisory Committee (UK).

DPBV   Disabled-passenger boarding vehicle.

DPC   *1* Defence Planning Committee (NATO).

2 Defence Production Committee (J).

3 Departure control.

4 Digital phase coding.

DPCM   Differential, or digital, pulse-code modulation.

DPD   Data-processing device.

DPDS   Distributed processing and display system (maritime sensors).

DPDT   Double-pole, double-throw.

DPDU   Data protocol data, or display, unit.

DPE   *1* Designated pilot examiner.

2 Department of Public Enterprise [runs Irish civil aviation].

3 Duration of present emergency, the notional period of service for members of the armed services conscripted or volunteered in WW2 (UK).

DPEE   Directorate of Proof and Experimental Establishments (UK, MoD formerly).

DPELS   Dual-pack Evolved Sea Sparrow launch system.

DPEM   Direct-purchased equipment maintenance.

DPEWS, D-pews   Design-to-price electronic-warfare suite.

DPF   Data-processing facility.

DPFG   Data-processing functional group.

DPG   *1* Data-processor, or processing, group.

2 Defense Planning Guidance, policy directive underlying the budget process (DoD).

DPH   Dew-point hygrometer.

DPI   Direct petrol-injection.

dpi   Dots per inch.

DPICM   Dual-purpose improved conventional munition.

DPKO   Department of PeaceKeeping Operations (UN).

DPL   *1* Disabled-passenger lift.

2 Dewpoint line.

DPLA   RPV or UAV (R).

DPLL   Digital phase-lock loop.

DPM   *1* Digital pulse-modulation.

2 Digital processing module.

3 Digital plotter map.

4 Development program manual.

DPMA   Data Processing Management Association (US).

DPMAA   Direction du Personnel Militaire de l'Armée de l'Air (F).

DPMC   Digital-plotter map computer.

DP/MC   Display processor and mission computer.

DPNG   Deepening.

DPOI   Desired point of impact, or interest.

DPP   *1* Deferred-payment plan.

2 Development and production phase.

DPPDB   Digital point-positioning database (NIMA DoD).

DPR   *1* Dual-port RAM(1).

2 Aeronautical Association (N. Korea).

DPRAM   See preceding.

DPRE   Designated parachute rigger/examiner.

DPS   *1* Differential phase-shift.

2 Dynamic pressure sensor (brake pedals).

3 Descent propulsion system (LM or planetary lander).

4 Deorbit propulsion stage.

5 Data-processing system.

DPSK   Digital, or differential, or differentially coherent, phase-shift keying.

DPT   *1* Durability proof test.

2 Depth.

DPTAC   Disabled Persons Transport Advisory Committee (DTLR).

DPTCA   Director, project technical cost-analysis.

DPU   *1* Display processor unit.

2 Digital, or data, processing unit.

DQ   Design qualification, demo of required reliability.

DQA   Directorate of Quality Assurance; TS adds Technical Support.

DQAB   Defence Quality Assurance Board; E adds Executive (UK).

DQ&R   Durability, quality and reliability.

DQAR   Digital quick-access recorder.

DQI   Digital-quality inertial.

DR   *1* Dead reckoning.

2 Dispense rate (mass/time) of chaff or aerosol (ECM).

3 Dispatch reliability.

4 Deck run.

5 Direct.

6 Data record[ing], or receptacle.

Dr   Drift.

d$_r$ *1* Increment of radius, e.g. helicopter rotor blade.

2 Radar resolution in range.

DRA   *1* Defence Research Agency (part of Qinetic/DSTL, previously DERA).

2 Direct radar access.

3 Dual-rail adapter.

4 Dual-row airdrop.

dracone   Large inflatable fluid (eg fuel, water) container, towable through sea and usable on soft land (strictly, Dracone).

Drads, DRADS   Degradation of radar defence systems (active jamming by MIRVs).

drag   Retarding force acting upon body in relative motion through fluid, parallel to direction of motion. Sum of all retarding forces acting on body, such as induced *, profile *. Basic equation is $D = C_D \frac{1}{2} \rho V^2 S$ where C_D is drag coefficient, ρ fluid density, V relative speed (e.g., TAS) and S total area, or total wing area.

drag area   Area of hypothetical surface having absolute drag coefficient of 1.0.

drag axis   Straight line through centre of gravity parallel to direction of relative fluid flow.

drag bracing   Internal bracing commonly used in fabric-covered wings to resist drag forces; may consist of adjustable wires, rod or tubes between front and rear spars or between compression ribs.

drag chute Parachute streamed from aircraft to reduce landing run, or steepen diving angle. Also called brake or braking parachute, deceleration parachute drogue parachute or parabrake.

drag-chute limit Maximum EAS at which drag chute may be deployed.

drag coefficient *1* Non-dimensional coefficient equal to total drag divided by $\frac{1}{2}\rho V^2 S$ where ρ is fluid density, V relative speed and S a representative area of the body, all units being compatible.

2 Coefficient representing drag on given body expressed in pounds on one square foot of area travelling at speed of one mile per hour (arch.).

drag creep The [undefinable] point at which drag starts to rise prematurely before M_{DD} is reached. The usual cause is formation of small shocks round a blunt leading edge, cured by root extension.

drag curve Plot of lift coefficient against drag coefficient, also known as drag polar.

drag-divergence Mach number Loosely, that at which shock formation becomes significant from viewpoints of drag, buffet and control. A common precise definition is M at which $dD/dM = 0.05$; NASA Langley chooses 0.1, while another definition is simply M 1.2. Also called drag-rise M, M_{DD}, and M_{div}.

drag hinge Approximately vertical hinge at root of helicopter main-rotor blade, allowing limited freedom to pivot to rear in plane of rotation.

drag index Usually means profile drag at 100 ft/s divided by total wetted area.

drag link Structural tie bracing a body, such as landing gear, against drag forces.

drag manoeuvre Air-combat manoeuvre in which one of a pair draws hostile aircraft into a firing position for his partner; can be used as verb.

drag member Structural component whose purpose is to react drag forces.

Dragon Deployable ram-air glider with on-board navigation (UAV).

drag parachute *1* See *drag chute*.

2 See *drogue (3)*.

drag polar Plot of C_D against $C_L{}^2$.

drag rib See *Compression rib*.

drag rise Sudden increase in wing drag on formation of shockwaves.

drag rope Thrown overboard from balloon to act as brake or variable ballast when landing; also called trail rope or guide rope.

drag rudder Wing-tip surface capable of imparting drag [e.g. by splitting into upper/lower halves] on tailless aircraft.

drag strut Strut reacting drag forces, esp. one incorporated in a wing.

drag-weight ratio Ratio of total drag at burnout to total weight of missile or rocket.

drag wires *1* Wires inside or outside wing(s) to react drag.

2 Wires led forward from car or other nacelle of airship to hull or envelope to react drag.

drain mast Pipe, usually telescopic through which liquid [fuel, water from cabin environmental system, grey water] can be extracted. No relevance to toilet servicing.

DRAM, Dram Dynamic random-access memory.

draping code Virtual processing tool [simulation] for composite structure in which dry preform[s] are wrapped around smooth tool.

Drapo Dessin et réalisation d'avions par ordinateur (F).

drawing introduction schedule Checks details for compliance with authorities.

DRB Defense Resources Board (US).

DRC *1* Defense Review Committee (NATO).

2 Data-recording cartridge.

DRD *1* Dry-runway distance.

2 Digital radar display.

DRDB Dual-redundant data bus.

DRDC Defense Research and Development Canada (Valcartier, Quebec).

DRDF VHF/UHF radio D/F (UK).

DRDL Defence R & D Laboratory (Hyderabad) or Laboratories (India).

DRDO Defense R & D Organization (US, India).

DRDPS Digital radar data-processing system.

DREA Defence Research Establishment Atlantic (Canada).

DRED Ducted-rocket engine development.

Dreem Drone radar electronic enhancement mechanism.

Drem lighting Visual landing approach aid with red, amber and green lights in accurately inclined tubes pointing up glidepath (from RAF Drem, Scotland, 1937).

DREO/P/S Defense Research Establishments: Ottawa/Pacific/Suffield (Canada).

DRER Designated Radio Engineering Representative (FAA).

dressed Equipped with all externally attached accessories, piping and control systems, esp. of an aircraft engine or accessory gearbox.

DRET Direction des Recherches, Etudes et Techniques (F).

DREV Defense Research Establishment, Valcartier (Canada).

DRF *1* Dual-role fighter.

2 Deutsche Rettungsflugwacht eV, air rescue service (G).

3 Data-recording facility.

DRFM Digital RF memory; S adds system[s], TG techniques generator.

DRFT Drift

DRG During.

DRIC Defence Research Information Centre (UK, Glasgow).

DRI/DRO Dolly roll-in to dolly roll-out; engine-change elapsed time.

drift *1* Lateral component of vehicle motion due to crosswind or to gyroscopic action of spinning projectile.

2 Slow unidirectional error movement of instrument pointer or other marker.

3 Slow unidirectional change in frequency of radio transmitter.

4 Angular deviation of spin axis of gyro away from fixed reference in space.

5 In semiconductors, movement of carriers in electric field.

6 Drag (until 1915).

7 Outward flow of boundary layer over swept wing, drawn towards tips by peak suction.

drift angle   Angle between heading (course) and track made good.

drift climb   Gentle climb after takeoff through noise-sensitive area at power just sufficient for ROC to be positive.

drift correction   Angular correction to track made good to obtain correct track (for navigation, bombing, survey etc).

drift-down   Gradual en route descent from top of flight profile (now rare).

drift error   Change in output of instrument over period of time, caused by random wander.

drift indicator   See *drift meter*.

drift meter   Instrument indicating drift angle; in simple optical form a hair line is rotated until objects on ground travel parallel with it.

drift sight   See *drift meter*.

drill   *1* Correct procedure to be followed meticulously, eg in particular phase of flight such as takeoff.

2 Training flight by a formation, including formation changes.

drill card   Checklist for correct procedures to be followed in particular phase of flight.

drill round   Dummy missile or gun ammunition used for training.

D-ring   *1* Ring in shape of capital D to which suspension ropes from balloon or other lighter-than-air craft are attached.

2 Handle for pulling parachute ripcord.

drink [the]   Open sea or ocean (RAF colloq., equivalent to *oggin*).

drip flap   Strip of fabric secured by one edge to envelope or outer cover of balloon or other lighter-than-air craft to deflect rain from surface below it and prevent it from dripping into basket or car; also helps to keep suspension ropes dry and non-conducting. Sometimes called drip band or drip strip.

drip loop   Inserted in wiring loom to allow for future extra length, and to direct condensation to drip in harmless place.

dripshield   Any tray for collecting fluid under machinery.

drip strip   See *drip flap*.

DRIR   Direct-readout infra-red.

DRIRU   Dry-rotor inertial reference unit.

DRISS   Digital read-in sub-system.

Drive   Documentation review into video entry.

driver, airframe   Pilot, especially in tanker/transport community (RAF colloq.).

drive surface   In NC machining or GPP, real or imaginary surface that defines direction of cutter travel.

driving band   Band of soft metal around projectile fired from rifled gun which deforms into barrel rifling to impart spin.

Drivmatic riveter   Patented power-driven riveter which closes aircraft rivets at high speed; can drill, countersink, insert rivet, close and mill head in sequence.

DRIW   Data redundancy in information warfare (USAF).

drizzle   In international weather code, precipitation from stratus or fog consisting of small water droplets.

DRK Flugdienst   Red Cross flying service (G).

DRL   Data-reduction laboratory.

DRLMS   Digital radar land-mass simulation (for TFR, GM, Nav-weaps, EW).

DRM   *1* Ducted rocket motor (USAF).

2 Digital recording module.

DRME   Direction des Recherches et Moyens d'Essais (F).

DRN   Document release notice.

DRO   *1* Daily routine orders.

2 Drone Recovery Officer.

drogue   *1* Conical funnel at end of in-flight refuelling hose used to draw hose out and stabilize it, and guide probe of receiver.

2 Fabric cone used as windsock, or towed behind aircraft as target for firing practice, or as sea anchor by seaplanes.

3 Conical parachute attached to aircraft, weapon or other body to slow it in flight, to extract larger parachute or cargo from aircraft hold, or for stabilizing the towing mass such as a re-entry body or ejection seat.

4 Part of connector on a spacecraft (eg Apollo lunar module) into which a docking probe fits.

drogue parachute   Drogue (3), tautological.

drogue recovery   Recovery system for spacecraft in which one or more small drogues are deployed to reduce aerodynamic heating and stabilize vehicle so that large recovery parachutes can be safely deployed.

drone   *1* Pre-programmed pilotless aircraft, usually employed as airborne target; either pilotless version of obsolete combat aircraft or smaller aircraft designed as a target. Totally different species from RPVs.

2 Loosely and unfortunately used as synonym for RPV or UAV.

droneway   Runway dedicated to UAV operations.

droop   *1* Downward curvature of leading edge of aerofoil to provide increased camber.

2 See *droop leading edge*.

3 Limited downward movement under gravity of door or access panel on underside, sufficient to have measurable effect on total aircraft drag.

droop balk   Mechanical interlock prohibiting (a) selection of all engines at takeoff power with droops (2) up, or (b) selection of droops up in flight at above a specified angle of attack.

drooping ailerons   Ailerons arranged to droop about 15° when flaps are lowered to increase lift while preserving lateral control.

droop leading edge   Wing leading edge hinged and rotated down to negative angle relative to wing for high-lift low-speed flight (esp. takeoff and landing); colloquially called droops.

droop nose   Nose designed to hinge down for low-speed flight and landing of slender delta aircraft and to provide crew forward vision at high angles of attack. Also called droop snoot.

droops   Droop leading-edge sections.

droop snoot   *1* Droop nose (colloq.).

2 Aircraft (esp. modification of familiar type) having extended down-sloping nose.

droop stop   Buffer incorporated in helicopter rotor hub to limit downward sag of blades at rest.

drop   *1* Dropping of airborne troops, equipment or supplies on specified * zone.

2 Correction used by airborne artillery observer or spotter to indicate desired decrease in range along spotting line.

191

drop altitude   *1* Altitude above MSL at which air drop is executed (DoD, NATO, CENTO).

2 Altitude of aircraft above ground at time of drop (see *drop height*).

drop forging   Forcing of metal or other materials in hot and plastic state to flow under pressure of blow(s) from drop hammer into mould or die to form parts of accurate shape.

drop height   Vertical distance between drop zone and aircraft (in SEATO, drop altitude).

drop interval   Time interval between drops (1).

drop line   Rope by which ground crew can walk balloon or other aerostat to new location.

dropmaster   *1* Person qualified to prepare, perform acceptance inspection, load, lash and eject items of cargo for air drop. Also called air despatcher.

2 Air crew member who, during drop, will relay required information between pilot and jumpmaster (USAF).

drop message   Written message dropped from aircraft to ground or surface unit (probably arch.).

drop model   Aerodynamically and dynamically correct model of fixed-wing a/c, originally balsa, dropped in still air or vertical tunnel.

drop-out   *1* Discrete variation in signal level during reproduction of recorded data which results in data-reduction error.

2 Of systems, automatic off-line disconnection following fault condition.

3 Disconnection of non-faulty autopilot in severe turbulence.

4 Oxygen mask which drops out automatically for passengers and crew following sudden loss of cabin pressure.

dropping angle   In level bombing, the angle between the line of sight to the target at the moment of release and the local vertical.

drops   Drop tanks [more than one, carried by one a/c].

dropsonde   Radiosonde dropped by parachute from high-flying aircraft to transmit atmospheric and weather measures at all heights as it descends. Used over water or other areas devoid of ground stations.

drop tank   Auxiliary fuel tank carried externally and designed to be jettisoned in flight when empty.

drop test   *1* Test of landing gear by dropping it from a height under various loads, with wheels at landing rpm.

2 Of models of future aircraft, free drop from balloon, helicopter or in spinning tunnel to check spinning or other characteristics.

3 Structural test of (usually unconventional) airframe by dropping on to hard surface.

4 Test of aerodynamics of spaceplane by drop from aircraft, eg NASA NB-52.

drop-test rig   Rig designed to drop-test (1) any structure.

drop zone   Specified area upon which airborne troops, equipment and supplies are dropped by parachute (abb. DZ).

DRP   Design review panel.

DR plot   Chart on which successive fixes, winds, positions and courses of an aircraft calculated by dead reckoning are shown.

DR position   Position of aircraft calculated by dead reckoning; symbol, dot in small square.

DRPS   Data receiving and processing station (Satcoms).

DRR   Data refresh rate.

DRS   *1* Detection and ranging set (radar, US).

2 Dead-reckoning subsystem.

3 Data-relay satellite.

4 Data-registration system (air cargo).

5 Design requirement specification.

DRS/CS   Digital range safety/command system.

DRSN   Drifting snow.

DRSS   Doppler radar scoring system.

DRTS   Detecting, ranging and tracking system (IR + laser).

DRU   *1* Data-retrieval unit.

2 Direct-reporting unit.

3 DGNSS reference unit.

drum   *1* Cylinder or series of discs upon which rotating blades of axial compressor are mounted; rotor minus its blades.

2 Portion of fuselage of broadly circular section, bounded by frames. Usually synonymous with plug or barrel section.

drum-and-disc construction   Axial compressor rotor in which some stages of blading are held in single-piece multi-stage drum and remainder in discrete discs.

drum test   Pressure test of fuselage section [or similar structure], normally in water tank.

DRV   Deutscher Reisebüro-Verband eV, travel-agents association (G).

DRVS   Doppler radar (or radial) velocity system.

DRVSM   Domestic RVSM, less than international.

DRX   Data reconstruction.

dry   *1* Without afterburner/augmentor in use.

2 Without charge for fuel (aircraft hire).

dry adiabatic lapse rate   Rate of decrease of atmospheric temperature with height, approximately 1°C per 100 m (1.8°F per 328 ft). This is close to rate at which ascending body of unsaturated air (clean air) cools due to adiabatic expansion.

dry bay   Structural compartment which, for a specific reason, is not used as an integral tank [it is normally surrounded by integral tankage].

dry-bulb thermometer   Ordinary thermometer to determine air temperatures, as distinct from wet-bulb type; has glass capillary of uniform bore and bulb partially filled with a fluid (usually mercury) and sealed with vacuum above.

dry contact   ARR hook-up between tanker and receiver without transfer of fuel.

Dryden   See *DFRC*.

dry cranking   Rotating main shaft[s] of gas turbine by external power, fuel system inoperative.

dry emplacement   Rocket or missile launch emplacement without provision for water cooling during launch.

dry film   Traditional film for optical camera, not digital smart card or video [originally to distinguish from wet film].

dry filter   One on which filtration is effected by a dry matrix, without oil.

dry fog   Haze due to dust or smoke.

dry fuel   See *solid propellant*.

dry gyro   Rotor has mechanical bearings and is not floated in fluid.

dry hub   Rotor hub with elastomeric bearings needing no lubrication.

dry hydrogen bomb   One without heavy water or other deuterium compounds.

dry ice   Solid CO_2, MPt -78.5C.

dry lease   Transport-aircraft lease by one airline to (usually) another, without flight or cabin crew or supporting services. Sometimes called barehull charter.

dry power   See *dry rating*.

dry rating   Of engine fitted with afterburner or augmentor, maximum thrust without these being ignited; usually same as MIL, maximum cold thrust.

dry run   Pre-firing operation and actuation of engine (esp. liquid-propellant rocket motor) control circuits and mechanical systems without causing propellants to flow or combustion to take place. This enables instrumentation and control circuits to be checked.

dry squeeze   Simulated operation of a system or device (from ** of gun trigger to operate combat camera only).

dry stores   Food and non-alcoholic beverages.

dry sump   Engine lubrication system in which oil does not remain in crankcase but is pumped out as fast as it collects and passed to outside tank or reservoir.

dry thrust   *Dry rating*.

dry weight   *1* Weight of engine exclusive of fuel, oil and liquid coolant, and including only those accessories essential to its running. [There are many more complex definitions].

2 Weight of liquid rocket vehicle without fuel, sometimes including payload.

Drzewiecki theory   Treatment of propeller blade as infinite number of chordwise elements; idealised, assumes each blade has no losses and meets undisturbed air.

DRZL   Drizzle.

DS   *1* Data sheet.

2 Directionally solidified alloy.

3 Documented sample.

4 Dynamic simulator.

5 Dust storm (ICAO).

6 Drone, anti-submarine (USN 1958–62).

7 Decision support.

8 Discarding sabot.

Ds   Maximum shoulder diameter (aircraft tyre/tire).

DS1   Digitization Stage 1 (UK MoD).

DSA   *1* Distributed system architecture.

2 Defense Supply Agency (US).

3 Disposal[s] Sales Agency (US DoD and UK MoD).

DSAC   Defence Scientific Advisory Council (UK).

DSAD   Digital service access device.

DSARC   Defense Systems Acquisition Review Council [or cycle] (US).

DSASO   Deputy senior air staff officer (RAF).

DSB   Defense Science Board (US).

DSB, d.s.b.   Double sideband.

DSC   *1* Defect-survival capability.

2 Distinguished Service Cross (UK).

3 Digital selective calling.

4 Digital scan convertor.

5 Disturbed-state concept.

DSCA   Defense Security [and] Co-operation Agency (US, approves weapons for export).

DSCE   Display select control equipment.

DSCS   Defense Satellite Communications System (US).

DSD   *1* Defence Support Division (NATO).

2 Data signal display [U adds unit].

DSDC   Digital signal data convertor.

DSEAD   Distributed suppression of enemy air defence.

DSEDM   Departure-sequenced engineering development model (FAA).

DSDU   Data signal display unit.

DSEO   Defense Systems Evaluation Office (US).

DSES   Defense Systems Evaluation Squadron (simulates hostile bombers attempting to penetrate Norad/ADC).

DSF   *1* Domestic supply flight (RAF).

2 Display systems function.

DSFC   Direct side-force control.

DSG   Design service goal [flight hours].

Dsg   Maximum shoulder diameter growth of tyre (tire).

DSIF   Deep Space Instrumentation Facility, stations at Cape Kennedy, in California (2) at Goldstone, in S Africa (2) and Woomera (NASA/JPL).

DSIR   Department of Scientific and Industrial Research (NZ, UK).

DSIS   Digital software integration station (fighter radar).

DSL   *1* Digital simulation language.

2 Depressed sightline.

3 Deutsche Schätzstelle für Luftfahrtzeuge eV (G).

4 Digital subscriber line.

DSLC   Dynamic-scattering liquid crystal.

DSM   *1* Dynamo situation modelling.

2 Departure-slot monitor.

3 Digital scene-matching.

4 Director of Support Management (RAF).

DSMAC   Digital scene-matching area-correlation, or correlator.

DSMC   *1* Direct seat-mile cost.

2 Direct simulation Monte Carlo.

DSM/YOF   Distinguished Staffel Member/Ye Olde Fokker (US).

DSMS   Digital stores-management system.

DSN   *1* Deep-Space Network, comprising DSIF and SFOF.

2 Defense switched network.

DSNS   Division of Space Nuclear Systems (AEC).

DSNT   Distant, = 30+ miles.

DSO   *1* Defence Sales Organization (UK).

2 Defense-, or defensive-, systems operator, or officer (US).

3 Distinguished Service Order (UK).

DSP   *1* Defense support program; O adds office (DoD).

2 Digital signal processor, or processing.

3 Departure spacing, or sequencing, program.

4 Display select panel.

5 Day surveillance payload.

6 Domain specific part.

DSPDRV   Display driver.

dsplcd   Displaced.

DSPS   Defense support program satellite[s].

DSR   *1* Director of Scientific Research.

2 Display-system replacement (FAA).

3 Directed-stick radiator (acoustics).

4 Dynamic super resolution.

DSRTK   Actual track between two waypoints.

DSS   *1* Department of Space Science (Estec).

2 Dipping sonar system.

3 Dynamic-signage system.

4 Decision support system.

5 Data-storage set.

6 Digital sequence/sequencer switch.

DSSP   Deep-Submergence System Program.

DSSS   Direct-sequence spread-spectrum.

DS³L, DSSSL   Document-style semantics and specification language.

DST   *1* Daylight-saving time.

    2 Dry specific thrust.

    3 Defense and space talks, or treaty (drafted 1988).

    4 Direction de la Surveillance du Territoire (F border security).

DSTL   Defence Science and Technology Laboratory (UK, created 2001 when US refused to share classified information).

DSTN   Double super-twist nematic.

DSTO   Defence Science & Technology Organization (Australia).

DSTS   Data storage and transfer set.

DSTU   Digital-signal transfer unit.

DS200   A directionally solidified HPT material.

DSU   *1* Direct Support Unit (USAF).

    2 Dynamic-sensor unit; gas-bearing yaw gyro and normal and lateral accelerometers.

    3 Digital switching unit.

    4 Data-storage unit; R adds receptacle.

    5 Defense-systems upgrade; P adds Program (USAF).

    6 Domain service unit.

    7 Data, or digital, signalling unit.

DSVT   Digital secure, or subscriber, voice terminal.

DSWA   Defense Special Weapons Agency.

DT   *1* Displaced threshold.

    2 Dual-tandem landing gear.

    3 Development test.

    4 Day TV (TADS).

    5 Damage-tolerant, or tolerance.

D$_T$ *1* Pressure drag.

    2 Trim drag.

D/T, D-T   Deuterium/tritium (NW fusion material).

DTA   *1* Direction des Transports Aériens, now Service des Transports Aériens (F).

    2 Dynamic, or drop, test article.

    3 Design and Technology Association (UK).

    4 Directorate of Technical Airworthiness (Canada).

    5 Deep-target attack.

DTAD   Demonstrated technology availability date.

DTAM   Descend to and maintain.

DT&E   Development, test and evaluation.

DTASS   Digital terrain-aided survival system.

DTAT   Direction Technique des Armements Terrestres (F).

DTC   *1* Design to cost.

    2 Distortion-tolerant control.

    3 Data-transfer cartridge.

    4 Defence Technology Centre[s] (UK MoD think-tank).

DTCA   Direction Technique de Constructions Aéronautiques (F).

DTCN   Direction Technique de Constructions Navales (F).

DTCS   Drone tracking and control system.

DTD   *1* Directorate of Technical Development (UK).

    2 Data-transfer device.

    3 Damage-tolerant design.

    4 Doppler turbulence detection.

    5 Data terminal display.

6 Document type definition.

DTDMA   Distributed (or distributive) time-division multiple access.

DTE   *1* Data terminal, or transfer, equipment.

    2 Digital target extractor.

DTEC   Direct thermal to electric conversion.

DTED   Digital terrain elevation data.

DTE/MM   Data-transfer equipment, mass memory.

DTEn   Direction Technique des Engins (F).

DTEO   Defence Test and Evaluation Organization (Boscombe Down, UK).

DTF   *1* Data transfer facility (software).

    2 Digital tape format.

DTG   *1* Distance to go.

    2 Dynamically tuned gyro.

    3 Date/time group.

DTH   Direct to home.

DTI   *1* Department of Trade and Industry (UK).

    2 Digital timebase interval.

    3 Distance to incident.

    4 Dial test indicator.

    5 Data-transfer interface; M adds module, U unit.

    6 Digital targeting information.

DTIC   Defense Technology Information Center (US).

DTIG   Deployed theater information grid.

DTK   Desired track.

DTL   Diode/transistor logic.

DTLCC   Design to life-cycle cost.

DTLR   Department for Transport, Local Government and the Regions (UK, includes civil aviation and aviation safety, became DETR).

DTM   *1* Data-transfer module.

    2 Digital terrain-management; -D adds and display.

    3 Digital terrain modelling.

    4 Demonstration test milestone.

DTM pt   Decision-to-miss point.

DTMC   Direct ton[ne]-mile cost.

DTMF   Dual-tone multi-frequency.

DTMMM, DTM³   Data-transfer module/mass-memory.

DTN   Data-transfer network.

DTO   *1* Desired test objective.

    2 Defense-technology objective.

DTOA   Difference in, or differential, or delta, time of arrival.

DT/OT   Development test[ing]/operational test[ing].

DTP   Design to price.

DTPC   Defense Technical Procedures Committee.

DTR   *1* Damage-tolerant rating.

    2 Defense Training Review (UK).

DTRA   Defense Threat-Reduction Agency (DoD, Dulles).

DTRD   Two-shaft turbojet; F adds with afterburner (USSR, R).

DTRI   Direction des Télécommunications du Réseau International (F).

DTRM   Dual-thrust rocket motor.

DTRS   Department of Transport and Regional Services (Australia).

DTRT   Deteriorate, deteriorating.

DTS   *1* Data-transfer system, or set.

    2 Dynamic test station.

    3 Desk-top simulator.

    4 Digital terrain system.

5 Data terminal set.

6 Diplomatic telecommunications system.

7 Development and Trials Squadron (AAC, UK).

DTSA   Defense Technology Security Administration (DoD).

DTSC   Digital Transmissions Standards Committee (UK).

DTSI   Defense-Trade security initiative (US).

DTSP   *1* Democratization transition systems program.

2 Division TUAV Sigint Program (USA).

DTU   *1* Display terminal unit.

2 Data terminal, transfer or transmission unit.

DTUC   *1* Design to unit cost.

2 Data-transfer-unit cartridge.

DTUPC   Design to unit production cost.

DTV   *1* Day, daylight or daytime TV.

2 Development[al] test vehicle.

3 Damage-tolerance verification.

DTVOR   Doppler-terminal VOR.

DTW, d.t.w.   Dual tandem wheels; usually means bogie, DT is more common.

DTWA   Dual trailing-wire antenna[s].

DU   *1* Depleted uranium.

2 Display unit.

3 Dust (ICAO).

Duad, DUAD   Dual air defence, battery consisting of only two fire units able to operate completely independently.

dual   *1* Duplication of cockpit controls permitting either pilot or co-pilot, instructor or pupil, to fly manually.

2 Pilot instruction in dual aircraft.

3 Airtime logged under * instruction.

dual-alloy disc   Turbine rotor comprising powder-metallurgy disc diffusion-bonded to outer cast blade ring.

dual-annular combustor   Gas-turbine combustion chamber comprising two coaxial elements at different radii, each with its own ring of fuel nozzles; one is used only for takeoff or other occasions demanding high power.

dual architecture   FCS(1) in which each surface is driven by its own local hydraulic systems and EHAs.

dual-axis rate transducer   Gyro based on electrically spun sphere of mercury, in which floats cruciform crystal sensor with 98% buoyancy; future rate sensor capable of operating at up to $\pm$ 60 g.

dual-capable   *1* Equally good at fighter and attack missions.

2 Can carry both conventional and nuclear bombs.

dual-combustion   Air-breathing engine, usually ramjet, which normally operates on two different fuels at different times. DCR feeds hot jet to main combustor burning JP-10.

dual compressor   Compressor split into low-pressure and high-pressure parts, each driven by separate turbine; also known as two-shaft or two-spool compressor.

dual-control aircraft   Provided with two sets of [usually interconnected] flight-control inceptors, usually for instructor and pupil.

dual designation   Right of two or more airlines of same nation to compete with each other on international route(s).

dual ignition   Piston engine ignition system employing two wholly duplicate means of igniting mixture.

dual-lane slide   Escape slide able to convey two separated streams of passengers simultaneously.

dual magneto   Single magneto with dual HT outputs to duplicate harness and plugs.

dual-mode DME   Can process /N or /P signals.

dual-mode EPU   Emergency power unit plus bleed from main engine.

dual-mode radar   Usually means MTI + SAR.

dual-mode seeker   Able to operate at either IR or RF frequencies.

dual persistence   CRT coated with phosphors giving bright display and dim previous indications.

dual-plane separation   Space vehicle staging in which interstage(s) falls away separately.

dual propellant   Ambiguous, can mean liquid bipropellant, double-base solid, or different propellants in two stages.

dual-purpose   Designed or intended to achieve two purposes, eg gun or missile (esp. on surface warship) intended to destroy surface or aerial targets.

dual-rail adapter   Interface enabling two AAMs to be carried on one pylon.

dual-rotation   Two-spool engine in which LP and HP shafts rotate in opposite directions.

dual-row airdrop   Use of slight (usually 4°) aircraft tilt to despatch pallets side-by-side.

dual-thrust motor   Solid rocket motor giving two levels of thrust by use of two propellant grains. In single-chamber unit boost grain may be bonded to sustainer grain, with thrust regulated by mechanically changing nozzle throat area or by using different compositions or configurations of grain. In dual-chamber unit, chambers may be in tandem or disposed concentrically.

dual-use technology   Military but can have civil applications [increasingly, perhaps, the other way round].

DUAT   Direct-user access terminal [S adds service].

dubler   Second prototype [doubler] (R).

DUC   *1* Dense upper cloud.

2 Distinguished unit citation (US).

Ducat   Duct acoustics and radiation.

DUCK, Duck   DCMS universal configuration key.

duck soup   Easy, no problem (US flying colloq.).

duct   *1* Passage or tube that confines and conducts fluid.

2 Channel or passage in airframe through which electric cables are run.

3 Portion of atmosphere (see * *propagation*).

ducted cooling   System in which cooling air is constrained to flow in duct(s) to or from power plant.

ducted-fan engine   Aircraft propulsor incorporating fan or propeller enclosed in duct; more especially jet engine in which enclosed fan is used to ingest ambient air to augment propulsive jet, thus providing greater mass flow. Most forms are better known today as turbofans.

ducted propulsor   Multi-bladed fan (usually with 5 to 12 variable-pitch blades) running in propulsive duct integrated with engine and airframe to give efficient and quiet low-speed propulsion. Source of power, piston engine or gas turbine.

ducted radiator   One installed in duct, esp. to give propulsive thrust.

ducted rocket   See *ram-rocket*.

duct heater   Burning in the bypass flow [turbofan].

ductility   Property of material which enables it to undergo plastic elongation under stress.

duct propagation Stratum of troposphere bounded above and below by layers having different refractive indeces which confines and propagates abnormally high proportion of VHF and UHF radiation, giving freak long-distance communications. Can be on surface or at height up to 16,000 ft (5 km); thickness seldom greater than 330 ft (100 m).

DUEO Deputy Unit Executive Officer (UK Police helos).

Dufaylite Airframe sandwich material with core of light plastics honeycomb.

duff gen Unreliable or incorrect information (RAF, WW2, colloq.).

dull switch Cockpit push-switch with dull appearance because never used.

DUM Dominant unstable mode, degree of freedom of aeroelastic or aeracoustic instability.

dumb Of munition, unguided or "iron"; not smart or brilliant.

dumb-bell *1* Manoeuvre in vertical plane resembling *, usually by helicopter, in making repeated low-level passes over same point.

2 Sign [usually white] in signals area denoting 'use paved areas only'. When ends have black bars = no restriction on taxiing but use runway for landing or TO.

dumb bogie Unimpressive opponent in air combat.

dummy deck Facsimile of aircraft carrier deck marked out on land.

dummy round Projectile, usually air-to-air or air-to-ground missile, fitted with dummy warhead for test or practice firing.

dummy run *1* Simulated firing practice, esp. air-to-ground gunnery or dive-bombing approach without release of a bomb. Also called dry run.

2 Trial approach to land, to release or pick up a load, as practice before actually performing operation.

dump *1* Emergency deactivation of whole system.

2 Of computer operation, to destroy, accidentally or intentionally, stored information, or to transfer all or part of contents of one section into another.

3 To jettison part of aircraft's load for safety or operational reasons, eg * fuel.

4 Of spacecraft, overboard release of waste water, astronaut waste products, etc.

5 Temporary storage area for bombs, ammunition, equipment or supplies.

6 To open * vavle of balloon.

dump and burn Awesome airshow entertainment in which a fighter makes a flypast in full afterburner and jettisoned fuel ignites in the hot jet[s].

dumper Lift dumper.

dump door Quick-acting hatch under hopper or tank for emergency release of chemicals, or under water tank for firefighting.

dump valve *1* Automatic valve which rapidly drains fuel manifold when fuel pressure falls below predetermined value.

2 Large-capacity valve fitted to any fluid system to empty it quickly for emergency or operational reasons.

3 Pilot-controlled valve which releases hot air or gas from balloon in controlled manner.

dunk To lower [e.g. sonobuoy] into water on tether or communication cable.

dunker training How to escape from helicopter underwater.

duo Co-ordinated aerobatics by two aircraft.

duo-tone *1* Colour of camouflage scheme with two main hues.

2 Two notes emitted by variometer.

Dupe Duplicate (Message).

duplex Circuit or channel which permits telegraphic communication in both directions simultaneously.

duplex burner *1* Alternative fuel entries and single exit orifice.

2 Small primary nozzle, used continuously, plus larger [often annular, surrounding primary] used only at T-O or other high-power regime.

duplexer Device which permits single antenna to be used for both transmitting and receiving, with minimal losses. (See *diplexer*).

Düppel Chaff (G, WW2).

DUR Duration, during [C adds climb, D descent, also expressed as **DURGC, DURGD**].

Durabond Metallic-composite adhesives for joining aluminium or stainless steel.

duralumin Wrought alloy containing 3–4.5 per cent copper, 0.4–1.0 per cent magnesium, up to 0.7 per cent manganese and the rest aluminium. Originally trade name.

Duramold Low-density structural material of thermosetting plastic-bonded wood (Clark, Fairchild, 1936).

duration *1* Maximum time aircraft can remain in air.

2 Time in seconds of operation of rocket engine.

duration model Traditional rubber-powered model aeroplane designed for flight endurance.

Durestos Composite structural material comprising asbestos fibres bonded with adhesive, usually formed by moulding under pressure.

dustbin *1* Gun turret lowered through aircraft floor to provide ventral defence (colloq., arch.).

2 Similar retractable container for sensor, esp. radar.

dust devil Small local whirlwind, dangerous to light aircraft.

dusting Controlled spreading of powder insecticide, fertiliser or other chemical by agricultural aircraft.

Dustoff Dedicated, unhesitating service to our fighting forces (US).

DUT Delft University of Technology.

Dutch roll Lateral oscillation with both rolling and yawing components; fault of early swept-wing aircraft. Especially dangerous at high altitude, when damping is insignificant. Can be stable [self-decaying], neutral, or dangerously divergent [unstable].

DUTE Digital universal test equipment (computer-guided probes check circuit boards).

duty crew Crew detailed to fly specific mission.

duty cycle Ratio of pulse duration time to pulse repetition time.

duty factor *1* In EDP (1), ratio of active time to total time.

2 In carrier composed of pulses that recur at regular intervals, product of pulse duration and PRF.

3 Several meanings in electronic jamming, usually fractions of unit time set is emitting or receiving.

duty ratio In radar, ratio of average to peak pulse power.

duty runway Runway designated for use by aircraft landing or taking off.

DUV Data under voice.

DV *1* Distinguished visitor (US).

2 Direct-vision (window).

3 Digital voltmeter.

4 Demonstrator/validation [also Demval].

DVA *1* Design verification article.

2 Doppler velocimeter/altimeter; often DV/A.

3 Diverse vector area [prescribed route not required].

DVB Digital video broadcast.

DVC *1* Direct voice control.

2 Digital voice communication[s].

3 Digital video conversion.

DVD *1* Direct-vendor delivery.

2 Direct-view display.

3 Digital versatile disk.

DVE Design verification engine.

DVF Demonstration and validation facility.

DVFR Defense Visual Flight Rules; procedures for use in an ADIZ (US).

DVG Display video generator.

D/VHF Doppler/VHF.

DVI Direct voice input.

DVI/O Direct voice input/output.

DVIT Digital video image [or imaging] transmission [S adds system].

DVK Deutsche Versuchsanstalt für Kraftfahrzeug und Fahrzeugmotoren (G).

DVL *1* Deutsche Versuchsanstalt für Luft- und Raumfahrt.

2 Data/voice logger.

DVLP Develop.

DVLS Digital voice-logging system.

DVM Digital voltmeter.

DVMC Digital video map computer.

DVMS Direct-voice management system.

DVO *1* Direct-view optics (or optical or option).

2 Direct voice output, synthesized speech confirming correct acceptance of DVI, and sometimes giving spoken warnings.

DVOF Digital vertical obstruction file, for terrain-referenced nav.

DVOR Doppler VOR, hence DVortac.

DVR Direct-view radar [D adds display, I indicator].

DVRS *1* Display, or digital, video recording system.

2 Digital voice recorder, or recording system.

DVS *1* Doppler velocity sensor.

2 Deutscher Verband für Schweisstechnik (G).

DVST Direct-vision storage tube.

DVT *1* Deep-vein thrombosis.

2 Delta voice terminal.

DW *1* Double wedge (aerofoil).

2 Drop weight.

3 Dual wheels.

D/W *1* Depth-to-width ratio in electron-beam hole drilling.

2 Drag-to-weight ratio.

DWan Direct wide-area network.

dwarf sonobuoy Smallest size, same 124 mm diameter as A-size but length only 305 mm (12 in).

DWC Distributed weapons co-ordination (CC&D).

DWD Deutscher Wetterdienst (meteorological service, G).

DWDF Delta-wing dual-fuselage.

DWDM Dense wavelength division multiplexing.

DWE Doppler wind experiment.

dwell *1* Brief rest period at end(s) of sinusoidal or other oscillatory motion, eg spot scanning CRT tube, bottom of drop-forging stroke, TDC and BDC in piston engine.

2 Period when scanning [eg, radar] beam remains looking at a particular target, to enhance resolution.

dwell illumination region Operator-selectable region of sky on which a surveillance radar can concentrate, ignoring remaining coverage.

dwell mark Caused by skin-mill, routing or similar tool remaining too long over one spot.

dwell time Among other meanings, the time spent in US between overseas tours (US armed forces).

DWG Digital word generator.

DWI Installation for generating giant magnetic pulses to explode mines (from cover name Directional Wireless Installation).

DWN Downdraughts.

DWP Digital wave processing.

DWPNT Dewpoint.

DWR Drag/weight ratio.

DWS Dispenser weapon system.

DWT Deutsche Gesellschaft für Wehrtechnik eV (G).

DX *1* See *duplex*.

2 Distance.

dyadics Sets of generalised vectors of aircraft motion [with not the usual three but nine components].

Dycoms Dynamic coherent measuring system.

DYN Dynamic pressure (EDP code).

dyn see *dyne*.

dynamic Frequently refreshed [computer memory].

dynamic amplifier Audio amplifier whose gain is proportional to average intensity of audio signal.

dynamic aquaplaning Landplane tyres running at high speed over shallow standing water and riding up out of contact with runway.

dynamic balance Rotating body in which all rotating masses are balanced within themselves so that no vibration is produced.

dynamic component See *dynamics*.

dynamic damper Device intended to damp out vibration by setting up forces opposing every motion.

dynamic directional stability $Cnß_{dyn}$, refinement of divergence treatment of a/c, esp. supersonic fighters, by Moul/Paulson 1958.

dynamic factor Ratio of load carried by structural part in accelerated flight to corresponding basic load (see *load factor*).

dynamic flight simulator Normally, flight simulator whose cabin is mounted on a centrifuge.

dynamic head See *dynamic pressure*; also called kinetic head.

dynamic heating Heating of the surface of an aircraft by virtue of its motion through the air. Heat is generated because the air at the surface is brought to rest relative to the aircraft, either by direct impact in the stagnation region or by the action of viscosity elsewhere. For practical purposes, synonymous with *aerodynamic heating* and kinetic heating.

dynamic hot bench Test facility enabling items such as avionics or auxiliary power systems to be connected as if they were in the aircraft.

dynamic lift Aerodynamic lift due to the movement of the air relative to a lighter-than-air aircraft.

dynamic load   A load imposed by dynamic action, as distinct from a static load. Specifically, with respect to aircraft, rockets or spacecraft, a load due to acceleration as imposed by manoeuvring, landing, gusts, firing guns, etc.

dynamic meteorology   The branch of meteorology that treats of the motions of the atmosphere and their relations to other meteorological phenomena.

dynamic model   A model of an aircraft (or other object) in which linear dimensions, mass and inertia are so represented as to make the motion of the model correspond to that of the full-scale aircraft.

dynamic particle filter   One which separates particulate solids from moving fluid by making them move relative to fluid; most common are centrifugal filter and momentum separation.

dynamic pressure   *1* Pressure of a fluid resulting from its motion when brought to rest on a surface, given by $q = \frac{1}{2}\rho V^2$; where ρ is density and V free-stream velocity; in incompressible flow, difference between total pressure and static pressure.

2 Pressure exerted on stagnation point(s) on a body by virtue of its motion through a fluid.

dynamic RAM   Constructed of periodically refreshed capacitor elements.

dynamics   Main rotating parts of helicopter airframe.

dynamic sampling   Concentration by flight recorder on one particular aspect of aircraft behaviour, esp. a malfunctioning channel.

dynamic scale   Scale of flow about a model relative to a flow about full-scale body; if two flows have same Reynolds number, both are at same **.

dynamic sidelobe level   That exceeded for 3 per cent of time on each main-beam scan.

dynamic similarity   *1* Relationship between model and full-scale body when, by virtue of similarity between dimensions, mass distributions, or elastic characteristics, aeroelastic motions are similar.

2 Similarity between fluid flows about a model and full-scale body when both have same Reynolds number.

dynamic-situation modelling   Program which seeks to create a framework for producing a real-time representation of the battlespace.

dynamic soaring   Soaring by making use of kinetic energy of air movements.

dynamic stability   Characteristic of a body that causes it, when disturbed from steady motion, to damp oscillations set up and gradually return to original state; used esp. of helicopters.

dynamic storage   In EDP (1), storage in which information is continuously changing position, as in delay line or magnetic drum.

dynamic system   See *dynamics*.

dynamic thrust   Work done by fan or propeller in imparting forward motion, equal to mass of air handled per second multiplied by (V_s–V), where V_s is slipstream velocity and V aircraft speed.

dynamic viscosity   Shear stress in a medium divided by shear velocity gradient, in each case between notional adjacent layers. Symbol μ, $= \upsilon\rho$ (kinematic viscosity multiplied by density), units $ML^{-1}T^{-1}$. SI unit Nsm^{-2} has no name, but see *poise*, = SI unit × 10.

dynamometer   Device designed to measure shaft power; torque or transmission * measures torque or torsional deflection of shaft; electric and hydraulic * measure power by electrical or fluid resistance; fan brake * uses air friction, and prony brake * applies torque by mechanical friction.

dynamotor   A d.c. generator and electric motor having common set of field poles and two or more armature windings; one armature is wound for low-voltage d.c. and serves as motor; other serves as generator for high-voltage d.c.

Dynarohr   Sound-suppressing multicell lining of engine fan cases and ducts (Rohr).

dyne   Abb. dyn, unit of force sufficient to accelerate 1 g at 1 cm s^{-2}; = 10^{-5}N or $\frac{1}{981}$ g or 2.248 × 10^{-6} lb.

dynode   Electrode of electron multiplier which emits secondary electrons when bombarded by electrons.

dysbarism   Decompression sickness, usually caused by release of nitrogen bubbles into blood at very low pressure.

DZ   *1* Drop (or dropping) zone.

2 light drizzle.

DZAA   Drop-zone assembly aid.

Dzus fastener   Countersunk screw with slotted shank anchored in removable or hinged panel, which hooks with half-turn into wire anchor on airframe.

Dzus rail   Mounting rail for LRUs.

E

E *1* Energy (but work often W).

2 Electric field strength.

3 Electromotive force.

4 Prefix exa = 10^{18}.

5 Young's modulus of elasticity.

6 US DoD role prefix, Special Electronic (from 1962).

7 UK role prefix, ECM training.

8 Illumination.

9 Endurance (usually safe endurance).

10 Estimated.

11 Sleet.

12 General controlled airspace (ICAO from 1990, US from September 1993).

13 East, or eastern.

14 Excellence (US defense contractors, WW2).

15 Emergency.

16 End (of reported weather).

17 See *e, E*.

18 Effective (machine performance).

e *1* Strain rate.

2 Induced drag.

3 Base of natural logarithms, c2.71828.

4 Emitter.

5 Electron charge, [strictly, preceded by minus sign] = 1.6021×10^{-19}C.

6 Eigen vector; suffix pm phugoid mode, sp short-period.

7 Elevator or slab tailplane.

8 Offset from hub of helicopter flapping hinge.

9 Subscript, exit.

10 Eccentricity of orbit or ellipse.

e, E Expansion ratio of rocket nozzle.

E_1 See *E-layer*.

E^2COTS See *EECOTS*.

E^2I Endoatmospheric/exoatmospheric interceptor.

E^3 *1* Energy-efficient engine (NASA).

2 End-to-end encryption (networks).

E-6B Analog dead-reckoning navigation computer (US, 1941).

E36 LRD Potassium acetate liquid runway dispenser.

e-bomb Renders electronic threat dumb, blind and incapable of retaliation.

e-check Electronic passenger-recognition system, usually biometric facial scan or iris scan

E-display Radar display with az/el target on rectangular x/y axes.

E-glass Ultra-high-strength, used as reinforcement in advanced structures (not in commercial GRP).

E-hitch Standard connector between tug and baggage trolley, with vertical pin.

E-layer Ionised layer of ionosphere typically 100–120 km, most pronounced in daytime; also called Heaviside or Kennelly-Heaviside or E_1 layer; some evidence of higher layer called E_2.

E-Pirep[s] Electronic pilot report[s].

E-plane Plane of antenna containing electric field; principal * is direction of maximum radiation.

E-scan Electronic scanning.

e-tag Electronic data tag.

EA *1* Enemy aircraft.

2 Electronic Attack, US aircraft category from 1962.

3 Engine-attributable.

4 Epoxy asphalt.

5 Engineering authority.

6 Environmentally acceptable.

7 l'Espace Affaires business class (F).

EAA *1* Experimental Aircraft Association (US, from 1953, office Oshkosh, Wisconsin).

2 East Anglian Aircraft Preservation Society (UK).

3 Export Administration Act (US).

EAAI European Association of Aerospace Industries.

EA&SD Evolutionary acquisition and spiral development (EW).

EAAP European Association for Aviation Psychology (office in Zurich).

EAAPS European Association of Airline Pilots' Schools.

EAAS Empire Air Armament School (RAF Manby, November 1944 to June 1949).

EAC *1* Expected approach clearance time.

2 Experimental-apparatus container.

3 Equipment Approvals Committee (UK MoD).

EACC European Airlift Control Cell (Eindhoven).

EACS Electronic automatic chart system.

EACSO East Africa Common Services Organization.

EADB Elevator-angle deviation bar.

EADE Extended air defense (US).

EADI Electronic attitude director, or *display*, *indicator*.

EADRCC Euro-Atlantic Disaster Response Co-ordination Centre.

EADS *1* Extended air-defense simulation.

2 Enterprise architecture and decision support (NSA1).

EAEM European Airlines Electronics Meetings[s].

EAF *1* Earned award fee.

2 Expeditionary Aerospace Force [= 10 AEFs] (USAF).

EAFAS European Academy for Aviation Safety, non-profit organization providing permanent training (Int.).

EAG European Air Group, air forces which share facilities (Belgium, France, Germany, Italy, Netherlands, Spain, UK).

EAGA European Advisory Group on Aerospace (EU).

EAGL European Association for Grey Literature.

Eagle *1* Extended, or early, airborne global launch evaluation, or evaluator (Awacs).

2 Evolutionary aerospace global laser engagement (AFRL).

Eagle Squadrons Three RAF units whose pilots were American volunteers (1940–42).

EAGS Expeditionary arresting-gear system.

EAI *1* Engine anti-ice.

2 Enterprise application integration.

EAISD European AIS(1) database.

EAL *1* Elevated approach light.

2 Evaluation assurance level.

EALS *1* Ejector augmented lift system.

2 Electromagnetic aircraft launch system.

EAM Emergency action message (NCA).

EAMDS European Airlines Medical Directors Society.

EAME   Europe/Africa/Middle East theatre (December 1941 to November 1945).

E&E   Escape and evasion.

E&M   Engineering and maintenance.

E&MD   Engineering and manufacturing development.

E&S   Executive and support (ATC services).

E&ST   Employment and Suitability Test Program (USAF).

EANPG   European Air Navigation Planning Group.

EANS   Empire Air Navigation School (UK).

EAP   *1* Experimental aircraft programme.

2 Engine-alert processor.

3 Emergency audio panel.

4 Educator Astronaut program.

EAPAS   Enhanced airworthiness program for airplane systems, concerned particularly with electric wiring (FAA).

EAPS   Engine air particle separator.

EAR   *1* Electronically agile radar.

2 Espace Aérien Réglementé (= danger, restricted or prohibited areas) (F).

EARB   European Airlines Research Bureau.

EARC   *1* Extraordinary Administration Radio Conference (UIT).

2 Elimination of ambiguity in R/T callsigns.

EARDA   European Armaments R&D Agency (EU objective, yet to be created).

early operational capability   First inventory aircraft, usually to lower standard than successors, to gain service experience.

early turn   Turn made ahead of fix in airway to ensure that aircraft does not, because of speed, course change required, wind, etc, fly outside airway boundaries.

early warning   Given by radar surveillance to LOS limit.

earnings   Profit (US).

EAROM, Earom   Electrical alterable read-only memory.

EARS   *1* Electromagnetic aircraft recovery system.

2 Environment analysis and reporting system.

EARSEL, Earsel   European Association of Remote Sensing Laboratories (Paris).

earth   Grounded side of electrical circuit or device.

Earth funnel   Funnel-like shape of lines of force above magnetic poles.

Earth inductor compass   Compass whose indication depends on current generated in coil revolving in Earth's magnetic field.

earthing   Connecting to earth; also adjective, as in * tyre (electrically conductive), * wire.

Earth/Moon system   Regarded as two-planet system whose centre of rotation is on line joining the centres.

Earth pendulum   See *Schuler pendulum*.

Earth-rate correction   Command rate applied to gyro to compensate for apparent precession caused by Earth's rotation.

earth return   Electrically bonded part of aircraft structure used in electrical circuit.

earthshine   Illumination of dark part of Moon produced by sunlight reflected from Earth's surface and atmosphere. Also called earthlight.

Earth station   Comsat, esp. Inmarsat, ground station, fixed or mobile.

earth station   Point where a conductor can be connected to earth.

earth system   All metallic parts of aircraft interconnected to form low-resistance network for safe distribution of electric currents and charges.

EARTS   *1* Enhanced aircraft radar test station.

2 En-route automated radar tracking system.

EAS   *1* Equivalent airspeed (see *airspeed*).

2 Espace Aérien Supérieur (upper airspace), (UIR).

3 Emergency avionics system.

4 Essential air service[s] (US).

EASA   *1* European Aviation Safety Agency, replacing JAAs (Int. 2002–).

2 European Aviation Suppliers Association (Int.).

Easie   Enhanced ATM and Mode-S implementation in Europe, see *EATMS(2)*.

EASS   *1* European Aviation Safety Seminar.

2 Evaluation and source-selection.

Eastern Test Range   Test range extending from Cape Canaveral across Atlantic into Indian Ocean and Antarctic.

EASU   Engine analyser and synchrophase unit.

Easy   Enhanced ADS(5) system.

EAT   *1* Estimated (or expected) approach time.

2 Electronic angle tracking.

Eatchip   European ATC harmonization and integration programme, from 1991, three phases, still active 2002 (ECAC/Eurocontrol).

EATMP   European air-traffic management programme.

eating irons   Knife, fork, spoon, issued to airmen (UK, colloq.).

EATMS   *1* European Air-Traffic Management System (Int.).

2 Enhanced air-traffic management and Mode-S Implementation in Europe (Int.).

EAU   *1* European accounting unit; see *International Accounting Unit*.

2 Engine analyser unit.

EB   *1* Essential bus.

2 Electron beam.

3 End-bend.

4 Executive Board (JAA).

5 US DoD role prefix for standoff jamming platform.

EBA   Engine bleed air.

EBAA   European Business Aviation Association (Int.).

EBACE   European Business Aviation Conference, or Convention, and Exhibition (Int.).

EBA/H   EBA plus hydrazine.

EBAPS, Ebaps   Electronically bombarded active-pixel sensor.

EBDM   Extended Buhr design method.

EBF   Externally blown flap.

EBHA   Electrical back-up hydraulic actuator.

EBIC   Electron-bombardment-induced conductivity.

EBI   Eyeballs in.

EBIT   Earnings before interest and taxes; DA adds depreciation and amortisation.

EBL   Electron-beam lithography.

EBM   *1* Electron-beam machining (or micro-analysis).

2 Electronic battle management.

EBO   *Effects-based operations.*

ebonite   Hard, brittle substance composed of sulphur and hard black rubber; possessing high inductive and insulating properties.

EBP   Electron-beam perforated.

EBPVD   Electron-beam physical vapour deposition.

EBR   Electron-beam recording.

EBRD   European Bank for Reconstruction and Development.

EBRM   Electronic bearing and range marker.

EBS   *1* Electronic beam squint tracking.

   2 Export baseline standard.

EBSC   European Bird Strike Committee.

Ebsicon   Standard NATO word for all SIT image tubes.

EBSV   Engine-bleed shutoff valve.

EBU   *1* European Broadcasting Union.

   2 Engine build-up or build unit.

ebullism   Formation of bubble, esp. in liquid rocket propellant or in biological or body fluids, caused by reduced ambient pressure.

e-business   Electronic business, usually means Internet.

EBW   Electron-beam welding.

EC   *1* Eddy current.

   2 Environmental control (system).

   3 Escadre de Chasse (fighter wing) (F).

   4 Elliptic-cubic (wing profiles).

   5 Electronic combat.

   6 Engine-caused.

   7 European Community, or Commission.

   8 Event criterion.

   9 Electronic checklist.

   10 Electronic commerce.

   11 Earth coverage (Satcoms).

E_c Compressive (bearing) strain.

ECA   Electronic control amplifier.

ECAC   *1* Pronounced E-kak, European Civil Aviation Conference [since 1956, now 36 countries]; RL adds reference level (ATC).

   2 European Civil Aviation Council [36 countries].

   3 Electromagnetic-compatibility analysis centre.

ECAM   *1* Electronic centralized aircraft monitor (presents all information on two CRTs in FFCC).

   2 Electronic caution alert module.

ECAP   *1* Electronic-combat adaptive processing.

   2 European capability action plan.

Ecarda   European coherent approach to R&D in ATM(7). (Int.).

E-Cars   Enhanced airline communications and reporting system.

ECB   *1* Electronic control box.

   2 Economical cruise boost.

ECBA   Electronic-combat battle management.

ECC   See *ECCM*.

ECCA   Engine-condition classification analysis.

ECCD   Electric cockpit-control device.

eccentricity   *1* Deviation from common centre or central point of application of load.

   2 Of any conic, ratio of length of radius vector through point on conic to distance of point from directrix.

   3 Of ellipse, ratio of distance between centre and focus to semimajor axis. Also called numerical *.

   4 Also of ellipse, distance between centre and focus. Also called linear *.

   5 Distance measured chordwise between a wing's aerodynamic centre and its elastic [torsional] axis.

ECCM   Electronic counter-countermeasures.

Eccosorb   Important family of commercially available SFAs (RAM).

ECD   *1* Excusable contract delay (no penalty).

   2 Equipment Capability Directorate[s] (MoD).

ECDES   Electronic combat digital evaluation system (USAF).

ECDIS   Electronic charts and data-information system[s].

ECDU   Enhanced control and display unit.

ECE   Economic Commission for Europe (UN).

ECEF   Earth-centred, Earth-fixed.

ECF   Enhanced connective facility (SNA).

ECFS   Empire Central Flying School.

ECG   Electrochemically assisted grinding.

ECGD   Export Credits Guarantee Department (UK).

ECH   Electrochemically assisted honing.

echelon   *1* Aircraft formation in which each member is above, behind, and to left or right of predecessor; such formation is said to be in * to port or starboard.

   2 Subdivision of headquarters, forward or rear.

   3 Level of command.

   4 Servicing unit detailed to provide ground support and maintenance facilities.

Echo, ECHO   Enhanced C^4ISR for homeland security operations, an attempted synthesis of GCCS, Adocs (2) and CATS(4) (US, 2003).

echo   *1* Pulse of reflected RF energy, esp. that reaching the receiver.

   2 Appearance on radar display of such energy returned from target; also called blip.

ECI   Electronic commerce infrastructure.

ECIF   Electronic Components Industry Federation (UK).

ECIM   Electronics computer-integrated manufacturing, ie, CIM of electronics.

ECIPS   Electronic-combat integrated pylon system.

ECIT   Enhanced communications interface transceiver.

ECL   *1* Emitter coupled logic.

   2 Electro-generated chemiluminescence.

   3 Engine-condition lever (CAA).

ecliptic   Apparent path of Sun among stars because of Earth's annual revolution; intersection of plane of Earth's orbit with celestial sphere, inclined at about 23° 27' to celestial equator.

ECLSS   Environmental control and life-support system [or subsystems].

ECM   *1* Electronic countermeasures.

   2 Electrochemical machining.

   3 Engine-condition monitoring.

   4 Electronic control module.

ECMJ   Escadrille de chasse multiplace de jour (multiseat day fighter squadron) (F).

ECMO   ECM officer (aircrew).

ECMS   Electronic component management system.

ECMT   European Conference of Ministers of Transport = CEMT.

ECN   Escadrille de chasse de nuit (night fighter squadron) (F).

ECNI   Enhanced CNI.

ECNP   Export control and non-proliferation (UK).

ECO   *1* Electron-coupled oscillator.

   2 Engineering change order.

ECOC   Enhanced Combat Operations Center.

Ecogas   European Council of GA Support (Int.).

ECOM   *1* Earth centre of mass.

   2 Electronic Command (USA).

Econ   Economy.

economical cruise mixture Piston engine mixture with which AMPG is maximum.

economiser Reservoir in continuous-flow oxygen system in which oxygen exhaled by user is collected for recirculation.

economiser valve Assists in regulating fuel flow through piston engine carburettor, opened by increased airflow.

economy Originally a passenger fare cheaper than first class, with less luxurious standards of cabin service, meals, seat pitch etc. IATA airlines introduced * class over North Atlantic in April 1958.

economy-class syndrome Normally means DVT (1).

ECOP Electronic copilot [colloq.].

ECP *1* Engineering change proposal, for introducing modification.

2 Etablissement Cinématographique et Photographique des Armées (F).

3 Effective candlepower (non-SI).

4 Eicas control panel.

ECPNL Equivalent Continuous Perceived Noise Level (see *noise*).

ECPP Effective critical parts plan.

ECPS Environmentally compatible propulsion system.

ECR *1* Electronic combat and reconnaissance.

2 Embedded computer resources.

ECS *1* Environmental control system.

2 Electronic Combat Squadron.

3 Engagement control station.

4 European company statute.

5 Engine-consumed spares.

6 Engineering compiler system.

7 Event-criterion sub-field.

8 Electronic chart system.

ECSL, ECSM, ECSR Respectively ECS(1) plus left card, miscellaneous, right card.

ECST Electronic-combat systems-tester (USAF).

ECSVR Engine-caused shop-visit rate.

ECT Enterprise caching technology.

ECTM Engine-condition trend-monitoring.

ECU *1* European Currency Unit (pronounced Ekyu, commonly called Euro).

2 Engine-change unit (complete bolt-on piston engine powerplant with cowl).

3 Environmental, or engine, or electronic, or Eicas, control unit.

4 Exercise Control Unit (a military formation).

5 External-compensation unit.

EC^{UK} Engineering Council (UK).

ECVS Emergency communications voice system, or switch.

ECW *1* Electronic Combat Wing.

2 Enhanced compressed wavelet.

ECWL Effective combat wing loading.

ED *1* Emergency distance (or distress signal).

2 Engineering development (part of progress schedule).

3 End of descent (Lockheed uses 'EoD').

4 Explosive device.

5 Environmental damage.

6 Eicas display.

E/D End of descent.

EDA *1* Effective disc area (helicopter).

2 Electronic design automation.

3 Excess defense article, available for sale (US DoD).

EdA Ejercito del Aire [Air Force, Spain].

EDAC See EDC(4).

EDAU Engine, or extended, data-acquisition unit.

EDB Extruded double-base.

EDC *1* European Defence Community.

2 Early display configuration.

3 Eros data centre.

4 Error detection and correction [often EDAC].

Edcars Engineering data computer-assisted retrieval system.

Edcas Equipment designers' cost analysis system.

EDCT Expected departure clearance time, issued to a flight as part of traffic-management program (FAA).

EDD Electronic data display (ATC flight data, tabular callsigns, heights, tracks and position information).

EDDS *1* Explosive-device detection system.

2 electronic document distribution service.

eddy *1* Local random fluid circulation drawing energy from flow on much larger scale and brought about by pressure irregularities, eg from passage of unstreamlined body.

2 In meteorology, developed vortex constituting local irregularity in wind producing gusts and lulls.

eddy current Generated in conductor by varying magnetic field; to reduce ** cores are built up of insulated laminations, iron dust or magnetic ferrite.

eddy damping Automatic damping by eddy currents generated by moving conductor.

eddy Mach wave radiation One of three major sources of jet-engine noise, associated with supersonically convecting disturbances.

EDG Electrical-discharge grinding.

edge alignment Distance, parallel to chord of propeller section, from centreline of blade to leading edge at any station.

edge effect Distortion of eddy-current pattern when testing for cracks near edge of material.

edge enhancement Increasing the contrast at the periphery of an image, to render it easier to distinguish [important in recon. and baggage screening].

edge flare Rim of abnormal brightness around edge of video picture.

edge keys Buttons around electronic display.

edge management Strict discipline of maintaining optimum LE of wing, tail [and pylons, if present] for aerodynamics and radar signature.

edge elevator Deck-edge elevator (carrier).

EDI *1* Electronic-data interchange, or interface, between single computers or groups; F adds function.

2 Electronic design information; L adds library.

3 Electron-drift instrument.

4 Engine-data interface; F adds function, U unit.

Edifact Electronic-data interchange for administration, commerce and transport.

EDIG European Defence Industries Group (Int., office Belgium).

E-Dircm Escort directional IR countermeasures (USAF).

EDIU Engine-data interface unit.

EDL *1* Engage/disengage logic.

2 Electrical-discharge laser.

3 Entry, descent and landing.

Edlar European data-link for aerial reconnaissance (Int.).

EDM *1* Electrical-discharge machining.

2 Engineering development model.

3 Evasive defence manoeuvres.

4 Engine-data multiplexer.

EDMS   Electronic data-management system[s].

EDO   Extended-duration orbiter.

EDP   *1* Electronic data-processor, or processing.

2 Engine-driven pump.

3 Experimental data-processor (Eurocontrol).

4 Engineering-development pallet.

EDS   *1* Explosive[s], or electronic, detection system.

2 European Distribution System; A adds aircraft (USAF).

EDSF   Electronic-data standard exchange.

EDSS   Explosives-detection security system.

EDT   *1* Eastern Daylight Time (US).

2 Expanded data-transfer; M adds module, S system.

3 Electronic drop tube.

EDU   *1* Enhanced, or engine, diagnostics unit.

2 Electronic display unit.

eductor   Duct-fed ejector[s] for powered VTOL lift.

Edwards   California [Mojave desert] AFB, site of AFFTC and NASA DFRC, previously called Muroc.

EE   *1* Emergency equipment, or egress.

2 Electronic[s] equipment [bay or compartment].

E/E   Electrical/electronic.

EEA   *1* Electronic Engineering Association (UK).

2 European Environment Agency (Int.).

EEC   *1* European Economic Community.

2 Engine electronic, or electronic engine, control; U adds unit.

3 Extendable exit cone.

EECots   Extended-environment commercial off the shelf.

EECS   Electrical/electronics cooling system.

EED   *1* Electromagnetic expulsive deicing.

2 Electro-explosive device.

EEE   Energy-efficient engine, also E3 or E^3.

EEFAE   Efficient and environmentally friendly aero engine; P adds program.

EEGS   Emergency electric-generating system.

EEI   *1* EFIS/EICAS interface.

2 Essential elements of information (reconnaissance).

3 Electrical-engineering instruction.

4 Electronic engine instrument[s].

EEL.3   Pioneer ester-based lubricant for gas-turbine engines (Esso).

EELV   Evolved expendable launch vehicle.

EEMAC, Eemac   Electrical & Electronic Manufacturers Association of Canada.

EEMP   Enhanced electromagnetic pulse.

EEMS   Electrostatic engine-monitoring system.

EEOC   En-route Expeditionary Operations Center (USAF).

EEOS   European Earth-observing satellite.

EEP   Experimental electronics package.

EEPGS   Enhanced EPGS, typically ½-ATR boxes and ½-volume.

EEProm, EEPROM, E^2Prom   Electronically-erasable programmable read-only memory.

EER   Extended echo-ranging.

EERM   Etablissement d'Etudes et de Recherches Météorologiques (F).

EEPSG   European Equipment Producers Support Group (Int.).

EES   *1* Electronically-enhanced sensing, or sensor.

2 Electrical Engineering Squadron (RAF).

EET   *1* Estimated elapsed time.

2 Escadron d'Expérimentation et de Transport (F).

EETC   Enhanced equipment trust certificate (leasing).

E-Etops   Initial E can mean early or EIS.

EEU   Elms [electrical load-management system] electronic unit.

EEVIP   Early extended-range twin operations validation and integration.

EEW   Equipped empty weight.

EEZ   Exclusive [coastal] economic zone; IG adds industry group.

EF   Evaluator Flight (RAF).

EFA   End-fire array (radar).

Efams   External fuel, armament and management system.

EFAS, Efas   *1* En-route flight advisory service.

2 Electronic-flash approach-light system.

Efato, EFATO   Engine failure at [or soon after] takeoff.

EFB   Electronic flight bag.

EFC   *1* Expected further clearance [time].

2 Elevator feel computer.

3 Engine-failure compensation mode.

EFCC   Enhanced fire-control computer; C adds configuration.

EFCS   Electrical [FBW] flight-control system.

efctv   Effective.

EFCU   Electronic fuel-, or flight-, control unit.

EFD   Electronic flight display.

Efdars, EFDARS   Expandable flight-data acquisition and recording system (FAA).

EFDC   Early-failure detection centre.

EFDPMA   Educational Foundation of DPMA (US).

EFDR   Expanded flight-data recorder.

EFDS   Electronic flight-data system.

EFE   Emitter feature extractor, an Elint tool.

EFEO   European Flight Engineers Organization (Int., merged into IFEO).

EFF   *1* Explosively formed fragment.

2 Effective.

3 Enhanced forward funding.

EFFE   European Federation of Flight Engineers, later EFEO.

effective angle of attack   Angle at which aerofoil produces a given lift coefficient in two-dimensional flow, also called AOA for infinite aspect ratio.

effective angle of incidence   See *effective angle of attack*.

effective aspect ratio   That of aerofoil of elliptical planform that, for same lift coefficient, has same induced-drag coefficient as aerofoil, or combination of aerofoils, in question.

effective atmosphere   That part of planetary atmosphere which measurably influences particular process of motion. For an Earth satellite limit is 120 miles, 193 km (see *mechanical border, sensible atmosphere*).

effective cover[age]   Region within which a navaid provides accurate and reliable guidance.

effective current   Difference between impressed current and counter-current.

effective exhaust velocity   Velocity of rocket jet after effects of friction, heat transfer, non-axially directed flow, etc.

effective helix angle   Angle of helix described by point on propeller blade in flight through still air measured relative to Earth.

effective horsepower   Power delivered to propeller.

effective pitch   Distance aircraft advances along flight-path for one revolution of propeller.

effective pitch radio   Basic propeller characteristic V/nd, where V is airspeed, n propeller rpm and d diameter, units being compatible.

effective profile drag   Difference between total wing drag and induced drag of wing with same aspect ratio but elliptically loaded.

effective propeller thrust   Net propulsive force; propeller thrust minus increase in drag due to slipstream.

effective range   Maximum distance at which weapon may be expected to strike target.

effective sortie   One which crosses the enemy frontier [see *sortie*].

effective span   Span minus correction for tip losses; usually defined as horizontal distance between tip chords.

effective terrestrial radiation   Amount by which IR radiation from Earth exceeds counter-radiation from atmosphere. Also called effective radiation or nocturnal radiation.

effective velocity ratio   Based on dynamic pressures

$$\sqrt{\frac{q}{q_j}}$$ where q_j is jet impingement stagnation pressure

(jet-lift ground effect).

effective wavelength   That corresponding to effective propagation velocity.

effector   Any device used to manoeuvre a vehicle in flight, now becoming popular in US as alternative to inceptor.

effects-based operations   Selection of a series of targets in a particular order, to achieve a specific final result.

efficiency   Ratio of output to input, usually expressed in percentage form.

efficiency of catch   Proportion of total water droplets in path of aircraft which actually strike it.

efflux   Total composition of gas or other fluid flowing out from a device, except that in an engine with a propulsive jet * excludes flows from auxiliary devices such as turbogenerators, heat exchangers and breathers.

effusion   Flow of gas through holes sufficiently large for velocity to be approximately proportional to square root of pressure difference.

EFH   *1* Earth far horizon.

  2 Engine flight hours.

  3 Equivalent flight hours [fatigue test].

EFI   Electronic flight instrumentation; S adds system; 8×8 colour CRTs.

EFIC   Electronic flight-instrument controller.

EFIDS   European flight-information display system.

EFIP   Electronic flight-instrument processor; CP adds control panel.

EFIS   See *EFI*; CP adds control panel.

EFL   *1* Emitter function logic.

  2 External-finance limits.

ERM   Enhanced fighter manoeuvrability, e.g. with TVC and RCFAM.

EFMCS   Enhanced flight-management computer system.

EFMS   Experimental flight-management system (Phare).

EFOGS   Enhanced fibre-optic-gyro missile.

EFP   Explosively formed penetrator, or projectile.

EFPS   Electronic flight-progress strip; D adds data.

EFT   *1* Elementary flying, or flight, training; E adds exercise, P programme, S school.

  2 Electronic funds transfer; S adds system.

EFVS   Enhanced flight vision system[s]. Allows aircraft below MDA and DH when not on Cat. II or III straight-in approach (FAA).

EFW   Electric field and wave.

EFX   Expeditionary forces experiment (USAF).

EGA   Exhaust-gas analyser.

EGAC   Enhanced general avionics computer.

Egads   Electronic ground automatic destruct sequencer button.

EGAS   European guaranteed access to space (five-year 2003–07 plan requiring €1 billion).

EGASF   European General Aviation Safety Foundation.

EGATS   European Guild of Air Traffic Services.

EGBU   Enhanced glide-bomb unit.

EGCU   Electrical-generator control unit.

EGDN   Ethylene-glycol dinitrate (a powerful explosive).

eggbeater   Helicopter with intermeshing rotors.

EGI   Embedded GPS/INS.

EGIHO   Expedited ground-initiated handoff.

EGIU   Electric[al] generator interface unit.

Eglin   Florida, largest AFB, home of many facilities including former APGC (USAF).

EGME   Ethylene-glycol monomethyl ether.

EGNOS, Egnos   European geostationary new, or navigation, overlay service, or system.

EGP   Exterior-gateway protocol.

EGPWS, EGPS   Ground-proximity warning system prefix E originally embedded, now enhanced; now called TAWS.

EGR   *1* Engine ground run[ning].

  2 Embedded GPS receiver.

egress   *1* Procedure for getting out of spacecraft in orbit or after planetary or lunar landing, whether for working in space or any other reason. Begins with putting on spacesuits, and includes depressurizing and opening hatch.

  2 Departure of combat aircraft from target area.

egress handle   Handle which fires ejection seat.

EGS   *1* Elementary gliding school.

  2 Exfoliation galvanic stress.

EGSE   Electrical ground support [or station] equipment.

EGT   Exhaust-gas temperature, measured immediately downstream of turbine[s] or exhaust valve.

EGTP   External ground test program.

EGW   Ethylene glycol and water.

EH   Edge enhancement.

E_h   Total energy at given speed and height.

EHA   *1* European Helicopter Association (Int.).

  2 Electro-hydrostatic, or -hydraulic, actuator, or actuation.

EHAC   En-route high-altitude chart.

EHAS   Electro-hydrostatic actuation system.

EHBS   Enhanced high-band subsystem.

EHD   Electro-hydrodynamic.

EHDD   Electronic head-down display.

e.h.f., EHF   Extra-, or extremely, high frequency, see Appendix 2.

EHL   Environmental health laboratory (USAF).

EHM   Engine health monitoring (or monitor).

EHOC   European Helicopter Operators' Committee (Int.).

ehp, e.h.p.   Equivalent horsepower. Usually total equivalent shaft horsepower.

EHR   Engine history recorder.

EHS   Enhanced surveillance.

EHSI   Electronic horizontal-situation indicator.

EHT   Electrothermal hydrazine thruster.

eht   Extra high tension (volts).

EHUM   Engine health and usage monitor.

EHV   Electro-hydraulic valve.

EI   *1* Earth (atmosphere) interface.
   2 Entry interface.
   3 Emissions index.
   4 Electronic intelligence, prefix to SEAD.

EIA   *1* Electronic Industries Alliance [Originally Association] (US).
   2 Environmental impact assessment.
   3 Element imaging array.
   4 Enhanced imagery analysis [W adds workstation].

EIANS   Eurocontrol Institute of Air Navigation Services.

EIB   European Investment Bank.

EICAS, Eicas   Engine indication, or instrument, (and) crew-alert[ing]-system; C adds control.

EICMS   Engine in-flight condition-monitoring system.

EID   *1* Electro-impulse deicing.
   2 Emitter identification.

EIDS   Engine-instrument display system.

Eiffel-type tunnel   Open-jet, non-return-flow wind tunnel in which whole working section is open.

eigen values   Discrete values of undetermined parameter involved in coefficient of differential equation, such that solution, with associated boundary conditions, exists only for these values; also called characteristic values or principal values.

eight   Flight manoeuvre in which aircraft flying horizontally follows track like large figure eight (see *Cuban**, *lazy* *).

eight-ball   Artificial horizon or attitude indicator (colloq., US).

eight-point roll   Roll executed in eight stages, with aircraft held momentarily after each roll increment of 45°.

eight pylon   Manoeuvre used in air racing in which aircraft is flown around pylons so that wingtip appears to pivot on pylon.

802-M, -11B   Leading wireless cabin system [in 2002] for use by individual passengers.

EIMS   European innovation monitoring system.

EIOTEC   Engineering, integration, operational test and evaluation contract.

EIP   *1* Enhanced industry participation.
   2 Environmental-impact parameter.
   3 École d'Initiation Pilotage (F).

EIPI   Extended initial protocol identifier.

EIRA   Ente Italiano Rilievi Aerofotogrammetrici.

EIRP   *1* Effective [or equivalent] isotropically radiated power.
   2 Earth incident radiated power.

EIS   *1* Entry into service.

   2 Environmental impact statement.
   3 Ejection initiation subsystem.
   4 Electronic instrument(ation) system.
   5 Engine indication [or instrument] system.

EISA   Extended industry-standard architecture.

EISF   Engine initial spares factor.

EISW   Equivalent isolated single-wheel load (LCN).

EIT   Exoatmospheric interceptor technology.

EITB   Engineering-Industry Training Board.

EIU   Interface unit prefixed by equipment, engine, electronic[s], Efis, Eicas or emergency.

ejectable   Able to be ejected from aircraft, esp. capsule, crew seat, sonobuoy, dropsonde or flight recorder.

ejection   Escape from aircraft by ejection seat.

ejection angle   Angle at which ejection seat leaves, measured relative to aircraft.

ejection capsule   *1* Detachable compartment serving as cockpit or cabin, which may be ejected as unit and parachuted to ground.
   2 Box containing recording instruments or data ejected and recovered by parachute or other device.

ejection chute   Parachute(s) used to decelerate ejection seat or capsule; often ballute or drogue.

ejection seat   Seat capable of being ejected in emergency to carry occupant clear of aircraft.

ejector   Device comprising nozzle, mixing tube and diffuser, utilising kinetic energy of fluid stream to pump another fluid from low-pressure region.

ejector augmented lift   *Ejector lift.*

ejector exhaust   Piston engine pipe(s) disposed or shaped to produce forward thrust, not necessarily incorporating an ejector.

ejector lift   Method of powered lift in which high-energy flow of hot gas (rarely, HP bleed air) from jet engine is expelled downwards through arrays of nozzles in large profiled vertical duct to entrain much greater flow of free air.

ejector nozzle   Propulsive nozzle for engine of supersonic aircraft whose jet can entrain a large surrounding airflow.

ejector ramjet   See *ram-rocket.*

ejector seat   See *ejection seat.*

EJS   Enhanced JTIDS.

EK   Equatorial air mass.

EKG   Electrocardiograph.

Ekman layer   Transition between surface boundary layer and free atmosphere.

EKP   Electronic knee-pad.

EKV   Exatmospheric kill vehicle.

ekW   Equivalent shaft power of turboprop, measured in kW. See *equivalent power.*

EL   *1* Electroluminescent.
   2 Ejector (augmented) lift.
   3 Emitter locator (or location).
   4 Elevation [or el].
   5 Electronic logbook; also see *ELB.*

Elac, ELAC   *1* En-route low-altitude chart.
   2 Elevator and aileron computer.

Elass, E-LASS   Enhanced low-altitude surveillance system.

elastance   Inability to hold electrostatic charge.

elastic axis   Spanwise line along cantilever wing along which load will produce bending but not torsion.

elastic centre   *1* Point within wing section at which

application of concentrated load will cause wing to deflect without rotation.

2 Point within wing section about which rotation will occur when wing is subjected to twist.

elastic collision   Collision between two particles in which no change occurs in their internal energy or in sum of their kinetic energies.

elastic instability   Condition in which compression member will fail in bending before failing compressive strength of material is reached.

elasticity   Property of material which enables a body deformed by stress to regain original dimensions when stress is removed.

elasticiser   Elastic substance or fuel used in solid rocket propellant to prevent cracking of grain and bind it to case.

elastic limit   Maximum stress withstood by material without causing permanent set or deformation. Hooke's Law asserts that within ** ratio of stress to strain is constant.

elastic model   Linear dimensions, mass distribution and stiffness are so represented that aero-elastic behaviour of model can be correlated with that of full-scale aircraft.

elastic modulus   Ratio of stress to strain [up to elastic limit].

elastic stability   Able to bear compressive yield stress of materials without buckling.

elastic stop nut   Nut in which self-locking is ensured by ring of fibre in which threads are formed as nut is screwed down.

elastivity   See *specific elastance*.

elastomeric bearing   Bearing in which angular (and some linear) relative motion is permitted by distortion of flexible blocks bonded to the two parts. Needs no maintenance.

elastomers   Rubber-like compounds used as pliable components in tyres, seals, gaskets etc.

elasto-optical effect   Variation in length and refractive index of fibre optics when subjected to tensile stress.

elastoplasticity   Theory of finite deformations.

el-az   Elevation/azimuth.

ELB   *1* Emergency locator beacon [A adds aircraft].

2 Extended [or extension of the] littoral battlespace.

3 Electronic logbook; FCG adds fault-correction guide, ISE in-service evaluation.

ELBA   Emergency locator beacon, aircraft.

elbow   *1* Angled section of piping used where change of direction is necessary.

2 Hollow fixture used for joining two lengths of electric conduit at an angle.

ELC   Engine-life computer.

ELCU   Electrical control unit (CAA).

ELD   *1* Electroluminescent display.

2 Earth leakage detector.

ELDO   European Launcher Development Organization (1960, now defunct).

ElectRelease   Patented epoxy adhesive, rapidly disbonded by application of low voltage.

electrical-discharge machining   Shaping hard metals by making the workpiece the anode in an electric circuit and eroding it by a shaped cathode tool, all submerged in ionised electrolyte.

electrical engine   Rocket in which propellant is accelerated by electrical device; also called electric rocket (see *electric propulsion*).

electrical interference   Undesirable and unintended effects on equipment due to electrical phenomena associated with other apparatus, cables, materials or meteorological conditions.

electrical load management   Supervises links between generators/alternators on main engines and APU, batteries and ground power supplies and on-board loads.

electric altimeter   Indicates height by variation of electrical capacitance. Also called electrostatic or capacity altimeter.

electric bonding   Interconnection of metallic parts for safe distribution of electrical charges.

electric energy   Product of current and time, 1MJ = 0.277 kWh, 1J = 1Ws.

electric field strength   Electric potential per unit distance across field, symbol E, units volts per metre.

electric flux density   Also called dielectric flux density, D $= 4\pi \times$ displacement current, units coulombs/metre^2.

electric gyro   One whose rotor is driven electrically.

electric propeller   Pitch-change mechanism is actuated electrically.

electric propulsion   General term describing all types of propulsion in which propellant consists of charged electrical particles accelerated by electric or magnetic fields or both; eg electrostatic, electromagnetic or electrothermal.

electric starter   Electric motor used to crank engine for starting.

electric steel   Steel made in electric furnace (induction or arc-type) which possesses uniform quality and higher strength than open-hearth steel of same carbon content.

electric tachometer   See *tachogenerator*.

electric welding   Welding by electric arc or passing large current through material.

electric wind   Emission of negative charge from sharp corner or point of conductor carrying high potential current. Also known as electric breeze.

electrochemical machining   Range of processes in which large direct current is passed through workpiece via shaped electrode in conductive electrolyte.

electrochemical treatment   Process involving application of electrical energy to produce chemical change in surface of material to be treated, such as anodization of aluminium alloys.

electrode   *1* Terminal at which electricity passes from one medium into another; positive is called anode and negative cathode.

2 Semiconductor element that performs one or more of the functions of emitting or collecting electrons or ions, or of controlling their movements by electric field.

3 In electron tube, conducting element that performs one or more of the functions of emitting, collecting or controlling, by electro-magnetic field, movement of electrons or ions.

electrodynamics   Science dealing with forces and energy transformation of electric currents, and associated magnetic fields.

electroforming   Building up a metal part of complex but thin form as an electroplated layer on a substrate, eg nickel on expanded polystyrene.

electro-hydraulic   Synonymous with electro-hydrostatic; both are abbreviated EHA.

electro-hydrostatic   Using hydraulic power to provide output force in localised system with all command and power provided by multi-redundant electric channels, which are much lighter than hydraulic piping.

electroimpulse deicing Mechanical method involving repeated [small] surface deformations caused by electric shocks.

electrojet Current sheet or stream moving in ionised layer in upper atmosphere; * move around Equator following sub-solar point and around polar regions, where they give rise to auroral phenomena.

electrokinetics Science dealing with electricity in motion, as distinguished from electrostatics. Electrokinetic potential symbol is ζ.

electroluminescence Emission of light caused by electric fields; gas light is emitted when kinetic energy of electrons or ions accelerated in field is transferred to atoms or molecules of gas.

electrolysis Chemical decomposition or change in chemical state produced by electric current.

electrolyte Liquid or paste conductor in electrolytic cell or battery; when acid, base or salt is dissolved in water dissolved material ionises, so that solution has electric potential and, when current is passed, will have different potential from metal immersed in it; solution used for anodizing aluminium and alloys, sulphuric or chromic acids being most common.

electrolytic corrosion Corrosion resulting from electrochemical action of dissimilar metals in presence of electrolyte.

electromagnet Magnet whose flux is produced by current in coil which encircles ferromagnetic core; temporarily magnetised while current flows.

electromagnetic Pertaining to magnetic field created by current; combined magnetic and electric fields accompanying movements of electrons through conductor. Abb. EM.

electromagnetic compatibility All aircraft systems can work simultaneously with no mutual interference.

electromagnetic expulsive deicing Sends intermittent giant pulses of EM energy which impart skin shocks which, though small amplitude, throw ice off.

electromagnetic focusing Control and concentration of electrons in narrow beam by magnetic fields.

electromagnetic frequency bands For administrative purposes various EM bands allotted letters (see Appendix 2).

electromagnetic induction Establishment of current in conductor cutting flux of electromagnet; principle of rotary electrical machines and transformers.

electromagnetic intrusion Intentional insertion of EM energy into transmission paths with object of causing confusion.

electromagnetic radiation Radiation made up of oscillating electric and magnetic fields and propagated in a vacuum at 299,792,456 m [983,571,007 ft]/s; includes gamma radiation, X-rays, ultra-violet, visible light, infrared radiation, radio and radar waves.

electromagnetic riveting Closing rivets by violent EM pulse.

electromagnetic rocket See *electrical engine, plasma rocket*.

electromagnetic spectrum EM radiation extending from gamma rays down through broadcast band and long radio waves.

electromagnetic units Several related systems of units [e.g. featuring abampere, abcoulomb, maxwell] now superseded by SI.

electromagnetic waves Waves associated with EM field, with electric and magnetic fields perpendicular to each other. Also known as electric waves, radio waves, light, X-rays, and by other names.

electromechanical *1* Using electricity as sole source of power and of command/control functions. Such systems are expected to displace hydraulics and other secondary power services, partly because of rare-earth magnets.
2 Control of engine fuel system by electrical signals.

electrometallurgy Use of electricity for smelting, refining, welding, annealing and other processes, and for electrolytic separation of metals and deposition from solutions.

electromotive force External electrical pressure (measured at source) which tends to produce flow of electrons in conducting medium; volt is ** required to maintain current of one ampere through resistance of one ohm.

electron Subatomic particle that possesses smallest negative charge, and which is so-called "fundamental particle" assumed to be building block of the Universe; mass at rest $m_c = 9.109 \times 10^{-28}$ g, negative charge 1.602×10^{-19} coulombs; charge/mass ratio $e/m_c = 1.7588 \times 10^{11}$ C kg^{-1}.

electron beam Stream of electrons focused by magnetic or electrostatic field and used for neutralisation of positively charged ion beam and to melt or weld materials with high melting points. Also called cathode ray.

electron-beam lithography 'Writing' parts of an integrated circuit (microchip) by means of beam of electrons.

electron-beam welding Use of powerful focused beam of electrons to make precision weld on workpiece in vacuum.

electron charge Unit, symbol e, -1.602×10^{-19} C.

electron gun Electrode structure which produces and may control one or more electron beams to produce TV picture or weld material.

electronic charge Electron charge.

electronic cloth Rapidly growing range of microelectronics based on low-cost flexible substrates.

electronic combat See *electronic warfare*.

electronic counter-countermeasures Subdivision of EW; actions to ensure effective use of electromagnetic radiation despite enemy use of countermeasures.

electronic countermeasures Subdivision of EW; actions to reduce or exploit effectiveness of enemy electromagnetic radiation.

electronic data-processing System using electronic computer(s) and other devices in gathering, transmission, processing and presentation of information.

electronic deception Deliberate radiation, reradiation, alteration, absorption or reflection of electromagnetic radiation, to mislead enemy in interpretation of data or present false indications; manipulative ** is alteration or simulation of friendly electromagnetic radiations to accomplish deception; imitative ** is introduction into enemy channels of radiation which imitates his own emissions.

electronic defence evaluation Mutual evaluation of radar(s) and aircraft by means of aircraft trying to penetrate radar through ECM.

electronic drop tube A multistation flight-strip manager.

electronic flight bag Software and data-services solution

207

to digitize logbooks, charts and other flight documents to achieve paperless cockpit.

electronic flight-control unit   Computer controlling surfaces used as spoilers and airbrakes, with or without roll-control function.

electronic flight instrument system   Replaces traditional flight instruments by full-colour CRT displays (typically three 200 × 200 nm, 8 × 8 in, for each pilot) each re-programmable to operate in different modes and giving high redundancy.

electronic interference   Disturbance that causes undesirable response in electronic equipment.

electronic intelligence   Detection, recording, analysis and cataloguing (where possible, linking with particular emitters) of all unfriendly EM emissions.

electronic jamming   Deliberate radiation, reradiation or reflection of electromagnetic signals with object of impairing use of electronic devices by enemy.

electronic line of sight   Path traversed by electromagnetic waves not subject to reflection or refraction by atmosphere.

electronics   Branch of physics concerned with emission, transmission, behaviour and effects of electrons.

electronic scanning   Scanning by cathode-ray tube, or sequenced emission from larger planar antenna array, instead of by mechanical means.

electronic warfare (also **electronic combat**)   Use of electromagnetic emissions as a weapon or a source of intelligence.

electron multiplier   Electron tube which delivers more electrons at output than it receives at input, because of secondary emission.

electron tube   Gas-filled tube having anode, cathode and sometimes other electrodes for controlling flow of electrons.

electron-volt   See *eV*.

electro-optical guidance   EO guidance makes use of visible (optical) contrast patterns of target or surrounding area to effect seeker lock-on and terminal homing. Three such systems are contrast edge tracker (Mk 84 EOGB and Walleye); contrast centroid tracker (Maverick); and optical area correlator, which scans contrast patterns in large area surrounding target.

electro-optic converter   Device which converts electricity into laser pulses for fibre-optic sensors.

electro-optics   Electronics involving visible or near-visible light, eg TV.

electroplating   Coating metal with deposit removed from electrode and carried by electrolyte in which object to be coated is immersed.

Electropult   Patented assisted-takeoff device, in effect a d.c. motor "unrolled" (US c1940).

electrostatic capacity   Measure of ability to hold electric charge, unit Farad, symbol F.

electrostatic deflection   Bending of electron beam during passage through electric field between two parallel flat electrodes; beam is deflected towards positive electrode.

electrostatic focusing   Use of electric field to focus stream of electrons to small beam.

electrostatic precipitation   Use of high voltages (large potential gradients) to remove particulate matter from gas flow, smoke or other volumes.

electrostatic rocket   See *ion rocket*, *ion engine*.

electrostatics   Study of electricity (charges) at rest.

electrostatic storage   Storage of information as electrostatic charges.

electrostatic unit, ESU   Unit of electric charge, amount of charge which repels similar charge in vacuum with force of one dyne; a statcoulomb.

Elektron   Magnesium alloys with 3–12% aluminium, 0.2–0.4% manganese and often 0.3–3.5% zinc.

element   *1* In electron tube, constituent part that contributes to electrical operation.

    2 In circuit, electrical device such as inductor, resistor, capacitor, generator, line, electrode or electron tube.

    3 In semiconductor device, integral part that contributes to its operation.

    4 Parameters defining orbit of body attracted by central, inverse-square force: longitude of ascending node, inclination of orbit plane, argument of perigee, eccentricity, semimajor axis, mean anomaly and epoch.

    5 Flight of two or three aircraft (US) or basic fighting unit of two aircraft (UK).

    6 Component parts of aircraft sufficiently distinctive and specific in type, shape or purpose as to be of major importance in design.

elementary charge   Electron charge.

elementary trainer   Ab initio, also known as primary trainer.

element leader   Lead aircraft or pilot of element or flight.

elephant ear   *1* Thick plate on rocket or missile used to reinforce hatch or aperture.

    2 Air intake consisting of twin inlets, one on each side of fuselage.

    3 Quasi-circular balancing area ahead of hinge axis of flight-control surface [rare after 1920].

Elev, elev   Elevation.

elevation   *1* Side or front view as drawn in orthographic projection.

    2 Vertical distance of point or level, measured from mean sea level.

    3 Height of airfield above mean sea level.

    4 Angle in vertical plane between local horizontal and line of sight to object.

elevation rudder   Elevator (arch.).

elevator   *1* Movable control surface for governing aircraft in pitch.

    2 Effectiveness of pitch control, as in expression "to run out of *"".

    3 In air intercept, code meaning 'take altitude indicated (in thousands of feet), calling off each 5,000 ft increment' (DoD).

elevator angle   Angle between chord of elevator and that of either the tailplane or aircraft longitudinal axis.

elevator tab   Trim (or other) tab attached to elevator.

11-9   Date of 2001 terrorist attacks on US.

elevons   Wing control surfaces combining functions of ailerons and elevators, esp. on delta-wing or 'tailless' aircraft.

elex   Electronics (colloq.).

e.l.f., ELF   *1* Extremely low frequency, see Appendix 2.

    2 Electronic location-finder.

    3 Aerosports federation (Estonia).

Elfin   ATR racking and module for housing instrument, electronic unit or other equipment.

ELG   Emergency landing ground.

ELGB   Emergency Loan Guarantee Board.

Elint   Electronic intelligence.

Elinvar   Trade name for an *invar* of steel character.
Elios   Elint identification and operating system.
Elips   Electronic integrated protection shield.
ELJ   External-load jettison.
elliptical orbit   Orbit of space object about primary body having form of ellipse. Nearest/furthest points pericentre/apocentre.
elliptic loading   Ideal form of spanwise loading of wing, lift vectors forming semi-ellipse seen in front elevation.
ELM   *1* Extended-length message.
2 Electrical load management [S adds system].
elongation   *1* Increase in length of hardware under tension.
2 Angle at Earth between lines to Sun and another celestial body of the solar system.
ELP   Electroluminescent panel.
ELQA   Extended link quality analysis (TADIL).
ELR   *1* Environmental lapse rate.
2 Extra-long-range.
ELS   *1* Emitter location system.
2 Emergency landing strip.
3 (Electron) energy-loss spectroscopy.
4 Electronic library system.
5 Elementary surveillance.
Elsa   Electronic lobe-switching antenna.
Elsec   Electronic security.
ELSS   Environmental life-support system.
ELSSE   Electronic sky screen equipment; indicates departure of rocket from predetermined trajectory.
ELT   *1* Emergency locator transponder [or transmitter].
2 Enforcement of laws and treaties.
3 Electronic light table, for EO reconnaissance.
4 Emergency landing technique.
ELV   Expendable launch vehicle.
Elvis   Enhanced linked virtual information systems.
EL/VT, ELVT   Ejector lift, vectored thrust.
EM   *1* Electromagnetic.
2 Energy manoeuvrability.
3 Electron microscope.
4 Element manager.
e.m.   Electromagnet.
e/m   Electron charge/mass ratio.
EMA   *1* Electromechanical actuator, or actuation.
2 Electron microprobe analysis.
3 External mounting assembly (helicopter).
4 Electronic missile acquisition.
EMAA   Etat-Major de l'Armée de l'Air (Chief of Staff, F).
EMAD   Engine-mounted accessory drive.
EMADS   Euromux management and data sheets.
EMAGR, E-MAGR   Enhanced miniaturized airborne GPS receiver.
Emals, EMALS   Electromagnetic aircraft launch system (catapult).
Emars   Electromagnetic aircraft recovery system (carrier).
EMARSSH, E-marsh   Europe Middle East route [structure] south of the Himalayas.
EMAS   *1* Electromechanical actuation system.
2 Environmentally modified airfield surface.
3 Engineered-material arresting system (ESCO).
EMAT   Electromagnetic acoustic transducer.
EMB   Extended MAD boom.
embedded   *1* Computer or other processor forming inte-

gral part of device or subsystem and thus unable to communicate directly with bus or highway or to be used for any other purpose.
2 Mixed clouds, usually Cu embedded in other types.
embedded optical databus   Plastic fibre-optic conductors printed on airframe structure, replacing looms of cables.
embedded training   Simulated threat data are fed to the avionics of a real airborne aircraft; can include audio and ground control.
embodiment loan   Loan of government property to private industry, research organization or individual, usually to enable recipient to fulfil government contract.
Embratel   Empresa Brasileira de Telecommunicacoes SA.
EMC   *1* Electromagnetic compatibility, or capability.
2 Entertainment multiplexer controller.
Emcat   Electromagnetic catapult.
EMCDB   Elastomer-modified cast double-base propellant.
Emcon   Emissions, or emission-monitor, control.
EMCS   Energy monitoring and control system.
EMD   *1* Emergency distance.
2 Eidgenossische Militärdepartment (Switz.).
3 Energy-management display.
4 Engine or engineering, model derivative.
5 Engineering and manufacturing development.
EMDa   Emergency distance available.
EMDM   Enhanced multiplex-demultiplex unit.
EMDP   Engine model derivative program (US).
EMDr   Emergency distance required.
EMDU   Enhanced main display unit (AEW aircraft).
EMEC   Enhanced master events controller.
EMEDI, Emedi   Electromagnetic-expulsion de-icing.
emer   Emergency.
Emerald   Emerging Research and Technology Department activities of relevance to ATM(7) concept definition (Euret).
emergency air   Compressed air for energizing hydraulic or pneumatic circuit in event of failure of normal power supply.
emergency cartridge   Provides combustion products to energize hydraulic or pneumatic circuit in event of failure of normal power supply.
emergency ceiling   Highest altitude for multi-engined aircraft at which best rate of climb is 50 ft per minute with throttle of one engine closed; also known as usable ceiling.
emergency combat capability   Condition exclusive of primary alert status whereby elements essential to combat-launch an ICBM are present and can effect launch under conditions of strategic warning (USAF).
emergency descent   Premature descent from operating altitude because of in-flight emergency.
emergency distance   Distance sufficient for all takeoff or landing emergencies, such as critical-engine failure at V_1, met by runway plus stopway and possibly clearway.
emergency exit   Door or window designed to be opened after emergency landing or aborted takeoff for passenger and crew evacuation.
emergency flotation gear   Inflatables fitted to aircraft in emergency to provide water buoyancy.
emergency landing   Landing made as result of inflight emergency.
emergency locator/transmitter   Radio beacon giving

position of crashed aircraft; fixed ***, portable ***, and survival *** (armoured and can float).

emergency parachute　Second stand-by parachute.

emergency power unit　On-board source of electrical and/or hydraulic power sufficient to continue controlled gliding flight following loss of main engines; commonly self-contained package using hydrazine monofuel (hence *MEPU*).

emergency rating　*1* Special rating of remaining helicopter engine[s] following failure of one; time-limited, typically to 30s; also called super-contingency.

　2 Piston engine rating for emergency sprint periods, with aid of high boost, water/methanol injection, etc.

emergency scramble　Aircraft carrier CAP launch of all available fighter aircraft; if smaller number required, numerals and/or type may be added (DoD).

emery　Hard abrasives based on corundum Al_2O_3.

EMF　Embarked military force.

emf, e.m.f.　Electromotive force.

EMG　Electromagnetic gun.

EMGFA　Armed forces general staff (Portugal).

EMI　*1* Electromagnetic induction, or inductor, or interference, or impulse[s].

　2 Environmental message interchange.

EMIH　EMI (1, 2) hardening.

EMI/HIRF　EMI (1) high intensity radio frequency.

EMIO　Egyptian Military Industrialization Organization.

emission　*1* Process by which body emits EM radiation as consequence of temperature only.

　2 Sending out of charged particles from surface for electrical propulsion.

　3 Loosely, any release from solid surface of electrical signal.

emissions control　Combat environment in which all detectable emissions are, as far as possible, prohibited. Thus, shipboard aircraft must use autonomous landing aids.

emissivity　Ratio of radiation emitted by body (if necessary in specified band of EM wavelengths) to that of perfect black body under same conditions; only luminescent can exceed 1, value for black body.

emitter　Device releasing radiation, usually in usable optical, IR or RF wavelengths.

EML　*1* Emergency medical link.

　2 Electromagnetic launcher.

Emma, EMMA　Engineering mock-up and manufacturing assembly.

eMMP　Electronic maintenance-management planning.

EMMU　Engine monitor multiplexer unit.

EMP　*1* Electromagnetic pulse (nuclear).

　2 Electric motor pump.

　3 Engine monitor panel.

　4 Engine motor pump [on ground, flight controls].

Empar, EMPAR　European multifunction phased-array radar.

EMPASS　Electromagnetic performance of air and ship system (USN).

empennage　Complete tail unit.

empirical　Based on observation and experiment rather than on theory; used esp. of mathematical formulae.

employment　Tactical usage of aircraft in desired area of operation; in airlift, movement of forces into a combat zone, usually in assault phase (USAF).

empty tunnel　No model in test section.

empty weight　Measured weight of individual aircraft less non-mandatory removable equipment and disposable load. OEW is preferred.

EMR　*1* Electromagnetic radiation.

　2 Electromagnetic resonance.

　3 Electromagnetic riveting.

EMRP　Effective monopole radiated power.

EMRS　Electromagnetic remote sensing.

EMRU　Electromechanical (or electromagnetic) release unit.

EMS　*1* Emergency medical service (usually helicopter).

　2 Entry monitor system.

　3 Equipment Maintenance Squadron (USAF).

　4 Engine management, or monitoring, system; see EMSC.

　5 Environmental management system (AEW radar).

　6 Electromagnetic-pulse shielding (hardening).

EMSC　Engine-monitoring system computer.

EMSG　European maintenance system guide.

EMSP　Enhanced modular signal processor.

EMT　*1* Equivalent megatons.

　2 Error-management training.

　3 Electronic maintenance trainer.

　4 Enhanced moving target; I adds indicator.

　5 Expert missile tracker.

EMTA　Engineering & Marine Training Authority (UK).

Emtas　Eco-management and audit scheme.

EMTE　Electromagnetic test environment.

EMU　*1* Extravehicular mobility unit; suit for exploring lunar surface.

　2 Engine maintenance, or monitoring, unit.

　3 Electronic mockup.

　4 Environment monitoring unit.

emu, e.m.u.　Electromagnetic unit[s].

EMUT　Enhanced manpack UHF terminal.

EMUX, Emux　Electrical multiplexing.

EMWR　Eddy Mach-wave radiation.

ENA　*1* Escuela Nacional de Aeronáutica (Arg.).

　2 Extended network addressing.

　3 Exhaust nozzle area.

　4 Exercise notification area.

ENAC　*1* École Nationale de l'Aviation Civile (F).

　2 Ente Nazionale per l'Aviazione Civile (Italy, certification).

ENAV　ATC authority (Italy).

ENB　Enhanced neutron bomb.

ENC　Electronic noise-cancelling.

encastré　Structural beam ends are not pinned but fixed.

enclosed cockpit　Provided with an overhead structure, either integral with the fuselage or a separate hinged or sliding canopy.

encoder　Analog-to-digital converter, eg converting linear or angular displacement, temperature or other variable to digital signals.

encoding altimeter　Presents usual display but in addition incorporates digitized output to transponder for transmission to ATC.

encounter　Time-continuous action between airborne friendly and hostile aircraft.

end-bend blading　Gas-turbine compressor blading whose ends (root and tip) are progressively given 3-D curvature to compensate for relatively sluggish flow over the inner and outer walls of the duct.

end-burning grain   Solid-propellant charge which burns only on transverse surface at one end, usually facing nozzle.

end effects   Aerodynamic effects due to fact wing span is finite.

end-fire   Linear aerial array whose direction of maximum radiation is along axis.

end game, endgame   In failed interception by AAM, time when missile runs out of V and energy.

end instrument   Converts data into electrical output for telemetry. Also called end organ or pickup.

end item   End-product ready for use.

endo-atmospheric   Within an atmosphere.

endothermic   Absorbing heat.

Endox Q-576   Alkaline soak added to water to form ultrasonic-cleaning fluid (Enthone).

endplate[s]   *1* Small auxiliary fins at or near tips of tailplane.

    2 * effect, aerodynamic effect of T-tail on fin, or of tanks, pods, missiles or fairings on wingtips.

end play   Unwanted axial movement of shaft.

end speed   Speed of aircraft relative to carrier at release from catapult.

end thrust   Thrust along axis of shaft.

endurance   Maximum time aircraft can continue flying under given conditions without refuelling.

endurance limit   Highest structural stress that permits indefinite repetition or reversal of loading; always less than yield stress (see *fatigue limit*).

endurance on station   Maximum time maritime aircraft can patrol in designated areas.

ENEC   Extendable nozzle exit cone.

Enema   Establissement National pour l'Exploitation Météorologique et Aéronautique (Algeria).

energy   Capacity to do work. SI unit = joule, or [more usefully] MJ = 0.3725 hp-h, 0.277 kWh; 1 kWh = 3.6000 MJ; 1hp-h = 2.68452 MJ. At any time * of a flying vehicle is given by $E_h = W (h + V^2/2_g)$ where h is height above MSL and W is instantaneous mass.

energy absorption test   See *drop test*.

energy conversion efficiency   Ratio of kinetic energy of jet leaving nozzle to that of hypothetical ideal jet leaving ideal nozzle using same fluid under same conditions.

energy density   Sound energy per unit volume (usual unit is non-SI: ergs/cc).

energy height   A measure of kinetic and potential energy of an air vehicle; $h_e = h + V^2/2_g$ where h is altitude above MSL and V is TAS expressed as a velocity.

energy level   Any specific value of energy which a particle may adopt; during transitions from one level to another, quanta or radiant energy are emitted or absorbed, frequencies depending on difference between levels.

energy management   Monitoring to minimise fuel expenditure for trajectory control, navigation, environmental control, etc.

energy manoeuvrability   Flight manoeuvres in which full use is made of kinetic energy of aircraft, normally in trading speed for altitude.

energy state   Total kinetic plus potential energy possessed by aircraft, particularly a fighter; normally expressed as altitude from SL reached (without propulsion) if all such energy were converted to potential (height) energy.

energy weapons   See *directed-energy*.

ENG   *1* Electronic news-gathering.

    2 Engine.

Engage   Armed position of some arrester hooks, extended or hinged down prior to landing

engage   *1* In air interception, order to attack designated contact (DoD usage).

    2 To contact arrester wire or barrier.

engagement   Encounter which involves hostile action by at least one participant.

engagement control   Exercised over functions of air-defence unit related to detection, identification, engagement and destruction of hostile targets.

engaging speed   Speed of aircraft relative to arrester wire at engagement.

engin   Missile (F).

engine altimeter   Indicates altitude corresponding to manifold pressure of supercharged engine.

engine-attributable   Caused by fault in an engine.

engine car   Airship car wholly or mainly devoted to propulsive machinery.

engine change unit   Aircraft piston engine removable as single unit with all accessories, cooling and oil systems.

engine cowling   Hinged or removable covering around aircraft engine shaped to keep drag to minimum and optimise flow of cooling air.

engineered material   Cellular concrete for overrun areas.

engineering   *1* Department responsible for detail design and development.

    2 Hardware design and development.

engineering mock-up   Full-scale replica of new aircraft or major part thereof, made [usually in metal] with high precision, partly in hard tooling, to check three-dimensional geometry of structure, systems, and equipment.

engineering time   Number of man-hours required to complete engineering task.

engineering units   Pre-SI (suggested obsolete) system of units for expressing lift and drag of wing or component part in lb/sq ft at 1 mph at specified angle of attack.

engine icing   A problem with all engines, but especially with piston engine with a choke-tube carburettor, where temperature is sharply reduced.

engine mounting   Structure by which engine is attached to airframe.

engine-out   Condition in which one engine of multi-engined aircraft gives no propulsive thrust.

engine-plus-fuel weight   A criterion of propulsive efficiency, heavy engines generally burning less fuel.

engine pod   See *pod*.

engine positioner   Dolly or trailer designed to carry engine, especially large turbofan, on cradle provided with hydraulic, or electrohydraulic, lateral, vertical, fore/aft, roll and pitch movement.

engine pressure ratio   Pressure ratio across complete compression system [possibly fan, booster and LP, IP and HP compressors]. In 1950 an axial spool of 15 stages achieved * of about 6; today this number of stages can exceed 50.

engine rating   Power permitted by regulations for specified use; maximum takeoff, combat, maximum continuous, weak mixture etc.

engine speed   Revolutions per minute of main or other specified rotor assembly.

English bias Missile aiming error at launch, and temporary guidance commands to overcome it.

ENH Earth near horizon.

enhanced GPWS Uses aircraft flight data to calculate envelope along projected flight path and compare this with internal terrain data base. Potential conflict gives ≤60 s aural/visual warning [in addition to normal GPWS output] and also displays terrain map showing clearance ahead.

enhanced vision system *1* Uses dual-band IR camera to project conformal image of scene ahead on to HUD, allowing approach to continue from 200 to 100 ft (30 m) decision height.

2 Another provides HUD-system to input Flir and/or MWR(1).

ENJJPT Euro-NATO Joint Jet Pilot Training.

ENK, ENNK Endo-atmospheric non-nuclear kill.

ENNA Enterprise Nationale de la Navigation Aérienne (Algeria).

ENOC Engineering network operation center.

enplanement Boarding by one passenger (US).

Enq Enquire.

ENR En-route, also ENRT.

ENRI Electronic Navigation Research Institute (Sendai, Japan).

enrichment *1* Adjustment by piston engine mixture control to produce richer mixture.

2 Artificial increase in percentage of isotope; thus, enriched uranium contains more than natural 0.75% of fissile U235.

en route *1* Between point of departure and destination.

2 Portion of flight on airways or desired track, excluding initial departure and approach phases.

en-route automated radar tracking system A step beyond ARTS IIIA with improved digital-display radars and fail-safe features (FAA).

en route base Air base between origin and destination of air force mission which has capability of supporting aircraft operating route.

en route clearance Valid to destination, either to joining stack or coming under approach control.

en route climb Climb to designated FL or cruising altitude on desired track.

en route height See *cruise altitude*.

en route support team Selected personnel, skills, equipment and supplies necessary to service and perform limited specialised maintenance on tactical aircraft at en route base (USAF).

en route time Time en route (1), normally measured from initial cruise altitude to TOD.

en route traffic control service Provided generally by ATC centres, to aircraft on IFR flight plan operating between departure and destination terminal areas.

ENS Euler/Navier-Stokes.

ENSA École Nationale Supérieure de l'Aéronautique (F).

ENSAE New designation of ENSA, with addition of "et de l'Espace".

ENSCE, or Ensce Enemy situation correlation element (US, intelligence).

ENSIP, Ensip Engine structural integrity program (USAF).

ENSMA Ecole Nationale Supérieure de Mécanique et d'Aéronautique (F).

ENSO El Niño southern oscillation.

Ensolite Very wide range of closed-cell foams made chiefly from VN (vinyl/nitrile PVC NBR rubber) or Neoprene; many applications (Uniroyal).

Entel Empresa Nacional de Telecommunicaciones (Argentina, Chile, etc).

Enterprise caching technology Combines VDA, APC(10) and selective caching in order to prevent superfluous data from being sent over electronic communications.

ENTG Euro/NATO Training Group.

enthalpy Total energy (heat content) of system or substance undergoing change from one stage to another under constant pressure, expressed as $H = E + PV$, where E is internal energy, P pressure and V volume; another expression is $Q = V + pV$.

entity Radar-detected aircraft seen on screen.

entomopter Flying machine based on insect aerodynamics.

ENTR Entire.

entrainment Sucking-in of induced fluid flow by high-velocity jet through duct.

entrance cone Portion of Eiffel-type tunnel upstream of working section.

entropy *1* In physics and thermodynamics, measure of unavailability of energy; symbol ϕ, or S, a measure of energy per unit temperature J/K. Specific *, symbol s, is * per unit mass kJ/kgK = 0.238846 Btu/lb°R; the reciprocal is 4.186798. Thus, in irreversible process, such as occurs in any real engine, * always increases. Any system or process having constant * is said to be isentropic.

2 In communications theory, measure of information disorder.

entry *1* Penetration of planetary atmosphere by spacecraft or other body travelling from outer space.

2 Fore part of body or aerofoil, esp. wing.

entry corridor Limits of route through atmosphere which returning spacecraft must follow. With too steep a trajectory spacecraft would burn up; with too shallow, spacecraft would bounce off atmosphere and be unable to return.

entry fix Precise reporting point at entry to FIR or control area, see next.

entry gate Point(s) of entry for incoming airways traffic to TMA.

entry interface Point during re-entry at which returning spacecraft encounters sensible atmosphere. Traditionally (non-SI) = 400,000 ft.

entry level Not precisely defined, but most common meaning is to describe small business jet, first to be bought by customer, who may later change to more costly replacement.

entry point *1* Ground position at which aircraft entering control zone crosses boundary.

2 Where supersonic track crosses coast inbound.

envelope *1* Of variable, curve which bounds values but does not consider possible simultaneous occurrences or correlations between different values.

2 Curve drawn through peaks of family of curves or through all limiting valves.

3 Glass or metal casing of electronic tube.

4 Hot air or gas container of non-rigid aerostat.

5 Outer cover of airship.

6 Volume of airspace bounded by limits of effective use of weapon.

envelope diameter *1* Diameter of circle encompassing engine or other irregular object.

2 Diameter of airship envelope.

ENVG Enhanced night-vision goggle[s] combining thermal imager with image intensifier.

environmental chamber Chamber in which humidity, temperature, pressure, solar radiation, noise and other variables may be controlled to simulate different environments.

environmental control system, ECS Produces environment in which human beings and equipment can work satisfactorily.

environmental lapse rate Measured rate of decrease of temperature with height; determined by vertical distribution of temperature at given time and place, and distinguished from process lapse rate of individual parcel of air.

environmental mock-up Mock-up cabin intended to assist design of ECS.

environmental stress screening Test procedure similar to burn-through for promoting reliability with growth.

environmental system Environmental control system.

EO *1* Electro-optical; used thus in subsequent definitions.

2 Engineering [or Executive] order.

3 Earth observation.

E/O Engine out.

EOAR European Office of Aerospace Research (USAF).

EOB Electronic order of battle.

EOBT Estimated off-block time.

EOC Electro-optic convertor.

EOCCM, EOC²M EO counter-countermeasures.

EOCM EO countermeasures.

EOD *1* Explosive-ordnance disposal, or demolition; S adds system, T reaining.

2 Electro-optical device [S adds system, T training].

3 End of day.

4 Enhanced operating database.

5 Embedded optical databus.

6 Erasable optical disc.

EODAP, Eodap Earth and ocean dynamic applications program.

EODC Earth-Observation Data Centre (UK).

EOE Elasto-optical effect.

EOEM Electronic original equipment manufacturer.

EOFC EO fire control.

EOGB Electro-optical[ly] guided bomb.

EO guidance Electro-optical guidance.

EOI Expression of interest, requested by potential customer from supplier; if answered, could lead to ITT (2).

EOIS EO imaging system.

EOIVS EO IR viewing system.

EOL *1* Engine-off landing.

2 Edge of light, use of reflected light at grazing angle to highlight surface imperfections.

Eolia European pre-operational data-link applications (ATC/Euret).

EOL power End-of-life power.

EOM *1* Earth observation mission.

2 End of message.

Eonnex Aircraft fabrics; name registered by Eonair Inc.

EOP *1* Engine oil pressure.

2 Engine operating point.

3 Enhanced operational capability.

EOQ Economic order quantity.

EOQC European Organization for Quality Control (Int.).

EOR Extend/off/retract.

EOS *1* Electrophoresis operations in space.

2 EO sensor/system/subsystem/surveillance. Thus EOSDS is surveillance and detection system, EOSS is sensor system.

3 Earth observation satellite, or observing system; DIS adds data and information system.

Eorsat Elint ocean-reconnaissance satellite.

EOSP EO signal processor.

EOT *1* Eo target/tracking/threat: S adds system, EOTADS target acquisition and detection system and EOTWD threat-warning development.

2 End of test.

3 End of text.

EOVL Engine-out vertical landing.

EOVS EO viewing system.

EOW EO warfare.

EP *1* External [or electric] power.

2 Environmental protection, [or plot], or processor [see *EPX*].

3 Engineering project.

4 Extended performance (target).

5 Electronic protection.

6 l'Espace Première = 1st class (F).

EPA *1* Environmental Protection Agency (US).

2 Epoxy polyamide.

3 Extended planning annex.

4 Economic price adjustment (US contracting).

5 Experimental (or European) prototype aircraft.

EPAD Electrically powered actuation design.

EPAF European participating [rarely, partner] air forces.

EPAM, Epam Electronic pilot activity and alertness monitor.

EPAS Expert process advisory set.

epaulettes In aviation, shoulder-mounted bages of rank or seniority, in civil aviation the most senior having four bars denoting captain of aircraft.

EPC *1* External power contactor (or connector).

2 Equipment Policy Committee (UK MoD).

3 Elementary Pilot Certificate.

EPCA Energy Policy and Conservation Act (US).

EPCO EANPG Co-ordination meeting[s] (ICAO).

EPD *1* Exhaust-plume dilution.

2 Electronic product definition.

3 Electric-power distribution; A adds assembly, S system.

Epera Extractor-parachute emergency-release assembly.

EPFA The European Property Flying Association [registered Wales, promotes aircraft in construction industry].

EPG European participating governments (or groups).

EPGS Electric[al] power generation, or generating, system.

ephemeris Periodical publication tabulating future positions of satellites or daily positions of celestial bodies and other astronomical data (plural = ephemerides).

ephemeris time   Uniform time defined by laws of dynamics, determined in principle by observed orbital motions of Earth and other planets (see *universal time*).

EPI   *1* Engine performance indicator.

2 Engineering process improvement.

3 Electronic-protection initiatives (AFRL).

4 Elevator position indicator.

EPIA   European Photovoltaic Industry Association (Int.).

EPIC, Epic   *1* Epitaxial passivated integrated circuit.

2 Engineering and product information control (management team).

3 Emergency procedures information centre (BAA).

4 Electronic Privacy Information Center (DC-based watchdog).

EPIRB, Epirb   Emergency position-indicating radio beacon, operating on 406 MHz in link with Sarsat.

EPL   Engine power lever, ie throttle.

EPLD   Electrically programmable logic device.

EPLRS   Enhanced position-location reporting system.

EPM   Electronic protection measures.

EPMaRV   Earth-penetrating manoeuvring re-entry vehicle; does not penetrate planet, only its atmosphere (USAF).

EPMS   *1* Engine performance monitoring system.

2 Electrical power management system.

EPN   European participating nations.

EPNdB   Equivalent Perceived Noise Decibel; unit of EPNL (see *noise*).

Epner   Ecole du Personnel Navigant Centre d'Essais et de Réception (F).

EPNL   Equivalent perceived noise level; measure of effect of noise on average human beings which takes into account sound pressure level (intensity), frequency, tonal value and duration.

EPO   Earth parking orbit.

epoch   Time when a satellite is established in orbit.

epoxy resin   Complex organic adhesive and electrical insulating material; addition of hardeners, plasticiers and fillers tailors its properties.

EPP   *1* Emergency power package.

2 Enhanced parallel port.

EPPIC, Epic   Enhanced precise positioning integrated capability (satellite).

Eppler   Family of wing sections for competition sailplanes; tailored to small R, high IAS for penetration.

EPR   *1* Engine pressure ratio.

2 External power receptacle.

3 Ethylene/propylene/rubber.

EPRL   Engine pressure ratio limit.

Eprom   Erasable programmable read-only message.

EPRT   Engine pressure-ratio transmitter.

EPS   *1* Emergency, or [confusing] electrical, power system (or supply).

2 Enhanced propulsion system.

3 Earning[s] per share.

EPSA   Emirates Parachute Sport Association.

EPSG   Equipment product supply group.

Epsilam   Copper-coated flexible substrate of ceramic-filled Teflon.

EPSRC   Engineering and Physical Science Research Council (UK, 1994–).

EPSU   European Public Service Union (Int.).

EPT   Egress procedures trainer, initially for the F-22 but with wide future possibilites.

EPTA   European Pultrusion Trade Association.

EPU   *1* Emergency power unit.

2 Electronic processing unit.

EPUU   EPLRS user unit [MLS can be suffix].

EPV   *1* Estimated programme value.

2 École du Personnel Volant (F).

EPW   Earth-penetrating warhead.

EPX   Environmental processor, military extension.

EQAR   Extended-storage [or expanded] quick-access recorder.

EQD   Electrical Quality-assurance Directorate (UK MoD).

EQPT   Equipment.

equal deflections   Principle used in analysis of statically indeterminate structure: two members rigidly attached must deflect an equal amount at point(s) of attachment under load.

equaliser   *1* Filter network which compensates over-specified frequency band for distortion introduced by variation of attenuation with frequency.

2 Connection between generators in parallel to equalise current and voltage.

equalising pulses   Signals sent before and after vertical synchronizing pulses to obtain correct start of lines in iconoscope, vidicon and display tubes.

equal taper   The same on LE and TE.

equation of time   Before 1965, difference between mean time and apparent time, usually labelled + or − to obtain apparent time. After 1965, correction applied to 12 hours + local mean time (LMT) to obtain local hour angle (LHA) of Sun.

equations of motion   Give information regarding motion of a body or point as a function of time when initial position and velocity are known.

equator   Primary great circle of sphere or spheroid, such as Earth, perpendicular to polar axis.

equatorial bulge   Excess of Earth's equatorial diameter over polar diameter.

equatorial satellite   One whose orbit plane coincides, or almost coincides, with Earth's equatorial plane.

equi-axed   Descriptive of traditional crystalline cast metal items.

equilibrium flow   Fluid flow in which energy is constant along streamlines, and composition at any point is not time-dependent.

equilibrium glide   Hypersonic gliding flight in which sum of vertical components of aerodynamic lift and centrifugal force is equal to weight at that height.

equilibrium height   At which, under given conditions, equilibrium is established between lift and weight of free aerostat without power.

equilibrium vapour   Vapour pressure of system in which two or more phases coexist in equilibrium; in meteorology reference is to water unless otherwise specified.

equinox   *1* Instant that Sun occupies one equinoctial point.

2 One of two points of intersection of eliptic and celestial equator, occupied by Sun when declination is 0°; also called equinoctial point.

equi-period transfer orbit   Orbit differing from first but having same period, eg that of lunar module following separation from command module.

equipment   Type or class of aircraft used or to be used on particular air-transport route(s).

equipment configuration report   Real-time all CMC, P/N S/N and DB/N.

equipment interchange   Agreement allowing aircraft to fly long routes over sectors of two or more carriers, crew being changed so that each carrier flies its own sectors.

equipment operationally ready   Weapon system is capable of safe use and all subsystems necessary for primary mission are ready (USAF).

equipped empty weight   Measured weight of individual aircraft including removable and other equipment but less disposable load.

equi-signal zone   Zone within which aircraft receives equal signals from left and right intersecting lobes, giving continuous on-track signal.

equivalence ratio   Ratio of stoichiometric to experimental air-fuel ratios.

equivalent airspeed   See *airspeed*.

equivalent brake horsepower   See *equivalent horsepower*.

equivalent circuit   Theoretical circuit diagram electrically equivalent to practical circuit or device.

equivalent drag area   See *equivalent flat-plate area*.

equivalent flat-plate area   Area of square flat plate, normal to free-stream relative airflow, which experiences same drag as the body or bodies under consideration.

equivalent horsepower   In turboprop, sum of horsepower, usually measured as brake hp, available at propeller shaft plus equivalent power derived from jet thrust by applying numerical factor to measure of thrust (abb. ehp). See *equivalent power*.

equivalent isotropically radiated power   Product of power to antenna multiplied by antenna gain in a particular direction relative to that from isotropic antenna.

equivalent kilowatt[s]   SI measure of power of turboprop, abb. ekW, see *equivalent power*.

equivalent monoplane   Monoplane wing having same lift and drag properties as combination of two or more wings under consideration.

equivalent monoplane aspect ratio   Wings and tip vortices of biplane mutually interfere; Prandtl showed increase in induced drag of each wing is: $\Delta D_i = \dfrac{\sigma L_1 L_2}{\tfrac{1}{2}\rho V^2 \pi b_1 b_2}$ where σ is Prandtl interference factor, L wing lifts, b spans, and $\tfrac{1}{2}\rho V^2$ dynamic head. Total added induced drag is twice that of single wing, so $**** = \dfrac{b_1^2}{S}\left[\dfrac{\mu^2(1+r)^2}{\mu + 2\sigma\mu r + r^2}\right]$ where b_1 is longer span, S total area, μ ratio $\dfrac{\text{shorter span}}{\text{longer span}}$, and r ratio $\dfrac{\text{shorter-span lift}}{\text{longer-span lift}}$

equivalent pendulum   Freely gimballed platform usually incorporating gyros and accelerometers, which has same period of oscillation as simple pendulum of particular length.

equivalent perceived noise level   LPNeq, $= L_E - 10 \log T/t_0$ where L_E is aircraft exposure level, T is total period of noise and t_0 is (usually) 1s (see *noise*).

equivalent potential temperature   Temperature given sample of air would have if brought adiabatically to top of atmosphere (ie to zero pressure) so that all water vapour is condensed and precipitated, remaining dry air then being compressed adiabatically to 1,000 millibars.

*** is therefore determined by absolute temperature, pressure and humidity.

equivalent power   See *equivalent horsepower*; in SI units power is measured in W or multiples thereof; to a first-order approximation ekW = kW + 68F$_n$ where F$_n$ is residual jet thrust in kN. In Imperial units jet thrust (lb force) is typically multiplied by 0.3846 [reciprocal 2.6] before being added to shaft power.

equivalent shaft horsepower   See *equivalent horsepower*.

equivalent single-wheel load   Mass which, supported by single wheel of size just large enough not to sink significantly into surface, causes same peak bending moment in airfield pavement as particular truck, bogie or other multi-wheel gear of actual aircraft.

equivalent temperature   Temperature particle of air would have if brought adiabatically to top of atmosphere (ie to zero pressure) so that all water vapour is condensed and precipitated, remaining dry air then being compressed adiabatically to original pressure.

equivalent wing   In stress analysis, same span as actual wing, but with chord at each section reduced in proportion to ratio of average beam load at that section to average beam load at section taken as standard.

ER   *1* Extended-range.
   2 Enhanced radiation.
   3 Echo reply.

E/R   Extend/retract.

Er   *Erbium*.

ERA   *1* European Regional Airline Association.
   2 Elastic recoil analysis, for hydrogen content.
   3 En-route [radar] array.
   4 Explosive reactive armour.
   5 Employment relations act (UK 1999).

ER-AAM   Extended-range air-to-air missile.

ERAAS   Extended-range autonomous attack system (UAV).

eradiation   See *Earth radiation*.

ERAM   *1* Extended-range anti-tank mine (or anti-armour munition).
   2 En-route automation modernization (FAA).

ERAQ   European Regional Airline Organization (Int.).

ERAP   Earth-resources aircraft program (US).

ERAPDS   Enhanced recognised air picture dissemination system.

ERAPS, Eraps   Expendable reliable-acoustic-path sonobuoy.

erase   In EDP (1) to expunge stored information, usually without affecting storage medium.

ERASL   Enhanced recognition and sensing lidar.

Erast   Environmental research aircraft and sensor technology (NASA).

E-Rast   Expendable Rast.

Erat   En-route absorption of (expected) terminal delay.

ERATS, Erats   En route advanced, or automated, tracking system.

ERAU   Embry-Riddle Aeronautical University (Daytona Beach, US).

ERB   *1* Executive, and also Engineering, Review Board.
   2 Executive responsibility budget.
   3. Earth radiation budget.

erbium   Bright silver metal, Er, density 9.1, MPt 1,529°C, important in optics [especially optical fibres] and in eye-safe Er-glass lasers on 2.9 μ.

ERBM   Electronic range/bearing marker.

ERBS　Earth radiation budget system, later satellite.

ERC　*1* Electronics Research Center, NASA, Cambridge, Massachusetts.

2 Extended runway centreline.

3 Engine-related causes.

4 En route chart.

ERCA　Etablissments Régional du Commissariat de l'Air (F).

ERCC　*1* En-route control centre.

2 Engine Requirements Co-ordinating Committee (CAA).

ERCS　*1* Emergency rocket communications system.

2 Enhanced radar cross-section (UAV decoy).

ERCE　Escadrille de Réception et de Convoyage Equipe [crew ferry flight unit] (F).

ERD　End-routing domain.

ERDA　Energy Research and Development Administration (US).

ERDE　Explosives Research and Development Establishment (formerly at Waltham Abbey, UK).

ERDI　ERD infrastructure.

ER/DL　Extended-range, data link.

ERE　External roll extrusion.

EREA　European research establishments in aeronautics, or for aerospace, launched 2001 with seven members.

erect　*1* Not inverted, vertical acceleration +1g.

2 To restore a horizon or standby horizon to give correct indication after upset.

erection　*1* Assembly and rigging of aircraft from component parts or from dismantled state; eg after crated shipment.

2 Of gyro, acceleration from rest to operating speed with axis in desired alignment. (Thus re-*, to restore proper axis alignment after being toppled.)

erector transporter　Vehicle used to convey ballistic rocket, elevate it for firing and act as launcher; also known as transporter erector.

E-region　Region of ionosphere in which E-layers and Sporadic E-layer tend to form.

EREL　Elevated runway-edge light.

ERFA　Conference on Economics of Route Air Navigation Facilities and Airports.

ERFCS　Extended-range fuel containment system [= tank].

ErG　Erbium-glass.

erg　Unit of energy in CGS (not SI) system; work done by force of one dyne acting through distance of 1 cm = 10^{-7}J.

ERGM　Extended-range guided munition (USA, USN).

ergometer exerciser　Device for exercising astronauts on long missions and measuring muscular work.

ERGP　Extended-range guided projectile.

ERHAC　En-route high-altitude chart.

Erint　Extended range-interceptor.

ERIS, Eris　Exoatmospheric re-entry vehicle interceptor system.

ERJ　External-combustion ramjet; one in which airflow and combustion are outside vehicle with profiled exterior surface.

erk　RAF slang (WW2) for airman ground crew possessing minimal skills and lowest rank (AC2 or ACH).

ERL　*1* Environmental Research Laboratories (NOAA).

2 Electronics Research Laboratory (Australia).

ERLAC　En-route low-altitude chart.

ERMA, Erma　Extended-red multi-alkali (Gen II image intensifiers).

ER/MP　Extended-range, multipurpose.

EROC　En-route obstacle clearance study group.

Erom　Erasable ROM.

Erops, EROPS　Extended-range operations, same as Etops.

EROS, Eros　*1* Earth Resources Observation Systems (US Geological Survey).

2 Earth-resources orbiting satellite.

3 Earth remote observing system.

erosion gauge　Instrument for measuring erosion by dust and micrometeorites on materials exposed to space environment.

ERP　*1* Effective radiated power.

2 Eye reference point.

3 Excitation [or exciter] receiver processor.

4 Enterprise resource planning, or process.

ER-PDU　Echo reply protocol data unit.

ER-PGB　Extended-range precision guided bomb.

ERPM　Engine rpm [normally of helicopter].

Erprobungsstelle　Proving (test) centre (G).

ERQ-PDU　Echo request protocol data unit.

ERRB　Enhanced radiation, reduced blast.

error　*1* In mathematics, difference between true value and calculated or observed value.

2 In EDP (1), incorrect step, process or result, whether due to machine malfunction or human intervention.

3 In air/ground bombing or photography, various definitions mainly concerned with linear miss-distance.

error band　Error value, usually expressed in per cent of full-scale, which defines maximum allowable error permitted for specified combination of transducer parameters.

error signal　Voltage proportional to difference between actual and desired condition. Thus, in radar, ** obtained from selsyns and AGC circuits and used to control servo to correct error.

ERS　*1* Earth-resources satellite.

2 Error-recovery service message (networks).

3 En-route supplement.

4 Emergency radio switching [system].

5 Electronic resource system.

ersatz　Substitute material (G).

ERSDS　En-route software development and support (FAA).

ERSU　Environmental remote-sensing unit.

ERT　*1* Elevator rigging tool.

2 Extended-range tank.

3 Earth-receive time [signal from planet].

ERTS　Earth-resources technology satellite.

ERU　*1* Ejector release unit for external stores.

2 Emergency reaction unit (USAFSS).

3 Engine relay unit.

ERV　Expendable rocket vehicle.

Ervis　Exoatmospheric re-entry vehicle interception (or interceptor) system (SDI).

ERW　Enhanced-radiation (neutron) weapon.

ER-WCMD　Extended-range wind-corrected munitions dispenser.

ERWE　Enhanced radar-warning equipment.

ES　*1* Escape slide.

2 Expert systems (artificial intelligence).

3 Electronically scanned.

4 Electrostatic.

5 Electronic support.

6 End system.

7 Extended squitter.

E$_s$ Specific energy, h + $v^2/2g$.

ES1, ES2   Radar antenna with electronic scanning about 1 or 2 axes, respectively.

ESA   *1* European Space Agency.

2 Enhanced signal average.

3 Electronic signature authentication.

4 Engineering source approval.

5 Electronically scanned array.

6 Enhanced [or electronic] situation[al] awareness [S adds system].

7 Embedded/special application, also E/SA.

ESAA   Electronic-scanning array antenna.

ES&DF   Electronic support and direction finding.

ESAS   *1* Electronically steerable antenna system.

2 See *ESA (6)*.

ESASC   EEA/SBAC Avionics Systems Committee.

ESB   Elevating sliding bridge, simpler than apron-drive type of airbridge.

ESC   *1* European Space Conference.

2 Executive steering committee.

3 Engine supervisory control.

4 Energy storage [and] control.

Escadre   Wing (military unit, F).

Escadrille   Flight (military unit, F).

Escadron   Squadron (F).

escalation   *1* Increase in scope, violence or weapons of conflict.

2 Increase in cost due to incorrect cost estimation, inflation, advances in technology, changes in specification or other factors.

escape   To achieve velocity and flightpath outward from primary body sufficient neither to fall back nor orbit.

escape capsule   See *ejection capsule*.

escape chute   Near-vertical chute forming part of structure, entered by opening pressure door in floor of crew compartment. Not to be confused with escape slide.

escape hatch   Hatch in aircraft, usually jettisonable, intended for use in abandoning aircraft; ventral for use in flight, dorsal after belly landing or ditching.

escape manoeuvre   *1* Several predetermined manoeuvres to evade hostile triple-A.

2 Maximum-rate manoeuvres to avoid CFIT.

3 Trajectory of spacecraft departing from Earth or evading planet [not normal usage].

escape orbit   Any of several paths body escaping from central force field must follow in order to escape.

escape rocket   Small rocket used to accelerate and separate payload near pad following launch-vehicle malfunction.

escape slide   Rapid-inflation pneumatic channel extended (usually from doors) from transport aircraft to enable passengers and crew to evacuate quickly in emergency. (Possible confusion with escape chute.)

escape spoiler   Aerodynamic baffle extended upstream of crew escape door or chute.

escape tower   Connects escape rocket(s) to vehicle; separated if ascent is normal.

escape velocity   Speed body must attain to escape from gravitational field. Earth 25,022 mph, 11.186 kms^{-1}, 36,700 ft/s, Moon 7,800 ft/s, Mars 16,700 ft/s and Jupiter 197,000 ft/s.

ESCC   European space components co-ordination (Estec).

Esces   Experimental Satellite-Communication Earth Station (India).

ESCS   *1* Emergency satcom system.

2 Electrical-system controller subsystem.

ESD   *1* Electronic Systems Division (USAF Systems Command).

2 European Security [or Strategy] and Defence [A adds Agency, I adds identity, P policy]; proposed EU task force.

3 Electrostatic discharge.

4 See *ESSD*.

Esdac   European Space Data Centre (now DIH).

ESDP   European Security and Defence Policy.

ESE   Earth-science enterprise (NASA).

ESF   European Science Foundation.

ESFC   Emergency surgery flying centre (helicopter).

ESG   *1* Electrostatically suspended gyro.

2 Extended-service goal.

3 Electronic Security Group (USAF).

ESGM   ESG monitor.

ESH   End system hello.

ESHE   École de Spécialisation sur Hélicoptères Embarqués (F).

eshp   Equivalent shaft horsepower, ehp.

ESI   *1* Engineering staff instruction.

2 Engine and system indication [D adds display, S system].

ESIC   Environmental Science Information Center (NOAA).

ESID   *1* Electrical-storm identification device.

2 Engine and system indication display.

ESIID   Embedded-system ionosphere interperability demonstration.

ESIL   Eye-safe IR laser.

ESIP   Engine structural integrity program (US).

ESIS   *1* Electronic standby instrument system.

2 See *ESI(2)*.

ESJ   Equivalent single jet.

ESKE   Enhanced station-keeping equipment.

ESL   *1* Earth-Sciences Laboratories (NOAA).

2 Eye-safe laser [R adds ranger].

Eslab   European Space Laboratory; now DSS of Estec.

ESLE   Electronic survivor-location equipment.

ESLR   Electronically scanned laser radar.

ESM   *1* Electronic support measures (UK).

2 Electronic surveillance measures, or measurement (US).

3 Electronic surveillance monitoring.

4 Enhanced space multiprocessor.

ESMB   Electrically-steered multi-beam.

ESMC   Eastern Space and Missile Center (USAF Patrick AFB).

ESMO, Esmo   ESM operator.

ESMR   Electronically scanned microwave radiometer.

ESO   Engineering standards order (FAA).

Esoc, ESOC   European Space Operations Centre, Darmstadt (Int.).

ESP   *1* External starting power.

2 Extended-service programme.

3 Elastically suspended pendulum.

4 Electrical standard practice[s].

5 En-route spacing program.

6 Expandable signal [or system] processor.

7 Expendable system programmes.

8 Engine surge protection.

ESPA　Electronically scanned phased-array.

ESPI　European Space Policy Institute.

Esprit　*1* European strategic programme for research into information technology.

2 Eye-slaved projected raster inset (Singer Link-Miles).

ESQAR　Extended-storage quick-access recorder.

ESR　*1* Electro-slag refined (or remelt).

2 European staff requirement (NATO).

3 Energy storage.

4 Emergency Sun reacquisition.

Esrange　Former European (now Swedish) space launch range, Kiruna.

ESRO　European Space Research Organization, now part of ESA.

ESRC　Engineering and Sciences Research Council (UK).

ESRDA　European Safety, Reliability and Data Association.

ESRP　European supersonic research programme [also called **PERS** (F)].

ESRRD　E-scope radar repeater display.

ESS　*1* Environmental stress screening.

2 Experiment support system (spacecraft).

3 ESM subsystem.

4 Electronic switching system.

5 Exercise support system.

6 Electronic Security Squadron (USAF).

ESSA　Environmental Science Services Administration (now NOAA).

ESSD　Electrostatic sensitive device[s].

essential bus　Electrical bus (bus-bar) on which are grouped nothing but essential electrical loads.

ESSL　Emergency speed select lever.

ESSS, ES³　*1* External stores support system.

2 Electronic sensors and systems sector.

Esswacs　Electronic solid-state wide-angle camera system.

EST　*1* Eastern Standard Time (US).

2 En-route support team (USAF).

3 See *E&ST*.

4 Elevation, slope, temperature.

5 Estimate[d].

ESTA　Electronically scanned tacan antenna.

establish　To achieve a steady state. In particular see next.

established　Aircraft confirmed as being stable at a prescribed flight condition, notably at a given FL or on a particular glidepath.

ESTAe　École Spéciale de Travaux Aéronautiques (F).

ESTC　European Space Tribology Centre.

Estec　European Space Research and Technology Centre.

Esteem　Elaboration of a strategy for the transition from Eatchip Phase III to EATMS (Euret).

Ester　EO sensor technology and evaluation research.

ester　Compound which reacts with water, acid or alkali to give an alcohol plus acid; important in many aerospace lubricants and other materials.

ESTL　European Space Tribology Laboratory (ESRO).

ESTOL, EStol　Extremely STOL.

ESU　*1* Electronic storage unit.

2 Emergency supply unit.

e.s.u.　Electrostatic unit.

ESV　Enhanced synthetic vision [S adds system].

ESVN　Executive secure-voice network (US civil govt.).

ESWL　Equivalent single-wheel load (of multi-wheel landing gear).

ET　*1* Emergency technology.

2 Extraterrestrial.

3 External tank.

e$_t$　*1* Tensile strain.

2 Thermal efficiency.

3 Environmental sensor unit.

ETA　*1* Estimated time of arrival.

2 Estimated time of acquisition.

3 Ejector thrust augmentation.

4 Effective turn angle.

ETAC　*1* Enlisted tactical air controller[s].

2 Engin tactique anti-chars (F).

ETACCS　European theatre air command and control study.

ETADS　Enhanced transportation automated data system.

Etalon　Small interferometer which reflects/refracts laser light to form interference pattern giving unique signature, rejecting all other sources.

ET&E　European test and evaluation (USAF).

ETAP　European technology acquisition plan, or programme.

eta patch　Fan-shaped patch of fabric and webbing secured to aerostat envelope.

e-Taws　Early, or embedded, terrain-awareness warning system.

ETB　*1* Engineering and Technology Board (UK Engineering Council).

2 End of block (ASCII).

3 Engineering test band.

ETBE　Ethyl-tertiary butyl ether.

ETBS　Etablissement Technique de Bourges (F).

ETC　*1* Environmental Test Centre (Foulness, UK).

2 Electro-thermal chemical.

3 Erroneous track change (FDR).

ETD　*1* Estimated time of departure.

2 Explosive[s] trace detection.

3 Expendable towed decay.

ETE　*1* Estimated time en route.

2 Environmental test and evaluation.

ETEB　Engineering Test and Evaluation Board (US).

ETEC　Expendable turbine engine concept.

ETES　Exotic threat-emitter system.

eTES　Enhanced total entertainment system.

ETF　*1* Electronic time fuze.

2 Enhanced tactical fighter.

3 Engine test facility.

4 Engineering task force.

ETG　*1* European Tripartite Group.

2 Electronic target generator.

Ethernet　Ether-net, yet uses coaxial cable or twisted pair of wires to link IEEE-802 radar images or data at <10 Mbps.

ethyl alcohol　Alcohol prepared from organic compound such as grain, starch or sugar; withstands high compression ratios but compared with conventional fuel costs more, has lower heat value and vapour pressure, and affinity for water, basically C_2H_5OH.

ethylene glycol   Principal additive in cooling systems of liquid-cooled engines, composed of saturated solution of ethylene oxide and water ($C_2H_6O_2$), BPt 197C.

ethylene oxide   Petroleum-derived gas used in FAE devices.

ETI   *1* Elapsed-time indicator.

2 Engine-technology improvement (US).

ETICS, Etics   Embedded tactical internet control system.

Etips   Electrothermal ice-protection system.

ETL   Elevated threshold light.

ETM   Elapsed-time measure[ment].

ETMP   Enhanced terrain-masked penetration.

ETMS   Enhanced traffic-management system (FAA).

ETNAS   Electro-level theodolite naval alignment system.

ETO   *1* Estimated time over, or overhead.

2 European theatre of operations (WW2).

Etops   Extended-range twin (engine) operations. Said to translate as: engines turning or passengers swimming.

ETOW   Engine time on wing.

ETP   *1* Equal time point.

2 Estimated time of penetration.

ETPS   Empire Test Pilots' School (originally Farnborough, now Boscombe Down, UK).

ETPU   Engine transient-pressure unit.

ETR   Eastern Test Range.

Etrac   Enhanced tactical radar correlator.

E-Tras   Electromechanical thrust-reverser actuation system.

ETRC   Expected taxi ramp clearance.

ETS   *1* Experimental test site.

2 External tank system.

3 Electronic systems test [S adds site].

4 Engineering test station.

5 Emitter targeting system.

ETSC   European Transport Safety Council (Int.).

ETSI   European Telecommunications Standards Institute.

ETSS   Enterprise targeting and strike system.

ETU   *1* External transmitter unit (IRCM).

2 Engineering Test Unit.

ETV   *1* Elevating [cargo] transfer vehicle.

2 [missile] Eject test vehicle.

ETVS   Enhanced terminal voice switch.

ETW   European transonic windtunnel.

ETX   End of transmission.

EU   *1* European Union.

2 Ejector unit (stores carrier).

3 Electronic[s] unit (many applications).

Eu   Europium.

EUAFS   enhanced upper-air forecast system.

Eucare   European confidential aviation reporting network (Int.).

EulG   K iron garnet.

Euler formula   Maximum load W of strut or long column,

$$W = \frac{kEI}{l^2},$$ where E is modulus of elasticity, I moment of inertia of strut section, k constant, and l^2 square of length between supports.

Eulerian angles   Systems of three angles which uniquely define with reference to one co-ordinate system (Earth axes) orientation of a second (body axes); orientation of second system is obtainable from first by rotation through each angle of turn, sequence being important. A singularity at 90° became significant with the latest [Russian] fighters.

Eulerian co-ordinators   System in which properties of fluid are assigned to points in space at each time, without attempting to identify individual parcels from one time to next.

EUM, Eumed   European–Mediterranean Air Navigation Region (ICAO).

Eumetsat   European meteorological satellite.

EUMS   Engine-usage monitoring system; EULMS adds 'life'.

EUPS   External uninterruptible power supply.

Eur   Eureka.

Eurac   *1* European aircraft-cost formula (includes landing, navigation and interest charges).

2 European Air Chiefs.

Euraca   European Air Carrier Assembly (Int., office Belg.).

Euram   European research on advanced materials (EC7).

EURANP   European air-navigation plan (ICAO).

EURATN   European ATN(1).

Eureca   European retrievable carrier.

Eureka   Ground beacon responding to Rebecca radar homing and distance-measuring system.

Eureka piece   Fragment of wreckage showing cause of catastrophe.

Euresco   European research conference[s].

Euret   European Research Programme in Transport.

EURFCB   European frequency-coordinating body (ICAO).

Euricas   European Research Institute for Civil Aviation Safety.

EuroCAE   European Organisation [spelt thus] for Civil-Aviation Electronics.

Eurocard   Standard single-sided PCB, 160 × 100 mm.

Eurocontrol   The European Organization for the Safety of Air Navigation, comprising Belgium, Netherlands [joint head office and 25 other states.

Eurogrid, Euro Grid   Digital map with terrain overlain by pilot-selected graphics (initially for military helicopters).

Eurogroup   Informal group of European defence ministers (NATO).

Euromep   European mission equipment package (helo night vision etc, F/G).

Europa   European undertakings for research organization, programmes and activities, an umbrella MoU (Int.).

European Air Chiefs   Free-ranging conference held twice per year since 1993 to promote air-power co-operation.

Europilote   European organization of airline pilots' associations.

europium   Symbol Eu, soft silvery metal, a lanthanide; density 5.243, MPt 822°C, many uses in phosphors, screen coatings, semiconducting alloys and lasers.

Europol   Intra-European air-transport policy (ECAC).

EUR-TFG   European Traffic-Forecast Group (Eurocontrol).

eutectic point   Lowest temperature at which mixture can be maintained in liquid phase; lowest melting or freezing point of alloy.

Eutelsat European Telecommunications Satellite Organization (Int.).

EUV Extreme ultra-violet, just beyond FUV (ie shorter wavelength); E adds Explorer.

EUVSA European Unmanned Vehicle Systems Association (Int.).

EV *1* Enhanced vision.

2 EAS (F).

3 Earned value.

eV Electron-volt, gain in energy acquired by electron gaining one volt in potential, 1.60219×10^{-19} J.

EVA *1* Extravehicular activity; carried on outside spacecraft or on lunar surface.

2 Equipement vocal pour l'aéronef, (cockpit human voice control) (F).

3 Equipe de Voltige Aérien (F, = team).

4 Economic value added.

evaluation *1* Appraisal of information in terms of credibility, reliability, pertinency and accuracy; for US and NATO letters A–F indicate reliability and numbers 1–6 accuracy; thus B-2 indicates probably true from usually reliable source, while E-means improbable from unreliable source.

2 Process of assessing proposal, design or hardware, usually on comparative basis in course of commercial or military procurement.

evaporative cooling Cooling system which uses latent heat of evaporation by allowing coolant to boil, then condensing and recycling it.

evaporative ice Ice formed in engine induction system on surface cooled by evaporation; can form from water or vapour at air temperatures up to 25°C.

EVAS Emergency vision assurance system = smoke goggles.

Evasion Ensemble de Visualisation et d'Affichage au Service de l'Instructeur à Organisation Numérique.

evasive action Flight manoeuvre performed by aircraft to evade defending forces, esp. AAA fire.

EVC Embedded visual computer, or computing.

EVCS Extravehicular communications system.

EVD *1* Elementary vortex distribution.

2 Explosive-vapour detector.

EVED Eidg. Verkehrs und Energiewirtschafts Departement (Switz.).

event At an airport, either a takeoff or a landing.

event marker Time-dependent indicator in HUD.

Everel propeller One of the few single-bladed propellers to have achieved any commercial success, the blade being counterbalanced by a lead cylinder (US c1930–40).

Everling number $\mathrm{No} = \dfrac{\mathrm{n}}{\mathrm{C_D}} = \dfrac{\mathrm{V_C}^3}{96{,}000\sqrt{\sigma}} \cdot \dfrac{\mathrm{W_o}/\mathrm{p}}{\mathrm{W_o}/\mathrm{S}}$

where n is propulsion efficiency, $\mathrm{C_D}$ total drag coefficient, $\mathrm{V_C}$ max level KEAS, σ relative density, $\mathrm{W_o}/\mathrm{P}$ power loading and $\mathrm{W_o}/\mathrm{S}$ wing loading.

Evett's Field Airfield serving WRE (Australia).

EVF Enter visual fix.

EVG Electrostatically supported vacuum gyro.

EVIR Enhanced-vision IR.

EVM *1* Engine-vibration monitor.

2 Error-vector magnitude.

3 Earth-viewing module.

EVO Hellenic arms industry.

EVR Electronic video recording.

EVS 1 Electro-optical viewing system.

2 Enhanced vision sensor, or system.

3 Electronic voice-switching [S adds system].

EVT *1* Extravehicular transfer.

2 Educational and vocational training (for return to civilian life).

EW *1* Electronic warfare.

2 Early warning.

3 Equatorial [warmer] air.

4 Elliptical waveguide.

5 Examining Wing.

EWAAS End-state wide-area augmentation system.

EWAC *1* Electronic-Warfare Aircraft Commander (USN).

2 Electronic-warfare analysis centre.

EWACS, Ewacs *1* Electronic wide-angle camera system.

2 Early warning and control system.

EWAISF Electronic warfare avionics integrated support facility.

EWAM Extended-window addressable memory.

EW&C Early warning and control.

EWAP Electronic-warfare AGE [access on ground equipment] panel.

EWAS Electronic-warfare analysis system.

EWAT Electronic-warfare advanced technology.

EWAU Electronic-warfare avionics unit (RAF).

EWCC *1* Electronic-Warfare Combat Co-ordinator.

2 Electronic-warfare co-ordination cell (NATO).

EWCS Electronic-warfare coordination system, or command station.

EWCU Electronic-warfare computer unit.

EWEDS Electronic-warfare evaluation display system.

EWEP Electronic-warfare evaluation program (USAF).

EWES Electronic-warfare evaluation system.

EWG Executive working group.

EW/GCI Early warning and ground-controlled intercept.

EWM Electronic-warfare management [S adds system, U unit].

EWMC Electronic-Warfare Mission Commander.

EWO *1* Electronic-Warfare Officer.

2 Emergency war order.

EWOP Electronic-Warfare Operator (USN).

EWOSE Electronic Warfare Operational Support Establishment (UK).

EWPA European Women Pilots' Association.

EWPI Electronic-warfare prime indicator.

EWR Early-warning radar.

EWS External weapon station.

EWSM Electronic-warfare (or early-warning) support measures (or surveillance measures).

EWSP Electronic-warfare self-protection.

EWTS Electronic-warfare training system.

Ex Expect[ed].

e_x Longitudinal distance between lift-jet centre-lines.

exa Prefix, multiply by 10^{18} [million million million], symbol E.

Exactor Mechanical remote-control system giving precise position.

exact orbit That Earth satellite must follow if exact sought-after data are to be obtained.

exact-point symbology   That showing a point rather than a locus; eg, in HUD, point where one projectile would have hit, rather than tracer line.

Excap   Expanded capability (= better).

exceedence   Single event, recordable on all HUM systems, in which engine or other device suffers an excursion in operating regime beyond allowable limits.

excess power   Difference between horsepower available and horsepower required; determines rate of climb. When horsepower available and required are equal, rate of climb falls to zero and absolute ceiling has been reached (see *SEP*).

exchange rates   Conversion factors used in calculating influence of variables in aircraft performance, most being guesses or assumptions; thus there are ** linking engine s.f.c., engine weight, parts cost, engine price and similar factors; ** will vary with stage length, operating conditions, costing formulae etc.

exciter   *1* Source of small current, such as battery or d.c. rotary generator, which supplies current for field windings of large electrical machine.

2 Oscillator which supplies carrier voltage to drive subsequent frequency-multiplying and amplifying circuits of transmitter.

3 Source of light used to stimulate photo-emissive cell.

exclusion zone   Airspace prohibited to aircraft, e.g. over National Monuments (US).

EXCM   Expendable [or external] countermeasures.

Excom   Extended communications search [SAR].

EXCP   Except[ed].

excursion   Undesired short-term variation of variable, such as instrument reading or flight path, away from correct value.

excursion fare   Promotional fare offered by airlines to stimulate traffic; usually applicable only to round trips, with limits on season, days available and/or trip duration.

excursion level   *1* In glidepath, maximum vertical or angular variation of centreline voltage/signal.

2 In glidepath, lowest safe angle of centreline voltage/signal.

educer   Outlet from diffuser of centrifugal compressor.

Exec   Executive.

execute missed approach   A mandatory ATC instruction.

exercise option   To convert option(s) into firm order(s).

exfoliation corrosion   Surface sheds thin flakes or layers.

exhaust branch   Short pipe from cylinder to exhaust manifold.

exhaust collector ring   Circular duct into which exhaust from radial engine is discharged.

exhaust cone   Assembly of outer pipe and inner cone which leads gas from turbine to jetpipe.

exhaust-driven supercharger   Turbocharger.

exhaust duct   *1* Tunnel through which gas is expelled from underground missile launcher.

2 Fan duct of aft-fan engine.

exhaust flame-damper   Expanding and shrouded pipes designed to prevent exhaust gas or stacks being seen at night.

exhaust gas analyser   Electrical instrument for indicating proportion of carbon monoxide and so indicating efficiency of combustion and correctness of fuel/air mixture.

exhaust manifold   Duct into which gas is led from number of cylinders.

exhaust plug   Streamlined body in exhaust nozzle for adjusting backpressure and giving propulsive thrust.

exhaust reheater   See *afterburner*.

exhaust stack   Exhaust pipe.

exhaust stator blades   Whole or partial ring of blades behind turbine to remove residual whirl from gas.

exhaust stroke   Fourth stroke in four-stroke cylinder, in which piston moves up to expel burnt gases.

exhaust stub   Short pipe linking cylinder direct with atmosphere.

exhaust turbocharger   See *turbocharger*.

exhaust velocity   Mean velocity of jet from rocket measured in plane of nozzle exit, v_e.

Eximbank   Export-Import Bank (US).

exit   Departure from battlefield [helo or fixed-wing]. Thus * **criteria**, required capabilities in flight performance and avionics to achieve this.

exit cone   Portion of wind tunnel into which air flows from working section.

exit fix   Reporting point at which aircraft leaves control area or FIR.

exo-atmospheric   Beyond the atmosphere.

exosphere   Outermost layer of atmosphere where collisions between molecular particles are so rare that only gravity will return escaping molecules; lower boundary is critical level of escape (region of escape) at 500–1,000 km.

exotic fuel   Any unusual fuel for air-breathing engine intended to produce greater thrust.

exotic material   Structural material seldom used in conventional applications; esp. one with melting point above 1,800°C.

expandable structure   One packaged in space vehicle and erected to full size and shape outside atmosphere.

expanded foam   Low-density material, usually rigid but of low mechanical strength, produced by chemical reaction in liquid state; often formed inside hollow metal airframe part.

expanding balloon   Kite balloon encircled by rubber cords or other devices to control shape when not full of gas; also known as dilatable balloon.

expanding brake   One whose segments are forced radially against drum by flexible sac.

expanding reamer   One with slotted flutes expanded by tapered pin.

expansion-deflection nozzle   Rocket nozzle in which jet enters top of bell-type nozzle moving radially outwards through an annular throat.

expansion joint   Pipe joint so constructed as to allow limited axial movement between sections held together.

expansion ratio   Ratio of cross-sectional area of rocket nozzle exit to area of nozzle throat.

expansion stroke   See *power stroke*.

expansion wave   Simple wave or progressive disturbance in compressible fluid, such that pressure and density decrease on crossing wave in direction of its motion; also known as rarefaction wave.

expansive corner   On supersonic body, convex corner [makes flow expand and accelerate].

expected approach clearance, EAC   Time at which arriving aircraft should be cleared to begin approach for landing; also known as expected approach time (EAT).

expected further clearance, EFC   Time at which it is expected additional clearance will be issued to aircraft.

expedite   ATC request: hurry up.

expendable construction   Rocket propellant tanks divided into sections jettisoned in sequence.

expendables   Missiles, RPVs, drones, and stores and materials consumed in action or in flight, esp. in space.

experimental aircraft   Aircraft whose objectives are fundamental research, or development of hardware having general application to many types of aircraft.

experimental mean pitch   Distance through which propeller advances along its axis during one revolution when giving no thrust.

exploding bridgewire   Metal wire which melts at high temperature, produced by large electrical impulse.

explosion turbine   Turbine rotated by gas from intermittent combustion process taking place in constant-volume chamber.

explosive bolt   One incorporating explosive charge so that, when detonated, whatever it secures in position is released.

explosive cladding   Use of explosive welding to clad one material with another.

explosive decompression   Rapid reduction of pressure caused by catastrophic leak in pressure cabin (eg loss of window).

explosive forming   High-energy-rate forming of sheet metal by using controlled explosive energy to blow workpiece against die.

explosive rivet   Blind rivet with partially hollow shank charged with black gunpowder which, when detonated, causes shank to bulge.

explosive welding   Effecting near-perfect bond between dissimilar metals by using explosion to drive them together under such pressure that joint melts and sweeps away previous surface impurities.

exposure level L_E, $= k \log \Sigma \ 10 L_{EPN} + 10$; can be amplified using L_{EPNi}/k where L_{EPNi} is i'th event and k is usually 10, with additions of $10T_o/t_o$ where t_o usually 1s and T_o may be 10s (see *noise*)

express   Property transported under air express tariffs filed with CAB; conducted on basis of agreement between Railways Express Agency and airlines.

Ext   Extension of runway.

EXTD   Extended.

extendable nozzle   Rocket exit cone retracted or extended to alter area ratio; also called extendable exit cone.

extended air defence   Defence against aircraft, UAVs and TBMs.

extended centreline   Centreline of runway extended in either direction indefinitely.

extended overwater operation   As defined by US FAR (Pt 1), an operation over water at horizontal distance more than 50 nm from nearest shore.

extended-range Dovap, Extradop   Baseline extension of Dovap to provide coherent reference to ground transmitter and all Dovap receivers located beyond line of sight.

extended-range operations   Modern engines are so reliable that twin-engined aircraft [large jets] can be certificated for Etops routes taking them 60, 90, 180, or 240 minutes away from nearest suitable airport at engine-out cruise speed.

extended-root blade   *1* Gas-turbine rotor blade in which aerofoil is carried on long platform in disc of reduced diameter.

2 Propeller blade with root extended in chord.

extension contract   Industrial contract formed as extension of previous contract in either scope or timing.

extension flap   See *area-increasing flap*.

extensometer   Instrument for measuring small amounts of deformation.

external aileron   Aileron mounted clear of wing surfaces but deflected conventionally.

external augmentor   Generalised description for arrangements which use high-energy primary flow to entrain ambient airflow remote from source of power, as in ejector lift.

external energiser   Portable motor used to supply motive power to engine inertia starters.

external gearbox   That attached outside casing of a gas-turbine engine, providing drives for accessories, usually from HP shaft, and connector for hand-turning.

external input   System input from source outside system.

externally blown flap   Flap in wake of main engine[s] when deflected, thus having greatly enhanced effect esp. in increasing lift.

externals   External inspection of aircraft carried out by pilot before boarding.

external storage   EDP (1) storage media separate from machine but capable of retaining information in form acceptable to it.

external supercharger   Impeller (manifold-pressure booster) located upstream of carburettor.

extinction   *1* Attenuation of light through absorption and scattering.

2 Cessation of combustion (see *flameout*).

extinction coefficient   In meteorology, space rate of diminution of transmitted light; attenuation coefficient applied to visible radiation.

extraction   *1* To recover friendly troops from hostile location.

2 To take shaft power from engine for performance-measuring purposes.

extraction parachute   Extracts cargo from aircraft and deploys main parachutes.

extraction zone   Specified ground area upon which supplies are delivered by extraction technique from low-flying aircraft.

extractor   *1* Part of firearm which engages rim or base of cartridge to pull it from chamber.

2 Computer-controlled device for automatic initiation and maintenance of all desirable radar contacts in ATC or air-defence system.

extra section   Extra flight by airline to take overflow of fully booked flight.

extravehicular activity   See *EVA*.

extremely high frequency   30–300 GHz.

EXTRM   Extreme.

extrusion   Hot or cold forming of metals, rubbers and plastics by forcing through die of appropriate cross-sectional shape.

ExW   Explosive welding.

e_y   Lateral distance between jet centrelines.

eye   In centrifugal compressor, that portion through which fluid enters.

eyeball   *1* Passenger-controlled spherical valve outlet for fresh air, usually overhead.

2 To search visually, or keep eyes on a target (colloq.).

eyeball design   Design by eye, without calculation.

eyeballs down   Jargon for severe positive acceleration.

eyeball/shooter   Manoeuvre in which lead fighter flies across to identify target visually, while wingman (shooter) remains able to fire BVR.

eyeballs in   Acceleration from behind when subject is seated upright, or below when prone [best].

eyeballs out   Decceleration when subject is seated upright [worst].

eyeballs up   Negative-g, downward acceleration.

eyebrow panel   Panel of instruments or controls in flight deck roof, above and behind windscreen.

eyebrow window   In roof of flight deck, also called VIT window.

eyelids   Jet-engine reverser or afterburner nozzle halves similar to eyelid in appearance and action.

eye relief   Distance from eyeball to NVG eyepiece or image of HMD.

F

F *1* Fahrenheit (contrary to SI).

2 Fighter aircraft category (USN since 1922, USAF since June 1948), UK prefix since 1942.

3 Flap angle.

4 Force, especially net propulsive force, thrust.

5 Farad.

6 Sonobuoy size, 0.3 m (1 ft) long.

7 Photographic category (USAS, USAAC, USAAF, 1924–47).

8 First class (seating).

9 Fuel mass.

10 Fuel, with suffix 12 to 44, thus F18 =100LL and F22 = 115 Grade (NATO).

11 Flashing [sequenced] light.

12 Fog.

13 Area forecast.

14 Magnetomotive force [also M].

15 Luminous flux [usually Φ].

F Faraday constant.

f *1* Frequency.

2 Frictional force.

3 Acceleration.

4 Equivalent parasitic area of aircraft.

5 Symbol meaning a function of [rarely, F].

6 Subscript, usually fuel, flap or fountain.

7 Normal stress.

8 Femto, 10^{-15}.

F^2T^2EA Find, fix, track, target, engage and assess (AFRL).

F^3 *1* Form/fit/function, called F-cubed.

2 Full flight-envelope flight, control law and display system.

3 Free-form fabrication.

F4 Airfield subgrade, standard asphalt.

F-class Restricted or advisory airspace (ICAO).

F-code In flight plan, aircraft has 4096-code transponder and approved R-nav.

F-display Target centred in rectangle, blip gives az/el aiming errors.

F-factor Dimensionless number interpreting vertical/horizontal strengths of windshear in terms of quantified reduction in climb performance.

F-layer One layer of ionosphere, at 150–300 km divided into F_1 and F_2 layers, F_2 being always present and having higher electron density.

f-pole Firing point of AAM which maximises aircraft/target separation at missile impact.

f-pole line Avionics limit which keeps fighter nose pointing within limits of radar gimbal boundary.

FA *1* Frequency agility, or frequency-agile.

2 Frontal (tactical) aviation (USSR, R)

3 Flight attendant, or assistant.

4 Free-air (tunnel).

5 Final approach.

F/A Fix/attack (display).

F_A Flaperon [rarely, flap] angle.

FAA *1* Federal Aviation Administration (US, since 1967 part of Department of Commerce, said to mean 'federal acronym association').

2 Federal Aviation Agency (US, 1958–67, independent body, previously CAA).

3 Fleet Air Arm (formed 1924 as part of RAF, from 1939 part of Royal Navy, until 1953 officialy called Naval Aviation.

4 Fuerza Aerea Argentina.

5 Foreign Airlines Association (UK, Japan).

6 Functional analysis and allocation.

7 Flasher and audio (alarm).

FAAA Flight Attendants Association of Australia.

FAAD Forward-area air defense; C_2I can be added, S adds system.

FAAM *1* Family of AAMs.

2 Fleet Air Arm Museum (Yeovilton, UK).

3 Facility for Airborne Atmospheric Measurements (Met. Office and NERC, UK).

FAAN Fight Against Aircraft Nuisance (UK).

FAAR Forward-area alerting radar, against low-flying aircraft.

FAAS Focal-area aerial surveillance.

FAATC *1* FAA Technical Center, Atlantic City, NJ, until 1981 called Nafec.

FAAWC Force Anti-Air-Warfare Commander.

FAB *1* General-purpose HE bomb (USSR, R).

2 Flight-authorization book.

3 Força Aerea Brasileira.

fabric Cloth or linen material of two main types: biased, multi-ply with one or more plies cut so that threads are transverse or diagonal; and parallel, with warp threads of all plies parallel.

fabricated Usually means welded.

fabrication *1* Generally, manufacture of hardware.

2 Specific, assembly by welding.

Fab-T, FAB-T Family of advanced beyond-LOS terminals (USAF).

FAC *1* Name of several air forces, usually Spanish- or Portugese-speaking.

2 Forward Air Control[ler].

3 Flight-augmentation computer.

4 Federal Airports Corporation (Australia).

5 Federal Aviation Commission (US, 1934–36).

6 Farnborough Aerospace Consortium (UK, 550+ companies).

7 Fast attack craft (marine).

FACCE See FAC^2E.

face *1* Any exposed quasi-flat surface, such as main area of turbine disk.

2 Any surface for mating with another.

3 Open end of duct to be joined to another, including front of gas-turbine-engine inlet.

4 Either surface of propeller or helicopter-rotor.

face alignment Distance perpendicular to chord from propeller or rotor blade chord centreline to flat face of blade at any station.

faceblind firing Method of firing ejection seat in which occupant pulls roller blind at top of seat down over his face, thus shielding latter from airstream on leaving aircraft.

faceplate *1* Disc mounted on nose of lathe spindle for

rotating work between centres or for gripping asymmetric item of short length.

2 Accessory mounting pad.

3 Transparent front of pressurized helmet.

Facet   Fault-assisted circuits for electronic training.

facet   Panel forming part of external visuals of simulator. The F-22 FMT has nine.

faceted aircraft   One whose external surface is made up of flat 2-D panels. Such an aircraft is theoretically invisible to hostile radars except for brief instants when one face is precisely normal to the incident signal. Such aircraft have severe flight-control limitations, and increased computer power now enables LO aircraft to have better aerodynamics.

Facets   Future anti-air concepts experimental technology (also terminal) seeker.

facility   *1* Physical plant, buildings and equipment (previously US usage).

2 Any part of adjunct of a physical plant or installation which is an operating entity.

3 An activity or installation which provides specific operating assistance to military or civil air operations.

facility availability   Actual/specified operating times, usually as percentage.

facility performance category   See *categories (3)*.

Faco, FACO   Final assembly and check-out.

FACP   Forward air control post.

FACS   Fully automatic compensation system.

facsimile   Telecommunications process in which picture of image is scanned and signals used locally or remotely, sent by telephone or TV, to reproduce * or likeness of subject image.

Factar   Follow-up action on accident reports.

Factor   Development of functional concepts from EATMS operational requirements (Euret).

factored field lengths   Any distance relative to CTOL operations (TOR, EMD, TOD etc) multiplied by factor to take account of engine failure at V_1, slippery surface or any other hazard.

factoring   Process of selecting and applying appropriate factors (of safety) in such areas as design and stress calculations, performance estimates etc.

factor of safety   *1* Factor by which limit load is multiplied to produce load used in design of aircraft or part; intended to provide margin of strength against loads greater than limit load, and against uncertainties in materials, construction, load estimation and stress analysis.

2 Ratio of ultimate strength to actual working stress or maximum permissible stress in use of material component.

factory loaded   Propellant charge or explosive filling added in plant before delivery.

factory remanufactured   Product, usually an engine, indistinguishable from new.

Facts   *1* FLIR-augmented Cobra Tow sight.

2 Fighter-aircraft-control training system.

FAC²E   Fighter air command and control enhancement.

FAD   *1* Fleet air defense (US).

2 Fast-action device.

3 Fighter aerodynamics development.

4 Feature analysis data.

5 Flexible-aircraft dynamics.

6 Funding authorization document.

7 Fuel advisory departure.

8 Forsvars & Aerospaceindustrien i Danmark.

FADA   Federación Argentina de Aeroclubes.

FADD   Fatigue and damage data.

fade   Decrease in received signal strength without change of receiver controls.

Fadec   Full-authority, or fully authoritative, digital engine (or electronic) control.

faded   Radio word meaning 'air-intercept contact has disappeared from reporing station's scope, and further information is estimated' (DoD).

fade-out   Fading in which received signal strength is reduced below noise level of receiver. Also known as radio fade-out, Dellinger effect, Mogel-Dellinger effect (see *blackout*).

fading   Variation of radio field strength caused by change in transmission medium.

FADR   Fixed [-site] air-defence radar.

FADS, Fads   Flush, or flexible air-data system.

FAE   *1* Fuel/air explosive; large class of ordnance devices.

2 Federación de Aeronáutica Española.

Faeshed   FAE store, helicopter delivery.

FAEI   Federation of Aerospace Enterprises in Ireland.

FAF   *1* Final-approach fix.

2 Full and free [flight controls].

FAFC   Full-authority fuel controller.

FAFL   Forces Aériennes Françaises Libres (1940–45).

FAFT   Fore/aft fuselage tankage (LH$_2$).

FAGC   Fast automatic gain control.

FAGr   Fernaufklärungsgruppe, long-range reconnaissance wing (G).

Fagsa, FAGSA   Federation of Airline General Sales Agents (UK).

Fahrenheit   Temperature scale, contrary to SI, in which ice point is 32° and boiling point of water 212°; thus to convert to °C subtract 32 and multiply by $^5/_9$; to convert to °K add 459.67 and multiply by $^5/_9$.

FAI   *1* Fédération Aéronautique Internationale, the supreme body ratifying aeronautical records, (office Lausanne, established 14 October 1905).

2 Fatal-accident inquiry.

3 First-article inspection.

fail-active   Quality of a dynamic functional system of remaining correctly operational after any single failure.

fail-hard   *1* Describes part or component, notable primary structure or other unduplicated load path, fracture of which would be catastrophic.

2 Describes system component whose failure renders system immediately misleading, incorrect or dangerous.

failing load   That which, when applied, will just cause structural member to fail.

fail link   Deliberate weak link to prevent overload damage to costly structure.

fail-operational   Any single system failure has no (or limited) effect on operation, though warning is given.

fail-passive   Failure inactivates system, thus preventing dangerous spurious signal or hardover output.

fail-safe   Normally structural design, rather than system, technique in which no crack can cause catastrophic failure of whole structure but is allowed to occur and be detected.

fail-soft   System failure does not inhibit operation but authority or limits of travel are reduced.

FAIP   First-assignment instructor pilot.

fairing   Secondary structure whose function is to reduce drag; eg streamlined cover, or junction between two parts.

fairing wire   Wire provided as point of attachment for outer cover to maintain contour of airship envelope.

fairlead   Streamlined tube through which trailing aerial or other cable exits aircraft.

fair over   To reduce drag of excrescence by fitting fairing over or around it.

faker   Strike aircraft engaged in air-defence exercise (DoD).

FAL   Facilitation of air transport (ICAO AIP).

Falac   Forward-area liaison and control.

Falcon   *1* Frequency-agile low coverage netted.

2 Force application and launch from the Continental US.

fallaway section   Part of rocket vehicle that separates; one that falls back to Earth.

fallback   *1* Immediate return of malfunctioning ballistic vehicle after vertical launch.

2 Material carried into air by nuclear explosion that ultimately drops back to Earth.

fallback area   Area to which personnel retire once missile (large, surface-launched) is ready for firing.

fallback programme   Second project undertaken as insurance against failure of first.

falling leaf   Aerobatic manoeuvre in which aircraft is stalled and then forced into spin; as soon as spin develops, controls are reversed; process is repeated, resulting in oscillations from side to side with little apparent change in heading.

fall off on a wing   See *stall turn*.

fallout   *1* Rain of radioactive particulate matter from nuclear explosion. Local * settles on surface within 24 hr; tropospheric * is deposited in narrow bands around Earth at about latitude of injection; stratospheric * falls slowly over much of Earth's surface.

2 See *spin-off*.

fallout contours   Lines joining points of equal radiation intensity.

fallout mission   Alternative or secondary combat mission, primary being impossible of accomplishment.

fallout pattern   Distribution as portrayed by fallout contours.

fallout prediction   Estimate before and immediately after nuclear detonation of location and intensity of militarily significant fallout.

fallout safe height   Altitude of detonation above which no militarily significant fallout will be produced.

fallout wind plot   Wind vector diagram from surface to highest altitude affecting fallout pattern.

false alarm   Appearance on a radar display of what appears to be a valid target but is caused by something else.

false cirrus   Cirrus-like clouds in advance of and at summit of thunder cloud, lacking feathery texture. Also known as thunderstorm cirrus.

false-colour film   *1* Has at least one emulsion layer sensitive to radiation outside visible spectrum (eg infra-red), in which representation of colours is deliberately altered.

2 Modified film whose dye layers produce assigned colours rather than natural ones.

false cone of silence   Radio-range phenomenon similar to cone of silence above transmitting station and likely to occur over mountains, ore deposits or other factor that causes dead spot in reception; lacks four characteristics of true counterpart; build-up, deadspot, surge and fade.

false glidepath   Loci of points in vertical plane containing runway centreline at which DDM is zero, other than those forming ILS glidepath.

false heat   Emitted by flares and other IR decoys.

false lift   Additional aerostat lift caused by positive superheat, temperature difference between gas and surrounding air.

false nosing   Built up of nose-rib formers and D-skin, attached to front spar to form leading edge.

false ogive   Rounded fairing added to nose of vehicle to improve streamlining. Also known as ballistic cap.

false ribs   Auxiliary nose ribs between main ribs forward of front spar to support fabric covering and improve contour.

false spar   Secondary spar not attached to fuselage, used as mounting for movable surfaces.

false start   Gas-turbine starting cycle which fails to achieve stable light-up; ability to survive is certification requirement.

FALW   Family of air-launched weapons.

FAM   *1* Fighter attack manoeuvring.

2 Federal Air Marshal [P adds program].

3 Final-approach mode.

Fame   Full-sky astrometric mapping explorer.

FAMG   Field artillery missile group (USA).

familiarisation   Training to acquaint technical personnel with specific system.

Famis   Full-aircraft management and inertial system.

famished   Air-intercept code: 'Have you any instructions for me?' (DoD).

Famos   Floating-gate avalanche-injection MOS.

FAMS   Family of air missile systems.

FAN   *1* Forward air navigator.

2 False-alarm normalization [normally holds CFAR to 10^{-6}].

fan   *1* Vaned rotary device for producing airflow.

2 Multi-bladed rotor, usually with single stage, serving as first stage of blading in turbofan [last stage in aft-* engine] and handling much greater airflow than core.

3 Propeller, when function is moving air rather than providing thrust.

4 Assembly of three or more reconnaissance or mapping cameras at such angles to each other as to provide wide lateral coverage with overlapping images.

fan blade off   Most severe turbofan certification requirement, ability to contain and survive severance of one entire fan rotor blade at redline speed without danger to aircraft.

F&E   Facilities and equipment.

F&F   Fire and forget.

F&R   Function and reliability.

F&U   Fire and update.

fan duct   Annular duct [in B-52H twin C-ducts] through which air compressed by fan of turbofan engine is delivered. Can be short, ending in annular propulsive nozzle surrounding core casing, or extend to rear where there may be a mixer. Almost always incorporates a reverser.

fan engine   *1* See *turbofan*.

2 Three-cylinder engine with one cylinder vertical and others at about 45° to it.

fan exit case   Casing surrounding fan (2) carrying reverser and often accessory gearbox.

fan-failure clutch　When necessary, disconnects engine fan from transmission of tilt-rotor.

Fang　Federation of Anti-Noise Groups (charity, UK, 1973–).

fan jet　Turbofan, or aircraft powered thereby (colloq.).

fan lift　Jet V/STOL system using large axial fans inside wings and fuselage covered by shutters above and below which are opened only in hovering mode.

fan mapping　Aerial survey using fan of cameras.

fan marker　Radio position-fix beacon radiating in vertical, fan-shaped [ellipse or dumbell] pattern, keyed for identification (see *radio beacon*, *Z-marker*, *FM-marker*).

fanned-beam antenna　Unidirectional antenna so designed that transverse cross-sections of major lobe are approximately elliptical.

fanning beam　Radiant-energy beam (eg radar) which sweeps back and forth over a limited arc (see *scan*).

Fanpac　Fan-noise prediction and control.

fan ramjet　See *augmented turbofan*.

FANS, Fans　Future air navigation system[s] (ICAO).

fan straightener　Radial vanes in front of and/or behind fan in wind tunnel to introduce or remove flow rotation usually counteracting that of fan.

fan stream burning　Thrust boosting by burning fuel in airflow downstream of fan; in some vectored engines same as PCB, in ejector lift and RALS after travel along large pipe.

FAO　Fabrication assistée par ordinateur (F).

FAOR　Fighter area of responsibility.

FAP　*1* Fleet average performance.
2 Force Aérienne de Projection (F).
3 Fuel-adjusted profit.
4 Frangible armour-piercing; DS adds discarding-sabot.
5 Forward attendant panel.
6 Final approach, or final-approach point.
7 Federation of Australian Pilots.
8 Fluorinated aluminium powder.

FAPA　*1* Future Aviation Professionals of America.
2 First Air Pilots Association.

FAQ　Frequently-asked question[s].

FAR　*1* Federal Aviation Regulation[s]; eg FAR-23 [also called Part 23] defines flight performance of private and taxi a/c ≤12,500 lb [5670 kg] MTOW, FAR-25 covers a/c above this limit, Pt 36 is concerned with noise and FAR-103 with single-seat ultralights, for example.
2 False-alarm rate.
3 Fighter/attack/recon. (pilot grading, USAF).
4 Field assessment (or functional area) review (US).
5 Force d'Action Rapide (F).
6 Forward-area rearm [or rearm/refuel, P adds point].
7 Federal Acquisition Regulations.
8 Federatia Aeronautica Romana.
9 Fatal-accident rate.

FARA　Formula Air Racing Association (UK).

farad　SI unit of electrical capacity, Symbol F; capacity of condenser (capacitor), which has potential difference of 1V when charged with 1C. More commonly used: micro-farad and picofarad.

faraday　Symbol F; non-SI unit of electric charge carried by 1 mole of singly-charged carbon-12 ions = $9.6487×10^4$C.

Faradex　Functional architecture reference for ATM (7) systems and data exchange (Euret).

Faraway　Fusion of radar and ADS (5) data through two-way data-link (Euret).

fare dilution　Dilution of airline revenue yield by excursion, affinity, group, seasonal or off-peak and other types of promotional fares, and by discounted or free travel to employees, or passengers on particular sectors.

fare structure　Complete range of airline fares, either approved by licensing authority such as CAB for domestic use or agreed at IATA traffic conferences for international use.

far-field boom　Supersonic N-wave boom after long travel has changed form, esp. by reducing rate of change of pressure at front and rear.

far-field noise　Noise, especially from jet engine, at considerable distance (typically 100+ metres) where higher frequencies are attenuated.

far IR, far infra-red　Wavelengths longer than 6 μ.

farm　Compact group of large number of aerials (antennas), especially protruding from aircraft.

farm-gate operations　Operational assistance and specialised tactical training provided to friendly foreign air force by United States armed forces; includes, under specifed conditions, flying of operational missions by combined US and foreign aircrew as part of training when such missions are beyond recipient's capability.

farm strip　Private airfield, usually with no facilities except hangar.

Farnborough　Location of Qinetic/Royal Aerospace Establishment (RAE), originally Royal Aircraft Factory (UK).

Farnborough indicator　Pioneer indicator for continuously recording pressure cycles in cylinder of piston engine.

Farnham roll　Large powered machine for two-dimensional bending of sheet metal.

FARRP　Forward-area rearming and refuelling point (or **FARP**, forward arming and refuelling point).

farval　Aerobatic manoeuvre in which two aircraft perform routine with one inverted above the other (thus a half-roll results in the pair changing places). See *double* *, (USN 1929, relaunched by Blue Angels 1962).

FAS　*1* Frequency-agile subsystem.
2 Flight-attendant station.
3 Forward acquisition sensor.
4 Federation for Air Sport (USSR).
5 Flare-augmentation system.
6 Fuel-advisory system.
7 Fore-and-aft scanner; S adds system.
8 Forces Aériennes Stratégiques (F, note plural, unlike FAT).
9 Future antenna suite.
10 Federal air surgeon (US).
11 Final-approach segment.

FASA　Friendly aircraft simulating aggressors.

Fasat　Future anti-satellite (weapon).

FASGW　Future air/surface guided weapon (UK).

FASH　Future amphibious support helicopter.

FASI　Federation of air sports (Indonesia).

Fasid, FASID　Facilities and services implementation document.

FASM　*1* Forward air-support munition.
2 Farnborough Air Sciences Museum (2003–).

Fasotragru　Officially written in capitals, Fleet Aviation

Specialized Operational Training Group; DET adds Detachment (USN).

FASS   Fore and-aft scanner system.

FAST, Fast   *1* Fan and supersonic turbine.

2 Fuel and sensor, tactical (clip-on pack).

3 Future aviation safety team (EC, JAA).

4 Flying-ambulance surgical trauma.

5 Fuselage automated, or automatic, splicing tool.

6 Fast-acting stabilizing [reefed drogue].

7 Final-approach spacing tool.

8 Fleet-aircrew simulation training.

9 Forecasting and assessment of science and technology.

10 Fly-away satellite terminal.

11 Forward-area support team.

12 Flight-advisory service test [of civil/military ground radar].

13 Fully automatic scoring target.

14 Flexible acquisition and sustainment tool (USAF).

15 Fuze air-to-surface technology (USAF).

FASTA   Farnborough Air Sciences Trust Association (UK).

Fasta   Flugzeugabwehrstartanlage (air-defence launcher, G).

Fast-action device   Thyristor switch which brings battery on line upon failure of generator or TRU.

Fastar   Forward-area surveillance and target-acquisition radar.

FASTC   Foreign Aerospace Science & Technology Directorate Center (USAF).

Fast CAP, Fastcap   Combat air patrol by fast jet.

Fastec, FASTec   Foundation for Advancing Science & Technology Education (US).

fast erection   Provision for super-rapid [usually elecrical] acceleration of gyro[s].

Fast FAC   Forward air controller in fast jet.

Fastjam   Flow analysis for selective, or selected, target jamming (Darpa).

fast jet   Generic title for ATC purposes of any aircraft with typical jet speed.

fast mover   Jet combat aircraft, especially in FAC role.

fast prototyping   Techniques for getting first flight article airborne at earliest possible date, ignoring deficiencies and making maximum use of simulation.

fast-reaction weapons demonstration   Ongoing research into optimum methods of dispensing multiple miniature smart submunitions (USAF).

Fastt   Flight-strip automation system for towers and Tracons.

FASU   Federation of Aeronautical Sports of Ukraine.

FAT   *1* Flechette anti-tank.

2 Factory acceptance test.

fat   *1* Overweight.

2 Material that can be removed to meet less-severe requirement, as in civil derivative of military engine.

Fatac   Force Aérienne Tactique (F).

fatal accident   One in which at least one occupant is killed; casualties on ground do not qualify.

Fatca   Federal ATC authority (Yugoslavia).

Fate   *1* Fuzing, arming, test and evaluation.

2 Factory acceptance test equipment.

FATG   Fixed air-to-ground, ie against non-moving target.

fathom   Nautical unit of sea depth, 6 ft, 1.8288 m.

fatigue   *1* Weakening or deterioration of metal or other material under load, esp. repeated cyclic load; causes cracks and ultimately failure.

2 Progressive decline in human ability to carry out appointed task apparent through lack of enthusiasm, inaccuracy, lassitude or other symptoms.

fatigue index   Arbitrary scale of airframe structure life terminating at 100, but capable of being extended to higher values by modification.

fatigue life   Minimum time, expressed in thousands of hours or specified number of load cycles, that structure is designed to operate without fatigue failure.

fatigue strength   Maximum stress that can be sustained for specified number of cycles without failure. Also known as fatigue limit.

fatigue test   Test in which specimen is subjected to known reversals of stress, such as alternate tension and compression, or cycle of known loads repeatedly applied and released.

Fatmi   Finnish air-traffic management integration.

FATO, Fato   Final approach and takeoff [determines size of heliport].

Future   Federation of Air Transport User Representatives in Europe.

FAU   Forward antenna unit.

FAUSST   Franco-Anglo-US SST Committee (1964–66).

fav   Fuel available.

favourable unbalanced field   A T–O with excess power available, so that surplus acc/stop distance can give screen height greater than 35 ft.

FAW   Fighter, all-weather (role prefix, UK).

FAWS   Future airborne weapon system.

faying surface   Overlapping of adjoining skin surfaces with edges exposed to airstream or to water.

FB   *1* Fighter-bomber (role prefix, UK). Also major subdivision of undergraduate pilots, as alternative to TTT (USAF).

2 Fingerprint biometrics.

3 Flare block [of cartridges].

F_B Aerodynamic loading due to buffeting pressure field.

F_b Burn time at average thrust (rocket).

FBC   *1* Fighter-bomber clear-weather day/night.

2 Fan/booster/compressor [module].

3 Fly by cable = mechanically signalled manual FCS.

FBE   Fleet battle experiment (USN).

FBG   Fighter/Bomber Group (USAAF).

FBI   Frequency and bias injection.

FBL   *1* Fly-by-light.

2 FIATA combined transport bill of lading.

FBM   *1* Fleet ballistic missile (S adds 'submarine' or 'system').

FBO   *1* Fixed-base operator.

2 Flights between overhauls.

3 Federal budget outlays.

4 Fan-blade off.

FBPAR   Fixed-base precision approach radar.

FBR   Fuji bearingless [main] rotor.

FBS   *1* *Fly-by-speech.*

2 *Fixed-base simulator.*

3 Forward-based system.

4 Flash/bang/smoke.

FBT   Fixed-base tower.

FBVL   Fédération Belge de Vol Libre.

FBW *Fly-by-wire.*
FC *1* First class.
 2 Fuel cell.
 3 Funnel cloud, tornado or waterspout.
 4 Foot-candle[s].
 5 Flight crew.
F$_c$ Fracto-cumulus.
f$_c$ Carrier frequency.
F/C Forecast.
FCA *1* Future cycle accumulation (engine).
 2 Functional configuration audit (software).
 3 Flight-control, or -critical, avionics.
 4 Fully controllable array.
FCAS Future combat air system.
FCB Frequency co-ordinating body.
FCBA *1* Fédération des Clubs Belges d'Aviation.
 2 Future carrier-based, or -borne, aircraft (UK, replaced by FJCA).
FCC *1* Federal Communications Commission (US, from 1934).
 2 Flight-control computer.
 3 Flight-connection centre (DAIS.2).
 4 Flat conductor cable.
FCCC, FC3 Framework Convention on Climate Change.
FC cost Flight-crew cost.
FCDA *1* Federal Civil-Defense Administration (US).
 2 Federación Colombiana de Deportes [sport] Aereos.
FCDC *1* Flight-control digital computer.
 2 Flight-control data concentrator.
 3 Flight-critical direct current.
 4 Flexible confidant detonating cord.
FCDM Flow-control decision message.
FCDS Flight-control display system.
FCE *1* Flight-control electronics.
 2 Flight-crew environment.
 3 Full cockpit emulator.
FCEM Flow-control execution message.
FC-ECY, FC/ECY First-class and economy.
FCES Flight-control electronics system.
FCF Functional check, or checkout, flight.
FCG Fatigue-crack growth.
FCGMS Fuel and c.g. management system.
FCI *1* Fuel-consumed indicator.
 2 Flight-command indicator.
FCL Flight-crew licensing (CAA).
FCLP Field carrier-landing practice.
FCLT Freeze calculated landing time (FAA).
FCMC Fuel control and monitoring computer.
FCNP Fire-control navigation panel.
FCNS Fiber-channel network switch (USN).
FCNU Flight-control navigation unit (UAV).
FCO *1* Formal change order (contract).
 2 Fire-control operator.
FCOC Fuel-cooled oil cooler.
FCOM Flight crew operating, or operations, manual.
FCP *1* Fuel cell powerplant.
 2 Flight-control panel, or processor.
FCPC Flight-control primary computer.
FCR Fire-control radar.
FCRC Federal Contract Research Center (US).
FCRS Flight-crew record system.
FCS *1* Flight control system.
 2 Fire control system, for management of weapons.
 3 Failure combat system.
 4 Future combat system (USA).
 5 Frame check sequence.
FCSC Flight-control secondary computer (DoD).
FCSS Fire-control sight system.
FCST Federal Council for Science & Technology (US).
Fcst, FCST Forecast.
FCT *1* First configuration test.
 2 Foreign comparative test[ing] (USAF).
 3 Flight-crew training [RM adds reference manual].
 4 Friction coefficient.
FCTP Flight-control technology program (VTOL, at WPAFB).
FCTR Fan/core thrust ratio.
FCTS Flight-controller training simulator.
FCU Feathering, fighter, flight or fuel control unit.
FD *1* Flight director.
 2 Frequency duplex.
 3 Frequency domain.
 4 Flight, or final, data.
 5 Flight deck [D adds documentation].
 6 Full development.
 7 Flight dynamics.
F$_d$ Takeoff-distance correction factor for slush, = TOD÷TOD for dry runway.
f$_d$ Doppler frequency.
FDA Flight-data acquisition; F adds function, S system and U unit.
FDAC Flight demonstration of ASTA (2) concepts.
FDAD Full digital Arts (1) display.
FDAF See *FDA.*
FDAI Flight-director attitude indicator.
FD&E Forces development and evaluation (USAF).
FDAU See *FDA.*
FDB Flight[plan] data bank.
FDC *1* Frequency-to-digital converter.
 2 Flight-director computer, or coupler.
 3 Flight-data concentrator, or centre.
 4 Fire direction centre; OPC adds operations planning cell.
FDCC Forward-deployed communication[s] center (USAF).
FDD Flight-data display.
FDDI Fibre-optics distributed, or fibre distribution, data interface.
FDDP Full-digital design process.
FDE *1* Fire detection and extinguishing.
 2 Fault detection and exclusion.
 3 Force development evaluation.
FDECU Field-deployable environmental-control unit.
FDEP Flight-data entry and print-out (or FDE panel).
FDF *1* Føreningen Danske Flyvere (Danish pilots' association).
 2 Fachverband Deutsche Flugdatenbearbeiter.
FDFF Føreningen af Danske Fabrikanter af Fly-material (Danish industry assoc.).
FDFM Flight-data and flow management (ICAO group).
FDH Flight-deck handset.
FDI *1* Flight director indicator.
 2 Flight-data interface [MU adds management units, U unit].
 3 Fault detection and isolation.
FDIO Flight-data input/output.

FDL   *1* Full-drawn line (symbology right across display).

2 Flight Dynamics Laboratory.

3 Fast deployment logistic (ship).

4 Fighter data link.

FDM   *1* Frequency-division multiple, or multiplex.

2 Fused deposition modelling.

FDMA   Frequency-division multiple access.

FDMS   Flight deflection measurement system.

FDMT   Flight-data monitoring tool.

FDMU   Flight-data management unit.

FDO   Flight-deck officer.

FDOA   Frequency difference of arrival.

FDP   *1* Flight [-plan] data processing, or processor; R adds replacement, S adds system, or service.

2 Flight-duty period.

3 Floating-deck pulser.

4 Funded delivery period.

FDR   *1* Flight-data recorder [A adds analysis, S system].

2 Flat-deck runway.

3 Flight-deck reporting.

FDS   *1* Flight director, or data, or display, system.

2 Fence disturbance sensor.

3 Flight-deck [of aircraft] simulator.

4 Field-deployable simulator.

FDSC   Future defence supply chain.

FDSO   Full dispersed-site operations.

FDSS   Flight display subsystem.

FDSU   Flight-data storage unit.

FDT   Flight-deck, or -data, terminal[s].

FDTE   Force development test and experimentation.

FDU   *1* Flight-data unit.

2 Flux detector unit.

FDVLO   First-day vertical liftoff.

FE   *1* Flight engineer [or examiner].

2 Ferroelectric.

3 Fan exit.

Fe   Iron.

F$_e$ Static thrust per engine at sea level.

FEA   *1* Federal Energy Administration (US).

2 Finite-element analysis.

FE(A)   Flight Examiner (Aeroplanes).

FEAF   Far East Air Force[s] (WW2 and Korea).

FEAR   Failure-effect[s] analysis report.

feasibility study   Determines whether plan is within capacity of, or makes best use of, resources available.

feathering   *1* Turning propeller blades to feathering angle, following engine failure or apparent malfunction, to minimise drag and prevent further damage.

2 Of helicopter, cyclic pitch.

feathering angle   See *feathering pitch*.

feathering button   Used to feather propeller; protected by hinged cover.

feathering hinge   Helicopter rotor-blade pivot which allows blade angle to be varied.

feathering pitch   Angular setting giving zero windmilling torque for stopped propeller (opposite ends of blades cancelling out), thus minimum drag.

feathering pump   After stoppage or failure of engine, provides hydraulic pressure to feather propeller.

feathers   Wing movables: slats, Krügers, droops, flaps, ailerons, spoilers (colloq.).

FEATMS   [sometimes **Feats**] Future European air-traffic management system.

feature console   In passenger cabin, clock, TAS readout, phone, fax etc.

feature-line overlap   Series of overlapping air photographs which follow ground feature such as river or road.

FEBA   Forward edge of battle area (replaced by FLOT).

FEC   Forward error correction.

fecal canister   Sealable container for human solid wastes in spaceflight.

FED   *1* Field-effect display.

2 Field emission display.

fed   Of radio, supplied with RF oscillations.

Federal Air Marshal   Armed guard carried [as ordinary passenger] on US commercial flights to deter terrorism (FAA).

Federal Flight-Deck Officer   Captain or copilot trained to carry a gun (TSA).

federated   Traditional arrangement of avionics in which each suite provides its own processor and a separate unit [usually called mission computer] distributes workload and output.

Fedix   Federal information exchange, online.

FEDN   Fondation pour les Etudes de Défense Nationale (F).

Feds   The FAA (1) [colloq.].

feed   *1* To provide signal.

2 Point at which signal enters circuit or device.

3 Signal entering circuit or device; input.

4 Means of supplying ammunition to gun, or chaff through dispenser.

feedback   *1* Return of portion of output to input; positive * adds to input, negative * subtracts from it.

2 Information on progress, results, field performance, returned to originating source.

3 Transmission of aerodynamic forces on control surfaces or rotor blades to cockpit controls; also forces so transmitted.

feedback control loop   Closed transmission path containing active transducer, forward path, feedback path, and one or more mixing points arranged to maintain prescribed relationship between input and output signals.

feedback control system   One or more feedback control loops to combine functions of controlled signals with functions of commands to tend to maintain prescribed relationships between them.

feedback path   Transmission path from loop output signal to loop feedback signal.

feeder   *1* Transmission line which connects aerial to transmitter or receiver.

2 Air route or service that feeds traffic to major domestic or international routes (see *commuter*, *third-level*).

feederliner   Transport aircraft used to operate feeder, commuter or third-level services.

feeder route   Links en-route to initial approach fix.

feed pipe   Pipe supplying any liquid.

feed tank   Small tank drawing fuel from main tankage and transferring it under pressure to an engine.

feel   Subjective pilot assessment of aircraft response to flight-control commands, stability, attitudes and other factors influencing his opinion.

feeler aileron   Small manual aileron whose primary purpose is to impart feel.

feel system   Mechanism in which control feel is augmented, improved or simulated artificially rather than provided only by aerodynamic forces on control surfaces (see *artificial feel*).

feet dry   Code: "I am, or contact designated is, over land" (DoD).

feet per second   Ft/s, = 0.3048 ms^{-1}, 1.09728 km/h.

feet wet   Code: "I am, or contact designated is, over water" (DoD).

FEFA   Future European Fighter Aircraft (project).

FEFI   Flight engineers fault isolation (technique and handbook).

FEGV   Fan-exit guide vane.

FE(H)   Flight Examiner [Helicopter].

FEI   *1* Federation of the Electronics Industry (UK, 300+ members).

     2 Field engineering instructions (NATS).

FEIA   Flight Engineers' International Association (US, merged into IFEO).

FEL   Free-electron laser.

FEl, Fel   Fibre-elastomer.

FELC   Field-effect liquid crystal.

FELD   Forward electrical load center (EP-3).

FELT, Felt   Free-electron laser technology; IE adds integration experiment (SDI).

felt   Non-woven materials used when properties of uni-directional fibre-reinforced plastics are not required; built up from fibres or whiskers of carbon, glass, formerly asbestos, etc.

felt strip   See *moleskin*.

FEM   *1* Force effectiveness measure.

     2 Finite-element model/mesh/method.

FEMA   Federal Emergency Management Agency (US).

femto   Prefix: multiplied by 10^{-15}, one thousandth of a millionth of a millionth; see *fermi*.

fence   *1* Line of readout or tracking stations for communication with satellite.

     2 Line or network of radar stations, on land or round periphery of surface fleet, for detecting enemy aircraft or missiles.

     3 Wall-like plate mounted on upper surface of wing, often continuing around leading edge, substantially parallel to airstream and used to prevent spanwise flow, esp. over swept wing at transonic speeds.

Fenda   Federación Nacional de los Desportes Aéros (Spain).

FENE   Fixed exit nozzle engine.

Fenestron   Helicopter tail rotor with numerous blades rotating in short duct inset into fin.

FEO   Federal Energy Office (US).

FEP   Front-end processor.

fermi   Unit of length, = 10^{-15}m.

Ferpic   Ferroelectric photoconductive image ceramic.

FEPS   Flight-envelope protection system.

ferret   Aircraft, ship or other platform equipped for detection, location, recording and analysis of hostile EM radiation (Elint mission).

Ferris scheme   Carefully designed paint scheme using two shades of colours to make it difficult to ascertain aircraft attitude (secondarily, aircraft type and direction of travel).

ferrite   Magnetic ceramics composed of salts of iron and another divalent metal; because of low eddy-current losses, cores constructed of sintered powders of these materials are widely used for rod aerials and cores of inductors for RF and video.

ferrite paint   See *iron paint*.

ferritic   Of ferrite.

ferrous   Derived from iron.

Ferroxcube   Proprietary non-metallic insulating magnetic materials which have extremely high resistivity and low eddy-current losses but do not become permanently magnetised.

ferrule   Small metal fitting or wire wrapping used to prevent loosening of wire terminal.

ferry flight   Flight whose purpose is to reposition aircraft at a different place.

ferry pilot   One responsible for delivering aircraft from one place to another; eg from manufacturer to customer.

ferry range   Distance unladen aircraft can be ferried; specified with or without ferry tanks.

ferry tank   Extra fuel tank for ferry flight over range greater than normal limit.

fertile material   Not itself fissile by thermal neutrons, can be converted into fissile material by irradiation; two are U-238 and Th-232, partially converted into Pu-239 and U-233.

FES   Flexible elastomer skin.

FESC   Forward electrical/electronic service centre.

Fescolizing   Patented electroplating of Cd, Cr or Ni.

FESG   Forecasting and Economic-Support Group.

FEST   Foreign emergency support team (USAF).

FET   Field-effect transistor.

FETAP   Fédération Européenne des Transports Aériens Privés.

FETT   First engine to test, date of first run of first complete engine of new type; confusingly sometimes said to mean first engine Type-Test, which might be years later.

FEW   *1* Fighter Escort Wing.

     2 Few clouds, usual = 2 oktas.

The Few   Collectively, the figher pilots defending the UK between 10 July and 31 October 1940.

few   Up to 7 hostile aircraft (DoD).

FEWP   Federation of European Women Pilots.

FEWSG   Fleet EW Support Group.

FF   Final fix.

FF, f.f.   Fuel flow.

F/F   First flight.

f_f Frequency to which digital filter is tuned.

FFA   *1* Flying Farmers Association (UK).

     2 Foam-filled aluminium.

     3 Flygtekniska Fösöksanstalten; aeronautical research institue, merged 2001 into FOI (Sweden).

FFAM   Fédération Française d'Aéro-Modélisme.

FFAR   *1* Folding-fin aircraft rocket (2.75-in calibre).

     2 Rarely, free-flight, or forward-firing, aircraft rocket.

     3 Feel forces/stick angle relationship.

FFATC   Free-flight [phase] air-traffic control.

FFBW   Fully fly-by-wire.

FFC   *1* Fan-failure clutch.

     2 For further clearance.

FFCC   Forward-facing crew cockpit.

FFCS   *1* Formation-flight control system.

     2 Fly-by-wire [primary] flight-control system.

     3 Free-fall control system (air-dropped ICBM).

FFD   FMS (1) flight data.

FFDO   Federal Flight-Deck Officer.

FFF *1* Film-forming foam (extinguishants).
2 Free-form fabrication.
FF/FU Fuel flow/fuel used (panel instrument).
FFG Code: guided-missile frigate (USN).
FFH For further headings.
FFI *1* Freedom from infection.
2 Forsvarets Forskningsinstittut (defence research, Norway).
FFIS Formation-flight instrumentation system.
FFK Full-function keyboard.
FFM *1* True-mass fuel flowmeter.
2 Far-field monitor.
FFMRRR, F^{2}MR^3 Folded fibreglass-mat rapid runway repair system.
FFN Far-field noise.
FFNAC Fédération Française des Navigants de l'Aviation Civile.
FFO *1* Furnace fuel oil.
2 Fixed-frequency oscillator.
3 Full fuzing option.
FFP *1* Firm fixed price.
2 FOV/focus/polarity.
3 Flight fine pitch.
4 Fédération Française de Parachutisme (F).
5 Frequent-flyer programme.
6 Free Flight Phase, followed by -1 or -2 (FAA).
7 Free-fall parachuting.
FFPB Free-fall practice bomb.
FFPS FFP(7) site.
FFPVL Fédération Française de Parachutisme Vol Libre (hang gliding).
FFR *1* Fuel-flow regulator.
2 Full flight regime, ie operative throughout each flight.
3 Full flight release (engine certification).
FFRAT Full-flight-regime autothrottle.
FFRDC Federally Funded R&D Center (FAA).
FFS *1* Full flight simulator.
2 Fee for service.
3 Formation Flight System (Honeywell).
FFSP Full-function signal processor.
FFT *1* Fast Fourier transform.
2 Full-scale fatigue test [S adds specimen].
FFTTEA See *F^{2}T^{2}EA*.
FFTx Fuel-flowmeter transmitter.
FFVC Forward-facing video camera.
FFVV Fédération Française de Vol à Voile (gliding).
FG *1* Fighter Group (USAAF, USAF).
2 Fog, defined as visibility ⩽⅝ mile.
3 Fuel-carrying glider (USAAF, 1944–47).
F$_g$ Gross thrust.
FGA Fighter, ground attack (role prefix, UK).
FGB Functional payload [or cargo] block (R).
FGC Flight-guidance computer.
FGCP Flight-guidance control panel.
FGCS Flight guidance and control system[s].
FGIH Federal government in-house.
FGM Flux-gate magnetometer.
FGMDSS Future global maritime distress and safety sytem, integrated with aviation satellites.
FGPA Field-gate programmable array.
FGR Fighter, ground attack, reconnaissance (role prefix, UK).
FGS Fine-guidance sensor.
F$_{gs}$ Gross thrust corrected to standard weight.

FGV Field-gradient voltage.
FH *1* Flight hour(s).
2 Frequency-hopping.
FHA *1* Fleet hour agreement (engine support).
2 Functional hazard assessment.
FHC Fluorinated hydrocarbon.
f.h.p. Friction horsepower [lost].
FHS Flight-hardware simulator.
FHSS Frequency-hopped spread spectrum.
FHU Force helicopter unit (helicopter portion of detached autonomous force).
FHV Fuel-heating value.
FHW Fault-history word.
FI *1* Flying Instructor.
2 Fault isolation.
3 Fluid injection (TVC).
4 Fatigue index.
F/I Flight idle engine power.
FIA *1* Fédération Internationale d'Astronautique.
2 Future imagery architecture (NIMA).
FIAS Formation Internationale Aéronautique et Spatiale.
Fiasts Fully integrated aircrew synthetic-training service (RAF).
FIAT *1* First installed article test.
2 Field Information Agency, Technical (US Group Control Council 1945–6).
FIATA Fédération Internationale des Associations de Transitaires ou Assimilés (International Federation of Forwarding Agents' Associations, office Zurich).
FIB Forwarding information base.
FIBDATD "Fix it but don't alter the drawings".
Fiberloy Family of composite materials based on boron fibres bonded in various resinous or plastics adhesives (Dow Chemical).
fibre Word used loosely of FBL and other systems employing optical fibres for all data transmission [US **fiber**].
Fibredux Family of CFRP and hybrid prepreg resins (Ciba-Geigy).
Fibrefrax H High-temperature ceramic-fibre insulating material, available as bulk fibre, blanket, felt or paper.
Fibreglass Glass-fibre, either raw fibre (many forms) or bonded into matrix and moulded or otherwise formed [capital F in UK, in US **fiberglass**].
Fibrelam Plastic honeycomb sandwich panel mechanically resistant to spike heels and not affected by galley or other spillage (Ciba/Geigy).
fibre optics Branch of optics concerned with propagation of light along thin fibres each comprising core and sheath of different glasses or other transparent material; light entering one end is transmitted by successive internal reflections. In practice extremely fine fibres a few microns in diameter are made up into bundles of 100,000 or more.
fibre-optics gyro Instrument (not a gyro at all) for measuring rotations by means of coherent light passed simultaneously both ways around a loop (typically 300–500 m long) of monomode optical fibre, rotation being measured instantly and precisely by phase shift at output.
fibrescope Fibre-optic borescope.
Fibs, FIBS Flight information billing system.
Fibua Fighting in built-up areas.
FIC *1* Flight information center, or centre.

2 Finance committee (ICAO).

3 Film integrated circuit.

4 Flying [or flight] instructor course.

5 Frequency/identification/course.

6 Flight inspection computer, compares aircraft position with that derived from navaids.

Fick's law   Basic law of gaseous diffusion: mass flux j = diffusion coefficient D times differential dC_1/dy where C_1 is concentration of gas 1 and y is distance from surface.

Ficon   Fighter conveyor, fighter carried to target by large bomber, to offer protection (USAF).

Fidag   Federazione Italiana Dipendenti Aviazione Generale.

fidelity   *1* Accuracy with which electronic or other system reproduces at output essential characteristics of input signal.

2 Handling * is degree to which flight simulator replicates handling of real aircraft.

FIDO   *1* Flight dynamics officer.

2 Fog Investigation and Dispersal Operation, UK method of dispersing fog in WW2 by burning fuel along runway edges.

3 Field integrated design and operations (Mars vehicle).

Fids, FIDS   *1* Fault identification and detection system.

2 Flight information display set [ATC radar] or system [for passenger information].

3 Fire detection and suppression system.

fiducial marks   Index marks on camera which form images on negative to determine position of optical centre or principal point of imagery; collimating marks.

field   *1* Airfield, as in * length.

2 Region of space within which each point has definite value; examples are gravitational, magnetic, electric, pressure, temperature, etc. If quantity specified at each point is vector, field is said to be vector *.

3 Customer service, thus * service, * rep, * report.

4 Operation at advanced base with austere facilities (military), thus * maintenance.

field alignment error   In ground DF station, error introduced by incorrect orientation of aerial elements.

field coils   Two fixed coils of DF goniometer at right angles to each other and connected to two halves of aerial system.

field extension   Organizational element performing operating functions that must be retained under direct control of parent staff office (USAF).

field-handling frame   Portable frame attached to airship on ground to afford grasp to large handling crew.

field inventory   Portfolio of used aircraft, parked and immediately available.

field length   Distance required for takeoff and landing, accelerate/stop, RTO and other operations as specified in flight manual (see *balanced* *).

field maintenance   That authorised and performed in field (4) in direct support of operational squadrons and other units; normally limited to replacement of unserviceable items.

field modification   One made in field (4), usually by FMK.

field of regard   Total angular coverage of sensor; with fixed installation same as FOV (2, next), but if gimballed depends on FOV plus slewing and elevation limits.

field of view   *1* Angle between two rays passing through perspective centre (rear nodal point) of camera lens to two opposite sides of format. Not to be confused with angle of view.

2 Total solid angle available when looking through sight, HUD or other optical system.

field operation   From forward airfield, esp. with unpaved runways.

field performance   That associated with takeoff and landing, esp. in context of certification.

field site   Completely unprepared stretch of terrain used in Harrier training.

field strength   *1* Flux density, intensity or gradient; also called field intensity, although this term does not follow strict radiometric definition of intensity (flux per unit solid angle).

2 Electric field strength, units Vm^{-1}.

3 Signal strength; magnitude of electric or magnetic component in direction of polarization.

4 See *magnetic* *.

field takeoff   From airfield, not ship or catapult [naval a/c].

field traffic   Surface vehicles on airfield.

field training detachment   Established to provide maintenance-orientated technical training, at operational location, on new systems and their aerospace ground equipment (USAF).

FIES   Factor of initial engine spares.

FIF   Fluorescent inspection fluid.

f_{IF}   Intermediate frequency (superhet. receiver).

FIFO   *1* Fail-isolated/fail-operative.

2 First in, first out.

FIFOR, Fifor   Flight forecast (Int.).

15-3-3-3   Alloy 76 Ti, 15 Va, 3 each Al, Cr, Sn.

fifth-freedom traffic   Picked up by airline of country A from country B and flown to country C (see *freedoms*).

FIG   *1* Fighter interceptor group.

2 Flight-idle gate.

fighter   Aircraft designed primarily to intercept and destroy other aircraft.

fighter affiliation   Training exercise carried out by bombers, other heavy aircraft, ground or naval forces, in co-operation with fighters.

fighter-bomber   Fighter able to carry air-to-surface weapons for ground attack and interdiction.

fighter controller   Officer on staff of tactical air controller charged with co-ordination and evaluation of air warning reports and operational control of aircraft allocated to him. Also known as fighter director (see also *air controller, tactical air controller, tactical air director*).

fighter cover   Patrol of fighter aircraft over specified area or force for purpose of repelling hostile aircraft.

fighter-direction aircraft   Equipped and manned for directing fighter operations.

fighter escort   Force of fighters detailed to protect other aircraft from attack by enemy aircraft.

fighter sweep   Offensive mission by fighter aircraft to seek out and destroy enemy aircraft or targets of opportunity in allotted area.

fighting harness   Seat harness [fighter and similar a/c 1920s].

fighting kite   Used in sport [originally China] in which objective is to cut rival's control cords.

fighting top   Cockpit box for gunner(s) on upper wing of large early bombers, accessed by ladder from fuselage.

fighting wing   Combat formation which allows wingman

to provide optimum coverage and maintain manoeuvrability during max-performance manoeuvres.

FIGS, Figs   Formation integrated gateway subsystem. Integrates radars, com. and airport systems gateways with VME-bus and LAN connections.

Figure-9 loop   Self-explanatory, aircraft progressively reducing [vertical-plane] turn radius to describe a 9.

figure of merit   Single numerical value describing quality of real system as percentage or decimal fraction of ideal or theoretical ideal.

FIH   Flight information handbook.

FIJPAé   Fédération Internationale des Journalistes Professionels de l'Aéronautique.

FIKI   Flight into known icing.

FIL   Fountain-induced lift.

FILA, Fila   Fighting intruder[s] at low altitude.

Filac   Federazione Italiana Lavoratori Aviazione Civile.

filament winding   Manufacture of pressure vessel (eg rocket-motor case) by winding continous high-strength filament on mandrel, bonded by adhesive.

File   Feature identification and location element (OSTA).

filed flightplan   That filed by pilot or his designated representative, without any subsequent changes.

FILG   Filling.

fill, filling   Threads in fabric which run perpendicular to selvage; weft.

fillers   *1* Paste or liquid used for filling pores of wood prior to applyng paint or varnish.

2 Pulse pairs generated by random noise in unsaturated DME beacon to maintain 2,700/s.

fillet   *1* Aerodynamic fairing giving radius at junction of two surfaces.

2 Fill which traditional weld makes at intersection of two parts.

3 Increased area of pavement at junctions of taxiways and runways to facilitate high-speed turn-offs and other manoeuvres.

filling   Increase in pressure in centre of low (meteorological); opposite of deepening.

filling sleeve   See *inflation sleeve*.

film chip   One incorporating thin or thick-film technology.

film cooling   Cooling of body by maintaining thin fluid (liquid, vapour or gas) layer over surface.

filmed   IIT coated with ion-barrier film to prevent feedback damaging delicate photocathode.

film-return satellite   Reconnaissance satellite which [possibly in a constellation of sensors] includes a camera using physical film, returned to Earth.

FILS, Fils   Fault-isolation and location system, integrates Bite with other systems.

filter   *1* See *centrifuge* *, *momentum separation* *, *dynamic particle* *.

2 Capacitance and/or inductance and resistance designed to pass given band of RF only. High-pass, low-pass, band-pass and band-stop * pass frequencies respectively above, below, between and outside desired frequencies. Frequencies at which attentuation falls by more than 3 dB are termed cut-off frequencies.

3 To study air warning information and eliminate any not of interest.

filter centre   Location in aircraft control and warning system at which information from observation posts is filtered (3) for further dissemination to air-defence control and direction centres (DoD).

filter crystal   Quartz crystal resonator used to control filter characteristics.

filter element   Cleansing medium in filter (1) with dry matrix or liquid (often oil) film.

filtering   *1* Analysis of signal into harmonic components.

2 Separation of wanted component of time series from unwanted residue (noise).

3 Suppression or attentuation of unwanted frequencies.

4 Cleansing of fluid flow of solid particles.

5 Process of interpreting reported information on vehicle movements to determine probable true tracks and, where applicable, heights or depths.

Filur   Flying innovative low-observable unmanned research.

FIM   *1* Fault-isolation monitoring [or manual].

2 Field-ion microscope.

FIN   Functional identification [or item] number.

fin   *1* Vertical or inclined aerofoil, usually at rear or on wingtip to increase directional stability.

2 Projecting flat plate to increase surface available to reject unwanted heat.

3 Those parts of stabilizers of kite balloon providing stability in pitch.

final   Inbound to active runway, called verbally by pilot when 4 nm from visible threshold.

final approach   *1* IFR, flightpath inbound, beginning at ** fix and extending to runway or to point where missed-approach procedure is executed.

2 VFR, flightpath in direction of landing along extended runway centreline from base leg to runway; hence "on finals".

final-approach altitude   Height at start of final approach.

final-approach fix   That from or over which published final IFR approach is executed.

final-approach gate   Position on extended runway centreline above which landing aircraft is required to pass at time assigned by approach control.

final-approach point   Start of final-approach segment of non-precision approach.

final-approach segment   Final approach (1).

final assembly   Assembly of major structural and sub-units which form completed aircraft; erection.

final-assembly drawing   Undimensioned drawing calling out all major installations on aircraft; complete index to particular model or sub-type (see *callout notes*).

final controller   Radar controller employed in transmission of PAR (previously GCA) talk-down instructions, and in passing monitoring information to pilots not using PAR.

final mass   Mass of rocket after burnout or cutoff.

final monitor aid   Program for management of parallel runways (FAA).

final procedure turn   Links base leg to approach.

finals   Final approach (colloq.).

final trim   Exact adjustment of ballistic missile or space launcher to desired cutoff velocity.

fin carrier   Frame laced to channel patches on aerostat to distribute loads from fin.

fine data channel   Channel of trajectory-measuring system delivering accurate but ambiguous data; coarse channel resolves ambiguity.

fineness ratio   Ratio of length of streamlined body to

maximum diameter, or some equivalent transverse dimension.

fine pitch　Governed propeller-blade angle most suitable for take-off and low-speed flight, between ground fine and range of coarser cruising settings.

fine-pitch stop　Sets limit of blade rotation into fine pitch.

fin flash　Rectilinear marking on fin, usually comprising stripes in national colours.

finger-bar controller　Pilot flight-control input in which fingers rotate cylinder forward or backward for pitch control and rock sideways for roll.

finger four　See *fingertip*.

finger lift　Finger-operated latch on front or rear of throttle lever to prevent inadvertent selection of afterburner or reverser.

finger patch　Aerostat envelope patch having radial 'fingers' to distribute load into fabric.

fingers　Long corridors projecting at about 90° from airport terminal to provide sufficient length for large number of gates.

finger-tight　Assembled so that item can readily be part-dismantled or stripped; usual state of prototype engine stored after cancellation.

fingertip　Formation in which four aircraft occupy positions suggested by fingertips of hand held horizontally.

finger twizzle　Twirl by finger signifying 'start engines'.

fin girder　Main vertical fin member in rigid airship.

finish　*1* External coating or covering of aircraft or part.

2 General appraisal by eye or touch of external surface quality of aircraft or part, esp. of all-metal construction.

finite-amplitude wave　Shockwave generated at front or rear of supersonic body of finite dimensions.

finite-displacement stick　Pilot's control column which transmits movements (even if small) and not electrical signals generated by force transducers.

finite wing　Wing having tips, thus all real wings other than annular.

Finnegan　Exercises involving detachment of NW-armed bomber[s] to dispersal base (RAF 1959–70).

Finrae　Ferranti inertial-navigation rapid-alignment equipment.

FINS, Fins　Fixed-imagery navigation sensor (Lantirn).

FIP　*1* Flight instruction program (AFROTC).

2 False-image projection, test of operator alertness.

3 Full intermediate power.

FIPS　Flight-information processing system.

FIR, Fir　*1* Flight Information Region.

2 Finite impulsive response.

3 Flight-incident recorder.

Firams, FIRAMS　Flight-incident recorder and monitor system.

FIRC　Flight-instructor refresher clinic.

Fire　Flammes infra-rouge embarquées, IR payloads (F).

fire　*1* To ignite rocket engine; start of main-chamber burn.

2 To launch rocket.

fire access door　Hinged flap, usually spring-loaded, through which fire extinguisher can be aimed when a/c parked.

fire-and-forget missile　One with IR seeker or other self-homing capability.

fire axe　Carried to enable crew to escape after crash or belly landing while on fire, normal exits unavailable.

fireball　*1* Luminous sphere formed a few millionths of a second after detonation of nuclear weapon.

2 Meteor with luminosity which equals or exceeds that of brightest planets.

fireblocker　Furnishing fabric meeting specific requirements as barrier to fire.

fire channel　Single data highway for C^3I, esp. of SAM system (eg Hawk has 2, Patriot has 10).

fire classification　Class A, wood, paper etc; B, petrol (gasoline), oil, other fuels, except, C, butane, propane, hydrogen etc.

fire-control radar　One providing target-information inputs to a weapon fire-control system.

fire-control system　System including radar(s) mounted on land, sea or air platform to provide exact data on target position and velocity before engagement with guns, missiles or other weapons.

fire deluge system　Remotely controlled pipes, hoses and spray outlets, situated throughout launch-pad area of large missile or space launching site, which operate if there is a fire or explosion in the area.

fired out　Fighter which has launched all its AAMs.

fire floor　Essentially horizontal floor or other sheet designed to be fireproof [at least for significant time].

firegate　In effect, the tap that, usually under computer control, governs dispensation rate of retardant in fire-fighting tanker.

fire point　Temperature at which material will give off vapour that will burn continuously after ignition (see *flashpoint*).

fireproof　Rules include 'at least as well as steel'.

fire pulse　Signal for remote control of fire (1); for fire (2) ususally called launch pulse.

fire resistant　Rules include 'at least as well as aluminium alloy'.

Fires　Firefighters' integrated response equipment system (USAF).

Firetex　Fire-blocking material, a viscose carbonised fabric reinforced with aramid fibres.

fire tunnel　Test facility for engine bay or other device for investigating temperatures, airflows, insulation, fuel leaks and fire suppression, etc.

fire unit　Basic subdivision of large SAM system (ie not infantry-operated), usually with four to 12 launchers at one location.

fire up　To start engine, especially first test of new type previously subjected only to motoring tests without combustion.

firewall　*1* Fire-resistant bulkhead designed to isolate engine from rest of aircraft.

2 Internet or Aeronet security barrier.

firing chamber　Test cell for static firing of small horizontally mounted rocket or missile.

firing console　Human interface with rocket engine or vehicle launch.

firing envelope　For any given airspeed and aerial target, the 3-D box of sky within which a fighter can launch a guided missile and achieve interception.

firing order　Sequence in which piston engine cylinders fire, invariably 1-3-4-2 or 1-5-3-6-2-4.

firing pass　Flight of combat aircraft towards air or ground target in which weapons are fired.

firing pit　Encloses rocket test stand on all sides except nozzle.

firing test   Static operation of rocket.

firing time   *1* Time between application of d.c. voltage to vacuum tube or solid-state device and start of current flow.

   2 Time between RF power and RF output in radar.

   3 Burn time.

Firms   Alternative to Firams.

FIRMU   Flight-incident recorder memory unit.

First   Flexible independent radar skills trainer.

first-angle projection   Side and plan show nearest faces of front elevation (US).

first buy   Lockheed term for firm order.

first day   First day of theoretical war, with part-loaded aircraft making vertical liftoffs. Later, defences are worn down, and aircraft make STOs with full weapons load.

first hop   Briefly airborne during high-speed taxi test prior to first flight.

first law of thermodynamics   Statement of conservation of energy for thermodynamic systems (not necessarily in equilibrium); fundamental form requires that heat absorbed serves either to raise internal energy or do external work.

first-line life   *1* Operational life of hardware in first-line service.

   2 Time between delivery of missile or RPV and its destruction or withdrawal. For planning purposes usually five years.

first-line service   Active operation in original design role with combat or training unit, or revenue service with air carrier.

first motion   First visible motion of vehicle at start of mission.

first officer   Civil airline rank = second pilot or copilot.

first pilot   See *PIC*.

first-run attack   Made immediately on a surface target not seen previously.

first stage   Lowest of two- or multi-stage launch vehicle, first to be fired.

first-tier customer   Purchaser [or lessee] of new aircraft.

first-tier supplier   One supplying direct to prime contractor.

firtree root   Usual gas-turbine rotor-blade root of tapered form with broached serrated edges providing multiple load-bearing faces.

FIS   *1* Flight Information Service [A adds automated, B adds broadcast, O adds Officer].

   2 Federal Inspection Services.

   3 Flight instructors' school.

   4 Fighter Interceptor Squadron.

   5 Flight-instrument system.

FISA   *1* Fédération Internationale des Sociétés Aérophilatéliques (office London).

   2 Foreign Intelligence Surveillance Act (US).

FIS-A   Flight information service, automated.

FIS-B   Flight information service, broadcast.

fiscal year   Financial year; for US, 1 October to 30 September; for Britain, 1 April to 31 March; for France and Germany, calendar year.

fishbone antenna   Coplanar elements in colinear pairs, coupled to balanced transmission line.

fish-head   Naval aircrew (RAF, WW2).

Fishpond   Active bomber radar giving bearing/distance of hostile aircraft (RAF 1943).

fishtailing   *1* Using coarse rudder to swing tail from one side to another in repeated S-turns, an alternative manoeuvre to sideslipping to steepen approach.

   2 Many aircraft [the author remembers the Oxford] proceeded in a mild zig-zag, which could not be arrested.

fishtail nozzle   Ends in triangular portion with narrow slit nozzle.

FISO   Flight Information Service Officer.

fissile   Fissionable by slow neutrons.

fission   Splitting of heavy nucleus into two approximately equal parts (nuclei of lighter elements), accompanied by release of large amount of energy and generally one or more neutrons.

fissionable   See *fissile*.

fission yield   Amount of energy released by fission in thermonuclear explosion as distinct from that released by fusion. The fission yield ratio, frequently expressed in per cent, is ratio of yield derived from fission to total.

Fist   *1* Fire support team.

   2 Flight-instrument and subsystem tasks.

   3 Future integrated supply team.

Fista   Flying IR signature technology aircraft.

FIT   Floating-input transistor.

fit   *1* Desired clearance, if any, between mating surfaces; eg push *, shrink *, force *.

   2 Total complement of avionic equipment in aircraft.

FITAP   Fédération Internationale des Transports Aériens Privés.

FITE   Fusion interfaces for tactical environment.

FITO   Forward indium-tin oxide.

FITS   *1* Flir internal targeting system.

   2 Fully integrated tactical system.

fitter   Skilled metalworking tradesman.

fitting   Assembly of parts mating with specified fit (1).

FITVC   Fluid-injection thrust-vector control.

five by five   Radio reception loud and clear. First figure denotes volume and second intelligibility, so "five by three" means loud but not very clear.

Five-Power Defence Agreement   UK/Australia/Malaysia/NZ/Singapore.

fix   *1* Aircraft position established by any independent means unrelated to a previous position.

   2 In particular, aircraft position established by intersection of two position lines. See *absolute* *, *outer* *, *running* *.

   3 Solution, possibly temporary, to technical problem (colloq.).

fixed-area nozzle   One whose cross-section cannot be adjusted.

fixed-base operator   Business operation at American airport usually including flying school, charter flights, sales agency for particular light aircraft and accessories, fuel/oil, maintenance and overhaul facilities and, sometimes, third-level or commuter airline.

fixed-base simulator   One in which cab does not move.

fixed-displacement pump   Fluid pump handling uniform volume on each repetitive cycle.

fixed distance marker   Located 300 m (1,000 ft) from threshold to provide marker for jet aircraft on other than precision instrument runway.

fixed-geometry   *1* Aircraft that does not have variable-sweep wings.

   2 Engine that does not have variable inlet or nozzle.

fixed gun   Gun fixed in aircraft to fire in one direction, usually forwards.

fixed-gun mode One for close-range snapshooting in which sightline is boresighted to gun line.
fixed landing gear One not designated to retract.
fixed light Constant luminous intensity when observed from fixed point.
fixed-loop aerial Not rotating relative to aircraft.
fixed munition *1* One used against a fixed target.
2 Gun ammunition in which projectiles are held in propellant cases.
fixed-pitch propeller One that has no provision for changing pitch of blades, and hence efficient at only one flight speed.
fixed point *1* Positional notation in computer operations in which corresponding places in different quantities are occupied by coefficients of same power of base (see *floating point*).
2 Notation in which base point is assumed to remain fixed with respect to one end of numeric expressions.
fixed-price contract One which either provides for firm price, or under appropriate circumstances may provide for adjustable price; several types, designed to facilitate proper pricing under varying circumstances (DoD).
fixed-price incentive contract Has provision for adjustment by formula based on relationship which final negotiated total cost bears to negotiated target cost as adjusted by approved changes (DoD).
fixed satellite See *geostationary*.
fixed slat Forward portion of aerofoil ahead of fixed slot built into structure.
fixed station Telecommunication station in aeronautical fixed service.
fixed target One that does not move relative to local Earth's surface.
fixed weight Total mass of aerostat in flying order without fuel, oil, dischargeable weight or payload.
fix end End of holding pattern flown over fix (1) at which aircraft enters pattern.
fixer network Radio or radar direction-finding stations which, operating in conjunction, plot positions of aircraft; fixer system (obs.).
fixture Small jig for detail subassembly.
fizzer To be on the * = to be on a disciplinary charge (RAF, WW2).
FJ *1* Fuel jettison.
2 Fast jet; DIC adds directed IR countermeasures, MAWS missile-approach warning system, PT pilot training, TS test squadron.
FJA Forward jamming antenna.
FJCA Future joint carrier [or combat] aircraft (UK).
F(jω) Fourier transform of a function of time.
FJS Fast-Jet Squadron (A&AEE).
FJTF Fast-Jet Training Fleet (RAF).
FKR Cosmonautics federation (R).
FL *1* Flight level, usually expressed in hundreds of feet; thus FL96 = 9,600 ft.
2 Foot-lambert[s] (see *luminance*).
3 Fan lift.
4 Flight line.
5 Flashing light.
FLA *1* Future large aircraft [aeroplane].
2 Foot-launched aircraft.
flade Fan blade.
Flag, FLAG *1* Floor level above ground.
2 Flemish Aerospace Group (Belg.).

3 Four-mode laser gyro, see *Flagship*.
flag Small brightly coloured plate, often Day-Glo red or orange, or diagonally striped yellow/black, which flicks into view in panel instrument, spacesuit instrument or any other subsystem or device, to give visual warning of fault or impending difficulty such as loss of electric power or low fuel level.
flag carrier *1* Airline designated as part of bilateral agreement to fly international route(s).
2 National state airline.
Flage, FLAGE Flexible lightweight agile guided experiment (anti-missile).
flagman *1* Person carrying chequered flag (formerly) employed at US airports, and still seen in Russia and other CIS, to direct arrivals to signalman or airport marshaller for parking.
2 Person carrying bright flag on tall mast to guide ag-aviation pilot towards end of each run.
flag operator *Flag carrier*.
flagship *1* Normal meaning can apply to airline service or individual aircraft, or [American, 1930–59] whole fleet.
2 Four-mode laser gyro software/hardware implemented partitioning, comprising Adiru, ADM, CDU, GNSSU and MSU.
flag stop Special unscheduled stop by scheduled airlift mission aircraft to load or unload traffic (USAF).
flag tracking Method of tracking helicopter rotor by holding fabric flag against blade tips coated with wet paint.
flak AAA fire.
flakship Small German warship tasked with defending other vesels against air attack (WW2).
flak-suppression fire Air-to-ground fire used to suppress AA defences immediately before and during air attack on surface targets (DoD). This definition should be amended to allow use of ASMs, cluster bombs etc.
FLAM Future land-attack missile.
flame attenuation Attenuation of radio signal by ionisation in rocket exhaust.
flame bucket See *flame deflector*.
flame chute Concrete and metal duct carrying flame and gas from bottom of silo or test pit to surface.
flame damper *1* Pre-radar shroud or extension to piston engine exhaust pipe to prevent visual detection at night.
2 See *flame trap*.
flame deflector Deflects hot gas of vertical-launch rocket engine from ground or from launching structure.
flame float Pyrotechnic marker that burns on water surface.
flame front Boundary of burning zone progressing through combustible mixture.
flame hardening Hardening metal surface by flame.
flameholder Body mounted in high-velocity combustible flow to create local region of turbulence and low velocity in which flame is stabilized.
flameout Cessation of combustion in gas turbine or other air-breather from cause other than fuel shutoff.
flamer Aircraft on fire in air, especially in air combat.
flame-resistant Not able to propagate flame after ignition source is removed.
flame stabilizer Flameholder.
flame trap Filter in piston engine induction system to prevent passage of flame upstream after blow-back or backfire.

flame tube *1* Perforated tube designed for mixing of fuel and air, in which fuel is burnt in gas turbine; usually inserted as inner liner in combustion chamber for diluting and cooling flame (UK).

2 Interconnector between combustors or between afterburner gutters (US).

flanging machine Metal-forming machine with high-speed plunger which bends up successive small portions of flange on moving workpiece.

flank *1* Lower side of fuselage or other aerodynamic body.

2 Sides of lower (inner radius) end of compressor or turbine blade.

FLAP Frankfurt, London, Amsterdam, Paris.

flap *1* Movable surface forming part of leading or trailing edge of aerofoil, esp. of wing, able to hinge downwards, swing down and forwards, translate aft on tracks or in some other way alter wing camber, cross-section and area in order to exert powerful effect on low-speed lift and drag. See following types: *double-slotted*, *dive-recovery*, *Fowler*, *Gouge*, *Junkers*, *Krüger*, *leading edge*, *manoeuvre*, *plain*, *slotted*, *split*, *triple-slotted*, *Youngman* and *Zapp*.

2 Side walls of thrust-augmenting ejector in powered-lift system, in fighters part of a retractable structure.

3 Hinged segment forming part of primary or secondary nozzle of afterburner.

4 Urgent activity (UK colloq.).

flap angle Angle between chord of flap and that of wing.

flap blowing Discharge of HP compressor bleed air over lowered flaps to prevent airflow breakaway. Normally air issues at about sonic speed through slit facing across flap upper surface, flow attaching to flap through Coanda effect. Also called Attinello flap (see *BLC*, *super-circulation*).

flaperon Surface combining roll-control function of aileron with increased lift and drag function of flap; can be differentially operated.

flap fan Experimental concept in which flaps carry small fans driven by engine bleed air (perhaps eight fans on each flap) to maintain attachment and provide powered lift.

flaplet *1* Loosely, any small flap.

2 Narrow-chord flap with circular-arc LE and flat top/bottom forming TE of Coanda CCW.

flappery Flaps, especially if prominent on Stol aircraft (colloq.).

flapping Angular oscillation of helicopter rotor blade about flapping hinge.

flapping angle Angle between tip-path plane and plane normal to axis.

flapping hinge Sensibly horizontal pivot on helicopter main-rotor hub which allows blade tip to rise and fall.

flapping plane Plane normal to plane of each flapping hinge axis.

flap-retraction-height Variable but always over 1,000 ft (305 m) with aircraft at or above FUSS.

Flaps *1* Force-level automated planning experiment (AAFCE).

2 Flat-aperture parabolic surface (antenna or mirror).

flap setting Predetermined angle of flap (1) for takeoff, landing or other flight condition.

flaps-extended speed The highest speed permissible with flaps in a prescribed extended position.

flaps-up safety speed Minimum TAS at which aircraft maintains positive ROC with flaps retracted.

flap-type control The common type of flight control surface.

FLAR Federatsii Lyubitelei Aviatsii Rossii, Federation of aviation amateurs (R).

Flair Fixed low-altitude intermediate-range, surveillance radar.

flare *1* Final nose-up pitch of landing aeroplane to reduce rate of descent approximately to zero at touchdown.

2 Distance sides of planing bottom of marine float or hull flare out from centreline.

3 Pyrotechnic aerial device for signalling or illumination; parachute * illuminates large area when released at altitude; wingtip * illuminates ground when landing.

4 Inverse taper (ie opening out) at tail of cylindrical body, as at base of rocket vehicle.

5 Eruptions from Sun's chromosphere, which may appear within minutes and fade within an hour; eject high-energy protons, cause radio fadeouts and magnetic disturbances on Earth.

6 Fixed source of ground or water illumination, of several types, usually burning kerosene or related fuel (generally obs.).

flare-augmentation system Electronic feedback on fixed-wing STOL to achieve minimum landing field length.

flare demand Coded Autoland signal commanding flare (1).

flare dud Nuclear weapon which detonates with anticipated yield but at altitude appreciably greater than intended; a dud in its effects on target (DoD).

flare out See *flare (1)*.

flare path Line of flares (6) or lights down one side or both sides of runway to provide illumination (generally obs.).

flare-path dinghy Attends flare path laid over water for marine aircraft.

Flash Folding lightweight acoustic sonar [or system] for helicopter.

flash *1* Basically rectangular pattern of vertical bars in national colours painted on military aircraft, usually covering portion of fin.

2 White semicircular or circular badge worn in headgear of aircrew cadet.

flashback Sudden upstream travel of flame in flow of combustible mixture in enclosed system.

flash/bang/smoke Signifies training target disabled.

flash burn Caused by radiation from nuclear explosion.

flasher unit Regular make/break switch in circuit of light which flashes rather than rotates.

flashing light Intermittent aeronautical surface light in which light periods are clearly shorter than dark, with repeated cycle. Usually has published frequency.

flashing off Drying of surface of film or finish until safe to gentle touch.

flash lobe Sudden peak in radar signal caused by radome [also one word].

flashover *1* Phenomenon in which material exposed to intense radiation suddenly ignites over its entire surface.

2 Sudden discharge between conductors with very large potential difference.

flashpoint Temperature at which vapour of substance, such as fuel, will flash or ignite momentarily. Lower than fire point.

flash suppressor Attachment to muzzle of gun which reduces or eliminates visible light emitted.

flash welding Electric welding by partial resistance welding under low pressure, heating by electric arc, and application of large compressive pressure forcing surplus weld out of joint.

FLAT Flight-plan aid tracking.

flat With horizontally opposed cylinders.

flatbed Transport aircraft (so far only conceptual) for conveying ISO containers on flatbed fuselage.

flat diameter Diameter of circle enclosing canopy of flat parachute when spread out on plane surface.

flat four Four-cylinder, horizontally opposed piston engine.

flat-H Piston engine with two superimposed rows of opposed cylinders.

flathatting Flying at lowest possible safe height (US colloq.).

flat-head rivet Thin-headed rivet used internally or where round or countersunk heads are for any reason not advisable.

flat pad Ship-mounted ballistic-missile launcher isolated from ship motion (colloq.).

flat-panel instrument With LCD display.

flat parachute Parachute whose canopy consists of triangular gores forming regular polygon when laid out flat.

flat pitch Ground fine pitch.

flat-rated *1* Engine throttled or otherwise restricted in output at low altitudes and thus able to give constant predictable power at all FLs up to given limit, shown as kink point on plot of power: altitude.

2 Engine restricted in ouput in cold ambient conditions and thus able to give constant predictable power in all air temperatures up to given published limit; eg * to 28.9°C.

3 Usually * combines (1) and (2). Hence, flat rating.

flat riser VTOL aircraft able to take off with fuselage substantially horizontal (term not normally applied to helicopters and other rotorcraft).

Flats Flight-line automatic test set (Varo).

flat sequence Display aerobatic routine imposed by low cloudbase.

flat six Six-cylinder, horizontally opposed piston engine.

flat spin Spin at large mean angle of attack but with longitudinal axis of the aircraft nearly horizontal; recovery difficult and prolonged because, with aircraft fully stalled, ailerons, elevators and rudder are ineffective; nose-up position and rotation carries slipstream away from tail. Anti-spin parachute can assist positive recovery.

flat template Representation on two-dimensional material of dimensions, areas and other characteristics of curved part; also known as layout template.

flattened-X Delta-3.

flattening out *1* "In alighting, the transition between the approach glide and the horizontal motion before making contact with the earth" (BSI).

2 Many authorities add recovery from a dive to level flight.

flat-top Aircraft carrier (colloq.).

flat zone Zone within indicated course sector or ILS glidepath sector in which slope of sector characteristic curve is zero.

FLB Association of airline operators (Austria).

FLC *1* Fuel and limitation(s) computer.

2 Flight-level change.

FLCH *1* Flechette(s), for piercing armour.

2 Change in flight level.

FLD Fault location device (wiring).

fld Field.

FLEEP Flying lunar-excursion experimental platform. One-man rocket-powered platform intended to enable astronaut to make quick hops on lunar surface.

fleet All aircraft of one type used by same operator.

fleet ballistic missile submarine Submarine designed to launch ballistic missiles.

fleet carrier CV, large surface vessel equipped to launch, recover and maintain powerful fixed-wing aircraft in any theatre.

fleet leader Aircraft in fleet having greatest flight time.

fleet noise level Average noise level throughout fleet.

fleet performance monitoring Confidential use of quick-access recorders to inspect actual way pilots fly [airline] aircraft, and in particular whether airspeeds, rates of descent, etc, are within limits.

Flem Fly-by landing excursion mode.

Flettner Aileron tab.

Flettner control See *servo tab*.

Flettner rotor Cylinder spinning on axis normal to airstream, generating transverse thrust/lift.

Flexadyne Proprietary (Rocketdyne) formulation of solid propellant.

Flexar Flexible adaptive radar.

flexbeam Torsionally compliant spar, typically of CFRP laminates, attaching helicopter rotor blade to hub and flexing to accommodate varying blade pitch and coning angles.

flexibility Ability of hardware, including aircraft, to operate efficiently over wide range of conditions; eg long or short sectors, high or low level.

flexibility factor Used in helicopter rotor stress calculations to make up for structure's flexibility.

flexible air-data system Versatile microprocessor-based DADC outputting MIL-1553B, Arinc 429, analog and IFF transponder.

flexible blade *1* Helicopter rotor blade with trailing-edge or balance tabs.

2 Helicopter rotor blade in which pivots are replaced by flexible structure.

flexible elastomer skin Reduces RCS by hiding joints and discontinuities.

flexible flight deck Post-WW2 concept of aircraft carrier whose aircraft would need no landing gear.

flexible gun Clearly a nonsensical idea: what is meant is pivoted, ball-mounted or in any other way manually aimed independently of the aircraft.

flexible takeoff, FTO Takeoff technique in which for TOW below MTOW, less than maximum engine thrust is selected. For given WAT condition, this thrust is computed by intersection of TOW and aircraft performance to comply with regulations, giving a theoretical "ambient temperature" Tf. Thrust selected is that which would be available at full power at Tf. FTO saves engine costs, reduces noise and extends engine life.

flexible tank Bag-type tank.

flexible [flex] targeting Mission is launched by local

commander with choice of targets and weapons to attack suddenly seen target.

flexible wall   Used in wind tunnels, engine air inlets and other ducts subject to large range of flow and Mach number; may be perforated to extract boundary layer.

flexural axis   Locus of flexural centres, points at which applied load produces pure bending without twist (note, on swept wing pure bending results in apparent twist, ie loss of incidence).

flexural wash-out   Apparent reduction in angle of incidence from root to tip as swept wing deflects upward under load.

flexure   Bending under load.

flex-wing   Foldable or collapsible single-surface wing for micro or hang glider.

FLG   *1* Flashing light.
   2 Falling (also Flg).

FLI   Fighter lead-in.

flicker   Subjective sensation resulting from periodic fluctuation in intensity of light at rates less than about 25/30 times a second, preventing complete continuity of images.

flicker control   See *bang-bang control.*

flicker marking   Various black-white schemes to render rotating propeller blades more visible on ground.

flicker rate   Refresh rate for CRT information below which flicker becomes noticeable; dependent on eye's persistence of vision and persistence of CRT phosphor.

flicker vertigo   Caused by light occulting (flickering) at frequencies from four to 20 per second (eg with single-propeller aircraft headed towards Sun at low rpm), producing nausea, dizziness or unconsciousness.

flick roll   *1* Essentially a horizontal spin, made by slowing to spin-entry speed with engine throttled back and then applying hard back stick and full rudder. Result should be a controlled very rapid 360° roll.
   2 "A rapidly executed roll" (B.S., 1940).

FLID   Flight identification.

Flidras   Flight-data relay and analysis system.

flight   *1* Movement of object through atmosphere or space sustained by aerodynamic, aerostatic or reaction forces, or by orbital speed.
   2 An instance of such movement.
   3 Specified group of aircraft engaged in common mission.
   4 Basic tactical unit of three or four aircraft.
   5 Flight sergeant (colloq., abb.).
   6 Radio call sign, Flight Directory (NASA).
   7 Particular scheduled air-carrier service, with three or four-figure identifying numbers, either routinely or on particular day.
   8 Fighting formation comprising two elements each of two aircraft (US, 1981 onwards).

flight advisory   Message giving advice or information broadcast to airborne aircraft or interested ground stations.

flight assist   Provision of maximum assistance to aircraft lost or in distress by flight-service stations, towers and centres (US).

flight attitude   *1* Defined by inclination of three vehicle axes to relative wind.
   2 Defined relative to Earth, ie local vertical.

flight augmentation computer   Main AFCS component providing yaw damping, pitch trim and flight-envelope monitoring and protection.

flight bag   *1* For personal equipment of flight crew member.
   2 Hang-glider body bag.

flight cap   Legally imposed arbitrary limit on number of flights (arrivals or departures) at airport, either throughout year or between particular [night] hours.

flight characteristic   Feature of handling, feel or performance exhibited by aircraft type or individual example.

flight check   Callsign prefix for navaid or other calibration/inspection (FAA).

flight compartment   See *flight deck (1).*

flight computer   Computer (2).

flight control   Specifically, control of trajectory.

flight-control data concentrator   Transmits info, such as control-surface position, to FDIU and displays, indicates failure status to FWCs and displays, and memorizes failures for central maintenance computer.

flight-control primary computer   In charge of generating control laws and controlling surfaces.

flight controls   Those governing trajectory of aircraft in flight.

flight-control secondary computer   Controls flight-control power system, spoilers and rudder trim and limits; if FCPC fails, also takes over Elac and yaw damper.

flight control system   See *control system.*

flight control unit   Primary flight-deck interface with AFCS providing autopilot modes, SPD/MACH, HDG SEL, ALT SEL and similar functions; some include autothrust.

flight coupon   Actual ticket or ticket book issued by air carrier to passenger.

flight crew   Personnel assigned to operate aircraft.

flight cycle   Sequence of operations and conditions, different for airframe, propulsion and each system or equipment item, which together make up one flight.

flight data recorder   See *flight recorder.*

flight deck   *1* Compartment in large aircraft occupied by flight crew.
   2 Upper deck of aircraft carrier.

flight despatcher   See *despatcher.*

flight director   *1* Flight instrument generally similar to attitude director giving information on pitch, roll and related parameters.
   2 Panel controller for autopilot.
   3 Most senior member of large wide-body cabin crew.

flight-director attitude indicator   Manned-spacecraft display indicating attitude, attitude error and rate of pitch, yaw and roll.

flight duty period   See *crew duty time.*

flight dynamics   General subject of motion of aerodyne and laws which govern it.

flight engineer   Aircrew member responsible for power-plant, systems and fuel management, and also sometimes for supervising turnround servicing. Today rare except R and military.

flight envelope   *1* See *gust envelope.*
   2 Curves of speed plotted against altitude or other variable defining performance limits and conditions within which equipment must work.

flight envelope monitoring   AFCS function providing computation of V_{MIN}, V_{MAX} and angle-of-attack limits.

flight-envelope protection   System in FBW [Airbus] aircraft which cannot normally be over-ridden by [eg hijacker] pilot which automatically commands climb or turn to avoid hitting an obstacle.

flight fine pitch   Finest propeller blade angle available in flight. Weight on wheels may remove * stop, enabling drag to be increased.

flight flutter kit   Installation, together with instrumentation, of 'bonkers' or other devices to induce flutter in flight-test aircraft.

flight-following   Maintaining contact with specified aircraft to determine en route progress.

flight idle   Lowest engine speed available in flight, set by ** stop, mechanical limit released to ground-idle position at touchdown.

flight indicator   Instrument combining lateral inclinometer, fore-and-aft inclinometer and turn indicator (obs.).

flight indicator board   Display in airport terminal showing arrivals and departures of airline flights.

flight information centre   Unit established to provide flight information service and alerting service.

flight information region, FIR   Airspace of defined dimensions within which flight information and alerting services are provided by air traffic control centre.

flight information service   Service giving advice and information useful for safe and efficient conduct of flights. In good weather provides listening watch only.

flight inspection   *1* By specially equipped aircraft, of accuracy of navaids.

2 Periodic examination of flight crew and ATC controllers.

flight instruments   Those used by pilot(s) to fly aircraft, esp. those providing basic information on flight attitude, speed and trajectory.

flight integrity   Close relationship between two friendly combat aircraft manoeuvring for mutual support.

flight level, FL   Level of surface of constant atmospheric pressure related to datum of 1013.25 mb (29.92 in mercury), expressed in hundreds of feet; thus FL 255 indicates 25,500 ft (see *QFE, QNH*).

flight-line   *1* Ramp area of airfield, where aircraft are parked and serviced.

2 In reconnaissance mission, prescribed ground path across targets.

flight Mach number   Free-stream Mach number measured in flight.

flight management system   Automatic computer-controlled system with autothrottle, possible Mach-hold, and complete control of navigation, including SIDs and STARs. Offers "menus" for minimum cost, minimum fuel burn or other objectives. Relieves workload, increases precision.

flight manual   Book prepared by aircraft manufacturer and carried on board, setting out recommended operating techniques, speeds, power settings, etc, necessary for flying particular type of aircraft. Known to airlines as operations manual.

flight mechanic   Pre-1935 title of flight engineer.

flight mechanics   One of the two components of flight dynamics, whose role is to establish the right balance between stability, manoeuvrability and control power.

flight number   See *flight (7)*.

flight office   On a GA airfield, centre for booking pleasure flights and carrying out domestic business, but not concerned with ATC or visiting pilots.

flight panel   Accepted definition: panel grouping all instruments necessary for continued flight without external references. Preferable: panel grouping available flight instruments.

flightpath   Trajectory of centre of gravity of vehicle referred to Earth or other fixed reference. In following five definitions H signifies * in horizontal plane and V in vertical plane.

flightpath angle   Acute angle between flightpath (V) and local horizontal, shown on FPA display by resolving G/S and V/S.

flightpath computer   See *course-line computer* (H).

flight-path controller   Digital system for Coast Guard (UK) helicopters comprising Doppler, 'radalt', attitude gyro and accelerometers to control low-altitude hover in absence of visual reference. Modes include autotransition, up/down, ability to overfly target.

flightpath deviation   Angular or linear difference between track and course of an aircraft (H).

flightpath recorder   Instrument for recording angle of flightpath (V) to horizontal.

flightpath sight   In HUD, direct aiming point showing distant point through which aircraft will pass (V, H).

flightpath vector   Prediction of future flightpath which replaces traditional flight director in advanced EFIS, especially to protect against windshear.

flight plan   *1* Specified information relating to whole or portion of intended flight (2); filed orally or in writing with air traffic control facility.

2 Common working document, both in spacecraft and at all ground stations during manned or unmanned space flight. Separate ** issued for lunar or planetary surface operations.

flight-plan correlation   Means of identifying aircraft by association with known flight plans.

flight platform   See *helipad*.

flight profile   Plot of complete flight (2) in vertical plane, usually altitude plotted against track distance.

flight process board   In ATC centre, displays all current FP strips.

flight-progress strip   ATC aide-memoire: paper strip typically 25 mm × 200 mm, coloured for traffic direction, giving one flight's c/s, FL, ETA as amended; slid into FP board until passed to colleague at handover.

flight-proximity demonstration   Flight in which a receiver aircraft formates behind a tanker even though neither may be equipped [yet] to supply or receive fuel.

flight rating test   One in which member of flight crew demonstrates ability to comply with requirements of particular licence or rating.

flight-readiness firing   Short-duration test of in-service space launcher, or other rocket vehicle, on launcher.

flight recorder   Device for automatically recording information on aircraft operation. Main type is flight data recorder (FDR), also colloquially called crash recorder or 'black box'. Records 50 or more parameters, including following mandatory channels: altitude, airspeed, vertical acceleration, heading, elapsed time at 1s intervals (UK also requires pitch, and usually control-surface positions, high-lift surface positions, engine speeds and flight-crew speech are also included). Such recorders are designed to survive crash accelerations, impacts, crushing and fire, and often carry underwater transponders or beacons. Normally recording medium, eg multi-track steel tape, is recycled every 25 h. Cockpit voice recorder (CVR) stores all speech on flight deck or cockpit, including intercom and radio. Maintenance recorders, eg AIDS, are linked by serial data highways to hundreds of transducers and other inputs recording many kinds of information (temperatures, vibrations, pressures and electronic parameters) to yield advance information of impending fault conditions or failures and improve system operation and economy. Highways lead to various logic and acquisition units, some for quick-look and others long-term; separate highway to protected FDR often provided.

flight reference card Carried in cockpit to provide quick detailed list of vital actions in event of all system failures or emergencies commonly encountered, with recommendations, suggestions and prohibitions.

flight regime State of being airborne, governing many systems and modes unavailable on ground.

flight release certificate Issued for aircraft with Permit to Fly (UK CAA).

flights Always plural, the offices [usually on the air side of hangars] and nearby aircraft parking areas of operational units on a permanent RAF station (becoming archaic usage).

flight-safety information management system Proposed unclassified database of accidents, incidents and malfunctions reported by participating air forces to improve safety of military aircraft.

flight schedule monitor Shared by FAA and user community, is the decision-making tool that forms basis of current flight information and air-traffic demand at each airport.

flight service station Facility providing flight assistance service (FAA).

flight shed Traditional British term for hangar in which prototypes are completed and readied for flight and subsequently are kept. Not normally associated with series production.

flight simulator Electronic device that can simulate entire flight characteristics of particular type of aircraft, with faithful reproduction of flight deck; used to test and check out flight crews, esp. in coping with emergencies, and (military) in completing combat missions according to role; or as design and engineering tool during aircraft development.

flight sister Female nursing officer trained for aeromedical duties.

flight space Space above and beyond Earth available for atmospheric or space flight.

Flight Standards District Office Handles all matters within assigned geographic area (FAA).

flight station *1* Flight crew position away from flight deck.

2 Base for marine aircraft (WW1).

flight status Indication of whether a given aircraft requires special handling by air traffic services.

flight strip *1* Auxiliary airfield on private property, farmland or adjacent to highway.

2 See *flight progress strip*.

flight structural mode excitation Allows pilot to command deterministic signals, such as swept-frequency sine waves, from FCC to excite all aircraft's flexural modes.

flight suit One-piece garment with various pockets, zips and velcros.

flight surgeon Physician (invariably not surgeon) trained in aeromedical practice whose primary duty is medical examination and care of aircrew on ground.

flight test *1* Test of vehicle by actual flight to achieve specific objectives.

2 Test of component mounted on or in carrier vehicle to subject it to conditions of flight.

3 Flight rating test.

flight test vehicle Special aircraft, missile or other vehicle for conduct of flight tests to explore either its own capabilities or those of equipment or component parts.

flight time *1* Elapsed time from moment aircraft first moves under its own power until moment it comes to rest at end of flight. For flying boats and seaplanes, buoy-to-buoy time (see *block time*).

2 For gliders and sailplanes, time from start of takeoff until end of landing.

3 For vehicles released in flight from parent carrier, measured from moment of release.

4 Aggregate of ** of all flights made by same basic structure or other hardware item.

flight vector Direction of travel; except in still air, not the same as azimuth of longitudinal axis. Essentially = track.

flight visibility Average forward horizontal distance from cockpit (assumed at typical light-aircraft FL) at which prominent unlighted object may be seen and identified by day and prominent lighted object may be identified by night.

flightway The airspace immediately beyond the end of a runway or other takeoff path; this concept has fallen into disuse.

flight weight Similar to production item; not a battleship test construction.

flightworthy Ready for flight; for aircraft, airworthy.

Flight Zero Brief unplanned flight, caused by sudden event during fast taxi or other testing prior to Flight 1.

flimsy orders Printed on onion-skin paper, can be eaten.

flinger ring Uses centrifugal force to inject W/M [in some Turbomeca engines, fuel] from entry to engine compressor.

FLIP, Flip *1* Flight information publication.

2 Floated lightweight inertial platform.

flip-flop *1* Bistable multivibrator; device having two stable states and two input signals each corresponding with one state; remains in either state until caused to flip or flop to other.

2 Bistable device with input which allows it to act as single-stage binary counter.

flipper *1* Elevator (US colloq.).

2 Before turning on final, flaps, pitch, power [some add roll].

FLIR, Flir Forward-looking IR.

Flit Fighter lead-in training (USAF, Holloman).

FLL *1* Flight line (two words) level (UK/NATO).

2 Fördergesellschaft für Luftschiffbau und Luftschiffahrt eV (G).

FLM *1* Flightline maintenance, or mechanic (US).

2 Foot-launched microlight.

FLMTS Flight line maintenance test set (Kollsman).

FLO *1* Flow control [of traffic]; -E adds East, -W West (ICAO).

2 Defence logistics organization (Norway); /Luft adds Air.

Flo Floodlights available on landing.

f_{LO} Local-oscillator frequency.

float *1* Horizontal distance travelled between flare and landing or alighting (see *ground effect*).

2 Watertight body with planing bottom forming alighting gear of * seaplane. Also wingtip *.

3 Ability of control surfaces to trail freely in airstream except when commanded by input; reckoned negative when surface deflected away from relative wind and positive when (because of overbalance ahead of hinge axis) surface moves against it.

4 Buoyant capsule in carburettor [carbureter].

floatation See *flotation*.

float displacement Mass of water displaced by totally submerged seaplane float.

floated gyro Floating gyro.

floated position That assumed by flight-control surface [esp. manual] in absence of pilot input.

float gear Floats (2) applied as modification to land-plane.

floating ailerons Designed to float (3).

floating gudgeon pin Free to rotate in both piston and connecting rod.

floating gyro Mass supported by hydrostatic force of surrounding liquid.

floating lines In photogrammetry, lines connecting same two points of detail on each print of stereo pair; used to determine whether or not points are intervisible, and drawn directly on prints or superimposed by transparent strips.

floating mark Mark or dot seen as occupying position in three-dimensional space formed by stereoscopic pair, used as reference in stereoscopy.

floating point EDP (1) positional notation in which corresponding places in different quantities are not necessarily occupied by coefficients of same power; eg 186,000 can be represented as 1.86×10^5. By shifting ** so number of signficant digits does not exceed machine capacity, widely varying quantities can be handled.

floating reticle One whose image can be moved within FOV.

float light See *flare (3)*.

floatplane See *float seaplane*.

float seaplane Aeroplane supported on water by separate floats (2), usually 2 or 3.

float-type carburettor Head of fuel supplied to jet is controlled by float (4) and needle valve.

float valve Fluid valve regulated by float acting on level in container.

float volume Ratio of seaplane gross weight to mass of unit volume (traditionally 1 ft^3) of water.

FLOLS Fresnel-lens optical landing system.

flood To overfill float chamber of carburettor, hence flooded.

flood flow *1* Unrestricted supply of hot high-pressure air to cockpit either for demist/deicing, or in emergency, or by auto switch triggered by excessive cabin altitude.

2 Has similar (various) meanings in oxygen systems.

floodlight Light providing general illumination over particular area.

flood valve Controls flow of fire extinguishant.

floor-loaded Aircraft is in factory for refurbishment or rework.

floor loading *1* Actual number of aircraft being modified.

2 Maximum [or actual] number of passengers.

floor locks Rows of attachments which interface with seating or VLDs.

floor vents Pass used cabin air to pressurized or non-pressurized lower fuselage.

flops Floating-point operations per second, hence M-*, Giga-*.

FLOT, Flot Forward line of own troops (formerly called Feba, forward edge of battle area).

flotation Quality of a wheel landing gear of operating from soft ground.

flotation bags, collars Inflatables used to provide buoyancy and stability for sea-recovered spacecraft.

flotation gear *1* Inflatable bags carried inside RPV, target or missile test vehicle to provide buoyancy after ditching.

2 Emergency inflatable bags surrounding landing gear of shipboard helicopter.

Flo Trak Patented arrangement of large-area plates fitting around landing-wheel tyre for enhanced flotation, esp. to permit combat aircraft to use soft surface.

flow augmenter Usually means ejector, but also applied to inducer at entry to centrifugal pump.

flowback Runback of water from wing leading edge in icing conditions.

flow chart Graphical symbolic representation of sequence of operations.

flow control Measures designed to maintain even flow of traffic into airspace or along route. Chief feature is acceptance of each aircraft into pre-booked slot at entry to controlled airspace, at agreed gate time, to provide orderly ATC service which does not become overloaded.

flow control system One form of main control for gas-turbine engine: fuel-pump delivery pressure is function of rpm, output being controlled to maintain set dP across throttle valve for any set air-inlet condition; ancillaries take care of transients and limitations. See *proportional *.

flow disrupter Small hinged or retractable plate intended to promote intentional stall.

flow fence Kevlar fabric shield surrounding top and sides of ejection-seat occupant.

flowmeter Instrument which measures fluid (gas or liquid) flow; numerous types based on venturi pressure drop, speed of free-spinning turbine, pitot pressure and many other principles; measure can be velocity at point, near-average velocity or, with density input, mass flow.

flow rake See *rake*.

flow regime Particular type of fluid flow (see *continuum *, free-molecule *, laminar *, slip *, turbulent *.

fl oz Fluid ounce.

FLPFM Foot-launched powered flying machine, eg, wheelless micro or engined parachute.

FLPP Future Launcher[s] Preparatory Programme (2002, ESA).

FLR Forward-looking radar.

FLRE Flare.

FLS Foreign Liaison Staff (MoD).

FLT, Flt Flight (unit).

FLTA Forward-looking terrain-avoidance.

FLTCAL Flight calibration.

FLTCK Flight check.

Flt Ctrl Flight control.

FLTP Future launcher technology programme (ESA, 1999, never implemented).

FLTR　Flightline tape-reader.

FLTS　Flight Test Squadron (USAF).

FltSatCom　Fleet satellite communication system; also clumsily written all capitals (USN).

FLTWO　Flight watch outlet.

FLUC　Fluctuating.

fluerics　See *fluidics*.

fluid　Liquid or gas.

fluid dynamics　Study of fluid motion.

fluid element　Second or supporting element in *fluid four*.

fluid four　Tactical formation in which second element is loosely spread in both vertical and horizontal planes to enhance manoeuvrability, look-out and mutual support.

fluidic nozzle　Jet-engine nozzle in which fluidic control is used to vary vectoring, profile and area.

fluidics　Branch of technology akin to electronics but using instead of electrons air or other fluid flowing at low pressure through pipes, valves and gates for control of external systems; one advantage is relaxed upper temperature limit.

fluidity　Reciprocal of viscosity.

fluidized bed　Container of finely divided solid particles supported in liquid-like state by upcurrent of air or other gas.

fluid mechanics　Study of static or moving fluids and reactions on bodies (includes aerodynamics, aerostatics, hydrodynamics, hydrostatics).

fluid ounce　Non-SI unit of volume, $1/_{160}$ Imp. gal. = 28.4131 cc.

fluid resistance　See *drag*.

fluorescein　Proprietary chemical supplied as solid block attached by lanyard to aircrew dinghy. When dropped into sea stains surface fluorescent greenish yellow.

fluorescence　Emission of photons, esp. visible light, during absorption of radiation of different wavelength from other source; photoluminescence (see *luminescence*, *phosphorescence*, *scintillation*).

fluorescent testing　Examination of item coated in fluorescent ink by UV light, to reveal crack as a bright line.

fluorine　Reactive yellow-green gas, used as liquid (BPt –188°C) oxidant in rockets (Isp 410 with LH_2) or in many cryogenic compounds such as oxygen difluoride (OF_2); * $^{-18}$ is radionuclide of half-life 110 min.

fluorocarbons　Generally resemble hydrocarbons, but F instead of H makes them more stable; many uses.

fluoroscope　Instrument with fluorescent screen supplied by processed signals from X-ray tube, used for immediate indirect viewing inside metal or composite structures.

fluoroscopy　X-ray TV.

flush antenna　One conforming with external shape of vehicle.

flush deck　Whole ship upper deck at same level.

flush intake　Not protruding, orifice in skin of vehicle.

flush on warning　Take off immediately radar evidence suggests hostile missile attack so that, when airfield is hit, aircraft are just out of dangerous radius of thermonuclear warhead.

flush rivet　Head is flush with surface into which it is countersunk.

flush weld　Plug or butt weld which leaves no weld material on surfaces.

fluted　Skin stiffened by evenly spaced parallel semi-

circular channels. External * cannot be aligned with complex slipstream.

flutter　*1* High-frequency oscillation of structure under interaction of aerodynamic and aeroelastic forces; basic mechanism is that aerodynamic load causes deflection of structure in bending and/or twist, which itself increases imposed aerodynamic load, structure overshooting neutral position on each cycle to cause load in opposite direction. Distinguished by number of degrees of freedom (bending and torsion of wing, aileron and other components are considered separately), symmetry across aircraft centreline, and other variables. When heating involved subject becomes aerothermoelasticity (see *classical* *, *hard* *, *soft* *).

2 Radio beat distortion when receiving two signals of almost same frequency.

flutter model　Flexible model with mass distribution, flexure and other features designed so that flutter qualities simulate those of full-scale aircraft.

flutter speed　Lowest EAS at which flutter occurs.

fluvial　Adjective meaning that seaplane [widest meaning] is intended to operate from calm water only. Opposite = maritime.

flux　*1* Generally, quantity proportional to surface integral of normal (90°) field (eg, magnetic) intensity over given cross-section.

2 Volume, mass or number of fluid elements or particles passing in given time through unit area of cross-section; eg luminous *, measured in lumens (abb. lm).

3 Magnetic * can be thought of as number of lines of force passing through particular coil or other closed figure; symbol Φ, unit weber Wb.)

4 Materials used in welding, brazing and soldering to clean mating surfaces, and/or form slag, which helps separate out oxides and impurities by flotation and exclude oxygen.

flux density　*1* Unit of magnetic ** is Wb/m^2, or tesla (T).

2 Neutron **, and particle physics generally, is particles per unit cross-section per second multiplied by velocity, $\Delta ø/\Delta t = nv$.

fluxgate　Sensitive detector giving electrical signal proportional to intensity of external magnetic field acting along its axis, used as sensing element of most remote-indicating compasses; also called fluxvalve.

fluxgate compass　Uses fluxgate to indicate, subject to corrections, direction of magnetic meridian.

fluxvalve　See *fluxgate*.

FLW　*1* Forward-looking [i.e. predictive] windshear radar.

2 Follow[s].

fly a desk　Retirement from professional flying (e.g., senior officer].

fly-away cost　Published retail price of GA aircraft, with specified avionics fit, ignoring spares, training or support.

fly-away disconnects　Launch-vehicle umbilicals on rigid arms which swing clear under power.

flyback　Controlled descent through atmosphere of returning aerospace-plane.

fly-back period　That during which CRT spot returns from end of one line to start of next when in raster-scan mode.

fly-back time　Time, usually ns or μs, for each fly-back period.

fly-back vehicle　Space vehicle intended to be reusable.

flybar   Flying by auditory reference.

fly before buy   Philosophy of flight evaluation of new aircraft type, esp. by military (government) customer for combat aircraft.

fly-by   *1* Interplanetary mission in which TV and instrumented spacecraft passes close to target planet but does not impact or orbit it.

2 Slow flight past tower to verify aircraft configuration or possible damage.

fly-by-cable   Mechanical links join cockpit to PFCUs.

fly-by-light   Flight-control system with signalling by optical fibres.

fly-by-speech   Flight-control system with input signalled by voice of pilot (various research programmes).

fly-by-wire   Flight-control system with electric signalling.

flyco   *1* Abbreviation for Wing Commander, Flying, at an RAF station.

2 Position aboard aircraft carrier from which all aircraft launches and recoveries are controlled.

Flygtekniska Föreningen   Society of Aeronautics (Sweden).

Flygtekniska Försökanstalten   Aeronautical Research Institute (Sweden).

fly-in   Informal gathering of private and club aircraft at particular airfield, usually with a relaxed programme of events and competitions.

flying boat   Seaplane whose main body is a hull with planing bottom; US = boat seaplane.

flying cable   Connects captive or kite balloon to winch.

flying controls   See *control system*.

flying diameter   Overall diameter of circular parachute canopy in normal operational descent.

Flying Farmers   Association of over 400 airstrip owners [not necessarily farmers] (UK).

flying machine   Powered aerodyne; common pre-1914, today humorous or derogatory.

flying order book   Set of rules governing the flying of aircraft owned by a club (UK usage), or by club members.

flying position   Attitude of aircraft when lateral and longitudinal axes are level or in flight attitude; esp. when aircraft on ground is supported in this attitude. Note: flight attitude varies with airspeed.

flying qualities   Loosely, stability and control as perceived by the pilot.

flying rigging   Distributes loads into balloon from flying cable.

flying roundup   Event, usually annual, at which passengers, usually handicapped children, are given flights, often in vintage transports (US).

flying shears   Rotary system for cutting long web of sheet metal or other material moving at high speed.

flying speed   *1* Loosely, minimum airspeed at which aeroplane can maintain level flight (preferably, positive climb) in specified configuration.

2 Another definition: speed reached on takeoff at which pilot has full control [but only for climbing straight ahead]. In modern terms V_2.

flying spot   Rapidly moving spot of light, usually generated by CRT, used to scan surface containing visual information.

flying stovepipe   Ramjet (colloq.).

flying-tab control   See *servo tab, Flettner*.

flying tail   Use of whole horizontal tail as primary control surface.

flying testbed   Aircraft or other vehicle used to carry new engine or other device for purpose of flight testing.

flying the ball   *1* Loosely, flying IFR, from traditional turn/slip indicator.

2 Correctly flown VFR carrier approach using mirror sight.

flying the needle   Navigating along airways by VOR.

flying time   See *flight time*.

flying weight   See *flight weight* (engine).

flying wing   Aeroplane consisting almost solely of wing, reflecting idealised concept of pure aerodynamic body providing lift but virtually devoid of drag-producing excrescences.

flying wires   Diagonal cables/wires placed under tension in 1g flight and used to join lower anchor (low on fuselage or within biplane cellule) to higher anchor further outboard on wing; also known as lift wires.

fly-off   *1* Competitive in-flight demonstration of performance and other qualities between two or more rival aircraft built to same requirement to determine which will be chosen for procurement.

2 Without hyphen, to take off from a ship, esp. by free takeoff.

fly-over noise   Noise made by aircraft over particular point, usually near airport on inbound/outbound track, chosen for noise measurements. *Approach/takeoff noise*.

fly space   Simulated volume of sky, especially above a terrain board.

FM   *1* Frequency modulation or modulated; instantaneous frequency of EM carrier wave is varied by amount proportional to instantaneous frequency of modulating (intelligence-carrying) signal, amplitude and modulated power remaining constant.

2 Frequency measurement.

3 Fan marker.

4 Facilities management.

5 Figure of merit.

6 Fissile material.

7 From.

8 Fighter, multiplace (USAAC 1936–41).

f_M Mode frequency [FCS airspeed].

f_m Modulating frequency.

FMA   *1* Flight-mode annunciator or annunciation.

2 Fleet Management, or Manager, agreement.

3 Final monitor aid.

4 Federacion Mexicana de Aeronáutica.

FM/AM   *1* Amplitude modulation of carrier by frequency-modulated subcarrier(s).

2 Alternate FM and AM operation.

F_{max} *1* Maximum thrust.

2 Peak thrust of rocket engine.

FMC   *1* Flight management computer [F adds function, DL data-link, S system, U unit].

2 Fully mission-capable (USAF).

3 Forward motion compensator, or compensation.

4 Fatigue-monitoring computer.

FMCS   FMC (1) or (4) system.

FMCT   Fissile-Material[s] Cutoff Treaty (repeatedly postponed).

FMCW   Frequency-modulated continuous [rarely, carrier] wave.

FMD   Flight management display.

FME   Failure modes and effects [A adds analysis, CA criticality analysis, TA task analysis], all self-explanatory tools for risk assessment.
FMEA   See previous.
FMEP   Friction mean effective pressure, torque measured on calibrated brake.
FMF   *1* Foreign military financing.
  2 Flight-management function.
FMG   Flight management guidance (C adds computer, EC envelope computer, S system).
FMGC   Flight management and guidance control (system).
FMI   Functional management inspection.
FMICW   FM (1) intermittent [or interrupted] continuous wave.
FMK   *1* Flyvematerielkommandoen (Denmark).
  2 Field modification kit.
FML   Frequency memory-loop.
FM marker   Fan marker, transmits fan-shaped pattern of coded identity signals upwards, usually across one leg of radio range station.
FMMWRA   Forward-looking MMW radar altimeter.
FM-9   Fuel modifier No 9, which with a carrier fluid forms Avgard.
FMOC   Flight Maneuver Operations Center (NASA).
FMOF   First manned orbital flight.
FMOP   Frequency modulation on pulse.
FMP   *1* Full [or flight] mode panel.
  2 Flow management position.
  3 Foreign military program.
FMPG   Flow Management Planning Group (ICAO).
FMS   *1* Flight-management system.
  2 Field maintenance squadron (USAF).
  3 Frequency-multiplexed subcarrier.
  4 Foreign military sales (DoD).
  5 Federation of materials societies.
  6 Fuel-management system.
  7 Flexible manufacturing systems.
  8 Full-mission simulator.
FMSC   Frequency Management Sub-Committee (NATO).
FMT   Full-mission trainer.
FMTAG   Foreign Military Training Affairs Group (US).
FMTI   Future missile-technology initiative (USA).
F/MTI   Fixed/moving target indicator.
FMU   *1* Flight-management unit.
  2 Fuel-metering, or management, unit.
  3 Flow [of air traffic] management unit.
FMV   Försvarets, Materielverk (Defence Materiel Administration, Sweden).
F_N Nozzle drag correction.
F_n *1* Net thrust.
  2 Receiver noise.
FNA   *1* Fédération Nationale Aéronautique (parent body of flying clubs, F).
  2 Final approach.
FNBA   Fédération Nationale Belge d'Aviation.
FNC   Favoured-nation clause.
FNCP   Flight navigation control panel.
FNL   Fleet noise level.
FNLN   Fine line, on radar indicating significant turbulence.

FNMOC   Fleet Numerical Meteorology and Oceanography Center.
fnp   Fusion point.
FNPT   Flight navigation procedure[s] trainer.
FNS   *1* Strategic nuclear forces (F).
  2 Fortified Navier-Stokes algorithm.
FNT   Front [+GNS = frontogenesis = front forming; LYS = frontolysis = decaying].
FO   *1* Foreign object.
  2 Fibre optics (resulting in numerous other FO acronyms).
  3 Fail operational, or operative (see *FOS*, *FOOS*).
  4 First Officer.
  5 Flag Officer.
  6 Fractional ownership.
  7 Forward observer.
F/O   *1* Flying Officer (RAF).
  2 First Officer (civil).
  3 Fuel/oil.
f_o Frequency, prior to Doppler shift.
FOA   *1* Field Operating Agency, unit of USAF distinct from major command.
  2 Follow-on attack (no precise definition).
  3 Försvarets Forskningsanstalt, Swedish defence research establishment (merged 2001 into FOI).
  4 Future offensive aircraft (C adds capability, S system).
FOAC   *1* Flag Officer, Aircraft Carriers.
  2 Future offensive air capability.
FOAEW   Future organic airborne early-warning.
foam carpet   Layer of foam put down on runway or other space by fire tenders to cushion impact of aircraft making wheels-up landing.
foamed plastics   Foaming agent provides minute voids to create low-density material used for insulation (thermal, mechanical shock etc), or to increase structural rigidity; often foamed in place within structure.
foam monitor   Turret-mounted foam gun on crash-fire vehicle.
foaming space   Free vapour volume above fuel in tank.
foam strip   Foam carpet.
FOARC   Fractional Ownership Aviation Rulemaking Committee.
FOAS   Future Offensive Air System [more recently, Support] (UK).
FOB   *1* Forward operating base.
  2 Fuel on board (suggest undesirable usage, confusion with established non-aero acronym free on board).
FOBS   Fractional-orbit bombardment system.
FOC   *1* Foreign-object check.
  2 Full (or final) operational capability.
  3 Flares/off/chaff.
  4 Faint-object camera.
  5 Fibre-optic control, or cable, or computer.
  6 Final operational clearance.
  7 Fuel/oil cooler.
  8 Flight-operational commonality, making possible mixed-fleet flying.
FOCA   Federal Office of Civil Aviation (Swiss).
focal length   Distance from optical centre of lens or surface of mirror to principal focus.
focal plane   That parallel to plane of lens or mirror and passing through focus.
focal point   *1* See *focus*.

2 Air Staff agency or individual designated as central source of information or guidance on specific programme or project requiring co-ordinated action by two or more Air Staff agencies (USAF).

Focas, FOCAS *1* Fibre-optic communications for aerospace systems (USAF).

2 Flag Officer Carriers and Amphibious Ships (RN).

3 Focused ordnance controller with aimpoint selection (USAF).

FOCR Final operational clearance recommendation.

FOCSI, Focsi Fibre-optic control-system integration.

focus *1* Point at which parallel rays of light meet after being refracted by lens or reflected by mirror.

2 Point having specific significance relative to geometrical figure such as ellipse, hyperbola or parabola.

FOD *1* Foreign-object damage [or debris].

2 Fibre-optic[s] data [B adds bus, C control, H highway, L link, MS multiplex systems and S system].

3 Flight Operations Department (UK CAA).

4 First operational delivery.

FODCS Fibre-optic[s] digital control system.

FOD plod Duty of removing from taxiways and runways anything that might cause FOD.

F/ODS, Fods Fire/overheat detection system.

FODT Fibre-optic[s] data transmission.

FOES Fibre-optic engine sensor.

FOFA Follow-on forces attack.

FO/FO Flame-on/flame-out.

FOF3 Flag Officer 3rd Flotilla, responsible for all naval aviation (UK, RN).

FOG *1* Fibre-optic[s] guidance.

2 Fibre-optic[s] gyro.

3 Flight-operations group.

fog *1* Form of cloud in surface layers of atmosphere caused by suspended particles of condensed moisture or smoke, reducing visibility to less than 1 km. Advection * results from arrival of warm humid air over cold surface; radiation * from cooling of water vapour created by evaporation during day by cold ground on clear night; sea * by condensation of moisture in warm air over cold sea (essentially advection).

2 See cabin *.

FOG-M, Fog-M Fibre-optic[s], guided missile (USA).

fog of war Notional idea invented to explain 'own goals' in recent conflicts (US).

Föhn, foehn Dry wind with strong downward component, warm for season, characteristic of mountainous regions.

FOI *1* Follow-on interceptor (USAF).

2 Totalförsvarets Forskningsinstitut [defence/aerospace research, Sweden].

FOIA Freedom of Information Act (US).

foil *1* Lifting surface of hydrofoil landing gear.

2 Spanwise trailing-edge members, one upper and one lower, forming integral part of compound wing.

FOL *1* Forward operating location.

2 First-order logic.

fold Joint in wing of carrier aircraft enabling it to be folded. Incorporates one or more horizontal, skewed or vertical hinges.

folded combustor Gas-turbine annular combustion chamber in which, to reduce engine overall length, the flow makes two 180° turns, leaving at a reduced chamber diameter to match high-speed turbine.

folded dipole Two parallel, closely spaced dipole antennas connected at ends, with one fed at centre.

folded fell seam Fabric seam which has both edges folded and located in same place between fabric layers.

folding fin *1* Aircraft or missile fin hinged axially at base to lie flat prior to launch.

2 Rocket or missile fin hinged transversely to emerge from housing slot, as in FFAR.

folding wing One hinged outboard of root to enable overall dimensions to be reduced when aircraft in hangar, esp. on carrier.

foliage penetration Fopen, challenging objective of wide range of EM-radiating systems, especially with MTI; other acronyms Lobstar, TUT.

follow-on Anything considered to be second or subsequent generation in development, esp. within same manufacturing team.

follow-on contract One whose terms repeat those for earlier stage of same programme.

follow-on development tests Tests during acquisition phase after completion of formal Cat II tests; consist of updating changes or additions not available previously (USAF).

follow-on production Serial production immediately subsequent to completion of development or previous production batch.

Folta Forward operating location training area.

FOM *1* Figure of merit.

2 Federation object model [simulation].

FOMC Fibre-optic micro-cable.

FONA Flag Officer, Naval Aviation (UK).

fone Telephone (FAA).

FOOS Fail-operational, fail-operational, fail-safe.

foot Non-SI unit of length, = 0.3048 m by definition.

foot-candle Non-SI unit of luminance = 1 lm/sq ft = 10.76 lux.

foot-lambert See *lambert*.

foot-launched Micro or powered parachute strapped to pilot's back.

foot motor Foot-operated hydraulic motor used to energise wheel brakes.

foot-pound Unit of work or energy, = 1.35582 J.

foot-poundal Unit of work or energy, = 0.042140 J.

foot-pound-second system Non-SI system of units still used in US; also called Imperial.

footprint *1* Area around airport enclosed by selected contour for LPN, EPNL, NNI or other noise measure.

2 Possible recovery area for spacecraft plotted from re-entry point.

foot-thumper Stall-warning device that triggers oscillating plunger in rudder pedal when stall imminent.

foot-to-head acceleration Ambiguous, means accelerating force acting on body from head to feet, negative g.

footwell Recess in outer skin of [usually combat] aircraft, often covered by spring-loaded flap, to enable crew to board without ladder.

FOP *1* Forward operating paid.

2 Fired outside parameters.

Fopen Foliage penetration, or penetrating (SAR, TUT).

FOQA Fleet [airline] or flight-operations, or operational, quality assurance (FAA).

FOQNH Forecast regional QNH.

FOR *1* Fail-operative redundant.

2 Field of regard.

Foracs   Fleet operational readiness and calibration system.

Forcap   Force CAP, combat air patrol maintained overhead a task force (USN).

force   SI unit is Newton (N), which gives mass of 1 kg acceleration of 1 ms^{-2} = 0.101968 kgf, 0.22482 lbf, 7.230658 pdl, 3.597 ozf, 10^5 dynes.

force balance transducer   Output from sensing member is amplified and fed back to element which causes force-summing member to return to rest position.

force coefficients   Aerodynamic forces, eg lift and drag, divided by dynamic pressure $\frac{1}{2}\rho V^2 S$.

force combat air patrol   Patrol of fighters maintained to protect task force against enemy aircraft.

forced convection   Process by which heat is transported by mechanical movement of air (cooling systems, meteorology).

force diagram   Vector presentation of force(s) acting on object, length and direction of each vector representing magnitude and direction of one force. If diagram forms closed polygon, forces are in equilibrium. If diagram fails to close, gap indicates unbalanced force. Hence, force polygon.

forced landing   Made when aircraft can no longer be kept airborne, for whatever reason.

forced oscillation   One in which response is imposed by excitation; if excitation is periodic and continuing, oscillation is steady-state.

forced vibration   See preceding entry.

force fit   Mating parts in which male dimension exceeds female (see *fit*).

force gradient   Relationship between pilot input force and aircraft response, e.g., degrees roll per second per pound of lateral force on stick.

force majeure   Literally, no choice; reason for crossing a frontier at other than a designated point of entry (now archaic).

force rendezvous   Navigational checkpoint at which formation of aircraft or ships joins main force.

force-sensing controller   Pilot's primary flight-control input (stick/pedals) which senses applied force without noticeable movement.

force structure   Currently effective operational inventory (US).

force vector   Line in force diagram representing force magnitude, direction and point of application.

Fords   Fleet [airline] operational reliability data system.

fore-and-aft level   Gravity-controlled indicator of pitch attitude (arch.).

forebody   *1* The front portion of a body in atmospheric flight; in * strake can mean front half of fuselage.

2 Planing bottom of float or hull upstream of step.

forebody strake   Low-aspect-ratio extensions of wing at root along sides of fuselage; like LERX and glove [which taper more sharply] they generate powerful vortices at high AOA to improve handling in extreme positive-acceleration manoeuvres.

forecast   Statement of expected meteorological conditions at given place during specified period; air-navigation * includes wind velocity at selected heights, cloud, visibility, precipitation, ice formation, and barometric pressures at airfields and sea level.

foreflap   Leading member of double or triple-slotted flap.

foreign air carrier   One registered in foreign country, except in case of multinational carriers (eg SAS or Air Afrique) in collaborating countries.

foreign military sales   Portion of United States military assistance authorised by Foreign Assistance Act (1961 as amended); differs from Military Assistance Program Grant Aid in that it is purchased by recipient country.

foreplane   Horizontal aerofoil mounted on nose or forward fuselage to improve take-off and low-speed handling, esp. of delta aircraft where wing lift is lost because of upward movement of elevons. * can be fixed or retractable, fixed-incidence or rotating, and have slats, flaps or elevators.

forerudder   Rudder at front of aircraft.

forging   Shaping metal softened by heating by slow, rapid or repeated blows, with or without a shaped female die or male/female dies.

forked rod   Piston engine connecting rod having forked bearing on crankshaft, fitting over big end of matching blade-type rod.

forklift capacity   Of cargo item, provided with an integral pallet, tineways or forklift entries.

formability   Unquantified measure of ease with which material can be shaped through plastic deformation.

Formac   Fibre-optic medium-access controller.

format   *1* Size and shape of map, chart or photo negative or print.

2 One of several selectable types of presentation for instrument (eg ADI or HSI) or display, such as moving map, radar map, alphanumerics, attitude indication, flight planning or en route.

3 Fortran matrix abstraction technique.

formation   Ordered arrangement of two or more vehicles proceeding together, especially in a geometric pattern.

formation flight   More than one aircraft which, by prior arrangement between pilots, navigate and report as single aircraft; FAA formation limits are no more than one mile laterally or longitudinally and within 100 ft vertically from leader.

formation-flight control system   Developed at UCLA to perfect autonomous formation flight.

formation light(s)   Fitted to aircraft to enable other aircraft to formate on it at night.

formatted   Electronically processed to be compatible with particular format (2).

form block   Block or die usually made of wood, zinc, steel or aluminium, over or into which sheet metal is formed.

form die   One which performs bending and sometimes light drawing operations upon flat blank.

form drag   Pressure drag minus induced drag.

former   Light secondary structure added to maintain or improve external shape; eg around basic box fuselage to give curved cross-section, or extra false wing ribs ahead of front spar to maintain profile where curvature is sharp.

form factor   Physical overall dimensions of a body, especially one carried externally, taken into account not so much aerodynamically as to avoid hardware conflicts.

forming   Forcing flat material to assume desired contours and curves, esp. compound curvatures, by such means as drop hammer, flanging machine, hydraulic press, stretch press, hand forming etc.

forming roll   Bends sheet metal into cylinders of various diameters and other single-curvature shapes.

Form 700 In many English-speaking air forces, document signed by captain on taking over aircraft to signify that he is satisfied that every pre-flight maintenance inspection has been completed.

formula costs Direct operating costs for transport aircraft as calculated to ATA formula, which assumes indirect costs as 100 per cent of direct and standardized values for passenger, freight and mail revenues.

Formula 1 Air-racing rules, laid down by FARA, for uniform class of racers (includes engine not over 200 cubic inches, wing not less than 66 square feet, empty weight not less than 500 lb, fixed landing gear and carefully defined cockpit/windshield geometry).

Formula V Lightweight FARA class with VW-derived engines.

FORS *1* Foreign airlines representatives (Sweden).

2 Fibre-optic rate sensor.

Forte Fast on-orbit recording of transient events.

Fortran Formula translator; computer language.

Forty-five line Trajectory at 45° to horizontal, flown in either direction, thus up-* line and down-* line (advanced aerobatics).

forward aeromedical evacuation Provides airlift for patients between battlefield and initial and subsequent points of treatment within combat zone.

forward air controller Member of tactical air control party who, from forward ground or airborne position, controls aircraft engaged in close air support of ground troops (DoD).

forward air control post Mobile tactical air control radar used to extend coverage and control in forward combat area.

forward lock Prevents selection of reverser except after landing; disabled by oleo deflection or, if possible, bogie-beam tilt.

forward oblique Oblique photograph of terrain directly ahead.

forward operating base Airfield used indefinitely to support tactical operations without establishing full support facilities.

forward operating location Forward operating base.

forward scatter Scattering of radiant energy into hemisphere of space bounded by plane normal to incident radiation and lying on side toward which radiation was advancing; opposite of backward scatter.

forward slip See *sideslip*.

forward speed Component of speed in horizontal plane.

forward stagnation point Point on leading edge of body in airstream which marks demarcation for airflow on either side; boundary-layer air is stationary.

forward supply point En route or turnaround station at which selected aircraft spares are prepositioned for support of assigned mission(s).

forward sweep Opposite of sweepback, wing tips being further forward than roots.

forward tilt Of helicopter rotor, forward angular deviation of locus of centroid of blade sections from plane of rotation.

forward tilt wing See *slew-wing aeroplane*.

FOS *1* Fail-operational, fail-safe.

2 Faint-object spectrograph.

3 Fibre-optics sensor (or sensing, or system).

4 Family of systems.

5 Fleet operations standards.

6 Forward observer system.

FOSA Fondation des Oeuvres Sociales de l'Air (F).

FOSC Fibre-optic satellite communications.

FOSI *1* Formatting output specifications instance.

2 Family of systems integration, or integrator (AFRL).

FOSS, Foss Fibre-optic sensor, or sensing, systems.

FOST Flag Officer, Sea Training.

foster parent Company contracted to provide service support, possibly extending to major modification and update, for an imported type of military aircraft.

FOT *1* Frequency of optimum transmission.

2 Fibre-optic twister for making inverted image upright.

3 Flight-operations Telex.

4 Follow-on operational test (USAF).

5 Future-oriented technologies.

FOT&E Follow-on test and evaluation (continues after entry to service, USAF).

FOT&FOE Inserts final operational.

FOTD Fibre-optic towed decoy.

FOTM Flight operations training manager.

fouling Unwanted deposition on points of piston engine spark plug.

fount Range of characters for electronic display.

fountain Vertical rising column of hot air, usually plus entrained debris, formed by jets of VTO jet aircraft hovering at low level; on impact with fuselage can exert undesired suckdown effect; alternatively, can be made to increase lift.

fountain sheet Vertically rising wall between adjacent jets in hovering VTO.

four-bank eight Flight training manoeuvre similar to figure of eight except that outer portions of loops are not circular but consist of two 45° turns linked by short, straight flightpath.

four-course beacon See *radio range*.

4-D 3-D plus time.

4-D R-nav Terminal guidance sufficiently accurate to put arrival's wheels on runway within guaranteed window of ± 10 s.

four greens Gear locked down and landing flap setting.

Fourier expansion Expansion of waveform or other oscillation in terms of fundamental and harmonics.

Fourier integral Representation of (x) for all values of x in terms of infinite integrals.

Fourier series Representation of function f(x) in interval (– L, L) by series consisting of sines and cosines with common period 2L; when f(x) is even, only cosine terms appear, when odd, only sine terms appear.

4midable Requirement/benefit definition study leading to 4-D meteorological databases (Euret).

four-minute turn See *standard turn*.

four-poster Jet engine for VTOL with two front fan nozzles and two rear core-jet nozzles, which in VTOL mode provides lift from four jets in rectangular pattern; or an aircraft equipped with such an engine.

four-wing configuration See *cruciform wing*.

FOV Field of view; EA adds eye angle relative to head.

FOVE Fleet operations versatile environment [cockpit interface].

FOW *1* Family of weapons.

2 Fibre-optic wire.

Fowler flap Special form of split flap that moves at first rearwards and then downwards along tracks, thus producing initial large increase in lift and at full deflection giving high lift and drag for landing.

fox away   Verbal code: "AAM has been fired or released" (DoD).

Fox Code   Air-combat numeral code, usually: 1, SARH mssile selected; 2, IR missile selected; 3, guns; 4, fired outside parameters to distract enemy. There is an alternative, used to report method used to dispose of adversary: 1, radar, AAM; 2, IR AAM; 3, gun(s); 4, made opponent fly into ground.

FP   *1* Flight progress.
  2 Fluoro-polyamide.
  3 Flight path.
  4 Full potential (aerodynamic flow).
  5 Fixed-price.
  6 Weather forecast (US states).
  7 Fuel, petroleum.

f$_p$ Freezing point.

FPA   *1* Fire Protection Association (UK).
  2 Flight-path angle.
  3 Flying Physicians' Association (US).
  4 Flight-path accelerometer.
  5 Focal-plane array.
  6 Forward pitch amplifier (SAAHS).
  7 Flat-plate antenna.

FPAC   Flight-path acceleration.

FP area   Food-preparation area.

FPAS   *1* Focal-plane array seeker.
  2 Flight-profile advisory system.

FPASS   Force-protection airborne surveillance system.

FPB   *1* Fuel preburner.
  2 Functional payload block.

FPC   *1* Flight-path control, or controller, or command.
  2 Flight-profile comparator.

FPCC   Flight propulsion control coupling.

FPCD   Flat-panel colour display.

FPCS   Flight-path control system, or (F-15 S/MTD) set.

FPC2   Force protection command and control.

FPD   *1* Flight planning document.
  2 Flat-panel display.
  3 Fee per departure.
  4 Freezing-point depressant.
  5 Flight-proximity demonstration.

FPDA   Five-Power Defence Agreement [which see].

FPDS   Feasibility pre-definition study.

FPDZ   Free-fall parachute drop zone; A adds activity (CAA, UK).

FPE   Fédération des Pilotes Européennes (federation of European women pilots).

FPEEPM   Floor proximity emergency-escape-path marking.

FPEPA   Fixed-price with economic price adjustment [i.e., not fixed].

FPF   Fixed-price firm.

FPG   Force per g.

FPGA   Field-programmable gate array.

FPI   *1* Fluoropolyimide.
  2 Fixed-price incentive.
  3 Fluorescent penetrant inspection.

FPIF   Fixed-price incentive firm.

FPIS   *1* Fixed-price incentive contract with successive targets.
  2 Forward propagation by ionospheric scatter.

FPIWA   First-pass in-weather attack.

FPL   *1* Flight-plan message, ie filed flight plan.
  2 Full performance level (FAA 1 air-traffic controllers).
  3 Fluctuating pressure (rarely, power) level(s).
  4 See *f-pole line).*

FPLN   Flight plan.

FPM   *1* Flightpath miles.
  2 Feet per minute (ft/min preferred).
  3 Fleet performance monitoring [or monitor].
  4 Flight-performance module (AFMSS).
  5 Fibre/fiber placement mandrel.

FPMU   Fuel-pump monitoring unit.

FPN   Fixed-pattern noise.

FPNM   Feet [altitude] per nautical mile.

FPOV   Fuel preburner oxidiser valve.

FPP   *1* Fixed-pitch propeller.
  2 Ferry Pilots' Pool.

FPPS   Flight-plan processing system.

FPR   *1* Fan pressure ratio.
  2 Fixed price, redeterminable.
  3 Federal Procurement Regulation (US).
  4 Flight-plan route.
  5 Flat-plate radiometer.

FPRA   FPR (2) of A-type.

FPRM   *1* Flight phase related mode.
  2 Fuel-pipe repair manual.

FPS, fps   *1* Foot-pound-second system.
  2 Feet per second.
  3 Fine-pitch stop.
  4 Fuel production and storage [O adds offshore].
  5 Flight-progress strip.
  6 Fast packet switch.

FP70   Low-expansion firefighting foam, hydrolized protein plus perfluorocarbon surfactant (Angus).

FPSP   Flight-progress-strip printer.

FPSR   Foot-pound-second Rankine, traditional British system of engineering units.

FP strip   *Flight progress.*

FPTS   Forward propagation by tropospheric scatter.

FPU   Fin processor unit.

FPV   Flight-path vector.

FQGS   Fuel-quantity gauging system.

FQHE   Fractional quantum Hall effect.

FQI   Fuel-quantity indicator (or indication).

FQIS   FQI switch (or system).

FQMS   Fuel-quantity management system.

FQP   Flow quality probe.

FQPSK   Feher-patented phase-shift keying.

FQPU   Fuel-quantity processor unit.

FQR   Formal Qualfication Review.

FQT   *1* Formal qualification testing.
  2 Frequent.

FQTI   Fuel-quantity totaliser indicator.

FR   *1* Flight refuelling.
  2 Flight recorder.
  3 Fighter reconnaissance.
  4 Forward relay (ATC).
  5 Fuel remaining.
  6 Frame.
  7 Falling rapidly.
  8 From.
  9 Federal Register (US).

F/R   *1* Function reliability.
  2 Final run.

F$_R$ *1* Aerodynamic loading (force) due to Rth mode.

2 Frictional force.

fr   Fuel required.

f_r   Radar p.r.f. [sometimes F_r].

FRA   *1* First-run attack.

2 Fuel-risk aversion.

3 Flap-retraction altitude.

4 See *fracto*.

fractional   Fractionally owned [adjective and noun].

fractional distillation   Heating crude petroleum to moderate temperature at atmospheric pressure and condensing different fractions separately.

fractional orbit bombardment system   ICBM initially launched into low Earth orbit, enabling warhead[s] to arrive on target from any direction.

fractional ownership   Aircraft, usually bizjet, is registered to prearranged [usually small] group, each having equal allocation of annual flight time.

fractionation   Use of large numbers of small (nuclear) warheads, esp. in counterforce attack; hence to fractionate.

fracto-   Prefix, clouds broken up into irregular, ragged fragments.

FRAD   Frame relay access device.

FRADU, Fradu   Fleet Requirements and Air Direction Unit (UK).

FRAG   Fatigue Research Advisory Group.

fragmentary order   Abbreviated form of operation order which eliminates need for restating information contained in basic operation order (DoD). Usually a daily supplement to SOOs governing conduct of specific mission.

fragmentation bomb   Charge contained in heavy case designed to hurl optimum fragments on bursting.

frag order   Fragmentary order.

FRAIS   Functional requirements for airport ground-movement control and management interconnection system.

FRAM, Fram   *1* Fleet rehabilitation and modernization.

2 Ferroelectric RAM (1).

F_{ram}   Ram drag.

frame   *1* In photography, single exposure within continuous sequence.

2 Transverse structural member of fuselage, hull, nacelle or pod, following its periphery.

3 Of engine, structural transverse diaphragm carrying main shaft bearing.

4 Picture area scanned in radar, TV, video or CRT (US = field).

frameless canopy   Single blown transparency in metal peripheral frame.

frame relay   New standard for handling small packets of high-speed burst data over WANs.

frame time   Symbol t_f, time required to scan one frame in radar search.

framing pulses   Transmitted in TV/video, SSR, IFF, ATCRBS and similar systems to synchronize transmitted and received timebases.

franchise   Legal right of a carrier to fly for hire or reward on a given route, without obligation to do so.

frangible   Designed to, or likely to, shatter on impact.

Frap   Fragmenting-payload, gun [initially 28-mm Mauser] projectile comprising 180 slender (1-g) tungsten-alloy needles.

FRAS   *1* Free-rocket anti-submarine.

2 Fuel-resources analysis system.

FRAT   Flight-readiness acceptance test.

fratricide   Shooting down friendly, esp. in air/air engagement.

fraudulent echo   Radar echo produced by DECM.

frax   Fractional ownership (colloq.).

FRC   *1* Flight reference card.

2 Fibre-reinforced [or fibrous refractory] composite (or concrete).

3 Future rotorcraft.

4 Federal Radio Commission (US).

5 Full route clearance.

FRCI   *1* Flexible reusable carbon insulation.

2 Fibrous refractory composite insulation.

freak   Verbal code meaning frequency in MHz (DoD).

FRED, Fred   Flexible routine[s] for engine development.

freddie   Air-intercept controlling unit (DoD).

free-air anomaly   Difference between observed and theoretical gravity computed for latitude and corrected for elevation above or below geoid.

free-air overpressure   Unreflected pressure in excess of ambient created by blast wave from explosion.

free airport   One at which people and goods may be trans-shipped without customs charges or examination.

free-air tunnel   Aerodynamic or other test with specimen moved through atmosphere, eg on aircraft or rail flatcar or rocket sled.

free atmosphere   That portion of Earth's atmosphere, above planetary boundary layer, in which effect of Earth's surface friction on air motion is negligible, and in which air is usually treated (dynamically) as ideal fluid. Base usually taken as geostrophic wind level.

free balloon   One floating untethered.

free-balloon concentration ring   Ring to which are attached ropes suspending basket and to which net is secured; also known as load ring.

free-balloon net   Distributes basket load over upper surface of envelope.

free-body principle   Stress-analysis procedure that involves isolating structure, considering it to be held in equilibrium by loads acting upon it.

free-call   Change frequency [non-mandatory ATC advisory].

free canopy   Cockpit canopy that is non-jettisonable.

freedoms   Five basic freedoms relating to air traffic negotiated in bilateral air agreements between governments of pairs of countries. First freedom is right to fly across territory of other country with whom agreement is made. Second is right to make technical or non-traffic stop in that country. Third is right to set down in that country traffic emplaned in home country of airline. Fourth is right to pick up traffic in that country destined for airline's home country. Fifth is right to pick up traffic in other country and carry it to any third state.

free drop   Dropping packaged equipment or supplies without parachute.

free electron   Not bound to an atom.

free fall   *1* Fall of body without guidance, thrust or braking device.

2 Free motion along Keplerian trajectory, in which force of gravity is counterbalanced by force of inertia.

3 Parachute jump in which parachute is manually activated at discretion of parachutist, or automatically at pre-set altitude.

4 Acceleration g under standard conditions, = 9.80665 ms^{-2}.

free-fall altimeter　　Worn by parachutist in free-fall jump to deploy parachute at pre-set altitude.

free-fall(ing) bomb　　Bomb without guidance.

free-fall landing gear　　Designed to be lowered by force of gravity and wind load.

free-fall model　　Unpowered model for spinning or other tests intended to fall freely after release.

free fan　　Name promoted by Boeing for UDF and propfan propulsion to avoid image of 'propeller'.

Free Flight　　Programme 'to provide controllers and NAS users with tangible functional enhancements' (FAA). Broadly, it puts decisions regarding commercial traffic over the US in the hands of pilots, monitored from the ground.

free flight　　*1* Without guidance except, possibly, simple stabilizing autopilot.

2 In air-carrier operations, captain can select his own FL and flight path, ATC intervening only to prevent conflict.

free-flight model　　One normally ground-launched and self-propelled, usually instrumented for aerodynamic or flutter measures.

free-flight tunnel　　One in which model can be observed in free flight, ie unmounted or fired from gun.

free-form fabrication　　Flexible manufacturing method for rapid development of prototype parts.

free gun　　Capable of being aimed independently of the aircraft.

free gyro　　Two degrees of freedom, spin axis may be oriented in any specified attitude and not provided with erection system.

free jet　　Fluid jet after emission from nozzle.

free-jet test　　In an open hypersonic airflow, rather than having air pumped into inlet.

freelance　　Verbal code: "Self-control of air-intercept aircraft is being employed" (DoD).

free-molecule flow　　*1* Flow regime in which mean free path is at least ten times typical dimension of flying body, as at height above 180 km.

2 Flow about body in which collisions between molecules of fluid are negligible compared with collisions between these and body.

free parachute　　Deployed manually by parachutist, not by static line.

free-piston engine　　Any of several types in which hot gas is generated between pistons oscillating freely in linear or toroidal cylnders.

FREER　　Free-route experimental encounter resolution.

free radical　　Atom or group of atoms broken away from stable compound by application of external energy; often highly reactive.

free-return trajectory　　One in which crippled spacecraft would fall back to Earth.

free rocket　　Unguided rocket.

free scan　　DME operating mode providing distance to all DMEs within LOS.

free space　　Ideal homogeneous medium possessing dielectric constant of unity and in which there is nothing to reflect, refract or absorb energy.

free-spinning tunnel　　Vertical wind tunnel used to test spinning characteristics using free models.

free-standing propellant　　Not case-bonded.

free stream　　Fluid outside region affected by aircraft or other body.

free-stream capture area　　Cross-sectional area of column of air ingested by jet engine, esp. ramjet.

free streamline　　One passing well away from moving body. Streamline separating fluid in motion from fluid at rest.

free-stream Mach number　　Mach number of body measured in free stream, unaccelerated by body's presence.

freestyle　　Aerobatic routine selected by pilot, not preset. Similarly * skydiving, not following a preset sequence.

free takeoff　　From aircraft carrier, not using catapult.

free turbine　　One that drives output shaft to propeller or helicopter rotors, and is not connected to compressor.

free-vortex compressor　　Axial compressor designed to impart tangential velocities inversely proportional to radius from axis.

free-vortex flow　　Persisting in fluid remote from source or solid surface, eg tornado.

freewheel　　Sprag clutch or other device which permits helicopter rotor system to continue to rotate even if input shaft from main gearbox is arrested.

freewing　　Wing freely able to rotate about spanwise axis, thus having incidence determined by relative wind while AOA remains constant [the reverse of conventional aeroplane].

freeze　　*1* To arrest dynamic operations, eg in simulator training.

2 Radar mode which, once commanded, permits one more scan; emissions then cease and display remains active but frozen until * button is pushed a second time.

freeze-out　　Method of controlling humidity by condensing water vapour, and possibly carbon dioxide, over cold surface.

freezing　　*1* Stage in design when all major features are irrevocably settled, thus enabling detail design to start.

2 Manually arresting input to display, leaving static (prior) situation for study.

free zone　　Customs-free area.

freight　　Cargo, including mail and unaccompanied baggage but excluding express.

freight consolidating　　Process of receiving shipments of less than carload/truckload size and assembling them into carload/truckload lots for onward movement to ultimate consignee or break-bulk point.

freight container　　See *container.*

freight doors　　Designed to take freight, vehicles or containers.

French landing　　With plenty of power on.

Frensor　　Freezing-point sensor.

Freon　　Trade name for family of halogenated hydrocarbons containing one or more fluorine atoms, including CFCs, widely used as refrigerant medium (eg in vapour-cycle air conditioning), as fire extinguishant and as aerosol propellant.

freq　　Frequency.

frequency　　*1* Reciprocal of primitive period of time-periodic function, symbol f, units (SI) Hz, (fps units) cycles per second.

2 Number of services operated by airline per day or per week over particular route.

frequency agility　　The ability to generate an output whose frequency is variable. This is a basic ECCM technique. In the case of radars the magnetron or other waveform generator can be tuned to give a different frequency on each scan. There are many other FA methods for communications and other emissions. The objective is invariably to prevent hostile jamming.

frequency band   Continuous range of frequencies extending between two limiting values; EM bands are listed in Appendix 2.

frequency bias   Constant frequency added to signal to prevent its frequency falling to zero.

frequency channel   *1* Band of frequencies which must be handled by carrier system to transmit specific quantity of information.

2 Band of frequencies within which station must maintain modulated carrier frequency to prevent interference with stations on adjacent channels.

3 Any telecommunications circuit over which telephone, telegraph or other signals may be sent.

frequency departure   Variation of carrier frequency or centre frequency from assigned value.

frequency deviation   *1* Maximum difference between IF (1) of FM wave and frequency of carrier.

2 In CW or AM transmission, variation of carrier frequency from assigned value.

frequency distortion   Produced by unequal amplification of signals or reproduction of sounds of different frequency; usual criterion for high-fidelity reproduction is level amplification over 20-15,000 Hz.

frequency division multiplex   Telecommunications allowing two or more signals to travel on one network simultaneously by modulating separate subcarriers, suitably spaced in frequency to prevent interference.

frequency drift   Slow change in frequency of oscillator or transmitter with time.

frequency equation   Relates phase speed to wavelength and to physical parameters of system in linear oscillation; also known as dispersion equation.

frequency hopping   Unpredictable continual and rapid changes of frequency of radar or other military electronics to defeat hostile ECM.

frequency modulated radar   One in which range is measured by interference beat frequencies between transmitted and received FM waves.

frequency modulation, FM   Instantaneous frequency of modulated wave differs from carrier by amount proportional to instantaneous value of modulating wave. Combination of phase and frequency modulation commonly referred to as FM.

frequency monitor   Stabilized receiver giving audible or visual indication of any departure of transmitter from assigned frequency.

frequency pairing   Association of each VOR frequency with a specific DME channel to reduce workload and errors.

frequency parameter   Ratio of airspeed to product of frequency of oscillation and representative length of oscillating system.

frequency response   *1* Portion of EM spectrum sensed within specified limits of error.

2 Response as function of excitation frequency.

frequency scanning   Pattern of radar-antenna movement formed by successive beams having different azimuth and frequency.

frequency-selective   Response only to narrow frequency band.

frequency-selective surface   Large family of frequency filters made up of band-pass or band-stop designs, eg loaded slots of various geometries, or doubly periodic arrays of metal elements or apertures in a conductive frame creating a plane-wave transmission with properties which are frequency-dependent. Fighter radomes are becoming treated with *, on the outside [fragile] because internal * could be destroyed by lightning strike.

frequency shift-keying   System of telegraph signalling in which keyed signal imposes small frequency shift on carrier; frequency changes of received signal are converted to amplitude changes.

frequency swing   Peak difference between maximum and minimum frequencies of FM signal.

frequency tolerance   Extent to which carrier (or centre of emission bandwidth) is permitted to depart from authorized frequency because of instability.

frequency-wild   Electric power system whose frequency is not stabilized but varies with rotational speed of generators.

frequent flyer   Fare-paying passenger rewarded for being good customer. [Varies with carrier].

FRES   Future rapid-effects system.

Frescan   Frequency scanning.

Fresnel lens   Form of echelon lens for generating parallel beam [familiarly used in lighthouses].

Fresnel mirror   Two planar mirrors joined at one edge, angle between them being almost 180°, for generating interference fringes.

Fresnel zone   Any spatial surface between transmitter and receiver, or radar and target, over which increase in distance over straightline path is equal to integer multiple of half wavelength.

Frespid   Frequency response identification (software).

FRET, Fret   First round effect[s] on target.

fretting   Rubbing together of solid surfaces, esp. slight movement but high contact pressure.

FRF   *1* Frequency-response function.

2 Flight-readiness firing.

FRFC   Full-range flow control.

FRFI   Fuel-related fare increase.

FRGN   Foreign.

FRH   Flap-retraction height.

friction   Force generated between solids, liquids or gases opposing relative motion.

friction coefficient   Friction (static, on point of relative motion, or dynamic) divided by perpendicular load pressing surfaces together.

friction horsepower   Indicated horsepower minus brake horsepower.

friction layer   Planetary boundary layer.

friction lock   Device in which friction is used to prevent unwanted movement; eg of throttle levers. Usually adjustable up to a positive lock.

friction range   For a given longitudinal trim setting, the range of airspeeds that can be flown stick-free due to FCS friction.

friction wake   That downstream of streamlined non-lifting body.

friction welding   Welding by rotating two parts together under load until surfaces are on point of melting and then forcing together to squeeze out joint material, simultaneously arresting rotation.

Friendly Functional requirement identification development methodology (Euret).

friendly Not hostile, contact positively identified as such (DoD).

FRIG Floated rate-integrated (or integrating) gyro.

Frise aileron Aileron having inset hinges and bevelled along upper leading edge; when lowered forms continuation of wing upper surface but when raised nose protrudes below wing, increasing drag and equalising aileron drag in banked turn.

FRL Fuselage reference line.

FRM *1* Failure-related mode.

2 Fault-report[ing] manual.

FRMALS Fédération Royale Marocaine de l'Aviation Légére et Sportive.

FRMG Forming (weather report).

FRMN Formation (weather report).

FRMR Frame reject[ion].

FRN Federal Radionavigation Plan, to phase out non-satellite navaids from 2010 (US DoT).

FRNG Firing.

FRO Failure requiring overhaul.

FROD Functionally related observable differences.

FROG Free rocket over ground.

frog Free-fall sport-parachuting position with body horizontal face-down, arms stretched parallel horizontal and lower legs vertical.

front *1* Boundary at Earth's surface between two contrasting air masses; usually associated with belt of cloud and precipitation, and more or less sharp change in wind (see *cold* *, *occluded* *, *stationary* *, *warm* *).

2 Occupant of front cockpit of tandem-seat aircraft, especially if aircraft commander (colloq., usually US/NATO).

frontal area Projected cross-section area of body viewed from front.

frontal attack Air intercept which terminates with heading crossing angle greater than 135° (DoD).

frontal weather To be expected in front; clouds, rain, temperature variation and other phenomena.

front course sector That situated on same side of localizer as runway.

front end Various meanings, including (1) system location (actual geometry immaterial) where EDP program is developed, and editing, compiling and peripheral access take place. (2) Start of major design process, long before appearance of hardware.

front enders Flight crew members, esp. in military transport (colloq.).

front ignition Solid rocket motor with igniter at end of filling or grain furthest from nozzle.

frontogenesis Process which produces discontinuity in atmosphere or increases intensity of existing front, generally caused by horizontal convergence of air currents possessing widely different properties.

frontolysis Process which tends to weaken or destroy a front.

front stagnation point See *forward stagnation point*.

Fropa Frontal passage.

Frost Fast read-out optical storage technology.

frost Small drops of dew which freeze upon contact with object colder than 0°C such as aircraft passing from cold air into warmer humid air. Glazed * (rain ice) is layer of smooth ice formed by rain falling on object at below 0°C. Hoar * is white semi-crystalline coating.

frost point Temperature of air at which frost forms on solid surface at same temperature.

Froude efficiency Basic element of propulsive efficiency: $\frac{2U}{V+U}$ where U is TAS and V is velocity of propulsive jet or propeller slipstream relative to the aircraft.

Froude number Non-dimensional ratio of inertial force to force of gravity for fluid flow; reciprocal of Reech number; Nfr = v^2/lg, where v is characteristic velocity and 1 characteristic length (may be given as square root of this ratio).

frozen No further design changes permitted.

frozen smoke Acoustic/thermal insulation [colloq. airogel said to be lowest-density semi-rigid material, able to support $10^3 \times$ own mass].

frozen stick Terrifying situation in which pilot cannot overcome stick force needed to recover from steep dive.

FRP *1* Fibre-reinforced plastic[s] or plywood.

2 Flight-refuelling probe.

3 Fuselage replacement programme.

4 Federal radio navigational plan.

5 Fares and Rates Panel.

6 Full-rate production.

FRPA Fixed reception pattern antenna.

FRQ Frequent, frequency.

FRR Flight-readiness review.

FRS *1* Fan rotation speed (N_1).

2 Fighter, reconnaissance, strike (role prefix, UK).

3 Fleet Replacement Squadron (USN, USMC).

4 Flying Refresher School (RAF).

5 Field Repair Squadron (RAF).

6 Flammability reduction system.

FRSI Flexible reusable silica insulation.

FRST Frost.

FRTO Flight radio-telephony operator.

FRTOL *1* Flight radio temporary operating license.

2 Flight R/T operator licence (UK).

FRTV Forward repair and test vehicle.

FRU *1* Forward Repair Unit (RAF).

2 Fleet Requirements Unit (RN).

fruiting SSR responses of aircraft to interrogation by other stations or responses of different aircraft to interrogation by other stations, in either case not synchronized with desired response by aircraft interrogated by own station; hence defruiting usually simple.

fruit salad Impressive rows of medal ribbons (UK, colloq.).

FRUSA Flexible rolled-up solar array.

FrW Friction welding.

FRWD Fast-reaction weapons demonstration.

FRZ Freeze (information or display).

FS *1* Fuselage station.

2 Frame station.

3 Fail-safe.

4 Fighter Squadron.

5 Front spar.

6 Flight simulator.

7 Feasibility study.

8 Full-supercharge gear.

9 Factor[s] of safety.

Fs Fractostratus.

f$_s$ Signal frequency.

f/s Not acceptable for ft/s.

FSA *1* Final squint angle.

2 Force-structure aircraft (US).

3 Fuel-savings advisory; S adds system and see *FSA/CAS*.

4 Frequency-spectrum availability.

5 Future strike aircraft (USAF).

FSAA Flight simulator for advanced aircraft (NASA).

FSA/CAS FSA (3) cockpit avionics system.

FSAF Future surface-to-air family (title often in French) (Int.).

FSAGA First sortie after ground alert.

FSAS *1* Fuel-savings advisory system.

2 Flight-service automation system.

FSAT *1* Full-scale aerial target.

2 Federal service of air transport (R).

FSB *1* Fan-stream burning.

2 Fasten seat-belts.

3 Flight Standardization Board (US).

4 Federal security service (R).

FSC *1* Force-sensing control[ler].

2 Fuel-saving[s] computer.

3 Forward shaped-charge warhead.

4 Flight Safety Committee (UK).

5 Flight-safety critical.

FSCL Fire-support co-ordinating line.

FSCM Federal supplier [or supply] code of manufacturer, or for manufacturers (US).

FSCP Flap/slat control processor.

FSCRS Flight-safety confidential reporting system [not same as Chirp].

FSCTE Fire Service Central Training Establishment (UK).

FSD *1* Full-scale deflection.

2 Full-scale development.

3 Federal Security Director (TSA).

FSDO Flight Standards District Office (FAA).

FSDPS Flight service data-processing system (FAA).

FSE *1* Fleet-supportability evaluation.

2 Field-service engineer, or evaluation.

FSED Full-scale engineering development.

FSEU Flap/slat electronics unit.

FSF *1* Flight Safety Foundation (US, 'provides leadership to over 600 member organisations in 75 countries').

2 Svenska Flygsportförbundet (Sweden).

3 Full-scale fatigue; S adds specimen, T test.

4 Forward Supply Flight (RAF).

FSG *1* Fluid-sphere gyro.

2 Fans stakeholder group.

3 Flight Simulation Group (RAeS, C adds Committee).

FSI Field service instruction.

FSII Fuel-system icing inhibitor.

F-Sims, FSIMS Flight-safety information-management system.

FSK Frequency-shift keying.

FSL *1* Full-stop landing, aircraft brought to halt.

2 Forecast Systems Laboratory (NOAA).

3 Fast serial link.

FSM Flight schedule monitor.

FSME Flight structural-mode excitation.

FSN *1* Factory serial number.

2 Federal stock number.

3 Field-service notice.

4 French-speaking nations.

FSOL Flight-safety occurrence list.

FSOV Fuel shut-off valve.

FSP *1* Flight [progress] strip printer.

2 Flexible sustainment program.

3 Fragment-simulator projectile (test of armour).

4 Function signalling panel.

5 Flir-stabilized payload.

6 Flight-screening program (USAF).

FSPF Military production authority (Iraq).

F$_{spil}$ Inlet spillage drag correction.

FSPM Flight-strip printing module.

FSP/UPT Flight-screening program and undergraduate pilot training (USAF).

FSR *1* Frequency set-on receive; S adds system.

2 Field-service representative.

3 Flight safety reporting.

4 Further special refit.

FSRI Florida Space Research Institute (Federal funding, US).

FSRS *1* Frequency-selective receiver system.

2 Flight-safety recording system.

FSS *1* Flight Service[s] Station (FAA).

2 Flight Standards Service (FAA).

3 Flying Selection, or Support, Squadron (RAF).

4 Frequency-selective surface[s].

5 Fixed satellite service.

6 Micro class, single-seat.

7 Fire-suppression system.

8 Flow-separation suppression; E adds element.

9 Flight-support system.

FSSP Forward-scattering spectrometer probe.

FSSR Federal Safety Standard Regulations.

FST *1* Flight-simulation test.

2 Fuel-spike test.

3 Full-scale tunnel.

4 Fuel-system trainer (F-22).

FSTA Future strategic tanker aircraft.

FSTP Full-spectrum threat protection.

FSU Flir sensor unit.

FSVL Fédération Suisse du Vol Libre (Switzerland).

FSVT Federal service of air transport; equivalent to FAA (R).

FSW *1* Forward-swept wing.

2 Friction-stir welding.

FT *1* Fault-tolerant.

2 Fast-track.

3 Fourier transform.

4 Functional test.

5 Terminal forecast.

6 False target; G adds generator, I imagery, LO lock-on, R rejection.

F-T Flight-time (diagram).

ft Foot, feet.

f(t) Function of time.

FTA *1* Fatigue-test article.

2 Fast tactical attack.

3 Fire-track area.

FTAAS Fast-time acoustic-analysis system.

FT-ADIR Fault-tolerant air-data inertial reference; S adds system, U unit.

FTAE First-time-around echo.

FTAJ Frequency/time ambiguity jamming.

FTB Flying, or flight, testbed.

FTC, ftc　*1* Fast time constant.
　2 Federal [and also Fair] Trade Commission (US).
FTCA　Future Tactical Combat Aircraft.
FTCCP　Flight-training candidates checks program (US Dept. of Justice).
FTD　*1* Field Training Detachment (USAF).
　2 Foreign Technology Division (USAF).
　3 Flight-training device[s].
FTDC　Fault-tolerant distributed computing.
FTE　*1* Flight-test engineer.
　2 Flight technical error.
　3 Full-time equivalent (personnel).
FTEP　Full-time error protection.
FTF　Flygtekniska Föreningen (Sweden).
FTG　*1* False-target generator.
　2 Fitting.
FTH　*1* Full-throttle horsepower.
　2 Further.
FTI　*1* Fixed time interval.
　2 Fixed target imagery, or indicator, or indication.
　3 Fast tactical imagery.
　4 Flight test instrumentation.
　5 Fictional threat image.
FTIR　Fourier transform of IR [S adds spectroscopy].
FTIT　Fan-turbine inlet temperature.
FTK　Freight tonne-kilometre(s).
FtL　Foot-lambert (non-SI unit of luminance or brightness).
FTLB　Flight-Time Limitation Board (UK).
FTLO　*1* Fast-tuned local oscillator.
　2 False-target lock-on.
FTM　*1* Frequency/time modulation.
　2 Flight-test mission objective(s).
ft/min　Feet [climb or descent] per minute.
FTNW　Future tactical nuclear weapon.
FTO　*1* Flexible take-off.
　2 Flying training organization.
FTP　*1* Functional test, procedure.
FTPP　Fault-tolerant power panel.
　2 File transfer protocol.
FTR　*1* Failed to return.
　2 Flat-tyre radius.
　3 False-target rejection.
　4 Future transport rotorcraft.
Ftr　Fighter.
FTRG　Fleet Tactical Readiness Group (USN).
FTS　*1* Flying Training School.
　2 Fatigue-test specimen.
　3 Federal Telecommunications System.
　4 Flexibe turret, or track, system.
　5 Flight termination system (UAV).
ft/s　Feet per second, which see for conversions.
FTSA　*1* Fault-tolerant system architecture.
　2 Fédération Tunisienne des Sports Aériens.
FTSC　Fault-tolerant spacecraft (or spaceborne) computer.
FTTU　Fliegender Technologie-trager unbemannt = unmanned technology testbed (ETAP).
FTU　Forward transmitter unit.
F$_{tu}$, F$_{TU}$　Ultimate tensile strength.
FTV　Flight test vehicle.
FTW　Follow-through warhead.
FTX　Field training exercise.
FTZ　Foreign trade zone, secure duty-free area at airport

legally outside territory administered by local customs authority.
FU　*1* Fuel uplifted.
　2 Fire unit.
　3 Forecast upper wind.
FU, Fu　Smoke (ICAO).
fu　Fuel used.
FUCE　Far-ultraviolet camera experiment.
fudge factor　Multiplying factor to allow for unknowns (colloq.).
fuel　Substance used to produce heat by chemical or nuclear reaction; usual chemical reaction is combustion with oxygen from atmosphere or from rocket oxidant (see *petrol*, *propellent*).
fuel accumulator　Container for storing fuel expelled during starting cycle to augment flow momentarily at predetermined fuel pressure.
fuel additive　Any material or substance added to a fuel to give it some desired quality; eg tetraethyl lead added as an anti-detonation ('knocking') agent.
fuel advisory departure　Procedure to save fuel by holding aircraft prior to engine start, rather than at destination stack.
fuel-air explosive　Selected liquid fuels which on warhead impact are scattered in fine cloud of large volume alongside target; this is then detonated (combining with atmospheric oxygen) less than 1 s after impact with blast effects generally greater than that of same mass of conventional explosives.
fuel-air mixture analyser　Measures piston engine air-to-fuel ratio. Chemical *** measures absorption of CO_2 by substance such as sodium or potassium hydroxide; physical (electro-chemical ***) measures difference in electrical resistance between two sampling cells, one open to air and other to exhaust.
fuel blending facility　One having authority to mix hydrocarbon fractions to produce piston or turbine fuels to specification, with correct addititives.
fuel burn　*Fuel consumption.*
fuel bypass　Maintains fuel pressure in carburettor float chamber of supercharged piston engine at fixed level above carburettor air pressure.
fuel capacity　Unless otherwise stated means actual volume of tank(s), not of system, and has no relevance to usable capacity.
fuel cell　Device which converts chemical energy directly into electricity; differs from storage battery in that reactants are supplied at rate determined by electrical load.
fuel chop　Sudden cutoff of supply, for whatever reason.
fuel consumption　Measured on volume and mass basis; 1 mpg = 0.354 km/1 (UK), 0.425 km/1 (US); 1 gal/mile = 2.825 1/km (UK), 2.353 1/km (US); including payload, 1 tonne-km l^{-1} = 2.868 UK ton-miles/UK gal; reciprocal is 0.348.
fuel control and monitoring computer　Measures and indicates fuel volume and mass, controls refuelling to selected levels whilst monitoring pumps and valves, monitors and controls tank temperatures, and controls tank utilization sequence to minimize wing bending moment and control c.g.
fuel control unit　Governs engine fuel supply in accordance with pilot demand, ambient conditions and engine limitations.

fuel-cooled   Cooled by fuel, either en route to engine or recirculated back to tank (see *regenerative cooling*).

fuel cut-off   Device for cutting off the supply of metered fuel to cylinders or a combustion chamber.

fuel dipper   Automatic adjustment of fuel flow to turbine engine, usually triggered to reduce flow at specific time, eg when firing guns or when any kind of upset disturbs inlet airflow to cause rapid rise in TGT.

fuel dumping   Release of fuel in flight to bring weight down to MLW in emergency.

fuel-flow regulator   Central element of gas-turbine engine CASC, driven by engine and incorporating two governors, pressure-drop and speed control, and two variable-orifice sliding valves, the VMO and the PDC (6).

fuel grade   Quality of piston engine petrol (gasoline), esp. as defined by anti-knock rating; previously called octane number. Usual grades are 80 (dyed red), 100 (green), 100L (green or blue, lower TEL content) and 115 (purple).

fuel injection   Inevitable in diesel or compression-ignition engines; term normally refers to Otto-cycle piston engine with mechanical injection of measured quantities of fuel either into carburation induction system or directly to cylinders. Unlike carburettor, ** unaffected by aerobatic or evasive manoeuvres and not prone to icing or fire hazards.

fuel jettison   Rapid discharge of fuel from aircraft in emergency.

fuel-jettison time   That required to reduce weight from MTO to MLW.

fuel lag   Deliberate short delay in injection of one propellant into rocket thrust chamber to establish particular ignition sequence.

fuel manifold   Peripheral main pipe with branch pipes distributing fuel to all burners of gas turbine.

fuel nozzle   In a gas turbine engine these atomise or vaporise the fuel to ensure very rapid burning in a short linear distance; see *Simplex*, *Duplex*, *Lubbock*, *airspray*, *vaporizing*.

fuel-pressure switch   Ensures full current is not applied to electric starter until fuel pressure has reached predetermined level.

fuel savings advisory   Provides command to pilots using flight-path optimization algorithms and Flight Manual data for L/D and thrust. An aditional mode can be preset for T-O/ldg parameters.

fuel shift   Gross long-term movement of fuel [eg, caused by prolonged steep climb], can be very dangerous.

fuel shut-off   See *cut-off*.

fuel sink   Total heat capacity of fuel as receptacle for surplus onboard energy; unit should be J.

fuel sloshing   Gross short-term movements of fuel or oxidant in part-empty tank.

fuel spike test   Tests engine surge margin.

fuel starvation   Fuel in tank[s] is for some reason prevented from reaching engine[s].

fuel state   The precise quantity of fuel in the tank[s] of an aircraft; quality of the fuel is irrelevant.

fuel tank   See *bag tank*, *integral tank*, *rigid tank*, *drop tank*.

fuel trimmer   An adjustment on gas-turbine fuel control to achieve a precise value of N_2 (NPC), adjusted to ISA sea-level for a particular engine, repeated after major overhaul and recorded on data plate.

fuel vent   Small pipe used to equalize pressure inside and outside tank.

FUF   Favourable unbalanced field.

fufo, FUFO   Full-fuzing options (NW).

FuG   Radio equipment (G).

fuh   Fuel mass at given height (usually rocket).

Führer weather   Sunshine, blue sky (WW2 Luftwaffe).

Fu L, FuL   Fuhrungsstab der Luftwaffe (G).

full annealing   Heating above critical temperature, or crystalline-transformation point, followed by slow cooling through range of transformation; results in relief of residual stress, greater ductility, increased toughness but lower strength.

full-authority control   Control system, today usually electronic but could be fluidic, which provides complete management function [of engine(s), system(s), or complete aircraft], with pilot serving as passive observer. Early example is Fadec, from which FAFC differs in lacking transient control intelligence, eg to control compressor airflow.

full flight regime   Designed to operate throughout each flight, eg FFR autothrottle.

full-flow oil system   Gas-turbine [usually advanced turbofan] lubrication system without a pressure-relief valve, filters and coolers being protected against extreme pressures by bypasses.

full fuzing options   For NW, air drop, air burst, ground burst, contact burst, time-delay after parachute-retarded, laydown retarded or free-fall.

full house   No parking stand left at terminal.

full lateral   Describes FMC or FMS (1) exercising total authority via AFCS from just after takeoff until final approach.

full mission trainer   Simulator capable of duplicating complete mission [initially for F-22] with any malfunction; can network with others.

full performance level   Category of most highly qualified air traffic controllers (FAA 1).

full-pressure suit   Completely enclosing wearer's body and able to sustain internal gas pressure convenient to human functions (see *partial pressure suit*, *water suit*, *spacesuit*).

full rudder, aileron or elevator   Hard-over demand signal by pilot or flight-control system moving surface to limit of travel.

full-scale development   Period when system or equipment and items for its support are tested and evaluated; intended output is pre-production system which closely approximates final product, documentation for production phase, and test results which demonstrate product will meet requirments (USAF).

full-scale tunnel   Wind tunnel for testing complete aircraft.

full-wave rectifier   Two elements in split circuit rectify both positive and negative halves of waveform, currents combining unidirectionally at output.

fully active   See *active*.

fully developed flow   Flow of viscous fluid over solid surface on which boundary layer has reached full thickness and velocity distribution remains constant downstream.

fully expanded nozzle   Normally, nozzle of jet engine, esp. rocket, whose supersonic divergent portion expands jet to ambient pressure at exit plane.

fully factored   Multiplied by ultimate factor of safety.

fully FBW   No mechanical backup, because aircraft could not be flown manually.

fully feathering   See *feathering*.

fully grown dimension   Dimension of part after process that results in enlargement, esp. high-temperature creep under tensile load.

fully ionised plasma   All neutral particles have lost at least one electron. With hydrogen no further ionisation is possible; other atoms can be further excited.

Fulta   Federazione Unitaria Lavoratori Trasporte Aereo (I).

Fulton system   Folding recovery system on nose of aircraft for retrieving space capsules and other parachuted loads in mid-air; used in modified form for repeated pick-up of persons or other loads from ground.

fumble factor   Factor of safety.

functionally shaped controls   Manual controls shaped to confirm their function, eg terminating in miniature wheel or flap.

fundamental frequency   *1* Of periodic quantity, lowest component frequency of sinusoidal quantity which has same period.

2 Of oscillating system, lowest natural frequency; normal associated mode of vibration is fundamental mode.

3 Reciprocal of period of wave.

4 Lowest resonant antenna frequency without added inductance or capacity.

fundamental units   Mass, length and time, from which all other units are derived. Main systems in use are SI, foot-pound-second, centimetre-gramme-second and metre-kilogramme-second.

funding   Process of finding and securing political approval for sums of money needed for military programmes (US).

fungus   Generalised term for life-forms growing at fuel/air interface in tank.

funicular polygon   See *force diagram*.

funk   General term for com. radio (G).

funnel   *1* Safe approach area centred on ILS glidepath and normally considered to extend 0.5° above and below and 2° left and right; aircraft inside * should be able to land.

2 Pre-1950 term for final approach area to runway, esp. any lighting or other guidance system for channelling aircraft to runway.

fu₀   Fuel mass at liftoff.

FUR   Future utility rotorcraft.

furaline   French rocket fuel; aniline with traces of anti-corrosive additive.

furlong   Non-SI unit of length, = 201.168 m.

furlough   *1* Leave of absence (holiday with pay) from armed forces (US).

2 Laying off (holiday without pay) from airline during traffic downturn or from similar cash-strapped employer (US).

furnishing   Non-structural cabin interior trim, seats, galley, toilet, baggage racks and associated features in commercial aircraft.

furnishing mock-up   Whole or part of cabin prepared for particular customer's evaluation and refinement.

FUS   Far-UV spectrometer.

FUSE   Far-UV Spectroscopic Explorer.

fuse   Linkage in operating system or structure so designed that, if the system or structure becomes overloaded, it fails at this place. Without a suffix, a weak link in an electrical circuit, usually a conductor with a low MPt. Not to be confused with *fuze*.

fuse bolt, fuse pin   Mechanically weak link at a point of high stress in a structural system, such as attachment for engine or landing gear.

fusee   Pyrotechnic squib installed in solid-propellant case to ignite charge over whole length.

fuselage   Main body of an aerodyne, absent in all-wing designs; when tail is attached to booms, called nacelle; with planing bottom called hull.

fuselage number   Identity of aircraft (R, USSR).

fuselage reference line   Straight line used as reference from which basic dimensions are laid out and major components located; usually along plane of symmetry and at convenient height.

fuse pin   Strong non-redundant pin attaching engine, or engine pylon, to airframe.

fusible plug   Several types, most important being fitted in main landing wheels; if temperature rises to dangerous value (at which tyre could burst) after excessive braking, ** blows and releases pressure in controled manner.

fusion   *1* So-called thermonuclear process whereby nuclei of light elements combine to form nucleus of heavier element, releasing very large amounts of energy; can be controlled and sustained.

2 Collection and integration of outputs from all of a range of sensors and warning systems, hence *IDF*.

3 Generally, data processing.

fusion bomb   One using energy of thermonuclear fusion.

fusion power density   Power generated per unit volume in controlled thermonuclear plasma; using deuterium at 10^{16} particles/ml and 60 kV, *** is about 1 kW, as kinetic energy of reaction products.

fusion reactor   Reactor in which thermonuclear fusion takes place.

fusion welding   Fusing edges of two base metals by using a welding flame to melt edges of metals, and a welding rod, which is similar in composition, to fuse or weld two metals together.

FUSS   Flaps-up safety speed.

future cycle accumulation   Number of LCF or operating (mission) cycles operator plans to accumulate on given hardware.

FUV   Far ultra-violet.

fuze   Device or mechanism designed to start detonation of high explosive under proper conditions of heat, impact, sound, elapsed time, proximity, external command, passage of electric current or other means, and usually without danger of detonation before weapon is armed.

fuzzy logic   Programming based on instructions not precisely specified numerically but on best estimates between 0 (false) and 1 (true).

FV   *1* Flight visibility.

2 Prefix to three-figure designators of alloys [mainly stainless steels] developed by Firth-Vickers (UK).

FVA   Federación Venezolana de Aeroclubes (Venezuela).

V$_{VAC}$ Vacuum thrust.

FVC   Forebody vortex control [T adds technology].

FVD   Fluorescent vacuum display.

FW   *1* Fiscal week.

2 Fixed wheel (sailplane).

3 Failure warning.

4 Filament-wound, see *filament winding*.

5 Fighter Wing (USAAF, USAF).

6 Frequency-wild.

f.w.   Full wave.

FWA   Flight watch area.

FW&A   Fraud, waste and abuse.

FWC   *1* Filament-wound cylinder, or case.

2 Flight, or fault, warning computer.

3 Flight watch centre.

FWCS   Flight-watch control station.

FWD   Falling-weight deflectometer.

fwd   Forward.

FWE   *1* Foreign weapons evaluation (US).

2 Fighter Wing Equivalent.

FWETE   Foreign weapons equipment technology evaluation (US DoD).

FWF   Firewall forward, the engine compartment of single-engine lightplane.

FWHM   Full wave, half modulation.

FWOC   *1* Forward wing operations centre (RAF).

2 Fleet Weather & Oceanographics Centre (UK).

FWS   *1* Fighter Weapons School (USAF).

2 Flight warning system.

3 Fire warning system.

4 Filter wedge spectrometer.

FWW   Fighter Weapons Wing.

FWWS   Food, water and waste subsystems.

FX   Fuel type unspecified.

FY   Fiscal year; US government runs 1 Oct. to 30 Sept.

F/Y   First-class and economy or tourist.

FYDP   *1* Five-year defence plan.

2 Future years defense program(US).

FYDS   Flight director/yaw-damper system.

F/Y ratio   Fission/yield ratio.

FZ, FZG   Freezing; **FZDZ** freezing drizzle, **FZFG** fog, **FZRN** rain.

fZ   Transverse VSI (magnetic) component.

FZP   Fresnel zone plate.

G

G *1* Giga, multiplied by 10^9.

2 Geostrophic force.

3 Geared (US piston engines).

4 Universal *gravitational constant* = 6.6705 × 10^{-11} Nm^2kg^{-2}.

5 Stress imposed on body due to applied force causing acceleration.

6 Gun (US DoD).

7 Shear modulus, rigidity.

8 Aircraft category, single-engine transport (USN 1939–42).

9 Aircraft category, glider (USA 1919–26 and USAF 1948–55).

10 Aircraft category, gyroplane [autogiro] (USAAC 1935–39).

11 Role prefix, air-refuelling tanker (USN 1958–62).

12 Role prefix, parasite carrier (USAF 1949–55).

13 Status prefix, permanently grounded (US from 1924).

14 Guard (suffix to serial on secret aircraft, UK, or to radio frequency).

15 Uncontrolled (airspace).

16 Green.

17 Gust[s].

18 Ground control.

19 Gain (radio).

20 Group.

21 Geschwader (G).

22 Sonobuoy size 419-mm long.

23 Conductance (also ℧).

24 Graphite.

25 Gauss.

g *1* Acceleration due to Earth gravity, international standard value being 9.80665 m/s^{-2}, assumed at standard sea level in atmosphere; often *g*, italic.

2 Gram[me]s.

3 Grid.

4 Suffix, gauge pressure.

5 Suffix, height AGL.

G^3 Gadolinium-gallium garnet.

g-break Sudden change of aircraft trajectory away from previous straight line in lateral or upward direction, eg following fast low-level run over airfield before landing.

G-display Rectangular, target is at centre (when radar aerial is aimed at it) and grows lateral 'wings' as range is closed; aiming errors result in appropriate displacements of blip from centre.

g-force Inertial force, that needed to accelerate mass, usually expressed in multiples of gravitational acceleration.

G-layer Layer of free electrons in ionosphere occasionally observed above F_2 layer.

g-loc Pilot blackout and LOC induced by severe and sustained vertical acceleration.

g-meter Indicates acceleration, usually in vertical plane.

G-seat, g-seat Seat simulating Z-axis acceleration, occupant normally wearing g-suit.

G-Star A spatial temporal anti-jam receiver for GPS-guided weapons.

g-suit Worn by pilot or astronaut; exerts pressure on abdomen and lower parts of body to prevent or retard collection of blood below chest under positive acceleration. Not necessarily pressure suit.

g-tolerance Tolerance of subject, human or device, to acceleration of specified level and duration. In case of human, usually vertical [head/toe direction].

GA *1* General aviation.

2 Gas analysis.

3 Gimbal angle.

4 Group accounting.

5 See *general-arrangement drawing*.

6 Go-around.

7 Military code for lethal nerve gas developed (G, 1937) as Tabun.

8 Ground attack aircraft category (USA 1919–24).

9 Goggle autoejector.

10 Grazhdanska Aviatsiya, civil aviation (R).

11 Grid amplifier, or array.

G/A Ground to air.

GAA *1* General Aviation Association (Australia).

2 Grid-array amplifier.

GAAC *1* General Aviation Awareness Council (UK).

2 General Aviation Airports Coalition (US).

GAAP Generally accepted accounting principles.

GAACC General Aviation Airworthiness Consultative Committee (UK).

GaAs Gallium arsenide.

GAATS Gander automated air-traffic system.

GAC *1* Gust-alleviation control.

2 Go-around computer.

GACC *1* Ground-attack control capability.

2 General Aviation Consultative Committee (CAA, UK, from 1997).

GACS Generic ATN com. service.

GACW Gust above constant wind.

gadget Code, radar equipment; type of equipment indicated by letter, followed by colour to indicate state of jamming: green *, clear; amber *, sector partially jammed; red *, sector completely jammed; blue *, completely jammed (DoD).

GADO General Aviation District Office (FAA).

gadolinium Lanthanide metal, Gd, isotopes are alpha-emitters.

GADS Generic aircraft display system.

G/A/G Ground to air to ground.

gage Gauge (US).

gaggle Group of aircraft (say, five to 20) flying together but with no semblance of formation.

GAIN, Gain Global aviation [or analysis and] information network (US, collects safety information).

gain *1* General term for increase in signal power in transmission, usually expressed in dB.

2 Increase or amplification; antenna * (* factor) is ratio of power transmitted along beam axis to that of isotropic radiator transmitting same total power; receiver * (video *) is amplification by receiver. Calculation for radars is usually equal to area of sphere of unit radius divided by area subtended on that sphere by solid angle equal to the 3-dB beam.

Gains, GAINS GPS air-data [or aided] laser inertial navigation system.

GAIS General ATC information system [mainly Wx].

GAIT *1* General aviation infrastructure tariff (Australia).

2 Ground-based augmentation and integrity.

gaiter Fireproof flexible cover over a pipe or tube, e.g. to prevent ingress of abrasive dust.

Gal, gal Non-SI unit of small acceleration, gal (Galileo) = 10^{-2}ms^{-2} = 1,000 mgal.

gal Gallon, type not specified.

GALAT, Galat Groupement Aviation Légère de l'Armée de Terre (F).

GALCIT Guggenheim Aeronautical Laboratory, Caltech.

galena Lead sulphide, PbS, used in IR cells.

gallery *1* Gas-turbine fuel manifold.

2 Fluid conduit formed within three-dimensional volume of material, eg drilled through body of pump.

galley Aircraft kitchen with provision for heating prepacked meals.

galling Pitting or marring of finished surface, esp. bearing surface, because of fretting.

gallium White metal, Ga, density 5.9, MPt 29.78°C, *-68 radionuclide of half-life 68 min.

gallon Non SI unit of liquid volume. Imperial * = 4.546087 litres = 277.42 cu in = 1.20095 US *; US* = 3.785412 litres = 231 cu in by definition = 0.83267 Imperial *.

galvanic corrosion Electrolytic action caused by contact of dissimilar metals or formation of oxygen cell in contact with metal.

galvanising Coating of metal, esp. steel sheet, by dipping in bath of molten zinc; protects from galvanic corrosion.

galvanometer Instrument which measures electric current passed through pivoted coil in magnetic field. With shunted external resistance, used as ammeter; with series resistance, voltmeter.

GAM *1* GPS-aided munition[s].

2 Ground-attack missile.

3 Ground-to-air missile, = SAM.

GAMA General Aviation Manufacturers' Association (US, formed 1970).

GAMAA Gate management and airport analysis (software).

Gambit General anti-material [US materiel] bomblet with improved terminal effects.

GAME Generalised automated maintenance environment (USN).

Game GPS approach-minima estimator.

Games GPS anomalies monitoring equipment suite.

GAMM Generalised air (or airlift) mobility model, baseline study using real military cargo spectrum.

gamma *1* γ, ratio of specific heats of gas.

2 Logarithmic function; ratio of contrast of transmitted scene to that of received display.

3 Flight path angle.

4 SI unit of very small changes in magnetic flux density; * (not abbr.) = 1nT.

gamma$_r$, γ_r Ratio of viscous damping to critical damping.

gamma plot Flight path angle plotted against airspeed.

gamma rays High-energy, short-wavelength EM radia-tion; very penetrating, similar to X-rays but nuclear in origin and usually more energetic. Symbol γ.

GAMS General airline management simulation.

Gamta, GAMTA General Aviation Manufacturers and Traders Association (UK).

GAN Global area network.

GaN Gallium nitride.

G&C Guidance and control.

gang drill Series of drill presses in close proximity on one bed, or drills operated simultaneously through interconnected chucks.

gang start All [multiple] engines start simultaneously.

gantry Large crane for erection and servicing of launch vehicles; * straddles vehicle and runs on tracks crossing launcher and work area.

GANTT General Agreement on National Trade and Tariffs.

GAO *1* General Accounting Office (US).

2 Groupe Aérien d'Observation (F).

GAP/Gap *1* General-aviation propulsion (NASA/industry).

2 GPS airborne pseudolite.

3 Ground-accident prevention (FSF).

4 Gas analysis package.

gap Distance between chords of any two adjacent superimposed wings of same aircraft, measured perpendicular to chord of upper wing at any point on its leading edge.

GAPA *1* Ground-to-air pilotless aircraft.

2 General Aviation Pilots' Association (US).

GAPAN, Gapan Guild of Air Pilots and Air Navigators [1929, livery company of City of London 1936].

gap coding Precise gaps in radio transmissions, or intervals of silence, used to represent letters, words or phrases.

GAPCU Ground and auxiliary power control unit, combines GPCU/APU-GCU.

GAPE, Gape General-aviation pilot education (FAA).

gap-filler Radar used to supplement long-range surveillance radar in area where coverage is inadequate.

gaping Discontinuity of skin profile caused by distortion or deflection of landing-gear doors, access doors and other separate panels intended to lie flush.

gapless ice guard Fitted within intake mouth; used in conjunction with automatic alternative inlet.

GAPP Geometric/arithmetic parallel processor, or processing.

gapped ice guard Mounted forward of air intake to provide a gap which does not ice up.

gapping Gap [increased clearance, intended or otherwise] between TE of compressor rotor blades and LE of next stator.

gap-squaring shears Tool used for squaring and cutting sheet metal; similar to squaring shears but able to slit long sheet.

GAR Ground abort rate.

GARA General Aviation Revitalisation Act, passed by US Congress 1994 to limit manufacturers' product liability, especially for aircraft over 18 years old.

garbage *1* Miscellaneous hardware in orbit or deep space; usually material ejected or broken away from launch vehicle or satellite.

2 EDP (1) software or program which has been degraded or in any other way rendered imperfect or unreliable, or output therefrom.

garbling Indecipherable responses from two aircraft

both at exactly same slant range of about 4 km (corresponding to 20.3 µs pulse interval) to SSR interrogation; caused by fact both sets of reply pulse trains are synchronised and superimposed (see *degarbling*).

garboard strake   That section of plating on a seaplane hull which extends fore and aft and is adjacent to the keel.

GARC   Groupe Aérien Régionale de Chasse (F).

GARD   General address reading device.

gardening   Aerial minelaying (RAF code name, WW2).

Gardian   General area-defense integrated anti-missile laser.

GAR/I   Ground acquisition receiver/interrogator.

GARP   Global Atmospheric Research Programme.

GARS   Gyrocompassing attitude reference system.

Garteur   Group for Aeronautical Research and Technology in Europe.

GAS   Global-positioning adaptive-antenna system.

gas bag   Flexible gas container of rigid airship; also known as gas cell.

gas-bag alarm   Indicates when predetermined gas-bag pressure has been reached.

gas-bag net   Mesh of cordage or wire to retain bag in position.

gas-bag wiring   Mesh of circumferential and longitudinal wiring enclosing each bag to take pressure and transmit lift.

gas bearing   Bearing for rotating assembly, usually high-speed, in which all forces are reacted by dynamic forces generated in contained volume of dry filtered gas, often He or N.

gas bleedoff   Bleeding-off of hot combustion gas from rocket engine for pressurizing or driving turbomachinery.

GASC   Ground Air Support Command (USAAF).

gas cap   Gas immediately in front of hypersonic body in atmosphere; compressed and heated, and if speed is sufficiently high becomes incandescent.

Gasco   *1* General-Aviation Safety Council (UK, formed 1965 as GAS Committee).

2 Ground Air Support Command (GASC was preferred).

gascolator   Filter fitted at lowest point of fuel system [archaic US usage].

gas constant   Constant factor R in equation of state for perfect gas; kJ/kg K = 185.863 ft-lbf/lb °R; 8,314.34 J/kmol/K for particular gas; specific ** r = R/m where m is molecular weight (see *Boltzmann*).

gas dynamics   Study of gases under high velocity, temperature, ionisation and other extreme conditions prohibiting compliance with aerodynamic laws.

gaseous electronics   Study of conduction of electricity through gases, involving Townsend, glow and arc discharges, and collision phenomena on atomic scale.

gaseous fuels   Those stored in aircraft in gaseous form, GH_2 being possible example (see *Blaugas*).

gaseous rocket   Bipropellant or monopropellant rocket utilising gaseous fuel and/or oxidiser.

gas film   Boundary layer on inner surface of combustion chamber.

gas generator   *1* Device for producing gases, hot or cold, under pressure. In some tip-drive helicopters * comprised gas-turbine core engines.

2 Gas-turbine core engine, eg turbofan minus LP turbine and fan or turboprop minus LP turbine, gearbox and propeller.

3 Major component supplying working fluid for turbopump of liquid rocket engine.

gash   Surplus, or pilfered (RAF, WW2).

gas hood   Cowl or ports in outer cover of airship through which gas can escape from inside hull.

gasifier   Machine for producing flow of hot gas, normally by combustion of fuel compressed by mechanism extracting energy from flow; free-piston and other systems but not gas turbine.

Gasil   General-aviation safety information leaflet (UK, CAA).

gas laser   One in which lasing medium is gaseous, eg HeNe, CO_2, Ar.

gas laws   Thermodynamic laws applying to perfect gases: Boyle-Mariotte, Charles, Dalton, equation of state. Also called perfect or ideal **.

gas main   Fabric hose running length of airship and having branches to gas bags for inflation.

gas misalignment   See *jet misalignment*.

gas motor   Mechanical drive system energised by high-pressure gas from combustion of solid fuel, usually single-shot.

gas oil   Residual hydrocarbon fuel left after distillation of gasolines and kerosenes from particular petroleum fraction. Can yield diesel oil and, by cracking, gasolines.

gasoline   Blends of hydrocarbon liquids, almost all petroleum products boiling at 32°/220°C, used as piston engine fuel. UK name *petrol*.

gas-operated   Automatic weapon in which initial rearward motion of moving parts, unlocking breech, is caused by piston moved by gas bled from barrel.

GASP   General Aviation Strategic Plan, being developed for 2004–08 (TSA).

gasper system   Low-pressure fresh-air supply piped to individual controllable outputs above each passenger or serving each crew-member.

gas pressurization   Feeding liquid propellant for rocket engine by piping high-pressure gas to dispel it from tank, with or without use of flexible liner; eliminates need for turbopump.

gas producer   See *gas generator (1)* or *(3)*.

gas ring   Spring ring for maintaining gastight seal between piston and cylinder.

gas rudder   See *gas vane*.

GASS   *1* Garde Aérienne Suisse de Sauvetage (Switz.), same initials in Italian but SRFW in German.

2 General air and surface situation (NATO).

gassing   Replenishing gas balloon with fresh gas to increase purity or make up for losses.

gassing factor   Quantity of gas required by aerostat over period of one year, ordinarily expressed as percentage of gas volume.

gas starter   Early (1920 onwards) method of starting multi-engined aircraft using airborne compressor set supplying compressed air or over-rich mixture to cylinders of each engine in correct sequence.

gas temperature   TET is usually measured between first stators and rotor, TGT at a point between stages and EGT can be same as JPT.

Gaston   Génération d'APT (auto-programmed tool) standard pour trajectoires d'outils normalisées (F).

gas triode   See *thyratron*.

gas trunk   Duct between gas-bag valve and gas hood.

gas tube   Electronic tube (valve) containing gas under very low pressure.

gas turbine   Engine incorporating turbine rotated by expanding hot gas. In usual form consists essentially of rotary air compressor, combustion chamber(s), and turbine driving compressor.

gas-turbine numerology   Stations within an engine are designated by numbers used as inferior suffixes which vary with engine configuration; in a simple turbojet the numbers are: 1, entrance to inlet duct; 2, entrance to compressor; 3, compressor delivery; 4, turbine entry; 5, turbine exit; 6, entry to nozzle; 7, in plane of nozzle. A two-spool engine with afterburner has numbers 1 to 10. A different set of suffixes identifies the shafts in a multishaft engine, 1 being LP, 2 HP or IP, and 3 HP or free-turbine in a three-shaft engine.

gas valves   Both rigid and non-rigid airships require valves to enable excess pressure in the gas bags or ballonets to be relieved. Lifting gas is released [automatically or manually] at the top, and air from valves on the underside.

gas vane   Aerodynamic TVC in jet of rocket (see *jet vane*).

gas volume   Volume of aerostat gas at SL.

gas welding   Fusion welding with hot gas flame; eg oxyacetylene on steels and oxy-hydrogen on aluminium.

GAT   *1* General air traffic.

2 General aviation terminal.

3 Greenwich apparent time.

Gatco, GATCO   Guild of Air Traffic Control Officers (UK, 1954–).

GATE, Gate   German Airport Technology and Equipment ev (association, 34 members).

gate   1 Point of passenger emplanement at airport.

2 Point at which commercial flight starts, or enters new sector of controlled airspace.

3 Removable lock to limit maximum travel of control lever (eg throttle) under normal conditions.

4 Position(s) on extended runway centreline above which inbound aircraft are required to pass at time assigned by approach control.

5 In air intercept: 'Fly at maximum possible speed for limited period' (DoD).

6 Control electrode of any device, esp. input connection to FET.

7 To control passage of signal in electronic circuits.

8 Circuit having output and input so designed that output is energised only when required input conditions are met, eg AND-*, OR-*, NOT-*

9 Circuit designed to receive signals in small fraction of principal time interval in radar or control system.

10 Range of fuel/air ratios through which combustion can be started.

11 Of trajectory, eg speed-record run, specified transverse apertures in space through which aircraft must pass to comply with regulations.

12 Of space mission, transverse aperture(s) defined in width and height at particular time or distance from liftoff or related to other body in space, through which vehicle must pass if mission is to accomplish objectives.

gated   *1* EM pulse permitted to function only under control of another pulse, usually synchronized.

2 Limited by gate (3).

gate guardian   Aircraft publicly displayed at entrance to major establishment, esp. air force base.

gate hold   Departure held at gate, common if delay exceeds 5 min.

gate position   Particular gate (1) numerically assigned to flight (7).

gate reader   Rapid checker of ticket and boarding pass located at gate (1).

gate time   *1* Agreed time at which flight enters new sector of controlled airspace.

2 Time at which aircraft passes particular point on track.

gate-to-gate   ATC is usually considered to be effective between gates (2).

gate valve   Valve controlling fluid flow by flat plate having linear motion across flow channel.

gateway   Customs airport, through which pax/cargo can enter country.

gathered parasheet   Parasheet whose periphery is constrained by hem cord.

gathering   Process of bringing guided missile into narrow pencil beam for subsequent guidance.

gating   *1* Process of selecting portions of EM wave which exist during one or more selected time intervals or which have magnitudes between selected limits.

2 Use of Q-switching or other control to permit laser to emit only during exact specified time intervals, esp. when used in rangefinding mode.

3 Imposing mechanical stop on piston engine throttle below selected pressure altitude.

gatling   Originally (1861) Gatling, automatic rapid-fire gun having a rotating assembly of several parallel barrels brought in succession in front of a single breech.

GATM   Global air-traffic management.

Gator   GPS ability to overcome resistance.

Gatorizing   Isothermal forging process for high-nickel turbine rotor blades and other super-alloys.

GATR   Ground/air transmit/receive.

GATS   GPS-aided targeting system; /GAM adds GPS-aided munition[s].

GATSS   Global air-transportation systems and services (R, CIS).

GATT   *1* Gate-assisted turnoff thyristor.

2 General agreement on tariffs and trade.

GAU   Gun, aircraft unit.

GAVC   GA Users Committee (LGK, LHR).

gauge (US gage)   *1* Any pressure-measuring instrument.

2 Hand comparator for GO/NO GO check on an exact dimension or screwthread.

3 Standard measures of sheet and wire thickness.

gauged fuel   Sum of fuel-gauge readings; fuel state, including unusable.

gauge pressure   Indicator reading showing amount by which system pressure exceeds atmospheric.

gauss   Non-SI unit of magnetic induction (flux density), $= 10^{-4}$T.

GAvA   Guild of Aviation Artists (UK).

GAVC   Ground/air visual code (FAA).

GAVRS   Gyrocompassing attitude and velocity reference system.

GAWG   General Aviation Working Group (Natmac).

GAZ   State aviation factory [almost 1,000] (USSR).

GB   *1* Gain/bandwidth.

2 Groupe de Bombardement (F).

3 Aircraft category, glide bomb (USAAF 1942–47).

4 Lethal nerve gas first produced as Sarin (G 1938) and later by US.

Gb   Gigabyte[s].

g.b.   Grid bias.

GBA   Groupe de Bombardement d'Assaut (F).

GBAD   Ground-based air defence; BC adds bridging capability, WS adds weapon system[s].

GBCS   Ground-based common sensors.

GBDM   Ground-based data management; S adds system.

GBI   Ground-based interceptor[s].

GBIB   Ground-based integrity broadcast.

Gbits/s   Gigabits (10^9) per second, not to be confused with Gb.

GBL   *1* Ground-based laboratory.

2 Ground-based laser.

3 Government bill of lading.

GBMD   Global ballistic-missile defence.

G/BMI   Graphite/bismaleimide composite.

GBM levels   Gough, Beard and McEvoy pioneered investigation of upper limits of force a pilot could be expected to exert on particular flight-control inputs (NACA).

GBP   Great Britain pound[s] Sterling.

GBR   Ground-based radar; P adds prototype (US NMD).

GBRAS   Ground-based regional augmentation system.

GBS   *1* Global broadcast service.

2 Ground-based software; T adds tool.

3 Ground-based sensor[s].

GBTA   Guild of British Travel Agents (1967–).

GBTS   Ground-based training system.

Gbyte   *Gb.*

GBU   Glide-bomb unit.

GC   *1* Groupe de Chasse (fighter wing, F).

2 Great circle.

3 Goggles-compatible (NVG).

4 Gyrocompass.

5 Ground control.

GCA   *1* Ground-controlled approach.

2 General controlled airspace, from 1990 called Class E (FAA).

GCAM   Ground collision-avoidance module.

GCAS, G-cas   Ground collision-avoidance system.

GCB   *1* Generator circuit-breaker.

2 Gun-control box (helicopter).

3 See next.

GC brg   Great-circle bearing.

GCC   *1* Graduated combat capability.

2 Goggles- (ie, NVG) compatible cockpit.

3 Gulf Cooperation Council (Inmarsat).

4 Global climate change; I adds initiative[s].

5 Ground-cluster controller (Acars).

GCCS   Global command and control system (DoD).

GCF   Ground-conditioning fan.

GCHQ   Government Communications HQ (Cheltenham, UK).

GCI   Ground-controlled interception.

GC/IMS   Gas chromatography/ion mobility spectrometry.

gcm^{-3}   SI unit of *density*.

GCMS   Gas-chromatograph mass spectrometer.

GCOS   Global climate observing system.

GCP   General conditions of purchase.

GCR   *1* Generator-control relay.

2 Ground-clutter reduction.

CGS   *1* Ground-control station[s].

2 Ground-clutter suppression.

Gc/s   Gigacycles = GHz.

GCSS   Global combat-support system.

GCT   *1* Government competitive test.

2 Greenwich Civil Time.

GCU   1 Generator control [and protection] unit.

2 Ground control unit [UAV].

GCV   Ground check vehicle [navaids].

GD   *1* Lethal nerve gas first produced (G, 1940) as Sarin (also GB).

2 General Duties branch, which includes aircrew (RAF).

GDB   Ground data-bus.

GDC   Gyro display coupler.

GDE   *1* Gas-discharge element, light source.

2 Graphics differential engine.

3 Ground demonstrator engine.

GDI   Get-down-itis, dangerous desperation to land for whatever reason.

GDL   *1* Gas-dynamics laboratory, pioneer rocket research centre (USSR).

2 Gas dynamic laser.

3 Ground data-link; P adds processor.

GDM   Generalized development model.

GDOP   Geometric[al] *dilution of precision*, also expressed as geometrical degradation of performance.

GDP   *1* See GDOP.

2 Graphics drawing processor.

3 Ground delay program.

GDS   *1* Ground debriefing station.

2 Global decision support, controls worldwide airlift; S adds system (USAF).

3 Global distribution system (Sita).

4 *Goldstone.*

GDT   Ground data terminal.

GE   Groupe d'entraînement (F).

G/E   Graphite/epoxy.

Ge   Germanium.

G$_e$ Gain, especially of radar jammer system.

GEANS   Gimballed electrostatic aircraft navigation system.

gear   Landing gear (colloq.).

geared engine   Piston engine with reduction gear to turn propeller more slowly than crankshaft, to enable power to be increased whilst not overspeeding propeller.

geared fan   Fan (2) driven through reduction gear.

geared propeller   See *geared engine.*

geared supercharger   Piston engine supercharger driven through friction clutch and step-up (speed-increasing) gears.

geared tab   Balance tab mechanically linked to control surface so that its angular movement is determined by that of main surface.

gear-type pump   Two intermeshing gears in close-fitting casing which pump fluid round outside of each gear in spaces between successive teeth and outer casing.

GEASA   Group of experts on aviation safety and assistance.

Gebecoma   Groupement Belge des Constructeurs de Matériel Aérospatiale.

GEC   Graphite epoxy composite.

GeCu   Germanium/copper (IR detector).

GEE   Ground exploitation, or evaluation, equipment.

Gee   Pioneer precision navaid, using interrelated VHF pulses transmitted from ground stations; position of aircraft determined by observing intervals between pulses from pairs of stations and plotting on hyperbolic map.

Gee-H   Secondary-radar navaid which enabled aircraft to determine position with precision by simultaneously measuring distances from two beacons.

GEEIA   Ground Electronics Engineering Installation Agency (USAF).

GEF   General Engineering Flight (RAF).

Gefra   Group of experts on the future regulatory arrangements [for international air transport].

gegenschein   Faint light area of sky opposite Sun and celestial sphere; believed to be reflection of sunlight from particles beyond Earth's orbit.

GEH   Graphite-epoxy honeycomb.

GEI   Groupement d'Economique Interêt (see *GIE*, which is more common).

Geiger counter   Geiger-Müller gas-filled tube containing electrodes; ionising radiation releases short pulse of current from negative to positive electrode, frequency of pulses indicating intensity of radiation.

GEJ   Group of Experts on Jurisprudence (ICAO).

GEL   Graphite-epoxy laminate.

GELIS, Gelis   Ground-emitter location and identification system.

GEM   *1* Ground-effect machine (ACV).

  2 Generic electronics module.

  3 Graphite epoxy (solid rocket) motor.

  4 Ground-environment material.

  5 Generalized emulation of microcircuits.

  6 Guided, or guidance, enhanced missile.

  7 Graphic [piston-] engine monitor, typically pictures CHT and EGT.

  8 GPS-embedded module.

GEMS, Gems   *1* Grouped engine monitoring systems.

  2 Global environment management system (noise).

  3 Global expeditionary medical system (USAF).

  4 Generic Earth-station management system.

GEN   *1* Generator.

  2 General [not the rank].

  3 General information (AIP).

Gen, gen   *1* Information, latest knowledge (colloq. RAF WW2).

  2 Generation, phase of development.

Gen2, Gen 3   New 'generations' of helicopter visionics.

genav   General aviation.

general air traffic   All traffic excluding OAT (3), special military (eg lo-level training) and local pleasure flights not notified for ATC purposes.

general-arrangement drawing   Usually three-view (front, side, plan) outline, in some cases with addition of dimensions, ground line, and broken lines giving additional information.

general aviation   All civil aviation except air transport for hire or reward; largest sectors are private (including company transport), agricultural and aerial work. Term introduced in US by CAA in 1951.

general cargo   Loose items, excluding large unit loads, not containerized or palletized.

general inference   General meteorological situation and future forecast.

General Flying Test   Taken at different levels by candidates for PPL, BCPL and CPL.

general-purpose aircraft   Military aircraft intended to fill multiplicity of roles; nearest modern equivalent is armed utility.

generate   To get a combat aircraft airborne and thus * a sortie.

generation   Family of all examples of particular hardware species designed at same time to meet similar requirements; after first * hard to define, and word generally used to impress audience of prior experience of particular manufacturer in field concerned.

generator   *1* Machine for generating direct current. Invariably used incorrectly to mean alternator.

  2 Device for producing clearly visible smoke, e.g. for aerobatics or for wind-tunnel.

generator line contactor   Main circuit-breaker between generator (ie, alternator) and AC bus.

generic   A particular meaning: using same region of EM spectrum for land, sea and air stations.

Genova   General overall validation for ATM (7) (Euret).

gentle turn   Primary flight manoeuvre in which bank angle does not exceed 25°.

Gen-X   Generic expendable[s].

GEO   Geostationary Earth orbit.

GEOAP   Ground equivalent onboard attitude processor.

geocentric   Related to centre of Earth.

geodesic line   Shortest line between two points on mathematically derived surface; ** on Earth called geodetic line.

geodesic radome   Spherical enclosure for large surface or ship radar whose structural members are geodetics and whose panels are perpendicular to transmissions.

geodesy   Science which deals mathematically with size and shape of Earth and its gravitational field, esp. with surveys of such precision that these measures must be taken into consideration.

geodetic construction   Methods of making curved space frames in which members follow geodesics along surface, each experiencing either tension or compression; resulting basketlike framework does not need stress-bearing covering.

GEODSS   Ground-based electro-optical deep-space surveillance system.

geographical envelope protection   Database which prevents the aircraft in which it is installed from entering a defined keep-out zone, eg overriding pilot's attempt to steer into a building.

geographical mile   One minute of arc at Equator, defined as 6,087.08 ft.

geographical position   Where line from centre of Earth to a heavenly body cuts Earth's surface (astronav).

geographic information system   Inputs land use in Gems (2), and other studies of airport noise.

geographic poles   North or south points of intersection of Earth's surface with axis of rotation, where all meridians meet.

geoid   Earth as defined by that geopotential surface which most nearly coincides with MSL.

geolocation   Finding where fixed targets are on the land surface, if possible near-instantaneously.

geomagnetic cavity Volume moving through solar wind occupied by Earth and surrounding magnetic field (magnetosphere).

geomagnetic co-ordinates System of spherical co-ordinates based on best fit of centred dipole to actual terrestrial magnetic field.

geomagnetic dipole Hypothetical magnetic dipole (bar magnet) located within Earth in such position as to give rise to actual terrestrial field.

geomagnetic equator Terrestrial great circle everywhere 90° from geomagnetic poles (should not be confused with magnetic equator).

geomagnetic poles North and south antipodal points marking intersection of Earth's surface with extended axis of geomagnetic dipole; north ** is 78½°N, 69°W, and south ** is 78½°S, 111°E. Should not be confused with magnetic pole.

geomagnetism *1* Magnetic phenomena exhibited by Earth and surrounding interplanetary space.

2 Study of magnetic field of Earth.

geometric dilution of precision See *dilution of precision*.

geometric pitch Distance propeller-blade element would advance in one revolution when moving along helix to which line defining blade angle of that element is tangential; ** of fixed-pitch propeller at standard radius is pitch of that propeller, and is marked on it.

geometric transition absorber Family of RAM (2) structures, most common being pyramidal type. Others include cones and sine waves.

geometric twist Variation along span of aerofoil of angle between chord and a fixed datum (see *aerodynamic twist*).

geometry-limited Restriction placed on aircraft attitude or configuration by geometric considerations; eg scraping tail on take-off or (variable-geometry aircraft) underwing stores fouling tailplane at max sweep.

geophones *1* Sensitive acoustic sensors for detecting sound transmitted through Earth's crust.

2 Seismic sensors buried along edge of runway to measure point of touchdown, severity of impact and bounces on landing (see *ALMS*).

geopotential height Height above Earth in units proportional to potential energy of unit mass (geopotential) relative to sea level. For most meteorological purposes same as geometric height, but geopotential metre = 0.98 metre; ** used under WMO convention for all aerological reports.

Georef World Geographic Reference System.

George Automatic pilot (colloq., arch.).

GEOS Geodynamic experimental ocean satellite[s] (NASA).

GeoSAR Geographic synthetic-aperture radar (NIMA).

geosphere Solid and liquid portions of Earth's lithosphere plus hydrosphere. Above * lies atmosphere, and at interface is found almost all biosphere zone of life.

geostationary altitude That at which body is in geostationary orbit. Accepted value is c35,880 km, 22,300 miles.

geostationary orbit Orbit in which satellite remains over same point on surface of Earth. Thus, period = 24 h; V is 3.07 kms^{-1}, 1.91 miles/s.

geostationary satellite Satellite in geostationary orbit; also called synchronous satellite.

geostrophic wind Wind the direction of which is determined by deflective force due to Earth's rotation.

geostrophic wind speed Calculated from pressure gradient, air density, rotational velocity of Earth and latitude, but neglecting curvature of wind's path.

geosynchronous Revolving at same angular speed as Earth, generally synonymous with geostationary.

GEP *1* Glassfibre/epoxy/PVC foam sandwich.

2 Geographical envelope protection.

GEPA, Gepa Groupement d'Etudes de Phénomenes Aériens (N adds 'Non-identifies') (F).

Gepna, GEPNA Groupe Européen de Planification de la Navigation Aérienne (Int.).

Gepta, GEPTA Group of experts on air-transport policy (Int.).

GERB Geostat Earth radiation budget.

germanium Silver-white, hard, brittle element possessing properties of both metals and non-metals; used in transistors, LEDs and IR windows (8–14 microns), but being superseded. Symbol Ge.

gerotor Generator-rotor.

gerotor pump Gear pump in which spur gear having n teeth rotates inside internal ring gear having n + 1 teeth.

GES *1* Ground Earth station.

2 Ground exploitation station [S adds subsystem].

Geschwader Luftwaffe formation equivalent to RAF group or US wing; also naval squadron (G).

GET *1* Ground elapsed time.

2 Ground entry terminal [Elint].

get To remove gas from vacuum system by sorption.

GETA Graphite-epoxy/titanium/aluminium.

get-away speed Airspeed at which seaplane or flying boat becomes entirely airborne.

GETI Ground elapsed time of ignition.

Gets GPS-enhanced theater support (US).

getter Material or device for removing gas by sorption.

GEU Guidance electronics unit.

GEV Ground-effect vehicle (usually = ACV).

GEW Ground-effect wing.

GEWP Generic electronic-warfare platform [pilotless aircraft for testing missile countermeasures].

GF *1* Gold film (transparency anti-icing).

2 Ground fog.

3 Geospacial Force (US).

GFA Gliding Federation of Australia.

GFAC *1* Ground forward air controller, ie not airborne.

2 Government furnished active countermeasures.

GFAE Government-furnished avionic (or aeronautical or aerospace) equipment (US).

GFDEP, GFDep Depth [thickness] of fog, estimated in feet.

GFE Government-furnished equipment; supplied by DoD to industrial contractor for incorporation in aircraft or other large product.

GFET, G-fet G-force environmental training.

GFH Groupement Française de l'Hélicoptère (F).

GFI *1* Government-furnished information (US).

2 Ground fault interrupter.

3 General format identifier.

GFK, GfK Glass-reinforced plastics (G).

GFL Gesellschaft zur Förderung der Luftschiffahrt (G).

G-Flops Billions of Flops.

GFM Government-furnished materiel (US).

GFP Ground fine pitch.

GFPT   Geospacial Force planning tool (US).

GFRP   Glass-fibre reinforced plastic[s].

GFS   *1* Gesellschaft zur Förderung der Segelflug-forschung (G).

2 Global fire-support system (Litton).

GFSK   Gaussian frequency-shift keying.

GFT   *1* General flight (or flying) test.

2 Generalized fast transform.

GfW   Gesellschaft für Weltraumforschung (West German national space society).

GF-X   Global Freight Exchange (London office).

GG   *1* Gravity gradient.

2 Gas generator.

3 Gyro-angling gain.

4 Graphics generator.

5 Ground-to-ground.

GGG   Gadolinium gallium garnet.

GGP   GPS guidance pack [age].

GGR   Ground-to-ground router.

GGS   *1* Gyro gunsight.

2 Gravity-gradient satellite.

3 GPS ground station.

GGTFM   Ground/ground traffic-flow management.

G/GV   Glove and glove vane.

GH   General [or ground] handling.

GH₂   Gaseous hydrogen.

GHA   Greenwich hour-angle, difference in longitude between heavenly body and 0° meridian. D adds diffrence, read off Godsave tables.

ghost   Extra images or blips on radar, TV or other display caused by signal reflection from hills, buildings or other objects; usually to right of primary image at distance proportional to reflected and direct path lengths.

GHQ   General headquarters.

GHW   Ground-handling wheels.

GHz   Gigahertz, billions of cycles per second.

GI   *1* Gum inhibitor.

2 Government issue (US).

3 Group identifier.

GIA   Glideslope intercept altitude.

GIB   *1* Guy in back, ie navigator or other back-seater in tandem two-seat military aircraft.

2 GNSS integrity broadcast.

GIBEA, Gibea   Guilde Belge des Electroniciens de l'Aviation (Belg.).

Gibson criteria   A series of assessments, such as attitude gain or frequency response plotted against phase angle, in an attempt to avoid PIOs.

GIC   GNSS, or GPS, or GPS/WAAS, integrity channel.

GICB   Ground-initiated comm-B, radio plus DME.

GID   Government Inspection Division (CAA).

GIDEP, Gidep   Government/industry data exchange program (US).

GIE   Groupement d'Interêt Economique.

GIEL   Groupement des Industries Electroniques (F).

GIES   Ground imagery exploitation station.

GIF   *1* Guy in front (see GIB).

2 Graphic[s] interchange file, or format.

Gifas, GIFAS   Groupement des Industries Français Aéronautiques et Spatiales; aerospace industries trade association (F).

Gift   Geosynchronous imaging Fourier transform spec-trometer.

GIG   *1* GPS integration guidelines.

2 Global information grid (DoD).

giga   Prefix, symbol G, $= \times 10^9$.

GIG-BE   GIG(2) bandwidth expansion, links centres with 10 Gbits/s fibre optics.

GIGO, Gigo   Garbage in, garbage out [EDP truism].

GIHO   Ground-initiated handoff.

GII   Global information infrastructure.

GIITS   General imagery intelligence training system.

Gilham Code   Gray code.

gill   Non-SI unit of liquid measure. UK * = 1.42065 × 10^{-4} m³; US * = 1.18294 × 10^{-4} m³.

gills   Hinged flaps at rear of engine cowling or other compartment to control cooling airflow.

Gimads   Generic integrated maintenance diagnostics (USAF).

gimbal   *1* Mounting with at least two, and usually three, mutually perpendicular and intersecting axes of rotation.

2 Gyro support which provides spin axis with degree of freedom.

3 To pivot propulsion engine for TVC.

4 To mount on *.

gimbal freedom   Maximum angular displacement about gyro output axis, expressed in degrees or in equivalent angular input.

gimballed chamber   Rocket-engine thrust chamber mounted on gimbal (3) so that it can swivel about one axis (or two perpendicular axes).

gimbal lock   Condition of two-degrees-of-freedom gyro wherein alignment of wheel spin axis with axis of freedom removes degree of freedom, rendering gyro useless.

Gins   *1* GPS/INS.

2 Gravimetric INS.

GIP   *1* Ground instructor pilot.

2 Generic interface processor.

3 Government/industry partnership.

GIPS   Geospatial information production system (NIMA).

GIRA   Groupe d'Instruction des Reserves de l'Air(F).

GIRD, Gird   Group for study of reaction (ie, rocket) engines (USSR, 1932–34).

GIRTS   Generic IR training system (ASD).

GIS   *1* Geographic[al] information system[s] (civil, US).

2 Graphic[al] information system.

Gismo   Globally integrated satellite mobile operating system.

GISS   Goddard Institute for Space Studies (NASA, New York City).

GIT   General interface terminal.

GITC   Guns in the cockpit.

GIUK   Greenland/Iceland UK, supposed air-defence gap.

GIVS   Groupe Interministeriel des Vols Sensibles, charged with security of commercial flights (F).

GKAP   State committee on aviation industry (USSR, R).

GKAT   State committee on aviation technology (USSR, R).

GKO   State defence committee (USSR, R).

GKS   Graphic[al] kernel system (WMO).

GL   *1* Ground level.

2 Group length.

GLA   Gust load alleviator.

GLAADS   Gun low-altitude air-defense system (USA).

GLAM   Groupe de Liaisons Aériennes Ministérielles (F).

gland   Short tube fitted to airship's envelope or gas bag through which rope may slide without leakage.

Glare   Glass-fibre prepreg tape reinforced aluminium alloy [usually multi-ply].

glareshield   Overhanging lip above instrument panel to protect pilot's night vision from bright reflections on windshield.

G-Lars   Guided launch and recovery system; /PLS adds precision landing system.

GLAS   Gust-load-alleviation system.

glass aircraft   One with high proportion of GRP in airframe, including skin.

glass cockpit   One featuring electronics displays in place of traditional instrument (colloq.).

glass-fibre   Produced by melting glass and spinning on revolving drum, fibres being typically 0.025 mm diameter. Fibreglass is registered name. Produced in many forms for structural or optical properties.

glass floor   Zero accidents or reportable incidents.

Glasshouse   Detention centre, military prison (UK, colloq.).

glass wool   Produced by forcing molten glass through orifices of approximately 1 µ diameter.

Glast   Gamma-ray large-area space telescope.

Glauert factor   Increase in lift coefficient due to fluid being compressible, $= (1 - M^2)^{-1/2}$.

Glavkosmos   Chief Administration of Space Launch Services (R).

Glavkoavia   Chief Administration of Aviation (USSR).

glazed frost   Rain ice, layer of smooth ice formed by fine rain falling on sub-zero surface.

glaze finish   Vitreous enamel coating on metal.

glaze ice   Transparent or translucent coating with glassy surface formed by contact with rain; part freezes on impact, most flowing back and freezing over surface.

GLC   Generator line contactor.

GLCM   Ground-launched cruise missile.

GLCS   Global launch control system (DoD).

GLD   Glider (ICAO).

glid   See *GLLD*.

glide   *1* Controlled descent by aerodyne, esp. aeroplane, under little or no engine thrust in which forward motion is maintained by gravity and vertical descent is controlled by lift forces. Rate of descent is given by $\upsilon_a = -(D/L)\upsilon$ where D is drag, L lift and V TAS.

2 Flightpath of *.

3 To descend in *.

glide bomb   Missile without propulsion but with aerofoils to provide lift and guidance; released from aircraft.

glide landing   No-flare landing.

glide mode   Flight-control system mode in which aircraft is automatically held to centre of glideslope.

glidepath   *1* Flightpath of aircraft in glide, esp. when making ILS landing.

2 Glideslope.

glidepath angle   That between local horizontal and straight line representing mean of glideslope.

glidepath beacon   ILS outer, middle or inner marker.

glidepath bend   Aberration in electronic glidepath.

glidepath indicator   ILS panel instrument.

glidepath localizer   Contradiction in terms (see *localizer*).

glidepath sector   Sector in vertical plane containing glideslope and extended runway centreline, limited by loci of points at which DDM is 0.175.

glider   Fixed-wing aerodyne designed to glide, ordinarily having no internal propulsion (see *sailplane*).

glide ratio   Ratio of horizontal distance travelled to height lost; TAS ÷ Vs (in same units).

glider flight time   Includes time on tow.

glider train   Two or more gliders towed in tandem behind one tug.

glider tug   Aircraft used to tow gliders.

glideslope   Radio beam in ILS providing vertical guidance (see *ILS*).

gliding angle   *1* Angle between local horizontal and glidepath. Traditionally $\gamma = \tan^{-1} D/L$ where D is drag and L lift.

2 Shallowest possible * of sailplane.

gliding range   Maximum distance that can be reached from given height in normal glide; also known as gliding distance.

gliding turn   Spiral flight manoeuvre consisting of sustained turn during glide; also known as spiral glide.

glim lamp   Source of illumination dim and local enough for use during blackout, esp. airfield lighting.

Glint   *1* Geostationary Earth orbit light-imaging national testbed (US).

2 Gated-laser illumination for night TV.

glint   Pulse-to-pulse change in amplitude of reflected radar signals, caused by reflection from object whose radar cross-section is rapidly changing.

GLIT, Glit   Chief State flight-test centre, Akhtyubinsk (R).

glitch   Small voltage surge affecting sensitive device; later general colloq. for technical problem.

GLLD   Ground laser locator-designator.

GLM   Gear limiting speed [usually means 'extended', rather than 'cycling'].

GLN   GPS landing and navigation; S adds system, U unit.

GLO   Ground liaison officer.

Global Area Network   Pioneer worldwide satcom service providing 64 kbit/s (Inmarsat 1991–).

Global Command and Control System   Overall electronic system tracking, and to some degree controlling, combat operations worldwide (DoD).

Global positioning and com   Gives GPS/TDMA for numerous linked users, esp. for station-keeping.

Global Positioning System   Worldwide system in which users derive their location by interrogating four satellites from total net of 24. Originally US military, which [2002] reserves to itself the greatest [centimetric] accuracy.

Globmet   Global meteorological service.

GLOC   See *g-loc*.

Glonass   Global navigation-satellite system (USSR, R).

glory   Aircraft in humid atmosphere viewed from above or below at centre of rings in spectral [rainbow] colours.

glove   *1* Fixed leading portion of wing root, esp. of variable-sweep wing.

2 Additional aerofoil profile added around normal wing, usually over limited span, for flight-test purposes.

GLOW   Gross lift-off weight (not spoken as word).

glow-discharge anemometer   Sensitive method of measuring gas velocity, esp. at low speeds or in turbulence, using cathode discharge between two pointed electrodes about 0.1 mm apart.

glow plug   Electric heating element, used in semi-diesel engines, which aids starting or, in one type of turboprop, provides inflight relight after flame-extinction.

GLS   *1* GPS- or GNSS-based landing system.

2 Gunlaying system.

3 Glider launching site.

GLTD   Ground-based laser target-designation.

GLU   GPS landing unit.

GLUAV   Gun-launched unmanned aerial vehicle.

glycerin   Glycerol, compound of C/H/O, soluble in water/alcohol, constituent of antifreezes.

glycol   See ethylene glycol.

glyptal   Synthetic resin made from glycerin and phthalic acid or phthalic anhydride.

GM   *1* Ground map mode (airborne radar).

2 Guaranteed minimum.

3 Gain [FCS feedback] margin.

4 Guided missile, usually with prefix.

5 Guidance material.

G/M   Gun/missiles selector switch on control column.

GMADS   Ground-based maintenance aid and diagnostic system.

GmbH   Gesellschaft mit beschränk Haftung, incorporated company (G).

GMC   Ground movement control, or controller.

GMCS   Ground-manoeuvring camera system.

GMD   Ground-based midcourse defense; S adds segment or system; in 2002 this replaced NIM as planned national defence against ballistic missiles (US).

GMDSS   Global maritime distress and safety system[s].

GMES   Global monitoring for [or of] environment and security (EC).

GMFSC   Ground mobile forces satellite communications.

GMI   Goddard management instruction.

GML   Gross moving load.

GMLA   Guided missiles and launch assemblies.

GMLRS   Guided multi-launch rocket system.

GMLS   Guided-missile launch system.

GMLTS   Guided-missile launcher test set.

GM-1   Nitrous oxide, piston-engine boost system (G, WW2).

GMP   Ground-movement planning.

GMR   *1* Ground-mapping radar.

2 Giant magnetoresistance.

GMRP   Guided-missile round pack.

GMS   *1* Geostationary meteorological satellite.

2 Groupement des Missiles Stratégiques (F).

3 Ground-based midcourse system.

4 Ground monitoring station.

GMSP   Global multi-mission support platform.

GMT   Greenwich mean time, or *Zulu,* now replaced by *UTC.*

GMTI   Ground moving-target indicator (or indication).

GMU   GPS measuring, or monitoring unit.

GMVLS   Guided-missile vertical launch system.

GN   See GN$_2$.

Gn   Green.

g$_n$ Standard value for gravitational acceleration.

GNA   Global network architecture.

G-nav   Navigation direct from A to B not on promulgated airway but crossing radials yet still using VOR/DME; name from graphic navigation, using computer-produced charts or hand-held equipment to give pilot a picture derived from VOR/DME inputs. Basic method of cross-radial navigation.

GNC   *1* Graphic numerical control.

2 General navigation computer.

3 Global navigation chart.

GNCS   Guidance navigation control system.

GND   Ground [CK adds check, CON control, FG fog].

GNE   Gross navigational error.

GNLS   GPS navigation and landing system.

GNLU   GPS navigation and landing unit.

gnomonic projection   Created by projecting from centre of Earth surface features on plane tangent to surface; distortion severe except near origin (point where plane touches Earth) but great circles are straight lines.

GNR   Global navigation receiver.

GNS   *1* Global navigation system.

2 Global navigation satellite [P adds panel, S system, SP system panel, SU sensor unit]; general term for all such spacecraft.

GNT   Gross nozzle thrust.

GN$_2$ Gaseous nitrogen.

GO   *1* Geared, opposed (US piston-engine designation).

2 General Order (military).

3 Groupe d'Observation (F).

GO$_2$ Gaseous oxygen.

go-ahead   Point in government programme at which prime contractor receives written authorization to proceed with full-scale development. Not an official term.

go-around   *Overshoot*; see *going around.*

go-around mode   Terminates aircraft approach and commands climb; also known as auto overshoot.

Goco, GOCO   Government owned, contractor-operated (US).

GODAE   Global ocean data assimilation experiment.

Goddard Space Flight Center   Greenbelt, Maryland, centre for NASA tracking and communications network.

Godsave   See GHA.

Goes, GOES   *1* Geostationary, or geosynchronous-orbit, or global, operational environmental satellite, suffixes East or West.

2 Gyrostabilized opto-electronic system.

go for broke   To fire all weapons in one pass of target.

go gauge   Dimensional gauge which must fit close, but without being forced, on or in the part for which it is intended.

go-home mode   Emergency RPV flight-control mode used following loss of navigation or command link.

going around: *1* Overshoot straight ahead (UK civil).

2 Make another circcuit (RAF).

Gold   General on-line diagnostic.

gold   Au, malleable metal with density 19.3, MPt 1,064°C, aerospace use mainly thermal-reflective coatings.

goldbeaters' fabric   Layer of cloth fabric cemented to one or more layers of goldbeaters' skin, making it gastight.

Gold C   Gliding certificate second only to Diamond C, requiring flight of ⩾300 km and other achievements.

golden arm   Supposed attribute of pilot whose ability and experience master all simulations and are acknowledged by peers.

golden handcuffs   Large cash sum to induce military pilot to extend period of service (UK, colloq.).

Goldfish Club   Club open to aircrew whose lives have been saved by dinghy made by UK company RFD Ltd.

goldie   Verbal code: "Aircraft automatic flight-control system and ground-control bombing system are engaged and awaiting electronic ground commands" (DoD).

goldie lock   Verbal code: "Ground controller has electronic control of aircraft" (DoD).

gold plating   Introduction of what (generally ignorant or partisan) politicians claim to be costly and unnecessary features in weapon systems (US).

Goldstone   DSIF stations NE of Barstow, CA.

golf ball   Turbulence control structure.

GoMats   Gulf of Mexico advanced traffic surveillance.

gondola   Car of airship.

gong   Medal or decoration (RAF colloq.).

gonio   VHF/DF (F).

goniometer   *1* Instrument for measuring angles between reflecting surfaces of crystal or prism.

2 Electrical transformer used with fixed and rotating aerials for determining bearing of radio station.

3 Motor-driven instrument used with four stationary aerials to deliver rotating signal field for VOR.

go/no-go   Step-by-step basis on which manned spaceflights are flown, with flight crew and mission control jointly making positive decision whether to continue into each new phase of mission.

go/no-go check list   Written guide for flight crews to determine go/no-go situation on any given subsystem deficiency (USAF).

go/no-go gauge   Dimensional gauge for checking whether part is within upper and lower tolerance limits.

go/no-go test equipment   Provides only one of two alternative answers to any question, eg whether given signal is in or out of tolerance.

good engine   One that continues to operate after other[s] failed.

Goodman diagrams   Various graphical plots used to determine parts life under repeated cyclic loads, most common having per cent alternating stress/endurance strength as ordinate and per cent mean stress/rupture strength as abscissa.

Goodrich de-icer   Original patented pulsating rubber de-icer for leading edges, intermittently inflated with air to break up ice.

Goodrich rivnut   See *Rivnut*.

goofers   Audience on island of carrier.

goolie chit   Written promise of reward if downed aircrew member is returned intact [UK Imperial, esp. North West Frontier, 1920–50].

goon   Guard at PoW Stalag Luft (RAF WW2).

gooseneck flare   Type of runway flare mounted on slender stem designed to bend easily if struck by aircraft.

GOR   *1* General Operational Requirement.

2 Guy on the right, in side-by-side military aircraft, normally navigator or electronic-warfare officer.

3 Ground occurrence report.

Gorac   Ground collision-avoidance system operational requirements and certification (Euret).

gore   *1* Shaped sector of parachute canopy normally bounded by two adjacent rigging lines.

2 Shaped section of airship envelope or gas bag, or balloon envelope.

3 Radial panel in airframe, esp. pressure bulkhead, hence * panel, * diaphragm.

GOS   *1* General operator station.

2 Grade of Service.

3 Gate-operating system.

4 Global observing system.

GOSC   General Officer Steering Committee.

Gosip   Government open-systems interconnection profile.

GosNII   State scientific research institute (USSR, R); -A or Aeronavigatsiya adds ATC/navigation/landing aids; -AS adds avionics; -GA adds all aspects of civil aviation; PAS adds ground test of aircraft systems.

Gospar   State commission for space research (R).

Gosport tube   Flexible speaking tube used in tandem open-cockpit trainers connecting instructor's mouthpiece with pupil's helmet or vice versa.

Gost   State research institute for fuels and lubricants (USSR, R).

Gothic delta   Wing whose basic triangular shape is modified to resemble Gothic window; also known as ogival delta.

GOTS   Government off the shelf.

Göttingen-type tunnel   Wind tunnel with return-flow circuit but open working section.

Gouge flap   Flap whose upper surface forms part of cylindrical surface; thus as flap rotates immediate movement is rearwards to increase area.

gox   Gaseous oxygen.

GoXML   Universal meta-language converting almost any data format into XML and back.

GP   *1* General purpose (bomb, or former RAF squadron role prefix).

2 Glove pylon.

3 Geographical position.

4 Glidepath.

Gp   Group.

GPa   Gigapascal.

GPADIRS   Global positioning air-data inertial reference system [U adds unit or replaces system, **GPIRS** omits AD].

GPADS   Guided-parafoil aerial [or air or airborne] delivery system.

GPALS   Global protection against limited strikes.

GP&C   Global positioning and communications.

GPB   Ground-power breaker.

GPBC   Gold-plated beryllium copper.

GPC   Government Procurement Code (US).

GPCDU   General-purpose control and display unit.

GPCU   Ground-power control unit.

GPDC   General-purpose digital computer.

GPEP   Global-positioning experiments program.

GPES   Ground-proximity extraction system.

GpFL   Group flashing light.

gph, GPH   Gallons per hour.

GPI   *1* Ground-positioning indicator.

2 Ground point of interception.

3 Glide-path indicator.

4 Global positioning inertial; N adds navigation, RS reference system, RU reference unit, SS sensor system.

5 Gas-penetrant imaging, or inspection.

GPIAA   Accident-investigation authority (Portugal).

GPIB   General-purpose instrument bus, or interface board.

GPIIA   Groupement Professionel des Industriels Importateurs de l'Aéronef (F).

GPIN   Global positioning [laser]-inertial navigation.

GPIRS   See *GPADIRS*.

GPM   Glass polycarbonate mix.

gpm   Gallons per minute (also gal/min).

GPMG   General-purpose machine gun.

GPP   *1* Graphic part-programming; technique for communicating with computer by words and diagrams, conveying pictures of shape required and operations necessary to produce it.

2 Generative process planning, basis for implementing FMS (7).

3 General-purpose processor.

GPPE   General-purpose processing element.

GPPPA, GP³A   Groupement Pour la Préservation du Patrimoine Aérien (F).

GPR   Glider Pilot Regiment (UK 1942–57).

GPRA   Glider Pilot Regimental Association.

GPRS   General packet radio services; see packet (3).

GPS   Global positioning system, or satellite (Navstar); ANT adds antenna, L1/L2/L5 see these entries.

GPSCS   General-purpose satellite communications system.

GPS-HMU   GPS height measuring unit.

GPSI   GPS interferometer.

GPSS   General-purpose simulation software.

GPSSU   GPS sensor unit.

GPT   Glidepath tracking.

GPTE   General-purpose test equipment.

GPU   *1* Ground power unit.

2 Gun pod unit.

GPVI   Graphic[s]-processor video interface.

GPW   Ground-proximity warning [C adds computer, S system, SU sensor unit].

GR, G/R   *1* Green run.

2 Ground attack, reconnaissance (role prefix, UK current).

3 General reconnaissance, ie Coastal Command (role prefix, UK, WW2).

4 Ground relay, or router.

5 Groupe de Reconnaissance (F).

6 Hail.

Gr   *1* Net climb gradient.

2 Graphite.

3 Grashof number.

grab   Tendency of wheel brakes to increase power suddenly without pilot input.

grabbit   Long boathook carried on large marine aircraft.

grab line   See *handling line.*

Grace   Gravity recovery and climate experiment (US/G).

GRAD   Gradient.

grad   Non-SI unit of plane angle, $= 0.9° = 1.5708 × 10^{-2}$ rad.

grade   *1* Of fuel, see *fuel grade.*

2 Unit of plane angle, defined as 0.9°.

Grade-A   Standard aircraft cotton fabric, long staple with 80 threads per inch across both warp and weft (US = fill).

graded fibre   Standard form of reinforcing-fibre raw material supplied according to diameter, length or other variable.

gradient   *1* Of net flightpath, has normal meaning, h/D%; note runway * = slope.

2 Space rate of decrease of function; if in three dimensions, vector normal to surfaces of constant value directed towards decreasing values. Ascendent is negative of *.

3 Loosely, magnitude of either * or ascendent.

4 Rate of change of quantity, or slope of curve when plotted graphically.

gradient distance   Linear distance from encounter with gust to point of peak intensity.

gradient of climb   See *climb gradient.*

gradient wind   Along isobars with velocity exactly balancing pressure gradient; equilibrium between force directed towards region of low pressure and centrifugal forces.

gradient wind speed   Calculated as for geostrophic but taking into account curvature of trajectory.

grading curve   In determining propeller performance by Drzwiecki theory, forces on infinitely small blade element are determined; curve of these forces (as the ordinate) against blade radius is **, from spinner or root and reaching maximum between 70%–90% tip radius.

Gradu   Gradual[ly].

Graetz number, Gz   Heat-transfer measure = Cp (specific heat at constant pressure) times mass flow divided by thermal conductivity and a length characteristic of body concerned.

Grafil   Registered name (Courtaulds) for carbon-fibre raw materials.

grain   *1* Entire case or extruded charge for solid rocket motor.

2 Particle of granular solid propellant, usually in gun ammunition.

3 Particle of metallic silver remaining in photographic emulsion after developing and fixing; these form dark area of image.

4 Non-SI unit of weight $= 0.0648$ g $= \dfrac{1}{7,000}$ lb.

grain orientation   Direction of solidification of metal.

GRAM, Gram   GPS receiver application module.

gramme   Fundamental SI unit of mass [gram in US], abb, g.

gramme-molecule   Mass in grammes of substance numerically equal to its molecular weight.

gramophone grooving   Close-pitch grooves in female part to form abradable seal round high-speed rotating member.

Grandfather rights   Permanent certificates for their existing route networks awarded US domestic airlines by CAB on its formation in July 1940. Hence Grandfather routes. Today loosely extended to all nations on basis 'If you've had this right in the past, you'll probably succeed in a fresh application'.

Grand Slam   RAF 22,000 lb [9,979 kg] deep penetration bomb of 1944.

grand slam   Verbal code: "All hostile aircraft sighted have been shot down" (DoD).

Grand Tour   Planned unmanned exploration of series of outer planets with same spacecraft using planets' gravitational fields to turn spacecraft from one to another; possible only once in each 180 years.

granularity   General measure of structure of very large EDP system based on number of processors used.

granular snow   Precipitation from stratus clouds (frozen drizzle) of small opaque grains 1 mm or less in diameter.

GRAP   Ground recognised air picture; -IOC adds initial operating capability.

grape   Purple-suited refuelling crewman on carrier (USN).

graphic part programming   Translation of three-dimensional co-ordinates of workpiece into computer program for NC machining, invariably using computer graphic displays as human interfaces.

graphics   Visual displays of any kind, esp. electronics displays forming part of EDP (1), EW or similar system (eg on CAD), and designed written matter and symbology inside and on skin of aircraft.

graphics differential engine   Uses *DTED* and algebraic/polynomial calculation to produce perspective digital landmass for *DMR*.

graphics drawing processor   An ASIC using subpixel addressing and anti-aliasing algorithms – eg, giving smooth dynamics, avoiding stair-stepping – to generate complex display formats at higher than 30 Hz.

graphic solution   Using geometric construction to solve problem; eg calculating point of no-return.

graphite   Soft naturally occurring allotropic form of carbon, also produced artificially and recently in form of strong fibres with perfect hexagonal crystalline structure. Large family commonly called carbon fibres includes many which in fact are graphite.

graphite bomb   Filled with filaments which short-out hostile electronics.

Grapioc   Ground recognised air picture initial operating capability (UK Army/RAF).

Grashof number, Gr   Heat-transfer parameter = $1^3 g\,(T_2\text{-}T_0)/v^2 T_0$ where 1 is a length, g is gravitational acceleration, T_1 and T_0 are temperatures, and v is kinematic viscosity.

grass   Random spikes projecting from timebase of CRT, radar or other electronic display caused by noise or deliberate jamming.

grasshopper   Safety-pin type of clip used to fasten cowl and other panels to perforated stud or similar anchor.

graticule   Any array of lines used as a reference for aiming, measurement or determining spatial relationships, esp. one of straight lines crossing at right angles on chart, map, CRT or other display, HUD or other human interface.

grating lobes   Undesirable radar-emissions caused by overlarge spacing between array elements which could reveal fighter's position.

grating spectrum   Produced by diffraction grating.

GRAU   State rocket and artillery directorate (USSR, R).

graunch   To damage aircraft or vehicle (UK, colloq.).

graveyard dive   One entered too close to the ground.

graveyard spiral   Without blind-flying instruments most simple aircraft, on entering cloud, can enter increasingly steep spiral, pilot under 1g and wings apparently level.

Graviner   Maker of fire extinguishers, became term for an extinguisher (RAF WW2).

gravipause   Point between two bodies where their gravity fields are equal and opposite.

gravireceptors   All sensors in human body for attitude, gravity and acceleration.

gravitation   Assumed universal property of all masses of attracting all other masses with force GMm/r² where G is universal * constant, M and m are two masses and r is mutual distance apart.

gravitational constant, G   Also called Newtonian constant, = 6.6732×10^{-11} Nm ²kg⁻²; other published values include 6.664, 6.669, 6.670 and 6.6705, in each case $\times 10^{-11}$.

graviton   Hypothetical elementary unit of gravitation.

gravity   Attraction experienced in vicinity of a mass, especially Earth. Standard value for terrestrial acceleration g = 9.80665 ms⁻² = 32.1740 ft/s.

gravity drop   Departure of inert projectile from initial trajectory.

gravity drop angle   Angle in vertical plane between gun line at moment of firing and straight line to a future projectile position.

gravity feed   Relying on fact liquids tend to flow downhill, unassisted by pump.

gravity seat   Simulator seat giving sensation of 'pulling-g', see next.

gravity suit   Aircrew suit, closely related to g-suit, with elements inflated/deflated by external system to give sensation of flight manoeuvres.

gravity tank   Container relying on gravity for feed, hence may be inoperable when inverted.

gray   *1* Grey (US spelling).
  2 Derived SI unit of absorbed dose of ionising radiation, equal to 1J/kg.

gray Code   Binary code used to transmit altitude data interleaved between transponder framing pulses, changing one digit at a time in Mode C.

gray scale   Grey scale.

grazing   Almost tangent to a curved surface, eg Sun-limb sensor or target-ranging system in low-level attack.

grazing angle   That between aircraft axis or sensor LOS and local Earth's surface.

GRB   Gamma-ray burst; CN adds co-ordinates network.

GRBL   Green-raster brightness level.

GRBM   Gamma-ray burst monitor.

GRC   *1* Glenn Research Center (NASA, Cleveland, Ohio).
  2 Glass-reinforced concrete.

GRCS   Guardrail common sensor.

GRDC   Gulf Range drone control; US adds update system.

GRDS   *1* Ground-roll director system, based on PVD.
  2 Generic radar display system.

GRE   *1* Ground readout equipment.
  2 Ground runup enclosure.

grease   *1* Lubricants based on hydrocarbon soaps emulsified in petroleum oils.
  2 To make a greaser.

greaser   Landing so smooth touchdown is imperceptible.

great circle   *1* Circle (usually small portion) on surface of sphere whose plane passes through sphere's centre.
  2 Intersection of Earth's surface and plane passing through Earth's centre.

great-circle chart   One on which all GCs are straight lines.

great-circle course   See next.

great-circle route   Shorter of two great circles linking all pairs of points on Earth's surface, giving minimum distance to fly; GC course is a misnomer because except along Equator or meridians course (hdg) is constantly changing.

great-circle track   See *great-circle route*.

Greatrex nozzle   Pioneer noise-reducing jet nozzle having several (typically six to eight) radial petal-like segments to increase length of periphery.

green   *1* Signal to proceed given by Aldis or similar lamp aimed at aircraft.

2 Friendly.

3 Coloured light[s] on instrument panel, esp. 3 * = landing gear down and locked.

green aircraft Flyable but still lacking interior furnishing and customer avionics, and still in * protective surface coat, awaiting painting.

green airway One running essentially E–W.

Green channel Airport route for arriving 'nothing to declare' passengers without dutiable possessions.

green density That of compacted powder prior to sintering.

green endorsement Written in logbook of aircrew member in green ink, showing exceptional ability, esp. for landing crippled or dangerous aircraft.

greenfield site Site considered for new airport or other facility where no structures exist at present.

Green Flag Tac-air war exercises strongly emphasising EW (USAF).

green flag In signals area = right-hand circuit.

greenhouse Long glazed canopy over tandem cockpits (colloq.).

greenhouse effect Filtering and reflective effect of Earth's atmosphere on solar and other radiation akin to that of glass panes; part of incoming spectrum penetrates to Earth, where it heats surface and causes reradiation of longer wavelengths, some of which are absorbed by atmospheric water vapour and again reradiated.

Greenie Technical air groundcrew (RN).

green run First run of new or overhauled engine or other item.

green suit[er] Soldier (USA).

green tube Unfurnished passenger aircraft.

Greenwich Earth's prime (0°) meridian, hence * apparent time (GAT), * hour angle (GHA), * mean time (GMT) and * sidereal time (GST).

green zone Traditionally, intersection between green and crossing airway at which it is traffic on crossing route that has responsibility for ensuring height separation.

Gremlin Family of mischievous imps responsible for faults (RAF, WW2).

Gretel Gramma-ray Eureca telescope.

grey body Unknown hypothetical body absorbing constant fraction of all wavelengths of incident EM radiation.

grey code Gray Code.

grey literature Technical documents produced by universities, laboratories and professional and government bodies, not normally available to public.

greyout Blurred vision under high positive acceleration less than that producing blackout.

grey scale, grey shades Standard series of achromatic tones linking black to white, typically 64 on modern display.

grey water Waste from handbasin; this can be fed to drain mast, unlike waste water.

grey wedge Standard filter whose opacity increases in known fashion across width, usually L to R; used in determining pulse distribution and other variables on CRT and other displays.

grf, g.r.f. Group repetition frequency.

GRG Ground-roll guidance.

GRI Group repetition interval.

Grib Gridded-binary data [chart of forecast weather].

grid *1* Perforated electrode between cathode and anode

of thermionic valve controlling flow of electrons into fine beam.

2 Metal cylinder at negative potential in CRT designed to concentrate electrons.

3 System of two sets of parallel lines crossing at 90° to form pattern of squares each identified by number and/or letters in margins; superimposed on maps, charts, photographs and multi-sensor outputs so that any point can be located by letter/number code. Usually also permits accurate measures of distance and direction. Often called military *, though most are civil.

grid bearing Direction of one point from another measured clockwise from grid (3) north.

grid bias Constant potential in series with input circuit between grid (1) and cathode to hold operation to one part of characteristic curve.

grid convergence Angle between true north and grid (3) north.

grid co-ordinates Rectilinear measures about two axes in flat plane of grid (3) facilitating conversion of lat/long and other Earth measures on to flat sheet by routine plane surveying.

grid heading Aircraft heading measured relative to grid (3) north.

grid leak Resistor allowing grid (1) charge to drain to cathode.

grid magnetic angle Angle between magnetic north and grid (3) north, measured E/W from latter; also known as grivation (= grid variation).

grid modulation AM achieved by applying modulating signal to grid (1).

grid north Zero datum of grid (3), close to true north.

grid ring Round top of traditional magnetic compass, rotated by hand when setting course.

grid ticks Small marks on neatline or along grid (3) lines showing alternative grid system(s).

grid variation See *grid magnetic angle*.

griff Reliable news or information.

Griffith wing Subsonic wing of very deep section with powerful suction slit on upper surface at about 70% chord to induce airflow to follow discontinuity between upper surface ahead of slit and thin trailing edge. Never successfully used.

grip range Range of thickness of material joinable by particular blind rivet or other fastener.

GRIS Global reconnaissance information system.

GRM Ground-roll monitor.

GRMS Ground reference and monitor station (DGPS).

GRND Ground.

GRO Gamma-ray observatory.

grommet *1* Rigid or reinforcing eyelet closed on to flexible surface.

2 Flexible ring set into rigid surface, often by peripheral groove matched with sheet thickness, providing bearing surface for pipe, cable or other line (1, 2) or control cable.

grooved runway One whose surface is traversed by one of four standards of shallow grooves tailored to climate, crossfall and other factors, along which water can escape even in heavy rain and strong wind to make critical aquaplaning depth extremely unusual.

groover Machine with large wheel, usually diamond-dressed saw, for cutting runway grooves.

GROS, Gros Civil Experimental Aeroplane Construction Organization (USSR).

gross altitude scale   Presentation of total altimeter operating range on one fixed scale (ASCC).

gross area   Area of projected surface of aerofoil, edges being assumed continuous through nacelles, fuselages, pods or other protuberances. Where tapered wing meets fuselage, edges projected in to meet at centreline, except in case where angle is extreme (eg, with glove, Lerx, strake), where end of root is taken across at 90°.

gross ceiling   Altitude at which gross climb gradient (see *gross performance*) is zero.

gross dry weight   Traditional measures of powerplant weight which included propeller hub (metal hub on which wooden propeller was mounted), all starters, primers, exhaust systems, fluid filters, air inlets and accessories, but excluding cooling system, fluid tanks and supply systems and instruments.

gross flightpath   Gross profile in climb-out segment.

gross flight performance   See *gross performance*.

gross height   Height of any point on gross flightpath.

gross lift   Buoyancy in ISA (1) of aerostat under standard conditions of inflation and with allowance for humidity.

gross moving load   Total moving mass of simulator, including upper baseplate and actuators.

gross performance   That actually measured on one aircraft of type, adjusted by small factor to reflect guaranteed rating and fleet minimum performance.

gross profile   Side elevation of aircraft trajectory, esp. following takeoff, corresponding to gross performance.

gross thrust   That developed by propulsion system in ideal conditions, not allowing for inlet momentum drag, inlet shock losses, duct losses, tailpipe losses, cooling drag, propeller slipstream drag, torque effects or any other effects.

gross upset   Major uncommanded departure in AOA/V/altitude/attitude.

gross weight   Traditional measure usually defined as maximum flying weight permitted; today MTOW.

gross wing area   *Gross area.*

ground   *1* US = earth.

2 To declare object or person unfit for flight.

3 Personnel on apron connected to aircraft by interphone cord.

ground-adjustable propeller   One whose pitch can be changed only by ground crew.

ground air vehicle   One designed for ground mobility but which can fly for short periods (ASCC).

ground alert   Status of aircraft fuelled and armed and crews able to take off within specified period, usually 15 minutes.

ground angle   *1* That between local horizontal and major axis of parked fuselage.

2 Maximum usable nose-up angle on landing, limited by tail scrape.

ground board   Flat surface representing the ground in wind tunnel.

ground clearance   *1* Vertical distance between airfield or deck and tips of helicopter main rotor blades in no-lift position.

2 Vertical distance between airfield or deck and specified part of aircraft or external stores.

ground clutter   Unwanted returns on radar display caused by direct reflection from ground.

ground collision avoidance system   To prevent airborne

aircraft from flying into the ground, not for preventing taxiing accidents.

ground contact   Glimpse of Earth sufficient to assist navigation.

ground control   Control tower position or other authority assigned to control all vehicles, including taxiing aircraft, on airfield movement area.

ground-controlled approach, GCA   Ground radar installation able to watch approaching aircraft and direct them to safe landing by radio (so-called talkdown) in bad visibility; and landing thus directed.

ground-controlled interception, GCI   Interception (1) controlled by ground radar and radio (usually voice-plain-language) advice.

ground crew   *1* Personnel assigned to cleaning, replenishment, servicing or maintenance of aircraft at turnround, between missions or in other routine situation.

2 Personnel assigned to manoeuvre aerostat on ground (see *landing crew*).

ground cushion   *1* Region of increased pressure beneath landing aeroplane caused by forward motion, proximity of ground and trapping of air ahead of flaps and under fuselage (can affect flow over tail and, for this and other reasons, cause pronounced pitching moment).

2 Region of increased lift under helicopter or jet V/STOL in low-altitude hovering mode caused by reflection of downwash, jets, entrained air and possibly entrained solids or liquids from ground.

ground delay program   Implemented to control traffic to airport where acceptance rate is reduced [expected to last a significant time, e.g. because of severe weather or an accident] by prohibiting flights to that airport to depart until a delayed EDCT.

ground Earth station   Aeronautical ground station.

grounded   Legally prohibited from flying.

ground effect   *1* Increased wing lift when flying in close proximity to ground, especially with low-wing aircraft.

2 Increased lift caused by interaction of powered lift system and ground, as with ground cushion (2), used in ACV (GEM).

3 All effects, invariably unwanted, caused by interference of ground on radars, radio navaids and other EM systems.

ground elapsed time, GET   Time measured from liftoff of major space mission, beginning with countup and continuing to provide one index of elapsed time unvarying with Earth time zone.

ground engineer   Skilled member of armed force or employee of MRO with power to certify work.

ground environment   *1* Environment experienced by ground equipment (no definition except to meet particular specifications which are variable).

2 Electronic environment created by ground stations, esp. for air-defence purposes.

ground equipment   *1* All non-flying portions of aerial weapon system.

2 All hardware retained on ground needed to support flight operations. Appears to be no clear definition; most authorities agree every item intimately associated with flight operations but exclude those concerned with training, design/development, marketing or other peripheral areas, and never include consumables.

ground fine pitch   Special ultra-fine pitch available after landing to increase drag on non-reversing installation; use of *** known as discing (pronounced disking).

ground-fine-pitch stop   Mechanical lock on hub released by compression of landing gear or other signal.

ground fire   Gunfire from ground directed against aircraft (most authorities exclude all but small-arms fire).

ground fog   Shallow fog caused by radiation chilling of surface at night.

ground half-coupling   That part attached to GSE affording direct connection with mating half in aircraft.

ground handling equipment   Ground equipment for lifting or moving large items, such as wings, missiles, spacecraft etc.

ground hold   Hold (1) for ATC purposes taken on ground before starting engines.

ground horizon   *1* Theoretical distance of horizon from sea level (see *horizon*).

2 Actual horizon seen from particular location.

ground idle   Governed running speed for engine with throttle fully closed; lower rpm than flight idle.

ground-imagery exploitation station   Each GIES comprises an IIW, an MD/RWW and an RRW (RAF reconnaissance).

ground lag   See *lag*.

ground liaison   Officer specially trained in offensive air support (DoD) and/or air reconnaissance (NATO, CENTO, IADB); organized as member of team under ground commander for liaison with air and/or navy.

ground loiter   Helicopter saving fuel by resting on ground between particular military tasks, in friendly or hostile territory.

ground loop   Involuntary uncontrolled turn while moving on ground, esp. during takeoff or landing, common on tailwheel aeroplanes with large ground angle, caused by directional instability; if at high speed, landing gear would normally collapse before turn had reached 180°.

ground marks   ICAN and other bodies decreed what information should be written [usually in letters/numbers 6.09m (20ft) high] on the ground or on buildings to aid pilots.

ground movement control   Military unit assigned to control of transport by land, esp. of air forces.

ground moving target indication   Separation of ground moving targets from clutter background by using their different Doppler shift, especially when looking ahead at small angles from track.

ground nadir   Point on ground vertically beneath perspective centre of camera lens when exposure was made; coincides with principal point in vertical photo.

ground observer   Trained person forming part of organization providing (DoD) visual and aural information on aircraft movements over defended area, (UK) information on fallout after nuclear attack.

ground occurrence report   Monitors failures [ground or inflight] traced to lapses by engineers.

ground-performance aircraft   One able to move itself on ground without using flight propulsion system (ASCC).

ground plane   Earthed system of conductors forming horizontal layer (mesh, sheet, radial rods etc) surrounding ground navaid.

ground plot   A calculated ground position.

ground position   Point on Earth vertically below aircraft.

ground-position indicator   Device fed with data from compass, ASI etc and giving continuous readout of DR position (obs.).

ground power unit   Source of power, usually electric and possibly pneumatic/hydraulic/shaft, supplied to parked aircraft.

ground-proximity extraction system   Standard technique for low-level airdrop of palletized cargo using shock-absorbing ground coupling which engages with hook suspended from pallet.

ground-proximity warning system   Uses forward-looking radar and sensitive altimeter[s] to give aural and/or visual warning, and in most systems, if ignored, to command violent pull-up to [typically] 30° climb.

ground radar aerial delivery   Method of air-dropping cargo, usually in A-22 (US) containers, from high altitude to avoid hostile fire, mountains or other hazards, with full parachute deployment delayed to increase accuracy.

ground readiness   Status of aircraft serviceable and crews standing by so that arming, briefing etc can be completed within any specified period (longer than 15 min of ground alert).

ground resonance   Dangerous natural vibration of helicopter on ground caused by stiffness and frequency of landing-gear legs amplifying primary frequency of main rotor; potentially catastrophic unless designed out, and even with certificated helicopter can occur as a result of severe landing shock.

ground return   See *ground clutter*.

ground roll   Distance travelled from point of touchdown to runway turnoff, stopping or other point marking end of landing.

ground run   Distance from brake-release to unstick, not same at TOR (see *takeoff*).

ground safety lock   Retraction lock.

ground sheet   Radial-wall flow of hot gas along ground beneath VTO [esp. jet-lift] hovering in ground effect.

ground signals   Bold visual symbols displayed in signal area.

groundspeed, G/S   Aircraft speed relative to local Earth.

groundspeed mode   Flight-system mode holding constant G/S.

ground spoiler   Spoiler available only after landing, usually as lift dumper.

ground start   Supply of propellants to large rocket vehicle from ground during ignition and hold-down so that at liftoff main-stage tanks are still full.

ground stop   Holds flight [usually scheduled, but in any case already cleared] at departure. Reasons might be closure of destination or to allow for implementation of longer-term solution to a destination problem, such as a *GDP*.

ground strafing   Attack by aircraft on tactical surface target, esp. by gunfire.

ground support   *1* Air power deployed for immediate assistance of friendly army, ie close air support; hence designation * aircraft.

2 Hardware needed to facilitate operation of aircraft, eg ladders, chocks, refuelling, replenishing and rearming equipment, loaders, tie-downs, blanks (4) and ground conditioning and power supplies; and use thereof.

ground support equipment, GSE   Ground equipment required for operation of aircraft [especially military], RPV or missile.

ground swing envelope   Plot of ground where obstructions would foul nose or tail of longest aircraft in most extreme positions on curves of taxiways or apron.

ground test Test on ground of equipment or system normally used in air.

ground test coupling Connections enabling airborne system to be tested on ground for fluid pressure and functioning, supply voltage or any other variable.

ground trace Ground track of satellite.

ground track Path on Earth's surface vertically below aircraft or satellite.

ground upset Accident caused to light aircraft or other vehicle by jet blast or large propeller slipstream.

ground visibility Prevailing visibility along Earth's surface as reported by accredited observer or measured by RVR.

groundwash Outward flow of wake turbulence from engines or wingtips of large aircraft on ground.

ground wave Radio or other EM waves taking direct path from ground transmitter to ground receiver (in practice mix of ground, ground-reflected and surface waves); subject to refraction in ducts in troposphere.

ground wire *1* US term for earthing wire.

2 Winched cable emerging from top of mooring mast and connected to airship mooring cable; US = mast line.

ground zero, GZ Point on Earth nearest to centre of nuclear detonation (which may be below, at or above GZ).

group *1* Military air formation consisting of two or more squadrons (DoD), or two or more wings (RAF).

2 Several sub-carrier oscillators in telemetry system.

3 Major portion of aircraft (eg. wing*) assigned to * (4).

4 Team of engineers assigned to design, stress, develop and possibly cost major portion of aircraft, often remaining intact to work on same part on successive programmes; common in US, where * titles are wing, fuselage, tail/controls, weight, electrical, hydraulic, armament and often others.

group flashing light, GpFL Ground light with regular emission of two or more flashes or Morse letter(s).

group technology General term for philosophy that links CAD with CAM to give CIM, based on recognising similarities between discrete parts.

group velocity Symbol U, that of entire disturbance of waves, equal to phase speed c minus wavelength 1 times dc/dl.

growl Missile tone heard in pilot headset indicating IR head locked on to target.

growler *1* Test equipment for short circuits in electrical machines (colloq.).

2 ECM aircraft, or a member of its crew.

growth Development to increase performance, hence * engine; this may or may not be physically larger.

GRP *1* Glass-reinforced plastics.

2 Geographic reference point.

GRR Glycol recovery and recycling.

GRS *1* Government rubber synthetic, Buna-S type.

2 Global reconnaissance strike (US).

3 Gamma-ray spectrometer.

GRSF Ground Radio Servicing Flight (RAF).

GRT Gross registered tonnage, measure of capacity of ship, = 100 ft^3 = 2.832 m^3.

GRU Main intelligence directorate of General Staff (USSR).

grunt manoeuvre One involving high g (colloq.).

Gruppe Group (G), equivalent to RAF wing.

GRV Glycol recovery vehicle.

GRVD Grooved runway.

GRVL Gravel runway or surface.

Gryphon FBMS/shore communications system.

GS *1* Ground speed.

2 Glideslope.

3 Ground plus station (costs).

4 Ground supply (usually electrical).

5 General schedule.

6 Galley service vehicle.

7 Ground stop.

8 Gliding School.

G/S *1* Ground speed.

2 Glideslope.

Gs Small hail or snow pellets.

GSA Gunsight, surface-to-air.

GSARS Ground-surveillance airborne radar system.

GSC *1* Ground switching centre.

2 Ground-station controller.

GSD *1* Graphics system design.

2 Grey- [gray-] scale definition.

3 Ground sample distance.

GSDI Ground speed and drift indicator.

GSE *1* Ground support equipment.

2 Ground swing envelope.

GSF Gross square feet [undesirable].

GSFC Goddard Space Flight Center, Greenbelt, MD (NASA).

GSFG Group of Soviet Forces in Germany (NATO name).

GSGA State service of civil aviation (R).

GSGG Gadolinium scandium gallium garnet.

GSI *1* Grand-scale integration (microelec.).

2 Government source inspection.

3 Glideslope indicator.

GSIF Ground-station information frame.

GSLV Geostationary [or geosynchronous] satellite launch vehicle (India).

GSM *1* Ground-station module, or mobile.

2 Global-station module.

3 GPS sensor module.

4 Global-systems mobile.

GSMC Global system for mobile communications.

GSMS Ground-station management system.

GSN Guidance unit (R).

GSO Geostationary orbit.

GSOC German Space Operations Centre, Oberpfaffenhofen.

GSP *1* Ground service plug (= socket).

2 Glareshield panel.

GSQA Government source quality assurance (US).

GSR *1* Ground surveillance radar.

2 General Staff requirement (UK, Army).

GSS Ground (or group) support system.

GSSS Gyrostabilized sight system.

GST *1* Greenwich sidereal time.

2 General Staff Target.

3 General skills test (proposed for NPPL).

GSTF Global Strike Task Force (USAF).

GSTRS Ground safety tracking and reporting system.

GSTS Ground-based surveillance and tracking system (SDI).

GSU Group Support Unit.

GSV   Gray-scale voltage[s].

G-switch   Activated by severe acceleration or impact.

GT   *1* Group technology.

   2 Rate, eg kg/h, of fuel consumption (USSR).

   3 Gas temperature.

   4 Greater than.

   5 Gain/thermal noise ratio, also G/T.

   6 Aircraft category, glider torpedo (USAAF 1942–47).

G$_t$ Gain of radar aerial (dB).

GTA   General terms agreement.

GTACS   *1* Ground-target attack control system.

   2 Ground-theater air control system (JFACC).

GTAW   Gas tungsten-arc welding.

GTC   *1* In IFF, group time cycle.

   2 Gyro time constant.

   3 Ground terminal computer of data-link.

GTDS   Ground tracking data station.

GTF   Ground test facility.

GTN   Global Transportation Network (web-based control system, to be upgraded to * 21 US DoD).

GTO   Geosynchronous (or geostationary) transfer orbit.

GTPE   Gun time per engagement.

GTR   *1* Gulf Test Range.

   2 General technical requirements.

   3 Greater.

GTRE   Gas Turbine Research Establishment (India).

GTRI   Georgia Tech. Research Institute.

GTS   *1* Gas-turbine starter.

   2 Glider Training School.

GTSIO   Geared, turbocharged, direct injection, opposed.

GTSS   Ground-target sensor surveillance.

GTT   Ground test time.

GTV   *1* Ground-test vehicle (helicopter).

   2 Guidance test vehicle (missile).

   3 Glide test vehicle.

g.u.   Gravity unit, standard unit for geophysical and MAD calculations, $= 10^{-6} \text{ ms}^{-2}$.

GUAP, Guap   Chief Administration of Aviation Industry (USSR).

guaranteed rating   Minimum power or thrust which manufacturer guarantees every engine of type will reach.

guard   Emergency VHF channel usually monitored as a secondary frequency by all air and ground stations in geographical area.

guarded switch   One protected against inadvertent operation by hinged cover or shroud.

guard frequency   Guard.

guardroom   Police post at entrance to RAF airfield or other military establishment (UK).

guardship   *1* Armed escort helicopter.

   2 Planeguard helicopter.

gudgeon pin   Links piston to connecting rod (US, wrist pin).

GUGVF   Chief Administration of Civil Air Fleet, of which Aeroflot is operating branch (USSR, R).

GUH   Get-U-Home.

GUI, G/UI   Graphics, or graphic[al], user interface: point and click, or retriever.

guidance   Control of vehicle trajectory, esp. that of unmanned, or of manned but according to external inputs (see *active homing, beam-rider, command *, electro-*

optical *, inertial *, IR *, laser *, midcourse *, passive homing, radar command *, semi-active homing, wire *).

guidance radar   One dedicated to providing pencil beam for beam-rider or radar command guidance or illumination beam for semi-active homing.

guidance system   Complete system providing guidance signals to flight-control system which steers vehicle.

guide ailerons   Small wing-tip ailerons providing normal feel on aircraft with plug-type spoiler ailerons.

guided bomb   Free-fall missile with guidance, esp. modified bomb.

guided missile   Vehicle able to deliver warhead to target; normally not including those travelling over land surface or entirely through water (torpedo) but including all with some form of aerial trajectory.

guided weapon   Guided missile (UK).

guide rope   See *drag rope*.

guide-surface canopy   Any of several families of parachute deployed from pack but able to be steered through air with translational motion.

guide vane   *1* See *stator blade*.

   2 Radial aerofoil struts at gas-turbine inlet designed to add or reduce swirl to airflow.

Guidonia   Large aeronautical research centre formerly (pre-1944) run by Italian defence ministry.

Guinea Pig Club   Members of Allied air forces in WW2 who had been critically burnt or injured and operated on semi-experimentally.

GULF, Gulf   Graphical user interface load-control facility.

gull wing   One having pronounced dihedral from root to c15–20% semi-span, then little dihedral or even anhedral to tip.

gull-wing canopy   In left/right halves, opened along centreline.

gull-wing door   One having pronounced curvature, concave on outer face, hinged parallel to aircraft longitudinal axis.

gully   Deep axial channel, eg. between two separated engines in fuselage of twin-jet aircraft.

gum   General term for viscous residues formed in gasolines (petrols) and to lesser extent other hydrocarbon fuels, mainly by slow oxidation.

gum inhibitor   Now called anti-oxidant additive.

Gumo, GUMO   Main and central directorates, each with a number (R, MoD).

Gump   Gas, undercarriage, mixture, propeller(s) (US arch.).

gun   *1* Good general term for airborne rifled weapons of all calibres, including recoilless installations; no clear definition at what low muzzle velocity * becomes projector.

   2 Piston engine throttle; hence to cut * = to close throttle, and to * engine = to apply full power (colloq., suggest arch.).

gunbore line   Projected axis of bore.

gun cross   HUD symbol indicating gun is armed, ready to fire.

gun gas   Emitted from muzzle, mix of initially incandescent gases from propellant deficient in unburned oxygen which if ingested by engine suddenly alters operating conditions.

gun jump   Angle between gunbore line at firing and projectile trajectory as it leaves muzzle.

Gunk Registered commercial solvent for oils and greases.

gunlaying radar Early AI radar with mode for assisting attack with fixed guns on target seen only on display.

Gunn oscillator Major family of GaAs diodes generating microwave outputs on application of small bias voltage.

gun pack Quickly replaceable unit comprising one or more fixed guns (sometimes with barrels remaining installed in aircraft), feed systems and ammunition tanks, either in streamlined pod or contained within aircraft.

gun perfection coefficient $\frac{\text{T-}m}{60\text{-M}}$ where m is mass of projectile, M mass of gun and T shots per minute.

gunship *1* Specially designed helicopter with slim two-seat fuselage, extensive protection and wide range of armament for roles in land warfare.

2 Large transport aircraft equipped with night sensors and guns for use against poorly defended ground targets.

gunsight line LOS to aiming point through gunsight fixed optics.

gun time per engagement Usually firing duration in seconds, aggregate of separate bursts, against one aerial target.

gun-type weapon Nuclear weapon triggered by firing together at maximum velocity two or more subcritical fissile masses.

gunwales Pronounced gunnels, the upper edge of the sides of a marine-aircraft hull or float [with a rounded top, hardly applicable].

Guppy Aircraft with grossly swollen or bulged fuselage, eg, for conveyance of space-launcher stages and wide-body components (colloq.).

Gusem Generic unified systems engineering metamodel.

gusset Small flat member used to reinforce joints and angles.

gust *1* Sudden increase in velocity of horizontal wind (see *gustiness factor*).

2 Suddenly encountered region of rising or falling air, causing moving aerodyne to experience sudden increase or decrease in angle of attack, = gust velocity u ÷ airspeed v. Vertical gust can theoretically be sharp-edged (instantaneous change from zero to maximum u) but normal design/airworthiness based on l-cosine (gradual) gust curve to which gust-alleviation factor applied.

gust alleviation Dynamic system for reducing effect of vertical gust on aeroplane (rarely, other aircraft) (see *active ailerons, Softride*).

gust-alleviation factor As aeroplane encounters gust it pitches (depending on wing/tail or foreplane geometry) and wing does not generate full extra lift until it has travelled several chord lengths into gust, both of which reduce sudden structure load below instantaneous encounter, BCAR assumes *** 0.61, ie assumptions are based on 61% of true sharp-edged gust.

gust curve Assumed plot of gust (invariably 2) velocity relative to surrounding air mass against horizontal distance from undisturbed air to position of peak u.

gust envelope Basic aircraft design plot, vertical axis being structural load factor (1) and horizontal axis airspeed; normal boundaries are positive-stall curve, peak positive gust (normal non-SI = 50 ft/s) to V_c, line to meet gust of half this strength ± 25 ft/s at V_D, then vertical V_D to negative half-strength gust, line to – 50 ft/s gust at V_c, and straight line at this negative gust value to meet posi-

tive stall at point less than 1 g. Recently new boundaries have been established at V_B at ±66ft/s.

gustiness factor Measure of gust (1), = difference between maximum gust and lull expressed as percentage of mean wind.

gust loading Increased structural loads caused by gust (1, 2).

gust locks Particular *control locks* preventing movement of flight controls of parked aircraft.

gust response Aircraft encountering gust (1, 2) experiences vertical acceleration made more severe by high speed, low wing loading (esp. large span, discounting flexure effect of wing) and some other factors. Normal measure of ** is number of 0.5 g vertical accelerations experienced by pilot's seat per minute under specified conditions at high (Mach 0.9) speed at low level.

Guti Rare clag in Zimbabwe.

gutter Afterburner flameholder having cross-section generally in form of V, open side to rear, to create strong turbulence sufficient to keep flame attached; see vapour *.

Guttman Original scaling technique used to assess community noise response assuming that any positive answer implies positive answer to all questions of lower order; final Guttman scale is normally: no action; sign petition; attend meeting; contact officials; visit officials; help organize action group.

GUVVF Chief Administration of Air Fleet (USSR).

GVC Girls Venture Corps; -AC or (AC) adds Air Cadets (UK 1939, incorporates WJAC).

GVE Graphics vector engine.

GVF Civil Air Fleet (USSR, R).

GVI General visual inspection.

G/VLLD Ground/vehicle laser locator designator.

GVLS Ground vortex length scale.

GVPF Geared variable-pitch fan.

GVRC GPS volume receiver card.

GVS *1* Ground velocity subsystem.

2 Global voice service.

GVSC Generic VHSIC spaceborne computer (USAF).

GVT Ground vibration test(s).

GVW Gross vehicle weight.

GW *1* Guided weapon.

2 Groundwave.

3 Gateway.

GWEN, Gwen Groundwave emergency network.

GWJ Garnet water jet for high-rate cutting of hard metals.

GWM Guam missile/space station.

GWS *1* Guided weapon system (RN).

2 Graphical weather service.

GWT Gross weight.

GWVSS Ground wind vortex sensing system.

G_x Gain of transponder RF amplifier.

gyro Gyroscope.

gyro angling gain CG = H/c, H sense (7).

gyrocompass Compass based upon space-rigidity of gyroscope; no true long-term instrument exists but see *directional gyro* and *Gyrosyn*, and sensing element of flux-gate compass is gyro-stabilized.

gyrodyne Aerodyne having engine power transmitted to lifting rotor(s) and propeller(s) used for thrust; convertiplane has wing in addition.

gyrograph Graphical plot of gyro drift against time.

gyro gunsight Sight for fixed guns using one or more

gyros (and RAE-developed Hooke's joint with two degrees of freedom) to provide automatic lead computation by measuring rates of sightline spin while remaining insensitive to rotation about sight axis itself caused by roll of host aircraft).

gyrohorizon   See *artificial horizon*.

gyro log   Form used to calculate and record gyro drift and drift rate (ASCC).

gyromagnetic compass   Directional gyro whose azimuth datum is maintained aligned with magnetic meridian by precession torquing from magnetic detector.

gyropilot   See *autopilot*.

gyroplane   Becoming a common US term for an *autogyro*.

gyrostabilized   Held in fixed attitude relative to space, subject to precession and wander.

gyrostat   Hughes-developed technique for satellites of great length spinning about minor axis.

Gyrosyn   Registered name for gyrosynchronized compass comprising DI (2) slaved to magnetic meridian by fluxgate.

gyro time constant   GTC = J/c.

gyro vertical   Local vertical indicated by vertical gyro.

GZ   Ground zero.

H

H   *1* Henry.

2 Total pressure.

3 Enthalpy.

4 High (synoptic chart).

5 High-altitude-class Vortac/Tacan or Route Chart.

6 NDB 50-1,999 W.

7 Angular momentum.

8 Helicopter mission category, USAF since 1948, USN since 1962.

9 US military aircraft modified mission prefix, search/rescue and aerial recovery (DoD).

10 Magnetizing force; horizontal component of Earth's field.

11 Stored in silo but raised to surface for launch (DoD, ICBM).

12 Hard temper (light-alloy suffix).

13 G/S home from CP.

14 Airway or map prefix, helicopter route.

15 Ambulance [hospital] category (USN 1929-31, 1942-44).

16 Transfer function.

17 Health (facility or RAF).

18 Piston engine with two crankshafts and parallel opposed cylinders.

19 Hard surface.

20 Haze.

21 Homing [beacon].

22 Hold, followed by direction.

23 Heavy.

24 Hazard.

25 Heliport.

26 Helicopter (PPL).

27 Hydrogen [see H_2].

28 Maximum section height (tyre).

29 Propeller pitch [P more common].

h   *1* Hour[s].

2 Prefix hecto = 10^2.

3 Hexode, heptode (ambiguous).

4 Hangarage available.

5 High (synoptic chart).

6 Heater (electronics).

7 Height above MSL, or height difference in flight trajectory.

8 Specific enthalpy.

9 Planck constant, = 6.62559×10^{-34}Js.

10 Height of blade CP above flapping axis.

11 Operator, 120° (electrical).

12 CC blowing jet slit height (also hj).

H+, H-   Hours plus or minus minutes, eg related to H-hour.

h^1, $\dot{h}$ Vertical velocity. Suffixes for glide-slope (GS), flare trajectory (FL) and reference trajectory (REF).

H_2   Gaseous hydrogen.

H24, H_{24}   Continuous-service airfield or facility.

H_2S   Original PPI mapping radar (UK, WW2).

H_2X   Development of H_sS at shorter wavelength in US (see *mickey, BTO*).

H-83282   Highly stable non-inflammable synthetic hydraulic fluid (USN).

H-bomb   *Hydrogen bomb.*

H-display   B-display with elevation angle indicated; target appears as bright line with slope proportional to sin elevation.

h-dot   See h^1, $\dot{h}$

H-engine   Piston engine with left and right rows of vertical opposed cylinders, two crankshafts geared to central output.

H-hour   Start of war, esp. time first landing aircraft reaches LZ, or similar clearly defined action.

H-film   Kapton hi-temperature polyimide.

H-plane   Plane of antenna's magnetic field, normal to E-plane.

H-Pres   Pressure altitude.

H-tail   One having twin fins on tips of tailplane.

HA   *1* Height of apogee.

2 Hour angle.

3 High altitude.

4 Housing allowance.

ha   Hectare[s], = 10^4 m^2.

HAA   *1* Helicopter Association of Australia.

2 Height above airport.

3 Historic Aircraft Association (UK).

4 Heavy anti-aircraft [gun, or fire].

5 High-altitude airship.

haar   Wet sea fog (UK, North Sea).

HAARS   High-altitude airdrop resupply system.

HAB   Heliport acquisition beacon.

HABM   Hypervelocity air-breathing missile.

haboob   Severe dust storm.

HABV   Hypersonic air-breathing vehicle.

HAC   *1* House Appropriations Committee (US Congress).

2 Hover/approach coupler.

3 High-acceleration cockpit.

4 Hélicoptère anti-char (F).

5 Helicopter aircraft controller.

6 Helicopter active control.

7 Heading alignment cone.

8 Helicopter Association of Canada.

Hacienda   Office of Aerospace Research (USAF, colloq.).

hack   *1* Aircraft informally used as general transport and utility vehicle by military unit (often captured from enemy or retired from combat duty).

2 To be able to accomplish (military/RAF, transative, colloq.).

3 To penetrate private network, especially a secure LAN.

HACP   High-altitude communications platform [unmanned airships].

HACS   Helicopter armoured crashworthy seat.

HACT   Helicopter active-control technology.

HAD   *1* Hybride analog/digital.

2 Hélicoptère d'Appui-Destruction (F).

3 Hardware architecture document.

Hadas   Helmet airborne display and sight.

Hadec   Highly-adaptive digital engine control.

HADS   *1* High-accuracy digital sensor.

2 Helicopter air-data system.

HAE   High-altitude endurance.

Hafnium   Hf, corrosion-resistant silver metal, density 13.3, MPt 2,230°C, * carbide and * nitride MPts are well beyond 3,000°C.

HAGB   Helicopter Association of Great Britain.

Hagen-Poiseuille   Law for velocity or pressure drop per unit length for pure streamline (laminar) flow in constant-section pipe. Equal to $u = \dfrac{P}{4\mu}(a^2 - r^2)$

where u is local velocity, P pressure drop per unit length, μ coefficient of viscosity, a radius of tube, and r radius at point concerned. Equation denotes parabolic velocity distribution.

HAGR   High-gain advanced GPS receiver.

HAH   Hot and high.

HAHO, Haho   High-altitude high-opening.

HAHST   High-altitude high-speed target.

HAHV, HaHv   Haute altitude, haute vitesse.

HAI   Helicopter Association International.

HAIL   Highlands and Islands Airports plc (UK).

hail   Precipitation in form of hard or soft ice pellets, varying size; maximum for certification test is usually 1 in (25.4 mm) sphere.

HAILSS   Helicopter aircrew integrated lift-support system.

HAINS, Hains   High-accuracy INS.

Hair   High-altitude IR.

hair absorber   Family of RAM (2) coatings consisting of dense mats of hair with conductive coatings, eg carbon-black in neoprene.

hairline crack   One in which the two sides are in close contact, hence inconspicuous to eye.

HAISS   High-altitude IR sensor system.

HAL   Height above landing [on helicopter pad].

Hale, HALE   High-altitude, long-endurance.

half-brightness life   Used in several senses, eg time from cutoff of stimulating radiation of phosphor to luminescence falling to half peak, time of cutoff of current to LED to same reduction in brightness, and various characteristics of storage tubes and displays.

half-Cuban   Aerobatic manoeuvre consisting of up-45 line, half-roll followed by remainder of loop [many variations and additions].

half glidepath   ILS glidepath within points at which DDM is 0.0875.

half-life   Time required for decomposition of half original mass or number of atoms of radio-active material.

half-mil   Aeronautical chart series on scale 1:500,000 (ICAO).

half-period zone   See *Fresnel zone*.

half-power points, rings   Points, whose locus is normally closed ring, where radiated power from antenna is half lobe maximum.

half-power width   Total angle at antenna between two opposite half-power points measured in plane containing lobe peak.

half-residence time   Time for quantity of delayed fallout (weapon debris) deposited in particular part of atmosphere to decrease to half original value.

half-reverse Cuban   Aerobatic manoeuvre: two left rolls up up-45 line, two flick rolls to right, increase trajectory to vertical to zero airspeed, tailslide followed by other manoeuvres on down-45.

half-roll   Rotation of aircraft sensibly about longitudinal axis through 180°, usually from upright to inverted attitude or vice versa; can be half of barrel roll or of slow roll.

half-thickness   Thickness of absorbing medium which transmits half intensity of radiation incident upon it.

halftone screen   Fine opaque grating usually scribed on glass to break up photographic image into halftone dot pattern, for printing or digitising purposes.

halftone tube   CRT containing conventional gun for writing separated from screen by fine-mesh electrode and storage plate.

half-view   Drawing showing half a symmetrical object.

half-wave rectification   Use of single-phase rectifier which passes only half of each alternate wave from input.

Hall effect   When current-carrying material is subjected to magnetic field (or when conductor is moved through magnetic field) potential difference is set up perpendicular to both current (or motion) and field; small in metals but important in semiconductor systems and in study of electricity in ionosphere. Since 1990 the operating principle of some space thrusters.

Halo   *1* High-altitude, low-opening (paradrop system).

  2 High-altitude large optics.

  3 High altitude [or agility], low observable.

  4 Hypersonic air-launched option.

  5 High altitude, long operation.

halo   *1* Any of several species of part-circular phenomena caused by ice crystals in upper atmosphere, chief of which is 22° radius around Sun or Moon.

  2 Coloured ring or disc seen on cloud in direction away from Sun, ie with aircraft shadow at centre; also called pilot's *.

  3 Reflection of cockpit instrument seen in canopy at night.

halo effect   Ability of SST service to generate or attract additional first-class subsonic traffic to same carrier on same route(s).

Halon   Family of halogen-based fluids, mainly BCF, stored as liquid under pressure and used as fire extinguishants and for inerting space above fuel in tanks.

Halsol   High-altitude solar [-powered aircraft].

HALT, Halt   Highly accelerated life test[s].

halteres   Twin vibrating prongs used in certain time-keeping systems.

hammer   Sudden violent excursions in pressure caused by reflected shockwaves in closed fluid (esp. hydraulic) system.

hammerhead   Large circular paved area at end of runway to facilitate turning.

hammer stall   Extreme stall turn in which aircraft rotates within c 5° of vertical plane; depending on aircraft, power retained until point of stall at apex. Also called hammerhead stall.

Hamots   High-altitude multiple-object tracking system.

HAMS   Hot-air management system.

handbook problem   One requiring in-flight consultation of flight manual for numerical answer.

hand bumping   Use of hand tools and backing dollies to shape sheet metal.

hand controller   Human interface to automatic or semi-automatic system, eg to HUD, multi-mode radar or large display; usually incorporates stick, rolling ball or triggers.

hand-crafted   Still used to mean IC (4) is custom-designed for particular application, even though design is entirely via computer graphics.

hand cranking Direct mechanical connection between crank and piston engine or small gas turbine (in latter case via step-up gears) for starting.

handed Items on left of aircraft are mirror-image of those on right; see next.

handed propellers Left and right ** rotate in opposite directions.

hand flying Piloting autopilot-equipped aircraft in fully manual mode.

hand forging Forging by hand tools.

hand forming Shaping ductile sheet with hand tools and accurate form blocks tailored to flanging, beading etc.

hand geometry Basis of one branch of biometrics concerned with linear and rotary measures, but not forces.

hand-held microphone Self-powered (battery) microphone for use in light aircraft following total electrical failure, usually connected to external antenna at 121.5 MHz.

hand inertia starter Hand crank spins flywheel which, when at full speed, is suddenly clutched to start piston engine.

handlebar moustache Wide with upturned ends (RAF fashion, WW2).

handler Apron loader of cargo.

Handley Page slat, slot See *slat*.

handline Firefighting hosepipe stowed on CFR vehicle, unreeled and deployed manually.

handling *1* Subjective impressions of response to controls of particular aircraft; hence * *squadron*.

2 Manoeuvring of aerostat by ground crew or landing party.

3 Providing full service for operator at airport where that operator has no staff. Customer is usually an air-carrier.

handling line *1* Primary line used by ground crew for handling aerostat.

2 Single or twin cables attached above c.g. of seaplane for use when hoisting by crane into or out of water.

handling pilot Pilot actually flying the aircraft.

handling squadron Assigned assessment of new aircraft, and compilation of Pilot's Notes.

handoff Transfer of control or surveillance from one ground radar controller to another.

handover Handoff (military).

hands-off *1* Condition in which non-autopilot aircraft, usually aeroplane, flies by itself in perfect trim with pilot(s) not touching controls.

2 Flight on autopilot.

3 Ground party release balloon basket, permitting ascent.

hands-on *1* With human(s) interfacing with system, which can be EDP (1) or aircraft in flight; appropriate only to complex autopilot aircraft in which * is rare in cruising flight. Hence, * training = practice in hand flying, or interfacing manually with military sensors and weapon systems.

2 Ground party hold down balloon basket.

hand starter Arrangement for starting engine by hand other than by swinging propeller.

hand-starter magneto Separate hand-controlled auxiliary magneto carried to aircraft and used to supply powerful spark when starting piston engine.

hand-turning gear Connection for a hand-crank, eg on same shaft as centrifugal breather to give maximum mechanical advantage. In piston engine can be same as manual or hand starter.

hangar Shelter for housing aircraft on ground.

hangarette *1* Hangar tailored to single aircraft, esp. hard shelter.

2 Weatherproof cover over missile launcher or similar installation.

3 Pre-fabricated hangar flown to FOL and rapidly erected.

hangar flying Social chat about flying by those involved in it, esp. pilots.

hangar queen Particular aircraft notoriously prone to unserviceability requiring major maintenance.

hangar rat Young enthusiast [self-explanatory].

hangfire Fault condition in which rocket missile fails to fire; vehicle or missile thus affected.

hang glider Large class of simple ultra-light aerodynes, broadly divided into those with flexible wings (most of Rogallo type) and those having wings with preformed aerofoil section (called rigids); majority have no controls and manoeuvre by translation of pilot mass to shift c.g. Can be monoplane or biplane, and may have rear tail, canard or auxiliary engine. Demarcation line with glider or ultra-light aeroplane becoming blurred.

hang up *1* Externally or internally carried store which fails to release when thus commanded.

2 Gas-turbine engine which starts but fails to spool-up; also called hung start.

h$_{ant}$ Height of ILS or MLS antenna above ground plane.

HANZ Helicopter Association of New Zealand.

HAP Hélicoptère d'appui et de protection (F).

Hapdar Hardpoint demonstration array radar.

HAPI, Hapi Helicopter approach-path indicator.

Happ High-altitude powered platform, lightweight, solar-powered, unmanned aircraft able to hold station for months to years.

Haps *1* Helmet-angle position sensor.

2 Helicopter acoustic processing system.

HAR Helicopter, air rescue (role prefix, UK).

harass *1* Air attack on any target in area of land battle not connected with CAS or interdiction, with objective of reducing enemy's combat effectiveness.

2 To interfere with progress of another aircraft by making repeated close passes against it.

Harc jet Hall-effect arc jet, using magnetic field on hydrogen plasma jet (see *Hall*).

Harco Hyperbolic area-coverage.

HARD Helicopter and aeroplane radar detection.

hard copy Immediately readable output, eg printed pages, as distinct from tape, microfilm or software.

hard data Remains in memory when power switched off.

hard deck The ground, especially in air combat at low altitude.

hardened *1* Protected against blast, ground shock, over-pressure, EMP, radiation and possibly other effects of nuclear explosion, and (DoD only) likely to be protected against chemical, biological or radiological attack (see *hardness*).

2 More recently, protected against terrorist attack, eg cockpit or LD3 container.

3 Of avionics, protected against EMP and any other

powerful external EM effects which in particular would normally degrade or destroy most memory cores.

4 Of metal, physically * by precipitation, quenching or cold working.

hard flutter   Normally well damped but extremely violent over one narrow range of conditions.

hard-iron magnetism   That induced into all magnetic parts of aircraft during manufacture, esp. by hammering or riveting, orientation and polarity depending on assembly heading and terrestrial latitude.

hard lander   Spacecraft designed to free-fall to surface of heavenly body.

hard landing   *1* Conventional aircraft landing with excessive rate of descent, esp. that results in damage or overstressing.

2 Arrival of hard lander on lunar/planetary surface.

hardness   *1* Various measures of physical * such as Moh's scale for non-engineering (eg geological) materials, and Rockwell, Vickers, Brinell and many other standard tests for precise quantified readings.

2 Any of five measures of nuclear hardening; eg resistance to overpressure, which varies from c 2 lb/sq in (c 14 kPa) for aircraft parked in open to over 11,000 lb/sq in (c 75 MPa) for hardest concrete silo.

hardover runaway   Sudden unwanted operation of system, esp. flight-control channel, to extreme limit of travel.

hardover signal   Fault condition resulting in full demand for unidirectional system operation unrestrained by normal feedback.

hardpoint   Anchorage built into aircraft structure for heavy external load, usually via intermediate pylon, MER or launcher.

hard radiation   High penetrating power from very short-wavelength; usual definition is ability to pass through 100 mm of lead.

hard recovery   Landing under difficulties, eg barrier crash into net in bad weather with tower out of action (probably colloq.).

hardstanding   Paved parking area (US, often hardstand).

hard target   *1* One that is hardened (1).

2 In air-to-air or surface-to-air firing practice, a rigid drone, as distinct from a sleeve or banner.

hard-target functional defeat   Use of air/ground weapons to disable deeply buried installations without necessarily killing humans.

hard temper   Modification of light-alloy properties, eg by cold working, denoted by scale such as 3/4H (= 75% fully hard); increased tensile strength is usually accompanied by reduced ductility, which in airframes tends to be equated with shorter fatigue life.

hardtop   Paved with permanent all-weather surface.

hard turn   Planned turn in air combat at rate governed by angle-off and range.

hard vacuum   High vacuum, pressure below 10^{-9} Nm^{-2} (10^{-7} torr).

hardwall hose   Does not collapse under atmospheric pressure when used for suction; not necessarily armoured.

hardware   *1* Originally introduced to distinguish mechanical parts of EDP (1) systems from software, today useful to imply manufactured items of any kind that exist, as distinct from software, system concepts, paper designs and proposals, simulations, capabilities and functions.

2 In narrow sense, small fasteners and similar small parts.

hardware in the loop   Flight-motion simulation system incorporating portions of actual aircraft dynamics.

hard wing   *1* One with simplified leading edge compared with particular previous wings, eg no slat.

2 Any wing with fixed-geometry leading edge.

3 In combat mission, wing man locked-in to follow leader in (almost) all circumstances.

Harm, HARM   High-speed anti-radiation missile.

Harmattan   Dry, dust-filled NE wind (W Africa).

harmonic   Component of sinusoidal or complex waveform or tone whose frequency is integer multiple of fundamental frequency.

harmonic analyser   Device, typically variable-frequency filter, which can resolve waveform into harmonic constituents.

harmonization   *1* Boresighting of all guns fixed to fire in same basic direction (eg ahead) so that all are correctly aligned with respect to aircraft axes, usually so that all gun axes converge at specified distance from aircraft.

2 Adjustment of flight-control system so that effect of controls about each axis matches that about others at all airspeeds; in particular so that handling about each axis in terms of rate of pitch, roll or yaw, and load (real or synthetic) experienced by pilot appear to be in harmony.

harness   *1* Assembly of straps with which member of flight crew can be secured to seat.

2 Assembly of straps with which parachutist can be attached to parachute, which may or may not be permanently attached to *.

3 System of straps and other restraining members with which cargo pallet or container is secured to cargo floor in cases where there is no inbuilt anchorage. Not restraint net.

4 Weatherproof, screened assembly of HT leads connecting ignition source to piston engine plugs or gas-turbine igniters.

HARP, Harp   *1* Helicopter Airworthiness Review Panel.

2 Helicopter advanced rotor program.

3 High-altitude research program.

Harper rivet   Extremely shallow convex head.

harpoon system   Any helicopter hold-down system based on firing barbed anchor into grid or other fixed base on ship platform.

Harry Clampers   Clamp (colloq.).

HARS, Hars   *1* Heading/attitude reference system.

2 High-altitude route system.

harsh environment recorder   Tape recorder with up to 28 tracks meeting most severe specification for temperature, vibration/shock and sustained acceleration.

hartley   Standard unit of information content generally taken as equal to $\log_2 10 = 3.219$ bits or 1 decimal digit.

Hartmann generator   Common electrically powered source of intense noise, usually with variable spectrum.

Hartridge smoke unit, HSU   One of most popular measures of smoke blackness; smoke is fed through chamber of known length and light meter measures light received from calibrated source at other end. Accurate for combustion engineers but for subjective assessment against sky jet diameter must be taken into account. PSU is similar (see *smoke*).

HAS   *1* Hardened aircraft shelter.

2 Hover-augmentation system (reduced pilot workload by using sensitive accelerometers in hover).

3 Hood, aircrew survival.

4 Helicopter anti-submarine (UK).

5 Heading and attitude (reference) system (or heading/attitude sensors).

6 Hazard awareness system.

HASA   Helicopter Association of Southern Africa.

HASC   House Armed Services Committee (US).

Hasell check   Height, airframe, security, engine(s), location, lookout; prior to spin or other harsh manoeuvre.

HASG   Helicopter Airworthiness Study Group (CAA).

HASI   Huygens atmospheric-structure instrument.

Hasp   High-altitude sampling program.

Hass, HASS   Highly accelerated stress screening.

HAST   High-altitude supersonic target.

Haste   Helicopter ambulance service to emergencies (US).

Hastelloy   Family of US Ni-Cr-Mo-Fe alloys combining strength at high temperatures with resistance to oxidation or corrosion.

HAT   *1* Height above touchdown (or threshold) (FAA).

2 Harbour acceptance trial(s) (Missile).

hat   *1* Section commonly used for stiffeners and other structural members, formerly called top-hat and resembling latter's outline.

2 Coolie *, flat cone multiway thumb switch on top of stick, usually controlling trim and radar elevation angle.

hatchback   Transport whose rear cabin area is Combi.

HATM   Hypervelocity anti-tank missile.

HATOD   Hill airtasking order defragger.

Hatol   Horizontal-attitude takeoff and landing.

HATR   *1* High-temperature attenuated total reflectance.

2 Hazardous air traffic report(ing).

HATS   *1* Helicopter automatic targeting system.

2 Heavy Aircraft Test Squadron (UK, WW2).

hatted   Appointed as a military commander (US colloq.).

Have   Permanent first word in code names of AFSC projects (USAF).

have numbers   Radio code: "I have received and understood wind and runway information for my inbound flight".

HAW   *1* Hypersonic aerodynamic weapon.

2 Hawaii space tracking station.

HAWC   Homing and warning computer.

Hawfcar   Helicopter adverse-weather fire-control acquisition radar.

Hawk   Homing all the way killer.

hawking the deck   To fly closely past carrier on right (starboard) side to check deck state before landing on.

Hawtads   Helicopter all-weather target-acquisition and destruction system.

Haybox   Helicopter jetpipe incorporating IRCM (colloq.).

hazard alert   Broadcast to all affected operators of existing or impending failure in hardware of nature likely to imperil safety of flight, usually as result of fault discovered on inspection; fault can be structural or in system operation, but invariably affects part of aircraft.

hazard beacon   Warns of permanent danger to air navigation.

Hazcam   Hazard [on planetary surface] camera.

hazmat   Hazardous material[s].

HB   *1* Brinell hardness.

2 Aircraft category, heavy bombardment (USAAC 1925-27).

3 Height of burst.

HBA   Hybrid buoyant aircraft.

HBAW   Handbook bulletin for airworthiness (FAA).

HBC   *1* Hot-bonding controller.

2 Heave by cable.

HBD   Hollow-bladed disk.

HBK   Handbook.

HBN   Hazard beacon.

HBP   High-band prototype.

HBPR   High bypass ratio.

HBR   *1* Human-behaviour representation.

2 High bit-rate.

HBS   *1* Hot ball and socket.

2 High-band subsystem [DU adds demonstration unit].

3 Hold-baggage screening.

HBT   Heterojunction bipolar transistor.

HC   *1* Hexachloroethane, ECM aerosol also usually containing ZnO and Al powder.

2 Hand controller.

3 Helicopter, cargo (UK) and USA (1959–62, became CH).

4 Critical height [ch preferred].

5 High-capacity [bombs, WW2].

6 Hydrocarbon[s].

7 Crane helicopter (USN 1952–55).

Hc   Height change.

HCA   *1* High-cycle aircraft, specimen which has completed more flights than any other of same type.

2 Hot compressed air (airfield snow clearance).

3 Historic cost accounting.

4 Helicopter Club of America.

HCC   High-thermal conductivity composite.

HCCS   Human-centred control system[s].

HCDC   House of Commons Defence Committee (UK).

HCDR   High channel density receiver.

HCF   *1* High-cycle fatigue.

2 Hollow carbon fibre.

HCGB   Helicopter Club of Great Britain.

HCHE   High-capacity HE (ammunition).

HCI   *1* Human/computer interface.

2 Helicopter Club of Ireland.

HCMM   Heat-capacity mapping mission (spacecraft).

HCMOS   High-density CMOS.

HCP   *1* Head-up control panel.

2 Hélicoptère de combat polyvalent [= multirole] (F).

HCS   Host computer system.

HCT   Helicopter control trainer.

HCTS   Helicopter collective training system.

HCU   *1* Heavy Conversion Unit (RAF, WW2).

2 Hydraulic control unit.

HCV   Hypervelocity, or hypersonic, cruise vehicle.

HCW   Heavily cold-worked [seamless tube].

HD   *1* Height difference, usually in attack mission between IP and target.

2 Distilled mustard gas code (USA).

3 Hydrogen decrepitated powder, for sintered and polymer-bonded magnets.

4 High-drag [bomb].

5 Hourly difference.

6 Home Defence [i.e., of UK, WW2, esp. civilian].

HDA   High-density acid (usually nitric).

HDAC   House [of Representatives] Defense Appropriations Committee (US).

HDAS   Hypersonic deep attack system.

HDB   High-density bombing, rebuild of bomber designed for nuclear warfare to carry heavier loads of conventional bombs.

HDBK   Handbook.

HDBT   Hard [and] deeply buried target [DC adds defeat capability].

HDD   Head-down display, ie a display inside cockpit, usually CRT raster plus symbology and TV overlay.

HDDR   Head-down display radar, or recorder.

HDEP   Haze depth.

HDF   *1* Hot-drape forming.
    2 High-frequency direction-finding (facility).

HDFPA   High-density focal-plane array.

Hdg   Heading.

Hdg C   Compass heading.

Hdg Sel   Heading select.

HDI   Homeland defense interceptor.

HDIP   Hazardous-duty incentive pay (US).

HDK   Hard disk.

HDL   Hybrid datalink.

HDLC   High-level datalink control; – B adds balanced..

HDLMS   Hybrid datalink management system.

HDMR   High-density multitrack recording.

HDOC   Hourly direct operating cost.

HDOP   Horizontal dilution of precision.

HDP   Hardware development plan.

HDPE   High-density polyethylene.

HDR   High data rate.

HDS   *1* Helicopter delivery service.
    2 Hard-disk subsystem.

HDT   High damage tolerance.

HDTA   High-density traffic airport (FAR 93).

HDTV   High-definition TV.

HDU   *1* Hose-drum unit for in-flight refuelling.
    2 Horizontal display unit.

HDV   Hydrant dispenser vehicle.

HE   *1* High explosive [43 suffix acronyms].
    2 High-energy [ignition].

He   Helium.

He, H$_e$ Altitude error.

he   Energy height, $h + v^2/2g$.

head   *1* Complete hub of helicopter rotor (main or tail) including flight-control linkage and all auxiliaries (nitrogen pressure signal, anti-icing connection, lights etc).
    2 Cutter and positioning system in most machine tools, esp. those in which workpiece is stationary.
    3 Loosely, downwind end of runway.

head-down   Looking into cockpit as distinct from outside aircraft.

head-down display   See *HDD*.

heading   Angle between horizontal reference datum and longitudinal axis of aircraft expressed as three-figure group 000°-360°; in UK also called course. Datum can be compass north, magnetic north or true north. Not to be confused with track [one dictionary actually calls * 'the direction of an aircraft path', precisely what it is not].

heading alignment cone   Imaginary cone of 18,000-ft upper diameter serving to guide Orbiter or STA(6) to runway.

heading hold   Flight-control mode which maintains selected heading.

heading-orientated map   One held in hand so that heading is towards top of sheet.

heading select   Flight-control mode in which aircraft automatically turns to and holds any inserted *.

head-level display   Immediately below HUD.

head moment   Total turning moment transmitted through head(1) in helicopter manoeuvre.

head-on   Flying directly towards other aircraft on reciprocal track.

heads down   Control of RPV or UAV without having it in view.

headset   Receive/transmit interface with radio communications or other system (eg aural warning or missile launch tone), either worn separately or built into helmet. Normally includes earpieces or earshells, noise-cancelling microphone on boom and binaural cable.

heads up   *1* Airborne intercept code: "Hostile force, whole or in part, got through defences" or "I am not in position to engage" (DoD).
    2 Occasionally used to mean head-up [only one human head can use a HUD at any one time].

head-up display   See *HUD*.

health   Generally everyday meaning but applied to hardware; thus * monitoring, can be equated with inspection, AIDS output etc.

HEAO   High-energy astronomical observatory.

heap cloud   Cloud with pronounced vertical development, e.g., cumulus family.

Heart   Health evaluation and risk tabulation.

heart-cut distillate   Particular family of kerosene fuels, in particular JP-6.

HEAT   *1* High-enthalpy ablation test.
    2 High-explosive anti-tank, shaped-charge warhead.
    3 Helicopter electric-actuation technology.
    4 High-energy advanced trainer.

Heat   Armament panel switch selection for IR AAMs.

heat   Measure of atomic/molecular kinetic energy, but not to be confused with temperature. Symbol Q, unit joule, $J = 0.238846$ cal$_{IT}$.

heat barrier   Supposed barrier to increasing flight Mach number due to kinetic heating.

heat-capacity   Quality of high-power lasers which permits sustained firing followed by brief cooling period.

heat-engine   Prime mover in which energy is extracted from thermodynamic system in which gas passes through cycle from which closed PV or T-entropy diagram can be plotted. Perfect ** has reversible (Carnot) cycle; all practical ** have lower-efficiency irreversible cycle. Nearly all aviation ** are constant-pressure gas turbines or Otto-cycle piston engine.

heater   Afterburner (colloq.).

heat-exchanger   Radiator in which two fluids are brought into close contact (eg, cold air and hot oil) so that one can reject heat to other.

heat flow rate   SI unit is watt, $W = 3.41214$ Btu/h $= 0.85985$ kcal/h.

heat of ablation   Measure of value of ablating material, rate of heat input divided by rate of mass loss.

heat pipe   Contains fluid which is alternately evaporated and condensed, transferring heat [in spacecraft, to space].

heat pulse   Total heat to be absorbed, dissipated, radiated or otherwise transferred from original kinetic energy of body on re-entry.

heat-seeker   Sensitive IR detector and homing guidance system.

heat shield   Usually non-structural layer protecting primary structure from high-temperature environment which would degrade strength, such as around aeroplane afterburners, at base of large rocket vehicle and, esp., over upstream face of body on re-entry (later is invariably ablative).

heat sink   *1* Any location to which heat may be removed from thermodynamic system.

2 Specifically, mass of metal, fuel, oil or other material which can accept unwanted heat.

heat treatment   Heating metals above specific temperatures followed by slow or rapid cooling to improve mechanical properties (see *carburising, case hardening, nitriding, precipitation **, solution ***).

heave   *1* Vertical movement without rotation (simulator).

2 Motion of taxiing aircraft [esp. marine] in vertical plane, fuselage [hull] remaining level.

heavier than air   Having density much greater than that of air, thus aerostatic buoyancy ignored in calculating lift, which is assumed generated wholly by aerodynamic or propulsive forces; aerodyne.

Heaviside layer   D, E, F, and F_1, layers in upper atmosphere (ionosphere) where UV solar radiation ionises gas molecules, allowing conduction of electricity.

heavy   *1* Flight control; difficult to move.

2 Maintenance, major and prolonged.

3 Warning to ATC of wake turbulence, aircraft MTO $\geqslant$300,000 lb (136,080 kg).

4 US civil radio traffic, any large transport, esp. large jet.

5 AA gun, >40-mm calibre (WW2); also, of course, can mean intense fire from many guns of all calibres.

6 Propeller aircraft, $\geqslant$5,700 kg [12,566 lb] MTOW.

7 Air in balloon envelope close to density of local atmosphere.

8 Large firefighting tanker, as distinct from Seat.

heavy alloy   One tailored to have exceptionally high density; most consist chiefly of tungsten, but osmium, iridium and depleted uranium are important.

heavy bomber   Aircraft designed to deliver heavy load of conventional ordnance to targets in enemy heartlands (arch.).

heavy dropping   Delivery by system of parachutes of exceptionally bulky or heavy load, suitably packaged and cushioned.

Heavy Fuel   In US usage, all DFs (diesel fuels) plus JP-5 and JP-8.

heavy landing   See *hard landing (1)*.

heavy maintenance   Maintenance taking 30 days or longer.

heavy metal   Classic warbird, esp. large and powerful (colloq.).

Heavyside   E-layer.

Heavy Wagon   Designated routes for lo military flights within Conus; 300 series for various nav/electronic systems evaluation and 400 Series for nav. training and weather evaluation at 500 ft (152 m) down to MOCA, normally not over 500 kt (USAF, USN).

HECS   High-performance engine control system.

hectare   Non-SI unit of large areas = 100 are = $10^4 m^2$.

hecto   Prefix, multiplied by 10^2, symbol h.

hectopascal   Pressure of 100 Nm^{-2} = 1 mb, symbol hPa.

hedge-hopping   Lo(2) flight by light GA aircraft (colloq.).

Hedi, HEDI   High-endoatmospheric defense interceptor (SDI).

HEDP   Ammunition, high-explosive, dual-purpose.

HEDR   High-energy dynamic radiography.

heeling   Roll due to turn while taxiing.

Heels   Helicopter emergency-egress lighting system.

HEF   High-expansion foam (anti-fire) usable at 1.5% concentration.

HEH   Hershey, Eberhardt, Hottel, basic analysis and charts for idealised piston engine cylinder performance.

HEI   *1* Ammunition, high-explosive incendiary.

2 Hot-end inspection.

height   *1* In performance, Ht defined as true vertical clearance between lowest part of aircraft (assumed aeroplane) in unbanked attitude and relevant datum.

2 General common definition: h, vertical distance of level, point or object considered as point, measured from specified datum (normally associated with QFE).

3 Vertical distance from reference ground level to specified upper extremity (close to but not necessarily highest point, which may be small whip aerial, static wick or other flexible part) of aircraft at specified weight (usually MTOW) and tyre inflation pressure when parked on flat horizontal surface.

height above airport, HAA   Height of MDA above published airfield elevation.

height above touchdown, HAT   Height of DH or MDA above highest obstruction in touchdown zone.

heightfinder radar, HFR   One whose primary output is precise height of distant target; has nodding aerial giving multi-lobe beam.

height lock   Function of autopilot or flight system in holding (after in some cases capturing) selected height h.

height ring   Visible on most displays of weather/mapping/AW/cloud-collision warning and other forward-looking radars as bright ring beyond zero-range ring formed by direct reflection from Earth's surface.

height/velocity curve   Fundamental plot of IAS against altitude included in helicopter flight manual; indicates region(s) from which safe autorotative descent is possible, normally assuming zero wind, sea level, MTOW.

Heim joint   Universal coupling in torque tube.

Heine mat   Flexible mat towed by support vessel on to which marine aircraft taxies before being hoisted on board.

HEIPT   Helicopter engines integrated product team.

HEI-SAP   High-explosive incendiary, semi-armour-piercing.

HEI-T   Ammunition, HE incendiary, tracer.

HEJOA   Heathrow Executive Jet Operators' Association,

HEL   High-energy laser [A adds applications, also see HELCM, Helex, HELJTO, Hels, HELSTF and Heltads].

Helarm   Helicopter, armed (sortie).

HELCM   HEL countermeasures.

HELD   Helicopter [or high-energy] laser designator.

Helex   HEL experiment[al].

Heliarc   Inert (helium) gas welding technique patented by Northrop 1940.

helical compressor   Diagonal-flow compressor.

helical gear   Spur gear with teeth arranged diagonally; contact begins at one end of tooth and proceeds diagonally across width, two teeth, normally transmit load at any time. Double * (herringbone) gear eliminates axial load.

helical tip speed   Actual airspeed at tip of propeller, taking account of vehicle's airspeed.

Heli-Coil   Patented hard-steel thread inserted in soft metal or to repair stripped female thread.

helicopter   Rotorcraft deriving both lift and control from one or more power-driven rotors rotating about substantially vertical axes. Rotors are driven as long as engines operate above certain minimum input speed, and airflow through rotor(s) is downwards except in autorotative engine-failure mode.

helicopter approach-path indicator   Hapi, ground optical aid giving steady white for correct [0.25-0.75°] glidepath, with green and flashing green above and red and flashing red below. Signal width 24°.

helicopter lane   Safety air corridor reserved for helicopters (DoD, NATO).

helideck   Operating platform for helicopters, e.g. on oil rig.

helidrop   Discharge of passengers/cargo/weapons while hovering.

heliocentric   Centred on the Sun.

heliopause   Frontier beyond which Sun has little or no influence.

Helios   Helicopter instrument and operational procedures simulator.

heliosphere   Region of solar influence.

heliosynchronous   Satellite orbit inclined at 90° to give complete Earth coverage, each pass having the same illumination angle.

helipad   Prepared area designated as take-off/landing area for helicopters; need have no facilities other than painted markings.

heliport   Facility for operating, basing, housing and maintaining helicopters; if civil, includes passenger facilities and usually mail/cargo channels.

heliport acquisition beacon   Marks landing pad[s] by white flashing Morse H.

helistat   Aerostat with added helicopter rotor(s).

helistop   Helipad served by civil helicopter.

helitanker   Large firefighting helicopter.

helitow   Towing by helicopter, normally applies to minesweeping.

helium   Inert gaseous element, He, density 0.0001785 (0.1785 gl^{-1}), 0.0112 lb/ft^3, BPt 3.2°, 4.2° K (two isotopes), helium 3 (three atomic nuclei) being important as lunar-source material for fusion power.

helix angle   *1* Of gear, measured between line tangent to tooth helix at PCD and shaft axis.

     2 Of propeller, see *effective**.

HELJTO   High-energy laser joint technology office.

Hellas   Helicopter laser [radar].

helmet   Individual fairing over piston-engine cylinder, hence helmeted cowl.

helmet sight   Any of several systems which attempt to interface between human aircrew looking at surface (or other) target and aiming of armament system; usually wearer's head is in some way accurately aligned with rigid flight helmet whose orientation is then automatically

monitored to give output signals fed to weapon-aiming system.

Helmholtz resonator   Hollow volume connecting with outside via small orifice, single-frequency output.

Helms   Helicopter malfunction system.

helo   Helicopter (colloq.).

HELP, Help   Hybrid electronic lightweight packaging.

Helps   Helicopter protection and support.

Helraps   Heliborne long-range active [acoustic-path] sonar.

Helras   Helicopter long-range sonar.

HELS, Hels   High-energy-laser system [TF adds test facility] (USA).

HELSTF   High-energy-laser system test facility, at WSMR (USA).

Heltads   High-energy-laser tactical air-defense system (USA).

Helweps   High-energy laser weapon system (TRW/USN).

HEMA   Heavy maintenance.

Hem-bird   Helicopter electromagnetic probe, suspended magnetometer for measuring thickness and other parameters of surface layers.

HEMC   High-explosive medium-capacity.

Hemes   Helicopter Hospital Emergency Medical Evacuation Service (USAF, USA).

hemispherical engine   Piston engine whose combustion spaces at TDC are hemispherical.

hemispherical radar   Mounted at nose and tail of aerial platform to provide complete 360° (2 × 180°) azimuth coverage.

hemispheric rule   For IFR, and VFR between FL30-290, altitude assignments are determined by magnetic bearing.

Hemloc   Helicopter emitter/locator countermeasures.

Hemp   Name of khaki-ochre colour [BS.4800-10B-21] of large patrol and tanker aircraft (RAF).

Hems   Helicopter emergency medisal service.

HEMT   High electron mobility transition.

HEND   High-energy-neutron detector.

He-Ne   Helium/neon laser.

henry, H   SI unit of inductance, that of closed circuit in which 1 V is generated when current varies at steady 1 A/s; in other units H=kg.m^{2}s^{-2}A^{-2}; plural henrys.

HEOB   Hostile electronic order of battle.

HEO   Highly elliptical [Earth] orbit; P adds payload.

HEP   High-explosive plastic.

HEPA   High-efficiency particulate air.

HEPL   High-energy-pulse laser.

heptane   Basic member of straight-chain (alkane) hydrocarbons (see *paraffin series*) with seven carbon atoms and thus 16 of hydrogen; zero-octane reference fuel.

HER   *1* Harsh [or high] environment recorder.

     2 Helicopter experimental radar.

Herald   Helicopter equipment for radar and laser detection.

Herbst manoeuvre   Post-stall [70° AOA] 180° turn using vectored thrust (X-31).

Herid   High-energy railgun integration demo, (USAF).

Hermes   Helicopter energy/rotor management system; automatically computes limiting payload.

Hermit   Harsh-environment robust micromechanical technology.

HERO, Hero   Historical evaluation and research organization.

herringbone gear   Double-helical (see *helical*).

HERT   Headquarters emergency relocation team.

Hertz, Hz   SI unit of frequency, = cycles per second; hence kHz, mHz, GHz etc.

Hesh   High-explosive squash-head.

hesitation roll   Aerobatic rolling manoeuvre, normally based on slow roll, in form of succession of quick aileron applications between which rate of roll is suddenly arrested; positions of arrest called points, and common ** are 4-point, 8-point or 16-point, 16 being difficult and rare.

HESSI, Hessi   High-energy solar spectroscopic imager.

HeSTOR   Helicopter simulator for technology operations and research.

HET   *1* Helicopter environmental technique, to increase safety and reduce disturbance by avoiding dwellings, and certainly urban areas.
    2 Hall-effect thruster.

HETE   High-energy transient Explorer.

Hete   High-energy transient experiment.

Heterodyne   Mixing of two alternating currents (eg radio signals) to generate third equal to sum or difference of their frequencies; verb or adjective.

heterojunction II logic   Microcircuits with inverted gates in which transistor is embedded (integrated injection) in expitaxial GaAs layer.

heterosphere   Earth atmosphere above c100km (62 miles).

HEU   *1* Highly enriched uranium.
    2 HUD electronics unit.

heuristic   Leading towards solution by trial and error (see *algorithm*).

HEUS   High(er)-energy upper stage.

HEW   *1* Heliborne, or helicopter, early warning.
    2 Health, Education and Welfare (US).

HEWS   Helicopter electronic-warfare suite, or system.

Hexal   Hexogen/aluminium bomb filling.

Hexapod   Usual mounting for cabin of flight simulator: two superimposed triangles (both horizontal when at rest) displaced spatially by 60° with corners joined by six actuating jacks to give 6-DoF motion.

Hexcel   Range of proprietary honeycombs, and honeycomb sandwiches, most with 24ST skins bonded by epoxy resin to glassfibre core [US company].

Hexogen   *RDX.*

Hexotonal   Mix of RDX, TNT and powdered aluminium.

HF   *1* Human factors; see HFE, HFR(2), HFRT, HFSG.
    2 Hydrogen fluoride; C adds cleaning.
    3 High-altitude fighter (UK role prefix, WW2).
    4 Hybrid fan.
    5 Heavy fuel, includes all DFs plus other blends.
    6 High-frequency, see *Appendix 2.*

H$_F$, H$_f$ Wheel height at flare initiation.

Hf   *Hafnium.*

h.f., HF   High-frequency, see *Appendix 2.*

HFAC   Helicopter flight-advisory computer.

HFAJ   High-frequency anti-jam[mer].

HfB$_2$ Hafnium boride.

HFCS   Helicopter flight-, or fire-, control system.

HFD   *1* Head-up flight display, S adds system.
    2 High-frequency data; CR adds communications radio, L link, R radio, RS radio system, U unit.

HFDF   *1* High-frequency direction-finding.
    2 Hydrogen fluoride/deuterium fluoride.

HFDL   HF datalink.

HFDM   HF data modem.

HFDR   HF data radio.

HFDS   Head-up flight display system.

HFE   *1* Human-factors engineering.
    2 Heavy-fuel engine.

HF-GWR   H.f. ground-wave [OTH] radar.

HFI   Helicopter Foundation International.

HFIP   H.f. improvement programme (NATO).

HFISA   Helicopter flight into Service Area.

HFK   Hochgeschwindigkeits Flugkörper, high-speed vehicle (G).

HFL   Hydrogen fluoride laser.

HFNPDU   HF network protocol data unit.

HFOL   Hydrogen fluoride overtone laser.

HFOR   Human-factors open reporting.

HFP   High-fragmentation projectile.

HFPD   High-frequency pulse detonation.

HFR   *1* Height-finder, or -finding, radar.
    2 Human-factors reporting.

HFRT   Human Factors Research and Technology Division (NASA).

HFS   High-frequency system[s].

HFSG   Human-factors steering, or study, group.

HFSNL   HF sub-network layer.

HFSWR   High-frequency surface-wave radar.

HFVT   Hybrid-fan vectored thrust.

HG   Head-up guidance; D adds display, S system from LOC/GS capture to touchdown.

Hg   *1* Mercury.
    2 Barometric, or boost, pressure in inches of (1).

HGA   High-gain antenna; S adds system.

HGAA   High-gain active antenna.

HGC   Head-up guidance computer.

HgCdTe   Mercury cadmium telluride, IR detector for 8-14μ.

HGI   Hot gas ingestion (ejector lift).

HGL   Hot gas leak.

HGM   Hot-gas manifold.

HGOPA   High-gain open planar array.

HGPADS, HGPads   High-glide precision air-delivery system (USA).

HGR   *1* Hot-gas recirculation [or reingestion].
    2 Hypervelocity guided rocket.

HGRI   Hot-gas reingestion.

HGS   *1* Holographic guidance system.
    2 Head-up guidance system.

H$_{gs}$ Height of glide-slope referenced to depressed surface.

HGSI   Hot-gas secondary injection TVC.

Hgt   Height.

HGU   Horizon gyro unit.

HGW   Higher gross weight.

HH   NDB class, over 2 kW.

HHC   Higher harmonic control.

HHLD   Heading hold.

HHMD   Hand-held metal detector.

HHS   Helicopter [shipboard] handling system.

HHT   Hand-held terminal [readout unit].

HHTI   Hand-held thermal imager.

HHTR   Hand-held tactical radar.

HHV   Higher heating value.

HHUD   Holographic HUD.

HHX   Hydrogen heat exchanger.

HI   *1* High-intensity light.

　　2 Heading indicator [direction indicator].

　　3 High, high-altitude, high-intensity, high band.

hi   Flight at tropopause or above, adopted by gas-turbine combat aircraft flying for range outside hostile environments.

HIA   Held in abeyance.

HIADC   High-integration air-data computer.

HIAL   High-intensity approach lighting; S adds system.

HIAPER, Hiaper   High-performance instrumented [or instruments] airborne platform for environmental research (NCAR).

HIB   Helicopter identification by [AT adds acoustic techniques *Hibat*, IRD adds IR detection *Hibird*, RAD adds radar detection *Hibrad*].

hibernate   To remain on station in space or in orbit with as many subsystems as possible switched off until reactivated by Earth command.

Hibird   Helicopter identification by IR detection.

Hibrad   Helicopter identification by radar detection.

HIC   Head impact criteria.

Hi-camp   Highly calibrated aircraft measurements program (Darpa).

Hicar   Acrilonitrile to DTD.5509.

Hi-CAT   CAT above 55,000 ft (16.8 km).

HICT   Hand-held imaging communications terminal.

HICU   HIPSS interface control unit.

Hidar   High instantaneous dynamic acoustic range.

Hidas   Helicopter integrated defensive-aids system, or suite.

hide   Locally constructed aircraft shelter of posts, net and camouflage.

Hidec   Highly integrated digital engine [or electronic] control.

HIDSS   Helmet integrated display sighting [or and sight] system.

Hiduminium   Family of duralumin-type alloys, including RR (Rolls-Royce) formulations, often also containing nickel and iron; developed for piston engine pistons and similar applications, also used in supersonic airframes (in Concorde also known by French designations such as AU2GN).

Hidyne   See *Hydyne*.

HIE   Helicopter-installed equipment.

Hi-Ex   High-expansion [typically volume factor of 1,000] firefighting foam.

HIF   Horizontal integration facility (space launchers).

HIFR   Helicopter in-flight refuelling from ship, also rendered hover in-flight refuelling.

HIG, Hig   Hermetic integrating gyroscope.

HIGE   Hovering in ground effect.

Higger   High-integrity GPS guidance enhanced receiver (blind STOVL).

high   Region of high atmospheric pressure, anticyclone.

high altitude   *1* Above 10 km (32,800 ft) (NATO).

　　2 Between 25,000 and 50,000 ft (7.6–15.2 km) (DoD).

High-altitude airship   To remain geostationary at 70,000 ft (21.34 km) for six months (US).

high-altitude bombing   Level bombing with release at over 15,000 ft (4.57 km) (DoD).

high-altitude burst   Nuclear weapon explosion at over 100,000 ft (30.5 km).

high blower   See *high supercharger*.

high boost   Piston engine operated at high inlet-manifold pressure, esp. one well above SL atmospheric pressure for takeoff.

high boss   Variable stator vane outer bearing.

high BPR   Bypass ratio exceeding 5.

high-capacity bomb   Large thin-case bomb for demolition of major soft target.

high cloud   Extremely loose classification, typical band being 6–8 km (20,000–60,000 ft) in tropics, 5–13 km (16,000–43,000 ft) in temperate, 3–8 km (10,000–26,000 ft) polar. Always mainly ice crystals, Ci, Ce, Cs.

high compressor   HP compressor (US).

high-cycle fatigue   That due to high-frequency vibrations, flexure or rotation of machinery, typically at rate many times per second.

high-density focal-plane array   EO sensor with thousands of 2-D IR elements integral in substrate with CCD readout and processing circuitry.

high-density seating   Normally, that giving greatest practical number of passengers, more even than all-economy, tourist or other classifications.

high-density tunnel   One having closed circuit filled with air or other gas under pressure, power required for given Mach/Reynolds combination being inversely proportional to density.

high-energy laser   No lasting definition; varies with family, whether continuous, pulse or gated, and with rapid progress, which by 1980 had made GW output not uncommon.

higher harmonic control   Reduction of helicopter stress, noise and vibration by using dynamic absorption methods.

higher heating value   Gross calorific value of fuel.

high-flotation gear   Landing gear having geometry and other characteristics suitable for operation from soft soils, sand and similar surfaces.

high-flotation tyre   One of low inflation pressure and very large contact area; not necessarily used on high-flotation gear.

high frequency   See *Appendix 2.*

high-frequency pulse detonation   Proposed propulsion systems based on pulse detonation at frequencies not less than 2 kHz, pioneered by Moscow State University.

high-gain aerial   Usually means strongly directional, designed to transmit or receive along single axis.

high gate   Different meaning in different programmes (lunar landing, air speed record runs, etc) but always a specified rectangle in exact relation to surface of Earth or other body through which vehicle must pass, often at precise time, for mission to be successful (see *gate [10, 11]*).

high gear   High supercharger gear.

high-incidence stall   Unaccelerated symmetric stall at highest angle of attack, normally with high-lift configuration.

high-inclination mission   Usually means satellite inclination from about 63° to 99° to Equator.

high-lift   System, device, configuration or mode giving lift greater than in clean or cruise configuration. Normal * devices include leading-edge flaps, droops, slats, Krügers, trailing-edge flaps, flap blowing and variable-sweep; other forms include BLC, EBF, USB, Jet Flap, vectored thrust and tilt-wing.

highly blown engine Piston engine in which supercharger is used to achieve high inlet manifold pressure at low altitudes. Terminology can be ambiguous; high and low-blown can mean high and low rated altitudes (critical heights).

highly supersonic Mach 4 to Mach 6.

high-Mach buffet Experienced when critical Mach number is exceeded on aircraft not designed for transonic flight; can affect control surfaces and primary structure, and cause trim changes which automatically either remove or intensify condition.

high-Mach flight Flight at high subsonic Mach number.

high-Mach trimmer See *Mach trimmer*.

high oblique Oblique reconnaissance or other aerial photograph in which portion of apparent horizon is visible.

high-octane *Fuel grade.* Not defined, but typically over 100.

high pitch Coarse-pitch.

high-pressure area Region of atmosphere where pressure significantly exceeds that of surroundings, esp. one bounded by enclosed isobar rings; anticyclone.

high-pressure compressor Applicable only to an engine with two or more compressors on separate shafts. High-pressure turbine likewise.

high-ratio engine Engine having high BPR.

high route Area-navigation route extending from 18,000 ft AMSL to FL450 (FAR. 1.).

high rudder See *top rudder*.

High-Shear rivet Patented close-tolerance steel threadless bolt held by swaged ring closed around groove.

high-speed alloys Brasses, steels, light alloys and other metals which can either be machined easily at high linear speed, or which retain strength at high temperature and can edge cutting tools (two almost opposite meanings for same term).

high-speed exit *High-speed turnoff.*

high-speed stall Accelerated stall in which stalling angle of attack is reached at relatively high airspeed; height well below aircraft ceiling is implied.

high-speed tunnel Traditionally, one in which effects of compressibility can be observed.

high-speed turnoff Taxiway forming transition curve from landing runway enabling aircraft to leave runway at earliest possible moment.

high-speed warning ADC-triggered subsystem giving aural and visual indication (usually duplicated) of inadvertent speed excursion beyond threshold usually set at just above Mмо; can be overridden to demonstrate MDF/VDF but aircraft thus equipped is not airworthy without it.

high-subsonic In the range of Mach numbers where sonic speed is locally exceeded at peak suctions, and compressibility effects, if any, are manifest; in older aircraft can be Mach 0.8, today higher values can be reached with no significant change in handling or trim (beyond 0.9 there may be major trim changes automatically countered by Mach trimmer). Hence * aircraft, general category of jet aircraft other than those designed for supersonic performance.

high supercharger gear In older high-power piston engine with mechanically driven supercharger it was common to have two drive ratios, high gear being automatically clutched in by aneroid at preselected height,

thus giving power/altitude curve with two major discontinuities; high gear unavailable at lower levels.

high-tailed aircraft Aeroplane with horizontal tail on top of fin; T-tail.

high-test peroxide, HTP Not aqueous solution but almost pure hydrogen peroxide, used in MEPUs, rocket engines and other applications as monofuel rapidly decomposed by silver-plated nickel into superheated steam and free oxygen in which kerosene or other fuel is also often burned.

high-time item That particular engine, aircraft or other hardware item that has flown more hours or completed more operating cycles than any other of same design.

high turbine HP turbine (US).

high vacuum Pressure less than 10^{-9} Nm^{-2}(10^{-7} torr).

high-velocity drop Airdrop in which conventional parachute system is not used, and speed of descent lies between 30 ft/s (low-velocity) and free-fall; requires retarding means plus stabilization to ensure impact on cushioned face.

highway Generalised term for channel for data in EDP (1) or other dynamic system.

high wing One attached at top of fuselage; see parasol, shoulder.

HIIL Heterojunction integrated injection logic.

HIK Heading index knob.

hikodan Wing (= UK group) (J).

HIL Horizontal integrity limit.

Hill's mirror Instrument for measuring wind-speed by observing smoke or other particulate matter.

Hilo 2 Hi/lo No 2, one of two worldwide programs for microcircuit design.

HILS Hardware-in-the-loop simulator.

HIM High-inclination mission.

Himad High to medium (altitude) air defence [s adds system].

Himat, HiMAT Highly manoeuvrable aircraft technology (research programme).

Himes, HIMES Highly manoeuvrable experimental space [vehicle] (J).

hinged wing General term for wing not always rigidly fixed but able to fold, vary sweep or incidence, slew or rotate in any other way relative to fuselage.

hinge moment Force required to rotate aerodynamic surface or resist incident air load on it, C_H = total aerodynamic load on surface (varies as square of airspeed) multiplied by distance from hinge axis to cp.

HIOC Hourly indirect operating cost.

Hip Hot isostatic pressing.

Hipar *1* High-power acquisition radar. *2* High-performance precision approach radar.

Hipas High-performance active sonar.

Hipcor High-power coherent radar.

HiPEP, Hipep High-power electric propulsion.

HIPO, Hipo Hierarchical input process output.

HIPPAG, Hippag High-pressure pure-air generator.

Hips Highly interactive problem solver.

HIPSS Helicopter integrated power and switching system.

HIR *1* Harmful interference to radio [ICAO study]. *2* Higher.

Hiran High-precision shoran.

HIRES *1* High resolution, pronounced hi-rez. *2* High-resolution Earth station.

HIRF   High-intensity radiated field, or radio frequency.
HIRL   High-intensity runway light/lights/lighting.
HIRM   High-incidence research model.
HIRNS, Hirns   Helicopter IR navigation system.
HIRS   Helicopter IR system.
HIRSS   Hover IR suppressor, or suppression, system.
Hirt   High-Reynolds-number tunnel.
Hirta, HIRTA   High-intensity radio transmission area.
Hirth coupling   A method of precisely locating the mating faces of two rotating assemblies, such as a turbine disc and a hollow driveshaft, by high-precision radial grooves which maintain perfect centering at all rotational speeds.
HIS   Hyperspectral imaging system.
Hisar   Hughes integrated surveillance and reconnaissance.
Hisat, HISAT   High-speed air-to-air target.
HISL   High-intensity strobe light.
Hisos   Helicopter integrated sonar system.
Hi-Spot   High-altitude surveillance platform for over-the-horizon targeting (airship, Lockheed).
Hiss   ASW helo. (USN colloq.).
historical   Based on past values.
HISU   Hybrid inertial sensor unit.
HIT, Hit   *1* Homing-intercept technology.
  2 Hybrid/inertial technology.
hit band   Plot of initial velocity against initial path angle for vehicle to make successful impact (hard/soft) on lunar or planetary surface; usually very narrow and curved.
hit dispersion   See *dispersion*.
HITL   Human in the loop.
Hitmore   Helicopter installed TV monitor recorder.
Hitron   Helicopter Interdiction Tactical Squadron (USCG).
HITS, Hits   Hostile identification/targeting system.
HITT   Holland Institute of Traffic Technology (Neth.).
hit the silk   Abandon aircraft by parachute in emergency (colloq.).
Hittile   Direct-hitting guided missile, ie one designed to explode inside target.
Hiwas   Hazardous in-flight weather advisory service.
HJ, H$_J$ *1* Sunrise to sunset (ICAO).
  2 Helicopter, utility (USN, 1944–49).
hj   CC blowing jet slit height.
HJV   Hot-jet velocity.
HKAOA   Hong Kong Aircrew Officers' Association.
Hkwr   Hookwire (available).
HL   *1* High-level airway, eg in North American VOR system, other than Jet Routes.
  2 Heavy lifter (airship class).
  3 Height loss.
HLA   *1* High-level architecture.
  2 Helicopter Listing Association (US).
HLBA   High-level bus analyser.
HLCS   High-lift control system.
HLD   Head-level display
HLDG   Holding.
HLDC   High-level data-link control.
HLE   Higher-layer entity.
HLF   Hybrid laminar flow [N adds nacelle].
HLH   Heavy-lift helicopter.
HLL   High-level language.
HLLV   Heavy-lift launch vehicle.
HLMD   Hull loss per million departures.

HLO   Helideck Landing Officer (oil rigs).
HLP   Hybrid linear potentiometer.
HLS   Helicopter landing site.
HLSI   Hybrid LSI.
HLSTO   Hailstones.
HLTF   High-level task force (MBFR treaty).
HLU   Hybrid linear unit.
HLUAV   Hand-launched UAV.
HLV   Heavy-lift vehicle (space launch).
HLWE   Helicopter laser warning equipment.
HLYR   Haze layer aloft.
HM   Heavy maintenance.
HMA   *1* Helicopter maritime attack.
  2 His/Her Majesty's Aircraft (UK civil airliner individual name prefix).
HMAFV   Her Majesty's Air Force Vessel (UK).
HMAS   Her Majesty's Australian Ship.
HMC   Health-monitoring computer.
HMC&E   Her Majesty's Customs and Excise.
HMCS   *1* Helmet-mounted cueing system.
  2 Her Majesty's Canadian Ship.
HMCU   Hydraulic monitoring computer unit.
HMD   *1* Helmet-mounted display [D adds device, SS and sight system].
  2 Helicopter mine dispenser.
  3 Hand-held metal detector.
HMDS   Hexamethyldisylazane.
HMEP   Helmet-mounted equipment platform.
HMF   Health maintenance facility, monitors health of space-station human crew.
HMFU   Hydromechanical fuel unit.
HMG   *1* Her/His Majesty's Government.
  2 Hydromechanical governor.
  3 Heavy machine gun.
  4 Hydraulically motored generator.
HMGP   Heavy machine-gun pod.
HMH   *1* Helmet-mounted HUD.
  2 Heavy helicopter squadron (USMC).
HMI   Human/machine interface.
HML   Hard mobile launcher.
HMLA   Helicopter squadron, light attack (USMC).
HMM   Helicopter squadron, medium (USMC).
HMOB   Hardened main operating base.
HMOS   High-density MOS.
HMOSP   Helicopter [or multimission] optronic stabilized payload.
HMP   HMG (3) pod.
HMR   *1* Homer.
  2 Health-monitor[ing] recorder.
  3 Helicopter main route.
HMRS   Hotline mobile repair shop.
HMS   *1* Health-monitoring system.
  2 High-modulus sheet (CFRP).
  3 Helmet-mounted sight; /D adds display, S system.
  4 His/Her Majesty's Ship, name bestowed on ships and airfields of the Royal Navy. The latter were named after seabirds; thus, RNAS Henstridge was HMS *Dipper*.
HMSO   Her/His Majesty's Stationery Office.
HMST   Helmet-mounted sensory technologies [PO adds program office].
HMT   Helicopter Training Squadron (USMC).
HMTAS   Helmet-mounted target-acquisition sight.
HMTDS   Helmet-mounted target designation system.

HMTIS   Helmet-mounted targeting and indication system (R).
HMU   *1* Hydromechanical unit.
 2 Helmet-mounted unit.
 3 Height-monitoring [or measuring] unit, for RVSM.
h$_\mu$, h$_{mu}$ Jet thickness or height.
HMV   Heavy-maintenance visit.
HMX   *1* Cyclotetramethylenetetranitramine; shock-sensitive explosive, most powerful in general use [see HNIW].
 2 Helicopter squadron, development (USMC).
HN, H$_N$ Sunset to sunrise (ICAO).
H$_n$ C.g. margin.
HN3   Nitrogen mustard gas (code USA).
HNIW   Hexanitrohexazaisowurtzitane, most powerful conventional explosive.
HNM   Helicopter noise model.
HNMO   Host-nation management office.
HNP   Hypersonics National Plan.
HNS   Host-nation support.
HNSC   House [of Representatives] National Security Committee (US).
HNVS   Helicopter night-vision system.
HO   *1* Service available to meet operational requirements (ICAO).
 2 Hardover.
 3 Helicopter, observation (USA, USN, 1948–62).
 4 Handoff.
HoA   *1* Heads of agreement.
 2 Head[s] of Agency.
hoar frost   White semi-crystalline coating deposited in clear air on surface colder than frost point (see *frost*).
HOB   Height of burst.
Hobbs meter   Logs engine operating time.
Hobo, HOBO   Homing bomb [S adds system].
HOBS   High off-boresight (missile aiming).
Hocac   Hands on cyclic and collective.
Hocas   Hands on collective and stick.
HOCSR   Host and oceanic computer system replacement (FAA).
HOD   Head-out display, HUD type equipment mounted on helmet and always in wearer's LOS.
HOE   *1* Homing overlay experiment, to kill RVs above atmosphere.
 2 Holographic optical element.
Hoerner tip   Sailplane wingtip of curved downturned shape; popular 1950–60.
Hofin   Hostile fire indicator; detects shockwaves caused by projectiles.
HOG   Hermann Oberth Gesellschaft eV (G).
HOGE   Hovering out of ground effect.
hogging   *1* Machining from solid, esp. of large part out of heavy forging.
 2 Deformation of airship such that both ends droop.
 3 Stressing condition of fuselage caused by sag of overhanging nose or tail in flight or on ground.
 4 Stressing condition of seaplane float or boat hull when weight wholly supported by large wave at midlength.
Hohmann   Describes minimum-energy transfer orbit between two bodies, using their gravitational attraction.
Hohner   Sailplane wingtip (see *Hoerner tip*).
HOJ   Homing on jamming; fundamental ECM technique.

HOL   *1* High[er] order language.
 2 Holiday.
hold   *1* To stop countdown until fault has been cleared.
 2 Period of * (1).
 3 To retain data in EDP (1) store after it has been copied into another.
 4 To fly standard pattern as requested until given further clearance (see *holding fix*).
 5 To wait on airfield at any time after arrival and before departure under tower instructions.
 6 Underfloor cargo compartment, identified as container * or bulk *.
 7 Above-floor compartment in all-cargo aircraft.
 8 Manual adjustment for vertical/horizontal synchronization of raster display.
holdback link   Strong link between carrier aircraft and deck, severed in accelerated takeoff; hence, any such link in launch of missile or RPV.
hold-down test   One carried out while vehicle, rocket motor or other device is mounted on test stand.
holding   Hold [4,5]; hence * course, * reciprocal, * side, non-* side, * fix end, * outer (or outbound) end, etc.
holding area   Region of holding fix.
holding bay   Airfield area where aircraft in motion (taxiing or towed) can be held (5) to facilitate ground movement or accommodate queue for takeoff.
holding fix   Navaid on which holding area is based, such as VOR radial intersection, Vortac, ILS outer marker etc.
holding off   In landing an aeroplane, the final moments of flight when the pilot keeps increasing pull on control column to postpone landing/alighting in order to do so either at lower speed or (tailwheel aircraft) in three-point attitude.
holding pattern   Racetrack pattern, two parallel legs joined by turns at 3°/s or at 30° bank.
holding point   *1* See *holding area*.
 2 Designated point for holding on airfield, especially before entering active runway.
holding room   Airport lounge.
holding side   That side of holding course on which pattern is flown.
hold short   Land and stop before reaching any runway intersection.
hole   *1* Vacant space left by electron missing from crystal lattice, hence any positive charge (fixed or mobile, in latter case flow of * cloud through p-type semiconductor).
 2 Air pocket (colloq.).
hollow charge   See *shaped charge*.
hollow spinner   Propeller spinner in form of annular ring through which passes airflow for engine, cooling or other purpose.
hollow-stem valve   Piston engine exhaust valve stem in form of tube filled with sodium or other material of high specific heat for conducting heat from head.
holography   Photographic record produced by interference between two sets of coherent waves. Many unique properties, one of which is 3-D nature of image.
HOM   *1* High-orbit mission.
 2 Harmony order management (software).
home   *1* To use matched-wavelength seeker and error-sensing flight-control system in order to fly automatically towards source of radiation.
 2 To steer aircraft manually towards particular point navaid or other emitter.

homebuilt   Difficult to define, but generally accepted as powered aircraft designed for construction by amateur. Majority are designed professionally, with plans approved by EAA or other authority and then duplicated and sold to homebuilders. In some cases successful * is put into commercial manufacture for sale, removing it from category. Excluded; modifications of existing types, replicas, restorations or copies of commercial designs.

Homeland Security   Effective October 2002, US Department exceeded in size only by Defense, with c170,000 personnel integrated from Coast Guard, federal Emergency, Customs, Immigration & Naturalization, Border Patrol and Secret Service.

home-in   See *home (2)*.

home plate   Base airfield to which aircraft are to recover after mission.

homer   VHF/DF station.

homing   *1* Procedures for bringing two radio or other EM stations, at least one airborne, together.
2 Guidance which causes a vehicle [e.g. aircraft or missile] automatically to fly towards a particular source of radiation.

homing beacon   One on which pilot can home (2).

homing guidance   System enabling unmanned vehicle to home (1), typically based on use of IR, IIR, EO/TV, ARH, SARH, or SALH.

homing receiver   Radio receiver indicating aurally/visually when heading (rather than track) is not towards transmitter.

homodyne   Reception of DSB in which local oscillator generates output synchronized with original carrier.

homogenous   Having uniform radar reflectivity at all points on surface (function of material substrate, coating[s] and orientation).

homosphere   Earth atmosphere below a height of about 100 km (62 miles), composition being almost constant.

homotron   Soft failure.

HOMP   Helicopter operations monitoring program.

HOMS   Homing optical missile system.

Honcho   Big chief, hence fighter ace or high-ranking military or company officer (US, colloq.).

honeycomb   *1* Low-density structural technique and materials based on hexagon-cell honeycomb sandwiched between two sheets too thin for stability alone; can be all-metal, or any of the three components can be of various other materials, joined by various adhesives. Can be supplied as standard sheet, or parts can be made individually with any form. Nid d'abeilles, NdA (F).
2 Grid of thin intersecting aerofoils or flat plates across tunnel or other duct intended to remove turbulence from fluid flow.

honking   Being airsick, hence *honk bag* (cabin-staff term).

HOO   Helicopter offshore operations.

hood   see *canopy*.

hoodoo   Hose-drum unit (RAF, colloq.).

hook   *1* Procedure used by air controller to direct EDP (1) of semi-automatic command and control system to take specified action on particular blip, signal or portion of display.
2 Aerobatic manoeuvre resembling a lateral Cobra.
3 Arrester hook.

Hooke's law   Up to elastic limit, strain set up in elastic body is proportional to applied stress.

hookman   Member of carrier deck party charged with unhooking aircraft after recovery.

hooks   Interfaces available for future avionics, weapons or other items not yet invented (colloquial).

Hoover highway   Command flight-path display (colloq.).

hop   *1* Very short flight; eg by aeroplane on fast taxi test or by aerodyne only marginally capable of flight.
2 Full circuit in advance of official first flight.
3 Jump from one EM frequency to another by ECCM subsystem.

Hope H_2 orbiting plane experiment [not same as next] (Japanese NAL).

Hope - X   High-orbiting plane experiment.

hopper   *1* Small receptacle in oil tank for hot or diluted oil to assist cold-weather start.
2 Container in ag-aircraft for dust (powdered chemical).

Hops   Helmet optical position sensor.

Horizon   Hélicoptère d'Observation Radar et d'Investigation sur Zone (F).

horizon   *1* Actual boundary where sky and planetary body appear to meet: visible *.
2 Apparent boundary as modified by atmospheric refraction, terrain, fog or other influence: apparent *.
3 Great circle on celestial sphere at all points 90° from zenith and nadir: celestial *.
4 Line resembling apparent * but above or below: false *.
5 Locus of points at which direct rays from terrestrial radio transmitter become tangent to Earth: radio *.
6 Artificial horizon (instrument).

horizon and pitch scale   HUD symbology comprising a horizon line, parallel lines above and below (±2° or 5°) for pitch information, and heading marks at 5° intervals.

horizon bar   Fixed horizontal reference in most types of horizon (6), HSI and attitude-displays.

horizon sensor   Radiometer or other sensitive passive receiver used to align one axis of spacecraft or satellite with apparent horizon of Earth or other body; also called *-seeker.

horizontal   Used in Hatol to mean short or vertical takeoff with fuselage substantially level; used in Hotol to mean takeoff along runway (CTOL) instead of straight up in vertical attitude.

horizontal error   Error in range, deflection or miss-radius which weapon system exceeds on half of all occasions when firing on target on horizontal plane. When trajectory at arrival is near-vertical ** is CEP; where trajectory slanting, giving elliptical dispersion, ** is probable error.

horizontally opposed engine   Piston engine with left and right rows of horizontal cylinders and central crankshaft.

horizontal parallax   Geocentric parallax of body on observer's horizon, = angle subtended at body by Earth radius at equator.

horizontal plane   Usually that through longitudinal [OX] axis and perpendicular to vertical.

horizontal project   One in which hierarchy in design/project team is minimised and each member works full-time on single project.

horizontal scanning   Scanning in azimuth only at near-0° elevation; rare (see *scanning*).

horizontal situation indicator, HSI   Standard pilot panel instrument for all except small GA aircraft; includes Hdg (T, M), angular deviation from VOR, INS or other track,

Tacan/DME or INS display, and alphanumeric readout of G/S and possibly other data.

horizontal stabilizer   Tailplane.

horizontal technology integration   Using common hardware, such as sensors, scanners/windows, stabilization and software shared between two or more systems, such as FLIR, IRST, LST, etc.

horn   *1* Operating arm of simple manual flight-control surface to which cable is attached.

2 Microwave aerial coupling waveguide to free air to give directional pattern; three basic forms are pyramid, sectoral and biconical, first two having rectilinear funnel appearance.

3 Acoustic emitter tube whose varying cross-section and final area control acoustic impedance and directivity.

4 Small area of control surface ahead of hinge axis, usually at tip, sometimes shielded at low deflection angles by fixed surface upstream and usually housing mass-balance; when surface deflected provides aerodynamic force assisting deflection.

horn aerial   See *horn (2)*.

horn balance   See *horn (4)*.

horn check   Verify correct functioning of landing-gear warning horn.

Hornet   Hazardous-ordnance engagement toolkit, to defeat terrorist SAM (US).

horsal   Horizontal ventral fin (originally on XF10F).

horsepower   Non-SI unit of power; traditional hp defined as 550 ft-lb/s = 0.745700 kW; metric * [called ch or PS] defined as 75 kg-m/s = 0.7355 kW. There is also electric * defined exactly as 0.746 kW. Brake * is measured at output shaft by applying known retarding torque; indicated * calculated from area of indicator diagram and rpm; frictional * is indicated minus brake *; equivalent * is turboprop bhp plus addition factored from residual jet thrust. Aircraft engines also have various * ratings, as do some nations' pilot certificates.

horsepower loading   See *power loading*.

horseshoe   Revetment having this shape in plan.

horseshoe vortex   *1* Combination of finite wing plus trailing vortex from each tip, forming circulation of approximate square-cornered U-shape in plan.

2 Semi-toroidal flow of ground sheet ahead of VTO hovering in headwind.

Horus   Hypersonic orbital return upper stage.

HOS   *1* Human operator simulator.

2 Helitow observation system.

hose   *1* Flexible conduit for fluid (liquid or gaseous).

2 Verb, to fire long burst of gunfire at or ahead of hostile or challenged aircraft.

HOSG   Helicopter Operations Study Group (UK).

host aircraft   That acting as carrier to test installation, or parent to UAV/drone.

host computer   One to which remote terminals or I/O devices are connected.

hostile   Target known or assumed to be enemy, esp. one seen on remote display.

hostile track   One which, based upon established criteria, is determined to be enemy airborne, ballistic or orbiting threat.

host nation   One whose territory houses infrastructure forming part of an international system or aircraft of a friendly air force.

Hot   High-subsonic, optically tracked, teleguided.

hot   *1* Fast-landing (colloq.).

2 With afterburner in operation.

3 With combustion in operation (pressure-jet tip drive or HTP/hydrocarbon rocket).

4 Strongly contaminated by fallout.

5 Hazardous to vicinity (see * *mission*).

6 Including propellant ignition (see * *test*).

7 Ready for firing (launch or static test).

8 High-temperature parts of gas turbine (see * *end*).

9 Thermally anti-iced, thus * prop.

10 In ground refuelling, with engine(s) running.

Hotac   Hotel! accommodation to provide sleep for civil flight crews.

hot and high   Airfield or helicopter platform where high altitude above MSL is combined with high ambient temperatures. These both reduce engine power and the lift from a wing or rotor.

Hotas   Hands on throttle and stick, design of cockpit of air-combat fighter so that pilot has every control switch, button or trigger needed in any combat on these two handholds [A adds aid].

hot bird   Fully functioning satellite.

hot box   Stolen item of avionics, for intelligence or commercial profit.

hot bucket   Turbine rotor blade (colloq., US).

hot chaff   Pyrophoric material for IRCM dispensed in covert [not emitting at visible wavelengths] manner.

hot day   Standard ISA condition for engine rating, aircraft performance certification and other temperature-dependent lawmaking.

hot dimpling   Dimpling of holes pre-heated to avoid cracking.

hotel mode   Aircraft on ground with full air-conditioning refrigeration.

hot end   Portion of gas-turbine subjected to high temperatures from combustion, normally all parts to rear of compressor (which itself can be hot enough to require special refractory alloys in final stages but is never included in this definition); called hot section in US.

hot fire   *Hot test.*

hot-gas recirculation   Any mechanism by which hot gas from a propulsion or lift jet can return to engine inlet.

hot gas system   One energized by gas bled from combustion of solid fuel, or from operating solid rocket motor.

hot gas valve   One used to control hot gas pressure for TVC purposes.

hot gun   Aircraft scrambled with loaded gun[s].

hot isostatic pressing   Temperature/pressure cycle for compacting sintered ceramic or encapsulated powders close to precise net shape.

hot leg   Presentation, or flypast, on which target simulates IR of hostile jet aircraft.

hot mike   Microphone continuously on transmit.

hot mission   Particular test (flight or otherwise) which involves hazards precluding other activity in same area.

Hotol   Horizontal takeoff and landing (spacecraft launch).

hot pit   Air refuelling in which fuel is transferred (UK colloq.).

hot rock   Inexperienced pilot eager to show off (colloq.).

hot rocket   One in which fuel is burned.

hot round   Rocket vehicle equipped with operative propulsion, in test programme where many are not.

HOTS   Hands on throttle and stick; Hotas is more usual.

hot section   Total of all parts of gas-turbine engine subjected to high temperatures from combustion of fuel, in modular engine exactly defined and normally including associated external dressing which in fact remains relatively cool.

hot shot   Method of igniting afterburner fuel [which would not otherwise ignite reliably, especially at high altitude] by spraying extra dose of fuel into engine combustion chamber to create hot flame which passes through turbine to afterburner; * lasts too short a time to damage turbine.

hot spot   *1* Place much hotter than environment, showing on 3–5 micron IR.

2 Local area on radar target giving intense return.

hot standby   Satellite (Comsat, navsat) available for instant use should operating satellite malfunction.

hot start   *1* Attempted start of gas turbine abandoned because of overtemperature indication.

2 Start, or attempted starts, of piston engine soon after previous run, often thwarted by vapour lock.

hot streak   Method of igniting afterburner by injecting fuel from special (normally inoperative) nozzle in main combustor, causing long flame to pass through turbine into afterburner primary zone. Can be synonymous with hot shot.

hot-stream nozzle   That from core engine in turbofan without mixer.

hot test   Static test of rocket engine in which actual firing takes place.

hotwell   Tank or portion of larger tank in which hot liquid collects.

hot winchback   Aircraft pulled into HAS with engines running.

hot-wire ammeter   One measuring current by I^2R heating of fine wire; hence * galvanometer, * voltmeter.

hot-wire anemometer   Measures airspeed or wind speed down to 10 mm/s by heating platinum wire to about 1,000°C and either measuring current I for constant T or resistance at constant I.

hot-wire ignition   Use of suddenly heated (but not exploded) resistance wire to set off rocket engine or gun ammunition.

hot-wire probe   Hot-wire anemometer.

hot-wire transducer   Detects and measures sound waves by change in resistance of heated wire.

hour   Mean solar = 3,600 s; sidereal * = 3,590.1704 s.

hour angle   Bearing of object on celestial sphere; angle at pole between hour circles of observer and object, abb HA (see *local ***, *Greenwich ***, *sidereal ***).

hour circle   Great circle of celestial sphere formed by projecting Earth meridian.

house aircraft   One used for research or development and property of user, normally a manufacturer.

HOV   Autopilot mode maintaining zero lateral/longitudinal groundspeed.

hover   *1* To fly (usually at low altitude) stationary relative to Earth, airspeed being that of local wind.

2 Exceptionally, to fly with zero airspeed, carried along at speed of wind in horizontal plane (while holding height constant).

3 Uncommon use, to be on station in geostationary orbit.

4 To operate or travel in air-cushion vehicle; obvious contradiction in terms.

Hovercraft   The original [registered] name of pioneer air-cushion vehicles, still popularly used without initial capital.

hovering ceiling   Greatest altitude at which helicopter under specified conditions can hover (2), normally defined as IGE (in ground effect) or OGE (out of ground effect).

hovering point   Location where V/STOL aircraft picks up or sets down load without landing.

hovering rig   Free-flying rig comprising open spaceframe in which are mounted jet lift engines and supporting systems planned for future aircraft to facilitate development of control systems in low hovering flight. These qualify as aircraft.

hover pit   Test facility for jet VTO or STOVL aircraft.

HOW   Handover word.

HOWD   Helicopter obstacle warning device.

howdah   Open gunner's cockpit above upper wing of large aircraft pre-1930.

Howe truss   One having upper and lower chords joined by verticals plus diagonals inclined outwards from bottom to top on each side of mid-point.

howgozit   Basic graphical plot of particular flight for planning purposes, always on basis of distance and with vertical scale flight level; includes weights, W/V and ambient T, and is amended as flight progresses (often arch.).

howler alert   Warns interphone handset is not properly seated.

Howls   Hostile-weapons location system; versatile phased-array radar.

HOWS   Helicopter obstacle warning system.

HP   *1* High pressure; suffix number (eg HP_8) denotes stage from which bleed air is taken.

2 Horsepower (undesirable).

3 Host processor.

4 Holding position, or pattern.

hp   *1* Horsepower.

2 High-pass filter.

HPA   *1* Human-powered aircraft; G adds London-based Group.

2 High-power amplifier or amplification.

hPa   Hectopascal.

HPAL   Human performance and limitations.

H-PAPI   Helicopter PAPI.

HPBB   High-power broadband.

HPBW   Half-power beamwidth.

HPC   *1* High-pressure compressor.

2 Calibrated pressure altitude.

HPD   *1* High-power discrimination, or discriminating.

2 High-PRF pulse-Doppler.

HPDA   Heavy propeller-driven aircraft.

HPEL   Hold[ing]-position edge light[s].

HPF   High-pass filter.

HPFL   High-power fiber/fibre laser (JTO, DoD).

HPFOTD   High-power fiber-optic towed decoy.

HPI   *1* High probability of intercept.

2 High-power illuminating, or illuminator (radar).

HPIR   HPI(1) receiver.

HPL   High-power laser (B adds blinding).

HPM   High-power microwave[s].

HPOC   High power offload to Cass (USN).

HPOT   High-pressure oxidiser turbopump.

HPOX   High-pressure oxygen.
HPR   *1* High-performance rescue (vehicle).
   2 High-power relay.
HPres, HPRES   Pressure altitude.
HPRF   *1* High pulse-repetition frequency.
   2 High-power radio frequency.
HPRL   Human Performance Research Laboratory (NASA).
HPRP   High-power reporting point (radar).
HPS   *1* Helmet pointing system.
   2 Horizon and pitch scale.
   3 Hardened personnel shelter.
   4 High-pressure sodium.
HPSC   House Permanent Select Committee; I adds on Intelligence (US).
HPSOV   High-pressure shut-off valve.
HPSS   High-pressure single-spool.
HPT   *1* High-pressure turbine.
   2 High-precision thermostat.
   3 High-power transmission.
HPTC   High-performance technical computing.
HPTE   High-performance turbine engine[s], or engined.
HPTIC   High-performance thermal-imaging committee (DERA/industry).
HPTS   High-power transmit[ter] set.
HPW   High-pressure water (snow removal).
HPX   HP-spool shaft-horsepower extraction.
HPZ   Helicopter protected zone.
hpz   Hectopièze[s], non-SI unit of pressure = 10^5 Pa.
HQ   Headquarters; -C^2 adds Command and Control, MSS mission-support system.
HQR   Handling-qualities rating.
Hr   Hour[s].
HR   *1* Helicopter route.
   2 Helicopter, transport (USN category 1944–62).
   3 Radar altitude.
   4 Hear, here or hours.
HRA   Highlands [of Scotland] restricted area.
H$_{RAD}$ Radar altitude measured in landing phase, = wheel height.
HRB   Horizon reference bar.
HRC   *1* Historical-records container.
   2 Helmet-release connector.
HRD   Home-routing domain.
HRDF   High-resolution direction-finding, or finder.
HRE   Hypersonic ramjet engine.
HRG   High-resolution geometric.
HRGM   High-resolution ground map.
HRIR   High-resolution IR radiometer.
HRL   Human Resources Laboratory (USAF).
HRN   High-resolution navigation mode.
HROD   High rate of descent.
HRP   Headset-receptacle panel.
HRR   *1* High-resolution radar.
   2 High-range resolution; GMTI adds ground MTI.
HRS   *1* Horizon-reference system.
   2 High-resolution stereoscopic.
   3 Hemispheric radar system.
hrs   Hours.
HRSAR   High-resolution SAR(2).
HRSCMR   High-resolution surface-composition mapping radiometer.
HRSI   High-temperature reusable surface, or silica,

insulation; Si-fibre tiles protected by fretted borosilicate glass with pigment for desired absorption/emission ratio.
HRST   Hyperspectral remote-sensing technology.
HRT   High-resolution textured [simulator].
HRTS   High-resolution telescope and spectrograph.
HRZ   Helicopter restricted zone.
HRZN   Horizon.
HS   *1* Scheduled hours only.
   2 Anti-submarine helicopter designation prefix (USN 1951–62).
   3 Helicopter Squadron.
   4 Hold short (command).
Hs   Maximum shoulder height of tyre [tire].
HSA   *1* Horizontal-stabilizer actuator.
   2 Homeland Security Agency (US, 2002).
HSAB   *1* Heavy-store[s] adapter beam.
   2 NDB with automatic weather broadcasts.
HSAC   Helicopter Safety Advisory Committee/Conference (US).
HSACE   HSA control electronics.
HSAD   High-speed anti-radiation demonstration.
HSARD   High-speed anti-radiation demonstration.
HSAS   Homeland security advisory system.
HSCS   High-specific creep-strength; M adds material[s].
HSCT   High-speed commercial transport; usually = M2.2-M5.
HSCU   Horizontal-stabilizer control unit.
HSD   *1* Hard surface, dry (runway).
   2 Horizontal-situation display.
   3 Homeland Security and Defense (US 2001).
HSDB   High-speed data bus.
HSDC   High-speed data channel; E adds extension.
HSE   Health and Safety Executive (UK).
HSEC   High-stability engine control.
HSF   Half sampling frequency.
HSFD   High-speed flight demonstration.
HSFRS   High-Speed Flight Research Station (NACA Muroc, became NASA Dryden).
HSF-RSAD   High-speed [film] frame, relay service access device.
HSI   *1* Horizontal-situation indicator.
   2 Hot-section inspection.
   3 Hours since inspection.
   4 Hyperspectral imagery.
HSIC   High-speed integrated circuit.
HSIT   Hardware/software integration test.
HSL   Heading select.
HSLA   High-strength low alloy steel.
HSLLADS   High-speed low-level air-delivery system.
HSLV   High-speed low-voltage (LSI).
HSM   *1* Hard-structure munition, conventional munition intended to have maximum effect on hardened targets.
   2 High-speed machining.
   3 Hardware security module.
HSMF   High-strength modified fluoropolymer.
HSN   Hot-stream nozzle.
HSO   Homeland Security Office (counter-terrorism, UK).
HSOS   Helicopter stabilized optical sight.
HSP   *1* High-speed [1,300 m/s] penetrator, or penetration; T adds technology.
   2 Head-shrinkable polyolefin.
   3 Hard spark[ing] plug.

HSR *1* High-speed research.
  2 High-stability reference.
  3 High sink-rate.
HSRP Hot-standby routing protocol.
HSS *1* Helicopter support ship.
  2 Helicopter stabiblized sight.
  3 Homeland security sensor, or suite.
  4 Hypersonic sound.
HSSL Helicopter self-screening launcher.
HSSM High-speed [Mach 8] strike missile.
HSSS Helicopter secure-speech system.
HST *1* Hypersonic transport.
  2 Heat-shrinkable thermoplastics.
  3 High-speed tunnel, or turnoff.
HSTA Horizontal-stabilizer trim actuator.
HSTF High-strength toughened fluoropolymer.
HSU Hartridge smoke unit.
HSV Highly supersonic vehicle.
HSVD Horizontal-situation video display.
HSWT High-speed wind tunnel.
HT *1* High turbine (US for HP turbine).
  2 Hard time, thus * life.
  3 Helicopter, training (USN category 1948–62, role prefix, UK).
  4 History tape.
H/T Hub-tip ratio.
ht Height.
h.t. High tension (electrical).
HTA *1* High-time aircraft of type.
  2 Horizontal target attack.
  3 Heavier than air.
  4 Hand-entered terrain altitude.
HTAD Hard-Target Attack Division (USAF).
HTADS Helmet target-acquisition and designation system.
HTC *1* Hard-target capability.
  2 See HTCU.
  3 Human to computer.
  4 Highest two-way channel.
HTCU Hover trim control unit.
HTD Hard-target defeat.
H_{TD} Wheel height at touchdown referenced to depressed surface.
HTDU High-temperature demonstration unit.
HTFD Hard-target functional defeat.
HTI Horizontal technology integration.
HTIFT Hard-target influence-fuze technology.
HTK Hit to kill.
HTKP Hard-target kill probability.
HTL High-threshold logic.
HTLC High-temperature load calibration.
HTMPFG One of earliest of at least 35 English-language safety acronyms = hood/harness, throttle/trim, mixture, pitch, fuel/friction-nut/flaps, gills/gyros/possibly goggles.
HTNS Helicopter tactical navigation system.
HTOVL Horizontal (conventional) takeoff, vertical landing.
HTP *1* High-test peroxide.
  2 High-tensile polyimide.
  3 Horizontal tailplane.
HTPB Hydroxyl-terminated polybutadiene, rocket propellant.
HTS *1* High-tensile sheet (CFRP).

  2 Harm targeting system.
  3 Hydraulic test stand.
  4 High-temperature superconductor.
HTSC High-temperature superconductor.
HTSF Hard-target smart fuze.
HTTB High-technology testbed (Lockheed).
HTU Hand-held terminal unit.
HTUFT Helicopter[s] taken up from trade, ie commandeered [also Hetuft].
HTZ Helicopter traffic zone.
HU *1* Helicopter, utility (USN, USA, 1950-62).
  2 Helmet unit, part of HMD(1).
hub *1* Strictly, those structural members required to hold together blades of propeller or rotor; in practice meaning has come to include all central portions, including pitch and other mechanisms, de-icing and instrumentation, but not spinner or other fairing.
  2 Missile body section containing attachments for cruciform wings.
  3 Chief airport of major city.
  4 Head office or largest plant of company.
hubbed, hubbing Parked at hub (3).
hub drive Driven by shaft input rather than tip jet reaction.
hub/tip ratio, H/T Ratio of radius at root of gas turbine fan, compressor or turbine blade, to radius at tip (usually inside tip shroud if fitted).
Hucks starter Device, often locally made, for starting piston engines. On a car or truck chassis a drive is taken from the engine to a long shaft whose height and inclination can be varied. This terminates in lateral pins which engage with a bayonet clutch on the end of the propeller shaft (suggest obs.).
HUCP Highest useful compression pressure.
HUCR Highest useful compression ratio.
Hud, HUD Head-up display; C adds camera, or computer.
Hudwac HUD weapon-aiming computer.
Hudwass HUD weapon-aiming subsystem.
Hufford Family of large machine tools which grip ductile sheet at each end in jaws on hydraulic rams mounted on pivoting platforms and stretch it over male die itself thrust forward on rams.
HUGS Head-up guidance system.
HULD Hardened unit-load device.
hull *1* Fuselage of flying boat.
  2 Main body of rigid airship.
  3 For insurance purposes, complete aircraft as defined in policy, ignoring any load or persons on board.
hull insurance That covering capital value (assigned at operative time, neither original first cost nor replacement cost) of aircraft.
hull lines Set of plan views of LWLs of marine aircraft or float.
hull loss Aircraft written off.
Hultec Hull to emitter correlation, EW/ESM software programs.
HUM, Hum Health and usage monitor, or monitoring; C adds computer, S adds system or and sensing, U unit.
humidity Wetness of gas, esp. amount of water vapour present in air; absolute * is mass of vapour in unit volume g/m^3, specific * is mass of vapour in unit mass of dry air g/kg, relative is percentage degree of saturation = 100 times actual vapour pressure divided by SVP.

humidity mixing ratio   Specific humidity.

Humint   Human intelligence, ie countermeasures against human decision-taking.

hump   Region where, in takeoff of marine aircraft, T/W –R/W is least; T is total thrust, W weight and R total resistance.

hump speed   That of marine aircraft (seaplane, flying boat) or ACV at which total resistance of water is greatest; in case of aircraft this speed [in US called V_H] is typically about half V_{us}.

Hums   Health and usage (and performance) monitoring system.

hundred per cent aircraft   One that exactly measures up to published performance.

hundredweight   Non-SI unit of mass, abb. cwt = 112lb, 50.8023kg; in US [called short *, sh.cwt.] = 100 lb, 45.3592 kg.

hung   Hung start.

hung round   Rocket or other missile which fails to release from aircraft.

hung stall   After compressor stall, engine fails to recover immediately.

hung start   Starting of main turbine engine which, for any reason, automatically or under manual control is arrested after ignition but before self-sustaining speed is reached.

Hunter   Head-up navigation and targeting equipment for retrofit.

hunter/killer   *1* Aircraft or other platform able to seek out and kill submarines unaided.

2 Pair of ASW aircraft, one with sensors, the other with weapons.

3 Co-ordinated task force comprising aircraft and surface forces combining in ASW mission.

hunting   Continual steady oscillation about neutral point; governed speed, governed position, desired flight attitude or other target regime; manifest in cyclic phugoids in aircraft pitch or yaw (rare in roll), shuttling of hydraulic spool valve, ceaseless rising and falling of speed of rotating machine or visible oscillation of instrument needle.

hunting tooth   Gear ratios which ensure that each tooth engages between a different pair of teeth on each revolution, to even out wear.

Huntsville   Alabama location of MSFC, USSRC.

HURCN   Hurricane.

hurricane   One name for tropical revolving storm, usually defined as one with winds whose mean speed exceeds 64 kt, 119 km/h.

hush house   Noise-suppressing testbed for jet engines (colloq.).

hushkit   Supplied by manufacturer to operator to quieten engine, invariably turbojet or turbofan already in service; often includes inlet acoustic liners, liners for tailpipe and new nozzle with longer periphery and faster mixing.

HV, h.v.   *1* High voltage.

2 High vacuum.

H/V   Height/velocity (plotted curve).

HVA   *1* Horizontal viewing arc.

2 High-value asset[s].

HVAC   Heating, ventilating and air-conditioning.

HVAP   High-velocity armour-piercing.

HVD   *1* Helmet visor display.

2 Helicopter video downlink.

HVDF   Hf/vhf D/F.

HVG   Hypervelocity gun.

HVI   Helmet/vehicle interface.

HVM   Hypervelocity missile.

HVML   High-volume minelayer.

HVOF   High-velocity oxygen/fuel plasma spray.

HVOR   High-altitude VOR.

HVPS   *1* High-voltage power supply [U adds unit].

2 High-volume precipitation sensor.

HVT   High-value target; A adds acquisition.

HVU   High-value unit [surface target].

HVY   Heavy.

HVZL   Hauptverwaltung der Zivilen Luftfahrt (DDR, East Germany).

HW   Hardware.

h.w.   Half-wave.

HWCI   Hardware configuration item.

HWD   Heavy-weight deflectometer.

HWIL   Hardware in the loop.

HWT   Hypersonic weapons technology.

HWVR   However.

HX, H_x   *1* Airfield has variable working hours.

2 Heat exchanger.

Hyblum   Long-established Al-Mg-Si alloys.

hybrid bearing   Combining bearing with rolling elements (ball, roller, needle) encased in free-rotating journal supported by fluid film. Typically Sinide elements in steel race.

hybrid buoyant aircraft   Those built so far, or planned, are vectored-thrust airships.

hybrid chip   Combining digital and analog functions.

hybrid composite   Matrix reinforced by fibres of two different types.

hybrid computer   One using digital techniques for large and precise arithmetic and logic beyond scope of analog machines, and analog wherever possible for highest computational speed; invariably a good compromise for large simulations and other specialized tasks [Ed.'s opinion].

hybrid display   One in which alphanumerics and symbology are cursively written on top of a raster background.

hybrid electronics   Combination in single integrated circuit of epitaxial monolithic or thin-film with one or more discrete devices.

hybrid fan   Refined form of tandem-fan lift/cruise engine which in lift mode blows entire fan airflow through twin vectored forward nozzles and separately induced core flow through aft vectored nozzle.

hybrid FCS   Flight-control system with digital outer loop and analog inner loop.

hybrid helicopter   Aerodyne with VTOL capability conferred by helicopter rotor[s] but which cruises (flies in translation mode) as an aeroplane.

hybrid IC, hybrid package   Hybrid electronics.

hybrid laminar flow   Combination of aerodynamic design and suction.

hybrid propulsion   *1* Aircraft propelled by two or more dissimilar species of prime mover; eg turboprop plus jet, or turbojet plus rocket.

2 Single propulsion engine capable of operating in two distinct modes; eg turborocket or ram rocket.

hybrid RAM   RAM(2), especially forming integral part

of airframe, combining two or more techniques to give broader bandwidth in thinner layer, eg magnetic/CA, graded dielectric/CA, etc.

hybrid rocket One using both liquid and solid propellants simultaneously; usual arrangement is solid fuel and liquid oxidant.

hybrid solar array Part folding, part flexible.

hybrid trajectory Any space trajectory intermediate between that for minimum energy or minimum time and alternatives offering greater payloads, longer launch windows or other advantages.

hybrid wave EM wave in waveguide having both magnetic and electric components in plane of propagation.

Hycatrol One of several trade names (FPT Industries) for rubber/metal bonded structures.

Hycorder Hyperspectral covered-lantern optical recognition device recorder.

hyd Hydraulic[s].

Hydim Hydraulic interface module.

hydrant dispenser Installation under apron or finger/gate area stand for refuelling aircraft without need for tanker vehicles.

hydrant pit Below-ground compartment, normally covered, housing connections and controls for fuel, hydraulic, lube oil or other liquid supply.

hydraulic catapult *Catapult*.

hydraulicing Abnormal resistance to movement of machine, esp. piston engine, caused by hydraulic lock in lower cylinders part-full of essentially incompressible oil; can cause serious damage.

hydraulic lock Use, frequently inadvertent, of essentially incompressible liquid to prevent movement of mechanical part.

hydraulic motor Source of mechanical power, usually rotary, driven hydraulically.

hydraulic power unit Source of power to energize hydraulic system, eg when main engines inoperative.

hydraulics Science of liquids either in motion (hydrodynamics) or as media for transmitting forces (hydrostatics). In aerospace generally science of nearly incompressible liquids enclosed in closed-circuit pipe systems at high pressure and used both to apply forces, with little fluid motion, and supply power, with large fluid motion. Media originally mineral oils, today also phosphate esters, cholorinated silicones, silicate esters and (supersonic and missiles) alkyl silicate esters.

hydraulic seal Total seal between two annular spaces in engine formed by ring-fins or flanges projecting into ring of oil created by centrifugal force.

hydraulic starting Used in small [eg, missile and UAV] jet engines: external supply feeds pressure to hydraulic motor; once engine is started, motor functions as hydraulic pump.

hydraulic system Complete aircraft installation comprising closed circuits of piping, engine-driven pumps, accumulators, valves, heat exchangers, filters and, usually, emergency input such as RAT or MEPU; normally divided into at least two systems with maximum degree of independence. Each system is assigned task of driving selected items by linear actuators, motors or other output devices.

hydrazine Family of chemicals, mostly colourless liquids, often corrosive; basic member is *, $(NH_2)_2$;

common rocket fuel is unsymmetrical dimethyl *, $NH_2N(CH_3)_2$; another is monomethyl *, $NH_2HH.CH_3$.

hydrobooster Hydraulic power unit used in boosted (not fully powered) flight-control system (colloq.).

hydrocarbon Compound of hydrogen and oxygen only; some millions are known, including all derivatives of petroleum, which in product form often have other elements added for specific purposes. Many aviation fuels are alkanes (paraffins), which are open chains with carbon atoms having single-valence bonds; first six members, with one to six carbon atoms respectively, are methane, ethane, propane, butane, pentane and hexane. These are prefixed n (normal), distinguished from prefix iso of more reactive branched alkanes. Alkenes have one carbon with double bond, based on ethylene $(CH_2)_2$. Aromatic * series are based on hexagon ring of benzene C_6H_6 with three double bonds.

hydrodynamics Science of fluid motion.

hydroflap Water rudder on flying boat.

hydrofoil Lifting surface operating in water. As well as being vehicles in their own right, surface-piercing and ladder * have been used on marine aircraft.

hydroforming Shaping parts by fluid pressure, esp. of thin-foil items.

hydrogen Least-dense element, comprising 88 per cent of atoms in Universe, LH_2 (liquid *) has density 77.0 gl^{-1}, 4.806 lb/ft^3 at 13.8K (the triple point) and 70.8 gl^{-1} at BPt of 20.28K, −253°C, where it becomes GH_2 (gaseous *) with density 0.0008988 (0.08988 gl^{-1}) or 0.00561 lb/ft^3; isotopes are bivalent deuterium, trivalent tritium.

hydrogenation Causing to combine with hydrogen, esp. at high pressure and in presence of catalyst such as nickel or platinum, in conversion of crude petroleum distillates into tailored fuels and other products. * of coal also important to future aviation.

hydrogen bomb So-called H-bomb, or thermonuclear weapon TN or TNW; comprises NW surrounded by lithium deuteride (LiD, the lithium being isotope Li-6) and a little tritium T. Triggering the NW emits neutrons which instantly convert the Li-6 into H+He-3+T. The He-3 and T then combine with remaining D to form more He and more neutrons, which also convert the U-238 bomb case into Pu-239, causing an additional (fission) reaction.

hydrogen bus Airport airside buses are among the first vehicles to be powered by hydrogen, usually GH_2 stored on vehicle roof.

hydrogen economy Hypothetical future in which Earth's limited reserves of petroleum are replaced by gaseous and liquid hydrogen. No new technology is needed, but see next.

hydrogen fusion Essentially limitless power could be unlocked if mankind could emulate the Sun and build a facility which continuously converted hydrogen into helium. Conversion would yield 630,000,000,000J of energy per gram.

hydroglider Glider with marine alighting gear.

hydrograph Hydrometer hard-copy output.

hydrokinetics Science of liquids in motion.

hydrolapse Rate of decrease of atmospheric water vapour with altitude.

hydromagnetics See *MHD*.

hydromechanical logic Performed with mechanical elements wherein information exists as hydraulic flows/pressures.

hydrometer   Instrument for determining density of liquids.

hydrometeor   *1* Any atmospheric water in solid or liquid form; all precipitation, fog, dew, frost etc.

    2 Phenomena dependent upon (1).

hydrophobic   Rain-repellant.

hydrophilic   Affinity for water, so acts as dessicant (various spellings).

hydroplane   Light boat which skims water surface on planing bottom when at high speed; erroneously misused for seaplane and/or hydrofoil.

hydroport   Inland airport for marine aircraft.

hydropress   Diverse family of hydraulic presses widely used in aerospace with large-area platen on which are mounted large or multiple tools around which sheet is shaped by rubber pad or mating tools.

hydroskis   Planing surface, usually in left/right pair and retractable, used for takeoff and landing of certain marine aircraft; have little buoyancy, so ski aircraft rests on water like flying boat when at rest.

hydrosphere   Water resting on or within crust of Earth or other body, excluding that in atmosphere.

hydrostatic bearing   Spinning shaft, sphere or other mass is supported (usually radially and axially) by filtered gas or liquid dynamic reaction, eliminating contact between fixed and moving solid surfaces.

hydrostatic drive   Transmits power, usually between rotating shafts, by pumping hydraulic fluid round closed circuit; both pump and motor usually have stroke or output infinitely variable down to zero, giving perfectly flexible dynamic link from zero to maximum output power.

hydrostatic equation   Applies when secondary effects (Earth curvature, friction, coriolis etc) ignored, leaving $dp/dz = -\rho g$ where p is pressure, z geometric height and ρ density.

hydrostatic extrusion   Advanced technique for extrusion of steels and other materials, usually at room temperature by forcing through die under extreme hydraulic pressure.

hydrostatic fuze   Triggered by depth-dependent water pressure.

hydrostatic test   Test of container (fuselage, solid rocket case) under high pressure using water or other liquid to minimize stored energy.

Hydrus   FBMS ship/shore communications system.

Hydulignum   Trade name (Hordern-Richmond) for densified wood.

Hydyne   Storable rocket fuel in various formulations based upon 60% UDMH and 40% diethylenetriamine.

hyetograph   *1* Recording rain gauge.

    2 Annual rainfall chart.

Hyfil   Trade name for family of CFRP raw materials marketed in standard forms or used for inhouse production (Rolls-Royce).

Hyflex   Hypersonic flight experiment.

HyFly   Programme of Mach-6 research (Darpa/ONR).

hygrograph   *1* Output from a recording hygrometer.

    2 Instrument for recording humidity, traditionally with pen positioned by bundle of stretched human hair.

hygrometer   Determines atmospheric humidity (strictly, wet and dry bulb * is psychrometer).

hygroscopic   Eager to absorb moisture.

Hy-Jet/I,II,III,IV   Ester-based non-inflammable hydraulic fluids (Chevron).

Hy-Lite   Hyperspectral long-wave imager for the tactical environment.

Hylite   British titanium alloys (+ Sn, Zr, Al); creep-resistant.

hyperabrupt   Device, eg diode, tailored to most rapid possible action.

hyperacoustic zone   Region above 100 km (62 miles) where mean free path approximates to wavelengths of sound; limit of sound propagation.

hyperbaric   *1* Having atmospheric pressure or oxygen concentration greater than normal sea-level.

    2 Internal body pressures greater than ambient.

hyperbola   Conic section obtained by plane cutting both nappes (normal right circular cone and its mirror-image inverted above); locus of points whose distances from two foci have constant difference, standard equation $x^2/a^2 - y^2/b^2 = 1$. Because of constant difference in distance from two foci, possible to base families of navaids on keyed radio emissions from two fixed stations.

hyperbolic navaids   Based upon synchronized emissions from fixed ground stations, often called master and slave(s), which are received by aircraft at time differences which yield lines of constant time (ie range) difference in form of hyperbolic position lines. First were Gee, Loran and Decca, later developed to give instantaneous readout of position or moving-map display.

hyperbolic error   That due to assumption that waves received at all antennas of an interferometer baseline are travelling in parallel directions.

hyperbolic frequencies   Measured in several tens of GHz.

hyperbolic re-entry   At hyperbolic speed.

hyperbolic speed   Sufficient to escape from Solar System; on Earth trajectory away from Sun about 40,597 km/h, 25,226 mph.

hypergolic   Of rocket propellants, those which ignite spontaneously when mixed.

hypermetropia   Long-sightedness.

hypermixing nozzle   In ejector-lift system, row of nozzles which alternately deflect jets in opposite directions to create large vortices promoting rapid mixing.

hyperoxia   Excess oxygen in the blood.

hypersat   Loose term for advanced small satellite[s].

hypersonic   Having Mach number exceeding 5 [another authority, M8 to M12].

Hypersonic National Plan   Co-ordinates DoD, NASA, industry and academia (US).

hypersound   Frequency greater than 10^9 Hz.

hyperspectral   Operating in several electromagnetic bands, dividing each colour into a separate channel to give a unique signature of absorption and emission.

hypertension   High blood pressure.

hyperventilation   Overbreathing; specif. reduced CO_2 causing * syndrome, dizziness, fainting, convulsions.

hypobaric   Having atmospheric pressure much less than normal at sea level.

hypocapnia CO_2 deficiency in blood.

hypoid   Bevel or helical gears transmitting power with some tooth-sliding action between shafts neither parallel not intersecting.

hypoventilation   Underbreathing.

hypoxaemia   Condition resulting from hypoxia.

hypoxia O_2 deficiency in blood, from whatever cause.

hypsometric tints   Colour gradations chosen for contrast by natural or artificial illumination (eg. for contour bands on topographic map).

HYR   Higher.

Hyrat   Hydraulic (pump driven by) ram-air turbine.

HySET   Hydrocarbon-fuel scramjet engine technology.

H$_Y$SID   Hypersonic systems integrated demonstrator.

hysteresis   *1* Generally, condition exhibited by system whose state results from previous history, specif. one whose instantaneous values lag behind prediction.

2 In ferromagnetic and some other materials, lagging of magnetic flux density T behind magnetic field strength A/m causing it (see * *loop*).

3 Effects caused by internal friction in elastic material undergoing varying (esp. rapidly oscillating) stress.

4 Lag in instrument indication; eg that of barometric altimeter in dive or climb.

hysteresis loop   Plot of magnetising field strength against magnetic flux density for ferromagnetic material; traditionally called B/H curve because flux density was called induction, symbol B, measured in gauss, and field strength (formerly in oersteds) was identified by symbol H; today's units are (field strength) A/m and (flux density) T.

HYT   High year of tenure (waiver programme).

Hytech   Hypersonic technology.

Hythe   Ciné camera mounted on Scarff ring or similar base for gunnery instruction; full name * camera gun.

Hytral   Noise-absorbent panels of sandwich construction.

Hyways   Hybrids with advanced yield for surveillance.

HZ   Haze (ICAO).

Hz   Hertz, SI unit of frequency, = cycles per second.

HZ Anlage   System of using a piston-engine solely to drive supercharger feeding propulsion engines [G, WW2].

HZS   Hrvatski Zrakoplovni Savez, aeronautical sport federation (Croatia).

I

I *1* Electric current.

2 Moment of inertia. suffixes include XX roll, YY pitch, ZZ yaw, R rotor [of engine or helicopter etc], ω angular momentum.

3 Prefix, direct-injection engine (US).

4 Luminous intensity.

5 Total heat content.

6 Total impulse, usually non-dimensional.

7 Intensity of turbulence.

8 Immigration.

9 Initial approach.

10 Instrument.

11 In-line.

12 Interrupted.

i *1* Intensity of rainfall.

2 'Square root of minus 1'.

I-band EM radiation, 37.5–30 mm, 8–10 GHz.

I-beam One of I section.

I-display When radar aerial pointed at target latter appears as circle at radius proportional to range; when aerial points away from target latter appears as segment showing magnitude/ direction of error.

I-local Index of local warnings [weather].

I-section Structural beam with vertical web and flat upper and lower booms.

I² Image-intensifying.

I²F Intelligent influence fuze.

I²S *1* Infra-red imaging system.

2 Integrated information system, cabin wireless LAN for passengers based on 802.11 and ARINC–763.

IA *1* Initial approach (FAA).

2 Inspection authorization (FAA).

3 Input axis.

4 Imagery analysis.

5 Initial attack (firefighting).

Ia Anode current.

IAA *1* International Academy of Astronautics.

2 International Aerospace Abstracts (AIAA).

3 Irish Aviation Authority (1994).

4 See *IAA (E)* and *IAA(O)*.

IAAA International Airforwarder and Agents Association.

IAAC *1* International Agricultural Aviation Centre (HQ in UK).

2 International Association of Aircargo Consolidators.

3 Information Assurance Advisory Council (UK).

IAAE Institution of Automotive & Aeronautical Engineers (Australia).

IAA(E) Inspector of Air Accidents (Engineering) [DETR, UK].

IAAFA Inter-American Air Forces Academy (US).

IAAH International Association of Aviation Historians.

IAAI *1* International Airports Authority of India.

2 Indonesian Aeronautical and Astronautical Institute.

IAA(O) Inspector of Air Accidents (Operations) [DETR, UK]

IAARC International Administrative Aeronautical Radio Conference.

IAASM International Academy of Aviation & Space Medicine.

IAATC International Association of Air Training Centers.

IAB Investment Appraisals Board (senior procurement body, MoD, UK).

IABA International Association of Aircraft Brokers and Agents.

IABCS Integrated aircraft brake control system.

IABG Industrieanlagen Betriebs GmbH (G).

IAC *1* Intelligence Analysis Center (US).

2 International Aerobatics Club (office Oshkosh, Wisconsin).

3 Instrument-approach chart.

4 Integrated avionics computer.

5 See *TIACA*.

6 Instituto de Aviação Civil (Brazil).

7 Irish Aviation Council (Greystones, Co. Wicklow).

IACA International Air Carrier Association (office UK).

IACAC International Association of Civil Aviation Chaplains (office JFK airport, NY).

IAC/AV Interstate Aviation Committee/Aviation Register (R, CIS).

IACC Inter-agency Air Cartographic Committee (US).

IACD Intelligent adviser [not advisor] capability demonstrator.

IACES International Air Cushion Engineering Society.

IACG Inter-agency consultative group (US, USSR, Europe, Japan, space science).

IACP Independent Association of Continental Pilots.

IACS Integrated avionics [also air-traffic] control system.

IACSP International Aeronautical Communications Service Provider.

IACS Integrated avionics [also air-traffic] control system.

IACZ Inter Airline Club Zurich (Int.).

IAD *1* Fighter Division (R).

2 Integrated antenna detector.

IADB Inter-American Defense Board.

IADS Integrated air-defence system.

IAE *1* Instituto de Aeronautica e Espaco, previously Instituto de Atividades Espaciais (Brazil).

2 Institute for Advancement in Engineering (US).

IAEA *1* Indian Air Engineers' Association.

2 International Atomic Energy Agency.

IAEM Instituto dos Altos Estudos Militares (Port.).

IAF *1* International Astronautical Federation.

2 Initial approach fix.

IAFA International Airfreight Forwarders' Association (J).

IA5 International Alphabet No. 5.

IAFU Improved assault fire unit (USA).

IAGC Instantaneous automatic gain control.

IAGS *1* Inter-American Geodetic Survey.

2 Integrated Arinc ground station.

IAHA International Air Handling Association.

IAHFR Improved airborne high-frequency radio: /NOE adds nap-of-the-Earth.

IAIN International Association of Institutes of Navigation.

IAIS Industrial Aerodynamics Information Service.

IAL International Airtraffic League.

IALA International Association of Lighthouse Authorities.

IALCE International Airlift Control Element (NATO).

IALPA Irish ALPA.

IAM *1* International Association of Machinists (US). See *IAMAW*.

2 Institute of Aviation Medicine (UK).

3 Inertially-aided munition, free-fall device (Northrop).

4 Instrument approach minima.

5 Institute for Advanced Materials (Petten, Neth.).

6 Initial approach mode.

IAMAW International Association of Machinists and Aerospace Workers (US).

IAMS Integrated armament management system.

IANA Internet Assigned Number Authority.

IANC *1* International Airlines Navigators Council.

2 International Air Navigation Convention (from 1919).

I&C, I&CO Installation and checkout.

I&E Installations and Environement (US DoD).

I&M Improvement and modernization.

I&T Integration and tape.

I&W Indications and warnings.

IAO *1* In and out of cloud.

2 Information Awareness Office (Darpa).

IAOA Indicated angle of attack.

IAOPA International Council of AOPAs (1962 –, office Frederick, MD).

IAP *1* Imagery architecture plan.

2 Fighter aviation regiment (USSR, R).

3 Instrument approach procedure (see * *chart*).

4 International airport.

5 Integrated actuation pack[age], usually an electrically driven pump.

6 Institution of Analysts and Programmers (UK).

IAPA *1* International Airline Passenger Association (Croydon, UK).

2 International Aviation Photographers' Associations.

3 Instrument-approach procedures automation.

IAPC *1* International Airport Planning Consortium (UK).

2 Instrument approach procedure chart.

IAPS *1* Ion auxiliary propulsion system.

2 Integrated avionics processing system.

IA-PVO Fighter aviation, air defence of the homeland (USSR, R).

IAR *1* Idle area reset.

2 Institute for Aerospace Research (Canada).

3 Inspection, or intersection, of air routes.

IARO International Air Rail Organisation (office Heathrow, UK).

IARP Inverse-address resolution protocol.

IAS *1* Institut Aéronautique et Spatial (F).

2 Indicated airspeed (see *airspeed*).

3 Integrated acoustic structure.

4 Interplanetary automated shuttle.

5 Impact attentuation system.

6 Ideal aerofoil/airfoil shape.

7 Integrated airport systems.

8 Institute of the Aeronautical Sciences, changed to Aerospace, and in 1962 to AIAA.

IASA *1* International Air Shipping Association.

2 International aviation safety assessment (FAA).

IASB Institut d'Aéronomie Spatiale de Belgique.

IASC *CASI (1)*.

IASM Institute of Aerospace Safety and Management (U of S California 1953, became ISSM).

IASS International air-safety seminars.

IAT *1* International atomic time.

2 International Association of Touristic managers.

IATA International Air Transport Association, 275 members, offices Montreal/Geneva, since 1945.

IATF International Airline Trust Fund.

IATP *1* International Airline Technical pool.

2 Inter-American Training Plan 1941–46.

IATS Intermediate automatic test system.

IATSC International Aeronautical Telecommunications Switching Centre.

IAU *1* International Accounting Unit.

2 International Astronomical Union.

3 Interface adaptor unit.

4 Integrated avionics unit.

IAW, i.a.w. In accordance with.

IAWA International Aviation Women's Association.

IAWG Industrial Avionics Working Group (UK).

IB *1* Inbound.

2 Incendiary bomb.

3 Ion beam, thus * erosion, * engine.

4 Interconnecting box.

I_b Burning-time impulse.

IBA *1* Inbound boom avoidance.

2 Fighter/bomber aviation (R).

3 International Bureau of Aviation (Europe).

IBAA International Business Aircraft Association (Europe).

IBAC International Business Aviation Council (Int.).

IBC *1* Individual blade control.

2 Intelligent bandwidth compression.

3 Integrated broadband communications.

IBCOS In-built checkout system.

IBCT Interim Brigade Combat Team (USA).

IBE Indirect battlefield effect.

Iberlant Iberia–Atlantic area (NATO).

IBF Internally blown flap.

IBIS, Ibis *1* ICAO birdstrike information system.

2 Israeli boost-intercept system.

IBKV Integrated helicopter avionics system (R).

IBLS Integrity beacon landing system.

IBN Image-based navigation.

IBn Identity (or identification) beacon.

IBP Iron-ball paint.

IBR *1* Integrally bladed rotor (gas turbine).

2 Intra-base radio.

3 Integrated baseline review.

4 Involuntary boarding refusal.

IBRD *1* Inflated ballute retarding device.

2 International Bank for Reconstruction and Development.

Ibris Interactive baggage reconciliation information system.

IBS *1* Integrated bridge system.

2 Integrated broadcast service, for correlating theatre intelligence for all US forces.

IBU Independent back-up unit.

IC *1* Internal combustion.

2 Interceptor controller.

3 Indirect cycle (sometimes i.c.).

4 Integrated circuit.

5 Intelligence community.

6 Ice crystals.

7 Inter-Cabinet.

8 Incident commander (aerial firefighting).

I/C, I & C *1* Installation and checkout.

2 Interface and control.

i.c. *1* Integrated circuit.

2 Internal combustion (IC more common).

i/c *1* In charge (UK usage).

2 Intercom.

ICA *1* International Committee of Aerospace Activities.

2 Initial cruise altitude.

3 International Cartographic Association.

ICAA *1* International Civil Airports Association.

2 International Committtee of Aerospace Activities.

ICAAS Integrated controls and avionics for air superiority (USAF).

ICADS, Icads Individual combat-aircrew display system (laptop training device).

ICAEA International Civil Aviation English [language] Association.

ICAF *1* International Committee on Aeronautical Fatigue.

2 Industrial College of the Armed Forces (US).

ICAI Intelligent computer-assisted instruction.

ICAM Integrated computer-aided manufacturing.

ICAN International Commission on Air Navigation (rue George Bizet, Paris, from 1921).

ICAO International Civil Aviation Organization (from 1947, 186 member-states, office Montreal).

ICAOPA See *IAOPA.*

Icaotam ICAO Technical Assistance Mission.

ICAP, I-Cap Improved capability.

Icarus Complex laptop combining flight/black-box data with real-time graphics (Qinetiq).

ICAS *1* International Council of the Aeronautical Sciences.

2 International Council of Air Shows.

ICAT International Center for Air Transportation (MIT).

Icats Integration command and telemetry system.

ICAU International Civil Aviation University, office Melbourne.

ICB International competitive bidding.

ICBM Intercontinental ballistic missile.

ICC *1* International Control Commission.

2 International Chamber of Commerce.

3 International Code of Conduct (use of space).

4 Integrated command and control.

5 Initial CAOC capability.

6 Information coordination circular.

7 Interstate Commerce Commission (US 1887–1996).

8 Integrated cargo carrier.

9 IAPS(2) card cage.

10 Interface and configuration cartridge.

ICCABMP International code of conduct against ballistic-missile proliferation.

ICCAIA International Coordinating Council of Aerospace Industries Associations.

ICCD Intensified charge-coupled device.

ICCP Integrated communications control panel (in aircraft).

ICCS *1* Integrated command and control subsystem.

2 Integrated communications control system (on ground, unrelated to ICCP).

ICD *1* Installation, or interface, control drawing.

2 Interface control document, or device.

ICDOC Interchangeability condition direct operating cost.

ICDS Integrated control and display system.

ICDU Intelligent control and display unit.

ICE *1* Interference cancellation equipment.

2 Internal-combustion engine.

3 In-circuit emulator, simulates portions of external hardware during debugging of control software.

4 Improved combat efficiency [or effectiveness].

5 Independent cost estimate.

6 Institution of Civil Engineers (UK, 1818–).

7 Innovative control effector[s].

ice Naturally occurring forms include crystals, fog, needles and pellets, self-explanatory.

icebox rivet One (eg 2024 alloy) which must be kept below 0°C until use.

ice frost Forms on cryogenic tank; if on rocket, easily shaken off at launch.

ice guard For piston engine, usually a mesh screen; see *gapless *, gapped *.*

IceHawk Proprietary [Goodrich] ice detection based on polarized IR.

ice ingestion Class of tests to determine ability of engine to swallow various cubes of ice, ice-water slush and sometimes hailstones at specified flow rates. See *ice rod*, *ice slab.*

ICEM Intergovernmental Committee for European Migrations.

IC engine Internal combustion, usually means reciprocating: Otto, Diesel, Stirling, Rankine etc.

icephobic "Lethal to ice", a preventative coating.

ice plate Strong plate on fuselage skin in plane of propellers.

ice point NTP equilibrium temperature for ice/water mixture.

ice rod Standard ingestion-test size: 1.25 in diameter × 12 in.

Icerun Ice on runway.

ICEsat Ice, cloud and [land] elevation satellite.

ice slab Standard ingestion-test size 12 in × 12 in × 0.5 in (305 × 305 × 12.7 mm). Is permitted to break before entering engine.

ice zapper System for removing accreted ice by giant electrical pulses (colloq.).

ICF *1* Initial contact frequency (ATC).

2 Inertial confinement fusion.

ICFP Inverse-Cassegrain flat plate.

ICG Icing.

ICGIC Icing in cloud[s].

ICGIP Icing in precipitation.

ICH Interline Club Holland (Int.).

ICHE Intercooler heat-exchanger.

ICHTHUS Integrated and coherent technology for aircraft health and usage support systems (Smiths).

ICI *1* International Commission for Illumination.

2 Initial capability inspection.

ICIAP Inter-agency Committee on International Aviation Policy (established 1963).

icing Accretion of ice or related material on aircraft. Occurs on ground in freezing fog and in flight through supercooled water droplets.

icing indicator Any of four families of device quantifying presence of icing and rate of accretion.

icing limits Usually upper and lower temperature limits, corresponding to one (sometimes two) flight levels.

icing meter Instrument giving rate of accretion, buildup thickness and cloud liquid-water content.

icing tanker Aircraft equipped to cause severe icing on another following close behind.

ICLECS Integrated closed-loop ECS.

ICM *1* Intercontinental missile.

2 Improved conventional munition.

3 Inter-console marker.

4 Integrated collection management.

5 Interim control module.

6 Interline communications manual.

ICMP Internet control message protocol.

ICMS *1* Integrated conventional-stores management system.

2 Integrated countermeasures system, or suite.

I-CMS Index of central maintenance system fault messages.

ICNI Integrated com, nav, IFF.

ICNIA Integrated communications, navigation and identification avionics (USAF).

ICNIS Integrated CNI set.

ICNS Integrated com/nav system.

ICO *1* Ignition cutoff.

2 Idle cutoff.

3 Intermediate circular orbit, 10,000–15,000 km.

ICOC International Code of Conduct (ICBMs; ABMP adds against BM proliferation).

Icon Integration contract.

Icons Integrated control and operations network system.

ICP *1* Integrated core processor [or processing; TD adds technology demonstration].

2 Integrated control panel.

3 Initial conflict probe.

ICPA Indian Commercial Pilots' Association.

ICR *1* In-commission rate.

2 Integrated cassette recorder, for voice/video to LAN/WAN.

ICRC International Committee of the Red Cross.

ICS *1* Improved composite structure.

2 Inverse conical scan(ning) (ECM).

3 Improved (or integrated) communications system.

4 Intercom switch, or system, or set.

5 Internal countermeasures system, or set.

6 Interim contractor support.

7 Inter-cockpit communication system.

8 Intelligent control system.

ICSA International Centre for Security Analysis.

ICSAR, Icsar Inter-agency committee on search and rescue.

ICSC International Communications Satellite Corporation.

ICSM *1* Integrated conventional-stores management; GPS can be a suffix.

2 Integrated communication signalling and monitoring.

ICSMA Integrated Communications System Management Agency.

ICSMS See *ICMS*.

ICSS Integrated communications switching system.

ICSU International Council of Scientific Unions.

ICT *1* Ice-contaminated tailplane [S adds stall].

2 In-country test[ing].

ICU *1* Interface computer, or converter, or control unit.

2 Interstation control unit.

3 Instrument comparator unit.

ICW *1* Independent carrier wave, or see *i.c.w.*

2 Interpersonal communications workshop.

3 Intermittent [or interrupted] continuous wave.

i.c.w. Interrupted continuous wave.

ICWAR, Icwar Improved continuous-wave acquisition radar.

ICWI Interrupted continuous-wave illumination.

ICY *1* Interchangeable Y-axis.

2 International Co-operation Year (1965).

ICZ Interchangeable Z-axis.

ID *1* Internal or inner diameter.

2 Inadvertent disconnect (flight refuel).

3 Identification, identity, identifier.

4 Inverse dark (video characters).

IDA *1* Istituto di Diritto Aeronautico (I).

2 Intermediate dialect of Atlas.

3 Integrated digital avionics.

4 Intelligence/decision/action.

5 Initial design activity.

6 Integrated digital audio; CS adds control system.

Idaflieg Interessengemeinschaft Deutscher Akademischer Fliegergruppen eV (G).

IDAP Integrated defensive-avionics program.

IDAS *1* Integrated defensive aids system (RWR plus jammer).

2 Integrated design automation system.

IDC *1* Imperial Defence College (UK).

2 Interactive design centre.

3 Inner dead centre [= upper in inverted engine].

4 Indicator display control.

5 Image-dissector camera.

IDCAOC Interim deployable CAOC.

IDCSP Initial Defense Com Sat Program (DoD).

IDD Interim deployment device.

IDE Integrated development environment.

IDEA Instituto de Experimentaciones Astronauticas (Argentina).

ideal fluid Perfect, inviscid fluid; forces are perpendicular to small-parcel boundaries, no kinetic energy can be degraded to heat, boundary layer absent.

ideal profile Flight profile and path for lowest fuel burn.

ideal rocket Theoretical rocket with perfect operation, eg no heat transfer, no turbulence, no friction etc.

Ideas International data exchange for aviation safety (ICAO).

IDECM Integrated defensive electronic countermeasures.

IDefy, IDEFY Indefinitely.

ident   *1* Special feature in ATCRBS and I/P in SIF to distinguish one displayed select code from other codes (FAA).

   2 ATC request; transponder sends extra pulse plus * code.

identification   *1* Proclamation of identity, eg by SSR or squawk.

   2 Visual recognition of aircraft type.

identification cable colour   Each cable or pipeline in modern aircraft is colour coded to indicate function.

identification feature   Characteristics built into each selected code in SSR, ATCRBS and military radars for ident purposes.

identification friend or foe   Automatic interrogation and response, by coded transmission from transponder, to proclaim friendly status or identify flight on SSR.

identification light   Pilot-controlled white lights visible from above or below for broadcasting identity by keying.

identification manoeuvre   In primitive radar GCA, manoeuvre commanded by controller to establish positive identity of customer on radar.

Idex   Imagery dissemination and exploitation [system].

IDF   *1* Intelligent data fusion, most efficient way of using large amounts of data from many sources.

   2 Indigenous Defence Fighter.

   3 IR decoy flare.

   4 Instantaneous direction-finding.

Idflieg   Inspektion der Fliegertruppen (G).

IDG   Integrated-drive generator.

IDH   Intelligent, or intelligence, data handling; S adds system.

IDI   Initial domain identifier.

IDIS   *1* Interactive distributive information and support [on-line manuals].

   2 Intelligent display and information system.

IDL   Intraflight datalink.

idle area reset   Open divergent afterburner nozzle with throttle closed [reduces temperatures].

idle cutoff   Position of piston engine mixture control that cuts off fuel supply, thus stopping engine.

idle descent   To bleed off height with engine[s] at flight idle.

idler   Gearwheel or shaft whose sole function is to transmit drive between two others.

idles   Repeated cycling of engine from idling to specified higher power.

IDLH   Immediately dangerous to life or health.

idling   Running at governed low speed consistent with reliable smooth operation, in most engines well above minimum sustaining rpm: usually two regimes, ground * being lower N_1 than flight * and obtained only when oleos compressed.

IDLS   International Data Link Society (2003–).

IDM   *1* Improved data modem.

   2 Inductive debris monitor.

IDP   *1* Individual development programme.

   2 Integrated data-processing.

   3 Imagery display processor.

   4 Initial domain port.

IDPA   International Deaf Pilots' Association.

IDPM   Institute of Data Processing Management (UK).

IDPS   Interface data-processing segment (NPOESS).

IDPU   Incursion and display processing unit.

IDR   Initial design review.

IDRF   Impact-Dynamics Research Facility (crash tests, NASA Langley).

IDRP   Inter-domain routing protocol.

IDS   *1* Interdiction/strike.

   2 Integrated display set [or system] (USAF).

   3 Improved data set (USAF).

   4 Integrated dynamic system (helo).

   5 Infra-red detection set.

   6 Integrated diagnostic system.

   7 Ice-detection system.

   8 IATA distribution services.

   9 Intrusion detection system.

   10 Information display system.

   11 Integrated deepwater system (USCG).

IDSCP   Initial Defense Satellite Communications Program.

IDST   Integrated decision support tool.

IDTC   Inter-deployment training cycle (USN).

IDU   Interactive display unit.

IE   *1* Institution of Electronics (UK).

   2 Instrument error.

   3 Initial equipment (RAF).

   4 Incremental ejection.

IEA   International Ergonomics Association.

IEAA   International EAA, UK based.

IEB   Institut für Extraterrestrische Biologie (G).

IEC   *1* International Electrotechnical Commission.

   2 IAPS (2) environmental control module.

   3 Inertial electrostatic confinement.

IECC   International Express Carriers Conference.

IECMS   In-flight engine-condition monitoring system.

IED   *1* Improvised explosive device; D adds disposal.

   2 Insertion/extraction device.

IEDD   *IED* (1) disposal.

IEE   The Institution of Electrical Engineers (UK, 1871–).

IEEE   Institute of Electrical and Electronic Engineers (US).

IEER   Improved extended echo ranging.

IEF   Interpretive execution facility (software).

IEM   Interpretive/explanatory material (JARs).

IEMats   Improved emergency message auto transmission system.

IEN   Internal engineering notice.

IEP   Interim Earth penetrator.

IEPG   Independent European Programme Group (NATO).

IEPR   Integrated engine pressure ratio.

IERE   Institution of Electronic and Radio Engineers (UK).

IERW   Initial entry rotary-wing [ITS adds integrated training system].

IES   *1* Image enhancement system.

   2 Imagery exploitation system.

   3 Interface editor system (ATC).

IES   Illuminating Engineering Society; AC adds Aviation Committee (US).

IESI   Integrated electronic standby instrument.

IESSG   Institute of Engineering, Surveying and Space Geodesy.

IET   Initial entry training.

IETF   Internet engineering task force.

IETM   Interactive electronic technical manual[s].

IEU   Interface electronics unit.

IEVS   IR enhanced vision system.

IEW   *1* Integrated electronic warfare; S adds system, UAV adds unmanned air vehicle.

2 Intelligence/electronic warfare; CS adds common sensor, UAV adds unmanned air vehicle.

IEWS   Information and electronic-warfare system.

IF   *1* Intermediate frequency (often i.f.).

2 Instrument flight.

3 Intensive flying (UK, RN).

4 Independent Force (RAF, 1918).

5 Ice fog.

6 Intermediate fix.

I/F   Interface.

I/F module   Inlet and fan module.

IFA   International Federation of Airworthiness; 120 members in 47 countries engaged in manufacturing, operating, insuring, etc., office UK.

IFAA   International Flight Attendants' Association.

Ifalpa, IFALPA   International Federation of Airline Pilots' Associations (95 members, office UK).

IFANS, Ifans   International Federation for the application of standards.

Ifapa, IFAPA   International Foundation of Airline Passengers' Associations.

IFast   Integrated flexibility (or facility) for avionics system test (USAF).

Ifatca, IFATCA   Internationl Federation of Air Traffic Controllers' Associations.

IFATE   International Federation of Airworthiness Technology and Engineering.

Ifatsea, IFATSEA   International Federation of Air-Traffic Safety Electronics Associations.

IFB   Invitation for bid.

IFBP   In-flight best procedure[s].

IFBS   Individual flexible barrier system.

IFC   Incentive-fee contract(ing).

IFCA   International Flight Catering Association (office UK).

IFCNC   Integrated flight-control and navigation computer.

IFCS   *1* Integrated, or intelligent, flight-control system.

2 Integrated fire-control system.

IFDAPS   Integrated flight-data processing system.

IFDFS   In-flight duty-free shop.

IFDL   Inter-/intra-flight data link.

IFE   *1* In-flight emergency.

2 In-flight entertainment [N adds network, S system].

IFEO   International Flight Engineers' Organization (office UK).

IFESS   Integrated flight entertainment and services system.

IFF   *1* Identification friend or foe.

2 International Flying Farmers (US, office Wichita).

3 Institute of Freight Forwarders (Int.).

IFFAA   International Federation of Forwarding Agent Associations (now IFFFA).

IFFCP   IFF (1) control panel.

IF/FCS   Integrated fire/flight control system.

IFFFA   International Federation of Air Freight Forwarders' Associations.

IFGR   Information for global research (USAF).

IFHA   International Federation of Helicopter Associations (office, US).

IFI   International Friction Index.

IFIAT   See *FITAP*.

Ifics   In-flight interceptor communications [or control] system.

I-file   Intelligence (surveillance computer).

IFIM   International Flight Information Manual (US).

IFIS   *1* Independent flight inspection system (for ILS, Vortac, etc).

2 Independent frequency-isolation system.

IFM   *1* Instantaneous frequency measurement.

2 In-flight monitor.

3 International Formula Midget [= Formula 1].

IFMA   In-flight mission abort.

IFME   In-flight medical emergency.

IFMIS   Integrated force management info system.

IFMP   Integrated financial management plan (NASA).

IFMR   IFM (1) receiver.

IFM/SHR   IFM (1) superheterodyne receiver.

IFMU   Integrated flight-management unit (UAV).

IFMW   Information for mobile warfare (USAF).

IFN   Institut Français de Navigation.

IFO   International field office.

IFOBL   In-flight-operable bomb lock.

IFOG   Interferometric fibre-optic gyro.

IFOP   Intensive flight-operations program.

IFOR   Implementation Force (NATO).

IFOSTP   International follow-on structural test programme.

IFOV   Instantaneous field of view.

IFP   *1* Initial flightpath.

2 In-flight performance [computer program].

3 In-flight phone.

4 Integrated flight planner.

IFPA   *1* International Fighter Pilots' Academy (Slovak Republic).

2 IFPS (1) area.

IFPC   Integrated flight and propulsion control (STOL); S adds system.

IFPG   *1* Intelligent flight-path guidance.

2 International Frequency Planning Group.

IFPL   *1* In-flight power loss.

2 ICAO flight plan.

IFPM   In-flight performance monitor.

IFPS   *1* Integrated, or initial, flight-plan processing system.

2 Intra-formation positioning system.

IFPTE   International Federation of Professional and Technical Engineers (Int.).

IFPZ   IFPS (1) zone.

IFR   *1* Instrument flight rules.

2 In-flight repair.

3 In-flight refuelling.

4 Initial flight release (of engine).

5 Internationaler Förderkreis für Raumfahrt (G).

IFRA   In-flight refuelling area, designated airspace.

IFRB   International Frequency Registration Board.

Ifrep, IFREP   In-flight report [reconnaissance].

IFRU   Interference rejection unit.

IFS   *1* Institut für Segelflugforschung (G).

2 Inspectorate, or Inspector, of Flight Safety (USAF, RAF).

3 Instrument flight simulator.

4 Inner fixed structure of reverser.

IFSA   In-Flight Service Association.

IFSAR, Ifsar   Interferometric SAR (2) for digital terrain [E adds elevations].

IFSD   In-flight shutdown [rate].

IFSS   *1* International flight service station.

   2 Instrumentation and flight safety system.

IFT   *1* Intelligent flight trainer.

   2 Integrated flight test, or training.

   3 Intercept flight test (NMD).

i.f.t.   Intermediate-frequency transformer.

IFTE   Integrated family of test equipment.

IFTO   International Federation of Tour Operators (office UK).

IFTS   Internal Flir [and] targeting system.

IFTU   Intensive Flying Trials Unit (UK).

IFU   Interface unit.

IFURTA, Ifurta   Institut de Formation Universitaire et de Recherche du Transport Aérien (F).

IFW   In-flight weight.

IG   *1* Imperial gallon (sometimes I.g.).

   2 Inspector-General.

   3 Image generation, or generator.

Ig   Grid current; hence Ig2 screen-grid current.

IGA   *1* Inner gimbal angle.

   2 Informazione geotropographiche aeronautiche (I).

   3 Irish Gliding Association.

IGAA   Inspection Générale de l'Armée de l'Air (F).

IGAC, Igacem   Inspection Générale de l'Aviation Civile et de la Météorologique (F).

Igan   Indium/gallium/arsenic/nitride.

IGB   Integral, or intermediate, gearbox.

IGBAD   Integrated ground-based air defence.

IGC   *1* International Gliding Commission.

   2 Intergovernmental Conference (US/EU).

IGDS   Integrated graduate development scheme.

IGE   *1* In ground effect.

   2 Instrument[ation] graphics environment.

IGEB   Inter-agency GPS Executive Board.

IGES   *1* International graphics exchange [file] standard.

   2 Interim ground Earth station.

IGFET   Insulated-gate field-effect transistor.

IGI   Instrument Ground-Instructor.

IGIA   Interagency Group on International Aviation [from 1960].

igloo   *1* Air cargo container in form of rigid box sized for above-floor loading.

   2 Small pressurized experiment container designed for use with Spacelab orbital laboratory.

IGN   Ignition.

igniter   Pyrotechnic (rocket) or high-energy discharge device for starting combustion of solid or liquid fuel.

igniter plug   Electric device for starting combustion of gas-turbine engine. Most common types are surface discharge, shunted surface discharge, and air-gap.

ignition delay time   Several definitions for elapsed period between ignition signal and stable or other desired combustion in piston engine, solid rocket etc. For solid motors, the elapsed time from ignition firing voltage to chamber pressure reaching 10% of maximum value, symbol t_d.

ignition harness   Complete screened assembly of h.t. cables serving spark plugs of piston engine.

ignition interference   Disturbance to radio communication caused by faulty ignition screening.

ignition loop   Plot of gas-turbine combustion stability on axes air/fuel ratio and air mass flow.

ignition screening   Surrounding layer of conductive braid to prevent signal emission from h.t. harness.

ignition servo unit   Automatically controls ignition timing.

IGO   *1* Prefix, US piston engine, direct injection, geared, opposed.

   2 Inter-governmental organization[s].

IGOR   Intercept ground optical recorder; long-focal-length tele-camera.

IGOS   Integrated Global Observing Strategy, or System (Int.).

IGPM   Imperial gallons per minute.

IGS   *1* Internal gun system.

   2 Instrument guidance system.

IGSO   IGO plus supercharged.

IGT   Industrial gas turbine.

IGTI   International Gas Turbine Institute (ASME, office Atlanta, GA).

Ig2   Screen-grid current.

IGV   Inlet guide vane.

IGW   Increased gross weight.

IGY   International Geophysical Year [1957–58].

IH   Inhibition height (GPWS).

I_h   *1* Mean horizontal candlepower.

   2 Heater current.

IH   Aviation Historical Association (Finland).

IHADSS   Integrated helmet and display sight[ing] system.

I-Hards   Improved high-altitude radiation-detection system.

IHAS   *1* Integrated helicopter attack system.

   2 Integrated hazard-awareness, or avoidance, system.

IHC   *1* Integrated hand control.

   2 Industry Harmonization Conference [rule making].

IHDSS   Integrated helmet display and sight[ing] system.

IHDTV   Intensified high-definition TV.

IHE   Insensitive, or improved, high explosive.

IHEC   Integrated helicopter emissions control.

IHEWS   Integrated helicopter EW suite.

IHFA   International Helicopter Firefighters Association (office in US).

IHMD   Integrated helmet-mounted display.

IHO   International Hydrographic Organization.

IHOC   International Helicopter Operations Committee.

i.h.p.   Indicated horsepower.

IHPA   Irish Hang-gliding and Paragliding Association.

IHPTET   Integrated high-performance turbine-engine technology, principally focused on propulsion of supersonic-cruise aircraft: takes off with maximum BPR, meeting civil noise legislation with simple nozzle, accelerates with BPR ≤1, cruises with BPR near 3, lands at high BPR (GE/USAF/Darpa).

IHTTET   Improved high-temperature turbine-engine technology (DoD).

IHTU   Inter-Service Hovercraft Trials Unit (UK, from 1962).

IHU   Integrated helmet unit.

IHUMS, I-Hums   Integrated Hums.

II   Image intensifier.

IIAE   Instituto de Investigaciones Aeronauticas y Espaciales (Argentina).

IIASA   International Institute for Applied Systems Analysis.

IIC   Image isocon camera.

IID   Integrated instrument display [S adds system].

IIDS   *1* Istituto Italiano di Diritto Spaziale.

     2 Integrated instrument display system.

IIE   Institution of Incorporated Engineers (UK).

IIF   Inserted in flight (data, target, destination etc).

IIMS   Initial implementation of Mode-S [ES adds enhanced surveillance].

IIN   *1* Istituto Italiano di Navigazione.

     2 Information infrastructure network (RAF).

IIP   Instrument, or instrumentation, incubator program (NASA).

IIR   *1* Imaging infra-red.

     2 Infra-red imaging radar (US).

     3 Infinite impulse response (signal processing).

     4 Incident [usually involving aircraft damage] investigation report.

IIR   Institute for International Research (US).

IIRA   Integrated inertial reference assembly.

IIRS   *1* Instrument inertial reference set.

     2 Imagery interpretability rating scale.

IIS   *1* Infra-red imaging system.

     2 Integrated instrument system.

IISA   Integrated inertial sensor assembly.

IISL   International Institute of Space Law.

IISS   International Institute for Strategic Studies (London, plus DC office).

IIT   Image-intensifier tube.

IITS   Infra-theatre imagery transmission system.

IITV   Image-intensified TV.

IIW   Image interpretation workstation, portion of GIES tasked with target location and selection.

IIWD, I²WD   Intelligence and Information Warfare Directorate (USA).

IJMS   Interim JTIDS message standard, or structure, or system.

IK   Club of Aeronautical Engineers (Finland).

IKAT   Interactive keyboard and terminal.

IKBS   Intelligent knowledge-based systems.

IKI   Space research institute (Soviet academy of sciences).

IKPT   Initial key personnel training.

IKSANO   English rendition of Information co-ordination council on air-navigation charges debts (R).

IKW   Intercept and kill weapon.

IL   *1* Infantry liaison (aircraft category, USA, 1919–25).

     2 Instytut Lotnictwa [aviation institute] (Poland).

ILA   *1* International Law Association (office London).

     2 Internationales Luftfahrt-Archiv (G).

     3 Image light amplifier.

ILAA   Integrated landing and approach aid.

ILAAS   Integrated low-altitude attack subsystem.

ILAC   Intake-lip acoustic liner.

ILAD   Inner-layer air defence.

ILAF   Identical location of accelerometer and force.

ILAS   Improved limb atmospheric spectrometer.

ILC   *1* Integrated laminating centre (or center).

     2 Increased-life core (engine).

ILCA   International Legal Committee for Aviation (from 1909, office Paris).

IL-check   C-check plus more detailed inspection, repair and update of systems and furnishing.

ILD   Injection laser diode.

ILF   In-line filter.

ILFPS   Integrated lift-fan propulsion system.

ILGH   Interessengemeinschaft Luftfahrtgeräte-Handel (G).

ILL   Internationale Luftverkehrsliga.

ILLF   Initial long-lead funding.

illuminance   Intensity of illumination, luminous flux per unit area, symbol E, unit lux, $lx = lm/m^2$.

illumination   *1* Illuminance.

     2 Lighting of target by radar or other signals, esp. to make it an emitter for SARH missile.

ILM   *1* Independent landing monitor.

     2 Intermediate-level maintenance.

ILO   International Labour [Labor] Organization.

ILS   *1* Instrument landing system.

     2 Integrated logistic (or logistics) support.

     3 Integrated [or intelligent] library system.

     4 Initial launch services.

     5 Integrated, or intelligent, [missile] launch system.

     6 Inventory Locator Service (US).

ILS integrity   Trust which can be placed in correctness of information supplied by ILS (1) facility (ICAO).

ILSMT   Integrated logistic support management team.

ILS Point A   On extended runway centreline 4 nm from threshold.

ILS Point B   On extended runway centreline 1,050 m (3,500 ft) from threshold.

ILS Point C   Intersection of straight line representing nominal (mean) glideslope and horizontal plane 30 m above threshold.

ILS Point D   6 m above centreline, 600 m upwind (ie towards localizer) of threshold.

ILS reference datum   Point at specified height vertically above intersection of centreline and threshold through which passes straight line representing nominal (mean) glideslope.

ILS reliability   *1* Facility: probability its signals are within specified tolerances.

     2 Signals: probability signal in space of specified characteristics is available to aircraft.

ILS-S   Integrated logistics system-supply (USAF).

ILVSI   Instantaneous-lag [or lead] VSI.

ILWS   International living with a star (NASA).

IM   *1* Inner marker.

     2 Intermediate maintenance.

     3 Inventory management.

     4 Intra-mural.

     5 Insensitive munition[s].

IMA   *1* Institut Médical de l'Aviation (Switzerland).

     2 Intermediate maintenance activity.

     3 Individual mobilization augmentee.

     4 Integrated modular avionics.

     5 Integrated multifunction apertures.

     6 International Museum of Airlines (Rockville, MD).

IMAA   Irish Microlight Aircraft Association.

IMAAWS   Infantry man-portable anti-armour/assault weapon system.

IMAC   Integrated microwave amplifier converter.

Imacs   Integrated manufacturing control systems.

IMAD   Integrated multisensor airborne display (Elint).

Image   Instrument for the measurement of air-traffic flow using ground environment.

image convertor   Converts image from invisible to visible wavelengths.

image degradation   That due to error in sensor operation, processing procedure or other fault by user.

image intensifier   Any of large family of electron tubes which multiply electron flow due to signal while ignoring noise; hence IIT, ** tube.

image-motion compensation   Synchronization of target image with recording sensor in vehicle, esp. low-level reconnaissance aircraft.

imagery   Representation of objects reproduced by optical or electronic means.

imagery collateral   Reference materials supporting imagery interpretation function (ASCC).

imagery correlation   Mutual relationship between different signatures on imagery of same object from different sensors.

imagery data-recording   Auto record of sensor speed, height, tilt, geographical position, time and possibly other parameters on to sensor matrix block at moment of imagery acquisition (ASCC).

imagery exploitation   Entire process from acquiring imagery to final dissemination of information.

imagery interpretation   Process of location, recognition, identification and description of objects visible on imagery (DoD, NATO).

imagery interpretation key   Diagrams, examples, charts, tables etc, which aid interpreters in rapid identification of objects.

imagery pack   Assembly of all records from different sensors covering common target area (ASCC).

imagery sortie   One flight by one aircraft for acquiring imagery (DoD, NATO, Cento).

IMAS   Integrated mission-avionics system.

IMBP   State institute of biomedical problems (R).

IMC   _1_ Instrument meteorological conditions (UK, see _IWR_).

  _2_ Image movement compensator (reconnaissance).

  _3_ Intermetallic-matrix composite.

  _4_ Indirect maintenance cost.

  _5_ See _Inst MC._

IMCC   _1_ Integrated mesoscopic cooler circuit.

  _2_ International Military Control Commission.

IMCPU   Improved master-controlling processor unit.

IMD   _1_ Indian Meteorological Department.

  _2_ Integrated mechanical diagnostics.

IMDS   Integrated maintenance data system[s] (USAF).

IMDT   Immediate.

IME   _1_ Indirect manufacturing expense.

  _2_ Integrated modelling environment.

IMEA   Integrated munitions effects assessment, tool for selecting aim points.

IMechE   Institution of Mechanical Engineers (UK, 1847–).

IMEP   International materiel evaluation program (US).

i.m.e.p.   Indicated mean effective pressure.

IMET   International Military Education and Training (NATO).

Imets   Integrated meteorological system.

IMEWS   Integrated missile early warning system, or satellite.

IMF   _1_ International Monetary Fund (UN agency).

  _2_ International Metalworkers Federation (trade union, office Geneva).

IMFCA   Institut de Mécanique des Fluides et Constructions Aérospatiales (Romania).

IMFIS, Imfis   Interoperability of military forces and information systems (Canada).

IMG   _1_ Immigration.

  _2_ Implementation management group.

IMI   _1_ Intermediate maintenance instruction.

  _2_ Improved manned interceptor.

  _3_ Initial maintenance interval.

  _4_ Interactive multimedia instruction.

  _5_ Imbedded [=embedded] message identifier.

Imint   Imagery intelligence.

IMIS   Integrated maintenance information system.

IMK   Increased maneuverability kit (US).

IML   _1_ International micro-gravity laboratory (Spacelab).

  _2_ Inner mould line.

IMLS   Interim MLS.

IMM   _1_ Intelligent menu-management.

  _2_ Interacting multiple model.

IMMC   Integrated mission-management computer.

immediate air support   That meeting specific request during battle and which cannot be planned in advance.

immediate award   Decoration awarded without the usual assessment process.

Immelmann   Air-combat manoeuvre. Two definitions current: (a) first half of loop followed by half-roll, resulting in 180° change of heading and gain in height; (b) steep climbing turn to bring guns to bear on target again following dive attack from front or rear.

immersed pump   Booster pump.

IMN   Indicated Mach number.

IMO   _1_ International Maritime Organization.

  _2_ International Meteorological Organization (1872–).

I_{mo} Motor specific impulse (solid propellant).

IMOK   I am OK.

IMP   _1_ Interactive microprogrammable.

  _2_ Indication of microwave propagation (anaprop) to exploit gaps in hostile radars and assess friendly radars in defence.

  _3_ Incremental modernization program [simulators].

  _4_ Inter-modulation product.

  _5_ Interplanetary monitoring platform.

Impact   Integrated multistatic passive/active concept testbed (sonobuoys).

impact area   Area having designated boundaries within which all ordnance is to hit ground.

impact microphone   Sensitive to small vibrations, as of micrometeoroid impact.

impact point   _1_ Point on drop zone where first parachutist or air-dropped cargo should land (NATO, IADB).

  _2_ Point at which projectile, bomb or re-entry vehicle impacts or is expected to impact (DoD).

  _3_ Reference point of accident from which all surveys are plotted, easily identifiable and either central or having highest density of contamination (USAF); this need not be actual point of impact.

impact pressure   See _stagnation pressure_.

impact tests   Charpy, Izod and similar tests to measure resistance of material to suddenly imposed stress.

Impatt   Impact-ionisation avalanche transit time (solid-state oscillator).

IMPD   Interactive multipurpose display.

IMPDS   Improved missile point-defence system.

impedance   Resistance to a.c. (1), symbol $Z = \sqrt{(R^2+X^2)}$ where R is ohmic resistance and X is reactance.

impeller   Single-stage radial-flow compressor rotor.

Imperial   System of units previously standardized in UK, such as foot (ft) and pound (lb).

Imperial gallon   See *gallon*.

impervious canopy   One through which ejection is prohibited.

impingement   Impact of high-velocity air or gas on structure (eg in reverse-thrust mode).

impingement cooling   Cooling of material by high-velocity air jets directed on to surface (usually internal surface of hollow blade or vane).

impingement injector   Liquid fuel and oxidant jets impact on each other to cause swift breakup and mixing.

implosion   Detonation of spherical array of inward-facing shaped charges (eg to crush fissile core of nuclear weapon).

IMPR   Improving.

impress   To commandeer a civil aircraft into government service in time of emergency.

improved climb T-O   Take-off at increased weight allowed by raising V_2 where second segment is limiting factor and runway distance is available.

improved conventional munition   Usually means fitted with electronic time fuze or RF proximity fuze.

IMPT   Important.

impulse   Rocket burn-time multiplied by burn-time average thrust; total energy imparted to vehicle.

impulse magneto   One whose drive incorporates stops and spring-loaded coupling (bypassed by centrifugal clutch in normal running) to give series of sudden rotations and thus hot sparks during starting.

impulse/reaction   Common form of gas turbine, combining impulse and reaction techniques.

impulse starter   Incorporated in impulse magneto.

impulse turbine   One whose working fluid enters at lowest pressure and maximum velocity, expands through diverging passages, and leaves at similar pressure and low velocity.

IMR   Imaging microwave radiometer.

IMRO   Industrial Marketing Research Organization (UK).

Imron   Range of du Pont poly enamels.

IMRS   Integrated maintenance recording system.

IMS   *1* Information management system.

2 Integrated management system.

3 Intermediate maintenance squadron.

4 International Military Services (MoD, UK) or Staff (NATO).

5 Integrated multiplex[ing] system, ties nav, weapons, air data and FBW.

6 Inertial measurement system.

7 Integrated mission system.

8 Inventory management system.

9 Ion mobility spectroscopy, or spectrometer (explosives/narcotics detection).

I'm safe   One of 35 English-language safety-related mnemonics: illness, medication, stress, alcohol [+ drugs], fatigue, emotional problems.

IM-6   Structural graphite/epoxy composite material (Hercules).

IMSS   Integrated multisensor system.

IMT   *1* International mobile telecommunications.

2 Immediate[ly].

IMTA   Intensive military training area.

IMTS   Integrated maintenance training system.

IMU   Inertial measurement unit.

IN   *1* Inertial navigator (or navigation).

2 Instrument navigator (arch.).

in   Inch, inches.

INA   *1* International Navigation Association Inc. (Charlotte Hall, MD, US).

2 Initial approach.

INACP   Integrated navigation-aids control panel.

inactivate   Of military unit, withdraw all personnel and transfer to inactive list.

INAS   Integrated, or inertial, navigation/attack system.

InAs   Indium arsenide.

InAsSb   Indium antimony arsenide.

INB   Iron-neodymium-boron magnetic alloy.

INBD, inbd   *1* Inbound.

2 Inboard.

inboard aileron   Aileron situated on inner wing between, or in place of, flaps.

inboard profile   Side-elevation drawing showing internal systems and equipment, sometimes as true section along centreline.

inboard quadrant   Inner selectable position of power (throttle) lever on left side of cockpit giving operative afterburner.

inbound   Approaching destination, thus * traffic, * controller etc.

inbound bearing   Normally QDM, not QDR or VOR radial.

INC   *1* Insertable nuclear component.

2 Interchangeability code.

3 In cloud.

4 Increase, increasing.

Inc   Incorporated (US company).

Inca   *1* Intelligent correlation agent (data fusion).

2 Integrated nuclear communications assessment.

3 Initiative en combustion avancée, future engines (F).

Incas   Integrated navigation and collision-avoidance system.

incendiary bomb   Bomb designed to ignite enemy infrastructures.

inceptor   Cockpit control forming interface between pilot and major change in trajectory, eg stick/yoke, throttle, pedal, cyclic, collective or nozzle angle lever.

Incerfa, INCERFA   Code: phase of uncertainty (ICAO).

inch   *1* To command powered actuator in rapid succession of small cycles to achieve target condition.

2 Non-SI unit of length, = 25.4 mm exactly.

inches   Traditional US measure of piston engine manifold pressure, = * Hg (mercury), see next.

In Hg   Non-SI unit of pressure, = (at 0°C] 3.38639×10^{-3} Nm^{-2}.

incidence   *1* Angle between chord of wing at centreline and OX axis.

2 Generally, the angular setting of any aerofoil or other plate-like surface to a reference axis.

3 Widely and incorrectly used to mean angle of attack.

incidence instability   Divergent aerofoil load caused by wing flexure simultaneously resulting in increased incidence.

incidence wires   Diagonal bracing wires in plane of biplane interplane struts.

incident report   Normal report of incident; this falls short of an accident, takes place on ground or in air during flying operations, and usually stems from human error.

incl   Inconclusive, include[d].

inclination   Angle between isobar and wind or airflow at given point.

inclined shock   Shockwave generated by body in airflow at Mach number significantly greater than 1, with angle such as to turn flow parallel to surface of body; in air-breathing inlet generated by centrebody or sharp-edged plate and focused on lip.

inclinometer   Instrument for measuring inclination; many forms, one being spirit level on pivoted arm [end of arm shows degrees on protractor while slight curvature of sliding spirit level gives minutes]. Some authorities call this a *clinometer*, which see.

included angle   That between longitudinal axis of body and free-stream vector.

incoherent backscatter[ing]   Random backscatter of a signal by individual electrons in the ionosphere.

Incomap   Family of mechanically alloyed Al alloys, esp. Al-Mg-Li-C-O.

Inconel   High-nickel chromium-iron refractory alloys (Int. Nickel and Mond).

Incos   Integrated control system [Hartman ASW].

Incospar   Indian National Committe for Space Research.

Incr   Increase.

incremental airbrakes   Capable of being controlled to intermediate positions.

incremental ejection   Dispensing chaff in discrete bundles.

incremental sensitivity   Change in received signal per unit displacement of ILS receiver from mean glidescope.

INCRSG   Increasing.

incursion   *1* Conflict, especially between two aircraft, on runway or elsewhere on airfield [FAA adds 'with active control tower'].

2 Any entry by aircraft into forbidden area, either on ground or in flight.

Ind   Indicator (for wind or landing direction).

INDAE   Instituto Nacional de Derecho Aeronutica y Espacial (Argentina).

indefinite callsign   C/s assigned to individual units/facilities etc and to large groupings.

INDEFLY   Indefinitely.

independence   One meaning is Busemann's principle that aerodynamic forces on a wing of high aspect ratio are independent of any V component in the spanwise direction.

independent   *1* Military unit with complete authority over tasking [e.g. UK * Air Force 1918].

2 R&D or programme not relying on external funding.

indexed wing/fins   Aerodynamic surfaces are at same angular setting to [usually missile] body, measured in transverse plane, eg all at 45° to horizontal or two vertical and two horizontal.

index error   That caused by misalignment of measurement mechanism of instrument (ASCC).

Indian   Hostile aircraft, esp. fighter.

indicated airspeed   See *airspeed*.

indicated altitude   That shown by altimeter set to latest known QNH.

indicated course   Locus of points in any horizontal plane at which ILS Loc needle is centred.

indicated course sector   Sector in any horizontal plane between loci of points at which ILS Loc needle is at FSD left or right.

indicated glidepath   Locus of points in vertical plane through runway centreline at which G/S needle is centred.

indicated hold   Autopilot mode maintaining present IAS.

indicated horsepower   See *horsepower*.

indicated Mach number   That shown on Machmeter.

indicator   Identifying 4-letter code for every airfield [in England and Wales beginning EG].

indicator diagram   Plot of piston engine cylinder pressure against piston position, often drawn by instrument attached to cylinder.

indirect air support   Given friendly land/sea forces by action other than in tactical battle area, eg by interdiction and air superiority.

indirect cycle   Nuclear propulsion with primary circuit, heat exchanger and secondary circuit.

indirect damage assessment   Revised target assessment based on new data such as actual weapon yield and ground zero.

indirect wave   One arriving by indirect path caused by reflection/refraction.

indium   In, soft silver-white metal, density 7.28, MPt 156.4°C.

individual controls   Control surfaces not attached to fixed surface but cantilevered from body (guided weapons).

induced drag   Drag due to component of wing resultant force along line of flight; drag due to lift.

induced flow   Fluid flow drawn in and accelerated by a high-velocity jet.

induced force   Usual aircraft design consideration is that caused by air entering engine inlet.

induced velocity   That due to wing vortex system and downwash, normally considered proportional to lift.

inducer   *1* Booster vanes at entry to centrifugal impeller, esp. rocket-engine turbopump.

2 Bleed ejector to induce cooling airflow.

inductance   Property of electric circuit to resist change in current as result of opposing magnetic linkage; see *self* *, *mutual* *.

induction compass   Based on induction coil pivoted to rotate in Earth's field.

induction heating   Heating electrically conductive material by h.f. field.

induction manifold   Pipe system conveying mixture to piston engine cylinders.

induction period   Specified delay between adding catalyst or hardener and applying or spraying material (coating, adhesive or thermo-setting structure).

induction phase   In pulsejet, portion of cycle when air is admitted.

induction stroke   In piston engine, portion of cycle when air or mixture is admitted.

induction system   In piston engine, entire flow path from combustion-air inlet to cylinder.

induction tunnel   Wind tunnel driven by jet engine(s) or compressed air via ejector system.

inductive coupling   *1* Mechanical shaft drive relying on magnetic linkage.

2 Magnetic coupling between primary and secondary coils, eg of transformer.

inductive reactance   Impedance due to inductance.

inelastic collision   Theoretical impact with no deformation or energy loss.

inert gas   Gas incapable of chemical reaction.

inertial anti-icer   Free-spinning vanes or other device for imparting rotation to engine airflow, ice or snow being flung out away from engine inlet.

inertial coupling   See *coupling (1)*.

inertial flight   No propulsion, controls locked central or free.

inertial guidance   Guidance by INS.

inertial gyro   Gyro of characteristics and quality to meet INS requirements.

inertial navigation system, INS   Assembly of super-accurate gyros to stabilize a gimballed platform on which is mounted a group of super-accurate accelerometers – typically one for each of the three rectilinear axes – to measure all accelerations imparted, which with one automatic time integration gives a continuous readout of velocity, and with a second time integration gives a readout of present position related to that at the start.

inertial orbit   Trajectory when coasting.

inertial platform   Rigid frame stabilized by gyro(s) to carry accelerometer(s).

inertia starter   Rotary-drive starter whose energy is stored in flywheel.

inertia welding   Welding by rapid rotation and pressure between mating surfaces.

inerting   Filling the space above the fuel in a tank with inert gas, usually nitrogen [USSR in WW2 used cooled engine exhaust, which contains oxygen].

inert round   Missile or ammunition partly or entirely dummy, and lacking propulsion.

INES   IUE newly extracted spectra.

INET   Inertial navigation equipment tester.

INEWS, Inews   Integrated electronic-warfare system [or suite].

INF   Intermediate-range nuclear force(s).

infant mortality   Failure at start of life-cycle.

Infco   Standing Committee for Science and Technology Information (ISO).

inferior planet   Planet with orbit inside that of Earth.

infiltration   Manufacturing stage with FRM (fibre-reinforced metal) components in which molten metal is used to fill gaps between compressed metal-coated fibres.

infinite aspect ratio   Many aerofoil calculations ignore effects at the ends (tips) and accordingly are true only for a wing of infinite aspect ratio (ie, it goes on forever or touches both walls of tunnel).

infinity   Symbol ∞, used as subscript to denote free-stream values.

Infis   Inertial-navigation flight inspection system.

inflatable aircraft   One whose airframe is of flexible fabric, stabilized by internal gas pressure. Term usually applied to aerodynes.

inflatable de-icer   One whose action is the repeated inflation and deflation of a flexible surface, thus breaking off accreted ice.

inflation manifold   Links several sources of gas to gas-filled aerostat.

inflation sleeve   Large thin-wall tube to which inflation manifold is connected.

inflator   Trolley-mounted powered fan to begin inflation of hot-air balloon.

inflection   Point at which curve reverses direction.

in-flight advisory   Sigmet/airmet broadcast to enroute pilots notifying conditions not anticipated at preflight briefing.

in-flight weight   Maximum authorized weight after in-flight refuelling, can exceed MTOW.

inflow   *1* The component of velocity through a helicopter main rotor normal to the tip-path plane.

2 There are two more [rare and confusing] meanings: increase in relative speed as air is sucked into a propeller; and inwards radial velocity as air is sucked into a propeller.

inflow ratio   Ratio of rotorcraft TAS and peripheral velocity $r\omega$ at tips of blades.

influence line   Graphical plot of shear, bending moment and other variables as point load moves along structure.

Info   Information frame.

informatics   Word gaining some ground in English from transliteration "informatique" (F) = EDP (1).

information dominance   Being quicker than the enemy in assessing combat situation and launching weapons.

information operations   Central method of waging war, including EW, psyops, comint and defence against cyber attack.

information pulses   Those repeated parts of the transmission in SSR or IFF that convey information (eg identity, flight level).

information technology management   The ground-based portion of an ADMS.

Infowar   Information warfare.

Infraero   Airports authority (Brazil).

infra-red, IR   Portion of EM spectrum with wavelength longer than deep red light, thus not visible, but sensed as heat. Near * wavelengths $0.75–1.5\mu$, intermediate * $1.5–20\mu$, far * $20–1,000\mu$.

infrasound   Sound, especially at very high power, at very low frequency [<c 20 Hz] and very large amplitude.

infrastructure   Fixed installations needed for activity (eg airfield, hangars, control tower, communications, fuel pipelines).

infrastructure authority   Those responsible for airfields, communications, radar and other ground installations.

infringement   Violation of air traffic control or other rules regarding operation of aircraft.

ingestion   Swallowing of foreign matter by engine (usually gas-turbine engine), including birds, ice, snow/slush/water, sand, rocket gas, catapult steam and metal parts; hence * test, * certification.

ingress   To re-enter spacecraft after EVA, including expeditions on lunar or planetary surface.

ingress route   Attack aircraft track from base to initial point.

inherent stability   In-flight stability achieved by basic shape of aircraft.

in Hg   Inches of mercury, non-SI unit of pressure = $3,386.39\ \mathrm{Nm^{-2}}$.

inhibiting   *1* To spray interior of machine or other item with anti-corrosion material before storage.

2 To coat inner surfaces of rocket solid-propellant grain to prohibit burning over treated areas.

3 The worst case in flight performance situations, eg instrument flight, failed engine and other adversities.

inhibitor *1* Additive to fuel or other liquid, eg Methyl Cellosolve (often +0.4% glycerine) to protect against ice and against formation of gumming residues, corrosion or fuel oxidation.

2 Refractory inert coating to control burning of solid grain.

3 Anti-corrosion oil for long-term storage.

INI Instituto Nacional de Industria (Spain).

INIT Initialization.

initial altitude Altitude(s) prescribed for IA (1) segment (FAA).

initial approach *1* Segment of standard instrument arrival or STAR between IA (1) fix and intermediate fix or point where aircraft established on intermediate approach course (FAA).

2 Portion of flight immediately before arrival over destination airfield or over reporting point from which final approach is commenced (Seato, IADB).

initial-approach area See *initial area*.

initial area Ground area of defined width between last preceding fix or DR position and either intersection of ILS, facility to be used in instrument approach or other point marking end of IA (1).

initial attack Tanker effort to contain a fire until ground firefighters can reach it (USFS).

initial contact frequency That used for ATC communication as aircraft enters new sector of controlled airspace.

initial heading That at start of rating period while using astro-gyro control (ASCC).

initial mass *1* That of rocket or rocket vehicle at launch.

2 That of fissile or other nuclear material before reaction.

initial operational capability Time at which particular hardware (eg weapon system) can first be employed effectively by trained and supported troops (USAF). Particular parts of the system may still be lacking.

initial point *1* Well defined fixed surface feature usable visually and/or electronically as starting point for attack on surface target (most existing definitions state 'starting point for bomb run'.)

2 Similar surface point where aircraft make final correction of course to pass over drop zone or other surface target.

3 Air-control point near landing zone from which helicopters are directed to landing sites.

4 First point at which moving target is located on plotting board or display system.

initial radiation That emitted from fireball within 60 s of nuclear burst, mainly neutrons and gamma rays.

initial surface That at start of solid-rocket burn; its area.

initial surface/throat area Fundamental non-dimensional characteristic of solid motor.

initial throat area Cross-section area of unused solid-motor throat.

initial-value problem One which, from given state, determines state of dynamic system at any future time.

initiation *1* Starting sequences leading to nuclear explosion.

2 Birth of new inbound track on air-defence system.

initiation phase In autoland, 10 nm from threshold, 2,500 ft (usually 205 kt, 15° flap).

initiator *1* Person assigned to initiate tracks in air-defence system.

2 Starts automated sequence in modern crew ejection.

INJ Injection.

injection Insertion of satellite into orbit, or spacecraft on to desired trajectory.

injection carburettor One delivering metred flow to nozzle in induction system, usually without venturi and with insensitivity to flight attitude.

injection flow Primary flow in ejector pump or induced augmenter.

injection point That at which satellite enters orbit.

injection pressure Pressure in system between turbo-pump and injector.

injection pump *1* Ejector.

2 Injection carburettor.

3 More usually, multi-plunger pump in direct-injection engine.

injection velocity That at injection point.

injector Point at which one or both propellants in liquid rocket enters chamber (usually many are used, distributed around chamber).

inlet Usually air intake at upstream end of air-breathing system.

inlet blocker Fixed or movable screen inside inlet duct[s] to engine[s] of combat aircraft to prevent hostile radar seeing front of engine.

inlet distortion *1* Departure from ideal airflow through inlet, esp. in yawed or turbulent conditions.

2 Plot of flow velocities across plane normal to axis of air-breathing engine, eg at entry to fan, entry to combustor or entry to turbine.

inlet duct Air passage linking inlet (intake) to engine.

inlet guide vane Fixed or variable-incidence stators preceding compressor rotor or turbine.

inlet particle separator Dynamic (usually centrifugal) device that removes solid particles from airflow entering engine.

inlet unstart Sudden gross disruption of airflow through supersonic engine inlet causing near-total loss of thrust; hard-to-cure hazard at Mach 3 and above (see *unstart*).

inline engine *1* Piston engine with all cylinders in single linear row; loosely (probably incorrectly) applied to engines with two or more rows such as the vee or opposed configurations.

2 Turbojet and ramjet mounted in tandem, sharing same airflow (usually with variable-geometry ducting).

Inmarsat International Maritime Satellite Organization.

INMG Instituto Nacional de Meteo e Geofisica (Portugal).

InN Indium nitride.

inner Applied to aeroplane, = closer to centreline.

inner horizontal surface Specified portion of local horizontal plane located above airfield and immediate surroundings.

inner loop Control loop via pilot, not autopilot.

inner marker ILS marker rarely installed near threshold, 3,000 Hz dots and white panel light.

inner mould line Low-cost technique for composite-structure tooling.

inner planets Four nearest Sun: Mercury, Venus, Earth, Mars.

inner space Within Solar System.

inner wing *1* Loosely, inboard part.

2 That on the inside of a turn, usually with reduced airspeed.

innovative controls   Many forms, e.g. replacing rudder and ailerons by rotating wingtips or slot/spoilers.

INO   Indian Ocean region (ICAO).

IN100   HPT material, high-Ni alloy, 1975.

INOP   Inoperative.

inoperative   Deliberately taken out of use (thus not a failed engine or other device).

INP   If not possible.

InP   Indium phosphide, semiconductor.

INPE   Instituto de Pesquisas Espaciais (space research) (Brazil).

inph   Interphone.

in-phase   Occurring at the same point in each of a series of phugoids or other SHM or repeated cycles, but caused by external stimulus.

in-plane bleeder   Sustained max-rate turn without change of height.

inplant   Done within factory instead of subcontracted, or relative to the factory (thus * facilities, * modification).

INPR   In progress.

input   *1* Point at which signal, data, energy or material enters system.

2 Signal, data, energy or material entering system.

input axis   Axis normal to gyro spin axis about which rotation of base causes maximum output.

INREQ   Request for information.

INRIA   Institution Nationale de Recherches d'Informatique et Automation (F).

INS   *1* Inertial navigation system.

2 Ion neutralization spectroscopy.

3 Immigration and Naturalization Services; PASS adds passenger accelerated service system (US).

4 Information network system.

Insacs   Interstate airway communication station (US).

In-Sap, IN-SAP   Intelligence special-access program.

InSb   Indium antimonide, IR detector 3–5μ.

insensitive munition   One impossible to detonate except by its own triggering system.

insert   *1* Small D-section body fixed inside propelling nozzle to trim area.

2 To place spacecraft in desired trajectory.

3 To convey friendly force (usually small and covert) to point on ground deep in enemy territory, usually by helicopter; hence insertion.

inserted blades   Not forming an integral [monolithic] part of the disc.

inset balance   Mass located within movable surface.

inset hinge   On conventional control surface one whose axis is to rear of leading edge.

inset light   One flush with airport pavement.

inside wing   That pointing towards centre of a turn.

insolation   Solar radiation received, usually at Earth or by spacecraft.

INSP   Inspection.

inspectability   Unquantifiable measure of extent to which fault, esp. structural, may escape notice of inspectors because of inaccessibility or any other reason.

inspect and repair as necessary   Not to pre-ordained schedule.

inspection   Search of hardware for evidence of existing or impending fault condition.

inspin yaw   Yawing moment holding or accelerating aircraft into spin.

INST, inst   Instrument.

instability   *1* Structural condition in which strut, web or other member buckles under compressive load.

2 Aerodynamic condition in which slightest disturbance triggers gross disruption or flight of body.

3 Meteorological, normal meaning.

installation envelope   Overall three-view dimensioned outline drawings.

instantaneous readout   System with zero lag between sensor and display.

instantaneous VSI   One giving instantaneous readout, with accelerometer air pumps to counteract lag. Also called instant-lead.

INSTL   Installed, installation.

InstMC   Institute of Measurement and Control (UK, 1944–).

instruction to proceed   Informal document accepted as guaranteeing payment in advance of contract.

instrument approach   Made under non-visual guidance, normally from aids on the ground.

instrument approach area   Volume of sky in which non-visual landing aid operates.

instrument approach runway   One providing non-visual directional guidance for straight-in approach.

instrument error   Difference between indicated and true value.

instrument flight   Using instruments in place of external cues.

instrument flight rules, IFR   Rules applied in cloud or whenever external cues are below VFR minima which prohibit non-IFR pilots/aircraft.

instrument landing   Ambiguous but usually IMC landing without ground aids (thus, non-ILS).

instrument landing system, ILS   Standard ground aid to landing comprising two radio guidance beams (localizer for direction in horizontal plane and glideslope for vertical plane with usual inclination 3°) and two markers for linear guidance. See headings beginning *glidepath*, and *categories*.

instrument meteorological conditions   Conditions less than minima for VFR.

instrument monitor   Various meanings from traditional flight-test panel camera(s) to automatic systems for notifying faults, ensuring majority vote or isolating failed instrument (rare).

instrument rating   Endorsement to pilot's licence allowing flight in IMC.

instrument runway   Instrument-approach or precision-approach runway.

insulators   Poor conductors of electricity, heat, noise or other forms of energy.

Int   *1* Intersection (FAA).

2 Intercom.

Inta   Instituto Nacional de Tecnica Aeroespacial Esteban Terradas (Spain).

intake   *1* Air inlet to propulsion or internal system.

2 Narrow-chord leading member of upper flap in CCW augmentor wing, upstream of shroud and separated by a slot.

intake duct   See *inlet duct*.

intake stroke   See *induction stroke*.

intcp   Intercept[or].

integral construction   Made from single slab of metal by machining and/or etching, or by forging and machining, to finished shape.

integral stator Fabricating a section [typically containing six to eight vanes] of the turbine HP stator ring as a monolithic extension of a flame tube.

integral tank Tank formed by coating aircraft structure with sealant.

integrated acoustic structure Noise-absorbent structures forming load-bearing part of main structure.

integrated aeroplane One in which single shape (eg Gothic delta with no separate fuselage) serves all functions, with no demarcation between parts.

integrated cargo carrier Unpressurized payload carrier fitting cargo bay of Shuttle.

integrated circuit Microelectronic device fabricated by successive etching, doping etc of single-crystal (usually silicon) substrate.

integrated communications control Switches audio paths, controls and displays 40 channels of NVM, formats synthesized alerts and provides for antenna selection and Bite.

integrated decision support tool Used in CDM to bring critical National Airspace System status and traffic management data together, graphics offering what-if capability.

integrated drive generator Electric generator (alternator) made as one unit with CSD to give constant-frequency output.

integrated dynamic system Combination of helicopter rotor hub, main transmission, swash plates, control system and hydraulic servos into single unit (MBB).

integrated electronic standby instrument Solid-state replacement for pneumatic airspeed, altitude and horizon.

integrated flight system Computer-linked FCS and panel displays which to a large degree relieve pilot of need to exercise judgement [arguable definition].

integrated flight test Current [1990–2020] meaning is to demonstrate identification and tracking of ICBM targets in space.

integrated flight training One meaning is ab initio use of flight instruments.

integrated instrumentation display Combines engine, transmission, accessory systems, rotor track/balance and vibration monitoring plus caution/warning.

integrated power unit APU plus IDG.

integrated servoactuator Flight-control power unit with integral failure-correction and only electric [or other] valve inputs, devoid of mechanical input from pilot.

integrating accelerometer Accelerometer whose output signal is first or second-order integration with respect to time (viz velocity or position).

integrating circuit Electronic circuit whose function is to integrate (mathematically) one variable with respect to another (usually time).

integration Assembly of stages, boosters and payloads of spacecraft (post-* normally means after mating of payloads).

integrator *1* Device, usually digital, for giving numerical approximation to integration (mathematical).

2 Mechanical latch for devices such as an escape slide or brake chute enabling ground engineers to open up for routine maintenance.

integrity Validity of structure or system, functioning in design role after suffering damage; usually, but not always, a mechanical quality associated with avoiding mechanical breakup. Loosely, resistance to failure.

integrity beacon landing system Combines DGPS with ground-based pseudolites to give accuracy ± 10 cm (ESTOL).

intelligent missile Vague popular concept normally taken to mean self-homing.

Intelsat International Telecommunications Satellite Organization.

intensive flying Purpose is to log flight hours on new equipment at maximum rate under operational conditions.

intensive student area Regions of US airspace where IFR flight is restricted.

Inter Intermittent, also *INTMT*.

interaction parameter Basic measure of relative dominance of fluid motion or magnetic field in MHD and plasma physics.

interactive computer One capable of progressive dialogue with human operator, via displays and lightpen or other method.

intercalation Insertion of chemical compounds in plasma layers between planes of base material such as graphite to enhance electrical conductivity.

interception *1* Flight manoeuvre to effect closure upon another aircraft or spacecraft.

2 To capture and hold desired flight condition (eg, VOR radial or ILS).

interceptor *1* Aircraft or spacecraft designed to intercept, and if necessary destroy, others.

2 Small hinged strip to block local airflow, esp. between slat and wing or immediately to rear of slat, operated on one wing only by applying aileron.

intercept point Computed location in space towards which vehicle (eg interceptor aircraft or spacecraft) is vectored.

interchanger Variable gearbox in one axis of powered flight-control system.

intercom Communication system within aircraft using crew headsets or loudspeakers but without any radio emission.

interconnectors Tubes conveying flame from each gas-turbine combustor primary zone to neighbour [in US called flame tube, which has different meaning in UK].

intercontinental ballistic missile Land-based, range over 5,500 nm (6,325 miles) (USAF).

intercooler Radiator for rejecting excess heat in enclosed fluid system.

intercostal Short longitudinal structural member (stringer) joining adjacent frames or ribs, usually to support access door or equipment.

intercrystalline corrosion Originating and propagating between crystals of alloy.

interdiction See *air interdiction, battlefield air interdiction*.

interface Boundary between mating portions of system; can be mechanical (eg inlet duct and engine) or electronic (eg central computer and navigation display).

interference *1* Mutual interaction between solid bodies in fluid flow, eg upper and lower biplane wings (see *Prandtl* *).

2 Disruption of radio communications by unwanted emission on same wavelength, as from unscreened ignition.

interference drag Drag caused by aerodynamic interference.

interference fit   Fit between parts where male dimension exceeds female (several exact definitions).

interference foul   Physical conflict between fixed and moving parts, esp. between control column/wheel/yoke and an obstruction.

interference strut   Obstructs cockpit if flight controls locked.

interferogram   Display or photograph of interferometer patterns for precise measurement or aerodynamic research.

interferometer   Optical measuring system using divided light beam later rejoined to give phase interference seen as light/dark fringes.

interferometer array   Aerial (antenna) able to emit or receive simultaneously at large number of accurately related locations.

interior ballistics   Branch of ballistics concerned with bodies under propulsion.

interlacing   *1* Scanning technique for raster display in which all odd-number lines are scanned and then all even.

2 IFF technique in which pulse trains of different modes are transmitted sequentially (to achieve enough transponder returns per scan it is rare to interlace more than three modes).

interline   Between different air carriers, hence interlining.

intermediate approach   That part of approach from arrival at first navigational facility or pre-determined fix to beginning of final approach.

intermediate case   Gas-turbine casing; several meanings, eg between two compressor spools or over compressor spool downstream of fan.

Intermediate contingency   *1* Turboshaft power rating below emergency (max. contingency) level, usually allowed for 30 [sometimes 60] min.

2 Of afterburning turbofan or turbojet, usually means maximum cold thrust.

intermediate frequency   That to which signal is shifted between its reception and transmission.

intermediate maintenance   At least six meanings, the most common being maintenance performed on a military user airbase in a specialized workshop.

intermediate-pressure   Compressor/turbine spool between LP and HP in three-shaft engine, abb. IP.

intermediate range   Traditional figure for ballistic missiles is 1,500 nm, about 1,727 miles, 2,780 km (originally USAF).

intermediate rating   Intermediate contingency.

intermediate shop   Flightline fault-isolating system, esp. for avionics, synthesising all forms of EM signal for HUD, radar, etc.

intermetallics   Compounds or alloys in which atoms of two or more metals are arranged in complex structures in fixed ratios. Some are semiconductors, but immediate work is concentrated on aluminides of refractory metals.

intermittent duct   Resonant air-breathing engine; also known as intermittent jet, but more common term is pulsejet.

intermodal   Capable of use by more than one form of transport, ideally by air, rail and truck.

internal air system   All airflows and pressure differences having no direct effect on engine thrust or power.

internal balance   By control-surface area ahead of hinge

[called the shelf] fitting in vented chamber in fixed structure. See *compound shelf*.

internal burning   Solid propellant rocket grain whose exposed surface is along centreline [e.g., star centred].

internal combustion   Originally described prime mover whose fuel was burned in the cylinder, unlike steam engine; in gas turbine and nuclear era meaning is blurred.

internal efficiency   Rocket thermal efficiency.

internal engineering notice   Issued by manufacturer, esp. to customers.

internal fuel   Contained in aircraft or spacecraft, as distinct from fuel in removable or jettisonable tanks.

internal gearbox   In a two-shaft engine comprises bevel gears from adjacent ends of LP and HP shafts from which drive is taken to external gearbox[es].

internally blown flap   *1* Usually, large conventional flap through which main propulsion jet(s) or bleed air can be ducted.

2 Rarely, jet flap.

internal power   Generated in the aircraft (electrical, hydraulic, etc).

internals   Items inside cockpit.

internal service   Air route wholly within one country.

internal starter/generator   Built into centre of engine as part of it.

internal supercharger   Downstream of carburettor.

international   As applied to flight, service or route, one passing through airspace of more than one country (though departure and destination may be in same country).

International Accounting Unit   Used by NATO and in other multinational infrastructure programmes, originally equal to £ sterling prior to 1967 devaluation.

international airport   Designated as airport of entry and departure for international traffic, and provided with necessary extra facilities (typically 12).

international altitude   That shown by ISA-calibrated pressure altimeter set to 1013.25 mb.

international boost   Piston engine boost control set to ISA.

International knot   See *knot*.

International power   The b.h.p. a piston engine is rated to develop at full throttle at International rpm at specified altitude.

International rpm   Highest crankshaft speed permitted in climbing flight for period exceeding 5 min.

International Standard Atmosphere, ISA   That agreed by ICAO and still used as common standard; defines pressure (1013.25 mb at MSL, about 29.92 in Hg), temperature (15°C at MSL) and relative density up to tropopause (see *atmosphere*).

InterNIC   Internet network information centre.

inter-ocular   Between eyes.

interphone   Intercom serving crew stations, service areas and ground-crew jacks (2).

interplane strut   Joins superimposed [biplane] wings.

interplanetary   Between planets [assumed, within plane of their orbits].

interpolation   Process of calculating or approximating values of function between known values.

interpulse period   PRI.

interrogate   To transmit IFF, SSR or ATC signal coded to trigger transponders.

interrogation mode   Any mode in which signals include

code to trigger transponder; eg Modes A, B and D for ATC transponders.

interrogator   Radio transceiver scanning in synchronism with primary radar or SSR requesting replies from all co-operative airborne transponders to reply; replies sent to video displays.

interrupted Bite   Ground test facility initiated manually via on-board control panel to enhance detection/location capability of C-Bite.

interscan   Brief time between scans on a timebase.

Intersect   Strictly, intesect, intelligent sensors for control technologies.

intersection   *1* Point where centrelines of runways coincide, hence * departure starts T-O at *.

2 Point on Earth, and vertical line through this point, where centrelines of airways cross.

interservices   Linking turbofan core to aircraft, thus * strut, * couplings, * interface.

Intersputnik   Soviet Bloc comsat system.

interstage   Space launcher airframe section between stages designed as major assembly housing guidance for other systems but without propulsion.

inter-tropical front, ITF   The assumed giant front where tradewinds meet N and S of Equator, also called ITCZ; not a normal front.

inter-turbine burner   Compact combustion system between HP and LP turbines in reheat-cycle engine.

intervalometer   Mechanical or electronic system controlling spacing of events (eg reconnaissance photographs).

in the blind   Without external cues, ie no visual reference outside aircraft.

in the groove   In the desired flight condition, esp. correctly aligned on the approach.

in the slot   Correctly lined up for landing.

intl   International.

INTMT   Intermittent.

Into   Intelligence officer.

into the mission   Measured from T-O at end of countdown, thus 30 s ***.

INTPS   Integrated navigation and tactical plotting system.

INTR   Interior.

intra-flight data-link   Secure communications between pilots and their sensors.

intraformation positioning system   Allows aircraft to fly close formation in blind conditions.

intramarket   Market for civil air transport in particular geographical region, which can be a single large country.

intrascope   See *borescope*.

intravehicular   Between two spacecraft.

intrepid birdman   Jokey reference to any pilot.

INTRG   Interrogated, interrogator.

INTRP   Interrupted, interruption.

intruder   *1* Aircraft engaged on interdiction, esp. against hostile aircraft and airfields (not necessarily by night).

2 An altitude-reporting aircraft considered to be a potential threat and processed by TCAS threat-detection logic.

intst   Intensity.

INTXN, intxn   Intersection.

INU   *1* Inertial navigation unit.

2 Inertial nav/attack unit.

Invar   Alloys formulated for near-zero coefficient of thermal expansion; *36 [36 per cent Ni] widely used for large CFRP moulds.

inventory   Complete list of hardware, esp. of items assigned to military units.

inventory service   Assigned to operational unit, including training units, but excluding those in evaluation, development, research and other non-operational status.

inverse monopulse   Missile or other guidance radar in which target Doppler shift is detected early in RF amplifier chain instead of at late stage.

inverse-square law   States point-source radiation and most other emissions fall off in intensity as square of distance from emitter.

inverse synthetic-aperture radar   Use of SAR/DBS technique but using a fixed (or moving) radar to integrate successive echoes from a moving target.

inversion   Local region of atmosphere where lapse rates are negative, ie temperature increases with height.

inversion point   Height at which inversion ceases and normal lapse rate begins.

inverted bipolar   Inverted gate.

inverted engine   Piston engine with crankshaft above cylinders.

inverted-flight valve   Commonest form is fitted in delivery from fuel tank to maintain feed under negative-g.

inverted gate   Bipolar with emitter downwards.

inverted gull wing   Seen in front elevation, slopes down from body with pronounced anhedral and then (suddenly or gradually) slopes up with dihedral to tip.

inverted loop   See next two entries. BS: "A complete revolution in flight in a vertical plane about lateral axis with upper surface on outside of curved flight path." Some insist * must be started with aircraft inverted, thus from bottom of manoeuvre.

inverted normal loop   Normal loop begun from inverted position.

inverted outside loop   See *bunt*.

inverted spin   Spin in inverted position. Strangely, BS definition is: "A spin with negative mean angle of incidence" [AOA is meant].

inverted vee-engine   Two inclined banks of cylinders below crankshaft.

inverter   Electrical machine or static rectifier that inverts polarity of each alternate AC sine wave to give DC output.

investment casting   Casting complex shapes in ceramic moulds formed as coatings on wax patterns which are then melted and run out, hence term lost-wax casting.

investment prediction   Assessment of required inventory of spares and GSE.

INVH   Integrated night-vision helmet.

inviscid   Without viscosity.

invocon   Innovative concepts in systems engineering.

INVOF   In the vicinity of.

involuntary boarding refusal   Passenger denied access to aircraft, for whatever reason.

involuntary retirement   Of aircraft, caused solely by legislation, esp. environmental non-compliance.

INVR   Institute of Noise and Vibration Research (UK).

INVRN   Inversion [weather].

inwales   Longitudinal members at junction of flying-boat topsides and deck. Often = gunwales.

in-weather   Flying in conditions that are wet and/or icy, but not poor visibility (USAF).

In-WX   In weather.

IO   *1* Identification officer (air-defence systems).

2 Direct-injection, opposed cylinders (US piston engines).

3 Image orthicon.

4 Information officer (US).

5 Information operations (US).

6 Integrated optics.

I/O   Input/output; C adds computer.

Io   Mean spherical candle power.

IOA   Initial operational assessment.

IoA   Institute of Acoustics.

IOAT   Indicated outside air temperature.

IOB   International Operations Bulletin.

IOBP   Imagery on-board processor.

I-obs   Index of crew and maintenance observations.

IOC   *1* International Oceanographic Commission (UN).

2 Initial operational [or operating] capability [or clearance].

3 Indirect operating cost.

4 International order of characters.

I/OC   Input/output concentrator, or controller.

I/OCE   Input/output control element.

IOCU   International Organization of Consumers' Unions.

IOD   *1* Inflight-opening doors (Stovl engine).

2 Internal- [as distinct from foreign] object damage.

iodine   Shiny black solid element, symbol I, density 4.9, sublimes to purple vapour, see silver iodide.

IOE   *1* Initial operating experience.

2 In-orbit experience.

IOFP   Intensive operational flying programme.

IOI   Item of interest/importance.

IOIS   Integrated operational intelligence system.

I-omega, Iω   Angular momentum.

ION   Institute of Navigation.

ion   Electrically charged atom or group of atoms; can be in solid or solution or free; charge can be positive (missing electron) or negative (usually through extra electron).

IONDS   Integrated operational nuclear detection system.

ionic propulsion   See *ion rocket*.

ionization   Conversion of atoms to ions.

ionization potential   Work measured in eV necessary to remove or add electron in ionization.

ionization screen   Barrier to charged particles which would otherwise damage human tissue.

ionogram   Plot of radio frequency against pulse round-trip time, ie electron density level for reflection (approx. equal to altitude of reflective layers).

ionopause   Ill-defined base of ionosphere; also known as D-region.

ionosphere   Entire ionized region of Earth's atmosphere (Kennelly-Heaviside, Appleton, E and F layers). Not Van Allen Belts.

ion rocket   Propulsor, usually small thruster, generating high-velocity jet of ions in electrostatic field.

IOP   *1* Institute of Petroleum.

2 Information operations planning [S adds system].

I/OP   Input/output processor.

IOR   *1* Indian Oceanic Region.

2 Immediate operational requirement.

IOS   *1* Instructor operating [or operated] station [or system].

2 Internal operating system.

IOSA   *1* Integrated optical spectrum analyser.

2 Integrated overhead Sigint architecture.

IOT   *1* In-orbit test.

2 Initial officer training.

IOTE, IOT&E   Initial operational test and evaluation.

IOVC   In overcast.

IP   *1* Initial point.

2 Intermediate-pressure.

3 Instructor (rarely, instrument) pilot.

4 Identification pulse (SSR).

5 Identification position (IFF).

6 Intellectual property (company law), R adds right[s].

7 Initial provisioning of spares.

8 Industrial participation.

9 Internet protocol.

10 Instrumented prototype [A adds aircraft].

11 Ice pellets.

12 Intercept point (radio dBm).

13 Initial production.

I/P   Identification position (US).

IPA   *1* Independent Pilots' Association (UK).

2 Instrumented production aircraft.

IPACG   International Pacific ATC Co-ordinating Group.

IPAD   *1* Integrated programs for aerospace [vehicle] design.

2 Improved processing and display [S adds system].

IPARS, Ipars   International programmed airlines reservation system.

IPAS   Integrated pressure air system.

IPAT   Inertial pointing-aided tracking.

IPB   *1* Illustrated parts breakdown.

2 Intelligence preparation of the battlespace, one of the pillars of PBA.

IPC   *1* Intermittent positive control (ATC backup, IPS used in US).

2 IP compressor.

3 Illustrated parts catalogue.

4 Integrated processing cabinet.

IPC-ASA   IPC (1) automatic separation assurance; advises conflicts to DABS/Adsel aircraft.

IPCC   International Panel on Climate Change.

IPCS   *1* Institution of Professional Civil Servants (UK).

2 Intelligent power-control system.

IPD   *1* Instituto de Pesquisas e Desenvolvimento (Brazil).

2 Improved point defence [MS adds missile system].

3 Initial production delivery.

4 Integrated product design [by prime plus suppliers], or delivery.

5 Imagery processing and dissemination; S adds system.

IPE   *1* Institution of Production Engineers (UK).

2 Increased- (or improved-) performance engine.

3 Interconnect, passive and electro-mechanical.

IPEC   *1* Inflight passenger entertainment and communications; C adds conference, S systems.

2 Integrated planning and execution center.

IPES   Individual passenger entertainment system.

Ipex   Immediate-purchase excursion; low-price fare without advance booking and without guarantee of seat.

IPF Integrated, or integration and processing facility.

I/PF Identification position feature.

IPFA Inspection des Programmes et Fabrications de l'Armement (F).

IPG Indian Pilots' Guide.

IPI *1* Intercept pattern for identification.
2 Inertial position insertion.
3 Initial protocol identifier.

IPID *1* IR perimeter intrusion detection.
2 Indefinite quantity, indefinite delivery.

IPK International Prototype Kilogramme, a platinum body kept at Sèvres.

IPL *1* Illustrated parts list.
2 Image Product Library (NIMA).

IPM *1* Interplanetary medium.
2 Immediate past Master (GAPAN).

IPMS *1* International Plastic (singular) Modelers (US spelling) Society.
2 Integrated platform management system.
3 Institution of Professionals, Managers and Specialists (UK).

IPN Iso-propyl nitrate.

IPNVG Integrated panoramic NVGs.

IPO *1* Initial public offering.
2 Integrated product ownership.

IPP *1* Institut für Plasma-Physik (G).
2 Information-processing panel.
3 Industrial preparedness planning (US).
4 Impact-point prediction.

IPPS Integrated power plant system.

IPR *1* Intellectual property rights, has particular relevance to software.
2 Inches per revolution.
3 Internet protocol router.

IPRA Independent precision radar approach.

IPS *1* Information presentation or processing system.
2 Instrument-pointing system.
3 Inlet particle separator.
4 Intelligence and planning squadron (RAF).
5 Intermediate pitch stop (helicopter).
6 Interactive pilot station.
7 Intermittent positive control (ATC).
8 Integrated power system.

i.p.s. Instructions per second.

IPSA Infirmières-Pilotes-Secouristes de l'Air (F).

IPSE Integrated product (or project) support environment (software).

IPT *1* Integrated project team; L adds leader (UK).
2 Integrated product team (US).
3 Intermediate-pressure turbine.
4 Integrated physiological trainer [a simulator].
5 Intelligent power terminal.

IPU *1* Interface processor unit.
2 Integrated power unit [usually electric].
3 Image processing unit.

IPV *1* Improve.
2 Icy runway, also IR.

IPW *1* Institut für Physikalische Weltraumforschung (G).
2 Ice-pellet shower.

IP/XML Internet protocol, extensible markup language.

IQA The Institute of Quality Assurance (UK).

IQSY International Quiet Sun Years [1964–65].

IQT Initial qualification training, or testing.

IR *1* Infra-red.
2 Instrument rating [or route, or rules].
3 Inspection report.
4 Incident report.
5 Initial reserve (RAF).
6 Ice on runway.

Ir Rotor-system inertia.

IRA *1* Inertial reference assembly.
2 Infra-red astronomy.
3 Initiated by requesting authority.

IRAC *1* Interdepartmental Radio Advisory Committee (US).
2 IR array camera.

IRAD *1* Investigate, research and define.
2 Independent research and development.
3 IR acquisition and designation [S adds system].

IRAN Inspect and repair as necessary.

IR&D Internal research and development, ie company funded.

IRAS *1* Interdiction/reconnaissance attack system.
2 IR astronomy satellite.

iraser IR laser (not recommended).

Irasi, IRASI IR atmospheric sounding interferometer.

IR augmenter Flare or otherwise intense IR source carried by vehicle to facilitate IR tracking.

IRAWS IR attack weapon system.

IRB *1* Industrial revenue bonds.
2 Integrated Requirements Board.

IRBM Intermediate-range ballistic missile.

IRC Industrial Reorganisation Corporation (UK).

IRCA Integrated real time in the cockpit/Real-time out of the cockpit for combat aircraft (AFRL).

IRCAM IR camera acquisition module.

IRCC International Radio Consultative Committee.

IRCCD IR-sensitive charge-coupled device.

IRCCM IR counter-countermeasures.

IRCM IR countermeasures (sometimes rendered IRC). Measures adopted to minimise IR emissions or render them misleading (eg by ejecting flares which it is hoped hostile IR-homing weapons will prefer instead of one's own aircraft).

IRCS Intrusion-resistant communications system.

IRD *1* Interoperability requirements document.
2 Independent (or internal) research and development (company-funded).
3 Inlet ram drag.
4 Integrated receiver/decoder.

IR decoy Intense heat source intended to distract IR-homing missiles.

IR detector Device containing cell sensitive to IR which raises electrons (cryogenically cooled to reduce noise) to higher energy level.

IRDF IR direction-finding.

IRDS IR detecting set.

IRDT Inflatable re-entry and descent technology.

IRDU IR detection unit.

IRE *1* Institution of Radio Engineers (UK).
2 IFF reply evaluator.
3 Instrument rating examiner.
4 Internal roll extrusion.

IR/EO IR electro-optical.

Ireps Integrated refractive effects prediction system (anaprop).

IREW IR electronic warfare.
IRF *1* Immediate Reaction Force (NATO).
 2 Institute of space physics (Sweden).
IRFFE Intelligent RF front end.
IRFI International Runway Friction Index.
IRFIS Inertial-referenced flight inspection system.
IRFITS IR fault-isolation test system.
IRFNA Inhibited red fuming nitric acid.
IRFPA IR focal-plane array.
IR guidance Use of IR seeker cell and combined optics and flight control to make vehicle home on suitable heat source.
IRI Inadvertent runway incursion.
IRIA Institut de Recherche Informatique et d'Automatique (F).
iridium Ir, hard, inert silvery metal, density 22.562, MPt 2,410°C.
IRIG Inter-range instrumentation group.
Iris *1* IR [or integrated radar] imaging system.
 2 Inferential retrieval index system.
 3 IR interferometer spectrometer.
iris scanner Looks into human eye to give unequivocal confirmation [or not] of identity.
IRJ IR jammer.
IRLS *1* IR linescan.
 2 Interrogation, recording and location system.
IRM *1* Information resources management.
 2 Intelligent robotics manufacturing.
 3 Ion-release module.
IRMP Inertial reference mode panel.
IRMS Integrated radio management system.
IRMW IR missile warning [S adds system or subsystem].
I-RNAV Boeing term for RNP-RNAV, meaning integrated-RNAV.
IROC International Requirements Oversight Council (US).
iron *1* Metallic element, Fe, density about 7.86, MPt 1,539°C.
 2 All magnetic parts of compass, aircraft etc, of whatever material.
 3 Colloquially, aircraft (especially warbird).
iron ball General term for a common magnetic RAM, powders of ferrites or carbonyl iron embedded in flexible matrix or sprayed on, see *iron paint*.
iron bird Ground rig to test major aircraft system[s].
iron bomb Simple free-fall bomb (colloq.).
iron compass Railway used as navaid (US colloq.).
ironless Constructed of non-magnetic materials.
iron mike Autopilot (colloq., obs.).
iron paint RAM (2) sprayed-on coating comprising ferrite magnetic balls of microscopic size contained in epoxy binder. Multiple layers give desired thickness.
iron reserve Fuel for taxiing at destination.
irons See *eating irons*.
Irotis, IR-OTIS IR optronic tracking and identification system.
IRP *1* Intermediate rated power.
 2 Interphone receptacle panel.
 3 Integrated refuelling panel.
 4 Intruder role player.
IRPG IR plume generator.
IRR *1* Integral rocket/ramjet.
 2 Infra-red radiation, or radiometer.

IRRCA Integrated real-time info into the cockpit/real-time info out of the cockpit for combat aircraft.
irreversible control Position of output governed only by input (thus, * flight control is unaffected by air load on surface).
irreversible screwjack Thread chosen so that linear position governed solely by rotation, not by operating load.
IRROLA Inflatable radar-reflective optical location aid.
irrotational flow Individual elements or parcels do not rotate about own axes.
IRRS IR reconnaissance system.
IRS *1* Inertial reference system.
 2 Improved radar simulator.
 3 Indian remote satellite.
 4 IR signature, or spectrograph.
 5 Internal Revenue Service (US).
 6 Interface requirements specification.
IR seeker See *IR detector*.
IR signature Complete plot of IR emissions from given source or vehicle, usually in form of intensity plotted against wavelength.
IRSM IR surveillance measures.
IR telescope camera system Hand-held, on long wand, for inspection [or search for explosives or drugs] in dark places.
IRSP Instrumentation radar support programme.
IRST IR search and track; S adds system.
IRT *1* Instrument rating test.
 2 Incident response team.
IRTH IR terminal homing.
IRTS *1* Initial radar training simulator.
 2 IR threat simulator.
IRTT IR tow target.
IRU Inertial reference unit.
IR/UV Infra-red [and] ultra-violet.
IRV Infra-red vision.
IRVAT Infra-red video automatic tracking.
Irvin Trade name of Irving Air Chute (parachutes).
Irving flap Split flap pivoted to links near mid-chord and lowered by actuator thrust on the leading edge.
IRVR Instrumented runway visual range.
IRWR IR warning receiver.
IS *1* Institut für Segelforschung (G).
 2 Implementation staff.
ISA *1* International standard atmosphere.
 2 International Stewardess, Hostess and Airliner Association, now ISAA (2).
 3 Instrument Society of America.
 4 Instruction set architecture.
 5 Inter-service agencies.
 6 Industry standard architecture.
 7 Integrated servoactuator.
ISA+21 International Society of Women Airline Pilots [ISA, international social affiliation].
ISAA *1* Israel Society of Aeronautics & Astronautics.
 2 International Steward/ess & Airliner Association.
ISA/AMPE Inter-service agencies/automated message processing exchange.
ISABE International Society for Air-Breathing Engines.
ISAC Institute of Space and Aeronautical Science (J).
ISADS Integrated strapdown air-data system.

ISAE   International Society of Airbreathing Engineers.

ISAF   International Security Assistance Force.

ISAG   International Simulation Advisory Group.

ISAHRS   Improved standard AHRS.

ISAL   Inverse synthetic-aperture lidar (or ladar).

isallobar   Line joining all places where, over given period (typically 3 h), barometric pressure at surface changes by same amount.

ISAMP   International Society of Aviation Maintenance Professionals (1996).

ISAN   Integrity of satellite navigation (G, research programme).

IS&S   Integrated systems and solutions, merger of command and control intelligence and data fusion (DoD).

ISAR   *1* Inverse synthetic-aperture radar.

2 Integrated, or intelligence, surveillance and reconnaissance; C adds cell.

ISAS   *1* Institute of Space and Aeronautical Science (India).

2 Institute of Space and Astronautical Science (J).

ISASI   International Society of Air Safety Investigators (office Sterling, VA, US).

ISA-SL   ISA (1) at sea level.

ISAT   Integrated site acceptance test.

Isatis   Integrated system for ATIS (1 or 2) (F).

ISAW   International Society of Aviation Writers.

ISB, isb   Independent sideband.

ISBA   Inertial-sensor-based avionics.

IS-BAO   International standard[s] for business aviation, or aircraft, operations.

ISC   *1* Integrated semiconductor circuit.

2 Instrumentation system[s] coupler.

3 Intercom set control.

4 Integrated systems controller.

5 Industry Steering Committee (Int.).

ISCA   Internatonal Steering Committee for Consumer Affairs.

ISCC, ISC²   Integrated space command and control.

ISCS   Integrated sensor control system.

ISD   *1* In-service, or initial service, date.

2 Interim situation display (FAA).

3 Integrated systems development.

ISDN   Integrated services digital [or data] network.

ISDOS   Information system design and optimization system.

ISDS   *1* Image switching and distribution system.

2 IRCM, or improved, self-defence system.

ISDU   Inertial system display unit.

IS/DX   Integrated services, digital exchanges.

ISE   *1* Interconnected stabilizer/elevator.

2 Intelligent synthesis environment.

3 In-service evaluation.

Iseman rivet   Hollow blind rivet set by driving drift pin from manufactured-head end.

isenergic   Without change in total energy per unit mass of fluid (in bulk or along a streamline).

isentropic   Without change in entropy (usually along streamline, with respect to time).

ISF   *1* Industrial space facility, two modules, one remaining in space and the other shuttling raw materials and completed products.

2 Integrated support facility.

ISG   *1* Inflatable survival gear (esp. liferafts, escape slides).

2 Internal starter/generator.

ISH   Intermediate system hello.

ISHM   Internal Security and Hazardous Materials office (FAA).

ISI   Institute for Scientific Information (international corporation, Philadelphia).

ISIS   *1* Integrated spar inspection system.

2 Integrated strike and interception system.

3 Integrated standby instrument system.

4 International satellites for ionospheric studies.

ISIT   Intensified silicon intensified target [*sic*].

ISJTA   Intensive student jet training area (US armed forces).

ISL   Inactive-status list.

island   *1* Enclosed area on 2-D graph or plot.

2 Superstructure above aircraft-carrier deck.

ISLN   Isolation.

ISLS   Interrogation, or interrogator, side-lobe suppression.

ISM   *1* Institut Suisse de Météorologie (Switzerland).

2 Internationale Segelflugzeugmesse (G).

ISMLS   Interim-standard MLS.

ISN   Inter-simulator network.

ISNS   International satellite-navigation system (FAA).

ISO   *1* International Organization for Standardization. Numerous aerospace documents, eg ISO 9000, International standards for quality; ISO. 9002, 19-part qualification of companies engaged in aerospace logistics, maintenance and training.

2 Also **iso**, isolation, isolated.

iso   Prefix, equal.

isobar   *1* Line on map or chart joining points of equal atmospheric pressure, usually reduced to MSL by ISA law.

2 Line on tunnel model or free-flight vehicle joining points of equal aerodynamic (surface stagnation) pressure.

isobaric range   Band of flight altitudes over which cabin pressure can be held constant.

isobar sweep   Angle between transverse axis and local direction of isobar (2).

isocentre   Intersection of interior bisector of camera tilt angle with film plane.

isochoric   Without change in volume.

isochrone   Line joining points of equal time difference in reception of radio signals.

isocline, isoclinic line   Line joining points of equal magnetic dip.

isoclinic wing   One maintaining constant angle of incidence while flexing under load.

isocontour   On weather radar = contour.

isodoppler contour   Joins all points of equal Doppler velocity.

isodynamic line   Line joining points of equal horizontal magnetic field intensity.

iso-echo   Radar display mode in which cloud turbulence centres appear dark or coloured, width of surround indicating rain/turbulence gradient; also called contour mode.

isogonal, isogonic line   Line joining points of equal magnetic declination (variation).

isogram   See *isopleth* (also specif. line joining places where meteorology event has same frequency of occurrence).

isogrid Pattern of reinforcing webs stiffening thin sheet (machined, chem-milled etc), usually based on equilateral or isoceles triangles.

isgriv Line joining points of equal angular difference between grid and magnetic north (grid magnetic angle).

isohyet Line joining places of equal rainfall.

Isolane PU/AP + al (F).

isolated drag Drag of a body considered in isolation; thus * of a jet-engine pod may be significantly less than its drag when hung on the aircraft.

ISOL[D] Isolated.

isolux Line joining points of equal light intensity.

isomers Compounds having same chemical formula but different structure.

isometric Drawing projection in which verticals are vertical, other two axes are equally inclined, and distances along all three axes are correct.

isometric switch Governs lock-on to aerial radar target.

iso-octane Hydrocarbon of paraffin series used as reference index for anti-knock ratings, normally C_8H_{18}.

iso-opinion charts Graphic attempts to portray general consensus of pilot evaluations, eg plotting undamped natural frequency against short-period damping.

ISO-PA ISO(1) protocol architecture.

isopleth Line joining points of a constant value of a variable with respect to space or time.

isopod Patented multi-mode packaging system.

iso-propyl nitrate Monopropellant for rockets and fuel for starters.

isopycnic Of unchanging density, or of equal densities, with respect to space or time.

ISOR Initial statement of requirements.

ISOS Inertially stabilized optronic sensor.

isoshear Joins places of equal wind shear.

isostatic Under equal pressure from all sides.

isotach Line joining points of equal wind speed, irrespective of direction.

isotherm Line joining points of equal temperature.

isothermal At constant temperature.

isothermal atmosphere Hypothetical atmosphere in hydrostatic equilibrium with constant temperature, also called exponential.

isothermal change Change of volume-pressure humidity and possibly other variables of perfect gas or gas mixture at constant temperature.

isothermal layer See *stratosphere*.

isotope Radioactive form of (normally non-radioactive) element.

isotope inspection X-ray using isotope as energy source.

isotope power Using isotope radiation as source of heat energy in closed fluid cycle.

Isotran Isolator transition, isolates Gunn diode from load mismatches and couples to waveguide and coaxial conductor.

isotropic Having same properties in all directions.

ISP *1* Interim support plan (FAA).
2 Internet service provider.
3 Integrated switching panel.

Isp Specific impulse.

ISPA International Society of Parametric Analysts.

ISQC Intersound quality-control facility [checks videocassettes].

ISR *1* Initial Service Release.
2 Intelligence, surveillance, reconnaissance.
3 Integrated signature reduction.
4 Image storage and retrieval.
5 Interrupt[ed] service routine.

Israc, ISRAC Intelligence, surveillance and reconnaissance cell (NATO).

ISRBM Intelligence, surveillance and reconnaissance battle manager (USAF).

ISRC International Search and Rescue Convention.

ISRO Indian Space Research Organization.

ISS *1* Inertial, or integrated, sensor system.
2 Instrument subsystem.
3 Indonesian Space Society.
4 International Space Station [MCETF adds management and cost evaluation task force] (US + R).
5 Information support system (ATC).
6 Integrated Satellite System.

ISSC Integrated system support contract.

ISSI International Space Science Institute (Berne).

ISSM Institute of Safety and Systems management, (U of S California, Pasadena).

ISSN Intermediate system subnetwork.

ISSR Independent secondary surveillance radar.

ISST ICBM silo superhardening technology.

IST Institute for Simulation and Training (University of Central Florida).

Istar *1* Integrated system test of an air-breathing rocket (NASA).
2 Intelligence, surveillance, target acquisition and reconnaissance.

ISTAT International Society of Transport Aircraft Trading.

ISTD Integrated space-technology demonstration (USAF).

ISTN Interfacility Satellite Telecommunications Network.

ISTP Integrated Space Transportation Plan (NASA).

ISTR Integrated systems test rig.

Istres Location of chief French Government flight-test establishment, the Centre des Essais en Vol (CEV), whose other locations are at Brétigny and Cazaux.

ISTS Integrated space transport system.

ISU *1* Intelligent, or inertial, or integrated, sensor unit.
2 Ignition servo unit.
3 Initial signal unit.
4 Intercommunication set [control] unit.
5 International Space University (Strasbourg).

ISURSS Interim small unit remote sensing system[s].

ISV *1* Intensified silicon target vidicon.
2 In-seat video [TV screen].

ISVR Institute of Sound and Vibration Research (Southampton, UK).

ISWL Isolated single-wheel load.

IT *1* Inclusive tour.
2 Information technology.
3 Interactive touchscreen.
4 Integral terminal, an HMI.
5 Independently targetable.

$\mathbf{I_T}$ Total impulse.

ITA *1* Institut du Transport Aérien (Int.).
2 Instituto Tecnologico de Aeronautica (Brazil).
3 International trade in arms/armaments.

$\mathbf{I_{TA}}$ Motor specific impulse.

ITAA *1* Inspection Technique de l'Armée de l'Air (F).
2 Information Technology Association of America.
ITAC Integrated tactical-aircraft control.
ITACS Integrated tactical air control system (USAF).
ITALD Improved theater [or tactical] air-launched decoy.
ITAM Institute for Theoretical and Applied Magnetics (R).
ITAP Interim track analysis program (US).
ITAR *1* International traffic[king] in Arms Regulations.
2 Integrated terrain access and retrieval system.
Itars Integrated terrain access and retrieval system.
ITAS *1* Improved tracking adjunct system.
2 Integrated tactical avionics system.
3 Integrated tower/approach system.
4 Improved target-acquisition system.
ITAV Ispettorato Telecommunicazioni e Assistenza al Volo (I).
ITB *1* Integrated testbed.
2 Inter-turbine burner.
I$_{TB}$ Burn time impulse.
ITC *1* Inclusive-tour charter.
2 Investment tax credit.
ITCI Interhost through check-in.
ITCM Integrated tactical countermeasures (USA, USAF).
ITCS *1* Integrated track and control system for RPV.
2 Integrated target control system.
3 IR telescope camera system.
ITCZ Inter-tropical convergence zone (see *inter-tropical front*).
ITD Interactive-touchscreen display.
ITE *1* Integral throat entrance.
2 Involute throat and exit [rocket nozzle].
3 Impact to egress [time spent on planetary surface].
ITEA International Test and Evaluation Association (office Fairfax, VA, US).
ITEC *1* Involute throat and exit cone (rocket).
2 Interoperability through European co-operation (Int.).
ITEM Integrated test and maintenance system.
Items Integtrated turbine-engine monitoring system.
ITER Improved triple ejector rack.
ITF *1* Inter-tropical front.
2 Intelligence task force (US).
3 International Transport Workers Federation (125 countries, office London).
ITGS Integrated track guidance system.
ITM *1* Information technology management, ground-based ADMS/EDMS.
2 Institute of Travel Managers in Industry & Commerce (UK).
ITMS Ion-trap mobility spectrometry.
ITO *1* Independent test organization (software V&V).
2 Indium-tin oxide.
3 Instrument takeoff.
ITOC Integrated tasking and operations centre.
ITOS Improved Tiros operational system.
ITP *1* Instruction [or intention] to proceed (UK).
2 Initial technical proposal (US).
3 Invitation to propose.
4 Intent[ion] to purchase.
ITPR IR temperature-profile radiometer.

ITPS International Test Pilots' School (UK).
ITR Integrated-technology rotor (USA).
ITRR Imaging-time resolved receiver.
ITS *1* Initial Training Squadron (rarely, school) (RAF).
2 Integrated test, trajectory or tower, system.
ITSE Integrated test and support environment.
ITSO Integrated training suite operations.
ITSS Integrated tactical surveillance system (USN).
ITT *1* Inter-stage turbine temperature.
2 Invitation to tender.
3 Innovation and technology transfer.
4 Institute of Travel and Tourism (UK).
ITTCC International Telegraph & Telephone Consultative Committee.
ITTS *1* Instrumentation targets and threat simulations.
2 Integrated tower and terminal system.
IT21 Info technology for 21st Century.
ITU International Telecommunications Union (formed as I Telegraph U in 1865).
IT-UAV IT (1) for UAVs.
ITV *1* Instrument test vehicle.
2 Inert test vehicle.
3 Inclination to vertical (parachute).
4 Idling throttle valve.
5 Independently targeted vehicle.
ITW Initial Training Wing (RAF).
ITWS Integrated terminal weather system (US, under development).
ITX Inclusive-tour excursion.
IU Input, or interface, or instruments, unit.
IUAI International Union of Aviation Insurers (office London).
IUE International UV Explorer.
IUKAdge Improved UK Air Defence Ground Environment.
IUMS Integrated utilities management system.
IUOTO International Union of Tourist Organizations.
IUPS Internal uninterruptible power supply.
IUS Inertial upper stage.
IUT Instructor under training (US).
IV *1* Isolation valve.
2 Interactive voice [M adds module, R response].
3 Inverted-vee piston engine.
4 Intelligent vehicle.
I/V Instrument/visual controlled airspace.
IVA *1* Input video amplifier.
2 Istituto del Valore Aeronautico (I).
Invadize Patented process of ion-vapour deposition of aluminium on steel.
IV&V Independent verification and validation (software).
IVD *1* Interactive video disk.
2 Integrated voice and data; M adds modem, N network.
IVDU Intelligent visual display unit.
IVH Integrated vehicle health [M adds management, MS monitoring system].
IVI Interchangeable virtual instrumentation (sets software standards).
IVKDF Institut von Kármán de Dynamique des Fluides (Int., also called VKIFD).
IVMMS Integrated vehicle mission-management system.

IVMS   Integrated vehicle management system.

IVR   Instrumented visual range.

IVRI   Intelligent vehicle research initiative (NASA).

IVRS   Interim voice-response system.

IVS   *1* Intelligent vehicle system.

   2 Interactive video system.

IVSC   Integrated vehicle support, or subsystem, control[s].

IVSI   Instantaneous VSI.

IVSN   Initial voice-switched network (NATO).

IVV   Instantaneous vertical velocity.

IW   *1* Individual weapon.

   2 Imminent warning (UK, WW2).

   3 Information warfare [also InfoWar], D adds defense [UK = defence], O offensive, or officer.

IWAAS   Initial wide-area augmentation system.

IWAC   Integrated weapon-aiming computer.

IWASM   Interntional Women's Air and Space Museum (Cleveland, OH).

IWBP   Independent wide-band repeater.

IWC   Integrated weapon complex.

IWFN   Inhibited white fuming nitric; A adds acid.

IWG   Internet Working Group.

IWI   Interceptor weapons instructor.

IWIDS   Integrated weather information display system.

IWM   *1* Imperial War Museum (London, UK).

   2 Institution of Works Managers (UK).

I-W/M   Index of warnings and malfunctions.

IWR   Instrument weather rating, proposed by JAA to replace UK's IMC.

IWS   Integrated weapons system.

IYQS   International Years of the Quiet Sun.

Izod   Standard test for impact strength of notched specimen subjected to transverse blow[s] of known energy.

IZS   Satellite (R).

J

J *1* Turbojet (US military engine designation).

2 Jet route in Class-A airspace (civil airways).

3 Joule[s].

4 General EM signal power.

5 RPV direction aircraft (DoD aircraft designation prefix).

6 Special test, temporary (DoD aircraft designation prefix).

7 Polar moment of inertia.

8 Aircraft category, transport (1926–31), utility (1931–55) (USN).

9 Life jackets.

10 Current density.

11 Propeller advance/diameter ratio.

j *1* Mass flux (gaseous diffusion).

2 Square root of minus 1.

3 Usually as subscript, fully expanded jet.

4 Operator, 90° (electrical).

J2 Diesel fuel (G, WW2).

J3E Single-sideband suppressed-carrier mode, ie HF SSB.

J9 Joint Experimental Directorate (USA).

J-band EM radiation 30–15 mm, 10–20 GHz.

J-curves Atmospheric post-stall manoeuvres by thrust-vectoring aircraft [X-31].

J-dinghy Circular, for crew of 6 (RAF, WW2).

J-display Time base is ring near edge of display, echo blip position varies with range; main use radar altimeters.

J-nose Fixed leading-edge structure.

J-turn Cobra followed immediately by rapid [30°–40°/s] yaw to reverse direction of flight.

JA Judge-advocate.

JAA *1* Japan Aeronautic Association.

2 Joint Aviation Authorities (ECAC, office at Hoofddorp, Netherlands)

JAAA Japan Ag-Aviation Association.

JAATT Joint air attack team tactics (outcome of Jaws [1]); hence JAAT.

Jabo Fighter-bomber (G).

JaboG Fighter-bomber wing (NATO, Luftwaffe).

JAC *1* Joint airworthiness code (AICMA).

2 Junta de Aeronautica Civil (Chile).

3 Joint Air Component [HQ adds headquarters] (UK).

JACIS *1* Japan Association of Air Cargo Information Systems.

2 Joint applications command information system (multiservice, UK).

jack *1* Powered linear actuator.

2 Jack box.

jack box Socket for plugging in communications headset.

jackpad Strong plate, usually square, distributing support of lifting jack into surrounding airframe.

jackscrew Screwthread converting rotary power to linear (jack) output, usually irreversible (see *ball* *).

jack stall Aerodynamic load on the surface overcomes force applied by powered flight-control unit.

jackstay Hinged strut for bracing cowl or other large panel when in open position.

JACMAS Joint Approach Control Meteorological Advisory Service.

Jacobs-Relf Original equations for calculation of M_{crit} and related C_p.

Jacola Joint Airports Committee Of Local Authorities (UK).

JACTS Joint Air Combat Training Squadron (USAF/USN).

JAFE Joint advanced fighter engine (US).

Jafna Joint Air Force/NASA (US).

Jafü Jagdführer, fighter leader (G, WW2).

JAG Judge-Advocate-General.

Jagd Hunt (G), hence Jagdflieger [fighter pilot], Jagdflugzeug [fighter], Jagdgeschwader [fighter group (US = wing)], Jagdgruppe [fighter wing (US = group)].

JAGO Joint air/ground operations.

Jaguar Joint air/ground operations unified adaptive replanning [or preplanning] (USAF).

JAIC Japanese Aircraft Industry Council.

jam Obliterate hostile EM transmission (esp. radar) by powerful emission on same wavelength(s).

Jamac Joint aeronautical materials activity (USAF).

James Air-damped pendulum driving deceleration (g) pointer carried in braked truck to give friction measure on runway (see *JBD, JBI*).

jammer High-power emitter to jam specific wavelengths.

jammer support receiver Automatically scans through programmed range of frequencies of interest.

jamming *1* Noise * obliterates by sheer power.

2 Deception * attempts to mislead enemy by causing false indications on his equipment.

jam nut Thin lock-nut.

Janet flights Contractor flights bringing staff to classified Nellis/Tonopah sites.

Janis, janis Joint Army/Navy intelligence studies (US).

jankers Punishment, confined to camp (RAF, colloq.).

JAN-TX Joint Army-Navy technical documentation (US).

Janus system Doppler radar technique in which frequency shift is measured as difference between beams to front and rear.

JAOC Joint Aerospace Operations Center (USAF).

JAOPA Jamaica Aviators, Operators and Pilots' Association.

JAPHAR, Japhar Joint airbreathing propulsion for hypersonic applications research.

JAPNMS Jtids air-platform network management system (RAF).

JAR Joint Aviation Requirement[s] [eg Pt 23 light aircraft, 25 based on FAR-25 + national variants, 66 training for maintenance engineers, 145 approval for maintenance organizations; – AWO adds all-weather operations, E Europe, FCL flight-crew licensing, OPS aircraft operation, VLA very light aircraft] (EC).

jaric, JARIC Joint Air Reconnaissance [&] Intelligence Centre [now part of DGIIA] (UK).

JART Joint assessment and ranking team.

JAS *1* Joint Airmiss Section, now Airprox (UK, civil/military).

   2 Fighter, attack, reconnaissance (Sweden).

JASA Joint airborne, or avionics, Sigint architecture.

JASC Joint Airmiss Steering Committee [ECAC].

JASDF Japanese Air Self-Defence Force = air force.

JASIF Joint airborne signal intelligence family.

JASO Joint Air-Support Organization (UK).

Jaspo Joint Aircraft-Survivability Program Office (US).

JASS *1* Joint Anti-Submarine School (UK).

   2 Joint airborne Sigint system.

JASSM Joint air-to-surface standoff missile (US).

JAST Joint advanced strike technology.

JASU Jet aircraft start unit (US).

JAT Joint affordability team (DoD).

JATCCCS Joint advanced tactical command control/com system (US tri-service).

JATCRU Joint Air Traffic Control Radar Unit (UK).

JATE Joint Air Transport Establishment (UK).

JATEU Joint Air Transportation Evaluation Unit (despite the US word transportation, an RAF formation).

Jato Jet (ie, rocket) assisted take-off.

Jats, JATS *1* Jamming analysis and transmission selection (ECM).

   2 Joint Air Traffic Services.

Jaumann absorber Multilayer RAM (2) absorber consisting of sandwich of Salisbury screens graded from high resistivity at front to low resistivity at back.

JAWG Joint Airprox [previously Airmiss] Working Group (ICAO and UK NATS).

Jaws *1* Joint attack weapon systems (fixed and rotary-wing against armour hostile to USA).

   2 Jamming and warning system (ECM).

   3 Joint airport weather studies.

JAXA Japan Aerospace Exploration Agency (from October 2003, incorporates ISAS, Nasda).

Jazz Rock Joint Survival System/Regional Operational Control Center (Adcom).

JB *1* Jet-propelled bomb category (USAAF, WW2).

   2 Jet barrier.

J-bar Jet runway barrier (FAA).

JBD *1* James brake decelerometer.

   2 Jet blast deflector.

JBG JagtBombergeschwader [also JaboG] (G).

JBI James brake index.

JBPDS Joint biological point detection system.

JCAB Japanese Civil Aviation Bureau.

JCAP Joint Committee on Aviation Pathology.

JCAS Joint close air support.

JCASR Joint Committee on Avionics Systems Research.

J-catch Joint countering attack helicopters (US programme against hostile battlefield helicopters).

JCC Jorn Control, or Co-ordination, Centre.

JCCDC Joint CCD(4) Center.

JCDP Joint conceptual definition phase.

JCEWS Joint combat electronic-warfare system.

JCIT Joint combat information terminal (USA).

JCMP Joint Common Missile Program (US, could involve UK).

JCMPO Joint Cruise Missiles Project Office (USAF/USN).

JCMT Joint collection management tools.

JCP Joint Certification Procedures [WG adds Working Group] (JAA).

JCR (Not JRC) Joint Research Centre of EC7 based in Ispra and co-ordinating some 50 laboratories on remote sensing and related problems.

JCS *1* Joint Chiefs of Staff (US).

   2 Jet-capable ship.

JCU Joint common user.

JDA *1* Japan Defence Agency.

   2 Joint Deployment Agency.

JDAM Joint direct-attack munition (US).

JDCC Joint Doctrine and Concept Centre (UK).

JDCU Jamming detection control unit.

JDP *1* Joint defensive planner (USAF).

   2 Joint definition phase.

JEA Joint endeavour agreement.

JEC Joint Economic Committee (US Congress).

JEDEC Joint experimental (sometimes engineering) development of electronic components.

Jeep Escort carrier or CVE (WW2).

JEF Joint Expeditionary Force[s]; E adds experiment, X exercise, but [2002] also means experiment (USAF + others).

JEFS Joint Elementary Flying School (UK).

JEFTS Joint Elementary Flying Training School (RAF Barkston Heath, since 2003 called DEFTS).

JEM Jet-effects model.

jempers JEMPRS (colloq.).

JEMPRS Joint en-route mission planning rehearsal system [typically 10 laptops, Ethernet network and palletized EMS terminal] (US DoD).

Jemtos Jet-engine maintenance task-oriented system.

JENA Jet-engine neural analyser.

Jengo Junior engineering officer (RAF).

Jenisys Joint-effects network-integrated system solution.

JEPES Joint engineer planning and execution system.

Jeppesen Widely used commercially sold database of airway and airport charts and information, named for Ebroy Jeppesen, whose 1926–40 notes formed basis.

jerk Rate of change of acceleration, V (esp. in ECM target or input motion).

jerk offload Extraction of large unit load from airlifter by sudden pull through ramp door (by parachute, ground extraction system etc).

Jesus nut Holds main rotor to rotor shaft on helicopter or, especially, light autogyro.

JET Joint estimate team.

jet *1* High-velocity gas flow discharged from nozzle (eg from * 2 or from open-* wind tunnel).

   2 Turbojet engine (loosely, turbofan).

   3 Aircraft powered by 2 (loosely, also by rocket).

   4 Calibrated orifice(s) in carburettor.

   5 To travel by 3.

Jet A Turbine kerosene similar to JP-5; freezes about -40°C; available US only.

Jet A-1 Turbine kerosene; freezes below -50°C, flash above 37.8°C; standard commercial fuel.

jet advisory area Specific regions along jet routes extending 14 nm each side of route segment from FL240–410 (radar) or FL270–310 + FL370–410 (non-radar).

jet advisory service That offered to certain civil jets in jet advisory areas; IFR separation or (in radar areas) other facilities (FAA).

jet age Loosely, era in which jet (3) aircraft became dominant.

jetavator See *jetevator*.

Jet B Wide-range distillate fuel similar to JP-4; freezes below -60°C, vapour pressure 2–3 lb/sq in.

jet blast Disturbance caused by ground running jet engine, hence need for blast fence.

JETDS Joint electronics type designation system (NATO).

jet-edge shear Rate of change of velocity per unit radial distance at outer boundary of jet (1), including on large scale the velocity profile at edge of atmospheric jetstream.

jet engine Any propulsion system whose reaction is generated by a jet (1), thus a turbofan, turbojet or ramjet. Generally used to mean air-breathing, esp. turbofan/turbojet, and thus arguably not a rocket, and certainly not applied to space thrusters of ES, ion, plasma and similar types.

jetevator Small power-actuated flap, spoiler or ring on skirt or nozzle exit or rocket for TVC.

jet flap Flap through which passes high-energy gas or air flow, discharged along trailing edge.

jet-fuel starter Main-engine starter burning main-engine fuel.

Jethete Proprietary refractory sheet metal (high-nickel).

Jeti Jet-engine test instrumentation.

jet-impingement stagnation pressure That which would be achieved if a high-velocity jet were to be brought precisely to rest on a fixed surface.

jet lag Mild temporary symptoms produced in human beings by fast travel through large meridian difference, ie through five or more time zones.

jet lift Using jet-engine thrust to support V/STOL aircraft.

jetpipe Pipe carrying hot gas from engine core [in a turbojet, turboshaft or turboprop, from whole engine] to propulsive nozzle.

jet propulsion Aircraft propulsion by jet engine(s).

jet route High-altitude route system for aircraft with high-altitude navaids, normally extending from 18,000 ft AMSL to FL450 (FAA).

jet sheet High-velocity fluid flow of essentially two-dimensional nature.

jet shoes Astronaut shoes for weightless walking.

jetstream Quasi-horizontal wind exceeding 80 kt (148 km/hr) in warm air at sharp boundary with cold, high troposphere or stratosphere, mid latitudes, predominantly westerly.

jet tab See *jetevator*.

jettison Discard fuel, canopy, external stores or other mass to reduce weight or remove a hazard; hence * pipe(s), * pump(s), * handle or switch.

JETTS Joint effects tactical targeting system.

jet vane See *jetevator*.

jet velocity In turbojet or turbofan core, proportional to square root of absolute temperature. In rocket, proportional to square root of absolute temperature divided by mean molecular weight of jet.

jetway See *bridge*.

JEWC Joint Electronic Warfare Center (US).

Jezebel Passive acoustic ASW search system (see *Julie*) (US).

JFACC Joint Force[s] Air Component Commander.

JFACTSU Joint FAC Training and Standardization Unit (UK).

JFASCC JFACC plus space (US).

JFC *1* Jet fuel control.

   2 Joint Force, or Field, Commander.

JFCOM Joint Forces Command (US).

JFDP Joint force development process, broad initiative at inter-service co-operation (US).

JFET Junction field-effect transistor.

JFI Joint fires initiative.

JFHQ Joint Forces HQ.

JFN Joint fires network (mission planning).

J-FOCSS, J-fox Joint-force command support system.

JFS Jet-fuel starter.

JFTO Joint flight-test organization.

JG *1* Jagdgeschwader, fighter unit equivalent to UK group or US wing (G).

   2 Junior Grade (USN).

JGr Jagdgruppe, equivalent to UK wing or US group (G).

JGSDF Japanese Ground Self-Defence Force (= army).

JHC Joint Helicopter Command (UK).

JHCU Jamming-head control unit; P adds processor.

JHL Joint Heavy Lift, proposed multiservice logistics aircraft (US).

JHMCS Joint helmet-mounted cueing system.

JHSU Joint Helicopter Support Unit (UK).

JHU Johns Hopkins University.

JIAAC Junta for civil air accidents (Argentina).

JIADS Joint integrated air-defense system.

JIAWG Joint Integrated Avionics Working Group (US).

JIC *1* Jet-induced circulation.

   2 Joint industrial company, prime contractor interface in typical collaborative programme with members assigned by all participating nations.

   3 Just in case.

   4 Joint intelligence centre.

JICF Jet in cross-flow.

Jico, JICO Joint interoperability, or interface, control officer.

JIES Joint interoperability evaluation system.

Jifdats Joint-services in-flight data transmission system (US).

jig Hard tooling; any rigid frame in which a part (eg of airframe) is assembled.

JIGI Joint-Stars imagery geolocation improvement.

jigsaw Multi-lobed rotor forming moving element of wheel brake.

JILL Jet-induced lift loss.

Jimpacs Joint improved multi-mission payload aerial surveillance, combat-survivable.

jink To take sharp avoiding action, eg against AAA fire.

Jintaccs Joint interoperability of tactical command and control systems.

JIO Joint Intelligence Organisation (Australia).

JIOC Joint Information Operations Center (USSC).

JIP Joint Interface Program, between different systems or services (US).

JIRD Joint initial, or interim, requirements document.

J/IST Joint, or JSF, integrated subsystems technology.

JIT production Just in time.

jitter *1* Transmission of DME pulses with random

spacing, to avoid locking-on to another aircraft interrogating same beacon.

2 ECCM technique in which p.r.f. is made to vary unpredictably.

jitterbug   Hand-held rotary buffer/polisher.

jitters   JTRS.

Jiva, JIVA   Joint intelligence virtual architecture.

JJPTP   Joint jet-pilot training programme.

J-Lars, JLARS   Joint-liaison advanced radio system (US civil government).

J-lens, JLENS   Joint land-attack missile/cruise missile defense elevated netted sensor system.

JM   JTIDS module[s].

JMCC   Johnson Missile Control Center (NASA JSC).

JMCIS   Joint Maritime Command information system (US).

JMD   Joint manufacturing demonstration.

JMEM   Joint munitions effectiveness manual (US armed forces).

JMG   Joint Meteorological Group.

JMIC   Joint Military Intelligence College (Washington DC).

JMOD   JSAF(1) Block Modernization Program.

JMPS, Jumps   Joint mission planning system.

JMR   Jason microwave radiometer.

JMRC   Joint mobile relay centre.

JMSDF   Japanese Maritime Self-Defence Force, = navy.

JMSNS   Justification of major-system new start.

JMSPO   Joint Meteorological System Program Office (DoD).

JMTSS   Joint multichannel trunking and switching system.

JNC   Jet navigation chart.

JNLWD   Joint Non-Lethal Weapons Directorate (DoD).

JNR   Jamming-to-noise ratio.

JNT   Joint.

JNWPU   Joint Numerical Weather Prediction Unit (US 1954–).

JOAC   Junior Officers Air Course (RN).

Joanna   Joint airborne night navigation and attack.

JOAP   Joint oil analysis program (USAF).

Jo-bolt   Patented internally threaded three-part rivet.

JOC   *1* Joint operations centre.

2 Jet orientation course.

jock, jockey   Combat aircraft pilot (colloq.).

JOCS   Joint operations command system.

joggle   Small vertical offset along edge of sheet or strip to allow it to overlap adjacent component.

joggling   Local squeezing of sheet-metal parts to improve abutment of mating surfaces.

Johanssen block   Super-accurate metal block for reference of various dimensions and surface finishes.

Johnson noise   RF thermal noise.

joined wing[s]   *Diamond wing.*

Joint Aerospace Operations Center   Carried in MC2A-X to control global task force (USAF).

Joint Aviation Authorities   Regulations adopted by Austria, Belgium, Cyprus, Czech Rep., Denmark, Finland, France, Germany, Greece, Hungary, Iceland, Ireland, Italy, Luxembourg, Malta, Monaco, Netherlands, Norway, Poland, Portugal, Slovenia, Spain, Sweden, Switzerland and UK.

jointery   Willing co-operation between armed forces (UK colloq.).

jointship   Vague buzzword meaning air/land/sea [space?] is all one conflict.

Joint-Stars   Joint surveillance and target-attack radar system (USAF).

joint-use restricted area   Restricted area in which, when not in use by using agency (eg USAF), FR or VFR clearance can be given for FAA traffic.

joint-wing   *Diamond wing.*

Joker   Fuel planning level selected to warn of imminent approach of Bingo.

JONA   Joint Office of Noise Abatement (DoT/NASA).

JOP   Junior officer pilot (RAF).

JOPES   Joint operation[s] planning and execution system.

JOpsC   Joint Operations Command [being prepared by Joint Forces Command 2003 for major combat and stability operations] (US).

JORD   Joint Operational Requirements Document.

JORN, Jorn   Jindalee OTH radar network (Australia).

Josephson junction   Formed by weakly linking superconductors at below 4°K, offering immense possibilities 0–1,000 GHz.

JOSS   Joint overseas switchboard (DoD).

Jostle   Powerful RAF active jammer, 1942–45.

JOTS   Joint operational tactical system.

JOTT   Junior officer tactics team (USN).

joule   SI unit of energy [potential, heat or kinetic], J = Nm = Ws = 10^7erg; MJ = 0.37251 hp-h.

Joule constant   Mechanical equivalent of heat, = 4.1858 J/15°C calorie, used in definition of 15° calorie.

Joule/Kelvin effect   Expansion of gas from high pressure through throttling orifice or porous plug; also called Joule/Thomson effect, see next.

Joule/Thomson coefficient   Rate of change $\left(\dfrac{dT}{dP} \right)$ H where T = temp, P = fluid pressure and H = enthalpy in reversible flow through porous plug.

Joule/Thomson cooling   Achieved by passing compressed gas through porous plug or small aperture.

Jovial   A real-time language for embedded computers [replaced by Ada].

joy   Success, satisfaction; thus "no *" = it failed to work (RAF).

joystick   Control column (suggest archaic).

JP   *1* Jamming pulse.

2 Jet propulsion/propelled/propellant.

JP-1   Original kerosene, Avtur (Jet Propulsion -1) fuel, replaced by Jet A-1, NATO F35.

JP-2   Improved kerosene, soon obsolete.

JP-3   Original wide-cut, including gasolines, kerosene and gas oil fractions.

JP-4   Wide-range distillate, Avtag, FS II, NATO F40, commonly available but reduces pump life and increases fire hazard; equivalent is Jet B fuel.

JP-5   Avcat, NATO F44, denser high-flash kerosene.

JP-6   'Heart-cut distillate', close control, good thermal stability, experimental use only.

JP-7   Special fuel for Mach 3 extreme-altitude aircraft (SR-71).

JP-8   Narrow-cut distillate, Jet A-1, F34, Avtur FS II.

JP-9   Starter fuel for cruise missiles [start at high altitude].

JP-10 High-density fuel developed to extend range of cruise missiles.

J-Pads Joint precision airdrop system.

J-Pals, JPALS Joint precision approach [and] landing system (USAF/USN).

J-Pass, JPASS Joint precision advanced strike system.

J-Pats, JPATS Joint primary aircraft training system (USAF/USN).

JPC J-tids portable capability (RAF).

JPDO Joint Planning and Development Office (FAA, DoC, DoD, DHS, NASA).

JPF Joint programmable fuze; PO adds Program Office.

JPI Joint precision interdiction (NATO).

JPITL Joint prioritized integrated target list.

JPL Jet Propulsion Laboratory, at Pasadena, a major facility of Caltech.

JPO *1* Joint program(me) office or organization.
2 Joint planning and development office for the transformation of the US aviation system (White House).

JPS Journal processing system.

JPSD Joint precision strike demonstration [PO adds project office] (USA).

j.p.t., JPT Jet-pipe temperature.

JPTS Jet propulsion, thermally stable, fuel for ultra-high altitude aircraft [originally for the U-2].

JR Utility transport aircraft class (USN 1935–55).

JRB Joint Reserve Base (US).

JRBA Joint Review Board Advisory Committee.

JRC *1* Joint Research Centre (Euratom Belgium/Germany/Italy/Netherlands).
2 Joint Resources Council (FAA).

JRCC *1* Joint Rescue Co-ordinating Centre (UK).
2 Joint Requirements Oversight Council (DoD).

JRDF Joint Rapid Deployment Force (UK).

JRDOD Joint research and development objective document.

JRIA Japan Rocket Industry Association.

JRMB Joint Requirements and Management Board (Office of JCS, US).

JROC Joint Requirements Oversight Council (US) or Committee (UK).

JRP *1* Joint Robotics Program (DoD).
2 Joint reconnaissance pod (RAF).

JRPG Joint Radar Planning Group (USAF/CAA, from 1956).

JRRF Joint Rapid Reaction Force (NATO).

JRS Japan Rocket Society.

JRSC Jam-resistant secure communications programme.

JRTC Joint Readiness Training Center (USA, USAF).

JS Job sheet.

J/S Jam-to-signal ratio.

JSA *1* Jet standard atmosphere.
2 Joint security area.

JSAF *1* Joint Sigint airborne family.
2 Joint Sigint avionics facility; LBSS adds low-band subsystem.

JSAM Joint Service aircrew mask.

JSAS Japanese institute of Space and Astronomical Sciences.

JSAT *1* Joint safety analysis team.
2 Joint system acceptance test.

JSC *1* Lyndon B. Johnson Spaceflight Center (NASA Houston).
2 Joint Security Commission (DoD).
3 Joint Steering Committee (ECAC-JAA and JARs).
4 Joint stock company.

JSCM Joint supersonic cruise missile (USN).

JSCMPO Joint-Service Cruise-Missile Program Office (USAF/USN).

JSD *1* Jackson system development (Ada).
2 Joint-Service[s] Designation.

JSDC Joint Service Defence College (Greenwich, UK).

JSDF Japan Self-Defence Force [=army; A adds Agency].

J-Sei Joint second-echelon interdiction (USA/USAF).

JSESPO Joint Surface-Effect Ship Program Office (US).

JSEW Joint-Service EW; S adds School.

JSF Joint Strike Fighter [40+ possible suffixes].

JSG Jump-strut [landing] gear.

J-Ship, JSHIP Joint shipboard helicopter integration process.

JSIAP Joint signal intelligence avionics program.

JSIC Joint Security Industry Council.

JSIES Joint-Services imagery exploitation system (UK).

JSims Joint simulation system.

JSIPS, J-sips Joint-Services imagery process[ing] system (US).

JSLO Joint-Services Liaison Organization (G).

JSME The Japan Society of Mechanical Engineers.

JSMRC Joint Service Medical Rehabilitation Centre (UK).

JSOA Joint Special-Operations Agency (US).

JSOR *1* Joint Strategic-Objective Plan (US).
2 Joint Services Operational Requirement (UK).

JSOW Joint-service stand-off weapon (US).

JSP Joint Services Publication, notably *318, Military Flying Regulations, esp. instrument, approach and departure procedures (UK).

JSPI Joint School of Photo Interpretation.

JSR *1* Joint Staff Requirement.
2 Jammer saturation range.
3 Jammer support receiver.
4 Jamming-to-signal ratio.

JSRC Joint Service Rescue Centre (UK).

JSRR Jam/signal ratio required.

JSS Joint Surveillance System (replaced Sage/BUIC).

JSSC Joint Services Staff Course (UK).

JSSEE Joint-Service software engineering environment[s].

J-Stars *Joint-Stars.*

JSTPS Joint strategic target planning staff.

JSTU Joint Service Trials Unit (UK).

JSWDL Joint-service weapon datalink.

JSWS Joint-Stars workstation.

JTA Joint technical architecture.

JTACMS Joint tactical missile system (USA/USAF).

JTAG Joint Test Action Group.

JTAGG Joint turbine advanced gas-generator (USA).

JTAGS, j-tags Joint tactical air/ground station, using IR data from DSP sensors to provide missile defence (USA).

JTAMD Joint theater air and missile defense; O adds organisation or office (USA).

JT&E   Joint test and evaluation.

JTAO   Joint tactical air order[s].

JTAV   Joint total-asset visibility.

JTC   Joint targeting cycle.

JTC³A   Joint Tactical Command, Control and Communications Agency (US).

JTCGAS   Joint Technical Coordinating Group for Aircraft Survivability (USN, industry).

JTDE   Joint technology-demonstrator engine (APSI).

JTEGG, j-tegg   Joint turbine-engine gas-generator.

JTF   *1* Joint task force.

    2 Joint tactical fusion; P adds program (US).

    3 Joint test facility.

JTGS   Joint Tactical Ground Station (US).

JTIDS, j-tids   Joint tactical information distribution system (all US Services, most of NATO).

JTR   Joint transport rotorcraft.

JTRS   Joint tactical radio system[s]; SCA adds software communications architecture.

JTRU   Joint Tropical Research Unit.

JTSO   Joint technical standing order.

JTSTR   Jetstream.

JTT   *1* Joint tactical terminal.

    2 Joint Trials Team.

JTTRE   Joint Tropical Trials and Research Establishment (Australia).

JTUAV   Joint [-service] tactical UAV.

JTW   Joint targeting workstation.

JU   Joint undertaking.

J-Ucas   Jount unmanned combat air system.

Judy   *1* Interceptor has contact and is assuming control of engagement (US).

    2 Target in practice interception: I have been hit (UK).

    3 See *Contact**.

JUEP   Joint UAV experimental programme (UK).

JUG   Joint Users Group.

jug   *1* Piston-engine cylinder.

    2 Drop tank (both meanings colloq.).

JUKL   ATC association (Yugoslavia, Serbia).

Julie/Jezebel   ASW system based on Julie plotting target position from echoes from ≤60 explosive charges detected by Jezebel.

Jumbo, jumbo jet   Boeing 747.

jumper   Cable temporarily attached to terminals to bypass part of electric power circuit.

jump jet   Jet VTOL aircraft (colloq.).

jumpliner   STOL transport (colloq.).

jumpmaster   Person in command of stick of parachutists.

Jump seat, jumpseat   Extra seat in cockpit or on flight deck not required by flight crew, but possibly occupied by authorized member of aircraft crew such as loadmaster.

jumpseater   Anyone occupying jump seat, eg passenger invited to flight deck (rare after '9-11').

jump strut gear   Main or (usually) nose landing gear capable of forcible extension to reduce TO ground roll of STOL aircraft.

jump takeoff   *1* Autogyro takeoff using stored rotor energy to achieve initial lift at zero airspeed.

    2 Jet VTO.

    3 Launch of anti-tank missile with initial jump to operating height.

junction   Mating surface between different types (eg p and n) of semiconductor material.

jungle penetrator   SAR device lowered into dense jungle by helicopter.

Jungly   Assault-transport helicopter, especially crew-member thereof (UK).

Junkers double wing   Plain flap mounted entirely aft of wing trailing edge.

junkhead   Head of sleeve-valve cylinder.

junkhead ring   Gas sealing ring between head and sleeve.

JURA   Joint-use restricted area.

jury strut   *1* Additional or temporary strut for particular short-term purpose.

    2 Short strut joining mid-point of main wing strut to wing.

Jusmag   Joint US Military Advisory Group.

juste retour   Division of workshares according to a partner-nation's investment, or commitment to purchase. The Eurofighter programme is the first in which this philosophy is no longer being applied.

Juxco   Joint unexploded ordance co-ordinating office.

JV   *1* Joint venture.

    2 Jettison vehicle [missile test].

JVC   Jet-vane control.

JVMF   Joint variable message format.

JVX   Joint-services advanced vertical lift aircraft (US).

JWA   Jointed-wing aircraft (microlight).

J-Warn   Joint warning and reporting network.

JWC   Joint Warfighting Center (JFCOM, Suffolk, VA).

JWE   Joint Warfare Establishment, Old Sarum [closed].

JWI   Joint Warrior, or warfare, interoperability; D adds demonstration.

JWIS   Joint Weather Impacts System (USAF).

K

K *1* Kelvin, used without degree symbol.

2 Telemetry, computing (JETDS).

3 Prefix (non-SI or metric) = 1,000, thus K-bit or K-byte, 32K, K-lb or Kp (kilopounds).

4 Factor of wing planform efficiency, or gain.

5 Tanker (UK and US aircraft category, US being a role prefix).

6 Airfield subgrade 300 lb/cu ft, ie dense concrete.

7 See *Knudsen number*.

8 Various ratios, eg surface area of solid grain to nozzle throat.

9 Circulation round aerofoil.

10 Potassium.

11 Km/h (flight plan).

12 Thermocouple of chrome/alumel type.

13 Invitation to transmit.

14 Knots (deprecated).

15 Smoke, obstruction to vision.

16 Bulk modulus.

k *1* Prefix kilo (× 1,000).

2 Cathode.

3 Cold (air mass), or colder than surface.

4 Boltzmann constant.

5 Radius of gyration.

6 Various time-constants, identified by suffix.

7 Thermal conductivity.

K-band EM radiation, 15-75 mm, 20-40 GHz.

k-bit 1,000 bits.

K-chart List of multipliers for calculating setback for bends other than 90° in metal or other elastic sheet.

K-dinghy For one man (RAF, WW2).

K-display Horizontal timebase shows two blips from target which vary in height if aerial azimuth direction is incorrect.

K-index, K-number Combined measure of moisture content and lapse rate over a range of pressure altitudes.

K-ration Standard field rations (USA, WW2, Korea).

K-loader Standard US military cargo aircraft loader elevating to side-door sill.

K-site Dummy airfield (UK, WW2).

K-wing Usually combination of canard plus slender delta.

K-words Kilo-words.

KA Message prefix (Morse).

KAB Guided bomb (R).

KADS, Kads Knowledge acquisition data system.

Kagohl Heavy-bomber unit (G, WW1).

KAI Kazan Aviation Institute (USSR, R).

Kai Kaizo (modification) suffix (J Army 1932-45).

Kaizen Meetings at which all concerned plan and then implement work flow [clean manufacturing].

Kalman filter Powerful software routine for combining multiple inputs (eg INS output and Doppler radar) to give most accurate single answer.

Kampf Battle, hence * geschwader = battle group [US wing] = bomber group; * Zerstörer = battle destroyer = heavy fighter (G).

kanat Underground aqueduct with surface breather tubes (NATO).

Kan-ban Use of coloured balls, cards or similar symbols to identify exact location of production-line hold-up (J).

Kapse Kernel APSE (software).

Kapton Polyimide materials, esp. age-resistant plastic sheet used as substrate for solar arrays or with gold coating as thermal insulation for spacecraft (DuPont).

Kapustin Yar Soviet ICBM/space 'cosmodrome' on flat territory near Caspian.

karabiner Steel D-ring on harness for clipping to 'dog lead' to prevent photographer, despatcher or other aviator from falling out of aircraft.

Karldap Administration centre for Karlsruhe ATC area.

Kármán-Moore Classic (1932) theory for aerodynamics of slender body of revolution travelling nose-first in supersonic gas flow.

Kármán street Endless succession of vortices, alternate left/right and clockwise/counterclockwise rotation, behind vibrating wire or strut in airflow.

Kármán-Tsien Classic (1939 and 1941) theory for compressibility effects on aerofoils, especially variation of Cp with M; most useful form is

$$Cp_M = \frac{Cp_0}{(1 - M^2)^{1/2} + \frac{1}{2}Cp_0(1 - [1 - M^2]^{1/2})}$$

KAS Killed on active service.

KASA Kenya Air sports Association.

katabatic wind Cold air flowing down mountain slope at night (keeps airfield on slope fog-free).

Katie Killer alert threat identification and evasion.

KAZ Air-refuelling unit (R).

KB *1* Construction (design) bureau (USSR).

2 Kite balloon (WW1).

kb Kilobyte[s].

KBAS Design bureau for automatic systems (R).

KBF Kalman-Bucy filter.

kbit 10^3 bits.

KBO *1* Weapon-system officer (G).

2 Kuiper-belt objects.

kb/s, KBPS Kilobytes per second.

KBS Knowledge-based systems.

KBU Keyboard unit.

KC Kill chains.

KCAB Korean (South) Civil Aviation Bureau.

kcal Kilogramme-calorie.

KCAS Knots calibrated airspeed.

KCK Composite sandwich, Kevlar/carbon/Kevlar.

kc/s Kilocycles, ie KHz.

KDAR Kleindrohne anti-radar = small harassment drone (G).

KDC Knock-down components.

KDD K-band dual digital.

KDEP Depth of smoke in thousands of feet.

KDR Kill/detection ratio (ASW).

KDS Keyboard display station.

KE Kinetic energy.

Ke Dielectric constant.

KEAS Knots equivalent airspeed.

KE-ASAT Kinetic-energy anti-satellite.

KEBPI Kinetic-energy boost-phase intercept.

keel	*1* Principal longitudinal member along ventral centreline of flying-boat hull or seaplane float.

2 Underside of flying-boat hull or seaplane float.

3 Principal centreline structural member of wing of microlight and similar aircraft.

keel angle	That between transverse line joining keel and chine and horizontal.

keel area	Strictly, projected plan area of keel (2).

2 Loosely, total side area of aircraft below OX axis.

keel section	The most-used part of runway, along centreline.

keel slider	Attachment of pilot or load to keel (3), with ability to slide fore/aft for stability and control.

keelson	I-beam forming structural backbone of marine hull.

KEK	Kinetic-energy kill; V adds vehicle.

Kelvin	Absolute (SI) temperature scale, degree symbol ° not used; absolute zero is 0K, triple point of water [0°C] is 273.16K, units same as °C. Thus $t_k = t_c + 273.16$, $= {}^5/_9$ ($t_f + 459.67$).

Kennelly-Heaviside layer	Original name for part of ionosphere, E-layer.

Kentucky windage	Deflection shooting by rule-of-thumb.

KEP	*1* Key emitter parameters (PD, PRF, frequency, etc).

2 Kinetic-energy penetrator.

KEPD	Kinetic-energy penetrator and destroyer.

Keplerian orbit	Satellite orbit linking two occasions when observer sees satellite against same point in space.

kerosene	Spelling used by Webster, van Nostrand and most other US dictionaries, by Oxford English Dictionary and by General Electric and Pratt & Whitney; **kerosine**, chosen by Chambers "**ene** is older spelling"; Encarta suggests "**ine** is US, Can, A, NZ", also preferred by Rolls-Royce; former UK name **paraffin**. Wide range of petroleum-derived fuels, primarily used for aviation turbine engines. Homologous hydrocarbon series with general formula C_nH_{2n+2}; first of this series (methane/ethane/propane/butane) are gases at STP, but next 11 are liquids (BPt 150-310°C) and form basis of jet fuels. Over 90 national designations, thus Jet A-1 is (among other things) DEngRD.2494 in UK, DEF(Aust) 240A, AIR 3405C (Fr), VTL 9130 (G), Gost-1027-67 and T-1 (R), FSD-M0754 (Sweden), 3-GP-23h (Canada), BA-PF-3 (Belgium), D1655-70/A-1 (ASTM), F-34 (NATO) and Avtur/FSII (International Service designation). For this fuel, density 0.796, thus 0.796 kg/l (c9.6 lb/Imp gal, 8.0 lb/US gal). Other major fuels include Avcat narrow-BPt fuel with high flash point (originally for carrier aircraft) and Avtag (turbine aviation gasoline, arguably not a kerosene), originally called JP-4 and in civil use JET B.

Kerr cell	Extremely fast electro-optical shutter based on glass container of nitrobenzene in electric field.

KET	Krypton evaluation technique.

kette	Section of three fighter aircraft (G, WW2).

Kevlar	Fibre-reinforced composite material with fibre properties superior to those of most glasses; called aramid fibre from resemblance to spider web (trade name, DuPont).

KEW	Kinetic-energy weapon[s], especially for SDI.

key	Position at start of letdown to landing, usually hi- * 15,000ft, lo- * 9,000 ft (US).

keyed emission	CW signal interrupted to convey intelligence.

keying	Interrupting current or signal by make/break switch, eg Morse key.

keying solution	Applied to give strong surface bond prior to application of dope, adhesive, corrosion protection or other coating.

keys	*1 Piano keys.*

2 On electronic display, edge keys.

keystroke	*1* Input from an edge key.

2 Basic element in electronic-display writing.

KF	Kinetic fires.

KFD	Key-fill device.

KFI	Krüger flaps indicator.

KfR	Kommission für Raumfahrttechnik (G).

KFS	Kerosene first stage.

KG	Kampfgeschwader, bomber wing (G, WW2).

K/G	Kevlar/graphite [IPH adds inter-ply hybrid].

kg	Kilogramme[s].

KGB	State committee for security (USSR, previously NKVD).

KGC	Japanese Science and Technology Agency.

kgf	Kilogrammes force or thrust (non-SI).

kgp	Kilogrammes force or thrust (poids, pond, puissance).

KGSK	Japanese Aeronautical Council.

KGV	Kryptographic variable (panel).

kGW, KGW	Thousands of pounds gross weight (highly ambiguous).

KH	Key Hole series of covert spy satellites (CIA/USAF).

Kh	Western rendition of Russian X, one use of which is to designate ASMs.

k_h, K_h	Factor for effect of forward speed on suckdown (jet lift).

KhAI	Kharkov Aviation Institute (USSR, Ukraine).

Khe Sanh	Extremely steep approach and departure (colloq., noun or verb).

KHTT	Know-how transfer and training.

kHz	Kilohertz, thousands of cycles per second.

Ki	Kitai (airframe type) number (J Army 1932-45).

KIA	Killed in action.

KIAS	Knots indicated airspeed.

Kic	Fracture toughness.

kick	Final impulse given by small upper-stage motor to space payload to achieve exact trajectory. Hence * motor, * stage, apogee * motor.

kickback	Bribe offered in large-scale contracting.

kicker	Direction needle [SBA, ILS, colloq.].

kick-off drift	In autolanding, separate control signal inserted before touchdown to yaw aircraft parallel to runway (but too late for crosswind to move aircraft laterally from centreline).

KIFA	Killed in flying accident.

KIFIS, Kifis	Instantaneous vertical speed indicator.

Ki-Gas	Piston-engine hand-priming system drawing fuel from main tanks.

Kilfrost	British liquid and paste deicer materials including propylene and other glycols.

kill	*1* Confirmed victory in air combat.

2 Destruction of missile or RV in flight.

kill box	Predesignated volume of sky, usually rectilinear, in which air or ground targets are sought.

kill chain　Process linking discovery of target by sensor to lock-on by shooter, also called *S2S* and *TCT*.

kill/loss ratio　Actual or claimed ratio of kills confirmed to losses suffered by a particular unit in a specified period; see kill ratio.

kill probability　Mathematical likelihood, based on experience, that a particular missile or attack will destroy its target.

kill ratio　*1* For a particular type of aircraft, total of its air victories divided by its own losses in air combat.
　2 Confirmed victories divided by number of hostile AAMs or SAMs launched.

kilo　Prefix, multiplied by 1,000; symbol k, Confusingly, often loosely used to mean kilogramme[s] or even kilometre[s].

kilobyte　Not 1,000 but 1,024 bytes, abb. kb.

kilocalorie　Non-SI unit of energy, often defined as 1/860 kWh, but see *calorie*.

kilocycle　Measure of frequency, = 1,000 Hz.

kilogramme　Measure of mass, = 1,000 grammes, in US kilogram, abb. kg, = 2.20462 lb. See *IPK*.

kilogramme-metre　Unit of work in the gravitational system, = 9.80665 Nm or 98,066,500 ergs.

kilohertz　SI unit of frequency, = 1,000 Hz, abb. kHz.

kilojoule　Si unit of energy, = 0.94786 Btu.

kilometre　In US often kilometer, unit of length = 1,000 m, 0.6214 statute mile, 0.5399568 Int. n.m., 0.5396129 UK n.m.

kilometric　Having a wavelength in the order of kilometre[s].

kilonewton　SI unit of force or thrust, abb. kN, = 224.80455 lb st.

kilopond　Kilogramme force, falling into disuse.

kiloton　Measure of NW explosive power, = that of 1,000 short tons of TNT (incorrectly, tonnes or long tons).

kilovolt　Measure of electric potential, = 1,000 V, abb. kV.

kilovolt-ampere　Measure of power of a.c. electrical machines, abb. kVA, numerically usually loosely = kW.

kilowatt　SI unit of power, abb. kW = 1,000 W, = 1.341 hp.

kilowatt-hour　Non-SI unit of electrical energy, kWh = 3.6 MJ.

kiloword　Unit of memory storage holding 1,024 words.

kinematic ranging　Aiming ahead of target correct lead angle to allow for target relative motion.

kinematic viscosity　Fluid viscosity divided by density, $^\mu/_\rho$, symbol *v*. unit $m^2 s^{-1}$ (no name) $= 10^4$ Stokes $= 10.7643 \, ft^2 \, s^{-1}$.

kinetheodolite　Tracking and recording instrument comprising high-speed camera whose frames bear az/el measures.

kinetic energy　That due to motion: for linear motion $E = \frac{1}{2} m V^2$; for rotary $E = \frac{1}{2} I \omega^2$ where I is moment of inertia and ω is angular velocity. Unit = joule. Note: symbols T and W are also used.

kinetic heating　Heating of boundary layer and surface beneath due to passage of body through gas, closely proportional to square of airspeed or Mach. For practical purposes synonymous with *aerodynamic heating*.

kinetic kill　Space interception technique using weapon whose kinetic energy alone provides its direct-impact destructive effect.

kinetic pressure　See *dynamic pressure*.

kinetic valve　Gas-turbine fuel-system valve in which two jets, one pump delivery pressure, the other pump servo pressure, point directly at each other with a variable interrupter blade where they meet.

kinetic weapons　All forms of projectile including bombs.

kingpost　One or more strong vertical struts above (and sometimes below) aircraft centreline providing attachment for primary flying and landing wires or struts, rare post-1916.

kink point　Any sharp corners on a graphical plot, e.g. of payload/range.

kip　Kilopound, 1,000 lbf (half a short ton), = 4,448.221615N. Not to be confused with *kilopond*.

Kips　Confusingly, in view of above, thousands of impulses, or instructions, per second.

KIR, Kir　Kinematic infra-red (flare).

Kirksite　Zinc alloy used for large airframe dies.

Kirkwood gaps　Gaps in the asteroid belts.

Kiruna　Swedish space launch and communications stations.

KIS　Kick-in step.

kiss landing　Touch at near-zero rate of descent.

kitbuilt　Constructed by customer from factory-supplied kit.

kite　*1* Aerodyne without propulsion tethered to semi-fixed point and sustained by wind.
　2 Aeroplane (UK colloq. usage, 1913-c50).

kite balloon　Balloon tethered to Earth or vehicle and shaped to derive stability (sometimes lift) from relative wind.

kitplane　Aeroplane assembled by customer from factory-built kit.

kitting　One of many aerospace meanings is to furnish item (engine, surface power unit, etc) with fresh consumables (filter elements, harnesses, circlips, gaskets) before reinstallation.

kJ　Kilojoule.

KJT　Japan Air Self-Defence Force.

KKV　Kinetic kill vehicle.

KKW　Kinetic kill weapon.

KLB, Klb　Pounds (lb) $\times 10^3$ [suggest not used without explanation].

KLD　Killed in line of duty, but not in action (US usage).

Klégécel　GRP/foam low-density sandwich materials (Kléber-Colombes).

klick(s)　Kilometre(s) (colloq.).

Klips, KLIPS　Thousands of logical inferences per second.

KLV　Copenhagen airport authority.

KLYR　Smoke layer aloft.

klystron　Velocity-modulated electron-tube UHF oscillator.

KM　Kriegministerium (war ministry, G).

kM　Rare prefix, kilomega (giga).

km　Kilometre[s]; hence km/h, kilometres per hour.

K Monel　Ni-Cr alloy; strong and corrosion-resistant, non-magnetic.

KMR　Kwajalein Missile Range, in mid-Pacific (USA).

KMW　Kuratorium der Mensch und der Weltraumfahrt eV (G).

Kn　*1* Solid rocket motor ratio of initial surface area to throat area.
　2 Knudsen number.
　3 Static margin.

kN Kilonewton; SI unit of force standard for most aerospace propulsion, = 224.809 lbf.

KnAPI Komsomolsk-on-Amur Polytechnic Institute.

knee Upper right corner on payload/range curve, or graph of similar shape.

kneeboard Notepad, stopwatch and other items strapped as unit to test pilot's leg above knee.

knee line Transverse line across [usually fighter] cockpit through front of pilot's knees.

kneeling Some landplanes have * landing gear to tilt nose-down to reduce space (aboard carrier), adjust cargo floor to truck bed, or alter wing angle of attack for catapult launch. Hence * parking.

knee panel Instrument or control panel at knee level, below side console.

knee window Low on side of cockpit [rare except helicopters].

Knickebein Navaid derived from Lorenz beam approach (G, WW2).

knife fight Close combat, usually one-on-one, (US term).

KNMI Royal Netherlands Meteorological Institute.

knock-down factor Arithmetical factor reducing allowable stresses imposed by certifying authority on uncertain structure [e.g. casting with porosity].

knockdown kit Complete aircraft packaged for shipment in component parts for assembly by foreign customer (usually also a licensee).

knock-down path The route followed by fire-fighters to a crashed aircraft, esp. the final few metres in which a path is blown through flames.

knocking Detonation (1).

knockout panel Portion of skin near pilot or other crewmember, often a window, which can be forced open from inside for emergency escape [esp. after belly landing].

knock rating Standard scale of resistance to knock (detonation) of piston-engine fuels, measured relative to iso-octane (100) (see *fuel grade*).

knot *1* Speed of 1 nautical mile per hour, International and in US = 6,076.12 ft/h = 1.15078 mph = 1.852 km/h = 1.6878 ft/s = 0.5144 ms^{-1}; in UK = 6,080 ft/h = 1.151 mph = 1.853184 km/h = 1.68 ft/s = 0.51477 ms^{-1}.

2 Nautical mile = 6,080 ft, 1,853.184 m.

knuckled Original adjective for levered-suspension landing gear.

knuckle pin See *wrist pin*.

Knudsen flow Gas flow in long tube at such near-zero pressure that mean free path is greater than tube radius.

Knudsen number, Kn Mean free path divided by characteristic length of body.

KNVvL Royal Netherlands Aeronautical Association.

KO Has been used to mean both kerosene and kerosene/oxygen.

Koch fitting Steel latches over wearer's chest connecting g-suit and harness to ejection seat and parachute.

KOCTY Smoke over city.

KOD Kick-off drift.

KOH Potassium hydroxide, used as standard in determining acidity and saponification of fuels, lubricants and hydraulic oils.

Koku Kantai Air fleet (J Navy, WW2).

Kokutai Complete air corps (J Navy, WW2).

Kollsman number Traditionally, altimeter setting, usually QFE or QFF.

kombi Release on glider suitable for winch launch or aerotow.

Komsomol League of Young Communists (USSR).

Komta Committee for heavy aviation (USSR, obs.).

Kops Thousands of operations per second.

Korex Family of aramid/phenolic honeycombs.

Koronas-F Solar observatory launched 2001 (R-Ukraine plus others).

KOSOS Department of experimental aeroplane construction (USSR, obs.).

Kourou *CSG*.

Kovar Fe/Ni/Co alloy whose coefficient of thermal expansion is close to that of glass.

Kp Kilopound, also called kip.

kp Kilopond, 1 kgf.

kPa Kilopascal[s], 1,000 N/m^2, SI unit of pressure.

kph Incorrect abbreviation for kilometres per hour.

KPP Key performance parameters.

KPS *1* Kilobytes per second.

2 Kills per sortie.

3 Knowledge-processing system.

KPU Keyboard printer unit.

k$_Q$ *Torque coefficient*.

Kraken Knowledge-rich acquisition of knowledge from experts who are non-logicians (USAF).

Kriegsmarine Navy (G, 1933-45).

KRL Khan Research Laboratories (Pakistan).

KRs & ACIs King's Regulations and Air Council Instructions (RAF 'bible' to 1952).

Krueger Not Krüger, leading-edge flap normally flush with undersurface, hinged down and forward to give bluff leading edge on high-speed profile.

krypton Kr, inert gas, density 3.7x10^{-3}, BPt -52°C.

Ks Factor of takeoff distance.

KSA Royal Swedish aero club (186 clubs).

KSC *1* John F Kennedy Space Center, includes Capes Canaveral and Kennedy, Merritt Island and the ETR.

2 Most confusing, kg/sq cm.

KSFC Thousands of simulated flight-cycles.

KSI, ksi Most confusing, thousands of pounds per square inch.

KT Geometric stress-concentration factor.

kT Kiloton[nes].

k$_T$ *Thrust coefficient*.

kt Knot[s].

KTAS Knots true airspeed.

KTD Key-transfer device.

kts See *kt*.

Küchemann carrot See *shock body*.

Küchemann 'Coke bottle' Wasp-waisted fuselage of high-subsonic aircraft with sides following local streamlines past wing. Not area ruled.

Küchemann tip Low-drag wingtip following outward-curved streamlines, with large-radius curve from leading edge to corner at trailing edge.

kulbit *1* A circle (R).

2 Somersault [Frolou], stopping motionless in vertical nose-up attitude whilst using vectoring to rotate in azimuth 360°.

3 Less dramatic, in horizontal or vertical flights point nose around circle centred on longitudinal axis.

KUR Key user requirement[s].

Kutney bump Drag-reducing bulge on inner side of pylon/wing junction.

Kutta-Zhukovsky Formula giving lift per unit span as KρV (circulation × density × velocity) for two-dimensional irrotational inviscid flow.

KV Kill vehicle.

kV Kilovolt[s].

kVA Kilovolt-ampere[s].

KVAR kVA-reactive, measure of reactive power.

KVDT Keyboard visual display terminal.

KW *1* Kinetic warhead.

    2 Key word; IC and OC add in, or out of context.

kW Kilowatt[s].

kWh Kilowatt-hour[s].

KWS Kampfwertsteigerung (G).

Kx, Ky 'Engineering' drag and lift coefficients.

kymograph *1* See *barograph*.

    2 Smoked-paper chart recording aircraft angular movements (arch.).

kytoon Combination kite + balloon or kite-shaped balloon.

kZ Vertical component of VSI (1).

KZB Drop tank (G).

KZO Mini-RPV for target detection and location (G).

k×k 1,000 × 1,000 pixels.

kΩ Kilohm.

L

L *1* Characteristic length of body.

2 Total lift.

3 Sound pressure level, see *Noise*.

4 Inductance.

5 Luminance, brightness.

6 Distance to applied load.

7 Low (navaid category), under 18,000 ft and En Route Low-altitude chart (FAA).

8 Low (synoptic chart).

9 Rolling moment.

10 Countermeasures (JETDS).

11 Lighted (airfield).

12 Code: IFR aircraft has DME and transponder.

13 Fighter designation prefix, low-altitude (UK, WW2).

14 Aircraft category, glider (prefix, USN, 1941–45), liaison (USAF from 1942, USN from 1962).

15 Modified-mission, cold-weather (prefix USAF, USN), searchlight carrier (USN until 1962).

16 Left (ident of parallel runway).

17 Lower wing.

18 Angular momentum.

19 In-line (piston engine).

20 Drizzle (ICAO).

21 Latitude.

22 Light (turbulence).

23 Locator.

24 Licensed airport.

25 Cleared to land.

26 Local.

27 Launch time of mission.

l *1* Litre[s].

2 Aerofoil section lift.

3 Aircraft overall length.

4 Distance from c.g. of aeroplane to c.p. of horizontal tail.

5 Stagnation line.

L̄ Average peak sound pressure level.

(L) Lower-airspace radar service, preceded by frequency.

L1 L-band carrier 1,575.42 MHz (GPS).

L2 L-band carrier, 1,227.6 MHz (GPS).

L2CS Two GPS codes, one with navigation data, for improved accuracy for civil users.

L3TV, L³TV Low-light-level TV.

L5 Third civil GPS frequency at 1,176 MHz, to be introduced 2005.

L$_{300/600}$ Sound pressure level for octave band 300/600 Hz, see *noise*.

L-band See Appendix 2.

L-display Central diametral timebase (usually vertical) on which appear transmitter and target blips at position giving range and offset according to pointing error.

LA *1* Lighter than air.

2 Launch azimuth.

3 Loiter altitude.

4 Limited authority.

5 Lietuvos Aeroklubas (Lithuania).

La Lanthanum.

L$_A$ A-weighted sound pressure level, see *noise*.

LAA *1* Light anti-aircraft gun or gunfire.

2 Low-altitude airspace.

3 Local-airport advisory.

4 Lowest acceptable altitude.

5 Low-altitude alert [S adds system].

6 Laboratory of Applied Anthropology.

7 Lateral accelerometer.

LAAAS, LA³S Low-altitude airfield-attack system.

LAAD Low-altitude air defence; S adds system.

LAADR Low-altitude airway departure route.

LAA/DR Low-altitude arrival/departure route.

LAANC Local Authorities Aircraft Noise Council (UK).

LAAP Low-altitude autopilot.

LAAS *1* Laboratoire d'Automatique et d'Analyse des Systèmes (F).

2 Low-altitude attack, or alert, system.

3 Local-area augmentation system, or scheme, proposed as GPS-based successor to ILS (FAA).

LAASH Litef analytical air-data system for helicopters.

LAAT Laser-augmented airborne TOW.

LAAV Light airborne ASW vehicle.

LABRV Large advanced ballistic re-entry vehicle.

Labs Low-altitude bombing system (guidance for upward toss of NW).

labyrinth seal Gas seal between fixed and moving parts comprising series of chambers which, though their sides do not quite touch, reduce gas escape close to zero.

LAC *1* Leading Aircraftman (RAF, obs.).

2 Low-altitude en route chart.

3 List of assessed contractors.

LACAC Latin American Civil Aviation Commission [from 1969].

Lace *1* Laser airborne communications experiment.

2 Liquid air cycle engine.

3 Low-power atmospheric compensation experiment.

4 Local adpative contrast enhancement.

Lacie Large-area crop inventory experiment.

lacing *1* Process of intermittent wrapping to join wire bundle into tight loom.

2 Stitching fabric to aircraft structure.

3 Threading wire through holes drilled at same location in every blade of a fan or turbine rotor to damp vibration.

lacking moral fibre Condition, often unfairly diagnosed, which resulted in aircrew being instantly removed from operations, and in most cases losing brevet and rank (RAF – WW2).

LACM Land-attack cruise missile.

Lacta Light-air-cushion triphibious aircraft.

LACW Leading Aircraftwoman (WRAF).

LAD *1* Laser acquisition device.

2 Large-area display.

3 Low-altitude dispenser.

4 Local-area differential [GPS, GNSS can be suffixed].

5 Launch-assist device.

ladar See *lidar*.

LADD Low-altitude (or angle) drogue delivery.

ladder network   Cascade of electrical sub-circuits each controlled by its predecessor (central to test equipment).

LADF   Lift-augmented ducted fan.

LADGNSS   Local-area differential GNSS.

LADGPS   Local-area differential GPS.

lading   Placing load on aircraft, including bar stocks and food trollies, excluding passengers and fuel.

LADS, Lads   *1* Lightweight air-defence systems.

2 Laser airborne depth sounder, for charting sea-bottom depth to assist in-shore ASW.

LAE   *1* Low-altitude extraction (para-drop cargo).

2 Licensed aircraft engineer (SLAET).

LAEO   Low-altitude electro-optics, or optical.

LAES   Landing Aids Experiment Station (Arcata, CA, from 1946).

LAF   Load-alleviation function.

LAFT   Light-aircraft flying task.

LAFTS   Laser and FLIR test set.

lag   *1* Angular crankshaft movement between a reference position (TDC, BDC) and open/closure of a valve.

2 Angular movement of electrical vector between reference position and current vector or other waveform.

3 Angular movement between helicopter main-rotor hub and (temporarily slower) blade; hence * hinge, * angle.

4 In instruments, eg VSI (1), normal meaning of delay.

5 In level bombing, horizontal distance between impact point [or an intermediate air position] and that achieved in vacuum.

Lageos   Laser geodynamic satellite.

lag-plane damping   Damping of fundamental mode of rotor system; critical for suppression of various instabilities, esp. ground and air resonance; expressed as % critical damping in non-rotating condition.

Lagrangian co-ordinates   Identify fluid parcels by assigning each a series of time-invariant co-ordinates such as transient spatial position. Constant-pressure meteorological balloon observations are Lagrangian.

Lagrangian points   Five positions in space where free body could maintain station with respect to satellite in existing two-body (eg Earth/Moon) system.

LAH   Light attack helicopter.

Lahat   Laser-homing anti-tank.

Lahaws, LAHAWS   Laser homing and warning system.

LAHC   Lincolnshire Aviation Heritage Centre.

LAHRS   Low-cost attitude/heading reference system.

LAHS   *1* Low-altitude high-speed route; military VFR training (US).

2 Land and hold short; O adds Operation.

LAI   Lean aerospace [or aircraft] initiative.

LAIRCM   Large-aircraft IRCM.

Lairs   *1* Large-aperture IR sensor.

2 Lightweight advanced imaging radar system.

3 Light-aircraft reconnaissance system.

LAL   Launch and leave.

Lα   Lift due to angle of attack (aircraft attitude), hence L_αWB = lift of blended wing/body due to α.

LALS   Linkless ammunition loading system.

LAM   *1* Long aerial mine (hence codename Mutton).

2 Loiter[ing] attack missile [-A adds aviation].

3 Logical acknowledgement message.

4 Laser additive manufacturing.

LAMA   Light Aircraft Manufacturers' Association (office Pleasanton, CA, US).

Lamars   Large-amplitude multi-mode aerospace research simulator.

lambda foot   Bifurcated shockwave near solid surface, resembling Greek letter.

lambert   Former unit of luminance, $\frac{1}{\pi}$ cd/cm^2; thus 1 foot-lambert = 3.426 cd/m^2.

Lambert projection   Map projection of modified conical type in which cone intersects Earth round two 'standard parallels' of latitude.

LAME, Lame   Line (or licensed) aircraft maintenance engineer.

Lamilloy   HP turbine cooling technology (former Allison co.).

lamina   Elementary (ie infinitely thin) slice.

laminar boundary layer   Comprises successive laminar layers, that adjacent to surface having zero relative velocity and successive layers adding velocity out to the free stream.

laminar flow   Fluid flow in which streamlines are invariant and maintain uniform separation, with perfect non-turbulent sliding between layers.

laminated   Made of thin layers bonded together.

LAMP, Lamp   *1* Lockheed adaptive modular platform, or payload,

2 Large advanced mirror program for space laser (SDI).

LAMPS, Lamps   *1* Low-altitude multi-purpose system (USAF).

2 Light airborne multi-purpose system (USN).

Lams, LAMS   *1* Load-alleviation and mode stabilization.

2 Light-aircraft maintenance schedule, or scheme, for aeroplanes ≤2,730 lb (1,238 kg) with 3-year C of A (CAA, UK).

3 Local-area missile system (ship defence).

4 Local-area mission system.

LAMV   Light aerial multipurpose vehicle.

LAN   Local area network.

lan   Inland.

LANA   Low-altitude night attack.

Lance   *1* Line algorithm for navigation in combat environment.

2 Launch and network control equipment.

land   Return to Earth or planetary surface of land-based vehicle; marine touchdown, prefer 'alight'.

land and hold short   Instruction by controller to landing aircraft to stop before first intersection or designated hold-short point to avoid conflict; recipient must tell ATC immediately if this clearance cannot be accepted.

land arm   Display signifying system is functioning in autoland mode.

land breeze   Offshore wind, towards sea.

L&D   Lateral and directional stability test.

lander   Spacecraft designed for soft landing on Moon or planet.

landing angle   Usually means angle between OX axis and ground at moment of touchdown.

landing area   Area of unpaved airfield reserved for landing and takeoff.

landing beacon or beam   Not recognized terms.

landing charge   Tax levied as part of airport charge.

landing circuit   Term used by Royal Navy in carrier operations.

landing compass   Precision magnetic compass on bubble-levelled tripod used as master when swinging aircraft.

landing crew   Large team(s) of handlers used in airship operations to hold ropes and manoeuvre ship to mast or walk it into hangar.

landing direction indicator   Visual device [eg tee, tetrahedron] at uncontrolled airfield.

landing distance, LD   Distance from runway threshold to aircraft stopping point.

landing distance available   That declared to be available by airfield authority.

landing flare   Released by aircraft over unlit airfield immediately before landing.

landing forecast   Met. forecast for destination at ETA, or nearest equivalent.

landing fuel allowance, LDG, ldg, LFA   Fuel mass required for theoretical circuit (go-around), landing and taxi at destination.

landing gear   Any portions of aircraft or spacecraft whose function is to enable a landing to be made; this includes wheels/skis/floats and attachments, and hook, but not flaps or lift-dumpers. Maximum speeds are usually prescribed for flight with * extended, as well as a lower maximum for cycling [act of retraction or extension], also called *LGOS*.

landing gross weight   Total weight of aircraft at point of touchdown in a particular landing, normally less than MLW.

landing ground   See *airfield*.

landing light   Forward-facing aircraft headlight, usually retractable or on nose or in leading edge, formerly to illuminate airfield and now used mainly as anti-collision or anti-bird beacon and for illuminating surface taxied over.

landing loads   Loads acting through structure of aircraft or spacecraft in design ultimate severe landing (or arrested landing).

landing long   Landing to touch down far down the runway.

landing mat   Various definitions, including flexible mat to reduce erosion or ingested debris in jet VTOL over unpaved surface.

landing minima   Worst weather, especially in terms of visibility, for legal landing; see *weather minima*. Traditionally involves DH and MDH.

landing on, landing-on   Recovery of aircraft on carrier, or helicopter on deck, by normal landing.

landing party   Landing crew (UK).

landing point   Intended or achieved point of MLG touchdown.

landing radar   Radio altimeter used by soft-landing spacecraft (rarely, by aircraft).

landing run   Actual achieved distance from touchdown point to stopping point. Also called roll or rollout.

landing runway   Runway assigned to arrivals.

landing site   Target for soft-landing spacecraft.

landing speed   Generally, TAS (sometimes defined as minimum TAS) at touchdown; plays no part in normal operation, which is based on V$_{\text{MCT}}$ and V$_{\text{AT}}$.

landing surface   Those areas declared by airfield authority to be available for landings.

landing tee   Large T-shaped sign in signals area of simple lightplane field, rotated to show wind direction.

landing weight   Predicted total mass of aircraft at landing.

landing wires   Bracing wires used to bear landing loads (eg sag of wings); also known as anti-lift wires.

landing zone   Area surrounding airfield where allowable heights of obstructions are limited (arch.).

land mine   In 1940 Germany discovered magnetic mines could be countered, so existing stocks were fitted with direct-impact fuzes and dropped (retaining the parachute) on British cities, receiving this name by recipients.

land on   To land an aircraft on ship, especially carrier.

landplane   Aircraft designed to operate from land, including ice or snow on skis.

Landsat   Earth-resources satellite.

LANE, Lane   Low-altitude navigation equipment.

lane   *1* One channel of a fly-by-wire or autopilot system.

    2 One hyperbolic track in early Gee or Decca navigation.

    3 One passenger guideway down escape slide.

lanes   Furrows on sea surface indicative of wind direction.

Langley   NASA research centre near Hampton, VA. abb. LaRC.

Langmuir-Blodgett   Film one molecule thick deposited on substrate in manufacture of very-high-speed devices.

LANL   Los Alamos National Laboratory (New Mexico, US), see *LASL*.

Lannion   French space communications ground station.

Lans, LANS   Land navigation system.

Lansu   Local air-navigation service unit (ATC).

lanthanum   La, soft silver metal, density 6.1, MPt 921°C.

Lantirn   Lo-altitude navigation[al] targeting IR for night (autonomous pod-mounted fire-control linked to HUD).

LAP   *1* Large-scale advanced propeller or propfan.

    2 Load, assemble and pack.

lap   *1* To fit two mating metal surfaces by rubbing together with fine abrasive.

    2 Crankshaft angular movement with inlet and exhaust valves open.

    3 Overlap (air reconnaissance).

    4 See *laps*.

Lapads   Lightweight acoustic processing and display system.

Lapam   Low-altitude penetrating attack missile.

Lapan   National Institute for Aeronautics and Space (Indonesia).

LAPB   Link access protocol, balanced.

LAPCB   Live Animals and Perishable Cargo Board (IATA).

Lapes   Low-altitude parachute extraction system.

lap joint   One sheet edge overlapping its neighbour.

Laplace   Name given to chief class of integral transforms, to elliptic partial differential equation and basic theorem on probabilities.

lap pack   Parachute pack carried on seated wearer's lap.

Laps   Local analysis and prediction system.

laps   Flaws in surface of steel castings.

lapse rate   Rate of reduction of temperature with height in atmosphere (see *dry adiabatic ***, *wet adiabatic ****).

LAPSS   Laser airborne photographic scanning system.

lap strap   Primitive seat harness with single belt across lap, almost universl in civil transport aircraft.

LAR   Live-animals requirements (IATA).

LARA   *1* Light armed reconnaissance aircraft.

    2 Low-altitude radar altimeter.

LARC   Low-altitude ride control (Softride); system for reducing vertical acceleration on crew compartment in high-speed flight through gusts.

LaRC   Langley Research Center (NASA, Hampton, VA, from 1920).

large aircraft   Over 12,500 lb (5,670 kg) MTOW (US).

large bird   Mass 4 lb, 1.8 kg.

large-scale integration   Typically c1,000 circuits, gates or logic functions on each chip; accepted upper limit is 16 kbit, above which is VLSI.

large space structure   Assembly of beams and girders having overall dimensions in range of hundreds of metres, posing unique problems of vibration and attitude control.

Larmor precession   Motion of charged particle attracted to fixed point in overall weak magnetic field.

LARS   *1* Lower-airspace radar advisory service (UK NATS).

2 Large-amplitude resonance simulator.

LAS   *1* Large astronomical satellite.

2 Lower airspace service.

Las   Landing aid system (Boeing).

Lasar   Light assault/attack reconfigurable.

LASC   Lead-angle steering command (air-to-air HUD mode).

Lascom   Laser communication (system).

Lascr   Light-activated silicon-controlled rectifier.

LASE   Lidar atmospheric sensing experiment.

laser   Any of many families of device for emitting coherent light (visible or outside visual spectrum); from light amplification by stimulated emission of radiation.

laser altimeter   Measures time for laser pulse to return from ground directly beneath aircraft.

laser centreline localizer   Amber on centreline, red or green, ultimately flashing, L or R.

laser cladding   Use of laser to deposit layer of [usually exotic] material.

Lasercom   Laser communications.

laser gyro   Any system, usually triangular, which senses rotation by measuring frequency shift of laser light trapped in closed circuit in horizontal plane.

laser inertial reference system   IRS using laser gyro(s).

laser ranging   Radar technique but using laser to illuminate target, range being function of time of return journey to target.

laser reference system   Gives attitude/heading, more precise than AHRS.

laser trimming   Use of laser mounted on travelling microscope to cut metal from specific sites in microelectronic circuit or device.

laser welding   Uses high-power (usually CO_2) laser for micro-precision welding; technique related to EBW.

LASH, Lash   Littoral airborne sensor hyperspectral, senses anomalies in light patterns to "see" through foliage or water.

LASL   Los Alamos Scientific Laboratory, centre of nuclear weapon and nuclear rocket research (University of New Mexico).

LASM   Land-attack Standard Missile.

Lasos   Lasers and space optical systems.

Laspac   Landing-gear avionics systems package.

LASR   Low-altitude surveillance radar.

Lass   Low-altitude surveillance system.

Lassi   Laser air speed sensor instrument.

Lassie, LASSIE   Low airspeed sensing and indicating equipment (helicopter and V/STOL).

last, LAST   Low-altitude supersonic target.

last-chance area   Parking spot near military runway for final check of armament status, seat pins, etc.

last-chance filter or screen   Final point for removal of foreign particles before oil [rarely, fuel] enters engine.

Laste   Low-altitude safety and target enhancement.

last form   Large-scale tooling for rapid manufacturing (acronym).

LASW   Littoral ASW; FNC adds future naval capabilities (US).

Lat, lat   Latitude.

Latar   Laser-augmented target acquisition and recognition.

Latas   Laser TAS system.

LATCC   London Air Traffic Control Centre (West Drayton, UK).

latching indicator   Instrument giving visual indication of security of door locks and pressure seals.

late-arming switch   Sub-circuit in aircraft weapon-control system, enabling arming of weapons or firing circuits to be left until moment of firing.

latency   *1* Delay in response of flight-simulator motion platform.

2 Delay between gathering intelligence and using it.

latent heat   Absorbed or emitted when material changes physical state.

lateral axis   Transverse OY axis, axis of pitch rotation.

lateral clinometer   See *cross-level*.

lateral-control criterion   Helix angle, described by wingtip in max-rate aileron roll: pb/2V where p is roll rate, b span and V airspeed.

lateral-control departure parameter   One method of studying loss or reversal of control in roll, for adverse yaw or other reason.

lateral datum   The transverse line passing through c.g. = lateral axis.

lateral deviation   Error in radio D/F caused by reflections or refractions.

lateral force   Forces acting parallel to line joining wingtips.

lateral gain   Width of fresh ground covered by each photo-reconnaissance run over area.

lateral qualities   Behaviour in roll.

lateral oscillation   One involving periodic roll, invariably with yaw and sideslip.

lateral-rotor helicopter   Main lifting rotors side-by-side.

lateral separation   Distance between parallel tracks of aircraft at same FL.

lateral stability   Stability in roll (secondarily, yaw and sideslip), measured by studying phugoid oscillation (stick free or fixed) following roll disturbance.

lateral translation   Manoeuvre possible with direct-force-control aircraft in which lateral velocity is commanded without change in heading.

lateral velocity   Speed component, usually relative to surrounding air, along line parallel to OY axis.

lateral tell   Communication of air surveillance and air target data sideways to other units along front.

late turn   One initiated at or after fix passage at waypoint.

Latex   Laser assocoié à une tourelle expérimentale (F).

Latis   Lightweight airborne thermal imaging system.

Latisha   Laser analysis and testing for intelligence, surveillance and hybrid applications (USAF).

latitude band　Between two parallels of latitude around Earth.

latitude nut　Screwed in or out on directional gyro (DI) to correct drift due to Earth rotation N or S of Equator.

LATN　Low-altitude tactical navigation area.

latr　Compass locator.

LATS　Launcher automatic test set (Varo).

lattice fin　A misnomer, not a fin but a powered control surface featuring a rectilinear criss-cross of flat surfaces. Also called trellis control.

launch　In addition to obvious, also take-off of manned combat mission.

launch bar　Towing link between catapult and nose leg.

launch complex　Entire ground facilities for launch of large space vehicle, probably including facilities for integration.

launch control centre　Manned room in launch complex from which countdown and launch, and possibly whole mission, are monitored and controlled.

launch cost　*1* Sum charged for placing customer's payload in desired orbit.

2 Nominal sum estimated, but not necessarily available, for design, development, construction and test of new major aircraft or engine; usually to certification in country of origin.

launch cycle　Typically 105 min, average time between launch and recovery for carrier aircraft; AEW/ASW can be launched for a double cycle.

launcher　*1* Interface unit between aircraft and externally or internally carried store, not necessarily with propulsion.

2 Container of tubes for firing unguided rockets, carried as external store or as retractable box.

3 Pad or other structure for land- or ship-based missile, space vehicle, RPV or other unmanned free-flight device.

launch escape　Ability of human crew to escape from slow-acceleration ballistic vehicle during countdown or in first seconds of flight, thus ** tower, ** motors, ** signal.

launch opportunity　Period in which all factors, including launch window, local weather and serviceability of all participating systems, is favourable.

launch pad　Platform with GSE for launch of ballistic vehicle; normally a fixed installation.

launch reliability　Percentage of planned missions on which combat aircraft took off on time.

launch vehicle　Vehicle providing propulsion for space payload or, rarely, atmospheric free-flight device; may be winged or ballistic but must lift off from Earth and impart nearly all impulse required.

launch window　Exactly defined period during which relative positions and velocities of Earth and other bodies are such that a particular interplanetary mission can be launched, may last minutes to days, and may be a unique opportunity or repeated at intervals.

LAV　Least absolute value.

LAW　Light anti-tank (or anti-armour) weapon.

LAWM　Lashenden Air Warfare Museum (UK).

LAWRS　Limited aviation weather reporting station.

LAWS　*1* Light aircraft warning system (UK Met. Office).

2 Lightweight aerial warning system (US).

LAX　*1* Limited-area automatic extraction.

2 Single noise event; more precisely L_{AX}; see *noise*.

lay　*1* Adjust aim of weapon in azimuth, elevation or both (obs.).

2 Spread aerial smokescreen.

3 Calculate or project course (obs.).

laydown　Release free-fall bombs in level flight at low altitude.

layer　Either of two ionised shells around Earth, called E and F, which see.

layered defence　*1* System for protecting fixed-base ICBMs by providing separate sensor/weapon systems for interception of hostile RVs at different altitudes.

2 More generally, any air defence system designed to assign different types of weapon to threats approaching in different height bands.

lay off　To redraw engineering part to full scale (has other meanings concerned with aiming.

layoff　Off-loading temporarily surplus employees (US).

layout　*1* Gross spatial arrangement of parts of aircraft (see *configuration* [*1*].

2 Arrangement of above-floor payload accommodation, eg one-class *.

3 Arrangement of drawings on sheet of paper; hence * draughtsman.

4 Geometrically correct drawing of sheet-metal part allowing for all bends, setbacks and joggles.

lay-up　*1* Basic assembly of parts for FRC structure before bonding under pressure and possibly heat.

2 To withdraw aircraft from service for modification or rebuild.

lazy eight　Flight manoeuvre in which nose describes figure 8, upper half above horizon and lower half below.

LB　*1* Light bomber.

2 Glider, bomb-carrying (USN, 1941–45).

3 Light bombardment aircraft category, USAAC 1924–32.

4 Laser beam.

5 Free balloon.

lb　Pound[s] mass, from Latin libra; as the plural is librae it is nonsense to write lbs.

LB film　Langmuir-Blodgett.

LBA　*1* Luftfahrt Bundesamt (office of civil aviation, G).

2 Local boarding application, host boarding gate control.

LBC　Linear block (digital) code.

LBCM　Locator back-course marker.

lbf　Pounds force.

LBH　Light battlefield helicopter.

LBI　*1* Low-band interrogator.

2 Long-baseline interferometry.

LBJ　Low-band jammer; A adds antenna, T transmitter.

LBL　*1* Left buttock line.

2 Länder-Behörden der Luftfahrt (G).

LBNL　Lawrence Berkeley National Laboratory, CA.

LBO　Leveraged buyout.

LBPR　Low (under 1.5) bypass ratio.

LBR　*1* Local base rescue.

2 Low bit rate.

3 See next.

LBRG, lbrg　Laser beam-riding guidance.

LBSD　Land-based strategic deterrent, to replace existing ICBMs by 2018 (USAF).

LBSS　Low-band subsystem.

lb st　Pound[s] force, static thrust.

LBTI　Long-burning target indicator.

LBVDS　Lightweight broadband variable-depth sonar.

LBW　*1* Laser-beam welding.

2 Learn by wire.

LC$_{50}$ CBW measure, lethal concentration in atmosphere required to kill 50% of exposed population.

LC   *1* Cargo aircraft, cold-weather operation (USAF, USN).

2 Local call, or control (FAA).

3 Inductance/capacitance.

4 Letter contract.

5 Least-cost.

LCA   *1* Light combat aircraft.

2 Layered component architecture.

LCAAS   Low-cost autonomous attack system.

LCAC   Landing craft, air-cushion.

LCAS   Light close air support.

LCB   *1* Line of constant bearing.

2 Liquid-cooled brake.

3 Lowest compliant bidder (NATO).

LCC   *1* Life-cycle cost.

2 Launch-control centre.

3 Linear cutting cord (canopy).

4 Load-carrying composite.

5 Leadless ceramic chip-carrier.

6 Lateral-control criterion.

7 Loran-C chart.

8 Local command centre.

LCCA   Lateral-control central actuator[s].

LCCC   Launch-control-centre computer.

LCCDU   Liquid-crystal crew display unit.

LCCG   Low-cost core guidance.

LCCMD   Low-cost cruise-missile defense (Darpa).

LCCP   Launch-control [SAM] computer program.

LCD   Liquid-crystal display.

LCDP   Lateral-control departure parameter.

L-CES   Limited-capability Earth station.

LCF   *1* Low-cycle fatigue; C adds counter, D damage, M meter.

2 Launch-control facility.

3 Link control field.

LCFPD   Liquid-crystal flat-panel display.

LCG   *1* Load classification group (I–VII, corresponding to LCN 120 – ⩾10).

2 Liquid-cooled garment.

LCH   Light combat helicopter.

LCH$_4$ Liquid methane.

LCI   *1* Low-cost inertial.

2 Low-cost interceptor (Darpa).

3 Logic[al] channel identifier.

LCL   *1* Local (FAA).

2 Lifting condensation level.

3 Laser centreline localizer.

LC/LO   Numerical percentage cost of least-cost compared with lo-observables.

LCLU   Landing control logic unit.

LCLV   Liquid-crystal light valve.

LCM   *1* Laser countermeasures.

2 Landing craft, medium.

3 Late change message.

4 Linear chirp modulation.

5 Lance-cartouches modulaire.

6 Logic control module.

7 Lithium/carbon monofluoride.

LCMS   *1* Low-cost missile system.

2 Local control and monitoring system.

LCN   *1* Load classification number; scale of values for

paved surfaces indicating ability to support loads without cracking or permanent deformation.

2 Logistics control numbers.

3 Local communications network.

LCO   *1* Life-cycle-oriented.

2 Launch control officer.

3 Limit-cycle oscillation.

4 Low-cost operation[s].

LCOS   Lead-computing optical sight; S adds system.

LCP   *1* Leachable chromate primer.

2 Launch control, or command, post.

3 Landing craft personnel.

4 Lighting control panel.

5 Last clicked position.

LCPK   Low-cost precision kill.

LCR   Link-connection refusal.

LCS   *1* Life-cycle cost, ie over whole useful life.

2 Liquid-crystal shutter.

LCSS   *1* Land combat support system.

2 Laser communications spacecraft (or satellite) system.

3 Liquid cooling sub-system.

LCSTB   Low-cost simulation testbed.

LCT   *1* Longitudinal cyclic trim.

2 Local civil time.

3 Landing craft, tank (WW2).

LCTD   Located.

Lctn   Location (FAA).

LCTR   Locator, suffixes M, O = middle or outer marker.

LCTS   Low-cost targeting system.

LCTV   Linac control and transit vehicle.

LCU   Laser code unit.

LCV   Landing craft vehicles [p adds personnel].

LCWDS   Low-cost weapon-delivery system.

LCZ, LCZR   Localizer.

LD   *1* Landing distance.

2 Lunar day.

3 Load device, prefix for designations of standard family of cargo containers and pallets, each of particular dimensions and with certificated permissible load.

4 Lower data.

5 Low-drag.

6 Lower deck.

L$_D$ D (daytime) weighted sound pressure level.

L/D   Lift/drag ratio.

LD$_{50}$ CBW measure of lethal dose; that which kills 50% of exposed population.

LDA   Localizer-type directional aid only.

LDA, LD$_a$ Landing distance available [H adds helicopter].

LDB   Launch data-bus.

LDC   *1* Less-developed countries (ICAO).

2 Lower-deck container.

LDCC   *1* Leaded chip carrier.

2 Lower-deck cargo compartment.

LDCM   Landsat data continuity mission.

LDCS   Local-departure control system, complete passenger-handling for non-hosted carriers.

LDDC   London Docklands Development Corporation (Stolport).

LDDI   Less-developed defence industries.

LDEF   Long-duration exposure facility (Shuttle).

L$_\delta$, L$_\Delta$　 Lift due to deflection (aeroelastic or surface rotation), thus L$_{\delta_T}$ = lift of tail due to deflection.

L$_{Den}$　 Noise level density, noise from all sources summed through each 24h (EC proposal).

LDG　 *1* Landing gear.

　2 Landing.

LDGP　 Low-drag general-purpose (bomb).

LDGPS　 Local-area differential GPS.

LDHD　 Low-density high demand.

LDI　 Landing direction indicator.

LDIN　 Lead-in [light system].

LDL　 Lower-deck lavatory.

LDM　 *1* Linear delta modulation.

　2 Lift/drag meter.

L/D$_{max}$　 Maximum attainable L/D.

LDMCRC　 Lower-deck mobile crew-rest container.

LDMX　 Local digital message exchange (secure terminals).

L$_{DN}$　 Duration of a noise.

LDNS　 Laser (or lightweight) Doppler navigation system.

LDO　 *1* Limited-duty officer (USN).

　2 Lease, develop, operate.

LDOC　 Long-distance operational control; F adds facility (HF radio).

LDP　 *1* Laser designator pod.

　2 Landing decision point (helicopter operations from small platforms).

LDPU　 Link and display processing unit (ATC).

LDR　 Low data-rate.

LDr　 Landing distance required.

L/Dr　 L/D for maximum range.

LDRF　 Laser designator rangefinder.

LDRU　 Light-duty release unit.

LDS　 *1* Layered-defence system.

　2 Lithium-doped silicon.

　3 Laser detecting set.

　4 Laser dazzle sight.

LD/SD　 Look down, shoot down.

LD-SVR　 Landing slant-visibility meter.

LDT　 *1* Lateral dispersion at touchdown (generally 10 ft/sec).

　2 Laser detector and tracker; SCAM adds strike camera.

　3 Local daylight time.

LDU　 *1* Lamp driver unit.

　2 Launcher decoder unit.

LDV　 *1* Limiting descent velocity.

　2 Laser Doppler velocimeter.

　3 Local Defence Volunteers (UK 1940).

ldw　 Landing weight.

LE　 *1* Leading edge; now has confusing additional meaning arising from expression '* of technology', signifying the very latest advances into unknown fields.

　2 Life extension.

　3 Link establish[ed].

Le　 Lewis number.

LEA　 Leurre [lure] electromagnétique actif (F).

LEAA　 Law-Enforcement Assistance Administration (US).

lead　 *1* Angular measurement of many variables (eg crankshaft motion between opening of exhaust value and TDC, or AC vectors related to zero-lead reference).

2 Angular distance between sightline to moving target and direction of aim to hit it.

3 First aircraft in element, or first element in large formation.

4 Dominant member of formation aerobatic display duo or team; role is to fly sequence precisely, without looking at No 2.

5 Different pronunciation, Pb, soft ductile metal, density 11.4, MPt 334°C.

lead aircraft　 *1* Aircraft with greater flight time than any other of similar type or using similar airframe.

2 Obviously, that leading a formation or group; see *leadplane*.

lead angle　 See *lead* (*2*).

lead azide　 Explosive triggered by mechanical deformation, used in detonators.

lead-computing sight　 Gyro or other sight sensitive to flight manoeuvres and providing a direct aiming mark to be superimposed over the target.

leaded fuel　 Containing small percentage TEL as anti-knock additive.

leader cable　 Electrically conductive cable buried along centreline of runway and taxiway to provide ground guidance in zero visibility.

lead-in　 *1* Formerly ground facilities and features between outer marker and threshold.

2 Tube through which aerial or towed MAD bird cable enters aircraft.

lead-in fighter　 Advanced jet trainer with which pupil can practise fighter missions, with sensors and weapons.

leading edge　 *1* Front edge of wing, rotor, tail or other aerofoil. Not precisely defined and, especially when made as detachable unit, extends to rear of 0% chord.

2 Rising slope of electronic pulse, esp. one on precise timebase, as in CRT, IFF, video etc.

3 Frontier of knowledge (see comment under LE).

leading-edge flap　 Any hinged high-lift surface attached to the leading edge but not forming the leading edge itself (ie, not a droop).

leading-edge root extension　 Sharp increase of wing chord at LE root, often almost flat and projecting ahead of wing profile proper, to cause strong vortex at high AOA and enhance lift, control and manoeuvrability. In extreme (long-chord form) becomes a large strake.

leading-edge sweep　 Angle between local (or, sometimes, mean) leading edge and OY axis.

leading panel　 The FSW in an oblique [slew] wing aircraft.

leadplane　 That guiding fire tankers to the retardant drop zone, orders sequence and approach path, watches for conflicts and relays altimeter setting (USFS).

lead pole　 Connects cable to tow banner.

lead-pursuit　 Traditional air-to-air attack using fixed guns, approach from rear and aiming ahead of crossing target.

lead-replacement petrol　 UK term for piston engine gasolines in which lead is replaced by VSR additives; in 2002 not yet approved for aviation.

LEADS, Leads　 Law-enforcement agencies data system (airport com. systems).

lead ship　 Prominently marked aircraft on which large day bomber formations formed up before setting course.

lead time　 Time between (a) placing order for bought-out item, or (b) starting fabrication of major airframe part or even (c) receiving heavy plate or other raw material, and emergence of finished aircraft. Expression also, incorrectly, used for time between ordering aircraft and its delivery.

leaf brake   Power tool for making radiused straight bends in sheet.

leakage drag   That due to local flows between fixed [eg, wing, tailplane] and movable parts of aircraft.

leaky turbojet   Turbofan of very low BPR (under 0.5).

lean   *1* Of fuel/air mixture, below stoichiometric, lacking fuel.

2 Linear distance at tip between position of backwards-leaning rotor blade (usually of gas-turbine compressor or helicopter) and position it would occupy if truly radial; the lean is sometimes along the tip-path plane and sometimes along chord line at tip.

Lean Aerospace Initiative   Programme by the SBAC and six UK universities to adapt the best practices in lean tools and processes (Toyota) to the aerospace industry. A major difference is that, unlike the motor industry, aerospace involves a great deal of non-recurring activity.

lean manufacturing   Keeping production line flowing with smallest possible inventory of components and work in progress, and elimination of muda [waste]. (Toyota).

lean mixture octane   At present this means fuel with TEL giving octane rating of 100. Essential for supercharged piston engine engines, replacements for TEL are being sought.

Leans (the)   Vertigo.

LEAP   Lightweight exo-atmospheric projectile.

leapfrog   To delay one ranging pulse train from radar to avoid two targets being superimposed.

learner cost   Extra element of direct-labour cost when work is unfamiliar.

learning curve   Fundamental curve portraying fall in manufacturing time or cost with increasing familiarity; abscissa is number of aircraft completed (often log scale) and ordinate is total direct labour cost, or total manufacturing man-hours or total manufacturing cost including raw materials and bought-out parts; usually an idealised curve not allowing for inflation.

Leasat   Leased satellite, or space bus hired out for different payloads.

leasing   Possession without title.

Lecos   Light (ie optical) electronic control system.

LEC   Locally employed civilian.

LECP   *1* Life-extension and capabilities program (US and ARRC).

2 Low-energy charged particle.

LED   *1* Light-emitting diode; - RHA adds recording-head assembly.

2 Leading-edge down (surface angular movement).

3 Leading-edge device(s).

4 Low endoatmospheric defence [I adds interceptor].

LEDDM   LED (1) dot matrix.

LEED   Low-energy electron diffraction.

lee wave   See *rotor cloud*.

LEF   *1* Leading-edge flap.

2 Light-emitting film.

left-hand circuit   Rectilinear circuit (1) with turns to left, anti-clockwise seen from above. Almost universal.

left-hand rotation   Anti-clockwise, viewed from rear.

left/right needle   Needle pivoted at top or bottom of panel instrument giving steering indication; pilot steers to keep needle vertical.

left seat   That of captain of aircraft; thus, ** time.

left-seater   Pilot in command, usually.

leg   *1* Main strut of landing gear.

2 Part of flight at constant heading between two waypoints.

3 Beam of radio range station, identified by particular flight as inbound * or outbound *.

legacy systems   *1* Those which a nation cannot afford to replace.

2 In general, those we use today, as distinct from the much better ones we can envisage. In the course of time everything becomes a *.

3 Specifically, the previous version.

legend   *1* Any fixed printed notice in cockpit.

2 Explanatory written matter on engineering drawing.

leg restraint   Strong belt automatically tightened round occupant's legs as ejection seat fires.

Lehar   Long-endurance high-altitude rotorcraft (USA).

LEIP   Leading-edge image process (auto map displays).

Lelfas   Long-endurance low-frequency active surveillance (ASW).

LEM   *1* Lunar excursion module.

2 Lean-enterprise model.

3 Linear electric motor.

Lemac   Leading edge of mean aerodynamic chord.

LEMF   Leading-edge manoeuvre flap.

Lemonnier   Class of resonant valveless pulsejets, named for inventor.

LEN   Low-entry networking.

len   Length.

length   Aeroplanes normally measured in flight attitude along OX axis with perpendiculars aligned with extremities of fixed airframe, normally including pitot or instrument booms; helicopters, must specify whether fuselage only or 'rotors turning', the latter being distance between perpendiculars to OX axis through periphery of rotor discs. Main cause of confusion is that measure is frequently taken from *Station Zero* at a location (often an arbitrary distance) in front of nose of aircraft. NATO measure is always major body * ignoring nose probes or booms, guns, FR probes, inlet centrebodies, rudder or tailplane overhang or any other projection or rotor blade. SI unit is metre m = 3.28084 ft = 39.37008 in.

Lens   Laser-engineered net shape.

lenticular   Having shape resembling side elevation of double-convex lens, with two arcs of large curvature meeting at pointed ends; thus * blade, a supersonic fan or compressor profile, and * cloud, found at tops of waves in lee of hills.

Lenz's law   Current induced in circuit moving relative to magnetic field will generate its own field opposing motion.

LEO   Low Earth orbit, parameters of which depend upon satellite mass, density, lifetime and other variables.

Leosat   Low Earth-orbiting satellite.

LEP   Light-emitting polymer.

L_{EPN} Effective perceived noise level, with tone/duration correction. See *noise*.

L_{EQ}, L_{eq} Energy-average sound pressure level.

LER   *1* Leading-edge radius.

2 Long-endurance rotorcraft.

3 Laser event recorder.

LeRC   Lewis Research Center (NASA).

LERX, Lerx   Leading-edge root extension, pronounced 'lurks'.

LES   *1* Leading-edge slat.
 2 Large-eddy simulation.
 3 Land Earth station.
 4 Launch escape system.

LESA   Lightweight electronically scanned array.

LESM   Lightweight ESM (1).

LESO, Leso   LES (3) operator.

LESS   Leading-edge subsystem (Space Shuttle).

LET   Launch and escape time (strategic bomber, cruise missile).

let-down   Complete procedure from TOD at end of cruise through the approach to landing; term concerned mainly with controlled adoption of successively lower flight levels rather than with the landing; thus * procedure.

lethal envelope   Volume, often spherical, within which parameters can be met for successful employment of particular munition.

letter boxing   Becoming squeezed between cloud layer and rising ground.

letter-box inlet   Large semi-rectangular air inlet along part of wing (or other) leading edge.

letter-box slot   Fixed slot at about 8% chord, usually ahead of aileron; further aft than slot formed by open slat.

letter of intent   Formal letter serving as notice by customer of intention to purchase, before negotiation of contract.

LEU   Leading edge up (surface angular movement).

leurre   Decoy [lure] (F).

l$_e$v   Tip chord of vertical tail.

level   Air intercept code: "Contact is at your angels".

level bust   Failure by dangerous margin [usually ± 300 ft] to fly at assigned FL.

level landing   Tail-up landing by tailwheel aeroplane; also known as a wheeler.

levelling circuit   AC filter circuit used to smooth out variation in bias voltage.

level of escape   Base of exosphere at which upward-moving particle has probability 1/e of colliding with another on way out of atmosphere.

level off   To pull out of dive or gentle let-down and hold height constant.

levels of similarity   Quantified lists of differences in aerodynamics and systems between early prototypes.

leverage   *1* Ratio between variables (eg, if Δ DOC due to Δ sfc is 8 times the cost of engines and spares to achieve Δ sfc then * of improved sfc is 8).
 2 Ratio of effect of destroying target to its own intrinsic value.

leveraged lease   Lease of aircraft on any of several forms of sliding scale.

levered suspension   Landing gear wheel(s) carried on arm pivoted to bottom of leg such that vertical travel of wheel is greater than that of shock strut.

LEVL   Leading-edge vortex lift.

LEW   Large eye/wheel distance, ie pilot must allow for his height above wheels at touchdown.

Lewis   NASA research centre for aeronautics, Cleveland, Ohio, Abb. LeRC.

Lewis aerial   Othogonal radar scanning sawtooth profile generated by electromechanical means to give flapping beams.

Lewis number   Le = Pr (Prandtl)/Sc (Schmidt), used in hypersonics.

LEWK   Loitering electronic-warfare killer (UAV).

LEWP   Line echo wave pattern.

LEX   Leading-edge extension (US terminology).

Lexan   Commercially produced polycarbonate plastic, usually transparent.

LF   *1* Load factor (structural).
 2 Load factor (traffic).
 3 Local forces, or landing force.
 4 Launch facility.

l.f., LF   Low frequency (see Appendix 2).

L + F   Leather and fabric.

LFA   *1* Landing fuel allowance.
 2 Low-flying area.
 3 Luftfahrtforschungsanstalt (G).
 4 Low-frequency active.
 5 Lawyers' Flying Association (UK).

LFAC   Ligne Française d'Aéronefs de Collection (aircraft preservation, F).

LFADS   Low-frequency active dipping sonar.

LFAS   Low-frequency active sonar.

LFATS   Low-frequency active towed sonar.

LFBB   Liquid-fuel flyback booster.

LFC   *1* Laminar flow control (see *BLC*).
 2 Longitudinal friction coefficient.
 3 Level of free convection.

LFD   *1* Lamp failure detector.
 2 Large freight door.

LFH   Lunar far horizon.

LFI   Light tactical fighter (R).

LFICS   Landing force integrated communications system.

LFL   Lower flammability limit.

LFM   *1* Low-powered fan marker.
 2 Laminated, or limited, fine mesh [weather model].
 3 Low-powered frequency modulation.

LFP   Loaded flank pitch (fir-tree blade root).

LFR   *1* Local flight regulations.
 2 Low/medium-frequency radio range.

LFRED   Liquid-fuelled ramjet engine development.

LFRJ   Liquid-fuelled ramjet (in solid rocket case).

LFR, LFRR   Low-frequency radio range.

LFS   Low-flying system (UK military).

LFSMS   Logistic force-structure management system(s).

LFV   Civil aviation board (Sweden).

LFW   Linear friction welding.

LFX   Limited-output full-area automatic extraction.

LG   *1* Landing gear; also, for F-22, * trainer (ground support item).
 2 Laser gyro.
 3 Landing ground.
 4 Lehrgeschwader, instructional group (G).

LGA   *1* Low-gain antenna.
 2 Local Government Association (UK, has two aviation groups).

LGB   Laser-guided bomb.

LGCIU   Landing-gear control and interface unit; typically controls LG and doors, monitors cargo door locks, senses flap/slat position and interfaces with ECAM, MRS and BITE (Dowty).

LGDM   Laser-guided dispenser munition.

LGE speed   Speed at which landing gear may be extended. See *landing gear*.

LGI   Laser glide slope indicator.

LGM   US weapon category, silo-launched missile.

LGOS   Landing-gear operating speed.

LGR   Laser guidance receiver.

LGS   *1* Laser gunfire simulator.

  2 Laser gyro strapdown.

LGSC   Linear glide-slope capture.

LGSM   Light ground-station module.

LGT   Landing-gear tread.

lgt   Light, lighting.

Lgtd   Lighted.

LGTR   Laser-guided training round.

LGW   Landing gross weight.

LGWB   Landing gear wheelbase.

LH   *1* Left-hand.

  2 Light helicopter.

L/H   Local horizontal.

LHA   *1* US Navy ship category, large helicopter assault carrier.

  2 Local hour angle.

LHC   *1* Light helicopter cycle (standard cycle for US turboshaft engine testing).

  2 Left-hand circuit.

  3 Lower-hold cargo.

LHe   Liquid helium.

LHF   Liquid-cooled, heavy fuel.

LHM   Laser-hardened materials.

LHN   Long-haul network.

LHOX   Low and high-pressure oxygen.

LHP   Lightning *HIRF* protection.

LHS   Left-hand side.

LH₂   Liquid hydrogen.

LHV   Fuel lower heating value, formerly measured in BTU/lb.

LHW   Laser-homing weapon.

LHWR   Lightning-hazard warning radar.

LHX   Light helicopter, experimental.

LI   *1* Lane identification (early Decca).

  2 Laser interrogator (or interrogation).

  3 Lithium-iron [LiFe preferred].

  4 Letter of intent.

  5 Lift index; numerically positive, negative or zero if atmosphere stable, unstable or neutral.

  6 Low-intensity light[s].

L$_i$ Maximum weighted noise level over series of i noise events.

LIB   *1* Left inboard.

  2 Loudspeaker intercom box.

libration   Small long-period oscillation, esp. that of Moon's aspect from Earth.

LIC   Low-intensity conflict [A adds aircraft, S system, hence Licas].

licence   US = license, document authorising holder to carry out functions specified; see *rating* (3), *validation*.

LID   *1* Lift-improvement device (jet V/STOL).

  2 Luftfahrt Information Dienst (DDR).

  3 Large integrated display.

  4 Laser irradiation detector.

  5 Liquid-interface diffusion (bonding).

lidar   Light detection and ranging, laser counterpart of radar.

LIF   Lead-in fighter.

LiFe   Lithium-iron.

life   *1* Allowable total period of operation of hardware item.

  2 To assign such a period; hence, a lifed part.

lifeboat   Transport vehicle for rescuing crew from spacecraft, usually parafoil Earth landing.

liferaft   Correct term for inflatable emergency 'dinghy'.

life-support system   Provides environment to sustain human life in space, including during EVA.

LIFM   Linear instantaneous frequency measurement.

Lifmop   Linear frequency-modulated pulse.

Lifo   Last in, first out.

LIFP   Low-inertia flat-plate (antenna).

LIFR   Low-altitude instrument flight rules.

LIFT   Lead-in fighter training, or team.

lift   *1* Total lifting force from a wing (component of resultant force along lift axis), aerostat envelope or other source excluding engine thrust. Normally, force supporting aircraft. Traditionally $L = C_L \frac{1}{2} \rho V^2 S$, where C_L is lift coefficient, ρ density, V velocity and S area.

  2 Any element of such lift, acting through particular point.

  3 Whole or part of an airborne operation, thus second * means second force to be airlifted.

  4 Aircraft-carrier elevator (British terminology).

  5 Total traffic capability of fleet of transport aircraft [esp. military].

lift axis   Line through c.g. perpendicular to relative wind in plane of symmetry.

lift coefficient C_L, dimensionless measure of lift of surface; actual lift divided by free-stream dynamic pressure $\frac{1}{2}\rho V^2$ and surface's area S.

lift/cruise engine   Turbofan or turbojet with vectoring to give jet lift or thrust.

lift curve   Plot of lift coefficient against angle of attack ($C_L:\alpha$).

lift curve slope   Inclination of lift curve at any point, rate of change $dC_L/d\alpha$.

lift-dependent drag   See *lift-induced drag*.

lift/drag ratio, L/D   Ratio of total lift to total drag, fundamental measure of efficiency of aircraft; L is normally constant and equal to weight but drag varies approx as square of airspeed; thus L/D plot is curve with peak at one particular airspeed for each aircraft, L/D$_{max}$.

lift dumper   Flat plate, usually long span and short chord, raised by powered system (rendered operative by weight on MLG) from upper surface of wing (usually inboard and at about 60% chord) after landing to destroy lift and improve wheel-brake traction. Usually synonymous with ground spoiler.

lift fan   *1* Turbofan of HBPR installed only for lift thrust.

  2 Free-running fan driven by tip turbine from external gas supply installed only for lift (note: 1 and 2 may have exit vanes to give a diagonal lift/thrust component).

lift-improvement device   Any aerodynamic strake, dam, flap or other fixed or movable surface to assist jet VTO by reducing hot-gas reingestion, suckdown or other undesirable effects.

lift index   Air stability expressed as positive number if stable, zero neutral and negative unstable.

lift-induced drag   For all practical purposes, the same as lift-dependent drag or drag due to lift, the rearwards component of the total [resultant] force vector on a wing.

Purists could say lift-dependent drag is the difference between drag at a given C_L and that at a datum C_L. They could also argue * is not synonymous with trailing-vortex drag, because the latter can exist in an inviscid flow.

lifting body   Aircraft whose chief or sole lift is generated by its body; usually hypersonic aircraft or spacecraft.

lifting re-entry   One in which aerodynamic lift forces play a significant role.

lift jet   Ultra-lightweight turbojet or turbofan installed only for upward thrust.

lift-lift/cruise   Equipped with both lift jet(s) and vectored-thrust engine.

lift motor   Engine driving vertical-axis prop/rotor on airship.

lift off   *1* Separation of any aircraft or other flight or space vehicle from ground or (eg Space Shuttle atmospheric tests) a parent vehicle. Hence * speed, V<small>LOF</small>. For aircraft, synonymous with unstick.

2 Undesirable gap between an eddy-current crack tester and the inspected surface.

lift strut   Bears tensile (rarely compression) load due to wing lift.

LIFTT   Leaders in flight-test training, includes [2002] ETPS, Epner, CCA, DUT and IAS.

lift/thrust   Ratio of lift to thrust of vectored-thrust engine, usually varies from unity to zero over range of nozzle movement; also see *L/T*.

lift vector   Vector drawn through point at which lift force acts, with angle showing direction (usually normal to chord or OX axis, irrespective of aircraft attitude) and length showing magnitude.

lift wire   Bears tensile load due to lift of wing.

LIG   *1* Laser image generator.

2 Lithium/iron gel.

light   Visible * extends from about 0.4 μ [red] to 0.75 μ [violet]. Velocity in vacuum = 299,792,456 m [983,571,007 ft]/s.

light aircraft   One having MTOW less than 12,500 lb (5,670 kg).

light alloy   One whose principal constituent is aluminium; some authorities add 'or magnesium' but these are usually described as magnesium alloys.

light anti-aircraft   Guns ≤ 40 mm.

light bomber   Today meaningless, and never universally defined.

light-emitting bar   Vertical bar of three (rarely, more) Si LEDs.

light-emitting diode   Solid-state diode emitting visible light when stimulated by electronic input, giving quick-reacting shaped light source.

light-emitting strip   Horizontal rectangular strip display made up of number of light-emitting bars, often used to give analog lateral-position readout.

lightening hole   Cut-out in relatively unstressed region of structural sheet part to save weight.

lighter than air   Buoyant in atmosphere (see *aerostat*).

light fighter   Unusually small fighter intended chiefly for close air-combat role.

light flight control   Easy to move, esp. when adjacent flight controls are heavy.

light gun   Aldis lamp or other projector of visible pencil beam, usually selectable white, red or green.

light ice   Traditionally, can be ignored for up to 1 h.

light machine gun   Not greater than rifle calibre.

light-microsecond   Almost exactly 300 m, 984 ft.

lightning   Any natural electrical discharge between clouds or between cloud and ground.

Lightning Bolt   Procedures enabling existing procurement to be streamlined, and commercially available items to be bought when appropriate (USAF).

lightoff, light off   *1* Ignition followed by acceleration of gas turbine.

2 Ignition of afterburner.

light pen   Fibre-optic device for interfacing and accessing computer via visual display.

light pipe   Single or bundle of optical fibres.

lightplane   See *light aircraft*.

light propeller aircraft   ≤5,700 kg MTOW.

Light Series   Carrier for four 20-lb practice bombs (UK, 1922-c60).

light turboprop   Aircraft category MTOW 7 t (15,432 lb).

light valve   Photoconductive layer controlling areas of liquid crystal illuminated in large display.

light water   *1* Water, as distinct from heavy water.

2 Trade name for *AFFF*.

lightweight fighter, LWF   Despite USAF competition 1972–75, never defined.

LIH/LIL/LIM   Light intensity high/low/medium.

LII   *1* Light image intensifier.

2 Flight research institute (USSR).

LIIPS   Leningrad institute for sail and communications engineers (USSR).

like on like   Liquid rocket with streams of fuel impinging on each other from some injectors and streams of oxidant on oxidant from others.

Lilo   Last in, last out.

lily pad   Forward operating base (USAF, esp. PACAF).

LIM   *1* Low-inclination mission.

2 Locator inner marker.

3 Light intensity medium.

4 Limit.

Lima   Laser ionisation mass analyser.

Limaçon   Quartic curve, r = a cos θ + b.

Limar   Laser imaging and ranging.

limb   Visible edge of heavenly body, esp. the Sun.

limit altitudes   Angles of pitch or bank which FCS prevents being exceeded.

limit-cycle oscillation   Sustained vibration at a fixed frequency and limited amplitude.

limited panel   Pilot instruction with key flight instruments obliterated and external cues absent (originally meant gyro instruments obliterated, and always horizon; today depends on panel).

limited remote communications outlet   Unmanned satellite air/ground com. facility operated as LRCO-A, VOR voice channel plus receiver, or LRCO-B, separate facility with transmit and receive capability, extending FSS service area (FAA).

limited-route concept   Operator, captain or whoever else prepares flight-plan, is offered very limited choice of routes through a controlled airspace.

limiter   One meaning is control device attached to transducer to prevent critical or threshold value being exceeded.

limiter spiral   Manoeuvre in which aircraft makes g-loaded roll on AOA limit, a form of corkscrew with stick in fully back L or R corner.

limiting load factor   See *design load factor*.

limiting Mach number   Maximum permitted for type of aircraft, usually before onset of buffet.

limiting runway   One whose length, altitude or temperature necessitates take-off below MTOW.

limiting speed   *1* Maximum IAS permitted in particular aircraft configuration, eg landing gear down.

2 Speed in any flight condition in which longitudinal acceleration is zero.

limiting velocity   Terminal velocity at specified angle to horizontal [not normal term].

limit load   Greatest anticipated stress on structural member, unfactored, from authorized ground and flight operation.

limit of proportionality   Tensile (rarely, other) stress at which material begins to suffer plastic deformation, acquiring permanent set.

limits   *1* Weather minima permitted for particular pilot or flight.

2 Boundaries of flight regimes, eg IAS or g in particular configurations.

Limnatran   Limited Atlantic regional air navigation.

LiMnO₂   Lithium manganese dioxide electric battery.

LIMSS   Logistics information management support system [hence LIMS, management system].

Linac   Linear accelerator for X-radiography.

Linas   Laser-inertial nav/attack system.

Lincs   *1* Leased-products interfacility national air space communications system; digital net connecting remote radar and Wx sites to ATC centres (FAA).

2 Long-haul interfacility com. system.

Lindberg detector   Fire detector with sealed network of stainless-steel tubing containing material which above set temperature emits gas, raising pressure.

Lindholme gear   Air/sea rescue equipment dropped to survivors; the original form (1942) was packaged in 10 buoyant containers.

line   *1* Single pipe in fluid system.

2 Single cable in electrical system.

3 Horizontal scan on raster display.

4 Cable or rope anchored to aerostat with other end free.

5 Flight-line.

6 Adjective, in revenue service with air carrier.

7 Future path of target.

8 Personal boast, from 'shooting a *' (RAF).

linear accelerator   In theory, any assisted-takeoff device. In practice, restricted to an "unrolled" electric motor.

linear aerial array   Yagi or other array of dipoles on straight axis.

linear aerospike   Rocket with two-dimensional expansive nozzle.

linear building   One in which operations take place in sequence from one end to other.

linear configuration   Vehicle assembled from separable stages arranged end to end in one line.

linear friction welding   Workpieces are rubbed together to reach welding temperature, giving a perfect bond by a solid-state process not involving melting.

linear hold   Usually, to delay landing by intercepting extended runway centreline far from airport, advising when 1,000 m from threshold.

linear motion   See *heave, surge, sway*.

linear optical sensor   Transducer in fibre-optic sensor system which, by splitting and reflecting laser pulses whose phase-displacement is then measured, translates mechanical movement (eg, of aileron) into a decodable output.

linear-scale instrument   Vertical or horizontal straight-line display, either tape or video, giving quantified output.

linear shaped charge   Explosive cord whose cross-section is that of hollow (shaped) charge, for unidirectional cutting.

line book   Written and witnessed record of lines (8) shot by members of Mess (RAF).

line check   Examination of crew qualified on type but proving ability to fly new route.

line inspection   Usually vague, but generally a special check not calling for aircraft to be moved to maintenance or engineering area or enter hangar.

lineman   Engineer, marshaller or other flight-line worker, esp. in general aviation.

line-mounted   Usually, supported entirely by a pipe or cable, thus * valve.

line of position   See *position line*.

line of reference   The intersection of the planes of reference and of symmetry.

line oriented flight training   Training (in air or simulator) of commercial aircrew flying as a crew and using SIDs and STARs and other regular procedures, esp. on routes of pupils' airline.

line-oriented safety audit   Collection of safety data and flight-crew performance as diagnostic tool.

liner   Sheet or sprayed-on heat insulator in some (non-case-bonded) solid motors.

line-replaceable   Capable of being removed from aircraft parked on flight-line and replaced by different example of same item.

linescan   IR graphics using raster display to generate picture.

line search   *1* To examine one strip of film from straight reconnaissance run.

2 In sea reconnaissance, to search on constant heading at maximum height at which target is identifiable.

line service   In revenue operation with air carrier.

Linesman   British attempt at combined air defence and ATC system (see *Mediator*).

line speed   Predicted take-off ASIR.

line squall   Violent cold front characterised by sudden drop in temperature, rise in pressure, thunderstorms and, especially, severe vertical and other gusts.

line up   To position aircraft on downwind end of runway, pointing along centreline.

line vortex   One in which vorticity is concentrated in a line.

liney   Apron marshaller (RAF, colloq.).

linkbelt   Ammunition feed using rigid inter-round links.

link chute   Discharges used ammunition links overboard.

link route   Authorised sector joining airways but not itself an airway.

Link's turbidity factor   See *turbidity factor*.

Link trainer   Traditional primitive electropneumatic flight (pilot training) simulator, not representative of aircraft type.

Link translator   Provides translation and forwarding between Tadil, Link 11, NATO Link 1 and other friendly communications.

LINS, Lins   Laser/inertial navigation system.

LION, Lion   Link interoperability network (UK MoD).

LIP   *1* Laboratory identification prototype.

  2 Limited-installation program.

  3 Lithium/iron polymer; E adds electrolyte.

lip   Leading edge of air inlet (other than a bodyside splitter plate).

lip microphone   For use, pivoted to be almost touching the mouth.

Lips, LIPS   Logical inferences per second (A12).

liquid-cooled   Loosely, any engine cooled by liquid, including water, but preferably restricted to cooling by water/alcohol or glycol mix.

liquid crystal   Organic liquids with elongated molecules which in electric fields arrange themselves to give controllable appearances.

liquid-film technique   Traditional method of coating surface with volatile oil to show demarcation between laminar/turbulent boundary layer and some details of flow direction.

liquid-fuel starter   Burning one or more liquids unlike that for main engine.

liquid inertia vibration eliminator   Heavy liquid, damps helicopter rotor vibration.

liquid injection TVC   Use of volatile fluid pumped into one side of rocket nozzle to create shockwave and deflect jet.

liquid oxygen   See *oxygen.*

liquid petroleum gas   Butane, heptane and similar gaseous hydrocarbon fuels stored as liquids under high pressure.

liquid propellant   Liquid fuel, monofuel or oxidant used in rocket.

liquid rocket   Rocket burning one or more liquid propellants.

Liquid Spring   Dowty shockstrut filled with liquid with deformable large molecules absorbing energy internally.

LIR   Laser intercept receiver.

LIRA, Lira   Low-intensity (ie limited war) reconnaissance aircraft, with simple optical/IR suite.

LIRCM   Large-aircraft IRCM.

LIRL   Low-intensity runway light(s).

LIRS   Laser inertial reference system.

LIRU   Laser inertial reference unit.

LIS   *1* Localizer inertial smoother.

  2 Lightning image shelter (NASA).

LISA, Lisa   Logistics Information Systems Agency.

LISB   Low-intensity [or less-intense] sonic boom.

LISE, Lise   Laser integrated space experiment (SDI).

LISN   Line-impedance stabilization network.

LiSO Cl$_2$ Lithium thionyl chloride.

listening out   Ready to receive broadcast transmissions on wavelength in use (US = listening watch, and predictably the latter is becoming standard).

listening post   *1* Installation under landing or takeoff climbout paths of airport for measuring and recording noise of all traffic.

  2 Installation, with or without sound-locator, giving warning of approach of possibly hostile aircraft (1917–45).

LIT   *1* Lead-in training (US).

  2 Light intratheatre transport.

lit   Litres (SI unit).

Lital   Medium- and high-strength Al-Li alloys (Alcan).

Litas, LITAS   Low-intensity two-colour approach slope system.

LITDL   Link-16 interoperable tactical data link.

Lite   Laser illuminator targeting equipment.

lithium   Extremely light (density 0.534) white metal, MPt 186°C, used in Al-Li alloys and as isotope Li-6 in NW.

lithium-drift detector   Ionising-radiation detector using semiconductor doped with lithium as n-type ions.

lithium tantalate   LiTaO$_3$ for modulating lasers.

lithometeor   Finely divided solid particles suspended in atmosphere.

lithosphere   Earth land mass, as distinct from atmosphere, hydrosphere.

litre   Metric unit of volume [in SI strictly called dm^3] = 10^{-3}m^3, 10^3 cm^3 = 0.219969 Imp. gal. = 0.264172 US gal. = 61.02361 in^3.

litres per kilometre   Measure of fuel burn = 0.3541 Imp. gal., 0.4252 US gal./statute mile; reciprocals respectively 2.82406, 2.3518.

LITS   Logistics Information Technology Strategy, or System [said to mean lost in time and space] (RAF, UK).

litter   Stretcher (medical).

Little F   Lt-Cdr (Flying), (RN).

littoral warfare   Coastal, shallow water.

LITVC   Liquid-injection thrust-vector control.

LIU   LAN interface unit.

l$_{iv}$ Root chord of vertical tail.

LIVE, Live   Liquid inertial vibration eliminator.

live drop   Release from aircraft of operative device, eg missile with propulsion and guidance and possibly warhead, as distinct from inert equivalent.

live engine   Operative engine(s) in aircraft with one or more real or simulated failures.

live flight deck   Subject to motion at sea, rather than in port (aircraft carrier).

live mail   Real air mail, as distinct from dummy loads.

live nut   Driven by rotary power unit along thread to give [usually irreversible] linear output.

livre[s]   Pounds (lb) avoirdupois (F).

lizards   Short lengths of rope, often with pulleyblock on free end, for ground-handling kite- or barrage balloon.

LJAO   London Joint Area Organization.

LJS   Laser jamming system.

LK   Product support (G).

LKLY   Likely.

LKN   Last known position [rarely, *LKP*].

LKR   Low-kiloton range.

LKS   Lakes.

LKV   LuchtvaartKundige Vereeniging (SA).

LL   *1* Low-level.

  2 Limit load,

  3 Flying laboratory, ie research aircraft (USSR, R).

  4 Low-lead [fuel].

  5 Long lead [time].

L$_L$ Loudness level (Stevens) in phons.

L-L   Line to line (AC voltage).

L/L   Latitude and longitude.

LLAD   Low-level air defence; S adds system.

Llanbedr   Airfield on Welsh coast serving Aberporth with targets (RAE, now Qinetiq).

LLAPI   Low-level air-picture interface.

LLC   *1* Lift-lift/cruise.

  2 Logic link control.

LLDF   Low-level discomfort factor.

LLDIN   Long-lead-in light system, can cross major city.

LLF   *1* Low-level fan of reconnaissance cameras.
  2 Long-lead funding.

LLGB   Launch-and-leave guided bomb.

LLH   Light liaison helicopter.

LLHK   Low-level height keeper.

LLLGB   Low-level laser-guided bomb.

LLLTV   Low-light-level television.

LLM   *1* Long-lead [time] material.
  2 Launcher loading module.

LLMS   Liquid-level measurement system.

LLNL   Lawrence Livermore National Laboratory, California.

LLP   *1* Low-level parachute.
  2 Left lower plug [all similar entries = avionics boxes].
  3 Limited-liability partnership.
  4 Life-limited, or limited-life, part[s].

LLS   Lightning location system.

LLT   Long lead time.

LLTOW   Landing limiting (or limited) takeoff weight.

LLTV   Low-light television.

LLV   Lower limit of video (HUD).

LLWAS   Low-level windshear alert system (sometimes LLWSAS).

LLWC   Low-level weather chart.

LLWD   Low-level weapons delivery.

LLWS   Low-level windshear.

LLZ   Localizer.

LM   *1* Lunar, or landing, module.
  2 Last-minute (cargo).
  3 Laser machining.
  4 Little movement.
  5 Locator, middle.
  6 Laser module.
  7 Line maintenance.

L/M   *1* Ratio of direct labour to material cost.
  2 List of materials.

lm   Lumen.

LMAE   Lunar module ascent engine.

LMAL   Langley Memorial Aeronautical Laboratory (NACA), became NASA LaRC.

L_{max} Peak sound level.

LMC   *1* A last-minute check-in.
  2 Life-monitoring computer.

LMD   *1* Laboratoire Météorologique Dynamique (F).
  2 Lithium manganese dioxide.

LMDE   Lunar module descent engine.

LME   *1* Line maintenance engineer.
  2 Link management entity.

LMF   *1* Liquid methane (or methanol) fuel.
  2 Lacking moral fibre (RAF, 1939–45).

L/MF   Low/medium frequency.

LMG   *1* Liquid methane gas.
  2 Light machine gun.

LMI   Logical management interface.

LMIT   Laser materials interaction testing.

LML   Lightweight multiple launcher.

LMLF   Limit manoeuvre load factor.

LM/LO   Liquid methane, liquid oxygen.

LMM   Compass locator at middle marker.

LMN   Local Mach number.

LMO   Lean-mixture octane [rating].

LMP   *1* Lunar module pilot.
  2 Left middle plug.

LMRS   London Military Radar Services.

LMS   *1* Least mean square.
  2 Land mobile service.
  3 Local maintenance system (navaid).
  4 Light monitor and switch.
  5 Learning management system [on-line].
  6 Line-maintenance service[s].

LMSJ   Lightweight modular support jammer.

LMSS   *1* Light mission support system (RAF).
  2 Land mobile satellite service.

LMST   Lightweight multiband satellite terminal.

LMT   *1* Local mean time.
  2 Locally manufactured tools, made to exact specification of an OEM.
  3 Limit.

LMTR   Laser marker and target ranger.

LMU   Line monitor unit.

LN   Glider, training (USN, 1941–45).

L_N *1* N (night) weighted sound pressure level.
  2 Confusingly, also the noise level exceeded for $_N$ % of each 24 h.

L-N   Line to neutral (AC volts).

LN_2 Liquid nitrogen.

LNA   Low-noise amplifier [DPL adds diplexer].

LNAV, L-Nav   Lateral navigation.

LNC   Loran [not necessarily Loran-C] chart.

LNDG   Landing.

LNG   *1* Liquefied natural gas.
  2 Long.

LNH   Lunar near horizon.

LNO   *1* Limited nuclear option.
  2 Liaison officer.

LNP, L_{NP}   Noise pollution level, equal to L_{eq} + 2.56 dB standard deviation.

LNSF   Light Night Striking Force.

LNTWA   Low-noise travelling-wave amplifier.

LO   *1* Low observables.
  2 Local oscillator.
  3 Compass locator at outer marker, also LOM.
  4 Longitude.
  5 Low band.
  6 See next.

lo   *1* Low level, variously interpreted as 60 m and 200 ft; minimum practical safe height for transonic attack.
  2 Minimum safe height to avoid obstructions, generally proportional to speed.

L/O   *1* Lift-off.
  2 Light off.

l.o.   Local oscillator.

LOA   *1* Letter of offer and acceptance; sometimes rendered as 'letter of agreement'.
  2 Launch on assessment.
  3 Letter of authorization.
  4 Line of attack.

LOAD, Load   Low-altitude defence (of ICBMs).

load cell   *1* Fluid-filled device for generating large forces accurately, eg in weighing large aircraft.
  2 Capsule containing strain gauge or other force transducer used, eg, in weighing aircraft.

Load classification number   *LCN*.

loadeo   Loading of explosive ordnance, structured procedures also used as basis for inter-unit competition (USAF).

loader   Loads computer main memory, esp. from transit tape.

load factor   *1* Vertical acceleration in g.

2 Stress applied to structural part as multiple of that in 1 g flight (not necessarily same definition as 1).

3 Ratio of failing load to assumed 1 g load in component.

4 Number of passenger seats occupied as percentage of those available.

5 Revenue ton-miles (or tonne-km) performed as percentage of RTM available.

6 Percentage of engine's maximum power needed for aircraft to fly (pre-1914).

load history   Crucially, a record of the number of times a particular load [stress, or vertical acceleration] has been exceeded by particular part.

loading   *1* Total aircraft mass divided by wing area (wing *) or total installed power (power *).

2 Volume fraction of composite (FRC) material occupied by strong fibres.

loading bridge   See *jetway*.

loading chart   Chart or diagram displaying correct locations for airborne loads in transport aircraft, esp. cargo.

loading coil   Inserted inductance, eg to increase electrical length of aerial [antenna] system.

loading diagram   *1* Standard graphical plot of forces in structural part or assembly.

2 Document with detailed plan of cargo floor and underfloor holds on which responsible official marks positions and masses of all cargo and final c.g. position.

loading loop   Graphical plot of transport aircraft weights against % SMC or MAC or distance from datum with fuel and payload forming closed figures.

loading up   Rich extinction of idling piston engine.

load manifest   Detailed inventory of cargo on commercial or military flight.

loadmaster   Member of military transport flight crew in charge of loading and unloading, and para-dropping etc if undertaken (but not of paratroops).

loadmeter   Ammeter, especially on light helicopter.

loadout   Total mission load (US).

loadpath   Sequence of structural elements carrying a load. Thus in F-22 * with missile-bay doors open is different from normal.

load programmer   Person in charge of structural fatigue test.

load ring   Rigid hoop to which balloon net and basket suspension are attached.

LOADS, Loads   Low-altitude air defence system.

load sheet   See *load manifest*.

load spreader   Rigid pallet for distributing dense loads over larger floor area.

load threshold   Notional maximum movements per hour ATC can accept.

load waterline   Waterline of marine aircraft at MTO weight.

LOAL   Lock-on after launch.

LOAS   Line operations assessment support.

LO-Axi   Low observable axisymmetric [N adds nozzle].

LOB   *1* Left outboard.

2 Launch operations building.

3 Line of bearing.

lobe   *1* One of two, four or more sub-beams that form directional radar beam from aerial [antenna] with reflector.

2 One of the (usually symmetrical) plan systems of regions of most intense noise from jet engine.

3 Eccentric profile of cam.

lobe nozzle   Jet-engine nozzle with mixing promoted by long multi-lobe periphery, resembling petals.

lobe switching   Radio D/F by time-variant switching of aerial [antenna] radiation pattern.

LOBL   Lock-on before launch.

Lobstar   Low-band structural array, principally for [radar] foliage penetration.

LOC   *1* Line of communication.

2 Letter[s] of credit.

3 Logistics operations centre.

4 Limited operational capability.

5 Level of capability.

6 Location, locator (ICAO).

7 Local-operations console.

8 Limiting oxygen content, minimum to sustain combustion.

9 See next.

loc   Localizer.

LOCAAS, Locaas   *1* Lo-observable comprehensive autonomous attack system.

2 Low-cost anti-armor submunition (USA).

3 Low-cost autonomous attack system (USN).

local area network   Bus, ring, PABX (telecommunications) or other transmission links for communication and EDP within an office, factory or other establishment.

local elastic instability   Small region of buckling in otherwise almost undistorted structure.

local hour angle, LHA   Hour angle of observed body measured relative to observer's meridian.

localizer   ILS aerial [antenna] and beam giving directional guidance.

localizer course   Locus of points in any horizontal plane at which DDM is zero.

localizer needle   Directional steering needle on ILS display.

localizer protected area   No aircraft or vehicles permitted to enter.

local Mach number   Actual Mach number at a point just outside boundary layer of aircraft or other vehicle. See local-surface airspeed.

local magnetic effects   Those peculiar to a region (eg ore deposits) which distort terrestrial field.

local mean time, LMT   Angle at celestial pole between local meridian and that of mean Sun; time elapsed since mean Sun's transit of observer's anti-meridian.

local meridian   That passing through a particular place.

local oscillator   Radio circuit generating RF with which received waves are combined.

local stress concentration   Intensification caused by shape of stressed part (eg at end of a crack).

local-surface airspeed   TAS+V_L where V_L is increment. [usually positive] induced by body's shape.

local traffic   Visible from tower.

local velocity   Relative speed of fluid flow over small area of body, essentially = local-surface airspeed.

local vertical   Line from centre of Earth through a particular place.

Locap   Lo combat air patrol.

Locat   Low-cost aerial target, or trainer.

Locate   Loran/Omega course and tracking equipment.

location bearings   Ball or tapered-roller bearings which determine the axial position of a shaft.

location indicator   Identification code [usually four letters] of aeronautical fixed station (ICAO).

locator   *1* L/MF NDB used as fix for final approach.

2 Portable radio beacon 121.5/243 MHz, carried on person or in parachute harness, sometimes with voice facility.

LOC-BC, loc-BC   Localizer backcourse.

LOCC   Launch, or lander, operations control centre.

LOCD   Low-cost dispenser.

LOCE   *1* Limited operational capability [for] Europe, see next.

2 Linked operations and intelligence centres, Europe

3 Large optical communications experiment.

Locid   Location identifier.

LOC inertia smoothing   Extra AFCS function, usually customer option, added in FCCs to alleviate effect of ILS localizer noise. Typically Setting 1 smoothes approach and provides survival after LOC failure below 100 ft/30 m; Setting 2 reduces minima on Cat II or III ILS.

Lockclad   Conventional 7 × 7 control cable covered with swaged aluminium envelope.

locked   *1* Gun bolt or breech block mechanically locked to barrel at moment of firing.

2 Flight controls mechanically locked to prevent damage by wind when parked; must be unlocked for flight.

3 Overbalanced flight-control surface driven to limit of deflection and not (or not readily) recoverable.

4 Carrier flight deck is for any reason not usable for flying.

locked-in-condition   Aircraft in dangerous flight condition (eg superstall) with airflow over controls inadequate for recovery to be possible.

locking wire   Fatigue-resistant wire pulled through ad hoc holes in series of nuts or other rotary fasteners and finally tightened and twisted off with inspector's stamp on soft metal seal at free ends.

lock nut   One that cannot loosen once tightened.

lock-on   1 Operating mode of many radars and other sensor systems in which pencil beam, having searched and found a target, thereafter remains pointing at that target.

2 When DME receives replies to 50+ per cent of interrogations.

lock time   Time from release of gun sear to firing of primer or detonator.

lock up   Lock on. Used esp. in airborne self-test of radar against other member of formation.

lock washer   Tightened under nut, prevents nut working loose (by biting into nut or by a tab manually turned up beside a flat on the nut).

lockwiring   Preventing rotation of nuts or bolts by threading fatigue-free wire through their heads to apply torque opposing movement.

loclad   Low-cost low-altitude dispenser.

Locom   Low-cost manufacturing [AMS adds of advanced metal structures].

LOC1   First level of capability (ACCS).

locpod   Low-cost powered off-boresight dispenser.

LOCR   Low-observables combat readiness (USAF).

Locus   Laser obstacle-cable unmasking system (helo).

locus   Path traced out by moving object, esp. one rotating in complex repeated orbit, eg tip of helicopter rotor blade.

Locusp   Low-cost uncooled sensor prototype.

LOD   Light-off detector.

Lodals   Long Odals.

LODE   Large optics demo experiment.

LOE   *1* Level of effort.

2 Loft- and line-oriented evaluation.

LOEC, LO ExCom   Low-observables Executive Committee, decides release of knowledge to partners/customers (US).

Lo-Erode   Concrete with surface reinforced with Meltex 19-11 stainless-steel fibres.

LOF   *1* Lift-off.

2 Line of flight.

Lofaads   Lo-altitude forward-area air defense system (USA).

Lofar   Low-frequency omnidirectional acoustic frequency analysis and recording (ASW).

Lo-Flyte   Low-observable flight-test experimental.

LOFT   Line-oriented flight training.

loft bombing   Low-level NW delivery also called toss bombing, etc (see *low-angle* *).

loft floor   Floor on which lofting is carried out.

lofting   Plotting full-size exact shapes of airframe, from which master templates, jigs, tooling, forging and stamping dies and other large parts can be constructed and NC tapes prepared.

LOG   Liquid oxygen/gasoline.

Log, log   Logarithm[ic].

log   Large ground-burning target marker (RAF, WW2).

LOGAIR   Air Logistics Command (USAF).

Logair   Logistics air network.

logarithmic decrement   Natural log of ratio of two successive amplitudes in damped harmonic or other oscillatory response.

logbook   Master history of member of aircrew, aircraft or other important functioning system in which are recorded times, events and occurrences.

Logholdair   Air logistics message (NATO).

logic   Electronic circuits and subcircuits constructed to obey mathematical laws.

Logmars   Logistics applications of [bar code] marking and reading symbols.

Logo   Limitation of government obligation.

Logspark   Logistic(al) park, housing 7 days' fuel, weapons and maintenance supplies, serving several STOVL hides.

LOH   Light observation helicopter.

Lohmannising   Metal dipped in amalgamating salt, pickled and then plated with two or more protective alloy coatings.

LOI   *1* Lunar orbit insertion.

2 Letter of intent to purchase.

loiter   *1* To fly for maximum endurance.

2 To fly a standing patrol.

loiter plate   Place where helicopters can practise hovering down to ground level (RAF).

Lola   Low-level [windshear] alert.

Lolex   Low-level extraction of parachute-retarded cargo.

LOM   *1* Compass locator at outer marker.

2 Localizer outer marker.

Lomads   Low-altitude missile air-defence system.

Lombard Original design of 'bonedome', 1947.

Lomcovak Unlimited flight manoeuvre in which aircraft tumbles about transverse axis whilst travelling sideways at near-zero airspeed [often mis-spelt Lomcevak].

LOMEZ Low-altitude missile engagement zone.

LOMS Line operations management, or monitoring, system.

LON Longitude (FAA).

Lone Ranger Detachments by single aircraft throughout non-Communist world (RAF 1959→).

long Longitude.

long-dated Long lead time, ie must be ordered months to years in advance of aircraft completion.

longeron Principal longitudinal structural members in fuselage, nacelle, airship, etc.

longitudinal axis OX axis, from nose to tail; roll axis.

longitudinal bulkhead Major full-depth web in plane of OX axis.

longitudinal dihedral Angular difference between incidence of wing and of horizontal tail (latter normally being less).

longitudinal force coefficient Component of resultant wing force resolved along chord; $C_X = C_D \cos \propto - C_L \sin \propto$.

longitudinal oscillation Vibration along longitudinal axis (chiefly in ballistic rocket vehicles); also known as pogo effect.

longitudinal plane Any plane of which OX axis is a part.

longitudinal separation Minimum distance or time between aircraft cleared to same track at same FL.

longitudinal short-period One of the five classic modes of aeroplane motion: near-constant airspeed, heavily damped by wing and tailplane/canard.

longitudinal stability Ability to recover automatically to level flight after sharp dive or climb command; generally, all stabilities in plane of symmetry. Static * is defined as tendency to return to trimmed airspeed and AOA following any mild disturbance, throttle fixed throughout.

longitudinal velocity "The component velocity along the longtitudianl axis relative to the air", (B.S., 1940). This need not be synonymous with true airspeed.

longitudinal wave One devoid of lateral components (eg sound).

long lead time Must be ordered months to years before aircraft completion; sum of contractual delays for item, delays in delivery (heavy forging may be years), processing time at sub-contractor or in plant, and time from when finished part joins aircraft to completion of aircraft.

long-range No valid modern definition except DoD * bomber operational radius over 2,500 nm.

long-range operations Philosophy possible with modern aircraft, where reliability is near-perfect. It ignores number of engines, and aims to achieve autonomous operation, with crew [two pilots] able to correct any fault or mishap.

long-wave Usually means wavelength over 1 km, but not a normal aerospace term.

Lons, LONS Local on-line network system.

look angle Angular limits of vision of EO or IR seeker.

look-down angle Limiting inclination of main beam in AWACS-type aircraft, strongly dependent on aerial [antenna] geometry.

look-down shoot-down Ability to destroy low-level hostile aircraft from high altitude against land or clutter background.

loom Tightly laced bundle of electric cables, instrumentation leads or other flexible wires.

loop *1* Flight manoeuvre in which aircraft rotates nose-up through 360° whilst keeping lateral axis horizontal; many variations but normal loop restores level flight on original heading but at slightly higher altitude. See inverted *, outside *, bunt.

 2 Conductive coil in vertical plane rotating about vertical axis to give bearing to ground radio station; a D/F loop (obs.).

loop detector Conductive loop buried in runway or taxiway to sense passage of aircraft and activate a display or airfield lights.

loop heat pipe In spacecraft, keeps vapour and cooled liquid separate by circulation through porous wick.

Loose Deuce Fighter combat tactic, 2 to 4 friendly aircraft manoeuvre to provide mutual support and increased firepower.

LOP Line of position [or positioning].

LOPC Lander operations planning center.

lopro Low probe; military aircraft mission.

LOR *1* Lunar orbital rendezvous.

 2 Launch off RWR target data.

Lora, LORA *1* Level of repair analysis.

 2 Low-frequency radar.

Loraas Long-range airborne ASW system.

Lorac Long-range accuracy, also called Loran-C or Cytac; Loran derivative.

Lorads Long-range radar and display system [now Lorads II].

Lorag Loads research advisory group.

Loran Long-range navigation, early [1941] but much developed hyperbolic navaid, using various onboard systems to translate time difference of reception of pulse-type transmissions from two or more fixed ground stations. In 1980 Loran-A [1,850/1,900/1,950 kHz] was replaced by Loran-C [100–110 kHz].

LORAS, Loras Linear omnidirectional resolving airspeed system.

Lord, LORD Laser obstacle ranging and display, for helicopters.

Lord mount Large family of patented anti-vibration mounts, usually metal/rubber.

Lorentz force, F$_L$ That on charged particle moving in magnetic field, = q(VB) where q is particle charge, V velocity and B magnetic induction (flux density).

Lorentz system Pioneer beam-approach landing system.

Lores, LORES Low-resolution.

LORO Lobe-on receive only.

Loroc *1* Long-range optical camera.

 2 Long-range offboard chaff.

Lorop Long-range oblique [or optical] photography [S adds system].

Lorv Low-observable re-entry vehicle, characterized by reduced radar cross-section.

LOS *1* Line of sight; I adds indicator.

 2 Loss of signal.

 3 Linear optical sensor.

 4 Line-oriented simulation.

 5 Level of service.

LOSA　Line Oriented, or Operations, Safety Audit (U. of Texas).

losas　Low-cost Scout acoustic system.

Losat　Line of sight, anti-tank.

Loschmidt number　Number of molecules of ideal gas per unit volume, $= 2.687 \times 10^{19}$ per cm^3.

Loss　Large-object salvage system.

lost-wax　Technique for casting intricate precision shapes, derived from Benvenuto Cellini c1550 but modified for modern refractory alloys. See *investment casting*.

LOT　Life of type [E adds extension].

lot　A particular meaning is one batch of production missiles.

Lotaws　Laser obstacle terrain-avoidance warning system.

Lotex　Life-of-type extension.

LO₂　Liquid oxygen.

lounge　Waiting area at airport for departure passengers between processing and gates.

LOV　Loss of vision due to opaque frames and other obstructions.

LOW　Launch on warning.

low　Geographical region of low atmospheric pressure.

low airburst　Fallout safe height of NW burst for maximum effect on surface target.

low altitude　US military traditional, 500/2,000 ft; today see *lo (1)*.

low-altitude airway　<18,000 ft [5,486 m] [see *low route*] (FAA).

low-altitude airway departure route　Provides operators with method to access under-utilized LAA (1) when upper airspace constrained, asking www.fly.faa.gov for procedures.

low-altitude bombing system　Early weapon-aiming electronics for tossing nuclear weapons in low attack, the high bomb trajectory giving aircraft time to escape explosion.

low-angle loft bombing　Free-fall loft bombing where release angle is within 35° of horizontal.

low approach　Premeditated overshoot.

low bidder　Manufacturer offering lowest price in industry competition.

low blower　See *low supercharger gear*.

low blown　Piston engine supercharged for maximum powers at low altitudes at expense of poor performance at height.

low boss　VSV inner bearing.

low cloud　Cumulus, cumulonimbus, stratocumulus, stratus and nimbostratus; base generally below 1,800 m, 5,900 ft.

low-cycle fatigue　Fatigue caused by changes in material stress resulting from changes in speed of rotating machines; from idling to take-off power and back to cruise rpm could represent single completed cycle.

low density, high demand　Assets [e.g. aircraft] available only in small numbers but needed in all theatres.

low Earth orbit　Below 2,000 km (1,243 miles), see *low orbit*.

lower airspace　No single definition; FAA usually below 14,500 ft AMSL.

lower airspace radar advisory service　Provided on request to local (say, within 50 km, 30 miles) uncontrolled traffic up to FL95 (CAA, UK).

lower-deck container　ULD shaped to fit underfloor space, either full- or half-width.

lower heating value　Net calorific value of fuel.

lower rotating ventral door　Chief pivoted member forming underside of D (or 2-D) nozzle.

lower sideband　Difference in frequency between modulation signal and AM carrier.

Lowest　Light occupant weight ejection-seat test.

low flying　Flight at minimum safe (sometimes unsafe) altitude for training or sport.

low-flying area　Geographical region within which low flying is authorized for training.

low frequency　Generally defined as 30-300 kHz.

low gate　Mechanical stop on piston-engine throttle box beyond which further opening inadvisable below rated height or other altitude limit.

low inclination mission　Satellite inclination less than 30° to Equator (eg 28.6°).

low-level discomfort factor　Usually measured as number of vertical accelerations exceeding specified level (0.5 g) experienced by occupants per minute.

low-level parachute　Rapid automatic opening for precision delivery of fragile loads.

low-light TV　Vidicon tube with multiplier tubes giving useful picture in near-darkness.

low-light-level TV　Generally used for same EO devices as preceding entry.

low/mid wing　Mounted about one-third way up body.

low oblique　Photography from lo altitude at oblique angles to either side, not vertical or ahead.

low observables　Stealth.

low orbit　Nominally, period 90 min or less.

low-pass filter　Designed to cut off all signals above given frequency.

low pitch　See *coarse pitch*.

low-pressure compressor　The first compressor downstream of the inlet [or fan, if fitted] in an engine with two or more shafts.

low route　Area-navigation routes not dependent on navaid-based airways, for low-level traffic, MAA (3) 4,000 ft AMSL (FAA).

low rudder　In tight turn, or other occasion with wings near-vertical, depressing pedal nearest ground to lower nose.

low silhouette　Squat aircraft, esp. in frontal aspect, easy to hide on battlefield (usually anti-tank helicopter).

low situational awareness　Criticism of combat pilot, failure to correlate several kinds of simultaneous input.

low-speed aerodynamics　Not defined; below 100 ft/s has been suggested.

low-speed aircraft　Several attempted definitions; UK CAA and FAR.91 do not define.

low-speed stall　Normal 1 g stall.

low supercharger gear　Lower of two gears in two-speed supercharger drive.

low vacuum　Pressure below 101.247 kPa (760 torr) and above some lower level usually agreed as 3.33 kPa (26 torr).

low-velocity drop　Paradrop or other drop with velocity below 30 ft/s (DoD, NATO).

low-visibility procedures　Adopted by airport controllers in visual control room [vary according to airport and RVR].

low wing　Mounted low on body, usually so that undersurfaces approximately coincide.

lox　Liquid oxygen.

loxodrome See *rhumb line*.

loxygen Liquid oxygen (arch.).

loz Liquid ozone.

LP *1* Low pressure.
2 Licensing panel.
3 Low-pass filter (often l.p.).
4 Liquid propellant.
5 Launch platform.
6 Low-power.
7 Linear polarization.

Lp Sound pressure level, usually measured over ⅓-octave or at a discrete frequency.

l̄p Rolling moment coefficient due to rolling.

LPA *1* Laboratoire de Physique de l'Atmosphère (F).
2 Linear power amplifier.

LPAR Large phased-array radar (Soviet ABM).

LPATS Lightning position and tracking system.

LPB Loss Prevention Bulletin; each issue lists over 20,000 stolen airline ticket numbers (IATA).

LPBA Lawyer Pilots' Bar Association (US).

LPC *1* Luftfahrt Presse-Club (G).
2 LP compressor.
3 Linear predictive coding.
4 Less paper in the cockpit.

LPCBA LP compressor bleed actuator.

LPCR Low-power colour radar.

LPD *1* Labelled plan display.
2 Low-prf pulse Doppler.
3 Log periodic dipole antenna, or array.
4 Low probability of detection.

LPDA Light propeller-driven aircraft.

LPDS Landing platform docking ship (USN).

LPDT Low-power distress transmitter.

LPDU Link-protocol data unit.

LPET Low-pressure elevated temperature glasscloth.

LPFI Logiky perspektive frontovoy istrebitel [lightweight future fighter] (R).

LPFT Low-pressure fuel turbopump.

LPG Liquid, or liquefied, petroleum gas.

LPH Landing platform, helicopter ship (USN).

LPHUD Low-profile HUD.

LPI *1* Low probability of interception.
2 Liquid-penetrant inspection.

LPIR LPI radar, multi-beam broadband coded waveform with very small sidelobes.

LPL Linear polarized laser.

LPLC, LPL/C Lift plus lift/cruise.

LPM *1* Looks per minute.
2 Landing path monitor.

l.p.m. Litres per minute.

LPN L𝐏𝐍 maximum value.

L𝐏𝐍 Perceived noise level.

L̄𝐏𝐍 Average peak outdoor L𝐏𝐍 at individual's residence.

LPNVGs Low-profile night-vision goggles.

LPO *1* Lunar parking orbit.
2 Lithium phospho-olivine.
3 Launch-panel operator.

LPOT Low-pressure oxidiser turbopump.

LPOX Low-pressure oxygen available.

LPP *1* Launch-point prediction.
2 Lean premixed/prevaporised.

LPPS Low-pressure plasma spray.

LPRE Low pulse-repetition frequency.

LPS Launch processing system.

LPSC Liquid Propulsion Systems Centre (Mahendragiri, ISRO).

LPT LP turbine.

LPTSW Linearly polarized transverse shear wave[s].

LPTV Low-profile transfer vehicle [for pax].

LPU *1* Logical program unit.
2 Line processor, or processing, unit.

LPX Extraction of shaft power from LP spool/shaft.

LQA Link quality analysis.

LQG Linear quadratic Gaussian.

LR *1* Long range.
2 Launch reliability of aircraft.
3 Line-replaceable.
4 Lead radial (VOR).
5 Prefix: the last message I received was . . .
6 Glider transport (USN, 1941-45).

l̄r Rolling moment coefficient due to yaw.

LRA *1* Line-replaceable assembly.
2 Landing-rights airport.
3 Laser retroreflector array.
4 Low-range radar altimeter.

LRAACA Long-range air anti-submarine capability aircraft.

LRAAS Long-range airborne ASW system.

LRAS Low-range airspeed system; ASI for V/STOL, helicopters.

LRAT Large radar array technology.

LRB Liquid-rocket booster (J).

LRBA Laboratoire des Recherches Balistiques et Aérodynamiques, Vernon (F).

LRBM Long-range ballistic missile (2,500 nm, 2,880 miles).

LRC *1* Long-range cruise.
2 Logistics Readiness Center (USAF).
3 Light reflective capacitor.

LRCA Long-range combat aircraft.

LRCO *1* Limited remote communications outlet.
2 Lead range control officer.

LRCSOW Long-range conventional stand-off weapon.

LRCU Landing rollout control unit.

LRD *1* Labelled radar display.
2 Laser ranger/designator.
3 Liquid runway deicer, usually glycol or PAF.

LRE *1* Launch and recovery element (UAV).
2 List of radioactive and hazardous elements.

LRF Laser rangefinder; D adds designator.

LRG Long-range.

LRGB Long-range glide bomb.

LRI *1* Long-range interceptor.
2 Air-traffic and airport administration (Hungary).
3 Line-replaceable item.

LRINF Long-range intermediate nuclear force(s).

LRIP Low-rate initial production.

LRIST Long-range IR search and track.

LRL *1* Lunar Receiving Laboratory.
2 Lightweight rocket launcher.

LRLS Laser radar landing system.

LRM *1* Long-range air-to-air missile.
2 Launching/reeling machine (towed MAD).
3 Line-replaceable module.

LRMP Long-range maritime patrol.

LRMTR Laser ranger and marked-target receiver.

LRMTS Laser ranger and marked-target seeker.

LRN   Long-range navigation.
LRNF   Low-Reynolds-number flight.
LRNS   Long-range navigation system.
LROPS   Long-range operations.
LRP   Lead-replacement petrol.
LRPA   Long-range patrol aircraft.
LRQA   Lloyd's Register Quality Assurance (UK).
LRR   Long-range [surveillance] radar.
LRRA   Low-range radar altimeter.
LRRR   Laser ranging retro-reflector.
LRS   *1* Load relief system.
   2 Laser reference system.
LRSI   Low-temperature reusable surface insulation (usually 99% pure silica-fibre felted tiles).
LRSO   Long-range stand-off; M adds missile.
LR^3   Laser ranging retro-reflector.
LRTS   Long-range tactical surveillance.
LRU   Line-replaceable unit.
LRV   *1* Lunar rover vehicle.
   2 Launch and recovery vehicle.
LRVD   Lower rotating ventral door.
LS   *1* Loudspeaker.
   2 Landmarks subsystem.
   3 Lavatory service vehicle.
   4 Prefix: the last message I sent . . .
   5 Lecture series.
   6 Loiter speed.
   7 Left side.
   8 Legacy software.
LSA   *1* Logistics support analysis [R adds records].
   2 Local-surface airspeed.
   3 Low situational awareness.
   4 Low-sidelobe antenna.
LSAB   London Society of Air-Britain.
LSALS   Lost survivor and asset locating system, detects chemical plume in ocean from crashed aircraft.
LSA, LSALT   Lowest safe altitude.
LSAR   Logistics support analysis records.
LSAS   Longitudinal-stability augmentation system.
LSAT   Logistic shelter, air-transportable.
LSB   *1* Lower sideband.
   2 Least significant bit.
LSC   Logistics support costs.
LSD   *1* Large-screen display.
   2 Least significant digit.
   3 Lithium sulphur dioxide.
LSEV   Less-stealthy export version.
LSF   Load-sheet fuel.
LSFFAR   Low-speed folding-fin aircraft rocket.
LS-FR   Low-speed frame relay; AD adds access device, SAD service access device.
LSI   *1* Large-scale integration (microelectronics).
   2 Lead system[s]-integrator.
LSIC   Large-scale integrated circuit.
LSIRR   Limb-scanning IR radiometer.
LSJ   Lifesaving jacket.
LSJSPO   Lethal Strike Joint Systems Project Office (DoD).
LSK   *1* Line-select key.
   2 Luftstreitkräfte, East German [DDR] air force 1949–90.
LSL   Landing ship, logistic.
LSLT   Line-of-sight link terminal (UAV).
LSM   Linear synchronous motor.

LSMU   Laser-communications space measurement unit.
LSN   Local sub-network.
LSO   *1* Landing safety officer (manages aircraft-carrier projector sight).
   2 Landing signals officer; T adds trainer.
   3 Limited strategic option.
LSP   *1* Large-screen projection.
   2 Logistics support plan.
   3 Longitudinal short-period.
   4 Link state PDU [protocol data unit].
   5 Locus of subvehicle points.
   6 Laser shock-peening.
LSQ   Line squall.
LSR   Loose snow on runway.
LSS   *1* Local speed of sound.
   2 Large space structure.
   3 Lighting sensor system.
   4 Land survey system [at crash site].
   5 Linear-scanning seeker.
   6 Laboratory-system specification (stealth).
LSSAS   Longitudinal static-stability augmentation system.
LSSM   *1* Local scientific survey module.
   2 Large-scale shared memory.
LSSS   Lightweight SHF satcom system.
LSST   *1* Lightweight secure Satcom terminal.
   2 Laser spot seeker/tracker.
LST   *1* Laser spot tracker.
   2 Lightweight satellite terminal.
   3 Lavatory servicing trailer.
   4 Local sidereal [or standard] time.
   5 Local Solar Time [on planet].
LSTAT   Life support for trauma and transport.
LSU   *1* Lavatory servicing unit.
   2 Legacy software upgrade.
LSV   *1* Laser speckle velocimetry.
   2 Lavatory servicing vehicle.
LT   *1* Low-pressure turbine.
   2 Low temperature.
   3 Lithium tantalate.
   4 Turn left after takeoff.
   5 Local time.
   6 Launch time.
L/T   Ratio of lift to thrust in jet-lift aircraft hovering or moving slowly in ground effect; subscripts, v ground vortex, h hover suckdown, w jet wake and f fountain.
L$_T$ Tail moment arm.
lt   Light, lighted.
l.t.   Low tension (electrical).
LTA   *1* Lighter than air.
   2 Lighter-Than-Air Society (US).
   3 Light transport aircraft.
   4 Light tactical aircraft.
   5 Long-term agreement.
   6 Limited test article.
LTAC   Light tactical airlift capability.
LT&E   Logistics testing and evaluation.
LTAS   See *The LTAS*.
LTB   Aircraft maintenance facility (G).
LTBT   Limited Test-Ban Treaty.
LTC   *1* Long-term costing, or contract.
   2 Lithium thionyl chloride (electric battery).
   3 Limited Type Certificate.

4 Lowest two-way channel.

LTCC   Low-temperature co-fired ceramic.

LTD   Laser target designator.

Ltd   Limited company (UK).

LTDP   Long-term defence plan (NATO).

LTD/R   Laser target designator/ranger.

LTDS   Laser target designator set.

LTE   *1* Loss of tail-rotor effectiveness.

2 Landline telephony.

3 Laser-tracker equipment, of aircraft in checking ILS.

LTF   Learn[ing] to fly.

LTFRS   Lantirn TFR system.

LTG   Lightning [CA adds cloud-to air, CC cloud-to-cloud, CCCG cloud/cloud/ground, CG cloud-to-ground, CW cloud-to-water, IC in clouds].

Lt Ho   Lighthouse.

LTIT   Low [-pressure] turbine inlet temperature.

LTKh   Flying qualities (ICAO).

LTL, ltl   Little.

LTM   *1* Load ton-mile.

2 Laser target marker.

3 Landsat thematic mapper.

4 Livestock transportation manual.

LTMA   London terminal control area.

LTMR   Laser target marker and receiver.

LTO   Landing and takeoff.

LTOF   Low-temperature optical facility.

LTP   *1* Laboratorio de Tecnologia de la Propulsión (Spain).

2 Longitudinal touchdown point.

3 Left top plug.

LTPG   Long-term planning guidelines (NATO).

LTPN   Tone-corrected perceived noise level, now judged of doubtful value, to account for intense tones, eg due to rotating fan and compressor blades.

LTPT   Low-turbulence pressure tunnel.

LTR   *1* Loop transfer recovery (EDP methodology).

2 Later.

LTS   *1* Load and trim sheet.

2 Landing threshold speed.

3 Lantirn targeting system.

4 Lights.

5 Link translator system.

LTSS   Long-term software support [P adds programme].

LTT   Landline teletypewriter.

LTV   Load threshold value.

Lt V   Light vessel.

L³TV   Low-light-level TV.

LTWA   Long trailing wire aerial, or antenna.

LTWTA   Linear, or linearized, travelling-wave tube amplifier.

LU   Logic(al) unit.

LUA   Launch under attack.

lubber line   Reference index, usually parallel to aircraft longitudinal axis, eg on compass, denoting aircraft heading.

Lubbock burner   Pioneer [1940] atomising burner for turbojets, devised by Isaac Lubbock; featured a sliding piston which controlled inlet ports to a swirl chamber.

lube   Lubricating oil (US, colloq.).

Lucero   Ground beacon keyed to 1.5 m AI or ASV radars, 1942-57.

Lucid   Software package for processing image data from all forms of visual sensor.

Lucite   Resin produced from methyl methacrylate, widely used for transparent plastics.

LUF   Lowest usable frequency.

Lufbery circle   Military ring formation in which all aircraft fly gentle turn following that in front.

Luftfartsverket   CAA (Norway), met. services (Sweden).

luhf   Lowest usable high frequency.

lumen   SI unit of luminous flux; lm = cd.sr (candela-steradian).

luminance   Brightness; intrinsic luminous intensity; illuminance on unit surface normal to radiation divided by subtended solid angle $L = cd\ m^{-2}$.

luminescence   Non-thermal emission of light, ie not incandescence but electro-*, phosphorescence, chemi-* and photo-* (fluorescence).

luminous flux   Light emitted in solid angle of 1 steradian by point source of luminous intensity 1 cd, symbol Φ, unit lumen; thus lm = cd.sr.

luminous intensity   Luminous energy per unit solid angle per second; unit candela, cd. A basic SI unit.

lunar boot   Astronaut footwear tailored to surface of Moon.

lunar orbit   Orbit round the Moon.

lunar orbital rendezvous   Lunar exploration by descent stage detached from larger spacecraft left in lunar orbit and which is rejoined for return to Earth.

Luneberg lens   Device, often spherical, designed for maximum reflectivity of radar energy back along incident path, tailored to wavelength; an enhancing corner reflector.

lusec   Lumen-second, quantity of luminous energy.

LUT   *1* Local-user-terminal [SAR satellites].

2 Limited user test.

3 Launch[er] umbilical tower.

lux   SI unit of illuminance; $lx = lm/m^2$.

LV   *1* Local vertical.

2 Lower [sideband] voice.

3 Low volume (crop spraying).

4 Light and variable.

5 Launch vehicle.

l̄v   Moment arm of vertical tail, usually measured from 25% MAC.

L/V   *1* Local vertical.

2 Lead/vinyl.

lv   Rolling moment coefficient due to sideslip.

L+V   Leather and vinyl.

LVA   *1* Large vertical aperture (radar).

2 Log video amplifier.

LVB   Luftverkeersbeveiliging, ATC authority (Netherlands).

LVD   Low-velocity detection.

LVDT   *1* Linear voltage differential transducer.

2 Linear variable differential (or displacement) transformer.

LVE   Leave, leaving.

LVER   Low-voltage electromagnetic riveter.

LVFE   Long variable-flap ejector (supersonic nozzle).

LVFR   Low-visibility flight rules.

LVIS   Low-voltage ignition system.

LVL   Level.

LVLCH   Change in FL.

LVNL   Civil ATC (Netherlands).

LVOR   Low-altitude VOR.

LVP   Low-visibility procedures.

LVPS   Low-voltage power supply.

LVT   Low-volume terminal (MIDS/JTIDS).

LVTO   Low-visibility take off.

LW   *1* Littoral warfare.
2 Landing weight [needs explanation].
3 Long-wave.

L$_w$ Sound power level; usually measured in ⅓-octave bands and can be measured in dB or W.

l.w.   Long wave.

LWA   Laser warning analyser.

LWABTJ   Lightweight afterburning turbojet.

LWAD   Littoral warfare advanced development.

L$_{WBT}$ Total lift of wing/body/tail.

LWC   *1* Light-water concentrate (firefighting foam).
2 Liquid-water content.

LWCCU   Lightweight common control unit.

LWD   Lowered.

LWF   Light-weight fighter.

LWIR   Long-wavelength IR.

LWL   Load waterline.

LWLD   Lightweight laser designator.

LWR   *1* Luftwaffenring eV (G).
2 Laser warning receiver.
3 Lower, or lowering.

LWS   Laser warning system.

LWSS   Lightweight sound system [helo sonar].

LWTR   Licence with type rating.

lx   Lux.

Lychgate   Multimedia data system linking RAF, MoD and other services (UK).

Lyman-Alpha   Radiation emitted by hydrogen at 12.16 pm (1,216 Å), penetrates Earth atmosphere to base of D-region (90 km, 55 miles).

LYR   Layer.

LYRD   Layered.

LYRS   Layers.

LZ   Landing zone (assault in land battle).

LZE   Luminous-zone emissivity (flare, IRCM).

LZS   Aeronautical association (Slovenia).

M

M   *1* Prefix mega, 10^6.
  2 Mass, except BS decrees m.
  3 Magnetic heading/course/bearing.
  4 Mach number [also M_n].
  5 Prefix minus (wind component).
  6 Maxwell.
  7 Dynamics, moment, esp. in pitch, with numerous suffixes.
  8 Meteorological (JETDS).
  9 Mutual inductance.
  10 Molecular weight.
  11 Structural bending moment, and generalized symbol for moment.
  12 Mandatory (NASA).
  13 Aircraft type designation: equipped to launch guided missiles (USN suffix 1955–62, prefix 1962–68).
  14 Telecom code: 'IFR aircraft has Tacan and transponder with no code capability' (FAA and others).
  15 Mean anomaly of orbit.
  16 Most ambiguously, thousand (ASA).
  17 Prefix maximum.
  18 Main.
  19 Maintain.
  20 Maritime air.
  21 Measured.
  22 Moderate.
  23 Multi-mission (US role prefix).
  24 MATZ penetration service [or (M)].
  25 Missing.
  26 Master station (Loran).
  27 Magnetization intensity; magnetic polarization.

(M)   *1* Torque, turning moment, also T.
  2 See *24* above.

m   *1* Metre[s].
  2 Prefix milli, 10^{-3}.
  3 Superplasticity.
  4 Modular ratio.
  5 Bypass ratio (USSR, R).
  6 Mass, esp. of electron.
  7 Minute[s].

m̊   Mass flow rate; also $\mathring{m} = kg\ s^{-1}$.

M1C   Cargo container half width of wide-body main deck.

m^2   Square metres.

M^2F^2   Multi-mode fire and forget.

M2M   Machine to machine [architecture].

M3P   Mini-mutes modification program.

M3R   Mobile multifunctional modular radar.

M-band   EM radiation, 5–3 mm, 60–100 GHz.

M-carcinotron   Backward-wave oscillator in which high power is possible by electron beam travelling between slow-wave structure and negative sole plate.

M-day   Day on which mobilization is to begin.

M-display   Has horizontal timebase along which target blip moves; operator moves second blip to line up on target by control graduated to indicate target range.

M-generator   Main generator (Gripen).

M-marker   *1* Middle marker.
  2 Low power NDB.

M-stoff   Methanol (G).

M-wave   EM millimetric wavelengths.

M-wing   Wing studied for low supersonic speeds with inner portions swept forward, outers swept back.

MA   *1* Mission abort.
  2 Naval aviation (USSR, R).
  3 Mobilization augmentee (US).
  4 Minor airfields.
  5 Meteorological authority (ICAO).
  6 Missed aproach.

M/A   Mach/airspeed.

mA   Milliampere[s].

ma, Ma   Mass flow (airflow); eg that passing through engine.

MAA   *1* Maximum authorized IFR altitude.
  2 Monitoring angle of attack.
  3 Missed-approach action.
  4 Mission-area analysis.
  5 Manufacturers' Aircraft Association [initially provided aircraft from a pool, 1917] (US).

MAAC   Medical Air Ambulance Company (USA).

MAAF   Mediterranean Allied Air Force (1942–45).

MAAG   Military Assistance Advisory Group (US).

MAAH   Minimum asymmetric approach height.

MAAM   Mid-Atlantic Air Museum (Reading, PA).

MAAS   Military air accident summary.

MAATS   Military automated air-traffic system (Canada).

MAAWLR   Miniature autonomous attack weapon, long-range.

MAB   *1* Modular-array basing (ICBM).
  2 Malaysia Airports Behrad (37 airports).

MABES   Manufacturing agent-based emulation system.

MAC   *1* Mean aeroydynamic chord, $\bar{\bar{c}}$.
  2 Military Airlift Command (USAF).
  3 Maintenance Analysis Center (FAA).
  4 Maintenance allocation chart.
  5 Master air control (UK, CAA).
  6 Medium-access controller.
  7 Multi-activity contract.
  8 Air defence command (Spain).
  9 Message act cancelled, or cancellation.
  10 Multiple all-up round canister, converts SSBN to fire cruise missiles.
  11 Mid-air collision.
  12 Mediterranean Air Command (US/UK, WW2).

Mac   Pitching moment of aircraft about aerodynamic centre.

MACA   Military assistance to civil authorities (US).

Macas   Mid-air collision-avoidance system.

MACC   *1* Military area control centre.
  2 Multi-application control computer.
  3 Modified Air Control Center.

MACCS   Marine air command and control system (DoD).

Mace, MACE   *1* Minimum-area crutchless ejector.
  2 Multinational alliance for criminal emergencies.

3 Multiple adaptive combat environment [for BVR training].

4 Multistatic ASW capability enhancement.

Mach   Mach number.

Mach angle   Angle between weak (ie point-source) shockwave and freestream flow; theoretically 90° at Mach 1, thereafter $\alpha = \sin^{-1} \frac{1}{M}$.

Mach-buster   Anyone who has flown faster than sound, esp. in pre-Concorde era when accomplishment had element of exclusivity.

Mach cone   Conical shock front from point source moving at supersonic speed relative to surrounding fluid; locus of Mach lines. Semi-vertex angle θ is given by cosec θ = M, where M is free-stream Mach number.

Mach disc   Visible disc at point of minimum diameter between jet shock and tail shock in supersonic jet, ie between adjacent shock diamonds.

Mach front   Mach stem.

Mach hold   Autopilot of AFCS mode holding Mach number at preset value.

machine   Flying machine, normally aeroplane (colloq., arch.).

machine gun   Magazine-fed automatic weapon using rifle ammunition.

machine language   Normally compatible with particular computer but unintelligible to humans.

machine screw   For pre-threaded holes and/or nuts, loosely = bolt.

machining centre   Major group of (usually NC) machine tools performing large number of operations on work-piece held either stationary or under positive control throughout.

Mach intersection   Junction of two or more shockwaves.

Mach-limited   Boundaries of flight performance set by restriction on permissible Mach number, not by thrust or other limitation, esp. * ceiling.

Mach line   *1* Weak (infinitesimal amplitude) shockwave.

2 Line on surface of body (ignoring boundary layer) at which accelerating free-stream flow reaches relative Mach 1.

Mach lock   See *Mach hold.*

Machmeter   Instrument giving near-instantaneous readout of Mach number.

Mach NO/YES   Air-intercept code: 'I have reached maximum speed and am not/am closing on target'.

Mach number   Ratio of true airspeed to speed of sound in surrounding fluid (which varies as square root of absolute temperature). Symbol M=V/a.

Mach reflection   Attenuated shockwave reflected from solid surface, eg walls of tunnel or Earth's surface. Reflection and some other effects approximate laws of optics.

Mach stem   Shock front (Mach front) formed by fusion of incident and ground-reflected blast waves from explosion, esp. from NW.

Mach trim coupler   Electronic subsystem of Mach trim system, which through analog (to 1970) or digital computational chain controls aircraft pitch-trim servo as function of Mach number; also contains switching, logic, monitoring and Bite.

Mach trimmer   Electronic/mechanical system for relieving pilot of task of correcting progressive deficiency in aircraft pitch trim and longitudinal stability at high Mach numbers; sensitive to Mach number and vertical

acceleration and automatically feeds primary pitch-trim demand to keep aircraft level or in desired attitude while leaving pilot authority to feed manual trim. In US called pitch trim compensator.

Mach tuck   Uncommanded, and possibly violent, nose-down pitching moment at high [usually just subsonic] Mach numbers.

MAC I   Model Aeronautics Council of Ireland.

MACIMS   Military Airlift Command integrated management system.

mackerel sky   Large area of high cloud giving dappled or banded effect (alto-cu or cirro-cu).

Maclaurin series   Special case of Taylor series where f(x) and all derivatives remain finite at x=0.

Macman   McDonnell anthropometric computerized man.

Macom, MACOM   Air-combat command (Spanish air force, EdeA).

Macro   Block of software capable of repeated use.

macro level   Preliminary design of software.

macroscopic   Generally, large enough to be visible to naked eye.

MACS   *1* Marine air control squadron (DoD).

2 Multiple-applications control system.

3 Mobile approach control system.

4 Modular airborne computer system (Gripen).

Mac-ship   Merchant aircraft carrier, merchant ship with flight deck [but nothing else] added (RN, 1942–44).

MAD   *1* Magnetic-anomaly detection, or detector.

2 Magnetic-azimuth detector.

3 Mutual, or massive, assured destruction.

4 Maintenance access door.

5 Mass air delivery (attack weapons).

6 Multiwavelength anomalous diffraction.

Madap   Maastricht automatic data-processing (Eurocontrol).

madapolam   Fabric woven from long-staple cotton, originally from Madapollam (single 1 originally erroneous).

Madar   Maintenance [or malfunction] analysis, detection and recording.

MADC   Miniature, or micro, air-data computer.

Maddls   Mirror-assisted dummy deck landings.

MADF   Missile assembly/disassembly facility.

Madge   *1* Microwave aircraft digital guidance equipment.

2 Malaysian air-defence ground environment.

3 Mascot design generator, enables software tools to be integrated.

MADLS   Mobile air-defence launching system.

MADM   Modified atmosphere density model.

MADME   Management and distribution system for more electric aircraft.

M-ADS, Mads   *1* Modified automatic dependent surveillance.

2 Modified air-defense system.

MAE   *1* Mean area of effectiveness (USAF).

2 Medium-altitude endurance (UAV).

3 Modular avionics emulator.

MAEO   *1* Master air electronics officer (RAF).

2 Medium-altitude electro-optics.

Maestro   Modular avionics enhancement system targeted for retrofit operations.

Mae West   Inflatable aircrew lifejacket tied round upper torso (WW2).

MAF *1* Mixed-amine fuel.

2 Marine amphibious force.

3 Michoud Assembly Facility (NASA, New Orleans).

4 Missionary Aviation Fellowship (A, PNG, UK), helped form (5).

5 Mission Aviation Fellowship (Int., numerous branches worldwide).

6 Military air forces (R).

MAFC Micro-adaptive flow control.

Maffs Modular airborne firefighting systems.

Mafis *1* Mobile automated field instrumentation system.

2 Multi-application fuzing initiation system.

Maflir Modified advanced forward-looking infra-red.

MAFT Major [or main] airframe fatigue test.

MAG *1* Military assistance, or advisory, group.

2 Marine air group.

3 Mobile arresting gear.

4 Machine gun.

5 Micromachined accelerometer gyro.

mag *1* Magnetic (FAA).

2 Magneto (eg * drop).

3 Magazine (camera) (NASA).

Magat Military assistance grant-aid training.

Magcom Guidance technique similar to Tercom but using variations in terrestrial magnetic field.

mag drop Reduction in rpm when either ignition source of dual-ignition piston engine is switched off; always checked before take-off to confirm both sources operative (from magneto rpm-drop).

Magerd MRCA AGE Requirement Document.

Magfet Magnetic Mosfet.

Maggs Modular advanced-graphics generation system, for simulator external visuals.

Magic *1* Microprocessor application of graphics with interactive communications (USAF).

2 Multiple-action global interactive control (Thomson-CSF).

3 Multiple-aircraft GPS integrated command and control (Herley-Vega).

magic eye Tuning system using miniature CRT with radial illumination around sector whose size varies with strength of received signal; important in pre-crystal tuning era; also called magic-T.

Magics Modular architecture for graphics and image console, or control, system.

Magiic Pronounced magic, Mobile Army Ground Imagery Interpretation Center (USA).

Magis Marine (or mobile) air/ground intelligence system.

maglev Magnetic levitation.

Magnaflux Non-destructive test for magnetic material using magnetic field and fluorescent ink; trade name.

Magnamite Family of CFRP composites (Hercules Inc.).

magnesium Mg, low-density (1.74) white metal, MPt 649°C, used pure and as alloy (Elektron, Dowmetal etc) for structural parts and incendiary bombs.

Magnesyn Patented remote-indicating system using induction between permanent-magnet rotor and saturable coil.

Magnet Multi-modal approach for GNSS 1 in European transport, implemented in two (A, B) concurrent forms (Euret).

Mag Net Mobile arresting gear using fast-erecting net to catch aircraft lacking serviceable tailhook.

magnetically anchored rate damper One in which gyro spin axis is restrained magnetically.

magnetic amplifier Various arrangements of saturable reactors such that small control current governs large output load/power/voltage.

magnetic anomaly Local irregularity in terrestrial magnetic field caused by presence of magnetic material such as submerged submarine or ore deposit.

magnetic azimuth detector Essentially, a compass.

magnetic bearing Direction of a fixed object measured clockwise from magnetic north.

magnetic compass Traditional compass indicating local horizontal direction of Earth's magnetic field.

magnetic core Doughnut-shaped ferrite ring storing either 1 or 0 in either of two stable magnetic states.

magnetic course Course (heading) indicated by simple magnetic compass after correction for deviation.

magnetic crack detection *Magnetic particle.*

magnetic crotchet Sudden change in numerical values of Earth's field usually ascribed to alteration in conductivity of ionosphere.

magnetic damping Use of eddy currents or other induced magnetic field to oppose oscillatory or vibratory motion.

magnetic declination Horizontal angle between terrestrial field and true N, ie between magnetic and geographic meridians; also known as variation.

magnetic deviation Errors in magnetic compass indication caused by local disturbance to field, esp. by 'iron' in aircraft.

magnetic dip Angle between local terrestrial field and local horizontal.

magnetic disc Computer storage in magnetic-oxide coating on surface of high-speed rotating disc.

magnetic drag cup Aluminium or copper cup surrounding rotating core in simple tachometer, rotated by generated eddy current.

magnetic equator Line joining points where angle of dip is zero.

magnetic field intensity Magnetizing force exerted on unit pole; also called field strength, symbol H, units Atm^{-1} [ampere-turns per metre].

magnetic flux Product of area and of field intensity perpendicular to it (in effect number of lines of force); $kg.m^2s^{-2}A^{-1}=V_s$; symbol Φ, unit is weber Wb.

magnetic flux density Measure of magnetic induction; $kg.s^{-2}A^{-1}=Vs.m^{-2}$; symbol B, unit is tesla $T = Wb/m^2$.

magnetic induction Induced magnetism in magnetic material, see previous entry.

magnetic inspection Many NDT methods which attempt to detect imperfections in magnetic-metal parts by the anomalies they cause in strong fields.

magnetic meridian Direction of horizontal component of terrestrial field near surface.

magnetic north North as indicated by the magnetic meridian.

magnetic orange pipe Brightly painted magnetized iron pipe filled with Styrofoam for aerial sweeping of mag-influence mines in shallow water.

magnetic particle inspection NDT for ferrous parts which when magnetized form N/S poles across cracks rendered visible by iron oxide powder viewed under UV.

magnetic permeability See *permeability.*

magnetic plug   Removable *chip detector*.

magnetic RAM   Any variety of RAM (2) which relies on magnetic materials such as very finely divided sintered nickel zinc ferrite, applied in layers on outside of aircraft skin; see *iron ball*.

magnetic refrigerator   Cryogenic cooler (eg for IR materials) operating by magneto-caloric repeated magnetization and demagnetization of suitable materials.

magnetic storm   Transient major disturbance in Earth's field.

magnetic surface wave   Travels across substrate, esp. ferromagnetic garnet.

magnetic tachometer   Most common speed-measuring system in aero engines measures frequency signal from magnetic and electrical interaction between toothed wheel and fixed sensor.

magnetic tape   Storage system in which information is recorded on and read off long strip coated with magnetic oxide.

magnetic turning error   Northerly turning error.

magnetic variation   Angle between magnetic meridian and true north, varies throughout globe and with time; thus used as ± correction to °M to give °T.

magnetizing force   Field strength, symbol H.

magneto-aerodynamics   Aerodynamics at high hypersonic speed and near-vacuum conditions, where magnetic and aerodynamic forces are similar.

magneto drop   See *mag drop*.

magnetohydrodynamics, MHD   Science of inter-action between magnetic fields and electrically conductive fluids, especially plasmas.

magnetometer   Instrument for measuring magnetic field intensity and direction.

magnetometer navaid   Precision position-fixing system based upon measurement of local terrestrial field, eg MAHRS.

magnetomotive force   Product of flux and reluctance (resistance), work needed to move unit pole against field; symbols F or M but unit ampere A.

magnetopause   Outer boundary of geomagnetic cavity.

magnetosphere   Region around Earth, from T-layer at 350 km to about 15 Earth radii, where magnetic field and ionised gas are dominant.

magnetostriction   Change in dimensions when magnetic material is magnetized; most pronounced in nickel.

magnetron   Pioneer resonant-cavity generator of high-power microwaves, with spinning electron beam deflected by transverse field.

magnitude   Apparent brightness of stars, planets, on scale of relative luminosity where × 100 brightness reduces magnitude by 5; hence brightest bodies have negative *.

Magnum   Missile[s] launched [at surface target].

Magnus effect   Force produced perpendicular to airflow past spinning body; basis of Flettner rotor and swerving golfball.

MAGR   Miniature airborne GPS receiver.

MAGS   *1* Miniaturized airborne GPS receiver [MAGR has also been used for this].

     2 Mode-S airport ground system.

MAGTF   Marine air/ground task force (USMC).

MAHA   Mid-Atlantic Helicopter Association.

MAHL   Maximum aircraft hook load (ASCC).

mahogany bomber   Non-agile fighter [colloq.].

MAHP   Missed-approach holding point.

MAHR   Microflex attitude and heading reference for light combat aeroplanes and helicopters, using strapdown three-axes magnetometer [S adds system, U unit].

MAHWP   Missed-approach holding waypoint.

MAI   *1* Moscow Aviation Institute.

     2 Mach/airpseed indicator.

Maica   Modelling and analysis of the impact of changes in ATM (7).

Maid/Miles   Magnetic anti-intrusion detector/magnetic intrusion line sensor.

MAIM   Méthodes Avancées en Ingénière Méchanique (F 2003, focused on aerospace).

main airfield   Permanent peacetime military airbase offering at least potential for all facilities.

main bangs   Transmitted pulses in radar (colloq.).

main beam   Principal beam from a radar; usually all others are unwanted.

main-beam clutter   Caused by main beam intersecting the ground.

main-beam killing   ECCM technique in which, while continuing to operate side lobes, main radar beam is suddenly cut off.

main bearing   Supports main rotating assembly in gas turbine, crankshaft in piston engine.

main bridle   Steel cable with eye at each end for mooring aircraft on water.

main float   Central float in three-float seaplane.

mainframe   Central computer in large or geographically dispersed EDP (1) system.

Main Gate   Crucial point of supposed no return in large programme, esp. in defence procurement (UK).

main gear   Main landing gear.

main landing gear   Each or all units of landing gear supporting nearly all weight of aircraft; other units are nosewheel, or tailwheel, or lateral outriggers.

main line   *1* Mooring line (airship).

     2 Data highway, excluding side branches attached at nodes.

     3 Cable from electric generator to aircraft electrical system.

main lobe   That on axis of transmitting dish antenna.

main parachute   Canopy supporting load, as distinct from drogues, extraction chute, etc.

mainplane   Wing, as distinct from tailplane or canard.

main rotor   Helicopter rotor providing lift and propulsion.

main runway   Airfield's principal runway, normally in use (usually longest).

main spar   Principal spar of wing, having modulus greater than others.

mainstage   *1* In multistage rocket vehicle, that stage having greatest thrust, excepting short-duration boosters (see [3]).

     2 In single-stage vehicle, main propulsion as distinct from verniers, roll-control and other motors.

     3 In smaller, or non-ballistic vehicles, sustainer propulsion after separation of boost motors(s).

     4 In large ballistic vehicle, period during which first-stage propulsion is delivering 90% or more of maximum thrust.

main step   Seaplane step below c.g.

maint   Maintenance.

maintenance   Work required, scheduled and otherwise,

to keep aircraft serviceable, other than repair. Term normally refers to minor tasks on flightline, but can also include major operations normally called overhaul.

maintenance access terminal Major connection through which engineers can interrogate systems and perform Bite checks.

maintenance burden Usually total maintenance cost, calculated as 1.8 times total maintenance labour cost.

maintenance data panel Electroluminescent or liquid-crystal display which collates, logs and displays any faults detected on avionics data bus.

maintenance dock Large structure fixed inside hangar with hinged or separable sections which can be closed tightly around large aircraft undergoing maintenance, providing staircases to platforms at different levels equipped with electric power and, usually, hydraulic and pneumatic power and water supply. Some are configured for particular type of aircraft.

maintenance lift Maintenance platform.

maintenance platform Mobile platform with scissors elevation to maximum height up to c10 m (c33 ft); some provide electric and other supplies.

maintenance recorder Wire or tape recorder for flight time, engine operation and variable number of other parameters.

Maintenance Review Board Establishes maintenance schedule for Transport Category aircraft.

maintenance schedule Prearranged plan for all maintenance required through life of item [but subject to revision].

maintenance status Non-operating condition deliberately imposed.

maintenance unit Military formation at fixed airbase able to store, modify, overhaul, flight-test and scrap aircraft.

main transverse Major frame of rigid airship joining all longitudinals.

main undercarriage Main landing gear.

mainwheel Wheel of main landing gear.

Mair Maritime air (NATO).

MAJCOM Major command (USAF).

Majiic Multi-sensor aerospace ground joint ISR interoperability coalition.

major aircraft review Search for items that can be cancelled (UK).

major axis Principal axis through solid body, usually along largest dimension or chief moment of inertia; where possible, axis of symmetry.

majority rule Philosophy whereby two or more operative channels (eg AFCS) always 'out-vote' single failed channel.

majority voting system Redundant system wherein outputs of three or more active channels are summed and output is fed back to each channel. When failure of one channel occurs, feedback causes all unfailed channels to act to offset failure; hence immediate shut-off of failure is not necessary.

major join Erection of aircraft, esp. mating fuselage to wing.

MAK *1* Interstate aviation committee (R, CIS).
 2 Maximal Arbeitsplatz Konzentration, standard for air pollution (Int.).

mAk Maritime Arctic air mass, colder than surface.

MAL Maximum allowable, hence *TOGW.

Malch, MALCH Multiple-access laser communications head (LEO satellite)

Mald, MALD Miniature air-launched decoy; -J adds jammer.

MALDT Mean administrative and logistics delay time.

MALE Medium-altitude long-endurance.

MALF Mostly aloft, precipitation not reaching ground.

MALJ Miniature air-launched jammer.

Mallory See *mercury battery*.

MALM Master Air Loadmaster.

MALS Medium-intensity approach light system; F adds sequenced flashing, R adds with runway-alignment lights.

Malta Microprocessor aircraft landing training aid.

MALV Miniature air-launched vehicle, unrelated to MALD/MALJ.

MAM Mission adapter module.

Mamba Mobile artillery-monitoring battlefield radar.

Mamis Mandatory aircraft modifications and inspections summary (FAA).

mammatus Cumuliform cloud with pendulous bulged undersurface.

Mammut Luftwaffe surveillance radar, 1944.

MAMS Mobile air movements squadron (RAF).

MAMT Multi-axial, or -axis, materials technology.

manacle Mounting in form of calipers surrounding and gripping circular section hardware, eg generator.

Manclos Manual command to line of sight.

mandatory To be complied with, often as AOG modification.

M&ES Magnetic and electromagnetic silencing.

M&O *1* Maintenance and overhaul.
 2 Maintenance and Operations.

M&R Maintenance and repair.

Mandrel High-power radar jammer (RAF, 1943).

mandrel Centrebody, usually circular section, around which tubular or female part is hand-forged, extruded or formed.

M&S *1* Marred and scarred (US category).
 2 Modelling and simulation.

maneton Pinch-fit coupling for female part tightened by bolt around (usually smooth-surface) pin, eg crankshaft web or big end.

manganese Mn, hard silver-white metal, density 7.3, MPt 1,244°C, important as alloying element in tough corrosion-resistant steels and with light alloys, brasses and bronzes.

manganese bronze Golden alloy, resistant to marine corrosion, sometimes used for compressor blading (DTD.197).

Manicom Manned information and communication facility.

manicured Beautifully kept [grass airfield].

manifest List of all passengers and cargo for one flight.

manifold Fluid pipe system distributing from single input to multiple outputs or vice versa (thus piston engine inlet *, exhaust *).

manifold pressure Pressure in inlet manifold of piston engine, normally local atmospheric plus boost (US traditionally inches Hg).

manipulator *1* Mechaical device resembling enlarged and more powerful human arm for positioning items in space.

2 Flat-plate aerofoil projections through aircraft boundary layer to cause major re-energization.

manning the rail   Crew of warship, esp. carrier, line main-deck periphery on entering/leaving harbour.

manoeuvrability   Measure of the maximum rate of change of magnitude and direction of the velocity vector. US maneuverability.

manoeuvrable re-entry vehicle   One which is capable of performing accurate preplanned flight manoeuvres during re-entry (DoD).

manoeuvre, maneuver   Any deliberate departure from straight-level flight or existing space trajectory.

manoeuvre ceiling   Maximum height at which aircraft, at given weight, can sustain specified load factor after onset of buffet; highest point of each buffet-boundary line.

manoeuvre diagram   Usually means V-n diagram or manoeuvring envelope.

manoeuvre flap   Usually small forward-hinged plate depressed under power from underside of inner wing at 20–40% chord to cause powerful nose-up trim change.

manoeuvre enhancement   Capability of a direct-force-control aircraft to achieve more rapid normal acceleration (upon pilot selection), typically by use of symmetric wing flaps for immediate increase in lift to eliminate difference between g (A_n) commanded and achieved.

manoeuvre induced error   Instrument error, usually pressure, gyro or magnetic, caused by flight manoeuvre.

manoeuvre load factor   Load factor (2).

manoeuvre margins   Two distances (stick-fixed and free) measured as % SMC from manoeuvre points to c.g. position.

manoeuvre motion   Real or simulated motion of cockpit resulting from pilot's control input, unlike disturbance motion.

manoeuvre point   Point around which aircraft rotates about all axes. Stick-fixed ** is c.g. position at which stick movement per g is zero; stick-free ** is c.g. position at which stick force per g is zero.

manoeuvring area   Area of airfield used for takeoffs, landings and associated manoeuvres (NATO).

manoeuvring ballistic re-entry vehicle   For practical purposes synonymous with manoeuvrable re-entry vehicle.

manoeuvring control force   Stick force per g.

manoeuvring envelope   Basic design envelope in which permissible speed (EAS) is plotted against load factor. From the origin the positive stall line extends to design limit load factor, thence to V_D, back to limit negative load factor at V_c and thence horizontally to intersect the negative stall line.

manoeuvring factor   See *load factor (1)*.

manometer   Linked twin or single vertical fluid tubes giving indication of pressure piped from remote source.

manometer bank   Large array of manometers side-by-side.

manometric lock   Autopilot function for capturing and holding constant pressure-based flight parameters, esp. IAS, altitude, vertical speed, M.

Manot   Missing-aircraft notice.

manpack   General term for astronaut/cosmonaut load carriers, designed according to local (eg lunar) gravity.

Manpads   Man-portable air-defense system (USA).

manprint   Manpower personnel integration.

man-rated   Sufficiently reliable to form part of manned spacecraft or launcher.

MANS, Mans   Missile and nudet surveillance.

man space   Assumed to include individual equipment and currently defined as 250 lb/13.5 cu ft (DoD).

Manta   Multi-axis no-tail aircraft.

Mantea   Management of surface traffic at European airports (Euret).

Mantech   Manufacturing technology (AFSC).

man-tended free-flyer   Autonomous space laboratory accessible to astronauts but not normally manned.

manual   Performed by hands, thus * control, * flying = hand-flying aircraft fitted with autopilot or AFCS.

manual D/F   Obtaining bearing by hand-rotation of loop and visual or aural judgement of signal.

manual feedback   Force experienced at pilot from manual FCS, or (rarely) other system.

manual override   Condition in which pilot physically overcomes AFCS through cable and/or linkage connec-tions and exerts flight control in excess of AFCS authority or in opposition to AFCS command.

manual reversion   Ability to switch from autopilot or AFCS to hand-flying (can be automatic in event of AFCS failure) wherein pilot's forces are transmitted to control surfaces.

manufactured head   Rivet head preformed when rivet made.

Manvis   Map and aviator's night-vision system (heli-copters).

many   Air intercept code: more than eight (DoD).

MAO   Mature aircraft objective.

MAOT   Mobile air operations team (British Army).

MAP   *1* Ministry of Aircraft Production (UK, 1940 –April 1946).

    2 Military Assistance Program (DoD).

    3 Machine assembly program.

    4 Missed-approach point, or procedure.

    5 Ministry of Aviation Industry (USSR, R).

    6 Manufacturing automation protocol.

    7 Manifold absolute (or air) pressure.

    8 Multiple aim point (USAF).

    9 Municipal airport (US).

    10 Maximum a priori probability.

    11 Modular airborne processor.

    12 Military Airport Program (FAA).

    13 Mode-annunciator panel.

map   Two-dimensional plot of two [sometimes several] parameters, usually based on X and Y axes, generally ≡ graphical plot.

MAPA   Malaysian Airlines' Pilots' Association.

Maple Flag   Canadian Red Flag.

map-matching   Navigation (usually of RPV or missile) by auto-correlation of terrain with stored strip of film; based on appearance, unlike Tercom.

Mapp   Methyl acetylene-propadiene/propane (FAE).

mapping radar   Producing pictorial display showing Earth's surface in detail (usually in sector ahead, some-times PPI 360° all round aircraft).

map reading   Navigating by comparing terrain with map, generally called contact flying.

MAPS, Maps   *1* Mobile aerial port squadron (MAC).

    2 Measurement of air pollution sensor (Shuttle), also rendered as measurement of atmospheric pollution from satellites.

3 Mission analysis and planning system, several models (USAF).

4 Military-aircraft planning system.

5 Meteorological and aeronautical presentation system (1976).

Mapse   Minimal APSE (Ada programming support environment).

MAPt   Missed-approach point.

MAPTFH   Mission aborts per thousand flight hours.

MAR   *1* March (ICAO).

2 At sea (ICAO).

3 Multiple- (or multi-function) array radar.

4 Mission abort rate.

5 Minimally attended radar.

6 Military Aircraft Release (UK).

7 Major aircraft review.

8 Multi-access receiver.

9 Multiple-access recorder.

MARA, Mara   Modular architecture for real-time applications.

maraging steel   High-alloy (Ni, Co, Mo) steels aged in martensitic condition for maximum tensile strength.

MARAIRMED   Maritime Air Forces, Mediterranean (NATO).

Maras, MARAS   Middle airspace radar advisory service.

MARC   Multi-access remote-control.

MARCS, Marcs   Military Airlift [Command] reaction communication system.

MARE   Miniature analog recording avionics.

Marendaz flap   Split flap hinged under wing entirely forward of TE.

Mareng tank   Flexible fuel cell (from Martin Engineering, US).

mare's tail sky   Cirrus.

marginal performance   Barely able to comply with airworthiness requirements, or to fly safely.

marginal weather   Only just good enough for safe or legal flight, actual conditions depending on whether flight is IFR or VFR. DoD definition: 'Sufficiently adverse to military operation so [*sic*] as to require imposition of procedural limitation'.

margin of lift   *1* Aerostat buoyancy (gross lift) minus mass.

2 Various meaningless definitions for aerodynes.

margin of safety   Percentage of ratio by which ultimate failing load of component exceeds design limit load.

maria   Flat areas on Moon once thought to be seas.

Marie   Martian radiation environment experiment.

marine aircraft   Designed to operate from water. Hence, marine aerodrome or airport.

maritime aircraft   Designed to operate over sea areas.

maritime PDNES   PDNES having extremely short pulses to reduce sea-clutter.

maritime SAR   Region within which USCG exercises SAR co-ordination function (FAA).

MARK   Material accountability and robotic kitting system (DoD).

Mark   Air-control agency code for commanded point of weapon release, usually preceded by word 'Standby' (DoD).

mark   *1* Abb. Mk, British word meaning version or sub-type to distinguish each variant of a basic aircraft design, whether produced in series or not. See * number.

2 Verb, to illuminate a target by flares, TIs, markers or laser.

marked target   One illuminated by (2) above to enable it to be attacked visually or by laser-homing weapons.

marker   *1* Distinctive visual (usually pyrotechnic), electronic or other device dropped on surface location.

2 Visual or electronic navigation aid indicating a fixed position, see * *beacon*.

3 Aircraft detailed to mark a target by air-dropped stores or to designate it by laser.

marker beacon   Any beacon (course-indicating, fan, outer, middle, inner) giving substantially vertical radiation, usually at 75 MHz, which gives aural and visual signal in cockpit (see *outer* *, *middle* * etc).

marker template   One having profile of common stringer or similar standard section, for marking end-cuts and cut-outs.

market   *1* The world total of available potential customers.

2 To promote one's product in (1).

marketable range or radius   Maximum sector distance sold to customers by air carrier; eg Dash-7 * 400 miles, while 'operational range' is 700 miles.

marking panel   Sheet of material displayed by ground troops for signalling to friendly aircraft.

marking team   Troops dropped into tactical area to establish navaids and possibly other electronic facilities.

marking up   Putting bright paint line round each airframe crack or blemish.

mark number   Suffix numeral identifying each *mark*; Roman to 1946, Arabic subsequently.

Marmon clamp   Patented ring joint for pipe sections, with wedge action.

MARRES   Manual radar-reconnaissance exploitation system.

marrying up   Offering up major airframe or other large items in erection of aircraft or space launcher, esp. when likelihood of imperfect fit.

Mars   *1* Mid-air recovery (USN), or retrieval (USAF) system.

2 Military affiliate radio system (DoD).

3 Multi-access retrieval system.

4 Modular airborne recorder series, or system.

5 Minimally attended radar station.

6 Meteorological and AIS (1) retrieval system.

7 Multi-applications recording, or reproducer, system.

8 Military Archive and Research Services (UK).

9 Mobile automatic reporting station.

10 Modular adaptable radar simulator.

11 Medium-altitude reconnaissance system.

Marsa   Military accepts responsibility for separation of aircraft (DoD/FAA).

marshal   *1* Person giving visual and/or aural signals to sporting aircraft or sailplanes to ensure line-up in correct take-off sequence.

2 See *Air Marshal*, *Sky marshal*.

marshaller   Person on apron or flightline giving visual signals or radio guidance to aircraft on ground, eg directions where to park; hence, * frequency, * van.

marshalling point   Place on airfield where aircraft come under control of a marshal (sporting, out-bound) or marshaller (inbound).

marshalling wand   Combined black flag and illuminated wand normally used one in each hand by marshaller.

MARSS   Microwave airborne radiometer and scanning system.

MART, Mart   *1* Modular air/ground remote terminal.

2 Mini-avion de Reconnaissance Télécommandé = RPV (F).

3 Mean active repair time.

Martacs   Maritime tactics.

martensite   Family of rapidly quenched [hence hard but brittle] steels.

Martha   Maillage anti-aérien des radars tactiques contre hélicoptères et avions.

Martlesham   Martlesham Heath, pre-1939 centre of military aircraft test and evaluation in UK, hence * figures (= indisputable).

MARV, Marv   Manoeuvrable, or manoeuvring [maneuvering], re-entry vehicle.

MAS   *1* Military Agency for Standardization (NATO).

2 Military area services.

3 Military assistance sales (DoD).

4 Middle airspace service.

5 Micro autonomous system.

6 Morphing aircraft structure.

MASC   Maritime airborne surveillance and control.

mascon   One of the mass concentrations scattered over lunar surface which distort orbits near the Moon.

MASDC   Military Aircraft Storage and Disposition Center (USAF Davis-Monthan AFB)).

MASE   Multi-axis seat ejection.

maser   Microwave amplification by stimulated emission of radiation, devices which pump electrons to higher energies and build up amplification of EM signals at microwave frequencies (see *laser*).

MASF   *1* Mobile aeromedical staging flight[s].

2 Military assistance service fund (DoD).

MASH, Mash   *1* Mobile Army Surgical Hospital (USA).

2 Mobile avionics screening handler.

MASI, Masi   Mach/ASI.

Masint   Measuring [or measurements] and signals [or signatures] intelligence.

masking   *1* Layer protecting substrate against applied process, eg metal foil or special paper when painting aircraft, or hi-resist. surface pattern when doping or etching microelectronic chip.

2 Deliberate use of additional emitters to hide or conceal true purpose of particular EM radiation.

3 Hiding battlefield helicopter behind natural cover.

4 Time taken for (3), typically ⩽2 s.

Masonite   Commercial formulation of compressed fibreboard, moulded to shape with accurate glassy finish on at least one surface.

MASP   Minimum aviation systems performance, A adds avionics, S standards or specification.

MASR   Multiple antenna surveillance radar.

MASS, Mass   *1* Marine air support squadron (DoD).

2 Mission avionics sensor synergism (data fusion).

3 Military approach and surveillance system.

4 Maritime air surveillance sortie.

5 Magnetic-array sensor system.

6 Multi-ammunition soft-kill system.

7 Military airborne surveillance system.

mass   Applied force on body divided by resulting accel-

eration; in practice gravitational attraction of body related to standard kilogramme.

mass axis   Line joining c.g. of all elements of part, esp. wing.

mass balance   Mass attached to flight-control surface, typically ahead of hinge axis, to reduce or eliminate inertial coupling with airframe flutter modes.

mass flow   *1* Quantity of fluid passing through closed system per unit time.

2 Specifically, mass of air passing through an engine or any of its components per second.

mass fraction   See *mass ratio (1)*.

mass injected pre-compressor cooled   Propulsion for initial part of Rascal, turbojet[s] with injection of water plus liquid air or LO_2.

massive ordnance   Conventional warhead weighing 9,072 kg (20,000 lb) or more (DoD).

mass law   Transmission of sound through a solid wall is very approximately inversely proportional to both frequency and mass per unit area.

mass parameter   Ratio of aircraft density [mass divided by wing area and mean chord] to air density multiplied by 2 and divided by lift slope; if necessary corrected for Mach number.

mass per unit area $1\ \mathrm{kg\,m^{-2}} = 3.27705\ \mathrm{oz\,ft^{-2}}$; $1\ \mathrm{lb\,ft^{-2}} = 4.88243\ \mathrm{kg\,m^{-2}}$.

mass per unit length $1\ \mathrm{kg\,m^{-1}} = 0.05600\ \mathrm{lb\,in^{-1}}$; $1\ \mathrm{lb\,in^{-1}} = 17.8580\ \mathrm{kg\,m^{-1}}$.

mass-properties engineering   Discipline dealing with weight, balance, moment of inertia, centre of gravity, stability and mission dynamics.

mass ratio   *1* Mass of vehicle, normally ballistic rocket, at liftoff divided by mass at all-burnt or other later condition.

2 Rarely and ambiguously, mass of propellants as fraction of total liftoff.

mass spectrograph   Instrument for converting molecules to ions and then separating these to determine exact isotopic weights.

mass taper   Rate of, or graphical plot of, reduction of cross-section-percentage material density from root to tip of blade (eg helicopter rotor), latter usually being externally untapered.

mass values   For average passengers [except charters]: male 88 kg, female 70, child 35.

MAST   *1* Military anti-shock trousers; pressurized, astronaut garb.

2 Military assistance to safety and traffic.

3 Married airmen sharing together (USAF).

4 Multimission airborne surveillance technology (USAF).

5 Multiple aircraft simulation terminal.

6 Maintenance, analysis, safety and training.

mast   *1* Tube projecting from underside of aircraft so that liquid can drain well away from airframe.

2 Rigid pillar for avionics antenna (aerial).

3 Rigid mooring structure for airship.

4 Pillar above centre of helicopter rotor for MMS (2).

Mastacs   Manoeuvrability-augmentation system for tactical air combat simulation (USN).

Master   Military aircraft satcom[s] terminal.

Master Bomber   Experienced crew-member tasked with staying over target throughout attack issuing instructions on [eg] which markers to use and which to avoid (RAF, WW2).

master caption panel   Flight-deck panel giving initial warning information and/or indication of AFCS mode, navigation mode etc.

master caution signal   Indication that at least one caution signal has been activated.

master connecting rod   See *master rod.*

master contour template   Full-size flat template showing all mould-line contours, rib contours or other related shapes to common reference.

master diversion airfield   Large and well-equipped airfield able to handle all kinds of aircraft whose home bases are shut by weather or enemy action, and promulgated as such.

master oscillator   That defining timebase and waveform for radio or radar.

master rating   In certain air forces, highest level of aircrew proficiency in each trade.

master rod   Connecting rod with big end and with attachments for all other rods in same plane.

master router template   Flat template, usually same size as metal sheet or slab, used to locate pilot and pin holes and set up sheet for routing or skin-milling by non-NC control.

master station   Various meanings, esp. that housing timebase and emitting original signals in R-Nav hyperbolic systems, other stations being slaves keyed to it.

master switch   Makes or breaks generator *main line*, must be ON to start engine[s].

master template   Reference for any form, usually two-dimensional, in manufacture; used to make jigs and tooling.

Mastif   Multiple-axis spin-test inertia facility.

Mastiff   Modular automated system to IFF.

mast-mounted sight   Group of sensors, typically optical, FLIR, laser, mounted on anti-vibration mast above helicopter rotor hub to allow crew to see targets without exposing helicopter.

MASU   *1* Multiple acceleration sensor unit.

2 Military Aviation Storage Unit (DARA Fleetlands, UK).

MAT   *1* Modification aproval test.

2 Moscow aviation technical high school.

3 Modular advanced test (Raytheon).

4 Maintenance-access terminal.

mat   *1* Flexible membrane laid on unprepared ground for limited number of V/STOL operations.

2 Area paved with quick-assembled metal mesh or planking.

3 Floating support towed by ship for recovery of marine aircraft.

Matac, MATAC   Tactical air command (Spanish air force, EdeA).

MATC   Mobile ATC (1).

Matcals   Marine (or Mobile) air-traffic control and all-weather landing system (USMC).

Match   Manned [or medium] anti-submarine troop- [or torpedo-] carrying helicopter.

matching   *1* Any prolonged process of developing dynamic parts to work properly together; eg inlet/engine/propelling nozzle over wide range of Mach numbers.

2 Achieving maximum energy transfer between electric circuits or devices by equalizing impedances and resistive/reactive components.

Matcon   Mobile air-traffic control unit (civil).

Matcu   Marine air-traffic control unit (USMC).

MATDSS   Maritime aircraft tactical decision support system.

Mate   *1* Modular automatic test equipment (Grumman).

2 Modular Autodin terminal equipment (Astronautics Corp.).

3 Man-portable aerial terrain explorer.

Matelo   Maritime air telecommunications organization (UK).

materiel   US term for hardware, logistic supplies, esp. military (but specif. excluding ships and naval aircraft); accent is on last syllable.

Matias   Magyar air-traffic integrated and automated system (Hungary).

mating   Offering up major portions of aircraft and joining together; the mass-production equivalent of marrying up. Thus * jig.

MATO, Mato   Military air-traffic operations (UK).

MATP   Military-aircraft tractor positioner, for confined spaces.

Matra, MATRA   Air-transport command (Spanish air force, EdeA).

Matrix   Multi-service automatic target-recognition imagery exploitation.

matrix   *1* Computer memory core of rectilinear array configuration, eg 2-D ferrite three-wire type.

2 Crystalline grain structure of metal.

3 Skeletal basis on which structure is formed, esp. by moulding.

4 Any 2-D rectilinear logic network.

MATS   *1* Military Air Transport Service (USAF/USN 1 June 1948, on 1 June 1966 became MAC).

2 Multi-altitude transponder system.

3 Manufacturing assembly tracking system.

4 Military-aircraft target system.

5 Miniature aerial target system.

MATSS   *1* Mobile aerosat [not aerostat] tracking and surveillance system.

2 Maritime aerostat tracking and surveillance system.

MATT   Multimission advanced tactical terminal (US Services).

matt   Non-reflective, external finish (usually paint) tending to diffuse EM radiation incident on it.

MATTS   Multiple airborne target trajectory system.

maturation   US term for growth of maturity, arguably = ageing.

maturity   Vague condition after impressive number of flight hours in which major problems naively thought unlikely. Supposed methods exist for quantifying degree of * for both new and derived equipment.

maturity factor   GDP growth × yield growth ÷ traffic growth.

MATV   Multi-axis thrust vectoring.

MATZ   Military air (or airfield or airport) traffic zone.

MAU   *1* Marine amphibious unit (USMC).

2 Million accounting units (EC).

3 Modular avionics unit[s].

MAUW   Maximum all-up weight (tantological).

MAV   *1* Micro air vehicle.

2 Mars ascent vehicle.

MAVACTD   Micro air-vehicle advanced concept technology demonstration.

Mavar   Parametric amplifier (from mixed amplification by variable reactance).

Mavus   Maritime VTOL UAV system.

MAW   *1* Mission-adaptive wing.
2 Missile approach warning, defensive system linked to RWR giving definite warning of a locked-on missile's approach.
3 Military Airlift Wing (USAF).
4 Marine Air Wing (USMC).

MAWP   Missed-approach waypoint.

MAWS   *1* Modular automated weather system.
2 Missile approach, or attack, warning system.

MAWTS   Marine Aviation Weapons and Tactics Squadron.

max   Maximum; CLB, CRZ add climb, cruise.

Maxaret   Pioneer (Dunlop) anti-skid system for wheel brakes.

max-des   Maximum rate of descent mode.

maxed out   Total overload of brain of jet pilot in combat.

maxi   Top RPV class, on basis of weight, altitude and speed; all are high-altitude jets.

Maxi decoy   ECM payload, originally bronze model aircraft with radar cross-section identical to attack aircraft.

maximum authorized altitude   Highest altitude on airway jet route or any other direct route for which MEA (1) is designated in FAR.95 at which navaid reception is assured (FAA).

maximum boom-free speed   Mach limit at high altitude to avoid shock reaching ground, variable with atmosphere in region of M 1.12.

maximum chamber pressure   Peak pressure in complete firing cycle of solid-propellant motor, symbol Pc max.

maximum C_L   Value at peak of curve where $dC_L/dx=0$, hence * angle of attack.

maximum climb thrust   For most jet engines, achieved by advancing throttle to obtain a predetermined EPR.

maximum cold thrust   Highest thrust without using (available) afterburner.

maximum contingency   Highest output turboshaft (rarely, other type of engine) can deliver or is authorized to deliver, usually for $2\frac{1}{2}$ min. following failure of other engine of multi-engine aircraft; usually requires subsequent inspection.

maximum continuous   Thrust or rpm limit available for unlimited period, usually using rich mixture if piston engine and all augmentation for gas turbine; often available for certification or emergency only.

maximum cruise rating   Highest power without augmentation or [piston engine] rich mixture; often this is highest available for normal continuous operation.

maximum cruise thrust   Same as max climb procedure.

maximum cruising speed   Highest speed permitted (usually in flight manual) for sustained operation; normally expressed as EAS or CAS, occasionally only as Mach.

maximum demonstrated   Precise limit [of whatever parameter] available to test pilot when aircraft is being certificated.

maximum diving speed   Highest speed demonstrated in certification (see V_{DF}).

maximum effort   Using every aircraft that can be made

serviceable, though usually without addition to list of allowable deficiencies.

maximum except takeoff   Highest available piston engine power other than takeoff rating.

maximum expected operating pressure   Peak (or 1.1 times peak) of pressure throughout burn of solid motor.

maximum gross weight   US term = *MTOW*.

maximum landing weight   MLW, certified value above which fuel must be jettisoned or burned off if landing becomes urgently necessary and possible structural damage is to be avoided.

maximum likelihood   A standard method of deriving stability and control plus cost function [difference between measured and estimated response, summed over time].

maximum loaded weight   To avoid confusion, now generally replaced by MRW.

maximum mean camber   Maximum camber of median line of wing sections from root to tip.

maximum net weight   Allowable payload of baggage trolley or other surface vehicle = maximum gross minus tare.

maximum on ground   Overall size of 'spot' on ground needed for landing, unloading and take-off of airlifter, probably in haste (under enemy fire). Most studies demand so many MOG Spots that STOVL capability is essential.

maximum operating Mach number   Self-explanatory, but still less than M_{NE}.

maximum payload   Limit for transport as defined in certification. For cargo aircraft usually same as *maximum structural payload*; for all-pax configuration, usually established by seat configuration at much lower value.

maximum performance takeoff   That in which energy transfer is at highest possible rate, usually limited by sliding of locked tyres and avoidance of striking tail on runway or exceeding gear-down EAS.

maximum power   For propulsion engine, power under ISA/SL conditions with engine operating at authorized limits of rpm, pressures and temperatures.

maximum power altitude   Lowest altitude at which full throttle is permissible (some definitions add: at maximum rpm for level flight); for supercharged piston engine, highest altitude at which maximum boost pressure can be maintained.

maximum q   Highest dynamic pressure attainable; for given aircraft, function of dive limits, atmospheric pressure and structural strength.

maximum ramp weight   The highest of all certified weight, = MTOW/MTWA plus fuel allowance for start-up and taxi.

maximum reading accelerometer   Records maximum reached.

maximum rpm   In shaft-drive engines, invariably a transient overspeed condition permitted by lag in propeller or engine-speed control system.

maximum speed   Highest TAS attainable in level flight, at best altitude.

maximum structural payload   Combined weight [or capacity] of pax, baggage and cargo certificated to be carried on main deck and in belly holds.

maximum takeoff rating   In jet engine, achieved by selecting water [if available] and advancing throttle to give TO EPR.

maximum takeoff weight   See *MTOW*.

maximum turn rate   That giving maximum number of deg/s rate of change of heading, ie rotation of longitudinal axis; unrelated to change in trajectory.

maximum undistorted output   Peak signal strength consistent with intelligible speech in early radio.

maximum usable frequency   Highest for shortwave communications between fixed points using ionospheric reflection (time-variant).

maximum vertical speed, Vm   Rare flight limitation demonstrated in VTOL aircraft, including civil-certificated helicopters.

maximum weight   See *MTOW, MLW, MZFW* etc.

maximum weak mixture   Highest power normally available from piston engine throughout flight; rich-mixture ratings can reduce engine life and aircraft range.

maximum zero-fuel weight   The maximum weight certificated with no usable fuel, probably less than MTOW minus usable fuel because wing fuel relieves the wing bending moment. Put another way, adding fuel may allow payload to be increased above that at MZFW.

Maxpac   Multi-axis pintle attitude control.

Maxwell   Non-Si unit of magnetic flux (conceptualised as "single line" of flux), $= 10^{-8}$Wb.

Mayday   International call for urgent assistance, from French "m'aidez!" Hence, to declare a *, to go *; usually sent on 121.5 MHz.

MB   *1* Magnetic bearing.
  2 Marker beacon.
  3 Mercury bromide (laser).
  4 Megabits.

Mb   Mbyte (megabyte), one million bytes.

mb   Millibar[s].

MBA   Main battle area.

MBAR, M-bar   Multi-beam acquisition radar.

MBAT   Multibeam array transmitter.

MBC   *1* Main-beam clutter.
  2 Microbiological corrosion.
  3 Military Budget Committee.
  4 Missile boresight correlator.
  5 Maritime Battle Center (NWDC).

MBCS   Manoeuvre-boost control system; senses aircraft response and adjusts flight controls for fastest completion of demand.

MBD   Molecular-beam deposition.

MBDOE   Million barrels per day oil equivalent.

MBE   *1* Molecular-beam epitaxy.
  2 Multiple-bit error.

MBF   Multi-body freighter.

MBFR   Mutual and balanced force reductions.

MBFS   Maximum boom-free speed.

Mbit   Megabit $= 2^{20} = 1,048,576$ bits; /S adds per second.

MBITR   Multi-band inter/intra team radio.

MBK   Missing, believed killed.

MBM   Magnetic bubble memory.

MBMMR   Multi-band multi-mode radio.

MBO   Management by objectives.

MBOH   Minimum break-off height.

MBP   *1* Maximum takeoff regime (USSR, Cyrillic characters actually represent MVR).
  2 Minimum burner pressure; V adds valve.

MBPS, MBps, MB/s   Megabits/second.

Mbps   Megabytes/second.

MBR   Marker-beacon receiver.

MBRGW   MBRW (adds redundant 'gross').

MBRV   Manoeuvrable ballistic re-entry vehicle.

MBRW   Maximum brake-release weight, at start of takeoff run.

MBS   *1* Millibars (on alphanumeric readout).
  2 Mobile base system (ISS robotic interface).
  3 Modular bomb set, for making IEDs for training.

MBTH   Material by the hour.

MBX   Management by exception.

Mbytes   Megabytes, $= 2^{20} = 1,048,576$ bytes.

MBZ   Mandatory broadcast zone.

MC   *1* Marine Corps (USMC).
  2 Medium-case, or capacity (bomb).
  3 Magnetic course.
  4 Multi-combiner (HUD).
  5 Maximum certificated (altitude).
  6 Manufacturing cost.
  7 Mission-capable.
  8 Motion cueing.
  9 Master change.
  10 Military characteristics (of NW).
  11 Materials consortium.
  12 Main, or mission, or monitoring, or multifunction, computer.
  13 Megacycles (incorrect usage).

M$_c$ Design cruise Mach number.

m/c   *1* Machined.
  2 Machine, colloquial = aeroplane.

m.c.   Moving-coil.

MCA   *1* Ministry of Civil Aviation.
  2 Minimum crossing altitude.
  3 Military/civil action.
  4 ISA in Cyrillic (R) characters, actually reading MSA.
  5 Military C. of A.
  6 Minimum controllable airspeed.
  7 Maritime and Coastguard Agency (UK).
  8 Maintenance-centre analyser (a HUMS).
  9 Major component assembly [hall].

MCAD   Mechanical computer-aided design.

MC&G   Mapping, Charting and Geodesy (USAF).

MCAS   Marine Corps Air Station (USMC).

MCASD   Military/civil air-safety day (RAF).

MCASMP   Marine Corps aviation simulator master plan (USMC).

MCB   Microwave circuit board.

MCBF   Mean cycles between failures.

MCC   *1* Mission control center (NASA).
  2 Midcourse correction.
  3 Meteorological communications centre.
  4 Mobile command centre.
  5 Main combustion chamber.
  6 Mission-control console.
  7 Mission crew commander.
  8 Manual control centre.
  9 Multi-crew co-operation.
  10 Mission computer cluster.
  11 Maintenance control computer, or centre.
  12 Management command and control.
  13 Main communications control; P adds panel.
  14 Manual control and counter.
  15 Maritime command and control.
  16 Mobile computer core.

MCCA, MC²A   Multi-sensor command and control [aircraft] (USAF).

MCCC, MC²C Multi-sensor command and control constellation (USAF).

MCCIS Maritime command and control information system (NATO).

MCCS Multi-function command and control system.

MCD *1* Magnetic chip-detector.

2 Marine Craft Detachment (RAF).

3 Multi-colour display.

MCDP Maintenance control and display panel.

M_CDR Critical drag-rise Mach number.

MCDU Multi-function controller/display unit; in cockpit, enables line maintenance to identify faulty LRU and replace without documentation.

MCDW Minimum-collateral-damage weapon.

MCE *1* Mission, or modular, control element, or module control equipment.

2 Microcircuit engineering; P adds program (NASA).

3 Military comunications electronics; B adds board, WG working Group (US).

MCEP Maneuver-criteria evaluation program (US).

MCF *1* Manoeuvring control force.

2 Military computer family.

mcf Thousands (not millions) of cubic feet.

MC/FD Modular chaff/flare dispenser.

MCG *1* MLS calibration generator.

2 Millimetre-wave contrast guidance.

3 Maintenance-cost guarantee; P adds plan.

M_CG Pitching moment about c.g.

McGee tube High-speed photo image tube, samples 'sausage' of electrons slice by slice.

MCGS Microwave command-guidance system.

MCI Moving-coil instrument, or indicator.

MCID *1* Manufacturing change in design.

2 Modular common inlet duct.

MCK Mission control keyboard.

McKinnon Wood bubble Shape of plot of pressure distribution on upper surface of wing upstream of shockwave at subsonic M exceeding Mcrit.

MCL Microcomputer compiler language.

MCLOS Manual command to line of sight.

MCLPM Mixed conventional loads pattern management.

MCM *1* Mine countermeasures [F adds Force].

2 1,000 circular mils.

MCMP Modular countermeasures pod.

MCN Manufacturing contract, or control, number.

MCNDC Missile capture network defense command (US).

MCO *1* Confusingly used to mean maximum continuous thrust.

2 Mars climate orbiter.

3 Mode-coupled oscillation.

M/CO Methane/castor oil.

MCOP Multilateral Crew Operations Panel (ISS crews).

MCops Millions of computer operations per second.

MCP *1* Maximum continuous power.

2 Maximum climb power.

3 Multi- or micro-channel plate (image intensifier).

4 Mode [or maintenance] control panel.

5 Military construction program (USAF).

6 Missile control panel.

7 Modular countermeasures pod.

MCPH Maintenance cost per hour.

MCR *1* Maximum cruising rate.

2 Maximum control range.

3 Mission capable [or completion] rate.

4 Modular-cabin crew rest.

Mcrit Critical Mach number.

Mcrit D Mcrit at which subsonic C_D rises by 0.002 at constant angle of attack.

MCRC Mobile control and reporting centre.

MCS *1* Minimum control speed.

2 Miniature control system (ATC).

3 Missile control system, on fighter.

4 Mixed-class seating.

5 Material [or materiel] certification statement (FAA).

6 Modular cooling system.

7 Main-computer, or mission-control, software.

8 Manoeuvre[s] control system.

9 Multifunctional control surface.

Mc/s Megacycles per second, = MHz.

MCSB Mission-control station backup (USAF).

MCSR Mission-completion success rate.

MCT *1* Mercury cadmium telluride; UK = CMT.

2 Maximum continuous thrust.

3 Confusingly, military combat thrust.

4 Mobile cable test[er], of wiring.

5 MOS (1)-controlled thyristor.

MCTA Militarily-critical technology agreement, signed by all SDI contractors.

MCTL Militarily-critical technologies list (US).

MCTOW Maximum certificated takeoff weight.

MC2A, MC²A Multirole command and control aircraft; X adds Experimental.

MC2C, MC²C Multimission, or multisensor, command and control constellation (USAF).

MCU *1* Matrix-character unit (display symbols).

2 Management-control unit (flight-data recorder).

3 Modular-concept unit, = ⅛-ATR.

4 Marine Craft Unit (RAF).

5 Mission-computer upgrade.

6 Multifunction, or modular, or mission, or master, or missile, or mobile, or monitor, or autothrottle motor, control unit.

M³A 'M-cubed', see MMMA.

MCV Mobile containment vessel, for explosions ≤15 kg TNT.

MCW *1* Modulated carrier-, or continuous-, wave; A2 emission.

2 Mid-cruise weight.

MCWB Multi-channel wideband.

MCWL Marine Corps Warfighting Laboratory (USMC).

MCXO Microprocessor-controlled crystal oscillator.

MD *1* Manoeuvre-demand (control loop).

2 Military District.

3 Or m.d., managing director.

4 Methyldichloroarsine (CW).

M/D Miscellaneous data (display).

M_d Design Mach number.

m_d Mass flow of dry gas.

MDA *1* Minimum descent altitude.

2 Master diversion airfield.

3 Multiple docking adapter.

4 Multilayer dielectric absorber (stealth).

5 Metal deactivator.

6 Missile Defense Agency (US, formed 2002 to manage GMD).

MDAP *1* Mutual Defense Assistance Program (US, NATO, June 1952).

2 Major defense-acquisition program (US).

MDAS Multifunction defense avionics system.

MDAU Modular data analysis unit.

MDBIC Missile defense/battle-integration center (USA).

MDBR Multifunction dextrous boro-robot.

MDC *1* Minimum-displacement [i.e., force-sensing] control[er].

2 Main display console.

3 Maintenance Data Centre (RAF).

4 Miniature (also micro, which is smaller) detonating cord.

5 Motor-driven compressor.

6 Multiple drone control.

7 Main-deck cargo (V adds volume).

8 Maintenance diagnostic computer.

MDCRS Meteorological data collection and reporting system, or service.

MDD *1* Meteorological data distribution.

2 Mission data debrief.

M_{DD} See *drag-divergence Mach number.*

MDDS MDD(2) software.

MDDU Multipurpose disk-drive unit [-O adds onboard, for BITE, CMC etc].

MDE Manual data entry.

M_{DET} Mach number at which shockwave is just detached.

MDEWS Modular digital EW suite.

MDF *1* MF D/F station (ICAO).

2 Mission degradation factor (USAF).

3 Mission data file.

M_{DF} Maximum demonstrated diving flight Mach number.

MDFDR Multiple dislocated flight-data recorder; S adds system.

MDFN Molybdenum-disulphide-filled nylon.

MDFY Modify.

MDG Map display generator.

MDGT Pronounced midget, mission data [or mission debriefing] ground terminal.

MDH Minimum descent height.

MDI *1* Miss-distance indicator; hence MDIS = * system.

2 Multifunction display indicator.

M_{DIV} Drag-divergence Mach number, which see

MDL *1* Mission data-loader.

2 Multipurpose data-link.

MDLC Manoeuvring direct-lift control.

MDLT Mobile data-link terminal.

MDM Multiplexer-demultiplexer.

MDNS Managed data network services.

MDNT Missile Defense National Team (US).

MDO Multidisciplinary design organization.

MDOF Multiple degrees of freedom.

MDP *1* Motor-driven pump.

2 Maintenance data panel (1553B data-bus).

3 Maximum dry power.

4 Modular display processor.

5 Multi-designation protocol.

MDPC Mission and display processing computer.

MDPS Metric data-processing system.

MDR *1* Mandatory defect reporting.

2 Magnetic-field dependent resistor.

3 Medium data-rate.

4 Micro-data recorder, linking surveillance radars with computers.

5 Mission data recorder.

M_{DR} Drag-rise Mach number.

MDRC Material Development and Readiness Command (USA).

MDRI Multipurpose display repeater indicator.

MD/RWW Map display and report-writing workstation, part of GIES.

MDS *1* Minimum discernible, or detectable, signal.

2 Metal-dielectric semiconductor.

3 Matériels de servitude, ie GSE (F).

4 Mission/design/series, basic system of designators of military aircraft (US).

5 Multilevel database system.

6 Measurement and debriefing station, or system [Tacts].

7 Manned destructive suppression.

8 Miss-distance sensor.

9 Medium-distance [EO] sensor.

10 Maintenance data station.

11 Mission distribution system.

12 Meteorological data system.

13 Modular dispenser system.

14 Mid-course defense segment [see *GMDS*].

15 Mobile data service.

16 Multi-static dependent surveillance.

MDT *1* Maintenance display terminal.

2 Mission-data tools, or terminal.

3 Mountain daylight time (US).

4 Moderate.

MDTC/P Mega-data-transfer cartridge, with processor.

MDTS *1* Mission data-transfer system.

2 Megabit digital troposcatter subsystem.

MDU *1* Microwave distribution unit.

2 Mine distributing unit.

MDWP Mutual Defense Weapons Program.

MD/WT Marine division/wing team (DoD).

ME *1* Manoeuvre enhancement mode.

2 Multi-engine, or main engine.

3 Mission-essential.

4 Maintenance error.

M€ Millions of Euros.

M_e Merkel number.

mE Maritime equatorial air.

m_e Electron rest-mass.

MEA *1* Minimum en-route IFR altitude.

2 Maintenance engineering analysis.

3 More-electric aircraft.

meacon To mislead enemy by receiving and instantly rebroadcasting radio-beacon signals from false positions.

Meads *1* MEA (2) data system.

2 Medium extended air-defense system (USA, G, I).

mean aerodynamic chord Chord of imaginary wing of constant section having same force vectors under all conditions as those of actual wing, symbol $\bar{\bar{c}}$ (in practice usually very close to mean chord $\bar{c}$).

mean blade width ratio Ratio of mean chord (width) of propeller blade to diameter.

mean chord Gross wing area divided by span, symbol $\bar{c}$.

mean day   Time between successive transits of mean Sun.

mean effective pressure   Mean pressure acting on piston during power stroke in Otto, diesel, two-stroke, Stirling and most other reciprocating IC engines.

mean fleet performance   Mean (sometimes average) flight performance (sometimes other parameters, eg engine IFSD rate) of entire airline fleet of one aircraft type.

mean free path   *1* Mean distance point could move in straight line without collision with surrounding fluid molecule; greater than mean distance between fluid molecules.

2 Mean distance a sound wave travels in an enclosed space before being reflected.

mean geometric pitch   Mean of geometric pitches of all elements from root to tip of propeller blade, symbol Pg.

mean line   Locus of points equidistant between nearest points of upper and lower surface from LE to TE of wing profile.

mean line of advance   Planned future track of centre of ships in convoy.

mean point of contact   Point whose co-ordinates are arithmetic means of co-ordinates of impacts of all weapons launched against same aiming point.

mean sea level   Average height of sea surface, usually calculated from hourly tide readings (in US, measured over 19-year period).

means of compliance   Methods and techniques formally demonstrated by manufacturer seeking certification of product from licensing authority.

mean solar day   Period between transits of mean Sun, 24 h 3 min 56.555 s of mean sidereal time (see *sidereal*).

mean square value   Arithmetic mean of squares of all values; * error equals sum of squares of errors divided by their number.

mean time between failures   For specified time interval, total operating time of population of materiel divided by total number of failures within population (USAF).

mean time to repair   For specified time interval, summation of active repair times divided by total number of malfunctions (USAF).

MEAP   More-electric aircraft project (DSB).

Mearts, M-EARTS   Micro en-route automated radar tracking system.

MEARW   Multi-engined advanced rotary-wing.

MEAS   Mechanical engineering aircraft squadron (RAF).

measured performance   That measured on one occasion under specified conditions; usually slightly higher than gross performance.

measured thrust   That measured on one occasion under specified conditions, usually by force transducers on testbed (see *installed/net/gross thrust*).

measured thrust coefficient   For solid-propellant rocket, thrust-versus-time integral over action-time interval divided by product of average throat area and integral of chamber pressure versus time over action-time interval; symbol C_f.

meatball   *1* Colloquial American for any guidance reference; thus, on the *, to hack the *.

2 Accurately flown final approach.

3 Carrier optical landing aid as seen by correctly aligned pilot.

4 Fresnel lens.

MEB   *1* Main elecronics box.

2 Marine Expeditionary Brigade.

3 Modular expendable block (countermeasures).

Mebul, MEBUL   Multiple engine build-up list.

MEC   *1* Main engine control.

2 Main equipment centre.

3 Modular electronics concept.

MECA   Missile electronics and computer assembly.

Mecaplex   Registered acrylic plastic sheet similar to Plexiglas.

mechanical de-icing   Using distortion (eg rubber boot), centrifugal force or other physical method to dislodge ice.

mechanical efficiency   Work delivered by a machine as percentage of work put in, difference being mainly frictional losses.

mechanical equivalent of heat   See *joule*.

Meco   Main-engine cutoff (large rocket vehicle).

MECV   Mobile explosive containment vessel.

MED   *1* Multifunction electronic display [S adds system, or subsystem].

2 Maximum-entropy discrimination.

3 Medium.

4 Medical.

MEDA   Military emergency diversion airfield.

Medcat   Medium-altitude CAT.

Medevac   Military airlift of sick or wounded.

median lethal dose   That over whole body which would be fatal to 50% of subjects.

median line   Line through aerofoil profile at all places equidistant from nearest points on upper and lower surfaces, = mean line.

median selection   Automatic choice of mean and rejection of extreme values; can also mean majority voting.

Mediator   Unsuccessful grand design for UK ATC (1) system embracing all air traffic.

medium aircraft   Meaningless, but in 1998 the 737 was cited as an example. An ICAO document once said, 7,000–136,000 kg MTOW.

medium altitude   Between 2,000 ft and 35,000 ft (DoD).

medium-altitude level bombing   Release between 8,000 ft and 15,000 ft (DoD).

medium-angle loft bombing   Release at 35° to 75° from horizontal.

medium bird   For impact or ingestion certification, one weighing 1.5 lb, 0.68 kg.

medium bomber   Former category defined quite differently by different air forces, either by bomb load or range, and so developed 1920–50 as to make numerical values meaningless.

medium cloud, CM   Cloud types prefixed by alto-; according to BSI with average height 8,000 ft to 20,000 ft (2,438–6,096 m).

medium frequency   EM radiation with superimposed carrier at 300 kHz–3 MHz.

medium-range ballistic missile   Operational range 600 to 1,500 nm (1,112–2,780 km); see *mid-range* ** (DoD).

medium-range transport   Full-payload range 1,500 to 3,500 nm (2,780–6,486 km).

medium-scale integration   Normally taken to mean 50–500 circuits or gates per chip.

medium-scale map   From 1:75,000 to 1:600,000 (DoD, IADB).

medium turn   Most authorities define as bank angle 25° to 45°.

MEDS, Meds Multifunctional electronic display subsystem, new (2002) guidance for Shuttle Orbiters and STA (6).

Medusa Multifunction electro-optics for defense of US aircraft (AFRL).

MEECN Minimum essential emergency communications network.

MEF *1* Minimum essential facilities.
 2 Marine Expeditionary Force.
 3 Maximum elevation figure.

MEFC Manual emergency fuel control.

mega *1* Prefix, $\times 10^6$; symbol M, thus MW = megawatt[s]; a omitted in megohm[s].
 2 In EDP = 2^{20} = 1,048,576.

megacycles Megahertz.

Megafloat Technology for floating offshore airports.

megahertz 10^6 cycles per second, MHz.

megaline 10^6 maxwells, = 10^{-2}Wb.

megaton, MT Explosive power equivalent to nominal 1,000,000 short tons of TNT.

MEGG Merging.

megger Universal electrical continuity and resistance tester (colloq., arch.).

megohm Ohm $\times 10^6$, symbol $M\Omega$.

MEHT Minimum eye-height over threshold.

MEHTF Multiple-event hard-target fuze.

MEI *1* Multi-engine instrument rating.
 2 Minimum-equipment item.
 3 Maintenance engineering inspection.
 4 Maintenance-error investigation.
 5 Missile ECM improvement.
 6 More-Electric Initiative.

MEIS Modular engine instrument system.

MEISR Minimum essential improvement in system reliability.

MEIT Multi-element integrated test.

MEK Methyl ethyl ketone solvent.

MEL *1* Multi-engine licence, or landplane.
 2 Minimum equipment list, list of ME (3) items.
 3 Missile ejector launcher.

MELC Main electrical load center (centre).

melée Confused close combat by numerous aircraft, opposite of one-on-one.

MELHS Mobile electroluminescent helipad [lighting] system.

Melios Minimum eyesafe laser IR observation.

MEM Module exchange magazine.

member Portion of structure bearing load.

memorandum of understanding Diplomatic agreement signed at ministerial level between governments agreed on collaborative programme, usually in advance of actual engineering.

MEMS *1* Maintenance-error management system (CAA, 2000–).
 2 Module-exchange mechanism system.

Mems Microelectromechanical systems, now with such prefixes as bio-, fluidic, optical, refractory and RF.

MENS Mission element need statement (DoD).

menu Range of variable inputs from human operator to EDP, fire-control system, display or other electronic system; files, codes, information, modes, formats, etc.

MEO *1* Mass in Earth orbit.
 2 Medium Earth orbit.

MEOP Maximum expected operating pressure (solid motors).

MEOTBF Mean engine operating time between failures.

MEP *1* Multi-engine pilot.
 2 Mission equipment passage.
 3 Marine environmental protection.
 4 Management engineering program(me).
 5 Member of European Parliament.
 6 Mean effective pressure.
 7 Mars exploration program[me].

Mephisto Multiple-effect penetrator high [or hard] sophisticated target optimized.

MEPT Multi-engine pilot trainer.

Mepu, MEPU Monofuel emergency power unit.

MER *1* Multiple ejector rack.
 2 Mission evaluation room.
 3 Minor equipment requirement [(A) adds Air] (UK).
 4 True height above MSL.
 5 Mars exploration rover.

Mera *1* Molecular-electronics radar.
 2 See *MER (3)*.

Mercator *1* Map projection: light at Earth centre projects map on to cylinder wrapped round Equator.
 2 Map electronic remote colour autonomous television output reader.

Mercier French aeroydynamicist with patents for close-fitting cowlings for radial engines and low-drag ailerons hinged diagonally across wingtip.

mercury Hg, the only metal liquid at NTP, MPt –39°C, BPt 357°C, density 13.5, used (rarely) as vapour in closed-circuit space power and as ion beam (thrusters).

mercury battery Dry cell, 1.2V, KOH electrolyte between mercuric oxide/graphite cathode and zinc anode.

Meredith effect Making a heat-rejection system, such as a radiator or oil cooler, give positive thrust.

merge *1* The coming together of two fighters about to engage in close combat.
 2 To fly close beside unknown aircraft in order to identify it visually.

meridian Great circle through any place on Earth and the poles.

meridian altitude Altitude of celestial body when on celestial meridian of observer (000° or 180° true).

meridian passage Time at which celestial body crosses observer's celestial meridian.

Merkel number Heat transfer equation Me=bS‴V/md where b is mass-transfer coefficient, S‴ transport surface per unit volume, V volume and md mass flow of dry gas.

Merlin *1* Multi-service extended-range low-cost interceptor (Darpa).
 2 Modular ejection-rated low-profile imaging for night.

Merod Message entry and readout device.

Merritt Island Site of largest KSC launch complexes (Saturn V vehicle).

Merto Maximum-energy rejected takeoff.

MES *1* Major equipment supplier (large projects).
 2 Main engine start(ing).
 3 Medium-energy source (laser).
 4 Multi-engine seaplane.
 5 Mobile Earth station.
 6 Multi-energy spectrum.

Mesa *1* Modular equipment stowage area (NASA).
 2 Multirole electronically scanned array [radar].

3 Minimum emergency safe altitude.

mesa   Raised portion[s] of microcircuit.

Mesar, MESAR   Multifunction electronically-scanned adaptive radar.

Mesas   Multirole electronically scanned aircraft, or airborne, system.

MESC   Mid electrical (or, UK CAA, equipment) service centre.

Mesfet   Metal semiconductor field-effect transistor.

MESG   Micro-electrostatically suspended gyro.

Mesh   Measuring engineering safety and health.

mesocline   Lower layer of mesosphere (1) where temperature rises to about 10°C at c60 km.

mesodecline   Upper layer of mesosphere (1) where temperature falls to about –90°C at mesopause.

mesometeorology   *1* Meteorology on scale between macro and micro.

2 Meteorology of mesosphere (1).

mesomorphic states   See *liquid crystal.*

meson   Family of elementary particles with energy 135–550 MeV and zero spin (possibly nine types, and corresponding anti-particles, but some may be resonances).

mesopause   Boundary between top of mesosphere (1) and thermosphere, 80–85 km.

mesopeak   Temperature at mesocline/mesodecline boundary.

mesosphere   *1* Thermal region of atmosphere between stratopause (c30 km) and mesopause (Chapman gives lower limit as 32 km).

2 Rarely, according to Wares, region between top of ionosphere (c400 km and supposed bottom of exosphere (500–1,000 km).

message pick up   Lowering hinged arm to clutch written message (RAF 1927–40).

Messir   Computerized AFTN message-switching system (F).

Met, MET   *1* Meteorology, meteorological.

2 Meteorological Office.

3 Meteorological service, or broadcast service.

4 Mission-event timer.

5 Multi-image exploitation tool.

6 Mission-essential task[s] (USAF).

Metag, METAG   Meteorological Advisory Group (ICAO).

metacentre   Intersection of line of buoyancy (of marine aircraft) and axis of symmetry.

metacentric height   Vertical distance from metacentre to c.g.

Metalastik   Patented family of metal/rubber bonded devices, mainly anti-vibration.

metal deactivator   Hydrocarbon fuel additive to reduce possibility of electrochemical action in tank or system, esp. catalytic effect of copper on fuel oxidation.

metal-injection moulding   Working material is mix of metal and plastic binder.

metallic fuel additive   *1* Various blends of high-energy (so-called "zip") turbine fuel incorporated boranes and metallic compounds.

2 Most high-energy solid rocket fuels incorporate finely divided aluminium powder.

metallics   See *metal matrix composites.*

metallization   Vacuum deposition of metal conductive paths in planar IC or other solid-state device. See next.

metallizing   Vast range of techniques involve coating with metal, usually by plasma-spraying with finely divided metal particles; typical example is electrical bonding of modern airframe, in which most major parts are isolated by non-conductive corrosion protection.

metal matrix composite   Various structural or refractory composites have a metal matrix, the most common being aluminium, magnesium, titanium and copper, and the fibres of whiskers silicon carbide or graphite.

metamic   Cermet (suggest undesirable).

Metar, METAR   Meteorological aeronautical radio code for routine reports.

metastable   Pseudo-equilibrium [e.g. supersaturated solution or supercooled liquid] needing small external disturbance to trigger violent change.

Metdial   Privately run telephone met. service (UK).

Météo-France   Meteorological service (Toulouse, F).

meteor   Small solid body in free fall through Earth's atmosphere; also called *-oid.

meteor bumper   Thin shield designed to protect spacecraft from penetration by meteors.

meteor burst communication   OTH radio technique in which short-burst signals are scattered by ionised trails left by meteors.

meteorite   Meteor large enough to strike Earth's surface.

meteorograph   Instrument for measuring and recording two or more meteorological measures, eg T, pressure, humidity.

meteorological rocket   Ballistic vehicle to loft payload to great height, there to free-fall or descend by ballute or parachute.

meteorological satellite   One designed to assist understanding of atmosphere and preparation/dissemination of actuals or forecasts.

meteorological wind   Forecast wind.

meteorology   Science of Earth's atmosphere (abb. met.).

metering   Control of flow, e.g. of air traffic entering terminal airspace.

metering orifice   Calibrated nozzle or tube passing exactly known flow at given pressure difference.

MetFAX   Fax-accessed weather service (UK Met. Office).

methane   Simplest hydrocarbon fuel CH_4; gas at NTP but cryogenic liquid.

methanol   Methyl alcohol, CH_3OH, BPt 64.6C, used in special blends of fuel for racing piston engine and as additive to water as anti-knock power boost fluid.

method of least work   Formula for analysis of statically indeterminate structures.

ME3   Refractory sintered nickel-based disc alloy.

methyl alcohol   See *methanol.*

methyl bromide CH_3Br, common filling for hand (and some other) fire extinguishers.

methyl cellosolve   Trade name for various strippers, esp. methylene chloride, CH_2Cl_2.

METI   Ministry of Economy, Trade and Industry [translation] (J).

Meto   Maximum except takeoff power.

Metoc   Meteorological and oceanographic (ASW).

METPS   Meteorological data processing system.

METR   Multiple-emitter targeting receiver.

metre   Fundamental SI unit of length, abb. m, 1,650,763.73 wavelengths in vacuum of radiation from transition $2p_{10}$ and $5d_5$ of Kr-86 atom.

Metrep　Meteorological report.

metric　Usually means MKS; aerospace is gradually standardizing on SI units.

metrics　Measurements, esp. of performance and other health measures normally falling under three headings: cost, quality and schedule-adherence.

Metro, metro　*1* Increasingly popular word meaning urban; applied to 'downtown' air services, commuter routes and aircraft.

2 Pilot-to-* voice call.

metrology　Science of precise measurement, esp. of linear dimensions.

metroplex routes　Selected recommended high-altitude IFR inner-city (US).

METS　Mobile engine test stand.

MetWEB　Internet-based aviation weather service (UK Met. Office).

MEU　Maritime Expeditionary Unit [SOC adds Special Operations Capable].

MeV　Mega-electron-volts, ev×10⁶.

MEW　*1* Manufactured empty weight.

2 Microwave early warning.

3 Ministry of Economic Warfare (UK, WW2).

4 Mean equivalent wind.

MEWS　*1* Modular EW simulator.

2 Microwave EW system.

3 Missile early-warning station.

MEWSG　Multiservice [or mobile] EW Support Group (NATO).

MEX　Microelectronics.

MEXE　Military Engineering Experimental Establishment (UK).

Mexepad　Air-portable ground cover for sustained jet-Stovl operations

MEZ　Missile (esp. AAM) engagement zone.

MF　*1* Main frame.

2 Main force.

3 Major field.

4 Multifunction.

5 Mandatory frequency.

m.f., MF　Medium frequency (see Appendix 2).

M²F²　Multimode fire and forget.

M_f　Fuel mass flow.

MFA　*1* Minimum flight altitude.

2 Multi-furnace assembly.

3 Military flying area.

MFAR　Multifunction array radar.

MFBF　Multifunction bomb fuze.

MFC　*1* Main fuel control, or controller.

2 Maximum fuel capacity.

3 Multi-frequency code, or coding.

4 Miniature flight computer.

M_FC　Maximum Mach number for satisfactory flight control and stability.

MFCD　*1* Modular flare/chaff dispenser.

2 Multifunction colour [or control] display [P adds panel].

MFCS　*1* Missile, or modified, fire-control system.

2 Microprocessor flight-control system (drones, targets).

MFCU　Multifunction concept [or control] unit.

MFD　Multifunction display [S adds system, U unit].

mfd　Microfarad.

MFESAR　Multifunction electronically-scanned adaptive radar.

MFF　*1* Mixed fighter force, such as all-weather look-down BVR mixed with simple visual day aircraft [O adds operations].

2 Mixed-fleet flying [common pool of pilots for airline's aircraft].

3 Multisite fatigue fracture.

MFFC　Mixed fighter force concept (eg, Phantoms/Hawks, RAF).

MFFO　Mixed-fleet flight operations.

MFG　*1* Miniature flex gyro.

2 Manufacturing.

3 Master-frequency generator.

MFHBF　Mean flight-hours between failures.

MFHDD　Multifunction head-down display.

MFI　Multirole tactical fighter (R).

MFIT　Mean fault isolation time.

MFK　Multifunction keyboard.

MFL　Minimum field length, TO distance to clear standard (35 or 50 ft) screen.

MFLI　Magnetic fluid-level indicator.

MFLOPS, Mflops　Million of flops, floating-point operations per second.

MFM　*1* Maintenance-fault memory.

2 Multisensor fusion module.

MFMA　Multifunction microwave aperture.

MFMS　Military flight-management system.

MFMT　Material and fleet-management program.

MFN　*1* Mode-forming network.

2 Most favoured nation.

MFO　Multinational force and observers.

MFOP　Maintenance-free operating period.

MFP　*1* Mean free path.

2 Main fuel pump.

MFPD　Multimode-flowpath propulsion demonstrator (scramjet).

MFPU　Mobile field-processing unit (photo-recon).

MFQ　Mouvement Français de Qualité.

MFR　*1* Multifunction radar, or receiver.

2 Medium-range forecast [weather].

Mfr　Manufacturer.

MFS　*1* Mexepad for Stovls.

2 Manned flight simulator.

3 Media file server.

M_FS　Free-stream Mach number.

MFSK　Multiple-frequency shift-keying.

MFSP　Multifunction Sigint payload.

MFSPP　Mediterranean forecasting system pilot project.

MFT　*1* Multi-radar fusion tracking.

2 Multifunction trainer, or training.

3 Meter fix time (US usage).

MFTS　Military Flying Training System, or Service (UK MoD).

MFV　*1* Main fuel valve.

2 Minimum forward visibility.

MG　*1* Master gauge.

2 Machine gun.

3 Motor glider.

Mg　*1* Magnesium.

2 Megagramme, ie 1,000 kg or 1 tonne.

mg　*1* Milligram[me]s.

2 Machine gun.

MGA　　*1* Middle gimbal angle.
　　2 Ministry of civil aviation (R).
MGARJS　Mobile ground/air radar jamming system.
MGCS　*1* Missile guidance and control system.
　　2 Mobile ground control station.
　　3 Meteosat ground computer system.
MGD　Magnetogasdynamic.
MGDA　Maintenance Group Defence Agency (UK, now part of DARA).
MGF　Metallized glass-fibre (chaff).
MgF₂　MgF_2 Magnesium fluoride (IR seeker domes).
MGGB　Modular guided glide bomb.
MGIR　Motor glider instrument, or instructor, rating.
MGM　Former US code, mobile-launcher surface-attack guided missile.
MGOS　Metal glass oxide silicon.
MGR　Miniature GPS receiver.
MGRS　Military grid-reference system.
MGS　*1* Mobile ground station (RPV).
　　2 Minimum groundspeed system.
　　3 Mars global surveyor.
　　4 Main-gear steering [CU adds control unit].
　　5 Microgravity science.
MGSM　Medium ground station module.
MGSS　Maintenance and ground-support system.
MGT　*1* Motor gas temperature (solid rocket).
　　2 Module ground terminal.
MGTOW　See *MTOW*.
MGTP　Main-gear touchdown point.
MGU　Midcourse guidance unit.
MGVF　Ministry of Civil Aviation (USSR, R).
MGW　Maximum gross weight (MTOW and MRW are more explicit).
MH　*1* NDB, less than 50 W.
　　2 Magnetic heading.
mH　Millihenry.
MHA　*1* Minimum holding altitude.
　　2 Maintenance hazard analysis.
MHD　*1* Magnetohydrodynamic(s).
　　2 Magnetic hard drive.
MHDD　Multifunction head-down display.
MHDF　Co-located MF and HF D/F (ICAO).
M/HFE　Maintainability and human factors engineering.
MHI　Missile [or manual] hit indicator.
mho　Unit of conductance, reciprocal ohm, symbol ℧, SI unit siemens.
MHP　Main hydraulic pump.
mhp　Muzzle horsepower, measure of automatic weapon.
MHR　Monopropellant hydrazine rocket.
MHRS　Magnetic heading-reference system.
MHS　*1* Message-handling system.
　　2 Masint hyperspectral study.
MHV　Miniature homing vehicle.
MHVDF　Co-located MF, HF and VHF/DF (ICAO).
MHz　Megahertz.
MI　*1* Model improvement.
　　2 Medium intensity.
　　3 Miles (not recommended).
　　4 Military intelligence.
　　5 Shallow (ICAO).
mi　Statute mile[s].
m.i., M/I　Minimum impulse.

MIA　*1* Missing in action.
　　2 Minimum IFR altitude.
Miacs　Multisensor integrated airborne command system.
MIAG　Modular integrated avionics group (UAV).
Mials　Medium-intensity approach-light system.
Miami　Microwave ice-accretion measurement instrument.
MIATS　Memory interrogation and test system.
MIB　*1* Minimum-impulse bit (rocket).
　　2 Management information base.
　　3 Mishap Investigation Board (NASA).
MIC　*1* Mineral insulated cable.
　　2 Microwave integrated circuit.
mic　microphone.
Mica　*1* Missile d'interception et de combat aérien (F).
　　2 Met [meteorological] improvements for controller aids (Euret).
Micad　Multipurpose integrated chemical-agent alarm device.
Micap　Mission-capable.
Micarta　Trade name for phenolic insulator and small-part material made from resin-impregnated cloth.
Micas　Miniature integrated camera spectrometer, or spectroscopy.
Mice　Microwave integrated checkout equipment.
mice　Small inserts fitted by hand to tailor cross-section area of fluid flow path, esp. gas-turbine jetpipe nozzle.
Michigan height　That at which an air-dropped store, esp. nuclear, is activated by its radar altimeter.
Mickey　H_2X (colloq.).
Mickey Finn　Major exercises involving numerous NW-carrying bombers from dispersed bases (RAF, 1960s).
MICNS　Modular integrated communications and navigation subsystem (Sotas).
MICOM, Micom　Missile Command (USA).
Micos　Multifunctional IR coherent optical scanner.
micrad　Microwave radiometer.
micro　*1* Prefix one-millionth, 10^{-6}, symbol μ.
　　2 Abb., microlight.
micro-adaptive flow control　High-frequency alternate blowing and sucking generated electromagnetically to preserve attachment of airflow to surface.
micro-adjuster　Small permanent magnets arranged to correct magnetic compass for coefficients P, Q, Cz, Fz.
Micro-aids　Micro-aircraft integrated data system.
microballoons　Microscopic hollow spheres used to increase bulk of fillers and potting compounds.
microbarograph　Records very small transient changes in atmospheric pressure at ground station.
microbiological corrosion　Eating away by micro-organisms living at fuel/air interfaces and on surface of virtually all aeronautical materials.
microbolometer　Bolometer able to measure microscopic regions (no numerical definition).
microburst　Most lethal form of vertical gust, in which core up to 2.5 km (1.5 miles) diameter forms vertical jet below convective cloud with downward velocity up to 20 m/s (4,000 ft/min), an almost instantaneous velocity difference of 80 kt, down to very low levels.
microchannel plate　Multiplies [× c10^5] electrons passing through and routes them to phosphor screen which converts to green light.

microcircuit Basic element of microelectronics, with numerous * fabricated on each epitaxial chip.

Micro-Earts Micro en route automated radar tracking system.

microelectronics Electronics based on solid-state devices of microscopic [down to nanometre] dimensions.

microfabrication Manufacture of microscopic [to nanometre] devices.

microfarad Farad $\times 10^{-6}$, symbol μF.

microfiche Common system of storing written or visual information with 24 to 288 microfilmed pages on each 100 mm $\times$ 150 mm (or 4 in $\times$ 6 in) sheet of film.

microflume Microfluidic molecular system.

microflyer Surveillance aircraft with no dimension exceeding 6 in (152.4 mm), TOW 4 oz (113.4 g).

microinch One millionth of an inch.

micro level Critical final design of software.

microlight Originally aeroplane with OWE $\leqslant$150 kg (330.7 lb), next definition was MTOW $\leqslant$390 kg, $\leqslant$50 litres fuel; latest definition: V_{so} $\leqslant$65 km/h, MTOW, 1-seat $\leqslant$300 kg [land], 330 kg [sea/amph]; 2-seat $\leqslant$450 kg [land], 495 kg [sea/amph]. (JAA, FAA). See *ultralight*.

micrometeorite Microscopic solid particle, many species throughout explored space.

micrometer Instrument for precise dimensional measurement, usually mechanical but sometimes based on fluid escape through small clearance.

micrometre One-millionth of metre, 10^{-6}m, symbol μ.

micron Previous name for micrometre, still common in US to avoid confusion with micrometer.

micronavigation Guidance of aircraft or missile in relation to nearby target (usually air/ground).

micronic Concerned with micron dimensions, hence * filter.

micronlitre Quantity of gas: 1 litre at pressure 1 μHg.

microprobe Instrument for investigating regions of micron size, hence laser-pulse * analyser.

microprogram Small sub-program for defining computer instructions in terms of other basic elemental operations.

microsatellite Mass 10–100 kg.

microsecond 10^{-6}s.

microshaving Precision-machining heads of driven countersunk rivets.

microstrip Microwave guide comprising conductor supported above ground plane, generally fabricated by printed circuitry.

microswitch Miniature electric switch governing system(s) according to external movement such as compression of MLG.

microtacan Microelectronic Tacan.

microtechnology Has come to mean manufacture at the molecular level, including Mems, microfabrication, nanotechnology, microfluidics and even DNA arrrays.

Microvision All-weather landing aid (Bendix) with pulsed beacons along runway energising pattern on HUD.

microwave background Found throughout universe, corresponding to temperature of 2.7K.

microwave landing system Post-1995 successor to ILS, TRSB (US) being political choice. Guides over ±40° sector landing or overshoot.

microwaves Electromagnetic radiation between RF and far-IR, normally 1–300 GHz.

MICS Manned interactive control station.

MID *1* Multifunction information distribution.

2 Maritime identification digits.

3 Middle.

4 Middle East (ICAO).

Mida Message interchange distributed application.

midair Midair collision.

Midas *1* Missile-defense alarm system (USA, USAF).

2 Manufacturing information distribution and acquisition system.

3 Multiplexer integration and digital [comsat subsystem] automation system.

4 Multifunction IR distributed-aperture system.

5 Multifunction integrated defensive avionics system.

6 Mobile instrumented data-acquisition system.

MiDASH, Midash Modular integrated display and sight helmet.

midcourse *1* Space trajectory linking departure from neighbourhood of Earth and arrival near destination, corresponding to atmospheric flight cruise; hence * guidance.

2 From missile boost burnout to start of terminal homing.

midcourse guidance Applies to midcourse (1), sometimes to (2).

midcruise weight Gross weight at midpoint of mission.

middle airspace Several national definitions, eg FL 180–290, FL 145–250.

Middleman Air-intercept code: VHF or UHF radio relay (DoD).

middle marker ILS marker on extended runway centre-line at ILS Point B, usually 1,067 m (3,500 ft) from threshold.

Midds Meteorology and oceanography integrated data-display system.

mid-flap Intermediate portion of triple-slotted flap.

mid-high wing Set about three-quarters way up fuselage.

midi Intermediate RPV class, on basis of weight, altitude and speed.

Midis Multifunction integrated defensive information system.

Midl Modular inter-operable data-link.

mid-life update Major refurbishment of costly military [rarely, large commercial] aircraft which in bygone days would have been replaced by a new design.

MIDN Midnight.

Midnight Air-intercept code: 'Change from close to broadcast control' (DoD).

MIDR Maintenance (or mandatory) incident and defect report.

mid-range ballistic missile One having range of 500–3,000 nm (927–5,560 km); see *medium-range* * (DoD).

MIDS, Mids *1* Management information and decision support.

2 Multifunction[al] or multiuser or multiple, information distribution system [FDL adds fighter data link] (NATO).

3 Multi-information display system (weather).

Midst Multiple interferometer determination of trajectories.

mid tank Usually, the centre tank in a wing, between the inner and outer.

MIDU *1* Missile-ignition delay unit.

2 Multipurpose interactive display unit.

mid-upper   *Dorsal turret.*

mid-value logic   Redundant system having odd number of active channels wherein system output is always that of intermediate-value channel, thus eliminating wild values.

mid-wave IR   This usually means wavelength 3–5 μ.

mid-wing aircraft   This usually means an aeroplane with its wing[s] in the so-called mid position, half-way down the fuselage, but see next.

mid-wing pylon   The centre pylon, or hardpoint, of three under each wing, between the inner- and outer-wing pylons; by itself confusing and in any case should be mid wing-pylon.

MIE   *1* Manoeuvre-induced error.

2 Managing (or management) in inflationary economy.

3 Minimum ignition energy.

MIES   Multi- [or modernised] imagery exploitation system.

MIF   Make it fly.

Mifass   Marine integrated fire and air support system.

MIFF   Anti-tank mine dispensed from MW-1 system (G).

MIFG   Shallow fog (ICAO).

Mifir, MIFIR   Microwave instantaneous-frequency indication receiver (ECM).

Mig, MIG   Miniature integrating gyro.

MiGCAP   Combat air patrol directed specifically against MiG aircraft, hence MiG screen.

Might, MIGHT   Malaysian Industry/Government Group for High Technology.

Migits   Miniature integrated GPS/INS tactical system.

MiG magnet   Photoflash (US, colloq.).

migration   *1* Movement at molecular level of one solid into another, eg of vinyl plasticiser into polyethylene core of electrical cable.

2 Movement of secondhand aircraft causing significant variation in demand for new aircraft.

Migrator   Microwave guidance radar beacon interrogator.

MIIRS   Modular imagery interpretation and reporting system.

MIITE   Multi-intelligence and information technology exploitation.

mike mike   Millimetres.

MIL   *1* Military (ICAO, US); hence * specifications, common to all US armed services; * rating, high-power engine ratings, usually close to max. cold or Meto.

2 Combined USAF/USN specifications, eg MIL-F-8785 covering basic flying qualities of aeroplanes.

3 Merritt Island.

mil   *1* One-thousandth of inch, 0.0254 mm; hence circular * = area of 1 mil circle = $506.7\mu^2$.

2 Angular measure, $^1/_{6400}$ of circle = 3.375′.

3 Military (UK, CAA).

Milcon   Military construction (USAF).

MILD   Magnetic-intrusion line detector.

Milds, MILDS   Missile [esp. ICBM] launch detection system.

mile   Imperial unit of length 1,609.344 m; aviation more commonly uses nautical mile, 1,853.184 m [US 1,852 exactly], except when reporting visibility.

Miles   Multiple integrated laser engagement system (USA).

miles in trail   Longitudinal spacing of en-route military aircraft in loose procession.

miles per hour   Statute mph = 1.609344 km/h, 0.86842 kt, 88 ft/min, 0.44704 ft/s; reciprocals 0.621371, 1.15152, 0.011364, 2.2369.

milestones   Usually I programme launch, II full development, III production.

miligraphic display   Large (19 in, 483 mm) software-driven colour display of radar overlaid on map.

military aerodrome traffic zone   Protected airspace typically extending 5 nm (9.26 km) round airfield periphery and from surface to 3,000 ft (914 m) AAL. Approach to main runway may be protected by an extended stub. In UK MATZ recognition by civil traffic is not mandatory.

military aircraft   Any operated by armed service, legal or insurrectionary, no matter what aircraft type; combat aircraft retired and privately owned is not military (see *paramilitary*).

Military Aircraft Release   Statement of operating envelope, build standard, limitations and procedures within which airworthiness has been established, which, if accepted, transfers responsibility to the Procurement Executive. This would apply were there to be a new British military aircraft.

military emergency diversion airfield   Available 24-h, 365 days, to aircraft in distress, offering radar vectoring (UK).

military power   Normally maximum cold (unaugmented) thrust; for piston engine METO.

military productivity   Several forms, including flying rate per squadron or wing, aircraft utilization, ordnance work rates, etc.

military qualification test, MQT   Final hurdle hardware must pass before entering US military service; schedule varies depending on item.

military spaceplane   MSP, family of sytems providing 'rapid global presence . . . including precision strike on terrestrial targets'. Main element will be SOV, which will probably carry CAV, MIS (3), OTV and SMV (USAF).

mill   *1* To machine fixed workpiece with rotary cutter.

2 Machine tool, often very large, for this purpose; hence skin *, spar *.

3 Large historic piston engine (US colloq.).

4 Large propeller (US colloq.).

5 Jet engine (US colloq.).

milled block circuit   Circuit (eg fluidic) machined from 3-D block.

Mill file   Single-cut tapered hand file for fine metalworking.

milli   Prefix one-thousandth, 10^{-3}, symbol m, thus milliampere = mA.

millibar   Unit of atmospheric pressure = 10^{-3} bar = $100 Nm^{-2}$, = $9.87×10^{-4}$ atm, $\simeq$ 29.53 in Hg [contrary to SI, = 100 Pa].

milligal   Former unit of gravitational attraction, = $10^{-5} ms^{-2}$ = 10 g.u.

millimetric   Concerned with magnitudes of the order of one millimetre; hence * waves, EM radiation of this order of wavelength.

milling machine   See *mill (2)*.

milliradian   Small angular measure = 3′26.25″.

MIL-1553B   Standard requirements for airborne digital data bus (US, now Int.).

MILS, M-ILS   See *MLS*.

Milsatcom   Military satellite communications.

Milspec Military specification.

Milstamp Military standard transportation and movement procedure (DoD).

Milstan Military standard.

Milstar Military strategic and tactical relay.

MIL-STD Military Standard(s) (US) eg, *-882 is that for safety requirements, *-1760 is protocol for air-dropped conventional weapons.

Milstrip Military standard requisitioning and issue procedure (DoD).

Miltracs Military air traffic control system(s).

MILTS Modern intermediate-level test station (USAF).

MILU Missile interface and logic unit (Smiths).

Milvan Military-owned ISO container.

MIM *1* Mobile-launched SAM (former US category).
  2 Minimum.

MIMD Multiple instruction, multiple data stream.

Mimic Microwave/millimetre-wave monolithic integrated circuit(s).

Mims Multispectral MWS(1).

MIMU Multisensor inertial-measurement unit.

min *1* Minimum.
  2 Minute (of time; abbreviation for minute of angular measure is ´).

M_{ind} Indicated Mach number.

mine countermeasures All methods of reducing damage or danger from land or sea mines.

mineral oil Lubricating oil of mineral (normally North American crude) base, thus not a synthetic turbine oil nor vegetable oil such as castor oil used in rotary engines; dyed red.

mineral wool Fibrous heat insulator made by blowing steam through molten slag with additives.

Miner's hypothesis If alternating stresses S_1, S_2 . . . are applied for n_1, n_2 . . . cycles, and N_1, N_2 . . . are the cycles to failure at constant S_1, S_2 . . ., fatigue damage will begin when $\Sigma \left(\frac{n}{N}\right) = 1$.

minfap Minimum facilities project; Eurocontrol, Maastricht.

Mings Micro-internetted unattended ground sensor[s].

mini Smallest RPV class, on basis of weight, altitude and speed; resemble large model aircraft.

miniature flex gyro DTG is mounted on flex shaft with opposing strut producing destabilizing force in proportion to rotor angular displacement. At tuned speed $(=f[N]^2)$ spring force and torques cancel.

Minigun GE multi-barrel belt-fed weapons of rifle calibre, usually 7.62 mm or 5.56 mm.

mini-M Single-channel voice, fax or data [small GA aircraft].

minima Lower limit of weather [esp. visibility] for particular aircraft and type of flight operation, esp. landing.

minimal flight path That for shortest time en route (ASCC).

minimally manned Aircraft available for flight as RPV but sometimes flown with safety pilot.

minimum airplane Concept, most active in US, of smallest/lightest/simplest aeroplane. By 1985 being accepted as MTOW not exceeding 100 kg, or 120 kg for two-seater.

minimum altitude Normally undefined except in same terms as lo.

minimum-altitude bombing Horizontal or glide bombing

(ie not toss) with release height below 900 ft (274 m) (DoD, IADB).

minimum break-off height For practical purposes, minimum descent height.

minimum burner pressure That below which combustion in idling engine may be extinguished, maintained by * valve.

minimum-control speed V_{MC} Lowest IAS at which aeroplane can always be flown safely (eg after sudden worst-case engine failure); specified as such in flight manual. See V_{MCA}, V_{MCG}, V_{MCL}.

minimum crossing altitude Lowest altitudes at certain radio fixes at which aircraft may cross en route to higher IFR MEA (FAA).

minimum decision altitude Minimum descent altitude (NESN, NFSN).

minimum descent altitude *1* MDA, lowest altitude, expressed in feet above MSL, without sight of runway to which descent is authorized on final approach or during circle-to-land manoeuvre in execution of standard instrument approach where no electronic glideslope is provided such as ILS, PAR (FAA). See *MD height*.
  2 Decision height (UK and others).

minimum descent height, MDH Height above touchdown, at which pilot making a non-precision instrument approach must either see to land or initiate overshoot, based entirely on topography and characteristics (including sink) of aircraft. Not used in US.

minimum design weight Not normally used: existing definitions are generally similar to operating weight.

minimum-energy orbit See *Hohmann orbit*.

minimum en route altitude That between radio fixes which meets obstruction clearances and assures good radio reception or navaid signal coverage; MEAs apply to entire width of all airways or other direct routes (FAA).

minimum equipment item Device whose failure does not delay departure, also called allowable deficiency, despatch deviation.

minimum flying speed Lowest TAS at which aeroplane or autogyro can maintain height; often well below V_{MCA} or angle of attack at which operative stall-warning system would trigger.

minimum fuel *1* Smallest quantity of fuel with which aircraft may be authorized to fly.
  2 Smallest quantity of fuel with which c.g. can fall within permitted range, normally sufficient for 30 min at maximum continuous power.
  3 Lowest quantity of fuel necessary to assure safe landing in sequence with other traffic without ATC priority (USAF).

minimum glide path This has appeared in print but is surely meaningless. There is only one glideslope angle at each ILS.

minimum gliding speed Lowest TAS at which aircraft can fly without propulsion; below V_{MCA}, or best-range gliding speed.

minimum groundspeed system Subsystem in AFCS which continuously calculates correct approach speed using TAS, G/S and W/V (entered into FMS by pilot) to protect against windshear.

minimum holding altitude Lowest altitude prescribed for holding pattern which complies with obstruction clearance and assures good radio/navaid signal reception.

minimum human force At least three sets of measures in

use as basis for aircraft/spacecraft design, factored to allow for injury, fatigue, g, anoxia and other influences.

minimum IFR altitude   As published in FAR-95 and 97, 1,000 ft above highest obstruction [2,000 ft in designated mountainous areas] within 5 statute miles of track [FAR word is "course"], or as otherwise authorized by ATC.

minimum line of detection   Arbitrary line at which hostile aircraft must be detected if defending interceptors are to destroy them before they reach vital area; usually same at ATC (1) line.

minimum line of interception   Arbitrary line at which hostile aircraft should be intercepted by aircraft if friendly AAA and SAMs are to destroy all objects not thus intercepted.

minimum military requirement   Specification or description of infrastructure designed to meet immediate or obvious future need and no more, on 'no frills' basis (NATO).

minimum navigation performance specification, MNPS   Effective from October 1978 over ocean areas FL275–400, calling for certain standards of navigation and close adherence to flightplan, and permitting major reductions in separation.

minimum normal burst altitude   Height AGL below which air-defense nuclear warheads are not normally detonated (DoD).

minimum obstruction clearance altitude   Specified altitude between ratio fixes on VOR/LF airways, off-airway routes or segments, which meets obstruction clearances and ensures acceptable signal coverage only with 22 nm of each VOR (FAA).

minimum off-route altitude   MORA charts published by Jeppesen show details of terrain and obstruction clearance within 10 nm of track.

minimum operating strip   All-weather runway devoid of non-portable facilities, see BOP (USMC).

minimum reception altitude   Lowest altitude required to receive adequate signals to determine VOR/Tacan/Vortac fixes (FAA).

minimum rpm   Engine rotational speed normally governed; with throttle on rearmost stop the speed depends on a governor, IAS and possibly other factors, such as MLG microswitch allowing a ground condition with superfine propeller pitch.

minimum runway   Not defined, other than a limiting runway for particular aircraft or mission, with regard to weather, aids and surrounding terrain.

minimums   *Minima.*

minimum safe altitude warning, MSAW   Included in ARTS, monitors all controlled aircraft and alerts controller of potentially unsafe situations, usually 100 ft below MDA.

minimum safe distance   Sum of radius of safety and buffer distance.

minimum sector altitude   That which in emergency will provide 300 m (984.2 ft) clearance within specified sector within 25 nm radius of radio aid.

minimum-sink speed   Of glider, self-explanatory, but note that MSS is not the same as speed for best L/D or glide ratio.

minimum speed   See *minimum flying speed.*

minimum TAS   Corresponds to IAS for minimum flying speed, and below MSL or on cold day can be lower.

minimum vectoring altitude   Lowest altitude, expressed in feet AMSL, to which aircraft may be vectored by radar controller.

minimum warning time   Sum of personnel and system reaction times.

mini-NW   Commonly called a mini-nuke, or mininuke, weapon with yield ⩽5kT.

minisat   Satellite launch mass 100–500 kg.

Minitat   Family of helicopter armament systems, from minimum tactical aircraft turret.

Minitrack   See *Stadan.*

minor equipments   Simple hardware items such as fasteners and clips, bought out by major supplier but not shared in collaborative programme.

MINS   Minimums (ICAO).

Mint   *1* Multi-intelligence.

    2 Mutual interference.

Mintech   Ministry of Technology (UK, obs.).

minute   *1* One-sixtieth of hour; abb. min.

    2 One-sixtieth of degree of angular measure, ′.

MIO   Multiple input/output control system.

Mio   Million (G).

MIP   Missile impact predictor [S adds set].

Mipas   Michelson interferometer for passive atmospheric sounding.

MIPB   Mono-isopropyl biphenyl.

MIPCC   Mass-injected [or injection] pre-compressor cooled [or cooling].

MIPI   Material in process inventory.

MIPR   Military interdepartmental purchase request (US).

MIPS, Mips   *1* Million instructions per second.

    2 Maintenance information planning system.

    3 Midlife improvement project study.

    4 Multiband imaging photometer for SIRTF.

    5 See *MIP.*

MIR   *1* Multiple-target instrumentation radar.

    2 Modular integrated rack, or racking.

    3 Micropower impulse radar.

    4 MLS inspection receiver.

    5 Manipulated information rate.

Mira   *1* Miniature IR alarm.

    2 Medium [wavelength] IR array.

Miracl   Mid-IR advanced chemical laser.

Miradcom   Missile R&D Command (USA).

Mirage   Microelectronic indicator for radar ground equipment.

Miran   Missile ranging system interrogating at 600 MHz with beacon reply at 580 MHz.

Mirc, MIRC   Missile in-range computer.

MIRFS   Multifunction integrated RF system.

MIRL   Medium-intensity runway-edge lights.

Mirls   Miniature IR linescan[ner].

mirror   Aerobatic manoeuvres by two aircraft in tight formation [usually one above the other], roll attitude of one 180° from that of partner.

mirror sight   Optical device to assist fixed-wing recovery on aircraft carrier, giving pitch-stabilized light indications to approaching pilot.

MIRTS   Modular[ized] infra-red transmitting set, protects against IR-seeking missiles.

Mirv   Multiple independently targeted re-entry vehicle[s], hence to * an existing missile, mirved warheads, * Salt limits.

MIS   *1* Management information system.

2 Missing (ICAO).

3 Micro-inertial sensor[s].

4 Modular insertion stage.

5 Man in space.

6 Meteorological impact statement.

MISDS   Multiple instruction, single data stream.

mismatch   *1* Inability of gas-turbine (invariably supersonic) inlet system to supply engine with correct airflow in yawed flight or other disturbed conditions.

2 Upper and lower dies not perfectly aligned in closed die forging.

MISR   Multiangle imaging spectroradiometer.

Misrep   Mission report.

misrigging   Incorrect angular setting of neutral position of control surface or other aerodynamic surface [less often, of wing(s)].

MISS, Miss   *1* Model integrated suspension system.

2 Missile-intercept scoring system.

miss-distance scorer   Indicates minimum passing distance between munition and target but does not give any co-ordinate values (ASCC).

missed approach   *1* One aborted for any reason, followed by a go-around (overshoot).

2 Standard flight procedure to be followed in (1). Captain or pilot flying elects to overshoot upon reaching DH and not seeing runway, or upon command to do so from approach controller.

missed-approach point   Point on published ILS approach, expressed as distance or time from FAF or height on glideslope, at which, if runway or approach lights are not in sight, MA(6) procedure must be initiated.

missile   No existing definition satisfactory; general assumption is flight wholly or mainly through atmosphere, with/without propulsion, with/without guidance, with one or more warheads.

missile free   Usually voice command, authority to launch AAM unless target is identified as friendly.

missile range   Ambiguous, can mean range (1) or (2).

Missile Technology Control Regime   Prohibits transfer to rogue state of hardware or technology with range over 300 km, 165 nm.

missilry   Technology of (assumed guided) missiles (colloq.).

missing-man flyby   Restrained flypast by formation team configured to leave one obvious gap in formation, symbolism (variable) being known to audience.

mission   *1* Single military operation flown by assigned force of aircraft.

2 Sortie, ie military operation flown by one aircraft.

3 Rarely (incorrectly) special flight of non-military nature.

4 Basic function or capability of missile or rocket, as shown by MDS designator (DoD).

mission-adaptive wing   One whose section profile varies automatically to suit the requirements of each flight condition.

mission-capability rate   Time out of each 24 h period that aircraft is available to perform its mission, expressed as percentage.

mission control [center]   Term usually means control of spacecraft.

mission degradation factor   Variables affecting mission, eg abort, navigation error, misidentification of target, etc.

mission equipment   See *role equipment*.

mission-oriented items   Those required following numerical assessment of enemy capabilities or targets.

mission profile   Graphical or written plot of flight level from start to finish of mission, usually a succession of lo and/or hi.

mission radius   Practical radius of action for aircraft equipped and loaded for mission on given profile, with allowances varying in peace or war; invariably less than half range with same load.

mission review report   Intelligence report containing information on all targets covered by one reconnaissance sortie.

mission specialist   Engineer or mathematician responsible for planning spaceflight, assigning payloads and assisting integration.

Mist   *1* Miniature spaceplane technology.

2 Meteorological-information self-briefing terminal.

3 Modular interoperable satellite terminal.

4 Mosaic IR sensor technology.

mist   *1* Visibility reduced by water droplets to 1–10 km.

2 Mixture of gas (invariably air) and finely divided liquid (usually an oil).

3 Definition 'popular expression for drizzle' is erroneous.

Mistel   Code name for explosive-filled pilotless aircraft guided to target by a pilot in a fighter initially linked to it (G, WW2).

misting   *1* Obscuration of transparency caused by condensed water droplets.

2 Tendency of fuel to form easily ignited dispersion in crash.

Mists   Modular(ized) IR transmitting set.

mistuning   Introducing microscopic variation into the thickness of gas-turbine rotor blades (alternate blades, pairs or quadrants) to reduce stress due to flutter.

MIT   *1* Massachusetts Institute of Technology.

2 Moscow Institute for Thermotechnology.

3 Miles in trail.

MITAS, Mitas   Multi-sensor imaging technology for airborne surveillance.

Mite   Micro tactical[ly] expendable.

MITI   Ministry of International Trade and Industry (J).

MITO   Minimum-interval takeoff.

MITS   Multi-integrated TCPED systems.

MIU   Missile interface unit.

mix   See *mixer*.

mixed conventional loads pattern management   Software enabling B-2 [possibly other aircraft] to attack target[s] with several types of free-fall weapon in single pass.

mixed fleet flying   Qualification of all airline's pilots to fly two aircraft types on successive flights, e.g. a narrow - and a widebody, or a four-engine aircraft and a twin (Airbus).

mixed-flow augmentation   Turbofan whose core and bypass flows are mixed upstream of augmenter (afterburner, reheat jetpipe).

mixed-flow compressor   Axial followed by centrifugal.

mixed-flow nozzle   Single nozzle for fan and core. Not necessarily a mixer.

mixed-manned UCAV   Fleet of UCAVs under integrated control, also called MSIC.

mixed mode   *1* Combining two EM-wave propagation modes.

2 Takeoffs and landings on same runway; hence MM operations.

mixed power aircraft Equipped with more than one species of propulsion, notably turbojet and rocket.

mixed propellant See *dual propellant*.

mixed traffic Contrasting types of aircraft (eg jets and lightplanes) under control or otherwise in same airspace, esp. on approach.

mixer *1* In communications radio, first detector in superhet receiver, which combines received signal with locally generated oscillation to yield intermediate frequency.

2 In AFCS, pitch/roll proportioning device, either mechanical or electronic, for supplying required signals to different surface power units.

3 Annular array of chutes in turbofan jetpipe which causes rapid mixing of core and bypass [fan] flows.

mixer nozzle/ejector Proposed propulsion for SSBJ: low-BPR tubofan with con/di ejector nozzle mixing jet and bypass flows.

mixing box Mixer (2), occasionally purely mechanical for translating flight-control input into required surface deflections on two axes.

mixing length Most common meaning is linear distance for fuel droplet to be vaporised.

mixture Air/fuel vapour suitable for piston engine, other than direct-injection types; hence * control, * ratio, etc.

MJ Megajoule[s], 10^6 joules.

Mj Fully expanded jet exit Mach number.

MJP Magnetron jamming pod.

MJPO Milsatcom Joint Project Office (USAF).

MJS mission Mars, Jupiter, Saturn mission; hence MJSU also includes Uranus.

MK Machine cannon (G).

Mk Mark (UK designations).

MKB Machine-construction [i.e., engine design] bureau (R).

MKC Multiple-kill capability.

MKR, Mkr Fan marker.

MKS *1* Metre/kilogramme/second [A adds ampere]; outmoded "engineering" system of metric units.

2 International Space Station (R).

MKSA Metre/kilogramme/second/ampere.

MKU Translated: 'multi-round catapult installation', rotary missile launcher (R).

MKV Miniature kill vehicle.

ML *1* Missile launcher.

2 Minelayer.

3 Middle locator.

M$_L$ Local Mach number.

ml Millilitre = cm^3.

MLA *1* Manoeuvre load alleviation.

2 Maneuver limited altitude.

ML(A) Magic lantern adaptation.

MLB *1* Multi-layer board.

2 Main-lobe blanking (ECM).

MLBM Modern large ballistic missile.

MLC Main-lobe clutter.

MLCCM Modular life-cycle cost model.

MLCM Microprocessor logic control module.

MLCS Mobile launching, or launcher, and control system.

MLD Maintenance logic diagram.

MLDI Meter-list display interval (timed arrivals, US).

MLE Missile launch envelope.

MLF Multilateral nuclear force [never activated].

MLFS Modular lightweight Flir system.

MLG Main landing gear; sometimes m.l.g.

MLI Mid-life improvement.

MLM Multipurpose laboratory module (ISS4).

MLLV Medium-lift launch vehicle.

MLMS Multipurpose lightweight missile system.

MLND, Mlnd Maximum landing weight (usually MLW).

MLP Multi-level pegging.

MLPRF Modular low-power RF.

MLRR Mode-locked ring resonator.

MLRS Multiple launch rocket system.

MLS *1* Microwave landing system.

2 Multi-level secure (or security) computer operating system.

3 Mapping and localization system (UAV).

MLST Mars local solar time.

MLT *1* Munitions lift trailer.

2 Multi-line terminal.

MLTI, MLT/I Mesosphere and lower themosphere/ionosphere.

MLTLVL Melting level.

MLU, MLUD *1* Mid-life update.

2 Monitor and logic unit.

MLV *1* Mobile launch vehicle.

2 Medium launch vehicle (upgraded Delta).

ML/V Memory loader/verifier.

MLW Maximum certificated landing weight.

MLWA MLW authorized.

MLWS Missile launch warning system.

MLZ MLS receiver.

MM *1* Rare prefix mega-mega (tera), 10^{12}.

2 Middle marker (ICAO).

3 Missile prefix mer/mer (sea/sea) (F).

4 Mass memory.

5 Mission Manager.

mm Millimetre[s].

MMA *1* Multimission aircraft.

2 Mono-methyl aniline.

3 Multimission maritime aircraft.

MMACS Multimission aft crew station; cockpit occupied by either crew member or avionics.

MM&T Manufacturing methods and technology.

M-marker Beacon with Morse-coded emission (obs.).

MMARS Military Middle Airspace Radar Service, available to all aircraft FL100–245 (UK).

MMB International bank in Moscow.

MMC *1* Metallic-matrix composite.

2 Monitor Mach computer.

3 Monopolies & Mergers Commission (UK).

4 Miniaturized munitions capability.

5 Modular mission [or mission management] computer.

6 Multimedia controller.

7 Mission Management Center (USA, SWC, USSC, 2001).

MMCCC Multi-mission command and control constellation [abb. MC^{2}C].

MMCM Multi-mission common modular.

MMD Master monitor [or mission-management, moving-map or multi-mode] display.

MMDL Multi-mode data-link.

MMDP   Modular mission display processor; drives HUD and MFDs.

MMDR   Multi-mode Doppler radar.

MMDT   Master-model design tool.

MME   *1* Maritime-mission electronics.

2 Miscellaneous military equipment [R adds requirement] (UK).

3 Modular mounting enclosure.

MMEL   Master minimum equipment list.

MMF, mmf   Magnetomotive force.

mmf   Micro-microfarad = picofarad, pf.

MMFF   Multimode fire and forget.

MMH   *1* Monomethyl hydrazine.

2 Maintenance man-hour.

MMI   *1* Mandatory modification and inspection.

2 Man/machine interface, or integration.

MMIC   *1* Millimetre-wave integrated circuit.

2 Monolithic microwave integrated circuit.

MMLS   *1* Modular MLS.

2 Mobile MLS.

MMLSA   Military MLS avionics.

MMKV   Multiple MKV.

MMMA   Multi-mission maritime aircraft.

MMMEL   Master mandatory minimum equipment list.

MMMMM   Sudden-change special weather report (ICAO).

MMMP   Multimission mobile processor.

MMMS, M³S   Multimission management system.

MMO   *1* Main meteorological office.

2 Mixed-mode operations.

M_{MO} Maximum operating Mach number.

MMOU   Multinational memorandum of understanding.

MMP   *1* Metallic materials processor (company category).

2 Modular mission payload.

3 Maintenance monitoring panel.

MMPM   MEECN message processing mode.

MMR   *1* Machmeter reading.

2 Minimum military requirement.

3 Multi-mode receiver.

4 Multi-mode [or multimission] radar [S adds system].

MMRC   Modular multi-role computer.

MMRH/FH   Mean maintenance and repair hours per flight hour.

MMRPV   Multi-mission RPV.

MMS   *1* Missile-management system.

2 Mast-mounted sight (or sensor).

3 Multi-mission modular spacecraft.

4 Metal measuring set.

5 Magnetic minesweeping system.

6 Moisture, or maintenance, management system.

7 Magnetospheric multiscale mission (NASA).

8 Mission-management system.

MMSA   Multimission surveillance aircraft.

MMSE   Minimum mean square error.

MMSS, M²S²   Mobile mass-storage system.

MMT   *1* Multiple-mirror telescope.

2 Message/management terminal.

3 Multimission trainer.

MMTD   Miniaturized, or miniature, munition[s] technology demonstration.

MMTS   Multi-modal transport system (Airbus).

MMU   *1* Manned maneuvering unit (gas-powered flight control system for Shuttle astronauts).

2 Mobile meteorological unit (RAF).

MM/UCAV   Mixed manned UCAV.

MMVX   Medium-speed support aircraft (project).

MMW   Millimetre, or millimetric, wave; R adds radar.

MN   *1* Meganewton[s].

2 Model number.

3 Magnetic north.

M_n Mach number.

mN   Millinewton[s] = $N \times 10^{-3}$.

m_n Neutron rest-mass.

MNA   Maritime Notification Area.

MNB   Multiple narrow beam(s).

MNC   Major NATO command.

MNCID   Management of network category interaction diagram.

MND   *1* Mission need documents (NATO).

2 Multi-National Division (SFOR).

3 Minimum nuclear deterrent.

4 Ministry of National Defence (Poland).

MNE   Mixer nozzle/ejector.

M_{NE} Mach number never to be exceeded, and not even approached in normal flying, but demonstrated in civil certification.

mnemonic   Easily remembered sequence to assist aircrew with their pre-flight and other checks, eg HTMPFFG = hood/harness, trim, mixture, pitch, fuel, flaps, gills/gyro (largely archaic except in general aviation).

MNFP   Multinational fighter program (US).

MNIIRS   Moscow scientific-research institute of radio communication.

MNLC   Multi-National Logistic Command.

MNLD   Mainland.

Mnm   Minimum (ICAO).

MNO   Ministry of national defence (Czech).

M_{NO} Normal operating Mach number, now generally replaced by M_{MO}.

MNOS   MOS with silicon nitride on oxide.

MNP   Magnetic north pole.

MNPA   Minimum navigation performance airspace, allowing 60 nm lateral separation at FL275–400.

MNPS   Minimum navigation performance standards, or specification [A adds airspace].

MNR   Minimum-noise route.

MNS   Mission-need[s] statement.

MNT   Monitor[ing].

MNTN   Maintain.

MO   *1* Meteorological Office.

2 Magneto-optical.

3 Ministry of defence (R).

4 Massive ordnance.

m.o.   Master oscillator.

MOA   *1* Memorandum of agreement.

2 Military operations area.

3 Minimum operating altitude.

4 Military operational approach[es].

MoA   Ministry of Aviation (defunct in UK).

MOAB   *1* Missile optimised anti-ballistic.

2 Massive ordnance aerial bomb.

3 Massive ordnance air-burst, or [more often] air blast.

MoAS   Ministry of Aviation Supply (UK defunct).

MOB   *1* Main operational base (USAF).

2 Mobile offshore base (ONR).

MOBA Mobility operations for built-up areas (USA).

Mobidic Modular bird with dispensing container (MBB/Aérospatiale).

mobile air movements team Air Force team trained for deployment on air-movement traffic duties.

mobile lounge Lounge which also conveys passengers between terminal and aircraft and vice versa (eg Washington Dulles Airport).

Mobile Packet Data service Internet Protocol service charged not on length of connection but on megabits sent or downloaded.

mobile quarantine facility Set up on vehicle, usually helicopter-capable ship, for receiving special cargo, eg lunar rock.

mobile satellite system One designed to serve mobile (eg aeronautical) subscribers.

MOBSS Mobility support squadron.

MOC *1* Maintenance operational check.
   2 Minimum operational characteristics.
   3 Minimum obstacle clearance.
   4 Mars orbiter [or orbital] camera.
   5 Modular [or mission, or multifunction] operations centre [or console].

MOCA Minimum obstruction- [or obstacle-] clearance altitude.

mock-up Quickly built replica of aircraft or other product, usually full-scale, to solve various problems (see *customer* *, *engineering* *, *furnishing* *, *hard* *).

MOCM Multispectral ocean colour monitor.

MOCN NC machine tool (F).

MOCVD Metal organic chemical vapour deposition.

MOD *1* Magneto-optical drive.
   2 Meteoroids and orbital debris.

MoD Ministry of Defence.

mod *1* Modification.
   2 Modulation, or modulator.
   3 Moderate.

Modar Modular aviation radar.

Modas *1* Modular data-acquisition system.
   2 Modular defensive-aids suite (UK, RAF).

MOD-DIG, Mod-Dig Modular digital image-generator.

mode *1* The five classical * of aeroplane motion; see *motion*.
   2 Any of selectable methods of operation of device or system.
   3 Number or letter referring to specific pulse spacing of signal transmitted by interrogator (IFF, SIR, SSR).
   4 Each possible configuration of spatial variable, eg EM wave, flutter or aerodynamic phenomena.

Mode A Pulse format for interrogation of ATCRBS, displaying aircraft ident on SSR.

Mode B Optional for ATCRBS xpdr.

Mode C Pulse format which also adds aircraft altitude.

Modec Automatic reporting of altitude.

mode-coupled oscillation PIOs, assumed linear, which interact with flexible modes of structure [in complex and usually unique ways].

Mode D Optional or unassigned xpdr mode.

model Ambiguously used in aviation:
   1 Small-scale replica for testing characteristics of full-size aircraft.
   2 In general aviation, improved version marketed at (often annual) intervals.

   3 To reproduce a functioning system synthetically, eg in EDP program.

model atmosphere Mathematically exact numerical values closely approximating to an idealized real atmosphere.

model basin See *towing tank*.

model cart Often large and elaborate wheeled truck on which are mounted a balance or sting and aircraft model, the whole then installed in tunnel working section with large number of electrical and manometric connections.

model qualification test US military clearance of new item, esp. engine, for production.

model tank *1* See *towing tank*.
   2 Tank filled with fluid for flow exploration round model by electrical potential analogy.

model tunnel Ambiguous, term best reserved for any small-scale model of a future large wind tunnel.

modem Telecommunications or EDP (1) device: modulator + demodulator.

moderator In nuclear reactor, substance specifically present to slow down neutrons by collisions with nuclei.

Mode S Selective interrogation of SSR offering Mode C plus mode-select for discrete interrogation and data-link.

modex Three-numeral designator for quick identification of individual aircraft within CAG or other unit (USN).

modification Temporary or permanent change to either single aircraft (for particular purpose) or all aircraft of type (to rectify fault or shortcoming or offer improved capability); can be unique, to owner's requirement, or one of planned and controlled series throughout life of aircraft.

modified close control Interceptor is told only target-position information (USAF).

modified mission MDS(4) letter added to left of basic mission letter to indicate a permanent alteration to basic mission (DoD).

modified PAR Precision approach guidance for high-performance aircraft to flare point instead of to touchdown point (USAF).

Modils Modular ILS.

Modir Modulated IR jammer.

Modis Moderate-resolution imaging spectro-radiometer, or spectrometer.

Modmor Commercially produced (Morgan Crucible) family of carbon/graphite fibres.

Modos Mobile Doppler Sodar.

MoD(PE) MoD Procurement Executive (UK).

modular Designed in discrete series of major components for ease of inspection, overhaul or repair by module replacement without greatly disturbing neighbours; engine may comprise fan, LP compressor, HP compressor, combustor, turbine and accessories modules, while major avionic system may be series of * boxes.

modular insertion stage A low-cost expendable upper stage to the MSP that would be used to deploy mission payloads for rapid replenishment and other critical-time missions (USAF).

modular lounge Airport lounge built as unit capable of (1) being moved from place to place to add capacity where required or (2) being driven to/from aircraft (mobile lounge).

modulated augmentation Afterburner (reheat) with fuel

flow continuously variable to give smooth increase in thrust from max. cold to max. augmented.

modulated waves Electromagnetic waves on which is impressed information in form of variation in amplitude (AM) or frequency (FM).

module *1* One of assemblies of modular system, easily replaced but seldom itself torn down except at major base or by manufacturer; eg spacecraft command *, engine fan *.

2 Single box of electronic equipment replaceable as plug-in unit.

3 Any standard dimensions or standard size of container.

4 Standard-capacity building block of computer memory.

modulus of elasticity Ratio of unit stress to unit deformation of structural material stressed below elastic limit (limit of proportionality), symbol K.

modulus of rigidity Shearing modulus of elasticity, proportional to angle of distortion but measured in stress per unit area, symbol C or N.

modulus of rupture In beam loaded in bending to failing point $S_M = Mc/I$ [M bending moment, c distance from neutral axis]; in shaft or tube failing in torsion $S_S = T_c/J$.

MoE, MOE *1* Measure of effectivness.

2 Maintenance organization exposition.

Moffett Moffett Field, CA, home of NACA lab., now known as NASA Ames Research Center, and an NAS.

Mofle Mid and outer fixed leading edge.

MOG Maximum on ground, areas suitable for large number of airlift transports in front-line LZ at time of peak delivery of, or resupply of, ground force.

Mogas Motor gasoline, ie ordinary automotive petrol. Usually 91 or 93 octane.

Mogr Moderate or greater.

MOG spots Unloading spots for STOVL transports in battle area, typically along short length of highway.

MOH Major overhaul.

Mohol Mix of Mogas and alcohol.

MOI Mars orbit insertion.

MOKA Methodology of knowledge acquisition.

MOL Manned orbital laboratory.

mol See *mole* (2).

MOLA Mars orbiter laser altimeter.

mold Mould (US spelling).

Mole Molecular electronics, many suffixes, not listed.

mole *1* Self-contained intelligence sensor air-dropped over hostile territory.

2 SI unit for amount of substance containing as many units [eg, atom, ion, molecule] as in 0.012 kg of C^{12}, abb. mol.

mole dropper Usually a UAV, see previous.

moleskin strip Soft strip around bonedome to eliminate scratching canopy.

Mollier diagram Plot of enthalpy (ordinate) against entropy.

Molochite Precision casting mould material, a Ca-Mg-Al silicate.

moly Colloquial for molybdenum disulphide lubricant.

molybdenum Mo, hard white metal, density 10.2, MPt 2,620°C, important alloying element in steels and in 'moly'.

moly bolt Patented bolt which can be inserted or removed with access from one side only.

MOM *1* Metal-oxide-metal.

2 Methanol/oxygen mix.

3 Moment.

moment Turning effect about an axis: force multiplied by perpendicular distance from axis to force.

moment arm Distance from axis to force, eg from c.g. to aerodynamic centre of tailplane.

moment coefficient *1* Any moment reduced to non-dimensional form, usually by dividing by dynamic pressure.

2 Particular values such as C_m (section *), C_{mo} (pitching *), C_{mac} (about aerodynamic centre) and C_{mcg} (about c.g.).

moment distribution Analytical method for frameworks based on bending moments at rigidly attached joints between members.

moment index Divisor, typically 10^4, to reduce numerical values in balance calculations of heavy or large aircraft.

moment of area Not normally used, though term 'second ***' synonymous with moment of inertia.

moment of inertia Symbol I, sum of all values of mr^2 where m is elementary particle and r its distance from axis about which I is measured, or Mk^2 where M is total mass and k is radius of gyration; SI unit usually $kg\text{-}m^2 = 23.73$ $lb\text{-}ft^2 = 0.73756$ slug-ft^2.

moment of momentum See *angular momentum*.

momentum Mass multiplied by velocity, symbol p = mv, units $kg\ ms^{-1}$.

momentum coefficient For CC/BLC System in Vj/qS, symbol $C\mu$.

momentum drag Drag due to change in momentum of air entering lateral or other non-ram inlet; major factor in design of ACVs and small high-power jet-lift V/STOLs; $= \dfrac{WV}{g}$ where W is mass flow and V vehicle speed.

momentum separation Flow separation from convex duct wall or other surface where Coanda effect is inadequate.

momentum theory Idealized Froude treatment for propellers and other driven rotors by calculating momentum ahead of and behind disc exerting uniform pressure on uniform flow.

momentum thickness Measure of reduction in momentum of flow due to viscous forces in boundary layer.

momentum thrust Component of jet-engine thrust due to acceleration of jet $= \dfrac{WVj}{g}$ where W is mass flow and V_j is jet velocity. See *thrust*.

momentum transfer method Drag measurement by pitot traverse of wake.

momentum wheel Flywheel, eg in attitude stabilization.

moment weight Mass of gas-turbine or other blade multiplied by distance of its c.g. from axis of rotation.

Moms, MOMS *1* Modular opto-electronic multi-spectral scanner.

2 Mission Operations and Mission Services (NASA).

MON *1* Above mountains (ICAO).

2 Mixed oxides of nitrogen.

3 Motor octane number.

4 Month.

Mon Monitor (NASA).

Mona   Advanced hyperbolic/computer/display airborne navigator system (from modular navigation).

Monab   Mobile naval airbase, on shore (USN).

Monel   Range of corrosion-resistant Ni-Cu alloys.

monergol   Hydrazine (F), often capital M.

Money flare   Flame from petrol-soaked roll of asbestos in steel basket (RAF, 1918 to WW2).

Monica   Large series of active radars used by RAF bombers 1943–46 for rear warning of interception.

monitor   *1* To listen out on a particular frequency.

2 Pipe, usually on RIV or CFR vehicle, through which water/foam or other retardant is delivered.

monitored system   One continuously subjected to surveillance and if necessary corrective action (eg isolating failed channel) by separate avionic subsystem; feature of autoland and other highly reliable systems.

monkey strip   Strong fabric strip securing backseater [esp. when standing to fire gun] to aircraft (UK, colloq., 1916–40).

Monnex   Dry-powder fire extinguishant.

monoball mount   Primary structural joint, eg pylon to wing, where loads are transmitted radially across spherical surface.

monobloc   *1* Shaped from single block, usually by forging or casting.

2 Control surface in form of single aerofoil, not hinged downstream of fixed fin or tailplane; also called slab type or all-moving.

monochromatic   EM radiation of single frequency, or extremely narrow band.

monochromator   Various devices for filtering out unwanted wavelengths.

monocoque   Three-dimensional form, such as fuselage, having all strength in skin and immediate underlying frames and stringers, with no interior structure or bracing. Some purists argue there must be skin only.

monocrystal   Turbine blade or other item formed from metal grown as a single crystal, devoid of inter-grain boundaries which would weaken it mechanically. Adjective monocrystalline.

MONOCV, Mono-CV   Future single-hull aircraft carrier.

monofuel   Fuel used alone without need for air or other oxidant; eg HTP, MMH.

monohud   Monocular HUD, for one eye only.

monolithic   Usually means *monocrystal*, but see next.

monolithic rocket   Having solid grain cast or extruded in one piece.

monomer   Substance consisting of single molecules, esp. one which can be polymerised.

monomethyl hydrazine   Rocket propellant; hydrazine in which one hydrogen atom is replaced by methyl group $N_2H_3.CH_3$.

monoplane   Aerodyne having single wing.

monopropellant   Monofuel used in liquid rocket, eg HTP.

monopulse   Radar technique using four overlapping pencil beams, two for azimuth, two for elevation, with circuitry so arranged that, when target is at centre, output voltage vanishes.

monorail   Installed along ceiling of paratroop transport, to which static lines are hooked.

monoslabs   Standard heavy concrete castings set into

ground to form large grid, seeded with grass, in regions of airfield subjected to severe jet erosion.

monospar   Wing construction based on single spar, strong in bending and torsion.

Monroe effect   Formation of hot high-velocity jet by explosion of shaped charge.

monsoon   *1* Winds which reverse direction seasonally, blowing seaward in winter.

2 In India, seasonal rains.

Monte Carlo method   Computation based on computerized random sampling.

Montgolfière   Common term for early (pre-c1850) hot-air balloons.

mooring area   Water reserved for mooring of marine aircraft.

mooring band   Reinforces upper part of envelope in line with mooring lines.

mooring drag   See *trail rope*.

mooring guy   Rope or cable for securing aerostat.

mooring line   See *mooring guy*.

mooring mast   Rigid or cable-braced high mast to which airship can be fastened and up/down which all supplies and people can pass; also called mooring tower.

Moose   Manned orbital-operations safety equipment.

MOOTW   Military operations other than war.

MOP   *1* Ministry of public works (Spanish-speaking countries).

2 Magnetic orange pipe.

3 Modulation on pulse.

4 Minimum operational performance; see * RS.

5 Massive ordnance penetrator.

6 See next.

MOP costs   Maintenance and overhaul personnel costs.

Mopitt   Measures of pollution in the troposphere.

MOPRS   Minimum operational performance requirements standards.

Mops, MOPS   *1* Minimum operational performance standards.

2 Million operations per second.

Moptar   Multiple-object phase tracking and ranging; short-baseline CW phase-comparison.

MOQA   Maintenance operations Quality Assurance.

MOR   *1* Mandatory occurrence reporting, or report (CAA).

2 MF omni-range (365–415 kHz).

3 Military operational requirement(s).

4 Meteorological optical range.

MORA   *1* Minimum off-route altitude.

2 Memorandum of requalification agreement.

Morag   MOR (1) advisory group.

Morass   Modern ramjet systems synthesis.

MoRE   Modular receiver/exciter.

More-Electric Initiative   Major effort to use electric power for secondary power [such as HEL] and propulsion (DoD, NASA, industry).

morph   A morphable aircraft.

morphing   To change shape of aircraft in flight [in gross and significant manner, not just by lowering flaps, for example]. Effort at present mainly on UAVs.

Morrison shelter   Mild-steel 'table shelter', able to support collapsed house; issued to British civilians 1942–44.

MORS   MOR (*1*) scheme.

Morse　Code for transmitting messages by succession of dots and dashes.

Morse key　Hand-operated switch for sending short and long signals [dots and dashes].

Morsviazsputnik　National signatory to Inmarsat (USSR).

mortar　Large-calibre low-velocity gun for projecting payloads, eg Shuttle drag chute.

MOS　*1* Metal oxide silicon, family of semiconductor devices.

　2 Maritime observation satellite.

　3 Minimum operating strip.

　4 Multiprocessor operating system[s].

MoS　*1* Ministry of Supply (UK, April 1946, became MoA).

　2 Military occupational specialty (USA).

MoS$_2$　Molybdenum disulphide.

Mosaic　Multifunction on the move secure adaptive integrated communications (USA).

mosaic　Large assemblage of mating aerial photographs, from optical camera, vidicon, IR or other source, with adjacent boundaries accurately aligned, to cover whole area under surveillance (military reconnaissance or photogrammetric mapping and surveying).

Mosdap　MOS dynamic analysis program.

MOSFET, Mosfet　Metal oxide silicon, field-effect transistor.

MOSLK　MOS (3) lighting kit (USMC).

MOSLS　Minimum operating strip lighting system.

MOSP　Multimission optronic stabilized payload.

Most, MOST　Metal oxide silicon (or semiconductor) transistor.

MOS-VAO　All-union association for experimental aircraft (USSR).

MOT　Maximum overhaul time(s).

MoT　Ministry of Transport (many countries).

Mote, MOT&E　Multinational operational test and evaluation.

MoTE　Mobile threat emitter; S adds system.

moteur canon　Patented (Hispano Suiza) installation of cannon between banks of V-12 engine, firing through geared propeller hub.

mother　Used in parental capacity, such as RPV control aircraft (* ship) or * printed-circuit board.

motion　In aeroplane, can occur in five classical modes: phugoid, roll, spiral, Dutch roll and longitudinal short-period. Other masses can rotate about 2, 3 or 6 axes.

motion seat　Seat in simulator, fitted to impart vibration and small accelerations, tighten harness, and give other sensations, often along any axis.

Motis　Message-oriented text interchange system.

motivator　Device causing major change in trajectory, eg control surface, flap, vectored nozzle, etc.

Motne　Meteorological operational telecommunications network in Europe.

motor　Best restricted to solid rockets and hydraulic/electrical/air machines, though several non-English languages have to use it for most or all main propulsion. Acceptable as adjective = powered, eg * glider.

motorboating　Oscillations at sub-audio frequency in any device.

motor glider　Ultra-light aeroplane derived from glider; best not used for powered or assisted sailplanes which are competition sailplanes with auxiliary propulsion (retractable, if propeller-type) used intermittently.

motoring　Rotating main engine by means of starter for purpose other than starting (usually gas turbine).

motoriste　Engine manufacturer (F).

motor octane number　Preferred scale of measurement of gasoline [petrol] resistance to detonation. Typically unleaded fuels are 85–87 MON, whereas RON figures are 95–97.

motor specific impulse　For solid-propellant rocket, total impulse divided by total motor loaded weight, symbol I_{Ta}, unit kNs or MNs (in US, lbf-s).

MOTR　Multi-object tracking radar.

MOTS　*1* Modular [or modified] off the shelf.

　2 Mobile tower systems.

MOTU　Maritime operational training unit.

MOTV　Manned orbital transfer vehicle.

MOU　*1* Memorandum of understanding.

　2 Mountain landings (Swiss national licence).

mould　*1* 3-D template.

　2 3-D reference tool for fabrication of composite parts of thermosetting sheet (eg canopy).

　3 Enclosed cavity for cast parts.

mould dimension　Any dimension to or along a mould line.

mould line　Line formed by intersection of two surfaces, eg sheet-metal parts.

Mouldy　Air-launched torpedo (UK colloq. 1920–50).

Mount　Military operations in urban terrain.

mountain wave　Powerful air mass immediately downstream of transverse mountain range, rotating about horizontal axis. There can be a succession of such waves. Arguably = rotor (3).

mounting　*1* Vague term acceptable for most objects hung on vehicle but not for landing gear or strut attachments. Includes all linkage of structural nature, and any ad hoc parts of airframe such as pylon, but not control or accessory systems.

　2 Process of fitting a tire [tyre] to a landing-gear wheel.

mounting pad　Circular attachment on engine or ADG for shaft-driven accessory.

mouse　See *mice*.

moused　Completely wrapped in soft copper wire [in mooring flying boat].

moustaches　Small canard foreplanes, usually retractable and used only at high-alpha regimes by supersonic delta.

MOUT　See *Mount*.

mouth　*1* Inlet of circular parachute, esp. where this has diameter less than canopy maximum.

　2 Open bottom of hot-air-balloon envelope.

mouth lock　Reef.

MOV　*1* Metal-oxide varistor.

　2 Main oxidiser valve.

mov　Move, moving.

movables　All flight-control, high-lift or high-drag surfaces on wing.

movement　*1* One aircraft flight as recorded by particular ground location; for airport, transits overhead are not counted and a * is either an arrival or a departure.

　2 Single military airlift operation by one or more aircraft.

movement area　Region where aircraft may be found on ground proceeding under their own power.

move off blocks　Start of an aeroplane flight from parked position, esp. commercial transport; time entered in ATC and flight logs. Where parked nose-on, time is from start of backward push.

MOVG Moving.

moving-armature speaker Audio loudspeaker driven by oscillation of part of soft-iron (ferromagnetic) circuit.

moving-base simulator One whose flight deck or cockpit can move linearly as well as rotate, usually in any direction.

moving-coil speaker Dynamic loudspeaker, driven by forces developed by interaction of currents in conductor and surrounding field.

moving flight deck In 1943–49 the advantages of making a carrier flight deck as a powered belt, with an axial motion similar to, or opposite to, that of landing aircraft, was shown to offer no advantage.

moving-iron instrument Coil carrying current to be measured moves soft-iron armature connected to indicating pointer.

moving-map display One having topographical, radar, IR, target or other map projected optically on screen with aircraft position fixed (usually at centre).

moving target indication Radar which incorporates circuits automatically eliminating fixed echoes so that display shows only those moving with respect to Earth. For airborne radar, eg AWACS, aircraft platform motion is automatically subtracted.

moving wings Wings whose incidence is varied in flight as primary means of trajectory control.

Movlas Manually operated visual landing-aid system (carrier LSO).

MOVPE Metal organic vapour-phase epitaxy.

MOVTAS, Movtas Modified visual target-acquisition system.

MOX, Mox Mixed dioxides of Pu and U.

Mozaic Measurement of ozone by Airbus in-service aircraft.

MP *1* Midpoint between fixes or waypoints.
　　2 Manoeuvre programmer (RPV).
　　3 Manifold pressure.
　　4 Multiphase (or pulse).
　　5 Manual proportional (flight-control system).
　　6 Manpower and Personnel (USAF).
　　7 Maintenance period.
　　8 Middle plug.
　　9 Main, or maritime, processor.
　　10 Modification proposal.
　　11 Multipurpose.
　　12 Mission planner.
　　13 Maximum payload.
　　14 Multi-platform.
　　15 Montreal Protocol.

mP Maritime polar air.

m$_p$ Proton-rest mass.

MPA *1* Man-powered aircraft.
　　2 Management problems analysis.
　　3 Maritime-patrol aircraft.
　　4 Million passengers [per] annum.
　　5 Module performance analysis.

MPa Megapascal, SI unit for high pressures = 145.0376 lb/in^2.

MPAA Multi-beam (or multifunctional) phased-array antenna (or aerial).

MPAG Man-Powered Aircraft Group.

MPAR Multipurpose airport radar (FAA).

MPAV Multi-purpose air vehicle.

MPB *1* Multiple pencil beam.
　　2 Mobility procedures branch.

MPBA Multiple practice bomb adapter.

MPBE Multiple-platform boresight equipment [helicopter].

MPBW Ministry of Public Buildings and Works (UK, defunct).

MPC *1* Multi-purpose console.
　　2 Military personnel centre.
　　3 Multilayer printed circuit.
　　4 Message-processing centre.
　　5 Missile practice camp.

MPCD Multipurpose colour display.

MP-CDL Multi-platform common data-link (USAF).

MPCF Million parts/particles per cubic foot.

MPCS *1* Mission planning and control station.

MPCU Multiport protocol converter unit.
　　2 Maintenance/power control system.

MPD *1* Medium-prf pulse Doppler.
　　2 Maximum permitted dose (radiation).
　　3 Maintenance planning document.
　　4 Multi-purpose display.
　　5 Mobile packet data (Satcoms); S adds service.

MPDI Multi-purpose display indicator; (see *MDRI*).

MPDM Multi-purpose dextrous manipulator.

MPDR Monopulse Doppler radar.

MPDS *1* Missile-penetrating discarding sabot.
　　2 Multi-purpose display system.
　　3 Maintenance-planning data support.
　　4 Message processing and distribution system.
　　5 See *MPD (5)*.

MPE *1* Mass-properties engineering.
　　2 Mission-planning element.

MPED Multi-purpose electronic display.

MPEL Maximum permissible exposure level.

MPET Metallized polyethylene terephthalate.

MPFF Multi-purpose fighter facility.

MP Flt Mission-planning Flight (RAF).

MPG *1* Or **m.p.g.**, miles per gallon, rare in aviation.
　　2 Moulded propellant gain.

mph Statute miles per hour, see *mile* for conversions.

MPHD Multi-purpose head-down.

MPHT Missile potential hazard team.

MPI *1* Mission and Payload-Integration office (NASA).
　　2 Magnetic-particle inspection.
　　3 Mean point of impact.
　　4 Multiprotocol interface.
　　5 Major periodic inspection.
　　6 Moving plan indicator.

mpig Miles per Imperial gallon (not recommended).

mPk Maritime polar, colder than surface.

MPL *1* Multi-programming level.
　　2 Mid-Pacific landing (NASA).
　　3 Mars polar lander.

MPLH Multi-purpose light helicopter.

MPLM Multi-purpose logistics module.

MPM *1* Multi-purpose missile, or modem.
　　2 Microwave power module (radar xmtr).

MPMS *1* Maritime patrol-mission system.
　　2 Missile-performance monitoring system (inertial guidance).

MPNL Master parts-number list.

MPO *1* Mission Planning Officer (RAF).
2 Main production order.
3 Mission-payload operator.
MPOA Multi-protocol over ATM.
MPP *1* Massively parallel processor.
2 Maintenance programme proposal.
MPPA, mppa Million passengers per annum.
MPPM Microwave pulsed power module.
MPPS Multi-purpose pylon system.
MPPU Multi-purpose processor unit.
MPR *1* MIL (military) power reserve.
2 Medium-power radar.
3 Microburst-prediction radar.
MPRD Metal produce and recovery depot.
MPRF Modulated, or medium, pulse-repetition frequency.
MP-Rtip Multi-platform RTIP.
MPS *1* Mixed-propellant[s] system.
2 Main propulsion stage, or system.
3 Mid-platform ship (pad amidships).
4 Materials processing in space.
5 Mission-payload subsystem.
6 Multiple protective structure (ICBM).
7 Military postal system (US includes other services).
8 Minimum-payload subsystem.
9 Minimum-performance specification.
10 Mission-planning, or processing, system.
11 Maintenance processing system.
12 Maximum-pitch stop.
13 Motor/pump set.
14 Metres per second (ICAO, contrary to SI, which dictates m/s).
mps Metres per second (see *14* above).
MPSM Multi-purpose submunition.
MPSS Man-hours per ship-set.
MPT *1* Memory point track.
2 Multi-purpose tracer.
3 Missile procedure trainer.
MPt Melting point.
MPTA Main-propulsion test article.
MPTO Maximum-performance takeoff (helicopter).
MPU *1* Missile power unit.
2 Master processor unit.
3 Message pick-up.
4 Mission programming unit.
5 Multifunction process unit.
MPUAV Multipurpose UAV.
MPVI Memory, planes, video input.
MPVO Local air defence (USSR, R).
MPVT Multi-pulse vectored thrust.
MPVU Multi-plot variable update.
MPW *1* Mission-planning workstation.
2 Minimum pavement width (required for 180° turn).
mPw Maritime polar air, warmer than surface.
$\overline{\text{mq}}$ Pitching moment coefficient due to pitching.
MQAD Materials Quality Assurance Directorate (UK MoD, now Qinetiq).
MQF Mobile quarantine facility.
MQLF Mobile quick-look facility.
MQP Moulded quality phenolic.
MQT Model qualification test.
MR *1* Maritime reconnaissance.
2 Medium-range.
3 Marker ranger (laser).

4 Military Region.
5 Machine-readable.
6 Magneto-resistive.
7 Mission rehearsal.
8 Memo for record.
9 Magnetic resonance.
10 Microwave receiver (Awacs).
11 Morning report (US usage).
12 Manual rectification.
mR Milliroentgen[s]; also incorrectly used to mean milliradian[s].
MRA *1* Major replaceable assembly.
2 Maximum relight altitude.
3 Minimum reception altitude.
MRAAM Medium-range air/air missile.
mRad, M-rad Milliradian, = 3'6.25", just over 0.05°.
MRAF Marshal of the Royal Air Force.
MRAG Metallics Research Advisory Group (MRCC).
MRALS, Mrals Marine remote-area [and] landing system, also rendered as **MRAALS**.
MRAM Magnetoresistive random-access memory.
MR&R Mobile reclamation and repair.
MRASM Medium-range air-to-surface missile.
MRB *1* Materials, or Management, Review Board.
2 Maintenance review board.
3 Main rotor blade.
MRBF Mean rounds before failure.
MRBM Ambiguous, can be medium-range or mid-range ballistic missile.
MRBR Maintenance Review Board Report.
MRBS Mean rounds between stoppages.
MRC *1* Meteorological Research Committee (UK).
2 Major regional conflict [or contingency], for practical purposes = war.
MCRA Multi-role combat aircraft.
MRCC Materials Research Consultative Committee (UK).
MRCLOS Missile reference command to line of sight.
MRCS *1* Multiple-RPV control system.
2 Mobile reporting and control system.
MRD Main-rotor diameter.
MRDA Missile round design agent.
MRDAU Multiple recorder data-acquisition unit.
MRDEC Missile Research, Development and Engineering Center (USA).
MRD/FT Missile restraint device field tester.
MRDL Multirole datalink.
MRDP Multi-radar data processing.
MRE *1* Mean radial error (weapon delivery).
2 Multi-role endurance (UAV).
3 Medium-range Empire (UK 1947).
4 Mission Rehearsal Exercise.
mrem Milliröntgen, thus mrem/h.
MRES Mobile remote-emitter simulator.
MRF *1* Meteorological Research Flight.
2 Multi-role proximity fuze.
3 Medical red flag.
4 Multi-role fighter.
MRFD Modular rugged flat display.
MRG *1* Master reference gyro.
2 Medium-range (ICAO).
3 Main-rotor gearbox.
MRGB Main-rotor gearbox.
MRGL Marginal.

MRH   Maintain runway heading.

MRI   *1* Magnetic/rubber inspection.

2 Magnetic resonance inspection, or imaging.

MRIR   Medium-resolution infra-red radiometer.

MRIS   Medium-resolution spectroradiometer.

MRIU   Missile remote interface unit.

MRL   Modular rocket launcher.

MRLG   Monolithic ring-laser gyro.

MRM   *1* Medium-range missile.

2 Map rectification machine.

MRMS   Mission-ready management service.

MRO   *1* Maintenance and Repair Organization.

2 Maintenance repair and overhaul.

3 Mars reconnaisance orbiter.

MROC   *1* Multi-role operations cabin (air defence).

2 Minimum [rarely maximum] rate of climb.

M-Rose   Multi-tasking real-time operating-system executive.

MRP   *1* Medium-angle rocket projectile (air/ground sight setting).

2 Multi-annual research programme (EC).

3 Materials requirement[s] planning.

4 Manufacturing resources planning.

5 Mobile repair party.

6 Machine-readable passport.

7 Modular reconnaisance pod.

8 Mission replay.

9 Meteorological reporting point.

MRPC   Mercury-Rankine power conversion.

MRPV   Mini-RPV.

MRR   *1* Mission-reliability rate.

2 Multimode radar; in Norwegian air defence, radio.

3 Manufacturing revision request.

4 Mechanical-reliability report.

5 Maritime radar reconnaisance.

6 Missile, or medium, range-recovery.

MRRV   *1* Manoeuvring (British spelling), or maneuverable, re-entry vehicle.

2 Multi-role reusable vehicle.

MRS   *1* Master, or military, radar station, or site.

2 Maintenance recording system.

3 Mobile radio service.

4 Medical record system.

5 Monitoring and ranging station.

6 Meteorological radar station, service, or system.

7 Mobility requirements study.

8 Mountain Rescue Service.

9 Magyar Repulo Szovetseg, sport-aviation association (Hungary).

MRS   Generalized mass, esp. in co-ordinates of dynamic model.

MRSA   Mandatory radar service area.

MRSI   Multiple-round[s] simultaneous impact.

MRSP   Multifunction radar-signal processor.

MRSR   Multi-role survivable radar.

MRT   *1* Maximum reheat thrust.

2 Military rated thrust.

3 Multi-role turret.

4 Miniature receive terminal.

5 Maintenance release tag.

6 Mountain rescue team.

7 Multi-radar tracker, or tracking.

8 Multiple remote terminal.

MRTA   Medium-range tactical aircraft.

MRTD   Minimum-resolvable temperature [rarely, time] difference.

MRTF   Mean rounds to failure (gun).

MRTFB   Major range and test-facility base (USAF).

MRTM   Maritime.

MRTS   Microwave repeater test set.

MRTT   *1* Multi-role tanker/transport.

2 Multi-role towed target.

MRTU   Multiple remote terminal unit.

MRU   *1* Mountain rescue unit.

2 Mobile radio, or radar, or reporting, unit.

3 Mach repeater unit.

4 Mobile receiver unit (UAV).

5 Maintenance recorder, or recording, unit.

6 Medical Rehabilitation Unit (RAF, Headley Court).

MRUASTAS   Medium-range unmanned aerial surveillance and target acquisition system.

MRV   *1* Multiple re-entry vehicles.

2 Machine-readable visa.

MRW   Maximum ramp weight.

MRWS   Manned remote work-station (for space-station construction).

MS   *1* Minus (ICAO).

2 Military standard (US).

3 Mild steel.

4 Maintenance schedule.

5 Medium supercharge[d].

6 Mossbauer spectrometer.

ms   Millisecond = 10^{-3}s.

m/s   Metres per second.

MSA   *1* Mission system(s) avionics.

2 Monolithic stretched acrylic.

3 Minimum safe, or separation, or sector, altitude.

4 Maritime Safety Agency(J).

MSAD   Multisatellite attitude determination.

MSAFQ   Minimum speed for acceptable flying qualities.

MSAM   Mobile (or, UK, medium) surface-to-air missile.

MSAR   Miniature synthetic-aperture radar.

MSAS   MTSAT satellite-based augmentation system.

MSAT   Multi-sensor aided targeting [-Air adds airborne].

MSAW   Minimum safe altitude warning.

MSB   *1* Most significant bit (EDP [1]).

2 Mandatory service bulletin.

MSBS   Mer/sol balistique stratégique; submarine-based strategic missile (F).

MSC   *1* Manned Spacecraft Center (NASA, now Johnson Space Center, Houston, TX).

2 Major subordinate command (NATO).

3 Manpower Services Commission (UK).

4 Military Sealift Command (US).

5 Multi-scan correlation.

6 Master server computer.

MSCADC   Miniature standard CADC.

MSCP   Mean spherical candlepower (non-SI).

MSD   *1* Mass storage device.

2 Minimum separation distance.

3 Multisensor display.

4 Multimode silent digital (radar).

5 Most significant digit.

6 Multiple-site damage [airframe].

7 Mid-span damper.

8 Map storage display.

MSDF   Manual spinning direcreton finding.

MSDPS   Military survey digital production system.

MSDS   Mode-selector damping service.

MSE   *1* Minimum single-engine speed.

   2 Manned spacecraft engineer.

   3 Mean square error (signal processing).

msec   Millisecond.

MSER   Multiple stores ejector rack.

MSET   Multi-sensor exploitations testbed.

MSF   *1* Militarily significant fallout.

   2 Multi-sensor fusion.

   3 Mission support facility.

   4 Mobile strike force.

MSFC   George C Marshall Space Flight Center (NASA, Huntsville, Alabama).

MSFN   Manned Space Flight Network (NASA).

MSFP   Mosaic-staring focal plane (early-warning sensors).

MSFSG   Manned spaceflight support group (USAF).

MSFU   Merchant ship fighter unit (WW2).

MSG   *1* Maintenance steering group (inter-airline).

   2 Message (ICAO).

   3 Microgravity science glovebox, for space experiments.

MSH   *1* Metastable helium.

   2 Medium support helicopter; ATF adds aircrew training facility.

MSHE   Multispectral hostile environment.

MSI   *1* Medium-scale integration.

   2 Moon's sphere of influence (NASA).

   3 Mission-systems integration.

   4 Multi-source integration or integrator.

MSIAC   Modeling and Simulation Information Analysis Center (DoD).

MSIC   Multi-ship integrated control.

MSIP   *1* Multinational [or multi-] staged improvement program.

   2 Multispectral imagery processor.

MSIS   Multisensor stabilized integrated system.

MSK   Modulation, or minimum, shift keying.

MSL   *1* Mean sea level.

   2 Mapping Sciences Laboratory (NASA).

msl   Missile.

MSLC   Maintenance stock level case.

MSLV   Microsatellite launch vehicle.

MSMA   Macro sensor management application.

MSMM   Multisensor, multimission.

MSMS, MS²   Maritime-surveillance mission system.

MSn   Manufacturer's serial number.

MSO   *1* Manager, shop operations.

   2 Molecular-sieve oxygen (S adds system).

MSOGS   Molecular-sieve oxygen generation system.

MSOSA   Modelling and simulation operational support activity.

MSOV   Modular standoff vehicle.

MSOW   Modular standoff weapon.

MSP   *1* Mach sweep programmer.

   2 Maintenance service plan.

   3 Magnetic speed probe.

   4 Mosaic sensor program(me).

   5 Maximum structural payload.

   6 Military spaceplane, also military spaceplane program (USAF).

   7 Multisensor processor.

   8 Mode-S protocol.

   9 Magnetic south pole.

   10 Mission-system processor.

MSPFE   Multi-sensor programmable feature extraction.

MSPS   Modular self-protection system.

MSPT   Marine silent power-transmission system (ASW).

MSR   *1* Multi-sensor reconnaissance.

   2 Missile-site radar.

   3 Modular strain recorder.

   4 Missile simulated round.

   5 Main supply route.

   6 Modular survivable radar.

   7 Maximum-speed range.

   8 Misrouted.

   9 Mars sample return.

MSRF   Microwave Space Research Facility (USN).

MSRS   Miniature-sonobuoy receiver system.

MSS   *1* Multi-spectral scanner, or satellite.

   2 Missile-sight subsystem.

   3 Mobile satellite service, or system.

   4 Model support strut.

   5 Mission-support system[s].

   6 Multi-sensor system.

   7 Maritime surveillance system.

   8 Minimum sinking speed.

MSSC   Maritime surface-surveillance capability.

MSSL   Mullard Space Science Laboratory (UK).

MSSMP   Multipurpose security and surveillance monitoring platform.

MSSP   *1* Mobile satellite service provider.

   2 Mode-S specific protocol.

MSSR   *1* Monopulse SSR.

   2 Mars surface-sample return.

   3 Multi-spectral scanning radiometer.

MSSS   *1* Man/seat separation system.

   2 Mode-S specific services.

MST   *1* Multisensor tracking, or turret.

   2 Moving surface target; E adds engagement.

   3 Mountain Standard Time.

   4 Microprocessor simulation technology.

   5 Missile surveillance technology.

   6 Microsystems technology.

   7 Mobile service tower.

   8 Mechanical-systems trainer.

MST&E   Multi-service test and evaluation.

M-Star, MSTAR   *1* Moving and stationary target acquisition and recognition.

   2 Man-portable surveillance and target-acquisition radar.

   3 MLRS smart tactical rocket.

MStart, M-Start   Missile system to attack relocatable targets.

MSTCS   Multiservice target control system.

MSTI   Miniature sensor technology integration.

MSTR   Moisture.

MSTRS   Miniature-satellite threat-reporting system.

MSTS   Multiservice tactical system.

MSTSAP   Medium-speed tactical-support aircraft program.

MSU   *1* Mobile satcom unit.

   2 Mode-selection unit.

   3 Master switching unit.

4 Mass storage unit.

5 Maintenance station unit.

MSV   Mobile [explosion-] suppression vessel.

mSV   Milli-solar volts.

MSW   *1* Maximum STOL weight.

2 Magnetic surface wave.

MSWS   Multisensor warning system.

MSX   Mid-course space experiment.

MSZ   Magnesium-stabilized zirconia.

MT   *1* Megaton[s], usually means 10^6 short tons.

2 Metric tonne[s], confusing.

3 Motor [surface] transport.

4 Mean, or minimum, time.

5 Message, or mobile, terminal.

6 Multi-frequency transducer.

7 Maximum-top (weather).

8 Mountain (ICAO).

mT   Maritime tropical air mass.

mt   Tonne[s].

MTA   *1* Military training area.

2 Multithread architecture.

3 Maintenance-task analysis.

MTACS, Mtacs   Marine tactical air control system.

MTAD   Multi-trace analysis display (EW).

MTADS   Modernized target-acquisition designation sight.

MTAE   Multiple-time-around echoes (radar and EW).

MTAP   Mine/torpedo aviation regiment (USSR, WW2).

MTAS   Millimetric, or modular, target-acquisition system.

MTAT   Mean turn-around time.

MTBAA   Mean time between avionics anomalies.

MTBCF   Mean time between critical failures.

MTBCM   Mean time between corrective maintenance.

MTBE   Methyl tertiary butyl-ether.

MTBF   Mean time between failures.

MTBF$_c$ Mean time between component failure (also written MTBCF).

MTBFRO, MTBFro   Mean time between failures requiring overhaul.

MTBI   Mean time between incidents.

MTBIE   Mean time between interruption, or instability, events.

MTBMA   *1* Mean time between maintenance action[s].

2 Mean time between mission aborts.

MTBO   Mean time between overhauls or outages.

MTBR   Mean time between repair, or removals, or replacement.

MTBSF   Mean time between significant failure.

MTBUER   Mean time between unscheduled engine removal.

MTBUM   Mean time between unscheduled maintenance.

MTBUR   Mean time between unscheduled, or unit, removals, or replacement, or repair.

MTC   *1* Mach-trim compensator, or computer.

2 Mission Training Center (DMT).

3 Maintenance terminal cabinet.

4 Military-Technical Co-operation (R).

5 Measured-term contract (UK, MoD).

MTCA   *1* Ministry of Transport and Civil Aviation (UK, Oct. 1951, became MCA).

2 Minimum terrain-clearance altitude.

3 Military terminal control area [MTMA more common].

MTCD   Medium-term conflict detection.

MTCR   Missile Technology Control Regime.

MTD   *1* Mounted.

2 Maintenace terminal display.

3 Maintenance training device.

4 Moving-target detection, or detector (primary radar).

MTDA   Mean time to dispatch alert.

MTDAS   Mobile telemetry data-acquisition system.

MTDS   *1* Marine tactical data system (USMC).

2 Mission training through distributed systems, or simulation.

MTE   *1* Megatons equivalent, unit of area destruction.

2 Modular threat emitter.

3 Moving-target exploitation.

4 Magnetic turning error.

MTEL   Methyl-triethyl lead.

MTF   *1* Modulation transfer function.

2 Mississippi Test Facility (NASA, now SSC 4).

3 Mid-tandem fan.

4 Maintenance terminal function.

MTFF   Man-tended free flyer.

MTG   Meeting.

MTGW   Maximum taxi gross weight.

MTHEL   Mobile tactical high-energy laser.

M3R   Multimission modular radar.

MTI   *1* Moving target indication, or indicator.

2 Multiple-target interception.

3 Multispectral thermal imager.

4 Marked temperature inversion.

MTIRA   Machine Tool Industry Research Association (UK).

MTIS   Modular thermal-imaging sight.

mTk   Maritime tropical air, colder than surface.

MTL   *1* Mean (rarely maximum) transmitter level.

2 Minimum triggering level [transponder].

3 Magnetic-tape loader.

MTLA   Minimum takeoff and landing area.

MTLM   Major throttle-lever movement.

MTM   *1* Mission and traffic model; co-ordinates Space Shuttle payloads and customer requirements.

2 Maximum takeoff mass.

3 Ministry of transport machine construction (USSR).

4 Million ton-miles (short tons).

5 Module test and maintenance [bus interface].

MTMA   Military terminal manoeuvring area.

MTMC   Military Traffic Management Command.

MTM/D   Million ton-miles per day (short tons).

MTMIU   Module test and maintenance bus interface unit.

MTMT   Multiple target and missile tracker.

MTN   Mountain[s].

MTO   See *MTOW*.

MTOD   Mean (or maximum) time on deck.

MTOE   *1* Mid-term operations estimate.

2 Modified table of organization and equipment.

MTOW   Maximum takeoff weight [MTOGW adds gross], a certificated value exceeded only during certification flight testing.

MTP   *1* Mandatory technical publications.

2 Maintenance test panel.

MTPA   Mobile transponder performance analyser.

MTPS   Maintenance Test Pilots' School (formerly RNAS Brawdy).
MTR   *1* Missile tracking radar.
   2 Marked-target receiver.
   3 Military training route.
   4 Main and tail rotor (helo configuration).
MTRA   Megafloat Technological Research Association.
MTRE   Missile test and readiness equipment.
MTRIV   Mission-tape recorder interface unit.
MTS   *1* Marked-target seeker.
   2 Mobile training set.
   3 Mobile test set.
   4 Motion/time survey.
   5 Maintenance training simulator.
   6 Multispectral, or multisensor, targeting system(s).
   7 Mobile telephone service.
   8 Mountains.
   9 Manned tactical simulator, or simulation.
MTSAT   Multifunctional transport satellite.
MTT   *1* Multiple-target tracking.
   2 Maintenance-training tutorial.
   3 Maximum turbine temperature.
MTTA   Machine Tool Technologies Association (UK).
MTTDA   Mean time to dispatch alert.
MTTF   Mean time to (next) failure [differs from MTBF in that no credit is given for items that have not failed].
MTTFSF   Mean time to first system failure.
MTTM   Mean time to maintenance [A adds alert].
MTTR   *1* Mean time to repair.
   2 Multiple-target tracking radar.
MTTRS   Mean time to restore service.
MTTS   Multi-task training system, versatile simulators which can be configured for particular aircraft types.
MTTUR   *MTUR*.
MTU   *1* Use metric units (ICAO).
   2 Mobile training unit.
   3 Magnetic-tape unit.
MTUR   Mean time to unscheduled removal, or replacement [differs from MTBUR in that no allowance is made for items not removed].
MTVC   *1* Motor thrust-vector control.
   2 Manual thrust-vector control.
MTW   *1* Major theatre [of] war.
   2 Mountain wave[s].
mTw   Maritime tropical air, warmer than surface.
MTWA   Maximum takeoff [or total] weight authorized, = *MTOW*.
M-type marker   See *M-marker*.
MTZT   Multiple time-zone travel.
MU   *1* Maintenance unit.
   2 Management unit [usually Acars].
   3 See μ [Greek letters].
MUAV   Micro, or maritime, UAV.
Mucels   Multiple communications emitter location system.
mud   Pilot of *-mover.
Mudas   Modular universal data-acquisition system.
mud-mover   Low-level close-support or attack aircraft (colloq.).
MUE   Modernized user equipment (GPS).
MUF, muf   Maximum usable frequency.
muff   Exhaust heat exchanger, usually for cabin heating.
muffler   Silencer (US).

Mufids   Multi-user flight-information display system.
μFORS   Mu (micro) fibre-optic rate sensor.
mufti   Civilian dress worn by Serviceman off duty (UK usage, Hindustani origin).
MUGS, Mugs   Multiple universal gunner system.
mule   *1* Refuelling bowser (US term).
   2 Hydraulic test rig (colloq.).
   3 Modular universal laser equipment.
Mullite   Commercially produced ablation material.
mult   Multiple, or multiplier.
Multack   Multiple target attack program (software).
multibank engine   Piston engine with several linear banks.
multi-body freighter   Hypothetical aircraft with payload in detachable bodies all carried on one wing.
multi-bogey   Air-combat situation in which there are many enemy aircraft.
multi-burn   See *multi-impulse*.
multi-cell   ACM engagement in which two or more DA cells participate.
multicellular foam   Material of low density with air or gas-filled closed cells such as expanded polystyrene; term not used for honeycombs.
multichannel receiver   Usual meaning is ability simultaneously to track several GPS signals.
multichannel selector   Manual controller for preselected communications channels.
Multicom   Multiple communications.
multicombiner   Optical system for projecting several sets of displays all focused at infinity on HUD screen.
multicommunications service   Mobile private communications service on 122.9 MHz for such activities as ag-aviation and forest firefighting.
multicolour system   Guidance, tracking or other target-oriented system which operates alternately or continuously on several EM wavelengths, not necessarily in visible range.
multicoupler   Device for making single aerial (antenna) serve several receivers.
multi-energy spectrum   Gives output in contrasting colours depending on average atomic number (baggage screening).
multi-foil wing   Wing of extremely thick (up to t/c 30%) supercritical profile comprising a main fixed portion and upper/lower rear hinged foils (up to 35% chord), with or without blowing between the foils.
multifunctional control surface   Flight-control system in which lateral and direct-lift control is effected by spoilers and full-span Fowler flaps and variable-camber tabs, one advantage being fast letdown and steeper approach.
multifunction display   EFIS display offering selectable weather radar, navigation maps, checklists and other information other than that on PFD.
multihead towbar   Fitted with head[s] for attachment to many types of NLG.
multi-impulse   Capable of repeated cutoff and restart to meet propulsion demands of mission (rocket).
multi-lane airway   For reasons of traffic, control, noise, disturbance or safety, airway divided into two (rarely three) parallel lanes, denoted by North/South or East/West, with Centre where three; lateral separation up to 20 km.
multilateral   Agreement on air-carrier rights between regional groups of sovereign states.

multilayer board   Any printed-circuit, hybrid or related electronic assembly fabricated as stack of subcircuits in single board.

multi-media classroom   Equipped with audio-visual aids, overhead projection, tape/slide systems, blackboard and student dialogue buttons.

multimission   Able to perform different military roles, esp. air-combat/interception and ground attack/interdiction.

multi-mode radar   Designed to function in several operating modes with quick switching between each, eg (for MMR for attack aircraft) ground map, ground map spoiled, etc, each mode requiring a different waveform.

multi-mode receiver   Usually means compatible with VOR, ILS, MLS, DGPS.

multipath effect   Anomalies in radar target range and position, and many other optical/radar situations, caused by receipt of reflected radiation from land or sea surface and other reflectors as well as direct signal from target.

multiplane   *1* Aeroplane with several sets of main wings, normally superimposed.

    2 Adjective: tail unit or other assembly with several superimposed horizontal surfaces.

multiple-access receiver   Lasercom to make uplink/downlink by sensing beacons to refine beam aiming.

multiple courses   Narrow false courses heard on radio range, esp. among mountains.

multiple echo[es]   Reception of more than one transmission of a single signal because of refractions and reflections.

multiple ejection rack, MER   Interface allowing several stores to be carried on single external-stores station, with positive ejector rams to ensure clean separation.

multiple independently targeted RV   Delivery system containing several missile re-entry vehicles each having own guidance system to separate targets.

multiple independent RV   Broader concept than above; system delivers several warheads which may free-fall or be independently targeted.

multiple kill capability   Usual meaning is that platform has several types of armament, eg gun plus two types of missile.

multiple options   Force employment alternatives depending on flexibility in tactical/strategic situations, retargeting and availability of conventional or nuclear weapons.

multiple RV   Re-entry vehicle containing number of separate warheads which scatter slightly but fall on same area target.

multiple-time-around echoes   Those from targets at exact multiples of the radar unambiguous range.

multiplexing   *1* Act of combining signals from many sources into common channel, requiring frequency, phase and time-division.

    2 In electrical wiring of whole vehicle, providing redundant pathways or alternative routes with auto switching triggered by battle damage or other discontinuity.

multiplexer, multiplexor   Device, often stored-program computer, which handles I/O (input/output) functions of on-line EDP (1) system with multiple communications channels.

multi-ply   Material, eg wood, fabric, metal or composite, assembled from several laminar layers.

multiplying valve, multiplier   Device whose output is exact product of several inputs.

multiprocessor   Multiprogrammed processor.

multiprogramming   Technique allowing single computer to run several programs simultaneously.

multi-row engine   Piston engine having more than two banks or radial rows of cylinders.

multi-sensor   Using more than one type of signal to gather information; eg optical camera, radar, IR linescan, passive elint etc.

multiservice connector   Personal coupling for suit pressure, electric heating, oxygen, radio and possibly other services.

multisite fracture   Main break was preceded by numerous micro-cracks.

multispectral   Capable of responding usefully to wide range of EM wavelengths including all visible hues, IR and possibly UV.

multi-stage rocket   Vehicle with several stages each fired and staged in succession.

multi-step   See *multi-stage rocket*.

multithread architecture   Simple parallel programming.

multitube launcher   External store filled with parallel tubes for air-launched rockets.

multivibrator   Oscillator having two cross-coupled valves or transistors operating alternately, each input coming from other stage's output; can be bistable (flip/flop), monostable or astable.

MULTS, Mults   Mobile universal Link-11 translator system.

Mumacs   Multiple unmanned aircraft control system.

mu-metal   High-permeability alloy used to screen equipment (eg CRT) from stray magnetic influences.

mu-meter   Rolling truck with pair of toed-out tyres, measures side force and hence coefficient of friction. From μ (mu), coefficient of sliding friction.

Mumps   Multi-user MEMS process.

MUN, MUNI, MUNL   Municipal.

Munk factor   Formula for performance of biplanes based on ratios of spans, lifts, gap and interference.

MUOS   Mobile- or, confusingly, multiple-, user objective system (DoD).

Mupsow   Multi-purpose stand-off weapon.

Murlin   Multiband research laser IR.

Muroc   Dry California lake, site of USAF flight-test centre of same name (now Edwards AFB).

MUS   Minimum-use specification.

MUSA, Musa   *1* Multiple-element steerable array.

    2 Frag submunition for semi-hard targets, dispensed by MW-1 system (G).

muscle pressure   Pneumatic pressure for applying force on a shaft bearing.

MUSE, Muse   *1* Multi-user system [and] environment [sometimes Musent], includes LBA, L-DCS, BVIS and other functions.

    2 Monitor of UV solar energy.

mush   To increase angle of attack suddenly but without immediate corresponding vertical acceleration, resulting from momentum along original path.

mush-head   See next.

mushroom rivet   One having thin convex head with sharp edge, for thin sheet.

Music   *1* Multiple signal classification.

2 Multi-spectral IRCM.

music   EW emissions, esp. jamming (colloq.).

Musical   Prefix to marking techniques denoting Oboe guidance (RAF 1943–45).

MUSPA   Area-denial mine, dispensed by MW-1 system (G).

Must   Multimission u.h.f. satcom terminal.

Mustrs, MUSTRS   Multisensor target-recognition system.

Muta, MUTA   Military upper [-level] traffic control area.

Mutes   Multiple-threat emitter system; simulates many hostile emissions simultaneously.

mute switch   Disconnects headset, microphone (term also often used for PA disconnect).

mutual inductance   The e.m.f. in a circuit caused by rapid change in surrounding magnetic field, unit = henry, symbol M, also = Wb per ampere.

mutual interception   By two friendly interceptors on each other.

mux   Multiplex[er].

muzzle brake   Any of variety of gun-muzzle gas deflector units to reduce recoil, or blast on surrounding structure.

muzzle cap   Frangible closure on gun muzzle to reduce drag and ingestion of precipitation.

muzzle energy   Kinetic energy of each projectile as it leaves gun, measured relative to gun.

muzzle horsepower   Standard non-SI measure of gun power, esp. automatic weapons; muzzle energy multiplied by rate of fire (units being compatible).

MV   *1* Muzzle velocity.

2 Miniature vehicle.

3 Mass value.

mV   Millivolt[s].

MVA   *1* Minimum vectoring altitude.

2 Multivariate analysis.

3 Megavolt-amperes, basic unit of large AC powers.

MVAR   Magnetic variation.

MVDF   MF and VHF D/F facilities co-located (ICAO).

MVEE   Military Vehicles Experimental Establishment (UK).

MVFR   Marginal VFR.

MVG   Moving.

MVIS   Microgravity vibration-isolating system.

MVL, mvl   Mid-value logic.

MVM   Muzzle-velocity measure(ment).

MVME   Modular virtual memory environment.

MVMT   Movement.

MVPS   Multiple vertical protective shelter (ICBM).

MVS   *1* Minimum vector speed (light pen).

2 Multi-vendor system (distributed networks).

MVSRF   Man/vehicle systems research facility (NASA).

MVT   Multi-view tomography.

MVTU   Moscow higher technical school.

MVW   *1* Maximum VTOL weight.

2 Maintenance virtual workplace.

MW   *1* Medium wave.

2 Methanol/water.

3 Megawatt[s].

4 Microwave.

5 Mine warfare.

M_W Bending moment at wing root.

mW   Milliwatt[s].

MWA   *1* Multiple weapons adaptor.

2 Momentum-wheel assembly.

MWARA   Major world air-route area.

MWCS   Multiple weapons carrier system.

MWD   Military working dog.

MWDP   *1* Mutual weapons development program.

2 Master warning display panel.

MWE   *1* Manufactured (or manufacturer's) weight empty.

2 Maximum weight empty.

MW(E)   Megawatts electrical.

MW50   Methanol 50%, water 49.5%, inhibitor 0.5%.

MWIR   Mid-, or medium-, wave IR.

MWL   *1* Minimum takeoff distance on water to clear standard (35 or 50 ft) obstacle.

2 Maintain wings level.

MWLD   Man-worn laser detector.

MWO   *1* Meteorological Watch Office (ICAO).

2 Modification work order.

MWP   *1* Museum of Women Pilots (US).

2 Meteorologist weather processor (FAA).

MWR   *1* Millimetre-wave radar.

2 Microwave radiometer.

MWS   *1* Master, or missile, warning system.

2 Missile Warning Squadron (USAF).

3 Multiple-warhead system[s].

4 Modular workstation.

MW(T)   Megawatts (thermal).

MWVS   Mission weapon visionics system.

MX   Mixed types of icing, white and clear (ICAO).

M_x Bending moment at any station x.

MXD   Mixed (ICAO).

MY   Multi-year [C adds contract(ing)].

Mylar   Tough transparent film, of terephthalate polyester family, usable down to extreme thinness (hence light weight per unit area).

Mynapak   Integrated circuit with up to three ceramic substrates carrying devices connected by gold wires, the whole being N_2-filled (BAe).

Myopia   Short sight, near-parallel rays being focused in front of the retina.

MYP   Multi-year procurement, or programme.

myriametre   Non-SI term for 10^4 metres, hence myriametric.

MZFW   Maximum zero-fuel weight.

N

N *1* Newton[s].

2 Shaft rotation speed, with suffix number (N$_1$, N$_2$ etc) for each shaft (see *gas-turbine numerology*).

3 Yawing moment.

4 Code: operates by sound in air (JETDS).

5 Code: navaid (ICAO).

6 Prefix, amount of cloud.

7 Number of turns in a coil, or of load cycles, usually per second.

8 North (ICAO).

9 Synoptic chart code: air mass has had characteristics changed.

10 Telecom code: aircraft.

11 Avogadro's number.

12 Nitrogen.

13 STOL aircraft (FAI).

14 Prefix, night, negative, nose, no.

15 Prefix, nuclear (USSR).

16 Knots [flight plan].

17 Nicrosil-nisil thermocouple.

18 IFR weather.

19 Non-scheduled civil transport flight.

20 Permanent special test (USAF/USN aircraft designation prefix).

21 Trainer (USN 1922–60).

22 Noise.

23 Modified mission suffix, night or all-weather (USN 1950–62).

n *1* Prefix, nano (10^{-9}).

2 Generalized symbol for an aircraft equipped with Tacan only and 64-code transponder.

3 Normal acceleration in g, load factor.

4 Number in a sample, any integer.

5 Frequency, esp. of rotation.

6 Refractive index.

7 Negative, hence n-type semiconductor.

$\dot{n}$ Angular acceleration.

N$_1$ Fan or LP compressor speed.

N$_2$ *1* Nitrogen.

2 IP compressor speed (CAA).

3 HP compressor speed (2-shaft engine).

N$_3$ HP compressor speed (CAA).

$\dot{N}'$, N° Rate of change of N(2) on spool-up or rundown.

$\bar{N}$ Vector representing integrated noise energy.

N-code ICAO code, amount of cloud.

N-display Target forms two blips on horizontal time-base (as in K-display), lateral positions giving range and relative amplitudes bearing.

N-layer N is set for any layer name [link, net, etc] or for the initial [open system architecture].

N-sector Sector of radio range in which Morse N is heard.

N-strut Arrangement of interplane struts or cabane struts resembling N.

n-type semiconductor One in which charge carriers are nearly all electrons.

N-wave Shockwave remote from source; far-field boom signature, when profile of pressure/linear distance has settled down to N-like profile.

NA *1* Noise abatement (procedure).

2 Numerical aperture.

N/A *1* Not applicable, not available, not authorized, not approved.

2 Navigation/attack.

N$_A$ Avogadro constant.

Na Sodium.

NAA *1* National Aeronautic Association of the US, founded 1905 as Aero Club of America (office Arlington VA).

2 National Aviation Academy (US).

3 National Aviation Authorities (European countries).

4 National Airport Authority (India).

NAAA National Agricultural Aviation Association (US), or National Aerial Applicators Association (US).

NAAAS National Association of Air Ambulance Services (UK).

NAAC *1* National Aviation Associations Coalition (US)

2 National Association of Agricultural Contractors (UK, promotes ag-aviation).

3 See *sodium acetate*.

NAADC North American Aerospace Defense Command.

NAAFI Pronounced "naffy", Navy, Army and Air Force Institutes [civilian organization in support of Other Ranks] (UK).

NAAP Netherlands Agency for Aerospace Promotion (NIVR).

NAAQS National ambient-air quality standard[s].

NAAS *1* National Association of Aerospace Sub-contractors (US).

2 Naval Auxiliary Air Station (USN).

NAATS National Association of Air Traffic Specialists (US).

NAAWS NATO anti-war warfare system.

NAB Navy Air Base (USN).

Nabs NATO air base Satcom.

NAC *1* North Atlantic Council, of NATO ambassadors.

2 Noise Advisory Council (UK).

3 National Air Communications (UK, 1939–40).

4 National Aviation Club (US).

5 Naval Air Command (UK).

6 Naval Avionics Center (USN).

7 Non-airline carrier.

8 Network access controller.

9 NASA Advisory Council.

NACA *1* National Advisory Committee for Aeronautics (US, 1915–58, became NASA).

2 National Air Carrier Association (US, from 1962).

NACA cowling Drag-reducing annular cowl for radial engines with aerofoil-profile leading edge and cylindrical main section.

NACA section Any of numerous aerofoil sections designed by NACA.

NACA standard atmosphere Original idealized atmosphere, published in 1925; later superseded by ICAO, ARDC and others.

NACEL   Naval air crew equipment laboratory.

nacelle   Streamlined body sized according to what it contains, which may be an engine, landing gear, human occupants etc; when tail carried on separate booms * takes place of fuselage.

NACES, Naces   Navy aircrew common ejection seat (USN).

Nacisa, NACISA   NATO Communications and Informations Systems Agency.

Nacma   NATO Air Command and control systems Management Agency.

Nacoss   National Approval Council for Security Systems (UK).

NACP   Noise-abatement climb procedure.

nacreous cloud   High layer cloud with iridescent appearance; also called mother-of-pearl cloud.

Nacsi   NATO Advisory Committee on Signals Intelligence.

NAD   *1* National Armament Director[s] (NATO nations).

2 Navy area defense (SAM, USN).

NADB   Netherlands Aircraft Development Board.

NADC   *1* US Naval Air Development Center.

2 Nuclear Affairs Defence Council (NATO).

3 NATO Air-Defence Committee.

Nadcap   National aerospace and defense contractors accreditation program (US).

NADEEC   NATO Air Defence Electronic Environment Committee (NATO).

NAD83   North American datum of 1983, precise geographic co-ordinates.

Nadep   Naval Aviation [or Air] Depot (USN).

Nadge   NATO air-defence ground environment; multinational programme for unified air-defence system of radars, computers, displays and communications from North Cape to eastern Turkey.

Nadgeco   Multinational company formed to implement Nadge plan.

Nadgemo   Nadge Management Office; formed within NATO to act as unified customer.

Nadin   *1* National airspace data-interchange network (FAA).

2 North American data-interchange network.

Nadir   Point on celestial sphere vertically below observer, ie 180° from zenith.

NADL   US Navy Avionics Development Laboratory.

NADS   *1* US Naval Air Development Station (from 1947).

2 NATO Armaments Directors.

3 Next available delivery slot.

NAE   *1* National Aeronautical Establishment (Canada).

2 National Academy of Engineering (US).

3 US Naval Air Engineering, concerned with aerospace installations in ships [F adds Facility 1956–62, L adds Laboratory from 1962].

NAEC   *1* US Naval Air Engineering Center (Philadelphia), NAMC renamed 1962.

2 National Aerospace Education Council (US).

Naegis, NAEGIS   NATO airborne early warning ground environment integration segment.

NAEL(SI)   Naval Air Engineering Laboratory (Ship Installations), from 1962.

NAES   *1* Naval Aviation Engineering Services (US); plus U = unit.

2 US Naval Air Experimental Station (1943–57).

NAEW   NATO airborne early warning; F adds force.

NAF   *1* Norsk Astronautisk Førening (Norway).

2 Non-appropriated fund(s).

3 Naval air facility (US).

Nafag   NATO Air Force Armaments Group.

Nafdu   Naval Air Fighting Development Unit.

NAFEC, Nafec   National Aviation Facilities Experimental Center; Atlantic City, NJ (US 1958, in 1981 became FAATC).

NAFI   National Association of Flight Instructors (US).

NAFIN, Nafin   Netherlands armed-forces integrated network.

NAFP   National aeronautical facilities program(me).

Nagara   Japanese word in Western use: a critical-path technique linking manufacturing operations, and balancing times to reduce lead times and WIP inventory.

NAGr   Short-range reconnaissance wing (G, WW2).

NAGS, Nags   NATO alliance ground surveillance.

NAGTE   Non-aircraft gas-turbine engine.

Nahema   NATO Helicopter Management Agency (Aix-en-Provence).

NAI   *1* Negro Airmen International (US).

2 Netherlands Aerospace Industries (member AECMA, title in English).

3 National Aerospace Initiative (US).

NAIA   National Aerospace Intelligence Agency (USAF).

Nails   National airspace integrated logistics support (FAA).

NAIU   Naval Accident Investigation Unit.

NAK   *1* National aero club (R).

2 Negative acknowledgement.

NaK   Generalised term for all mixtures of sodium and potassium; used eg as liquid-metal heat-transfer media.

NAKA   National aerospace agency (Kazakhstan).

naked   See *clean (1, 2)*.

Nallads   Norweigian Army low-level air-defence system.

NAM   *1* National Association of Manufacturers (US).

2 National Atomic Museum (Albuquerque, NM).

3 NATO Air Meet (annual).

4 North American region (ICAO).

nam, n.a.m.   Nautical air miles; hence */lb of fuel.

NAMAS   National Measurement and Accreditation Service (UK).

NAMC   US Naval Air Material Center (1943, in 1962 became NAEC).

Nameadsma   NATO medium extended air-defence system management agency.

NAMFI, Namfi   NATO Missile Firing Installation (Crete).

NAMI   Scientific auto-motor institute (Sweden).

Namis   NATO automated meteorological information system.

NAML   Naval Aircraft Materials Laboratory (UK).

NAMMA   NATO MRCA Management Agency.

NAMRL   Naval Aerospace Medical Research Laboratory (USN).

NAMS   NATO Maintenance and Supply [A adds Agency, O Organization].

NAMTD   Naval Air Maintenance Training Detachment (USN).

NAMU   US Naval Aircraft Modification Unit, (1943–47).

NAND NOT + AND logic device, retains output until voltage at all inputs, then goes to 0.

nano Prefix, $\times 10^{-9}$, hence many technologies eg, * electronics, * engineering, * physics, * structures.

nanosatellite Mass $1 \leqslant 10$ kg.

NAO National Audit Office (US).

NAOAL National Aviation Officer for Airworthiness and Logistics (US Fire Service).

NAOC National Airborne Operations Center (E-4B platform for NCA).

NAOS North Atlantic Ocean station.

NAP *1* Noise-abatement procedure.
  2 Normal acceleration point (SST).
  3 National airport plan (US).

nap *1* Local profile of land surface; hence * of Earth, * of the Earth flying (as close to ground as possible).
  2 Short fibre ends along edge of fabric.

napalm *1* Mixture of naphtha and palm oil (hence name), usually with additives, used as incendiary material.
  2 Air-dropped ordnance filled with * designed to burst and distribute flame over large area.

NAPC Naval Air Projects Co-ordination office (USN).

NAPCA, Napca National Air-Pollution Control Association (US, now APCC).

nape Use napalm against (colloq.).

Napgel Mix of ethylene/propylene glycols [de-icing fluid tradename].

naphtha Generalized name for inflammable oils distilled from coal tar and other sources.

NAPL National Air Photo Library (Canada).

NAPMA NATO AEW&C Programme Management Agency.

Napnoc No acceptable price, no contract.

NaPO NASA Pasadena Office.

Napol North Atlantic Policy [ECAC working group].

NAPP National Association of Priest Pilots (US).

NAPR NATO armaments planning review.

NAPTC Naval Air Propulsion Test Center (USN).

NAR National airspace review (US).

NARC Nexcom Aviation Rulemaking Committee (FAA).

NARF *1* Naval Air Reserve Force (USN).
  2 Naval Air Rework Facility; Pensacola (USN).

NARG Navaids and area-navigation working group (ICAO).

NARIM NAS (2) research and investment model.

NARO Naval Aircraft Repair Organization (Gosport, UK, now part of DARA).

narrow-body Commercial transport with fuselage width of approximately 10 ft (3 m) with single aisle between passenger seats.

narrow gate AAM operating mode permitting homing on target only within narrow limits of rate of closure.

NARS Navigaton and attitude reference system.

NARTEL National air radio telecommunications (UK).

NARUC National Association of Regulatory Utility Commissioners (US).

NAS *1* Naval Air Station (USN).
  2 National airspace system (FAA).
  3 National Academy of Sciences (US).
  4 Nozzle actuation system.
  5 National aerospace standards (FAA, CAA, UK).
  6 Numeric aerodynamic simulation.

7 Nav/attack system.

NASA National Aeronautics and Space Administration (US, from 1 October 1958 successor to NACA).

Nasad National Association of Sport Aircraft Designers (US).

NASAF North African Strategic Air Force (US/UK, WW2).

NASAO National Association of State Aviation Officials (US).

NASC *1* Naval Air Systems Command (USN).
  2 National Aeronautics and Space Council (US).
  3 National Association of Spotters' Clubs (UK, 1941–46).
  4 National Aerospace Standards Committee.
  5 National AIS (1) system centre (UK).

Nascom NASA Communications Network.

NASDA National Space Development Agency (J).

NASF Navigation and Attack Systems Flight (RAF).

NASH Nav/attack system for helicopters.

NASIP, Nasip National aviation safety inspection program (FAA).

NASM National Air and Space Museum (part of the Smithsonian Institution, Washington DC).

NASMDEF NATO anti-ship missile defence evaluation facility.

NASN National Air-Sampling Network (EPA).

NASO *1* Naval Aviation Supply Office (USN).
  2 Non-acoustic systems operator.

NASP *1* National aerospace plane (US).
  2 National airport plan (US).
  3 Navy airship program (USN).

Naspac NAS (2) performance-analysis capability.

Naspals NAS (2) precision approach and landing system (FAA).

NASPG North Atlantic Systems Planning Group.

Naspo National Airspace System Planning Office (FAA, from 1966).

NASR Naval and Air Staff Requirement.

Nasroc New anti-submarine rocket (J).

NASS *1* Naval Anti-Submarine School.
  2 Non-acoustic sensing system.

NASSA National AeroSpace Services Association (US).

Nassi NAS (2) status information, developed under FFP1 to provide such data as ARC, RVR and delays.

Nassim NAS (2) simulation.

Nasstat NAS (2) facility status database and display.

NAST Naval Air Staff Target.

Nastar Navier-Stokes analysis for arbitrary regime.

Nastran NASA structural analysis (GP finite-element program).

NASU Naval Air Support Unit (formerly RNAS Brawdy).

NASWDU Naval Air/Sea Warfare Development Unit (UK).

NAT *1* North Atlantic Region (ICAO).
  2 North Atlantic tracks.
  3 Netherlands Airport Technology (trade association, 32 members).

NATA *1* National Air Taxi Association (UK).
  2 National Air Transportation Association, Inc. (US, established 1940).
  3 National Aviation Trades Association (US).

Natar   NATO transatlantic advanced radar (Belgium, Canada, Denmark, Luxembourg, Norway, US).

NATC   *1* Naval Air Test Center (Patuxent River, Md).
2 Naval Air Training Command (US).
3 National Air Transportation Conferences (US).

Natca   National Air Traffic Controllers Association (US).

Natcapit   North Atlantic Capacity And Inclusive Tours Panel.

NATCC   National Air Transport Co-ordinating Committee (US).

NATCS   National Air Traffic Control Services (UK, now NATS).

NATF   US Naval Air Test Facility.

Natinad   NATO Integrated Air Defence [S adds system].

national airline   Designated flag carrier of sovereign state.

National Airspace System   That administered by the FAA, traditionally linking VORs.

National Air Traffic Services   Provides ATC over UK, division of CAA but from 2002 part-privatized.

National Flight Data Center   FAA office in DC which collates useful information on civil airspace and publishes it each weekday in the NFD Digest.

national responsibility   Part of collaborative programme wholly assigned to one country.

National Route Plan   Rules and procedures designed to increase the flexibility of user flight planning (FAA).

National Search and Rescue Plan   US inter-agency agreement to facilitate application of full national resources in emergencies.

Natlas   National Testing Laboratories Accreditation Scheme (NPL, UK).

Natmac   National air-traffic management advisory committee (CAA).

NATO   *1* North Atlantic Treaty Organization, formed 1949, always written thus yet spoken as Nato (Int.).
2 N African theatre of operations (WW2).

Natops   Naval Air Training and Operating Procedures Standardization (USN).

NATS   *1* National Air Traffic Services (UK).
2 Naval Air Transport Service (USN, 1941–62).
3 National Air Telecommunications Service [U adds Unit] (US).
4 North Atlantic track system.

Natsim   Network advanced training simulator: -ATC adds ATC (1); -Stars adds scenario, radar and target simulation.

NATSPG   North Atlantic Systems Planning Group (ICAO).

Natsu   Nominated air traffic service unit (CAA).

NATTC   *1* Naval Air Turbine Test Center (USN).
2 Naval Air Technical Training Command.

natural buffet   Buffet occuring automatically near stall as result of airflow turbulence, thus giving warning to pilot.

natural finish   Unpainted.

natural frequency   That at which a system oscillates if given one sudden perturbation and thereafter left to itself.

natural laminar flow   Laminar flow induced by quality of aircraft skin, notably by extremely fine parallel grooves or "sharkskin teeth".

natural language   That spoken by human beings (usually means English).

naturally aspirated   Not supercharged but left to draw in air at local atmospheric pressure.

natural satellite   Not man-made.

natural wavelength   That corresponding to natural frequency of tuned electronic circuit; that at which open aerial [antenna] will oscillate.

NAU   Network access unit.

nausea bag   Sickbag.

nautical mile   Standard unit of distance in air navigation, totally at variance with SI; aviation uses International *, 6,076.1 ft, 1,852 m; UK * is 6,080 ft, 1,853.18 m. A common aviation approximation is 6,000 ft (1,828.80 m). Abb. n.m. See *knot*.

nautical twilight   Period when Sun's upper limb is below visible horizon and Sun's centre is not more than 12° below celestial horizon.

nav   Navigation, navigator.

navaid   Navigation aid, esp. one of electronic nature located at fixed ground station.

Navair   US Naval Air Systems Command.

naval aircraft   *1* Loosely, one used by a navy.
2 One specially equipped for operation from aircraft carrier or other warship.

nav/attack system   One offering either pilot guidance or direct command of aircraft to ensure accurate navigation and weapon delivery against surface target.

nav/bomb   Crew-member combines functions of navigator and bomb aimer.

nav/com   Loosely, navigation and communications, or a single radio transceiver used for both functions.

navex   Navigation exercise.

Navhars   Navigation, heading and attitude reference system.

Navier-Stokes   Basic set of equations for motion of body or flow parcel in viscous fluid.

navigation aid   Any facility intended to assist takeoff, en route flight and landing.

navigation datacard   Portable holder of customized database.

navigation flare   Bright-burning pyrotechnic dropped over open country at night to provide fixed object for measurement of drift (obs.).

navigation float   Navigation aid in form of clearly visible float, with or without pyrotechnics, for drift measurement over sea; hence navigation flame-float, navigation smoke-float (obs).

navigation lights   Regulation wingtip lights (red on left, blue-green on right)) visible from ahead through 110° to side, and white light at tail visible each side of rear centre-line.

navigation satellite   Artificial satellite whose purpose is to assist navigation [not only of aircraft].

navigation smoke bomb   See *navigation flare*.

navigation stars   Those used in astro-navigation.

Navsat   Navigation satellite.

Navsep   Specialist [semi-permanent] panel on navigation and separation of aircraft.

Navspasur   Naval space surveillance system (USN).

Navspoc   Naval Space Operations Center (Dahlgren, VA).

Navstar   Pioneer GPS system based on 24 satellites in 63° orbits at 11,000 n.m.; acronym = Navigation system tracking [or time] and range.

Navwar Navigation warfare, eg by jamming enemy's GPS reception.

Navwass Navigation and weapon-aiming subsystem.

NAW Night/adverse weather.

Nawacs NATO Awacs.

Nawas National warning system (US).

NAWAU National Aviation Weather Advisory Unit (Kansas City).

NAWC Naval Air Warfare Center [AD adds Aircraft Division, WD adds Weapons Division] (USN).

NAWS Naval Air Weapons Station (USN China lake).

Nawtol Night/all-weather takeoff and landing.

Naxos German WW2 family of passive electronic systems tuned to home on H_2S.

NB *1* Enhanced-radiation weapon (colloq., from 'neutron bomb').

2 Northbound.

3 Night bombardment [L, S, added long/short range] (USA, 1919–26).

Nb₃Sn Nb_3Sn Niobium/tin superconductive alloy.

NBA *1* New-build aircraft.

2 Certification authority (Finland).

NBAA National Business Aviation Association (US).

NBC Nuclear, biological, chemical (warfare).

n.b.c. Noise-balancing control (radio).

NBCAP National beacon code-allocation plan (US).

n.b.f.m. Narrow-band frequency modulation.

NBFR Not before.

NBH Normal business hours (ICAO).

NBMD NATO business management directory.

NBP No-break [electrical] power; T adds transfer.

NBPA National Broadcast Pilots Association (US).

NBR Nitrile-based rubber.

NBS *1* National Bureau of Standards, now NIST (US).

2 Navigation/bombing system.

NBSV Narrow-band secure voice.

NC *1* Numerical control (machine tool).

2 Nitrocellulose (or Nc).

3 No change (ICAO).

4 Node, or nozzle, controller.

5 Normally closed.

6 No charge.

7 Narrow coverage (Satcoms).

8 Network centric, or netcentric.

N_c N_c Compressor speed in rpm.

N/C *1* New [installation] concept.

2 Numerical control.

NCA *1* National Command Authorities (US).

2 Nuclear-capable aircraft.

3 NATO conventional armaments; PC adds Planning Committee, RC review committee.

NCAA National Council of Aircraft Appraisers (US).

NCAE National Coalition for Aviation Education (US).

NCAGE NATO commercial and government entity code.

NCAP Night combat air patrol, also called nightcap.

NCAR National Center for Atmospheric Research (Boulder, Colorado).

NCAS Nomex core, aluminium skin.

NCC *1* Network control center/centre.

2 Nickel-coated carbon.

NCCAFB NASA Center for Computational Astrobiology and Fundamental Biology (at Ames).

NCCCA, NC3A NATO Consultation, Command and Control [or Communications] Agency.

NCCIS NATO Command, control and information system.

NCCT Netcentric, or network-centric, collaborative targeting.

NCD *1* No computed data.

2 Net control device.

NCDS Navigaton/checklist display system.

NCDU Navigation[al] control and display unit.

NCE Non-cooperative emitter.

NCGS Nomex core, GRP skin.

NCI *1* Navigation control indicator.

2 Not currently implemented.

NCIAP Networked communications/intelligence weapon data-link architecture program (USAF).

NCIS National Criminal Intelligence Service (UK).

NCISA NATO Communicatons and Information Systems Agency.

NCMA National Contract Management Association (US, now Int.).

NCMC Norad Cheyenne Mountain Complex.

NCMS Network channel-management system.

NC/NG Nitrocellulose/nitroglycerine.

NCO *1* Non-commissioned officer.

2 Numerically controlled oscillator.

3 Navigation/communications operator.

4 Network-centric operations.

NCPA National Center for Physical Acoustics (University of Mississippi).

NCPS Network-centric, or netcentric, precision strike.

NCQR National Centre for Quality and Reliability (UK).

NCRP Non-compulsory reporting point.

NCS *1* Numerical-Control Society (US).

2 Network control, or coordinating, station (Comsats).

3 NATO codification system.

NCSC National Cargo Security Council (US).

NCT *1* National Centre for Tribology, now ESTB (UK).

2 Non-cooperative, or non-cooperating, target; see NCTR.

3 NATO comparative, or cooperative, test.

4 Network control terminal.

5 National Commission on Terrorism (US).

NC3A NATO Consultation, Command and Control Agency.

NCTI Nav [not navigation] Canada Training Institute.

NCTR Non-cooperative target recognition.

NCU Navigation computer unit; R adds readout.

NCW Network-centric warfare.

NCWX No change in weather.

ND *1* Nose-down (trim control).

2 Unable to deliver message, notify originator (ICAO).

3 Navigation display.

4 No date.

5 Neutron detector.

6 Networks Directorate.

Nd Neodymium.

NDA National defense area (US, not DoD property but military or security interest).

NDAA National Defense Authorization Act (US 2002).

NDAC *1* Northern Defence Affairs Committee (NATO).

2 National Defense Advisory Commission (US, 1940).

NDB   *1* Non-directional beacon [L adds locator].

2 Navigation database, stored in FMC.

3 Nuclear depth bomb.

NDBC   Non-discrete beacon code.

NDCL   Nozzles-down cue line.

NDE   Non-destructive evaluation, or examination (NDT research).

NDEW   Nuclear directed-energy weapon.

NDH   No damage history.

NDHQ   Address for all Department of National Defence offices (Canada).

NDI   *1* Non-destructive inspection.

2 Non-development[al] item.

NDIA   National Defense Industrial Association (US).

NDIC   National Defence Industries Council (UK).

NDM   Noise Definition Manual (JARs).

Ndot   = ṅ, esp. rate of change of shaft speed.

NDP   *1* National Defense Panel, or planning (US).

2 National Disclosure Policy [C adds Committee].

3 Night and day payload (UAV).

NDPER   National designated pilot examiner registry (US).

ND point   Nominal deceleration point (SST).

NDRC   National Defense Research Committee (US, WW2).

NDS   Nuclear detection system.

NDT   Non-destructive testing.

NDTA   National Defense Transportation Association (US).

NDTS   Non-Destructive Testing Society (UK).

NDU   Navigation display unit.

NDV   Nuclear delivery vehicle.

NdYAG   Nd-doped YAG.

NE   *1* No echo [radar weather].

2 Network element.

N/E   *1* Neon/ethane mixture (for flushing payload compartments in space).

2 Northings and eastings.

Ne   Neon.

N$_e$ *1* Number of engines.

2 Shaft power (USSR, R).

NEA   *1* National Electronics Association (US).

2 Nuclear Energy Agency (OECD).

3 Nitrogen-enhanced, or enriched, air.

4 Near-Earth asteroid [R adds rendezvous, T tracking, TS tracking system].

NEAC   Noise and Emission Advisory Committee (IATA).

NEACP   National-emergency airborne command post (US).

NEA4   Nitrogen-enriched air with 4% oxygen and other constituents.

Neal-Smith   Criterion for assessing pitch control: lead/lag pilot model with time delay 0.3s [adjusted to aircraft] provides closed-loop resonance and phase lead for comparison with opinions of real pilots.

NEADS   NorthEast Air Defense Sector (US).

NEAM   New England Air Museum (Windsor Locks).

NEAN   North European ADS-B network.

Neap   Northern European CNS/ATM (7) application project (Euret).

NEAR   Near-Earth Asteroid Rendezvous (NASA).

near encounter   Close fly-by of planet or other body.

near field   *1* Shockwave region close to source (hence ** signature).

2 Sonic boom area closest to SST track.

near-field effects   With jet aircraft, include entrainment of free stream, blockage behind the nozzles and [STOVL] suck-down.

near-hit   Airmiss.

near-IR   Usually defined as 0.75-1.5μ.

near miss   DoD term for airmiss.

near space   Region between 100,000 and 120,000 ft [30.48 – 36.58 km], regarded by USAF as a "not used" region which could be exploited. Hence NSMV.

NEAT   North European Aerospace Test facility (Kiruna, Sweden).

Neat, NEAT   Near-Earth asteroid tracking.

neatlines   Parallels and meridians surrounding body of map (NATO); also called sheetlines.

NEB   *1* National Enterprise Board (UK).

2 National Energy Board (C).

3 Nuclear exo-atmospheric burst (DoD).

4 National Examiner Board (US).

NEC   Network-enabled capability.

NECI   Noise exposure computer/integrator.

neck   Lower tube-like portion of gas-balloon envelope.

necking   Local reduction in cross-section of member due to plastic flow under tension.

necking down   Reducing the diversity of types, e.g. on a carrier.

necklace vortex   Formed at a junction between a quasi-flat surface, such as the local skin of a fuselage, and a bluff projection, such as the leading edge of the wing. Unable to penetrate the adverse pressure-gradient, the fuselage boundary layer separates and forms a *, coiling above and below the wing. In most applications the same as a horse-shoe vortex.

neck moment   Bending moment on neck on entering slip-stream in ejection.

NEDO   English rendition of New Energy and Industrial Technology Organization (Japan).

NEDS   Narcotics eradication delivery system (US).

NEEC   Noise-excluding ear capsule.

needle   Rotary 'hand' of traditional dial-type instrument; but see plural.

needle and ball   See *turn and slip*.

needle beam   Extremely directional radio beam with suppressed sidelobes (difficult for enemy to detect).

needles   Generalized term for instrument (esp. flight instrument) readings; thus 'the * all dropped to zero'.

needle split   Divergence between helicopter engine and rotor speed indications, with normally superimposed needles.

needle valve   One offering fine adjustment of fluid flow by linear translation of tapering pointed rod centred in orifice.

NEF   Noise exposure forecast.

NEFC   NATO electronic-warfare fusion cell.

NEFD   Noise equivalent flux density.

Nefma, NEFMA   NATO European Fighter Management Agency (Int.).

NEG   Negative.

"negative"   Voice communications word meaning "no".

negative altitude   Angular distance below horizon; depression (ASCC).

negative area   *1* Area on tephigram enclosed between

path of rising particle at all times colder than environment and surrounding air.

2 Generally, vague area (volume) surrounding colder air that happens to be rising.

negative camber Usually interpreted as concave on upper surface.

negative feedback *1* Signal either reversed in sense or otherwise out of phase and thus tending to increase departure from original condition.

2 Transfer of part of amplifier output in reverse phase to input.

negative g Subject to acceleration in the vertical plane in the opposite-to-normal sense, eg aircraft in sustained inverted flight or in pushover from steep climb to steep dive; wings are bent 'downwards' (relative to aircraft attitude) and pilot can experience 'red-out'. This condition is intermittently inevitable in severe turbulence but is normally prohibited for non-aerobatic aircraft.

negative-g valve Inverted-flight valve.

negative image Apart from photographic meaning, transposition of blacks and whites in TV, EO or IR picture.

negative pole Cathode, or S-seeking pole of magnet.

negative pressure relief valve Prevents dP in pressurized aircraft becoming negative.

negative rolling moment Tending to rotate aircraft anticlockwise, seen from rear.

negative stability See *instability.*

negative stagger Backwards stagger; lower plane in advance of upper.

negative stall Stall under negative g; this regime provides lower left boundary of basic manoeuvre envelope.

negative sweep See *forward sweep.*

negative terminal That from which electrons flow; thus towards which 'current' flows.

negative-torque signal Indication of fault characterized by driven member (eg propeller) tending to drive driving input (eg turboprop).

negative yaw Rotating aircraft anticlockwise about z-axis seen from above.

NEGL Negligible.

negotiation Commercial discussion preceding contract.

negotiation threshold Point in escalating conflict at which either participant is likely to draw back and initiate negotiation.

NEH Code: 'I am connecting you to a station accepting traffic for station you request' (ICAO).

N18 Powder metallurgy nickel alloy, MPt 1,225–1,323°C.

NEL National Engineering Laboratory (UK).

NEMA National Electrical Manufacturers Association (US).

nematic See *liquid crystal.*

NEMO, Nemo Navy [or Naval] Earth Map Observer.

NEMP Nuclear electromagnetic pulse.

Nems, NEMS Nano-electromechanical system[s].

Nemspa National EMS Pilots' Association (US).

NEO *1* National energy outlook (US).

2 Near-Earth object[s].

3 Non-combatant evacuation, or extraction, operation.

neodymium Nd, silver metal, density 7.0, MPt 1,021°C, used in lasers and in permanent magnets of highest known energy level [see next].

Neomax Magnetic material $Nd_{15}Fe_{77}B_8$.

neon Ne, inert gas, density 0.9×10^{-3}, BPt –246°C.

Neoprene Family of synthetic rubbers (polymerized chloroprenes) resistant to hydrocarbon fluids.

NEP *1* Noise equivalent power.

2 Nuclear-electric propulsion.

Nepal drop Low-level airdrop of cargo, usually food, in which about 20 sacks are attached to strong plywood sheets by cords which break on impact.

neper Unit (Napier) expressing scalar ratio of two currents, $N = nat \log(I_1/I_2)$ nepers, $= 8.686$ dB. Applicable to all scalar ratios of like quantities.

nephelometer Measures light scattering by fine particles in suspension.

nepheloscope Measures temperature changes in gases rapidly compressed or expanded.

nephometer Convex mirror divided into one central and five radial parts for estimating cloud cover.

nephoscope Optical instrument for measuring direction and angular velocity of cloud motion.

NEPP Normal and emergency preflight procedures.

Neptco Process for manufacturing soft-skin composite in which main fibres are pultruded rods.

Nerc, NERC *1* New EnRoute Centre (Swanwick, NATS, UK).

2 Natural Environment Research Council (UK).

3 National Environmental Research Centre (EPA).

NERO *1* National Energy Resources Organization (US).

2 Nederlanse Vereniging voor Raketonderzoek.

Nerva Nuclear energy, or engine, for rocket-vehicle applications; May 1961.

NES Netcentric Enterprise Services, moves nets such as GCCS and GLCS into web-based environment.

NESC *1* Naval Electronics Systems Command (USN).

2 National Environmental Satellite Center (NOAA).

3 NASA Engineering & Safety Center (LaRC, 2003–).

NESDIS, Nesdis National environmental satellite data and information service or system (NOAA).

NESN NATO English-speaking nations.

Nest Non-expendable space transport [S adds system].

Nesterov loop Flying 360° circle in horizontal plane whilst rolling continuously (1913).

NET *1* Network entity title.

2 Network enabled technologies.

net *1* Tailored mesh forming structural link between traditional gas balloon envelope and useful load.

2 Electronic, optical or other telecommunications system(s) forming single service covering designated area accessible at any point.

net area *1* Traditional gross area (normally of wing) minus projected horizontal area of fuselages, booms, nacelles, pods, etc.

net dimensions Those of final shape, thus a * moulded core.

netcentricity Conduct of future warfare governed by an integrated ISR network using aircraft and UAVs; hence netcentric warfare, etc.

NETD Noise equivalent temperature difference.

net dry weight Basic weight of engine or other device calculated according to various rules but always excluding fluids (fuel, lubricant, coolant, etc) and usually all accessories, protective systems, instrumentation systems, etc, not essential for device to function.

net flightpath That followed by aircraft, esp. aeroplane, after application of factors (particularly for average-aircraft performance and average pilot skill); ie gross flightpath fully factored.

net height That at any point on net flightpath, esp. during takeoff and climb-out.

Netma NATO Eurofighter and Tornado Management Agency.

net performance Gross performance factored to take account of temporary variations (from whatever cause) and pilot handling skill.

net propulsive force See *net thrust*.

net radiation factor Percentage of radiant energy emitted by one surface or volume that is absorbed by another surface or volume.

nettage decrement Percentage difference between gross and net performance.

net thrust Fn, effective thrust of jet engine numerically equal to change in momentum of fluids (air/gas and fuel) passing through engine (plus, in engine operating with choked nozzle, extra aeroydynamic thrust generated in nozzle).

netting Electronically interlinking numerous related stations, such as SAM launchers or battlefield communications centres, which are dispersed randomly over a wide area and move relative to each other.

net weight Loosely empty weight, but usually excluded from aerospace usage.

net wing area Gross area minus projected plan area of fuselage, nacelles over wing and other non-aerofoil parts.

network System of communications linking computers and other management tools.

network centric warfare Radical new command, control and communications systems, with every post, vehicle [including airborne] and sensor interconnected. A core objective is to find targets using multiple—possibly widely dispersed—sensors and provide near-instaneous target data.

neural networks Loosely follow architecture of human neurons and their dendritic connections; *NNP* is a pioneer multiple instruction/multiple data neural processor.

neutral-angle intake An inlet whose mouth is shaped to minimise variation of ram pressure with airspeed.

neutral area "A strip of ground of specified width adjoining the sides of a runway" (B.S., 1940).

neutral axis Locus of points within structural member at which bending imposes neither tension nor compression.

neutral burning Combustion of solid propellant grain in which exposed surface [and thus thrust] remains almost constant over burn-time.

neutral engine Main propulsion engine devoid of dressing peculiar to either particular aircraft type or to left or right installation in multi-engined aircraft, and thus available for quick completion for desired installation.

neutral equilibrium Normally means that system will tend to stay in most recently commanded attitude or condition, without oscillation, unstable divergence or recovery of previous condition.

neutral flame Neither oxidizing nor reducing.

neutral hole Aperture cut from sheet, esp. in wall of internally pressurized container (usually taken to be cylindrical) shaped so that peak stress around periphery is minimised and stress in surrounding material is as if hole did not exist. Credited to E.H. Mansfield, RAE. Normally approximates to ellipse.

neutralized controls Usually taken to mean centralized.

neutralized track Air intercept code: target is ineffective or unusable.

neutral point *1* Location of aircraft, esp. aeroplane, c.g. at which stability would be neutral; rear extremity from which static margin is measured. More strictly, stick-fixed ** is c.g. position at which stick movement to trim a change in speed is zero; stick-free ** is c.g. position at which stick force to trim a change in speed is zero.

2 Lagrangian point.

3 Any sky direction where polarization of diffuse (ie not from specific source) radiation is zero.

neutral stability See *neutral equilibrium*.

neutrino Elusive small particle, rest mass 0, spin ½.

neutron Particle of atomic nucleus having no charge and mass 1.675×10^{-24} g (proton is 1.672×10^{-24} g).

neutron bomb Enhanced-radiation weapon.

neutron radiography NDT method similar to X-ray inspection and 'photography' but using beam of neutrons.

Nevatron Possible future gyro: magnetic field guides atoms at near-zero temperature round 20-mm ring.

NEW *1* NATO electronic warfare [AC adds advisory committee, TS training system].

2 Network-enabled warfare.

Newac NATO Electronic Warfare Advisory Committee.

new blue Recruit (USAF).

Newhaven Visual marking of ground target (RAF WW2).

newton SI unit of force, $= 1$ kg m s^{-2} $= 10^5$ dyn $= 7.233$ pdls $= 0.224809$ lbf $= 0.10197$ kgf; written without initial capital, but symbol N.

Newtonian flow That in extremely rarified gas where mean free path is in order of metres and hypersonic body is surrounded by incident arriving molecules passing between those bouncing off surface; also called free-molecule flow.

Newtonian mechanics Those based on Newton's laws of motion, in which mass and energy are unrelated.

Newtonian speed of sound Relation a $= \sqrt{p\rho}$ where p is pressure and ρ density.

Newtonian stress Fundamental shear stress in fluid, given by law $\tau = \mu \frac{du}{dy}$ where μ is fluid viscosity and u is fluid velocity at distance y from fixed surface.

Newton's laws Briefly: (1) body at rest remains at rest unless acted upon by outside force, (2) change in motion (momentum) is proportional to applied force, and (3) to every action (force or change in momentum) there is equal and opposite reaction.

Newts Naval electronic-warfare training system (USN).

Nex, NEX Next-generation (prefix, DoD).

Nexcom Next-generation air/ground communications [P adds program] (FAA).

Nexrad Next-generation radar.

NEXST Next-generation supersonic transport.

Next NASA evolutionary xenon thruster.

Nexwos Next-generation weather-observing system (FAA, 1995).

NEZ No-escape zone.

NF *1* French material specification prefix.

2 Night fighter.

3 Notched filter.

N_f, N_F *1* Fan rpm.

2 Relationship between power-turbine speed and gearbox governor.

nF   Nanofarad.

n.f.   *1* Noise factor (radio).

2 Negative feedback.

NFAC   National Full-scale Aerodynamics Complex (NASA Ames).

NFCS   Nuclear forces communications satellite.

NFCT   Non-Federal control tower (FAA).

NFD   National flight data [C adds Center, D Digest, PS processing system] (FAA).

NFE   Near-field effects.

NFF   No fault found.

NFG   The Newfoundland Group, airline labour watchdog (US).

NFH   NATO frigate helicopter.

NFIP   National foreign intelligence program (US).

NFIS   Navigational flight inspection system.

N5+   Refractory NiCo alloy used for monocrystal engine parts.

NFKK   Women's aero association (J).

NFLC   National Flying Laboratory Centre (Cranfield, UK).

n.f.m., NFM   Narrow-band FM.

NFMS   Navigation and flight-management system.

NFN   *1* Near-field noise.

2 Naval fires network (USN).

NFO   Naval flight officer (USN).

NFOV   Narrow field of view.

NFP   Net flightpath.

NFPA   National Fire Protection Association (US); many annexes, eg * 417 specifies resistance of ADB (2) to an external fire.

NFRL   Naval Facilities Research Laboratories (USN).

NFS   *1* Network file system.

2 National Fire Service (UK, WW2).

3 Near-field source.

NFSN   NATO French-speaking nations.

NFT   *1* Night-flying test.

2 Navigation flight test.

NFTC   NATO flying training in Canada.

NFTM   Noise and flight-track monitoring.

NFWS   Navy Fighter Weapons School (NAS Miramar).

NFZ   No-fly zone.

NG   Natural gas, hence LNG = liquid natural gas.

NG, Ng   Nitroglycerine.

N$_g$ Gas-generator rpm.

NGA   National Geospatial-intelligence Agency (US, was NIMA/Nima).

NGAGC   See *Nexcom.*

NGATM   New-generation air-traffic manager.

NGAUS   National Guard Association of the US.

NGB   National Guard Bureau (US).

NGC   Nylon/graphite composite.

NGCCS   Next-generation command and control system.

NGDC   National Geographic Data Center (NOAA).

NGDR   Next-generation [broadband] digital receiver.

NGE   Non-ground effect.

NGEA   Nouvelle génération école/appui [combat-aircraft trainer] (F).

NGIFF   New-generation IFF.

NGL   Natural-gas liquids.

NGLS   Next-generation launch system.

NGLT   Next-generation launcher technology.

NGM   Nested-grid model [weather computer program].

NGO   Non-government organization (US).

NGPS   Navstar global positioning system.

NGR   *1* Night-goggle readable (ie at IR level).

2 Next-generation GPS receiver.

3 Nitrogen gas reduction.

NGS   *1* Naval gunfire support.

2 National Geodetic Survey (NOAA).

NGSP   National geodetic-satellite program (NASA).

NGSST   Next-generation SST.

NGST   Next-generation space telescope, following Hubble.

NGT   *1* New-generation trainer.

2 Night.

NGTCS   Next-generation target control system.

NGTE   National Gas Turbine Establishment (formerly at Pyestock, UK).

NGV   Nozzle guide vane.

NGW   Nuclear gravity weapon.

N$_H$ HP rpm, N_2.

NHA   Naval Helicopter Association (office Coronado, CA, US).

NHC   *1* Navigator's hand controller.

2 National Hurricane Center (NWS).

NHCR   National Flying League (J).

NHE   Notes and helps editor.

NHGA   National Hang-Gliding Association (UK).

NHP   Non-handling pilot.

NHR   National Flying Association (J).

NHS   National hypersonics strategy (US).

NI   Noisiness Index (South Africa, Van Niekerk/Muller, 1969; see *noise*.

Ni   Nickel.

NIAC   National Infrastructure Advisory Council (US, counter-terrorism).

NIAG   NATO Industrial Advisory Group.

Nial   Nickel/aluminide diffusion coating.

NIAR   National Institute for Aviation Research (Wichita State University).

NIAST   National Institute for Aeronautics and Systems Technology (South Africa).

NIAT   *1* MAP-sponsored factory (USSR).

2 National institute for engineering research (R).

NIB   Neodymium, iron, boron, permanent-magnet material.

nib   *1* Any substantially axial fore or aft-pointing fairing, usually with concave surfaces.

2 Aft-pointing fairing between, and projecting behind, two closely spaced jetpipes.

3 Forward-pointing extension at inner end of fixed glove on VG aircraft.

nibbler   Machine tool for eating away edge of sheet by repeated local vertical shearing, in some cases with ability to impart lateral compression or joggling.

nibble   An approach to very limit of g-induced stall in air combat, also see next entry.

nibbles   Stall testing of aircraft in which AOA is increased in small increments at 1 g, culminating in fully developed stalls.

NIBS   Neutral industry booking system (IG adds interest group).

NIC   *1* Newly industrializing country.

2 New installation concept.

Nicalloy   Nickel-iron alloy, low initial but high maximum permeability for transformers, etc.

Nicap   National Investigation Committee on Aerial Phenomena (US).

Nicasil   Ni/Cd/Si.

Ni/Cd, Nicad   Nickel/cadmium electric battery.

NICE, Nice   *1* NAT(2) implementation manager, or management, cost/effectiveness.
  2 NATO internet cryptography equipment.

Nicerol   Widely used protein foam compound for fire-fighting.

Nichols diagram   Plots stability of rigid aeroplane showing open-loop frequency response in each axis following control-surface deflection.

Nichrome   American heat-resisting alloys (c 85% Ni, 15% Cr), eg for resistance wire.

nickel   Ni, silver-white magnetic metal important in corrosion-resistant alloys; density 8.9, MPt 1,453°C.

nickel/cadmium battery   Cell having KOH (potassium hydroxide) electrolyte, positive plates of nickel hydroxide and negative plates of cadmium hydroxide.

Nicmos   Near-IR camera and multi-object spectrometer.

Nicral   Nickel/chromium/aluminium plasma spray.

Nicrosil   Ni/Cr/Si.

NICS, Nics   NATO integrated communications system; O added "Organization", MA added 'Management Agency'; became NACISA.

NID   National interest determination.

Nidjam   Nav/ident deception jammer.

NIDTS   NATO integrated digital transmission system.

NIE   National Intelligence Estimate[s] (US).

NIF   National infrastructure forum (UK).

NIFA   National Intercollegiate Flying Association (US).

NIFC   National Inter-Agency Fire Center (US).

Ni/Fe, Nife   Nickel/iron electric battery.

night airglow   See *airglow*.

night and all-weather   Strictly, interceptor can be used at night or in any weather, seldom true at time this term was in use; more accurately meant night and rain or snow but with acceptable landing minima.

Night cap, NCAP   Night combat air patrol (DoD).

night effect   Phenomenon most noticeable near sunrise and sunset when directional radio signals (D/F, radio range, VOR) give false readings thought to be due to variations in ionosphere.

night fighter   Aircraft intended to intercept other aircraft at night; today implicit in term 'fighter' or 'interceptor'.

night-flying chart   Special editions of regional charts, usually 1:1,000,000, eliminating all detail unseen at night but emphasising lights, navaids and fields with night facilities (US).

Night Owl   Night ground-attack mission.

Night rating   Enables holder of PPL to fly at night as PIC with passenger[s].

night vision   Human seeing after eyes have had time fully to adapt to near-absence of light, with irises fully open.

NIH   Not invented here, rejection of foreign developments.

NIHL   Noise-induced hearing loss.

NiH$_2$ Nickel/hydrogen electric battery.

NII   *1* Scientific test institute (USSR, R; many, each covering one subject).
  2 National information infrastructure.

NIID   Defence manufacturers association (Netherlands).

NIIR   Radio engineering research institute (R).

NIIRS   National imagery interpretability rating scale (US, runs 0 to 9).

NIL   *1* Code: 'I have no message for you' (ICAO).
  2 National Information Library (NIMA).

NILE, Nile   NATO improved link 11.

NIM   National Imagery & Mapping; A adds Agency, C college (DoD, Bethesda, MD).

Nima   See previous; now *NGA*.

nimbostratus, Ns   Thick dark blanket cloud in low/middle band (2,000–6,000 m), mainly ice crystals and supercooled water, large horizontal extent, usually rain.

nimbus   Not normally used as cloud type but as adjective meaning rain-producing, as in Cb and Ns.

Nimby   Not in my back yard.

Nimocast   British casting alloys, composition akin to Nimonics.

Nimonic   Family of refractory and anticorrosive alloys based chiefly on nickel, originally Mond patents, important where creep-resistance essential; Inconel, Hastelloy, Udimet and Waspalloys related.

NIMS, Nims   National airspace system Infrastructure Management System (FAA).

90-minute rule   Certification requirement that twin-engined passenger aircraft may fly transoceanic sectors provided they are never more than 90 minutes from an emergency alternate.

The 99s   US women in aviation educational charity [sometimes spelt out].

NINST   Non-instrument [runway].

niobium   Nb, shiny grey metal, density 8.6, MPt 2,468°C.

NIOSH   National Institute for Occupational Safety and Health (US).

NIP   Network interface processor.

nip   *1* Local compression between adjacent components, esp. that used to secure a third part, eg compressor rotor between discs.
  2 Local compression caused by deflection under operating conditions, eg axial movement at periphery of conical compressor or turbine disc.

NIPC   National Infrastructure Protection Center (US).

Niprnet   Non-classified information protocol router network (DoD).

NIR   *1* Near infra-red.
  2 Network interface router; V adds VHF data-link.

NIS   *1* NATO identification system.
  2 Nose-in stand (airport ramp).
  3 Not in stock.
  4 Not in service.

Nisac   National Infrastructure Simulation and Analysis Center (LASL/Sandia, a response to 9-11).

Nisil   Ni/Si.

NIST   National Institute of Standards and Technology (US, previously NBS).

nit   Name, not normally used, of SI unit of luminance, cd/m^2.

NITE   Night imaging and threat evaluation.

Nitenol   Alloys of N$_i$ and T$_i$, variable properties.

NITEworks   Network, integration, test and experimentation works (UK, MoD).

NITF   National imagery transmission format.

Nite-Op   Night-imaging through electro-optics.

Nitralloy British steels for nitrided parts with small amounts of Cr, Al, Mn, C, Si and possibly Ni and Mo.

nitrate dope Aircraft fabric dope comprising cellulose fibres dissolved in nitric acid, plus pigment, thinner, etc.

nitriding Surface hardening of steels by prolonged heating in nitrogen-rich atmosphere.

nitrogen Generally unreactive gas forming 78.03% by mass of sea-level air, symbol N_2, BPt –195.8°C, density 1.25 gl^{-1}; dry gas important as inert purging medium, liquid LN_2 used as cryogenic heat-transfer fluid.

nitrogen desaturation Human condition caused by nitrogen deficiency.

nitrogen narcosis Human condition caused by apparent reaction between tissue fats and nitrogen under pressure.

nitrogen tetroxide N_2O_4, most common storable liquid oxidant, often called NTO, BPt 21°C, Isp 285 with UDMH, 290 with hydrazine.

nitroguanidine Major constituent of many gun propellants, commercial name Picrite.

nitromethane CH_3NO_2, oily liquid, monopropellant.

nitrous oxide N_2O 'laughing gas', used as source of oxygen in power-boosting piston engines in WW2.

NIU *1* Nitrogen inerting unit.
  2 Network, or navigation, interface unit.

NIVO, Nivo Dark green night-bomber paint, later RDM2 (RAF).

NIVR Netherlands Institute for Aerospace Development.

NIW Night and in-weather.

NJ Noise jamming.

NJE Nominal jet edge.

NJG Nachtjagdgeschwader, night-fighter group [US = wing] (G, WW2).

NJSK Private Pilots' Association (J).

NKAP State commissariat for aviation industry (USSR).

NKF Non-kinetic fires.

NKK Nihon Koku Kyokai, Aeronautical Association (J).

NKO State commissariat for defence (USSR).

NKSK Aero Engineers' Association (J).

NKTP State commissariat for heavy industry (USSR).

NKUGK Society of Aeronautical and Space Sciences (J).

NKVD State commissariat for internal affairs (now KGB).

N/kW Newtons per kilowatt, fundamental performance measure of Hall-effect thrusters.

NL *1* Natural language.
  2 Normenstelle Luftfahrt (G).

N_L *1* LP rpm [N_1 preferred].
  2 Normal load factor.

NLA *1* New large aeroplane study group.
  2 Noise-level analyser.

NLAW Next-generation light anti-armour weapon.

NLB Nose loader, or loading, bridge; see *bridge*.

NLC Noctilucent cloud.

NLCM Non-lethal countermeasures.

NLF *1* Natural laminar flow.
  2 Normal load factor.

NLG *1* Nose landing gear.
  2 Noise Liaison Group (UK).

NLL No load lubrication.

NLM Network-loadable module.

NLO No local, or live, operator, also called nulls.

NLOS Non-line of sight; CA adds combined arms.

NLP Network layer protocol.

NLR Nationaal Lucht- en Ruimtevaartlaboratorium (Netherlands).

NLRB National Labor Relations Board (US).

NLRGC Aeromedical research centre (Netherlands).

NLS New launch system (NASA/USAF).

NLT Not less than.

NLW Non-lethal weapon.

NM, n.m., nm *1* Nautical mile; nm preferred except by ICAO. Note confusion with nanometre.
  2 Network management.

Nm SI unit of torque or moment, Newton-metre = 0.73756 lbf-ft.

nm *1* Nanometre (10^{-9}m).
  2 Nautical mile[s], or n.m.

NMAC Near mid-air collision.

NMB National Mediation Board (US).

NMC *1* Naval Missile Center (Pt Mugu, CA).
  2 Not mission-capable.
  3 Naval Materiel Command (USN).
  4 Satellite network management centre.
  5 National Meteorological Center (NWS).
  6 Net[work] monitoring and control.

NMCC National Military Command Center (US).

NMCCD Network-management category class diagram.

NMCS National military command system (US).

NMD National missile defense [i.e., against ICBMs], LSI added lead systems integration; in 2002 replaced by GMD.

NMDPS Network-management data-processing.

NMEA National Marine Electronics Association (US).

NMF Network management function.

NMG Numerical master geometry.

NMH Nickel-metal hydride.

NMIC National Military Intelligence Center (US).

NMIRS Network management interface requirements specification.

NMKB Model Aeronautics Association (J).

NML Normal.

NMM National mission model (US).

NMO National military objectives (US).

NMOS Negative (n-type) metal-oxide semiconductor, or silicon.

NMP *1* Navigation microfilm projector.
  2 Network management plan.

nmpg Nautical miles per gallon [gallon not specified].

NMR Nuclear magnetic resonance.

NMRS Numerous.

NMS *1* Navigation, or network, management system.
  2 Noise monitoring system.
  3 Non-motion simulator.

Nms SI unit of angular momentum, Newton-metre-second.

NMSB Non-modification Service Bulletin.

NMSI National Museum of Science and Industry ["the Science Museum"], London.

NMT *1* Non-manoeuvring target.
  2 Not more than.
  3 Noise-monitoring terminal.

NMU Navigaton management unit.

NN Network node; see *NNSS*.

NNA　Neutral and non-aligned.

NNC　Non- [or not] noise certificated aircraft.

NNE　Noise and number exposure.

NNI　*1* Noise and number index (see *noise*).

2 National nanotechnology initiative (US).

NNK　Non-nuclear kill.

NNMSB　Non-nuclear munitions safety board.

NNP　Neural-network processor (SBIR/USN).

NNR　Hot-air Balloon League (J).

NNSA　National Nuclear Security Adminstration system (US DoE).

NNSS　Network-node switching system.

NNTS　Nevada Nuclear Test Site.

NO　*1* Nitric oxide, colourless gas.

2 Normally open.

3 Not available, not operative.

4 Notice to Airmen.

5 Night observation aircraft category (USA 1919–24).

N_o　Characteristic frequency, esp. centroid of power spectral density distribution.

NO_2　Nitrogen dioxide (peroxide), pungent brown gas.

N_2O_4　*Nitrogen tetroxide.*

NOA　*1* Non-operational aircraft.

2 Not organizationally assigned (aircraft stored for future use).

3 New obligation(al) authority (US).

NOAA　*1* National Oceanic and Atmospheric Administration (US).

2 New optimisation approaches for air-traffic flow management (Euret).

NOACT　National Overseas Air-Cargo Terminal (USN).

Noball　Code name for German flying bombs and rockets, hence * targets were mainly launch sites.

no brains　Aircraft handling always predictable, pilot can relax.

no-break supply　One whose emergency standby system comes on-line instantaneously, in theory without losing one waveform or pulse in coded train.

NOC　*1* Network operations centre.

2 Notice of change.

Nocar　North Atlantic oceanic concept and requirements.

Nocas　Night-operation[s]-capable avionics system.

Nocom　No communications (Acars).

noctilucent cloud　Appearing self-luminous at twilight in high [50+°] latitudes, caused by particulate matter at height 75+ km.

Nocus　North continental US (Loran chain).

NOD, Nod　Night observation device [LR adds long-range].

nodalisation　Equipment of helicopter with antivibration couplings between rotor head and fuselage.

Noda-Matic　Patented vibration-isolation system in which helicopter fuselage is suspended from rotor via arrangement of tuned vibrating masses which cancel out rotor vibrations (Bell).

nodding　Deflection under vertical acceleration of masses cantilevered ahead of or behind main structure, eg forward fuselage (in flight only) or engine on wing pylon well ahead of leading edge.

nodding aerial　One oscillating only, or principally, in vertical plane, eg HFR.

noddy cap　Protective cover for delicate (eg IR-homing) missile nose (RAF, colloq.).

NODE　National operational [ATC] display equipment.

node　*1* In structures, location of point where load variation causes only rotation but no linear deflection.

2 Point, line or surface in wave system where some major variable has zero amplitude.

3 In any network, terminal point or point where two channels branch.

4 Intersection of orbit of satellite with plane of orbit of primary.

5 Location in mobility system where movement is originated, processed or terminated (DoD).

NODLOR　Night observation device, long-range.

no-draft forging　One forged essentially to finish dimensions, thus needing little if any machining.

NODS　Night observation and detection system.

nodular cast iron　See *SG cast iron.*

NOE　Nap of the Earth, ie flight as low as posible over undulating terrain.

noed　Knot (F).

no-escape zone　In AAM engagement with fast target, often less than 0.25 maximum AAM range.

NOF　International Notam Office (ICAO).

no-feathering axis　Axis of swashplate, about which there is no feathering moment or first-harmonic variation of cyclic pitch.

no-flare landing　Aeroplane landing (rarely, other aircraft) in which approach trajectory is continued in essentially straight line until landing gear hits ground.

no-fly zone　Airspace prohibited to the aircraft of that country, and [usually] patrolled by aircraft of a hostile country to ensure compliance.

NOFORN　No foreign [dissemination of information] (US).

Nogaps　Navy operational global atmospheric prediction system (USN).

no-go gauge　One whose linear dimension (between faces, threads or diameter) is just below smallest permitted limit for part.

no-go item　One whose failure or absence from aircraft prohibits takeoff according to operating rules (though not necessarily rendering it unairworthy).

NOGS, Nogs　Night observation gunship system.

NOI　*1* Notice of intention.

2 Notice of Inquiry (US)

NOISE　National Organization to Insure [*sic*] a Sound-controlled Environment (US).

noise　*1* Noise in air. Basic unit is decibel, dB, 0.1 bel, measure of sound pressure above local atmosphere on logarithmic scale, usually related to starting reference pressure of 2×10^{-5} Nm^{-2}. Sound pressure level $L = 10 \log p^2/p_o^2 = 20 \log (p/p_o)$ where p_o is reference pressure and p actual measured pressure. Alternative is to use source power level $L_W = 10 \log (W/W_o)$ dB where W_o is reference power commonly taken to be 10^{-12} W (Watts). Pressure levels are more common, and log scale allows for million-fold increase in human perceived pressures, each 6 dB increment representing doubling of pressure level. Study of aircraft noise from 1952 led to many new measures in attempts to quantify noise nuisance. In 1953 CNR (Community Noise Rating) gave single-number scale based on public response to six generally quantifiable factors, and in 1957 NC (Noise Criteria) curves attempted to portray equal-loudness contours taking into account discrete tones, impulsive nature of some sounds and other variables. By this time many workers had tried to quantify human aural response to different frequencies and tones with mixed frequencies, and curves drawn in 1959 were labelled L_{PN} (Perceived Noise Level) in units of dB(PN), sometimes written PNdB. Despite its complexity this gained major foothold, and virtually eliminated traditional measures (phon, relating sound pressure level to standard 1 kHz tone, and sone, loudness corresponding to 40 phons). Various weighted dB measures were introduced for the measures taken by meters with scales adjusted to equal-loudness contours for different overall pressure levels, these being called by various letters (thus, A-weighted = dBA = L_A). In 1961 a series of surveys measured annoyance according to new measures, L_{PN50} or L_{PN90} (L_{PN} exceeded by 50 or 90 per cent of aircraft), D_{85} or D_{95} (10 log time in seconds when sound pressure exceeded 85 or 95 dB), and N (number of aircraft 'passing over', latter criterion not being defined); result was single value for location called NNI (Noise and Number Index). Another 1961 unit was derived by splitting noise into one-third-octave bands and assigning each band a Noy rating by comparing with subjective noisiness of random noise centred on 1 kHz; individual Noy figures then added by method allowing for masking of one band by others and presented in PNdB. Further work allowing for particular features – such as intense pure tones, as from compressor blading, in otherwise broadband jet sound – led to use of L_{EPN} (Effective Perceived Noise Level) measured in EPNdB (Effective Perceived Noise dB) in first-draft legislation in 1966, which led to FAR Pt 36 and subsequently closely similar ICAO Annex 16. By this time at least 20 national or local authorities had published research, including Australia's AI (Annoyance Index) = $\overline{L}_{PN}$ + 10 log N; German Störindex $\overline{Q}$ based on dBA; French R-index = $\overline{L}_{PN}$ + 10 log N-30; Dutch Total Noise Rating B based on log of summation of A-weighted pressure levels; American CNR (Community Noise Rating) based on many variables; California's CNEL (Community Noise Equivalent Level) using WECPNLs (Weighted Equivalent Continuous Perceived Noise Levels) varying with time of day and season; American NEF (Noise Exposure Forecast) = $\overline{L}_{EPN}$ + 10 log N-K where K is 88 by day and 76 by night; the European Community's L_{DEN} = noise from all sources [noise density] summed through each 24 h; South Africa's NI (Noisiness Index) = $\overline{L} \times$ 10 log N + 10 log T_a/T where T_a and T are times; British TNI (Traffic Noise Index) and resulting L_{eq} (Equivalent Average Sound Pressure Level), which led to L_{NP} (Noise Pollution Level) = L_{eq} + 2.56 σ where σ is standard deviation of dB fluctuations. Further measures include LAX or L_{AX}, also called Senel (Single-Event Noise Exposure Level), SIL (Speech Interference Level), L_{TPN} (Tone-Corrected Perceived Noise Level) and various octave-band measures such as $L_{300-600}$ (sound pressure level of band 300–600 Hz). Also see *Approach* *, *sideline* *, *takeoff* *.

2 Background noise, that present in electronic amplification, communication or recording system in absence of signal.

3 Thermal or Johnson noise caused by thermal agitation of charge carriers.

4 White (Gaussian) noise, constant energy per unit bandwidth.

5 Shot noise, fluctuation in charge-carrier current.

6 Random noise (eg white, shot), uniform energy versus frequency distribution.

noise (electronic)   *1* Effects of unwanted signals, including those generated within the system.

2 Unwanted signals themselves.

noise abatement   Deliberate procedures whose objective is minimization of noise perceived at ground. See next.

noise abatement climb procedure   Maximum power from brakes-release to reach maximum attainable height AGL at point where ground track crosses boundary of built-up area, or location of listening post[s], there cutting back power to predetermined value just sufficient to maintain positive rate of climb [or 2% gross gradient] at V_2 + 15 kt, until either built-up area is passed (some operators add 1 nm margin) or height AGL exceeds FL 50, where all-engines en-route climb is started.

noise-absorbing material   Wide range of materials, usually used as non-structural linings, containing precisely sized cells which convert impinging sound energy into heat. Most are honeycomb sandwiches whose facesheet is perforated.

Noise and Number Index   See *noise (1)*.

noise attenuation   Design and/or constructional features whose purpose is to minimise externally perceived noise. Techniques include addition of sound energy-absorbing linings, structural and aerodynamic features to change frequencies (eg of blades passing), mechanical design to reduce noise of bearings and gear teeth, and maximization of jet-nozzle periphery to increase rapidity of jet/atmosphere mixing.

noise carpet   Area along aircraft track subjected to significant noise nuisance.

noise certification   Certifying authorities in all but a few states require compliance with noise (and emissions in most cases) legislation for all new civil aircraft. Older aircraft have to comply at specified future dates.

noise contour   Locus of points on ground at which specified air traffic results in particular perceived noise level, NNI or other noise nuisance. Traffic may be an average of arrivals or departures, a weighted average, a 'noisiest' aircraft type or an NNI figure taking frequency of flights into account.

noise exposure   Not defined but related to noise levels, number of events (though Senel [see *noise (1)*] is one event) and time of day. Hence ** forecast.

noise factor   Ratio of audio to thermal noise at same frequency.

noise floor   Hypothetical minimum background noise level.

noise footprint   Outline – generally footprint-shaped – of enclosed region around runway bounded by particular noise contour (often 90 EPNL) resulting from one landing and one takeoff by particular aircraft type operating at MTOW in ISA with measures taken at standard reference points and other locations.

noise-reduction rating   Quantified measure of effectiveness of ear-defenders and aircraft headsets, unit is dB (EPA).

noise reference points   In civil aircraft certification, three locations at which noise measures are taken. See *approach noise*, *sideline noise* and *takeoff noise*.

noise-shield aircraft   One in which basic configuration, by design or as fortuitous bonus, places major portions of structure between main noise sources and ground.

noise shielding   Portions of aircraft which, by design or fortuitously, are interposed between noise sources and distant observers.

noise suppression   See *noise attenuation*.

noise suppressor   Jet nozzle configured to reduce noise by increasing periphery of nozzle(s) and speeding mixing.

no joy   Air intercept code: 'I have been unsuccessful, or have no information'.

NOK   Next of kin.

NOL   Naval Ordnance Laboratory (USN).

no lift   Stencilled instruction to ground personnel prohibiting application of lifting forces in local area of airframe.

no-lift angle   That between no-lift direction and chord.

no-lift direction   Angle of attack of two-dimensional aerofoil section at which lift is zero at low airspeeds. In practical wing ** varies from root to tip.

no-lift wire   One bracing aerofoil from above; also called anti-lift wire.

NOLO, nolo   No live, or local, operator, ie RPV is preprogrammed.

NOM   National Operations Manager (ATCCC).

Nomad   *1* Naval operations and maintenance aviation deck (USN).
   2 North Sea operations for mutual air defence (Raids).

Nomex   Family of nylon/phenolic honeycomb structures, core resin-impregnated or coated paper.

nominal acceleration point   That geographical location at which SST is to begin supersonic acceleration.

nominal deceleration point   Location, varying with flight level and pitch attitude, at which SST is to begin deceleration to subsonic regime.

nominal dimension   Various interpretations, typically that indicated on drawing before allowances, fits and tolerances.

nominal gas capacity   That of gas cells of aerostat under defined conditions of inflation, ambient pressure and flight attitude.

nominal jet edge   Boundary of discrete high-energy jet [eg from jet engine], conventionally taken as locus of points at which V is 10% of maximum.

nominal performance   Published, or according to brochure.

nominal pitch   See *standard pitch*.

nominal weapon   Nuclear weapon having yield of approximately 20 kT.

Nomos   Noise-monitoring system.

NOMSS   National operational meteorological satellite system (NOAA).

NON   Unmodulated NDB, transmitting no information.

non-co-operative scorer   One whose ammunition is not modified, or does not need modification, for scoring purposes (ASCC).

non-co-operative target   One without emissions, transponder or enhancement device.

non-destructive testing   Methods of testing structures for integrity, esp. absence of manufacturing flaws or cracks, that do not impair serviceability or future life.

non-developmental item   For practical purposes = off the shelf.

non-differential spoilers   Main feature is that in airbrake mode all spoilers remain extended even in demand for roll (see *spoilers*).

non-directional beacon   ADF ground station sending in 190–550 kHz range with keyed identification carrier.

non-effective sortie   Aircraft which for whatever reason fails to accomplish mission (DoD).

non-ferrous   Metals and alloys not based on iron; term usually also excludes aluminium alloys and generally means coppers and brasses.

non-fluff   Lint-free.

non-flying prototype   Essentially mock-up built to full flight standard but, for whatever reason, not cleared or intended for flight.

non-frangible wheel   Various techniques applied to design and fabrication of turbine disc to preclude possibility of rupture in overspeed or asymmetric condition.

non-galvanic corrosion   That due to causes other than formation of eleric cells; two important examples are fretting and microbiological.

non-handling pilot   Member of civil flight crew not actually flying the aircraft.

non-holding side   That on left side of holding course inbound towards holding fix.

non-instrument runway   No ILS.

non-interchangeable socket   Otherwise standard multipin or other sockets on device which ensure correct attachment of several connectors.

non-kinetic-energy weapons   Lasers, microwaves, radio and similar wave systems.

non-landing section   That length of runway from original threshold to displaced threshold.

non-operating active aircraft   Allowance, usually 10%, above UE level to make up for IRAN, modifications and heavy maintenance (USAF).

NONP   Non-precision approach runway.

non-precision approach   Without electronic glideslope.

non-precision instrument runway   Without ILS.

non-program aircraft   Those in inventory other than active or reserve, eg experimental or withdrawn (DoD).

non-radar   Self-explanatory, but can mean pilot is not using radar provided.

non-return-flow tunnel   Simple wind tunnel open at both ends.

non-rigid airship   Without rigid skeleton or stress-bearing covering around lifting cells.

non-sked   Non-scheduled, ie not operating to a timetable (colloq.).

non-structural   Other than primary or secondary structure; physical breakage of part would not imperil continued flight.

non-traffic stop   Stop by transport aircraft in public service planned in advance for reasons other than to pick up or set down.

non-volatile   Permanent memory.

NOO   Naval Oceanographic Office.

Nopac   North Pacific.

NOPR   Notice of proposed rulemaking (FAA).

NOPT, NoPT   No procedure turn required (FAA).

Nora   Not only radar, for post-2006 Gripen.

Norad   North American Air Defense Command (US/Canada, ratified 1958).

Norcote   Spacecraft coatings based on phenolic resin and powdered cork.

NORDA   No radio; also NORDO.

Norden sight   Complex but highly accurate optical bombsight for high-altitude level bombing (US, 1941–49).

Norden gear   Patented carrier-landing energy-absorption system.

NORDO   Alternative to NORDA, common in UK.

Norfab   Fire-blocking aluminised material incorporating polyamide binder and glassfibre.

NOR gate   Logic circuit usable as either AND or OR, depending on logic levels chosen to represent 0 or 1.

NORM, Norm   Not operationally ready, because need for maintenance.

normal   *1* Perpendicular to.

2 Maximum continuous, eg engine rating (R).

normal acceleration   Acceleration in vertical plane relative to aircraft, along OZ axis (eg as result of rotation about OY axis).

normal axis   Vertical axis (note: may not be vertical but must be at 90° to longitudinal axis in plane of symmetry); also called OZ axis. Positive direction is downward.

normal force   That measured on body in fluid flow at 90° to free-stream direction, symbol Z (rarely N).

normal force coefficient   Dimensionless coefficient C_Z derived from Z; also written $C_L \cos \alpha + C_D \sin \alpha$ where C_L and C_D are lift/drag coefficients and α is angle of attack.

normal glide   That at which glide ratio is maximum.

normal gross weight   Usually same as MTOW; excluding all overload, emergency or alternate gross weights.

normal horsepower   Not defined but generally same as rated hp.

normalizing   Stress-relieving heat treatment usually comprising heating to above critical temperature followed by cooling in atmosphere.

normal landing   For tailwheel aeroplane, three-point landing.

normal load factor   That measured along normal axis [the usual meaning].

normal loop   Loop, as distinct from inverted loop, starting and finishing in straight and level flight in upright attitude.

normally aspirated   Unsupercharged.

normal mode   Free vibration of undamped system.

normal outsize cargo   That having cross-section greater than 9 ft × 10 ft, which is C-130 or C-141 size (DoD).

normal pressure drag C_{Dp}, downwind resultant force coefficient.

normal propeller state   Usual condition for helicopter under power, with rotor thrust in opposition to flow direction through disc.

normal rating   Maximum continuous (R).

normal shock   Shockwave at 90° to fluid flow direction.

normal spin   Intentional spin entered from upright attitude and recoverable by centralizing controls or applying opposite rudder (US usually adds 'within two turns').

normal turn   Procedure turn through 360° in two minutes.

normal velocity   "The component velocity along the normal axis relative to the air", (B.S., 1940).

Normand theorem   On tephigram a dry-adiabatic line drawn through dry-bulb temperature, saturated-adiabatic through wet-bulb temperature, and dewpoint line through dewpoint temperature all meet at point which represents condensation level.

NORS, Nors   Not operationally ready, spare parts (or supply, as an order).

Norse   Nuclear, optical and radar [radiation] signature evaluation [or estimation].

northerly turning error   Transient errors in magnetic-compass reading caused by vertical component of magnetic field, at maximum when turning off northerly or southerly course. In N hemisphere compass is sluggish and lags behind when turning to L or R through northerly heading and races ahead when turning L or R through southerly. In US usually called magnetic turning error.

north mode   Display, eg Automap or moving-map, has N at top.

NOS   *1* Night observation sight, or surveillance.

2 National Ocean Service.

NOSC   NATO operations support cell.

nose   *1* Foremost part of vehicle, measured relative to direction of travel, excluding secondary structures or probes.

2 Leading portion of aerofoil, hence D-*, * rib.

nose art   Decorative painting on aircraft forward fuselage or nose.

nose battens   Radial stiffeners around nose of airship.

nose-cap   *1* Removable nose, usually body of revolution, forming forward extremity of larger forebody.

2 Small spinner not extending further aft than front of blade roots.

3 Bow cap.

nosecone   Essentially conical nose of high-speed vehicle, esp. fairing over re-entry vehicle.

nose-dive   Dive at very steep (near-vertical) angle.

nose down   *1* To push over from level flight into glide or dive.

2 To fly with fuselage in ** attitude, though not necessarily losing height.

nose drive   Shaft drive to auxiliaries taken off front of gas-turbine engine along axis of symmetry, eg ** generator.

nose entry   Shape of aircraft nose evaluated from aerodynamic and aesthetic viewpoints.

nose gear   Forward unit of tricycle landing gear, no matter how far location is from nose.

nose gearbox   Gearbox mounted on front (usually on centreline) of gas-turbine engine to drive auxiliaries, helicopter shafting or propeller (ie, turboprop); not mounted remotely on struts.

nose-heavy   Tending to rotate nose-down when controls released.

nose in   *1* To taxi and park facing terminal building or finger.

2 Aircraft thus parked (see *Agnis, Safeway, sidemarker*).

nose landing gear   *Nose gear.*

nose leg   Main leg of nose gear.

nose over   To overturn (eg after landing tailwheel-type aircraft on soft ground) by rotating tail-up to inverted position.

noseplane   Canard foreplane mounted at or ahead of nose; not applicable to conventional modern canards.

noseplate   Metal plate on centreline of hang glider linking leading-edge tubes.

nose radar   Radar whose aerials (antennas) are in nose of aircraft pointing ahead, esp. for use against targets ahead of aircraft.

nose ribs   Ribs along leading edge extending chordwise only as far as front spar.

nose slice   Maximum-rate yaw induced by rudder.

nose slots   Apertures in low-pressure region of high velocity around nose for discharge of fluid flow, eg cooling air.

nose tow   Standard US Navy method of catapult link for accelerated carrier takeoff by pulling on nose leg.

nose up   To rotate in pitch from level flight to climb.

nosewheel   Wheel(s) of nose gear.

no-show   Airline passenger who has booked ticket but fails to check in for flight.

Nosig   No significant meteorological change (ICAO).

no step   Stencilled warning on aircraft: do not put weight on this area.

NOT   Naczelna Organizacja Techniczna (Polish federation of engineering associations).

Notal   Not to all.

Notam, NOTAM   Notice to Airmen, identified as notice or as Airmen Advisory, disseminated by all means to give information on establishment, conditon or change in any aeronautical facility, service, procedure or hazard; suffix D distant [wide dissemination], L local (ICAO).

Notam code   Standard code for transmitting Notams; eg QAUED 3 MC 5813 142359 is interpreted as 'Met com operating frequency of 3 MHz will be changed to 5,813 kHz on 14th of this month at 23.59'.

Notar   No tail rotor, torque reaction supplied by offset thrust from air blown through slit in tail boom (Hughes, then McDonnell Douglas, now Boeing).

notch   Essentially chordwise or streamwise sawcut or groove over nose of aerofoil.

notch aerial   Formed by cutout in skin of vehicle, leaving aperture matched to wavelength (usually in HF com. band) and covered with dielectric skin to original profile.

notched cone nozzle   Promising primary nozzle in ejector-lift system in which primary flow is discharged through row(s) of fishtail (lozenge-section) nozzles generating plumes orthogonal to long axis of duct.

notched elevators   Cut away at trailing edge for rudder movement.

notch effect   Shortcoming of early (1960s) all-flying tailplanes or slab tails in which demands tended to be in noticeable increments [pilot often reverted to flying on trimmer].

notch flap   Leading-edge flap extended from fuselage.

notifiable accident   One which cannot legally go unreported, where any person suffers injury, third-party property is damaged or public are in any way put at risk; variable rules governing scale of damage to aircraft.

no-transgression zone   Airspace where aircraft under positive electronic IFR control must not penetrate, esp. region 900+ m/2,000+ ft wide between aircraft making ILS approaches on to parallel runways.

NOTS   Naval Ordnance Test Station (USN, China Lake, Inyokern etc, now NWC).

Notus   Notice to users (Arinc).

Nova   Networked open versatile architecture.

NOV-AB   Non-persistent Toxic-B gas bomb (USSR).

Novcam   Non-volatile charge-addressed memory.

Novoview   Range of CGI(2) visual systems, some textured (Rediffusion).

Novram   Non-volatile RAM.

Nowcast   Report on current weather.

no-wind position   Geographical position aircraft would have occupied had wind velocity been zero.

NOx, NOX   Nitrogen/oxygen breathing mixture (normally means supplied in absence of atmosphere).

NO_x   Shorthand for all oxides of nitrogen resulting from combustion of fuel in air.

noy, Noy   Subjective measure of noisiness in bandwidths of one-third octave (see *noise*).

NOZ   No operating zone.

nozpos   Nozzle position[s], ie angle.

nozzle   *1* In jet-propulsion or reaction-jet control, aperture through which fluid escapes from system to atmosphere and in which as much energy as possible is converted to kinetic energy (see *choked, con/di*).

2 Wind-tunnel section immediately upstream of working section.

3 Primary aperture through which fuel is injected into gas-turbine engine combustion chamber [US usage].

4 Section of fluid flow duct upstream of axial turbine, in which flow is controlled to enter turbine at favourable direction, pressure and velocity. Form depends on whether turbine is impulse/reaction or pure impulse.

5 Incorrectly, nozzle guide vane.

nozzle blade   See *nozzle guide vane*.

nozzle block   See *nozzle (2)*.

nozzle box   Assembly of nozzle guide vanes (all, or a group filling portion of periphery) and surrounding walls of gas duct.

nozzle bucket   See *nozzle guide vane*.

nozzle contraction ratio   Ratio of flow cross-section area at inlet or start of nozzle (esp. con/di or rocket) to area at throat.

nozzle diaphragm   Nozzle (4) as complete assembly.

nozzle efficiency   Usually ratio of actual change in kinetic energy across nozzle to ideal value for given inlet conditions.

nozzle exit area   Area of cross-section of flow at exit.

nozzle expansion ratio   In supersonic nozzle, ratio of exit area to throat area.

nozzle guide vane   Radial aerofoils upstream of axial gas turbine, convergent passages between which form nozzle (4) through which gas is directed on to turbine rotor blades. Also called turbine stator.

nozzle insert   Small blocking body fixed inside nozzle (1) to trim exit area; colloq. = mice.

nozzle ring   Complete 360° assembly of nozzle guide vanes.

nozzle throat   Region of con/di nozzle having smallest cross-section area.

nozzle thrust coefficient   Usually defined as actual achieved thrust divided by product of chamber pressure and throat area (rocket).

NP   *1* Noise-preferential.

2 North Pacific region (ICAO).

N_p   *1* Number of passengers.

2 Speed (rpm) of power turbine or, confusingly, propeller.

$\overline{np}$ Yawing moment coefficient due to rolling.

NPA   *1* Notice of proposed amendment (FAA, JARs).

2 National Packaging Authority (UK).

3 National Pilots' Association [trade union] (US).

4 Non-precision approach.

NPB   *1* Nuclear-powered (or propelled) bomber.

2 Nadge Policy Board.

NPD   Need and planning documents.

NPDU   Network protocol data unit.

NPE   Navy preliminary evaluation (USN).

NPG   *1* No performance group type aircraft (CAA).
2 Nuclear planning group.
3 Non-unit personnel generator.

NPI   Non-precision instrument.

NPIAS   National Plan of Integrated Airport Systems (US).

NPL   National Physical Laboratory (UK, 1899–).

NPLOs   NATO production and logistics organizations.

NPL tunnel   A closed-jet tunnel, original * = return flow, standard * = non-return flow.

NPO   Scientific production union (USSR, R).

NPOESS   National, later (2002) changed to New, polar-orbiting operational environmental satellite system (NASA, NOAA, USAF).

NPP   *1* Research and production enterprise (R).
2 NPOESS preparatory project.

NPPL   National Private Pilot's Licence [proposed, VFR only] (UK).

NPR   *1* No power recovery.
2 Nozzle pressure ratio.
3 Nuclear posture review (US).
4 Noise preferential route.

NPRM   Notice of proposed rulemaking (FAA).

NPRS   Non-persistent.

NPS   *1* Non-prior service.
2 Naval Post-graduate School (USN).
3 Nuclear profiled sortie.

NPT   Non-proliferation treaty, 1 July 1968.

NPTR   *1* National Parachute Test Range.
2 No procedure turn required.

NPU   Navigation processor unit.

NQA   National Quality Assurance [certification authority] (UK).

NQIS   Navigation-quality inertial sensor.

NQR   Nuclear quadrupole resonance.

NR, Nr   *1* Helicopter main-rotor rpm.
2 Network router.

$\overline{n_r}$ Yawing moment coefficient due to yawing.

N_r *1* Helicopter main-rotor rpm.
2 Number (FAA).

NRA   Nuclear reaction analysis, ion-beam technique for light elements.

NRAG   Non-metallics research advisory group (MRCC).

NRC   *1* National Research Council.
2 Nuclear reporting cell.
3 Nuclear Regulatory Commission (US, 1974 onwards).

NRCS   Normalized radar cross-section.

NRD   Network routing domain.

NRDC   *1* National Research Development Corporation (UK).
2 Natural Resources Defense Council (US).

NRDS   Nuclear rocket development station (Jackass Flats, US).

NRF   NATO Response Force (2002–).

NRH   No reply heard.

NRI   Net radio interface.

NRL   Naval Research Laboratory (USN).

NRMM   Navaids remote maintenance and monitoring.

NRO   National Reconnaissance Office (US).

NROSS, Nross   Navy remote ocean sensing system or satellite (spacecraft, USN).

NROTC   Naval Reserve Officer Training Corps (US).

NRP   *1* Normal rated power.
2 Narrow programmable receiver.
3 National Route Program, or Plan (US).
4 Non-return point.

NRPA   Net return on productive assets.

NRPB   National Radiological Protection Board.

NRR   Noise-reduction rating (EPA).

NRRC   Nuclear-Risk Reduction Centers (1987).

NRSC   National Remote Sensing Centre (in civil enclave at former RAE Farnborough, UK).

NRST   Nearest.

NRT   Near real time [DF adds data fusion, RAS resource allocation system].

NRTS   Not repairable this station (USAF).

nrv   Non-return valve.

NRZ   Non-return to zero.

NS   *1* Non-skid.
2 Network service.

Ns   *1* Nimbostratus.
2 Newton-second, also N-s.

ns   Nanosecond[s] (10^{-9}s).

n_s Maximum sustained normal acceleration.

NSA   *1* National Security Agency (US).
2 National Standards Association (US).

NSAP   Network service access point.

NSAR   North Sea air-combat manoeuvring instrumented range.

NSAWC   Naval Strike and Air Warfare Center (USN).

NSC   *1* National Security Council (US).
2 No significant cloud.
3 New Scottish [ATC] Centre, Prestwick.

NSCA   National Safety Council of America.

NSDA   National Space Development Agency (J).

NSDU   Network service data unit.

NSE   Near-synchronous equatorial.

NSEU   Neutron single-event upset.

NSF   National Science Foundation (US).

NSFAC   Non-Scheduled Flying Advisory Committee (CAA, 1945).

NSG   *1* NATO Standardization Group.
2 North-seeking gyrocompass.

NSGr   Night close-support group (G, WW2).

NSIA   National Security Industrial Association (US).

NSIU   Navigation switching interface unit.

NSM   Non-contact stress measurement.

NSMS   Non-intrusive stress-monitoring system.

NSMV   Near-space maneuvering vehicle (USAF).

NSN   National [or NATO] stock number[s].

NSNF   Non-strategic nuclear forces.

NSO   Navigation/systems operator.

NSOC   Naval Satellite Operations Center (Pt Mugu, CA).

NSP   *1* Normal sea-level power.
2 Night surveillance payload.
3 National Simulator Program; O adds Office (FAA).

NSPE   National Society of Professional Engineers (US).

NSPOL   Non-scheduled policy (ECAC).

NSR   *1* No scheduled removal.
2 Naval [or NATO] Staff Requirement.

NSRI   National Soil Resources Institute (U. of Cranfield).

NSRP   National Search and Rescue Plan (US).

NSS   *1* National seismic station (US).

    2 Near-source simulation.

    3 Navigaton subsystem.

    4 National Security Space (US).

NSSA   *1* National Safe Skies Alliance (US).

    2 National Space Society of Australia.

NSSC   Naval Ship Systems Command (USN).

NSSFC   National Severe Storms Forecast Center (Kansas City).

NSSI   National Security Space Integration; O adds Office (DoD).

NSSL   National Severe Storms Laboratory (US).

NST   *1* NATO Staff Target.

    2 Noise, spikes and transients.

NSTAC   National Security Telecommunications Advisory Council (US).

NSTAP   National strategic technology acquisition plan (UK).

NSTB   National satellite testbed.

NSTD   Non-standard.

NSTL   *1* National Space Technology Laboratories (NASA, previously MTF, now SSC (4)).

    2 National security threat list (US).

NSTO   Near single stage to orbit.

NSTP   National space technology programme (DTI, UK).

N-strut   Interplane or other bracing system having general shape of N.

NSVN   NATO secure voice network.

NSW   *1* Nominal specification weight.

    2 No significant weather [TAF or Metar].

NSWC   Naval Surface Warfare Center (USN).

NSWP   Non-Soviet Warsaw Pact.

nT   Nanotesla; terrestrial field varies from c 25,000 (magnetic equator) to c 70,000 nT (poles), vertical component usually being measured in geophysical prospecting and ASW.

nt   Nit (name not usually used).

NTAOCH   Notice to AOC holders (CAA).

NTAP   Notice to Airmen publication (USGPO).

NTAS   Norad tactical Autovon system (USAF).

NTAT   Near-term Acme technology.

NTB   *1* National test bed (US, SDI).

    2 Nuclear test ban; T adds Treaty.

NTC   *1* National Training Center (US DoD).

    2 Numerator time constant.

    3 Notice.

NTD   National Test Director (ISS).

NTDD   Normalized total departure delay.

NTDS   Naval tactical data, or distribution, system.

NTE   *1* Not to exceed.

    2 Northerly turning error.

NTF   *1* No trouble found.

    2 NATO task force.

NTFWTC   NATO Tactical Fighter and Weapons Training Centre.

NTG   Nachrichtentechnische Gesellschaft im VDE (G).

Nth country   Next power to possess nuclear weapons (DoD).

NTI   Next [runway] turn indicator.

NTIA   National Telecommunications and Information Administration (US).

NTIS   National Technical Information Service (US).

NTK   Scientific and technical committee (many in USSR, R).

NTM   *1* National technical means, of verification of MBFR and precise Earth mapping.

    2 NDT manual.

    3 NATO Tiger Meet.

NTMV   National technical means of verification.

NTNF   Royal Norwegian council for scientific and industrial research.

NTO   *1* *Nitrogen tetroxide.*

    2 Notice to operators.

    3 No technical objection[s].

NTOS   No time on station.

NTP   Normal (ie standard) temperature and pressure.

NTPD   NTP dry (gas bottle capacities).

NTS   *1* Negative-torque signal.

    2 Navigation technology satellite (USN).

    3 Night targeting system.

NTSB   National Transportation Safety Board (US, from 1966).

NTSC   National Television Standards Commitee (US).

NTT   Non-threat traffic.

NTTC   National Technology Transfer Center (NASA).

NTU   *1* State technical administration (USSR, R).

    2 Navigation training unit.

    3 New-threat upgrade.

    4 Not taken up [civil registration].

NTW   Navy theater-wide; /BMD adds ballistic-missile defense (USN).

N₂O₄   *Nitrogen tetroxide.* — N_2O_4 *Nitrogen tetroxide.*

NTWS   New threat-warning system.

NTZ   No-transgression zone.

NU   Nose-up.

Nu   *Nusselt number.*

nuclear airburst   Explosion at height AGL greater than maximum radius of fireball.

nuclear bomber   Ambiguous but generally taken to mean aircraft able to deliver nuclear weapons.

nuclear capable   Nuclear bomber.

nuclear cloud   All-inclusive term for volume of hot gas, dust, smoke, and other particulate matter from nuclear bomb and environment carried aloft with fireball (DoD, NATO).

nuclear column   Hollow cylinder of water and spray thrown up from underwater nuclear explosion through which hot high-pressure gases escape to atmosphere (DoD, NATO).

nuclear defence   Defence against attack by nuclear or radiological weapons.

nuclear dud   NW which after being triggered fails to provide any explosion of that portion designed to produce nuclear yield (DoD).

nuclear emulsion   Thick layer used on photo-type plates for recording tracks of energetic particles.

nuclear energy   That liberated by fission; more rarely, that liberated by fusion reaction, and some definitions include radioactive decay.

nuclear explosive   Material designed to achieve greatest uncontrolled fission or fission + fusion reaction.

nuclear/heater propulsion   Using nuclear reactor to heat working fluid for rocket propulsion.

nuclear incident   Unexpected event short of NWA but resulting in damage to NW or associated facilities or increase in possibility of explosion or radioactive contamination (DoD).

nuclear radiation   All EM and particulate radiations from nuclear processes.

nuclear reactor   Device for containing controlled nuclear fission (rarely, fusion) reaction.

nuclear rocket   Usually one whose working fluid is heated in nuclear reactor.

nuclear safety line   Line drawn (if possible through prominent topographic features) on map to serve as reference in describing levels of protective measures, degrees of damage and limits allowed for effects of friendly NW.

nuclear surface burst   One in which centre of fireball is below that height equal to maximum radius.

nuclear underground burst   One in which centre of detonation lies below original ground level.

nuclear weapon   One in which almost all released energy results from fission, fusion or both; abb. NW. Original 1945 designs used chain-reaction triggered by critical mass of uranium with unnaturally high concentration of isotope U-235. Second form was based on plutonium Pu-239. Loosely called atomic, or fission, bomb. See *hydrogen bomb*.

nuclear-weapon(s) accident   Unplanned occurrence resulting in loss of, or serious damage to, nuclear weapons or components resulting in actual or potential hazard to life or property (DoD, NATO).

nuclear-weapon degradation   Degeneration of NW to such extent that anticipated yield is reduced.

nuclear-weapon employment time   Reaction time of NW.

nuclear-weapon exercise   Exercises involving real NW but excluding launching or flying operations.

nuclear yield   Energy released in detonation of NW measured in terms of mass of TNT required to liberate same amount: categorized as very low (less than 1 kT), low (1–10 kT), medium (10–50 kT), high (50-500 kT) and very high (over 500 kT).

nucleating agent   The material released into the atmosphere in cloud seeding, such as silver iodide.

nucleon   Particle of atomic nucleus, eg neutron, proton.

nucleonics   Science of applications of nucleons and atomic emissions.

nucleus   *1* Particle upon which atmospheric water vapour can condense and/or freeze.

2 Positively charged core of atom.

nuclide   Member of family of atoms distinguished by having same nucleus (immediately decaying nuclear states are excluded).

Nudets   Nuclear detonation detection and reporting system; system for surveillance of friendly target areas and immediately providing place, ground zero, burst height and yield of nuclear explosions (DoD, NATO).

Nufas   NATO uhf frequency-assignment system.

Nugget   First-tour aviator (USN).

nugget   Obturator (US, colloq.).

NUGP   Nominal unit ground pressure.

NUI   Network-user identification.

nuisance malfunction   One caused by failure of safety system, eg auto-disconnect of AFCS, main system being serviceable throughout.

null   *1* Orientation of receiver aerial [antenna] (eg ADF) in which no signal is heard.

2 In DLC, angular setting (usually about 7°) about which spoilers oscillate.

3 Location in space where, at any moment, gravitational forces cancel out (eg five centres of libration near Earth/Moon system).

nullo   Flown without human on board [cockpit-equipped target or other RPV], from 'no live, or local, operator'.

null position   See *null (2)*.

NUM   Nuova unita maggiore, future aircraft carrier (Italy).

Numast   National Union of Marine, Aviation and Shipping Transport officers (UK).

No 1 engine   Port outer: engines numbered across aircraft port to starboard.

numbers   *1* Runway designator.

2 Piano keys, to land on the *.

3 Any numerical data needed for flight, as 'have *'.

numerator time constant   Controls rate at which attitude changes result in flight-path changes, symbol $T\theta_2$.

numerical control   Control of system, esp. machine tool, by series of commands expressed in numeric (digital) terms giving locations, directions and speeds, and secondary information (eg cutter speed) in correct sequence. Software usually punched paper or magnetic tape.

numerical weather prediction   Repeatedly refined methods in which powerful computers solve equations and simulate local atmosphere, hence numerical forecast, numerical model.

Nunio   Network universal input/output [local Ethernet].

NURK   Astronautics Society (J).

nurse balloon   Fabric gas container used as reservoir or to maintain constant inflation pressure in aerostat on ground.

Nusselt number   Non-dimensional parameter $Nu = qD/\lambda\delta T$ where q is quantity of heat, D is typical length, λ is thermal conductivity and δT is temperature difference.

Nut   Near-unity probability.

nutating feed   Microwave feed to tracking radar in which beam oscillates in one plane while plane of polarization (and usually aerial [antenna] reflector) remains fixed.

nutation   *1* Oscillation of axis of rotating body (eg gyro).

2 Irregularities in precession of equinoxes and other effects and of rotary precession of Earth's axis in period of 18.6 years (* period) with maximum displacement of 9.21 s (constant of *).

nutator   Drive mechanism causing dipole or aerial [antenna] feed horn to gyrate about focus of paraboloidal aerial without changing plane of polarization.

nutcracker   V/STOL aeroplane whose fuselage hinges near mid-length to vector main-engine thrust.

nut plate, nutplate   Nut, esp. elastic stop nut, provided with flat-plate base attached permanently to airframe to provide anchor into which attachment bolt can be screwed for securing access panel or other removable item.

nut runner   Large nut threaded on screwjack so that when either ** or screw is rotated a linear motion results (eg to drive tailplane).

NV   *1* Naamloze Vennootschap (Netherlands, Belgium, company constitution).

2 National Variants, to FAR-25 and similar.

3 Night vision.

4 Non-volatile.

NVCA   National Venture Capital Association (US).

NVEO   Night-vision electro-optics.

NVESD   Night Vision Enhanced [or and Electronic] Sensors Directorate (USA).

NVG   Night-vision goggles; T adds training.

NVIS   *1* Near-vertical incident skywave.

    2 Night-vision imaging system.

NVL   Netherlands air transport association.

NVLP   Netherlands Aerospace Writers Association.

NVM   Non-volatile memory.

NVPS   Night-vision pilotage [or piloting] system.

NVQ   National Vocational Qualification (UK).

NVR   Netherlands astronautical society.

NVRAM   Non-volatile RAM.

NVS   *1* Night-vision system.

    2 Noise and vibration suppression [or simulation].

nvt   Neutron volts, measure of ionising radiation.

NV thrust   Nominal vacuum thrust.

NVTS   Night-vision targeting sight.

NVvL   Nederelandse Vereniging voor Luchtvaarttechniek.

NW   *1* Nuclear weapon, or warfare.

    2 Nosewheel.

NWA   Nuclear-weapon(s) accident.

NWAA   North West Aerospace Alliance; 180+ companies (UK).

NWC   *1* Naval Weapons Center (China Lake, USN).

    2 National War College (US).

NWDC   Navy Warfare Development Command (USN).

NWDS   Navigation and weapons-delivery system.

NWEF   Naval Weapons Evaluation Facility (Albuquerque, USN).

NWP   Numerical weather prediction.

NWPRA   National Women's Pylon Racing Association (US, from 1964).

NWS   *1* National Weather Service (NOAA, US).

    2 Nosewheel steering.

    3 North[ern] Warning System (former Dew Line).

    4 NW (1) state, or status.

NWSSG   Nuclear weapon system safety group.

NWT   Natural work team.

NWTS   Naval Weapon Test Squadron (NAS Point Mugu).

NWV   New World Vistas (USAF).

NXT   Next [K adds track].

$\overline{n}_y$ Yawing moment coefficient due to sideslip.

Nycote   Nylon lacquer protective coating.

nylon   Long-chain synthetic polymer amide with recurrent amide groups distributed along chain (some definitions add 'capable of being drawn into filament whose structure is oriented along fibre'). Large family, now often used for 3-D mouldings and many structural purposes (formerly TM for Du Pont de Nemours).

nylon letdown   Parachute.

NYO   Not yet operating.

N_z *1* Normal load factor, ie along OZ axis.

    2 AMSU normal-acceleration ouput.

NZAF   New Zealand Aviation Federation Inc.

NZAPA   NZ Airline Pilots' Association.

NZAT   NZ Airport Technologies (trade association, 19 members).

NZCA   NZ College of Aviation.

NZDF   NZ Defence Force.

NZG   Near-zero growth (tyres).

NZGA   NZ College of Aviation.

NZMAA   NZ Model Aeronautical Association.

NZMS   NZ Meteorological service.

NZRA   NZ Rotorcraft Association.

NZSFA   NZ Space Flight Association Ltd.

O

O *1* Opposed configuration (US piston engine designation).

2 Operating cost over given period.

3 Ground speed 'on' (onwards from critical point).

4 Observation aircraft category (US services from 1922).

5 Hang-glider category (FAI).

6 Other meanings include odd, over, oxygen and optional.

O₂ Oxygen.

O₃ Ozone.

O-ring Flexible fluid-sealing ring having O-section in free state.

OA *1* Output axis.

2 Operational analysis.

3 Observation amphibian (USAAC, USAAF, 1925–47).

O/A On or about.

OAA Orient Airlines Association.

OAAN Organismo Autónomo Aeropuertos Nacionales (Spain).

OAB Outer air battle.

OAC *1* Operations Advisory Committee (UK CAA).

2 Oceanic Area Control [C adds Centre].

3 Austrian aero club.

OACI ICAO (F).

OACS Optically active coding system.

OAD Office of Aviation Development [CAA, 1949 on].

OADS Omnidirectonal air-data system (helicopter).

OAE Optimized after erosion (helicopter blade profile).

OAF Optical-alignment facility.

OAFS Open apron, free-standing (ie, no air-bridge).

OAFU Observers advanced flying unit.

OAG *1* Operational Advisory Group (BATA, USN).

2 Official Airline Guide.

OAIRMS Open-architecture integrated radio management system.

OALC Ogden Air Logistics Center (Hill AFB, Utah).

OALT Operationally acceptable level of traffic.

OAM Office of Aviation Medicine (FAA).

OAMCM Organic airborne mine countermeasures.

OAMP Optical airborne measurement platform.

OAMS Orbital attitude and manoeuvre system.

O&C Operations and checkout.

O&D Origin and destination.

O&I Operations and integration.

O&M Operations & maintenance [CM adds configuration management].

O&O Organizational and operational (RPV).

O&S *1* Operational and support (costs).

2 Operations & Sustainment, of hardware in service (USAF).

OANS Observers air navigation school.

OAO *1* Orbiting astronomical observatory.

2 Out-of-area operations.

OAOI On-and-off instruments.

OAP *1* Organizzazione dell'Aviazione Privata e d'Affari (I).

2 Offset aiming point.

3 On-board attitude processor.

OAPD Online airline product database (IATA).

OAPEC Organization of Arab Petroleum Exporting Countries.

OAPP Office of Aviation Policy and Plans (FAA 1).

OAPS Oblique air photograph strip.

OAR Office of Aerospace Research (USAF and FAA/DoT).

OARB Orient airlines research bureau.

OARF Outdoor Aerodynamic Research Facility (NASA).

OARN Off-airways R-nav.

OART Office of Advanced Research & Technology (NASA; now AST).

OAS *1* Offensive avionics system.

2 Offensive air support (air support directly linked to land operations, US).

3 Omnidirectional airspeed system (helo).

4 Open-access service.

5 Optimal aircraft scheduling.

6 Obstacle assessment surface.

7 Office of Aviation Safety (CAA).

8 On active service.

9 Oceanic Automation System (FAA/USN).

OASC Officer and Aircrew Selection Centre (RAF, Biggin Hill).

OASD Office of the Assistant SecDef (US).

OASF Orbiting astronomical support facility.

Oasis *1* Oceanic and atmospheric scientific information system (NOAA).

2 Operational application of special intelligence system (AAFCE).

3 Omnidirectional approach-slope indicator system.

4 Operational and supportability implementation system (USAF/FAA).

5 Organic airborne and surface-influence sweep.

6 Oceanic Area system improvement study.

7 On-board aircraft server and information system.

OASPL Overall sound pressure level.

OAST Office of Aeronautics and Space Technology (NASA), more usually AST.

Oasys Obstacle-avoidance, or awareness, system.

OAT *1* Outside air temperature.

2 Operational acceptance test.

3 Operational air traffic (ie military).

4 Oxide aligned transistor.

5 Operational, or optional, auxiliary terminal (Acars).

6 Operational airfield test set.

7 Orbit and attitude tracking; S adds system.

OAV Organic air vehicle.

OAVUK Society for aviation and gliding of Ukraine and Crimea.

OB *1* Outbound.

2 Balloon club (Austria).

3 Off-boresight.

4 Operations base (USA).

5 Operational Bulletin.

O/B *1* Outbound.

2 Outboard.

OBA   Outbound boom avoidance (SST).

OBAP   Organization of Black Airline Pilots (US).

OBC   *1* Optical barrel camera.

2 On-board computer.

OBCO(S)   On-board cargo operations (system).

Ob.dL   Oberkommando der Luftwaffe (G).

OBDMS   On-board data-monitoring systems.

OBE   *1* Off-board expendables.

2 Overtaken by events.

OBEWS   *1* On-board electronic warfare simulation.

2 On-board EW system(s).

OBFM   Offensive basic flight manoeuvres.

OBI   Omni-bearing indicator.

Obiggs   On-board inert gas generating system.

OBIS   On-board information system.

objective   *1* Normally, in optical, EO or IR system, first lens or lens group to receive incoming radiation.

2 Military target for capture or other action by surface forces; not used for target for aerial attack.

oblique   Oblique photograph.

oblique camera   One mounted in aircraft with axis between vertical and horizontal.

oblique photograph   One taken by oblique camera; subdivided into high * in which apparent horizon appears and low * in which it does not.

oblique projection   Map projection with axis inclined at oblique angle (say, 20° to 60°) to plane of Equator.

oblique-shock inlet   Inlet designed for use in supersonic vehicle and provided with centrebody, wedge or other projecting portion intended to focus oblique shockwave on opposite lip.

oblique shockwave   Inclined shock formed whenever supersonic flow has to turn through finite angle in compressive direction.

oblique wing   Wing arranged to pivot at mid-point as single unit so that, as one half is swept back, other half is swept forwards. Also called slew wing.

OBLMS   On-board life-monitoring system.

Oboe   WW2 precision navaid of SSR type with Cat station near Dover and Mouse station in Norfolk sending synchronized pulses which formed continuous note along correct flightpath over distant target; Mouse operators also sent signals to tell aircrew when to release bombs or TIs.

Obogs   On-board oxygen generation (or generating) system.

OBP   *1* On-board processor, or processing [satcom].

2 Omnidirectional ball panel.

3 Operational build plan.

OBPR   Office of Biological and Physical Research (NASA).

OBR   Optical beam rider, or riding.

OBRC   Operating budget review committee.

OBRM   On-board replacement module.

OBS   *1* Omni-bearing selector.

2 Observe/observed/observing (ICAO).

3 On-board simulation.

4 Organizational breakdown structure.

5 Optical bypass switch.

6 Observation-balloon system (tethered near major highway).

obs   *1* Obstruction lights.

2 Observe[d], observation.

3 Obstacle.

Obsc, OBSC   Obscure/obscured/obscuring (ICAO).

obscuration methods   Instrument techniques for measuring visible smoke (eg from jet engine) such as HSU and PSU where cutoff of calibrated light beam is measured over known distance.

observation   *1* Many military definitions, most agreeing that * platform is for gathering all possible tactical information about an enemy.

2 Complete set of meteorological readings at one place and time.

observation balloon   In bygone wars, tethered balloon carrying human who reported on fall of shot and on enemy activity.

observation balloon system   In US, usually tethered at airfield to monitor potentially conflicting road traffic, eg at runway crossing.

observation mirror   A horizontal mirror with superimposed graticule used in the same way as a camera obscura.

observed value   Measured, not calculated.

observer   Common title for second crew member in two-seat military (especially combat) aircraft whose functions may include navigation, systems management, electronic warfare, command guidance of weapons and other tasks; title is traditional and may bear no relation to actual duties.

OBST, obst   Obstruction, obstacle.

obstacle clearance height   Lowest height above runway threshold or aerodrome elevation used to establish compliance with obstacle clearance criteria in instrument approach.

obstn   Obstruction (FAA).

obstruction   Also called obstacle, a real or notional solid body forming hazard to aircraft on or near runway or flight path. For certification purposes has height of 10.7 mm (35 ft) or 15 m (50 ft).

obstruction angle   Angle between horizontal and line joining highest point of object in flightway [ie, approach path] to nearest point of appropriate runway.

obstruction clearance surface   Surface in form of plane or flat cone sloping at obstruction angle.

obstruction-free zone   Free from all fixed obstacles except light frangibly mounted navaids.

obstruction light   Red light visible 360° on top of object dangerous to moving aircraft in air or on ground.

obstruction marker   Object of approved shape or colour marking obstruction or boundary of hazardous surface on airfield.

Obtex   Off-board targeting experiment[s].

obturator   Rigid or flexible body tailored to preventing escape of gas under pressure from particular orifice or other leakage path.

obturator ring   Piston ring of L-section intended to be gastight.

OBW   *1* On-board wheelchair (for disabled pax).

2 Off-boresight weapon.

OC, O/C   *1* O'clock.

2 Officer commanding.

3 Over-current.

4 Obstacle clearance (ICAO panel).

5 Optical communications.

6 Obstruction chart.

7 Officer candidate.

8 Oil consumption.

9 On course.

OCA *1* Obstacle clearance altitude [suffix H_{fm}, height for finals and straight missed approach; H_{ps}, height for precision segment].

  2 Oceanic control area.

  3 Offensive counter-air.

OCALC Oklahoma City Air Logistics Center (Tinker AFB).

OCAMS, Ocams On-board checkout and monitoring system.

OCC *1* Operation[s] [or Oceanic] control centre.

  2 Occulting.

  3 Occupied [telephone line].

  4 Operator control console.

Occar Organisme Conjointe de Co-opération en Matière d'Armément (Int.).

occluded front Warm air mass forced aloft by overtaking cold front.

occlusion *1* Atmospheric region in which cold front has overtaken slow or stationary warm front and forced warm air mass upwards.

  2 Trapping of gas bubbles in solidifying material or by molecular adhesion on surface, esp. removal of gas in high-vacuum technology by a getter.

OCCM Optical counter-countermeasures.

occultation Complete disappearance of body, esp. source of illumination such as star or aircraft lights, behind another object of much larger apparent size.

occulting Flashing, but with illuminated periods clearly predominant.

OCD *1* Operational concept demonstration.

  2 Oceanic clearance delivery.

  3 Office of Civilian Defense (US, 1941).

oceanic airspace Controlled airspace over ocean areas.

oceanic clearance Clearance delivery to enter oceanic airspace.

oceanic navigation error report Filed when surveillance monitor observes aircraft exiting oceanic airspace seriously off track.

OCF Occluded frontal passage.

OCFNT Occluded front.

OCH Obstacle clearance height.

OCI *1* Outside of clearance indicator (MLS).

  2 Out-of-coverage indicator (navaids).

OCIG Oceanic communications improvement group.

OCIP Offensive capability improvement program.

OCL *1* Obstacle [or obstruction] clearance limit.

  2 Optimum cruising level.

OCLD Occlude.

OCLN Occlusion.

OCM *1* On-condition maintenance.

  2 Optical countermeasures.

OCNL Occasional [Y adds occasionally].

OCNR Office of the Chief of Naval Research (USN).

OCP *1* Onmnidirectional control pattern.

  2 Oceanic clearance processor.

OCR *1* Optical-character recognition, or reading.

  2 Oceanic control region.

  3 On-condition replacement.

  4 Office of Commitments and Requirements.

  5 Occur.

OCS *1* Optimum-cost speed.

  2 Officer Candidate School (US).

  3 Oceanic Control System (NZ).

  4 Obstacle clearance surface.

  5 Operational control segment.

  6 Officers Command School (RAF Henlow).

oct Octane (FAA).

octa Unit of visible sky area representing one-eighth of total area visible to celestial horizon, now *okta*.

Octagon Trade-name of a range of de-icing fluids, including potassium and sodium acetates, sodium formate and propylene glycol.

octal *1* Standard base for electronic device having eight connector pins.

  2 Counting system to base 8.

octane number Standard system for expressing resistance of hydrocarbon or other fuel to detonation in piston engine, ranging from 0 (equivalent to n-heptane) through 100 (iso-octane) upwards to about 150. Measures are taken with lean mixture and with rich, latter giving higher values. Also called knock rating (strictly, anti-knock) or performance number, esp. when above 100. See *MON*, *RON*.

octane rating See *octane number*.

octane test Standard test in which sample of fuel is used to run special variable-compression single-cylinder engine and compared with mixtures of n-heptane and iso-octane, or (above 100 performance number) with other reference fuels.

octant Bubble sextant able to measure angles to 90°.

octave Interval between any two frequencies having ratio 1:2.

octet Complete set of four pairs of electrons forming layers 2 and 3 (in all noble gases except He, outer shell).

Octol HE warhead filling, HMX/TNT.

Octopus Operational and certified takeoff and landing performance universal software.

Octu Officer Cadet Training Unit (UK).

OCU *1* Operational Conversion Unit (UK).

  2 Optical coupler, or control, unit.

  3 Operational capabilities upgrade.

  4 Operator control unit (IRCM).

oculogyral illusion Apparent movement of fixed objects seen under high-g.

OCV Organisme de Contrôle en Vol (F).

OCXO Oven-compensated crystal oscillator.

OD *1* Outside diameter.

  2 Ordnance delivery.

  3 Olive drab colour.

  4 Optical disk.

ODA Overseas Development Administration (UK).

Odads Omnidirectional air-data system.

Odals, ODALS Omnidirectional approach lighting system.

Odapi Omnidirectional approach-path indicator.

Odaps Oceanic display and processing [or planning] system.

ODC Office of Defense Co-operation (US DoD office in Germany).

ODF Operational degradation factor.

ODFDMU Optical disk flight-data management unit.

ODID, Odid Operational display and input development.

Odin Operational data interface.

Odis Operational data interface system.

ODL Optical, or Oceanic, data-link.

ODM *1* Ministry of Overseas Development (UK).

  2 Operating data manual.

3 Operational development model.

4 Ordnance deployment manager.

5 Optical driver modem.

6 Original design manufacturer.

ODMS   *1* Oil-debris monitoring system.

2 Operational data-management system (ATC).

odometer   Digital readout of numerical quantity, esp. distance, as in DME.

ODP   Office of Defense Plants (US 1945).

ODR   *1* Overland downlook radar.

2 Operator difference requirements.

3 Overnight defect rectification.

ODS   *1* Oxide-dispersion strengthened.

2 Operational debrief station.

3 Optical-disk system.

4 Ozone-depleting substance[s].

5 Operator, or operator input [and], display system (also OIDS).

6 Office of Defense Supplies (US).

7 Orbit-determination system.

ODT   *1* Overland downlook technology.

2 Omnidirectional transmitter or transmission.

3 Office of Defense Transportation (US).

ODTC   Office of Defence Trade Controls (US).

ODU   Optical display unit.

ODVF   Society of friends of the air fleet (USSR).

ODW   Optimum-drag windmilling.

OE   *1* Opto-electronic (not necessarily synonymous with EO).

2 Over-excitation.

OEANC   Optical and electronic aiming and navigation complex.

OEB   Operations engineering bulletin.

OEC   Operational emergency clearance.

OECD   Organization for Economic Co-operation and Development (19 member countries).

OED   *1* Operational evaluation demonstration.

2 Operator-workstation embedded disk.

3 Operational-employment date (UK).

OEI   One engine inoperative.

OEIM   Organisme Egyptien pour l'Industrie Militaire.

OELD   Organic electroluminescent display.

OEM   *1* Original equipment manufacturer.

2 Office of Emergency Management (US).

OEO   Operational equipment objective.

OEP   *1* Office of Emergency Preparedness (US).

2 Operational evolution plan (FAA).

OER   *1* Operational effectiveness rate.

2 Officer effectiveness report.

oersted   Non-SI unit of magnetic field strength, = 79.577 A/m, exact conversion is $1,000/4\pi$.

OEST   Outline European staff target.

OEU   *1* Operational Evaluation Unit (RAF).

2 Overhead electronics unit[s].

OEW   Operating empty weight [OWE is preferred].

OEX   Orbiter experiment; SS adds support system[s] (NASA).

OF   *1* Over-frequency.

2 Objective Force, major long-term program (USA).

3 Ocean Flight [tower system]; DVPS adds data visual and processing system.

O/F   Oxidizer (oxidant)-to-fuel ratio.

OF₂   Oxygen difluoride rocket propellant.

OFA   Office Fédéral de l'Air (Switzerland).

OFAB   HE fragmentation (bomb, USSR, R).

OFAM   Office Fédéral des Aérodromes Militaires (Switzerland).

OFCM   Office of the Federal Co-ordinator for Meteorological Services and Supporting Research (US Dept. of Commerce).

OFCR   Overhead flight-crew rest, in normal wide-body unused above-ceiling volume.

OFDM   *1* Operational flight-data monitoring.

2 Orthogonal frequency-division multiplexing.

OFDPS   Oceanic flight-data processing system.

OFEMA, Ofema   Office Français d'Exportation des Matériels Aéronautiques.

off-block time   When aircraft leaves gate [probably travelling backwards] or starts to taxi.

offboard   Released or ejected from the platform [aircraft or ship], usually as decoy.

offboard data   Supplied by sensors outside the aircraft.

off-design   *1* Any operating condition other than that or those for which equipment was intended.

2 Less common definition: at or near stall or buffet boundary, or in severe turbulence.

offensive avionics   Those carried in attacking aircraft to assist fulfilment of mission, eg by providing navigation, target sensing and weapon guidance.

offensive sweep   Low-level flight by fighters over enemy territory looking for targets of opportunity.

office   Cockpit or flight deck (colloq.).

off-line   *1* Connected to computer but not forming part of dedicated controlled sytem.

2 Computer peripheral not directly communicating with central processor.

3 Not on airline's route network (eg Western Airlines has an * office in Washington DC).

4 Computer or EDP (4) installation which stores input data for processing when commanded.

5 No longer supplying power [usually electric] to system.

off-mounted   SSR aerial [antenna] not mounted on primary radar but on its own turning gear, and can thus be either synchronized to primary radar or rotated at predetermined data rate.

offset   *1* In major international purchase, eg of quantity of combat aircraft, agreement in reverse direction in which purchasing country is awarded one or more contracts for products which may or may not be connected with original hardware. This kind of * may be (1) purely window dressing to render costly import less unpalatable, (2) important commercial deal to benefit original importing country, (3) designed to bolster ailing home industry, or (4) important vehicle for transfer of advanced technology to importer.

2 Linear distances, usually small, measured from baseline to joggled or tapered edge, bevel, flange at angle other than 90°, or similar part dimension.

3 Precisely defined point on ground in surface attack (see * *bombing*) or in space in air interception (see * *point*).

4 Linear or angular difference beween major axis (eg of aircraft or engine) and axis of drive shaft from gearbox.

5 In fir-tree root, distance between teeth measured parallel to major axis of blade.

6 Length of normal common to landing-gear castor axis and wheel axle [can be zero].

7 Lateral and vertical distances from desired G/S and

runway centreline of (a) an actual landing aircraft, or (b) ILS beam (locus of peak signal strength).

8 Bearing/distance to runway threshold from point navaid, eg Tacan.

9 See * *factor)*.

10 See next.

offset angle   *1* Angle in vertical plane between horizontal through tug and line joining tug to glider, tow-target or other towed body.

2 Offset (4).

offset attack   See *offset bombing*.

offset bombing   Any surface-attack procedure which employs aiming or reference point other than target.

offset distance   *1* See *offset (2)*.

2 Distance from desired ground zero or actual ground zero to point target or centre of area target (DoD, NATO).

offset factor   In parafoil delivery, distance divided by drop altitude.

offset frequency simplex   Two stations transmit to each other on slightly different frequencies.

offset frontal   Meeting between two DA (3) aircraft and two hostiles (7–10 miles ahead) on reciprocal, such that no aircraft will pass between opposing pair.

offset hinge   Helicopter main-rotor hinge at substantial radial distance from axis of rotation.

offshore procurement   Procurement by direct obligation of MAP funds of materiel outside US territory; may include common items financed by other appropriations (DoD). In other words, US taxpayers paid for aircraft made in Europe, to assist industrial recovery 1951–55.

offst   Offset track.

offtake   Route through which power-source physical output is extracted, notably pipe for hot bleed air or shaft for rotary power.

off the shelf   Already fully developed to military or commercial standards and available from industry for procurement without change to meet military requirement.

off-wing   With the engine removed from the aircraft [not necessarily an under-wing installation].

off-wing slide   Emergency escape slide well ahead of or behind wing.

OFIS, Ofis   Operational Flight Information Service (ICAO).

OF/NT   Operational flight/navigation trainer (USN).

OFO   Orbiting frog otolith [programme].

OFP   *1* Operational flight programme.

2 Occluded frontal passage.

OFPP   Office of Procurement and Policy (DoD).

OFS   Operational flying school (RN).

OFSHR   Offshore.

OFT   *1* Operational flight trainer.

2 Operational flight test.

3 Operational flying training.

4 Orbital flight test.

5 Office of Fair Trading (UK).

6 Outer-fix time.

7 Office of Force Transformation (DoD).

OFTS   Operational flight and tactics simulator.

OFU   Overseas Ferry Unit (RAF, WW2).

OFV   Outflow valve.

OFZ   Obstacle-free zone.

OG   Observation group (USAAC, USAAF).

OGA   *1* Outer gimbal angle.

2 Office Général de l'Air (F, export distribution).

OGE   *1* Out of ground effect; supported by lifting rotor(s) in free air with no land surface in proximity.

2 Operational ground equipment.

ogee   See *ogive*.

oggin, the   The sea (RN colloq.).

ogive   *1* Loosely, any shape formed by planar curve whose radius increases until it becomes straight line.

2 Wing plan having approximate form of curve becoming straight line parallel to longitudinal axis on each side of centreline; so-called 'Gothic window' shape.

3 Body of revolution formed by rotation of curve as in (1) about axis parallel to line on same side of line as curve.

4 Any of various other shapes such as conical * (body formed by rotation of two straight lines forming cone/cylinder), tangent * (circular arc tangent to line) or secant * (circular arc meeting line at angle).

5 Common definition is 'surface of revolution generated when circular arc and line segment are rotated about axis parallel to line'; this is incorrect since curve need not be circular arc and shape can be planar and not body of revolution.

OGO   Orbiting geophysical observatory.

OGV   Outlet guide vane, or straightener vane.

OGW   Space agency (Austria).

OH   *1* Overhaul.

2 Overhead.

O/H   Overheat.

OHA   Operating hazard analysis.

OHAR   Overhead attendant rest (above ceiling of widebody centre fuselage).

OHC   *1* Operating hours counter.

2 Overhauled condition.

3 Overhead camshaft.

OHD   Overhead.

OHM   Office of Hazardous Materials (US Dept of Transportation).

ohm   *1* SI unit of electric resistance, defined as that of conductor across which potential difference of 1 V produces current of 1 A.

2 By analogy with (1), mechanical measure of resistance derived by dividing applied force by velocity; has dimensions of g/s.

Ohmist, OHMIST   Offshore helicopter meteorological information self-briefing terminal.

ohmmeter   Usually, compact instrument for giving quick approximate indication of resistance of circuit to direct current.

OHS   Office of Homeland Security (US, DoD, created 2001).

OHU, Ohud   Optical head, or overhead, unit of HUD.

OHV   Overhead valve[s].

OI   *1* Operational interruption.

2 Office of Information.

3 On instruments.

4 Operating instruction.

OIA   Office of International Aviation (US).

OIB   Orbit-insertion burn.

OIC   *1* Organized immigration crime.

2 Officer in charge.

OICRs   Operational intelligence collection requirements.

OID   *1* Optical incremental digitiser.

2 Outline installation drawing.

3 Operator input and display; S adds system.

OIDT　Operator interactive display terminal.

OIG　Office of Inspector-General (US, eg NASA).

oil　Vague term which could mean crude petroleum, hydrocarbon-based lubricant or other 'oil-like' materials. Even ester-based synthetic lubricants are often marketed as turbine oils.

oil bottle　Container of lubricating oil fed by air/gas pressure or even spring-loaded plunger, to short-life engine, eg of cruise missile.

Oil Burner routes　Published routes within continental US along which USAF, USN and USMC conduct high-speed training missions at low level in VFR and IFR.

oil-can　Noun and verb, portion of metal skin where compressive stress imposed in manufacture has resulted in slight local bulge between rows of rivets or other attachment which, when subjected to pressure difference or perpendicular force at centre, can suddenly spring inwards noisily; potential fatigue hazard with thin skin.

oil control ring(s)　Piston rings whose main purpose is to prevent loss of lubricating oil from cylinder wall up into combustion chamber; usually one or two, below main compression rings.

oil cooler　Heat exchanger whose purpose is to remove heat from lubricating-oil circuit; usually cooled by air or fuel flow.

oil coring　See *coring*.

oil dilution　Mixing petrol with lubricating oil to reduce viscosity to facilitate starting large piston engine at very low temperature.

oil drive　Hydraulic drive, usually signifying infinitely variable ratio.

oil radiator　See *oil cooler*.

oil ring　Scraper ring.

OIML　Organisation Internationale de la Métrologie Légale (Int.).

OIN　*1* Organisation Internationale de Normalisation (Int.).

2 Overhaul information notice.

OIP　Optimum implementation plan.

OIPC　Organisation Internationale de Protection Civile (Int.).

OIRI　French for CCIR.

OIS　*1* Obstacle identification surface.

2 On-board information system.

OIT　Operator-information telex.

OITS　Organización Internacional de Telecomunicaciones por Satelites (Int.).

OIU　*1* Orientation/introduction unit.

2 Operator interface unit.

OJCS　Organization (or Office) of the Joint Chiefs of Staff (US).

OJT　On the job training (DoD).

OKB　Experimental construction bureau, where aircraft are designed and developed (USSR, R).

OKC　Over [a named radio station].

OKO　Experimental construction section (various, USSR).

okta　Unit of sky area equal to one-eighth of total sky visible to celestial horizon [previously octa].

OL　Operating location.

OLAN　On-board local area network.

OLBR　Operational laser-beam recorder.

OLC　Operational level of capability, marked by achievement of different Phases.

OLDI　On-line data interchange.

OLDP　On-line data processing (or processor).

OLED　Organic light-emitting diode.

OLEEA　On-line exhaust emissions analysis.

oleo　Telescopic strut which absorbs energy of compressive loads by allowing hydraulic oil or other fluid to escape under pressure through small restrictor orifice, usually controlled by one-way valve to reduce rebound.

oleo-pneumatic　Absorbing shock by combination of air compression and forcing liquid through an orifice.

OLF　Outlying field.

OLGS　Operational-level ground station [OLTGS adds Tactical].

olive　Small ellipsoid resembling an olive threaded on cable either to act as bearing surface in tubular guideway, or in continuous row separated by small spheres located in recesses in olive ends, to provide two-way push/pull mechanical interconnection (eg in Bloctube system).

Olive Branch route　Lo-Level routes for B-52 training in continental US.

OLM　*1* On-line, or operational loads, monitoring, or measurement.

2 Off-line mapper.

3 Organizational-level maintenance.

4 Octane lean mixture.

Ologs　Open-loop oxygen generating system.

OLOS, Olos　Out of line of sight.

OLP　Open-loop phase.

OLR　Offload route[s].

OLRT　On-line, real time.

OLS　*1* Optical landing system.

2 Operational linescan, or linescan system.

OLTF　Open-loop transfer function.

OLTS　Optical-line termination equipment.

OM　*1* Outer marker.

2 Organizational maintenance.

3 Operation & Maintenance (USAF).

4 Other management.

5 Over maximum (often O/M).

6 Occupational medicine.

7 Operations module.

8 Out [of use] for maintenance.

OMAG　Independent naval aviation group (USSR).

OMAR　Optical microwave approach and ranging (UAV all-weather landing).

OMB　*1* Office of Management & Budget (US, White House).

2 Oilless magnetic bearing.

OMC　Organic-matrix composite.

Omcads　Oil movement control and distribution system, manages fuel [not lubricating oil] installation at airport including separation of contaminants.

OMCFP　Optimized MAC computer flight plan.

OMCM　Operational and maintenance configuration management (software).

OMD　On-board maintenance documentation.

OME　*1* Operating mass empty, usually same as OWE.

2 Operational mission environment.

Omega　Accurate long-range radio navaid of VLF hyperbolic type, covering entire Earth from eight ground stations and usable down to SL or underwater.

omega (ω)　*1* Angular velocity (SI unit is rad/s).

2 Angular frequency.

3 Any generalized frequency, thus ω_n is natural frequency and ω_r is undamped natural frequency of rth mode.

OMEV　Independent naval helicopter squadron, Bulgaria's naval air unit, previously OPLEV.

OMG　Operational manoeuvre group (USSR).

OMI　Omnibearing magnetic indicator.

OMIS　Operations-management information system.

OMM　*1* Organisation Météorologique Mondiale = WMO (Int.).

2 Oxygen-mask microphone.

OMMS　Oxygen-mask-mounted sight.

Omni　VOR (colloq.).

omni-axial nozzle　Rocket motor nozzle capable of being vectored to any angle within prescribed limits about pivot point defined by intersection of axes of symmetry of propelling system and nozzle.

omni-bearing indicator　VOR panel instrument.

omni-bearing selector, OBS　Knob on most VOR/ILS indicators which is turned to each required VOR bearing, which appears in a three-digit window display, left/right needle thereafter showing difference from required heading.

omnidirectional aerial　Antenna emitting to all points of compass (assumed equal signal strength throughout 360°).

omnidirectional ball panel　Removable floor panel containing free-running balls on which cargo containers are moved.

omnidirectional beacon　Fixed ground radio station giving non-directional signals for airborne DF receivers (obs).

omniflash beacon　Airborne flashing (strobe) light visible equally through 360° in azimuth.

omni-range　See *VOR*.

omnivision　Uninterrupted view through 360°.

OMNTS　Over mountains; also OMTNS.

OMOS　Department of experimental marine aircraft construction (USSR, defunct).

OMPS　Ozone-mapping and profiler sensor, or suite.

OMS　*1* Crew chief (USAF).

2 Orbital manoeuvring system, or subsystem.

3 Organisation Mondiale de la Santé (Int.).

4 On-board maintenance system.

5 Order, or operating, or operational, management system.

6 On-board monitoring system.

OMT　*1* Other military targets.

2 On-board maintenance terminal.

3 Object modelling [US modeling] technique.

4 Optical modular technology.

OMV　Orbital manoeuvring vehicle.

ON　*1* Omega navigation.

2 Octane number.

ONA　*1* Office of Noise Abatement (FAA).

2 Operational net assessment.

ONAC　Office of Noise Abatement and Control (EPA, US).

on-block time　Parked at gate.

on-board aircraft server　Integrates flight deck, cabin, maintenance and ground-based operations into seamless system, each aircraft being a LAN.

on-board cargo operations system　Microprocessor-based system which automatically positions each item for correct c.g. and best structural integrity.

on-board EW simulator　Internally [ie, passively] simulates external threats as seen by on-board sensors [all likely wavelengths].

on-board oxygen generation　Usually removes nitrogen from bleed air by Zeolite molecular sieve.

ONC　Operational navigation chart[s].

on call　Ready for prearranged mission to be requested.

on-call target　Planned nuclear target to be attacked not at specific time but on request.

on-condition maintenance　Performed only when condition of item demands, instead of at scheduled intervals.

on-course signal　Electronic signal indicating receiver is on desired course, usually in form of steady note created by two superimposed coded signals.

OND　Optical neural device.

1½-stage　Vehicle configuration, esp. ballistic rocket, in which single set of propellant tanks feed two or more thrust chambers, one or more of which is jettisoned in flight.

one-dimensional flow　Flow through duct in which all parameters are, or are assumed to be, constant throughout any section normal to the direction of flow.

one-dot error　One scale divison from centre on ILS/VOR needle or similar-type instrument, usually equal to 0.5°.

1553B　Standard databus highway (US/NATO).

180°-approach　Any of several standard approach procedures involving downwind leg and 180° turn; variations include 180° accuracy landing, 180° spot landing, 180° U-turn approach and 180° semicircular approach.

180° error　Misreading compass by flying reciprocal of desired course, formerly avoided by rule 'red on red'.

1-min turn　Procedure turn through 360° (rarely, 180°) taking one minute to complete; bank angle is about 30°.

one-off　Aircraft of which only a single example is constructed.

one-on-one　Classical air combat by two opposing aircraft.

ONER　Oceanic navigation error report.

ONERA　Office National d'Etudes et de Recherches Aérospatiales (F).

ONFA　Opera Nazionale per i Figli degli Aviatori (I).

onglet　Leading-edge root fillet.

on-line　*1* Of computer of EDP (1) system, automatically receiving and instantly processing information and sending output data to required destination.

2 Of EDP (1) peripherals, operating under direct control of central processor.

on-mounted　Indicates that SSR aerial [antenna], ECM aerial or similar ancillary device is mounted on, and oscillates or rotates with, main radar.

ONN　Optical neural net.

ONR　Office of Naval Research (USN).

ONS　Omega navigation system.

ONSHR　Onshore.

ONST　Outline NATO staff target.

on the beam　Correctly aligned on ILS glidepath or other guidance beam. Today has come to mean 'on the right track' in most general sense.

on the deck　At minimum safe altitude (colloq.).

on the fizzer, on the hooks　On a charge, facing disciplinary action (RAF colloq.).

on top *1* Above unbroken cloud, with CAVU environment.

2 Directly above operating sonobuoy or indicated submarine position.

Ontrac II American VLF navaid of rho-rho type.

ONVL Over-the-nose vision line, lower limit in side elevation of pilot's FOV.

on-wing With the engine installed in the aircraft [not necessarily hung on a wing].

OOA *1* Object-oriented analysis.

2 Out of area.

3 Offshore Operators' Association (UK).

OOB Order of battle.

OOD *1* Object-oriented design, or development.

2 Officer of the day.

OODA Observe[r]-orientation-decision-action; hence Ooda-loop.

OODB Object-oriented data-base; MS adds management system.

OOF Out of flatness (machined plate).

OOK Department of special constructions (USSR, obs).

OOOI Out [from gate], off [T-O], on [landed], in [at gate] (Acars).

OOP Object-oriented programming.

OOR Out of region.

OORA Object-oriented requirements analysis.

OOS *1* Department of special aircraft construction (USSR, obs).

2 Out of service.

OOSA Office for Outer Space Affairs (UN).

OOTW Operations other than war.

OOV Objects of verification.

OOW Out of window, simulator depicts external scene.

OP *1* Observation post, position or point.

2 Oil pressure.

3 Operating procedure.

op An operational flight, hence ops.

OPA *1* Component of GB Sarin, isopropylamine plus isopropylalcohol.

2 Office of Price Administration (US).

3 Opaque (icing).

4 Open planar array.

OPAC Operations of aircraft [ECAC working group].

OPACI, Opaci Opera Pionieri e Anziani dell'Aviazione Civile Italiana.

opacity In optical systems or photographic film, reciprocal of transmittance (log O = density d); term absorbance is now recommended.

OPAL, Opal *1* Order processing automated line.

2 Orbiting picosat automated launcher.

opaque plasma One through which EM signals cannot pass; generally plasma is opaque for EM frequencies below plasma frequency.

opaque rime White icing of porous form caused by rapid freezing of small droplets.

OPAS, Opas *1* Overhead-panel Arinc-629 system.

2 Operational assignment (ICAO).

OPAT Office des Ports Aériens de Tunisie.

Opats Object-position and tracking sensor.

OPB Oxygen preburner.

OPBC Overhead-panel bus controller.

OPC *1* Operational control (ICAO).

2 Optical-phase conjugation.

Opcom Operational command.

OPCW Organizaton for the Prohibition of Chemical Weapons.

Opdef Operational defect.

OPEC, Opec Organisation of Petroleum Exporting Countries.

open *1* Aircraft has * cockpit with no canopy.

2 With normal control path disconnected; such failure interrupts or seriously distorts signal passing along that channel.

open angle Angle of mating part slightly greater than 90° or other angle of edge of metal structure such as angle section; fault condition in sheet metalwork.

open architecture Easily added to or modified.

open-bladed Not enclosed in a shroud or duct.

open-centre system Hydraulic system without an accumulator.

open circuit *1* Electrical, circuit interrupted.

2 Wind tunnel, one having no return path.

Open Class FAI/CIVV categories for competition sailplanes, in one case with Standard Class span of 15 m and alternatively with span unrestricted but usually 17 to 20 m. In each case all refinements such as flaps are permitted.

open cockpit Not provided with a canopy, leaving occupant's head [and possibly upper torso] in slipstream.

open-coil armature Ends of each coil connected to different bars of commutator.

open competition Industrial competition in which all proposals, promises or offers are communicated to all participants.

open delta Three-phase transformer comprising two single-phase transformers linking three lines.

open-ended Spaceflight continued until either all possible information has been gained, and mission objectives met, or spacecraft systems run low or become faulty.

open-gore Parachute in which fabric is absent from one gore to assist trajectory control.

open-hearth Principal method of high-quality steelmaking in many countries, using shallow regenerative furnace with or without oxygen.

open improved site Open-air site for military storage whose surface has been graded and surfaced with topping or hard paving.

open items *1* Work done on sections of airframe at location away from assembly line prior to major join.

2 The sections worked on in [1].

3 Parts or procedures inadvertently omitted during manufacture.

open-jaw ticket Generally, airline booking out by one route and return by another or over return route to different destination.

open-jet tunnel Wind tunnel in which working section is open and has no tunnel walls.

open-link ammunition Rounds are held in clips that do not encircle each case, which can be pushed out diagonally instead of removed only to rear.

Open List List of military aircraft on which nothing was secret (RAF 1918 – c1960).

open loop Servomechanism or other system comprising control path only, with no measurement of result or feedback to give self-correcting action.

open production line Inactive, but in a state ready to resume production should further aircraft be required.

open propfan   Propfan or UDF without a surrounding shroud.

open rate   Situation in which fares on particular air-carrier route are fixed by each operator independently.

open rotor   *Open propfan.*

open section   Structural member devoid of closed and thus uninspectable surface.

Open Skies   Country allows unlimited traffic rights to foreign carriers, usually reciprocal.

open system   Using stored food/oxygen and discarding all body wastes.

oper   Operate (FAA).

operand   EDP (1) quantity entering or arising in instruction; can assume many forms, from argument to address code; generally a word on which operation is to be performed.

operating active aircraft   Those for which funds are provided, as distinct from non-operating active aircraft (DoD).

operating envelope   Plot of extreme boundaries of conditions to which hardware is subject, eg acceleration/temperature, or (in case of aircraft) altitude/Mach or V/n gust envelope.

operating expenditure   Airline's total costs of generating ATKs.

operating ratio   Airline operating revenue divided by operating expenditure.

operating revenue   Gross income from an airline's air-carrier operations, normally excluding other activities and any state subsidy.

operating speed   Traditionally 87.5 per cent of rated rpm for light piston engine (US, obs.).

operating weight empty   Total mass of aircraft ready for flight, including oil, water, food and bar stocks, passenger consumable stores, flight and cabin crews and their baggage and, according to most definitions, empty baggage containers where appropriate, but excluding fuel and payload. In case of military aircraft, ADI fluid and drop tanks are excluded but EW pods are included.

operation   *1* A military action, or carrying out of mission.

2 Operational flight, recorded by individual's logbook; if enemy opposition was weak, a completed * might count as only ½ in assessing individual's total.

operational   Ready to accomplish mission.

operational agility   'The ability to adapt and respond rapidly and precisely with safety and poise to maximise mission effectiveness', also defined as 'airframe agility + systems agility + weapons agility'.

operational aircraft   In British usage, one intended and ready for combat missions, as distinct from transport or training.

operational air traffic   Generally, that which cannot conform to requirements of flights within airways or controlled airspace.

Operational Bulletin   Issued when necessary, eg following difficulty or technical failure, giving advice, especially to flight crews [but not giving mandatory instructions].

operational characteristics   Those numerically specified parameters describing system performance; system can be aircraft, radar, etc.

operational command   Normally synonymous with operational control, and covers all functions involving composion of forces, assignment of tasks, designation of objectives and direction of mission; does not include such matters as administration, discipline or training.

operational control   *1* Authority granted a commander to direct forces to accomplish missions (NATO, etc).

2 Exercise of authority over initiation, continuation, diversion and termination of a flight (ICAO).

operational control communications   Communications required for operational control (2) between aircraft and operating agencies.

operational control segment   That portion of communications link of GPS or Navstar used to transmit commands.

Operational Conversion Unit   Turned qualified pilots and other crew members into crews fit for operations (RAF WW2 and later).

operational development model   Almost same as a prototype, but to evaluate hardware for a new or modified mission, eg an AEW/AWACS airship.

operational effect rate   Usually, flying rate.

operational evaluation   Test and analysis of system under operational conditions, with consideration of capability offered, manning and cost, potential enemy accomplishments and alternatives, to form basis for decision on quantity production. In practice ** is often not accomplished until long after production decision.

operational factors   Those exercising constraints on flightplan, notably ATC (1), pilot workload, available aids, specified refuelling locations, etc.

operational flight   In the face of the enemy, carrying weapons or cameras or troops or food parcels or towing glider, etc, and underlined in red in logbook. US = combat mission.

operational flightplan   Operator's plan for safe conduct of particular flight.

operational interchangeability   Ability to substitute one item for another of different composition or origin without loss of effectiveness or performance.

operational interruption   Period during which an item is unserviceable.

operational load measurement   On-going structural audit to establish safe fatigue life, esp. of fighter or military trainer.

operational loss or damage   Loss or damage to military item caused by reason other than combat.

operational manoeuvrability ceiling   Maximum height at which, for any given weight, aeroplane can sustain specified load factor (vertical acceleration) at onset of buffet.

operational meal   Special treat for aircrew returned from an operational mission; key element: a fried egg (RAF, WW2).

operational mission   Many definitions, all faulty; suggested: a miltary flight in furtherance of armed conflict. Aircraft need not be armed, and need not encounter enemy.

operational mode   Any of selectable alternative methods of operation for functioning system, eg computer-controlled, automatic with feedback, automatic open-loop, semi-automatic and manual.

operational net assessment   Study of a hostile country from economic, military, political and diplomatic viewpoints, and potential for fomenting revolution (US).

operational performance categories   Categories I to IIIC defining ILS and blind (automatic) landing installation performance. See *Categories (3)*.

operational phase   Period from acceptance by first user to elimination from inventory.

operational readiness   Capability of unit or hardware to perform assigned missions or functions. Hence ** inspection.

operational readiness platform   Ramp where aircraft at various ready states (including alert states) are parked, with all required connections (eg telebrief) in place.

operational readiness training   Consolidated instructional period wherein qualified personnel for operational units are given integrated training in operational mode.

operational research   Generally, analytical study of problems to provide numerical (some defininitions say "scientific") basis for decision-taking; in US often called operations research.

operational test and evaluation   Test and analysis of system under operational conditions, promulgation of associated doctrines and procedures, and continuing evaluation against new threats or changed environment or circumstances.

operations   Engaged in operational flying.

operations manual   Usually, definitive handbook for user of system, as distinct from engineering, repair or design manuals; for aircraft called flight manual.

operative   *1* Able to operate; in true fail-* system there is no loss in performance (though there may be in integrity) after failure.

2 Employee engaged in routine operation of tool or machine on production work.

operator   *1* Person, organization or enterprise engaged in or offering to engage in aircraft operation (ICAO).

2 Licensed air carrier.

3 American term for employee other than foreman or manager.

operator's local representative   Agent located to obtain meteorological information for operational purposes and to provide operational information to local met. office.

Opeval   Operational evaluation.

OPF   Orbiter processing facility.

Opfor   The opposing force[s] [tactical simulation].

Op.Hr.   Operating hours.

OPIAR   Association of aerospace companies (Romania).

OPLE   Omega position-locating equipment; synchronous satellite providing control beyond LOS.

OPLEV   See *OMEV*.

OPM   *1* Office of Personnel Management.

2 Operation performance monitor.

3 Operations Policy Manual (JARs).

Opmet   Operational meteorological information or service.

OPN   Optimised-profile navigation.

opn   Operation, open, opening.

Opos   Open purchase order summary.

OPOV   Oxygen preburner oxidiser valve.

opportunity servicing   Servicing carried out at any convenient time, within specified flight or operating-time intervals.

opportunity target   See *target of opportunity*.

opposed engine   Piston engine having left and right rows of (usually horizontal) cylinders either exactly or approx-

imately opposite to each other, with crankshaft on centre-line.

opposed-piston engine   Piston engine, usually of two-cycle compression-ignition type, in which each cylinder has central combustion space and two pistons working in opposition and driving crankshafts at opposite extremities.

opposition manoeuvres   Two aircraft perform in unison, one flying the mirror image of the other's manoeuvres.

OPR   *1* Overall engine pressure ratio.

2 Office of Primary Responsibility (DoD).

3 Operate, operation[s], operating, operator, operational, etc.

4 Operational preference.

5 Once per revolution [or o.p.r.].

OPS   *1* Operational performance standards.

2 Operations per second [or o.p.s.].

Ops, ops   Operations; an individual might complete 25 * and still be on * (RAF WW2).

opsec   Operational security.

OPSP   Operations Panel (ICAO).

Opspecs   Operational specifications.

OPT   *1* Optional.

2 Optimum.

optical barrel camera   Precise long-focus camera for SR-71 aircraft.

optical communications   Systems using coherent (laser) light, usually channelled along optical fibres; owing to extremely high frequencies such systems have enormous information-rate capacity.

optical countermeasures   ECM in visible range of EM spectrum.

optical display unit   The optical element of a HUD, also called optical head unit.

optical electronics   One meaning is use of optics to focus IR on sensitive detector in search for electronic emitter.

optical fibre   Filament drawn from two kinds of glass having different refractive indices, in form of core and sheath; light entering at one end is reflected each time it encounters core/sheath boundary until it emerges at other end.

optical guidance   Usually, passive guidance based on sensitive photocells which sense presence of target and cause vehicle to home on it.

optical gyro   Fibre-optic gyro.

optical head unit   See *optical display*.

optical instrumentation   Use of optical systems (telescope, theodolite, precision graticules and marking on vehicle) in conjunction with cameras and TV to provide stored information of location in three dimensions, velocities and other parameters of hardware or phenomena.

optical landing system   Refined form of mirror sight in which gyrostabilized light beams indicate to approaching pilot deviation from glidepath (USN, USMC).

optical line of sight   LOS through atmosphere, which in precision measures is not to be regarded as straight.

optical linescan   Term used for various line-by-line scanning at optical wavelengths, including RBV (video).

optically flat panel   Window used in visual sighting systems through which light passes with near-zero distortion.

optical maser   See *laser*.

optical Mems   Mems operating at optical wavelengths.

optical pyrometer   Instrument giving numerical readout of temperature of hot surface by measuring incandescent brightness.

optical reference point   Index mark on canopy or elsewhere providing pilot with sightline guidance in vertical or near-vertical landing.

optical turbulence   Fluctuating distortion of light rays caused by time-varying gradients of refractive index.

optimized profile   Flight trajectory calculated to minimize energy consumption throughout flight (but for operational reasons seldom attainable).

optimum angle of attack   That giving maximum L/D and thus highest cruise efficiency.

optimum angle of glide   That at which *glide ratio* is a maximum.

optimum coupling   Matched radio impedance in which load equals output of amplifier or transmission line, for maximum power transfer.

optimum height   Height of air burst or conventional explosion producing maximum effect against given target.

option   *1* Acquisition of particular place in aircraft or other major equipment item production programme, with or without deposit and without agreement to purchase.

2 Variation in standard of product offering customer choice, eg avionic fit, long-range tankage, de-icers etc.

3 Alternative decisions open to commander engaged in battle.

optional, or optimal, yield management   Determining traffic mix by class code and fare type to maximize revenue.

opto-electronic   Combining optical and electronic subsystems in single device; term appears to be used synonymously with EO.

optronics   Technology of systems using wavelengths from optical to electronic. *Appendix 4*.

optrotheodolite   Doubtful designation for precision position-correlating instrument combining laser and video camera.

OPU   Overspeed protection unit.

OPV   Optionally [or operationally] piloted vehicle; some of this class are helicopter UAVs.

OQAR   Optical quick-access recorder.

OQPSK   Offset quadrature phase-shift keyed, or keying.

OR   *1* Operational research.

2 Operational requirement[s] (UK).

3 Operational readiness; also (1985) operationally ready (DoD).

4 See *OR gate*.

5 Open rotor.

6 Other ranks.

7 On request.

O/R   On request.

Oracle   Operational research and critical-link evaluation.

Orads   Optical ranging and detecting systems (USA).

Oralloy   Classified metal alloy, not necessarily based on U-235, used as preferred primary fissile material in NW, with or without tritium.

Orange   Could be friendly or hostile, uncertain.

Orange Force   Simulated hostile force during exercise (NATO, IADB).

oranges   Weather (RAF, WW2).

Oranges Sour   Air-intercept code: weather unsuitable for mission (DoD).

Oranges Sweet   Air-intercept code: weather suitable for mission (DoD).

Orasis   Optical real-time adaptive spectral identification system.

ORB, Orb   *1* Operations record book (RAF).

2 Omnidirectional radio beacon.

orbat   Order of battle.

orbit   *1* Closed elliptical or quasi-circular path of body revolving around source of gravitational attraction balanced by its own centrifugal force.

2 Closed pattern, usually circular or racetrack, followed repeatedly by aircraft, eg in air-intercept loiter or when holding.

3 To follow an * as in (1) or (2).

orbit[al] decay   *Decay (1)*.

orbital elements   See *orbital parameters*.

orbital manoeuvring vehicle   Used from Shuttle and later from space station to position and retrieve satellites about 1,000 miles distant (NASA).

orbital parameters   Basic numerical values describing orbit, such as apogee, perigee, inclination and period.

orbital period   Time to complete one revolution (relative to space, ignoring such irrelevancies as rotation of primary body); symbol P, whose square is always proportional to cube of semi-major axis of orbit (distance from centre of primary to apogee).

orbital rendezvous   See *rendezvous*.

orbital transfer vehicle   *1* The first * was a reusable tug shuttling between a planetary or lunar orbiter and higher orbits.

2 Part of the planned MSP, a vehicle for orbit transfers and servicing, potentially reducing spacecraft weight and extending on-orbit life.

orbit determination   Calculation of orbital parameters of unknown satellite.

orbiter   *1* Portion of spaceflight system designed to orbit, as distinct from booster stages, tanks, etc.

2 In particular, bus to inspect planetary body from a distance, with or without probes.

orbit point   Geographically or electronically defined location above Earth's surface used as reference in stationing orbiting aircraft.

ORC   Oversight and review committee.

Orca   Optimized raster chart analyzer [*sic*].

Orchidée   Observatoire radar cohérent heliporté, d'investigation des éléments ennemis; name means orchid (F).

Orcon   Fire-resistant insulation batting using RK carbon fibre.

ORD   *1* Order.

2 Operational, or operations, Requirement[s] Document.

3 Operational readiness demonstration.

4 Optional retirement date (RAF).

Ordalt   Ordnance alteration.

order book   Real or symbolic book containing list of customers for major equipment item, esp. commercial transport, broken down into firm orders, options, leases, etc, and fading from importance with programme's maturity and second-hand customers.

ordinate   *1* Vertical axis on graph or other geometric plot.

2 Vertical distances measured from datum, eg from chord line to upper/lower surfaces of aerofoil.

ordnance   Explosives, chemicals and pyrotechnics for use against an enemy, including nuclear, together with directly associated hardware such as guns, rocket launchers, etc.

ordnance alteration   Change [usually last-minute] to load carried by attack aircraft.

ordnance devices   Explosive or pyrotechnic parts of spacecraft, RPV or other [especially non-weapon] system.

ordnance work rate   Various measures intended to offer single numerical value of rate of application of ordnance in particular theatre or against one target.

Orel   Omnidirectional runway-edge lights/lighting.

Orex   Orbital re-entry experiment.

ORGALIME, Orgalime   Organization for liaison between European electrical and mechanical engineering industries.

organic   *1* Originally, living material; today broadened to cover all compounds of carbon and hydrogen, and often with other elements.

2 Assigned to, and forming essential part of, military formation.

3 In many applications involving heat exchange (eg * reactor, * Rankine cycle), use of organic (1) fluid, often mixture of polyphenyls.

organic air vehicle   Organic(2) aircraft; but the first meaning is a small UAV controlled by a squad of infantry (SOF, USA).

organizational-level maintenance   Performed on a squadron or other operator unit, eg on an installed or backed-off engine.

organometallic   Compound having metal atom attached to organic radical.

OR gate   Logic device which, whenever a voltage (eg a 1) appears at any input, responds with voltage at output).

ORHE   On-ramp handling equipment.

ORI   Operational readiness inspection.

orientation   *1* Determination of approximate position or attitude by external visual reference.

2 Turning of instrument or map until datum point or meridian is aligned with datum point or true meridian on Earth (ASCC).

3 Direction of fibres in lay-up, prepreg or finished FRC part.

orifice meter   Fluid-flow measuring technique where pressure is measured upstream and downstream of transverse plate with calibrated hole.

origin   *1* Point 0,0 on cartesian graph.

2 Base from which map projection is drawn.

originator   Command by whose authority a message is sent.

ORLA   Optimum repair-level analysis.

ornithopter   Aerodyne intended to fly by means of wings flapping or oscillating about any axis, but not rotating.

Oroca   Off-route obstruction-clearance altitude.

orographic uplift   Large-scale uplift and cooling of air mass caused by mountains; hence * rain or precipitation. Generally synonymous with mountain-wave turbulence.

OROS, Oros   Optical read-only storage.

ORP   *1* Optical reference point.

2 On-request reporting point.

3 Operational readiness platform.

ORS   *1* Off-range site.

2 Offensive radar system.

3 Occurrence reporting system.

4 Operational Research Section.

ORT   *1* Optical relay tube (TADS).

2 Optical resonance transfer.

3 Operational readiness test.

ORTA   Office of Research and Technology Assessment.

orthicon   Camera tube with secondary emission reduced or eliminated by using low-velocity electrons to scan two-sided sensitive mosaic, one side scanned and reverse face illuminated, resulting in stored charges on mosaic being removed by beam and generating output signal.

orthodrome   Great circle, hence orthodromic course.

orthogonal   Originally meant perpendicular; today blurred, as shown below.

orthogonal aerial   Pair of transmitting and receiving aerials [antennas], or single T/R aerial, designed to measure difference in polarization between radiated signal and echo from target.

orthogonal biplane   One without stagger.

orthogonal scanning   Use of combined axial and lateral magnetic fields to control low-velocity electron beam in camera tubes.

orthographic   Rectilinear projection of solid objects on two-dimensional sheet by use of parallel rays, in contrast to conical perspective rays; result is six possible views (left, right, front, rear, top, bottom), of which four (front, top, left, right) are normally used (in different relative positions in Europe and N America, see *third-angle projection*), each representing appearance of object as seen from infinitely large distance.

orthomorphic   Map projection in which meridians and parallels cross at right angles, all angles are correct and distortion of scale is equal in all directions from any point; several variations are in use, chief being Lambert's conformal and Mercator's.

ORTL   Optical resonance transfer laser.

ORU   Orbital replacement unit.

OS   *1* Observation squadron.

2 Operating system(s).

3 Observation scout, ship-based aircraft category (USN 1935–45).

4 Also O/S, ordnance server [UCAVs and simulation].

O/S   Overspeed.

OSA   *1* Operational support aircraft.

2 Official Secrets Act.

3 Open-systems architecture.

OSACOM   Written thus though said as a word, Operational Support Airlift Command (USA).

OSAD   Outer-Space Affairs Division (UN, Int.).

OSAF   Office of the Secretary of the Air Force (US).

OSAMC   Open-systems architecture mission computer.

OSBV   One-shot bonding verfahren (G).

OSC   *1* Overhead stowage compartment(s) for carry-on baggage.

2 Optical sensor converter.

Oscar   *1* Open-systems core-, or commercial-, avionics requirements.

2 Orbiting satellite[s] carrying amateur radio.

OSCE   Organization for Security and Co-operation in Europe (Vienna, many Permanent Missions and delegations).

oscillation   Repeated fluctuation of quantity on each side of mean value.

oscillator Device for converting direct current into alternating current to give control or carrier frequency, using not rotating mechanism but transistor or thermionic valves.

oscillogram Hard-copy (eg photographic) record of oscilloscope output.

oscillograph Sensitive electromechanical device translating varying electrical quantities into visible trace on paper or film (generally superseded by oscilloscope).

Oscillogyro Patented (Ferranti) gyro having unique property of measuring displacements about two axes.

oscilloscope Device translating varying electric quantities into visible traces on screen of CRT or similar display.

OSD *1* Office of the Secretary of Defense (US).
   2 Out of service date.
   3 On-screen display.

OSDBU Office of Small and Disadvantaged Business Utilization (NASA).

OSDS Oceanic system development support (FAA).

OSE *1* Optical shaft encoder.
   2 Office of Systems Engineering (DoT, US).
   3 Operations support equipment.

Oseen flow See *viscous flow*.

OSEM Office of Systems Engineering and Management (FAA, US).

OSF *1* Optronique secteur frontale, fighter EO sensor (F).
   2 Open-systems foundation.

OSHA Occupational Safety and Health Act [or Administration] (US).

OSI Open systems, or standard, interconnection, or interface.

OSIA On-Site Inspection Agency (INF Treaty).

OSIE OSI environment.

OSI-RM OSI reference model.

OSK Department of special construction (USSR, obs.).

OSM Optical support measures, eg laser threat warning.

osmium Hard, heavy, corrosion-resistant silvery metal, MPt 3,054°C, densest element (at 20°C = 22.588), symbol Os.

OSO Orbiting solar observatory [Nos 1–8].

Osoaviakhim Society for the support of defence, aviation and chemical industries (USSR).

OSP *1* Offshore procurement.
   2 Orbital/suborbital program (USAF).
   3 Optical surveillance platform.
   4 Operational special program (JTIDS).
   5 Orbital Space Plane (NASA).

OSPF Open shortest path first.

OSR *1* Optical solar reflector.
   2 On-site repair.
   3 Order status report.

OSRD Office of Scientific Research and Development (US, from 1941).

OSS *1* Office of Strategic Services (US 1942, became CIA 1947).
   2 Department of experimental landplane construction (USSR, defunct).
   3 Omega sensor system.
   4 On-site support.
   5 Office of Space Sciences (US).

OSSAI Organisation Scientifique et Sportive d'Aérostation Internationale.

OS-SAP Operations and support special-access program.

OST *1* Outline staff target (UK).
   2 Operational suitability test (DoD).
   3 Office of Strategic Trade (US).
   4 Outer Space Treaty (1967).
   5 Office of the Secretary of Transportation (US).

OSTA Office of Space and Terrestrial Applications (NASA).

OSTD Office of SST Development (DoT, US).

OSTIV Organisation Scientifique et Technique Internationale du Vol à Voile (Int., office Delft).

OSTM Ocean-surface topography mission.

OSTP Office of Science and Technology Policy (US).

OSU Omega/vlf sensor unit.

OSV Ocean station vessel.

Oswatitsch inlet Supersonic inlet/diffuser in form of body of revolution having pointed conical centrebody of progressively increasing semi-angle, thus generating succession of inclined shocks at increasingly coarse angles all focused on peripheral lip.

OT *1* Oil temperature.
   2 Other times, or over time.
   3 Operational test.

OTA *1* Office of Technology Assessment (Congress, US, from 1972).
   2 Orbital transfer assembly.
   3 Over the [or overflight] top attack.
   4 Other transaction agreement[s].

OTAA Office of Trade Adjustment Assistance (US).

OTAEF Operational Test & Evaluation Force (but see OTE, OT&E).

OTAES Optical technology Apollo extension system.

OTAN NATO (F).

OT&E Operational test and evaluation.

OTAR Over-the-air re-keying.

OTAS On top and smooth.

OTC *1* Operational training course (USAF).
   2 One-stop tour charter.
   3 Overseas Telecommunications Commission of Australia.
   4 Operational [or official] test centre.
   5 Officer in Tactical Command.

OTCIXS Officer in tactical command information exchange system.

OTD *1* Origin to destination.
   2 Optimal terminal descent.
   3 Optical-transient detector.

OTDA Office of Tracking and Data-Acquisition (NASA).

OTDF Outlet temperature distribution factor.

OTE National telecommunications organisation (Greece).

OTE, OT&E Operational test and evaluation [A adds Agency, C adds Command] (USA).

OTER Over-temperature emergency rating.

Otevfor Operational test and evaluation for operational requirements (USN).

OTFP Operational traffic flow planning.

OTH *1* Over the horizon; hence * radar.
   2 Other.

OTH-B Over the horizon, backscatter.

other ranks Soldiers who are not yet NCOs, army equivalent of airmen (UK).

OTH-R   Over the horizon, radar.

OTH-SW   Over the horizon, surface wave.

OTH-T, OTHT   Over the horizon, target[ting].

OTIS   Optronic tracking and identification system.

OTLK   Outlook.

otolith   Organ of inner ear sensitive to attitude and acceleration.

OTP   *1* Office of Telecommunications Policy (White House, US).

2 On-top position [ie, above submerged submarine].

3 Operation[al] test programme [S adds set].

OTPI   On-top position indicator.

OTPS   Oceanic traffic-planning system (FAA).

OTQ   Over-temperature qualification.

OTR   *1* Operational turn-round.

2 Oceanic transition route.

3 Other.

4 Overberg Test Range (S Africa).

5 Operational Test and Readiness; R adds Review.

OTS   *1* Over the shoulder.

2 Orbital test satellite.

3 Out of service.

4 Organized track system[s].

5 Off the shelf.

6 One-man ticketing system, or structure.

OTSI   Operating time since inspection.

Otto cycle   General name for four-stroke spark-ignition cycle for piston engine.

OTU   Operational training unit.

OTV   *1* Orbital transfer vehicle.

2 Obstruction to vision.

OTW   *1* On the wing; hence * replacement of engine module.

2 Over the wing; B adds bridge.

OTWD   Out-the-window display (simulator).

OU   Operations unit.

OUBD   Outbound.

OUE   Operational utility evaluation.

ounce   Imperial mass, 28.3495 g, abb. oz.

out   *1* Radio procedure word: conversation is completed.

2 Unserviceable.

outage   *1* Loosely, failure to function, especially of communications.

2 Period during which communications station is faulty.

outboard   *1* In direction from centreline to wingtip.

2 Further from centreline (eg * engine, in relation to other[s] on same side).

3 Carried outside main structure.

outbound   *1* Towards destination; hence VOR * radial.

2 Away from Conus (US military usage).

3 In holding pattern, side opposite to inbound (which heads towards holding fix).

outbound bearing   That of outbound leg of holding pattern or of VOR radial en route to destination or next waypoint.

outbound radial   That linking fix or beacon to next waypoint.

outburst   Lateral spread across surface of air arriving in downburst.

outer   Further from aircraft centreline, but see * wing.

outer air battle   Taking place 100 nm or more from surface fleet Battle Force, beyond radius of ship-to-air weapons.

outer cover   External skin of rigid airship.

outer fix   Fix in destination terminal area, other than approach fix, to which aircraft are normally cleared by ATC and from which they are cleared to approach for final approach fix (FAA).

outer loop   FCS in autopilot mode, human pilot being inner loop.

outer marker   ILS marker normally on approach centre-line approximately 4½ nm (8.3 km) from threshold; usually identified by 400 Hz aural dashes and synchronized blue panel light.

outer panel   Left or right wing outboard of centre section.

outer wing   Apart from obvious meaning: in a turn, that pointing away from centre of turn, experiencing increased airspeed.

outfall valve   Usually synonymous with outflow valve.

outflow   Violent low-level wind radiating out from downburst, or created beneath hovering VTOL aircraft.

outflow pattern   Isobar pattern above wing at positive angles of attack.

outflow valve   That incorporated in isobaric cabin-pressure regulator through which air is allowed to escape to atmosphere during climb to isobaric altitude.

outgassing   Emission of gas from metals and other materials in high-vacuum conditions.

outlet temperature distribution factor   Combustion chamber outlet peak temperature (T) minus outlet mean T divided by mean combustion chamber T rise.

out of alignment   In case of propeller or rotor blade, having incorrect sweep (bent forward or back in plane of rotation).

out-of-control lights   Two superimposed red lights 6 ft (2 m) apart hoisted at night on mast of marine aircraft whose engines have failed and which is not anchored.

out of phase   Not synchronized; two or more cycles or wavetrains which have same frequencies but pass through maxima and minima at different times.

out of pitch   In case of propeller or rotor blade, having blade angle different from that of remaining blades.

out of step   Not synchronized; two or more series of digital pulse trains or other discrete phenomena having essentially same frequency but occurring at different times.

out of track   In case of propeller or helicopter rotor blade, having incorrect tilt so that particular tip does not lie on tip-path plane.

output   *1* Total product or yield from system, eg tonne-km for airline.

2 Delivery from computer, as hard copy or graphics display.

3 One of five basic functional sections of most EDP(1) systems.

4 Signal from transducer or other instrumentation.

5 Power developed by engine or other prime mover.

6 Shaft at which power is extracted from shaft-drive engine.

7 Verb, to deliver an * (2), hence outputting.

outriggers   *1* Primary structural members carrying tail of aeroplane with short fuselage, nacelle or boat hull, also called booms.

2 Ancillary landing gear near wingtip or at other location well outboard from centreline on aircraft with bicycle or other type of centreline landing gear.

outside air temperature *1* Free-air static temperature (UK).

2 The uncorrected reading of the OAT instrument, requiring corrections to obtain true static temperature (US).

outside loop Inverted loop, performed with aircraft upper surface outwards and thus under negative g.

outside roll One started from negative g, esp. from inverted attitude.

outside wing That on outer side of spin, having greater airspeed than inside wing.

outsourcing Subcontractor brings in someone else to help.

outspin On outer side of spin, hence * aileron or rudder.

outspin yaw Yaw inevitably present in normal spin, usually function of pitch attitude.

outyears Period during which item is in operational service (USAF, later general use).

OUV Oskar-Ursinus Vereinigung, association of amateur aircraft constructors (G).

OV *1* Over-voltage.

2 Aircraft mission code, observation V/STOL (DoD).

3 Outside vendor.

4 Location (ICAO).

5 Orbiting vehicle.

ovalisation Structural distortion such that circular parts become ovals, esp. on pylon-mounted turbofans.

oval office The President of the United States; thus * green = his authority to proceed.

OVC Overcast.

ovenise To bake electronic device to simulate overload test condition.

over Radio procedure word: 'My transmission is ended and I expect a response from you'.

overall efficiency In case of aircraft propulsion system, usually thermal efficiency multiplied by propulsive (Froude) efficiency.

overall length Most definitions for aircraft stipulate in flying attitude; distance between local perpendiculars aligned with extremities of basic airframe (but usually excluding pitot or instrument probes, static wicks etc). See *length*.

overall pressure ratio OPR, ratio of total pressure at delivery to combustion chamber to that at engine inlet [the latter taken as unity, so OPR 26:1 can be written as 26]; in a turbofan OPR = FPR × compressor PR.

overbalanced Provided with excessive balancing mass, aerodynamic area or other source of control-surface moment assisting pilot, so that control surface has little feel (and in extreme case can move by itself to hard-over deflection).

overboard bleed From compressor to assist starting, not used in normal running.

overbooking Practice of selling more tickets for a particular flight than there are seats, to compensate for (1) no-shows and (2) individual pax or agents making multiple reservations for one journey.

overbuilt Made stronger than necessary, usually to allow for future development.

overburn Operation of rocket or other space propulsion system for too long a period (0.1 s can be significant), resulting in incorrect cutoff velocity and/or trajectory.

overcast Means 0.9+ cloud cover except IWC = $\frac{7}{8}$+ [= 0.875+].

overcontrol To apply more control deflection than necessary, commonly resulting in succession of excursions on each side of desired flight condition.

overdesigned *1* Stronger than necessary to meet requirements.

2 More refined than necessary, usually resulting in lighter structure at expense of high cost and complexity.

overexpanded nozzle One in which jet leaves at below ambient pressure.

overfin Fin added above high [T-type] horizontal tail.

overflight Flight over particular route, esp. across unfriendly territory.

overflight top attack Class of missiles which detonate shaped charges downwards as they pass over armoured vehicle.

overfly To fly over.

overhang *1* Spanwise distance from junction of wing and bracing struts to tip; length of cantilever mainplane.

2 Spanwise distance to tip of upper wing of biplane from point vertically above tip of lower wing.

3 See next.

overhang balance Mass distributed along leading edge of control surface, as distinct from horn balance.

overhaul Regularly scheduled procedure of dismantling (the extent varying with which *), inspection, replacement of parts if necessary and return to service. For large aircraft a major * can take weeks and involve refurnishing and exterior repaint.

overhaul period Also called TBO, the time between consecutive overhauls. For aircraft or engine or other functioning item * is actual operating time in hours. For missile and some other items such as ejection seat it is total elapsed time.

overhead Apart from normal meanings, involving the use of one or more satellites.

overhead approach Landing procedure in which pilot flies downwind over point of touchdown and then makes descending circular pattern centred on extended landing centreline; also called 360° landing.

overhead camshaft In upright cylinder, above the head; thus, beyond horizontal cylinder, below an inverted cylinder.

overhead stick Control column pivoted to cockpit roof.

overhead stowage Provision in passenger transport for hand baggage and other rigid or heavy items to be stowed in latched bins above seats.

overhead suspension Docked rigid airship hung from roof of shed.

overhung *1* Cantilevered beyond last point of support, esp. applied to fan, compressor or turbine stages mounted at extremity of shaft system beyond nearest bearing.

2 Projecting beyond tip of fixed surface, adjective normally applied to aileron or elevator.

overlap *1* Percentage of photograph duplicated on next frame of same film (see *lap*).

2 See *overlap zone*.

overlap tell Transfer of information to adjacent air-defence facility of (normally hostile or unidentified) tracks in latter's region.

overlap zone Designated area on each side of boundary between adjacent tactical air control or defence regions wherein co-ordination and interaction are required.

overlay Upper (exoatmospheric) layer of layered defence system (SDI).

overload report   Filed by ATC controller when workload could lead to safety being compromised.

overload response   Procedures designed to enable an inadequate ATC system to handle traffic at peak periods (UK).

overload weight   *1* Alternate MTO weight authorized only in unusual circumstances.

2 Any takeoff weight exceeding permitted maximum.

overpressure   *1* Limiting values reached above and below normal atmospheric pressure during passage of blast wave of explosion (nuclear or conventional).

2 Peak positive pressure reached during passage of boom from supersonic aircraft.

override   Facility for bypassing or overcoming normal limit on command action to obtain exceptional response, usually in emergency (eg * boost).

override boost   Emergency higher boost pressure available in exceptional circumstances such as overload takeoff or in combat.

overrun   *1* To fail to stop within available landing distance.

2 Control facility causing camera to continue operating for preset number of frames or seconds after cutoff of control.

overrun area   Paved area beyond end of runway which for any reason is not in use; usually marked with herring-bone pattern.

overrun barrier   Layer of material beyond end of runway provided to arrest overrunning aircraft with minimal damage; PFA is preferred to gravel.

overrunning clutch   One allowing driven member (eg helicopter rotor) to run ahead of drive system.

overseer   Common term for software used in managerial role in large system.

overshoe   Externally attached de-icer of pulsating-rubber mechanical type; also called boot.

overshoot   *1* To abandon landing and make fresh approach; in US called missed approach or go-around.

2 To land too far across available landing area and make uncontrolled excursion into region beyond.

3 To exceed desired IAS, altitude or other flight condition.

4 To pass through enemy aircraft's future flight path in plane of symmetry.

overshoot area   Designated area of semi-prepared ground available to overshooting (2) aircraft (usually GA only) from which they may taxi relatively undamaged.

overspeed condition   Normally applied to rotating machinery, and permissible for brief specified period and within specified limit. An exceptional and usually undesired condition, and in no way synonymous (as claimed in one source) with takeoff rpm.

overstow position   Inoperative position of target-type reverser with geometry which precludes inadvertent operation in flight.

overstress   Application of load which could cause structural failure [particularly positive or negative pitch input which could break wing].

oversweep   Ultra-acute setting of VG pivoted wings [eg, to make carrier-based F-14 aircraft more compact when parked].

overswing   Undesired azimuth excursion by aircraft with long, heavy body during fast turn while taxiing, esp. on slippery surface.

overtake velocity   Excess speed over that of aerial target ahead.

over the fence   Immediately before touchdown (colloq., normally GA).

over the horizon   Beyond the visible or LOS horizon; region accessible only to particular species of radar.

over-the-shoulder delivery   See *toss bombing*.

over the top   *On top (1)*.

over the weather   At pressurized cruising levels; basic meaning is free from turbulence except CAT, but cloud may still be present and cu-nims must be avoided.

over the wing   Aeroplane configuration in which turbofan is located above wing to give USB powered lift but without surface scrubbing losses.

overturn boundary   Limiting sea state for safe operation of float or pontoon-mounted helicopter.

overturn pylon   Strong structure to protect occupants of small aircraft in overturn on ground.

overwater   Particular provisions apply to commercial transports for use on sea crossings where power-off glide to land is not possible. Adjective also applied to extended-range versions, irrespective of actual intended routes.

OVHD   Overhead.

OVHT   Overheat.

OVI   Department for war inventions (of RKKA, USSR, defunct).

OVR, ovr   Over.

ovrd   Override.

ovrn   Overrun.

OVS   Overhead video system.

OVTR   Optical videotape recorder.

OW   Oblique (slew) wing.

OWB   Over-the-wing bridge (passenger loading).

OWE   Operating weight empty.

OWF   Optimum working frequency.

OWL   *1* On-wing life (main engine).

2 Obstacle-warning ladar.

3 Over water and land.

4 On-target weapon, long-range.

OWLD   Obstacle warning, location and detection.

OWM   One-way mission.

OWRM   Other war reserve materiel (US).

OWS   *1* Orbital workshop.

2 Obstacle warning system (helo).

3 Optical weapon sight.

OWSF   Oblique-wing single fuselage.

OWTF   Oblique-wing twin fuselage.

OWWS   Operational windshear warning system.

OX, Ox   *1* Longitudinal axis about which aircraft rolls.

2 Oxygen: OX1 HP gaseous, OX2 LP gaseous, OX3 HP replacement bottles, OX4 LP bottles, no code for liquid.

oxidant   Chemical carried as rocket propellant for combination with fuel; examples are lox, nitrogen tetroxide, nitric acid and mixtures of fluorine and oxygen (flox).

oxidation   *1* Removal of electron from atom or atomic group.

2 Combination with oxygen, as in burning or rusting.

oxidiser   Oxidant (N American usage).

oxidising flame   In gas welding or cutting, one having excess oxygen.

oximeter   Instrument giving numerical readout of blood oxygen concentration.

OXRB   Oxygen replacement bottles available.

oxy   Oxygen.

oxygen O_2, odourless gas, vital to life, density 1.43 gl^{-1}; liquid [lox] density 1.1, MPt –218°C, BPt –183°C. For aircraft breathing purposes, Grade A mandatory, 99.5% pure, many other requirements including water ≤0.005 mg/l. In US OX1/2/3/4 are respectively HP gas, LP gas, HP replacement bottle and LP replacement bottle.

oxygen bottle   Container for gaseous oxygen under pressure.

oxygen converter   Liquid oxygen converter (ie boiler) supplying gox to breathing system.

oxygen mask   Face mask for supplying gaseous oxygen to wearer, usually on demand, and separating exhaled breath.

oxygen microphone   Microphone fitted to oxygen mask (probably arch.).

OY   Option Year.

OY, Oy   Lateral axis about which aircraft pitches.

OYM   Optional yield management.

OZ, Oz   Vertical axis about which aircraft yaws.

oz   Ounce, $^1/_{16}$ lb = 28.35 g.

OZB   Civil aviation directory, under Ministry of Works (Austria).

ozone O_3, allotropic form of oxygen present in minute quantities in air, liquid fainly bluish, very reactive, BPt –182.97°C.

ozone layer   See *ozonosphere*.

ozonosphere   Layer of upper atmosphere, roughly at 20–40 km altitude, where ozone concentration is much higher than at SL and ozone plays major role in establishing radiation balance.

P

P *1* Generalized symbol for force.

2 Power, esp. bhp of shaft engine or peak power of transmitted signal.

3 Aircraft category, pursuit (USA, USAS, USAAC, USAAF, 1925–47, USN 1923), and patrol (USN from 1923, USAF from 1962).

4 Prohibited area.

5 Poise.

6 Period, eg orbital.

7 Aircraft designation suffix, photographic (USN, to 1962).

8 Pressure; normally p but gas-turbine pressures normally written P_1, P_2 etc.

9 Pulsed, or pulse (eg with suffix numeral P1, P2 etc).

10 Pilot, with numerical suffix 1, 2, 3 etc to show hierarchical ranking of pilots in same crew.

11 Prefix, plus (wind component).

12 Production cost.

13 Ratio of full-scale length to model length, esp. in radio aerials and related fields.

14 Probability.

15 Telecom code: aircraft has Tacan only and 4096-code transponder.

16 Radar (JETDS).

17 Missile launch environment: soft pad (DoD).

18 Phosphorus.

19 Provisional.

20 Aerospace craft (FAI category).

21 Primary, primary frequency, or winding.

22 Precision (DME, GPS, MLS).

23 Packet (TDM mode).

24 Paved surface.

25 Prefix peta = 10^{15}, thousand million million.

26 Airport with instrument-approach procedure.

27 Number of persons on board.

28 Polar air.

29 Positive.

30 Precipitation.

31 Prototype.

32 Propeller *pitch*.

33 Polarization [electric].

p *1* Pico, 10^{-12}.

2 Generalized symbol for pressure.

3 Structural pitch.

4 Structural stress.

5 Pound mass (ambiguous, eg psi, see next).

6 Per (ambiguous, eg lb psi).

7 Parts (eg ppm, parts per million).

8 Port (ie, left).

9 Pentode.

10 Pulse-modulated radio emission.

11 Plate (electrical).

12 Generalized symbol for positive or positive values, eg *-type semiconductor.

13 Angular velocity in roll; roll-rate output (AMSU).

14 Momentum.

p̄ *1* Average pressure difference across upper/lower surfaces of aerofoil.

2 Specific weight (air-breathing jet engine).

3 Average rocket-motor chamber pressure.

4 Generalized symbol for averaged pressures.

P_0 Total static pressure head; stagnation pressure.

P_1 *1* Compressor inlet total pressure.

2 First pilot, captain of aircraft.

P_{12}, $P_{\frac{1}{2}}$ Fan inlet pressure.

P_2 *1* Compressor delivery pressure (single-spool) or fan delivery.

2 Copilot.

3 Two PAPIs.

P2P Confusingly, peer-to-peer.

P_3 Compressor delivery pressure (two-spool).

P^3 *1* Public/private partnership.

2 Pulse-pair processing.

P^3I Pre-planned product [or program for] improvement (pronounced P-cubed I).

P4 Runway has 4 PAPIs.

P.4, P.6 Standard magnetic compasses (UK, 1930–50).

P-alt Pressure altitude.

P-band Frequency band, 225–390 MHz (Appendix 2).

P-channel Packet-mode channel, in two forms, *Pd* and *Psmc*.

P-charge Projectile-charge warhead (SAM).

P-clamp For ordinary electric/electronic cables.

P-clip For convoluted conduit with raised rib to prevent lateral movement.

P-code Precise or protected navigation code, billions of pseudo-random numbers on 10.23 MHz repeating each 267 days, each 168-h (one week) segment unique to particular GPS satellite.

P-display Radar display of PPI type.

P-factor *1* Asymmetric propeller loading.

2 Effect on propeller aircraft of spiral slipstream[s].

P-lead Primary (switch) cable of magneto.

P-line Electric cross-country power line.

P-time Proposed departure time.

p-type semiconductor One in which majority of charge-carriers are holes (electron absences).

PA *1* Public, or passenger, address.

2 Performance appraisal.

3 Public affairs.

4 Proposal architecture.

5 Product assurance (especially software).

6 Precision attitude.

7 Pilot's associate, or assistant.

8 Porte avions, aircraft carrier (F).

9 Pressure altitude.

10 Pursuit, aircooled (USA 1919–24).

11 Power[ed] approach [esp. to carrier].

12 Power amplifier.

13 Precision approach, suffixes 1/2/3 for Categories I/II/III.

14 Point Arguello, CA, US.

15 Pad abort.

P/A Payload/altitude curve (sounding rocket).

Pa *1* Pascal, SI unit of pressure.

2 Polar Atlantic.

P_a Action-time average chamber pressure; sometimes abb. Pc.

pa, p.a. Power amplifier.

PAA *1* Partial air alert.

2 Primary aircraft authorization, or authorized.

3 Passenger-address amplifier.

4 Phased-array antenna.

PAAMS Principal anti-air missile system.

PA&E Program analysis and evaluation (DoD).

PABST Primary adhesively bonded structure technology.

PABX Private automatic branch exchange.

PAC *1* Parachute and cable (rocket-launched 'SAM', WW2).

2 Public Accounts Committee (UK, House of Commons).

3 Programmable armament control.

4 Photometrical airfield calibration.

5 Path attenuation compensation.

6 Pacific Region (ICAO).

7 Penetration-aid[s] carrier.

8 Precision attitude control.

9 Public-address control.

10 Poste aérienne de commande (F).

11 Presidential Advisory Committee (US).

12 Pulley and cable [flight control].

PACA *1* Precision attitude control augmentation; /IAGSS adds improved air/ground sight system.

2 Programmed adaptive clutter attenuator.

PACAF, Pacaf Pacific Air Forces (USAF).

PACCS Post-attack command and control system.

Pacer Portable aircraft condition evaluation recorder.

PA/CI Passenger address and cabin [or communications] interphone [S adds system].

pacing item Task whose time for accomplishment determines overall time for programme; any major item on critical path.

PACIS Pilot aid and close-in surveillance.

pack *1* Bag in which parachute is packed.

2 Complete system in demountable package which may be attached inside or outside aircraft, eg for multi-sensor reconnaissance, rocket thrust-boost, gun and ammunition, or air-conditioning.

3 In particular, ECS air-conditioning unit comprising bootstrap ACM(1) plus air/air heat exchanger.

package *1* See *pack (1)*.

2 Complete offer of international contract covering either collaborative development or manufacture of aircraft complete with associated offsets, loans, training and other attractions, and with agreed share to each partner of overall effort and/or cost.

3 Complete offer of international contract covering sale of aircraft with associated offers of offsets, training, service support, construction of support facilities, etc.

4 Assembly of attack aircraft, tankers, EW jammers, etc, to carry out a mission.

package aircraft Subject of package (2, 3) deal.

packaged *1* Of petroleum products – eg POL, jet fuel – contained in drums having capacity not greater than 55 US gal (208.2 l).

2 Of electric generating plant, nuclear reactor, etc, mounted on skids and transportable thereon as operative unit.

packaged propellant Rocket motor fed by liquid propellant(s) sealed in container forming single unit with thrust chamber.

package gun One mounted in fairing outside aircraft fuselage, usually fixed.

package programme Timetable for package (2, 3).

packet *1* Single pulse of radar, digitized signal or other coded EM emission.

2 Prepacked quantity of chaff, either pre-cut or as roving, for loading into dispenser.

3 Discrete parcel of data sent through multiple network channels and reassembled at point of delivery.

packet mode TDM mode for signalling, control and data com. See *Pd* and *Psmc* (GPS).

pack hardening Nitriding by heating inside loose pack of carbonaceous material.

PACS *1* Passive attitude control system, eg gravity gradient.

2 Pitch-augmentation control system.

PACT *1* Portable automatic calibration tracker.

2 Precision aircraft control technology (CCV).

Pactec Partners in aviation and communications technology, non-profit humanitarian organization working in developing countries.

Pacts Parliamentary Advisory Council for Transport Security (UK, 2002).

PAD *1* Picture assembly device.

2 Point air defence; T adds trainer.

3 Persistent area-denial, or dominance.

4 Precision aerial delivery (USAF).

5 Packet assembler/disassembler.

6 Primary Awacs display.

7 Pad-abort demonstrator, or demonstration.

PADA Phased-array downlink antenna.

pad *1* Platform for launch of vehicle, usually of ballistic or unmanned nature (colloq. and becoming arch.).

2 Platform for operation of helicopter, esp. on surface vessel.

3 Attachment face on engine or other energy source for accessory, usually with central shaft drive.

4 Network of resistances or impedances to couple to transmission lines effectively.

pad abort Abort before launch from pad (1).

padder See *padding capacitor*.

padding capacitor In series with superhet. local-oscillator tuning capacitor to match exactly to frequency.

paddle In a rate gyro, damping drag surfaces which generate angular difference proportional to first-order rate term.

paddle blade Propeller blade having unusually wide chord maintained to tip.

paddleplane See *cyclogiro*.

paddle switch Ambiguously used to denote electrical switches operated by dynamic air pressure, dynamic water pressure in ditching and various other activation sources.

PADIS, Padis Procedures for analysing the design of interactive solutions (fire-control software).

Padlocked "I have my vision fixed on bogies/bandits and will not look away".

PADM Persistant area-dominance munition.

pad output Power rating of shaft at each pad (3) location on engine, APU or MEPU.

PADR Product-assurance design review.

PADS, Pads *1* Passive advanced sonobuoy.

2 Position and azimuth determining system.

3 Precision automatic dependent surveillance.

PADUS   Principal Assistant Deputy Under-Secretary (US).

PAF   *1* Parabolic aircraft flight.

2 Police de l'Air et des Frontières (F).

3 Potassium acetate fluid (de-icer).

PAFAM, Pafam   Performance and failure assessment monitor (independent flight guidance with flight-deck displays during autolanding).

PAFU   Pilots' advanced flying unit.

PAG   *1* Programmable automatic gauge, for NC dimensional checking.

2 Portable arrester gear.

Pageos   Passive GEOS [Geodynamic experimental ocean satellite(s)].

PAH   *1* Anti-tank helicopter (G).

2 Polycyclic aromatic hydrocarbon.

3 Production approval holder (FAA).

PAI   Parachute Association of Ireland Ltd.

Paid, PAID   Parked-aircraft intrusion detector.

paint   To create blip on radar display, esp. one giving position of aircraft or other object.

paintless aircraft   Instead of usual finishes, covered with protective film of appliqué materials, saving weight, cost and support requirements.

paint-on bag   Flexible vacuum bag formed in place to cover layup of composite material fabrication and seal tool in prototype moulding of large parts.

paint-stripe loading   Vertical stripes inside cargo aircraft to assist loaders in achieving correct load distribution, esp. with regard to c.g.

PAIR, Pair   Precision-approach interferometer radar.

pair   Section of two fighters operating together.

paired channels   Selecting Vortac or ILS frequency automatically also selects a DME.

PaK   Anti-armour cannon (G).

PAK-FA   Future air complex for tactical aviation, loosely = next-generation fighter (R).

PAL   *1* Passive augmenting (or augmentation) lens, eg Luneberg.

2 Standard German TV/video system; from phase alternation line.

3 Permissive action link.

4 Portable airfield lighting; S adds set.

5 Programmable array logic.

6 Propulseur d'appoint liquide (F).

7 Preprocessor assignment logic.

PALC   Point Arguello Launch Complex.

Palea   Philippine Airlines Employees Association.

PALF   Partly aloft (precipitation not reaching ground).

palisade   Large hinged flight-deck barrier (RN carriers 1919–32).

palladium   Pd, silvery malleable metal, density 12.0, MPt 1,552°C.

pallet   *1* Standard platform for air freight, eg (Imperial measures are still used) 88 in × 125 in, 88 in × 108 in and 96 in × 125 in.

2 Flat base, not necessarily of above dimensions, for combining stores or carrying single item to form unit load, in some cases retained as temporary or permanent base in operation.

3 Standard-dimension base on which can be assembled experiments, ancillaries and other payloads, eg for Spacelab.

4 Standard-dimension base carrying work-pieces in mechanized and automated manufacturing.

5 Configured (shape-matching) payload carrier for attachment close against airframe exterior, eg fuel * (FAST Pack).

pallette   Subsystems.

palletized bladder   Flexible container for liquid mounted on pallet (1) for airfreighting.

palletrolley   Pallet (2) with wheels for ground transport, eg for Sea Skua missile.

Palmachim   On coast south of Tel Aviv, base for spaceflight, IRBM, ABM (Israel).

Palmer scan   Radar scanning technique in which conical scan is superimposed on some other, eg Palmer-sector (beam oscillates to and fro over azimuth sector while continuously making small-angle conical scan), Palmer-raster, Palmer-circular etc.

palm tree   Upward bomb burst (formation aerobatics).

Palnut   Thin nut, usually pressed steel, screwed tightly above ordinary nut to prevent loosening of latter.

PALS, Pals   *1* Precision approach and landing system.

2 Portable airfield light set (USAF).

3 Positioning and locating system.

palsar   Phased-array L-band SAR.

PAM, p.a.m.   *1* Pulse-amplitude modulation; also, confusingly, phase-amplitude modulation.

2 Payload assist module.

3 Picture assembly multiplexer.

4 Plasma-arc machining.

5 Precision attack missile.

6 Peripheral adapter module.

7 Portable automated mesonets (weather).

PAMA   Professional Aviation Maintenance (formerly Mechanics) Association (US).

PAMC   Provisional acceptable means of compliance (ICAO).

Pamela   Process abstraction method for embedded large applications.

Pamir   *1* Passive airborne modular IR.

2 Phased-array multifunction imaging radar.

PAMR   Public-access mobile radio.

PAMS   Point anti-missile system.

PAN   *1* Polyacrilonitrile, precursor of carbon fibre in most processes.

2 Originating station has urgent message to transmit concerning safety of vehicle or occupant(s) (DoD); see Pan.

3 Porte avions nucléaires (F).

Pan   Radio code indicating uncertainty or alert, general broadcast to widest area but not yet at level of Mayday.

pan   *1* Base of seat tailored to pilot with seat-type pack.

2 Paved dispersal point for single aircraft.

3 Carrier for air-drop parachuted load.

pancake   *1* Vague term supposedly indicating landing at abnormally low forward speed and high sink rate (arch.).

2 Air-intercept code: "I wish to land", usually followed by word giving reason, eg * fuel (DoD).

3 Verb, to land (arch.).

4 Flat-topped fuselage regarded as a lifting surface.

pancam   Panoramic camera (planetary lander).

P&A   Priorities and allocations.

Panda   Personnel and administration.

P&ES   Personnel and equipment supply.

P&F   Particles and fields.

P&I   Paint and interior.

P&O   Plant and operations.

P&P   *1* Plans and programs (DoD).

2 Payments and progress (NATO).

P&SS   Provost and Security Service (RAF).

P&T   Priorities and traffic.

P&TC   Personnel and Training Command (RAF).

panel   *1* Single piece of aircraft skin, eg fuselage *.

2 Major section of wing as finished component, eg outer *.

3 Essentially planar base carrying instruments.

4 Complete portion of stressed-skin structure, sandwich or, rarely, other constructional form, used for fatigue or static test purposes; not necessarily portion of actual aircraft.

5 Subdivision of parachute canopy, either complete gore or portion thereof.

6 Body of experts, eg examining *.

7 Marking panel.

panel code   Standard code for visual ground/air communication (IADB).

panel elements   Items mounted on panel (3) such as instruments, switches and displays.

panel fill factor   The percentage of the complete instrument panel facing the pilot that is occupied by multifunction displays.

panelled   Equipped with instruments and avionics, thus well * (colloq.).

panhandle   Aircraft dispersal of frying-pan shape.

pan-handle   Ejection-seat firing handle on seat pan.

pan-head screw   One having thin, large head with slightly convex upper surface.

panic button   Gives emergency recovery to unaccelerated horizontal flight with wings level (USAF).

PANNI   Psychological assessment of noise and number index [see noise].

pannier   Small 1-, 2- or 4-tube rocket launcher added to CBLS or other stores carrier.

panoramic camera   One which by means of system of oscillating or rotating optics or mirrors scans wide strip of terrain usually from horizon to horizon. Mounted vertically or obliquely to scan across or along line of flight.

PANS, Pans   Procedures for Air-Navigation Services; /OPS adds aircraft operations, /RAC adds radio com. procedures.

PANS-ABS   PANS abbreviations and codes.

Pantera   Precision attack navigation and targeting with extended range acquisition.

panting   In/out springy movement of thin skin under compressive stress; generally similar to oil-can effect.

pantobase   Landing gear designed for land, water, snow and possibly other surfaces.

pantry   Aircraft kitchen without provision for heating prepacked meals.

pants   Fixed fairing over landing gear leg and, esp., wheels; generally similar to spats but chiefly US usage.

pants duct   Large-diameter Y-piece forming junction or bifurcation in duct system.

PANYNJ   Port Authority of New York and New Jersey.

PAO   Poly-alpha olefin.

PAP   *1* Precision approach path.

2 Plug and play (transport a/c fuselage).

3 Paintless aircraft programme.

4 Propulsion [or propulseur] d'appoint à poudre [= solid booster rocket].

5 Pierced, or perforated, aluminium plank[ing].

Papa   Parallax aircraft-parking aid.

paperless   Totally on computer; thus * cockpit, in which all paper is replaced by EFB.

paper returns   Written lists of [military] aircraft at readiness, serviceable, unserviceable and written off, possibly adjusted for political reasons.

PAPI, Papi   Precision approach path indicator.

PAPM, p.a.p.m.   Pulse amplitude and phase modification; narrows signal bandwidth to quadruple carrying capacity per channel.

Paprica   Passenger relief in co-operation with airlines.

PAPS, Paps   *1* Periodic armaments planning system (NATO).

2 Phased armaments programme systems.

3 Precision-airdrop planning software (USAF).

PAR   *1* Precision approach radar.

2 Progressive aircraft rework (USAF, USN).

3 Phased-array radar, ie electronically steered instead of mechanically scanned.

4 Perimeter-acquisition radar (ABM).

5 Pulse-acquisition radar (SAM).

6 Program(me) appraisal and review.

7 Preferential, or preferred, arrival route.

8 Power analyser and recorder.

9 Parachute/airbag recovery.

10 Parallel.

11 Physiological ageing rating.

12 Performance and reliability.

13 Private pilot, airplane, recreational.

Par   Precision aircraft reference (Lear-Siegler twin-gyro platform).

parabola   Conic section made by cutting right circular cone parallel to any of its elements; locus of point which moves so that distance from line (directrix) equals distance from point (focus), so eccentricity = 1.

parabolic aerial   One whose reflecting surface forms portion of parabola, thus converting plane waves into spherical waves or vice versa, and either emitting pencil beam from point feeder or focusing incoming radiation to single-point receiver. Also called parabolic antenna/mirror/reflector.

parabolic trajectory   As eccentricity is unity it represents least eccentricity for escape from attracting body.

paraboloid   Shape in 3-D formed by rotating parabola about major axis.

parabrake   Braking parachute.

Parac   PAR(4) attack characterization, system travelling-wave tube.

parachute   *1* Any device comprising flexible drag or drag + lift surface from which load is suspended by shroud lines. Originally canopy was umbrella-shaped but today may be of Rogallo or other semi-winged types offering precision control of landing within large area. Distinction between * and hang glider is blurred, but chief features distinguishing * are rapid deployment from packed condition and suspension of load well below canopy.

2 Verb, to use (1).

3 Verb, to deploy an aerodynamic-drag system (UAV).

parachute deployment height   Height AGL at which canopy is fully deployed.

parachute flare Illuminating flare equipped with parachute to prolong descent.

parachute gore See *gore*.

parachute harness That to which personal parachute is attached.

parachute pack Bag containing packed parachute.

parachute tower *1* Tower or mast from which parachute descents are made for sport or instruction.

2 High-ceiling part of parachute section of airbase wherein parachutes are hung to dry after use.

parachute tray Rigid base to pack of parachute used in heavy dropping.

parachute vent Aperture in top of canopy or left by blank gore(s) to ensure stable descent.

para-circular Near-circular orbit (intended to be circular).

paradrag drop Ultra-low airdrop of cargo using drag of arrester parachute to extract and halt payload (NATO).

paradrop Delivery by parachute (from height significantly greater than paradrag drop) of personnel or cargo from aircraft (NATO).

paraffin *Kerosene.*

parafoil Parachute able to fly as aerodyne (L/D about 3) and stall, but with canopy instantly deployed from packed condition and attached by normal harness.

parafoveal vision Used at night to see extremely dim sources; eye is oriented so that what light there is falls on area of retina populated mainly with rods, not in central (foveal) region but surrounding it.

paraglider *1* Inflatable hypersonic spacecraft having form of metallised-fabric kite of paper-dart shape.

2 Also formerly applied to various flexible-wing gliders, towed kites and parafoils but today no longer used; each device must be either a kite, hang glider, parachute or parafoil.

Paralkatone Protective sealant film sprayed on aluminium alloy aircraft, usually after completion or during erection.

parallax aircraft parking aid Built into airport gate, gives precise optical guidance for each type of aircraft.

parallax error Caused by viewing bi-planar display [eg dial instrument] obliquely.

parallel *1* Great circles parallel to Equator (* of latitude).

2 Circle on celstial sphere parallel to celestial equator (* of declination).

3 Connected so that current flow, signal, etc divides and passes through all components simultaneously, thereafter recombining.

4 Software connection so that each signal has its own wire; hence a 16-bit wire has 16 conductor paths.

parallel actuators Two or more in parallel to drive single load; usually physically separated and tied by load in force or torque-summing fashion (thus providing rip-stop design).

parallel aerofoil Having constant section from tip to tip; two-dimensional.

parallel burning Solid grain which ignites at centre over whole length and burns purely radially outwards to case.

parallel double-wedge Aerofoil whose section is that of flat hexagon with sharp wedges at leading and trailing edges and parallel upper and lower surfaces; poor at subsonic speeds but acceptable supersonic shape, esp. for missiles.

parallel-heading square Training manoeuvre in which helicopter is flown at low level around a square, each of the four legs being flown with helicopter aligned with that side of square.

parallel ILS Serving parallel runways, with minimum lateral separation of 1,500 m, 5,000 ft.

parallel-line design Aircraft whose wing LE/TE, tips and sharp edges are all parallel. This concentrates radar signature into very narrow beams, easily missed by enemy. B-2 is example.

parallel offset Airway R-Nav route running alongside designated airway.

parallel of origin Parallel of latitude used as an origin (2).

parallel redundancy Addition of channels in parallel (3) mode purely to increase redundancy (1).

parallel resonant circuit Tuned circuit with inductive and capacitive elements in parallel (3).

parallel runways Runways at same airport on same alignment and with sufficient lateral separation to permit simultaneous use, including parallel ILS approaches. Designators have suffix L or R [left or right] thus 28L is south of 28R.

parallel servo One located in control system so that servo output drives in parallel with major input; usually arranged to drive both cockpit controls and flight-control system and thus performing as alternative to pilot.

parallel warfare Use of a sensor network and EBO to render an enemy helpless (USAF).

parallel yaw damper One connected direct into pilot/rudder control loop and, in some cases, resulting in movement of cockpit pedals. Normally superseded by series damper.

paramagnetic Possessing magnetic permeability above 1 and permanent magnetic moment; atoms tend to align in direction of external field giving resultant magnetic moment and tend to move to strongest part of field.

Paramatta Code for RAF WW2 blind target-marking by PFF using precision-aimed TIs of sophisticated types repeated at intervals; usual form with electronic aids called Musical *.

parameter Basic definable characteristic or quality, esp. one that can be expressed numerically; specialized meanings in maths, EDP (1) and statistics.

parametric amplifier Reactance amplifier dependent on time-varying reactance (fed by signal and a pump source) forming part of tuned circuit; variable-capacitance usually Varactor or crystal diode. Also called Mavar (mixed amplification by variable reactance).

parametric study Study based on numerical values of system variables.

parametric take-off number Product of (landing wing loading × 1.11) ×(MTOW ÷ [total installed thrust × C_{L2}]).

paramotor Engine/propeller and seat unit for attachment to hang glider or parafoil.

paramp See *parametric amplifier*.

pararescue team Trained personnel who reach site of incident by parachute and render aid.

parasheet Parachute in form of regular polygon with rigging lines attached to apices; subdivided into gathered * (periphery constrained by hem cord) and ungathered *.

parasite Aircraft, invariably aeroplane, which for portion of flight relies on another for propulsion and possibly also for lift. Can ride on back of parent, or be attached elsewhere externally or internally, or be towed.

parasite drag   Sum of all drag components from all non-lifting parts of body, usually defined as total drag minus induced drag. Not normally used (see *profile drag*).

parasitic element   Resonant element of directional aerial excited to produce directional pattern; eg reflector added to Yagi aerial.

parasitic material   RAM added [usually by adhesive] to exterior skin to reduce RCS.

parasitic oscillation   Generated by stray inductances and capacitances and eliminated by parasitic stopper.

parasiting   Technique of using parasite.

paraski   Sport combining downhill ski with lifting surface such as parachute, parafoil or hang glider; competitions test landing precision.

parasol wing   One from which rest of aircraft is suspended by ties.; hence, parasol monoplane.

paravane   *1* Hydrodynamic body towed (e.g., by helicopter) on end of minesweeping cable.
2 Aerodynamic body towed on end of cable from aircraft to keep line taut and steady, eg in mid-air recovery.

para-visual director   Attention-getting non-quantitative flight instrument giving bold linear indication of excursions in pitch or azimuth, generated by rotation of striped 'barber pole' display; usually mounted on flight-deck coaming; indicator in GRDS is AMLCD (Smiths Industries).

parawing   Variation of basic Rogallo flexible wing developed at NASA Langley for payload recovery and marketed by Northrop; not current term in hang gliding.

PARC   Particle-Astronomy Research Council (UK).

parcel   Any small volume of gas, esp. of atmosphere.

Parcs   Perimeter acquisition radar characterization system.

PARD   Pilotless Aircraft Research Division (NACA, from 1946).

Pardop   Passive ranging Doppler.

parent   *1* Main aircraft in composite or carrier of parasite.
2 Aircraft carrying one or more RPVs over which it exercises control following release.

parent body   Primary around which satellite orbits.

PARES   Passive-radar ESM system.

parity   *1* Symmetric property of wave function or other phenomenon; value is 1 if function is unchanged by inversion of its co-ordinate system.
2 Precise keying of two or more sets of data [eg EDP (1) software, tapes, reconnaissance pictures or wire recordings] so that they run to common time-base. System parities are assigned individual tracks in such devices as reconnaissance signal recorders.

park   To establish synchronous satellite at its operating position, where it is parked.

parked   Aircraft, esp. commercial transport, not in active use over extended period.

Parkerizing   Anti-corrosion treatment in which metal part is boiled in solution of phosphoric acid and manganese dioxide and then dipped in oil.

Parker-Kalon, PK   Patented family of self-tapping screws for sheet.

parking brake   That applied continuously after aircraft is parked, to prevent subsequent rolling; also (US) called park brake.

parking catch   Fitted to normal handbrake (rarely, toe brake) to convert to parking brake.

parking orbit   Temporary spacecraft orbit for such purposes as waiting for correct timing or for delivery of components, spacecraft or station assembly, or rectification of fault.

PARL   Parallel.

Parm, PARM   *1* Persistent anti-radiation missile.
2 Precision approach runway monitor, = PRM (2).

Parmod   Progressive aircraft rework modification (USAF, USN).

parrot   Code for IFF transponder (DoD).

PARS, Pars   *1* Programmed airline reservation system.
2 Primary attitude-reference system.

parsec   Unit of length on cosmological scale equal to distance at which object has heliocentric parallax of 1" (one second of arc) = 206,265 times semi-major axis of Earth's orbit = 3.26 light-years = 3.0857×10^{16}m, abb. pc. Multiples are kpc, kilo-* and Mpc, mega-*.

Parsecs   Program for astronomical research and scientific effects concerning space.

Part   Chapter in rules governing civil aviation; e.g. * 91 governs business aviation and * 121 scheduled air carriers (FAA).

part-charter   Particular flight, scheduled according to timetable.

partial-admission turbine   One in which working fluid enters around only part of periphery.

partial pressure   That exerted by each gas in gaseous mixture; Dalton's law states value is same as if that gas occupied whole volume occupied by mixture at same temperature.

partial-pressure suit   Skintight suit covering body except head, hands and feet and inflatable (usually by stitched-in tubing) to press on wearer and oppose internal pressures.

partial priority   Unusual status accorded special traffic (SST was intended) by ATC services in special circumstances short of emergency.

partial shielding   *1* Protection against micrometeorites incident from one direction only.
2 Radiation shielding of crew of nuclear-propelled aircraft whilst leaving radiation free to escape in other directions.

particle separator   Device for cleaning solid (and possibly liquid) particles from air entering engine or other device, typically by centrifuging effect in vortex.

particle static   Radio-communication static thought to be due to accumulation or shedding of particulate matter in flight (not defined and suggested arch.).

parting out   Procedure for dismantling aircraft to yield certifiable parts.

Part-Publication   Category of military aircraft whose characteristics could be openly published except for such items as performance, fuel capacity and other sensitive details (UK).

part-span shroud   Snubber [clapper] near mid-length of fan [rarely, compressor] blade; abuts on neighbours to damp out flutter or other vibration.

part surface   Surface being cut in GPP.

part-throttle reheat   Augmentation of engine by afterburner brought in at less than maximum cold thrust, giving smoothly increased thrust and avoiding sudden augmentation by selecting afterburner at maximum rpm.

Afterburner stays lit as throttle is closed, extinguished only at low fuel flow at low thrust level.

Parylene   Conformal (vapour-deposited) barrier coating used to protect precision parts, especially electronics, from hostile environments (Union Carbide).

PAS   *1* Performance advisory system (Simmonds subsystem offering minimum fuel and maximum propulsive efficiency).

   2 Public-, or passenger-, address system.

   3 Power available, shaft (UK, CAA = spindle).

   4 Polyarylene sulphide.

   5 Projector approach sight.

   6 Paralkatone adhesive sealant [I adds inhibiting, TP temperature protection].

   7 Police Air Support (UK).

   8 Prime and suppliers.

   9 Perch and stare (UAV).

   10 Primary alerting system.

   11 Pulse-analysis system.

   12 Pseudo aircraft system.

   13 Precision-attack system[s].

PAS-2   A pioneer PAS (4) processed at relatively low temperatures for advanced airframes (Phillips Petroleum).

Pasar   Podded advanced synthetic-aperture radar; S adds system.

PASC   Pacific Area Standards Congress.

Pascal   Software language, becoming disused.

pascal   SI unit of pressure, Pa, = N/m^2 = 0.02088 lb/ft^2, 0.000145038 lb/in^2. As it is such a small unit kPa or MPa are more common.

Pascal's law   In fluid at rest, all pressures acting on given point are equal in magnitude in each direction (neglects acceleration due to gravity).

PASI   Pre-application statement of intent.

Paspo   Precision-attack systems program office.

Pass, PASS   *1* Parked-aircraft security system.

   2 Passive and active sensor subsystem.

   3 Passive aircraft surveillance system.

   4 Primary avionics subsystem.

   5 Pylon-accommodated self-protection suite, or system.

   6 Personnel-access security system.

   7 Professional Airways Systems Specialists (US trade union).

pass   *1* Single run by aircraft past point on ground or other object such as aircraft in flight on same heading at markedly lower speed.

   2 Short tactical run or dive by aircraft at target; single sweep through or within firing range of enemy air formation (DoD, IADB).

   3 Single orbit by satellite, starting at point Equator is crossed northbound.

   4 Single passage by satellite overhead.

   5 Single period of time during which satellite is within radio contact of control or data-acquisition station.

passenger   In addition to basic meaning of humans carried in vehicle other than vehicle's crew, normal meaning in air-carrier terminology is FPP (fare-paying *), normally, excluding company employees on cheap rate. Latter, however, are normally included in traffic statistics.

passenger boarding bridge   Passenger loading bridge.

passenger breathing equipment   *1* Drop-down oxygen mask.

   2 Smoke hood.

passenger loading bridge   Covered walkway linking terminal and parked aircraft; see *bridge*.

passenger profile   Distribution of pax at one airport through 24-h period.

passenger service charge   Fee levied by airport authority on each departure passenger to help cover its costs.

passenger service unit   Originally comprised overhead punkah outlet, now adds light(s), call button and possibly other interfaces.

passivating   Coating, esp. metal surface, with inert film to prevent electrochemical corrosion.

passive   *1* Receiving but not emitting.

   2 System, eg PFCS, which can only fail open.

passive air defence   Includes such activities as cover and concealment, camouflage, deception, dispersion and shelters.

passive countermeasures   Detection and analysis of hostile emissions (see *ESM*).

passive guidance   Use of received signals from whatever source in order to navigate.

passive homing   Tending to fly towards source of emission from target, usually IR.

passive jamming   Use of chaff and confusion reflectors.

passive landing gear   One utilizing optimised self-adaptive damping devices.

passive missile   One equippped with guidance able to home on emissions generated by the target.

passive mode   Use of airborne radar in receive-only mode, eg AWACS.

passive munition   *1* Delivered safe and then triggered either by timing system or radio signal.

   2 Inert and undisturbed, eg mine.

passive paralleling   Simplest and most common type of redundancy wherein two parallel functional devices are utilized such that, if one fails, second is still available. Limited to simple elements of control system which can only fail passive, such as springs and linkages. When failure of one element occurs, there is an acceptable change in performance or capability.

passive pilot   One who, on long-haul, is relaxing or taking a nap.

passive radar   Misnomer; normally interpreted as passive mode.

passive ranging   Trajectory-measuring systems that do not rquire a transmitter or transponder in vehicle, eg Pardop. See next.

passive ranging subsystem   Doppler plus phase rate of change to give direction and range.

passive sonar   Listens for emissions from submarine.

passive thermal control   Changing attitude of spacecraft to even-out incident solar flux, esp. by roll through 180°; so-called barbecue manoeuvre.

pass off   To test and clear production item.

PAT   *1* Process action team (AFLC).

   2 Pilot access, or applications, terminal.

   3 Private pilot, airplane, recreational, transition.

PATA   Polish air-traffic agency.

PATC   Professional, administrative, technical and clerical.

Patca   Plurilateral agreement on trade in civil aircraft.

Patch   Precision approach to coupled hover.

patch   *1* Strong fabric attachment cemented to aerostat envelope (see *eta* *, *channel* *, *split* *, *rigging* *).

2 Embroidered or printed badge worn on working (especially flying) clothing.

3 Area of ground illuminated by airborne radar.

PATCO, Patco   Professional Air Traffic Controllers Organization (US, lost status 1981).

Patec   Portable automatic test equipment calibrator.

path   *1* Track of aircraft under control of TMA.

2 Projection on Earth's surface of satellite orbital plane; same as track.

3 Loosely, aircraft trajectory in 3-D.

Pathfinder   *1* Software with immense capacity to extract intelligence from vast amounts of data.

2 Passive thermal Flir for navigation, detection and enhanced resolution.

pathfinder aircraft   One with special crew plus drop-zone/landing-zone marking teams, markers and/or electronic navaids to prepare DZ/LZ for main force (NATO).

pathfinder drop-zone control   Communication and operation centre from which pathfinders exercise guidance (DoD, LADB).

Pathfinder Force   Elite sub-force within RAF Bomber Command 1943–45 charged with marking targets for main-force attack, abb. PFF. Pathfinder Association supports needy survivors.

pathfinders   Four meanings, all defined as plural:

1 Experienced aircrew who lead formation to DZ, RP or target.

2 Teams dropped or air-landed at objective to operate navaids.

3 Radar or other navaid used to facilitate homing on objective, esp. in bad visibility.

4 Teams air-delivered into enemy territory to determine best approach/withdrawal lanes, LZs and sites for heliborne forces (all DoD, LADB).

path-stretching   Deliberately routeing incoming traffic over longer path (1) to achieve correct spacing on approach.

Patio   Platform for ATM (7) tools integration up to pre-operation (Euret).

PATN   See *Pro-ATN*.

Patrick AFB   USAF base at Cape Canaveral supporting AMR.

Patriot   *1* Provide appropriate tools required to intercept and obstruct terrorism (US).

2 Phased-array tracking to intercept of target.

patrol aircraft   *1* Generally accepted term for aircraft engaged in offshore duties such as search/rescue, customs/immigration and enforcement of marine laws.

2 US designation for large ocean combat aircraft engaged in ASW, maritime reconnaissance and mining.

PATS, Pats   *1* Precision automated tracking station.

2 Prototype automatic target screener.

3 Primary aircraft training system.

4 Precision-attack targeting system.

patter   Flying instructor's well-rehearsed flow of verbal instructions and comments.

pattern   *1* Replica of part used in constructing casting mould.

2 Radiation of transmitting aerial as plotted on diagram of field strength for each bearing.

3 Circuit (1) (US).

4 Shape traced out on ground by track of aircraft, esp.

in circuit, making procedure turns, in holding stack or other circumstance demanding accurate geometry.

5 Authorized flightpath (1) to point of touch-down.

pattern aircraft   One supplied to participant in production programme, eg licensee or co-producer, to serve as master to instruct engineers and solve arguments, and possibly facilitate local change orders.

pattern bombing   Systematic covering of target area with carpet of bombs uniformly distributed according to plan.

pattern generator   Signal generator whose output provides TV-type pattern for testing TV, video, EO or visual-display systems.

PATTS   Programmed auto trim/test system.

PAT3   Parallel advanced tactical targeting technology.

PATU   Pan-African Telecommunications Union.

Patuxent River   US Naval Air Test Center.

PATWAS, Patwas   Pilots' automatic telephone weather answering service (FSS, US).

PAU   Passenger-address unit.

PAUC   Program-acquisition unit cost.

PAV   *1* Power assurance valve (CAA).

2 Prototype air vehicle.

Pave   With this prefix the USAF designated numerous tactical electronic systems.

pavement   Airfield paved area, including runways, taxiways, aprons and possibly dispersals.

pavement condition index   A subjective scale based on heave, cracking and surface spalling.

P_{avg}   Average power of pulsed-radar signal.

PAW   Plasma-arc welding.

PAWES   Performance assessment and workload evaluation system.

Paws   *1* Phased-array warning system.

2 Passive airborne, or all-threat, warning system.

3 Portable analyst, or ASAS, workstation.

4 Passive approach warning sensor.

pax   Passenger[s].

Pax River   See *Patuxent River*.

Paxsim   Models passenger flow through all public areas of terminal, assisting design.

payback   Aircraft added to procurement contract to replace one diverted, eg to another customer.

payload   *1* That part of useful load from which revenue is derived (BSI). See *Maximum* *.

2 Load expressed in short tons, US gal or number of passengers vehicle is designed to transport under specified conditions (DoD etc).

3 In missile, warhead section (DoD), plus container and activating devices (NATO); both definitions need modifying in light of MIRV and similar techniques.

4 For spaceflight, not yet defined; * for Saturn V launch vehicle was complete Apollo spacecraft, whose LM ascent stage had its own * in terms of two crew plus lunar rock.

5 For ECM, all discrete devices released or ejected, eg chaff, flares and jammers in prepackaged * form.

6 Maximum load that can be positioned on upper base-plate (simulator).

payload integration   Matching payload (4) to spacecraft, including power supply, environmental system and software.

payload of opportunity   One put aboard space launcher because mass/volume is available.

payload/range   Fundamental measure of capability of transport aircraft usually presented as graphical plot of payload (1) against range with specified operating conditions and reserves.

payload specialist   Engineer skilled in designing and integrating payloads (4), esp. those of academic-research or commercial nature.

PB   *1* Payload bay.

   2 Passenger bridge.

   3 Aircraft category, patrol bomber (USN, 1935–62); pursuit, biplace (USAAC 1935–41).

   4 Pre-brief[ed].

   5 Particle beam.

P_B   *1* Roll-rate induced angle of attack.

   2 Fuel-burner pressure.

Pb   *Lead*.

P_b   *1* Burning-time average chamber pressure (solid rocket motor).

   2 Gas-turbine combustion-chamber pressure.

   3 Polar modified air.

PBA   *1* Pitch bias actuator.

   2 Plastic blasting media.

   3 Predictive battlespace awareness.

PBAA   Polybutadiene acrylic acid (solid rocket propellant).

PBAN   Polybutadiene acrylonitrile (solid rocket propellant).

PBATS   Portable battlefield attack system, all-weather all-aspect SAM.

PBB   Passenger boarding bridge, see *bridge*.

PBC   Practice bomb carrier.

PBCPI   Permanent bar-code parts identification.

PBCT   Proposed boundary crossing time.

PBD   Program budget decision, or document (US DoD).

PB/D   Place, bearing and distance (waypoint).

PBDI   Position, bearing and distance indicator.

PBE   *1* Protective, or passenger, breathing equipment [smokehood].

   2 Protein-based electronics.

PBF   *1* Personnel (occasionally pilot) briefing facility.

   2 Polymer-based film [appliqué].

PBG   Private pilot, free balloon, gas.

PBH   Private pilot, free balloon, hot air.

PBI   *1* Polybenzinidazole, fire-blocking chemical incorporated into felts and other fabrics (Celanese).

   2 Push-button indicator.

PBID   Post-burn-in data.

PBIL   Predicted, or projected, bomb-impact line.

PBJ   Partial-band jammer.

PBL   *1* Probable.

   2 Performance-based logistics.

PBM   Pulse-bias modulation.

PBO   Performance-based organization.

PbO   Lead oxide.

PBP&E   Professional books, papers and equipment.

PBPS   Post-boost propulsion system.

PBS   *1* Prefabricated bituminous surface.

   2 Plasma-based stealth.

PbS   Lead sulphide.

PBSI   Pushbutton selector/indicator.

PBT   Polymer-based thermosetting.

PBTH   Power by the hour.

PBV   Post-boost vehicle.

PBW   *1* Particle-beam [or plasma-based] weapon.

   2 Power by wire.

PBX   *1* Atmospheric pressure (USSR, R).

   2 Family of special explosives used to trigger NW devices; PBX-9505 is related to, and used with, Cyclotrol; PBX-9404 is still in general use; PBX-9502 is IHE.

   3 Private branch exchange.

PC   *1* Production, or positive, control.

   2 Printed circuit.

   3 Physical conditioning.

   4 Permanent Commission (RAF).

   5 Pulse compression.

   6 Personal computer.

   7 Pilot, or production, certificate.

   8 Pilotage chart.

   9 Programmable controller.

   10 Power converter.

   11 Potential conflict.

Pc   *1* Chamber pressure (any rocket).

   2 Polar continental, or Canadian, air mass.

   3 Cumulative probability of detection (radar).

P_c   Burn-time or action-time average chamber pressure.

pc   Parsec[s].

PC1, PC2 etc   Powered flight-control systems.

PCA   *1* Polar-cap absorption.

   2 Positive-controlled airspace, or area.

   3 Physical configuration audit (software).

   4 Photovoltaic concentrator array.

   5 Pre-conditioned air.

   6 Propulsion-controlled airplane (no aerodynamic controls).

   7 Presidency of Civil Aviation (Saudi Arabia).

PCAS   Pitch-control augmentation system.

PCASP   Passive-cavity aerosol spectrometer probe.

PCATD   Personal-computer aviation-training device[s].

PCB   *1* Printed-circuit board.

   2 Plenum-chamber burning.

   3 Polychlorinated biphenyl(s).

   4 Pilot control bay (UAV-GCS).

   5 Publications Clearance Branch (UK MoD).

PCBW   Provisional combat bomb wing (USAAF).

PCC   *1* Parts-consumption cost.

   2 Pin-cushion correction.

   3 Portland-cement concrete.

   4 Pilot/control[ler] communication[s].

   5 Prague Capabilities Commitment (NATO).

PCCB   Program-configuration control board.

PCD   Proceed.

PCDN   Process-change design notice.

PCDU   Portable [electric] cable diagnostic unit.

PCE   *1* Phase control electronics.

   2 Professional continuing education (USAF).

PCF   *1* Pulse-compression filter.

   2 Passenger-cum-freight (CAA).

   3 Protein crystallization facility.

   4 Pounds per cubic foot [not recommended].

   5 Photonic crystal fibre.

PCFTD   Personal-computer flight-training device.

PCG   Projectile common guidance.

PC(G)   Missile-armed patrol craft.

PCI   *1* Pavement-condition index, scale based on heave, cracking, etc.

   2 Protocol control information.

   3 Pattern of cockpit indications [failure method].

   4 Physical-configuration inspection.

   5 Pounds per cubic inch [not recommended].

pciBIRD   A 6-DOF magnetic tracker with processor card.

PCID   Preliminary change in design.

PCIDM   Personal-control improved data modem.

PCIG   Personal-computer image generator.

PCIP   Precipitation.

PCIPB   President's Critical-Infrastructure Protection Board (US).

PCL   *1* Power control lever [not flight controls].

2 Pilot-controlled [cabin] lighting.

3 Passive coherent location.

PCLL   Persistent cell [storm stays in same place].

PCM   *1* Pulse-code modulation, or modulated.

2 Pyrotechnic countermeasures.

3 Phase-control module.

4 Post-crash management.

5 Parametric cost modelling.

6 Prearranged contact mode.

7 Power-converter module.

Pc max   Maximum chamber pressure (solid rocket motor).

PCMCIA   Personal-computer memory-card interface association.

PCN   *1* Pavement Classification Number; part of proposed ICAO standard ACN-PCN system for relating aircraft footprints to pavement strengths.

2 Personal communications network.

PCO   *1* Photochemical oxidant, = smog.

2 Procuring (or procurement) contracting officer (DoD).

PCOA   Phase-control-only array.

PCP   *1* Platoon command post (SAM).

2 Program change proposal.

PCPN   Precipitation amount.

PCRT   Projection CRT.

PCS   *1* Performance command system.

2 Portable, or piloting, control station (RPV).

3 Permanent change of station (military).

4 Pitch (or powerplant) control system.

5 Power conditioning system.

PCSB   Pulse-coded scanning beam.

PCSC   Precision-cast single-crystal.

PCSV   Pilot-to-controller service.

PCT   *1* Power control test.

2 Portable common tool [software; E adds environment, I interface].

PCU   *1* Powered-control unit (flight controls).

2 Pilot's, or passenger, or parachute, or portable, or propeller, or power (engine), control unit.

3 Pod conditioning unit.

4 Pitch-change unit (helo).

PCV   Pneumatic control valve.

PCZ   Positive-control zone.

PD   *1* Pulse Doppler.

2 Powered, or profile, descent.

3 Power-doubler (radio).

4 Preliminary design.

5 Project definition (abb. results in confusion with [4]).

6 Period (full stop) in message.

7 Pre-digital.

8 Presidential decision (US).

9 Procurement document.

10 Probability of detection.

11 Pictorial display [/CLC adds course-line computer].

12 Point detonating.

13 Programme Director.

Pd   Packet mode, data. GPS com. link used for signalling and ground/air messages.

P_d   Probability of detection in single look or sweep (radar).

p.d.   Potential difference.

PDA, p.d.a.   *1* Post-deflection modulation.

2 Photon detector assembly (space telescope).

3 Problem-detection audit.

4 Personal digital assistant.

5 Premature-descent alert.

PDADS   Passenger digital-activated display system.

PDAF   Probabilistic data association filtering.

PDAM   Precision direct-attack munition.

PDAR   Preferential departure and arrival route.

PDAT   Portable data access terminal.

PDB   *1* Precision digital barometer.

2 Performance data-base.

PDBS   Pilot data, or direct, broadcast system.

PDC   *1* Performance data computer.

2 Public dividend capital.

3 Personnel despatch centre.

4 Programme development card; /PMM adds performance-monitor module.

5 Power distribution centre.

6 Pre-departure clearance.

7 Pressure-drop control.

PDCG   Pressure-drop control governor.

PDCO   Pressure-drop control orifice.

PDCS   Performance-data computer system (Lear-Siegler/Boeing).

PDCU   Panel data concentration unit.

PDCV   Pressure-drop control valve.

PDD   *1* Post-delivery development.

2 Package design document.

PDDI   Product definition data interface.

PDE   Pulse detonator engine.

PDES   *1* Pulse-Doppler elevation scan; PDNES plus electronic scanning in vertical plane to give target height.

2 Product-data exchange specification.

PDEW   [Passengers] per day each way, measure of total flow.

PDF   *1* Precision direction finding.

2 Primary display function.

3 Portable document file [F adds format].

PDFN   Phase-distortion at first null.

PDG   *1* Precision-drop glider.

2 Président directeur-général (F).

3 Programmable display generator.

PDI   *1* Powered descent insertion.

2 Primary direction indicator.

PDID   Pulse-doppler identification.

PDL   Program design [or description] language.

PDM   *1* Pulse-duration modulation.

2 Primary development model.

3 Programmed depot maintenance.

4 Presidential decision memorandum.

5 Product-data management.

6 Pilot decision-making.

P-DME   Precision DME.

PDMF   Programmable digital matched filter(s).

PDMM   Pulse-Doppler map matching.

PDMS   Point-defense missile system.

PDN   *1* Public data network.

2 Pulse-Doppler navigation.

PDNES   Pulse-Doppler non-elevation scan; surveillance down to surface, without indication of target height.

PDO   Pendulum dynamic observer.

PDOP   *1* Phase degradation of performance.

2 Position dilution of precision (GPS).

PDOS   Powered-door operating [or opening] system.

PDP   *1* Program(me) decision package.

2 Plasma, or performance, display panel.

3 Portable data processor [S adds system].

4 Project definition phase.

5 Polar-diagram plotter.

PDQ   *1* Photo-data quantiser (or quantifier).

2 Pre-defined qualities.

PDR   *1* Preliminary design review.

2 Pulse-Doppler radar.

3 Pilot's display recorder (Ferranti).

4 Predetermined routeing.

5 Primary defect rate.

6 Preferential departure route.

7 Programmable digital radio.

8 Pressure-drop regulator.

9 Program-development review.

PDRC   Pressure-drop ratio control.

PDRJ   Pulse-Doppler radar jammer.

PDRR   Product/program project definition and risk reduction.

PDS   *1* Passive detection system.

2 Prestocked dispersal site.

3 Pulse-Doppler search mode.

4 Passenger distribution system.

5 Post-design service(s), or support.

6 Primary display system.

7 Project definition study.

8 Portable data store.

PDSTT   Pulse-Doppler single-target track mode.

PDT   *1* Pyrotechnic door [or deployment] thruster.

2 Pacific Daylight Time.

3 Pliable display technology.

PDU   *1* Pilot['s] display unit.

2 Power drive, or distribution, unit.

3 Pneumatic drive unit.

4 Protocol data unit.

PDV   *1* Pressurizing and dump valve.

2 Parafoil-delivered, or -deployed, vehicle, for MOP(5).

PDVOR   Precision DVOR.

PDW   *1* Pulse-detonation wave [rocket].

2 Priority delayed weather.

PDWC   Post-departure weather change.

PDZC   Pathfinder drop-zone control.

PE   *1* Procurement Executive (UK, MoD).

2 Position error, also p.e.

3 Pilot error.

4 Piston engine.

5 Program, or processing, element (DoD).

6 Professional engineer (US).

7 Permanent echo.

8 Ice pellets.

9 Pre-emptive.

10 Precision engagement.

P_E Pressure in core-engine jetpipe.

Pe   See *Péclet number*.

P_e Total installed power of shaft-drive engine[s].

PEAD   PE(10) assessment definition.

peak   *1* In production programme, time or rate of maximum output.

2 In aerospace vehicle, highest points in sine-wave flight, typically 200,000 ft.

peaking circuit   Extends frequency response of video amplifier at highest frequencies.

peak suction   Lowest pressure on upper surface of wing or on 2-D aerofoil profile; rarely, lowest pressure on other convex surface of aerodynamic body.

peak/trough ratio   Peak (usually summer) traffic rate divided by lowest (usually winter) traffic rate.

peaky   Aerofoil section (profile) of traditional type causing large acceleration of flow over leading edge and hence very low pressure (large peak suction) over narrow strip at 8–15% chord; opposite to supercritical or roof-top.

PEAT   Procedural event analysis tool.

PEC   *1* Personal equipment connector.

2 Pressure, or position, error correction.

3 Pulsed eddy current.

PECHS   Percussion/electric conversion hardware system [ammunition].

pecked line   Broken or dashed line in graphics or artwork.

pecking   Touching ground with propeller tips, esp. on takeoff.

Péclet number $Pe = \dfrac{Vl}{\lambda}$ where V is fluid flow velocity, l a length and λ thermal conductivity; applies in heat transfer at low airspeeds.

PECM   Passive electronic countermeasures.

PECP   Primary entry control point, to hostile airspace.

PECT   Peer entity contact table.

Pectenometers   Class of aerodynes weighing c6.5g, 0.23 oz, which fly like, e.g., butterflies, by clapping aerofoils together, ejecting air rearwards. On separation, air is sucked in from the front (NRL).

PED   *1* Program(mme) element description.

2 Post-exit deflector.

3 Passenger [or portable, or personal] electronic device[s].

pedal turn   Changing heading (azimuth) of hovering helicopter or Stovl using pedals only.

pedestal   *1* Raised box between pilots on flight deck carrying numerous control and system interfaces.

2 Pillar supporting aircraft on ski (and, it is suggested, float).

Pedoba   Portable electronic device[s] on board aircraft.

PEDS   Portal explosive detection system.

PEE   Photoelectron emission.

PEEK   Polyetheretherketone.

peel off   To roll away from straight-and-level flight, esp. from a formation, and dive away; normally manoeuvre performed in sequence by all aircraft of formation.

peen   *1* To cold-work metal surface by repeated light blows of ball-pane hammer [peen forming] or bombardment with hard balls [shot peening].

2 To deform metal part by series of hammer blows, eg to set rivet or burr end of bolt to prevent loss of nut.

PEG   Polyethylene glycol.

Pegasys   Precision and extended-glide airdrop system (USA).

PEI   *1* Polyetherimide, important family of resins.

2 Professional engineering institutions.

PEL *1* Personnel licensing.

2 Precision elastic-limit.

PELS Personal equipment life-support system, a smokehood, with decompression protection.

PELSS Precision emitter location strike system.

Peltier effect Generation of current by making circuit containing two different metals and keeping two junctions at different temperatures; some definitions refer to generation or absorption of heat at junctions upon passage of current.

PELTP Personnel Licensing and Training Panel (ICAO).

PEM *1* Parametric estimation model.

2 Program element monitor (DoD).

3 Plastic-encapsulated microcircuit, hence plural Pems.

PEMB Procurement Executive Management Board (UK).

PEN Photonic exchange network.

penaids See *penetration aids.*

penalty Deficiency in one aspect of aircraft design or performance in return for improvement in another; eg adding thermocouple in jetpipe to indicate temperature assists pilot but creates drag in jet reflected in reduced cruising speed, range or both.

pencil beam Narrow, strongly directional beam with minimal sidelobes.

pendant *1* Arrester wire (deck cable).

2 Flag indicating centre of arrester wire.

pendulum damper Crankshaft counterweight vibration damper in form of pivoted mass.

pendulum stability That inherent in large vertical distance between high centre of wing lift and low c.g.

pendulum valves Gravity-operated flaps covering ports in housing of air-operated gyro which sense tilt and control airflow to cause restoring (erecting) precessive force.

penetrant dye Method of detecting cracks and pores by washing part with coloured liquid and then with white developer.

Penetrate Passive enhanced navigation with terrain-referenced avionics: radar altimeter and INS plus 3D terrain model in database (Ferranti).

penetration *1* Flight into hostile, esp. defended, airspace.

2 Ability of sailplane to keep going by trading kinetic energy for distance with minimal loss of height or speed while heading for next thermal.

3 Flight deep into cloud, esp. one with large vertical extent and severe turbulence.

4 Flight into eye of tropical revolving storm.

5 Success in previously unattacked market.

6 Progress into flight manoeuvre, esp. one posing potential problem or danger, hence stall-*.

7 Portion of high-altitude instrument approach which prescribes descent to start of approach (DoD).

penetration aids Devices and systems assisting vehicle to accomplish penetration (1), eg jammers, chaff, flares, decoys and warning systems, low (lo) flight level, reduced RCS and improved vehicle hardness.

penetration area That within which enemy defences are to be neutralized to degree assisting succeeding aircraft to reach targets.

penetration factor Reciprocal of amplification factor of valve.

penetration fighter One intended to fly deep into hostile territory (term now rarely used).

penetrator *1* Tool, usually powered, for quickly cutting apertures in side of aircraft for speedy rescue of occupants.

2 Aircraft designed for penetration (1).

3 High-density projectile designed to pierce armour by kinetic energy.

penetrometer Measures ability of airfield surface to support static or moving aircraft.

penguin Officer devoid of brevet (RAF).

pennant Mooring and haul-down wire for captive balloon.

penny-farthing Helicopter of normal configuration (colloq.).

Pensky Martin Standard apparatus for determination of flashpoint; used in two forms, closed or open.

penthouse roof Top of cylinder combustion space having sloping sides.

pentode Thermionic valve having three grids.

PEO Program Executive Officer, or Office (US).

PEP, p.e.p. *1* Peak envelope power.

2 Pre-engine production.

3 Productivity enhancement programme.

4 Pulse envelope programming.

5 Predesignated ejection point.

6 Pulsed-energy projector.

PEPE Parallel-element processing ensemble (or element).

PEPP Planetary-entry parachute program (Martin Marietta/NASA).

PEPS, Peps *1* Positive-expulsion propellant system.

2 Pintle escape propulsion system.

PER Performance.

PERA Production Engineering Research Association (Melton Mowbray, UK).

perceived noise level See *noise.*

perch position In pattern (3) with touchdown point 45° behind.

percussion cap Detonator/igniter for ammunition activated by sudden deformation.

percussion gun Device for cheaply making loud bangs to scare birds.

perfect fluid Usually implies inviscid and incompressible, as well as homogeneous.

perfect gas One exactly obeying Joule and Boyle laws so that $PV = nRT$; also obeys Charles, Gay-Lussac and several other laws.

perforated Pierced by holes, in case of aerodynamic surface typically removing 25–30% of projected area; perforation of flap or airbrake can reduce actuation power requirement and increase drag of surface, esp. when at large angle to airflow, but destroys value of flap as lifting surface and thus confined to divebrakes and airbrakes.

performance *1* From operational viewpoint, ability of system to perform, esp. expressed in numerical values.

2 From flight-safety viewpoint, ability of aircraft to perform required functions and manoeuvres, esp. in degraded condition and under adverse circumstances, again expressed numerically. Subdivided into three categories: 1, measured *, that actually recorded for particular aircraft at particular time; 2, gross *, factored from measured to allow for poorest aircraft in fleet, guaranteed-minimum instead of average thrust propulsion and possibly other degrading factors; and 3, net *, factored from gross to allow for further possible temporary variations (airframe damage or icing or certain non-critical fault conditions) and minimum standard of pilot skill and experience.

3 Narrow interpretation of numerical values of aircraft flight limits such as speeds, altitudes and payload/ranges.

performance-command system   Microprocessor-based crew advisory system in commercial transports which optimises performance and saves fuel.

performance factors   Those deriving from aircraft performance which affect airline traffic achieved, as distinct from commercial and operational factors.

performance groups   UK aeroplane categories: those whose first C of A was issued before 1951 are NPG (no performance group); those built since 1951 are Group A, large multi-engined; B, spare; C, light multi-engined; D, single-engined; X, foreign multi-engined imported before particular dates.

performance index   Non-dimensional comparator for radars, derived from xmtr peak power, prf, beam width, pulse width, antenna gain and receiver noise.

performance limiting conditions   Those demanding flight to near boundaries of flight performance or point at which auto-ignition, stick-shaker or other subsystem is triggered.

performance management system   See *flight management system*.

performance number   Knock rating; below 100 called octane number.

performance reduction   Historically, calculations to reduce measured aircraft performance to standard values at a chosen weight and in particular atmospheric pressure and temperature.

performance-type glider   Sailplane, esp. competition sailplane.

periapsis   Pericentre of orbit.

pericentre   Point on orbit closest to primary.

pericynthian   Point in spacecraft trajectory closest to Moon.

perifocus   See *pericentre*.

perigee   Pericentre of Earth orbit.

perihelion   Pericentre of solar orbit.

perilune   Pericentre of lunar orbit.

perimeter   *1* Periphery of airfield flying area.

2 Boundary of defended area.

perimeter track   Taxi track linking ends of runways, revetments or dispersals and main hangar/apron area.

period   *1* Time interval between successive passages through particular point in same direction of SHM or other wave motion $= \dfrac{1}{f}$ or $\dfrac{1}{f_t}$.

2 Time interval between successive passages through particular bounding plane in same direction of satellite, usually time between successive northbound transits of Equator (orbital *); see *sidereal* *, *synodic* *.

periodic inspection   Inspection, with or without tear-down, according to published schedule of time intervals irrespective of performance (1) of device.

periodic reservation   Several aircraft book slots on same Satcom T-channel.

period of performance   Timeframe in which work is done.

peripheral hem   Leading edge of parachute canopy.

peripheral VLS   Distributes SAM launchers along ship to enhance survivability.

peri track   Perimeter.

PERM   Permanent.

Permalloy   Family of American Fe/Ni alloys which are magnetically soft, with high permeability at low magnetizing forces.

permanent echo   Terrestrial features as seen on radar at fixed site.

permanent magnet   One which retains its magnetism in absence of strong demagnetizing field.

Permaswage   Patented fluid-tight method of connecting pipes.

permatron   Gas tube similar to thyratron with magnetic-field control.

PERME, Perme   Propellants, Explosives and Rocket Motor Establishment (Westcott, Waltham Abbey, UK).

permeability   *1* In magnet, ratio B/H where B is magnetic flux induction and H is magnetizing force, symbol μ, this is divided into absolute and relative or specific *, μ_r, except in emu system, where * of free space is defined as unity; measured in henry/metre.

2 In aerostat, measure of rate at which gas at STP can pass through fabric, usually expressed in litres per 24 h.

3 In amphibian, ratio of volume of landing-gear compartments that can be occupied by water to whole buoyant volume.

permeability tuning   Radio tuning by varying permeability of inductor core, usually with translating bar of ferrite.

Perminvar   Family of low-hysteresis alloys of Fe/Ni/Co.

permissive action link   Highly secure, jam-proof data links (various categories) which enable bomber crew to trigger weapon live from cockpit when bomber is at safe distance (hi-alt laydown only).

permittivity   Dielectric constant: relative * is ratio of electric flux density in medium to that which same force would produce in vacuum, symbol ϵ_r, unit farads per metre; absolute * ϵ_o is in vacuum, often called free space (see *Coulomb law*).

Permit to fly   Issued to categories, such as homebuilts, warbirds, classic GA aircraft and micros, that do not qualify for a C of A but may be flown with restrictions. (UK CAA).

Permly   Permanently (FAA).

PERP   Peak effective radiated power.

perpendicular-heading square   Training manoeuvre in which helicopter is flown around square, crabbing at 90° sideways along each of the four sides.

persistent area-denial   Use of loitering robotic systems [ie, UAVs] to provide non-stop surveillance/attack over battlefield for period of days to months.

personal equipment connector   Quick-make/break multi-channel coupling for aircrew oxygen, R/T, intercom, g-suit and EC. Not necessarily synonymous with AEA.

personnel locator beacon   Miniature transmitter sending coded signal and worn on flying clothing.

personnel reaction time   Time from nuclear warning to all defensive measures taken.

Perspex   Family of methyl methacrylate plastics important as aircraft transparent material (ICI).

Pert, PERT   Programme Evaluation and Review Technique; critical path method.

PES   *1* Passenger entertainment system, audio plus film.

2 Polyethersulphone.

3 Pre-entry [to inlet] streamline.

PESO   Product engineering services office.

PET   *1* Piston-engine time.

2 Positron emission tomography.

3 Polyethylene terephthalate.

4 Point of equal time.

5 Pacific engineering trials.

PETA, Peta   Pulsed-ejector thrust augmentor.

peta   Prefix, $\times 10^{15}$, symbol P.

Petal   Preliminary Eurocontrol test of air/ground data link; II adds Phase II, IIe adds Phase II Extension.

petal cowling   One divided into large hinged segments opening like petals of flower.

PETN   Pentaerythritol tetranitrate (explosive).

petrol   Fuel or solvent tailored from hydrocarbons [eg heptane, hexane, octane] plus additives. Density 0.708–0.72, giving typical mass 3.93 kg (8.65 lb) gal^{-1}, 3.27 kg (7.2 lb) US gal^{-1}, different fractions boil at from 41°C. US = gasoline.

Petroseal   A film-forming foam fire-extinguishant.

Pett, PETT   Project engineers & technologists for tomorrow.

petticoat   In airship gas duct, pleated sleeve able to seal duct yet leaving clear passage when released.

Petty valve   Operated manually after fighter mission to release pressure in gun-firing pneumatic circuit (pre-1950).

PEU   Pod, power or processing electronics unit.

PF   *1* Pilot flying (others being PNF).

2 Plastics factor.

3 Preformed fragment.

4 See *p.f.*

PF, P/F   *1* Powder forging (or forged).

2 Primary/final talkdown.

P$_f$ Pressure in fan duct of turbofan.

pF   Picofarad, 10^{-12} farad.

p.f.   Power factor.

PFA   *1* Popular Flying Association (UK, formed 1946 as ULAA).

2 Pulverised fuel ash.

3 Porous friction asphalt.

4 Probability of false alarms.

PFB   Preliminary flying badge.

PFC   *1* Primary flight control, or computer.

2 Powered flight (or flying) control(s).

3 Pre-flight console.

4 Passenger facility charge[s].

5 Porous friction course (see *PFA (3)*).

PFCES   Primary flight-control electronic system.

PFCS   Primary flight-control, or computer, system.

PFCU   Powered flying-control unit, surface power unit.

PFD   *1* Primary flight display, or director; /ND adds navigation display, S subsystem.

2 Planned flight data.

3 Personal flotation device.

PFE   *1* Purchaser-furnished equipment.

2 Path-following error.

Pfenninger wing   Strut-braced, yet high-subsonic wing with exceptional aspect ratio and reduced thickness, yielding L/D of up to 40.

PFF   *1* Pathfinder Force (RAF, WW2).

2 Perspective [prospective is meant] frontal fighter (US).

3 Panel fill factor.

4 Preformed fragment.

PFFT   Parallel fast-Fourier transform.

PFH   Per flight (or flying) hour.

PFHE   Prefragmented high explosive.

PFI   *1* Post-flight inspection (US = after-flight).

2 Private [or project] finance initiative.

PFIA   Professional Flight Instructors Association (UK).

PFIAB   President's Foreign Intelligence Advisory Board (US).

PFIS   *1* Portable flight-inspection system.

2 Passenger flight-information system.

PFL   Practice forced landing (A adds 'area').

PFLD   Pilot's fault-list display.

PFM   *1* Pulse-frequency modulation.

2 Pre-flight message.

PFMA   Post-flight mission analysis.

PFMGO   Pre-flight message-generating officer (RAF).

PFN   Pulse-forming network.

PFNA   Pulsed fast-neutron analysis.

PFO   Port-facing oblique.

PFP   *1* Partnership for peace (NATO).

2 Proximity-fuze programmer.

3 Primary flight permit.

PFPS   Portable flight-planning system.

PFQT   Preliminary flight qualification test.

PFR   *1* Permitted flying route (CAA).

2 Passenger flow rate.

3 Post-flight report.

PFRT   Preliminary flight rating test.

PFS   *1* Primary flying squadron.

2 Product file sets.

PFSV   Pilot-to-forecaster service.

PFTA   Payload flight-test article.

PFTI   Phototelesis fast tactical imagery.

PFTS   *1* Production flight-test schedule.

2 Permanent field training site.

PG, p.g.   *1* Processing gain.

2 Plastic/gas.

3 Peelable graphics, esp. airline livery or logo on aircraft.

4 Program management assistance group (USAF).

5 Photographic group (USAAF).

6 Aircraft category pursuit, ground attack (USA 1919–24); powered glider (USAAF 1943–47).

PGA   *1* Pressure-garment assembly.

2 Pin grid array.

PGB   *1* Precision-guided bomb.

2 Power gearbox.

PG bearing   Plastic/gas, avoiding solid/solid contact.

PGCS   Portable ground-control station (UAV).

PGEN   Program generator (software).

PGL   Private pilot, glider.

PGM   *1* Precision-guided munition.

2 Guided missile (ICBM) launched from soft pad (DoD, obs.).

3 Precision ground map[ping].

PGMPS   PGM (1) planning software.

PGN   Passenger-generated noise.

PGO   Foreplane, canard (R).

PGRV   *1* Post-boost guided re-entry vehicle (has IMU and TPU).

2 Precision-guided re-entry vehicle (DoD).

PGS Pilot-guard system.

PGSC Personnel guide surface canopy.

PGSE Peculiar (ie, to type) ground-support equipment.

PGSM Precision-guided submunition.

PGT Private pilot, gyroplane [autogyro], transition.

PH *1* Hit probability (also **P$_H$**).

2 Porte hélicoptères (F).

3 Public holiday.

4 Pacific, Hawaii (ICAO).

PHA *1* Preliminary hazard analysis.

2 Polymerized healing agent (composites).

PHAK *Pilot's Handbook of Aeronautical Knowledge* (US Government).

phantom beacon R-Nav waypoint where no beacon actually exists.

phantom contract See *phantom order*.

phantom drawing One using phantom lines.

phantom lines Geometrically accurate but incomplete lines merely giving location of item or alternative positions thereof, eg to show avionic equipment in structural airframe drawing.

phantom member Non-existent member added in pre-computer era to assist solution of structural analysis.

phantom order Draft contract with manufacturer with provision for preplanning immediate production in time of crisis or conflict (DoD).

PHAR Program for harmonized air-traffic-management research [E adds in Europe, or Eurocontrol].

Pharos Plan-handling and radar operating system.

Pharus Phased-array universal synthetic-array radar.

phase *1* In any periodic cycle, fraction of period (1) measured from any defined reference.

2 In reactive circuit, relationship between current and voltage.

3 In physical chemistry, distinct homogeneous physical states separated by sharp boundaries, eg liquid/solid, solid/solid, or immiscible liquids.

4 Periodic variation in solar illumination of Moon as seen from Earth.

5 Normal meaning often applied to programme planning, amphibious assault, establishment of military government etc. In DoD, Phase 1 is concept, 2 is proof of concept, 3 is downselect and demonstration.

phase-advance Subsystem which senses aeroplane rate of change of pitch and triggers stick-pusher progressively earlier as rate increases, so that pitching momentum never takes AOA beyond prescribed limiting value.

phase angle *1* Phase (1) difference between two sets of periodic phenomena expressed in angular measure.

2 Angle between current and voltage in rotating-vector plot of alternating current.

3 Angle between sightlines to Sun and Earth measured at remote locations, eg other celestial body.

phased array Physically fixed antenna (aerial) scanned electronically, usually in both x and y (horizontal and vertical) axes.

phase difference Measure of phase angle (1) from any VOR radial related to that on bearing 000°.

phase discriminator Detector of phase modulation.

phase inverter Radio or other signal-processing stage with unity gain whose output is reciprocal of input (not synonymous with half-wave rectifier).

phase modulation Carrier phase angle (1) varies from carrier angle by amount proportional to instantaneous amplitude of modulating signal and at a rate proportional to modulation frequency, PhM.

phase out, phaseout Progressive withdrawal from production or active service.

phase shift Phase difference (not necessarily VOR); change of phase angle (1).

phase shifter Circuits which steer the beam emitted by a planar-array radar antenna.

phase velocity That of equiphase surface of travelling plane wave along wave normal; also called wave speed/velocity.

Phasst Programmable high-altitude single-soldier transport.

PHB Pilot's handbook (USGPO).

PHD Pilot's horizontal display.

PHE Penetrating high-altitude endurance.

PHEI Penetrator HE incendiary.

p-HEMT Pseudomorphic high-electron-mobility transistor.

phenolic/epoxy Family of resins and adhesives much used in composites derived from phenol (carbolic acid) and characterized by oxygen bridges linking hydrocarbon radicals.

phenolics Large family of synthetic polymers (plastics/resins/adhesives) dating from 1907 and mainly unmodified phenol-formaldehydes or (esp. in case of adhesives) resorcinol-formaldehydes.

PH-15-7Mo Refractory stainless steel, primary structure of B-70.

PHI *1* Position and homing indicator.

2 Pitot-head inoperative.

Phibuf Performance buffet-limit.

Phigs Prgrammer's hierarchical international graphic standard, or system.

Phillips entry Shape of leading edge of typical modern wing (Horatio Phillips, 1886).

Phills Portable hyperspectral imaging low-light spectrometer.

Phinom Nominal bank angle.

PHLD Powered high-lift device.

PHM *1* Proportional hazards modelling.

2 Prognostic[s] and health monitoring, or management.

phon See *noise*; not used in modern work.

phonetic alphabet See Appendix 4.

phoney war Northern France, 3 September 1939 to 10 May 1940.

phonic wheel Sensor for tachometer, transmits signal at frequency proportional to shaft speed.

phosphate esters Fire-resistant hydraulic fluids based on esters of P(18) acids.

phosphor Substance which is luminescent; those in radars/CRT/TV etc are commonly zinc sulphide/zinc and selenide/copper compounds, but cadmium and rare earths are common. Those for printing on opaque substrates are unrelated.

phosphorescence Luminescence which continues more than 10^{-8} s after cutoff of excitation, usually being visible to eye for days thereafter.

phosphorus P, three main forms, esp. white * soft non-metal, spontaneously flammable, MPt 44°C, density 1.8.

phot Non-Si unit of illuminance = $lm/cm^2 = 10^4$ lx.

photint Photographic intelligence.

photoactivated Activated by light.

photocathode Electrode for photoelectric emission.

photochemical Involving chemical change and emission/absorption of radiation.

photochromic Having colour, transmittance or other optical property changed by variation in incident light.

photochromy Colour photography.

photoconductive Having electrical resistance varied by illumination.

photodiode Diode converting light into electricity; hence * array yields signals which when processed analyse incident light pattern.

photodrafting See *photographic lofting*.

photoelectric Involving light and electricity, usually by absorbing photons and emitting electrons.

photoelectric cell Transducer converting EM radiation in visible, IR or UV wavelengths into electricity; abb. photocell.

photoelectron Electron ejected, eg from metal surface, by impact of energetic (short wavelength) photon.

photoelectronics Involving electrons and photons (many devices).

photoemissive Emitting electrons (ie electric current) when illuminated.

photoflash Pyrotechnic cartridge producing brief but intense illumination, esp. for lo-level night reconnaissance.

photogrammetry Making accurate measurements and drawings, esp. surveying and mapmaking, by photographic means.

photographic layout drawing Photographic lofting.

photographic lofting Lofting entirely with photographs.

photographic reconnaissance *Photo reconnaissance*.

photographic transmission density Log of opacity (base 10), thus perfectly transparent film has *** of zero, while one transmitting only 10% has *** = 1.

photometer Instrument for measuring luminous intensity, luminance or illuminance.

photometrical calibration Regular measurement of output of all airfield lighting, especially on Cat II, III runways.

photometry Science or technology of measuring luminous flux, luminous intensity, luminance and illuminance.

photomultiplier Tube containing photocathode, several intermediate electrodes (dynodes) and output electrode; also called multiplier phototube.

photon Elementary parcel of EM energy emitted by transition of single electron, with energy $h\nu$ (h = Planck constant, ν frequency) and momentum $h\nu/c$ where c is velocity of light.

photonic material Material designed to manipulate light, as distinct from electrons, with properties based on an arrangement of atoms on an artificial structure.

photon rocket Theoretically achievable rocket whose working fluid is light, ie stream of photons; small thrust but in deep space very high I_{sp}, and vehicle velocity could be significant fraction of that of light.

photopic vision Using retinal cones, hence colours distinguishable.

photo reconnaissance Military mission to bring back images of scenes in enemy territory, such as buildings and structures, troop movements, ships and results of previous attacks. Can be high vertical, low oblique, stereo, overlapping strip, etc.

photosensitivity Degree to which substance changes chemical or electrical state when light falls on it.

photosensor Device operating by photoconductivity, eg light valve.

photosmoke method One of two techniques for measuring smokiness of jet by direct determination of optical density (other is Hartridge); gives output in PSU.

photosphere Intensely hot, bright outer layer of Sun's atmosphere.

phototheodolite Instrument comprising camera whose azimuth and elevation are precisely recorded (usually on its own film).

phototransistor Solid-state device, originally Ge wafer, generating holes by light absorption and multiplying this photocurrent by transistor action at collector.

phototube Electron tube (vacuum tube) containing photoemissive cathode and collecting anode (usually plus other sub-devices).

photovoltaic cell Transducer which, like photoelectric cell, converts EM radiation in visible or near-visible wavelengths into electricity; unlike photocell its purpose is to generate usable current instead of merely giving signal or serving other purpose calling for very low power; example is solar cell.

phraseology Accepted forms of speech, codes phonetic alphabet, etc, used to facilitate telecommunications, usually by voice.

PHS Precision hover sensor.

PHT *1* Private pilot, helicopter, recreational, transition. *2* High-temperature platform.

phugoid One of the five classical modes of aeroplane motion, a long-period oscillation of pitch axis, perpetually hunting about level attitude and trimmed speed, a switchback trajectory at almost unvarying AOA; noun and also adj, eg * oscillation.

phut-phut *Put-put.*

PHY Physical interface (device).

P_{hvd} Fluid system pressure, esp. hydraulic test pressure of rocket motor case.

PI *1* Point of interception (navigation plot).
 2 Photographic interpreter (or interpretation).
 3 Process (or program) instruction.
 4 Practice interception.
 5 Program introduction (D adds 'document').
 6 Production investment.
 7 Production installation, of new equipment in Service aircraft.
 8 Product improvement.
 9 Precipitation identification.
 10 Parameter identifier.
 11 Pipeline inspection.
 12 Principal investigator.

Pi Input power, esp. of jammer.

PIA *1* Pilots' International Association (US).
 2 Pilot-interpreted approach.
 3 Proprietary Industries Association (US).
 4 Performance integrity and availability.

PIAC, Piac Peak instantaneous airborne count[s].

PIAG, Piag Propulsion Installation Advisory Group (Int.).

Pianeg, Pianet Planning the implementation of an improved AFS/AFTN network (ICAO).

piano hinge One continuous along edge of hinged item.

piano keys Black/white runway end markings (colloq.).

piano wire Finest steel wire normally produced; 0.8–0.95% C, very high uts, accurate dimensionally.

PIB   Preflight information bulletin.

Pibal   Pilot-balloon aloft (observation).

PIC   *1* Pilot in command.

2 Prime integration contract.

3 Price-improvement curve.

4 Potential icing [category].

Picao, PICAO   Provisional ICAO (1945–47).

Picasso   Predicted ionograms correlated against segmented swept output (ionospheric analysis).

PICC   Processor interface controller and communication.

piccolo actuator   PFCU or other actuator whose output is generated by row of parallel jacks fitting within thin aerofoil.

piccolo tube   Tube perforated by (usually linear) row of holes from which hot deicing air is blown, usually to impinge on inside surface of a leading edge.

pick-a-back   *1* See *composite aircraft*.

2 Superimposed printed-circuit boards.

picket   *1* AEW or AWACS aircraft.

2 Instrumented oceangoing ship on missile range.

picketing   Securing aircraft against movement when parked in open, normally by attachment to heavy masses or spiral rods screwed into ground.

picketing anchor   Spiral rod with eye at upper end.

pickle   Tactical air code: moment of manual triggering of system, esp. release of ordnance on surface target.

pickle button   That commanding release of airdropped stores.

pickled facility   Warm long-term storage, esp. for NW.

pickling   Soaking in dulute acid solutions to remove oxides or other surface films or inter-crystalline carbides and surface scale. Principal acids are HCl, H_2SO_4, HNO_3 and HFl.

pick-off   Sensor of angular motion or position; many types, eg electric potentiometer, angular digitizer, photocell, magnetic coil moving-iron reluctance bridge, or fluidic valve or gate.

pickoff excitation   Normally a frequency.

pickoff sensitivity   Usually signal voltage per unit angular travel.

pickup   A fault or omission noticed and corrected later.

PICL, Picl   Pool-item candidate list.

pico   Prefix 10^{-12}; hence one picosecond (1 ps) is one millionth of one millionth of a second.

picocell   Small radio tower on passenger aircraft to instruct handsets to communicate with it exclusively and at lowest power.

picosatellite   Mass $\leqslant$ 0.5 lb, 0.2268 kg.

picric acid   *Trinitrophenol.*

Pics, PICS   *1* Photogrammetric integrated control system.

2 Protocol implementation conformance statement[s].

picture manoeuvre   Manoeuvre made by large aerobatic team involving wide separation of aircraft to fill large part of display area, eg bomb-burst.

PID   *1* Program introduction document (DoD).

2 Passive identification device.

3 Photo-ionization detector.

4 Post-impact delay.

5 Portable intruder detector.

6 Parameter, process or primitive identifier.

PIDP   Programmable indicator data-processor (USAF).

PIDS, Pids   *1* Prime-item development specification.

2 Pylon integrated dispenser system.

3 Positive identification system.

4 Perimeter, or portable, intrusion detection system.

piece of cake   A task posing no problems (RAF colloq., WW2).

pier   Long corridor, usually two-level, connecting airport terminal with gates.

pierce   To cut part from sheet; hence large family of presswork dies such as * and cut off, * and form, * and trim.

pierced-steel planking   Standard (mainly WW2 to 1950) unit of prefabricated airfield surface; mild steel plates measuring 119.75 in × 16 in and weighing 65 lb (29.5 kg) with interlocking edges.

pie-shaped   NLG steering, or other, inceptor having shape of segment of disc.

pièze   Non-Si unit of pressure used in French legal system = 1 sn/m^2 = 1 kN/m^2 = 1kPA = 0.14503 lb/in^2.

piezoelectric   Relationship exhibited by certain crystalline substances, esp. single crystals, between electric potential difference and mechanical stress; eg applying voltage (DC or AC) across opposite faces results in expansion/ contraction or vibration, while applying stress or vibration results in potential difference.

PIF   *1* Photo-interpretation facility.

2 Pilot's Information File (US, WW2).

3 Pilotage in force [control by thrusters or thrust-vectoring].

pif/paf   Missile control system combining lateral thrusters at c.g. [pif] with aerodynamic surfaces [paf].

Pifet   Piezoelectric field-effect transistor.

PIG, Pig   *1* Pendulous integrating gyro.

2 Pilot's Information Guide.

Piga   PIG accelerometer.

Pigeon, pigeons   Air-intercept code: "Your base bears X° and is Y miles away".

piggyback   Composite aircraft, or aircraft carrying large vehicle superimposed.

Pigma   Pressurized-inert-gas metal arc.

pigtail   *1* Projecting rigid pipe, usually with 90° bend and threaded connection for attachment to fluid system.

2 Short length of any other kind of cable or transmission line projecting from device for attachment to system.

PIHM   Protective integrated hood mask.

pillow tank   Dracone or similar flexible fluid storage.

Pilot   *1* Piloted low-speed test (ambiguous).

2 Pod integrated localization, observation, transmission.

pilot   Person designated as *. Previous definitions involved operation of particular controls (in one case 'mechanisms') or guidance of aircraft in 3-D flight, none of which need be done in advanced aircraft, though * required to monitor. In case of RPV * may be in other aircraft or on ground. For command-guided missiles preferred term is operator.

pilotage   Contact flying, navigating by visible surface landmarks.

pilot assister   Qualified pilot in right-hand seat of aircraft training navigators or other crew members [generally = copilot].

pilot balloon   Meteorological balloon; alternatively, small free balloon devoid of instrumentation, observation of which from ground enables wind at different heights to be caclulated.

pilot canopy   Small auxiliary canopy, ejection of which pulls out main canopy (personal, cargo and braking parachutes).

pilot certificate   In many countries, title of document licensing pilot according to five to 11 categories. In UK and many other countries called licence.

pilot chute   *Pilot canopy.*

pilot control bay   Location of flight trajectory and navigation interface in UAV GCS.

piloted   Supervised by human beings, usually on board, playing active and direct role in control of vehicle.

pilot flying   In multi-crew operation, pilot actually flying the aircraft, also called handling pilot.

pilot hole   Small but precisely located hole serving as guide to subsequent larger drilling.

pilot in command   Person responsible for aircraft in flight.

pilot induced oscillations   Potentially dangerous or even catastrophic pitch oscillations caused by pilot trying to stop them. Cause may be oversensitive system with very light input forces, or restricted hydraulic flow rates in PFCUs so that pilot is always making late corrections with ever-greater magnitude.

pilot-interpreted system   One, eg early AI radar, requiring skill and judgement on part of operator, in contrast to modern digital readout and unambiguous indications. Note: early systems were often interpreted by other members of crew but no term exists.

pilotless aircraft   Ambiguous: aircraft whose pilot has departed or aircraft designed to fly unmanned (arch.).

pilot opinion rating   Subjective assessment of aircraft stability and handling, measured according to Cooper scale.

pilot parachute   *1* (See *pilot canopy*).

   2 Parachute worn by pilot, eg with seat-type pack or forming part of ejection seat.

pilot plane   Auxiliary surface mounted ahead of main surface (some definitions add 'and free to take up position in line with wind').

pilot pushing   Unlawfully urging aircrew to work excessive hours.

Pilot's Associate   Artificial intelligence aid for fighter pilots combining software, hardware and advanced pilot/vehicle interfaces (McDonnell Douglas, Texas Instruments).

pilot's automatic telephone weather answering service   Weather advisory continuously available by telephone (US).

pilot's discretion   ATC has given pilot freedom to choose timing/place/rate of climb or descent.

pilot's notes   Handbook providing operating instructions, helpful advice and all significant images and numerical data for pilot of particular type or sub-type of aircraft.

pilot's preference kit   Small bag of allowed personal effects (NASA).

pilot's reference eye position   Assumed position of eyes of normal pilot looking ahead (as for landing) in particular type of aircraft, esp. in designing cockpit.

pilot's trace   Rough overlay to map made by pilot of reconnaissance aircraft immediately after sortie showing

locations, directions, number and order of sensing runs together with sensors used on each.

pilot's view   Working section of tunnel as seen by notional pilot of vehicle under test.

PIM, Pim   *1* Previous intended movement, of aircraft carrier.

   2 Processor in memory.

Pimaws   Passive IR missile-approach warning system.

PIMPF   Programmable intelligent multi-purpose fuze.

PIN, p-i-n   Semiconductor p-n junction diode with interleaved layer of intrinsic semiconductor (from 'positive-intrinsic-negative').

pinch hitter   Safety pilot, esp. one who is unlicensed, and frequently the spouse or partner of the PIC (US colloq.).

PIND   Particle impact noise detection, test carried out by acoustically sensing foreign particles in electronic devices.

PINE, Pine   Passive infra-red night equipment.

pinger   *1* Acoustic transducer array or other source of ASW underwater signals.

   2 Operating sonobuoy.

   3 Crew member in charge of ASW sonics.

pingly   ASW helicopter, or a crew-member thereof (UK, colloq.).

ping-pong snow   Loose aggregations similar in size to table-tennis ball.

ping-pong test   Pressure test in which air is used and volume is filled with ping-pong balls to reduce stored energy.

pin joint   Joint between structural members where link is pivot, thus no bending moment can be transmitted and members of structure entirely pin-jointed must all be in pure tension or compression.

pink and green   Uniform of former USAAF (colloq.).

pinked   Cut with zig-zag edge (with pinking shears); almost universal with fabric coverings.

pinking   See *knocking*, *detonation*.

pinking shears   Shears or scissors which make a zig-zag cut.

Pinlite   Miniature light source of discrete-device type.

pink slip   Piece of paper telling employee he/she has been laid off (US).

pinpoint   *1* Precise fix.

   2 Small positively identified ground feature providing fix.

PINS   Pipeline Inspection Notification System, warns low-flying military of low-flying GA, esp. PI(11) helicopters and ag-aircraft.

pins   *1* Palletized INS.

   2 Pipeline inspection notification system (UK).

pin stowage   Authorized and clearly visible attachments for safety pins removed from ejection seat.

pint   Non-SI measure of capacity, pt = (UK) 0.568261 l (cm^3); US liquid * = 0.473176 l (cm^3), dry * = 0.550610 cm^3.

pintle   Word used in normal sense, as cantilever pivot-pin, in types of gun mount, esp. on underside or in doorway of helicopters.

PIO   *1* Pilot-induced oscillation(s).

   2 Processor input/output.

   3 Public Information Officer (UK).

PIP   *1* Product-improvement programme.

   2 Predicted impact point (NASA).

   3 Pulse-interval processor.

4 Program initialization parameter.

5 Production investment phase.

pip   Small blip on CRT, especially one used (usually as one of series) as timing mark.

PIPA, Pipa   Pulse integrating pendulous accelerometer.

Pipals   Pilot-interpreted precision-approach [and] landing system.

Piperack   Advanced jammer for AI radars (RAF 192, 214 Sqns 1944).

pipper   Aiming mark, typically 2-mil-diameter dot on HUD or other sight system.

Pip pin   Patented family of connecting pins having one end headed (often with knurled drum) and other chamfered and provided with two spring-loaded round-head plungers 180° apart which keep pin in position; usually steel and not normally used as permanent fixture.

PIPS, Pips   Pilot internet practice [not practise] service.

PIR   *1* Precision instrument runway.

2 Pilot incident report.

3 Property irregularity report, dealing with loss/damage to baggage and other pax possessions.

4 Priority intelligence requirement(s).

5 Passive IR, see next.

6 Passive IR radiometry.

7 Parachute Infantry Regiment (USA).

piracy   Used in traditional sense for unauthorized appropriation of aircraft, ie hijacking.

Pirani gauge   Measures vacuum by Wheatstone bridge and resistance wire.

Pirate   Passive IR airborne tracking eequipment.

Piraz   Positive-identification and radar advisory zone.

PIRC   Pre-emptive IR countermeasure.

PIRE   Pipe internal roll extrusion.

Pirep   Automatic pilot report programme; pilot reports actual weather on discrete frequency to chosen VOR nearby where message is taped and rebroadcast by VOR until an amending Pirep is received (FAA, from 1960).

Pirsa   Portable ILS receiver-signal analyser.

Pisa, PISA   *1* Pilot's IR sighting ability.

2 Portable ILS/VOR signal analyser.

piston   Aircraft, esp. newly built, with piston engine[s] (US colloq.).

piston engine   One in which working fluid yields energy by expanding and driving piston along cylinder, specif. IC engine of Otto (by far most common in aviation), diesel or Stirling type.

piston ring   Precision-ground abrasion-resistant ring fitted in groove around piston to make spring-tight fit against cylinder (see *gas ring*, *junk ring*, *obturator ring*, *oil-scraper ring*).

piston-ring seal   One of hard metal pressed against cylinder wall by its own elastic stress.

PIT   *1* Pilot instructor training.

2 Prioritized image-transmission.

3 Telecommunications research institute (Poland).

pit   Location, usually referenced to ground, of air-refuelling contact; a * stop (colloq.).

pitch   *1* Angular displacement (rotation) about lateral (OY) axis.

2 Arguably, angular displacement about that axis which, at any moment, is perpendicular to both vehicle longitudinal (OX) axis and local vertical; thus in vertical bank, according to this widespread definition, * = yaw.

3 See *propeller **.

4 In case of ballistic vehicle, rotation about axis perpendicular to vertical plane containing vehicle's longitudinal axis.

5 Angular setting, measured at defined station, of helicopter main- or tail-rotor blade relative to axis of rotation.

6 Uniform distance between evenly spaced objects in row, eg rivets, bolt threads or passenger seats (in each case measured from same reference point in each object).

7 Rotation of camera about axis parallel to vehicle lateral axis; also known as tip.

8 See *porpoise (1)*.

9 Rotation of main landing-gear bogie beam about transverse axis.

pitch attitude   Angle between vehicle longitudinal (OX) axis and defined reference plane, eg local horizontal.

pitch axis   See *lateral axis*.

pitch bucking   Repeated sequence in which canard foreplane stalls, nose drops, canard unstalls, aircraft pitches up and repeats cycle.

pitch circle   See *pitch curves*.

pitch cones   Contacting cones of bevel gears on which normal pressure angles are equal.

pitch control   *1* That giving manual control of propeller pitch (3), eg by moving datum of CSU.

2 That, normally effected by stick, giving control in pitch (1) of VTOL aircraft at zero or low airspeed.

3 That giving control of pitch attitude of spacecraft.

4 Combined cyclic/collective systems of helicopter (in this case, rotor pitch).

pitch curves   Intersection of tooth surfaces in pitch cones.

pitch cylinder   Notional cylinder conaining all points of contact between teeth of spur or helical gears.

pitch diameter   That of pitch line in circular wheel.

pitching   See *pitch (1)*.

pitching moment   One causing pitch (1), measured as positive when nose-up or tail-heavy. Basic equation $M = C_M \tfrac{1}{2} \rho V^2 Sc$, where C_M is total moment coefficient, ρ density, V velocity, S wing area and c wing chord.

pitching tank   Towing tank in which tendency of marine aircraft to *porpoise* (1) can be studied.

pitch jet   RCJ providing low-airspeed control in pitch (1).

pitch line   Locus of points at which centres, contact points or pitch (6) of gearwheel teeth or bolt threads are measured.

pitchover   Pronounced departure from upwards-pointing attitude, eg at point where ballistic vehicle is programmed to pitch (4) away from vertical, aircraft in stall-turn pitches at highest point and aircraft attempting absolute-altitude record runs out of kinetic energy.

pitch plane   That common to both pitch cylinders of helical or spur gears, pitch cones of mating bevel gears, or both pitch cylinders of wormwheel (on which axial and transverse pitches are equal) and pitch line of mating gear.

pitch point   Point of contact of two pitch circles.

pitch pointing   Advanced FCS mode giving ability to vary pitch (1) and thus AOA at constant flightpath angle.

pitch range   Angular range of travel of blade, esp. of propeller.

pitch ratio   Ratio of pitch (3) to diameter.

pitch setting   *1* Act of setting up propeller or helicopter rotor so that all blades have correct pitch according to that commanded.

2 Actual pitch of blade(s) at particular time, in case of propeller usually at 0.75 radius.

pitch speed   Product of mean geometric pitch and number of revolutions made in unit time (latter is s or h depending on unit used to express answer).

pitch trim compensator   See *Mach trimmer*.

pitch trimmer   Scissors link connecting bottom of main-leg outer casing and one end of landing gear bogie beam.

pitch-up   Uncommanded positive pitch (1) experienced by some aeroplanes at high subsonic Mach numbers or (tip stall on swept wing or tailplane in wing downwash) at large AOA.

PITL   Pilot in the loop [of the flight-control system].

pitot bomb   Pitot head carried on free-weathercocking mass towed on cable.

pitot comb   Row of pitot tubes, eg in vertical row behind wing.

pitot head   Sensing head for pitot/static system. In case where static pressure is taken from skin vent, pitot pressure only, thus essentially = pitot tube.

pitot pressure   That sensed by pitot head, intended to be close to stagnation pressure.

pitot rake   See *pitot comb*.

pitot/static system   Instrumentation system fed by combination of pitot pressure and local static pressure, difference giving dynamic head and thus ASIR.

pitot traverse   Taking successive measures of pitot pressure under same conditions but at different places, esp. along vertical (less commonly horizontal) line in wake of wing or other body; result indicates fluid momentum transfer and thus drag.

pitot tube   Open-ended tube facing forwards into fluid flow, thus generating internal pressure equal to stagnation pressure (in case of supersonic flow, that downstream of normal shock).

PITS, Pits   Passive identification and targeting system.

PIU   *1* Pilot-induced undulation (PIO is better).

2 Processor, or pylon, or programmable, interface unit.

3 Intermediate contingency power (F).

4 Plasma ignition unit.

PIV, p.i.v.   *1* Pressure isolating valve.

2 Peak inverse voltage.

Piver   Programmation et interprétation des vols d'engins de reconnaissance (F).

PIXE   Pronounced pixie, proton-induced X-ray emission, ion-beam technique.

pixel   Picture element, from which electronically transmitted picture is assembled.

PJ   Parajumper, in helo rescue crew.

Pj   *1* Radar received-power from jammer.

2 Jetpipe pressure.

PJBD   Permanent Joint Board on, or of, Defence (Canada/US).

PJC   Permanent Joint Council (Russia-NATO).

PJE   Parachute-jumping exercise.

PJF   Partially jet-borne flight.

PJH   PLRS/JTIDS hybrid.

PJI   Parachute-jumping instructor.

PK, P$_k$ *1* Kill probability.

2 Peak.

PKD   *1* Path of known delay.

2 Parts knock-down, aircraft or other product supplied unassembled.

PKE   Pluto Kuiper Express (NASA).

PKI   Public key infrastructure.

PKM   Perigee kick motor.

PKO   Cosmic (space) defence forces (USSR).

PKP   *1* Passenger-kilometres performed.

2 Predicted kill point.

PK screw   *Parker-Kalon*.

PL   *1* Position line.

2 Plain language (often P/L).

3 Pulse, or parameter, length.

4 Pilote de ligne (F).

5 Parts list.

6 Powered lift.

7 Primary lighting.

8 Public Law (US).

P/L   *1* Payload.

2 Plain language.

PLA   *1* Programmable logic array.

2 Power-lever angle.

3 Power-lift aircraft.

4 Pre-launch activites, for new aircraft.

5 Post-launch autonomy.

6 Practice low approach.

7 Plain-language address.

8 Private pilot, lighter-than-air, airship.

PLAB   Napalm (USSR, R).

placard value   Published numerical values of aircraft performance, esp. those concerned with safety or limiting speeds and often displayed on placard (small plate) fixed in cockpit.

place   Seat; hence '4-* ship' = four-seat aircraft (US usage).

PLACO, Placo   Planning committee (ISO).

Plaid   Precision location and identification.

plain bearing   One in which rotating shaft is simply run in surrounding fixed support, usually lined with bearing metal, without needles, rollers, balls or dynamic pressure from air or gas, but with interposed oil film.

plain flap   Simple flap in which trailing edge of wing is hinged.

plain language   Message not coded for security.

plan   One of three basic orthogonal views, that showing object from above; hence *-form.

planar   Essentially lying in one plane, 2-D; hence * technology or * electronics include solid-state devices constructed as various deposited layers, with etching, metallization and other layer-modification.

planar-array radar   One whose aerial comprises numerous (normally identical) elements in flat array; probably electronically scanned and probably synonymous with phased-array.

Planck constant   Symbol h, = 6.626196×10^{-34} Js.

Planck Law   Fundamental law of quantum theory: E = hv where E is value of quantum in units of energy, h is *Planck constant* and v is frequency.

plane   *1* Aeroplane or airplane (colloq., rarely used in professional aerospace).

2 A wing, either left or right or complete tip-to-tip.

3 To move over water at speed sufficiently high for hydrodynamic and aerodynamic lift to predominate over buoyancy.

plane angle   Angular measure in 2D, planar, unit rad [radian] = 57.2958°.

plane-change engine   Small rocket, usually MHR, whose thrust alters orbital plane of satellite.

plane flying   Navigation without electronic aids over short distances such that curvature of Earth is neglectable (in Editor's view, nonsense).

plane-guard   Routine duty of aircraft (today helicopter) stationed off port (left) quarter (towards stern) of carrier while flying operations are in progress; rescues ditched aircrew and performs other tasks.

planemaker   Aircraft manufacturer (colloq.).

plane of reference   That, perpendicular to plane of symmetry and in front of [or possibly touching] the nose, from which all nose-to-tail stations are measured.

plane of rotation   That in which tips of blades of rotating object travel; in case of helicopter main rotor synonymous with tip-path plane, and thus seldom perpendicular to shaft axis.

plane of symmetry   That containing OX and OZ axes, dividing aircraft into (usually mirror-image) left/right halves.

plane-polarized   EM radiation, eg light, in which electric force and direction of propagation remain in one plane.

planer   Machine tool, often large, whose workpiece is cut by linear motion past fixed tool. Skin mill has revolving cutter(s).

planetary boundary layer   From planet surface to geostrophic wind level, including Ekman layer.

planetary gear   Reduction gear in which driven sun-wheel turns planet-wheels engaging with fixed outer annulus; any gearwheel whose centre describes circular path around another.

planetary lander   Spacecraft designed to (usually soft-) land on planet.

planetocentric   Related to planet's centre, eg * orbit.

plane wave   One whose front is normal to propagation direction.

planform   Geometric shape in plan, esp. of wings and other aerofoils.

planimeter   Instrument for mechanically measuring area on plane surface.

planing bottom   Faired smooth surface on underside of float or hull (BSI); this omits to note need for deadrise, chine, step etc, needed to plane (3).

plank wing   Traditional wing, as distinct from swept or other modern planform (colloq.).

plan-label display   Radar display on which SSR alphanumeric information can be written in association with positional echo or symbol; usually fast synthetic, or mixed-phosphor (hard for high-refresh alphanumerics and soft for raw position symbol).

planned flight   One for which flightplan is filed and which has specific purpose, ie not air experience or joyride.

planned load   One made up in advance and tailored to cargo-aircraft type and mission.

planning-programming-budgeting   Integrated system for management of DOD budget and Five-Year Defense Program (USAF).

planometer   Surface plate.

plan-position indicator   P-type display in which scene appears in plan with observer, radar or other sensor at centre; objects at radial distance giving range (usually linear scale, often selectable to several values) and with correct bearing (000° usually at top or 12 o'clock position); offset PPI moves sensor to position away from centre, typically to 6 o'clock margin. Expanded-centre PPI has zero range at ring surrounding centre.

plan range   In air reconnaissance, horizontal distance from sub-aircraft point (that where local vertical through aircraft intersects surface) and ground object.

PLAP, Plap   Power-lever angle prime, throttle in UFC (1) to which engine responds irrespective of pilot demand.

PLASI, Plasi   Pulse-light approach slope indicator.

plasma   Assembly of neutral atoms, ions, electrons and possibly molecules in which particle motion is determined by EM interactions; electrically conductive, hence responds to magnetic field. Study called MHD or hydromagnetics.

plasma antenna   Glass or ceramic container of ionized gas.

plasma-based stealth   Making aircraft morwe or less invisible to hostile radars by enveloping them in an intense EM field.

plasma engine   See *plasma rocket*.

plasma ignition   Source of 1,000 J/s at 5,000K for solids, liquids or gases.

plasma jet   Jet of plasma produced by MHD.

plasma panel   Electronic display of gas-discharge type, usually AC, usually orange (Ne), but many other colours with different gases.

plasma plating   Deposition of refractory, abrasion-resistant or anti-corrosive coating by means of intensely hot (c16,600°C) plasma jet moving at supersonic speed into which coat material is introduced as powder.

plasma rocket   One whose working fluid is a plasma, accelerated by intense EM field (ie, plasma jet).

plasma sheath   That surrounding re-entry vehicle or spacecraft, serving as barrier to radio communications.

plasma wind tunnel   One capable of simulating spacecraft re-entry to Earth atmosphere.

plastic   Not elastic, tending to remain in deformed shape or position (see *plastics*).

plastic effect   Electronic display shows relief but little tonal value.

Plasticele   Transparent plastic which, unlike Pyralin, is non-inflammable and does not discolour with age.

plastic flow   That caused by stress beyond elastic limit and remaining when stress is removed.

plastic gyro   Wheel assembled from moulded plastics components.

plastic instability   Column failure due to plastic flow in compression rather than to bending.

plasticity   Ability to be deformed to new permanent shape.

plasticizer   Substance added to polymer to change properties to improve mouldability or other useful properties; usually liquid of high boiling point. In solid propellants used chiefly to increase flexibility, strengthen bonding and eliminate cracking.

plastics   General terms for vast range of synthetic materials made by mixing constituents of which prime members are polymers, for distribution as liquid, fibre, granules or sheet subsequently moulded (see *thermoplastic*, *thermosetting*) or used as reinforcement with adhesive bonding. Properties range from rubbers to highly crystalline fibres. Singular 'plastic' is adjective; in describing part or finished product preferred usage is 'plastics' or, if possible, name material, eg PTFE, PVC, GRP or CFRP.

plastic factor   Additional factor, typically 1.2–1.5, applied in designing primary structure in fibre-reinforced composite; this 'factor of ignorance' is being relaxed as experience is gained.

PLAT   Pilot's landing-aid TV.

Plate   *1* Sheet thicker than 0.25 in (6.35 mm); in airframes invariably machined or chem-milled. Not to be confused with sheet.

2 Principal anode of vacuum tube.

3 Pocket-size sheet of paper, plastics or aluminium on which are printed details of facilities, aids and approach data for one airport.

plate brake   Mechanical brake for rotating shaft, eg landing wheel or helicopter rotor, where retarded moving member is ring fabricated from heavy plate (steel, titanium or beryllium, or CFRP-based) often stacked in parallel; essential difference from disc brake is that ring is gripped from both sides.

plate-wired memory   Advanced and highly compact memory woven on loom from coated wire giving non-volatile storage and low-nanosec speeds.

platelet   Small plate, esp. one which is perforated by, or whose surface contains precision channels usually produced by, photo-etching, for fluidic or rocket-injector system.

platelet injector   Rocket injector assembled from large stack of platelets to give optimum multiple paths for (usually two) liquid propellants.

platform   *1* Vehicle carrying sensors and/or weapons, eg aircraft or spacecraft.

2 Extended root of turbine (rarely, other) blade linking root attachment to outer aerofoil.

3 Raised operating area for helicopter or V/STOL, esp. on surface vessel, also called pad.

4 See *airdrop platform*.

platform drop   Drop of loaded platform (4) from rear-loading aircraft with roller conveyor.

platform dynamics   Those resulting from motion of platform (1), esp. as they affect ECM/ESM, eg range, range-rate (velocity), acceleration and acceleration-rate (jerk); can cause receiver to lose lock or sychronization.

platform face   That forming inner end of aerofoil portion of turbine rotor blade and part of inner wall of gas duct.

platform strength   Number of aircraft available.

Platinizing   Coating steel with Zn.

platinum   Costly non-corroding metal, density 21.5, MPt 1,773°C, symbol Pt.

Plato   *1* Program logic for automatic teaching operations.

2 Pilot low-altitude terrain overlay.

platypus   Flat 2-D jet-engine propulsive nozzle.

playing area   Area of operations possible with digital ATC (1) simulator, typically, 256, 512 or 1,024 miles square.

PLB   *1* Personnel (or personal) locator beacon.

2 Passenger loading bridge.

PLC   Programmable logic controller.

plc   Public limited company.

PLCU   Primary-lighting control unit.

PLD   *1* Pulse-length discrimination (in MTI circuits eliminates fast-moving clouds and other moving objects whose size precludes their being targets).

2 Precision laser designator.

3 Proportional lift-dump [mode].

4 Programmable logic device.

pleasure flight   One made by private pilot landing back at the point of departure.

plenum chamber   Airtight chamber, esp. one containing fluid-flow sink such as operative air-breathing engine; essential for gas turbine having double-entry or reverse-flow compressor with ingestion all round periphery.

plenum-chamber burning   Boosting thrust of vectored-thrust turbofan by burning additional fuel in the 'cold' nozzles downstream of the fan.

plenum-chamber door   Blow-in door to increase airflow into chamber when internal depression falls below selected level, eg on take-off.

Plesetsk   Soviet, now Russian, ICBM base and launch establishment for Cosmos and many other large ballistic systems; in Leningrad military district at 62.9°N 40.1°E.

Plexiglas   Registered name (Rohm & Haas) of family of acrylic-acid resin plastics, esp. transparent, widely used for blown mouldings; essentially US counterpart of Perspex, though different material.

Plezit   PLZT.

PLF   *1* Precise local fix.

2 Parachute landfall.

3 Powered-lift facility.

4 Passenger load factor.

PLGR   Precise, or precision, lightweight, or location, GPS receiver.

PLH   Propeller load horsepower.

PLI   Pre-load indicator (projects from head of structural bolt).

pliss   See *PLSS (2)* (colloq.).

PLL   Phase-lock[ed] loop.

PLM   *1* Pulse-length modulation.

2 Pulse-length monitor.

3 Product life-cycle management.

PLN   Flight plan.

P-LOCAAS   Powered derivative of the low-cost autonomous attack submunition.

Plod   Passenger landed on deck (carrier onboard delivery).

PLOG   Pilot's [flight-planning] log.

Plog   Pilot-log record[s], no formal definition.

Plonk   AC2 or ACH/GD, lowest form of life (RAF, WW2).

plot   *1* Graphical representation of two or more variables on 2-D surface.

2 Graphical construction for solving navigation problems, eg triangle of velocities.

3 Map, chart or graph representing data of any sort (DoD).

4 Visual display, eg on radar, of aerial object at particular time; hence * extraction.

5 Portion of map or overlay showing outlines of areas covered by reconnaissance or survey photographs.

plot extraction   Translating radar plot (4) into quantified target position information, formerly done manually.

plot extractor   Electronic system which detects replies from primary or secondary radar and, after making validity check, digitizes information ready for transmission over narrow-band link equipment or high-grade telephone line; where input is SSR basic range/bearing information can be supplemented by identity and height. 'Extraction' derives from fact system eliminates information not needed.

plotting board   Large horizontal (rarely vertical) surface upon which positions of moving objects are shown with respect to co-ordinates or fixed reference points.

plotting chart   Chart designed for graphical methods of navigation.

ploughing, plowing   Taxiing marine aircraft at below planing speed.

PLP   *1* Pipeline patrol.

2 Parallel-line planform.

PLRO   Plain-language readout.

PLRS   Precision (or position) location (and) reporting system.

PLS   *1* Precision landing system; R adds receiver.

2 Palletized loading system.

3 Personnel locator, or location, system.

4 Plasma subsystem, to measure solar wind.

PLSS   *1* Precision location strike system.

2 Portable life-support system.

PLU   *1* Position (or preservation of) location uncertainty.

2 Program load unit.

plug   *1* Extra section, usually of constant cross-section, added in front of or behind wing when *stretching* fuselage.

2 Air-refuelling contact, hence wet *, dry *.

3 As verb, one meaning is to blank over passenger window with metal skin.

plug aileron   Has form of curved sheet forming segment of cylinder, extended on pivoted brackets from curved slot in wing.

plug door   One so designed, eg with inward/upward travel or with retractable upper and lower portions, that it is larger than doorway, two mating with thick tapered edges to increase security of pressurized fuselage. Pressurization load merely forces door more tightly against frame.

plug gauge   Male-type gauge, not always of circular cross-section and often tapered or threaded, for checking dimensions of holes and internal threads.

plug inlet   One form of inlet for air-breathing propulsion at Mach 3.5–6 in which axisymmetric duct tapers from sharp lip to rear, and contains large plug (spike) which when translated fully aft can seal flow; in most forms rearward spike travel renders shock-on-lip operation impossible.

plug nozzle   Proposed for rocket engines: combustion chamber is annular toroidal form discharging around central cone with curved profile which converts initial inwards radial component into pure axial flow. Also called spike nozzle.

plug ring   Translating ring surrounding rear of piston engine cowl to control cooling airflow; not necessarily provided with hinged flaps or shutters (ie gills).

plug tap   Final non-bevelled or bottoming tap for completing threaded hole.

plug weld   One made by drilling through part of structure, eg boom splice or skin/stringer, and welding through hole to increase strength of joint.

plug window   Overlarge window with tapered periphery (as plug door) used as emergency exit in pressurized fuselage.

plumber[s]   Ground crew (RAF colloq.).

Plumbicon   Photoconductive camera tube similar to vidicon but using semiconductor PbO target doped to behave as reverse-biased PIN.

plumbing   Pipework for liquid systems, eg hydraulics, lube oil, Lox etc (colloq.).

plume   Originally having specific applications, today a general word meaning entire wake from jet (airbreathing or rocket) bounded at periphery and at more vague downstream extremity by envelope enclosing all parts having significant effects on environment, eg thermal, aerodynamic, acoustic, or as IR source.

plume target   Aerial target emitting plume simulating that from hostile jet aircraft.

plus count   Forward count begun at lift-off of spaceflight, continued throughout mission to provide GET reference.

plutonium   Silvery metal, many isotopes, all radioactive and toxic, density 19.8, MPt (typical) 641°C, most important constituent of NW devices, alone or with an alloy, symbol Pu.

pluviometer   Rain gauge.

PLV   Payload launch vehicle (ABV).

PLVL   Present level.

PLW   Ploughed (US).

PLWS   Precision lighting warning system.

Plymetal   Plywood/aluminium sandwiches, invariably non-structural.

PLZT   *1* Lead lanthanum zirconate titanate.

2 Polarized lead/zinc titanate.

PM   *1* Pulse modulation.

2 Phase modulation, or modulated.

3 Permanent magnet.

4 Program(me) manager, or management.

5 Powder metallurgy.

6 Pressurized module.

7 Poly medialite.

8 Power management.

9 Phase margin [coupling between FCS and structure].

10 Pilot monitoring.

P^m Polar maritime.

P/M   Presentation/manoeuvring (simulator).

PMA   *1* Projected map assembly.

2 Parts manufacturing [or manufacturer] approval, or authority.

3 Permanent-magnet alternator.

4 Pressurized mating adapter (docking).

5 Portable maintenance aid.

6 Positive mental attitude (US).

7 Propagation management and assessment [algorithms].

8 Post-mission analysis.

PMA/A   Probable missed approach per arrival.

PMADS   Pedestal-mounted air-defense [missile] system.

PMAS   Performance measurement analysis system, for assessing contractors and management (DoD).

PMAT   Portable maintenance access terminal.

PMAWS   Passive missile approach warning system.

P_{max} *1* Maximum power (electronic).

2 Maximum pressure (usually MEOP).

PMB   Phare Management Board.

PMC   *1* Plastic/metal composite sandwich, usually two metal skins bonded to low-density core.

2 President of Mess Committee (RAF).

3 Maximum continuous power (F).

4 Polymer/matrix composite.

5 Power management control.

6 Performance management computer (S adds "system").

7 Personal multimedia communications.

8 Provisional memory cover.

9 Private military [including air] companies.

10 PCI mezzanine card.

PMCS　Portable mission control station (UAV).

PMD　*1* Projected map display.

2 Maximum take-off power (F).

3 Panel-mounted display.

4 Programme management directive.

PMDB　Production management[s] data base.

PMDS　Projected-map display set.

PME　*1* Professional military education (US).

2 Precision-measurement equipment [L adds laboratory] (USAF).

PMF　Processeur militaire Français.

PMFT　Post-maintenance flight test.

PMG　Permanent-magnet generator.

PMH　Patrol missile hydrofoil (US Navy).

PMI　*1* Payload margin indicator.

2 Performance management indicator.

PMIRR　Pressure-modulated IR radiometer.

PMM　Performance monitor module.

PMMA　Poly-methylmethacrylate.

PMN　Lead/magnesium niobate [-PT adds lead titanate].

PMO　*1* Program(me) management office.

2 Principal medical officer.

3 Prime maintenance organization[s].

PMOP　Phase modulation on pulse.

PMOS　Positive (p-type) metal-oxide silicon, or semiconductor.

PMP　*1* Program management proposal.

2 Propulsion modernization program.

3 Premodulation processor.

PMPS　Portable mission-planning system (USAF).

PMQC　Purchase-material quality control (UK, quality assurance).

PMR　*1* Pacific Missile Range, from Vandenberg (WTR) and Pt Mugu (US); F adds Facility, at Kauai, Hawaii, and Kwajalein Atoll.

2 Proton magnetic resonance.

3 Portable MLS receiver.

4 Private mobile radio.

PMRAFNS　Princess Mary's RAF Nursing Service (UK).

PMRT　Program management responsibility transfer (DoD).

PMS　*1* Performance/power/programme management system.

2 Projected map system (or subsystem).

3 Personnel Management Squadrons (RAF).

4 Performance measurement system (of programme).

5 Process and materials specifications (manual).

6 Poor-man solution.

PMSP　Parallel-module signal processor.

PMST　Project-management support team.

PMSV　Pilot-to-metro service.

PMTC　Pacific Missile Test Center.

PMU　Maximum contingency power (F).

PN　*1* See *pseudonoise*.

2 Performance number.

3 Prior notice.

4 Pursuit, night (USA 1919–24).

P/N　Part number.

PNA　Point of no alternate.

PNB　Pilot/navigator/bomb aimer (former RAF aircrew grade, higher than SEG).

PNC　Pneumatic nozzle control.

PNCP　Peripheral node control point (SNA).

PNCS　Performance and navigation computer system.

PND　*1* Pilot numerical display, including distance to go, G/S.

2 Primary navigation display.

PNdB　Perceived noise decibels (see *noise*).

PNDC　Pakistan National Development Complex.

pneud, pneudraulic　Combined hydraulic and pneumatic operation.

pneumatic　*1* Air-operated; term usually reserved for services taking very small flow at high pressure energized by shaft-driven compressor or one-shot bottle. Ambiguously also often used for services taking very large flow at low pressure to drive turbines, air motors and cabin environmental controllers.

2 Also describes panel instruments driven by air pressure or air-driven gyro.

pneumatic altimeter　Traditional barometric altimeter.

pneumatic bearing　Externally pressurized gas bearing.

pneumatic deicing　Removal of ice accretion by alternate inflation and deflation of flexible tubes along [wing or tail] LE.

pneumatic logic　That used in fluidics.

pneumatic power module　Stores energy for extremely rapid high-power release.

pneumodynamic　System supplies air, or other gas, in rapidly time-variant controlled manner.

PNF　Pilot not flying, the non-handling pilot.

PNG　*1* Pseudonoise generator.

2 Passive night (-vision) goggles.

PNI　Pictorial navigation indicator.

PNJ　Pulse(d) noise jamming.

PNL　Perceived noise level (see *noise*).

PNLT　Tone-corrected PNL.

pnm　Passenger nautical mile.

PNP　Programmed numerical path.

PNR　*1* Point of no return.

2 Prior notice required.

3 Part number.

4 Passenger name record, full details on database (US).

PNS　Pictorial navigation system.

PNTR　Permanent normal trade relations.

PNU　Precision navigation upgrade.

PNVG　Panoramic night-vision goggles.

PNVS　Pilot's night vision system, or sensor (attack helicopters).

PNWA　Prevention of nuclear war agreement (1973).

PO　*1* Purchase order, or option.

2 Preposition operation(s).

3 Dust devil (from French).

POA　*1* Position of advantage (air combat).

2 Plain old Acars.

POAP　Photoconductor on active pixel.

POB　*1* Persons, or pilot, on board.

2 Polymer optical backplane.

POBA　Project for on-board autonomy (orbiter).

POC　*1* Point of contact.

2 Parts obsolescence cost.

3 Professional Officer Course (ROTC).

4 Payload Operations Center (MSFC).

POC, PoC   Proof of concept.

POCA   Parting-out candidate aircraft.

POCC   Payload operations control center, now POC (4).

pocket   Short spanwise length of light trailing edge attached to rear of helicopter rotor-blade spar, many * being required for each blade.

POCU   Pre-OCU.

POD   *1* Preliminary orbit determination.

2 Probability of detection.

pod   Streamlined container carried on pylon, strut or other attachment entirely outside airframe and housing propulsion system, reconnaissance sensors, flight-refu-elling hosereel or similar devices.

PODA   Pre-operational data-link applications.

Podas   Portable data-acquisition system.

podded   Accommodated in pod.

podding   Philosophy and technique of using pods, esp. for accommodation of main engines.

pod formation   Formation of friendly aircraft disposed so that ECM assets give maximum mutual protection.

podium   Desk for boarding agent at airport gate.

PODS, Pods   *1* Portable data store.

2 Portable digitizer subsystem.

POE   *1* Probability of error.

2 Port [includes airfields] of embarkation.

Poems   Pre-operational European Mode-S (ATC).

POES   Polar-orbiting environmental satellite.

Poet   *1* Portable opto-electronic tracker.

2 Primed oscillator expendable transponder.

POF   Plastic optical fibre.

POFM   Petrol/oil fuel mixture.

Pogo   *1* See *pogo effect*.

2 Tail-standing VTOL (colloq.).

3 Local below-airways ATC linking Paris airports.

4 Air-intercept code: "Switch to preceding channel or, if unable to establish communications, to next channel after *" (DoD).

5 Mechanical link in PFCS which in case of any jammed components automatically overrides drive cam.

6 Precision on-board GPS optimization (missile guidance).

pogo effect   Longitudinal oscillation or vibration, esp. of vehicle having high ratio of thrust to mass and whose propulsive thrust may suffer short-term variations; characterized by significant and uncomfortable axial accelerations and, in large ballistic rocket, severe propellant sloshing.

pogo-stick   Precisely calibrated penetrometer for measuring strength of paved or unpaved surfaces.

POH   *1* Pilot operator's [or pilot's operating] handbook.

2 Put on hold.

Pohwaro   Pulsated overheated water rocket (FFA, Switzerland).

POI   *1* Programme of instruction.

2 Probability of intercept.

3 Point of interest.

Point Arguello   At south Vandenberg, Western Test Range (US).

point defence   *1* Defence of specified geographical areas, cities and vital installations; distinguishing feature is that missile guidance radars are near launch sites (USAF).

2 Defence of single surface ship, esp. against incoming missile.

point-designation grid   Grid drawn on map or photograph whose sole purpose is to assist location of small features.

point discharge   Gaseous electrical discharge from surface of small radius (point, or tips of static wick) at markedly different potential from surrounding; unlike corona discharge, ** is silent and non-luminous.

point light   Luminous signal without perceptible length (ICAO).

point mass   Simplifying equations of motion by assuming aircraft has no dimensions, thus eliminating torques and moments.

Point Mugu   Location of Naval Missile Center and Naval Missile Range (US).

point navaid   Electronic navigation aid located at a single site, e.g. NDB, VOR.

point of attachment   Where balloon rigging cable joins flying cable.

point of entry   Where aircraft enters control zone.

point of equal time   Same time to reach destination or return to start.

point of interest   Airfield, or point navaid.

point of inversion   Height at which lapse rate at last passes through zero; where temperature begins to fall.

point of no alternate   Geographical position on track or time at which fuel remaining becomes insufficient to reach declared alternate.

point of no return   Geographical position on track or time at which fuel remaining becomes insufficient for aircraft to return to starting point (DoD, NATO wording 'to its own or some other associated base').

point parallel   Standard form of rendezvous for boom-type tanker and large receiver, in which aircraft fly reciprocal tracks to ARCP, tanker then turning 180° to come up 3.5 miles ahead of receiver at roughly same height, heading and speed.

point target   *1* One requiring accurate placement of conventional ordnance in order to neutralize or destroy it (DoD).

2 With NW, one in which target radius is not greater than one-fifth radius of damage.

point to point   Linear motion of tool between NC-instructed commands.

point vortex   Section of straight-line vortex in 2-D motion.

Poise   Pointing and stabilization platform element (USA, RPVs).

poise   Non-SI unit of dynamic viscosity, defined as 1 $dyn.s.m^{-2} = 0.1\ Ns.m^{-2} = 0.067197\ lb.s/ft^2$.

Poiseuille equation   Relates flow through tube (defined as 'long and thin'; elsewhere as 'capillary') to variables: $Q = \pi Pr^4/8l\mu$ where Q is volume per unit time (seconds), P is pressure difference across ends of tube, r is radius, l is length and μ is viscosity.

Poiseuille flow   *1* Viscous laminar flow in circular-section pipe.

2 Viscous laminar flow between close bounding planes.

Poisson's equation   In stressed material $\sigma = (E/2n)-1$ where σ is Poisson's ratio, E is Young's modulus and n is modulus of rigidity.

Poisson's ratio   Ratio of lateral contraction (in absence of local waisting) per unit breadth to longitudinal extension per unit length for material stretched within elastic limit, symbol σ.

Poits Payload orientation and instrumented tracking system.

poka yoke Japanese words meaning foolproof, ie eliminating disruption caused by faulty work.

poke Propulsive thrust or power (colloq.), hence 'pokier', etc.

poke welding Similar to spot welding but using single 'poked' electrode, the other being clamped to any convenient point on workpiece.

POL Petrol, or petroleum, oil and lubricant.

Pol, pol Polarity (but P in FFP).

Polar Precision over-the-horizon land attack rocket.

polar *1* Air mass supposedly originating near pole, hence cold and usually dry: thus * Atlantic, * continental, * front, * maritime.

2 Parameter plotted on polar co-ordinates.

3 Basic performance curve of sailplane in which sink speed is plotted against EAS (units of two scales differ, traditionally ft/s against kt but today m/s: km/h).

polar-cap absorption Radio blackout by HF absorption in ionospheric storms.

polar continental Typically extremely cold, dry and stable air mass; abb. P_c.

polar control Twist-and-steer flight control.

polar co-ordinates Those defining locations by means of angle of a radius vector, measured relative to agreed direction, and vector length; similar to rho-theta navigation.

polar diagram One giving values round 360° from a point, eg noise, IR or radio emission, stress or temperature.

polar distance Angular distance from celestial pole.

polar front That separating polar air mass from contrasting mass.

polarimeter Instrument for determining degree of polarization of EM radiation, esp. light.

polarimetry Particular meaning in aerospace is examination of changes in polarization of radar returns from different types of target surface.

polariscope Instrument for detecting polarized radiation.

polarity *1* Of line segment, having both ends distinguishable.

2 Hence, of physical system, having two contrasting points, specif. oppositely marked terminals of electric cell or plus/minus characteristics of ions.

polarization Many meanings but chiefly associated with EM radiation in which * can be plane, elliptical or circular. Plane * rotates all wave motion so that all E (electric) vibrations take place in one plane (plane of vibration) and all H (magnetic) vibration takes place at 90° to this (plane of *).

polarization diversity Having ability to swtich from plane-polarized to circular-polarized (radar).

polar maritime Air mass that is cold and, though absolute humidity and dewpoint temperature low, relative humidity is high; abb. P_m.

polar moment of inertia Moment of inertia of area about axis perpendicular to its plane. Traditional symbol not I but J.

polar navigation Navigation at high latitudes, distinguished in bygone days on account of unreliability of magnetic compass, possible electrical/radio inteference

and other problems such as rapid change of meridians and map-projection difficulties.

polar orbit Orbit passing over, or close to, poles of primary body.

polar Pacific Air mass originating over N Pacific or N America, seasonal characteristics; abb. P_P.

polar plot Locating a point, eg target, by polar co-ordinates.

Polar stereographic Map projection, that of high-latitude region projected on flat sheet touching Earth at Pole by light source at opposite pole. Parallels expand at sec^2 co-lat/2.

polar triangle One formed by three intersecting great circles.

pole *1* Origin of polar co-ordinate system.

2 Point of concentration of magnetic charge (magnetic *).

3 Point of concentration of electric charge (dipole and sought-after monopole).

4 Intersection of Earth's surface and axis of rotation (geographic *).

5 For any circle on spherical surface, intersection of surface and normal line through centre of circle.

6 Parts of surface of magnet through which magnetic flux emanates or enters (theoretical but necessary concept).

7 Terminal of battery.

pole model Miniature or full-scale model, eg complete aircraft, mounted on tall pole to measure RCS (2) from all aspects.

pole piece That part of core of electromagnet which terminates at air gap.

poll *1* To ask specific questions of number of sensors; questions are normally asked sequentially, and answers constitute an update of information in system.

2 Technique used in data transmission whereby several terminals share communication channels, particular channel chosen for given terminal being determined by testing each to find one free, or to locate channel on which incoming data are present. Also used to call for transmissions from remote terminals by signal from central terminal; method used for avoiding contention.

polled ACARS mode in which airborne system transmits only in response to received uplink message.

poll the room To obtain consensus in solving problem in manned space mission.

polonium Po, grey semi-metal, energy source for spacecraft, density 9.3, MPt 254°C.

polyanilines Range of electrically conducting plastics made by oxidising aniline and then polymerizing with various acids; basis of smart skins.

polybutadiene acrylic acid, PBAA Important solid rocket propellant and monofuel.

polybutadiene acrylonitrile, PBAN Important solid rocket propellant and monofuel.

polyconic Mapmaking by projecting latitude bands on the succession of cones, each centred on Earth's axis and each touching surface along parallel passing through centre of map, subsequent strips on single-curvature conical surface being unrolled.

Polydol Alcohol-resistant foam liquid for firefighting made from protein hydrolysate and an organometallic complex; mixes with fresh or sea-water.

polygon warhead One having several (typically 8–16)

radial faces from which are projected tailored fragments, steels balls, shaped-charge jets or other damage mechanisms. Common type of warhead for SAMs and AAMs, radial blasts being so timed that their plane passes through target.

polymerization   Basic processes for making large (high-polymer) molecules from small ones, normally without chemical change; can be by addition, condensation, rearrangement or other methods.

polymer optical backplane   Airframe structure on which are printed fibre-optic conductors.

polymorph   Aircraft capable of gross change of shape.

polyphase coding   Pulse-compression technique for radar, esp. for fighters, in which successive phases are rotated by fixed angles such as 90°.

polyvalence, polyvalent   Multirole (F).

POM   *1* Printer output microfilm.

2 Program objective(s) memorandum (DoD).

Pomcus   Prepositioned overseas materiel configured in unit sets (ie, grouped by user units).

POMO, Pomo   Production oriented maintenance organization (USAF).

Pomros   Power-off minimum rate of sink.

POMS   Production-oriented maintenance system.

pond   Reservoir for cooling water below testbed for large rockets.

ponding   Settlement of runway subsoil leading to formation of standing-water pond in rain.

pongo   Member of friendly army (RAF colloq.).

PONO   Project office nominated official.

pontoon   *1* Inflatable buoyancy bag used as permanent alighting gear for helicopter.

2 Rigid float, of circular or rectangular section, used on early water-based aircraft.

POO   *1* Payload of opportunity.

2 Pronounced poo, to clear manned-spacecraft computer display prior to solving fresh problem, from P-zero-zero (colloq.).

Pooley's   Flight guide to UK and Ireland, commercially published annually.

pool fire   Burning pool of fuel surrounding crashed aircraft.

poopy suit   Flight-crew overwater survival suit (colloq.).

POP   *1* Probability of precipitation.

2 Product optimization programme.

3 Plug-in optronic payload.

4 Period of performance.

5 Point of presence.

pop   Sudden rise by target, from ground cover or out of clutter on radar.

Popeye   Air-intercept code: "I am in cloud or reduced visibility".

popouts   Fast-inflating buoyancy bags for flotation in emergency.

poppet valve   Common mushroom-type valve of piston engine.

Pop rivet   Pioneer form of tubular rivet closed by withdrawing central mandrel which forms integral part of each rivet and is gripped by jaws of tension tool.

POPS, Pops   *1* Position and orientation propulsion system.

2 Pyrotechnic optical-plume simulator.

Popular Flying Association   Represents sport aviation, amateur builders and group-operated aircraft (UK).

pop-up   Manoeuvre made by attack aircraft in transition from lo penetration to altitude from which target can be identified and attacked; eliminated by BFPA (blind first-pass attack).

pop-up missile   *1* One ejected, often by gas generator/piston or other device not forming part of missile propulsion, from launch system in vertical direction, subsequently making fast pitch-over on target heading. Launcher need not move in azimuth or elevation, thus minimizing reaction time.

2 Missile making pop-up manoeuvre as it nears previously located target on which it then dives.

pop-up test   Test of pop-up missile (1) launch system.

POR   *1* Pilot opinion rating.

2 Pacific Oceanic Region.

Poroly   Porous metal produced by incomplete sintering, mainly used as filter.

porous friction asphalt   Asphalt whose constituents and lay-down process are tailored to give high-μ surface which will not permit standing water to remain. Preferred top layer to runway.

porpoise   *1* Undulatory (near-phugoid) motion of marine aircraft and some landplanes with bicycle landing gear characterized by pitch oscillation and limited-amplitude vertical travel; normally problem only at particular speed(s).

2 In absence of radio contact, deliberate roller-coaster flight to indicate to pilot of friendly interceptor (more rarely, tower) that aircraft is in distress.

port   *1* Left side or direction, aircraft viewed from rear.

2 Aperture in solid rocket motor case opened for thrust termination.

3 Aperture[s] in volume under pressure regularly opened and closed by valves, such as in piston engine or reciprocating-compressor cylinder.

portable data store   Computerized flight-planning data carried by member of flight crew and plugged in before take-off.

portal   *1* Air/land interface: airfield or heliport.

2 Airline website.

Porteous loop   Loop with 360° aileron roll added at zenith.

portfolio   Owner's catalogue of aircraft [usually used] for lease or sale.

POS   *1* Peacetime operating stocks.

2 Position.

3 Positive.

posigrade   In direction of travel, hence * rocket increases speed of satellite in order to reach higher orbit. Opposite of retrograde.

POS-Init   Position initialization.

position   *1* For celestial body, bearing and altitude.

2 Location of crew member, esp. in military aircraft.

3 Location of manually aimed defensive gun(s) in military aircraft (today archaic); gun *.

4 Location of AAA defending land target; gun *.

5 To fly aircraft to where it is needed; hence * or positioning flight.

position error   That induced in ASI system by fact that stagnation pressure sensed is seldom that of true free stream (see *airspeed*).

position light   See *navigation light*.

position line   Line along which vehicle is known to be at particular moment, eg by taking VOR bearing. Two PLs are needed for fix.

position stabilized   Held to linear trajectory, eg LOS, in absence of commanding signal (eg antitank missiles).

positive acceleration   Acceleration upwards along OZ axis, to right along OY and forwards (ie to increase speed) along OX. Axes are always vehicle related.

positive area   That enclosed on tephigram between path of rising particle and surrounding air when particle is throughout warmer than surroundings.

positive coarse pitch   Locked minimum-drag setting of non-feathering propeller after engine failure.

positive control   *1* Operation of air traffic in radar/non-radar ground control environment in which positive identification, tracking and direction of all aircraft in airspace is conducted by authorized agency (DoD).

2 Command/control and release procedures, and operational procedures that provide acceptable level of assurance against misuse of nuclear warheads, when these warheads are part of a weapon system (USAF).

positive coupling   Mutually inductive coupling such that increase in one coil induces rising voltage in other similar in sense to that caused by increasing current in other.

positive-displacement pump   One delivering fluid in discrete parcels in irreversible way, eg pumps using pistons, vanes or gearwheels in contrast to centrifugal blower.

positive-driven supercharger   Mechanical drive as distinct from turbo.

positive-expulsion propellant system   One in which liquid propellants are forced from container by gas pressure, esp. acting on flexible bag containing propellant(s).

positive feedback   Feedback that results in increasing amplification; also called regenerative.

positive g   See *positive acceleration*.

positive-identification and radar advisory zone   Specified area established for identification and flight-following of aircraft in vicinity of fleet defended area.

positive ion   One deficient in one or more electrons.

positive pitch   *1* Nose-up.

2 Propeller set for forward flight as distinct from braking.

positive pressure cabin   Arch., see *pressure cabin*.

positive rolling moment   Tending to roll right-wing-down.

positive stability   Aeroplane tends of own accord to resume original condition, esp. level flight, after disturbance (upset or gust) in pitch, without pilot action. Term applies only to motion in vertical plane, along OZ axis.

positive stagger   Leading edge of upper wing is ahead of that of lower.

positive stall   Stall under 1 g flight or positive acceleration, ie not from inverted attitude or negative g.

positive yaw   Tending to rotate anticlockwise about OZ axis seen from above, ie nose to left.

positron   Short-lived particle equal in mass to electron but positive in sign.

positron emission tomography   Technique for measuring fluid flow in which positrons are used as labels, generating 511 keV gamma rays which can be read from outside the structure of an engine or other container.

Posix   *1* Portable operating system interface for computer environments.

2 Also translated as Portable operating system IX.

POSN   Position.

POS-Ref   Position reference.

POSS, Poss   *1* Power-off stalling speed.

2 Precipitation-occurrence sensor system.

3 Possible (ICAO).

POST   *1* Portable optical sensor testbed (hardened sites).

2 Passive optical seeker technique (or technology).

3 Production-oriented scheduling technique.

4 Point-of-sale terminal.

post   *1* Vertical primary structure, eg fin *, king-*.

2 Main landing leg, thus MLG of YC-14 described as of four-* type.

post-attack period   Between termination of nuclear exchange and cessation of hostilities.

post-boost   After cut-off of mainstage propulsion.

post-boost vehicle   Vehicle and payload after cut-off of mainstage propulsion, and hence posing new problems to SDI acquisition and tracking systems.

post-exit deflector   Powered flap(s) downstream of engine nozzle, able to vector entire thrust for STOVL or air combat.

post-flight report   Basically comprises ECAM warnings and maintenance status, printed on demand.

post-integration   Occurring after assembly of complete space launch system but before lift-off.

post-pass   After (usually soon after) satellite has passed overhead.

post-stall gyration   Uncontrolled motions about one or more axes, usually involving large excursions in AOA, following departure (3).

post-strike   Immediately after attack on surface target, hence * reconnaissance provides information for damage assessment.

posture   Military strength, disposition and readiness as it affects capabilities (DoD).

POT, pot   Potentiometer.

pot   Piston engine cylinder (colloq.).

potassium   K (from kalium), reactive silvery metal, MPt 64°C, density 0.86.

potential   *1* In electric field, work done in bringing unit positive charge to that point from infinite distance.

2 Value atmospheric thermodynamic variable would have if expanded or compressed adiabatically to 1,000 mb (100 kPa).

3 Specialized meanings in thermodynamics, geodesy and celestial mechanics.

potential density   That which parcel of air would have if adjusted adiabatically to 1,000 mb (100 kPa), given by $\rho' = p/R\theta$ where ρ' is **, p is pressure at 100 kPa, R is gas constant and θ is potential temperature.

potential difference   Measured between any two points in conductive circuit, in volts; if between terminals of battery, when no current is flowing.

potential energy   That possessed by mass by virtue of position in gravitational field; can yield ** by 'falling', ie moving towards region of lower **.

potential flow   Fluid motion in which vorticity is everywhere zero.

potential gradient   Local space (linear) rate of change of potential.

potential refractive index   That so formulated that variation with height in adiabatic atmosphere is zero.

potential temperature That which parcel of dry air would have if adiabatically brought to pressure of 100 kPa.

potentiometer *1* Variable resistance (rheostat), esp. one of precise type giving, eg, accurate radar pointing information.

2 Instrument for measuring EMF (potential differnce, esp. DC), usually without drawing current from circuit measured.

POTS *1* Production off the shelf.

2 Plain old telephone service, or system.

potted device Electronic component encapsulated in resin, mainly to give mechanical protection.

pouce Inch (F).

Pounce Air-intercept code: "I am in position to engage target".

pound Non-SI unit of mass, abb. lb, = 0.45359237 kg. Note, plural also lb. Strictly, not the same as force or weight of 1 lbf.

poundal Non-SI unit of force, abb pdl, = 0.138255 N = 0.31081 lbf.

pound force Abb. lbf, see *pound weight*.

pounds per square inch Non-SI unit of pressure, abb. lbf/sq in (psi or lb psi not recommended), = 6,894.76 Pa =6.89476 kPa; = 14.696 ata = 2.03596 in Hg.

pound weight Attraction of standard Earth gravity for mass of 1 lb, abb. lbf (lb force), = 4.44822 N. See *pound*.

pour point Temperature established by standard pour test as lowest at which liquid, esp. fuel or lube oil, will flow.

POV Pigmented oil varnish.

POW Prisoner of war.

Powdered Tactical code: enemy aircraft broke up in the air.

powder metallurgy Production of finished or near-finished parts by fusion under heat and pressure of metal in form of finely divided powder. Usually synonymous with sintering.

power Rate of doing work; SI unit is watt (W) = J/s; kW = 1.34012 hp = 1.359621 cv, PS; hp = 0.7457 kW; cv, PS = 0.7355 kW.

power amplification Ratio of AC power at output to AC power at input circuit.

power amplifier Amplifier designed to deliver large output current into low impedance to obtain power gain as distinct from voltage gain.

power-assisted flight control FCS in which power inputs assist pilot by overcoming major part of hinge moments while leaving pilot to move surfaces directly and experience direct feedback, and with difficulty control aircraft in event of system failure.

power brake US term for sheet-metal press, esp. one using mating male/female dies or rubber.

power bumping Sheet-metal forming on bumping hammer.

power by the hour Contract for total support of customer's engines at agreed rate per hour flown. Contractor may be O&M or OEM.

power centroid Point which will be selected by seeker (radar or IR) as centre of target; in case where both real target and decoy (eg flare) are visible, ** likely to be in space between them.

power coefficient In calculating propeller performance by the Drzewiecki method, a grading curve to torque component plotted against blade radius may be drawn for each pitch angle to arrive at function K_Q, constant for each value of advance ratio J; ** equals K_QJ^2, and has symbol Cp. Total power absorbed is $Cp\rho$ n^3 D^5, where ρ is air density, n rotational speed and D tip diameter.

power convertor Apron vehicle or trailer converting 50 or 60-Hz current to 400 Hz.

power density *1* Power per unit volume (electronic device, nuclear reactor).

2 Power per unit area of beam cross-section (radar).

power dive Steep dive with engine at full throttle, suggest no longer relevant in jet era.

power drive unit *1* In floor of hold, usually electric, for positioning cargo.

2 Source of mechanical power for secondary power system, eg to actuate landing gear or flight-control surface, a recent usage.

3 More specifically, device for converting electric, hydraulic or pneumatic power into rotary or linear mechanical power.

powered approach One in which aircraft propulsion is giving significant thrust.

powered ascent That from lunar or other non-Earth surface.

powered controls Powered flight-control system.

powered descent That of Lunar Module or other soft-landing device on to surface other than Earth.

powered flight Flight of vehicle while self-propelled.

powered flight-control system Vehicle flight-control system which uses irreversible actuation such that no manual reversion is possible, though pilot may have manual override over AFCS at input.

powered high-lift system High-lift system in which energy drawn from main propulsion plays direct role, either by flow augmentation, flap-blowing, USB or other technique other than jet lift or use of rotors.

powered lift Lift resulting wholly (or in case of CC/BLC aerofoil, partly) from engine power. Almost all PL aircraft can fly at very low airspeed or hover.

powered-lift aircraft In 2002 FAA was defining rules for what were once called convertiplanes, with VTOL capability but [most] cruising in aeroplane mode.

powered-lift regime Flight regime in which sustained flight at below POSS is possible because lift and control moments are derived from installed powerplants.

powered slat One positively driven to extended position instead of being moved by its own aerodynamic lift.

power egg Complete ECU, usually without propeller (normally piston engine).

power factor *1* That by which product of alternating current and voltage must be multiplied to obtain true load, = cos ϕ where ϕ is phase angle between current and voltage. [θ is also common]; p.f. = P [watts]/VI.

2 Measure of dielectric loss of capacitor.

3 In wind tunnel, ratio of driving power to kinetic energy multiplied by mass flow in working section.

power feedback Feedback in which significant amount of power (electrical, electronic or acoustic) is transferred.

power folding Folding wings by power actuation, usually controlled from cockpit and sometimes by external protected control.

power gain *1* Ratio of power, usually expressed in dB, delivered by amplifier or other transducer to power absorbed by input.

2 In any direction, $4\pi \times$ ratio of radiation intensity sent out by aerial to RF power input; with strongly directional (eg pencil beam) aerial close to zero except on axis of main lobe.

power gearbox   In a turboprop engine, that which drives the propeller.

power head   On large machine tool, cutter and associated drive.

power in, power out   Aircraft parks at airport and departs without use of towing vehicle.

power in, push out   Tug is used to push aircraft from nose-in stand to taxilane.

power jet   In traditional carburettor, fuel jet which comes into operation only when throttle advances beyond maximum-cruise position.

power lever   Throttle, esp. for gas turbine of any kind.

power loading   *1* For propeller-driven aeroplane, W/P_e, total mass (usually taken as MTOW) divided by total installed horsepower, ehp in case of turboprop; units lb/hp = 0.60864 kg kW^{-1}, reciprocal 1.643. Jet equivalent = thrust loading.

2 Input shaft power divided by projected area of propeller [not propeller disc].

power-on spin   One entered from power-on stall.

power-on stall   Normal stall in which propulsive power is maintained at significant (eg normal cruising) level throughout; nose is raised higher to lose speed at acceptable rate and slipstream usually adds to lift, generally resulting in more extreme attitude and more violent pitchover when stall finally occurs.

power overlap   Overlap of firing strokes in multi-cylinder piston engine; for four-stroke engine more than four cylinders are needed to make this concept of continuous power effective.

power performance index   Computerized index of helicopter engine power corrected for atmospheric pressure/temperature, altitude, power-turbine temperature and drive torque.

power plant, powerplant   *1* Those permanently installed prime movers responsible for propulsion, including their number; thus * of L-1011 TriStar is three RB.211 turbofans (some authorities would add 'plus ST6 APU').

2 One prime mover, of any type, in complete form plus accessories, silenced nozzle, propeller and associated subsystem and, in some cases, surrounding cowling. This is a different meaning from (1).

power port   Socket beside passenger seat for laptop or to charge switched-off mobile.

power press   Any press whose actuation force does not derive from human muscle; need not have vertical motion.

power processing unit   Converts electricity from solar panels into power for a Hall-effect thruster.

power rating   Rating (1).

power-recovery turbine   Driven by exhaust gas from piston engine to put extra power into crankshaft or other output.

power section   Gas-generator part of engine, especially where two share a common output gearbox and drive.

power shear   Shear for cutting heavy sheet or plate, with hydraulic ram(s) moving blade.

power spectrum   Plot of Sa(f), signal amplitude against frequency, normally symmetrical about peak at fc and with form varying with modulation.

power/speed coefficient   *1* Function in propeller performance calculation, introduced by F.E. Weick;

$$C_s = \frac{J}{2C_p{}^{0.2}} = \frac{\rho^{0.2}V}{P^{0.2}n^{0.4}}$$

where J is advance ratio, C_P power coefficient, ρ density, V slipstream velocity, P power consumption and n rotational speed.

2 Variation of installed power or thrust with airspeed, esp. for jet engine; normally plotted at constant air density.

power take-off   Shaft from which power is available to drive accessory or other item.

power train   Assembly of gearwheels, shafts, clutches and possibly other items for transmitting shaft power. Generally synonymous with transmission.

power transfer unit   Interconnection between otherwise totally separated hydraulic systems, eg Green and Yellow, which enables power to be transfered from one to the other.

power turbine   Mechanically independent turbine (ie connected to engine only by shaft bearings and gas path) driving propeller gearbox of turboprop or output shaft of turboshaft.

power unit   *1* Power plant (2) (not recommended).

2 Device which either generates electrical or EM radiative power or produces required currents or signals from raw AC or DC input.

3 Source of energy, other than propulsion, for missile, RPV or spacecraft where there is no shaft-driven alternator or other supply.

4 Prime mover of APU or EPU/MEPU.

power venturi   Venturi used to operate air-driven instruments or other devices relying on suction.

PP   *1* Descent through cloud (ICAO code).

2 Present position.

3 Pilot production.

4 Peak-to-peak, also p.p.

5 See *P/P*.

6 Pre-production aircraft.

7 Probability percentage.

P/P   *1* Private, or pupil, pilot.

2 Push/pull.

P$_P$ Polar Pacific.

p.p.   Peak-to-peak.

PPA   *1* Pre-planned attack.

2 Passengers per annum.

3 Pre-production aircraft.

PPAC   Product performance agreement center (US).

PPARC   Particle Physics & Astronomy Research Council (UK).

P/PATM   Passengers per air transport movement.

PPB   *1* Program/planning/budgeting; ES adds evaluation system, S adds system (US).

2 Parts per billion; (M) adds by mass, (V) by volume.

3 Passenger protective breathing; E adds equipment.

PPC   *1* Production possibilities curve.

2 Pulse position code [uplink/downlink].

PPD   Proximity/point detonating.

PPDU   Physical-layer protocol data unit.

PPE   Passengers (per year) per employee.

PPF   *1* Production-phase financing.

2 Payload processing facility.

PPFRT   Prototype preliminary flight rating test.

P/PFRT   Pupil-pilot flight rating test.
PPG   *1* Planning policy guidance.
   2 Powered para-glider.
   3 Pacific Proving Ground[s] (USA).
PPH, pph   Pounds or, less commonly, pints, per hour.
PPhA   Passive phased-array radar.
PPI   *1* Plan-position indicator.
   2 Power performance index.
   3 Photo production and interpretation.
PPIF   Precision photographic interpretation facility (pre-1980, photo processing [or production] and interpretation facility).
PPINA   PPI (1) not available.
PPINE   PPI (1) normal, no echoes observed.
PPIOM   PPI (1) out for maintenance, or inoperative.
PPL   *1* Private Pilot's Licence; (A) adds aeroplane, (AS) airship, (B) balloon, (G) gyroplane/autogyro, (GR) ground examiner, (H) helicopter, (IR) instrument rating, (microlight), (R) examiner, (SLMG) self-launching motor-glider, (X, followed by appropriate suffix) examiner authorized to conduct flight and ground tests (CAA).
   2 Polypropylene.
   3 State airports enterprise (Poland).
   4 Pulsed plasma thruster.
   5 Processor-to-processor link.
PPLI   Precise participant location information.
PPM   *1* Pulse-position modulation.
   2 Pounds, or pages, per minute.
   3 Production performance measurement.
   4 Performance programs manual.
   5 Pre-processor module.
   6 Programmable processing module.
   7 Periodic permanent magnet.
   8 Pneumatic power module.
ppm   Parts per million.
PPMF   Permanent periodic magnetic focusing.
PPMFD, P²MFD   Projection primary multi-function display.
PPMS   Precision power-measurement system (EW).
ppn   Precipitation.
PPO   *1* Prior permission only.
   2 President's Pilot Office.
   3 Position pick-off; U adds unit.
PPOS   Present position.
PPP   *1* Synoptic chart: pressure referred to normal mean sea level.
   2 Public/private partnership (UK).
   3 Point-to-point protocol (computers).
   4 Pulse-pair processing.
PPR   *1* Prior permission required.
   2 Periodic performance report.
   3 Prospective price redetermination.
   4 Plans, programmes, [and] requirements.
PPRM   Power plant recording and monitoring.
PPS   *1* Passenger processing system.
   2 Photovoltaic power system.
   3 Pilot's performance system.
   4 Polyphenylene sulphide.
   5 Parliamentary Private Secretary (UK).
   6 Post-production support.
   7 Precise position[ing] service (GPS).
   8 Provisional project structure.
   9 Precision pointing system.
   10 Policy planning staff.

   11 Propulsion pod system (J-Stars).
   12 Packets per second (see next).
   13 Propulsion prognosis system.
pps   *1* Pulses per second.
   2 Pixels per second.
PPSI   *1* Probe-powered speed indicator.
   2 Pounds per square inch [not recommended].
PPSN   *1* Public packet-switching network.
   2 Present position.
PPT   *1* Perspective-pole track (HUD).
   2 Pulsed plasma thruster.
PPU   Power processing unit.
PPV   Pre-production verification.
PQ   Category, man-carrying target (USAAF 1942–47).
PQAR   Product quality action request.
PQC   Poor-quality cost.
PQE   Product-quality engineer.
PQO   Principal Quality Officer (UK).
PQT   Production qualification and test.
PR   *1* Photo-reconnaissance.
   2 Pitch rate.
   3 Pressure ratio.
   4 Polysulphide rubber.
   5 Ply rating (tyre).
   6 Purchase request.
   7 Periodic reservation.
   8 Precipitation, or primary, radar.
   9 Partial.
   10 Procurement Regulation (US).
   11 Public relations.
Pr   *1* Prandtl number.
   2 Radiation pressure.
PRA   Popular Rotorcraft Association (US, office Mentone, IN).
PRADS, Prads   Parachute/retrorocket air drop system.
PRAIM, Praim   Predictive receiver autonomous integrity monitoring.
PRAM, Pram   *1* Productivity, reliability, availability and maintainability.
   2 Pre-recorded announcement machine [for passengers].
Prandtl-Glauert equation   States that lift, drag or pressure coefficients at high subsonic speeds, where compressibility must be taken into account, are equal to those at lower speeds factored by Glauert factor $(1-M^2)^{-\frac{1}{2}}$ where M is Mach number, ie

$$C_L = \frac{C_L \,[\text{incompressible}]}{\sqrt{1-M^2}} \quad.$$

Prandtl-Glauert law   That describing effect of compressibility on lift of 2-D wing.
Prandtl interference factor   Dimensionless factor σ used in determining interference between wings or biplane; dependent upon ratios

$\dfrac{\text{gap}}{\text{mean span}}$ and $\dfrac{\text{upper span}}{\text{lower span}}$; where both wings have same span, *** $\simeq 0.5$ for gap/span ratio of 0.2 and about 0.375 for gap/span ratio of 0.3.

Prandtl-Meyer expansion   Original treatment of supersonic expansion round corner, in which entropy remains constant but pressure, density, temperature and refractive index all fall. Prandtl solved 2-D case in 1907.
Prandtl number   Ratio of momentum diffusivity to thermal diffusivity, $Pr = \mu \, C_P/\lambda = \nu\alpha$, where μ is viscosity,

C_P is specific heat at constant pressure, λ is thermal conductivity, ν is kinematic viscosity and α is angle of attack.

Prandtl-Schlichting   Original treatment of transition flow between laminar and turbulent boundary layers forming link curve on plot of skin-friction coefficient against Reynolds number.

prang   To have accident, esp. to crash aircraft; also derived noun (RAF, WW2, colloq.). If written on aircraft (PRANG), Puerto Rico Air National Guard.

PRAT, Prat   Production reliability acceptance test.

Pratt truss   Basic braced monoplane.

PRAWS   Pitch/roll attitude warning system.

prayer mat   Portable anti-erosion mat for jet V/STOL operation.

PRC   *1* Program(me) review committee.
2 Performance Review Commission (Eurocontrol).

PRCA   Pitch/roll control assembly.

PRCR   Preliminary request for customer response.

PRCS   Pitch reaction-control system.

PRCTN   Precaution.

PRD   *1* Program review data.
2 Primary radar data.

PRDA   Program research/development announcement (US).

PRDS   Processed radar display system.

pre-balanced assembly   Gas-turbine module held in store until needed.

precautionary flight zone   That beyond left boundary of helicopter h/V (height/velocity) curve, where safe autorotative descent is unlikely or impossible.

precautionary landing   Practice forced landing; various techniques but invariably objective is to arrive at correct point with correct speed for minimum ground run, then overshoot without touching.

precautionary launch   Launch of aircraft loaded with NW from airbase or carrier under immediate threat of nuclear attack, not necessarily on mission to enemy target but to preclude friendly-aircraft (ie, its own) destruction.

precession   Rotation of axis of spinning body, eg gyro, when acted upon by external torque, such that spinning body tries to rotate in same plane and in same direction as impressed torque; ie axis tends to rotate about a line (axis of *) normal to both original axis and that of impressed torque. Horizontal component called drift; vertical called topple.

precession cone   That described by longitudinal axis of slender body, eg ballistic rocket, devoid of attitude or roll stabilization; two cones each having apex at vehicle c.g.

precipitation   Moisture released from atmosphere, esp. that in large enough particles to fall sensibly, ie excluding fog and mist; examples are rain, snow, hail, sleet and drizzle.

precipitation attenuation   Loss of RF signal due to presence of precipitation.

precipitation drag   That caused by slush or standing water on aircraft wheels; snow not normally included in this term.

precipitation hardening   Usually synonymous with ageing (aluminium alloys).

precipitation heat treatment   Artificial ageing, usually preceded by solution heat treatment.

precipitation interference   Static discharge, normally considered to be caused by precipitation and atmospheric dust, which increases ambient noise level, making some reception (eg NDB) difficult.

precipitation static   See previous.

precise encrypted   Accurate GPS using not only L1 (1,575 MHz) but also L2 (1,227 MHz), available only to US and Allied armed forces; abb. P(Y).

precise local fix   An offset (3) near inconspicuous ground target.

precise positioning service   Self-explanatory, based on dual-frequency P-code of GPS.

precision approach   *1* One in which pilot is provided by ground PAR controller with accurate guidance to extended centreline, informed of glidepath interception and thereafter provided with precise az/el guidance together with distance from touchdown at intervals not greater that 1 mile (rarely, 1 km). Hence, * radar.
2 One in which ILS is used.

precision-approach interferometer radar   Low-power ground interrogator near runway triggers replies from airborne SSR transponder received at three azimuth dishes (readout within 0.02°) and four elevation dishes (within 0.03°).

precision-approach monitoring   Use of monopulse *ESPA* radar with fast (1s) refresh rate to give track information accurate enough for landing on parallel runways 700 ft apart.

precision approach path-indicator   Optical aid to holding glidepath. Banks of coloured lights are precisely aligned towards incoming aircraft, red showing too low and white too high [some put green between the two].

precision approach radar, PAR   Primary radar designed specifically to provide accurate azimuth, elevation and range information on aircraft from range of at least 10 miles (16 km) on extended centreline to threshold.

precision bombing   Level bombing directed at specific point target.

precision casting   One in which main surface areas are to finish dimensions, needing only drilling or trimming and surface treatment.

precision-drop glider   Glider, usually of Rogallo type, for delivery of cargo to point target after release from nearby aircraft.

precision instrument runway   One served by operative non-visual precision approach aid, esp. ILS; specially marked with non-precision instrument runway indications plus TDZ markers, fixed-distance markers and side stripes. See *Categories (3)*.

precision landing system   Precise blind landing system based on GPS with civil/military interoperability. (Raytheon).

precision runway monitor   Installation permits simultaneous landings on close parallel runways.

precision spin   Accurately performed spin completed as training manoeuvre: entry must be positive, spin must be fully developed within quarter-turn and spin recovery must take place after prescribed number (or number and fraction) of turns.

precision tool   One not held in hand whose positions and attitudes in relation to workpiece are established to limits closer than engineering tolerances.

precision turn   Training manoeuvre often conducted above straight road or similar reference in which smoothness of entry, quick roll to correct bank attitude, absence of skid, correct rate of turn and proper exit on desired heading are mandatory.

precision velocity update　Use of Doppler tracking of radar ground clutter to determine exact ground speed and drift.

precombustion chamber　Rare ancillary chamber used in start cycle of rocket or ramjet where burning fuel is produced to act as ignition source for main chamber. Also feature of certain diesel or oil engines.

Precomm　Preliminary communications search, start of SAR process.

precursor　*1* Material such as PAN yarn from which by carbonization process carbon or graphite fibre is made.

2 Air pressure wave ahead of main blast wave in nuclear explosion of appropriate yield over heat-absorbing or dusty surface.

predatory pricing　Price structure established by airline on particular route[s], or by manufacturer of hardware, alleged by rivals to be aimed at elimination of competition; such prices are often held to be below direct costs.

predesignated ejection point　Location where pilot can depart in knowledge that aircraft will crash harmlessly.

prediction　Predicted numerical value.

prediction angle　That between LOS to target and gun line when properly pointed for hit on future position.

predictive battlespace awareness　"A commander-driven process to predict and pre-empt adversary actions when and where we choose" (USAF).

predictive gate　Radar range-gate sensitive to hostile frequency-hopping and ECCM.

predictive maintenance　Not mere remote diagnostics but the synthesis of massive amounts of fleet data, overlain with AI plus the remote diagnostics from each item [eg, engine] to forecast and thus prevent any future malfunction or unscheduled maintenance.

predictor　Mechanical computer used (1926–45) to predict future positions of target for AAA.

predominant height　Height of 51% or more of structures within area of similar surface material subject to air reconnaissance.

pre-emptive attack　Attack initiated on basis of incontrovertible evidence that enemy attack is imminent (DoD). Word 'nuclear' absent.

preferential routes　Within an ARTCC preferential arrival (PAR) and departure (PDR) routes link an airport to an en-route point and simplify flight planning.

preferential runway　One suggested or offered for departure to minimize local noise nuisance and afford route away from urban areas even though not optimum from viewpoint of departing traffic.

preferred IFR route　In effect aerial motorways listed to guide long-distance pilots in US and facilitate ATC; divided into high- and low-altitude routes, some involving SIDs or STARs and one list being unidirectional (FAA).

preflight　*1* Pre-flight inspection, and verb to accomplish same.

2 After completion of new aircraft when systems are checked and other work done prior to first flight.

pre-flight inspection　That carried out on aircraft before each flight by flight crew, especially PIC, to broadly standard procedure which varies from type to type. Obvious tasks are to remove removables, check for fluid leaks or superficial damage, inspect tyres and landing gear and have cleared objects that could be blown by slipstream.

On entering aircraft it is then necessary to * cockpit (task sometimes performed by ground crew).

pre-flight line　Parking line for newly completed aircraft where work is carried out prior to first flight; hence pre-flight hangar.

preformed cable　For flight control, one whose individual wires are preformed to final spiral, or spiral-spiral, shape; thus they do not splay out should cable be cut.

prefragmented warhead　One containing large number of high-density steel balls or similar penetrative objects. Term not to be used for continuous-rod or heavy casing scored by grooves.

pre-ignition　Fault of IC piston (esp. Otto) engine in which, due to incandescent carbon or other cause, mixture ignites before spark, and may run after ignition has been switched off.

pre-launch period　Time between receipt of alarm and take-off of manned aircraft, esp. NW carrier.

Prelim　Preliminary data.

preliminary flight rating test　First assessment of pupil's aptitude.

premate preparation　Tasks accomplished before vehicle integration, eg before Shuttle has SRBs and tank added.

Premir　Precise multi-intelligence registration, to fuse multiple images into one display (AFRL).

pre-owned　Used, secondhand.

pre-oxygenation　Breathing pure O_2 prior to high-altitude or space flight, period depending on duration and altitude of flight.

PREP, Prep　*1* Pilot's reference eye position.

2 Part-reliability enhancement programme.

3 Prepare, preparation.

4 Primary-radar engineering program.

prepaid ticket advice　Paid for (usually in local currency) in country of destination and sent as valid ticket to passenger in country of origin. Price structure usually same as in country of origin.

Prepha　Programme for research and study into hypersonic propulsion (F).

preplanned air support　Air support in accordance with a program(me) planned in advance of operations.

preplanned mission request　Request for an air strike or reconnaissance mission on target which can be anticipated sufficiently in advance to permit detailed mission co-ordination and planning.

prepositioned　In place prior to hostilities to ensure timely support of initial operations.

prepreg　CFRP raw material in form of oriented fibres, tows or other reinforcement impregnated with resin and supplied as sheet, strip or other form ready for final cutting and moulding, usually as number of plies with different orientation, to make product. Not relevant to product made by filament winding. Note: terminology bound to spread into boron/epoxy and many other forms of FRC construction.

pre-production aircraft　Incorporates modifications shown to be necessary by testing prototype[s] but still short of final build-standard. Usually not sold to customer.

pre-rotation gear　Landing gear whose wheels are spun-up before touchdown.

pre-rotation vanes　Upstream of fan in wind tunnel to impart swirl which is eliminated by fan.

PRES   Pressure.

Preselect   Autopilot mode to capture desired condition, usually altitude.

presentation   *1* Geometric form, character and style of display, esp. one of electronic nature; eg PPI plus alphanumerics, line plot plus cursive writing or TV picture plus alphanumerics.

  2 Presentation flight.

  3 Putting up a target to be shot at with guns, SAMs or by fighters; each * can be racetrack, snaketrack, straight line or other form.

présentation   The merge (F).

presentation flight   First showing of new type of aircraft to Press/customers/public, usually [unlike demo flight] in sedate flypasts.

preset guidance   Pre-programmed autopilot determining mission in advance.

PRESFR   Pressure falling rapidly.

pre-simulator training   Training on Technamation and similar animated displays, with mainly dummy controls, to ensure familiarity with systems and procedures.

PRESRR   Pressure rising rapidly.

Press   *1* Pacific Range electromagnetic signature study (USAF, DARPA).

  2 Project review, evaluation and scheduling system.

  3 Pressure.

Presselswitch   Pressure selector with manual input.

press fit   Loose term which usually means interference fit.

press tooling   Tooling for use in presses, such as rubber or mating male/female dies for making finished or near-finished parts in sheet; not usually applied to tools for applying curvature to plate.

press to transmit   Pushbutton on control wheel or other inceptor enabling pilot to use R/T whilst flying manually.

press-ups   Jet STOVL training: vertical takeoffs and landings repeated on same spot.

pressure   Force per unit area. The SI unit is the pascal, $1\ Pa = Nm^{-2} = 0.02089\ lbf/ft^{-2}$; $1\ lbf/ft^2 = 47.8303\ Pa$; $1\ kPa = 0.29530\ in\ Hg$; $1\ in\ Hg = 3.38639\ kPa$; $1\ MPa = 0.0675\ UK\ tonf/in^2 = 151.2\ lbf/in^2$; $1\ lbf/in^2 = 6.89476\ kNm^{-2} = 2.03602\ in\ Hg$; $1\ in\ Hg = 0.49115\ lbf/in^2 = 3.38639\ kNm^{-2}$; $1\ in\ H_2O = 249.089\ Nm^{-2} = 5.2023\ lb/ft^2$; $1\ bar = 10^5\ Nm^{-2} = 750.08\ mm = 29.5307\ in\ Hg$; $1\ ata = 101.325\ kNm^{-2} = 1.01325\ bar = 14.6959\ lb/in^{-2} = 760.01856\ mm = 29.922\ in\ Hg$.

pressure accumulator   Device for storing energy in form of compressed fluid, typically by gas (nitrogen or air) trapped above liquid (eg hydraulic system).

pressure altimeter   Conventional altimeter driven by aneroid capsule(s) and measuring not height but local atmospheric pressure.

pressure altitude   *1* Height in atmosphere measured as vertical distance above standard sealevel reference plane defined in pressure terms, invariably 1013.2 mb; thus height indicated by pressure altimeter set to QNE and corrected for IE (2) and PE (2).

  2 Altitude in standard atmosphere corresponding to atmospheric pressure in real atmosphere.

  3 That simulated in pressure (vacuum) environmental chamber.

  4 That at which gas cell(s) of aerostat become full.

pressure-balance seal   Disc carried by gas-turbine shaft

acted upon by internal air pressure to maintain positive forward load on shaft location bearing.

pressure breathing   Respiration of gas (eg O_2 or mixture) at pressure greater than ambient surrounding wearer of mask; hence * mask. Converse of demand breathing.

pressure bulkhead   One sealed to serve as boundary of pressure cabin or pressurized section of fuselage.

pressure cabin   Volume in aircraft occupied by human beings in which pressure is always maintained at or above selected level (eg equivalent to atmospheric height of 2,500 m or 8,000 ft) for comfort of occupants no matter how high aircraft may ascend. Term derives from early (c1930–40) usage when ** was entity installed in fuselage. Today inappropriate for civil transports where entire fuselage is pressurized except for nose tip, extreme tail and cut-out for wing; preferable simply to use adjective pressurized.

pressure chamber   Strictly, one in which environmental pressure is raised above atmospheric; decompression chamber is preferred.

pressure coefficient   Local pressure (eg measured on surface of body such as wing) divided by dynamic pressure, thus $Cp = P/\frac{1}{2}\rho V^2$.

pressure cooling   Cooling of heat-generating device (eg piston engine) by liquid maintained under pressure to raise boiling point.

pressure-demand system   Demand oxygen system supplying at above wearer's local pressure.

pressure differential   Difference in pressure between two volumes, eg pressurized fuselage and surrounding atmosphere, dP.

pressure door   Sealed to form part of boundary of pressure cabin or pressurized cargo compartment.

pressure drag   Drag due to integral (summation) of all forces normal to surface resolved along free-stream direction; most wings in cruising flight have small ** because adverse pressures on and immediately below leading edge are largely countered by helpful pressures over rear portion.

pressure-drop control   Called a * valve, * orifice or * system, this turbofan fuel-control unit comprises sliding variable-aperture orifices moved by a centrifugal governor controlling transmission of primary to main fuel pressure.

pressure error   Instruments using a pitot/static system suffer from *, made up of compressibility error and position error.

pressure face   Side of propeller or helicopter-rotor blade formed by lower surfaces of aerofoil elements, over which pressure is usually greater than atmospheric.

pressure fatigue   Structural fatigue induced in pressurized fuselage by repeated reversals of pressurization stress.

pressure flap   Large fabric inwards- (rarely, outwards-) relief valve(s) in skin of airship to allow air to flow in during descent so that internal pressure shall never be significantly below that of surrounding atmosphere.

pressure garment assembly   NASA terminology for types of space suit, without PLSS.

pressure gradient   Rate of change of pressure along any line normal to local isobar direction, eg in atmosphere or on surface of lifting wing.

pressure head   Combined pitot/static tubes.

pressure height   *1* See *pressure altitude (4)*.

  2 That, related to standard atmosphere, at which gas

cells reach predetermined super-pressure (BSI). Thus, that at which gas release valves open.

pressure helmet   See *pressure suit*.

pressure instruments   Flight instruments operated by air pressure difference, eg ASI, VSI and simple altimeter.

pressure jet   Helicopter tip-drive unit comprising propulsive jet on or near tip of main-rotor blade fed with compressed air supplied along interior of blade. May or may not have provision for combustion of fuel, which if used results in combustion-* or tip-burning.

pressure lubrication   Lube oil supplied under pressure, usually from gear pump, and ducted through drillings and pipes to main bearings and other parts (see *pressure-relief valve system*).

pressure manometer   Liquid-filled U-tube at foot of aerostat gas cell.

pressure-pattern flying   Technique of long-range navigation (1944–60) in which isobar patterns provided basis for flight-planning to make maximum use of following winds.

pressure pump   That supplying fluid under pressure to closed system, eg lube oil (less often, hydraulics).

pressure ratio   Ratio of fluid pressures between two points in flow, notably between stages of axial compressor (PRPS, per stage) or between inlet and delivery of compressor spool or entire compressor system (eg fan, IP compressor, HP compressor). For piston engines, term is compression ratio.

pressure recovery   Also called recovered pressure, that generated in flow through duct by conversion of kinetic energy, esp. after addition of heat as in cooling radiator. When heat added can exceed stagnation pressure.

pressure-relief valve system   Oil supply, eg for gas-turbine engine, in which delivery oil pressure is opposed by relief valve backed by spring plus atmospheric pressure via oil tank; as engine rpm increases, delivery pressure eventually forces relief valve open, spilling excess back to tank, thus holding steady pressure and flow to bearings at all flight engine speeds.

pressure rigid airship   One combining features of rigid and non-rigid design to maintain shape and skin tension.

pressure seal   Inflatable seal between rigid members which is pressure-tight when inflated but when deflated offers no resistance to relative motion of parts; universal in pressurized doors and canopies.

pressure stabilized   Having structural rigidity wholly dependent on maintenance of positive internal pressure, eg Atlas missile.

pressure structure   See *inflatables*.

pressure suit   Suit which, when rigid-facepiece helmet is attached, completely encloses wearer's body and within which gas pressure is maintained at level suitable for maintenance of bodily function (apart from addition of comment on necessity of adding helmet, this is DoD, NATO). Also called full-pressure suit. Some US definitions call this a pressurized suit, to be distinguished from ** because, according to these authorities, ** merely 'provides pressure upon the body . . . a pressure suit is distinguished from a pressurized suit, which inflates, although it may be fitted with inflating parts that tighten the garment as ambient pressure decreases'. This is partial-**, distinction must also be drawn between **, g-suit and water suit.

pressure surface   Curved and irregular surface corresponding to particular atmospheric pressure at any time;

usually plotted for 1,000 (often part below Earth's surface), 900, 800, 700, 500, 300 and 100 mb (see *contour chart*).

pressure switch   One actuated by reaching preset pressure, esp. that in starter circuit triggered by fuel pressure.

pressure test   Invariably means response of operative device or, esp., inflated structure, eg fuselage; objective not to measure pressure.

pressure thrust   Product of pressure difference between rocket exhaust pressure and ambient pressure multiplied by area of nozzle; value may range from negative at SL to positive at high altitude. Rocket thrust = ** plus momentum thrust.

pressure transducer   Device, based on at least eight principles, for generating output current or signal proportional to fluid pressure or pressure difference.

pressure tube   Tube fitted to aerostat envelope or gas cell to which pressure gauge may be coupled.

pressure vessel   Container for gas under (usually high) pressure.

pressure waistcoat   Pressurized covering over torso, as much to resist g as to facilitate breathing at high altitude.

pressurization   Inflation: increasing pressure in closed system. Term usually means pressurized fuselage; see *pressure cabin*.

pressurized blade   Patented (Sikorsky) technique for early warning of crack in helicopter main-rotor blade primary structure by inflating with dry nitrogen or other gas and sealing; loss of pressure signals cockpit lamp or other alarm.

pressurized feed   Supply of liquid rocket propellants under gas pressure.

pressurized ignition   Piston-engine ignition system sealed and held close to sea-level pressure to reduce arcing and electrode wear at high altitude.

pressurized tank   Tank for liquid expelled by pressurizing gas.

Pressurs   Pre-strike surveillance/reconnaissance system.

pre-stage   Intermediate stage of operation in start-up of large liquid-propellant rocket in which propellants are fed to main chamber by main turbopumps and ignited, but with propellant flow less than 25% of maximum; once proper combustion has been confirmed, main-stage operation is signalled (ie allowing full flow).

Prestal   Specially formulated aluminium alloy with exceptional elongation and malleability for large-deformation presswork.

pre-stall buffet   Aerodynamic buffet induced by turbulence over wing and/or control surfaces or fixed tail giving warning of imminent stall.

pre-stocked site   Storage of POL/ordnance at remote site which might in emergency (crisis or war) be operating base for V/STOL aircraft.

Prestone   Trade name of family of coloured and colourless ethylene-glycol coolants.

prestretched cable   Control cable loaded to 60% uts for 3 min. immediately before installation.

prestrike recon   Mission undertaken to obtain complete information on previously known targets.

pre-swirl nozzles   Ring of curved stators which impart swirl to cooling air fed to turbine disc and reduce temperature and pressure.

pre-tensioning   Tightening a bolt to such a degree that it is always under tension, reducing stress fluctuation in service.

pre-TR Rapid-switching cell which protects radar transmit/receive tube from undesirable input.

prev Previous.

prevailing visibility Horizontal distance at which targets of known distance are visible over at least half horizon; determined by viewing dark objects against horizon sky by day and moderate-intensity unfocused lights by night. Not necessarily RVR.

preventive advisory Resolution advisory instructing pilot to avoid certain deviations from established vertical velocity (TCAS).

preventive maintenance Systematic inspection, detection and correction of incipient failures either before they occur or before they develop into major defects (DoD).

preventive perimeter Outer perimeter formed during emergency security operations by stationing aerospace security forces at key vantage points and avenues of approach to vital areas of base (USAF).

preview Brief series of tests by military customer of new type of aircraft to assess potential. No longer common.

PRF *1* Pulse-repetition, or recurrence, frequency.
2 Primary reference fuel.

PRFD PRF (1) distribution.

PRFJ PRF (1) jitter (2).

PRFS PRF (1) stagger.

PRG *1* Program review group (US).
2 Photo-Reconnaissance Group.
3 Private pilot, rotorcraft, gyroplane (FAA).

PRH Private pilot, rotorcraft, helicopter (FAA).

PRI *1* Pulse-repetition interval, ie gap between pulses.
2 Primary.
3 Primary rate interface.
4 Performance Review Institute (Int.).

Pride Pulse-repetition, or recognition, interval, de-inter-leaving.

PriFly Aircraft-carrier Air Commander in Primary Flying Control.

PRIM, Prim Flight-control primary computer.

primary *1* Primary body.
2 Primary glider.

primary air/airflow *1* That mixed with fuel in primary combustion zone.
2 Airflow through engine core.

primary aircraft authorization *1* Total number of particular aircraft type procured in current FY (DoD).
2 Number of aircraft assigned to inventory of squadron or other unit (USAF).

primary alerting system Leased-circuit voice communications network formerly linking SAC HQ to operating squadrons of missiles or bombers.

primary area Published area along airway within which obstacle clearance is provided.

primary body That around which satellite orbits or to/from which spacecraft escapes or is attracted.

primary circulation Atmospheric circulation on gross planetary scale.

primary combustion That immediately surrounding burning fuel, using only minor fraction of total airflow; in gas turbine most air is used in secondary region for diluting and cooling.

primary configuration That in which weapon system is

delivered or in which its primary mission capability is contained (USAF)

primary control See *primary flight controls*.

primary cosmic rays Those reaching Earth from outer space.

primary defect rate Failures per unit time, usually 10^5 flight hours.

primary effects Of NW: blast, radiation, heat, EM pulse.

primary flight computers Usually three, for redundancy, link inceptors to surface power units via computers/ACE (10) units to implement control laws, protect aircraft and minimize pilot workload.

primary flight controls Those providing control of trajectory, as distinct from trimmers, drag-increasers and high-lift devices. Conventionally ailerons/spoilers (where latter are used for roll), elevators and/or tailplane (or fore-plane) and rudder. DLC usually excluded.

primary flight display Electronic cockpit instrument telling pilot everything he needs to know in vertical plane, to enable him to fly the aircraft. Partnered by ND, for horizontal plane.

primary frequency R/T frequency assigned to aircraft as first choice for air/ground communications in an R/T network.

primary glider Strong and simple training glider in which no attempt is made to achieve soaring capability. Often consists of a skid on which is a lattice-frame linking wing and tail.

primary heat-exchanger That which removes heat from source and rejects it to atmosphere or to a secondary circuit.

primary holes Those through which primary air enters gas-turbine combustor.

primary inspection Minimum scheduled periodic lubricating and servicing check applied to aircraft and its removable equipment, including examination for defects and simple functional testing of systems such as flight controls, radio and electrics (ASCC, qualified by 'UK usage').

primary instability Failure of strut or other compression member through buckling (transverse movement near centre).

primary instruments Those giving basic information on flight trajectory and airspeed, eg traditionally horizon, turn/slip, ASI, VSI and altimeter. Heading information not included.

primary lighting Usually that essential for safe flight; cockpit lighting, navigation, rotating beacon, icing, formation.

primary member Any part of primary structure, though not ususally applied to sheet or machined skin.

primary modulus of elasticity That for sandwich or other multi-component material below yield point (if applicable) of weaker component.

primary nozzle *1* That through which main flow of fuel passes in fuel burner or injector.
2 That for primary air to combustor or airflow through pipe of vaporizing burner.

primary radar One using plain reflection of transmitted radiation from target, as distinct from retransmission on same or different wavelength.

primary runway Used in preference to others whenever conditions permit.

primary space launch One starting from Earth.

primary stress Basic applied tension or compression, as distinct from stresses induced by deflection of structures such as buckling.

primary structure That whose failure at any point would imperil safety.

primary surveillance radar In ATC system, determines aircraft range and azimuth in controlled airspace.

primary target Main target of surface-attack mission.

primary trainer That on which pupil pilot begins flight instruction, also called ab initio, elementary (but not basic).

primary zone See *primary combustion*.

PRIME, Prime *1* Precision recovery including manoeuvrable re-entry.

2 Products comprising interdependent mechanical and electronic parts.

prime See *prime contractor* or *priming*.

prime airlift Number of aircraft of force that can be continuously maintained in cycle: home base, on-load base, off-load base, recycle.

Prime Beef Worldwide Base Engineer Emergency Force for direct combat support or to assist recovery from natural disasters (USAF).

prime contact That for entire responsibility for design, development and (usually) assembly and test of complete functioning system, eg aircraft, missile etc. Manufacture, esp. in production, may be assigned by prime contractor to other companies or shared by consortium. An essential is control of programme management. (Strict DoD definition: any contract entered into directly by DoD procuring activity).

prime contractor That awarded prime contract.

prime crew *1* Crew of strategic bomber/tanker/ICBM flight with long experience of combat duty.

2 That assigned to fly space mission (see *back-up crew*) (NASA).

prime maintenance organization A maintenance partner to shoulder airline's engineering responsibilities.

prime meridian Longitude 0°.

prime mover *1* Source of mechanical energy, eg engine.

2 In military sense, surface vehicle designed chiefly for towing and normally accommodating crew, ammunition etc.

primer Subsystem, usually energized by hand-pump in cockpit, for spraying fuel into piston-engine inlet manifold (rarely, elsewhere) to facilitate starting from cold.

priming *1* Use of primer.

2 Sensitive high explosive for detonating main charge.

priming pump Primer.

principal axes *1* Traditionally, those orthogonal axes which eliminate product of inertia terms from equations of motion. Modern computers make these redundant.

2 Axis of relative wind.

3 Principal inertia axis.

4 Rectilinear axes in plane of cross-section of structural member about which moment of inertia is maximum and minimum.

5 Loosely, any reference axes OX/OY/OZ.

principal axis of symmetry In most aircraft, the longitudinal axis, OX or X-axis.

principal inertia axis That line passing through c.g. and usually in place of symmetry about which long, slender body tends to rotate when rolling; in case of aircraft actual axis of rotation normally lies between this and wind axis.

principal parallel On oblique photograph, line through principal point parallel to true horizon.

principal plane On oblique photograph, vertical plane containing principal point, perspective centre of lens and ground nadir.

principal point On oblique photograph, foot of perpendicular to photo plane through perspective centre, usually determined by joining opposite collimating or fiducial marks.

principal site concept All prototypes of new combat aircraft do entire flight-test programme at one customer central site under close customer control (USN).

principal tensile stress For cutout in pressurized cylinder, that along axis where stress is maximum, hence *** factor, = maximum ***/hoop stress. For neutral hole (ellipse in proportions $2/\sqrt{2}$) *** = hoop stress.

principal vertical On oblique photograph, line through principal point perpendicular to true horizon.

Prind Present indication.

printed circuit Electric or electronic circuit formed by deposition of conductive and/or semiconductive material on insulating base; many techniques and many substrates, eg thin/thick film, foil etc.

printed-circuit board Printed circuit on rigid substrate provided with multiple plug-in or other contact terminals and on which are mounted discrete devices.

printed communications Telecommunications network providing printout at each terminal of all messages.

print reference Identifying reference of each air-reconnaissance photograph.

print-through Unwanted transfer of strong signals from one part of magnetic tape to another pressed against it on spool.

priority designated Two-digit number from 01 to 20 resulting from combination of assigned force/activity designator and local urgency-of-need designator (USAF).

priority induction Immediate transfer to USAF of Air Force Reserve members who fail to participate in training.

prior permission *1* That which must be granted by appropriate national authority before start of flight(s) landing in or flying over territory of nation concerned.

2 Specific permission to land at particular airfield within particular times.

Prism *1* Programmed real-time information system for management (Northrop/USA).

2 Panchromatic remote-sensing instrument for stereo mapping.

3 Photo-reconnaissance intelligence strike module.

prisoner nut Lock nut.

private aircraft One owned by private pilot.

private flight One made by private pilot from one airfield to another [see *pleasure*].

private pilot One licensed by national authority to fly particular type, class or group of aircraft without payment, and precluded from carrying fare-paying passengers.

private venture Major product designed, developed and tested at company risk, esp. in case of military aircraft not requested by government.

PRJMP Pressure jump.

PRKG Parking.

PRM *1* Presidential review memorandum.

2 Precision runway monitor[ing].

3 Proposed rule-making.

PRMD   Pilot's repeater map display.

PRN   Pseudo-random noise.

P-RNAV   Precision RNAV, accurate $\geq$ 1 nm (1.8 km), adequate for TMAs.

PrNK   Pritsino navigatsionniy kompleks. Nav/attack system (USSR, R).

PRNSA   Pseudo-random noise signal assembly.

PRO   *1* Anti-rocket (ie anti-ICBM/SLBM) forces (USSR).

2 Procedures and Requirements Overview (ICAO).

3 Public Record Office (Kew, UK).

PROAR   Area forecast, height indicated in pressure units (ICAO).

Pro-ATN   Prototype aeronautical telecoms network (Euret).

Proavia   Paris-based international airport trade organization, 39 members.

PROB   Probable, probability, of weather.

Proba   Project for on-board autonomy.

probability percentage   Likelihood of prediction in Met. report; only two are used, 30% and 40%.

probable   *1* See *probably destroyed*.

2 Qualifying term in photo interpretation where facts point to object's identity without much doubt (ASCC).

probable errors   In range, deflection, height of burst etc, those which are exceeded as often as not (DoD).

probably destroyed   Assessment on enemy aircraft seen to break off combat in circumstances which lead to conclusion it must be a loss, though not actually seen to crash (DoD, NATO).

Probag   Installation for heat-shrink-wrapping baggage.

probe   *1* Instrument boom; see * *errors*, * *parameters*.

2 Rigid receiving tube for fuel passed by flight-refuelling drogue.

3 Any device used to obtain information (esp. quantified) about environment, esp. unmanned instrument-carrying spacecraft.

4 In particular, a sensor which leaves an orbiter and descends to planetary surface.

5 Loop or straight wire for coupling to waveguide and extracting energy from electric or magnetic component of radiation.

6 Transducer which converts shaft speed, sensed as sinusoidal magnetic field from toothed wheel on shaft into alternating signal current; also called MSP.

probe and drogue   British (Flight Refuelling) method of refuelling aircraft in flight.

probe errors   Those originating in probe (1), as far as possible corrected by ADC.

probe parameters   Subject to particular installation can include AOA, yaw/sideslip, total pressure, total temperature and static pressure.

PROC   *1* Procedure.

2 Procurement.

procedure alpha   Ceremonial manning of flight deck of carrier, eg when entering harbour, often with crew arranged to spell vessel's name or a slogan.

procedure manoeuvre   Accurately flown and/or timed flight manoeuvre for identifying or ATC purpose.

procedure turn   Flight manoeuvre in which turn is made at constant rate away from track followed by constant-rate turn in opposite direction to enable aircraft to capture and hold reciprocal; precision manoeuvre used in radio range, holding over point fix prior to joining stack, joining

ILS wtihout radar vectoring, or in simulator training of traditional (Link) type. Designated left or right depending on first turn direction.

process annealing   Heating ferrous alloys to below critical temperature followed by cooling in air or other medium.

Procon   Protocol convertor.

Procod   Production cost by drawing number.

Procru, ProCru   Procedure oriented crew-station model.

procurement   Process of obtaining personnel, services, supplies and equipment (DoD).

procurement lead time   Interval in months between initiating procurement action and receipt into supply system of production article; composed of sum of administrative lead time and production lead time (DoD).

prod   Make probe/drogue inflight-refuelling contact (colloq.).

product definition data interface   Standardizes digital descriptions of part properties and configurations required by manufacturer.

product improvement   Significant change in design of hardware to improve desirable features, and marketed as such; usually initiated to meet market need, either because of competition or to rectify deficiency.

production   Ideal * is manufacture of successive items which are identical, also called series *. With complete aircraft successive items tend to differ in furnishing, equipment and even engine type.

production base   Total national production capacity available for manufcture of items to meet material needs (DoD).

production investment   Funding for tooling, or the cost of same.

production lead time   Time between placing contract and receipt of hardware; subdivided according to whether contract is initial or reorder.

production phase   Period between production approval until last item is delivered and accepted.

productive potential   Payload multiplied by range.

productivity   In airline performance, traffic units (eg LTM) generated per hour per aircraft of particular type or per employee.

product support   Assistance provided by manufacturer to all customers throughout period of product's use in form of training, publications, spare parts, modification kits, product-improvement and immediate response to difficulties.

PROF   Profile.

Profi   Product, or project, financing (multinational).

proficiency student   One on refresher course.

proficiency training   Flight training for desk-bound flight personnel.

profile   Outline of body in side elevation; many submeanings. Common use is to describe shape of cross-section of wing or other aerofoil section. Outboard * is external shape of body, esp. aircraft fuselage. Inboard * is longitudinal cross-section showing how interior is utilized. Flight * is orthogonal projection of flightpath on vertical surface containing nominal track showing variation in height (either AGL or AMSL) along straight bottom axis representing track.

Profiledata   Library of NC-machining software (Ferranti).

profile descent   Uninterrupted from TOD (1) to interception of glide-path.

profile drag   Total drag minus induced drag; sum of form drag and surface-friction drag. One interpretation suggests 'drag of wing with camber and twist removed'.

profile-drag power loss   That expended in overcoming total profile drag of propeller blades.

profile line   Profile of terrain, eg near airport drawn as map inset.

profile milling   Milling variable profile (surface levels) in plate, eg in machined or integrally stiffened skin panel.

profile template   Template for hand-guided machine tool, eg router.

profile thickness   Maximum distance between upper and lower contours of aerofoil profile, each measured normal to mean line; essentially wing thickness.

profilometer   Instrument for measuring surface roughness.

PROG   *1* Prognosis, prognostic, progress.

2 In particular, progress page on MCDU.

prognostic chart   Forecast of meteorological elements for specified time and location depicted graphically.

prognostics and health monitoring   System assigned to notice small change in component performance and trigger maintenance action to prevent failure.

prograde   Direct, or progressive orbit; satellite launched into Earth orbit with inclination from 001° to 179°.

program   *1* Vast topic, covered in EDP (1) dictionaries; simplest definition is group of related instructions which when followed by computer will solve a given problem.

2 Programme (US).

program acquisition unit cost   Includes average procurement unit cost plus the unit slice of the total R&D cost since program inception. These are sunk costs, and do not affect price paid for production aircraft (DoD).

Program Aircraft   Total of active and reserve aircraft (USAF).

programmable   Capable of being controlled by different programs by change of software.

programmable display generator   Generator of 3-colour raster formats plus calligraphic symbology, all under variable software control.

programme   Life history of major project, typified by such * milestones as definition of requirement, feasibility study, project definition, engineering design, hardware manufacture/development/flight test, flight development, service clearance (eg qualification or certification), production, modification and product-improvement, fault-rectification, phaseout.

program(me) element   Portion of giant system (eg, SDI) which is broken down to facilitate assignment of tasks to companies or participating nations.

programmer   Human being engaged in programming.

programmer comparator   Versatile automatic testing station providing serial evaluation of all kinds of analog and digital signals, eg in checking avionic systems in aircraft, missiles or spacecraft.

programming   Art of preparing set of terms and instructions which EDP (1) machine can understand and obey and which when followed by that machine will result in solution to problem for which program was written.

progressive burning   See *progressive propellant*.

progressive die   One which performs series of operations (usually on sheet metal) at successive strokes of press.

progressive feel   Artificial feel proportional to dynamic pressure.

progressive orbit   See *prograde*.

progressive powder   Solid fuel, usually not gunpowder, which burns increasingly fast as combustion pressure increases.

progressive propellant   Solid rocket-motor grain so shaped and ignited that area of combustion, and hence speed of burning, rate of consumption and thrust, all increase throughout period of burn. Any radial burning technique tends to be * unless original grain has deep star centre such that combustion area is constant (neutural burning).

progressive servicing   Servicing performed at military airbase where major tasks are subdivided into sections performed at times fitting in best with operational readiness requirements.

progressive stall   Ideal stall quality where breakdown of wing flow occurs gradually, with well-signalled symptoms; to obtain it a breaker strip or fence may be needed.

progressive strip   See *Flight progress strip*.

prohibited area   Airspace of defined dimensions identified by area on surface within which flight by aircraft is prohibited, usually for reason of national security or to safeguard wildlife. Height ranges from surface to published value such as 4,000 ft or 18,000 ft (FAA).

Proj   Projection.

project   Planned undertaking of something to be accomplished (DoD).

Project Blue Book   Official dossier on UFOs (USAF).

project cycle   Life history of platform or weapon system [concept, not individual examples].

project design   Programme phase in which design is refined by evaluating alternative choices, making performance/capability/cost trade-offs and ultimately arriving at optimized configuration on which engineering design can begin. Work possible in this stage includes tunnel testing, cockpit mock-up improvement and basic systems design, but excludes stressing (detail engineering design).

projected area   Area projected from 3-D surface to plane, or from one plane to another.

projected blade area   Area of propeller or other blade projected on to plane normal to axis of rotation; solidity is not based on this but on total area.

projected flightpath   That which aircraft, esp. aeroplane, will follow in immediate future in absence of further disturbance.

project engineer   Engineer assigned to oversee design and technical management of specific project, reporting to chief engineer or v-p engineering.

projectile velocity   Resultant of muzzle and aircraft velocities.

projection   In cartography, any systematic arrangement of parallels and meridians portraying quasi-spherical planetary surface on plane of map.

project officer   Military or civilian individual responsible for accomplishment of project; usually limited-duration appointment and not one already established within organizational and supervisory channels (USAF).

projector   *1* Illuminating source sending out pencil beam of visible light, eg vertically up at cloudbase.

2 Long-dash broken line to show projection of line or

surface from one plane to another in engineering drawing (drafting).

projector (approach) sight　Mirror sight or similar landing guidance optical system on carrier.

proliferation　Spread of NWs to additional nations.

PROM, Prom　Programmable read-only memory.

promethium　Pm, radioactive metal, among other things used in small batteries, MPt 1,168°C, density 7.2.

prominent target　One which predominates over chaff and other decoys.

Promis　Procurement management information system.

promulgated　Published openly; eg * in ACIs, or VOR beacon * range, in latter case * in Air Pilot and various flight guides.

prone　Lying down; invariably * pilot positions are supported mainly on front of torso and thighs at angle of about 20° with head up to look ahead and toes supported in rear pedals.

Pronto　*1* Code: as quickly as possible (DoD).
2 Program for NC tool operation, acronym.

prontour　Chart used to forecast future pressure-surface contours.

prony brake　Simple mechanical peripheral-band brake whose torque, multiplied by shaft speed, gives brake horsepower.

proof factor　Arbitrarily assigned factor of safety imposed above proof load; for UK civil aircraft * is 1, thus proof and limit loads are same; for UK military aircraft * is 1.125, thus proof load is $1/_8$ higher than limit load. See *ultimate factor*.

proof load　Design limit load × proof factor. Maximum which primary structure is designed to bear whilst remaining serviceable. This vague definition survives in official publications, yet does not mention fatigue effect of repeated loading.

proof of concept article　Prototype.

proof positive/negative　Two sets of proof loads established for particular aircraft type and demonstrated in static test.

proof strength　That required to survive proof loads (pos/neg).

proof stress　Stress at yield point.

prop　Propeller, or aeroplane with propeller[s] (colloq.).

propaganda balloon　Free balloon carrying propaganda leaflets scattered at timed intervals when prevailing wind is expected to carry it over enemy cities.

propagation constant　Complex quantity of plane wave; real part is attenuation constant (nepers/unit length) and imaginary part is phase constant (radians/unit length).

propagation error　In ranging system, algebraic sum of propagation velocity error and (important at long ranges and low angles) curved-path error.

propagation rate　Linear velocity of structural crack.

propagation ratio　Between two points in path of plane wave, ratio of complex electric field strength.

propagation velocity　For EM wave (light, radio) in vacuum taken to be 2.997925×10^8 m/s.

propagation velocity error　Difference between assumed and effective velocities over ray path.

propane　Gaseous hydrocarbon of paraffin series, $CH_3Ch_2CH_3$, BPt −45°C.

prop banners　Sleeves, usually bright Day-Glo colour, announcing (eg) FOR SALE, FOR RENT, slipped over blades.

propellant　Medium used for propulsion, as in * charge of gun ammunition or material burned to form jet of rocket. Rocket * can be solid, liquid or gas, or combination. Where two are mixed rocket is bi-*, common mixture being fuel plus oxidant (oxidiser). Where catalyst is consumed and adds to jet this also is *. In uncommon case where single liquid is used rocket is mono-*. In rockets and thrusters where no chemical combustion takes place preferable to use term 'working fluid'.

propellant mass fraction　See *mass ratio*.

propellant specific impulse　See *specific impulse*.

propellant volume　Total volume occupied by propellant, esp. solid grain, Vp.

propeller　Rotating hub with helical radial blades converting shaft power into aerodynamic thrust. Shaft power provided by human or prime mover. Tip-drive * possible (would require modified definition). Most existing definitions state 'power-driven', implying use of engine. Left and right-hand rotation respectively mean anti-clockwise and clockwise seen from behind. Can be pusher or tractor, latter at one time often being called airscrew (now arch.). Types of propellers (co-axial, reverse-pitch etc) are covered separately (see also *propulsor*). Similar screw for converting energy of slipstream into shaft power is windmill or RAT (ram-air turbine). Note : in 1930–50 UK usage resulted in a 1939 glossary having the entry 'Propeller : colloquial term meaning "airscrew".'

propeller area　Usually means total area of blades obtained by integrating total of areas of elementary chordwise slices, taking blades as having no thickness; ie each slice is projected in plane of its local chord. Essentially same as outline area of blades with twist removed.

propeller balance stand　Trestle having two horizontal and parallel steel knife-edges (about 1–3 mm radius) on which propeller can be balanced on short slave-shaft.

propeller bar　Handtool used to loosen or tighten main retaining nut on many lightplane propellers.

propeller blade　Thrust-generating aerofoil of propeller.

propeller blade angle　Except in feathered position (when close to 90°) acute angle β between chord line of blade measured at standard radius and plane of rotation, latter being normal to axis of rotation.

propeller blade-width ratio　Ratio of widest chord to propeller diameter.

propeller brake　Brake to stop rotation of propeller after engine shut-down, either to speed passenger disembarkation or, in case of free-turbine engine, to prevent prolonged windmilling with aircraft parked.

propeller camber ratio　Ratio of blade maximum thickness to chord at any station.

propeller cavitation　Generation of near-vacuum on suction face near tip at high Mach numbers (not necessarily at high flight speed).

propeller characteristic　Fundamental curve of V/nD (velocity of advance divided by rpm × diameter) plotted against Cc (speed/power coefficient) (see *Weick*).

propeller disc　Circular area swept out by propeller.

propeller efficiency　Useful work expressed as thrust imparted divided by power input; thrust hp/shp = thrust × slipstream velocity divided by $2\pi nQ$ where n is rpm and Q is drive torque. Some authorities cite two values of **, one called net (net thrust hp/shp) and the other propulsive (propulsive thrust hp/shp).

propeller governor   Usually simple centrifugal governor which keeps shuttle valve oscillating about null position feeding oil to increase or decrease rpm and thus hold rotational speed constant irrespective of aircraft forward speed.

propeller hub   Central portion of propeller carried on drive-shaft, usually made as separate unit into which blades (with simple fixed-pitch propeller, whole propeller) are inserted.

propeller interference   Aerodynamic effects, mainly drag, of bluff bodies immediately downstream, eg radiator or cylinders.

propeller pitch   The angular setting of the blades of a propeller. The *blade angle* (θ) is the angle between the chord at any element (station) on the blade and the axis of rotation. The *helix angle* (ϕ), also called the *angle of advance*, is the angle between the actual direction of motion (velocity) at any element relative to Earth and the actual direction of motion relative to the aeroplane; it is equal to the blade angle minus the angle of attack (α). The *geometric pitch* (p or P, rarely H) is a linear dimension equal to the distance any blade element would move forward in one revolution in absence of any slip ($= 2\pi \tan \phi$); it is likely to be almost constant from root to tip. *Experimental pitch*, also called *ideal pitch*, is the distance the propeller would move forward in one revolution when giving neither thrust nor drag. Actual pitch, also called *effective pitch*, or *practical pitch* or *advance per revolution*, is the distance travelled forward relative to the atmosphere (not to the slipstream). Standard pitch, also called nominal pitch, is the pitch angle at *standard radius*. The *pitch ratio* is the ratio of geometric pitch divided by the circular distance travelled by the blade tip (circumference). The *effective pitch ratio* is the ratio of effective pitch divided by the tip circumference, often expressed as $V/\pi nd$ where V is TAS and n rotational speed in compatible units. The ratio of effective pitch divided by ideal pitch, expressed as a percentage, is called *slip*. Other characteristics include *power coefficient* $P/\rho n^3 d^5$, where P is shaft power and ρ air density, and *thrust coefficient* $T/\rho n^2 d^4$, where T is thrust. All the above apply to a fixed-pitch propeller. When pitch is variable, *fine pitch* enables maximum power to be achieved for takeoff, *coarse pitch* enables high TAS to be achieved with fuel economy in cruising flight, *reverse pitch* sets the blades at a negative angle of attack to shorten the landing run, with the engine delivering high power, and *feathering pitch* sets the blades edge-on to the oncoming air so that, with the engine inoperative, torque imparted by the air to the inner part of each blade exactly neutralizes that over the outer part, thus stopping rotation.

propeller rake   See *rake (3, 4)*.

propeller root   On a simple fixed-pitch propeller, where a blade joins the hub; on a propeller with separate inserted blades, where the blade joins the shank.

propeller rotational speed   Angular velocity, denoted by n [usually rps] or ω[rad/s].

propeller shaft   That on which propeller is mounted.

propeller shank   The innermost part of each blade, normally inside the spinner, gripped by the structure of the hub and incorporating the drive for changing pitch.

propeller solidity   See *solidity*.

propeller speed   *Propeller rotational speed*, but often rpm.

propeller standard radius   In US by custom, 75% of tip radius; elsewhere, usually 66.6%, symbol P_s.

propeller state   Normal condition of helicopter main rotor in which thrust is in opposite direction (upwards) to flow both through and outside rotor disc.

propeller thrust   That imparted by the aerodynamic foce on the blades through the hub to propel the aircraft, symbol T, normally = $P\eta/V$ ($\times$ 550 in Imperial measures), where P is shaft power, η propeller efficiency and V true airspeed.

propeller tipping   Metal skin on tip and outer leading edge of soft (eg wood) blade.

propeller torque   Torque imparted by propeller drive shaft and reacted by aerodynamic rolling moment of aircraft (usually to some degree in-built) or asymmetric load on left/right main gears on ground; symbol Q.

propeller turbine   See *turboprop*.

propeller wash   Slipstream, esp. on ground; also called prop blast; see *propwash*.

propeller width ratio   Product of blade-width ratio and number of blades; akin to solidity.

propelling nozzle   That at exit from fluid jet system used for propulsion, esp. turbojet or ramjet; for supersonic flight variable in shape (con-di) and area.

propfan   *1* Advanced propeller for use at high Mach numbers, characterized by having six to 12 blades each with thin, sharp-edged lenticular profile and curved scimitar shape, overall solidity exceeding unity and loading being high. Can be tractor or pusher, shrouded or open, and for highest efficiency at about Mach 0.8 has two contra-rotating units.

2 Complete engine whose thrust is generated by (1).

propjet   See *turboprop* (colloq.).

proportional control   Effect at output, eg surface movement, is proportional to input, eg stick movement; opposite of flicker or bang-bang.

proportional flow control   A fuel control system for large turbofans in which main flow is adjusted by precision control of a small parallel flow [often incorporating a pressure-drop control connected to a kinetic valve].

proportional navigation   Control of trajectory in order to home on target by changing course by several times (typically 3.5) rate of change of sightline to target; thus angular rate of velocity vector is proportional to angular rate of line of sight to target.

proposal   Formal and comprehensive document in which manufacturer sets out before government procurement officials complete technical specification and performance of proposed item, including timing and prices of development and, possibly, production programme.

proprioceptive   Pertaining to stimuli produced within body, esp. human body, by proprioceptors.

proprioceptor   Internal receptor for stimuli, such as tendon tension, originating in somatic organ. Body balance maintained by * plus eyes and ear laybrinths.

proprotor   *1* Large propeller[s] of VTOL aircraft whose axis can be rotated 90° to give lift.

2 Convertiplane with fixed-axis propeller[s] and lifting rotor.

prop strike   Tips of propeller[s] hitting ground, eg on takeoff.

propulseur d'appoint   Strap-on-booster (F).

propulsion efficiency   Term not recommended (see *propeller efficiency*, *propulsive efficiency*).

propulsion integration   The aerodynamic integration of the propulsion system into the air vehicle, ignoring physical [structural] considerations.

propulsion system   Sum of all components which are required to propel vehicle, eg engine, accessories and engine-control system, fuel system, protection devices, inlets and cooling systems.

propulsive duct   Not recommended; usually means pulsejet or subsonic ramjet.

propulsive efficiency   *1* Broadly, energy imparted to vehicle as percentage of energy imparted to jet (from propeller or other prime-moving device) or expended in burning fuel. Basic equation is Froude efficiency $\eta_F = 1/(1 + \delta V/2V)$ where V is velocity of vehicle and δV is total increase in velocity of air in jet measured as difference between air well ahead of vehicle and that at fastest-moving part of jet (with propeller this is well behind plane of blades but with turbojet probably in plane of nozzle exit). Theoretically but not practically δV could be zero and jet would remain stationary with respect to free-stream; η_F then is 100%. Another expression for same relationship is $\eta_F = 2/(R + 1)$ where R is ratio of jet velocity relative to vehicle velocity, which in theoretical perfect case becomes unity. Another treatment is

$$* = \frac{\text{work done on aircraft}}{\text{energy imparted to jet}} \text{ which reduces to } \frac{2V}{V+V_j},$$

where V is aircraft speed and V_j is jet.

2 Some authorities insist definition is input energy from burning fuel divided by energy of jet, allowing for all engine losses, expressed as percentage.

propulsor   *1* Multi-bladed fan, usually with variable pitch or constant-speed control, used as superior alternative to propeller; in most cases surrounded by profiled duct.

2   Since 1980 US usage has introduced a different meaning: the core of a turbofan engine, comprising compressor, combustor and HP turbine.

propwash   Airflow caused by propeller alone, usually helical but velocities are measured in axial directions only.

propylene oxide   Stable liquid used in rapidly dispersed form as fuel/air explosive.

proration   Actual yield of fares to carrier.

PRORO, Proro   AFS code: route forecast with height indication in pressure units.

PROSAB   Parachute flare (USSR, R).

Prosat   Promotional satellite.

pro-spin   Tending to initiate, sustain or accelerate spin.

protected angle   Selected (usually solid) angular limits of DOA within which received signals are accepted and amplified.

protected range   Limits (eg of radius and altitude) within which ground navaid or landing aid is protected against interference.

protected system   One incorporating maximum security to allow electrical transmission of classified plain language.

protection ratio   Ratio of wanted to unwanted received signal strength, eg in VOR, ADF, NDB etc.

protective breathing equipment   Filtering mask donned by airline cabin crew in event of fire in cabin.

protocol   Set of rules for format and content of messages between communicating processors or processes.

protocol control information   Exchanged between peer [open-systems] members to co-ordinate joint information.

protocol data unit   The N-PDU combines the N-PCI with the N-UD or N-SDU, the total information transferred between peer network members.

protoflight model   Qualification model of spacecraft that is later actually flown in space.

proton   Positively charged elementary particle of mass number 1 forming part of every complete atomic nucleus; charge magnitude equal but opposite to that of electron e.

proton-magnetometer   Family of devices usually used in space instrumentation in which field strength is measured by investigating effect upon atomic nuclei (eg NMR and Larmor precession).

prototype   First example(s) built of item intended for production; as far as possible representative of definitive article but usually inevitably deficient in many respects. In case of modern aircraft traditional * now rare; with civil programmes production is initiated in parallel with engineering design/development and even first examples are likely to be sold. With military aircraft first batch may be termed development aircraft. Aerodynamic * merely has correct shape. Breadboard or brassboard * is used to develop avionics. Purpose of * is to assist development; secondary role is to permit customer evaluation.

PROV   Provisional (ICAO).

proverse rudder   Application of rudder to augment roll commanded by lateral control system.

proving flight   Unscheduled flight by new type of commercial transport over intended routes by crew at least partly provided by customer to establish compatibility with sectors, aids, airfields and, esp., terminals and airline's ground equipment and staff.

proving ground   Military area dedicated to testing of ordnance, esp. of new types.

provisioning   Precisely calculated schedule of necessary spare parts, types, numbers, prices, dates and locations to support operation of functioning system, eg commercial transport, fighter, radar or computer.

PROX   Proximity, or GPWS.

proximate splitter   Term coined (Pratt & Whitney) for aerodynamic flow splitter added to eliminate afterburner light-up pulsations in afterburning turbofan from reaching compressor.

proximity fuse   Fuze which initiates itself by remotely sensing presence, distance and/or direction of target or associated environment by means of signal generated by fuze or emitted by target or by detecting disturbance in natural field surrounding target (DoD). Can be radio (radar), IR, visual (EO), acoustic or magnetic.

proximity scorer   Hit/miss device triggered by entry of munition into spherical volume with scorer at centre; indicates only that munition entered this volume, without giving miss-distance.

proximity switch   Switch today used in place of microswitch in exposed locations (MLG, trim tab etc) to signal mechanical position; usually variable-reactance sensor feeding micro-electronic module. Normal principle is proximity of target plate of high-permeability metal to sensor, giving varying voltage across bridge.

PRP   *1* Premature-removal period.

2 Pulse recurrence (or repetition) period = 1/PRF).

3 Parent rule point.

4 Power-deployed reserve parachute.

5 Personnel reliability program (DoD).

6 Propulsion replacement program (ICBM, USAF).

PRPA Professional Race, or Racing, Pilots' Association (affiliate of NAA).

PRPS Pressure rise per stage.

PRR *1* Premature-removal rate.

2 Power ready relay.

3 Production, or processing, readiness review.

4 Pulse-repetition rate.

PRRC Pitch/roll rate changer assembly.

PRRFC Planar randomly reinforced fibre composite.

PRS *1* Pressure-ratio sensor(s).

2 Public regulated service (satcom).

PRSD Power reactant storage and distribution subsystem, LO_2 and LH_2 for fuel cells generating electric power (Space Shuttle).

PRSG Pulse-rebalanced strapdown gyro; INS gyro whose dynamic error is reduced by compensating re-balance loop electronically.

PRSOV Pressure regulating and shut-off valve.

PRSS Passive ranging subsystem.

PRST Persist [ent].

P/RST Press to reset.

PRT *1* Pulse recurrence (or repetition) time.

2 Pulse rise time.

3 Propeller research tunnel.

4 Power-recovery turbine.

5 Processing remote terminal.

$\mathbf{P_{rt}}$, $\mathbf{P_{RT}}$ Aggregate perimeter of all jets of aircraft.

PRTB Mobile ICBM repair technical base (R).

PRTCS Portable radar tracking [and] control system (UAVs).

PRTR Printer.

PRTV Production representative, or readiness, test vehicle.

PRU *1* Photo-reconnaissance unit (RAF, WW2).

2 Performance reference unit.

prudent limit of endurance Time during which aircraft can remain airborne and retain given safety margin of fuel (NATO).

prudent limit of patrol Time at which aircraft must depart from its operational area in order to return to base and arrive there with given margin (usually 20%) of fuel.

Prune Accident-prone person, esp. pilot, from mythical Pilot Officer Percy Prune, (RAF, WW2).

PRV Pressure regulating (or relief) valve.

PRW Passive radar warning.

PS *1* Pferdestärken = cv = 0.98632 hp = 0.7355 kW.

2 Pitot/static.

3 Passenger-service costs.

4 Procurement, or performance, specification.

5 Photoemission scintillation.

6 Photo squadron.

7 Payload specialist (spaceflight).

8 Primary/search director.

9 Power supply [A adds assembly].

10 Plus.

11 Positive.

12 See *P/S*.

$\mathbf{P_s}$ *1* Static pressure.

2 Radar power received from target.

3 Time rate-of-change of specific energy.

4 Standard pitch of propeller.

5 Specific excess power.

6 Probability of survival. (Suffix bare, lone unsupported aircraft; Suffix E or enhanced, all possible on-board avionics and EW/ECM systems).

7 Stagnation pressure.

ps Picosecond (10^{-12}s).

P/S, P-S *1* Pitot/static.

2 Primary/search director.

3 Pressure-sensitive.

4 Primary to secondary (airports).

PSA *1* Prefabricated surface, or surfacing, aluminium, or semi-permanent airfield [meanings can be same].

2 Provisional site acceptance.

3 Pressure-swing absorption.

4 Pilot's associate.

5 Power-spindle angle (TLA is preferred).

6 Lb/in^2 absolute (deprecated).

7 Problem-statement analyser.

8 Precision Strike Association (US).

9 Plasma stealth antenna.

p.s.a. Passed staff college [air] (RAF).

PSAC Presidential Science Advisory Committee (US).

PSAI Public Safety Aviation Institute (US).

PSAS Pitch, or primary, stability augmentation system.

PSB Plough, sweeper and blower [snow].

PSBL Possible.

PSBR Public-sector borrowing requirement.

PSC *1* Principal site concept (USN).

2 Product-support committee.

3 Program support contract (NASA).

4 Performance-seeking control.

5 Photo safety chase [TB adds testbed].

6 Plasma-sprayed ceramic.

7 Public-sector comparator.

8 Photographic sensor control; S adds system.

PSD *1* Power spectral density.

2 Physiological Support Division (SR-71 ops).

3 Port-sharing device.

PSDN Packet-, or public-, switched data, or digital, network.

PSDP Programmable-signal data-processor.

PSE *1* Passenger service equipment (reading lights, call button etc).

2 Passive seismic experiment.

3 Project support environment.

4 Phase-shifting element.

5 Packet switching exchange.

PSEU Proximity slat, or sensor, electronics unit.

pseudo-adiabatic Process by which saturated air parcel undergoes adiabatic transformation, water being assumed to fall out as condensed.

ps aircraft system Hi-fi multi-aircraft real-time simulation environment to support ATC research (NASA Ames).

pseudo analog Electronic display which simulates traditional instrument, eg by using fixed LED matrix to form 'dial' plus computer-driven LEDs to generate 'pointer' and alphanumerics.

pseudo fly-by-wire Flight control system in which at least one axis of control is normally, at one point at least, electrical; * systems have capability for manual reversion or override.

Pseudolite   A particular scheme using surrogate satellites in a HALE UAV to counter hostile GPS jamming.

pseudolite   Transmitter of differential GPS signals from ground to improve acquisition and accuracy [from 'pseudo-satellite'].

pseudonoise   Technique in which PN code generated sends out wideband noise-like signal which is then viewed as carrier on which message is imposed; usually direct-sequence modulation approach is used in which PN code directly balance-modulates carrier.

pseudopursuit navigation   Homing method in which missile is directed towards target instantaneous position in azimuth while pursuit navigation in elevation is delayed until more favourable attack angle (note: this is not the same as AOA) on target is achieved (DoD).

pseudo radar   ADS (5) with display.

pseudo-random noise   See *pseudonoise*.

pseudorange   Satellite-derived range uncorrected for errors in synchronization between the two clocks.

pseudo stereo   *1* False impression of stereoscopic relief (ASCC).

2 Common audio meaning, simulating stereophonic by using two speakers, one via brief delay, from single channel.

pseudo-3-D   Large family of techniques for generating subjective impression of three dimensions on (usually) planar display. One is psuedo stereo (1); various methods used in PPI displays and several techniques in computer-driven OTW displays for simulators, some coming under heading of CGI.

PSF   *1* Phosphosilicate fibre (optical fibre).

2 Polystyrene foam.

3 Personnel services flight (RAF).

psf, PSF   Pounds per square foot (non-SI and in any case strongly discouraged).

PSFP   Pre-simulator familiarization panel.

PSG   *1* Post-stall gyration.

2 Passing, passage.

PSGR   Passenger[s].

PSH   Prime standard and handbook.

psi, PSI   *1* Pounds per square inch.

2 Permanent staff instructor.

3 Pulsing, or pulsating, slope indicator; -L or -R shows which side of runway.

4 Proliferation security initiatve.

PSIA   Pounds per square inch, absolute.

PSID   Pounds per square inch, differential.

PSIG, psig   Pounds per square inch, gauge.

PSK   *1* Phase [rarely, pulse] shift keying; differentially used in satcom terminals.

2 Prospect[s].

PSL   *1* Polystyrene latex.

2 Problem statement language.

3 Physical Science Laboratory (NM State Univ.).

PSLO   Product-support logistic operation.

PSLV   Polar-satellite launch vehicle.

PSM   *1* Passenger statute-mile.

2 Post-stall manoeuvring, or mode.

3 Personal safety monitor.

4 Power-system, or supply, module.

PSMC   Preselected manual control.

Psmc   GPS packet-mode channel for system management and control. Continuously broadcast by each satellite to inform system configuration and status, plus time and frequency information needed by an AES seeking to log on.

PSMK   Personal-sensor moding key.

PSN   *1* Potassium/sodium niobate.

2 Position.

3 Packet switch[ed] network.

PSO   *1* Pilot systems officer, GIB in dual-control combat aircraft (USAF).

2 Protective service operations (mission).

3 Peace support operations.

4 Program[me] support office.

PSOM   Polar strap-on motors.

psophometer   Instrument which attempts to measure perceived noise level.

PSP   *1* Pierced steel planking.

2 Programmable signal processor.

3 Personal (or personnel) survival pack.

4 Product support programme.

5 Primary special pay (DoD).

6 Pressure-sensitive paint.

7 Production software package.

PSPL   Preferred standard parts list.

PSR   *1* Primary surveillance radar.

2 Precision secondary radar.

3 Pulsar.

4 Post-strike reconnaissance.

5 Packed snow on runway.

6 Point of safe return.

7 Parts status report.

PSRE   Propulsion-system rocket engine.

PSRU   Piston-engine propeller-speed-reducing unit (= gearbox).

PSS   *1* Precision slab synchro.

2 Proximity sensor system.

3 Primary [flight-control] surface servo.

4 Product-support services.

5 Payload specialist station.

PSSA   Pilot-stick sensor assembly.

P$_{ssk}$ SSKP.

PSSPO   Precision Strike Systems Project Office (USAF).

PSSSU   Power supply and system selector unit.

PST   *1* Pacific Standard Time.

2 Propeller STOL transport.

3 Lead scandium tantalate.

PSTB   Propulsion-system testbed.

PSTN   Public switched telephone network.

PSTS   *1* Public switched telephone service.

2 Precision Sigint tracking system.

3 Propulsion-system test stand.

PSU   *1* Photosmoke unit (not same as Hartridge).

2 Power supply, or switching, unit (electronic).

3 Passenger service unit.

4 Program-setting unit (IRCM).

PSV   Public service vehicle.

psychological warfare   Planned use of propaganda and other psychological actions having primary purpose of influencing opinions, emotions, attitudes and behaviour of hostile foreign groups (DoD); (NATO substitutes for 'hostile foreign' 'enemy, neutral or friendly').

psychrometer   Instrument for measuring atmospheric humidity, usually comprising dry and wet-bulb thermometers.

psyops Psychological operations, psywar plus political, military, economic and ideological ops.

psywar See *psychological warfare*.

PSZ *1* Ceramic, mix of tetragonal zirconia in cubic zirconia.

   2 Public safety zone.

PT *1* Power turbine.

   2 Primary trainer (USA, USAS, USAAC, 1925–47).

   3 Public transport.

   4 Pesawat Terbang [company constitution, Indonesia].

   5 Procedure turn.

   6 Point.

Pt Platinum.

P$_t$ *1* Rocket chamber pressure at termination (solid propellant).

   2 Total pressure.

   3 Pitot pressure.

   4 Radar power transmitted.

   5 Pennant number.

pt Pint.

p$_t$ Tensile stress.

P$_{t2}$ etc Gas-turbine total pressures at usual numbered locations, ending with P$_{t7}$ for nozzle exit.

PTA *1* Prepaid ticket advice in foreign currency.

   2 Polskie Towarzystwo Astronautyczne (Poland).

   3 Propfan test assessment.

   4 Pilotless target aircraft.

   5 Part-throttle afterburning (see part-throttle reheat).

   6 Practice target area.

PTAB Small hollow-charge anti-armour bomblet, usually followed by figure giving weight in kg (USSR, R).

PTAG Portable tactical aircraft guidance.

PTAN Precision terrain-aided navigation.

PTAS Pilotless target aircraft squadron.

PTB *1* External tank, or drop tank (USSR, R).

   2 Patrol torpedo-bomber (USN 1937).

PTC *1* Programming & test centre.

   2 Part-through crack.

   3 Pitch trim compensator.

   4 Passive thermal control.

   5 Personnel and Training Command (RAF).

   6 Personal Technical Certificate (NATS).

   7 Pack temperature controller.

PTCHY Patchy.

PTCS Portable tracking and control station.

PTCV Primary temperature control valve.

PTD *1* Provisional technical document.

   2 Performance test domain.

PTE Performance test engine.

PTEH Per thousand engine hours.

PTF Permit to fly.

PTFCE, ptfce Polytrifluorochlorethylene.

ptfe Polytetrafluoroethylene.

PTH Path.

PTI Packet-type identifier.

PTID Programmable tactical information display.

PTIR Precision tracking and illuminating radar.

PTIT Power-turbine inlet temperature.

PTK New hollow-charge bomblet (USSR).

PTL *1* Primary technical leaflet.

   2 Prioritized target list.

PTLY Partly.

PTM *1* Pulse-time modulation.

   2 Pressure transducer module.

   3 Peripheral transition module (SBC).

   4 Power thermal management.

PTMU Pressure/temperature measurement unit.

PTN *1* Procedure turn.

   2 Position.

   3 Public telephone network.

   4 Pattern [S plural].

   5 Portion [S plural].

PTO *1* Power take-off (shaft output).

   2 Permeability tuned oscillator.

   3 Participating test organization.

   4 Overhaul by the customer (R).

   5 Part-time operation.

   6 Pacific theatre of operations (WW2).

   7 Personal ticket office (software).

P to F *1* Permit to fly.

   2 Permission to fire.

P-tots Portable transparency optical test system (for rainbowing).

PTP *1* Paper-tape punch.

   2 Programmable touch panel.

   3 Programming and test panel (CIDS).

   4 Purchase to payment (electronic delivery).

PTPS Passive thermal protection system.

PTR *1* Part-throttle reheat.

   2 Paper-tape reader.

   3 Power-turbine rotor.

   4 Peak/trough ratio.

   5 Production test requirements.

   6 Program Trouble Report (Stars).

PTRP Propfan technology readiness programme.

PTRS Program tracking and reporting system (ATOS).

PTS *1* Pilot Training Squadron (US).

   2 Photogrammetric target system.

   3 Parachute Training School.

   4 Polar track structure.

   5 Pre-training screen.

   6 Problem-tracking system.

PTSA Prior to sample-approval.

PTSD Production test specification document.

PtSi Platinum/silicide.

PTSN Public telephone switching network.

PTT *1* Part-task trainer (simulator).

   2 Post, telegraph, telephone (many European nations).

   3 Press, or push, to transmit, or talk, or test.

PTTEM Preliminary tactical technical economical requirement, following UTTEM and leading to proto-type(s) (Sweden).

PTTS Pressure/temperature test set.

PTU Power transfer unit; transmits power but not fluid between hydraulic systems. Can provide vital back-up following failure of an engine, esp. left engine.

PTV *1* Propulsion technology validation.

   2 Propulsion test vehicle.

PTW Precision targeting workstation.

P2P See PTP(4).

Pty Proprietary [company constitution, RSA].

PU, p.u. *1* Pick-up.

   2 Propellant-utilization system.

   3 Physical unit (SNA).

   4 Polyurethane.

Pu Plutonium.

public-address system, PA Interphone voice circuit used by captain or cabin crew to address passengers.

477

public aircraft   *1* Not public, ie used exclusively in government service (US).

2 Used by the public, including for hire or reward (UK).

public dividend capital   Strange term for direct gift of taxpayer's money to national airline (UK).

public safety zone   Area adjacent to end of runway where development, such as housing, is restricted.

published route   One for which an IFR altitude has been published.

puck   Replacement pad for plate or disc brake.

pucker   Local buckling of sheet metal in compression, eg on flange around inside of bend.

PUD   Power unit de-icing (CAA).

puddle welding   Blind attachment [e.g., of stringers to skin, or of sheet to a core] by Argon-arc torch having access to one side of the workpiece only.

PUDT   Portable user data terminal.

puff pipe   Pipe taking bleed air to an RCS control valve in jet-lift aircraft. Hence puffer jet (colloq.).

Pugs   Propellant utilization and gauging system (NASA).

pukka   True, thus * Gen = reliable info (from Hindustani, RAF WW2).

pull   *1* To operate, eg to * spoilers (colloq.).

2 To engage arrester wire.

3 To engage arrester wire, causing damage or breakage to arrester system.

pulled   Arrester wire whose anchorage or shock-absorbing system has been damaged.

pulling through   Rotating propeller by at least 2 to 3 blades by hand, for whatever reason.

pull lead   To pull nose of aircraft further round to aim correct distance ahead of target.

pull-off   *1* Practice parachute jump in which slipstream pulls wearer of opened parachute off wing or other suitable part of aircraft.

2 Weak link in anti-spin parachute, calibrated to break at given IAS.

pull out   *1* To recover from dive to level flight or zoom.

2 To extend arrester wire or airfield barrier near limits.

pull-out area   Carrier deckspace kept clear for decelerating arrivals.

pull-out distance   Distance travelled by hook between engaging wire and coming to stop; also called run-out.

pull-ring   Parachute operating handle or D-ring.

pull up   Short sudden climb from level flight, normally trading speed for height (usually general aviation or tactical attack).

pull-up point   Geographical point at which aircraft must pull up from lo approach to gain sufficient height to make attack or execute retirement (DoD).

pulsating rubber   De-icer boot.

pulsator   Engine instrument showing both engine speed and oil circulation by pulsations of oil in glass dome (rare after 1917).

Pulse   Precision up-shot laser-steerable equipment.

pulse   *1* Transient phenomenon, esp. in radio or other EM signal, characteriszd by rise, brief finite duration and decay.

2 Single solid-propellant grain, two or more of which are contained in rocket motor casing, hence two-* rocket can give two impulses, separated by selectable interval.

pulse-amplitude modulation   Signal is broken down into bits, amplitude of each being measured to give series of discrete values.

pulse-bias modulation   Following a skid [loss of wheel traction], determines how long before full pilot pedal pressure is restored.

pulse code   Sequences of pulses conveying information.

pulse-code modulation   Modulation involving pulse codes, esp. that which translates continuously modulating signal into stream of digital pulses all of uniform height (amplitude), information being conveyed by spacing/duration. PCM output is compatible with all digital EDP (1) and virtually eliminates transmission errors.

pulse compression   Radar techniques for increasing pulse amplitude and reducing length, see chirp, binary phase modulation, polyphase coding.

pulse decay time   Time pulse takes to fall from high to low value, normally from 90% peak to 10%.

pulse-detonation engine   Aircraft propulsion jet engine in which fuel is burned in a high-frequency series of detonations, without moving parts; suitable for up to Mach 5. See *HFPD*.

pulse Doppler, PD   Radar mode using pulse trains and Doppler processing in which received signals are examined by mixers and band-pass filters which eliminate everything except genuine targets or objects of interest. Doppler technique injects information on relative range-rates as well as eliminating non-targets.

pulsed production line   Whole line moves regularly to new position at [quite long] intervals.

pulse duration   Time that single pulse exceeds stated value, usually 10%, of peak; ICAO selects time over 50%.

pulse-duration modulation   Also called pulse-time modulation, pulse-width modulation, translates CW signal into succession of constant-amplitude pulses of varying width (time duration).

pulse envelope programming   Surveillance-radar mode in which one beam searches for high-altitude hostile aircraft and a second in PD mode searches for low-level aircraft and missiles; possibly a third searches for surface targets.

pulse-forming line   Radar circuit generating short high-voltage pulses.

pulse-frequency modulation   More precisely called PRF modulation; CW signal is translated into succession of constant-amplitude pulses transmitted at frequency proportional to amplitude of original signal.

pulse interval   Time between consecutive pulses both measured at same point.

pulsejet   Air-breathing jet engine in which air is intermittently induced or allowed to enter, mixed with fuel, ignited (by electric discharge, residual combustion products or other method) and expelled as single expanded charge of hot gas giving pulse of thrust. Cyclic operation (typical frequency 30–60 Hz) may be inherent in aerothermodynamics of duct or imposed by sprung flap-valves or other oscillating one-way valving at inlet.

pulse length   Physical length of radar pulse, irrespective of time.

pulse-light approach slope indicator   Pilot sees red/white lights which pulse with a frequency proportional to deviation from glidepath.

pulse limiting rate   Highest PRF allowing time for echo to reach receiver in gap between pulses.

pulse modulation   Variety of methods of translating CW

signals into digital pulses to reduce bandwidth required, eliminate errors and, where possible, improve signal/noise ratio (see separate entries).

pulse packet   Concept of radar signal pulse as physical entity occupying particular 3-D volume, esp. particular length.

pulse-phase modulation   See *pulse-position modulation.*

pulse-position modulation   CW signal is translated into succession of constant-amplitude pulses whose position (ie time from start of each frame or timebase period) is proportional to amplitude of original wave at corresponding point.

pulse radar   Most common type, in which signals are in form of pulses: also called pulsed radar.

pulse-repetition frequency   Average number of radar pulses transmitted in unit time.

pulse-repetition period   Average elapsed time between a point [eg, peak] on radar pulse and same point on next.

pulse rise time   Time required for EM pulse to rise from a low to a high value, normally from 10 to 90% of peak but occasionally from 5 to 95 or from 1 to 99.

pulse separator   Receiver circuit which removes imposed (regular) pulse train.

pulse sorter   ECM device which selects one pulse from many for detailed measurement.

pulse spike   Erroneous sharp super-peak superimposed on pulse.

pulse-time modulation   CW carrier is modulated by pulse train of lower frequency which in turn is modulated with variable characteristic, which may be amplitude, duration, PRF or position.

pulse train   Succession of pulses.

pulse width   Unlike duration, this measure of CW signal is normally defined as width (time) between half-power points.

pulse-width modulation   CW signal is translated into train of constant-amplitude pulses whose widths are each proportional to corresponding amplitude of original signal. (Width in this context has no relevance to half-power points).

pulsometer   Visual indicator of fluid flow, calibrated 08-18 [trad. engineering meaning not used in aerospace].

pultrude   Fabrication process combining pulling through die and extrusion under back-pressure (eg Grafil CFC).

pultrusion   Raw material or finished section made by a pultrude-type process.

pulverized fuel ash   From coal-burning power stations [utilities], brings aircraft to a stop in overrun area with little damage.

Puma   Parallel unstructured maritime aerodynamics; program to generate flow data, eg to predict noise from helicopter landing gear and other complex shapes.

PUMP   Pre-upgrade maintenance programme (RAF).

pump-up time   Time taken to inflate gas storage in blow-down tunnel.

PUN   ICAO code: prepare new perforated tape for message.

punch, punch out   *1* To eject.
2 Confusingly, to start engine(s).

punch welding   *Poke welding.*

pundit   *1* Aerodrome beacon with Morse identifying sequence (arch.).

2 Portable ultrasonic non-destructive digital indicating tester.

punkah, punkah louvre   Fresh-air jets in passenger cabin, of whatever geometry and location.

PUP   *1* Pitch-up point in pop-up delivery.
2 Performance update programme.
3 Principal-user processor.

pupil pilot   General British term for student pilot, other terms being (US) undergraduate pilot, trainee and PUT. Pilots seeking higher qualifications are not *.

purchase   Single grip on yoke, spectacles or other aileron input; thus full aileron may be two-* task.

purchase cable   That connecting pendant to arresting gear under flight deck.

Pure-clad   Trade name (Reynolds) for Alclad-type material.

pure jet   Not defined, but usually meant turbojet as distinct from turboprop (arch.).

pure pursuit-course lead   Course in which velocity vector of attacking aircraft is always directed towards instantaneous position of target (ASCC).

pure research   In case of aircraft, one whose purpose is to obtain knowledge of general application; aircraft employed may be specially designed or modified version of familiar type.

Purex   Plutonium/uranium extraction.

purge   *1* To clean and flush device, eg liquid-propellant rocket, by high-rate pumping of inert gas, eg dry nitrogen. This removes potentially dangerous propellants and helps preserve hardware; secondary function may be to trigger various valves and leave inert gas occupying internal chambers and piping. Term also sometimes used to mean inhibiting.

2 To prevent admission of air to space above fuel in aircraft tank by continuously pumping in inert gas (usually dry nitrogen).

purity   Volume percentage of lifting gas in aerostat gas cell.

Purolator   Patented lubricating-oil filter of conventional form.

Purple   Air-intercept code: unit is suspected of carrying nuclear weapons (DoD).

Purple Airway   Special temporary airway established and promulgated in Notams for Royal flight(s) in UK and certain other areas.

pursuit aircraft   Interceptor (US, term faded from use in WW2).

pursuit course   Course in which attacker must maintain a lead angle over velocity vector of target to predicted point in space at which gun- or rocket-fire would intercept target (ASCC) (see *pure pursuit*).

pursuit missile   One which can be fired only from astern of air target, eg because IR homing head is unreliable from any other aspect.

PUS   Permanent Under-Secretary of State (UK).

pushback   Transfer from gate to clear area or taxiway by tractor [AM or tug] attached to NLG, hence * clearance, * crew, * time, * tug.

pushboom   Very long rigid sleeve ahead of aircraft nose, carrying surface radiating elements.

pushbutton indicator   Pushbutton which when depressed illuminates.

pushdown effect   Generalized rule that new commercial transports enter service on densest routes, pushing older types down to lesser markets; new type is itself then pushed down over 10 to 20 years.

pusher   See *stick-pusher*.

pusher aircraft   One with pusher propeller(s) only.

pusher propeller   One mounted behind engine so that drive shaft is in compression.

push fit   Fit just requiring light force to assemble.

push-off drift   See *kick-off drift*.

pushover   Nose-down manoeuvre commanded by stick-pusher; in effect same as pitchover.

push/pull   *1* Throttle, eg on lightplanes, having linear motion sliding through panel.

2 Amplifier having two similar valves or transistors connected in anti-phase and with I/O circuits combined about earthed centre.

3 Installation of tractor and pusher propellers in tandem, posibly separated by entire fuselage.

push-push actuator   Linear actuator having uni-directional output interleaved by weak or slow return stroke.

pushrod   Rod transmitting cam motion of valve gear to rocker or other drive to poppet valve(s).

PUT   *1* Pop-up test of ICBM, SLBM, ABM or other launch system of externally energized cold-launch type.

2 Pilot under training, also Pu/t.

put-put   Small underpowered aeroplane [onomatopoeia].

PV   *1* Pressure/volume or pressure × volume.

2 Private-venture aircraft.

3 Product verification (formerly MQT).

4 Prevailing visibility.

5 Prime vendor.

6 Positive vetting.

7 Post vacant.

8 Parameter value.

PVA   *1* Polyvinyl acetate (or pva).

2 Plane view area, normally of precision forging.

PVASI   Para, or pulsating, visual approach slope indicator.

PVB   Polyvinyl butyral (or pvb).

PVC   *1* Polyvinyl chloride (or pvc).

2 Permanent virtual circuit.

PVD   *1* Para-visual director.

2 Plan-view display; -E adds emulator.

3 Peripheral-vision display.

4 Air-data sensor, esp. pitot, pitch, yaw (USSR, R).

PVDF   Polyvinylidene fluoride (or pvdf).

PVDU   Portable VOR deviation unit.

PVF   Polyvinyl fluoride (or pvf).

PVI   *1* Pilot/vehicle interface.

2 Para-visual indicator.

PVL   Prevail.

PVLS   Peripheral vertical-launch system.

PVM   Primary visual marker.

PVO   Air defence of the homeland, made up of IA (manned interceptors) and ZR (zenith rockets, ie SAMs) (USSR, R).

PVOR   Precision VOR.

PVO-SV   Troops of air defence of ground forces (USSR, R).

PVR   Premature voluntary retirement.

PVRD   Ramjet (USSR, R).

PVS   Pilot's vision system, or visual subsystem.

Pvs   Pitch of tunnel vane set (distance between vanes).

PVT   *1* Product verification test.

2 Private [operator].

3 Position/velocity/time.

4 Personal verifier terminal.

PVTOS   Physical vapour transport of organic solids, for space manufacturing.

PVU   *1* Portable ventilator unit.

2 Precision velocity update.

PVV   Proof vertical velocity (MLG demo case).

PW   *1* Pulse width.

2 Plated wire (memory).

3 Potable water.

4 Pursuit, water-cooled (USA 1919–24).

PWA   Public Works Administration (US from 1933).

PWB   Pilot['s] weather briefing.

PWBH   Pilot's Weight and Balance Handbook (US).

PWD   Present-weather detector (visibility, precipitation, snow depth).

PWG   Planning Working Group (Eurocontrol).

PWHQ   Primary war headquarters.

PWI   *1* Proximity, or pilot, warning indicator.

2 Preliminary warning instruction.

PWIN   Prototype WWMCCS intercomputer network.

PWM   Pulse-width modulation; hence PWMI = * inverter.

PWP   *1* Pylon weight plug.

2 Plasticized white phosphorus.

PWQT   Potable-water quantity transmitter.

PWR   *1* Power.

2 Passive warning radar.

PWRS   Prepositioned war reserve stocks.

PWS   *1* Proximity warning system (generally helicopter applications).

2 Performance work statement.

3 Potable water supply.

PWSC   Preferred weapon-system concept.

PWSDE   Effective power delivered at shaft of turbo-prop = ehp.

PWT   *1* Propulsion wind tunnel.

2 Plasma wind tunnel.

PWW   Predictive windshear warning.

PX   Post Exchange, became BX and today AAFEX, which see.

PY   *1* Program(me) year.

2 Spray.

P(Y)   Precise encrypted (GPS).

PYBBN   Pitch yaw balanced-beam nozzle.

pylon   *1* Rigid pillar-like structure projecting upwards to carry load (eg engine) or protect occupants in overturn (crash *).

2 Streamline-section structure transmitting stress from external load to airframe, eg engine pod, ordnance, drop tank etc. Can extend above or below wing or horizontally or at other angle from fuselage. For engine pod often extended to *strut.

3 Object on surface used as landmark for race turning point.

4 Object on surface used as reference for pilots performing flight manoeuvres.

5 Aircraft [usually private or club] tipped on nose, tail in air (colloq.).

pylon select   I/O device linking crew to WCS (weapon control system) or Navwass, enabling specific pylons, ie particular portions of load, to be selected for attack on particular target.

pylon spar Principal structural member of engine pylon normally (in case of wing engine) linking engine mounts direct to wing box.

pylon strut Pylon (2) linking engine pod to wing or rear fuselage.

Pyralin Cellulose-base transparent plastic formerly used for windows.

pyramidal absorber Family of absorbers of EM radiation (ie RAM 2) and noise, characterized by entire surface of structure being covered by long pointed pyramids with apices pointing towards source of radiation. Similar geometry is found in airframe underlying structure of low-observables aircraft.

pyranometer Actinometer which measures combined solar and diffuse sky radiation, sensor viewing entire visible sky. Also called solarimeter.

Pyrene Trade name for fire exitinguishers and for carbon tetrachloride filling.

pyrgeometer Actinometer which measures terrestrial radiation.

pyrheliometer Actinometer which measures only direct solar radiation.

pyroelectric detector Sensitive detector of IR; radiation enters via precision window of Ge and light pipe directs it to microscopic flake of TGS which rises in temperature, changing polarization and generating surface charge amplified in low-noise electronics.

pyrogen Pyrotechnic generator; ignition squib for solid-propellant motors comprising electric resistance or bridgewire which ignites hotburning powder charge.

pyrolitic carbon Allotrope of carbon derived from controlled pyrolysis of char-yielding resin (eg phenolic resoles or Novolaks) to form matrix used in carbon/carbon composite, with reinforcement by carbon or graphite fibre. Used in rocket nozzle liners, high-temperature wheel brakes, etc.

pyrolysis Chemical decomposition by heating.

pyromechanical actuator One energized by solid fuel or other combustible charge.

pyrometer Instrument for measuring high temperatures; optical, electrical resistance, thermocouple, radiation, etc.

pyron Non-SI unit of EM radiant intensity (no abb.), = calories (Int.)/cm^2/min.

pyrophoric Igniting spontaneously on contact with air.

pyrotechnic Today includes not only visual 'firework' devices but also precision igniters for large solid motors, single-shot actuators, hot-gas generators and IR flares giving accurately controlled decoy wavelength.

pz See *pièze*.

PZD Petrolatum/zinc dust.

PZRK Portable rocket (missile) air-defence system (R).

PZT Lead zirconate titanate.

Q

Q *1* Quantity of electricity, esp. electric charge.

2 Applied shear force [see q2].

3 JETDS code: sonar.

4 JETDS code: special, or combination of purposes.

5 Probability of failure.

6 Quantity of light, heat or other EM radiation.

7 Static moment of area about any axis.

8 Aircraft category, target (USAF 1948–62).

9 Modified-mission prefix: drone or RPV (USAF, USN, since 1962).

10 Modified-mission suffix: electronic counter-measures (USN 1945–62).

11 Generalized symbol for torque.

12 Common prefix 'quiet'.

13 Quadrature, and quadrature component of coherent video (radar) signal.

14 Squall[s].

15 Enthalpy (H is more common).

16 Volume of fluid (gas or liquid).

17 Reactive power.

18 See *Q-factor*.

q *1* Dynamic pressure.

2 Shear stress [τ is preferred].

3 Tetrode.

4 Heat flux (rate of flow).

5 Angular velocity (rate of change) in pitch, i.e. pitch rate.

6 Generalized symbol for rate of flow, eg volume, energy etc.

7 Range (Breguet formula).

8 AMSU pitch-rate output.

Q̄ Störindex; German measure of annoyance-weighted sound pressure level (see *noise*).

Q-aerial Combination of dipole plus quarter-wavelength of twin-wire line to match feeder impedance to dipole.

Q-alpha, QA Free-stream dynamic pressure.

Q-ball, q-ball Spherical or hemispherical-nosed instrument package on nose of spacecraft or aircraft sensing q, AOA, AOY, total temperature and other parameters.

Q-band Obsolete EM radiation band with limits 33–50 GHz (US), 26.5–40 GHz (UK), now covered by K, L bands.

Q-bay Heated and pressurized compartment for reconnaissance sensors, including large camera looking through quartz window.

Q-code Basic telecommunications code of three-letter groups in three sections: QAA–QNZ are limits of Aeronautical Code; QOA–QQZ is Maritime; QRA–QUZ is for all services. Many of the entries which follow are from this code, of which nine are still in use.

Q-correction That applied to observed altitudes of star Polaris (because not quite at N celestial pole).

Q-dinghy Oval, for a crew of up to 8 (RAF, WW2).

Q-factor *1* Figure of merit of inductance, ratio of reactance to resistance.

2 Ratio of energy stored to energy dissipated per radian in electrical or mechanical system.

3 Generally, sharpness of resonance or frequency selec-

tivity of vibratory system having one degree of freedom, mechanical or electronic.

Q-fan Quiet fan.

q-feel Flight-control feel synthetically made to resemble natural feedback from aerodynamic loads by making it approximately proportional to dynamic pressure.

Q-meter Instrument for measuring Q (1).

Q-series propellants Patented (Thiokol) slow-burning solid fuels for gas generation with LL-521 coolant keeping flame below 1,093°C.

Q-Shed Proposed QRA hangar for three RAF airfields in UK.

Q-site Dummy airfield with simulated flarepath and other lights (UK, WW2).

q, Q-spring Mechanical connection with stiffness proportional to dynamic pressure.

q-stops Mechanical limits on flight-control system response, and thus surface movement, commanded by q-system.

Q-switching Extremely rapid switching of laser by means of Kerr cell or similar opto-electronic device; essential for shortest-duration high energy bursts lasting only a few ns.

q-system That sensing dynamic pressure, eg drawing processed signal from ADC, and feeding it to flight-control and other q-sensitive systems or devices.

QA *1* Quality assurance.

2 Quasi-analog.

QAA Quality Assurance Agency for Higher Education (UK).

QAAC Quebec Association of Air Carriers.

QAB *1* Quality Assurance Board (MoD-PE, UK).

2 Code: "May I have clearance for – from – to – at FL –?"

QAC *1* Quality action case.

2 Civil aircraft qualification (F).

QAD Quick attach/detach.

QADDM Quasi-analog DDM(1).

QAE Quality assurance evaluator.

QAF Code: "Will you advise me when you are/were at –?"

QAG Code: "Arrange your flight to arrive at/over at –?"

QAH Code: "What is your height above –?"

QAI Code: "What is essential traffic regarding my aircraft?"

QAK Code: "Is there risk of collision?"

QAL Code: "Are you going to land at –?"

QAM *1* Code: "What is latest met.?"

2 Quadrature amplitude modulation.

QAN Code: "What is surface wind?"

QAO *1* Code: "What is wind at your location at different FLs?"

2 Quality assurance office (DoD).

QAR *1* Quick-access recorder.

2 Quality-assurance representative.

QA/RM Quality assurance and risk management.

Qasar Quality assurance systems analysis review (FAA from 1971).

QAT Qualified for all three.

QAVC Quiescent automatic volume control.

QB Quiet [deceased] Birdmen (US society from 1921).

QBA Code: "What is horizontal visibility at –?"

QBAR Dynamic pressure.

QBB Code: "What is amount, type and height above field of cloudbase?"

QBI Code: "Is flight under IFR compulsory?" Hence, QBI conditions = bad weather. Colloquially said to mean 'quite bloody impossible'.

QC *1* Quality control.

2 Quick change, ie from pax to cargo configuration.

3 Quiet, clean (as prefix).

4 Quality circle (usually plural).

5 Quota count (noise).

6 Quantum computing.

7 See * *card.*

QCA Qualifications and Curriculum Authority (UK).

QC card Quadrantal correction.

QCDA Quickened climb/dive angle.

QCDP Quality-control development programme.

QCE Quality-control engineer.

QCG Qualifiable Code Generator, embedded graphics tool.

QCGAT Quiet, clean general-aviation turbofan.

QCI Qualified crewman instructor.

QCIP Quality-control inspection procedure.

QCPSK Quadrature coherent phase-shift keying.

QCS Query control station (ECM).

QCSEE, QC-see Quiet, clean short-haul experimental engine (NASA).

QD *1* Quantity distance (explosives).

2 Quick dump.

3 Quick disconnect.

4 Quantum diode.

QDB Quick-disconnect button.

QDG Quick-draw graphics.

QDL Code: "I intend to ask for series of bearings."

QDM *1* Code: "Will you indicate magnetic heading for me to steer towards you, with no wind?"

2 Quick-donning mask.

QDR *1* Code: "What is my magnetic bearing from you?"

2 Quality-control deficiency report.

3 Quadrennial Defense Review (US).

QE Quality engineering.

QEC *1* Quick engine change unit, ie ready-to-install powerplant.

2 Quadrantal error correction, or corrector.

QEP Quality-enhancement program.

QET Quick engine test.

QFA Meteorological forecast prefix.

QFE *1* Quiet, fuel-efficient.

2 Code: "To what should I set my altimeter to obtain height above your location?" Usually requests airfield pressure. Thus, altimeter reads zero on landing there.

QFF Code: "What is present atmospheric pressure converted to MSL at your location?"

QFI Qualified flying instructor.

Q-Flow Quota flow control.

QFU Code: "What is [magnetic] direction/designation of runway to be used?"

QGH Code: "May I land using – procedure?" Requests letdown procedure using radio aids.

QH Queen's Honorary, followed by third [possibly fourth] letter denoting Chaplain, Dental Surgeon, Nursing Sister, Physician, Surgeon, all military appointments.

QHI Qualified helicopter instructor.

QHNI Qualified helicopter navigation instructor.

QI Quality improvement.

QinetiQ Supposed PPP created 2001 by renaming DERA; when US refused to share classified projects with a PPP, 25% was split off to form DSTL. In fact by 2003 no private shares had been announced, but flotation was still the eventual objective (UK).

QIS Quality information system.

Qitars Qualitative intratheatre airlift requirements study.

QK Quick-flashing.

QL Ethyl-2 (di-isopropylamino) ethylmethylphosphonite, component of VX nerve gas.

QM *1* Quartermaster; G adds -General, S -Sergeant.

2 Quality management.

QMAC *1* Quarter-orbit magnetic altitude control.

2 Questionnaire on the method of allocating cost[s].

QMDR Quality-monitoring deficiency report.

QMP Quality management plan.

QMS Quarterly manning statistics.

QMW Code: "What is/are freezing levels?"

QN Quiet nacelle.

q_n Jet dynamic pressure.

QNE Code: "What will my altimeter read on landing at – if set to 1013.2 mb?" Note: answer is pressure height of airfield. Used by all aircraft over FL180 in US.

QNH Code: "To what should I set my altimeter to read your airfield height on arrival?" Assuming ISA throughout, answer is equivalent MSL pressure as calculated by destination ATC. Regional *, or lowest-forecast *, is value below which actual * is predicted not to fall in given period and location; gives supposedly safe terrain clearance.

QNM Quantized normal/MTI (video).

QNY Code: "What is present weather at your location?"

q_o Freestream dynamic pressure.

QOC Quality officer in charge (UK).

QOP Quality operating procedure[s].

QOR Qualitative operational requirement.

QoS Quality of service.

QOT&E Qualification operational test and evaluation.

QPD Quality procedural document.

QPL Qualified products list.

qpp Quiescent push/pull.

QPR *1* Quality problem report.

2 Quality procedural requirement.

QPSK Quadrature, or quadrative, phase-shift, or pulse-shift.

QR *1* Quiet radar.

2 Quadrantal receiver.

3 Quadrupole resonance.

q_r Generalized co-ordinate.

QRA *1* Code: "What is name of your station?"

2 Quick-reaction alert (RAF), originally with suffixes (I) or (B) for interceptor or bomber.

3 Quiet reconnaissance airplane (USA).

QRB Quick-release buckle.

Q_{rb} Generalized force of rth mode due to buffeting pressure field.

QRC Quick-reaction capability, or [IRCM] contract.
QRF *1* Quick-release fitting.
   2 Quick-reaction force.
Q_{rf} Generalized forcing function.
QRGA Quadrupole residual gas analyser, instrument for measuring ultrahigh vacuum.
QRH Quick-reference handbook.
QRI Quick-reaction interceptor.
QRMC Quadrennial review of military compensation.
QRP Quick-reaction package.
QRS Quality requirements systems.
Q_{rs} Generalized force in rth mode due to sth-mode aerodynamic excitation.
QRT Code: "Shall I stop (or please stop) sending?"
QRTOL Quiet RTOL.
QS Quick scan.
QSEG Qualification systems-engineering group.
QSH Quiet short-haul.
QSJ Quiet supersonic jet.
QSP Quiet supersonic platform, research programme by NASA, Darpa, Northrop Grumman.
QSR *1* Quick strike reconnaissance.
   2 Quarterly service report.
QSRA Quiet short-haul research aircraft.
QSS Quality, safety and security.
QSTNRY Quasi-stationary.
QSTOL Quiet STOL.
QSY Code: "Change radio frequency now to –".
QT Quart, quarter.
QT&E Qualification test and evaluation.
QTE 1 Code: "What is my true bearing from you?"; today often changed to "line of position" = PL.
   2 Qualification test and evaluation.
Q-tech Quality technology.
QTF 1 Code: "Will you give me position of my station according to bearings taken by D/F stations you control?" (Note: many countries, including UK, can no longer provide D/F fix.)
   2 Quarter-turn fastener.
Q-tip Propeller with sweptback tips.
QTM Quality technical memorandum.
QTOL Quiet takeoff and landing.
QTR *1* Quiet tail rotor.
   2 Quality technical requirement.
   3 Quad tilt-rotor.
qts Dynamic pressure in wind-tunnel test section.
QTY Quantity.
quad *1* Group of four attitude-control thrusters indexed at 000°/090°/180°/270° relative to vehicle major axis.
   2 Quadrant (ICAO).
   3 Loosely, any group of four, especially of air-launched missiles fired from one container or pylon.
quad actuator Among other things, central summing/proportioning/switching unit between triply redundant systems to ensure that maximum redundancy and maximum authority are preserved.
quaded cable Telecom cable with two twisted pairs (four channels).
quadrant *1* Radio range area between equisignal zones (actually much more than 90°).
   2 Circular-arc operating lever on control surface of airship or large, slow aeroplane (probably obs.).
   3 Circular-arc mounting for throttle, pitch or other operating levers in cockpit.

quadrantal cruising levels Quadrantal heights.
quadrantal error That caused by presence of metal structure (eg receiving aircraft) distorting radio signal.
quadrantal heights Specified flight levels assigned to traffic on heading in each of four 90° quadrants; intention is that traffic on conflicting headings shall be separated in height. See *Quadrantal Rule.*
quadrantal points Intercardinal headings: 045°, 135°, 225° and 315°.
Quadrantal Rule Quadrantal heights allocated in UK uncontrolled airspace below FL250.
quadrature Quarter-phase difference between two wave trains of same frequency, ie displacement of 90°.
quadrex Quadruply redundant.
quadricycle *1* Landing gear with four wheels or wheel groups disposed at corners of rectangle in plan.
   2 Loosely (suggest incorrectly) any aircraft with four landing gears even if three are in line.
quadruplane Aeroplane with four superimposed wings.
quadruple register Instrument recording wind speed/direction, sunshine and rain (not snow).
quadruplex Any fourfold system, ie four parallel channels.
quadruplex system Dynamic system that is quadruply redundant and thus provides multiple failure capability, eg SFO/FS, DFO and DFO/FS.
quadrupole Idealized noise source made up of four equidistant sources, each diametrically opposite pair emitting positive pressure peaks while the other pair are at peak negative (all having same frequency).
quadrupole resonance Method of detecting small quantities of specific material by irradiating with RF tuned to that material's molecules.
quad tilt-rotor Aircraft lifted and propelled by four tilt rotors.
qualification Clearance for service; thus also for production.
qualitative limitation In arms control, eg SALT II, concerned only with weapon-system capability.
quality control That management function by which conformance to established standards is assured, performance is measured, and, in event of defects, corrective action is initiated (USAF).
quantified Expressed in numerical terms.
quantitative limitation One concerned solely with numbers, eg of weapon systems in SALT II.
quantized Restricted to particular set of discrete values.
quantum bit Atomic electron or photon.
quantum dot Semiconductor device with overall dimensions of a few (one definition, ≤10) nanometres.
quarantine Secure bonded store for parts which, for whatever reason, are illegal or accompanied by incorrect documentation.
quart Non-SI measure of capacity, qt = (UK) 1.137 litres, (US) 0.946 litres.
quarter-chord Locus of all points lying at 25% chord (of wing or other aerofoil), each measurement being in plane parallel to longitudinal axis of aircraft. Normal interpretation is that measures are for wing in clean condition, ignoring ancillary items such as stall-breaker strips and anything else attached to wing, but including dogtooth and similar basic modifications to outline.
quarterlights Windows at oblique angle between front windshield (UK windscreen) and side.

Quarter-Million   Aeronautical charts [rarely, maps] published on scale of 1:250,000 (ICAO).

quarter-phase   Electrically 90° (see *quadrature*).

quarter-turn fastener   Cowl or panel fastener released by 90° anticlockwise turn.

quarter-wave aerial   One quarter-wavelength long and resonating at slightly less than 4l where l is length; hence quarter-wave line is same length of co-axial or twin-wire transmission line.

quartz   Natural and synthesized mineral, SiO_2, with many electrical, electronic and refractory structural uses but noted for piezoelectric properties in perfect-crystal form.

Quattrocopter   Helicopter lifted and controlled by contra-rotating two-blade rotors on vertical axes at each corner of a square or X-shaped base; largest would be manned, smallest 500 g (1.1 lb).

Qubit   Quantum bit [pronounced cubit].

Queen Bee   Senior WAAF or WRAF officer on station (colloq.).

Queen Mary   Articulated road vehicle for transporting dismantled or crashed aircraft (RAF).

quenching   *1* Sudden cooling of hot metal in water, oil or other medium to obtain desired crystalline properties.

2 Blanketing solid grain of rocket in mid-operation, by various related techniques, to obtain variable-pulse operation.

Questol   Quiet experimental STOL.

QUI   Quito, Ecuador, space tracking-station.

quick cam   Flap-track profile giving long rearwards travel to increase area/lift and with sharp downwards travel at end which, when full flap selected, rotates flap to landing angle giving high drag.

quick-change aircraft   One whose interior is configured for passenger or freight operation and whose seat units, passenger-service units, pantries, toilets and often trim can be removed in minutes and replaced by freight restraints (attached to original floor rails) and protective wall panels.

quick-connect parachute   Chest-type pack which user can clip to harness almost instantly.

quick-disconnect couplings   Mating fluid-pipe couplings incorporating self-sealing shut-off valves to allow disconnection under pressure without loss of fluid.

quick-donning mask   Simple oxygen mask, usually of drop-down type but sometimes carried in separate pack, which in theory can be put on with one hand in a few seconds.

quick dump   Switch for getting rid of entire load of rockets, normally by jettisoning launchers.

quick-engine-change unit   See *engine-change unit*.

quickie GCA   GCA or PAR approach conducted in circumstances calling for priority (possibly notified emergency) and eliminating procedure or identifying turns.

quick-reaction alert   In 1960s it was calculated that Western Europe would have 4 minutes warning of a Soviet strike by nuclear missiles, so RAF bomber squadrons practised getting retaliatory V-bombers airborne within 4 min. of an 'out of the blue' warning at any time of day or night; USAF SAC had parallel exercises.

quick-release box   Parachute-harness latch released by 90° rotation and blow from hand.

quick-search procedure   One in which double normal number of aircraft are used and entire area is searched on outbound leg (NATO).

Quicktrans   Long term contract airlift within Conus in support of USN, USMC and DoD agencies (US).

quiescent current   Flow cathode/anode in absence of input signal.

quiescent flow   Small flow of air maintained throughout no-load period to keep bleed-air turbine running; eliminated in modern systems.

quiescent modulation   Amplitude modulation in which carrier is radiated only during modulation.

quiet automatic gain control   Automatic variation of bias of one or more amplifying stages preceding detector, in which output is suppressed for all signals too weak to trigger control.

quiet radar   Scans scene continuously with several thousand very narrow beams in rapid random FH sequence.

quiet Sun   Free from significant sunspots or unusual radiation.

quiet supersonic aircraft   Specifically means that any sonic boom generated does not reach the ground, or does so innocuously. See QSP, SSBD.

quill shaft   Slim drive shaft or driven shaft projecting as cantilever and terminating in splined coupling inserted into mating female portion. Requires no key, tolerates small misalignment, and absorbs torsional vibration. Can be keyed at both ends.

quilted blanket   Thermal insulating blanket with insulator, eg rock-wool, sealed in stainless-steel foil layers joined along criss-cross bonds; tailored to particular application.

quota flow control   Various methods of metering traffic to congested airports.

QUJ   Code: "Will you indicate true track to reach you?"; ie zero-wind heading.

QV   Quantitative visualization.

QVI   Quasi-vertical incidence, sounder for monitoring ionosphere.

QWI   Qualified weapons instructor.

QWIP   Quantum-well IR photodetector.

QW mechanism   Quick-wind (reconnaissance camera).

485

R

R *1* Generalized term for range [aircraft, missile, radar or other signal].

2 Generalized term for radius, from sheet-metal work to aircraft mission.

3 Resistance (electrical, fluid flow, marine aircraft on water), or resistor.

4 Reynolds number (also Re, R$_N$).

5 Resultant.

6 Gas constant.

7 US piston-engine code: radial.

8 Aircraft category, racer (USA 1919–24, USN 1922–28); transport (USN 1931–62); helicopter (USAAF 1941–47), reconnaissance (USAF 1948–62).

9 US Navy modified-mission prefix reconnaissance, suffix transport conversion (to 1962).

10 Generalized term for reliability, and probability thereof.

11 Moment of resistance (also M).

12 Radiance.

13 JETDS code: radio.

14 JETDS code: receiving only, ie passive detection.

15 Revenue.

16 Restricted area (ICAO).

17 Received, or receive only (ICAO).

18 Repair facilities available (FAA).

19 Right (runway designation with parallel pair).

20 Suffix: radial (thus 234°R = VOR bearing).

21 Generalized term for rate (eg code rate, data rate, rate of roll or turn).

22 Total rainfall.

23 Modulus of rupture.

24 US missile code for vehicle type; unguided rocket.

25 US missile code for launch environment: ship.

26 Rüstsatz, field conversion kit (G, WW2).

27 Microlight aircraft category (FAI).

28 Reject, rejected (EFIS or nav display).

29 Suffix, area navigation plus altitude-encoding transponder.

30 Red.

31 Airport surveillance radar.

32 Route, or route-tuned (navaid).

33 Reluctance [also denoted as S].

r *1* Radius, or radius of rotation.

2 Suffix: required.

3 Rotational speed (eg rpm); angular velocity in roll.

4 Ratio.

5 Rocket burn rate.

6 AMSU yaw-rate output.

7 Resistivity [ρ is preferred].

8 Spherical, or cylindrical, co-ordinate.

R* Universal gas constant, 8.31432 J k mol^{-1} K^{-1}.

R* Radar range at which probability of detection is * per cent.

Ṙ Range rate.

°R *1* Rankine, = $^5/_9$ K.

2 Réaumur.

R^{2}P Repair and return packaging.

R^{2}W Robot[ic] rotary wingman.

R^3 Reduced runway reliance, or reduced reliance on runways.

R-12 Refrigerant 12.

R-channel Though sometimes said to mean radio, R here means random-access. Used for air/ground signalling and data, see Rd and R$_{smc}$.

R-display Extended A-display enabling particular radar echo to be magnified for close examination.

R-stoff Mixture of 57% crude oxide monoxylidene and 43% triethylamine (G).

RA *1* Rain (ICAO).

2 Receiver attenuation.

3 Research association(s) (UK).

4 Right ascension.

5 Reliability analysis.

6 Radio altimeter.

7 Restricted article.

8 Runway/final approach.

9 Resolution advisory; see **TA*, **/VSI*.

10 Research announcement (Darpa).

11 Research Author (NACA, NASA).

12 Risk analysis.

13 Regional augmentation.

14 Reference axis, or [STA] altitude.

15 Rocket-assisted.

16 Relay assembly.

17 Radar summary map.

18 Radius of action.

19 Rate alarm [lightning frequency].

20 Region Aérienne (F, defence; NE, Atlantic, Mediterranean).

21 Routing area.

22 Reportable accident.

RAA *1* Regional Airline Association (US).

2 Roll-augmentation actuator.

3 Regional Airport Authority (UK).

RAAA Regional Airline Association of Australia.

RAAKS Russian association of aviation and space insurance companies.

RAAWS Radar altimeter and altitude warning system.

RAB *1* Registrar Accreditation Board (US).

2 Régiment Aérienne de Bombardement (F).

rabbit Video display of beacon response to two un-synchronized [e.g., alien] interrogating radars; also called running rabbits, or rabbit tracks.

rabbit lights Sequentially flashing lead-in approach lights (colloq.).

Rabdart Rapid aircraft battle-damage augmentation repair team.

Rabfac Radar beacon, forward air control.

RAC *1* Rules of the air and traffic control services (UK, and all AIPs).

2 Radiometric area correlation (guidance).

3 Radar-absorbing chaff.

4 Radar-aiming complex (R, Anglicized form).

5 Rulemaking Advisory Committee (FAA).

6 Régiment Aérienne de Chasse (fighter) (F).

RACE Real Aero Club de España.

race compound Secure park where aircraft are scrutinized by handicappers.

racecourse Race pattern.

race pattern Closed flightpath comprising two semi-circles joined by two parallel straight legs; one FL of stack.

race-pattern hold Normally selected for hold of longer than 2 min., latter being time for 360° turn; straight legs adjusted to give required hold unless pattern is established and published.

race rotation Gross rotation of spiral nature imparted by propeller to slipstream.

racetrack See *race pattern*.

raceway Major conduit or linear attachment along which are installed multiple cable looms and possibly fluid pipe-runs.

RACGAT Russian-American Co-ordination Group for ATC.

RACH Remotely actuated cargo hook.

Rachel Reperage acoustique d'hélicoptères (detection system, F).

racing the count Inaccuracy when using Consol or related navaid when flying across position lines close to station.

rack *1* Attachment for air-dropped store (see *hardpoint, pylon, ejector rack, station*).

2 Framework inside aircraft for accommodation of avionic equipments.

rack control Powered flight control with surface actuation by linear rack/pinion drive (suggested obs.).

racking Arrangement of installation of avionics according to boxes and attachment racks of standard dimensions published by ATR, Elfin etc.

RACO *1* Roll attitude cut-out.

2 Radar/video convertor.

racon Radar beacon transponder carried in aircraft for interrogation by ground station, eg SSR.

RACP Revolutionary Aeropropulsion Concepts Program (NASA).

RACR³A Royal Aero Club Records, Racing and Rally Association (UK).

RACS Redundant attitude-control system.

RACT Royal Aero Club Trust (UK).

RAD *1* Radar approach aid (FAA).

2 Radiation accumulated dose.

3 Rigid-aircraft dynamics.

4 Rapid application development.

5 Rapid-access data.

rad *1* VOR radial.

2 Radian.

3 Radiation dose absorbed, unit of radiant energy received: non-SI, = 0.01 J/kg.

RA&D Requirements analysis and design.

RADA Random-access discrete address.

Radag Radar aimpoint [or area] and guidance.

Radalt The reading on a radio altimeter.

Radan Radiation direction and nature (ECM).

radar Use of reflected EM radiation, normally with wavelength in RF spectrum between 30 m and 3 mm, to give information on distant target. Information may include range, range rate, bearing, height and relative velocity, or may be pictorial for reconnaissance purposes. (Originally US acronym, from radio detection and range, or ranging.)

radar-absorbent material Range of surface coatings, those in use security-classified, which greatly reduce strength of RF energy reflected back along incident path. It is not known if RAMs extend to substrates and structures.

radar advisory Message providing advice and information based on radar observation.

radar advisory service Outside controlled airspace, controller provides heading, distance (and, if known, FL) of conflicting non-participating traffic, together with any advice necessary.

radar aerial Portion of radar system used to radiate or intercept signals; US, = antenna.

radar altimeter See *radio altimeter*.

radar altimetry area Large and comparatively level terrain with defined elevation which can be used in determining altitude of airborne equipment by means of radar (DoD, NATO).

radar altitude Automatic FCS mode in which height AGL is maintained constant by autopilot slaved to radio altimeter.

radar approach Approach executed under direction of radar controller.

radar approach control Facility providing approach control service by means of ASR and PAR (USAF).

radar beacon Transponder carried by aircraft, missile or spacecraft which, when it receives correct pulse code from ground radar, immediately retransmits identifying code on same or different wavelength.

radar beam Energy emitted highly directionally because of antenna geometry, eg centimetric dish, decimetric yagi.

radar blip See *blip*.

radar bombing Level bombing using radar bombsight.

radar bomb scoring Aircraft transmits a signal at moment of simulated release of free-fall bomb. Plotters on ground determine precisely where an actual bomb would have fallen.

radar bombsight Sight for level bombing in which, irrespective of whether target can be viewed optically, numerical data on target relative position and velocity are fed by radar carried in aircraft.

radar boresight line That along axis of aerial.

radar calibration *1* Use of radar direction and distance information to check another system.

2 More often, use of accurately positioned aerial target to check accuracy of ground radar.

radar camouflage Use of radar-absorbent or reflective materials to change radar-echoing properties of object's surface; does not include dispensed countermeasures, jamming or other active technique.

radar clear range Bombing or firing range which accepts responsibility for avoiding danger to aircraft straying into it.

radar clutter See *clutter*.

radar command Command guidance in which target and missile positions and velocities are continuously determined by radar.

radar contact Identification of echo on radar display, esp. as that sought. When thus advised, in civil ATC, aircraft ceases normal reporting.

radar control Control of air traffic based upon position/height information supplied by radars.

radar-controlled gun AAA gun, aircraft turret or other gun system whose aim is controlled by radar and computer which feeds all information necessary (sightline spin, lead etc).

radar controller Air traffic controller whose positional information is provided by radar displays, and holding radar rating appropriate to assigned functions.

radar countermeasures See *countermeasures, ECM*.

radar coverage Limits within which objects can be effectively detected by radar(s) at given site or installation; may be angular, polar diagram or in other terms.

radar cross-section, RCS Apparent size of target as judged by its displayed echo, determined (in absence of ECM activity) by true size, range, aspect, geometric shape, materials, surface texture and treatment and other factors including intervening dust and precipitation. Normally defined by ratio P_r/P_s where P_r is radar power received at target and P_s is power reflected, plotted as polar (1) in horizontal plane.

radar display Visual electronic display of radar information.

radar distance Distance to target and return, thus 1 radar-ft = 2 ft.

radar echo Signal indication of object which has reflected energy back to radar.

radar element Radar as portion of large system. e.g. PAR as part of overall GCA.

radar fire Gunfire guided by radar, or against radar-tracked target.

radar fix Obtained from PPI radar map display.

radar flight-following General observation of progress of identified aircraft targets sufficiently to retain their identity or observation of specific radar targets (FAA).

radar foot *Radar distance*.

radar fuzing Comprehensive duplicated radar altimeters are installed in many NW to detonate device at selected height above surface.

radar gunlaying Aiming using radar target position/velocity and, usually, radar input of range; can be manual, automatic with manual override or fully auto.

radar handover Transfer of control using radar.

radar horizon At any location, line along which direct radar rays are tangential to Earth's surface; in practice, usually same as radio horizon.

radar identification Use of transponder or procedure manoeuvre to establish positive identification of object seen on radar.

radar imagery Imagery produced by recording radar waves reflected from target surface in air/surface reconnaissance.

radar indicator Radar display.

radar information service Controller informs pilot of heading, distance (and, if known, FL) of conflicting traffic, but does not offer advice.

radar integration Automatic integration of information from air-defence, naval and other primary and secondary radars.

radar intelligence item Feature which is radar-significant but which cannot be identified exactly at moment of its appearance.

radar map Cartographical information superimposed on radar display.

radar mapping Cartography based upon radar, esp. SLAR.

radar mile Unit of time equal to 10.75 µs; time for EM radiation to reach target 1 statute mile distant and return. Rarely, 12.369 µs [also given as 12.359] for nautical mile.

radar monitoring One of three types of radar service afforded by civil controllers; radar flight-following of aircraft navigated by own pilots or crews, and advice on deviations from track, possible conflicting traffic or progress on instrument approach from final approach fix to runway.

radar navigation guidance One of three types of radar service, ground vectoring of aircraft to provide course guidance.

radar netting Linking of several radars (surveillance, HFR or similar air-defence radars) to single centre to provide integrated target information.

radar netting station Centre which can receive data from radars and exchange these among other radar stations, thus forming netting system.

radar netting unit Optional electronic equipments converting air-defence fire-distribution system command centre to radar netting station.

radar performance Usually means peak power divided by minimum detectable signal power, but there are other meanings.

radar picket Ship, aircraft or vehicle stationed at distance from surface force for purpose of increasing radar detection range. Hence, ** combat air patrol.

radar picket escort Surface vessel dedicated to ESM, ECM and electronic search facilities. USN designation DER.

radar position symbol Computer generated.

radar prediction Graphic portrayal of estimated radar intensity, persistence and shape of cultural and natural features of specific area (USAF).

radar range Commonly means distance at which particular object can be detected with (1) 100% reliability or (2, more usual definition) 50% reliability.

radar ranging Use of radar to obtain continuous input of target range.

radar receiver Ambiguous term when used alone; can mean receiver of radar that also transmits, or passive equipment used in ESM, ECM or radar warning.

radar reconnaissance Reconnaissance using radar(s) to determine nature of terrain and enemy activity.

radar reflectivity As in optics, fraction of incident radiation reflected by target, normally measured on unit area perpendicular to radiation. Modified by radar camouflage and RAM (2).

radar response Visual indication on radar display of signal transmitted by target following radar interrogation.

radar return See *radar echo*.

radar scan See *scan*.

radar scanner Radar aerial (antenna) able to scan.

radarscope Radar display in which information is presented visually for human assessment.

radarscope overlay Transparent overlay for comparison and identification of radar returns.

radarscope photography Film record of radar display [cassette giving way to CD].

radar screen Radar display.

radar separation Radar service in which air traffic is spaced in accord with established minima.

radar service Monitoring, navigation guidance or separation.

radar signal film Film on which are recorded all signals acquired by a coherent radar and viewed or processed through optical correlator to permit interpretation.

radar signature See *signature*.

radar silence Imposed discipline prohibiting transmission by radar on some or all frequencies.

radarsonde, radar-sonde Meteorological facility in which balloon carries instrumentation recording temperature, pressure and other atmospheric data which are transmitted when triggered by secondary radar whose observation of balloon position when related to time and known rate of ascent gives wind velocities at different heights.

radar surveillance Radar observation of given geographical area or airspace for purpose of performing radar function, eg traffic control or defence.

radar target designator control Automatically moves acquisition symbology to bracket target prior to lock-on.

radar tracking station Radar facility with capability of tracking moving targets.

radar track position Extrapolation of aircraft future position by computer based upon radar information and used by computer for tracking.

radar vector Heading (course to steer) issued as part of radar navigation guidance.

radar weather echo intensity Scale of six levels of intensity giving rough idea of likely turbulence (NWS).

Radat Freezing-level data.

RAD/BAR Radio/pressure altimeter, or selector switch between both.

RADC Rome Air Development Center, New York state (USAF).

Radcon Rapid detection of concealed time-critical targets.

RADE Receive antenna distribution equipment.

Radel Trade name for resins and thermosetting plastics (Amoco).

Radex *1* Rapid-deployment exercise.

2 Ground-based calibrated transmitter/receiver for testing navaids.

radiac Adjective meaning radioactivity detection, indication and computation; applied to radiological instruments and equipment.

radiac dosimeter Measures aggregate ionising radiation received by that instrument.

radial *1* Piston engine whose cylinders are arranged radially, like spokes of wheel; unlike rotary, cylinders fixed, propeller driven from crankshaft.

2 Magnetic bearing extending from point-source navaid, eg VOR, Tacan, Vortac; usually QDR.

3 Tyre [tire] construction in which rubber casing is coated with ply cords arranged transversely and extending to but not under the bead, stabilized by stiff circumferential belt.

radial cancellation Methods of reducing propeller or propfan noise by sweeping [usually back] the blades.

radial compressor Centrifugal compressor.

radial displacement Distortion of tall buildings in low-level reconnaissance photos.

radial drill Machine tool having drill chuck carried on pivoted radial arm of variable length.

radial engine See *Radial (1)*.

radial error Distance between desired point of impact of munitions and actual point, both points projected and measured on imaginary plane perpendicular to munition flightpath.

radial error probable That circle drawn on radial error plane through which 50% of actual munitions pass, with centre at projected target position.

radial flow Fluid flow inwards or outwards along substantially radial path, usually outwards in supercharger or centrifugal compressor and inwards in * turbine.

radial flyability Unquantifiable measure of ease with which pilot can accurately hold radial (2), esp. when near station; varies with terrain-induced errors and other factors which distort or otherwise influence signals.

radial GSI, RGSI Replaced DME in many aircraft; essentially DME panel instrument with inbuilt wind correction facilitating choice of FL giving best ground speed.

radial staging *1* Gas-turbine or reheat combustor with fuel fed selectively to two or more rings of burners at different radii from engine axis.

2 Rocket vehicle able to shed stages at different radii from major axis.

radial struts Those connecting inner and outer ridge main joints of airship transverse frames.

radial temperature distribution factor Circumferentially measured combustor outlet peak gas temperature minus outlet mean temperature divided by mean combustor temperature rise.

radial velocity Velocity of approach or recession between two bodies, ie component of relative velocity along line connecting them.

radial wall jet Outward flow along ground beneath jet-lift aircraft or helicopter hovering in ground effect.

radial-wing configuration Use of several, eg four, wings mounted radially to permit flight manoeuvre instantaneously along any plane containing longitudinal axis without prior need for roll.

radial wires Join vertices of airship's main transverse frames to central fitting or to those diametrically opposite.

radian SI unit of plane angle; angle subtended at centre of circle by arc equal in length to diameter, rad = 57.2958° = 57°17'44.8".

radiant energy EM radiation, eg heat (IR), light, radio and radar. Arguably, also occasionally used for other energy, esp. acoustic.

radiant-energy density Instantaneous value for amount of energy in unit volume of propagating medium, symbol u. With pulse radars depends on pulse length and position.

radiant-energy thermometer Instrument which determines black-body temperature; emitter need be 'black' only over range of wavelengths studied.

radiant flux Time rate of flow of radiant energy, ø.

radiant-flux density Radiant flux per unit area; when applied to source, called radiant emittance, radiancy, symbol W; when applied to receiver, called irradiance or (not recommended) irradiancy, symbol H.

radiant intensity Radiant flux per unit solid angle, measured in given direction, SI unit W/sr.

radiant temperature That recorded by total-radiation pyrometer; when sighted on non-black body is less than true temperature.

radiating element Any portion of radar aerial, esp. one of electronically scanned type, which emits or reflects transmitted energy.

radiation   *1* Process by which EM waves are propagated.

2 Process by which other forms of energy are propagated, eg heat from solid body, kinetic by ocean waves and sound waves through atmosphere or other medium.

3 Other forms of energy propagation such as nuclear *, high-energy ionising *, radioactivity.

radiation area   Place where human being could receive 5+ mrem/h or total of 150 mrem in 5 consecutive days.

radiation belts   Belts of charged particles trapped within planetary magnetic field, esp. Van Allen belts.

radiation burn   Damage to skin caused by ionising radiation.

radiation constants   Two physical constants: First ** = $2\pi hc^2 = 3.7418 \times 10^{-16}$ Wm^2; Second ** = hc/k = 1.4388 × 10^{-2} mK.

radiation cooling   Cooling by direct radiation from surface, normally implying high temperature (eg rocket thrust-chamber skirt).

radiation dose   Amount of ionising radiation absorbed by substance, esp. over short period of time (see *rad* [3], *roentgen, rem*).

radiation dose rate   Dose per unit time.

radiation efficiency   Ratio of power radiated to power supplied to transmitting aerial at given frequency.

radiation field   Volume occupied by radiation (1), esp. that around conductor carrying AC or RF, comprising electrical and magnetic (inductive) components.

radiation fog   Usually shallow fog caused by radiative cooling (often at night) of ground to below dewpoint, combined with gentle mixing, saturation and condensation.

radiation hardening   Gradual hardening and embrittlement of most metals exposed to intense nuclear radiation, resulting in limited choice of metals for high-integrity structures in such environments.

radiation illness   Disorders, some fatal, caused by excessive exposure to ionising radiation.

radiation intelligence   That derived from collection and analysis of non-information-bearing radiation unintentionally emitted by foreign devices, excluding that generated by detonation of NW.

radiation intensity   Radiation dose rate, normally that measured in air. RI-3, eg, is value 3 h after NW burst.

radiation laws   Those describing black-body radiation: Stefan-Boltzmann, Planck, Kirchoff and Wien.

radiation lobe   See *lobe*.

radiation medicine   Branch dealing with radiation and human beings.

radiation pattern   Graphical representation of radiation (1) of device, esp. radio, radar or similar emitting aerial, plotted as field strength as function of direction. Normally plotted as plan-view polar or as vertical cross-section, but can be in plane of magnetic or electrical polarization of waves. Free-space ** depends on wavelength, feed system and reflector. Field** takes account of real situation in which waves are reflected from ground or other objects so that direct and reflected waves interfere with each other. Also called coverage diagram, aerial (antenna) pattern, lobe pattern.

radiation pressure   That exerted on solid body by incident radiation (1), symbol P_r.

radiation pyrometer   Pyrometer measuring light wavelengths and giving readout in terms of temperature.

radiation scattering   Diversion of radiation (EM, thermal or nuclear) caused by collision or interaction with atoms, molecules or large particles between source (esp. NW explosion) and remote site; thus radiation is received from many directions.

radiation sickness   See *radiation illness*.

radiation situation map   One showing actual and/or predicted radiation situation, usually intensity, in particular ground area.

radiation source   Generally a man-made, portable, sealed source of radioactivity.

radiator   *1* Source of radiant energy, esp. EM or RF, eg hostile operating radar.

2 Heat exchanger, esp. for rejecting unwanted heat to a sink. Common usage restricts term to devices that dump heat overboard, eg to atmosphere, and to call those that use heat sink on board (eg fuel) heat-exchangers.

radiator header   Tank in which liquid coolant is received from heat source, eg engine, and distributed to cooling elements, ie radiator(s).

Radic   Rapidly deployable integrated command and control system, links navies to Nadge.

Radil   ROCC/Awacs digital information link.

radio   *1* Use of EM radiation between about 5 kHz and 3 THz to convey information.

2 Qualifying adjective denoting that a height above ground has been measured by radio altimeter, thus '50 ft *'.

radioactive ionization gauge   Measuring device in which ions produced by radiation (usually alpha particles) from radioactive source discharge a capacitor.

radioactivity   *1* Spontaneous disintegration of nuclei of unstable isotope yielding alpha and/or beta particles, often accompanied by gamma radiation.

2 Number of spontaneous disintegrations per unit mass per unit time, usually measured in curies.

radioactivity concentration guide   Published values (DoD, NATO etc) of quantities of listed radioisotopes permissible per unit volume of air and water for continuous consumption.

radio altimeter   Instrument, invariably of CW FM type, giving readout of height AGL by time-varying frequency and measuring difference in frequency of received waves, this being proportional to time and hence to height. Sometimes called radar altimeter.

radio approach aids   Those which assist landing in bad visibility, notably ILS, MLS; also called radio or electronic landing aids.

radio astronomy   Study of radio emissions received by Earth, esp. those which can be associated with source of EM emissions on visible, X-ray or other wavelengths.

radio bands   Artificially divided segments of the EM spectrum, listed in Appendix 2.

radio beacon   Fixed ground station emitting RF signals, esp. those containing identifying information, which enable mobile stations to determine their position relative to it.

radio beam   Transmitted by directional antenna to maximize radiated power at long range.

radio bearing   Usually, angle between apparent direction of fixed station and a reference direction, eg, true or magnetic N. Hence true **, magnetic **.

radio biology Often written as one word, study of effects of radiation on life.

radio channel One band of frequencies sufficient for practical radio communication; sum of emission bandwidth, sideband spread (interference guard bands) and tolerance for frequency variation.

radio check Request to ground station to transmit to confirm audibility [readability].

radiochemistry Chemistry of radioactive materials.

radio command Command guidance using a radio link.

radio compass Originally, fixed-loop receiver with which aircraft could home on to any selected fixed station. Later superseded by ADF and other navaids.

radio control Vague, but generally means control of vehicle trajectory with commands transmitted over radio link.

radio countermeasures Those branches or activities of ECM concerned with telecommunications.

radio coupling box Inputs ADF, VOR, ILS, etc, to autopilot.

radio deception Use of radio to deceive enemy; includes sending false despatches, using deceptive headings and employing enemy callsigns (DoD, IADB).

radio detection Detection of object's presence by radio, without information on position.

radio determination satellite system Satellite system which enables receiver stations to determine position, velocity or other characteristics by propagation of radio waves.

radio direction-finding See *direction finding*.

radio direction-finding database Aggregate of information, provided by air and surface means, necessary to support radio D/F operations to produce fixes on target transmitters/emitters.

radio duct Shallow quasi-horizontal layer(s) in atmosphere wherein temperature and moisture gradients result in abnormally high refraction lapse rate, causing RF signals to become trapped within layer (see *anomalous propagation, skip effect*).

radio energy That which propagates at radio frequencies.

radio facility chart Original series of US airway maps based on radio range; name still common for modern air maps showing all facilities, airways, control zones, TMAs, etc.

radio fix *1* Fix of mobile station, eg aircraft, obtained by use of radio navaid, esp. by traditional crossings of position lines.

2 Geographical location of friendly or, esp., enemy emitter obtained by various ESM and D/F techniques.

radio frequencies Abb. RF, those EM frequencies used for radio or related purposes. Common-use subdivisions are: VLF, below 30 kHz; LF, 30–300 kHz; MF, 300 kHz–3 MHz; HF, 3–30 MHz; VHF, 30–300 MHz; UHF, 300 MHz–3 GHz; SHF, 3–30 GHz; EHF, 30–300 GHz; unnamed, 300 GHz–3 THz. See Appendix 2.

radio goniometer See *direction-finder*.

radiography Photography using X-rays, gamma rays or other ionizing radiation; important NDT method, often using radiation source inside test object and film outside.

radio guard Radio station, eg aircraft, which listens out on assigned frequencies and handles traffic and records transmissions.

radio guidance Guidance or navigation system using

radio waves, eg point-source aids, area coverage (R-Nav), global (Omega) and command methods.

radio height Height above ground measured by radio altimeter.

radio hole Direction of propagation suffering abnormal attenuation or fading, usually caused by local refraction.

radio horizon At any location, line along which direct rays from RF transmitter become tangential to Earth's surface; extends beyond visual horizon because of atmospheric refraction, and varies according to whether propagation is sub-, normal or super-standard.

radio-inertial guidance Various systems combining inertial and radio tracking and/or command (probably obs.).

radio interferometer Interferometer operating at RF.

radioisotope Unstable isotope that decays spontaneously, emitting radiation.

radioisotope thermoelectric generator Self-contained power system in which a radioisotope is used to heat one junction in a circuit containing dissimilar metals and thus generate sustained electricity.

radiolocation Original UK name for radar.

radiological defence Measures taken against radiation hazard resulting from use of NW and RW.

radiological survey Directed effort to determine distribution and dose rates of radiation in an area. Hence ** flight altitude.

radiological weapons Established forms of radioactive materials or radiation-producing devices, including intentional employment of NW fallout, to cause casualties or restrict use of terrain; abb. RW.

radiology Science of ionizing radiation in treatment of disease.

radio loop D/F loop aerial.

radioluminescence Emission in visible EM range, typically with characteristic hues, in radiation from radioactive materials.

radio-magnetic indicator Magnetic and radio panel instrument in which card (dial) rotates so that heading appears opposite index mark at top of case while needle rotates to show magnetic bearing (QDM) of tuned beacon.

radio marker beacon See *marker*.

radio mast Mast projecting above or below aircraft, esp. older aircraft, usually as one anchor for MF/HF wire aerial. Modern VHF blade seldom thus termed.

radiometeorograph Meteorograph transmitting readings by radio (see *radio-sonde*).

radiometer Instrument for detecting, and usually also measuring, IR and closely related radiation (see *bolometer, actinometer, photometer*).

radiometric resolution This is usually measured in bits.

radio navigation All uses of radio to determine location, obtain heading information and warn of obstructions or hazards.

radionuclide Nuclide spontaneously emitting radiation.

radiophotoluminescence Photoluminescence exhibited by substance after irradiation by radionuclide or other source of beta or gamma rays.

radio proximity fuze See *proximity fuze*.

radio range Original radio navaid (US), a land (rarely, ship) fixed station of aeronautical radio service broadcasting continuous coded signals which on one side of an airway are heard as a Morse N (-·) and on other as A (·-); in a less common system signals are I (··) and A. In the equisignal zone along centre of airway both signals combine, A and N forming continuous note and I and A cancelling to send no signal to pair of reeds which, as soon as aircraft strays from centreline, vibrate visually and aurally. Now replaced by VOR and later navaids.

radio rangefinding   Use of radio (essentially in this case not radar) to determine range.

radio-range orientation   Technique needed when flying radio range of finding and identifying station and then accurately flying correct leg to next station or destination.

radio recognition   Determination by radio means of another station's identity; DoD wording is 'friendly or enemy character, or individually, of another'.

radioresistance/radiosensitivity   Measures of resistance or sensitivity of living cells to injurious radiation.

radio silence   Period during which some or all RF emitters (eg of military force) are kept inoperative.

radio-sonde   Instrumentation for measurement of atmospheric data, usually temperature, pressure and humidity, carried aloft by balloon together with electronics for converting answers into code for RF transmission to ground station at intervals (see *radarsonde*).

radio sonobuoy   Sonobuoy which transmits information to friendly receivers.

radio telegraphy   Transmission of telegraphic codes by radio; often one word.

radio telephony   Transmission of speech by radio, R/T, often one word.

radio-telephony network   Integrated group of aeronautical fixed stations which use and guard frequencies from same family and co-operate to maximize reliability of air/ground communications.

radio telescope   Radio receiver capable of greatly amplifying, recording and determining source-direction of radio waves received on Earth from outer space.

radio waves   EM radiation with wavelengths from tens of kilometres (VLF) down to tens of microns (above EHF, so-called decimillimetric waves). See Appendix 2.

radius   See *radius of action*.

radius block   Steel block with edge of accurate radius for sheet-metalwork.

radius dimpling   Pressing dimples in thin sheet with hemispheric or cone-shaped mating tools.

radius gauge   Hand instrument for measuring inside or outside-bend radii.

radius of action   Maximum distance aircraft can travel away from its base along given course with normal combat load and return without refuelling, allowing for all safety and operating factors. Today, esp. with gas-turbine propulsion, *** varies greatly with mission profile, fuel consumption being several times higher in lo regime with afterburner lit.

radius of damage   Radius from ground zero within which there is 50% probability of achieving desired damage.

radius of gyration   Distance from body's centre of gyration to selected axis; square root of ratio of moment of inertia, about selected axis, divided by mass.

radius of integration   Radius from ground zero within which effects of NW and conventional weapons are to be integrated.

radius of safety   Horizontal distance from ground zero within which NW effects on friendly troops are un-acceptable.

radius of turn   For conventional aeroplane $R = V^2/_g \tan \theta$ where V is TAS and θ is angle of bank.

radius rod   Major bracing strut in retracting landing gear; normally pivoted near lower end of main leg and upper end pinned to backlink, actuator or sliding block, but many arrangements all with same name.

radius vector   Vector connecting body with object which may have relative motion; specif., vector joining primary body to satellite, as in Earth/Moon system.

radix point   In any number system, index separating numbers having negative powers from those having positive; eg decimal point, binary point.

Radnet   Radar-data interchange network [Eurocontrol].

radome   Protective covering (and in aircraft aerodynamic fairing) over radar or other aerial, esp. one with mechanical scanning; made of dielectric material selected according to operating wavelength and other factors.

Radop   Radar/optical tracking methods.

Radot   Recording automatic digital optical tracker.

Radpac   Radar package (Panavia); versatile software to extend radar capability, esp. in air-to-air missions.

rad/s   Radians per second.

RADU   Range and azimuth display unit.

RAE   Royal Aerospace (until 1988 Aircraft) Establishment (MoD PE), later called DERA Farnborough, now defunct. Britain's greatest aeronautical research centre, at Farnborough, Hampshire; in 1912 called Royal Aircraft Factory, 1918 Royal Aircraft Establishment, in 1988 Royal Aerospace Establishment, now closed; an outstation opened as RAE Bedford in 1955.

RAeC   Royal Aero Club (UK, founded 29 October 1901).

RAeS   Royal Aeronautical Society (UK, founded 1866).

Raevam   RAE variable-aerofoil mechanism; infinitely adjustable flexible leading edge.

RAF   *1* Royal Aircraft Factory (UK, 1912–18); also Royal Air Force, but military services as such are excluded from this dictionary, as are airlines.

   2 Recursive adaptive filter.

   3 Requirements analysis folder.

RAFA   The Royal Air Forces Association (1943–).

RAFBF   Royal Air Force Benevolent Fund; E adds Enterprises (Fund founded October 1919).

RAFC   Regional area forecast center/centre.

RAFCS   Redesignated AFCS.

RAFES   RAF Escaping Society, charity for assisting those who helped Allied aircrew evade or escape capture in WW2.

Rafic   Radar and flight-information capture (noise-monitoring software).

RAFL   Rainfall.

RAFO   Reserve of Air Force officers (UK).

RAFSEE   RAF Signals Engineering Establishment (Henlow).

RAFSPA   RAF Sport Parachute Association.

Rafts   Reconnaissance/attack/fighter training system (USAF).

RAG   *1* Replacement Air Group, supplying aircrew to carriers (USN).

   2 Runway arrester gear.

   3 Ragged cloud.

   4 Range/azimuth gating.

rag and tube   Constructed from tubing, with fabric covering (US, colloq.).

RAGS   Mixed rain and hail.

rag-wing   Fabric-covered aircraft (US, colloq.).

RAHE   Ram-air heat-exchanger.

RAHR   Royal Artillery Hebrides Range (UK).

RAI   *1* Registro Aeronautico Italiano; national licensing authority.

2 Runway alignment indicator (ICAO).

3 Remote attitude indicator.

4 Regional Airspace Initiative.

5 Radio-altitude [or altimeter] indication [or indicator].

6 Reconnaissance/attack interface.

7 Ram-air inflation.

RAID   *1* Redundant array of independent disks.

2 Rapid alerting and identification display.

3 Rapid aerostat initial deployment (USA).

Raider   Replacement advanced intelligent dual ejector rack.

Raidrs   Rapid attack identification detection and reporting system.

Raids, RAIDS   *1* Rangeless airborne instrumented debriefing system, see Urits.

2 Radar airborne intrusion detection system.

RAIL   Runway alignment indicator light; S adds system (FAA).

rail   Launcher on ground or aircraft mount for missile requiring support and/or guidance while accelerating.

rail-drive airbridge   Can move in/out telescopically, and in vertical plane, but not in azimuth.

rail garrison basing   Deployment of ICBM (LGM-118A) units wholly contained within US railroad trains.

railgun   General term for family of actual or possible space weapons whose projectiles are accelerated by EM effects.

RAILS   See *RAIL*.

RAIM   Receiver autonomous integrity monitor(ing).

Rain   Reduction of airframe and installation noise.

rain   Precipitation in form of water droplets making noticeable individual impact, diameter roughly 1 to 5.5 mm.

rain band   Spiral cloud area of tropical revolving storm where there is heavy rain.

rainbow(ing)   Appearance in laminated transparency of multicoloured (spectral) stress patterns.

rainfall   Term for radioactive precipitation from base-surge clouds after underwater NW burst (DoD).

rain gauge   Various forms of precipitation gauge designed primarily for measuring fall of water droplets, usually by funnelling fall over known area into calibrated container.

rain ice   Most dangerous airframe icing; similar to glaze ice and caused by supercooled raindrops which take sufficiently long to freeze for flowback to be extensive.

rain loop   See *drip flap*.

rainmaking   See *seeding (2)*.

rain-out   *1* Removal of solid particulate pollution by rain.

2 Radioactive material in atmosphere brought down by precipitation and present on surface (DoD).

rain static   Radio interference believed to be caused by rain bearing electrostatic charges.

RAIS   Redundant array of inexpensive systems [architecture].

RAISD   Research and Acquisition Information Systems Division (Andrews AFB).

RAJ   Ring-around jammer (ECM).

Rajpo   Range Joint Project Office.

rake   *1* Angle between local vertical and swivel or castor axis of swivel-mounted wheel, eg tailwheel.

2 Angle between quasi-straight edge of wingtip and aircraft longitudinal axis, called positive * when leading edge is shorter than trailing edge and negative * when trailing edge is shorter.

3 Distance, measured parallel to aircraft longitudinal axis, between front of propeller blade at tip and plane of rotation through axial mid-point of hub (simple fixed-pitch propeller).

4 Acute angle between line joining centroids of propeller blade from root to tip and plane of rotation (often zero and possibly synonymous with 3).

5 Comb, linear or other array of pitot heads.

raked tip   Sharp sweepback on extreme outer section of wing, propeller/helicopter blade or other aerofoil.

raking shot   Gunfire almost aligned with hostile aircraft longitudinal axis.

RAL   Rutherford Appleton Laboratory (UK, SERC, now part of CLRC).

Ralacs   Radio-altimeter low-altitude control system (RPV).

RAL beacon   Downwind of threshold, shows runway alignment.

RALS, Rals   Remote augmented lift system.

RALT   Radar altitude.

RAM   *1* Random-access memory.

2 Radar-absorbing [or absorbent, or attenuating] material(s).

3 Research and applications module (Space Shuttle).

4 Rapid area maintenance (AFLC).

5 Raid-assessment mode (radar).

6 Reliable/available/maintainable.

7 Rolling-airframe missile.

8 Route-adherence monitoring.

ram   *1* Increase in pressure in forward-facing tube, duct, inlet etc as result of vehicle speed through atmosphere: if fluid flow were brought to rest in duct pressure would be q, dynamic pressure. Hence * inlet, * pressure, * -jet, * air, * effect.

2 Main movable portion of hydropress; term not encouraged for most hydraulic devices having linear force output, for which actuator is preferred.

ram air   Air rammed in at forward-facing inlet.

ram-air parafoil   Flexible double-membrane wing inflated for rigidity by ram air.

ram-air temperature   That of ram air brought to rest with respect to vehicle; stagnation temperature.

ram-air turbine   Small windmill extended into slipstream to provide shaft power for essential services [eg., flight control or electric power] following total engine or electrical failure.

Ramana   Role of agile management in aerospace.

Raman effect   Scattering, with change in polarization and wavelength, of light passing through transparent solid, liquid or gas.

ramark   Fixed radio beacon continuously transmitting (sometimes coded signal) to cause radial line on PPI radars.

ramburner   Turbofan which at Mach 2.5–4 closes off core to become a ramjet.

RAMCC Regional Air-Movement Control Center (US).

ram compression See *ram (1)*.

Ramdi, RAMDI Radioactive miss-distance indicator.

ram drag See *momentum drag*.

Ramics Rapid airborne mine-clearance system (USN).

ram inlet Forward-facing inlet designed to achieve maximum ram recovery, also called ram [or ram-air] intake.

ramjet Air-breathing jet engine similar to turbojet but without mechanical compressor or turbine; compression is accomplished entirely by ram (1) and is thus sensitive to vehicle forward speed and non-existent at rest (hence * cannot start from rest). Inefficient at Mach numbers below 3 but extremely important for unmanned vehicles, esp. in conjunction with rocket (eg ramrocket). Also called athodyd, Lorin duct; not to be confused with pulsejet or resonant ducts.

RAM net Camouflage net made of RAM (2).

Ramos Remote automatic meteorological observing station.

RAMP, Ramp *1* Robotic applications for modular payloads (USA).

2 Radar modernization program.

3 Reconnaissance avionics maintainability program (USAF).

ramp *1* Main aircraft parking area at airport, airfield, airbase.

2 Sharp-edged wedge with sloping wall forming inner wall of supersonic inlet duct to create oblique shock(s) and improve pressure recovery, esp. at supersonic speeds; usually has variable geometry.

3 Inclined track for launch of target, RPV, UAV or missile under moderate acceleration.

ramp capacity Number of aircraft, of specified general size class, for which ramp (1) is designed, including nose-in and off-terminal parking.

ramp check Visual external inspection of aircraft plus replenishing hydraulic fluid, oil, water and other consumables, plus checking tyres and brakes.

ramp extension Increase in size of ramp (1) to augment capacity or allow for larger aircraft with wide turning circles.

ram pressure Ram (1).

Ramps Resource allocation and multipath scheduling.

ramp services All services needed by civil aircraft on transit stop or turnround, normally excluding mechanized freight handling and supplies (eg pantry/bar stocks) brought from terminal.

ramp status Accorded a new prototype after it has gone 'out the door' and is being readied for taxi tests.

ramp-to-ramp See *block time*.

ramp up To increase (US).

ramp weight, MRW Maximum weight permissible for aircraft, equal to MTOW plus fuel allowance for main engines and APU for start, run-up and taxi.

ram recovery Pressure actually achieved in ram inlet, expressed as absolute value or as percentage of total available dynamic pressure.

ramrocket Important species of propulsion system for unmanned vehicles, comprising rocket (solid, liquid or hybrid) and integral ramjet propulsion. Vehicle is launched by rocket, at conclusion of whose burn at supersonic speed nozzle is jettisoned, leaving larger con-di

nozzle, air inlet is extended or opened and ramjet operation takes place with combustion of liquid or hybrid type in original solid motor case and chamber (see *ducted rocket*).

Ramrod Day attack by bombers escorted by fighters with primary objective destruction of target (WW2).

RAMS, Rams *1* Remote automatic multipurpose station.

2 Removable auxiliary memory set.

3 Rapid assembly of munitions system, refined procedures for assembling munitions over sustained period (USAF).

4 Reliability, availability, maintainability and safety.

5 Reorganized ATC mathematical simulator (Eurocontrol).

6 Regional atmospheric modelling system.

Ramsbottom A standard procedure for determining carbon residue left after combustion of lubricating oils.

Ramses Radar mode-S evaluation system.

ram's horn *1* Pilot flight-control yoke generally resembling ram's horn, more upright than spectacles.

2 Microwave aerial of spiral (usually exponential) form.

RAMU Removable auxiliary memory unit.

ram void pressure Achieved pressure (pressure recovery) inside inlet duct, esp. at supersonic speed; ratio **/freestream total pressure is function controlling dump doors on SSTs.

ram wing Vehicle designed to fly in ground effect; arguably synonymous with ekranoplan.

ram yoke Carrier, usually one of a pair, for AAM, powered by cartridge to separate missile under peak negative-g conditions.

RAN Regional Air Navigation panel (ICAO).

RANC Radar attenuation, noise and clutter (US, DNA).

R&A Report and accounts.

R&C Rolled and coined.

R&D Research and development.

R&E Radio and electronic, as distinct from A&E.

R&M *1* Reports and memoranda (UK).

2 Reliability and maintainability (US).

R&O Repair and overhaul.

random access Ability of computer memory to remember contents and addresses of all memory stores immediately; access time is independent of location of preceding record.

random-demand planning Planning for supplies necessitated by in-service failures.

random energy That of fluid particles in disordered motion, rapidly degrading to heat, eg downstream of shock.

random error Unpredictable, caused by short-period disturbances in system or in measuring method, normally having Gaussian distribution over period; excludes major failures, human errors and errors of systematic nature.

random flight Unplanned local flight by light aircraft, esp. one without radio.

random R-nav routes Direct routes making full use of R-nav capability.

random scatter See *scatter (1)*.

R&QA Reliability and quality assurance.

R&R *1* Routeing and record.

2 Rest and recreation, or Rest and recuperation.

3 Recovery and repair.

4 Repair and return.

Rands Range, nose, distance, speed.

R&SU Repair and Servicing Unit (RAF).

R&T Research and technology [A adds acquisition].

range *1* Distance aircraft can travel, under given conditions, without refuelling in flight. By itself has little meaning, except for very small, simple aircraft. Maximum-fuel * normally taken to mean IFR reserves for multi-engine aircraft, VFR for single. Calculations for military aircraft traditionally assume external tank(s) retained, ammunition not fired, includes distance during climb but not fuel for warm-up, takeoff or reserve. Definitions of what constitute short- *, ultra-long * etc, have never ceased to proliferate, but may soon firm up. One formula is $R=W_fV/_{c'}F$ where W_f is mass of fuel [in practice, of usable fuel], V is TAS, c' thrust sfc and F net propulsive force. See Breguet.

2 Limiting distance, over intercontinental * measured as great circle, missile, RPV or other unmanned vehicle can travel with specified load and following specified flight profiles.

3 Distance between observer or weapon launch point and target.

4 Land and/or water area equipped and designated for vehicle testing, esp. missile, UAV or RPV, or testing ordnance or practice shooting at targets.

5 Difference between upper and lower limits for variable, eg frequency or wavelength coverage of receiver, pitch of propeller blades or many other variables.

6 To organize aircraft on carrier flight deck into desired sequence with closest packing.

range ambiguity In several early radio navaids, eg radio range, possibility of flying reciprocal or obtaining misleading distance indication.

range analog bar Horizontal bar appearing in some optical, HUD or radar sight displays once full lock-on has been achieved, its length showing range.

range and bearing launch Missile has these parameters preset, flies on bearing and at last possible moment switches on radar seeker to acquire target.

range attenuation Inverse-square decrease in radar power density with range.

range/azimuth display unit Various instruments, esp. HSI giving VOR/DME information.

range bar Bold, usually horizontal, failure warning flag for panel instruments.

range bin Store location in SSR plot extractor; each range increment on given azimuth has store location from range 0 to range limit in which detected targets are stored until end of scan, when each is extracted and, together with scan azimuth, passed out as a plot.

range error Distance measured from imaginary line drawn through desired impact point and one drawn through actual impact point, both parallel and perpendicular to axis of attack (USAF); distance means perpendicular distance.

range error probable Distance between two parallel lines drawn perpendicular to axis of attack and equidistant from desired mean point of impact between which fall 50% of impact points of independently aimed weapons.

range gate pull-off Basic ECM jamming technique usually used to pull hostile radar off target and thus provide infinite JSR for angle jamming; JSRR depends on

many variables and is seldom effective technique when radar is manually controlled.

range gating Limiting radar or laser to detect targets only within upper and lower (often narrow) range limits.

range lights Row of green lights marking each end of usable runway at simple airfield.

range markers *1* Parallel lines, concentric circles or other fixed graticule on display giving indication of range.

2 Single synthetic echo(es) injected into radar display timebase to give sharp blip, circle or other clear indication at selected range.

3 Two upright markers, illuminated at night, placed so that, when aligned, they assist piloting or in beaching amphibious craft.

range mean pairs Continuous analysis of peaks and troughs of variable function.

range octagon Computer-generated octagon around HUDS sight target which unwinds at rate of one side lost for each 100 m closure.

range-only tracking System for accurately measuring vehicle range by phase-comparison technique; vehicle transponder is interrogated by transmitter and replies are recorded by three or more widely separated receivers. In US operated on 387 and 417 MHz.

Ranger Deep penetration of hostile territory looking for air or surface targets (RAF, WW2).

range rate $\dot{R}$, rate at which range changes, $\dfrac{dR}{dT}$ [range being to a target, or of radar signal, etc].

range resolution Ability of radar to discriminate (separate) two targets on same bearing but with small range separation; determined mainly by pulse length.

range ring Any circle or arc on PPI display indicating range.

range safety officer Person charged with supervising safety to personnel on range (4) and ensuring that no object travels beyond range boundary.

range strobe See *range markers (2)*.

range tracking element Radar subcircuit which monitors received-echo times and operates range gate immediately before each return is received.

ranging Process of establishing target distance, by radar, optics/lasers, echo, intermittent, manual, gunfire, explosive-echo or navigational means.

ranging time Time taken by EM (e.g., radar) signal to travel to target and return, approximately 6.7 µs per km, 10.7 per mile, 12.4 per n.m. (nautical mile).

RANK Replacement alphanumeric keyboard.

rank Position in airline fleet: e.g. * 7 is seventh of type to be delivered.

Rankine Absolute Fahrenheit scale of temperature, °R = °F + 459.67, hence K = 5/9R.

Rankine cycle Thermodynamic cycle forming basis of modern steam plant, including many vapour-cycle machines for aerospace, all having closed circuit: boiler/prime mover/condenser/pump/boiler. In space power systems working fluid is usually metallic vapour, eg Hg, Cs.

Rankine-Hugoniot Relationship between pressure and density on each side of plane shockwave (Rankine 1870), from which p, V and γ for inclined shocks can be deduced.

Ranntac Reduction of aircraft noise by nacelle treatment and active control.

Ranrap Random-range program (ECM).

Ransac Range-surveillance aircraft.

Ransu   Regional air-navigation service unit.
Rant   Re-entry antenna test (ABRES).
RAO   Régiment Aérienne d'Observation (F).
Raob   Radio-sonde observation.
RAOC   Regional air operations centre.
RAOD   Ram-air overboard dump.
RAP   *1* Reliable acoustic path (sonar).
   2 Resource allocation plan.
   3 Reliability analysis program(me).
   4 Rocket-assisted projectile (ECM).
   5 Radar-absorbent (or absorbing) paint.
   6 Radio, or random, access point.
   7 Recognized air picture.
   8 Rack and pinion.
   9 Replacement acoustic processor.
   10 Air reconnaissance regiment (R).
Rapcon, RAPCON   Radar approach control (FAA).
RAPD   Recognized air picture display.
Rapec   Rocket-assisted personnel ejection catapult (Martin-Baker).
Raphael   Radar de photographie aérienne electronique; -TH adds transmission herzienne (F).
RAPID   Met. change [esp. significant] expected within 30 minutes.
Rapid   *1* Real-time acquisition program of inflight data (Sikorsky).
   2 Retrorocket-assisted parachute inflight delivery.
   3 Rugged[ized] advanced pathogen identification device.
   4 Radar passive identification.
rapid area supply support   Ability to deliver supply/ transport/packaging teams where needed (USAF AFLC).
rapid-bloom   Chaff or other ECM dispensed payload which quickly (within 0.1 s) assumes dimensions or emission properties resembling those of actual aircraft.
rapid deployment force   Military force, usually trained in amphibious, urban and peacekeeping operations, ready for near-immediate deployment to distant trouble spot. American RDF includes armour, air and naval power.
Rapide   Reliability and performance in demonstrated engines.
rapid engineer deployment   Quick-reaction civil-engineer squadrons that provide heavy construction and repair capability for theatre commander (USAF).
rapid-extraction parachute   One designed to deploy in less than 0.5 s from initial mechanical or electrical signal, eg for lo-alt delivery or crew escape at minimum altitude.
rapid-fracture surface   That left by high-rate crack propagation, showing as dark (often arrowhead) zones separated by bright zones of slow fatigue failure.
rapid-intervention vehicle   Fast off-road vehicle, first at scene of crash on or near [usually military] airfield, equipped to rescue crew.
rapid prototyping   Standard life-cycle method of software development.
rapid pucker-factor take-off   One on a favourable unbalanced field.
rapid-reload capability   Ability of single ICBM silo or SLBM tube system to fire at rapid rate; not defined in SALT II but generally taken as more than one round per 24 h. Note: in West, no reload missiles exist.
Rapids   *1* Radar passive identification system.
   2 Real-time acquisition and processing integrated data system [satellite ground station].

Rapnet   Regional air-traffic service packet-switched network.
RAPP   Recognized air-picture production.
Rapport   Rapid alert programmed power-management of radar targets.
RAPPS   Remote-area precision positioning system.
RAPS, Raps   *1* RPV advanced payload system(s) (USAF RPV ECM).
   2 Radar prediction, or protection, system.
   3 Recording, analysis, playback and simulation [radar data].
   4 RPV autoland position sensor.
Rapsat   Ranging and processing satellite.
RAPT, Rapt   *1* Radar procedures trainer.
   2 Recognized air-picture troop.
Raptor   Responsive aircraft program for theater operations [USAF UAVs, not connected with name of F-22].
RAR   *1* Request radar blip identification (RBI) message.
   2 Radar arrival route.
Rara, RARA   Reusable active RF augmentation.
RARDE   Royal Armament Research & Development Establishment, (Fort Halstead, Sevenoaks and Chertsey, UK).
rare-earth   General adjective for electric machines using ** magnetic materials, notably SmCo.
rarefaction   Supersonic flow through diverging duct, in which pressure decreases while velocity increases.
Rareps   Radar [weather] reports.
RARF   Radar, antenna and RF, integrated system in which multiple radar functions are performed with single electrically scanned aerial on time-shared non-interference basis (Emerson).
RARO   Remote aerial refuelling operation.
RARS   Radar recording system.
RAS   *1* Rectified airspeed, see *airspeed.*
   2 Rough air speed, for flight through gust (2).
   3 Replenishment at sea.
   4 Radar-absorbing, or absorbent, structure.
   5 Remote active spectrometer.
   6 Radar Advisory Service; A adds Area.
   7 Row address strobe.
   8 Runway alert system.
RASA   Russian Aviation and Space Agency.
RASC   Regional AIS system centre.
Rascal   *1* Responsive-access, small-cargo, affordable launch (Darpa).
   2 Ramjet, small-calibre.
   3 Radar do scoperta e controllo aereo locale (short-range air surveillance) (I).
RASD   Requirements and system definition.
RASE   Rapid automatic sweep equipment (ECM).
RASER, Raser   Revolutionary aerospace engine research (NASA).
RASH   Rain showers.
RASI, Rasi   Radar and navaid simulator.
RA/SI   Rate alarm/storm intensity (thunderstorms).
RASM   Revenue per aircraft, or available, seat-mile.
RASN   Rain and snow (ICAO).
Rasp, RASP   *1* Recognized air and surface picture.
   2 Radar Applications Specialist Panel (Eurocontrol).
   3 Rapid-acquisition spectrum processor.
Raspberry ripple   Colour scheme [red/white] of training and research aircraft (RAF/RN/Qinetiq).

RASS   *1* Rapid area supply/support (USAF).

  2 Radio acoustic sounding system.

  3 Radar analysis support system.

Rast, RAST   *1* Recovery assist, secure and traverse (shipboard helicopter).

  2 Radar-augmented sub-target.

  3 Replacement aerial subsonic target.

Rastas   Radiating site and target acquisition system (airborne ECM).

raster   Generation of large-area display, eg TV screen, by close-spaced horizontal lines scanned either alternately or in sequence.

raster scan   See *scan*.

Rasur   Radio surveillance for intelligence purposes.

RASV   Reusable aerospace vehicle.

RAT   *1* Ram-air turbine.

  2 Ram-air temperature.

  3 Rock abrasion tool (Martian exploration).

  4 Radar active target.

Rat   Hostile intruder aircraft flying alone.

Rata, RA/TA   Resolution advisory/traffic advisory.

RATC   Radar air-traffic control [C adds centre, F facility].

Ratchet   Rapid ATC HMI evaluation tool.

ratcheting   See roll *.

rate   Rate of change, first derivative of variable with respect to time; thus angle *, * gyro. Symbol is dot placed centrally above value, thus if a is acceleration along flight-path $\dot{a}$ is rate of change of acceleration, or acceleration*.

rated   Qualified for specific flight duty, especially to fly particular aircraft type.

rated altitude   That at which piston engine gives maximum power, for supercharged engine usually at height considerably above S/L; lowest altitude at which full throttle is permissible (or, usually, obtainable) or maximum boost pressure can be maintained at maximum rpm in level flight.

rated coverage   Area within which strength of NDB vertical field of ground wave exceeds minimum value specified for geographical area where situated.

rated power   Any of several specified limits to gas-turbine power, eg take-off, 2½ min. contingency, maximum continuous.

rate gyro   Gyro whose indication gives a rate term, rate of change of attitude; single degree-of-freedom gyro with primarily elastic restraint of spin axis about output axis.

rate integrating gyro   Single-degree-of-freedom gyro whose output axis is linked to spin axis by viscous restraint so that angular displacement is proportional to integral of angular rate of change of attitude. Abb. Rig (not recommended).

rate of climb   Rate of gain of height, vertical component of airspeed of aircraft (normally aerodyne) established in climbing flight at quasi-constant airspeed (ie not in zoom trading speed for height). For helicopter, two values; maximum *** and maximum vertical ***.

rate-of-climb indicator   See *VSI*.

rate of temperature-rise indicator   Fire-warning system; does not respond to temperature but to positive rate of change.

rate of turn   Rate of change of heading, proportional to bank angle θ; R (turn radius), for constant speed, is exactly proportional to tan θ. See *turn rate*.

rates of exchange   Tradeoff multipliers, eg numerical value linking takeoff field length or MTOW for unchanged takeoff field length per °C change in ambient temperature.

Rate 1 turn   Sustained $3°s^{-1}$, = $180°\ min^{-1}$. See *turn rate*.

rate structure   Comprehensive and in general internationally agreed prices for air carriage of unit mass of all commodities.

RATG   Ram-air turbine generator.

rating   *1* Authorized operating regimes, with limiting numerical values, for engine or other functional device or system. For gas-turbine engine can include takeoff, maximum climb, maximum cruise, OEI contingency, etc; for piston engine can include METO, maximum weak mixture, etc.

  2 Anti-knock value of piston-engine fuel.

  3 Endorsements, additional qualifications, privileges and limitations added to airmen certificate (US) or pilot's licence (UK), eg night *, IMC *, seaplane *, multi-engine *, instrument * and flying instructor *. See *Type*.

rating spring   Spring, usually with linear output, whose change in length is accurately proportional to applied force.

ratio changer   Device for varying the response of a flight-control surface, especially rudder, to input demand, usually in proportion to q (dynamic pressure).

ratio of specific heats   For a gas, ratio of specific heat at constant pressure divided by specific heat at constant volume; for air $\gamma = C_p/C_v = 1.401$.

Rato, RATO   Rocket-assisted takeoff; G or g adds gear, meaning not landing gear but equipment.

Ratrace   Radar transmitter waveguide shape facilitating use of common aerial for transmitting and receiving.

RATS, Rats   *1* Rapid area transportation support (USAF).

  2 Rescue advanced-tactics school.

  3 Roving aerodynamic test system.

Ratscat   Radar-target scatter; more fully, radar-target test scatter facility.

RATT   Radio teletype.

Rattlr, RATTLR   Revolutionary approach to time-critical long range (USN).

RAU   Radio access unit.

RAV   Robotic air vehicle [also called Rave].

Rave   *1* Rapidly adjustable variable exhaust.

  2 Reconfigurable advanced visualization environment, virtual-reality display.

raven   Electronic-warfare officer, specialist flight crew member (US).

RA/VSI   Resolution advisory/vertical speed indicator.

raw   Versatile word meaning ready for next stage of processing; thus * AC is ready for precise frequency control before being supplied to avionics; * data can be of many forms (instrument readings, reconnaissance photos, tabular statistics); and * material is in fact anything but * yet still unmanufactured into product.

rawin   Wind velocity at different heights computed by radar tracking of balloon (usually with transponder).

rawinsonde   See *radar-sonde*.

RAWS   *1* Radar altitude warning system.

  2 Radar attack (and) warning system.

  3 Remote-area weather station.

  4 Role-adaptable weapon system.

Rayleigh atmosphere   Ideal atmosphere devoid of all

particles larger than about 0.1 wavelength of incident radiation.

Rayleigh flow   First (1876) theory since Newton for lift of inclined-plane wing; streamlines travel direct to surface, where they are brought to rest, losing all relative energy.

Rayleigh formula   Another Rayleigh equation gives loss of pitot pressure caused by presence of normal shock, in terms only of Mach numbers, pressures and γ (ratio of specific heats of gas).

Rayleigh limit   One-quarter of an EM wave, maximum difference in optical path.

Rayleigh number   Non-dimensional ratio Nra = $\dfrac{gd^3 \propto (\theta_2 - \theta_1)}{v\text{k}}$ where g is acceleration due to gravity, d is vertical separation of two horizontal planes in fluid system (eg atmosphere), $\theta_2 - \theta_1$ is temperature difference across planes, $\propto$ is coefficient of thermal expansion, v is coefficient of kinetic viscosity and k is thermal conductivity.

Rayleigh scattering   Normal scattering of radiation by particles whose ruling size is 0.1 or less that of radiant wavelength; explains why sky is blue above and red/orange at sunset.

Rayleigh wave   2-D barotropic fluid disturbance or wave propagated along free solid surface.

RB   *1* Rescue boat (ICAO).
2 Rapid-bloom (ECM).
3 Radar-blip identification message.
4 Reduced blast (NW).
5 Relative bearing.
6 Readback.
7 Rain began [time].
8 See * *switch*).
9 Rudder boost.

$\mathbf{R_B}$ See *Rockwell*.

Rb   Robot, = guided missile (Sweden).

$\mathbf{R_b, r_b}$ Solid rocket motor burn rate; also r.

RBC   Rapid-bloom countermeasures (or chaff).

RBCC   Rocket-based combined cycle, also called RBC2.

RBCI   Radio-based combat identity, or identification.

RBD   Radar-beacon digitizer.

RB/ER   Reduced blast, enhanced radiation (neutron bomb).

RBF   Remove before flight.

RBGM   Radar-beam ground-map operating mode.

RBI   *1* Relative-bearing indicator.
2 Radar-blip identification (ICAO).

RBIM   Rudder-boost internal monitor.

RBL   Range and bearing launch.

RBM   *1* Wing-root bending moment.
2 Real-time batch monitor.

RBN   Radio beacon.

RBOC   Rapid-bloom offboard countermeasures.

RBPS   Rotor-blade protection system.

RBS   *1* Radar beacon system (FAA).
2 Rutherford back-scattering.
3 Radar bomb scoring.

RBSJ   Recirculating-ball screwjack.

RB switch   Selects radio or barometric altitude, eg on HUD.

RBT   *1* Remote batch terminal.
2 Resistance-bulb thermometer.

RBV   Return-beam vidicon.

RC   *1* Rotating-combustion, engine of Wankel type.
2 Reaction control.
3 Resistance/capacitance (also R-C).
4 Rotating components of gas-turbine engine.
5 Rate of climb at MTOW, or r/c.
6 Radio-controlled, or R/C.
7 Rounds counter.
8 Reverse course [= back course].
9 Radar computer.
10 Rotary coupler (AWACS).
11 Remote command.

R-C   Resistor/capacitor network.

$\mathbf{R_C}$ See *Rockwell*.

r/c   Rate of climb.

RC-1   Rate of climb, one engine inoperative.

RC-2   Rate of climb, all engines operating.

RCA   Reach cruising altitude (ICAO).

RCAG   *1* Remote center air/ground (FAA).
2 Rescue, or remote, communications air/ground.

RC&W   Rack connectors and wiring.

RCB   Remelted cast bar.

RCC   *1* Reinforced carbon/carbon.
2 Rescue co-ordination [or control] centre.
3 Resistance/capacitance coupling (often r.c.c.).
4 Research Consultative Committee (UK).
5 Repetitive chime clacker.
6 Remote charge converter.
7 Regional control centre.

RCCB   Reverse-current circuit-breaker.

RCDI   Rate of climb/descent indicator, = VSI.

RCDS   *1* Royal College of Defence Studies (London, UK).
2 Radio-controlled destruction system.

RCDU   Radar control display unit (UK E-3D).

RCE   *1* Rotating-combustion engine.
2 Radio control equipment.

RCF   *1* Radio communications failure message (ICAO).
2 Remote-control facility.

RCFAM   Roll-coupled fuselage aiming mode.

RCFCA   Royal Canadian Flying Clubs Association.

RCFWT   Radio-controlled fixed-wing target.

RCG   Reaction-cured glass, covers thermal-protection tiles (Space Shuttle).

RCH   Reach, reaching.

RCJ   Reaction control jet.

RCL   *1* Runway centreline; L adds lights.
2 Radio communications link.

RCLM   Runway centreline marking (FAA).

RCLR   Recleared.

RCLS   Runway centreline light system (FAA).

RCM   *1* Radio countermeasures.
2 Reliability-centred maintenance.
3 Rollercoaster manoeuvre.
4 Restricted corrosive material.
5 Reciprocating chemical muscle.
6 Requirements criteria and methods (ICAO).

RCMAT   Radio-controlled miniature aerial target.

RC/MC   Rounds counter, magazine controller.

RCMG   Rifle-calibre machine gun.

RCMP   Radar control and maintenance panel.

RCMS   *1* Resonator-controlled microwave source.
2 Remote control [and] monitoring system.

RCMU   Remote control and monitoring unit.

RCO   *1* Remote communications outlet.
  2 Range control officer.
  3 Remote control officer (or operator).
RC/OI   Reaction control/orbital insertion subsystem.
RCP   *1* Role-change package.
  2 Radio control panel.
  3 Required communications performance.
RCR   *1* Runway condition reading.
  2 Route contingency reserve (fuel).
  3 Routeing and circuit restoral.
R-CRS   Report on course.
RCS   *1* Reaction-control system.
  2 Radar cross-section.
  3 Ride-control system.
  4 Range control station.
  5 Remote control system.
RCSM   Resident customer-support manager.
RCSR   Radar cross-section reduction.
RCSS   *1* Remote-controlled signals-intelligence system.
  2 Remotely controlled surveillance system.
RCSU   Remote-control and status unit (navaids).
RCT   *1* Reply code train (IFF, SSR).
  2 Royal Corps of Transport (UK).
  3 Reverse-conducting thyristor.
  4 Rear-crew trainer.
RCTC   Rear-crew trainer cabin.
RC2P   Rehosted command and control processor.
RCU   *1* Range converter unit.
  2 Reaction, or receiver, or rudder, or recorder, or remote control unit.
RCV   *1* Reaction-control valve.
  2 Robotic command (or remotely controlled) vehicle.
rcv   Receive [**rcvr** receiver, **rcvs** receives].
RCW   Roll-control wheel.
RD   *1* Ramp door, thus * uplock.
  2 Random discontinuous fibre.
  3 Rigged dry, for dusting rather than spraying (ag-aviation).
  4 Report departure, or departing.
  5 Relative density.
  6 Restricted data.
  7 Long-range (R)
  8 Radar ranging (R).
Rd   Radio channel for data.
r/d   Rate of descent.
RDA   *1* Runway de-icing agent.
  2 Requirements, development and analysis.
RD&A   Research, development and acquisition.
RD&E   Research, development and engineering.
RDARA   Regional domestic air-route area.
RDAS   Reconnaissance data annotation set.
RDAU   Remote data acquisition unit.
RDB   *1* Requirements data bank (originally AFLC).
  2 Raw-data buffer.
RDBM   Relational database management (more often RDM); S adds system.
RDC   *1* Rate-of-descent computer.
  2 Regional dissemination center (NASA, technology transfer).
  3 Radar data converter, or correlator.
  4 Ramp door control.
  5 Remote data concentrator.
  6 Routeing-domain confederation.
  7 Resolver-to-digital convertor.

RDCE   Radio distribution and control equipment.
RDD   *1* Routine dynamic display (maritime/ASW navigation as distinct from tactical situation).
  2 Radar data display.
RDDBS   Redundant distributed data-base system.
RDDMI   Remote (or radio) digital direction magnetic indicator.
RDDU   Remote demand and display unit.
RDE, R&DE   Research and development engineering (or evaluation).
RDEC   Rapid data-entry cassette.
RDF   *1* Radio direction-finding (security cover for radar, 1936–42).
  2 Rapid Deployment Force (US).
  3 Routeing-domain format.
RDG   Ridge of high pressure.
RDH   Reference datum height of ILS.
RDI   *1* Radar Doppler à impulsions (F).
  2 Reference designated indicator.
  3 Routeing-domain identifier.
RDIDS   Rapid-deployment intrusion-detection system.
RDJTF   Rapid Deployment Joint Task Force (US).
RDL, rdl   *1* Radial (bearing, heading).
  2 Rapid-deployment launcher.
RDM   *1* Random-deflection monitor.
  2 Relative-distance measurement.
  3 Radar Doppler multifunction (F).
RDMI   Radio directional (or distance) magnetic indicator.
RDMS   Relational database management systems.
RDMSS   Redmiss.
RDO   Redistribution order.
rdo   Radio (FAA).
RDP   *1* Radar data, or disk, processor, or processing; S adds system, U unit.
  2 Range Doppler profile, or profiling.
RDPS   Radar data processing system.
RDPU   Radar display processing unit.
RDR   *1* Radar.
  2 Radar departure route.
RDS   *1* Rudder-disconnect speed (Autoland).
  2 Remote diagnostics server.
RDSS   *1* Rapidly deployable surveillance system.
  2 Radio determination satellite service (Inmarsat).
  3 Replacement data-storage system.
RDT   Radar-data transfer.
RDT&E   Research, development, test and engineering (or evaluation).
RDU   *1* Remote display unit.
  2 Receiver/decoder unit.
RDV   Requirements development and validation.
RDVS   Rapid-deployment voice switch (FAA).
RDX   *1* Powerful explosive ($CH_2N.$ $NO_2)_3$; also called Hexogen, Cyclonite. Now family of explosives still used as bomb, mine and other fillings, though under other names and designations.
  2 Radar-data extractor.
RE   *1* Recent, ie qualifying as 'actual' met data.
  2 Rain erosion, hence * resistance.
  3 Re-engined.
  4 Role equipment.
  5 Random energy.
  6 Research establishment.
  7 Random echo.

8 Rain ended (time).

9 Range error.

R/E   Receiver/exciter.

R_e, R_E   Reynolds number, if for any reason R is unavailable.

REA   Radar echoing area (suggest = RCS [2]).

REAs   Responsible engineering activities.

Reach   Realisable integrated modular avionics common access helicopters.

reach   Length of threaded portion of sparking plug.

reachback   Having immediate access to many sources of information and intelligence located far [back] from scene of conflict.

React   *1* Reliability evaluation and control technique.

2 Rain echo attenuation compensation technology.

3 Rapid execution and combat targeting (Minuteman upgrade).

reactance   Component of impedance in AC circuits due to inductance or capacitance; value is $2\pi fL$ for inductance and $1/2\pi fC$ for capacitance where f is frequency, L is henrys and C is farads.

reactant   Any substance consumed in reaction, esp. in rocket motor combustion or other generation of working fluid.

reactant ratio   Ratio of mass flow of oxidant to fuel.

reacted pressure   See *stalled pressure*.

reaction   See *positive feedback, regeneration (2)*.

reaction chamber   One in which chemical reaction takes place, specifically that in which reactants produce gas to drive turbopump or to power MEPU.

reaction-control jet   Thruster or RCV jet.

reaction-control system   Primary attitude/trajectory control system of spacecraft, or of V/STOL aircraft at speeds too low for conventional controls to be effective. Control moments are imparted by *thrusters* or *reaction-control valves*.

reaction-control valve   Small nozzles supplied with hot, high-pressure main-engine bleed air at extremities of V/STOL aircraft; at low airspeed pilot operates * to control attitude and trajectory of aircraft.

reaction engine, reaction propulsion   Vague; all aerospace propulsion is by Newtonian reaction.

reaction sphere   Spacecraft orientation control system comprising free-running dense sphere surrounded by three magnetic coils, one for each attitude axis. Also called free sphere.

reaction time   Elapsed time between stimulus and response, eg between command to launch missile and actual launch, between first receipt of warning and dispatch of weapon system, or between pilot seeing conflicting aircraft and initiation of manoeuvre to avert collision.

reaction turbine   One in which main tangential driving force comes from acceleration of working fluid through converging passages between rotor blades; rare in gas turbines, which are usually impulse/reaction or impulse.

Reaction Wheel   Patented advanced gyro for spacecraft stabilization (Honeywell).

reactive employment   Use of device in retaliation to appearance of hostile threat, esp. of ARM in response to operative and threatening emitter; opposite of pre-emptive.

reactive-material warhead   Classified metal/plastics assembly which, upon impacting a target, triggers a reaction which releases intense pressure and heat.

reactive mission plan   Taking account of previous errors, losses or failures.

reactor   *1* Choke or high-inductance coil.

2 Core in which nuclear fission (thermal or fast) or fusion reaction takes place.

read   *1* To obtain information from EDP (1) storage and transfer it to another device or address.

2 Of humans, to receive and understand telecommunications message.

readability   Numerical scale [1 is best] of audibility of voice radio.

read across   To have direct relevance to quite different programme or problem; eg use of LH_2 in space propulsion solved problems which ** to LH_2 in future commercial airline operation.

readback   Procedure whereby receiving station repeats all or part of message to originator to verify accuracy.

readout   *1* Data played back from tape or other store, eg from flight recorder.

2 To broadcast data either from storage or as received.

3 In EDP (1), to transfer word(s) from memory to another location.

Reads, READS   Re-entry air-data system.

read/write memory   Each cell is selected by appropriate signal, and stored data may be sensed at output or by changes in response to other inputs.

ready   Weapon system is available for use with no delay save reaction time.

ready CAP   Combat air patrol aircraft on standby status; ready for take-off.

ready light   Indicates particular munition is ready for use.

ready position   Designated place where stick (4) waits for helicopter or for order to emplane.

ready room   Where hardware is prepared for use, esp. missiles, RPVs and other unmanned vehicles and esp. aboard warship.

real fluid   One exhibiting viscosity.

real precession   That induced in gyro by applied force, eg friction or imbalance, and not by Earth's rotation.

real time   *1* In EDP (1), operation in which event times are controlled by portions of system other than computer and cannot be modified for convenience in processing; not necessarily simultaneous with time in everyday sense.

2 Absence of delay, except for time required for transmission by EM energy, between occurrence of event or transmission of data, and knowledge of events or reception of data at some other location (DoD). Neither definition has anything to do with time computer is processing as distinct from warm-up time or rest periods, which incorrectly figure in some popular definitions.

real-time clock   One indicating actual time.

real-time reconnaissance   Sensing, recording and relaying information 'as it happens'.

real wander   See *real precession*.

rear area   That behind combat and forward areas in battle.

rear bearing   In single-shaft gas turbine, that supporting turbine.

rear-box wing   Usually means three-spar.

rear cover   Aft closure of piston-engine [esp. radial]

crankcase, incorporating drives and mounting faces for accessories.

rear echelon   Elements of air-transport force not required in objective area.

rear-loading   Provided with door(s) and ramp extending across full cross-section of fuselage interior for loading of vehicles and other bulky cargo, and for air dropping.

rearming   Replenishment of consumed ordnance between missions.

rear pit   Back seat in tandem cockpit, hence rear-pitter (military colloq.).

rearplane   Rear wing of tandem-wing aircraft. Not justified when foreplane is clearly for control rather than lift but correct term where front and rear wings are comparable in size.

rear port   Aperture, usually circular, in solid rocket motor case at end or on face opposite to nozzle which can be opened to reduce internal pressure for thrust cutoff.

rear stagnation point   That at which fluid is at rest on surface of body on downstream side and from which a streamline emanates.

rear step   That at trailing edge of seaplane (flying boat) afterbody, where planing bottom terminates; in some designs a point, but usually either vee-step or a sharp-edge discontinuity sufficient to break hydrodynamic attachment to skin.

rear view display   HUD system in which information is fed from behind pilot's helmet via optical fibres to collimating lens and combiner glass close in front of eyes.

Réaumur   Non-SI temperature scale on which ice point (1 atmosphere) 0° and water boiling point 80°, symbol R.

ReB   Re-entry body.

rebated blade   Solid light-alloy (rarely, wood laminate or GRP) propeller blade whose leading edge is cut away to accommodate encapsulated electrothermal anticing element.

Rebecca   Pioneer DME used by airborne interrogation of Eureka beacon (1942). Still in production in 1970 as * 12 for airborne forces and used in conjunction with MR 343 air-dropped beacon.

rebreather   Closed-circuit oxygen system from which CO_2 and water are continuously removed, pressure being maintained by adding fresh oxygen (see * *bag*).

rebreather bag   Gastight sac in oxygen line near mask so that incoming gas is diluted with exhaled breath; normally feature of airline passenger system.

Rebro   Relay broadcasting.

REC   *1* Received, receiving (ICAO).
   2 Radio-electronic combat (USSR, R).
   3 Recommend.

recalescence point   Temperature, exhibited only by iron and some other ferromagnetic materials, at which on cooling from white heat exothermic change in crystalline structure halts cooling and can cause momentary rise in temperature.

Recap   Reliability evaluation and corrective-action program.

recapture   Regain of revenue from spilt passengers (*spill effect*) who rebook on later flight by same carrier.

Recat   Reduced-energy for civil air transport.

recce   Reconnaissance (UK, colloq., pronounced recky).

RECD   Received.

Receiver autonomous integrity monitoring   Can detect a satellite malfunction if five satellites are in view, and can

identify the failed satellite if six are in view, thus meeting requirement for sole means of oceanic navigation.

reception   All arrangements for receipt of air drop.

recession   Linear rate at which ablating material is eroded, normally measured normal to local surface.

Rechlin   Location near Neubrandenburg of largest and oldest German aircraft/air armament test establishment before 1945.

reciprocal   Heading 180° from previous heading, or from that intended.

reciprocal bearing   Bearing of observer from remote station.

reciprocal latching   Use of phase-shifters to generate computer-controlled sequence and distribution pattern for electronically scanned radar transmission.

reciprocating engine   One in which piston(s) oscillate to and fro in cylinder, eg Otto, diesel, Stirling, etc.

recirculating ball system   Mechanical friction-reducing technique in which contact between spiral thread of screwjack and surrounding nut or runner is transmitted via no-gap stream of bearing balls with rolling contact only, which after reaching end of nut are returned via external tube to start.

reclaiming   Repairing or otherwise modifying a part or tool or material to fit it for further use.

Reclama   Please reconsider proposed action or decision (DoD).

RECMF   Radio and Electronic Components Manufacturers' Federation (UK).

recognition   In imagery interpretation, determination of type or class of object without positive identification (ASCC).

recoil   For automatic weapons in aircraft usual measures are average, peak and counter-*; all are reduced by fitting muzzle brake.

recommended practices   Not mandatory but published by ICAO.

recon   Abb., reconnaissance.

reconciliation   Reclaim of baggage by the correct individual, ie the owner. System has many uses, such as automatic offload if passenger does not board.

reconfiguration   Change in aerodynamics and external form of aircraft, especially while in flight.

reconnaissance by fire   Firing on suspected enemy to draw retaliatory fire.

reconnaissance exploitation report   Written, accompanies imagery.

reconnaissance pallet   Multi-sensor pallet (1) installed in or under tactical aircraft in lieu of other load.

reconnaissance pod   Pod housing reconnaissance sensors carried externally.

reconnaissance reference point   Conspicuous geographic location from which reconnaissance objectives can be found.

reconnaissance slipper   Fairing housing reconnaissance sensors carried scabbed against exterior skin.

recontouring   Improving leading edge of fan or compressor blade[s] after erosion in service.

record as target   Code for listing target for reference or future engagement (DoD).

record card   Sheet of card, fitting into filing box, issued for each aircraft, on which were recorded all significant modifications or repairs to Service aircraft prior to about 1955, when newer systems became available.

recorder   Device translating data into hard-copy record on magnetic tape, wire, film, punched paper tape or other medium.

recorder data package   Integrated airborne systems providing record of subsystem performance in flight, eg arming and fuzing of Mirved warheads, for recovery and ground analysis.

recording accelerometer   Counts and stores vertical acceleration.

recording altimeter   Barograph providing permanent [originally paper] print-out.

recording storage tube   Electronic tube which accepts CRT or other display picture, stores for period (eg 12 h) and reads out as often as required for analysis, monitor activation or various conversion processes.

recovered pressure   That measured inside duct downstream of ram inlet.

recovery   1 Retrieval, in air or on surface, of part or whole of used RPV, target, test missile, spacecraft, instrument capsule or other inert body.

2 Retrieval of glider (sailplane) that has landed away from own airfield.

3 Return to base and safe landing (on land or aircraft carrier) of aircraft.

4 Return of combat aircraft from post-strike base to home base or designated recycle airfield.

5 Retrieval, normally by crane helicopter, of crashed or shot-down aircraft, in friendly or enemy territory.

6 Retrieval, by special trailers and vehicles, plus airbags or jacks, of belly-landed or disabled aircraft.

7 Completion of flight manoeuvre and resumption of straight/level flight.

8 Conversion of kinetic to pressure (potential) energy in fluid flow (see *ram* *).

9 Returning late programme to schedule.

recovery airfield   One at which aircraft might land post-H-hour but from which combat missions are not expected to be conducted (DoD).

recovery base   Rear-area airfield used for maintenance and servicing to eliminate need for such services at airfields in combat zone (USAF).

recovery capsule   Capsule containing reconnaissance pictures, instrumentation records or other data designed to separate from satellite, ICBM or other carrier and survive re-entry at preplanned location.

recovery footprint   Area within which recovery capsule, returning manned spacecraft or other object is expected to fall.

recovery guidance system   Displays on flight-director command bars pitch guidance for recovery during continuous windshear; also for TOGA to maintain familiarity.

recovery initiation window   Small block of airspace within ASW(4) from which UAV can be guided to touchdown point.

recovery package   Contains devices to assist recovery and retrieval of re-entry body, eg brightly coloured buoyant balloons, radio beacons and coloured pyrotechnics.

recovery station   Occupied by carrier deck crew who run out to attend each landing aircraft.

recovery temperature   See *adiabatic recovery temperature*.

recovery time   Time for gas-discharge tube to return to neutral under grid control.

REC/PLB   Recording and playback.

recreational   Self-explanatory; * pilot certificate is issued by FAA in US for pilot of * vehicle.

RECT   Rectangular.

rectangular co-ordinates   See *cartesian*.

rectangular input   Input, eg to FCS, which instantaneously jumps from null to a maximum commanded value, to give a rate command.

rectenna   Rectifying aerial (antenna), especially of directionally beamed microwave power.

rectification   1 Process of projecting tilted or oblique photograph on to horizontal reference plane.

2 See *rectifying*.

rectified airspeed   See *airspeed*.

rectified altitude   Sextant altitude corrected only for inaccuracies in reference level (dip, coriolis) and reading (index, instrument and personal).

rectifier   Static device exhibiting strongly unidirectional conduction properties such that it can convert a.c. to d.c.

rectifying   1 Elimination of errors which convert compass heading into track, eg deviation, variation and wind.

2 Converting a.c. to d.c., by any means.

rectifying antenna   Receives radar wave and separates its two components, the electric field and magnetic field, the former being conducted into the hot jet[s] from the engine[s], which become[s] ionized and cooled, and the magnetic component being dissipated in the hot engine and degaussed.

rectifying valve   Thermionic valve which rectifies by virtue of unilateral conductivity of cathode/anode path.

rectilinear flight   Straight and level, and 0g imposed and thus in aeronautical sense called + 1g.

rectilinear propagation   Sent out in straight lines (ignoring relativistic effects).

rectilinear scanning   TV raster (see *scan*).

recuperation   Recovering heat from engine [turboprop or piston engine] exhaust and inserting it at useful place in operating cycle, hence recuperator.

recurrent training   Regular review of human performance in normal, abnormal and emergency situations.

recycle   1 To stop countdown and re-enter count at earlier point.

2 To return to start of EDP (1) program without entering fresh data.

3 To give completely new checkout to missile or other device (USAF).

4 To remove part or complete module from engine, avionic device or other hardware and put through remanufacturing or inspection sequence to clear it for reuse with undiminished life.

recycle airfield   One from which combat aircraft can be prepared for reuse away from home base.

RED   1 Reconnaissance Engineering Directorate (US AFSC).

2 Reference engineering drawing or data.

3 Retractable ejector duct.

red   Hostile. Thus, colour of threat on radar, or of any team in exercise or R&D effort acting part of enemy.

Red airway   One running more or less east/west.

Redar   Range engineering data acquisition and reduction.

Redars   Reference engineering data automated retrieval system.

Redcap   Real-time electromagnetic digitally controlled analyser processor.

Red channel   Airport route for arrivals declaring the possession of dutiable goods.

REDCOM   Readiness Command (USA).

Red endorsement   Manuscript endorsement in flying logbook confirming bad airmanship or other [usually potentially dangerous] act or failure to act.

redeployment airfield   One not occupied in peacetime but available upon outbreak of war for use by units redeployed from peacetime locations; substantially same facilities as main airfield (DoD, NATO).

redeye   Long overnight flight.

Red Flag   Major tac-air exercises based on Nellis AFB, taking place several times per year over instrumented air/air and air/ground ranges under various 'real war' scenarios.

Red Force   Hostile force in exercise.

Red Horse   Rapid engineering deployment and heavy operations repair squadron, engineering (USAF).

REDL   Runway edge lights.

red-line value   One never to be exceeded, eg V_{NE} or * fan speed.

Redmiss   Remotely deployable mission support system.

red-on-red   Traditional rule for setting up simple magnetic compass and avoiding mistakenly flying reciprocal.

red-out   Loss of vision in powerful and sustained negative acceleration, ie where subject is restrained in seat only by harness over shoulders. Vision is replaced by primarily blood-coloured input.

red pole   North-seeking.

redrive   Propeller drive incorporating speed-reducing gears or belt (homebuilts).

reduced frequency   Ratio of product of frequency of oscillation and representative length of oscillating system to airspeed (nl/V); dimensionless parameter determining flutter amplitude.

reduced modulus of elasticity   Theoretical value expressing relationship between modulus of elasticity and tangent modulus beyond limit of proportionality.

reduced vertical separation minima   Progressive introduction of reduced minima [initially, 1,000 ft, 305 m] for aircraft whose altitude measuring, maintaining and reporting systems meet RVSM numerical standards. Introduced on N Atlantic between FL310–FL390, and extended 2002 to include all airspace of ECAC member states above FL290.

reducing mask   Metal-sheet stamping enabling small instrument to be mounted in space on panel for larger one.

reducing valve   Valve which reduces pressure in fluid system to precise lower value at * output.

reduction   *1* Removal of oxygen or other electronegative atom or group; hence reducing flame.

2 Conversion of raw measured values of flight performance or met. observations into standard forms, limiting values and, esp., graphical plots and derived values for comparative purposes.

reduction factor   Large number, usually power of 10, used as divisor of all values in calculations to reduce size of whole numbers involved.

reduction gear   Speed-reducing gear, ie output turns slower than input, torque being increased in the same ratio.

redundancy   *1* Provision of two or more means of accomplishing task where one alone would suffice in absence of failure. In parallel * several usually similar systems all operate together so that failure of one either leaves remainder operative or, in majority-rule *, can always be over-powered by remainder. Standby * has primary systems automatically switched in by malfunction-detection system.

2 Design of primary structure so that even after failure of any component there will remain enough load paths to carry all expected loads with adequate margin of safety.

3 In EDP (1) or information handling, amount by which logarithm of number of symbols available at source exceeds average information content per symbol.

redundant attitude-control system   Cold-gas jet control of roll during powered flight, three-axis control during periods coasting.

redundant structure   Basically, one possessing too many members; in practice one not amenable to simple stress analysis, eg because it has joints that are fixed instead of pinned or more than one member sharing load in way calling for elegant analytical solution. Not synonymous with fail-safe structure.

Redux   Family of adhesives for most structural materials, including metals, normally applied in form of sheet, powder or liquid and cured (bonded) under heat and pressure (CIBA).

Redwood   Together with * Admiralty, a traditional instrument for measuring kinematic viscosity; rare in aerospace.

REDZ   Recent drizzle.

Red zone   Zone of intersection on Red airway (US) in which Red traffic maintains height and conflicting traffic procedure is published.

Reed & Prince screw   Recess-head crosspoint screw driven by tool with single taper.

reed valve   Leaf valve, usually of thin spring steel, giving unidirectional flow of fluid, eg in air-conditioning system or pulsejet.

reef   *1* To make sudden deliberate maximum-rate departure from straight flight [normally fighter combat].

2 To fit restraining cord round parachute, see next.

reefed parachute   One in which canopy is (usually temporarily) restrained against full deployment by encircling cord.

reefing sequence   Systematic timed deployment of parachute, first in reefed condition and later to full deployment (eg Apollo CM recovery).

reeling   Paying out or drawing in cable for a tow-target; hence RM = * machine and RMCS * machine control system, offering: out, stop (preselected length), in, and other functions.

re-entrant angle   Sudden change in direction of external surface (of aircraft skin, structural member etc) such that angle measured externally is less than 180°; in case of ** on surface of body, one causing local acceleration of airflow and, in supersonic flight, local attached shockwave.

re-entry   Process of travelling [returning is implied] from outer space into and through planetary (esp. Earth's) atmosphere, in case of Earth proceeding down to planetary surface where *recovery* takes place. Strictly the term should be entry.

re-entry body   Body designed to survive extreme aerodynamic heating, high temperatures and large sustained deceleration of re-entry, eg ICBM RV, manned spacecraft, instrument or reconnaissance capsule.

re-entry corridor   Optimum trajectory through atmosphere for lifting body returning to Earth.

re-entry plasma   Plasma inevitably formed around all bodies arriving in Earth's atmosphere from outer space due to kinetic heating and ionization, forming barrier to radio signals.

re-entry system   That portion of ballistic missile designed to place one or more RVs on terminal trajectories so as to arrive at selected targets. Includes penaids, spacers, deployment modules and associated programming, sensing and control devices.

re-entry vehicle   That part of a strategic ballistic missile or space vehicle designed to re-enter Earth's atmosphere in terminal portion of its trajectory; can be manoeuvrable **, or one of several multiple ** or multiple independently targeted **.

REF   *1* Reference.

2 Refuge.

refan   To replace original fan or LP compressor of turbojet, bypass turbojet or turbofan with fan having larger diameter but fewer stages, for greater economy and less noise; hence refanned engine.

reference area   Area used in components or coefficients of aerodynamic forces acting on a body.

reference axes   Those in relation to which a body's attitude and motion are described; in case of aircraft they normally include body axes, all normal to each other and passing through c.g., called O (OX fore/aft, longitudinal; OY laterally, lateral or axis of pitch; OZ vertically, vertical or axis of yaw); wind axis (direction of free-stream relative wind); and principal inertia axis passing through c.g. in plane of symmetry and at small angle to OX.

reference datum   *1* Arbitrary location at or beyond extremities of structure, eg aircraft, or on centreline or other major axis, from which all distances are measured and station numbers derived. Normally ** remains throughout life of aircraft even if stretching of fuselage or alteration of span of wing changes actual distances to established stations.

2 Imaginary vertical plane at or near nose of aircraft from which all horizontal distances are measured for balance purposes; called balance station zero (NATO).

reference designated indicator   Display which enables aircraft [with or without input by ground engineers] instantly to identify any electronic or mechanical fault.

reference eye position   Typical position of pilot's eyes used in design of cockpit or flight deck.

reference fix   Known position inserted into INS at start of flight.

reference fuel   Piston-engine (Otto) fuel of known anti-knock rating used as reference to fuel whose octane or performance number is to be established.

reference humidity   That specified for mandatory performance information; whichever is lesser of 70% RH to 33°C or 35 mb vapour pressure.

reference landing distance   Abb. RLD, BCAR procedure for calculating landing distance on wet runways [especially jet transports] assuming 3° approach at constant thrust and 15 kt excess speed at 30 ft, excessive [usually 7s] float and slow application of decelerating devices.

reference line   *1* Convenient line on surface used by observer, eg FAC, as line to which spots (4) are related.

2 Single horizontal level showing correct operation by group(s) of systems or measured parameters indicated on vertical-tape instruments; thus a malfunction immediately stands out as a discontinuity.

reference meridian   That selected to establish grid north or local time (ASCC).

reference phase   Non-directional signal emitted by VOR having constant phase through 360° azimuth.

reference plane   See *datum plane*.

reference plate   Minimum-size plate cut from cheap material or scrap containing witness holes or slots cut from every tool in NC program and incorporating part of every sub-routine in program.

reference point   Fixed datum near centre of airfield landing area (obs. except general aviation).

reference pressure $\frac{1}{2}\rho V^2$ where ρ is fluid density and V is relative velocity.

reference pressure ratio   Pressure ratio published for engine, and specialized meanings.

reference section   Traditional definition: a section of structure, displacements of which are taken as co-ordinates in a semi-rigid representation.

reference signal   That against which telemetry data signals are compared to check differences in time, phase etc.

reference sound   Commonly a random noise of one octave bandwidth centred on 1,000 Hz presented frontally.

reference speed   On its own, too vague. V_{REF} is a reference speed used for comparative purposes in takeoff and landing modes; commonly stall speed V_s or target threshold speed V_{AT}, but needs to be defined whenever used [for example, whether clean or full flap].

reference wet hard surface   Numerically specified slippery runway surface for determining braking standards.

reference zero   Datum point from which horizontal and vertical distances are measured to each point on takeoff net flightpath.

refire time   Time required after initial launch to fire second missile (ICBM is assumed) from same silo or other launcher (see *rapid-reload*).

REFL   Reflection.

reflectance   Ratio of reflected to incident radiant flux, symbol ρ.

reflected interrogation   A misnomer, as reply is fresh signal (SSR).

reflected memory bus   High-speed parallel data bus connecting all nodes or subsystems in a large EDP system.

reflected shockwave   That reflected when a shock-wave strikes a boundary between its original medium and one of greater density; part of energy generates shock which continues through denser medium but remainder is reflected in original medium, important case being in boundary between tunnel wall and tunnel working fluid.

reflection coefficient   Measure of mismatch between two impedances; ** a = $(Z_1 - Z_2) / (Z_1 + Z_2)$ where Z_1, Z_2 are impedances.

reflection-interference waves   Intermittent peaks in sea state caused by reinforcement of incident and reflected waves near cliff or sea wall.

reflection interval   Time for radar pulse to reach target and return.

reflection-plane model   Tunnel model comprising one half of model aircraft sliced down axis of symmetry which rests on floor of tunnel.

reflection suppression   False echoes (ghost aircraft) in radars, esp. SSR, are usually suppressed by a suppression transmitter feeding a separate Yagi which, at moment main interrogator scans reflecting surface, sends two pulses, first larger than second; this suppression pair take direct path to aircraft and suppress its transponder before arrival of interrogation signal via reflector; thus no spurious reply is received.

reflectivity factor   Measure of the fraction of incident radar energy reflected by a target, symbol Z.

reflectometer   Instrument for measuring transmission-line reflection coefficient.

reflector   *1* Reflecting surface, usually copper gauze, so sited and shaped as to reflect radiation from primary radiator (eg of radar) in correct phase relationship to reinforce forward and reduce backward radiation; usually paraboloidal but flat in modern fighter radars.

2 Parasitic element located near primary radiator to reduce emission in all directions other than main lobe.

3 Repeller electrode of reflex klystron and similar tubes.

4 Material of high scattering cross-section surrounding core of nuclear reactor.

reflector sight   Gunsight (rarely, for anti-tank missiles) in which reticle aiming mark(s) are projected as bright points on glass screen through which pilot or other aimer views target. Lead angle (aim-off) is assessed by gyro-electronics that measure rate of sightline spin and, in conjunction with range set by pilot, adjust aiming mark so that rounds should hit if reticle is superimposed on target.

reflex camber   Shape of aerofoil in which mean line curves upwards towards trailing edge; eg to provide download well aft of c.g., increasing with airspeed, in tail-less aircraft.

reflex ratio   Measure of structural flexibility of propeller blade tested as cantilever beam (loaded in direction parallel to axis of rotation; US standard is that all blades of same propeller must exhibit tip movement uniform within 0.4 in [10 mm]). Test not applicable to certain blade types, eg hollow steel.

Reflex   Detachments on ground alert [with NW] to overseas base[s] (SAC).

reflex sight   ASCC term for reflector sight.

reflex trailing edge   See *reflex camber*.

refly   To make second test flight to clear snag [verb or noun].

REFRA   Recent freezing rain.

refraction   Change in direction of travel of supposed linear radiation, eg EM radiation or sound, due to variation in properties (eg refractive index, air temperature) of transmitting medium. Can be gradual over a distance or instantaneous at boundary between two media; for radio/radar important forms are atmospheric * (low-altitude temperature, pressure, humidity; also responsible for errors in apparent altitude of celestial bodies), coastal * (change in propagation path at land/sea boundary) and ionospheric * (change of direction in passage through ionized layer).

refractive index   Ratio of phase velocity of EM radiation in free space (vacuum) to phase velocity in medium considered; normally related to air, though in fact ** for air is not unity, common S/L value being 1.003. For radio normally refined to modified ** (n + h/a) where n is ** at height h and a is radius of Earth.

refractory   Resistant to high temperatures; normally implies able to retain precise structural shape and bear appreciable stress at temperatures up to 2,200° C (4,000°F) without significant long-term change.

refresh rate   Rate at which data on electronic display, eg radar, are resupplied in order to maintain bright picture free from flicker; early radar had no refresh and picture was generated only once on each scan; today data are updated at each revolution but are refreshed 100 to 200 times between updates. Panel-instrument * is usually 40–80 Hz.

refrigerant   Working fluid pumped round closed circuit to extract heat, usually by repeated evaporation/condensation.

Refrigerant 12   Difluorodichloromethane (DDM).

refrigerant injection   Water injection into gas turbine.

refrigeration capacity   Common unit is 'tons', normally a heat-extraction rate sufficient to convert one short ton of water at 0°C to ice every 24 h, equivalent to 3.517 kW = 4.715 hp.

refrigeration icing   Caused by sudden drop in temperature of airflow, notably by depression in choke tube and/or evaporation of fuel.

refrigeration tunnel   Wind tunnel used for icing tests.

REFS   Reconfigurable engineering flight simulator.

refusal speed   That at which a takeoff can no longer be aborted, equivalent to V_1 (RAF).

REG   *1* Registration, registered.

2 Regular.

3 Regulation.

Rega, REGA   Swiss air ambulance association. Not an acronym, but French and Italian counterparts GASS are.

Regal   Range and elevation guidance for approach and landing (FAA test 1960).

regard   Total solid angle of 'vision' of a sensor trainable to point in different directions; FOV may be much less.

regeneration   *1* Introduction of closed-circuit operation in gas-turbine plant in which by various methods part of exhaust heat is extracted and used to increase temperature of incoming airflow; hence regenerative gas turbine.

2 Increase in radio detector sensitivity by positive feedback.

3 Favourable heat transfer between rocket thrust chamber nozzle and propellant (see *regenerative cooling*).

4 In EDP (1) rewriting to prevent memory deterioration.

regenerative cooling   Use of a cool incoming liquid, eg rocket-engine propellant, to remove heat from hot hardware, eg rocket nozzle skirt and exit cone. Essential feature is that heat transfer is beneficial to both cooled item and coolant.

RegFD   Regulation fair disclosure (US Securities and Exchange Commission).

regime   One defined mode of operation, clearly distinguished from other types of operation of same device; eg USAF fighter flight-suit pressure regulator has lo * maintaining 3 lb/sq in and hi * maintaining 6.5.

region   See *flight information region*.

regional airport   One serving a number of (usually modest-sized) local communities.

Regional augmentation   Navsat transponders broadcast ground-derived GPS integrity assurance and accuracy enhancement.

regional carrier   Civil operator whose route network covers only a minor part of a large country; often same as third-level.

regional QNH   See *QNH*.

register   Small array of bistable circuits storing one EDP (1) word.

registering balloon   Small free balloon carrying recording meteorological instruments; balloonsonde.

registration   Entry of civil aircraft into records of national certification authority, with allocation of letter/number code displayed on aircraft and * certificate which in most countries must be displayed inside aircraft.

REGL   Regional.

REGR   Recent hail.

regression   Precession of nodes; eg Moon completes revolution in 18.6 years.

regression rate   Linear rate at which solid-propellant grain burns, measured normal to local surface; in 1970 typically 0.25 mm/s, in 2002 over 5 mm/s.

regressive burning   Solid-motor combustion in which burn surface area and hence chamber pressure and thrust all fall throughout period of burn.

regressive orbit   See *retrograde*.

regroup airfield   Military or civil airfield at which post-H-hour, aircraft would reassemble for rearming, refuelling and resumption of armed alert, overseas deployment or further combat missions (DoD, IADB).

REGS   Regulations.

regular airfield   One which may be listed in flight plan as intended destination.

regular airport   That at which scheduled service calls (UK usage).

regulated take-off weight   See *WAT-limited*.

REH   Rapid-erect hangar.

reheat   Original [UK] term for *afterburning*.

REHM   Recording engine health monitor.

REI   *1* Repair engineering instruction.
  2 Reusable external insulation

Reid vapour pressure   Absolute pressure of liquid [usually hydrocarbon fuel] in enclosed volume at given temperature, usually 100°F (37.8C).

REIL   Runway-end identification or identifier lights (FAA); S adds system.

reinforced carbon/carbon   Composite material comprising high-strength carbon fibres bonded in matrix of pyrolitic carbon; or can be thought of as pyrolitic carbon reinforced with CF.

reingestion stall   Gas-turbine compressor stall induced by reingestion of hot gas during reverse-thrust mode.

Reins   Radar-equipped inertial navigation system (Autonetics 1956).

Rejac   Receiver/jammer capability.

reject   To dump heat out of system into supposed sink, eg atmosphere.

rejected take-off   One aborted after it has begun, ie between brakes-release and decision point.

rejector   Inductance/capacitance in parallel to reject one resonant frequency.

Rejex   Soil-barrier protection for painted surfaces.

rejoining aircraft   One returned to service after [for whatever reason] having been parked for long period.

REL   *1* Relative direction.
  2 Runway edge light[s], lighting.

relateral tell   Relay of air-defence information between facilities via third; appropriate between automated centres in degraded communications environment.

relative altitude   See *vertical separation*.

relative bearing   Normally means bearing of surface feature or other aircraft relative to current heading.

relative-bearing indicator, RBI   Shows bearing of tuned fixed station related to aircraft longitudinal axis; unlike RMI, does not show heading.

relative density   *1* Density at height in atmosphere related to that at S/L, ρ/ρ_o, symbol σ.
  2 Specific gravity.

relative efficiency of biplane   Ratio of wing loadings of upper and lower wings.

relative humidity   Water content of unit volume of atmosphere expressed as percentage of saturation water content at same temperature.

relative inclinometer   Flight instrument indicating attitude with respect to apparent gravity (resultant of gravity and applied acceleration).

relative permittivity   See *permittivity*.

relative scatter intensity   Ratio of radiant intensity scattered in given direction to that in direction of incident beam; symbol F(ø), relative scattering function.

relative target altitude   Vertical difference between altitudes of interceptor and target.

relative wind   Velocity of free-stream air measured with respect to body in flight, in case of aircraft normally same as true airspeed but seldom exactly aligned with longitudinal axis.

relaxation time   *1* Elapsed time between removal of disturbance and restoration of equilibrium conditions among molecules of fluid (eg air), operative parts of system or other dynamic components of system.
  2 Time for exponentially decaying quantity to decrease in amplitude by 1/e = 0.36788.

relaxed static stability   CCV-derived manoeuvre enhancement for air-combat fighter, involving rear c.g., longer nose, smaller fin etc, with artificial stability imparted by AFCS.

relay   Device in which small control signal, usually electrical, is made not only to operate at a distance but also to control large and possibly high-power devices; most serve switching functions or to protect devices against supply faults.

relay time   Elapsed time between instant message is completely received and that when it is completely transmitted.

release   Clearance of aircraft for line service, eg after original development or overhaul.

release altitude   Altitude of aircraft AGL at actual time of release of ordnance, tow target etc (DoD).

release blade   That selected in FBO test.

released   *1* Of aircraft to unit, clearance for inventory service.
  2 Of drawing, approval for transmission to manufacturing or production department.
  3 Of air-defence unit, crews and/or weapons no longer needed at readiness; when * they will be informed when state of readiness will be resumed (NESN).

released, available   Ready at 20 min. notice (FAA).

release point   Point on ground directly above which first paratroop or cargo item is air-dropped (NATO).

release time   Departure time issued by ATC to avoid conflicts.

reliability   Probability that hardware will operate without failure for specified time; in practice also a result of operations already accomplished (see *despatch* *, *MTBF*).

reliability coefficient   Percentage probability that aircraft can fly route on particular day with x% load factor and y% likelihood of diversion. Note: not a measure of hardware reliability.

relief hole   Drilled in metal sheet to allow intersecting bends to be made to that point without buckling.

relief on station   Two UAVs simultaneously worked by same ground control station.

relief tube   Personal urinal pipe normally discharging overboard.

relief valve   Fluid system valve which releases pressure at preset value.

relight   *1* To restart combustion in gas turbine after mid-air flameout.

    2 To use ridge lift to prolong glide of sailplane.

relight envelope   Published diagram of permissible limits of TAS and height outside which engine relight should be attempted only in emergency. ** forms part of flight manual of civil gas turbine aircraft, and for military aircraft is given for main combustion and afterburner.

relighting altitude   That up to which safe and reliable restarting of power unit (gas turbine is implied) is possible (CAA).

reluctance   Opposition to magnetic flux, property of magnetic circuit akin to resistance in electric circuit which limits value of flux for given MMF; symbol R or S, numerically = l/μ A where 1 is length of magnetic path, μ is permeability and A cross-sectional area.

Relvel   Relative velocity.

REM   *1* Rocket engine module.

    2 Remaining.

rem   Quantity of ionizing radiation which, when absorbed by body, produces same physiological effect as 1 roentgen of X-ray or gamma radiation; from roentgen-equivalent man.

remanence   Flux density remaining (residual) in material after removal of magnetizing force. Also called retentivity.

remanufactured   Aircraft completely stripped and inspected, usually by original builder, and with new structure added where necessary before reassembly and delivery with clearance for specified long life free from fatigue. Usually opportunity is taken to update systems and equipment also.

Remap   Research maximization and prioritization (NASA).

Rembass   Remotely monitored battlefield-area surveillance (or sensor) system.

Remco   Reference Materials Committee (ISO).

Remdeg   Reliability military data exchange guide.

Remis   Reliability and maintainability information system[s].

Remoco   Remote monitoring and control.

remote aerial refuelling operation   With boom-equipped aircraft, using CCTV from flight deck in absence of boom operator.

remote augmented lift system   Arrangement for providing enhanced jet lift for V/STOL on demand in which large airflow bled from engine(s) is piped to auxiliary combustion chamber, usually near nose of aircraft, to provide additional high-energy lift jet. Latter may have means for modulation and linked vectoring, and RALS always requires large pilot-controlled diverter valve(s).

remote augmentor   *Remote nozzle.*

remote communications outlet   Remotely controlled unmanned air/ground com. station providing UHF and VHF transmit-and-receive capability to extend range of FSS (FAA).

remote fan   Mechanically independent fan driven by tip-driven turbine blades fed by hot gas from main engines via switch-in deflectors; used to provide jet lift for V/STOL or, rarely, to increase thrust at low speeds.

remote frequency display   Shows details and status of all on-board radios.

remote indicating compass   Magnetic compass whose sensing element is installed in extremity of aircraft where deviation is minimal, with transmitter system serving repeater dial(s) facing crew.

remotely piloted vehicle, RPV   Aerodyne usually of aeroplane type whose pilot does not fly with it but controls it from another aircraft or from station on surface. Authority of remote pilot is usually absolute, though to ease pilot workload some RPVs have choice of preselected auto pilot/computer programs for at least part of each mission.

remote mass balance   Connected by linkage to surface where * cannot be accommodated.

remote nozzle   Substantially vertical nozzle in RALS fed with air from main-engine fan and burning fuel to generate about 45% of total lift thrust.

remote receiving station   Friendly station remote from UAV or RPV control station at which its transmission (eg reconnaissance information) can be received.

remote-source lighting   Use of [possibly very long] fibre optics.

remoting   Transmission of ATC, SSR or air defence radar display by landline, microwave link or other means to distant centre.

remous   Air turbulence (F, commonly used by English-speaking aviators pre-WW1).

removables   All items flight crew must remove from outside of aircraft before take-off.

REMP   Replacement and modernization program[me].

Remro   Remote radar operator.

REMS   Remote sensor.

REMSA   Requirements for emergency and safety airborne equipment, training and procedures (ECAC).

Remsevs   Remote-control secure-voice system.

rendering   Th acceptance by a Contracting State of another's C of A or licence.

rendezvous   Meeting of two aircraft or spacecraft at preplanned place and time. For aircraft can include air-refuelling hook-up but for spacecraft physical connection is docking.

rendezvous orbit insertion   Establishment by one spacecraft of orbit almost identical to that of another before actual rendezvous.

rendezvous radar   Esp. in spacecraft, small ranging and range-rate radar carried to facilitate rendezvous and subsequent docking.

René alloys   Family of American high-temperature alloys with Ni base plus Cr, Mo, Co, Ti, Fe, Al, and small amounts of C, Bo.

reneg   To go back on previous agreement (US, colloq.).

renegotiation   Procedure not uncommon in US where manufacturer is required (usually by government) to re-negotiate an existing active contract; usual cause is allegation of excessive profit.

renewal   Procedure, usually annual, for inspection of civil aircraft for certification * and of civil pilot * of certificate or licence. Hence * rate, total of fees payable, and * inspection, by authority-approved organization.

R/EO   Radar/electro-optical.

REP   *1* Reporting point (ICAO).
    2 Reference eye position (or point).
    3 ECM (R).
    4 Replacement.
    5 Representative.
    6 Range error probable.

repair-cycle aircraft   Those in active inventory in or awaiting depot maintenance (DoD).

repairman certificate   Issued by FAA to skilled tradesman engaged in repair or maintenance of aircraft or parts.

Repairnet   Reconstitution post-attack interoperable radio network.

repeater jammer   ECM receiver and transmitter which receives hostile signals and amplifies, muliplies and retransmits them for purposes of deception or jamming. Basic deception mode is to give false indication of range, azimuth or number of targets.

repetition rate   *1* In radar, number of pulses per second.
    2 In automatic gun, cyclic rate.

repetitive chime clacker   Gives aural warning indicating fault in aircraft configuration.

repetitive flight plan   Kept on file for frequent [eg. scheduled] identical flights.

Repin (G)   Replacement inertial navigation unit with GPS added.

replaceable panel   Aircraft maintenance access panel which has to be replaced; not an interchangeable panel, which can be opened in about one-tenth the time.

replacement factor   Estimated percentage of hardware items that over given period will need replacing from all causes except accidents.

replenishing   Refilling of aircraft with consumables such as fuel, oil, liquid oxygen and compressed gases to authorized pressure; rearming is excluded (NATO).

replenishing phase   Part of operating cycle of pulsejet in which depression in duct induces fresh charge.

reply code   That repeated series of pulses transmitted by SSR transponder in aircraft when interrogated; typically up to 12 information pulses between two framing pulses 20.3 µs apart.

reply-code evaluator   Automatic avionic subsystem which reads reply code and determines if valid, what identify and in case of military IFF whether friendly or hostile.

reply efficiency   A transponder's valid responses as percentage of interrogations.

reporting point   Geographical point in relation to which aircraft position is reported.

repositioning   Moving aerobatic aircraft from end of one manoeuvre to start of next, at correct height/speed.

repressurant   Material, eg compressed air, oxygen and water vapour, stored in spacecraft outside pressurized volume and fed in to refill interior after depressurized operations.

req, REQ   On request.

REQMT   Requirement.

requalification   To requalify approved system to meet more severe demand or environment.

request for proposals   Document sent by central government to one or more industrial contractors outlining future requirement stated by armed force(s) and inviting suggestions on how this should best be met; RFP calls for analysis of problem by manufacturer and for submission of general scheme for hardware which in case of aircraft and most other equipment includes three-view drawing, basic description, estimated weights and performance, timescale and costs.

required flightpath   That necessary to satisfy immediate task of pilot; not a recognised performance parameter but general objective weighed against projected flightpath.

required navigation performance   Measure of accuracy of a particular segment or block of airspace.

required track   Path aircraft commander wishes to follow; refers always to future intention. Often a ruled line joining two waypoints.

requirement   Predicted future need spelt out by armed force, usually after long process of refinement and assessment of alternatives.

RER   *1* Reconnaissance exploitation report.
    2 Radar electrical rack.
    3 Residual error rate.

RERA   Recent rain.

RERP   Reliability enhancement and re-engining program.

RERTE   Re-route.

RES   *1* Radio emission surveillance.
    2 Reserved.
    3 Reservoir.
    4 Radar emitter, or environment, simulator.
    5 Remote, or reprogrammable, emitter simulator.

Resa, RESA   *1* Research, evaluation and systems analysis.
    2 Runway-end safety area.
    3 Rotating electronically scanned array.

Rescap   Rescue combat air patrol.

rescue basket   Lowered from helicopter to rescue person unconscious or incapacitated.

rescue co-ordination centre   Initiates, manages and terminates rescue efforts within particular area by all branches offering help.

research and development   Generalized term covering process of development of specific items of hardware; research involved is usually minimal and always applied, main effort being directed at solving engineering problems which are invariably unpredicted and occasionally require new fundamental knowledge; abb. R&D, but US favours RDT&E, which is unnecessarily clumsy.

research coupling   Disseminating research results to maximize early and widespread use.

research octane number   Usual scale of measurement for

anti-knock value of automotive fuels. On this basis unleaded gasolines are 95–97 RON, while leaded might have rich-mixture RON of 130. Aeronautical gasolines are commonly MON-rated.

research rocket   Usually unguided ballistic rocket whose purpose is to lift scientific payload to high altitude for free-fall or parachute descent. Very occasionally purpose is to advance technology of rockets.

research vehicle   Atmospheric or space vehicle, manned or otherwise, whose purpose is to provide answers to research problems; not normally prototype of production article but often associated with specific programme.

reseau   *1* Group of met. stations operating under common direction; hence international *.

2 Grid of fine parallel lines used in image analysis.

Réseau du Sport de l'Air   Association of homebuilders (F).

reserve aircraft   Those accumulated in excess of immediate needs for active aircraft and retained in inventory against possible future needs (DoD).

reserve buoyancy   Additional mass or applied force needed to immerse completely floats or hull of seaplane or flying boat already at specified (usually MTO) weight.

reserve factor   Ratio of actual strength of structure to minimum required for a specified condition.

reserve fuel   See *reserves*.

reserve parachute   Second, standby parachute usually worn by professional parachutists and many others who make frequent deliberate descents; 7.3 m/24 ft ** mandatory for sport parachuting.

reserve power   See *specific excess power*.

reserves   Quantities of consumables, esp. fuel, planned to be unconsumed when aircraft arrives at destination and available for holding (stacking), go-arounds, diversions and other contingencies.

reservoir   Storage (not header) tank in fluid system, eg hydraulics.

reset   To restore device, eg bistable gate, memory address or fire-warning system, to original untriggered state.

RESH   Recent showers.

residence time   *1* Time fuel droplet or gas particle remains in either gas-turbine combustor or afterburner.

2 Time, usually expressed as a halftime (not to be confused with half-life) radioactive material, eg fallout, remains in atmosphere after NW detonation.

residual magnetism   See *remanence*.

residual propulsive force   Net propulsive force minus total aircraft drag, F–D.

residuals   *1* Any fluid left in spacecraft tanks after use.

2 Difference (plus/minus) between intended δV (velocity increment) for a burn and that achieved (NASA).

residual stress   That in structural component in absence of applied load, due to heat treatment, fabrication or other internal source.

residual thrust   *1* That produced by jet engine (turbojet or turbofan) at flight or ground-idle setting.

2 That produced by any other propulsion engine after deliberate shutdown, or cutoff.

resilience   Measure of energy which must be expended in distorting material to elastic limit.

resin   Vast profusion of natural and, increasingly, synthetic materials used as adhesives, fillers, binders and for insulation. Various types used in nearly all fibre-reinforced composites.

resin lamp   Small lamp with filament bulb used not for illumination but to indicate position, e.g. of wingtips of aircraft in night formation.

resistance   Opposition to flow of electric current, R = V/I, unit ohm[s], symbol Ω.

resistance derivatives   Quantities, generally dimensionless, which express variation in forces and moments acting on aircraft after upset. In general case there are 18 translatory and 18 rotary.

resistance welding   Using internal resistance of metal workpiece to very large electric current to produce heating which, under pressure, forms weld (see *spot weld, seam weld*).

resistivity   Specific resistance; electrical resistance of unit length of material of unit cross-section. Convenient unit ohm/cm^3, though not strictly SI.

Resistojet   Various patented (eg Avco, Marquardt) space thrusters using liquid ammonia or other working fluid, including biowastes, accelerated by solar-powered electrothermal chamber with de Laval nozzle.

resistor   Electrical device offering accurately specified resistance.

reslams   Repeated throttle slams for engine test.

RESM   Radar electronic support measures.

RESN   Recent snow.

resojet   Usually means resonant pulsejet.

resolution   *1* Measure of ability of optical system, radar, video/TV or other EO system, photographic film or other scene-reproducing method to reveal two closely spaced objects as separate bodies; normally defined in terms of angle at receiver subtended by two objects which can just be distinguished as separate.

2 Ability of device, as in (1), to render barely distinguishable pattern of black/white lines; expressed in number of lines per mm which can just be distinguished from flat grey tone. Both (1) and (2) also called resolving power.

3 Measure of response of gyro to small change at input; minimum change that will cause detectable change in output for inputs greater than threshold, expressed as % of half input range.

4 Separation of vector quantity into vertical/horizontal components.

resolution advisory   Verbal or display indication recommending increased vertical separation relative to an intruding aircraft.

resolver   Subsystem in spacecraft INS which measures changes of attitude, esp. rotation about longitudinal axis, and informs guidance computers. In general a rotary digitiser converting small angular movements to digital signals.

resolving power   *1* See *resolution (1, 2)*.

2 Ability of radar set or camera to form distinguishable images (ASCC); term unhelpful and indistinguishable in practice from resolution (1, 2).

3 Reciprocal of unidirectional aerial beamwidth measured in degrees (not synonymous with resolution [1], which is affected by other factors).

resonance   *1* Condition in which oscillating system such as free wave or aircraft structure oscillates under forcing input at natural frequency, such that any change in frequency of impressed excitation causes decrease in response.

2 Condition of AC circuit when, at given frequency, inductive and capacitive reactances are equal.

3 In specific case of rotary-wing aircraft, particularly helicopter, condition in which natural frequency of landing gear corresponds with main-rotor rpm.

resonance test   Structural exploration of natural frequencies by excitation over slowly varying wide range of frequencies.

resonant duct   See *resonant pulsejet*.

resonant pulsejet   Pulsejet in which intermittent operation occurs at natural frequency of operating air/fuel-burning duct system, without need for one-way flap valves.

resonating cavity   Closed hollow space of precise geometry having electrically conductive walls in which microwaves are generated when excited by EM field or electron beam; examples are magnetron, klystron, rhumbatron; also called resonant cavity.

resonator   *1* See *resonating cavity*.

2 Magnetostrictive ferromagnetic rod excited to respond at several distinct frequencies.

3 Lecher wire.

4 Piezoelectric crystal.

5 Acoustic enclosure having single-frequency response.

Resound   Reduction of engine source noise through understanding and novel design.

RESP   Receiver/exciter/synchronizer processor.

responder   Receiving unit in transponder.

response   A specialized aerospace meaning is rejection of chaff or flare decoy by a missile seeker, which re-acquires real target.

responsor   Electronic device used to receive electronic challenge and display a reply thereto (DoD).

RESS   Radar emission, or environmental, simulator system.

ressource   "Sudden cabrage following a dive" [i.e., pull out]; between the wars it was held there was no English equivalent.

REST, Rest   *1* Radar electronic scan technique; spherical/planar lens array aerial, RF angle-error sampling circuits, auto search/detection/confirmation, and display/recording.

2 Re-entry environmental and systems technology.

rest angle   Angular position of hinged surface when not in use. May correctly incorporate slight offset, droop or other departure from housed or neutral.

restart   Start of burn after previous cutoff.

restart time   Time between completion of adjustment of tunnel model and taking first readings.

rest-EVA period   Scheduled period in manned space mission when in absence of assigned duties person may rest or conduct EVA.

restitution   Determination of true planimetric position from reconnaissance photographs (NATO).

rest mass   Mass of body when at rest (absolute in cosmological term); other masses $m = m_o \sqrt{1 - (v^2/c^2)}$ where m_o is **, v is velocity and c is speed of light.

restoring couple   Couple producing restoring moment.

restoring moment   Moment generated by upset, ie rotary excursion from original or desired condition, which tends to restore original condition.

rest period   Time on duty on ground during which flight crew is relieved of all duties.

REST   Radar electronic scan[ning] technique.

RESTR   Restrict[ion].

restrained aircraft   One undergoing dynamic structural test or analysis, eg flutter, with one or more parts anchored.

restraint   *1* Standard series of tiedowns, webbing and nylon-cord nets to prevent movement of bulk cargo.

2 Process of binding, lashing and wedging items into one unit on to or into transport to ensure immobility during transit (DoD), NATO).

3 Mechanical latching of container, pallet or other ULD.

restricted airspace   See *restricted area*.

restricted area   *1* Airspace above surface area of published dimensions within which flight of aircraft is subject to restrictions caused, eg, by 'unusual and often invisible hazards' such as AAA or SAM activity; in US published in FAR 73, in UK by CAA and in all Notams, charts and commercial publications.

2 Area where restrictions are in force to minimise interference between friendly forces.

restricted burning   See *restricted propellant*.

restricted data   Those pertaining to design, manufacture or use of NW, or special nuclear material.

restricted fire plan   Safety measure for friendly aircraft which establishes airspace 'reasonably safe from friendly surface-delivered non-nuclear fires' (DoD).

restricted propellant   Solid-propellant grain whose surface is only partly available for ignition, remainder being protected by restrictor.

restrictor   Layer of solid-propellant fuel containing no oxidant (oxidizer) or of non-combustible material bonded firmly to inner surface of grain to prevent that part being ignited except by flame travelling within propellant under *.

restructurable   Able to be rapidly modified [redesigned] in flight to overcome dangerous problem, such as hard-over runaway (flight-control system).

resultant   Sum of two or more vectors. Hence * action, * force, * lift, * velocity.

RESYNC   Resynchronize, resynchronizing.

RET   *1* Rapid-exit taxiway.

2 Retired.

3 Reliability evaluation test.

ret   NASA code for time between routine events.

retard   *1* To cause piston engine ignition to occur later in each cycle (normally still before TDC).

2 AT/SC mode in which throttles bleed off at programmed rate during landing flare.

3 Displayed instruction to chop power to Flight Idle.

4 Mechanical block on forward movement of throttle lever[s].

retardant   Tailored additive to water dropped by fire-bomber [several trade names].

retardation probe   Tapering rod thrust into hydraulic aperture to offer increasingly great resistance to carrier arrester-wire pull-out.

retarded bomb   Free-fall bomb with airbrakes, drogue, parachute or other high-drag device deployed automatically on release.

retarder parachute   Small auxiliary parachute to pull main-parachute rigging lines out in advance of canopy.

retention   Maintenance of full design performance over long period of service.

retention area   Highly loaded parts of turbine or compressor disc around blade roots.

reticle   *1* Any kind of mark, such as black ring, illuminated cross or ring of bright diamonds (to give three examples) used to assist any form of optical aiming, eg aerial gunnery, spacecraft docking or airdrop on marker.

2 In photogrammetry, cross or system of lines in image plane of viewing apparatus used singly as reference mark in monocular instruments or in pair to form floating mark in certain stereoscopes (ASCC).

reticulated   Having form of fine network; hence * plastic or * foam are 3-D volumes of low-density fire-resistant foam which can be foamed in place inside or outside fuel tanks and other items to prevent build-up of fuel/air mix and, even in presence of severe combat damage or post-crash rupture, prohibit explosion or swift spread of fire.

Retimet   Patented (Dunlop) reticulated metal, low-density 3-D mesh of various metal strips or filaments.

retinal scanning display   Mounted on pilot's head and projects pixels [of flight data] into pilot's eye, focused at infinity.

retirement   Withdrawal of serviceable aircraft on grounds of age, see next.

retirement life   Aggregate of running time of engine or other device at which decision is taken, usually by operator but sometimes suggested by manufacturer, that further overhauls are uneconomic; main reasons are obsolescence of design or onset of fatigue problems.

retrace   American term for flyback.

retractable   Capable of being withdrawn into aircraft so that it no longer protrudes, or protrudes only partially; applies to many devices which are * into all parts of aircraft, such as landing gear, inlet spikes, hooks, MAD gear, Fowler flaps, spoilers, sensor pods, radars and, formerly, gun turrets.

retractable ejector duct   Two-position jet-engine reverser for use in flight.

retraction lock   Mechanical device to prevent inadvertent retraction of landing gear; today is removable but backed up by second lock actuated by compression of undercarriage oleos.

retread crew   Military flight crew returned to OCU after tour of combat duty.

retreating blade   That on side of a lifting rotor, eg on helicopter, moving relative to aircraft in same sense as slipstream; thus its airspeed is difference between its own speed and true airspeed, which is normally positive at tip, zero at a particular part-span radius and negative near root. Inboard of zero-airspeed radius, airflow is from trailing to leading edge.

retreating-blade stall   Stall of retreating blade at high helicopter forward speeds, when angle of attack of retreating blade is excessive, especially towards tip; exceeding flight-manual forward speed causes stall over near-rectangular area of disc whose centre is behind c.g. and thus effect is to cause nose-up roll towards retreating side.

retrieval   Mid-air snatch of parachuted load, eg spacecraft.

retrieve   Task of following sailplane on cross-country to goal, dismantling it and bringing it back in trailer.

retrievers   Controller workstations in Comfile ATC

system which pass data and radar from Ethernet to operator consoles.

retrimming   *1* Adjustment of flight-control trim for different flight condition [major error on takeoff can be catastrophic].

2 Adjustment of engine trim to allow for deterioration caused by deposits, erosion, birdstrikes and other factors.

retro   Usually means retrofire or retrorocket.

retrofire   To fire retrorocket.

retrofit   Modification, esp. involving addition of new or improved equipment, to item already in service; hence * action, * mod.

retrograde   *1* Orbit in direction different from normal; eg of a planet apparently moving westward against fixed stars.

2 Orbit in direction opposite to rotation of primary body, eg Triton around Neptune or Earth satellite launched at inclination from 180° through 270° to 360/000°.

3 In traffic direction opposite to normal, eg cargo of military logistic type moving towards United States, or military command away from enemy.

retroreflection   Reflection, eg of EM radiation, parallel to incident rays; hence retroreflector, device for accomplishing same, eg corner reflector, Luneberg lens.

retrorocket   Rocket fitted to vehicle to oppose forward motion, eg to bring satellite out of orbit and back to Earth. Loosely used to mean what are more precisely called separation or staging motors or, on aircraft, braking rocket.

retrosequence   Event sequence before, during and after retrofire.

retrothrust   Thrust opposing motion; can be used for aircraft reverse thrust.

Rets, RETS   *1* Remote target system, or remote-equipment target system.

2 Recent thunderstorm[s].

3 Radar extractor and tracking system.

return   Echo (radar), esp. that due to clutter sources; often plural.

return-flow   Combustor in which incoming air and issuing gas travel parallel in opposite directions; also called reverse-flow.

return-flow tunnel   Wind tunnel in which air is contained in enclosed circuit.

return grab   Returns shuttle of catapult (deck accelerator) to start position.

return line   *1* That traced on CRT by flyback.

2 Fluid pipe bringing fluid back from device to pump.

return load   *1* That transmitted to aircraft by stopping forward motion of action in gun, in direction opposite to recoil.

2 Personnel and/or cargo to be transported by returning carrier (DoD); 'carrier' means any vehicle, not aircraft carrier.

return oil   Scavenge oil; hence instrument showing * temperature.

returns   See *return*.

return to base   Code: proceed to point indicated by displayed information which is being used as point from which aircraft can return to place where they can land; command heading, speed and altitude may be used (DoD).

REU   Remote electronic[s] unit.

reusability   Extent to which space hardware can be used for repeated launches, especially if recovery is made from desert or ocean.

REV   Revision message.

Reveal   Real-time electronic video enhancement at long range.

Reven   Reverser/variable exhaust nozzle.

revenue   Air-carrier income from traffic sales.

revenue yield   Rate per unit of traffic, eg cents per ton-mile.

reverberation   Persistence of sound in enclosed space as result of continued multiple reflections, with or without continued emission by source.

reverberation time   Time between cut-off of source and diminution of sound, measured as time-average of acoustic energy density, to fall to 10^{-6} of original.

reversal   *1* Half a cycle of oscillating applied load, ie from maximum load in one direction to maximum load in opposite direction.
   2 Control reversal (see *reversed controls*).

reversal parameter   Sign ($\pm$) of partial derivative of pitch-loop bandwidth to pilot's pitch control gain, considered in design of carrier aircraft.

reversal speed   Lowest EAS at which control reversal is manifest.

reversal temperature   That at which characteristic spectral lines of incandescent gas disappear against black-body spectrum.

reversal zone   Zone within ILS glideslope or course sector in which slope of sector characteristic curve is negative.

reverse bias   That which reduces current.

reverse blindness   Obscuration of flight-deck vision by snow or other material in reverse-thrust mode.

reverse breakdown voltage   That at which reverse current across p/n junction increases rapidly with little increase in reverse voltage.

reversed controls   Flight-control axis about which, in particular severe conditions, application of pilot input demand causes aircraft response in opposite sense, normally due to aeroelastic distortion of structure. Usual axis is roll, where under very high EAS (ideally not within limits of flight manual) large aileron deflection causes opposite twist of wing which more than neutralizes rolling moment due to aileron; in effect aileron acts as tab and wing as aileron.

reverse dihedral   Destabilizing rolling moment in sideslip at high AOA when slipstream increases lift of leeward (trailing) wing. Also called negative dihedral.

reversed lobsterback   A Phase II combustor heat shield [Rolls-Royce, superseded by Phase-V].

reversed rolling moment   That due to reversed control in rolling plane; also called roll reversal.

reverse engineering   Process of studying a finished product [precise geometry, materials and surface finish] so that it can be copied. Classic case is Tupolev's * of B-29 to create Tu-4.

reverse-flow combustor   One in which air enters at front, travels to rear-mounted fuel burners and then returns as hot gas within flame tube to leave radially inward from front; also called folded combustor, return-flow.

reverse-flow engine   Gas turbine incorporating axial compressor which draws in air around rear end and compresses it in forwards direction, before turning flow radially outwards (often by added centrifugal stage) to flow back to rear through combustor(s).

reverse-flow region   *1* Quasi-circular region near hub of helicopter main rotor disc, on retreating side, within which relative airflow is from trailing to leading edge, ie helicopter airspeed is greater than blade speed due to rotation.
   2 Any region in turbulent boundary layer in which there is a majority-flow reversal.

reverse idle   Power-lever setting at which engine is at idle (usually flight-idle because prior to touchdown on committed landing) with reverser buckets in reverse-thrust mode.

reverse launch   In direction opposing Earth's rotation.

reverse localizer   Back course.

reversement   See *reverse turn*.

reverse origami   Unfolding of spacecraft, or its aerial[s].

reverse pitch   Special ground-only setting available on some propellers and ducted propulsors, including several variable-pitch turbofans, in which blades accelerate air forwards, creating retrothrust proportional to engine power without change in direction of rotation.

reverser   Device for deflecting some or all of efflux from jet engine to give reverse thrust (retrothrust); can take form of pivoting clamshell buckets, blocker doors and peripheral cascades, or other forms, and may include turbofan core or fan exit only. Angle through which jet is turned seldom exceeds 135°.

reverse thrust   Operating mode for jet engine equipped with reverser, obtainable only by overcoming gate or detent which may be locked until weight is on oleos for specified period, eg 1.5 s; normally obtained by moving power levers past idle down to ** mode, further movement in this direction opening throttle to full power to give maximum retrothrust.

reverse torque   Any situation in which the driven member [e.g., propeller] drives the prime mover. Could be dangerous, as in a Viscount which was dived at high IAS when one propeller ran away fully fine.

reverse turn   Opposite of Immelmann: half-roll followed by half loop. Also called reversement.

reverse-velocity rotor   Main lifting rotor of compound or hybrid helicopter which behaves as an aeroplane in cruising flight, with a large portion in high-speed reversed relative airflow.

reversible propeller   One in which reverse pitch may be selected.

reversing layer   Thin lower part of Sun's atmosphere; cooler than photosphere and source of Fraunhofer lines.

reversion   Change of operating mode (eg, but not exclusively, of flight-control system from normal powered to a degraded or manual mode).

reversionary   Available following failure of primary system.

reversionary facility   Facility for changing operating mode, either automatically or upon human command, esp. one following failure or degradation of existing channel or subsystem.

reversionary lane   Back-up or standby channel.

reversionary mode   Normally means advanced integrated flight system is available for pilot input of selected navigation mode from choice of several unrelated systems, eg INS, local R-Nav, Doppler, VOR/DME or Omega.

reverted rubber   Rubber heated beyond critical point at which it loses basic mechanical properties, esp. elasticity, and becomes sticky and permanently deformable. In one of the three aquaplaning modes, lack of anti-skid system causes locked wheel(s), reverted rubber in contact with runway covered in standing water, rapid steam generation and aquaplaning on steam layer above water.

revetment   Area protected on three sides by blast-resistant wall of concrete, sandbags, compacted earth or other material, either to protect occupants and parked combat aircraft or other stores against external attack or to protect occupants, eg launch crew, against hazardous rocket or similar tests.

Revi   Reflector sight (G).

Revise   Research vehicle for inflight submunition ejection.

revival   Restart of spacecraft systems after period of rest or shutdown, eg after long mid-course en route to planet when * is commanded by characteristic telecom signal.

rev/min   SI abb. for revolutions per minute.

revolution   *1* In engineering terms, one rotation of shaft or rotary system.

2 In spaceflight, motion of body about axis remote from itself, eg of planet around Sun.

3 One complete orbit starting and finishing at same point. In practice this is unattainable concept because of rotation of Earth, Earth's revolution (2) around Sun, Sun's motion through local galaxy, etc. Time for Earth-satellite revolution, as distinct from period (see *sidereal* and *synodic orbit*), is of meaning only in specif. case where inclination is 090° or 270° along Equator when, because of Earth rotation, time is shortened or extended by 6 min. over orbital period. At all other inclinations satellite follows fresh track on each orbit and never makes * in this sense.

Revolutionary Turbine Accelerator   Unconventional scheme for turbofan for flight Mach No of 4.25 (NASA).

Revolution in Military Affairs   Owning the sky and space over the battlespace; giving unchallengeable command and control.

revolver cannon   One whose ammunition is fed to chamber via rotary cylinder driven by main action and in whose several chambers successive rounds pass through complete firing sequence, in one position being fired down a single fixed barrel. See *rotary cannon*.

REW   Radio-electronic warfare.

Reward   Reporting working and reliability data.

REWTS   Responsive electronic-warfare training system.

Reynolds number   Most important dimensionless coefficient used as indication of scale of fluid flow, and fundamental to all viscous fluids; $R = \rho Vl/\mu$ where ρ is density, V velocity, l a characteristic length (eg chord of wing) and μ viscosity $= Vl/v$ where v is kinematic viscosity. Expression is ratio of inertia to viscous forces. It shows, eg, that for dimensional similarity model tests in tunnels should be run at pressures greater than atmospheric.

Reynolds stress   *1* Shear stress in laminar boundary layer in viscous fluid (see *skin friction*).

2 Term(s) representing momentum transfer due to turbulence.

RF   *1* Radio frequency or facility.

2 Aircraft designation prefix, reconnaissance/fighter (DoD).

3 Regional forces.

4 Royal Flight; thus, issue of an * Notam.

RFA   *1* Request for alteration.

2 Royal Fleet Auxiliary; ship usually with helicopter pad.

RFACA   Royal Federation of Aero Clubs of Australia.

RFAS   *1* Russian Federation and Associated States.

2 Reaction Force Air Staff (NATO).

RF-ATE   RF (1) auto test equipment.

RFC   *1* Reinforced fibre composite.

2 RF choke or communication[s].

3 Royal Flying Corps (UK, 1912–18).

4 Request for change, or comments.

5 Retirement for cause (USAF).

6 Radio facilities chart.

7 Reconstruction Finance Corporation (US 1932–56).

RFCM   RF countermeasures.

RFCP   RF Costs Panel (ICAO).

RFCS   RF com. set.

RFD   Remote frequency display.

RFDIU   RF/digital interface unit.

RFDS   Royal Flying Doctor Service of Australia.

RFDU   RF distribution unit.

RFEG   RF environment generator, to test whole aircraft.

RFF   *1* Research Flight Facility (NOAA).

2 Rescue and firefighting.

RFFE   RF front end.

RFFP   Rescue and Fire Fighting Panel (ICAO).

RFFS   Rescue and firefighting service (CAA).

RFGS   RF generation subsystem (ECM).

RFI   *1* Request for information.

2 RF interference, or radar-frequency interferometer.

3 Resin-film insulation, or infusion.

RFID, RF/ID   RF identification (of passenger or baggage).

RFIS   Receiver fire-control computer-interface software.

RJF, RF/J   RF jammer.

RFL   *1* Restricted flammable liquid (cargo).

2 Rocket flare launcher.

RFLG   Refuelling.

RFM   RF module.

RF-Mems   RF microelectronic/mechanical system[s].

RFNA   Red fuming nitric acid.

RFP   *1* Request for proposal[s].

2 Remote front panel.

RFPB   Reserve Forces Policy Board (DoD).

RFPI   Rapid force projection initiative.

RFPU   Recorder/film-processor unit.

RFQ   Request for quote/quotation[s].

RFR   *1* Request for revision.

2 Restriction of Flying Regulation[s].

RFS   *1* Reserve Flying School (UK).

2 RF surveillance; /ECM adds and ECM.

3 Restricted flammable solids (cargo).

RFSC   Rubberized friction and seal coat (runway).

RFSP   Replacement flight-strip printer.

RFSS   *1* RF surveillance system.

2 Remote flight-service station.

RF surveillance   Maintaining continuous monitor and record of all hostile RFs.

RFT   *1* Ready for training (US).

2 Request for tender.

3 Right first time.

RFTDL Rangefinder target-designator laser.

RFTP Request for technical proposal[s].

RFTR Ring-fin tail rotor.

RFTS *1* Robot flexible transfer system.

2 Reserve Flying Training School.

RFU Radar, or radio, frequency unit.

RFV Request for visit [by accredited foreign staff, eg attaché].

RFY Reduced fission yield.

RG *1* Retractable gear.

2 Range (lights, ICAO).

3 Reconnaissance group.

4 Recombinant gas (accumulator).

5 Rotorcraft/gyroplane.

6 Regular General Aviation use.

RGB *1* Reduction-gearbox module.

2 Red/green/blue (systems identification and colour TV tube).

3 Rail garrison basing.

RGC Radar graphics computer.

RGCS Review of the general concepts of separation (ATC); P adds panel.

RGD Ragged.

RGF Range-gate filter.

RGL Runway guard light[s].

RGM US designation prefix: ship-launched surface-attack missile.

rgn Region (ICAO).

RGO Royal Greenwich Observatory.

RGPI/RGPO Random combination of RGPO and range-gate pull-in; ECM technique calling for predictive gate but effective against manual operator. Essentially *Ranrap*.

RGPO *Range gate pull-off*.

RGPS Relative GPS.

RGS Recovery guidance system.

RGSI Radial groundspeed indicator (DME).

RGSN Homing radar system or beacon (R).

RGT *1* Remote ground terminal.

2 Reliability growth test.

rgt Right.

RGV Rotating guide vane.

RGWS Radar-guided weapon system.

RH, rh, r.h. *1* Right-hand.

2 Relative humidity.

3 Reheat.

4 Rotorcraft, helicopter.

5 Radio handler.

6 Rolled homogenous, see next.

RHA *1* Rolled homogenous armour.

2 Recording-head assembly.

RHAG Rotary hydraulic arresting gear.

RHAW Radar homing and warning; S adds system.

RHC Right-hand circuit.

Rhea, RHEA Role of the human in the evolution of ATM (7) systems (Euret).

rheostat Infinitely variable resistance for control of current.

RHI *1* Range/height indicator (radar).

2 Relative-height indicator.

Rhino, RHINO Range/height indicator mode not operating.

RHL Rudder hinge line.

RHLP Rotatable horizontal log periodic.

RHO *1* Density of air or other medium (EDP), from Greek.

2 Response on handoff.

rho Generalized term for radio-derived distance, usually a radial distance from fixed station.

rhodium Rh, hard silvery metal, one use is electrode for spark plugs, MPt 1,966°C, density 12.4.

rhombic aerial Short-wave directional aerial comprising two dipoles forming horizontal rhombus emitting travelling wave, thus having reflector at one end and non-inductive resistance at other.

rhombus wing One whose section is symmetrical double wedge, ie for supersonic flight only.

rho-rho Radio navaid giving distances from two fixed stations, thus a radio fix.

rho-rho-rho Radio navaid, eg Omega, giving simultaneous distances from three fixed stations.

rho-theta Radio navaid providing fix by one distance from fixed station and also bearing from that station; not normally written $\rho\theta$.

RHP Radar head processor.

RHS Right-hand seat, or side.

RHSM Reduced horizontal-separation minima.

Rhubarb Small-scale attack by fighters or fighter-bombers on enemy surface targets (esp. airfield[s]), using cloud etc to achieve surprise (RAF, WW2).

rhumbatron Common type of resonant cavity.

rhumb line Line drawn on Earth cutting all meridians at same angle.

rhumb-line course One flown at constant heading.

RHWR Radar homing and warning receiver.

RHWS Radar homing and warning system; basic part of ECM kit of penetrating aircraft; also written RHAWS.

rhythmic light See *code light*, *flashing beacon*, *occulting light*, [all now rare].

RI *1* Radar interrogator.

2 Remote indicator.

3 Radio-inertial (often R/I).

4 Root insert [747 wing].

5 Risk index.

6 Radiation intensity.

RIA *1* Range instrumentation aircraft.

2 Rapid inertial alignment.

3 Regulatory impact assessment (UK Cabinet Office), examines effect of new legislation.

4 Radio interface adaptor (TRV).

RIAP Revised instrument approach procedure (usually plural).

RIB *1* Rigid inflatable boat (RAF).

2 Right inboard.

3 Routeing information base.

rib *1* Primary structural member running across wing or other aerofoil essentially in chordwise direction; in highly swept wing axis may occasionally be aligned more with aircraft longitudinal axis but essential feature is that * joins leading and trailing edges and maintains correct section profile.

2 Light peripheral member not part of primary structure whose purpose is to maintain profile of aerofoil and support fabric or thin wood covering (see *compression* *, *nose* *).

ribbon heater Electrothermal tape coiled around pipe to prevent freezing, sections joined by approved connectors.

ribbon microphone   Comprises thin corrugated strip of aluminium alloy suspended between poles of permanent magnet; output is signals generated by strip vibrating perpendicular to field.

ribbon parachute   One whose canopy is formed from rings (rarely, spiral) of ribbon, giving high porosity but reduced opening shock and good stability.

ribbon spray   Water flung sideways by planing bottom at high speed; caused by first contact of hull or float with water and leaves at high speed at shallow angle; also called velocity spray.

riblet   *1* Portion of rib, eg extending only from front spar to LE.

2 Carefully profiled microgroove, no larger than fine scratch, which, repeated millions of times to cover entire non-laminar part of aircraft skin, can reduce drag up to c 3 per cent.

RIBS   Readiness in base services.

RIC   *1* Reconnaissance Intelligence Centre (RAF Marham).

2 Reconnaissance interpretation center.

rich   Having excess of fuel (well above stoichiometric) for given flow of air or other oxidant. Hence * mixture.

Richardson effect   See *thermionic emission*.

Richardson number   Ri, non-dimensional quality in study of vertical shear in atmosphere.

rich cut   Sudden loss of piston engine power caused by over-rich mixture, notably caused by flooding of float-chamber carburettor under negative g.

rich extinction   Failure of combustion caused by excessively rich mixture.

rich mixture   Piston engine fuel/air mixtures significantly above stoichiometric, among other things reducing combustion temperature and enabling higher boost pressure to be used without detonation. Thus 100-octane Avgas has an RMO rating of 130.

RICS   Rubber-impregnated chopped strands.

RIDE, Ride   Radio communications intercept and D/F equipment.

ride control   Automatically commanded aerodynamic control system which reduces, and attempts to eliminate, vertical accelerations caused by flight through gusts, esp. by penetrating aircraft at high (possibly transonic) speed at lo level. Typically includes sensitive g-sensors, computer and foreplanes (possibly augmented by fore-rudder or section of main rudder) to minimize vertical acceleration of crew compartment. In B-1 called LARC, later SMCS.

ridge   Narrow extended portion of anticyclone or other high.

ridge girder   Structural member forming part of stiff-joined main transverse frame of airship, usually qualified as inner or outer and separated by main radial struts. Each ** links two longitudinals.

ridge lift   Provided by air on upwind side of ridge.

ridge lines   Bright lines overlaid on HUD along summits, ridges and edges.

riding lights   Those displayed by marine aircraft moored or at anchor.

riding the controls   Not definable, but tendency of pilot to keep making small unnecessary control movements.

RIDR   Runway-incursion detection radar.

RIDS, Rids   *1* Radio information distribution system (digital airborne CNI systems).

2 Ramp-information display system.

RIF   *1* Reduction in force (military); hence, personnel can be riffed (USAF).

2 Reclearance in flight.

RIG, Rig   Rate integrating gyro.

rig   *1* To adjust wing angular setting, wash-in wash-out, dihedral, control-surface neutral positions and other aerodynamic shape determinants to obtain desired flight characteristics; normally applied only to light GA aircraft, in which it is possible to * by adjusting tensions of bracing wires and even alter shape of fuselage.

2 To prepare a load for airdrop (NATO).

3 Purpose-designed test installation for development of jet-lift V/STOL aircraft (in which case * may fly) or complete aircraft system, eg fuel, hydraulics, environmental, landing gear or propulsion. Usually full-scale and non-flying and often incorporating flight-quality hardware; eg fuel-system * can test entire aircraft fuel system in extreme attitudes and under abnormal environmental conditions.

rigger   *1* Historically, ground engineer responsible for rigging (1, 2).

2 Today, engineer responsible for fine adjustment of flight controls, flaps, airbrakes and certain operative systems.

rigging   *1* See *rig (1)*; esp. adjustment of flight control system, even in modern powered system, so that all surfaces have exactly correct rest angles and system responses.

2 Complete system of wires, cables and cords by which aerostat (esp. kite balloon or other moored type) is secured to main cable(s) or handling guys, and by which crew operate valves etc.

3 Equipment for dusting (dry *) or spraying (wet *) on ag-aircraft.

rigging angle of incidence   See *angle of incidence*.

rigging band   Strong tape band around kite balloon or other moored aerostat envelope to which all rigging (2) and payload are attached.

rigging lines   Those connecting parachute canopy to load.

rigging patch   Patch connecting rigging (2) to aerostat, in place of rigging band.

rigging position   Aircraft attitude in which lateral axis and an arbitrary longitudinal axis (possibly actual longitudinal axis) are both horizontal.

rigging tab   Ground-adjustable tab.

right ascension   Angular distance measured E from vernal equinox; arc of celestial equator, or angle at celestial pole, of given celestial body measured in hours (h) or degrees (°).

right-hand circuit   Circuit (1) with turns to right, clockwise seen from above (unusual).

right-hand rotation   Clockwise; in case of engine/propeller, seen from behind. No ruling exists for pusher engines, one school claiming that ** engine is unchanged when installed as pusher and another claiming that ** means seen from direction in which propeller is beyond engine.

right wind   One blowing on aircraft from right, causing drift to left (suggest arch.).

rigid airship   One having rigid framework or envelope to maintain desired shape at all times.

rigid frame One having fixed joints, and thus statically indeterminate.

rigidity Property of gyro of maintaining axis pointing in fixed direction; also called gyroscopic inertia.

rigid rotor *1* Rotor of helicopter (in theory also auto-gyro) whose blades have no lead/lag [drag] or flapping hinges but can rotate to change pitch.

2 Bearingless helicopter lifting rotor, in which all control input is reacted by root attachment flexible in bending and torsion.

RIIA Royal Institute of International Affairs (UK).

RIIS Route integration instrumentation system (SAC 1).

rill Deep, narrow depression across lunar surface.

Rilsa Resident integrated logistic support activity.

RIM US guided-missile designation prefix, ship-launched for aerial interception.

Rimcas Runway-incursion monitoring and conflict-alert subsystem.

rime Icing type; rough, milky and opaque, formed by instantaneous freezing of super-cooled water droplets. Definition of BS, under heading 'frost': 'ice of feathery nature on windward side of exposed objects when frost and fog occur together'.

Rimpatt Read impatt diode.

RIMS Replacement inertial measurement system [or set].

RIN The Royal Institute of Navigation (UK, 1947–).

ring *1* Common structural part of most gas turbines.

2 Frame of fuselage of circular or oval section.

ring and bead sight Rudimentary gunsight in which aimer aligns target with bead foresight and correct part of ring backsight for required aim-off.

ring around Self-interrogation of beacon due to insufficient isolation between transmitter and receiver, or to triggering of airborne transponders at close range by unwanted sidelobes; in either case result is disastrous ring around display origin. Suppressed by ISLS.

ring counter Three or more bistable devices and gates so interconnected that only one is ON at one time; input pulses advance ON around ring.

ring cowling Simple ring, usually with concave inner profile (in longitudinal plane) and convex outer, surrounding radial engine (see *Townend ring*, from which stemmed long-chord NACA cowl).

Ringelman Basic method of assessing jet smoke by subjective opinion, in use since 1896 for matching smoke plumes against grey (gray) tones; hence * number, numbered scale from white to black. Said to be free from subjective variation but affected by angle, eg if four exhaust plumes are seen from the side.

ring-fin tail rotor Helicopter tail rotor mounted in profiled duct in centre of tail fin (vertical stabilizer).

ring-frame Main transverse structural frame(s) of rigid airship.

ring gauge Hand gauge for checking outside diameter of circular work, in some cases threaded; can be go/no-go type [US = gage].

ring laser gyro Measures rotation, and rate of rotation, by sending laser light in both directions round closed circuit [in parked aircraft, outputs rotation of Earth].

ring rolling Forming rings, eg for engine, by rolling on specially set-up machine which produces required radius.

rings Black/white stripes on cuffs or shoulders denoting rank (RAF).

ring-sail parachute One having construction based not on gores but on rings, usually arranged in form of upper disc and large skirt ring with separating gap. Unlike ribbon parachute, there is usually only one gap (sometimes two).

ring seal Shaft seal with small freedom to move in close-fitting static housing.

ring-slot parachute Family of parachutes with basic ring construction with concentric slots, merging into ribbon form.

ring spring Spring assembled from stack of steel rings with mating chamfered edges whose friction on compression reduces rebound.

Ringstone Ceramic or jewel pivot for instrument arbor (shaft).

ringworm Cracked dope on fabric, especially in concentric circles, caused by local pressure or impact.

Rint Radiation intelligence.

RIO Radar intercept officer (USN aircrew).

riometer Relative ionospheric opacity meter.

RIP *1* Resin impregnation, or resin-impregnated plastics.

2 Remote instrument package.

ripcord *1* Cord which deploys parachute, pulled by wearer, static line or automatic (eg barostat control) system.

2 Cord, usually manually pulled, for tearing open aerostat rip panel.

RIPP Radar-information processing post.

rip panel, ripping panel Panel or patch on aerostat envelope, usually near top, which for emergency deflation can be ripped open: parachute ** right round, sealable in flight; velcro ** ¾ way round, not resealable in flight.

ripple *1* To fire large battery of rockets in timed sequence, interval typically 0.01 s.

2 Residual small alternating component superimposed on DC output.

RIPS Rotor ice protection system.

ripster Carrier offering competitive fare (colloq.).

rip-stop Structural design technique prohibiting unchecked growth of crack, eg by making part in parallel sections. Extends throughout airframe and also systems, eg to prevent single crack from affecting two hydraulic systems.

RIR Range instrumentation radar.

RIRP Retractable inflight refuelling probe.

RIS *1* Range information system (ECM).

2 Range instrumentation system.

3 Reconnaissance interface system.

4 Radar information service.

5 Radar integration system.

RISC Reduced instruction-set computer (or computing), 10,000-plus active devices on single GaAs chip.

riser *1* Quasi-vertical channel in casting mould either for admitting poured metal or for escape of air.

2 VTOL aircraft, esp. aerodyne (colloq.).

3 Main duct conveying ECS fresh air to top of cabin or flight-deck.

rise time *1* Time taken for a pulse, waveform or other repeated phenomenon to increase from a minimum to a maximum (usually measured from 10 to 90 per cent of peak value).

2 Time taken for flare to reach 90 per cent peak radiant power, usually measured from start of emission.

3 Time for large abrupt demand in pitch [e.g., stick back] to reach peak rotation rate; effective * subtracts effective time delay [intersection of maximum-slope tangent with origin].

rising-sun magnetron   One in which resonators for two frequencies are arranged alternately for mode separation.

RISLS   Receiver interrogation sidelobe suppression.

RISS   Real-time IR/EO scene simulator [or simulation].

Rista   Reconnaissance intelligence [rarely, IR] surveillance and target acquisition.

RIT   *1* Rotor inlet temperature.

2 Remote interactive terminal, plural RITs.

3 Radio interface terminal.

4 Remote image transceiver.

RITA, Rita   *1* Rapid imagery transmission to aircraft.

2 RF thruster assembly.

3 Réseau integre de transmission automatique (F).

RITE   Right, direction of turn (ICAO).

RIU   Radio interface unit.

RIV   Rapid-intervention vehicle.

rivet   Essential feature is deformation of shank or surrounding collar to make permanent joint, but even this no longer true of some complex types, a few of which can be reused. Definition needed.

rivet gun   Simple hand tool, usually pneumatic, with cup-ended striker which closes rivet by repeated blows.

rivet hammer   Plain hand hammer with small flat steel face.

riveting machine   Numerous forms of machine tool, most of which are fixed and through which work is fed; most close rivet between powered tool and anvil and some can drill, dimple/countersink and insert rivet, close it and then mill head.

rivet mandrel   Rod passing through tubular rivet closed on withdrawal.

rivet rash   Paint refuses to adhere to rivet heads because of coating of lubricant required for interference fit.

rivet set   Hand tool having female shape which forms correct head as rivet is closed.

rivet snap   See *rivet set*.

rivet squeeze(r)   Hand tool which heads solid rivets by single quiet deformation, usually by mechanical leverage from operator.

Rivnut   Patented (B.F. Goodrich) tubular blind rivet with threaded shank which, after closure, acts as nut (eg for panel fasteners).

RIW   *1* Repairable in works.

2 Reliability interim (or improvement) warranty.

3 Recovery initiation window.

RJ   *1* Ramjet.

2 Regional jet (aircraft class).

RJ-5   Conventional-type jet fuel for US expendable engines; high energy per unit volume.

RJT   *1* Technical rejection message (ICAO).

2 Ground Self-Defence Forces (J).

RK   *1* Range known.

2 Russian meanings include variable-area wing and reconnaissance/artillery correction.

RKA   Russian space agency Rosaviakosmos.

RKIIGA   Riga [now Latvia] Red-banner institute of civil-aviation engineers (USSR).

RL   *1* Rhumb line.

2 Runway edge lights.

3 Report when leaving.

4 Rocket launcher.

5 Radioluminescence.

R/L   Redline.

RLA   *1* Repair level analysis.

2 Relay to.

3 Railway (not railroad) Labor Act; formerly governed negotiations of air-carrier unions in US.

RLB   Reversed lobsterback (Rolls-Royce).

RLC   Request level (FL) change; E adds en-route.

RLD   *1* Rijksluchtvaartdienst (Netherlands certification authority).

2 Reference landing distance.

RLE   Response on link establishment.

RLG   *1* Ring laser gyro.

2 Relief landing ground.

RLGM   Remote loop group multiplexer.

RLLS   Runway lead-in lighting system.

RLM   Reichsluftfahrtministerium (German air ministry to 1945).

RLNA   Requested [flight] level not available.

RLPCR   Air-navigation services of Czech Republic.

RLS   *1* Reservoir level sensor.

2 Raster-line structure.

3 Radius of landing site (lunar).

4 Reliable link source.

5 Remote light sensor.

RLT   Rolling liquid transporter, eg Dracone.

RLV   Recoverable, or reusable, launch vehicle.

RLW   *1* Regie der Luchtwegen (CAA, Belgium, see *RVA*).

2 Raising/lowering winch.

RLY   Relay.

RM   *1* Reference materials in calibration of R&D measures.

2 Radio maintenance.

3 Reflected memory.

4 Reeling machine.

5 Risk management.

6 Research memorandum.

7 Remarks.

8 Raksha Mantralaya (MoD, India).

RMA   *1* Reliability, maintainability and availability.

2 Royal Mail Aircraft (followed by individual name of UK airliner, now obs).

3 Rear maintenance area.

4 Revolution in military affairs (US).

RM&A   Reliability, maintainability and availability.

RM&T   Reliability, maintainability and testability.

R$_{max}$ Maximum range, especially of weapon system or radar.

RMCB   Remote-control circuit-breaker.

RMCC   Remote monitoring and control console.

RMCDE   Radio-message conversion and distribution equipment.

RMCS   *1* Remote monitoring and control system (LADGPS).

2 Royal Military College of Science (UK).

RMD   Radar monitoring display.

RMDI   Radio (or remote) magnetic direction indicator.

RME   *1* Rocket Motor Executive (UK, PERME).

2 Resonant multiphoton excitation.

RMEF   Reconnaissance mobile exploitation facility.

RMetS   Royal Meteorological Society [1850, Royal 1866–].

RMF   Reconfigurable modular family.

RMG   Resource Management Group (USAF).

RMHK   Re-engine modification hardware kit.

RMI   *1* Remote, or radio, magnetic indicator.

2 Radio magnetic interference.

RMK   Remark[s].

RML   *1* Radar microwave link.

2 Recirculating memory loop.

RM/L   Reeling machine/launcher.

RMM   *1* Read-mostly memory.

2 Remote maintenance monitor, or remote monitoring and maintenance.

RMMC   *1* Remote maintenance and monitoring configuration.

2 Remote maintenance monitoring and control.

RMMS   *1* Remote maintenance-monitoring system, or status.

2 Radio-magnetic management system.

RMN   Remain.

RMNDR   Remainder.

RMO   Rich-mixture octane [rating].

RMOS   Refractory MOS.

RMP   *1* Root mean power.

2 Range mean pairs.

3 Reprogrammable microprocessor.

4 Radio management panel.

5 Remote maintenance panel.

6 Risk miniaturization program.

RMPA   Replacement maritime-patrol aircraft.

RMR   Remote map-reader.

RMS   *1* Root mean square.

2 Route, or reconnaissance, or rocket, management system.

3 Range measurement system.

4 Remote manipulator system (Shuttle).

5 Roof-mounted sight.

6 Reusable multipurpose spacecraft.

7 Rotary mirror scanner.

8 Recurring manufacturing support.

9 Remote monitoring system.

10 Radiation meteoroid satellite.

RMSE   Root-mean-square error.

RMT   Reliability, maintainability, testability.

RMU   Radio management unit.

RMV   Remotely manned (or managed, or manipulated) vehicle.

RMVL   Removal.

RMVP   Reliability maintenance and validation program(me).

RMW   Reactive-material warhead, which see.

RN   Recovery net (RPV).

R$_N$, R$_n$ Reynolds number, if for any reason R is unavailable.

RNAC   Reinforced North Atlantic Council.

RNARY   Royal Naval Aircraft Repair Yard.

RNAS   Royal Naval Air Station; shore airbase, also known by ship name (prefaced HMS), usually name of seabird. Previously (1912–18) Royal Naval Air Service.

R-Nav, RNAV, R-nav   Area navigation, can have *RNPC* added.

RNAY   Royal Naval Aircraft Yard.

RNDZ   Rendezvous.

RNEFTS   Royal Navy Elementary Flying Training Squadron.

RNEP   Robust nuclear earth penetrator (DoD).

RNF   Radio navigation facilities.

RNG   Radio range (ICAO, FAA).

RNGA   Range arc.

RNGSA   Royal Naval Gliding and Soaring Association.

RNHF   Royal Navy Historic Flight.

RNI   Russian research institute, IKP adds space-instrument making.

RNII   Reaction-engine scientific research institute (USSR, R).

RNIP   Ring-laser gyro navigation improvement programme.

RNMP   Replacement of the nautical-mile panel (ICAO).

RNP   Required navigation performance, defined as maximum expected en-route error; /ANP adds actual navigation performance, C or -C adds capability, -5 means accuracy ≤5 nm (9.3 km) 95% of time, -RNAV see above.

RNPU   Radar-navigation processing unit.

RNR   Receive[r] not ready.

RNS   *1* Radar-navigation service, or system.

2 Regular use by non-scheduled carriers.

RNTP   Radio-navigation tuning panel.

rnwy   Runway.

RNZAC   Royal New Zealand Aero Club.

RO   *1* Range-only memory.

2 Radio, or radar, operator.

3 Royal Ordnance (UK).

4 Rollout.

5 Report when over runway.

6 Routeing organization.

7 Receive only.

8 Rotorcycle category (USN, 1954–59).

R/O   Radio officer.

R$_o$ *1* Radar range at which signal/noise ratio is 1.

2 Range to a patch or mapped swath.

ROA   *1* Return on assets.

2 Remotely-operated aircraft.

3 Radius of action.

4 Recognised operating agency.

roach   Wake of marine aircraft when deep in water; characterized by dense, almost vertical up-flow immediately behind float or hull.

ROAD   Retired on active duty.

roadable aircraft   One which, usually after some alteration, can be driven on public highway as a car.

roadmap   Graphical plot, usually with time as X-axis, giving pictorial overview of a programme, input efforts, available technologies or other variables.

Roadstead   Code name for attacks on enemy ships at sea (RAF, WW2).

roaks   Cavities in steel castings caused by carbon monoxide.

ROB   *1* Right outboard.

2 Radar order of battle.

rob   See next entry, derived verb.

robbery action   Procedure whereby serviceable part is

taken from aircraft or assembly in order to get the latter back in service (not cannibalization because * is temporary).

Robe, ROBE   Roll-on beyond line-of-sight enhancement, or extension.

Robeps   Radar is operating below performance standard[s].

Robex   Regional Opmet bulletin exchange.

Robinson anemometer   Standard three or four-cup, with worm drive.

robot   In addition to usual meaning, guided missile (Sweden).

robotics   Use of intelligent machines to replace human beings in routine mechanical tasks.

robot rotary wingman   UAV helicopter able to fly fast at low level accompanying manned helicopter, see *VCAR* (Darpa).

robustness   Unquantifiable quality of a flight-control system to continue to function in face of uncertainties or failure in actuator, airframe or sensor input, especially despite dangerous conditions such as uncommanded hardover surface deflection.

ROC   *1* Rate of climb.

2 Required, or rate of, operational capability (pronounced rock).

3 Royal Observer Corps (UK, civilian NW/fallout monitoring organization).

4 Receiver operating characteristic (ASW).

ROCC   Region(al) (or range) Operations Control Center (Norad, SSS).

Roche's limit   Critical radius from primary within which planetary satellite cannot form; equal to 2.44 times radius of primary.

rock   *1* Uncommanded excursion or oscillation in roll. See *SIWR*.

2 Rotary movement of wing[s] relative to fuselage, usually in plane perpendicular to longitudinal axis, caused by worn root anchor pins and various other faults (usually old GA aeroplanes).

3 To move throttle lever[s] laterally.

rockair   Small high-altitude sounding-rocket technique involving launch from aircraft, eg jet fighter, in vertical attitude at high altitude (probably arch.).

Rock apes   RAF Regiment (colloq.).

Rockbestos   Synthetic fire/electricity insulator.

rocker arm   See-saw arm pivoted to piston engine cylinder head transmitting push-rod actuation to valve(s). Hence rocker box, enclosing valve gear.

rocket   *1* Propulsion system containing all ingredients needed to form its jet, or vehicle thus propelled. Most involve combustion, but examples which do not are monopropellants (eg HTP) and pressurized boiling water (eg Pohwaro).

2 Verbal or written reprimand (UK, colloq.).

rocket ammunition   Rocket-powered projectiles of relatively small size fired from aircraft or other platform, mobile or stationary (USAF). Unguided, normally finned and/or spin-stabilized.

rocket artillery   Artillery in which projectiles are propelled by rocket power but given guidance only during launch by amount of thrust and direction of take-off (USAF). This definition is meant to cover tube-type airborne launchers.

rocket-assisted take-off   Take-off in which horizontal

acceleration, and usually vertical component of thrust, are augmented by one or more rockets which may form part of aircraft or be jettisoned when spent. Ambiguously given US term JATO.

rocket/athodyd   See *ram rocket*.

rocket-based combined cycle   Rocket with valve-controlled duct to scramjet.

rocket cluster   Large group of parallel rockets normally all fired together.

rocket engine   Term best reserved for rocket using liquid propellants, esp. with pump feed and control system.

rocket fuel   Ambiguous; can mean fuel as distinct from oxidant, or a solid propellant.

rocket grain   See *grain*.

rocket igniter   Device for starting combustion in any form of rocket; usually an electrically fired pyrotechnic, but many contrasting varieties.

rocket launcher   Anything for launching rockets, specif. aircraft pod, pylon or rail from which rocket ammunition is fired.

rocket loop   Flight manoeuvre in which loop is made with excess initial speed, traded in loop for height.

rocket motor   Term preferred for all solid-propellant rockets, of any size, together with hybrids of modest complexity.

rocket on rotor   Method of temporarily greatly boosting helicopter lifting power by switching on tip-drive jets fed with HTP from tank above hub.

rocket power   See *rocket propulsion*.

rocket propellant   See *propellant*.

rocket propulsion   Propulsion of vehicle by rocket, excluding rocket drive of gyro wheels, turbopumps and other shaft-output devices.

rocketry   Technology of rockets (colloq.).

rocket sled   Sled guided by straight track and propelled by rocket(s) to high (often supersonic) speed for use as test platform.

rocketsonde   Launched by rocket, operates on parachute descent.

rocket thrust   Measured either for sea-level or vacuum conditions, latter being about 26% greater.

rocking   Uncommanded lateral and directional oscillation, chiefly in roll.

rockoon   Sounding rocket fired from helium balloon.

rockover   Nose-down rotation on landing, normally by tailwheel aircraft, in extreme case by 180°.

Rockwell number   Measure of hardness, esp. of metals, derived by impressing tool (for soft metals Type B, $1/16$-in/1.5875 mm steel ball, result being R_B number; for very hard Type C, 120° diamond cone, result R_C number) first with 10 kg load and then with 90 or 140 kg, and measuring change in indent area.

Rococa   Rate of change of cabin altitude.

ROD   *1* Rate of descent.

2 Repair on demand.

rod   Photoreceptor in retina for scotopic (night) vision and detection of movement.

rod aerial   Aerial in form of rigid tube or bar conductor.

Rodar   Helicopter rotor-blade radar (Ferranti).

Rodeo   Code name for sweep over enemy territory by fighters, on preplanned and opportunity targets (RAF, WW2).

Rodnet   Reliable on-board data network.

Rods, RODS   *1* Robust optical data system.

2 Ruggedized optical-disk system.

ROE   *1* Rules of engagement.

2 Return on equity.

ROEA   Read-only, electrically alterable memory.

roentgen   Unit of exposure to ionizing radiation, quantity producing 1 e.s.u. in 1 cm^3 dry air at 0°C; non-SI, = 2.58×10^{-4} C/kg (C = coulombs).

roentgen equivalent man   See *rem*.

ROF   *1* Royal Ordnance Factories (UK, at one time various locations).

2 Rate of fire.

3 Rollover force, imposed by aircraft wheel.

Rofor, ROFOR   Route forecast.

Rogallo   Originator of patented family of flexible-wing aircraft characterized by delta wing plan with three rigid members in form of arrow-head joined by flexible fabric which inflates upwards to arch under flight loads; originally paragliders with poor L/D, later developed to include powered aircraft.

Roger   Voice code: "I have received and understood all of your last transmission".

rogue   Aircraft which displays flight characteristics and/or performance markedly inferior to others of same type or, in particularly dangerous specimens, which is unpredictable.

rogue state   Country considered by the USA/UK to pose a possible NW threat.

Rohacell   Low-density cellular plastics material of thermosetting type used to stabilize sandwich structures, usually with CFRP skins.

Rohrbond   Patented metal adhesive bonding, especially for noise-absorbent sandwich.

ROI   Return on investment.

Roink   See *Ronk* (USN).

ROL   Roll-on landing.

Role, ROLE   Receive only link eleven (Link 11, US/NATO).

role-change package   Removable equipment fit (eg flight-refuelling probe installation) stored available for use when required; does not necessarily change role.

role equipment   Equipment attached to or carried in aircraft for particular duties or mission(s) which can be removed subsequently; eg helicopter can offload ASW gear or anti-tank weapon/sight system and take on winch, buoyancy bags and furnishing for rescuees.

roleur   Non-flying taxi trainer (F, suggest obs].

roll   *1* Rotation about longitudinal (OX) axis; one of the five classical modes of aeroplane motion.

2 To move across surface on takeoff.

3 The length of * (2).

rollback   *1* Pushback.

2 Engine rundown or spooldown.

roll bar, roll cage   See *roll pylon*.

roll cloud   See *rotor cloud*.

roll coupling   See *inertia coupling*.

roll damper   Flight-control damper operating on ailerons or differential spoilers, either to preclude Dutch roll or because aircraft is difficult to hold wings-level in turbulence.

roll electrodes   Freely rotating mating discs used for resistance-welding continuous seams or evenly spaced spots.

roller   *1* Landing by tailwheel aircraft with fuselage substantially level, eg because three-pointer is difficult or risks severe bounce, or because of gusty conditions.

2 Touch-and-go (RAF).

roller/ball transfer   Movement of container[s] or pallet[s] over ballmats.

roller-blind instrument   Panel instrument whose display makes use, usually as backdrop, of roller blind (eg giving black/white sky/Earth indication of horizon). Not synonymous with tape instrument, which is analog linear scale.

roller cloud   See *rotor cloud*.

roller drive   Patented (TRW) reduction or stepup gear in which gear teeth and most bearings are eliminated, smooth preloaded rollers (Sun and two-step rings of planets) transmitting torque without lubrication.

roller-map display   Navigation display based on moving map (usually air topographic printed on film at reduced scale) projected by optics on circular screen on which heading appears as vertical central line (usually with present position as small ring near 6 o'clock position). In advanced forms, eg Ferranti Comed, combined with electronic displays and information readouts.

rolleron   Flight control serving as primary for pitch or pitch/yaw and roll, esp. in radial-winged missiles, which have four * to handle all manoeuvres.

rollgang   Any of many makes of roller system for facilitating movement of cargo, especially in removable floor panels of standard size.

rolling   Engaged in roll (1, 2).

rolling balance   Tunnel balance measuring forces and moments while model rolls about axis parallel to airflow.

rolling ball   Compact two-axis human input to dynamic system, eg ATC or SSR radar, air-defence plot or computer display, in which partially recessed sphere drives two potentiometers giving rotary output proportional to rotation of ball about each of two perpendicular horizontal axes; output can be analog voltage or digital pulses. Smallest types fit on end of cockpit levers to command panel displays or HUD. In US ATC called track ball.

rolling engagement   Air-combat training in which one aircraft (eg Aggressor role) takes over when predecessor is at bingo.

rolling instability   Lateral instability; depending whether neutral or positive instability, an upset about OX will fail to be restored or become divergent.

rolling moment   Component about longitudinal (OX) axis of couple due to relative airflow; measured positive if rotates right-wing down, negative if vice versa.

rolling plane   Vertical plane normal to OX axis, ie containing line through both wingtips.

rolling radius   Effective radius of landing wheels allowing for deflection (function of aircraft weight, speed and tyre inflation pressure).

rolling tailplane   Taileron.

rolling take-off   Any take-off by helicopter, tiltrotor or other VTO aircraft which, for any reason, accelerates forwards before leaving ground.

rolling vertical take-off   Take-off with vectored nozzles aft for short distance (USAF 50 ft, in practice more helps), at which point nozzles are vectored to 60° or thereabouts for near-vertical ascent. Often shortened to rolling take-off.

roll-in point   Point in space where aircraft enters diving attack.

roll jet   Reaction-control jet for roll.

roll off   Uncommanded roll when at high AOA (eg due to climb at low speed) due to contamination by ice or other matter).

roll off the top   See *Immelmann*.

roll-on landing   Helicopter landing with substantial forward speed, eg following loss of tail rotor.

roll on top   Half loop followed by 360° roll followed by remaining half of loop.

rollout   *1* Ground roll after landing, esp. when continued to later turnoff to ease brake wear.

2 Emergence of new aircraft, especially first of new type, from factory, often carefully staged ceremony.

3 Termination of flight manoeuvre designed to place (normally combat-type) aircraft in optimal position for completion of intended activity (USAF).

rollout RVR   Readout values from RVR equipment located nearest runway end (FAA).

rollover   Apart from generalized meaning in business investment, replacement of airline line equipment by larger aircraft, in response to or anticipation of increased traffic.

roll-over stand   Maintenance stand for engine [rarely, other items] providing hand-crank gearbox to rotate unit for all-round access.

Rollpin   Press-fit pin made of roll of spring steel, forming self-retaining bearing pivot, axle or hinge-pin.

roll post   Strange US term for the reaction-control jets needed for lateral control on jet STOVL aircraft, esp. the F-35B. Each comprises a downward-pointing nozzle with an electrically powered valve controlling high-power bleed air from the main engine (maximum 6% of main-engine airflow per nozzle). The previously existing terms were puffer jet and RCJ.

roll pylon   Strong structure on ag-aircraft to give protection to pilot's head in event of aircraft becoming inverted on ground.

roll ratcheting   Uncommanded oscillation of lateral controls, typically at 2–3 Hz, caused by human neuro-muscular input [but emphatically not PIO], force-sensing stick gain and command prefiltering.

roll rate   Measures of lateral-control power, notably instantaneous * and sustained *.

roll-rate gyro   A single gyroscope used to measure rate of roll of its cage, or aircraft.

roll reversal   Causes, apart from basic aeroelastic reversal of control, include jack stall, spoiler blowback, adverse yaw and loss of lift of accelerated wing due to increased compressibility.

Rollspring   Patented (Lockheed Georgia) mechanical drive using belt of thin spring steel.

roll-stabilized   Prevented, eg by autopilot or vertical gyro, from rolling; common characteristic of cruciform-winged missiles as an alternative to deliberate roll at known rate.

roll whiskers   Short diagonal lines showing bank-angle limits (HUD symbology).

rolometer   Extra-sensitive bolometer (ASCC).

ROM   *1* Read-only memory.

2 Rough order of magnitude (prefix).

Romag   Remote map generator.

RON   *1* Receive, receiving, only (ICAO).

2 Remain overnight (FAA).

3 Research octane number (fuel).

Ronchi   Modification of Schlieren technique to give quantitative results, with second slit removed and grid of parallel wires added to give light/dark shadow planes.

Roncz   Family of [originally sailplane] aerofoils, designations prefixed RQW for wing, RQHT/RQVT for horizontal or vertical tail.

Ronk   Resident officer in charge (USN, colloq.).

Ronly   Receiver only.

röntgen   See *roentgen*.

roofline   Generalized term for that level in passenger cabin at which lights, fresh-air (punkah) louvres, call buttons and drop-down oxygen are located. Hence * locker = overhead stowage bin.

roof rat   Carrier deck crew (USN colloq.).

rooftop   Aerofoil profile which at typical cruise AOA generates lift more or less uniformly across large middle part of chord (chordwise plot has semi-elliptic form). Not precisely defined and generally synonymous with super-critical.

root   Junction of aerofoil with fuselage or similar supporting structure, including unspecified small inner-most portion of aerofoil itself.

root bending moment   Bending moment at the wing root, strongly influenced by lift generated near the tip, especially by an increase in span.

root mean square   Square root of arithmetic mean of squares of all possible values of given function.

root-mean-square error   Square root of arithmetic mean of squares of deviations from arithmetic mean of whole; standard deviation σ.

Roots   One of many types of positive-displacement fluid compressors (eg superchargers, cabin blowers) in which two mating rotors are driven by external gears to revolve together inside casing; rotors in this case are identical and resemble fat figure-eights.

root section   Profile of aerofoil at root; in both wings and propellers often significantly different from remainder of surface.

ROP   *1* Rotor overspeed protection.

2 Runway observation post, for human guess at RVR.

ROPA   Reserve Officers Personnel Act (ROPMA adds 'Management') (US).

Ropar   Regional operators' program for airframe reliability; co-operative effort by US carriers.

rope   Element of chaff consisting of long roll of metallic foil or wire designed for broad low-frequency response.

rope chaff   Chaff which contains one or more rope elements (DoD).

ROPW   Read-only, permanently woven memory.

ropy   Or *ropey*, general critical/derogatory adjective and adverb (RAF WW2).

ROR   *1* Range-only radar.

2 Rocket on rotor.

3 Red on red.

Ro/Ro   Roll-on roll-off, ship configured for transport of vehicles driven on or off under own power. See *T-AKX*.

Rorsat   Radar ocean-reconnaissance satellite.

ROS   *1* Rotor outside stator (electrical machines).

2 Read-only storage.

3 Relief on station (UAVs).

4 Rosman.

Rosa   Runway, or road and runway, surface analyser.

rose   Polar plot, normally subdivided from 000° to 360° clockwise but often including only quadrantal, cardinal and possibly other points and occasionally only radial lines with variable angular intervals. Term applies to any such diagram, eg in panel instrument (ADF, RMI, HSI etc), compass or wind *.

rosette   *1* Standard series of patterns, generally resembling petals of flower, for applying strain gauges to signal stress in all directions, usually in sheet.

2 Similar-shaped hand weld made in hole through sheet or tube to secure second member inside.

Rosman   See *Stadan.*

Rossby chart   Basic diagram assisting in establishing character of air masses and in met. forecasting generally; energy diagram in which specific humidity is plotted against equivalent potential temperature, result being named characteristic curve.

Rossby number   Ratio of inertial forces to Coriolis forces in a rotating fluid.

Rosto, ROSTO   Paramilitary sports and technical society (R).

ROT   Reserve officer training.

Rotachute   Patented (Hafner/ML Aviation) free-running personal rotor used WW2 instead of parachute for accurate paratroop descent.

rotaglider   Generalized term for gliders with free-running rotary wing in place of fixed wing.

rotaplane   Flying machine whose support in flight is derived from reaction of air on one or more rotors which normally rotate freely on substantially vertical axes (BSI); today called autogyro, making this term redundant.

rotary atomizer   Rapidly spinning drum or disk from which fuel enters gas-turbine combustion chamber, notably on Turbomeca engines.

rotary bombdoor   Door to aircraft internal weapon bay made as single stress-bearing beam pivoted at extremities and itself loaded with ordnance or, alternatively, fuel tank or reconnaissance sensors; to drop ordnance, door is rotated 180°, thereafter normally reversing direction to close.

rotary cannon   Gun with several (eg six) barrels which in operation rotate at high speed around the gun centreline, at any moment each barrel being in different point in operating cycle. In many respects similar to Gatling (1862).

rotary converter   A.c. input drives generator for d.c. output.

rotary derivatives   Stability derivatives expressing variation in forces and moments resulting from changes in aircraft rate of rotation about all axes.

rotary engine   *1* Historically, piston engine (usually Otto but occasionally two-stroke) with radial cylinders which drive themselves round a fixed crankshaft; propeller is thus attached to crankcase.

2 Today, an RC (1) engine.

rotary loads   The aerodynamic loads corresponding to rotation about the three axes, roll/pitch/yaw.

rotary scan   See *scan.*

rotary shears   Hand or power tool used for slitting sheet along straight lines or curves of variable radius.

rotary-wing aircraft   See *rotorcraft.*

rotate   *1* Voice command, usually P2 to P1, at V_R.

2 To cycle personnel, especially military, through repetitive tour of duty.

3 See *rotation.*

rotated out   Removed from combat duty after tour.

rotating beacon   *1* Bright aircraft-mounted light steady or flashing while rotating continuously in azimuth, usually synonymous with strobe; today tends to be illuminated whenever engine run on ground.

2 Ground transmitter having directional radiation pattern rotated continuously at predetermined rate (BSI); suggested arch. term overtaken by more precise names, eg VOR.

rotating-combustion engine   IC engine of positive-displacement type in which main moving parts are not reciprocating but rotary; usually based on Otto cycle. According to best-known worker in this field, Felix Wankel, 864 possible configurations divided into single-rotation machines (SIM), planetary-rotation machines (PLM), SIM-type rotating-piston machines (SROM) and PLM-type rotating-piston machines (PROM). Unfortunately acronyms confuse with equally important EDP (1) meaning of PROM etc.

rotating-cylinder flap   High-lift wing flap in which circulation is assisted by high-speed rotary cylinder mounted along flap leading edge (Coanda and Flettner relevant), eg in NASA YOV-10A.

rotating guide vane   Curved inner leading edge of centrifugal (rarely, other types) of compressor or impeller rotor, main purpose of which is to smooth change of direction of flow from axial to radial.

rotating stall   There is no brief single form describing stall in axial compressors and other rotating-blade machines.

rotating-wing aircraft   See *rotorcraft.*

rotation   *1* Positive, ie nose-up, rotation of aeroplane about lateral (pitch) axis immediately before becoming airborne; in transports commanded at V_R.

2 One round trip to destination, especially by scheduled transport aircraft.

rotational speed   Number of rotations in unit time, measured as rpm (per minute) or rps (per second) or radians per second; 1 rpm = 0.104720 rad/s; 1 rad/s = 9.54927 rpm.

ROTC   Reserve Officers' Training Corps (USAF).

Rotherham pump   Windmill fuel pump (1914–c35).

ROTHR   Relocatable over-the-horizon radar; -B adds backscatter.

rotochute   Free-rotor airbrake fitted to certain air-dropped sonobuoys.

rotodome   Slowly rotating radome of discus shape used to fair in antenna of AWACS-type aircraft, usually housing main and IFF/link antennas across diameter, back-to-back.

rotor   *1* System of rotating aerofoils (ASCC); this includes propeller, so should be qualified by adding 'whose primary purpose is lift'.

2 Main rotating part of machine, eg gas-turbine engine, turbopump or alternator.

3 Local air mass rotating about substantially horizontal axis; when downstream of mountain ridge can be exceedingly dangerous.

rotor cloud   Unusual cloud usually of Ac type and often dangerously turbulent; found in rotor flow, normally in lee of mountain range or ridge. Also called roll cloud.

rotorcraft   Aerodyne deriving lift from rotor(s).

rotorcraft load   See *slung load.*

rotorcraft pilot's associate   Digital terrain map to assist plotting masked routes.

rotor damping   Damping of blades about any pivot axis in helicopter main rotor (see *lag-plane, soft in plane*).

rotor disc   *1* Circular area swept by blades of helicopter or autogyro rotor.

2 Structural disc holding compressor or turbine rotor blades.

rotor flow   Large-scale rotary flow of atmosphere about substantially horizontal axis in lee of mountain or sharp ridge, which when wind very strong is turbulent to point of being dangerous, with vertical gusts of 30 m (100 ft)/s (see *rotor streaming*).

rotor force   Resultant imparted by lifting rotor, analysed into lift and drag, or thrust (perpendicular to disc) and in-plane (H-) force.

rotor governing mode   Control mode in which rotor (1) speed is held constant.

rotor head   Complete assembly of rotating components at centre of lifting rotor of helicopter or autogyro, including hub, blade attachments and pivot bearings, if any, and complete control mechanism, as well as such adjuncts as anti-icing, electrics and pressurization instrument leads rotating with hub.

rotor hinge   In a non-rigid [articulated] rotor, the drag and flapping hinges.

rotor hub   Primary structure connecting blades of helicopter or autogyro rotor. Some authorities define hub to mean same as head, which is unhelpful.

rotor incidence   Angle between plane normal to axis of rotation and relative wind (BSI).

rotor inlet temperature   Assumed equal to SOT (gas turbine).

rotor kite   Towed engineless autogyro.

rotor mast   Pylon carrying rotor in small rotorcraft.

rotorplane   See *rotorcraft.*

rotor slap   Noise, often almost explosive, caused by interaction between each helicopter main rotor blade and the vortex from its predecessor.

rotor streaming   Shedding of turbulent rotors (3) downstream, often near ground and very dangerous to aircraft.

rotor tip drive, tip jet   *Tip jet.*

rotor torque   That imparted to airframe by helicopter rotor, esp. main lifting rotor of hub-driven type, which has to be countered by tail rotor or use of two main rotors.

ROTR   Rate of temperature rise.

Rotte   Fighter aircraft in loose pair (G).

rough-air speed   Recommended speed for flight in turbulence, V_{RA} (hence rough-air Mach, M_{RA}); lies between V_A and V_C and usually near V_B.

rough field   Defined by standard categories of surface profile, including post-attack damage repair: A, single blister up to 1.5 in H (height) over aircraft-travel distance of 80 ft; B, H up to 3 in; C/D, two 3 in blisters within 80 ft; E, two up to 4.5 in; 2E, two up to 9 in.

roughness   Criterion affecting surface finish: irregularities that are closely spaced (see next entry).

roughness factor   Rayleigh criterion $\Delta_h = \lambda/8 \sin \psi$ where λ wavelength and ψ angle of incidence.

roughness width cutoff   Maximum width of surface irregularities to be included in measure of surface height;

in Imperial measure usually 0.03 in, but occasionally 0.003, 0.010 or 0.10.

roulement   Rotation of aircraft through front-line squadrons (RAF).

rouleur   Ground trainer with wing too small for flight (F).

round   *1* Single munition, missile or device to be loaded on or in delivery platform (USAF).

2 Parachute with circular canopy [smokejumper term].

round-down   Rear terminator of aircraft carrier flight deck, usually normal to landing direction (term derives from older ships where deck fell away over stern in large-radius curve).

rounded trailing edge   Aerofoil designed for flow attachment by Coanda-effect blowing to give very high (10+) lift coefficients.

roundel   National marking for military aircraft in form of concentric rings.

round-head rivet   Usual rivet for thin sheet where countersinking is not required; head OD larger than button-head.

round-trip time   For radar, see *ranging time.*

route   *1* Defined path, consisting of one or more courses, which aircraft traverses in horizontal plane over surface of Earth (FAA).

2 Published route linking traffic points of air carrier and used in traffic and rights negotiations.

3 Path followed by channel of AFTN network.

route flight   *1* One from A to B.

2 Military mission, usually transport, over established route.

route package   Geographical division of enemy landmass for purposes of air-strike targeting.

router   Rhyming with doubter, machine tool having high-speed cutter, usually rotating about vertical axis, capable of being positioned anywhere over horizontal work; cutter can be side or end-cutting and in large machines may be NC.

router   Pronounced rooter, seamlessly connects wireless networks on land/sea/air cellular [Wi Fi] satellite systems.

route sector   Route (1) between two traffic points.

route segment   Route (1) between two way-points or [ICAO] consecutive significant points.

route speed   En route speed, usually synonymous with block speed.

route stage   *Stage* (4).

routine   *1* Series of step-by-step EDP (1) instructions forming part of program; hence portions of same are subroutines.

2 Complete sequence of aerobatic manoeuvres planned by competitor or airshow participant.

routing   Itinerary followed by message in AFTN.

routing list   That showing AFTN centre which outgoing circuit to use for each addressee.

ROV   Remotely operated vehicle.

Rover   Armed reconnaissance, usually against shipping (RAF, WW2).

rover   Self-propelled explorer of planetary or lunar surface.

roving   Traditional meaning was slightly twisted hank of textile fibre; modern meaning in aerospace is continuous fibre (eg carbon, graphite, boron, glass) for filament winding, or continuous raw material for ECM chaff to be cut to required response length in dispenser.

roving aerodynamic test system   System for measuring flow quality and other parameters at various points (eg vane sets) in wind tunnel.

ROW   *1* Rest of world.
  2 Right of way.

row   Group of cylinders of radial engine in one plane, all driving one crankpin (strictly, not a row but planar array); hence two-* engine has two planes one behind the other.

row section   Group of consecutive seat rows called to board at the same time.

Roydazide   Hot-isostatic-pressed sinide.

Royal Flight   Single mission notified by Notam of flight carrying at least one member of the Royal Family [v. rarely, other VIP] (UK).

Royal Naval Aircraft Yard   Major maintenance facility for RN aircraft.

Royal Observer Corps   Civilian body established to keep watch on sky [and surface if necessary] and report occurrences (UK, 1925 –).

ROZ   RON (G).

RP   *1* Reporting point, or post.
  2 Rocket projectile.
  3 Rocket propellant (fuel designation prefix).
  4 Route package.
  5 Rapid prototyping.
  6 Report [when you are] passing.
  7 Reticulated polyurethane.
  8 Routeing protocol.

R/P   *1* Rocket projectile (WW2 usage).
  2 Receiver processor.

RP-1   *1* Common kerosene fuel for rockets and ramjets.
  2 One-minute racetrack pattern for holding.

RPA   *1* Remotely piloted aircraft.
  2 Recreational pilot, airplane.
  3 Rotorcraft pilot's associate.

RPB   Retarded practice bomb.

RPC   *1* Recreational Pilot['s] Certificate (FAA).
  2 Remote procedure call[s].

RPCC   Rotating-parts cycle count.

RPCM   Reply processing and channel management.

RPD   *1* Rapid, or rapid inertial alignment.
  2 Random pulse discrimination.

RPDE   Reliability and performance in demonstrator engine[s].

RPDL   Receiver pilot director light.

RPDLY   Rapidly.

RPE   Rocket Propulsion Establishment (Westcott, UK).

RPEH   Rate per engine hour.

RPF   Reticulated-plastics foam.

RPFS   Radio position-fixing system.

RPFT   Rapid pucker factor take-off.

RPG   *1* Rounds [of ammunition] per gun.
  2 Regional Planning Group (ICAO).
  3 Radar product generator.
  4 Receiver/processor group.
  5 Recreational pilot, gyroplane.

RPGT   Radar procedures ground trainer.

RPH   *1* Remotely piloted helicopter.
  2 Recreational pilot, helicopter.

RPI   *1* Runway point of intersection.
  2 Rapid process improvement.

RPK   *1* Revenue passenger-kilometres.

  2 Fragmentation bomblet (as well as a series of Kalashnikov LMGs) (USSR, R).

RPL   *1* Radiophotoluminescence.
  2 Rated power level (electronic).
  3 Runway planning length.
  4 Repetitive flight plan; s adds system.
  5 Regional Plans (ICAO Standing Group).

RPLC   Replace[d].

RPM   *1* Revenue passenger-mile[s].
  2 Rounds per minute (often rds/min or spm).
  3 Reliability prediction manual.
  4 Research and program(me) management.
  5 Remotely piloted munition.
  6 Radar performance monitor.

rpm, r.p.m.   Revolutions per minute; SI dictates rev/min.

RPMB   Remotely piloted mini-blimp.

RPMD   Repeater projected map display.

RPO   *1* Rotorcraft program office.
  2 Resident project officer.

RPOA   Recognised private-operation agency (CCITT).

RPOADS   Remotely piloted observation aircraft designator system (USA).

RPOW   Relative power, one-way (SSR and other radars).

RPPG   Radar Planning and Policy Group (ICAO).

RPRT   Report.

RPRV   Remotely piloted research vehicle.

RPS   *1* Radar position symbol.
  2 Regional pressure setting.
  3 Robot passive sonar.
  4 Recording and playback system [voice/radar/LAN].

rps   *1* Radians per second (contrary to SI and confusing, see next).
  2 Revolutions per second.

RPSTL   Repair parts and special tool list.

RPT, Rpt   *1* Repeat, or I repeat.
  2 Revenue passenger transport.

R/Pt   Reporting point.

RPU   *1* Receiver processor Unit.
  2 Radar processing unit.
  3 Remote processing unit.

RPV   Remotely piloted vehicle, term normally confined to fixed-wing aerodynes; US preference for UAV is making * unfashionable.

RPX   Radar-data processing executive.

R/P/Y   Roll, pitch and yaw.

RQ   Designation prefixes for Roncz aerofoil.

RQ   *1* Preface: request (ICAO).
  2 Request for quotation.
  3 Role designation prefix: surveillance UAV (US).
  4 Designation prefix for Roncz airfoils.

RQD   Required.

RQL   Rich burn, quick quench, lean burn.

RQMNT[S]   Requirement[s].

RQP   Request, flight plan.

RQRD   Required.

RQS   *1* Request, supplementary flight plan.
  2 Rescue Squadron (USAF).

RR   *1* LF or MF radio range (FAA).
  2 Rendezvous radar.
  3 Rain-repellant liquid.
  4 Rain area.
  5 Radiometric resolution.
  6 Report [upon] reaching.

7 Rising rapidly.

8 Receiver ready.

R/R Remove and replace.

RRA *1* Radar recording and analysis.

2 Radar reflective area.

RRAB Cluster-dispensed incendiary (USSR, R).

RRAM Rapid-response aerospace manufacture.

RRC Regional radar center.

RRCC Rotorcraft Requirements Co-ordinating Committee (CAA).

RRCM Rudder-ratio control module.

RRCS *1* Remote radio control system [control of airfield lighting].

2 Remote radar control system.

RRCSR Rapid response to critical system requirements (USA).

RRE *1* Royal Radar Establishment, became RSRE.

2 Risk-reduction effort.

RRF Ready Reserve Force[s] (US).

RRG Roll-rate gyro.

RRI Router [pronounced rooter] reference implementation (ATN).

RRL Runway-remaining lights.

RRM Risk-reduction measures.

RRP *1* Runway reference point.

2 Risk-reduction plan.

RRPS Ready-reinforcement personnel section.

RRR, R³ *1* Rapid runway repair[s].

2 Rapid-reinforcement plan.

3 Radar-data recording and replay system.

RRS *1* Remote receiving station (UAV).

2 Risk and revenue sharing; P adds partner.

RRTC Roll-response time constant; subjective assessment from 0.1–10 s of apparent lag in response to large roll demand suddenly applied.

RRTES Reconnaissance real-time exploitation system.

RRTS *1* Radar real-time simulator.

2 Remote radar tracking station, or system.

RRU Remote readout unit.

RRW *1* Robot rotary [-winged] wingman.

2 Recce [reconnaissance] report workstation, for creating and disseminating reports in various formats.

RRWD Radar remote weather display; S adds system.

RRZ Radar regulation zone.

RS *1* Reconnaissance/strike (USAF) or Reconnaissance Squadron.

2 Reserve Squadron (RFC).

3 Re-entry system.

4 Rear spar.

5 Rapidly solidified; MMC adds metal-matrix composite.

6 Regular use by scheduled carriers.

7 Remain well over to right side of runway.

8 Radio set.

9 Receiver segment.

10 Record Special, ground observation of Sigmet.

11 Ring-slot [parachute]; HV adds high-velocity.

12 Remote sensing.

R_s Separation radius, of ground sheet.

RSA *1* Rate sensor assembly.

2 Réseau du Sport de l'Air (F).

3 Range standardization and automation.

4 Runway safety area.

5 Reference-station antenna (GPS + datalink).

RSAF Royal Small Arms Factories (UK, defunct).

RSAOC Region/sector air operations center[s].

RSB *1* Recovery speed brake.

2 Rescue/security boat.

RSC *1* Rescue sub-centre (ICAO).

2 Radar-scattering camouflage.

3 Runway surface condition.

4 Remote switching control.

5 Radar scan converter.

RSCAA, Rscaal Remote-sensing chemical-agent alarm.

RSCD RSC (3).

RSCU Ramp spill control unit.

RSD *1* Retinal scanning display.

2 Raster-scan display.

3 Rapid securing device (shipboard helicopter).

RSDP Reliable sequencing delivery [confirmation] protocol.

RSDU Radar/sonar display unit.

RSG, rsg Rising.

RSH Reservoirs souples héliportables (F).

RSI *1* Reusable surface insulation.

2 Rationalization, standardization, interoperability (NATO).

3 Remote status indicator.

RSIP Radar sensitivity [or system] improvement programme.

RSIS Rotorcraft systems integration simulator (NASA).

RSITA Regulations, SITA.

RSIU Radar-set interface unit.

RSL *1* Range-safety launch (RAF).

2 Remote-source lighting.

RSLS Receiver side-lobe suppression.

RSM Runway surface and markings.

R_{smc} Random-access channel for air/ground signals.

RSMMC Rapidly solidified metal-matrix composite.

RSN Regional subnetwork.

RSO *1* Reconnaissance Systems Officer, or operator.

2 Regional Safety Officer.

3 Runway Supervisory Officer.

RSP *1* Radar start point.

2 Responder beacon, or response (ICAO).

3 Radar signal processor.

4 Revenue-sharing participant.

5 Risk-sharing partner.

6 Reversion Select Panel.

7 Required surveillance performance.

8 Responsive space program (AFRL).

RSPD Rapid-solidification plasma deposition.

RSPL Recommended spare-parts list.

RSPT Report [when] starting procedure turn.

RSR *1* En-route surveillance radar.

2 Radar service request.

3 Rapid solidification rate; P adds process.

RSRA Rotor-systems research aircraft.

RSRE Royal Signals and Radar Establishment, Great Malvern.

RSRM Reusable solid rocket motor; not multipulse, but refurbished after each mission.

RSRS Radio and Space Research Station, now Appleton Laboratory.

RSS *1* Relaxed static stability.

2 Reflection suppressor system (SSR).

3 Radar subsystem.

4 Reliability and Safety Society (UK).

5 Remote surveillance system.

6 Rosette scanning seeker.

7 Root/sum/square.

8 Received signal strength; I adds indicator.

9 Remote Sensing Society (UK).

RSSK　Rigid seat survival kit, opened during descent.

RSSN　Reaction-sintered silicon nitride.

RSSP　Radar Systems Specialist Panel (ICAO).

RST　*1* Recording storage tube.

2 Reheat specific thrust.

3 Rapidly solidified titanium.

RSTA　Reconnaissance, surveillance and target acquisition.

RSTD, rstd　Restricted.

RSTE　Reynolds-stress transport equation[s].

RSTM　Reynolds-stress transport model.

RSTR　Restricted.

RSU　*1* Runway supervisory unit.

2 Rate sensor unit.

3 Remote sampling unit.

4 Relay switching unit.

RSV　*1* Reserve [fuel].

2 Reparto Sperimentale di Volo (I).

RSVP　*1* Restartable solid variable-pulse [rocket].

2 Rotating surveillance-vehicle platform.

RT　*1* Real time, often followed by control, display, environment, interface, language, management, processor, system, etc.

2 Reaction time.

3 Remote terminal.

4 Resistance training.

5 Replaceable tile[s], or RSI (1).

6 Right turn after takeoff.

R/T　*1* Radio/telephone, or telephony, or telecommunications.

2 Receiver/transmitter (beacon).

RTA　*1* Real, or required, time of arrival.

2 Real-time acquisition.

3 Rotation target altitude.

4 Receiver/transmitter antenna.

5 Research and Technology Agency, supports RTO(6) (NATO).

6 Revolutionary Turbine Accelerator.

RT&BTL　Radar tracking and beacon tracking level (ARTS).

RTB　*1* Return, or returned, to base.

2 Research and Technology Board, decides policy (NATO).

3 Rocket [ICBM] repair technical base, each numbered (R).

R/TBDA　Real-time battle-damage assessment.

RTC　*1* Real-time control; AF adds autonomous flight.

2 Rotorcraft.

3 Resident Training Center (USAF).

4 Real-time clock.

RTCA　Radio Technical Commission for Aeronautics (US, from 1935).

RTD　*1* Radar-target designator control.

2 Delayed, or routine time delayed.

3 Real-Time display.

4 Research and technology development.

5 Routed.

RTDC　Real-time damage computation.

RTDF　Radial temperature distribution factor.

RTDS　Real-time dissemination shelter.

RTE　Route.

RTEL　Runway threshold and end lights [co-located].

RTES　Real-time embedded system.

RTF　*1* Radiotelephone or radiotelephony (ICAO), also rendered rtf.

2 Round-trip flight.

3 Remote test facility.

4 Real-time fusion; I adds of information.

5 Also RT/F, revisit time/frequency (satellite).

RTG　*1* Radiotelegraph[y].

2 Real target gate.

3 Radioisotope thermoelectric generator.

R-θ, R-theta　Radio navaid family giving distance and bearing from fixed station.

RTHL　Runway threshold lights.

RTI　*1* Radar-target interrogator, or indicator.

2 Real-time interrogate.

3 Run-time infrastructure.

RTIC　*1* Real time in the cockpit.

2 Real-time information in the cockpit.

RTIL　Runway-threshold identification light[s].

RTIP　Radar-technology insertion program.

RTIRL　Real-time IR linescan.

RTK　Revenue tonne-kilometres.

RTLS　Return to launch site.

RTM　*1* Resin transfer moulding.

2 Revenue ton-mile.

3 Remote telemetry, or radio-transmission, module.

RTMM　Removable transportable media module.

RTMR　Real-time mission radius.

RTN, RTNG　Return[ing], returned.

RTO　*1* Rejected take-off.

2 Resident Technical Officer, of government in manufacturer's plant (UK).

3 Responsible test organization.

4 Range training officer (ACMI).

5 Runway turnoff lights.

6 Research and Technology Organization (NATO).

RTOAA　Rejected take-off area available.

RTOC　Real time out of the cockpit, often preceded by RTIC (1).

RTODA　Rejected take-off distance available; H adds helicopter.

RTOG　Regulated take-off graph, plot on which actual weight is assessed against T°, wind etc.

RTOL　Reduced (or, sometimes, restricted) take-off and landing, conventional transport with field length from 900 m (3,000 ft) to 1,500 m (5,000 ft).

RTOR　Right turn on red.

RTOS　Real-, or run-, time operating system[s].

RTOW　Regulated takeoff weight.

RTP　*1* Reporting and turn point.

2 Routine technical publication.

3 Radio tuning panel.

4 Reliability test plan.

5 Research and technology projects.

6 Receiver/transmitter/processor.

7 Research Technology Programme (Agard).

RtP　Real-time perspective.

RTPA Room-temperature parametric amplifier.

RTPTR Real-time precision-targeting radar.

RTQC Real-time quality control.

RTR *1* Real-time reconnaissance.

2 Reserve thrust rating.

3 Remote transmitter/receiver site.

RTRCDS Real-time reconnaissance cockpit display system.

RTRD Retard.

RTRN Return.

RTS *1* Radar target simulator.

2 Replacement training squadron.

3 Radar test subsystem.

4 Remote tracking station[s], AFSCF global network.

5 Returned [or release] to service (ICAO).

6 Request to send.

7 Regional turbine shop.

8 Rapid targeting system.

RTSS Red-telephone switching system.

RTT *1* Reserve take-off thrust.

2 Radio-teletypewriter, ie, teleprinter (ICAO).

3 Radio telephone terminal.

4 Radar target teacher.

5 Radio telemetry theodolite (US calibration).

RTTC Regional Technology Transfer Center (NASA).

RTTI Run-time type identification.

RTTL Range target-towing launch (RAF).

RTTP Real-time tracking and positioning.

RTTS *1* Robotic target training system.

2 RAF transportable telecommunications system.

RTTY, rtty Radio-teletype.

RTU *1* Replacement training unit (USAF).

2 Radio tuning unit.

3 Remote terminal unit.

RTV *1* Rocket test vehicle (UK).

2 Rapid terrain visualization (JPSD).

RTW Return to works (UK).

R²CSR See *RRSCR*

RTY Raw total yield (megatons).

RTZ Runway touchdown zone; L adds lights.

RU *1* Range unknown.

2 Rack unit.

R_u Maximum unambiguous radar range.

RUA Release unit adapter (torpedo).

rubber aircraft Aircraft in early project stage when gross variation in design is still possible (colloq.).

rubber at the ramp Airpower (US, colloq.).

rubber-base propellant Solid rocket propellant in which fuel is related to synthetic-rubber latex (wide range of resins, plasticizers and other additives) mixed with oxidant often of AP or other perchlorate type, finally cured into rubber-like grain. Examples include PBAA, PBAN.

rubber boot See *pneumatic deicing*.

rubberdraulics Rubber presswork technology, especially when flexible medium flows like liquid.

rubber press Press for forming sheet-metal parts by forcing thick pad of rubber down under high pressure on to cut-out flat parts (or, if blanking strips added, uncut sheet) arranged above male dies against which parts are forced with minimal wrinkling, rubber acting sideways as well as downwards.

rubber slick Stripe of rubber melted off tyre at touch-down and adhering to runway. Resulting blackened region shows touchdown zone preferred by actual traffic.

rubbing seal Gastight seal between fixed and moving members actually in contact; usually at least one mating face is carbon.

rubbing strip Numerous non-structural parts whose purpose is to accept impact from doors, ground equipment (eg steps and vehicles) and abrasion by inlet blanking plates or rescue winch cable.

rub indicator Sensor giving cockpit indication of eccentric running of rotating assembly, eg engine shaft.

rub rail Mount and launch rail for missile shipped and launched from canister; sometimes four, each locating a wingtip.

RUC Rapid update cycle (weather).

rudder Primary control surface in yaw; when nose-mounted, prefaced by nose- or fore-. Term also includes fixed fin of kite balloon, usually ventral, providing weathercock stability.

rudder bar Centre-pivoted bar providing pilot rudder input in simplest ultralights and historic aircraft. Traditional term for rudder input even when linear pedals are fitted.

rudder-bias strut Simple engine-out device comprising piston in rudder circuit with engine-bleed air piped to each side; failure of either engine causes immediate application of rudder.

rudder lock Potentially dangerous flight condition with rudder locked at maximum deflection. Caused by reversal of aerodynamic moment at large sideslip angle [suggest simple manual rudder only].

rudder pedal Left/right pedals for pilot's feet acting as manual input to rudder, and in modern aircraft to wheel-brakes.

rudder post Traditional (suggested arch.) term for leading-edge member of rudder carrying hinges; with modern inset hinges place taken by internal spar.

rudder reversal Roll reversal using rudder only, usually in maximum-performance high-alpha air-combat manoeuvres.

rudder roll Unwanted roll produced solely by coarse use of rudder.

rudder torque Twisting moment exerted by rudder on rear fuselage.

ruddervator Movable flight-control surface of butterfly tail, combining duties of rudder and elevator. Sometimes 'ruddevator'.

ruddervon Control surface, usually of traditional trailing-edge form, able to serve as rudder, aileron or elevator on so-called tailless aircraft. In most applications, outboard of elevon[s].

RUF Rough.

ruling dimension Basic measure, almost impossible to alter, e.g. diameter of fuselage.

ruling material That used for most of airframe structure.

RUM Ship-launched anti-submarine missile (USN code).

rumble *1* Rocket combustion instability audibly obvious from low growl or *.

2 Unstable pulsing at low frequency [300–700 Hz] in jet engine or afterburner.

3 Prolonged flat landing approach under power, to * in.

rumble seat   Occasional seat, eg for flight-deck observer or for stewardess on landing/take-off.

run   *1* Ground or distance traversed by wheels, or water by floats/hull, on takeoff and landing.

2 Number of production articles built to common type, or elapsed time to produce same.

3 Single flight over target (also called pass) for assessment, release of ordnance or operation of reconnaissance sensors.

4 Single flight past designated point(s), eg in attempt on speed record.

runaway   Undesired operation of device, eg PFCU, when not commanded; in dangerous extreme case continuing to limit of travel, giving hard-over condition.

runback ice   Very dangerous accretion of ice which forms on LE, is melted by deicing system or kinetic [ram-energy] heating, and refreezes further back.

rundown   *1* Fall off to zero rpm after normal closure of shut-off valve or HP cocks, or flameout or engine failure (also called spool-down).

2 Decay in production rate due to falling demand or imminent termination.

rundown time   Time for engine to come to rest, giving rough indication of internal rubbing.

run-flat tyre   Various aircraft tyres, pioneered by Goodrich, designed for high-speed landings after deflation caused by take-off blowout, fire or combat damage.

running fit   Slight clearance between mating parts allowing rotation or other relative motion.

running fix   Approximate fix obtained by taking a bearing of fixed station, or in any other way obtaining PL, then obtaining second PL and adjusting to common time.

running in   Act of running newly built or completely overhauled engine or other machine to ensure parts run together under controlled gentle operating conditions.

running landing   *1* Helicopter landing made into wind with groundspeed and/or translational lift at touchdown; with skid gear demands careful collective after touchdown to avoid abrupt stop.

2 Jet V/STOL landing with significant forward speed.

running mate   New transport aircraft type, usually smaller, to accompany trunk-route type already in service and offering same advanced-technology appeal.

running order   Traditional condition for measuring mass of piston engine: including radiator, coolant, internal oil, external pipes and controls, but excluding fuel, oil, tanks, reserve coolant, exhaust tailpipes and instruments.

running rigging   Rigging for kite balloon or other aerial object which by system of vee-lines and pulleys automatically adjusts to direction of pull.

running take-off   *1* Started without lineup and hold, speed never slackening on arrival at runway in use.

2 By helicopter or other VTO aircraft making preliminary ground run.

runoff area   Strip or pad beside runway where arrivals can clear quickly for following traffic (also run-off, run off).

run-on landing   Roll-on landing.

run-out   *1* Distance travelled by carrier aircraft between engaging hook and coming to stop, also called pull-out.

2 Distance travelled by aircraft after encountering runway barrier, drag wire, decelerating bed [such as EMAS] or other arresting system.

3 Distance travelled by gun or barrel, on recoil stroke.

run-up   *1* To accelerate engine under own power.

2 Portion of flight immediately preceding target run.

3 To test piston engine, briefly at high power and to check dual ignition, before takeoff.

run-up area   Portion of airfield near taxiway designated for run-up (3).

run-up drag   *1* Drag caused by windmilling propeller in air-start.

2 Drag caused by need to accelerate landing wheels on touchdown.

runway   Paved surface, usually rectangular and of defined extent, available and suitable for aeroplane take-off and landing. Equipped * includes stopway, clearway, surface markings and designators. All-weather * includes lighting. Instrument * adds electronic aid, eg ILS, MLS. Unpaved * = airstrip.

runway alignment   Direction of runway centreline, published as first two digits, in both directions, eg 13/31; also called direction number. See *runway designator*.

runway alert system   Passively monitors ASDE and other sources to detect potential RWI.

runway alignment factor   Maximum angular departure from alignment admissible for straight-in approach, normally 30°.

runway alignment indicator   Group of flush lights offering directional guidance on takeoff (to some degree on landing) in bad visibility.

runway basic length   That length selected for aerodrome planning purposes required for takeoff or landing in ISA for zero wind, elevation and slope.

runway capacity   Frequency of landings and/or takeoffs, or mixture, possible or permissible with minimal approach spacing, published for IFR and VFR. In complete airport varies greatly with number and arrangement of runways, and conflicts of crossing traffic.

runway condition   Numerical output from braking [decelerometer] tests.

runway controller   ATC controller stationed, usually in caravan with checkerboard markings, close beside downwind end of runway in use; equipped with signal pistol, Aldis and telephone to tower (rare since WW2).

runway designator   Numerical alignment plus qualifier left/right if necessary; thus New York JFK offers 4R/22L, 4L/22R, 13R/31L and 13L/31R.

runway direction numbers   Numerical values of alignment.

runway-edge lights   White lights grouped according to intensity: fixed-intensity LIRL (low-intensity runway lights) and variable MIRL and HIRL. Designed to withstand flying stones.

runway end   End of runway in use, identified by markings; beyond may be similar-surfaced blast pad, overrun (RESA) or stopway and (unpaved) clearway.

runway-end lights, REIL   Often pair of flashing white lights, at major airport continuous transverse row of bi-directional lights showing green towards approach and red towards runway.

runway-end safety area   UK term for overrun area; area adjoining runway in use, symmetrical about extended centreline, intended to minimize damage to undershoot or overrun aircraft.

runway floodlight   Appears incorrectly in some dictionaries; does not exist.

runway gradient   Not used; correct word is slope.

runway guard   Flashing yellow lights where taxiway meets runway.

runway incursion   Crossing runway in use, or taxiing beyond takeoff holding point, without ATC clearance. Abb. RWI.

runway localizer   See *localizer*.

runway markings   Basic, 2-digit direction number and centreline; Instrument adds threshold; All-weather (precision) adds side strips and zones.

runway occupancy   Elapsed time particular aircraft is on runway, on arrival or departure.

runway profile descent   Published procedure for complete IFR controlled arrival from en-route to glide-path.

runway separation   Time and distance intervals between arrivals and/or departures on one runway. Distance between parallel runways is spacing.

runway spacing   Perpendicular distance between centre-lines of parallel runways at same airport.

runway strip   Defined area including runway and stopway intended to reduce risk of damage to aircraft that run off runway in any direction and to protect aircraft flying over it during takeoff or landing.

runway threshold   Threshold (1).

runway turnoff lights   Fixed wide-beam white lights on each side of the forward fuselage giving lateral illumination.

runway visibility value, RVV   Determined for particular runway by transmissometer with readout in tower; generally being replaced by RVR.

runway visual range, RVR   Value representing horizontal distance pilot will see centreline or edge lights or runway markings down runway from approach end. Once recorded by an observer 76 m from centreline, now by RVR system.

RUR   Ship-launched anti-submarine rocket (USN code).

RUSA   Roll-up solar array.

RUSI   Royal United Services Institute for Defence Studies (UK).

ruslick   Anti-corrosive coating for bright metal in saline (ocean) environment.

RUT   Standard regional route transmitting frequencies (ICAO).

Ruticon   Family of photoconductive/liquid-crystal devices used in large-screen projection, typically with potential across photoconductor and elastomer with mirror surface for readout, scanned on opposite face by CRT input.

RV   *1* Radar vector.
   2 Re-entry vehicle.
   3 Residual value.
   4 Rescue vessel (RAF, ICAO).
   5 Rescue vehicle.
   6 Rendezvous.

RVA   *1* Radar vectoring area.
   2 Régie des Voies Aériennes (Belgium, see *RLW*).

RvA   National accreditation body (Netherlands).

RVC   Radar video corridor.

RVD   *1* Radar video.
   2 Rear-view display.

RVDP   Radar video data processor.

RVDT   *1* Rotating variable- [or voltage-] differential transducer (servocontrol position sensor).
   2 Rotating voltage-displacement transformer.

R/VGPO   Synchronized range and velocity deception, see *RGPO*, *VGPO*.

RVL   Rolling vertical landing.

RVO   Runway visibility by observer.

RVP   *1* Reid vapour pressure.
   2 Ramp void pressure.

RV-PVO   Radiotekhnicheski-Voiska PVO, radio-technical air-defence troops (R).

RVR   *1* Runway visual range; C adds centre, R rollout, T touchdown area.
   2 Rear-view radar.
   3 Reverse-velocity rotor.

RVR system   Various electro-optical instruments for measuring RVR without subjective interpretation, usually by calibrated light source and transmissometer receiver separated by distance great enough to avoid too much error from local smoke.

RVS   Reduced vertical separation[s]; M adds minima [US often minimums], MK minima kit.

RVSD   Revised.

RVSM   Reduced vertical-separation minima, or minimum.

RVSN   Strategic rocket forces (USSR).

RVSP   Radar video signal processor.

RVT   Remote video terminal.

RVV   Runway visibility value.

RV/WH   Re-entry vehicle/warhead.

RW   *1* Radiological weapon(s).
   2 Retractable wheel (sailplanes).
   3 Rigged wet, ie for spraying (ag-aircraft).
   4 Reconnaissance wing.
   5 Rain shower.
   6 Runway.

R/W   *1* Marine aircraft on water total resistance divided by weight, equivalent to an L/D ratio.
   2 Runway.
   3 Read/write.

RWA   *1* Reaction wheel assembly.
   2 Rhomboid-wing aircraft.
   3 Rotating-Wing Aviation, RU adds Research Unit (USA).

RW/D   Rigged wet and dry, ie for both spraying and dusting (ag-aircraft).

RWE   Radar warning equipment.

RWG   Rotorcraft Working Group (UK/US).

RWI   Runway incursion.

RWIS   Runway weather information system.

RWM   *1* Code for SAR helicopter (ICAO).
   2 Read-only wire (or composite-wire) memory.
   3 Read/write memory.

RWR   *1* Rear-warning radar.
   2 Radar warning receiver.

RWRT   Real world, real time.

RWS   *1* Radar warning system.
   2 Range while search.
   3 Remote workstation.

RW+S   Rigged wet plus spreader (ag-aircraft).

RWT   Radar warning trainer.

RWTS   Rotary-Wing Training School (RAF, WW2).

RWU   Rain shower, intensity unknown.

RWY, rwy   Runway (ICAO); -TDZ adds touchdown zone.

RX   *1* Report [when you are] crossing [specified point].
   2 Receive.

Rx   Receiver, or reception only.

R(X)   Correlation function.

Ryton   A trade-name for PPS (4), with high opacity to EMI(1) (Phillips Petroleum).

RZ   *1* Recovery zone.
   2 Reconnaissance zone.
   3 Return to zero.

RZI   Real-zero interpolation.

S

S *1* Generalized symbol for area, eg gross wing area.

2 Entropy (not UK and some other countries).

3 US piston engines, supercharged (hence, TS = turbocharged).

4 Aircraft category, scout (USN 1922-46), sonic test (USAAF 1946-47), ASW (USN from 1946).

5 Aircraft modified mission, ASW (USN pre-1962 suffix, post-1962 prefix).

6 Aircraft category, strike (RAF, RN).

7 Siemens, normally written siemens.

8 JETDS code, special, or detection/range-bearing/search.

9 South, southern latitudes (ICAO).

10 Section modulus (alternatively Z).

11 Surface, common missile code for both launch location and target.

12 Single [especially single wheel on each landing-gear leg, and runway bearing capacity for such wheel].

13 Distance between contact areas of dual wheels.

14 Serviceable (CAA).

15 Secondary [and secondary airport].

16 Saturated [especially traffic between city pair].

17 VLA (BCAR section).

18 Supplementary [frequency].

19 Snow, or squalls.

20 Stoke[s].

21 Signal or radar power, or energy.

22 Scheduled flight.

23 Superior (warm air mass).

24 Suffix, light-alloys = wrought.

25 Strouhal number [St is preferred].

26 Reluctance.

27 Apparent power (electrical).

28 Sulphur.

s *1* Second[s] (time).

2 Stress.

3 Generalized symbol for linear measure, length, distance; esp. used for aerofoil semi-span (b/2).

4 Starboard.

5 Square (eg in rms).

6 Spherical (eg in scp).

7 Static [not total] pressure.

8 Laplace variable.

9 Specific entropy.

$\vec{S}$ Vector: integrated signal energy.

S0, S_0 Segment zero.

S1 Segment 1.

S2S Sensor-to-shooter.

S^3 *1* S-cubed, stick-shaker speed.

2 Step stress screening.

S4 Special-services switching system.

S-band Former common-use radar band originally 19.33-5.77 cm, 1.55-5.2 GHz, later rationalized to 15-7.5 cm, 2-4 GHz; now occupied by E and F.

S-code IFR flightplan code aircraft has 64-code transponder and approved R-Nav.

S-duct Curved duct supplying air to centre engine in trijets.

S-gear Full supercharge.

S-ing Series of S-turns, especially in taxiing.

S-manoeuvre To weave in horizontal plane.

S-mode Aircraft transponder provides data-link capability, e.g. altitude, bearing, range.

S-pattern Wavy track resulting from S-ing.

S-Stoff Rocket fuel: 90-97% RFNA, 10-3% sulphuric acid (G).

S-turn To describe S in horizontal plane.

SA *1* Situational awareness.

2 Sand or dust storm (ICAO).

3 Standby altimeter.

4 Shaft angle.

5 Surface-to-air.

6 Submerged-arc (weld).

7 Spin axis.

8 Single-aisle (passenger transport aircraft internal layout).

9 Structural audit.

10 Structured analysis (software).

11 Stand-alone.

12 Safety altitude.

13 Standard Atmosphere.

14 Safety action (FAA).

15 Société Anonyme (F), and Sociedad Anónimo (Spain).

16 Special access [black].

17 Selective availability, of GPS.

18 Simple approach [runway lighting].

19 Arsine, CW agent.

20 Scientific Advisor (UK).

21 Ship's Airplane (USN 1917-19).

22 Surface aviation scheduled weather report.

23 Search/attack (USN).

SAA *1* Society of Airline Analysts (US).

2 Safety and arming.

3 Swiss Aerobatic Association.

4 Systems application architecture.

5 Service access area.

6 Supersonic adversary aircraft [also SSA] (USN).

7 Cargo container for main deck (code).

8 Suphuric-acid anodized [surface finish].

9 School of Army Aviation (UK).

10 South Atlantic anomaly.

SAAA Sport Aircraft Association of Australia.

SAAATS, SA3TS South African advanced ATC system.

SAAC *1* Simulator for air-to-air combat.

2 Society of Amateur Aircraft Constructors (Ireland, 1978–).

SAAF Small austere airfield.

SAAFA South African Air Force Association.

SAAFI South African Association of Flying Instructors.

SAAHS Stability-augmentation/attitude-hold system.

SAALC San Antonio Air Logistics Center (Kelly AFB).

SAAM *1* Special-assignment airlift mission.

2 Surface [or sol (F)] -air anti-missile.

SAAPA South African Airways Pilots' Association.

SAARU Secondary attitude and air-data reference unit.

SA/AS, SAAS   Selectively available [or selective-availability] anti-spoofing; M adds module (GPS).
SAB   *1* Scientific Advisory Board (USAF).
   2 Self-aligning bearing.
SABA, Saba   Often pronounced 'Sabre', small agile battlefield aircraft.
SABAR   Satellites, balloons and rockets.
Sabatier   *1* Reversal phenomenon in photo processing occurring when developed image is exposed to diffuse light and redeveloped (not encountered with X-rays).
   2 Fully developed reaction process for recovery of pure oxygen from human exhalation/excretion with by-products such as methane and carbon dioxide.
Saber   Simplified acquisition of base engineering requirements.
SABH   NDB providing automatic weather broadcasts.
Sabin   Unit of acoustic absorption equal to 1 ft^2 of surface absorbing all sound energy falling on it. Appears to be no SI unit available.
Sabmis   Surface-to-air ballistic-missile interception system.
sabot   Annular driving mechanism to enable subcalibre projectile to be fired at very high muzzle velocity from gun; usually has form of drum open at one end to receive projectile and divided radially to separate into sections beyond muzzle.
SABR   Support, amphibious and battlefield rotorcraft (RAF/RN).
SAC   *1* Strategic Air Command, formed 1946 by USAAF, principal 'deterrence' of USAF until formation of Air Combat Command in 1992, initials now mean Space Analysis Center (AFSPC).
   2 Space Applications Centre (India).
   3 Space Activities Commission (Japan).
   4 Seldom used for Soaring Association of Canada.
   5 Supplemental air carrier.
   6 Subsecretaria de Aviación Civil (Spain).
   7 Standing advisory committee.
   8 Standard arbitration clause.
   9 Sectional aeronautical chart.
   10 Single annular combustor.
   11 Stealthy affordable capsule.
   12 Surface-analysis chart (weather).
SACA   Service Administrative de la Commissariat de l'Air (F).
SACC   Society of Air Cargo Correspondents (UK, 1979–).
SACCA   Scottish Advisory Council for Civil Aviation.
SACCS   SAC (1) automated command and control system.
Sacdin   SAC digital information network (USAF).
Saceur   Supreme Allied Commander, Europe.
Sacintnet   SAC intelligence network (from 1990).
SACL   Strobe anti-collision light.
Saclame   Standing Advisory Committee for Licensed Aircraft Mechanical Engineers (UK CAA).
Saclant   Supreme Allied Commander, Atlantic.
Saclos   Semi-automatic command to line of sight.
SaCo   Samarium cobalt (magnet material).
SACP   *1* Surface/air courte portée; short-range SAM (F).
   2 Standing Advisory Committee on Pilot licensing (CAA, UK).
sacrificial corrosion   Metal is protected against corrosion

by being coated with metal less noble than itself, which is attacked preferentially.
Sacru   Semi-automatic cargo release unit.
SACS   *1* Speed/attitude control system (takeoff).
   2 Secondary attitude and compass system.
   3 Small air-capable ship.
   4 Secure-access control system.
SACT   *1* Signal acquisition conditioning team, or terminal.
   2 Supreme Allied Command Transformation (NATO, Norfolk Va).
SACU   Stand-alone communications unit.
SAD   *1* Submarine anomaly detector.
   2 Spares advanced data.
   3 Self-adhesive decal.
   4 Solar-array drive.
   5 Situational-awareness display.
   6 Surface-to-air defence.
   7 Standard advanced Dewar [A adds assembly, D display].
SADA   Semi-automatic air-defence system (Spain).
Sadarm   Sense (or search) and destroy armour (USA/Aerojet).
SADC   Secondary air-data computer.
saddle   Shaped wooden former on which keel of marine aircraft rests during manufacture.
Sadis   Satellite distribution system.
SADL   Situation[al] awareness data-link.
SADM   Special atomic demolition munition.
SAdO   Station Administration Officer (RAF).
Sadral   Système autodéfense rapproche anti-aérien légèr (F).
Sadram   Seek and destroy radar-assisted mission; locks on for long period after hostile emitter silent.
SADRG   Semi-active Doppler radar guidance (SADRH substitutes 'homing').
SADS   Satellite distribution service.
SADT   Structured analysis and design techniques.
SAE   *1* Society of Automotive Engineers (US).
   2 Servicing Appraisal Executive (RAF).
   3 Semi-actuator ejector.
   4 Sender alignment equipment [beam-rider].
   5 Service Acquisition Executive[s] (USAF/USN).
SAE ratings   Lube-oil ratings (10 to 70) based on Saybolt viscosity.
SAF   *1* Specified approach funnel.
   2 Secretary of the Air Force (US).
   3 Simulation and analysis facility (USAF).
SAFA   Safety assessment of foreign aircraft (USAF).
SAFCS   Standard automatic flight-control system.
SAFE   *1* Formerly Survival & Flight Equipment Association, now simply SAFE Association (US).
   2 Soil airfield fighter environment (DoD).
   3 Simple architecture for full electrical.
Safe-bar   Safeland barrier.
safe burst height   That above which damage/fallout is locally acceptable.
Safecon   Safety and Flight Evaluation Conference (NIFA).
Safe Flight kit   AOA-sensitive stall-warning system normally displaying green arrows on glareshield of GA aircraft.
safe life   Basic design and certification philosophy for primary structure; whether or not there are redundant

load paths or fail-safe provisions, a total acceptable life (flight time) is published in flight manual and relevant structure must then be replaced, even if no crack is visible.

Safer   *1* System for aircrew flight extension and return; lifting/propelled ejection capsule.

    2 Special aviation fire and explosion reduction (FAA panel, US, from 1977).

Safegrip   Proprietary anti-icing/deicing fluid, based on potassium acetate.

safety   As verb, see *safetying*.

safety advisory   Issued by ATC to aircraft under its control to warn of (1) terrain obstruction and (2) aircraft conflict, in the judgement of the controller (FAA).

safety altitude   Loosely, one at which collision with surface is unlikely at approximate location; not accepted term (see *DH, MDA*, etc).

safety area   Designated area around helicopter, ICAO = 0.25 overall length [minimum 3 m] and must be load bearing, FAA = 0.33 main-rotor diameter [minimum 6 m], need not be load bearing (downwash only).

safety barrier   Emergency arresting barrier across carrier flight deck (US often barricade net).

safety belt   For passenger, normally called seat belt or lap strap; for flight (occasional cabin) crew, seat harness.

safety disc   Disc of accurately known strength sealing fluid system, eg cartridge-operated, serving as safety valve.

safety equipment   Vast range of personal items, of which the most obvious is a parachute, others being helmet, goggles, armour, whistle, dinghy (with fluorescein dye), Sarbe (or similar) beacon, torch, mirror, pocket GPS, Very pistol and in some circumstances anti-NW flash protection or a handgun. Does not include normal flying clothing, oxygen mask or microphone, but does also include aircraft equipment, such as seat harness, axe, fire extinguisher and escape slide.

safety equippers   Specialized ground staff charged with maintenance of safety equipment.

safety factor   See *factor of safety*.

safety height   Not accepted term (see *DH, MDA*, etc). One authority: lowest at which safe to fly on instruments.

safety imagination   Ability to extrapolate from a real near-accident to hypothetical accident.

safetying   *1* Rendering explosive or pyrotechnic device safe by positive means. See *safing*.

    2 Installing locking wire or other device which prevents an attachment from becoming loose.

safety net   One designed to catch flying object with minimal damage, eg ejection seat on test model in tunnel (esp. spinning tunnel) or flying rotor blade in rotor test rig.

safety pilot   One present in cockpit to prevent accident, eg to radio-controlled aircraft or RPV, or with pupil on instrument practice.

safety pins   Those inserted to disarm ejection seat except in flight.

safety plug   Blow-out plug serving as safety valve in case of excess pressure (eg JATO bottle) or excess temperature (eg tyre after prolonged braking).

safety speed   That above which aircraft is safe to fly with given load and configuration; formerly several, eg FUSS, but most important today is V_2.

safety thread   That connecting D-ring (ripcord) to parachute pack.

safety wire   Locking wire passed through holes in nuts,

turnbuckle barrels and other fasteners in such a way that they cannot loosen subsequently.

safety zone   Area reserved for non-combat operations by friendly forces (DoD).

Safeway   Proprietary deicers: * SD is a solid, * KA liquid.

Safeway Pneutronic   Guidance installation for nose-in parking; pavement pressure pads sense nosewheel(s) and illuminate progress lights facing flight deck while arrows, lights and other displays give guidance.

Saffire   Synthetic-aperture fully focused imaging radar equipment.

SAFI   Semi-automatic flight inspection (FAA, from 1962); S adds system, = Safis.

safing   Process of rendering potentially dangerous device or system inoperative; eg complex procedure in Space Shuttle Orbiter post-landing.

Safire   Scanning airborne filter radiometer.

Safoc   Semi-automatic flight operations centre.

SAF/OI   Secretary of the Air Force/Office of Information (USAF).

SAFP   Slotted-array flat plate; A adds antenna.

SAF/PA   As above, Public Affairs.

SAFR   Schweizerische Arbeitsgemeinschaft für Raumfahrt (previously Raketentechnik) (Switzerland).

SAFU   Safety, arming and fuzing (sometimes functioning) unit.

SAG   *1* Support air group (Royal Navy).

    2 Scientific advisory group (US).

    3 Surface action group (USN, concerned with ship targets).

    4 Survivability analysis group (DoD).

    5 Semi-active guidance.

sag   See *sagging*.

SAGA   *1* Studies Analysis and Gaming Agency (DoD).

    2 System of azimuth guidance for approach; simple optical ILS.

    3 Statistics of accidents in General Aviation.

SAGE, Sage   *1* Semi-Automatic Ground Environment; pioneer (from 1953, US) computer-controlled radar and communications system for defence of large airspace and management of interceptors and SAMs.

    2 Stratospheric aerosol and gas experiment.

sagging   *1* Distortion of airship (any kind) caused by upward loads near ends or lack of lift at centre.

    2 Bending stress on seaplane float or flying-boat hull caused by water support concentrated near ends due to swell or waves.

SAGr   Maritime reconnaissance group (G. WW2).

Sags   Semi-active gravity-gradient stabilization.

SAGW   Surface-to-air guided weapon (UK usage).

SAH   *1* Semi-active homing.

    2 Sample and hold (EDP [1]).

    3 Select and hold.

    4 School of Aircraft Handling.

SAHIS, Sahis   Standby attitude, heading and rate of turn indicating system.

SAHR   Semi-active homing radar (SAHRG adds 'guidance').

SAHRS   Standby, or standard, or secondary, attitude/heading reference system.

SAI   *1* Standby airspeed indicator.

    2 Spherical attitude indicator.

    3 Single-aperture interferometer [or interferometry].

4 System architecture and interface.
SAIA Swedish Aerospace Industries Association.
SAIF Standard avionics integrated fuzing (sets dispenser payload fuzes milliseconds before release).
SAIG Single-axis integrated gyro.
SAIL Shuttle-Avionics Integration Laboratories.
sail *1* Flat surface pointed towards Sun or other celestial object and attached to spacecraft, eg carrying solar cells.
2 Very large lightweight reflective surface proposed for space propulsion by pressure of sunlight. Potential for $6×10^5$ km/h (380,000 mph).
3 To navigate seaplane (see *sailing [2]*).
4 Projecting structure above hull of submarine; also called fin, bridgefin or, formerly, conning tower.
5 (Usually plural) Small winglets arranged at different angles around wingtip, typically four disposed from front to rear.
6 Loosely and ambiguously, any winglet.
7 Recent usage, upper or lower surface of fabric-covered rigid wing, thus upper-*, lower-*.
sailing *1* Undesired rotation of helicopter rotors or aeroplane propeller in high wind.
2 Navigation of seaplane on water, esp. in conditions of wind and current.
sailplane Glider designed for soaring.
sails See *sail (5)*.
sailwing Aerodyne whose wing assumes lifting (near-aerofoil) profile only in presence of suitable relative wind; class includes parawings but normally has flexible surface(s) restrained by rigid periphery.
Saint *1* Surveillance, acquisition, identification, notification and tracking.
2 Satellite inspection technique.
Saint Elmo's fire Brush discharge caused by build-up of electrostatic potential, notably on propeller blades; luminous and often audible.
SAIP Semi-automated imagery [or Imint] processing [or processor].
SAIRS, Sairs Standardized advanced IR sender [or system] (Martin Marietta).
SAIRST Situational awareness IR search and track.
SAKh Written CAX, mean aerodynamic chord (R).
SAL *1* Strategic arms limitation.
2 Security access level.
3 Selected altitude layer decoder.
4 Semi-active laser.
Salbei See *SV-stoff*.
SALC Sacramento Air Logistics Center (McLellan AFB).
Salisbury screen Oldest and simplest RAM (2), comprising resonant absorber created by placing resistive sheet on spacer of low dielectric constant in front of metal plate (eg aircraft skin).
Salkit Supplemental airfield lighting kit.
salmon *1* Streamlined bulge forming tip to wing of aircraft having centreline gear and no outrigger (eg sailplane); designed to withstand rubbing on ground.
2 Streamlined fairing on wingtip for purpose other than housing fuel (applied to Su-25 where * incorporates airbrakes; not used for EW tip pods so far).
Salomon damper Dynamic damper for crankshaft balance weights to remove oscillatory loads.
SALR Saturated adiabatic lapse-rate.

SALS *1* Short approach-light system; F adds with flashing lights (FAA).
2 Separate-access landing system.
3 Shipborne aircraft landing system.
4 Service de l'Aviation Légère et Sportive (F).
SALT Strategic Arms Limitation Treaty (* I 26 May 1972, * II 18 June 1979).
salted weapon NW which has, in addition to normal components, extra elements or isotopes which capture neutrons at time of explosion and produce additional radioactive products; generally, opposite to clean NW.
Salthorse Commander of a carrier despite being a non-aviator (RN, colloq.).
Salti Synthetic-aperture lidar for tactical imaging.
Salto di Quirra NATO air firing range, including AAA and SAMs, in Sardinia.
salvage *1* To scrap complete aircraft or other equipment but recover parts or material.
2 To retrieve aircraft after landing away from any airfield.
3 To retrieve potentially dangerous flight situation, eg high rate of sink.
salvo *1* Simultaneous group of ECM bursts, esp. from dispensed payloads; not necessarily all of same species, eg could be two chaff, one IR, one jammer.
2 In close air support/interdiction, method of delivery in which all weapons of specific type are released or fired simultaneously (DoD).
3 See *salvos*.
salvos Air-intercept code: "I am about to open fire"; can add further word specifying weapon(s), eg * mushroom = "I am about to fire special weapon", ie Genie (DoD).
SAM *1* Surface-to-air missile.
2 School of Aerospace Medicine (USAF).
3 Sound-absorbing material(s).
4 Standard (or standardized) assembly module.
5 Société Aérostatique et Météorologique, 1865 (F).
6 Special Air Missions squadron (USAF).
7 Standard avionic module.
8 Structural analysis and maintenance.
9 South American region, / SAT adds South Atlantic (ICAO).
10 Situational-awareness mode.
11 Structural-anomaly mapping.
SAMA *1* Small Aircraft Manufacturers' Association (US).
2 Systema aeronautico modulare anti-cendio (I).
3 Semi-automated manoeuvre analysis.
samarium Sm, rare-earth metal, density 7.52, MPt 1,077°C, alloying element in magnets.
SAMD Stratospheric aerosol measurement device.
SAMF Société Aérostatique et Météorologique de France (1865).
SAMI System-acquisition management inspection.
Samir Système d'alerte missile IR (F).
Samoa Système aérotransportable mobile pour les opérations aériennes (F).
Samoc SAM operations center.
Samos, SAMOS *1* Stacked-gate avalanche-injection MOS.
2 Satellite and missile observation system.
SAMP Sol-air [or système d'autodéfense] moyenne portée [short-range SAM]; / N adds navale, T terrestre [land] (F).

Sampe, SAMPE Originally Society of Aerospace Material and Process Engineers, now Society of Advanced Material and Process Engineering (US).

sampled recording Automatic switching so that many ($<10^6$) parameters can be measured each second.

sample length Lenth of specimen studied in examination of variable, eg surface finish.

sample rate Rate per unit time for flight-test programme or other variable.

Sams, SAMS *1* Six-axis motion system (simulator).

2 Software automated management support.

3 Spare assembly, maintenance and servicing (spacecraft).

SAMSO Space and Missile Systems Organization, Air Force Systems Command (USAF).

Samson *1* Strategic automatic message-switching operational network (Burroughs).

2 Special avionics mission strap-on, now a self-powered FLIR pod (Lockheed).

SAMT *1* Simulated aircraft maintenance trainer.

2 State-of-the-art medium terminal (Satcom).

SAMTEC Space and Missile Test Center (DoD).

SAN *1* Satellite access node[s]; / AP adds air-portable.

2 Sanitary.

3 Storage area network.

S&A Safety and arming.

sand and spinach Green/brown camouflage (colloq.).

sandbag Passsenger carried free by airline to make up necessary ballast.

sandbag bumping American term for hand-shaping sheet metal by hammering against tough sandbag.

sandbag line Rope joining sandbag loops to prevent wear.

sandbag loops Cord loops over aerostat (esp. kite balloon) envelope carrying sandbag at each end; also called sandbag bridle.

sandbag ring Cable or rope round balloon basket from which sandbags are (suggest were) hung, forming easily jettisoned load.

sandblasting Scouring material surface with high-velocity jet of sand or other abrasive for various purposes; not common in modern aerospace, replaced by tailored steel shot or glass beads.

sand casting Casting metal parts in sand mould.

S&E Scientists and engineers.

SANDF South African National Defence Force.

S&I Safety and initiating.

sanding coat Heavy-bodied paint or dope coat which fills irregularities in surface, leaving smooth base for top coats.

S&M Supply and movements (RAF).

Sandow cord See *bungee*.

sand pillar *Dust devil.*

S&S Sensors and shooters.

SANDT School of Applied Non-Destructive Testing.

S&R Search and rescue.

S&TI Scientific and technical intelligence.

sandwich aircraft Positioned horizontally or vertically between two enemy aircraft.

sandwich construction Large family of constructional methods, most of them patented, in which two load-bearing skins are joined to stabilizing low-density core. In most types both skins are locally flat and parallel, but most can be shaped to single or compound curvature

before or after bonding three components together, and a few are tailored so that * thickness varies to meet requirement of part. Cores can be metal or paper/plastics honeycombs, foamed plastics, balsa or many other choices, including dimpled or corrugated sheet.

sandwich manoeuvre One fighter, usually DA partner, turns away and accelerates while other falls in behind enemy.

sandwich moulding Polymer A injected inside polymer B in mould.

sandwich plate Plate, part of primary structure, located between two other members, eg on centreline between left/right keel girders.

Sandy Search and rescue aircraft (colloq.).

San Marco Italian-operated space launch platform on island off Kenya.

SANS Small-angle neutron scattering.

Santal Système anti-aérien légèr (F).

SAO *1* Special access only (high security classification, also called black).

2 Special Activities Office (DoD).

3 Surface aviation observation.

SAOCS Submarine/aircraft optical communication system.

SAOEU Strike/Attack Operational Evaluation Unit.

SAOWP Structural analysis and optimization working party.

SAP *1* Seaborne air platform, eg for jet V/STOL.

2 Semi-armour piercing [many possible suffixes, such as, HE, high-explosive, API; armour-piercing incendiary].

3 Silicon avalanche photodiode.

4 Simulated attack profile.

5 Special-access programme [OC adds oversight committee].

6 Survivable adaptive planning [E adds experiment].

7 Soon as possible.

8 Sensing and processing.

9 Service access point[s].

SAPC South African Parachute Club.

SAPE Survivable adaptive planning experiment.

Saphyre Swerve aero-propulsion hypersonic research experiment (NASA).

SAPM Synchronized air-power management (USAF).

SAPO Standby arrangement for peacekeeper operations (UN).

SAR *1* Search and rescue.

2 Synthetic-aperture [or array] radar; FTI adds fixed-target indication.

3 Selected acquisition report (US one each FY).

4 Stand-alone radar.

5 Semi-active radar.

6 Signal-acquisition remote.

7 Special access required.

8 Starter-assisted relight.

Sara Selective-adhesion release agent.

Sarah Search and rescue and homing; personal radio beacon (Ultra).

Sarbe Search and rescue beacon equipment; military/civil (Burndept).

SARC Systems Acquisition Review Council (USA).

Sarcap, SarCAP SAR (1) combat air patrol.

Sarda State and regional defense airlift (US).

Sardam Sonar-array remote detection of aircraft and

management, enables submerged submarine to detect and track distant aircraft.

SARG   Semi-active radar guidance.

Sarge   Surveillance and reconnaissance ground equipment.

SARH   *1* Semi-active radar homing.
  2 Society of Air Racing Historians (Int.).

Sarie   Semi-automatic radar identification equipment; part of Abbey Hill (UK).

Sarin   Toxic nerve gas.

Saris   *1* Synthetic-aperture radar interpretation system.
  2 Semi-active radar imaging seeker.

SARLupe   Radar surveillance satellites (G).

SARO   Supply aero-engine record office (UK).

Saros   Search and rescue for Open Skies.

SARP, sarp   *1* Semi-automatic radar plotting.
  2 Signal auto radar-processing system.
  3 See next.

SARPS, Sarps   Standards and recommended practices (ICAO).

SARS   *1* Support and restraint system.
  2 Static automatic reporting system.
  3 Severe acute respiratory syndrome.

Sarsat   *1* Search and rescue satellite.
  2 Search and rescue satellite-aided tracking.

SART, Sart   *1* Semi-active [artificial] radar target.
  2 Search and rescue transponder.
  3 Structural airframe repair technician.
  4 Self-activating reactive target.

Sartaf   SAR (1) task force.

Sartor, SARTOR   Standards and routes to registration (EC[UK]).

SAS   *1* Stability-augmentation system.
  2 Satellite Applications Section (NOAA).
  3 Small-angle scattering.
  4 Stall-avoidance subsystem.
  5 Staring-array seeker.
  6 Single audio system.
  7 Special Air Service (UK).
  8 Survival avionics system.
  9 Sensors and Avionic Systems (UK Qinetiq).
  10 Signature augmentation subsystem.
  11 Support analysis software.
  12 Société par Actions Simplifiée (F, joint stock co).
  13 Station address set.
  14 School of Aviation Safety (USN, USMC).
  15 Small Astronomy Satellite[s].

SASC   Senate Armed Services Committee (US).

SAS/CSS   SAS (1) plus control-stick steering.

SASE   Semi-automatic support equipment.

Sashlite   Illuminated under bomber at instant of bomb release and plotted on camera obscura (RAF training, 1930-45).

SASI   The Society of Air Safety Investigators (Canada).

SASIG   Strategic Aviation Special Interest Group (UK, airport location).

SASO   *1* Pronounced sasso, Senior Air Staff Officer (RAF).
  2 Stability and support operations.

SASP   Single advanced signal processor.

SASRS   Satellite-aided search and rescue system.

SASS   Strategic airborne, or small aerostat, surveillance system.

SASSR   Small-aperture SSR.

SAST   *1* Strategic analyses in science and technology.
  2 Shanghai Academy of Spaceflight Technology.

SAT   *1* Software audit team.
  2 Static air temperature.
  3 Situational-awareness technology.
  4 Strategic action team (FAA).

SATA, Sata   Small-aperture telescope augmentation.

Sataf   Site-activation task force.

Satair   Sea acceptance trials, air.

SATCC   Southern African Transportation Co-ordinating Commission.

Satco, SATCO   Senior air traffic control officer.

satcom[s]   Generalized term for satellite communications.

Satcoma   Satellite Communications Agency (US).

SATCP   Sol-air très courte portée (F).

satellite   *1* Body revolving in equilibrium orbit around primary, natural or man-made.
  2 Man-made device intended to become * (1).
  3 Military airfield auxiliary to nearby main airfield and relying on latter for admin. and most services.
  4 Sub-terminal at airport to disperse processing and bring passengers nearer relevant gates.

satellite landing system   Based on a DGPS ground station, currently provides Cat.1 to equipped aircraft [DGNSSU plus interfaces] to all runways within 30 nm [56 km]; later growth to Cat.III.

satellite/pier layout   Airport terminal is connected to some satellites (4) and some piers.

satellite telephone intermediate unit   Cabin interface between satcom and terrestrial telephone [or telecommunications] avionics with CTU (2).

satelloid   Satellite whose orbit is within planetary atmosphere and thus requires continuous or intermittent thrust.

SATF   *1* Strike and terrain-following (radar).
  2 Shuttle activation task force (also rendered Sataf).

Satin   *1* SAC automated total-information network (USAF).
  2 Survivability augmentation for transport installation-now (Lockheed-Georgia ECM).

Satka   Surveillance, acquisition, tracking and kill assessment.

Satnav   Satellite navigation, ie satellite-assisted.

Satnet   Satellite network.

SATO   Scheduled airlines ticket office.

satphone   Satellite telephone, especially in aircraft.

Satrack   Satellite tracking, first GPS system.

SATS   *1* Small airfield for tactical support (USMC).
  2 Small-arms target system.
  3 Shuttle avionics test set.
  4 Small-aircraft technology [changed from transportation] system, to relieve pressure on hubs (NASA).

SATSAR   Satellite-aided SAR study group (ICAO).

Satsim   Saturation countermeasures simulator (USAF).

SATT   Small-aircraft training target.

saturable reactor   Soft-core inductor control for pulsed radars and magnetic amplifiers.

saturated adiabatic lapse-rate   Rate of decrease of temperature with height for parcel of saturated air.

saturated air   Air containing greatest possible density of water vapour, such that in any given period number of molecular break-ups equals number of recombinations; RH (2) = 100%.

saturated beacon   Ground transponder beacon, eg

DME, interrogated by so many aircraft (usually 100 simultaneously is limit) that its AGC cuts out replies except to 100 strongest interrogators.

saturation   Measure of airport traffic: current movements [usually per hour] ÷ maximum allowable.

saturation diving   Undersea submergence at depths in order of 300 m for prolonged period when blood is saturated with chosen breathing mixture.

saturation vapour pressure   Vapour pressure of particular substance, variable with temperature, which at given temperature is in equilibrium with plane surface of same substance in liquid or solid phase.

Saturn   *1* Second-generation anti-jam tactical uhf radios for NATO.

   2 The interoperable waveform for (1).

SAU   *1* Safety/arming unit.

   2 Signal acquisition unit.

   3 Surface-attack unit.

saunter   Air-intercept code: "Fly for best endurance".

sausage   *1* Kite balloon (colloq., arch.).

   2 According to one authority, a windsock.

saumon   Salmon (F).

Sauter mean diameter   Diameter of droplet having same surface : volume ratio as an entire liquid spray, measured in μ.

SAvA   Society of Aviation Artists (UK, 1953–).

SAVAC   Simulates, analyses, visualizes activated circuitry (Chrysler/USAF).

Savasi   Simplified abbreviated Vasi.

save   Rescue of a downed pilot, esp. behind hostile lines.

save-list item   Item of equipment to be salvaged from aircraft at 'boneyard' site.

SAVR   Strapdown attitude/velocity reference.

SAW   *1* Surface acoustic wave; / CAD adds chemical-agent detection.

   2 Society of Aviation Writers (several countries).

   3 Submerged-arc welding.

   4 Special Air Warfare (USAF).

sawcut   Chordwise slot in leading edge of aerofoil to promote chordwise flow, energize boundary layer or serve other aerodynamic function.

SAWE   *1* Society of Allied [previously Aeronautical] Weight Engineers (US).

   2 Simulated area-weapons effects; can have suffixes NBC or RF.

SAWHQ   Shape alternative war HQ.

SAWOS   Semi-automatic weather observation system.

SAWRS   Supplementary, or supplemental, aviation weather-reporting station (NOAA).

SAWS   Satellite, or silent, attack warning system.

SAWSS   Shipboard aircraft weight subsystem.

sawtooth   *1* Voltage or timebase which when plotted has appearance of saw edge with series of linear-rate climbs and near-vertical descents.

   2 See *dogtooth*.

   3 Flight profile of motor-glider.

sawtooth nozzle   Propulsive nozzle of jet engine, especially turbofan fan duct and/or core, terminating in a zig-zag edge. This typically gices c3 dB reduction in takeoff noise. Also called sawtooth mixer.

say again   Please repeat last bit of message.

Saybolt   Standard test for viscosity of liquid, esp. lubricating oil, in which sample heated to known temperature is poured through calibrated orifice and time in seconds recorded for 60 ml (cm³); hence *SUS*.

SB   *1* Service bulletin.

   2 Sideband.

   3 Scrieve board.

   4 Sonic boom.

   5 Speedbrake(s).

   6 Scout Bomber (USN aircraft category 1934-46).

   7 Snow began, followed by time.

   8 Southbound.

   9 Spot beam.

   10 Side of body (casting).

sb   Stilb[s].

SBA   *1* Standard beam approach, pioneer electronic landing aid providing lateral (azimuth) and distance (marker-beacon) guidance; led to ILS.

   2 Small Business Administration (US, from 1953).

   3 Smaller Businesses Association (US, from 1953).

   4 Spot-beam antenna.

   5 Serial-bus analyser.

   6 Space-based assets.

SBAC   Society of British Aerospace Companies (1916-1970 The Society of British Aircraft Constructors).

SBAMS   Sea-based air master study.

SBB   Single-beam blanking.

SBC   *1* Senate Budget Committee (US).

   2 Single-board computer.

   3 Sonic-Boom Committee (ICAO).

   4 Small bayonet cap.

SB-comp   Service-bulletin computerization.

SBD   Schematic block diagram.

SBE   Single-bit error.

SBF   Svenska Ballong Federationen (Sweden).

SBG   Stand-by gyro.

SBGS   Sonic-boom ground signature.

SBH   Support by the hour.

SBHN   Solar-blade heliogyro nanosat.

SBI   Space-based interceptor, for disabling hostile BV, PBV, RV, Asat, etc (SDI).

SBIR   Small-business innovation, or innovative, research.

SBIRS   Space-based IR system; HP adds Hi-Program, L adds Low [replaced by STSS].

SBJ   Supersonic business jet.

SBKEWS   Space-based kinetic-energy weapon system.

SBKKV   Space-based kinetic-kill vehicle.

SBL   *1* Space-based laser; BMD adds ballistic-missile defense.

   2 Scanned-beam laminography.

SBM   *1* Space battle management.

   2 Sinter-bonded mesh.

   3 Scheduled base maintenance (DARA).

SBMC3, SBMC³   SBM(1) Command, control and communications.

SBMCS   Sea-based mid-course system [ballistic-missile defence].

SBN   Strontium barium niobate.

SBNPB   Space-based neutral-particle beam.

SBO   *1* Sideband[s] only.

   2 Specific behavioural objective.

SBR   *1* Space-based radar.

   2 Signal-to-backround ratio.

SBS   *1* Smart/small/space-based bomb system.

   2 Satellite business systems.

3 Self-contained booster stage.

SBSS   *1* Standard-base supply system (USAF).

2 Space-based soldier system (USA).

3 Space-based surveillance system (USAF).

SBT   Self-briefing terminal.

SBTC   Sino-British Trade Council.

SBTDS   Sea-based terminal-defence system (USN).

SBUV   Solar backscatter UV.

SBW   *1* Steer [nosewheel] by wire.

2 Search bubble window, giving view vertically downwards.

SBX   Sea-based X-band radar, crucial element in BMD system testbed.

SBY   Standby.

SC   *1* Single-crystal.

2 Speed-control (system).

3 Short circuit (also s.c.).

4 Subgrade code (ICAO).

5 Service ceiling.

6 Stratified [or shaped] charge.

7 Statement of capability.

8 Structured coupling.

9 Single-card.

10 Special committee, or category.

11 Surface combatant, = warship.

Sc   *1* Stratocumulus.

2 Schmidt number.

S/C, S/c   *1* Spacecraft.

2 Step[ped] climb.

3 Supercharged.

S$_c$ Unit compressive stress.

SC-1   One engine inoperative.

SC-2   All engines operating.

SCA   *1* Single-crystal alloy.

2 Services de la Circulation Aérienne = ATC (F).

3 Simulation control area.

4 Supercritical compound aerofoil [or airfoil].

5 Short-circuit analysis.

6 Self-contained approach [blind landing aid].

7 Software communications architecture.

8 Strategic-capabilities assessment.

scab   External payload carried flush against pylon or aircraft skin; hence to *-on, scabbed.

SCAD, Scad   *1* Subsonic-cruise armed decoy.

2 Often ScAd, Scientific Advisor (NATO).

3 Single-channel amplifier/detector.

4 Stock control and distribution.

Scada   Supervisory control and data acquisition.

SCADC   Standard control (or, USAF, central) air-data computer.

Scads   *1* Shipboard containerized air-defence system (BAe).

2 Simultaneous calibration of air-data systems.

SCAF   Supply, control and accounts flight (RAF).

scalar   Quantity having magnitude only, as distinct from vector.

scale   To reproduce on different scale of size, esp. to produce gas-turbine engine larger or smaller than original but broadly similar aerodynamically.

scale altitude effect   Compressibility error, so called because at high speeds ASIR over-reads as if it were at lower altitude.

scaled   Basically unchanged in design but smaller or larger than original.

scale effect   Sum of all effects of change in size of body in fluid flow, keeping shape same; more specif., effect of alteration in Reynolds number.

scale factor   Output for given input; eg for accelerometer** is output current in mA per unit of applied acceleration g.

scale model   Model forming exact miniature of original.

scaler   Electronic counter producing one output pulse for given number of inputs; thus binary * has scaling factor of 2 and decade * has SF of 10.

scale strength   Relationship between actual structural strength of model and its size and same relationship for full-scale aircraft or other item; hence scale stiffness.

scaling factor   Number of input pulses for one scaler output pulse.

scaling law   Mathematical equations permitting effects of NW explosion of given yield to be determined as function of distance from GZ provided that corresponding effect is known for reference explosion, eg 1 kt.

scalloping   VOR bearing error due to distortion of propagation over uneven terrain; also called bends.

Scalp   Système de croisière conventionnel autonome à longue portée (F).

scalping   Rough-machining surface layers off ingot.

SCAM, Scam   Strike camera.

SCAMA, Scama   Switching, conferencing and monitoring arrangement (NASA).

SCAN, Scan   *1* Surface-condition analyser.

2 Simulated comprehensive air navigation.

3 Self-correcting automatic navigator.

scan   *1* Motion of electronic beam through space searching for target (see * *types*).

2 In EM or acoustics search, one rotation of sensor; this may determine a timebase (NATO).

3 Air intercept code: "Search sector indicated and report any contacts".

4 In TV and other video systems, process of continuously translating scene into picture elements and thus varying electrical signal(s).

5 To make one complete cycle or sweep of eyes across either selected flight instruments or external scene ahead.

SC&D   Stock control and distribution.

scandium   Sc, silvery metal, density 3.0, MPt 1,541°C, could become important in aerospace.

scanner   *1* Device which scans (usually 1 or 4), including electron beams in image tubes and CRTs, TV cameras and many electronic display systems, but esp. including radar transmitter/receiver aerials (antennas) which are scanned mechanically or electronically.

2 Radio receiver add-on which trawls the airband.

scanner column   Vertical member carrying two or more superimposed radar aerials, eg in nose of C-5B.

scanning   *1* Process by which radar aerial scans, either by physical rotary movement (usally driven hydraulically) or by electronic scanning.

2 Process by which electron beam scans, accomplished by electrostatic or electromagnetic plates or coils.

3 Action of keeping eyes sweeping over external scene and/or flight instruments and other internals.

4 In particular, searching sky ahead visually, either front-to-side or side-to-side, to avoid a mid-air.

5 Use of scanner (2).

scanning-beam MLS   See *MLS*.

scanning DME　DME that scans stations and locks on to strongest signal without pilot action.

scanning field　Area, usually rectangular, scanned by electron beam in TV camera, image tube, CRT or electronic display.

scanning generator　Timebase controlling scanning (2).

scanning sonobuoy　One whose acoustic sensors (either passive or active) scan to give directional information and/or to filter out noise from sources other than target.

scanning spot　Point of light where electron beam strikes face of CRT or other scanning field.

scan period　Time period of basic scan types other than conical and lobe-switching, or period of lowest repetitive cycle in more complex combinations; basic units in US are °/s, mils/s or s/cycle.

scan stealing　Appropriating a main scan for writing symbols and alphanumerics when interscan period is too brief; term not recommended.

scan transfer　Sudden switching of scanning (3) from external to internal scene or vice versa, esp. immediately before bad-visibility touchdown.

scan types　There are many varieties of radar scan (1) patterns. What follows is descriptive of motions of centre-line axis (boresight line) of main lobe only. Each * is associated with particular type(s) of display, to which aerial (antenna) az/el information is supplied by potentiometer or other servo system. Fixed-scan radar points in one direction only. Manually controlled, points where directed. Conical, traces out circular path forming small-angle cone with radar at apex; a variation, spiral, traces spiral path beginning and ending at centre of circle. Sector, scans through limited az angle (unidirectional, from L to R or R to L only, snapping back to start after each scan; bidirectional, L-R-L-R). Circular, rotates continuously in horizontal plane (common AWACS mode). Sector display is circular scan with long-persistence phosphor in one [important] sector. Helical, scans continuously in az while winding up and down elevation from 0° to 90° and back. Palmer is conical scan superimposed on another, eg Palmer-circular or Palmer-sector covers 360° periphery or arc respectively with conical scans. Raster is TV method of horizontal lines, usually interlaced; eg an air-intercept 6-bar raster might scan to R along line 1 (top), to L along 4, to R along 3, to L along 6, to R along 2, to L along 5 and back to start. Palmer-raster is another air-intercept scan with conical scan along two or more horizontal bars; thus 3-bar PR makes quick rings along top bar, next along middle and back along bottom, then back via middle to top. Track while scan (TWS) uses an az radar and an el radar simultaneously scanning in both planes.

SCAR, Scar　*1* Strike control and reconnaissance, co-in mission (USAF).

2 Strike co-ordinating armed reconnaissance.

3 Sistemi de Control de Armamento (I).

scarf　*1* Inclination in vertical plane of cutoff from rod, tube or other section.

2 Inclination in vertical plane of engine inlet or nozzle, thus zero-* = vertical.

scarf cloud　Thin cirrus draping summit or anvil of Cb.

scarfed　Cut off at an oblique angle; hence * joint, inlet, nozzle etc.

Scarff ring　Standard British cockpit mount for hand-aimed machine gun 1917-40 with ring-mounted elevating U-frame.

scarf joint　Structural joint, invariably in wood, in which mating members are given flat taper to give large glued/pinned area.

scarifying　Increasing coefficient of friction of airfield surface by cutting shallow grooves, simultaneously removing rubber and other unwanted residues.

Scarlet　Solar concentrator array with refractive linear element technology.

Scart　Syndicat des Constructeurs d'Appareils Radio-récepteurs et Téléviseurs (F).

scar weight　Weight penalty remaining [from brackets, cables etc] when mission-specific features are removed.

SCAS　Stability and control augmentation system.

Scat, SCAT　*1* Supersonic civil air transport.

2 Space communications and tracking.

3 Scout/attack helicopter.

4 Speed command of attitude and thrust.

5 Special-category [or single-contractor] aviation training.

6 Security control of air traffic; ANA adds and air navigation plan, ER adds and EM radiation (US DoD, 1952).

7 Satellite control of air traffic [oil rigs].

Scatana　Security control of air traffic and air-navigation aids; special provisions and instructions in time of defence emergency (FAA).

ScATCC　Scottish ATC Centre.

Scatha, SCATHA　Spacecraft charging at high altitude[s].

Scat-I　Special-category I, capability for Cat.I landings provided by SLS to all runways within radius of [usually] 30 nm, 56 km.

scatter　*1* Distribution, either ordered or, more usually, random of measured values about mean point (eg of 1,000 measures of wing span of similar type aircraft).

2 Distribution of impact points of projectiles aimed at same target.

3 See *scattering*.

scattered cloud　Seldom used; cloud amount reported only in octas, see *SCT (1)*.

scattering　*1* Diffusion of radiation in all directions caused by small particulate matter in atmosphere; effect varies according to ratio betwen wavelength and particle size; when this exceeds about 10 Rayleigh scattering occurs (see *back-*, *tropospheric **).

2 Trajectory changes of sub-atomic particles caused by collisions of various interactions; can be elastic or (if there is energy transfer) inelastic.

scattering loss　That part of transmission loss due to scattering (1) or to target's rough surface.

scattering power　Ratio of total radar power scattered by target to total power received at target; also called scattering cross-section.

scatterometer　Carried by satellite or aircraft to measure light reflected from ocean surface to give information on local wind.

scatter point　Geographic point where race competitors cease to be constrained to narrow take-off corridor.

scatter propagation　See *back-scatter*, *tropospheric scatter*.

scatter tolerance　That allowed on dimensions of die forging, often measured at random locations.

scatter weapon   One releasing or dispensing many bomblets or mines.

scavenge oil   Lubricating oil on its way back from the lubricated part, also called return oil.

scavenge pipe   Carries lube oil from machine, eg engine, to tank. In US, often scavenger.

scavenge pump   Pumps lube oil out of machine, in case of engine fitted with wet sump, from base of sump. Gas turbines generally have several scavenge (return) gears on same shaft as pressure gears. In US, often scavenger.

scavenge(r) system   Exit ducting from wind tunnel for removal of contaminants, eg from smoke apparatus, combustion products from burning tests or exhaust from combustion devices.

SCC   *1* Standing Consultative Commission (arms control, ABM treaty).

    2 Security consultative committee.

    3 Sector control, or command, centre.

SCCI   System controllers and cockpit indicators.

SCCOA   Système de commandement et de conduite des opérations aériennes (F).

SCCS   Source-code control system (software).

SCD   *1* Speed computing display; airspeed (often TAS) needle plus Mach counter.

    2 Signal command decoder.

    3 Specification control drawing.

    4 System category diagram.

SCDA   Software cost-driver attribute.

SCDDS   Sensor control-data display set.

SCDL   Surveillance and control data-link.

SCDU   *1* Selective control decode unit (IFF).

    2 Satellite control data unit.

SCE   *1* Signal conditioning equipment.

    2 Single corporate entity.

    3 Spacecraft command encoder.

scenaric computer   One able to assemble visual scenes, eg in Tepigen.

SCEPS   Stored chemical energy propulsion system.

Sceptr   Suitcase emergency procedures trainer (cheap erasable ROM).

SCF   *1* Satellite control facility.

    2 Single-configuration fleet.

    3 Stress-concentration factor.

SCFN   Spherical convergent flap nozzle.

SCG   Speed-control governor.

SCH   *1* Sonobuoy cable hold (autopilot selector).

    2 Simplified combined harness.

schedule   *1* Precisely controlled mechanical movements to meet system demands, carried out automatically by system usually provided with feedback; eg variable inlets and nozzles in supersonic airbreathing engine installation.

    2 Preplanned sequence of time events, eg aircraft inspections and overhauls or timetabled civil flights.

scheduled service   Air-carrier service for any kind of payload run to timetable.

scheduled speed   Any of type-specific speeds published in flight manual, eg V_S, V_{AT}; in no way connected with speed in commercial use, which is defined as block speed.

scheduling   Numerical, analog or graphical description of sequence of scheduled (1) movements, eg of turbojet inlet spike.

schedule inventory   List of all safety equipment and other removable items carried on board.

schematic diagram   Drawing which explains functions and general spatial relationships but which uses standard symbols and makes no attempt to portray visual appearance. Often includes only one subsystem, rest of machine, aircraft or other device being in outline or phantom line only. Alternatives are exploded drawing, block diagram.

Schlichting   Original (1936) theory treating of flat plate in supersonic flow.

schlieren   German word (nearest English equivalent is 'striations') for various shadowgraph-like techniques for optical investigations based on 1859 method of Foucault. Basic feature is small light source, parallel rays of light through region under investigation and opaque cut-off at focus of second lens projecting image on screen or photographic film (either point source and pinhole cut-off or line source and line cut-off). Variations in density, eg in flow through shock-waves, Prandtl-Meyer expansions and supersonic flow generally, are sharply visible as tonal gradations.

Schmidt camera   Elegant astronomical camera/telescope with objective in form of thin plate of glass and rear concave spherical mirror focusing on curved film. Objective plate has one surface figured (thicker and convex at centre, thin and concave around periphery) to correct mirror's aberration, its own chromatic aberration being slight because of small thickness.

Schmidt duct   Pioneer flap-valve pulsejet.

Schmidt number $Sc = \mu/\rho D_{12}$ where μ is viscosity, ρ is density and D_{12} diffusion coefficient; ratio of viscous and mass diffusivity, or kinematic viscosity divided by mass diffusivity.

Schottky defect   Atom missing from crystal lattice.

Schottky diode   Barrier-layer device based on rectification properties of contact between metal and semiconductor due to formation of barrier layer at point of contact.

Schottky effect   Small variation in electron current of thermionic valve caused by variation in anode voltage affecting work done by electrons in escaping.

schräge Musik   Oblique music = jazz, code name for night-fighter armament of upward-firing cannon (G, WW2).

Schuler pendulum   One whose length equals radius of Earth, and thus when carried in vehicle moving near Earth's surface always indicates local vertical. In practice any pendulum having same period of approximately 84 min., achieved at particular relationship between c.g. and pivot, such that centre of rotation of pendulum is always at centre of Earth. Used in stable platforms of INS.

Schuler tuning   Adjusting period of Schuler pendulum so that its centre of rotation exactly corresponds with centre of Earth.

Schultz-Grunow   Standard treatment for turbulent flow in viscid fluid at R from 10^6 to 10^{10}.

Schwarm   Two Rottes, fighters in two loose pairs (G).

SCI   *1* Smoke curtain installation, for laying smokescreen.

    2 Switched collector impedance.

    3 Secure [or sensitive] compartmented information (DoD).

    4 Scalable coherent interface.

    5 Serial communication interface.

SCIA   Spacecraft checkout and integration area.

SCID   Software configuration index drawing.

SCIDA System co-ordinating installation design authority.

SCIDM Single-cord improved data modem.

science pilot Experienced researcher in a scientific discipline subsequently qualified as pilot, eg of Space Shuttle (NASA).

Scimitar System for countering interdiction missiles and target-acquisition radar[s].

scimitar wing One whose planform is curved; usually means same as crescent wing but has been applied to early wings curved across whole span with 'sweep' at tips and zero sweep at root.

Scinda Scintillation network decision aid (USAF space command).

scintillation *1* See *glint* (radar).

2 Rapid and random variation in appearance of small light source viewed through atmosphere, esp. variation in luminance.

3 Brief light emission by single event (eg impingement on phosphor) (see * *counter*).

scintillation counter Instrument for measuring alpha, beta or gamma radiation by counting scintillations (3); also called scintillator or scintillating counter.

scintillation spectrometer Scintillation counter plus pulse-height analyser for radiation energy distribution.

scintillometer Photoelectric photometer for measurement of wind speed near tropopause by various Schlieren-type measures of stellar scintillation. Also called scintillation meter.

SCIRP Semiconductor IR photography.

SCISE Self-contained in-seat entertainment.

scissor Any flight manoeuvre made by coarse rudder at low airspeed. See *scissors*.

scissor lift Platform mounted on two pairs of pivoted arms giving true vertical motion without any tilt. Hence, scissor jack, scissor drive.

scissors Flight manoeuvre performed by two or more pairs of aircraft crossing at angle like two halves of *; basically a series of turn reversals intended to make following enemy overshoot.

scissor wing Wing made as single plane from tip to tip pivoted to fuselage at mid-point; usually synonymous with slew wing.

SCIU Code for radio-altitude indicator or reading.

SCL Space-charge limited (C adds 'current').

sclerometer Hand instrument measuring hardness by load needed to make scratch of standard depth with diamond point rotated along arc.

scleroscope Hand instrument measuring hardness by rebound height of hard-tipped (steel or diamond depending on pattern) rod-hammer dropped inside tube from standard height on to surface. Common commercial model is Shore *.

SCM *1* Single-crystal material (or metal).

2 Single-chip microprocessor.

3 Software configuration management.

4 Spoiler control module.

5 Surface contamination module.

6 Silicon carbide monofilament.

7 Self-contained munition.

8 Supply-chain management.

SCMR *1* Surface-composition mapping radiometer.

2 Surface-combatant maritime rotorcraft (UK RN).

SCN *1* Specification change notice.

2 Satellite control network.

3 Self-contained navigation [S adds system, U adds unit].

SCO Sub-carrier oscillator.

SCOB Scattered clouds or better.

Scoff Society for Conquest of the Fear of Flying (US).

SCOMP, Scomp Secure communications processor.

scooter bogie One with two wheels only, in tandem.

SCOPE Simple checkout program (language).

Scope Spacecraft operational-performance evaluation.

scope *1* Electronic display, esp. one supplying output of radar (colloq.).

2 Generalized term for optical viewing instrument, eg microscope, CRT, etc (colloq., vague).

3 According to Webster, distance within which missile carries; unknown in aerospace.

Scope Command From 2002 the replacement for all previous h.f. ground communications (USAF).

scopodromic Headed in direction of target.

Score *1* Signal communications by orbital relay experiment.

2 Stratified-charge omnivorous [ie, multifuel] rotary engine.

3 Supplier cost-reduction effort.

Score defect Damage caused by movement of penetrative item across softer surface.

scoring Apart from normal use (eg in military exercises), the process by which a customer numerically evaluates bidders' proposals.

Scot Satellite-communications on-board (ship) terminal.

scotopic Vision with retinal rods associated with extremely low light intensity and detection of gross movement.

Scott Single-channel objective tactical terminal.

scout Single-seat aircraft, armed for air combat, operating in patrol or reconnaissance role (WW1).

SCP, scp *1* Spherical candlepower.

2 System-concept paper.

3 Single-card processing.

4 Spacecraft control processor.

5 Supersonic camera pod (RAAF).

SCPC Single channel per carrier [or single carrier per channel] alternative to TDMA; mobile ISDN.

SCPI Supersonic-cruise propulsion integration.

SCPL Senior commercial pilot's licence (no longer issued).

SCPU Suite central processing unit.

SCR *1* Silicon-controlled rectifier.

2 Signal/clutter ratio.

3 Stratified-charge rotary (engine).

4 Single-channel radio.

5 Selective chopper radiometer.

SCRA *1* Single-channel radio access.

2 See *SCRIA*.

scramble *1* Take off as quickly as possible (usually followed by course and altitude instructions) (DoD). Any urgent call for military (usually combat) aircraft to take off and leave vicinity of base, either as training manoeuvre or for sudden operational reason.

2 To attempt to provide telecom security by rendering transmission unintelligble to third parties (ie by scrambling), eg by speech inversion.

scrambled egg   Gilt rim round peak of Service-dress cap of group captain and Air Ranks (RAF, colloq.).

scramble pan   Dispersal from which fighter can make immediate takeoff.

scramjet   Supersonic-combustion ramjet; one in which flow through combustor itself is still supersonic.

Scram/Lace   Supersonic combustion ramjet, liquid air cycle engine (NASA).

scrap   *1* Workpiece containing defect, even trivial, sufficient to cause it to fail an inspection.

2 To break up old unwanted aircraft or one beyond economic repair.

3 Residue of (2) other than produce or salvaged items.

scraper ring   *1* Spring piston-ring with sharp-angled lower periphery for removing oil from cylinder wall.

2 Middle ring on squadron-leader badge of rank (RAF, colloq.).

scrap view   Small inset showing detail (eg item from different viewpoint or in different configuration) added where room permits on main drawing.

scratchbuilt   Replica of historic aircraft containing no authentic parts.

SC/RC   Stratified-charge rotating-combustion engine.

SCRE   Syndicat des Constructeurs de Relais Electriques (F).

screaming   Undefined term descriptive of high-frequency combustion instability in rocket characterized by high-pitched noise (see *screeching*).

screeching   Undefined term describing high-frequency combustion instability in rocket or afterburner character-ized by harsh, shrill noise more irregular than screaming.

screeding   Levelling and smoothing resin adhesive in bonding operation; hence * tool.

screen   *1* Imaginary obstruction having form of level-top wall normal to flightpath which aeroplane would just clear on takeoff or landing with landing gear extended and wings level.

2 Arrangement of ships, submarines and aircraft for protection of ship(s) against attack (DoD).

3 Wire-mesh gauze or sieve (ASCC).

4 Electrically conducting or magnetically permeable enclosure which shields either contents or exterior against unwanted magnetic/electrical fields.

5 Face of electronic display, eg CRT, TV, projection system, etc.

screen burn   See *ion burn*.

screen captain   Senior training or supervisory pilot in role of examiner. May be employee of certification authority.

screened   Provided with screen (4), thus * ignition has enveloping earthed conductive covering to prevent escape of R/F interference.

screened horn   Balance surface entirely downstream of fixed surface.

screened pair   Twin electric cable incorporating earthed screen (4).

screener   Airport X-ray machine.

screen filter   Fluid filter whose element is a fine metal mesh screen.

screen grid   Screen (3) between anode and control grid in thermionic valve to reduce electrostatic influence of anode.

screen height   Height above ground of top of screen (1), normally 35 ft (10.67 m), occasionally 10 m and rarely 50

ft, on takeoff; on landing usually 30 ft (often interpreted as 10 m). Depends upon aircraft performance group and particular case considered.

screening   Meanings include: 1 Examining candidate as fit to handle classified information. 2 Airport security examination of passengers and baggage. 3 *Screen* (4).

screen navigator   Carried in navigator training aircraft to ensure safe return despite pupil errors.

screen speed   Speed at moment aeroplane passes over screen (see *screen height*), either assumed or as target value, on takeoff or landing. Does not have V-suffix abbreviation.

screw   Generalized term for threaded connector rotated into workpiece and not held by nut; many quick-fasteners have *-thread and no sharp dividing line is possible.

screw gauge   Hand instrument for measuring major diameter of screw, typically by graduated line scales forming small-angle notch.

screwjack   Actuator having rotary input and linear output obtained by screwthread, often with interposed recirculating balls.

screw pitch gauge   Hand instrument for checking thread on metalworking screw or bolt, usually with selection of blades each having one 'threaded' edge.

SCRIA   Supply-chain relationships in aerospace.

scrieve board   Portable board on which lofting lines are recorded either by scribing or some other undeformable method. Often used with locating blocks for actually assembling flight hardware, eg frames, thus becoming a jig. Becoming obsolete.

SCRJ   Supersonic combustion ramjet.

scroll   Any curved duct or guidance channel, especially leading air out of centrifugal compressor or cooling air across face of turbine disc.

SCRT   Single-channel receiver/transponder.

scrub   *1* To abandon project, esp. a planned flight or military mission.

2 To eliminate a pupil from course of instruction, following failure of * check.

scrubbing   *1* Lateral sliding of landing-gear tyre on hard pavement, eg of inner bogie wheels in sharp turn.

2 Significant contact between tips of rotating members, esp. blades, in gas turbine and casing; also called rubbing.

3 Rapid wear in piston engine caused by detonation.

scrubbing torque limit   Maximum permitted torque imposed on landing-gear leg by scrubbing (1), which usually determines limit of steering angle of nose gear, and hence minimum turn radius.

scrub check   Last-chance assessment of pupil pilot by senior instructor (usually CFI).

SCS   *1* Stabilization control system.

2 Speed control system.

3 Sea control ship.

4 Survivable control system.

5 Single-crystal sapphire (IR domes).

6 Society for Computer Simulation (US).

7 Signal Corps Set (USA, WW2).

8 Satellite communication system.

9 Single-channel simplex.

10 Slaved compass system.

11 Space Control Squadron.

SCSC   Strategic conventional standoff capability.

SCS-51   Signal Corps Set 51, original form of ILS.

SCSI   *1* Small computer system[s] interface.

2 Simulation Computer Society International [AC adds Advisory Council].

3 Single-card serial interface.

SCT   *1* Scattered cloud, also **SCTD**, CAA = 3 to 4 oktas, ICAO = ⅛-⅘ , FAA = 0.1-0.5.

2 Scanning telescope (NASA).

3 Surface-charge transistor.

4 Single-channel transponder.

5 Staff continuation training (CAA).

6 Seat and canopy trainer.

SCTI   Service Centrale des Télécommunications et de l'Informatique (F).

SCTR   Sector.

SCTV/GDHS   Spacecraft TV ground data-handling system.

SCU   *1* Signal conditioning unit.

2 Signal convertor unit (satcoms).

3 System[s], switching, stores, sensor, station, supplemental, or secondary control unit.

4 Satellite communications unit.

SCUC   Satellite Communications Users Conference (Int.).

Scud   Subsonic-cruise unarmed decoy.

scud   Shredded or fragmentary cloud, typically Fs, moving with apparent greater speed below solid layer of higher cloud.

scuff plate   Protects airframe against impacts from GSE.

scupper   Fuel-tight recess around gravity filler, usually with its own drain.

scuttle   *1* Hatch in top of fuselage (US usage).

2 According to Webster: 'An airport. *Brit.*'; unknown to author.

SCV   Sub-clutter visibility.

SCWA   Single-channel wire access.

SCWG   Satellite communications working group.

SD   *1* Shaft delivery, ie rotary output.

2 Shipping document.

3 Structured design (software).

4 Service dress (RAF).

5 Specification detail.

6 System display, or device.

7 Storm detection (NWS, ARTCC).

8 Side display.

9 Service deviation, or Standard deviation.

10 Self-destruct, or destroying.

11 Radar weather report (ICAO).

12 Surveillance drone.

S$_D$ Distance between centres of ground-contact areas of diagonally-opposite wheels of a bogie.

SDA   *1* System, or sister, design authority.

2 Strategic-defense architecture.

SDAC   System-data analog converter.

SDACS   Solid-fuel divert and attitude-control system.

SD&D   System development and demonstration (USAF).

SDAS   Source-data automation system[s].

SDAT   *1* Sector design and analysis tool (ATC).

2 Silicon-diode array target.

SDAU   Safety Data and Analysis Unit (CAA, UK).

SDB   *1* Small-diameter bomb.

2 Small disadvantaged business.

SDBY   Standby (ICAO).

SDC   *1* Synchro-to-digital converter.

2 Signal data converter, or computer.

3 Space Defense Center (Colorado Springs, USAF).

4 System data capture.

5 Shuttle Data Center.

6 Satellite data communications; S adds system.

7 Shaft-driven compressor.

8 Supersonic-dash capability.

9 Strategic Defense Command (USAF).

10 Situation-display console.

SDD   *1* Standard disk drive.

2 System design and development.

3 System development and demonstration [now SD&D].

4 Synthetic dynamic display.

5 Sensor data degradation.

6 Situation data display; -T adds Tower-Brite.

SDDM   SecDef decision memorandum (US).

SDE   *1* Spatial-database engine.

2 Software development environment[s].

3 Scatter detection enhancement.

SDF   *1* Single degree of freedom.

2 Simplified directional facility (FAA).

3 Self-destruct fuze.

4 Stepdown fix.

5 Strategic deterrent forces [previously RVSN] (R).

SDFOV   Simultaneous dual field of view.

SDG   Speed-decreasing gearbox.

SDI   *1* Strategic (sometimes rendered as Space) Defense Initiative (US); O adds Organization, P Program, PO Participation Office.

2 System discharge indicator.

3 Selective dissemination of information (telecommunications).

4 Source destination identifier.

SDIO   SDI Organization, or Office (US).

SDIP   SDI Program.

SDIPO   SDIP Office (UK).

SDIS   Small-diameter imaging seeker.

SDLC   Synchronous datalink controller.

SDLF   Shaft-driven lift fan.

SDLM   Standard depot-level maintenance.

SDLV   Shuttle-derivation launch vehicle.

SDM   *1* Scatter-drop mine (or munition); helicopter weapon.

2 Site-defense of Minuteman; anti-intruder system.

3 System definition manual.

4 Speaker drive module.

5 Ship-defense missile.

6 Sum in depth of modulation.

7 Space division multiplex[ing].

8 Structural dynamics model.

SDMA   Space-division multiple-access.

SDME   Software development and maintenance environment.

SDMI   Strategic distribution management initiative (USTC/DLA).

SDMS   *1* Software development maintenance system.

2 Shipboard data-multiplex system.

3 Support defense missile system.

4 Sensor to decision-maker to shooter.

SDN   System descriptive note.

Sdn Bhd   Sendirian Berhad (company constitution, Malaysia).

SDNRIU   Secure digital net radio interface unit.

SDO   Serial digital output.

SDOF   Single degree of freedom.

SDOM   Six degrees of motion (simulator).

SDP   *1* System-definition phase.

   2 Signal-data processor.

   3 Standard datum plane.

   4 Surveillance data processing.

SDPDS   Surveillance data-processing and distribution system[s].

SDPR   Federal directorate of supply and procurement (Jugoslavia).

SDPU   Sensor-data processor unit.

SDR   *1* System design report (or review, or responsibility).

   2 Service difficulty report.

   3 Signal data recorder.

   4 Special drawing rights; assist airlines in inter-line fare transactions.

   5 System development requirement.

   6 Strategic Defence Review (UK).

   7 Supplier data requirements [L adds list].

   8 Software-defined radio.

   9 Signal-to-distortion ratio.

SDRAM   Synchronous dynamic RAM.

SDRL   Subcontract, or supplier, data requirements list.

SDRS   Splash-detection radar system.

SDS   *1* Software development system.

   2 Satellite, or secondary, data system.

   3 Strategic defense system (US, SDI).

   4 Small digital switch.

   5 Short-distance sensor.

SDSAWG   Software Development Standards and Ada Working Group.

SDSMS   Self-defense surface missile system.

SDT   System development tool.

SDTI   Système de drones tactiques intérimaire (F).

SDU   *1* Selective decode unit (SSR).

   2 Safety data unit (CAA).

   3 System design utility.

   4 Satellite, or Service, data unit.

   5 Smart, or sensor, display unit.

   6 Signal [ATC] distribution unit.

   7 Secure-data unit (JTIDS).

SDVP   System demonstration validation program(me).

SDW   Symmetrical double-wedge aerofoil.

SE   *1* Support equipment.

   2 Systems, or safety, or sustaining, engineering.

   3 Servicing echelon.

   4 Snow ended.

   5 Storage element.

   6 Synthetic environment.

S/E   Single-engine[d].

Se   Selenium.

SE^2   Software engineeering environment.

SEA   *1* Single-engine asymmetry.

   2 Stored-energy actuator.

   3 Software engineering architecture.

   4 South-East Asia region (ICAO).

sea/air/land team   Group trained and equipped for unconventional and paramilitary operations including surveillance/reconnaissance in and from restricted waters, rivers and coastal areas (DoD).

Sea ALL   Sea airborne lead line, UAVs for naval perimeter defence (USN).

sea anchor   Anchor for marine aircraft in deep water, typically canvas or rubberized-fabric bucket or drogue trailed from stern.

Seabed Treaty   Prohibition of NW, 11 February 1971.

Seabee   Construction Battalion (USN).

sea bias   See *water bias*.

sea breeze   Onshore wind during day.

SEAC   *1* Support-Equipment Advisory Commitee (inter-airline).

   2 South-East Asia Command (WW2).

SEAD   Suppression of enemy air defence(s).

sea disturbance   See *sea state*.

seadrome   Floating aerodrome for refuelling transoceanic aeroplanes [unrealised idea 1931–39].

Seafac   Systems engineering avionics facility.

sea fog   Fog over sea usually caused by moist air over cold water.

SEAGA   Selective employment of ground and air alert (SAC).

Seagull   EDP for flight simulators using distributed computers linked by reflective-memory bus (Rediffusion/Gould).

SEAL   Sea/air/land.

sea lane   Area of water with fixed markers showing its use by marine aircraft.

sealant   Material tailored to particular duty of ensuring fluid-tight (liquid or gas) seal between mating materials; often applied as continuous layer, eg inside integral tank.

Sealdrum   Portable, collapsible rubber container for POL, water and other liquids (US Royal).

sealed   Officially closed, eg by lock wire with lead seal, until completion of particular flight or test programme.

sealed-balance control   Flight-control surface whose gap between the leading edge [ahead of the hinge] and the fixed surface is made airtight by flexible strip.

sea level   *1* Actual height of sea surface, used as local height reference.

   2 More often, height corresponding to pressure of 1,013.25 mb.

   3 In UK, referred to mean high-water at Newlyn, Cornwall.

sea-level corrections   See *STP (1)*.

seam   Joint in fabric, eg overlap or three fell-types.

sea marker   Anything dropped on sea to provide visible indication of drift.

sea-motion corrector   See *sea-surface correction*.

seam weld   Continuous, hopefully pressure-tight, weld made by roll electrodes; in theory can be accomplished by close repeated spot welding or by hand process, but unusual.

SE&I   Systems engineering and integration.

Seapac   Sea-activated parachute automatic crew release.

seaplane   *1* Float seaplane, marine aircraft having one, two or (rarely) more separate floats and conventional fuselage (UK).

   2 Marine aircraft of any fixed-wing type (US).

seaplane basin   *1* See *towing tank*.

   2 Some dictionaries define this as a place having sheltered water for seaplanes.

seaplane marker   Buoyant or bottom-fixed marker at marine aerodrome, also called taxi-channel marker.

seaplane rating   Endorsement on licence qualifying holder to fly or maintain seaplanes.

seaplane tank   See *pitching tank*, *towing tank*.

seaplane trim   Angle between mean water surface and

aircraft longitudinal axis or other reference when parked on water.

sear   Generalized term for catch, pawl or other latching device which holds breechblock or bolt of gun at open position.

sea rating   See *sea state*.

search   *1* Systematic reconnaissance of defined area such that all parts pass within visibility.

2 DME mode: emit interrogation pulse pair and scans whole operating area for reply.

search and rescue facility   One responsible for maintaining search and rescue service for persons and property in distress.

search jammer   Automatic jammer.

searchlight   Any powerful directional surface light for illumination of cloudbase or other aerial objects.

searchlight scanning   Sector scanning (see *scan types*).

search mission   Air reconnaissance to search for specific surface object(s).

search radar   Vague; generally means surveillance radar operating in search mode, and often applied to weather radar.

search rate   Reciprocal of average time to search one PN chip-width, ie chips/s (ECM).

sea return   See *clutter*.

Sears-Haack   Profile of body having minimum supersonic drag, normally body of revolution having pointed ends and fineness ratio appropriate to Mach number; complicated by addition of aerofoils.

SEARW   Single-engine advanced rotary wing.

seasat   Generalized name for oceanographic satellite.

sea smoke   See *steaming fog*.

sea state   Condition of sea surface as related to standard list of reference conditons, invariably Beaufort scale.

sea-surface correction   Electronically applied correction to Doppler output to remove false velocities due to motions of waves.

SEAT   *1* Site-equipment acceptance test.

2 Status evaluation and test.

Seat   Single-engine air tanker (USFS).

SEATO   Pronounced Seato, SE Asia Treaty Organization, 1954-76.

seat pack   Parachute pack worn in such a way that it forms seat cushion.

seat selection   Offering customer for particular commercial flight right to select seat from those remaining, subsequently shown on boarding pass.

seat-type parachute   One with seat pack.

SEB   *1* Staphylococcus enterotoxin B, BW agent.

2 System[s] electronics box.

Sebass   Spatially enhanced broadband Army spectrographic system (USA).

SEBRW   Single-engine basic rotary wing.

SEC   *1* Secondary emission cathode.

2 Secondary-electron conduction.

3 Securities and Exchange Commission (US).

4 Secondary (NASA).

5 Spoiler and elevator computer.

6 Security [problems] working group (ECAC).

7 Special-event charter.

8 Section, or sectional aeronautical chart.

9 Sector.

10 Survivable engine control.

11 Software-enabled control (AFRL-Darpa).

Sec   Seconds (contrary to SI but still ICAO).

Secad   SEC (10) algorithm development (NAWC).

Secam   Safety, efficiency and capacity of ATM (7) methodologies (Euret).

Secant   Separation and control of aircraft, non-synchronous techniques.

secant modulus   E, slope of stress/strain curve to any point; thus equal to modulus of elasticity up to limit of proportionality, thereafter varying with applied stress to point of rupture.

SecDef   Secretary of State for Defense (US).

SECL   Symmetrical emitter-coupled logic.

Seco, SECO   S-IVB engine cut-off (Apollo).

second   *1* SI unit of time, symbol s, = 9,192,631,770 transitions between two ground states of Cs-133 atom.

2 Unit of plane angular measure, symbol " = 4.848 × 10^{-6}.rad.

secondary   Small area of low pressure on periphery of large cyclone.

secondary airflow   That used to dilute and cool flame in gas-turbine combustor.

secondary airport   Smaller airport at city possessing hub airport.

secondary battery   Electric battery rechargeable by reverse DC current.

secondary bending   In a beam or column whose chief applied stress is transverse, that bending moment due to axial load.

secondary controls   See *secondary flight controls*.

secondary depression   See *secondary*.

secondary display   Simplified display of flight instruments enabling aircraft to be safely flown following failure of primary.

secondary electron emission   Flow of electrons from metal surface under bombardment by high-energy electrons or protons.

secondary emission   Ejection of subatomic particles and/or photons from atoms or particles subjected to primary radiation, eg cosmic rays.

secondary explosion   Explosion at surface target caused by air attack but additional to explosions of air-dropped ordnance.

secondary fan airflow   Airflow through fan which does not pass through engine core.

secondary flight controls   Those used intermittently to change lift or speed but not trajectory; eg flaps, slats, Krügers, airbrakes, droops and, except in DLC, spoilers.

secondary frequency   Assigned to an aircraft as standby in air/ground.

secondary front   One formed within an air mass.

secondary glider   Not defined, but training glider more advanced than primary and with enclosed cockpit. Generally any glider intermediate between primary and Standard Class sailplane.

secondary great circle   See *meridian*.

secondary heat-exchanger   That which rejects heat to atmosphere from secondary circuit heated by primary heat-exchanger.

secondary holes   Those admitting secondary air to combustor.

secondary instruments   *1* Those giving information unconnected with gross flight trajectory, ie not a primary flight instrument.

2 Those whose calibration is determined by comparison with an absolute instrument.

secondary members   See *secondary structure*.

secondary modulus   Modulus of elasticity for composite of other two-component material (esp. two-metal components) beyond point at which weaker material yields.

secondary nozzle   Annular nozzle surrounding primary nozzle of jet engine through which may pass cooling airflow, inlet excess and various other flows around engine.

secondary power system   Mechanical power system on board aircraft other than main engines, eg shaft-driven accessories, gearboxes, gas-turbine starter, APU, EPU or MEPU, inter-engine cross-shafting and major bleed ducting with air-turbine drives if fitted.

secondary radar   Radar in which interrogatory pulse is sent to distant transponder which is triggered to send back a different pulse code to originator. Examples are airborne DME triggering ground DME facility and ground SSR triggering airborne transponder.

secondary radiation   Usually synonymous with secondary emission.

secondary stall   Rotating stall [engine].

secondary stress   That resulting from deflection under load, eg of end-loaded column in bending.

secondary structure   Structural parts of airframe whose failure does not immediately imperil continued safe flight.

secondary surveillance radar   Ground radar which in-terrogates air traffic with identifiable codes of pulses, triggers distinctive response from each target, extracts plots and assigns identity to each, normally presented as flight number, altitude and other information beside target on radar display [Mode A provides coded target identiy, C altitude and S selective interrogation]. Aerial normally slaved in azimuth to main surveillance (primary) radar and may be on-mounted.

secondary winding   That of transformer, magneto or other electrical device from which output is supplied.

second buy   See *option (1)*; originally Lockheed term for option with paid deposit.

second day   In all-out war, assumption that enemey's long-range air defences are destroyed.

second dicky   Co-pilot (UK, colloq.).

second-line servicing   That carried out over planned period when aircraft is out of line service, sometimes at special off-base facility.

second moment of area   Moment of inertia of a structural section whose mass is unity; SI m^4 = 115.86183 ft^4, cm^4 = 0.024025 in^4.

second pilot   *1* Person designated as second pilot in flight crew to assist PIC; in commercial crew usually has rank of First Officer. In ASCC definition: 'not necessarily quali-fied on type'.

2 Unofficial term for passenger who has completed short [usually 8–10h] flying course to enable him or her to land light [⩽12,500lb] aircraft following incapacitation of pilot.

second segment   Second segment of normal takeoff for large or advanced aerooplanes beginning at gear retraction at V_2 and maintaining this speed in climb to top of climb at end of segment when aircraft is levelled out, or climbed less steeply, to accelerate to FUSS or for power cutback in noise-abatement procedure, which see.

second source   Manufacturer assigned by government to augment output of major hardware item, eg aircraft or missile, with assistance of original design company, which remains in production as first source; no royalty is normally payable, and ** has no commercial rights to design, nor permission to sell to third party.

second strike   Strategic attack with NW mounted after enemy's nuclear attack has taken place; objective of hardening is to confer a ** capability. DoD: 'The first counterblow of a war; generally associated with nuclear operations'.

second throat   That downstream of working section and upstream of exhaust diffuser in simple blow-down super-sonic tunnel. In operation traversed by weak normal shock.

second-trace return   Caused by large echoing target outside range scale.

Secop   Single-engine climb-out procedure.

Secor   Sequential collation of range. Usually * /DME, which collates range with distance derived from DME. A basic technique in global mapping with geodesic satellites.

secretress   GA stewardess also serving as office secretary in flight.

SECS   Sequential-events control system.

SECT, Sect   Simulator for electronic-combat training.

Sect   Sector.

section   *1* Cross-section.

2 Major portion of airframe, eg nose*, tail*, but becomes dangerously ambiguous with wing*, body* when meaning normally (1).

3 Small subdivision of military air unit, normally (DoD) two combat aircraft.

4 Raw material rolled or extruded to standard (often complex) cross-section, as distinct from sheet, billets or strip.

5 Major subdivision of missile or rocket vehicle, eg guidance*.

6 Subdivision into similar parts, eg six sections of slat.

sectional   Noun, VFR navigation chart, equivalent to ICAO 'half-million' (US).

section modulus   Moment of inertia of structural member cross-section divided by perpendicular distance from neutral axis to outermost surface of section, ie most highy stressed fibre.

sector   *1* Subdivision of air-defence frontier.

2 Limited range of azimuth, eg through which radar may scan (see * under *scan types*).

3 Subdivision of airspace by radio range characterized by letter A or N, also called quadrant.

4 Portion of commercial route between two traffic points.

sector controller   Air-defence controller in charge of sector (1).

sector display   See *scan types*.

sector distance   Length of air route sector (4).

sector fuel   That allowed for in flightplan for one sector (4).

sector management tool   Software providing a traffic-management unit with ability to maintain sector integrity through use of ground delays.

sector scan   See *scan types*.

sector temperature/wind   Those met values assumed in flight-planning one sector (4).

sector time   In commercial operation, scheduled or actual time for sector (4).

sector visibility   Within particular sector (2) of horizon seen from tower.

SECU   Spoiler[s] electronic control unit.

secular   Having a very long time period, eg a century or more.

secure   Proof against interference by enemy and esp. against information content of signals being deciphered by enemy. Various shades of meaning, eg * air refuelling is one beyond enemy radar range in which all communication is by lamp.

Security   Over 20 terms relating this topic to aerospace all appear to be self-explanatory.

SED   *1* Safe escape distance; minimum radius from airbase at which aircraft tail-on is assumed safe against hostile NW with GZ at airbase.

2 Scanning electron diffraction.

3 Sensor evolutionary development.

4 Systems engineering documentation.

5 Secondary Eicas display.

6 Society of Engineering Designers (UK).

SEDF   Surface-emitting distributed feedback.

SEDIS, Sedis   Surface-emitter detection identification system (ESM).

Sedris   Synthetic-environment data representation [and] interchange specification.

SEE   *1* Society of Environmental Engineers (UK).

2 Secondary electron emission.

see and avoid   Basic onus on pilots in VMC to maintain lookout.

Seebeck effect   Generation of EMF or current by dissimilar metals in circuit with junctions at different temperatures (see *Peltier*).

seeding   *1* Aerial dispensing, at controlled rate per unit time or unit distance, of ECM payload such as chaff, flare pellets or other dispersed medium including aerosols. Hence * rate, * distance.

2 Aerial dispensing, at controlled rate per unit time, of condensation nuclei such as crystals of silver iodide or dry ice (solid CO_2) to trigger precipitation in rainmaking.

seeing   *1* Colloq.uially used to mean ability of radar to reach highly reflective part[s] of target.

2 Quality of observability (astronomy).

seeker   Device able to sense radiation from target, lock-on and steer towards it, using radar (active or semi-active), optics (usually passive), laser or IR; rarely other methods. Normal sensor for terminal guidance of missiles and other guided ordnance.

Seenot   Air/sea rescue, prefix to longer words (G).

see-saw rotor   See *teetering*.

SEF   Stability enhancement functions.

Sefis, SEFIS   Simulated electronic flight-instruments system.

SEFT   Section d'Etudes et Fabrications des Télécommunications (F).

SEG   Signaller, engineer, gunner (RAF aircrew grade, WW2).

segment   *1* One of six standard subdivisions of take-off flightpath: 1, screen to gear-up at V_2; 2, gear-up to top of initial climb at V_2; 3, level acceleration to FUSS; 4, flaps-up to 5-min. power point at FUSS; 5, level acceleration to en route climb speed; 6, climb to 1,500 ft at ERCS; rest is en route climb.

2 A second definition is: one of two subdivisions of climbout, first-* from lift-off to location of noise listening

post or airport boundary, second-* from that point with power cut back in drift-climb until sensitive area has been overflown or a declared height is reached. See *noise-abatement climb*.

3 Portion of en route flight between two waypoints.

4 Portion of missile flight, eg boost, cruise (or midcourse) and terminal.

SEGS   *1* Selective (inclination) glideslope.

2 Standard-entry guidance system.

SEI   *1* Small-engine instruction.

2 Standby engine indicator (or instrument) (CAA). In other usage E is engineer.

3 Specific emitter identification.

4 Software Engineering Institute (US).

SEID   Système d'écartometrie IR différentielle (F missile guidance).

SEIFR   Pronounced safer, single-engine IFR.

SE-IMC   Single-engine instrument meteorological conditions.

SEIRS   Small-engine IR suppressor.

SEL   *1* Selcal (ICAO).

2 Space Environment Laboratory (NOAA).

3 Single-engine licence.

4 Single-engine, landplane.

5 Select, selector identifier.

6 Switchable eyesafe laser.

7 Sound exposure level, see *noise*.

8 Surface-emitting laser.

9 Sun/Earth Lagrange point; 1, 2, respectively denote near or dark side.

Selcal   Selective calling, system enabling ground controller to call a single aircraft, usually long-haul oceanic airline or bizjet, without the crew having to listen out on that frequency or any other aircraft being bothered. Selcal code for each aircraft is four-letter code using letters A through M, triggering light or loudspeaker. Airborne decoder usually accepts inputs from VHF or HF. See *Adsel*, *DABS*.

Selecta-vision   Registered name for holographic video recording.

select code   That code displayed when ground interrogator (SSR, or in US ATCRBS) and airborne transponder are operating on same mode and code simultaneously.

selected track   In traditional US-style airways structure, link route off-airways authorized to particular flight equipped with R-nav. Where R-nav systems (eg Decca) in use this term has little meaning because commercial and IFR pilots are not confined to airways and VOR radials.

selective address   See *selective interrogation*.

selective availability   Management of the GPS by the US DoD so that civilian receivers cannot have access to the [22-m, 72-ft] accuracy of P(Y).

selective fading   Distortion of signal caused by variation in ionospheric density which causes fading that varies with frequency.

selective feathering   Manual feathering using selector to pick out correct propeller.

selective fit   Non-standard and rare engineering fit selected by hand to achieve desired mating of parts.

selective identification facility (or feature), SIF   Airborne pulse-type transponder which provides automatic selective identification of aircraft in which installed (NATO);

early form of selective interrogation used with Mk X ATCRBS.

selective interrogation   With automated ATC systems, once aircraft acquired by computer, interrogation necessary only as radar scans exact sector containing that aircraft; if each transponder has its own (discrete) address code, computer can order interrogation on that code in particular sector, so transponders reply only when selected, number of replies cut by perhaps 99% and interference negligible (saturation avoided); two current systems are Adsel (UK) and Dabs (US).

selective jamming   Jamming on single frequencies; often synonymous with spot jamming but can cover several specif. frequencies.

selective loading   Arranging cargo load to facilitate unloading in desired sequence.

selective pitch   Early term for variable pitch.

selectivity   Measure of ability of radio receiver to discriminate between wanted signal and interference signals on adjacent frequencies.

selector   Manual demand input offering choice, eg * valve may have four or more fluid-flow positions, flap * may offer four or more settings, while landing-gear * normally offers only two.

selenium   Se, non-metallic crystalline element, density 4.8, MPt 217°C, high electrical resistance except when illuminated, thus used in photocells, resistance bridges and, with semiconductor properties, in rectifiers.

seleno-   Prefix, of the Moon: hence *-centric orbit, *-id (lunar satellite), *-logy (study of Moon).

self-aligning bearing   Ball (occasionally other forms) bearing with outer race having part-spheroidal track allowing variation in shaft attitude.

self-bias   See *automatic grid bias*.

self-contained night attack   Capability of single aircraft navigating by night to acquire and strike designated point target and return to base (USAF omits word 'point').

self-destroying fuze   One designed to explode projectile before end of its flight.

self-destruct   Ability of missile, RPV or similar air vehicle to explode into harmless fragments at particular time or geographical location.

self-erecting   Of gyro, capable of erecting automatically after being toppled or started from rest.

self-exciting   Generator's own current supplies magnet coils.

self-FAC   To carry out entire surface-attack-mission without help of FAC or other aircrft giving aiming directions.

self-forging fragment   Warhead which punches through thin top armour of AFVs: transverse disc of explosive converts disc of dense metal ahead of it into streamlined globule moving at over 9,000 ft (2,750 m)/s, its impact energy exceeding shear strength of armour.

self-generated multipath effect   Multipath radio/radar problems caused by change in aircraft configuration, eg carriage of large stores.

self-guided missile   One guided by means other than command.

self-induced vibration   Caused by internal conversion of non-oscillatory to oscillatory excitation; also called self-excited vibration.

self-induced wing rock   Sustained roll, with little or no yaw, in flight at high AOA by slender delta or aircraft with

wing of low aspect ratio, caused by asymmetric shedding of LE vortices.

self-inductance   Ratio of magnetic flux linking circuit to flux-producing current, ie e.m.f induced per unit rate of change of current; symbol L, unit henry, also webers per ampere; analogous to inertia in mechanical system.

self-manoeuvring stand   Airport apron arranged so that gate parking and departure are accomplished by aircraft taxiing under own power. Normally nose-in parking avoided, and gate parking is accomplished with fuselage approximately parallel to adjacent wall of finger or terminal.

self-noise   Internally generated noise within system, esp. within sonar.

Selfoc   Sheet electric-light focusing; patented (Nippon) two-component optical-fibre system.

self-repairing   Having inbuilt ability automatically to adjust itself to best possible condition (eg operative channels, gain, feedback, power management) following any malfunction or damage.

self-rescue system   Patented (Bell Aerospace) jet-propelled foldable parawing to enable combat aircrew to eject and fly to friendly area.

self-sealing tank   Fluid, esp. hydrocarbon fuel, tank constructed with layer of soft unvulcanized synthetic rubber sandwiched between main structural layers so that combat damage causes leak, bringing fuel into contact with rubber, which swells swiftly to block hole.

self stall   Self-induced stall brought about by inherent tendency when near stalling AOA to pitch-up, eg by stall well outboard on swept wing with root still lifting strongly, thus progressively increasing nose-up tendency.

self-sustaining speed   Gas-turbine rpm [typically 35% maximum] at which, during start cycle, external cranking is no longer needed.

self-test   Sequential test program performed by equipment (usually electronic) upon itself to determine whether it is operating correctly and which subsystems or circuits are faulty. In advanced form pinpoints each fault to particular device or PCB and stops.

selint   Selective interrogation; not a particular system but technique.

SELR   Switchable eyesafe laser ranger [or rangefinder]; D adds designator.

SELS   Severe local storm[s].

selscan   Selective-call scanning.

selsyn   From self-sychronous, patented (US General Electric) synchronization system for transmitting remote compass indication to cockpit; fluxvalve current fed to three 120° coils driving central coil carrying indicator needle.

selvedge   Woven edge of fabric from mill, does not unravel (US = salvage).

SEM   *1* Scanning electron microscope, microscopy.

   2 Space environment monitoring (or monitor).

   3 Service engineered [in UK, engineering] modification.

   4 Standard electronic [or system equipment] module.

   5 Superconducting electrical machine (or machinery).

   6 Society for Experimental Mechanics (US).

SEMA   *1* Special electronics mission aircraft (USAF).

   2 Station engineering management aid (RAF, computerized maintenance management).

   3 Smart electro-mechanical actuator.

4 Syndicat des Equipements et Matériels Aéronautiques (Paris).

semaphore TVC   Thrust-vector control using oscillating blades pivoted to move across propulsive jet transversely and by partial blocking of flow path cause angular deflection.

SEMC   Standard electronic-memory cartridge.

SEM-E   Standard electronic module Format E.

semi-active homing   Homing on to radiation reflected or scattered by target illuminated by radar forming part of system but mounted on fighter or surface platform, ie, not carried by the homing vehicle.

semi-active landing gear   One in which the damping forces are controlled in response to the aircraft motion.

semi-active missile   Missile whose guidance system includes a receiver only, to home on a target illuminated by a radar, laser, or other emitter located elsewhere.

semi-angled deck   Carrier deck whose axis is at maximum diagonal angle permitted by original hull and deck without addition of large overhanging portion.

semi-annular wing   Wing whose front elevation forms lower half of circle centred on fuselage or on each of two engine nacelles.

semi-armour-piercing   Important category of gun-launched projectiles with AP properties conferred by high muzzle velocity and choice of materials but not relying entirely upon KE for penetration and with internal explosive charge. Some have added incendiary or tracer.

semi-automatic gun   One which ejects spent case and reloads but fires only upon command; same as repetition of single-shot.

semi-automatic machine tool   Usually, one that must be commanded to begin cycle of operations but thereafter runs through that cycle to completion, thereupon awaiting next start order.

semi-cantilever   Not defined; loose term often meaning a braced cantilever, as in many light-plane wings.

semicircular separation   System of allocating cruise height in Airways and ⩾24,500 ft (7468 m) amsl outside controlled airspace in IFR (outside UK, in most European countries VFR also). Typically magnetic tracks 000°–180° have one set of assigned tracks and 181°–360° have the remainder. In US *Hemispheric Rule* applies.

semiconductor   Electronic conductor whose room-temperature resistivity lies between that of metals and insulators (say, in range 10^{-2}-10^9 ohm/cm) and which compared with metals has very few charge carriers, energy bands being full or empty (except for a few electrons or holes excited by heat in intrinsic * or provided by impurity doped in extrinsic *); n-type * has free-electron (negative) charge carriers and p-type has free-hole (positive) charge carriers.

semi-controlled mosaic   One made up of photographs on approximately same scale so arranged that major ground features match geographical co-ordinates.

semi-hardened   Given some protection against nuclear attack, eg buried in ground but without concrete, or enclosed in concrete but not buried; aircraft shelters in NATO are an example at bottom end of hardening scale.

semi-levered bogie   One whose rear bogie beam can be locked to the front half before takeoff to increase tail clearance on rotation.

semiotics   Imparting information by signs and symbols.

semi-monocoque   Structure in which loads are carried part by frame/stringer combination and part by skin. Almost all modern fuselages are of this type.

semi-ogive   Sliced in half longitudinally [see ogive].

semi-permanent runway   One paved with prefabricated metal by any of several standard methods.

semi-rigid airship   Non-rigid with a single longitudinal member to assist in maintaining shape of envelope and distribute loads.

semi-rigid rotor   Main lifting rotor of helicopter in which there are no hinges but flexible root behaves as leaf spring in vertical and horizontal planes to allow some freedom in flapping and lag modes.

semi-rigid theory   Approximate treatment for elastic structure in which only finite number of degrees of freedom are allowed, each with one fixed mode.

semi-sonic blading   Rotor blading (normally of axial compressor or fan) in which tips are just supersonic.

semi-span   Half wingspan, theoretically measured from centreline but occasionally measured from side of fuselage, esp. in calculating wing bending modes. Symbol b/2.

semi-stalled   Stalled over portion of total aerofoil while remainder continues with attached lifting flow.

semi-transparent photocathode   Radiation, eg light, on one side produces photoelectric emission on other.

SEMMS   Solar electric multimission spacecraft.

SEMP   System(s) engineering management plan.

Semtex   Plastic explosive, PETN plus syrene-butadiene copolymer.

Senap   Signal emulation of navigation and landing [translation from Czech].

Sender   Self-navigating drone, expendable/recoverable (USN).

SENEAM   CAA (1) Mexico.

Senel   Single-event noise exposure level (see *noise*).

Sengap   Small [jet] engine advanced program (Darpa).

Sengo, SENGO   Senior Engineering Officer (RAF).

senior pilot   Second-in-command of RN air squadron (UK).

sense aerial   Fixed vertical receiver aerial with output same as maximum obtained from D/F or ADF loop and in phase with loop only over 180° to resolve 180° ambiguity of basic radio D/F method.

sense indicator   Direct-reading to/from readout on VOR/ILS and similar panel instruments.

sensible atmosphere   That part offering measurable aerodynamic resistance.

sensible horizon   Circle of celestial sphere formed by its intersection with plane through eye of observer and perpendicular to local vertical.

sensing   Removing ambiguity by sense aerial.

sensitive altimeter   Pressure (aneroid) instrument more accurate than simple altimeter with aneroid stack, corrective adjustments and three-needle or counter-pointer readout; not as accurate as servo-assisted.

sensitivity   Generalized term for output divided by input, eg instrument or system response per unit stimulus.

sensitivity-level command   Instruction to TCAS [any type] to control threat volume.

sensitivity time control, STC   Automatically reduces gain of weather radar as aircraft approaches cloud to avoid near clouds appearing brilliant and distant clouds faint; normally operates within radius of 45 km/25 nm.

Senso   Sensor operator aboard ASW aircraft.

sensor   According to DoD: 'A technical means to extend

man's senses; equipment which detects and indicates terrain configuration, presence of military targets, and other man-made and natural objects and activities by means of energy emitted or reflected by such objects; energy may be nuclear, EM (visible and invisible portions of spectrum), chemical, biological, thermal or mechanical, including sound, blast and vibration'. Definition could be even broader; any transducer converting an input stimulus into a usable ouput. It is suggested there are three classes:

1 Input device for detecting and measuring vehicle motion and air data.

2 Device for graphically illustrating a remote object.

3 Device for detecting and precisely locating a target.

sensor fusion   Automatic combination of outputs from different types of sensor into a single display.

Sentac   Senso (tactical).

Sentai   *1* Division or flotilla comprising aircraft of one or (usually) two carriers (Japanese Navy, WW2).

2 Basic combat unit equivalent to UK wing or US group (Japan).

SEO   Station Engineering Officer (RAF).

SEOS   Stabilized electro-optical system.

SEP   *1* Specific excess power.

2 Single-engine performance.

3 Separate, separation.

4 Safety and emergency procedures.

5 Strategic equity partner.

6 Single-event phenomenon.

7 Spherical-error probability.

8 Secondary electric power.

Sepak   Suspension of expendable penaids by kite.

separated flow   Flow no longer attached to surface of immersed body.

separated lift   Vortex lift.

separation   *1* Breakdown of attached fluid flow round body into gross turbulence, occuring at particular time (stall) or place (* point); possible to have sustained equilibrium with attached (laminar or turbulent) flow upstream and complete * downstream.

2 Authorized lateral, longitudinal and vertical clearances (distances) between aircraft under positive control.

3 Severing of links between rocket stages or other fallaway sections, also called staging.

4 Time when (3) occurs.

5 Discharge, release from active duty (USAF).

6 Distance, along any axis or direction, between interceptor and target.

7 Periphery of ground sheet, where it lifts from the surface.

separation manoeuvre   Energy-gaining manoeuvre at low-alpha, high thrust, to close (reduce) or extend (increase) separation in air combat.

separation minima   Minimum longitudinal, lateral or vertical distances by which aircraft are spaced through application of ATC (1) procedures (FAA).

separation motor   Thruster to assist separation (3).

separation point   In 2-D flow, point at which velocity of boundary layer relative to body becomes zero and flow separates from surface.

separation standards   ICAO term for separation (2) minima.

separation test vehicle, STV   Air vehicle for assisting

development of separation (3), esp. of tandem or wrap-round boost motors.

SEPC   Secondary electric-power contactor.

SEPD   Standard for the exchange of product data.

SEPDS   Secondary electric power distribution system.

Sepla, SEPLA   Sindicato Español Pilotos Lineas Aéreas (Spain).

SEPM   Scanning electric-potential microscope [differs from SEM].

SEPP   Stress evaluation prediction program.

SEPS   *1* Solar electric propulsion system (or stage).

2 Supplemental electric power system.

SEPSTO   Single-engine protected short takeoff.

SEPT   Synthetic environmental procedures [or procedural] trainer.

SEQ   Sequence.

sequenced doors   Landing-gear doors close after gear has been extended.

sequenced ejection   *1* Automatic small delays are built in between events, eg canopy, stick, calf garters, seat, drogue, harness release etc.

2 Ejection from multiseat aircraft in which crew members are fired in close-spaced series, captain or aircraft commander last.

sequence valve   Fluid-flow controller scheduled to perform series of actions in sequence, each completion starting that following. Common US term: sequencer.

sequencing   Assignation by ATC or radar controllers of strict order in which aircraft under control are to proceed, eg by selecting arrivals from holding points and, with path-stretching if necessary, achieving correct time/distance separation as they join localizer.

sequential collation of range, Secor   Long-base-line system for determining vehicle trajectory by phase-comparison of responses of vehicle transponder to interrogation by three ground stations.

sequential computer   Connected in series with other equipment, eg SSR or air-defence radars, eg to predict conflicts and advise on courses of action.

SER   *1* Service, served, serving.

2 Stop [at] end of runway.

3 Serial number, or series.

4 Snap experimental reactor.

SERB   *1* Selective Early-Retirement Board.

2 Space-Experiments Review Board.

SERC   Science and Engineering Research Council (UK).

SERD   Support-equipment recommended data.

SERE   Survival, evasion, resistance and escape (US joint services).

SEREB, Sereb   Société pour l'Etude et la Réalisation d'Engins Balistiques (F).

Serf   Studies of the economics of route facilities (ICAO).

serial   *1* Element or group of elements within series given numerical or alphabetical designation; also that designation (DoD, NATO).

2 Numerical identity of particular hardware item, eg aircraft. Usually displayed on item concerned and recorded in all events concerning item (eg entered in pilot's log book, used as radio callsign in US and many other air forces).

3 In sequence as distinct from parallel; hence * data, * wiring.

serial data   Successive signals passed over single wire or channel.

serial number   See *serial (2)*.

serial rudders   Rudder made in front and rear portions, latter hinged to former and deflecting through greater angle (eg on Dash 7). Also called serially hinged.

series   *1* General term for subdivision or group within a larger related group, eg aircraft type Halifax II Series 1A, in this case corresponding to block number, modification state and other national terms.

2 In routine sequence as in manufacture of successive identical articles, eg * production, * aircraft; in this context often redundant word.

3 Mathematical expression with sequence of terms having form $a_1 + a_2 + \ldots a_n$.

4 Connected in succession on same line, wire or channel and thus all carrying same signal, current or flow. Thus a turbine bearing, rotor disc and OGV may be cooled by a single airflow in *.

series burn   Consecutive burns of single or multiple rocket motors, eg on Space Shuttle Orbiter.

series loading   Addition of inductance in series to increase electrical length of aerial and reduce natural frequency of system.

series modulation   Connection of modulator in series with amplifier.

series/parallel redundancy   Connection of fluid pipes or other lines to give particular item (eg control valves of LMAE) choice of series or parallel redundancy, either on command or automatic and switched by sensed failure.

series production   Manufacture of successive identical (or near-identical) articles.

series redundancy   Connection of two or more similar items, eg control valves, in series so that failure of one does not imperil functioning system.

series resonant circuit   One in which inductances and capacitances are connected in series.

series servo   Servo located in control sytem so that its output adds to that of a major input. Commonly used with SAS actuators to superimpose controls on primary commands without motion at major input.

series yaw damper   One connected into rudder circuit at PFCU, driving surface only but having no effect upstream and thus not felt at pedals; may be operative at all times, including takeoff and landing.

SERL   Services Electronics Research Laboratory (UK, MoD PE, Baldock).

SermeTel   Coating systems [notably Process 2000] comprising aluminium-filled ceramic basecoat [sacrificial] and inert glossy-ceramic top-coat.

SERN, Sern   Single-expansion ramp nozzle.

Serno   Serial number.

serpentine inlet   Shaped to prevent hostile radars from 'seeing' the engine.

Serrate   Family of similar passive receivers carried by night intruders and giving bearing of hostile night-fighter radars (RAF, WW2).

serrated skin joint   Having a sawtooth edge to minimize radar cross-section.

SERT   Space electric (or electrostatic) rocket test.

Service   See *service (2)*.

service   *1* Use, employment for design function; thus squadron *, line * (civil airline) etc.

2 Major branch of national armed forces.

3 To carry out routine maintenance and replenishment.

4 Facility offered to aviators, eg radar *, ATC *.

service area   *1* Geographical extent of coverage of radio navaid or other surface-based electronic system.

2 Part of airfield assigned to routine servicing.

3 Part of airfield dedicated to support services, eg crash/fire/rescue, transport vehicles, trolley and stairway parking, etc.

Service Bulletin   Advisory notice issued by manufacturer of aircraft, engine or equipment alerting operators to actual or predicted faults which require rectification, remedial maintenance or design modification. Some are prefaced mandatory, but, unlike ADs, SBs cannot be legally enforced.

service ceiling   Basic performance parameter for (usually military) aircraft; height which maximum rate of climb has fallen to lowest value practical for military operations, in UK and US traditionally equal to 100 ft/min.

Service Deviation   Temporary permitted deviation from MAR to meet urgent OR covering a modification which should eventually be subject to MAR procedures. Aircraft airworthiness under SD is responsibility of the user Service (UK forces).

service door   Door in aircraft outer skin covering a maintenance or control panel, eg for cargo loading.

service engineering   Function of determining integrity of materiel and services to measure and maintain operational reliability, approve design changes and assure conformance with approved specifications and standards (USAF).

service load   "The total weight of crew, removable armaments and equipment normally carried" (UK usage, 1920–40).

service loads   Structural loads actually met in service.

Serviceman   Member of the armed forces (UK origin); need for Servicewoman is suggested.

service module   Major element of spacecraft supplying secondary power and consumables.

service stand   Place assigned to a particular flight (7); can be gate position or marshalled location on distant apron. Some servicing is normally performed here before departure.

service tank   Fuel tank located near engine to which fuel from other tanks is pumped and from which fuel is supplied to engine (arch.).

Service Technique   See *STAé*.

service test   Test of hardware or technique under simulated or actual operational conditions to confirm satisfaction of military requirements.

service-test model   Model (full-scale item is implied) used to determine characteristics, capabilities and limitations under simulated or actual service operational conditions (ASCC).

service tower   Tower used to afford access to whole length of tall (eg ballistic or Shuttle) vehicle before liftoff; generally synonymous with gantry.

service transport unit   Installation conveying electric and hydraulic (in some cases pneumatic) power and a wide range of liquids across apron-drive bridges to aircraft.

servicing   To carry out service (3).

servicing appraisal exercise   Formal study of servicing of particular hardware item (eg combat aircraft) in simulated combat conditions, to yield job times, difficulties, conflicts and shortcomings and make recommendations (UK usage).

servicing instruction   Issued to remedy or prevent defect in military hardware item when action required may be urgent and recurrent (UK usage). May be issued when defect is suspected but not confirmed; roughly equivalent to Airworthiness Directive.

serving cord   Usually seven-strand machine cord, used for wrapping control-cable splices.

servo   Servomechanism, but now word in own right.

servoactuator   The actuator in a servomechanism.

servo-assisted altimeter   Pressure altimeter in which capsule movement is measured by sensitive EM pick-off whose output is amplified and used to drive motor geared to display.

servo-assisted controls   See *servocontrols*.

servocontrols   Not defined, and not recommended. BSI definition: a control devised to reinforce pilot's effort by a relay. A US definition: a * is practically identical to a trimming tab. Appears to be general vague term for primary flight controls (not mentioned in BSI) where surface deflection is produced by force other than, or additional to, that of muscles. Such added force may come from PFCU or various types of tab (see *servotab*).

Servodyne   Registered (Automotive Products/Lockheed UK) family of pioneer PFCUs with mechanical signalling and hydraulic output.

servoed   Operated by servo.

servo link   See *servo loop*.

servo loop   Conrol system in which human input is amplified by servomechanism provided with feedback so that, as desired output is attained, demand is cancelled.

servomechanism   Force-amplifying mechanism such that output accurately follows input, even when rapidly varying, but has much greater power. Motions can be rotary but usually linear, and can be controlled by input only (open-loop) or by follow-up feedback (closed-loop) forming servo loop. Essential feature is that * constantly compares demand with output, any difference generating an error signal which drives output in required direction to reduce error to zero.

servomotor   Rotary-output machine providing power locally; not necessarily part of servomechanism.

servo optical mechanical   Modelling program for IR ray tracing.

servo rudder   Auxilliary rudder driven directly by pilot's pedals and moving main rudder by twin cantilever beams attaching ** to trailing edge of main surface. Common 1922-38. Precursor of servo tab but not a servo-mechanism.

servo system   Servomechanism with feedback.

servotab   Tab in primary flight-control surface moved directly by pilot to generate aerodynamic force moving main surface.

SES   *1* Surface-effect ship.
2 Shuttle engineering simulator.
3 Software exploitation segment (JSIPS).
4 Single-event signal.
5 Secure equipment system.
6 Stored-energy system.
7 Support-equipment summary.
8 Société Européenne des Satellites (Luxembourg).
9 Single-engine seaplane.
10 Single European Sky.

11 Space environment simulator.

SESA   Society for Experimental Stress Analysis (US).

SESC   Special environmental sample container.

SESMA   Special-event search and master analysis.

SESP   Space Experiment Support Program (USAF).

Sespo   Support Equipment Systems Project Office.

sesquiplane   Biplane whose lower wing has less than half area of upper.

SESS   *1* Space environmental support system.
2 Session.

Sessia   Société d'Etudes de constructions de Souffleries, Simulateurs et Instruments Aérodynamiques (F).

SET   *1* Split engine transportation (fan and core travel as two items).
2 The Space Education Trust (RAeS).
3 Science, engineering and technology.

set   *1* Drift (US, suggest arch.).
2 To place storage device, binary cell or other bistable switch or gate in particular state; condition thus obtained.
3 Complete kit of special tools and/or parts for particular job, eg field modification; also called shop *.
4 Permanent deflection imparted by straining beyond elastic limit, esp. lateral bending of saw teeth.
5 Hand tool/hammer for closing rivets (rare in aerospace).

SETA, SE/TA   Systems [or scientific and] engineering technical assistance.

Setac   German augmented sector-Tacan system used as precision approach aid.

setback   Distance from mould line (edge) and bend tangent line to allow for radius of bend in sheet metalwork.

SETD   Scheduled ETD.

SETE   Supersonic expendable turbine engine.

SETI   Search for extraterrestrial intelligence.

SETL   Single-event threshold level.

SETO   Single-engine takeoff.

SETOLS   Surface-effect takeoff and landing system.

set-on   Assignment of offensive electronics, especially ECM jammer, to counter a particular threat, specifying frequency, signal modulation and, if possible, direction.

SETP   Society of Experimental Test Pilots (Int. but US-based, 2,100 + members, 1955–).

SETR   Specific Equipment Type Rating (NATS engineers).

SETS   *1* Seeker evaluation test system.
2 Severe-environment tape system.

setting   Angle of incidence of wing, flap, tailplane (or trimmer) or other surface.

setting hammer   See *peening*.

setting the hook   Preventing roll reversal in air combat by centralizing stick and using rudder.

settling   Sink of helicopter on takeoff (if pilot does not increase collective) caused by lift ceasing to exceed weight as rotor rises out of ground effect.

settling chamber   Section of wind tunnel upstream of working section in which large increase in cross-section results in great reduction in flow velocity and allows turbulence to die out before acceleration into working-section throat.

SEU   *1* Sensor, or sight, or seat, or system, electronics unit.
2 Stores ejector unit.
3 Single-event upset.

SEV *1* Surface-effect vehicle, usually synonymous with ACV.

2 Severe.

Seval, SEVAL Sensor EW Tactical Evaluator (USN).

seven-bar format Standard format for presenting alphanumeric numerals in electronic displays, all numerals being created by illuminating some of four vertical and three horizontal bars (eg LEDs).

7 by 19 Standard high-strength steel cable made up of seven strands each of 19 twisted wires.

7 by 7 Standard steel cable made up of seven strands each of seven twisted wires.

seven flyings Seven types of flight by dead reckoning, two plane (plane and traverse) and five spherical (composite, great circle, Mercator, middle-altitude and parallel) (US usage, arch.).

720° precision turn Standard US training manoeuvre; two complete circles flown at full power as near as possible constant height at bank angle of 60°.

1760 Standard interface for air-launched stores and other external loads (DoD).

7500 International transponder code 'I am being hijacked'.

Severe Weather Avoidance Plan Approved plan to minimize ATC disruption caused by occasional need to re-route traffic to/through impacted terminal and/or ARTCC areas (FAA).

SEVIRI, Seviri Spinning enhanced visible IR imager.

SEVVA Security Evaluation, Validation and Verification Agency (UK).

Sewaco Sensor weapon command and control system.

SEWS Satellite early-warning system.

SEWT Simulator for electronic-warfare training.

sextant Optical instrument for measuring altitude of celestial bodies.

Seybolt Hand tester of aircraft fabric which measures force required to punch a controlled hole.

SEZ Selector engagement zone.

SF *1* Signal frequency (often s.f.).

2 Scheduled freight.

3 Stick force.

4 Secondary/final talkdown.

5 Special Forces (USA).

6 Standard form.

7 Sampling frequency.

S/F Ratio of system operating time divided by flight hours.

SFA *1* Société Française d'Astronautique (F).

2 Sous-direction de la Formation Aéronautique (F).

3 Sintered ferrite absorber (RAM).

4 Single-frequency approach.

SFACT Service de la Formation Aéronautique et du Contrôle Technique; responsible for civil aircraft and aircrew licensing (F).

SFAR Special Federal Aviation Regulation[s] (US).

SFC *1* Simulated flight cycle[s].

2 Surface (ICAO).

3 Side-force control.

sfc, s.f.c. Specific fuel consumption.

SFCA Service des Fabrications du Commissariat de l'Air (F).

SFCC *1* Side-facing crew cockpit, ie has flight engineer panel.

2 Slat/flap control computer.

SFCS *1* Survivable flight-control system, USAF/McDonnell FBW programme of 1968.

2 Safety flight-control system; protects aircraft (eg Concorde) against excessive AOA or jammed control column.

3 Secondary flight-control system.

4 Simplified fire-control system.

SFD *1* Simple formattable document, immediately ready for message transmission.

2 Special Forces Directorate (UK).

SFDAS Société Française de Droit Aérien et Spatial (F).

SFDB Superplastic forming and diffusion bonding; important manufacturing technique in which structure is welded from sheet into gas-tight envelope and then inflated in heated mould until diffusion-bonded into desired shape.

SFDC Satellite Field Distribution Center (NMC).

SFDF Subsystem fault-detection function.

SFDR Standard flight-data recorder.

SFDS Secondary, or standby, flight-display, or flight-data, system.

SFE *1* Supplier-furnished equipment.

2 Sensor front end; / GA adds gimbal assembly.

SFEA Survival and Flight Equipment Association (UK).

sferics Study of atmospheric radio interference, esp. from met. point of view; sometimes spelt spherics.

SFF *1* Self-forging, or forming, fragment.

2 Svensk Flughistorik Forening (Sweden).

SFFL Standard foreign fare level (IATA).

SF/g Stick force per g.

SFI Special flying instruction (CAA).

SFIRR Solid-fuel integrated rocket/ramjet.

SFK Aramid (spider-fibre) reinforced plastics composite (G).

SFL *1* Safe fatigue life.

2 Sequenced flashing light[s].

SFLOC, SFloc Synoptic filing of location, of sources of atmospherics.

SFM *1* Sensor-fuzed munition.

2 Self-forging munition.

3 Surface feet per minute [machining].

SFO Simulated flameout.

SFOC Space flight operations center.

SFOCS Single fibre optical communications system.

SFOF Space Flight Operations Facility, part of DSN.

SFOR Stabilization Force (NATO).

SFOV Sensor field of view.

SFPA Staring focal-plane array; hence SFPAS adds seeker.

SFPCA Society for the Preservation of Commercial Aircraft (US).

SFPD Smart flat-panel display.

SFPM Surface feet per minute; linear speed measure for machining or grinding.

SFPMAC Société Française de Physiologie et de Médécine Aéronautique et Cosmonautique (F).

SFQL Structured full-text query language.

SFR Stepped-frequency radar.

SFRJ Solid-fuel ramjet.

SFRM Total hours since factory remanufacture.

SFS *1* Side-force surface.

2 Simulator/fallback system.

SFSO   Station Flight-Safety Officer (RAF).

SFSS   Satellite field service station.

SFT   *1* Satellite field terminal.

2 Standard food trolley (loaded, 136 kg, 300 lb).

3 Surface friction tester.

SFTE   Society of Flight Test Engineers; US, with a London-based European Chapter.

SFTS   *1* Service flying training school.

2 Spaceflight telecommunications system.

3 Synthetic flight training system.

SF21   Safe Flight 21, three-year [1999-2001] programme to demonstrate Free Flight, using ADS-B and TIS-B (FAA).

SFU   Suitable for upgrade.

SFUV   Self-filtered ultra-violet.

SFW   Sensor-fuzed weapon(s).

SG   *1* Specific gravity, or s.g.

2 Spheroidal-graphite (cast iron).

3 Screen grid.

4 Shell gun, = cannon.

5 Sortie generation, or sorties generated.

6 Schlachtgeschwader, close-support attack wing (G, WW2).

7 Symbol, or signal, generator.

8 Snow grains.

9 Study group.

s.g.   Specific gravity.

SGA   Silicon gate array.

SGAC   Secrétariat Général à l'Aviation Civile (C adds 'Commerciale') (F).

SGAD   Supersonic global attack demonstrator (US AFRL).

SGC   *1* Symbol-generator computing.

2 Swept gain control.

3 Smoke-generator cartridge.

SGCAS   Study Group on Certification of Automatic Systems.

SGCI   SG cast iron.

SGD   Synthesized, or smart, graphic display[s].

SGDF   Shaft- and gear-driven fan.

SGDN   Secrétariat Général de las Défense Nationale (F).

SGDP   Selected ground delay program(me); departures causing overload at congested arrival fix are manually assigned chosen later departure time.

SGDU   Smart graphic display unit.

SGE   USAF Support Group Europe (RAF Kemble).

SGEMP   System-generated electromagnetic pulse.

SGF   Second-generation Flir.

SGFNT   Significant.

SGI   Silicon graphics image [generator].

SGIT   Special-group inclusive tour.

SGL   *1* Signal (ICAO).

2 Static ground line.

SGLS   Space ground link subsystem (USAF).

SGLV   Second-generation launch vehicle.

SGME   Self-generated multipath effect.

SGML   Standard generalized markup language.

SGN   Standing Group NATO.

SGP   Smart graphics processor.

SGR   Sortie-generation rate.

SG Rep   Standing Group Representative.

SGS   *1* Surface guidance system.

2 Sub-grid scale.

3 Satellite ground system, or station.

SGSI   Stabilized glideslope indicator.

SGT   Satellite ground terminal.

SGU   Signal-generator unit.

SH   *1* Showers (ICAO).

2 Support helicopter.

S/H   Sample and hold; maintains present analog velocity until next sampling.

SHA   *1* System hazard analysis.

2 Sidereal hour angle.

3 Swiss Helicopter Association.

shack   Verb, to score direct hit (colloq.).

shackle   *1* Loosely used in conventional sense to mean link attaching dropped stores to carrier or rack; not recommended.

2 To swap places to enable a pair to exploit tactical situation.

Shade   Shared data environment.

Shadow   Subsonic hovering armament direction and observation window (UAV).

shadow   *1* Wingman ordered to stick close to leader in all circumstances.

2 To duplicate all functions of a manufacturing plant to provide exact second source.

shadow box   Compartmented container for kit of parts (lean manufacturing).

shadow factor   Multiplication factor derived from Sun's declination, target latitude and time in determining object heights from reconnaissance picture shadows; also called tan alt.

shadow factory   Manufacturing plant built and owned by government but managed by selected industrial company to duplicate production of urgently required weapon or product.

shadowgraph   Technique, or photograph made by it, in which point-source light is focused parallel through tunnel working section and on to film; density gradients are visible as changed tonal values, proportional to second derivative of refractive index (Schlieren = first derivative).

shadow-mask tube   Three-colour TV tube with three guns projecting red/green/blue beams through mask with about 500,000 holes.

shadow region   Region where EM signals, eg radio or radar, are poorly received, usually because of LOS difficulties.

shadow shading   Aircraft camouflage (UK, 1936-39).

shadow squadron   Identity which a flying-training unit would assume in war or national emergency (RAF).

Shaef   Supreme HQ Allied Expeditionary Force (NW Europe 1944-45).

Shaft   Smart hard-target attack fuzing technology.

shaft   Transmitter of torque joining two rotating assemblies, such as a turbine and compressor. In a 3-* engine the LP [fan] * passes down the engine centreline inside the IP *, which in turn is surrounded for part of its length by the HP *.

shaft horsepower   Horsepower measured at an engine output shaft, ignoring potentially useful energy in the efflux. Also called torque hp and brake hp. Numerically $2\pi nQ$ where n is rpm and Q drive torque.

shaft power   Power available from rotating shaft, = torque × rpm.

shaft speed   Rate of rotation, rpm or rad s^{-1}.

shaft turbine   Turboshaft engine, gas turbine providing power at an output shaft, in some cases driven by free turbine.

shaker   See *vibration generator*.

shake-table test   Any of various standard test schedules for delicate items conducted on vibration generator to simulate vibration in service (according to USAF 'during launch of missile or other vehicle').

shale fuel   Aviation jet fuel, notably JP-4, derived from oil-bearing shales.

shall   Shipborne [or shipboard] helicopter approach and landing lighting.

Shanicle   Radio guidance system of hyperbolic type used on TM-61B cruise missile 1954: name from short-range navigation vehicle.

shank   Inner portion of some propeller blades where section is not aerofoil but circular.

Shape   Supreme HQ Allied Powers Europe (Int., B-7010 Belgium).

shape   Appearance of electronic (eg radar) signal pulse when plotted in form of amplitude against time. This shape is not related to electronic shaping.

shaped-beam aerial   One emitting beam whose main lobe is shaped by electronic phasing.

shaped charge   Hollow charge; warhead whose target-facing surface has form of re-entrant cone to generate armour-piercing jet.

shaped-charge accelerator   Propulsion system using a shaped-charge propellant, a near-explosive, to achieve highest possible speed of man-made object in atmosphere after prior acceleration by rockets.

shaped sonic boom   Wavefront tailored for rapid attenuation, so that it does not extend to Earth's surface.

shaper   Machine tool having single unidirectional cutter with horizontal (occasionally vertical) reciprocating action.

shaping   Particular electronic meaning is tailoring shape of pulses or signal waveform, eg to assist manual command guidance of RPV or missile.

SHAR, Shar   Shriharikota Range (India).

Sharc   Swedish highly advanced research configuration (UAV).

shared aerial   One used by several receivers.

Shares   Shared airline reservations system (LA-based).

Shark   Silent hard-kill [effective against switched-off radars].

Sharp, SHARP   *1* Standard hardware acquisition and reliability program, which has addressed all military electronics packaging, including shift from DIN to NAC connector system (US).

   2 Strapdown heading and attitude-reference platform.

   3 Shared reconnaissance pod.

   4 System-oriented high-range-resolution automatic target-recognition program (USAF).

sharp-edged gust   Gust characterized by high rate of change of vertical air velocity per unit horizontal distance at particular place, thus aircraft encounters full gust vertical velocity almost instantaneously.

Shars   Strapdown heading and attitude reference system.

shaving   Finishing machine-tool cut removing very small amount of metal; hence shave die, female die with cutting edges for finish-cutting cams etc.

SHBL   Solid hydrogen, boron lattice.

SHC   Synthesized hydrocarbons.

shear   Stress in which parallel planes of loaded material tend to slide past each other; hence also deliberate cutting action by cutting edges which slide past each other.

shear centre   Of a wing, axis about which wing deflects in torsion, also called flexural centre of elastic axis.

shearing   Cutting workpiece or material by shears without formation of chips.

shear lag   Structural stress diffusion in which lag of longitudinal displacement of one part of longitudinal section relative to another is result of shear applied parallel to length of structure.

shear load   Load (force) applied in shear, eg of engine pod on wing spar.

shear modulus E_s, modulus of rigidity, approximately half modulus of elasticity.

shear neck   Local reduction in diameter of a shaft to ensure that, in event of sudden increase in load [eg from something jamming gearteeth], failure will occur at this point.

shear nut   Thin nut used on bolt where load is entirely in shear, merely to retain bolt.

shearout   An interconnection [eg in powered flight control] designed to break if overloaded.

shear plan   Lofting plan of body, eg fuselage, flying-boat hull, showing half-sections as numerous transverse planes.

shear rate   Vertical wind gradient, often measured in kt per 1,000 feet.

shear slide   Free-sliding piston moved by pressurant along length of propellant tank to force liquid propellant into (usually rocket) engine.

shear spinning   Method of forming solid rocket case from preformed thick tubular billet by rolling against rotating mandrel, using two rollers 180° apart, normally performed at room temperature.

shear strength   Stress required to produce fracture in plane of cross-section by two opposed forces with small offset.

shear stress   Component of any stress lying in plane of area where stress is measured; for fluid, equal to $\tau, \mu \dfrac{du}{dy}$ (see *Newton's laws*). Existence of ** in fluid is evidence of viscosity.

shear wave   Wave in elastic medium causing any element of medium to change shape but not volume; in isotropic medium a transverse wave, mathematically one whose velocity field has zero divergence.

sheath   *1* Metal tip, and often leading edge, to soft-blade propeller; also called tipping.

   2 Envelope of plasma surrounding re-entry body.

sheathing   See *sheath (1)*.

SHEB   Solid hydrogen, embedded boron.

shed   Traditional term for shelter (hangar) for aerostats, esp. airship.

shedding   Action for removal of ice from aircraft in flight, rain from windshield (windscreen) and non-vaporised material separated from ablating surface on re-entry.

sheep dipping   Process whereby CIA pilots were given fully documented false professional backgrounds.

sheer lines   Outlines of vertical sections of fuselage [or, especially, hull or float] parallel to longitudinal axis.

sheet   Standard form of raw material; in case of metal, uniform sheet not over ⅛ in (0.125 in, 3.175 mm) thick; thicker metal = plate.

sheet moulding compound　2-D fibre-reinforced plastics not needing complex laying-up procedure.

Shelf　Super-hard, extremely low frequency; military communications system.

shelf　*1* Figurative location where items are stored before use; thus * life, published maximum period during which item will not deteriorate in suitable storage; off-the-*, standard commercial product already available.

2 Spanwise strip[s] hinged to leading edge of movable surface and to trailing edge of fixed structure [eg elevator/tailplane]; see compound *.

3 Longitudinal beams outboard of fighter engine[s] carrying tailplanes.

shell　*1* Bare monocoque structure, eg fuselage or nacelle; shade of meaning includes thin-skinned and deformable, thus engine carcase excluded. Some definitions state 'curved'.

2 Ordnance projectile launched from gun or other tube, with or without own propulsion, and containing explosive, incendiary or other active filling; calibre normally greater than 20 mm. Word also applicable to AP projectiles of such calibres.

3 Supposed hollow spheres at different radii from atomic nucleus occupied by electrons, 2 in innermost *, 8 in next, 18 in next etc, all electrons in each * sharing similar energy level.

shellac　Naturally derived resinous varnish.

shell curve　Plot of control effectiveness [X-axis] against control damping [Y].

Shelldyne　Family of related synthetic fuels developed by Shell for USAF expendable turbojets, mainly characterized by high energy per unit volume.

shelter　*1* Unhardened (generally recessed or underground) accommodation for civilians faced with air attack, in some cases attempting to offer some protection against nuclear attack (fallout *).

2 Unhardened reinforced-concrete structure accommodating (usually single) combat aircraft at dispersal and offering protection against conventional attack, eg Tab-Vee.

shelter marshal　Officer in charge of security and movements within HAS, HPS or PBF.

Sheradizing　Anti-corrosion treatment similar to case hardening but employing Zn dust.

Sheridan tool　Family of large stretch-presses often able to apply double curvature to thick plate.

SHF　Support helicopter force.

s.h.f., SHF　Super-high frequency (see Appendix 2).

SHFE　*1* Sustained hypersonic flight experiment (UK).

2 Small heavy-fuel engine.

SHFT　Shift.

SHGR　Hail shower.

SHGS　Small hail or snow pellets.

SHI　Standby horizon indicator.

Shi　Experimental number, with numerical prefix for year of Emperor's reign; thus 16-* = 1941 (Japanese Navy, 1931-45).

shield　See *shielding (1)*.

shielded bearing　Ball/roller/needle race with metal ring on each side to reduce ingress of dirt.

shielded cable　See *screened*.

shielded configuration　Aircraft deliberately designed so

that parts of major structure, eg wing, are often interposed between sources (of noise or IR radiation) and ground observers or defences.

shielding　*1* Material of suitable thickness and physical characteristics used to protect personnel from radiation during manufacture, handling and transport of radio-active and fissionable materials (DoD, NATO).

2 Obstructions which tend to protect personnel or materials from effects of NW (DoD).

3 Design philosophy of installing crucial parts of primary structure, whether damage-tolerant or otherwise, as far as possible behind others or in some other way geometrically protected from in-service damage.

4 See *screen (4)* (US usage).

shift　*1* Ability to move origin of radar P-type display away from centre of display; limit of * usually to periphery.

2 See *fuel* *.

shim　Thin spacer, from piece of paper (usually unacceptable) to large precision part tailored to specific application, to fill gap or adjust separation between parts; examples, to obtain exact rig (1) neutral and full-deflection settings of powered flight control surface, and to adjust separation of two ends of recirculating ball screwjack so that when bolted together balls have no play and no friction.

shim, shimming　To remove small amount of material to improve fit between mating surfaces [opposite of previous].

shimless assembly　Maintenance of such manufacturing accuracy as to eliminate need for shims [objective in V-22 programme].

shimmy　Rapid lateral angular oscillation of a trailing castoring wheel running over surface where coefficient of friction exceeds critical value, cured by twin-tread tyre or proper design of landing gear; usually affects unsteered nosewheel or tailwheel (or supermarket trolley).

shimmy damper　Add-on damper, usually with pneumatic/hydraulic dashpot, resistant to rapid variation of castor angle.

shingles　Refractory skin panels able to oversail each other at the edges as they expand [eg on Earth re-entry].

shingling　When the clapper of a fan blade overrides its neighbour.

shiny switch　Cockpit button, or switch in constant use (colloq.).

Ship　Software/hardware implemented partitioning.

ship　Aeroplane (US, colloq., suggest archaic).

shipboard aircraft　Aircraft designed to operate from surface vessel or submarine, including marine aircraft and rotorcraft (eg rotor-kite). Some definitions equate term with land aeroplane based on carrier.

shipborne aircraft landing system　Tracks helicopter from ship, and cockpit display guides pilot in radio silence.

shipment　Complete consignment of hardware (probably not conveyed by ship), eg from manufacturer to operator or logistic base to user unit.

ship-plane　Imprecise; most definitions restrict term to land aeroplane for operation from deck of carrier but USN includes catapulted seaplanes formerly used from surface warships.

ship-set　Complete inventory of particular items for one aircraft.

shirtsleeve environment　Popular and often true description of desired environment in high-flying aircraft and spacecraft in which human performance is improved if special clothing does not have to be worn.

SHK   Space hit-to-kill (NMD).

SHLD   Shaped-hole laser drilling.

SHLW   Shallow.

SHM   Simple harmonic motion.

SHNKUK   Roman initials of Society of Japanese Aerospace Companies.

SHNMO   Shape host-nation management office (NATO).

shoals   Scanning hydrographic operational airborne lidar survey.

Shoc   Standoff high-speed option [or operation] for counterproliferation.

shock   *1* Shockwave.

2 Single large-energy pressure wave (see *shock front (2)*).

3 Often used to mean impact, single large externally applied impulse causing acceleration.

shock-absorber   Device for dissipating energy by resisting vertical movements between landing gear (wheels or floats) and aircraft when running across surface, usually with unidirectional quality to reduce rebound and bounce (see *oleo*); other methods include simple steel or composite leaf springs, steel ring springs, rubber blocks in compression and bungee in tension.

shock body   Streamlined volume added (eg on rear of wing) to improve area-rule distribution; also called Whitcomb body, Küchemann carrot, speed bump, etc.

shock cloud   Localized cloud caused by violent changes in flow conditions in close proximity to supersonic aircraft, notably in Prandtl-Meyer expansions, eg over wing and canopy.

shock compression   Fluid flow compression occurring virtually instantaneously in passage through shockwave; for normal shock, ratio of pressures $p_1/p_2 = 7M^2/6 - 1/6$ where M is initial Mach.

shock cord   See *bungee*.

shock diamonds   Approximately diamond-shaped reflections, brilliantly luminous in hot jet (eg from rocket or afterburner), caused by reflection of internal inclined shocks from edge of jet at boundary with atmosphere.

shock drag   That drag associated with a shockwave (which always causes loss in total or static pressure), normally varying as fourth power of velocity or pressure amplitude.

shock excitation   Generation of oscillations in circuit at natural frequency by external pulses, eg for sawtooth generation.

shock expulsion   Faulty operating condition of inlet to supersonic airbreathing engine in which, for various reasons, gross flow breakdown occurs and inlet shock system is expelled forwards; accompanying large and possibly dangerous increase in drag. Often used synonymously with inlet unstart.

shock front   *1* See *shockwave*.

2 Boundary between pressure disturbance created by explosion (in air, water or earth) and ambient surrounding medium.

shock isolator   Device, usually mechanical and assembled from solid parts such as deformable rubbers/plastics or metal deflecting well within elastic limit, which absorbs input movements (eg to accommodate input vibration while keeping output still) or cushions large impacts (by permitting output to travel over a distance which absorbs energy within permitted limits of acceleration). Thus, some absorb vibration, usually of small amplitude, while others absorb shock, which in case of ICBM suspended in silo may require travel in order of 1 m.

shock mount   Shock isolator on which delicate object is mounted.

shock softening   Reduction of linear rate of pressure rise through shockwave, esp. in proximity to subsonic boundary layer; ie increase in thickness of shock (1) from about 10^{-3} mm by several orders of magnitude.

shock spectrum   Plot of peak amplitude of response of single-degree-of-freedom system to various single applied shocks (3).

shock stall   Gross breakdown of flow behind shockwave on wing (esp. one of large t/c ratio or for any other reason causing large airflow acceleration) at about critical Mach number, causing symptoms of loss of lift and turbulent wake resembling stall, but at normal AOA.

shock strut   Main energy-absorbing member of landing gear; may or may not be main structural member but (unlike shock absorber) is always part of structure.

shock tube   Wind tunnel for hypersonic studies in which fluid at high pressures, usually involving rapid combustion to increase energy, is released by rupturing diaphragm and accelerates through evacuated working section containing model. Many varieties, most having stoichiometric gas mixture as driver and large-expansion-ratio (over 200) supersonic nozzle upstream of working section, giving M up to 30 and T around 18,000° K.

shockwave   *1* Surface of discontinuity between free-stream fluid and that affected by body moving at relative velocity greater than speed of sound in surrounding fluid. As fluid accelerates round body, if it eventually reaches local Mach 1 a weak shock forms perpendicular to flow, called a normal shock. Pressure difference $(p_1-p_0)/p_0$ is zero and flow downstream is subsonic. As M increases, shock leans back, becoming an inclined shock, at angle $\alpha = \sin^{-1} 1/M$, while pressure ratio and velocity of propagation V_0 increase according to M and angle of deflection, a property of geometry of body; for 15° deflection (ie wedge or cone of 15° semi-angle) at Mach 3 static and $(p_1-p_0)/p_0$ and $V_0/a_0 = 2.1$, ie * moves at twice speed of sound.

2 Continuously propagated pressure pulse formed by blast from explosion in air by air blast, underwater by water blast and underground by earth blast (DoD, NATO).

Shodop   Short-range Doppler.

shoe   Detachable interface between pylon or hardpoint and store, often specific to latter.

SHOL   Ship/helicopter operational limit.

shoot bolt   Linear bolt type of panel latch.

shooter   Aircraft detailed to attack a target, as distinct from one whose task is to mark or designate.

shooting the breeze   Engaging in casual shoptalk (US).

shoot up   *1* To attack a surface target with gunfire.

2 To simulate this at an airshow.

shop head   End of rivet upset when rivet is used.

shop visit   Removal of item from aircraft for repair or other attention in specially equipped workshop, usually of customer.

shop-visit rate   Frequency, measured on occasions per

unit of flight-time [eg, per 10^5 h] with which particular item [eg, engine] is removed from aircraft for repair or overhaul; often a global fleet average.

Shorad Short-range air-defense; S adds system (USA).

Shoran From short-range navigation, precision radio navaid based on timing pulsed transmissions from two or more fixed stations; in conjunction with suitable computer used for blind bombing.

shore Strut supporting airship during manufacture.

shoreline Line drawn straight across all inlets less than 55.6 km (30 nm) wide (ICAO).

short-distance navaid One usable within 320 km/200 miles (NATO).

short field Limiting field or runway demanding special takeoff procedure.

short finals *1* Last part of approach, usually defined as that commencing at inner marker.

2 Radio call made from aircraft 2 n.m. (3,706 m) from threshold, or on final approach from shortened circuit.

short-haul Several definitions, eg maximum-payload range (knee of graph) 1,609 km (1,000 statute miles) or less; see also *short-range transport*.

short hundredweight US unit of mass = 100 lb = 45.3592 kg.

short-life engine One designed for single flight or any other purpose not requiring prolonged use, and normally qualified for running time of 50 h.

short lift Use of STO to enable powered-lift aircraft to carry enhanced payload.

short-lift rating Thrust rating permitted for [usually] 15 s for VTO or VL.

short period In assessment of factors such as lateral-control damping of fighters, usually means ≤1.5 s.

short-range attack missile ASM launched at range not exposing launch aircraft to terminal defences (USAF).

short-range ballistic missile Up to about 600 nm (1,112 km, 691 miles) (DoD).

short-range clearance Authorizes IFR departure to proceed to a fix short of destination pending further clearance.

short-range Doppler Trajectory measurement using Dovap plus Elsse.

short-range transport Range at normal cruising conditions not to exceed 1,200 nm (2,224 km, 1,382 miles).

short round *1* Round of ammunition deficient in length (DoD 'in which projectile has been seated too deeply'), causing stoppage.

2 Ordnance delivered on friendly troops.

short stacks Briefest form of piston engine exhaust for cowled engine.

short takeoff and landing, STOL Usually defined as able to take off or land over 50 ft screen (note, not 35 ft) within total distance of 1,500 ft (457 m).

short ton US ton of 2,000 lb, = 907.185 kg.

short trail Towing position for sleeve (presumably other forms) of target in which target is immediately astern of towing aircraft.

short wave *1* Not defined and rare in aerospace: traditional radio meaning is decametric (10-100 m) corresponding to 30-3 MHz; FAA meaning is frequencies 7.7-2.8 MHz; scientific is 0.4-1μ wavelengths.

2 In spectrometry, band 2.5-45 μ.

shot *1* Commercial lead * for shotguns, normally used as cheap variable mass.

2 Tailored hardened steel balls of graded sizes.

3 Solid-projectile ammunition, eg for air-firing practice or AP type.

4 Report indicating a gun has been fired (DoD).

5 Single flight of unguided ballistic rocket, eg probe.

shotgun wind Appearing to come from all points of the compass.

shot-peening Bombarding metal surface with air-propelled shot (2), usually to harden and relieve internal stress.

shoulder *1* Area immediately beyond edge of pavement, such as a runway or parking apron, so prepared as to provide transition between pavement and adjacent surface.

2 See * *season*.

shoulder bolt Thread of smaller diameter than shank; for attaching plastics parts where over-tightening must be avoided.

shoulder cowl Usually means cowling panel(s) hinged upwards near top of sides.

shoulder fare That charged for period between standard and off-peak.

shoulder harness Seat harness including straps passing over shoulders to prevent body jack-knifing forward.

shoulder pylon Auxiliary pylons [usually for AIM-9 or similar missiles] on sides of main pylon for tank or other heavy store.

shoulder season Intermediate demand between low and peak, or intermediate time of day, in determining passenger fare structure.

shoulder wing Wing attached between mid and high positions. Original German *Schulterdecker* implied wing depth more than half that of fuselage, with blended wing/body junction.

show Preplanned air operation, especially over hostile territory (RAF, WW2, colloq.).

shower(s) Precipitation from convective cloud characterized by sudden onset, rapid variation in intensity and sudden stop, with intervening periods of part-clear sky.

showerhead Liquid-propellant rocket injector in which numerous fuel and oxidant sprays are distributed (with various forms of impingement) over flat or curved surface.

show finish Glass-like finish (on homebuilts, usually) achieved by repeated doping and rubbing down.

shp, s.h.p. Shaft horsepower [sometimes written SHP].

SHR *1* Shear (weather).

2 Superheterodyne receiver.

SHRA Heavy rain showers.

shrimpboat Small marker of clear plastic on which controller writes flight identity, FL and other information, subsequently moved by hand to remain adjacent to blip on display.

shrinkage Natural reduction in dimensions of most castings on cooling (see *shrink rule*).

shrink fit Force for interference fit between two metal parts obtained by heating outer, cooling inner, or both.

shrink rule Casting mould made ×0.010 (linear) oversize to allow for shrinkage.

shroud *1* Plate formed integrally with gas-turbine fan, compressor or turbine blade usually in plane perpendicular to blade major axis. In fan and upstream compressor blades usually as part-span, * being formed in halves on each side of blade and mating with those adjacent to damp vibration. In turbine rotor invariably on tip, serving as ring minimizing gas leakage around periphery.

2 Circular duct surrounding propeller or propulsive fan.

3 Heat-resistant aerodynamic fairing over space payload or any forward-facing projection on space launch vehicle, ICBM or other hypersonic vehicle.

4 Extensions of fixed surface of aerofoil (eg wing, tailplane) projecting behind hinge line of movable surface (eg flap, aileron) to reduce drag or improve flight control.

5 Main upper hinged flap at rear of augmentor CCW wing, downstream of intake; can have trailing tab.

6 Covering plate on face of centrifugal impeller enclosing flow passages and preventing leakage.

shroud coolant   Refrigerant cooling volume in which cryogenic cooling takes place, eg LH$_2$ serves as ** surrounding liquefaction of helium.

shroud balance   Leading edge of control surface enclosed within trailing edge of fixed structure.

shrouded blade   Blade fitted with shroud (1).

shrouded impeller   Centrifugal impeller, eg of super-charger, enclosed by shroud (2).

shrouded insulator   Radio aerial insulator (eg HF wire) fitted with overlying shroud (normal meaning of word) to prevent ice connection to metal structure.

shroud lines   Main suspension cords of parachute connecting load to canopy.

SHRS   Stabilized horizon-reference system.

SHS   Since hot section (inspection flight hours).

SHSS   Short-haul system simulation.

Shud, S-HUD   Smart HUD.

shunt connection   Bypass circuit, usually taking most of current.

shunt excitation   Feeding mast radiator (aerial) about 0.25 of way up, earthed at base.

Shup   Silo-hardness upgrade program (USAF).

shutdown   *1* For rocket engine, cut-off.

2 For conventional aircraft-propulsion engine, reducing power to zero and rendering inactive, eg by turning off HP cocks. If in flight becomes IFSD. Normally follows obvious or signalled failure.

3 Event in which (2) occurs.

shuttered fuze   Inadvertent initiation of detonator will not initiate booster or burst charge.

shuttle   *1* Generalized term for reusable space launch vehicle recovered by aeroplane-type flight.

2 High-frequency trunk-route service characterized by no-reservations, payment on board and, in some cases, aircraft always boarding, and departing if full before announced time.

3 Sliding drive member of flight-deck accelerator (catapult).

shuttle bombing   Use of two bases, aircraft making one-way bombing missions between them.

shuttle valve   Fluid-flow valve of bistable type which passes flow from one line and isolates other or vice versa.

SHWR   Shower.

SHYFE, Shyfe   Sustained hypersonic flight experiment (UK).

SI   *1* Système International d'Unités; standardized system of units adopted (but not yet fully implemented) by all industrialized nations.

2 Servicing instruction.

3 Straight-in (approach).

4 Spark-ignited, or ignition.

5 Single [or spark].

6 Standby instrument[s].

7 Supporting interrogator.

8 Surveillance [and] intelligence.

9 Suomen Ilmaluliitto, aeronautical association (Finland).

10 Selective interrogator.

Si   *1* Silicon.

2 On an oil analysis = dirt, foreign matter.

Si$_3$N$_4$ Silicon nitride.

SIA   *1* Structural-integrity audit.

2 Service de l'Information Aéronautique (F).

3 Semiconductor Industry Association (US).

SIAE   Salons Internationaux de l'Aéronautique et de l'Espace (F).

SIAG   Salons Internationaux de l'Aviation Générale (F).

Sialon   Silicon-aluminium-oxinitride.

Siam   Self-initiated anti-aircraft missile; carries out IFF interrogation and handles subsequent interception automatically.

siamese   To join two similar items into single paired unit or to bifurcate duct into two equal parts; hence siamesed, adj.

SIAP   Systems integration and assurance phase, of procurement process (UK).

Siap[s]   Straight-in approach procedure[s].

SIAR   Service de Surveillance Industrielle de l'Armement (F).

SIAT   Service instructor aircrew training.

SIB   *1* Special Investigation Branch (UK).

2 Service Information Bulletin.

3 Subject indicator box [message sent by signal].

sibilant filter   One removing hissing frequencies from speech on R/T.

SIC   *1* Steady initial climb, ie V$_4$.

2 Standards Information Center (NBS).

3 Service instruction circular.

4 Second in command.

SiC   Silicon carbide.

Sicas   SSR improvements and collision-avoidance system [P adds panel].

SICBM   Small ICBM.

SICM   Small intercontinental missile.

SID   *1* Standard instrument departure.

2 System integration demonstration.

3 Spray-impingement drag (marine aircraft).

4 Supplemental inspection document.

5 Switch-in deflector.

6 Situation information display.

7 Sensor-image display.

Sidcot   One-piece flying suit with numerous pockets and zips and high fur collar, widely used from 1929 including by RAF (from designer Sidney Cotton).

SIDE   Suprathermal ion detector experiment.

side-arm controller   Primary flight-control input in form of miniature control column at side of cockpit (of combat aircraft) on console incorporating armrest to facilitate accurate flight under conditions of large applied acceleration.

sideband   Band of frequencies produced above and below carrier frequency by modulation; sum and difference products are called upper * and lower *.

side-by-side *1* Two-seat aircraft in which seats are in same transverse plane at same level.

2 Piston engine in which connecting rods are both same, with big ends side-by-side on crankpin (thus, cylinders of opposed banks are not quite in line); constructional form of most modern light aircraft engines.

side car, sidecar *1* Airship car suspended away from centreline plane.

2 A parallel emitter, notably an LWIR boresighted alongside a laser.

side direction Normal to plane of symmetry.

side elevation Portrayed as seen from side, in case of drawing as seen from infinite distance, ie orthographically.

side fence[s] Fences above wing and/or flap of USB aircraft to restrain lateral spread of main engine jet.

side float Usually means sponson.

side force Force acting normal to plane of symmetry.

side-force control Aircraft flight-control system capable of exerting lateral [transverse] force, normally by vertical surfaces in front of as well as behind c.g. Aircraft with ** can almost instantly change track by flying diagonally, without need to roll or change fuselage axis; make immediate lateral corrections to line of fire of fixed gun; move laterally out of hostile gunfire without prior roll.

side-force surface One designed to generate transverse force acting almost through c.g., as on NASA TIFS C-131.

side frequencies Carrier plus or minus audio frequency.

sideline noise Measured beside takeoff run at distance from centreline of 450 m (ICAO Annex 16) or 0.35 nm (FAR 36 CAN 5, 4-engined aircraft) or 0.25 nm (2-, 3-engined). See *noise*.

sidelobe Lobes of aerial radiation propagated at angle to main lobe, normally unwanted and cause of clutter or false returns, eg obscuring actual location of sender.

sidelobe clutter Echoes caused by intersection of radar sidelobes with the ground.

sidelobe suppression Various techniques for eliminating not presence of sidelobes but their effects.

side-looking Scanning to either side of aircraft track; hence SLAR, radar whose output is detailed picture of terrain near track, either all on one side or equally on both sides, depending on aerial arrangement.

side marker board Display beside airport gate arranged for nose-in parking giving indicator marks, usually vertical white bars on black board; when aligned with captain's left shoulder, airbridge is aligned with passenger door.

Sident Site identification.

side number Bold three-digit Modex number (US Navy).

side oblique Photograph taken with camera oblique and perpendicular to longitudinal axis of aircraft.

sidereal Pertaining to stars, but see following entries. Rhymes with material.

sidereal day Time for one rotation of Earth as defined by period between successive transits of vernal equinox (in ASCC wording, alternative name, First Point of Aries) over upper branch of any chosen meridian, equal to 24 h of mean sidereal time or 23 h 56 min 4.09054 s of mean solar time.

sidereal hour angle Angular distance west of vernal equinox; arc of celestial equator or angle at celestial pole between vernal equinox and hour circle of observer (see *right ascension*).

sidereal month Average period of revolution of Moon with respect to stars: 27 days 7 h 43 min 11.5 s.

sidereal period *1* Time taken by planet to complete revolution around primary as seen from primary and referred to fixed stars.

2 Interval between two successive returns of Earth satellite to same geocentric right ascension.

sidereal time Time measured from rotation of Earth related to vernal equinox, called local time or Greenwich time depending on choice of meridian; when adjusted for nutation inaccuracy called mean time.

sidereal year Period of Earth's rotation around Sun, related to stars; in 2002 equal to 365 days 6 h 9 min 9.55454 s and increasing at about 0.000095 s per year (see *tropical year*).

sideslip Flight manoeuvre in which controls are deliberately crossed, eg to * to left aeroplane is banked to left while right rudder is applied; result is not much change in track but flight path inclined downwards, ie steady loss of height without significant change in airspeed and with longitudinal axis markedly displaced from flightpath. Angle of * is angle between plane of symmetry and direction of motion (flightpath, or relative wind). Rate of * is component of velocity along lateral axis.

sidestep Following an instrument approach, clearance to land on a parallel runway not more than 1,200 ft away laterally.

sidestick Small control column on cockpit side panel, usually on R, often input force with almost no noticeable movement.

sidetone Reproduction of sound in a speaker or headset from speaker's own transmitter, thus hearing own voice.

sidewalk Chordwise walkway above wing root.

sidetracking skate See *skate*.

sidewall treatment Addition of sound-absorbent material along sides of passenger cabin.

sidewash Sideways deflection of free stream behind wing in sideslip or yawed flight, dominated at tail by vortex flow.

sideways translational tendency Characteristic of single-rotor helicopter to drift to L or R under thrust of anti-torque tail rotor, unless main rotor tip-path plane is tilted in opposition to neutralize this thrust.

SIDs Standard instrument departures; SIDS, standard instrument departure system.

SIE Self-initiated elmination, = dropout.

siemens Not Siemens, SI unit of conductance, reciprocal of ohm, S = $^{1}/\Omega$; also SI unit of admittance and susceptance.

SIERE Syndicat des Industries Electroniques de Reproduction et d'Enregistrement (F).

Sierra Prefix, supersonic (esp. SST) airway (ICAO).

Sierracote Family of patented glass and/or acrylic transparencies which may include anti-icing heating by transparent film.

SIF *1* Selective identification facility (or feature).

2 Standard interchange format.

3 Spares investment forecast.

4 System interrogation facility.

Sifbronze British alloys of copper, silicon and zinc for low-temperature gas welding.

SIFF Successor IFF.

SIFL   Standard industry fare level (US).

Sifet   Società Italiana Fotogrammetria e Topografia (I).

SIFTA   Sistema Interamericano de Telecommunicaciones para las Fuerzas Aereas (Int.).

SIFV   Sensor instantaneous field of view.

SIG   *1* Signature (DoD, ICAO).

  2 Significant (ICAO).

  3 Stellar/inertial guidance.

  4 Simplified inertial guidance.

  5 Special industry group.

sight   *1* Optical device for measuring (eg drift *) or aiming (gun *), often incorporating magnification or combined with HUD (Hud-*).

  2 To take observation with sextant.

sight gauge, sight glass   Graduated vertical window in fluid container indicating level of contents; alternative to dipstick.

sighting   Visual contact; does not include other forms of contact, eg radar, sonar.

sighting angle   Angle between LOS to aiming point and local vertical; at time of bomb release, same as dropping angle (ASCC).

sighting out of wind   Rigging by eye (arch.).

sighting pendant   Vertical plumb wire ahead of airship control car to aid in steering and drift estimation.

sighting station   Crew station from which sight (2) taken or other optical measures made.

sight tracking line   LOS from a computing gunsight reticle image to target.

SIGI   Space integrated GPS/INS.

Sigint   See *signal intelligence*.

Sigma   *1* Scale of statistically measuring products and services by counting rate of defects; thus 4Σ is average and 6Σ near-perfect.

  2 Originally in French, system for interception, goniometry, monitoring and analysis.

Sigmet   Weather advisory service to warn of potentially hazardous (significant) extreme meteorological conditions dangerous to most aircraft, eg extreme turbulence, severe icing, squall lines, dense fog.

signal   *1* Anything conveying information, eg by visible, audible or tactile means.

  2 Pulse(s) of EM radiation conveying information over communications system.

  3 Information conveyed by (1, 2).

  4 Any electronic carrier of information, as distinct from noise.

  5 Standard visual symbols displayed on simple airfield.

signal area   Plot of ground at simple airfield, usually adjacent to tower, set aside and equipped for display of ground-to-air signals (5) for informing pilots of aircraft not equipped with radio, eg of circuit direction, local hazards, prohibitions, etc.

signal data   General term for elint output or for any recorded (if possible quantified) information on received signals (2).

signal flare   Flare pyrotechnic whose use conveys known meaning, eg two-star red.

signal frequency   Frequency of transmitted carrier as distinct from component parts, or received carrier as distinct from intermediate frequency to which it is converted.

signal generator   Versatile oscillator capable of outputting any desired RF or AF at any selected ampli-

tude and with output modulated to simulate actual transmission.

signal intelligence   General term for communications intelligence plus electronic intelligence; art of detecting, recording, analysing and interpreting all unknown or hostile signals.

signalman   Authorized person using hands, wands and/or lights to marshal aircraft on apron or elsewhere on airport manoeuvring area.

signal pistol   Projector of pyrotechnics, eg Very pistol.

signal star   Pyrotechnic of distinctive character emitted from cartridge or signal pistol.

signal-to-noise ratio   Ratio of amplitude of desired signal to amplitude of noise signals at a given point in time (DoD). Ratio, at selected point in circuit, of signal power to total circuit noise power (ASCC). Number of dB by which level of fully modulated signal at maximum output exceeds noise level, all values being rms.

signature   Characteristic pattern of target displayed by detection and identification equipment (DoD, NATO etc). Like thumb-print, * analysed with sufficient accuracy can identify source as to type and even to specific example of emitter. Can be EM (radar, IR, optical), acoustic (eg SST boom *) or velocity (Doppler *).

significant obstacle   Posing potential threat to aircraft.

significant tracks   Tracks of aircraft or missiles which behave in an unusual manner which warrants attention and could pose a threat to a defended area (DoD, NATO).

significant turn   Change of heading large enough for explicit account to be taken operationally of reduction of climb gradient.

significant weather   Potentially dangerous to aviation.

Sigsec   Signals security (USA).

SIGWX   Significant weather [also Sigwr].

SIIDAS   Sensor independent integrated defensive aids suite.

s(i,k)   Slope of sound pressure level; change between ⅓-octave SPLs at i-th band at k-th moment in time. Hence $\Delta s(i,k)$ = change in SPL slope, $s'(i,k)$ is adjusted slope between adjacent adjusted bands, and $-\bar{s}(i,k)$ is average slope (see *noise*).

SIL   *1* Suomen Ilmailu Litto (Finnish Aeronautical Society).

  2 Site d'intégration lanceurs (F); launcher assembly site.

  3 System[s] integration laboratory.

  4 Service information letter.

Silastic   Proprietary range of silicone rubbers and sealants including many resistant to hydrocarbon fuels.

SILC   *1* Space-object identification in living colour.

  2 Semiconductor IR laser countermeasure[s].

Silence   Significantly lower community exposure to aircraft noise and emissions.

Silencer   Significantly lower aircraft noise for community in Europe research (EC).

silent target   Non-emitting, ie no radio, radar, IR, laser, TV or other detectable radiation.

silica   Silicon dioxide, quartz, SiO_2.

silica gel   Numerous stable sols of colloidal particles, some important as drying agents.

silicon   Si, abundant versatile solid element with many forms, eg dark crystal or brown powder; important in alloys, in vast range of compounds and in pure crystalline form as semiconductor; basis of major part of microelectronics. Density 2.3, MPt 1,410°C.

silicon chip Popular name for microcircuit fabricated in chip scribed from slice of single-crystal epitaxial silicon. Formerly each chip contained devices and subcircuits, today can be a complete equipment requiring only packaging and human interfaces.

silicone Generalized term for polymeric organo-siloxanes of form $(R_2SiO)_n$, including elastomers, rubbers, plastics, water-repellants and finishes, eg resins, lacquers and paints.

silicone cork Relatively cheap ablator and heat insulator, eg on Shuttle external tank.

silicon nitride Refractory ceramic, Si_3N_4; increases strength slightly to 1,200°C, good thermal shock resistance becauase of small thermal expansion, and resists many forms of chemical attack.

silk Parachute (colloq.), hence hit the *, take to the *.

SILL Strategic illuminator laser.

SilMU Silicon micro-machined electromechanical system inertial measurement unit.

silo Missile shelter that consists of hardened vertical hole in ground with facilities either for lifting missile to a launch position or for direct launch from shelter. Normally closed by hardened lid and provided with shock-isolating missile supports.

silver Metallic element, Ag, density 10.5, MPt 962°C, important in electronics, photography, joining metals, cloud seeding and disposal of explosives and nerve gases.

silver ball Inceptor controlling Stovl engine nozzle[s] (colloq.).

Silver bullet Procurement process for highly classified items (US).

Silver C Intermediate certificate and badge for gliding requiring flight of at least five hours, gain in height of at least 1,000 m (3,281 ft) and straight flight of at least 50 km (31 miles).

silver cell Silver/zinc.

Silver Flag Exercises training combat support forces to operate in a hostile bare/austere base environment.

silver iodide AgI, sprayed into clouds to promote rainfall [seeding].

silver solder Jointing alloy of silver, copper and nickel.

silver-strip indicator Inserted in fuel filter, detects any abnormal concentration of sulphur in fuel.

silver tux Astronaut suit (US, colloq.).

silver/zinc Ag/Zn elecric dry battery common in applications calling for single-shot high-power electric supply.

SIM *1* SAM (1) intercept missile; bomber defence.

2 System improvement modification.

3 Space interferometry mission.

4 Security identification module.

Sim Simulation, simulator.

SIMA Scientific Instruments Manufacturers Association of GB (UK).

Simaf Simulation and Analysis Facility (USAF).

SIMD Single instruction, multiple data stream.

Simmonds-Corsey Mechanical remote-control system in which push/pull commands are transmitted by flexible cable in conduit on which are threaded mating tubes and 'olives' giving bidirectional control.

Simmonds nut Pioneer stop nut incorporating tightly held fibre washer.

Simop Simulataneous operation of co-located radios.

Simos Simulator orthogon system.

simple architecture for full electrical Landing-gear brake-by-wire system in which both main and alternate [alternative] systems are electric, mainly or wholly eliminating hydraulics.

simple flap Hinged wing trailing edge, with or without shroud (4) but without intervening slot.

simple harmonic motion Regular oscillation as exemplified by alternating current or drag-free swinging pendulum; projection on any axis in same plane of point moving round circle with constant angular velocity; expressed by $y = a \cos(2\pi nt + b)$ where y is distance from origin (at centre) at time t, n frequency and b a phase constant such that at $t = 0$, $y = a \cos b$. A plot of SHM is a characteristic wavy line, passing through the origin (displacement zero) with peak velocity, rising to maximum displacement where velocity passes through zero and reverses, returning through the origin to describe a precise mirror-image terminating at the start of the next cycle. The sum of positive and negative displacements [peaks plus troughs] is called the amplitude, the time between successive passes through the same point in the cycle the period, and the reciprocal of the period [or number of cycles per unit time] the frequency.

simple stress Either pure tension, pure compression or pure shear.

simplex *1* With no provision for redundancy.

2 Communication on a single channel which is unidirectional in operation; thus, when receiving, cannot transmit.

simplex burner Simple gas-turbine fuel burner fed by single pipe leading to nozzle surrounded by air swirl vanes and with flow proportional to square root of supply pressure.

simplex communications Communications technique in which signals pass in one direction only at any one time; can be single or double channel and switched by press-to-speak, manual T/R switch or voice-operated.

simplified directional facility ILS localizer (108.1-111.9 MHz) with aerial offset from runway and emitting beam usually not exactly aligned with it nor providing G/S information.

simplified passenger travel Attempt from 1999 to streamline 'repetitive identity checks at airports'. Since 11 September 2001 emphasis has included new identification measures (ICAO).

simply supported beam Pin-jointed at both ends.

Sims *1* Secondary-ion mass spectrometry.

2 Signal-identification mobile system.

SIM2 Use of three sets of axes (wind, tunnel stability and body) for force and moment equations (NASA 1980 onwards).

Simu Single-input multi-unit.

SIMUL Simultaneously (ICAO).

simulated forced landing Includes all actions except landing.

simulated-operations test Operational test needed to support statements of new requirements and support positions and programmes (USAF).

simulated attack profile Typically, lo mission in which pilot is tasked to find and attack several point targets.

simulator Dynamic device which attempts to reproduce behaviour of another dynamic device, eg aircraft or missile, under static and controlled conditions for research, engineering design, detail development or personnel training. Those used for research are similar in appearance to other large electronic items; those for training often incorporate a complete cockpit or flight deck carried on hydraulic rams giving motion about all three linear axes and three rotary axes (eg exactly reproducing asymmetric swing on take-off engine failure or buffet at approach to stall with correct fuselage pitch attitudes), and model-form or electronically generated external scene for flight in neighbourhood of particular selected airport(s).

simulator sickness   Caused by conflict between sensations, control inputs and visual cues.

simultaneous approach   Two aircraft land on parallel runways at the same place.

simultaneous dual field of view   Split-screen system generated by two LOS telescopes focusing IR on single detector array, converted to formatted electronic picture, half wide FOV and half magnified narrow FOV (Hughes/Eltro).

simultaneous engagement   Concurrent engagement of target by interceptors and SAMs (DoD).

simultaneous pitch control   See *collective*.

simultaneous range   Radio range which simultaneously broadcasts voice messages.

SIN   Significant-item number.

Sinaga   Sindicato Nazionale Gente dell'Aria (I).

SINCGARS, Sincgars   Single-channel ground and airborne, or ground/air, radio subsystem; SIP adds system-improvement programme and V adds vhf.

sine curve   Obtained by plotting sine on linear axis, graphically identical to plot of SHM.

sine wave   Wave of SHM form, eg EM radiation.

sine-wave flight   Repeated SHM in vertical plane, especially spacecraft skipping in and out of sensible atmosphere.

sine-wave spar   Structural member whose web has a sine-wave profile.

S-ing   Performing succession of S-turns.

single   A single-engine aircraft, term normally used for GA aeroplanes.

single-acting   Actuator pushing in one direction only, with spring return; push-push.

single-aisle aircraft   Narrow-body, having twin or triple seats on each side of one axial aisle.

single-axis autopilot   One offering stability or control about one aircraft axis only, eg pitch, roll or yaw.

single-axis head   Homing head whose sensor scans only in one plane.

single-base propellant   Traditional term originating in so-called smokeless powders based on either Nc or Ng alone (see *double-base*).

single-bay   Having only one set of interplane struts joining wings of biplane on each side of centreline.

single-channel simplex   Same frequency in both directions.

single-configuration fleet   Not only are all aircraft in airline fleet of same type but all have same build-standard, avionics, cockpit and furnishing.

single-crystal alloy   Complete workpiece formed in piece of metal grown as single crystal, possibly containing occasional atomic imperfections but devoid of gross intercrystalline joints and as far as possible with lattice orientation selected to increase strength in direction of greatest applied stress.

single curvature   Curved only in one plane, as surface of regular cylinder.

single-direction route   Self-explanatory, usually high-altitude IFR.

single-entry compressor   Radial or centrifugal impeller with vanes on one side only.

Single European Sky   Overdue plan for unified ATC, adopted by EC December 2002, still years from implementation.

single-expansion ramp nozzle   One form of non-axisymmetric jet engine nozzle in which supersonic jet is accelerated along sloping wall (expansive flow) on one side only. Has attractions for SSTOVL.

single-face repair   Repair to sandwich structure involving core and one face only.

single-flare joint   End of rigid pipe flared out but not turned back on itself.

single-float seaplane   Large central float and small stabilizing floats outboard.

single ignition   One coil of magneto, feeding one plug per piston engine cylinder.

single-pass heat-exchanger   Each fluid passes once through without turning.

single-regime engine   One designed to operate always under same conditions, eg jet VTOL.

single-rotation   *1* Composed of one unit (propeller or propfan), as distinct from two equal units rotating in opposite directions.

2 Flap hinged about a single fixed axis well below the wing, and thus moving along a circular arc.

single-row engine   Radial engine with all cylinders in same plane driving one crankpin.

single-shaft engine   Gas turbine in which all compressor stages are connected to the same turbine. There may be an independent shaft linking a free turbine to a shaft output.

single-sideband   Reduction of bandwidth by transmitting only one sideband and suppressing other (and usually carrier also); receiver heterodynes at original carrier frequency.

single-sideband suppressed carrier   Band of audio-intelligence frequencies translated to radio frequencies with no distortion of signal.

single-sink flow   Fluid flow capable of exact representaton using a single sink.

Single Sky   Wide range of measures intended to consolidate Upper Airspace throughout Europe by end of 2004 (EU).

single-spar wing   Wing in which primary flight loads are borne by one spar (as distinct from a box), possibly made with two booms on at least one edge or having U or circular section. Does not preclude secondary spanwise member(s) for trailing-edge surface loads.

single-spring flexure   Tunnel balance in which model is supported by single (usually vertical transverse) member locally thinnned at one transverse point to serve as pivot sensitive to forces or couples about that axis only.

single-stage compressor   Compressor achieving total overall pressure ratio in one operation, eg by centrifugal impeller or single row of axial blades.

single-stage turbine   Turbine having only one set of axial rotor blades or inward radial vanes.

single-stage vehicle   Aerospace vehicle (aircraft, RPV, missile, rocket) with only one propulsion system. It is hoped eventually to create a * to orbit, or to inter-planetary space.

single-surface rudder Normal leading edge back to spar [near maximum thickness], aft of which skin is on centre-line with half-ribs on each side. Also called splitter-plate rudder or tadpole rudder.

single-surface wing Wing having upper [lifting] surface only, characsistic of pre-1914 aeroplanes and many hang gliders and microlights.

single-target track Traditional fighter radar tracking mode, usually with phase-comparison monopulse.

single-tipping Sheath (1) made in one piece.

single-up Competition in which winner takes all.

single-wedge aerofoil Cross-section has sharp leading edge, flat sides and blunt or square trailing edge, eg vertical tail of X-15. Extremely inefficient at low speeds.

single-wheel gear Main landing gear of large aircraft with one wheel on each shock strut.

sinide Silicon nitride.

sink *1* Theoretical point in fluid flow at which fluid is consumed at constant rate.

2 Large mass to which waste heat can be rejected.

sinking Rapid uncontrolled increase in rate of descent with little change in horizontal attitude, usually caused by increasing AOA at approach to stall.

sinking speed Vertical component of velocity of aircraft without propulsive or sustaining power in still air; for glider or engine-out aeroplane, = TAS × sin gliding angle = TAS ÷ glide ratio.

sink rate Rate of descent of free-fall unpowered lifting-body, especially glider at best L/D. Usually same as sinking speed.

SINS, Sins Ship's inertial navigation system (FBMS).

Sintac Système integre d'identification, de navigation, de contrôle du trafic aérien, anti-collision et de communication (F).

sintering Bonding powder or granules under heat and pressure; no melting takes place and product may be ceramic, cermet, metal or many other types of material, and compact or having any desired degree of controlled porosity. Related terms: diffusion bonding, hot isostatic pressing, powder metallurgy.

sinusoidal Having form of sine waves; characteristic of SHM.

SIO, SI/O Special input/output.

SIOE Space and Information Operations Element (USSPACECOM).

SIOP Single integrated operational, or operating, plan (US use of NW).

SIOU Serial input/output unit (threat warning).

SIP *1* Surface-impact, or impulsion, propulsion.

2 Self-improver pilot.

3 Single inline package.

4 Stockpile, or system, improvement program.

5 Structural, or system[s] or subsystem[s] integrity program.

Sipac Sindicato Italiano Piloti Aviazione Civile (I).

Sipaer Serviço de Investigação e Prevenção de Acidentes Aeronauticos (Brazil).

Sipri Stockholm International Peace Research Institute.

Siprnet Secure international protocol router network (DoD).

Sips, SIPS *1* Survey information processing system.

2 Small integrated propulsion system.

3 Software-intensive projects.

4 Structural-integration program[me] system.

SIPU Super-integrated power unit, engine designed to run on many fuels including MOM and MEPU.

SIR *1* Strip, inspect and rebuild.

2 Snow and ice on runway.

3 Search/interrogation radar.

4 Shuttle-imaging radar.

5 System integrated receiver (TJS).

6 Screening information request.

SIRA Sensor IR/acoustic.

SIRE Satellite IR experiment.

siren noise Caused by escape of air or gas under pressure between fixed and rapidly moving apertures, such as engine blading, successive pulses giving frequency.

SIRFC Suite of integrated RF countermeasures, pronounced Surfac, sometimes written SIRCM or SIRFCM.

Sirias Synergistic integrated receiver techniques for interference adaptation and suppression, for M-code GPS (AFRL).

Sirpa Service d'Information et de Relations Publiques des Armées (F).

Sirs, SIRS *1* Signal-intelligence receiving system.

2 Satellite IR spectrometer.

SIRST Surveillance IR search and track.

SIRTF Space [originally Shuttle] IR telescope facility.

SIRU Space inertial-reference unit.

SIS *1* Stall-identification [rarely, inhibition] system.

2 Standard instruction sheet.

3 Software-interface specification.

4 Semiconductor/insulator/semiconductor.

5 Signal in space.

6 Superheterodyne IFM subsystem.

7 Secret Intelligence Service.

8 Satellite interceptor system.

Sisal Single-assignment language.

SISC Standard Information Systems Center, part of AFCAC (Gunter AFB).

SISCM Suite of integrated sensors and counter-measures.

SISD *1* Space Information Systems Division (USAF Colorado Springs).

2 Single instruction, single data-stream.

SISO Simulation Interoperability Standards Organization (Int.).

SIT *1* Silicon-intensified target.

2 Selective-identification transponder.

3 Surplus-inventory tag.

4 Spontaneous-ignition temperature.

5 System integration and test.

Sita Société Internationale de Téléecommunications Aéronautiques (Int., office Geneva, serves 350 air carriers).

site error Radio navaid inaccuracy caused by radiation reaching destination by indirect, ie reflected, routes; eg NDB or VOR radiation reflected from building near beacon.

Sitelesc Syndicat des Industries de Tubes Electroniques et Semi-Conducteurs (F).

SiTF See *STF*.

SITP System-integration test plan; /D adds description.

SITR System-integration test report.

SITREP, Sitrep   Situation report.

SITS   Systems integrated test station.

situational awareness   A buzz-word in air-combat theory; unquantifiable ability of pilot to keep abreast of what is happening to all friends and all foes. Not related to experience.

SIU   *1* Sensor, or server, or systems, or satellite, or Standard, or Sidewinder, interface unit.

2 Supportability and infrastructure upgrade[s].

SIV   Separation integration vehicle.

Sivam   System for vigilance of the Amazon.

SiVSG   Silicon vibrating-structure gyro.

SIWL   Single isolated-wheel load.

SIWR   Self-induced wing rock.

six   Six-o'clock position, ie directly behind one's own aircraft; hence "check *!".

6 by 7 cable   Flexible aircraft cable with cotton core surrounded by six cables each twisted from seven wires.

six-component balance   Wind-tunnel balance that simultaneously measures forces and moments (couples) about all three axes.

Six Cs   A checklist for an emergency: confess (problem), climb, communicate, conserve, comply, consult.

6-DOF   Six degrees of freedom; two directions in each of three dimensions.

Six-sigma   Method of quantifying and measuring products and services to achieve 'world class' performance; common denominator is number of defects per unit or task.

666   IFR currency requirements: previous calendar months, number of approaches, hours.

16-g seat   Strength requirement of airline seat (current FAR).

sized fibre   Virgin carbon fibre sized with resin binder.

SJ   Ski-jump.

SJAC   Society of Japanese Aircraft Constructors; Japanese-language acronym = SHNKUK.

SJB   Semi-jetborne (jet VTOL).

SJC   Standard job card.

SJF   Standing Joint Force.

SJI   Stores jettison indicator.

SJR   Single jet-driven rotor (helo).

Skad, SKAD   Survival kit, air-droppable.

skate   Platform(s) with castoring wheels or air-cushion pads for moving large aircraft on ground.

SKB   Student construction (design) bureau (USSR).

SKC   Sky clear (ICAO = ≤⅛ cloud).

SKD   Semi-knocked-down.

SKE   Station-keeping equipment.

sked   Scheduled (ICAO).

skeg   *1* Small fixed fin at rear of afterbody step of marine aircraft to improve stability when taxiing.

2 ACV sidewall.

3 Ventral strip along seaplane afterbody serving as support on land.

skew aileron   One whose hinge axis is markedly not parallel to transverse axis but diagonally across tip.

skew angle   Angle between principal axes of fuselage and wing of oblique-wing aircraft.

skew compressor   Fluid compressor intermediate between centrifugal (radial) and axial.

SKF   Superkritischer Flügel; supercritical wing (G).

SKG   Schnellkampfgeschwader, fast bomber wing (G, WW2).

Skiatron   Dark-trace CRT or related display.

skidding   *1* Sliding outward in turn because of insufficient bank or excess rudder, opposite of slip.

2 Incorrect operation of ball or roller bearing in which sliding friction occurs instead of pure rolling for various reasons.

Skiddometer   Towed runway-friction measurer with 17%-slip braked centre wheel.

skid fin   Fixed fin mounted high above c.g., eg above upper wing of biplane, to reduce skidding (1).

skid landing gear   *1* In early aeroplanes, rigid ski-shaped member projecting ahead of landing gear to prevent nosing-over.

2 In helicopters, fixed tubular landing gear, often provided with small auxiliary wheels (eg winched down by hand) to confer ground mobility.

skid-out   See *skidding (1)*.

skid transducer   Input sensor of anti-skid wheel brake system, able to sense any sudden variation in wheel rotational speed.

skiing glider   Skier also wearing lifting aerofoil or parafoil.

ski-jump   Take-off over a * ramp.

ski-jump ramp   Curved ramp terminating at (ideally) about 12° to horizontal providing large benefits to rolling take-off by vectored-thrust STOVL aircraft, including shorter run and/or greater weight of fuel/weapons, and increased safety (particularly off ship) in event of failure of engine or nozzles.

ski landing gear   Designed for ice or compacted snow, often with heating to prevent adhesion.

skimmer   *1* Missile, eg anti-ship category, programmed to fly just above crests of waves.

2 ACV (colloq.).

skin   *1* Outer covering of air vehicle, ACV or spacecraft, except that in case of vehicle covered with ablative layer or thermal-insulating tiles * is underlying structural layer. Can be made of any material including fabric or Mylar.

2 Outer component of sandwich.

skin Doppler   Determination of air-vehicle velocity by radar.

skin drag   See *surface-friction drag*.

skin echo   Popular term for object, especially aircraft, as seen on radar.

skin effect   Concentration of AC (electron flow) towards surface of conductor.

skin friction   See *surface-friction drag*.

skin-friction coefficient   Non-dimensional form of skin-friction drag on body immersed in a laminar, viscous, incompressible flow, $\gamma = \dfrac{\tau_o}{\frac{1}{2}\,\rho\,\mathrm{Um}^2}$ where τ_o is shearing stress at solid surface, ρ density and U_m mean flow velocity $\left(= \dfrac{16}{R} \text{ where R is Reynolds No} \right)$.

skin mill   Large machine tool with revolving cutter(s) under which passes workpiece with linear motion; cutter axis can vary from horizontal to vertical. Often NC machine and able to sculpt complete wing skin.

skin paint   *1* =Radar indication caused by reflected radar signal from object (DoD); ie blip.

2 Fix obtained by ground radar on aerial target.

skin temperature   Temperature of outer surface of body, esp. in sustained supersonic flight.

skin tracking   Tracking of object by means of a skin paint (1) (DoD).

skip   See *skip re-entry*.

skip altitude   Lowest point of a trough.

ski pad   Large-area pad attached to helicopter landing gear for operations from snow, tundra, muskeg, swamp and sand.

skip bombing   Method of aerial bombing in which bomb is released from such a low altitude that it slides or glances along surface of water or ground and strikes target at or above water or ground level (DoD).

skip distance   Distance from transmitter at which first reflected sky wave can be received, increasing with frequency.

skip/glide   See *boost/glide*.

skip it   Air intercept code: "Do not attack, cease interception".

skipping   Sine-wave flight.

skip re-entry   Atmospheric entry by lifting-body spacecraft in which energy is lost in penetrating atmosphere in curving trajectory reminiscent of stone skipping on pond, possible only with very accurate trajectory control if first skip is not to result in permanent departure from Earth (see *lifting re-entry*).

skirt   *1* Lowest part of body of large ballistic vehicle surrounding rocket engine(s).

2 Lowest part of parachute canopy.

3 Lowest part of envelope of hot-air balloon.

4 Flexible structure surrounding and containing cushion of amphibious and many other types of ACV enabling vehicle to run over waves or rough ground.

skirt fog   Steam cloud during launch from wet pad.

ski-toe locus   Imaginary object shaped like front of ski which TFR runs along terrain by electronic means; size and form are often variable to give soft to hard ride.

SKRVT   State Commission for Development and Co-ordination of Science and Technology (Czech Republic).

SKSV   National certification authority (Jugoslavia).

SKT   Specialty knowledge tests (USAF).

sky compass   Instrument for determining azimuth of Sun from polarization of sunlight.

sky conditions   Amount of cloud in oktas.

sky diver   Sport parachutist.

Skydrol   Synthetic non-flammable phosphate ester-based hydraulic fluids (Monsanto).

Sky Guards   Trained professionals flying as passengers to protect against any airborne threat [first introduced by ANA, Japan].

Skyhook   Plastic balloon for meteorological observation and rocket launch (see *rockoon*).

skyjack   Aerial hijack (colloq.).

skyline   Verb, to get low enemy aircraft visibly above horizon with sky background, or to expose own aircraft in same way.

Skymarshal   Armed Federal security officer riding on US commercial flights to deter hijacking 1970-73, and reintroduced 2001 (US Customs Service).

Sky Miles   Free travel assigned by airline to frequent flyer, or offered as competition prize.

skyquake   Sonic boom from large hypersonic aircraft.

sky screen   Simple optical (camera obscura) device showing range safety officer if vehicle departs from safe trajectory; often one for track and another for vertical profile.

sky shouting   Use of aircraft-borne loudspeaker; in most countries prohibited for private use.

sky wave   That portion of a radiated wave that travels in space and is returned to Earth by refraction in ionosphere (ASCC). Several other authorities use word 'reflection'. Also called ionospheric or indirect.

sky-wave correction   Factor to be applied to some hyperbolic navaids, eg Loran, if sky waves are used instead of ground waves; varies with relative distances to master and slave(s).

sky writing   Writing, if possible against blue-sky background, using oil added to exhaust or other system and forming characters by accurately flying along their outlines, or using tight formation to switch smoke under computer/radio control while flying straight along words.

SL   *1* Sea level, also S/L.

2 Space-limited, or launcher.

3 Service letter.

4 Schätzstellen für Luftfahrtzeuge (G, Austria).

5 Short landing.

6 Sensitivity level.

7 Start line.

8 Standing lenticular.

9 Sound level.

S/L   *1* Shoot/lock.

2 Sub-level.

3 Sea level.

SLA   *1* Small light aeroplane.

2 Self-launching aircraft.

3 Service-level agreement.

4 Spacecraft/Lunar-module adaptor.

Slab   Sealed lead-acid battery.

slab-sided   Having essentially flat (usually near-vertical) sides.

slab tailplane   Horizontal tail formed as single pivoted surface and used as primary flight control; no fixed tailplane or hinged elevators (US: horizontal stabilizer). Called taileron when installed in left/right halves capable of being driven in opposite directions.

slab trailing edge   Blunt railing edge, esp. squared off normal to line of flight.

SLAE   Now SLAET.

SLAEA   The Society of Licensed Aircraft Engineers Australia.

SLAET   The Society of Licensed Aircraft Engineers & Technologists (UK, 1943, reconstituted 1962).

slag refining   See *electro-slag refining*.

SLAM   *1* Supersonic low-altitude missile.

2 Standoff land attack missile [ER adds extended range or expanded response].

3 Scanning laser acoustic microscopy.

Slam   RAAM surface-launched advanced medium-range AAM, ie SL-Amraam. Suggest confusion with Slammer.

slam acceleration   Most rapid possible acceleration of engine, esp. gas turbine, typified by violent forward movement of power lever to limit.

Slammer   Amraam, or especially aircraft armed with live Amraams.

slamming   Impact of front or rear wheels (depending on design geometry) of bogie on ground at vertical velocity greater than that of aircraft, caused by added velocity imparted by rotation of bogie beam.

Slammr   Sideways-looking airborne multimission radar.

Slam-R   Small lightweight airborne MTI-radar.
Slams   Surface look-alike mine systems.
slant course line   Intersection of course surface and plane of nominal ILS glidepath.
slanted deck   Angled deck.
slant range   LOS range between aircraft (aerial target) and fixed ground station; not same as range plotted on map, hence ** correction to radio navaid distances which is small until aircraft height is greater than 20% of **.
slant visual range   *SVR (1)*.
SLAP   *1* Saboted light-armour penetrator.
   2 Slot-allocation procedure.
   3 Service-life assessment program.
slap   See *rotor* *.
SLAR   Side[ways]-looking airborne (or aircraft) radar.
slash   Radar beacon reply presented as a short line on display.
slash mark   Oblique stroke /.
slash rating   *1* For electric machines and other accessories, normally 150% of base load.
   2 More generally, any special increased rating of an engine or other machine, printed to right of a slash mark.
SLAT, Slat   *1* Slow, low, airborne target.
   2 Supersonic low-altitude target.
   3 Ship-launched air [or aerial] target.
slat   *1* Movable portion of leading edge of aerofoil, esp. wing, which in cruising flight is recessed against main surface and forms part of profile; at high angle of attack either lifts away under its own aerodynamic load or is driven under power to move forward and down and leave intervening slot.
   2 Fixed leading-edge portion of aerofoil, either wing or tailplane (in latter case often inverted, lying along underside), forming slot ahead of main surface. Both (1, 2) postpone flow breakaway at high AOA and thus delay stall.
Slate   Small lightweight altitude-transmitting equipment, beacon for GA (FAA 1961).
slatted   Fitted with a slat (1, 2).
slave   *1* See *slaving*.
   2 See *slave station*.
   3 Adjective descriptive of any item installed in an aircraft purely to check the dimensions and interface connections; hence * engine, * APU, not intended to be operated.
Slave actuator   Actuator, usually ballscrew, forming one of a number transmitting motion originating at a remote power source. Distributed along wing to move flaps, and around engine to move reverser blocker doors.
slave aerial   Mechanically scanned aerial slaved to another, eg SSR slaved to surveillance radar but off-mounted.
slaved gyro   One whose spin axis is maintained in alignment with an external direction, eg magnetic N or local vertical.
slave landing gear   Temporary landing gear used in factory to move incomplete aircraft.
slave shaft   Short shaft on which rotating item is temporarily mounted (possibly loosely) when perfecting balance.
slave station   Radio station whose emissions are controlled in exact synchronization or phase with a master station at different geographical location.

Slavianoff   System of arc welding using metal wire or rod as positive electrode.
slaving   *1* To constrain a body to maintain an attitude in exact alignment with another.
   2 To key a transmitter to radiate in exact phase or synchronization with a master.
SLB   *1* System link budget.
   2 Sidelobe blanking.
SLBM   Submarine-launched ballistic missile.
SLC   *1* Sidelobe clutter, or cancellation.
   2 Software life cycle.
   3 Space launch complex (pronounced slick).
   4 Submarine laser communications.
   5 Source lines of code.
   6 Sonobuoy launch container.
   7 Sensitivity-level command (TCAS).
   8 Synchronous-link control.
SLCM   *1* Submarine- [or sea- , or ship-] launched cruise missile.
   2 Survivable low communications system.
SLD   *1* Short lift, dry.
   2 Supercooled large droplets.
   3 Solid sky cover.
S/LD   Spoiler(s) and lift dumper(s).
Sleaford Tech   Derogatory term for RAF College, Cranwell.
SLECR   Software-loadable equipment configuration report.
sled   Track-mounted wheelless vehicle accelerated to high [often supersonic] speed by rocket[s] on which test devices [such as fighter forward fuselage with ejection seat] can be mounted.
Sleec   Slender lifting-entry emergency craft.
sleet   Precipitation of rain/snow mix or partially melted snow. Two special US usages: frozen rain in form of clear drops of ice, and glaze ice covering surface objects; both highly ambiguous.
sleeve   *1* Sleeve target.
   2 Plastics cylinder used as colour-coded electrical cable marker.
   3 Valve mating with bore of sleeve-valve cylinder.
   4 Fabric tube for filling gas aerostat.
sleeve target   Tapered tube of flexible fabric, open both ends and towed large-end first; can incorporate reflective prism, mesh, MDI or other enhancement.
sleeve valve   Any of various techniques for piston-engine valve gear using one or two concentric sleeves between piston and cylinder with suitably shaped ports in walls lining up intermittently with inlet/exhaust connections on cylinder; usual is Burt-McCollum single sleeve.
slender body   One of such large slenderness ratio that squares and higher powers of disturbances can be ignored.
slender delta   Aeroplane whose wing has ogival delta plan with very low aspect ratio such that at Mach numbers exceeding 2 entire wing lies within conical shockwave from nose.
slenderness ratio   Length/diameter of fuselage or other slender body. Generally, synonymous with fineness ratio.
slender wing   Not defined: any wing of very low aspect ratio.
SLEP   Service-life extension programme.
SLES   Spacecraft life-extension system.
SLEW   Single-tone link-11 waveform (Int.).
slew   *1* To rotate in azimuth.

2 To offset centre of P-type or similar display laterally, eg to study air traffic or surface feature off edge.

3 To rotate gyro spin axis by applied torque at 90°.

slewed flight   Yawed, eg with applied rudder while holding height and with wings level.

slewing   Slew; also defined as changing scale on radar display (not recommended).

slew-wing aeroplane   One whose wing is pivoted as one unit about mid-point, thus as one tip moves forward opposite tip moves aft. In some forms there is no fuselage, and wing obliquity to airflow is determined solely by tip fin(s) and engine pod angles.

SLF   Shuttle landing facility.

SLFCS   Survivable LF communications system (SAC, USAF).

SLFP   Suction-lift fuel pump.

SLG   Satellite landing ground, ie auxiliary field.

SLGPS   Small lightweight GPS.

SLGT   Slight; CHC adds chance = <20% likelihood.

SLH   System-level health.

SLI   *1* Staatliche Luftfahrtinspektion (DDR).

2 Space launch initiative (NASA).

3 System-level interface.

SLIC   Submarine-launched intercontinental missile.

Slice   Internationally agreed subdivision of international funds, usually allocated in * groups or as 1-year * for infrastructure (NATO).

slice   *1* Possibly violent uncontrolled departure in yaw, usually at extreme AOA.

2 Intended rapid yaw.

slice(d)   Maximum-performance hard nose-down turn, over 90° bank.

slice weight   Maximum mass of bird material between consecutive fan blades.

slick   Any streamlined free-fall store, especially GP bomb.

slick wing   One with no provision for pylons or hardpoints.

SLICS   Safe-lane indicator computing system.

slide raft   Escape slide which can be detached and used subsequently as fully equipped life raft.

slidewire   Wire carrying transport trolley providing emergency escape from top of space-launch service tower.

sliding carpet   Moving aircraft-carrier flight deck, never actually tested.

sliding window   Figurative (electronic) window in SSR which looks into each range bin in turn, feeds any traffic or other reply found there to plot extractor and usually also defruits.

SLIM   *1* Surface-launched interceptor missile (concept).

2 Software life-cycle management.

3 Simplified logistics and improved maintenance.

slime light   External low-voltage strip light to facilitate night formation flying.

slim jet   See *single-aisle, narrow-body*.

slinger ring   Channel or pipe around propeller hub (inside spinner if fitted) to which controlled supply of deicing fluid can be fed for centrifugal distribution along blades.

slinging point   Clearly indicated location on airframe or major assembly around which sling of crane can be passed as loop for hoisting.

slip   *1* Sliding towards inside of turn as result of excessive bank.

2 Loosely, any yawed flight causing indication towards centre of turn or lower wing on turn/* indicator, eg forward *, side-*; in particular, controlled flight of helicopter in direction not in line with fore/aft axis.

3 Measure of loss of propulsive power defined as difference between geometric and effective pitch, see *propeller pitch*.

4 Slurry composed of finely divided ceramic or glass suspended in liquid, eg for coating surfaces in precision casting techniques.

5 Difference between speed of induction motor under load and synchronized speed, expressed as percentage.

6 Crystalline defect characterized by (usually local) displacement of atoms in one plane by one atomic space.

7 Launch slipway for marine aircraft.

8 Shackle for bomb or other dropped store (becoming arch.).

9 To change flight crews at one stopping place on airline route.

slip bands   Microscopic parallel lines visible on polished metal stressed beyond yield point.

slip cover   Fabric cover previously cut to shape and sewn, available from store tailored to aircraft type.

slip crew   Airline flight or cabin crew who leave or join as operating crew at intermediate point in multi-sector flight. In some cases crew may continue on same aircraft but off-duty.

slip flow   Flow in extremely rarefied fluid where mean free path is comparable with dimensions of body (see *free-molecular flow, Newtonian flow*).

slip function   Basic propeller parameter, also called effective pitch ratio, see *propeller pitch*.

slip gauge   Extremely accurate wafers of steel of known thicknesses which can be stacked to build blocks accurate to within 0.25 μ (10^{-5} in).

slip-in   See *slip (1)*.

slip joint   One permitting axial sliding, eg in exhaust manifold to allow for expansion.

slippage mark   White rectangle painted on wheel/tyre to show relative rotation.

slip pattern   Planned arrangement for slipping crews detailed on crew roster.

slipper   *1* Adjective describing drop tank or other air-dropped or externally carried store shaped to fit underside and leading edge of wing.

2 Generalized term for precision part designed to slide over another, eg in air or other fluid bearings, or in con-rod big end which mates on periphery of master rod and held in place by rings, there being no ring-type big end but only a driving *.

slippery   Having large momentum and little drag (colloq.).

slipping torque   Torque at which piston-engine starter clutch will slip.

slipping turn   Turn with slip (1).

slipring   Conducting ring rotating with rotor of electrical machine to transfer current without commutating.

slip speed   Supercharger rpm rquired to maintain given pressure differential between inlet and delivery manifold when no air is being delivered.

slipstream   *1* Airflow immediately surrounding aircraft; if behind propeller * is propwash. Velocities are measured relative to aircraft.

2 To follow direction of streamlines, eg in case of freely hinged nozzle tailfeathers or elevator.

3 To follow in wake of another aircraft. Note: (2, 3) are verbs.

slipstream factor   Usually means ratio of mean speed of propeller slipstream relative to aircraft divided by true airspeed.

slipstreaming   Slipstream (2).

slip tank   A tank, eg fuel, water, oil, which can be jettisoned; used in airships and a few aeroplanes but now overtaken by *drop tank*.

slipway   Sloping ramp along which marine aircraft can enter or leave water.

SLIR   Sideways-looking IR.

SLIRBM   Submarine-launched IRBM (proposal, USN).

SL-ISA   Sea level, international standard atmosphere.

slitting shear   Hand or powered shears of lever type used for heavy or very wide sheet (not plate).

sliver fraction   Volume of slivers remaining in case at web burnout divided by total propellant volume, symbol λ_s.

slivers   Fragments of solid propellant which are left unconsumed after rocket-motor burnout.

SLKT   Survivability, lethality and key technologies.

SLM   *1* Standard-length message.

2 Spatial light modulator.

SLMG   Self-launching motor glider.

SLMM   Sea-launched mobile mine.

SLO   Slow.

SLOA   Special Letter of Authorization.

SLOC   *1* Sea lines of communication.

2 Source lines of code [also SLoC].

Slomar   Space logistics, maintenance and rescue.

slope   *1* Glideslope angle to horizontal, usually 3° but steeper for STOL or noise-reduction approaches.

2 Rate of change of one variable with respect to another, tangent from origin to any point on curve plotting y against x or dy/dx.

3 Of runway, mean inclincation, Δh/L expressed as percentage.

slope angle   Acute angle measured in vertical plane between flightpath and local horizontal.

slope line system   Approach-light system giving vertical guidance by appearance of ground lights (UK 1946-50).

slope of lift curve   Unit is increment of lift per radian change in AOA.

SLOR   Swept local-oscillator receiver.

SLOS   *1* Star line of sight.

2 Stabilized long-range observation system, or optical sight.

sloshing   Gross oscillatory motion of liquid in tank sufficient to impose severe structural stress or affect vehicle trajectory; one cause of pogo effect; * is short-term, unlike fuel shift.

sloshing baffles   Transverse perforated bulkheads, usually part of tank structure, to curb sloshing; strictly anti-**.

Slot   Sequential logic tester.

slot   *1* Suitably profiled gap between main aerofoil, esp. wing or tailplane, and slat or other leading-edge portion through which airflow is accelerated at high AOA to prevent breakaway; usually curves up and back to direct air over upper surface but on tailplane often inverted.

2 Gap between wing and hinged trailing-edge surface,

eg flap or aileron, through which air flows attached across movable upper surface.

3 Particular allotted time for using facility (eg gunnery range), for space launch (also called window) or for controlled aircraft departure or arrival, esp. at busy airport; hence to secure a *, to miss one's *.

4 Physical aperture for * aerial.

5 Particular band of aircraft weight or flight performance.

6 Figurative situation or position, esp. a target situation, eg in the * = correctly set up for landing.

7 In carrier flying, to enter landing pattern by flying up ship's starboard side followed by break downwind.

8 Amplifying (7), window in sky about 300 ft to right (stbd) of ship's bows and 600 ft above sea.

slot aerial   Aerial (antenna), eg for DME, in form of slot cut in metal skin, often backed by reflective cavity and aerodynamically faired by dielectric; normally 0.5 wavelength long and 0.05 wide, with polarization usually 90° to plane of slot.

slot/spoiler control   Lateral, and alternatively multi-axis, flight control combining powered variable slot [or leading-edge flap] and upper-surface spoiler [also serving as airbrake when used symmetrically].

slotted aerofoil   One incorporating a fixed slot (1); if slot results from motion of a slat correct adjective is slatted.

slotted aileron   Aileron separated from wing by slot (2).

slotted flap   Flap, usually not translating but simply hinged, forming whole of local trailing edge and separated from wing by slot (2).

slottery   *1* Allocation of capped number of slots (3) by lottery.

2 Arrangement of slots (1), especially when complex (colloq.).

slow-blow fuse   Cartridge designed to withstand brief overload (electrical).

slow-CAP   Combat air patrol to protect slow-flying aircraft.

slow roll   Precision flight manoeuvre in which aircraft, usually fixed-wing, is rolled through 360° by ailerons (using rudder as necessary) while keeping longitudinal axis sensibly constant on original heading; unlike barrel roll imparts -1 g in inverted attitude. Can be performed with longitudinal axis at any inclination in vertical plane. In US called aileron roll.

slow-running cut-out   Pilot-operated valve which stops piston engine by turning off supply of metered fuel (carburettor engine only).

slow-running jet   Fine carburettor jet which alone supplies fuel to piston engine mixture when throttle is at idling position.

SLP   *1* Space-limited payload.

2 Survivor-locator package (USAF).

3 Sequential linear programming.

4 Slope; SLPG sloping.

5 Sea-level pressure.

6 Speed-limiting point.

SLR   *1* Side-looking radar.

2 Slush on runway.

3 Standard lapse rate.

SLRR   Side-looking reconnaissance radar.

SLRS   Space lift range system.

SLS   *1* Side-lobe suppression.

2 Sea-level static [or standard].

3 Self-launching sailplane.

4 Strained-layer superlattice.

5 Satellite landing system.

SLSAR   Side-looking synthetic-aperture radar.

SLSL   Space Life Science Laboratory (KSC, 2003–).

SLST   Sea-level static thrust.

SLS-TO   Sea-level static, takeoff.

SLT   Sleet.

SLTO   Sea-level takeoff.

SLU   *1* Surface-launched unit.

2 Stabilized laser unit.

3 Switching logic unit.

SLUC   System-level use case.

sludge   Viscous slimy deposit gradually formed in lubricating oil by oxidation, water contamination and other reactions.

sludge chamber   Cavity, tube or other region in crankshaft web or crankpin, supercharger drive gear or other rotary component in which sludge is deliberately trapped by centrifugal force.

SLUFAE   Surface-launched unit fuel/air explosive.

slug   *1* Non-SI (UK only) unit of mass = g lbf =14.5939 kg.

2 Pre-rivet forming feedstock for Drivmatic or similar riveting machine.

3 Ferritic cylinder for varying coil permeability or inductance.

4 Metal or dielectric cylinder for waveguide impedance transforming.

5 Slab of alloy from which SFF warhead is formed.

6 Body of water or other contaminant in tank of fuel or other liquid.

slugging   Malfunction in vapour-cycle ECS in which the compressor pumps liquid refrigerant.

slung load   Payload carried below helicopter on single cable. Class A, does not extend below landing gear, is fixed and cannot be jettisoned; Class B is jesttisonable; Class C remains touching land or water.

Slurs   Shoulder-launched unmanned reconnaissance system.

slurry   Suspension of finely divided solid particles in liquid, usually capable of being pumped; physical form of experimental fuels, particles often being metal.

slush   Mixture of snow and water with SG 0.5 to 0.8; below these values is called wet snow, above is called standing water.

slush fund   Money allocated for bribery, usually of persons able to influence potential customers.

slush hydrogen   High-energy H_2 slurry, typically with 116 per cent density of LH_2, in some forms gelatinous and in others a pumpable mix of solid, liquid and gas.

SLV   *1* Satellite, or small, launch vehicle.

2 Space-launched vehicle.

3 Synchronization lock valve.

4 Service-level verifier.

SLW   *1* Short-lift wet [T adds thrust].

2 Supercooled liquid water [C adds content].

3 Slow.

SM   *1* Statute mile (often s.m. or mile).

2 Static margin.

3 Service, or simulation, module.

4 Sandwich moulding.

5 Strategic missile (former DoD designation prefix).

6 Standard missile (USN).

7 Special mission (US).

8 Standards manual.

9 Stockpile memo (NW).

10 Smoke.

sm, s.m.   *1* Statute mile[s].

2 Short emission.

SMA   *1* Squadron maintenance area.

2 Stato Maggiore Aeronautica (I).

3 Surplus military aircraft.

4 Shape-memory alloy.

5 Surface-movement advisor (ATC).

6 Signal message address.

7 Service de la Maintenance Aéronautique, part of DGA (F).

SMAC   Scene-matching area-correlator.

SMAAC   Structural maintenance of ageing aircraft (EU).

SMAE   The Society of Model Aeronautical Engineers (UK, 1922).

small aircraft   *1* One of below 12,500 lb (5,670 kg) MTOW (US).

2 Between 17 and 40 tonnes MTOW (UK wake turbulence).

small arms   All arms up to and including 0.6 in (15.24 mm) calibre (DoD). Despite 'all arms', term normally means guns.

small business   One with fewer than 500 employees.

small circle   That described on spherical or spheroidal surface by intersection of plane not passing through body's centre.

small end   End of connecting rod pin-jointed to piston.

small light aeroplane   Two groups seek definition.

small/medium enterprise   Less than £30m annual turnover, ≤250 employees, ≤25% owned by voting rights in another company (EU).

small perturbation   One for which 2nd and higher-order terms are ignored.

smallsat   Small satellite: FAA ≤ 2,000 lb; ESA ≤ 400 kg (882 lb), ≤ €15 million.

small-scale integration   Usually 10 or fewer gates or other functions per IC.

SM&E   Semiconductor materials and equipment.

SMAP, Smap   *1* Systems-management analysis project (AFSC).

2 Simultaneous MAP (10).

Smart   *1* Secure mobile anti-jam reliable [tactical] terminal.

2 Scalable multiprocessor architecture for real time [U-* = ultra].

3 Spurt message alphanumeric radio terminal.

4 Small-firms merit award for research and technology (UK).

5 Supersonic military aerospace research track.

6 Situation-monitoring analysis and reporting tablet.

7 Smart munition advanced rocket.

8 Small missions for advanced research in, or and, technology (ESA).

smart   *1* Capable of being guided, by self-homing or external command, to achieve direct hit on point target.

2 Generalized term for clever, eg smart jammer listens for hostile emission and then jams on correct wavelength.

smart actuator   Containing an embedded processor.

smart bogey   Formidable opponent in air combat.

smart display unit   Combines functions of mission

computer, colour-graphics processor and display/control panel.

smart electromechanical actuator   Based on brushless motors using rare-earth magnets, with position feedback using Hall-effect sensors.

smart fuze   Fuze incorporating linear accelerometer and processor chip to measure decelerations after first contact with target and detonate warhead at a predetermined point.

smart graphics processor   Locally generates display imagery on which overlays merge external video.

smartlet   Smart bomblet.

smart skin   *1* External skin incorporating microstructures in micron range of size which can gang together like phased arrays to allow transmission, reception and processing of EM information in skin surface.

2 Loosely, any electrically conductive skin.

smart-T   Secure mobile anti-jam reliable tactical terminal.

Smash   Southeast Asia multisensor armament system, helicopter.

Smatcals   Signature-managed ATC approach and landing system (USN).

SMAU   Stop-motion aim-point upgrade.

SMAW   Shielded metal-arc welding.

SMB   Side marker board, for airport parking guidance.

SMC   *1* Standard mean chord.

2 Surface movement control (ICAO).

3 Space and Missile-systems Center (USAF, Los Angeles AFB).

4 System management and communication.

SmCo   Samarium cobalt, chief rare-earth magnetic material accepting 20 to 30 times normal current for short overload periods.

SMCS   *1* Spoiler mode control system.

2 Structural mode control system.

3 Survivable missile control station (USAF).

SMD   *1* Shop modification drawing.

2 System management directive.

3 Surface-mounted device (electronics).

4 Sauter mean diameter.

5 System maintenance diagnostics.

SMDARS   Sea-based mid-course defense advanced radar suite.

SMDC   *1* Shielded mild detonating cord.

2 Space & Missile Defense Command (USA).

SMDI   Smart-motion de-interlacing.

SMDP   Standardized military drawing program, for microcircuits (US).

SMDPS   Strategic-missile defense and planning system.

SMDS   Switched multi-megabit data service.

SMDU   Strapdown magnetic detector unit.

SME   *1* Small/medium enterprise.

2 Special mission equipment.

3 System-management entity.

4 Solar mesosphere explorer.

5 Society of Manufacturing Engineers (US).

smear   Degraded radio reception due to another transmission on same frequency or degraded TV picture due to ghost image closely following primary image.

smear camera   See *streak camera*.

smearer   Subcircuit to eliminate pulse-amplification overshoot.

smear metal   Metal melted by high-speed machining or welding and deposited on workpiece.

SMEAT   Skylab medical experiments altitude test.

SMEC   Strategic Missiles Evaluation Committee (USAF, formerly).

smectic phase   Liquid-crystal phase having layered structure with constant preferred direction; flow is abnormal and X-ray diffraction pattern is obtained from one direction only.

SMED   Single-minute exchange of dies [to eliminate extended set-up times].

SMEI   Solar mass ejection imager.

SMER   Smart multiple ejector rack.

SMES   *1* Strategic Missile Evaluation Squadron.

2 Superconducting magnetic-energy storage.

SMET   Simulated mission endurance testing.

SMEU   Switchable main electronic unit.

SMF   Sintered metal fibre.

SMFA   Service du Matériel de la formation Aéronautique (F).

SMG   *1* Sync/message/guard time-slot.

2 Spinning-mass gyro.

3 System Management Group (ICAO AAG).

SMGCS   Surface-movement guidance and control system, pronounced smigs.

SMHMS   Standardized magnetic helmet-mounted sight (USN).

SMI   *1* Structural merit index.

2 Standard message identifiers.

3 San Marco Island.

SMILS   Sonobuoy missile-impact location system.

S_{min} Minimum detectable signal power.

Smith-Barry   Pioneer formalized system of flying training, 1914.

Smith diagram   Standard plot for solving electrical transmission-line problems.

SMK   Smoke.

SML   Small.

SMLS   *1* Interim standard MLS.

2 Seamless (pipe, tube).

SMLV   Standard memory-load verification.

SMM   *1* Space manufacturing module.

2 Solar maximum mission.

3 Service Météorologique Métropolitain (F).

SMMC   System maintenance monitoring console.

SMMR   Scanning multi-frequency microwave radiometer.

SMO   *1* Supplementary meteorological office.

2 Synchronized modulated oscillator.

3 Shelter management office.

smog   Fog contaminated by liquid and/or solid industrial pollutants, particularly smoke.

SMOH   Since major overhaul; suffixes LE, RE = left/right engine.

SMOLED   Often pronounced smo-led, small-molecule organic LED.

smoke apparatus   Aircraft installation for leaving either a smoke screen or, today more often, smoke trail of desired colour for display purposes.

smoke bomb   Air-dropped pyrotechnic, able to float, for indicating wind velocity.

smoke box   Container of slow-burning fuel for producing wind-indicating smoke trail.

smoke float   See *smoke bomb*.

smokehood   Light but fire-resistant transparent bag enveloping wearer's head to offer short-term protection against smoke and toxic fumes; must withstand decompression.

smokejumper   Firefighter who parachutes, abseils, or [less often] is air-landed, on burning area.

smoke pot   Remotely triggered device fired to indicate a hit on ground target.

smoke tunnel   Not precisely defined; wind tunnel in which either general recirculating smoke or discrete streams give visible flow indication.

SMP   *1* Self-maintenance period (aircraft carrier).

2 System management processor (1553B data bus).

3 Sintered metal powder.

4 System management and performance [testing].

5 Stores-management processor.

6 System main processor (SMGCS).

SMPS   Switched-mode power supply.

SMR   *1* Surface-movement radar (ICAO).

2 Svenska Mekanisters Riksförening (Sweden).

3 Selective message routing.

4 Stores management and release; S adds system.

SMRD   Spin motor detector (rate gyro).

SMRS   Stores management and release system.

SMRT   Soldier metabolic remote telemonitor (AFRL).

SMS   *1* Strategic Missile Squadron (USAF).

2 Stores-management set, or system.

3 Suspended manoeuvring system.

4 Supply and movements squadron (RAF).

5 Space mission simulator.

6 Sensor monitoring set.

7 Smart materials and structures.

8 Synchronous meteorological satellite.

9 Setting mini-station.

10 Signal-measurement system.

11 Short message, or messaging, service.

12 Spectrum monitoring system.

13 Safety management system (airline).

14 Stratosphere, or stratospheric, and mesosphere, or mesospheric, sounder.

SMT   *1* Shadow-mask tube.

2 Système modulair thermique, IR common module (F).

3 Sector management tool.

4 Surface-mount[ed] technology.

5 Static/mobile/transportable.

6 Square-mesh track.

7 Servo-mount [elevator, or aileron/rudder].

8 Station management.

9 Standard-message text.

SMTC   Space and Missiles Test Center (Vandenberg AFB, USAF).

SMTD   Stol/manoeuvring technology demonstrator.

SMTH   Smooth.

SMTI   Selective moving-target indicator.

SMTO   Space & Missile Test Organization (AFSC).

SMTP   Standard mail transfer protocol.

SMTS   *1* Store (4) management test set.

2 Space and missile tracking system.

SMU   System[s]-, or sensor, management unit.

SMUD   Standoff munitions disrupter, for destroying unexploded anti-airfield munitions.

Smurf   Side-mounted under-root fin, small curved surface ahead of and below LE root of horizontal tail to eliminate pitch-down in low-airspeed manoeuvres.

SMV   Space maneuver vehicle (US).

SMW   Strategic Missile Wing (USAF).

SMWHT   Somewhat.

SMWP   Standby master warning panel.

SN   *1* Snow (ICAO).

2 Since new (often S/N).

3 Shipping notice.

4 Scout trainer (USN aircraft category 1939-48).

5 Secretary of the Navy (US).

6 Strategic navigation.

7 Subnetwork.

S/N   *1* Stress against number of alternating load cycles to failure; S is normally ratio of alternating load to ultimate strength, so that for S = 1 N = 1, while for S = 0.8 N may be 10^4.

2 Serial number.

3 Signal/noise ratio.

Sn   Tin.

sn   Sthène.

SNA   *1* Sindicator Nacional dos Aeronautas (Brazil).

2 System network analysis, or architecture.

3 National aero club (Slovakia).

SNABV   Syndicat National des Agences et Bureaux de Voyages (F).

SNAC   Subnetwork access; P adds protocol.

SNAEC, Snaec   Special notice to aircraft and engine contractors (UK, MoD).

snag   *1* Fault condition or impediment to progress (UK, colloq.), hence * list.

2 Dogtooth or other abrupt discontinuity in leading edge.

Snagfa   Syndicat National des Agents et Groupeurs de Fret [freight] aérien (F).

snake drill   Hand drill with tool bit driven by flexible connection.

snake mode   Control mode in which pursuing aircraft flies preprogrammed weaving path to allow time to accomplish identification functions (DoD).

snaketrack   Weaving flight path, under pilot control or preprogrammed, in case of aerial target to provide greater challenge to defences.

snaking   Natural oscillation in yaw at approximately constant amplitude.

SNAP, Snap   *1* Systems for, or supplementary, nuclear auxiliary power; major programme for space electric power generation using RTG and similar methods.

2 Synchronous numeric array processor, an add-on to enable small computers to act as simulators.

3 Steerable null antenna processor.

Snapac   Sindacarto Nazionale Autonomo Personale Aviazione Civile (I).

snap-action   *1* Positive full-range movement of bistable device, esp. mechanical, eg spring-loaded two-pole switch or valve.

2 Various meanings in electronics, esp. abrupt jump in output of magnetic amplifier with large positive feedback.

snap-down   Ability to see, engage and destroy target at much lower level, esp. one very close to surface; hence * missile. Crucial factor is ability of radar or other sensor to see target against ground clutter.

snap gauge   C-frame go/no-go gauge for shaft measures, usually with one anvil adjustable.

snap in Of combat aircrew, to enter and connect up to aircraft prior to mission.

snap report Preliminary report by aircrew of observations, prior to compilation of mission report (ASCC etc). Term not to be used (DoD).

snap ring Sprung fastener which locks into peripheral groove on either inside or outside diameter.

snap roll See *flick roll*.

snap-shoot Traditionally, quickly aimed shot; in modern air combat, shooting with fixed gun without need for prior tracking using correctly interpreted HUD sight symbology, usually providing a tracer line.

snapshot 'Photograph' of all input parameters recorded at particular points in time by HUM to log steady-state conditions for future long-term analysis.

snap start Prelaunch AAM condition in which weapon is pre-tuned to guidance radar and then returned to passive mode but ready for instant launch.

snap-up Rapid maximum-performance pull-up to engage target at higher altitude. US aerodynamicists have called the Cobra manoeuvre a snapup [one word].

snap-up missile AAM capable of engaging and destroying target aircraft at much greater height than launch platform.

SNavO Senior, or station, navigation officer (RAF).

SNAW School of Naval Air Warfare (St. Merryn, UK).

SNC *1* Standard navigation computer; small leg-strapped box giving continuous moving-map readout showing aircraft position.

2 Strategic Nuclear Command (India).

SNCO Senior NCO.

SNCR Subnetwork connection reference.

SNCTA Syndicat National des Contrôleurs du Trafic Aérien (F).

SNCTAA Syndicat National des Cadres et Techniciens de l'Aéronautique et de l'Astronautique (F).

SND Secondary navigaton display.

SNDC Subnetwork dependent convergence; F adds function, P protocol.

SNDV Strategic nuclear delivery vehicle[s].

SNEA Sindicato Nacional das Empresas Aeroviarias (Brazilian Air Transport Association).

Snell law Law of index of refraction, n, $\sin \theta = n_2 \sin \theta_2$ where n are refractive indices of two media.

SNF Short-range nuclear force(s).

SNFLK Snowflakes.

SNG Synthetic (or synthesized, or substitute) natural gas.

SNI *1* Signal/noise index (Omega).

2 SNA network interconnection.

3 Stand number indicator.

SNICF Subnetwork independent convergence function.

snifter(s) Small spin motors (often one motor with tangential nozzles) for small (eg anti-tank) missiles.

SNII Aeroplane scientific test institute of GVF (USSR, R).

Snipag Syndicat National des Industriels et Professionnels de l'Aviation Générale (F).

snips Hand shears for cutting sheet metal.

Snirfag Syndicat National des Industriels Réparateurs et Fournisseurs de l'Aviation Générale (F).

SNL Sandia National Laboratory (US).

SNLE *1* Sous-marin nucléaire lanceur engins (missile-firing submarine, F).

2 Subnetwork link establishment.

SNM Special nuclear material.

SNMP Simple network management protocol.

SNO Snow.

SNOE Smart noise operation equipment (ECM).

SNOINCR Snow depth increase in past hour.

Snomac Syndicat National des Officiers Mécaniciens de l'Aviation Civile (F).

Snorac Syndicat National des Officiers Radios de l'Aviation Civile (F).

snorkel Pipe through which a helitanker can refill its tank[s] while hovering over a source.

Snort Supersonic naval ordnance rocket track.

snort Submarine schnorkel pipe, esp. tip seen on radar above ocean.

Snorun Snow on runway.

snout In a gas-turbine engine with one or more drum-like combustion chambers, the entrance for primary air upstream of the burner, also called primary-air scoop. Absent from annular chamber.

snow *1* Precipitation in form of feathery ice crystals (BSI); better definition is: small (under 1 mm) grains, granular *; long (2 + mm) grains, ice needles; large agglomerations in form of flakes, *. Dry * is SG 0.2 to 0.35; wet * is 0.35 to 0.5.

2 Speckled interference on electronic display.

3 Air-intercept code: sweep jamming (ie display looks like *).

Snowdrop Service Policeman (USAAF, from white helmet; RAF term was snoop).

snow gauge Combination of rain gauge and vertical measuring stick to determine snow moisture content.

snow lights Specially designed runway-edge lights which stand above level of any snow yet snap off if struck by aircraft.

snow pellets Small white opaque pellets of water/ice, softer than hail.

snowplough mode Use of canards as airbrakes after landing.

snow static Severe R/F interference caused by snow.

Snowtam Special-series Notam announcing presence or removal of hazardous conditions due to snow, ice, slush or standing water in association with these on movement area.

SNPA Subnetwork point of attachment.

SNPC National civil protection (rescue) service (I).

SNPDU Subnetwork protocol data unit.

SNPL Syndicat National des Pilotes de Ligne (F).

SNPNAC Syndicat National du Personnel Navigant de l'Aéronautique Civile (F).

SNPO Space Nuclear-Propulsion Office (AEC, NASA).

SNR, S/NR Signal to noise ratio.

SNRS Sunrise.

SNS Secure network server.

SNSDU Subnetwork service data unit.

SNSH Snow showers.

SNST Sunset.

SNTA Syndicat National des Transporteurs Aériens (F).

snubber *1* Device which greatly increases stiffness of elastic system whenever deflection or travel exceeds given limiting value, eg rubber block to arrest travel of shock suspension and special features (often hydraulic) to arrest travel of actuator at full stroke.

2 Very loosely used to mean part-span shroud on fan blade.

SNVQ   Scottish National-Vocational Qualification.

SNVS   Stabilized night-vision system.

SNW   Snow.

SNWFL   Snowfall.

SNZF   Sintered nickel/zinc ferrite (RAM).

SO   *1* Second Officer.
2 Scout observation (USN aircraft category 1934-46).
3 Special order.

SOA   *1* Spectrometric oil analysis.
2 State of the art.
3 Special Operations aircraft.
4 Space-surveillance network optical augmentation.
5 Separate operating agency.

SOACMS   Special-operations aircraft combat-mission simulator.

SOAG   Special Operations Aviation Group (USA).

SoAG   School of Air Gunnery.

soakdown   Period after engine shutdown when heat is dissipated.

SOAP, Soap   Spectrometric oil-analysis programme.

SOAR   *1* Shuttle Orbiter applications and requirements.
2 Special Operations Aviation Regiment (USA).

soar   To prolong sailplane flight by seeking upcurrents, especially thermals.

Soarex   Sub-orbital   aerodynamic   re-entry experiment[s].

SOAU   Special-operations avionics upgrade.

SOAWS   Satellite on-board attack warning system.

SOB   *1* Souls on board, traditional maritime count of everyone on board.
2 Stand-off bomb.

SOC   *1* Struck off charge, no longer on unit strength.
2 Shut-off cock.
3 Sector Operations Centre (RAF).
4 Satellite, or systems, operations complex.
5 Start of climb.
6 Single overhead camshaft.
7 State of charge.
8 System on a chip.
9 See *Socom*.

SOCC   *1* Sector Operations Control Centre.
2 Space Operations Control Center (NOAA, Suitland, MD).

SOCJ   Standoff communications jammer.

socked-in   Airfield closed by weather, especially by fog (colloq.).

Socom   Special Operations Command (USA).

Socrates   Sensor for optically characterizing remote turbulence emanating sound (FAA), i.e. laser detection and plotting of turbulent wakes.

Socus   South Continental US Loran chain.

SOD   Satellite Operations Directorate.

Soda   Statement of demonstrated ability.

Sodals   Simplified omnidirectional approach-light system.

Sodar   Sonic detection and ranging, usually for wind velocity and turbulence.

sodium   Na [natrium], silvery reactive metal, density 0.97, MPt 98°C, vast range of compounds and used as heat-transfer medium in piston-engine exhaust valves.

sodium acetate   Soluble solid used as solid or liquid deicer, principally for airfields.

sodium light   Deep yellow, wavelength 589.6 nm, approach lighting loosely called sodiums, same wavelength for * line-reversal pyrometer.

SODP   Start-of-deceleration point.

SOE   The Society of Engineers (UK, 1854–).

SOF   *1* Service Officiel Français (F).
2 Special Operations Force(s) (USAF).
3 Stand-off flare (IRCM).
4 Strategic offensive forces.
5 Supervisor of flying.
6 Safety of flight [I adds issue].
7 Satellite Operations Facility (UK).

SOFA   Status of forces agreement.

SOFAG   Special-ops force assistance group (USAF).

Sofar   Sound fixing and ranging, technique for fixing position at sea by time-difference measures of sound from explosion (eg of depth charge, usually at considerable depth) or impact of spacecraft on surface. Hence * bomb, special sound-producing bomb.

Sofats   Special Operations Forces aircrew training system.

SoFC   School of Flying Control.

SOFI   *1* Sprayed-on foam insulation.
2 See *SOF (6)*.

Sofia   Stratospheric observatory for IR astronomy.

Soflam   Special operations forces laser marker.

Sofnet   Solar observing and forecasting network (DoD, 1963-72).

Sofprep   Special Operations Forces, planning, rehearsal and execution preparation.

S of S   Secretary of State, = minister (UK).

SOFT   Site operational functional test.

Soft   See *soft keys*.

soft   Not hardened against NW explosion.

soft-blade propeller   Made of wood, glass-fibre or other abradable material.

soft bomb   Bomb dropped for purpose other than causing damage or casualties.

soft failure   *1* Usual interpretation is cessation of function without any incorrect function (eg no hardover signal).
2 In EDP, short-term transient failure followed by return to normal; believed caused by alpha particles.

soft flutter   Flutter that is possibly severe but non-divergent, and confined within apparently safe amplitude limits.

soft hail   See *snow pellets*.

soft-in-plane   Semi-rigid helicopter main rotor of subcritical type, fundamental lag frequency being less than N_1 (rotor rpm) at normal operating speed.

soft iron   Iron containing little carbon, as distinct from steel; loses nearly all magnetism when external field is removed.

soft keys   MFD keys which give direct pilot control of individual system of other hardware items via databus. Function of each is displayed by caption.

soft landing   Gentle landing as distinct from hard (free fall); term applies to arrival on surfaces other than Earth.

soft obstacle   *1* Conceptual obstacle, eg 35 ft screen.
2 One physically present, eg ILS localizer, but whose engineering design minimizes damage to colliding aircraft.

soft radiation   Unable to penetrate more than 100 mm of lead.

soft ride   Ride, esp. in lo-flying aircraft at high-subsonic speed, judged comfortable and in no way rough enough to impair crew functions. Normally defined as fewer than two 0.5 g bumps per minute (see *LLDF*). Occasionally selected mode in TFR system.

soft skin   Composite structure which removes major loads from outer skin and concentrates stresses or underlying [usually under/over woven] fibre stiffeners.

soft target   *1* Not hardened or armoured, such as house or merchant ship.

2 In air-to-air firing practice, a sleeve or banner.

soft tooling   Tooling whose dimensions can be adjusted, normally within small limits (eg could not accommodate different 707 fin sizes, which required new tooling).

soft undocking   One that does not influence subsequent trajectory, eg by not using separating thrusters.

soft valve   Thermionic valve into which some air has leaked.

software   *1* All programs and component parts of programs used in EDP (1), eg routines, assemblers, compilers and narrators. Divided into two parts. Basic *, usually provided by equipment manufacturer, is machine-oriented and is essential to permit or extend use of particular hardware; examples are diagnostic programs, compilers, I/O conversion routines and programs for file or data-management. Application * is normally user-oriented and often compiled by user or a subcontracted * house, to enable machine to handle specific tasks; may include GP packages, eg NC tool, payroll or airline booking, or locally created programs for highly specific tasks, eg exploring flight characteristics of unbuilt aircraft.

2 Ambiguously, also used for parachute or drogue packs, esp. those installed in aircraft. This usage is potentially misleading.

software-enabled control   Systems giving intelligent UAVs ability to respond autonomously to external threats and internal faults.

SOG   *1* Special Operations Group (US).

2 Singlet oxygen generator, supplies excited oxygen for high-power COIL.

SoGR   School of General Reconnaissance.

SOH   *1* Since overhaul.

2 Start of header.

Soho   Solar and heliospheric observatory.

SOI   *1* Space object identification (Norad).

2 Silicon-on-insulator (Mosfet).

3 System operator instructions.

SOIA   Simultaneous offset instrument approach.

SOIC   Small outline integrated circuit.

SOICA   The State Organization for Iraqi Civil Aviation.

SOIR   Study Group, operations on parallel instrument runways.

SOIS   *1* Silicon on insulating substrate.

2 Space-object identification system.

SOIT   Satellite operational implementation team (FAA).

SOJ   Stand-off jammer, or jamming.

SOLAP   Shop-order location and reporting.

Solar   Shared on-line automated reservation system.

solar apex   Point on celestial sphere towards which Sun is moving.

solar array   Large assembly of solar cells, on rigid frame, folding, in roll-up sheet or other geometric form.

solar atmospheric tides   Cyclic variations in atmospheric pressure ascribed to Sun's gravitation, with primary 12-h component (about ± 1.5 mb at Equator, 0.5 in mid-latitudes) and much smaller 6-h and 8-h effects.

solar battery   See *solar cell*.

solar cell   Photoelectric (photovoltaic) device converting sunlight direct into electricity, usually by liberating electrons and holes in silicon p-n junction.

solar chamber   Test chamber in which is simulated solar radiation outside atmosphere.

solar constant   Rate at which solar radiation is received outside atmosphere on unit area normal to solar radiation at Earth's mean distance from Sun, about $1.38769 \, \text{kW/m}^2$.

solar cycle   Approx 11-year cycle in sunspot frequency.

solar day   *1* Time between two successive solar transits of same meridian, ie time for Earth to rotate once on its axis with respect to Sun (mean or apparent), see *sidereal year*, *solar year*.

2 Time for Sun to rotate on its axis with respect to fixed stars.

solar flare   See *flare (2)*.

Solar Happ   Solar high-altitude powered platform.

solari board   Usual type of large electromechanical display at airports indicating flight arrivals and departures.

solar noise   Solar radiation at RF frequencies.

solar paddle   Solar-cell array on fixed frame resembling paddle.

solar panel   Any fixed planar solar array.

solar-particle alert network   Global observation system to warn astronauts of solar flares.

solar propulsion   Loose term sometimes used for rocket systems based on electric power, which can be derived from solar cells, but best restricted to solar sailing using a sail (2).

solar radiation   Solar constant.

solar simulator   Device for simulating solar radiation outside Earth's atmosphere.

solar wind   Plasma radiating from Sun, assumed equally in all directions, which in vicinity of Earth has T about $200,000°K$, V about 400 km/s and density of 3.5 particles/cm^3; grossly distorts terrestrial magnetic field, causing upstream shock front; another effect is to blow comet tails downstream from Sun.

Solas   Safety of life at sea.

solar year   365 days 5 h 48 min 45.5 s.

solderless splice   Joint made by crimping or machine-wrapping.

SOLE   *1* Start-of-life efficiency of thermoelectric module or other progressively degraded power-conversion device.

2 Society of Logistics Engineers (US).

solenoid   *1* Range of simple electromagnetic devices in which current in a coil (usually of cylindrical form) moves iron core, eg to operate a switch.

2 Tube formed in space by intersection of two surfaces at which a particular quantity (eg pressure, temperature) is everywhere equal.

Solic   Special-operations low-intensity combat.

solid   Apparently immovable [flight controls, especially ailerons].

solid angle   Portion of space viewed from given point and bounded by cone whose vertex is at that point, measured by area of sphere of unit radius centred at same point cut by bounding cone. SI unit is steradian.

solid conductor   One containing single wire.

solid fuel   Preferably reserved for fuel not used as propellant but as energy source, eg for EPU. For rocket, see *solid propellant*.

solidity   *1* Ratio of total area (not projected area but integral of chord lengths across length of blade) of propeller or rotor to disc area. Basic measure of proportion of disc occupied by blades.

2 At standard radius, ratio of sum of blade chords to circumference, which is not same as (1) since blades of different rotors or propellers are not all same plan shape.

solid motor   Rocket filled with solid propellant.

solid nose   Term to distinguish aircraft nose with metal skin or radome from others of same aircraft type which are glazed (colloq.).

solid propellant   Rocket propellant containing all ingredients for propulsive jet in solid form, either in cast, extruded or otherwise prepared grain or in granular, powder, multiple-rod or other form. Some definitions questionably exclude non-monolithic forms.

solids   In ag-aviation, non-liquid chemicals, eg powders, dusts and granules.

solid-shaft engine   Not free turbine.

solid-state devices   Electronic devices using properties of solids, especially semiconductors.

solid-state oxygen   Alkali-metal chlorates which can readily be made to yield free oxygen.

solid surface   RF reflector, esp. for very large aerial, whose reflective surface is not wire mesh but aluminium sheet.

solid target   Not banner or drogue but towed aerodyne.

solid wire   Single tinned or galvanised steel wire.

SOLL, Soll   Special operations, lo-level.

solo   According to some definitions, pilot flying unaccompanied by instructor, but this could admit passengers; invariably means pilot is only human occupant of aircraft.

solstice   Either of two points on ecliptic furthest from celestial equator, direction of Sun's centre at maximum declination; N hemisphere is summer *, about 22 June; S hemisphere is winter *, about 23 December.

solstitial colure   Celestial great circle through poles and solstices.

solution heat treatment   First stage of heat treatment of certain light alloys in which salt bath (often $NaNO_3/KNO_3$) is used for accurate heating followed by room-temperature cooling or quenching.

solvent extraction   Various processes in which solvents, often hydrocarbons, eg propane, are used to separate lube-oil products from pipe-still distillates. Other meanings in processing of coal.

SOM   *1* Stand-off missile.

2 Simulation object model.

3 Search on the move [sensor].

4 Servo optical mechanical.

5 System operator manual.

6 Side oblique mode.

somatic   Affecting exposed individual only, as distinct from offspring.

somatogravic   Relating the human body to acceleration; * illusion is dangerous feeling that overshooting aircraft

(especially jet) is entering steep climb when it is actually very close to ground.

Sommerfield matting   Mass-produced airfield pavement in form of 75 ft (22.86 m) rolls of 13 SWG wire mesh reinforced at 8 in (203 mm) intervals by steel rod with hooked end linking to adjacent strip (UK, WW2).

SON   *1* Statement of operational need; regarded as a statement of deficiency.

2 Silicon-on-nothing, Mosfet supported at edges only.

sonar   From sound navigation and ranging. Use of word as method of communication under water, detection of surface or submerged targets and measurement of range, and in some cases bearing and relative speed. Analogous to radar, and similarly may be active (emitting high-energy sound waves of tailored form, eg from * transducer) and working with reflections from all submerged objects, or passive, in which receivers listen for sounds emitted from targets.

sonar capsule   Device giving enhanced echoes to sonar to assist location of marine object, eg floating RV or space payload.

sonar transducer   Translates electrical energy into high-intensity sound, normally used in multiple to form a sonar stave (typically radiating 500-1,000 W), which in turn is used in multiple to form 360° or directional array.

SoNC   School of Naval Co-operation.

sonde   Airborne telemetry system, to transmit meteorological or other atmospheric data.

sone   Primitive unit of perceived noise equal to that from simple 1 kHz tone 40 dB above listener's threshold. Subjective judgement of any sound enables it to be expressed in sones (0.001 * = millisone); useless for noise investigation (see *noise*).

sonic   *1* Pertaining to local speed of sound.

2 Approximately at local speed of sound.

sonic bang   Noise heard as shockwave(s) from supersonic object pass hearer's ears; small object at close range generates sharp crack, normal aircraft at close range one or more loud bangs (resembling close thunder) and distant SST dull boom(s) resembling distant thunder (see *boom signature*). Crack of a whip and natural thunder are both examples.

sonic barrier   See *sound barrier*.

sonic boom   See *boom, sonic bang*.

sonic drilling   See *ultrasonic machining*.

sonic erosion   See *ultrasonic machining*.

sonic fatigue   Suffered by structure, especially thin sheet, subject to intense sound.

sonic line   Curved surface above or below wing or other body which has accelerated flow beyond Mach 1, at which M = 1, enclosing region of supersonic flow terminated at rear by shockwave.

sonics   *1* Aggregate of installed sonars, sonobuoys and displays in platform, eg in aircraft.

2 Technology of applying sound to functions other than those related to hearing.

sonic soldering   See *ultrasonic bonding*.

sonic speed   Local speed of sound, symbol a.

sonic venturi   Venturi in which sonic speed is reached at throat, thereby automatically limiting maximum flow (eg in bleed systems).

Sonne   Pioneer German long-range navaid, developed from radio range, became Consolan.

sonobuoy   Discrete sonar devices immersed or dropped

into water; can be active (emitting) or passive, directional or non-directional, and except when dunked by helicopter normally provide readout by radio, usually upon command.

sonodunking   Action of dunking permanently attached sonobuoy.

SOO   Standard operations (or operational) orders.

SOON, Soon   Solar-observatory optical network (USAF).

SOP   *1* Standard operating, or operational, procedure, or platform, or practice.

2 Confusingly, special operating procedure[s].

SOPA   Standard operating procedure amplified.

SOPC   Shuttle Operations and Planning Complex (USAF).

Sopemea   Société pour le Perfectionnement des Matériels et Equipements Aérospatiaux (F).

SOR   *1* Struck off records.

2 Specific, or statement of, operational requirement.

3 State of readiness.

4 Specific Operational Requirement (USAF).

sorb   To acquire gas by sorption; hence sorbent material.

Soreas   Syndicat des Fabricants d'Organes et d'Equipements Aéronautiques et Spatiaux (F).

SORO   Scan on, receive only.

sorption   Taking up of gas by absorption, adsorption, chemisorption or any combination of these.

SORT   *1* Structures for orbiting radio telescopes.

2 Simulated optical range tester.

sortation   Process of reading airline baggage bar code and online * messages and directing item accordingly; must handle airline's own code, IATA 10-digit, Code 39 [old USPS] and Code 93 [new USPS].

sortie   An operational flight by one aircraft. It is generated when the aircraft takes off. Some authorities insist mission must be offensive against surface target. Thus an effective * crosses enemy frontier or front line, an accredited * places bomb[s] on target.

sortie capacity   Maximum number of sorties mounted by unit or other airpower source (eg one airfield) in stated period, usually 24 h.

sortie generation   Ability of combat unit to put its aircraft in air, especially around clock.

sortie number   Reference identifying all images secured by all sensors on one air reconnaissance sortie.

sortie plot   Map overlay representing area(s) covered by imagery during one sortie.

sortie rate   Number of combat missions actually performed by unit (eg squadron or polk) in 24 h period.

sortie reference   See *sortie number*.

SOS   *1* International distress signal.

2 Silicon on sapphire.

3 Special, or Space, Operations Squadron (USAF).

4 Sidewall overhead stowage in passenger airliner.

5 Stabilized optical sight.

6 Squadron Officer School (AU).

Sostar   Stand-off surveillance and target acquisition radar (F, G, Spain, Netherlands).

Sosus   Sound surveillance system.

SOT   *1* Stator outlet temperature.

2 Small outline transistor.

3 Specific operational test.

4 Stand-off tactical [countermeasures evaluation trainer].

5 Solar optical telescope.

Sotas   Stand-off target acquisition system.

SOTD   Stabilized optical tracking device.

SoTT   School of Technical Training.

SOTV   Solar orbit transfer vehicle.

sound   Longitudinal pressure waves transmitted through elastic medium. In atmosphere velocity is a =

$$\sqrt{\frac{\gamma p}{\rho}}$$ = c332.2 ms^{-1} at 0°C and 344 at room temperature; in fresh water 1,410 ms^{-1}, in sea-water 1,540. Audible to ear at frequencies c20 Hz to 20 kHz. See *noise*.

sound attenuation   Reduction in sound intensity, esp. through deliberate conversion to other energy forms, eg heat, in absorbent or other layers of material.

sound barrier   Conceptual barrier to manned flight at supersonic speed when this was extremely difficult, ie before about 1952.

sound energy   Measure of either total emitted energy (for brief sound, eg explosion) or sustained rate of energy transfer for prolonged sound, in latter case measured in watts.

sounding   *1* Any penetration of natural environment for observation or measurement.

2 Complete set of measures taken in and of upper atmosphere for met or other purpose. Hence * balloon, unmanned free balloon carrying upper-atmosphere instruments; * rocket, stabilized but usually unguided rocket carrying upper-atmosphere instruments.

sound intensity   Average sound power passing at given point through unit area normal to propagation, expressed either in W/cm^2 or as sound level.

sound level   Ratio of sound power to a zero reference, expressed in dB (see *noise*).

sound locator   Device for concentrating incident noise from aircraft, usually by one or more large exponential horns rotated in azimuth and elevation until intensity is maximum.

sound power   Sound energy (rate for sustained sound) in watts.

sound pressure   Total instantaneous pressure at point at given time minus static pressure; unit is N/m^2.

sound pressure level   20 log SPL/reference pressure (see *noise*).

sound probe   Instrument responding to sound, eg sound pressure, without significantly altering sound field.

sound ranging   Determining location of sound source by measuring times of arrival at different locations.

sound suppressor   See *suppressor*.

sound wave   Disturbance conveying sound in form of longitudinal alternate compressions/rarefactions through any medium. Frequency spread may be much greater than human aural range, and extreme-energy case is blast wave, which initially (like shockwave) travels faster than sound.

souped up   Tuned to generate maximum possible power [engine of racing aircraft] (colloq.).

source   *1* Contractor for entire article, eg aircraft or missile (see *second* *). More recently, a * of small items or material for restoration.

2 Origin of noise.

3 Origin of fluid in fluid flow, or of large uniform air mass in atmosphere.

4 Solid-state electrode connection corresponding to cathode.

5 Verb, to assign a * 1.

source noise　Generated noise, that emitted by source in all directions as distinct from that received by observer.

Sourdine　Study of optimization procedures for decreasing the impact of noise around airports.

souris　Inlet centrebody shock-cone (F, literally 'mouse').

SOV　*1* Shut-off valve.

2 Simulated operational vehicle.

3 Space operations [or operating, or observation] vehicle, part of MSP.

SOV-AB　Replacement persistent Toxic-B lethal gas and dispenser system (USSR).

sovereignty　ICAN Rule　1 decrees every state has "complete and exclusive * over the airspace above its territory."

SOW　*1* Stand-off weapon.

2 Statement of work.

3 Special Operations Wing.

SOWG　Science Operations Working Group (JPL).

SP　*1* Stabilized platform.

2 Speed brake.

3 Scheduled passenger.

4 Staging post.

5 Self-propelled.

6 Single-phase.

7 Schedule planning.

8 Software protocol.

9 Solar-powered.

10 South Pacific ocean (ICAO).

11 Snow pellets.

12 Special, or special performance.

13 Self-protect (ARM mode).

14 Space.

15 Service provider.

S/P　*1* Speed/power measurement point, ie 1 g, level flight, constant V.

2 Serial/parallel.

SPA　*1* Solar-powered aircraft.

2 Seaplane Pilots Association (US).

3 Special-purpose aircraft, usually RPV.

4 Surplus Property Administration (US).

5 Schedules planning and analysis.

6 Special-Purpose Audit.

7 SkyTeam Pilot Alliance (Delta/Air France).

SP/A　Smart procurement, acquisition.

SpA　Società per Azioni [company, I].

SPAAG　Self-propelled anti-aircraft gun.

Spacdar　Specialist Panel on Automatic Conflict Detection And Resolution (ICAO).

Space　Software productivty and cost estimation.

space　Various precise definitions, but loosely volume in which celestial bodies move and esp. local portion of solar system outside Earth's atmosphere.

space age　Conceptual period in which human beings first learned to operate in space, beginning 1957 (first artificial satellite) or 1961 (first manned space flight).

space/air vehicle　See *aerospace vehicle*.

space biology　See *bioastronautics*.

spaceborne　Travelling through space; suggest unnecessary word.

space capsule　Environmentally controlled container in

which device or living organism flies in space; suggest not used for human occupation.

space charge　Negative charge carried by cathode electrons which unless continuously accelerated away bar further emission.

space-charge region　See *depletion layer*.

SpaceCom　Space Command (USAF).

spacecraft　Self-contained space vehicle, manned or unmanned.

spaced armour　Fitted in layers with sufficient gaps to defeat HEAT or hollow-charge weapons.

spaced-diversity　Radio communications technique which avoids fading by using three or more receiver aerials spaced 10 or more wavelengths apart, all feeding separate amplifying channels.

space defence　All measures designed to destroy attacking enemy vehicles, including missiles, while in space, or to nullify or reduce effectiveness of such attack (DoD).

space equivalent　Region in atmosphere where one particular parameter is similar to that in space.

space-erectable　Capable of being assembled in space, eg radiation shield.

space fabrication　Manufacturing or building operations in space.

spacefighter　Aerospaceplane with sufficient endurance and manoeuvrability to survey and if necessary disable hostile satellites.

space-fixed reference　3-D cartesian co-ordinate system related to fixed stars.

spaceflight　*1* Journey through space of man-made object.

2 The technology required for (1).

spaceframe　3-D framed structure assembled from simple tubes or girders with pinned or fixed joints, usually built up from succession of triangulated assemblies for rigidity. Note: may have nothing to do with space.

space gyro　Gyro having complete freedom about all axes, as distinct from rate, tied or Earth gyros.

spacelab　Laboratory for operations in space; with capital S, ESA/NASA programme.

Space Launch Initiative　Far-ranging study of Shuttle-replacement options (NASA).

space medicine　Branch of aerospace medicine concerned with health of human beings before, during and after spaceflight.

Spacemetal　Range of corrugated-core sandwich structures in stainless steels patented in 1957–60 by North American Aviation [initially for Navaho missile].

space operations vehicle　Primary component of MSP, reusable launch vehicle [probably T^{3}O] carrying various upper mission stages.

space probe　See *probe (3)*.

space qualified　Certificated for prolonged use in space environment, especially outside a satellite or space station.

spacer　See *shim*.

space segment　That in space forming part of large system with stations on Earth.

space simulator　Simulator wherein is reproduced one or more parameters of space environment; arguably impossible to simulate all.

space station　Permanent structure, probably manned,

established in space; probably in Earth orbit, and with routine crew replacement and import of materials.

spacesuit   Pressurized suit for EVA and other operations in space or on lunar surface.

Space tasking order   Describes the configuration of constellations required to support a specific mission (US).

Spacetrack   Global system of radar, optical and radiometric sensors linked to computation/analysis centre at Norad for detection, tracking and cataloguing of all man-made objects in Earth orbit. USAF portion of Spadats (DoD).

space tug   Propulsion vehicle for attachment to space materials, capsules, payloads and laboratories delivered to local area by Shuttle for onward exact positioning.

space wave   Combined direct wave and ground-reflected wave from transmitter to receiver, as distinct from surface wave or ionosphere-refracted wave.

space zones   Loose subdivision of local space into translunar (between or near Earth and Moon); interplanetary; interstellar.

SPAD   Signal processing and display (sonar).

Spadats   Space detection and tracking system; reports orbital parameters of all satellites and debris to central control facility (DoD).

SPADCCS   Space command and control system.

SPADE, Spade   *1* Single programmable access demand exchange; for small comsat-users.

2 Space acquisitions defence experiment.

spade   *1* Term for several detail features of aircraft, notably fixed or retractable blades projecting into propulsive jet in attempt to enlarge and break up periphery, promote mixing and reduce noise.

2 In particular, a small horizontal plate or aerofoil mounted on a miniature pylon under an aileron to give area ahead of hinge axis.

Spades   Small parafoil autonomous delivery system.

Spadoc   Space Defense Operations Center (USAF).

SPAé   Service de la Production Aéronautique (F).

SPAF   Svenska Privat och Affärs Flygföreningen (Swedish AOPA).

spaghetti   *1* Complex masses of electrical, hydraulic or other pipes or cables (colloq.).

2 Insulating or colour-identification tubing slipped over wires (colloq.).

spallation   Particular forms of spalling resulting from various separation mechanisms, notably between coatings and base material.

spalling   *1* Separation of pieces of armour from inner face of sheet after warhead impact.

2 Failure mode of surface of concrete under impingement of hot jet.

SPAM, Spam   Special-purpose aircraft modification.

Spamcan   All-metal light aircraft (US derogatory colloq. c1945–60).

SPAN, span   *1* Solar-particle (or proton) alert network (NOAA).

2 Stored-program alphanumerics, ATC system tested at Indianapolis ARTCC 1965.

3 Spacecraft analysis (NASA).

span   *1* Distance between extremities of wingtips. Term not normally applied to rotorcraft and often a nominal dimension which may or may not take into account tip tanks, ECM pods, winglets and similar extras. When folding wings are folded term is width. With variable sweep or slew wing, angular position must be quoted. Symbol b.

2 For a quasi-vertical winglet, the height.

3 Operative radial distance from root to tip of rotating aerofoil (eg of helicopter, gas-turbine compressor blade) discounting any inner porton not of aerofoil section. Not defined whether should include tip appendage, eg shroud or tip-drive propulsion. Note: effectively radius, not diameter.

4 Elapsed time within which NW should detonate.

5 Axial distance between centres of bearings supporting a shaft, or between supports of a beam or truss.

spanloader   Aeroplane (conceivably, glider) carrying payload distributed across most or all of span (1), normally in ISO containers fitting within wing profile.

span loading   Weight of aeroplane or glider divided by square of span (1); W/b^2.

Spanwar   See *Spawar*.

spanwise   In a lateral (transverse) direction, esp. along wing towards tip.

spanwise lift distribution   Plot in front elevation of actual wing lift for elemental chordwise sections of wing from tip to tip, normally (ie without active ailerons or DLC) having semi-elliptic form falling to zero at tips.

SPAR, Spar   *1* Semi-permanent airfield runway.

2 Solid-state phased-array radar.

3 Special progressive aircraft rework.

4 Special problem areas report.

5 Super-precision approach radar.

spar   *1* Major structural member of slender form projecting out from one end (which may be pinned or fixed).

2 Specifically, main spanwise structural member(s) of wing or rotorcraft rotor. Wing may have one to many discrete *, or two may be made into single strong box-* (often integral tank) to which secondary leading and trailing structures added. D-* is box structure formed by thick leading edge and a * forming upright of D. Usually formed from web and two or more booms.

spare   Item certified as suitable to replace one that is faulty.

spareable   Item capable of being supplied, esp. from stock, as spare.

spar frame   Particularly strong frame or bulkhead to which a spar (2) is attached.

Spark   Solid-propellant advanced ramjet kinetic-energy missile.

spark discharge   Electrical discharge, usually brief, resulting in very large electron flow linking points of high potential difference along narrow and brilliantly luminous path.

spark erosion   See *spark machining*.

spark-ignition engine   Piston engine, eg of Otto type, in which hot electrical spark is used to ignite mixture before each power stroke; thus, not diesel.

sparking plug   Term (spark plug in N America) reserved for plugs for spark-ignition engine and those few jet engines where ordinary commercial plug is used; term for gas turbines generally is igniter plug.

spark machining   Precision machining of extremely hard or otherwise difficult material by minute stream of HF sparks struck between anode tool (often of particular shape) and cathode workpiece; usually action is purely thermal.

579

spark photography *1* Simplest and oldest method of high-speed photography, in which shutter is left open and scene is illuminated by single point-source brilliant spark at exact time.

2 Various techniques, eg in tunnels, in which hot spark is used to create local airstream of contrasting refractive index.

sparks Wireless operator (UK traditional colloq.).

Spars Women's arm of USCG (from semper paratus).

Sparta Special anti-missile research tests, Australia (US programme).

SPAS *1* Shuttle pallet satellite.

2 Safety performance analysis system (ATOS).

SPASC System planning and system control.

Spasm Self-propelled air-to-surface missile.

Spasur Space-surveillance system with mission of detecting and establishing orbital parameters of every man-made object in Earth orbit, using fan of CW, across Conus; US Navy portion of Spadats.

Spasyn Space synchro.

Spate Special-purpose automatic test equipment.

spatial Relating to space, not in sense of cosmonautics but 3-D volume of any (possibly very small) size; thus, concerned with geometric position. Normal dictionary entries need revision.

spatial awareness Suggest same as situational awareness.

spatial disorientation Colloq.ually, not knowing which way is up; eg after losing control of aircraft in cloud.

spatial light modulator Hybrid optoelectronic module combining speed and parallelism of optics with integration of electronic chips.

spatial resolution Ability of sensor to distinguish between two very close distant objects; thus an angular measure (see *resolution*).

spatiography Mapping ('geography') of space.

Spato Single-pilot air-taxi operator (US).

spats Aerodynamic fairings over fixed landing wheels; purists do not allow term to encompass trousers.

Spawar Space and Naval Warfare Systems Command (USN).

Spawn Space protection and warning (Darpa).

SPB Seaplane base.

SPC *1* Synthetic particulate chaff.

2 Software productivity consortium.

3 Special-purpose company.

4 Statistical process control.

5 Stored-program control.

SPCD Space Communications Division (USAF).

SPCV Special-purpose corporate vehicle.

SPD *1* Spectral power distribution.

2 Speed.

3 System programme director.

4 Spool positioning device (umbilical lock).

SPDA Secondary-power distribution assembly.

SPDM Solid-propellant divert motor[s] (MKV).

SPE *1* Solar-particle event.

2 Solid-polymer electrolyte.

3 Seller-purchased equipment.

Spear *1* Selectable precision effects at range (RAF).

2 Support programme for evaluation of activities in research.

3 Spontaneous protection enabling accelerated response.

Spears Screener performance evaluation and reporting system (FAA).

SPEC Standard for professional engineering competence (UK).

Spec Special meteorological report, pronounced speck.

spec Specification, pronounced speck.

SPECI Special, especially special Metar.

special access See *SAO*.

special air mission One conducted by 89 Military Airlift Wing, whether or not President is aboard (USAF).

special-assignment airlift Airlift which for any reason cannot be accommodated by channel airlift (DoD).

special cargo Item requiring unusual handling, eg detonators, precision instruments.

special flight One set up to move a specific load.

special flying instruction Warning issued to pilots regarding handling of particular aircraft type [usually temporary].

specialized undergraduate pilot training After division into FB or TTT (USAF).

special material Nuclear, esp. fissile, material not naturally occurring, eg Pu-239, U-235 or enriched uranium.

special pilot ratings Instrument rating and instructor rating (US).

special reconnaissance Flight made covertly, ie without hostile detection.

special rules zone Protected airspace surrounding minor airfield which does not justify a control zone; extends from surface to published (usually low) FL.

special summary drawing Prepared for each production aircraft in GA and certain other civil categories listing all equipment fits, avionic fits, furnishing fabrics/colours, seat types and similar customer choices.

special technical instruction Commands urgent non-recurrent action to remedy serious defect (UK military).

Special Traffic Management Program Reservation programme implemented to regulate arrivals and departures at airports serving special events attracting heavy traffic (FAA).

Special-use airspace Too numerous to define; includes any precisely defined area that either cannot be entered or cannot be entered except by special aircraft, or with prior permission.

Special VFR Particular weather minima below those for normal VFR but which permit VFR flight; hence ** operations = flight within control zone under Special VFR clearance.

specific air range See *specific range*.

specification *1* Numerical and descriptive statement of capabilities required of new hardware item, eg future type of aircraft.

2 Concise list of basic numerical measures describing existing hardware item.

specific consumption See *specific fuel consumption*.

specific energy Energy per unit mass, whether released by chemical combustion or degradation of KE, units (SI) $J\,kg^{-1}$.

specific entropy Entropy per unit mass, see *entropy*. Units, $1\,kJ\,kg^{-1} = 0.4299\,Btu/lb$; $1\,Btu/lb = 2,326.00\,J\,kg^{-1}$.

specific excess power Propulsion power available over and above that needed to propel aircraft in level flight at given reference speed, and thus available for climb or manoeuvres at sustained high speed. With streamlined aeroplane (fighter) drag is small in relation to thrust at most reference speeds, so SEP is broadly governed by thrust/weight ratio T/W.

specific fuel consumption Symbol c′, rate of consumption of fuel for unit power or thrust, and thus basic measure of efficiency of prime mover; term confined to air-breathing engines (equivalent for rocket is specific impulse). SI unit for jet engines (turbojet with/without afterburner, turbofan, ramjet or pulsejet) is mg/Ns (milligrammes per Newton-second); traditional Imperial measure is lb/h/lb thrust, measure being for SL static ISA condition unless otherwise specified. For shaft engines (turboshaft, turboprop, piston) SI unit is μg/J (microgrammes per joule); traditional Imperial unit is lb/h/hp, hp being qualified for turboprop as shaft or equivalent. Conversons, jet engines, 1 mg/Ns = 0.0353 lb/h/lb st; 1 lb/h/lb st = 28.325 mg/Ns; 1 kg/h/kN = 0.0098 lb/h/lb st; 1 lb/h/lb st = 102.04 kg/h/kN; shaft engines, 1 μg/J = 0.00592 lb/h/shp; 1 lb/h/shp = 169.0 μg/J; 1 kg/h/cv = 2.2352 lb/h/shp; 1 lb/h/shp = 0.4474 kg/h/cv.

specific gravity, SG Density of material expressed as decimal fraction (less than or greater than unity) of density of water at 4°C.

specific heat Quantity of heat required to raise unit mass of material by unit temperature, usually from 0° to 1°C. Traditional measure is calories per gramme, the SI unit is kJ/kg K, = 0.238846 Btu/lb°F, CHU/lb°C, Cal/g°C. For gases thermodynamic process must be stated; C_p (** at constant pressure) and C_v (** at constant volume) are not same.

specific humidity Ratio of mass of water vapour to total mass of moist gas (dimensionless).

specific impulse I_{sp} basic performance parameter of rocket, = thrust divided by rate of consumption of propellants in compatible units, = total impulse divided by total mass of propellants (see also *motor* **, = total impulse divided by total loaded mass of solid motor), in general = effective jet velocity divided by g; unit = s (seconds) derived from force × seconds divided by mass (strictly, in SI Newtons force cannot be divided by kilogrammes mass).

specific optical density Numerical scale of atmospheric opacity, on which dense smoke = 800.

specific power Not defined; often used for thrust/weight ratio of prime mover, esp. including electrical battery or fuel cell. For air-breathing engines see *power/* or *thrust/weight ratio*.

specific propellant consumption Reciprocal of I_{sp}, mass flow of propellants or rate of burning to generate unit thrust in rocket.

specific range Air distance flown for unit consumption of fuel; traditional unit is nam/lb (NAMP), while SI would be air-km/kg.

specific search Reconnaissance of limited number of points for specific information.

specific speed Basic performance parameter of hydraulic turbine; trad. unit is rpm at which 1 hp is generated with head of 1 ft. SI unit needed for modern hydraulic motors and rocket turbopumps.

specific stiffness Stiffness (usually Young's modulus) divided by density, E/ρ.

specific strength Ratio of ultimate (tensile/compresive/shear/bending) strength to density.

specific tasking Planning phase in which commanders designate actual units to fill force list of operation plan.

specific thrust *1* Net thrust of jet engine divided by total inlet mass flow.

2 Confusingly, propulsive thrust divided by engine mass (not normally applicable to rockets).

specific volume Volume per unit mass, reciprocal of density; 1 cm^3 [cc]/kg = 0.02768 in ^{3}/lb.

specific weight *1* Symbol p̄, engine mass divided by net thrust (air-breathing jet engines).

2 Symbol ω̄, engine mass divided by net thrust (stub) power (piston engines). Strictly, mass should include cowl and propeller.

3 Dynamic equivalent of density, force [not mass] divided by volume; 1 kN/m^3 = 6.365858 lbf/ft^3.

specified approach funnel Funnel extending about 0.5° above and below PAR gildepath and ±2° of PAR centre-line within which aircraft is not advised by PAR controller monitoring ILS approach.

specimen performance Performance, esp. of civil transport, worked out and presented as guide to correct procedures and methods of compliance.

specklegram Photograph taken by laser speckle interferometry.

speckle interferometry Technique for measurement of almost vanishingly small angles.

spectacles *1* Main control wheel comprising single left/right curved handlebars, suggest synonymous with ram's horn; yoke is generalized term for all configurations.

2 Figure-eight manoeuvre in the vertical plane.

Spectra High-modulus composite material for body armour to protect against explosions and mechanical impacts (trade name).

spectral line Indication of single frequency, or very narrow band, in continuous spectrum (2), denoting presence of identifiable atoms or molecules.

spectrograph Spectroscope with camera or other recorder.

spectroheliograph Takes pictures of Sun in monochromatic light; spectrohelioscope is for direct viewing.

spectrometer Instrument for analysing spectrum to read out wavelengths and/or energies.

spectrometric analysis Usual method is to analyse spectrum of light as test substance is burned in electric arc.

spectrophotometer Photometer which measures variation of radiant intensity with wavelength.

spectropyrheliometer Measures variation of intensity of direct solar radiation with wavelength.

spectroradiometer See *spectrophotometer*.

spectroscope Instrument for dispersing light into spectrum.

spectrum *1* Visual display, EDP (1) printout, photo record or other presentation of variation of radiation intensity (sometimes other parameters, eg sound pressure level) with wavelength/frequency or other variables.

2 Continuous range of electromagnetic wavelength/frequencies within which radiation forms common grouping, eg visible *, IR *.

3 Stylised colours of visible *.

4 Various specialized terms in maths, mechanics etc.

spectrum line *Spectral line*.

specular Offering a smooth reflecting surface; rule-of-thumb demarcation from diffuse scattering surface is roughness factor. In low-level attack with radar/laser, reflection from calm sea is * up to grazing angles near 20°;

steeper attack gives diffuse. Hence importance in stealth aircraft of non-* RAM.

SPED   *1* Small-parcel explosive detection; S adds system.

  2 Supersonic planetary entry decelerator.

SPEEA   Society of Professional Engineering Employees in Aerospace (US, office Seattle, WA).

speech inversion   Simple form of scrambling in which frequencies relative to a reference value are inverted.

speed   Scalar quantity, as distinct from velocity; see *airspeed*; also used for rotational *, see *shaft* *, not used for other rate-type quantities such as *frequency* (Appendix 2). SI unit is ms^{-1} = 2.23694 mph = 1.94386 kt = 3.6 [exact] km/h; km/h = 0.621373 mph = 0.53961 kt; kt = 1.60934 mph = 1.85318 km/h = 0.514773 ms^{-1}.

speed brake   See *airbrake*.

speed bulge, speed bump   See *shock body*.

speed capsule   See *shock body*.

speed control computer   Accepts flight data from aircraft sensors and provides flight-management outputs for throttle servoactuators and, probably, flight director and autopilot to maintain selected TAS.

speed control governor   Engine control device invariably based on centrifugal flyweights adjusting a fuel orifice, in some engines with datum reset for emergencies or W/M injection.

speed course   Accurately surveyed course for attempts on speed records; previously also used for true G/S measures (now arch.).

speed generator   See *engine* *.

speed jeans   G-suit (colloq.).

speed lock   Autopilot sub-mode in which TAS (in some systems G/S) is held constant.

speed of rotation   *Rotational speed*.

speed of sound or light   See *sound, light*.

speed/power   See *power/speed coefficient*.

speed probe   Basic sensor for rpm of shaft, with magnetic pole piece held close to teeth of existing gear or special toothed disc to generate emf whose frequency is transmitted. Has virtually replaced tachometer except in GA aircraft.

speed range   Aircraft maximum level speed minus minimum speed, eg V_{mc}.

speed reference system   Chief meaning is subsystem on advanced transports (Airbus) which provides flight director visual guidance on how far pilot should haul back in windshear without triggering stick shaker.

speed rotor   Main input sensor of anti-skid brake, whose angular velocity is held against sudden reductions.

speed stability   Condition such that aircraft tends to return to preset speed following any excursion, thrust remaining constant; condition not obtained below V_{md}.

Speedtape   Commercially available thin aluminium sheet, resembling stout foil, with adhesive backing revealed by peeling off skin; for rapid repair, which can be surveyed as permanent.

speed trend   Additional protection against windshear, shows what speed/AOA will be 10 s hence with no action by pilot.

spelk   Unusable bundles of short lengths of surplus reinforcing fibres.

spent fuel   Fissile fuel whose allotted life has been consumed, though material still fissile.

SPER, Sper   *1* Syndicat des Industries de Matériel Professionnel Electronique et Radio-Electrique (F).

  2 Strategic-planning executive review.

SPERT, Spert   Scheduled (or simplified) programmed evaluation and review technique.

SPET   Solid-propellant electric thruster.

SPETC   Solid-propellant electrothermal-chemical (gun).

SPEW, Spew   Small-platform electronic warfare.

Spews   Self-protection electronic-warfare system.

SPF   *1* Svensk Pilotförening (Sweden).

  2 Superplastic forming.

SPFDB   See *SFDB*.

SPGG   Solid-propellant gas generator.

SPGR   Special-purpose GPS receiver.

sphere of influence   Volume around body in space within which small particle is attracted to body; in case of Earth not true sphere.

Spheric   System for protection of helicopters by radar and IR countermeasures.

spherical angle   Angle between two great circles.

spherical convergent flap nozzle   Propelling nozzle of augmented turbofan providing not only variable area and profile but also flow blocking and reversal and ±20° vectoring in any plane.

spherical data system   Long-range navigation system, eg for maritime patrol aircraft, related to spherical Earth surface.

spherical triangle   Formed by arcs of three great circles.

spherics   See *sferics*.

spherodizing   Hot soaking of irons and steels close to (usually just below) critical temperature followed by slow cooling.

spherometer   Instrument with three legs and central micrometer leg for measuring radius of convex or concave spherical surfaces.

SPHVM   Self-propelled high-velocity missile.

sphygmomanometer   Instrument for measuring blood pressure.

SPI   *1* Surface position indicator.

  2 Short-pulse insertion (SSR).

  3 Spike-position indicator.

  4 Scatter-plate interferometer.

  5 Special position identification (pulse).

  6 Smart procurement initiative.

  7 Symbolic pictorial indicator.

  8 Software process improvement.

Spice   Smart precise-impact cost/effective.

Spicsy   Standard protocols to support intra-centre communications between air-traffic management system components.

spicules   Long, bright filaments briefly extended from chromosphere.

spider   *1* Structural heart of propeller or helicopter rotor in form of hub integral with radial members which bear all stresses from attached blades.

  2 Multi-finger plate securing structural members grouped at a common joint, each finger being aligned with and secured to one of members, all being in same plane.

spider beam   Spaceframe-type rocket interstage structure.

SPIE   *1* Society of Photo-optical Instrumentation Engineers (US).

2 Special insertion and extraction [of troops, usually by abseiling in, helo out].

spigot See *sprag*.

spike *1* Conical inlet centrebody of supersonic airbreathing engine, usually designed to translate.

2 Short-duration transient (in signal, current, radar display or any other oscillating variable) in which amplitude makes large excursion beyond normal.

3 Centrebody of * nozzle.

4 As verb, to designate by laser.

5 Long tapered tube ahead of nose of SLBM to generate conical shock and reduce drag of bluff nose during climb up through atmosphere.

Spikebuoy Tactical warfare sensor dropped from air to land on spike (eg in jungle), thereafter transmitting on command ground tremors which can be attributed to human beings or vehicles.

spike inlet 2-D airbreathing inlet in form of body of revolution with central spike (1).

spike nozzle Rocket nozzle in which gas escapes through ring around centrebody in form of concave-profile cone.

spill To cause spilling.

spillage *1* Amount by which mass flow into airbreathing inlet is less than datum flow (in UK called intake flow).

2 The actual airflow [eg, spilt around nacelle] which fails to enter inlet.

3 Flow of air from below wing to above at tip.

spillage drag Difference between drag at given engine airflow and drag at datum flow.

spill burner Gas-turbine burner in which fuel is supplied at constant high pressure, giving good swirl and atomization, excess over requirement being 'spilt' back through second pipe for reuse.

spill door Auxiliary door usually spring-loaded to open outwards, through which excess engine airflow (spillage) escapes with minimum drag, eg in high-speed cruise or on letdown.

spill effect Loss of revenue, esp. pax, at certain times on highest-density routes because seats are not available; this offsets cost benefit of not using larger aircraft.

spilling Escape of air at one part of parachute canopy periphery, either through instability or for directional control of trajectory.

spillover Airflow deflected to pass outside airbreathing inlet in supersonic flight with detached shock.

SPILS, Spils Spin- [or stall-] prevention and incidence (AOA is meant) limiting system.

spin *1* Sustained spiral descent of fixed-wing aerodyne with AOA beyond stalling angle; in most cases a stable autorotation (see *flat* *, *inverted* *, *upward* *.

2 To shape by spinning.

spin avoidance system Seldom seen as a separate system, detects AOA and possibly other factors such as airspeed and emits aural or visual warning.

spin axis Axis of rotation of gyro wheel.

spin box Primitive flight-test device comprising pen, paper-tape drive, altimeter and two stop-watches.

spin chute Anti-spin parachute.

spin dimpling Dimpling by coldworking tool (eg 60° cone) around hole for rivet or other fastener under pressure, without cutting.

spindle Structural frame in rigid airship to which mooring cone is attached.

spindling Generalized term for machining of wooden

parts, esp. to effect particular desired uniform cross-section to spar, longeron or other structural member. * machine basically resembles spar mill or router.

spine Non-structural fairing along dorsal centreline of aircraft or along outside of ballistic vehicle (occasionally in other locations but always parallel to longitudinal axis) covering pipes, controls or other services. In some cases merely drag-reducing fairing linking canopy to fin.

spine hood Hinged or removable cover over equipment in spine.

spin motor Rocket(s) imparting spin (rotation about longitudinal axis) to missile or other vehicle.

spinner Streamlined fairing over propeller hub; not used for similar fairing over helicopter rotor hub.

spinning *1* Sheet-metal shaping by forcing against spinning die of desired profile (suitable only for bodies of revolution).

2 To engage in spin (1).

spinning nose dive Flight manoeuvre in which aeroplane or glider is rolled with ailerons while in steep unstalled dive (US usage).

spinning test *1* To explore spin characteristics of aircraft.

2 Basic strength test of propeller or helicopter rotor, or (often at very high speed in low-drag environment) other rotating assemblies, eg turbine or fan rotor.

spinning tunnel See *spin tunnel*.

spin-off *1* Predicted or unexpected advances in one technology caused by transfer of technical solutions from another. Also called fallout.

2 In particular, transfer from defence to civil.

spin-on Technology transfer from civil to defence.

spin parachute See *anti-spin parachute*.

spin-recovery parachute See *anti-spin parachute*.

spin rocket See *spin motor*.

spin stabilized Given gyroscopic directional pointing stability by high-speed rotation about longitudinal axis.

spin table Large disc on which objects, including human beings, can be rapidly rotated about vertical axis for varous test and research purposes; ** in turn may be mounted on arm of centrifuge to give additional sustained unidirectional lateral acceleration.

Spintcom Special-intelligence communications.

spintronics Technology of devices whose operation depends on electron or nuclear spin.

spin tunnel Wind tunnel in which flow through working section is vertically upwards, thus free model supported by airflow can be examined for spinning characteristics.

SPIP Transponder identification pulse.

SPIR Single-pilot instrument rating.

SP(IR) Satellite picture, IR.

spiral *1* One of the five basic modes of aeroplane motion, slow divergence or convergence in level flight with gentle banks L/R persisting long enough for large heading changes.

2 Flight manoeuvre in which at least 360° change in heading is effected while in glide or shallow dive (chiefly US usage).

spiral aerial (antenna) Aerial in form of single conductor wound as spiral on conical dielectric support; common as passive RWR receiver.

spiral angle Angle between pitch cone generator of bevel gear and tangent to tooth trace; positive for right-hand gear.

spiral beval gear   Crown gear has pitch curves inclined to pitch element and usually circular arcs.

spiral dive   Extremely dangerous flight manoeuvre in which aircraft, invariably fixed-wing aerodyne, is unwittingly in spiral descent with neither turn nor slip indicating and in absence of external cues few indications other than horizon instrument if fitted.

spiral divergence   Spiral dive with vertical acceleration ± 1 g.

spiral glide   Sustained gliding turn.

spiral instability   Faulty aeroplane flight characteristic in which there is an inherent tendency to depart from straight and level flight into oscillating sideslip and bank, latter always being too great for tendency to turn. (Long-established UK definition unrelated to Dutch roll, which is a high-subsonic phenomenon; suggest arch.).

spiral scan   See *scan types*.

spiral stability   Desired aeroplane characteristic in which, in co-ordinated turn, it automatically resumes straight and level flight on release of flight controls. Note that no slip is present to assist recovery, and that large fin (needed for other reasons) exerts adverse influence.

Spirit   Spectral IR rocket-borne interferometer telescope.

spiroid gear   Patented gear in which conical worm engages with many teeth simultaneously of face-type gear, possibly with exceptionally large reduction ratio.

SPIT, Spit   Smart procurement implementation team.

spitting   Air intercept code: "I am about to lay sonobuoys and may be out of radio contact (because very low) for few minutes" (DoD).

SPJ   Self-protection jammer.

SPKL   Sprinkle (rain).

SPKR   Speaker.

SPL   *1* Sound pressure level, see *noise*.

2 Standard parts list(ing).

3 Supplementary flightplan message.

4 Sun-pumped laser.

5 Signature and Propagation Laboratory (USA).

6 Special.

7 Student Pilot's Licence [in UK replaced by CAA medical certificate].

splash   *1* Code word sent to observer 5 s before estimated impact of weapon(s).

2 Target destruction verified by visual or radar means.

3 Generalized term for action of destroying aerial target.

splash code   Letter/number identity of navigation beacon (WW2).

splash cooling strip   Welded around interior of flame tube of annular chamber so that cooling air entering through perforations is converted into thin sheet moving acrosss inner surface.

splash-detection radar   Pinpoints impact of vehicle with ocean to facilitate positive scoring and vehicle recovery (eg in Nike X ABM tests).

splashdown   End of space mission in which spacecraft, capsule or other recoverable object impacts ocean surface; defined as either a time or a location.

splashed   Air-intercept code: enemy aircraft shot down (followed by number and type) (DoD).

splash lubrication   Use of small lips or vanes on connecting rods or crankpins to splash oil inside crankcase; rare in aviation.

splatter   *1* Adjacent-channel interference in pulsed transmissions, measured as amount of spectrum energy that can appear in adjacent channel; varies greatly with different pulse waveforms.

2 Cloud of canopy fragments after MDC detonation.

splice   Structural joint made by plate overlapping both members, plus doubler.

spline   *1* Axial groove in shaft for meshing with driven member; hence splined shaft has entire periphery formed into splines, invariably of rectangular section.

2 Flexible non-structural strip (various materials) bent to required curvature in construction of fairing.

split   Verb, to divide an engine into modules, notably into fan and core, hence *-engine shipping, etc.

split-altitude profile   Flight-profile has two main flight levels.

split-axle gear   Landing gear on simple low-performance aircraft in which there is no axle or other linking transverse member. Also called divided gear.

split basing   Division of tactical-aircraft unit's resources between two operating bases.

split cameras   Two or more cameras fixed so that imagery of one overlaps that of neighbour(s) by selected amount.

split charter   Commercial flight flown on charter to two companies, each providing part of payload.

split-compressor engine   Gas turbine in which compression is performed by two separate rotating assemblies running at different speeds; can be axial + centrifugal, while term two-spool suggests axial + axial.

split courses   See *multiple courses*.

split distance   Two friendly fighters fly apart for mutual interception practice, usually head-on.

split flap   Flap formed from only underside of aerofoil, depressed with plain hinge leaving upper surface unaltered; gives high drag but little extra lift.

split-flow engine   Turbofan in which fan air is diverted to blow wing or flaps and core jet is used for propulsion (and, if vectoring is added, for lift).

split gear   Landing-gear trucks (B-52) attempt to steer some left and some right simultaneously.

split landing gear   See *split-axle gear*.

split line   Small flash-projecting ridge round surface of die forging.

split load   Drop of firefighting retardant at two locations in same mission.

split mission   Several meanings, including (1) profile part hi and part lo, (2) task includes recon and attack, (3) task includes two surface targets, (4) transport flight carries loads for two destinations, and (5) part subsonic and part supersonic dash.

split needles   Various flight-instrument indications for aeroplane (eg dual engine-speed indicator with one engine at flight idle) or helicopter [eg, gross disharmony between speeds of engine and rotor system].

split pair   See *split vertical photography*.

split patch   Patch for reinforcing end of surface (eg fabric air inlet) at junction with airship envelope.

split-plane manoeuvring   Air-combat manoeuvring for mutual support following a defensive split.

split-S   Flight manoeuvre comprising half flick (snap) roll followed by second half of loop, resulting in loss of height and 180° change in heading.

split-surface control   Powered flight-control surface

subdivided into two or three portions, each of same area and each driven by its own independent power unit (thus, VC10 has four elevators and three rudders).

splitter   *1* Fixed or laterally movable surface dividing fluid flow in duct, eg to feed two engines.

2 Machine which divides or apportions signals, shaft power or other services among selected recipients.

splitter box   Divides one rotary input among two outputs in variable proportions, eg to control either or both of two-roll-control or DLC spoilers.

splitter gearbox   *1* Splitter box.

2 Gearbox dividing input along two channels, eg from reverser PDU (3) to left/right half-rings of slave actuators.

splitter panels   Sound-absorbing panels, usually radial struts or concentric rings, to reduce noise from jet-engine inlet.

splitter plate   On centreline of single-surface rudder.

split vertical photography   Simultaneously triggered reconnaissance cameras, each tilted same angle to left or right of centreline, with small centreline overlap.

split wheel   Made in inner and outer halves to facilitate changing tyre.

split-work blade   Gas-turbine fan blade comprising inner (compressor supercharging) portion separated from outer (fan) portion by part-span shroud.

SPM   *1* System planning manual.

2 Solar-proton (or particle) monitor.

3 Solid propellant motor.

4 Security protection module (networks).

5 Superalloy powder metallurgy.

6 Scanning-probe microscope, or microscopy.

7 Surface-position monitor.

8 Stabilizer position module[s].

spm   Shots per minute.

SPMS   Special-purpose manipulator system.

SPN/GEANS   Standard precision navigator, gimballed electrostatically, aircraft navigation system.

SPO   *1* System program office.

2 Safety petty officer (carrier catapult).

3 Single-pilot operation.

4 Strategic plan of operation [see *SPT*].

SPOC   Single point of contact (Sarsat).

SPOH   Since part, or propeller, or partial, overhaul.

spoil   To reduce, not necessarily to zero (eg lift or thrust).

spoiler   *1* Hinged or otherwise movable surface on upper rear surface of wing which when open reduces lift, and usually also increases drag. Most are essentially flat plates hinged at or beyond leading edge and power-driven to open upwards, either symmetrically to incline flight path downwards or differentially to command or augment roll. Supersonic combat aircraft often have no other roll control, while others use * only at low speeds or high speeds. * operation on commercial transports is invariably linked with position of speed brakes, response depending on whether differential or not. DLC * is primary flight control, esp. during landing. Many * used as lift-dumpers after touchdown.

2 Movable deflector used to kill most if not all residual thrust from engine or from turbofan core; much simpler and lighter than reverser.

3 Comb, flap or other device extended on command to break up local airstream, eg ahead of open weapons bay.

spoiler/elevator computer   Multiplexes input control signals to provide roll and speedbrake control and, if fitted, gust-alleviation; following failure of pitch computer also provides back-up main and trim control in pitch.

spoiler initiation angle   Rotary deflection of pilot's roll control (spectacles, handwheel, etc) at which spoiler(s) start to open (from wings-level flight with speedbrake fully closed).

spoiler-mode control system   Governs spoilers as DLC in flight and auto dumper and speedbrake on touchdown.

spoileron   Small spoiler either augmenting ailerons with large input, or serving as primary control in roll.

spoking   *1* Regular or erratic flashing of rotating time-base on PPI or other radial display.

2 Any PPI display which radiates out from a central origin.

SPOL   Solar-powered obstruction light[s].

sponginess, spongy   Descriptive but vague term for flight controls where response appears unduly delayed or uncertain, usually due to cable stretch, play in pully supports and mechanical wear.

sponsons   *1* Symmetric projections low on each side of flying-boat hull in form of short thick wings to provide stability on water in place of outer-wing floats.

2 Projections from helicopter fuselage in form of short thick wings to provide attachment for main landing gear (and in some cases retraction stowage space).

3 Short, thin wing-like structures projecting from aeroplane or helicopter fuselage to carry weapons, external tanks, guns and possibly other devices.

spoof   *1* To copy hostile IFF reply code.

2 See spoofing.

spoofer   Air-intercept code: contact is employing electronic or tactical deception measures (DoD).

spoofing   Acting the part of hostile forces, especially in EW.

spook   Adjective or noun: aircraft, manned or unmanned, flying EW missions, especially clandestine.

spool   *1* One complete axial compressor rotor, in case of multi-shaft engine forming LP, IP or HP portion of complete compressor. Some authorities include the drive-shaft and turbine.

2 Attachment anchor, usually one of L/R pair, for catapult bridle of carrier aircraft.

3 To open throttles (colloq.).

spool down   To allow gas-turbine engine rpm to decay to zero, eg after closing shut-off valve or HP cocks; normally two words, or one as adjective, hence spooldown time. Also called *rundown*.

spool duct   Joins engine to augmentor and UAA nozzle.

spool up   To accelerate engine rpm, esp. to TO power or at least to much higher level than previously; normally turbofan or turbojet.

sporadic E   Irregular radio-reception and disturbance caused by abnormal variation in E-layer.

sportplane   No definition, but genreally taken to mean small GA aircraft in which flight performance dominates other qualities.

SPOT   *1* Spot wind (ICAO).

2 Speed, position, track.

spot   *1* To form up aircraft in close ranks on carrier deck ready for free or catapult takeoffs.

2 Designated place on airfield where landing is to be made.

3 Bright region where electrons strike fluorescent tube face in CRT and many other displays or image converters.

4 To determine, by observation, deviations of ordnance from target for purpose of supplying necessary information for adjustment of fire (DoD).

5 To search from ground for hostile aircraft in own airspace (UK civilian usage, WW2; hence spotter, spotting).

6 Code, spot wind.

spot annealing   Annealing local area of hard steel, eg to drill and tap fixing hole.

spot beam   Electromagnetic beam made as parallel as possible to maximize power at great distance.

spot elevation   See *spot height* (US usage).

spot facing   Local surface-machining round hole or other point, to improve surface finish, adjust dimensions or provide square-on surface for bolt head.

spot fuel   Uplifted and paid for on the spot, as distinct from part of ongoing contract.

spot height   Height of point, esp. mountain peak or other high point, marked on map or chart.

spot hover   Helicopter training manoeuvre in which machine is hovered at low level over point, turning through four successive headings 90° apart.

spot jamming   Jamming of specific frequency or channel.

spot landing   Aeroplane landing made from specified position and height AGL on to spot (2); form of accuracy landing.

spotlight   A DBS radar operating mode using very narrow beam with highest possible resolution, steered to dwell on targets of high interest so that numerous echoes can be integrated.

spot net   Com net used for spot (4) information.

spot report   Sent from attack or reconnaissance aircraft stating passage overhead specific target.

spot size   Diameter of spot (3).

spotter   Person assigned to task of watching for and identifying hostile aircraft; hence raid *, official responsible for immediate warning of imminent attack to high-priority establishment whose personnel would remain at work throughout air raids (UK civilian usage, WW2).

spotting   Act of arranging aircraft on flight deck (see *spot [1]*).

spotting factor   Ease with which particular aircraft type can be spot (1) positioned on deck; not quantified but takes into account overall folded dimensions and turn radius and possibly stability and laden weight.

spot weld   Local, usually circular, weld quickly made by electrical resistance jaws working on sheet; tool can be point or roll electrodes.

spot wind   Wind measured at one geographical location.

spot wobble   Technique for imposing small SHM waveform on each line of TV or similar raster display to blur separate lines.

SPR   *1* Solid-propellant rocket.

2 A commercial rain-repellant for transparencies.

3 Secretarial program review (US).

4 Synchronization phase reversal [Mode-S xpdr].

5 Secure packet radio.

SPRA   Special-purpose reconnaissance aircraft; one designed to cross hostile frontier covertly.

Sprag   Water-spray type arresting gear.

sprag   *1* Pivoting and lockable projection in cargo-floor

guiderails to prevent movement of pallets or platforms. Occasionally (not DoD) called spigot.

2 Pawl or spigot-type mechanical lock for a rotary member, as in next two entries.

sprag brake   Positive mechanical brake for locking wheels of helicopter parked in confined area on rolling deck.

sprag clutch   Positive mechanical lock for failed engine and drive of helicopter, while permitting continued overrunning or autorotation of rotor system.

sprashpot   Sprag and dashpot in series.

spray   To dispense liquid from aircraft for agricultural, defoliant or ECM purposes.

spray bar   Bar for spraying, esp. for ag purposes, arranged spanwise below trailing edge and usually freely swinging to knock upward on impact with obstruction.

spray dam   Strip projecting along forebody chine of marine aircraft, as far as possible following streamline in cruising flight, to deflect water spray downward.

spray dome   Mound of water thrown into atmosphere when shockwave from underwater NW reaches surface.

Spraymat   Patented (Napier, now Lucas) electrothermal anti-icing and de-icing mats featuring sprayed-on metallic layers.

spray strip   See *spray dam*.

SPRD   *1* Spread.

2 Solid-propellant rocket motor (R).

spreader   *1* Ag aircraft rigged for solids.

2 Mechanical (usually centrifugal) dispenser of ag solids.

spreader bar   *1* Horizontal member(s) separating and joining floats of twin-float seaplane.

2 Horizontal axle-like member joining left/right landing gears of early or light aircraft, other than true live axle.

spread spectrum   Vast and growing technology forming complete divison of electronics, esp. military and avionics, fundamental of which is use of PN, FH, TH or any combination of these to modulate signal whose bandwidth is much wider than that of plain message. Latter is conventional (eg biphase digital or PDM analog) and is merged digitally with ** modulation to generate emitted signal. Advantages are very great anti-jam capability, military security, multiple access, low detectability, Selcal capability and auto transmitter ident, multipath tolerance, and inherent precision-nav capability.

springback   Angular distance through which metal bent to new shape springs back after bending force is removed; allowed for in making tooling or in hand operations.

spring bow   Supports arrester wire above carrier deck.

spring drive   Coupling inserted between piston-engine crankshaft and supercharger [rarely, propeller or reduction gear] to prevent transmission of cyclic vibrations. See Bibby.

spring feel   Simplest form of artificial feel, in which force is applied to pilot's flight control (eg stick or yoke in pitch) by linear spring, thus within limits exactly proportional to deflection and unvarying with dynamic pressure.

spring sheet-holder   Small lock in form of cylinder and spring-loaded rod with locking end inserted by pliers through holes in sheets to hold location during riveting.

spring strut   Mechanical link imposing absolute limit on force transmitted.

spring tab   Servo tab whose deflection relative to surface is resisted by spring, usually torsion bar, which is often preloaded so that at gentle inputs pilot moves surface and tab unaided; at higher input spring is overcome and tab deflected to assist.

springy tab   Not same as previous entry, tab moved upwards only by spring, having powerful effect at low IAS, making pilot pull back on yoke giving stick-free stability. Also called Vee tab.

Sprite   *1* Surveillance, patrol, reconnaissance, intelligence-gathering, target-designation and EW (UK).

2 Signal processing in the element.

sprites   Transient visual phenomena, typically luminous streaks, seen in ionosphere above giant thunderstorms.

SPRL   Spiral.

sprog   Totally inexperienced [noun and adjective] (RAF, WW2).

SPS   *1* Secondary, or solar, power system.

2 Samples per second.

3 Blown flap (USSR, R).

4 Self-protection system, or subsystem.

5 Simplified processing station.

6 Service (module) propulsion system.

7 Standard position[ing] system, or service, part of GPS Navstar.

8 Signal processing system.

9 Sensor processing subsystem.

10 Standard Procurement System (DoD).

SPSS   Statistical package for social sciences (CAA).

SPT   *1* Strategic Planning Team (FAA, NAS).

2 Shop processing time.

3 Simplified passenger travel.

4 Signal-processing tools.

SPT-B   Selectable-performance target, ballistic.

SPU   *1* Short power-up (SAAHS self-test).

2 Intercom (USSR, R).

3 Subsystem power unit.

4 Signal processing unit.

5 Stores power unit.

Spur   Space-power unit reactor.

spur gear   Gearwheel with straight teeth round periphery; many dictionaries add 'parallel to axis of rotation', not so, most aero-engines contain *helical* *.

Sputnik   Russian word for 'fellow traveller', name of first artificial satellite, launched 4 October 1957; later colloq. for any satellite or even any spacecraft.

sputtering   Ejection of metal atoms from cold cathode by evaporation or ion bombardment, either as nuisance, to form fine coating on substrate or to form colloidal metal solution.

SPW   *1* Self-protection weapon (tactical attack aircraft AAM, not necessarily fired in forwards direction).

2 Secondary power.

SQ   *1* Squall (ICAO).

2 Software, or service, quality.

3 Squelch, squawk.

4 Super-quick [fuze].

5 Square.

SQA   *1* Software quality assurance; / CM adds /configuration management.

2 Service quality agreement.

SQAL   Squall.

SQAN   Squall line.

SQB   Service-quality billing; P adds processor.

SQC   Statistical quality control.

SQD   Service-quality data.

sq ft   Square feet.

SQL   Structured query language.

SQLN   Squall line.

SQM   *1* Software quality metrics, quantified measures of SQ.

2 Software quality management.

sq m   Square metres, m^2 preferred.

Sqn, sqn   Squadron.

squadron   Any of many types of military or naval unit, including common administrative unit of combat aircraft; according to many services 'consisting of two or more flights'. Foreign-language equivalents include Staffel (G), escadron (F), escuadron (S), eskadrilya (USSR, R).

squall   Strong but intermittent wind; gust whose effect lasts minutes rather than seconds and extends a kilometre/mile or more horizontally.

squall line   Line of established or developing thunderstorms.

squarco   Radar beam squinting plus area coverage.

square   *1* See *signal area*.

2 Large square marked on remote part of airfield, to be accurately flown by pupil helicopter pilot, especially in presence of wind.

square bashing   Drill, esp. on barrack square (UK).

square course   Airfield circuit (US usage); does not mean literally square.

square/cube law   Basic geometric law: areas of similar-shaped solid bodies are proportional to squares of linear dimensions, and volumes (ie for equal densities, masses) to cubes. Thus if two aeroplanes are of same shape but one has twice linear dimensions, it will have four times wing area and eight times weight, hence W/S is doubled.

square engine   Piston engine whose bore equals stroke.

square foot $0.0929m^2$.

square inch $645.16 \ mm^2$.

square mile $2.58999 \ km^2$.

square parachute   One whose canopy is approximately square when laid out flat.

square search   Various standard air/surface search patterns in form of overlapping rectangles (usually squares or near-squares) so that after a period aircraft has examined a large strip or rectangle with minimal duplication.

square stall   Stall with wings level throughout.

square thread   Thread with vertical faces (unlike Acme) used for transmitting power as linear thrust in either direction.

square wave   EM or other periodic wave which alternates between steady positive and steady negative values in time extremely brief by comparison with steady periods.

square-wing biplane   Upper and lower spans equal; according to some authorities, also without stagger.

squaring shears   Hand or power shears for cutting sheet positioned on marked-out platform.

squash head   Warhead for use against hard targets, esp. those with single thick metallic armour layer; basic principle is transmission of intense shockwave through armour, causing transverse acceleration high enough to spall pieces off inside face.

squat(ting)   Downward movement of marine aircraft c.g. (eg due to trough) in essentially level attitude while running on water.

squat switch   Bistable switch triggered by sustained

(usually 2.5 s) compression of main or nose landing-gear struts on touchdown, to operate lift dumpers, reversers and/or other devices. Term also applicable when input is rotation of MLG bogie beam.

squawk   Generalized word for airborne transponder or IFF operation and keying; when used alone usually a ground-radar ATC command to switch to normal or to directed mode. See following entries.

squawk alt   Switch to active Mode C with auto altitude-reporting.

squawk Charlie   See *squawk alt*.

squawk flash   Operate IFF I/P switch.

squawk ident   Engage ident feature; civil counterpart of flash.

squawking   Air/ground code: showing IFF/transponder in mode/code indicated.

squawk low   Switch to low sensitivity.

squawk Mayday   Switch to emergency position; for civil transponder Mode A, Code 7700; for mil IFF Mode 3, Code 7700 plus emergency feature.

squawk mike   Operate IFF MIC switch and key transmitter as directed.

squawk normal   Switch to normal sensitivity.

squawk (number)   Operate in Mode A/3 on designated code.

squawk (number) and ident   Operate on specified code in Mode A/3 and engage ident (mil IFF I/P).

squawk standby   Switch to standby position.

squeaker   *1* Fitted to friendly aircraft to warn of nearby balloon cables (WW2).

2 Perfect touchdown, synonymous with greaser.

squeeze-film bearing   Provided with small annular space between outer track and housing filled with pressure-feed lube oil which cushions dynamic radial loads, thus reducing engine vibraton and possible fatigue.

squeeze riveting   Rivet closure by single sustained force instead of blows.

squelch   *1* Subcircuit in communications receiver which holds down volume to reduce output noise until a signal is received.

2 Pilot control of volume or signal/noise ratio.

squib   Any small pyrotechnic used as source of hot gas, eg to fire igniter of rocket or fuel in some (eg missile/RPV) gas turbines.

squib valve   Ambiguous, has been used to mean (1) solenoid valve for controlling thrusters (Apollo RCS) or (2) squib-actuated valve.

Squid   Trade name, see (1) next.

squid   *1* Superconducting quantum interference device; pair of Josephson junctions.

2 Semi-stable part-opened regime of parachute canopy, normally encountered only at extreme airspeeds.

squidding   Operation of parachute in squid position.

squint   *1* Small angular error between actual and theoretical direction of main-lobe axis or direction of maximum radiation of radar or other directional emitter (see * *angle*).

2 Operating mode of forward-looking radar or SAR in which beam is steered to image object obliquely, ahead of or behind platform.

squint angle   *1* Maximum angle away from missile axis at which homing head (IR, radar, EO, etc) can acquire and lock on to normal emitting point target at significant distance.

2 Angle of squint.

Squippers   Safety-equipment fitters (colloq.).

squirrel cage   *1* Air-combat dogfight (colloq.).

2 Induction motor whose rotor comprises axial bars joined to rings at each end, pulled round by rotating field.

squirt   Jet aircraft (colloq., c 1942-48).

squitter   *1* Filler pulses transmitted by transponder between interrogations.

2 Spontaneous transmission generated once per second, without interrogation.

3 Random pulse pairs generated as fillers.

SQL   Structured query language.

Sql   Squelch.

SQP   Signal-quality parameter.

SR   *1* Short-range.

2 Search rate.

3 Sunlight-readable.

4 Special rules.

5 Single rotation.

6 Sortie rate.

7 Strategic reconnaissance, role designator (US), originally transposition of RS (recon/strike).

8 Sunrise.

9 Service report, or request.

10 Solid rocket.

11 Slow-speed route (FAA).

12 Switched reluctance (see SRMG, SRSG).

13 Shear rate.

14 Staff requirement [(A) adds Air].

15 Stopped-rotor [adjective].

sr   Steradian.

SRA   *1* Special-rules area, or airspace.

2 Shop-replaceable, or repair, assemblies.

3 Surveillance radar approach (CAA).

4 Spin reference axis.

5 Specialized repair activity (DoD).

6 Surveillance and reconnaissance aircraft.

SRAA   Short-range air-to-air, M adds missile, W warfare; but see next.

SRAAW   Short-range anti-armour weapon (UAV).

SRADD   Software requirements and design description.

SRALT   Short-range air-launch target (USN).

SRAM   *1* Short-range attack missile.

2 Static, or strategic, random-access memory.

SRARM   Short-range anti-radiation (or radar) missile.

SRB   Solid-rocket booster.

SRBM   Short-range ballistic missile.

SRBOC   Super-rapid-blooming offboard chaff.

SRBS   Skeletal reference baseline simulator (SDI).

SRC   *1* Science Research Council.

2 Secondary radar code.

3 Sample return container, or capsule [spaceflight].

4 Surveillance-radar computer.

SRCC   *1* Structures Research Consultative Committee (SBAC).

2 Standard radar-control console (NATO).

SRCU   Secure remote-control unit.

SRD   *1* Short-range diversion.

2 Service-revealed difficulty.

3 Systems requirements document.

SRDE   *1* Signals Research & Development Establishment (Christchurch, UK, now closed).

2 Search-radar data extractor.

SRDS Systems Research and Development Service (FAA).

SRE Surveillance radar element; portion of GCA.

S_{ref} Reference area.

SREJ Selective reject.

SREM Software requirements engineering methodology.

SRF State-rate feedback [IMF adds implicit mode following].

SRFCS Self-repairing flight-control system.

SRFP Sunlight-readable flat panel; D adds display.

SRFW Schweizerische Rettungsflugwacht (= GASS) (Switzerland).

SRG *1* Safety Regulation Group (UK CAA).
2 Short-range.

SRHit, SRHIT Short-range homing intercept technology (SDI).

SRI Short-range insert.

SRINF Short-range intermediate nuclear force[s].

SRL Società Responsibilita Limitata (I).

SRLD Small rocket lift device.

SRM *1* Solid-rocket motor [U adds upgrade].
2 Short-range missiles (HUD selection).
3 System resource manager (software).
4 Speech-recognition module.
5 Structural repair manual point.
6 Selective-reject mode.

SRMG Switched-reluctance motor-generator.

SR-30 30 minutes before sunrise.

SRO *1* Station routine orders.
2 Senior ranking officer.
3 Space Research Organization (India), usually called ISRO.
4 Sensitive reconnaissance operation[s].
5 Superintendent of Range Operations.

SROB Short-range omnidirectional beacon.

SRP *1* Steep rocket projectile (GGS selection, WW2).
2 Software rapid prototyping.
3 Slot [ATC] reference.
4 Computer (R).
5 Service resource planning.
6 Stabilization reference package; / PDS adds position-determining system.
7 Sustained readiness program.
8 Selected reference point.

SRPG Strain-range pair counter.

SRR *1* Search/rescue region (ICAO).
2 System requirements review.
3 Software requirements review.
4 Short-range recovery.
5 Strategic resources review (NASA).
6 Satellite recognition receiver.

SRS *1* Sonobuoy reference system.
2 Speed reference system.
3 Survival radio set.
4 Smoke-repellant system.
5 Strategic Reconnaissance Squadron (USAF).

Srs Series.

SRSG Switched-reluctance starter-generator.

SRSK Short-range station-keeping.

SR-SS Sunrise to sunset.

SRST Sunlight-readable see-through; HW adds head-wearable.

SRT *1* System readiness test[s].

2 Syllabus for recurrent training.
3 Standard remote terminal.
4 Satellite receiver/transmitter.

SRTM Shuttle-radar topography mission.

SRTO Supervisory Resident Technical Officer.

SRTS Short-range thermal sight.

SRU *1* Shop-replaceable unit.
2 Scanner receiver unit.

SRV *1* Surveillance.
2 Surrogate research vehicle.

SRW Strategic Reconnaissance Wing (USAF).

SRWBR Short-range wideband radio.

SRY Secondary.

SRZ *1* Special-rules zero.
2 Surveillance-radar zone.

SS *1* Sunset.
2 Spread spectrum.
3 Single-slot[ted].
4 Surface-to-surface.
5 Sandstorm (ICAO).
6 System status.
7 Sliding scale.
8 Source-substantiation.
9 System[s] simulation.
10 Sector search (MTI).
11 SAR (2) spotlight (radar mode).

S_s Torsional stress.

SSA *1* The Soaring Society of America.
2 Self-Soar Association (US).
3 Static-stability augmentation.
4 Stick sensor assembly.
5 Supersonic adversary; A adds aircraft.
6 Special security agreement.
7 Safe sector altitude.
8 Soviet strategic aviation, transliteration of ADD.
9 Solid-state amplifier.

SSAC *1* Source-Selection Advisory Council.
2 Solid-state aircooled.

SSADP System station annunciator display panel.

SSAE Society of Senior Aerospace Engineers (US).

SSAI Solid-state attitude indicator.

SSAL Simplied short approach lights; F adds sequence flashing lights, R adds RAIL, S system.

SS-Anars Space sextant autonomous navigation and reference system.

SSAP Survival stabilator actuation package.

SSAR Spotlight SAR.

SSAT Subsonic subscale aerial target.

SSB *1* Single-sideband.
2 Space Science Board (US, NAS).
3 Split system breakers.
4 Ballistic-missile-firing conventionally powered submarine.
5 Supersonic bomber.
6 Small smart bomb; REX adds range-extension.
7 Shaped sonic boom. See next.

SSBD Shaped sonic-boom demonstration, which see.

SSBJ Supersonic business jet.

SSBN Ship, submersible, ballistic [missile], nuclear [powered] (USN).

SSBS Sol/sol balistique stratégique (=MRBM, F).

SSC *1* Short-service commission (RAF).
2 Sidestick controller.
3 Solid-state scanner, or scanning.

4 Stennis Space Center (NASA, St Louis, Mississippi).

5 Slot/spoiler control.

SSCP System services control point.

SSCVFDR Solid-state cockpit voice and flight-data recorder; SSCVDR omits flight-.

SSCVR Solid-state cockpit voice recorder.

SSD Systems and Simulation Division (USAF).

SSDC *1* Space and Strategic Defense Command (USAF).

2 Solid-state data-carrier.

SSDS Ship self-defence system.

SSE *1* Software support environment.

2 Site security enhancement.

3 Simulation/stimulation equipment (ASW).

SSEB Source-Selection Evaluation Board.

SSEC *1* Static-source error correction.

2 Systems and Software Engineering Centre (Qinetiq, Malvern).

SSF *1* Station Services Flight (RAF).

2 Special sensor laser threat detector.

SSFDP Solid-state floating-deck pulser.

SSFDR Solid-state flight-data recorder.

SSFM Steerable sensor-fuzed munition.

SSG Former USN code for cruise-missile submarine; N adds nuclear.

SSGW Surface-to-surface guided weapon.

SSHCL Solid-state heat [originally high] - capacity laser, sustained 100 kW.

SSI *1* Small-scale integration.

2 Silver-strip integrator.

3 Significant structural, or structurally significant, item[s].

4 Sandwich speckle interferometry.

5 Sensitive security information.

6 Single-system image.

SSIA Service de Surveillance Industrielle de l'Armément (F).

SSICA Stick-sensor and interface control assembly.

SSID *1* Solid-state ice detector.

2 Supplementary structural-inspection document.

SSIES Special sensor ionospheric electron scintillation.

SSIM Standard schedules information manual, or manager.

SSIP *1* Supplemental structural inspection program.

2 SubSystems integration project (NICS).

SSIXS Submarine satellite information-exchange system.

SSJ Serrated skin joint.

SSJ5 Special sensor electron/ion spectrometer.

SSK Single-shot kill; P adds probability.

SSL *1* Speed select lever.

2 Single site location.

SSM *1* Surface-to-surface missile.

2 Sea-skimming, or - skimmer, missile.

3 Seat statute-mile, non-SI traffic unit.

4 Special-shape market (advert-shape aerostats).

5 Standard schedule message.

6 Sign status matrix.

7 Software sizing model.

8 Sector search mode (J-Stars).

9 Stick-shaker margin.

10 Special-sensor magnetometer.

SSMA Spread-spectrum multiple access.

SSMC Single-source maintenance contract.

SSME Space Shuttle main engine.

SSMI Special sensor microwave imager; S adds sounder.

SSMO Summary of synoptic meteorological observations.

SSMT Simulated system maintenance trainer.

SSN *1* Attack submarine, nuclear powered.

2 Specification serial number.

SSNR Mobile missile guidance station (USSR, R).

SSO *1* Sun-synchronous orbit.

2 Special-service officer (Australia).

3 Special-systems operator.

4 Surveillance-sensor operator.

SSP *1* Sensor select panel.

2 Source selection panel (or procedure) (NATO).

3 Second-source producer.

4 Solid (unpierced) steel planking.

5 Surface-science package.

SSPA Solid-state phased array, or power amplifer.

SSPAR Solid-state phased-array radar [S adds system].

SSPI Sighting system passive/IR.

SS+30 30 minutes after sunset.

SSPM Space-station propulsion module.

SSPS Satellite solar power system (or station).

SSQ Station sick quarters.

SSQAR Solid-state quick-access recorder.

SSR *1* Secondary surveillance radar (/ RPG adds Regional Planning Group).

2 Solid-state relay, or recorder.

SSRF Shell/shield replacement fabric.

SSRMS Space-station remote manipulator system.

SSRT Mobile detection and designation radar (USSR, R).

SSR video Video (raw radar) signal processed in SSR computer to exclude unwanted information and leave graphical display of targets, data and other displayed information in correct display positions.

SSS *1* Space surveillance system (USN).

2 Small scientific satellite.

3 Strategic satellite system; also survivable strategic satellite.

4 Stick-shaker speed.

5 Simulator-specific software.

6 System support segment [JSIPS].

SSSC Single-sideband suppressed carrier.

SS-SR Sunset to sunrise.

SSST *1* Solid-state star tracker.

2 Supersonic sea-skimming target.

SST *1* Supersonic transport.

2 Static-strength test(ing).

3 Static storage tank.

4 Standard serviceability test.

5 Single-subscriber terminal.

6 Solid-state transmitter [also SSTX].

SSTDMA Satellite system TDMA.

SSTI Stabilized steerable thermal imaging.

SSTO Single stage to orbit.

S-stoff RFNA (G).

SSTOL Super-short, or supersonic and short, takeoff and landing.

SSTS Space-based surveillance and tracking system, to acquire and track PBVs, RVs and ASATs (SDI).

SSU *1* Signal summing unit.

2 Semiconductor storage unit.

3 Sensor surveying unit.

4 Subsequent signal unit.

SSULI   Special sensor UV limb-imager.

SSUP   Space station users panel (Int.).

SSUS   Spinning solid upper stage.

SSUSI   Special sensor UV spectrographic imager.

SSV   Standard service volume (radio).

SSVT   Satcom secure voice terminal.

SSW   Swept square wave (ECM).

ST   *1* Standard time.

2 Stairway (passenger, not powered).

3 Sharp-transition (VASI).

4 Single tandem [landing gear, eg C-130].

5 Strategic transport, air.

6 Static thrust; st is preferred.

7 Statistics.

S_T *1* Area of horizontal tail.

2 Distance between centres of contact areas of tandem wheels.

St   *1* Stratus.

2 Stokes.

3 Stanton number.

4 Strouhal number (also S).

5 Static thrust.

S_t Unit tensile stress.

st   *1* Static thrust.

2 Stone.

STA   *1* Service des Transports Aériens (F).

2 Station (also Sta), eg fuselage STA 307.8.

3 Supersonic transport aircraft (duplicates SST).

4 Static test article.

5 Straight-in approach (ICAO).

6 Shuttle Training Aircraft, STA prefix for other acronyms.

7 Structural test airframe (or article).

8 Section Technique de l'Armée.

9 Scheduled time of arrival.

10 Surveillance and target acquisition.

11 Satcom terminal assemblage.

Sta   Station.

STAARTE   Scientific training and access to aircraft for atmospheric research throughout Europe.

STAB   Stabilizer.

Stab   *1* Staff, especially staff flight of four aircraft (G).

2 Steered agile beam.

stabbing   Assembly process in which finished gas-turbine stator blade (IGV) is fired at high speed through unbroken ring to form tight-gripping joint.

stabbing band   Narrow projecting band around rotating part, eg turbine disc, from which metal can be removed for dynamic balancing (now generally superseded).

STABE   Second-time-around beacon echo.

stabilator   Slab horizontal tail used as single primary control (stabilizer/elevator). Normally pitch axis only, but occasionally used to mean taileron (US usage).

stability   Generally, quality of resisting disturbance from existing condition and tendency to restore or return to that condition when disturbance is removed. For aircraft, meaning is confined to basic flight control and defined as tendency to resume original (normally straight/level but not necessarily) attitude after upset (rotation about any axis); qualified according to axis and whether stick-fixed or stick-free. See *motion*. For atmosphere, temperature distribution such that particle tends to stay at original

level. For structure, ability to develop internal forces resisting those externally applied. For materials other than structural, usually ability to withstand harsh environment (eg high temperature) without even gradual physical or chemical change.

stability augmentation   *1* Various species of auxiliary subsystem added to primary flight-control system (usually of helicopter, advanced aeroplane or spacecraft) to achieve desired vehicle characteristics by selection of variable gains in feedback loops from surfaces. In some forms surfaces are commanded, eg yaw damper. Modern fighters with relaxed longitudinal stability would be dangerously unstable without **. Usually has limited authority and does not move pilot's controls.

2 Some authorities insist ** artificially improves stability while retaining control in the hands of the human pilot.

stability axes   Introduced c 1939 to simplify calculation, X-axis aligned with relative wind and remaining axes fixed relative to body throughout subsequent disturbance. Made redundant by computers.

stability derivatives   See *derivatives*.

stability factor   Ratio of change in transistor collector current to change in I_{co} (DC collector current for zero emitted current).

stability limits   *1* Forward and aft c.g. limits.

2 Extreme angles of incidence to which taxiing seaplane can be trimmed without porpoising.

stability loop   Plot of limits of gas-turbine combustion, one too fuel-rich and the other too weak, on ordinate air/fuel ratio and abscissa mass flow.

stability margin   See *static margin*.

stabilization   *1* Positive action to maintain stability, esp. of spacecraft or payload, when term invariably refers to attitude. Passive * is any method requiring no sensing, logic or power, eg gravity-gradient, spin and solar-wind/aerodynamic pressure. Semi-passive requires stored momentum, eg gravity-gradient plus CMG. Semi-active introduces limited thrust/torqueing, eg on one axis. Active features sensing, logic and control about all axes. Hybrid are systems with more than three degrees of freedom, eg to control despun/gimballed/independent secondary devices such as aerials or telescopes.

2 On starting engines of multi-engined aircraft, attainment of steady rpm, EGT and other parameters on all engines, usually prior to taxi.

stablized approach   On glidepath at correct airspeed, correctly configured, all checklists and paperwork complete.

stabilized gyro   Usually means aligned with a desired direction, eg Earth centre, true N or magnetic meridian.

stabilized platform   Invariably, platform maintained always horizontal at any place on or near Earth, ie perpendicular to local vertical.

stabilizer   *1* Tailplane or slab horizontal tail (US).

2 Loosely, any fixed tail surface, but fin normally prefaced by 'vertical' (US).

3 Low-density core (eg foamed-in-place plastics, balsa, honeycomb) filling interior of secondary structure, control surface, flap, door or similar structure.

4 Additives to retard chemical reactions.

5 Gyro subsystem to stabilize pivoted or gimballed device, eg radar aerial.

6 Flame *.

stabilizing altitude Altitude at which actual rate of climb is zero.

stabilizing floats Small seaplane-type floats mounted well outboard under wing of flying boat or three-float seaplane to provide roll stability when afloat.

stabilizing gears Small landing gear carried well outboard (eg near or at tips of wing) of landplane with centreline main gears to provide lateral stability on ground, esp. when turning. Also called outriggers.

stabilizing parachute Used to stabilize fall of otherwise unstable paradropped load.

Stabimatic Simple modular autopilot for GA, buildable from wing-leveller using vacuum aileron input to fully coupled 3-axis system capable of capturing desired FL.

stable An air force's total front-line inventory of one aircraft type [rarely, entire available inventory of all types].

stable aerofoil Complete wing whose CP travel is very small.

stable air Air mass in which actual lapse rate is less than adiabatic lapse rate (dry or saturated, depending on humidity), in extreme cases becoming negative (ie inversion).

stable-base film Reconnaissance or scientific film of extremely high dimensional stability.

stable equilibrium Body returns to original location after being displaced.

stable oscillation *1* Oscillation whose amplitude is constant.

2 Oscillation whose amplitude decreases (BS).

stable platform See *stabilized platform*.

stable spread Standard attitude for minimum rate of descent in sport parachuting: face-down, arms and legs spread widely.

Stabo Anti-runway munition (G).

STAC Supersonic Transport Aircraft Committee (UK, 1956-62).

stack *1* Superimposed series of holding patterns, each at assigned FL.

2 To assign to hold in *.

3 Piston-engine exhaust pipe, of any length or configuration (US).

4 To assemble multi-stage launch vehicle (colloq.), and vehicle thus assembled.

Staco Standing Committee for the Study of principles of Standardization (ISO).

STacSAR Small tactical SAR (2).

Stadan Space tracking and data network. Previously called Minitrack, there are fixed linear aerials at College, AK; St John's, Newfoundland; Goldstone, CA; E Grand Forks, MN; Blossom Pt, MD; Ft Myers, FL; Quito, Ecuador; Lima, Peru; Santiago, Chile; Winkfield, England; Johannesburg, S Africa; and Woomera, Australia. Large dishes are located at Fairbanks, AK; Rosman, NC; and Canberra, Australia.

stadiametric aiming Optical aiming using lead angle calculated from apparent size of target and aspect, using various methods including subjective judgement (suggest arch.).

stadiametric ranging Estimating target range from knowledge of its true size.

stadiametric warning Based on range-closure derived from apparent size of other body.

STAé Service Technique Aéronautique (F).

Staff, STAFF *1* Smart target-activated fire-and-forget.

2 Spatio-temporal analysis of field fluctuation.

Staffel Squadron (G).

Staff-Pak Four interlinked laboratory modules designed for installation in transport (C-130) to provide electrically noise-free environment.

staff pilot Experienced military pilot assigned to special duties, ie not with operational or training unit.

Stag Simultaneous telemetry and graphics.

stag Stagnation.

stage *1* One complete element of propulsion, jettisoned (staged) when propellants are consumed (normally applied to rocket). In a multi-* vehicle each * fires in sequence following separation of predecessor to reduce mass remaining.

2 One complete element of multi-* process, normally compression or expansion, through which fluid is passed. Passage through a single long diffuser, venturi or other tapering duct is not * but use of several in succession causes each to become one *.

3 One complete element of fluid-flow compressing or expanding (eg power-extracting) device; eg one planar assembly of compressor rotor blades and associated ring of stator blades.

4 Sector (4) or, military, portion of air route between two staging units; sometimes, for flight planning, one point on route.

5 Various meanings in electronics, EDP (1) and other disciplines.

stage cost Direct operating cost of flying one (mean, or one specific) stage (4).

stage count Simple list of total number of stages of blading in fan, compressors and turbines, thus GP7200 * reads 1-5-9-2-6.

staged combustion Or * combustor, fuel is supplied to groups of burners arranged in rings of different radius or in axially spaced rings or radial arms, ignited successively.

staged crew Prepositioned at staging unit to take over incoming flight.

stage flight One flight forming part of longer multi-stage journey.

stage fuel Fuel burned in flying one stage (4); hence ** carpet, plot of variables for flight-planning purposes.

stage length Air-route distance between two staging points; in commercial use normally synonymous with sector distance.

stage sheet Completed for each stage of maintenance listing all configuration changes and parts replaced.

stage time *1* Planned or actual time at which stage (1) takes place.

2 Sector time.

Stagg Small turbine advanced gas-generator.

stagger *1* Distance measured parallel to aircraft longitudinal axis between biplane lower-wing leading edge and vertical projection on to lower-wing extended chord line of upper-wing leading edge at same spanwise station (UK). Negative when upper plane is aft of lower.

2 Acute angle measured in vertical plane parallel to aircraft longitudinal axis between leading edges of lower and upper planes at same spanwise location (US). Negative when upper plane is aft of lower.

3 PRF variation by various means involving interleaving trains separated by offset interval; alternative EW technique to PRF jitter.

stagger angle   *1* In biplane, acute angle in plane parallel to aircraft longitudinal axis between line joining points equidistant from centreline on upper and lower LE (see *stagger* [1]) and local vertical.

2 In rotor blade of gas-turbine axial compressor turbine rotor, angle between principal chord at any radius and [usually] plane through axis of rotation; chord line may be drawn as tangent to LE/TE, or even through front of aerofoil only.

stagger tuning   Increasing pass-bandwidth of RF receiver by tuning different output stages (one meaning of stage [5]) slightly above or below central frequency.

stagger-wing   Biplane of any make with negative stagger.

stagger wire   Diagonal wire joining lower and upper wings of biplane and lying approximately in plane parallel to axis of symmetry. (US term; UK = incidence wire).

staging   *1* Separation of one stage (1) from next.

2 Time at which * (1) is scheduled or actually occurs.

3 Flying by separate stages (4), with or without changes of crew.

staging area   Geographical area between mounting area of exercise and objective, esp. for airborne or amphibious operation.

staging base   Landing and takeoff area with minimum servicing, supply and shelter provided for temporary occupancy of military aircraft during course of movement from one location to another (DoD).

staging point, unit   Place or organization linking two stages (4).

stagnation line   Locus joining stagnation points, eg boundary between radial-wall jets under hovering VTOL.

stagnation point   Point on surface of body in viscous fluid flow (one facing upstream and one down) where fluid is at rest with respect to body, flow in boundary layer on each side of ** being in opposite directions.

stagnation pressure   Pressure at stagnation point, normally same as total head, total pressure or pitot pressure, = sum of local atmospheric plus dynamic pressures.

stagnation region   Region close to upstream stagnation point.

stagnation stall   *1* Several related afflictions of afterburning turbofans normally occurring on afterburner light-up at high altitude at modest airspeed, in most cases with rapid pressure pulses (say, seven per second) in fan duct causing oscillating stall of fan and then core engine.

2 In flight of aircraft, any stall that is not self-correcting.

stagnation streamline   That which in any representaton of 2D flow passes through front and rear stagnation points on immersed body.

stagnation temperature   That at stagnation point, when all relative kinetic energy has been converted isentropically to heat.

stainless steel   Generally, steels with 12-20 per cent chromium. Most common is 18-8, these being % chromium and nickel.

Stairs   Sensor technology for affordable IR systems.

stair-stepping   Step cruising.

STAJ   Short-term anti-jam.

stake   Anvil-type bench tool for sheet.

stake out   To picket aircraft.

staking   Swageing terminal on to electrical conductor.

Stalag Luft   Prison camp for captured aircrew (G, WW2).

STALD   Standoff tactical air-launched decoy.

stale track   Shown on radar display at last known position, even though it has not appeared on susbequent updates.

stall   *1* Gross change in fluid flow around aerofoil, usually occurring suddenly at any 2-D section aligned with flow, at AOA just beyond limit for attached flow (at which lift coefficient is maximum); characterized by complete separation of boundary layer from upper surface and large reduction in lift. Traditional wings normally * at AOA near 16°-18°, which can be attained at any airspeed depending on applied vertical acceleration. AOA for * is increased by slat, and some highly swept (variable-sweep at max. sweep) wings exhibit no * even at AOA beyond 60°.

2 Sudden breakdown in fluid flow previously attached to solid surface, caused by changed angle of either surface or flow, violent pressure pulse in flow (esp. travelling upstream) or other severe disturbance, eg flutter.

3 Point at which opposing linear force or torque overcomes that of driving member (eg PFCU, hydraulic motor, airbrake or tailplane actuator), causing a commanded movement to be arrested.

stalled pressure   Delivery pressure at which delivery from variable-stroke fluid pump, centrifugal compressor or certain other pumps falls to zero; also called reacted pressure.

stall fence   Fence whose purpose is primarily to improve behaviour at stall.

stalling angle   AOA at which flow suddenly separates from upper surface; probably that at which C_L is a maximum

stalling flutter   *1* Flutter in one or more degrees of freedom near angle of stall (1).

2 In particular, flutter of any stalled aerofoil [eg, compressor blade] drawing energy from surrounding flow.

stalling speed   Any speed at which stalling AOA is reached, esp. that at 1 g when ** is at lowest value (when depends on aircraft weight, aircraft configuration and air density, among other variables). Usually assumptions include SL ISA, gear/flaps down, power off.

stall line   Boundary between acceptable operating conditions for gas turbine and stall zone for any given altitude as plotted on compressor map.

stall margin   Difference, normally expressed as available spread of rpm, between gas-turbine operating line (at any altitude and for transient slam accelerations, etc) and stall line.

stall out   To stall as result of attempting too steep a climb or for any other reason, esp. when chasing opponent, thus leaving manoeuvre incomplete or failing to get into firing position.

stall protection system   Aeroplane flight-control subsystem sensitive to AOA (sometimes sensed at points on either side of centreline to cater for rapid-roll AOAs) which at given value triggers positive action to prevent stall; obvious example is stick-pusher, but stick-shaker is sometimes considered for inclusion.

stall quality   Pilot's subjective opinion of behaviour of aircraft (normally fixed-wing aerodyne) at stall, assessed for all types of stall (eg accelerated) and configurations (eg dirty).

stall strip   Transverse ridge or other projection added to skin of aeroplane, usually in spanwise direction along leading edge, to serve as stall-promoter, create warning buffet and ensure stall (1) occurs first at that point.

stall tolerance   Generally non-quantified quality of gas-turbine fan and/or compressor to accept distorted airflow or other disturbance (eg gun gas, ingested jet gas, birds, hail or pressure pulses moving upstream) without stalling.

stall turn   Flight manoeuvre in which aircraft (aeroplane or glider) is pulled up into very steep climb, usually with engine cut well back, until on point of stall full rudder is applied to cause rapid rotation in yaw, with wings rotating in near-vertical plane; ends in dive and pullout on to desired heading (generally on to reciprocal). In US hammerhead stall (also see *wingover*).

stall warning   Anything giving pilot warning of impending stall, eg natural buffet, inbuilt ** system sensing AOA and giving visible or aural warning, or stick-shaker with or without knocker.

stall warning and identification system   SWIS, system commanded by AOA vane (12) whose signals are analysed for AOA and rate; because of natural lag or hysteresis in system, trigger is fired progressively earlier as rate is increased by building in a phase advance giving protection at all rates.

stall zone   Region beyond stall line of gas turbine at any altitude where attempted pressure ratio is too great for rpm and airflow.

stalo   Stabilized local oscillator.

Staloc   Self-tracking automatic lock-on circuit.

stamo   Stabilized master oscillator.

Stamp   *1* Small tactical aerial mobility platform (USMC).
2 Single-tube auto multipoint.

STAN   Sum total and nose gear.

Stanag, STANAG   Standard NATO, or Standard-ization, agreement.

stand   *1* Place for parking one aircraft, especially at airport terminal.
2 Fixed or mobile mounting for item undergoing inspection, test, maintenance or repair.

stand-alone   Generalized term meaning equipment, eg radar, is not integrated directly into existing system of radars or other sensors, computers and com. network. Increasingly being used for fixed-based weapon systems and even vehicle-mounted (eg airborne) equipments. Thus * ASW helicopter operates autonomously, without needing senors or other platforms.

standard acceleration due to gravity   See *gravity*.

standard aerodrome   "An aerodrome suitable for the operation of regular day and night services" [BS.1940]. Now arch.

standard air munitions package   Conventional air/ground munitions required for 30-day support of one aircraft of specific type, air-transportable in three equal increments.

standard atmosphere   Model atmosphere defined in terms of pressure, density and temperature for all heights, assuming perfect gas, devoid of any form of water or suspended matter; approximates to real atmosphere and taken as reference for aircraft performance and all other quantitative measures. First NACA 1925 (see *model*

atmosphere), later refined 12 times, current 1980 ICAO Doc 7488. Physical constants; P_0 10,132.5 Pa; T_0 288.15°K; M_0 2.89644× 10^{-4} kg/mole; ρ_0 1.2250 kg/m³; R * 8.31432 J k-mol^{-1} K^{-1}; temperature gradient from -5,000 m (5 km below SL) to altitude (11,000 m) at which T is -56.5°C is -0.0065°C per standard geopotential metre; from 11 to 20 km temperature gradient is zero; from 20 to 32 km temperature gradient is +0.0010°C per standard geopotential metre.

Standard Beam Approach   Pioneer landing aid providing lateral guidance and series of marker beacons.

standard bird   Two (occasionally three) specifications for birds used as inert (sometimes frozen) bodies fired into gas turbine in ingestion testing.

standard body   Not wide-body, ie width in region of 3 m. Loosely synonymous with narrow-body, slim jet, single-aisle.

Standard Class   Sailplane competition class limiting span to 15 m and prohibiting high-lift flaps and certain other features.

standard conditions   Standard temperature and pressure.

standard data message   NATO message format for digital communications between national or international units or facilities; an example of a Stanag result.

Standard Day   At ISA pressure and temperature.

standard day of supply   Total amount of supplies needed for average day as defined by NATO Standing Group rates.

standard design memo, SDM   International standardized proforma for hard-copy communication of information affecting hardware design, often as computer printout.

standard deviation   Quantification of dispersion of data points about mean value: square root of average of all squares of variances (amount by which each point differs from mean); symbol σ.

standard DME arrival   Arrival routes based on DME distances.

standard electronic module   See *ATR* or *MCU*.

standard empty weight   No longer definable, most precise equivalent is APS.

standard gravity   See *gravity*.

standard industry fare level   Hypothetical revenue rate per mile invented by CAB as guide to IRS in taxing GA aircraft on non-company business (US).

standard instrument departure   Preplanned, coded ATC IFR departure routing, preprinted for pilot use in textual form often (at major traffic points) supplemented by graphics; abb. SID.

standardization   Objective of achieving interoperability through use of either uniform or at least compatible hard-ware; significant that definition of word by DoD, NATO, SEATO, CENTO and IADB is in each case different, while definition 'standardization agreement' differs for NATO, SEATO and CENTO.

standardized product   One conforming to specifications resulting from same technical requirements (NATO, CENTO, IADB).

standard load   One preplanned as to dimensions, weight and balance, and designated by a number or other classi-fication (NATO).

standard mean chord   Gross wing area divided by span; position defined by co-ordinates of quarter-chord point and an inclination found by integrating (three methods). Also called geometric mean chord, symbol c̄. Numerically equal to chord of rectangular wing of same span and gross area.

standard NPL tunnel   Closed-jet, no return flow.

standard of build   Precise description of which of various options were followed in construction and equipment of aircraft, esp. of prototype or development aircraft where *** changes between one aircraft and next.

standard of preparation   Defines list of equipment installed.

standard operating platform   An agreed build-standard for newly constructed airfields, mainly in Western Germany (NATO 1951-54).

standard option   Choice of build standard, engine, avionics, finish colours, furnishing or other variables offered to all customers (eg improved stopping on Advanced 727).

standard parallel   Parallel on map or chart along which scale is as stated.

standard pitch   See *propeller* *.

standard pressure   *1* Standard SL atmospheric pressure of 10,132.5 Pa.

2 In meteorology, usually 1,000 mb = 10^5 Pa.

standard propagation   Assumes smooth spherical Earth of uniform dielectric constant and conductivity under standard atmospheric refraction decreasing uniformly with height.

standard radio atmosphere   One having excess modified refractive index (also see *standard refraction*).

standard-rate turn   Usually heading changes 3°/s or 2 min for 360°.

standard refraction   Idealized ratio refraction decreasing uniformly with height at 39×10^{-6} units per km; included in groundwave calculations by enlarging Earth radius to 8.5×10^6 m.

standard structure   Not normally used in aerospace; elsewhere often structure whose dimensions are everywhere mid-way between tolerance limits.

standard temperature   Value upon which a temperature scale is based, in physics normally 273.15°K (0°C), but for practical (eg gas-turbine rating) purposes normally that at zero height in standard atmosphere, 288.15°K.

standard terminal arrival route   Preplanned coded ATC IFR arrival routing, preprinted for pilot use in textual form often (at major traffic points) supplemented by graphics; abb. Star.

standard time, ST   Universally adopted time for all countries, based on zone time but modified to suit country's longitude span and if necesssary zoned such that difference between ST and GMT is always divisible exactly by 0.5 h.

standard turn   See *standard-rate turn*.

standard weight   *1* Term formerly used in FAA certification as being certificated gross weight.

2 Also used to mean assumed mass of such loads as adult passenger, parachute, and unit volumes (eg litre or US gal) of fluids.

standby, stand-by   *1* Generalized term for being available at short notice, in some cases (eg redundant flight-control channel) instantaneous and in others (interceptor on *) at minutes.

2 R/T code: "I must pause for a few seconds"; if followed by "out" means pause may be much longer but channel must be kept clear for resumption, the meaning

becoming "other stations please do not transmit on this frequency".

3 Able to board flight if seat available.

standby item   One duplicating another in function and used following failure of that normally operative.

standby mode   One of several basic operating modes for equipment, eg radar, normally characterized by receivers shut down but transmitters warmed and ready for immediate power, or DME powered up but not transmitting.

standby redundancy   System design such that redundant duplicative channels do not normally operate (as in parallel) but are switched on following failure of those normally operative.

stand-down   *1* Particular aircraft, though serviceable, remains for long period on ground, for whatever reason.

2 Base or unit is deactivated.

stand fix   Most accurate of all fixes, obtained with aircraft parked on surveyed location.

standing detachment   Semi-permanent deployment (RAF).

Standing Group   The permanent body of NATO.

standing water   Defined as mean depth exceeding 12.7 mm (0.5 in).

standing wave   *1* Oscillatory motion in vertical plane of air downwind of steep hill or mountain face in which troughs and peaks (latter usually marked by cloud at various levels) remain roughly stationary; at lowest level cloud often rotor, high levels usually lenticular. Strong ** often associated with jetstream.

2 Stationary wave formed in transmission line by superposition of travelling wave reflected back from point of impedance-change of forward wave.

3 Stationary wave or wave-pattern formed in vibrating body, eg turbine disc, by reflection.

standing wire   That length of cable consumed in making splice.

stand-off   *1* Distance from target surface to reference point on hollow-charge warhead (usually apex of cone).

2 To remain outside airfield circuit or pattern, normally following command or positive decision to do so, eg following landing-gear failure or obstruction on runway.

3 To have to park too far from terminal to use airbridges.

4 To remain outside effective range of enemy defences, esp. when making an attack with * missile.

stand-off ability   Capability of forcing, eg by outgunning, similar enemy vehicles to remain beyond their own firing range.

stand-off armour   Armour fixed on outside of existing armour with sufficient spacing to protect against hollow-charge piercing weapons. Unlike spaced armour, normally only two layers in all.

stand-off bomb   ASM (1) launched beyond enemy defence perimeter. Odd UK term of 1950s usually synonymous with cruise missile.

stand-off flare   *1* IRCM payload dispensed at sufficient distance to protect against enemy heat-homing missiles.

2 Rarely, illuminating flare dropped on parachute beside rather than over surface target, usually in helicopter ASV attack.

stand-off missile   One which may be launched at a distance (from target) sufficient to allow attacking personnel to evade defensive fire from target area (USAF).

stand-off steps Passenger/crew stairways kept at terminal equipped with loading bridges at gates to cater for overflow traffic that has to stand-off (3).

stand up To become operational at commissioning ceremony of new squadron or other unit (US).

Staneval Standards evaluation.

stang fairing Fairing on sides of jet engine installation, especially pod, covering reverser bucket hinges and rams.

Stanly Statistics and analysis, of ATC data.

Stano Surveillance, target acquisition and night observation; C adds counter-intelligence center.

Stansit Statistics on non-scheduled air transport (ECAC).

Stanton number Non-dimensional number defining heat transfer through surface; $St = -q/\rho \, VC_p \, dT$ where q is total quantity of heat, ρ is density of fluid (eg air), V is relative velocity, C_p is specific heat at constant pressure and dT is recovery temperature minus wall temperature.

STAP *1* Statistics panel (ICAO).

2 Space/time adaptive processor.

STAPL Ship-tethered aerial platform (Kaman).

STAR, Star *1* Standard terminal arrival route[s].

2 Ship tactical airborne RPV.

3 Surface-to-air recovery (or, USAF, retrieval[s]).

4 Satellite de télécommunications, d'applications et de recherche (F) [see *Star (2)*].

5 Space thermionic advanced reactor.

6 Strategic and tactical airbone reconnaissance (Thomson-CSF), or recovery.

7 Star tracking using ambient radio.

8 Supersonic tailless-aircraft research.

9 System-threat assessment report.

10 Subsonic-transport acoustic research.

11 Studies, tests and applied research (Eurocontrol).

12 Strategic-arms reduction.

13 Signal threat analysis and recording.

14 Simulation trainer for ATC and radar.

15 Surveillance and threat-alert radar.

Star *1* Starboard.

2 Satellite for telecom, applications and research.

star *1* Star-shaped empty space along centre of solid rocket propellant grain.

2 Formation formed all on same level by wrist-linked team of free-fall parachutists.

3 Helicopter main-rotor control radial arms, located under hub; one fixed, one rotating.

Staran Association processor in ARTS-II (FAR).

starboard Naval-derived term for right, right-hand or towards right, seen from behind. Thus, from front, * is on left.

STAR-C STAR (5)-compact.

star-centred Cast or extruded with star (1).

stardust Air-impingement haze in acrylic finish.

staring focal-plane array Important class of sensors which operate like human eye, with focal plane of input optics covered with dense micro-mosaic of 2-D receptors which 'look' continuously, direction of target being determined from knowledge of which detector(s) see it.

STAR-M STAR (5)-mid range (over 30 kW).

starplates Upper and lower spiders each formed from one piece of composite material or metal (titanium) forging and forming structural basis of modern non-artic-

ulated helicopter rotor hub. Not to be confused with star (3).

Stars *1* Silent tactical-attack/reconnaissance system.

2 Plural of STAR, Star (1).

3 Standard terminal and arrival reporting system (CAA).

4 Small transportable (or tethered) aerostat relocatable system.

5 Surveillance and, or stand-off, target-attack radar system (USAF).

6 Standard terminal automation-replacement system (FAA, on-going 2003).

7 Software technology for adaptable reliable system.

8 Space-based telemetry and range safety.

START Strategic Arms Reduction Treaty; *1, 31 July 1991; *2, 3 January 1993; *3, not yet ratified 2003; *4, being discussed.

Start *1* Spacecraft technology and advanced re-entry test.

2 Solid-state angular rate transducer.

3 Shf tri-band advanced range-extension terminal.

4 Special threat-awareness receiver/transmitter.

starter Device for cranking any prime mover of rotary type during starting: 14 basic species.

starter exhaust Overboard discharge of exhaust from starter of cartridge, monofuel, fuel/air, bipropellant or air-bleed types; potentially dangerous.

starter gearbox Box containing gear train through which starter cranks engine; may be reduction or step-up gears.

starter/generator Single electrical machine, usually DC, serving both as electrical starter and electrical generator (see *CSDS*).

starter magneto Various forms of magneto designed to provide powerful spark during start, eg hand-cranked, impulse starter and LT boosting energy transfer.

Startex Start of exercise (UK).

starting chamber *1* Combustion chamber in multi-chamber (multi-can) gas turbine in which igniter is fitted, flame thereafter being carried round by inter-chamber pipes.

2 Liquid-rocket precombustion chamber.

starting coil Auxiliary induction coil used as HT booster when starting piston engine; alternatively used as energy-transfer LT booster.

starting pressure Minimum rocket combustion pressure at which nozzle exit plane is shock-free.

starting transients Temporary variations in pressures, flows, velocities and temperatures during complete start sequence of rocket.

starting vortex Transverse vortex left by lifting wing at start of motion providing essential link behind subsequent vortices from wingtips (part theoretical concept, since at start wing may not be lifting and ** has zero strength, but necessary because trailing tip vortices cannot have free ends).

star tracker Optical or opto-electronic sensor which automatically locks on to preselected celestial body or bodies to provide input to astro or astro-inertial nav system.

Starts Strategic arms reduction treaties.

startup Launch of new OEM.

start-up airline Commonly taken to mean first six months' operations.

start-up costs Those incurred in launching new type of aircraft. Precise definition lacking; general opinion is that it covers all costs up to certification, excluding related engine and systems.

Star-21 Strategic aerospace review for 21st Century (EU).

star washer Hard steel washer with multiple twisted radial projections which bite into superimposed nut and prevent it becoming loose.

STAS Space transportation architecture studies.

STAT *1* Statistical.

2 Statute (US).

state *1* Readiness condition of combat aircraft, from cockpit-alert to unserviceable (different national sub-divisions).

2 In connection with runway, condition of surface or traffic occupancy.

3 For rotorcraft main rotor, usually three subdivisions: propeller *, vortex ring * and windmill-brake *.

4 Amount of main propulsion fuel remaining, esp. when running low.

State chicken Air-intercept code: fuel state requires recovery, tanker or diversion (DoD).

State lamb Air-intercept code: "I do not have enough fuel for intercept plus reserve required for carrier recovery".

state of occurrence State in which an incident, such as an accident, takes place.

state of the art Level to which technology and science have at any designated cut-off time been developed in any given industry or group of industries (USAF).

State tiger Air-intercept code: "I have fuel for completion of mission".

Statfor, STATFOR Panel on Statistics and Forecasts (Eurocontrol).

static *1* At rest, or at rest relative to solid surface or local atmosphere [ram pressure zero].

2 Structural test with application of single increasing load.

3 Radio and other com. interference, esp. that due to discharge of * electricity.

static air temperature Static temperature.

static balance *1* Aircraft condition in which there is no resulting moment about any axis.

2 Control-surface condition in which in absence of any applied torque surface is freely balanced about hinge axis, either because c.g. lies on that axis or because mass balance has been added.

3 Propeller condition in which when supported on rod on knife-edge it rests in any position.

static bomb Tube-mounted instrument outputting TAS.

static cable Longitudinal cable supported at each end along interior of transport to which parachute strops of troops and airdropped loads are attached. Also called strop line, anchor cable.

static ceiling Altitude at which airship in ISA (in some definitions, without forward speed) neither gains nor loses altitude after all ballast has been dumped.

static characteristic Basic plot of thermionic valve, grid volts against anode current.

static conversion Conversion of energy from one form to another without use of moving parts; common example is solid-state AC/DC converter.

static discharger Device for harmless distribution of static electricity, static wick.

static electricity Electric charge built up on non-conductive surface or insulated body by deposition of electrons or positive charges, eg by friction between air and aircraft or between fuel and hose, ultimately reaching very large potential difference.

static firing Firing test of rocket motor, engine or vehicle while attached to test stand.

static flux Magnetic field through magneto frame and inductor with latter stationary.

static friction Force required to initiate relative movement between surfaces in contact.

static gearing ratio Ratio of angular deflection of vehicle (esp. missile or RPV) control surface to rotation of vehicle axis which caused surface to deflect.

static ground line Connection of aircraft earthing (grounding) system to earth.

static head Pitot head measuring ambient static pressure.

static instability Unlikely design fault in which aircraft, once disturbed from straight/level flight, suffers increasing upset in absence of aerodynamic inputs, eg due to high c.g.

static inverter Non-rotating device for converting a.c. to d.c.: two forms, either transistor or gas tube [eg thyratron].

static lift Difference in mass between gas contained in aerostat at rest and air displaced by whole aerostat.

static line *1* Links parachute with static cable so that as parachute leaves aircraft it is opened automatically. Hence ** jump.

2 Line of parked aircraft on static display at airshow.

static load Applied force is unidirectional, either held constant or increasing in programmed way from zero to maximum.

static longitudinal stability Static stability (1).

static margin Basic measure of aeroplane static stability (primarily in pitch), normally defined as distance of c.g. ahead of neutral point expressed as %MAC. Measured stick-fixed or stick-free; former is %MAC proportional to stick displacement with percentage change of speed from trimmed value, called positive when direction backward for lower speed; stick-free is %MAC proportional to rate of change of applied stick force with percentage change of speed from trimmed speed, positive when pull needed by lower speed.

static marking Flaws on imagery caused by static-electricty discharge.

static moment Product of area and its distance (measured from centroid) from reference axis.

Staticon Photoconductive type of TV camera tube.

static pin Wire fitting, usually multi-pin, pulled from parachute by static line.

static port See *static vent*.

static positioning GPS output made possible by fact receiver is fixed relative to Earth.

static power installation Electrical ring mains on airfield.

static pressure That exerted by fluid on surface of body moving with it; alternatively on a surface exactly parallel to local free-stream velocity vector and which does not affect local flow.

static pressure gradient Possible progressive increase in static pressure along constant-section duct, eg tunnel working section, resulting from progressive thickening of boundary layer.

static pressure head   See *static head*.

static propeller thrust   That generated by propeller restrained against movement.

static radar   One whose scanning is electronic, not mechanical.

static radial   Normal radial engine, as contrasted with rotary.

static rail   Rigid form of static cable.

static RAM   Constructed of bistable transistor elements, not needing refreshing.

static sag   Angular or vertical-linear distance through which a structure sags at rest compared with some other reference condition, esp. wingtip at rest compared with 1 g flight.

static soaring   Soaring (US term, now arch., contrasting normal soaring with so-called dynamic soaring).

static-source error   Correction applied to measured static pressure [of particular aircraft] to give correct reading over range of altitudes and Mach numbers.

static-source selector   Pilot switch connecting static system with chosen sensor.

static stability   *1* Measure of tendency of aeroplane to return to or diverge from initial trimmed condition following upset, with throttle[s] fixed.

2 Study of applied moments imposed on aeroplane following upset of AOA or speed from initial trimmed condition.

3 Quality of having positive static margin.

4 Stability attained by action of weight, esp. by having low c.g. (common US definition entirely unrelated to normal meanings).

static-stability compensator   Limited-authority subsystem which ensures aeroplane has stick-free static stability by applying small nose-down trim force immediately before stalling AOA is reached, esp. to avoid self-stall; in UK called static-stability augmenter.

static system   Barometric-ASI plumbing incorporating pitot/static head of static vent and connections to ASI, VSI and possibly other devices; usually includes alternative sources.

static temperature T_s, temperature measured by thermometer moving with fluid [or measured after airspeed is reduced to zero] and thus measuring true temperature due to random motion of fluid molecules. Numerically equal to total air temperature corrected for Mach effect.

static test   For aeroplane or other structure, test under static loads; hence constructor may build ** specimen as well as fatigue specimen. For some vehicles, usually means test while fixed in one place.

static 3-D radar   Large surface radar with fixed aerial scanned electronically.

static thrust   Thrust of jet engine restrained against forward motion, especially measured under ISA sea-level conditions.

static tube   That transmitting pressure from static head.

static turn indicator   Rare instrument driven by pressure difference from two modified static heads mounted transversely at wingtips.

static vent   Carefully designed opening in plate aligned with skin of aircraft which under most flight conditions senses true static pressure.

static wick   Device for discharging static electricity from fine (very small radius) tips of thousands of conductive wires, braid or graphite particles built into flexible wick projecting behind trailing edge.

statimeter   Instrument for measuring static thrust.

station   *1* Dimension measured from * origin locating all planes along fuselage normal to longitudinal axis, or along wing normal to transverse axis. Every structural part, bolt hole, equipment item, door frame and every other dimensional reference is in form of * number (in US in inches, elsewhere in mm). Corresponding term for vertical distances is WL, waterlevel).

2 Fixed base for military, naval, air or research operations.

3 Airport (esp. staff, facilities and costs); airline usage, usually adjective only.

4 Planned or actual position of geostationary satellite.

5 Position relative to other vehicles of one vehicle in group (esp. of aircraft and ships).

6 Location of radio transmitter.

7 Location in gas turbine engine. There is no standard system. Pratt & Whitney uses AM [ambient] followed by 1 [entry to fan] leading to [eg] 4 for HPT inlet and 4.5 for HPT exit.

stationary front   Front without significant motion over Earth's surface.

stationary orbit   See *geostationary*.

stationary reservation   Altitude reservation over a fixed area.

stationary shockwave   Shockwave at rest relative to solid surface.

station box   Human interface with intercom and/or remote radio transmitter/receiver, comprising jack socket, tuning and other controls; in large aircraft links up to 12 audio (crew) inputs to intercom and com. systems.

station capacity   Maximum number of aircraft that can be handled simultaneously by radio (esp. DME) station.

station costs   Airline costs attributed to staff and facilities at airports.

station ident   Ident feature keyed into ILS localizer transmission at regular intervals, either three (military, two) letter code or voice.

stationkeeping   Ability of vehicle, esp. satellite, to maintain prescribed orbit or station (most geosynchronous satellites follow small figure-eight shapes).

station-keeping equipment   Small 360° radar scanning horizontally and used to assist several aircraft to hold exact formation up to several miles apart (eg for precisely sequenced air-drops).

stationmaster   CO of station, from railway usage, usually Grp Capt (RAF, colloq.).

station passage   Flight directly overhead a station (6), esp. a point navaid, eg VOR, when 'to' becomes 'from' and bearings become reciprocals.

station set   Selected items of mission-support equipment prepositioned at designated locations for support of war or emergency ops (USAF).

station time   Time at which crews, pax and cargo are to be on board military transport ready for take-off (DoD).

station zero   Origin of aircraft measurement system [station (1)] along each axis; for longitudinal axis may be at or in front of nose.

statistical accelerometer   Instrument recording number

of times particular acceleration is exceeded; apparently synonymous with recording accelerometer.

stator   Fixed part of rotary machine, eg electrical generator or motor, hydraulic pump, gas turbine compressor or turbine, RC engine casing or brake discs keyed to landing-gear axle.

stator blade   Fixed (except for variable incidence) blade attached to axial compressor stator casing in radial row between each stage of rotor blades. US = stator vane.

stator casing   That enclosing axial compressor.

stator outlet temperature   In most gas turbines gas path expands through turbine[s] and SOT is higher than TET. See *turbine temperatures*.

statorless turbine   Designs in which contra-rotating rotors eliminate need for stators.

stator vane   Stator blade, usually abbreviated to vane (US).

statoscope   Instrument (now arch.) for detecting small excursions from preset pressure altitude. Comprised sealed U-tube containing drop of red liquid reposing normally at centre, connected to thermos flask.

Status   MFD key which returns pilot to basic menu for complete change of function.

status   Assessment of target, esp. one seen only on radar, as friend, foe or unknown.

statute mile   Non-SI unit of length = 1609.344 m.

Stavka   Supreme-command staff (USSR).

stay   Lateral strut or bracing member, esp. for landing gear.

STB   *1* Stop bars.

   2 Systems testbed.

STBA   Service Technique des Bases Aériennes (F).

STBL   Stable.

STBY   Standby, standby instruments..

STC   *1* Sensitivity, or swept, time constant, or control.

   2 Self-test capability.

   3 Supplemental Type Certificate (US).

   4 Satellite test centre.

   5 Short-term confict (A adds alert).

   6 Strike Command (RAF).

STCA   *1* Short-term collision-avoidance.

   2 Short-term conflict-alert.

STCICS   Strike Command integrated communications system (RAF).

STCM   Stabilizer-trim control module.

STCR   System test and checkout report.

STD   *1* System technology demonstration, or description.

   2 Scheduled time of departure.

   3 Standard, as in next.

STD bus   Traditional US-developed 8-bit bus for computers and related EDP.

STDBY   Standby (alternative).

STDMA   Space, or synchronized, time-division multiple access.

STDN   Space- [flight] tracking and data network.

STDY   Steady.

STE   *1* Sun-tracking error.

   2 Synthetic training equipment.

   3 Scheduled time en-route.

   4 Send, then encrypt.

STEADES, Steades   Safety trend evaluation analysis and data-exchange system (IATA).

Steady   Air-intercept code: "I am on prescribed heading", or "Straighten out on present heading".

steady flow   See *time-invariant flow*.

steady initial-climb speed   See V_4.

steady-state condition   One that is time-invariant, as applied to signal or flutter amplitude, physical or chemical properties or any other variable.

stealth   Technology for making tangible objects, initially aircraft, as invisible and undetectable as possible. It covers all EM wavelengths as well as sound, and is increasingly essential for survival in defended airspace. Also adjective.

steam bombing   Visual manual attack on target of opportunity using free-fall bombs, especially by advanced automated aircraft (colloq.).

steam catapult   *Catapult*.

steam cooling   Cooling piston engine by allowing slightly pressurized water to boil in cooling jackets, to be condensed in dragless double-skin radiator. Also called evaporative cooling.

steam gauge   Traditional dial instrument in a modern cockpit (colloq.).

steam(ing) fog   Forms when supersaturated freezing air with inversion moves over warm water; also called Arctic smoke or sea smoke.

Stears   Stand-off tactical electronic airborne reconnaissance system.

STE bus   New international standard 8-bit computer bus originally developed for Eurocards which has rendered STD obsolete.

STEC   Solar/thermal energy conversion.

STED   Space Test and Evaluation Directorate (USAF).

steel drag chute   Reverser (colloq.).

steep approach   *1* That adopted by helicopter pilot descending into obstructed, eg urban, heliport, begun at 9 m/200 ft above selected landing spot and made straight-in at close to 50° from downwind.

   2 For aeroplanes with limited STOL capability, on an individual airport basis; usually 5.5°.

steepest-descent method   Basic method of optimization in which all contour lines (each representing a plotted variable) are crossed perpendicularly.

steep gliding turn   Steep turn performed in glide, if continued resulting in tight spiral (more common in US).

steep turn   Various definitions with bank angles: over 50°; 45°-70°; over 60°.

Stefan-Boltzmann constant   That in Stefan-Boltzmann law, $\sigma = 5.66961 \times 10^{-8} \text{W/m}^2\text{K}^4$.

Stefan-Boltzmann law   Basic law of thermal radiation: total radiation from black body is proportional to 4th power of absolute temperature, $E = \sigma T^4$.

STEI   Service Technique de l'Electronique et de l'Informatique (F).

stellar guidance   See *astronavigation*.

stellar/inertial guidance   Inertial navigation intermittently updated and refined by astro.

Stellite   Large family of hard alloys of Co (30-80%), Cr (10-40%), W (0.25-14%) and Mo (0.1-5%), and in one case with 30% Ni and 5% Fe. Some cannot be machined; common use is piston-engine valve heads and seats.

St Elmo's fire   See *corona discharge*.

Stem   *1* Shaped-tube electrolytic machining, see * *drilling*.

   2 System (or spaceflight) trainer and exercise module.

   3 Space-to-Earth missile concept.

stem Strong quasi-vertical member at bow of marine aircraft.

stem drilling Use of titanium tool [cathode] feeding 20% HNO_3 into workpiece; can produce holes $\geqslant$0.5 mm, 0.02 in, diameter and length 120 mm, 4.7 in, see *capillary*.

Step, STEP *1* Software test and evaluation project, concerned with entire software life cycle.

2 Standard equipment package.

3 MFD key which pages through all available formats, usually at 1 Hz.

4 Standard for the exchange of product model data.

step *1* Segment of climb from one FL to next, each normally begun either on ATC clearance or upon arriving at suitable gross weight from burning fuel.

2 Sharp or angled discontinuity in planing bottom (float or hull) to improve planing characteristics and ease takeoff.

3 Stage (1), latter being preferable.

step-aside gearbox One in which input and output shafts are not co-axial.

step climb Gaining altitude in a series of steps, each accurately flown to minimize fuel burn and comply with ATC. Abb. STEPCLB.

step cruise Protracted stepped climb covering most of flight.

step pad Secondary structure built externally on top of fuselage, esp. of helicopter, for ground crew or other persons.

stepped climb Climb in series of steps (1), separated by slow (drift-up) climb or level flight.

stepped formation One in which successive aircraft or elements are at higher or lower level.

stepped solvents Solvents in liquid (eg aircraft finish) which evaporate at very different rates.

step taxi To taxi marine aircraft fast enough to ride on step (2).

step-up gear One in which output shaft speed is higher than input, torque being reduced in same ratio; opposite of reduction gear.

steradian SI unit of solid angle, abb. sr: that solid angle which, having as its vertex centre of sphere, cuts off area equal to that of square whose sides are equal to radius of sphere.

stereogram Stereoscopic set of graphics or imagery arranged for viewing.

stereographic See *polar stereographic*.

stereographic coverage Air reconnaissance cover by overlapping imagery to provide 3-D picture; 53% overlap is minimum and 60% normal.

stereoscopic cover NATO term for stereographic cover.

stereoscopic pair Two images for stereographic (stereoscopic) viewing of same scene.

sterile areas Parts of airfield between aircraft manoeuvring areas.

sterilization-proof Solid rocket propellants (fuels and, esp., binders) are often degraded by sterilization (4) heating, and * binders have had to be developed.

sterilize *1* To mark off portion of runway as unusable, for any reason.

2 To blank off portion of instrument panel, eg for equipment not yet available.

3 To prohibit unauthorized access, esp. to entire airside of civil airport for security reasons.

4 Normal meaning for spacecraft either departing for or arriving from other planet.

5 To deactivate device, eg mine, eg after preset period.

6 To render unusable or unavailable; eg F-111 MLG * underside of fuselage for weapon carriage.

stern attack Air-intercept attack which terminates with crossing angle of 45° or less (DoD).

stern-droop Structural sag of rear end of airship.

sternheaviness Tailheaviness of airship.

sternpost Single vertical member marking rear termination of fuselage, hull or float (BS adds 'not to be confused with rudder post', but often same member).

sternpost angle Acute angle between horizontal (usually same as longitudinal axis or underside of keel ahead of step) and line joining top of step on centreline and afterbody terminator, bottom of sternpost or rearstep (heel) of flying-boat hull.

stern wave Formed at low speeds as swell ahead of stern of taxiing marine aircraft.

stero route Named and established, in most respects synonymous with Airway.

Stevenson screen Standard Meteorological Office slatted box for ground instruments.

Stevi Sperry turbine-engine vibration indicator.

STEW, Stew *1* Surface-threat electronic warfare [S adds system].

2 Stewardess [confusing, now rare].

Stex Space technology experiment satellite.

STF *1* Self-test facility.

2 Special trials flight.

StF Stratiform cloud.

StFra Stratus fractus.

StFrm Stratiform.

STFV Sensor total field of view.

STG Strong.

StG Stukageschwader, dive-bomber group (US wing] (G, WW2).

STGR Search, track and guidance radar.

sthène Non-SI unit of force formerly standard in French legal system, that giving 1 tonne acceleration of 1 m/s^2; sn = 1,000 N = 1 kN.

S^3T See *SSST (2)*.

STI Standard (or special) technical instruction.

stick *1* Control column (colloq.).

2 Any primary pilot input in pitch, in figurative sense, eg *-free static stability.

3 Succession of missiles (ordnance items would be better) fired or released separately at predetermined intervals from single aircarft (ASCC).

4 Number of parachutists who jump from one aperture of aircraft during one run over DZ.

5 A pilot, especially fighter or aerobatic (colloq.).

stick and string Generalized term for construction of pre-WW1 aeroplanes (colloq.).

stick-canceller Electronic box in FCS whose output is a stick movement, eg in ASW hover.

sticker Usually on windscreen, displays selling price [GA aircraft, usually used].

stick-fixed With elevator or tailplane trimmed to hold level flight and thereafter held in this position.

stick-fixed static stability See *static margin*.

stick force per g Pilot's applied force in direct fore/aft movement divided by vertical acceleration resulting.

stick-free With elevator or tailplane trimmed to hold

level flight and thereafter left free (not applicable with irreversible flight controls).

sticking   Tendency of planing bottom of marine aircraft to adhere to water on takeoff.

stick knocker   Stall-warning device added to stick shaker to give loud knocking aural warning.

stick movement per g   Linear fore/aft movement measured at top of stick divided by resulting vertical acceleration.

stick pusher   Positive stall-prevention system which when triggered (by sensed AOA or AOA factored by a rate-of-increase term) forces stick forward, commanding aircraft to rotate from climb to shallow dive.

stick shaker   Stall-warning system which when triggered by AOA passing preset value (occasionally factored to allow for high rate of increase) applies large oscillating force which shakes stick (2) (normally a large yoke) rapidly through small angle in fore/aft plane.

stick-shaker speed   SSS or S-cubed, really a misnomer as system is triggered by AOA, not a particular speed.

stick spacing   Linear distance on ground between ordnance items dropped in a stick (3).

stick time   Logged pilot time, esp. as PIC.

stick travel   Total range of travel of stick in either fore/aft or side-to-side direction, normally measured not as angle but as linear distance.

stick travel per g   See *stick movement per g*.

stiction   See *static friction*.

stiffener   Normally a strip or beam attached to sheet to resist load normal to surface. An integral * is formed in skin itself, usually as pressed channel parallel to relative wind.

stiffening bead   Integral stiffener (see above).

stiffness   Ability of system to resist a prescribed deviation. In structural member within elastic limit, ratio of steady applied force to resulting displacement or ratio of applied torque to resulting angular deflection. For many dynamic parts such as servo-actuators there is static * and dynamic *. Static * is stiffness characteristic exhibited in steady-state condition, normally trying to approach infinity (discounting compliance within actuator and deflection of surrounding structure) up to stall thrust. Dynamic * is an apparent value dependent on ability of flow power of servo valve to hold output against oscillating load, up to critical frequency (varying with magnitude of applied load) at which servo contribution becomes negligible to give a degraded 'infinite-frequency *'.

stiffness criterion   Relationship between stiffness and other properties of structure which when satisfied ensures prevention of flutter or other type of instability or loss of control (BS).

stiff nut   Nut provided with means for gripping male thread to provide sustained torque resisting rotation once tightened.

stiff pavement   One able to accept any input aircraft bending moment; stiffness measure is tonnes vertical load to produce 1 cm deflection under point of application.

stiff wing   Fixed-wing, ie not helicopter (colloq.).

stilb   Non-Si unit of luminance, sb = 10^4 cd/m^2; plural stilbs, sb.

Stile   Sensor technologies integrated laboratory environment.

Stiletto   Becoming generalized term for anti-radar D/F and passive-ranging systems.

stiletto criterion   Basic design case for passenger floors: intensity of * heel loading of 100 kg/cm^2, or over 100 times limit for distributed heavy cargo.

stiletto weld   One whose cross-section resembles exclamation mark without stop.

stillage   A specific meaning in aerospace is small wheeled dolly for ground movement of external store, eg bomb. Usual meaning is static warehouse storage.

still-air range   Not a normal performance or flight manual figure; in general a vague estimate of distance under ideal still-air cruise conditions aircraft could fly without air refuelling, ignoring ATC constraints, reserves or any other factor. Unlike ferry range, usually assumes some (occasionally maximum) payload.

stilling chamber   Large volume in which fluid flow eddies and gross turbulence are brought to rest; generally synonymous with settling chamber.

ST-IN   Straight in.

Stinfo   Scientific and technical information (data management).

sting   *1* Long cantilever tube projecting upstream in tunnel to which model is attached with minimum interference from mount.

2 Long cantilever tube projecting directly ahead of nose of aircraft to carry instrumentation with minimum interference from mount or following aircraft; also called probe, instrumentation boom.

sting hook   Normal form of arrester hook in form of single strong tube (term introduced WW2 to distinguish from A-frame).

Stings   Stellar-inertial guidance system.

sting switch   Activated by springy rod projecting below RPV (or other aircraft), triggered on landing.

Stir   *1* Surveillance target-indicating radar.

2 Separate tracking and illuminating radar[s].

Stirs   Strapdown inertial reference system.

Stirling cycle   Heat-engine cycle in which heat is added at CV followed by isothermal expansion with heat addition, heat is then rejected at CV followed by isothermal compression with heat rejection; very efficient, esp. with regenerator, but mechanical problems (Philips use patented Rollsock to seal reciprocating parts). So far used in aerospace mainly for space power generation or cryogenic cooling.

STIS   *1* Stabilized thermal-imaging sight [S adds system].

2 Space telescope imaging spectrograph.

STIU   Satellite telephone [or telecoms] intermediate unit.

STK   Satellite toolkit, for software analysis.

STLO   *1* Science and technology (or scientific and technical) liaison office.

2 Simulation and Training Liaison Officer (RAF).

STM   *1* Short-term memory.

2 Supersonic tactical missile.

3 Significant technical milestone(s).

4 Storm.

5 Serial transition module.

STMP   Special Traffic Management Program (NAS).

STN, Stn   Station.

STNA   Service Technique de la Navigation Aérienne (F).

Stn No   Stanton number (or St).

STNPA   Système technique de neutralization des pirates de l'air (F).

STNR   Stationary.

STO   *1* Short take-off.

  2 System test objective.

  3 Station telecom officer.

  4 Council (Soviet) for labour and defence (USSR).

  5 Science and technology objective.

  6 Signature technology office (USAF).

  7 Space tasking order.

S/TO   Search/track operator.

STOAL   Short takeoff and arrested landing.

Stobal   Short takeoff but arrested landing.

Stobar   Short takeoff but arrested recovery.

STOC   Special Tactical Operations Center (US NMCC).

STOCC   Space Telescope Operations Control Center.

stochastic   Implying presence of unknown or random variable; thus, * process is ordered set of observations, each a sample from a probability distribution.

stock   *1* Raw material preformed to standard dimensions as sheet, strip, tube etc.

  2 Material surplus to a part's finished dimensions, to be removed during manufacture.

stockpile/target sequence, STS   Order of events in removing NW from storage and assembling, testing, transporting and delivering to target (DoD).

stocks   Shaped supports on which flying-boat hull is built, but in no sense tooling.

stock template, ST   One developed by trial and error, esp. for parts undergoing severe deformation.

Stoddard   Common naphtha-like hydrocarbon solvent.

STOGW   Short takeoff gross weight.

STOH   Since top overhaul.

stoichiometric   Provided in exact proportions required for complete chemical combination, esp. of fuel/air mixtures.

stoke   Non-SI unit of kinematic viscosity, 1 St = 10^{-4} m^2/s.

Stokes law   Terminal velocity of sphere [density ρ, radius r] falling through fluid [density ρ_o, dynamic viscosity η] is V = $\frac{2gr}{9\eta}$ $(\rho-\rho_o)$.

Stokes litter   Litter [UK = stretcher] designed for helicopter recovery of injured casualty.

Stol, STOL   Short [rarely, slow] take-off and landing.

Stoland   Digital system permitting fully automatic landings into STOL airfield (NASA).

Stolport   Airport, esp. urban (metropolitan), configured and designated for Stol operations.

Stol runway   Runway, normally 900 m (2,000 ft), specifically designated and marked for Stol operations (FAA). Letters STOL at threshold and a TDAP.

STOM   Ship-to-objective manoeuvre.

stone   Non-SI unit of weight, usually applied to humans, = 14 lb = 6.3503 kg.

stoneguards   Various mesh screens or deflectors; undefineable.

Stoner Mudge   Patented technique of coating integral tanks with rubbery sealant.

stonk   To destroy a surface target [verb or noun], (colloq.).

stooge   To fly aimlessly in order to stay in particular area (eg awaiting orders, or in hope of encountering enemy).

stooping   Atmospheric refractive phenomenon in which image (mirage) of distant object is vertically foreshortened.

STOP   Structural/thermal/optical program.

stop(s)   Mechanical limiter(s) to permissible travel of flying-control or other mechanism.

stop alt squawk   Turn off altitude-reporting switch and continue Mode C framing pulses.

stop-and-go   See *circuit*.

stop countersink   Fitted with collar to limit penetration.

stop drill   *1* See *stop hole*.

  2 Drill with collar to limit penetration.

stop hole   Hole drilled in end of fatigue crack to provide larger radius and halt further spread.

stop nut   Nut which stops in place without further action (such as wiring or bending up tabs).

stopover   *1* Stop by pax at intermediate airport authorized by ticket for prescribed period.

  2 One-night stay away from base by slip crew (civil or military).

stopped-rotor aircraft   See *stowed rotor aircraft*.

stopping   Loosely, sealer for cracks, esp. in pressurized riveted joint.

stop squawk   Switch off transponder, or a particular mode.

stopway   *1* Defined rectangular area on ground at end of runway in direction of takeoff (ie beyond upwind end, symmetrical about extended CL) prepared as a suitable area in which aircraft can be stopped in case of abandoned takeoff (ICAO).

  2 Area beyond takeoff runway, no less wide and centred upon extended CL, able to support aircraft during aborted takeoff without causing structural damage to it and thus designated for use (FAA).

stop weld, stop-weld   Material, usually in form of paint, which has so high a melting point that it is not fused by any conventional welding process. Inserted as applied layer along line followed in seam-welding, results in treated area remaining unwelded.

storage   Generalized term for any kind of memory in EDP (1), display technology, EW and similar disciplines. Not normally used in aircraft for fuel or other consumables, ordnance and ammunition or cargo/baggage.

storage CRT   Various families able to write information into a storage surface by adding/subtracting from an initial potential; most configurations have symmetrical layout with write gun at one end and read gun at other.

storage oil   Intended as corrosion preventative, for engines [especially piston engines], usually mineral or other oil plus corrosion inhibitor.

storage system   In electronics and displays, basically comprising direct-view storage tubes and those in which storage is entirely separate; invariably concerned with vector inputs, converted into analog (beam-deflection) signals and varying luminance.

storage tube   See *storage CRT*.

Storc   Satellite tracking of RV convoys.

store   *1* Basic element of EDP (1) storage; normally a bistable device accommodating one bit.

  2 Generalized adjective for storage, hence * address, etc.

  3 To place information in memory for future reading.

4 Generalized term for any mission-related payload carried by combat aircraft in form of discrete streamlined device either carried externally or released from internal bay; anything that occupies a pylon or ERU, whether intended for air-dropping or not.

stores   Domestic supplies and consumables for large aircraft on long flight.

stores inventory display   Cockpit readout, often illuminated on command, showing locations and identities of all stores (4) remaining on board.

stores pylon   Pylon for carrying, and if necessary releasing, a store or stores (4).

Storm   *1* Sensor tactical-operations range module[s].

  2 Standard stores management [S adds system].

storm cell   Central region of most intense turbulence, indicated in cockpit radar display by black hole or red colour.

Stormfest   Stormscale operational research meteorological fronts experimental systems test (FAA 1992).

Storms   Stores management system.

storm scope   Weather radar.

storm-warning radar   See *airborne weather radar*.

stovepipe   Ramjet (colloq.).

Stovie   Pilot of Stovl aircraft (colloq.).

Stovl, STOVL   Short take-off, vertical landing; common operating mode of Harrier family.

STOW   Synthetic theatre of war.

Stow   System for takeoff weight and c.g., also called IWBS, with display readout driven by sensors on all landing gears.

stowed-rotor aircraft   Rotorcraft, normally helicopter in vertical mode, whose lifting rotor(s) can be stopped and retracted in wingborne cruising flight. Not yet achieved.

stow position   Travelling or inoperative az/el attitude of large radar or other steerable aerial, normally pointing to zenith.

STP   *1* Standard temperature and pressure.

  2 Space-test Program (DoD).

  3 Space Technology Program (Darpa).

  4 Systems Technology Program (DoD).

  5 Short-term planning.

  6 Status test panel.

  7 Sensor track processor.

  8 Standard training package.

  9 System-test package procedure.

  10 Solar terrestrial probe[s], or receiver.

STPAé   Service Technique des Programmes Aéronautiques (F).

STPD   Standard temperature and pressure, dry.

STPE   Service Technique des Poudres et Explosifs (F).

STR   *1* Service trials report.

  2 Sidetone ranging modulation; EW (spread-spectrum) acquisition assistance technique.

  3 Sustained turn rate (usually means maximum).

  4 Software trouble report.

  5 Systems technology radar.

  6 Standard test rack.

  7 Same type rating.

  8 Sonar transmitter/receiver.

  9 Satellite transceiver.

  10 Solar terrestrial receiver.

STRA   Simultaneous turnround actions.

Stradographe   Runway friction measurer with braked or toed-in wheels and EDP (1) for nine variables.

strafe   To rake with fire, eg from automatic guns; spelt as shown (not straff) and rhyming with chafe. US usage is in favour of 'straff', rhyming with chaff, and 'straffing' is becoming predominant.

straight   Applied to fuel, lubricant, etc, = without additives.

straighteners   Vanes, cascades, or, in tunnel, transverse flat-plate honeycomb, to remove swirl or turbulence from flow.

straight-flow   Gas turbine of normal, ie not reverse-flow, layout.

straight-in approach   In IFR, an instrument approach wherein final approach is begun without a prior procedure turn. In VFR, entry of traffic pattern by interception of extended runway CL without executing any other portion of traffic pattern (FAA).

straight leading edge   Having no taper; likewise for trailing edge.

straight-pass attack   One using on-board weapon-aiming system to hit point surface target without search or visual acquisition, thus in straight run at highest speed at lo level.

straight roller bearing   Not tapered, thus no axial load expected.

straight-run   Hydrocarbon distillate, eg from original crude, representing all products separating out between specified upper/lower temperature limits; not normally used as aviation fuel.

straight spur gear   Straight teeth parallel to shaft.

straight stall   One performed with minimal yaw, using rudder if necessary to hold heading.

straight-through duct   Inlet duct to trijet centre engine in which flow from inlet to nozzle is essentially straight, as distinct from S-duct.

straight wing   Of traditional planform, specif. not swept.

strain   Deformation under stress expressed as a percentage of an original dimension; length, area or volume.

strained silicon   Perfect lattices which permit electrons either to move faster or to travel with less power.

strain energy   Elastic energy recovered from body by removing stress.

straingauge   Device for transducing strain into electrical signal, usually by extremely accurate measurement of change of resistance of conductor.

strain hardening   Increase in hardness and reduction in ductility caused by strain, esp. by cold-working (eg rolling); the only way to harden some wrought light alloys. Introduces a strain exponent (expressed as n) into stress/strain equations.

strain rate   Strain per unit time, normally under uniform stress and often synonymous with creep.

strain viewer   Instrument giving pictorial strain pattern using polarized light.

strake   *1* Long but shallow surface normal to skin and aligned with local airflow; extremely low-aspect-ratio fin.

  2 Fin[s] mounted on upper part of underwing engine pod, normal to surface, to generate vortex passing over wing.

  3 One row from stem to stern of single plates cladding marine aircraft.

stranded conductor   Electric cable containing numerous conductive wires twisted together within single insulating sheath.

Stranger   Air-intercept code: unidentified target. Normally ground/air message followed by bearing, distance, altitude, in that order.

strangle   General term meaning please switch off a particular emitter (military usage).

Strangle Parrot   Ground/air code: switch off IFF.

Strap, STRAP   *1* Sonobuoy thinned random-array project. Acoustic process or in aircraft correlates signals from 15 to 20 buoys, four of which emit low-power signals to fix precise position of each buoy.

2 Straight-through repeater antenna program, or performance.

strapdown   Generalized adjective for device mounted so that its attitude changes with that of aircraft or spacecraft; specif., one not gimballed about three axes.

strapdown INS   Simplied INS using strapdown platform.

strapdown platform   Platform for INS on which sensing gyros and accelerometers are fastened without relative motion; usually there are three mutually perpendicular gyro/accelerometer units, and in some cases an element of redundancy from a fourth mounted at 45° to all three others.

strapping   *1* Interconnection of resonant chambers of cavity magnetron or related oscillator give one stable preferred mode.

2 Calibration of storage tank so that measurement of contents depth can be related to actual volume.

3 Metal or other straps, wire or other ties around palletized, igloo or other loose cargo.

Strata   Siimulated training research advanced testbed for avionics.

Stratcom   Strategic Command (USAF).

strategic   Concerned with broad politico-military objectives and enemy's warmaking potential.

strategic aeromedical evacuation   Airflift of patients out of theatre of operations to main support area.

strategic airlift   In support of all arms between area commands or between home state and overseas area.

strategic air transport   According to DoD, one in accord with strategic plan; according to NATO, movement between theatres by scheduled service, special flight, air logistic support or medevac.

strategic air warfare   Air operations designed to effect progressive destruction and disintegration of enemy's warmaking capacity (NATO; DoD is much longer definition which adds nothing).

strategic attack   Aerospace attack on selected vital targets of enemy nation to destroy warmaking capacity or will to fight.

strategic bomber   Delivery-system aeroplane for strategic attack.

Strategic Defense Initiative   Plan by Reagan administration to enhance US ability to conduct warfare in space, eg by using direct-impact missiles or super-power lasers to incapacitate satellites.

strategic plan   Plan for overall conduct of war.

Strategic Planning Team   Focal point for development of future NAS (FAA).

strategic psywar   Actions designed to undermine enemy's will to fight.

strategic transport   Aircraft for transport between theatres.

strategic warning   Notification that enemy-initiated

hostilities may be imminent. Hence ** lead-time, time elapsing between ** receipt and beginning of hostilities. May include two action periods, ** predecision time and ** postdecision time in which national commander takes a decision to respond positively to **.

stratified-charge engine   Piston engine in which mixture strength is varied in controlled manner during induction stroke, with minimum turbulence, if possible to leave layered charge in cylinder with highest density near source of ignition. Now especially important in RC (1) engines.

stratiform cloud   Sheets in stable thin layers.

stratocumulus, Sc   Layer of connected cloudlets at low-cloud level often arranged in aligned rows.

stratopause   Atmospheric layer at top of stratosphere where inversion ceases at 270.65°K [though such accuracy is pointless].

stratosphere   Atmospheric region between tropopause and stratopause within which temperature remains essentially constant and then, at upper level, rises with altitude.

stratus   Uniform layer of low (usually grey) cloud, well clear of surface.

straw man   Pilot of well below average competencce.

straw qualities   Possessed by aircraft safe enough to be routinely flown by straw man.

stray   *1* Naturally occurring EM signals, eg static.

2 Errant marker or TI put in wrong place.

STRC   Strategic training route complex.

streak   *1* Horizontal smear, usually white, following moving image on TV or other raster display.

2 Long flame in airbreathing engine normally denoting abnormal combustion at one point.

streak camera   Family of cameras for ultrafast photography in which changing scene is viewed through slit perpendicular to main image variation and optics sweep image along fixed arc of film. Basic type has rotating mirror, normally projecting through array of biconvex lenses.

streaking   *1* Unwanted manifestation of streak (1).

2 Unwanted manifestation of streak (2) in gas turbine, usually resulting in reduced life or damage to NGVs.

stream   *1* To release parachute retarding horizontal motion, eg braking parachute.

2 To dispense chaff as solid, at random intervals or as bursts (DoD).

3 Jet stream (colloq.).

4 Shower of meteoroids with similar orbits and timing.

streamer   Anything that follows or indicates streamlines, in particular *windsock*.

streamering   *1* Visible brush discharge.

2 Unreefed parachute canopy opens but fails to deploy fully; hence streamered.

stream function   Basic parameter of 2-D non-divergent fluid flow with value (symbol ψ) constant along each streamline related to velocities along each axis by u = d ψ/dy, v = d ψ/dx.

stream landing   Landings by group of aircraft in quick succession.

streamline   Line marking path of particle of fluid in homogenous flow; esp. in streamline flow; line whose tangent is everywhere parallel to instantaneous velocity at that point.

streamlined   3-D body shaped such that fluid drag is a minimum.

streamline flow   Fluid flow that is laminar and time-invariant, and in which each streamline is devoid of a closed curve or sudden change in direction.

streamline position   That in which a hinged or pivoted body, eg control surface or pylon on a variable-sweep wing, is aligned with relative wind.

streamline wire   One whose section is streamlined, though seldom optimum (usually two intersecting circular arcs).

Streams   Surface-traffic enhancement and automation support system.

stream surface   2-D sheet made up of streamlines.

stream take-off   Take-off by group of aircraft in quick succession with departure in trail formation.

stream thrust   Total of pressure force and time rate of momentum flow across any cross-section in fluid flow, $F = PA + \rho AV^2$.

stream tube   In a laminar fluid flow, volume of flow enclosed by streamlines passing through upstream and downstream closed loops (not necessarily circular) placed normal to flow. At any point velocity is inversely proportional to ** cross-section area.

streamwise tip   Wingtip of high-subsonic or transonic aeroplane in which leading edge is curved progressively back parallel to local airflow to eliminate outward sweep of isobars; one form is *Küchemann tip*.

street   Regular procession of straightline vortices shed alternately from above and below body, eg naturally oscillating cylinder or wire, each vortex following same path as next-but-one predecessor. Normally if D is body diameter, two half-streets are separated by 1.2-D and spacing between vortices in each half-street is 4.3 D. Also called Kármán * or von Kármán *.

streetcar   STCR (colloq.).

strength   *1* Physical * is ability to withstand stress without rupture; normally subdivided into compressive, shear and tensile, the latter usually being stress at the yield point.
2 In radio and related fields, signal amplitude in W or dB.
3 Dielectric * is maximum potential gradient in V/mm.

strength test   See *static test*.

strength/weight ratio   For material, ultimate tensile strength/density; for a structural member, breaking stress/weight.

stress   *1* Condition within elastic material caused by applied load, temperature gradient or any other force-producing mechanism, measured as force divided by area. Unit * is force per unit area normal to direction of force. It is this force that resists externally applied loads. Common units are $MNm^{-2} = 0.06475$ UK tons/in^2; $kNm^{-2} = 0.14504$ lb/in^2; $kg\ mm^{-2} = 0.63497$ UK tons/in^2.
2 Generalized term for psychological, physiological or mental load on organism, esp. human, which reduces proficiency.
3 Measure of resistance of viscous fluid to shear between adjacent layers (see *viscosity*, *Newton's laws*).

stress analysis   Determination of all loads borne by all elements of structure in all flight conditions, external reaction points of application and direction, and allowable and actual stresses in each member.

stress-bearing   Required to resist applied load(s).

stress concentration   Localized region of increased stress caused by sudden changes in section, poor design and

manufacturing imperfections, eg tooling marks (see *stress raiser*).

stress concentration factor   Peak actual local stress divided by stress for member calculated by any standard method without presence of stress raisers, such as sharp-corner apertures or external surfaces. Neutral-hole * is unity.

stress corrosion   *1* Metal cracking due to residual stress from manufacturing processes or concentrated stresses caused by flight loads and/or poor design.
2 Exfoliation corrosion.

stress cycle   Complete cycle of variation of stress with time, repeated more or less identically (very numerous for piston-engine conrod, less for wing LCF).

stress distribution   Variation of stress across cross-section of member.

stressed skin   Form of semi-monocoque construction in which skin, nearly always metal, bears significant proportion of flight loads, and makes principal contribution to stiffness.

stress-free stock   Selection of stock material for primary structure in which presence of residual internal stress results in rejection or return for further treatment; important in heavy plate for wing skins, etc.

stressing   Stress analysis of structural members, usually while altering their design to attain optimized structure.

Stresskin   Patented metal (esp. stainless) sandwich panels requiring no supporting structure over surface.

stress raiser   Local abrupt change in section resulting in stress concentration; severity varies inversely with radius, so that a single scratch (eg from emery particle) can over a period initiate a fatigue crack that will eventually prove catastrophic.

stress ratio   Maximum to minimum ratio in one stress cycle.

stress-relief annealing   Heating to beyond critical temperature, and slow cooling to relieve internal stress, eg after cold working or welding.

stress/strain curve   Plot of strain resulting from all stresses from zero to yield point and on to rupture. Normally linear over most of plot to yield point (limit of proportionality where elastic deformation gives way to plastic).

stress wave   Sonic pulse propagated through various devices, eg magnetostrictive-tablet display; also called strain wave.

stress wrinkle   Visible wrinkling of skin caused by applied load, esp. in secondary structure, eg sagging of rear fuselage of B-52.

stretch   *1* Increase in capacity of transport by adding plugs to fuselage, normally both in front of and behind wing. Noun and verb.
2 To apply tensile stress exceeding elastic limit.

stretchability   Potential for stretch (1).

stretching   *1* Process of introducing a stretch (1).
2 See *stretch-wrap forming*.

stretching press   See *stretch press*.

stretch-levelling   Stretching sheet or plate just beyond elastic limit (typically 1.5% elongation) to remove all irregularities; also called stress-levelling.

stretchout   Agreement, initiated by contractor or customer, to reduce rate of production without altering quantity to be built.

stretch point   Fuselage station at which plug is to be

added for stretch (1). Possibly allowed for in original design for stretch planned far in future.

stretch press   Any of several families of press, mainly hydraulic (eg Hufford, Sheridan), in which sheet is pulled beyond elastic limit over a 3-D die or tool of correct profile; invariably both ends are pulled equally over die at centre.

stretch-wrap forming   Use of stretch press of various types (eg Hufford) in which, as rams operate, their axes simultaneously rotate to wrap workpiece around tool.

Streuwaffen   Cluster bomblets or scatter weapons (G).

strew   To lay down sonobuoys in prescribed sonar pattern.

stricam   Structure-integrated camouflage.

stricken   Subjected to strike (5). Normally applied to military aircraft which are reduced to produce.

Stricom   Simulation, Training and Instrumentation Command (USA).

Strict   Structurally integrated inlet control technology, uses air jets to keep engine airflow attached to duct wall.

Strida   Système de Traitement et de Représentations des Informations de Défense Aérienne (air-defence system, now * II, F).

strike   *1* An attack designed to inflict damage on, seize or destroy an objective (DoD, NATO).

2 As (1) is suggestive of target within reach of surface forces, better definition is a tactical close-support or inter-diction attack on surface target, with conventional or nuclear weapons.

3 Significant impact(s) with foreign objects, esp. with bird(s) while on takeoff run or airborne.

4 Verb, from above.

5 Verb, to remove aircraft from active inventory.

6 Verb, to fold wings of carrier-based aircraft and move down to hangar [also * down].

strike aircraft   Aircraft, normally aeroplane, for carrying out strike (2); definition which restricts term to naval carrier-based aircraft is obsolete, though such aircraft are not excluded.

strike camera   Camera, usually optical wavelengths but can include radar and/or IR, operated automatically or on command during process of carrying out strike to record fall of ordnance and give preliminary indication of likely results.

strike-CAP   Fighter role in which strike task is predominant; offensive ordnance is jettisoned only under direct attack.

strike control and reconnaissance   Mission flown for primary purpose of acquiring and reporting air-interdiction targets and controlling air strikes against such targets (USAF).

strike down   To fold aircraft on deck and transfer to hangar(s) aboard carrier; term comes from traditional verb to secure items so that they cannot move relative to deck.

strike force   Composed of units capable of conducting strikes (1), attack or assault operations (DoD); not necessarily aviation.

strike photography   Imagery secured during air strike.

striker   Various meanings in connection with firing ammunition and ordnance devices by impact on percussion cap, in some cases synonymous with firing pin, and in others a hammer for hitting firing pin or intermediate between pin and percussion device. Hence * pin.

striking voltage   Critical potential difference across gas-discharge tube and certain other devices at which discharge occurs and current flows.

string   *1* Generalized term for assembly of devices in essentially linear sequence through which signal or object passes.

2 Literally piece of string or wool used as crude indication of relative wind, esp. of yaw (probably obsolete).

3 In modern usage, operating channel in electronic device, such as autopilot.

stringer   Longitudinal member (ie in fuselage more or less aligned with longitudinal axis and in wing and tail surfaces more or less perpendicular to this axis) which gives airframe its shape and provides basis for skin. In fuselage they link frames and in aerofoils they link ribs. Most existing definitions are obsolete, describing * as light auxiliary or fill-in member; modern transport fuselage has no other longitudinals apart from possible underfloor keel on centreline. Integral skin removes need for * except in some structures where integral stiffeners are used plus * at 90° different orientation.

strip   *1* To dismantle, also called teardown.

2 See *strip stock*.

3 See *stripping*.

4 Farm strip; loosely, any private airfield.

strip and digit   Instrument based on roller-blind tape (usually with variable indices) plus alphanumeric window.

strip antenna   One or more laminate dipoles flush with skin surface.

Stripline   Microwave transmission line formed from two close strip conductors face-to-face or single strip close to conductive surface; also called Microstrip and other registered names.

strip map   Various forms (eg folded paper or film for projection) of topographical map in strip form, either covering entire flightplan track or sector(s) between WPs or other intermediate points.

stripper   *1* Regular user of unlicensed airfield, such as a farmer's own land.

2 Material, usually liquid, used for *stripping*.

stripping   One meaning is controlled removal of paint, transfers (decals) and other layers from aircraft skin. Wet * employs various solvents, which can combine with the removed materials. Dry * uses high-velocity blast from air at 0.5-5 bars to direct non-aggressive abrasive such as various plastics, starches or sodium bicarbonate; often a magnetic screen is used to collect ferrous dust and paint particles.

strip plot   Portion of map or overlay upon which is delineated coverage of air-reconnaissance imagery without indicating outlines of individual prints.

strip stock   Standard forms of metallic raw material in long, narrow strips, often coiled; suggest term confined to flat sections, only.

strobe   *1* Originally abb. of stroboscope, high-intensity flashing light source.

2 Continuous high-intensity light source rotating about vertical axis to point repeatedly to each azimuth.

3 See *strobe marker*.

4 See *strobe unit*.

5 To select particular portion of waveform timebase or other time subdivision in cyclic phenomenon.

6 Stroboscope.

strobe marker   Small bright spot, short gap or other discontinuity on radar timebase or other portion of cyclically scanned display to indicate portion receiving special attention.

strobe pulse   See *strobe marker*.

strobe timebase   Small section of timebase containing target blip, Loran signal or other object of interest which is extracted by strobe marker and expanded to fill original timebase width. In some cases process can be repeated, giving expanded **.

strobe unit   In missile guidance system, generates strobing pulses which search and lock-on to target reflections and supply target range and relative velocity to guidance computer.

Strobokerr   Camera for ultrafast photography in which incoming scene passes through Kerr cell, which divides it into pulses of light so brief that clear non-smeared images are projected on to film revolving at high speed inside drum.

stroboscope   Instrument for apparently bringing rotating or oscillating objects to rest by intermittent phased illumination, eg for inspection of deflection under high-speed operation or to determine rpm or Hz.

stroke   *1* Linear distance moved by piston of piston engine from TDC to BDC or vice versa.

2 One complete translation of piston from TDC to BDC or vice versa, performing particular operation; thus inlet or induction *, compression *, power * and exhaust *.

3 Linear or angular distance travelled by electron or other beam forming timebase or any other electronic display.

4 Linear or angular distance travelled by output of actuator.

5 One major 'flash' making up flash of lightning, normally repeated every 38-45 ms until discharge complete.

6 Basic element of *-generated writing.

stroke font   Total repertoire of alphanumeric characters possible with particular stroke-generated writing system.

stroke-generated writing   Major branch of alphanumerics for electronic displays in which each character is assembled from one or more straight-line strokes. Simplest format is 7-segment, also called DHW (double-hung window), all segments of which are used in figure 8 (this is common in LED and LCD displays for minimum cost, eg watches); another common system is 14 or 16-bit starburst, and best results are yielded by multi-stroke systems using five lengths and up to 40 orientations.

stroking   Correct cycling of pump plungers over full stroke, eg in variable-displacement pump with swash drive from one end.

strongback   Structural member added over large surface to provide rigidity in bending and torsion, eg across large rotodomes.

strongpoint   See *hardpoint*.

strontium   Sr, soft, reactive, silvery metal, many uses as element and compounds, Sr-90 dangerously radioactive. Density 2.5, MPt 769°C.

strop   *1* Flexible loop connecting deck accelerator (catapult) to aircraft not fitted with nose towbar.

2 Length of webbing connecting static line of airdrop load to anchor cable (DoD).

3 Also see *suspension* *.

Strouhal number   Constant for particular bodies in fluid flows giving rise to street of vortices and from which frequency of shedding can be derived; $S = fD/V$ where f is frequency, D is diameter of cylinder or wire and V is velocity; for regular cylinder S is 0.2 for R between 10^3 and 2×10^5.

structural aspect ratio   See *aspect ratio*.

structural damping   Total damping of assembled structure.

structural design   Total task of designing structure.

structural element   Subdivisons of structure such as a spar, rib or frame, which may themselves be assembled from smaller structural members.

structural factor   Seldom used ratio of structure mass to sum of structural plus propellant mass for rocket vehicle; mass ratio more common.

structural failure   Breakage under load.

structural limit   Greatest weight (mass) at which aircraft can show compliance with structural requirements, usually MRW.

structural machine screw   HT-steel machine screw (term often synonymous with bolt) with unthreaded shank portion below head.

structural member   Portion of structure, esp. any part important in bearing loads; can be single piece of material or assembly, and demarcation with structural element is blurred.

structural merit index   Ratio E/ρ, modulus divided by density.

structural-mode control system   Flight-control subsystem whose purpose is to reduce stresses in structure during flight through gusts and other turbulence, esp. in dense air, or to reduce vertical accelerations experienced by crew compartment. Requires vertical-acceleration sensor and separate horizontal (occasionally inclined or vertical) surfaces to apply required countering forces to airframe, usually symmetric about c.g. to avoid applying pitch moment. Often similar in principle to active controls.

structural placard   Structurally significant airspeeds, eg maximum gear-down or flap-extension IAS, displayed in cockpit.

structural RAM   RAM (2) material, often of hybrid types such as CA overlying magnetic RAM, bearing some or all of local skin structural loads.

structural section   Cross-section not necessarily unvarying, of structural materials or finished parts of high fineness ratio, produced mainly by extrusion or rolling between shaped dies but often machined to incorporate tapers or discontinuities. Standard * appear in priced catalogues.

strut   *1* Externally mounted structural member intended to bear compressive loads, and usually of streamline section.

2 Ambiguously, stub wing or pylon * attaching engine pod to fuselage or wing (loads mainly tensile and bending).

3 Loosely, any major bracing member or portion of truss even if load is tensile (suggest incorrect usage), thus a lift * on braced highwing monoplane is a misnomer [it is a tie].

strutjet   RBCC engine using multiple struts in air inlet.

strut skin   Load-bearing additional skin, eg bonded reinforcement.

STRV   Space technology research vehicle.
STS   *1* Space transportation system; usually means Shuttle.
   2 Space Tourism Society.
   3 Stockpile to target sequence.
   4 Stable time subfield.
   5 Status.
   6 Support and test station.
   7 SAR (2) transit stripmap (radar mode).
   8 Sensor to shooter (usually S2S).
sts   Stones (non-SI unit of weight).
STSA   Short-term strategic aircraft (UK).
STSC   Sequentially triggered shaped charge [=BWA].
STSMT   Service Technique des Systèmes de Missiles Tactiques (F).
STSS   Space tracking and surveillance system [formerly SBIRS-Low].
STSSS   *ST3S*.
STST   *1* Strategic transportable satellite terminal.
   2 Sensor-to-shooter time.
STT   *1* Solution-treatment temperature.
   2 Single-target track[ing].
STTA   Service Technique des Télécommunications de l'Air (F).
STTE   *1* Special-to-type test equipment.
   2 Alternative to STTEA.
STTEA   Service Technique des Télécommunications et des Equipements de l'Aéronautique (F).
STTEM   Final tactical technical economical requirement, follows PTTEM and leads to series production (Sweden).
ST3S   Service Technique des Systèmes Stratégiques et Spatiaux (F).
STU   *1* Service trials unit (UK).
   2 Sensor transmitter unit.
   3 Satellite terminal unit.
   4 Secure terminal, or telephone, unit.
   5 Service transport unit.
stub   *1* Portion of transmission line up to ¼-wavelength long connected in shunt with dipole feeder to match impedances.
   2 Short straight exhaust stack carrying gas from one piston-engine cylinder direct to atmosphere just clear of cowl or any downstream structure.
stub antenna/aerial   Stub (1), usually for DME, IFF and beacon transponders.
stub exhaust   See *stub (2)*.
stub float   See *sponson*.
stub plane   Short length of wing projecting from fuselage or hull to which main wings and/or main landing gear are attached (BS). See *stub wing*.
stub power   Net power available for propulsion.
stub runway   Short section of runway projecting on far side of an intersection which may be specifically designated for STOL takeoffs which avoid all conflicts with CTOLS on ground and in air. Not normally used for landings.
stub shaft   *1* Projects from driving or driven item just far enough for gearwheel and bearing.
   2 Short shaft running in its own bearings with bevel drive to an accessory [demanding longitudinal location] driven by splines [permitting axial movement by driver].
stub thrust   Net thrust available for propulsion.
stub wing   Short aerofoil projecting approximately hori-

zontally from fuselage serving either as a structural bracing member (eg linking lift struts and main gears) or as a beam supporting external stores racks or other loads.
stub-wing stabilizer   See *sponson*.
stud   *1* Headless bolt threaded at both ends but not in mid-portion used for attaching items to threaded hole in casting or forging.
   2 Projecting pin used as fulcrum, pivot, locating dowel or for other purpose.
   3 Projecting push-button human interface.
student pilot   Person authorized and licensed to receive flying instruction from rated flight instructor but who has not yet attained any licence.
study   Formal investigation, with or without research or any kind of testing, of possible solutions to a stated future requirement; hence * phase, * program, * contract.
stuffed   Large component, eg wing panel or section of fuselage, manufactured complete with internal systems and equipment. In case of transport fuselage, not necessarily furnished. In case of nacelle, contains equipped engine.
Stuka   Dive bomber (G).
stunting   Performing aerobatics (colloq., becoming obs.).
S-turn   Flight manoeuvre in form of S in horizontal plane, often carried out across road, railway, fence or similar axis.
STV   *1* Structural, or systems, or supersonic, or separation, test vehicle.
   2 Satellite transfer vehicle.
STVA   Self-tuning vibration analyser.
STVS   Small-tower voice switch.
STW   Special-to-weapon equipment, eg in WAS.
STWA   Short(er) trailing-wire antenna.
STW&ARS   Satellite threat warning and attack reporting system (USAF).
STWL   Stopway lights.
S2S   Sensor to shooter.
STX   Start of text.
STY   Standby.
styrene   Liquid hydrocarbon ($C_6H_5CH{:}CH_2$) polymerised into polystyrene for use as basis of important resins, low-density foams (expanded polystyrene) and many other products.
Styrofoam   Most widely used of the styrene-based foams, used as sandwich structure filler.
SU   *1* Signal[s] unit.
   2 Submersible unit (sonar).
   3 Servicing unit.
SUA   Special-use airspace.
SUAV   Small unmanned [or unit] air vehicle.
Suave   Small UAV engine.
sub   Smaller contractor, partner chosen in *teaming down*; can be thought of as short for subordinate or sub-contractor.
subassembly   Assembly (eg structural, electronic or for some other system) forming part of a larger item.
subcarrier   Subsidiary modulated carrier which in turn modulates primary RF.
subchannel   Intermittent fraction of telemetry channel conveying one repeatedly sensed measurement.
sub-cloud car   Car suspended below above-cloud aerostat giving view of ground.
subcommutation   Commutation of additional telemetry

channels, output of which is fed to primary commutator; called synchronous if commutation frequency is multiple of that of primary (which it usually is).

subcontract Agreement other than prime contract to perform work on same programme. Normally between subcontractor and prime contractor for work assisting latter to complete task. Does not necessarily confer design authority or responsibility.

subcritical mass Mass of fissile material inadequate in magnitude or configuration to sustain chain reaction.

subcritical rotor Rotor, esp. main lifting rotor of helicopter, whose fundamental flapping, lag-plane or other resonant frequency is less than normal operating frequency, latter being frequency with which one blade passes given angular position.

subcritical wing Wing designed not to exceed M_{crit} at any point.

subgrade Soil underlying airport pavement.

subgrade code ICAO coding for strength of subgrade: A, 150 MN m^{-3}; B, 80; C, 40; D, 20.

subgravity Condition in which apparent vertical acceleration is between zero and + 1 g.

subharmonic SHM or other sinusoidal waveform whose frequency is exact multiple of related fundamental waveform.

sub-idle Operating speed or control selection below idling, occasionally provided with main engines for ground operation.

SUBJ Subject [to . . .].

subjective Dependent upon personal opinion; in such fields as aircraft noise and visible smoke unquantified * measures play a major role, usually as the result of seeking views of statistically significant population.

sub-kiloton weapon NW whose yield is below 1 kT.

sublimation Direct transition, in either direction, between solid and vapour state.

sublimator Solid material designed to reject waste heat by sublimation to space environment, usually a porous plate and in some cases not a true * but medium for evaporating liquid heat-transfer medium.

submarine Verb, to slide down out of seat harness in violent deceleration [impossible with 5-point harness].

submarine missile Missile launched by submarine, eg against distant city target.

submarine rocket Submarine-launched rocket (Subroc) weapon for use against other submarines.

submarine striking force Force of submarines having guided or ballistic missile capabilities and formed to launch offensive nuclear strike (DoD).

submodulator AF power amplifier preceding main RF modulator.

submunition Self-contained warhead carried in multiple (tens or hundreds) inside single delivery vehicle, eg air/ground store.

subprogram See *subroutine*.

Subroc Submarine-launched [guided, nuclear-warhead] rocket.

subroutine Set of instructions necessary to enable EDP (1) computer to carry out well defined mathematical or logical operation, forming part of complete program; unit of routine for specific sub-task, usually written in relative or symbolic coding even if full routine is not, and in closed * entered by jump path from main routine and reverting to main routine at completion (one sequence of instruc-

tions can be * and at same time a main routine with respect to its own *).

subsatellite Satellite of Moon.

subsatellite point Point on Earth's surface directly below Earth satellite, ie on local vertical through satellite at any time.

subscale prototyping Not defined. The A380 conceivably might be assisted by a balsa chuck-glider.

subsidence *1* Extensive sinking of air mass, eg in polar high, in which air forced to descend is warmed by compression, increasing stability.

2 Disturbance to aircraft flight which dies away without oscillation (UK usage only and unusual).

subsonic Slower than speed of sound in surrounding medium. According to one lexicographer, Mach numbers less than 0.7.

subsonic flow Flow whose velocity is less than speed of sound within it; in contracting duct accelerates and rarefies slightly and in expanding duct decelerates and compresses slightly.

substantiation Formal demonstration of compliance, eg with design fatigue-life requirement.

substrate In various kinds of planar technology, structural layer upon which operative layers and/or devices are formed; in most microelectronics has low resistance, while in solid-state circuitry and solar cells * is normally an insulator.

substratosphere Imprecise, generally taken to mean upper layer of troposphere.

subsystem There is no clear demarcation between system and * though it is simple to give examples of dynamic organized groupings of devices that can be seen to be one or other. It could be argued, eg, that because it is part of a flight-control system a stability-augmentation system is a subsystem. Again, a BITE can be considerd an integral part of a system or as a *.

suckdown Downwards pull on jet-lift aircraft when hovering in ground effect, usually caused by upward flow around fuselage from fountain.

suck-in Traditional verbal command in hand-starting piston engine, meaning 'Do not energize ignition (cylinders are being filled with mixture)'.

suck-out Deliberate or undesired characteristic of some vertically translating airbrakes, esp. on sailplanes, of being pulled open by local depression under certain flight conditions.

Sucsede Successful user-centered systems engineering development and environment (software tools).

suction Withdrawal of fluid through a region of local depression.

suction-cup gun Spraygun using ejector effect to withdraw medium.

suction face That side of propeller blade, normally facing forward, formed from upper surface of its aerofoil.

suction flap Not normal term; in some cases means blown flap.

suction gauge Instrument measuring pressure below atmospheric, usually by aneroid or bourdon tube.

suction gyro Gyro whose rotor is driven by atmospheric air jets trying to fill evacuated case.

suction stroke Induction or inlet stroke.

suction wing *1* One whose boundary layer is continually sucked away by powered suction system.

2 Wing of deep section with upper rear discontinuity

from which large airflow is removed by suction to maintain attached flow (rare after 1949).

SUD   Stretched upper deck.

sudden ionospheric disturbance   Abnormal behaviour of ionosphere following passage of radiation travelling at speed of light from source of solar flare, affecting Earth's sunlit face. Gradual return over following hour or more.

sudden pull-up   Basic stressing case in which severe positive g is applied by violent nose-up elevator command (US term).

sudden stoppage   Inflight stoppage of piston engine apparently within about one turn of propeller, or of turbine within about one second; indicative of severe internal damage.

Sudec   Supervisory digital electronic control, less authority than Fadec.

SUEM   Spin-up and ejection mechanism (Mars lander).

suite   Aggregate of all equipments, not necessarily integrated or forming a common system, of similar general type carried in vehicle. Especially favoured in connection with ECM, for which as many as 14 separate electronic equipments may be carried (and in theory operated simultaneously) though most are linked only through common power supplies and possibly cockpit displays. For GA avionics * is supplanting 'fit'.

SUL   Yaw-damper servo-actuator.

sulfate   Salt or ester of sulfuric acid.

sulfated   Condition of lead/acid battery after prolonged discharge; lead sulfate plates cannot be restored by charging.

sulfidation corrosion   Accelerated metallic corrosion due to sea salts in atmosphere.

sulfide   Compound of sulfur with element or radical.

sulfur   S, former UK names brimstone and then sulphur, important in many kinds of compound in several forms, commonest being yellow crystalline, density 2.1, MPt 113°C.

SUM   *1* Surface-to-underwater missile.

  2 Structural-usage monitor.

  3 System user module.

  4 Scheduled unit maintenance (DARA).

sumerian cobalt   One of many cobalt-steel materials used for permanent magnets.

summer solstice   Point on ecliptic, or time thereof, at which Sun reaches maximum N declination; about 21 June.

summing gear   Differential gears which add or subtract motions of two members.

summing unit   Device whose output is sum of inputs; can be mechanical, fluid flow (including pneumatic logic) or electronic.

sump   Low region of fluid (liquid) system where liquid tends to collect by gravity. In piston engine bottom of crankcase, which in wet-* engine also serves as oil tank. In fuel tank lowest point with tank in normal attitude, where water collects and may be drained.

sump jar   Container in vent line from battery box in which alkaline chemicals neutralize battery-charging gas.

Sums, SUMS   *1* Shallow underwater missile system.

  2 Structural-usage monitoring system.

Sun   Apparent diameter (optical) 13.92×10^8 m, mean density 1.41 g/cm^3, mass c 1.99×10^{30} kg, surface gravity 274 ms^{-2} (26.9g), rotational period c 25.4 days at Equator (33 at 75°N/S lat), radiating surface temperature 5,800K, chromosphere up to c 10^6K, emits various corpuscular and EM radiation at total energy c 3.39×10^{26} Js^{-1}; moving with solar system through interstellar medium in local arm of galaxy at c 20 km s^{-1}. See *solar wind*.

Sun compass   Compass based on az/el of Sun, formerly used near magnetic poles where magnetic compass unreliable.

Sundstrand drive   Infinitely variable hydraulic CSD.

sun gear   Central gearwheel in planetary reduction gear.

sunk costs   Costs incurred in development which are paid off in another programme.

sunlight-readable   Readable in illumination (illuminance) of 10^4 lx; basic requirement of military cockpit warning panels and displays.

SunRez   Proprietary resin activated [set hard] by UV light.

Sun-synchronous   See heliosynchronous.

SUO   *1* Small unit operations.

  2 Aileron/elevator/rudder servoactuator.

SUP   *1* Smart upgrade procurement.

  2 Supplement.

  3 Suspected unapproved part[s] (FAA).

Suparco   Space and Upper-Atmosphere Research Commission (Pakistan).

Sup Aéro   See *ENSAé*.

superadiabatic lapse rate   Greater than DALR, such that potential temperature decreases with height.

superaerodynamics   Aerodynamics involving such high relative velocities and such low densities that body has passed before air molecules can collide with others and exchange energy; also called free-molecule flow and Newtonian aerodynamics, and akin to MHD flow.

superalloy   Any alloy designed for extremely severe conditions, esp. at very high temperatures.

superaugmentation   Either of two methods of achieving gains in flight performance by imparting artificially stable flight characteristics to aeroplanes that are inherently unstable.

superblock   Major section (20%) of CVF, each assigned to different yard because of lack of national capability to construct large ship (UK).

superboom   Boom from SST or other aircraft which, because of reflection and/or refraction in atmosphere, is heard up to 250 km from flightpath.

supercharged core engine   Turbofan in which fan is regarded as 'supercharger' for core; no need for this concept.

supercharged harness   Pressurized harness to reduce ignition arcing.

supercharged turboprop   Powerful turboprop derated and matched with lower-capacity gearbox and propeller, thus giving same power at all airfield heights and temperatures; essentially synonymous with flat-rated.

supercharger   Compressor driven by crankshaft step-up gears or by exhaust turbine which increases density of air or mixture supplied to cylinders of piston engine, either to boost power or, more often, to assist in maintaining power at high altitudes. Virtually all are single-stage centrifugal, in some cases with choice of gear ratios and formerly (WW2) with two consecutive stages and intercooler. Term also (suggest formerly) used for cabin blower.

supercirculation   Increase of wing lift by increasing circulation by positive power-consuming means; secondary gains include postponement of stall, reduction of drag (both by improving flow and by enabling wing and other aerofoils to be smaller) and as means towards realizing laminar flow. Commonest form is blown flap, with more ambitious schemes discharging supersonic bleed air along upper part of leading edge or other places; used facing to rear to accelerate boundary layer and as by-product to impart thrust.

supercompression   Piston engine with such a high compression ratio that it must not be operated at full throttle below a given height (today unusual).

superconductivity   Near-zero electrical resistance exhibited by some metals, esp. particular mixtures, as 0°K is approached. Extremely powerful currents and magnetic fields are possible eg making possible frictionless gyroscope. One branch of cryogenics.

supercooled   Vapour and finely dispersed water droplets can exist as vapour and liquid at below 0°C, freezing immediately on contact with solid object.

supercooled large droplet   Diameter ⩾300 μm.

supercritical   Loosely, any flow involving regions where M>1.

supercritical wing   Aerofoil designed to cruise at above M_{crit}, characterized by bluff leading edge, flattish top, bulged underside and downcurved trailing edge; by reducing peak suction maximum acceleration and shock formation are delayed and wing can be deeper, have less sweep, house more fuel and weigh less than conventional wing for same cruise M.

supercruise   Sustained flight at supersonic speed with engine[s] in dry thrust, without afterburner.

supercruiser   Aircraft designed to cruise at supersonic speed, usually Mach 1.5 to 2.

superheat   *1* Temperature difference between aerostat gas or hot air and surrounding atmosphere; called positive if gas is warmer than atmosphere.

2 Heat energy added to gas or vapour after evaporation has been completed.

superheated vapour   Vapour heated above its boiling point for given pressure.

superheterodyne   Radio receiver in which received signal is mixed (heterodyned) with local oscillatory frequency to give intermediate frequency which is then amplified with various advantages.

superhigh frequency   See *frequency, radio*.

superior planets   Those further out than Earth, ie Mars to Pluto.

superluminal   Velocity greater than that of light, by expansion/contraction of spacetime.

supermanoeuvrability   Ability to perform controlled supermanoeuvres.

supermanoeuvre   *1* A sustained manoeuvre which increases AOA beyond the 1-g stall.

2 A sustained manoeuvre which increases AOA beyond the actual accelerated stall.

3 A dynamic manoeuvre in which angular momentum in the pitching plane momentarily increases AOA to a peak beyond stall.

superplasticity   Property of flowing like hot glass at elevated temperatures under modest applied pressures with no tendency to necking or fracture; possessed by many alloys, eg Prestal at 250°-260°C.

superposition   *1* Principle in stress analysis that aggregate of all strains caused by a load system may be considered to be the sum of all individual strains experienced by each member taken in isolation.

2 Identical principle for algebraic sum of currents or voltages in linear network.

3 Ability of subatomic particle [many] to exist in more than one place, or one state, at same time.

superpressure   Pressure difference between gas in aerostat at any point and surrounding atmosphere at same height; called positive if gas pressure is greater (as it usually is).

superpressure balloon   Unvented envelope strong enough not to burst in long-duration voyage at constant pressure height.

superrefraction   Warm air over cold sea, extends radio/radar ranges.

super search mode   Radar scans entire HUD field of view.

supersonic   *1* Faster than speed of sound in surrounding medium. One lexicographer says Mach 1.2+.

2 R/T callsign, suffix when actually cruising at * speed.

supersonic-combustion ramjet   Ramjet whose combustion system is (very unusually) designed to function at supersonic speed; abb. SCRJ or scramjet (even Mach-6 ramjet vehicles invariably burn fuel in subsonic airflow).

supersonic compressor   Axial compressor in which fluid velocity is supersonic relative to whole length of rotor blades, stator blades or both, with oblique shocks giving greatest possible pressure rise per stage. (Some axial and centrifugal compressors not classed as * do in fact have local flow over Mach 1 at periphery at maximum rpm).

supersonic dash capability   Ability to fly safely at over Mach 1. Originally this was to enable a high-subsonic attack aircraft to exceed Mach 1 for brief period to escape defences. Now it is seen as a possibility for Sonic Cruise type aircraft to exceed Mach 1 in recovering from upsets.

supersonic diffuser   Contracting duct (see *supersonic flow*).

supersonic flow   Flow which relative to immersed body or surrounding walls is supersonic. In contracting duct decelerates and compresses; in expanding duct accelerates and rarefies.

supersonic inlet   Air inlet designed for supersonic flow both past and through it for at least part of flight; ideally has centrebody or side wedge to create attached oblique shock and various forms of variable geometry and auxiliary doors.

supersonic jet   *1* Propulsive jet from rocket, ramjet or afterburner whose velocity relative to source is supersonic.

2 Supersonic aeroplane (colloq.).

supersonic nozzle   Propulsive or wind-tunnel nozzle through which relative flow velocity is supersonic. Ideally of con/di form with variable profile and area.

supersonic propeller   Propeller whose blades are designed to operate with supersonic relative velocity over major portion of surface. Noise problem appears insoluble.

supersonic tunnel   Wind tunnel capable of supersonic speed in working section (either brief or sustained).

supersonic turbine   Turbine of any kind designed to operate with flow velocity relative to rotor blades supersonic (rare).

superstall   Progressive stall attainable by certain aeroplanes, eg T-tail with rear-mounted engines, in which (partly because at low speeds drag increases faster than lift when pilot pulls nose up) decay of speed and increasing AOA leads to stable condition in which aeroplane descends in approximately constant attitude (not far removed from level flight) but with decaying speed and AOA increasing continuously so that after long period it approaches 90°. Root cause is combination of nose-up pitching moment plus immersion of horizontal tail in wing wake, destroying effectiveness. If not recoverable, called locked-in stall.

superstandard propagation　Propagation with super-refraction.

superstructure　*1* Secondary structure built above main fuselage or other part of aircraft (rare; eg not used for AWACS radar aerial).

2 Secondary fairing structure to streamline box-like truss.

SUPP　Supplemental.

supplemental carrier　US air carrier operating under supplemental certificate, normally authorizing services of various kinds other than scheduled.

Supplemental Type Certificate, STC　Authorizes alteration to aircraft, engine or other item operating under approved type certificate (US).

supplementary aerodrome　One designated for use by aircraft unable to reach its regular or alternate aerodrome (BS, suggest arch.).

supply balloon　Flexible container for storing gas at low pressure ready for aerostats; normally too heavy to fly even if free. Hence supply main, supply tube, links * with aerostat needing supply of gas.

supply chain　The complete network of subcontractors to a major company or programme, especially arranged hierarchically.

support　*1* All services and material needed or provided to assist operator after delivery (see * *items*).

2 Action taken to assist friendly unit in battle.

3 Part of force or unit held back at start of action as reserve.

4 Underpinning of new programme by R&D effort to ensure answers are available to technical problems.

supporting aircraft　All active aircraft other than unit aircraft (DoD).

supporting surfaces　Those aerofoil surfaces whose chief function is to provide lift for aerodyne; can be fixed or rotating.

support items　For support (1) typically publications, training, simulator and instructional rigs, auxiliary ground equipment, spare parts, testing, warranty provisions and field modification kits.

support zone　Designated surface area for airlanding or other operations in direct support of battle (no longer used DoD, NATO).

suppressant　Active ingredient for suppressing an action, normally fire, esp. one automatically released by sensitive pressure sensors in fuel tanks or similar regions; passive suppressing methods, eg reticulated foam, are not *.

suppressed　*1* Installed so that item does not project beyond skin of aircraft; thus * aerial is synonymous with flush aerial.

2 Emitter, especially engine(s), designed or installed to minimize emissions; for civil aircraft noise predominates, for military aircraft IR radiation.

suppressive　Intended to suppress hostile defensive fire by offensive action, eg direct attack with weapons and

offensive ECM; hence * attack, * support, * weapons (eg ARMs).

suppressor　*1* Jet-engine nozzle either configured for minimum noise or shielded by additional surrounding duct.

2 Various additions to electrical or electronic devices or circuits to reduce unwanted leakages, emissions or other phenomena, eg extra grids in thermionic valves (tubes) to stop secondary emission and large series resistors in HT circuits to eliminate sparking.

suppressor pulse　Sent out to disable L-band [original meaning, 0.39-1.55 GHz] avionics during transmitting period of other equipment on similar wavelength, to prevent interference or damage.

SUPPS　Regional supplementary procedures.

Supra　Support for the use of presently unserved airspace (Euret).

supra-aural　Fitting over the ear.

Supral　Superplastic aluminium alloy marketed for SPFDB applications (British Alcan).

suprathermal ion detector　One of ALSEP experiments left on Moon; measures energetic ions impacting suface to determine solar-wind energies.

SUPT　Specialized undergraduate pilot training.

SURE, Sure　Sensor update and refurbishment effort (USN).

surf　Verb, see *surfing*.

surface　*1* Aerofoil, esp. large, eg wing (not small rotating, eg compressor blade).

2 Exterior of aircraft, eg * friction drag.

3 2-D layer corresponding to particular pressure altitude.

4 2-D layer in any plane corresponding to particular electronic radiation pattern or time difference.

5 Hinged or extendable area for flight control, lift augmentation or drag augmentation.

6 Generalized term for Earth's *, hence * target = one on land or water.

surface acoustic wave　Travelling across polished piezo-electric substrate at controllable microwave frequencies.

surface actuator　Device which physically moves a surface, eg control surface; need not embody any form of control function.

surface boundary layer　Atmosphere in contact with Earth's surface, extending up to base of Ekman layer (anemometer level).

surface burning　Combustion of fuel for propulsion on outside of aeroplane, proposed for variable-geometry aircraft for Mach numbers of about 5, using variable body profile for ramjet effect.

surface cooler　See *surface radiator*.

surface corrosion　Galvanic (non-mechanical) attack on surface, eg by salt spray, often under paint film.

surface discharge　Most common type of gas-turbine igniter, in which semiconductor [usually SiC] permits leak from tungsten electrode to body, ionizing path for main high-energy flashover.

surface effect　Effects on air-supported vehicle of close horizontal surface beneath (synonymous with helicopter ground effect), hence ** vehicle (= air-cushion vehicle), ** ship (US usage in latter cases).

surface-friction drag　Drag due to all forces tangential to surface, notably shearing of boundary layer; added to form drag makes profile drag; added to pressure drag makes total drag.

surface gauge   Gauge (US, gage) in form of precision stand moved about on surface plate carrying adjustable scriber for transfer of exact height measures, eg in marking out or in checking finished workpieces.

surface inversion   Atmospheric inversion with base at Earth's surface.

surface loading   Mean normal force per unit area carried by a particular aerofoil under specified aerodynamic conditions (BS); in case of wing term is wing loading; suggest few cases where ** needed.

surface management system   Looks into near future to manage departures and avoid congestion or other problems.

surface movement   One vehicle (eg aircraft, or even bicycle), in motion on airfield movement area. Hence ** indicator, usually a PPI radar.

surface of discontinuity   Sloping demarcation between warm and cold air masses.

surface operations   In the US civil community, any movement on airport/airfield surface.

surface plate   Steel table with extremely flat and smooth surface.

surface power unit   Surface actuator embodying control functions, eg control valves, feedback inputs, summing units and possibly redundancy provisions.

surface sampler   Device for scooping up specimen of planetary or other surface for analysis or other study (eg to investigate for presence of life).

surface tape   Pinked-edge strips of fabric doped over all seams, rib-stitching and edges of fabric covering. Also called finishing tape.

surface target   One on land or sea.

surface tension   Tendency of a liquid/gas surface to minimize its area [because all the attractive liquid molecules are on one side of the surface], symbol γ.

surface-to-air missile   Missile launched from surface (6) against target above surface (DoD). Hence ** envelope, that airspace within kill capabilities of particular SAM system; ** installation, a ** site with system installed; ** site, prepared plot of ground designated for but not occupied by SAM system.

surface-to-surface missile   Surface (6)-launched missile designed to operate against target on surface (6), including those underground.

surface visibility   At eye level.

surface wave   *1* Radio wave travelling round surface (6); most effective propagation mode of LW/LF.

2 Acoustic wave in surface-wave device.

surface-wave device   New family of electronic devices based on surface acoustic waves sent across piezoelectric slab or other substrate; originally (1970s) used for delay lines and now for complex signal processing.

surface wind   Generalized term for wind measured at surface (6); in US gradually switching from 20 ft (6.5 m) anemometer level to ICAO 10 m (32.8 ft) level.

surface zero   See *ground zero*.

surfacing   Improving low-drag quality of vehicle surface (in general task called stopping and *), eg by perfecting flatness and smoothness.

surfactant   Surface active agent, material (usually liquid or particulate) which alters surface tension and/or

performs other tasks at boundaries between dissimilar materials, eg detergent.

surfing   Riding with enhanced L/D ratio on the field of increased pressure created by a hypersonic vehicle's own shockwaves.

surge   *1* Gross breakdown of airflow through compressor, normally of axial type, resulting from local stall and usually characterized by muffled bang and sudden increase in turbine temperature; hence * line, * point. Often used synonymously with stall.

2 Various abnormally large currents, signal amplitudes or voltages in electrics or radio, eg on first switching on or caused by lightning or static discharge.

3 Planned large increase in flying rate of military unit, eg to explore ultimate potential of personnel and hardware over short or longer term under crisis conditions.

4 Unplanned transient increase in flow of fuel in aerial refuelling, sometimes causing a disconnect.

5 General change in atmospheric pressure at surface apparently superimposed on predicted diurnal or cyclonic change.

6 Fore/aft linear motion (simulator).

surge box   Term used for various kinds of device in aircraft fuel system to reduce pressure/flow excursions caused by fuel momentum either in tanks (sloshing) or in pipelines, esp. during high-rate refuelling.

surge diverter   Protective semiconductor device having negative resistance/temperature coefficient to earth voltage surges.

surge line   Boundary between gas-turbine operating region and region where surge of compressor is certain; locus of all surge points. Generally same as stall line.

surge point   Any combination of airflow and pressure ratio for gas turbine at which surge occurs.

surging   *1* Occurrence of surge (1).

2 Fault in wind tunnel characterized by erratic or low-frequency pulsations in velocity, flow and pressure.

3 See *sloshing*.

surpic   Surface picture.

surrogate factory   One engaged in assembly of kit-built aircraft.

Surtass   Surveillance towed-array sonar system.

surveillance   Systematic observation of aerospace, surface or subsurface objects by any kind(s) of sensor.

surveillance approach   Instrument approach conducted in accordance with directions issued by ground controller referring to a surveillance radar display (DoD).

surveillance radar   *1* Primary radar scanning in azimuth, often through 360°, supplying P-type display (PPI). Not normally giving elevation or height of aerial targets.

2 Specif., primary radar whose purpose is to determine az/el position, track and (with SSR) identity of all aerial targets, and to provide radar separation, navigational assistance, storm warning and vectoring for final approach (but not normally to handle complete radar approach).

surveillance radar element, SRE   Portion of GCA system which vectors incoming traffic until established on ILS and handover to PAR.

surveillance system   Any means of surveillance not contained wholly in one vehicle or site, eg RPV, electronic communications and guidance, digital sensor data-link, and control/receive ground station.

survey   *1* To examine damaged vehicle, eg crashed

aircraft, often on behalf of insurers or underwriters, to establish damage, possibility of salvage (eg to fly out from crash site) and best course of action.

2 Normal meaning in photogrammetry and mapping.

3 Examination by surveyor (see below).

surveyor Technically qualified and designated official empowered to collaborate with aerospace design staff on behalf of national certification authority and examine subsequent hardware and design software to establish compliance with airworthiness requirements; usually concerned with particular design aspect, eg fluid systems, or with particular class of aircraft.

survivability Capability of a system to withstand a man-made hostile environment without suffering an abortive impairment of its ability to perform its designated mission (USAF). Refers specif. to various effects of NW attack, eg degradation of volatile memory.

survival capsule Detachable crew compartment, normally of military aircraft, capable of separation in emergency and soft landing on land or water, thereafter serving as shelter.

survival kit Man-portable package containing items to help sustain life remote from other human beings.

survival radio Self-contained, portable, shockproof, floating radio emitting homing (and possibly voice) signals on 121.5 and 243 MHz.

Survsat Survivable sat-com system.

SUS *1* Saybolt universal second[s].

2 Signal, underwater sound.

susceptance Reciprocal of reactance, defined as ratio of current quadrature component to voltage for same frequency, $B = 1/x$, unit siemens.

susceptibility Degree to which any hardware is open to an attack as a result of inherent weakness (DoD).

suspended underwing unit Dispenser.

suspension *1* Linkage between aerostat and load, hence * band (fabric band linking envelope to * lines), * bar connecting suspension lines to basket ropes in balloon, * line, main connections between envelope and basket or suspended car, and * winch connecting kite balloon to surface (6).

2 System of particles dispersed through fluid, including atmosphere.

suspension strop Webbing or wire rope connecting helicopter and cargo sling.

sustained flight Time-invariant flight, ie steady lift and airspeed.

sustained readiness program Keeping fleet operational by replacing structural parts damaged by fatigue or corrosion.

sustainer Propulsion, either rocket or airbreathing, that provides power for sustained flight following short high-thrust acceleration period under power of boost motor(s). Not normally applied to any stage of large or small multi-stage vehicle; must be long-duration propulsion system handling entire mission after separation or burnout of booster (in ramrockets and many related systems may in fact use booster case for combustion).

SUT *1* Autothrottle servo.

2 Tailplane [stabilizer] trim servo.

3 Surface and underwater target.

4 System under test.

Sutherland law Gives temperature variation for viscosity of air.

Sutton harness Traditional (WW1) personal seat harness with two lapstraps, two shoulder straps and central pinned clip passing through all four.

SUU *1* Suspended underwing unit (US).

2 Secondary user unit.

SUVOS Semiconductor UV optical source[s].

SUW Surface warfare.

SV *1* Satellite, or space, vehicle.

2 State vector.

3 Static vent.

4 Simulation validation.

5 Shop visit.

6 Synthetic vision [S adds system].

SVA Security violation alert.

SVAS Shuttleworth Veteran Aeroplane Society (UK).

SVC *1* Service, service message.

2 Switched virtual circuit.

SVCBL Serviceable (ICAO).

SVCS Secure-voice communications system.

SVD *1* System verification diagram (software).

2 Space Vehicles Directorate (USAF).

SVF Schweizerische Vereinigung für Flugwissen-schaften (Switzerland).

SVFR Special VFR.

SVGA Superior video-graphics array, 800×600 pixels.

SVI Smoke volatility index.

SVIS Synthetic vision information system[s].

SVLR Schweizerische Vereinigung für Luft-und Raumrecht (Switzerland).

SVMS Space-vehicle motor simulator.

SVN Satellite vehicle number.

SVO Servo.

SVP Static-vent plate.

SVR *1* Slant visual range; attempt to give pilot on final approach idea of when he will acquire approach lighting, reported as either nominal contact height at which 150 m (500 ft) segment of one crossbar will become visible, or at top of shallow fog layer as minimum visible length of lighting.

2 Shop-visit rate.

3 Severe.

4 Service of external intelligence (R).

SVRL Several.

SVRR Service readiness review.

SVS *1* Synthetic-vision system.

2 Secure voice switch; E adds equipment.

SVSG Silicon vibrating-structure gyro.

SV-stoff Rocket propellant, typically 85% RFNA, 15% sulfuric acid (G).

SVT Servo throttle.

SVU Satellite voice unit.

SVUOM State institute for protection of materials (Czech Rep.).

SVVT VTOL (R).

SVWT Schweizerische Vereinigung für Weltraum-technik (Switzerland).

SW *1* Single wheel (MLG).

2 Short-wave.

3 Surface wave.

4 Strategic Wing (USAF).

5 Space Wing (USAF).

6 Secretary of War (US).

7 Skin waviness.

8 Single-wedge (aerofoil).

9 Snow shower.

10 Software.

S/W   *1* Software.

2 Surface wind.

Swaarm   Smart weapon, anti-armour.

Swafrap   Swedish air force rapid-reaction force.

swageing   Joining by cold-squeezing one member around another, eg electrical or control-cable terminal or end-fitting on to end of cable.

SWALAS   Shallow-water ASW localization and attack system.

swallowed shock   Position of shockwave across airbreathing engine or other inlet inside duct, when mass flow is ρVA but internal pressure is low (normally equated with zero thrust). Can be feature of plain pitot inlet or any other type. Resumption of correct engine operation restores shock to normal position at inlet.

swallowing capacity   Ability of inlet to handle large airflow, esp. over wide range of air densities and Mach numbers.

SWAP   *1* Severe-weather avoidance plan, or program (US).

2 System-worthiness analysis program (FAA 1966).

Swaps   Standing-wave acoustic parametric source.

Swarm   *1* Stabilized weapon and reconnaissance mount.

2 Small warfighter array of reconfigurable modules. See next.

swarming   Warfighting concept in which dozens to hundreds of small networked warfighting units stealthily coalesce to attack a target and then disperse, ready for the next 'pulse'.

swashplate   Disc rigidly or pivotally mounted on shaft as drive mechanism for plungers or rams arranged parallel to shaft; when disc is normal to shaft, plunger stroke is zero, increasing to maximum at maximum * obliquity. In hydraulic motor * is driven, not driving, member.

SWAT   Special weapons and tactics team (US police forces).

Swat, SWaT   Slotted-waveguide technology.

swath, swathe   The preferred spelling is the former, universal in the US. Among aerospace meanings are:

1 Area treated in each pass over field by AG-aircraft, usually without deliberate overlap.

2 Area covered in one pass over target by SAR, camera or similar sensor, invariably with overlap.

sway   Lateral movement without rotation (simulator).

sway braces, swaybraces   *1* See *crutches*.

2 Additional struts, not normally part of aircraft, required to brace particular large or winged store.

sway space   Clear space left around any shock-mounted item.

SWC   *1* Special Weapons Center (Kirtland AFB, NM; USAF).

2 Sky wave correction (Loran).

3 Solar wind composition.

4 Space Warfare Center (USAF, Schriever AFB, Colorado).

5 Strategic Warfare Center.

6 Spot-wind chart.

SWCL   Short-wavelength chemical laser.

SWD   Surface-wave device.

SWE   *1* Stress wave emission.

2 Software engineering.

sweat cooling   See *transpiration cooling*.

sweating, sweated joint   Joining two tinned members without additional solder or brazing metal.

sweep   *1* Sweepback or sweepforward.

2 Total angular movement of aerial, eg surveillance radar, oscillating in azimuth (sector scan).

3 Total movement, normally expressed in linear measure, of time-base spot scanning across CRT or other display.

4 One complete cycle of VG wing.

5 Angular deviation of locus of centroids of propeller blade sections from radial line tangential thereto at propeller axis projected on plane of rotation (BS).

6 Offensive tactical mission against surface targets, normally targets of opportunity (WW2).

7 To range over continuous (usually large) band of frequencies.

8 To employ technical means to uncover covert surveillance devices (DoD).

sweepback   Visibly obvious backwards inclination of aerofoil from root to tip so that leading edge meets relative wind obliquely. This is usually done to increase critical Mach number.

sweepback angle   Angle between normal to longitudinal OX axis (axis of symmetry in most aircraft) and reference line on aerofoil, normally 0.25 (one-quarter) chord line or, less often, leading edge; both normal line and reference line lie in same plane, which is usually that containing centroids of aerofoil sections from root to tip (thus for canted verticals, ** measured in plane of each surface).

sweepforward   Visibly obvious forwards inclination of aerofoil from root to tip so that leading edge meets relative wind obliquely; hence * angle, or forward-sweep angle.

sweeping   Modifying wing or tail to incorporate *sweep* (1).

sweeping check   Confirming that cockpit flight controls [inceptors] move freely over full range of travel.

sweep jamming   To emit narrow band of jamming able to sweep (7) back and forth over wide operating band of frequencies.

sweep oscillator   Signal generator whose frequency is varied periodically by fixed amount at constant amplitude above and below central fixed frequency; also called Wobbulator (UK), sweep generator (US) or scanning generator.

sweep-tip blade   Helicopter rotor (main or tail) blade whose locus of centroids is radial from root to near tip and then sharply inclines back.

sweet spot   Condition in which aircraft can maintain precise altitude, heading and speed without pilot input or autopilot.

swept   Incorporating sweepback (never used of forward sweep).

Swerve   Sandia winged energetic re-entry vehicle experiment.

SW/FR   Slow write, fast read.

SWG   *1* Standard wire gauge (UK); standard range of sheet thicknesses.

2 Square waffle grid (space structures).

Swift   *1* Standoff all-weather radar for inflight terrain surveillance.

2 Specification of working position in future ATC.

SWIM, Swim   System-wide integrity management (USAF).

Swims   Shallow-water influence minesweeping system.

swing   *1* Involuntary and often uncontrollable divergent excursion from desired track of tailwheel-type aeroplane running on ground.

2 To turn propeller by hand to start piston engine; if not engaged in starting engine, or with turboprop, term is to hand-turn or pull-through.

3 To calibrate compass deviation by recording its value at regular intervals, usually 15°, during 360° rotation of aircraft on compass base.

4 Distortion of radio range; also called night effect.

5 Sudden yaw of aeroplane consequent upon loss of power of engine mounted away from centreline.

swing-by   Close pass of planet or other celestial body by spacecraft on Grand Tour.

swing force   Aircraft or complete combat unit can fly air/air and air/ground in same mission.

swinging base   Compass base; also called deviation clock.

swinging compass   Magnetic compass used as standard for calibrating that in aircraft.

swing-piston engine   Various topological families of piston engine in which two, three or more pistons oscillate around toroidal cylinder alternately compressing mixture between them, being driven by firing strokes or, in some, acting as pumps. Most do not have mechanical drive but supply gas, eg to drive turbine.

swing-role   Often interpreted differently from swing force in that aircraft and crews can fly offensive missions or (on different occasions) defensive missions.

swing-wing aircraft   Aeroplane with variable sweep (1); also called VG aircraft (colloq.).

Swinter   Study of women in non-traditional environments and roles, which included military pilot training (Canada).

SWIP   Super weight-improvement program (US).

Swipe   Simulated weapon-impact predicting equipment.

SWIR   Short-wave IR.

swirl   *1* Gross rotation of flow about axis approximately aligned with flow direction, eg in propeller slipstream, upstream or downstream of turbine, downstream of gas turbine fuel nozzle, or induced by large drive fan in low-speed tunnel (removed by straighteners).

2 Rotation of air in whirling-arm room or other non-evacuated chamber containing high-speed rotating object.

swirl vanes   Fixed aerofoils for imparting *swirl* to a fluid flow. At inwards-radial entry to a centrifugal compressor they are ususally miniature wings parallel to axis of rotation. In turbine-disc cooling air and upstream of fuel burners they are radial.

SWIS   *1* Stall-warning and identification system.

2 Satellite weather information system.

swishtail   See *fishtail*.

SWIT   Software integration and test.

switch   A specialized aerospace meaning is the rejection of the target by a missile seeker, which instead locks on to a decoy; this may be followed by response.

switches off   Traditional verbal command in hand-starting piston engine to ensure that ignition is inoperative at start (actually switches normally closed, short-circuiting HT).

switches on   Seldom used; normal call is "contact".

switch-in deflector   For jet lift, shuts off normal jet nozzle and diverts flow through rotatable side cascade (1957-70).

switching system   Automatic switching in large network, eg military communications, airline reservations or nationwide computer link; normally electro-mechanical pre-1960 and electronic later, allowing for on-line, real-time messages, data transfer, storage or display.

switchology   Fluency in human interaction with operating systems (colloq.).

swivelling engine   Entire engine, or liquid-rocket thrust chamber, that is gimbal-mounted or pivoted so that thrust axes can rotate relative to vehicle.

Swizz   SWIS (colloq.).

SWL   Strategic-weapon launcher.

SWLAN   Secure wireless local area network.

SWMCM   Shallow-water mine countermeasures.

SWO   Station Warrant Officer (RAF).

Sword   *1* Stand-off all-weather observation and reconnaissance drone.

2 System for all-weather observation by radar on drone.

3 Short-range missile defence with optimised radar distribution (US).

SWOS   Synoptic weather observing (or observation) station.

SWPA   South-West Pacific Area (painted on many captured Japanese aircraft 1944-45).

SWPC   Small War Plants Corporation (US, 1942–50).

SWR   *1* Standing wave ratio.

2 Surface-wave radar.

SWS   *1* Standard warning system (CAA).

2 Strategic Weapons School.

SWSL   Supplemental Weather Service location.

SWTDL   Surface-wave tapped delay line.

SWTL   Surface-wave transmission line.

SWTRR   Software test readiness review.

SWU   Switching unit.

SWY   Stopway.

SX   Sheet explosive.

S_x Simplex.

SXGA   Super extra graphics array, 1,280×1,024 pixels.

SXT, Sxt   Sextant.

SXTF   Satellite X-ray test facility.

SYC   Statistical yield control.

SYCAF   Système de Couplage Automatique sur Faisceau (ILS coupler, F).

Sycep   Syndicat des Industries de Composants Electroniques Passifs (F).

Syco   Symbiotic communications (USAF/DARPA).

Syers, SYERS   Senior Year electro-optical relay [or reconnaissance] system [P^3 adds preplanned product-improvement program].

Sygong   System go/no-go.

sylphon   Stack of aneroid capsules; sometimes called * tube.

symbology   Symbols conveying meanings to human beings, the technology of their design and production and their incorporation in systems and displays; most important are alphanumerics, in various national languages, followed by more than 9,000 standard conventional symbols so far available for various technologies. About 50 different forms and variations have been agreed for HUDs, Hudsights and other weapon-aiming systems, eg simple cross or cross/ring reticles, range rings that unwind as range closes, and various aiming lines, wing bars and arrow or triangular markers.

symmetric aerofoil   Wing profile whose mean line is straight.

symmetric double-wedge   Wing profile in form of sharp-edged parallelogram, used mainly for supersonic missiles whose subsonic qualities are unimportant.

symmetric flight   Both left/right wings equally loaded.

symmetric flutter   Left/right symmetry in amplitude and direction.

symmetric immersion   Both flying-boat tip floats in water equally (rare).

symmetric instrumentation   Installation of experimental sensing equipment (such as pressure transducers) over the entire surface of aircraft such that for each sensing head on left half there is an exactly corresponding unit on right.

symmetric principal axis   See *principal axis of symmetry*.

symmetric pull-out   Pull-out from dive with wings level.

symmetric stall   Stall with wings level, longitudinal axis rotating within plane of symmetry.

symmetry check   Measurements to corrresponding L/R points from centreline.

Syname   Syndicat National de la Mesure Electrique et Electronique (F).

synchro   Generalized term for bipolar a.c. synchronous systems in which a master unit or sensor commands identical response (eg angular position) by one or more instruments or other receivers. An alternative to voltage signalling by potentiometer and digital signalling by encoder.

synchronization   *1* Commanding all aircraft engines to rotate at same speed.

2 Commanding automatic guns to fire at cyclic rate forming exact fraction of multiple of blade-passing frequency of propeller; not same as interrupter.

3 Process of adjusting timing (epoch), frequency and phase of spread-spectrum receiver's PN correlation to match those of received signal.

4 Process of preadjusting outputs of two or more control (eg FCS) channels to reduce dead-zone if operated together or switchover transient if operated separately.

synchronized aerobatics   Performed by two [possibly more] sections of same aerobatic team; see *synchro pair*.

synchronous corridor   Equatorial belt within which synchronous satellite must remain (normally describing small vertical figure-eights).

synchronous orbit   See *geostationary orbit*.

synchronous satellite   One whose rotation is synchronized with that of Earth; also called geostationary.

synchronous sighting   See *tachymetric aiming*.

synchro pair   Two aircraft which perform synchronized manoeuvres to entertain crowd while rest of team reposition.

synchrophasing   Commanding all propellers of multi-engine aircraft to rotate in step with propeller of master engine, with all blades instantaneously at same angular positions.

synchropter   Helicopter lifted by two or more rotors whose blades intermesh (suggest colloq.).

synchroscope   Instrument for giving visual indication of synchronization, or lack of it, between two or more frequencies or speeds.

syncom   Synchronous communications (satellite).

sync pulse/signal   Sync is generalized term for sychronization between TV camera and receiver, or between any raster-scan sensor and display or output, hence * is integral part of transmitted waveform to maintain lock on synchronization.

syncrude   Synthetic crude petroleum; starting point for various synthesized petroleum-type hydrocarbons.

synergic ascent   Following synergic curve.

synergic curve   Trajectory for departing spacecraft for minimum energy requirement, ie lowest propellant consumption for given position and velocity; on Earth initially vertical to leave denser atmosphere quickly and then curving to take advantage of Earth's rotation.

synergism   Favourable interrelationship between variables such that overall benefit of a change is greater than sum of individual gains; eg scaling down aircraft size has synergistic effects on structure weight, drag, engine size, fuel consumption and fuel mass for given range.

Synjet   Non-petroleum-based fuel produced to current (or broadened) Jet A specification.

synodic period   Interval of time between identical positions of celestial body in solar system measured with respect to Sun.

synodic satellite   Hypothetical Earth satellite located on Earth/Moon axis at 0.84 lunar distance from Earth.

SYNOP   Special meteorological report.

synoptic chart   Standardized map of weather, showing isobars, fronts and weather symbols, and covering large area for one particular time.

synoptic meteorology   Collection of meteorological information covering large area at one time (as near as possible to present), esp. with view to forecasting.

syntactic foam   Composite material consisting predominantly of premanufactured hollow microspheres embedded in resin; for radomes usually 30-140 μ spheres with 1.5 μ wall, in epoxy or polyimide matrix.

synthetic-aperture radar   Various methods of summing the returns from many locations [eg, at TAS 600 kt, spacing 1 kHz gives returns from 1 ft apart, so a block of 50 gives definition equal to 50-ft antenna].

synthetic lubricants   Post-1948 families of turbine oils originally based on esters of sebacic acid, especially dioctyl sebacate; later with complex thickeners added.

synthetic resins   Too numerous to outline, but mainly polymers or copolymers and often thermosetting; used as bases for many materials (eg plastics and paints) and as adhesives, including nearly all those for aerospace bonding and for fibre-reinforced composites.

synthetic rubber   Vast family of rubber-like materials originally (1917) based on isoprene and today nearly all based on copolymers of butadiene; includes many solid propellants.

synthetic training   All training that simulates, eg with simple Link, mimic boards, system rigs, air-combat simulators and, esp., flight simulators. Also generally held to include actual flight training when something, eg absence of external vision, is simulated.

SyOP   Security operating procedure[s].

Syrca   Système de restitution de combat aérien (F).

syrup   Generic term for accumulation of spilled beverages, water [and condensate], toilet and lavatory fluids and dust.

sys   System, or system identifier.

Syscat-B   System Category B (FAA format).

Sysci   System configuration item.
Sysco   System co-ordination (ATC).
Sysop   Systems operator.
SYSPO   Systems Command Program Office[r] (USAF).
system   *1* Generalized term for any dynamically functioning organization of man-made devices.

2 Portion of vehicle, eg aircraft, missile, etc, forming integral network of related and inter-controlled devices to accomplish set of specif. related functions.

3 Composite of equipment, skills and techniques capable of performing and/or supporting operational role (USAF).

4 Often used to mean (1) plus supporting equipment, documents, training devices and all other products and services, as in weapon *.

5 Incorrectly used to mean mere assemblage of mechanical parts, eg engine LP *.

system concept   Integrated approach to design, procurement or operation of system in sense (4).

system discharge indicator, SDI   Yellow disc or blow-out plug in aircraft skin to indicate fire-extinguishing system discharged for reason other than fire or overheat warning.

Système International   In full, SI d'Unités, system of unified units of measurement adopted by all principal industrialized countries since 1960 and gradually being implemented; seven base units, metre (m), kilogram(me) (kg), second (s), ampere (A), kelvin (K), candela (cd) and mole (mol).

system(s) engineering   Not briefly definable, but extremely broad discipline akin to operations research whose main objective is to apply broad overview of entire system (4) in order to advise customer and/or management on objectives and possibilities and refine and integrate all subsystems before start of hardware design.

system source selection   Selection by government of industrial source, known as system prime.

system turnover   Formal acceptance by customer of responsibility for system (4).

Systo   Systems Command program officer (USAF).

Systrid   Powerful 3-D CAD technique (Battelle).

syzygy   Point on orbit, esp. that of Moon, at which body is in conjunction or opposition.

SZH   Schweizerische Zentrale für Handelsförderung (Switzerland).

T

T *1* Temperature, esp. absolute.

2 Tesla[s], nT being more common.

3 Prefix tera, 10^{12}.

4 Aircraft designation, transport (USA 1919–24), torpedo [carrier] (USN 1922–35), transport (USN 1927–30), trainer (RAF, RN, since 1941, USN, USAF, since 1948).

5 Modified mission, suffix, two-seat trainer version (USN 1946–62), prefix, trainer version (USAAF, USAF, 1943 to date).

6 Time of day, or in countdown.

7 Transmitting, or transmits only.

8 Thrust (F preferred).

9 Torque.

10 Period.

11 Kinetic energy.

12 Aircraft has transponder with 64-code but no altitude encoding.

13 Short ton[s], usage usually confined to NW yield.

14 Navaid terminal frequency.

15 Designation suffix, turbocharged [piston engine].

16 Designation prefix: turboprop.

17 Tropical air mass.

18 Class: light autogyro (BCARs).

19 Twin-wheel landing gears.

20 Terrain-clearance altitude.

21 Airport terminal, followed by number.

22 Tracer.

23 Radar-target strength.

24 Electrochemical transport number (also t).

25 Threshold lights, lighting.

26 See basic T.

27 True [headings].

28 Designation suffix: fully heat-treated.

29 A small T-shaped hangar for one [usually small private] aircraft.

30 Training, also (T).

t *1* Elapsed time in seconds.

2 Tonne[s].

3 Thickness, esp. maximum of aerofoil.

4 Triode.

5 Turns.

6 One revolution (F), as in t/min = rpm.

7 Often, non-absolute temperature.

8 Threads, eg in tpi (per inch).

9 Trend landing forecast.

10 As subscript, total pressure.

T_0 *1* Free-stream total temperature.

2 Absolute temperature.

^0T True heading.

T+ Severe thunderstorm.

T_1 Compressor or fan inlet total temperature.

T^2 Total time.

T^2A Total-terrain avionics.

T^2CAS *1* Taws-TCAS; the box still fits a single LRU.

2 Traffic and terrain collision-avoidance system.

T^3 Tailplane trimming tank.

T_3 HP turbine inlet temperature (can have other meanings).

$T_{4,5}$ Common US usage for power-turbine inlet temperature.

T_6 or T_7 Total turbine exit (outlet) temperature.

T-bridge Telescopic, usually means apron-drive.

T-channel TDMA channel for air/ground data messages too long for R-channel (GPS).

T-layer Notional top of the ionosphere.

T-rail Standard longitudinal floor rail tailored to receive pax seats, cargo tie-down rings or stretcher racking.

T-section Shaped like T or inverted T in cross-section.

T-Stoff HTP plus a little oxyquinoline or phosphate stabilizer (G).

T-tail Aeroplane tail with horizontal surface mounted on top of fin (vertical stabilizer).

T-time Reference time in countdown, often that for start of engine ignition.

TA *1* Trunnion angle (INS).

2 Telescoped ammunition.

3 Twin-aisle (transport).

4 Target alert, or acquisition.

5 Terrain avoidance.

6 Tuition assistance.

7 Transition altitude (CAA).

8 True altitude.

9 Trainer, aircooled (USA 1919–24).

10 Ambient temperature.

11 Terrain-clearance altitude.

12 Traffic advisory.

13 Technical assessment.

14 Terminal automation.

15 Towed array.

16 Technischen Ausschuss (DAeC).

Ta Tantalum.

T_a *1* Actual temperature.

2 Radar antenna noise.

3 Tropical Atlantic.

t_a Action time.

TAA *1* Transport Association of America.

2 Technical Assistance Agreement.

TAAATS The Australian advanced air-traffic system.

TAAF Test, analyse and fix.

TAAM *1* Terminal-area altitude monitoring.

2 Tactical (or Tomahawk) airfield attack missile.

3 Total airspace and airport modeller.

TAAP Tethered-aerostat antenna program.

TAATD Target acquisition advanced technology demonstration.

TAB Technical Assistance Bureau (ICAO).

Tab Tabulation and insertion program.

tab *1* Hinged rear portion of flight-control surface used for trimming (trim *), to reduce hinge moment and increase control power (servo *, balance *, spring * etc) or to increase hinge moment and effectiveness of powered surface (anti-balance *, flap *).

2 Rarely, auxiliary aerofoil hinged to trailing edge of flight-control surface, usually of servo type.

3 Small hinged spoilers on inside of propelling nozzle or

rocket expansion nozzle for thrust vectoring or noise reduction.

4 Hinged panels along lower edges of side flaps (walls) of ejector-lift duct, deflected to give maximum augmentation.

5 Abb. tabulator, alphanumeric display.

tabbed flap   Flap, usually Fowler, whose trailing edge is hinged (without slot) and at maximum landing setting is pulled down by linkage to much greater angle than main surface.

tablock   Flat washer with rectangular tab bent up to lock nut (sometimes capital T).

TABMS   Tactical air battle management system.

Tabs   Telephone automated [also total avionics] briefing system, or service.

tabulated altitude   Altitude of celestial body read from a table.

Tabun   Lethal nerve gas (G, 1937; in US called GA).

TABV   Theatre airbase vulnerability.

Tab-Vee   Alternative form of TABV, common name for NATO hardened shelter for aircraft.

TAC   *1* Tactical Air Command (USAF, now part of Air Combat Command).

2 Trim augmentation computer.

3 Turbo-alternator compressor (Brayton cycle).

4 Thermosetting asbestos composite.

5 Terminal-area chart.

6 Test-access control.

7 Thrust-asymmetry compensation.

8 Total accumulated cycles (especially USAF).

9 True-airspeed computer.

Tac   Usually tactical.

Tacamo   Take charge and move out (USN airborne v.l.f. strategic communications system, mainly NCA to SSBNs).

Tacan   Tactical air navigation; UHF R-theta-type navaid giving bearing/distance of aircraft from an interrogated ground station.

Tacbe   Lightweight tactical beacon equipment; 2-channel ground/air voice (Burndept).

TACC   *1* Tactical Air Control Centre (RAF).

2 Theater Air Control Center (USAF TAC).

3 Tactical air combat cycle.

4 Tanker Airlift Command, or Control, Center (USAF).

Taccar   Time-average clutter coherent airborne radar.

Taccims   Theatre automated command and control information management system.

TACCO, Tacco   Tactical co-ordinator; flight-crew member of ASW aircraft responsible for overall mission management during search or attack operations.

TACCS   Tactical air command and control specialist.

Tacdew   Tactical combat direction and EW.

TACDS   Threat-adaptive countermeasures dispensing, or dispenser, system.

TA/CE   Technical analysis and cost estimate.

Taces   Tactical communications exploitation system.

Taceval   Tactical evaluation, hardware development, test and training effort on production air weapon systems (USAF).

Tacfax   Tactical digital facsimile.

TACG   Tactical air control group (USAF).

TacGA   Tactical ground-to-air.

tach   Tachometer, pronounced tack, as are following entries.

tachogenerator, tachometer generator   Tachometer of electric synchronous-motor type with shaft-driven generator feeding AC to one or more synchronized displays.

tachometer   Instrument for indicating speed of rotating shaft, in rpm and/or as percentage of normal maximum.

tachometer cable   Rotary drive of mechanical tachometer, two spiral layers of steel wire.

tachometric   Tachymetric.

tachymetric aiming   Aiming of gunfire, bomb or other weapon by continuously maintaining sightline on target, thus determining speed relative to surface target and in some cases track through target.

TACIU   Test access control interface unit.

Tacjam   Tactical jamming of UHF/VHF communications.

tack   Degree of stickiness in prepreg resin.

tack coat   Very thin coat of finish, usually dope, which precedes the full-density wet coat.

tack rag   Soft lint-free rag slightly damp with thinner.

tack weld   Small dab of weld metal making local link to hold parts in correct location (but capable of easy rupture if in error) while main weld is made.

TACL   Tactical all-weather collection at long range.

TACLS, Tacls   Tactical airborne combat laser system[s] (USAF).

Tacmet   Tactical countermeasures evaluation trainer.

TACMS   Tactical missile system (USA).

Tacnav   Tactical navigator, or navigation; mod or /mod adds modification.

Tacom   Tactical-area communications.

Tacon   Tactical control.

Taconis oscillation   Pulsating (c 1 Hz) resonance in cryogenic refrigerant.

Tacor   Threat-assessment and control receiver (ECM).

Tacos   Tactical airborne countermeasures or strike.

TACP   Theatre [or tactical] air control party.

TacR   Tactical reconnaissance.

TACS   *1* Tactical, or theater, air control system (USAF).

2 Less often, tactical air control squadron.

3 Thruster, attitude-control system.

Tacsatcom   Tactical satellite communications (DoD).

TACSI, Tacsi   TACS (1) improvements.

Tacsim   Tactical simulation.

TACT   Transonic aircraft technology (NASA).

tact   Tactical air-traffic flow management.

Tactas   Tactical towed-array sonar.

Tactass   Tactical towed acoustic-sensor system.

Tactec   Totally advanced communications technology (RCA).

tactical   Generalized term meaning concerned with warfare against directly opposing forces, usually involving air, land and sea forces together, and in limited theatre of operations.

tactical aeromedical evacuation   From combat zone to outside it, or between points in combat zone.

tactical air combat cycle   Standard fighter mission assumed in determining engine life.

tactical air command center   Theatre HQ of USMC air operations.

tactical air control centre   Principal centre, shore or ship-based, from which all tactical air is controlled.

tactical air co-ordinator   Directs, from aircraft, air close support of surface forces.

tactical aircraft shelter   Normally protects against conventional attack but may be extended to offer protection against NW blast, radiation and CBW.

tactical air officer (afloat)   Responsible under amphibious task-force commander for all supporting air operations until control is passed ashore.

tactical bomb line   See *bomb line*.

tactical code   Two-digit number in various colours on combat aircraft (R, CIS).

tactical finish   Camouflaged: can be all one colour.

tactical input segment   Subsystem for receiving EO and IR images in real time.

tactical intervention vehicle   Designed to rescue hostages from parked aircraft.

tactical laser weapon system   An array of mirrors aim powerful laser simultaneously at multiple munitions.

tactical targeting network technologies   Creates networks between airborne platforms passing data at ≤ 2 Mbit/s over distances ≤ 100 nm, 185.3 km (USAF).

Tactifs   Tactical integrated flight system.

tactile faceplate   Electronic display screen sensitive to fingertip touch for reprogramming, selecting from menu, changing scale or operating mode, or adjusting any variable.

tactile situational awareness system   A high-tech aircrew vest.

Tacts   Tactical aircrew combat training system (ACMI).

TAD   *1* Turbo-alternator drive.

   2 Target assembly data.

   3 Technology availability date (or data).

   4 Towed aerial decoy.

   5 Theater, or tactical, air defense.

   6 Trim-aid device.

   7 Target acquisition and designation.

   8 Terrain-awareness display.

TADC   Tactical air direction center.

Tadds   Target alert data display sets; part of FAAR.

tadec, TADEC   Totally automatic digital engine control [piston engines].

Tadil   *1* Tactical digital intelligence, or information, link (C adds command, J adds joint [service]).

   2 Tactical aircraft digital information link.

TADIRCM   Tactical aircraft directional, or directable, IRCM.

TADIXS-B   Tactical data information exchange system – broadcast.

Tadjet   Transport, airdrop, jettison.

TADMS   TR-1 Asars-2 data manipulation system.

Tadoc   Transportable, or tactical, air-defense operations center, possible confusion with Tradoc.

tadpole   Track of moving target on radar display presented with comet-like tail to show direction of travel. Most air-defence radars can select tadpoles on or off.

tadpole profile   Aerodynamic profile with conventional nose followed by single-surface construction downstream (eg fin of A-4 followed by single-skin rudder).

TADS, Tads   *1* Tactical air defense sight (US).

   2 Towed angular deception system.

   3 Target acquisition and designation sight, /PNVS adds pilot's night-vision system.

   4 Target airborne data system.

   5 Tactical laser and designation system.

TAE   Thrust-augmented entomopter.

TAEL   Turning-area edge light[s].

TAEM   Terminal-area energy management.

TAERS   Tactical aircrew eye respiratory system.

TAF   *1* Tactical air force.

   2 Terminal area , or aerodrome, forecast (international meteorological figure-code).

   3 Thermal acoustic fatigue.

TAFI   Turn-around fault isolation.

TAFIIS   TAF (1) integrated information system.

Tafim   Technical architecture framework for information management.

TAFS   Airfield meteorological forecast.

Tafseg   Tactical air force systems engineering group.

TAG   *1* Telegraphist/air gunner (Royal Navy, WW2).

   2 Thrust-alleviated gyroscope.

   3 Tactical Airlift Group (USAF).

   4 Telescoped-ammunition gun.

   5 Transport Air Group (USMC).

   6 Tailored air group.

   7 Target-adaptive guidance.

   8 Towed acoustic generator.

   9 Technical Advisory Group (USAF).

   10 Test analysis guide.

Taggart   Sometimes rendered Taggent, a tagging agent incorporated in a strike weapon and released on detonation for detecting and tracking biological aerosols.

TAG   Telegraphist/air gunner.

tagging   Attaching unmissable warning notice during maintenance to point out, e.g., that item has been switched off or disabled.

Tags   Technology for automated generation of systems.

Tagwes   Target weapons effects simulation, or simulator.

TAH   Transfer and hold.

TA/H   Twin altitude/height.

TAI   *1* Total active inventory.

   2 Thermal anti-icing.

TAIC   Transport Accident Investigation Commission (NZ).

tail   *1* Rear part of aircraft, where applicable.

   2 Assembly of aerofoils whose main purpose is stability and control, normally located at rear of aerodyne or airship.

   3 Trailing luminous area behind blip of moving target.

   4 Normal verb meaning in air-intercept shadowing from astern.

tail boom   Tubular cantilever(s) carrying tail (2) attached either above short fuselage nacelle or as L/R pair to wings.

tail bumper   Projecting or reinforced structure under tail designed to withstand impacts and scraping on runway.

tail chute   See *tail parachute*.

tailcone   Conical fairing of rear of body, esp. downstream of turbine disc in jetpipe.

tail damping power factor   Numerically the product of TDR [see next] and URVC.

tail damping ratio   A [suggested limited] measure of antispin quality based on side area under tailplane multiplied by distance to c.g.

tail drag   Restraining mass free to slide on ground to which moored airship stern is attached.

taildragger   Aircraft with tailwheel or tailskid (colloq.).

tailed delta   Aircraft with delta wing and horizontal tail.

tail-end Charlie *1* Formation of aircraft in single line, each behind the other.

2 Last aircraft in such a line.

3 Rear gunner in tail of large aircraft (1935–50).

taileron Single-piece horizontal tail surface, one of two forming tailplane whose left/right halves can operate in unison (as tailplane commanding pitch) or differentially (as ailerons commanding roll). Term preferable to ailevator or rolling tailplane. Elevon differs in that it is hinged to wing. US term stabilator is ambiguous and can mean * or slab tailplane.

tailets Small fixed fins on underside of tailplane near each tip.

tailfeathers Free-floating flaps forming periphery of supersonic airbreathing propulsive nozzle, usually as outer boundary of large secondary nozzle. These take up slipstreaming angular positions aligned with streamlines.

tail fin Fixed stabilizing fin at rear, 'tail' normally being redundant.

tail-first Aerodyne configuration in which the only auxiliary horizontal surface is ahead of the wing, commonly called a foreplane or canard.

tail float Float supporting tail of float seaplane (now arch.).

tail group Complete tail (2), considered as design task or as element of total aircraft mass.

tail guy Secures tail of moored airship, often to tail drag.

tailheaviness Condition in which aircraft rotates nose-up unless prevented.

tailhook *1* CTOL by carrier aircraft.

2 Naval pilot (colloq.).

tailless aircraft Normally applied to aeroplanes and gliders only and usually meaning that there is no separate horizontal stabilizing or control surface, though there may be a vertical tail (2). In extreme case there is no tail surface, and (esp. if fuselage vestigial or absent) this is more often called flying-wing aircraft).

tail load Vertical up or down thrust acting on tailplane.

tail logo Bold logo of operator displayed on tail; hence ** light, also valuable as anti-collision beacon.

tail number See *serial (2)*.

tailored fuel Synthesized to meet specific operational specification.

tail parachute Parachute attached to tail, normally for anti-spin or anti-superstall purpose. Not used as braking parachute.

tailpipe Exhaust pipe of turboprop or turboshaft; according to some, piston engine exhaust pipe downstream of collector or manifold.

tailplane *1* More or less horizontal aerofoil at tail of aerodyne (invariably fixed-wing) providing stability in pitch; fixed or adjustable only for trim, and carrying elevators (US = stabilizer).

2 Aerofoil pivoted at tail about horizontal axis and driven directly by pilot of fixed-wing aerodyne or rotorcraft as primary flight control in pitch in translational flight; forms complete surface without separate elevators (US = stabilizer).

tailplane tank Fuel tank, invariably integral, in horizontal tail, to increase system capacity and, esp., to control longitudinal trim without drag.

tail rotor Helicopter anti-torque rotor, rotating at tail about more or less horizontal axis. Not used for rear tandem rotor.

tail setting angle The acute angle between chord lines of wing and tailplane (1).

tailsitter VTOL aerodyne whose fuselage is approximately perpendicular at takeoff, in hovering mode and at landing, today preferably called Vatol.

tailskid Projection supporting tail of aerodyne on ground, esp. one whose c.g. is well aft of main landing gear.

tailskid shoe Replaceable pad on end of tailskid which slides on ground.

tailslide Transient flight condition of fixed-wing aerodyne in which relative wind is from astern, eg in stall from near-vertical climbing attitude.

tailspin Spin (arch.).

tailstander *Tailsitter*.

tailstrike Scraping rear fuselage on runway on rotation. Hence, * indicator, frangible foil which causes a bright flash on EICAS.

tailstrike protection Any of several systems which prevent a tailstrike, usually by limiting authority of horizontal tail.

tail surface Any aerofoil forming part of tail (2).

tail undercarriage Rearmost unit of tailwheel-type landing gear (rare, suggest arch.).

tail unit Complete tail (2) of horizontal, vertical and/or canted surfaces, often including ventral fins or strakes. Also called empennage.

tail view Tail-on view showing object from directly astern; not normal aspect for layout drawing.

tailwagging *1* Lateral flexure of fuselage.

2 Flat turns, esp. to steepen glide.

tail warning radar Aft-facing radar, usually of active type, intended to detect other aircraft (and possibly SAMs) intercepting from behind.

tailwheel *1* Rear wheel of * type landing gear, supporting tail on ground.

2 Auxiliary wheel under tail of aircraft with nosewheel-type landing gear (eg Albemarle); fitted in place of tail bumper.

tailwheel landing gear Landing gear comprising left/right main units ahead of c.g. and tailwheel at rear.

tailwind Wind blowing approximately from astern of aircraft and thus increasing groundspeed.

TAIMS Three-axis inertial measurement system.

TAINS, Tains Tercom and, or Tercom-aided, inertial navigation system.

TAIR Terminal-area instrumentation radar.

TAIRCW Tactical air control wing.

TAIS *1* Tactical air intelligence systems.

2 Technology application information system (SDI).

3 Tactical airspace integration system.

4 Thermal active intervention system.

Take 5 Traffic crossing airway must maintain prescribed separation of 5 nm horizontally and 5,000 ft vertically from any GAT track in airway.

takeoff *1* Procedure in which aerodyne becomes airborne; not normally used for launch of glider (except on aerotow) or high-acceleration launch of missile or RPV, and never for any ballistic vertical-liftoff vehicle. In author's opinion verb is best as two words, noun and adjective as single word without hyphen.

2 Moment or place at which aerodyne leaves ground or water.

3 Net flightpath from brakes-release to screen height.

4 Power * for extraction of shaft power.

take off   To perform a takeoff.

takeoff boost   Boost pressure permitted for takeoff, usually 2 minute limit.

takeoff cone   Airspace occupied by aircraft in first minutes of flight.

takeoff distance, TOD   Field length measured from brake-release to reference zero (at screen); can be longer than runway and extreme limit TOD_a = entire runway + stopway + clearway = $TOR_a \times 1.5$. For multi-engine aeroplanes usually factored according to number of operative engines, thus TOD_4 or TOD_3. TOD_1 = TOD required for particular aircraft and WAT, not normally to exceed 0.87 TOD_a.

takeoff distance available   Actual distance at particular time, not necessarily length of runway.

takeoff distance ratio   TOD into wind divided by TOD downwind [with tailwind], usually expressed as percentage.

takeoff/liftoff area   Heliport area, a square with side equal to main-rotor diameter (FAA).

takeoff limit   No general meaning.

takeoff mass   Not normal term; for rocket or space launcher usually liftoff mass or launch mass.

takeoff noise   Measured on extended runway centreline 3.5 nm (strictly 6,485.5 m, but taken as 6.5 km) from brakes-release. A second reference point, not used for certification, is at side or runway opposite supposed start of run 1 nm from centreline.

takeoff power   Power authorized for piston engine or turboprop for takeoff, usually 2½-minute rating for turbine engines. In case of turboshaft, a lower rating than 2½-minute contingency.

takeoff rating   *1* Boost/manifold pressure/rpm figures authorized for piston engine at takeoff.

2 Thrust published for turbojet or turbofan at takeoff, normally achieved by engine control system rather than set directly by pilot, and subject to ATR or FTO techniques.

takeoff rocket   See *rocket-assisted takeoff*.

takeoff run   *1* Loosely, distance travelled over land or water in aeroplane or aerotow-glider takeoff to point of becoming airborne.

2 TOR, field length measured from brake-release to end of ground run plus one-third of airborne distance to screen height. TOR_a = TOR available = length of runway; TOR_4, TOR_3 are factored for engine-out cases, and TOR_r = TOR required.

takeoff safety speed V_2, lowest speed at which aeroplane complies with required handling criteria for climb-out following engine failure at takeoff.

takeoff speed   Not defined but loosely = unstick speed.

takeoff thrust   Takeoff rating (2).

takeoff weight   *1* See *MTOW*.

2 Actual weight at takeoff (2) on particular occasion.

TAKEOVER   In HUD or as caption: autopilot has disconnected.

T-AKX   Ro/Ro ship commandeered in emergency for RDF.

TAL   Transatlantic abort landing (Shuttle).

talbot   MKS unit of luminous energy; 1*/s = 1 lm.

TALC   Tactical airlift center (USAF).

TALCE   Tanker airlift control element (USAF).

TALCM   Tactical air-launched cruise missile.

TALD   *1* Tactical air-launched decoy [vehicle or mission].

2 Tactical airborne laser designator.

Taleos   Terrain-aided localization using EO sensors.

talkdown   Landing, esp. in bad visibility, using GCA.

talk-through   Facility whereby two mobile radio stations communicate via a base station.

tall aircraft   One calling for LEW technique or experience.

tall-aircraft VASI   See *T-VASI*.

Tallboy   Armour-piercing bomb, 12,000 lb [5443 kg] (RAF WW2).

Tallboy torch   Turbinlite.

Tally   Visual sighting of air-to-air target (RAF).

Tally Ho   Air-intercept code: target visually sighted. Normally followed by Heads Up or Pounce (DoD).

Talon   Theater application launch on notice.

Talon(s)   Tactical airborne Loran (system).

TALT   Tactical arms limitation talks.

TAM   Technical acknowledgement message (ICAO).

Tamda   Tactical acoustics measurement and decision aid.

TAMF   Training Aircraft Maintenance Facility (DARA St Athan).

Tammac   Tactical airborne, or aircraft, moving-map capability.

Tamps   Tactical automated, or aircraft, mission-planning system.

Tams   *1* Total-airport management system.

2 Transportable automated meteorological station.

T&B   Turn and bank.

T&E   Test, or trial, and evaluation.

tandem actuator   Has two pistons or jacks on same axis, with linear output.

tandem bicycle gear   Two main landing gears on centreline.

tandem boost   Rocket boost motor(s) mounted directly behind main vehicle, staging rearwards at burnout.

tandem clapping aerial swimmer   Small (the first was 19.5 kg, 43 lb) electrically powered aircraft with four reverse-cambered aerofoils which clap against each other in alternate pairs (NRL).

tandem-fan engine   Gas turbine with single core driving front and rear fans on common shaft projecting ahead of engine; fans can have shared inlet for conventional flight or valved separate inlet and exits for V/STOL.

tandem main gears   Two or more similar main gears in tandem on left and on right, as on C-5 or C-17.

tandem rotors   Helicopter lifted by two (usually identical but handed) rotors, designated front and rear.

tandem seating   One behind the other, in combat aircraft usually with rear seat at higher level.

tandem vehicle   One assembled from portions, eg stages, assembled in tandem and staged axially, in contrast to strap-on, lateral or other configuration such as Space Shuttle.

tandem-wheel gear   Two or more similar wheels in tandem on one leg, ie not a bogie.

tandem-wing aircraft   One lifted by two wings in tandem, neither bearing more than 80% of total weight.

T&M   Time and material [contract].

T&O   Training and operations.

T&R   Training and Readiness (USN).

T&S   Turn and slip.

Tanegashima   Principal Japanese satellite launch facility.

tangential ellipse   See *Hohmann*.

tangential landing   Running landing by rotorcraft or VTOL.

tangent modulus   Slope of tangent to stress/strain curve at any point.

tangent of camber   In aerofoil profile, line drawn tangent to mean camber line at intersection with leading edge, in modern wings occasionally negative (sloping up to front).

TANGO, Tango   Technology application to the near-term business goals and objectives, 34 partners in 12 countries with part-EU funding.

Tango   *1* Standard ground position marker in shape of T (RAF).

2 Turbulent (Airmet advisory).

tank   Container of all fuel, liquid, propellant, lube oil, hydraulic fluid, anti-icing fluid or toilet chemical, and often gun ammunition; not used for containers of breathing Lox, air-conditioning refrigerants, potable water or supressant/extinguishant.

tankage   Aggregate capacity of all tanks for particular fluid.

tank circuit   Tuned RF circuit with capacitor and inductor in parallel.

tanker   Aircraft equipped for inflight refuelling of others.

tank farm   Cluster of storage tanks, usually for fuel, at airport.

tank pump   Booster pump.

tank sealer   Various thermoplastic liquids, resistant to hydrocarbon fuels, sloshed inside integral tankage to seal all interior surface; today superior methods of wet assembly and multiple coatings are used.

tank vent   *Vent* (1).

Tans, TANS   Tactical air navigation system; airborne computer storing many waypoints and fed by other inputs (eg Doppler, magnetic heading).

tantalum   Ta, shiny metal, MPt 2,996°C, density 16.7, important in refractory alloys. A carbide is harder than diamond.

TAOC   Tactical air operations centre (RAF/Army).

TAOM   Tactical air operations module.

TAOR   Tactical area of responsibility (UK).

TAP   *1* Terminal approach procedure.

2 Air-transport regiment (R).

3 Tactical autopilot, or technology.

4 Technical, or technology, assessment programme.

5 Terminal-area productivity.

tap   *1* To bleed; hence tapping, pipe for bleed air.

2 To cut threads in drilled hole; also tool for doing this.

3 Electrical power wire connected to main conductor at point along latter.

4 Engine throttle or power lever (colloq.).

TAPA   3-D antenna pattern analyser (USAF ECM).

Tape   Total airport performance and evaluation (Euret).

tape   *1* Main meaning in aerospace is as medium for software, usually magnetic or punched paper.

2 One form of CF or other reinforced-plastics prepreg, used for layups or moulding but seldom for filament winding.

3 Pinked-edge fabric strip used for surface finishing (surface *).

tape control   Automatic control, eg of machine tool, by tape (1).

tape instrument   Cockpit instrument whose presentation is based on linear tape driven over end spools, usually in conjunction with fixed and/or movable index pointers or bars. Usually vertical, as in VSFI.

tapelayer   Computer-controlled tool for laying-up prepreg tape in manufacture of composite parts; automatically positions, starts, stops and dumps material rejected during prior editing.

tape lay-up   The parts produced by the tapelayer.

tape mission   Reconnaissance of Elint type in which digital (eg signal) or digitized pictorial information is stored on 7-track magnetic tape from which whole mission profile can be assigned to exact ground track, with each hostile emitter or other target assigned to precise location and timing.

Taper   Turbulent air-pilot environment research (1960–65 NASA-FAA).

taper   For given wing section profile * equal in plan and thickness, usually defined as straight or compound; in some aerofoils * not equal in plan/thickness so section profile changes.

tapered sheet   Thickness varies (usually at uniform rate) along one axis.

taper ratio   Normally defined as ratio of tip chord C_t to either root chord or equivalent centreline chord C_c.

taper reamer   Used to smooth and true previously tapped hole.

taper tap   Hand-turned tap (2) to initiate thread cutting.

Tapley meter   Damped pendulum in heavy stable case whose limit of swing feeds record of instantaneous or maximum vehicle deceleration; not suitable for runway friction measures.

Taps, TAPS   *1* Tercom aircraft (or tracking and) positioning system.

2 Terminal applications processor system.

3 Target analysis and planning system.

4 Twin annular pre-swirl.

taps   Throttles (colloq.).

tap test   Crude search for delamination or other flaw in composite structure, typically with a coin.

TAR   *1* Terminal-area surveillance radar (ICAO).

2 Terminal approach radar.

3 Thrust auto reduce (SST).

4 Threat-avoidance receiver; passive ECM.

5 Trials ATN router.

6 Test action request.

Tara   Terminal and regional airspace.

Tarad   Tracking asynchronous radar data.

Taran   *1* Tactical attack radar and navigation.

2 Test and repair as necessary.

Taras   Tactical [digital] radio system (Sweden).

Tarasov-Bauer   Computer-based method of smoothing out judge's scores to eliminate highest and lowest (CIVA).

TARC   *1* Transport Aircraft Requirements Committee (UK 1956–62).

2 Tactical air reconnaissance center (USAF).

Tarcap   Target combat aircraft practice (practise).

TARE, Tare   *1* Tactical air reconnaissance equipment.

2 Telemetry, or telegraph, automatic relay equipment (NATO).

tare   Unladen, without load, crew or fuel; normally used only in connection with surface vehicles, except for ULDs, where * includes linings and fittings according to specification or registered with IATA.

tare effect Forces and moments on tunnel model caused by support-structure interference.

tare weight allowance Free allowance given by IATA to shippers for ULDs not owned by members.

TAREWS, Tarews Tactical air reconnaissance and electronic-warfare support; RPV (USAF).

TARG, Targ Telescoped ammunition revolver gun.

Target Training and rehearsal generation toolkit.

target *1* Objective of air-combat mission, either in air or on surface.

2 Objective of intelligence or Elint activity.

3 Any true echo (blip) seen or radar, and object causing it.

4 Objective of any missile.

5 To insert position co-ordinates of fixed surface * into guidance software of ballistic or cruise missile; also called targeting.

6 Unpiloted (towed or RPV) aerodyne serving as target for friendly fire.

7 Aircraft within surveillance range of TCAS.

target acquisition Detection, identification and location in sufficient detail for effective employment of weapons.

target alert EFIS warning of future turbulence.

target allocation In air-defence weapon assignment, process of assigning particular target or airspace to particular interceptor or SAM unit (NESN).

target approach point Navigation checkpoint, usually prominent land feature similar to initial point, over which final turn in to DZ or LZ is made.

target CAP Target combat air patrol; patrol of fighters over enemy target area to destroy hostile aircraft and cover friendly surface forces.

target capture To detect, identify and locate a target in flight.

target crossing speed Relative lateral velocity or sight-line spin (angular rate) or aerial target seen from interceptor.

target date Date on which particular planned event should take place.

target designation Marking or otherwise pointing out a target, or setting it into HUD or fire-control system.

target designation control Throttle thumbswitch for slewing sight (or HUD) brackets to contain a surface target.

target director post Positions friendly aircraft, in all weathers, over predetermined geographical positions, eg targets.

target discrimination See *discrimination*.

target dossier File of assembled intelligence information on target, normally including multisensor readouts and Elint.

target drone Pilotless target aircraft, today often an RPV.

target ensemble Region of sky occupied [or expected to be occupied] by multiple air/ground munitions.

target-following radar One locked on to target.

target indicator, TI Visible pyrotechnic, electronic homing beacon or other device air-dropped on surface target.

targeting *1* To target (5).

2 Distribution of targets assigned to weapons, esp. to ICBMs and SLBMs.

target marker Visible pyrotechnic dropped on surface target, or aircraft dropping same.

target of opportunity *1* Target visible to a sensor or observer and within range of weapons and against which fire has not been scheduled or requested (DoD).

2 Target which appears during combat and which can be reached by weapons and against which fire has not been scheduled (NATO).

Note: both the above can be ground or air.

3 NW target detected after operation begins that should be attacked as soon as possible within time limits for co-ordination and warning friendly forces.

target pattern Flightpath of aircraft (meaning is normally in plan view) during attack phase (DoD).

target price That hoped to be achieved, eg in incentive-type contract.

target recognition Positive identification of type of target (eg type of aircraft), by visual means or by high-resolution sensor giving jet modulation or prop/rotor reflection signature.

target reverser Jet-engine (turbojet or turbofan) thrust reverser comprising two deflectors (also called clamshells or buckets) which swing down to meet downstream of nozzle.

target strength $T = E - (S+2H)$ where S is source, E echo and H radar transmission loss; unit is dB.

target symbol Computer-generated on display.

target tape Basic software for programming missile guidance of inertial and certain other species.

target tug Manned aircraft or RPV towing target (6) for live air/air or surface/air firing.

tarmac Colloq. UK (esp. non-aviation people) for paved apron; US = hardstand.

Tarmos Tactical radio monitoring system.

Tarms Tactical aerial resource-management study (aerial firefighting).

TARN Telegraphic auto routing network (NATO, Litton).

Tarpol Tariff policy.

Tarpos Target positional data [attack on moving surface target].

Tarps Tactical-aircraft, or air, reconnaissance-pod system (USN).

TARS, Tars *1* Tethered aerostat radar system.

2 Theatre airborne, or tactical air, reconnaissance system.

3 Tactical Air Research and Survey Office (USAF, formerly).

Tarsp Tactical air radar signal processor.

TAS *1* True airspeed.

2 Training aggressor squadron.

3 Target-acquisition system.

4 Typed air station.

5 Thallium arsenic selenide.

6 Tactical acoustic system.

7 Towed-array sonar.

8 Tracking adjunct system (SAM).

9 Targeting avionics system.

10 Traffic-avoidance system (Japan).

11 Tactical Air Squadron (Poland).

12 Transportable aerosat system.

TASA Thai Aero Sport Association.

TASC *1* Touch-activated screen (or simulator) control.

2 Technical and Air Safety Committee of GAPAN.

TASD   Trajectory and signature data.

Tasdac   Tactical secure-data communications (USAF).

taser   Hand-held NLW which delivers temporarily incapacitating shock via probes fired by nitrogen gun.

Tases   Tactical airborne signal exploitation system.

Tasets   Tactical steerable emitter threat simulator.

Tasi   True airspeed indicator (pronounced 'tarzi').

task   *1* Specific assignment to one air vehicle, or any other military force, normally involving operational or simulated mission or particular training exercise or programme.

2 Specific assignment to competitors in sailplane championships, eg speed round triangle, distance or declared goal, not disclosed prior to day.

tasked   Required to fulfil certain tasks (1), either of variable operational or routine nature. Not available for other missions.

tasking   Process of assigning tasks to available units or individual aircraft or crews to fulfil all mission requirements.

TASM   *1* Total available seat-miles.

2 Tactical anti-ship, or air-to-surface, missile.

3 Top-attack submunition.

Tasmo   Tactical air support for maritime operations.

TASR   *1* Tactical automated situation receiver.

2 Terminal airport surveillance radar.

TASS   Tactical Air Support Squadron (USAF).

Tass   *1* Tactical automated security system.

2 Towed-array surveillance system.

3 Terminal-area surveillance system (1995 onwards).

TASST   Tentative airworthiness standards for [a future] SST (FAA).

Tasuma   Target and surveillance unmanned aircraft.

Tasval   Tactical-aircraft survivability against armour; post-Jaws (USA/USAF).

TAT   *1* Total, or true, air temperature.

2 Turnaround time.

3 Tactical aircraft turret (helicopter).

TATC   Terminal ATC; A adds automation.

Tatcof   Transportable ATC facility.

TATF   Terminal Automation Test Facility (FAA).

TATI, Tati   Trim and tailplane incidence (indicator) (pronounced 'tatty').

TATP   Triacetone triperoxide, high explosive.

TATS   *1* Tactical aircraft training system.

2 Tactical Aerial Targets Squadron (USAF).

TAU   *1* Target acquisition and tracking unit.

2 Terminal access unit.

3 Threat awareness unit.

TAV   *1* Transatmospheric vehicle.

2 Total asset visibility.

TAW   *1* Thrust-augmented wing.

2 Terrain-awareness warning.

3 Tactical Airlift Wing.

TAWC   Tactical Air Warfare Center.

TAWDS   Target-acquisition/weapon-delivery system, with Pave Mover.

Taws   *1* Terrain-awareness [or avoidance] warning system [previously EGPWS, now e-TAWS].

2 Theater airborne warning system.

tax   Taxiway lights (ICAO).

TAXI   Taxi and parking facilities airfield chart.

taxi   To move aircraft on surface (land or water) under its own power.

taxi channel   Defined path for marine aircraft.

taxi-holding position   Designated point at which all vehicles may be required to hold to provide adequate clearance for arrivals/departures on runway.

taxiing   Participle/gerund from taxi; note spelling.

taxilane   Path on large apron or other paved area to be followed by nose gear, marked by continuous white line.

taxitrack   Assigned taxiing route at land airfield, not necessarily paved. Most or all may be perimeter track.

taxiway   Assigned taxiing route at land airfield, paved.

Taylor diagram   Plot of dry and saturated adiabatic curves on axes of pressure and volume (reciprocal of density) showing loss of pitot pressure in moist air.

Taylor/Maccoll   Original more exact solution for pressure over unyawed circular cone in supersonic flow (1932).

Taylor Maclaurin   Mathematical expansion of $f(x)$ for values near $x = 0$.

Taylor recorder   Automatically counted number of times a preset vertical acceleration was exceeded (RAE, 1950).

Taylor series   Power series of $f(x)$ in ascending powers of $x - a$ where $f(x)$ and derivatives are continuous near $x = a$.

TB   *1* Turn/bank.

2 Timebase.

3 Torpedo-bomber (USN, 1935–46).

4 Terminal block.

5 Heavy bomber (USSR).

6 Turbulence.

t_b Burn time.

TBA   *1* Total (helicopter rotor) blade area.

2 Test-bed analysis.

T-bar system   See *T-VASI*.

TBB   Transfer-bus breaker.

TBC   *1* Toss-bombing computer.

2 Tailored-bloom chaff.

3 Tactical bombing competition.

4 Thermal barrier coating.

TBCC   Turbojet-, or turbine-, based combined cycle.

TBCP   Telebrief control panel.

TBD   *1* To be determined, or decided.

2 Time/bearing display.

3 Trail[ing] blade damage.

TBE   Timebase error.

TBH   *1* Truck-bed height.

2 Turbine- [or thrust-] bearing housing.

TBI   Turn/bank indicator = turn/slip.

TbIG   Terbium iron garnet.

TBL   Towbar-less, i.e. not fitted with towbar.

TBM   *1* Theater battle management; CS adds core system[s], S system.

2 Tactical, or theater, ballistic missile; D adds defense, DFS defense feasibility study, EWS early-warning system (USA).

TBO   Time between overhauls.

TBPA   Torso back protective armour.

TBR   Torpedo-bomber reconnaissance.

TBRP   Timebase recurrence period.

TBS   *1* To be supplied, or specified.

2 Tailored business stream (DCAC).

TBT   *1* Turbine-bearing temperature (also used to mean turbine temperature).

2 Transonic blowdown tunnel.

TBV   Tilt-body vehicle.

TBW   Throttle by wire.

tBX   Air temperature (USSR, R, Cyrillic characters).
TC   *1* Toilet cart.
   2 Time constant, or critical [see TCAIA, TCT].
   3 Turn (or twin) co-ordinator.
   4 Thermocouple.
   5 Test-complete (verb).
   6 Type, or Technical, or Transport, Certificate (US).
   7 Top of climb.
   8 Taxiway centreline (lights).
   9 Turbocharged.
   10 Tropical cyclone, or continental.
   11 Top of cylinder.
   12 True course.
   13 Transport Canada, with many suffixes, including N (data-processing network) and TSB (technical services branch).
   14 Transformational communications (USAF).
Tc   *1* Tropical continental air mass.
   2 PN code bit length, also called chip width.
   3 Adiabatic flame temperature of rocket.
   4 Superconducting critical temperature.
T/C   Top of climb.
t/c   Thickness/chord ratio of aerofoil.
TCA   *1* Terminal control area.
   2 Télécommande automatique; IR/optical + wire guidance for missile. Operator merely keeps sight on target.
   3 Time of closest approach.
   4 Track crossing angle.
   5 Temperature control amplifier.
   6 Technical collaboration agreement.
   7 Turbine cooling airflow.
   8 Technical, or technology, concept aircraft (SST).
   9 Tungsten carbide alloy.
   10 Traffic-collision avoidance; D adds device (for GA; 1980s).
   11 Throttle-control assembly.
   12 Transformational communications architecture.
TCAA   *1* Taiwan CAA.
   2 Transatlantic Common Aviation Area.
TCAC   Tactical Control and Analysis Center.
TCAIA   Time-critical automatic identification and attack.
TCAR   Transatlantic collaborative advanced radar.
TCAS   *1* Pronounced T-cass, traffic alert and collision-avoidance system [see entry]; -RA adds resolution advisory.
   2 Tandem clapping aerial swimmer.
TCB   *1* Turret control box (helicopter).
   2 Trusted computer, or computing, base.
TCBM   Transcontinental ballistic missile.
TCC   *1* Thermal-control coating.
   2 Thrust control computer.
   3 Telecommunications center(s).
   4 Titanium-coated carbon.
   5 Turbine-case cooling.
   6 Tactical co-ordination console.
   7 Tactical control center.
   8 Technical co-ordinating committee.
   9 Troop Carrier Command (USAAF, WW2).
   10 Temporary Council committee.
TC³, TCCC   Tower control computer complex; S adds system.
TCCF   *1* Tactical combat control facility (USAF).

   2 Technical communication control facility.
TCCP   Take-command control panel.
TCD   Time-critical data.
TCDD   Tower-cab digital display (ATC).
TCDL   Tactical common data-link.
TCDS   Type Certificate data sheet.
TCEA   Training Centre for Experimental Aero-dynamics (NATO, Brussels).
TCF   Terain clearance floor.
tcf   Trillion cubic feet.
TCG   Troop Carrier Group (USAAF).
TCH   *1* Threshold crossing height.
   2 See *TKP*.
TCI   *1* Tape-controlled inspection.
   2 Time-controlled item.
TCIM   Tactical communications interface modem, or module.
TCIR   Toxic-chemical inventory release.
TCJ   Tactical communications jamming.
TCL   Taxiway centreline light[s].
TCLT   Tentative calculated landing time.
TCM   *1* Trim-control module.
   2 Trellis coded modulation.
   3 Throttle clutch motor.
   4 Trajectory-correction manoeuvre.
   5 Technical co-ordination meeting.
   6 Transformational communications milsatcom, or military [also called TCS, TSAT].
TCMA   Time co-ordinated multiple access.
TCML   Target co-ordinate map locator.
TCMS   Test-content management system.
TCN   Tacan.
TCO   *1* Total cost of ownership.
   2 Tape-controlled oscillator.
   3 Tactical Control Officer.
   4 Tone cut-off (noise reduction).
TComSS   Telephonics communications management system.
TCP   *1* Transfer-of-control point.
   2 Transmission control program, or protocol (Autodin).
   3 Tri-cresyl phosphate.
   4 Thrust centre position (of gross thrust vector).
   5 Takeoff-chart computation program.
TCPA   Time to closest point of approach.
TCPED   Tasking, collection, processing, exploitation and dissemination.
TCP/IP   TCP (2) internet protocol.
TCQ   Throttle control quadrant.
TCR   *1* Terrain-closure rate.
   2 Thickness/chord ratio.
   3 Time-compliance requirements.
TCS   *1* Tilt-control switch (tilt-wing V/STOL).
   2 TV camera set (F-14).
   3 Tracking and communications subsystem (ACMI).
   4 Telemetry and command system (satcom).
   5 Turbulence-control structure.
   6 Tactical-control system, or squadron.
   7 Trusted computer system, meeting requirements for secure access.
   8 Tactical-command system.
   9 Touch-control steering.
   10 Target control set.
   11 Total-component support.

12 Troop Carrier Squadron (USAAF).

TC/s Teracycles per second, = THz.

TCSC Titanium-coated silicon carbide.

TCSEC Trusted-computer system evaluation criteria.

TCSS Terminal communications switching system.

TCT *1* Time-critical target, or targeting; A adds aid.

2 Tactical computer terminal.

3 Transverse-current tube.

4 Target-centred tracker.

5 Takeoff configuration test.

6 Turbomachinery and combustion technology.

7 Targeting-cycle timeline = S2S.

TCTO Time-compliant, or compliance, technical order.

TCTT Time-critical target technology.

TCU *1* Tracking control, or and communications, unit.

2 Thermal cueing, or control, unit.

3 Tracking and communication unit (UAV).

4 Take-control unit.

5 Tacan control unit.

6 Telephone conversion unit.

TCU, TCu Towering cumulus.

TCV *1* Total-containment vessel.

2 Terminal-configured vehicle.

TCW *1* Terminal controller workstation.

2 Tactical Communications Wing.

3 Tactics and Countermeasures Wing (RAF AWS).

4 Troop Carrier Wing (USAAF).

TCX Transfer-of-control cancellation message.

TCXO Temperature-controlled crystal oscillator.

TD *1* Target drone (USN category 1942–46).

2 Touchdown.

3 Transposition docking.

4 Time difference, or delay.

5 Tunnel diode.

6 Test directive.

7 Time duplex.

8 Tactical Director (USAF).

9 Tactical display.

10 Thrust decay; S adds system.

T_d, Td, T–D Dewpoint temperature.

t_d Ignition delay time of rocket.

T/D Top of descent.

TDA *1* Tunnel-diode amplifier.

2 Temporary danger area.

3 Trade and Development Agency (US).

4 Theater-defense architecture.

5 Today.

TD&E *1* Transposition, docking and LM ejection.

2 Tactics development and evaluation.

TDAP Touchdown aim point.

TDAR Tactical defence alerting radar.

TDAS Test, or tracking and, data acquisition system.

TDATS, T-Dats Target detection, acquisition and tracking system.

TDC *1* Top dead centre.

2 Through-deck cruiser, for Stovls.

3 Technical Development Center (FAA).

4 Target designator, or designation, control.

TDCP Tactical-data communications processor (USMC).

TDCS Traffic-data collection system.

TDD *1* Tactical-related data-dissemination system.

2 Target-detection device.

TDE *1* Target-data extractor.

2 Tactical-data equipment.

TDEC Technical Development and Evaluation Center (CAA Indianapolis from 1939).

TDEFS Technology demonstrator for enhancement and future systems.

TDEU Test and data-extractor unit.

TDF *1* Tactical digital facsimile.

2 Tactical-display framework (Awacs).

TDG *1* Triggered-discharge gauge.

2 Two-displacement gyro.

TDI *1* Triple-display indicator [fluid pressure, three dial scales].

2 Tapped delay input.

3 Trade-data interchange, part of Apex.

4 Time-delay and integration (TICM).

5 Time-of-day interface.

TDL *1* Tactical data-link; PS adds processing system.

2 Tactical data-loop; S adds system.

3 Trapped delay-line.

4 Truck dock lift.

TDLS Tower data-link services, such as pre-departure clearance and D-ATIS.

TDM *1* Time-division, or -domain, multiplex.

2 Tactical-data management, or modem.

TDMA Time-division, or domain, multiple access.

TDMMS Telemetry Doppler metric measurement system.

TDMS Test-documentation, or tactical-data, management system.

TDO Tornado.

TDoA Time-difference, or delay, of arrival.

TDOP Time-dilution of precision.

TDP *1* Touchdown point; D adds dispersion.

2 Target-data panel.

3 Technology-demonstration, or development, programme, or project.

4 Three-day planning; F/C adds forecast chart.

TDPF Tail-damping power factor.

TDPS Tracking, or test, data-processing system.

TDR *1* Tail-damping ratio.

2 Takeoff-distance ratio.

3 Technical-despatch reliability.

4 Transponder [XPDR more common].

5 Traffic-data record.

6 Terminal Doppler radar.

TDRE Tracking and data relay experiment.

TDRS *1* Tracking and data-relay satellite; S adds system.

2 Technology demonstration and [or for] risk-reduction.

TDS *1* Tactical-data system.

2 Time/distance/speed scale.

3 Tactical Drone Squadron.

4 Thermal diffuse scattering.

5 Training Depot Station (RFC).

6 Target-designation sight.

7 Threat deception system.

8 Threat-detection system; -FA adds fighter aircraft, -H helicopter.

9 Thrust-decay system.

10 Terminal display system.

TDST Tower data services terminal.

TDT *1* Tactical data terminal.

2 Transonic dynamics tunnel.

TDTG   Twin delta tandem [landing] gear.

TDTS   Tactical-data transfer system.

TDU   *1* Test, or TV, or terminal, display unit.

2 Torpedo Development Unit (RAF Gosport, 1938, became ATDU).

TDV   *1* Technology development vehicle.

2 Truck dock vehicle (cargo handling).

TDWR   Terminal Doppler weather radar.

TDX   Target-data extractor.

TDY   Temporary duty.

TDZ   Touchdown zone; CL adds centreline lighting, E adds elevation, L lights, M marking.

TDZ marking   White axial stripe on each side of runway.

TE, t.e.   *1* Trailing edge.

2 Tactical evaluation.

3 Taxiway edge (lighting).

4 Table of equipment.

T/E   Twin-engine[d].

Te   Tellurium.

TEA   *1* Tri-ethyl alcohol, hypergolic igniter.

2 Transferred-electron amplifier.

TEAM, Team   *1* Training equipment and maintenance (NATS).

2 Technology for efficient agile mixed-signal micro-systems.

teaming agreement   *1* Inter-company agreement for marketing purposes, not involving licensing or co-production.

2 Now coming to mean any inter-company agreement to assist penetration of markets.

teaming down   Getting into partnership with smaller companies.

teaming up   Getting into partnership with one or more giant companies, one of which may be eventual prime.

teampack   Packaged for carrying across rough terrain by team of 2 to 8 personnel.

Teams   *1* Tactical evaluation and monitoring system (Northrop).

2 Tactical electronic-aircraft [or EA-6B] mission support system (USN/USMC).

tearaway connector   Umbilical pull-off coupling.

teardown   Dismantling into component parts.

teardrop   Standard procedure flying pattern similar to racetrack but with one end having large-radius and the other small.

teardrop canopy   Of smoothly streamlined shape, usually moulded from one transparent sheet.

tear off a strip   To deliver spoken reprimand (RAF, colloq.).

tearoff cap   Lightly sewn fabric parachute cover torn off pack by static line.

tearstrap   Doubler fastened [if possible, bonded] to skin to arrest progress of tensile crack.

tease   Faulty operation of circuit-breaker in which snap-action is absent; hence *-free.

TEB   Tri-ethyl borane.

TEC   *1* Trans-Earth coast.

2 Thermal (or thermoelectric) energy converter.

3 Thermoelectric cooler, or cooling.

4 Tower en-route control.

tech   Adjective, to go * = unserviceable (colloq.).

TECEVAL   Technical evaluation (USN).

tech mod   Technology modernization.

Technamation   Technical animation, methods for training and educational displays giving illusion of motion, eg flow through pipes, rotation of shafts, etc.

Technical Assistance Agreements   Bilateral agreements permitting disclosure of sensitive items by the US to the UK, notably concerned with LO technology (10 negotiated by early 2003).

technical delay   Delay ascribed to fault in hardware, lasting longer than 5 (sometimes 15) min.

technical despatch reliability   Percentage of scheduled flights which are unaffected by any prior technical fault, but ignoring delays due to other causes.

technical electrics   All services other than commercial electrics.

technically closed   Problem has been solved.

Technical Standard Order   Establishes quality control for avionics and other equipment; thus TSO'd items bear higher price (FAA).

technical stop   Stop by commercial transport at airport for reasons other than traffic; not shown in timetable.

technical survey   Inspection for monitoring (bugging) systems (DoD).

Techroll   Patented (CSD) configuration for vectoring nozzle of solid-propellant rocket motor in which nozzle drive forces are reduced by fluid-filled constant-volume surround sealed by flexible diaphragm.

Tecmus   Tactical ECM upgrade system.

Tecom   Test and Evaluation Command (USA, APG).

Tecos   Terminal co-ordination system.

TECR   Technical reason (ICAO).

Tecstat   Nazionale Associazione Tecnici di Stato (I).

TED   *1* Transferred-electron device.

2 Tactical (or threat) evaluation display.

3 Trailing edge down, or device[s].

4 Threat-environment description.

5 Tool and equipment drawing.

6 Trace [of] explosives detector.

TEDA   Triethylenediamine.

Tedlar   Flexible PVF film for surface protection (registered name).

TEDS   Tactical expendable drone system, for ECM saturation jamming.

TEE   Tubular extendible element, produced by un-rolling steel tape.

tee   Air/ground wind-direction indicator in shape of large T in white, either placed on ground and occasionally rotated or pivoted to base (and in a few cases moved by weather-cocking). Cross-piece is at downwind end of upright.

TE-Ebaps   Transferred-electron electron-bombarded active pixel sensor.

tee connector   T-shaped plumbing connector.

Tee Emm   Training memoranda [and excellent periodical] (RAF).

tee gearbox   One rotary shaft geared to another at 90° at a point other than one end.

tee junction   T-shaped connection of two microwave waveguides.

teetering rotor   Helicopter main rotor with two blades freely pivoted as one unit about horizontal axis transverse to line joining blade tips.

TEF   Total environment facility, for processing reconnaissance data.

Teflon   Trade name (du Pont) for large family of fluoro-carbon-resin rubbers and plastics.

Tefzel   Trade name (du Pont) for EPTE products.

TEG   *1* Tactical exploitation group (satellites).

2 Thermo-electric generator.

tehp   Total equivalent horsepower, normally same as ehp.

TEI   *1* Trans-Earth insertion (or injection).

2 Text-element identifier[s].

3 Thermocouple engine instrument.

tektites   Small glassy bodies unrelated to surrounding Earth surface and believed of extraterrestrial origin.

TEL   *1* Tetraethyl lead.

2 Telebrief.

3 Telephonic (ICAO).

4 Transporter/erector/launcher; AR adds and radar.

Telar   TEL (4) and radar, on one vehicle.

TELATS   Tactical electronic locating and targeting system (USAF).

telebrief   Direct telephone link between ground personnel, eg air controller or ground crew, and military aircrew seated in aircraft on ground.

telecommunications   Transmission, emission or reception of signs, signals, writing, images or sounds by wire, radio, visual or other EM system; abb. telecom.

teleconference   Conference between participants linked by telecom system.

Teleflex   Mechanical remote-control systems in which push/pull commands are transmitted by tube-mounted cable with complex coiled overlayers [able to drive toothed wheel].

telegraph   Telecom using succession of identical electrical pulses.

teleguided   Not a normal expression; could mean a missile guided by radio command or by wires.

telematics   This word does not appear in normal English dictionaries. It appears to mean automatic control over wide areas encompassing several systems.

telemetry   Transmission of real-time data by radio link, eg from missile to ground station; today invariably digital and important in RPVs and unmanned reconnaissance systems. Data can be pressure, velocity, surface angular position or any other instrument output, or any form of reconnaissance output. Telemeter is verb; use as noun arch. Noun is * system or telemetering system.

teleoperator   Robot for performing mechanical tasks under remote control.

telephone   Transmission of sounds, signals or images by wire or other discrete-path link, eg microwave beam or optical link using free coherent beam or fibres.

telephone box   Figurative enclosure of aircraft whose energy has decayed in air combat to point where he is low and slow and has 'no place to go'.

telephotography   Photography of distant objects on Earth.

telephotometer   Visibility meter.

teleprinter   Telegraphy with keyed input and printed written output.

teleprocessing   EDP (1) by computer fed by telecom system.

teleran   Television radar air navigation; use of ground radar to feed airborne TV display.

Telesacs   Telematics for safety-critical systems, in particular co-ordination of ACAS, STCA and precision navigation (Euret).

telescience   Increasing output of space science experi-

ments by use of Internet and broadband satellite communications to involve ground-based researchers.

telescope   *1* In astronomy, instrument for collecting EM radiation (esp. light, radio, IR and X-ray) from extraterrestrial sources.

2 To reduce overall dimensions by folding or, esp., linear retraction, eg helicopter rotor.

3 To reduce propeller diameter by cropping tips.

telescoped ammunition   Rounds in which the projectile is carried largely within the case, reducing length and increasing propellant energy per unit overall volume.

telescope gauge   Precision rod sliding in tube and locked to measured dimension, eg hole diameter, subsequently measured by micrometer.

telescoping   To telescope (2, 3).

telescramble   To render telecom, usually telephone, conversation unintelligible by scrambling.

teletype   US term for teleprinter; hence *-writer; often capital T (registered name).

television   Transmission and reception of real-time imagery by electronic means. Link usually by radio but may be any other telecom form, and imagery usually keyed to sound channel, entire received signal also being recordable by receiver; abb. TV.

television command   See *television guidance*.

television guidance   Command guidance by radio link sending steering commands from operator watching TV picture taken by camera in nose of vehicle.

telint   Telemetry intelligence.

telling   See *track telling*.

telltale   An indicator of position external to cockpit, such as rods projecting through wing skin to show landing-gear position.

tellurium   Te, semi-metal, density 6.2, MPt 450°C, metal alloys, glass, ceramics, electronics.

Telops   Telemetry on-line processing system.

TELS   Turbine-engine loads simulator.

TEM   *1* Transmission electron microscope (or microscopy).

2 Technical error message (ICAO).

3 Thermally expanded metal.

4 Illustrated tool and equipment manual.

T/EMM   Thermal and energy-management module.

TEMP, Temp   *1* Temperature (ICAO).

2 Temporary.

3 Test and evaluation master plan (AFSC).

temper   Degree of hardness introduced to metal by heat treatment, cold-working or other process.

temperature   Property of material systems, commonly called intensity of heat, determining whether they are in thermodynamic equilibrium. Normally a measure of translation kinetic energy of atoms or molecules. SI unit is K, not necessarily written °K. Specified as reported *, a local actual value, or as forecast * or declared *, read from statistical tables.

temperature accountability   All factors, in aircraft design, operation and certification determined by reduction in propulsive thrust and wing lift caused by increase in atmospheric temperature.

temperature coefficient   Rates of change of variable per unit change in temperature.

temperature coefficient of pressure   For a given ratio of solid-rocket-motor surface to throat area, $\pi_k = \delta I_n\, P_c / \delta T$

where I_n is initial specific impulse, P_c is chamber pressure and T is temperature.

temperature correction   Correction applied to bring instrument reading to STP conditions.

temperature gradient   Rate of change of temperature with unit distance through material in direction normal to isotherm surfaces, esp. rate of change of temperature in atmosphere with unit increase in height.

temperature inversion   See *inversion*.

temperature lapse rate   *Lapse rate*.

temperature probe   Sensor protruding into air-stream, giving output requiring correction to give static temperature.

temperature recovery factor   Usually, equilibrium temperature of solid surface in high-speed flow, varying according to turbulence of boundary layer; usually T_w and given by several formulae.

temperature shear   Rapid change in atmospheric temperature with horizontal or vertical travel [can cause unacceptable change in Mach whilst holding airspeed/altitude constant].

temperature stress   Stress caused by temperature, esp. changes in temperature between different parts of mono-lithic body.

temperature traverse   Series of temperature (usually stagnation/total temperature) measures taken either over area or along straight line perpendicular to fluid flow, eg at exit from combustion chamber.

Tempest   Transient EMP emanation standard.

template   Simple pattern, usually planar, either cut to shape of a part or with shape and dimensions marked on surface, used as guide in repeated marking out of desired shape. In UK occasionally written templet.

Temple-Yarwood   Formula for pressure coefficient at low Mach as function of critical Mach M_c: $C = 1 - 0.522 (1 + 0.2 M_c^2)^3 / M_c^2 (1 - 0.05 M_c^2)^2$.

Tempo   *1* Technical military planning operation.

    2 Also TEMPO, temporary or temporarily.

temporary flight restriction   Order prohibiting un-authorized aircraft from airspace above major accident, natural disaster or other event.

temporary revision   Document printed on yellow paper which temporarily amends an item in a maintenance manual [now also issued electronically].

TEMS   Turbine-engine monitoring system.

TEN   Tactical environment network (USMC).

Tencap   Tactical exploitation of national capabilities (US).

TEND   Trend forecast.

tendency   Variation with respect to time, esp. change in atmospheric pressure in 3 h period prior to an obser-vation.

Tenley   Secure voice system for Tri-Tac, for NSA (US).

Tensabarrier   Seat-belt-type barrier to control people movements at airport (BAA); quickly closed or opened.

tensile strength   Tensile force per unit cross-section required to cause rupture.

tensile stress   That produced by two external forces acting in direct opposition tending to increase distance between their points of application.

tensiometer   Measures actual tensile stress in flexible cable, such as flight-control circuit; can be used for flexible bracing wires.

tensioner   Self-contained mechanism inserted into cable

carrying tensile load, eg in manual flight-control system, which maintains desired (usually constant) tension throughout; often in form of spring-loaded quadrants.

tension field   Surface within which tensile force acts, with direction parallel to forces. Hence * beam, eg wing spar, within which * acts diagonally in vertical plane, tending to pull upper and lower booms together.

tension regulator   See *tensioner*.

tent   Quasi-conical upper compartment of long-endurance balloon housing near-spherical helium cell, the main purpose being thermal insulation.

tenuity factor   In level bombing, correction for variation with height of atmospheric density.

TEO   Transferred-electron oscillator.

TEOC   *1* Tactical electro-optical camera.

    2 Technical-objective camera.

TEORS   Tactical electro-optical reconnaissance system.

TEOSS, Teoss   *1* Tactical emitter operational support system; EW locator (USAF).

    2 Tracking electro-optical sensor suite.

TEP   Tactical electronic plot.

tephigram   Graphical plot of atmospheric temperature and entropy on grid of intersecting isothermals and isentropic lines against vertical axis of height (decreasing pressure levels); also written Tϕ gram, for temperature and entropy. Pronounced tee-fie-gram.

Tepigen   Television picture generation (or generator).

Tepop   Tracking-error propagation and orbit prediction.

TER   *1* Triple ejector rack.

    2 Total-energy requirements.

    3 Terrain-following radar (TFR preferred).

TERA   Terminal effects research and analysis.

tera   Prefix = $\times 10^{12}$, symbol T.

terabit   One trillion bits/s.

teraflops   One trillion flops = 10^{12} operations per second.

TeraGrid   Most powerful computing system, created in US under auspices of NSF by linking 3,300+ processors to give speed of 13.6+ teraflops and storage of 450+ tril-lion bytes.

terbium   Tb, soft silvery metal, density 8.23, MPt 1,356°C, importance growing.

Tercom   Terrain comparison or terrain contour-matching; navigation technique in which vehicle guidance memory compares profiles of terrain below, sensed as unique sequences of digital height measures, with those already stored; hence, each match with terrain increases accuracy of refinement of basic (eg INS) guidance, whereas most systems degrade with time.

térébenthine   Refined turpentine used as rocket fuel (F).

Terec   Tactical electronic reconnaissance (Litton).

TERLS   Thumba Equatorial Rocket Launching Station (UN facility in India).

Term   Terminates.

terminal   *1* Building, either discrete or dispersed, at airport which links airside with landside and through which all passenger traffic passes, or through which all cargo traffic passes. Very seldom is there one * for pax and cargo, and at many major hub airports each airline or group of airlines has its own *.

    2 Downtown (city-centre) building at which passengers may check in for flights and from which they may be conveyed by public transport with baggage already checked to pass straight through a * (1).

3 Normal general meaning of being connecting point through which flow passes into or out of a system in electric circuits, air traffic, data processing and many other disciplines.

4 Final portion of flight of missile between midcourse and target.

terminal airport   That at which flight terminates. Also correctly used as point at which particular item of traffic leaves flight.

terminal alternate   Alternate named in flightplan as second-choice terminal (3) if for any reason normal destination unattainable.

terminal area   See *terminal control area*.

terminal ballistics   Behaviour of projectiles at impact with, and penetration of, target.

terminal building   See *terminal (1)*.

terminal clearance capacity   Maximum amount of cargo or personnel that can be moved through terminal (3) daily (DoD).

terminal communications   Communications services or facilities within terminal control area other than those used for approach or ground movement.

terminal control area   Airspace control area, or portion thereof, normally at confluence of airways or air traffic service routes in vicinity of one or more airports. Extends from surface or from higher FL to specified FL and within it all aircraft are subject to specified rules and requirements. Often a TMA.

terminal count   Final portion of countdown ending in lift-off.

terminal guidance   *1* That governing trajectory from end of midcourse to impact with or detonation beside target.

2 Electronic, mechanical, visual or other assistance given to aircraft pilot to facilitate arrival at, operation within or over, or departure from, air landing or airdrop facility (DoD).

terminal manoeuvring area, TMA   Controlled airspace region surrounding busiest airports (usually large city with many airfields); normally permanent IFR with other traffic by dispensation.

terminal nosedive   Vertical dive at full power (arch.).

terminal phase   *1* Portion of trajectory of ballistic missile between re-entry and target. Also other meanings in particular space missions, eg LM/CM redocking in lunar orbit.

2 Final part of missile trajectory after missile's own seeker has detected and locked on to target.

terminal radar service area   Primarily an electronic environment, not extending below a floor at medium FL, providing radar vectoring and sequencing of all IFR and VFR aircraft landing at primary airport, separation of all aircraft in TRSA service area, and advisories on all unidentified aircraft on a workload-permitting basis.

terminal velocity   *1* Highest speed of which aeroplane (rarely, other aircraft) is capable, reached at end of infinitely long vertical dive at full power through uniform atmosphere (suggest arch.).

2 Ultimate speed reached by inert body in free fall through particular prescribed atmosphere.

terminal VOR   VOR located at or near airport at which particular flight terminates and specified as navaid used in final approach clearance.

terminating bar lights   Red lights between final-approach lights and wing-bar lights.

terminator   *1* Solid-propellant rocket subsystem comprising signal input, squib or detonator and blow-out port(s) for causing immediate thrust termination.

2 Boundary between sunlit and dark sides of planetary body, eg Earth, Moon.

TERMM   Transportable emergency-response monitoring module.

Terms   Terminal management system.

Tern, TERN   Terminal and en route nav.

ternary   Device capable of three states, normally called 0,1, x.

terne plate   US term for tinned, or lead-coated, mild steel sheet (not plate).

Terp   *1* Turbine-engine reliability programme.

2 Terminal instrument-approach procedure [see Terps].

Terpes   Tactical electronic reconnaissance processing and evaluation system (primarily USN/USMC).

Terprom   Terrain profile matching, usually similar in principle to Tercom.

Terps   Terminal en-route procedures (FAA).

terrain-avoidance system   System, usually radar-based, providing pilot or other crew member with situation display of ground or obstacles ahead which project above either horizontal plane parallel to aircraft or plane containing aircraft pitch/roll axes so that pilot can manoeuvre aircraft laterally to avoid obstruction. Radar becomes primary flight instrument.

terrain board   Physical model of landscape formerly used in simulation of air activity.

terrain-clearance system   System, usually radar-based, providing pilot or autopilot with climb/dive signals such that aircraft maintains preset flight level while clearing peaks within selected height in vertical plane through flight vector. Unlike terrain-following, after each protruding peak aircraft levels out at prescribed FL.

terrain comparison   See *Tercom*.

terrain database   Comprises computer-stored 2-D grid of ground spot heights plus land culture information.

terrain-following system   System, usually radar-based, which provides pilot or autopilot with climb/dive signals such that aircraft will maintain as closely as possible a selected lo height above ground contour in vertical plane through flight vector. In effect system projects radar ski-toe locus which slides over terrain ahead to give minimum safe clearance.

terrain masking   Obscuration of aerial and other targets by hills or buildings, esp. as seen at acute grazing angles by overland downlook radar.

terrain orientation   Holding topographical map so that aircraft heading is at top of sheet or folded sheet.

terrain profile   Outline of profile of ground surface, usually with vertical scale × 5 (sometimes × 10) published on approach chart or other documents to assist pilots.

terrain-profile recorder   Airborne instrument, recording sensitive radar or laser altimeter, giving hard-copy readout for mapping and surveying.

terrain-referenced navigation   Terrestrial reference guidance.

terrestrial radiation   Earth IR radiation; also called eradiation.

terrestrial reference guidance   Any method providing steering intelligence from characteristics (usually stored as quantified digital measures) of surface being overflown, thereby achieving flight along a predetermined path without the need for emissions. One example is Tercom. Also called terrain-reference [or referenced] navigation.

terrestrial refraction   Refraction observed in light from source within Earth atmosphere; thus caused only by inhomogeneities of atmosphere itself.

terrestrial scintillation   Generalized term for scintillation effects observed in light from sources within Earth atmosphere; also called atmospheric boil, optical haze and shimmer.

Tersi   Series of EW jamming and aerial-pattern simulators.

tertiary airflow   That passing through tertiary holes.

tertiary holes   Apertures in gas-turbine flame tube or combustor downstream of secondary holes admitting air purely for dilution and cooling purposes to achieve desired uniform gas temperature across chamber exit plane.

TERTM   Thermal-expansion resin transfer machine.

TES   *1* Test and evaluation squadron.

2 Technology Experiment Satellite (India).

3 Tactical environment, or engagement, simulation [or system].

4 Thermal emission spectrometer.

5 Transportable Earth station.

6 Threat-emitter system.

7 Trials end system (ATN).

TESAC, Tesac   Training and evaluation system for active countermeasures.

Tesar   *1* Tactically enhanced synthetic-aperture radar.

2 Tactical-endurance synthetic-aperture radar (UAV) or search and rescue.

tesla   SI unit of magnetic flux density; $1T = 1$ Wb/m^2 = 10^4G.

tesla coil   Induction coil without iron core normally giving HF output.

TES-N   Tactical exploitation system – Navy (USN).

Tess   *1* Tactical engagement, or threat-emitter, simulation system.

2 Transport efficiency support system.

Test   Checklist at start of takeoff: time, engine instruments, strobe, transponder.

testbed   Mounting, either on ground or in form of aircraft, upon which item can be mounted or installed for test purposes. When an aircraft may be, but not necessarily, prefixed by 'flight' or 'flying'. In US normally called test stand.

test cell   Usually horizontal test stand, eg for rocket motor, surrounded except on operative side by protective shelter giving protection from weather and limited protection externally from explosion inside.

test chamber   Environmentally controlled sealed chamber in which test can take place; eg can simulate stratosphere or hard vacuum with space solar radiation.

test clip   Spring-steel clip for quick electrical connections to terminals.

test club   Club propeller making no pretence at aerofoil shape but merely having stubby projecting arms in correct balance.

test diamond   Region in supersonic tunnel working section within which model is placed and within which flow conditions are essentially constant at any one time.

test firing   Firing of rocket, of any type, while mounted on testbed (test stand).

test flight   Flight by aircraft, winged spacecraft or cruise

missile for purpose of evaluating or measuring performance, handling or system operation.

test pattern   Geometric pattern used in testing electronic displays.

test program[me] set   Small box which is brought to check out cockpit processors.

test rig   See *rig (3)*.

test section   *1* Tunnel working section.

2 Special glove aerofoil carried on flying testbed.

test set   Packaged equipment, either versatile or for testing specific system, of electronic, hydraulic, pneumatic, microwave/RF or any other character, which can readily be brought to aircraft or have device brought to it.

test vehicle   Air vehicle, normally unmanned, built to test major element of its design, construction or systems and thus prove new concept.

TET   *1* Turbine-entry temperature, see *turbine temperatures*.

2 Tolerable exposure time, esp. with reference to aircraft in high-speed flight in gusts.

3 Technical evaluation test.

TETA   Triethylene-triamine.

Tete   Total estimated time en route.

tethered satellite   One connected to a space station, Shuttle Orbiter or other parent body by a fine cable up to 100 km (62.1 miles) in length.

Tetra   *1* Turbine-engine transient response analyser (EDP code).

2 Terrestrial trunked radio.

tetraethyl lead   Liquid added to some petrols (gasolines) to improve resistance to detonation (anti-knock value or fuel grade); base material is Pb(C$_2$H$_5$)$_4$. Resulting fuel is called leaded.

tetrode   Thermionic valve (tube) containing cathode, plate and two other electrodes.

Tetwog   Turbine-engine testing working group.

TEU   Trailing edge up.

TEV   Test, evaluation and verification.

Tevi   Turbine-engine vibration indicator.

Tewa, TEWA   Threat-evaluation and weapon assignment.

TEWS, Tews   Tactical electronic-warfare suite, or electronic warning system.

TEWT   Pronounced tute, tactical exercise without troops.

Textolite   Obsolete 'plastics'; hot pressed canvas/resin laminates.

textured visuals   Visuals whose CGI tones are not plain grey shades but have texture corresponding to real life, giving enhanced spatial cues. Usually achieved by digital and photographic imagery combined.

TEZ   Total exclusion zone.

TF   *1* Trip fuel.

2 US military engine-designation prefix: turbofan.

3 US military aircraft designation prefix: fighter trainer; dual version of established fighter.

4 Turbine fuel is available.

5 Toroidal field.

6 Technology forecasting.

7 Torpedo (armed) fighter (UK, pre-1953).

8 Terrain-following.

9 Thin film.

10 Tandem fan.

11 Task force.

12 Toll free.

13 Time/frequency (or T/F).

T$_f$ *1* Temperature at flexible T-O rating.

2 Fuel temperature.

t$_f$ Radar frame time.

TFA   Transfer-function analyser.

t$_{fa}$ Average time between false alarms.

T-fast   Technology for frequency-agile digital synthe-sized transmitters.

TFB   Tower fly-by.

TFC   *1* Total final consumption.

2 Tactical fusion centre (AAFCE).

3 Tactical fire control.

4 Tactical flag command; C adds centre.

tfc   Traffic (FAA).

TFD   *1* Thin-film diode.

2 Time/frequency display.

TFE   *1* Terrain-following E-scope.

2 Therminioc fuel element.

3 Tetrafluoroethylene, major additive to magnesium IRCM.

TFEC   Tactical fighter electronic combat.

TFEL   Thin-film electroluminescent display, a CRT alternative.

TFF   Tri-fluid fuel.

TFG   *1* Thrust-floated gyro.

2 Tactical Fighter Group.

TFH   Thick-film hybrid.

TF/HF   Tandem fan, hybrid fan.

T-fix   Elapsed time since last update of position of moving surface target.

TFK   Trainer facility kit.

T-Flir   Targeting FLIR.

TFM   *1* Tactical flight management system.

2 Traffic flow management.

TFMS   Tactical frequency-management system.

TFOV   Total field of view, limited in HUD by head freedom and optical aberrations.

TFP   Technology forecast panel

TFPA   Torso front protective armour.

TFPRT   Thin-film platinum resistance thermometer.

TFR   *1* Terrain-following radar.

2 Total fuel remaining.

3 Temporary flight restriction area (FAA).

TFS   Tactical Fighter Squadron.

TFSB   Tungsten-filament seven-bar display.

TFSF   Time to first system failure.

TFSUSS   Task-force on scientific users of space station.

TFT   *1* Thin-film transistor.

2 Trim for takeoff.

TFTA^2   Terrain following, terrain avoidance, threat avoidance.

TFTP   Trivial file transfer protocol.

TFTS   *1* Tactical Fighter Training Squadron.

2 Terrestrial flight telecommunications, or telephone, system.

TFU   *1* Turret Flir unit.

2 Technical follow-up.

TFW   Tactical Fighter Wing.

TFWC   Tactical Fighter Weapons Center.

TG   *1* Transportgeschwader (G).

2 Tactical, or Task, Group.

3 Techniques generator (EW).

4 Transmission gate.

5 Training glider (USAAF, 1941–47).

6 Timer – VDL management entity.

7 Terminal guidance.

Tg   Tropical Gulf.

TGA   Target gate analysis.

TGAT   Tactical GPS anti-jam technology.

TGB   Transfer gearbox.

TGC   *1* Travel-group charter (US term, basically = UK ABC).

2 Turbulence gain control.

TGCR   Tactical generic cable replacement (FO trans-mission system).

TG4   Maximum time between GSIF's timer.

TGG   Third-generation gyro (Northrop).

TGIF   Transportable ground interface facility.

TGL   *1* Touch-and-go landing (ICAO).

2 Temporary guidance leaflet (FAA).

TGO   Thermally-grown oxide.

TGP   *1* Twin-gyro platform.

2 Terminally guided projectile.

TGS   *1* Triglycine sulphate; pyroelectric IR detector material.

2 Taxiing, or taxiway, guidance system, (ICAO) or sign.

3 Turreted gun system.

4 Maximum link-overlap timer.

TGSM   Terminally guided sub-muniton.

TGT   Turbine gas temperature.

2 Target.

3 Titanium/graphite/titanium.

Tgt Opp   Target of opportunity.

TG3   Ground-station's maximum time between trans-missions.

TGW   Terminally guided weapon.

TH   True heading.

T$_H$ Total temperature.

THAAD   Theater high-altitude air defense [S adds system] (USA).

THAD   Terminal-homing accuracy demonstrator.

Thagg   Tactical high-antijam GPS guidance.

THAR   Tyre height above runway.

THAWS, Thaws   Tactical homing and warning system (RCA).

THDG   True heading.

THDR   Thunder.

theatre   Geographical area of military operations in which commander of unified or specified command has complete responsibility; today used as adjective, often synonymous with tactical.

theatre range   Range of combat aircraft within a theatre, as distinct from deploy range.

Thel, THEL   *1* Tactical high-energy laser.

2 Theater high-energy laser (US, Israel).

Thelact   Tactical high-energy laser advanced-concept technology.

The LTAS   The Lighter-Than-Air Society (Akron, OH, US).

Themis   Thermal-emission imaging system.

Then Year   Actual funds voted or spent; must be factored for inflation to enable comparison to be made with 'now'.

theodolite   Optical sight or telescope whose az/el can be accurately read off angular scales.

theoretical gravity That at Earth's surface if Earth's mass was reshaped as perfect sphere.

theoretical thrust coefficient A thrust/time value for solid-propellant rockets computed from large equation involving an effective value and assumed conditions for various areas and pressures. Symbol C°_1.

therapeutic adaptor Coupled to continuous-flow oxygen mask, approximately triples flow rate; used for passengers with respiratory or heart problem.

therapeutic oxygen Administered primarily to treat ailment, eg pulmonary or cardiac faults.

therm Non-SI unit of energy = 10^5 BTU = 105.506 MJ.

thermal 1 Local column of rising air in atmosphere, usually caused by surface heat source.

2 To use (1) as energy input for soaring flight.

thermal acoustic fatigue Fatigue of structure caused by impingement or close proximity of hot gas jet.

thermal anticing Anticing by heating affected surface.

thermal barrier Notional barrier to further increase in some variable, eg flight speed in atmosphere or turbine entry temperature in engines, caused by inability of materials to withstand increased temperatures. Continually being eroded by new refractory materials.

thermal barrier coating Vast range of refractory materials, usually deposited by electron beam or plasma spray, based on zirconia, yttrium and similar exotics.

thermal battery Electrical cell stored inactive and activated chemically for one-shot high-power output.

thermal blooming See *blooming*.

thermal coefficient of expansion Increase of (1) length per unit length, or (2) area per unit area, or (3) volume per unit volume, caused by rise in temperature of 1°C (often defined as from 0° to 1°C, or from 15° to 16°C).

thermal conductance Rate of flow of heat per unit time through unit cross-section area; 1 BTU.ft^{-2}h.°F = 5.67826 Wm^{-2}h.°C; 1 Wm^{-2}h.°C = 0.17611 BTU.ft^{-2}.h.°F.

thermal conductivity Time rate of flow of heat through unit area normal to temperature gradient per unit T° difference. Symbol λ or k, rate given by Fourier's law. SI unit is Wm^{-1}K^{-1}; Imperial (obs.) might be BTU ft^{-1}s^{-1}F^{-1}.

thermal cueing unit Adjunct to FLIR-based attack system which puts marker boxes round all likely surface targets, picking them according to their high temperature, and which automatically feeds target co-ordinates to the attack system if any of these boxes is touched by the pilot on the HDD touch display.

thermal cycling Oscillating between low and high temperatures.

thermal de-icing De-icing by heating affected surface.

thermal diffusivity Measure of transfer of heat by diffusion analogous to viscous motion; symbol $\alpha = \lambda/\rho C_p$.

thermal diode Solid-state generator of electricity comprising layer of semiconductor at room temperature joined by thermal insulative layer to layer heated to 250–450°C.

thermal efficiency Basic efficiency parameter of heat engine, defined as percentage ratio of work done in given time to mechanical equivalent of heat energy burned in fuel supplied in same period.

thermal emission EM radiation solely due to body's temperature (which if hot enough contains strong visible radiation).

thermal excitation Acquisition of excess energy by atoms or molecules as result of collisions.

thermal expansion Increase in dimensions caused by increase in temperature.

thermal exposure Calories/cm^2 received by normal surface in course of complete NW detonation (DoD).

thermal fatigue Mechanical fatigue caused by stresses repeatedly imposed by thermal cycling.

thermal gradient See *temperature gradient*.

thermal gradiometer Airborne instrument for detecting thermals by thermocouples on wing-tips which, in presence of temperature difference, sends electrical signal to cockpit indicator.

thermal heating Tautological; kinetic heating is meant.

thermal imagery Produced by measuring and electronically recording thermal radiation from objects (NATO). Normally IR wavelengths only are implied. Hence thermal imaging, to produce pictorial displays or printouts showing variation of temperature over field of view.

thermal index A forecast value of the temperature difference between sinking and rising air.

thermal instability Any combination of temperature gradient, thermal conductivity and viscosity resulting in convective currents, eg wind in atmosphere.

thermal keel Generated by positioning engine jet nozzles well forward under the fuselage [helps reduce generation of sonic boom].

thermal lift 1 Lift due to thermal (1).

2 Lift imparted to air mass because of greater density of cold surrounding air, not quite synonymous with (1).

thermal load Imprecise term usually meaning temperature gradient or temperature stress.

thermally expanded metal Fabrication of parts from aluminium alloy sheets rolled together with intervening patterns of 'ink'; the latter prevents the sheets bonding and, on subsequent heating, expands to force the unbonded parts to fit a mould.

thermal neutron Neutron slowed, eg in moderator, to thermal equilibrium with surroundings at about 2,200 m/s (so-called slow neutron); * analysis is principal method used in detecting presence of explosives.

thermal noise RF noise caused by thermal agitation in dissipative body (any conductor or semiconductor), also called Johnson noise.

thermal picture synthesizer Matrix of heat-emitting thin-film resistors on Si substrate, each representing individually addressed pixel to give overall large picture rate of 50 Hz.

thermal protection Protection against kinetic heating during atmospheric entry (re-entry) of spacecraft structure, RV or other body, esp. one intended for repeated space missions.

thermal pulse Total IR emission from NW detonation, or plot of IR flux against time during complete burst and fireball climb.

thermal radiation 1 See *thermal emission*.

2 Total heat and light radiation produced by NW detonation (DoD).

thermal relief valve Safety valve in fluid system to guard against excessive pressure caused by overheating.

thermal runaway 1 Fault condition with element of danger affecting Ni/Cd batteries characterized by particular cells losing resistance (possibly because of high temperature) and thus taking increased current, lowering resistance still further in chain-reactive process.

2 Similar divergent overheating in current-carrying transistor.

thermal sensitivity   Of IR camera, quantified difference in temperature required to output different tonal value between black/white, typically 0.02–0.1°C.

thermal shock   Severe mechanical stress resulting from sudden extreme temperature gradient.

thermal soaring   See *soaring*.

thermal stress   See *temperature stress*.

thermal switch   Switch activated by temperature difference or particular temperature.

thermal thicket   Flight conditions in which kinetic heating (or other thermal problems) is a factor to be considered but does not yet impose a thermal barrier (colloq.).

thermal wind   Notional vector difference between winds at different heights, caused by horizontal variation of atmospheric temperature and hence pressure at all upper heights (note: not pressure surfaces).

thermal X-rays   EM radiation, mainly in soft (low energy) X-ray region, emitted by extremely hot NW debris.

thermel   Any device based on Seebeck, eg thermocouple, thermopile.

thermie   Non-SI unit of work (mechanical energy); 1 th = 4.1855 MJ.

thermionic   Involving electrons emitted from hot bodies.

thermionic converter   Electric generator powered by hot emitter and cold collector.

thermionic rectifier   Depends on unidirectional electron flow from cathode to anode.

thermionic tube   See *thermionic valve*.

thermionic valve   Evacuated capsule, usually glass, containing heated cathode emitting electrons attracted to anode, usually via one or more intervening control electrodes usually called grids. In US called vacuum tube.

thermistor   Protective resistor based on semiconductor having high negative temperature coefficient of resistance.

thermite   Mixture of finely divided magnesium and iron oxide used as heat source in welding and as incendiary filling; originally spelt with capital T.

thermobaric warhead   Creating both high-temperature and blast-wave effects [often said of FAE].

thermobarograph   Provides continuous readout of temperature and pressure.

thermochemistry   Branch of chemistry concerned with thermally induced reactions and relationship between chemical changes and heat.

thermochromic   LO technology in which appearance is changed by variation in temperature.

thermochromic tube   CRT with phosphor replaced by heat-sensitive layer.

thermocline   Sharp submarine temperature gradient.

thermocouple   Instrument based on Seebeck effect which measures temperature difference between pair of dissimilar-metal junctions; much used for high-temperature measures using refractory metals, and in common copper/constantan junction at room temperature, eg for met. observation.

thermodynamics   Science based upon heat flow and temperature changes, esp. those in moving fluids.

thermodynamic cycle   Operating cycle of any heat engine. In some, eg virtually all piston engines, one parcel of fluid at a time goes through complete ** in same enclosed (usually variable-size) volume; in others, eg gas turbines, continuous flow of fluid goes through ** by passing from one part of device to another, each component handling only one part of **. The working fluid may be recycled, continually changing state liquid/vapour.

thermodynamic efficiency   See *thermal efficiency*.

thermodynamic energy equations   Exact expressions of variation of pressure, volume and temperature in reversible processes in perfect gas.

thermodynamic equilibrium   Time-invariant state in which all processes are balanced by reverse process and entropy production vanishes.

thermoelectric cooling   Local cooling using Peltier and cooling 'hot' junction; 'cold' junction then falls to desired level at -20 to -30°C.

thermoelectric generator   Electric generator based on thermocouples using Seebeck, Thompson, Kelvin or Peltier effects; common spacecraft systems use nuclear reactor or radio-isotope to heat junction often based on Ge/Si alloy.

thermogram   *1* Single-line output of traditional thermograph.

2 Pictorial output of thermographic camera.

thermograph   Recording thermometer using pen/chart or light-spot trace on film. Output is a thermogram.

thermographic camera   IR camera, usually of IRLS type.

thermography   Translation of temperature changes in a scanned scene into visual picture, today important in military and civil aerial reconnaissance, industrial process control, medicine and many other fields. Either black/white (black = cold, white = hot) or colour.

thermohydrometer   Hydrometer with thermometer, giving two chart outputs.

thermometer   Instrument for measuring temperature.

thermometer screen   Louvred box screening thermometer from direct sunlight; usually contains other met. instruments and in US called instrument shelter.

thermonuclear   Processes in which extremely high temperatures are used to initiate fusion of light nuclei.

thermonuclear weapon   *Hydrogen bomb*.

thermopile   Thermoelectric generator comprising stack of thermocouples.

thermoplastic recording   Patented (GE) process for recording sound or video signals via electron beam direct on thermoplastic layer heated by microscopic currents induced in underlying conductive layer.

thermoplastics   Large class of synthetic polymers which may be repeatedly softened and remoulded by heating.

thermosetting plastics   Synthetic polymers that are chemically changed irreversibly by chemical action, for example a hardening agent, or by EM radiation, notably heat or UV irradiation, generally setting hard.

thermosphere   Outermost region of atmosphere from top of mesosphere outwards into space, characterized by more or less steadily increasing temperature with distance from Earth.

thermostat   Device for maintaining a desired temperature by taking action at preset limits of low and high temperature.

thermotropic model   Atmosphere used in forecasting one temperature and one pressure surface.

Thesh   Threshold, also Thld, THR.

theta   Greek letter θ, used for many parameters, including pitch angle (thus, θ = pitch rate) and azimuth (hence Rθ). See Appendix 1.

THI   Tactical hit indicator.

thickened fuel   Aircraft fuel designed to resist fine dispersion and instead to break down in crash into globules with near-zero surrounding vapour; generally synonymous with gelled fuel.

thick-film   Very diverse technology of electronics involving processing, high-current devices, current-generation (inc. solar cells) and many other topics, mainly using insulating substrates but often with semiconductor layer.

thickness   Of wing, maximum straight-line distance from external skin of upper surface to external skin of lower surface measured in plane of aerofoil profile and perpendicular to chord line.

thickness/chord   Ratio of thickness to chord of wing, both measured in plane of aerofoil profile at same station.

thickness distance   Distance aft of leading edge of maximum thickness of supersonic rhomboidal or double-wedge wing, expressed as % chord.

thickness gauge   See *feeler gauge*.

thickness lines   Lines joining points on chart where vertical distance between pressure surfaces is everywhere same.

thickness ratio   Wing t/c ratio.

thimble   *1* Pear-shaped eye around which end of control cable is spliced.

2 Ratchet turning knob of hand micrometer.

3 Pimple-like radome, especially on or under nose [usually adjective].

thin-case bomb   Conventional bomb for blast effect against soft target. Also called light-case (UK, WW2).

thindown   Progressive energy loss by primary cosmic rays in ionising surrounding medium.

thin-film circuit   Electrical or electronic circuit formed by depositing thin film on (usually insulating) substrate; normal manufacturing methods are vacuum deposition and cathode sputtering. Films may be conductive, semiconductor or insulating.

thin-film lubrication   Imperfect, with occasional metal/metal contact.

thin-film transistor   IGFET constructed by evaporating on to insulating substrate metal electrodes, semiconductor layer(s), insulating upper layer and metallic gate; abb. TFT.

think tank   Centralized group of people normally working for government or large corporation engaged in futures, forecasting, ultra-new technologies and other disciplines calling for visionary judgement.

thinner(s)   Solvents for paint, dope and other liquids to reduce viscosity.

thin route   Airline route, usually intercontinental, offering only modest traffic.

thin-tape system   Applied to aircraft skin to increase stealthiness of joints.

THIR   Temperature, humidity and IR radiometer.

third-angle projection   Convention in engineering drawing in which front view, side elevation and plan each show face nearest to it in adjacent view; traditional US arrangement becoming standard in European aerospace.

third-level carrier   Generalized term for 'third tier' of scheduled airline operations, also called feeder or commuter and often of radial nature serving single city hub. No clear demarcation separating from second-level (local-service or regional).

thixotropic   Becoming liquid when vibrated or stirred, setting after standing for a period.

THK   *1* Turk Hava Kurumu (national air-sport association, Turkey).

2 Thick.

THL   *1* Tailplane hinge line.

2 Tourelle hélicoptere leger (F).

Thld   Threshold.

THN   Thin.

ThO$_2$ Thorium oxide.

Thor, THOR   *1* Thermionic opening reactor (burst power up to GW range).

2 Terahertz operational reachback (Darpa).

thorium   Th, silvery radioactive metal, density 11.7, MPt 1,750°C.

Thornel   Tradename for carbon and graphite fibres.

thou   Thousandth of an inch, 25.4 µ.

THP   *1* Thrust horsepower, often thp.

2 Through-hole plated.

3 Turbo-hydraulic pump.

4 Total-head pressure.

THR   *1* Threshold, threshold lights.

2 Turboreacteur à hélice rapide = propfan (F).

3 Thrust.

thread chaser   Tool for removing contamination, eg paint or dirt, from thread.

thread filter   Long fine screwthread on outer surface of cartridge inserted tightly into surrounding unthreaded cylinder to filter fine fragments, typically as last-chance * before oil reaches vital feed jets.

thread gauge   Hand gauge with many specimen threads, one of which is matched with part.

threading the needle   Process of accurately flying through a small gate in airspace, eg in setting a speed record (colloq.).

thread insert   Steel helix screwed into soft (eg aluminium) hole.

threat   *1* Hostile anti-aircraft defences, especially air-defence radars, SAM systems, AAA and fighters.

2 A target that has satisfied the * -detection logic and therefore requires a traffic or resolution advisory (TCAS).

Threat awareness unit   Minimum time flight crew need to discern collision threat and take avoiding action; performance envelope of aircraft divided by closure rate of intruder.

threat circle   Projected on cockpit display showing computed region in which LO aircraft might expect to be detected by particular hostile radars.

threat cloud   Total collection of warheads, chaff and other penetration aids in ICBM attack.

threat evaluation   Process of detecting, analysing and classifying hostile offensive systems, either in warning of attack or during penetration of hostile territory when systems are surface-to-air.

threat library   Numerical characteristics of hostile threats, especially EM emitters, stored in friendly computer (eg of RWR receiver).

threat simulation   Simulation of hostile offensive systems, eg by add-ons to RPV target to include emissions, dispensed payloads and jamming.

three-axis autopilot   Has authority in pitch, roll and yaw.

3-bar VASI   Comprises VASI plus additional pair of upwind (210 m, 700 ft) wing bars symmetrically disposed about centreline each having at least two light units, for use by LEW aircraft.

three-bearing nozzle   Propulsive nozzle able to vector in two planes for STOVL.

three-body problem   Mechanics of motion of small body in gravity of two others.

three-control aeroplane   Conventional, with separate pilot input for each rotational axis.

3-D cam   Cam whose profile varies across its width and which moves axially as well as rotationally.

3-D flow   Fluid flow which cannot be represented fully in 2-D, eg flow over a real wing.

3-D radar   Radar enabling position of target to be determined in 3-D space, either by Cartesian methods or, more often, by az/el plus slant range.

3-D tool   Jig or fixture used to define exact shape of finished assembly, eg complex hydraulic piping or wiring loom.

3E   Environment, efficiency, economy.

three-float seaplane   Main float on the centreline and stabilizing float on left and right.

3GCS   Third-generation cellular system[s].

three greens   Landing gear is down and locked (colloq.).

3He   Helium, valency 3.

3LM   Third-level maintenance.

three-moment equation   For solving bending moments and other loads at ends of two adjoining spans of continuous beam.

3-P   Planning, production, progress.

three-phase current   Alternating electrical current made up of three phases, each with vector separation of 120°; carried by triple wire.

three-phase equilibrium   See *triple point*.

3φ   Three-phase current [3-phi].

three-pointer altimeter   Dial instrument with short needle for thousands (ft or m), mid-length for hundreds and longest for tens.

three-point landing   Correctly judged landing by tailwheel-type aeroplane in which main and tail wheels touch ground simultaneously with wing stalled.

three-point mooring   Mooring for aerostat in which three lines are run (often from single point, eg nose of airship) to three ground anchors, usually at apices of equilateral triangle.

three-point tanker   Equipped with two outer-wing HDUs and one at the tail.

3-pole switch   Opens and closes three conductors or circuits.

three-poster   STOVL or V/STOL vectored-thrust propulsion system having three jets; normally two cold fan jets and one hot core jet, but alternatively two main (rear) jets plus an auxiliary nose jet fed via a bleed air duct.

three-shaft engine   Gas turbine having LP, IP and HP shaft systems.

three-stream engine   *1* Turbofan (HBPR) in which fan thrust (probably VIGV modulated) and core jet are used for propulsion and LP compressor (core supercharger) is used for blowing purposes.

2 Any engine in which fan thrust, core thrust and lift thrust or bleed are used separately.

3 to 1 rule   Air distance 3 n.m. for each 1,000 ft lost in letdown.

3-view drawing   GA drawing, normally showing elevation (left side), front and plan.

3-way switch   Routes input along either of two outputs.

3-wire   Target of most carrier arrested landings, No 3 wire; hence ** landing.

3-wire circuit   Neutral wire between two outer wires, latter having potential difference from neutral equal to half that between them.

threshold   *1* Beginning of usable portion of runway, ie downwind end.

2 In automatic control systems, point at which response is first noticed, usually defined in terms of input displacement (see * level).

3 Flight condition when fixed-wing aerodyne is on point of stall.

4 Point at which sound just becomes audible (* of audibility or of hearing), normally 2×10^{-5} N/m^2.

5 EAS giving lowest comfortable cruising, possibly higher than that for minimum fuel.

6 See thresholds.

threshold contrast   Smallest contrast in luminance visible under given conditions.

threshold crossing height   Height of glideslope above threshold.

threshold curve   Plot of sound frequency against noise level in dB (or other noise measure) just audible against quiet background, eg anechoic chamber.

threshold displacement   Linear distance between end of full-strength runway pavement and displaced threshold, with latter shown on airfield charts as white bar across runway crossed by narrow black line, and expressed as minus quantity in certain navaid figures, eg Vorloc II $= -380$ ft.

threshold dose   Minimum quantity of radiation producing detectable biological effect.

threshold illuminance   Minimum value of illuminance eye can detect under given dark adaptation and target size; also called flux-density threshold.

threshold level   Threshold (2), esp. in rate gyro developing electrical output as function of rate of turn; that angular rate after rotational acceleration from rest at which there is first indication of output, or change in output; normal unit is °/s $\times 10^{-6}$.

threshold lights   If fitted, bidirectional units, showing green towards approach and red towards runway, in continuous row across threshold (rare at displaced threshold).

threshold limit value   Average airborne concentration of toxic substance[s] normal person can withstand 8 h per day 5 days per week, usually expressed as ppm or mg/m^3 at 25°C/760 mm Hg.

threshold marking   For simple runway, runway number in white, visible to pilot on approach; if displaced threshold, preceded by white transverse bar touched by four arrowheads pointing upwind and preceded by series of centreline arrows. For instrument runway, four bold white axial stripes in rectangular group on each side preceeding runway number.

thresholds   Limits on programme monetary changes imposed by US Defense Secretary.

threshold sampling time   Time since overhaul at which engines are removed and inspected in preparation for extension in TBO; * may be less or more than new TBO.

threshold speed V_T, V_{AT} and $V_{T\,max}$.

THR HOLD   Thrust, or throttle, hold.

THRFTR   Thereafter.

throat   Point of smallest cross-section in duct, especially that in con/di nozzle, supersonic tunnel upstream of working section, and rocket engine or motor thrust chamber and nozzle.

throatable   Jet or fluid flow controllable by changing shape or area of throat (unusual except in tunnels).

throat control   In gas turbines, system controlling flow through nozzle guide vanes upstream of turbine.

throatless chamber   Rocket thrust chamber without throat yet still achieving supersonic expansion, eg multi-chamber toroidal type.

throatless shear(s)   Power shear for cutting large sheet or plate which may be rotated during cut to leave curved edge.

throat microphone   Microphone held against skin of throat; better for deep or guttural voices or languages.

throttle   *1* Input control, usually hand lever rotating through arc, for main vehicle propulsion.

2 System responsible under pilot for varying engine power.

3 Valve in carburettor or fuel control which governs admission to engine of either air, fuel or (piston engine only) mixture.

4 To reduce power of engine, also called to * back.

5 To constrict fluid flow path and thus reduce mass flow.

throttle back   To reduce power.

throttle friction   Pilot-operated device which greatly increases resistance of throttle lever(s) to movement, effectively locking them in set position; also called friction lock.

throttle icing   Ice accretion in carburettor near or on partially closed throttle (3).

throttle lock   See *throttle friction*.

throttle push   Pilot action to increase power.

throttle sensitivity   Change in thrust or power per unit movement of throttle lever.

throttle tension   Locking resistance value of friction lock.

throttling capability   Range of thrust expressed as percentage to 100, over which liquid rocket (occasionally other type of engine or propulsion) is designed to operate.

through-deck ship   Generally, one with flight deck unobstructed by any full-width superstructure, even though not necessarily extending to bow.

through-stick feedback   Characteristic of some autopilots that, when engaged, pilots flight controls move.

through-thickness pinning   Repair of major damage to composite structure in which numerous fine pins are collapsed by a foam carrier.

throw   *1* Part of crankshaft to which conrod attached, comprising webs and crankpin.

2 Loose measurement of distance to which ECS fresh-air inlet projects, in absence of bulk cabin air movement.

thrower ring   Flange on rotating shaft which flings off leakage oil or other fluid.

throw weight   Total mass of payload carried by ballistic missile, in case of ICBM including warheads, RVs, decoys and other penaids, post-boost propulsion and terminal guidance systems.

THRP   Port throttle (caption).

THRS   Starboard throttle (caption).

THRU   *1* Through.

2 I am switching you to . . .

thrust   Force, esp. that imparting propulsion, SI unit is newton (N), conveniently multiplied in kN = 224.80 lb st; 1 lb st = 4.44483 N. Useful * of turbofan or turbojet is resultant of all * generated by fan (positive), front of combustor (positive), turbines (negative) and nozzles (negative), * at each location being $AP + \dfrac{WV}{g}$ where A is area of flow cross section, P is pressure, W is mass flow and V flow velocity. Overall net * is $A(P-Po) + \dfrac{W(V_j-V)}{g}$ where P is static pressure at nozzle, Po local atmospheric pressure, V_j is jet velocity and V velocity of aircraft.

thrust angle   Acute angle between axis of nozzle of canted solid motor and centreline axis of vehicle, measured in plane passing through both axes if possible.

thrust augmentation   Usually means afterburning, but also applied to water injection, and to piston engine ejector-exhaust schemes.

thrust-augmented wing   Aeroplane wing in which enhancement of circulation by powered-lift system also gives significant additional thrust (many arrangements, but augmentation of thrust invariably secondary objective).

thrust axis   Axis along which resultant propulsive thrust acts. With a turbofan this is resultant of fan and core jets, and with a turboprop that of propeller and (probably angled) jet. In multiengined aircraft * can oscillate because of engines outboard on flexible wing.

thrust bearing   Bearing, usually tapered roller, needle or ball, that resists axial shaft load due to propeller thrust.

thrust buildup   Sequence of programmed events in large rocket engine between ignition and liftoff.

thrust bump   Sudden uncommanded change [especially increase] in thrust.

thrust chamber   Complete thrust-producing portion of liquid rocket engine comprising combustion chamber and nozzle, often mounted on gimbals; not applicable to other types of engine.

thrust coefficient   *1* For propeller, basic performance calculation method based on Drzwiecki method of plotting grading curve of thrust against blade radius, yielding value k_T, constant for each value of advance ratio J; ** then equals $k_T J^2$ and has symbol C_T. This is also measured thrust divided by $\rho n^2 D^4$ where ρ is density, n rpm and D diameter.

2 For rocket motor, measured ** is thrust: time integral over action-time interval divided by product of average throat area and integral of chamber pressure: time over action-time interval, symbol C_f.

3 For CC/BLC blowing slit, T/qS.

thrust component   In propeller theory (Drzwiecki), force on one element parallel to axis of rotation, T_c; convenient to plot T_c as ordinate against blade radius, area under curve being measure of total thrust, $T = N\frac{1}{2}\rho V^2 \int_o^r T_c\, dr$ where N is number of blades, $\frac{1}{2}\rho V^2$ dynamic head and T_c integrated between axis of rotation (or, in practice, spinner diameter) and tip radius r.

thrust control computer   AFCS computer providing control of engine N_1 and thrust, computation of engine limit parameters, and autothrottle.

thrust cutoff   See *cutoff*.

thrust decay   Gradual falloff in thrust of solid motor, usually a slow fall from peak to cutoff or burnout followed by rapid ** over 2 to 8 s and to zero after perhaps 10–12 s.

thrust-decay system Idle-area reset (turbofan).

thrust deflector Various schemes for V/STOL or STOVL, see four-poster, switch-in deflector, vectored thrust, etc.

thrust equivalent horsepower See *thrust horsepower*.

thruster Small propulsor, normally any of many kinds of rocket, used for spacecraft attitude control or fine adjustment of velocity.

thrust face Side of propeller blade corresponding to underside of aerofoil.

thrust/frontal area Jet-engine thrust divided by engine's nominal or published frontal area; fair criterion in early days of jet propulsion but today meaningless. Important only in highly supersonic aircraft, in which area of propelling nozzle exceeds that of engine.

thrust horsepower Seldom-used measure attempting to determine power imparted to aircraft. For propeller aircraft normally engine bhp or shp multiplied by propeller efficiency (in case of turboprop plus a variable component due to exhaust thrust). For jet engines, basically thrust actually imparted to aircraft multiplied by TAS, keeping units compatible. See *equivalent horsepower*.

thrust lever Jet-engine throttle, or power lever.

thrust line Thrust axis.

thrust loading W/F, total mass (in this case, weight) of jet-propelled vehicle divided by aggregate thrust, usually calculated for SLS-TO condition; units lb/kN = 224.8 lb/lb st, reciprocal 0.004448; 1 lb/lb st = 102.04 kg kN^{-1}, reciprocal 0.0098.

thrust meter Instrument for measuring thrust, more commonly of jet engine.

thrust power Appears always to be synonymous with thrust horsepower.

thrust rating computer Central element in auto power management system (ATS).

thrust rating panel AFCS cockpit display of limiting and target values of engine parameters, and selectors for operating mode (climb, cruise, MCT or TO/GA) or FTO temperature(s).

thrust reverser See *reverser*.

thrust section Portion of vehicle, esp. slender rocket, containing propulsion.

thrust specific fuel consumption See *specific fuel consumption*.

thrust spoiler Pilot-controlled spoiler which when actuated diverts jet from jet engine (esp. from turbofan core) to reduce thrust close to zero. Lighter and simpler than a reverser and merely eliminates possibly embarrassing idling thrust.

thrust structure In large ballistic vehicle propelled by multiple rocket chambers, structure which transmits thrust from all chambers and diffuses it into airframe. Normally large tubular truss structure at rear but can include side structures for laterally attached motors, eg SRBs.

thrust terminator Any quick-acting device for terminating thrust of solid rocket motor, including blow-off ports, nozzle ejection and inert-liquid injection into case.

thrust time lag Time from abrupt throttle movement to reach stabilized thrust or power.

thrust-vectoring Control of vehicle trajectory by rotating thrust line, esp. that of rocket; may involve gimballed chamber, rotation of chamber about skewed axis, inert-liquid injection at nozzle-skirt periphery, jet tabs, spoilers, refractory vanes and other methods; abb. TVC.

thrust-weight ratio Basic measure of combat aeroplane performance: thrust (normally SLS-TO) divided by total mass of aircraft.

thrust wire Diagonal bracing wire transmitting airship propulsion thrust to envelope.

THRUT Throughout.

THSA Trimmable horizontal-stabilizer actuator.

THSD Thousand[s].

THT Transient heat transfer.

THUM, Thum Meteorological readings of temperature and humidity, hence * flight.

thumbprint Common meaning is aircraft T/W (thrust: weight ratio) plotted against W/S (wing loading).

thumbstick Pilot input controller, eg for RPV or anti-tank missile, in form of miniature stick operated by thumb, typically attached to pistol grip and with * pivots between vertical thumb and operator.

Thump Meteorological readings of temperature, humidity and pressure.

thunderstorm effect Error, possibly approaching 180°, of ADF in vicinity of thunderstorm; needle may point to nearby Cb or flick over, giving false indication of station passage.

thyratron Gas-discharge triode used as relay, switch or sawtooth generator.

thyristor Multilayer semiconductor device also called Si-controlled rectifier; bistable, in one state high-impedance in both directions, in other high-impedance in one direction only.

THz Terahertz.

TI *1* Target indicator.
 2 Thermal imager, or infra-red.
 3 Training [or tactics] instructor.
 4 Trial installation.
 5 Thermal index.

Ti Titanium; hence such alloys as Ti3Al2.5V, Ti6Al4V, Ti6Al2Sn4Zr2Mo and Ti10V2Fe3Al.

TIA *1* Type inspection and authorization; allows FAA to fly new aircraft.
 2 Telephone interface adaptor card (TRV).
 3 Telecommunications Industry Association (US).

TIAA Travel Industry Association of America.

TIACA The International Air Cargo Association

TIAD Tactical internet for air defence.

TiAl General symbol for titanium aluminides.

TIALD Thermal imaging airborne laser designator.

ti-aluminides Alloys of titanium and aluminium.

TIAS *1* Target identification and acquisition system (ARMs).
 2 True indicated airspeed.

TIB Technical Intelligence Bureau (former UK government department, still a title in many countries).

TIBA Traffic information broadcast by aircraft.

Tibs, TIBS *1* Tactical information broadcast service (USAF).
 2 Telephone information briefing service.

TiB$_2$ Titanium boride.

TIC *1* Technical information centre.

2 Tantalum integrated circuit.

3 Target-insertion controller.

4 Total inventory count.

5 Transport & Infrastructure Committee (US House of Reps.).

6 Technologies of information and communication (also F).

tic Visual marking pulse on telemetry readout indicating time intervals, often every 0.5 s (see *time* *).

TICC Technical Information and Communications Committee (ATA).

TICCS, TIC²S Target information command and control system.

tick Audible marking pulse serving as regular (often infrequent, eg each 10 s or 60 s) time signal.

ticket Pilot's licence (colloquial, especially pre-1914).

TICM Thermal-imaging common module(s).

TiCo Titanium-colombium.

Ticonal Magnetic alloy of Ni/Co plus a little Al/Cu.

tic-tac airplane Miniature free-flight aircraft for sonic-boom research.

TID *1* Tactical (or target) information display.

2 Touch input device.

3 Technical-interface description.

tiddleywinks effect Tendency of nose gear to project stones and other loose objects laterally.

TIDLS, Tidals Tactical information data-link system.

TIDP Telemetry and image data processing.

tie Structural member normally loaded in tension.

tie bar Filament-wound tension member connecting helicopter main-rotor blade to hub; fatigue-proof because of large number of load-bearing members. Also called dog-bone.

TIEC Tactical information exchange capability (RAF).

tied gyro Gyro whose rigidity is related to Earth rather than space; eg that in traditional horizon has axis tied by gravity aligned with local vertical.

tied on Air-intercept code: "Aircraft indicated is in formation with me."

tiedown *1* Picketing arrangement for aircraft left in open (US).

2 Cargo lashing.

tiedown diagram Drawing illustrating method of securing particular type or item of cargo in particular vehicle (DoD).

tiedown point Permanent attachment point for cargo provided on or in vehicle (DoD); hence * pattern.

tiedown test Rocket engine static test.

tie rod General term for tie of rod-like form, esp. with threaded ends.

tiers Different levels assigned to subcontractors in major programme; Tier 1 are usually assigned responsibility for design and test, as well as manufacture.

Ties Tactical information exchange system.

TIF, Tif *1* Takeoff inhibit function, temporarily suppresses all non-essential cockpit warnings.

2 Text interchange format.

3 Terminal interface function.

4 Tactical Imagery Intelligence Flight [not TIIF] (RAF).

TIFS Total in-flight simulator.

TIG *1* Tungsten inert-gas welding.

2 Time of ignition.

Tiger *1* Targeting by image georegistration.

2 Terrifically insensitive to ground effect radar.

tightening Tendency of aeroplane or glider trimmed for level flight to increase rate of a commanded turn or dive pull-out, demanding a push force on stick or yoke to hold constant g.

TIGO Prefix, US piston engine, turbocharged, direct-injection, geared, opposed cylinders.

TiGr Titanium/graphite composite.

TII Threshold inspection interval.

TiiMs Texas Institute for Intelligent Bio-Nano Materials and Structures for Aerospace Vehicles.

til Until.

tile *1* Thin-film or thick-film substrate; also occasionally used for substrate of solar cell.

2 Discrete unit of surface thermal-protection system for RV or large spacecraft, eg Space Shuttle, inspectable and replaceable.

Till Tracking [and] illuminating laser (ABL).

TILS Tactical ILS.

tilt *1* Angular deviation of locus of centroids of sections of helicopter main-rotor blade from plane of rotation (BSI). Measured as forward or backward though actually up/down.

2 Angular movement or offset of camera axis about aircraft longitudinal axis (NATO).

tilt angle Angle between axis of air camera and aircraft vertical (OZ) axis; normally angle at perspective centre between photograph perpendicular and plumb line (NATO).

tilt-body vehicle Usually synonymous with tilting fuselage, standing upright on its tail for VTOL. A totally different species has the wing and power (lift/propulsion) system able to rotate up to 90° with respect to the free-pivoted wing and tail; this family are usually STOLs.

tilting-duct VTOL VTOL aeroplane which in hovering mode is lifted by ducted propellers or fans rotated through approx. 90° for translational flight.

tilting-engine/jet/propeller/wing Same definition as above but for different pivoted component. Tilting-jet means entire engine is pivoted.

tilting fuselage Unusual class of VTOL aeroplanes in which fuselage can be pivoted near mid-length, in some cases complete with attached wing, in order for jet thrust to act vertically. Also called nutcracker aircraft.

tilting head *1* Rotorcraft head pivoted about lateral axis relative to supporting structure.

2 Machine-tool cutter and drive pivoted about horizontal axis.

tilting-nozzle VTOL Not used; term is jet-deflection, vectored-thrust, lift/cruise, vectoring or vectored-jet.

tilt-rotor VTOL aeroplane lifted in hovering mode by one or (usually) more rotors which are rotated through approx. 90° for translational flight.

tilt-wing VTOL aeroplane whose wing, carrying complete propulsion system, is pivoted upwards through approx. 90° in vertical mode, thrust then exceeding total weight.

TIM *1* Total inventory management.

2 Training integrated [or integration] management [S adds systems].

3 Target information module.

time Normally measured by subatomic frequency reference, eg crystal clock, but defined according to position of celestial reference point; depending on which point chosen

* called solar (Sun), lunar (Moon) or sidereal (vernal equinox), solar being subdivided into mean or apparent according to which Sun. Practical time designated GMT or according to designated longitude zone. SI unit is s, 3,600 to h, 86,400 to week.

timebase *1* Straight line traced by spot on CRT or other display of cartesian and several other types providing timescale for measurement, eg of target range.

2 Straight line, regularly incorporating time tic, on data readout.

time between overhauls, TBO Period recommended by manufacturer and beyond which all warranties become invalid and operation may be in violation of certification.

time box Small box, usually rectangular or square, which moves along cockpit display future track, according to flight plan, at selected groundspeed.

time-change item One whose operation is limited to number of operating hours, number of operating cycles or (rarely) passage of time, and which must be periodically replaced on this basis.

time circle Basic symbology of many HUDs and other attack systems in which bright circle starts at 60 s and unwinds anticlockwise to 180° at 30 s and to vanish at 0 s.

time-compliant technical order Mandatory instruction for modification or for retrofit of equipment.

time constant *1* Usually, time taken from start of input signal for instrument to indicate specified % final reading; for exponential response, eg thermometer, time to reach 63.2% final reading; also called relaxation time, lag coefficient. Same meaning in charge/discharge of electrical C/R circuit or current in L/R circuit.

2 Time taken for aeroplane to reach maximum angular velocity [any axis] after hard-over control input.

TIMED, Timed Thermosphere, ionosphere, mesosphere, energetics and dynamics satellite launched 2001 (NASA).

time dilation Apparent slowing-down of time as observer's speed reaches significant fraction of that of light; also called clock paradox or twin paradox.

time dilution of precision Measure of errors [usually in navigation] resulting from errors or variation in measured or calculated time.

time/distance/speed scale Simple written scale, either purchased (in which case of sliderule type) or prepared before flight, with which unknown distance or speed can be immediately read if other two factors are known.

time-division multiplex Dividing several continuous measures, eg in telemetry system, or several input signals, to form single continuous interlaced pulse train sent over single channel to multiple receivers.

time-division multiplex access When multiple transmitters are using a single carrier the carrier is time-shared to avoid messages being garbled at receiver.

time group Four digits denoting time in hours and minutes, such as 1730.

time hack Time at which a future event is scheduled, eg at which a particular squadron is to start engines (colloq. chiefly military).

time in service For maintenance time records, aircraft log and similar purposes, elapsed time from aircraft leaving surface until touching it again on landing (FAA).

time lag Any delay between stimulus and response, or cause and effect, esp. that between start of signal and full indication by instrument.

time mean bleed Short period of time during which large RCS bleeds are expected to be used, and beyond which thrust must be reduced.

time of flight Elapsed time from weapon launch, release or departure from gun muzzle to instant it strikes target or detonates.

time-of-flight spectrometer Instrument sorting particles, esp. neutrons, according to time to travel known distance.

time of origin Local time message is released for transmission.

time of useful consciousness See *time reserve*.

time on target *1* Time, either planned or actual, at which aircraft attacks or photographs target.

2 Time at which NW detonation is planned at specified GZ (DoD).

time over target Time at which aircraft arrive(s) over designated point for purpose of conducting an air mission on a target (USAF).

time pulse distributor EDP (1) circuit that generates timing pulses during machine cycle, gated by command generator to carry out commanded operations.

time reserve Time between sudden total loss of oxygen supply and time when human can no longer be relied upon to function normally or rationally.

time-response parameter Addition of input time delay to assessment of response to pilot input of pitch rate [rotation] and normal acceleration.

Time-Rite Patented indicator of piston position for timing (1).

time series Sequence of time-variant measures, either continuous (eg barograph trace) or discrete (eg hourly met. pressure readings).

time sharing *1* Use of one EDP (1) processor or computer, usually large and beyond means or requirements of each customer, by a number of customers or users whose programs are run in short bursts in time-division multiplexed form switched according to cyclic formula agreed between users (in simplest form, a round robin).

2 Planned allocation of time to external scanning [typically 18 s] and to looking around cockpit [typically 3 s].

time signal *1* Broadcast signal used as very accurate time reference.

2 Time reference mark along border of reconnaissance imagery or other film.

time/size plot Diagram whose ordinate is a measure of aircraft size, eg MTOW or pax seats, and abscissa is time in years.

time slot Slot (3).

time/speed scale Scale for given groundspeed used in conjunction with plotting chart or topographical map.

timeswitch Electrical switch activated by time of day or elapsed time from a start point.

time/temperature cycle recorder Records time engine spends at critically high TGT, to give realistic indication of hot-end life.

time tic Time reference mark along telemetry readout; usually small inverted V every second along straight timebase.

time tick Regular time signal of one or more audible brief sounds.

time to go In air intercept, time to fly to offset point from any other initial position; after offset point, time to fly to intercept point (DoD).

time zone   Regions of local standard time, esp. over sea areas, where they are exactly divided by 15° widths of longitude.

timing   *1* Angular positions of piston-engine crankshaft at which valves first rise from seats or touch them again, and at which spark occurs; also called valve *, ignition *.

2 In US, assessment of human pilot's ability to co-ordinate flight controls on correct time basis for smooth manoeuvres; not often regarded as a topic elsewhere unless demonstrably faulty.

timing consideration   Measure of time missile (or, possibly, other weapon such as aeroplane) is exposed on ground between withdrawal from hardened shelter and launch (probably arch.).

timing disc   Disc, engraved marking or other feature on piston engine to assist establishing exact crankshaft angular positions for timing purposes.

timing parallax   Film distance between time signal (2) and corresponding frame of imagery.

timing pulse   Pulse used as time reference in telemetry, radar and SSR and other electronic systems.

TiMMC   Titanium metal-matrix composite.

Timos   Total-implant MOS device or circuit.

Tims, TIMS   *1* Technology integration of missile subsystems.

2 Tactical information management system.

3 Training Integration Management System, flight scheduling and student records (USAF).

tin   *1* Soft white metal, density 7.31, MPt 231.85°C, symbol Sn (stannum).

2 To coat surface of mild steel sheet with tin to prevent corrosion.

3 To coat metal surface with solder before making joint.

4 Aircraft, not necessarily metal (US colloq.).

Tina   Thermal-imaging navigation aid.

tinfish   Torpedo (UK colloq.).

TINS, Tins   Thermal-imaging navigation system, or set.

Tinsel   Transmitter carried by bomber to jam ground instructions to German fighters (RAF WW2).

tin-strip   Metal prefabricated-plank airstrip for STOVL.

t$_{int}$ Integration time, especially radar filter integration time.

tin-tray game   Stewardess trolley race.

tin wing   Lightplane whose wings are metal-skinned.

TIO   US piston engine designation: turbocharged, direct injection, opposed.

TIOS   Two-in-one service (Satcoms).

TIP   *1* Message code: until past specified waypoint or other point (ICAO).

2 Tracking and impact prediction.

3 Technical information panel (Agard).

4 Test integration plan.

5 Tailored instruction program (US).

6 Threat image projection, to test X-ray baggage screeners.

7 Technical improvement program, or technically improved product (IFF).

8 Transit improvement program.

9 Tiros information processor.

tip   *1* Extremity of aerofoil.

2 Angle of rotation of reconnaissance camera about aircraft transverse axis; also called pitch.

3 Wing-tip fuel tank (DoD) (colloq., adjective).

tip aileron   Aileron forming most or all of tip of wing.

tip cargo   Special cargo, eg radioactive isotopes, carried in small compartment in wingtip of some transports.

tip chord   Chord at tip of aerofoil, esp. wing, normally measured parallel to plane or symmetry of wing (for variable-sweep, at minimum sweep angle) between points where straight leading/trailing edges meet curvature at tip. Where both edges have pronounced sweep at tip, or where they are joined by line not parallel to plane of symmetry (eg Lightning, Tornado) other definitions apply, often unique to type.

tip cropping   Cutting off at Mach angle.

tip dragger   *1* Spoiler above wingtip used asymmetrically to cause yaw.

2 Sailplane (colloq.).

tip drive   Rotation of main rotor(s) of rotocraft by thrust applied at or near tips.

tip droop   Downward folding of wingtips through large angle, usually 60°–80°, to move forward aerodynamic centre of wing at supersonic speed and decrease trim drag; in some aircraft (XB-70) also generated compression lift.

tip float   See *stabilizing float*.

tip generator   Wingtip vortex generator.

TIPI   Tactical information processing and interpretation system (USAF).

tip in   To bank steeply away from takeoff flight path.

tip jet   Any system providing propulsive thrust at the tip of a helicopter main-rotor blade: pressure jet, cold [compressed-air] jet, ramjet, pulsejet, rocket or turbojet.

tip loss   Inefficiency of tip of aerofoil in lifting mode caused by spanwise deflection of isobars and relative wind, in some transonic cases approaching 90° and making tip mere dead weight.

tip loss factor   Correcting factor in calculating lift of rotorcraft lifting rotor to allow for tip loss.

tip-path plane   Plane containing path of tips of helicopter or other rotorcraft main lifting rotor, tilted in direction of travel or horizontal acceleration.

tipping   See *propeller tipping*.

tip pod   Streamlined container carried centred on or below tip of aerofoil.

tip radius   Usually synonymous with radius.

tip rake   See *rake*.

Tips, TIPS   *1* Total integrated pneumatic system (C-5).

2 Telemetry integrated processing system (AFSC).

3 Technical issue panels (FAA).

4 Transatlantic industrial proposal solution[s] (AGS6).

tipsail   See *winglet*.

tip shroud   *Shroud 1*.

tip speed   Tangential speed of rotating tip of propeller or rotor due solely to its rotation and ignoring superimposed vehicle airspeed; i.e., $V = r\varpi$, radius multiplied by angular velocity.

tip stall   Stall of tip of aerofoil, esp. wing, while remainder of surface remains unstalled; common condition caused mainly by higher lift coefficient at tip unless stall strip applied inboard.

tip tank   Fuel tank formed as streamlined body, jettisonable or otherwise, carried centred on or below wingtip.

tip trailing vortex   See *vortex*.

tip vortex   See *vortex*.

TIR   *1* Total indicator reading.

2 Target-illuminating radar.

3 Tracking and illuminating radar

4 Thermal-imaging radar.

5 Traffic information radar.

6 Thermal infra-red.

7 Twin intermeshing rotors (helicopter).

TIRC　Tactical IR countermeasure.

tire　UK spelling 'tyre' is used in this dictionary.

tiredness　General deterioration of airframe caused by long and intensive use, primarily manifest in repeated cyclic loading and successive severe gusts but also including superficial damage caused by impact of steps, ground vehicles, stones etc; no significant crack need be present but many structural parts will not be original and many boltholes will be oversized and re-reamed for bolts of increased diameter.

Tiros　TV/IR observation satellite[s].

Tirp　Terminal instrument radar procedure.

TIRS　Transverse-impulse rocket subsystem (planetary lander).

TIRSS　Theatre intelligence, reconnaissance and surveillance study (USAF).

TIS, tis　*1* Tracking information (or instrumentation) subsystem.

2 Thermal-imaging sensor, or system.

3 Tactical Intelligence Squadron.

4 Traffic information service(s) [aircraft-position datalink, ground or airborne receiver].

5 Tactical input segment (satellite).

Tisar　Terrestrial inverse SAR.

Tis-B　Traffic information service, broadcast (FAA).

TISD　Tactical Information Systems Division (Langley AFB).

Tiseo　Target-identification system, or sensor, electro-optical.

TISH, Tish　Thermal-imaging sensor head.

TISS, Tiss　Thermal-imaging security system, or surveillance system.

TIT　Turbine inlet temperature; see *turbine temperatures*.

tit　Any control button, especially to fire guns (UK, colloq., WW2).

Titan　Thunderstorm identification, analysis and 'nowcasting', under development from 1990s (USWB, NASA, FAA).

titanium　Ti, hard silvery metal, density 4.5, MPt 1,660°C, reactive but bulk metal passivated by oxide/nitride coating in atmosphere, vast range of aerospace uses, main tonnage Ti-Al-V alloys, see *Ti*.

titanium aluminides　Rapidly growing range of refractory (820°C) metals with properties marred only by poor toughness and ductility.

Tite　Tews intermediate test equipment.

title block　Standardized rectangular format on drawing, usually lower right corner, listing title, part numbers, mod states, names of draughtsmen/tracers etc, dates and other information.

titles　Name of owner or operator painted on commercial or GA aircraft, to be read from a distance.

TIU　Time insertion unit.

TIV　Tactical intervention vehicle.

TIVO　US piston-engine designation: turbocharged, direct injection, vertical crankshaft (for helicopter), opposed.

TIW　Total[ly] integrated warfare.

TIZ　Traffic information zone.

TJ　Turbojet.

TJAG　The Judge Advocate-General.

TJF　Transportable JTIDS facility (RAF).

TJRJ　Turbojet/ramjet or turboramjet.

TJS　Tactical jamming system.

TK　*1* Turbocharger (R, G).

2 Thermal keel.

Tk　Track, track angle.

TKE, TkE　Track-angle error.

TKF　*1* Tactical combat aircraft (G).

2 Takeoff, also TKO,Tkof.

TKM　Tonne-kilometres.

TKOF, tkof　Takeoff.

TKP　*1* Tonne-km performed; basic measure of airline traffic.

2 Transport clearing house (R).

TKS　Chemical de-icing pastes and pumped liquid (typically 60% aqueous solution of glycol), from Tecalemit/Kilfrost/Sheepbridge-Stokes.

TKT　Sandwich of Teflon/Kapton/Teflon, uniquely resistant even to electric arcing.

TL　*1* Thermoluminescence.

2 Transition level.

3 Transmission loss.

4 Until.

5 Terminal location (Acars/AFEPS).

6 Turbinen-luftstromtriebwerke = turbojet (G).

T/L　Top level.

TLA　*1* Towed linear-array sonar.

2 Throttle- [or thrust-] lever angle.

TLAR　*1* "That looks about right".

2 Top-level aircraft requirements [for reliability].

TLBR　Tactical laser beam recorder.

TLC　*1* Trans-lunar coast.

2 Ton[ne]s lifting capacity.

3 Through-life costs.

4 Takeoff and landing chart program.

5 Tender loving care.

TLD　*1* Technical-log defect.

2 Top-level domain.

TLDHS　Target location designation and hand-off system.

TLDM　Royal Malaysian navy.

TLE　*1* Type life extension.

2 Target-location error.

3 Treaty-limited equipment.

TLG　Tail landing gear.

TLI　Trans-lunar insertion.

TLLF　Tactical low-level flight, or flying.

TLM　Telemetry-word.

TLMC　Time limits and maintenance checks.

TLO　*1* Terminal learning objective.

2 Touchdown/, or takeoff/, liftoff area [also TLOF].

TLP　Tactical leadership program(me).

TLR　Target-locating radar.

TLS　*1* Tactical, or transponder, landing system.

2 Translunar shuttle.

3 Through-life support.

4 Target level of safety.

5 Training laser system (MoD, UK).

TLSI　Technical-log special inspection.

TLSS　Tactical life-support system (USAF flight suit).

TLTV　Towbarless tractor vehicle.

TLV　*1* Transition level.

2 Threshold limit value.

TLWD Tailwind.
TLWS Tactical laser weapon system.
TM *1* Training memoranda.
  2 Tactical missile.
  3 Trade mark.
  4 Ton-mile (seldom abb.).
  5 Transcendental meditation, relevant to aerospace.
  6 Transverse magnetic EM propagation mode.
  7 Telemetry.
  8 Technical manual, or memorandum.
  9 Thrust magnitude (of gross thrust vector).
  10 Time.
  11 Transmit manifold (Awacs).
  12 Timer/media (access control).
  13 Thermal model.
Tm Tropical maritime.
TMA *1* Terminal manoeuvring (or control) area, ie terminal airspace.
  2 Trimethylamine.
  3 Target-motion analysis.
  4 Traffic management advisor (FAA).
  5 Timer/media access.
TMAC Tactical medium-altitude camera.
TMB *1* Time mean bleed.
  2 Turbulent mixing boundary.
TMBACA Times microwave broadband airborne cable assembly.
TMC *1* Thrust-management computer [F adds function, S system].
  2 Titanium [or titanium-aluminide metal-] matrix composite.
  3 Terminal control.
TMCR Total maintenance-cost reduction.
TMCS Technical monitoring and control[ling] system.
TMD *1* Tactical munitions dispenser.
  2 Theatre missile defence.
  3 Test, measure and diagnose [or measurement and diagnostic].
  4 Tactical modular display.
TME Total mission energy, normally in non-SI kWh.
TMEL Trimethyl-ethyl lead.
TMET Tethered medium Earth terminal.
TMF *1* True-mass flowmeter.
  2 Thrust-management function.
TMG *1* Track made good.
  2 Thermal/meteoroid garment.
  3 Towing motor glider.
  4 Ton-miles per gallon.
T/MGS Transportable/mobile ground station.
TMIS Technicians maintenance information system.
TML *1* Tetramethyl lead.
  2 Terminal.
  3 TV microwave link.
TMM Tantalum manganese-oxide metal device.
TMMC Titanium/metal-matrix composite.
TMMS TOW mast-mounted sight.
TMN True Mach number.
TMO *1* Traffic management office (AFSC).
  2 Ten [nautical] miles out [from threshold].
TMP *1* Transverse-magnetized plasma.
  2 Twin machine-gun pod.
  3 Theatre mission planning; S adds system.
  4 Test-measurement program[me].
TMPA Traffic-management program alert.

tmpr, tmprly Temporarily.
TMRC Technical-manual reference card.
TMS *1* Thrust-management system.
  2 Test and monitoring station.
  3 Traffic, or technical, management system, or specialist.
  4 Tactical mission system (helicopters).
  5 Transformer mains supply.
TMSA *1* Trainer-mission simulator aircraft.
  2 Technical Marketing Society of America.
TMT Technology management team (ASTOVL).
TMU *1* Traffic management unit (FAA).
  2 Transducer matching unit (sonar).
TMW Tomorrow.
TMXO Tactical miniature crystal oscillator.
TN *1* Nuclear, thermonuclear (weapon prefix, USSR).
  2 Technology need.
  3 Technical note.
  4 True north.
TNA *1* Truth in Negotiations Act (US Congress).
  2 Training-needs analysis.
  3 Thermal- neutron analysis, or activation.
  4 Twin altitude.
TNAV, T-nav So-called four-dimensional navigation system commanding three spatial dimensions and time.
TNC Terminal node controller.
TND Trace narcotics detector.
TNDCY Tendency.
TNE Tungsten nuclear engine.
TNEL Total noise exposure level; see *noise*.
TNF Theatre nuclear forces (S^3 or S-cubed adds 'survivability, security and safety').
TNGT Tonight.
TNH Turn height.
TNI *1* Total noise index; see *noise*.
  2 Trusted network interpretation.
TNR Transfer of control message, non-radar.
TNS Technical news-sheet.
TNT *1* Trinitro-toluene; for * equivalent see *yield*.
  2 Tragflügel neue technologie, advanced supercritical wing (G).
TNW *1* Theatre nuclear weapon.
  2 Tactical nuclear warfare.
TO, T-O *1* Takeoff.
  2 Technical order.
  3 Table of organization.
TOA *1* Total obligation[al] authority, sum that may be obligated in coming FY for contracts possibly running for many years hence.
  2 Time of arrival, hence TOA/DME.
  3 Usually plural, transportation operating agencies (MAC, MSC and MTMC, US).
  4 Training options analysis [software tool].
TOAA Takeoff obstacle accountability areas (study).
TO&E Table of organization and equipment.
toboggan In-flight refuelling technique in which shallow dive is maintained to match speeds of fast tanker (if necessary with spoilers or airbrakes) and slow receiver.
TOC *1* Top of climb.
  2 Total operating cost (often t.o.c.).
  3 Travel order card.
  4 Tactical operations center (US).
  5 Transfer of communication[s].
Toca Theatre operational CIS(3) architecture.

TOCC   Tactical operations control centre.

TOCG   Takeoff c.g. position.

TOCS, Tocs   Terminal operations control system.

TOD   *1* Top of descent.

2 Takeoff distance.

3 Terrain/obstacle database.

4 Time of day, or of departure.

TOD$_a$, TODA   Takeoff distance available; H adds helicopter.

TO dist   Takeoff distance.

TODP   Takeoff decision point.

TOD$_r$, TODR   Takeoff distance required.

TODS, Tods   Tactical optical-disk system.

TOE   *1* Ton (usually tonne or short ton) of oil equivalent; measure of energy.

2 Table of organization and equipment.

toe   *1* Figurative forward extremity of ski shape whose contact with ground is commanded by TFR.

2 Any lateral extremity at foot of graphical plot.

toe brakes   See *wheelbrakes*.

toed in   Left/right (eg engines) have axes which in horizontal plane are inclined to meet aircraft centreline ahead of nose. Hence, **toed out**; axes meet centreline to rear, as in case of engines whose axes are perpendicular to tapered leading edge (eg Ju 52/3m).

toe out angle   Angle between major chord of winglet and OX axis, generating inward side force.

toe plates   Hinged tapered plates along outer edges of cargo-aircraft vehicle ramp.

TOEPR   Takeoff engine pressure-ratio.

TOF   *1* Takeoff fuel; quantity aboard at takeoff.

2 Time of flight.

3 Trigger on failure.

TOFL   Takeoff field length.

TO-FLX   Derated (flexible) take-off.

TOFP   Takeoff flightpath.

to/from   Indication of whether certain radio navaids are moving towards or away from ground station, either by caption window in instrument or by various switches or procedures; also called sense indication. Esp. applies to VOR.

TOFT   Tactical operational flight trainer [simulator] (USN).

TO/GA   Takeoff or go-around (overshoot).

Toga button   Automatically advances throttle levers to takeoff thrust.

toggling   Joggling.

TOGR   Takeoff ground roll.

TOGW   Takeoff gross weight, either published MTOW or that at one particular takeoff.

TOI   Time of intercept.

TOJ   Track on jam[ming].

TOL   Takeoff and landing; A adds analysis.

Told card   Takeoff and landing data, kept handy in cockpit.

tolerance   *1* Maximum departure permitted between dimension of an actual part and its nominal value; usually part may be either over or undersize (eg 653 ± 0.1 mm) but occasionally * is unilateral (eg 653 - 0.1 mm).

2 Maximum error permissible in calibration of instrument or other device.

3 Maximum quantity of harmful radiation which may be received by particular person with negligible results, also called * dose.

4 Ability of individual to withstand cumulative doses of drug.

toluene   Flammable liquid used as solvent and thinner; also called methyl benzene ($C_6H_5CH_3$) or toluol.

TOM   Target object map.

tombstone technology   Advances triggered by fatal accidents.

tome   X-ray slice through running engine or other subject. Generated by neutrons, gamma rays or other radiation [today usually PET (2)].

tomodromic   Heading to intersect a particular line, eg trajectory of another aircraft.

tomogram   Array of tomes scanned in sequence to provide 3-D picture.

TOMS   Total ozone mapping spectrometer.

ton   *1* Standard SI-related unit is tonne (t), = 1,000 kg = 1 Mg = 0.984207 long ton = 2,204.6236 lb. In Americas 2,000 lb, commonly called short * (not abb.), = 907.18474 kg. In UK and Commonwealth 2,240 lb, commonly called long * (not abb.), = 1,016.0469088 kg = 1.12 short *. In aerospace much confusion exists because of these three values, especially 12% difference between short and long * in aircraft payloads, airline traffic (usually short *), airfield pavements (mainly metric) and many other areas. When used incorrectly as a force [tonf] UK ton = 9.96402 kN, US ton = 8.89644 kN.

2 In air-conditioning and *refrigeration*, rate of removal of heat sufficient to freeze 1 short * of ice each 24 h = 3,140.05 W (if long * is basis, 3,516.85 W).

tonal balance   Can refer to audio frequencies (balance across pitches of sounds as heard by listener) or to white/grey/black or colour tones in radar or other electronic display.

tone   *1* Sound of one pitch containing no harmonics, usually synonymous with mono-*.

2 Specifically, in AAM launch, aural note which changes to singing or growling after IR lock-on.

tone localizer   Localizer whose L/R indications are received as contrasting tones heard on each side of glide-slope centreline.

tonf   Ton force, non-SI unit of force, = 9,964.02 N.

Tonka   See *R-Stoff*.

ton-mile   Unit of aircraft work; assuming long ton and statute mile = 1.5838 tonne-km (reciprocal 0.6314); * per Imp. gal = 0.3484 tonne-km/litre (reciprocal 2.8703).

tonne   Metric ton, 10^3 kg = 0.98433 long tons = 1.10231 US tons.

TO N1   Takeoff engine fan speed.

TOO   Target of opportunity.

tool   Though obviously normal meaning applies in aerospace, an added meaning is extension to include any device or construction facilitating manufacture or assembly, even when it plays no part in shaping workpiece. Examples include assembly structures of species in most cases preferably called jigs (more explicit), as well as temporary fixtures, struts and props, inflatable bags, rubber press-* and dies of all kinds, and devices for holding or locating during tests or other operations.

tool bit   Small cutting tool, usually from square steel bar with super-hard added tip, fixed in place on machine tool; not used for drills and millers.

tool design   Design of tools for particular programme, esp. design of all required jigs, fixtures, templates, gauges and special-purpose tools, eg for checking dimensions and alignment of large parts.

tooling　See *Tool*.

toolmaker　Skilled person, usually previously machinist, responsible for making many special-purpose in-plant tools (both jigs/fixtures and cutting tools) and in particular for setting up machines for semi-skilled operatives and minders, today often versed in NC.

toolroom　Originally room where cutting tool bits were kept, today clean (often in strict sense) environment for super-accurate measures, gauges and manufacturing operations calling for abnormal standards of accuracy.

tool steel　High-carbon steels retaining extreme hardness at elevated temperatures (note: bits are now usually carbides, cermet or other materials).

TOP　*1* Total obscuring power; basic measure of chaff or aerosol, in US expressed in non-SI units sq ft/lb (cross-section of sky per unit mass dispensed), for 80% opaqueness to hostile radar or other sensor.

2 Tube à ondes progressives = TWT (F).

3 Technical and office protocol, similar to CNMA and MAP 6 (US).

4 Takeoff power.

5 Top of cloud(s).

Topcap　Total objective plan for career airmen personnel (USAF).

top chord　Main transverse (end-to-end) upper member of truss.

top cover　Defending friendly fighters watching over bomber or attack aircraft from higher level, esp. while over hostile territory.

top dead centre　Instantaneous position of piston engine or reciprocating-pump crankpin in which centreline of crankshaft, crankpin and cylinder are all in line with piston at extreme top of stroke; hence also corresponding position of piston.

top dressing　Application of ag-chemical to land or growing crops from above; normally method of applying fertilizers rather than insecticides, for which technique may be to coat undersides of leaves also.

top-hat　Family of standard structural sections based on five straight surfaces, each at 90° to neighbour(s); resemble top hat in shape.

Topkat　Tele-operated precision kill and targeting.

top loading　Increasing apparent (effective) height of radiating aerial by adding metal plate, mesh or radial wires at extremity.

TOPM　Takeoff performance monitor; S adds system.

topocentric　Referred to observer's position; measures, usually linear distance or az/el, based on observer's position as origin.

topographic　Representing physical features of Earth's surface, both natural and man-made; hence * display, * map. DoD definition of * map: one which presents vertical position of features in measurable form as well as horizontal. Normally, essential feature is use of contour lines, as well as normal positional information.

top overhaul　Overhaul of piston-engine cylinders (valve grinding, ring replacement, decarbonization etc) without opening crankcase.

topping　Operating cycle of liquid-propellant turbopump for rocket engine in which cryogenic fuel is heated, producing high-pressure gas used to drive turbine(s); this gas then passes at lower pressure to combustion chamber

(different nozzles from main flow), where it burns. Hence * cycle, * engine.

topping off　Replacement of cryogenic propellant lost by boiloff.

topping up　Replenishment of gas-filled aerostat, eg after a flight.

topple　Real or apparent wander in vertical plane of gyro-axis (see *toppled*).

toppled　Gyro whose gimbals have for any reason ceased to maintain its correct axis in space, so that further rotation of mounting results in violent direct precession. Traditional gyro instruments can be * by aerobatics or any rotation of aircraft axes beyond defined limits, instrument then being useless as attitude reference until gyro has settled again into normal operation. New term is needed for either topple or toppled.

top rudder　Applying rudder towards the upper wing in a turn: thus, in a steeply banked L turn, pushing on R pedal [eg, to keep nose from dropping below horizon].

TOPS　*1* Thermoelectric outer-planet spacecraft.

2 Transfer orbit and payload-testing support.

Topsar　Topographical synthetic-aperture radar.

Topscene　Tactical operational scene (Tamps).

Top Secret　High grade of defence classification for material whose unauthorized disclosure might result in severance of diplomatic relations, war or collapse of defence planning.

Topsep　Targeting optimization for solar-electric propulsion.

top shock　Shockwave on upper surface of aerofoil.

topside　On carrier flight deck, esp. movement thereto by elevator; eg coming * as aircraft or other item appears level with deck (USN).

top-temperature control　Any subsystem limiting a temperature to a specified safe limit, esp. that for TET, TGT or equivalents.

TOR, T-OR　*1* Takeoff run.

2 Tentative operational requirement.

3 Terms of reference.

tor　Torr.

TOR$_a$, TORA　Takeoff run available, usually = TOR.

Toray　Trade name [from torched rayon] of carbon fibres of outstanding specific strength and modulus.

torch igniter　Combined igniter plug and fuel atomizer emitting jet of flame from burning fuel. Very rare in gas turbines but occasionally used in afterburners and a minority of liquid and solid rockets.

torching　*1* Faulty operation of gas turbine, esp. jet engine, in which unburned fuel travels past turbine and results in flames travelling down jetpipe, often expelled from exhaust nozzle.

2 Faulty operation of piston engine in which unburned fuel travels through exhaust valve and burns in exhaust pipe, often causing visible flame beyond exhaust nozzle.

3 Degassing in ultrahigh-vacuum technology by applying gas flame to walls.

toric　Having a surface described by a segment of a conic section.

toric combiner　Optical lens assembly used to combine a generated-information display with an image of real world.

Torlon　Heat-resistant resin used in graphite-fibre composites for high-temperature applications (Amoco).

tornado　Localized violent whirlwind east of Rockies in

US with such low pressure in core as to explode structures in its path, usually pendant under a Cb. Also used for Gulf of Guinea thunder squalls advancing westwards in line.

toroidal   Shaped like doughnut.

toroidal vanes   Rings of curved section guiding air to eye of centrifugal compressor.

torpedo director   Traditional optical sight for aerial torpedo attack; user sets target size/speed and receives azimuth guidance.

torque   For all practical purposes, synonymous with turning moment or couple [which see]. A rigorous definition is effectiveness of a force in setting a body into rotation, according to which trying to loosen a tight nut unsuccessfully or rotate free end of rod fixed at other is not application of * (though in second case it is torsion). Often invertedly defined as resistance to a twisting action. For propellers see * *component*.

torque box   Box-like structure, eg wing torsion box, designed to resist applied torque.

torque brake   Variable brake on rotating shaft, eg slat drive, triggered at particular point of system travel.

torque coefficient   Product of propeller torque divided by $\rho N^2 D^5$; $k_Q = Q/\rho N^2 D.^5$.

torque component Q_c, tangential force acting in plane of propeller rotation on any elementary chordwise lamina; thus total propeller torque $Q = N\frac{1}{2}\rho\ V^2 \times$ integral of Q_c from axis to tip with respect to radius.

torque dynamometer   Measures shaft power by measuring N (rpm) and torque.

torque effect   Reaction on vehicle of torque applied to propeller or rotor (** for rotodome usually ignored); in helicopter countered by tail rotor.

torque horsepower   Shaft horsepower, often same as brake horsepower. Use to be discouraged because of confusion with thp (thrust horsepower).

torque link   Pivoted links preventing relative rotation between cylinder and piston of oleo shockstrut; limiting factor with bogie main gears on allowable steering angle of nose gear. Also called scissors or nutcracker.

torquemeter   Device, either instrument or component part of engine, for measuring torque; in turboprop or some piston engines, usually oil-pressure system sensing axial load on reduction-gear planetary helical gears or, less often, tangential reactive load around annulus gear.

torquer   Device imparting torque to an axis of freedom of a gyro, usually in response to signal input.

torque roll   Performable only by aircraft with fast-responding (eg piston) engine on centreline giving very large torque in relation to aircraft weight; approach is made at flight idle at minimum safe flying speed, whereupon throttle is banged wide open to cause rapid roll in opposite direction as aircraft accelerates, pilot recovering to wings-level with aileron.

torque-set screw   Can be repeatedly unscrewed without losing original torque needed to release; used to latch long-MTBM panels.

torque stand   Test stand for engines, esp. aircraft piston engines.

torque tube   Tubular member designed to withstand torque, either one applied inevitably and to be resisted or one to be transmitted as part of drive system (eg in primary flight controls).

torque wrench   Hand tool with dial or other direct

readout of torque imparted, usually set to slip if overloaded.

torquing   Input to gyro from torquer for slaving, capturing, slewing, cageing etc.

torr   Non-SI unit of high-vacuum pressure, = 133.322 $N/m^2 = 0.0193368$ lbf/in^2. Originally (and still very nearly) 1 mm Hg.

torsion   Deflection, usually within elastic range, caused by twisting, ie applied torque (note: * is result, not an applied stress).

torsional divergence   Potentially lethal design fault in which wing's aerodynamic centre is ahead of shear centre or elastic axis, lurking unsuspected until a critical IAS is exceeded.

torsional instability   Characteristic of structural member such that, when loaded in compression or bending, it will twist before reaching ultimate compressive stress.

torsional load   One imparting turning moment or torque.

torsional stress   Stress resulting from applied torque; for torque tube $S_s = Tc/J$ where T is torque, c is radius and J polar moment of inertia.

torsion balance   Instrument containing light horizontal rod suspended by fine fibre for measuring weak forces, eg gravitation, radiation.

torsion-bar tab   See *spring tab*.

torsion box   Main structural basis of wing, comprising front and rear spars joined by strong upper and lower skins; also called wing box, inter-spar box.

torso harness   Normal seat harness of military pilot restraining torso over full length.

TOS   *1* Transfer orbit stage (eg Shuttle to geosynchronous).

   2 Time on station.

   3 Traffic orientation scheme.

   4 Tactical operations system.

   5 Tiros operational satellite[s].

TOSA   Takeoff space available.

TOSS   Television optical scoring system.

toss bombing   Method of attack on surface target with free-fall bomb, esp. NW, in which aircraft flies toward target, pulls up in vertical plane and releases bomb at angle that compensates for effect of gravity drop; similar to loft-bombing and unrestricted as to altitude but normally entered from lo. Two main varieties: forward **, in which bomb is released at angle short of 90° (usually about 70°), after which aircraft continues with Immelmann-type manoeuvre; and over-the-shoulder **, in which aircraft overflies target and releases at angle beyond 90°.

TOT   *1* Time on (or over) target.

   2 Turbine outlet temperature.

   3 Time-oriented task.

   4 Total.

   5 Transfer of title, on delivery of aircraft.

t_{ot} Time on target (radar).

total   Noun or adjective, damaged far beyond economic repair.

total air temperature   Temperature of air brought to rest including rise due to compressibility.

total blade-width ratio   Ratio of propeller diameter to product of number of blades and maximum blade chord; also known as total propeller-width ratio.

total conductivity   Sum of electrical conductivities of all free ions, positive and negative, in given volume of atmosphere.

total curvature   Change in direction of ray between object and observer.

total drag   Component of total aerodynamic force on unducted body parallel to free-stream direction = induced plus profile = pressure plus surface friction.

total-energy equation   Expression for sum of pressure, kinetic and potential energies of given volume of atmosphere as result of combining mechanical energy equation with thermodynamic.

total engagement training   Involving actual or simulated firing of weapons.

total equivalent horsepower   See *equivalent horsepower*.

total head   See *total pressure*.

total impulse   Basic measure of quantity of energy imparted to vehicle by rocket, = integral of thrust versus time over total operating time, abb. I^t, expressed in Ns, kNs or (US) lbf-s.

totalizer   Indicator showing quantity of variable (fuel, ammunition etc) that has passed sensing point (see *detotalising*).

totalled   Damaged beyond repair.

total lift   Component along lift axis of resultant force on aircraft.

total-loss lubrication   System in which oil is supplied, usually under cartridge or other stored gas pressure, and finally dumped overboard; common in target or cruise-missile propulsion.

total noise rating   See *noise*.

total obligational authority   Money for 5-year defence programme or any portion for a given FY (DoD).

total operating time   Time between ignition of solid-propellant rocket motor and time when thrust decays to zero; this is usually at least 15% longer than burn time and 10% longer than action time.

total-package procurement   Award of one very large prime contract for entire operative system from conceptual stage through R&D to engineering design, test and production.

total pressure   Pressure that would be reached in fluid moving past body if its relative velocity were to be brought to zero adiabatically and isentropically. For low speeds taken as p ½ ρV^2 and at high Mach numbers as $p(1 + [\gamma - 1] M^2/2)^{\gamma/\gamma-1}$. Also called total head or stagnation pressure, and usually same as pitot pressure, impact pressure.

total-pressure head   See *pitot head*.

total propeller width ratio   See *total blade-width ratio*.

total refraction   Curvature of radiation out of layer or medium.

total system   Entire system supplied to virgin site, including accommodation, civil engineering, power supply, refuse disposal etc, as turnkey contract.

total temperature   Temperature of particle of fluid at stagnation point or otherwise brought to rest adiabatically and isentropically. If T_s is static temperature, $T_H = T_s(1 + ½[\gamma - 1] M^2)^{-1}$.

total terrain avionics   Combine digital contour and map information to enhance sensors and displays for carefree flight at high speeds at low altitudes in bad weather without any high-energy or readily detectable emissions, thus facilitating stealth design.

TOTE, Tote   Tracker optical thermally enhanced.

tote board   Display board presenting written information in tabular form, esp. in ATC (1) flight-progress board or cockpit alphanumeric tab (4).

TOTS, tots   Tower operator training system (USN ATC).

touch-and-go   Practice landing in which aeroplane is permitted to touch runway briefly; in many cases flaps are moved to take-off setting while weight is on wheels.

touch-control steering   Small inputs by pilot to change flight path while in autopilot mode.

touchdown   *1* Moment, or location, of contact of aircraft with surface on landing or of soft-landing spacecraft with designated destination surface.

2 Intersection of glidepath with Earth's surface, not necessarily point of any actual landing.

touch down   To perform touchdown.

touchdown aim-point   Area of runway on which pilot intends to land. This is usually in touchdown zone, but on STOL runway a ** marker is provided in form of 90 m (200 ft) axial white strip on each runway edge projecting inwards from white edge strips.

touchdown point   That programmed into UCARS or other UAV recovery system.

touchdown ROD   Touchdown rate of descent; value shown by ** indicator, usually sensitive VSI based on radio altimeter or laser altimeter.

touchdown RVR   Touchdown runway visual range; RVR at time and place of landing.

touchdown zone   That portion of runway selected by most pilots as touchdown zone; on precision instrument runway marked by three close axial white bars 90 m (200 ft) long on each side between centreline and edge, beginning 150 m (500 ft) beyond threshold.

touch-screen technology   Ability of advanced displays to interface with humans by direct fingertip touch of part of display of interest, notably by touching particular line or word in alphanumeric readout.

touchwire   Human input to electronic display in form of matrix of fine wires, any of which, when touched, switches enlarged local region of display to fill entire area (or, in alphanumerics switches in amplified readout of that particular item).

toughness   Ability of structural material to absorb mechanical energy in plastic deformation without fracture.

tour   *1* Individual crew member's assigned total of combat missions. To qualify, mission must be effective. RAF * in WW2 was usually 30.

2 More generally, * of particular duty for military personnel.

touring aircraft   *1* Original meaning, aircraft designed for long pleasure flights.

2 Aircraft making appearances at successive air displays.

touring motor glider   Light aeroplane designed to cruise under power but to soar when conditions permit.

tourist   Originally (1949) special high-density airline accommodation usually synonymous with coach; today standard type of seating, denoted by symbol Y.

TOVC   Top of overcast.

TOVS   Tiros operational vertical sounder.

TOW   *1* Takeoff weight, usually meaning MTOW.

2 Tube-launched optically-tracked wire-guided (missile).

3 Time of week (GPS).

tow *1* Standard manufactured form of reinforcing fibre, eg carbon, graphite, as long unwoven staple.

2 Aero * for one flight of sailplane; hence on *, * release.

towbar Connects tug and nose gear for towing or pushing away from gate.

towbarless tractor One designed to lift NLG off ground on to tractor body.

tow dart Dart-type aerial target towed by RPV or target drone or, in some cases and on 900 m (2,000 ft) line, by manned aircraft.

towed body Remote sensing unit of helicopter MAD or airborne magnetometer.

towed glider Glider on aero tow.

towel rack Rail-like aerial (antenna) for HF com or Loran (colloq.).

tower Airport or airfield control tower, esp. service or facility based therein; hence * airport, * frequency, * controlled. Increasingly coming to mean seat of ATC even if no physical * exists.

tower fly-by Fly past tower at low level, eg for determination of position error or visual check on aircraft configuration.

towering Opposite of stooping, refraction phenomenon in atmosphere in which visual image of distant object appears extended vertically.

towering cumulus Building rapidly, so that height exceeds any lateral dimension.

towering takeoff Helicopter rises vertically under full power and goes ahead as rate of climb decays to zero.

tower shaft Radial shaft transmitting drive from engine spool to accessory gearbox or other unit.

tower-snag recovery Recovery of RPV or other winged vehicle by flying it to hook a line suspended between arms on tower built for this purpose; in most systems line imparts decelerative drag which stalls vehicle within distance significantly less than height of tower.

towhook Pilot-operated coupling release on glider for tow cable.

towing basin See *towing tank*.

towing eye Eye (structural ring) attached to nose gear or other part of aircraft for towing on ground.

towing sleeve Towed sleeve (drogue) target.

towing tank Long, narrow water tank for hydrodynamic tests, also called seaplane basin/tank, along which models of hull/float forms, skis, ACVs and other objects are towed. Seldom used for wavemaking.

Townend ring Pioneer ring-type cowling for radial engines, with chord seldom greater than external diameter of cylinders and no pretension at true aerofoil shape, though usually with tube around inner side of leading edge.

Townsend avalanche Cascade multiplication of ions in gas-filled counting-tube technology.

Townsend coefficients In DC gas-discharge, First ** = η = number of electron/ion pairs per volt; Second ** = γ = number of secondary electrons emitted from cathode per impacting positive ion.

Townsend discharge DC discharge between two electrodes immersed in gas and requiring cathode electron emission.

Townsend ionization coefficients Average number of ionizing collisions electron makes in drifting unit distance along applied field.

tow-reeling machine Powered winch for winding in cable towing aerial target or other device.

tow rope Connection between tug and glider, in WW2 typically 9 in [229 mm] or 10 in [254 mm] (circumference). Manila [hemp], later replaced by Nylon. See *cable*.

tow tractor Usually means prime mover for towing aircraft or baggage train.

Toxic-B Most common lethal air-dropped or dispensed war agent (USSR).

TP *1* Turning point.

  2 Thermoplastics.

  3 Test pilot, or point.

  4 Target, or training, practice.

  5 Teleprinter.

  6 Terminal processor.

  7 Traffic pattern; A adds altitude.

  8 Turbulence plot.

  9 Technical Publication[s]

  10 Tactics planner.

  11 Two-seat pursuit (USA 1919–24).

  12 Telecommunication [singular] processor.

Tp *1* Tailplane trim [actuator].

  2 Tropical Pacific.

TPA *1* Target-practice ammunition.

  2 Traffic-pattern altitude.

  3 Taildragger Pilots' Association (US).

  4 Trigger-pulse amplifier [D adds driver].

TPAR *1* Tactical penetration-aid rocket, carrying expendable ECM payloads.

  2 Trans-polar air route.

TPC *1* Total programme cost(s).

  2 Tactical pilotage chart.

  3 Temperature- [or thermally-] protective coating.

  4 Technical Partnership Canada.

TPCI Technical publications combined index.

TPD Tracking processing device.

TPDR Transponder (more often TXP or XPDR).

TPDU Transport Protocol data unit.

TPE *1* Tracking and pointing experiment.

  2 Thermoplastic elastomer.

TPED Tasking, processing, exploitation and dissemination (Imint).

TPF *1* Terminal phase, final.

  2 Technology performance financing.

  3 Terrestrial-planet finder.

TPFDD Time-phased force and deployment data.

TP4 Transport Protocol Class 4.

TPFP Target-practice frangible projectile [-T adds -tracer].

TPFT Tunable pipelined frequency transform.

TPG *1* TV picture generator, Tepigen.

  2 Technology planning guide.

  3 Topping.

T-phi, Tϕ T-S.

TPI Terminal phase initiation.

t.p.i. Turns (or threads) per inch.

TPIS Tyre-pressure indicating system.

TPL *1* Transmitted pulse length.

  2 Terminal permission list (Acars/Afeps).

TPM *1* Terrain profile matching.

  2 Technical performance management.

TPMU Tyre-pressure monitor unit.

TPN Technical procedure notice; (L) adds electronic.

TPP *1* Tip-path plane.

2 Total-package procurement (C adds concept).

3 Technology program plan (AFSC).

4 Terminal procedures publication.

5 Tri-phenyl phosphate, extreme-pressure anti-scuff oil additive.

TPPX   Target-practice proximity-fuzed.

TPR   *1* Terrain profile recorder.

2 Thermoplastic rubber.

3 Transponder [XPDR preferred].

4 Transient-phase restoration.

TPRM   Trusted protocol reference model.

TPS   *1* Thermal protection system.

2 Test program set.

3 Technical problem-solving.

4 Thermal picture synthesizer.

5 Test Pilot School (USAF).

TPSA   Technologies, processes and system attributes.

TPSRS   Terminal primary and secondary radar system(s).

TPT, TP/T   *1* Target practice, tracer.

2 Transonic pressure tunnel.

TPTA   Tailplane trim actuator.

TPTO   Temporary permission to operate.

TPU   *1* Terminal position update (RV).

2 Tactical, or transmitter, or transceiver, processing unit.

3 Turbine power unit (Gripen).

TPWG   Test planning working group.

TPX-42   Numeric decoder of aircraft beacons.

TQ   Total quality; C adds cost, E engineering, M management, PP planning and producibility and S supportability.

TQA   Throttle-quadrant assembly.

TR, Tr   *1* Track.

2 Thrust reverser.

3 Tracking rate.

4 Tactical reconnaissance (US, role prefix).

5 Torpedo reconnaissance (UK, defunct).

6 Technical report.

7 Temporary revision (ADRES, CAATS).

8 Braking parachute (R).

9 Total reaction.

10 Trace.

T/R   *1* Transmitter/receiver, communications radio.

2 Transformer/rectifier (or TR).

3 Thrust reverser.

t$_r$ Round-trip transit time, especially of radar signal.

TRA   *1* Track angle.

2 Radar transfer-of-control message.

3 Thrust-reverser aft (SST).

4 Temporarily reserved [or restricted] airspace [or area].

5 Terrain-referenced avionics.

6 Thrust-reduction altitude.

TRAAMS   Time-referenced angle-of-arrival measurement system.

TRAC, Trac   *1* Telescoping-rotor aircraft.

2 Trials recording and analysis console, giving immediate video-tape of fire-control system performance.

3 Tactical radar correlator.

4 Terminal radar approach control.

5 Tradoc Analysis Center.

6 Transit research and attitude control (satellite).

Traca, Trac-A   Total radar aperture-control antenna.

Tracals   Traffic-control approach and landing system (USAF).

Trace   *1* Test equipment for rapid automatic checkout and evaluation.

2 Taxiing and routing of aircraft co-ordination equipment (US 1960s).

trace   *1* Line on CRT and many other displays made by electron beam, successive sweeps being linked by retraces.

2 Line of data on any linear graphic printout visible to eye (thus, not applicable to magnetic tape).

3 EDP (1) diagnostic technique which analyses each instruction and writes it on an output device as each is executed.

trace ice   Rate of accretion just exceeds sublimation.

tracer   *1* Ammunition whose projectiles leave bright visible trails.

2 Substance added (usually in very small proportion) to main flow in order latter may be followed accurately through process, living organism etc; * may be physical, chemical or, often, radioactive.

tracer display   True historic display on Hudwac or similar sight system featuring tracer line and other symbology for snapshooting, normally with real target scene through combiner glass.

tracer line   Bright line on sight system showing locus of points where a projectile would now be had it been fired during preceding few seconds, ie where projectiles from continuous burst would now be. A range marker, usually a ring, is superimposed at actual target range. Pilot must then place this ring over target in order to hit it, or arrange for target to pass through ring.

track   *1* Path of aircraft over Earth's surface from take-off to touchdown.

2 At any time in flight, angle between a reference datum, and actual flightpath of aircraft over Earth's surface, measured clockwise from 000° round to 360°. Magnetic * is referred to magnetic N; true * is referred to true N and is * normally used in plotting; required * is that desired; * made good is that found by inspection to have been achieved; great-circle and rhumbline * are those which are thus represented on chart.

3 To observe or plot a * (1), eg by radar or on plotting board.

4 Series of related contacts on plotting board.

5 To display or record successive positions of moving object.

6 To lock on to source of radiation and obtain guidance therefrom.

7 Path traced by tips of propeller, rotor or similar rotating radial-arm assembly.

8 Distance measured as straight line between centre of contact area of left mainwheel, or geometric centre of left main-gear bogie, and corresponding centre on right; in case of aircraft with centreline gears and outriggers, measured between outriggers; if landing gears are skids, measured between lines of contact; if main gears are skis, measured between centrelines; if gears are inflatable pontoons, measured between centres of ground contact area; not normally applied to marine aircraft, but would presumably be distance between CLs of two floats.

9 As plural, rails along which travel area-increasing flaps or certain translating leading edge slats, carrying these surfaces out approx. in direction normal to leading or trailing edge.

10 To keep device aimed at moving target.

11 Conductive path on printed-circuit board.

12 DME mode after lock-on, when pulse-pair rate is reduced.

13 Position and velocity of aircraft estimated from correlated surveillance data (TCAS).

track angle   See *track (1)*.

track ball   Basic human interface with electronic displays, either for inputting data or calling up portion of display for any reason; comprises ball recessed into console rolled by operator's palm in any direction to generate either stream of digital pulses or analog voltage about two co-ordinate axes to achieve desired place, eg particular aircraft on display.

track beacon   See *NDB*.

track clearance   Clearance to fly stated track (1) as far as particular fix.

track correlation   Correlating track information using all available data, for identification purposes.

track crossing angle   *1* Angular difference between tracks of interceptor and target at time of intercept (DoD).

2 Generally, angle between two flight paths measured from tail of reference aircraft.

tracker   Hand-held electronic reader of coded information on parcel or letter, which is then automatically sent by radio (satellite if necessary) to management displays.

track handover   Process of transferring responsibility for production of air-defence track from one track-production area to next (NESN).

tracking   *1* Air intercept code: "By my evaluation, target is steering true course indicated" (DoD).

2 Precise and continuous position-finding of targets by radar, optical or other means (NESN). Hence synonymous with track (5).

3 Measure of correct rotation of separate blades of helicopter main rotor in that each should follow exactly behind its predecessor, ie all tips should lie in common tip-path plane.

4 Procedure to ensure * (3) by holding paper or fabric against painted blade tips and adjusting hub settings until a single spot results.

5 Correct holding of frequency relationships between all receiver circuits tuned from same shaft to maintain constant intermediate frequency in superhet or constant difference frequencies.

6 Keeping device, eg fighter aircraft, aimed at target; hence synonymous with track (10).

7 Flight path in horizontal plane, especially along ILS glidepath.

tracking station   Fixed station for tracking (2) objects in air or space.

track initiator   Person responsible for taking decision on appearance of unknown blip on air-defence or other surveillance radar that it represents a target whose track is to be determined and assigned an identity.

track intervals   Convenient time/distance divisions between checkpoints when navigating visually.

track lock lever   Hand lever locking flight-crew seat in desired position.

track made good   See *track (1)*.

track marker   Symbology on display indicating track, eg straight black or bright line, with or without arrowhead,

cross, ring or square, depending on type of display; absent when display is auto track-oriented.

track oriented   Aligned with current track at 12 o'clock position, eg hand-held map, projected map display, etc.

track prioritization   Order of threat priority can be manually or automatically assigned to several targets, usually on basis of TTG.

track production   Function of air-surveillance organization in which active and passive radar inputs are correlated into coherent position reports together with historic positions, identities, heights, strengths and direction of flight (NESN).

track-production area   Area in which tracks are produced by one radar station.

track repetition   Time between exact overflights of spot on Earth by satellite.

track separation   *1* Lateral distance between aircraft tracks imposed by ATC.

2 Distance (often at Equator) or angular longitude difference between successive passes of Earth satellite.

track symbology   Symbols used to display and identify tracks on radar or data-readout console or other electronic display.

track telling   Process of communicating air-defence, surveillance and tactical-data information between command and control systems and facilities: back tell, transfer from higher to lower echelon of command; cross tell, between facilities at same level; forward tell, to higher level; lateral tell, across front at same level; overlap tell, to adjacent facility concerning tracks detected in latter's area; and relateral tell; via third party.

track via missile   SAM or AAM guidance system based on multirole electronically scanned radar [eg, Patriot].

trackway   Standard prefabricated military track for land vehicles, quickly laid for recovery of force-landed aircraft or across infilled bomb crater on airfield, esp. to speed reopening of bombed runway.

track-while-scan   Radar/ECM scan produced by two unidirectional sector scans simultaneously scanning in two planes, usually one vertically and one horizontally, allowing target common to both to be accurately tracked in az/el as well as (medium PRF) range/V. Target is not alerted as subject to special interest.

Tracon   Terminal radar approach control (FAA).

Tracs   *1* Terminal radar and [or approach] control systems (Canada DND).

2 Test and repair control system, automated data retrieval (TRACS).

3 Tool[s] for rapid advances in cockpit simulation.

4 Transportable radar and communications simulator.

traction wave   Generated on tread surface of under-inflated tyre at high speed.

tractor   Adjective meaning pulling, hence * aeroplane is pulled by propellers, not pushed; * propeller is in front of engine and driven by shaft in tension. Converse is pusher.

tradcom   Transportation R&D Command (USAF, defunct).

trade   *1* Targets, eg a plurality of hostile aerial targets.

2 Of fighter, to encounter hostile aircraft.

tradeoff   Generalized term for fair exchange between inter-related variables; thus in aircraft design there are numerous and continuing examples of * between wing area, thrust, fuel consumption, gust response, structure weight and many other parameters; in aircraft flight management pilot can * (used as verb) speed for height, etc.

Tradoc   Training & Doctrine Command (USA); possible confusion with Tadoc.

traffic   *1* Quantity of vehicles, eg aircraft, in operation; measured as number in flight in region at one time, number under positive control, or general number in vicinity, or as number in given period. For control purposes includes * on movement area.

2 Number of landings and take-offs at airport in given period, eg one calendar year.

3 One aircraft in flight as reported to or noticed by another in vicinity.

4 Output of commercial or military air transport operator, measured in such units as number of pax or mass of cargo carried multiplied by mean distance each is transported, eg passenger miles or tonne-km (standard units compatible with SI are needed).

5 Number or frequency of messages on telecom system.

"traffic"   Repeated, aural warning of midair (TCAS).

traffic advisory   Information [without comment] sent to pilot about other traffic within ± 1,200 ft FL and [at existing closure speed] 45 s in time.

traffic alert and collision-avoidance system   As initially conceived, exists in two levels. TCAS I is the baseline system which merely senses potentially conflicting traffic and warns crew [by traffic advisory]. TCAS II additionally provides traffic information within c30 nm [55 km], and two conflicting equipped aircraft are manoeuvred apart.

traffic circuit   See *Circuit*.

traffic density   The number of xpdr-equipped aircraft [excluding one's own] within R nm [1.85R km] ÷ πR^2.

traffic information, radar   Information issued to alert aircraft to radar target observed on ground radar display which may be in such proximity to its position or intended route as to warrant its attention.

traffic lights   Any red/amber/green presentation, especially that by a radar altimeter referenced to a preselected low (minimum safe) height setting.

traffic pattern   See *circuit*; * usually used for tracks/profiles of arrivals and departures of non-GA traffic, ie military or commercial.

traffic situation display   TMS (3) tool for monitoring position of traffic to determine demand on airports and sensors.

TrAG   Training air group (RN, WW2).

trail   *1* Relative motion of dropped store, eg free-fall bomb, behind aircraft flying at constant V, broken down into * distance, cross-*, range component of *, * angle, cross-* angle, and range component of cross-*.

2 Distance between centre of tyre contact area with ground and intersection with ground of free castoring axis; not relevant to power-steered aircraft.

3 To shadow another aircraft or hostile ship(s).

4 Tendency of freely hinged (ie, not irreversible) control surface to align itself with relative wind. Normally negative, surface 'floating' in line with wind, but an overbalanced surface has positive * and unless restrained will be blown to limit of its deflection.

5 To fly a tanker behind potential receivers.

6 To extend tanker's hose.

trail angle   *1* Angle between vertical and line joining bomb impact to aircraft at time of impact.

2 Several angles in landing gear, including acute angle between bogie beam and aircraft horizontal plane (negative when front axle is lower than rear) and acute angle between local aircraft vertical and axis of main-gear oleo strut (often not relevant because of gear geometry).

trail blade   That immediately following in same stage of engine fan, compressor or turbine, especially following a blade that breaks off.

trail distance   Horizontal distance between point of bomb impact and point vertically below aircraft at time of bomb impact.

trailer   *1* Aircraft following and keeping under surveillance a designated airborne contact (DoD).

2 General US term for towed road vehicle, esp. for human occupancy.

trail formation   In direct line-ahead, each aircraft or element being directly in front of those following.

trailing aerial   Aerial in form of long wire, usually with weight or drogue on end, capable of being wound in or out from underside of aircraft.

trailing area   Area of flight control or other pivoted surface on aircraft downwind of hinge axis.

trailing blade   Trail blade (US usage).

trailing edge   *1* Rear edge of aerofoil or streamlined strut.

2 Outline of pulse as amplitude falls from peak to zero or minmum positive value.

trailing finger   Extra electrode in piston-engine magneto distributor which transmits large current from booster magneto to cylinder next in firing order when engine is started.

trailing flap   Not a normal term but could be applied to Junkers double wing.

trailing sweep   Sweep (5) when deviation is towards trailing edge.

trailing vortex   Vortex extending downstream from point on body.

trailing vortex drag   See *lift-induced drag*.

trail length   Length of cable connecting aircraft to braking parachute, anti-spin parachute or other drag device.

trail line   That between aircraft in level flight and bomb released from it. Projected, it reaches *trail point*.

trail point   Where trail line reaches Earth's surface.

trail rope   *1* Trailed by balloon over ground to reduce groundspeed and assist in regulating height.

2 Carried in airship for ground handling.

train   Bombs dropped in short intervals or sequence (DoD).

train bombing   Two or more bombs released at predetermined interval from one aircraft as result of single actuation of release mechanism (USAF).

trainer   Aircraft for training flight personnel, esp. pilots.

training aids   Items whose primary purpose is to assist instruction and growth of operator skill/familiarity, such as publications, tapes, films, mimic boards, systems rigs, procedure trainers and simulators.

training package   Self-contained arrangement to train personnel (eg of an air force or of purchaser of GA aircraft) for fixed fee or fixed outlay per month; may include design of trainer, construction of facilities as turnkey contract or part of larger programme.

trajectory   Flightpath in 3-D of any object, eg aeroplane or electron or other particle, with exception of orbits and other closed paths. Can be ballistic, acted on only by atmospheric drag and gravity, or controlled by various external forces.

trajectory band   Webbing strip round top of aerostat envelope to reduce distortion.

trajectory plotting   A particular meaning is using a wreckage field and knowledge of winds at all relevant altitudes to establish point in sky at which an aircraft broke up.

trajectory scorer   Instrument carried by aerial target which continuously defines position of intercepting missile in sphere whose centre coincides with origin of target's co-ordinate axes; readout is time-history record of missile range and angular position commensurate with scoring requirements.

trajectory shift   Distance or angular measure of deviation of missile from ballistic trajectory under influence of a thrust mechanism (ASCC).

Trakmat   UV-stabilized polypropylene overlain by mesh of galvanised-steel rods and wires.

TRAM, Tram   Target-recognition attack multi-sensor; DRS adds 'detection and ranging set'.

tram   Trammel bar, or as verb to use same (colloq.).

tramline pointer(s)   Twin parallel lines or bars between which is to be aligned instrument needle.

tramlines   Guidance lines on flight deck for V/STOL aircraft proceeding under own power.

trammel and adjust   Traditional procedure for rigging airframe c1910–35.

trammel bar   Hand gauge in form of straight bar, set-square, triangle or other shape provided with precision locating feet; used in checking dimensions, angles and alignments of large structures; abb. tram or tram bar.

tramming   Use of trammel bar.

tramping   *1* Uncommanded oscillation of rudder [less often, other control surfaces].

2 Oscillation or vibration of aircraft in vertical plane.

3 Zigzag flight path as result of (1).

tranche   Production batch, not all aircraft necessarily being to same standard (UK).

trans, TRANS   *1* Transmit, transmitter, transmitting.

2 Transition; ALT adds altitude, LEV level.

transatmospheric   Operating between upper atmosphere and sub-orbital regime.

transatmospheric vehicle   Also called aerospace plane, spacecraft capable of atmospheric flight with full propulsion, lift and control, and recovery at base similar to that of aeroplane. Launch may be either by vertical rocket or horizontal takeoff.

transattack period   From initiation of NW attack to its termination (DoD).

transborder   Crossing frontier, eg airborne pollution, fallout, virus etc.

transceiver   Radio transmitter and receiver sharing common case and subcircuits, precluding simultaneous transmission and reception.

transcowl   Translating (fore/aft-moving) structure of fan reverser.

transducer   Device for translating energy from one form to another, eg mechanical strain to electrical signal (straingauge), temperature to electrical signal (thermocouple), or electrical signal into sound (earphone or loudspeaker).

transducer gain   Ratio, usually expressed in dB, of power delivered to transducer load (output) to available power input.

transductor   Any magnetic device, eg saturable reactor or magnetic amplifier, in which non-linear characteristic controls circuit.

Transec   Transmission security.

transerter   Device for sampling unknown surface material, eg on planet other than Earth; hence * auger, tube containing spiral auger which carries sample material to various instruments and experiments.

transfer   Transport between airport and ultimate destination.

transfer duct   Air duct between front and rear fans in tandem-fan engine, containing shut-off valve and auxiliary inlet system for rear fan and core in lift mode.

transfer ellipse   See *transfer orbit*.

transfer loader   Wheeled or tracked vehicle with platform positioned at any convenient height and with horizontal adjustment, used in transfer of cargo or casualties between modes of transport.

transfer function   Mathematical treatment of ratio of output response to input signal, usually a Laplace transform, expressible as plot of frequency and in closed-loop systems controlling sensitivity of output to system error.

transfer of control   Action whereby responsibility for provision of separation of an aircraft is transferred from one controller to another (see *handover, handoff*).

transfer orbit   Elliptical orbit linking two other orbits, eg one round Moon and one round Earth; for minimum energy invariably tangential to both linked orbits (see *Hohmann*).

transfer punch   Centre punch for transferring positions of template holes to sheet beneath.

transferred position line   PL redrawn to slightly later time, parallel to original but displaced by calculated ground distance.

transformation   Methods (Laplace, Fourier) of simplifying solution of differential equations; hence Laplace or Fourier transform or inverse transform.

Transformation[al]   Revolution in application of armed force brought about by netcentric warfare.

transformer   Device for transferring energy from one electrical circuit to another, usually with change of voltage, by magnetic induction.

transformer/rectifier   Device for converting a.c. to d.c. at a different voltage; can be rotary machine or solid state.

transient   *1* Temporary surge or excursion of variable, eg on first switching on.

2 Short-duration electrical impulse having steep leading edge and repeated irregularly.

3 Awaiting orders, or staging through en route to another destination.

transient distortion   Inability of equipment to reproduce very brief signals.

transient peak ratio   Peak value of phugoid parameter divided by that of immediate predecessor.

transient performance   Air-combat performance sustainable for a few seconds only (eg by trading height for speed or vice versa, or allowing speed/energy to decay).

transient response   Response to sudden changes in demand, eg hydraulic system or liquid rocket engine, where this factor is significant.

transient trimmer   Short-duration input to longitudinal trim system to counter known disturbing moment, eg when extension of rocket pack under F-86D caused pitch-attitude changes affecting aiming.

Transire   Single-sheet document stamped by Customs on entering country, or different island within country.

transistor   Electric/electronic device for amplification or control consisting of semiconductor material to which are attached metal electrodes. Name comes from transfer resistor, and in simplest form one electrode is emitter, connected to p-type material, separated from other p-type electrode (called collector) by layer of n-type.

transistor amplifier   Amplifier employing one or more transistors arranged in any of several configurations, eg common-emitter, common-collector or common-base.

transit   *1* Passage of celestial body across meridian.

2 Passage of one aircraft through controlled airspace.

3 Instrument used to determine (1).

4 Apparent passage of celestial body across face of another.

5 Condition in which three points are aligned, eg observer and two objects on Earth's surface, prefaced by 'in' (ie said to be in *).

6 Period spent on ground by passenger between arriving on one flight and departing on another, hence * area, * trollies etc.

7 A passenger in transit (6).

8 Period spent on ground by aircraft, especially commercial transport, between flights; the most frequent interval written into schedules for maintenance. Also defined as turnaround stop enroute.

9 Motion of landing gear during retraction or extension.

transit bearing   Measuring time at which two surface features have same (measured) bearing from aircraft in flight.

transit mode   Configuration of mobile system, eg SAM missile, radars and support facilities, for moving on ground to new location with radars folded, missiles packed, launchers at 0° elevation and doors closed, etc.

transition   *1* One meaning in aerospace is change from jet-supported VTOL flight to wing-borne translational flight and vice versa.

2 Another is sudden switch from blind instrument approach to visual on first sighting ground, e.g. runway lights.

3 Another is SID to airway and thence to Star.

transitional surface   Specified surface sloping up and out from edge of approach surface and from line originating at end of inner edge of each approach area, drawn parallel to runway centreline in direction of landing (ICAO).

transition altitude   QNH, altitude in vicinity of airfield at or below which aircraft control is referred to true altitude (see *transition level*).

transition distance   Ground distance covered in transition (1 or 2).

transition down   Change in helicopter flight level to dunk sonobuoy in sea in ASW operation.

transition envelope   That portion of flight envelope in which trimmed controllable flight is possible in powered flight regime, bounded by airspeed, height, ROC, power, conversion angle, AOA, control margins, etc (USAF).

transition flight   Flight at TAS below power-off stall speed, where lift is derived from both wing and power-plant.

transition height   QFE, at or below which altitude is referred to that of airfield.

transition layer   Airspace between transition level and transition altitude (NESN).

transition level   QNE, lowest flight level available for use above transition altitude (DoD).

transition lift parameter   For jet VTOL, $L/T \times Aj/S$, where L is wing lift, T is jet lift, Aj is jet area (total) and S is wing area.

transition manoeuvre   Aeroplane manoeuvre linking two glidepaths or approach trajectories; unusual except at airports where approach has to be on instrument runway and actual landing on a parallel runway.

transition point   Point on 2-D aerofoil or other surface at which boundary layer changes from laminar to turbulent, extremely sensitive to surface roughness, temperature difference, steadiness of upstream flow and other factors, and difficult to locate accurately in model testing.

transition strip   Area of airfield adjacent to runway or taxiway suitably paved to allow aircraft to taxi across it in all weathers.

transition temperature   Many meanings in which particular temperature-dependent change takes place, but esp. temperature range in which metal ductility or fracture mode changes rapidly.

transition up   Change in helicopter flight level to pull sonobuoy out of water.

transition zone   *1* Narrow atmospheric region along front where characteristics change rapidly, values lying between those of dissimilar air masses on either side.

2 Short section of glidepath within which average pilot makes transition from IFR to visual.

transitron   Pentode oscillator with negative resistance and near-constant sum of anode/screen current.

transit time   *1* Elapsed time between instant of filing message with AFTN station for transmission and instant it is made available to addressee.

2 Elapsed time between electrodes in valve or other device for any electron.

translating   Moving in straight line relative to surroundings.

translating centrebody   Supersonic-inlet centrebody able to move linearly into or out of inlet under control of automatic control system.

translating nozzle   Jet engine nozzle which in reverse mode moves to rear, further from engine, opening gap in jetpipe for gas deflected by reverser clamshells.

translation   Motion in more or less straight line, from A to B, with no rotation about any axis.

translational flight   Flight at sensible airspeed, such that wing generates lift; loosely from A to B, moving under power from one place to another.

translational lift   Additional lift gained by helicopter in translational flight resulting from induced airflow through main rotor(s) gained from forward airspeed.

translation bearing   Mechanical bearing permitting sliding motion, eg air-cushion pad or oilfilm bearing.

translation rocket   Separation or staging rocket.

translatory resistance derivatives   Those expressing moments and forces caused by small changes in translational velocity.

transloader   Vehicle for transporting missiles [invariably SAMs] and loading them on or into launchers.

Transloc   Transportable (ground station) Loran-C.

translucent Permitting EM radiation, esp. light, to pass through but diffused in direction.

translucent rime See *glazed ice*.

translunar Different definitions; most authorities agree word means extending from Earth to just beyond Moon, but a minority claim it excludes all space inside lunar orbit.

transmission *1* Process by which EM radiation or any other radiated flux is propagated, esp. through tangible medium transparent to such radiation.

2 Process of sending signal via telecom network.

3 Signal or message thus sent.

4 Mechanism transmitting mechanical energy or power, eg between helicopter engine(s) and rotors.

transmission anomaly Deviation from inverse-square-law propagation, esp. in underwater sound; symbol A, = H – 20 log R where H is transmission loss in dB and R is horizontal range in m.

transmission coefficient Radiant energy which remains after passage through a layer, medium interface or other intervening material, relative to that incident upon it; expressed as fraction or percentage; symbol $\tau = e^{-\sigma}$.

transmission factor Ratio of dose inside shielding material or hard shelter to that received outside in NW attack.

transmission grating Diffraction grating ruled on transparent substrate.

transmission level Ratio, usually expressed in dB, of signal power at any point in network to power at reference point.

transmission limit Particular frequency or wavelength above or below which almost all power is absorbed or diffused by medium.

transmission line Conductor conveying electrical power or signal, esp. wire, coaxial cable or waveguide carrying information signals.

transmission loss Decrease in power, usually expressed in dB, between energy sent and that received; in radio propagation through space, ratio expressed in dB of power received by aerial to that sent out by identical transmitting aerial; in underwater sound, symbol H, = S – L where S is sound level (dB) and L is SPL in dB above reference rms value (traditionally 1 dyne/cm^2).

transmission modes Possible configurations of electric and magnetic field patterns in waveguide; TM (transverse magnetic) has magnetic vector perpendicular to direction of travel; TE (transverse electric) has electrical vector perpendicular to direction of travel; TEM has both vectors perpendicular to direction of travel. There are an infinite number of modes, but each cannot function below a particular cutoff frequency, and two parallel wires send nothing but DC.

transmission rate Of xpdr, average number of pulse pairs per second.

transmissivity Ratio of radiation transmitted through medium, or through unit distance of it, to that incident upon it, usually expressed as %. In some usages transmission coefficient is synonymous.

transmissometer Invariably, synonymous with visibility meter, telephotometer; instrument for measuring atmospheric extinction coefficient and determination of visual range.

transmittance *1* Ratio of EM radiation transmitted through medium to that incident upon it, $T = I/I_o$ (essentially synonymous with transmissivity).

2 Ratio of luminance of surface at which light leaves medium to illuminance of incident surface (provided units are compatible).

transmitter Equipment for converting code, sound or video signals into modulated RF signal and amplifying and broadcasting latter.

transmitter chain Klystron with TWT driver.

transmitter/receiver See *transceiver*.

transmutation Conversion of atoms into different element(s) by nuclear radiation.

transom Traditional name for near-vertical transverse bulkhead at stern of boat; occasionally present in seaplane float or flying-boat hull.

transonic General term for fluid flow in which relative velocity around immersed body or surrounding duct is subsonic in some places (seldom below M 0.8) and supersonic in others (seldom above 1.2). Thus * range depends on shape of body, being very narrow and close to M 1.0 for slender body with pointed nose of small semi-angle and very thin wings.

transonic blading Rotating blading whose surrounding fluid has subsonic relative velocity at root and supersonic at tip.

transonic transport Aircraft designed to cruise at about M 1.15 without producing sonic bang at ground level.

transonic tunnel Wind tunnel whose working section can operate at Mach numbers close to that of sound, say 0.8 to 1.2.

transosonde Balloon, normally for meteorological purposes, designed to maintain constant pressure level.

transparency *1* Portions of airframe optically transparent, eg windows, canopies, moulded noses, etc; also called glazing.

2 Imagery fixed on transparent base for viewing by transmitted light, often synonymous with diapositive.

transparent *1* One aerospace meaning is that no special maintenance is required.

2 Another is that making a major change, such as fitting a different type of engine, will be undetectable by either the pilot or the aircraft systems.

transparent plasma Plasma through which EM radiation, esp. that used by communications system, can pass; generally plasma is transparent to frequencies higher than its own.

transpiration Flow of fluid through passages very long in relation to diameter (eg interstices of porous solid) yet large enough for flow to depend chiefly on pressure difference and fluid viscosity.

transpiration cooling Cooling of hot solid by fluid, eg air, passed under pressure through its porous wall.

Transply Sandwich of two metal sheets each chemically etched to form passageways linking holes perforating both sheets (holes not in line, forcing cooling air to flow within sheet).

transponder Transmitter/responder; radio device which when triggered by correct received signal sends out precoded reply on same (rarely different) wavelength; received signal usually called interrogation, and reply usually coded pulse train. ATC allocates Modes A and B four-digit numbers to provide identification; Mode C gives auto reading from encoding altimeter. Used in many systems, eg telemetry, DME, SSR, IFF; abb. xponder, XPDR, TPDR.

transponder india   Code for ICAO SSR (DoD).

transponder landing system   Closed-loop approach aid which uses aircraft xpdr to give aircraft position and transmits guidance on normal G/S and localizer frequencies to ILS instrument.

transponder sierra   Code for IFF Mk X/SIF (DoD).

transponder tango   Code for IFF Mk X basic (DoD).

transport   Aircraft designed for carrying ten or more passengers or equivalent cargo and having MTOW greater than 12,500 lb (5,670 kg). Note: this was originally US usage, where * normal meaning is called transportation, which also means ticket.

Transportation Security Administration   Formed after 9–11 [2001] as part of DoT but separate from FAA (US).

transport equation   Complicated integral/differential equation by Boltzmann for distribution function in fluid, eg gas at low pressure, subject to flow and intermolecular collision.

transporter   *1* Land vehicle, usually large and for off-road use, for carrying large missile or other mobile system or system element.

2 Airside vehicle for ULD.

transporter/erector   Transporter for ballistic missile or large radar which also erects its load into firing or operating position. Hence **/launcher, which also fires missile.

transporter mast   See next.

transporter tower   Airship mooring mast on mobile base.

transport joint   Joint between major portions of structure, eg between centre section and outer panel, dismantled for transport (US transportation) in another vehicle.

transport wander   See *apparent wander*.

transputer   Transmitter plus computer with integrated signal-processing architecture.

trans-sonic   See *transonic*.

transuranium elements   Also called transuranic, those of atomic number higher than 92 (uranium); not occurring in nature but produced by nuclear reactions.

transverse   Though this means athwartships or sideways it is often related not to basic vehicle axes but to major axis of local part; thus in wing it is common to consider spar as longitudinal and ribs as *.

transverse axis   OY, pitch axis, parallel to line through wingtips.

transverse bulkhead   See *bulkhead*.

transverse electric   See *transmission modes*.

transverse-flow effect   In helicopter translational flight air passing through main rotor is initially at higher level and that passing through rear of rotor disc is accelerated as it passes across top of rotor; ** is differential lift caused by this difference in relative wind, causing blades in rear part of disc to flap upward. Note: in this usage transverse means longitudinal.

transverse load   One acting more or less normal to major axis of member, thus tending to bend it; thus weight of fuselage forms ** on wing.

transversely isotropic   Materials having uniform elastic properties in one plane, independent of axis of testing.

transverse magnetic   See *transmission modes*.

transverse member   Structural member running across from side to side; in wing and other aerofoils often interpreted as in chordwise direction.

transverse Mercator   Map projection in which meridian is used as false equator; map is that produced by light source at Earth centre projecting on to cylinder wrapped round Earth touching along selected meridian, if necessary passing across pole. Parallels near pole almost circular but become ellipses of increasing elongation until Equator is straight line; great circles are straight lines parallel to selected meridian, otherwise complex curves.

transverse pitch   Perpendicular distance between two rows of rivets.

transverse wave   Displacement direction of each particle is parallel to wave front and normal to direction of propagation; includes EM waves and water-surface waves.

Trap, TRAP   *1* Terminal radiation airborne measurements program; note, not Tramp.

2 Tactical related applications.

3 Tactical recovery, aircraft and personnel (USAF).

trap   *1* Radio receiver, subcircuit which absorbs unwanted signals.

2 In ultrahigh-vacuum technology, device which prevents vapour pressure of mercury or oil in diffusion pump from reaching evacuated region.

3 Rare filter in solid-propellant rocket to prevent escape through nozzle of unburned propellant.

4 Hollows in piston-engine rotating components in which oil sludge collects by centrifugal action.

5 See *flame trap*.

6 Verb, to make arrested landing on carrier.

7 Landing made as in (6).

Trapatt   Trapped-plasma avalanche-triggered transit device.

trapeze bar   *1* Transverse bar linking balloon basket suspension to riggings attached to envelope, permitting relative pitch but not yaw or roll; also called suspension bar.

2 Transverse bar on underside of airship or large aerodyne for attachment (including inflight release and recovery) of aeroplane or other aircraft.

trapeze beam   General name for beam pivoted to two swinging parallel arms; also called bifilar suspension.

trapezium distortion   Distortion of basic rectangular image on CRT or other display, eg caused by unbalanced deflection voltages.

trapezoidal modulation   Involves changing waveform from sinusoidal to near-trapezoidal, with 95% modulation straight-top.

trapezoidal section   Supersonic wing section looking like extremely flat shallow rhombus with flat top and bottom joined by wedge leading/trailing edges both on upper surface.

trapezoidal wing   Usually means one whose plan is *, not section; in most cases leading and trailing edges are both at 90° to flightpath, joined by tips with straight rake which is usually negative and at Mach angle.

trapped fuel   That fuel always remaining in tanks, in worst case on ground using booster pumps for defuelling and switching off immediately associated LP warning lights illuminate.

trapper   CFS examiner (RAF colloq.).

trapping   Process by which particles are caught in radiation belts.

trap weight Maximum weight permitted for arrested carrier landing.

TRASR Tactical remote assessment/surveillance radar [through-wall sensor].

travelling-wave aerial One in which sinusoidal waves travel from feeder to terminated end.

travelling-wave amplifier Microwave amplifier depending on interaction between slow-wave field (eg travelling along wire helix) and electron beam directed along axis.

travelling-wave magnetron Usual type of modern multi-cavity magnetron in which TWT-type amplification is used at high power.

travelling-wave tube Various species of microwave amplifiers in which interaction takes place between electron beam and RF field travelling in same or opposite direction or in other arrangement; abb. TWT.

travel pod Streamlined external pod carried on normal pylon or hardpoint of fighter for pilot or crew personal baggage [rarely, for aircraft support equipment].

traverse *1* To rotate in azimuth.

2 To take set of readings along line, either discrete readings (eg pitot pressure) at selected points or forming continuous plot, eg from point well below trailing edge of wing to point well above (in this case called pitot *).

3 Surveying method based on accurate distance/angle measures between fixed points.

4 Aircraft track made up of large number of straight legs linked by turns usually of 90° or 180°; used in search, patrol or EW duties (US term).

traverse flying Flying along traverse (4).

trawling Free-ranging search for targets of opportunity.

TR box Transmit-receive switch in system (eg radar) using one aerial for emitting and reception; prevents emitted signal from being passed to receiver.

TRC *1* True track.

2 Thrust rating computer.

3 Traffic counts.

TRCS Techniques for determining near-field radar cross-section of space vehicles, eg RVs.

TRCV Tri-colour VASI.

TRD *1* Western rendering of turbo-reaktivny dvigatel, Russian for turbojet.

2 Torsional resonance damper.

3 Test requirements document.

4 Towed radar decoy.

5 Transit routing domain.

TRDI Technical Research and Development Institute (JDA).

TRE *1* Has been used in classified ads to mean transmitter/receiver equipment, transponder equipment, tactical receive equipment.

2 Telecommunications Research Establishment, Malvern, later RSRE.

3 Type-rating examiner.

4 Target-rich environment.

tread *1* Track (8) (US usage).

2 Normal meaning for * of tyre.

Trecom Transportation Research Command (USA).

T-Recs, TRECS Tactical radar electronic combat system.

TREE, Tree *1* Transient radiation effects on electronics.

2 Test and repair of electronic equipment.

TREF *1* Transient Radiation Effects Laboratory (USAF).

2 Transportable reconnaissance exploitation facility (RAF).

trefoil Cluster of three parachutes.

Trek Telescience research kit, ground-based workstations.

trellis control See *lattice fin*.

trenched *1* Of rotating shaft, provided with multiple grooves forming labyrinth seal.

2 Of passenger-transport aisle, lower than rest of floor on which seats are mounted.

TREND Conditions [eg, for landing] in next two hours.

Trends Tilt-rotor engineering database system.

TRES Tactical radar and ESM system.

TRF *1* Tuned radio frequency.

2 Threat radar frequency (US adds 'spectrum utilization').

3 Tactical replay facility.

TRG *1* Tuned rate-gyro.

2 Training.

TRGB Tail-rotor gearbox.

TRIA *1* Tungsten-reinforced iron alloy.

2 Tracking range instrumented aircraft (USAF).

Triac *1* Test Resources Improvement Advisory Council (AFSC).

2 Without initial capital, semiconductor gate (switch) similar to silicon rectifier but triggered by either positive or negative pulse.

Triad *1* US deterrent concept based on simultaneous demonstration of hard land missiles, SLBMs and recallable bombers.

2 Triple air defense (USA).

3 Technique for reading an integrated air defense (US); S adds system.

triage Urgent investigation of casualties of NW attack to determine which need, and will respond to, medical treatment.

triagraph Three-digit (numeral/letter) callsign used by whole formation or squadron (often changed), each aircraft having two-number suffix.

triangle of velocities Basic triangle in DR navigation with sides representing heading (course) and TAS, track and G/S, and wind velocity.

triangular parachute One whose canopy is approximately triangular when laid out flat.

triangular pattern Regular repeated flight pattern flown by aircraft with radio failure: equilateral triangle with sides 1 min (jet) or 2 min (others); flown left or right-handed depending on whether transmitter and receiver failed or only transmitter.

triangulation Mensuration technique, eg in sheet metalwork, in which whole area is divided into equal adjoining triangles.

triangulation balloon Small balloon used as sighting mark in triangulation survey.

triangulation station Point on land whose position is determined by triangulation; also called trig point.

triboelectrification Electricity produced by frictional processes.

tribology Study of solid surfaces sliding over one another, with or without interposed fluid.

tribometer Instrument for measuring sliding friction,

usually small-scale between smooth surfaces (ie not of runway).

tri-camera photography   Simultaneous exposures by fan of three overlapping reconnaissance cameras.

trichlorethylene   Common solvent and cleaner, $CH_2 HCl_4$.

trichromatic   Three-colour, eg TV or electronic display.

tricycle   Though nearly all aeroplane landing gears support at three points (ignoring different number of wheels at each point) this adjective means use of nose-wheel instead of tailwheel. Becoming redundant, adj. being needed only to distinguish tailwheel-type aircraft.

Tridop   CW Doppler trajectory measurement using three fixed receivers.

tri-ethyl borane   TEB, volatile pyrophoric [spontaneously igniting on contact with air] ignition liquid.

Tri-fluid fuel   Rocket propellant comprising liquid hydrogen peroxide, decomposed peroxide and JP-7.

Trifom   Trilateral fibre-optic missile (France/Germany/Italy).

triform   Structural member, usually an extruded section, providing attachment faces along three planes passing through same line, e.g. a broad-arrow or a Y shape.

Trigatron   Pulse modulator having DC-charged hemispherical electrode discharged by pulse from hemispherical trigger in gas-filled envelope.

trigger   *1* Pulse used in electronic circuits to start or stop an operation.

2 To initiate action using pulse of EM energy, eg leading edge of first pulse in SSR pulse train begins * of transponder.

3 Sharp strip, usually on leading edge, to initiate stall.

triggering transformer   Extra-high-voltage transformer connected in series with high-energy igniter to ionise gap and allow triggering spark to occur.

trigger motor   Linear actuator in pressure circuit (usually pneumatic), or driven electromagnetically, to fire gun.

trijet   Aeroplane propelled by three jet engines.

trike   *1* Microlight using trike unit.

2 Trike unit.

trike unit   Three-wheel chassis, usually with engine and seat(s), forming basic body and landing gear of microlight or ULA.

trillion $10^{12} = 10,000,000,000,000$.

Trim   *1* Trail/roads interdiction multisensor.

2 Time-related instruction management.

trim   *1* Basic measure of any residual moments about aircraft c.g. in hands-off flight.

2 Condition in which sum of all such moments is zero.

3 Condition in which aircraft is in static balance in pitch (BSI).

4 To adjust trimmers or other devices to obtain desired hands-off aircraft attitude (according to BSI, in pitch only).

5 Angle between longitudinal axis (OX) and local horizontal, esp. of airship, marine aircraft or seaplane float on water.

6 To make fine adjustment to value of any variable, eg velocity at cutoff of ICBM, fuel flow to engine or capacitance deposited on circuit.

7 To make fine adjustments to flap, LG door, external access panel or other part of aircraft surface so that when closed there are no surface discontinuities.

trim aid device   Patented add-on to eliminate need for rudder to counter torque and gyro effects.

trim air   Hot bleed air added downstream of ECS pack(s) to achieve desired cabin air temperature.

trim angle   Trim (5), positive when bow is higher than stern.

trim cord   Short length of cord doped above or below trailing edge of control surface of simple aircraft to adjust trim.

trim curve   Plot of elevator or tailplane angle (ηT) against airspeed or Mach for each c.g. position, altitude etc.

trim die   Die which trims to final dimensions.

trim drag   *1* Sustained increment of induced drag caused by need to increase wing lift to counter download on tailplane, plus the induced drag of the tailplane itself, plus component parallel to downwash on tailplane.

2 Sustained increment of drag caused by need to deflect pivoted surface continuously in order to achieve required trim, eg to deflect elevons or tailplane up to counteract rearward shift in CP as aeroplane accelerates to supersonic cruise speed. Normally eliminated in SST by pumping fuel aft to shift c.g.

trimetric drawing   3-D perspective.

trimetrogon   Reconnaissance camera installation of three cameras in which one is vertical and others take high obliques at 90° to line of flight at inclination of 60° from vertical.

trim for takeoff   Automatic or manual trimming of flight-control system to correct settings for takeoff; abb. TFT.

trimmed   In correct trim; also called trimmed-out.

trimmer   Trimming system about any axis (as plural, about all axes); normally trim tab(s).

trimmeron   Small auxiliary irreversible surface used for roll trim.

trimming moment   Moment about reference point, usually c.g., exerted by trimming system or by seaplane float or hull when held in water at particular fixed trim angle.

trimming strip   Strip of metal (occasionally length of cord or wire) attached to trailing edge of control surface and adjustable on ground to achieve desired trim.

trimming system   Flight-control subsystem through which pilot inputs bias controls about all axes to obtain desired trim; in most cases ** operates via hinged trim tabs but in supersonic fighters inputs are separate irreversible actuators driving into surface power units and in simple lightplanes often a spring-loading device in cockpit.

trimming tab   See *trim tab*.

trimming tanks   Fuel tanks (occasionally for other liquids) located as far as possible from c.g. between which fuel can be pumped to achieve desired trim in pitch, eg in SST at subsonic or supersonic speed.

trimode scanner   Circular aerial (antenna) using three EM modes in microwave cavity to move amplitude distribution in azimuth to scan 360°.

trim size   Finished size of map or chart sheet.

trim speed   Precise value of N_2, adjusted to ISA sea level, for particular engine; recorded by many manufacturers on data plate. It can correspond to MIL, rated thrust or some other power.

trim tab   Small hinged portion of trailing edge of primary flight-control surface whose setting relative to surface is set by pilot via screwthread, powered trimmer actuator or other system preventing subsequent rotation under air loads and whose effect is to hold main surface in desired neutral position for trimmed flight.

trim tank   See *trimming tanks*.

trim template   Template used for marking finished shape of part already formed (eg in press) but not trimmed.

trinitrophenol   Yellow crystalline high explosive.

triode   Thermionic valve containing cathode, anode and control grid.

trip   *1* One complete flown sector; hence * time, * fuel.

2 To activate an electrical circuit-protection system, or action thus triggered, eg overvoltage *.

triplane   *1* Aeroplane with three main lifting surfaces, in practice superimposed but not required by definition.

2 In modern usage, aeroplane with foreplane and tailplane.

triple   Triple seat unit.

triple-A   Anti-aircraft gunfire or defences, from AA artillery (colloq.).

triple display indicator   Shows Mach, KEAS and altitude.

triple ejector rack   Rack for carrying and forcibly ejecting three dropped stores.

triple-H   Hot, high, humid (all reduce takeoff performance).

triple modular redundancy   Majority-vote redundancy by three parallel systems.

triple-output GPU   Normally means electrical supply offers three voltages, eg 28 V DC, 115 and 415 V AC.

triple point   *1* Unique thermodynamic equilibrium value of temperature/pressure at which substance can exist as solid, liquid and vapour.

2 In any diametral vertical section through NW air burst, intersection of incident, reflected and fused (Mach) shock fronts; height above surface (called Mach stem) increases with distance from GZ (DoD).

triple pressure gauge   On British aircraft 1932–60 showed pneumatic brake system pressure and actual pressure at each main wheel.

triple release   Release of three free-fall bombs at short intervals, with middle one intended to be on target (suggest arch.).

triple seat   Passenger seat unit for three persons side-by-side.

triple-tandem   Actuator of tandem type with three actuation sections.

triple torque indicator   Vertical scales on AMLCD show engine output and main/tail-rotor torque.

triplex   Among many meanings, use of three parallel systems with minimal or zero interconnection to provide majority-vote redundancy; such triple redundancy provides for continued operation after single failure (SFO/FS operation). In * detection/correction system one channel may be a model.

triplexer   Dual duplexer allowing one aerial to feed two radar receivers simultaneously, both isolated during transmission of each pulse.

trip line   Cable or thin rope attached to far end of sea-anchor drogue which when pulled spills out contents to allow drogue to be wound back on board.

tripod   Vectored-thrust turbofan with one hot and two cold nozzles.

TriSar   Triple-mode synthetic-aperture radar.

TRIT   Turbine (usually HP) rotor inlet temperature.

Tri-Tac   Tri-service tactical communications (DoD).

tritium   Unstable radioactive isotope of hydrogen of mass number 3, crucial component of NW fusion materials such as Pu, Li-6 and deuterium.

TRIXS   Tactical reconnaissance intelligence exchange system, or service.

TRJ   Tactical radar jammer.

TRK, Trk   Track.

TRL   *1* Tyre rolling limit; upper right boundary of aeroplane takeoff performance carpet.

2 Thrust-reduction limit in flexible takeoff.

3 Transition level.

TRM   *1* Technical requirements manual.

2 Training and readiness manual (US).

3 Team resource management.

4 Tow-reeling machine.

Tr(M)   Magnetic track.

TRML   Terminal.

TRMM   Tropical-rainfall measuring mission.

TRN   *1* Terrain-reference(d) navigation [S adds system].

2 Terminal radar numeric.

TRO   Technical resources operation.

trochoid   Path traced by point on circle which rolls along straight line. Commonest RC engine pistons are of related profiles.

TROF   Trough.

troland   Unit of retinal illuminance; that produced by viewing surface whose luminance is 1 cd/m^2 through artificial pupil of 1 mm^2 area centred on natural pupil. E (in *) = LA (in nits × mm^2).

trolley-acc   Pronounced trolley-ack, 12-V or 28-V accumulator on two-wheel mount for starting aircraft.

trolley dolly   Stewardess (colloq., suggest derogatory).

troll(ing)   Flying of random pattern by EW/Elint/ECM aircraft to try to trigger and detect hostile emissions.

trombone   Adjustable U-shaped length of waveguide or co-axial line used for phasing.

trombone lever   Cockpit lever with linear trombone-like movement, eg F-111 wing sweep.

tromboning   See *path stretching*.

TROP   Tropopause.

tropical air mass   Warm air originating at low latitudes, esp. in subtropical high-pressure system.

tropical conditions   Various standardized conditions normally including ambient temperature at least 30°C (usually higher) and 100% RH, ie hot and moist.

tropical continental   T$_c$, air mass characterized by high temperature and low humidity; usually unstable and associated with clear sky.

tropical cyclone   Rotating storm: tropical depression, winds <63 km/h; tropical storm, <118.5 km/h; typhoon, cyclone or hurricane, 120+ km/h. See *tropical revolving storm*.

tropical maritime   T$_m$, air mass characterized by high temperature and humidity.

Tropical Maximum Atmosphere   One of many artificially averaged atmospheres.

tropical revolving storm   Largest and most violent form of thermal depression, generally less than 800 km (500 miles) diameter but with pressure falling in centre to about 960 mb, originating in ITCZ in 5° to 15° N or S; called cyclone (Bay of Bengal, Arabian Sea), hurricane (S

Indian Ocean and W Indies), typhoon (China Sea) and willy willy (W Australia).

tropical trials   Trials under tropical conditions, for most equipments simulated exactly to specification but for aircraft by flying to actual hot/high airfield; for aircraft low atmospheric density more important than high humidity specified for, say, avionics.

tropical year   Period of revolution of Earth round Sun with respect to vernal equinox (with respect to stars, 359°9′ 59.7″); 365 d 5 h 48 min 45.85 s in 1990, increasing 0.0053 s per year.

tropopause   Boundary between troposphere and stratosphere characterized by abrupt change in lapse rate (except at high latitudes in winter when change almost undetectable); height fixed in ISA at 11 km (36,089.3 ft), at pressure about 230 mb and relative density about 30%. In practice much higher over Equator (about 15–20 km) than over poles (about 8–10 km).

troposphere   Lowest portion of atmosphere, extending from surface to stratosphere; characterized by lapse rate, humidity, vertical air movements and weather. Subdivided into surface boundary layer, Ekman layer and free atmosphere.

tropospheric scatter   OTH radio propagation by reflection or scattering from irregularly ionised regions of troposphere; using forward scatter at about 25–60 MHz, ranges of about 1,400 km (870 miles) are possible.

tropospheric wave   Radio wave propagated by reflection from place of rapid change in dielectric constant (high ionisation gradient).

troubleshooting   Process of investigating and detecting cause of hardware malfunction (US usage; UK = fault diagnosis).

Tro/Tri   Tactical reconnaissance optical, tactical reconnaissance IR day/night.

trough   *1* Long but narrow region of low atmospheric pressure, opposite of ridge, with curving isobars at apex and absence of recognizable front.

2 Point in space where gravitational fields (eg Earth/Moon) cancel out (colloq.); neutral point but has implication of vagueness in location.

3 Repeated low portions of sine-wave flight 115–132K ft.

trouser   Fairing for fixed landing gear in form of continuous streamline section round leg and upper part of wheel, usually tapering in chord and thickness.

Trowal   Trough of warm air aloft (suggest arch.).

TRP   *1* Threat-recognition processor (USN).

2 Tuition-refund program(me).

3 Thrust-rating panel.

4 Terminal rendezvous point (usually TRV).

5 Time-response parameter.

6 Timed reporting point [operational squadron = time on target].

7 Mode-S transponder.

TRR   *1* Total removal rate.

2 Tyre (tire) rolling radius.

3 Test, rejection and repair.

4 Throttleable ram-rocket.

5 Test readiness review.

TRRAP   Technology readiness risk assessment programme.

TrReq   Track required.

TRRN   Terrain.

TRRR, TR³   Trilateration range and range-rate.

TRS   *1* Tropical revolving storm.

2 Tactical reconnaissance system [or sensor, or squadron].

3 Teleoperator retrieval system.

4 Track reporting system.

5 Triple-redundant system.

6 Tail-rotor swashplate.

7 Training research simulator.

8 Terminal-radar simulator.

TRSA   Terminal radar service area.

TRSB   Time-referenced scanning beam MLS.

TRSR   Turbo-rogue space receiver; uses GPS to determine satellite position by triangulation.

TRT   Terec remote terminal.

Tr(T)   True track.

TRTD   Treated (runway).

TRTG   Tactical radar threat generator.

TRTO   Type-rating training organization.

TRTT   Tactical-record traffic teletypewriter.

TRU   *1* Transformer/rectifier unit, eg for converting raw a.c. into low-voltage d.c.

2 Transmitter/receiver unit.

3 True.

truck   Bogie of main landing gear (US usage); same word also used for four MLGs of B-52, which are twin-wheel, one axle.

truck-bed height   Commonly accepted average height of payload bed of road truck (UK, lorry), usually taken as 41 in (1,030 mm).

TRUD, trud   Time remaining until dive; count of time from launch of upper-atmosphere or other vertical ballistic rocket vehicle and time at which it reaches apogee and begins dive; hence * count.

true airspeed   See *airspeed*.

true altitude   *1* Actual height above SL; calibrated altitude corrected for air temperature.

2 Observed altitude of body.

true bearing   Angle between meridian plane (referred to true N) at observer and vertical plane through observer and observed point.

true course   Angle between aircraft longitudinal axis OX and plane of local meridian (referred to true N).

true heading   See *true course*.

true-historic display   One showing what would have happened, eg line showing locus of impact points where bullets would have hit had they been fired.

true meridian   Great circle through geographical poles.

true north   Direction towards N pole of meridian through observer.

true position   Position of celestial body or spacecraft computed from orbital parameters of Earth and body without allowance for flight time.

true power   Actual (I^2R) power of a.c. electrical circuit.

true prime vertical   Vertical circle through true E and W points of horizon.

true stress   That computed as force divided by true area normal to the load, as distinct from Engineering stress.

true Sun   Sun as it appears to Earth observer, as distinct from mean, dynamic mean.

true track   Angle between true N and aircraft path over ground.

true vertical   Local vertical, line passing through observer and centre of Earth.

Tru-Loc Flight-control cable end-fitting swaged on by machine with tensile rating equal to breaking strength of cable.

truncation error EDP (1) error resulting from use of only finite number of terms of infinite series, or from other simplifying techniques, eg calculus of finite differences.

trunk *1* Large-section lightweight air-conditioning duct, or duct between gasbag valve and gas hood.

2 Major route, with or without surrounding box conduit, for large number of controls, services, cables, wire looms and other lines.

3 * route or operator.

4 Compartment for baggage or general storage in GA aircraft (US usage derived from cars).

trunk route *1* Most important type of commercial route, eg between largest cities in country or countries, offering highest traffic. Hence * operator, domestic *, * traffic.

2 Established air route along which strategic moves of military forces can take place (NATO).

truss Rigid load-bearing planar structure of spaceframe type, usually comprising essentially horizontal upper and lower chords linked by various vertical and diagonal members.

trusted Generalized description of network elements meeting specific measures of security.

truth table EDP (1) technique where coded signals are allocated to particular addresses, comprising series of adjacent address codes and output codes.

TRV *1* Terminal rendezvous point (army helicopters).

2 Tower restoral vehicle (USAF).

TRVR Touchdown runway visual range.

TRW *1* Tactical Reconnaissance Wing (USAF).

2 Thundershower (ICAO).

try-again missile Conceptual AAM of 1950s which, finding initial interception was outside design manoeuvre limits, made programmed turn for second attempt.

tryptique See *carnet*.

TS *1* Thunderstorm (ICAO).

2 Transport, or Training, Squadron, or service.

3 Transattack survivability, i.e. a post-strike system.

4 Thunderstorm sensor.

5 Track system (UAV).

6 Torpedo-bomber/scout (USN 1943).

7 Two-stroke.

8 Transmitter segment.

9 Turbosupercharged (Satcom).

10 Time source.

Ts *1* Static temperature.

2 Note: Russian 'Ts' in this dictionary is rendered as C.

T$_S$ *1* Total noise of radar system.

2 System equivalent noise temperature.

3 Sampling interval time.

T-S Graphical plot of absolute temperature (ordinate)–against entropy per unit mass of fluid S. Fundamental diagram in thermodynamics.

TSA *1* Training systems acquisition.

2 Tail-strike assembly (USAF).

3 Transportation Security Administration (US, from 2001).

TSAAC TSA(3) Access Certificate, required by Pt 91 operators in order to fly to airports normally accessible only to scheduled carriers.

TsAGI Common Western rendition of Central Aero and Hydrodynamics (research) Institute (USSR, R).

TSAM Tri-Service Attack Missile (US).

TSAP Transport service access point.

TSAR, Tsar Tactical and strategic advanced reconnaissance.

TSAS Tactile situation-awareness system.

TSB Transportation Safety Board (Canada).

TSC *1* Transportation Systems Center (US DoT).

2 Triple store carrier.

3 Tactical Support Center (USN).

4 Term service commitment.

5 Training system contract.

TSCM *1* Technical surveillance countermeasure(s).

2 Tactical-strike coordination module (Tamps).

TSD *1* Tactical situation display.

2 Traffic situation display.

3 Time/speed/distance.

TS diagram Plot of temperature against entropy, a closed 4-sided figure.

TSDIU Transport service data-interface unit.

TSDS Telemetry storage and display system (UAV).

TSE Total system error.

Ts ENTROSPAS, TsentroSpas English renditions of ministry for civil defence, emergencies and natural disasters (R).

TSF *1* Technical Supply Flight (RAF).

2 Originally telegraphie sans fil, = radio, now Télécoms sans Frontières (F).

tsfc, TSFC Thrust specific fuel consumption; SFC of air-breathing jet engine.

TSFE Thermally stimulated field emission.

TSGR Thunderstorm plus hail (ICAO).

TSGT Transportable satellite ground terminal.

TSHWR Thundershower.

TSI *1* Turn and slip indicator.

2 Track-situation indicator.

3 Transportation Safety Institute (US, from 1971).

TsIAM Common Western rendition of Central (research) Institute for Aviation Motors (USSR, R).

TSIC Touch-screen interactive control.

TSIO US piston-engine code: turbosupercharged, direct injection, opposed; now called TIO.

TSIP Trimble Standard interface protocol.

TSIR Total system-integration responsibility (AFSC).

TsKB Common Western rendition of Central Construction (design) Bureau (USSR, closed).

TSM *1* Autothrottle servo mount.

2 Trouble-shooting, or technical-support, manual.

TSMO Time since major overhaul.

TSMP Terminal-segment master plan.

TSMT Transmit.

TSMTR Transmitter (alternative to XMTR).

TSN Time since new.

TsNII English rendition of CSRI, Central Science Research Insitute (R).

TSNT Transient.

TSO *1* Time since overhaul.

2 Technical service order (FAA).

3 Technical standard(s) order (CAA, from 1947).

TSOC Touch-screen operator console.

TSOR Tentative specific operational requirement.

TSP *1* Turret stabilized platform.

2 Transonic small perturbation.

3 Thermal scanning polygon [M adds motor].

4 Time/space position [I adds indication or information].

5 Total support package (industry/RAF).

6 Transmitted signal power.

7 Twisted shielded pair.

8 Transmitter/responder.

9 Tailstrike protection [which see].

TSPG   Training Systems Product Group (USAF).

TSPR   Total system-performance, or program, responsibility (AFSC).

TSQLS   Thundersqualls.

TSR   *1* Torpedo spotter reconnaissance.

2 Tactical strike/reconnaissance.

3 Twin side-by-side rotors (helicopter).

TSRA   Thunderstorm plus rain.

TSS   *1* SST (F).

2 Tunable solid-state (laser).

3 Tethered satellite system.

4 Tacco sub-system.

5 Tactical surveillance sonobuoy, or supervisor.

6 Tail-strike sensor.

7 Target sight system.

8 Tangential signal sensitivity.

9 Technology support and services.

TSSA   *1* Thunderstorm plus duststorm or sandstorm (ICAO).

2 Transport Salaried Staff Association (UK trade union).

TSSAM   Tri-Service stand-off attack missile.

TSSC   *1* Technical supply subcommittee (AEA).

2 Training System Support Center (F-22).

3 Technical and Safety Standing Committee.

TSSR   Total-systems support responsibility.

TSSTS   Tactical sigint system training simulator.

TST   *1* Transonic transport (usually M 1.1–1.2).

2 Threshold sampling time.

3 Technical support team.

4 Time-sensitive target, or targeting.

TSTA   Takeoff safety training aid.

TTSM   Time-source transition module.

TSTM[S]   Thunderstorm[s].

TSTO   Two stage to orbit.

TSU   Telebrief-[ing] switching unit.

TSUS   Tariff schedules of the US.

TSV   *1* Time since shop visit.

2 Through-sight video.

TSW   *1* Tactical Supply Wing (RAF).

2 Transverse shear-wave.

TT   *1* Total time.

2 Dry-bulb [total] temperature.

3 Teletypewriter (ICAO).

4 Turnround time.

5 Target towing (role prefix, UK, defunct).

6 All-weather (F).

7 Threat transmitter.

8 Twin-tandem landing gear.

9 Torpedo tube.

10 True track.

11 Test tools.

TTA   Total-terrain avionics.

TTAE   See *TTAF/E*.

TTAF   Total time on airframe.

TTAF/E, TTAF&E   Total time, airframe and engine.

TT&C   Telemetry, tracking and control, or command.

TTB   *1* Tanker/transport/bomber (USAF aircrew).

2 Target-triggered burst.

TTBT   Threshold Test-Ban Treaty.

TTBTS   TTB(1) training system.

TTC   *1* Tracking, telemetry and command.

2 Technical Training Center (USAF), or Command (RAF, formerly).

3 Tape-transport cartridge.

4 Top-temperature control.

TTCP   The Technical Co-operation Program[me] (US/UK).

TTCR   *1* Time/temperature cycle recorder, or recording.

2 Triangular trihedral corner reflector.

TTCS   Tactical terminal, or target-tracking, control system.

TTD   Tactical-threat display.

TTDF   Tip-turbine-driven fan.

TTE   Tooling and testing equivalency working group (AMC).

TTEMP   Temperature test and evaluation master plan.

TTF   *1* Time to first fix.

2 Tanker task force.

3 Threat training facility (US).

4 Target-towing flight.

TTFF   Time to first fix.

TTG   Time to go, range divided by closing speed.

$T_{\theta 2}$ T–theta–2, numerator time-constant.

TTI   Triple torque indicator.

TTL   *1* Transistor/transistor logic; IC adds integrated circuit.

2 Torpedo-tube launch.

TTLS   Transportable-transponder landing system.

TTM   Tape-transfer magazine.

TTNT   Tactical-targeting network technologies.

TTP   *1* Time to protection, measured from firing chaff, aerosol, flare or other dispensed payload to time aircraft can be judged protected.

2 Through-thickness pinning.

3 Time-triggered protocol.

TTR   *1* Target-tracking radar.

2 Twin tandem rotors (helicopter).

3 TCAS-II transmitter/receiver.

4 Tonopah Test Range, Utah.

TTS   *1* Time to station; time in minutes to fly to tuned DME station. Often a selectable mode on DME panel instrument.

2 Technology Transfer Society (US).

3 Thin-tape system.

4 Total technical service (MRO).

5 Total training system.

TTSMOH   Total time since major overhaul.

TTSN   Total time since new.

TTT   *1* Tactical technical requirement(s), basis of each military specification (USSR, R).

2 Tailplane trim[ming] tank.

3 Template-tracing technique.

TTTO, T³O   Two stage[s] to orbit.

TTTS   Tanker/transport training system.

TTU   *1* Triplex transducer unit feeding height, TAS and Mach to CSAS.

2 Torpedo Training Unit (RAF).

TTW   Time of tension to war, or transition to war.

T²CAS   Combined TCAS/TAWS.

TTWS   Terminal threat warning system.

TTY   Telephone/teletypewriter.

TU   Towed unit.

TUAV   Tactical unmanned air, or aerial, vehicle [R adds radar].

tub   Tub section has been used to describe lower rear half of main fuselage structure of pod/boom helicopters.

Tuballoy   Depleted uranium.

tube   *1* Vacuum tube (UK = thermionic valve).

2 Large modern jetliner (colloq.).

tube oil   Liquid primer for hot-coating interior of inaccessible workpieces.

tube yawmeter   Array of four (or five if one in centre) pitot tubes each inclined at about 45° to flow and radially spaced at 90° so that dP across each opposite pair gives flow inclination in that plane; provides complete picture of local flow direction, little affected by yaw up to 15°.

tubo-annular   Gas-turbine combustor of annular type within which are separate tubular flame tubes. Latter are sometimes linked into common turbine NGV ring.

tubular combustor   Gas-turbine combustion chamber whose upstream part is of essentially circular cross-section with central fuel burner. Engine may have one or more (early turbojets had as many as 16 because true annular technology unknown).

tubular rivet   Rivet whose shank is a tube, usually inserted on central mandrel which on withdrawal closes upset head.

tubulators   Numerous open tubes in surface of sailplane wing in hope of sustaining a laminar boundary-layer.

tuck, tuck in   *1* Tendency to tighten turn, without pilot command.

2 At high Mach numbers, uncommanded nose-down pitching moment, also called Mach *; could be violent to point of causing structural failure [the opposite of (1)].

tuck-under   Strong tendency to dive, especially at high Mach number.

tufting   Covering aircraft, model or portion thereof with tufts.

tufts   Short pieces of wool, thread or other very light, flexible and easily visible material which give qualitative picture of local airflow direction and (from steadiness or oscillatory motion or turbulence) vorticity or turbulence. Mounted on aircraft or model skin or at varying distances from it.

tug   *1* Self-contained propulsion system for long-term use in space tasked with taking payloads, satellites, spacecraft and portions of station structure, propellants and other supplies (eg delivered by Shuttle Orbiter) and placing them in desired locations or trajectories. Intention is to attach * to all types of payload, in various ways giving thrust without rotation, and to replenish propellants in space.

2 Aeroplane for towing glider, target or other unpowered object. Hence * pilot, * queue, * landing area.

tugging   Towing a target (colloq.).

tulip   Spray of fuel from gas-turbine burner at fuel pressure not high enough to atomize continuous film but too high for it to converge as bubble.

tulip valve   Piston-engine exhaust valve shaped in side elevation like tulip.

tumble   *1* To rotate about lateral axis, ie end over end; rare in aircraft but not uncommon in spacecraft.

2 To rotate metal parts in drum, often with powder

abrasive, to remove flash or burrs and obtain polished surface.

3 Of gyro wheel, to precess to limit after toppling.

tumblehome   Distance measured parallel to transverse axis from vertical line through widest extremity of seaplane or flying-boat hull to any point on skin above.

tumblehome line   Abrupt change of curvature of mould line, buttock line or water-level contour at end of contour [marine aircraft].

tumble limit   Angular displacement in pitch or roll at which traditional gyro instrument is on gimbal stops.

tumbler   Drum in which metal parts are tumbled (2).

tumbler switch   Snap-action electrical switch with short operating lever.

tumbling   To tumble (1, 2, 3); to be in that condition.

tunable beam approach   Pre-ILS landing system (BABS, SBA) in which pilot tuned receiver to particular airfield (arch.).

tuned-bandpass transformer   Typically, primary winding forms anode load of one stage and secondary drives grid of succeeding stage.

tuned circuit   Oscillatory circuit with capacitance and inductance selected or tuned to resonate at frequency of applied signal.

tuned-grid circuit   Parallel resonant circuit linking grid and cathode with maximum response at resonant frequency; similar resonant response for tuned anode circuit.

tuner   Subcircuit or device in RF receiver which selects desired frequency and rejects all others.

tungsten   Silver-grey metal, symbol W, density 19.3, MPt 3,407°C; one of densest, hardest, strongest and most refractory metals known.

tungsten inert-gas welding   DC current is passed through cathode of 4% thoriated tungsten and through workpiece with high-purity argon fed to both sides of weld and to form gas lens from torch.

Tungum   Corrosion-resistant alloy of copper with 14.6% zinc and small amounts of other metals.

tuning   Fine adjustment over continuous analog range of values to obtain that desired, eg RF frequency/wavelength or optimum operating condition of engine or other device.

tunnel   *1* See *wind tunnel*.

2 Axial fairing along body, eg to cover pipes, cables and other lines routed outside skin.

3 Channel along which baggage passes through screener.

tunnel diode   Semiconductor device having single p-n junction across which electrons flow by quantum tunnelling; when biased to centre of a negative-resistance mode can operate as amplifier, oscillator or switch.

tunnel shock   That occurring immediately downstream of supersonic working section in wind tunnel when Mach number falls below unity; intense if no second throat.

tunnel vision   Inability to perceive anything outside an extremely small angular range of FOV, as if one were in a tunnel; caused by disease or high g.

TUP   Technology utilization program (NASA, technology transfer).

turb, TURBC   Turbulence (ICAO).

turbidity   Any condition of atmosphere which reduces its transparency to radiation, esp. optical; term normally applied to cloud-free sky where * is caused by suspended matter, scintillation and other effects.

turbine *1* Gas-turbine shaft power, ie turboshaft or turboprop; eg helicopter can be said to have * power.

 2 Prime mover whose power is obtained by action of working fluid (water, steam, cold or hot gas etc) reacting on blades or shaped passages which rotate a shaft. One or more are source of power in all gas-* engines and in turbochargers, turbopumps and many other rotary machines.

turbine-airscrew unit   See *turboprop*.

turbine bearing   Bearing supporting turbine shaft.

turbine blade   Radial aerofoil mounted in edge of turbine disc whose tangential force rotates turbine rotor. Each turbine stage has many blades, occasionally fabricated in groups of two or more. Called bucket in US. Inwards-radial turbine does not have *.

turbine disc   Central member upon which turbine rotor blades are mounted; in many multistage turbines there are no flat discs, first and last stages being conical and intervening stages being rings gripped between them.

turbine entry temperature   See *turbine temperatures*.

turbine gas temperature   See *turbine temperatures*.

turbine rotor   Complete turbine rotating assembly with 1–6 stages; stators are excluded.

turbine shroud   Shroud around periphery of turbine rotor stage, formed by * section on tip of each blade.

turbine stage   Single turbine disc or ring with inserted blades; for completeness should be associated with preceding stator (IGV) stage.

turbine stator   Ring of fixed blades, also called inlet guide vanes, upstream of each turbine rotor stage, on to whose blades ** directs gas with optimum distribution of V and pressure for maximum turbine work and efficiency (eg changes radial pressure/V, previously uniform, so that with increasing radius pressure increases while V falls).

turbine temperatures   This entry describes temperatures in gas-turbine engines. Alphabetically, these are: CET, CIT, COT, EGT, RIT, SOT, TDT, TET, TGT and TIT. Following a particle through an engine, CIT is compressor inlet temperature, which is that of the ambient atmosphere corrected for any ram effect. COT is compressor outlet temperature, measured immediately behind the final stage of compression. CET, combustor exit temperature (also called combustion chamber outlet temperature, giving a second meaning for COT) is the temperature of the gas at the entry to the first-stage turbine stator. SOT is stator (first-stage) outlet temperature. The gas is then cooled 20°–120°C by injection of cooling air from the turbine disc (whose temperature TDT is cooled by the airflow) and other sources to give the RIT (rotor inlet temperature), also called TET (turbine entry temperature) and TIT (turbine inlet temperature. In a single-shaft engine the gas leaves the turbine at EGT (exhaust gas temperature). In a two-shaft engine the gas leaving the HP turbine upstream of the LP turbine first-stage stator is measured as TGT (turbine gas temperature). Having passed through the LP turbine, the gas is reduced to EGT.

turbine vane   US term for turbine stator blade.

turbine wheel   US usage for either one complete stage or complete turbine, possibly of several stages.

Turbinlite   Airborne searchlight for night interception experiments (RAF, WW2).

turbo *1* Turbocharger.

 2 Generalized prefix meaning driven by or associated with gas turbine, eg *-supercharger, *-ramjet.

turboblower   Air blower driven by exhaust-gas turbine to sweep burnt mixture from cylinders of two-stroke diesel. (Today also often used, esp. with diesels, to mean turbocharger).

turbocharger   Piston-engine supercharger driven by exhaust-gas turbine.

turbofan   Most important form of propulsion for all except slow aeroplanes (say, below 600 km/h, 375 mph); comprises gas-turbine core engine, essentially a simple turbojet, plus extra turbine stages (usually on separate LP shaft) driving large-diameter fan ducting very large propulsive airflow round core engine and generating most of thrust. For given fuel consumption generates much more takeoff thrust than turbojet, with many times less noise, but performance falls off more rapidly with forward speed. A few * engines have *aft fan* whose blades form outward extensions of those of the compressor turbine.

turbofan-prop   Turbofan driving propfan mounted ahead of inlet and acting as fan booster stage to give jet as well as shaft power.

turbojet   Simplest form of gas turbine, comprising compressor, combustion chamber and turbine, latter extracting only just enough energy from gas flow to drive compressor. Most of energy remains in gas, which is expanded to atmosphere at high velocity through constricting propelling nozzle. In supersonic aircraft often fitted with afterburner.

turbojet-based combined cycle   Propulsion system mating turbojet and scramjet, with separate [not valved] flow paths.

Turboline   FS100 fuel additive to improve high-temperature stability.

turboprop   Gas turbine similar to turbofan but with extra turbine power geared down to drive propeller. Difference between two forms of engine is of degree only; fan of turbofan is invariably shrouded, running inside profiled case, while propeller (but not propulsor) is always geared and normally operates unshrouded and unducted. In general * has much higher bypass ratio than turbofan, and is tailored to slower aircraft.

turbopump   Pump driven by turbine turned by gas, e.g. from rocket propellants.

turboramjet   Combination of turbojet and ramjet as integrated propulsion for supersonic aircraft. In theory afterburning turbojet or turbofan can be classed as *, but in practice true * is large ramjet within or upstream of which is turbojet for starting from rest and acceleration to ramjet lightup speed. In some forms large valves or duct-diversion doors are needed to change internal flows between turbojet and ramjet modes.

turborocket   Various combinations of gas turbine and rocket in one engine.

turboshaft   Gas turbine for delivering shaft power, eg to power helicopter, ACV or other non-flying vehicle. Essentially a turbofan or turboprop with fan or propeller removed. Often can deliver power at either end, and usually has at least one stage of speed-reducing gearbox.

turbostarter   Main-engine starter driven by turbine turned by gas from cartridge, IPN or other fuel.

turbosupercharger   See *turbocharger*.

TURBT   Turbulent.

turbulator   See *vortex generator*.

turbulence   Time-variant random motion of fluid in which velocity of any particle, or at any point, is characterized by wild and unpredictable fluctuations which are extremely effective in conveying heat, momentum and material from one part of fluid to others. Called isentropic if rms velocity is same in all directions. US NWS defines light * as wind varies 0–19 ft/s and moderate as 19–35 ft/s. $(5.79–10.67 \text{ ms}^1)$.

turbulence cloud   Cloud formed because of atmospheric turbulence, usually distinctive layer above condensation level about 30 mb (say, 300 m, 1,000 ft) thick in otherwise stable air.

turbulence control structure   Gigantic 'golf-ball' with porous walls attached to inlet of engine on outdoor test to eliminate effect of wind.

turbulence number   R (4) at which C_d of smooth sphere becomes 0.3.

turbulence plot   Term has been used for plotting wake turbulence behind aircraft, building or other bodies, and also for recording geographical locations of severe atmospheric turbulence, including CAT, over a long period.

turbulence screen   Screen across wind tunnel to reduce turbulence, usually rectilinear array of crossing sharp-edged strips.

turbulent boundary layer   One that is no longer laminar, characterized by gross random lateral motions, and Reynolds stresses much larger than viscous; all boundary layers become turbulent at R = 250,000+ unless surface unusually smooth, though apart from rise in skin-friction drag there should be no other significant effect and no separation.

turbulent bursts   Microscopic eruptions which occur constantly over aircraft (or other) surface, beginning at surface; responsible for most of skin-friction drag and nearly half total aerodynamic drag.

turbulent flow   Flow having turbulence superimposed on main movement, measured as velocity increments about all three axes expressed as fraction or % of mean flow velocity.

turkey   *1* Badly designed aircraft, especially aeroplane, with sluggish or dangerous handling.

2 Aircraft with performance so poor as to be useless.

turn   Angular change of track; thus 30° * does not mean 30° bank. See * *rate*.

turn and bank   Traditional flight instrument, at bottom right in the *Basic 6*: one centre-pivoted needle moves L/R around the upper arc to indicate slip, while a second moves round lower arc to indicate rate of turn; straight and level or at rest both needles are vertical.

turn and slip   Traditional flight instrument indicating rate of turn by a needle moving L/R around upper arc, and slip/skid by a ball in a lateral tube in lower arc. Alternatively, the needle is in the lower arc and the upper is a curved tube with a bubble.

turnaround   Elapsed time between aircraft parking at stopping point and moving off to continue flight or carry out fresh mission.

turnaround cycle   Comprises loading time at home base, time to/from destination, unloading/loading time at destination, unloading time at home, planned maintenance and where applicable, time awaiting facilities (DoD).

turnaround time   See *turnaround*.

turnback   Abandonment of scheduled sector and diversion to alternate or to starting point, 80% [2002] for reasons not related to the aircraft.

turnbuckle   Double-ended threaded barrel for adjusting wire/cable tension.

turn co-ordinator   Rate gyro which senses rotation about both roll and yaw axes (US).

turn errors   See *turning errors*.

turn indicator   Flight instrument indicating rate of turn about aircraft vertical axis, almost always combined with slip/skid.

turning   *1* General term for operation on a lathe in which workpiece is rotated.

2 Manoeuvre by which parachutist rotates to face direction of drift.

3 Rotating engine by means other than own power; also called cranking.

turning errors   Those due to instrument deficiencies, eg due to acceleration; see *acceleration error*.

turning moment   See *couple*.

turn-in point   Point in space at which aircraft starts to turn from approach direction to line of attack (DoD, NATO, becoming arch.).

turnoff   Point at which aeroplane leaves runway to taxi to parking place, normally also junction of runway and paved taxiway; high-speed * is configured with gentle turn radius to avoid overstressing landing gear laterally.

turnoff lights   Flush lights at 15 m (50 ft) intervals defining curved path from runway centreline to taxiway centreline.

turnover assembly   Added at tail of vertical-launch missile to rotate towards target, then jettisoned.

turnover structure   Strong structure intended to protect aircraft occupants in event of overturn on ground, eg crash arch., crash pylon.

turnover voltage   Reverse V of point-contact semiconductor device (corresponding to reverse breakdown V of junction device) beyond which control over reverse current is lost.

turn radius   Half lateral distance required to change heading 180°.

turn rate   By convention, aircraft change of heading rate is measured as Rate 1 = 360° in 2 min., Rate 2 in 1 min., Rate 3 in 32s, Rate 4 in 20s.

turnstile aerial   Crossed dipoles, equal quadrature-phase signals.

turpentine   One of terpenes, BPt 155°C; solvent, thinner and (in France) rocket fuel.

turret   Enclosure for airborne gun(s) able to rotate in azimuth and with provision for gun rotation in elevation; manned or remotely controlled, but in latter case merged indefinably into barbette. Later forms (c1955) exactly resembled sting of wasp and did not fit traditional configuration. Today found mainly on helicopters, always with remote control.

turret lathe   Equipped with rotating toolholder for six (occasionally more or less) tools, indexed to work in sequence; large capstan with power operation and turret on main bed-slides.

turtleback   Top of fuselage, esp. aft of cockpit.

TUSA   Tactical unmanned surveillance aircraft.

Tuslog   The US Logistics Group (USAFE).

TUT   Targets under trees (AC^2ISRC).

TUV   Tactical unmanned vehicle.

TV *1* Television.

2 Terminal velocity.

3 Theatre of war (USSR, R).

4 Thrust vectoring.

5 Transfer vehicle (cargo).

6 Test vehicle.

TVA *1* Thrust vector angle (of gross thrust vector).

2 Tuned vibration absorber.

3 Target vector analysis.

T-VASI See *VASI*.

TVAT Television air trainer; carried on external pylon.

TVBC Turbine vane and blade cooling.

TVBS Television broadcast satellite.

TVC *1* Thrust-vector control (S adds system).

2 Turbine vane cooling.

TV command Guidance by human operator watching TV picture from camera in nose of controlled vehicle.

TVCR Tower visual control room.

TVD *1* Turboprop (USSR, R).

2 Theatre(s) of military operation (USSR, R).

TVDS Tactical-video distribution system.

TVDU Television display unit.

TVE Total vertical [separation] error.

TVGS Text-to-voice generation system.

TV homing Automatic homing guidance by comparing TV picture with sequence stored in missile memory.

TVI TV interference.

TVIT, T-VIT Tactical video imaging terminal.

TVLA Tuned vertical line array (sonobuoy).

TVM Track-via-missile.

T-VOR, TVOR Terminal VOR.

TVPM Time-varying pulse manipulation.

TVR *1* Track-via-missile radar guidance.

2 Trajectory/velocity radar.

TVRS *1* Tactical voice-recognition system.

2 Tactical video receiving system.

TVS *1* Target value structure.

2 Thermal video system.

TVSU Television sight unit.

TVT TV tracker.

TV/TR Thrust vectoring, thrust reversing.

TW *1* Threat warning.

2 Training, water-cooled (USA 1919–24).

T/W Thrust/weight ratio.

T$_w$ Torque applied at wing root.

TW/AA Tactical warning and attack assessment (Norad).

TWACN Theatre-wide area communications network [Cormorant].

TWATN Theater-wide area telecommunications network, not same as above.

TWC Tungsten-wire composite.

TWCC Threat warning and countermeasures control.

TWD *1* *Touchwire display*.

2 Toward[s].

TWDL *1* Two-way data-link.

2 Terminal-weather data-link.

TWDR Terminal weather Doppler radar (TDWR is more usual).

TWE Threat-warning equipment.

TWEB Transcribed weather broadcast.

12-5 Rule limiting MTOW to 12,500 lb.

1250 Paybook for RAF airmen in which career details are recorded.

20-minute rating A-h rating of battery indicating current required to discharge from maximum to zero in 20 min.

TWF Tail warning function (ECM).

TWG *1* Treaty Working Group (JAA).

2 Technical Working Group (US/NATO Stanags).

TWI Threat- (or tail-) warning indicator.

Twids Taut-wire [perimeter] intrusion detection system.

twilight Designated civil *, nautical * or astronomical * at Sun zenith angles of 96°, 102°, 108°, respectively.

twilight band Strip, about one aircraft span in width, where radio range A/N just detected against steady note (obs.).

twilight effect Faulty indications of radio navaids of pre-1950 (occasionally, later) during twilight, ascribed to ionospheric distortions.

twilight zone Bi-signal radio range zone where only A or N heard (obs.).

twin Aircraft, usually aeroplane, powered by two similar engines; arguably, one powered by two different engines, eg jet plus prop.

twin-aisle aircraft Usually synonymous with widebody, transport whose passenger seating is divided by two axial aisles.

twin-array VASI One located ahead of desired touch-down point, the other beyond.

twin cable Plastic extrusion containing two side-by-side untwisted conductors.

twin-contact tyre See *twin-tread*.

twin-float seaplane One supported on water by two similar floats side-by-side.

twin-gyro platform Platform housing two gimballed gyros providing precision heading and attitude reference for military aircraft; additional input, eg Doppler, is needed for position readout.

twinkle roll Has been used to mean flick roll, but normally maximum-rate slow aileron roll performed by two or more formating aircraft simultaneously, each about own axis.

twin paradox See *time dilation*.

twin-row engine Piston engine of radial type with two rows each occupying one plane and each driving on one crankpin.

twin-shaft *Two-shaft engine*.

twin-spool *Two-spool*.

twin tail Conventional tail with twin vertical surfaces, often at or close to extremities of tailplane in slipstream from engines, esp. with single fuselage.

twin-tread tyre Contacting ground around two circular treads separated by deep groove; intended to reduce shimmy, esp. of tailwheels.

twin-tub aircraft Two-seater, especially with open cock-pits.

twin-wing aircraft *1* Term currently in use for diamond-wing aircraft; especially when tips are joined by vortex-generator endplates.

2 Since 2002 also used to mean a UAV biplane.

TWIP Terminal weather information for pilots.

twist *1* Variation in angle of incidence along aerofoil, always present in rotating blades. In wing, normally subdivided into aerodynamic (defined as variation in no-lift direction along span) and geometric (variation in angle between chord and fixed datum along span).

2 Roll about longitudinal axis, as in * and steer.

twist and steer Control method for fixed-wing aerodynes

in which it is necessary first to roll to (normally large) bank angle and then pull round on to desired heading. Only possible method on aircraft (eg missiles) with no control surfaces other than main wings, which pivot independently for twist and together for steer. Not applicable to conventional aeroplanes, which make simultaneous harmonized movements about all axes.

twister *1* Device twisting electric component of EM, especially coherent radar.

   2 Tornado (colloq.).

twitch factor Unquantified factor degrading flight-crew (esp. pilot) performance in terms of accuracy or incidence of errors, resulting from fear or high workload. In simplest case ** magnifies task of adding 2+2.

twizzle Various manoeuvres adopted by combat pilots, esp. of large aircraft, to throw fighter off aim or effect escape; always involved steep climb and/or dive, combined with limited rolling manoeuvres (WW2).

TWMS Tactical-weapons management system.

Twng Towering.

Two, 2 The partner in a formation-display duo; his role is to keep his eyes glued on Lead and position accordingly.

2½-axis machining Continuous-path machining (invariably with NC) on x, y axes, usually intermittent on z.

two-axis autopilot Simple autopilot having authority in roll and yaw only.

two-axis homing head Scans in az and el.

two-bay biplane One with wing assembly comprising two bays on each side of fuselage, each comprising rectangular cellule of inner and outer interplane struts with points of attachment linked by diagonal wires. Loosely, biplane with interplane struts at two different spanwise locations on each side of fuselage (discounting any struts near fuselage).

two-colour pyrometer High temperature is measured at two wavelengths.

two-control aircraft Aeroplane or glider with flight-control system in two axes only, invariably ailerons and elevator; thus no rudder or pedals.

two-crew Misnomer; what is meant is that flight crew comprises two persons; * operation is often matter between airline employers and ALPA, and largest transports frequently have flight crew of at least three.

two-cycle See *two-stroke.*

2-D *1* Two-dimensional.

   2 Position defined by lateral guidance only in Satnav.

2-DCD Two-dimensional convergent/divergent (nozzle).

2-D flow Flow which can be described completely in one plane, eg around wing of infinite span.

two-dimensional matrix Code for marking parts for identification, small area containing 100 times more information than bar code.

2-D inlet Two-dimensional inlets are in theory ideal for operation over wide range of Mach numbers but because of end and corner effects and mechanical complexity are in practice inferior to axi-symmetric.

2-DM Two-dimensional matrix.

2-D nozzle Jet nozzle of rectangular cross-section and constant longitudinal profile which can vector thrust in vertical plane. Studied for future US fighters. Also called platypus.

   2 **DOF** Two degrees of freedom.

2-D radar Radar giving target position in two dimen-

sions, eg az/el, or range/bearing; only a few can pinpoint body's location in 3-D space.

2-D wing Usually one of infinite span, whose tip effects can thus be ignored; for some purposes constant-section wing joined at each end to tunnel wall is approximation.

twofold flocking Turbofan test for birdstrikes over whole area of fan and spinner.

two-frequency Glidepath and/or localizer having two unrelated radiation patterns.

2 Gins Two-gimbal INS (colloq.).

two-inceptor control Fundamental control strategy for pilot of jet STOVL aircraft in which left hand controls speed only (including forward acceleration from hover) and right hand controls flight trajectory (including hover height).

two-level bridge (jetway) Passenger bridge whose landside end can be raised or lowered to mate with either arrival or departure floor level.

2LM Second-level maintenance.

two-man rule Philosophy under which no individual is allowed unaccompanied access to NW or certain designated components or associated system interfaces.

two-meal service Two meals served on one sector.

2-minute turn Rate 1 turn; also called standard turn or procedure turn.

two-moment equation Relates simple-beam loading to shear and BM.

2 on 1 injector Rocket liquid injector spraying two streams of liquid A on to one stream of liquid B, the three meeting at various angles depending on design.

2-place Two-seater.

2.5-D Describes advanced synthetic vision systems which give important information on 3-D shapes.

2.5-engine aircraft Twin-engine aircraft certificated to use APU [or other centreline auxiliary engine] to give propulsive thrust, especially at takeoff.

2-point landing See *wheeler.*

2-point suspension Bifilar, hung from two points at same level on two filaments or cables.

two-point tanker Fitted with two HDUs which can be used simultaneously.

2-pole switch Opens or closes both sides of same circuit or two separate circuits simultaneously.

two-position propeller One having only two settings, fine and coarse.

two-pulse rocket Motor, usually solid propellant, comprising two stages (eg, boost and sustain fired in series).

two-rate oleo Normally, first part of travel is at low rate (ie, large deflection per unit load); after loading to given limit, often close to static position at gross weight, high-rate law applies. Common on naval helicopters.

two-tow radial See *twin-row engine.*

two-segment approach Two angles, usually 6°/3°.

two-shaft engine Gas turbine with mechanically independent LP and HP sections each running at its own speed, or a free-turbine turboshaft or turboprop.

two-speed supercharger Driven by step-up gearbox which, usually automatically at given pressure height, changes ratio to increase impeller speed at high altitudes.

two-spool Gas turbine of two-shaft type, originally confined to engines having two axial compressors (LP and HP) in series. Also called split-compressor engine.

two-stage Adjective, achieved in two parts. A turboprop may have a * compressor through the parts of which the air flows consecutively, whereas the propeller is driven by a * gearbox in which both stages operate simultaneously.

two-stage amber Synthetic day/night pilot training aid in which pilot wears blue goggles and flies trainer with amber transparencies (or vice versa) which effectively eliminates external cues while allowing view of instruments.

two-stage igniter Generally, one that first ignites a local flame, burning main fuel, which in turn is used to ignite main combustion (gas turbine or rocket but now rare).

two-stage supercharger One having two impellers in series, either both driven by crankshaft or with first (LP impeller) driven by exhaust turbo.

2-star red Standard distress pyrotechnic, fired without pistol.

two-step supercharge Throttle is gated below rated altitude, thereafter being free to move to maximum position; used with or without two drive ratios.

2-stick Dual-control aircraft (colloq.).

two-stroke Piston-engine cycle in which every upstroke expels exhaust and compresses mixture and every downstroke provides power. Often abb. 2T. Not normally certificatable for unrestricted aviation use.

2-view drawing Orthographic projection drawing comprising two views, usually left-side elevation and plan.

TWP *1* Two-way programme.
2 Technical-work programme.

TWR *1* Threat-warning receiver (formerly sometimes rendered as tail-warning radar).
2 Tower (or twr); suffixes: INC, in cloud; INH, in haze; INK, in smoke; INP, in precipitation; INUN, in unknown obscuration.
3 Turbulence weather radar.

twr Tower; one definition = aerodrome control.

T/W ratio Thrust/weight ratio.

TWRG Towering.

TWS *1* Tail-warning set or system.
2 Track-while-scan.
3 Threat-warning system.
4 Through-wall surveillance.
5 Thermal weapon sight.
6 Tactical work station.
7 Terminal weather system.

TWSC Thin-wall steel case.

TWSRO Track-while-scan in receive mode only.

TWSS *1* TOW weapon subsystem.
2 Track-while-scan system.

TWT *1* Travelling-wave tube.
2 Transonic wind tunnel.

TWTA TWT amplifier.

TWU *1* Tactical Weapons Unit (RAF).
2 Transport Workers Union, now Transport Workers of America (US).

TWX *1* Theatre war exercise.
2 Teletypewriter exchange.

TWY, twy Taxiway.

TWYL Taxiway link.

TX, tx Transmitter, transmit, transmission.

T_x Horizontal component of thrust.

TXP Transponder (also TPDR, XPDR etc).

TXT Text.

TYD, TY$ Then-year dollars, ie based on a selected historic monetary value.

TYP Type of aircraft (ICAO).

type *1* Particular type of aircraft, ignoring marks or subdivisions.
2 Human being, especially adult male (as 'a good *', 'a quiet *', colloq. RAF usage, WW2).

type approval Issue of type certificate.

type certificate Legal document, in US issued by FAA, allowing manufacturer to offer item (eg aircraft, engine) for sale.

type certificate data sheets Official specifications to which each unit (aircraft, engine, propeller) commercially offered for sale must conform (FAA).

type conversion Clearance of qualified pilot to fly additional type of aircraft.

Typed Air Station Establishment assigned task of providing full support for particular aircraft type, even though based elsewhere or embarked (RN).

Type Rating Endorsement on licence specifically qualifying holder to fly or maintain particular type of aircraft. These are required if MTOW exceeds 12,500 lb (5,670 kg).

type record Various books or document dossiers, some of which have legal status, but basically complete record of all decisions made regarding particular type of hardware, esp. including all modifications, service difficulties, airworthiness directives and alterations made locally, eg on owner's initiative.

type test Basic government test clearing engine or other machine for production and acceptance by government customers (UK).

TYPH Typhoon.

typhoon *Tropical cyclone, revolving storm.*

typical In structures and structural components, arithmetical mean structure calculated by measuring sections of all members actually produced.

Tyranno A Japanese high-modulus ceramic fibre.

Ty-Rap Patented nylon strap for tying and identifying bundles (looms) of electrical wiring.

tyre Specifically for landing-gear wheels; US = tire.

tyre sizes Usual sequence of three numerical values is overall diameter, section width (undeflected) and (if present) inner diameter or bead size. Other measures include Types III and VII, radial and metric. There is a need for a uniform scheme.

Tyro Callsign prefix, when calling military or D&D: I am inexperienced.

Tyuratam Soviet test centre for ICBMs, various other missiles and FOBs, and launch centre for Cosmos military satellites; location 45.8°N, 63.4°E.

T_z Vertical component of thrust.

TZD True zenith distance.

TZM Titanium, zirconium, molybdenum.

TZP Ceramic comprising mix of tetragonal zirconia in fine-grained polycrystalline zirconia.

U

U *1* Overall heat-transfer coefficient.

2 Linear acceleration (archaic usage).

3 Internal or intrinsic energy.

4 Aircraft has 4096-code transponder with altitude encoding.

5 With appropriate suffix, flow velocity component [along orthogonal axes, radial or tangential], or vehicle speed, eg U_o.

6 Designation prefix, aircraft, unpiloted (USN 1946–55), RAF (from 1956), or utility (USAF from 1952), USN (from 1955).

7 Modified mission, prefix and suffix, utility (USN).

8 Designation first letter, missile, underwater-launched (DoD).

9 Designation second letter, missile, attack on under-water target (DoD).

10 Airfield is unlicensed.

11 Upravlayemaya, = guided (R).

12 *Umrüst*, = factory modification (G, WW2); -B or *Bausatz* added conversion kit.

13 Upper (wing).

14 Identity unknown.

15 Until.

16 Unicom.

17 Uranium.

18 Other meanings include unmanned, unwatched, unverified, upward and user-fee.

u *1* Force in structural member due to unit load.

2 Tangential velocity, or linear velocity of point in rotating structure, eg aeroplane in roll.

3 Also used for translational velocities.

4 Surface unpaved.

5 Unit of atomic mass.

6 Specific internal energy.

U-alpha Free-stream fluid velocity, usually written $U\alpha$.

U-code See *U (4)*.

U-index Monthly mean of differences between consecutive daily mean values of horizontal component of Earth's magnetic field.

u-index Value of U-index divided by sine of magnetic co-latitude multiplied by cos of angle between magnetic meridian and horizontal component.

U-joint Universal joint.

U-tail Aeroplane tail with twin verticals attached to fuselage (can be inclined outwards but not applicable to butterfly tail).

U-235 Fissile uranium, isotope typically 0.71 per cent of natural metal, nearly all the rest being U-238.

UA *1* Uncontrolled airspace.

2 Unit of account; standard accounting unit (EEC).

3 Until advised.

4 Airep, upper-air pirep.

5 Unnumbered acknowledgement.

UAA *1* Up and away (JSF).

2 University Aviation Association (US).

3 Upper advisory area.

UAAA Ultralight Aircraft Association of Australia.

UAB Until advised by (ICAO).

UAC *1* Upper-air computer, or center (FAA).

2 Upper-airspace control (CAA).

3 Upper-area control centre.

4 Universal avionics computer.

UAD *1* Unidirectional aligned discontinuous.

2 Upper advisory (route).

UADE Ultralightweight aerial diesel engine [usually Ulade].

UAI Union Astronomique Internationale (Int.).

UAJ Unattended jammer.

UAL *1* Unidirectional approach light[ing].

2 Unit authorization list.

Uα See *U-alpha*.

UALV Unmanned airlift vehicle.

UAM Underwater-to-air (missile).

UANC Upper-airspace navigation chart.

U&L Upper and lower (wings).

UAOS Unmanned aerial, or aeronautical, observation system.

UAPP Unified adapative planning (or preplanning) program.

UAR *1* Upper air route.

2 Unattended radar.

UARS *1* Unattended radar station.

2 Unmanned air reconnaissance system.

3 Upper-atmosphere research satellite [NASA, 1991–2001].

UART Universal asynchronous receiver/transmitter.

UAS *1* University Air Squadron(s). (UK).

2 Upper airspace.

3 Uninhabited air system.

UASA Upper-airspace service area.

UASTAS Unmanned airborne surveillance and target-acquisition system(s) (DND).

UAT Universal-access transceiver.

UATI Union des Associations Techniques Internationales (Int.).

UATP Universal air travel plan (Int., US-based).

UAV Unmanned, or uninhabited, air, or aerial, vehicle; B adds battlelab, S systems, SA Systems Association.

UB Utility bus.

UBE Ultra-bypass engine.

Ubee, ubie Alternatives to UBE.

UBEP UK bomb enhancement programme.

UBF Underground baggage facility.

UBI Uplink block identifier.

UBRE Unit bulk refuelling equipment, 4 × 4 truck.

U/c, u/c *1* Undercarriage, ie landing gear.

2 Under construction.

UCAR, Ucar Uninhabited, or unmanned, combat armed rotorcraft [previously called R^2W] (USA/Darpa).

Ucars Unmanned air vehicle common automatic, or automated, recovery system.

UCAV Unmanned, or uninhabited, combat air vehicle.

UCC Ultra-compact combustor.

UCCEGA Union des Chambres de Commerce et Établissements Gestionnaires d' Aéroports (F).

Uchinoura Rocket launch site, Kagoshima prefecture (J).

UCI *1* Unit construction index; measure of flotation

(ability of landplane to use soft airfields) based on MTOW, tyre pressure and gear geometry.

    2 User computer interface.

UCIIR   Uncooled imaging IR; SAL adds semi-active laser.

UCIMU   Unione Costruttori Italiani Macchine Utensili.

UCL   Up command link; opposite of DDL.

UCNI   Unified com/nav/ident.

UCP   Ultrasonic capacitance probe.

UCR   Uniform commercial rate[s].

UCS   *1* Utilities control system.

    2 Uniform chromaticity scale.

    3 Universal control station (NATO).

UCWA   Urgent center weather advisory (Sigmet).

UD   *1* User data.

    2 Upper deck.

    3 Unaligned discontinuous.

UDA   Upper advisory area.

UDACS   Universal display and control system.

UDCS   Universal drone control system.

UDDF   Up- and down-draughts.

u_{DE} Gust vertical speed.

Udet buoy   Open-sea tethered buoy to assist downed aircrew (G, WW2).

U-deta   Unsymmetrical diethyltriamine.

UDF   *1* U.h.f. direction-finding.

    2 Unducted fan engine (GE trademark).

    3 Unit development folder (software).

UDL   Universal or unclassified, data-link.

UDM   Universal docking module.

UDMH   Unsymmetrical dimethylhydrazine rocket fuel.

UDMU   Universal decoder memory unit (Gpats).

udometer   *Rain gauge.*

UDP   User Datagram protocol.

UDS   *1* Unidirectional solidification.

    2 Unsupported (or unstocked) dispersed (or dispersal) site.

UDT   Unidirectional transducer.

UDTE   Upgraded data-transfer equipment.

UE   Unit equipment, or establishment, list of aircraft or other items serving with combat units.

UE, u/e   Under-excitation.

UECNA   Union Européenne Contre les Nuisances des Avions (Int.).

UEDP   Uncontained engine debris pattern.

UEET   Ultra-efficient engine technology (NASA).

UEJ   Unattached expendable jammer.

UEO   Western European Union (F).

UER   Unscheduled engine removal.

UESA   U.h.f. electronically scanned array, or array antenna.

UEU   Universal exciter upgrade.

UEWR   Upgraded early-warning radar.

UF   Uplink format.

UFA   Until further advised.

UFC   *1* Unified fuel control.

    2 Up-front control; D adds display, P panel.

UFCM   Uncommanded flight-control movement.

UFD   *1* Up-front display.

    2 Ultra-flat display.

UFDR   Universal flight-data recorder.

UFIR   UAV-to-fighter imagery relay (USAF).

UFL   Upper flammability limit.

UFN   Until further notice.

UFO   *1* Unidentified flying object.

    2 U.h.f. follow-on (DoD satellites).

    3 United Flying Octogenarians (US).

UFPA   Uncooled focal-plane array.

UFS   Ultimate factor of safety.

UFTAS, Uftas   Uniform flight-test analysis system (AFFTC).

UGAI   Unione Giornalisti Aerospaziali Italiani. (I)

Ugine-Séjournet   Method of extruding steel using molten glass as lubricant.

UGM   US prefix, underwater-launched surface-attack missile.

UGS   *1* Upgraded silo.

    2 Unattended ground sensor(s).

UGV   Unmanned ground vehicle, usually controlled from air.

UH   UDMH

UHB   Ultra-high bypass engine.

UHC   Unburned hydrocarbons.

UHCA   Ultra-high-capacity aircraft.

UHD   Ultra-high density.

UHDT   Unable higher, due traffic.

u.h.f.   Ultra-high frequency, see Appendix 2.

UHPT   Undergraduate helicopter pilot training.

UHR   Ultra-high resolution.

UHS   Ultra-high-speed (logic).

UHV   Ultra-high vacuum.

UI   Unnumbered information.

UIC   Upper-airspace/information centre (ICAO).

UIF   Unfavourable information file.

UIL   User interface language.

UIP   Upgrading instructor pilots.

UIR   Upper-airspace flight-information region.

UIS   Upper-airspace/information service, or system.

UIT   *1* Union Internationale des Télécommunications.

    2 UV imaging telescope.

UJT   Unijunction transistor.

UK   Training centre (USSR, R).

UKAATS   UK Advanced Air Traffic System.

UKAB   UK Airprox Board (CAA).

UKACC   UK Air Cargo Club.

UKACCIS   UK air command and control information system.

UKAdge   UK air-defence ground environment.

UKADR   UK air-defence region.

UKAPE   The UK Association of Professional Engineers.

UKAS   UK Accreditation Service.

UKcEB   UK Council for Electronic Business.

UKF   Unbemanntes Kampfflugzeug = UAV (G).

UKFSC   UK Flight Safety Committee.

UKIRCM   UK IR countermeasures.

UKISC   UK Industrial Space Committee (FEI + SBAC).

UKLF   UK land forces.

UKMF   UK Mobile Force.

UKMSCS   UK military satellite communications system.

UKN   Unknown.

UKOOA   UK Offshore Operators' Association.

UKRAOC   UK Regional Air Operations Centre.

UKSCC   UK Satellite-navigation Co-ordinating Committee.

UKSEDS　UK Students for the Exploration and Development of Space (RAeS).

UKWMO　UK Warning and Monitoring Organization.

UL　*1* Ultralight.
　2 Unleaded.
　3 Uplink

ULA　*1* Uncommitted logic array.
　2 Ultra-light aircraft.
　3 Fairbanks, AK, tracking station.

ULAA　Ultra-Light Aircraft Association (Australia; in UK became PFA in 1952).

Ulade　Ultralight aeronautical diesel engine.

ULAIDS　Universal locator airborne integrated data system.

Ulana　Unified local-area network architecture.

ULB　Underwater locator beacon.

ULC　Unit load container.

ULCE　Unified life-cycle engineering.

ULCS　Unit-level circuit switch.

ULD　Unit load device.

ULDB　Ultra-long-duration balloon.

ULD Carrier　See *Dolly.*

ULEA　Ultra-long-endurance aircraft.

ULEV　Unmanned long-endurance vehicle.

ULH　Ultra-long-haul.

ULL　Ullage (NASA).

ULLA　Ultra-low-level airdrop.

ullage　Volume above liquid in a tank; occasionally misused to mean last dregs of liquid itself remaining in tank.

ullage engine　Rocket motor fired in ullage manoeuvre.

ullage manoeuvre　Applying axial thrust to space-launcher stage or other liquid propellant tank(s) that are nearly empty in order to collect remaining propellants around delivery pipe connection.

ullage motor　See *ullage engine.*

ullage space　See *ullage.*

ullage washing　Injecting gaseous nitrogen from ground supply before takeoff.

ULM　Ultra-leger motorisé = ultralight aircraft (F). MTOW (1 seat) 300 kg (661.4 lb), (2 seat) 450 kg (992 lb), stall ⩽ 35 kt (65 km/h).

ULMS　*1* Undersea-launch (or long-range) missile system.
　2 Unit-level message switch.

ULR　Ultra-long-range; loosely = ULH.

ULSA　Ultra-low-sidelobe antenna.

Ultem　Fire-blocking furnishing materials based on PEI, seen as replacement for Lexan (GE).

ultimate factor of safety　Number by which limit (or proof) load is multiplied to obtain ultimate load; purely arbitrary factor of safety, usually varying from 1.5 to 2, representing best humanly attainable compromise between economic structure weight and aircraft that will not break.

ultimate load　Greatest load that any structural member is required to carry without breaking; that at which it may legally be on verge of breaking, and permanently deformed. Usually, limit load × UFS.

ultimate strength　Strength required to bear ultimate loads, or product of greatest load considered possible in service multiplied by UFS. Gust requirements, in which structure oscillates, can be superimposed on maximum static load to demand even greater design strength in "flappable" parts of structure, esp. wing. Hence, ultimate bending/compressive/tensile/torsional strength.

ultimate stress　That in a member or piece of material at moment of fracture; in theory that in structural member loaded to ultimate strength.

Ult-Join　Unit-level trainer, joint operations integrated network.

ultra-bypass engine　One name of unducted propfan, with BPR from 20 to 50 (Boeing).

ultra-high bypass　UHB, alternative to above (McDonnell Douglas).

ultra-high-density seating　Seating configuration for maximum number of passengers (invariably greater than original certification limit) with minimal pitch and no galley.

ultra-high frequency　300–3,000 MHz, see Appendix 2.

ultra-high magnetic field　Generally, greater than 10 T (100 kG).

ultra-high-speed photography　Rate higher than 10^4 images/s.

ultra-high vacuum　Previously below 10^{-13} torr (not apparently yet defined in SI, but this is $13.33 \times 10^{-11} \text{N/m}^2$); density 3.2×10^3 mol/cc, mean free path (air 25°C) 3.8×10^{10}.

ultralight aircraft　Categories of small aeroplane. In USA, 1-seat with empty weight ⩽254 lb (115.2 kg); in Australia, 1 or 2 seats, MTOW 540 kg (1,190.5 lb). France, see *ULM*, UK, see *microlight*, *SLA.*

ultra-long-haul　FDP exceeding 16 hours.

ultra-low airdrop　Below 15 m (50 ft) AGL.

ultramicroscope　Instrument for observing extremely small particles by intense illumination causing diffraction rings on dark background.

ultra-short takeoff　Use of short forward run by aircraft capable of VTO, normally called merely STO. Can involve pilot-selectable operating modes.

ultrasonic　Mechanical vibrations, eg sound waves, of frequency too high to be audible to humans (opposite = infrasonic). Generally frequencies above 15 kHz, usually generated by electro-acoustic transducer and propagated through solids, liquids and gases.

ultrasonic bonding　Techniques of joining materials with ultrasonic energy, in some cases to cause local melting (ultrasonic welding) and in others to cause either enhanced local diffusion or merely heating to accelerate curing or setting of adhesive.

ultrasonic cleaning　The item to be cleaned is immersed in liquid [water or various proprietary fluids] and surrounded by small bubbles created by magnetostrictive transducers. The collapse of the bubbles [implosion] bombards the surface with shockwaves.

ultrasonic inspection　NDT method in which cracks are revealed by discontinuity in propagation of ultrasonic waves through metal.

ultrasonic machining　Techniques for shaping solids, esp. those too hard or for any other reason not machinable by conventional means. Most common method uses shaped tool oscillated vertically above work at frequency about 20 kHz, with space between tool/work fed with hard powder abrasive.

ultrasonic rolling　Transducer (typically 20 kHz) placed inside one (rarely both) of rolls in rolling mill with objective of reducing rolling energy needed, reducing required temperature of metal being rolled and rolling to thinner gauges.

ultrasonic welding　Techniques in which ultrasonic vibrations are used to join two metal (rarely, either or both is non-metal) surfaces brought into contact. Surfaces usually well mating, atomically clean and pressed together under pressure. In some techniques ultrasonic rearrangement of atoms is sufficient to cause melting at interface, while in others there is diffusion between the two faces.

ultrasonic wind sensor　Most have three sources exchanging pulses between each pair, times being precisely measured.

ultrasound　Sound of ultrasonic frequency.

ultra-violet　Band of EM radiation of frequency just too high (ie wavelength too short) to be visible; links visible violet at about 3.8 to 4×10^{-7} m to limiting wavelength put at from 1 to 1.36×10^{-8} m (next major part of spectrum is X-ray at appreciably shorter wavelength).

ultra-violet detector　Material, eg single-crystal silicon carbide, not triggered by sunlight or other incident radiation but instantly responsive to UV.

ultra-violet imagery　That produced by sensing UV reflected from target (DoD).

ULV　*1* Upper limit of video (HUD).

　2 Ultra-low volume (crop spraying).

UM　Rhymes with 'come', an unaccompanied minor [passenger under 16].

UMA　*1* Unmanned aircraft, = drone or RPV.

　2 Uffici Meteorologici Aeronautica (I).

　3 Unusual military activities.

Umark　Unit maintenance aerial recovery kit.

umbilical　*1* Connection between vehicle, esp. space launcher, missile (esp. large ballistic) or RPV, and ground, along which pass all necessary supplies and signals up to moment of liftoff or launch, eg computer data-link, telemetry, guidance programming, topping-up liquid propellants, electrical/hydraulic/pneumatic power and instrumentation. Generally grouped into one large multiple conduit with pull-off connector. Occasionally plural, when two serve one vehicle.

　2 Jetway, loading bridge (colloq.).

umbilical cord　See *umbilical (1)*.

umbilical tower　Tower carrying umbilical lines and connector to an upper stage of a ballistic vehicle; often called umbilical mast.

UMD　*1* Unrefuelled mission distance.

　2 Unit manning document.

UMG　Universal message generator.

umkehr　Anomaly of relative zenith intensities of scattered sunlight near UV as Sun nears horizon, often cap U.

UML　Unified modelling language.

UMLS　Universal MLS.

UMR　Unrefuelled mission radius.

umrust　Modified, modification (G).

UMS　Unified message switch.

UMT　*1* Universal military training.

　2 Universal mount.

UMTE　Unmanned threat emitter.

UMTS　Universal mobile telecommunications system, providing all previous facilities plus TV and video-conferencing on mobile devices (2004 on).

UN　United Nations, many suffixes.

UNA　Unable.

unaccelerated flight　Flight without imposed acceleration, but there are several conflicting definitions and term needs redefining to avoid ambiguity; (1) flight along straight-line trajectory at constant V, thus no imposed acceleration; (2) rectilinear flight (eg straight and level) with no imposed acceleration normal to flightpath other than Earth gravity; (3) weightless flight with not even gravitational acceleration, thus curvilinear around Earth. Definition (2) allows acceleration along line of flight, thus speed can vary.

unaccompanied baggage　Not carried on same aircraft with pax or crew to which it belongs; DoD definition adds 'not carried free on ticket used for personal travel'.

unanimity rule　Basic IATA principle that fare from A to B is same as from B to A and same for all carriers offering identical service; can be relaxed.

UNAP　Unable to approve.

unapproved　Not tested to established requirements, thus flown on special dispensation, eg VW engines with single ignition.

unapproved part　Can physically be installed but does not fulfil requirements. See *counterfeit*.

unaugmented　In case of turbojet or turbofan, not equipped with afterburner, or not using afterburner.

unavbl　Unavailable.

Unavia　Italian national organization for study and development of aircraft technology.

unbalanced cell　Cell of Ni/Cd battery which has discharged more than others; first step to thermal runaway.

unbalanced field　Any takeoff in which accelerate/stop distance is not the same as for normal takeoff to 35 ft.

unbalanced turn　One with slip or skid.

unblown　*1* STOL aircraft with USB, IBF, EBF or other powered blowing system in flight mode with blowing inoperative.

　2 Of piston engine, not equipped with supercharger.

unburnt hydrocarbons　Essentially unburnt fuel, contaminant emitted by engines and subject to emissions legislation.

UNC　United Nations command; followed by various force initials.

uncertain　Category of aircraft whose safety is not known, normally applied when 30 minutes have elapsed since arrival message or ETA and not answering radio call; hence uncertainty phase.

Uncertificated　Aircraft category; airworthiness not established.

UNCL　*1* Unified numerical-control language.

　2 Unclassified.

Unclassified　*1* Security category for official matter which requires no safeguards but may be controlled for other reasons.

　2 Performance category for aircraft, usually civil transports, in service prior to 1951 and thus not built to CAR.4(b), BCARs or SR.422A/B.

uncontrolled airspace　Airspace where no ATC service is provided.

uncontrolled mosaic　Made up of uncorrected images matched from print to print without ground control or other orientation, giving mosaic on which distances and bearings cannot be accurately measured.

uncooled　Descriptive of turbine blade devoid of internal or transpiration cooling.

uncooperative　Generalized adjective for vehicle of essentially passive nature, devoid of helpful emissions and not responding when electronically challenged; applied to aircraft, usually means not equipped with transponder.

uncoupled　Vibration mode wholly independent of others at same time.

Unctad　UN Conference on Trade and Development.

unctld　Uncontrolled airspace (FAA).

undamped　Vibration of free nature dependent only on internal mass and elastic and inertia forces.

underbreathing　See *hypoventilation.*

undercarriage　Landing gear (UK usage; original BSI definition included main wheels, skids or floats and support and explicitly excluded tail wheel or skid). Hence * door, indicator, same as landing gear door, indicator. Note: floats are usually called alighting gear.

undercast　Solid cloud cover seen from above.

underdeck spray　Pad deluge directed up at underside of launch pad (in some cases deliberately including underside of base of launched vehicle).

undergraduate　Aircrew (esp. pilot) pupil who has not yet qualified, ie won his/her wings.

underlay　Lowest stratum of layered defence.

undershoot　*1* Faulty approach by aerodyne, usually fixed wing, which if continued to ground results in landing short of desired area, eg before threshold; normally corrected by go-around (overshoot).

2 To fail to capture desired flight condition, eg altitude, IAS, by falling short when approaching value from below, normally through small lack of aircraft energy.

under the hood　Instrument flight training in which pupil is prevented from seeing outside aircraft, originally by unfolding opaque hood, later by two-stage amber or other method.

under the radar　Flight levels as close as possible (see *lo*) to ground in attempt to thwart hostile attempts to obtain positive track by defensive radars; becoming pious hope.

underwater missile　Launched below water surface.

under way　Marine aircraft, moving on water [some authorities: under weigh, from weighing the anchor].

underwedge bleed　Secondary airflow extracted from location on underside of variable wedge above supersonic airbreathing inlet, usually from point of maximum wedge depth at throat.

under wing　In service [engine possibly hung on rear fuselage], thus 30,000 h * = total time.

undevelopable　3-D curvature, which cannot be drawn accurately on flat sheet and can be made only by some type of sheet forming.

undk　Undock(ing).

undock　*1* To separate two vehicles in space previously joined and with intercommunication but not necessarily sharing common atmosphere.

2 To remove airship from hangar.

unducted fan　Engine in which gas-turbine core drives fan blades external to cowled engine. Blades can be mounted on ring of turbine blades in gas flow, or driven via gearbox (tractor or pusher) from separate turbine.

unduplicated　Generalized term meaning that in assessing airline or other route network each sector is counted once only, in one direction only, despite fact it is used in both directions and may form part of from 2 to 48 distinct routes; hence * route mileage, * network.

unfactored　Not multiplied by a factor, eg factor of safety; hence * load = limit load. See next.

unfactored performance　That expected from average aircraft, flown by average pilot, with no safety factors.

unfavourable unbalanced field　One whose clearway allows takeoff at increased weight at which TOD to 35 ft exceeds accelerate/stop distance.

unfeather　To restore propeller from feathered state to normal operative state with engine transmitting power and blades in positive pitch.

unfilmed　IIT not coated with protective ion barrier in order to improve resolution and SNR.

Unfo　Undergraduate Naval Flight Officer (USN).

UNGA　*1* United Nations General Assembly.

2 Unione Nazionale Giovanile Aeronautica (I).

UNICE, Unice　Union of the industrial federations of the EEC countries (English translation).

Unicom　Private radio communications service on five frequencies based at airports, heliports etc, with or without tower in addition; used for various advisories other than ATC purposes (US).

unidentified flying object　Something seen in the sky which, by virtue of shape or behaviour, cannot be identified as something known to humanity.

unidirectional aerial　Single well-defined direction of maximum gain.

unidirectional composite　All fibres are parallel, usually aligned with direction of applied load.

unidirectional current　Flowing in one direction only, eg signal pulses or d.c. with superimposed a.c.

unidirectional solidification　Techniques for obtaining strongly preferred direction of crystalline grains on solidification of alloy from melt such that each crystal forms long string or column aligned with maximum applied load, transverse intercrystalline joints being rare. In most applications, eg turbine rotor blades, a step on route to single-crystal material.

Unido　UN Industrial Development Organizations.

unified fuel control　Control system for supersonic airbreathing engine governing engine acceleration, pilot input response, Mach, nozzle and Plap.

unified system　Generalized term for TTC (tracking, telemetry and command) system handling all digital, video and voice signals in integrated system for larger or complex air vehicle or space payload. Also applied to simpler satellites, missiles and RPVs, eg unified-pulse systems in which each command pulse train comprises vehicle address, message ident and message, synchronized with radar ranging pulses; thus one ** handles all vehicles.

Unified thread　Standard 60° screwthread (US, UK, Canada).

uniform acceleration　Time-invariant, giving straight line on V/T plot.

uniformly distributed load　One imposing constant force on each unit area of horizontal floor.

uniformly varying load　Magnitude varies directly with distance along straight-line axis from reference point.

uniform photo-interpretation report　Third-phase machine-formatted intelligence report of particular objective containing detailed information extracted from photo sensor imagery (USAF).

uniform velocity　Time-invariant speed and direction.

Unihedd　Universal head-down display; standardized

cockpit display of information from many sources and sensors with high brightness.

unijunction transistor Bar of doped semiconductor with p-n junction near centre, normally forward-biased, giving sharp peak of emitter V for small emitter current; used in pulse generators and sweep circuits. Also called double-base diode.

unilateral instrumentation In unyawed flight each half-aircraft behaves as mirror image of other and, therefore, full information can be obtained by fitting pressure-sensing heads, transducers, and other experimental equipment on one half only; ** should not be used during yawed flight.

unipole Hypothetical aerial radiating equally in all directions.

Unishear Fixed or portable high-speed shear for thin sheet.

unit *1* Fundamental subdivision of any numerical measure. Nearly all industrial countries have in theory adopted SI system. Base * of seven fundamental quantities are: length, metre (m); mass, kilogram(me) (kg); time, second(s); electrical current, ampere (A); thermodynamic temperature, kelvin (K); luminous intensity, candela (cd); and amount of substance, mole (mol). SI * of electricity or charge is coulomb. For payment of electricity supply, kWh.

2 Military element whose structure is prescribed by component authority; normally applied only to small organizations at field level, eg squadron, or to element of airborne force, eg two or three aircraft.

3 One complete item from production, eg production *, * sales.

unit aircraft Those provided to an aircraft unit for performance of a flying mission (DoD).

unit area That equal to square of unit length on each side.

unitary Having a single warhead, as distinct from dispensed or clustered munitions.

unit cost *1* Airline's total costs divided by generated RPKs or RTKs.

2 See Average flyaway *, average procurement *, program acquisition *.

unit deformation Deformation divided by original length or other undistorted measure; ie deformation per unit of original length.

Uniter Secure survivable fixed telecom network, part of DFTS and particularly linking UKAdge (RAF).

unit hydraulic tail Power unit driving tailplane (horizontal stabilizer, taileron).

unit load Also called unitized load, any collection of cargo items packaged to fit a unit load device.

unit load device Platform or container for cargo, of standard ISO dimensions and interfacing with handling and restraint systems, e.g., LD3 container or 88×125 pallet.

unit of issue Measure in which commodity (esp. military) is issued, eg by number, dozens, metres, feet, US gal etc.

units of measurement See *unit (1)*. In most practical flying SI is still unattained, distances being in nautical miles, pressures in lb/sq in, atmos, bar and various other units (hardly ever Pa), acceleration in g, and volumes in litres, US gal or Imp. gal instead of m^3 (but fuel mass generally in kg).

unit stress Usually load divided by cross-section area normal to applied force direction.

unit training device Simulator.

universal gas constant See *gas constant*.

Universal Metrics A set of six performance measures designed to be applicable across the UK [and thus probably any other] aerospace industry, launched in 2002 by the SBAC Lean Aerospace Initiative. The objective is to enable companies to measure their performance and that of their supply chain.

universal motor Electric motor operative on a.c. or d.c.

universal polar stereographic Grid for regions from 80° latitude to poles.

universal receiver Operative on a.c. or d.c., with various protective devices.

Universal Safety Oversight Audit Inspection by ICAO-led team of safety and security measures taken by 187 member states to counter terrorism.

universal shunt Resistor in parallel with ammeter to increase range of currents measured as FSD is approached.

universal tester Usually, hand meter for measuring a.c./d.c. current, a.c./d.c. volts and resistances.

universal time Defined by rotation of Earth, and thus not absolutely uniform; ** 1, corrected for polar motion; ** 2, corrected for seasonal variation in rotation.

universal timing disc Disc graduated 000°/360° with associated pendulous pointer, attached to piston-engine crankshaft.

universal transmission function Attempts to describe mathematically IR propagation in atmosphere.

universal transverse Mercator Grid co-ordinate system from 84°N to 80°S.

UNK, unk, unkn Unknown.

unknown traffic Flight details not known to the ATSU with which you are in contact.

UNL Unlimited (altitude, ceiling).

unlgtd Unlighted (FAA).

Unlimited *1* Air race class for piston-engined aircraft: no restriction on engine capacity.

2 Aerobatic class: no restriction is imposed on flight manoeuvres apart from airframe strength limits.

unlimited ceiling Traditionally, less than 50% cloud, base 9,750+ft (2,972 m) AGL (FAA).

unload To reduce g (normal acceleration), usually to restore lost speed.

unlocked Automatic-gun action in which at moment of firing breech is not locked (eg to barrel or case) but is of sufficient mass (inertia) for pressure to have fallen to safe level before breech opens significantly; usually synonymous with blowback action.

unltd Unlimited.

unmanned Aircraft has no pilot on board [but may have one in another aircraft or on ground].

unmask Point at which vehicle becomes visible to defending surveillance systems.

unmkd Unmarked (eg obstruction) (FAA).

Unmovic UN Monitoring, Verification and Inspection Commission.

UN No Four-digit number identifying a particular dangerous material in cargo.

UNOSC UN Outer Space Committee.

unpaired channel DME channel without a corresponding VOR or ILS frequency.

unprepared airfield   Usual meaning is without permanent paved runway.

unpumpable fuel   That fuel deliberately designed to be unpumpable, through levels of sumps and booster-pump inlet sills, to avoid ingestion of any water.

unrefuelled range   Range on fuel carried at takeoff; usually applied to aircraft with provision for inflight refuelling.

unreinforced ablator   Without honeycomb filling or backing.

unrel   Unreliable.

unreliability coding   Series of (usually five) dots following transmissions made by unusable Tacan for calibration or test purposes, indicating not to be used.

unrotated projectile   Original name for unguided air and ground-launched rockets (UK, WW2).

UNRSTD   Unrestricted.

UNSA   Unmanned naval strike aircraft [projects].

unscheduled maintenance   Those unpredictable maintenance requirements not previously planned or programmed but which require prompt attention and must be added to, integrated with or substituted for previously scheduled workloads (USAF).

unsealed strip   Runway with no waterproof coating; normally means compacted earth, gravel or other substrate with or without top layer of rolled material or prefabricated mat of mesh or steel planking.

unshielded   A major meaning is that portion of aerodynamic surface in full slipstream, not in wake of another part of aircraft, especially that part of rudder not in wake from horizontal tail in established spin. The URVC is based on this multiplied by its moment arm.

unstable aerofoil   Generally means one with extensive CP travel.

unstable air   Air in which temperature decreases with height at rate greater than DALR or SALR; thus a parcel given a small vertical movement (up or down) will continue with increasing speed.

unstable aircraft   General meaning is one which, when diverted [even slightly] from straight/level flight, will diverge in uncontrolled manner.

unstart   Explosively violent breakdown of correct (ie started) airflow through supersonic inlet to airbreathing engine, notably with expulsion of shockwave(s) and temporary reversal of flow. All highly supersonic (M3+) engines have such a large contraction ratio and constricted throat that any yaw, spillage from neighbouring engine, gunfire or other disturbance can cause *.

UNSTBL   Unstable air mass.

UNSTDY   Unsteady.

unstick   Point at which fixed-wing aerodyne leaves surface of land or water; hence * run = ground or water run; * speed = that at which aircraft becomes airborne, usually about 25% of way from V_R to V_2 symbol V_{us}.

unstick-speed ratio V/V_{us}, ratio of aircraft speed to unstick speed either as % or as fraction. Usually plotted as abscissa on takeoff performance graph, esp. of marine aircraft.

unstressed   Not bearing significant external load.

unsupported site   Possible operating site for V/STOL aircraft but devoid of prestocked supplies, eg POL, ammunition etc.

Unsvc   Ground facility is unserviceable.

unsymmetrical   Generalized chemical description of molecular structure where left is not mirror image of right; eg in 1, 2 dimethylhydrazine each N has an H and a CH_3 attached (symmetrical), but 1, 1 dimethylhydrazine has left N joined to two H atoms and right N joined to two CH_3 methyl molecules and is unsymmetrical (UDMH).

unsymmetric flight   Condition in which aircraft (aeroplane or glider) is not balanced about longitudinal axis, due to roll, roll/yaw or other rotary manoeuvre causing gross alteration to normally symmetric wing lift. Hence unsymmetric load, that in unsymmetric flight.

unsymmetric thrust   Thrust with one failed engine away from centreline; normally called asymmetric.

UNT   Undergraduate navigator training (S adds system) (USAF).

UNTSO   UN Truce Supervisory Organization.

unusable fuel   Fuel that cannot be used in flight with wings level and at cruise AOA (or nose 3° up). *Trapped fuel* is that fuel remaining in worst case on ground using booster pumps for defuelling and switching off immediately associated LP warning lights illuminate. See *Unpumpable fuel.*

UNUSBL   Unusable.

unwarned exposed   Friendly troops are in open when NW detonates on near enemy.

UOES   User operational evaluation system.

UOR   Urgent Operational Requirement (UK).

UOS   UCAV operational system.

UP   *1* Unrotated projectile (1938–1944 name for British rockets, virtually all air/ground).

   2 Unguided projectile.

   3 Unruly passenger.

   4 Unknown precipitation.

   5 Universal platform.

up and away   *1* Descriptive of all powered-lift systems capable of giving VTO.

   2 The operative mode giving VTO, as distinct from (e.g.) STOVL.

UPB   Unruly passenger behaviour, also called *air rage.*

UPC   Unit production cost.

UPCF   Union des Pilotes Civils de France.

up-chaff   Normally, between chaff cloud or stream and target; hence * interception, with good radar view of target.

up-conversion   Move to a higher EM frequency band.

update   To refresh memory, radar picture or other electronic device with later information; hence * rate, rate (possibly as often as kHz) at which a system or input is scanned for new or changed values.

UPDFTS   Updraughts.

updraught carburettor   One fed by duct conveying air upwards from below; in US updraft carbureter.

UPDTS   Updates.

Uped   UV pre-initiation electrical discharge (laser).

up-45 line   Straight sustained climb at inclination of 45°.

up-front control   Single small panel in fighter cockpit giving complete control of all CNI functions.

up gear   US voice command or check for raising landing gear.

upgrade   To rebook passenger into higher class.

UPI   Undercarriage position indicator (UK).

Upkeep   Water-skipping dambusting bomb (UK, 1943).

upkeep   Generalized US term for all tasks aimed at

preventing deterioration of hardware, eg GA aircraft; less used for commercial and military.

uplift *1* Total disposable load of cargo aircraft, or cargo taken on board.

2 Fuel taken on board, esp. away from home base.

3 Fuel taken aboard from air-refuelling tanker aircraft.

uplink *1* Telemetry, command, data or other electronic link between Earth and spacecraft.

2 Com. link from ground to aircraft, especially telephone call to passenger.

upload Load acting vertically upwards, or vertical component of loads, eg airloads due to lift.

uplock Mechanical lock securing device, eg landing gear, in up or housed position.

UPM *1* Universal processor module.

2 Ultra-portable multiplexer.

upper air Portion of atmosphere above lower troposphere, normally (eg for synoptic purposes) that above pressure height 850 mb. Hence, * chart. See next three entries.

upper-air observation Observation of upper air, above effective range of surface measures; also called sounding.

upper airspace Normally all FLs above 250 (7,620 m/25,000 ft).

upper atmosphere Not strict term, normally interpreted as above tropopause, but also as above 30 km or 20 miles.

upper baseplate Triangular or hexagonal frame on which cabin of flight simulator is mounted.

upper branch That half of meridian or celestial meridian passing through observer or observer's zenith.

Upper Flight Information Region Same geographic areas as FIR but imposes special rules above 24,500 ft (7468 m) [in US, 25,000 ft, 7,620 m].

upper limb That half of limb of celestial body having greater altitude.

upper sideband USB = carrier frequency + modulation frequency; it carries the information.

upper stage Second or subsequent stage in multistage rocket, in most vehicles not fired in atmosphere.

upper-surface aileron Split surface forming part of upper surface of wing only, used for roll or, in some cases, as spoiler (rare).

upper-surface blowing Discharge of main propulsive jets (flattened laterally to cover more span) across top of wing; in high-lift mode deflected down by Coanda-effect attachment to upper surface of large trailing-edge flaps to give augmented lift. Can achieve C_L up to about 12 but reduces cruise efficiency.

upper-tier technology That required for exo-atmospheric ballistic-missile defence.

upper wing Top wing of biplane; hence ** pylon, pylon attached above top plane of biplane.

UPPL Undergraduate private pilot's licence.

UPR Upper airspace.

uprated Cleared to deliver more power, usually but not necessarily after incorporation of improvements; eg * gas turbine may operate at higher TGT or incorporate more efficient blades with reduced tip clearance; hence uprating, process of authorizing greater output from existing or improved machine.

uprig To adjust neutral position of spoiler or airbrakes (normally recessed in upper surface of wing) to provide up/down direct lift control.

UPRM Universal platform resource management.

UPS *1* Uninterruptable power supply, or supplies.

2 UV photoelectron spectroscopy.

upset *1* Sudden externally imposed or undesired disturbance to flightpath, eg by violent gust. Classed as moderate if not worse than zero-g, severe if vertical acceleration is strongly negative. Usual definition of severe includes speed varying anywhere from V_S to V_{DF}, bank $\pm60°$ and pitch $\pm30°$.

2 Metalworking process akin to forging in which rod or other slender workpiece is heated and placed under axial compression to reduce length and increase diameter, usually at particular place and to attain particular longitudinal profile.

upset rate Frequency of problems suffered by computer[s] due to radiation.

UPSLP Upslope.

UPSMS UPS(1) management system.

upstage In a direction towards nose of multistage ballistic vehicle; * and downstage can refer to positions of items within same stage or to items in different stages.

upstairs *1* Aloft, ie flying (colloq.).

2 High altitude, especially as distinct from low.

upstream injection Fuel is sprayed in opposite direction to engine airflow.

UPT Undergraduate pilot training (S adds system.)

uptime Time when equipment is available for use.

UPU Universal Postal Union (Int.).

upward Charlie Upward roll (UK colloq.).

upward ident Upward identification light; white light visible from hemisphere above aircraft and usually provided with manual keying from cockpit, eg to send aircraft callsign or other message.

upward roll Roll while in steep climb, usually zoom after fast low pass.

upwash *1* Upward movement of air around outside of trailing vortex behind wing giving positive lift, usually curving over to become downwash on inner side of vortex.

2 Upward movement of air ahead of leading edge of subsonic wing giving positive lift.

3 See fountain.

upwind Towards the direction from which the wind is coming: G/S = TAS - W/V.

UQ Ultra-quick (fuze).

UR Unsatisfactory report.

URA *1* Unrestricted article.

2 Ultra-reliable aircraft.

uranium U, silvery radioactive metal, density 19.0, MPt 1,132°C. See *nuclear weapon*, *depleted* *.

URAV Uninhabited, or unmanned, reconnaissance air vehicle.

URCS Unmanned radar and communications station.

URD User requirement[s] document.

URE Unintentional radiation exploitation.

urea $CO(NH_2)_2$, traditional runway deicer, also used in some plastics.

URET User-request evaluation tool; CCLD adds core-capability limited deployment (FAA/FFP-1).

Ureti University research, engineering and technology institutes (NASA-DoD).

URG Underway Replenishment Group.

URIPS, Urips Undersea radioisotope power supply.

Urits   USAF rangeless instrumentation [originally interim] training system, not compatible with Actions and Raids, which are not compatible with each other.

URR   Ultra-reliable radar.

URS   User-requirements specification.

Ursi, URSI   Union Radio Scientifique International (Int., pronounced as word).

URTA   Upset recovery training aid.

URV   Unmanned research vehicle.

URVC   Unshielded rudder volume coefficient.

US   *1* Under instruction.

2 Ultrasonic (sometimes U/S).

U/S   Unserviceable.

USA   *1* United States Army.

2 Useful screen area (HDD).

3 United Space Alliance, a linking of US industrial firms.

USAAAVS   USA Agency for Aviation Safety.

USAAC   USA Air Corps (1926–42).

USAADS   USA Air Defense School.

USAAF   USA Air Force (1941–47).

USAARL   USA Aeromedical Research Laboratory.

USAAS   USA Air Service (1918–26).

USAAVSCOM   USA Aviation Systems Command.

usability factor   Percentage of time a runway has cross-wind within published limits.

usable fuel   Not defined, but usually means fuel actually available, with no reserve, typically 95–98% of system capacity.

usable lift   *1* Thermal worth using (sailplane).

2 See *useful lift*.

USABMD   USA Ballistic Missile Defense Agency.

USAC   Urban Systems Inter-Agency Advisory Committee (US).

USAEC   USA Electronics Command.

USAETL   USA Engineering Topographical Laboratory.

USAF   US Air Force [see entries beginning AF].

USAFA   USAF Academy.

USAFE   USAF Europe.

USAFSS   USAF Security Service.

usage   Usually means hours flown.

USAID   US Agency for International Development.

US Aire   US Aerospace Industry Representatives in Europe (Paris, 70+ member companies).

USAMICOM   USA Missile Command.

USANCA   USA Nuclear and Chemical Agency.

US&RAeC   United Service and Royal Aero Club (UK, formerly).

USANG   US Air National Guard.

USAS   United States Air Service of the Army Signal Corps (1918–20, when it became USAAS).

USASF   USA Special Forces.

USASMDC   USA Space and Missile Defense Command.

USATA   US Air Tour Association (office Calverton MD).

USB   *1* Upper-surface blowing; T adds technology.

2 Upper sideband.

3 Universal serial bus.

4 Unified S-band.

USCG   US Coast Guard.

USCS   US Customs Service.

USD   *1* Unmanned surveillance drone.

2 US dollars.

USDAO   US defense attaché offices.

USDOC, Usdoc   US Department of Commerce.

USDR&E   Under-SecDef for Research & Engineering.

use   Workpiece metallurgically flawless and suitable for forging (UK trad.).

useful lift   Difference between fixed weight of aerostat and gross weight, available for fuel, oil, consumables and payload.

useful load   For any aircraft, difference between empty weight and laden weight; for certificated aircraft, difference between OEW and MTOW (or MRW if separately authorized). Note: OEW includes many items previously loosely included in **.

user data   N-user data may be transferred between peer network [OSI Model] members, as necessary.

USERS, Users   Unmanned [microgravity] space-experiment recovery system.

user segment   World population of GPS receivers.

USFE   Upper-stage flight experiment.

USFQIS   Ultrasonic fuel-quantity indication system.

USFS   US Forest Service.

USG, US gallon   Non-SI unit of volume = $3.785411784 \times 10^{-3}$ m^3 = 0.83267 Imp. gal.

USGPM   US gal/min.

USGPO   US Government Printing Office.

USGS   US Geological Survey.

USGW   Underwater-to-surface guided weapon.

USHGA   US Hang Gliding Association.

USIA   US Information Agency.

USIAS   Union Syndicale des Industries Aéronautiques et Spatiales, became Gifas (F).

USJFCOM   US Joint Forces Command.

USL   Underslung load.

USM   Utility-systems management.

USMC   US Marine Corps.

USMDA   US Missile Defense Agency.

USMS   Pronounced uz'ms, utility systems management system, software programmed into six microprocessors which control a complete fighter.

USMTF   US message text format[ted]; ACO adds airspace-control order[s].

USMV   Unmanned space maneuvering vehicle.

USN   US Navy.

USNI   US Naval Institute.

USNO   US Naval Observatory.

USNR   US Navy Reserve[s].

USNTPS   USN Test Pilot School, Patuxent River.

USOA   Universal Safety Oversight Audit [P adds Program] (ICAO).

USPA   US Parachute Association.

USPS   *1* US Postal Service.

2 US Planetary Society.

USR   Special work control (USSR).

USRA   Universities Space Research Association.

USS   United States [Naval] Ship.

USSC   United States Space Command, merged 2002 with Strategic Command.

USSF   United States Space Foundation (office Colorado Springs).

USSOCM   US Special Operations Command.

USSOF   US Special Operations Forces.

USSP   Universal sensor signal processor.

USSRC   US Space & Rocket Center, adjacent to MSFC.

USSTAF   US Strategic Air Forces.

UST   Upper-surface transition, location of shockwave.

USTB   Unstabilized.

USTC   United States Transportation Command.

USTDA   United States Trade and Development Agency.

USTO   Ultra-short takeoff.

US ton   Short ton (see *ton*).

USTS   Uhf satellite-tracking system.

USUA   US Ultralight Association.

USV   Überschall Verkehrsflugzeug = SST (G).

US/VTOL   Ultra-short or VTOL.

USW   *1* Undersea warfare.
    2 Ultrasonic welding.

USWB   US Weather Bureau.

UT   *1* Under training (often u/t).
    2 Universal time.
    3 Ultrasonic.
    4 Update time.

UTA   *1* Unit training assembly.
    2 Upper-airspace control area (CAA).

UTC   Universal time, co-ordinated [previously GMT, and see Zulu].

UTCS   *1* Universal tactical control system (aerial targets).
    2 Universal target control station.

UTD   Unit training device.

UTIAS   University of Toronto Institute for Aerospace Studies.

utillidor   Thermally insulated, heated conduit (above or below ground level) conveying water, steam, sewage and fire pipes between buildings (USAF).

utilities control system   In a complete reversal of meanings, the UCS connects flight-control system, landing-gear computers, fuel computers, right glareshields, computer symbol generators and power generators.

utility   *1* Generalized adjective for aircraft intended for, or assigned to, variety of missions mainly of transport nature but with limited payload; usually denotes odd-job status but occasionally equipped with mission interfaces, eg for photography, target towing or even surface attack in limited war.
    2 FAA = limited aerobatic category.
    3 * glider – secondary or training glider.

utility finish   Provides fabric with tautness and fill, but lacks gloss.

utility system   Usually loose term meaning that system serves routine domestic functions not crucial to safety of vehicle; occasionally means special system for specific purpose, and sometimes even a standby emergency system. Never involves flight control, navigation or weapon release, but see *utilities control system*.

utilization   *1* Proportion of total time an equipment is available for use or is actually used.

2 Number of hours per year something, eg aircraft, is used.

utilization rate   *1* For civil transports, utilization (2).
    2 For combat aircraft, flying rate, usually qualified according to conditions, eg normal ** based on (say) 40-h week; emergency ** based on maximum attainable on 6-day week; wartime ** based on 7-day week with wartime crew, maintenance and safety criteria.

UTLS   Upper troposphere [and] lower stratosphere.

UTM   Universal transverse Mercator.

UTP   *1* Unit test pilot.
    2 Unshielded twisted pair.

UTR   Universal [or uniform] temperature reference.

UTS   Ultimate tensile strength.

U_{TS} Velocity of tunnel test section (alternative to V_{TS}, sometimes preferred for empty tunnel).

UTT   Upper-tier technology.

UTTAS   Utility tactical-transport aircraft system.

UTTEM   Draft tactical technical economical requirement, start of each new programme (Sweden).

UUA   Urgent upper-air Pirep.

UUM   *1* Unification of Units of Measurement; P adds Panel (ICAO).
    2 Former US category, underwater launch, underwater target, attack missile.

UUPI   Ultrasonic undercarriage position indicator.

UUT   Unit under test.

UV   *1* Ultra-violet.
    2 Under-voltage.
    3 Unmanned vehicle.
    4 Upper sideband, voice.

UVAS   Unmanned vehicle for aerial surveillance.

UVDF   U.h.f./v.h.f. D/F.

UVDS   UV flame detector system.

Uverom   Ultra-violet erasable read-only memory.

Uveprom   UV erasable programmable ROM.

UVL   UV laser.

UVM   User view menu.

UVR   UV-resistant [paint].

UVS   *1* UV spectrometer, or spectrometry.
    2 Unmanned-vehicle system.

UVV   Ultimate vertical velocity; landing-gear design case.

UVVS   Administration of the air force (USSR, obs.).

UW   *1* Unconventional warfare.
    2 Unique word.

UWB   *1* Ultra-wideband [R adds radio].
    2 Underwater battlespace.

UWNDS   Upper-airspace wind[s].

UWR   Ungrooved wet runway.

UWS   Urgent weather Sigmet.

UWW   Underwater weapons.

UWY   Upper-airspace airway.

UXB   Unexploded bomb.

UXO   Unexploded ordnance.

V

V *1* Volts, potential, e.m.f.

2 Velocity, including TAS, EAS or ASIR.

3 US piston-engine designation prefix, vee-type.

4 US piston-engine designation prefix, vertical orientation, ie crankshaft vertical (currently in use).

5 JETDS code, visible light(s).

6 Volume, or volume-fraction.

7 Total shear stress.

8 Designation prefix: convertiplane (USAF 1954–62), V/STOL (USAF from 1954, USN from 1962).

9 US military-aircraft designation modifying prefix: staff/VIP.

10 Unit designation prefix: airplane (USN 1922–62).

11 Potential energy.

12 Vanadium.

13 Prefix, Victor airway.

14 Varying, variation, or varying between (Metar), or variable.

15 Ground/air visual code: require assistance.

16 Secondary station (Loran).

17 Visibility, visual or visual descent point.

18 Experimental (G).

v *1* Specific volume of gas.

2 Component of RMS velocity; phase velocity of EM wave.

3 Thermionic valve.

4 Linear [called lateral] velocity of point due to rotation of body in pitch.

5 Relative velocity between two moving bodies or points.

6 Propwash velocity relative to undisturbed air, ie 'true' V_p.

(V) suffix indicating item of electronic equipment can be configured to suit a number of platforms or system applications. Normally followed by numeral identifying which (JETDS).

$\bar{V}$ Horizontal tail volume coefficient ($l_t S_T / \bar{c} S$).

V' Radial inflow velocity, eg to eye of centrifugal compressor.

$\overset{\circ}{V}$ Volume rate of flow.

V_1 Decision speed; ASIR defining decision point on take-off at which, should critical engine fail, pilot can elect to abandon takeoff or continue. Calculated by WAT and runway friction index for each takeoff, never less than V_{MCG}. Also regarded as engine-failure recognition speed, made up of V_{EF} plus increment due to pilot thinking time.

V2 Two-box VASI, on either side of runway.

V_2 Takeoff safety speed; lowest ASIR at which aeroplane complies with those handling criteria associated with climb following one engine failure, normally obtained by factoring V_{MCA}, V_{MSL} and pre-stall buffet speed. Aeroplane should reach V_2 at screen after engine failure at V_1 and climb out to 120 m height without speed falling below V_2.

V_3 Normal screen with all engines operating, at which aeroplane is assumed to pass through screen height in normal takeoff; usually about $V_2 + 10$ kt.

V4 Four-box VASI.

V_4 Steady initial climb speed for first-segment noise-abatement climb with all engines operating.

V6 Six-box VASI.

V12 A V6 on each side of the runway.

V16 Two V4 VASIs on each side of the runway.

V90 Category of off-base military airstrips dispersed through countryside and usable by fighters (Sweden).

V-aerial Two rod conductors balance-fed at apex with geometry giving desired directional propagation.

V-band Original radar frequency band 46–56 GHz (obs.).

V-beam radar Uses an inclined and a vertical beam to determine target bearing, range and altitude.

V-belt Drive belt of tapering cross-section, often coming to narrow inner edge like V, mating with pulleys having inclined inner peripheral faces.

V-block Hardwood block with large V-notch used in hand sheet-metalwork.

V-bombers The UK's only strategic jet bombers (1951–90), named Valiant, Vulcan, Victor.

V-engine See *vee engine*.

V-tail See *butterfly tail*.

VA *1* Volts × amperes; basic measure of a.c. or reactive electrical power.

2 Visual aids (ICAO panel).

3 Unit prefix: fixed-wing attack squadron (USN).

4 Voice-activated.

5 Air army (USSR).

6 Veterans Administration (US).

7 Visual approach, and VASI.

8 Vortex advisory.

9 Volcanic ash.

V_A Design manoeuvring speed; on basic manoeuvring envelope speed at intersection of positive stall curve (assumed in cruise configuration) with n_1 (limiting positive manoeuvring load factor). Highest EAS at which limit load factor can be pulled.

V_a *1* Aquaplaning speed; speed (usually ASIR) at which wheels lose effective contact with runway covered with standing water.

2 Axial gas velocity.

V-A Volt-amperes.

VAA *1* Vintage Aircraft Association (US).

2 Vertical alert annunciator.

VAAC Vectored-thrust aircraft advanced [flight] control.

VAATE Versatile affordable advanced turbine engine (USAF programme).

VAB Vehicle (originally Vertical) Assembly Building, at KSC.

V_{AB} Resultant velocity of circulation of two particles A and B.

VABI, Vabi Variable-area bypass injector.

VAC *1* (Vintage Aeroplane Club, 1951–55; Vintage Aircraft Group 1964–74; Vintage Aircraft Club 1974–, all UK)

2 Valiant Air Command [Florida-based warbird centre].

3 Volts a.c.

4 Visual-approach chart.

VACA   The Vintage Aircraft Club of Australia.

vacancy   Unoccupied lattice site in crystal.

Vacbi   Video and computer-based instruction.

VACIS, Vacis   Vehicle and cargo inspection system.

VACS   *1* Variable-autonomy control system (UAVs).

2 Voice-activated control system.

VACT   Visual air-combat training.

vacuum casting   See *vacuum melting*.

vacuum evaporation   Loss of molecules from body's surface in space.

vacuum gauge   Instrument for measuring low fluid pressure, eg Pirani, Knudsen, ionization, McLeod.

vacuum melting   Almost self-explanatory, preparation of advanced steels and other alloys in a near-vacuum in order to reduce [almost eliminate] the unwanted formation of oxides, hydrides and other impurities.

vacuum orbit   Orbit of satellite round incomparably more massive body in complete absence of any atmosphere.

vacuum pump   Device for establishing unidirectional flow of gas molecules and thus of evacuating fixed enclosure; following gross evacuation by mechanical pump alternatives include vapour/diffusion pump, cryopump etc. Also, in aircraft, simple device for applying depression to air-driven flight instruments, eg venturi.

vacuum specific impulse   Specific impulse in vacuum operation.

vacuum system control   Sucks air from aircraft toilet WC.

vacuum thrust   Thrust of rocket in vacuum, typically about 25% greater than at sea level (actual thrust rises by approx. product of atmospheric pressure and rocket nozzle exit area).

vacuum tube   Electronic tube whose internal pressure is so low that residual gas or vapour atoms or molecules have no significant effect on operation; also called thermionic valve in most common form.

vacuum tunnel   Wind tunnel operated at much less than sea-level pressure.

VAD   *1* Velocity/azimuth display.

2 VHPIC applications demonstration (programme).

3 Visual approach and departure (usually helicopter) chart.

VADR   Voice and data recorder.

VADS   Vulcan air-defense system (USA).

VAE   Virtual air environment.

VAES   Voice-activated electronic system.

VAFA   Vintage Aircraft and Flying Association (UK).

VAFB   Vandenberg AFB (USAF), see Vandenberg.

Vaftad   Volcanic-ash forecast transport and distribution.

VAI   Voice-interactive avionics (not VIA).

VAI-IPR   Versatile affordable integrated-inlet fan for performance and reliability.

VAL   *1* Visual approach and landing [chart].

2 Variable approach light [3 intensities, 2°–8° azimuth].

3 In valleys [mist/fog].

Valid   Variability and life data.

validation   Generalized word for activity intended to re-validate licence or authority, esp. training, examinations and emergency-procedure practice of graduate pilot, instructor or ATC (1) officer. ATC * involves simulated crises worse than any normally encountered.

validation phase   Period when major-programme

characteristics are refined by study and test to validate alternatives and decide whether to proceed to FSD.

valise   Storage envelope for liferaft.

V∞

V-alpha, free-stream velocity.

VALPT   Variable-area LP turbine.

Value   Validated aircraft logistics utilization evaluation.

value engineering   Complete engineering discipline devoted to seeking ways to achieve desired hardware performance, quality and reliability at minimal total cost, eg by elimination of unnecessary items, changes in material and manufacturing method, and simplification of design.

valve   *1* Device for controlling fluid flow, eg into and out of piston-engine cylinder (inlet *, exhaust *) or into/out of aerostat, esp. airship.

2 Numerous other fluid-flow control devices in hydraulics, oxygen, propellants, hot-gas, bleed air, carburetion [US spelling] and other systems.

3 Vacuum tube (UK usage).

4 To release air or gas from aerostat into atmosphere.

valve clearance   Gap between end of stem and rocker arm.

valve duration   Time or angular crankshaft movement during which a valve remains open.

valve face   Mating edge of valve (1) bedded by grinding into seat.

valve gear   Mechanism driving valves of piston engine.

valve hood   Umbrella-like cowl protecting main airship gas valve against rain or icing.

valve lag   Angular motion of crankshaft between either maker's specified valve-closing position or TDC and point at which valve actually closes.

valve lead   'Leed', not 'led', angular motion of crankshaft between closure of valve and either maker's specified closing position or TDC.

valve lift   Total linear motion of poppet valve, ie cam stroke × mechanical advantage of valve gear.

valve line   Cord operating aerostat gas valve.

valve petticoat   See *petticoat*.

valve ports   Inlet and exhaust passages forming part of cylinder head.

valve rigging   Linkage inside aerostat (airship) envelope by means of which automatic valve is opened.

valve seat   Angled ring, usually made of hard erosion-resistant material, forming poppet-valve mating face in cylinder head.

valve-seat recession   Accelerating erosion of piston-engine poppet-valve seat caused by fragments of soft seat fuzing to valve [avoided by leaded fuel or LRP].

valve timing   Exact plot of crankshaft angular positions at which piston-engine valves open and close.

VAM   *1* Variable aerofoil mechanism; usually infinitely reprofilable leading edge.

2 Visual anamorphic movie; external-scene (OTW) add-on to simulator.

3 Visual approach monitor (wide-body HUDs).

Vamom   Visite d'aptitude à la mise en oeuvre et à la maintenance (Armée de l'Air, F).

Vamp   *1* Variable anamorphic motion picture; see *VAM (2)*.

2 VHSIC avionics modular processor.

VAMS   Vector airspeed measuring system.

Van, VAN   *1* Value-added, or visual-area, network.

2 Variable-area nozzle.

3 See next (2).

van *1* Generalized US term for air-conditioned towed vehicles for major support operations, eg strip, check and reassemble major avionic systems in field. Not called trailer.

2 Common term for [usually checkerboard painted] runway control vehicle, called VAN in code.

vanadium Hard silver-white metal, symbol V, density 6.1, MPt 1,887°C, important alloying element, esp. in steels.

Van Allen belt(s) Inner and outer zones of high-intensity particulate radiation trapped in Earth's magnetic field around Equator (inner mainly protons, outer mainly electrons) at radii from Earth's centre from 8,700 to 26,000 km.

Van de Graaff Registered name of high-voltage generator using rotating belt to convey electrical charges.

Vandenberg Originally Camp Cooke, CA, today main West Coast rocket test base, head of Pacific Missile Range, address Lompoc (USAF).

Van der Waals equation Best known equation describing behaviour of real (as distinct from perfect) gas: $(p + a/b^2)(v - b) = RT$, where p is pressure, v volume, R universal gas constant, T °K and a, b constants.

Van der Waals forces Interatomic or intermolecular attractive forces between interacting varying dipole moments; varies as seventh power of radius.

V&F Vinyl and fabric.

V&V Verification and validation (software).

Van Dyke Generalized theory of aerodynamics at Mach 5+ (hypersonics), where air can no longer be assumed perfect gas owing to molecular vibration, dissociation, electronic excitation and ionization.

vane *1* Generalized term for thin (flat or, usually, curved but not necessarily aerofoil section) aerodynamic surface either fixed in order to turn air or gas flow, or freely pivoted and thus aligning itself with fluid direction.

2 Stator blade of compressor or turbine, (US usage).

3 Gas guide surface at nozzle to turbine (US and UK); abb. IGV.

4 Strips, usually of circular-arc section, in cascade at corners of wind tunnel, or used as valves to control flow.

5 Radial strips around fuel burner of gas turbine imparting rotary vortex motion. Often called swirl *.

6 Curved surfaces in cascade at angle bends of airflow in many gas turbines; again called swirl * though rotation is not desired.

7 Curved forward extensions to radial arms of centrifugal compressor or supercharger impeller, called rotating guide *, RGV.

8 Alternative (unusual) term for fence.

9 Swinging retractable leading-edge flaps normally housed in fixed glove of swing-wing aircraft and extended in high-lift mode, called glove *.

10 Common term for slat fixed to leading edge of flap and for various other auxiliary aerofoils carried on flaps, ailerons, droops and leading-edge slats.

11 Normal meaning for weathercocking surface, eg to indicate wind direction or relative wind.

12 Particular application of (1) is sensor in SWIS.

vane pump Large family of fluid pumps in which flat surfaces oscillate and rotate inside chamber eccentrically arranged around drive shaft.

vane rate Angular velocity of vane in SWIS indicating rate at which aircraft is approaching stall.

vane set Row of vanes (4) across wind tunnel, either fixed and profiled to change flow direction or pivoted and ganged to rotate in unison to control or divert flow.

vanilla aircraft Baseline aircraft on which individual customer fits are incorporated.

Vanvis Visual and near-visual intercept system; visual refers to EM frequency used, not to pilot acquiring target visually.

Van Zelm Catcher installation for catapult bridles thrown from aircraft carriers.

VAP *1* Vortex avoidance procedure.

2 Visual-Aids Panel (ICAO).

3 Spray dispensing system for HCN or persistent Toxic-B (USSR).

4 Value-added processor.

VAPI Visual approach-path indicator; also known as vertical speed and approach-path indicator, suitable for fixed-wing and helicopter operation.

Vapo Velocity at apogee.

vaporiser Heat exchanger which converts Lox into breathable gas.

vaporising combustor Gas-turbine combustion system in which fuel is vaporised prior to passage through burner and ignition (usually in *walking stick*); hence vaporiser, in which fuel is vaporised, and vaporising burner.

vapour Substance in gaseous state but below critical temperature; thus can be converted by pressure alone to liquid or solid (US = vapor).

vapour cycle Closed-circuit refrigeration, eg for air-conditioning, in which heat is extracted by refrigerant alternately evaporated and condensed.

vapour degreasing Immersion in hot solvent vapour, eg trichlorethylene.

vapour gutter Assembly of rings and radial struts, usually of > section, to retain flame in afterburner; also called flameholder or stabilizer.

vapour lock Complete breakdown of supply of liquid, eg fuel from tank, because of blockage by bubble of vapour at high point in pipe.

vapour-phase inhibitor Nitrite-based chemical, often locked in paper or other solid, which protects metal parts against corrosion by preventing formation of vapour, esp. of water; usually white powder which volatilises and recrystallises on metal surface.

vapour pressure Pressure exerted by molecules of vapour on walls of container; with mixture, sum of partial pressures. Compared with kerosene, wide-cut fuels have high volatility and thus high *, hastening boiling and vapour lock.

vapour tension Maximum attainable vapour pressure exerted by plane liquid surface with vapour above, varies with temperature.

vapour trail See *condensation trail*.

vapour-type thermometer Needle is driven by Bourdon tube sensing pressure from vapour capsule whose temperature is that indicated.

V$_{APP}$ Approach speed.

VAPS, Vaps *1* Virtual avionics [or applications] prototyping system.

2 Visual approaches.

Vaptar Variable-parameter terrain-avoidance radar.

VAQ Unit prefix: airborne early-warning squadron (USN).

VAR *1* V.h.f. aural range, ie radio range.

2 Visual/aural range; radio range in which two airways are located by A/N signals and two by visual means, eg panel instrument (obs.).

3 Volt/ampere(s) reactive; unit of wattless (reactive) electrical power.

4 Vacuum-arc remelting.

5 Variation.

6 Volcanic-activity reporting.

var Variation (magnetic).

varactor Device employing p-n junction whose capacitance is varied by reverse voltage; important in parametric amplifiers.

VARI Vacuum-assisted resin injection.

variable When applied to IGV or stator ring, means variable-incidence; when applied to jet-engine inlet or nozzle, means variable profile and area.

variable-area nozzle Propelling nozzle whose cross-section area can be altered (usually, together with profile) to match changed Mach number and afterburner operation (airbreather) or atmospheric pressure from SL to vacuum (rocket).

variable-area wing Rare arrangements have included variable span and even retractable lower wing of biplane into upper, but modern flap systems (eg Fowler) give significant change.

variable camber *1* Apart from experimental aircraft, most important methods are hinged leading and trailing edges, eg F-16, which can each be pivoted slightly up, centred or pivoted fully down independently to suit desired flight condition. Tail surfaces, eg tailplane, rudder, are today often divided spanwise to change not just surface angle but also camber, for greater power.

2 Camber is varied by elastic deformation of surface, eg 747 Kruegers.

variable-camber flap *1* See *anti-balance tab*.

2 Flap, usually Krueger, whose profile changes on extension.

variable colour stripping Removal of part[s] of spectrum to indicate presence of organic material or explosive devices.

variable-cycle engine Jet engine in which path of working fluid can be altered by shutters/valves/doors, eg to convert from turbojet or turbofan to ramjet, hybrid rocket or other form in cruising flight. Some authorities claim adding an afterburner and variable nozzle results in a *.

variable-datum boost control Auto boost control for supercharged piston engine in which governed boost pressure increases as pilot's throttle lever is opened.

variable-delivery pump Fluid pump whose output can be varied independently of drive speed, usually by variable-angle swashplate driving stroked plungers (at 90° stroke is zero).

variable-density tunnel Wind tunnel whose pressure can be varied over wide range (usually not below atmospheric) while in operation; normally pressurized to achieve desired Reynolds number.

variable-diameter tilt-rotor Full diameter for takeoff, transition to reduced diameter for fast translational flight.

variable-discharge turbine Gas turbine, eg driven by piston-engine exhaust, whose throughput (mass flow) can

be controlled by valve (often called waste gate) to match turbine power to altitude and other variables.

variable-displacement pump Fluid pump whose output (mass flow) can be varied over wide range, often to zero, for any given input drive speed; often synonymous with variable-delivery, but ** explicitly implies reciprocating plungers or pistons whose stroke is variable.

variable-floor-level bridge Passenger bridge whose airside end can be raised and lowered to match aircraft sill height.

variable-flow ducted rocket Propulsion for long-range AAM and ASMs offering sustained supersonic [even hypersonic] speed from combining rocket with ramjet.

variable-geometry aircraft Aircraft whose shape can be varied in gross manner, ie more fundamentally than by retractable landing gear or flaps; term has come to mean variable wing sweep, so should not now be used in any other context, unless circumstances change or clear explanation of meaning is furnished.

variable-geometry engine Invariably refers to air-breathing engine for supersonic propulsion in which for reasonable efficiency it is essential to have not only fully variable inlet and nozzle but also variation in flow path in engine itself, eg to divert flow around HP compressor in supersonic mode or to convert engine into ramjet. Needs explanation when used.

variable-geometry inlet See *variable inlet*.

variable-incidence Pivotally mounted so that angle of incidence can be altered. * guide vane, stator blade or turbine inlet guide vane whose incidence is altered for best compromise between flow incident on leading edge and flow angle leaving trailing edge, invariably auto scheduled by engine control system. * tails, tailplane whose incidence is varied either for trimming, with elevators as primary flight control surfaces, or as primary flight control. * wing: wing pivoted on transverse axis so that over full (large) range of flight AOA fuselage can remain more or less level, eg to improve pilot view on approach or permit short landing gear.

variable inlet Variable-geometry airbreathing engine inlet whose area, lip/wedge/centrebody axial position and duct profile can all be adjusted to match required flight shock position and mass flow. Mere downstream auxiliary inlets or spill doors do not qualify.

variable-inlet guide vane Gas-turbine IGV whose incidence varies according to engine operating regime; very rare upstream of turbine but common upstream of first and often subsequent stages of axial compressor, to match airflow mass flow and whirl to rotor blade conditions and avoid stall; controlled by auto system always sensitive to rotor speed and inlet air temperature and occasionally to other variables.

variable load All variables aboard aircraft other than fuel and payload.

variable metering orifice In gas-turbine fuel system, key element in CASC comprising triangular orifice moved axially by stack of aneroid capsules and part-covered by sleeve moved by centrifugal SCG.

variable overhead Varies with number of particular item manufactured.

variable-pitch Synonymous with variable incidence. Normally confined to propellers, where incidence is called pitch. Usually means pitch can be varied on ground, or by pilot in flight, often only as choice of either coarse or fine pitch, without auto control such as constant-speed. Fine distinction between * and adjustable-pitch, latter explicitly meaning ground-adjustable only.

variable-ratio Two main applications: in shaft-drive gearbox * drive is usually synonymous with constant-speed drive, ie ratio is varied to hold output speed constant despite varying input; * bypass engine or turbofan (rare) has bleeds, two-position shutters or doors to change ratio of airflow between bypass duct and core.

variable-stability aircraft Aircraft, invariably aeroplane, whose flight-control surfaces, and possibly structure, can be acted upon in flight to effect gross change of stability and control characteristics, either for research or to mimic the behaviour of a totally different type. The inputs should be seamless, not noticed by pilot.

variable-stator Usually means gas turbine with not just one but several rows of variable-incidence stator blades (vanes) in axial compressor(s).

variable-stroke Though many reciprocating engines and machines patented with *, invariably means axial oscillating plunger liquid pump driven by variable-angle swashplate (see *variable-delivery*).

variable-sweep Aerofoil is pivoted so that sweep angle can be varied. Mainly applicable to main wings, where left/right wings are made separate from rest of structure and attached by large diameter fatigue-free pivots so that both surfaces can be scheduled (either auto or by pilot command) by actuator over wide range of sweep angles, symmetrical about axis of symmetry. Does not mean slew-wing. (Helicopter blade rotation about drag hinge is strictly * but is not called such). Also called variable-geometry (VG) and (colloq.) swing-wing.

variable-timing Feature of magneto drives enabling ignition to be advanced or retarded [there are many other aerospace meanings].

variance Mathematical average of square of deviations from mean value.

variance rate Difference between standard and actual wages (US hourly-paid workers).

variant Different version of same basic aircraft type.

variation *1* Horizontal angle between local magnetic and geographical meridians, expressed as E or W to indicate direction of magnetic pole from true. Also called declination (see *grid**).

2 Detailed schedule of change orders or special furnishings or equipment specific to one customer.

3 Small periodic change in astronomical latitude of Earth locations due to wandering of poles.

varicam Pioneer variable-camber aerofoil.

varicowl Generalized term for any variable-geometry inlet, esp. for main engine.

vario VSI (F, colloq.).

variocoupler RF transformer with fixed and moving windings.

variometer *1* Aneroid-type VSI (traditional term of gliding fraternity; * used to seek thermals).

2 Variable inductive coupler with fixed and moving coils or rotors for comparing magnetic fields, esp. Earth's field.

varistor Two-electrode semiconductor resistor characterized by resistance varying inversely with applied V, in either direction of current.

Varite Resistor characterized by negative temperature coefficient of resistance.

varnish *1* Solutions of resins, eg common gum or wood rosin, in drying oil, eg linseed.

2 Thick slimes in overheated lubricant.

Varsol Naptha-like petroleum solvent (trade name).

VARTM Vacuum-assisted resin transfer moulding [molding].

VAS *1* Visual augmentation system, or sleeve.

2 Voice-activated system.

3 Visible/IR spin-scan radiometer atmosphere sounder.

VASI, Vasi Visual approach slope indicator.

Vasis Vasi system (UK style).

Vast *1* Versatile avionics shop test(er); packaged laboratory for installation on Navy airfields or carriers; occasionally 'a' held to mean automatic.

2 Vibration analysis systems technique (GE).

Vastac Vector-assisted attack.

VAT *1* Value-added tax; applicable to light and sport aircraft (UK).

2 Vernier axial thruster.

3 Visually-augmented target.

4 Vertical-acceleration threshold.

V_{AT} Target threshold speed; scheduled speed for arrival at threshold at screen height of 10 m after steady stable approach at angle of descent not less than 3°. Usually about 1.3 V_s (see V_{Tmax}, V_{Tmin}). Subdivided: V_{AT_0}, all engines operating; V_{AT_1}, one engine failed; V_{AT_2}, two engines failed, etc.

VATAS Voice and tone annunciation system, synthesized voice warnings to helicopter pilot in NOE flight.

Vat-B Short-form weather report comprising word 'weather' and four numbers (DoD, from vis, amount [cloud], top [cloud] and base [cloud]).

Vatcas Very advanced ATC automation system (G).

VATE Versatile affordable turbine engine (NASA).

VATLS Visual airborne target-locator system; laser ranging device.

Vatol Vertical-attitude takeoff and landing, ie 'tail-standing'.

VATS, Vats Video-augmented tracking system.

VATT Visually augmented tow target.

VAU Voltage averaging unit.

VAW Volcanic-ash warning (ICAO study group).

VAWS Voice-alarm warning system.

VAWT Vertical-axis wind turbine.

VB *1* Vertical build [of engine]; A adds area.

2 Vacuum-bonded.

V_B Maximum speed at which specified gust (eg ±66 ft/s) can be withstood without airframe damage; hence speed at left lower and upper corners of basic gust envelope. Usually more than half V_c and less than half V_D. Plays role in establishing V_{RA}.

VBC *1* Velocity blast contour.

2 Video bandwidth compression.

V_{BE} Speed for best flight endurance.

VBI Verband Beratender Ingenieure (G).

VBM Volatile bulk memory.

VBMC Virtual battlefield management center.

VBO Velocity at burnout.

V_{BR} Speed for best range.

V_{BROC} ASIR for best rate of climb of helicopter, usually at or near SL.

VBS Visual bootstrapping subsystem.

VBV Variable-bypass valve.

VBW   *1* Vertikalbordwaffe, dispenser of attack payloads (G).

    2 Vertical ballistic weapon.

VC   *1* US military aircraft prefix, staff/VIP transport.

    2 Variable-camber.

    3 Vanadium carbide.

    4 Or V_c, velocity command.

    5 Variable-cycle; E adds engine.

    6 Vicinity (5–10 miles, or n.m., of airport).

    7 Voice communication [many suffixes].

    8 Vertical-cavity.

    9 Vereinigung Cockpit eV (ALPA, G).

    10 Virtual circuit.

    11 Video cassette.

V_c *1* Design cruising speed, usually one of speeds used in establishing structural strength.

    2 Relative closing speed between two aerial targets.

V_c *1* Rate of climb (note: not IAS or other horizontal speed in climb).

    2 Circular velocity, satellite speed in linear measure.

VCAA   Vintage and Classic Aeroplane Association.

VCAI   Viable combat-avionics initiative, attempt to study system life prior to procurement (USAF).

VCAS   *1* Vice-Chief of the Air Staff.

    2 Visual calibration augmentation system.

VCASS   Visually coupled airborne systems simulator (a helmet, Armstrong Lab.).

VCATS   Visually coupled acquisition and targeting system (HMD).

VCB   Virtual-circuit bridge.

VCC   *1* Vehicle-control centre.

    2 Colour CRT display (F).

    3 Voice communication control.

    4 Video control centre, in pax cabin.

VCCO   Voltage-controlled clock oscillator.

VCCS   Voice communication[s] control system, or switch.

VCD   *1* Variable-capacitance diode.

    2 Voltage-control, or -controlled, device.

VCE   *1* Variable-cycle engine.

    2 Virtual collaborative engineering.

VCFS   Visually coupled flight system.

VCID   Voice-controlled interactive device.

VCIR   Vistal caesium IR.

VC/L   Vehicle container/launcher (UAV).

VCM   Visual countermeasures (OCM more usual).

VC$_{max}$ Active maximum-control speed.

VC$_{min}$ Active minimum-control speed.

VCN   Visual computing network.

VCNTY   Vicinity.

VCO   *1* Variable clock, or controlled, oscillator.

    2 Voltage-controlled oscillator.

V_{con} Conventional flying qualities (STOVL aircraft).

VCOP   Virtual-cockpit optimization program (USA).

VCOS   *1* Vice-Chief of Staff.

VCO/S   Voltage controller oscillator/synthesizer.

VCR   *1* Video cassette recorder.

    2 Visual control room, in tower.

    3 Variable compression ratio.

    4 Voice-command recognition.

VCRG   V.h.f. Channel Requirements Group.

VCS   *1* Vertical cross-section.

    2 Vehicle control station, or system.

    3 Voice communication switch, or system.

    4 Voice-controlled system.

    5 Visually coupled system.

    6 Video-camera station, or system.

    7 Variable colour stripping.

VCSD   Very-close-in ship defence.

VCSEL   Vertical-cavity surface-emitting laser.

VCSELI   VCSE laser-based interconnect[s].

VCSS   *1* Vinyl-coated stainless steel.

    2 Voice-controlled, or communication, switching system.

VCSU   Video-code suppression unit.

VCTS   Variable-cockpit training system.

VCU   *1* Video conversion unit.

    2 VDL(2) control unit.

VCXI   Visual course cross-pointer indicator.

VCXO   Voltage-controlled crystal oscillator.

VCY   Vicinity.

VD   *1* Video detector, or disk.

    2 V.h.f. data.

    3 Visual display.

    4 Variable displacement.

V_D *1* Design diving speed; highest speed at which aircraft is normally permitted to fly, forming vertical right-hand boundary to both basic manoeuvring envelope and basic gust envelope. One of speeds used in establishing structural strength.

    2 Heading to a DME distance.

V_d Doppler velocity [range rate].

v_d Rate of descent (note: not airspeed during descent).

v.d.   Vapour density.

VDA   *1* Versatile drone autopilot.

    2 Vehicule de défense anti-aérienne (F).

    3 Vertical data analysis.

VDC   *1* Variable-datum control.

    2 Visual-display controller.

Vd.c.   Volts direct-current.

VDD   Version description document.

VDDL   Verband Deutscher Drachenfluglehrer eV (kite-flying instructors, G).

VDE   Verein Deutscher Elektrotechniker (G, technical society).

VDF   *1* V.h.f. D/F; most common ground D/F.

    2 Verein Deutscher Flugzeugführer (G, PIC society).

    3 Verband Deutscher Flugleiter (G).

V_{DF} Maximum demonstrated flight diving speed; highest IAS at which aircraft is ever flown (normally only during certification); associated with poor strength margins (inadequate for severe gust) and possibly poor handling, yet attainable at full power even in climb on many transports at low/medium levels.

VDFT   Verband Deutscher Flugsicherungs-Techniker eV (flight-safety technicians, G).

VDGS   Visual docking guidance system.

VDI   *1* Vertical displacement indicator.

    2 Variable-depth (sonar) output indicator.

    3 Verband Deutscher Ingenieure (G, technical society).

    4 Vertical-deviation indication, or indicator.

VDL   *1* Verband Deutscher Luftfahrt-Techniker (G, aerospace engineers' society).

    2 V.h.f. data-link (Arinc), or digital link.

VDLM   V.h.f. data-link Mode, followed by 1, 2, 3 or 4.

VDLU   Verband Deutscher Luftfahrt-Unternehmen (G).

VDM *1* Viscous damped mount.

  2 Visual display module.

  3 VSCS display module.

VDMA *1* Variable-destination multiple access.

  2 Verein Deutscher Maschinenbau-Anstalten (G).

VDNX, VDNKh Exhibition of national-economy achievement (USSR).

VDP *1* Validation demonstration phase.

  2 Visual descent point.

  3 Variable-displacement pump.

VDR, v.d.r. *1* Voltage-dependent resistor.

  2 V.h.f. data, or digital, radio.

  3 Variable-diameter rotor.

VDRS Vehicle data recorder system.

VDRT Visual display research tool.

VDS *1* Variable-depth sonar.

  2 Video-disc simulation.

VDSM Very deep sub-micron.

VDT *1* Variable-discharge turbine (driven by piston-engine exhaust).

  2 Vapour-deposited tungsten.

  3 Variable-deflection thruster.

  4 Variable-density [wind] tunnel.

VDTR Variable-diameter tilt-rotor.

VDU Visual or video display unit.

VDVP Variable-displacement vane pump.

VE *1* Value engineering.

  2 Visual exempt[ed].

V$_E$ Generalized symbol for a limiting speed at which something (eg flap, landing gear) may be selected and extended; suffixes, eg V$_{EI}$, V$_{EN}$, used for specific items and additional terms are used (more are needed) for limiting speeds for subsequent retraction. Main gears (usually not nose) are generally cleared to V$_{MO}$ when locked down, but V$_E$ always applies for selecting down or up.

V$_e$ *1* Equivalent airspeed.

  2 Effective velocity ratio Vα/V$_j$.

VEB Vehicle equipment bay.

Vebal Vertical ballistic (downward-firing anti-armour).

VECP Value-engineering change proposal.

VECTOR, Vector Vectoring Estol control tailless operational research (US/Germany).

vector *1* Adj, having both magnitude and direction; thus velocity is * quantity, speed is not.

  2 Line representing quantity's magnitude, direction and point of application, eg force on structure, or aircraft heading/TAS.

  3 A particular aircraft heading or track, eg that needed to arrive at destination.

  4 To issue headings to aircraft to provide navigation guidance (DoD definition adds 'by radar').

  5 On HUD, aircraft direction of travel, usually synonymous with track.

  6 To control trajectory by altering thrust axis of propulsion.

  7 In translation from Italian, rocket launch vehicle.

  8 Prefix to three-digit heading passed to interceptor engaged in interception (for recovery, corresponding word is steer).

vector computer Device for solving vector triangles, eg CSC.

vectored Capable of being pointed in chosen directions.

vectored attack Surface attack in which weapon carrier is vectored (4) to weapon-delivery point by unit which holds contact on target (DoD, NATO).

vectored thrust Propulsive thrust whose axis can be rotated to control vehicle trajectory; term normally applied to swivelling-nozzle jet engine of aeroplane, corresponding term for space and military rockets being usually TVC.

vector flight control Control of trajectory by vectored thrust.

vectoring *Vectored.*

vector force Resultant of wing's lift [or lift coefficient] and pitching moment [or its coefficient], acting through c.p.

vectoring in forward flight See *viff.*

vector quantity One that has magnitude and direction.

vector sight Traditional type of bomb sight incorporating mechanical representation of vectors of relevant vector triangle.

vector steering Control of trajectory by vectored thrust.

vector triangle Closed figure formed from three vectors, eg (1) heading/TAS, track/GS and W/V, or (2) lift, drag and resultant force on lifting wing.

vector velocity See *vorticity.* Tantological on its own, velocity implying direction.

Vectra Extremely strong liquid-crystal polymer, injection moulded and often strengthened with chopped graphite or glass (Hoechst Celanese).

VEDM Vehicle and engine display management.

vee-belt *V-belt.*

vee depression Vee-shaped low extending between two highs, usually with squall.

veeder counter Stepping digital counter, eg odometer; today often LED or LCD.

vee engine Piston engine whose cylinders are arranged in two inclined in-line rows (banks) in V form seen from either end, driving on common crankshaft; hence vee-12 (often called V-12) ** with six cylinders in each bank.

vee formation Aircraft formation in shape of horizontal V proceeding apex-first, for symmetry with odd number of aircraft.

Veep VIP.

veering Change of wind direction clockwise seen from above; now applies in either hemisphere.

vee tab Elevator spring tab with no-load up-deflection balanced by positive [typically + 10°] trim tab feature of some large aircraft in 1940s.

vee tail See *butterfly.*

V$_{EF}$ Speed at which critical engine failure occurs in accelerate/stop takeoff; defines V$_1$ as * plus speed gained with critical engine inoperative during time pilot takes to recognize situation and respond.

vegetable Mine laid by aircraft at sea (RAF, WW2), hence gardening.

vegetable oil Several, esp. castor, used for engine lubrication pre-1935.

VEGV Variable exit guide vane.

VEH Variable edge enhancement, see *EH.*

vehicle Self-propelled, boosted or towed conveyance for transporting a burden on land, sea or through air or space. This is DoD/NATO wording. Only possible word in most generalized contexts, and also only word covering aircraft, spacecraft, missiles, RPVs etc. Air * identifies flying portion of weapon or reconnaissance system that has extensive non-flying portions.

vehicle axes   Axes, usually cartesian, related to vehicle rather than to Earth or space.

vehicle correlator   Radar subsystem intended to eliminate clutter caused by detection of large numbers of road vehicles [e.g., fleeing civilians] whose speeds exceed lower limit for detection in look-down mode.

vehicle mass ratio   Ratio of final mass of vehicle, usually M_f, after cutoff or burnout of propulsion to initial mass, usually M_o. Normally applied to rocket vehicles.

veil cloud   Loose term meaning either Cs or cloud forming thin veil on mountain.

VEK   Equivalent airspeed (EDP).

VEL   *1* True airspeed (EDP).

     2 Velocity.

vela sensor   Usually measures velocity and angle of attack.

velocimeter   *1* Generalized term for velocity meter.

     2 CW-reflection Doppler system for measurement of radial velocity, ie speed of approach or recession relative to observer.

velocity   *1* Measure of motion; speed (linear or angular) in specified direction.

     2 Loosely (though common in fluid flow), speed. SI unit is m/s.

velocity blast contour   Plot of jet wake velocities immediately behind large jet as it proceeds from gate to takeoff point. Usual measure is 6 ft (1.83 m) above ground at distance 50 ft (15.24 m) behind tail.

velocity budget   Sum of all velocities in planning space-flight.

velocity factor   Ratio of speed of RF wave along conductor to its speed in free space, usually 0.6 to almost unity; symbol k. In typical co-axial cable value is about 0.66, often expressed as a percentage.

velocity gate   Basic ability of CW, Doppler and certain other tracking radars to sense and lock on to particular radial velocity characteristic of target. Hence ** pull-off, basic ECM technique to pull radar off target signal and thus give infinite JSRR.

velocity gradient   Rate of change of fluid speed per unit distance traversed perpendicular to streamlines, eg in boundary layer.

velocity head   Not accepted term; usually means pitot pressure but has even been used to mean kinetic energy of unit mass of fluid.

velocity jump   Angle between launch line and line of departure (ASCC).

velocity microphone   Electrical output = f(V) where V is mean speed of particles on which sound waves impact.

velocity modulation   Various techniques of modulating electron beams, eg by h.f. transverse field which impresses sinusoidal velocity contour causing corresponding variation in intensity of scanning spot.

velocity of advance   Airspeed past propeller blades ignoring speed due to rotation of blades; essentially synonymous with slipstream speed. Always greater than aircraft airspeed, provided engine is operating.

velocity of light   In vacuum 2.9979250×10^8 m/s, symbol c.

velocity of propagation   Speed at which EM wave travels along conductor, eg co-axial cable, waveguide; usually = kc, where k is velocity factor.

velocity of rotation   Rotational speed.

velocity of sound   See *sound*.

velocity potential   Integral of flow velocity parallel to surface, $= \phi$, $= Ux + Vy$ (rectilinear) $= \dfrac{\Gamma\theta}{2\pi}$ (vortex).

velocity profile   Plot of velocity of viscous fluid in traverse (2) perpendicular to flow direction; thus for laminar flow through small tube ** = parabolic curve.

velocity ratio   *1* Common meaning is $\sqrt{q/q_j}$ where q_j = dynamic pressure of jet.

     2 For jet-lift [STOVL], free-stream V divided by jet V.

     3 Generally, velocity of jet or propeller slipstream to free-stream relative velocity.

     4 Mechanical advantage.

velocity signature   Record of Doppler track of aerial or other moving target.

velocity transducer   Generates electrical output proportional to imparted V, symbol V_e.

velocity vector   Main reference on primary flight display showing desired flight path on which aircraft is held by FCS.

VEMD   Vehicle and engine multifunction, or management, display.

VEN   Variable-exhaust nozzle.

vendor   Supplier to a programme, almost always manufacturer of finished parts, devices and equipment though * list often includes suppliers of raw material and even services.

vendor audit   Survey of vendor profiles, eg before entering into discussion or negotiation.

vendor profile   Detailed standardized description of vendor companies for benefit of large-system prime contractors.

veneer   Thin sheet wood; when applied to plastics usually means sheet with simulated woodgrain.

vent   *1* Opening to atmosphere, eg from fuel tank, to equalize internal and external pressures.

     2 Opening in centre of parachute for stabilization.

     3 Precision aperture in aircraft skin sensing true local static pressure; called static *.

     4 Spanwise aperture on USB flap either to entrain air from below or to reduce Coanda attachment to upper surface.

vent cap   Vent (2) patch.

vent hem   Reinforced hem around vent (2).

ventilated shock   Shockwave at UST point which has been stabilized and weakened by a double slot system.

ventilated suit   Partial-pressure or pressure suit provided with ventilation system, eg by ventilation garment.

ventilated tunnel   Wind tunnel whose working section is perforated by holes or slots to prevent choking at Mach 0.8–1.2.

ventilated wet suit   Designed to protect downed aircrew against exposure.

ventilation garment   Light inner suit, forming part of pressure suit, through which dry air is pumped to control body surface temperature and evaporation.

venting capability   Ability of a closed volume to dissipate sudden increase in pressure.

venting cycle   Regular or automatic venting of Ni/Cd battery to avoid cell imbalance or thermal runaway.

vent patch   Piece of fabric covering vent (2) sewn to hem.

ventral   *1* On belly side of body; hence with horizontal fuselage, on underside, or, occasionally, firing or directed through underside.

     2 Underfin, vertical or inclined, fixed or hinged.

ventral container/pallet/pod/radar/tank Criterion for adjective 'ventral', as distinct from fuselage, centreline etc, is that item either forms underside of fuselage or is flush against it, eg Beech 99 baggage container, RF-111 reconnaissance pallet, Harrier gun pod, RA-5C SLAR and Attacker drop tank.

ventral fin Fixed or movable fin on underside of body, usually but not necessarily at tail.

ventral inlet Inlet on underside of body far enough from nose not to be called chin inlet.

ventral nozzle Plain downward facing nozzle in underside of jetpipe from ejector-lift engine, with pilot or auto-controlled valve.

venturi Duct for fluid flow which contracts to minimum cross-section at throat and then expands (usually to same area as inlet); pressure at throat falls to minimum value which can be used to drive vacuum instruments or as measure of airspeed; flow throughout always subsonic.

venturi meter Instrument for measuring fluid mass flow through calibrated horizontal venturi;

$$\text{flow } Q = A_2 \sqrt{\frac{p_1 - p_2}{\rho/2[1 - (A_2/A_1)^2]}}$$ where p_1, A_1 are pressure

and cross-sectional area at start of venturi, p_2, A_2 are values at throat and ρ is density. Assumes flow laminar and compressibility ignored.

venturi pitot Combination of pitot tube and venturi, one giving pressure above atmospheric, other below.

venturi tube See *venturi*.

Venus Visual-engine numeric spaces, produces smooth 3-D landscapes.

VEO-wing Vectored engine over, used in several RALS studies.

VER *1* Vertical ejector, or ejection, rack; VER-2 is twin-store *.

2 Vertical.

3 Version.

verb APPC command.

Verdan Versatile digital analyser; pioneer airborne computer used in Reins (Autonetics); colloq. translated as 'Very Effective Replacement for Dumb-Ass Navigators'.

Verey, verey See Very.

verge ring 000°–360° ring rotated to set traditional magnetic compass.

verification *1* To ensure meaning and phraseology of transmitted message conveys exact intention of originator.

2 Any action, including surveillance, inspection, identification, detection, to confirm compliance with arms-control agreement.

3 Confirmation of flightplan by inserting waypoints and destination in navigation system and causing system and displays to drive through to destination in accelerated timescale.

4 Re-examination of firing data (DoD).

vernal equinox Intersection of ecliptic and celestial equator occupied by Sun about March 21 as it passes through zero declination, changing from S declination to N.

vernier Aid to fine adjustment or refined measurement, eg linear scale placed adjacent to main scale, with which, by aligning two corresponding divisions, main scale can be read with enhanced accuracy.

vernier engine/motor/rocket Small rocket which remains operating or is started after shut-down of main propulsion

to provide accurate final adjustment of vehicle velocity; in most cases has thrust-vector control to refine exact direction of trajectory.

vertex *1* Highest point of trajectory.

2 Point on great circle nearest a pole

3 Node or branch point in network.

vertical *1* Vertical or slightly inclined fin on supersonic fighter-type aircraft, either at nose or tail, usually under body at nose but usually above at tail, often pivoted and power-driven to serve as primary flight-control surface. Twin canted * form U-tail.

2 See *vertical air photograph*.

vertical air photograph Taken with optical axis of camera perpendicular to Earth's surface (note: definition, agreed by DoD, NATO, does not say with axis vertical; over mountains, axis could be at 45° which is not intended).

vertical and/or short take-off and landing, V/STOL Usually means that aircraft is fixed-wing aerodyne with jet lift giving horizontal attitude VTOL capability at less than maximum weight but that in most missions STO is preferred; landing at light weight may be vertical, thus, STOVL. Some VTOLs do not have STO capability and thus cannot be called V/STOL.

vertical attitude Aircraft is rotated approximately 90° nose-up from normal flight attitude; thus a Vatol may have to operate from a special mobile or ship-mounted platform.

vertical axis *1* Vehicle-related ** is that axis perpendicular to both longitudinal and transverse horizontal axes; in aircraft called OZ, usually lying in plane of symmetry.

2 Local vertical.

vertical bank Aircraft is rolled through 90°, so that OZ axis is horizontal and OY axis is coincident with local vertical.

vertical camera Optical axis is perpendicular to Earth's surface; according to other authorities, axis is aligned with local vertical, no matter what local terrain may be. Need for clarification.

vertical-cavity SEL(8) Semiconductor laser whose optical modes are excited perpendicularly to laminate, i.e. vertically.

vertical circle Great circle of celestial sphere passing through zenith and nadir.

vertical clearance Height above ground.

vertical data analysis Identifies and separates out types of data in successive packets to reduce volume of header-style information (satcoms).

vertical development Depth of cloud from base to top.

vertical ejector rack Carries two superimposed external stores; much lower clean drag than TER.

vertical engine *1* Piston engine or gas turbine whose main rotating member rotates about vertical axis, usually for helicopter drive or jet lift.

2 Piston engine whose cylinders are vertical above or below crankshaft.

vertical envelopment Tactical manoeuvre in which troops, air-dropped or air-landed, attack flanks and rear of hostile surface force, effectively cutting off latter.

vertical fin Fin; traditional fin is fixed, and word 'vertical' alone usually means powered fin serving as flight control.

vertical force Vertical component of Earth's magnetic field.

vertical gust Gust; but in V-g recorder V signifies vehicle velocity.

vertical gyro Two-degrees-of-freedom gyro torqued on gimbal mounts to hold spin axis vertical, thus giving output signals proportional to rotation about two orthogonal axes, usually pitch/roll.

vertical interval Difference in height between two locations, eg between two targets or between observer and target.

vertical launch Launch of vehicle on initially vertical trajectory where such trajectory is not inevitable; eg non-ballistic vehicle or one launched from launcher of variable elevation.

vertical layout Traditional layout of project or design office in which an administrative or seniority hierarchy takes precedence over (1) project or programme, and (2) type of work.

vertical lift Lift force along local vertical generated by aerodyne wing, rotor or engine.

vertical navigation, VNav Guidance of flight trajectory in vertical plane, eg to minimize pilot workload in letdowns, holding patterns and during climb or descent to ATC cleared FLs along particular routes or on early stages of approach; provided by modern transport navigation systems, esp. those of energy-management type.

vertical overspill Vertical beam from weather radar reflected from surface below to give height ring on display; not present in mapping mode.

vertical pincer Any DA(3) engagement, one low and one high (relative to target).

vertical pressure gradient Change of atmospheric pressure per unit change in height (traditionally per 1,000 ft in UK/US).

vertical probable error Product of range probable error and slope of fall.

vertical reference Earth-related vertical axis, ie local vertical normally approximated by vertical gyro.

vertical reference gyro *Vertical gyro.*

vertical replenishment Use of VTOL (helicopter) or V/STOL aircraft for transfer of stores and/or ammunition from ship to ship or to shore (NESN).

vertical reverse Aerobatic manoeuvre related to half flick (snap) roll; begun from tight turn by pulling hard back and applying full top rudder to flick inverted, thereafter completing second half of loop to recover level flight; often called vertical reversement (US term).

vertical riser Flat-rising VTOL, ie jet-lift aircraft taking off with fuselage horizontal.

vertical rolling scissors Defensive descending manoeuvre in vertical plane in attempt to make enemy overshoot and fly into attacker's future flight path.

vertical separation Specified difference in FL between air traffic on conflicting courses; normally published for (1) tracks 000–179 and (2) tracks 180–359, and for FLs 0–180, 180–290 and 290 +.

vertical situation display Abb. VSD, flight instrument designed to avoid CFIT. It adds a large rectangle in the lower half of the ND showing a side profile of the flight path and the terrain, based on current track. This gives a valuable extra view supplementing TAWS (Boeing).

vertical speed *1* Helicopter autopilot mode, rate of change of pressure altitude.

2 Loosely, rate of change of height.

vertical speed indicator Panel instrument indicating vertical speed, ie rate of climb/descent; invariably one pointer zeroed at 9 o'clock.

vertical spin tunnel See *spinning tunnel.*

vertical stabilizer Fin (US).

vertical stiffeners Angle or other sections riveted or bonded at intervals along spar web or fuselage keel to resist buckling in vertical plane.

vertical strip Single flightline of overlapping vertical reconnaissance images, eg of beach or road.

vertical tail Traditionally, fin(s) and rudder(s); hence ** area, aggregate area in side elevation of fin(s) and rudder(s), together with dorsal fin and any ventral fin(s) but exclusive of fillets, fairings or bullets.

vertical tail length Distance from c.g. to aerodynamic centre of vertical tail.

vertical take-off and landing Aerodyne has capability of rising from surface without airspeed, hovering and returning to soft landing again without airspeed, generating lift greater than its weight by rotors, ducted fans, jets, deflected propulsion or other internally energized means.

vertical tape instrument Display has roller blind translating vertically, against which are read fixed and/or moving index markers; usually engine instruments are grouped in multi-engine aircraft so that in correct operation all similar readouts are at same levels.

vertical translation Motion of aeroplane in vertical plane, esp. under direct lift force, without change of pitch attitude; can be achieved, eg, by Harrier viff or by F-16 symmetric wing-flaperon deflection with scheduled flaperon/tailplane interconnect gain and with pitch hold engaged.

vertical tunnel See *spinning tunnel.*

vertical turn Turn with approx. 90° bank.

vertical virage Turn with approx. 90° bank (arch.).

vertical visibility Self-explanatory, can be looking down or up.

vertical wind tunnel See *spinning tunnel*

vertigo Subjective sensations caused by faults in inner ear semicircular canals: subjective * = external world is moving past sufferer; objective * = external world is rotating.

vertiplane VTOL aircraft having fixed wing with flaps powerful enough to lift aircraft at zero forward speed by deflecting propwash; FAI category E-4 but no records and (it would appear) no current flying examples.

vertrep See *vertical replenishment.*

Vervis Vertical visibility.

Very Patented signal pistol, standard Allied aviation from WW1; hence * light, * pistol etc.

very high Above FL 500 (DoD).

very high frequency 30-300MHz, see Appendix 2.

very high frequency omni-range see *VOR.*

very high speed photography Image rate 500 to 10^4/s. Faster = ultra.

very large aircraft No definition known to exist.

very-large-scale integration Commonly accepted as over 10^5 devices (some authorities, over 16 kbit) per chip.

very low frequency 3–30 kHz, see Appendix 2.

very light aircraft *1* MTOW ⩽750 kg, 1,653lb (FAA).

2 ⩽390 kg, 860 lb (BCAR).

V_{esc} Escape velocity = $\sqrt{2K/R}$ where K is a constant (universal gravitational constant × primary-body mass) and R is distance from centre of primary body.

vespel polyimide   Coating for bearing surfaces retaining low-friction qualities to high temperature.

Vesta   Vecteur à statoréacteur [long-burn ramjet] (F).

VF   *1* Voice, or variable, frequency.

     2 Unit prefix, fighter squadron (USN).

V_F Design flap limiting speed; replaced by V_{FE}.

V_f *1* Surface wind.

     2 Volume fraction of fibre or whisker reinforced composites; expressed as % volume occupied by fibre.

     3 Fuel flow (CAA).

VFB   Video-frame buffer.

V_{FC} Maximum speed for flight stability ($_{FC}$ = full control); usually synonymous with V_{MO}, little used outside US and suggest passing from use.

VFCT   Voice frequency carrier telegraph.

VFD   Vacuum fluorescent display.

VFDR   Variable-flow ducted rocket.

V_{FE} Maximum flaps-extended placard speed; usually an ASIR and in most flight manuals precise meaning is explained, eg whether limit is for landing setting or any lesser setting. Note: this is invariably a limit for an established flap setting; it does not allow for changed settings.

VFLB   Variable-floor-level bridge.

VFMED   Variable-format message entry device.

VFO   Variable-frequency oscillator.

VFOP   Visual flight [rules] operations panel.

VFR   Visual flight rules, G adds Group.

VfR   Verein für Raumschiffahrt, society for space travel (G).

VFR-OT, VFR/OT   VFR on top.

VFSS   Variable-frequency selection system.

V_{FTO} V_1 for flexible take-off, factored for full power in event of one engine failure at reduced rating.

VFW   Veterans of Foreign Wars (US).

$V_{FXR(R)}$ Maximum flap-retraction speed.

$V_{FXR(X)}$ Maximum flap-extension speed.

VG   *1* Variable-geometry.

     2 Vertical gyro.

     3 Vortex generator.

     4 See V_{GND}.

Vg   Geostrophic wind.

V-g   Aircraft speed and normal acceleration; hence * diagram, graphic plot of these parameters, * recorder, primitive instrument recording speed and applied loading in vertical (very rarely in other) plane.

VGA   *1* Variable graphics array.

     2 Video graphics array, 640 × 480 pixels.

     3 Video graphics adapter.

VG/DG   Vertical gyro/directional gyro.

VGG   Visual graphics generator.

Vgh   Aircraft speed, vertical acceleration and height; hence * recorder, continuous recording of these parameters on wire or tape.

VGI   Vertical gyro instrument.

VGK   *1* Supreme military command (USSR, R)

     2 A CFD method (Fortran) for predicting characteristics of a 2-D single-element aerofoil.

V_{GND} Velocity relative to the ground, suggest usually = G/S.

VGPO   Velocity-gate pull-off.

V_{grad} Gradient wind.

VGS   *1* Velocity-gate steal; usually synonymous with VGPO.

     2 Volunteer Gliding School[s] (UK ATC, now Air Cadets).

     3 Video guidance sensor.

     4 Visual guidance system.

VGTD   Gas-turbine APU (R).

VH   *1* Designation, very heavy aircraft or unit (SAC, formerly).

     2 Or V_H, velocity hold.

V_H Maximum speed in level flight with maximum continuous power; little used outside US. Definition should add that H = high altitude; this power at medium/low FLs would usually exceed V_{DF}.

V/H   Velocity/height ratio [in compatible units] in taking reconnaissance imagery and in sensor design (M0.9/200 ft gives * = 5).

V_h Hump speed.

VHDL   VHSIC hardware design [or description or descriptive] language.

v.h.f., VHF   Very high frequency, see Appendix 2; /DF, adds direction finder; /PTN = radio link to public-telephone network.

v.h.f. omni-range   See *VOR*.

VHFRT   V.h.f. R/T.

VHL   Very high level (software language).

VHPIC   Very-high performance (or power) integrated circuit.

VHRR   Very-high-resolution radiometer.

VHS   Very high speed [electronics].

VHSIC   Very-high-speed integrated circuit; Si or SOS, then GaAs, two to three orders of magnitude faster than MSI/LSI; -2 adds Phase 2.

VI   *1* Visual identification mode.

     2 Viscosity index.

     3 Video interface.

     4 Variable-intensity (eights).

     5 Heading [course] to intercept.

V_i *1* IAS (not ASIR).

     2 EAS.

     3 Velocity induced by winglet, normally inwards normal to direction of flight. Note: winglet lift has a useful thrust component.

VIA   Versatile integrated avionics.

VIAM   All-union (research) institute for aviation materials (USSR).

Viapure   Optical-grade polyurethane interlayer material.

VIAS   Indicated airspeed.

vibrating voltage regulator   Controller of d.c. machines which uses vibrating points to sense voltage and adjust resistance controlling field current.

vibration indicator   Instrument or sensor for either recording or indicating mechanical vibration either remote from crew or of too high a frequency to be obvious, esp. emanating from turbine engines. Modern turbofan engines are fitted with crystal transmitters whose signals are filtered to pinpoint the source.

vibration isolator   See *isolator*.

vibration meter   Instrument for recording vibration frequency; very rarely either amplitude or acceleration in addition.

vibration mode   Most physical objects, from wings to quartz crystals, can vibrate at a fundamental mode having lowest frequency, or (depending on dimensions, mounting, impressed forces, coupling and other factors) at any of numerous modes having different visual pattern and higher frequency. Fundamental mode usually longitudinal over length or thickness, flexural over width or thickness, and shear over thickness or face.

vibrator *1* Mechanical source of high-power sinusoidal vibration for test purposes; also called vibration generator.

2 Rapid-action switch for alternately reversing polarity of transformer primary fed from d.c. to give raw a.c. output; also called vibrator converter.

vibratory torque control Mechanical coupling for rotary output, eg from piston engine, which at low rpm locks drive into hydraulically stiff configuration but at higher rpm unlocks and allows drive to be taken through slender quill shaft.

vibrograph Seismic instrument giving record of vibration displacement/time.

vibro polishing Immersion in a vat of small abrasive particles vibrated at a selected frequency.

Vic, vic Formation of aircraft all in same horizontal plane having shape of V flying point-first; minimum number of aircraft 3.

vicious cycling Standard maintenance procedure for Ni/Cd battery in which charge is violently drained and replaced; also called deep cycling.

Vickers pyramid Hardness testing machine in which precise force drives pyramid-point diamond into specimen, hence * Number = measure of surface hardness.

Vicon Visual confirmation (of voice instruction or clearance, especially to take off).

Victor airway Airway linking VORs, thus virtually all airways in US and many other countries; identified by prefix V (from 1952).

VID *1* Visual identification, or VID required.

2 Virtual (confusingly, also visual) image display.

video Generalized adjective or noun for electronic transmission of visual information.

video compression Returning of video from primary radar so that each radial trace is briefly stored and then written more quickly; this allows time for cursive writing of synthetic data.

video detector Diode which demodulates video signal.

video display Electronic display which, whether or not it presents alphanumerics, symbology and other information, presents pictures.

video extractor System for analysing all signals, selecting all that form part of useful image (eg in TV or radar) and excluding all others; in SSR usually synonymous with plot extractor.

video link Telecom system conveying pictorial information.

video map(ping) Superimposition on radar display of fixed information or picture, usually derived from fine-grain photo plate scanned in synchronization with rotation of radar aerial or from computer memory, eg in GCA to show exact relationship of aircraft to surface obstructions.

video signal Telecom signal conveying video information.

Videotex System of computerized self-service information terminals giving complete information (usually without charge) to all users, eg passengers at airports.

vidicon Most common form of video (TV) picture (camera) tube, in which light pattern is stored on photoconductive surface; this is then scanned by electron beam, which deposits electrons to neutralize charge and thus generate output signal.

Vidissector Modern form of pioneer Farnsworth camera tube; used in space TV surveillance (ITT).

VIDS Visual integrated [or information] display set [or system].

VIE Video image exploitation.

Les Vieilles Racines French association of aerospace pioneers and professionals.

Les Vieilles Tiges French association of pioneer pilots.

Vierendeel Girder (truss) comprising upper/lower chords and verticals, without diagonals or shear web, designed for flexure.

view Opinion; thus 'to take a dim *' = to oppose or regret a decision or situation (RAF WW2).

VIEWS, Views *1* Vibration indicator early-warning system.

2 Virtual integrated EW simulator.

VIF Vertical integration facility (space launch vehicle).

viff, VIFF Vectoring in forward flight; pilot control of trajectory by direct control of propulsive thrust axis of jet-lift V/STOL aeroplane, selecting downwards for lift (normal acceleration) and forward of vertical for deceleration, thus performing combat manoeuvres unmatchable by any conventional aircraft. Hence verb to viff, viffing etc.

Vigil Vinten integrated IR linescan.

Vigo pad Small concrete pan preferably surrounded by concealing trees (STOVL).

VIGV Variable-incidence (or inlet or integral) guide vane[s].

VII Viscosity index improver.

VIIRS Visible/IR imaging radiometer suite.

VIM *1* Vacuum-induction melting.

2 Vendor information manual.

V_{IMD} Speed (IAS) for minimum drag.

V_{IMP} Speed (IAS) for minimum power, not necessarily same as above.

V_{i-mp} Speed corresponding with lowest power at which both height and speed can be maintained, ie minimum speed for continuous cruise (not current use).

$V\infty$ V-infinity, ie free-stream velocity.

V (Int)[2] Vehicle integrated intelligence.

vinyl ester Low-viscosity solventless liquid resin used as alternative to epoxy in wet lay-up of FRC materials.

VIO Violent, meaning heavy static or other radio interference, normal code VLNT for other meanings.

violet Route(s) into and out from target area on colour radar.

VIP *1* Value improvement programme.

2 Common meaning, very important person.

3 Video integrated processor, or presentation (see next entry).

4 Vehicle improvement programme.

5 Variable installation position (engines).

6 Voice over internet protocol.

VIP levels Those of video processor yielding weather echo intensity for precipitation, from * 1 (weak) to * 6 (extreme).

Viper Video-input encoder.

VIPPS Visual-imaging pass-production system, controlling personnel access.

virage   Tight turn (pre-WW1).

virga   Streaks of water or ice particles falling from cloud but evaporating before reaching surface.

virgin fibre   Continuous tow, long staple.

VIRSS   Visual and IR sensor systems.

virtual airline   Air carrier created only as legal entity to facilitate franchising agreement.

virtual cockpit   One offering no natural external view, only displays from sensors.

virtual collaborative engineering   Links users at remote locations into conference [eg, all can have input to a drawing].

virtual gravity   Terrestrial acceleration acting on parcel of atmosphere, reduced by centrifugal force due to parcel's relative motion; symbol g* = approx. 99.99% g.

virtual height   Apparent height of ionized atmospheric layer calculated from time for radio pulse to complete vertical round trip.

virtual image   One visible in mirror but not projectable on surface.

virtual image display   Small CRT binocular colour high-resolution image of surface target, which appears to be 935 mm (36.9 in) behind face of magnifying lens.

virtual inertia   That part of inertia forces acting on oscillating body due to surrounding fluid (eg air) and proportional to fluid density.

virtual level   Energy level of subatomic nuclear system for which excitation energy exceeds lowest nuclear-particle dissociated energy.

virtual manufacturing   Integration of available technologies to get right information to right people at right time to increase speed and accuracy of decisions.

virtual mass   Actual mass plus apparent mass.

virtual piston   Pumping effect caused by collapse of launch tube by explosive lens.

virtual star   Created by laser illuminating diffuse sodium c100 km above Earth to give continuous readout of atmospheric distortion of telescope optics.

virtual stress   See Reynolds stress.

virtual temperature   Temperature parcel of air would have had had it been entirely free of water vapour, symbol $T_v = (1 + 0.61 q) T$ where T is measured temperature and q is specific humidity.

VIS   Voice-interactive subsystem.

V_{IS} Lowest selectable airspeed.

vis   Invariably means visibility, but ambiguous.

Visa   Vertically interconnected sensor array.

viscoelasticity   Behaviour of material which has hereditary or prior stress-history memory and exhibits viscous and delayed elastic response to stress superimposed on normal instantaneous elastic strain.

viscosimeter   Instrument for measuring viscosity; Saybolt and Engler * are simple calibrated containers with narrow orifice, result being obtained by timing run-off; accurate (absolute) * include Stokes (falling speed of small sphere), rotating-cylinder (outer cylinder drives inner via fluid interface whose drive torque is measured), capillary tube (Poiseuille), and oscillating disc (parallel and close to plane surface).

viscosity   1 Dynamic * can be considered internal friction in fluid; property which enables fluid to generate tangential forces and offer dissipative resistance to flow, defined as ratio of shear stress to strain; in air almost unaffected by pressure but increases with temperature.

Symbol μ; unit $Ns/m^2 = 1,000$ cP; for air 1.78593×10^{-5} Ns/m^2 (in traditional units 3.73 slug/ft-s).

2 Kinematic * is μ/ρ where ρ is density; varies with pressure as well as temperature, units are $m^2/s = 10^6$ cSt; also called dynamic *. Symbol v.

viscosity coefficient   Synonymous with viscosity.

viscosity index   Usually synonymous with viscosity, or its variation with temperature.

viscosity-index improver   Long-chain waxy polymer[s] which stay thick at elevated temperatures.

viscosity manometer   Instrument for measuring very low fluid pressure by torque exerted on disc suspended on quartz fibre very close to spinning disc; examples are Dushman and Langmuir gauges.

viscosity valve   Liquid-system control valve controlled by viscosity of medium, eg in bypassing lube-oil cooler.

viscous aquaplaning   Occurs when the runway is merely damp, with a water film not penetrated by tyres; important on smooth surfaces, especially coated with deposited tyre rubber; persists to low speeds.

viscous damping   Energy dissipation in vibrating system in which motion is opposed by force proportional to relative velocity.

viscous flow   Flow in which viscosity is important; can be laminar or turbulent but criterion is that smallest cross-section of flow must be very large in relation to mean free path. At very low R inertia becomes unimportant and flow is governed by Stokes equations.

viscous fluid   One in which viscosity is significant.

viscous force   Force per unit mass or volume due to tangential shear in fluid.

viscous stress   Fluid shear stress, symbol $\tau = \mu du/dy$ (Newton's law) where u is velocity distant y from surface or other reference layer.

Visgard   Polyurethane conductive coating for transparencies, anti-fog, anti-scratch, anti-static.

visibility   Distance at which large dark object can just be seen against horizon sky in daylight (DoD, NATO, see RVR, RVV).

visibility meter   Instrument for measurement of visibility, visual range, eg telephotometer, transmissometer, nephelometer etc.

visibility prevailing   Distance at which known fixed objects can be seen round at least half horizon.

visibility value   Distance (see visibility) along runway, in miles and tenths.

visible horizon   Circle around observer where Earth and sky appear to meet (USAF). Also called natural horizon.

visible light   EM radiation to which human eye responds, typically with wavelength from 400 to 700 mm, 0.4 to 0.7μ.

visible line   Full line on drawing representing a line visible in assembled subject item.

visible radiation   See visible light.

visible spectrum   See visible light.

visionics   Collective term for optical and electronic devices, operating at many EM wavelengths, which enhance human vision, especially at night and in bad weather or smoke or other adverse conditions.

vision-in-turn window   Eyebrow window in roof of flight deck or cockpit.

visitor   Person entering a foreign state for period less than three months.

Vismir   Visible to medium infra-red.

Visol Butyl ether + 15% aniline; rocket fuel with Salbei (G).

visor *1* Pivoting or translating fairing for forward-facing cockpit windows of supersonic or hypersonic aircraft, in latter case with thermal-protection coating.

2 Hinged screen, transparent clear or tinted, protecting eyes against solar radiation and/or micrometeorites in space exploration.

Visrep Visual report by reconnaissance pilot.

VISSR Visible IR spin/scan radiometry (or radiometer).

Vista *1* Very intelligent surveillance and target acquisition.

2 Variable-stability inflight-simulator test aircraft.

3 Virtual integrated software testbed for avionics.

4 Visual imagery simulation training aid.

Visual Visual imaging system for approach and landing, based on a 3.5-μ Flir.

visual See *visual contact*.

visual acuity Human ability to see clearly, as tested by letter card.

visual approach Landing approach conducted by visual reference to surface, esp. by aircraft on IFR flightplan having received authorization.

visual approach path indicator Simple Vasi for private pilots with red/amber/green sectors (Lockheed-Georgia).

visual approach slope indicator Systems of visible lights arranged in pattern at landing end of runway providing descent (and limited lateral guidance) information. Basic * comprises downwind and upwind bars each of three lights on each side of runway (in US, total of 2, 4 or 12), each light projecting red lower segment and white upper. When correctly on glidepath, as close as possible to ILS glidepath, pilot sees pink. For long or LEW aircraft extra light bars are added in 3-bar *, 3-bar Avasi, T-Vasi and AT-Vasi. FAA-adopted 1970.

visual arrow Modern equivalent of finger four (*fingertip*), too tight for tactical use but common on recovery to airfield.

visual/aural range See *VAR*.

visual contact To catch sight of, or a sudden good view of, target or Earth's surface.

visual cue Visual contact of object outside vehicle, esp. Earth, sufficient to give pilot orientation and position information; eg first glimpse of ground through cloud during letdown in remote place far from electronic aids.

visual descent point Optional point on final approach course of non-precision landing, identified by a navaid, from which VASI should be visible.

visual envelope Plot of flight-deck window areas showing range of P1 vision.

Visual Flight Operations Panel Permanently reviews what is permissible in each class of airspace (ICAO).

visual flight rules Those rules as prescribed by national authority for visual flight, with corresponding relaxed requirements for flight instruments, radio etc; in US typically demands flight visibility not less than 3 st. miles (1 outside controlled airspace) with specified distances from any clouds; * *on top* gives a VFR clearance to an IFR flight above the weather or cloud.

visual head-up image Imagery created outside simulator, focused at infinity; alternative or addition to terrain model or CGI.

visual hold[ing] Over fixed feature easily recognized from altitude.

visual identification Control (ATC) mode in which aircraft follows a radar target and is automatically positioned to allow visual identification (NESN); ie interceptor pilot looks at other aircraft and, if possible, identifies type and registration or other identity.

visualization Making fluid flow visible, eg by injection of carbon tetrachloride or other liquid, smoke, tufting, schlieren, spark photo, hot wire, china clay on surface, liquid films etc.

visually coupled Night/all-weather sensor(s) are slaved to pilot's helmet-mounted system, so pilot appears to 'see' normally (field of regard and FOV impose limitations).

Visually Coupled System Specific system linked with Tiara, providing monochromatic overlain with aircraft performance data, on pilot's HMD.

visual meteorological conditions Generally UK counterpart to VFR, requiring visibility 5+ miles and 1,000 ft vertical and 1 n.m. horizontal from any cloud [variations if a/c below 3,000 ft a.m.s.l.].

visual/optical countermeasures Those directed against human eyesight, ranging from reduction of night adaptation to blinding.

visual photometry Luminous intensity judged by eye.

visual range Distance under day conditions at which apparent contrast between specified target and background becomes equal to contrast threshold of observer (there is also night *) (see *RVR, RVV*).

visual reference Earth's surface, esp. that clearly identified and thus giving geographical position as well as attitude and orientation guidance, used as reference in controlling flight trajectory, if necessary down to touchdown.

visual report Identical to hot report but based solely on aircrew observations; answers specific questions concerning target for which sortie was flown (ASCC). (ASCC adds 'not to be confused with inflight report', while DoD states 'not to be used; use inflight report').

visuals Lifelike scenes viewed through windows of simulator, almost always wholly synthetic and computer-generated (but in old simulators achieved by optics moved over a large model).

visual separation Basic method of avoiding collisions in TMAs or on ground by seeing and avoiding; normally involves conflicts between arrivals and departures and can be accomplished by pilot action or tower instruction.

VIT Vision in turn.

vital actions Rigorously learned sequences instilled into all pilots, either specific to type or, more often, as general good airmanship; necessary eg before entering aircraft, upon entering cockpit, on starting engine, before taxiing, before takeoff, etc. On simple aircraft generally remembered by mnemonics, eg Bumpf (check brakes, u/c, mixture, pitch, flaps) or HTMPFFG (hood/harness, trim/throttle friction, mixture, pitch, fuel cocks, flaps, gills/gyros).

vital area Designated area or installation to be defended by air-defense units (DoD).

VITC Vertical-interval time code.

Vitreloy Dense 'metallic glass' (proprietary, Howmet).

vitrifying Transformation of ceramic from crystalline phase into amorphous or glassy state.

Vitro-Lube Ceramic-bonded dry-film lubricant with wide temperature range.

VITS Video-image tracking system.

VIU   *1* Video interface unit.

   2 Voloclub Italiano Ultraleggeri (ultralights).

Vivid   Video verification and identification.

V_J Velocity of propulsive jet, normally measured relative to vehicle.

VKIFD   Von Kármán Institute for Fluid Dynamics (Int.).

VKS   Military space force[s] (R).

VL   *1* Vertical landing, or lift, or launch.

   2 Tunnel velocity × characteristic model length.

V/L   VOR/localizer.

V_L Relative velocity of flow on underside of aerofoil.

VLA   *1* Very light aircraft.

   2 Very large aircraft.

   3 Vertical line array.

   4 Variable lever arm (flight-control ratio changer).

VLAAS   Vertical-launch autonomous attack system.

VLAC   Vertical-Lift Aircraft Council (US Aerospace Industries Association).

VLAD   Vertical line-array Difar.

Vladimir   Very-large-array demonstration imager for IR.

VLAR   Vertical [attitude] launch and recovery.

VLBI   Very long baseline interferometry.

VLBTI   Very-long-burning target indicator.

VLC   Very low clearance.

VLCHV   Very low cost harrassment vehicle.

$V^L/_D$, $V^l/_d$ Best lift/drag speed.

VLE   Virtual learning environment.

V_{LE} Maximum speed with landing gear locked down (E = extended); flight manual specifies whether main gear only or all units. Always significantly higher than V_E values.

VLEA   Very-long-endurance aircraft, powered by solar cells charging fuel cell(s) for continuous flight of months or years.

VLED   Visible LED.

VLES   Very large eddy simulation.

VLF   Vectored-lift fighter; also see next.

v.l.f.   Very low frequency. See Appendix 2.

v.l.f. Omega   Global system of long-range navigation using v.l.f. radio.

V_{LG} No longer used; was vague maximum landing- gear speed, now replaced by V_{EI}, V_{LE}.

VLNT   Violent.

VLO   Very low observability.

V_{LO} US term, maximum speed for landing-gear operation; synonymous with UK V_E.

V_{LOF} Lift-off speed, at which aeroplane becomes airborne; suggest = V_T.

VLR   *1* Very long range [1942–45, ranges today considered modest].

   2 Telecom code for long-range search/rescue aircraft.

   3 Velocity/length Reynolds number.

   4 Very light rotorcraft.

VLS   *1* Visible light sensor.

   2 Vertical launch system.

   3 Veiculo Lancador de Satelittes (Brazil).

VLSI   Very-large-scale integration.

VLSIC   Very-large-scale integrated circuit.

VLSIPA   Very-large-scale integration photonics architecture.

VLV   Valve.

VLY   *1* Volley.

   2 Valley.

VM, v.m.   *1* Velocity modulation.

   2 Voltmeter (also V/M).

   3 Voter/monitor (also V/M).

   4 Visual [target] marker.

   5 Heading to a manual termination.

V_M Speed at which precipitation (esp. slush) drag is maximum; always well below aquaplaning speed.

V_m *1* Volume fraction of composite material occupied by matrix.

   2 Missile velocity.

V/M   See VM(2), (3).

VMA   Code, fixed-wing attack squadron (USMC).

VMAD   Vertically mounted accessory drive.

V_{max} Maximum CAS for clean aircraft.

VMC   *1* Visual meteorological conditions.

   2 Vehicle management computer.

V_{MC} Minimum-control speed; more precisely specified as following three entries:

V_{MCA} Minimum speed at which aeroplane can be controlled in air; defined as limiting speed above which it is possible to climb away with not more than 5° bank and with yaw arrested after suffering failure of critical engine in takeoff configuration, with engine windmilling and c.g. at aft limit. There are usually suggested limits of required rudder-pedal force and on absolute value of *.

V_{MCG} Minimum speed at which aeroplane can be controlled on ground; defined as that above which pilot can maintain directional control after failure of critical engine without applying more than 70 kg pedal force, without going off runway and if possible while holding centreline, with 7+ kt crosswind and wet surface.

V_{MCL} Minimum speed at which aeroplane can be controlled in the air in landing configuration, while applying maximum possible variations of power on remaling engine[s] after failure of critical engine.

V_{MCL2} As above but with any two engines inoperative.

V_{MCP} Speed, usually EAS, at maximum continuous power in level flight; V_H is more commonly used.

VMDI   Vector miss-distance indicator.

VME   *1* Virtual memory environment.

   2 V.h.f. management entity [bus].

   3 Versa module Eurocard [bus].

V_{ME} Airspeed of sailplane for maximum endurance.

VMECC   Versa module Eurocard card cage.

VMF   *1* Code, fixed-wing fighter squadron (USMC).

   2 Navy (USSR).

VMFA   Code, fixed-wing fighter/attack squadron (USMC).

V_{MG} Vertical main landing gear strut load at MRW and with full aft c.g.

V_{min} Minimum CAS for basic clean aircraft.

$V_{M(LO)}$ Minimum maneuver speed (US).

VMM   *1* Veille Météorologique Mondiale = World Weather Watch.

   2 Vehicle management module.

VM3, VM³, VMMM   Versatile mass media memory, 4 RTMMs.

VMO   Variable metering orifice.

V_{MO} Maximum permitted operating speed under any condition, higher than V_{NE} and less than V_{DF} but latter is wholly exceptional limit not intended to be reached except during certification flying.

VMS   *1* Vehicle management system.

2 Vehicle monitoring system (robotics).

3 Vertical motion simulator.

4 Vehicle motion sensor.

5 Variable metering sleeve.

V$_{MS}$ Minimum EAS observed during normal symmetric stall, usually less than V$_s$.

VMU *1* Voice management [or message] unit.

2 Velocity measurement unit.

3 Marine Air Group (USMC).

V$_{MU}$ Minimum demonstrated unstick speed, at which with all engines operating, and without regard to safety, noise-abatement or any other factor, aeroplane will leave ground and hold positive climb.

VN *1* Speed multiplied by (or plotted against) normal acceleration; more commonly V-n.

2 Vinyl/nitrile (PVC nitrile-based rubber).

V$_n$ Component of wind acting perpendicular to heading; also called normal wind, normal component or (loosely) crosswind component.

V-n Flight speed (usually EAS) multiplied by or plotted against normal acceleration; hence * diagram, of which two forms; basic manoeuvring envelope and basic gust envelope.

V$_{NA}$ Noise-abatement climb speed, usually synonymous with V$_4$ (1st segment) but also commonly used for 4th segment Fuss.

V-Nav, VNAV, V/NAV Vertical navigation; generalized topic of control of flight trajectory in vertical plane; now becoming automatic in transport-aircraft energy-managing flight systems. In a specific system, an add-on to LNAV giving glideslope guidance down to 350 ft AGL (WAAS).

V$_{NE}$ Never-exceed speed; an exceptional permitted maximum beyond V$_{MO}$ of which captain may avail himself in unusual circumstances. Implication is that * must be reported and explained.

VNII All-union research institute [EM adds electro-mechanics, RA radio engineering] (R).

V90 Category of off-base airstrips usable by fighters (Sweden).

VNIR Visible to near-IR.

V$_{NO}$ Maximum permitted normal-operating speed, generally replaced by V$_{MO}$; in smooth air can be exceeded 'with caution'.

VNR V.h.f. navigation receiver.

VNRT Very near real time.

VNTSC Volpe National Transportation Systems Center (US DoT).

VNV ALPA (Netherlands).

VO *1* Visual optics.

2 US piston engine code, vertical crankshaft, opposed.

V$_o$ Operational speed (airspeed in FCS calculations).

VOA *1* Velocity of arrival.

2 Vsesoyuznoe Obshchestvo Aviastroitelei (aeronautical society, USSR, R).

3 Volatile organic analyser.

VOB Vacuum optical bench.

VOC *1* Validation of concept.

2 Visual optical countermeasures.

3 Volatile organic compound[s].

Vocational Tax relief Granted by Inland Revenue against cost of some commercial-pilot training (UK).

Vocoder Voice coder; device responding to spoken input (usually previously stored) to generate synthetic speech output.

VOCRAD Voice radio.

VOCS Voice-operated carrier suppressor.

VOD *1* Vertical-on-board delivery; usually synonymous with vertrep.

2 Vertical obstruction data.

3 Video on demand.

Voder Device with keyboard input controlling generation of electronic sounds, esp. synthetic speech output.

VODR Has been used to mean VOR.

VOF Volume of fluid.

VÖFVL Verband Österreichischer Flugverkehrsleiter (Austria).

Vogad Voice-operated gain-adjusting device; auto volume compressor or expander.

Voice Voice optimal interrogator (USN).

voice frequency Normally taken as 25 Hz to 3 kHz for telecommunications, much greater range for hi-fi.

voice-grade channel Covers about 300–3,000 Hz, for speech, analog, digital or facsimile.

voice keying System enabling telecommunications to use common R/T transmit and receive sites and similar frequencies but with voice-operated carrier suppressor and delay network to switch outgoing signal to transmitter and incoming to receiver. Remote stations normally switch automatically to receive mode except when user is speaking.

voiceless homing Any electronic homing system not using speech; traditionally meant radio range (arch.).

voice message unit Software-controlled system providing voice or tone warnings of faults, sensor activity and other occurrences.

voice-operated relay See *voice keying*.

voice rotating beacon Short-range radio navaid transmitting stored-speech headings (usually QDMs) which differ from 000° round to 359°; a form of talking VOR, also called talking beacon, abb. VRB (arch.).

void Undesired gap in welded joint.

void fraction *1* Percentage of total frontal area of jet engine through which airflow passes.

2 Also several meanings in composite materials and structures.

Voigt effect Double refraction (associated with Zeeman) of light passing through vapour perpendicular to strong magnetic field.

VOIR Venus-orbiting imaging radar.

Voiska-PVO Troops of air defence of homeland (USSR, R).

Voispond Proposed Calsel function that would automatically identify an aircraft by a voice recording.

VOL *1* Vertical on-board landing.

2 Volume.

volatile *1* EDP (1) memory which dissipates stored information when electrical power is switched off; thus, next morning or after weekend all bits must be restored before computer operation. Also means electrical transients cause corruption (though this may be only temporary). Hence * data, * memory, volatility.

2 Having high *vapour pressure*, and thus low boiling or subliming temperature at SL pressure; hence volatility.

vol à voile(s) Gliding, soaring (F).

vol d'abeille Beeline, straight-line distance (F).

Volmet Routine ground-to-air broadcast of meteorological information (ICAO). Today such broadcasts are Metars and apply to a designated list of airfields.

Volocan Radar tracking/computing of flight paths to solve stacking problems (USAF 1953).

vol piqué Dive (F).

vol plané Planing (inclined) flight, ie glide by powered aeroplane (arch. except in F).

volt SI unit of EMF, = W/A.

voltage amplifier Delivers small current to high impedance to obtain voltage gain.

voltage-dependent resistor Ohms = f(V); resistance varies directly with applied voltage.

voltage drop PD across any impedance carrying current.

voltage-fed aerial (antenna) Fed from one end, where signal potential is maximum.

voltage gain Ratio of output/input voltages.

voltage standing-wave ratio Ratio maximum/minimum V along waveguide or coaxial.

volt-ampere SI unit of alternating-current power, symbol S, made up of power component P watts and reactive component Q; $S = \sqrt{P^2 + Q^2}$; see *power factor*.

voltmeter Instrument for measuring potential difference, ie V.

volume SI unit is m^3 (conversion factors for non-SI measures, from $ft^3 \times 0.02831684$, UK gal $\times 0.004546087$ and US gal $\times 0.003895411$); litre $(dm^3) = 0.035287\ ft^3$ $= 60.9756\ in^3$; cm^3 (cc) $= 0.06102\ in^3$; UK gal [Imp. gal] $= 1.20095$ US gal; US gal $= 0.83267$ UK gal.

volume fraction *1* Proportion, usually %, of reinforced composite (FRP) occupied by reinforcing fibres.

2 Generally, proportion of whole volume occupied by particular substance.

3 For aerostats see *air* *, *gas* *.

volumetric efficiency Volume of combustible mixture (in diesel, air) actually drawn into cylinder of piston engine on each operating cycle divided by capacity (swept volume) of cylinder, usually expressed as %. Symbol η_v or e_v.

volumetric loading Also called * density, total volume of solid rocket motor propellant divided by total volume of unloaded case, usually expressed as %. Symbol λ.

volume unit Measure of audio volume to be outputted by electrical current, expressed in dB equal to ratio of magnitude of electrical waves to magnitude of reference volume, usually 1 mW; abb. VU.

volute Spiral or planar helix; thus, spiral casing of centrifugal compressor or supercharger impeller.

Vom Volts/ohms/milliamps tester.

Vomit Comet Aircraft, e.g. KC-135, used for zero-gravity tests.

Von Brand Standard method of measuring jet smoke by passing measured gas volume through filter and then recording intensity of calibrated light reflected by filter pad. Gives quantified measure of particulate matter trapped by chosen filter. Hence * scale for visible smoke.

von Kármán street See *street* (also called Kármán street).

VOP Variation of price.

VOR V.h.f. omnidirectional radio range, announced by RCA in 1941 and forced through by US to become universal global [except USSR] radio navaid. Comprises fixed beacon emitting fixed circular horizontal radiation pattern at 108–118MHz on which is superimposed rotating directional pattern at 30 Hz giving output whose phase modulation is unique for each bearing from beacon. Thus airborne station can read from panel instrument bearing of aircraft from station, called inbound or outbound radial. Each fixed station identified by three-letter keyed intermittent transmission (sometimes voice). See Doppler *.

VOR/DME VOR steering guidance with DME distance information.

Vorgen Voltage generator.

VOR/ILS Linkage of VOR signals to aircraft ILS so that left/right steering guidance is given by ILS panel instrument. Latter often called * deviation indicator, or Vorloc.

Vorloc, VOR/Loc Panel instrument giving steering guidance from received VOR signals; can be complete ILS indicator or (esp. in light aircraft) simple VOR receiver with localizer needle only.

VORMB, VOR/MB VOR marker beacon.

Vormet Sends scripted pilot's weather reports to overflying aircraft.

Vorpostenboot Flakship (G).

Vortac Combination of VOR and Tacan (occasionally written VOR/TAC) offering from one fixed station VOR az, Tacan az and Tacan (DME) distance information; ident codes prove VOR and Tacan signals are both from same fixed station. Normally * is end-product of trying to integrate civil (VOR/DME) with military (Tacan) navaids, latter being u.h.f. and therefore inherently incompatible.

vortex Fluid in rotational motion (possessing vorticity), eg streamed behind wingtip or across leading edge of slender delta. See line *, point *, trailing *.

vortex breakdown Sudden separation of large vortex from leading edge of slender delta (naturally followed by its decay) at particular AOA (higher than stalling AOA for most wings); essentially represents stall of slender-delta wing.

vortex burst See *vortex breakdown*.

vortex dissipator Bleed-air jet(s) blown down below and ahead of jet-engine inlet to prevent ingestion of material from unpaved airfields or contaminated runways.

vortex drag Drag caused by vortex formation; not normally a recognized part of aircraft drag.

vortex filament Line along which intense (theoretically infinite, at R = 0) vorticity is concentrated; either closed loop or extending to infinity.

vortex flap Hinged along its leading edge just behind leading edge of wing on upper surface. Opened to 45° traps vortex to increase lift.

vortex flow Fluid flow combining rotation with translational motion.

vortex generator Small flat blade perpendicular to skin of aircraft or other body set at angle to airflow to cause vortex which stirs boundary layer, usually to increase relative speed of boundary layer and keep it attached to surface; also called turbulator.

vortex hazard Danger to aircraft, esp. light aircraft, from powerful vortices trailed behind wingtips of large aircraft; also called wake hazard, wake turbulence.

vortex lift Lift generated by slender delta or similar wing having sharp, acutely swept leading edge (subsonic relative velocity normal to leading edge): large and powerful vortex is shed evenly on left/right wings, adding major non-linear increment to lift; also postpones stall to lower speed and extreme AOA, but with high drag.

vortex line Line whose direction at every point coincides with rotation vector, all of whose tangents are parallel with local direction of vorticity. Must be closed curve or extend to infinity, or to edge of fluid or to a point on an infinitely intense vortex sheet.

vortex ring Vortex forming closed ring (eg smoke ring); collar vortex, and formed by helicopter as it slows to the hover.

vortex-ring state Operating state of rotorcraft (esp. helicopter) main rotor in which direction of flow through rotor is in opposite sense to relative vertical flow outside rotor disc and opposite to rotor thrust. Occurs in autorotative landing, and can occur with rotor under power if rate of descent equals rotor downwash velocity.

vortex separation Filtration of different types of particle from fluid by different centrifugal forces in vortex motion.

vortex sheet Theoretical infinitely thin layer of fluid characterized by infinite vorticity; in practice layer of finite thickness formed by large number of small vortices, eg as trailed behind lifting wing (where much of vorticity is quickly rolled up into two large tip vortices).

vortex street See *street*.

vortex strength Circulation round any body or other closed system, symbol Γ, constant at all points on a vortex filament.

vortex trail Visible (white) trail from wingtip, propeller tip etc, caused by intense vortex.

vortex tube Device devoid of moving parts in which pressure difference induces fluid flow through tangential slots into tube; violent vortex divides flow into surrounding warm flow and cold (about 40°C cooler) core.

vortex turbine Mounted in optimum location at wingtip to extract power from tip vortex.

vorticity Vector measure of local rotation in fluid; in uniformly rotating fluid proportional to angular velocity (in UK, exactly defined as twice angular velocity). Symbol $q = \vec{\nabla} V$ where $\vec{\nabla}$ is del (mathematical operator) and V is vector velocity (° curl V in US; often called rot. V, from rotation, in Europe).

vorticity component Circulation around elementary surface normal to direction of vorticity divided by area of surface; more strictly, limit of circulation as area of element approaches zero.

vortillon Name coined by McDonnell Douglas to describe fence around underside of DC-9 wing leading edge controlling boundary-layer direction.

VORW VOR without voice.

VOS *1* Velocity of sound.
2 Voice-operated switch.

VOT, Vot VOR test signal; ground facility for testing accuracy of VOR receivers.

voter Binary logic element or device which compares signal condition in two or more channels and changes state whenever a predetermined signal mismatch occurs, usually to exclude a minority 'outvoted' signal.

voter threshold Difference between signals at which voter is switched or triggered; normally difference between one selected signal and mid-value signal from all others in parallel system.

voting system System in which outputs of several parallel channels are sensed and compared by voter so that any single malfunctioning channel may be excluded.

Votol, VOTOL Vertical-only take-off and landing.

VOWS, Vows Valuation of weight saved; measure of financial reward (usually in increased annual earning power) from cutting each unit of mass (kg or lb) from empty weight; eg VOWS for Concorde in 1974 currency was £50/lb.

VOx Vanadium oxide.

VOX See *Vox*.

Vox Voice (communication, keying or activation).

VP *1* Variable-pitch.
2 Code: fixed-wing patrol squadron (USN); suffix B adds bomber.
3 Vector processor.
4 Video processor.
5 Combat helicopter regiment (R).
6 Validation parameter.

V_p *1* Propellant volume; that volume occupied by solid propellant in a rocket motor.
2 Propwash velocity, V + v.

v.p. Vapour pressure.

VPAC Vapour-phase aluminide coating.

V_{path} Vertical path.

VPC Vertical-path computer.

VPD Virtual product design.

VPDS Variable public display system.

$V_\%$ Best angle-of-climb speed (UK usage, US = Vx).

$V_{\%SE}$ Best angle-of-climb speed, single-engine (UK).

V_{peri} Perigee velocity, maximum speed of satellite

$$= \sqrt{K\,\frac{2}{r_o} - \frac{1}{a}}$$ where K is constant, r_o is distance to

centre of primary and a is semi-axis of elliptical orbit.

VPI *1* Vapour-phase inhibitor.
2 Vertical-position indicator.

vpm, v.p.m. Vibrations per minute.

VPN *1* Vickers pyramid number.
2 Variable primary nozzle.
3 Virtual private network.
4 Vendor part number.

VPR Voice position report.

VPS Vacuum plasma spray.

VPTAR See *Vaptar*.

VPTS Voice-processing training system.

VPU *1* Video, or voice, processing unit.
2 Vortac position unit.

VPU(D) Voice processor unit with data mode.

VPVO Air-defence troops (R).

VR *1* Resultant velocity.
2 Vernier radial (thruster).
3 Visual reconnaissance.
4 Volunteer Reserve.
5 Veer[ring].
6 Visual route, rules or range.
7 Vortex ring.
8 Virtual reality.
9 Voice recognition.

V_R *1* Rotation speed, at which PIC starts to pull back on yoke to rotate aeroplane in pitch; normally determined by one of following: not less than 1.05 V_{MCA}; not less than 1.1 V_{MS}; not less than 1.05 or 1.1 V_{MU}; and it must allow 1.1 V_{MCA} or 1.2 V_{MS} to be achieved at screen after one engine failure.
2 Radar velocity, i.e., of aircraft in which it is mounted.
3 Heading to a radial (US wording).

V_r Radial velocity, eg component of velocity along sightline to target (rate of change of range) or fluid speed along radial direction in centrifugal compressor measured relative to compressor (ie eliminating tangential component).

VRA   Variable-response [research] aircraft.

V_{RA} Rough-air speed; maximum recommended EAS for flight in turbulence.

VRAM   Video random-access memory.

VRB   Voice rotating beacon.

VRBL   Variable.

VRC   *1* Vendor reject crib.
    2 Value relay centre (EC).

VRD   Virtual retinal display.

V_{REF} *1* Loosely, any reference speed.
    2 In jet transports, a typical approach speed at about 1.35 V_{S0} (chiefly US).

VRF   Code: ferry squadron (USN).

VRG   *1* Vertical reference gyro, = vertical gyro.
    2 VDL(2) reference guide.

vrille   Spin (arch.).

VRMT   Virtual-reality maintenance trainer.

VROC   *1* Vertical rate of climb (helicopter).
    2 Validated rate of climb.

VRP   *1* Visual reporting post or point.
    2 Visual refrence point, giving a fix.

VRS   *1* Video recording system.
    2 Voice response system.
    3 VTOL recovery and surveillance [aircraft] (USCG).
    4 Vortex-ring state.

VRT   Virtual-reality toolkit [& S adds and simulator].

VRU   Vertical reference unit.

VS   *1* USN squadron code: fixed-wing ASW.
    2 Vestigial sideband.
    3 Velocity search (radar).
    4 Vane set.
    5 Vertical speed (also V/S, VSP).
    6 Versus.

V_s *1* Stalling speed; IAS at which aeroplane exhibits characteristics or behaviour accepted as defining stall (in US, FAA adds 'or minimum steady flight speed at which airplane is controllable', which is not same thing and is separately defined in UK as V_{MCA}).
    2 Velocity of slip (propeller).
    3 Slipstream velocity, relative to aircraft.

V/S   Vertical speed.

VSA   By visual reference to ground (ICAO code).

VSAM   Vestigial-sideband amplitude modulation.

VSAT   Very small aperture terminal.

VSB   *1* Vendor service bulletin.
    2 Visible.

vsby   Visibility.

VSC   *1* Video scan converter.
    2 Vacuum system control.

VSCAS   Variable-stability control-augmentation system.

VSCF   Variable-speed constant-frequency.

VSCS   *1* Voice switching and communications, or control, system (FAA).
    2 Vertical-stabilizer control system (Notar).

VSD   *1* Vertical situation display.
    2 Video symbology display.
    3 VDL(2) specific DTE address.

VSDR   Variable-speed digital recorder.

VSER   Vertical-speed and energy rate.

VSFI   Vertical-scale flight instrument.

VSG   Vibrating-structure gyro.

VShorad   Very short-range air defense (USA).

VSI   *1* Vertical-speed indicator, output is rate of climb or descent.
    2 Vertical soft-iron; component of Earth's field.
    3 Vapour space inhibitor = VPI.
    4 Vacuum superinsulation.
    5 Velocity and steering indicator.
    6 Variable-swath imagery.

V_{SI} Indicated stalling speed [avoid confusion with V_{s1}].

VSIP   Virtual-system implementation.

V6, V16   Vasi with 6 or 16 boxes, the 16-box being on both sides of the runway.

VSJ   Vazduhoplovni Savez Jugoslavije, aeronautical sport union of former Yugoslavia.

VSL   Vertical-speed limit [A adds advisory, which may be preventative or corrective].

VSLD   Velocity-, or vertical-, search lookdown.

V-sled, VSLED   Vibration, structural life and engine diagnostic system.

VSM   Vertical-separation minimum, or minima.

V_{s0} Stalling speed with flaps at landing setting, engine[s] idling.

V_{s1} Stalling speed in a specified configuration other than clean.

V_{s1g} Stalling speed under 1 g vertical (normal) acceleration; obtained from V_s by correcting for any imposed normal acceleration that may have been present during an actual measured stall; a 'pure' V_s not normally entering into performance calculations.

VSP   Vertical speed; VS more common.

VSR   *1* Very short range.
    2 Volume search radar.
    3 Valve-seat recession.

VSRA   V/STOL research aircraft (NASA).

VSRAD   Very short range air defence.

VSRS   Variable-speed rotor system.

VSS   *1* Video signal simulator.
    2 Variable-stability system.
    3 Vehicle systems simulator.

VSSA   Variable-stability simulator aircraft.

VSSC   Vikram Sarabhai Space Centre (India).

V_{SSE} Minimum speed, selected by manufacturer, for intentionally shutting down one engine in flight for pilot training; in UK and some other countries this is prohibited below 3,000 ft (907 m) AGL.

VSSG   Vertical Separation Study Group (Navsep).

VST   Variable-stability trainer; aeroplane with avionics and flight-control surfaces added to enable it precisely to duplicate flight characteristics of other types.

V_{st} One reference states 'stall or minimum flight speed, flaps up, no power'; not a normally recognised abbreviation.

V-Star   Variable search and track air-defence radar.

V/STOL   Vertical or short takeoff and landing.

VSTT   Variable-speed training target.

VSV   Variable-stator vane, or valve [A adds actuator, AS adds actuating system].

VSVT   CAA of lithuania (1992), in 1994 became DCA(2).

VSW   *1* Vertical speed and windshear; hence VSWI = * indicator.

2 Variable-sweep wing.

3 Verification software.

VSWE　Virtual strike warfare environment.

VSWR　Voltage standing-wave ratio.

VT　*1* Vernier thruster.

2 Voltage transient.

3 Vectored thrust.

4 Internal prison, eg for design teams (USSR, 1929–42).

5 Target speed.

6 Video tracker [SC adds system controller].

7 Variable time. or timing.

8 Validity time[s].

V_T　*1* Takeoff speed.

2 Confusingly, threshold speed, see V_{TDM}, V_{Tmax}, V_{Tmin}.

3 Velocity of target.

4 Threshold voltage, especially that established for automatic target detection.

V_t　True airspeed, in aerodynamics.

VTA　*1* Military transport aviation (USSR, R).

2 Vibration tuning amplifier.

3 Vertex time of arrival.

4 Voice terrain advisory.

VTAS　*1* Visual target acquisition system.

2 True airspeed.

3 Voice, throttle and stick.

VTC　*1* Vectored (or vectoring) thrust control.

2 Vernier thruster control.

3 Variable time-constant.

4 Vibratory torque control (Teledyne Continental).

V_{TDM}　Minimum threshold speed demonstrated.

V_{TDP}　Vectored-thrust ducted propeller.

VTK, V_{TK}　Vertical track distance.

VTL　Prefix to aviation fluid specification(6).

VTM　Voltage-tunable magnetron.

V_{Tmax}　Maximum threshold speed, above which risk of overrunning is judged unacceptable; usually V_{AT} + 15 kt.

V_{Tmin}　Minimum threshold speed, below which risk of stall (esp. in windshear) is judged unacceptable; usually V_{AT} - 5 kt.

VTO　*1* Vertical takeoff.

2 Varactor-tuned oscillator.

3 Volumetric top-off.

4 Visiting technical officer (UK).

VTOCL　Vertical takeoff, conventional landing.

VTO grid　Vertical takeoff grid designed to reduce erosion and reingestion problems in operating jet-lift aircraft from unprepared surfaces.

VTOGW　Vertical takeoff gross weight.

VTOL　Vertical takeoff and landing; VTOVL adds redundant second 'vertical'.

VTOSS　Takeoff safety speed (rotorcraft) (FAA).

VTPR　Vertical temperature-profile radiometer.

VTR　*1* Video tape recorder.

2 Vocational training, or tax, relief.

3 Variable takeoff rating.

VTRAT　Visual threat recognition and avoidance trainer.

V_{TRK}, **V/TRK**　Vertical track.

VTS　*1* Video target simulator.

2 Voice telecom system.

V_{TS}　Velocity of tunnel test section.

VU　Volume unit.

VTUAV　*1* VTOL UAV.

2 VTOL tactical UAV.

VTVL　Vertical takeoff, vertical landing [suggest = VTOL].

V_U　*1* Relative airspeed across upper (positively cambered or lifting) surface of aerofoil.

2 Tangential velocity, eg of flow leaving centrifugal compressor.

3 Utility speed (US usage).

VUAV　Vertical UAV.

V/u.h.f.　Very and ultra-high frequency; Appendix 2.

Vulcanoids　Hypothetical asteroids within the orbit of Mercury.

Vulkollan　Hard erosion-resistant thermosetting plastics material (trade name, F).

V_{us}　Unstick speed of marine aircraft.

VV　*1* Vertical visibility, measured (most countries) in hundreds of feet, or runway visibility value.

2 Valve voltmeter.

3 Validation and verification (sometimes V&V).

4 Velocity vector.

V/V　Vertical velocity.

VVA　*1* Zhukovskiy air force academy (USSR, R).

2 Vietnam Veterans of America; F adds Foundation.

3 Voltage variable attenuator.

VV&A　Verification, validation and accreditation.

VVI　Vertical velocity indicator, suggest = VSI.

VVIA　VVA engineering academy (USSR, R).

VVR　Voice and video recorder.

VVS, V-VS　Air forces (USSR, R).

VVT　Variable valve timing.

VW　Vortex wake; S adds spacing.

V_w　Tailwind, or tailwind component.

VWA　Virtual worktop architecture.

VWF　Verband der Wissenschaftler an Forschungs-instituten (G).

VWP　Visa-waiver program (Dept. of State).

VWRS　Vibrating-wire rate sensor.

VWS　*1* Vertical windshear.

2 Ventilated wet suit.

VX　*1* General code for nerve gases.

2 Test and Evaluation Squadron (USN).

V_x　Airspeed for best angle of climb, segment not specified (US usage).

VXE　Antarctic Development (ie, exploration) Squadron (USN).

VXO　Variable crystal oscillator.

V_{XSE}　Airspeed for best angle of climb, single-engine (US usage).

V_Y　Airspeed for best rate of climb (US usage).

VYRO, Vyro　Vertical yaw and roll.

V_{YSE}　Airspeed for best rate of climb, single-engine (US usage, also called blueline speed).

VZ　Aircraft designator, vertical-lift research (USA 1957–62).

V_{ZRC}　Airspeed for zero rate of climb, at which with one engine inoperative drag reduces gradient to zero.

W

W *1* Watt[s], and general symbol for power in SI countries.

2 Weight, loosely synonymous with mass, and mass flow, esp. through jet engine..

3 Force of applied load.

4 Energy [work]; E is preferred.

5 Tungsten [from wolfram].

6 Aircraft mission, prefix, electronic search or AEW (USN 1952–62).

7 Modified mission, suffix, AEW (USN 1944–62); prefix, weather reconnaissance (USAF from 1958, USN from 1962).

8 JETDS code: armament, automatic flight or remotely piloted.

9 Weather, and airport with NWS office (US).

10 West, western longitude.

11 Weapon.

12 Wave[s] or Mach-wave angle.

13 IFR flightplan; approved R-nav but no xpdr.

14 Wing [military unit].

15 Prefix, NW warhead.

16 White light.

17 Width, maximum tyre [tire] cross-section.

18 Warning.

19 Indefinite ceiling, sky obscured.

20 Secondary station (Loran).

21 See W-engine.

22 Without voice (radio).

23 Suffix, quenched in cold water.

w *1* Generalized symbol for special fluid velocities, eg vertical gust, wing downwash, propeller slipstream etc.

2 Warm (air mass).

3 Load per unit distance or per unit area.

4 Specific loading.

5 Linear velocity due to yaw.

6 Suffix, wing; thus Ww = wing weight.

7 Rate term for weight or mass, eg per unit time.

8 Generalized symbol for work.

9 Range, of values.

W-code W (13): approved R-Nav but no transponder.

W-engine Piston engine with three linear banks of cylinders about 50°–60° apart; also called broad-arrow.

W-wing Shaped like W in planform with sweepback inboard and forward sweep outboard.

WA *1* Work authorization.

2 Prefix: word after . . .

3 Airmet weather advisory.

W/A Weight per cross-section area (warhead).

W$_A$ Equipped airframe weight.

W$_a$ Air mass flow, eg passing through engine per second.

WAA War Assets Administration (US, 1946 –).

WAAF Women's Auxiliary Air Force (UK, 1939–49).

WAAM Wide-area anti-armour munition.

WAAS *1* Wide-area active surveillance (radar).

2 World airline accident summary (UK CAA).

3 Wide-area augmentation system (GPS, US counterpart to Egnos).

WAASA Women's Aviation Association of South Africa.

WAC *1* World Aerobatic Championships.

2 Weapon-aiming computer.

3 Wide-angle collimated; S adds system.

4 World Aeronautical Chart (1,000,000 scale).

5 Women's Army Corps (US, 1943 –).

WACA World Airlines Clubs Association (Int.).

WACCS Warning and caution computer system.

WACD Wide-area change detection (DDB).

WACRA, Wacra World Airline Customer Relations Association (Int.).

WAD *1* Workload assessment device.

2 Wide-angle differential (see *GNSS, *GPS).

WADC Wright Air Development Center (USAF).

WADD Wright Air Development Division.

WADDS Wind and altimeter [setting] digital-display system.

WADGNSS Wide-area differential global navsat system.

WADGPS Wide-area differential GPS.

WAEA World Airline Entertainment Association (Int.).

WAEO World Aerospace Education Organization (UK).

WAF Women in the (US) Air Force (1948 –).

WAFC World area forecast centre.

wafer Complete (near-circular) slice of single crystal (usually epitaxial) semiconductor material on which numerous elecronic devices are constructed, subsequently separated by scribing and cleavage to make chips.

waffle plate Thin metal sheet stabilized by impressed dimples, often parallel rectangles. Same name for more complex sandwich structures.

WAFS *1* Women's Auxiliary Ferrying Squadron (US, 1942).

2 World area forecast system.

WAG *1* World average growth.

2 World Air Games, held annually.

WAGE Wide-area GPS enhancement.

waggle Rapidly repeated bank to left and right [say, ±20°].

Wagner bar Pioneer spoiler-type flight control with bang/bang solenoid operation for radio command guidance of missiles (from 1937).

Wagner beam Idealized pure tension-field beam assumed to have zero compressive strength and thus to react loads as diagonal tensions; generally, a beam designed to buckle in operation.

wagon wheel See *wheel (6)*.

wagonwheel propellant Solid rocket motor propellant grain with cross-section having form [positive or negative] of wagon wheel.

WAGS Windshear alert and guidance system.

WAHS World Airline Historical Society (US).

WAI *1* Wing anti-ice, anti-icing.

2 Women in Aviation, International (1994 –, office in Ohio).

wailer   Unmistakeable warning triggered by autopilot disconnect without human response.

WAIN   Wide-area integrated network.

waist   *1* Amidships portion of fuselage or hull; thus * gunner, firing laterally (often from rear midships area).

2 Middle portion of gas-turbine engine, esp. where diameter here is less than elsewhere; thus * gearbox for accessories.

waisting   *1* Local reduction in diameter caused by plastic flow under tension, esp. at point of failure.

2 Local reduction of cross-sectional area of body, eg to conform to Area Rule.

waiting beacon   "A low-power omni-directional beacon used by aircraft waiting their turn to follow the radio track" (B.S., 1940).

waiver   Whereas usual meaning is permanent relinquishment, in air law or certification it is a postponement.

WAK   Wing adapter kit.

wake   *1* Fluid downstream of body where total head has been changed by body's presence; usually to some degree turbulent.

2 Wingtip vortices left in atmosphere behind aircraft whose wing is generating lift, in case of large/heavy aircraft very powerful and persistent, capable of destroying light aircraft.

wake-interaction noise   Generated by impingement of wake from moving blade on object downstream, eg stator; in gas turbine hundreds of such interactions can generate noises at blade-passing frequency of each stage of blading, plus harmonics.

wake separation   Divergence of wakes (2) behind large aircraft.

wake turbulence   Turbulence due to wakes (2), behind large aircraft with powerful downward motion.

wake turbulence classification   Applies to conventional aeroplanes: light aircraft $\leqslant 7$ t (15,432 lb); medium = 7–136 t (c 300,000 lb); heavy = 136+ t (ICAO). The UK has a definition 'small aircraft' = 17–40 t.

wake vortex   Wake (2).

walkaround oxygen   Bottle carried by aircrew when disconnected from system.

walkback   Return of deck arrester wire (pendant) to 'cocked' or ready position.

walking   *1* To advance engine throttles (power levers) in asymmetric steps, left/right/left etc.

2 To keep straight on takeoff with violent left/right rudder.

walking beam   Pivoted beam transmitting force or power, eg to retract landing gear; secondary meaning arising by chance from repeated usage is landing-gear retraction and bracing beam whose upper end is not pivoted direct to airframe but to a sliding or translating member, eg long jackscrew.

walking in/out   Movement of aerostat, esp. airship, in or out of hangar by large ground crew walking while holding tethering lines.

walking-stick   *1* Laser used to indicate ground clearance in descent through cloud with ceiling close to terrain (not colloq.; name derives from stick of blind person).

2 Gas-turbine vaporizing burner in which fuel is heated in tube with 180° bend looking like top of *.

walk-on service   Airline service in which anyone could walk directly on board, buying ticket from cabin crew [not after 9-11].

walkway   Catwalk or other narrow structure provided in airship (rarely, large aircraft of other type) to provide human access to other part; hence * girder.

wall   *1* Sides of tyre, between wheel and tread.

2 Internal boundaries of wind tunnel; not only vertical sides.

wall constraint   Distortion of flow round model in tunnel with closed working section arising from fact that walls are seldom aligned with streamlines.

wall energy   Energy per unit area of boundary between oppositely oriented magnetic domains.

Wallops Flight Facility   NASA island complex on US east coast [Hampton, VA] used for nearly all small rocket launches.

wallowing   Uncommanded motions about all three axes.

Walter   Emergency locator beacon emitting battery-powered pulses.

WAM   Window addressable memory.

WAMRS   Water-activated mask-release system.

Wams, WAMS   *1* Weapon-aiming mode selector.

2 Women Aircraft Mechanics Service (1942 –) (US).

WAN   Wide-area network (distributed systems).

wander   See *apparent precession*.

wandering   Slow and apparently steady uncommanded change in heading; may from time to time reverse itself.

Wanganui   Blind skymarking of target obscured by cloud: Main Force set zero wind and on a given heading bombed on parachute flare (RAF WW2).

WAP   Wireless application protocol [mobile telephones].

WAPS, Waps   *1* Weighted airman promotion system (US).

2 Wide-area precision surveillance.

3 Whole-airframe parachute system.

WAR   Warning.

War Air Service Program   American (DoD) plan for civil airline routes, equipment and services following withdrawal of CRAF aircraft.

warbird   Historic military aircraft, real or replica.

WARC   World Administrative Radio Conference [suffix MOB added mobile service, ST space telecommunications, and two numbers indicated last two digits of year, eg 1992] (ITU).

war consumables   All essential expendables directly related to hardware of a weapon/support system or combat/support activity (USAF).

War Executive   Officer in squadron responsible for detailed planning of all combat missions (RAF).

war game   Simulation, by whatever means, of warfare using rules intended to depict real life.

war gas   Chemical agent designed for use against human body directly; eg not used to contaminate water supply. Liquid, solid or vapour.

warhead   Portion of munition containing HE, nuclear, thermonuclear, CBR (2) or inert materials intended to inflict damage. DoD recognizes additional term, * section, as assembled * including appropriate skin sections and related components; thus * alone need not include fuzing, arming, safety and other subsystems.

Warhorse, War Horse   Wide-angle reconnaissance hyperspectral overhead real-time surveillance experiment.

warhud   Wide-angle raster HUD.

warload   Generally taken to include expendable weapons, external fuel and external EW.

Warloc   W-band advanced radar for low-observables control.

Warlord   War Executive (colloq.).

Warmaps   Wartime Manpower Planning System (US).

war materiel   All purchasable items required to support US and Allied forces after M-day (US).

warm front   Locus of points along Earth's surface at which advancing warm air leaves surface and rises over cold air.

warm gas thruster   Propulsive jet composed of HP stored gas heated (eg by main rocket combustion) before expulsion through separate nozzle; various configurations but common for vehicle roll control.

warm sector   Portion of depression, esp. recently formed, occupied by warm air.

warm-up   Generalized term for process, or necessary elapsed time, in which device is operated solely for purpose of bringing it to steady-state operating condition, with steady running speed, temperature, pressure or other variables. Examples: piston engine, gyro.

warm-up time   Published time for device, eg gyro, to reach specified performance from moment of energization.

Warn   Weather-analysis radar network.

warned exposed   Friendly forces are lying prone with all skin covered and wearing at least two-layer summer uniform.

warned protected   Friendly forces are in armoured vehicles or crouched in holes with improvised overhead shielding.

warning area   Airspace over international waters in which, because of military exercise, non-participating aircraft may be at risk.

warning indicator   Device intended to give visual or aural warning of hazard, eg fault condition or hostile attack.

warning in/out   Two books in which members record details of temporary duty elsewhere, leave and other absence (RAF Officers' Mess).

warning net   Designated telecom system for disseminating information on enemy activity to all commands.

warning order   Preliminary notice of friendly action to follow.

warning panel   Area of display, eg on aircraft flight deck, containing numerous designated warning indicators or captions, with or without attention-getting master *.

warning receiver   Passive EM receiver with primary function of warning user his unit/vehicle/location is being illuminated by an EM signal of interest.

warning red   Attack by hostile aircraft/missiles imminent or in progress.

warning streamer   Brightly coloured fabric strip, flexible in light breeze, drawing attention to protective cover or other item which must be removed from vehicle before flight or launch.

Warp   Weather and radar processor, a 1990s development (FAA).

warp   Threads in fabric parallel to selvage, continuous full length of material.

warping   Twisting wings asymmetrically to obtain lateral stability and control; usually imposed by diagonal downward pull on rear of wing near tip to twist (wash-in) and increase camber.

Warps   Wing air-refuelling pod system.

warp-sheet   Standard raw-material form of reinforcing fibre, esp. carbon/graphite, in which broad sheet is made up entirely of parallel fibres; usually used cut to shape in multiple laminate structure.

warpwheel   Pre-1914 lateral control wheel.

WARR   Wissenschaftliche Arbeitsgemeinschaft für Raketentechnik und Raumfahrt (G).

war-readiness spares kit   Prepared kit of spares and repair parts to sustain planned wartime or contingency operations of weapon system for specified period (USAF).

Warren   Structure in form of frame truss comprising upper/lower chords joined by symmetric diagonal members only; hence * bracing, * girder, * struts, * truss. * biplane in front view has only diagonal interplane struts, forming continuous zig-zag along wing.

Warren-Young   Wing of rhomboid form with sweptback front wing joined at tips to swept-forward rear wing.

war reserve   Inactive stocks, subdivided into nuclear and other, of all forms of supplies (US = materiel) drawn upon in event of war.

Warrior   Conceptual American all-arms fighting man able to project force anywhere on the globe.

WARS   West Atlantic route structure.

Warsaw Convention   Principal international agreement on carriage by air, signed 12 October 1929 and subsequently amended, notably at The Hague in 1955.

wartime rate   Maximum attainable flying rate based on seven-day week and with wartime maintenance/safety criteria.

warting   Pitting of metal surface, esp. in gas turbine, caused by combined actions of carbon and atmospheric salts plus thermal cycling.

WAS   *1* Weapon-aiming system.

    2 Wide-area search, or surveillance.

    3 Weapons avionics simulator.

WASAA   Wide-area search [and] autonomous-attack: M adds munition, MM miniature munition.

WASD   Wide-area surveillance and detection.

WASG   *1* Warranty and service guarantee.

    2 World Airline Suppliers' Guide.

wash   *1* Wake (colloq.).

    2 To play upon or around, as 'the hot jet can * tyres of aircraft astern'.

wash-in, washin   Inbuilt wing twist resulting in angle of incidence increasing towards tip.

washing fluid   For cleaning aircraft exterior, typically mains water, possibly with a little detergent; for engine compressors, distilled water plus 1–11 per cent solvent, plus [option] inhibiting oil.

washing machine   Trainer used by commanding officer or CFI and thus that in which failed pupil makes last flight (US colloq., arch.).

wash-out, washout   *1* Inbuilt wing twist resulting in angle of incidence reducing towards tips.

    2 To fail course of flight (pilot) instruction.

    3 Failed pupil pilot.

    4 Removal of particulate matter from atmosphere by rain.

    5 See next.

washout phase   Point at which flight simulator can no longer sustain sensation of acceleration.

wash primer   Self-etching primer to prepare surface of Al or Mg for subsequent priming or painting.

WASP, Wasp   *1* War Air Service Program (DoD).

2 Wide-area special [or surveillance] projectile (USAF).

3 Women Air Service Pilots (US, replaced WAFS 1942–44).

4 Weasel attack signal processor.

5 White alternate sector propeller [black/white to prevent deaths on the ground].

6 Windshear airborne sensors program (NASA/FAA).

Waspaloy Registered (Pratt & Whitney) nickel alloys for gas-turbine rotor blading and similar purposes, typically with about 19% Cr, 14% Co and also Mo, Ti, Al etc.

Wassar Wide-angle search synthetic-aperture radar.

wastage Those pupils who fail a course of instruction, esp. aircrew; hence * rate, % failing.

waste energy Energy not put to use, that in propulsive jet being $W(v_j–V)/2g$, where $(v_j–V)$ is *waste velocity*.

waste gate Controllable nozzle box for exhaust gas turbine of turbocharged piston engine; hence ** valve, which when open allows gas to bypass turbine but which gradually closes with height until at rated full-throttle or other selected height valve is closed completely.

wastes *1* Human body wastes, esp. fecal.

2 Surplus radioactive equipment and materials.

waste velocity In the propulsion of any aircraft, the difference between the speed of the propulsive jet (behind anything giving propulsion, e.g. a propeller or helicopter rotor) and the TAS, i.e. $(v_j–V)$. At the start of conventional takeoff * is 100%; it would be zero were it possible for a high-speed aircraft to leave its jet at rest with respect to surrounding air.

WAT *1* Weight/altitude/temperature; factors independent of runway which govern each takeoff and determine whether aeroplane can meet specified positive climb criteria after engine failure at V_1; pronounced 'watt', but invariably written all in capitals.

2 Western Atlantic (ICAO, RVSM).

WATA World Association of Travel Agents (Int.).

watch office Aerodrome air traffic control centre (UK 1918). Progressively replaced after c 1933 by control tower, but term still common in 1939–45.

WAT curve Graphical plot of WAT limitation for particular aircraft type; hence WAT limit, limiting value(s) of WAT at which performance is minimum for compliance with requirements.

water bag Polythene bag carrying water ballast.

water ballast Standard ballast carried by competitive sailplane.

water barrier *1* Runway overrun barrier using water as retarding material.

2 Notional barrier to prolonged spaceflight caused by fact that plants used for fresh food/oxygen continuously convert more material to water than they return in consumable form.

water bias See *sea bias*.

water bomber Aircraft designed for surface (eg forest) firefighting by dropping large masses of water; can be marine aircraft with means for quick on-water replenishment using ram inlet on planing bottom.

water cart Dispenser of water to aircraft on apron, with supplies of either or both demineralized water for engines or potable water for passengers.

water-collecting sump Low point in any system where water could collect, esp. fuel tank and tray under vapour-cycle air-conditioning coils, from which water can be extracted.

water-displacing fluid Commercial liquids (eg LPS-3) which preferentially attach themselves to metal surface in place of local droplets of moisture, thus arresting corrosion.

water equivalent depth Measure of depth of precipitation contamination on runway; WED = actual depth × density, thus 20 mm slush with SG 0.5 gives WED 10 mm. For water, *** = actual depth.

waterfall Basic model of software life cycle.

water flaps Surfaces hinged about near-vertical axis near afterbody keel of marine aircraft (esp. jet) which when under water are used differentially for steering and together for braking (usually also used as airbrakes).

water gauge Pressure expressed as height of column of water; 1 in $H_2O = 249.089 \text{ Nm}^{-2}$; 100 mm $= 980.66 \text{ Nm}^{-2}$.

water injection Injection of demineralized water, either pure or with 30–67% alcohol or (more commonly) 44–60% methanol, into cylinders of piston engine to cool charge and eliminate detonation at maximum BMEP, or into compressor inlet or combustor of gas turbine to cool air and thereby increase density and thus mass flow and power.

water jacket Container for cooling water around cylinder.

water level Generalized term in lofting and aerospace construction generally (except vehicles whose major axis is vertical, eg most space launchers) to denote measures in the vertical plane; thus ** 193 = 193 mm above aircraft reference datum for measures in vertical plane, which may be longitudinal axis OX or some other essentially horizontal reference; abb. WL. Note: in US unit is often still inches.

waterline *1* Intersection of body exterior profile and a horizontal plane; often used as synonymous with water level, thus WL 0 is lowest point of body and all subsequent slicing planes are parallel to prime longitudinal axis or other horizontal reference. Thus * view, * plot (all waterlines drawn on common axis of symmetry).

2 Any horizontal reference other than local Earth surface used in aircraft attitude instrument or HUD.

waterloop Inadvertent turn by marine aircraft on water, eg after dipping wingtip float at high speed (full ** rare because usually aircraft rolls in opposite direction through centripetal force).

water/methanol See *water injection*.

water recovery Recovery of usable water from propulsion exhaust, esp. aboard airship for use as ballast.

water resistance Drag caused by water to aircraft moving through it, made up of skin friction and wave-making.

water rudder Small surface usually hinged on centreline of marine aircraft to sternpost or rearstep heel; used for directional control on water.

waterspout Visible water-filled tornado over sea.

water suit Anti-g suit in which interlining is filled with water which automatically provides approx. required hydrostatic pressures under large normal accelerations.

water tunnel Similar to wind tunnel but using water as working fluid for large R at low V.

water twister Rotary liquid-turbine device which absorbs energy in MAG arrested landings.

WATOG, Watog World Airlines Technical Operations Glossary.

WATS Wide-area tracking system.

watt SI unit of power (not only electrical power), W = J/s. Conversion factors: hp 745.7 exactly, CV 735.499, Btu/min 17.5725, in each case to convert to * from unit stated.

watt-hour SI unit of energy = 3,600 J.

wattless power Reactive power VAR; also called wattless component.

wattmeter Instrument for measuring electrical power.

wave *1* Disturbance propagated in medium such that at any point displacement = f (time) and at any time displacement of point = f(position); any time-varying quantity that is also an f(position). This definition falls down for light and other EM radiation, which appears not to need a 'medium' for propagation (f = function of).

2 Formation of assault vehicles (land, sea or air) timed to hit hostile territory at about same time.

wave angle Angle between upstream free-stream direction and an oblique shock created by a real [large source] supersonic body, symbol α.

waveband Particular portion of EM spectrum in telecommunications frequency region assigned by national authority for specific purpose.

wave cloud Formed at crest of a lee wave.

wave crest Peak of waveform.

wave disturbance Discontinuity or distortion along a met. front.

wave drag Additional increment of aerodynamic drag caused by shockwave formation, made up of distribution of volume along length (longitudinal axis) and drag due to lift; symbol for coefficient of ** C_{DW}.

waveform Shape of a repetitive (eg sinusoidal) wave when plotted as amplitude against time-base or when displayed on CRT.

waveform generator Converts d.c. or raw a.c. into any desired waveform output, with any frequency, amplitude (both time-varying if required) or other characteristic, eg for testing airborne electronic systems.

wave front *1* Leading edge of shockwave group, or of blast wave from explosion.

2 In a repetitive wave, a surface formed by points which all have the same phase at a given time.

waveguide Conductor for EM radiation in reverse sense to normal conductor in that radiation travels through insulator (usually atmosphere) surrounded by metal walls, usually rectangular cross-section, along which waves propagate by multiple internal reflection. Rarer form is dielectric cylinder along whose outer surface EM radiation propagates.

waveguide modes See *propagation modes*.

waveguide mode suppressor Filter matching particular waveguide cross-section designed to suppress undesirable propagation modes.

wavelength Distance between successive wave crests; symbol $\lambda = v/f$ where v is velocity of EM radiation (usually close to speed of light) and f is frequency.

wavelet Small shockwave, usually present in large numbers in boundary layers and around surface of supersonic body. Sometimes called Mach wave or Mach *.

wave lift Lift on lee side of ridge or mountains.

wavemaking resistance Drag of taxiing marine aircraft caused by gross displacement of water in waves; reaches maximum at about 20–30% of unstick speed.

wave motion Oscillatory motion of particle(s) caused by passage of wave(s), usually involving little or no net trans-lation, ie particle resumes near-original position after wave has passed. Direction of ** varies with transverse waves (eg EM radiation), longitudinal waves (sound) or other forms, eg surface (water/air interface) waves.

wave number Reciprocal of wavelength $1/\lambda$ or (alternatively) $2\pi/\lambda$.

waveoff, wave-off Any landing prevented, for whatever reason, by a command from the ground terminal or carrier DLCO.

wave period Elapsed time between successive crests, 1/f, where f = frequency.

waverider Hypersonic aircraft designed to use shockwaves to increase L/D ratio.

Waves Women Accepted for Voluntary Emergency Service (USN, from 1942).

wave soaring Using wave lift.

wave trough Point of minimum, or maximum-negative, amplitude, usually half-way between crests.

waviness Surface irregularities with spacing greater than for roughness; height is mean difference between peaks and valleys and spacing is distance between peaks.

way Speed of marine aircraft relative to water surface, also called way on.

waybill Document listing description of each item of cargo, consignor, consignee, route, destination, flight number, date and other information.

Waymouth unit Capacitance-type fuel contents gauge (tradename).

waypoint *1* Predetermined and accurately known geographical position forming start or end of route segment.

2 In US, as (1) but with addition 'whose position is defined relative to a Vortac station' (FAA).

3 In military operations, a point or series of points in space to which an aircraft may be vectored.

WB *1* Weather Bureau (US, NOAA).

2 Prefix: word before . . .

W/B Weight and balance.

Wb Weber.

W-BAR Wing bar [runway lights].

WBC *1* Weight and balance computer.

2 Wideband convertor.

WBD Wideband data, or detector.

WBF Wing-borne flight (jet V/STOL).

WBG Wideband gapfiller.

WBL Wing buttock [or base] line.

WBM Weight and balance manual.

WBO Wien-bridge oscillator.

WBPT Wet-bulb potential temperature.

WBR Wideband receiver.

WBS *1* Work breakdown structure.

2 Weight and balance system.

WBSS Wideband switching system.

WBSV Wideband secure voice.

WBT *1* Web-based training.

2 Wideband transmitter.

WBTM Weather Bureau technical memoranda.

WBVTR Wideband video-tape recorder.

WC *1* Weather centre.

2 Wire-combed (runway surface).

3 Warnings and cautions (ECAM).

W/C Wavechange.

WCA Wind correction angle.

WCAN Wideband communications airborne network.

WCCS   Wireless control and communication system.

WCFB   Wide-chord fan blade.

WCG   Water-cooled garment.

WCM   Weapon control module.

WCMD   Wind-corrected munitions dispenser.

WCMS   *1* Wing-contamination [especially ice] monitoring system.

2 Weapon[s] control and management system.

WCNS   Weapon control and navigation system.

WCO   World Customs Organization (Brussels).

W/comp   Wind component.

WCP   *1* Working capital productivity.

2 Weapon[s] control panel.

3 WXR [weather radar] control panel[s].

WCQL   Worst-cycle quality level.

WCR   Weight/capacity ratio.

WCS   *1* Weapon control system.

2 Waveguide communications system.

3 Writable control store (EDP).

WCSPL   Waist-catapult safe parking line, aligned across deck midway between Cats 2 and 3 (USN).

WCTB   Wing carry-through box, joining pivots of variable-sweep aircraft.

WCTL   Worst-condition time-lag.

WD   *1* Wind direction.

2 Warning display.

3 Word or word group.

4 War Department (US, replaced 1947 by DoD).

WDA   Weather display adapter.

WDAU   Weapon-dispenser arming unit.

WDC   Weapon[s]-delivery computer.

WDD   Western Development Division (USAF 1954–62, later Samso).

WDDS   Weather-data display system.

WDEL   Weapons Development and Engineering Laboratories (US).

WDF   Water-displacing fluid.

WDI   Wind-direction indicator.

WDIP   Weapon-data input panel.

WDL   Weapon datalink; A adds archive.

WDLY   Widely.

WDM   Wave - [or wavelength-] division multiplexing; AOR adds area of responsibility.

WDNS   Weapon-delivery and navigation system.

WDS   Wavelength dispersive spectrometer.

WDSPRD   Widespread.

WDU   Wireless Development Unit (UK, WW2).

WDX   Weather-data extractor, or extraction.

WE   Weekend.

W_E   Mass of propulsion system (from weight of engine[s]).

WEA   Weather.

WEAA   Western European Airports Association (Int.).

WEAAC   West European Airport Authorities Conference (Int.).

WEAAP   West European Association for Aviation Psychology (Int.).

WEAG   Western European Armaments Group.

weakest maintained   Weakest fuel mixture at which under specified conditions maximum power can be maintained; also called WMMP.

weak link   Point at which structure, esp. hold-back tie (eg on aircraft about to be catapult-launched), is designed to break when normal operating load is applied. Catapult launch ** resists full thrust of aircraft engines but breaks when catapult thrust is added. Occasionally a safety feature fracturing only on overload.

weak mixture   Fuel/air ratio for piston engine below stoichiometric; economical for cruising but engine runs hot. Hence ** rating, maximum power permitted for specified conditions cruising with **; ** knock rating, fuel performance-number grade under economical-cruise conditions.

weak tie   Structural weak link designed to fail in normal operation (eg holdback on catapult takeoff).

weapon-aiming system   That governing launch trajectory of unguided weapon.

weapon bay   Internal compartment for carriage of weapons, esp. of varied types, eg AAMs, ASMs, NWs, free-fall bombs, guns, sensors, cruise missiles etc. If for one type of weapon preferable to be more explicit. Derived terms include ** door, ** fuel tank, ** hosereel pack.

weapon control system   Avionics and possibly other subsystems (eg optics) built into launching aircraft to manage weapons before release and release them at correct points along desired trajectories. Should not be used to mean radio command or other form of guidance system of missile.

weapon debris   Residue of NW after explosion; not usually well defined but generally means all solids (assumed recondensed from vapour) originally forming casing, fuzing and other parts, plus unexpended Pu, U-235 or other fissile material.

weapon delivery   Total action required to locate target, establish release conditions and maintain guidance to target if required (ASCC).

weaponeering   Process of determining quantity of specific weapon necessary for required degree of damage to particular (surface) target. Takes into account defences, errors, reliabilities etc.

weaponized   Modified to carry weapons (USAF UAVs).

weapon line   See **bomb line**.

weapon-replaceable assembly   Any item, not necessarily related to weapons, that can be quickly removed and replaced, such as a PCB.

weapons assignment   Process by which weapons are assigned to individual air weapons controllers for an assigned mission (DoD).

weapons of mass destruction   For arms-control purposes, strategic NW, C, B, R devices with potential of killing large numbers of people, but exclusive of delivery systems.

weapons recommendation sheet   Defines intention of attack and recommends nature of weapons, tonnage, fuzing, spacing, desired mean points of impact, intervals of reattack and expected damage.

weapons state of readiness   In DoD usage, lists of numbers of air-defence weapons and reaction times: 2 min, 5 min, 15 min, 30 min, 1 h, 3-h, and released from readiness.

weapon system   *1* A weapon and those components required for its operation (DoD, NATO). This could simply be a part of a manned aircraft, eg radar, HUD, WCS.

2 Composite of equipment, skills and techniques that form instrument of combat which usually, but not necessarily, has aerospace vehicle as its major operational element (USAF). As originally conceived in 1951, includes all type-specific GSE, training aids, publications and every other item necessary for sustained deployment.

weapon-systems physical security Concerned to protect aerospace operational resources against physical damage.

weapon/target line Sightline (straight line) from weapon to target.

weather Short-term variations in atmosphere, esp. lower atmosphere.

weather advisory Expression of hazardous weather likely to affect air traffic, not predicted when area forecast was made.

weather beam Emitted by radar operating in weather mode, conical pencil of approx 5° total angle projecting horizontally ahead (thus filling whole troposphere about 100 km ahead).

weather categories *1* Traditional **, eg US cat C (contact), N (instrument) and X (closed), common today in many countries.

2 Precise measures of DH/RVR as they affect arrivals; Cat 1, DH 60 m/200 ft or better, RVR 800 m/2,600 ft or more; Cat 2, 60–30 m/200–100 ft, 800–400 m/2,600–1,300 ft; Cat 3a, 0 along runway, 200 m/700 ft in final descent phase; Cat 3b, 0, 50 m/150 ft; Cat 3c, 0, 0, (visual taxiing impossible).

weather central Organization collecting, processing and outputting all local weather information.

weathercocking Tendency of aerodynamic vehicle to align longitudinal axis with relative wind; note that this affects pitch as well as yaw.

weathercock stability Basic directional stability of air vehicle or re-entering spacecraft; in CCV (eg modern fighter) this is degraded to ultimate degree and replaced by synethetic * applied by avionics linking sensors to flight controls.

weather forecast Prediction of weather within area, at point or along route for specified period.

weather map Shows weather prevailing, or predicted to prevail.

weather minima Worst weather under which flight operations may be conducted, subdivided into VFR and IFR; usually defined in terms of ceiling, visibility and specific hazards to flight.

weather radar Airborne radar (less often, surface radar) whose purpose is indication of weather along planned track; traditional ouput is picture of heavy precipitation, but modern * can indicate severe turbulence (in meaningful colours) even if precipitation absent.

weather reconnaissance Flight undertaken to take measurements (traditionally = thum = temp + humidity) at specified flight levels up to near aircraft ceiling; today rare but also includes all forms of weather research.

weather report *1* Broadcast * by national weather service, eg each hour.

2 An actual, transmitted by airborne flight crew.

weather satellite See *met. satellite*.

weathervane US term for weathercock (eg on building), and for weathercocking tendency of aircraft on ground to face into wind. Hence * effect of vertical tail, which progressively gives directional stability to VTOL aircraft as forward speed increases.

weathervane stability That provided in flight by fixed tail surfaces.

weave *1* To make continuous and smooth changes of direction and height while over a period following a desired track; weaving assigned to proportion of fighters escorting slower aircraft so that continuous watch could be kept astern and in other difficult areas. Hence, weaver.

2 Angular wander of spin axis, esp. of gyro, radar scanner, rotary mirror etc.

WEB Web effective burn (time).

web *1* Principal vertical member of a beam, spar or other primary structure running length of wing or fuselage, providing strength necessary to resist shear and keep upper and lower booms (chords) correct distance apart. Occasionally expanded to * member, * plate.

2 In solid-propellant rocket, distance through which propellant burning surface will advance from initial surface until * burnout as defined by two-tanget method; usually measured as linear distance perpendicular to initial surface, symbol τ_w.

3 Any material form resembling sheet, either as discrete pieces or continuous * unrolled from drum or coil; esp. sheet form of prepreg, supplied in standard widths.

web average burning-surface area Total volume of solid rocket propellant, excluding slivers, divided by web; thus has dimensions of an area, symbol A_s.

webbing Strong close-woven fabric strip produced to specified UTS, used eg for securing loose (bulk) cargo.

weber, Wb SI unit of magnetic flux; that flux which, linking circuit of one turn, produces EMF of 1 V as it is reduced to zero uniformly in 1 s. Symbol Φ. To convert from maxwell, multiply by 10^{-8}.

web fraction Web (2) divided by internal radius of motor case or chamber; symbol f, expressed as %.

web rib Rib fabricated from sheet or plate.

WEC World Energy Conference.

WECMC Wing electronic-combat management, or managers', course.

WECPNL Weighted equivalent continuous perceived noise level (see *noise*).

WED Water equivalent depth.

wedge *1* Air mass having wedge shape in plan, esp. such a mass of high pressure extending between two lows.

2 Sharp-edged essentially 2-D * forming one wall of 2-D inlet of supersonic airbreathing engine, extending ahead of inlet so that its shock may be focused on inlet lip and normally extending rearwards as a variable ramp. Hence * inlet.

3 Small * added above trailing edge of aileron [rarely, other control surface] giving blunt trailing edge; supersonic equivalent of aileron cord.

wedge aerofoil Sharp LE and blunt TE, useless except at supersonic speed where efficiency is high, so confined to use on missiles. See *parallel double-*.

WEDS Weapons effects display system.

weeds, in the *1* At lowest possible level, on TFR or manually (colloq.).

2 Location of [usually inadvertent] landing outside airfield boundary.

weeping wing Fitted with liquid-injection leading-edge de-icing.

wef With effect from.

Wefax Weather facsimile format; one selectable mode of data transmission between weather satellite and ground printout.

Weft Wings/engine(s)/fuselage/tail; most basic of mnemonics used in early aircraft-recognition instruction.

WEG Weapons Evaluation Group (USAF).

Wehnelt   Type of cathode whose emissivity is enhanced by coating of radioactive-metal oxides.

Weick   Coefficient for calculating propeller characteristic, also called speed/power coefficient, $C_s = V^5 \sqrt{\rho/PN^2}$, where V is velocity of advance, ρ is density, P is power and N is rpm.

weighing   Today almost all determination of aircraft weight is done by moving landing gears over platforms supported on load cells which measure forces by strain-gauges, whose output may be summed and displayed automatically. A few aircraft have landing-gear hydraulics giving a cockpit readout of weight and c.g. position.

weighing points   Locations, published in engineering documents and stencilled on aircraft, where jacks may be applied in weighing process.

weighing record   Hard copy updated each time aircraft is weighed; includes c.g. position.

weigh-off   Free ballooning of airship before casting off to refine trim.

weight   Force exerted on a mass by Earth's gravity; thus a figure unique to a particular location which by international agreement is any at which g (free-fall acceleration) is 9.80665 m/s. For modern aeroplane some measures include: empty, complete aircraft plus systems measured in accord with specification (eg in US military usage, MIL-STD-3374); CCDR, empty minus items listed in STD-25140A such as engine(s), starter(s), electrics and avionics where removable direct from racking, wheels/brakes/tyres/tubes; standard empty, bare aircraft plus unusable fuel, full oil and full operating fluids (thus excluding potable water); basic empty, standard empty plus optional equipment; structure, bare airframe without systems and equipment other than wing/tail movables and flight-control power units, flap actuation, landing gear and actuation, and equipped engine installation(s) minus engines; useful load, those items that when added to * empty (for transport, OEW) will add up to gross weight for design mission (transport, MTOW); operating weight (military), empty plus useful load minus expendable fuel (internal/external), ammunition and stores; operating empty (OEW), equipped empty + all consumables (fuel, lube, filled galleys and bonded stocks, toiletries etc) + removable furnishings, reading and entertainment materials, cutlery, flight and cabin crews and their baggage, ship's papers; zero-fuel (ZFW), (military) operating plus ammunition/missiles/stores, (transport) MTOW minus usable fuel; gross (military), MTOW (civil), allowable at moment of takeoff; ramp (MRW), (civil) allowable at moment of starting engines; basic flight design (military), takeoff with full internal fuel and useful load for primary mission; minimum flying gross (military), empty + minimum crew, 5% usable/unusable fuel (zero for flutter) and lube consistent with fuel; maximum design, military equivalent of MRW allowing for full internal/external fuel (in some cases extended to higher figure still after air refuelling); maximum landing (MLW, civil), figure specified for each type between ZFW and MTOW; landplane landing design gross, basic flight design gross plus empty external tanks and pylons minus 60% internal fuel; maximum landing design gross, maximum design minus dropped tanks, fuel expended in one go-around (overshoot) or 3 minutes (whichever is less) and any items routinely dropped immediately after

takeoff; bogey, also called target bogey, established 4% below specification * (in practice * tends to rise, and bogey is usually a pious hope); specification (military), that number written into original agreed specification; job-package target/bogey, series of targets for each * group parts-breakdown; current, also called current status, that representing best available information, obtained by adding to previously reported status all subsequent revisions.

weight and balance sheet   Document carried with transport (military/civil) recording distribution of weight and c.g. at takeoff and (military) landing.

weight breakdown   Subdivision of aircraft weight (usually a design gross) into broad headings: structure (itself divided into wing group, tail group, fuselage and landing gear), power-plant, equipment services, and disposable load (latter divided into fuel/consumable items and payload).

weight coefficients   Dimensionless ratios BF/TOW, BF/ZFW, RSV/LW, RSV/ZFW, TOW/LW and TOW/ZFW.

weight flow   See *mass flow*.

weight gradient   Required change in weight, eg MTOW, for unit change in temperature, usually expressed in kg/°C, in such corrections as QNH variation, air-conditioning and anti-icing.

weight in running order   Traditional measure of piston-engine weight in which radiator/coolant/pipes/controls were added, plus oil within engine, but excluding tanks/fuel/oil/reserve coolant/exhaust tailpipes/instruments.

weightless   Condition in which no observer within system can detect any gravitational acceleration; can be produced either in free fall near a massive attracting body, eg Earth satellite, or remote from any attractive body.

weight-limited   Payload that can be carried is limited by restriction on aircraft MTOW or ZFW and not by available space.

weight per horsepower   Usually incorrectly called power/weight ratio, basic measure of piston or other shaft-output engine; dry weight divided by maximum power (latter can be 2½-minute contingency).

weight per unit thrust   Dry weight (mass) of jet engine divided by a specified measure of thrust (for turbojets/turbofans usually SLS/takeoff); in SI it is not possible to divide a mass by a force, but a meaningful ratio is still obtainable provided units of both are compatible and specified.

weightshift control   Controlling aircraft [micro, hang glider, or similar] by pilot moving his/her c.g. laterally or longitudinally.

Weir tables   Azimuth diagram and tables for interpreting radio direction finding (obs.).

weld bead   Metal deposited along welded joint.

weld bonding   Combination of resistance spot-welding and adhesive bonding with properties superior to either alone.

weld continuity   Specified as tack, intermittent, continuous.

welded patch   Thin sheet-steel patch welded over local damage in steel tubular airframe.

welded wing   Pair of aircraft in unvarying side-by-side formation about 500 ft apart.

welded steel blade   Propeller blade assembled by edge-welding two shaped sheets of steel to form aerofoil, also called hollow-steel.

weld fusion zone　Width of bead.

welding　Joining metal parts by local melting, with or without addition of filler metal to increase strength of joint, using gas torch, electric arc, electrical resistance, friction, explosive, ultrasonic vibration and other methods, often with local atmosphere of inert gas. Techniques generally called diffusion bonding are closely allied but often require no heat or added metal and rely upon natural bonding of two clean surfaces in intimate contact.

welding flux　Material, eg provided as coating on welding rod, which melts and flows over joint, excluding oxygen.

welding jig　Fixture for holding parts to be welded in exact relative positions while joints are made.

welding machine　Invariably an electric machine welding workpieces by spot, roll or seam methods.

welding rod　Consumable rod of correct metal to act as joint filler which also conveys current to form arc struck against workpiece; diameter selected according to current and usually with flux coating.

well　*1* Generalized code word (including air intercept) = serviceable.

2 Internal space or compartment for retractable item such as landing gear, FLIR or radar.

Wellington boot　Radar viewing vizor.

WEM　Warning electronic module.

WEMA　Western Electronic Manufacturers Association (US).

WEP　*1* Weapon effect planning.

2 War emergency power.

WES　*1* Warning electronic system.

2 Weapons effects simulation, or system.

Westland-Irving　British name for an internally balanced flight-control surface, esp. aileron.

wet　*1* To come in contact with surface of body; hence wetted area.

2 With water injection.

3 Of station or pylon, plumbed for fuel, or carrying a tank.

4 Fuel included [hire cost per hour].

5 Structure is sealed to house fuel [see * *wing*, but adjective also applicable to fuselage, fin or horizontal tail, eg Airbuses].

wet adiabatic　See *saturated adiabatic*, *SALR*.

wet and dry bulb　See *pychrometer*.

wet assembly　Important technique for modern aerospace structures in which all primary components are not only given successive surface treatments but are put together and joined while their surfaces are still wet; eg each component would be anodised, then coated with primer and finally with Thiokol sealer, rivets or bolts also being coated with Thiokol except in case of interference-fit bolts (Taper-loks or radius-nose Hi-locks) which fill holes completely.

wet boost　Boost pressure permissible for piston engine with water injection in operation.

wet builder　Manufacturer employing wet-assembly techniques.

wet-bulb potential　Temperature air parcel would have if adiabatically cooled to 100% RH and then adiabatically brought to 1,000 mb level.

wet-bulb thermometer　Has sensitive element surrounded

by muslin kept moist by supply of water, hence reading gives indirect measure of relative humidity.

wet emplacement　Rocket test emplacement or vehicle launch pad whose flame deflector and nearby parts are cooled by deluge of water.

wet film　Film, usually large-format, used in traditional optical camera.

wet filter　Particles are retained by a liquid film on element surface.

WET FUR　Mnemonic for remembering aircraft components for recognition purposes: wings, engines, tail, fuselage, undercarriage, radiators (or radomes) (UK, WW2).

wet H-bomb　Thermonuclear device whose fusion material is liquid, cryogenic or otherwise not a dry solid.

wet layup　Fabrication of composite structure using reinforcements saturated with liquid resin.

wet lease　Hire of commercial transport from another carrier complete with crew (at least flight crew, but often not cabin crew) and in effect forming continuation of previous operation, with major servicing performed by owner, but with hirer's logo and insignia temporarily applied.

wet pad　Wet emplacement for launches.

wet point, wet pylon　Wet station.

wet rating　Power or thrust with water or water/methanol (rarely, water/alcohol) injection.

wet-run anti-icing　Surface kept continuously above temperature at which droplets freeze.

wet sensor　ASW sensor dropped or dunked into ocean, eg sonobuoy.

wet start　Faulty start of gas turbine in which unburned or burning fuel is ejected from tailpipe.

wet station　Plumbed for fuel [or other liquid] carried in external tank.

wet suit　Standard anti-exposure suit for working in sea; eg in winch rescue.

wet sump　Piston-engine sump which serves as container for entire supply of lube oil.

wet takeoff　Takeoff with water injection.

wetted area　Total area of surface of body over which fluid flow passes and on which boundary layer forms. In case of aircraft usually simplified to visible external skin, ignoring inner surfaces or air inlets, ducts, jetpipes and air-conditioning system.

wetted fibre　Fibre, eg carbon/graphite/boron/glass, coated with same resin as will be used to form matrix of finished part.

wetted surface　See *wetted area*.

wetting agent　Surface-active agent which, usually by destroying surface tension, causes liquid to spread quickly over entire surface of solid or be absorbed thereby.

wet weight　Weight of devices plus liquids normally present when operating.

wet wing　Wing whose structure forms integral fuel tank. Note: not merely a wing in which fuel tanks are housed.

wet workshop　Space workshop launched as operative rocket stage whose propellants must be consumed and removed prior to equipping as workshop in space.

WEU　*1* Assembly of Western European Union (1955 –).

2 Warning electronic unit.

WEWO, Wewo　Wing Electronic Warfare Officer (RAF).

WF US mission prefix, weather-reconnaissance fighter.

W_F Total weight of fuel.

W_f Mass flow rate of fuel.

WFA WXR [weather radar] flat-plate antenna.

WFC *1* Wallops Flight Center (NASA).

2 Wet-film camera.

3 War Finance Corporation (US, WW1).

WFD Widespread fatigue damage.

WFG Waveform generator.

WFNT Warm front.

WFO Weather forecast office.

WFOV Wide field of view.

WFP Warm front passage.

WFR Water/fuel ratio.

WFS Wide-field sensor.

WFU Withdrawn from use.

WFZ Weapons-free zone.

WG *1* Working group.

2 Water gauge.

3 Weapons-grade Pu.

Wg Maximum growth, section width of tyre [tire].

WGD Windshield (windscreen) guidance display, transport-aircraft HUD for Cat IIIB operations.

WGF Wideband gap-filler.

WGMD Wire-grid micrometeoroid detector.

WGN White gaussian noise.

WGS *1* Weapon guidance system (GPS).

2 Followed by 72 or 84, World Geodetic Survey [1972] or System [1984].

3 Wideband gapfiller satellite.

WGT Weight.

WGU Waveguide unit.

WH Hurricane advisory.

Wh *1* Watt-hour.

2 White.

WHCA White House Communications Agency.

Wheatstone bridge Circuit with known and variable resistances with which unknown resistance can be accurately measured.

wheel *1* Early aircraft commonly had such an input to lateral control system (cf ships, road vehicles); terminology is needed for all forms of stick/control column/handwheel/yoke/spectacles. No modern captain would say "take the *".

2 Complete turbine rotor, esp. single-stage or small size.

3 High-speed gyrostabilizing device (see *reaction* *, *internal* *, *gyro* *).

4 Verb, change of heading by formation of aircraft.

5 Verb, to introduce 3-D curvature to sheet (see *wheeling*).

6 Noun, a defensive formation of several aircraft circling in horizontal plane.

wheelbarrow *1* Manoeuvre for airshows in which aircraft proceeds with nosewheel[s] in contact with runway and tail high in air (only possible with certain types of aircraft).

2 To reposition tailwheel-type light aeroplane by lifting tail and walking.

wheelbarrowing Violent oscillation in pitch on ground, either because of undulating surface or harsh brake application and rebounds from nose gear (not applicable to tailwheel aircraft).

wheelbase Distance in side elevation between wheel centres of nose and main landing gears; where there are two sets of MLGs (eg 747, C-5A) measure is to point midway between mean points of contact of front and rear MLGs.

wheelbrakes Brakes acting on landing-gear wheels.

wheelcase Compartment containing train of drive gears to accessories which are usually mounted thereon, the whole tailored to fit against or around prime mover, eg turbofan. There is only a shade of difference between this and accessory gearbox.

wheel door(s) Covers landing wheels when retracted.

wheeler Tail-high landing by tailwheel aeroplane, subsequently sinking into three-point position, also called wheely or wheelie.

wheeling Sheet-metal forming (on * machine) by locally squeezing it between upper and lower rollers; workpiece hand-positioned to achieve desired 3-D curvatures.

wheel landing See *wheeler*.

wheel load Vertical force exerted by each landing wheel on ground; hence ** capacity, but this is imprecise in comparison with various soil-mechanics and civil-engineering measures, eg CBR.

wheel mode Satellite or position thereof rotates, often fairly slowly (5–30 rpm), for attitude stabilization.

wheel satellite Made in shape of wheel, normally rotating either for attitude stabilization or to impart artificial gravity to occupants.

wheel trimming By separate [usually small] wheel inceptor[s], requiring pilot to remove hand from flight control.

wheel well Compartment in which one unit of landing gear is housed when retracted.

which transponder R/T code: please state type of transponder fitted, IFF, SSR or ATCRBS (DoD).

Whidds War HQ information display and dissemination system (USA).

whifferdil Nose up 30°–60°, bank 90°, change heading 180° ending in dive.

whiffletree *1* Also rendered whippletree, originally horse-traction linkage, which equalized and distributed pull of many horses at one point; now used to distribute pull of one large hydraulic jack or other load applicator over an area of airframe via array of beams pulled at intermediate point (usually mid-point) and transmitting pull from both ends to other such beams.

2 Also used loosely for any pivoted beam or bellcrank (US).

whip One meaning is upward jump of carrier pendant or runway arrester wire after being depressed by passage across it of landing wheel.

whip aerial Flexible aerial (antenna), either quarter-wave or of arbitrary length and usually vertically polarized, projecting at about 90° from skin.

whip stall US term which appears ill-defined; existing definitions agree only on fact that * is complete, violent and involves large positive change in pitch attitude; some suggest it is entered with flick manoeuvre, others make this impossible by suggesting possible tail-slide at outset. One authority gives tail-slide as alternative meaning. Recent authorities equate * with stall turn.

whirl Rotational flow of fluid, esp. when in translational motion in duct, in which case streamlines follow spiral paths.

whirling arm Large family of instruments or installations in which object (usually under test) is whirled on end of balanced beam rotated about vertical axis; originally used for aerodynamic tests, eg of aerofoils, but now more often used to apply sustained high-g acceleration to human beings and devices, sometimes in evacuated chamber.

whirling mode   Vibratory mode of shaft in which elastic bending is suffered, giving severe out-of-straightness either at end or along length, pronounced only at certain critical rpm (critical whirling speeds).

whirl-mode flutter   Aeroelastic flutter, esp. of wing, in which input energy is derived, at least initially, from whirling of engine or propeller shaft (catastrophic on one type of turboprop transport).

whirltower   Installation for static testing of helicopter main rotor, including overspeed conditions. Peripheral net retains separated blades.

whirl velocity   Peripheral or tangential velocity, eg of flow in centrifugal compressor.

whirlwind   Small local tornado in dry air, without cloud or rain.

whirlybird   Helicopter (rarely, other rotorcraft) (colloq.).

whisker   *1* Small single crystal, usually lenticular, whose strength is very close to maximum theoretically attainable.

2 Sharpened contact pressed against semiconductor in point-contact transistor or solid-state diode; hence * resistance.

3 See next.

whiskers   Multiple weak shocks on surface of real body, through each of which flow is retarded slightly. Hence whiskering effect.

whisper-shout   A sequence of ATCRBS interrogations and suppressions of varying power levels transmitted by TCAS equipment to reduce severity of synchronous interference and multipath problems.

whistle   Annoying high-pitched note, usually slightly varying in pitch, caused by interference carrier in superhet reception.

whistler   RF signal, usually in form of falling note, generated by lightning; heard on Earth and Jupiter and in former case bounces to and fro along magnetic-field lines between N/S hemispheres.

Whitcomb body   Streamlined body added to aircraft (eg wing trailing edge) to improve Area Rule volume distribution; many other names, eg speed bump, Küchemann carrot.

White   *1* Not black, hence can be disclosed to public.

2 Friendly, as alternative to Blue.

3 Overseeing authority in war game.

white, white hot   TV/video mode giving positive (normal) picture, which in IR display renders hot as white and cold as dark.

white body   Hypothetical surface which does not absorb EM radiation at any wavelength, ie absorptivity always zero.

white level   Maximum permissible video signal, 100% + modulation or 0% −.

white noise   *1* Strictly, noise having constant energy per unit bandwidth (Hz).

2 Commonly used to mean spectrum of generally uniform level on constant-% bandwidth basis (so-called

broadband noise) without discrete-frequency components.

whiteout   *1* Loss of orientation with respect to horizon caused by overcast sky and sunlight reflecting off snow.

2 Zero visibility caused by what one authority calls ping-pong-ball snow.

White Rating   Holder of licence is authorized to land in cloud base ⩾ 400 ft, 122 m.

white room   *1* Super-clean room in which air is continuously filtered to eliminate micron-size and larger particles and special rules almost eliminate introduction of contaminants by human beings or objects (eg india-rubber, pencils, handkerchiefs, etc, prohibited).

2 Anechoic chamber.

3 Overseeing authority in war game, esp. in netcentric warfare.

White Sands   Chief USA missile range, large area of New Mexico; abb. WSMR.

white-scarf syndrome   Alleged antipathy towards UAVs by human pilots (US, esp. USAF).

white-tailed   *1* Not bearing markings of a commercial air carrier.

2 Available for wet or dry lease with peelable logo and name.

3 Completed but unsold.

4 Company demonstrator (recently including UAV).

White World   Ordinary [not black], especially in manufacturing industry.

Whitney punch   Hand-operated tool for punching holes of selected sizes in metal sheet.

Whitworth   Traditional UK screwthread with radiused crest and root and 55° angle.

whizzer   TV zoom lens.

whizzkid   Civilian analyst or adviser in DoD.

whizzo   WSO (colloq.).

WHO   World Health Organization (UN agency, HQ Switzerland).

whole-aircraft charter   Operator charters complete aircraft for one flight or for a period, in contrast to split charter.

whole-body counter   Nucleonic instrument for identifying and measuring body burden (whole-body received radiation) of human beings and other living organisms.

whole-range distance   Horizontal distance between point vertically below release point and whole-range point.

whole-range point   Point on surface vertically below aircraft at moment of impact of bomb released by it, assuming constant aircraft velocity (ASCC).

WI   *1* Welding Institute (BWRA).

2 Within.

3 Wallops Island.

WIA   *1* Wounded in action.

2 Women in Aviation (US).

WIAS   Weather information and display system.

WIC   *1* Warning information correlation.

2 Women, infants and children.

3 Weapons instructor course.

wick   *1* See *static wick*.

2 Throttle(s); eg to turn up * = increase propulsion power (colloq.). Invariably used for turbine engine(s), esp. jet(s).

WID   Width.

WIDE   Projection equipment used in simulator to

generate FOV of 150° × 40° (Redifon); WIDE II, five projectors for 200°.

Wide   Wide-angle infinity display equipment.

wide-area augmentation system   Under development from 1995 to improve GPS accuracy (FAA).

wideband amplifier   One offering uniform response over many decades of frequency.

wideband dipole   Large ratio diameter/length.

wideband ratio   Ratio of occupied frequency bandwidth to intelligence bandwidth.

wide body   See *wide-body aircraft*.

wide-body aircraft   Commercial transport with internal cabin width sufficient for normal passenger seating to be divided into three axial groups by two aisles; in practice this means not less than 4.72 m (15 ft 6 in) (B.767, narrowest **).

wide-cut   Generalized term for aviation turbine fuels assembled from wider range of hydrocarbon fractions than kerosene-type fuels; more accurate term is wide boiling-point range, for whereas little or no kerosene-type boils below 174°C, * begins to boil at 52–53° and fractions continue to boil off at up to about 220°. SG well below 0.76, compared with 0.79–0.8 for kerosene-type fuels. Widely held to be dangerous because of high volatility (arguable) but operational fuel of nearly all air forces and several airlines.

wide-deck   Lycoming term for cylinder having wide base held by hexagonal nuts on large-radius circle; no separate hold-down rings or plates.

widget   Any small cunning, fascinating device.

Widia   German range of sintered tungsten carbides with 3–13% Co.

WIDS   Weather information display system.

width   Maximum lateral dimension of lifting-body aircraft, equivalent to span.

width of sheaf   Lateral interval between centres of flak bursts or bomb impacts (DoD).

WIDU   Wireless Intelligence and Development Unit (RAF, from October 1940).

WIE   With immediate effect.

Wiedemann-Franz law   States ratio of electrical to thermal conductivity for all metals is proportional to °K (for most metals observed value is slightly higher than ** figure).

Wien bridge   Stabilized oscillator whose frequency is determined by circuits incorporating resistances, capacitances and two triodes.

Wien law   States wavelength of peak radiation from hot-body source is inversely proportional to °K.

WIFU   Weapons interface unit.

WIG   Wing in ground effect.

wigglystrip   Rolled [occasionally machined] refractory strip with endless square-corner up/down form [combustors and afterburners].

WII light   Wing ice inspection light.

Wilco   R/T code: "I will comply with your instruction".

wildfire   Forest fire out of control.

Wildhaber-Novikov   Best known form of conformal gears with mating profiles formed by convex/concave circular arcs whose centres of curvature are on or near pitch circles.

Wild Weasel   Though original name of specific programme, now generalized term for dedicated EW plat-form based on airframe of combat (eg attack or fighter) aircraft.

Williot   Diagram which graphically portrays deflections of all joints (panel points) of loaded planar truss; result contains small inaccuracies which are removed by Mohr correction diagram, result being called Williot-Mohr.

will not fire   Code sent to spotter or other requesting agency affirming that target will not be engaged by surface fire.

willy-willy   Tropical cyclone (Australia).

Wimet   British range of tungsten carbides.

WIMS, Wims   Worldwide intratheatre mobility study.

WIN   *1* WWMCCS intercomputer network.
2 Web industrial network.

winch launch   Launch of glider by winch, usually locally built on road vehicle and driven by latter's engine.

winchman   Member of aircrew of rescue helicopter in charge of winching payloads, eg rescuees, and who may hand winch to colleague and descend to organize pick-up of incapacitated rescuee from below.

winch suspension   Rigging joining kite balloon and flying cable.

wind across   Horizontal component of wind at 90° to catapult or centreline of (axial or angled) deck.

windage   Loss of rpm of rotating device caused by air drag.

windage jump   Vertical jump (up or down) of bullet trajectory caused by crosswind, eg firing laterally from bomber or gunship.

wind angle   Angle between wind direction and true course (heading), measured 000–180° L or R of course.

wind axes   *1* In UK usage, three rectilinear axes (u, v, w) with origin within aircraft, ususally at c.g., and directions each representing component of relative wind in longitudinal, lateral and vertical planes. Traditional names: lift axis, + upward; drag axis, + to rear; crosswind axis, + to left.
2 In US usage, X, Y, Z, each the exact opposite of drag, sideforce and lift and acting through c.g. to eliminate rotary and translational motions.

wind cone   See *windsock*.

wind correction angle   Difference between course and track.

wind diagram   US for triangle of velocities.

wind direction   That from which wind is coming, expressed as number from 000° to 359°.

wind down   Horizontal component of wind along axis of catapult or centreline of (axial or angled) deck.

Windee   Wind-tunnel data encoding and evaluation (EDP [1]).

wind factor   Net effect of wind on aircraft progress, expressed as ± knots (USAF).

wind-gauge sight   Drift meter; instrument which (BSI definition), by determining track on two or more courses, enables air, wind and ground speeds to be represented by vectors (arch.).

wind gradient   Rate of change of wind with unit increase in height AGL; usually factor of interest is component of wind along runway, thus direct headwind at 150 m/500 ft veering to crosswind at threshold is regarded as change equal to full wind speed even though actual wind speed does not alter. See windshear.

winding   See filament *.

WINDMG   Wind magnitude.

windmill   *1* Of inoperative engine, to be driven by propeller (piston or turboprop) or by ram airflow through it (turbojet or turbofan); hence windmilling, windmilling drag etc.

2 Small propeller-like * used on older aircraft to drive electrical generator and occasionally other machines; in principle like modern RAT but permanently operating.

3 Of free-turbine propeller, to spin idly in wind when aircraft parked (normally prevented by brake).

4 Propeller (colloq.).

windmill-brake state   Operating condition of helicopter rotor in which thrust, flow through disc and flow outside disc are all in same direction.

wind over deck   Vector sum of wind plus speed of carrier.

window   *1* Transparent area in skin for aircraft optics or IR (in latter case not apparently 'transparent').

2 WW2 code for frequency-cut metal reflective slivers, wire, foil etc that later became better known by US name chaff (ECM).

3 Launch opportunity, defined as unique and possibly brief time period in which spacecraft can be launched from its particular site and accomplish its mission; usually recurring after a matter of days to months.

4 Small band of wavelengths in EM spectrum to which Earth atmosphere is transparent; there are many such, though together they account for only small part of total spectrum, rest being blocked. Like (3) this meaning rests on * being a small transparent gap in a dark continuum.

5 Verb, to enlarge local part of drawing or graphic display, usually to show greater detail.

Windpads   Wind-profile precision air-delivery system (para-drop accuracy, USAF).

WINDR   Wind direction.

wind rose   Polar plot for fixed station showing frequency of winds and strengths over given (long) period from 000°–359°.

wind rotor   Multi-blade rotor for forced aircooling of landing-wheel brake, of which it forms part.

winds aloft   US term for upper winds.

windscreen   Windows through which pilot(s) look ahead, called windshield in US. Originally on open-cockpit aeroplanes complete assembly of frame and windows ahead of pilot's head. On modern flight deck less obvious and generally replaced by such term as flight deck windows/transparencies/fenestration, which includes side and roof (eyebrow) windows.

windscreen wiper   Term confined to mechanical devices with oscillating blades; rotary-disc and air-blast (eg for rain-shedding) excluded.

windshear   Exceptionally large local wind gradient. Originally defined as 'change of wind velocity with distance along an axis at right angles to wind direction, specified vertical or horizontal' (BSI), which is same as wind gradient. Today recognized as extremely dangerous phenomenon because encountered chiefly at low altitude (in squall or local frontal systems) in approach configuration at speed where * makes sudden and potentially disastrous difference to airspeed and thus lift. In practice pilot must take into account air movement in vertical plane (see *downburst*) because sudden encounter with downward gust is more serious than mere fall-off in headwind. Often accompanied by severe turbulence and precipitation which can make traditional ASI under-read.

windshear indicator   Modern electronic displays will show * situation well, but with traditional instruments an extra dial is avoided by adding an energy-rate pointer to VSI; this striped needle is driven by combined verticalspeed rate and rate of change of airspeed to show rate of change of aircraft energy. Pilot can readily work throttles to keep this needle coincident with VSI needle.

windshield   See *windscreen*.

windshield guidance display   Optical projector in glareshield giving HUD-type ground-roll guidance after blind (Cat IIIB) touchdown.

wind shift   Sudden change in wind direction.

wind sleeve   See next.

windsock   Traditional fabric sleeve hung from mast to give rough indication of local wind strength/direction; also called wind sleeve, wind cone.

wind star   Plot for determining wind by drawing drifts measured on two headings (US, arch.).

wind tee   White T-shaped indicator displayed in signals area to show pilots wind direction; also called wind T. See *tee*.

wind tetrahedron   US counterpart of wind T (which is also used in US): large pyramid shape indicating wind direction, rotated on pedestal.

wind tunnel   Any of family of devices in which fluid is pumped though duct to flow past object under test. Duct can be closed circuit or open at both ends. Working section, containing body under test, can be closed or open (called open jet). Fluid can be air at any temperature/pressure or various other gases or vapours. Operation can be continuous, intermittent or as brief as a millisec. Particular species is spinning tunnel.

wind-tunnel balance   Apparatus for measuring forces and moments on object tested in tunnel; originally included actual mounting for object but today usually electrical force-transducers built into sting.

wind-tunnel stability axes   Considered more helpful term than mere 'stability axes' for data from tunnels.

wind-up turn   Turn by winged aerodyne that becomes ever-tighter [rhymes with mind].

wind vane   Small pivoted blade which aligns itself with local airflow; usually drives via rotary viscous damper to one or more precision potentiometers to form a flow-angle transmitter for AOA (pivoting on horizontal axis) or yaw.

windy drill   Workshop tool driven by high-pressure air (colloquial).

wing   *1* Main supporting surface of fixed-* aerodyne; despite term 'rotating-* aircraft' not used in that context, usual word for rotating supporting aerofoils being blade. Normally taken to cover all parts within main aerofoil envelope, including all movable surfaces. Many terms, eg * jig, * structure, are judged self-explanatory.

2 Numerous types of military unit, eg in RAF basic combat administration organization comprising perhaps three squadrons, often all sharing same base, but in US usually a larger unit comprising one or more groups with support organizations, in Navy a self-contained unit for deployment of organic air power at sea (one * per carrier) or land, and in Marines those aviation elements for support of one division.

3 Extreme left or right of battlefront or aerial formation.

4 To be positioned on *, to be alongside in formation (see *wingman*).

5 To fly (colloq.).

wing area *1* Area of surface encompassed by planview outline drawn along leading and trailing edges, including all movable surfaces in cruise configuration, and including areas of fuselage, nacelles or other bodies enclosed by lines joining intersections of leading or trailing edges with such bodies. One expression is

$$S = \int_{-b/2}^{b/2} c.dy$$ where b is span, c chord at distance y across

lateral axis. Thus, pods hung below wing are excluded; wingtip tanks are normally excluded; winglets are included as they appear in plan view; in case where root meets fuselage or nacelle at sweep approaching 90° (eg with forebody strakes) most authorities include these portions and fuselage up to intersection of leading edge with fuselage. VG 'swing wing' is measured at minimum sweep. Foreplane for lifting canard is counted separately.

2 Net ** excludes projected areas of fuselage, nacelles etc. Note: US measures of gross * often omit all areas outside basic wing trapezium, such as exist if taper is greater at root.

wing arrangement *1* Basic configuration of aeroplane.

2 Height at which monoplane wing is mounted, eg low, mid etc; see *wing position*.

wing axis Locus of all aerodynamic centres.

wing bar Row of approach lights perpendicular to runway and starting beyond runway edge (Vasi).

wing bending moment Bending moment. Hence *** relief, reduction afforded by masses (fuel, engine pods) distributed across span instead of being located in or on fuselage.

wing bending torsion mode Aeroelastic deflection, sustained or flutter, of swept wing in which bending introduces twist.

wing/body fillet Fillet, possibly very large, filling in rear part of wing/body junction to prevent separation of flow and possibly reduce peak velocities and suctions.

wing box Primary structure of modern stressed-skin in which all loads are taken by cantilever beams comprising upper and lower (usually machined) skins joined to front and rear spars, plus small number of ribs (occasionally additional spar[s] within *). This strong structure, usually by far heaviest single piece of airframe, is usually sealed to form integral tank.

wing car Airship car suspended to L or R of centreline.

wing cell See *cell, cellule*.

wing chord See *chord*.

wing drag When lifting, induced plus profile drags.

wing drop Sudden loss of lift on one wing, eg near stalling AOA, causing rapid roll not recoverable by aileron.

wing fence See *fence*.

wing fillet See *fillet*; avoided if possible, but often introduced because of need to accommodate main landing gears and in some cases combined with air-conditioning ducts on underside.

wing flaps See *flap*.

wing flutter See *flutter*.

wing guns Guns mounted within or attached to wing.

wing heavy Tending to roll in one direction.

wing in ground effect Class of vehicles, arguably aircraft, supported by a wing riding close above Earth surface.

Unlike ACV (hovercraft) they rely for lift on forward speed.

wingless wonder *1* Officer without brevet (RAF colloq., derogatory).

2 Aircraft whose wings have been permanently removed, to end its days in ground test programmes [eg, catapult/arrest].

winglet *1* Upturned wingtip or added auxiliary aerofoil(s) above and/or below tip to increase efficiency of wing in cruise, usually by reducing tip vortex and thus recovering energy lost therein and improving circulation and lift of outer portion of wing.

2 Miniature wing mounted horizontally on fuselage (not at nose or tail), on interplane struts or elsewhere (eg nacelles), often not so much for lift as to carry external load or connect bracing struts or main gears to fuselage.

winglet lift This invariably means the resultant force on the winglet, which is usually perpendicular to the surface in the vertical plane and inclined forwards [giving a positive thrust component] in the horizontal plane.

wing leveller Simple single-axis autopilot with authority only in roll; often with heading lock and VOR/ADF coupler. Generally synonymous with aileron-centring device.

wing loading Gross weight or MTOW divided by wing area (1); 1 lb/sq ft = 4.88243 kg m^{-2}; reciprocal 0.204816.

wingman Second in element of two combat aircraft, esp. interceptors; term loosely applied to pilot or aircraft. Flies off wingtip of element leader except when required to perform manoeuvres, eg day/visual intercept of unidentified aircraft.

wingover US flight manoeuvre most briefly defined as climbing turn followed by diving turn; at apogee aircraft (usually trainer) is almost stalled, and rotation continues in pitch and roll so that recovery takes place at lower level by diving out on reciprocal. Almost = stall turn.

wing overhang See *overhang*.

wing panel See *panel*.

wing pivot That on which VG wing ('swing wing') is attached.

wing plan Shape of wing outline seen from above (see *plan*).

wing position Height at which wing (1) is mounted relative to fuselage, esp. as seen from front, eg low, shoulder, parasol etc.

wing profile See *profile*.

wing radiator Cooling radiator mounted on wing, esp. inside wing, fed by leading-edge inlet.

wing reactions Those forces applied to fuselage by wing.

wing rib See *rib*.

wing rider A wing walker [and more accurate term].

wing rock See *rock (2)*.

wing root Junction of wing with fuselage (not with nacelle or any other body). Some authorities include junction of wing with opposite wing, eg on centreline (upper wing of biplane). Hence ** chord, ** fillet, ** thickness.

Wings *1* Weather information and navigational graphics system (PC-based, overlays latest Wx on planned route by GA pilot).

2 Commander (Flying) on RN carrier.

wing section Appears to be synonymous with aerofoil section, wing profile; only ambiguity caused by erroneously using term to mean portion of wing, eg centre section, outer panel.

wingset   Left and right (port + starboard) wings for same aircraft.

wing setting   See *angle of incidence*.

wingside   More specific than airside: in immediate proximity of parked aircraft.

wing skid   Protective skid on underside near tip (rare since 1916).

wing skin   Usually refers to large stressed skin forming upper or lower surface of wing box; largest single piece of material in aircraft.

wing slot   Slot built into wing; extends through outer wing from lower to upper surface with profile generally similar to that left by open slat.

wing spar   Principal spanwise member of wing, in traditional or light aircraft usually isolated but in modern stressed-skin wing forming one face of wing box which itself behaves as a spar. Always extends full available depth, unlike stringers. In slender delta built perpendicular to longitudinal axis, thus in such wings there are many spars which terminate not at tip but at points along leading edge.

wing spread   See *span*.

wing strut   Primary structural links joining wings of biplane (interplane struts) or bracing high-wing monoplane diagonally to fuselage (such are actually ties in flight but struts on ground).

wing sweep   See *sweep*.

wing tanks   *1* Tanks, normally for fuel, either accommodated in wing or (integral) formed by sealing wing box.

2 Wing-mounted drop tanks where aircraft also has drop tanks elsewhere; thus ** may be dropped first to permit sweep to be increased.

wingtip   Outer extremity of wing; either extreme tip or general area.

wingtip aileron   Aileron forming entire tip of wing, either with chordwise inner end or extending inboard along trailing edge.

wingtip extension   Increase of span by adding at tip at same dihedral angle as wing.

wingtip fence   Winglet of very low aspect ratio.

wingtip flare   Pyrotechnic attached to wingtip and ignited by pilot to assist night landing (obs. technique).

wingtip float   Stabilizing float of flying boat, usually inboard from wingtip.

wingtip handler   Person walking beside sailplane being towed across airfield holding one tip so that wings are level.

wingtip rake   See *rake*.

wingtip sail   Winglet, especially prominent.

wingtip tank   External fuel tank, jettisonable or not, carried on wingtip.

wingtip vortex generator   Ducted windmill at wingtip providing shaft power. Does the opposite of generating vortex.

wingtip vortices   See *vortex*; always present off tips of conventional wings at lifting AOA, and when intense (eg tight turn) pressure at centre falls so low that moisture condenses to leave white visible trail.

wing truss   Wing plus all bracing struts and wires transmitting loads to or from fuselage.

wingunder   Bunt [suggest unpremeditated].

wing waggle   Waggle.

wing walk   Area marked in upper surface where maintenance engineers may walk in soft shoes.

wing walker   Passenger, invariably attractive girl, who seldom walks but rides standing securely attached to upper centre section of biplane at airshow.

wing warping   Lateral control by warping; called primary on lower wing of biplane, secondary on upper. Relies on wing torsional flexure (see also *warping*).

wing yawmeter   Yawmeter in form of miniature wing, aligned vertically or (for AOA) horizontally, with sensing holes at 0.15 or less chord.

WINN, Winn   Weather information network (NASA, Honeywell and others).

WINS, Wins   *1* Workshop in negotiating skills.

2 Wireless instrumentation station [sometimes WIS].

3 Wireless integrated network sensor.

WINTEM   Upper wind[s] and temperature[s].

winterization   Process of equipping aircraft for flight in Arctic-type environment; obviously includes full anti-icing and de-icing of airframe, engines, propellers and flight-deck windows and extends to landing gears, systems, fluid specifications and many items of GSE.

winter solstice   Point on ecliptic occupied by Sun at maximum southerly declination, around 22 December.

WIP   Work in progress (check if field is open).

wipe-off switch   Attached externally under belly or engine pod in most vulnerable place in belly landing; triggers safety system, eg tank inerting or fire extinguishers.

WIPO   World Intellectual Property Organization; handles copyright (eg EDP data); Berne Convention.

Wipps, WIPPS   Wideband integrated platform protection system.

Wire   Wide-field IR explorer.

wire   Misleadingly in common use for the multi-wire cables used to arrest aircraft on carriers.

wire braid   Woven covering over ignition cables to prevent escape of emissions causing radio interference.

wire bundle   Large group of electrical wires individually tagged, clipped together and then attached as a unit to structure, usually in wireway.

wired   Wind-tunnel integrated RTIC/RTOC experiments and demonstrations.

wired program   One employing wired storage; also called fixed program.

wire-drawing   *1* Manufacture of wire of required diameter by drawing through circular die.

2 Part-throttling fluid flow by passage through constriction (colloq.).

wired storage   EDP (1) storage which was originally literally wired in and could be erased only by physically removing it; today any indestructable storage (usually for ROM).

wire edge   Sharp burr along edge of sheet freshly cut by shear.

wire gauge   *1* Measure of diameter of wire or thickness of sheet; in UK by SWG measure based on Imperial (0.001–0.5 in).

2 Hand gauge for measurement of sheet thickness or wire diameter.

wire group   Several wires routed to common destination and tied by clips.

wire guidance   Command guidance of missile or other vehicle by electrical signals (bang/bang or analog) conveyed along fine wires unrolled behind vehicle and used to position surfaces governing trajectory.

wireless telegraphy Transmission of Morse by radio.

wire link Telemetry in which signals are conveyed by wire instead of by radio; also called hard-wire telemetry.

wire locking Tying a group of nuts with safety wire to prevent rotation.

wiresonde Meteorological balloon at low altitude transmitting data over fine wire(s).

wire-strike kit Measures taken to reduce lethality of impact between low-flying aircraft, usually small helicopter, and electric (eg national grid supply) cable. Simplest form is forward-sloping sharp-edged deflector. Guillotines [as in WW2] appear to be extinct.

wireway Large conduit forming secondary structure along which numerous electrical cicuits (occasionally also fluid lines) are routed; in civil aircraft under floor or behind trim.

wire-wound Constructed by wrapping ultra-high-strength (eg tungsten) wire on mandrel and bonding by plasma spray or other method. Usually final part, eg rocket case or nozzle, has several layers aligned in different directions; also called wire-wrapped.

wire-wrapped connection Connection between single electrical signal wire and terminal made not with solder but by crimp wrapping by machine designed for purpose.

wiring All internal wires in airship, divided into structural *, preserving cross-section and other dimensions, and gasbag *, distributing lift and preventing chafing.

WIS WWMCCS information system.

Wisp Waves in space plasma.

WIST Windshear and turbulence (ICAO study group).

witness *1* Hole, groove, recess or slot cut in reference plate by each tool used in complete NC program for producing a large machined workpiece; if each * is dimensionally correct this proves software and cutters and reference plate is stored as master for that program.

2 Marks on surface of structure or component showing where two surfaces rubbed, scored, impacted, jammed or became deformed prior to crash of aircraft.

WIU Weapon interface unit.

Wizard! Expression of approval (RAF WW2).

Wizzo WSO (colloq.uial).

WJAC Women's Junior Air Corps (1939–, UK, became GVC [AC]).

WJIS Wall-jet induced suckdown.

WK Week.

WK, WKN Weaken, weakening.

wksp Workshop.

WL *1* Water level, eg WL 365.8.

2 Weight-limited.

3 Will.

W_L Actual aircraft landing weight.

WL reference Horizontal axis used to define water levels, usually synonymous with static ground line (usually not synonymous with longitudinal axis).

WLC Whole-life cost[s].

WLD Welded [pipe or tube].

WLDP Warning-light display panel.

WLFL Wet landing field length.

WLG Wing-mounted landing gear.

WLOP Air-defence and aviation force (Poland).

WLR Weapon-locating radar.

WLR-Arbeitskreis Defence and aerospace trade union (G).

WLU Wireless LAN unit.

W/M, WM *1* Water/methanol.

2 Weak mixture.

3 WXR [weather radar] mount [also WM].

4 Warm.

WMA Weather-radar waveguide adapter and antenna pedestal.

WMAP Wilkinson microwave anisotropy probe.

WMC *1* Warfare management computer.

2 War Manpower Commission (US, WW2).

WMD *1* Weapons of mass destruction [-CST adds US National Guard civil support teams].

2 Walkthrough metal detector.

WM50 50/50 mix of water and methanol.

WMFNT Warm front.

WMI Weather-radar indicator mount.

WMMP Weak-mixture for maximum power.

WMO World Meteorological Organization (UN agency, 1947 –, HQ Switzerland).

W/MOD With modification of vertical profile.

WMS *1* World Magnetic Survey (ICSU).

2 Water mist system [fire protection].

3 Wide-area master station.

WMSC Weather-message switching centre; R adds replacement.

WMT Weather-radar mount.

WN Week number.

W_N The Nth root of unity used in expressing coefficients of Fourier transform in complex notation $= e^{-j2\pi/N}$.

W-N Wildhaber-Novikov.

WNA Wrought nickel alloy.

WNC Weapon[s] and navigation computer.

WND Wind.

WNS Wideband netted sensors.

WNW Wideband networking waveform[s].

WO *1* Work order.

2 Winch operator.

W/O Written off.

W_0 Gross weight.

W/o Without.

WOA Warbirds of America.

wobble plate See *swash plate*.

wobble pump Cockpit hand pump, eg for building up fuel pressure before starting piston engine.

wobbulator FM signal generator, varied at constant amplitude above/below central frequency.

WOC Wing Operations Center (US).

WOD Wind over deck (carrier).

WOFF Weight of fuel flow; usually means per unit time.

wolfram Tungsten.

Wolf trap Instrument with sample collector and numerous culture chambers for detecting life in interplanetary space.

Womble Wire-operated mobile bomb-lifting equipment.

Wong Weight on nose gear.

wooden bomb Hypothetical concept of weapon or device which has 100% reliability, infinite shelf life and requires no special storage, surveillance or handling (DoD).

wooden round Missile that fails to work when needed (colloq.).

Wood's metal Low-melting alloy (70°C), typically 50% Bi, 25% Pb, 12.5% Sn, 12.5% Cd.

woolmeter   Yawstring.

woolpack   Cumulus.

Woolworth carrier   CVE, escort carrier (WW2 colloq.).

Woomera   Location in South Australia of Weapons Research Establishment, head of missile range extending northwest to Indian Ocean.

WOP   Wet (integral-tank) outer panel.

Wop   Radio ('wireless') operator (colloq., arch.); hence * AG adds air gunner.

Word   Wind-oriented rocket deployment (Stencel seat).

word   Basic group of ordered characters, bits or digits handled in EDP (1) as one unit.

word rate   Frequency derived from elapsed time between end of one word and start of next.

words twice   Please repeat each phrase or sentence.

Worf   Window Observational Research Facility.

work   *1* Transfer of energy, defined as force multiplied by distance through which output moves.

2 Transitive verb, to use a particular ground station by airborne radio officer, hence working that station.

work function   *1* Thermodynamic: Helmholtz free energy $A = U - TS$ where U is internal energy, T $^{\circ}$K and S entropy.

2 Electronic: energy (usually measured in eV) supplied to electron at Fermi level in metal to remove it to infinite distance; governs thermionic emission.

work hardening   See *strain hardening*.

working   Productive flight by Ag-aircraft or RPH, hence * height, * speed, * endurance; last measure excludes time between base and * site.

working fluid   Fluid used as medium for transfer of energy within system or (in wind tunnel) for study of flow around body. Can be gas, vapour or liquid.

working line   On any graphical plot describing performance of a device, the line to which device is intended to perform. Prime example is gas-turbine compressor pressure ratio plotted against airflow, always a safe distance below surge line.

working load   That borne by structure or member in normal operation; in aircraft varies greatly with turbulence, manoeuvres and with aeroelastic forces, and maximum is usually taken. Hence working stress experienced.

working pressure/temperature   Value typical in normal operation.

working section   Where model or other body is placed in wind tunnel; may be of open-jet form, without bounding walls, in traditional low-speed tunnel.

working up   Process of turning newly formed squadron [or other unit] into state of complete combat-readiness.

work lights   Powerful airborne floodlights, directed forward or laterally, to enable agricultural aircraft to work in the dark.

work package   *1* Quantified series of operations required to make one part of airframe; eg one wing panel may be divided into 15 **, each of short time period and exactly costable.

2 Basic unit of manufacturing effort into which entire product (eg aeroplane) is divided for allocation of effort in agreed manner between programme partners.

work parameter   dH/T, change of enthalpy over total temperature, or $CP^{dT}/_T$, change in specific heat.

works   *1* A factory (UK, suggest colloq.).

2 A functioning mechanism (UK, colloq.).

workstation   Local interface to any major CAE system or computer network.

world fare   Averaged value(s) of IATA tariffs.

World Geographic Reference System   Better known as Georef, uniform position-fixing and designation system for control of aircraft, targeting of ICBMs and many other functions (USAF).

World Weather Watch   Scheme to share and disseminate weather-satellite information internationally; also called VMM (WMO).

Worm   Write once, read many, or multiple.

wormhole   Theoretically achievable short cut between two points [lightyears apart] by warping spacetime.

Wortmann   Family of wing profiles designed by Prof F.X. Wortmann of Stuttgart; tailored to R appropriate to sailplanes (ie small chord), with outstandingly low drag which, esp. for flapped sections, extends over wide range of C_L.

WOS   Weather observing station.

WOT   Wide-open throttle.

Wow, WOW   *1* Weight on [rarely, off] wheels, hence Wow switch.

2 Women ordnance workers (US, WW2).

WP   *1* Warsaw Pact (now defunct).

2 Working paper.

3 White phosphorus; -T adds tracer.

W$_p$ Payload.

W/P   *1* Waypoint; also WP or WPT.

2 Work in progress.

WPA   *1* Wheelchair Pilots Association (US).

2 Works Progress Administration (US 1935, for Federal relief).

3 Works Projects Administration (followed * 2 to develop US civil airfields for defense purposes, 1939–43).

wPa   Warm polar Atlantic.

WPAFB   Wright-Patterson AFB, Dayton, OH.

WPB   War Production Board (US 1942 –).

WPC   World plotting chart.

wPc   Warm polar continental.

WPFC   World Precision-Flying Championships.

WPM, wpm   Words per minute.

Wpn, wpn   Weapon.

wPp   Warm polar Pacific.

WPR   *1* Wideband programmable receiver.

2 Waypoint position report[s].

WPT   Waypoint.

WPU   Weapon programming, or processing, unit.

WR   Work request.

WRA   *1* Weapon replaceable assembly.

2 War Resources Administration (US, WW2).

WRAC   Women's Royal Army Corps (UK 1946–93).

WRAF   Women's Royal Air Force (UK, 1949–94).

WRALC   Warner-Robins Air Logistics Center (USAF).

wrap-around   Wrap-round.

wrapped connection   Not soldered but made by wire wrapping.

wrapping cord   See *serving cord*.

wrap-round boosts   Boost motors, usually four spaced at 90°, arranged around sides of body of missile or test vehicle and at burnout separated laterally.

wrap-round engine   Turboramjet comprising turbojet core surrounded by separate ramjet duct; differs from afterburning turbofan in that there is no fan, all burners are in roughly same axial plane, and in RJ mode core is shut off.

wrap-round fins Rocket or missile fins which, before launch, are recessed around the curved body; they open after launch on hinges parallel to the longitudinal axis.

wrap-round windscreen Manufactured as single blown or vacuum-formed moulding forming entire front of cabinm, or cockpit, extending from nose or engine cowl to behind front seat[s].

wrap-round wing skin Single sheet wrapped around leading edge and forming upper/lower wing surfaces back to rear spar.

WRB War Resources Board (US, WW2).

WRCP Weather-radar control panel.

WRCS Weapon-release computer set (USAF).

WRD War-reserve drop-tank (large quantities in store).

WRDA Weapons range danger area.

WRE Weapons Research Establishment (Woomera, Australia).

Wrebus WRE break-up system; radio-commanded method of explosively disintegrating errant vehicle.

wreckage field Total area of land or seabed considered to contain every part of crashed aircraft, especially one that broke up in flight.

wreckage trail Wreckage field from aircraft which broke up at high altitude.

Wren[s] See *WRNS*.

wring (wringing it) out To demonstrate one's piloting skill by flamboyant demonstration intended to push aircraft (usually training aeroplane) to limits; has overtones of poor airmanship.

wrinkle Visible buckle in skin.

wrist actimeter Gives alarm if pitch attitude changes, even slightly, with pilot dozing [interpreted as no wrist movement in 4 min]; can be set to alarm even with no aircraft rotation.

wrist pin *1* In UK: attaches radial-engine con-rod (articulated rod) to master rod; also called knuckle pin. Hence ** end of articulated rod.

2 US : attaches piston to con-rod [in UK called gudgeon pin].

write In EDP (1), to enter in memory, to record input information.

write off To damage an aircraft so severely it is not judged worth repairing.

write-off *1* An aircraft so severely damaged that repair is uneconomic.

2 Unobligated balance of funds removed from account involved.

written off Struck off charge, for reasons of obsolescence, unrepairable damage, total cannibalization or destruction in crash; does not include sale or MIA.

WRM *1* War-readiness (or reserve) materiel (USAF).

2 Warm.

WRMFNT Warm front.

WRNG Warning.

WRNS Women's Royal Naval Service.

WRO Western Range Operations; CI adds communications and information (VAFB).

wrought Generally, shaped by plastic deformation, eg by forging or other hot or cold working as distinct from casting, sintering or machining.

WRP Wing reference plane.

WRR Wide-range ramjet (from 1995, France-Russia).

WRS *1* Word-recognition system.

2 Wide-area reference station.

3 Weather Reconnaissance Squadron.

WRSK War-readiness spares (or supply) kit (USAF).

WRT Weather-radar receiver/transmitter.

WS *1* Weapon system.

2 Weather service (FAA).

3 Water service (cart).

4 Windshear (Sigmet warning).

5 Weather-advisory Sigmet.

6 Wireless School (RFC, RAF).

W/S *1* Warning/status (display).

2 Windscreen.

W$_s$ *1* Structure weight.

2 Maximum shoulder width of tyre [tire].

WS^3 See *WSSS*.

WSADS Wind-supported air, or aerial, delivery system.

WSC *1* Weapon-system controller.

2 World Space Congress (2002).

WSCP Warning-system control panel.

WSCS Weapon-system computational subsystem.

WSD Wind speed and direction (W/V is preferred).

WSDDM Weather support for deicing decision-making.

WSDL Weapon-system data-link.

WSDPS Weapon-system design and performance specification.

WSEG Weapon-systems evaluation group (DoD).

WSEP Weapon-system evaluation program (USAF).

WSF Weapon storage flight (RAF).

WSFO Weather Service Forecast Office (US NWS).

Wsg Maximum growth shoulder width of tyre.

WSHFT Windshift.

WSI *1* Windshear indicator.

2 Water-separometer index (fuel).

WSIM Water-separometer (or separation) index, modified.

WSIP Weapon[s] system improvement programme.

WSK Wire-strike kit.

WSL *1* Weapon-system level (programme lead responsibility).

2 Wide-spectrum language.

3 Workstation lift [cargo handling].

WSMC Western Space & Missile Center (Vandenberg, USAF).

WSMIS Weapon-system management information system.

WSMO Weather Service Meteorological Observatory.

WSMR White Sands Missile Range (USA, not same as WSTF).

WSMS Windshear monitor system.

WSO *1* Weapon-system operator (or weapon-systems officer).

2 Weather Service Office.

WSP *1* Weapon-system partnership.

2 Wx systems processor.

WSPO Weapon-system project (or program[me]) office.

WSPS *1* Weapon-system physics section.

2 Wire-strike protection system.

WSR *1* Weather Service Radar.

2 Wet snow on runway.

3 Warning and surveillance receiver.

WSRN   Western Satellite Research Network.
WSS   Weapon specific simulator.
WSSA   Weapon-system support activity.
WSSD   Weapon-system support development.
WSSG   Warning-system signal generator.
WSSN   Worldwide seismic sensor network.
WSSP   Weapon-system support program(me).
WSSS   Weapon storage and security system.
WST   *1* Weapon-system trainer.
   2 Weather Service advisory, convective Sigmet.
W/STEP   With step-change in altitude.
WSTF   White Sands Test Facility (NASA, Las Cruces, NM).
WSW   Windshear warning; / RG adds recovery guidance, / RGS adds system.
WT   *1* Waste-ticket (US manufacturing).
   2 Water twister.
   3 Wind tunnel.
   4 Water tank.
   5 Tornado watch.
W/T, w.t   Wireless telegraphy.
wt   Weight.
WTA   Water, turbine aircraft; demineralized water for injection with or without methanol.
wTa   Warm tropical Atlantic.
WTC   Windshield temperature controller.
WTD   Wehrtechnischen dienstelle, military test (G).
WTE   Wingtip extension.
wTg   Warm tropical Gulf [of Mexico].
WTI   Weapons and tactics instructor.
WTMD   Walk-through metal detector.
WTO   *1* World Trade Organization.
   2 Warsaw Treaty Organization.
W$_{To}$ Actual aircraft takeoff weight.
wTp   Warm tropical Pacific.
WTR   Western Test Range.
WTS   *1* Two-seat trike (microlight class).
   2 War Training Service (US, succeeded CPT Program 1942–44).
WTSPT   Waterspout.
WTSS   Wideband tactical surveillance system.
WTT   Weapons tactics, or tactical, trainer.
WTTC   World Travel and Tourism Council (Int.).

WTU   Wing-tip (electronic] unit.
W$_u$ Useful load.
WUI   Wildland/urban interface.
Würzburg   Standard German aerial-target tracking radar, 1940–45.
WV   *1* Warfighter visualization.
   2 Wave(s).
W/V   Wind velocity.
WVA   Wake-vortex avoidance [S adds system].
WVG   Wingtip vortex generator.
WVR   Within visual range.
WW   *1* Present weather (ICAO).
   2 White world.
   3 Windshear warning; / RGS adds recovery guidance system.
W/W   War weary (US, WW2, refers to hardware).
WWABNCP   Worldwide airborne national command post(s) (US).
WWACPS   Worldwide airborne command-post system (US).
WWMCCS   Worldwide military command and control system (US).
WWO   Wing Weapon Officer (RAF).
WWP   World Weather Programme.
WWS   Wild Weasel Squadron (USAF).
WWSSN   Worldwide Standardized Seismograph Network.
W$_{W+u}$ Weight of wing plus undercarriage (landing gear).
WWV   NBS exact-time service (US).
WWW   *1* World Weather Watch (WMO).
   2 World Wide Web.
W(WW)   Past weather (ICAO).
WX   *1* Weather.
   2 Airborne weather radar.
   3 Indefinite obscuration.
WX/C   Weather-radar controller.
WXI   Weather-radar indicator.
WXNIL   No significant weather.
WXP   Weather-radar panel.
WXR   Weather radar.
WXS   Weather-radar system.
WYPT   Waypoint altitude.

X

X *1* Longitudinal axis (more strictly OX); all measurements parallel to this direction, esp. force; or a distance along a streamline.

2 Generalized term for reactance.

3 US DoD aircraft designation mission prefix: research; and also modified-mission prefix: experimental.

4 JETDS code: (1) identification, recognition; (2) facsimile, TV.

5 Length of beam or other structural member.

6 IFR flight plan code; transponder, no code.

7 On request.

8 Closed or abandoned.

9 Ground-to-air visual code: require medical aid.

10 Weather: intense.

11 Sky obscured.

12 On request.

13 Prefix, piston engine with one crankshaft driven by four cylinders, or banks of cylinders in form of X when viewed from front.

14 Generalized prefix: expendable.

x *1* Generalized term for unknown quantity.

2 Horizontal axis (abscissa) of cartesian coordinates or graphical figure.

3 Any value measured parallel to (2) or any coordinate point measured along that axis.

4 Reactance (X is more common).

5 Longitudinal displacement; distances measured along OX axis.

6 Mole fraction.

X^1 Forward extent of VTO ground vortex.

x̄ *1* Position of c.g. as co-ordinate along OX axis.

2 Average of several values of x.

X-25 Istel communications network allowing dial-up access to 9.6 kb/s.

X-aerial Crossed rods, two longer (dipoles) and two shorter (director).

X-allocation First 126 paired (1–63, 64–126) DME interrogation frequencies [see X-channel].

X-axis X (1) or x (2).

X-band Former common-usage radar frequency band centred on wavelength of 3 cm, later amended to 2.73–5.77 cm (about 10.9–5.2 GHz), see *Appendix 2*.

x-bar Crossbar.

X-C Cross-country.

X-channel DME or Tacan channel associated or paired with another radio service on same frequency. There are 126, of which first 63 have ground/air 63 MHz below air/ground frequency and second 63 have ground/air 63 MHz above air/ground frequency.

X-Cty Cross-country; XC, X-C also common.

X-cut crystal Cut parallel to Z-axis, perpendicular to X-axis.

X-engine See X(13).

X-Geräte Pioneer beam-riding aid to navigation and bombing (G, WW2).

X-glider Expendable glider (ASW).

X-licences Range of licences for ground engineers, with endorsements for many disciplines and equipments (UK).

X-plates Vertically parallel deflection plates in CRT whose potential difference deflects beam horizontally and creates timebase.

X-ray Extremely short-wavelength EM radiation, with frequency higher than any other except gamma and nuclear radiations; typical wavelengths 10–1,000 pm.

X-ray analysis Based on diffraction of X-rays by crystalline solids.

X-ray astronomy Study of X-rays arriving at Earth from space (not possible at surface of Earth because of attenuation by atmosphere).

X-unit Non-SI unit of length = 10^{-11} cm = 0.1 pm = 100 fm.

X-wing Aircraft able to operate as helicopter or, with special four-blade rotor stopped with blades diagonal to airstream, as aeroplane.

XA *1* Extended architecture, giving (for example) relief from storage and I/O constraints.

2 ARINC code.

XB IATA code.

Xbar Crossbar (ICAO style).

XC Cross-country.

X$_c$ Capacitive reactance.

XCP Except.

X/C Percentage of aerofoil chord.

XCSRA Cross-Channel Special Rules Area (UK).

XCTR Exciter.

XCVR, Xcvr Transceiver.

X/d Distance along jet as multiple of diameter.

XDM Experimental development model.

XDR *1* Extended [ie, high] data-rate.

2 External data representation.

xducer Transducer.

Xe Xenon.

xenon Xe, inert gas, density 5.9 gl^{-1}, BPt $-107°$C, used in lasers and gas-discharge tubes; best fuel for ion propulsion.

xfer XFR.

XFR Transfer.

xfmr Transformer.

XFSS Auxiliary flight-service station.

XFV Exo-skeleton flying vehicle.

XG Centre of gravity (c.g. is preferred).

XGA Extra graphics array, 1,024 × 768 pixels.

XID Exchange identification.

XIPS Xenon-ion propulsion system.

X$_L$ Inductive reactance.

x$_L$ Moment of total lift of aerodyne about c.g.

XLTR Translator.

XM *1* Extra marker.

2 External master.

XMIT[S] Transmit[s].

XML Extensible markup language.

XMM X-ray multi-mirror [mission].

xmsn Tramnsmission.

xmt XMIT.

XMTR, xmtr Transmitter.

XN Runway intersection.

X$_n$ *1* Net thrust [at a specified condition].

2 The nth in-phase signal sample applied to digital filter during any one integration period.

XNG   Crossing.

XO   Executive officer (USN).

XOE   Operational electronic-warfare branch.

XPCD   Expected.

XPD   XPDR.

XPDR   ATC transponder.

xponder   XPDR.

XPR   XPDR.

XPS   X-ray photoelectron spectroscopy.

XR   *1* Code: Office of CoS Plans and Programs (AFSC).
2 Extended-range.

XRD   X-ray diffractometer or diffractometry.

XRED   X-ray energy dispersion analysis.

XRF/NAA   X-ray fluorescence neutron-activation analysis.

XRL   X-ray laser.

XRT   X-ray telescope.

XS   *1* Atmospherics (ICAO).
2 Code: SITA.

xsmn   Transmission (both radio and helicopter dynamic parts).

XSS   Expendable sonobuoy (or sonar or sonic) sensor(s).

XST   *1* Experimental survivable technology.
2 Experimental supersonic transport.

xtal   Crystal.

XTI   X/open transport interface.

XTK   Cross-track distance error.

XTP   Express transfer protocol.

XTRM   Extreme.

XUAV   Expendable UAV.

XUV   Experimental unmanned vehicle (USA).

X_{W+u}   Distance along OX axis of c.g. of wing and undercarriage (landing gear) from c.g. of complete aircraft.

XX   Heavy [stormy] weather.

XXX   CW transmission for uncertainty or alert (Int.).

Xydar   Liquid-crystal thermoplastic, good microwave transparency and often glass-reinforced (Dartco).

xylene   Family of toxic aromatic and inflammable hydrocarbons in some ways resembling benzene, with general formula C_8H_{10}, widely used as solvent; trade name Xylol.

Xylonite   Proprietary thermosetting material similar to Celluloid.

Y

Y *1* Yaw, yaw angle.

2 Lateral axis (strictly OY) or any measure or component along that axis, esp. lateral force, or cross-stream distance in wind tunnel.

3 Admittance.

4 IATA symbol, tourist class.

5 US DoD aircraft designation prefix; prototype/service-test quantity.

6 Yttrium.

7 Year (eg in FY, fiscal *).

8 Adiabatic factor.

9 Bessel function.

10 Induced pressure, eg caused by presence of jet.

11 Rules governing change from IFR to visual.

12 Other meanings include yes (affirmative) and yellow.

y *1* Vertical axis (ordinate) of cartesian coordinates or graphical figure.

2 Any value measured parallel to (1) or any co-ordinate point measured along that axis.

3 Admittance (Y is more common).

4 Perpendicular distance from extreme [surface] fibre to neutral axis.

ȳ Co-ordinate position of c.g. along OY axis (normally at or near origin).

Y-allocation Second group of 126 frequency-paired Tacan/civil DME channels.

Y-alloys British aluminium alloys originally developed by Rolls-Royce and HDA for pistons, typically with 4% Cu, 2% Ni, 1.5% Mg and other elements, which because of their retention of strength to 250–300°C were chosen as ruling materials for Concorde with names RR.58 (UK) and AU2GN (F).

Y-axis See *Y (2), y (1)*

Y-channel See *Y-allocation*.

Y-connection One end of all three coils of 3-phase electrical machine connected to common point while other ends constitute 3-phase line; alternative to delta, also called star.

Y-cut crystal Cut parallel to Z-axis, perpendicular to Y-axis; thus parallel to one face of hexagaon.

Y-duct Leads from two [usually lateral] inlets to single engine.

Y-loader Trolley with manually pumped hydraulic jack for loading bombs and other stores, picking them up from stillage.

Y-plates Horizontally parallel deflector plates whose potential difference positions electron beam vertically in CRT.

Y-scale Scale along line of principal vertical in oblique reconnaissance imagery, or along any other line which on ground area shown would be parallel to this.

Y-section Structural section resembling Y, often with flanged or beaded edges.

Y-service Electronic intelligence (UK, WW2).

Y-valve Lube-oil drain valve from dry sump.

Y-winding See *Y-connection*.

YAF Yesterday's Air Force, California (US).

YAG, Yag Yttrium aluminium garnet (laser).

Yagi aerial Directional aerial comprising driven dipole, reflector and one or more (usually linear array) parasitic dipole directors spaced at 0.15–0.25 wavelength; maximum radiation is projected parallel to long axis.

yard Traditional Imperial unit of length = 0.9144 m exactly.

yaw Rotation of aircraft about vertical (OZ) axis; positive = clockwise seen from above; symbol ψ.

yaw damper Automatic subsystem in aeroplane (usually jet) FCS which senses onset of yaw and immediately applies corrective rudder to eliminate it. Early types were called parallel because they operated whole circuit including pedals; modern series ** has no effect on FCS further forward than fin (though sensing gyro may be near c.g.) and its activity is unnoticed by pilot. Most aircraft are flyable throughout virtually whole flight envelope with ** inoperative.

yawed wing Wing proceeding obliquely to relative wind; slewed wing.

yaw guy Cable along ground under mooring airship for attachment of yaw-guy wires.

yaw-guy wires Ropes or cables dropped from bow of airship before mooring for securing to yaw guys; these stop nose from swinging.

yawhead Yaw sensor, eg angled pitots, on pivoted vane.

yawing moment Moment tending to rotate aircraft about vertical OZ axis, symbol usually $N = qS\bar{c}c_n$ where q is constant, S is wing area, $\bar{c}$ is mean chord and c_n ** coefficient; measured positive if clockwise seen from above.

yaw lines Yaw-guy wires (US).

yawmeter Instrument or sensor for detecting yaw; can be simple device, eg yaw string or two or more pitot tubes at different inclinations whose pressure difference is sensed, or electronic gyro-fed subsystem.

yaw pointing Additional flight-control mode for some modern fighters in which yaw can be controlled without changing flight trajectory by varying sideslip angle while holding zero lateral acceleration. Usually achieved by deflecting vertical canard while linked to rudder(s) and roll-control surfaces. Gives much quicker and better gun-aiming.

yaw string Crude yawmeter comprising string, wool or other filament allowed to align with relative wind, eg ahead of windscreen.

yaw vane Small vertical aerofoil on long pivoted arm giving weathercock indication of yaw.

Yb Ytterbium.

YBC Years between calibrations.

YBCO Yttrium barium copper oxide.

YbYAG Ytterbium/YAG.

YC *1* Tourist class.

2 Yaw computer.

YCV Ship class: aircraft transportation lighter (USN).

YCZ Yellow caution zone (runway lighting).

YD, Y/D Yaw damper.

yd Yards [rarely, yds].

YDA Yesterday.

YDC Yaw-damper computer.

yellow General colour for caution [in many airborne and ground systems], also called amber.

yellow arc Range on dial instrument indicating caution, or higher than normal.

yellow card *1* Formal warning to airshow participant following single serious breach of rules or potentially serious flying error.

2 Warning notice from captain to passenger[s].

yellow caution zone Region (of runway or glideslope angle) marked by yellow or amber lights.

yellow gear/stuff Aircraft GSE vehicles on airfield or carrier (service carts, tugs, dollies, handling and store-loading equipment).

yellow sector *1* Area on left of ILS centreline (on right if using back course).

2 Arc on traditional ASI which should not be entered in severe turbulence.

yield *1* Explosive power of NW, measured in TNT equivalent weights and usually given as: very low, under 1 kt; low, 1–10 kt; medium, 10–50 kt; high, 50–500 kt; very high, over 500 kt (0.5 Mt).

2 Revenue per traffic unit, eg per tonne-km, pax-mile, etc.

yield factor of safety Specified factor used in some airworthiness requirements to prevent permanent deformation of structures.

yield load Limit load × yield factor of safety.

yield point Unit tensile stress at which deformation continues (to breakage) without further increase in applied load. Measured by loading specimen to point where a permanent set occurs, typically 0.2%. This is slightly higher than elastic limit and not normally approached in practice.

yield strength Unit stress corresponding to a specified permanent elongation, for light alloys usually taken as 0.2%.

yield stress Ambiguously, stress at yield point, which may be higher than at yield strength; except in the case of ductile materials that experience strain hardening, the greatest that material can reach.

YIG Yttrium indium garnet.

YL Year lease (suffix to two digits of year).

YMS Yield management system.

Y_n The nth quadrature signal sample applied to a digital filter during any one integration period.

Yoder rolling Manufacture of complex sections by sequential precision rolling operations tailored to each section.

yoke *1* Control column of large aircraft in which roll input is by two laterally pivoted handgrips in form of a Y. Occasionally refers to a wheel or spectacles.

2 Main magnetic structure of electrical machine supporting poles and conveying flux round linkage on each side of armature.

3 Frame on which are wound CRT deflection coils, or case of high-permeability metal surrounding such coils.

4 Interconnecting cross-member or tie.

5 In particular, tie linking helicopter main-rotor blade to hub.

6 Forked mounting, eg passing both sides of nosewheel.

yokemount On centre of handwheel or spectacles (eg, for document or notepad).

YOS Years of service.

YOT "You over there," man in right-hand seat of F-111, A-6 etc (colloq.).

Young-Helmholtz Original theory of colour vision, based on receptors for red/green/blue.

Youngman flap Patented (Fairey) trailing-edge flap carried on struts below trailing edge and in addition to normal deflection also having a negative (usually –30°) setting for use as dive brake.

Young's modulus Basic measure of material strength under tension, ratio of normal stress (within limit of proportionality) to strain, ie ratio of tensile load per unit cross-section area to elongation per unit length, within elastic limit, symbol E. SI units kN/m^2 or MN/m^2.

yo-yo Family of air-combat manoeuvres in horizontal and vertical planes intended to reduce angle-off or hold nose/tail separation and thus prevent overshoot of defender's turn. Hi-speed * trades speed for height, lo-speed * opposite.

YP Yield point.

YPA Young Pilots' Association (UK).

YR Your.

yr Year.

YS Yield strength (or stress).

YSAS Yaw-stability augmentation system.

YSZ Yttria-stabilized zirconia.

YTD year to date.

YTS Youth training scheme (UK).

Y_{TS} Cross-stream distance across test section (tunnel).

ytterbium Yb, silvery metal, density 6.97, MPt 824°C, increasing use in electronics and steels.

yttrium Y, soft silvery metal, density 4.47, MPt 1,522°C, used in alloys, glasses, semiconductors and YAG, YIG.

Y2K 2000 AD.

Y_2O_3 Yttrium oxide.

yugging Uncommanded rapid unsteadiness, usually up/down and rocking movements, between aircraft in formation aerobatic team, usually caused by turbulence.

Yukawa potential Describes meson field about a nucleon.

Y_{VS} Cross-stream distance along vane set (tunnel).

Z

Z *1* Vertical (normal) axis (OZ), or any parameter measured along that axis or parallel to it, eg normal force (also called N by some authorities) or structural depth.

2 Impedance.

3 Zenith.

4 Section modulus (also called S).

5 Time suffix: GMT = Zulu.

6 US military aircraft designation prefix: obsolete (pre-1962).

7 US military aircraft modified mission designation prefix: planning (ie still a project).

8 USN aircraft modified mission suffix: administrative version (pre-1962).

9 USN aircraft designation prefix: airship (1922–62).

10 Fluid pressure-difference (dP) due to difference in level (in Earth gravity).

11 Viscosity relative to absolute viscosity of water.

12 Atomic number.

13 Reflectivity factor.

14 Tower.

15 Flight rules governing change visual to IFR.

16 Freezing rain.

17 See *Zwilling*.

z *1* Normal axis of cartesian figure (ie perpendicular to plane of paper).

2 Distance measured along normal axis (also Z).

3 Impedence (Z more common).

4 Zone.

5 Ion valency [anion negative].

z̄ Co-ordinate position of c.g. measured along OZ normal axis.

Z-axis See *Z (1)*.

Z-battery Fired unguided AA rockets (UK, WW2).

Z-beacon See *Z-marker*.

Z-correction Correction for coriolis applied by moving celestial fix or PL at 90° to track (ASCC).

Z-fibre Process for curing laminates through which are passed small-diameter pins of pultruded CF or other material.

Z-Hud, Z-HUD HUD whose main optical power is a reflector subtending <10° from optical axis, reducing aberrations, and forming exit pupil in front of pilot's eye.

Z-marker Also called station locator, former fan or cone marker filling radio-range cone of silence, typically with audible 3 MHz tone (suggest obsolete).

Z-meter Impedance meter.

Z-mill Sendzimir [steel rolling mill].

Z-optics Device in which optical path has shape of Z, as in Z-Hud.

Z-scale Scale used in calculating height of object in oblique reconnaissance image.

Z-section Structural section in general form resembling Z, often with flanged or beaded edges and all angles 90°.

Z-Stoff Aqueous solution of calcium (rarely sodium) permanganate (G, WW2).

Z-technology IR sensor staring focal-plane array using chips along Z-axis.

ZA *1* Zone aerial (antenna).

2 Zenitnaya artilleriya, AA gun (USSR, R).

3 Zinc/air [battery].

ZAB Incendiary bomb or warhead (USSR, R).

ZAD Zone Aérienne de Défense (west, NE and SE, under Cafda, F).

Zahn cup Container with calibrated hole for measurement of liquid viscosity.

Zamak Family of Zn-based die-casting and stamping alloys, some of them important for rubber-press dies.

ZAO Zakrytoe Aktsionernoye Obshchestvo, company constitution (R).

ZAP Box or dispenser of ZABs (USSR, R).

zap *1* To smash or render unserviceable (colloq.).

2 To adorn visiting friendly combat aircraft from competitor unit with unauthorized insignia or paint scheme.

3 To disable a (usually aerial) target in one pass.

Zapp flap Early form of trailing-edge flap resembling split type but with leading edge translated aft along guides and with pivoted arms holding surface at about mid-chord; thus small hinge moment or operating force.

ZAR *1* Zero-airspeed radius, at which retreating blade of helicopter is at rest with respect to local airflow.

2 Zeus acquisition radar.

ZAW Zentralausschuss der Werbewirtschaft (G).

Zbrozek corrections Refinements to alleviation factor.

ZBTA Zinc-base trial alloy; cheap material for NC reference plates.

ZCET, Z-Cet Zero CO_2 emissions.

ZD *1* Zero defects.

2 Zone de Défense (F, five departments in Pacific and Caribbean).

ZD Moment arm of total drag about c.g., ie vertical distance from drag resultant to c.g.

ZDA Zone défense aérienne (F, now RAs).

ZDS Zonal drying system.

ZEDS Zonal electrical distribution system of future CV.

Zeeman Effect in which line spectra, eg of incandescent gas, are each split up into two or more components by passage through magnetic field.

Zell, ZEL Zero-length launching, or to make such a launch; hence zelling.

Zelmal Zero-length launch and flexible-mat landing.

zener current That flowing in insulator in electrical field sufficiently intense to excite electrons direct from valence to conduction.

zener diode Si junction diode having specific peak inverse voltage, at which point it breaks down and suddenly allows flow.

zener voltage *1* Field strength needed to initiate zener current, about 10^6V/mm.

2 That associated with reverse-VA of semiconductor, more or less constant over wide range of current.

zenith Point on celestial sphere directly overhead. Strictly this is observer's *; astronomical * is where plumb-line would intersect celestial sphere, and geographical * is where line perpendicular to smooth Earth would do so. These are not all synonymous.

zenith attraction Effect of planetary gravity on free-falling body (originally, Earth on meteorite) of increasing velocity and moving radiant toward zenith.

zenith distance Angular distance from celestial body to zenith.

zenographic Of positions on surface of Jupiter, referred to planet's equator and specified reference meridian.

zeolite Natural hydrated silicate of Ca and Al.

ZEPL Self-forging fragment submunition (G).

zero defects Conceptual (sometimes attainable) goal of perfection, esp. in manufacturing industry (chiefly US).

zero-delivery pressure In delivery line of variable-displacement pump at normal operating speed with zero stroke.

zero-draft forging One without draft, ie to finish dimensions.

Zerodur Glass ceramic with very low coefficient of expansion; used for LINS.

zero-force separation Interstage separation in which all mechanical links have already been broken, including all electrical connectors, etc.

zero-fuel weight MTOW minus total usable-fuel weight; usually limiting case for wing bending moment because wings are empty and fuselage is full.

zero-g See *zero gravity*.

zero-gravity Free-fall, weightlessness; or in deep space remote from massive bodies.

zero gross gradient Altitude at which gross climb gradient (all engines operating, corrected for anti-icing) is zero.

Zerol Zero-spiral angle bevel gear.

zero lash Mechanism adjusted until there is no play or backlash, eg by using hydraulic valve lifters in piston engine.

zero-length launch Launching of vehicle from aircraft or surface launcher in such a way that it is free as soon as it begins to move; in case of aircraft rockets, hung on clips instead of being accelerated along tubes or rails; for surface launch, thrust away by rocket boost motors whose vertical component is greater than weight. Sometimes abb. to zero-launch; usually abb. Zell, * launch(ing).

zero-lifed Restored to new condition after having seen service; applies esp. to airframe or engine and follows meticulous inspection and rectification such that flight-time count may legally be restarted.

zero lift Angle of attack at which aerofoil generates neither positive nor negative lift, ie no force normal to airstream; usually an obvious negative AOA for tradi-tional cambered wing at below M_{crit} but for supersonic wing can be positive angle.

zero-lift line Drawn through trailing edge of aerofoil parallel to relative wind when lift is zero.

zero meridian Prime Meridian, 0° longitude.

zero point Location of centre of NW at moment of detonation; usually above or below ground zero.

zero-power transfer switch Automatically switches load when waveform is passing through zero.

zero-range ring Circular arc forming range origin of most weather radars, and many other polar-type displays (most unusual to have origin as point).

zero rate of climb speed TAS at which, for an established engine power [not necessarily maximum], drag reduces climb gradient to zero.

Zero Reader Pioneer (Sperry) flight instrument with cross-pointers driven by various selectable sensors to control trajectory (vertical and lateral displacement), pilot steering aircraft so that both needles cross at origin of display in method first used on ILS meter. In most respects similar to ILS meter that can accept inputs from en route navaids and attitude sensors.

zero stage Extra axial stage added on front of existing multi-stage axial compressor in new uprated version of gas-turbine engine; can be overhung ahead of front bearing and may or may not be preceded by inlet guide vanes. Hence, more rarely, zero-**, for a second additional upstream stage.

zero-thrust pitch Distance propeller advances in one revolution when operating at normal speed but moved through still air by external force so that it generates no thrust; also called exponential mean pitch.

zero-timed Engine overhauled to Service limits, less rigorous than factory-remanufactured.

zero-torque pitch Distance propeller advances in one revolution when moved through still air by external force at such a speed that its drive torque is zero; ie when wind-milling in frictionless bearings. Symbol P_a.

zero/zero landing Totally blind helicopter landing enveloped in snow.

zero/zero seat Ejection seat qualified for operation at zero height, zero airspeed; ie pilot can safely eject from parked aircraft.

zero-zero stage See *zero stage*.

Zerstörer Destroyer, large or long-range fighter (G, WW2).

ZF Zero force (separation).

z_F Moment arm about c.g. of net propulsive force; usually acts in purely vertical plane, ie in pitch.

ZFCG Zero-fuel c.g.

ZFT Zero flight time (of hardware, or pilot recurrent training or type-conversion entirely on simulator).

ZFW Zero-fuel weight.

ZG *1* Zero gravity (also Z-g).
2 Zerstörergeschwader (G, WW2).

ZGG Zero gross gradient.

ZHR Zenithal hourly rate (measure of meteor shower).

ZHUD See *Z-HUD*.

Zhukovsky theory Original (1907) description of airflow round wing in which viscous effects transmit disturbance far from solid surface and lift is direct function of airflow.

ZI Zone of the Interior (US).

zinc Zn, blue-white metal, density 6.92, MPt 907°C, cheap and important in alloys for casting (esp. die-casting) and dies, form blocks and other press tooling, see Prestal, and in surface galvanizing, and electric batteries (see next).

zinc/air Electrical battery in which KOH (potassium hydroxide) electrolyte is pumped through cell with Zn cathode (converted to oxide) and anode of porous Ni through which is pumped air (oxygen). Much higher energy density than lead/acid.

zinc chromate $ZnCrO_4$, yellow pigment used as basis for yellow-green * primer; mixed with alkyd resins to give strongly adhering anticorrosive treatment almost universal in metal aircraft construction whose chromate ions are released by moisture.

zinc sulfide ZnS, important phosphor in electronic screens and lighting.

zinger Snag (US colloq., especially in air combat).

zip fuel Exotic or high-energy fuel for airbreathing

engines, esp. ethyl diborane and other liquids based on boron compounds.

ZIPO Zone indicate [or indicating] position officer [helicopter landing].

zipped, zipped up Blast/radiation shields in place over all glazed areas of bomber after release of NW.

Zipper *1* Target CAP (combat air patrol) at dawn or dusk (DoD).

2 ZIPO.

ZIPS See *XIPS*.

zip strap Sharp-edge adhesive sealing strip covering gap or joint in LO aircraft, renewed after maintenance.

zipstring Something simple and cheap.

zirconia Zirconium oxide ZrO_2, important refractory (MPt 2,500+°C) ceramic and abrasive.

zirconium Zr, white, ductile metal, density 6.48, MPt 1,857°C; important as alloying element and in nuclear applications.

ZL Freezing drizzle.

ZLA Zero-lift angle.

ZLBH Zero-lifed bare hull.

ZLDI Zentralstelle für Luft- und Raumfahrt-dokumentation und Info (G).

ZM Z-marker, v.h.f. station location.

ZN Azimuth.

Zn Zinc.

ZNKJRK All-Japan Air Transport and Service Association.

ZnSe Zinc selenide.

Z_0 Drift (1).

ZOC Zone of convergence (Eurocontrol).

zodiac Band of celestial sphere centred on ecliptic extending 8° on either side and containing Sun, Moon and all planets used for navigation purposes (except, sometimes, Venus).

zodiacal counterglow See *gegenschein*.

zodiacal light Faint cone of light seen (esp. in tropics) pointing towards ecliptic after sunset or before sunrise.

ZOE Zone of exclusion.

ZOH Zero-order hold in FCS sampling.

ZOK Factory for experimental construction (USSR).

Zombie Soviet strategic-reconnaissance aircraft intruding [legally] into Western airspace.

ZOMP Weapon(s) of mass destruction (USSR).

zonal comfort system Use of evaporative cooling to manage moisture in passenger compartment (trademark, CTT).

zonal drying Removal of water condensate from thermal and acoustic insulation of transport aircraft (CTT).

zonal wind In N hemisphere, wind's westerly component.

Zone 1, 2 and 3 Surface skin areas whose smoothness and perfect profile are of high [1], medium [2] or low [3] sensitivity for causing aerodynamic drag.

zone *1* Administrative region of airspace, esp. controlled airspace.

2 Portion of drawing (see *zoning*).

3 Quadrant of radio range, portions of (early-type) Decca coverage and other navaid subdivisions.

4 Sector of Earth sharing common time, bounded by two standard meridians; there are 24.

5 Circular areas centred on NW explosion: I, within MSD (minimum safe distance), within which all friendly forces evacuated; II, all personnel maximum protection: III, minimum protection.

6 Regions of aircraft surface: * 3 combines thick boundary layer with modest local flow velocity; * 2 is intermediate; * 1 combines thin boundary layer with high local velocity, and is acutely sensitive to any disturbance [eg, caused by a rivet head].

7 Portion of aircraft/spacecraft with separately controllable ECS.

zone marker See *Z-marker*.

zone numbers Those locating an item on zoned drawing.

zone of intersection Portion of civil airway overlapping or lying within any other airway (US chiefly, eg CAR 60.104).

zone of protection Within cone of 45° total apex angle whose apex is top of lightning conductor (eg on airport building or tower).

zone signals Radio-range quadrant signals (see *zone [3]*).

zone time *1* Civil time of meridian passing through centre of a time zone.

2 Time kept in sea areas in 15° zone of longitude or multiple of 15° from prime meridian (ASCC).

zoning *1* Dividing large engineering drawing into numbered/lettered grid so that items can more quickly be located by assigning each a grid reference.

2 Division of parts of aircraft into *, esp. for fire-protection purposes.

zoom *1* Abnormally steep climb trading speed for height; applies chiefly to majority of aeroplanes whose T/W ratio is much less than 1 even at near SL; normally a manoeuvre in low-level display flying.

2 Optimized steep climb by high-performance jet (which at SL might have T/W greater than 1, and thus could make sustained climb at 90°) at high altitude, normally starting at maximum level Mach, trading speed for height in order to reach exceptional height far above sustained level ceiling.

3 To enlarge or reduce image of object in TV, video, camera, etc, using lens of continuously variable focal length (* lens).

4 In air-combat, unloaded low-alpha climb to gain maximum height for minimum dissipation of energy.

zoom boundary Limits of flight envelope (in which ordinate is altitude) attained by zooming (see *zoom [2]*).

zoom ceiling That attained with a zoom (2).

zoom climb Zoom (1); second word redundant.

Zorflex Actvated-charcoal cloth (CCI).

ZOS Zone of separation.

ZP Zone of protection.

ZPI Zone position indicator radar.

ZPU Hostile AAA using close-range automatic weapons (US colloq., from a Soviet 14.5 mm AA gun designation).

ZR *1* Freezing rain.

2 Zenitnaya raketnaya = SAM; K adds system, PVO has its own entry, SD adds medium-range, V troops (USSR, R).

Zr Zirconium.

ZrB_2 Zirconium boride.

ZRE Plus suffix numeral, British Mg-Zr alloys.

ZROC Zero rate of climb; hence * speed, that (on back of drag curve) at which slender delta can just hold height at full power near SL.

ZST   Zone standard time.

ZTA   Ceramic comprising tetragonal zirconia in alumina.

ZTC   Zero temperature controller.

ZTDL   Zero-thrust descent and landing.

ZTV   Zone trim valve, governs each zone (7) in controlling ECS.

Zulu, zulu   *1* Phonetic word for Z, thus * time = UTC, * flightplan = one wherein all times are UTC [previously GMT.

     2 Denotes QRA status (USAF).

     3 Airmet advisory: freezing, icing.

ZVEI   Zentralverband der Elektrotechnik-Industrie und Elektronikindustrie.

Zwilling   Twin or coupled, esp. twin-fuselage (G).

Zyglo   NDT process in which part is coated with penetrant dye, cleaned, coated with powder developer (which extracts dye from any cracks) and examined in UV when dye fluoresces.

Zytel   Nylon materials which remain flexible at extreme cryogenic temperatures (registered name).

Appendix 1:

Greek symbols

The following are some of the aerospace usages of letters of the Greek alphabet. In almost all cases the meaning is refined by a selection of suffixes or other additions.

α (alpha) *1* Angle of attack, or tilt angle of primary nozzle (α, AOA rate; α_{geo}, geometric AOA; α_o, difference in AOA between wing and horizontal tail).
2 Generalized symbol for acceleration, angular as well as linear.
3 Absorption factor.
4 Propeller or compressor axial inflow factor.
5 Decay or attenuation coefficient (radio).
6 Coefficient of linear thermal expansion.
7 Degree of dissociation.

β (beta) *1* Angle of sideslip ($\dot{\beta}$, sideslip rate).
2 Angle between fluid and tangentially moving blade; pitch angle.
3 Particular operating regime of propeller when pilot selects pitch directly (available on ground only).
4 Angle of incidence of tailplane relative to main wing.
5 Angle of yaw (airships).
6 Coefficient of cubic thermal expansion (γ preferred).
7 Luminance factor.
8 Diffuser angle.

γ (gamma) *1* Ratio of specific heats of gas (at constant pressure and constant volume).
2 Surface tension.
3 Dihedral angle.
4 Flight path (or gliding) angle, relative to horizontal: common suffixes are: c, climb; CMD, commanded by pilot; CON, commanded by an AFCS; D, d, dive or descent; DISP, displayed in cockpit; S, corrected to sea level.
5 Electrical conductivity.
6 Free-stream velocity (non-dimensional).
7 Magnetometer field strength, in MAD operations usually = 10^{-5} oersted.
8 Coefficient of cubic thermal expansion.
9 Shear strain (also ϕ).
10 Suffix TiAl, low-density titanium aluminide (<850°C).
11 Second Townsend coefficient.

Γ (capital gamma) *1* Circulation, vortex strength, rotation in fluid flow.
2 Gamma function (mathematics).
3 Momentum conservation along streamlines.

δ (delta) *1* Small increment.
2 Control-surface or jet-deflection angle (but see other letters for angles of individual surfaces). Suffix F or USB, flap deflection angle, TE flaperon.
3 Unit elongation or structural deflection.
4 Atmospheric pressure ratio, or relative pressure.

5 Slope of runway.
6 Boundary-layer thickness.
7 Blade elementary drag coefficient, suffixes A/E/F for aileron/elevator/flap.

Δ (capital delta) *1* Generalized prefix, difference, differential.
2 Deflection of panel point of truss.
3 Increment, not necessarily small.
4 Load on water, hydrostatic lift.
5 ∇, operator del (mathematics).
6 Offset from datum or desired value.
7 Suffix 3 or H, Diamond tail rotor.
8 Suffix t = time difference.

ϵ (epsilon) *1* Angle of downwash.
2 Permittivity (dielectric constant).
3 Eccentricity or (suffix) eccentric load.
4 Emissivity.
5 Strain, direct (also e).

ζ (zeta) *1* Rudder angle.
2 Damping.
3 Electrokinetic potential.

η (eta) *1* Generalized symbol for efficiency, with suffixes indicating which, eg η_a antenna aperture, η_m mechanical, η_o overall, η_p propulsive, η_r radiation, η_T thermal, or tailplane setting angle, η_v volumetric.
2 Elevator angle.
3 Ratio of wing lifts (biplane).
4 Electrolytic polarization.
5 Dynamic viscosity [μ is more common].
6 Normalized incremental coefficient of backscatter.
7 First Townsend coefficient.

θ (theta) *1* Generalized symbol for bearing, azimuth direction, angular distance, polar co-ordinate angle etc, denoted by suffixes a [azimuth], e [elevation] etc.
2 Angle of pitch (all meanings, eg of fuselage attitude, propeller or rotor blade etc); subscripts many, eg bl blade pitch angle, T, thrust-vector angle.
3 Angle of twist or torsional strain.
4 Phase displacement (also ϕ).
5 Temperature or temperature ratio.
6 Semi-vertex angle of Mach cone.
7 Spherical co-ordinate measures.

θ/R Bearing and range.

Θ (capital theta) *1* Absolute temperature, or temperature ratio.
2 Pitch attitude, usually in radians.
3 Often with subscript n, deflection of total gross thrust vector from aircraft datum.
4 Subscript a, scan width in azimuth.
5 Subscript e, scan width in elevation.

6 Subscript F, deflection of forward nozzle(s).

7 Subscript NN, null-to-null antenna beamwidth.

8 Subscript R, deflection of rear nozzles.

9 Subscript 3dB; 3dB antenna beamwidth.

10 Subscript XG, inclination of total gross thrust vector to flight path.

ι (iota)   No common meanings.

κ (kappa)   *1* Permittivity (dielectric constant) (also ϵ).

2 Compressibility.

3 Conductivity, esp. in electrochemistry.

λ (lambda)   *1* Wavelength.

2 Area ratio of wing S/b^2.

3 Damping coefficient.

4 Scale ratio (model: full size).

5 Thermal conductivity.

6 Longitude.

7 Solid rocket volumetric loading.

8 Eigenvalues, with suffixes such as ph = phugoid mode, sp = short-period.

Λ (capital lambda)   *1* Angle of sweepback, measured at 0.25 chord; if leading or trailing edge with subscripts LE, TE.

2 Permeance.

3 Ionic/molar conductance.

μ (mu)   *1* Generalized prefix, micro ($\times 10^{-6}$, one millionth), thus $\mu m = 10^{-6}$ metre.

2 Without any other symbol, micrometre or micron.

3 Coefficient of friction.

4 Permeability; μ_0, of vacuum.

5 Joule-Thomson coefficient.

6 Dynamic viscosity.

7 Aircraft velocity (non-dimensional).

8 Ratio of spans (biplane).

9 Amplification factor (radio).

10 μ_1, relative density of aircraft (aerodyne).

11 μ', specific fuel consumption of jet engine (normally air-breathing). Not to be confused with fact that sfc of shaft-drive engines is measured in $\mu g/J$ (microgrammes per joule).

12 Helicopter rotor-blade advance ratio.

13 Engine bypass ratio.

14 Subscript FORS, microfibre optic rate sensors.

15 μs, microsecond[s].

16 Refractive index.

ν (nu)   *1* Kinematic viscosity; suffix t, effective turbulent.

2 Wave, or phase, velocity.

3 Reluctance, reluctivity.

4 Number of molecules (stoichiometric); ν_c, molecular diffusivity.

5 Frequency in s^{-1}.

6 Poisson's ratio.

7 $\bar{\nu}$, wave number (also ν).

ξ (xi)   *1* Aileron angle.

2 Damping, for each mode; $\dot{\xi}$ is average.

o (omicron)   Too like o to be used separately.

π (pi)   *1* Ratio of circumference of circle to diameter, 3.141592653.

2 Solar parallax.

Π (capital pi)   *1* Generalized symbol for a product (math).

2 Often italic, osmotic pressure.

ρ (rho)   *1* Generalized symbol for density; ρ_0 air density at SL.

2 Radius of curvature.

3 Resistivity.

4 Reflection factor, reflectance.

5 Electric charge density (MHD, C/m^3).

σ (sigma)   *1* Relative density, eg to standard atmosphere, ρ/ρ_0.

2 Poisson's ratio (ν preferred).

3 Prandtl interference factor (biplane).

4 Stefan-Boltzmann constant.

5 Electrical conductivity.

6 Velocity error, eg in INS readout.

7 Standard deviation.

8 Normal stress (also f) and surface tension (γ more common).

9 Radar cross-section.

10 Subscript zero, incremental backscatter coefficient.

Σ (capital sigma)   Generalized symbol for a summation.

τ (tau)   *1* Temperature (also θ).

2 Transmission factor or coefficient.

3 Shear stress in fluid (also q).

4 Period (growth or decay), or time delay.

5 Radar pulse width [subscript comp, compressed pulse width].

υ (upsilon)   No aerospace meanings, loosely used interchangeably with v.

ϕ (phi)   *1* Generalized function symbol.

2 Angle of bank or roll.

3 Angle between flight path and local horizontal.

4 Slope of runway.

5 Rotor inflow angle.

6 Effective helix angle.

7 Latitude, terrestrial, celestial etc.

8 Shear strain (also γ, also called fluidity).

9 Polar co-ordinate angle.

10 Radiant [luminous] or magnetic flux (also capital).

11 Entropy.

12 Velocity potential.

13 Phase displacement (also θ); subscript m, phase margin.

14 Cos ϕ, electrical power factor.

Φ (capital phi)   *1* Flux, magnetic or luminous.

2 Thrust or lift augmentation ratio in ejector.

3 Phase angle between input signal and polar co-ordinates of digital filter.

4 Φ Frequency difference between input signal and digital filter.

χ (chi)   *1* Mass.

2 Magnetic susceptibility.

3 Symbol for probability.

ψ (psi)   *1* Stream function.

2 Angle of yaw.

3 Azimuth angle.

4 Electrostatic, dielectric or luminous flux.

5 Helmholtz free energy.

6 Sweep angle of fin (vertical stabilizer), often with suffix number giving % chord of point of measurement.

7 Helicopter rotor-blade azimuth position.

Ψ (capital psi)   Electric flux.

ω (omega)   *1* Frequency (suffix c, carrier; n, natural).

2 Angular velocity in rad/s.

3 $\bar{\omega}$, specific weight, esp. piston engines.

4 Vorticity.

Ω (capital omega)   *1* Ohms resistance.

2 Resultant velocity, or angular velocity.

3 Solid angle (also *ω*).

4 Angular velocity, eg of helicopter rotor.

℧   Conductance.

Appendix 2:

Electromagnetic frequency bands

For convenience, the EM frequency spectrum has been divided up into arbitrary 'bands', each of which has been alotted an identifying number (radio) or letter (radar):

Radio

The following terminology is that agreed by the CCIR: Band 4, v.l.f. (very low frequency), 3–30 kHz; Band 5, l.f., 30–300 kHz; Band 6, m.f. (medium), 300–3,000 kHz (3 MHz); Band 7, h.f. (high), 3–30 MHz; Band 8, v.h.f. (very high), 30–300 MHz; Band 9, u.h.f. (ultra-high), 300–3,000 MHz (3 GHz); Band 10, s.h.f. (super-high), 3–30 GHz; Band 11, e.h.f. (extremely high), 30–300 GHz; Band 12 (awaiting a name), 300–3,000 GHz (3 THz).

Radar

During World War 2, to assist security, radar wavebands were given arbitrary letters: P-band, 0.225–0.39 GHz; L, 0.39–1.55; S, 1.55–5.2; X, 5.2–10.9; K, 10.9–36; Q, 36–46; V, 46–56.

From 1946 several schemes proliferated. For radar, one authority divided the spectrum into convenient wavelengths, resulting in (figures rounded off): L, 1–2 GHz; S, 2–4; C, 4–8; X, 8–13 [in US sometmes 12.5]; Q (changed to Ku), 13–20; K, 20–30; Ka, 30–40.

European usage centred on: P, 80–390 MHz; L, 390–2,500 MHz (2.5 GHz); S, 2.5–4.1 GHz; C, 4.1–7.0 GHz; X, 7.0–11.5 GHz; J, 11.5–18.0 GHz; K, 18–33 GHz; O, 33–40 GHz; Q, 40–60 GHz; V, 60–90 GHz.

The ITU refined this to the following, now actually nominated for worldwide use: v.h.f., 138–144 and 216–225 MHz; u.h.f., 420–450 and 890–942 MHz; L, 1.215–1.4 GHz; S, 2.3–2.5 GHz; C, 5.25–5.925 GHz; X, 8.5–10.68 GHz; Ku, 13.4–14.0 and 15.7–17.7 GHz; K, 24.05–24.25 GHz; Ka, 33.4–36.0 GHz.

A different rationalized version has been adopted for space communications (not with GPS): L, 0.39–1.55 GHz; S, 1.55–5.2 GHz; C [overlapping S and X], 3.7–6.2 GHz; X, 5.2–10.9 GHz; Ku, 15.35–17.25 GHz; K, 10.9–36.0 GHz; Ka, 33–36 GHz.

In 1977 the US introduced a supposed definitive system covering an expanded range. This was adopted for electronic countermeasures, but the 'old' systems are still commonly used for military radar: h.f., 10–30 MHz; v.h.f., 30–100 MHz; A, 100–300 MHz; B, 300–500 MHz (0.5 GHz); C, 0.5–1 GHz; D, 1–2 GHz; E, 2–3 GHz; F, 3–4 GHz; G, 4–6 GHz; H, 6–8 GHz; I, 8-10 GHz; J, 10–20 GHz; K, 20–40 GHz; L, 40–60 GHz; M, 60–100 GHz.

This system was adopted by NATO, but with the longer wavelengths changed to: A, 0–250 MHz; B, 250–500 MHz.

Light

For completeness, at frequencies higher than those listed above, the microwaves of radar give way to IR (infra-red). From this point it is more common to cite wavelength, the reciprocal of frequency. IR covers a range of wavelengths roughly extending from $100\,\mu$ (10^{-4} m) down to $0.75\,\mu$. As wavelength is reduced further, the light becomes visible to the human eye, disappearing again into the UV (ultra-violet) at about $0.75\,\mu$.

Appendix 3:

FAI categories

For the purpose of homologating records the FAI groups all aircraft into the following categories:

A (Balloons)
AA Gas balloons,
AM Mixed (hot air plus gas),
AX Hot air. The following suffix numbers indicate envelope volume:

1 up to 250 m^3,	**2** 250–400,
3 400–600,	**4** 600–900,
5 900–1,200,	**6** 1,200–1,600,
7 1,600–2,200,	**8** 2,200–3,000,
9 3,000–4,000,	**10** 4,000–6,000,
11 6,000–9,000,	**12** 9,000–12,000,
13 12,000–16,000,	**14** 16,000–22,000,
15 22,000+.	

B (Dirigibles)
BX Hot-air airships, suffixes:

3 900–1,600 m^3,	**4** 1,600–3,000,
5 3,000–6,000,	**6** 6,000–12,000,
7 12,000–25,000,	**8** 25,000–50,000,
9 50,000–100,000,	**10** 100,000 +.

C (Aeroplanes)
Group I Piston-engined.
C-1-a/o Landplanes up to 300 kg empty weight,

C-1-a 300–500,	**b** 500–1,000,
c 1,000–1,750,	**d** 1,750–3,000,
e 3,000–6,000,	**f** 6,000–8,000.
C-2 Seaplanes.	**a** up to 600 kg,
b 600–1,200,	**c** 1,200–2,100,
d 2,100–3,400.	

C-3 Amphibians. Weight classes as C-2.

Group II Turboprops. As Group I plus

C-1-g 8,000–12,000 kg,	**h** 12,000–16,00
i 16,000–20,000,	**j** 20,000–25,000.

Group III Turbojets and turbofans. As above plus
C-1-k 25,000–35,000 kg,   **l** 35,000–45,000, **m** 45,000–55,000.

Group IV   Rocket aircraft.

D (Gliders):
D-1 Single-seat,   **D-2** Multi-seat.
DM (Motor-gliders):
DM-1 Single-seat,   **DM-2** multi-seat.

E (Rotorcraft)
E-1 Land rotorcraft. Weight categories as for Group I landplanes.
E-2 Convertiplanes, **E-3** Autogyros, **E-4** Vertiplanes.

F (Model aircraft): Various subsections.

G (Parachuting): Various subsections.

H (Jet-lift aircraft): This section is growing.

I (Man-powered aircraft)

K (Spacecraft): Various types of mission.

N (STOL aircraft): The FAI is having difficulty in drawing demarcation lines for this category.

O (Hang-gliders):
Various subsections.

P (Aerospace craft):
Category new in 1985.

R (Microlights):
Category new in 1985, various subsections.

Appendix 4:

Phonetic alphabets

Early radios suffered severely from interference, which often made messages almost impossible to understand. The meaning was greatly clarified by inventing a word to confirm each letter, as far as possible with no two words sounding similar. Even with modern clear electronic communications a phonetic alphabet is often helpful. The following alphabets are those used in English-language aviation:

UK usage 1912 to October 1942
Ack
Beer
Charlie
Dog
Emma, later Edward
Freddie
George
Harry
Ink
Johnny
King
London
Monkey
Nuts
Orange
Pip
Queen, or Queenie
Robert
Sugar
Toc
Uncle
Vic
William
X-ray
Yorker, later York
Zebra

US/UK Combined Phonetic Alphabet October 1942
Adam; from Nov 42 Able
Baker
Charlie
Dog
Easy
Fox
George
How
Item
Jig
King
Love
Mike
Negat (USAAF); from Nov 42 Nan
Oboe
Prep (USAAF); from Nov 42 Peter
Queen
Roger
Sugar
Tare
Uncle
Victor
William
X-ray
Yoke
Zed (USAAF); from Nov 42 Zebra

ICAO, from 1952
Alpha; in US often Alfa
Bravo
Coca or Coco, from 1953 Charlie
Delta
Echo
Foxtrot
Golf
Hotel
India
Juliet, in US often Juliett
Kilo
Lima
Metro, from 1953 Mike
Nectar, from 1953 November
Oscar
Papa
Quebec
Romeo
Sierra
Tango
Union, from 1953 Uniform
Victor
Whiskey
Extra or X-extra, from 1953 X-ray
Yankee
Zulu

There are many such phonetic alphabets. The following is one used by German-speakers:
Anton
Berta, or Bruno
Cäsar
Dora
Emil
Friedrich, or Fritz
Gustav
Heinrich
Ida
Josef
Kurfürst
Ludwig
Martha
Nordpol
Otto
Paula
Quelle
Richard
Siegfried
Toni
Ulrich
Viktor
Wilhelm
Xantippe
Ypern
Zeppelin

Appendix 5:

US military aircraft designations

Since 1962 all US military aircraft have been designated according to a common system which assigns a letter for the basic mission, followed after a hyphen by a number for the aircraft basic type. A simple example is B-1, signifying bomber type 1. Modifying letters are then added to give information on permanent changes to the basic mission, and occasionally a status prefix is added to show that the vehicle is 'not standard because of its test, modification, experimental, or prototype design'. Between the basic mission letter and the hyphen a further letter can be added to denote the following 'vehicle types': rotary-wing, V/STOL, glider, lighter-than-air. To the right of the number is added a series number, running consecutively from A (for the first production version) onwards, omitting I and O; the series letter is changed for each 'major modification that alters significantly the relationship of the aerospace vehicle to its non-expendable system components or changes its logistics support'. Finally, in the fullest form of each designation, a block number is added to identify identical aircraft forming one production 'block'; these numbers are usually multiples of 5, intermediate numbers then being assigned to identify later field modifications.

Status prefix

G	Permanently grounded
J	Special test (temporary)
N	Special test (permanent)
X	Experimental
Y	Prototype
Z	Planning

Modified mission

A	Attack
C	Transport
D	Director
E	Special electronic installation
F	Fighter
H	Search and rescue
K	Tanker
L	Cold weather
M	Multi-mission
O	Observation
P	Patrol
Q	Drone
R	Reconnaissance
S	Anti-submarine
T	Trainer
U	Utility
W	Weather

Basic mission

A	Attack
B	Bomber
C	Transport
E	Special electronic installation
F	Fighter
O	Observation
P	Patrol
R	Reconnaissance
S	Anti-submarine
T	Trainer
U	Utility
X	Research

Vehicle type

G	Glider
H	Helicopter
V	VTOL/STOL
Z	Lighter-than-air vehicle

As an example of how the system works, if there were a special-test version of the trainer variant of the US Marine Corps Harrier it would be the NTAV-8B (ignoring any block number) = N, special test; T, trainer; A, attack; V, V/STOL; 8, eighth V/STOL type; B, second production model.

Appendix 6:

US engine designations

Each US engine manufacturer has its own entirely individual designation system. Department of Defense designations for jet engines are governed by a common scheme, though the Navy still uses a strictly numerical sequence of Mk (Mark) numbers for its solid rocket motors. The following is the DoD scheme for jet and turbine engines:

Status prefix

J	Special test
X	Experimental
Y	Prototype

Engine category

F	Turbofan (current)
J	Turbojet
LR	Liquid rocket
RJ	Ramjet
SR	Solid rocket
T	Turboprop/turboshaft
TF	Turbofan (formerly)

Manufacturer code

A	Allison, now Rolls-Royce
AJ	Aerojet-General
GA	Garrett, now Honeywell
GE	General Electric
L	Textron Lycoming, Stratford, now Honeywell
LD	Textron Lycoming Williamsport
MA	Marquardt
NA	Rocketdyne (North American)
P	Pratt & Whitney (formerly)
PW	Pratt & Whitney (current)
RR	Rolls-Royce
T	Teledyne CAE
TC	Morton Thiokol
WR	Williams Research

Designations are completed by a suffix model number. In 1945 these began at 1 for AF numbers, using odd numbers only, and at 2 for the Navy, using even numbers. Thus the prototype C-130 had YT56-A-1 turboprops, while the first F-8 Crusader had a J57-P-12 turbojet. Today AF numbers start at 100 or 200 and Navy numbers at 400. Thus the F-15C has F100-PW-220 turbofans.

Piston engines are designated by a letter giving the geometrical configuration of the cylinders, followed by a number giving the cubic capacity (displacement, or swept volume) in cubic inches rounded off to the nearest multiple of 5. Prefix letters can then be added (if necessary in multiple) giving further information. Suffix letters (previously numbers) indicate successive models of the same basic design.

Status prefix

X	Experimental
Y	Prototype

Prefix letter

A	Aerobatic
G	Geared
H	Helicopter
I	Direct fuel injection
T	Turbosupercharged
V	Vertical mounting

Configuration letter

L	Inline (upright or inverted)
O	Horizontally opposed
R	Radial
RC	Rotating combustion (Wankel type)
V	Vee

Thus Textron Lycoming's TIGO-541-E is the fifth model in a family of opposed engines of 541.5 cu in capacity (in a new series distinguished by the number 541 from the original series rounded off to 540) with turbocharger, direct injection and geared drive.

Appendix 7:

US missile and RPV designations

US unmanned air vehicles have their own designation system. This works in the same way as that for military aircraft, though it should be noted that manned aircraft converted as remotely piloted drones or targets retain their original designation, with the drone prefix (Q) added. An exception is the Sperry (Convair) PQM–102, which is in accord with neither system.

Status prefix

C	Captive
D	Dummy
J	Special test (temporary)
M	Maintenance
N	Special test (permanent)
X	Experimental
Y	Prototype
Z	Planning

Launch environment

A	Air
B	Multiple
C	Coffin
F	Individual
G	Runway
H	Silo-stored
L	Silo-launched
M	Mobile
P	Soft pad
R	Ship
U	Underwater attack

Mission

D	Decoy
E	Special electronic installation
G	Surface attack
I	Aerial intercept
Q	Drone
T	Training
U	Underwater attack
W	Weather

Type

M	Guided missile/drone
N	Probe
R	Rocket

As an example, the original experimental Lockheed Aquila RPVs were designated XMQM-105: **X**, experimental; **M**, mobile launcher; **Q**, drone; **M**, guided (missile or drone); 105th guided type.

Appendix 8:

Joint electronics type designation system

US military electronic equipment is designated with the following series of letters and numbers reading left to right:

1 Prefix AN, indicating that the equipment is a formally designated military system.

2 Three-letter equipment indicator code. The first letter indicates the platform from which the equipment operates, the second its type and the third its function. Platform letters are: **A** Piloted aircraft, **B** Underwater mobile (submarine), **D** Pilotless carrier, **F** Fixed ground, **G** General ground use, **K** Amphibious, **M** Mobile (ground), **P** Portable, **S** Water (surface), **T** Ground-transportable, **U** General utility, **V** Vehicular (ground), **W** Water (surface and underwater combination), **Z** Piloted/pilotless airborne vehicle combination. Type indicators are: **A** Invisible light/heat radiation, **C** Carrier, **D** Radiac, **G** Telegraph or teletype, **I** Interphone/public address, **J** Electro-mechanical or inertial wire-covered, **K** Telemetry, **L** Countermeasures, **M** Meteorological, **N** Sound in air, **P** Radar, **Q** Sonar/underwater sound, **R** Radio, **S** Special or combination of types, **T** Telephone (wire), **V** Visual and visible light, **W** Armament, **X** Facsimile or television, **Y** Data-processing.

Function indicators are: **A** Attachment, **B** Bombing, **C** Communications, **D** Direction-finding, reconnaissance and/or surveillance, **E** Ejection and/or release, **G** Fire control or searchlight direction, **H** Recording and/or reproducing, **K** Computing, **M** Maintenance and/or test assembly, **N** Navigation aid, **Q** Special or combination of purposes, **R** Receiving/passive detection, **S** Detection and/or range and bearing search, **T** Transmitting, **W** Automatic flight/remote control, **X** Identification and recognition, **Y** Surveillance and control.

3 Number indicating place in the chronological sequence of all such systems to have entered service.

4 Designation modifying suffix giving additional information: **A**, **B**, **C** etc Successive major variants, **(V)** available in various configurations, **(V)**1, 2, 3 etc indicates the variant used in a particular installation, **(X)** under development, **( )** not yet formally designated.

As an example, AN/ALR-62(V)4 indicates the 62nd type of piloted-aircraft countermeasures receiver/passive-detection system; this variant, the fourth, was specific to the EF-111A electronic-warfare aircraft.

Appendix 9:

Civil aircraft registrations

AP	Pakistan	HZ	Saudi Arabia
A2	Botswana	H4	Solomon Islands
A3	Tonga	I	Italy
A4O	Oman	JA	Japan
A5	Bhutan	JY	Jordan
A6	United Arab Emirates	J2	Djibouti
A7	Qatar	J3	Grenada
A9C	Bahrain	J5	Guinea-Bissau
B	China, also Taiwan	J6	St Lucia
C, CF	Canada	J7	Dominica (not Dominican Republic)
CC	Chile	J8	St Vincent
CN	Morocco	LN	Norway
CP	Bolivia	LQ, LV	Argentina
CR, CS	Portugal	LX	Luxembourg
CS	Macau	LY	Lithuania
CU	Cuba	LZ	Bulgaria
CX	Uruguay	MT	Mongolia
C2	Nauru	N	United States of America and outlying territories
C5	The Gambia		
C6	Bahamas	OB	Peru
C9	Mozambique	OD	Lebanon, Palestine
D	Germany	OE	Austria
DQ	Fiji	OH	Finland
D2	Angola	OK	Czech Republic
D4	Cape Verde	OM	Slovakia
D6	Comoro, Republic of	OO	Belgium
EC	Spain	OY	Denmark
EI	Ireland	P	Korea, Democratic People's Republic of
EL	Liberia		
EP	Iran	PH	Netherlands, Kingdom of the
ER	Moldova	PJ	Bonaire (Netherlands Antilles)
ES	Estonia	PK	Indonesia
ET	Ethiopia	PP, PT	Brazil
EW	Belarus	PZ	Suriname
EX	Kygyzstan	P2	Papua New Guinea
EY	Tajikstan	P4	Aruba, Netherlands Caribbean
EZ	Turkmenistan	RA	Russian Federation
E3	Eritrea	RDPL	Lao, People's Democratic Republic of
F	France	RP	Philippines
F-O	French overseas departments and territories	SE	Sweden
		SP	Poland
G	United Kingdom	ST	Sudan
HA	Hungary	SU	Egypt
HB + national emblem	Switzerland	SX	Greece
		S2	Bangladesh
HC	Ecuador	S5	Slovenia
HH	Haiti	S7	Seychelles
HI	Dominican Republic	S9	São Tômé and Principe
HK	Colombia	TC	Turkey
HL	Korea, Republic of	TF	Iceland
HMAY	See MT	TG	Guatemala
HP	Panama	TI	Costa Rica
HR	Honduras	TJ	Cameroun
HS	Thailand	TL	Central African Republic

TN	Congo, People's Republic of the	ZK, ZL, ZM	New Zealand
TR	Gabon	ZP	Paraguay
TS	Tunisia	ZS, ZT, ZU	South Africa, Transkei,
TT	Chad		Boputhatswana
TU	Ivory Coast	3A	Monaco
TY	Benin	3B	Mauritius
TZ	Mali	3C	Equatorial Guinea
T2	Tuvalu	3D	Swaziland
T3	Kiribati	3X	Guinea, Republic of
T9	Bosnia-Herzegovina	4K	Azerbaijan
UK	Uzbekistan	4L	Georgia, Republic of
UN	Kazakhstan	4R	Sri Lanka
UR	Ukraine	4W	Yemen, Arab Republic of the
VH	Australia	4X	Israel
VN	Vietnam	5A	Libyan Arab Jamahiriya
VP-B	Bermuda	5B	Cyprus
VP-C	Cayman Islands	5H	Tanzania, United Republic of
VP-F	Falkland Islands	5N	Nigeria
VP-G	Gibraltar	5R	Malagasy Republic (Madagascar)
VP-LA	Leeward Islands	5T	Mauritania
VP-LP	British Virgin Islands	5U	Niger
VQ-H	St Helena	5V	Togo
VQ-T	Turks and Caicos Islands	5W	Western Samoa
VR-H, now		5X	Uganda
incorporated		5Y	Kenya
into B	Hong Kong	6O	Somalia
VT	India	6V, 6W	Senegal
V2	Antigua	6Y	Jamaica
V3	Belize	7O	Yemen, Democratic People's Republic
V4	Nevis		of the
V5	Namibia	7P	Lesotho
V6	Micronesia	7Q	Malawi
V7	Marshall Islands	7T	Algeria
V8	Brunei Darussalam	8P	Barbados
XA, XB, XC	Mexico	8Q	Maldives
XT	Burkina Faso (formerly Upper Volta).	8R	Guyana
XU	Democratic Kampuchea (Cambodia)	9A	Croatia
XV	Vietnam	9G	Ghana
XY, XZ	Myanmar	9H	Malta
YA	Afghanistan	9J	Zambia
YI	Iraq	9K	Kuwait
YJ	Vanuatu	9L	Sierra Leone
YK	Syrian Arab Republic	9M	Malaysia
YL	Latvia	9N	Nepal
YN	Nicaragua	9P	Barbados
YR	Romania	9Q	Zaïre
YS	El Salvador	9U	Burundi
YU	Yugoslavia, Serbia, Montenegro	9V	Singapore
YV	Venezuela	9XR	Rwanda
Z	Zimbabwe	9Y	Trinidad and Tobago
ZA	Albania		

Appendix 10:

NATO Reporting Names

Since WW2 there have been three successive series of 'type numbers' or 'reporting names'" assigned to Soviet aircraft by the NATO Air Standardization Co-ordinating Committee. When first assigned each new name was classified; this extended even to the suffix letters which identify important modifications of each basic design. For some reason the practice of assigning invented names has continued even though the correct designations are known. Bomber and reconnaissance names begin with B, transports with C, fighters with F, helicopters with H and other types with M. A single-syllable name (except for helicopters) denotes a propeller aircraft. Older aircraft are omitted from the following list.

Backfire	Tu-22M	**Firebar**	Yak-28P
Badger	Tu-16 (Tu-88 and Chinese H-6)	**Fishbed**	MiG-21 (single-seaters and Chinese J-7)
Beagle	Chinese H-5 (licence-built Il-28)	**Fishpot**	Su-11 (single-seater)
Bear	Tu-95, Tu-142	**Fitter**	Su-7, Su-17, Su-20 and Su-22 (including some two-seaters)
Blackjack	Tu-160	**Flagon**	Su-15 and Su-21 (including two-seaters)
Blinder	Tu-22 (Tu-105)	**Flanker**	Su-27, Su-30, Su-32, Su-33, Su-35, Su-37
Brewer	Most Yak-28 tactical versions	**Flogger**	MiG-23 and MiG-27 (including trainers)
Camber	Il-86	**Forger**	Yak-38
Candid	Il-76	**Foxbat**	MiG-25 (including reconnaissance and trainer versions)
Careless	Tu-154, Tu-164	**Foxhound**	MiG-31
Cash	An-28	**Frogfoot**	Su-25, Su-28
Clank	An-30	**Fulcrum**	MiG-29
Classic	Il-62	**Halo**	Mi-26
Cline	An-32	**Harke**	Mi-10
Clobber	Yak-42	**Havoc**	Mi-28
Coaler	An-72, An-74	**Haze**	Mi-14
Cock	An-22	**Helix**	Ka-27, Ka-28, Ka-29, Ka-31, Ka-32
Codling	Yak-40	**Hermit**	Mi-34
Coke	An-24 (Chinese Y-7)	**Hind**	Mi-24, Mi-25, Mi-35
Colt	An-2 (Chinese Y-5)	**Hip**	Mi-8, Mi-9, Mi-17
Condor	An-124	**Hokum**	Ka-50
Coot	Il-18 and variants	**Hoodlum**	Ka-26
Crate	Il-14	**Hook**	Mi-6, Mi-22
Crusty	Tu-134	**Hoplite**	Mi-2 (excluding PZL variants)
Cub	An-12 and variants	**Hormone**	Ka-25
Cuff	Be-32	**Hound**	Mi-4 (and Chinese Z-5)
Curl	An-26	**Madcap**	An-71
Farmer	MiG-19 (and Chinese J-6, including trainer)	**Maestro**	Yak-28U
Fencer	Su-24		
Fiddler	Tu-28P/128 (Tu-102)		

Mail	Be-12/M-12		**Moose**	Yak-11
Mainstay	Beriev A-50M		**Moss**	Tu-126
Mascot	Chinese HJ-5 (licence-built Il-28U)		**Moujik**	Su-7U (but not later trainer versions)
Max	Yak-18 (and most variants except Yak-50/52/55 and Chinese CJ-6)			

Chinese Aircraft   Most indigenous Chinese designs have not been assigned NATO reporting names. Two exceptions which have been published are:

May	Il-38
Midas	Il-78M
Midget	MiG-15UTI
Mongol	MiG-21U variants

Fantan	Nanchang Q-5
Finback	Shenyang J-8

Appendix 11:
Powers of 10

Y yotto	=	x 10^{24}		**d deci**	=	x 10^{-1}
Z zeta	=	x 10^{21}		**c centi**	=	x 10^{-2}
E exa	=	x 10^{18}		**m milli**	=	x 10^{-3}
P peta	=	x 10^{15}		**μ micro**	=	x 10^{-6}
T tera	=	x 10^{12}		**n nano**	=	x 10^{-9}
G giga	=	x 10^{9}		**p pico**	=	x 10^{-12}
M mega	=	x 10^{6}		**f femto**	=	x 10^{-15}
k kilo	=	x 10^{3}		**a atto**	=	x 10^{-18}
h hecto	=	x 10^{2}		**z zepto**	=	x 10^{-21}
da deca	=	x 10		**y yocto**	=	x 10^{-24}